PHILIP'S

WORLD
ATLAS

Philip's are grateful to the following for acting as specialist geography consultants on '*The World in Focus*' front section:

Professor D. Brunsden, Kings College, University of London, UK
Dr C. Clarke, Oxford University, UK
Dr I. S. Evans, Durham University, UK
Professor P. Haggett, University of Bristol, UK
Professor K. McLachlan, University of London, UK
Professor M. Monmonier, Syracuse University, New York, USA
Professor M-L. Hsu, University of Minnesota, Minnesota, USA
Professor M. J. Tooley, University of St Andrews, UK
Dr T. Unwin, Royal Holloway, University of London, UK

THE WORLD IN FOCUS
Cartography by Philip's

Picture Acknowledgements
Robin Scagell/Galaxy page 3

Illustrations: Stefan Chabluk

WORLD CITIES
Cartography by Philip's

Page 10, Dublin: The town plan of Dublin is based on Ordnance Survey Ireland by permission of the Government Permit Number 8621.
© Ordnance Survey Ireland and Government of Ireland.

Page 11, Edinburgh, and page 15, London:
This product includes mapping data licensed from Ordnance Survey® with the permission of the Controller of Her Majesty's Stationery Office. © Crown copyright 2010. All rights reserved. Licence number 100011710.

All satellite images in this section courtesy of Fugro NPA Ltd, Edenbridge, Kent, UK (www.satmaps.com).

Published in Great Britain in 2010 by Philip's,
a division of Octopus Publishing Group Limited
(www.octopusbooks.co.uk)
Endeavour House, 189 Shaftesbury Avenue, London WC2H 8JY
An Hachette UK Company (www.hachette.co.uk)

Copyright © 2010 Philip's

Cartography by Philip's

ISBN 978-1-84907-104-8

A CIP catalogue record for this book is available from the British Library.

Printed in Hong Kong

Details of other Philip's titles and services can be found on our website at:
www.philips-maps.co.uk

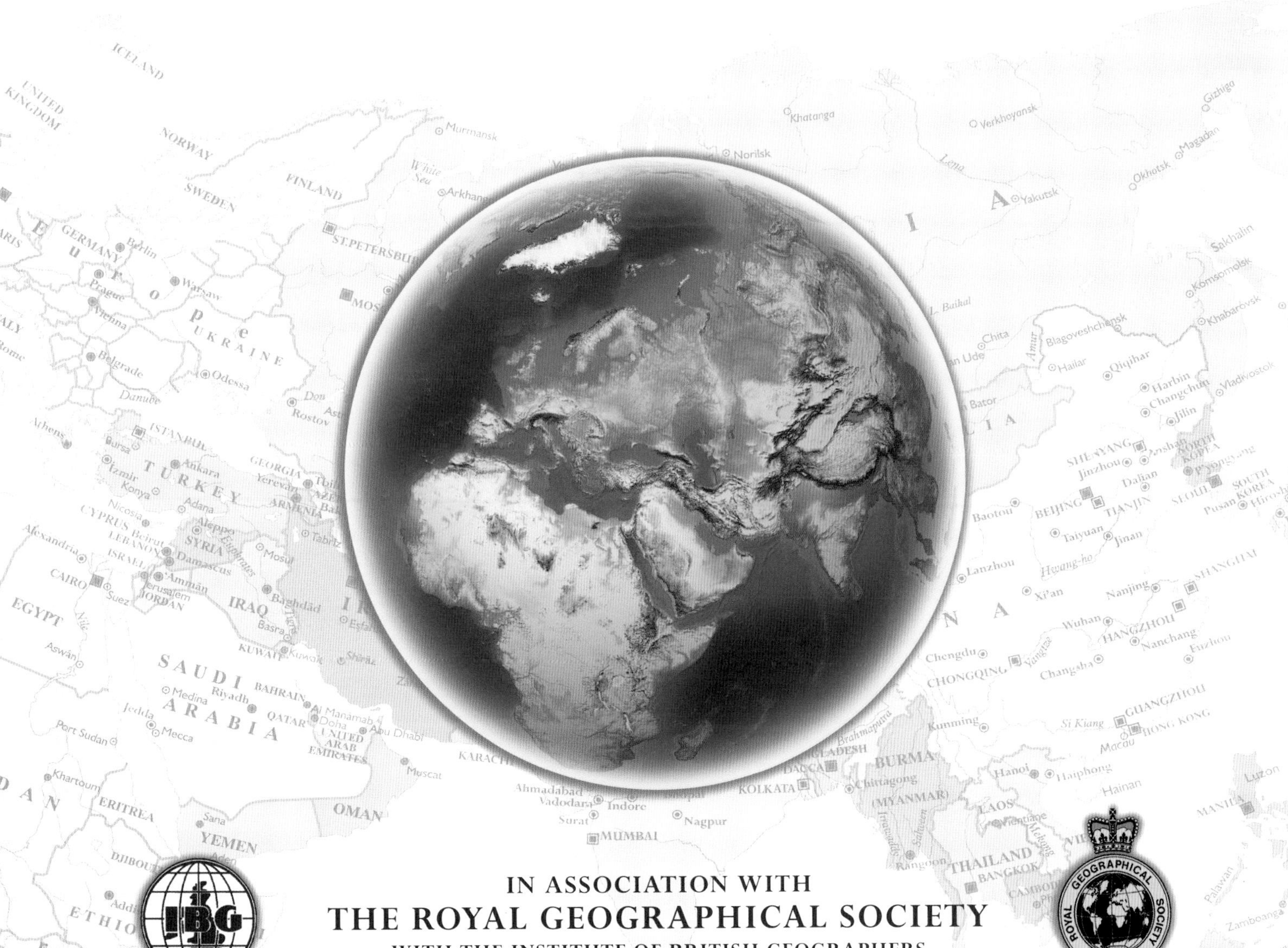

PHILIP'S

WORLD ATLAS

IN ASSOCIATION WITH
THE ROYAL GEOGRAPHICAL SOCIETY
WITH THE INSTITUTE OF BRITISH GEOGRAPHERS

Contents

World Statistics

Countries vi
Physical Dimensions vii
User Guide viii

The World in Focus

Planet Earth 2–3
Restless Earth 4–5
Landforms 6–7
Oceans 8–9
Climate 10–11
Water and Vegetation 12–13
Environment 14–15
Population 16–17
The Human Family 18–19
Wealth 20–21
Quality of Life 22–23
Energy 24–25
Production 26–27
Trade 28–29
Travel and Tourism 30–31

World Cities

Amsterdam 2
Athens 2
Atlanta 3
Baghdad 3
Bangkok 3
Barcelona 4
Beijing 4
Berlin 5
Boston 6
Brussels 6
Budapest 7
Buenos Aires 7
Cairo 7
Cape Town 8
Copenhagen 8
Chicago 9
Delhi 10
Dublin 10
Edinburgh 11
Guangzhou 11
Helsinki 11
Hong Kong 12
Istanbul 12
Jakarta 12
Jerusalem 13
Johannesburg 13
Karachi 13
Kolkata 14
Lagos 14
Las Vegas 14
Lima 14
London 15
Lisbon 16
Los Angeles 16
Madrid 17
Manila 17
Melbourne 18
Mexico City 18
Miami 18
Milan 19
Moscow 19
Montréal 20
Mumbai 20
Munich 21
New Orleans 21
New York 22
Orlando 23
Osaka 23
Oslo 23
Paris 24
Prague 25
Rio de Janeiro 25
Rome 26
San Francisco 26
St Petersburg 27
Santiago 27
São Paulo 27
Seoul 27
Shanghai 28
Singapore 28
Stockholm 29
Sydney 29
Tokyo 30
Tehran 31
Toronto 31
Vienna 32
Warsaw 32
Washington 33
Wellington 33
Index to City Maps 34–39
World Cities Satellite Images 40–48

World Maps

Map Symbols 1

The World: Political 1:95 000 000 2–3

Arctic Ocean 1:35 000 000 4

Antarctica 1:35 000 000 5

Europe

Europe: Physical 1:20 000 000 6

Europe: Political 1:20 000 000 7

Scandinavia 1:6 000 000
Iceland 1:6 000 000
Færoe Islands 1:6 000 000 8–9

Ireland 1:2 000 000 10

Scotland 1:2 000 000 11

England and Wales 1:2 000 000 12–13

British Isles 1:5 000 000 14

Netherlands, Belgium and Luxembourg 1:2 500 000 15

Central Europe 1:5 000 000 16–17

Eastern Europe and Turkey 1:10 000 000 18–19

France 1:5 000 000 20

Spain and Portugal 1:5 000 000 21

Italy and the Balkan States 1:5 000 000 22–23

The Balearics, the Canaries and Madeira 1:1 000 000 / 1:2 000 000 24

Madeira
Canary Islands
Balearic Islands

Malta, Crete, Corfu, Rhodes and Cyprus 1:1 000 000 / 1:1 300 000 25

Corfu
Malta
Rhodes
Crete
Cyprus

Asia

Asia: Physical 1:47 000 000 26

Asia: Political 1:47 000 000 27

Russia and Central Asia 1:20 000 000 28–29

Japan 1:5 000 000 30–31

China and the Far East 1:15 000 000
Hong Kong and Macau 1:1 000 000 32–33

Northern China and Korea 1:6 000 000 34–35

Indonesia and the Philippines 1:12 500 000
Java and Madura 1:7 500 000
Bali 1:2 000 000 36–37

Mainland South-east Asia 1:6 000 000
Ko Phuket 1:1 000 000
Ko Samui 1:1 000 000
Pinang 1:1 000 000
Singapore 1:1 000 000 38–39

South Asia 1:10 000 000 40–41

Northern India, Pakistan and Nepal
1:6 000 000
Jammu and Kashmir 1:6 000 000

42–43

The Middle East
1:7 000 000

44–45

The Near East 1:2 500 000

46

Arabia and the Horn of Africa 1:15 000 000

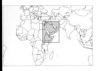

47

Africa

Africa: Physical
1:42 000 000   48

Africa: Political
1:42 000 000   49

Northern Africa
1:15 000 000
Azores 1:15 000 000
Cape Verde Islands 1:10 000 000

50–51

Central and Southern Africa
1:15 000 000
Seychelles 1:2 500 000
Réunion 1:2 500 000
Mauritius 1:2 500 000
Comoros 1:8 000 000

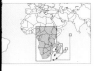

52–53

East Africa 1:8 000 000

54–55

Southern Africa
1:8 000 000
Madagascar 1:8 000 000

56–57

Australia and Oceania

Australia and Oceania: Physical and Political
1:50 000 000   58

New Zealand 1:6 000 000
Samoan Islands 1:6 000 000
Fiji 1:6 000 000
Tonga 1:6 000 000
Tahiti and Moorea 1:1 000 000

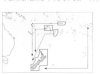

59

Western Australia
1:8 000 000

60–61

Eastern Australia
1:8 000 000
Torres Strait 1:8 000 000
Whitsunday Islands 1:2 500 000
Tasmania 1:8 000 000

62–63

Pacific Ocean
1:54 000 000   64–65

North America

North America: Physical
1:35 000 000   66

North America: Political
1:35 000 000   67

Canada 1:15 000 000

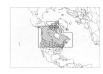

68–69

Western Canada 1:7 000 000

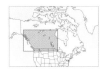

70–71

Eastern Canada 1:7 000 000

72–73

United States 1:15 000 000
Hawai'i 1:15 000 000
Alaska 1:15 000 000

74–75

United States – West
1:6 700 000

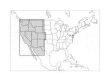

76–77

Central and Southern California and Western Washington
1:2 500 000

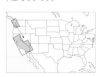

78–79

United States – Midwest and North-east
1:6 700 000

80–81

North-eastern United States
1:2 500 000

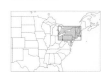

82–83

United States – South
1:6 700 000

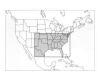

84–85

Mexico 1:8 000 000

86–87

Central America and the West Indies 1:8 000 000
Jamaica 1:3 000 000
Guadeloupe 1:2 000 000
Martinique 1:2 000 000
Puerto Rico 1:3 000 000
Virgin Islands 1:2 000 000
St Lucia 1:2 000 000
Barbados 1:2 000 000

88–89

South America

South America: Physical
1:35 000 000   90

South America: Political
1:35 000 000   91

South America – North
1:16 000 000
Trinidad and Tobago 1:2 500 000

92–93

Central South America
1:8 000 000

94–95

South America – South
1:16 000 000

96

Index to World Maps
97–176

World Statistics: Countries

This alphabetical list includes the principal countries and territories of the world. If a territory is not completely independent, the country it is associated with is named. The area figures give the total area of land, inland water and ice. The population figures are 2009 estimates where available. The annual income is the Gross Domestic Product per capita in US dollars. The figures are the latest available, usually 2009 estimates.

Country/Territory	Area km² Thousands	Area miles² Thousands	Population Thousands	Capital	Annual Income US $
Afghanistan	652	252	28,396	Kabul	800
Albania	28.7	11.1	3,639	Tirana	6,200
Algeria	2,382	920	34,178	Algiers	7,100
American Samoa (US)	0.20	0.08	66	Pago Pago	8,000
Andorra	0.47	0.18	84	Andorra La Vella	42,500
Angola	1,247	481	12,799	Luanda	8,800
Anguilla (UK)	0.10	0.04	14	The Valley	8,800
Antigua & Barbuda	0.44	0.17	86	St John's	18,100
Argentina	2,780	1,074	40,914	Buenos Aires	13,800
Armenia	29.8	11.5	2,967	Yerevan	5,900
Aruba (Netherlands)	0.19	0.07	103	Oranjestad	21,800
Australia	7,741	2,989	21,263	Canberra	38,500
Austria	83.9	32.4	8,210	Vienna	39,400
Azerbaijan	86.6	33.4	8,239	Baku	9,900
Azores (Portugal)	2.2	0.86	236	Ponta Delgada	15,000
Bahamas	13.9	5.4	308	Nassau	29,800
Bahrain	0.69	0.27	729	Manama	38,400
Bangladesh	144	55.6	156,051	Dhaka	1,600
Barbados	0.43	0.17	285	Bridgetown	18,500
Belarus	208	80.2	9,649	Minsk	11,600
Belgium	30.5	11.8	10,414	Brussels	36,600
Belize	23.0	8.9	308	Belmopan	8,200
Benin	113	43.5	8,792	Porto-Novo	1,500
Bermuda (UK)	0.05	0.02	68	Hamilton	69,900
Bhutan	47.0	18.1	691	Thimphu	6,200
Bolivia	1,099	424	9,775	La Paz/Sucre	4,600
Bosnia-Herzegovina	51.2	19.8	4,613	Sarajevo	6,300
Botswana	582	225	1,991	Gaborone	12,100
Brazil	8,514	3,287	198,739	Brasília	10,200
Brunei	5.8	2.2	388	Bandar Seri Begawan	50,100
Bulgaria	111	42.8	7,205	Sofia	12,600
Burkina Faso	274	106	15,746	Ouagadougou	1,200
Burma (Myanmar)	677	261	48,138	Rangoon/Naypyidaw	1,200
Burundi	27.8	10.7	9,511	Bujumbura	300
Cambodia	181	69.9	14,494	Phnom Penh	1,900
Cameroon	475	184	18,879	Yaoundé	2,300
Canada	9,971	3,850	33,487	Ottawa	38,400
Canary Is. (Spain)	7.2	2.8	1,682	Las Palmas/Santa Cruz	19,900
Cape Verde Is.	4.0	1.6	429	Praia	3,900
Cayman Is. (UK)	0.26	0.10	49	George Town	43,800
Central African Republic	623	241	4,511	Bangui	700
Chad	1,284	496	10,329	Ndjaména	1,500
Chile	757	292	16,602	Santiago	14,700
China	9,597	3,705	1,338,613	Beijing	6,500
Colombia	1,139	440	43,677	Bogotá	9,200
Comoros	2.2	0.86	752	Moroni	1,000
Congo	342	132	4,013	Brazzaville	4,200
Congo (Dem. Rep. of the)	2,345	905	68,693	Kinshasa	300
Cook Is. (NZ)	0.24	0.09	12	Avarua	9,100
Costa Rica	51.1	19.7	4,254	San José	11,300
Croatia	56.5	21.8	4,489	Zagreb	17,600
Cuba	111	42.8	11,452	Havana	9,700
Cyprus	9.3	3.6	1,085	Nicosia	21,200
Czech Republic	78.9	30.5	10,212	Prague	25,100
Denmark	43.1	16.6	5,501	Copenhagen	36,200
Djibouti	23.2	9.0	725	Djibouti	2,800
Dominica	0.75	0.29	73	Roseau	10,200
Dominican Republic	48.5	18.7	9,650	Santo Domingo	8,200
East Timor	14.9	5.7	1,132	Dili	2,400
Ecuador	284	109	14,573	Quito	7,300
Egypt	1,001	387	78,867	Cairo	6,000
El Salvador	21.0	8.1	7,185	San Salvador	6,000
Equatorial Guinea	28.1	10.8	633	Malabo	36,100
Eritrea	118	45.4	5,647	Asmara	700
Estonia	45.1	17.4	1,299	Tallinn	18,800
Ethiopia	1,104	426	85,237	Addis Ababa	900
Faroe Is. (Denmark)	1.4	0.54	49	Tórshavn	31,000
Fiji	18.3	7.1	945	Suva	3,800
Finland	338	131	5,250	Helsinki	34,900
France	552	213	64,058	Paris	32,800
French Guiana (France)	90.0	34.7	203	Cayenne	8,300
French Polynesia (France)	4.0	1.5	287	Papeete	18,000
Gabon	268	103	1,515	Libreville	13,700
Gambia, The	11.3	4.4	1,778	Banjul	1,300
Gaza Strip (OPT)*	0.36	0.14	1,552	–	3,100
Georgia	69.7	26.9	4,616	Tbilisi	4,500
Germany	357	138	82,330	Berlin	34,200
Ghana	239	92.1	23,888	Accra	1,500
Gibraltar (UK)	0.006	0.002	29	Gibraltar Town	38,200
Greece	132	50.9	10,737	Athens	32,100
Greenland (Denmark)	2,176	840	58	Nuuk	34,700
Grenada	0.34	0.13	91	St George's	12,700
Guadeloupe (France)	1.7	0.66	453	Basse-Terre	7,900
Guam (US)	0.55	0.21	178	Agana	15,000
Guatemala	109	42.0	13,277	Guatemala City	5,200
Guinea	246	94.9	10,058	Conakry	1,100
Guinea-Bissau	36.1	13.9	1,534	Bissau	600
Guyana	215	83.0	753	Georgetown	3,900
Haiti	27.8	10.7	9,036	Port-au-Prince	1,300
Honduras	112	43.3	7,834	Tegucigalpa	4,200
Hungary	93.0	35.9	9,906	Budapest	18,800
Iceland	103	39.8	307	Reykjavik	39,800
India	3,287	1,269	1,156,898	New Delhi	3,100
Indonesia	1,905	735	240,272	Jakarta	4,000
Iran	1,648	636	66,429	Tehran	12,900
Iraq	438	169	28,946	Baghdad	3,300
Ireland	70.3	27.1	4,203	Dublin	42,200
Israel	20.6	8.0	7,234	Jerusalem	28,400
Italy	301	116	58,126	Rome	30,200
Ivory Coast (Côte d'Ivoire)	322	125	20,617	Yamoussoukro	1,700
Jamaica	11.0	4.2	2,826	Kingston	8,300
Japan	378	146	127,079	Tokyo	32,600
Jordan	89.3	34.5	6,269	Amman	5,300
Kazakhstan	2,725	1,052	15,399	Astana	11,400
Kenya	580	224	39,003	Nairobi	1,600
Kiribati	0.73	0.28	113	Tarawa	5,300
Korea, North	121	46.5	22,665	Pyŏngyang	1,800
Korea, South	99.3	38.3	48,509	Seoul	27,700
Kosovo	10.9	4.2	1,805	Pristina	2,300
Kuwait	17.8	6.9	2,693	Kuwait City	55,800
Kyrgyzstan	200	77.2	5,432	Bishkek	2,100
Laos	237	91.4	6,834	Vientiane	2,100
Latvia	64.6	24.9	2,232	Riga	14,500
Lebanon	10.4	4.0	4,017	Beirut	11,500
Lesotho	30.4	11.7	2,131	Maseru	1,500
Liberia	111	43.0	3,442	Monrovia	500
Libya	1,760	679	6,324	Tripoli	14,600
Liechtenstein	0.16	0.06	35	Vaduz	122,100
Lithuania	65.2	25.2	3,555	Vilnius	15,000
Luxembourg	2.6	1.0	492	Luxembourg	77,600
Macedonia (FYROM)	25.7	9.9	2,067	Skopje	9,000
Madagascar	587	227	20,654	Antananarivo	1,000
Madeira (Portugal)	0.78	0.30	241	Funchal	22,700
Malawi	118	45.7	15,029	Lilongwe	900
Malaysia	330	127	25,716	Kuala Lumpur/Putrajaya	14,700
Maldives	0.30	0.12	396	Malé	4,200
Mali	1,240	479	13,443	Bamako	1,100
Malta	0.32	0.12	405	Valletta	23,800
Marshall Is.	0.18	0.07	65	Majuro	2,500
Martinique (France)	1.1	0.43	436	Fort-de-France	14,400
Mauritania	1,026	396	3,129	Nouakchott	2,100
Mauritius	2.0	0.79	1,284	Port Louis	12,400
Mayotte (France)	0.37	0.14	224	Mamoudzou	4,900
Mexico	1,958	756	111,212	Mexico City	13,200
Micronesia, Fed. States of	0.70	0.27	107	Palikir	2,200
Moldova	33.9	13.1	4,321	Kishinev	2,400
Monaco	0.001	0.0004	33	Monaco	30,000
Mongolia	1,567	605	3,041	Ulan Bator	3,400
Montenegro	14.0	5.4	672	Podgorica	9,800
Morocco	447	172	31,285	Rabat	4,600
Mozambique	802	309	21,669	Maputo	900
Namibia	824	318	2,109	Windhoek	6,400
Nauru	0.02	0.008	14	Yaren	5,000
Nepal	147	56.8	28,563	Katmandu	1,200
Netherlands	41.5	16.0	16,716	Amsterdam/The Hague	39,000
Netherlands Antilles (Neths)†	0.80	0.31	227	Willemstad	16,000
New Caledonia (France)	18.6	7.2	228	Nouméa	15,000
New Zealand	271	104	4,213	Wellington	27,700
Nicaragua	130	50.2	5,891	Managua	2,800
Niger	1,267	489	15,306	Niamey	700
Nigeria	924	357	149,229	Abuja	2,400
Northern Mariana Is. (US)	0.46	0.18	51	Saipan	12,500
Norway	324	125	4,661	Oslo	59,300
Oman	310	119	3,418	Muscat	20,300
Pakistan	796	307	174,579	Islamabad	2,600
Palau	0.46	0.18	21	Melekeok	8,100
Panama	75.5	29.2	3,360	Panamá	11,900
Papua New Guinea	463	179	5,941	Port Moresby	2,300
Paraguay	407	157	6,996	Asunción	4,100
Peru	1,285	496	29,547	Lima	8,600
Philippines	300	116	97,977	Manila	3,300
Poland	323	125	38,483	Warsaw	17,800
Portugal	88.8	34.3	10,708	Lisbon	21,700
Puerto Rico (US)	8.9	3.4	3,966	San Juan	17,100
Qatar	11.0	4.2	833	Doha	121,400
Réunion (France)	2.5	0.97	788	St-Denis	6,200
Romania	238	92.0	22,215	Bucharest	11,500
Russia	17,075	6,593	140,041	Moscow	15,200
Rwanda	26.3	10.2	10,746	Kigali	1,000
St Kitts & Nevis	0.26	0.10	40	Basseterre	18,800
St Lucia	0.54	0.21	160	Castries	10,900
St Vincent & Grenadines	0.39	0.15	105	Kingstown	18,100
Samoa	2.8	1.1	220	Apia	4,700
San Marino	0.06	0.02	30	San Marino	41,900
São Tomé & Príncipe	0.96	0.37	213	São Tomé	1,400
Saudi Arabia	2,150	830	28,687	Riyadh	20,300
Senegal	197	76.0	13,712	Dakar	1,700
Serbia	77.5	29.9	7,379	Belgrade	10,400
Seychelles	0.46	0.18	87	Victoria	19,400
Sierra Leone	71.7	27.7	5,132	Freetown	900
Singapore	0.68	0.26	4,658	Singapore City	50,300
Slovak Republic	49.0	18.9	5,463	Bratislava	21,100
Slovenia	20.3	7.8	2,006	Ljubljana	28,200
Solomon Is.	28.9	11.2	596	Honiara	2,600
Somalia	638	246	9,832	Mogadishu	600
South Africa	1,221	471	49,052	Cape Town/Pretoria	10,000
Spain	498	192	40,525	Madrid	33,700
Sri Lanka	65.6	25.3	21,325	Colombo	4,500
Sudan	2,506	967	41,088	Khartoum	2,300
Suriname	163	63.0	481	Paramaribo	8,800
Swaziland	17.4	6.7	1,337	Mbabane	4,400
Sweden	450	174	9,060	Stockholm	36,800
Switzerland	41.3	15.9	7,604	Bern	41,600
Syria	185	71.5	21,763	Damascus	4,700
Taiwan	36.0	13.9	22,974	Taipei	30,200
Tajikistan	143	55.3	7,349	Dushanbe	1,800
Tanzania	945	365	41,049	Dodoma	1,400
Thailand	513	198	65,998	Bangkok	8,100
Togo	56.8	21.9	6,032	Lomé	900
Tonga	0.65	0.25	121	Nuku'alofa	4,600
Trinidad & Tobago	5.1	2.0	1,230	Port of Spain	23,300
Tunisia	164	63.2	10,486	Tunis	8,000
Turkey	775	299	76,806	Ankara	11,200
Turkmenistan	488	188	4,885	Ashkhabad	6,700
Turks & Caicos Is. (UK)	0.43	0.17	23	Cockburn Town	11,500
Tuvalu	0.03	0.01	12	Fongafale	1,600
Uganda	241	93.1	32,370	Kampala	1,300
Ukraine	604	233	45,700	Kiev	6,400
United Arab Emirates	83.6	32.3	4,798	Abu Dhabi	41,800
United Kingdom	242	93.4	61,113	London	35,400
United States of America	9,629	3,718	307,212	Washington, DC	46,400
Uruguay	175	67.6	3,494	Montevideo	12,600
Uzbekistan	447	173	27,606	Tashkent	2,800
Vanuatu	12.2	4.7	219	Port-Vila	4,800
Venezuela	912	352	26,815	Caracas	13,200
Vietnam	332	128	88,577	Hanoi	2,900
Virgin Is. (UK)	0.15	0.06	24	Road Town	38,500
Virgin Is. (US)	0.35	0.13	110	Charlotte Amalie	14,500
Wallis & Futuna Is. (France)	0.20	0.08	15	Mata-Utu	3,800
West Bank (OPT)*	5.9	2.3	2,416	–	2,900
Western Sahara	266	103	405	El Aaiún	2,500
Yemen	528	204	22,858	Sana'	2,500
Zambia	753	291	11,863	Lusaka	1,500
Zimbabwe	391	151	11,393	Harare	200

*OPT = Occupied Palestinian Territory

† Plans have been announced to dissolve the Netherlands Antilles as a political entity in October 2010. The five islands will then each have a new constitutional status within the Kingdom of the Netherlands.

World Statistics: Physical Dimensions

Each topic list is divided into continents and within a continent the items are listed in order of size. The bottom part of many of the lists is selective in order to give examples from as many different countries as possible. The order of the continents is the same as in the atlas, beginning with Europe and ending with South America. The figures are rounded as appropriate.

World, Continents, Oceans

	km²	miles²	%
The World	509,450,000	196,672,000	–
Land	149,450,000	57,688,000	29.3
Water	360,000,000	138,984,000	70.7
Asia	44,500,000	17,177,000	29.8
Africa	30,302,000	11,697,000	20.3
North America	24,241,000	9,357,000	16.2
South America	17,793,000	6,868,000	11.9
Antarctica	14,100,000	5,443,000	9.4
Europe	9,957,000	3,843,000	6.7
Australia & Oceania	8,557,000	3,303,000	5.7
Pacific Ocean	155,557,000	60,061,000	46.4
Atlantic Ocean	76,762,000	29,638,000	22.9
Indian Ocean	68,556,000	26,470,000	20.4
Southern Ocean	20,327,000	7,848,000	6.1
Arctic Ocean	14,056,000	5,427,000	4.2

Ocean Depths

Atlantic Ocean	m	ft
Puerto Rico (Milwaukee) Deep	8,605	28,232
Cayman Trench	7,680	25,197
Gulf of Mexico	5,203	17,070
Mediterranean Sea	5,121	16,801
Black Sea	2,211	7,254
North Sea	660	2,165

Indian Ocean	m	ft
Java Trench	7,450	24,442
Red Sea	2,635	8,454

Pacific Ocean	m	ft
Mariana Trench	11,022	36,161
Tonga Trench	10,882	35,702
Japan Trench	10,554	34,626
Kuril Trench	10,542	34,587

Arctic Ocean	m	ft
Molloy Deep	5,608	18,399

Southern Ocean	m	ft
South Sandwich Trench	7,235	23,737

Mountains

Europe		m	ft
Elbrus	Russia	5,642	18,510
Dykh-Tau	Russia	5,205	17,076
Shkhara	Russia/Georgia	5,201	17,064
Koshtan-Tau	Russia	5,152	16,903
Kazbek	Russia/Georgia	5,047	16,558
Pushkin	Russia/Georgia	5,033	16,512
Katyn-Tau	Russia/Georgia	4,979	16,335
Shota Rustaveli	Russia/Georgia	4,860	15,945
Mont Blanc	France/Italy	4,808	15,774
Monte Rosa	Italy/Switzerland	4,634	15,203
Dom	Switzerland	4,545	14,911
Liskamm	Switzerland	4,527	14,852
Weisshorn	Switzerland	4,505	14,780
Taschorn	Switzerland	4,490	14,730
Matterhorn/Cervino	Italy/Switzerland	4,478	14,691
Grossglockner	Austria	3,797	12,457
Mulhacén	Spain	3,478	11,411
Zugspitze	Germany	2,962	9,718
Olympus	Greece	2,917	9,570
Galdhøpiggen	Norway	2,469	8,100
Ben Nevis	UK	1,342	4,403

Asia		m	ft
Everest	China/Nepal	8,850	29,035
K2 (Godwin Austen)	China/Kashmir	8,611	28,251
Kanchenjunga	India/Nepal	8,598	28,208
Lhotse	China/Nepal	8,516	27,939
Makalu	China/Nepal	8,481	27,824
Cho Oyu	China/Nepal	8,201	26,906
Dhaulagiri	Nepal	8,167	26,795
Manaslu	Nepal	8,156	26,758
Nanga Parbat	Kashmir	8,126	26,660
Annapurna	Nepal	8,078	26,502
Gasherbrum	China/Kashmir	8,068	26,469
Broad Peak	China/Kashmir	8,051	26,414
Xixabangma	China	8,012	26,286
Kangbachen	Nepal	7,858	25,781
Trivor	Pakistan	7,720	25,328
Pik Imeni Ismail Samani	Tajikistan	7,495	24,590
Demavend	Iran	5,604	18,386
Ararat	Turkey	5,165	16,945
Gunong Kinabalu	Malaysia (Borneo)	4,101	13,455
Fuji-San	Japan	3,776	12,388

Africa		m	ft
Kilimanjaro	Tanzania	5,895	19,340
Mt Kenya	Kenya	5,199	17,057
Ruwenzori (Margherita)	Ug./Congo (D.R.)	5,109	16,762
Meru	Tanzania	4,565	14,977
Ras Dashen	Ethiopia	4,533	14,872
Karisimbi	Rwanda/Congo (D.R.)	4,507	14,787
Mt Elgon	Kenya/Uganda	4,321	14,176
Batu	Ethiopia	4,307	14,130
Toubkal	Morocco	4,165	13,665
Mt Cameroun	Cameroon	4,070	13,353

Oceania		m	ft
Puncak Jaya	Indonesia	5,029	16,499
Puncak Trikora	Indonesia	4,730	15,518
Puncak Mandala	Indonesia	4,702	15,427
Mt Wilhelm	Papua New Guinea	4,508	14,790
Mauna Kea	USA (Hawai'i)	4,205	13,796
Mauna Loa	USA (Hawai'i)	4,169	13,681
Aoraki Mt Cook	New Zealand	3,753	12,313
Mt Kosciuszko	Australia	2,228	7,310

North America		m	ft
Mt McKinley (Denali)	USA (Alaska)	6,194	20,321
Mt Logan	Canada	5,959	19,551
Pico de Orizaba	Mexico	5,610	18,405
Mt St Elias	USA/Canada	5,489	18,008
Popocatépetl	Mexico	5,452	17,887
Mt Foraker	USA (Alaska)	5,304	17,401
Iztaccihuatl	Mexico	5,286	17,343
Mt Lucania	Canada	5,226	17,146
Mt Steele	Canada	5,073	16,644
Mt Bona	USA (Alaska)	5,005	16,420
Mt Whitney	USA	4,418	14,495
Tajumulco	Guatemala	4,220	13,845
Chirripó Grande	Costa Rica	3,837	12,589
Pico Duarte	Dominican Rep.	3,175	10,417

South America		m	ft
Aconcagua	Argentina	6,962	22,841
Bonete	Argentina	6,872	22,546
Ojos del Salado	Argentina/Chile	6,863	22,516
Pissis	Argentina	6,779	22,241
Mercedario	Argentina/Chile	6,770	22,211
Huascarán	Peru	6,768	22,204
Llullaillaco	Argentina/Chile	6,723	22,057
Nevado de Cachi	Argentina	6,720	22,047
Yerupaja	Peru	6,632	21,758
Sajama	Bolivia	6,520	21,391
Chimborazo	Ecuador	6,267	20,561
Pico Cristóbal Colón	Colombia	5,800	19,029
Pico Bolivar	Venezuela	5,007	16,427

Antarctica		m	ft
Vinson Massif		4,897	16,066
Mt Kirkpatrick		4,528	14,855

Rivers

Europe		km	miles
Volga	Caspian Sea	3,700	2,300
Danube	Black Sea	2,850	1,770
Ural	Caspian Sea	2,535	1,575
Dnepr (Dnipro)	Black Sea	2,285	1,420
Kama	Volga	2,030	1,260
Don	Black Sea	1,990	1,240
Petchora	Arctic Ocean	1,790	1,110
Oka	Volga	1,480	920
Dnister (Dniester)	Black Sea	1,400	870
Vyatka	Kama	1,370	850
Rhine	North Sea	1,320	820
N. Dvina	Arctic Ocean	1,290	800
Elbe	North Sea	1,145	710

Asia		km	miles
Yangtze	Pacific Ocean	6,380	3,960
Yenisey–Angara	Arctic Ocean	5,550	3,445
Huang He	Pacific Ocean	5,464	3,395
Ob–Irtysh	Arctic Ocean	5,410	3,360
Mekong	Pacific Ocean	4,500	2,795
Amur	Pacific Ocean	4,442	2,760
Lena	Arctic Ocean	4,402	2,735
Irtysh	Ob	4,250	2,640
Yenisey	Arctic Ocean	4,090	2,540
Ob	Arctic Ocean	3,680	2,285
Indus	Indian Ocean	3,100	1,925
Brahmaputra	Indian Ocean	2,900	1,800
Syrdarya	Aral Sea	2,860	1,775
Salween	Indian Ocean	2,800	1,740
Euphrates	Indian Ocean	2,700	1,675
Amudarya	Aral Sea	2,540	1,575

Africa		km	miles
Nile	Mediterranean	6,695	4,160
Congo	Atlantic Ocean	4,670	2,900
Niger	Atlantic Ocean	4,180	2,595
Zambezi	Indian Ocean	3,540	2,200
Oubangi/Uele	Congo (D.R.)	2,250	1,400
Kasai	Congo (D.R.)	1,950	1,210
Shaballe	Indian Ocean	1,930	1,200
Orange	Atlantic Ocean	1,860	1,155
Cubango	Okavango Delta	1,800	1,120
Limpopo	Indian Ocean	1,770	1,100
Senegal	Atlantic Ocean	1,640	1,020

Australia		km	miles
Murray–Darling	Southern Ocean	3,750	2,330
Darling	Murray	3,070	1,905
Murray	Southern Ocean	2,575	1,600
Murrumbidgee	Murray	1,690	1,050

North America		km	miles
Mississippi–Missouri	Gulf of Mexico	5,971	3,710
Mackenzie	Arctic Ocean	4,240	2,630
Missouri	Mississippi	4,088	2,540
Mississippi	Gulf of Mexico	3,782	2,350
Yukon	Pacific Ocean	3,185	1,980
Rio Grande	Gulf of Mexico	3,030	1,880
Arkansas	Mississippi	2,340	1,450
Colorado	Pacific Ocean	2,330	1,445
Red	Mississippi	2,040	1,270
Columbia	Pacific Ocean	1,950	1,210
Saskatchewan	Lake Winnipeg	1,940	1,205

South America		km	miles
Amazon	Atlantic Ocean	6,450	4,010
Paraná–Plate	Atlantic Ocean	4,500	2,800
Purus	Amazon	3,350	2,080
Madeira	Amazon	3,200	1,990
São Francisco	Atlantic Ocean	2,900	1,800
Paraná	Plate	2,800	1,740
Tocantins	Atlantic Ocean	2,750	1,710
Orinoco	Atlantic Ocean	2,740	1,700
Paraguay	Paraná	2,550	1,580
Pilcomayo	Paraná	2,500	1,550
Araguaia	Tocantins	2,250	1,400

Lakes

Europe		km²	miles²
Lake Ladoga	Russia	17,700	6,800
Lake Onega	Russia	9,700	3,700
Saimaa system	Finland	8,000	3,100
Vänern	Sweden	5,500	2,100

Asia		km²	miles²
Caspian Sea	Asia	371,000	143,000
Lake Baikal	Russia	30,500	11,780
Tonlé Sap	Cambodia	20,000	7,700
Lake Balqash	Kazakhstan	18,500	7,100
Aral Sea	Kazakhstan/Uzbekistan	17,160	6,625

Africa		km²	miles²
Lake Victoria	East Africa	68,000	26,300
Lake Tanganyika	Central Africa	33,000	13,000
Lake Malawi/Nyasa	East Africa	29,600	11,430
Lake Chad	Central Africa	25,000	9,700
Lake Bangweulu	Zambia	9,840	3,800
Lake Turkana	Ethiopia/Kenya	8,500	3,290

Australia		km²	miles²
Lake Eyre	Australia	8,900	3,400
Lake Torrens	Australia	5,800	2,200
Lake Gairdner	Australia	4,800	1,900

North America		km²	miles²
Lake Superior	Canada/USA	82,350	31,800
Lake Huron	Canada/USA	59,600	23,010
Lake Michigan	USA	58,000	22,400
Great Bear Lake	Canada	31,800	12,280
Great Slave Lake	Canada	28,500	11,000
Lake Erie	Canada/USA	25,700	9,900
Lake Winnipeg	Canada	24,400	9,400
Lake Ontario	Canada/USA	19,500	7,500
Lake Nicaragua	Nicaragua	8,200	3,200

South America		km²	miles²
Lake Titicaca	Bolivia/Peru	8,300	3,200
Lake Poopo	Bolivia	2,800	1,100

Islands

Europe		km²	miles²
Great Britain	UK	229,880	88,700
Iceland	Atlantic Ocean	103,000	39,800
Ireland	Ireland/UK	84,400	32,600
Novaya Zemlya (N.)	Russia	48,200	18,600
Sicily	Italy	25,500	9,800
Corsica	France	8,700	3,400

Asia		km²	miles²
Borneo	South-east Asia	744,360	287,400
Sumatra	Indonesia	473,600	182,860
Honshu	Japan	230,500	88,980
Sulawesi (Celebes)	Indonesia	189,000	73,000
Java	Indonesia	126,700	48,900
Luzon	Philippines	104,700	40,400
Hokkaido	Japan	78,400	30,300

Africa		km²	miles²
Madagascar	Indian Ocean	587,040	226,660
Socotra	Indian Ocean	3,600	1,400
Réunion	Indian Ocean	2,500	965

Oceania		km²	miles²
New Guinea	Indonesia/Papua NG	821,030	317,000
New Zealand (S.)	Pacific Ocean	150,500	58,100
New Zealand (N.)	Pacific Ocean	114,700	44,300
Tasmania	Australia	67,800	26,200
Hawai'i	Pacific Ocean	10,450	4,000

North America		km²	miles²
Greenland	Atlantic Ocean	2,175,600	839,800
Baffin Is.	Canada	508,000	196,100
Victoria Is.	Canada	212,200	81,900
Ellesmere Is.	Canada	212,000	81,800
Cuba	Caribbean Sea	110,860	42,800
Hispaniola	Dominican Rep./Haiti	76,200	29,400
Jamaica	Caribbean Sea	11,400	4,400
Puerto Rico	Atlantic Ocean	8,900	3,400

South America		km²	miles²
Tierra del Fuego	Argentina/Chile	47,000	18,100
Falkland Is. (E.)	Atlantic Ocean	6,800	2,600

User Guide

The reference maps which form the main body of this atlas have been prepared in accordance with the highest standards of international cartography to provide an accurate and detailed representation of the Earth. The scales and projections used have been carefully chosen to give balanced coverage of the world, while emphasizing the most densely populated and economically significant regions. A hallmark of Philip's mapping is the use of hill shading and relief colouring to create a graphic impression of landforms: this makes the maps exceptionally easy to read. However, knowledge of the key features employed in the construction and presentation of the maps will enable the reader to derive the fullest benefit from the atlas.

Map sequence

The atlas covers the Earth continent by continent: first Europe; then its land neighbour Asia (mapped north before south, in a clockwise sequence), then Africa, Australia and Oceania, North America and South America. This is the classic arrangement adopted by most cartographers since the 16th century. For each continent, there are maps at a variety of scales. First, physical relief and political maps of the whole continent; then a series of larger-scale maps of the regions within the continent, each followed, where required, by still larger-scale maps of the most important or densely populated areas. The governing principle is that by turning the pages of the atlas, the reader moves steadily from north to south through each continent, with each map overlapping its neighbours.

Map presentation

With very few exceptions (for example, for the Arctic and Antarctica), the maps are drawn with north at the top, regardless of whether they are presented upright or sideways on the page. In the borders will be found the map title; a locator diagram showing the area covered; continuation arrows showing the page numbers for maps of adjacent areas; the scale; the projection used; the degrees of latitude and longitude; and the letters and figures used in the index for locating place names and geographical features. Physical relief maps also have a height reference panel identifying the colours used for each layer of contouring.

Map symbols

Each map contains a vast amount of detail which can only be conveyed clearly and accurately by the use of symbols. Points and circles of varying sizes locate and identify the relative importance of towns and cities; different styles of type are employed for administrative, geographical and regional place names. A variety of pictorial symbols denote features such as glaciers and marshes, as well as man-made structures including roads, railways, airports and canals.

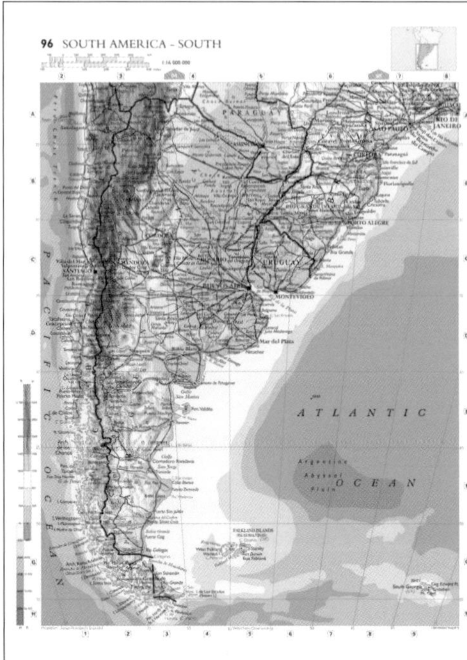

International borders are shown by red lines. Where neighbouring countries are in dispute, for example in the Middle East, the maps show the *de facto* boundary between nations, regardless of the legal or historical situation. The symbols are explained on the first page of the World Maps section of the atlas.

Map scales

The scale of each map is given in the numerical form known as the 'representative fraction'. The first figure is always one, signifying one unit of distance on the map; the second figure, usually in millions, is the number by which the map unit must be multiplied to give the equivalent distance on the Earth's surface. Calculations can easily be made in centimetres and kilometres, by dividing the Earth units figure by 100 000 (i.e. deleting the last five 0s). Thus 1:1 000 000 means 1 cm = 10 km. The calculation for inches and miles is more laborious, but 1 000 000 divided by 63 360 (the number of inches in a mile) shows that the ratio 1:1 000 000 means approximately 1 inch = 16 miles. The table below provides distance equivalents for scales down to 1:50 000 000.

LARGE SCALE		
1:1 000 000	1 cm = 10 km	1 inch = 16 miles
1:2 500 000	1 cm = 25 km	1 inch = 39.5 miles
1:5 000 000	1 cm = 50 km	1 inch = 79 miles
1:6 000 000	1 cm = 60 km	1 inch = 95 miles
1:8 000 000	1 cm = 80 km	1 inch = 126 miles
1:10 000 000	1 cm = 100 km	1 inch = 158 miles
1:15 000 000	1 cm = 150 km	1 inch = 237 miles
1:20 000 000	1 cm = 200 km	1 inch = 316 miles
1:50 000 000	1 cm = 500 km	1 inch = 790 miles
SMALL SCALE		

Measuring distances

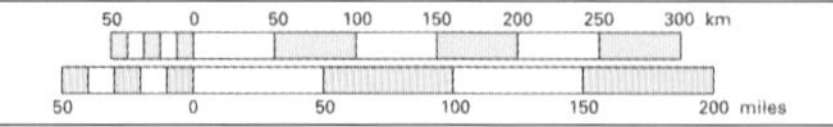

Although each map is accompanied by a scale bar, distances cannot always be measured with confidence because of the distortions involved in portraying the curved surface of the Earth on a flat page. As a general rule, the larger the map scale (i.e. the lower the number of Earth units in the representative fraction), the more accurate and reliable will be the distance measured. On small-scale maps such as those of the world and of entire continents, measurement may only be accurate along the 'standard parallels', or central axes, and should not be attempted without considering the map projection.

Latitude and longitude

Accurate positioning of individual points on the Earth's surface is made possible by reference to the geometrical system of latitude and longitude. Latitude *parallels* are drawn west-east around the Earth and numbered by degrees north and south of the Equator, which is designated 0° of latitude. Longitude *meridians* are drawn north–south and numbered by degrees east and west of the *prime meridian*, 0° of longitude, which passes through Greenwich in England. By referring to these co-ordinates and their subdivisions of minutes ($^1/60$th of a degree) and seconds ($^1/60$th of a minute), any place on Earth can be located to within a few hundred metres. Latitude and longitude are indicated by blue lines on the maps; they are straight or curved according to the projection employed. Reference to these lines is the easiest way of determining the relative positions of places on different maps, and for plotting compass directions.

Name forms

For ease of reference, both English and local name forms appear in the atlas. Oceans, seas and countries are shown in English throughout the atlas; country names may be abbreviated to their commonly accepted form (for example, Germany, not The Federal Republic of Germany). Conventional English forms are also used for place names on the smaller-scale maps of the continents. However, local name forms are used on all large-scale and regional maps, with the English form given in brackets only for important cities – the large-scale map of Russia and Central Asia thus shows Moskva (Moscow). For countries which do not use a Roman script, place names have been transcribed according to the systems adopted by the British and US Geographic Names Authorities. For China, the Pin Yin system has been used, with some more widely known forms appearing in brackets, as with Beijing (Peking). Both English and local names appear in the index, the English form being cross-referenced to the local form.

THE
WORLD
IN FOCUS

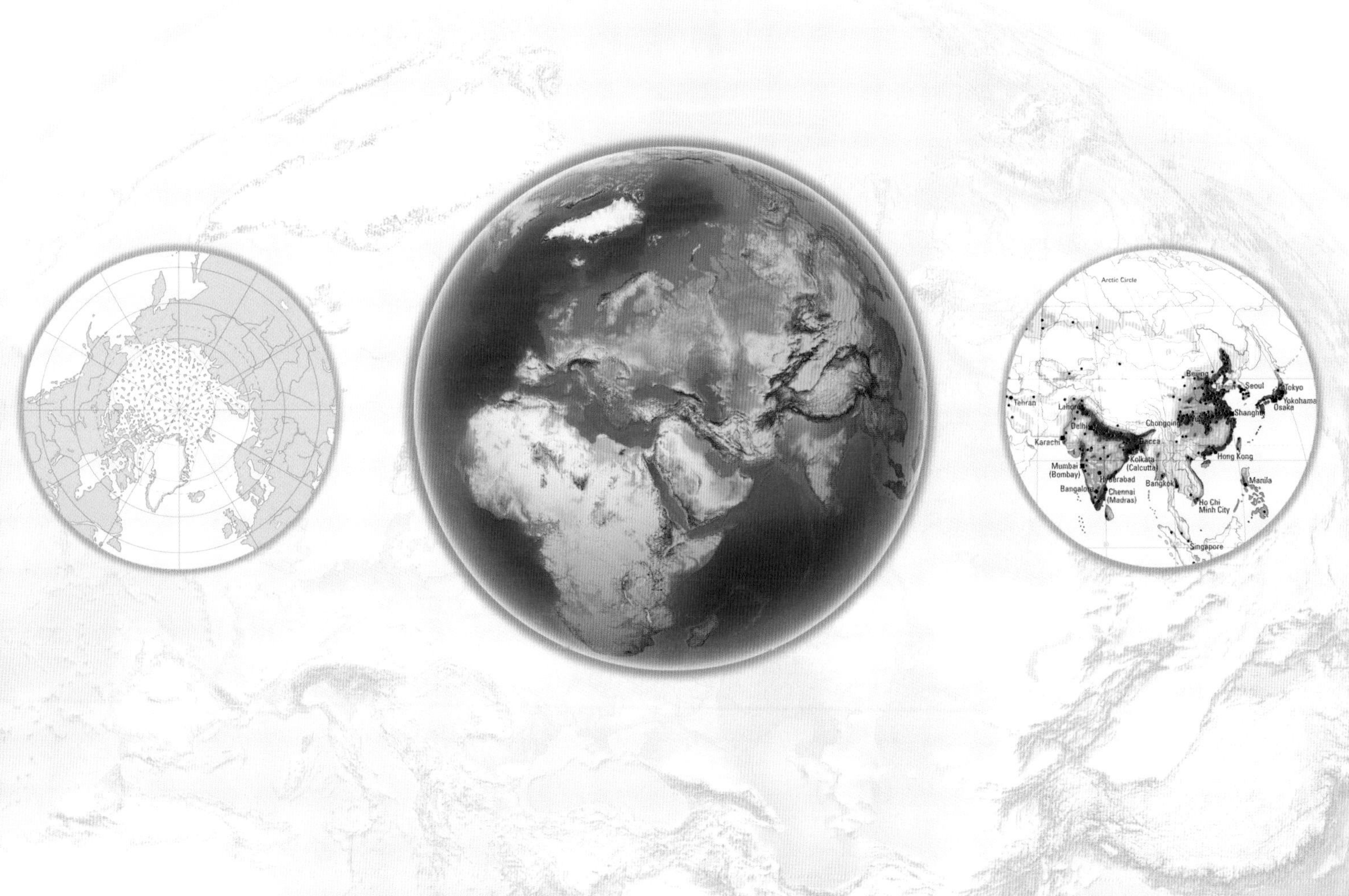

Planet Earth

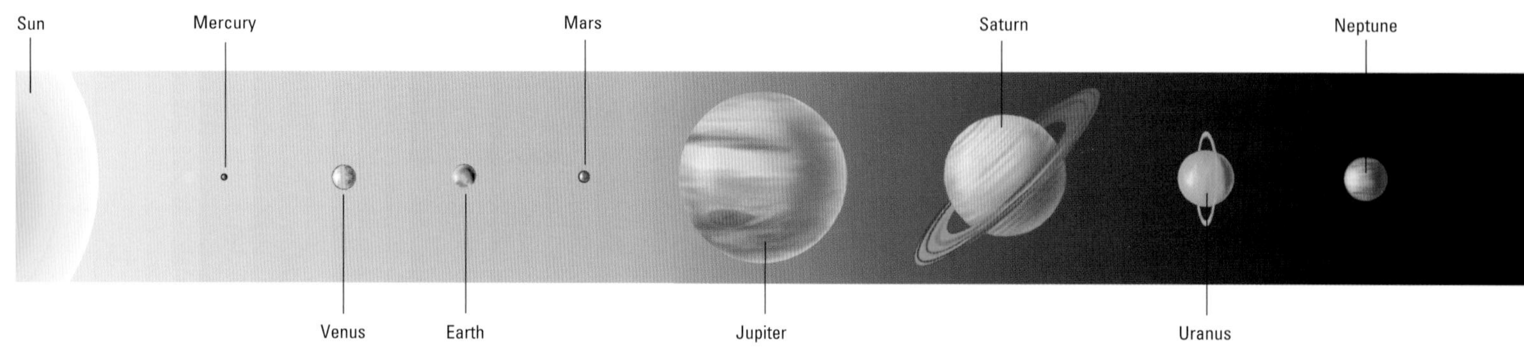

The Solar System

A minute part of one of the billions of galaxies (collections of stars) that populate the Universe, the Solar System lies about 26,000 light-years from the centre of our own Galaxy, the 'Milky Way'. Thought to be about 5 billion years old, it consists of a central Sun with eight planets and their moons revolving around it, attracted by its gravitational pull. The planets orbit the Sun in the same direction – anti-clockwise when viewed from above the Sun's north pole – and almost in the same plane. Their orbital distances, however, vary enormously.

The Sun's diameter is 109 times that of the Earth, and the temperature at its core – caused by continuous thermonuclear fusions of hydrogen into helium – is estimated to be 15 million degrees Celsius. It is the Solar System's only source of light and heat.

Profile of the Planets

	Mean distance from Sun (million km)	Mass (Earth = 1)	Period of orbit (Earth days/years)	Period of rotation (Earth days)	Equatorial diameter (km)	Number of known satellites*
Mercury	57.9	0.06	87.97 days	58.65	4,879	0
Venus	108.2	0.82	224.7 days	243.02	12,104	0
Earth	149.6	1.00	365.3 days	1.00	12,756	1
Mars	227.9	0.11	687.0 days	1.029	6,792	2
Jupiter	778	317.8	11.86 years	0.411	142,984	63
Saturn	1,427	95.2	29.45 years	0.428	120,536	62
Uranus	2,871	14.5	84.02 years	0.720	51,118	27
Neptune	4,498	17.2	164.8 years	0.673	49,528	13

* Number of known satellites at mid-2010

All planetary orbits are elliptical in form, but only Mercury follows a path that deviates noticeably from a circular one. In 2006, Pluto was demoted from its former status as a planet and is now regarded as a member of the Kuiper Belt of icy bodies at the fringes of the Solar System.

The Seasons

Seasons occur because the Earth's axis is tilted at an angle of approximately 23½°. When the northern hemisphere is tilted to a maximum extent towards the Sun, on 21 June, the Sun is overhead at the Tropic of Cancer (latitude 23½° North). This is midsummer, or the summer solstice, in the northern hemisphere.

On 22 or 23 September, the Sun is overhead at the Equator, and day and night are of equal length throughout the world. This is the autumnal equinox in the northern hemisphere. On 21 or 22 December, the Sun is overhead at the Tropic of Capricorn (23½° South), the winter solstice in the northern hemisphere. The overhead Sun then tracks north until, on 21 March, it is overhead at the Equator. This is the spring (vernal) equinox in the northern hemisphere.

In the southern hemisphere, the seasons are the reverse of those in the north.

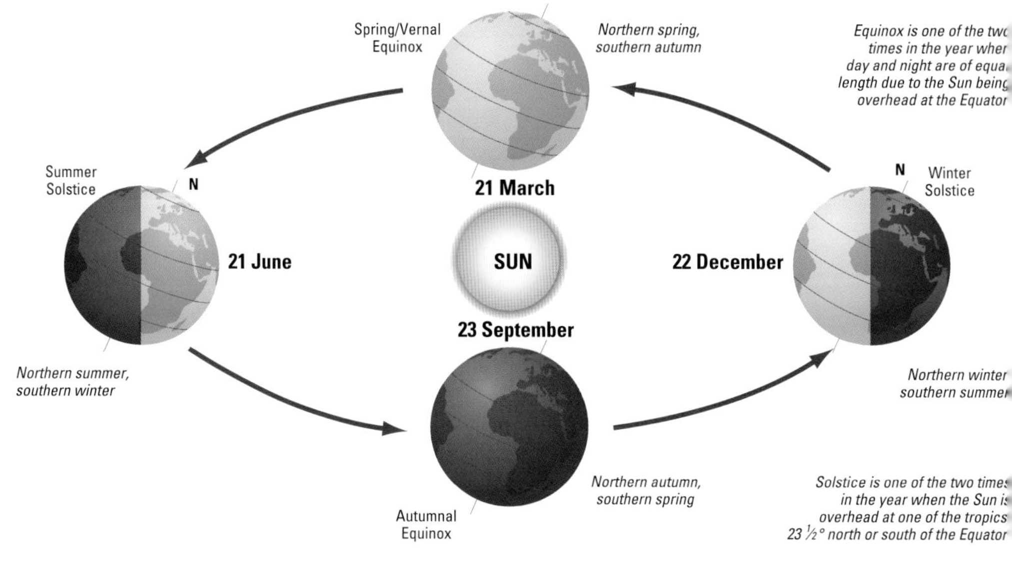

Day and Night

The Sun appears to rise in the east, reach its highest point at noon, and then set in the west, to be followed by night. In reality, it is not the Sun that is moving but the Earth rotating from west to east. The moment when the Sun's upper limb first appears above the horizon is termed sunrise; the moment when the Sun's upper limb disappears below the horizon is sunset.

At the summer solstice in the northern hemisphere (21 June), the Arctic has total daylight and the Antarctic total darkness. The opposite occurs at the winter solstice (21 or 22 December). At the Equator, the length of day and night are almost equal all year.

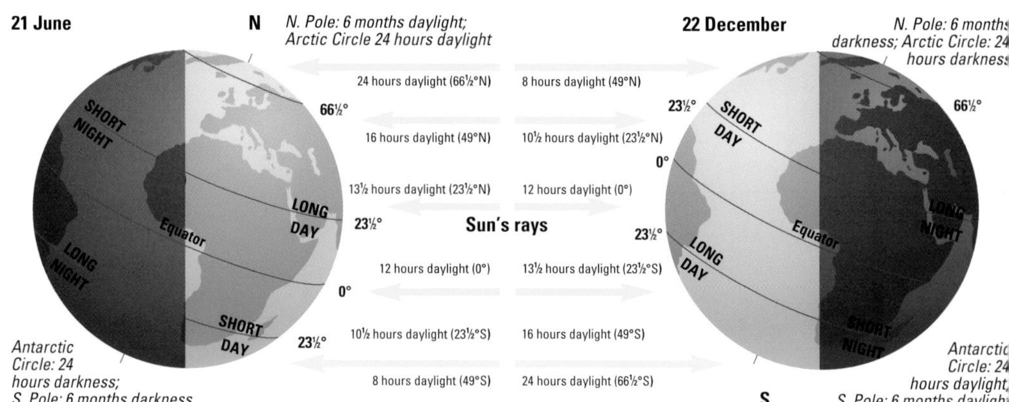

Time

Year: The time taken by the Earth to revolve around the Sun, or 365.24 days.

Leap Year: A calendar year of 366 days, 29 February being the additional day. It offsets the difference between the calendar and the solar year.

Month: The 12 calendar months of the year are approximately equal in length to a lunar month.

Week: An artificial period of 7 days, not based on astronomical time.

Day: The time taken by the Earth to complete one rotation on its axis.

Hour: 24 hours make one day. The day is divided into hours a.m. (ante meridiem or before noon) and p.m. (post meridiem or after noon), although most timetables now use the 24-hour system, from midnight to midnight.

Sunrise

Spring Equinox   Autumnal Equinox

Hours AM / Latitude

60°N / 40°N / 20°N / 0°(Equator) / 20°S / 40°S / 60°S

Months of the year
J F M A M J J A S O N D

Sunset

Spring Equinox   Autumnal Equinox

Hours PM / Latitude

60°S / 40°S / 20°S / 0°(Equator) / 20°N / 40°N / 60°N

Months of the year
J F M A M J J A S O N D

The Moon

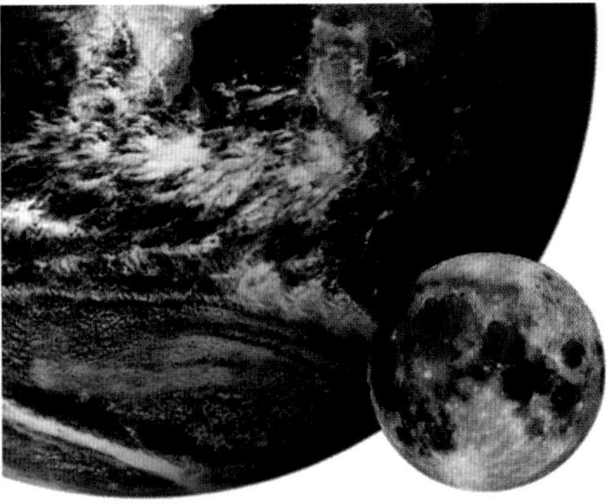

The Moon rotates more slowly than the Earth, taking just over 27 days to make one complete rotation on its axis. This corresponds to the Moon's orbital period around the Earth, and therefore the Moon always presents the same hemisphere towards us; some 41% of the Moon's far side is never visible from the Earth. The interval between one New Moon and the next is 29½ days – this is called a lunation, or lunar month.

The Moon shines only by reflected sunlight, and emits no light of its own. During each lunation the Moon displays a complete cycle of phases, caused by the changing angle of illumination from the Sun.

Phases of the Moon

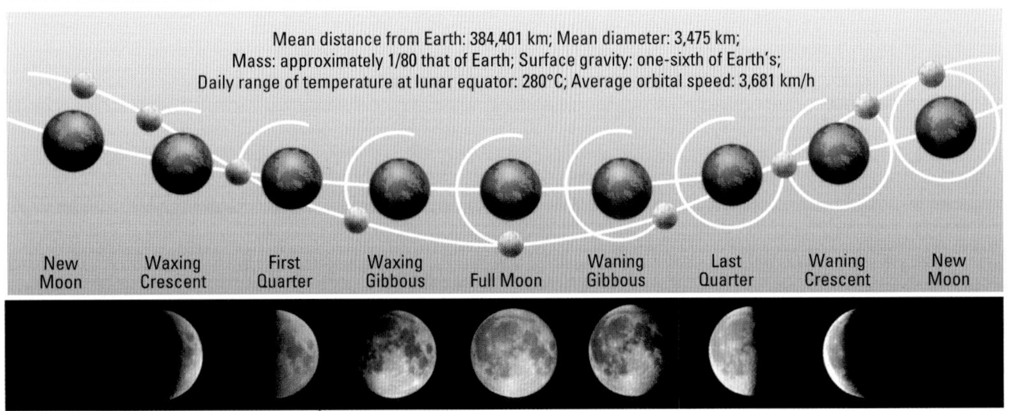

Mean distance from Earth: 384,401 km; Mean diameter: 3,475 km; Mass: approximately 1/80 that of Earth; Surface gravity: one-sixth of Earth's; Daily range of temperature at lunar equator: 280°C; Average orbital speed: 3,681 km/h

| New Moon | Waxing Crescent | First Quarter | Waxing Gibbous | Full Moon | Waning Gibbous | Last Quarter | Waning Crescent | New Moon |

Eclipses

When the Moon passes between the Sun and the Earth, the Sun becomes partially eclipsed (1). A partial eclipse becomes a total eclipse if the Moon proceeds to cover the Sun completely (2) and the dark central part of the lunar shadow touches the Earth. The broad geographical zone covered by the Moon's outer shadow (P), has only a very small central area (often less than 100 km wide) that experiences totality. Totality can never last for more than 7½ minutes at maximum, but is usually much briefer than this. Lunar eclipses take place when the Moon moves through the shadow of the Earth, and can be partial or total. Any single location on Earth can experience a maximum of four solar and three lunar eclipses in any single year, while a total solar eclipse occurs an average of once every 360 years for any given location.

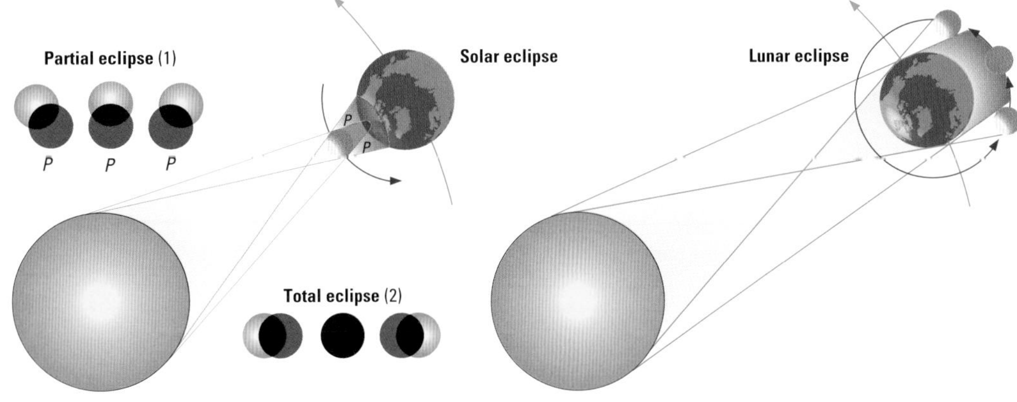

Partial eclipse (1)

Solar eclipse

Lunar eclipse

Total eclipse (2)

Tides

The daily rise and fall of the ocean's tides are the result of the gravitational pull of the Moon and that of the Sun, though the effect of the latter is not as strong as that of the Moon. This effect is greatest on the hemisphere facing the Moon and causes a tidal 'bulge'.

Spring tides occur when the Sun, Earth and Moon are aligned; high tides are at their highest, and low tides fall to their lowest. When the Moon and Sun are furthest out of line (near the Moon's First and Last Quarters), neap tides occur, producing the smallest range between high and low tides.

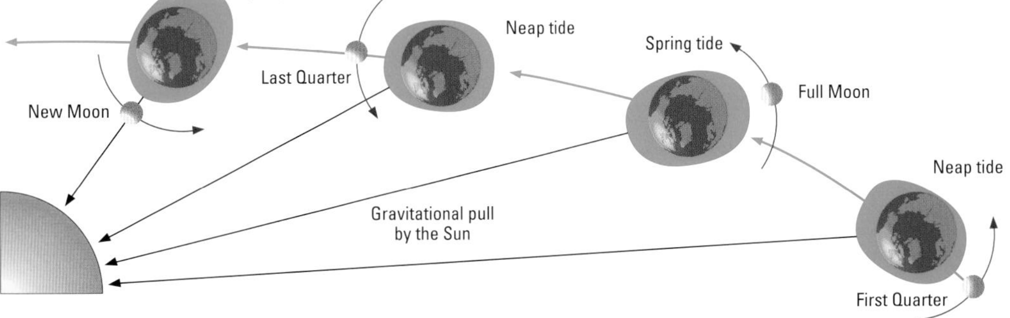

Spring tide

Neap tide

Last Quarter

Spring tide

New Moon

Full Moon

Neap tide

Gravitational pull by the Sun

First Quarter

Restless Earth

The Earth's Structure

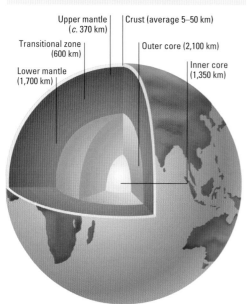

Upper mantle (c. 370 km)
Crust (average 5–50 km)
Transitional zone (600 km)
Outer core (2,100 km)
Lower mantle (1,700 km)
Inner core (1,350 km)

Continental Drift

About 200 million years ago the original Pangaea landmass began to split into two continental groups, which further separated over time to produce the present-day configuration.

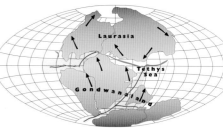

180 million years ago

135 million years ago

Present day

Trench
Rift
New ocean floor
Zones of slippage

Notable Earthquakes Since 1900

Year	Location	Richter Scale	Deaths
1906	San Francisco, USA	8.3	3,000
1906	Valparaiso, Chile	8.6	22,000
1908	Messina, Italy	7.5	83,000
1915	Avezzano, Italy	7.5	30,000
1920	Gansu (Kansu), China	8.6	180,000
1923	Yokohama, Japan	8.3	143,000
1927	Nan Shan, China	8.3	200,000
1932	Gansu (Kansu), China	7.6	70,000
1933	Sanriku, Japan	8.9	2,990
1934	Bihar, India/Nepal	8.4	10,700
1935	Quetta, India (now Pakistan)	7.5	60,000
1939	Chillan, Chile	8.3	28,000
1939	Erzincan, Turkey	7.9	30,000
1960	S. W. Chile	9.5	2,200
1960	Agadir, Morocco	5.8	12,000
1962	Khorasan, Iran	7.1	12,230
1964	Anchorage, USA	9.2	125
1968	N. E. Iran	7.4	12,000
1970	N. Peru	7.8	70,000
1972	Managua, Nicaragua	6.2	5,000
1974	N. Pakistan	6.3	5,200
1976	Guatemala	7.5	22,500
1976	Tangshan, China	8.2	255,000
1978	Tabas, Iran	7.7	25,000
1980	El Asnam, Algeria	7.3	20,000
1980	S. Italy	7.2	4,800
1985	Mexico City, Mexico	8.1	4,200
1988	N.W. Armenia	6.8	55,000
1990	N. Iran	7.7	36,000
1993	Maharashtra, India	6.4	30,000
1994	Los Angeles, USA	6.6	51
1995	Kobe, Japan	7.2	5,000
1995	Sakhalin Is., Russia	7.5	2,000
1997	N. E. Iran	7.1	2,400
1998	Takhar, Afghanistan	6.1	4,200
1998	Rostaq, Afghanistan	7.0	5,000
1999	Izmit, Turkey	7.4	15,000
1999	Taipei, Taiwan	7.6	1,700
2001	Gujarat, India	7.7	14,000
2002	Baghlan, Afghanistan	6.1	1,000
2003	Boumerdes, Algeria	6.8	2,200
2003	Bam, Iran	6.6	30,000
2004	Sumatra, Indonesia	9.0	250,000
2005	N. Pakistan	7.6	74,000
2006	Java, Indonesia	6.4	6,200
2007	S. Peru	8.0	600
2008	Sichuan, China	7.9	70,000
2010	Haiti	7.0	230,000

Earthquakes

Earthquake magnitude is usually rated according to either the Richter or the Modified Mercalli scale, both devised by seismologists in the 1930s. The Richter scale measures absolute earthquake power with mathematical precision: each step upwards represents a tenfold increase in shockwave amplitude. Theoretically, there is no upper limit, but most of the largest earthquakes measured have been rated at between 8.8 and 8.9. The 12–point Mercalli scale, based on observed effects, is often more meaningful, ranging from I (earthquakes noticed only by seismographs) to XII (total destruction); intermediate points include V (people awakened at night; unstable objects overturned), VII (collapse of ordinary buildings; chimneys and monuments fall), and IX (conspicuous cracks in ground; serious damage to reservoirs).

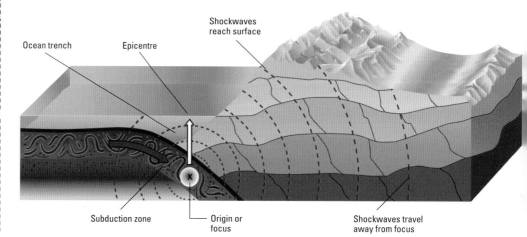

Shockwaves reach surface
Ocean trench
Epicentre
Subduction zone
Origin or focus
Shockwaves travel away from focus

Structure and Earthquakes

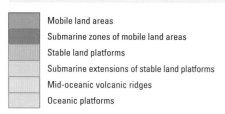

Mobile land areas
Submarine zones of mobile land areas
Stable land platforms
Submarine extensions of stable land platforms
Mid-oceanic volcanic ridges
Oceanic platforms

1976 ○ Principal earthquakes and dates (since 1900)

Earthquakes are a series of rapid vibrations originating from the slipping or faulting of parts of the Earth's crust when stresses within build up to breaking point. They usually happen at depths varying from 8 km to 30 km. Severe earthquakes cause extensive damage when they take place in populated areas, destroying structures and severing communications. Most initial loss of life occurs due to secondary causes such as falling masonry, fires and flooding.

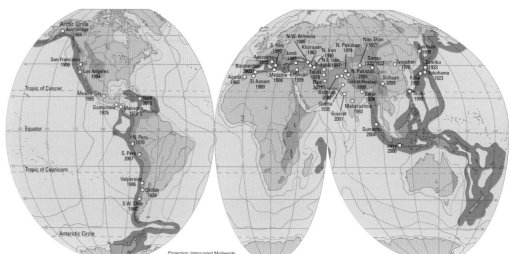

Projection: Interrupted Mollweide

Plate Tectonics

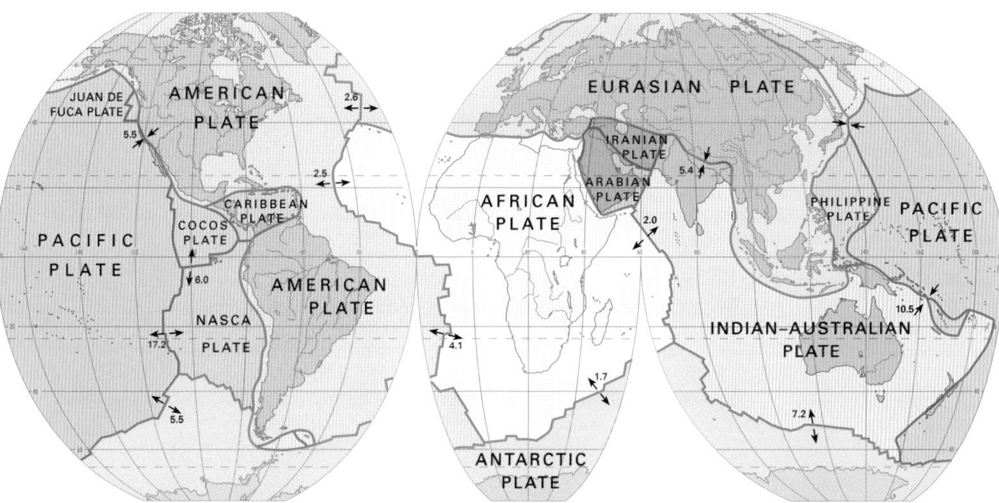

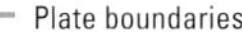

The drifting of the continents is a feature that is unique to planet Earth. The complementary, almost jigsaw-puzzle fit of the coastlines on each side of the Atlantic Ocean inspired Alfred Wegener's theory of continental drift in 1915. The theory suggested that the ancient supercontinent, which Wegener named Pangaea, incorporated all of the Earth's landmasses and gradually split up to form today's continents.

The original debate about continental drift was a prelude to a more radical idea: plate tectonics. The basic theory is that the Earth's crust is made up of a series of rigid plates which float on a soft layer of the mantle and are moved about by continental convection currents within the Earth's interior. These plates diverge and converge along margins marked by seismic activity. Plates diverge from mid-ocean ridges where molten lava pushes upwards and forces the plates apart at rates of up to 40 mm [1.6 in] a year.

The three diagrams, left, give some examples of plate boundaries from around the world. Diagram (a) shows sea-floor spreading at the Mid-Atlantic Ridge as the American and African plates slowly diverge. The same thing is happening in (b) where sea-floor spreading at the Mid-Indian Ocean Ridge is forcing the Indian–Australian plate to collide into the Eurasian plate. In (c) oceanic crust (sima) is being subducted beneath lighter continental crust (sial).

Volcanoes

Volcanoes occur when hot liquefied rock beneath the Earth's crust is pushed up by pressure to the surface as molten lava. Some volcanoes erupt in an explosive way, throwing out rocks and ash, whilst others are effusive and lava flows out of the vent. There are volcanoes which are both, such as Mount Fuji. An accumulation of lava and cinders creates cones of variable size and shape. As a result of many eruptions over centuries, Mount Etna in Sicily has a circumference of more than 120 km [75 miles].

Climatologists believe that volcanic ash, if ejected high into the atmosphere, can influence temperature and weather for several years afterwards. The 1991 eruption of Mount Pinatubo in the Philippines ejected more than 20 million tonnes of dust and ash 32 km [20 miles] into the atmosphere and is believed to have accelerated ozone depletion over a large part of the globe.

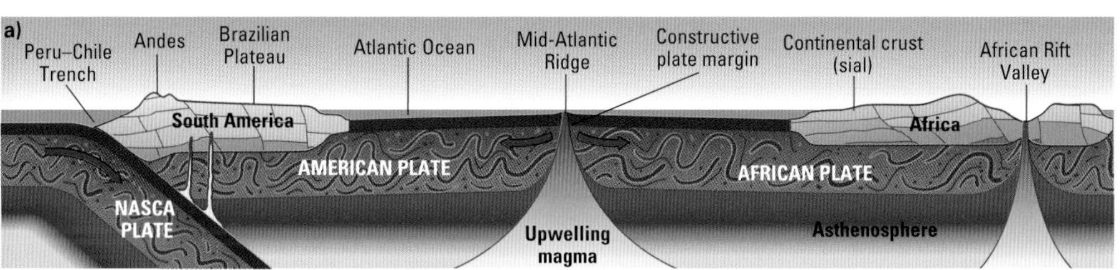

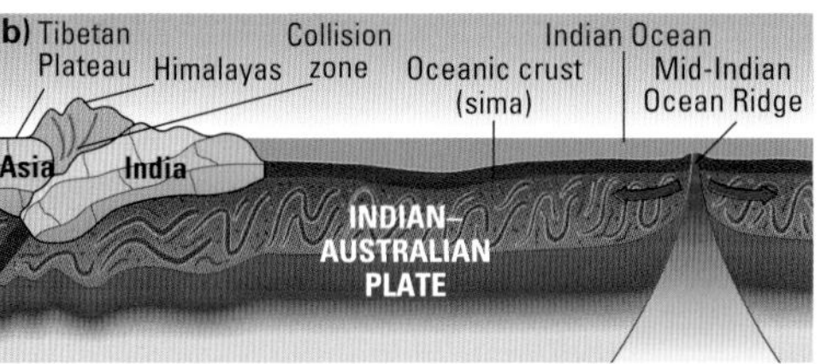

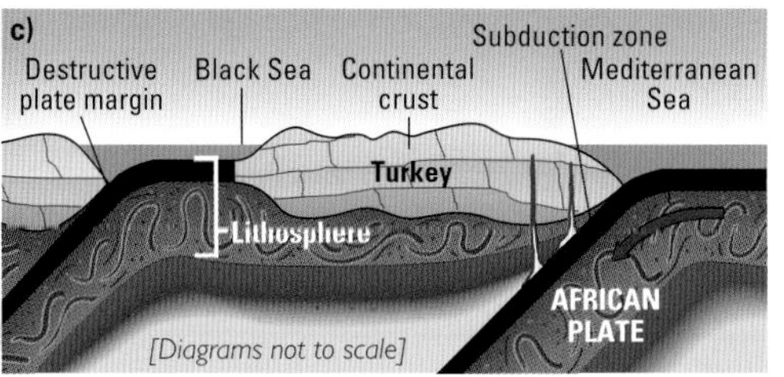

[Diagrams not to scale]

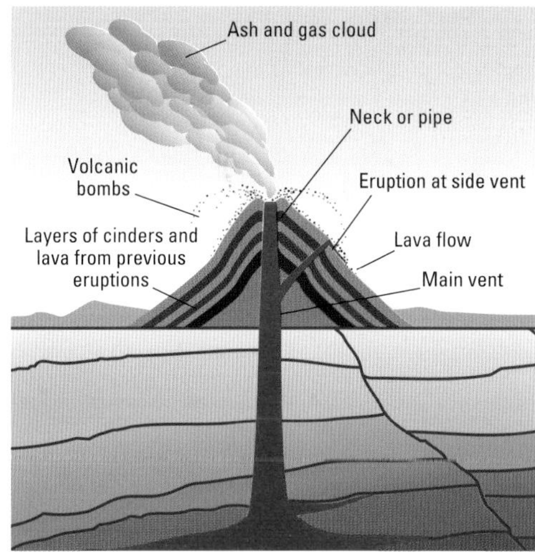

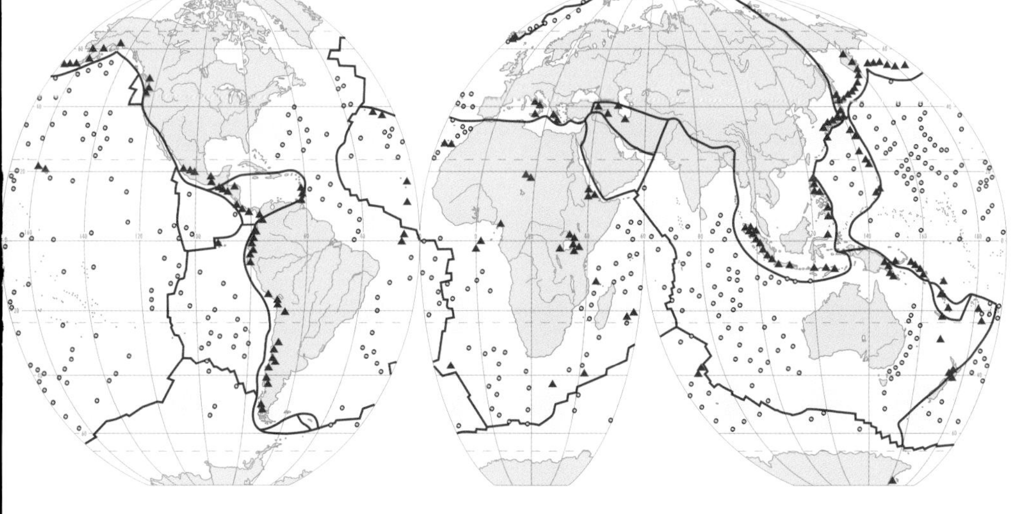

Distribution of Volcanoes

Volcanoes today may be the subject of considerable scientific study but they remain both dramatic and unpredictable: in 1991 Mount Pinatubo, 100 km [62 miles] north of the Philippines capital Manila, suddenly burst into life after lying dormant for more than six centuries. Most of the world's active volcanoes occur in a belt around the Pacific Ocean, on the edge of the Pacific plate, called the 'ring of fire'. Indonesia has the greatest concentration with 90 volcanoes, 12 of which are active. The most famous, Krakatoa, erupted in 1883 with such force that the resulting tidal wave killed 36,000 people, and tremors were felt as far away as Australia.

⊙ Submarine volcanoes

▲ Land volcanoes active since 1700

— Boundaries of tectonic plates

Landforms

The Rock Cycle

James Hutton first proposed the rock cycle in the late 1700s after he observed the slow but steady effects of erosion.

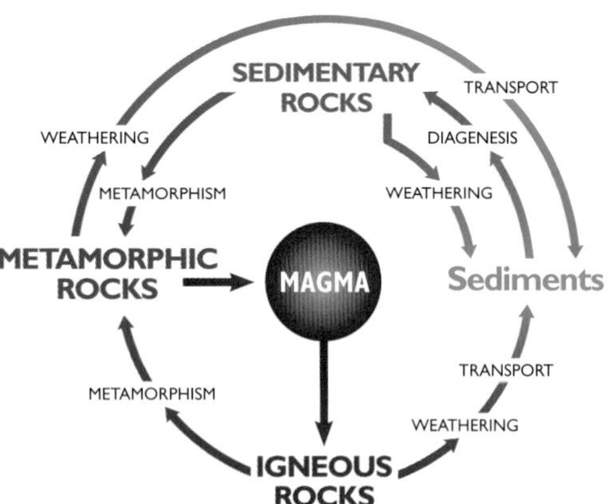

Above and below the surface of the oceans, the features of the Earth's crust are constantly changing. The phenomenal forces generated by convection currents in the molten core of our planet carry the vast segments or 'plates' of the crust across the globe in an endless cycle of creation and destruction. A continent may travel little more than 25 mm [1 in] per year, yet in the vast span of geological time this process throws up giant mountain ranges and creates new land.

Destruction of the landscape, however, begins as soon as it is formed. Wind, water, ice and sea, the main agents of erosion, mount a constant assault that even the most resistant rocks cannot withstand. Mountain peaks may dwindle by as little as a few millimetres each year, but if they are not uplifted by further movements of the crust they will eventually be reduced to rubble and transported away.

Water is the most powerful agent of erosion – it has been estimated that 100 billion tonnes of sediment are washed into the oceans every year.

Three Asian rivers account for 20% of this total: the Huang He, in China, and the Brahmaputra and the Ganges in Bangladesh.

Rivers and glaciers, like the sea itself, generate much of their effect through abrasion – pounding the land with the debris they carry with them. But as well as destroying they also create new landforms, many of them spectacular: vast deltas like those of the Mississippi and the Nile, or the deep fjords cut by glaciers in British Columbia, Norway and New Zealand.

Geologists once considered that landscapes evolved from 'young', newly uplifted mountainous areas, through a 'mature' hilly stage, to an 'old age' stage when the land was reduced to an almost flat plain, or peneplain. This theory, called the 'cycle of erosion', fell into disuse when it became evident that so many factors, including the effects of plate tectonics and climatic change, constantly interrupt the cycle, which takes no account of the highly complex interactions that shape the surface of our planet.

Mountain Building

Mountains are formed when pressures on the Earth's crust caused by continental drift become so intense that the surface buckles or cracks. This happens where oceanic crust is subducted by continental crust or, more dramatically, where two tectonic plates collide: the Rockies, Andes, Alps, Urals and Himalayas resulted from such impacts. These are all known as fold mountains because they were formed by the compression of the rocks, forcing the surface to bend and fold like a crumpled rug. The Himalayas were formed from the folded former sediments of the Tethys Sea, which was trapped in the collision zone between the Indian and Eurasian plates.

The other main mountain-building process occurs when the crust fractures to create faults, allowing rock to be forced upwards in large blocks; or when the pressure of magma within the crust forces the surface to bulge into a dome, or erupts to form a volcano. Large mountain ranges may reveal a combination of these features; the Alps, for example, have been compressed so violently that the folds are fragmented by numerous faults and intrusions of molten igneous rock.

Over millions of years, even the greatest mountain ranges can be reduced by the agents of erosion (most notably rivers) to a low rugged landscape known as a peneplain.

Types of faults: Faults occur where the crust is being stretched or compressed so violently that the rock strata break in a horizontal or vertical movement. They are classified by the direction in which the blocks of rock have moved. A normal fault results when a vertical movement causes the surface to break apart; compression causes a reverse fault. Horizontal movement causes shearing, known as a strike-slip fault. When the rock breaks in two places, the central block may be pushed up in a horst fault, or sink (creating a rift valley) in a graben fault.

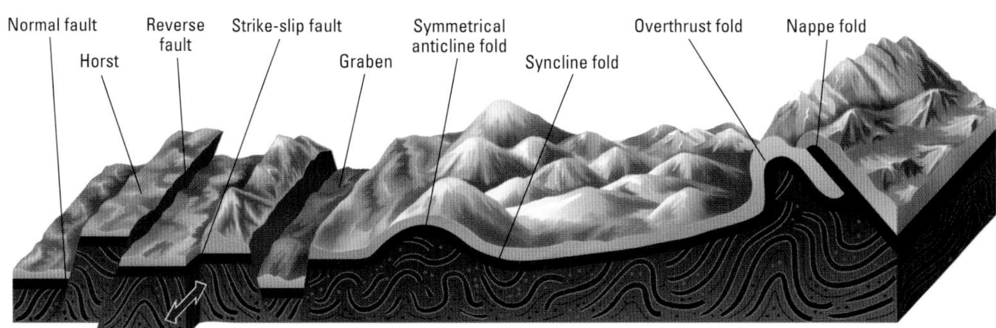

Types of fold: Folds occur when rock strata are squeezed and compressed. They are common, therefore, at destructive plate margins and where plates have collided, forcing the rocks to buckle into mountain ranges. Geographers give different names to the degrees of fold that result from continuing pressure on the rock. A simple fold may be symmetric, with even slopes on either side, but as the pressure builds up, one slope becomes steeper and the fold becomes asymmetric. Later, the ridge or 'anticline' at the top of the fold may slide over the lower ground or 'syncline' to form a recumbent fold. Eventually, the rock strata may break under the pressure to form an overthrust and finally a nappe fold.

Continental Glaciation

Ice sheets were at their greatest extent about 200,000 years ago. The maximum advance of the last Ice Age was about 18,000 years ago, when ice covered virtually all of Canada and reached as far south as the Bristol Channel in Britain.

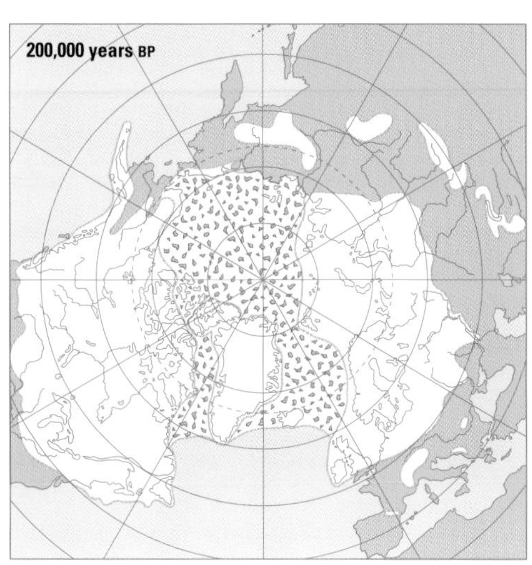

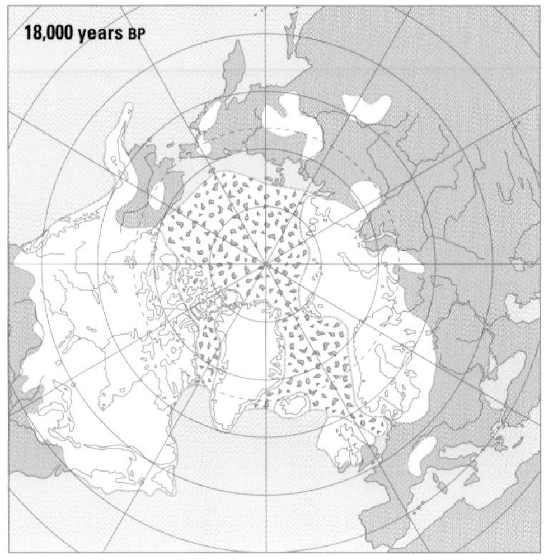

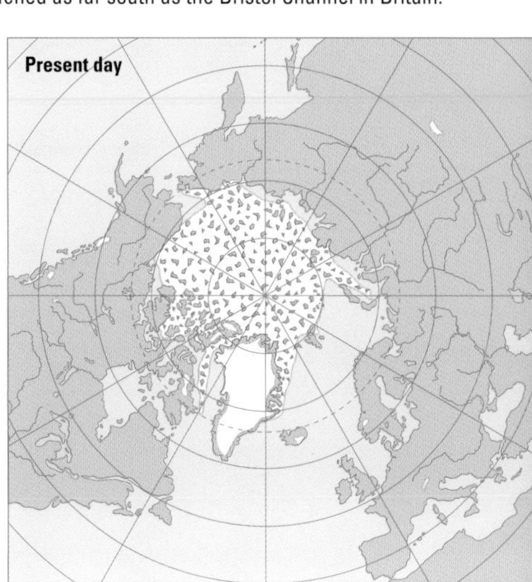

6

Natural Landforms

A stylized diagram to show some of the major natural landforms found in the mid-latitudes.

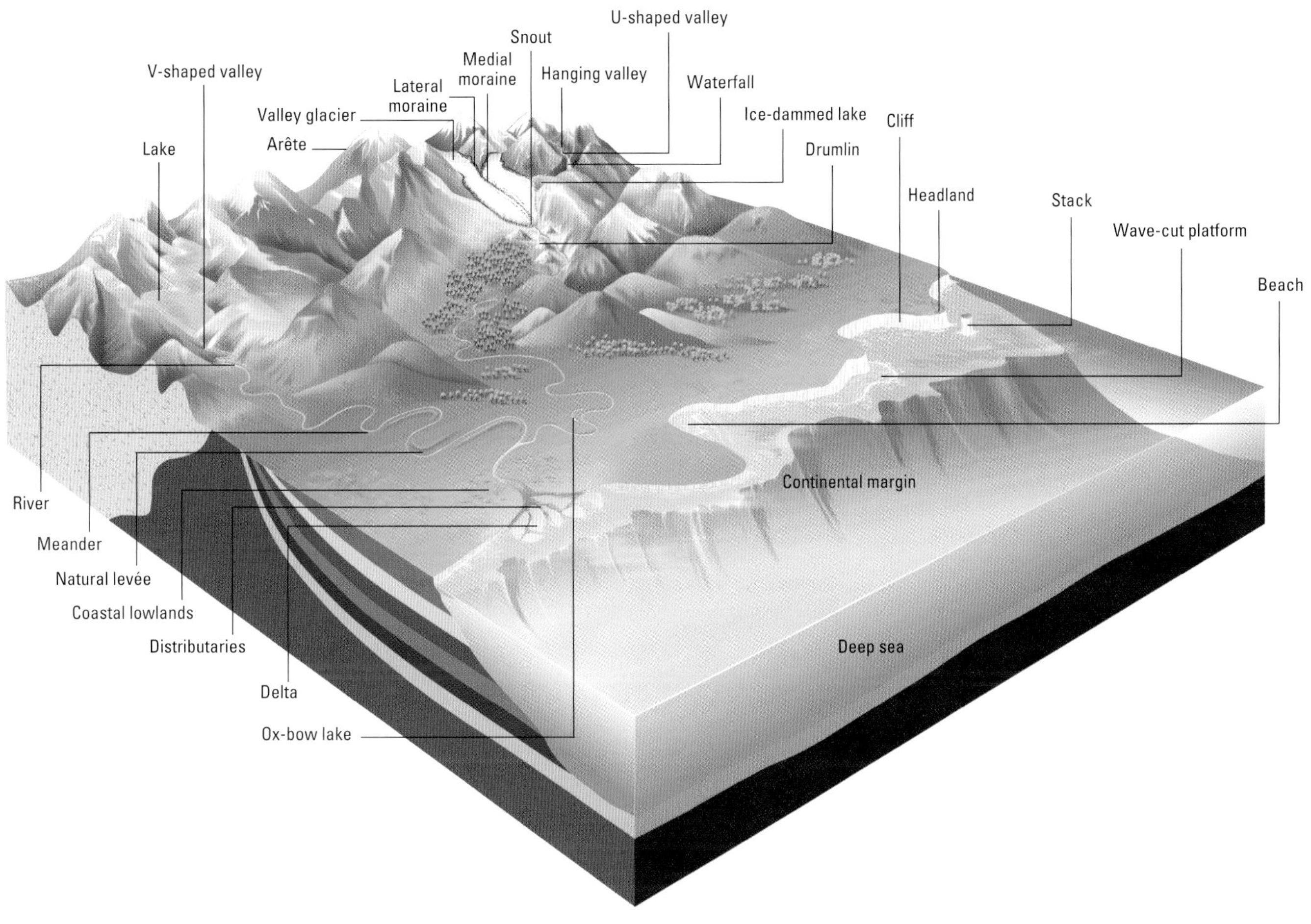

Desert Landscapes

The popular image that deserts are all huge expanses of sand is wrong. Despite harsh conditions, deserts contain some of the most varied and interesting landscapes in the world. They are also one of the most extensive environments – the hot and cold deserts together cover almost 40% of the Earth's surface.

The three types of hot desert are known by their Arabic names: sand desert, called *erg*, covers only about one-fifth of the world's desert; the rest is divided between *hammada* (areas of bare rock) and *reg* (broad plains covered by loose gravel or pebbles).

In areas of *erg*, such as the Namib Desert, the shape of the dunes reflects the character of local winds. Where winds are constant in direction, crescent-shaped *barchan* dunes form. In areas of bare rock, wind-blown sand is a major agent of erosion. The erosion is mainly confined to within 2 m [6.5 ft] of the surface, producing characteristic mushroom-shaped rocks.

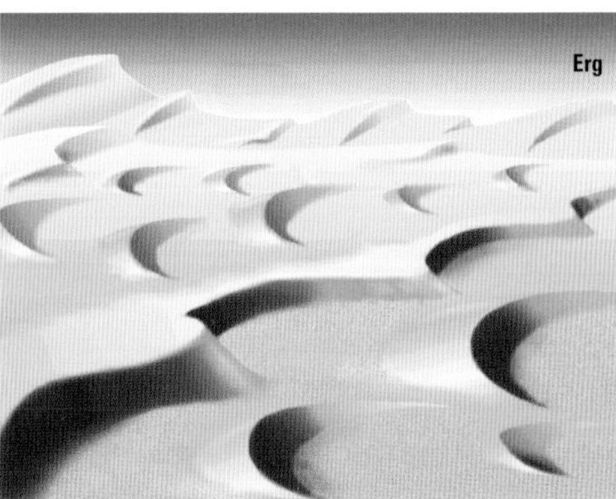
Erg

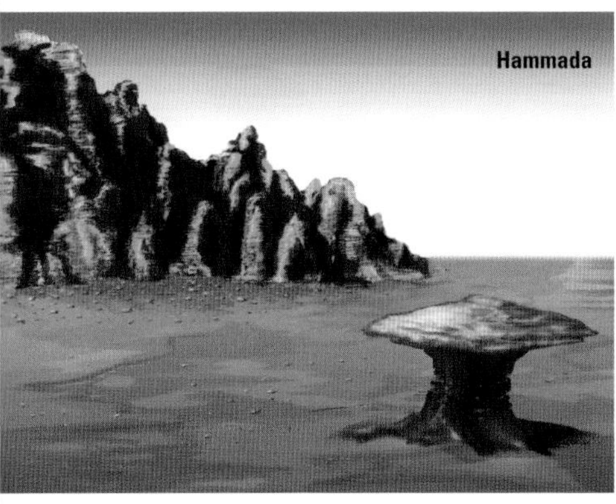
Hammada

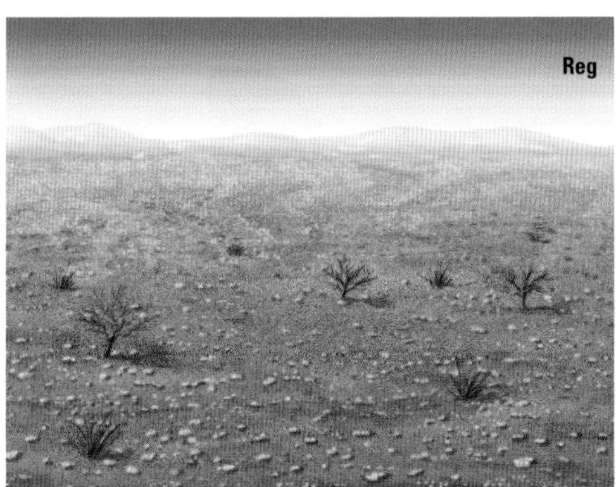
Reg

Surface Processes

Catastrophic changes to natural landforms are periodically caused by such phenomena as avalanches, landslides and volcanic eruptions, but most of the processes that shape the Earth's surface operate extremely slowly in human terms. One estimate, based on a study in the United States, suggested that 1 m [3 ft] of land was removed from the entire surface of the country, on average, every 29,500 years. However, the time-scale varies from 1,300 years to 154,200 years depending on the terrain and climate.

In hot, dry climates, mechanical weathering, a result of rapid temperature changes, causes the outer layers of rock to peel away, while in cold mountainous regions, boulders are prised apart when water freezes in cracks in rocks. Chemical weathering, at its greatest in warm, humid regions, is responsible for hollowing out limestone caves and decomposing granites.

The erosion of soil and rock is greatest on sloping land and the steeper the slope, the greater the tendency for mass wasting – the movement of soil and rock downhill under the influence of gravity. The mechanisms of mass wasting (ranging from very slow to very rapid) vary with the type of material, but the presence of water as a lubricant is usually an important factor.

Running water is the world's leading agent of erosion and transportation. The energy of a river depends on several factors, including its velocity and volume, and its erosive power is at its peak when it is in full flood. Sea waves also exert tremendous erosive power during storms when they hurl pebbles against the shore, undercutting cliffs and hollowing out caves.

Glacier ice forms in mountain hollows and spills out to form valley glaciers, which transport rocks shattered by frost action. As glaciers move, rocks embedded into the ice erode steep-sided, U-shaped valleys. Evidence of glaciation in mountain regions includes cirques, knife-edged ridges, or arêtes, and pyramidal peaks.

Oceans

The Great Oceans

Relative sizes of the world's oceans

Pacific
Atlantic
Indian
Southern
Arctic

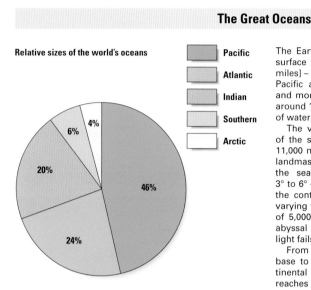

4%
6%
20%
46%
24%

From ancient times to about the 15th century, the legendary 'Seven Seas' comprised the Red Sea, Mediterranean Sea, Persian Gulf, Black Sea, Adriatic Sea, Caspian Sea and Indian Sea.

The Earth is a watery planet: more than 70% of its surface – over 360,000,000 sq km [140,000,000 sq miles] – is covered by the oceans and seas. The mighty Pacific alone accounts for nearly 36% of the total, and more than 46% of the sea area. Gravity holds in around 1,400 million cubic km [320 million cubic miles] of water, of which over 97% is saline.

The vast underwater world starts in the shallows of the seaside and plunges to depths of more than 11,000 m [36,000 ft]. The continental shelf, part of the landmass, drops gently to around 200 m [650 ft]; here the seabed falls away suddenly at an angle of 3° to 6° – the continental slope. The third stage, called the continental rise, is more gradual with gradients varying from 1 in 100 to 1 in 700. At an average depth of 5,000 m [16,500 ft] there begins the aptly-named abyssal plain – massive submarine depths where sunlight fails to penetrate and few creatures can survive.

From these plains rise volcanoes which, taken from base to top, rival and even surpass the tallest continental mountains in height. Mauna Kea, on Hawai'i, reaches a total of 10,203 m [33,400 ft], some 1,355 m [4,500 ft] higher than Mount Everest, though scarcely 40% is visible above sea level.

In addition, there are underwater mountain chains up to 1,000 km [600 miles] across, whose peaks sometimes appear above sea level as islands, such as Iceland and Tristan da Cunha.

The Ocean Depths

Average and maximum depths of the world's great oceans, in metres

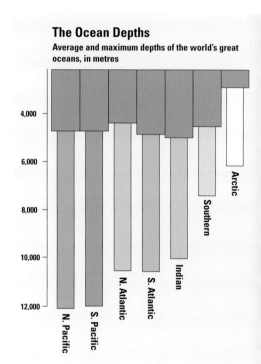

4,000
6,000
8,000
10,000
12,000

N. Pacific
S. Pacific
N. Atlantic
S. Atlantic
Indian
Southern
Arctic

Ocean Currents

January ocean currents (Northern Hemisphere: winter; Southern Hemisphere: summer)

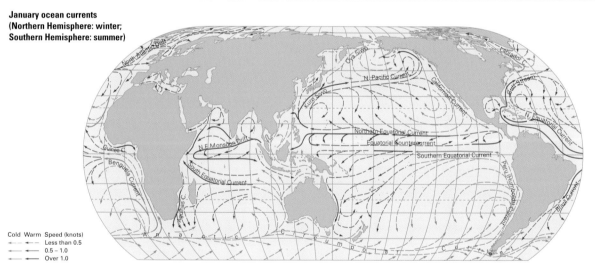

Cold Warm Speed (knots)
Less than 0.5
0.5 – 1.0
Over 1.0

July ocean currents (Northern Hemisphere: summer; Southern Hemisphere: winter)

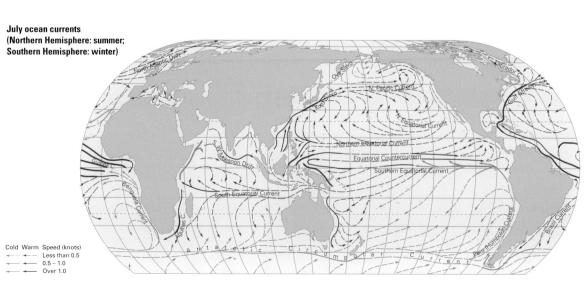

Cold Warm Speed (knots)
Less than 0.5
0.5 – 1.0
Over 1.0

Moving immense quantities of energy as well as billions of tonnes of water every hour, the ocean currents are a vital part of the great heat engine that drives the Earth's climate. They themselves are produced by a twofold mechanism. At the surface, winds push huge masses of water before them; in the deep ocean, below an abrupt temperature gradient that separates the churning surface waters from the still depths, density variations cause slow vertical movements.

The pattern of circulation of the great surface currents is determined by the displacement known as the Coriolis effect. As the Earth turns beneath a moving object – whether it is a tennis ball or a vast mass of water – it appears to be deflected to one side. The deflection is most obvious near the Equator, where the Earth's surface is spinning eastwards at 1,700 km/h [1,050 mph]; currents moving polewards are curved clockwise in the northern hemisphere and anti-clockwise in the southern.

The result is a system of spinning circles known as 'gyres'. The Coriolis effect piles up water on the left of each gyre, creating a narrow, fast-moving stream that is matched by a slower, broader returning current on the right. North and south of the Equator, the fastest currents are located in the west and in the east respectively. In each case, warm water moves from the Equator and cold water returns to it. Cold currents often bring an upwelling of nutrients with them, supporting the world's most economically important fisheries.

Depending on the prevailing winds, some currents on or near the Equator may reverse their direction in the course of the year – a seasonal variation on which Asian monsoon rains depend, and whose occasional failure can bring disaster to millions of people.

World Fishing Areas

Main commercial fishing areas (numbered FAO regions)

Catch by top marine fishing areas, million tonnes (2006)

1.	Pacific, NW	[61]	21.2	22.6%
2.	Pacific, SE	[87]	14.7	15.7%
3.	Pacific, WC	[71]	10.7	11.4%
4.	Atlantic, NE	[27]	9.7	10.3%
5.	Indian, E	[57]	5.6	6.0%
6.	Indian, W	[51]	4.4	4.7%
7.	Atlantic, EC	[34]	3.5	3.7%
8.	Pacific, NE	[67]	3.1	3.3%
9.	Atlantic, NW	[21]	2.2	2.3%
10.	Atlantic, SW	[41]	1.8	1.9%

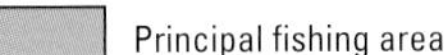

Principal fishing areas

Leading fishing nations

China 16.3% Peru 8% Indonesia 5.4% USA 5.3% Japan 4.7% India 4.4% Chile 4.2%

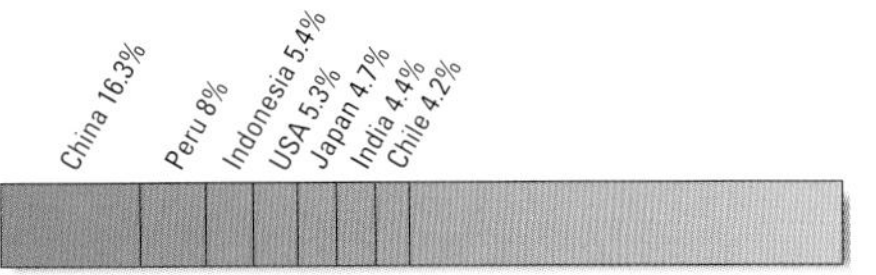

World total (2007): 90.1 million tonnes
(Marine catch 90.3% : Inland catch 9.7%)

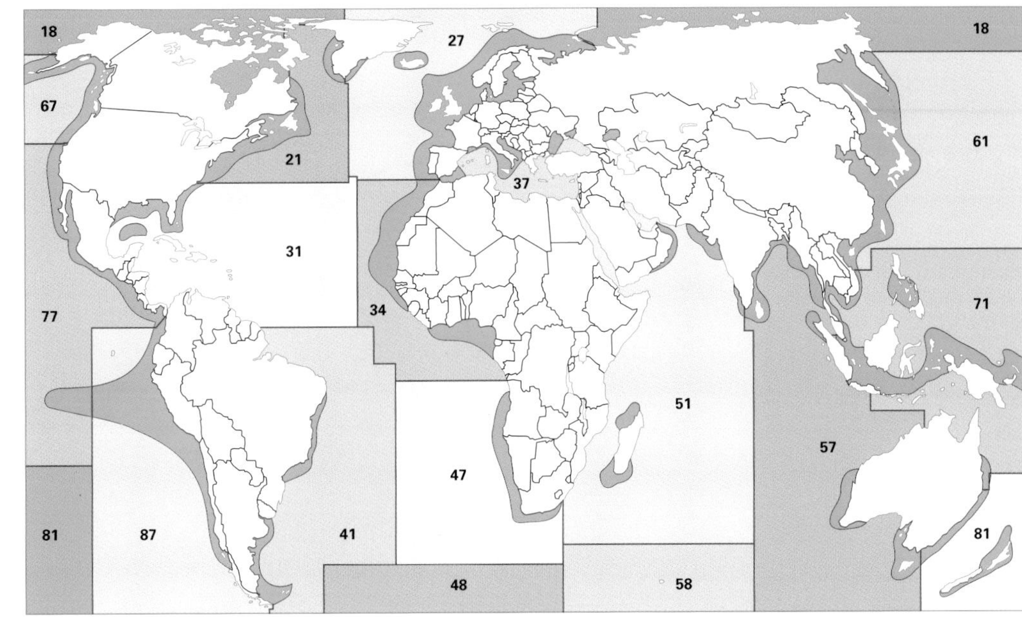

Marine Pollution

Sources of marine oil pollution

- Tanker operations
- Municipal wastes
- Tanker accidents
- Bilge and fuel oils
- Natural seeps
- Industrial waste
- Urban runoff
- Coastal oil refining
- Offshore oil rigs
- River runoffs
- Other

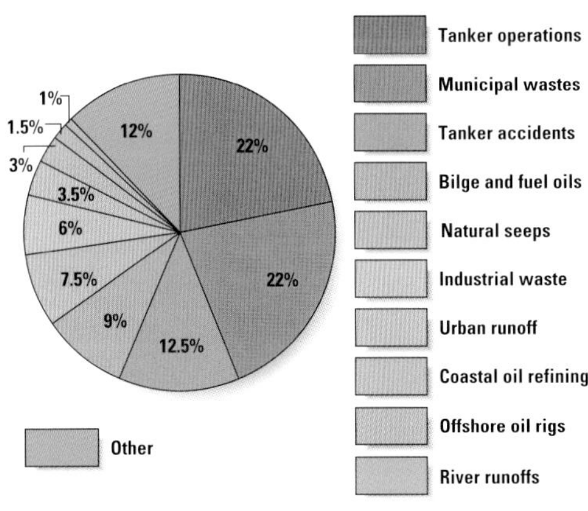

Oil Spills

Major oil spills from tankers and combined carriers

Year	Vessel	Location	Spill (barrels)*	Cause
1979	Atlantic Empress	West Indies	1,890,000	collision
1983	Castillo De Bellver	South Africa	1,760,000	fire
1978	Amoco Cadiz	France	1,628,000	grounding
1991	Haven	Italy	1,029,000	explosion
1988	Odyssey	Canada	1,000,000	fire
1967	Torrey Canyon	UK	909,000	grounding
1972	Sea Star	Gulf of Oman	902,250	collision
1977	Hawaiian Patriot	Hawaiian Is.	742,500	fire
1979	Independenta	Turkey	696,350	collision
1993	Braer	UK	625,000	grounding
1996	Sea Empress	UK	515,000	grounding
2002	Prestige	Spain	463,250	storm

Other sources of major oil spills

Year	Vessel	Location	Spill (barrels)	Cause
1983	Nowruz oilfield	Persian Gulf	4,250,000[†]	war
1979	Ixtoc 1 oilwell	Gulf of Mexico	4,200,000	blow-out
2010	Deepwater Horizon	Gulf of Mexico	3–8,700,000[†]	blow-out

* 1 barrel = 0.136 tonnes/159 lit./35 Imperial gal./42 US gal. [†] estimated

River Pollution

Sources of river pollution, USA

- Agriculture
- Mining
- Forestry
- Urban runoff
- Hydro-engineering
- Construction
- Land disposal
- Other

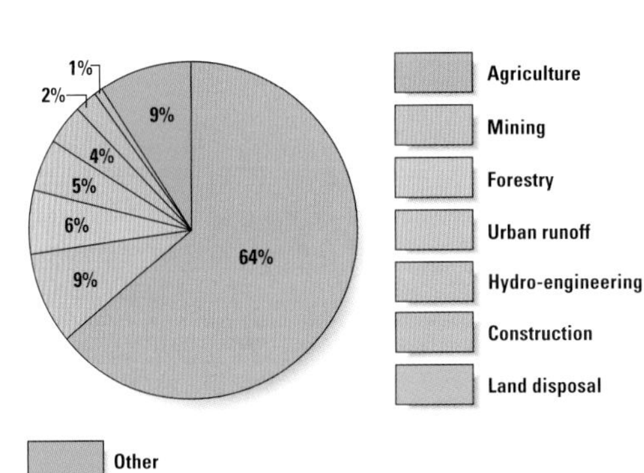

Water Pollution

- ■ Severely polluted sea areas and lakes
- ■ Polluted sea areas and lakes
- ■ Areas of frequent oil pollution by shipping
- ◣ Major oil tanker spills
- ▲ Major oil rig blow-outs
- ▼ Offshore dumpsites for industrial and municipal waste
- ── Severely polluted rivers and estuaries

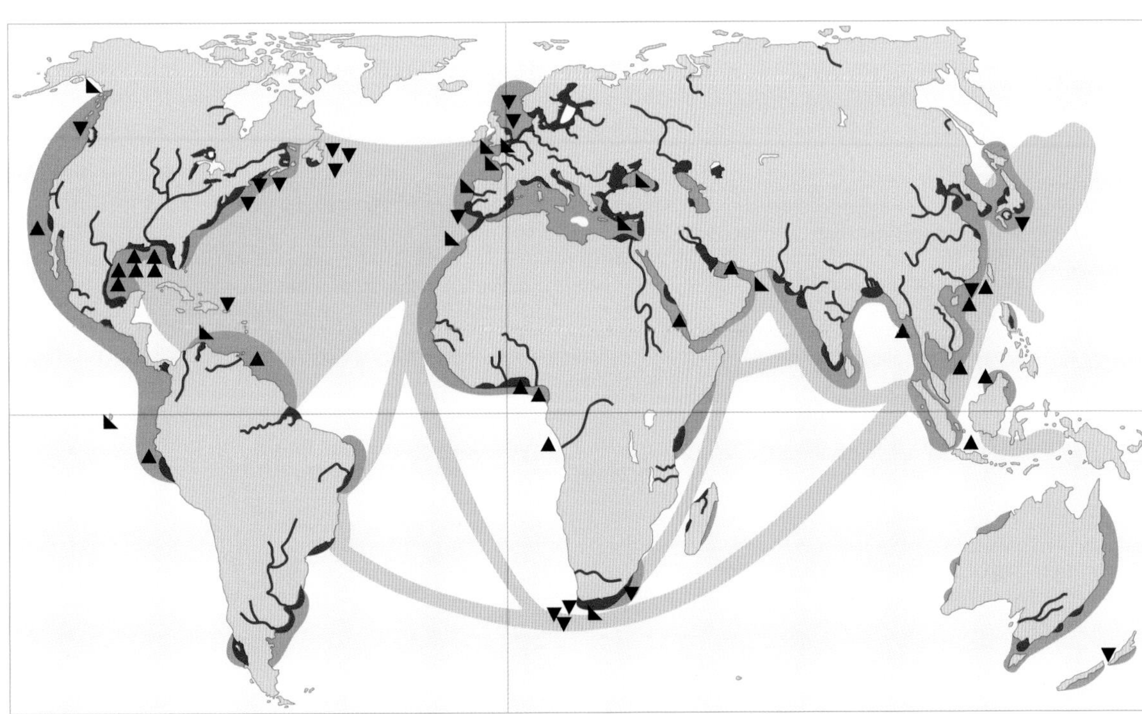

The most notorious oil rig blow-out of recent years occurred when the BP rig 'Deepwater Horizon' exploded on 20 April 2010. It is estimated that between 3,000,000 and 8,700,000 barrels of crude oil were spilled into the Gulf of Mexico, making it the world's second worst oil spill in history (after the Lakeview Gusher, California, of 1910).

Climate

Climate Regions

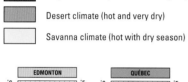

Tropical climate (hot with rain all year)

Desert climate (hot and very dry)

Savanna climate (hot with dry season)

Steppe climate (warm and dry)

Mild climate (warm and wet)

Continental climate (wet with cold winter)

Subarctic climate (very cold winter)

Polar climate (very cold and dry)

Mountainous climate (altitude affects climate)

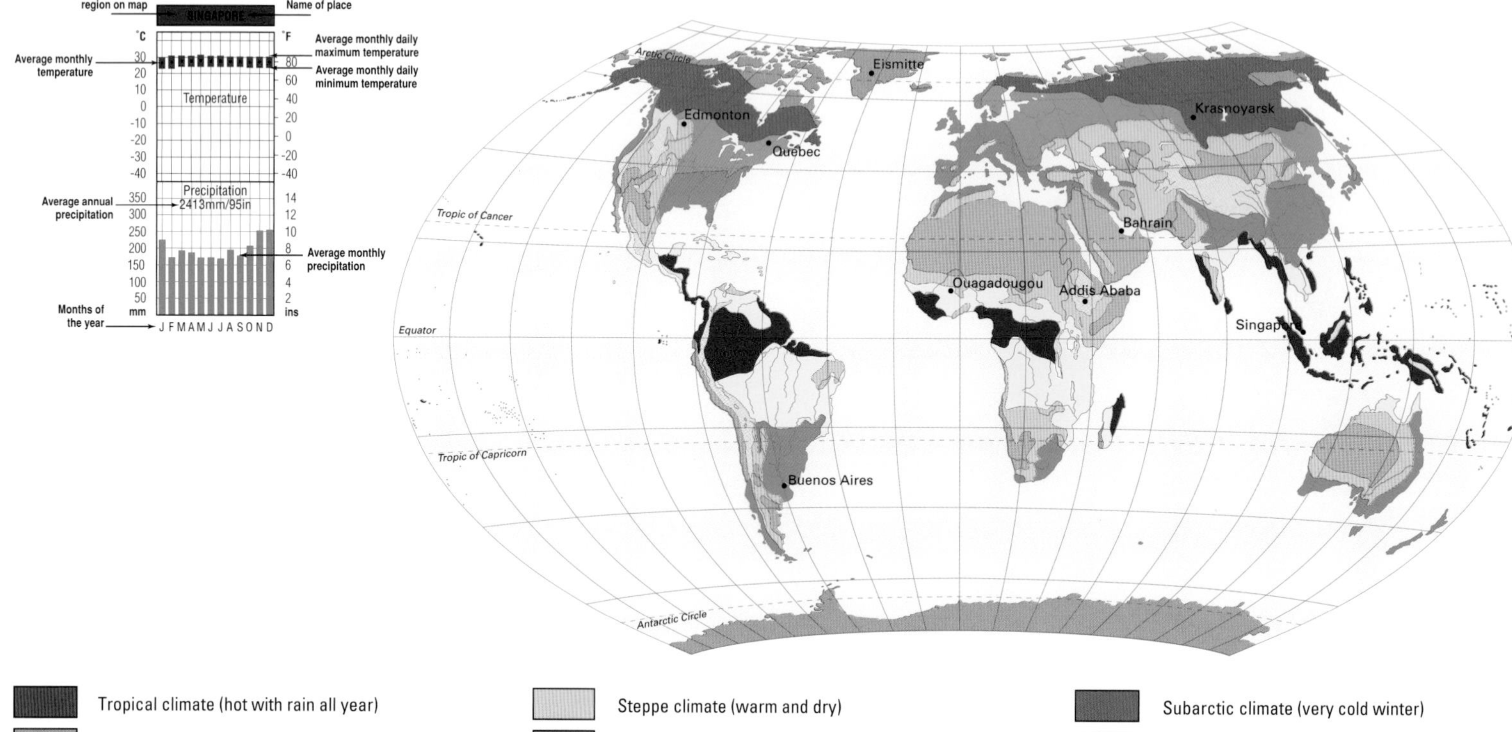

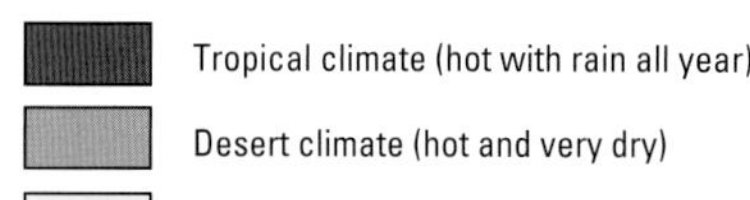

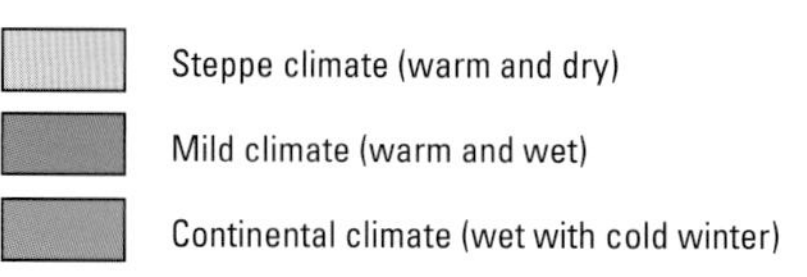

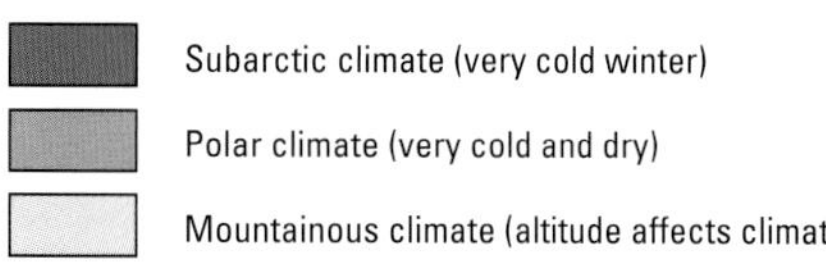

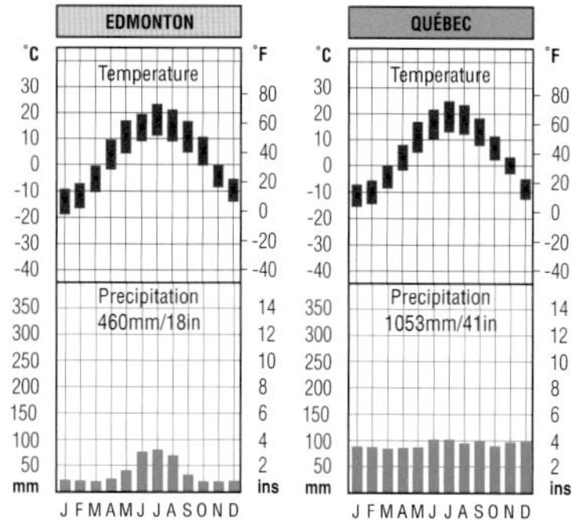

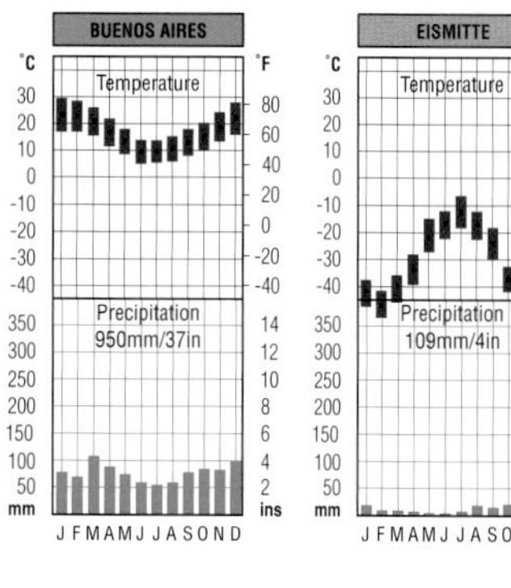

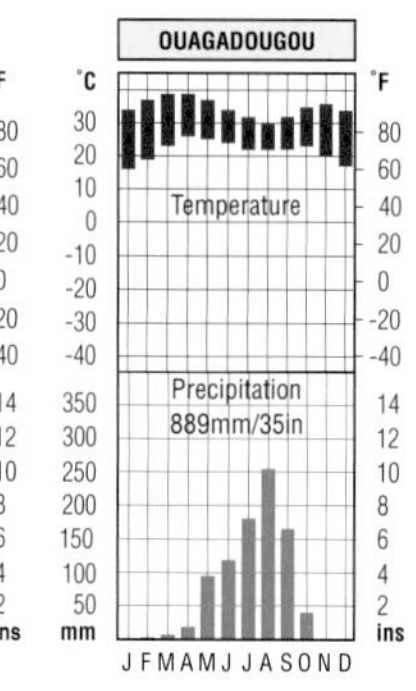

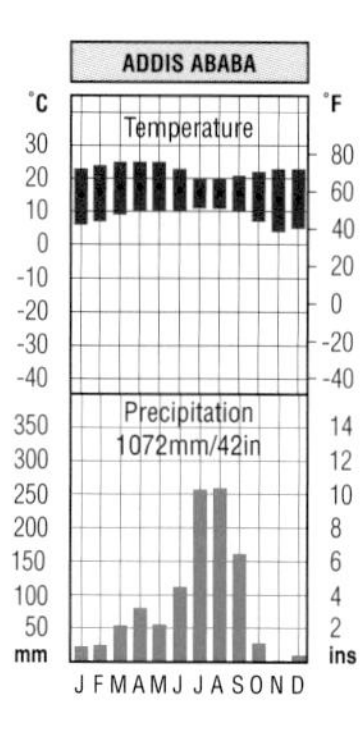

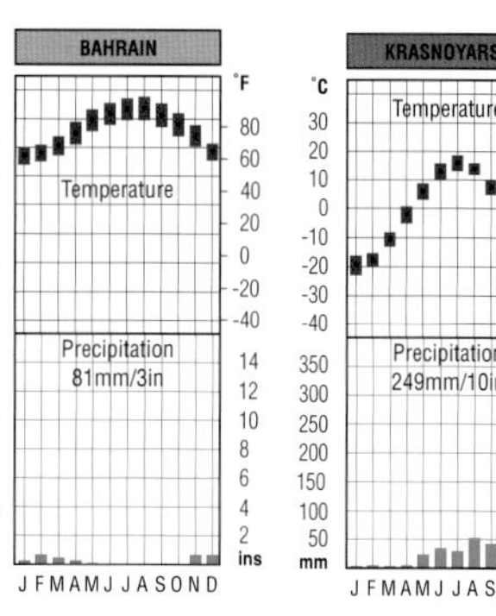

Climate Records

Temperature

Highest recorded shade temperature: Al Aziziyah, Libya, 57.7°C [135.9°F], 13 September 1922.

Highest mean annual temperature: Dallol, Ethiopia, 34.4°C [94°F], 1960–66.

Longest heatwave: Marble Bar, W. Australia, 162 days over 38°C [100°F], 23 October 1923 to 7 April 1924.

Lowest recorded temperature (outside poles): Verkhoyansk, Siberia, –68°C [–93.6°F], 7 February 1982.

Lowest mean annual temperature: Polus Nedostupnosti, Pole of Cold, Antarctica, –57.8°C [–72°F].

Precipitation

Driest place: Quillagua, Chile, mean annual rainfall 0.5 mm [0.02 in], 1964–2001.

Wettest place (average): Mt Wai-ale-ale, Hawai'i, USA, mean annual rainfall 11,680 mm [459.8 in].

Wettest place (12 months): Cherrapunji, Meghalaya, N. E. India, 26,461 mm [1,042 in], August 1860 to July 1861. Cherrapunji also holds the record for the most rainfall in one month: 2,930 mm [115 in], July 1861.

Wettest place (24 hours): Fac Fac, Réunion, Indian Ocean, 1,825 mm [71.9 in], 15–16 March 1952.

Heaviest hailstones: Gopalganj, Bangladesh, up to 1.02 kg [2.25 lb], 14 April 1986 (killed 92 people).

Heaviest snowfall (continuous): Bessans, Savoie, France, 1,730 mm [68 in] in 19 hours, 5–6 April 1969.

Heaviest snowfall (season/year): Mt Baker, Washington, USA, 28,956 mm [1,140 in], June 1998 to June 1999.

Pressure and winds

Highest barometric pressure: Agata, Siberia (at 262 m [862 ft] altitude), 1,083.8 mb, 31 December 1968.

Lowest barometric pressure: Typhoon Tip, Guam, Pacific Ocean, 870 mb, 12 October 1979.

Highest recorded wind speed: Bridge Creek, Oklahoma, USA, 512 km/h [318 mph], 3 May 1999. Measured by Doppler radar monitoring a tornado.

Windiest place: Port Martin, Antarctica, where winds of more than 64 km/h [40 mph] occur for not less than 100 days a year.

Climate

Climate is weather in the long term: the seasonal pattern of hot and cold, wet and dry, averaged over time (usually 30 years). At the simplest level, it is caused by the uneven heating of the Earth. Surplus heat at the Equator passes towards the poles, levelling out the energy differential. Its passage is marked by a ceaseless churning of the atmosphere and the oceans, further agitated by the Earth's diurnal spin and the motion it imparts to moving air and water. The heat's means of transport – by winds and ocean currents, by the continual evaporation and recondensation of water molecules – is the weather itself. There are four basic types of climate, each of which can be further subdivided: tropical, desert (dry), temperate and polar.

Composition of Dry Air

Nitrogen	78.09%	Sulphur dioxide	trace
Oxygen	20.95%	Nitrogen oxide	trace
Argon	0.93%	Methane	trace
Water vapour	0.2–4.0%	Dust	trace
Carbon dioxide	0.03%	Helium	trace
Ozone	0.00006%	Neon	trace

El Niño

In a normal year, south-easterly trade winds drive surface waters westwards off the coast of South America, drawing cold, nutrient-rich water up from below. In an El Niño year (which occurs every 2–7 years), warm water from the west Pacific suppresses up-welling in the east, depriving the region of nutrients. The water is warmed by as much as 7°C [12°F], disturbing the tropical atmospheric circulation. During an intense El Niño, the south-east trade winds change direction and become equatorial westerlies, resulting in climatic extremes in many regions of the world, such as drought in parts of Australia and India, and heavy rainfall in south-eastern USA. An intense El Niño occurred in 1997–8, with resultant freak weather conditions across the entire Pacific region.

Normal year

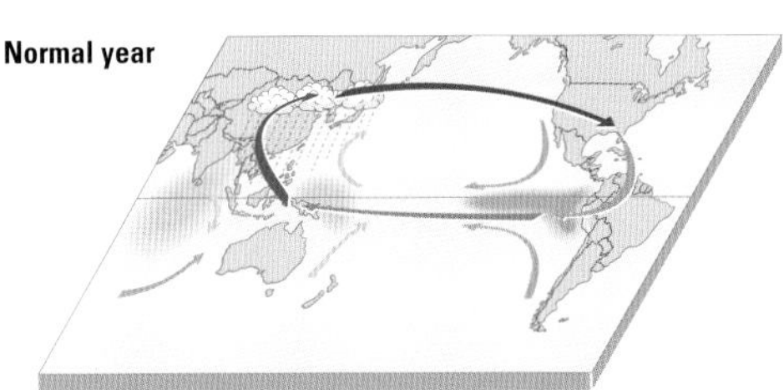

El Niño event

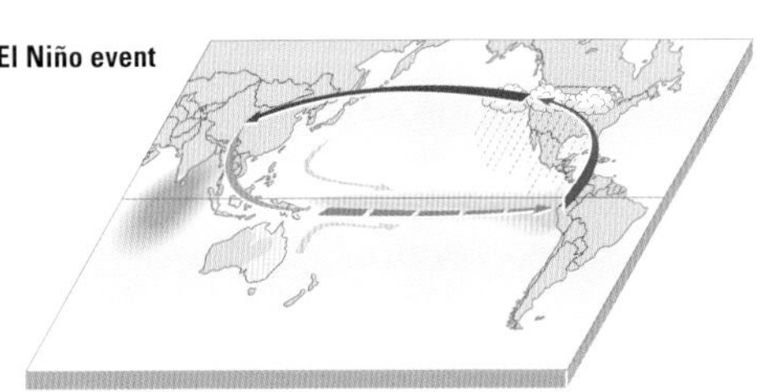

Beaufort Wind Scale

Named after the 19th-century British naval officer who devised it, the Beaufort Scale assesses wind speed according to its effects. It was originally designed as an aid for sailors, but has since been adapted for use on the land.

Scale	Wind speed km/h	mph	Effect
0	0–1	0–1	**Calm** Smoke rises vertically
1	1–5	1–3	**Light air** Wind direction shown only by smoke drift
2	6–11	4–7	**Light breeze** Wind felt on face; leaves rustle; vanes moved by wind
3	12–19	8–12	**Gentle breeze** Leaves and small twigs in constant motion; wind extends small flag
4	20–28	13–18	**Moderate** Raises dust and loose paper; small branches move
5	29–38	19–24	**Fresh** Small trees in leaf sway; wavelets on inland waters
6	39–49	25–31	**Strong** Large branches move; difficult to use umbrellas
7	50–61	32–38	**Near gale** Whole trees in motion; difficult to walk against wind
8	62–74	39–46	**Gale** Twigs break from trees; walking very difficult
9	75–88	47–54	**Strong gale** Slight structural damage
10	89–102	55–63	**Storm** Trees uprooted; serious structural damage
11	103–117	64–72	**Violent storm** Widespread damage
12	118+	73+	**Hurricane**

Conversions
°C = (°F − 32) × 5/9; °F = (°C × 9/5) + 32; 0°C = 32°F
1 in = 25.4 mm; 1 mm = 0.0394 in; 100 mm = 3.94 in

Temperature

Average temperature in January

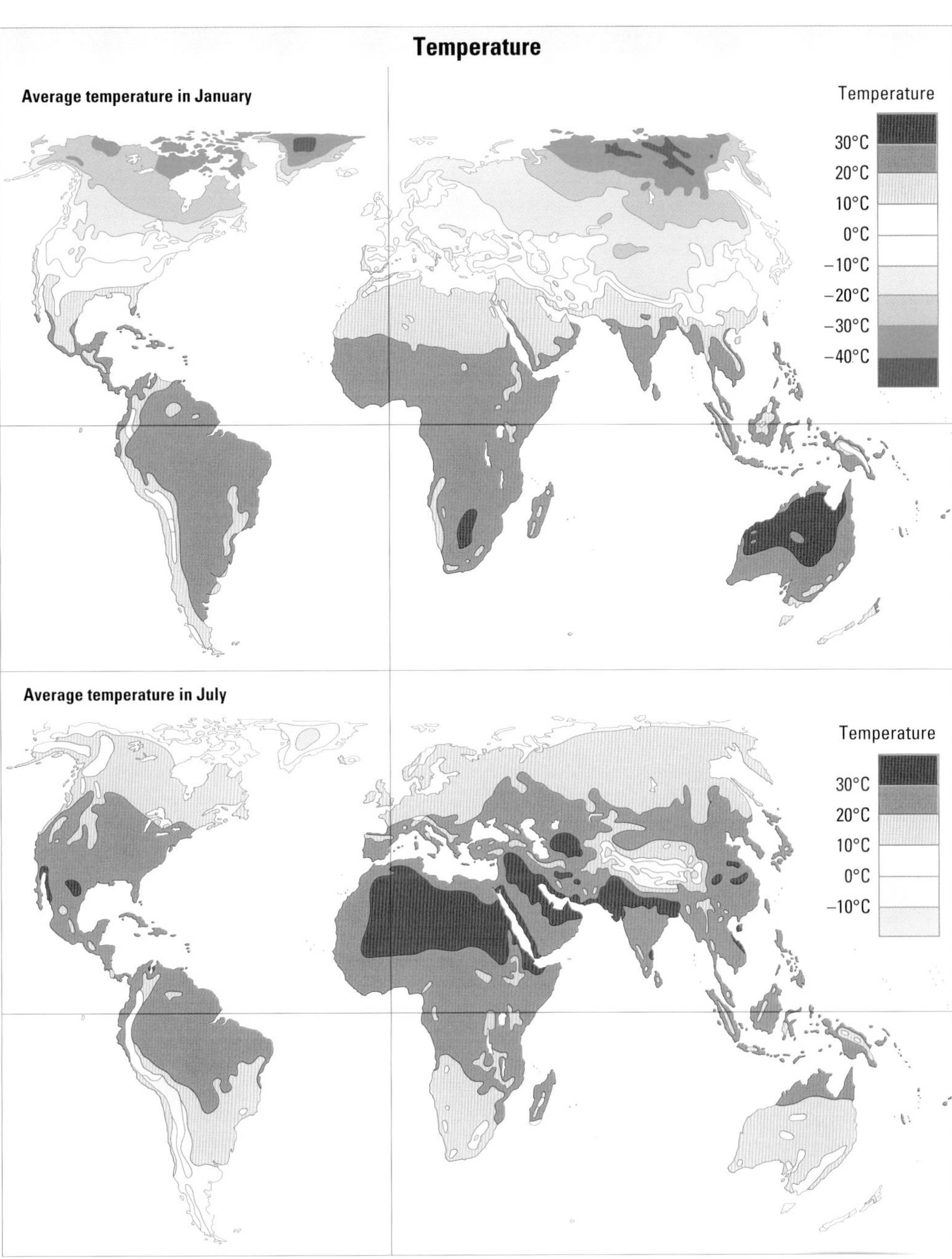

Temperature

- 30°C
- 20°C
- 10°C
- 0°C
- −10°C
- −20°C
- −30°C
- −40°C

Average temperature in July

Temperature

- 30°C
- 20°C
- 10°C
- 0°C
- −10°C

Precipitation (Rainfall and Snow)

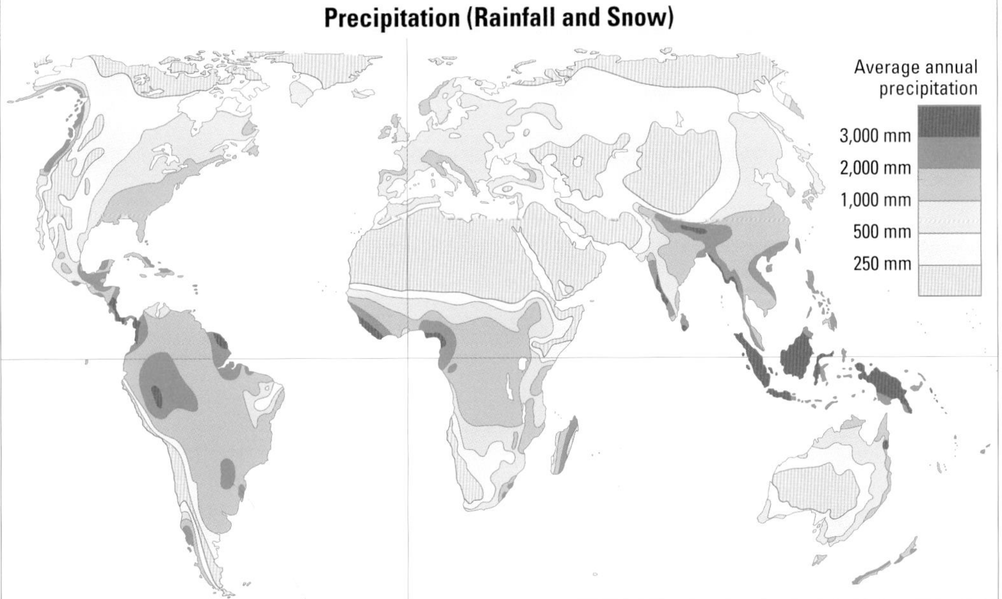

Average annual precipitation

- 3,000 mm
- 2,000 mm
- 1,000 mm
- 500 mm
- 250 mm

Water and Vegetation

The Hydrological Cycle

The world's water balance is regulated by the constant recycling of water between the oceans, atmosphere and land. The movement of water between these three reservoirs is known as the hydrological cycle. The oceans play a vital role in the hydrological cycle: 74% of the total precipitation falls over the oceans and 84% of the total evaporation comes from the oceans.

Water Distribution

The distribution of planetary water, by percentage. Oceans and ice caps together account for more than 99% of the total; the breakdown of the remainder is estimated.

All water

- 97.4% — Oceans
- 2.6% — Fresh water

Fresh water

- 76.6% — Ice caps and glaciers
- 22.7% — Groundwater
- 0.5% — Active water

Active water

- 52% — Lakes
- 36% — Soil moisture
- 7.1% — Atmosphere
- 3.5% — Rivers
- 1.4% — Living things

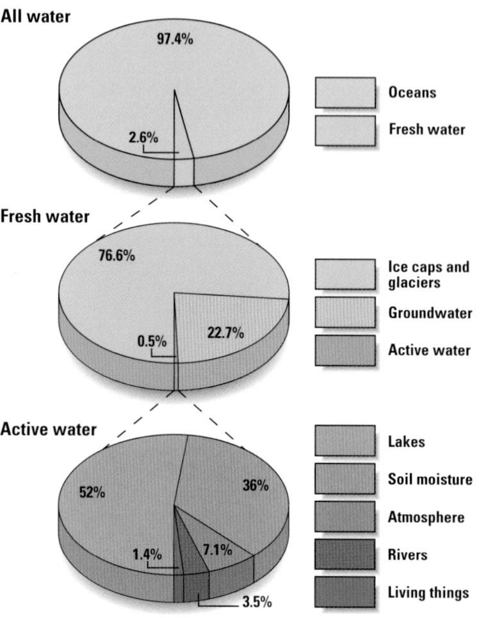

Water Utilization

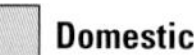

The percentage breakdown of water usage by sector, selected countries (2007)

Legend: Domestic | Industrial | Agriculture

Countries listed: Algeria, Australia, Egypt, France, Ghana, India, Mexico, Poland, Russia, Saudi Arabia, UK, USA

Scale: 0 — 20 — 40 — 60 — 80 — 100

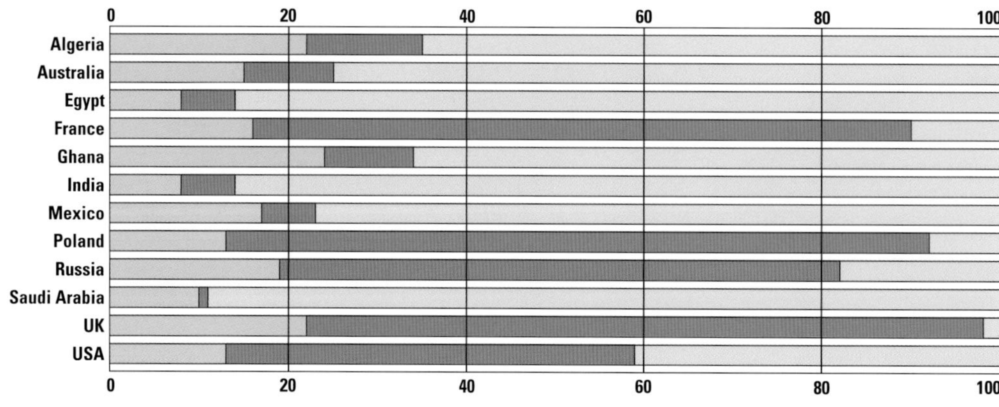

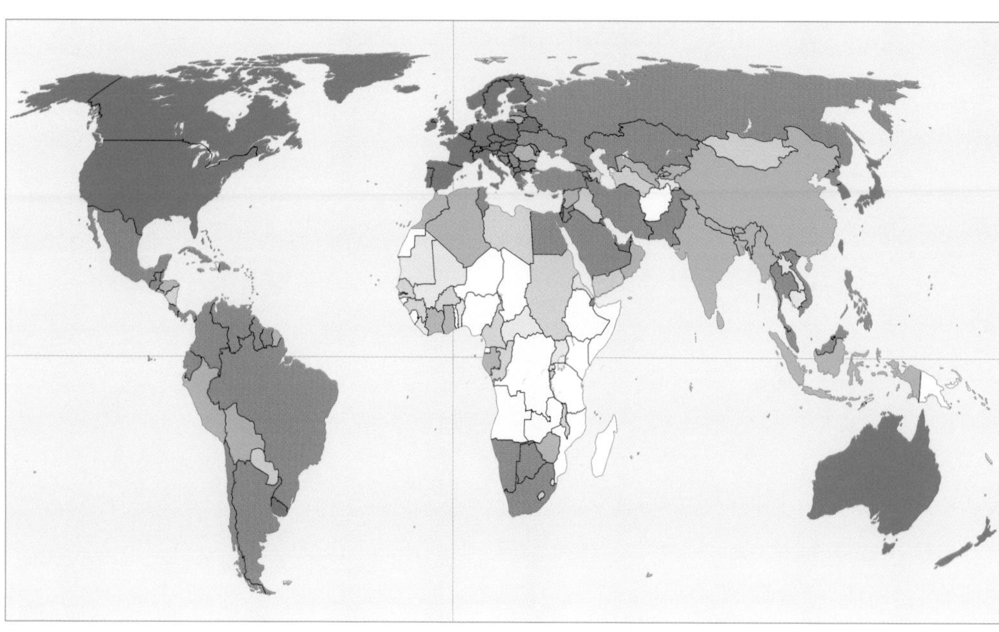

Water Usage

Almost all the world's water is 3,000 million years old, and all of it cycles endlessly through the hydrosphere, though at different rates. Water vapour circulates over days or even hours, deep ocean water circulates over millennia, and ice-cap water remains solid for millions of years.

Fresh water is essential to all terrestrial life. Humans cannot survive more than a few days without it, and even the hardiest desert plants and animals could not exist without some water. Agriculture requires huge quantities of fresh water: without large-scale irrigation most of the world's people would starve. In the USA, agriculture uses 41% and industry 46% of all water withdrawals.

According to the latest figures, the average North American uses 1.3 million litres of water per year. This is more than six times the average African, who uses just 186,000 litres of water each year. Europeans and Australians use 694,000 litres per year.

Water Supply

Percentage of total population with access to safe drinking water (2006)

- 100% with safe water
- 90 – 100% with safe water
- 80 – 90% with safe water
- 70 – 80% with safe water
- 60 – 70% with safe water
- Less than 60% with safe water
- No data available

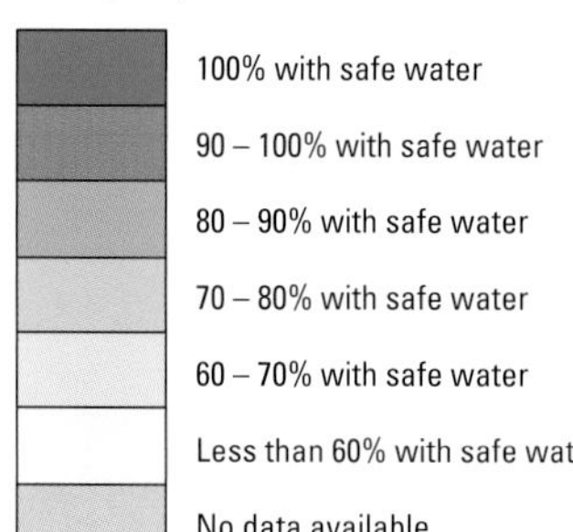

Least well-provided countries

Madagascar 47%	Mozambique 42%
Nigeria 47%	Niger 42%
Congo (Dem. Rep.) 46%	Papua New Guinea 40%
Equatorial Guinea 43%	Somalia 29%
Ethiopia 42%	Western Sahara 26%

Natural Vegetation

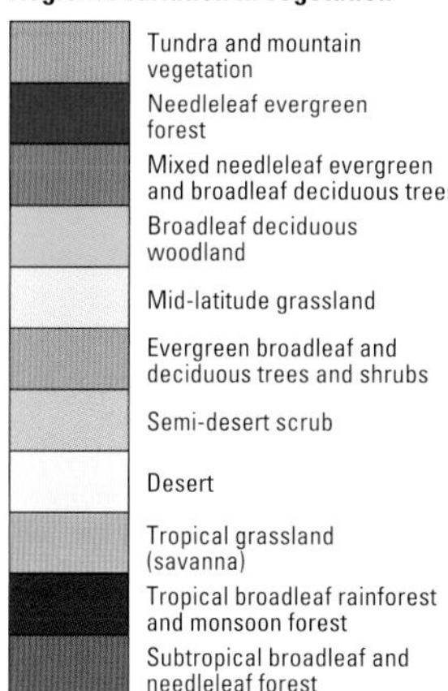

Regional variation in vegetation

- Tundra and mountain vegetation
- Needleleaf evergreen forest
- Mixed needleleaf evergreen and broadleaf deciduous trees
- Broadleaf deciduous woodland
- Mid-latitude grassland
- Evergreen broadleaf and deciduous trees and shrubs
- Semi-desert scrub
- Desert
- Tropical grassland (savanna)
- Tropical broadleaf rainforest and monsoon forest
- Subtropical broadleaf and needleleaf forest

The map shows the natural 'climax vegetation' of regions, as dictated by climate and topography. In most cases, however, agricultural activity has drastically altered the vegetation pattern. Western Europe, for example, lost most of its broadleaf forest many centuries ago, while irrigation has turned some natural semi-desert into productive land.

Land Use by Continent (2007)

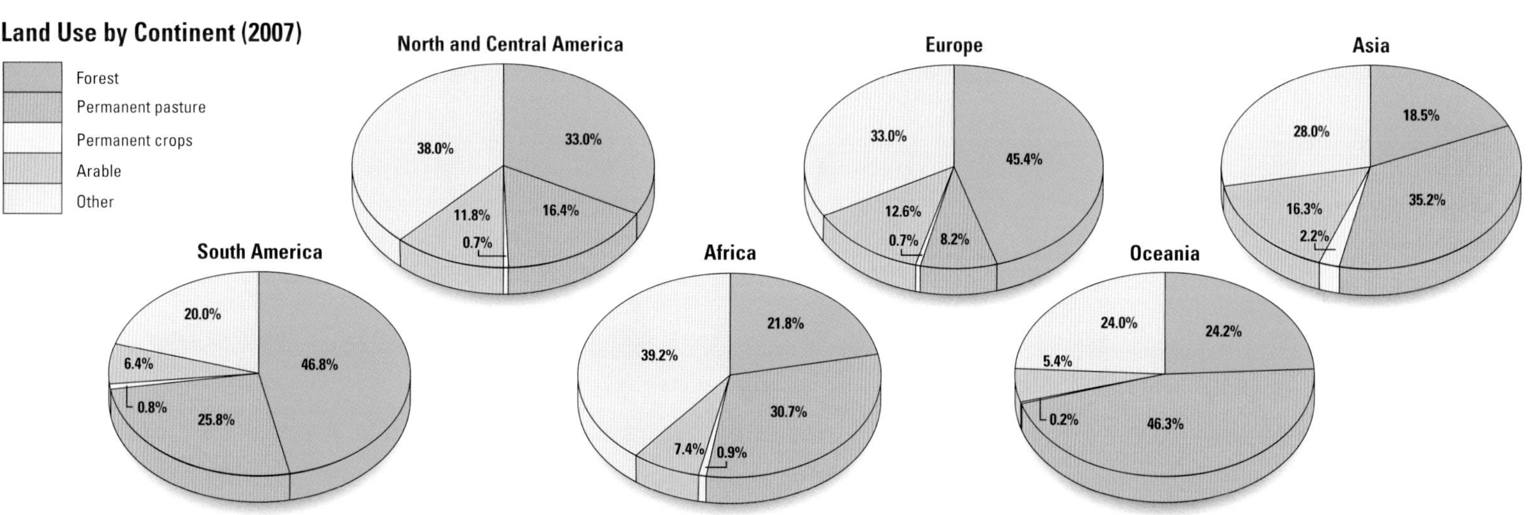

- Forest
- Permanent pasture
- Permanent crops
- Arable
- Other

North and Central America
33.0% / 16.4% / 0.7% / 11.8% / 38.0%

Europe
45.4% / 8.2% / 0.7% / 12.6% / 33.0%

Asia
18.5% / 35.2% / 2.2% / 16.3% / 28.0%

South America
46.8% / 25.8% / 0.8% / 6.4% / 20.0%

Africa
21.8% / 30.7% / 0.9% / 7.4% / 39.2%

Oceania
24.2% / 46.3% / 0.2% / 5.4% / 24.0%

Forestry: Production

Forest and woodland (million hectares)		Annual production (2008, million cubic metres)	
		Fuelwood	Industrial roundwood*
World	*3,869.5*	*1,891.5*	*1,556.7*
Europe	1,039.3	152.5	504.6
S. America	885.6	200.9	185.7
Africa	649.9	637.6	70.3
N. & C. America	549.3	131.5	500.3
Asia	547.8	753.7	243.3
Oceania	197.6	15.9	52.4

Paper and Board

Top producers (2008)**		Top exporters (2008)**	
China	.83,685	Germany	.13,254
USA	.80,178	Finland	.11,851
Japan	.28,360	USA	.11,707
Germany	.22,842	Canada	.10,910
Canada	.15,773	Sweden	.10,579

* roundwood is timber as it is felled
** in thousand tonnes

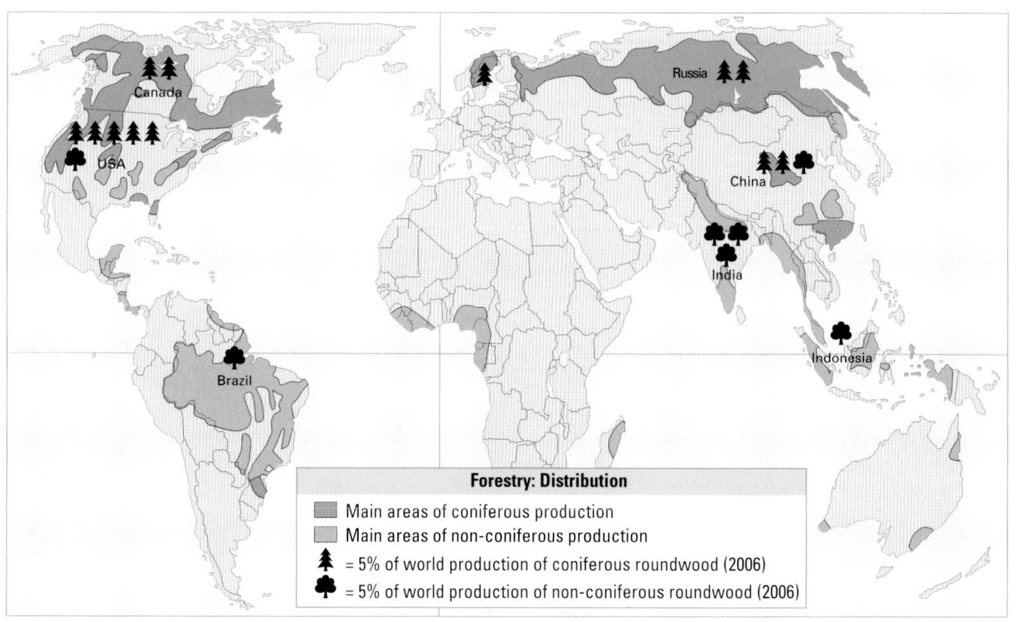

Forestry: Distribution

- Main areas of coniferous production
- Main areas of non-coniferous production
- 🌲 = 5% of world production of coniferous roundwood (2006)
- 🌳 = 5% of world production of non-coniferous roundwood (2006)

Environment

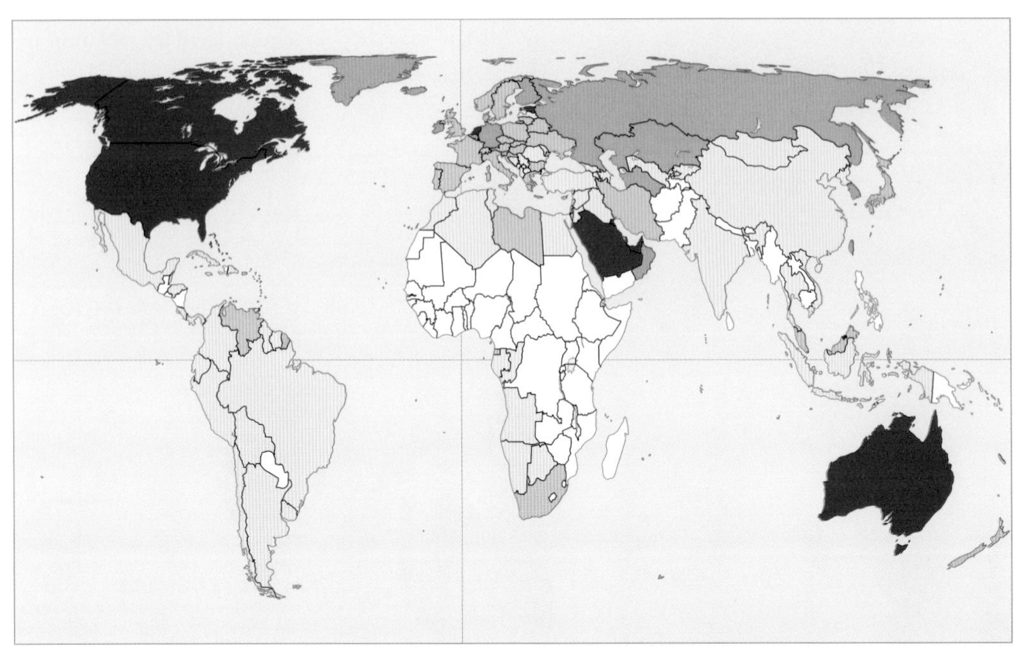

Global Warming

Carbon dioxide emissions in tonnes per capita (2008)

	Over 15
	10 – 15
	5 – 10
	1 – 5
	Under 1
	No data available

Carbon Dioxide

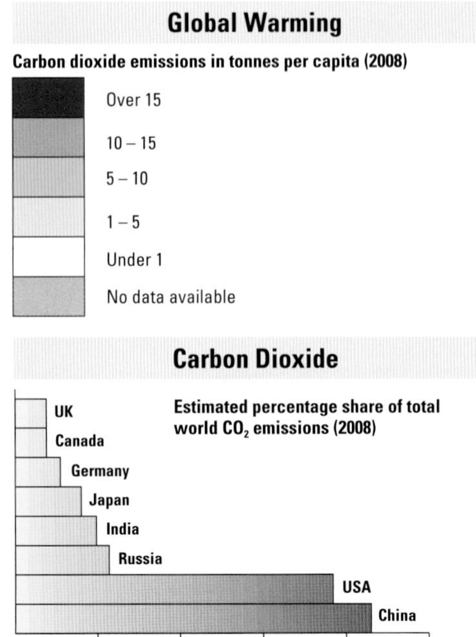

Estimated percentage share of total world CO_2 emissions (2008)

UK
Canada
Germany
Japan
India
Russia
USA
China

5%   10%   15%   20%   25%

Predicted Change in Precipitation

The difference between actual annual average precipitation, 1960–1990, and the predicted annual average precipitation, 2070–2100. It should be noted that these predicted annual mean changes mask quite significant seasonal detail.

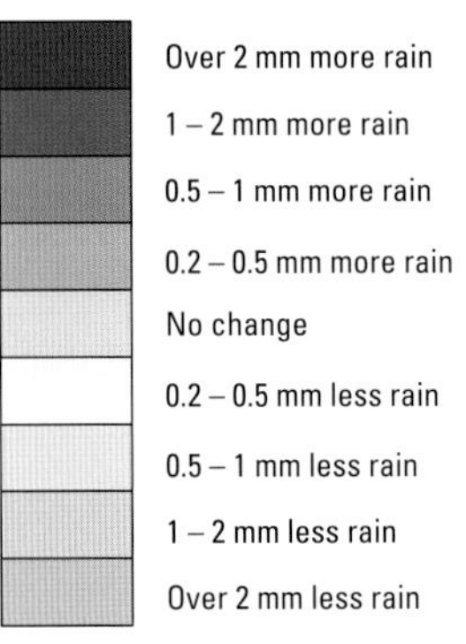

	Over 2 mm more rain
	1 – 2 mm more rain
	0.5 – 1 mm more rain
	0.2 – 0.5 mm more rain
	No change
	0.2 – 0.5 mm less rain
	0.5 – 1 mm less rain
	1 – 2 mm less rain
	Over 2 mm less rain

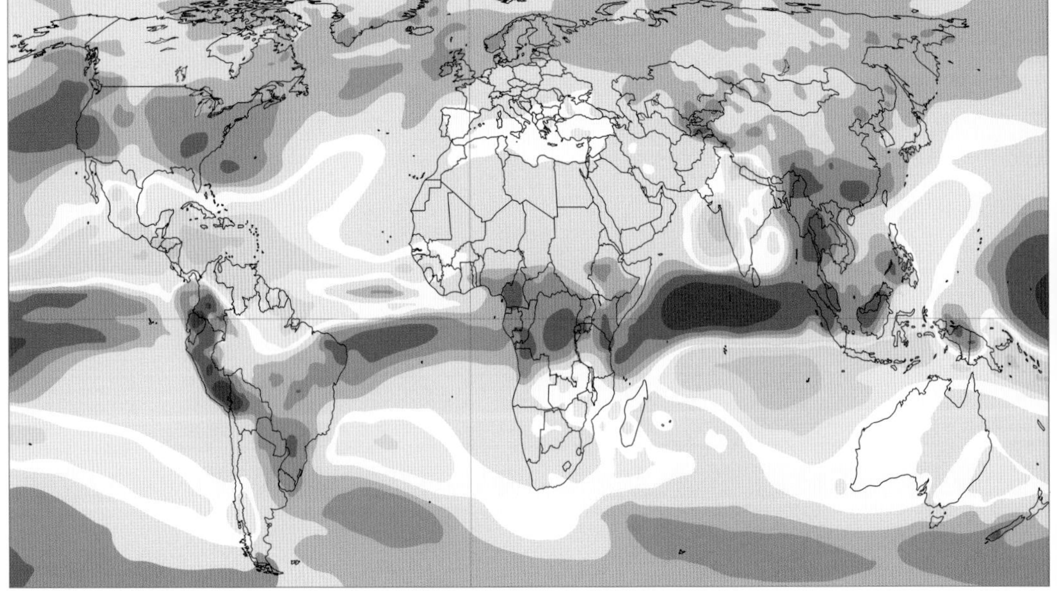

Predicted Change in Temperature

The difference between actual annual average surface air temperature, 1960–1990, and the predicted annual average surface air temperature, 2070–2100. This map shows the predicted increase, assuming a 'medium growth' of global economy and assuming that no measures are taken to combat the emission of greenhouse gases.

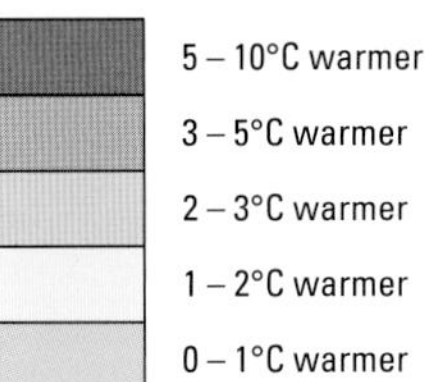

	5 – 10°C warmer
	3 – 5°C warmer
	2 – 3°C warmer
	1 – 2°C warmer
	0 – 1°C warmer

Source: The Hadley Centre of Climate Prediction and Research, The Met. Office

Projected Change in Global Warming

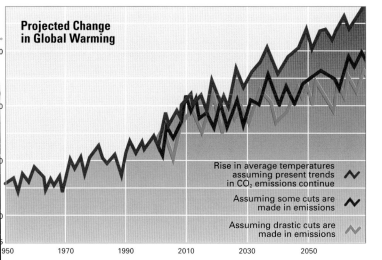

Rise in average temperatures assuming present trends in CO₂ emissions continue ∿

Assuming some cuts are made in emissions ∿

Assuming drastic cuts are made in emissions ∿

950   1970   1990   2010   2030   2050

Possible Effect of Sea Level Rise in Florida

Sea levels have risen worldwide by about 2 cm since 1900. If CO_2 emissions continue at the same rate, the sea level is expected to rise by 7.4 m by 2200. The map shows the dramatic effects that such a rise could have on the southern part of Florida in the USA.

Submerged land area if sea level rises 4.5 m

Submerged land area if sea level rises 7.4 m

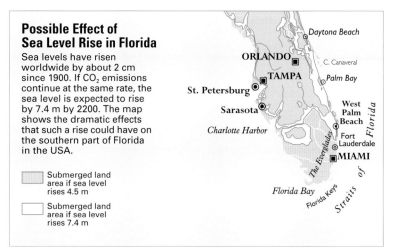

The Greenhouse Effect

Carbon dioxide is increased by burning fossil fuels and cutting forests

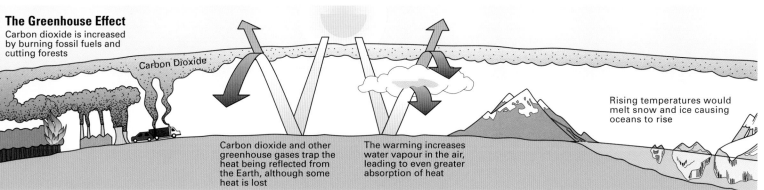

Carbon Dioxide

Carbon dioxide and other greenhouse gases trap the heat being reflected from the Earth, although some heat is lost

The warming increases water vapour in the air, leading to even greater absorption of heat

Rising temperatures would melt snow and ice causing oceans to rise

Desertification

Existing deserts

Areas with a high risk of desertification

Areas with a moderate risk of desertification

Former areas of rainforest

Existing rainforest

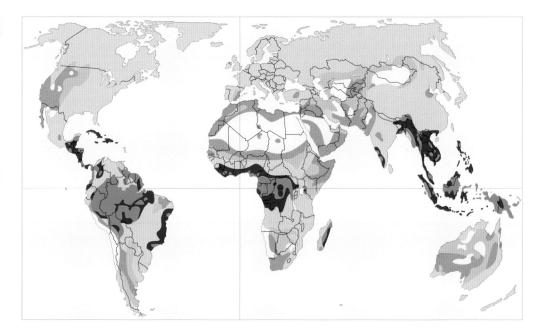

Forest Clearance

Thousands of hectares of forest cleared annually, tropical countries surveyed 1980–85, 1990–95 and 2000–05. Loss as a percentage of remaining stocks is shown in figures on each column. Gain is indicated as a minus figure.

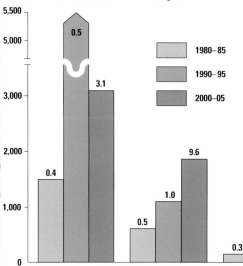

1980–85

1990–95

2000–05

Deforestation

The Earth's remaining forests are under attack from three directions: expanding agriculture, logging, and growing consumption of fuelwood, often in combination. Sometimes deforestation is the direct result of government policy, as in the efforts made to resettle the urban poor in some parts of Brazil; just as often, it comes about despite state attempts at conservation.

Loggers, licensed or unlicensed, blaze a trail into virgin forest, often destroying twice as many trees as they harvest. Landless farmers follow, burning away most of what remains to plant their crops, completing the destruction. However, some countries such as Vietnam and Costa Rica have successfully implemented reafforestation programmes.

Brazil	Indonesia	India	Burma	Thailand	Vietnam	Philippines	Costa Rica
0.4 / 0.5 / 3.1	0.5 / 1.0 / 9.6	0.3 / 0.0 / 0.7	0.3 / 1.4 / 4.7	2.4 / 2.6 / 2.0	0.7 / 1.4 / −12.2	1.0 / 3.5 / 4.2	4.0 / 3.0 / −0.6

Population

Developed nations such as the UK have populations evenly spread across the age groups and, usually, a growing proportion of elderly people. The great majority of the people in developing nations, however, are in the younger age groups, about to enter their most fertile years. In time, these population profiles should resemble the world profile (even Nigeria has made recent progress by reducing its birth rate), but the transition will come about only after a few more generations of rapid population growth.

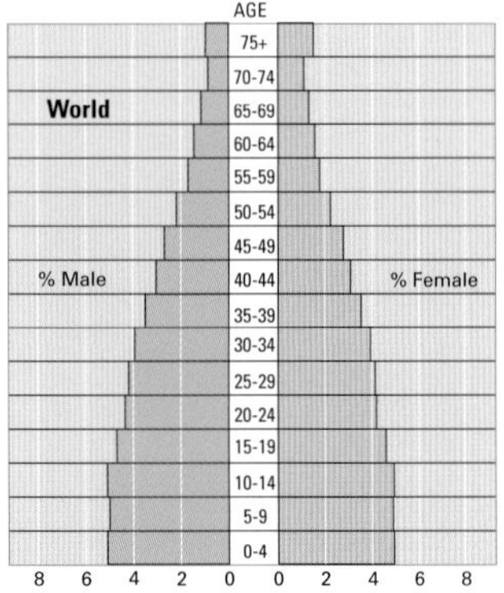

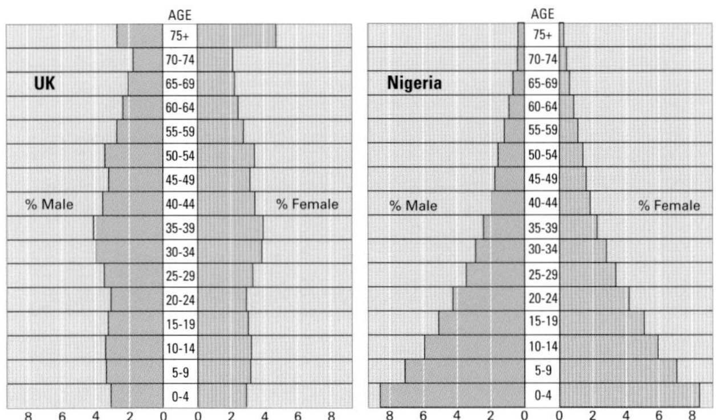

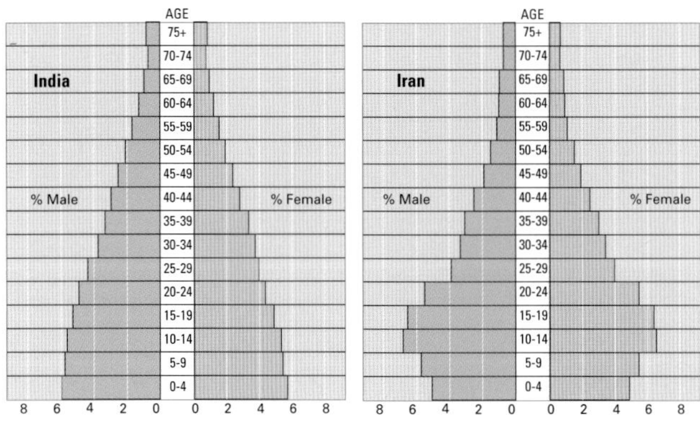

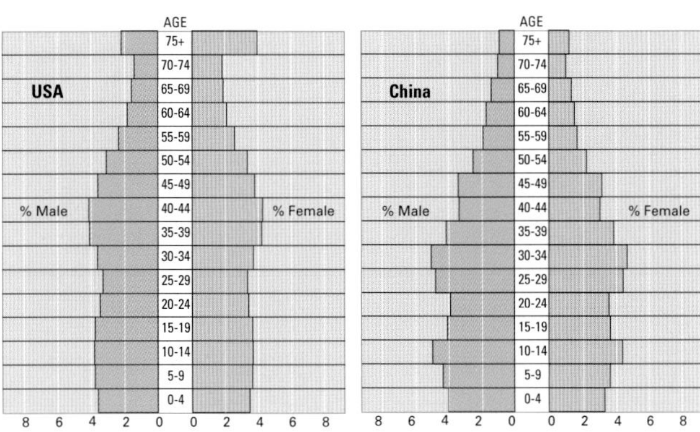

Most Populous Nations, in millions (2009 estimates)

1.	China	1,339	9. Russia	140	17. Turkey	77
2.	India	1,157	10. Japan	127	18. Congo (Dem. Rep.)	69
3.	USA	307	11. Mexico	111	19. Iran	66
4.	Indonesia	240	12. Philippines	98	20. Thailand	66
5.	Brazil	199	13. Vietnam	89	21. France	64
6.	Pakistan	175	14. Ethiopia	85	22. UK	61
7.	Bangladesh	156	15. Germany	82	23. Italy	58
8.	Nigeria	149	16. Egypt	79	24. South Africa	49

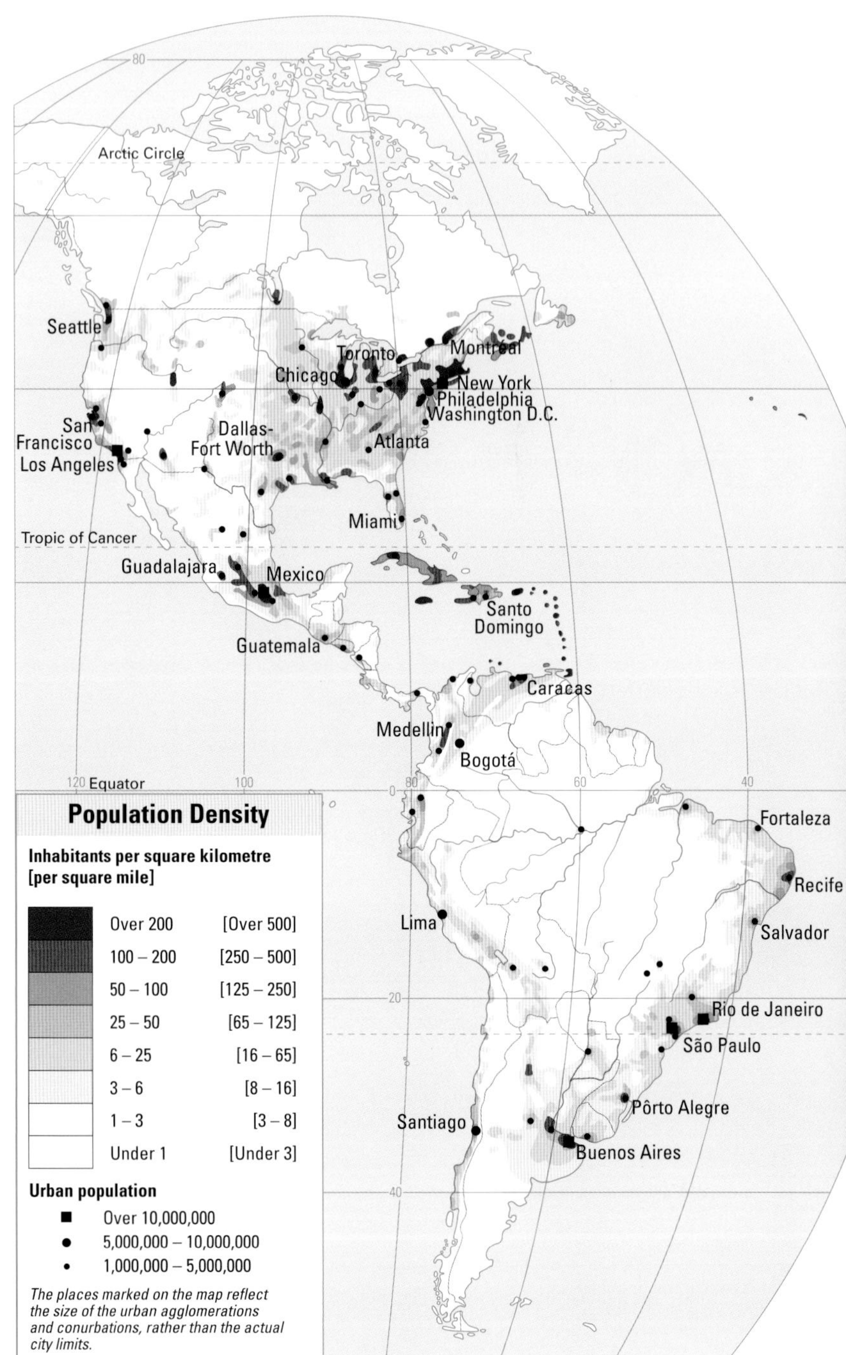

Population Density

Inhabitants per square kilometre
[per square mile]

Over 200	[Over 500]
100 – 200	[250 – 500]
50 – 100	[125 – 250]
25 – 50	[65 – 125]
6 – 25	[16 – 65]
3 – 6	[8 – 16]
1 – 3	[3 – 8]
Under 1	[Under 3]

Urban population

- ■ Over 10,000,000
- ● 5,000,000 – 10,000,000
- • 1,000,000 – 5,000,000

The places marked on the map reflect the size of the urban agglomerations and conurbations, rather than the actual city limits.

Continental Comparisons

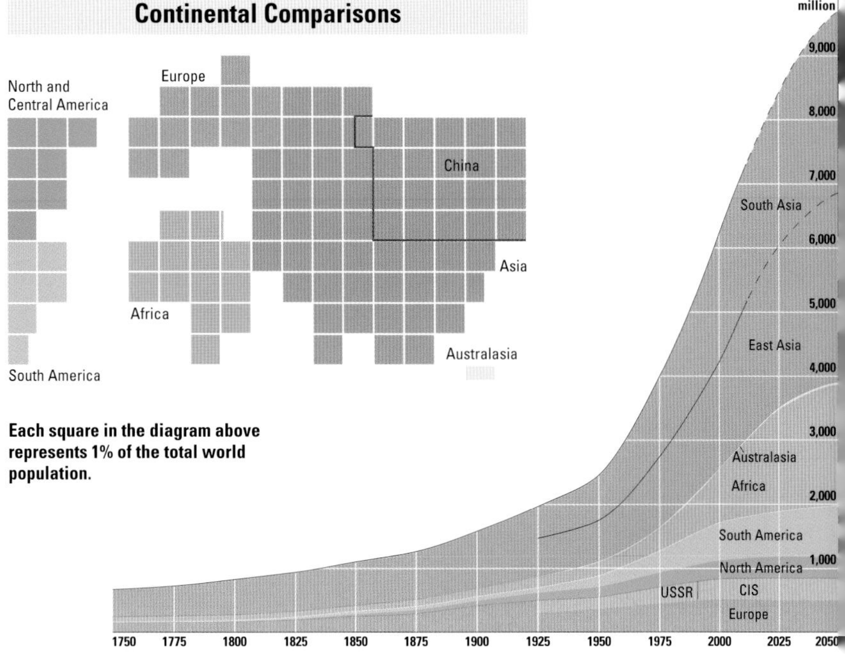

Each square in the diagram above represents 1% of the total world population.

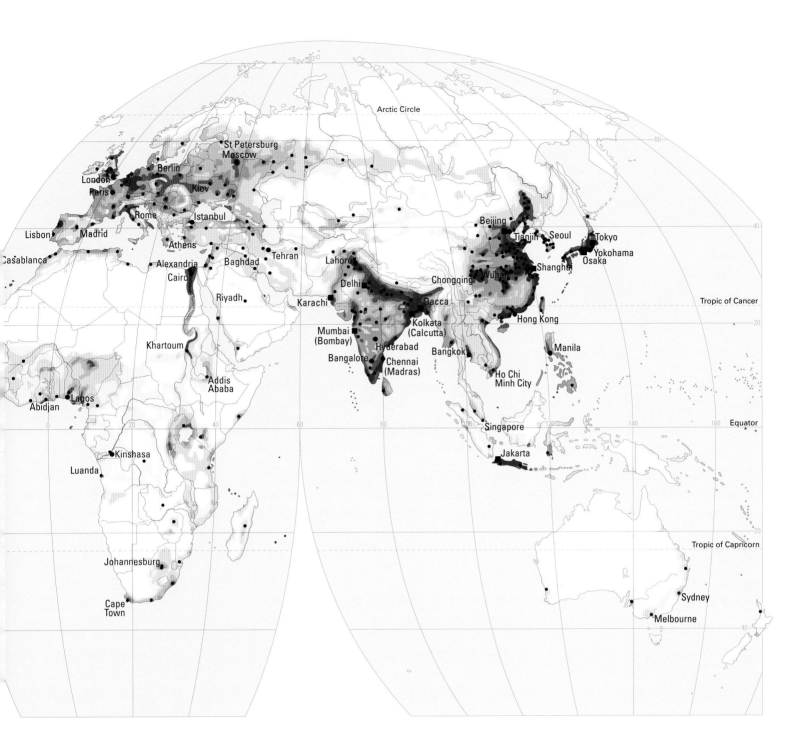

Arctic Circle

St Petersburg
Moscow
Berlin
London
Paris
Kiev
Rome
Istanbul
Lisbon
Madrid
Athens
Casablanca
Alexandria
Baghdad
Tehran
Cairo
Lahore
Riyadh
Karachi
Delhi
Dacca
Chongqing
Wuhan
Shanghai
Beijing
Tianjin
Seoul
Tokyo
Yokohama
Osaka
Hong Kong
Tropic of Cancer
Khartoum
Mumbai
(Bombay)
Kolkata
(Calcutta)
Hyderabad
Bangkok
Bangalore
Chennai
(Madras)
Manila
Addis
Ababa
Ho Chi
Minh City
Lagos
Abidjan
Singapore
Equator
Kinshasa
Jakarta
Luanda
Tropic of Capricorn
Johannesburg
Sydney
Cape
Town
Melbourne

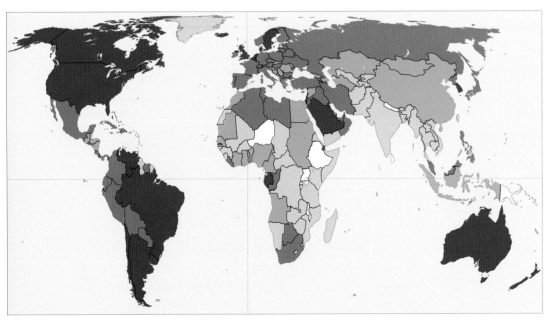

Urban Population

Percentage of total population living in towns and cities (2009)

	Over 80%
	60 – 80%
	40 – 60%
	20 – 40%
	Under 20%
	No data available

Most urbanized		Least urbanized	
Singapore	100%	Burundi	11%
Kuwait	98%	Papua New Guinea	13%
Belgium	97%	Uganda	13%
Qatar	96%	Trinidad & Tobago	14%
Malta	95%	Sri Lanka	15%

17

The Human Family

Predominant Languages

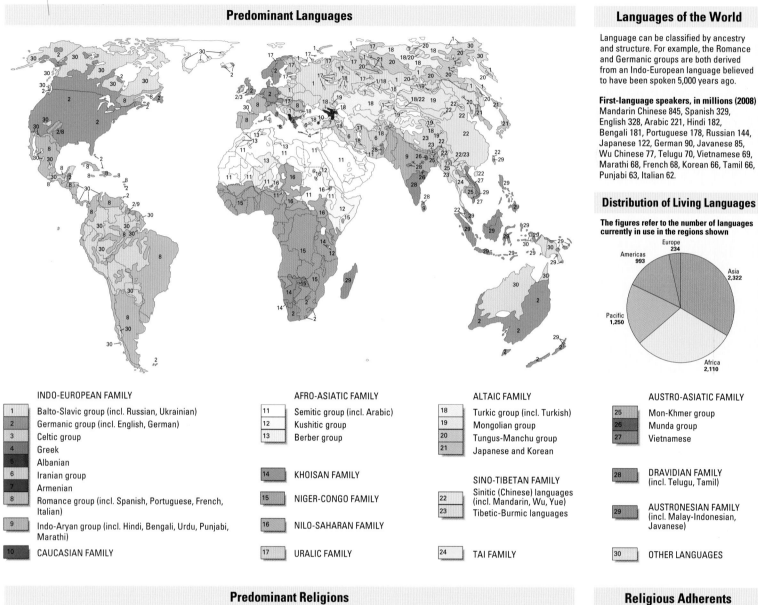

Languages of the World

Language can be classified by ancestry and structure. For example, the Romance and Germanic groups are both derived from an Indo-European language believed to have been spoken 5,000 years ago.

First-language speakers, in millions (2008)
Mandarin Chinese 845, Spanish 329, English 328, Arabic 221, Hindi 182, Bengali 181, Portuguese 178, Russian 144, Japanese 122, German 90, Javanese 85, Wu Chinese 77, Telugu 70, Vietnamese 69, Marathi 68, French 68, Korean 66, Tamil 66, Punjabi 63, Italian 62.

Distribution of Living Languages

The figures refer to the number of languages currently in use in the regions shown

Europe 234
Americas 993
Asia 2,322
Pacific 1,250
Africa 2,110

INDO-EUROPEAN FAMILY

1	Balto-Slavic group (incl. Russian, Ukrainian)
2	Germanic group (incl. English, German)
3	Celtic group
4	Greek
5	Albanian
6	Iranian group
7	Armenian
8	Romance group (incl. Spanish, Portuguese, French, Italian)
9	Indo-Aryan group (incl. Hindi, Bengali, Urdu, Punjabi, Marathi)
10	CAUCASIAN FAMILY

AFRO-ASIATIC FAMILY

11	Semitic group (incl. Arabic)
12	Kushitic group
13	Berber group
14	KHOISAN FAMILY
15	NIGER-CONGO FAMILY
16	NILO-SAHARAN FAMILY
17	URALIC FAMILY

ALTAIC FAMILY

18	Turkic group (incl. Turkish)
19	Mongolian group
20	Tungus-Manchu group
21	Japanese and Korean

SINO-TIBETAN FAMILY

22	Sinitic (Chinese) languages (incl. Mandarin, Wu, Yue)
23	Tibetic-Burmic languages
24	TAI FAMILY

AUSTRO-ASIATIC FAMILY

25	Mon-Khmer group
26	Munda group
27	Vietnamese
28	DRAVIDIAN FAMILY (incl. Telugu, Tamil)
29	AUSTRONESIAN FAMILY (incl. Malay-Indonesian, Javanese)
30	OTHER LANGUAGES

Predominant Religions

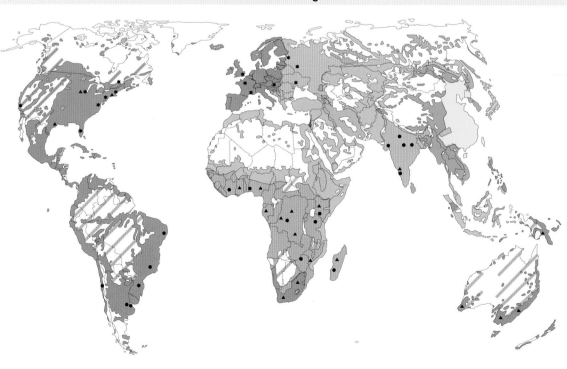

Religious Adherents

Religious adherents in millions (2006)

Christianity	2,100	Hindu	900
Roman Catholic	1,050	Chinese folk	394
Protestant	396	Buddhism	376
Orthodox	240	Ethnic religions	300
Anglican	73	New religions	103
Others	341	Sikhism	23
Islam	1,070	Spiritism	15
Sunni	940	Judaism	14
Shi'ite	120	Baha'i	7
Others	10	Confucianism	6
Non-religious/		Jainism	4
Agnostic/Atheist	1,100	Shintoism	4

- Roman Catholicism
- Orthodox and other Eastern Churches
- Protestantism
- Sunni Islam
- Shi'ite Islam
- Buddhism
- Hinduism
- Confucianism
- Judaism
- Shintoism
- Tribal Religions

United Nations

Created in 1945 to promote peace and co-operation, and based in New York, the United Nations is the world's largest international organization, with 192 members and an annual budget of US $4.2 billion (2008–9). Each member of the General Assembly has one vote, while the five permanent members of the 15-nation Security Council – China, France, Russia, UK and USA – each hold a veto. The Secretariat is the UN's principal administrative arm. The 54 members of the Economic and Social Council are responsible for economic, social, cultural, educational, health and related matters. The UN has 16 specialized agencies – based in Canada, France, Switzerland and Italy, as well as the USA – which help members in fields such as education (UNESCO), agriculture (FAO), medicine (WHO) and finance (IFC). By the end of 1994, all the original 11 trust territories of the Trusteeship Council had become independent.

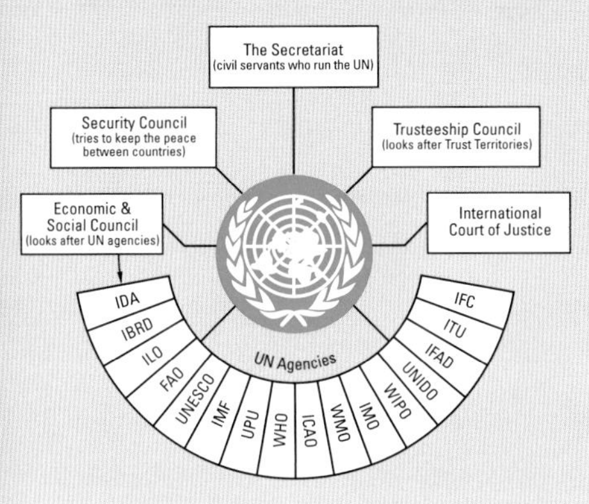

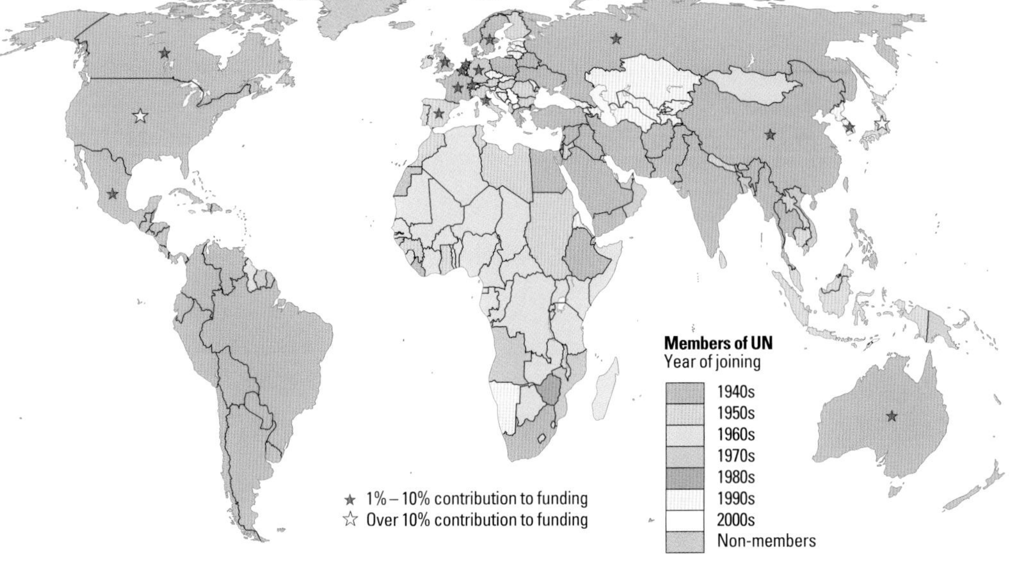

Members of UN
Year of joining

- 1940s
- 1950s
- 1960s
- 1970s
- 1980s
- 1990s
- 2000s
- Non-members

★ 1% – 10% contribution to funding
☆ Over 10% contribution to funding

MEMBERSHIP OF THE UN From the original 51, membership of the UN has now grown to 192. Recent additions include East Timor, Switzerland and Montenegro. There are only two independent states which are not members of the UN – Taiwan and the Vatican City. All the successor states of the former USSR had joined by the end of 1992. The official languages of the UN are Chinese, English, French, Russian, Spanish and Arabic.

FUNDING The UN budget for 2008–9 was US $4.2 billion. Contributions are assessed by the members' ability to pay, with the maximum 22% of the total (USA's share), and the minimum 0.001%. The 27-member EU pays nearly 39% of the budget.

PEACEKEEPING The UN has been involved in 64 peacekeeping operations worldwide since 1948.

International Organizations

ACP African-Caribbean-Pacific (formed in 1963). Members have economic ties with the EU.
APEC Asia-Pacific Economic Co-operation (formed in 1989). It aims to enhance economic growth and prosperity for the region and to strengthen the Asia-Pacific community. APEC is the only intergovernmental grouping in the world operating on the basis of non-binding commitments, open dialogue, and equal respect for the views of all participants. There are 21 member economies.
ARAB LEAGUE (formed in 1945). The League's aim is to promote economic, social, political and military co-operation. There are 22 member nations.
ASEAN Association of South-east Asian Nations (formed in 1967). Cambodia joined in 1999.
AU The African Union replaced the Organization of African Unity (formed in 1963) in 2002. Its 53 members represent over 94% of Africa's population. Arabic, English, French and Portuguese are recognized as working languages.
COLOMBO PLAN (formed in 1951). Its 25 members aim to promote economic and social development in Asia and the Pacific.
COMMONWEALTH The Commonwealth of Nations evolved from the British Empire. Pakistan was suspended in 1999, but reinstated in 2004. Zimbabwe was suspended in 2002 and, in response to its continued suspension, Zimbabwe left the Commonwealth in 2003. Fiji Islands was suspended in 2006 following a military coup. Rwanda joined the Commonwealth in 2009, as the 54th member state, becoming only the second country which was not formerly a British colony to be admitted to the group.
EU European Union (evolved from the European Community in 1993). Cyprus, the Czech Republic, Estonia, Hungary, Latvia, Lithuania, Malta, Poland, the Slovak Republic and Slovenia joined the EU in May 2004; Bulgaria and Romania joined in 2007. The other 15 members of the EU are Austria, Belgium, Denmark, Finland, France, Germany, Greece, Ireland, Italy, Luxembourg, Netherlands, Portugal, Spain, Sweden and the UK. Together, the 27 members aim to integrate economies, co-ordinate social developments and bring about political union.
LAIA Latin American Integration Association (1980). Its aim is to promote freer regional trade.
NATO North Atlantic Treaty Organization (formed in 1949). It continues despite the winding-up of the Warsaw Pact in 1991. Bulgaria, Estonia, Latvia, Lithuania, Romania, the Slovak Republic and Slovenia became members in 2004.

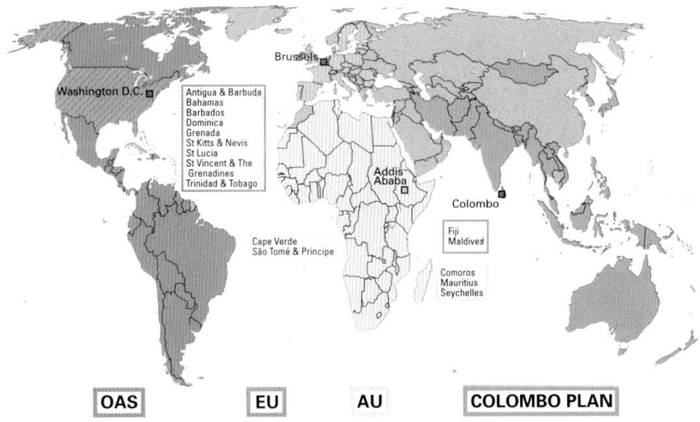

OAS   EU   AU   COLOMBO PLAN

OAS Organization of American States (formed in 1948). It aims to promote social and economic co-operation between countries in the developed North America and developing Latin America.
OECD Organization for Economic Co-operation and Development (formed in 1961). It comprises 30 major free-market economies. Poland, Hungary and South Korea joined in 1996, and the Slovak Republic in 2000. The 'G8' is its 'inner group' of leading industrial nations, comprising Canada, France, Germany, Italy, Japan, Russia, the UK and the USA.
OPEC Organization of Petroleum Exporting Countries (formed in 1960). It controls about three-quarters of the world's oil supply. Gabon formally withdrew from OPEC in August 1996.

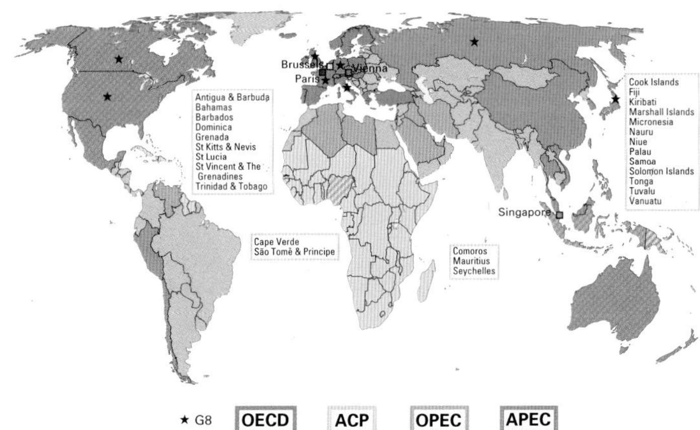

★ G8   OECD   ACP   OPEC   APEC

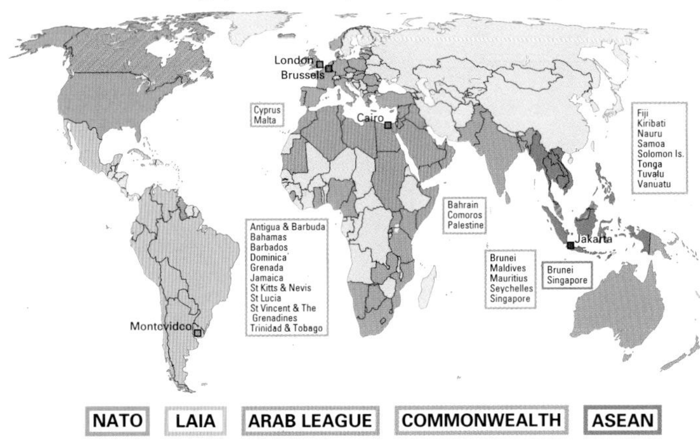

NATO   LAIA   ARAB LEAGUE   COMMONWEALTH   ASEAN

19

Wealth

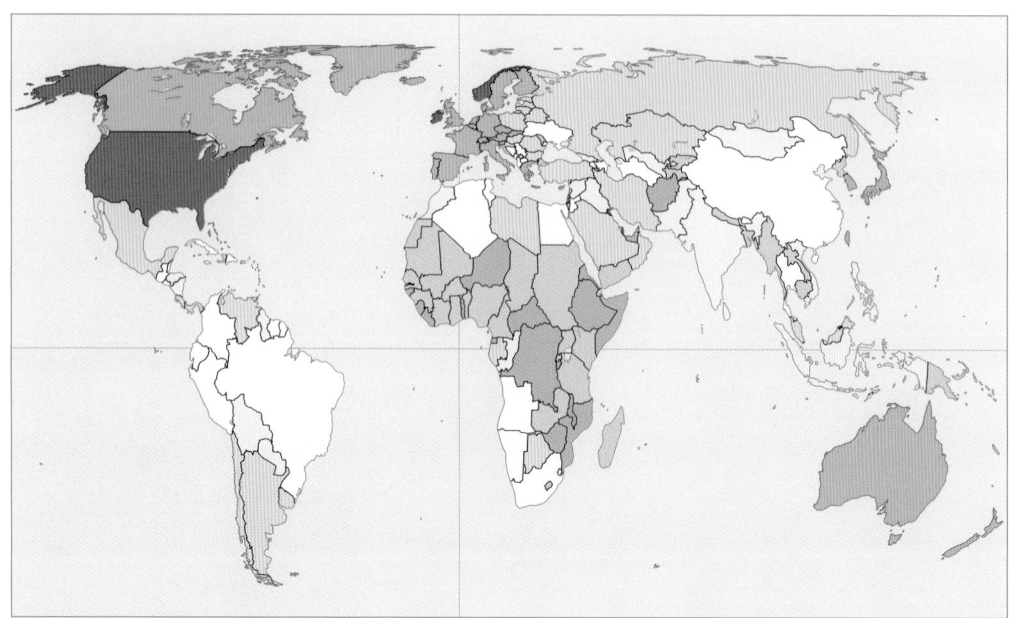

Levels of Income

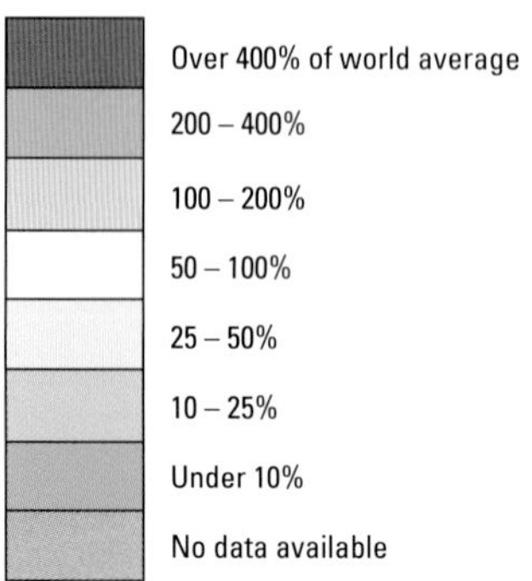

Gross National Income per capita: the value of total production divided by the population (2009)

	Over 400% of world average
	200 – 400%
	100 – 200%
	50 – 100%
	25 – 50%
	10 – 25%
	Under 10%
	No data available

Wealth Creation

The Gross National Income (GNI) of the world's largest economies, US $ million (2008)

1.	USA	14,466,100	21.	Indonesia	458,200
2.	Japan	4,879,200	22.	Poland	453,000
3.	China	3,899,300	23.	Norway	415,200
4.	Germany	3,485,700	24.	Austria	386,000
5.	UK	2,787,200	25.	Saudi Arabia	374,300
6.	France	2,702,200	26.	Denmark	325,100
7.	Italy	2,109,100	27.	Greece	322,000
8.	Spain	1,456,500	28.	Argentina	287,200
9.	Brazil	1,411,200	29.	South Africa	283,300
10.	Canada	1,390,000	30.	Venezuela	257,800
11.	Russia	1,364,500	31.	Finland	255,700
12.	India	1,215,500	32.	Iran	251,500
13.	Mexico	1,061,400	33.	Ireland	221,200
14.	South Korea	1,046,300	34.	Hong Kong	219,300
15.	Australia	862,500	35.	Portugal	218,400
16.	Netherlands	824,600	36.	Colombia	207,400
17.	Turkey	690,700	37.	Thailand	191,700
18.	Switzerland	498,500	38.	Malaysia	188,100
19.	Belgium	474,500	39.	Israel	180,500
20.	Sweden	469,700	40.	Nigeria	175,600

The Wealth Gap

The world's richest and poorest countries, by Gross National Income (GNI) per capita in US $ (2008)

Richest countries			Poorest countries		
1.	Luxembourg	64,320	1.	Congo (Dem. Rep.)	290
2.	Norway	58,500	2.	Liberia	300
3.	Kuwait	52,610	3.	Burundi	380
4.	Macau (China)	52,260	4.	Guinea-Bissau	530
5.	Brunei	50,200	5.	Eritrea	630
6.	Singapore	47,940	6.	Niger	680
7.	USA	46,970	7.	Central African Rep.	730
8.	Switzerland	46,460	8.	Sierra Leone	750
9.	Hong Kong (China)	43,960	9.	Mozambique	770
10.	Netherlands	41,670	10.	Togo	820
11.	Sweden	38,180	11.	Malawi	830
12.	Austria	37,680	12.	Ethiopia	870
13.	Ireland	37,350	13.	Rwanda	1,010
14.	Denmark	37,280	14.	Madagascar	1,040
15.	Canada	36,220	15.	Mali	1,090
16.	UK	36,130	16.	Nepal	1,120
17.	Germany	35,940	17.	Uganda	1,140
18.	Finland	35,660	18.	Burkina Faso	1,160
19.	Japan	35,220	=	Chad	1,160
20.	Belgium	34,760	20.	Comoros	1,170

Continental Shares

Shares of population and of wealth (GNI) by continent

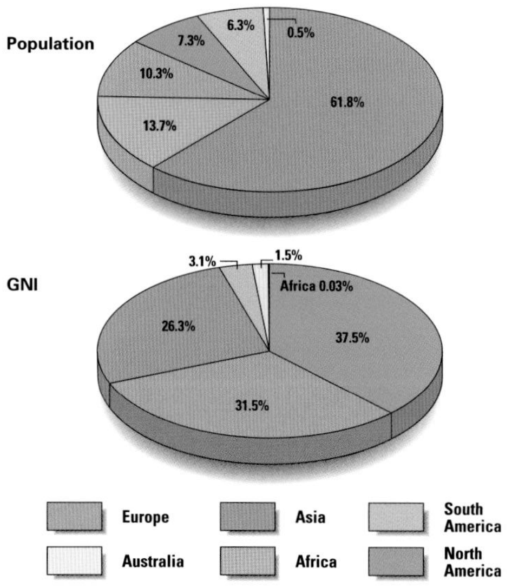

Population

GNI

	Europe		Asia		South America
	Australia		Africa		North America

Inflation

Average annual rate of inflation (2009)

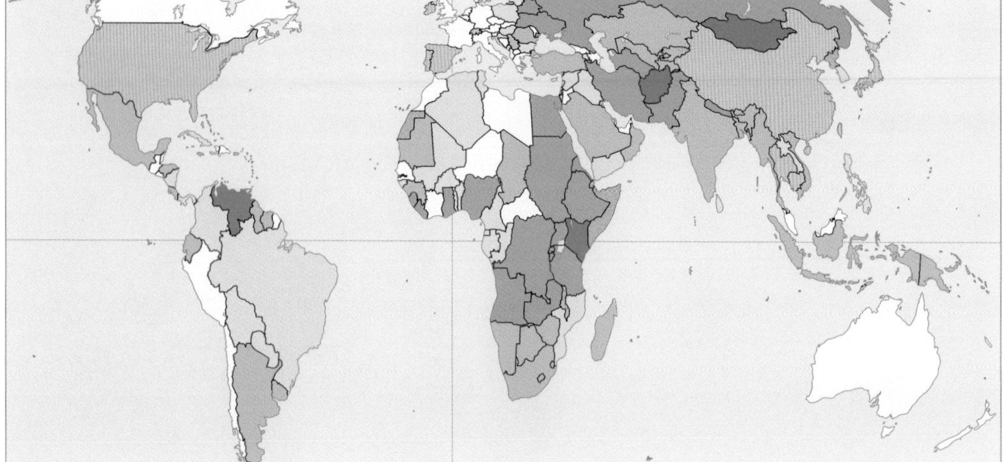

	Over 20%
	10% – 20%
	5% – 10%
	2.5% – 5%
	Under 2.5%
	Negative inflation
	No data available

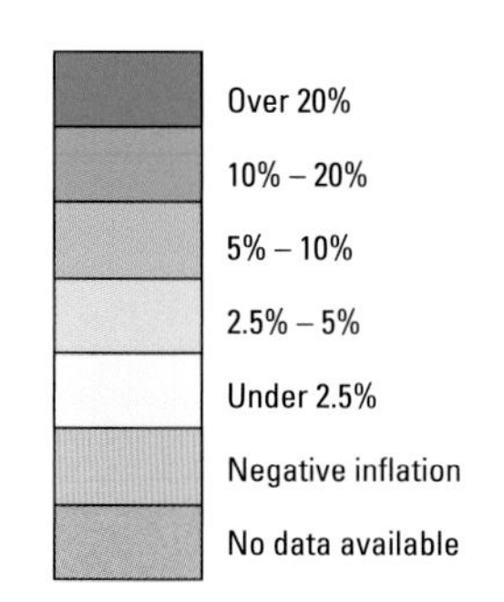

Highest average inflation

Seychelles	34%
Mongolia	28%
Venezuela	27%

Lowest average inflation

Qatar	–3.9%
Ireland	–3.9%
San Marino	–3.5%

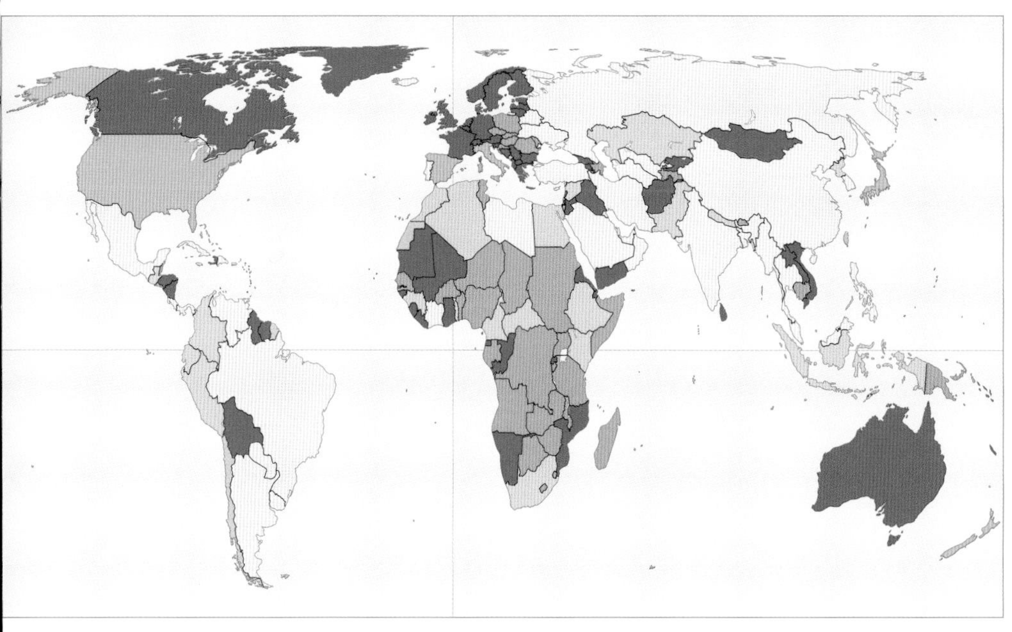

International Aid

Official Development Assistance (ODA) provided and received, per capita (2007)

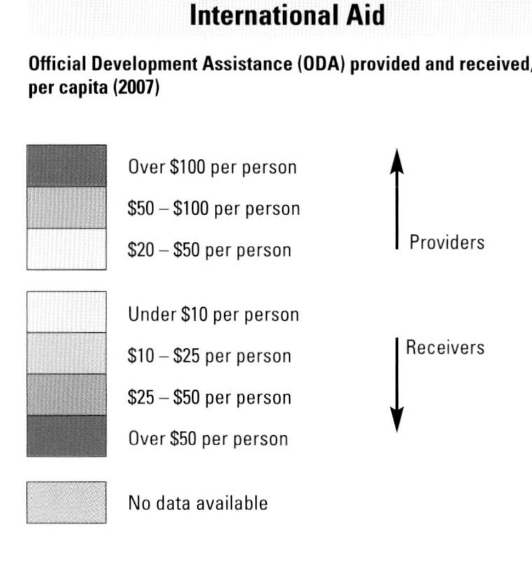

- Over $100 per person
- $50 – $100 per person
- $20 – $50 per person

Providers

- Under $10 per person
- $10 – $25 per person
- $25 – $50 per person
- Over $50 per person

Receivers

- No data available

Debt and Aid

International debtors and the aid they receive

Although aid grants make a vital contribution to many of the world's poorer countries, they are usually dwarfed by the burden of debt that the developing economies are expected to repay. It is estimated that the total debt burden of developing countries is US$523 billion.

- Debt, US $ per capita (2007)
- Aid, US $ per capita (2007)

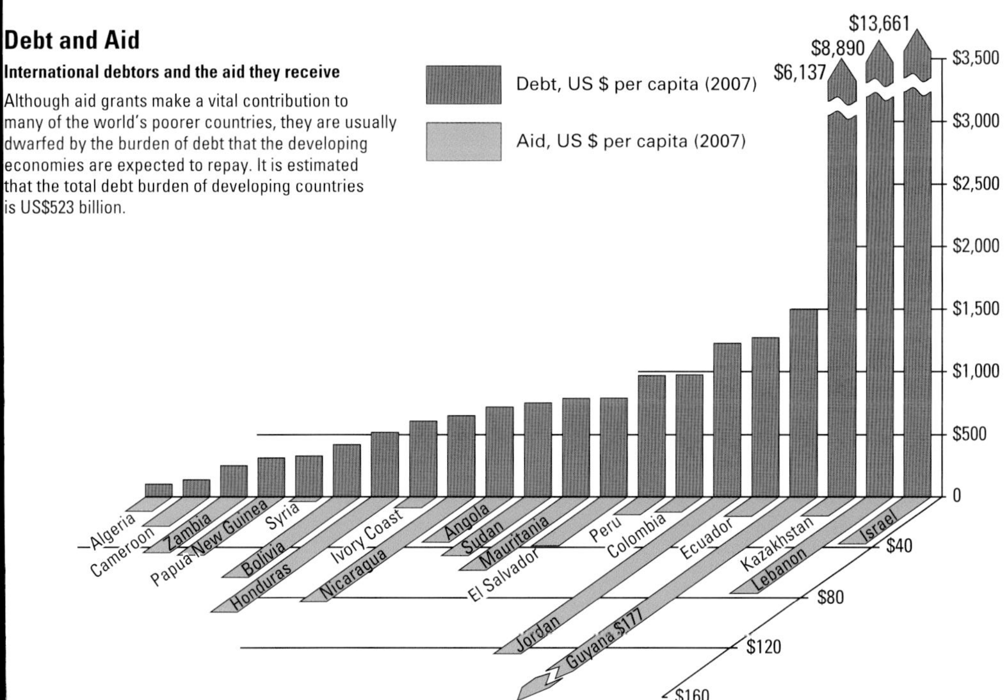

Distribution of Spending

Percentage share of household spending, selected countries

- Food
- Clothing
- Energy & Housing
- Medicine & Education
- Transport
- Other

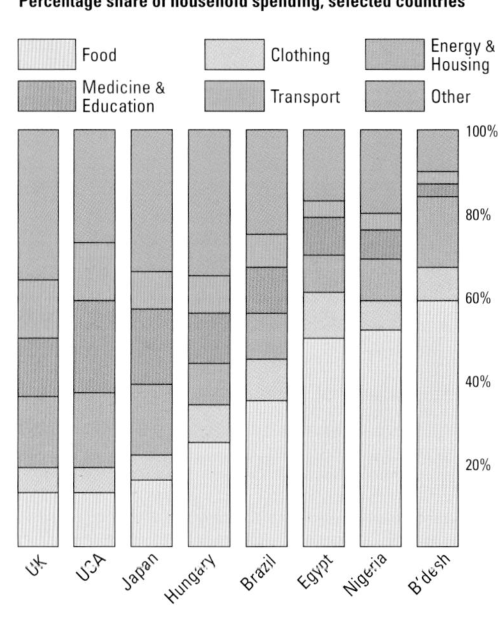

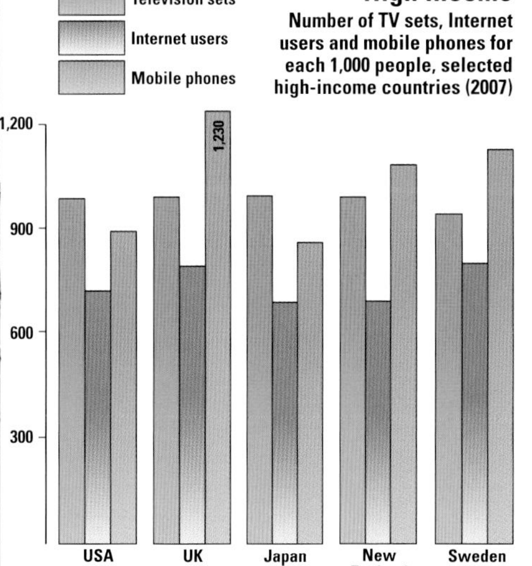

- Television sets
- Internet users
- Mobile phones

High Income
Number of TV sets, Internet users and mobile phones for each 1,000 people, selected high-income countries (2007)

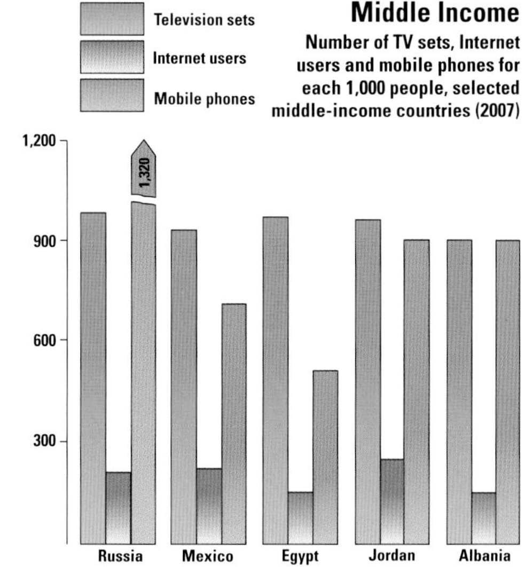

- Television sets
- Internet users
- Mobile phones

Middle Income
Number of TV sets, Internet users and mobile phones for each 1,000 people, selected middle-income countries (2007)

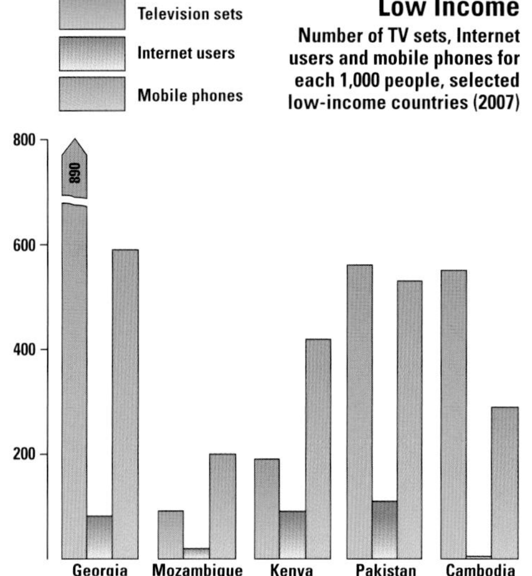

- Television sets
- Internet users
- Mobile phones

Low Income
Number of TV sets, Internet users and mobile phones for each 1,000 people, selected low-income countries (2007)

21

Quality of Life

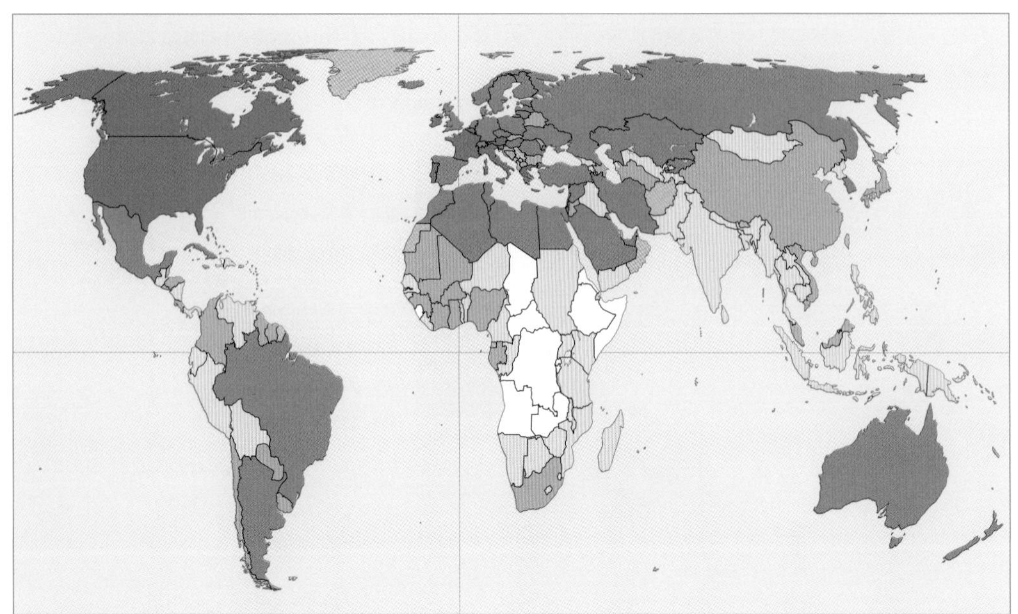

Daily Food Consumption

Average daily food intake in calories per person (2005)

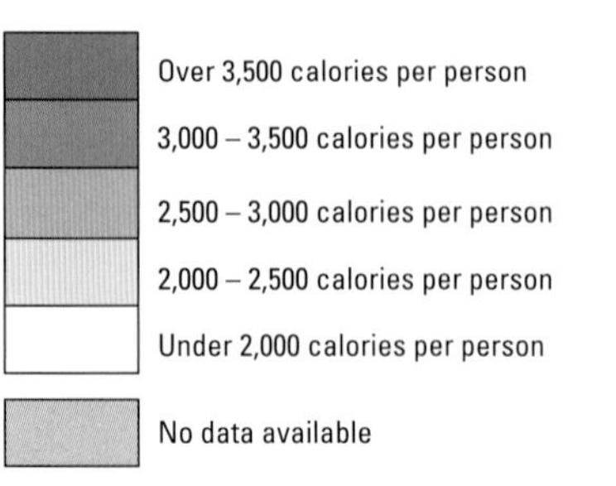

- Over 3,500 calories per person
- 3,000 – 3,500 calories per person
- 2,500 – 3,000 calories per person
- 2,000 – 2,500 calories per person
- Under 2,000 calories per person
- No data available

Hospital Capacity

Hospital beds available for each 1,000 people (2007)

Highest capacity		Lowest capacity	
Japan	14.1	Angola	0.1
Belarus	11.1	Cambodia	0.1
Russia	9.7	Malawi	0.1
Ukraine	8.7	Senegal	0.1
South Korea	8.6	Ethiopia	0.2
Czech Republic	8.4	Nepal	0.2
Germany	8.3	Bangladesh	0.3
Azerbaijan	8.1	Guinea	0.3
Lithuania	8.0	Madagascar	0.3
Hungary	7.9	Mali	0.3
Kazakhstan	7.8	Afghanistan	0.4
Austria	7.6	Chad	0.4
Latvia	7.6	Sierra Leone	0.4
Malta	7.6	Benin	0.5
Iceland	7.5	Nigeria	0.5

Although the ratio of people to hospital beds gives a good approximation of a country's health provision, it is not an absolute indicator. Raw numbers may mask inefficiency and other weaknesses: the high availability of beds in Belarus, for example, has not prevented infant mortality rates over three times as high as in the United Kingdom and the United States.

Life Expectancy

Years of life expectancy at birth, selected countries (2009)

The chart shows combined data for both sexes. On average, women live longer than men worldwide, even in developing countries with high maternal mortality rates. Overall, life expectancy is steadily rising, though the difference between rich and poor nations remains dramatic.

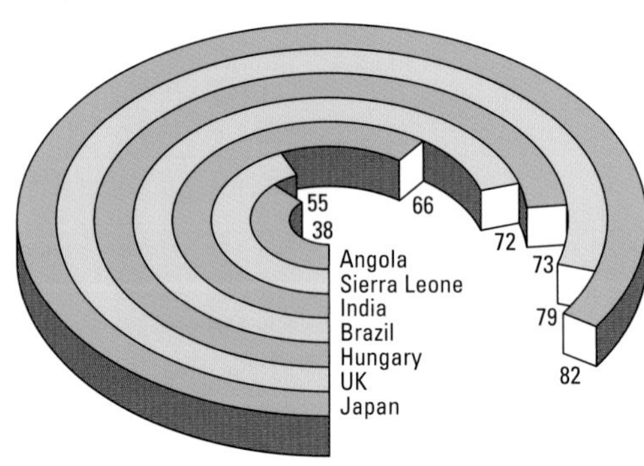

Angola 38
Sierra Leone 55
India 66
Brazil 72
Hungary 73
UK 79
Japan 82

Causes of Death

Causes of death for selected countries by percentage

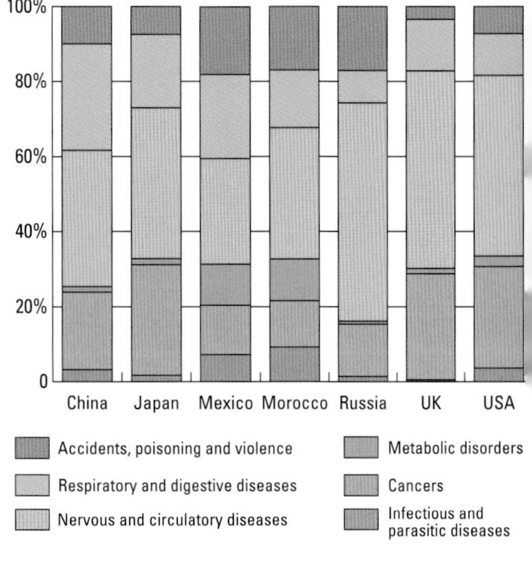

- Accidents, poisoning and violence
- Respiratory and digestive diseases
- Nervous and circulatory diseases
- Metabolic disorders
- Cancers
- Infectious and parasitic diseases

Infant Mortality

Number of babies who died under the age of one, per 1,000 live births (2009)

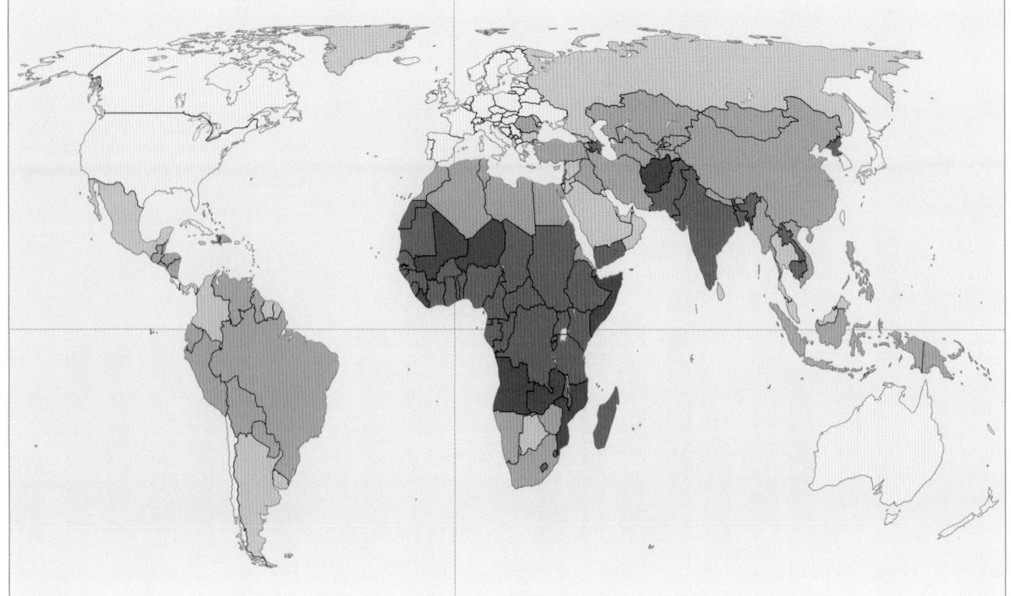

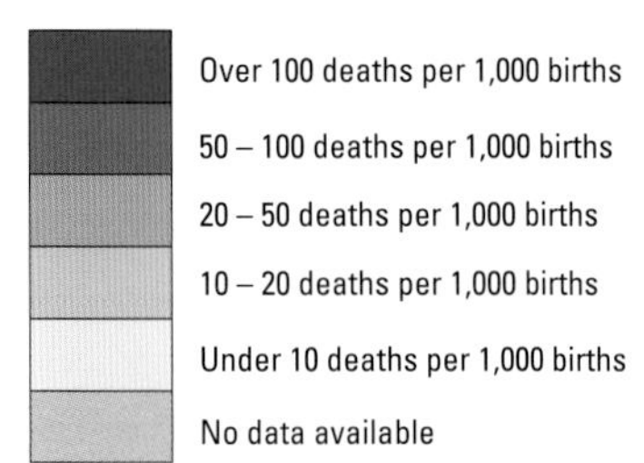

- Over 100 deaths per 1,000 births
- 50 – 100 deaths per 1,000 births
- 20 – 50 deaths per 1,000 births
- 10 – 20 deaths per 1,000 births
- Under 10 deaths per 1,000 births
- No data available

Highest infant mortality		Lowest infant mortality	
Angola	180 deaths	Japan	3 deaths
Afghanistan	153 deaths	Iceland	3 deaths
Liberia	138 deaths	France	3 deaths

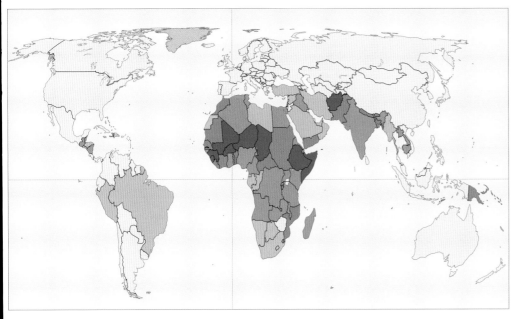

Illiteracy

Percentage of the total adult population unable to read or write (2007)

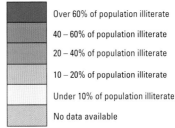

Over 60% of population illiterate

40 – 60% of population illiterate

20 – 40% of population illiterate

10 – 20% of population illiterate

Under 10% of population illiterate

No data available

Countries with the highest and lowest illiteracy rates

Highest		Lowest	
Burkina Faso	87	Australia	0
Niger	83	Denmark	0
Mali	81	Finland	0
Sierra Leone	69	Liechtenstein	0
Guinea	64	Luxembourg	0

Fertility and Education

Fertility rates compared with female education, selected countries (2008)

Percentage of females aged 12–17 in secondary education

Fertility rate: average number of children borne per woman

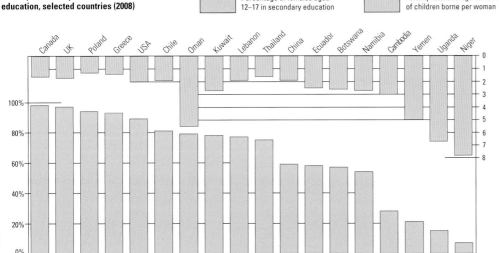

Living Standards

At first sight, most international contrasts in living standards are swamped by differences in wealth. The rich not only have more money, they have more of everything, including years of life. Those with only a little money are obliged to spend most of it on food and clothing, the basic maintenance costs of their existence; air travel and tourism are unlikely to feature on their expenditure lists. However, poverty and wealth are both relative: slum dwellers living on social security payments in an affluent industrial country have far more resources at their disposal than an average African peasant, but feel their own poverty nonetheless. A middle-class Indian lawyer cannot command a fraction of the earnings of a counterpart living in New York, London or Rome; nevertheless, he rightly sees himself as prosperous.

The rich not only live longer, on average, than the poor, they also die from different causes. Infectious and parasitic diseases, all but eliminated in the developed world, remain a scourge in the developing nations. On the other hand, more than two-thirds of the populations of OECD nations eventually succumb to cancer or circulatory disease.

Human Development Index

The Human Development Index (HDI), calculated by the UN Development Programme (UNDP), gives a value to countries using indicators of life expectancy, education and standards of living (2007). Higher values show more developed countries.

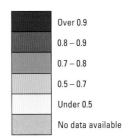

Over 0.9

0.8 – 0.9

0.7 – 0.8

0.5 – 0.7

Under 0.5

No data available

Highest values		Lowest values	
Norway	0.971	Niger	0.340
Australia	0.970	Afghanistan	0.352
Iceland	0.969	Sierra Leone	0.365
Canada	0.966	Central African Rep.	0.369
Ireland	0.965	Mali	0.371

Energy

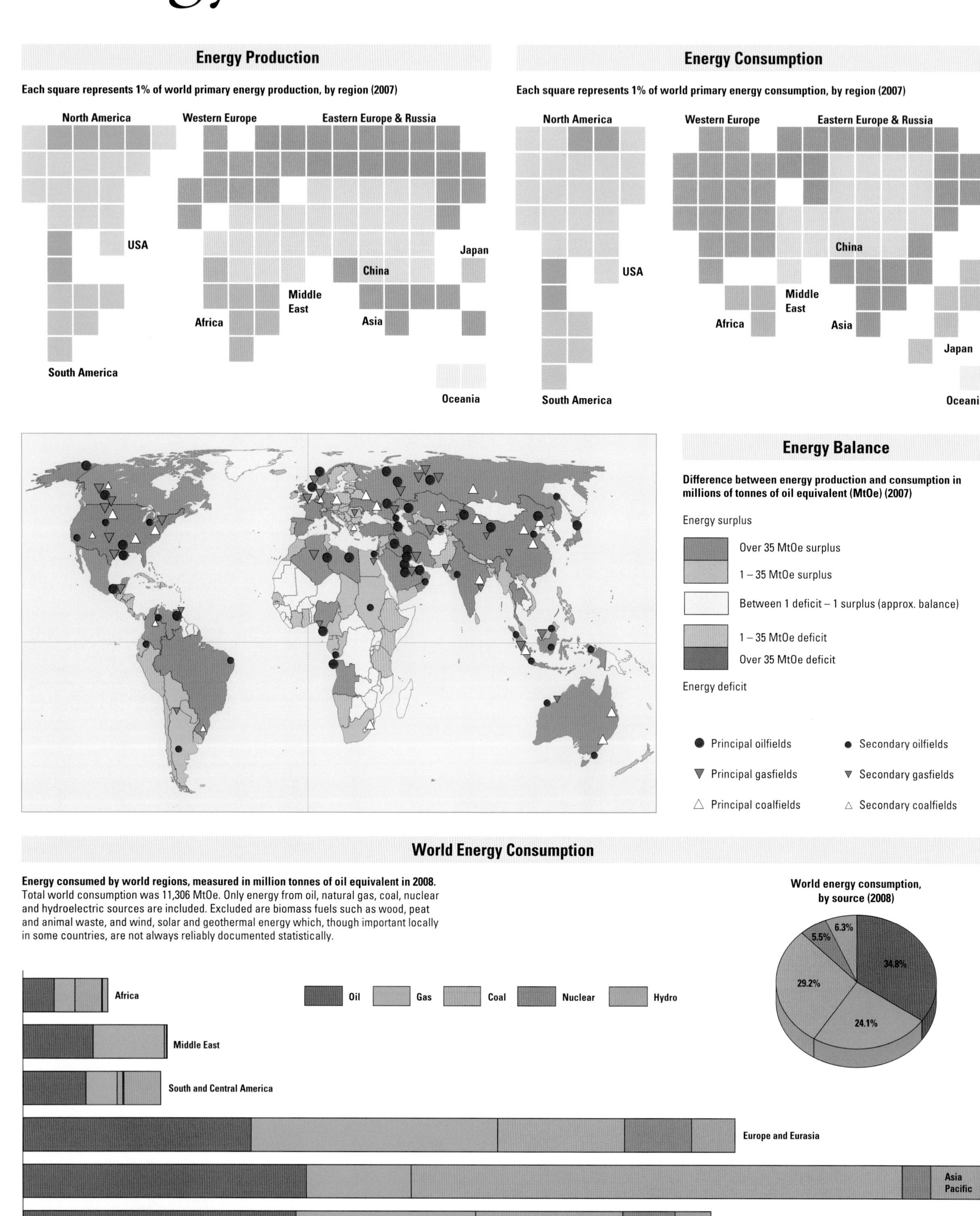

Energy Production

Each square represents 1% of world primary energy production, by region (2007)

North America
Western Europe
Eastern Europe & Russia
USA
Japan
China
Middle East
Africa
Asia
South America
Oceania

Energy Consumption

Each square represents 1% of world primary energy consumption, by region (2007)

North America
Western Europe
Eastern Europe & Russia
China
USA
Middle East
Africa
Asia
South America
Japan
Oceania

Energy Balance

Difference between energy production and consumption in millions of tonnes of oil equivalent (MtOe) (2007)

Energy surplus

Over 35 MtOe surplus

1 – 35 MtOe surplus

Between 1 deficit – 1 surplus (approx. balance)

1 – 35 MtOe deficit

Over 35 MtOe deficit

Energy deficit

● Principal oilfields   ● Secondary oilfields

▼ Principal gasfields   ▼ Secondary gasfields

△ Principal coalfields   △ Secondary coalfields

World Energy Consumption

Energy consumed by world regions, measured in million tonnes of oil equivalent in 2008.
Total world consumption was 11,306 MtOe. Only energy from oil, natural gas, coal, nuclear and hydroelectric sources are included. Excluded are biomass fuels such as wood, peat and animal waste, and wind, solar and geothermal energy which, though important locally in some countries, are not always reliably documented statistically.

World energy consumption, by source (2008)

6.3%
5.5%
34.8%
29.2%
24.1%

Oil   Gas   Coal   Nuclear   Hydro

Africa

Middle East

South and Central America

Europe and Eurasia

Asia Pacific

North America

500   1,000   1,500   2,000   2,500   3,000   3,500   4,0

million tonnes of oil equivalent

Source: BP Statistical Review of World Energy 2009

24

Energy

Energy is used to keep us warm or cool, fuel our industries and our transport systems, and even feed us; high-intensity agriculture, with its use of fertilizers, pesticides and machinery, is heavily energy-dependent. Although we live in a high-energy society, there are vast discrepancies between rich and poor; for example, a North American consumes six times as much energy as a Chinese person. But even developing nations have more power at their disposal than was imaginable a century ago.

The distribution of energy supplies, most importantly fossil fuels (coal, oil and natural gas), is very uneven. In addition, the diagrams and map opposite show that the largest producers of energy are not necessarily the largest consumers. The movement of energy supplies around the world is therefore an important component of international trade.

As the finite reserves of fossil fuels are depleted, renewable energy sources, such as solar, hydro-thermal, wind, tidal and biomass, will become increasingly important around the world.

Nuclear Power

Major producers by percentage of world total and by percentage of domestic electricity generation (2008)

Country	% of world total production	Country	% of nuclear as proportion of domestic electricity
1. USA	31.0%	1. France	77.5%
2. France	16.1%	2. Lithuania	75.6%
3. Japan	9.4%	3. Slovak Rep.	56.7%
4. Russia	5.9%	4. Belgium	55.4%
5. South Korea	5.5%	5. Ukraine	45.5%
6. Germany	5.4%	6. Slovenia	42.2%
7. Canada	3.4%	7. Sweden	41.9%
8. Ukraine	3.2%	8. Armenia	40.7%
9. China	2.5%	9. Switzerland	40.2%
10. Sweden	2.3%	10. Hungary	37.2%

Although the 1980s were a bad time for the nuclear power industry (fears of long-term environmental damage were heavily reinforced by the 1986 disaster at Chernobyl), the industry picked up in the early 1990s. Sixteen countries currently rely on nuclear power to supply over 25% of their electricity requirements. There are over 400 operating nuclear power stations worldwide, with over 100 more planned or under construction.

Hydroelectricity

Major producers by percentage of world total and by percentage of domestic electricity generation (2007)

Country	% of world total production	Country	% of hydroelectric as proportion of domestic electricity
1. China	14.3%	1. Lesotho	100%
2. Brazil	12.3%	= Bhutan	100%
3. Canada	12.2%	= Paraguay	100%
4. USA	8.3%	4. Mozambique	99.9%
5. Russia	5.8%	5. Congo (Rep. Dem.)	99.7%
6. Norway	4.4%	6. Nepal	99.5%
7. India	4.1%	7. Zambia	99.4%
8. Venezuela	2.8%	8. Norway	98.7%
9. Japan	2.4%	9. Tajikistan	97.9%
10. Sweden	2.2%	10. Burundi	97.8%

Countries heavily reliant on hydroelectricity are usually small and non-industrial: a high proportion of hydroelectric power more often reflects a modest energy budget than vast hydroelectric resources. The USA, for instance, produces only 6% of its power requirements from hydroelectricity; yet that 6% amounts to almost half the hydropower generated by most of Africa.

Fuel Exports

Fuels as a percentage of total value of exports (2007)

- Over 50%
- 10 – 50%
- 1 – 10%
- Under 1%
- No data available

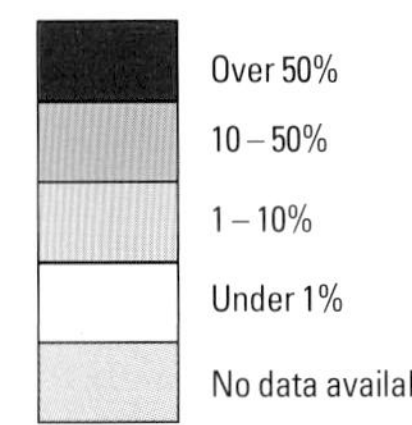

In the 1970s, oil exports became a political issue when OPEC sought to increase the influence of developing countries in world affairs by raising oil prices and restricting production. But its power was short-lived, following a fall in demand for oil in the 1980s, due to an increase in energy efficiency and development of alternative resources. However, with the heavy energy demands of the Asian economies early in the 21st century, both oil and gas prices have risen sharply.

Conversion Rates

1 barrel = 0.136 tonnes or 159 litres or 35 Imperial gallons or 42 US gallons

1 tonne = 7.33 barrels or 1,185 litres or 256 Imperial gallons or 261 US gallons

1 tonne oil = 1.5 tonnes hard coal or 3.0 tonnes lignite or 12,000 kWh

1 Imperial gallon = 1.201 US gallons or 4.546 litres or 277.4 cubic inches

Measurements

For historical reasons, oil is traded in 'barrels'. The weight and volume equivalents (shown right) are all based on average-density 'Arabian light' crude oil.

The energy equivalents given for a tonne of oil are also somewhat imprecise: oil and coal of different qualities will have varying energy contents, a fact usually reflected in their price on world markets.

Energy Reserves

World Oil Reserves

World oil reserves by region and country, billion tonnes (2008)

World total: 170.8 billion tonnes

Al:	Algeria	No:	Norway
Au:	Australia	Po:	Poland
Br:	Brazil	Ru:	Russia
Cn:	China	SA:	Saudi Arabia
In:	Indonesia	S Af:	South Africa
Iq:	Iraq	UAE:	United Arab Emirates
Ka:	Kazakhstan	Uk:	Ukraine
Li:	Libya	USA:	United States of America
Ni:	Nigeria	Ve:	Venezuela

World Gas Reserves

World natural gas reserves by region and country, billion tonnes of oil equivalent (2008)

World total: 169.9 billion tonnes of oil equivalent

World Coal Reserves

World coal reserves (including lignite) by region and country, billion tonnes (2008)

World total: 826.0 billion tonnes

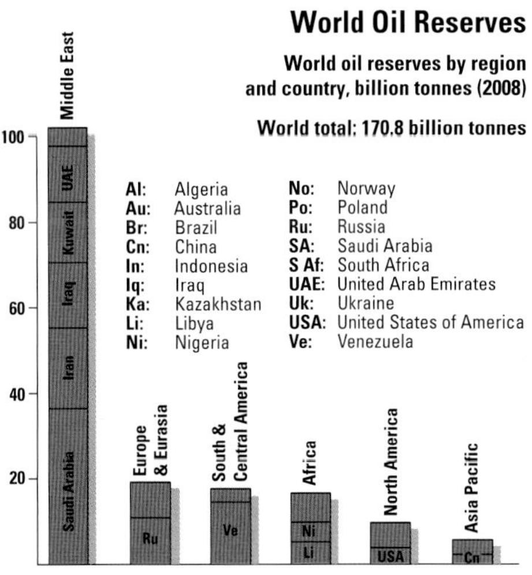

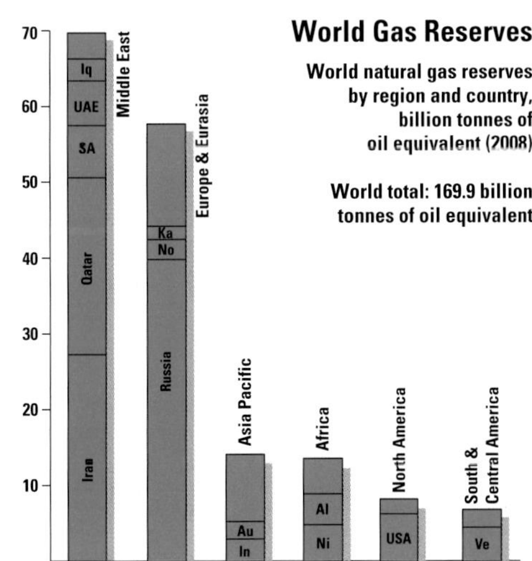

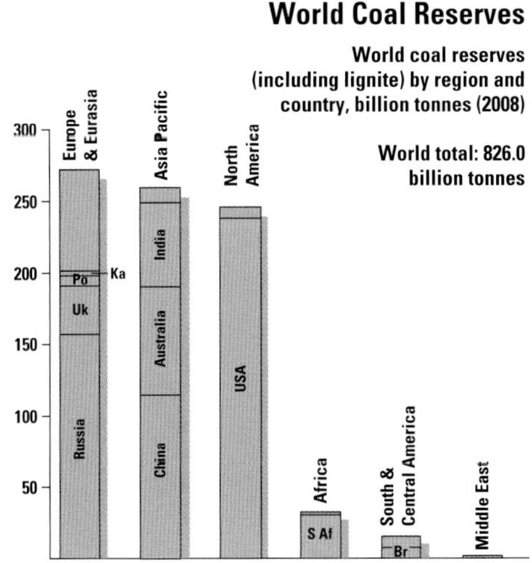

Production

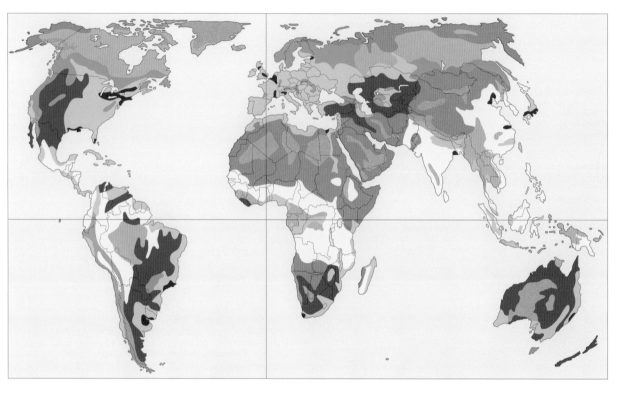

Agriculture

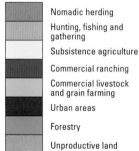

Predominant type of farming or land use

- Nomadic herding
- Hunting, fishing and gathering
- Subsistence agriculture
- Commercial ranching
- Commercial livestock and grain farming
- Urban areas
- Forestry
- Unproductive land

The development of agriculture has transformed human existence more than any other. The whole business of farming is constantly developing: due mainly to the new varieties of rice and wheat, world grain production has more than doubled since 1965. New machinery and modern agricultural techniques enable relatively few farmers to produce enough food for the world's 6 billion or so people.

Staple Crops

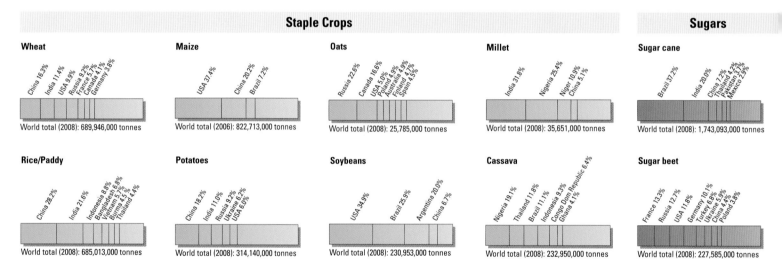

Wheat

China 16.3%, India 11.4%, USA 9.9%, Russia 9.2%, France 5.7%, Canada 4.1%, Germany 3.8%

World total (2008): 689,946,000 tonnes

Maize

USA 37.4%, China 20.2%, Brazil 7.2%

World total (2006): 822,713,000 tonnes

Oats

Russia 22.6%, Canada 16.6%, USA 5.0%, Poland 4.9%, Australia 4.9%, Finland 4.7%, Spain 4.5%

World total (2008): 25,785,000 tonnes

Millet

India 31.8%, Nigeria 25.4%, Niger 10.9%, China 5.1%

World total (2008): 35,651,000 tonnes

Rice/Paddy

China 28.2%, India 21.6%, Indonesia 8.8%, Bangladesh 6.6%, Vietnam 5.7%, Burma 4.5%, Thailand 4.4%

World total (2008): 685,013,000 tonnes

Potatoes

China 18.2%, India 11.0%, Russia 9.2%, Ukraine 6.2%, USA 6.0%

World total (2008): 314,140,000 tonnes

Soybeans

USA 34.9%, Brazil 25.9%, Argentina 20.0%, China 6.7%

World total (2008): 230,953,000 tonnes

Cassava

Nigeria 19.1%, Thailand 11.8%, Brazil 11.1%, Indonesia 9.3%, Congo Dom Republic 6.4%, Ghana 4.1%

World total (2008): 232,950,000 tonnes

Sugars

Sugar cane

Brazil 37.2%, India 20.0%, China 7.2%, Thailand 4.2%, Pakistan 3.3%, Mexico 2.9%

World total (2008): 1,743,093,000 tonnes

Sugar beet

France 13.3%, Russia 12.7%, USA 11.8%, Germany 10.1%, Turkey 6.8%, Ukraine 5.3%, Poland 3.8%

World total (2008): 227,585,000 tonnes

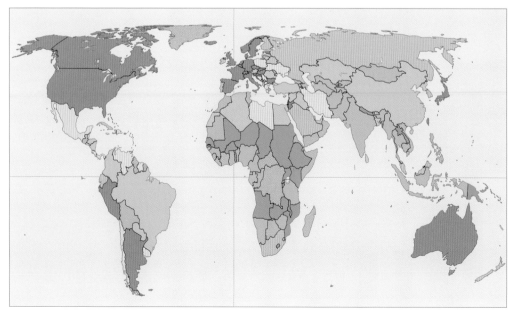

Employment

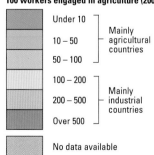

The number of workers employed in manufacturing for every 100 workers engaged in agriculture (2007)

- Under 10 — Mainly agricultural countries
- 10 – 50 — Mainly agricultural countries
- 50 – 100 — Mainly agricultural countries
- 100 – 200 — Mainly industrial countries
- 200 – 500 — Mainly industrial countries
- Over 500 — Mainly industrial countries
- No data available

Countries with the highest number of workers employed in manufacturing per 100 workers engaged in agriculture (2007)

1. Bahrain	7,900	6. Peru	2,400
2. USA	3,800	7. Argentina	2,300
3. San Marino	3,700	8. Singapore	2,200
4. Micronesia, Fed. States	3,400	9. Liechtenstein	2,150
5. Sweden	2,800	10. Andorra	2,000

Mineral Production

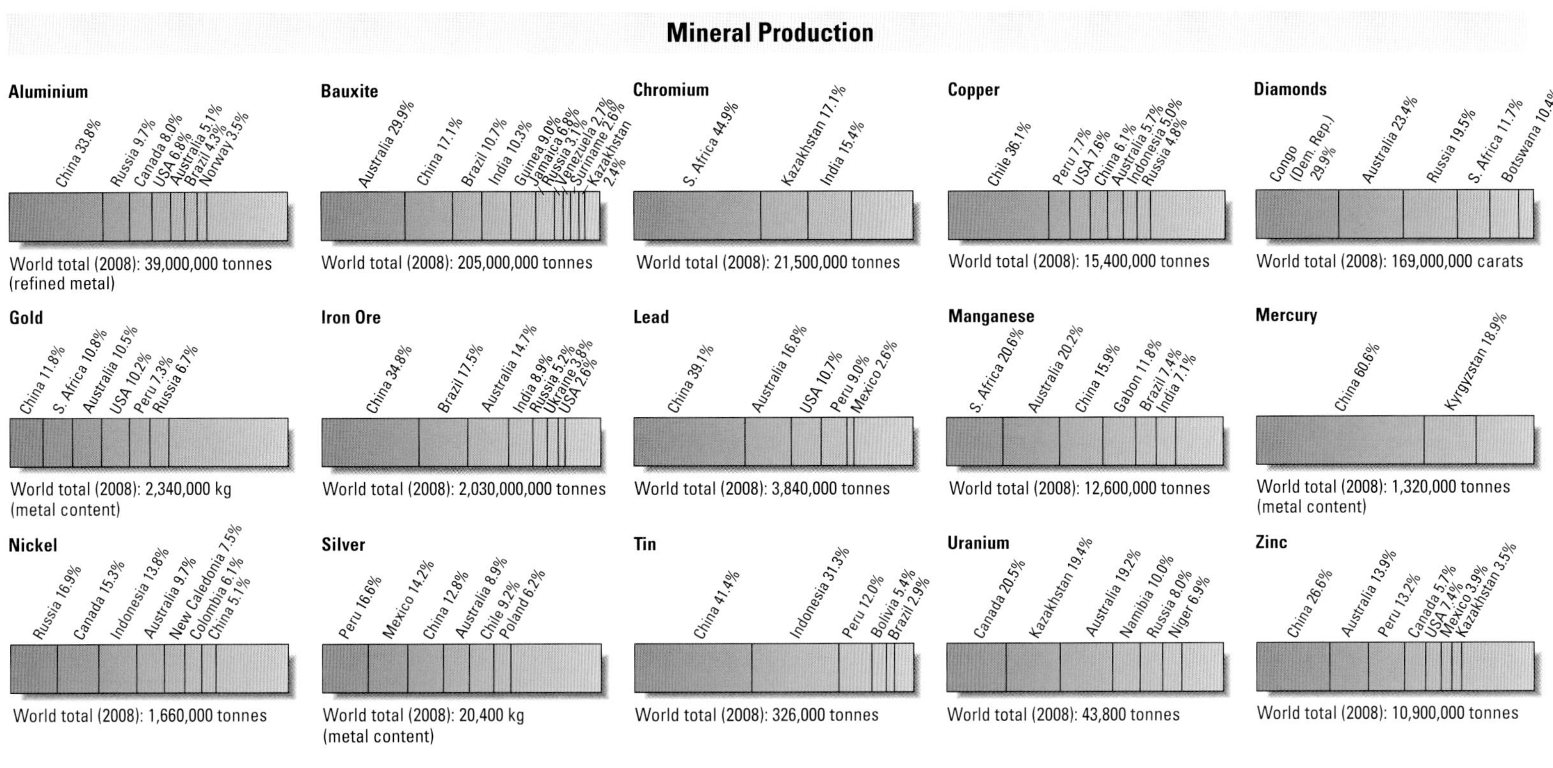

Aluminium

China 33.8% · Russia 9.7% · Canada 8.0% · USA 6.8% · Australia 5.1% · Brazil 4.3% · Norway 3.5%

World total (2008): 39,000,000 tonnes
(refined metal)

Bauxite

Australia 29.9% · China 17.1% · Brazil 10.7% · India 10.3% · Guinea 9.0% · Russia 6.8% · Jamaica 6.8% · Venezuela 2.7% · Suriname 2.6% · Kazakhstan 2.4%

World total (2008): 205,000,000 tonnes

Chromium

S. Africa 44.9% · Kazakhstan 17.1% · India 15.4%

World total (2008): 21,500,000 tonnes

Copper

Chile 36.1% · Peru 7.7% · USA 7.6% · China 6.1% · Australia 5.7% · Indonesia 5.0% · Russia 4.8%

World total (2008): 15,400,000 tonnes

Diamonds

Congo (Dem. Rep.) 29.9% · Australia 23.4% · Russia 19.5% · S. Africa 11.7% · Botswana 10.4%

World total (2008): 169,000,000 carats

Gold

China 11.8% · S. Africa 10.8% · Australia 10.5% · USA 10.2% · Peru 7.3% · Russia 6.7%

World total (2008): 2,340,000 kg
(metal content)

Iron Ore

China 34.8% · Brazil 17.5% · Australia 14.7% · India 8.9% · Russia 5.2% · Ukraine 3.8% · USA 2.6%

World total (2008): 2,030,000,000 tonnes

Lead

China 39.1% · Australia 16.8% · USA 10.7% · Peru 9.0% · Mexico 2.6%

World total (2008): 3,840,000 tonnes

Manganese

S. Africa 20.6% · Australia 20.2% · China 15.9% · Gabon 11.8% · Brazil 7.4% · India 7.1%

World total (2008): 12,600,000 tonnes

Mercury

China 60.6% · Kyrgyzstan 18.9%

World total (2008): 1,320,000 tonnes
(metal content)

Nickel

Russia 16.9% · Canada 15.3% · Indonesia 13.8% · Australia 9.7% · New Caledonia 7.5% · Colombia 6.1% · China 5.1%

World total (2008): 1,660,000 tonnes

Silver

Peru 16.6% · Mexico 14.2% · China 12.8% · Australia 8.5% · Chile 9.2% · Poland 6.2%

World total (2008): 20,400 kg
(metal content)

Tin

China 41.4% · Indonesia 31.3% · Peru 12.0% · Bolivia 5.4% · Brazil 2.9%

World total (2008): 326,000 tonnes

Uranium

Canada 20.5% · Kazakhstan 19.4% · Australia 19.2% · Namibia 10.0% · Russia 8.0% · Niger 6.9%

World total (2008): 43,800 tonnes

Zinc

China 26.6% · Australia 13.9% · Peru 13.2% · Canada 5.7% · USA 7.4% · Mexico 3.9% · Kazakhstan 3.5%

World total (2008): 10,900,000 tonnes

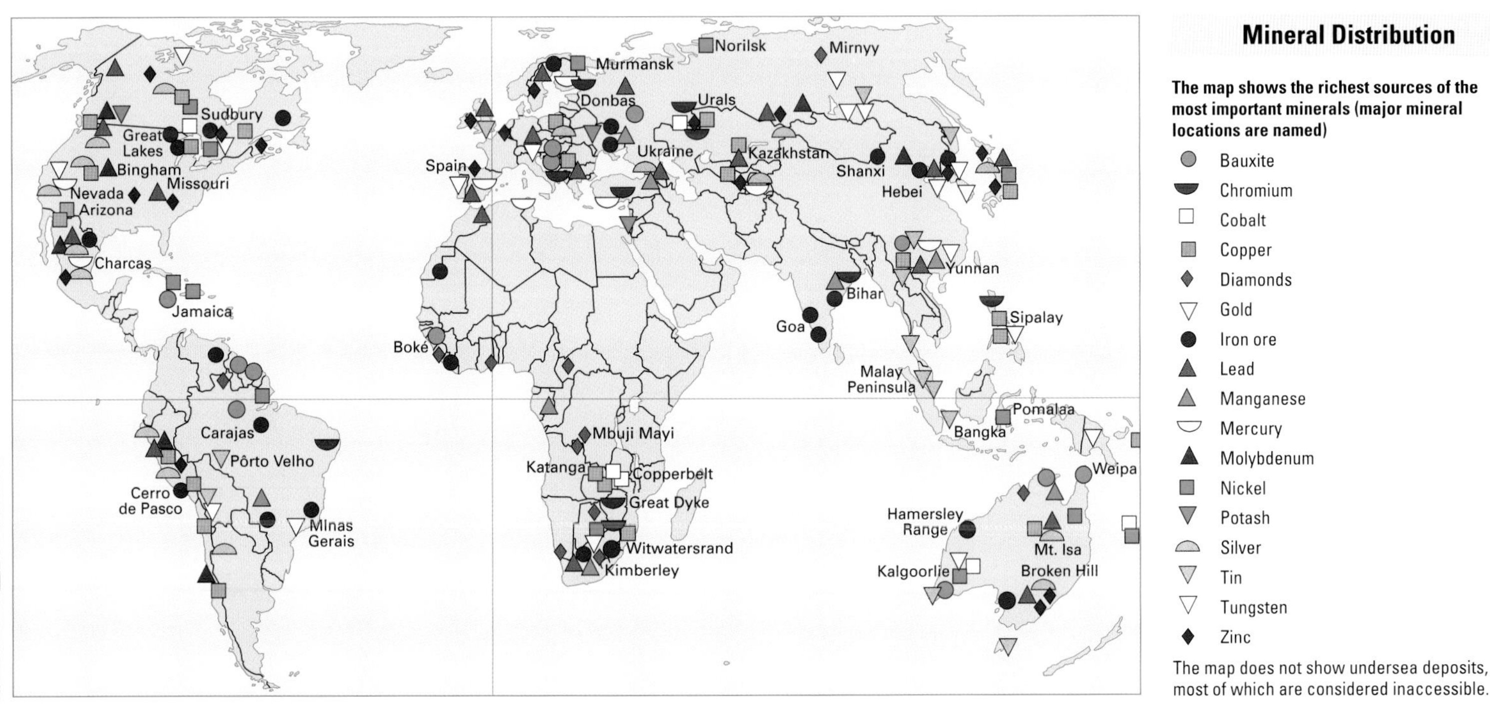

Mineral Distribution

The map shows the richest sources of the most important minerals (major mineral locations are named)

- ◯ Bauxite
- ◗ Chromium
- ☐ Cobalt
- ▢ Copper
- ◆ Diamonds
- ▽ Gold
- ● Iron ore
- ▲ Lead
- △ Manganese
- ▽ Mercury
- ▲ Molybdenum
- ▪ Nickel
- ▼ Potash
- ◠ Silver
- ▽ Tin
- ▽ Tungsten
- ◆ Zinc

The map does not show undersea deposits, most of which are considered inaccessible.

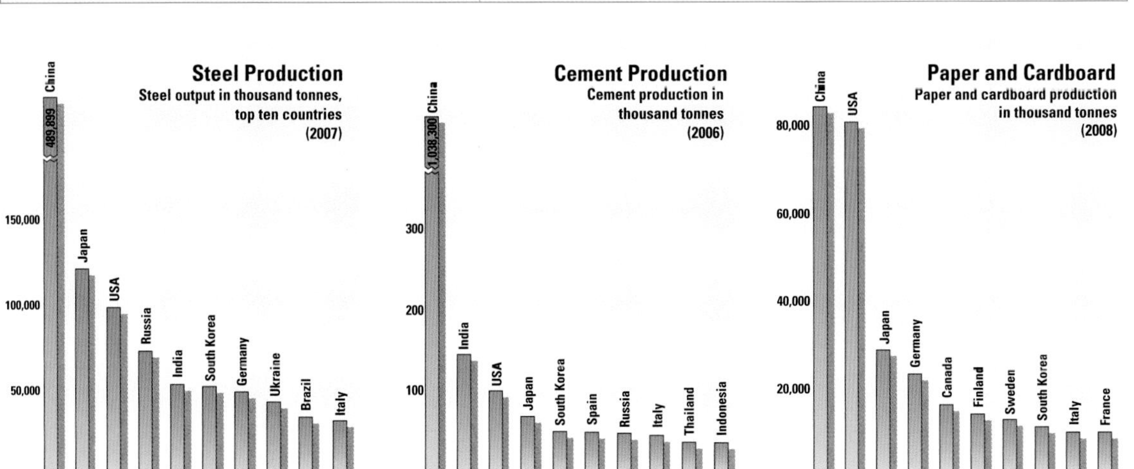

Steel Production
Steel output in thousand tonnes, top ten countries (2007)

China 489,898 · Japan · USA · Russia · India · South Korea · Germany · Ukraine · Brazil · Italy

Cement Production
Cement production in thousand tonnes (2006)

China 1,038,300 · India · USA · Japan · South Korea · Spain · Russia · Italy · Thailand · Indonesia

Paper and Cardboard
Paper and cardboard production in thousand tonnes (2008)

China · USA · Japan · Germany · Canada · Finland · Sweden · South Korea · Italy · France

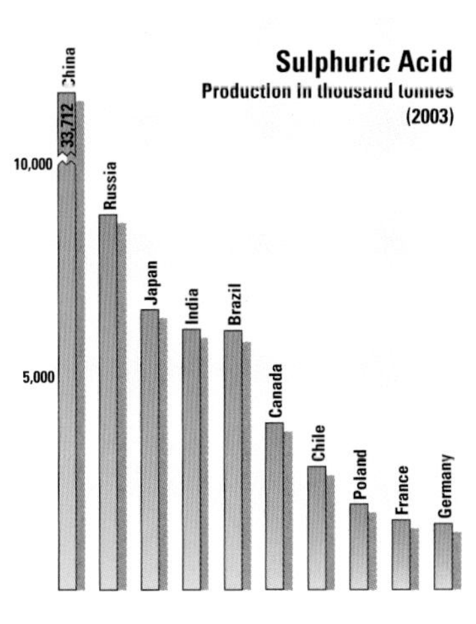

Sulphuric Acid
Production in thousand tonnes (2003)

China 33,712 · Russia · Japan · India · Brazil · Canada · Chile · Poland · France · Germany

Trade

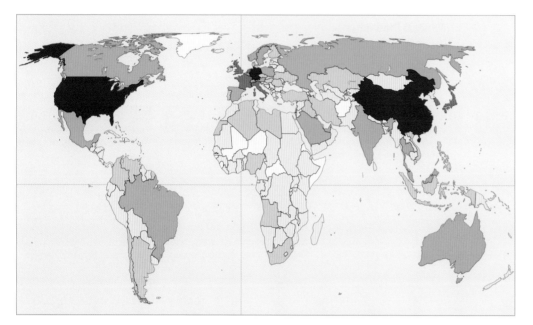

Share of World Trade

Percentage share of total world exports by value (2009)

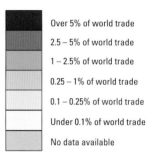

- Over 5% of world trade
- 2.5 – 5% of world trade
- 1 – 2.5% of world trade
- 0.25 – 1% of world trade
- 0.1 – 0.25% of world trade
- Under 0.1% of world trade
- No data available

Countries with the largest share of world trade

1. China	9.86%	6. Netherlands	3.28%
2. Germany	9.80%	7. Italy	3.05%
3. USA	8.21%	8. South Korea	2.93%
4. Japan	4.26%	9. UK	2.90%
5. France	3.77%	10. Hong Kong (China)	2.70%

The Main Trading Nations

The imports and exports of the top ten trading nations as a percentage of world trade (2006). Each country's trade in manufactured goods is shown in dark blue

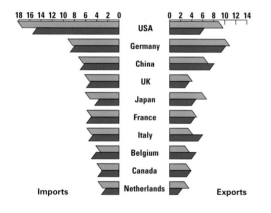

Imports — USA, Germany, China, UK, Japan, France, Italy, Belgium, Canada, Netherlands — Exports

Major Exports

Leading manufactured items and their exporters

Motor Vehicles
World total (2008): US$ 3,355,798 million

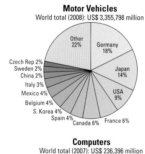

Germany 18%, Japan 14%, USA 9%, France 6%, Canada 6%, Spain 4%, S. Korea 4%, Belgium 4%, Mexico 4%, Italy 3%, China 2%, Sweden 2%, Czech Rep 2%, Other 22%, UK 5%

Telecommunications Gear
World total (2008): US$ 1,619,703 million

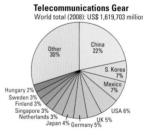

China 22%, S. Korea 7%, Mexico 7%, USA 6%, Germany 5%, Japan 4%, Netherlands 3%, Singapore 3%, Finland 3%, Sweden 3%, Hungary 2%, Other 30%

Petrol Products
World total (2008): US$ 1,819,371 million

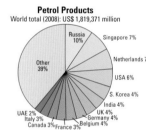

Russia 10%, Singapore 7%, Netherlands 7%, USA 6%, S. Korea 4%, India 4%, UK 4%, Germany 4%, Belgium 4%, France 3%, Canada 3%, Italy 3%, UAE 2%, Other 39%

Computers
World total (2007): US$ 236,396 million

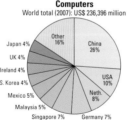

China 26%, USA 10%, Neth. 8%, Germany 7%, Singapore 7%, Malaysia 5%, Mexico 5%, S. Korea 4%, Ireland 4%, UK 4%, Japan 4%, Other 16%

Electrical Components
World total (2008): US$ 5,333,323 million

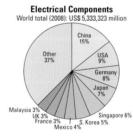

China 15%, USA 9%, Germany 8%, Japan 7%, Singapore 6%, S. Korea 5%, Mexico 4%, France 3%, UK 3%, Malaysia 3%, Other 37%

Pharmaceuticals
World total (2008): US$ 1,238,425 million

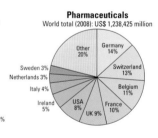

Germany 14%, Switzerland 13%, Belgium 11%, France 10%, UK 9%, USA 8%, Ireland 5%, Italy 4%, Netherlands 3%, Sweden 3%, Other 20%

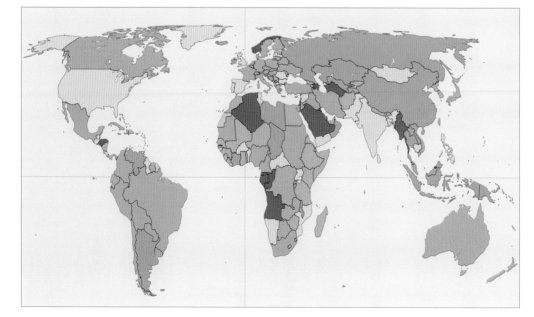

Balance of Trade

Value of exports in proportion to the value of imports (2009)

Imports exceed exports by:
- More than 40%
- 20 – 40%
- 20% either side
- 20 – 40%
- More than 40%
- No data available

Exports exceed imports by:

The total world trade balance should amount to zero, since exports must equal imports on a global scale. In practice, at least $100 billion in exports go unrecorded, leaving the world with an apparent deficit and many countries in a better position than public accounting reveals. However, a favourable trade balance is not necessarily a sign of prosperity: many poorer countries must maintain a high surplus in order to service debts, and do so by restricting imports below the levels needed to sustain successful economies.

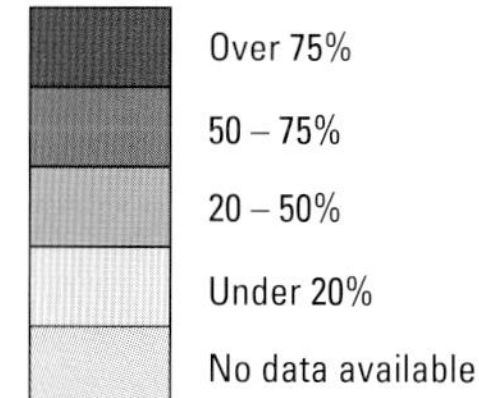

Trade in Primary Exports

Primary exports as a percentage of total export value (2008)

- Over 75%
- 50 – 75%
- 20 – 50%
- Under 20%
- No data available

Primary exports are raw materials or partly processed products that form the basis for manufacturing. They are the necessary requirements of industries and include agricultural products, minerals, fuels and timber, as well as many semi-manufactured goods such as cotton, which has been spun but not woven, wood pulp or flour. Many developed countries have few natural resources and rely on imports for the majority of their primary products. The countries of South-east Asia export hardwoods to the rest of the world, while many South American countries are heavily dependent on coffee exports.

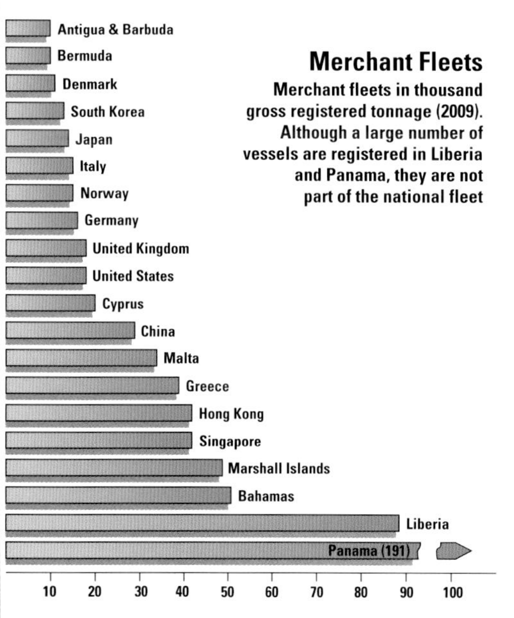

Merchant Fleets

Merchant fleets in thousand gross registered tonnage (2009). Although a large number of vessels are registered in Liberia and Panama, they are not part of the national fleet

Antigua & Barbuda
Bermuda
Denmark
South Korea
Japan
Italy
Norway
Germany
United Kingdom
United States
Cyprus
China
Malta
Greece
Hong Kong
Singapore
Marshall Islands
Bahamas
Liberia
Panama (191)

10   20   30   40   50   60   70   80   90   100

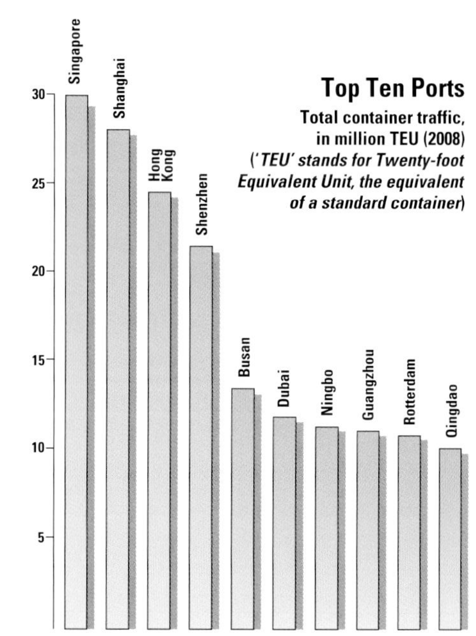

Top Ten Ports

Total container traffic, in million TEU (2008) ('TEU' stands for Twenty-foot Equivalent Unit, the equivalent of a standard container)

Singapore, Shanghai, Hong Kong, Shenzhen, Busan, Dubai, Ningbo, Guangzhou, Rotterdam, Qingdao

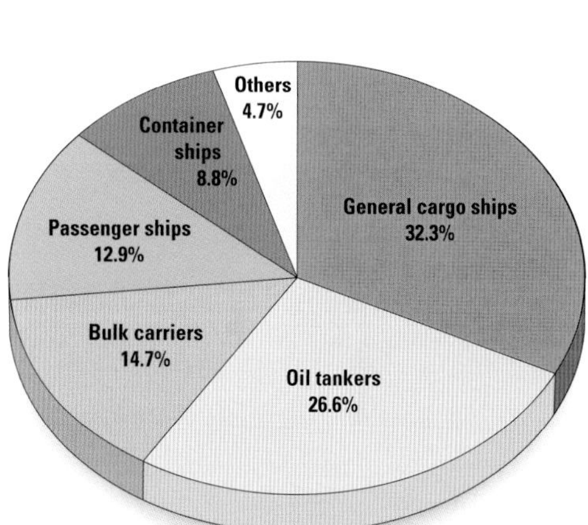

Types of Vessels

World fleet by type of vessel (2009)

- Others 4.7%
- Container ships 8.8%
- Passenger ships 12.9%
- Bulk carriers 14.7%
- General cargo ships 32.3%
- Oil tankers 26.6%

Exports Per Capita

Value of exports in US $, divided by total population (2009)

- Over 10,000
- 5,000 – 10,000
- 1,000 – 5,000
- 500 – 1,000
- 100 – 500
- Under 100
- No data available

Countries with highest exports per capita

San Marino	$153,413
Svalbard (Norway)	$93,573
Liechtenstein	$71,057
Singapore	$52,603
Hong Kong (China)	$46,335
Qatar	$44,919

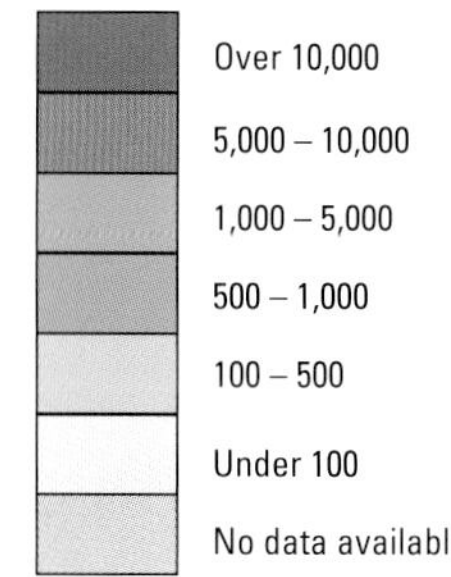

Travel and Tourism

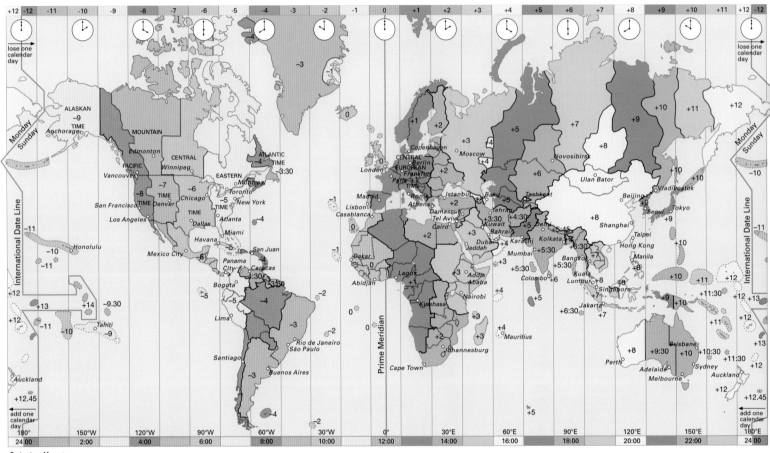

Projection: Mercator

Time Zones

Zones using UT (GMT)	Zones ahead of UT (GMT)	Certain time zones are affected by the incidence of daylight saving time in countries where it is adopted.
Zones behind UT (GMT)	Half-hour zones	
International boundaries	Time-zone boundaries	Actual solar time, when it is noon at Greenwich, is shown along the top of the map.
10 Hours fast or slow of UT or Co-ordinated Universal Time	International Date Line	

The world is divided into 24 time zones, each centred on meridians at 15° intervals, which is the longitudinal distance the sun travels every hour. The meridian running through Greenwich, London, passes through the middle of the first zone.

Rail and Road: The Leading Nations

Total rail network ('000 km)	Passenger km per head per year	Total road network ('000 km)	Vehicle km per head per year	Number of vehicles per km of roads
1. USA 233.8	Japan 1,891	USA 6,378.3	USA................12,505	Hong Kong 287
2. Russia 85.5	Switzerland 1,751	India 3,319.6	Luxembourg 7,989	Qatar...................284
3. Canada 73.2	Belarus..............1,334	China 1,765.2	Kuwait 7,251	UAE......................232
4. India 63.1	France 1,203	Brazil 1,724.9	France 7,142	Germany 195
5. China................60.5	Ukraine 1,100	Canada.............1,408.8	Sweden 6,991	Lebanon 191
6. Germany...........36.1	Russia................1,080	Japan 1,171.4	Germany 6,806	Macau.................172
7. Argentina 34.2	Austria...............1,008	France 893.1	Denmark 6,764	Singapore 167
8. France..............29.3	Denmark 999	Australia 811.6	Austria................6,518	South Korea 160
9. Mexico..............26.5	Netherlands 855	Spain 664.9	Netherlands 5,984	Kuwait 156
10. South Africa........22.7	Germany 842	Russia 537.3	UK 5,738	Taiwan 150
11. Brazil.................22.1	Italy 811	Italy 479.7	Canada................5,493	Israel 111
12. Ukraine 22.1	Belgium 795	UK 371.9	Italy.....................4,852	Malta 110

Air Travel

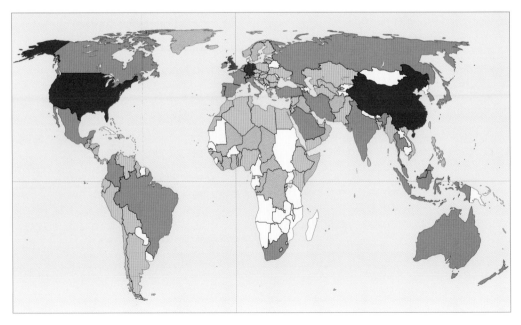

Number of air passengers carried (2008)

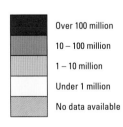

	Over 100 million
	10 – 100 million
	1 – 10 million
	Under 1 million
	No data available

World's busiest airports (2009) – total passengers	World's busiest airports (2009) – international passengers
1. Atlanta (Hartsfield Internat'l)	1. London (Heathrow)
2. London (Heathrow)	2. Paris (Charles de Gaulle)
3. Beijing (Capital Internat'l)	3. Amsterdam (Schipol)
4. Chicago (O'Hare Internat'l)	4. Hong Kong (International)
5. Tokyo (Haneda)	5. Frankfurt (International)

30

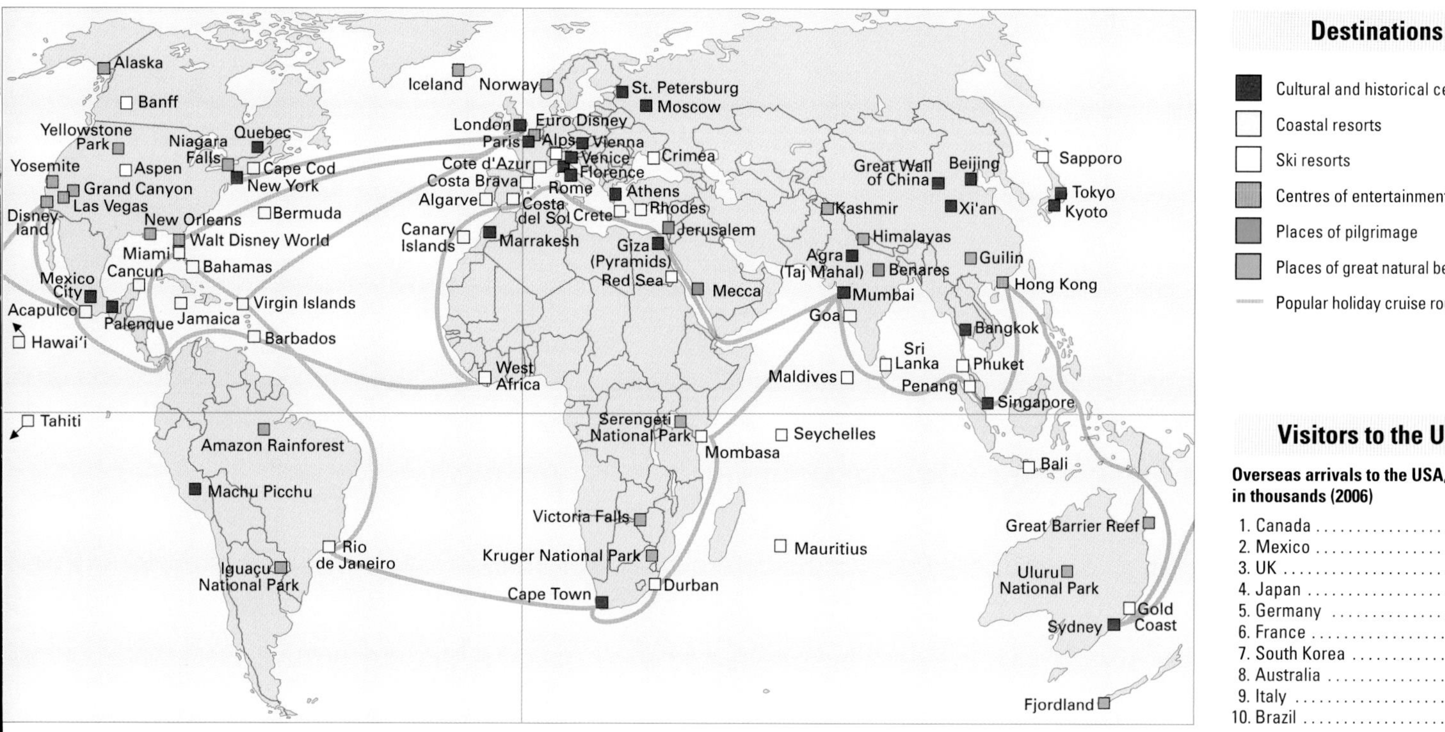

Destinations

- ■ Cultural and historical centres
- □ Coastal resorts
- □ Ski resorts
- ▨ Centres of entertainment
- ▨ Places of pilgrimage
- ▨ Places of great natural beauty
- — Popular holiday cruise routes

Visitors to the USA

Overseas arrivals to the USA, in thousands (2006)

1.	Canada	15,995
2.	Mexico	13,400
3.	UK	4,176
4.	Japan	3,673
5.	Germany	1,386
6.	France	790
7.	South Korea	758
8.	Australia	603
9.	Italy	533
10.	Brazil	525

Tourist Spending

Countries spending the most on overseas tourism, US $ million (2006)

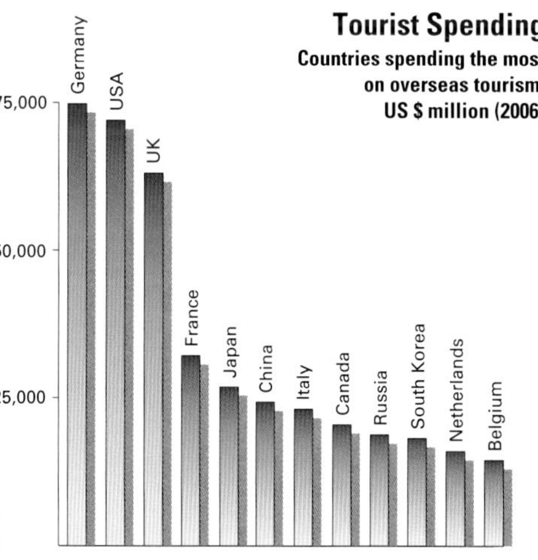

Importance of Tourism

		Arrivals from abroad (2006)	% of world total (2006)
1.	France	76,001,000	9.0%
2.	Spain	55,577,000	6.6%
3.	USA	46,085,000	5.4%
4.	China	41,761,000	4.9%
5.	Italy	36,513,000	4.3%
6.	UK	29,970,000	3.5%
7.	Germany	21,500,000	2.5%
8.	Mexico	20,617,000	2.4%
9.	Turkey	20,273,000	2.4%
10.	Austria	19,952,000	2.4%
11.	Russia	19,940,000	2.4%
12.	Canada	19,152,000	2.3%

The 846 million international arrivals in 2006 represented an additional 43 million over the 2005 level – making a new record year for the industry. Growth was common to all regions, but particularly strong in Asia and the Pacific, and in the Middle East.

Tourist Earnings

Countries receiving the most from overseas tourism, US $ million (2006)

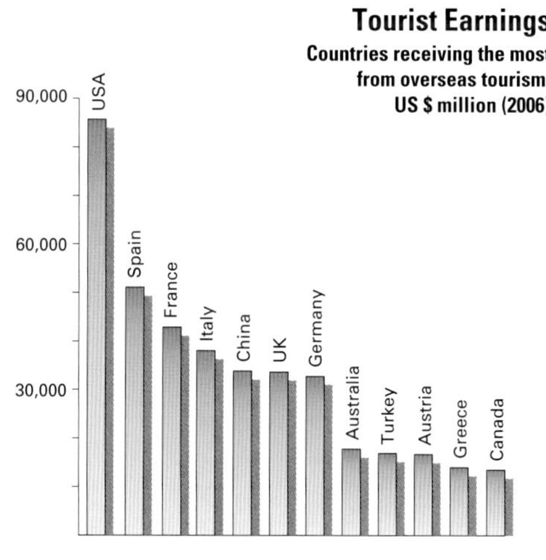

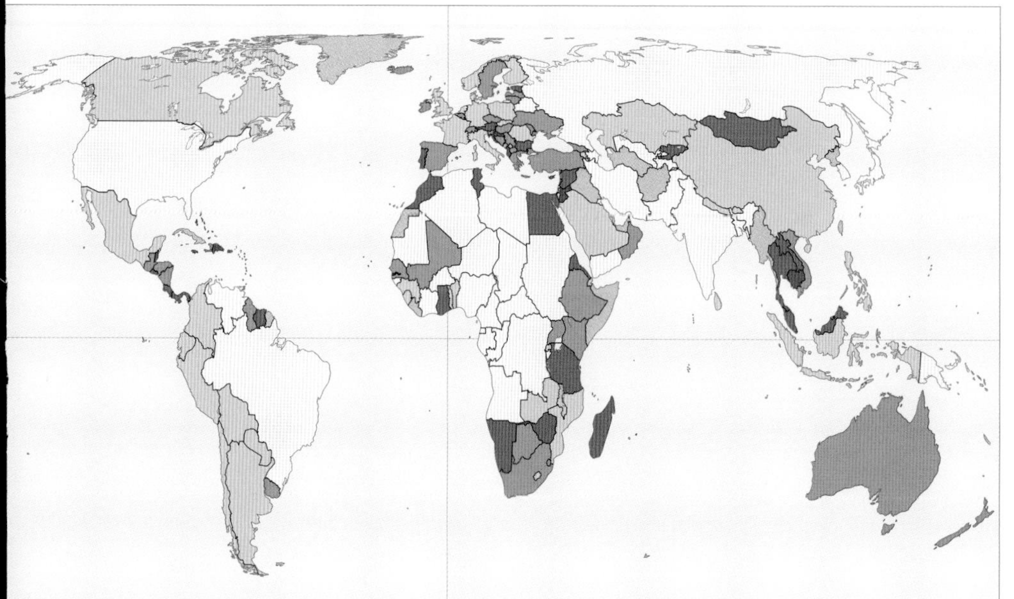

Tourism

Tourism receipts as a percentage of Gross National Income (2006)

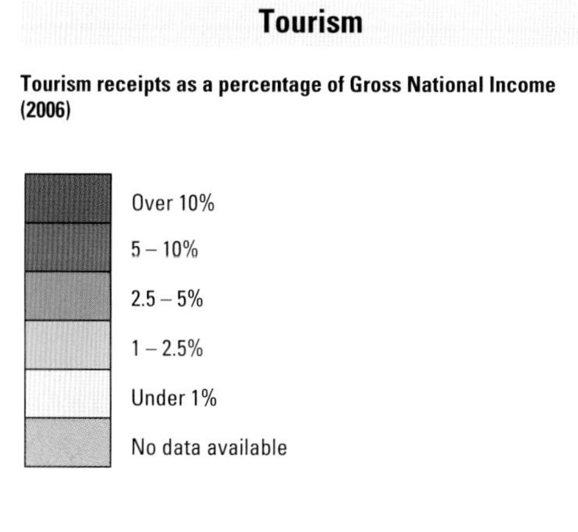

- Over 10%
- 5 – 10%
- 2.5 – 5%
- 1 – 2.5%
- Under 1%
- No data available

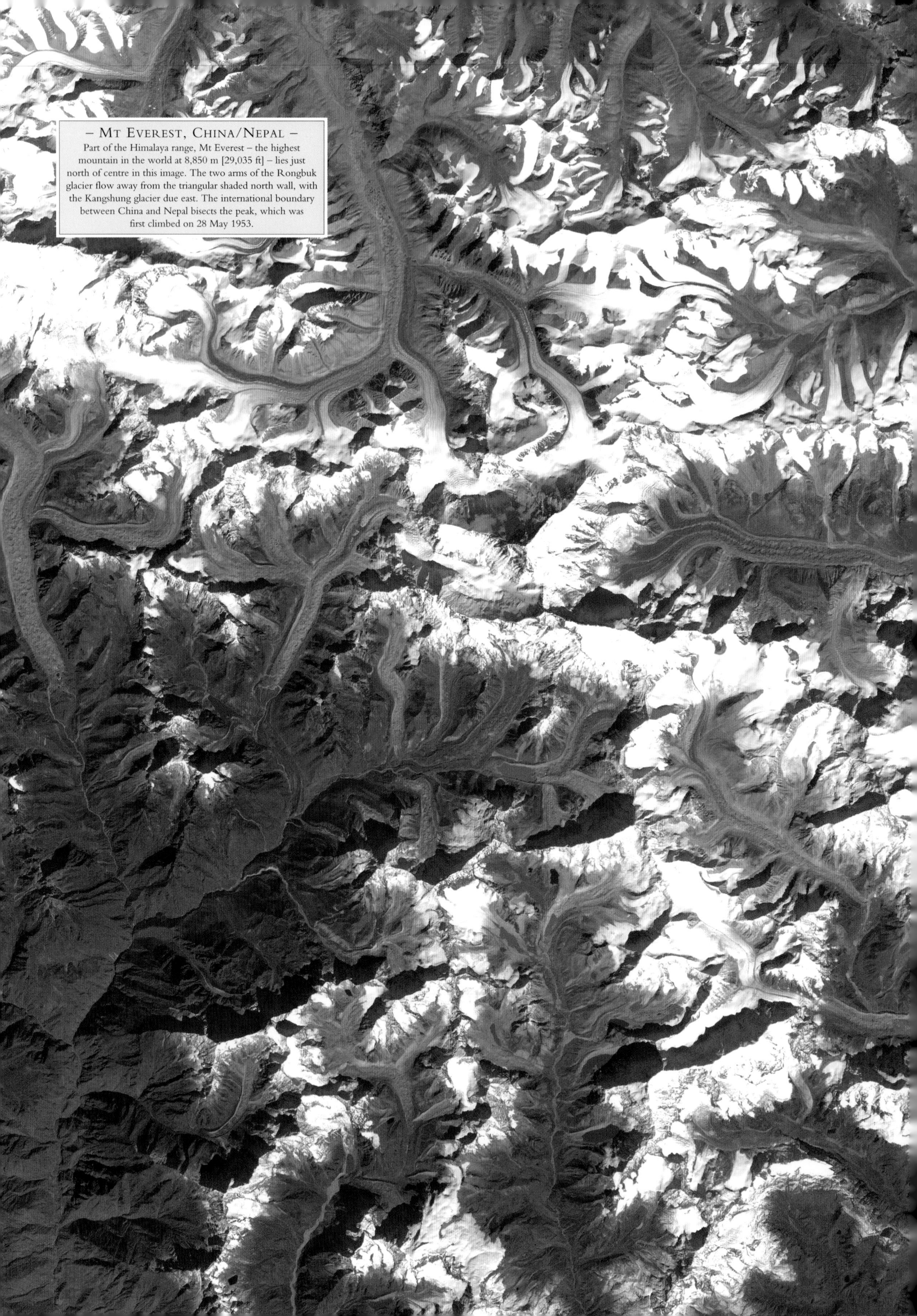

— MT EVEREST, CHINA/NEPAL —

Part of the Himalaya range, Mt Everest – the highest
mountain in the world at 8,850 m [29,035 ft] – lies just
north of centre in this image. The two arms of the Rongbuk
glacier flow away from the triangular shaded north wall, with
the Kangshung glacier due east. The international boundary
between China and Nepal bisects the peak, which was
first climbed on 28 May 1953.

WORLD CITIES

AMSTERDAM	2	HONG KONG	12	MUMBAI	20	TEHRAN	31
ATHENS	2	ISTANBUL	12	MUNICH	21	TORONTO	31
ATLANTA	3	JAKARTA	12	NEW ORLEANS	21	VIENNA	32
BAGHDAD	3	JERUSALEM	13	NEW YORK	22	WARSAW	32
BANGKOK	3	JOHANNESBURG	13	ORLANDO	23	WASHINGTON	33
BARCELONA	4	KARACHI	13	OSAKA	23	WELLINGTON	33
BEIJING	4	KOLKATA	14	OSLO	23		
BERLIN	5	LAGOS	14	PARIS	24	INDEX	34–39
BOSTON	6	LAS VEGAS	14	PRAGUE	25		
BRUSSELS	6	LIMA	14	RIO DE JANEIRO	25		
BUDAPEST	7	LONDON	15	ROME	26	WORLD CITIES	
BUENOS AIRES	7	LOS ANGELES	16	SAN FRANCISCO	26	SATELLITE IMAGES	
CAIRO	7	LISBON	16	ST PETERSBURG	27	CAPE TOWN	40
CAPE TOWN	8	MADRID	17	SANTIAGO	27	KARACHI	41
COPENHAGEN	8	MANILA	17	SÃO PAULO	27	LONDON	42
CHICAGO	9	MELBOURNE	18	SEOUL	27	MELBOURNE	43
DELHI	10	MEXICO CITY	18	SHANGHAI	28	NEW ORLEANS	44
DUBLIN	10	MIAMI	18	SINGAPORE	28	NEW YORK	45
EDINBURGH	11	MILAN	19	STOCKHOLM	29	SAN FRANCISCO	46
GUANGZHOU	11	MOSCOW	19	SYDNEY	29	SANTIAGO	47
HELSINKI	11	MONTRÉAL	20	TOKYO	30	TOKYO	48

CITY MAPS

Motorway, freeway, expressway – with road number	A10
Motorway, freeway, expressway – with European road number	E51
Road junction	
Under construction	
Tunnel	
Primary road – with road number dual carriageway	14
single carriageway	14
Secondary road – with road number dual carriageway	96
single carriageway	96
Other road	
Ferry	
Railroad	
Principal station	Estación del Norte
Height above sea level (m)	705 ▲
Airport	✈
Airfield	⊕
Central area coverage	
Urban area	
Woodlands and parks	

CENTRAL AREA MAPS

Motorway, freeway, expressway	
Through route	
Secondary road	
Dual carriageway	
Other road	
Tunnel	
Limited access/ pedestrian road	
Parks and open space	
Railroad	
Rail/bus station	
Underground, metro station	
Funicular	
Cable car	
Abbey, cathedral	✝
Church of interest	†
Synagogue	✡
Shrine, temple	
Mosque	
Public building	
Tourist information	i
Place of interest	Palace

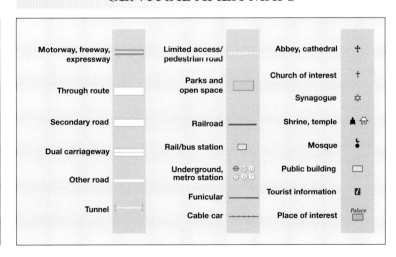

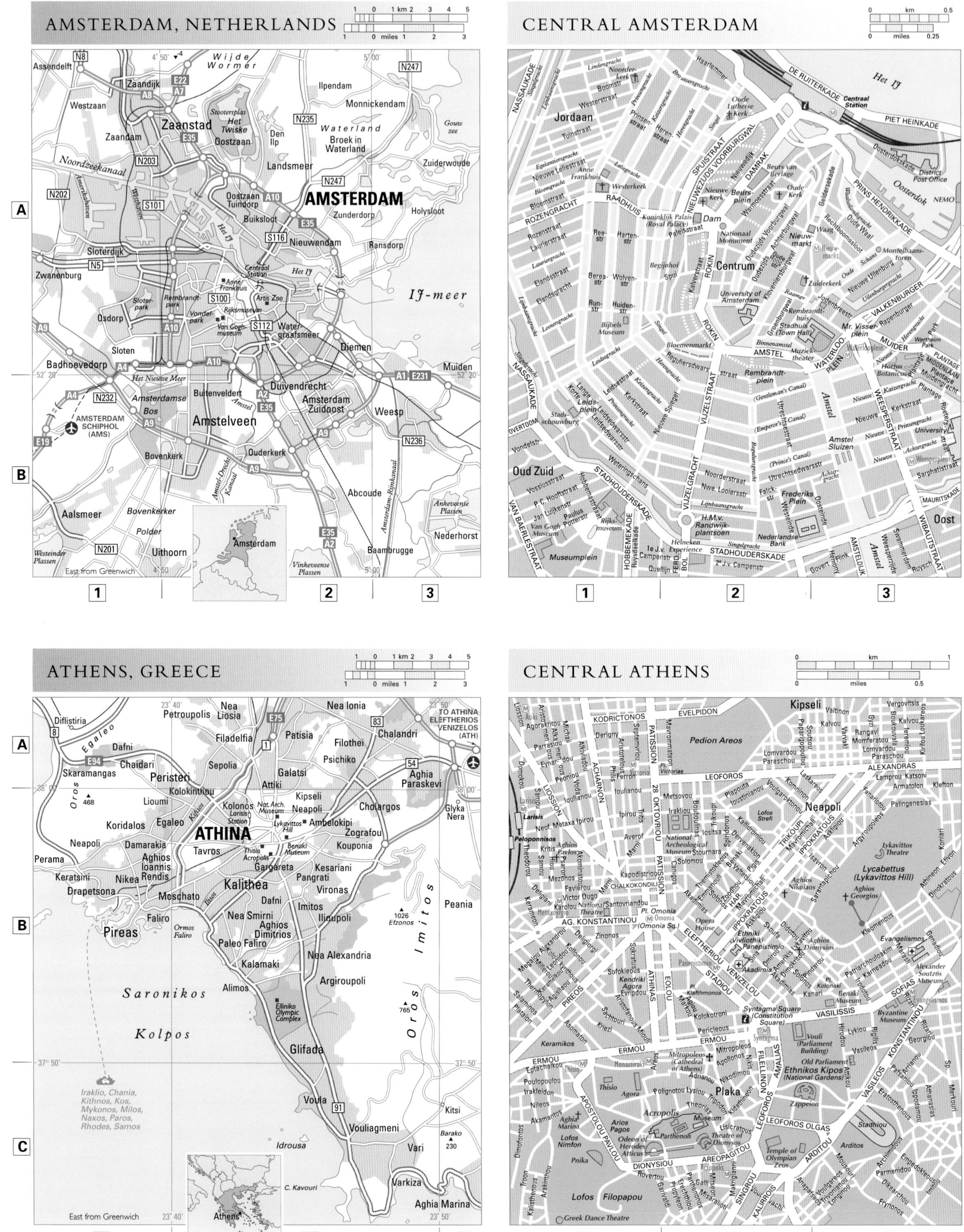
AMSTERDAM, NETHERLANDS
CENTRAL AMSTERDAM
ATHENS, GREECE
CENTRAL ATHENS
Assendelft
Zaandijk
Westzaan
Zaanstad
Zaandam
Oostzaan
Ilpendam
Monnickendam
Waterland
Broek in Waterland
Den Ilp
Landsmeer
Zuiderwoude
Noordzeekanaal
AMSTERDAM
Oostzaan Tuindorp
Buiksloot
Zunderdorp
Holysloot
Nieuwendam
Ransdorp
Sloterdijk
Zwanenburg
IJ-meer
Osdorp
Sloten
Badhoevedorp
Diemen
Muiden
Duivendrecht
Buitenveldert
Amsterdam Zuidoost
Weesp
Amstelveen
AMSTERDAM SCHIPHOL (AMS)
Bovenkerk
Ouderkerk
Abcoude
Aalsmeer
Bovenkerker Polder
Uithoorn
Nederhorst
Baambrugge
East from Greenwich
Jordaan
Het IJ
Centraal Station
PIET HEINKADE
Westerkerk
Dam
Centrum
Nieuwmarkt
Rembrandtplein
Amstel
Oud Zuid
Museumplein
Oost
Kipseli
Pedion Areos
Neapoli
Lycabettus (Lykavittos Hill)
Pl. Omonia (Omonia Sq.)
Syntagma Square (Constitution Square)
Plaka
Acropolis
Parthenon
Temple of Olympian Zeus
Lofos Filopapou
Greek Dance Theatre
Petroupolis
Nea Liosia
Nea Ionia
Chalandri
Filadelfia
Patisia
Filothei
Psichiko
Dafni
Chaidari
Skaramangas
Peristeri
Sepolia
Galatsi
Aghia Paraskevi
Kolokinthou
Attiki
Liouni
Kipseli
Cholargos
Glyka Nera
Koridalos
Egaleo
Kolonos
ATHINA
Ambelokipi
Zografou
Neapoli
Damarakia
Tavros
Kouponia
Perama
Aghios Ioannis Rendis
Gargareta
Kesariani
Keratsini
Nikea
Kalithea
Pangrati
Drapetsona
Moschato
Vironas
Peania
Dafni
Imitos
Faliro
Nea Smirni
Ilioupoli
Pireas
Aghios Dimitrios
Paleo Faliro
Kalamaki
Nea Alexandria
Saronikos Kolpos
Alimos
Argiroupoli
Glifada
Voula
Kitsi
Vouliagmeni
Vari
Varkiza
Aghia Marina
Iraklio, Chania, Kithnos, Kos, Mykonos, Milos, Naxos, Paros, Rhodes, Samos
Athens
East from Greenwich

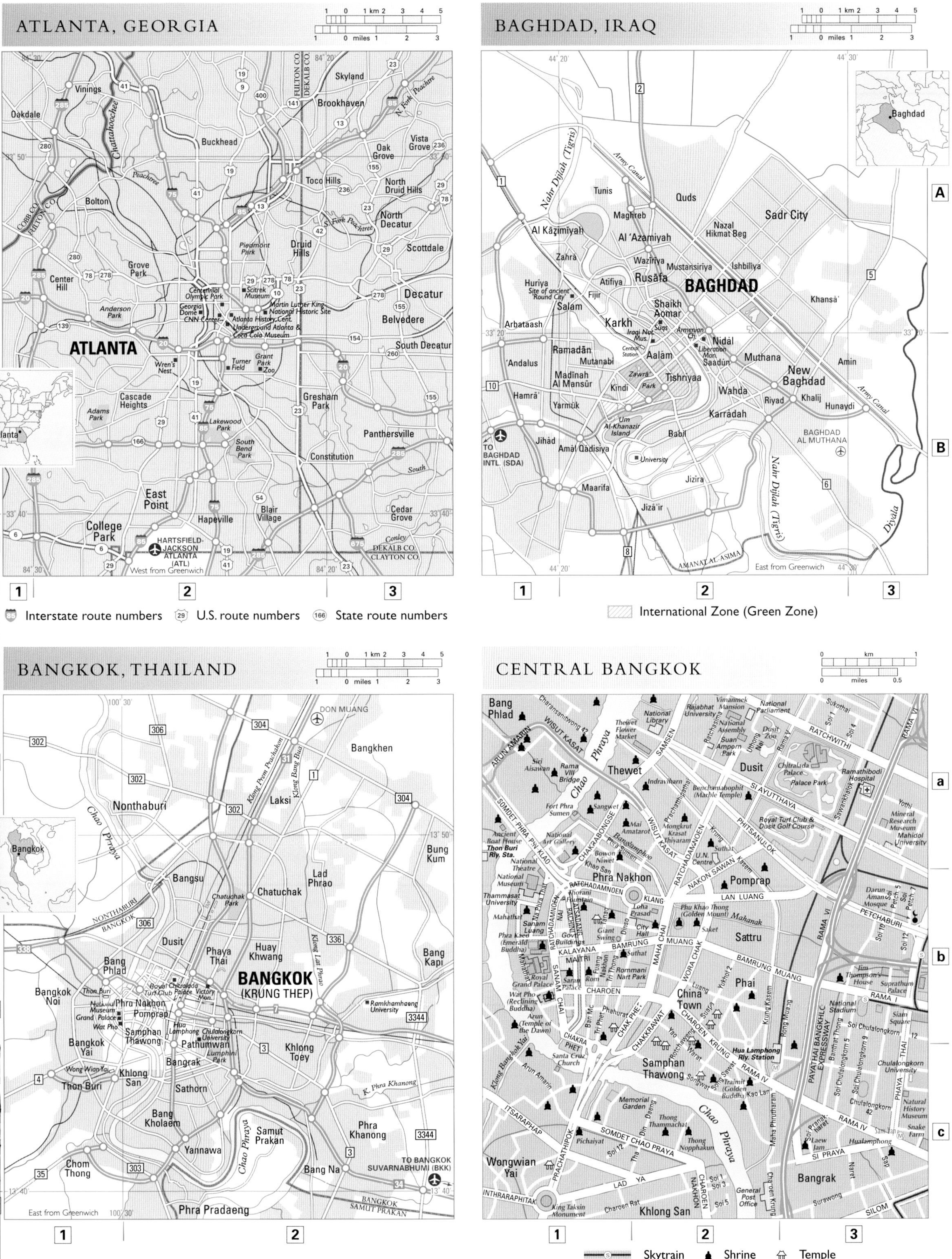

ATLANTA, GEORGIA

BAGHDAD, IRAQ

Interstate route numbers   U.S. route numbers   State route numbers

International Zone (Green Zone)

BANGKOK, THAILAND

CENTRAL BANGKOK

Skytrain   Shrine   Temple

COPYRIGHT PHILIP'S

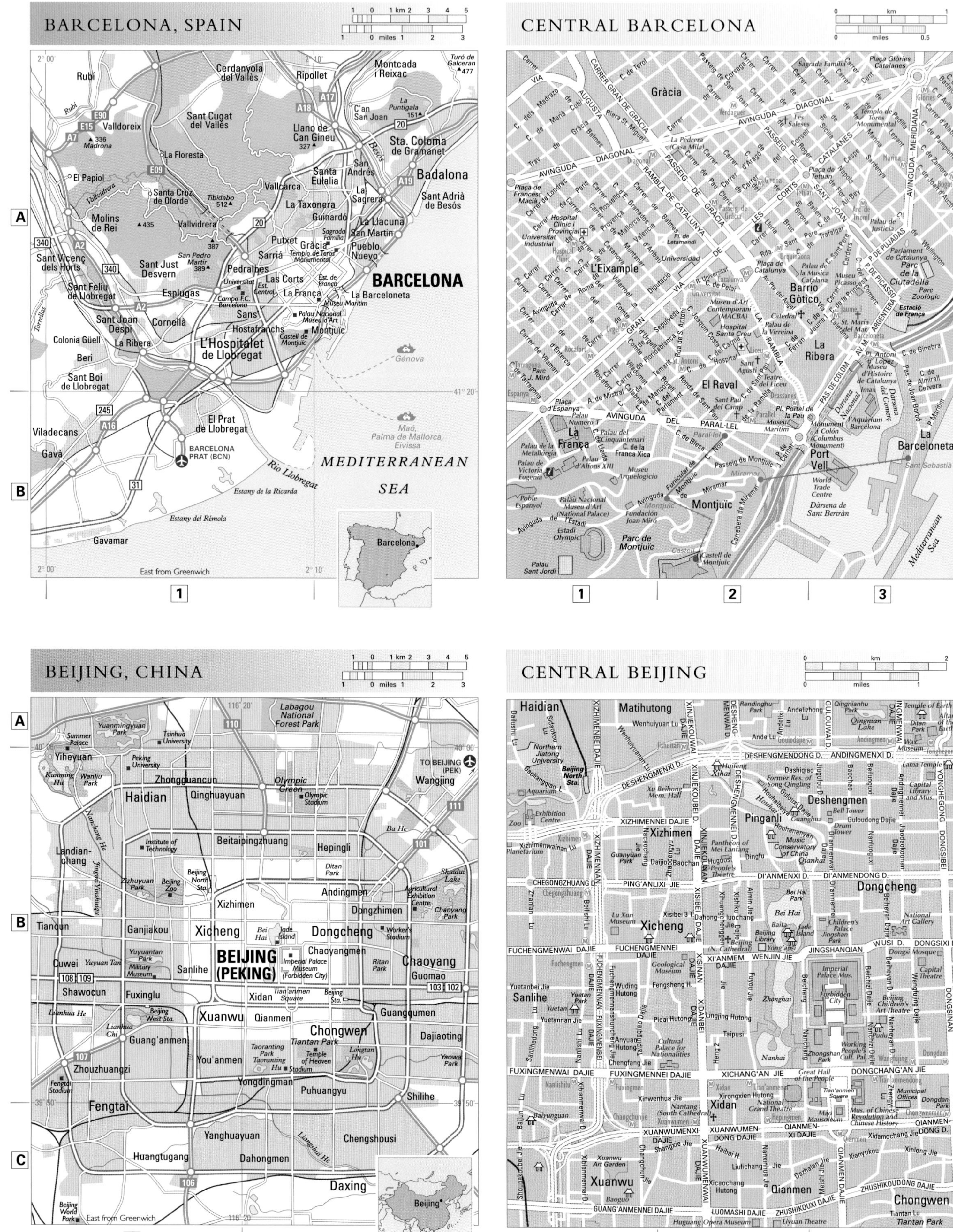

BARCELONA, SPAIN

CENTRAL BARCELONA

BEIJING, CHINA

CENTRAL BEIJING

⌂ Temple

BERLIN, GERMANY

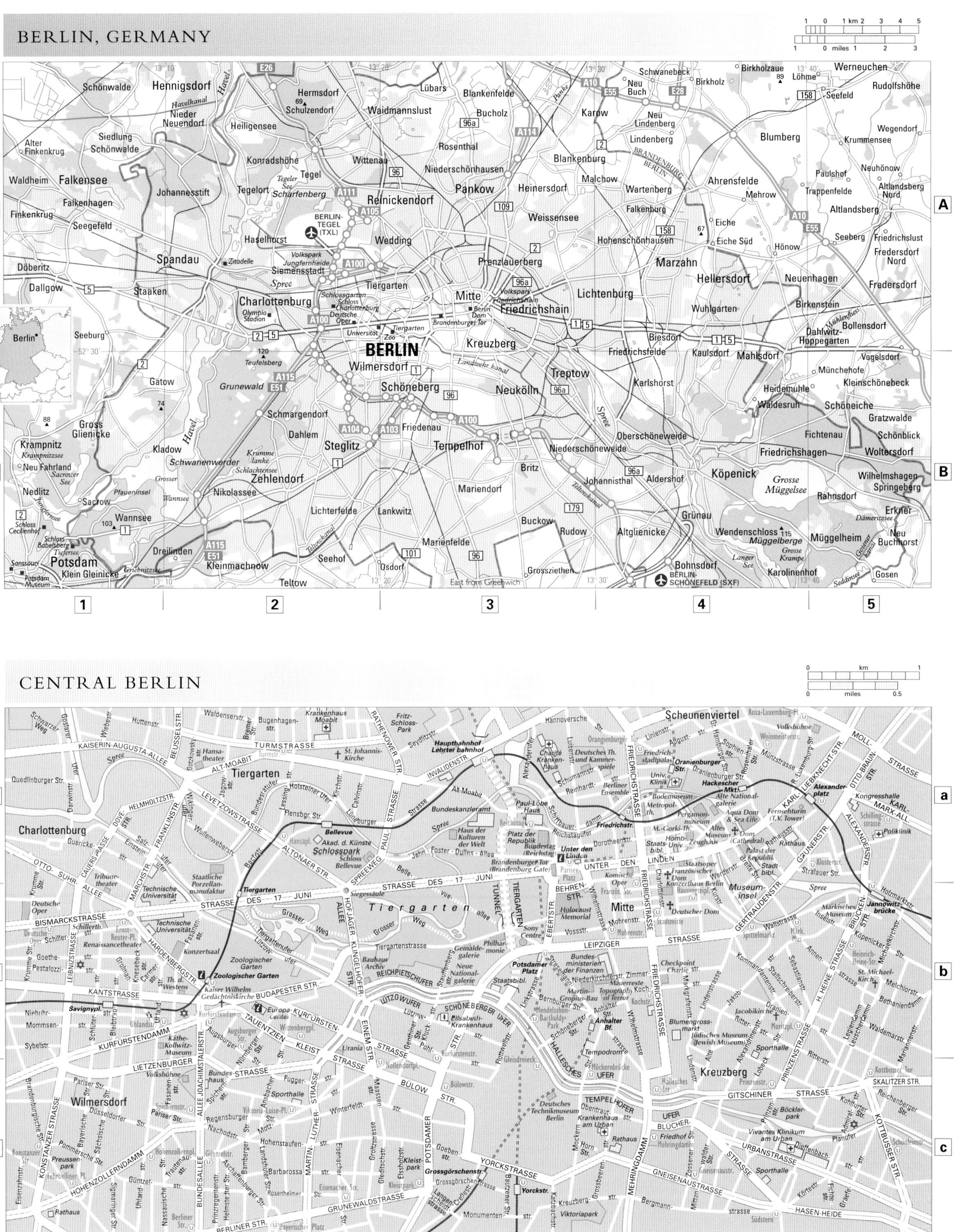

CENTRAL BERLIN

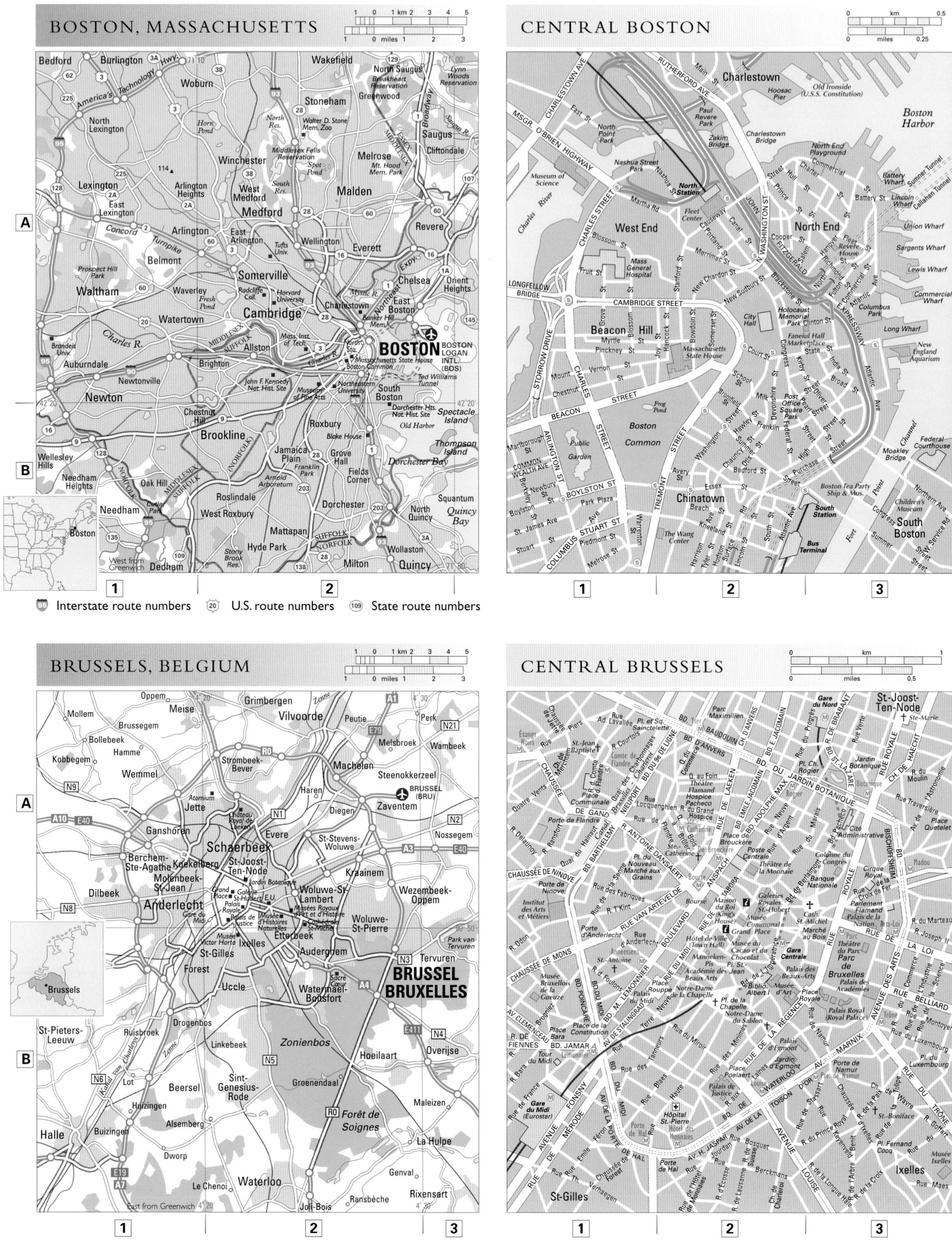

BOSTON, MASSACHUSETTS

CENTRAL BOSTON

Interstate route numbers
U.S. route numbers
State route numbers

BRUSSELS, BELGIUM

CENTRAL BRUSSELS

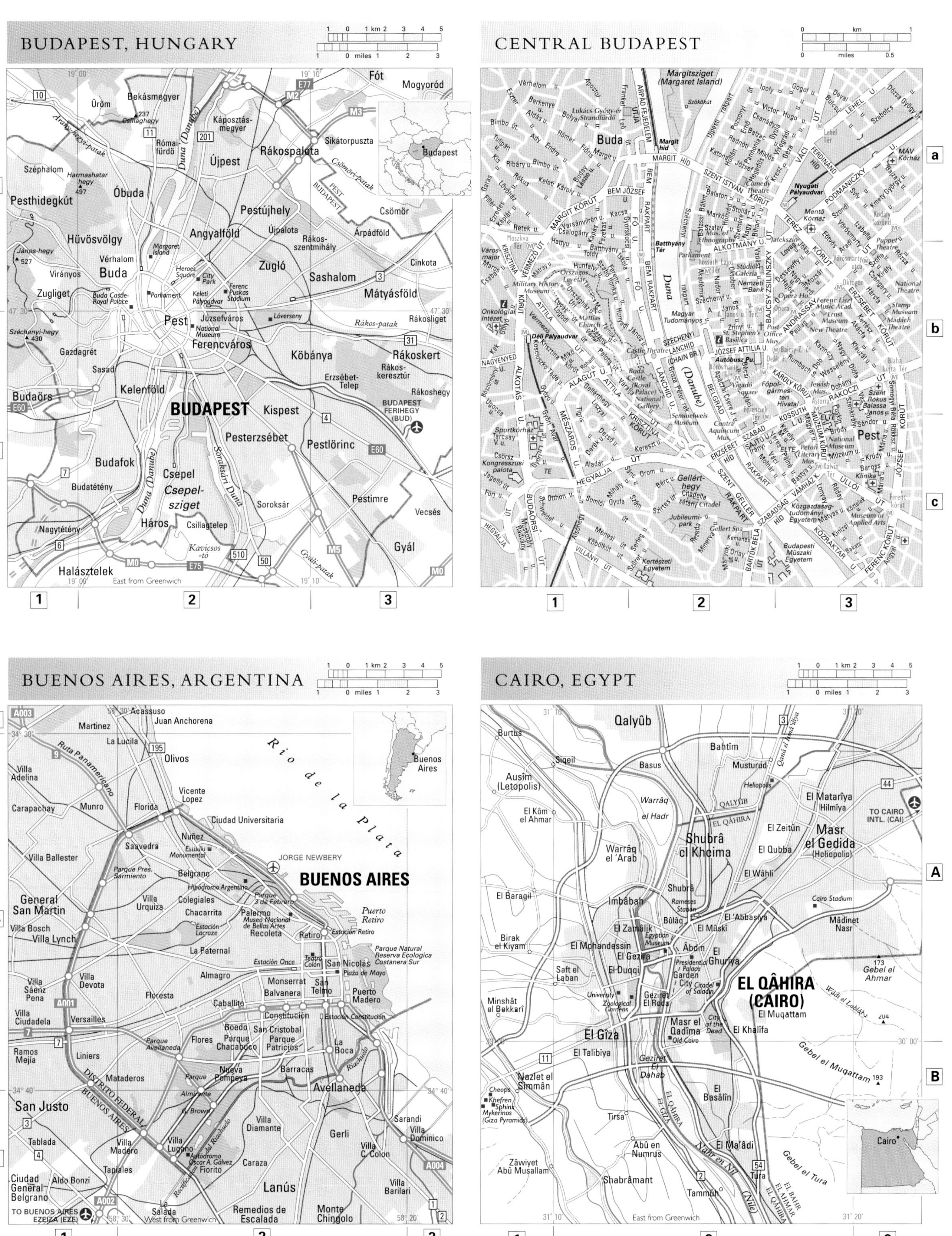
BUDAPEST, HUNGARY
CENTRAL BUDAPEST
BUENOS AIRES, ARGENTINA
CAIRO, EGYPT

CAPE TOWN, SOUTH AFRICA

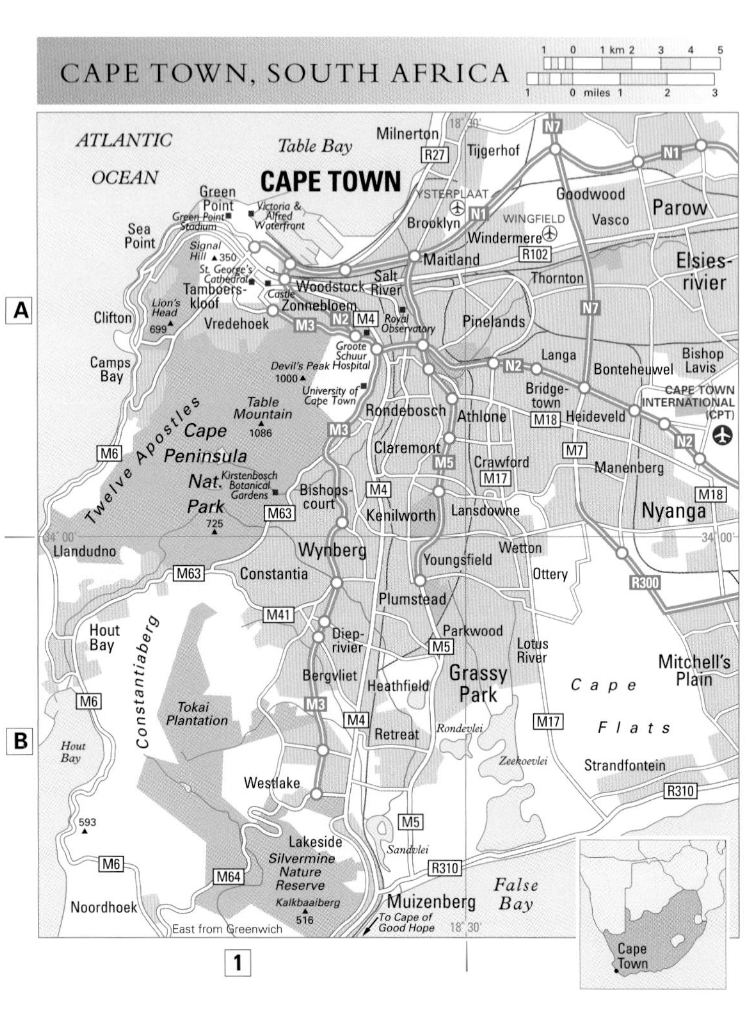

CENTRAL CAPE TOWN

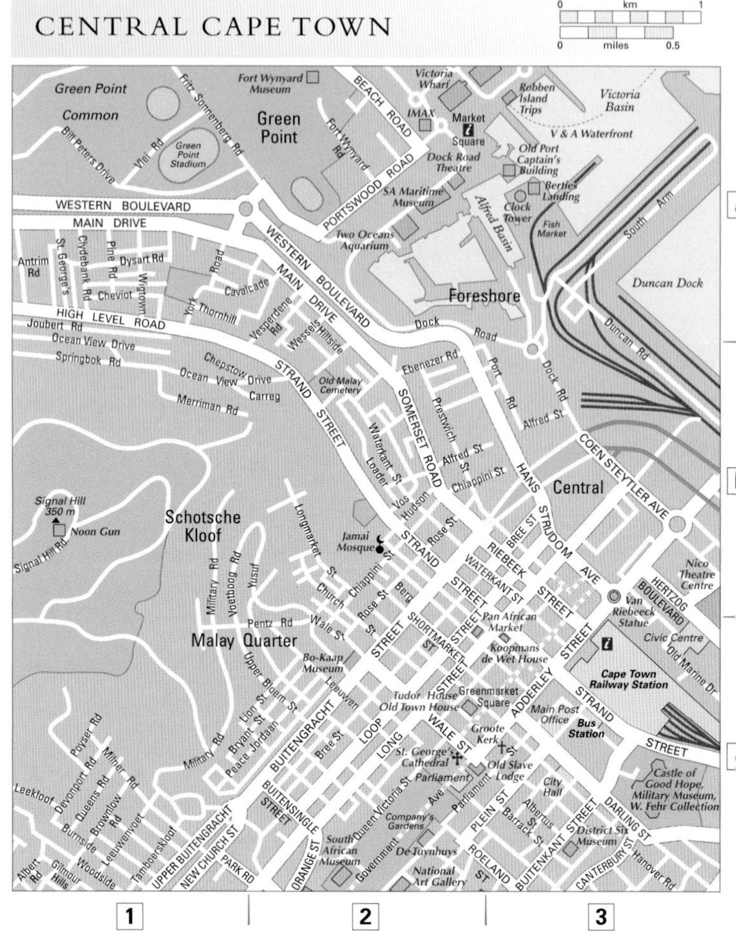

COPENHAGEN, DENMARK

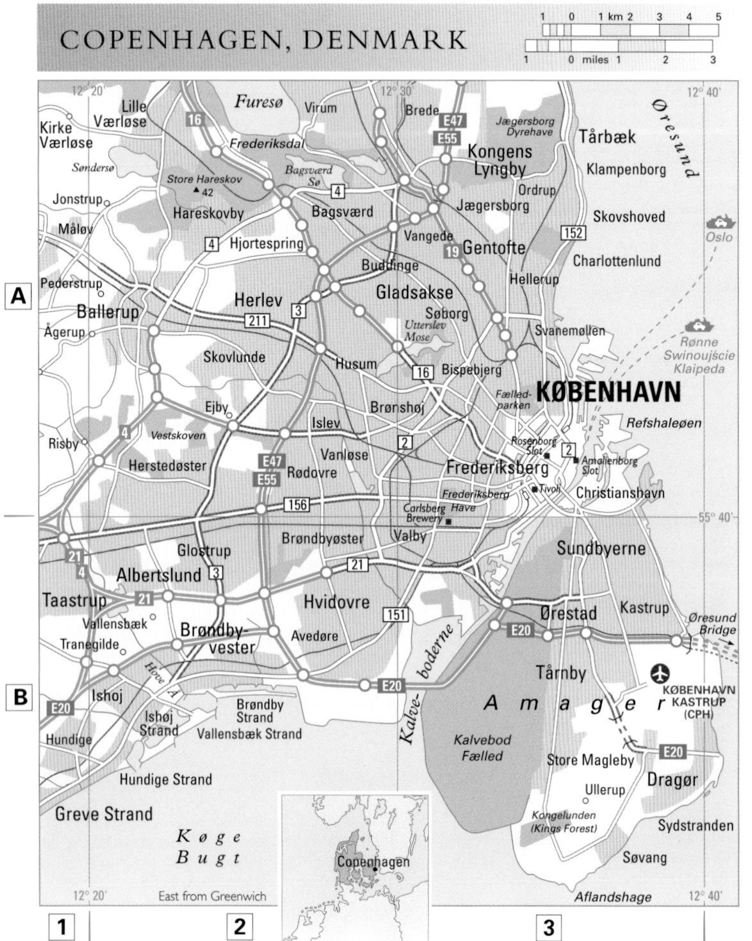

CENTRAL COPENHAGEN

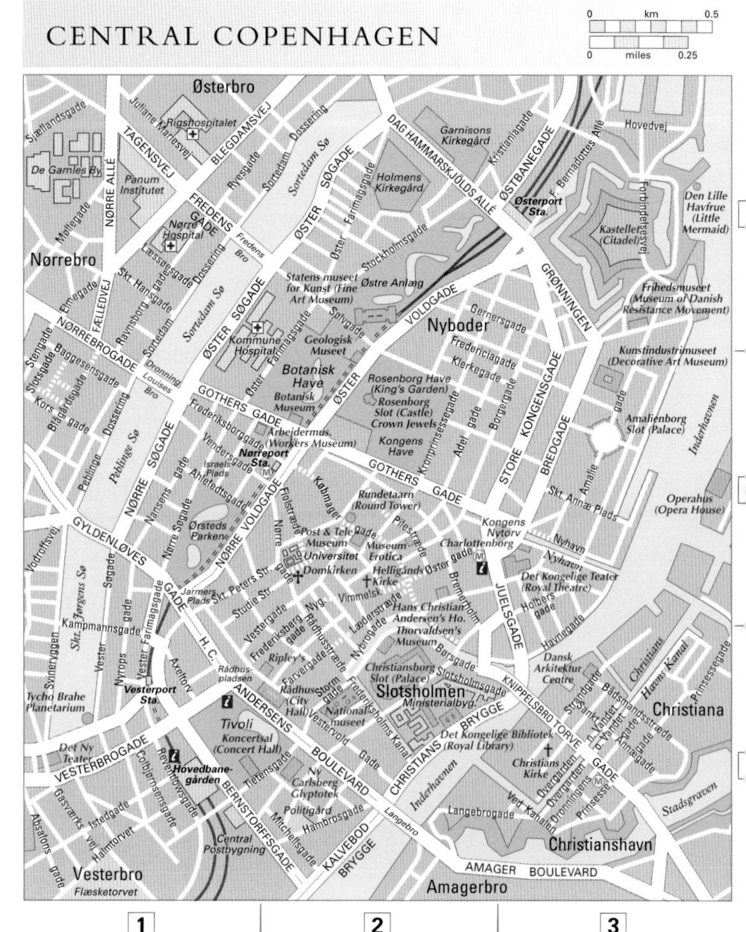

CHICAGO, ILLINOIS

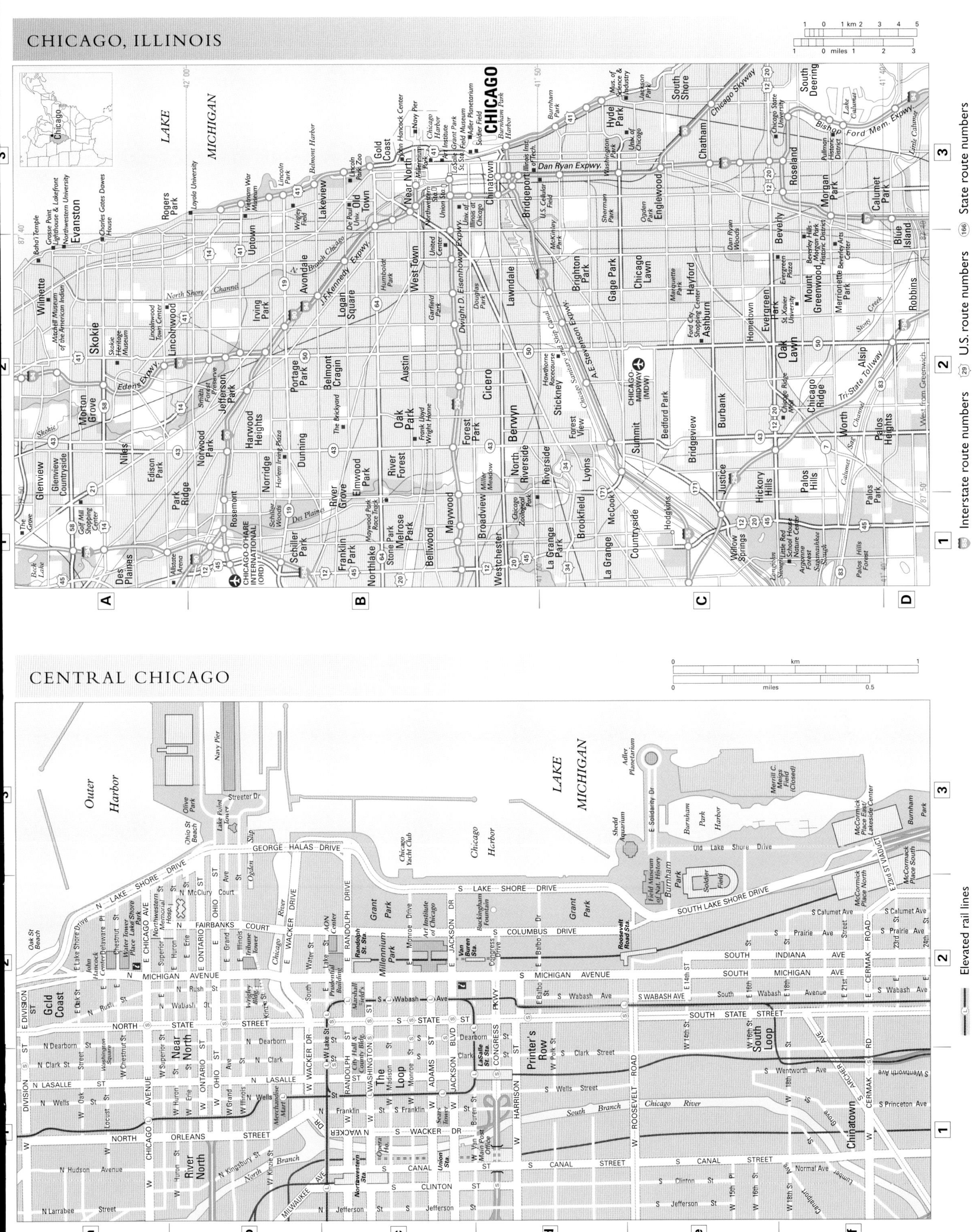

CENTRAL CHICAGO

Interstate route numbers

State route numbers

U.S. route numbers

Elevated rail lines

COPYRIGHT PHILIP'S

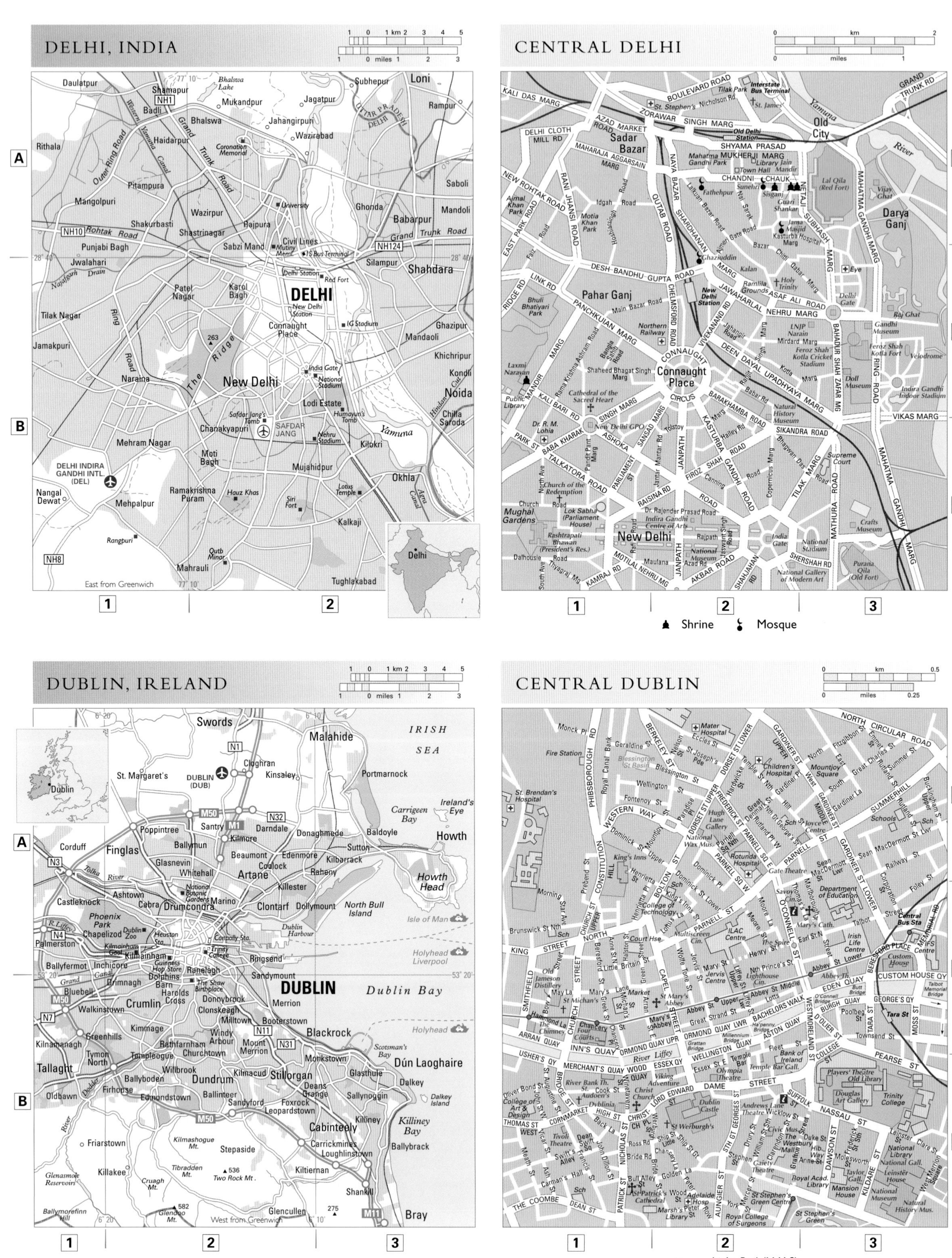

DELHI, INDIA

CENTRAL DELHI

▲ Shrine   ⚲ Mosque

DUBLIN, IRELAND

CENTRAL DUBLIN

— Light Rail (LUAS)

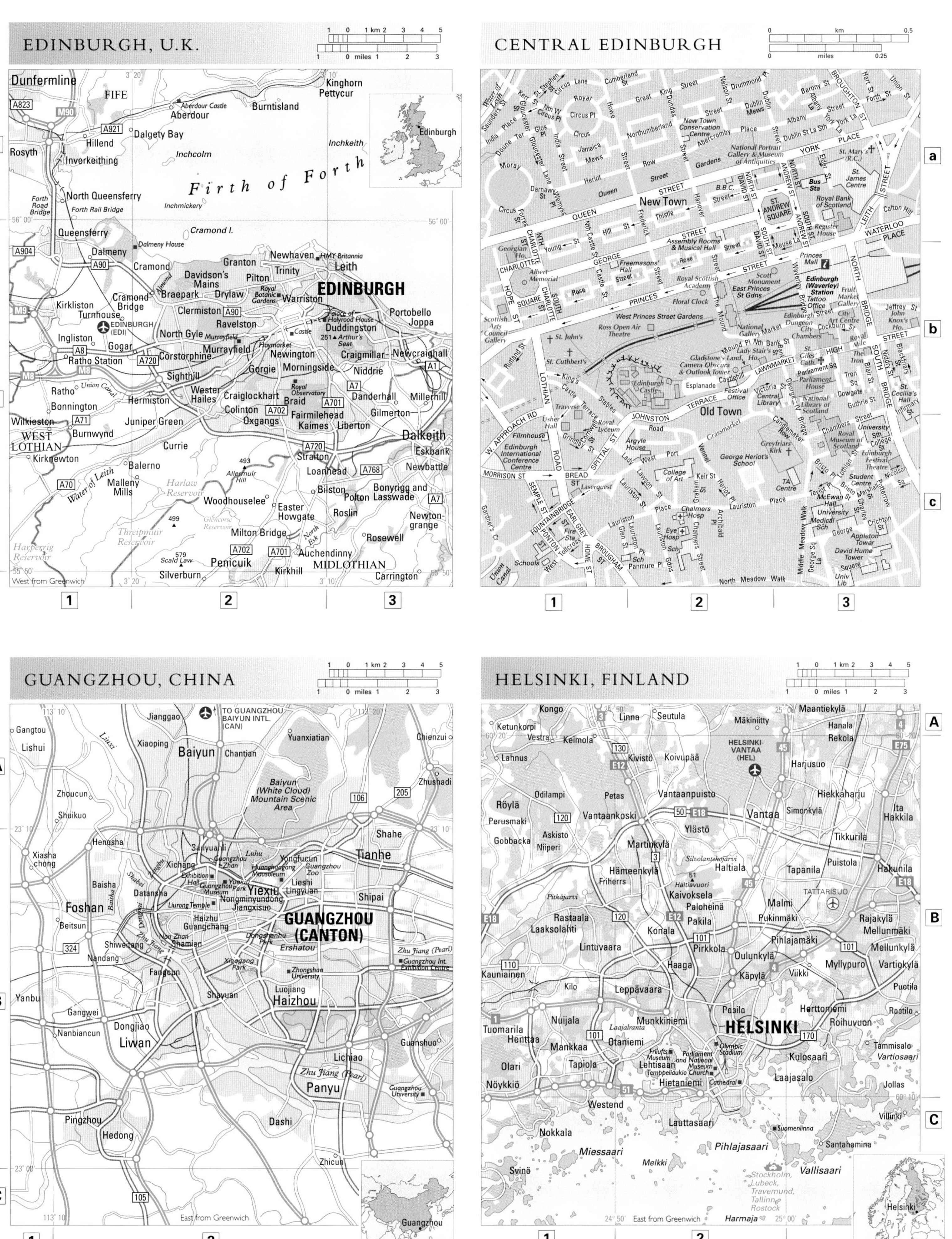
EDINBURGH, U.K.
1 0 1 km 2 3 4 5
1 0 miles 1 2 3
Dunfermline
FIFE
Aberdour Castle
Aberdour
Burntisland
Kinghorn
Pettycur
Dalgety Bay
Hillend
Rosyth
Inverkeithing
Inchcolm
Inchkeith
Edinburgh
Firth of Forth
Forth Road Bridge
North Queensferry
Forth Rail Bridge
Inchmickery
Queensferry
Cramond I.
Dalmeny House
Dalmeny
Cramond
Granton
Newhaven
HMY Britannia
Leith
Trinity
Davidson's Mains
Pilton
EDINBURGH
Kirkliston
Cramond Bridge
Braepark
Drylaw
Royal Botanic Gardens
Warriston
Turnhouse
Clermiston
Palace of Holyrood House
Portobello
Joppa
EDINBURGH (EDI)
Ravelston
Duddingston
Ingliston
North Gyle
Murrayfield
Castle
Arthur's Seat
Gogar
Corstorphine
Haymarket
Newington
Craigmillar
Newcraighall
Ratho Station
Murrayfield
Sighthill
Gorgie
Morningside
Niddrie
Ratho
Union Canal
Wester Hailes
Craiglockhart
Royal Observatory
Danderhall
Millerhill
Bonnington
Hermiston
Braid
Colinton
Fairmilehead
Gilmerton
Wilkieston
Juniper Green
Oxgangs
Kaimes
Liberton
WEST LOTHIAN
Burnwynd
Currie
Dalkeith
Straiton
Eskbank
Kirknewton
Balerno
Allermuir Hill
Loanhead
Newbattle
Water of Leith
Malleny Mills
Harlaw Reservoir
Woodhouselee
Bilston
Polton
Bonyrigg and Lasswade
Easter Howgate
Roslin
Newton-grange
Threipmuir Reservoir
Glencorse Reservoir
Milton Bridge
North Esk
Rosewell
Harperrig Reservoir
Scald Law
Penicuik
Auchendinny
MIDLOTHIAN
Silverburn
Kirkhill
Carrington
West from Greenwich
CENTRAL EDINBURGH
0 km 0.5
0 miles 0.25
Royal Circus
New Town Conservation Centre
National Portrait Gallery & Museum of Antiquities
Queen Street
New Town
St. Andrew Square
Bus Sta
Royal Bank of Scotland
Register House
George Street
Assembly Rooms & Musical Hall
Freemasons' Hall
Royal Scottish Academy
Scott Monument
East Princes St Gdns
Princes Mall
Edinburgh (Waverley) Station
Princes Street
Floral Clock
West Princes Street Gardens
Ross Open Air Theatre
National Gallery
Edinburgh Castle
Gladstone's Land
Camera Obscura & Outlook Tower
Lady Stair's Ho.
St. Giles' Cath.
Parliament Sq
Parliament House
High Street
Old Town
National Library of Scotland
Victoria St
Lawnmarket
Grassmarket
Greyfriars Kirk
George Heriot's School
Royal Museum of Scotland
Chambers Street
Festival Theatre
University
St. John's
St. Cuthbert's
Usher Hall
Traverse Theatre
Royal Lyceum
Filmhouse
Edinburgh International Conference Centre
Lothian Road
Johnston Terrace
College of Art
Lauriston Place
Chalmers Hosp
Eye Hosp
Meadow Walk
George Square
David Hume Tower
Appleton Tower
North Meadow Walk
Union Canal
Schools
Waterloo Place
Calton Hill
Jeffrey St
John Knox's Ho.
GUANGZHOU, CHINA
1 0 1 km 2 3 4 5
1 0 miles 1 2 3
TO GUANGZHOU BAIYUN INTL. (CAN)
Jianggao
Gangtou
Lishui
Xiaoping
Baiyun
Chantian
Yuanxiatian
Chienzui
Zhushadi
Baiyun (White Cloud) Mountain Scenic Area
Zhoucun
Shuikuo
Hengsha
Shahe
Tianhe
Sanyuanli
Xiasha chong
Xichang
Yongfucun
Guangzhou Zoo
Exhibition Hall
Yuexiu Park
Yiexiu
Lieshi Lingyuan
Baisha
Datansha
Shipai
Foshan
Liurong Temple
Nongminyundong Jiangxisuo
Haizhu Guangchang
GUANGZHOU (CANTON)
Beitsun
Shiweitang
Shamian
Ershatou
Zhu Jiang (Pearl)
Guangzhou Int. Exhibition Centre
Nandang
Fangcun
Zhongshan University
Luojiang
Haizhou
Yanbu
Shayuan
Gangwei
Nanbiancun
Dongjiao
Liwan
Guanshuo
Lichiao
Panyu
Guangzhou University
Pingzhou
Hedong
Dashi
Zhicun
Guangzhou
East from Greenwich
HELSINKI, FINLAND
1 0 1 km 2 3 4 5
1 0 miles 1 2 3
Kongo
Linna
Seutula
Mäkiniitty
Maantiekylä
Ketunkorpi
Vestra
Keimola
Hanala
Rekola
HELSINKI-VANTAA (HEL)
Lahnus
Kivistö
Koivupää
Harjusuo
Röylä
Odilampi
Petas
Vantaanpuisto
Hiekkaharju
Ita Hakkila
Perusmaki
Vantaankoski
Vantaa
Simonkylä
Gobbacka
Askisto
Niiperi
Ylästö
Tikkurila
Martinkylä
Silvolantekojärvi
Puistola
Hämeenkylä
Haltiala
Tapanila
Hakunila
Friherrs
Haltiavuori
Kaivoksela
TATTARISUO
Pitkäjärvi
Paloheinä
Malmi
Rastaala
Pakila
Pukinmäki
Rajakylä
Laaksolahti
Mellunmäki
Konala
Lintuvaara
Pihlajamäki
Pirkkola
Mellunkylä
Oulunkylä
Haaga
Myllypuro
Vartiokylä
Kauniainen
Käpylä
Viikki
Kilo
Leppävaara
Puotila
Pasila
Herttoniemi
Rostilo
Nuijala
Munkkiniemi
Tuomarila
HELSINKI
Roihuvuori
Henttaa
Laajalranta
Otaniemi
Mankkaa
Tammisalo
Vartiosaari
Olympic Stadium
Frilufts Museum
Parliament and National Museum
Tapiola
Lehtisaari
Temppeliaukio Church
Kulosaari
Olari
Cathedral
Laajasalo
Nöykkiö
Hietaniemi
Jollas
Westend
Villinki
Lauttasaari
Suomenlinna
Nokkala
Santahamina
Miessaari
Pihlajasaari
Melkki
Vallisaari
Svinö
Stockholm, Lubeck, Travemund, Tallinn, Rostock
Harmaja
East from Greenwich
Helsinki

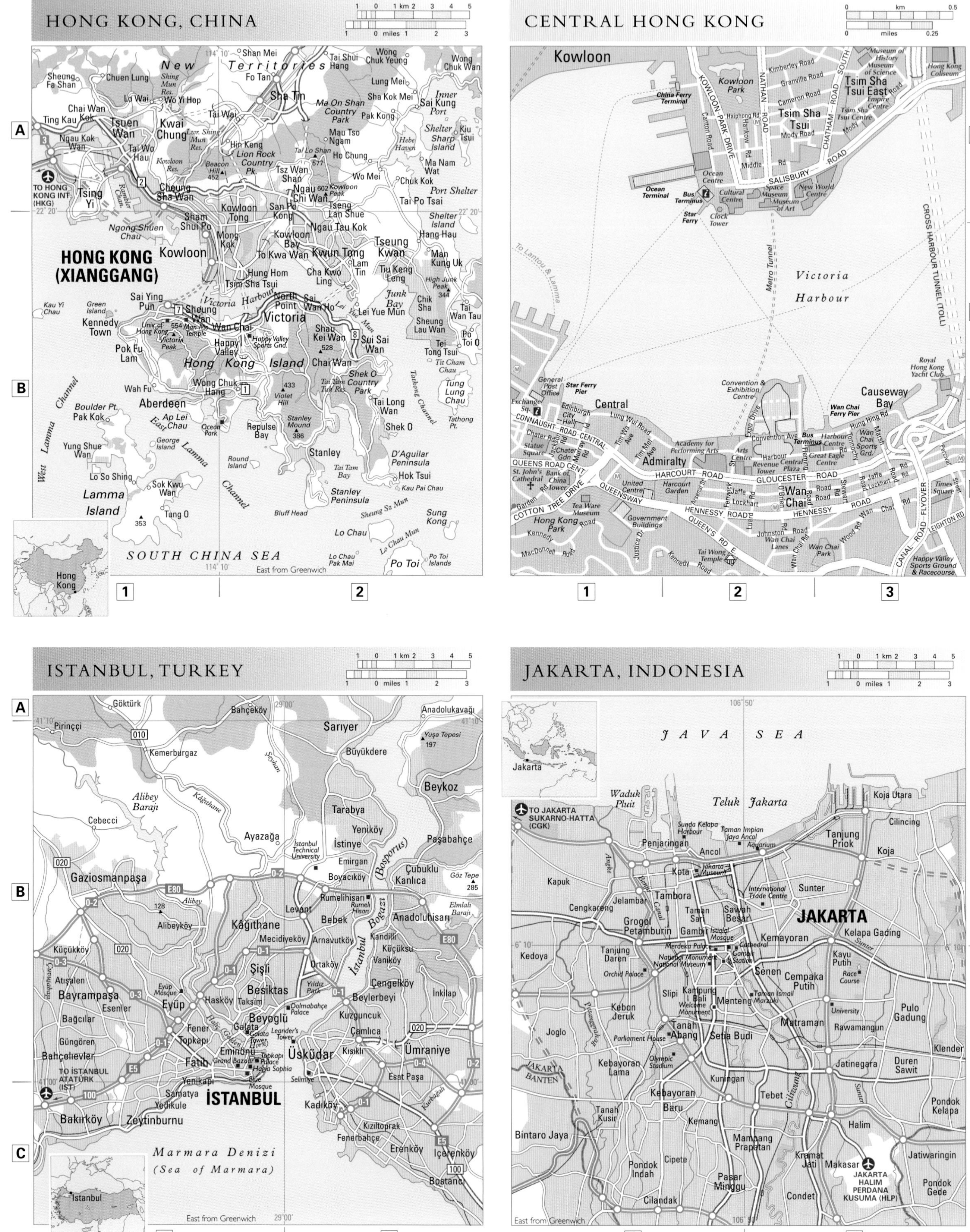

HONG KONG, CHINA
New Territories
HONG KONG (XIANGGANG)
Kowloon
Victoria
Hong Kong Island
Lamma Island
Sha Tin
Tsuen Wan
Kwai Chung
Tsing Yi
Aberdeen
Stanley
Victoria Harbour
SOUTH CHINA SEA
TO HONG KONG INT. (HKG)
East from Greenwich
CENTRAL HONG KONG
Kowloon
Tsim Sha Tsui
Tsim Sha Tsui East
Victoria Harbour
Central
Admiralty
Wan Chai
Causeway Bay
CROSS HARBOUR TUNNEL (TOLL)
ISTANBUL, TURKEY
İSTANBUL
Sarıyer
Beykoz
Kâğıthane
Şişli
Beşiktaş
Beyoğlu
Üsküdar
Kadıköy
Fatih
Eyüp
Bakırköy
Zeytinburnu
Ümraniye
Gaziosmanpaşa
Bayrampaşa
İstanbul Boğazı (Bosporus)
Marmara Denizi (Sea of Marmara)
TO İSTANBUL ATATÜRK (IST)
East from Greenwich
JAKARTA, INDONESIA
JAVA SEA
Teluk Jakarta
JAKARTA
Tanjung Priok
Kota
Kemayoran
Tanah Abang
Kebayoran Baru
Jatinegara
TO JAKARTA SUKARNO-HATTA (CGK)
JAKARTA HALIM PERDANA KUSUMA (HLP)
East from Greenwich

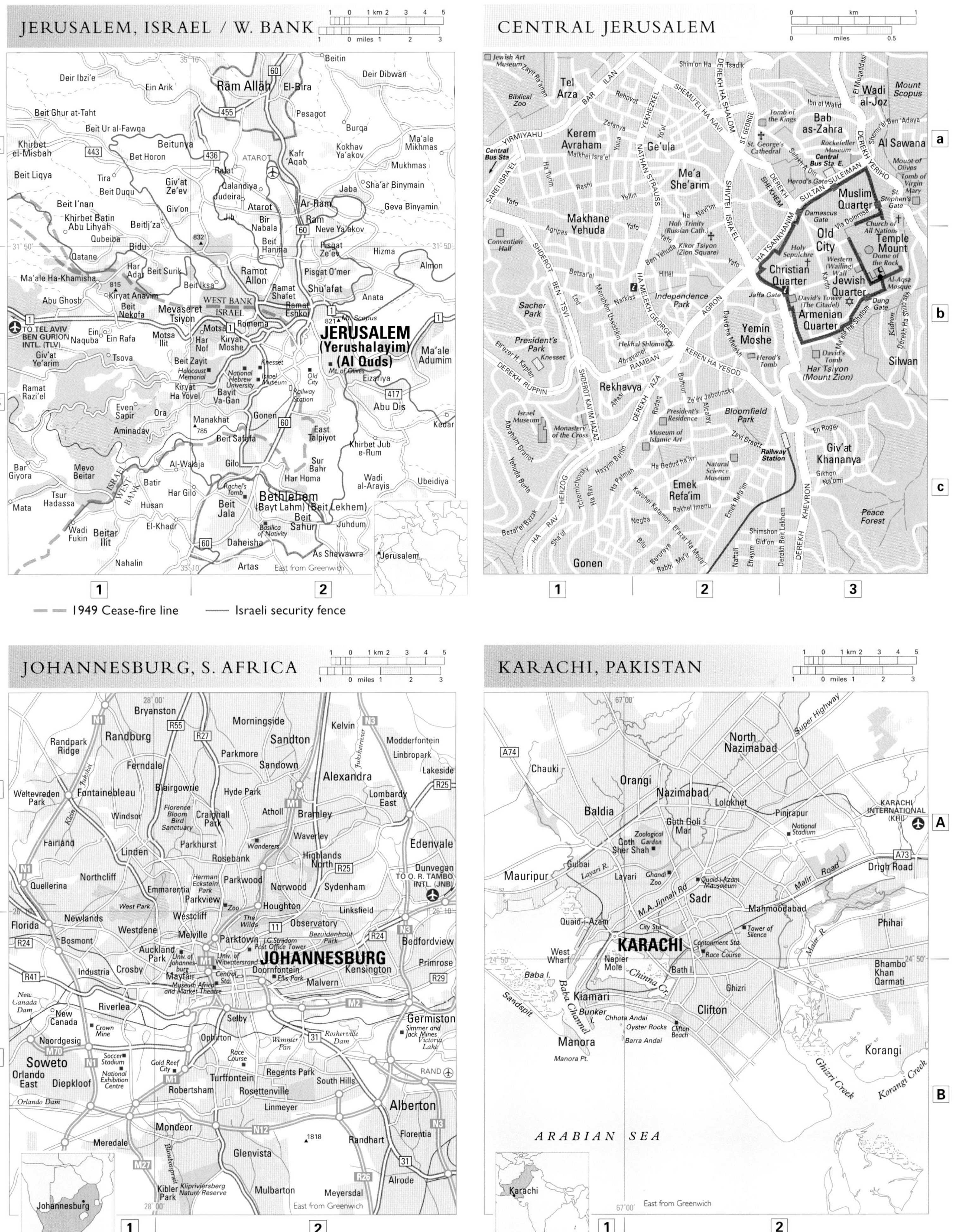
JERUSALEM, ISRAEL / W. BANK
Râm Allāh
El-Bira
Beitin
Deir Dibwan
Deir Ibzi'e
Ein Arik
Beit Ghur at-Taht
Beit Ur al-Fawqa
Pesagot
Burqa
Khirbet el-Misbah
Beitunya
Bet Horon
Kafr 'Aqab
Kokhav Ya'akov
Ma'ale Mikhmas
Mukhmas
Beit Liqya
Tira
Beit Duqu
Giv'at Ze'ev
Qalandiya
Jaba
Sha'ar Binyamin
Beit I'nan
Khirbet Batin
Abu Lihyah
Qubeiba
Giv'on
Atarot
Ar-Ram
Geva Binyamin
Bir Nabala
Neve Ya'akov
Beit Hanina
Pisgat Ze'ev
Hizma
Qatane
Biddu
Ma'ale Ha-Khamisha
Har Adar
Beit Surik
Ramot Allon
Pisgat O'mer
Almon
Beit Iksa
Shu'afat
Ramat Shafat
Ramat Eshkol
Abu Ghosh
Kiryat Anavim
Mevaseret Tsiyon
WEST BANK
ISRAEL
Anata
TO TEL AVIV BEN GURION INTL. (TLV)
Ein Naquba
Ein Rafa
Motsa Illit
Har Nof
Kiryat Moshe
Romema
Mt. Scopus
JERUSALEM
(Yerushalayim)
(Al Quds)
Ma'ale Adumim
Giv'at Ye'arim
Tsova
Beit Zayit
Holocaust Memorial
Knesset
National Hebrew University
Israel Museum
Old City
Mt. of Olives
Eizariya
Ramat Razi'el
Kiryat Ha Yovel
Bayit Va-Gan
Railway Station
Abu Dis
Even Sapir
Ora
Manakhat
Gonen
Aminadav
Beit Safafa
East Talpiyot
Kedar
Khirbet Jub e-Rum
Bar Giyora
Mevo Beitar
Al-Walaja
Gilo
Sur Bahr
Har Homa
Wadi al-Arayis
Ubeidiya
Tsur Hadassa
Batir
Har Gilo
Rachel's Tomb
Beit Jala
Bethlehem
(Bayt Lahm) (Beit Lekhem)
Beit Sahur
Mata
Husan
El-Khadr
Juhdum
Wadi Fukin
Beitar Illit
Basilica of Nativity
Daheisha
As Shawawra
Jerusalem
Nahalin
Artas
East from Greenwich
1949 Cease-fire line
Israeli security fence
CENTRAL JERUSALEM
Jewish Art Museum
Tel Arza
Shim'on Ha Tsadik
Wadi al-Joz
Mount Scopus
Biblical Zoo
Tomb of the Kings
Bab as-Zahra
Kerem Avraham
Ge'ula
St. George's Cathedral
Rockefeller Museum
Central Bus Sta.
Al Sawana
Mount of Olives
Me'a She'arim
Herod's Gate
Muslim Quarter
Tomb of Virgin Mary
St. Stephen's Gate
Makhane Yehuda
Damascus Gate
Old City
Church of All Nations
Temple Mount
Dome of the Rock
Convention Hall
Holy Trinity (Russian Cath.)
Kikar Tsiyon (Zion Square)
Holy Sepulchre
Western (Wailing) Wall
Christian Quarter
Jewish Quarter
Al-Aqsa Mosque
Sacher Park
Independence Park
Jaffa Gate
David's Tower (The Citadel)
Armenian Quarter
Dung Gate
President's Park
Knesset
Hekhal Shlomo
Yemin Moshe
Herod's Tomb
David's Tomb
Har Tsiyon (Mount Zion)
Silwan
Rekhavya
Israel Museum
Monastery of the Cross
President's Residence
Museum of Islamic Art
Bloomfield Park
Railway Station
Giv'at Khananya
Natural Science Museum
Emek Refa'im
Peace Forest
Gonen
JOHANNESBURG, S. AFRICA
Bryanston
Randburg
Morningside
Sandton
Kelvin
Randpark Ridge
Parkmore
Modderfontein
Linbropark
Ferndale
Sandown
Alexandra
Lakeside
Weltevreden Park
Fontainebleau
Blairgowrie
Hyde Park
Lombardy East
Windsor
Florence Bloom Bird Sanctuary
Craighall Park
Atholl
Bramley
Fairland
Parkhurst
Wanderers
Waverley
Edenvale
Linden
Rosebank
Highlands North
Dunvegan
Quellerina
Northcliff
Herman Eckstein Park
Parkwood
Norwood
Sydenham
TO O. R. TAMBO INTL. (JNB)
Emmarentia
Parkview
Zoo
Houghton
Newlands
West Park
Westcliff
The Wilds
Observatory
Linksfield
Florida
Westdene
Melville
Parktown
Bezuidenhout Park
Bedfordview
Bosmont
Auckland Park
Univ. of Johannesburg
J.G. Strijdom Post Office Tower
JOHANNESBURG
Univ. of Witwatersrand
Doornfontein
Kensington
Primrose
Industria
Crosby
Mayfair
Central Sta.
Ellis Park
Malvern
Museum Africa and Market Theatre
New Canada Dam
Riverlea
New Canada
Selby
Germiston
Crown Mine
Simmer and Jack Mines
Victoria Lake
Noordgesig
Ophirton
Wemmer Pan
Rosherville Dam
Soweto
Soccer Stadium
Race Course
Orlando East
Diepkloof
National Exhibition Centre
Gold Reef City
Turffontein
Regents Park
South Hills
RAND
Orlando Dam
Robertsham
Rosettenville
Linmeyer
Alberton
Mondeor
Meredale
Florentia
Glenvista
Randhart
Johannesburg
Kibler Park
Klipriviersberg Nature Reserve
Mulbarton
Meyersdal
Alrode
East from Greenwich
KARACHI, PAKISTAN
Super Highway
North Nazimabad
Chauki
Orangi
Nazimabad
Lolokhet
Baldia
Pinjrapur
KARACHI INTERNATIONAL (KHI)
National Stadium
Goth Goli Mar
Zoological Garden
Goth Sher Shah
Drigh Road
Mauripur
Gulbai
Layari R.
Layari
Ghandi Zoo
Quaid-i-Azam Mausoleum
Malir Road
Sadr
Mahmoodabad
Quaid-i-Azam
Phihai
Tower of Silence
City Sta.
Malir R.
West Wharf
KARACHI
Cantonment Sta.
Race Course
Napier Mole
Baba I.
Bath I.
Bhambo Khan Qarmati
Chinna Cr.
Baba Channel
Sandspit
Kiamari
Ghizri
Bunker I.
Chhota Andai
Clifton
Oyster Rocks
Clifton Beach
Manora
Barra Andai
Korangi
Manora Pt.
Ghizri Creek
Korangi Creek
ARABIAN SEA
Karachi
East from Greenwich

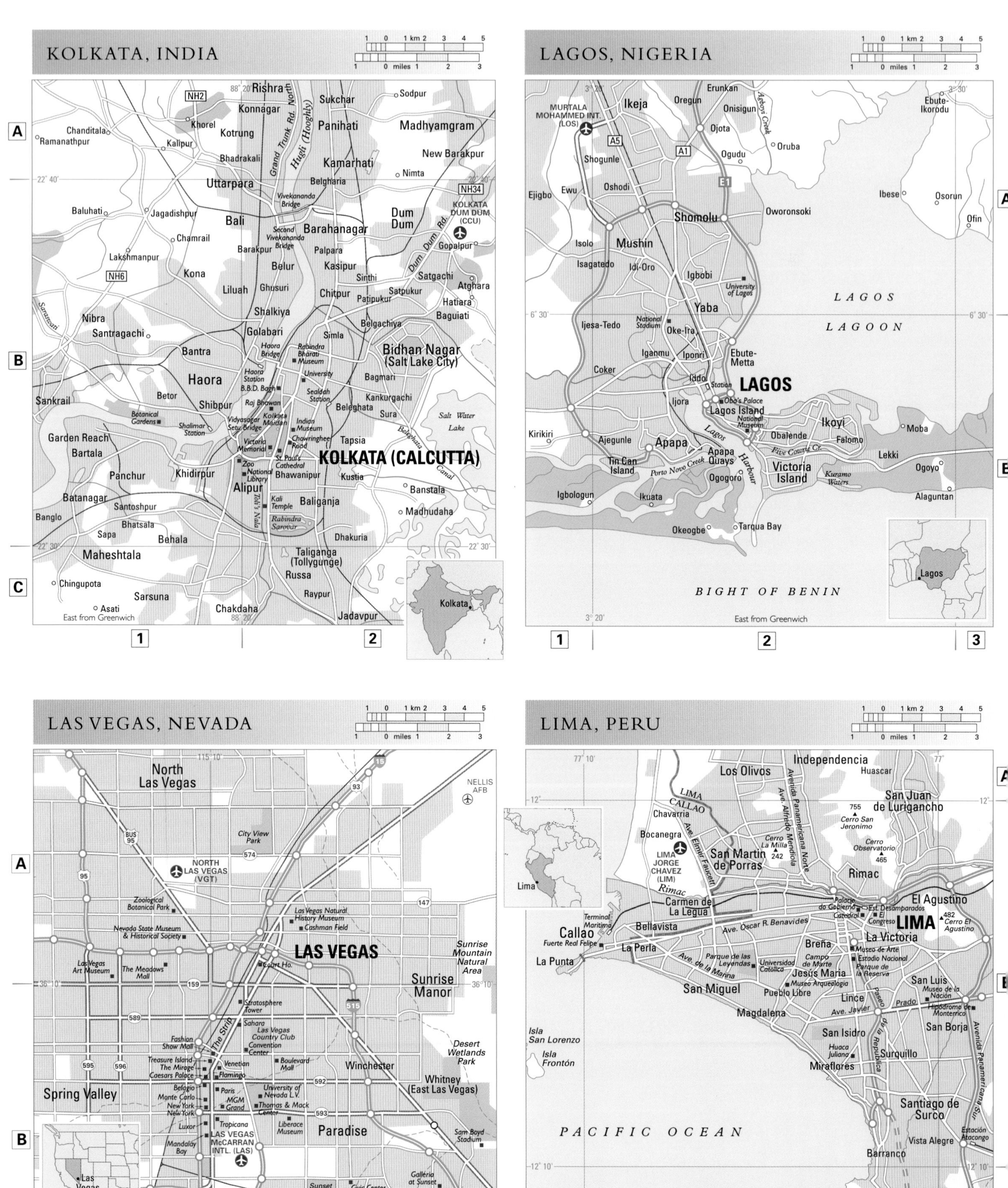

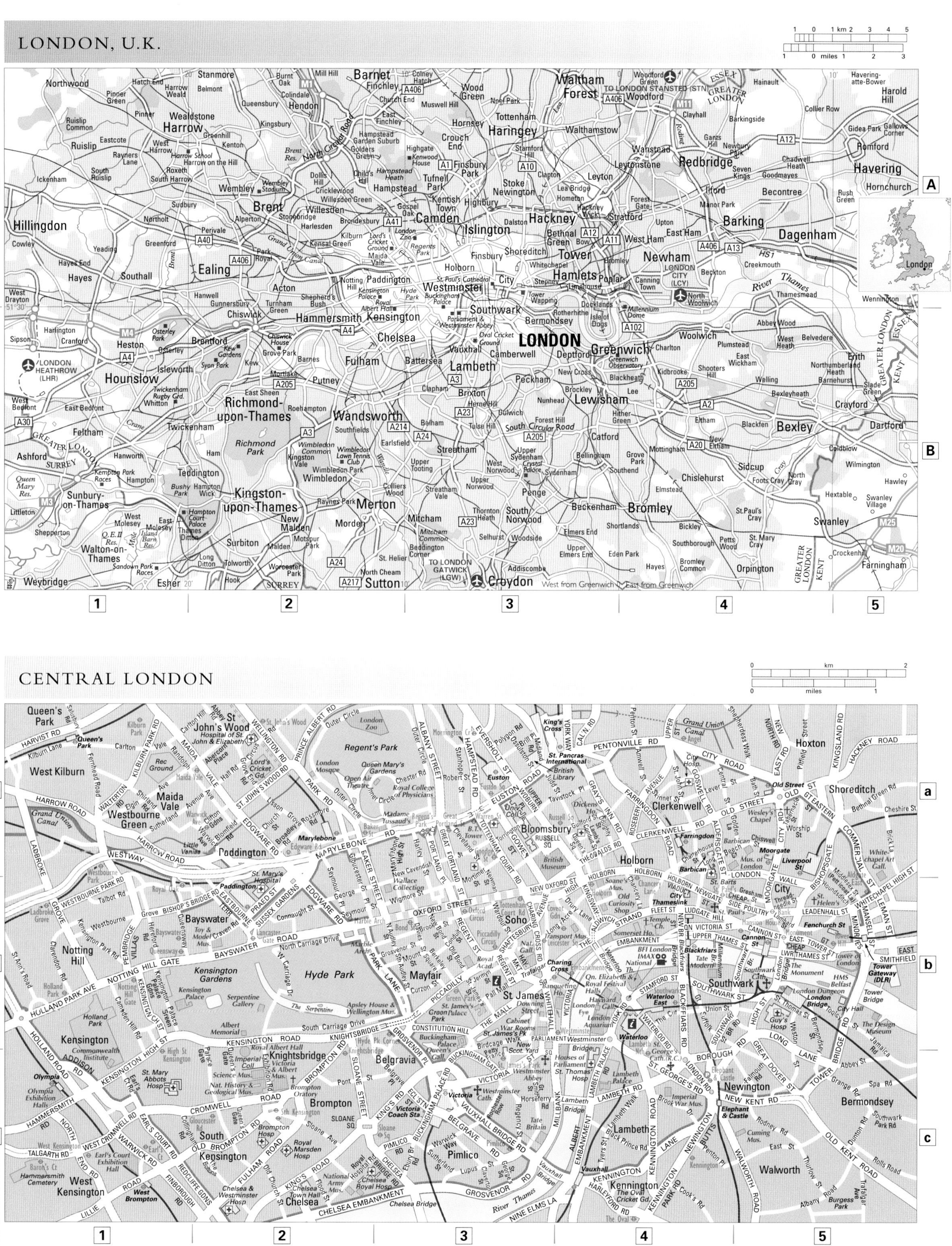

LONDON, U.K.

CENTRAL LONDON

Congestion Charging Zone

COPYRIGHT PHILIP'S

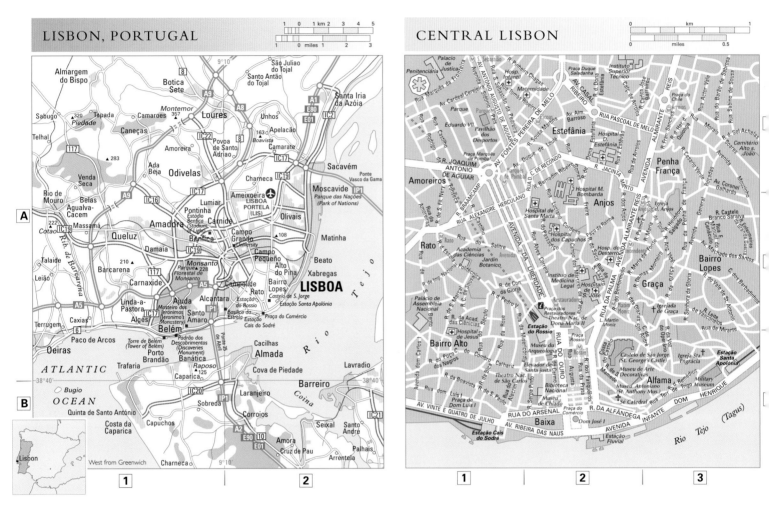

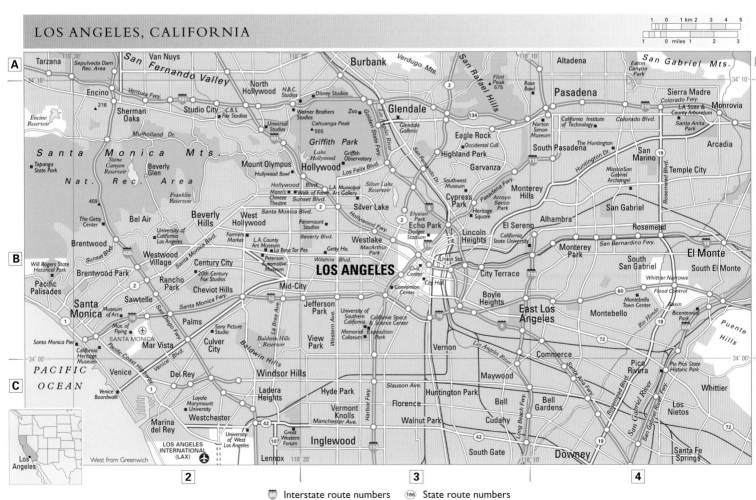

Interstate route numbers   State route numbers

COPYRIGHT PHILIP'S

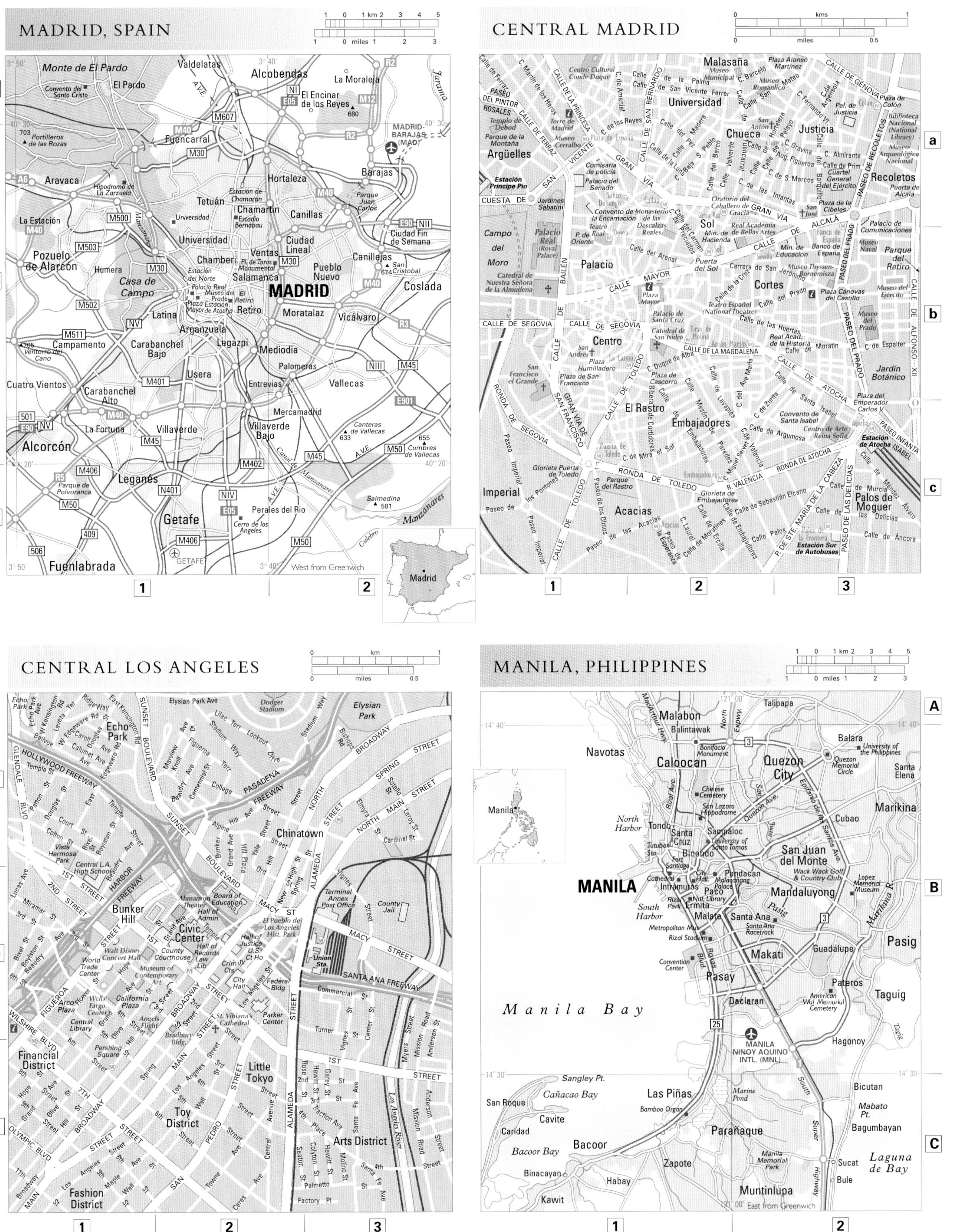
MADRID, SPAIN
Monte de El Pardo
El Pardo
Valdelatas
Alcobendas
La Moraleja
El Encinar de los Reyes
MADRID-BARAJAS (MAD)
Fuencarral
Aravaca
Barajas
Hortaleza
Tetuán
Chamartín
Canillas
La Estación
Pozuelo de Alarcón
Universidad
Ciudad Lineal
Ciudad Fin de Semana
Chamberí
Ventas
Canillejas
Humera
Casa de Campo
Salamanca
Pueblo Nuevo
Coslada
MADRID
Latina
Retiro
Moratalaz
Vicálvaro
Arganzuela
Campamento
Carabanchel Bajo
Legazpi
Mediodía
Usera
Palomeras
Cuatro Vientos
Carabanchel Alto
Entrevías
Vallecas
Mercamadrid
La Fortuna
Villaverde
Villaverde Bajo
Alcorcón
Leganés
Getafe
Perales del Río
Fuenlabrada
West from Greenwich
Madrid
CENTRAL MADRID
Malasaña
Universidad
Chueca
Justicia
Argüelles
Recoletos
Sol
Palacio
Cortes
Centro
El Rastro
Embajadores
Imperial
Acacias
Palos de Moguer
Jardín Botánico
Estación de Atocha
CENTRAL LOS ANGELES
Echo Park
Elysian Park
Dodger Stadium
Chinatown
Bunker Hill
Civic Center
Union Sta.
Financial District
Little Tokyo
Toy District
Arts District
Fashion District
MANILA, PHILIPPINES
Malabon
Navotas
Caloocan
Quezon City
Marikina
Tondo
Santa Cruz
Sampaloc
San Juan del Monte
Cubao
MANILA
Intramuros
Ermita
Malate
Paco
Mandaluyong
Santa Ana
Makati
Pasig
Pasay
Pateros
Taguig
Manila Bay
MANILA NINOY AQUINO INTL. (MNL)
Hagonoy
Las Piñas
Cavite
Bacoor
Parañaque
Zapote
Muntinlupa
Laguna de Bay
East from Greenwich
Manila

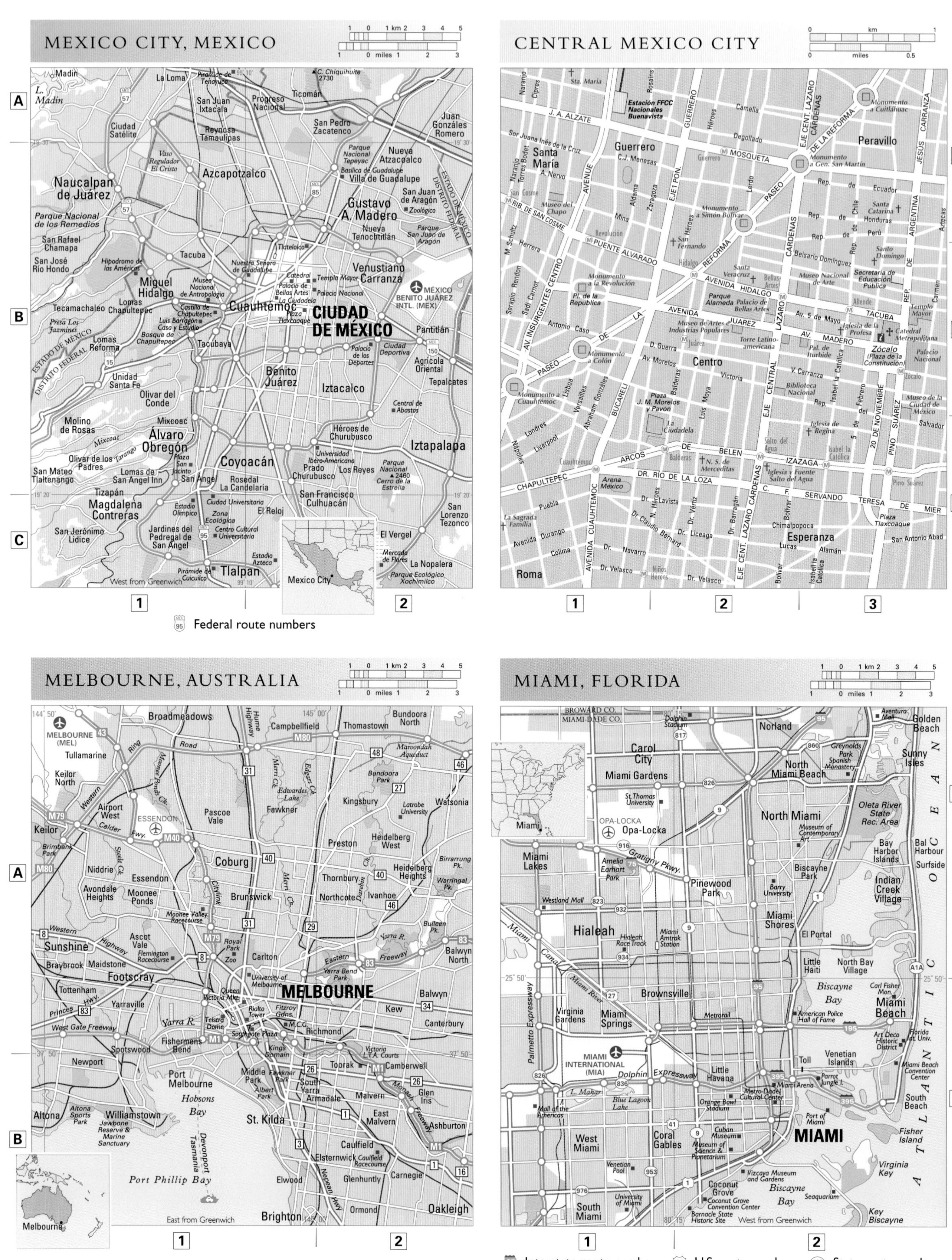

MEXICO CITY, MEXICO

1 0 1 km 2 3 4 5
1 0 miles 1 2 3

West from Greenwich

Mexico City

95 Federal route numbers

CENTRAL MEXICO CITY

km
0 miles 0.5

MELBOURNE, AUSTRALIA

1 0 1 km 2 3 4 5
1 0 miles 1 2 3

East from Greenwich

Melbourne

MIAMI, FLORIDA

1 0 1 km 2 3 4 5
1 0 miles 1 2 3

West from Greenwich

Miami

85 Interstate route numbers   29 U.S. route numbers   166 State route numbers

COPYRIGHT PHILIP'S

MILAN, ITALY

1 0 1 km 2 3 4 5
1 0 miles 1 2 3

Coronno
Cesate
Pertusella
Limbiate
Varedo
Muggiò
Autodromo
Concorezzo
Monza
Garbagnate Milanese
Parco Regionale
Can. Villoresi
Palazzolo
Incirano
Dugnano
Nova Milanese
Senago
Amata
Cassina Nuova
Paderno
Lainate
Valera
Cusano Milanino
Cinisello Balsamo
San Fruttuoso
Brughério
Bollate
Passirana
Arese
Terrazzano
Ospiate
Cormano
Bresso
San Maurizio al Lambro
Rho
Bruzzano
Parco Regionale
Affori
Sesto San Giovanni
Cologno Monzese
Novate Milanese
Bovisa
Precotto
Naviglio d. Martesana
Statale Padana superiore
Cornaredo
Pero
Greco
Crescenzago
Vimodrone
Pioltello
Musocco
Milano Due
Vighignolo
Figino
Olona
Trenno
Bolòinasco
Loreto
Parco Lambro
Segrate
Séttimo Milanese
St dio San Siro
MILANO
Stazione Centrale
Lambrate
Tangenziale Est
Seguro
San Siro
Fiera Camp.
Castello Sforzesco
Brera
La Scala
Ortica
Milano San Felice
Quinto Romano
Sta. Ferrovie Nord
Duomo
Basilica di Sant Ambr gio
Città degli Studi
Idroscalo
San Bóvio
Monzoro
Bággio
MILAN LINATE (LIN)
Assiano
Calvairate
Cusago
Cesano Boscone
San Cristoforo
Morivione
Gamboloita
Mezzate
Vigentino
Peschiera Borromeo
Quartiere Zingone
Naviglio Grande
Córsico
Triulzo
Romano Banco
Metanopoli
Trezzano sul Naviglio
Tangenziale Ovest
Assago
Chiaravalle Milanese
Gratosóglio
San Donato Milanese
Gaggiano
Buccinasco
Quinto de Stampi
Poasco
San Giuliano Milanese
San Novo
Sesto Ulteriano
Mediglia
San Pietro Cúsico
Gudo Gamb.
Mirasole
Zivido
Pontesesto
Opera
Via Emilia
Milan
Rozzano
Fizzonasco
San Brera
Zibido San Giacomo
Naviglio di Pavia
Statale dei Giovi
Tolcinasco
Mezzano
Locate di Triulzi
Zúnico
9° 10' East from Greenwich
45° 30'
527
35
36
233
E66
A4
A51
A8
A4
E35
A50
11
A7
415
494
412
9
E35
A50
A1
E35
A
B
1
2

CENTRAL MOSCOW

0 km 1
0 miles 0.5

SAD.-SAMOTECHNAYA
SAD.-SUHAREVSKAYA
SAD.-TRIUMFALNAYA ULITSA
Svetnoy Boulevard
Old Moscow Circus
Suharevskaya
SAD.-SPASSKAYA
Sergievskiy Per.
Mayakovskiy Ploshchad
Tchaikovsky Concert Hall
Mayakovskaya
CHEKHOVA U.
Russian Cinema
PETROVSKIY BOULEVARD
Trubnaya Pl.
'BOULEVARD RING' ROZHDESTVENSKIY BOULEVARD
Kostyanskiy Per.
U. SRETENKA
Youth Theatre
Pushkinskaya
Tverskaya
TVERSKAYA
STRASTNOY BLD.
Chekhovskaya
Convent of the Nativity of the Virgin
Ulitsa Rozhdestvenka
Sadovskikh Per.
Museum of the Revolution
Pushkin Ploshchad
PUSHKINSKAYA
PETROVKA
NEGLINNAYA
SVETNOY BOULEVARD
Varsonofevskiy Per.
ULITSA LUBYANKA
Turgenevskaya
Turgenevskaya Pl.
Chisty Prudy
'BOULEVARD RING' TVERSKOY BOULEVARD
Stoleshnikov
Petrovskiy Passage
Bolshoy Theatre
Kuznetskiy Most
Detskiy Theatre
U. MYASNITSKAYA
Gorky Theatre
Pereulok
Teatralnaya
Lubyanka
Komsomolskiy Bld.
MAL. BRONNAYA ULITSA
Chekhov Theatre
Central Post Office
Okhotniy Ryad
Theatre Square
TEATRALNIY PROJ.
Ploshchad Lubyanskaya
LUBYANSKIY
Gorky House Museum
Ulitsa Stanislavskovo
Ulitsa Nezhdanovoy
Ulitsa Ogaryova
Belinskogo Ul.
Ermolovoy Theatre
Slavanskiy Bazar
Ul. Nikolskaya
Bolshoy Per.
NOVAYA PL.
Polytechnic Museum
Kitai Gorod
Nogina
Revolution Square
Pl. Revolyutsiy
Devyatinskiy
GERSENA
Manezhnaya Ploshchad
Historical Museum
Lenin Museum
Gum Shopping Arcade
Ulitsa Ilinka
Vladimirova Pereulok
SLAVYANSKAYA PL.
PROSPEKT
NIKITSKIY BLD.
Moscow Conservatoire
ULITSA
Semazhko Ulitsa
University
Red Square
Central Exhibition Hall
Alexander Garden
ULITSA
Arsenal
Council of Ministers
Lenin Mausoleum
Arbatskaya Ploshchad
VOZDVIZHENKA U.
Museum of Russian Architecture
Aleksandrovsky Sad
Presidium of the Supreme Soviet
St. Basil's Cathedral
ULITSA VARVARKA
Arbatskaya
RYAD OKHOTNIY
MANEZHNAYA
Palace of Congress
Ivan Square
Kremlin
Central Concert Hall
ULITSA ARBAT
Lenin State Library
U. ZNAMENKA
Terem Palace
Cathedral Square
Archangel Cathedral
KITAISKI PERULOK
Armoury Palace
Kremlin Palace
MOSKVORETS. NAB.
Marx-Engels Ulitsa
Borovitskaya Ploshchad
KREMLEVSKAYA NABEREZHNAYA
Moskva (Moscow)
RAUSHSKAYA NAB.
GOGOLEVSKIY BOULEVARD
'BOULEVARD RING'
Pushkin Fine Arts Museum
VOLKHONKA ULITSA
SOFIYSKAYA NABEREZHNAYA
Ryleyev Ulitsa
Cathedral of Christ the Saviour
BOLSHOY KAMENIY MOST
BOLOTNAYA NAB.
Vodootvodny Kanal
SADOVNICHESKAYA
Kropotkinskaya
KADASHEVSKAYA NAB.
OVCHINNIKOVSKAYA
a
b
c
1
2
3

MOSCOW, RUSSIA

1 0 1 km 2 3 4 5
1 0 miles 1 2 3

Skhodnya
37° 20'
Putilkovo
Bratsevo
TO MOSCOW SHEREMETYEVO INTL. (SVO)
37° 30'
Degunino
Vladykino
37° 40'
Yaroslavskoye Prospekt
GOROD MOSKVA
MOSKVA OBLAST
37° 50'
Medvezhiy Ozyora
Medvezhiy Ozyora
Mitino
Khimki-Khovrino
Babushkin
Novonikolyskoye
157
Almazova
Chernyovo
Penyagino
M10
Likhoborka
Pekhra-Pokrovskoye
55° 50'
Tushino
Nikolskiy
Petrovsko-Razumovskoye
Dzerzhinskiy Park
M8
Losiny Ostrov National Park
Krasnogorsk
Timiryazev Park
Abramtsevo
Pavshino
Ostankino
Moskovskaya Koltsevaya Avtomobilnaya Doroga (MKAD) (Moscow Ring Road)
Golyevo
Myakinino
Leningradskiy Prospekt
Yauza
Vostochnyy
Balashikha
Strogino
Bogorodskoye
Sosenka
Galyanovo
140
M9
Pokrovsko-Sresnevo
Petrovskiy Park
Sokolniki Park
Novaya
Arkhangelskoye
Troitse-Lykovo
Sokolniki
Izmaylovo
Gorenki
M7
Pekhra-Yakovievskaya
Zakharkovo
Moskva
Frunze
Dzerzhinskiy
Serebryanka
Vishnyaki
Rublovo
Khorosovo
Sverdlov
Leningrad Station
Yaroslavl Station
Izmayloskiy Park
Nikolyskoye
Pekhorka
Mnevniki
MOSKVA
Kazan Station
Leportovo
150
Moscow
Tatarovo
Saltykovka
Krasno-Presnenskaya
Bolshoy Theatre
Bauman
Novogireyevo
Reutov
Razdory
Cherepkovo
Moskva (Moscow)
Red Square, St. Basil's Cath. Lenin Museum
Kursk Station
Serebryanka
Kutsino
MOSKVA OBLAST
GOROD MOSKVA
Krylatskoye
Kremlin
Perovo
Zheleznodorozhnyy
Barvikha
Setun
Fili-Mazilovo
Kiev Station
Tretyakov Art Gallery
Kuskovo
Romashkovo
Kuntsevo
Zhdanov
Plyushchevo
Vochnyaki
Rudnevka
Fenino
Poduskino
Novodevichy Convent
Pavelet Station
Davydkovo
Gorky Park
Vykhino
Kosino
Kozhukhovo
Temnikovo
Nemchinovka
Luzhniki Sports Centre, Lenin Stadium
Moskvoretskiy
94
B
Novoivanovskoye
Lomonosov Moscow State University
Volgogradskiy Prospekt
Zhulebino
Mikhelysona
Chornaya
Lochino
Aminyevo
Leninskiye Gory
Moscow Circus
Oktyabrskiy
Tekstilyshchik
Kuzyminki
Marusino
Mamonovo
150
(MKAD)
Bakovka
Setun
Zarechye
Ochakovo
Ramenki
Leninskiy Prospekt
Nogatino
Lyublino
Nekrasovka
Odintsovo
Cheryomushki
Lyubertsy
Korenevo
M1
Meshcherskiy
Yugo-Zarad
Dyakovo
Maryino
55° 40'
Nikulino
Moskva (Moscow)
Volkhonka-Zil
Tomilino
Kraskovo
Chobuty
Solntsevo
Troparevo
Zyuzino
Kuryanovo
Kotelyniki
Malakhovka
Peredelkino
Belyayevo Bogorodskoye
Bittsevsky Forest Park
Kapotnya
M5
Chkalova
Orlovo
Chertanovka
Brateyevo
Rasskazovka
Rumyantsevo
M3
250
M2
Lenino
M4
TO DOMODEDOVO INTL. (DME)
Tokarevo
Vnukovo
37° 30' East from Greenwich
Chertanovo
M6
Borisovo
Dzerzhinskiy
37° 50'
38°
A
B
C
1
2
3
4
5
6

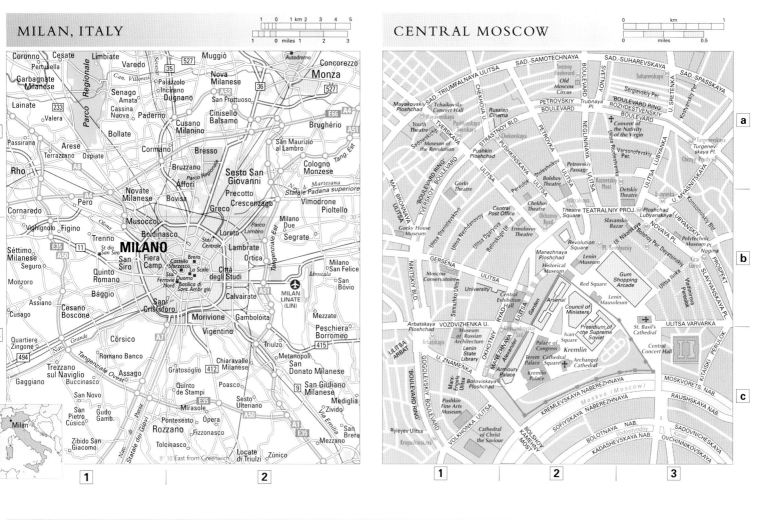

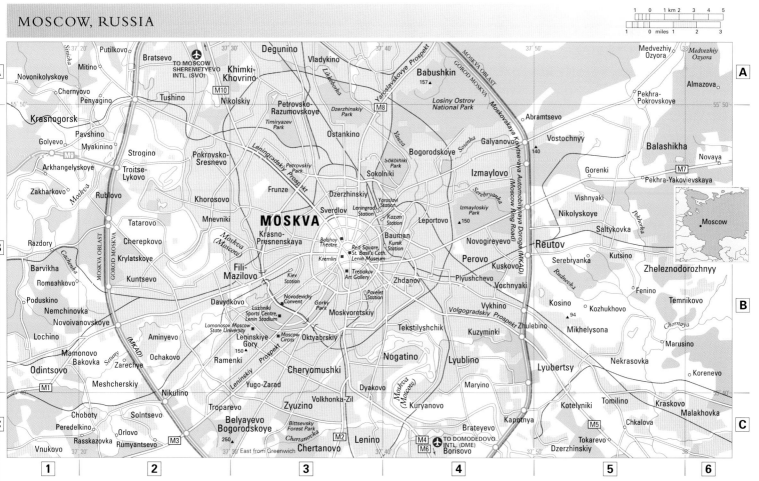

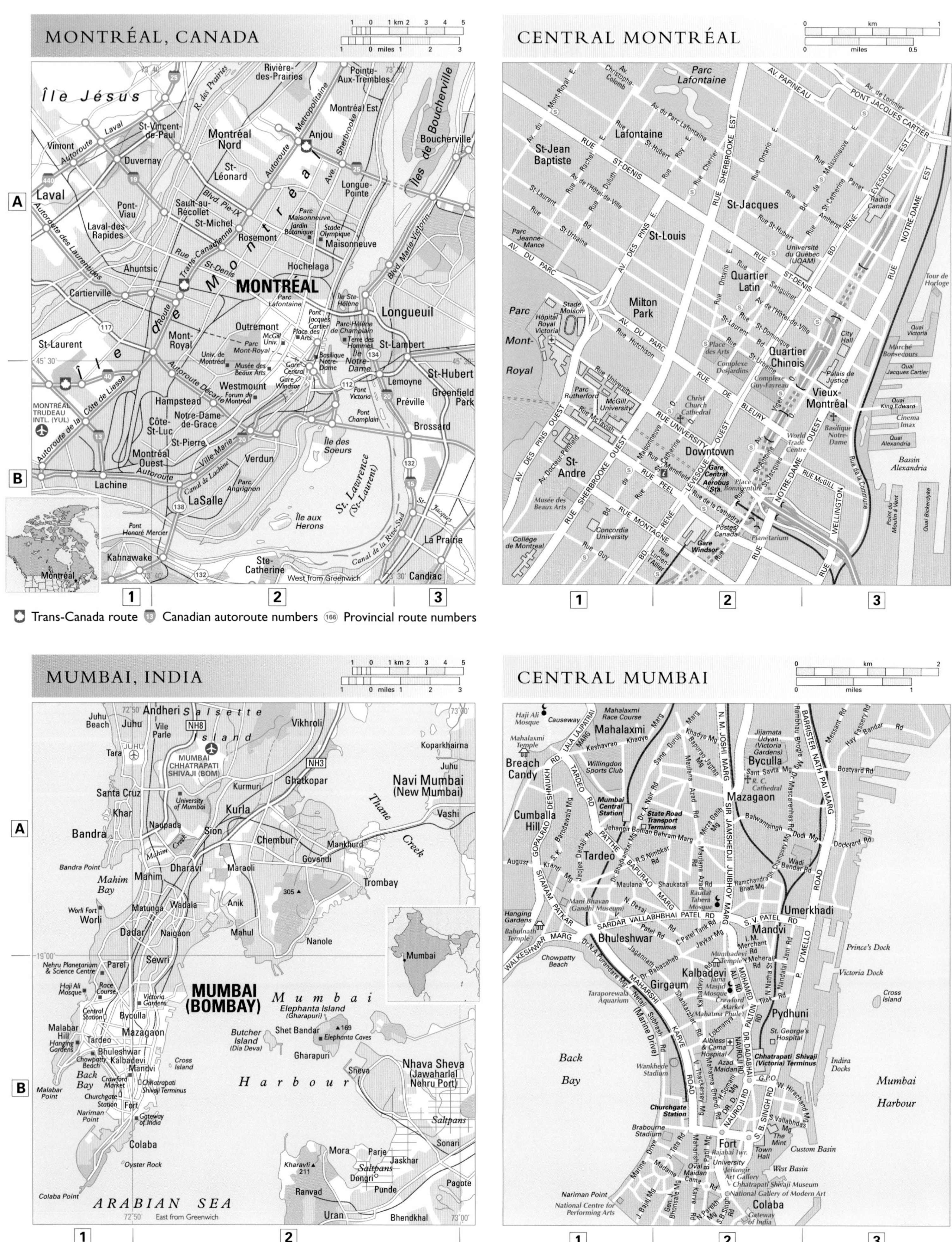

MONTRÉAL, CANADA

Île Jésus

Laval
Vimont
St-Vincent-de-Paul
Laval
Montréal Nord
Rivière-des-Prairies
Pointe-Aux-Trembles
Montréal Est
Boucherville
Anjou
Duvernay
St-Léonard
Sault-au-Récollet
St-Michel
Rosemont
Parc Maisonneuve Jardin Botanique
Stade Olympique
Maisonneuve
Longue-Pointe
Pont-Viau
Laval-des-Rapides
Ahuntsic
Cartierville
St-Laurent
MONTRÉAL
Hochelaga
Île Ste-Hélène
Parc Lafontaine
Outremont
Mont-Royal
McGill Univ.
Pont Jacques Cartier
Parc-Hélène de Champlain
Terre des Hommes
Île Notre-Dame
Longueuil
St-Lambert
Westmount
Univ. de Montréal
Musée des Beaux Arts
Place des Arts
Gare Central
Gare Windsor
Basilique Notre-Dame
Île Notre-Dame
St-Hubert
Lemoyne
Préville
Greenfield Park
Hampstead
Côte-St-Luc
Notre-Dame-de-Grace
St-Pierre
Forum de Montréal
Pont Victoria
Pont Champlain
Brossard
MONTRÉAL TRUDEAU INTL. (YUL)
Montréal Ouest
Ville Marie
Verdun
Île des Soeurs
Lachine
LaSalle
Canal de Lachine
Parc Angrignon
St. Lawrence (St-Laurent)
La Prairie
Kahnawake
Pont Honoré Mercier
Ste-Catherine
Île aux Herons
Candiac
West from Greenwich

Trans-Canada route   Canadian autoroute numbers   Provincial route numbers

CENTRAL MONTRÉAL

Parc Lafontaine
St-Jean Baptiste
Lafontaine
St-Jacques
Radio Canada
St-Louis
Université du Québec (UQAM)
Parc Jeanne-Mance
Quartier Latin
Tour de Horloge
Milton Park
Parc Mont-Royal
Stade Molson
Hôpital Royal Victoria
Place des Arts
Quartier Chinois
Quai Victoria
Marché Bonsecours
Quai Jacques Cartier
Parc Rutherford
McGill University
Christ Church Cathedral
Complexe Desjardins
Complexe Guy-Favreau
Vieux-Montréal
Palais de Justice
City Hall
World Trade Centre
Basilique Notre-Dame
Quai King Edward
Cinema Imax
Quai Alexandria
St-Andre
Musée des Beaux Arts
Downtown
Gare Central Aerobus Sta.
Place Bonaventure
Bassin Alexandria
Concordia University
Postes Canada
Gare Windsor
Planetarium
Point du Moulin à Vent
Quai Bickerdyke
Collège de Montréal

MUMBAI, INDIA

Andheri   Salsette Island
Juhu Beach
Juhu
Vile Parle
NH8
Vikhroli
Tara
MUMBAI CHHATRAPATI SHIVAJI (BOM)
NH3
Koparkhairna
Santa Cruz
Kurmuri
Ghatkopar
Juhu
Navi Mumbai (New Mumbai)
University of Mumbai
Khar
Naupada
Kurla
Chembur
Vashi
Bandra
Sion
Mankhurd
Thane Creek
Bandra Point
Mahim
Dharavi
Maraoli
Govandi
Trombay
Mahim Bay
Matunga
Wadala
Anik
305
Worli Fort
Worli
Dadar
Naigaon
Mahul
Nanole
Nehru Planetarium & Science Centre
Parel
Haji Ali Mosque
Race Course
Sewri
MUMBAI (BOMBAY)
Mumbai
Elephanta Island (Gharapuri)
Central Station
Victoria Gardens
Byculla
Butcher Island (Dia Deva)
Shet Bandar
169
Elephanta Caves
Malabar Hill
Hanging Gardens
Tardeo
Mazagaon
Gharapuri
Nhava Sheva (Jawaharlal Nehru Port)
Chowpatty Beach
Bhuleshwar
Kalbadevi
Mandvi
Cross Island
Sheva
Malabar Point
Back Bay
Crawford Market
Chhatrapati Shivaji Terminus
Harbour
Churchgate Station
Fort
Saltpans
Nariman Point
Gateway of India
Colaba
Mora
Parje
Jaskhar
Saltpans
Sonari
Colaba Point
Oyster Rock
Dongri
Punde
Pagote
ARABIAN SEA
Kharavli 211
Ranvad
East from Greenwich
Uran
Bhendkhal

CENTRAL MUMBAI

Haji Ali Mosque
Mahalaxmi Race Course
Causeway
Mahalaxmi Temple
Mahalaxmi
N. M. JOSHI MARG
Jijamata Udyan (Victoria Gardens)
Breach Candy
Willingdon Sports Club
Keshavrao Khadye
Mumbai Central Station
State Road Transport Terminus
Byculla
Mazagaon
Cumballa Hill
Tardeo
R.S. Nimbkar Rd
Umerkhadi
Hanging Gardens
Mani Bhavan (Gandhi Museum)
Raudat Tahera Mosque
Mandvi
Babulnath Temple
SARDAR VALLABHBHAI PATEL RD
Bhuleshwar
Mumbadevi Temple
Merchant
Chowpatty Beach
Jagannath
Kalbadevi
Girgaum
Prince's Dock
Taraporewala Aquarium
Crawford Market
Pydhuni
Victoria Dock
Back Bay
Albless & Cama Hospital
Azad Maidan
Chhatrapati Shivaji (Victoria) Terminus
St. George's Hospital
Cross Island
Indira Docks
Wankhede Stadium
G.P.O.
Churchgate Station
Mumbai Harbour
Brabourne Stadium
The Mint
Custom Basin
Fort
Town Hall
West Basin
Nariman Point
National Centre for Performing Arts
Rajabai Twr.
University
Jehangir Art Gallery
Oval Maidan
Chhatrapati Shivaji Museum
National Gallery of Modern Art
Colaba
Gateway of India

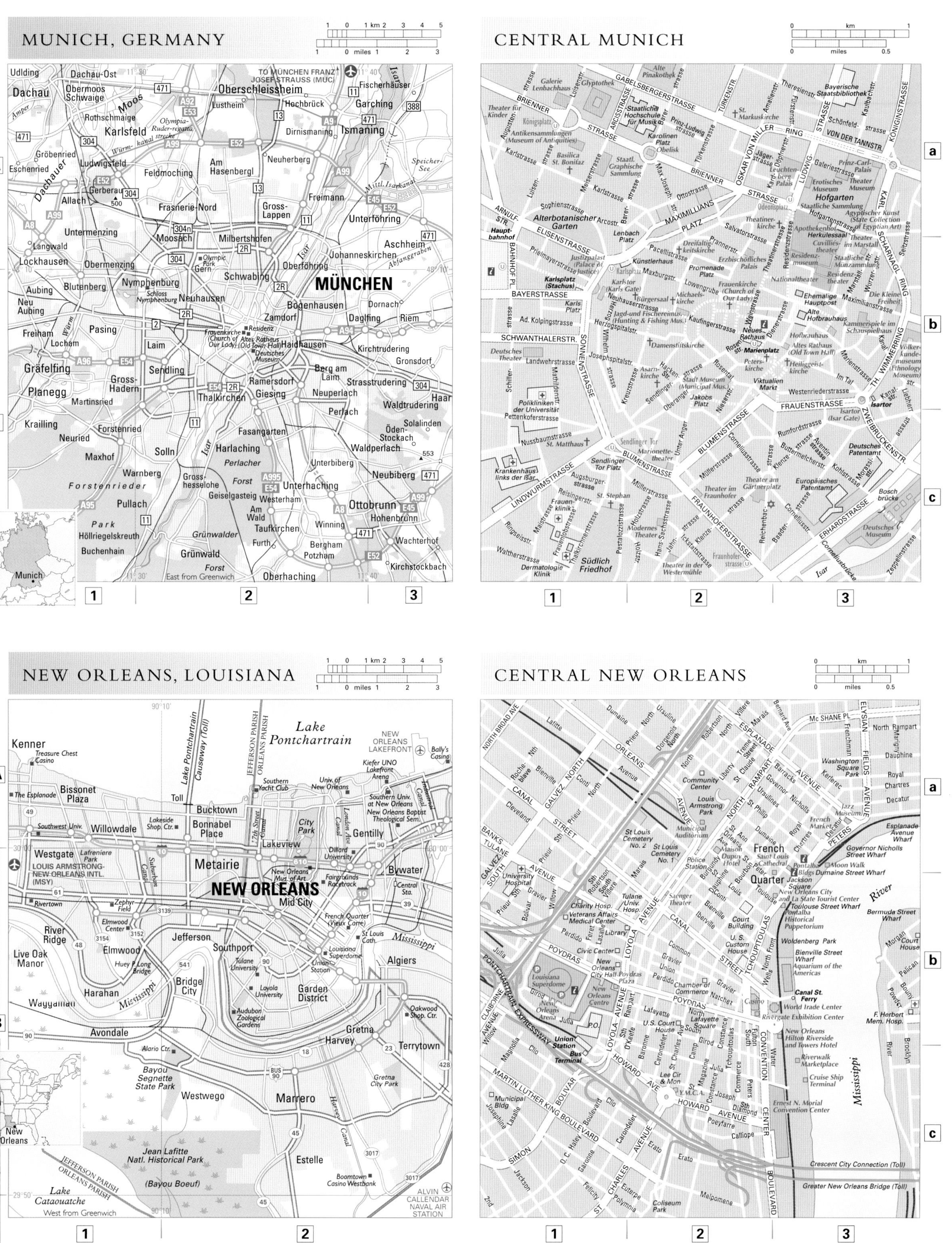

MUNICH, GERMANY

CENTRAL MUNICH

NEW ORLEANS, LOUISIANA

CENTRAL NEW ORLEANS

Interstate route numbers   ⑰ U.S. route numbers   ⑷₁₇ State route numbers

NEW YORK, NEW YORK

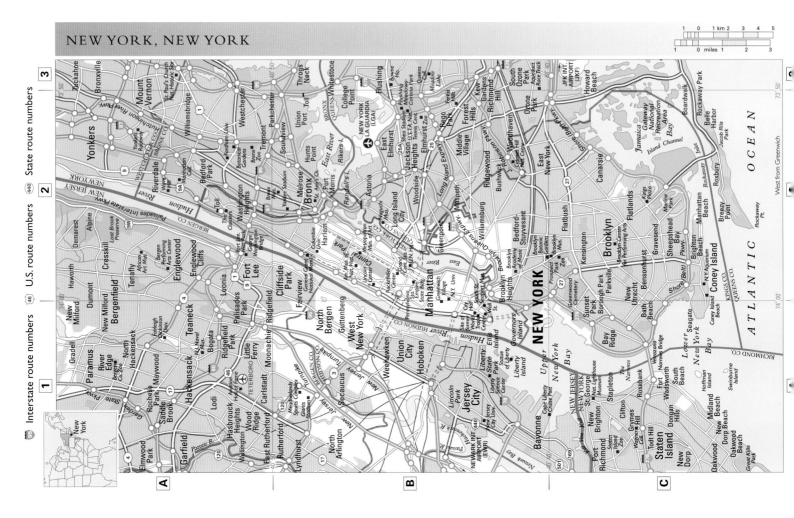

CENTRAL NEW YORK

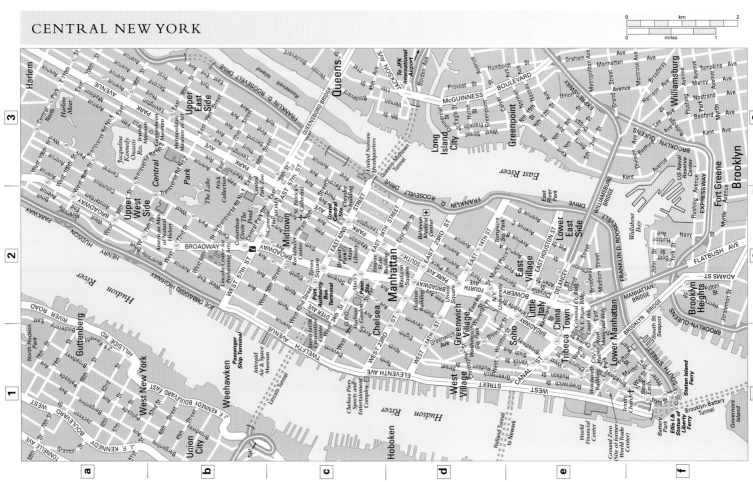

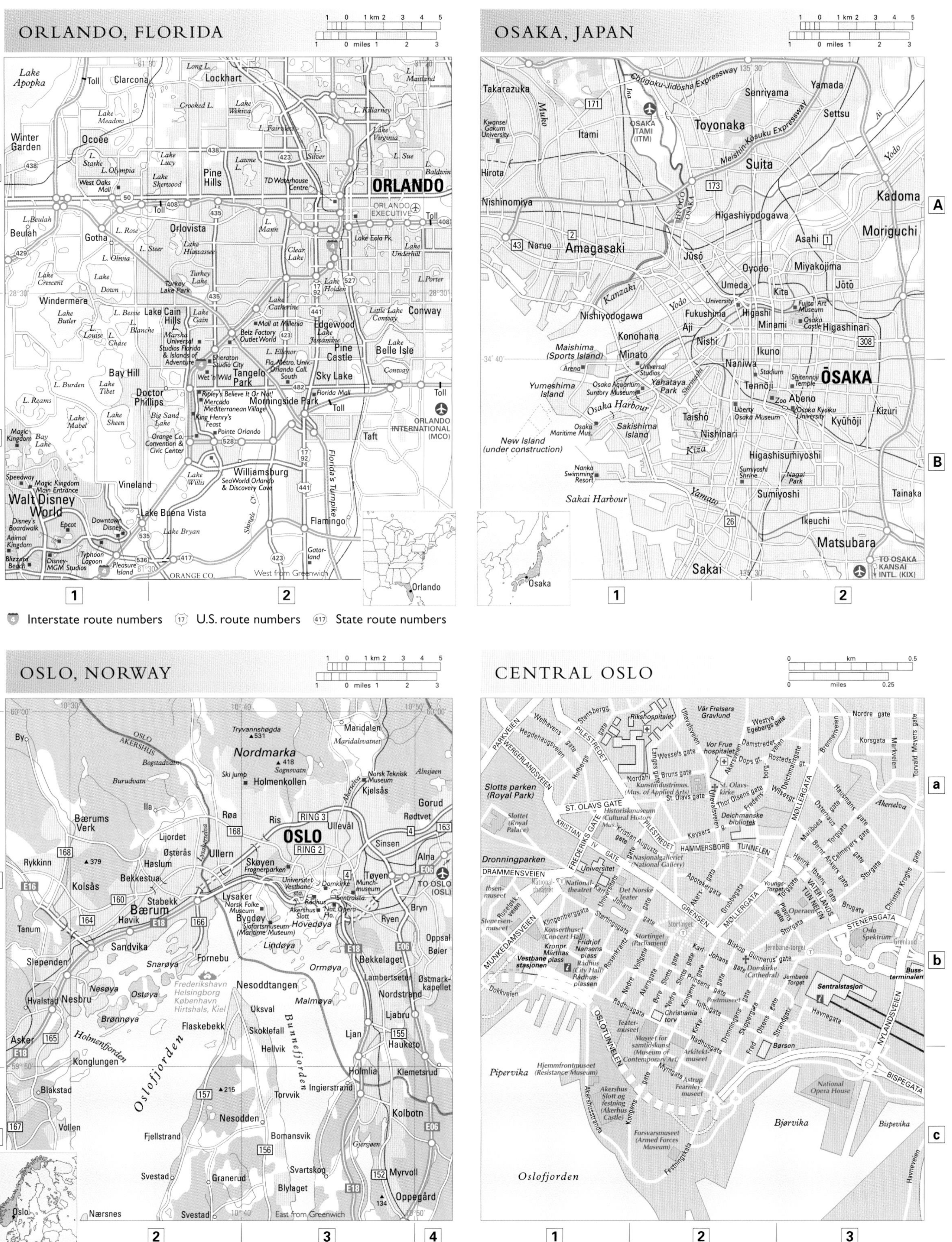
ORLANDO, FLORIDA
1 0 1 km 2 3 4 5
1 0 miles 1 2 3
Lake Apopka
Clarcona
Lockhart
Winter Garden
Ocoee
Pine Hills
ORLANDO
ORLANDO EXECUTIVE
Beulah
Gotha
Orlovista
Windermere
Lake Cain Hills
Conway
Edgewood
Belle Isle
Pine Castle
Sky Lake
Bay Hill
Tangelo Park
Doctor Phillips
Morningside Park
ORLANDO INTERNATIONAL (MCO)
Taft
Vineland
Williamsburg
Walt Disney World
Lake Buena Vista
Flamingo
Florida's Turnpike
Orlando
Interstate route numbers
U.S. route numbers
State route numbers
OSAKA, JAPAN
1 0 1 km 2 3 4 5
1 0 miles 1 2 3
Takarazuka
Itami
OSAKA ITAMI (ITM)
Toyonaka
Senriyama
Yamada
Settsu
Suita
Hirota
Nishinomiya
Kadoma
Higashiyodogawa
Moriguchi
Asahi
Amagasaki
Naruo
Jūsō
Oyodo
Miyakojima
Umeda
Kita
Jōtō
Nishiyodogawa
Fukushima
Higashi
Minami
Higashinari
Konohana
Nishi
Minato
Ikuno
Naniwa
ŌSAKA
Tennōji
Abeno
Taishō
Nishinari
Kizuri
Kyūhōji
Yumeshima Island
Sakishima Island
Osaka Harbour
New Island (under construction)
Higashisumiyoshi
Sumiyoshi
Sakai Harbour
Tainaka
Ikeuchi
Matsubara
Sakai
TO OSAKA KANSAI INTL. (KIX)
Osaka
OSLO, NORWAY
1 0 1 km 2 3 4 5
1 0 miles 1 2 3
Nordmarka
Maridalen
Holmenkollen
Kjelsås
Gorud
Rødtvet
Bærums Verk
Røa
Ris
OSLO
Ullevål
Sinsen
Rykkinn
Østerås
Ullern
Haslum
Skøyen
Alna
Kolsås
Bekkestua
Tøyen
TO OSLO (OSL)
Stabekk
Lysaker
Bærum
Bygdøy
Hovedøya
Bryn
Høvik
Ryen
Tanum
Sandvika
Fornebu
Oppsal
Bøler
Lindøya
Bekkelaget
Slependen
Ormøya
Østmarkkapellet
Nesoddtangen
Nordstrand
Hvalstad
Nesbru
Malmøya
Ljabru
Flaskebekk
Uksval
Asker
Bunnefjorden
Oslofjorden
Ljan
Hauketo
Skoklefall
Hellvik
Konglungen
Holmenfjorden
Holmlia
Klemetsrud
Blakstad
Torvvik
Ingierstrand
Kolbotn
Vollen
Nesodden
Fjellstrand
Bomansvik
Svestad
Granerud
Svartskog
Myrvoll
Blylaget
Oppegård
Nærsnes
East from Greenwich
CENTRAL OSLO
0 km 0.5
0 miles 0.25
Rikshospitalet
Vår Frelsers Gravlund
Slotts parken (Royal Park)
Slottet (Royal Palace)
Dronningparken
Universitet
Nationaltheatret
Stortinget (Parliament)
Domkirke (Cathedral)
Sentralstasjon
Rådhus (City Hall)
Akershus Slott og festning (Akershus Castle)
Oslo Spektrum
National Opera House
Pipervika
Bjørvika
Bispevika
Oslofjorden

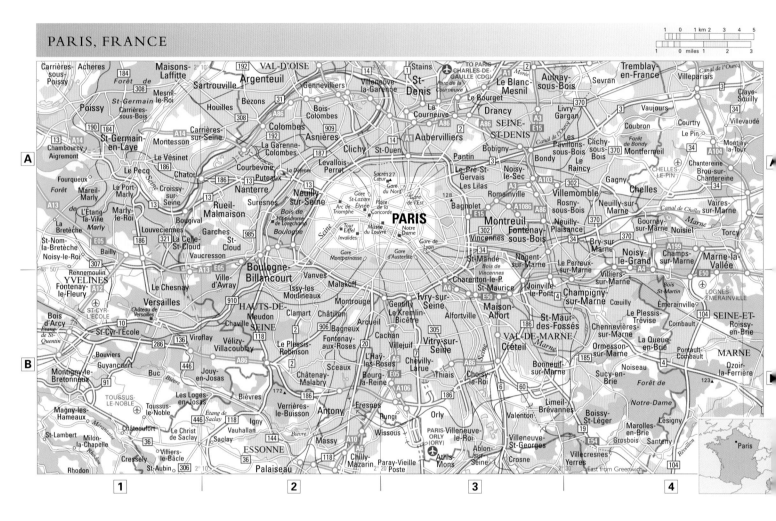

PARIS, FRANCE

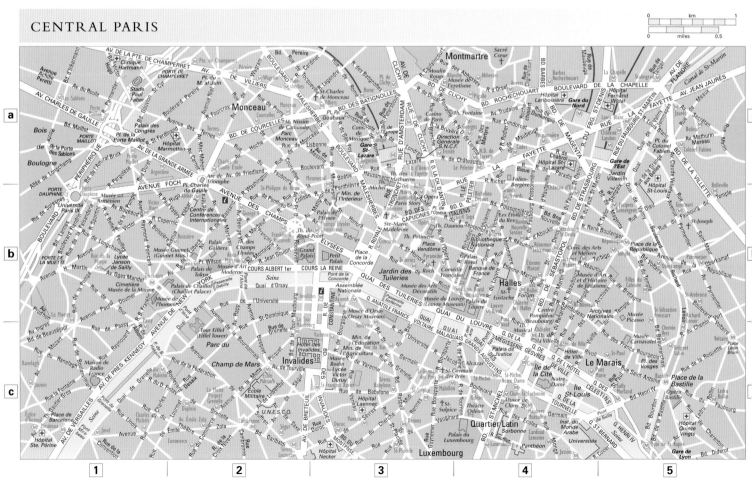

CENTRAL PARIS

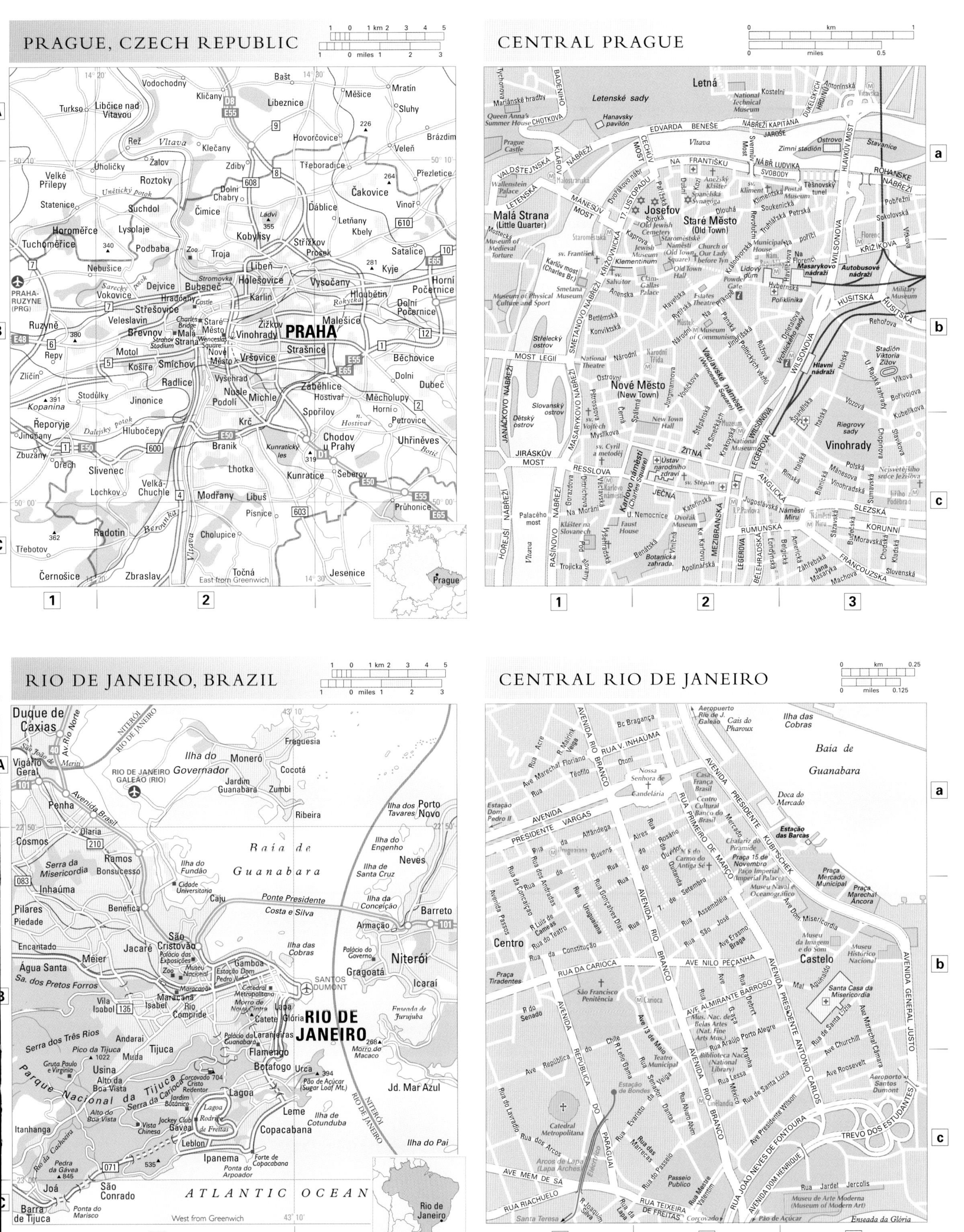
PRAGUE, CZECH REPUBLIC
CENTRAL PRAGUE
RIO DE JANEIRO, BRAZIL
CENTRAL RIO DE JANEIRO
PRAHA
Vltava
Letná
Malá Strana
(Little Quarter)
Josefov
Staré Město
(Old Town)
Nové Město
(New Town)
Vinohrady
Hlavní nádraží
Prague Castle
Karlův most
(Charles Br.)
Václavské náměstí
(Wenceslas Square)
Karlovo náměstí
(Charles Square)
RIO DE JANEIRO
Baía de Guanabara
Ilha do Governador
Niterói
Copacabana
Ipanema
Leblon
Botafogo
Flamengo
Lagoa
Barra de Tijuca
ATLANTIC OCEAN
Parque Nacional da Tijuca
Ponte Presidente Costa e Silva
Ilha das Cobras
Centro
Castelo
AVENIDA RIO BRANCO
AVENIDA PRESIDENTE VARGAS
Catedral Metropolitana
Arcos da Lapa
(Lapa Arches)
Museu de Arte Moderna
(Museum of Modern Art)
Enseada da Glória
East from Greenwich
West from Greenwich

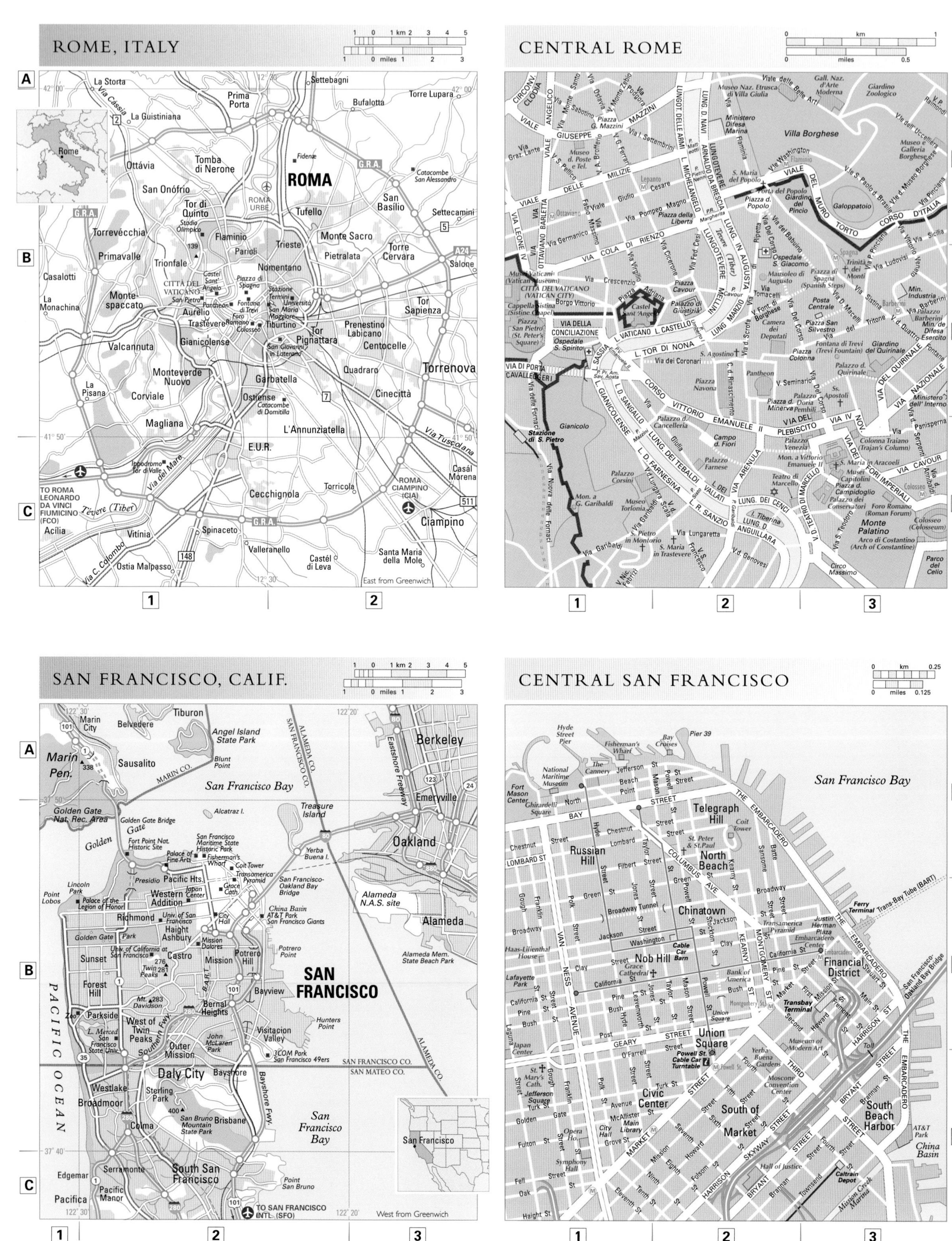

ROME, ITALY

CENTRAL ROME

SAN FRANCISCO, CALIF.

CENTRAL SAN FRANCISCO

Interstate route numbers   U.S. route numbers   State route numbers

—— Cable Car route

COPYRIGHT PHILIP'S

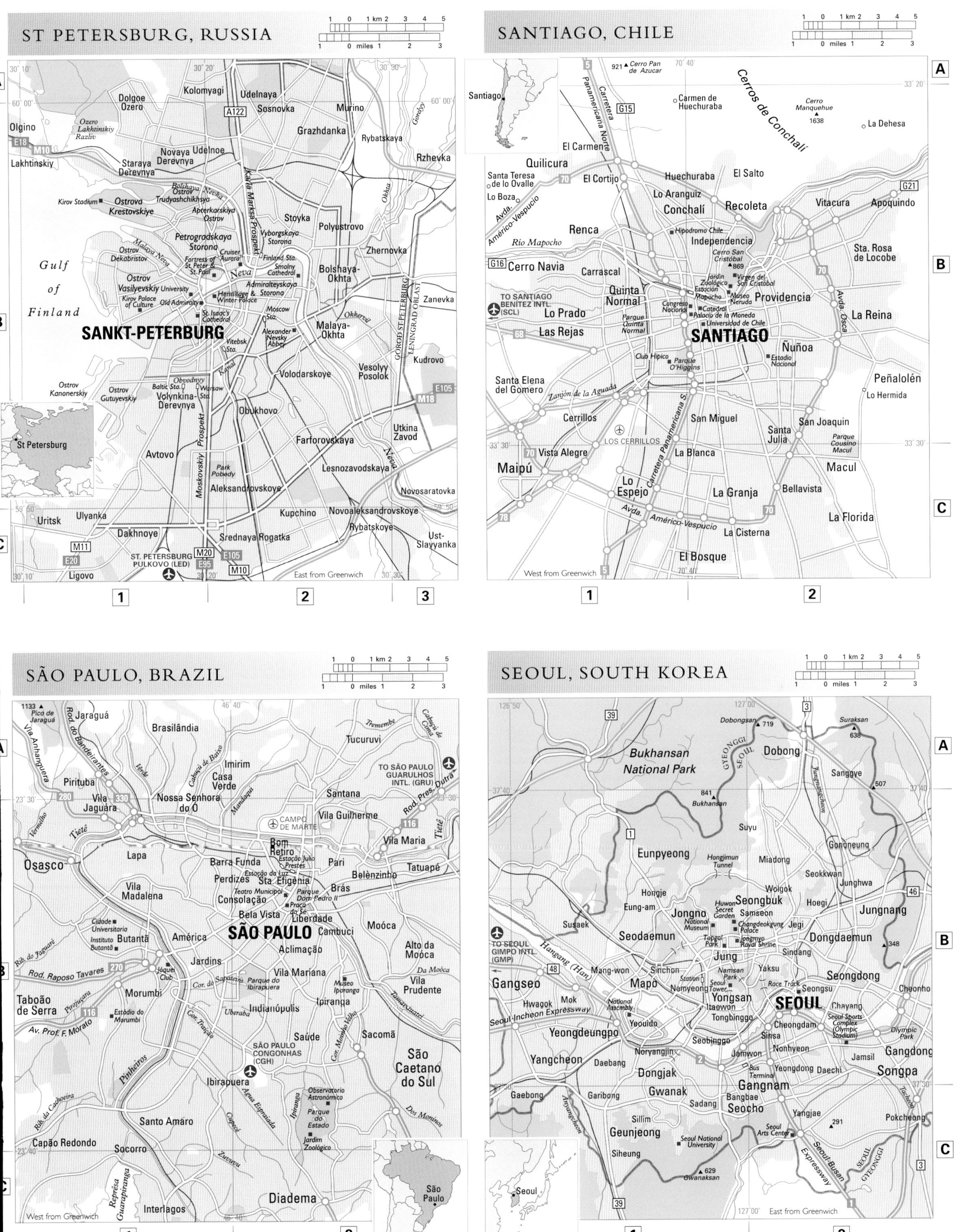
ST PETERSBURG, RUSSIA
Dolgoe Ozero
Kolomyagi
Udelnaya
Sosnovka
Murino
Olgino
Grazhdanka
Rybatskaya
Novaya Udelnoe
Staraya Derevnya
Lakhtinskiy
Rzhevka
Kirov Stadium
Ostrova Krestovskiye
Stoyka
Polyustrovo
Petrogradskaya Storona
Vyborgskaya Storona
Zhernovka
Gulf of Finland
Ostrov Dekabristov
Ostrov Vasilyevskiy
Bolshaya-Okhta
Zanevka
SANKT-PETERBURG
Malaya-Okhta
Kudrovo
Vesolyy Posolok
Volodarskoye
Ostrov Kanonerskiy
Ostrov Gutuyevskiy
Volynkina-Derevnya
Obukhovo
St Petersburg
Utkina Zavod
Farforovskaya
Avtovo
Lesnozavodskaya
Aleksandrovskoye
Novosaratovka
Uritsk
Ulyanka
Kupchino
Novoaleksandrovskoye
Dakhnoye
Rybatskoye
Srednaya Rogatka
Ust-Slavyanka
ST. PETERSBURG PULKOVO (LED)
Ligovo
East from Greenwich
SANTIAGO, CHILE
Cerro Pan de Azucar
Carmen de Huechuraba
Cerros de Conchalí
Cerro Manquehue 1638
La Dehesa
Santiago
El Carmen
Quilicura
Santa Teresa de lo Ovalle
El Cortijo
Huechuraba
El Salto
Lo Boza
Lo Aranguiz
Conchalí
Recoleta
Vitacura
Apoquindo
Renca
Independencia
Río Mapocho
Sta. Rosa de Locobe
Cerro Navia
Carrascal
Quinta Normal
Providencia
TO SANTIAGO BENITEZ INTL. (SCL)
Lo Prado
La Reina
Las Rejas
SANTIAGO
Ñuñoa
Peñalolén
Santa Elena del Gomero
Lo Hermida
Cerrillos
San Miguel
San Joaquin
Santa Julia
LOS CERRILLOS
Vista Alegre
La Blanca
Macul
Maipú
Lo Espejo
La Granja
Bellavista
La Florida
La Cisterna
El Bosque
West from Greenwich
SÃO PAULO, BRAZIL
Jaraguá
Brasilândia
Tucuruvi
Imirim
Casa Verde
TO SÃO PAULO GUARULHOS INTL. (GRU)
Pirituba
Vila Jaguára
Nossa Senhora do Ó
Santana
Vila Guilherme
CAMPO DE MARTE
Vila Maria
Lapa
Bom Retiro
Osasco
Barra Funda
Pari
Tatuapé
Belénzinho
Perdizes
Sta. Efigênia
Brás
Vila Madalena
Consolação
Liberdade
Moóca
Bela Vista
Butantã
América
SÃO PAULO
Cambuci
Alto da Moóca
Aclimação
Jardins
Vila Mariana
Vila Prudente
Taboão de Serra
Morumbi
Indianópolis
Ipiranga
Saúde
Sacomã
SÃO PAULO CONGONHAS (CGH)
São Caetano do Sul
Ibirapuera
Santo Amaro
Capão Redondo
Socorro
Diadema
Interlagos
São Paulo
West from Greenwich
SEOUL, SOUTH KOREA
Dobongsan 719
Suraksan 638
Bukhansan National Park
Dobong
Sanggye
Bukhansan 841
Suyu
Gongneung
Eunpyeong
Miadong
Seokkwan
Junghwa
Hongje
Seongbuk
Eung-am
Jongno
Hoegi
Jungnang
Susaek
Seodaemun
Dongdaemun
TO SEOUL GIMPO INTL. (GMP)
Jung
Sindang
Mang-won
Yaksu
Seongdong
Gangseo
Mapo
Yongsan
SEOUL
Cheonho
Hwagok
Mok
Seoul-Incheon Expressway
Yeouido
Tongbinggo
Cheongdam
Yeongdeungpo
Seobinggo
Sinsa
Gangdong
Noryangjin
Jamwon
Nonhyeon
Yangcheon
Daebang
Dongjak
Yeongdong
Daechi
Jamsil
Songpa
Gwanak
Gangnam
Gaebong
Garibong
Sadang
Bangbae
Seocho
Yangjae
Pokcheong
Sillim
Geunjeong
Seoul National University
Siheung
Gwanaksan 629
Seoul-Busan Expressway
Seoul
East from Greenwich

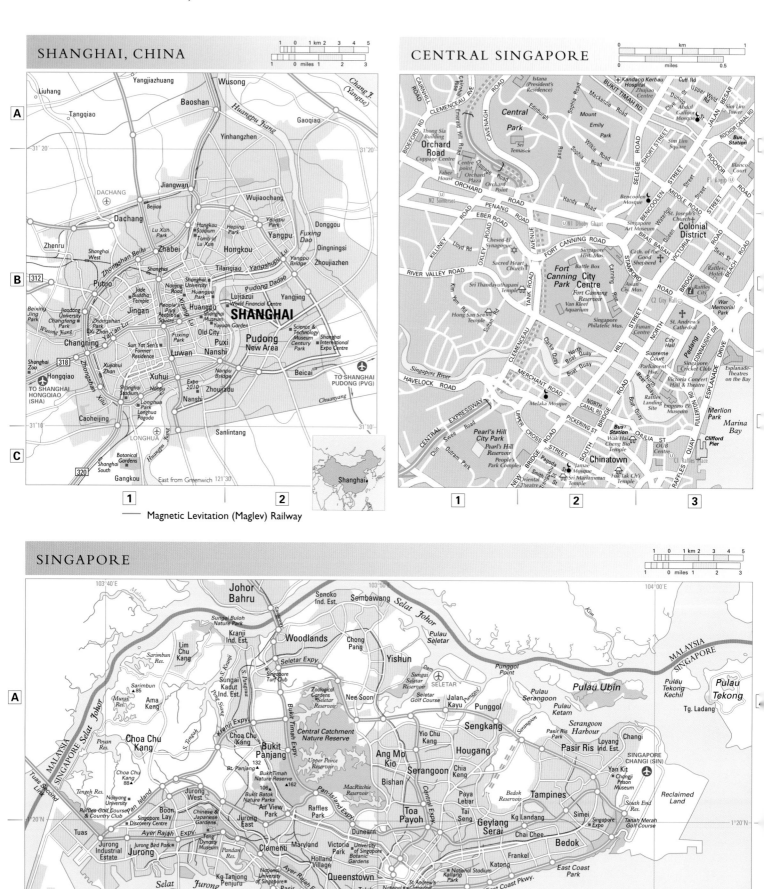

SHANGHAI, CHINA

CENTRAL SINGAPORE

Magnetic Levitation (Maglev) Railway

SINGAPORE

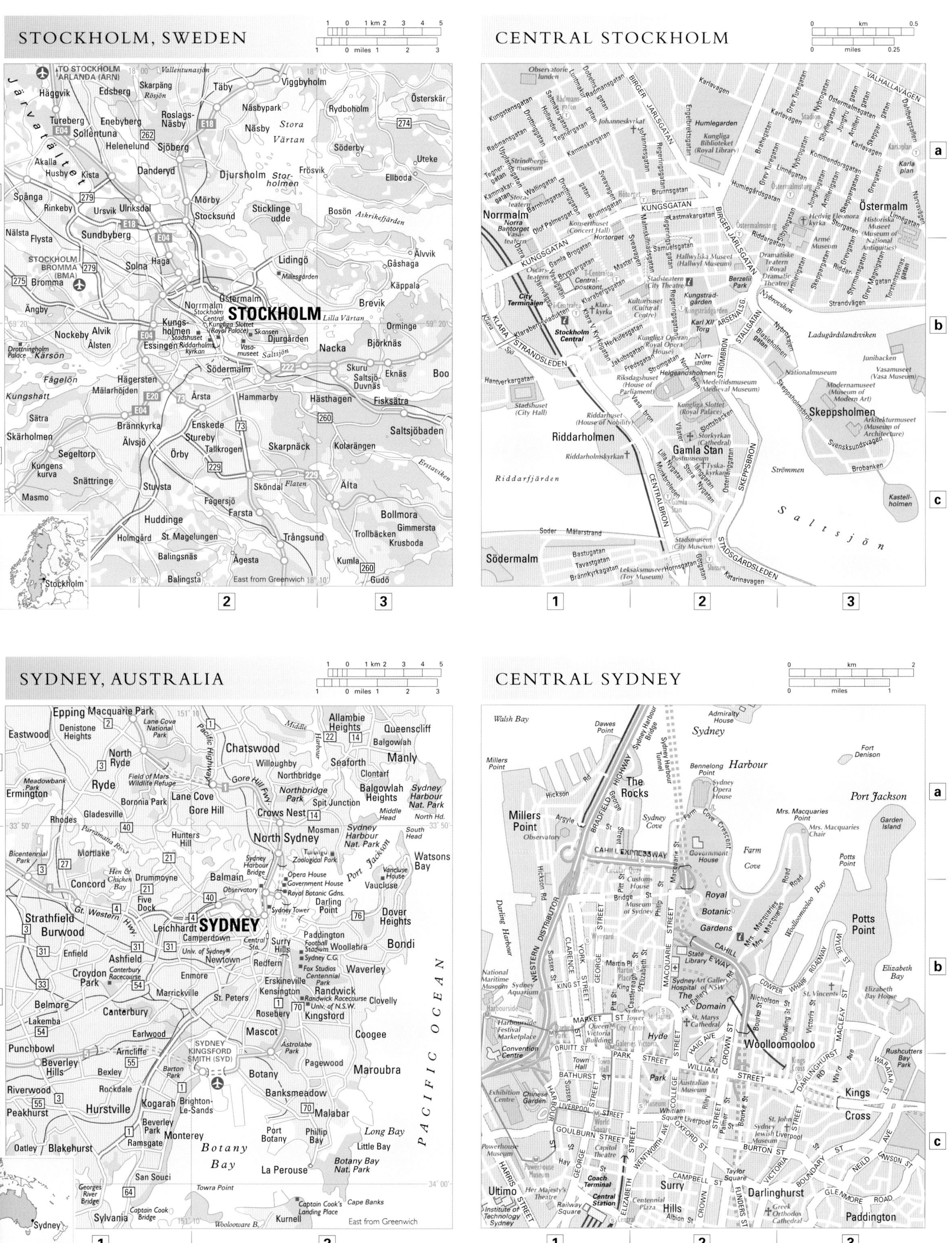

STOCKHOLM, SWEDEN

CENTRAL STOCKHOLM

SYDNEY, AUSTRALIA

CENTRAL SYDNEY

— ⓜ — Monorail

TOKYO, JAPAN

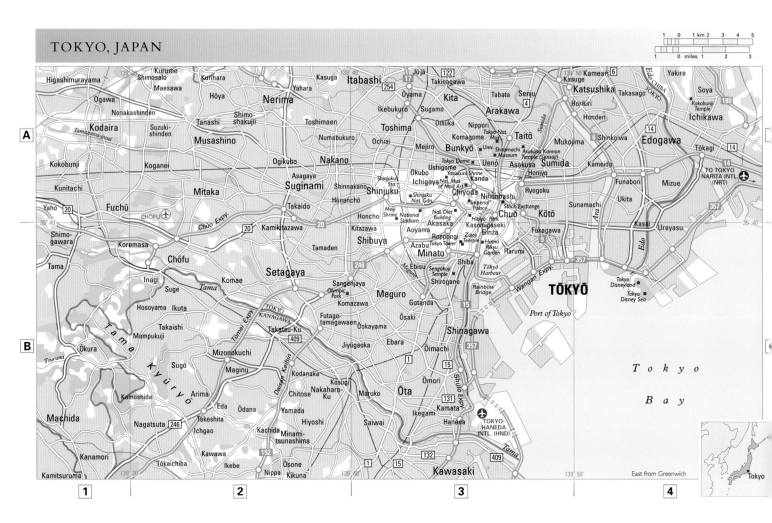

CENTRAL TOKYO

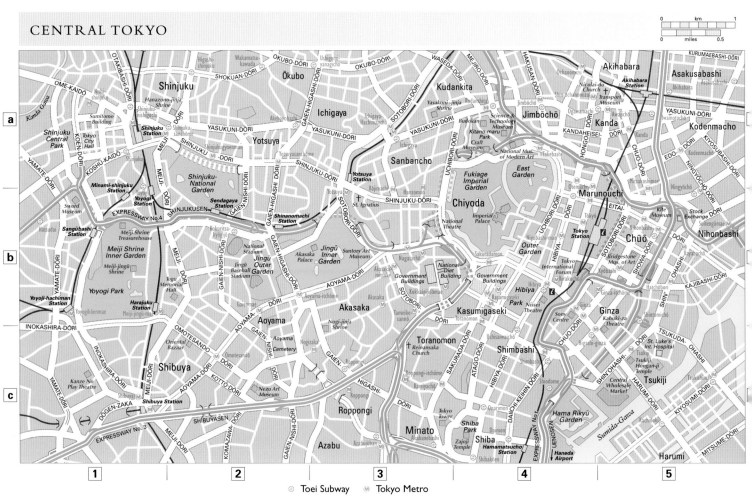

⊖ Toei Subway   Ⓜ Tokyo Metro

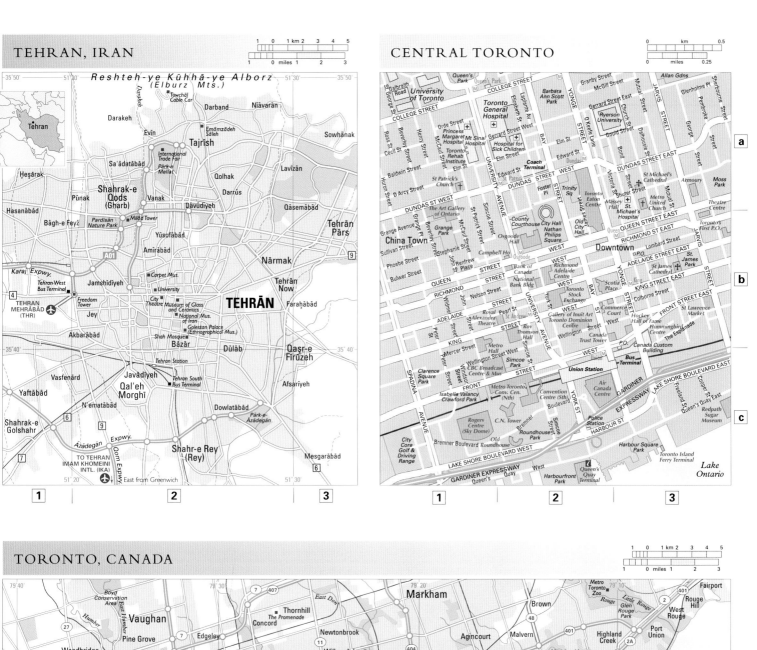

TEHRAN, IRAN

Reshteh-ye Kūhhā-ye Alborz
(Elburz Mts.)

Towchal Cable Car
Darakeh
Darband
Niāvarān
Ēvīn
Emāmzādeh Şāleh
Sowhānak
Tajrīsh
Lavīzān
Sa'ādatābād
Qolhak
Heşārak
Shahrak-e Qods (Gharb)
Vanak
Darrūs
Qāsemābād
Pūnak
Dāvūdīyeh
Bāgh-e Feyž
Pardisān Nature Park
Mīlād Tower
Hasanābād
Yūsofābād
Tehrān Pārs
Amīrābād
Nārmak
Jamshīdīyeh
Tehrān Now
TEHRĀN
Freedom Tower
City Theatre
University
Farahābād
TEHRAN MEHRĀBĀD (THR)
Jey
Museum of Glass and Ceramics
National Mus. of Iran
Akbarābād
Shah Mosque
Golestan Palace (Ethnographical Mus.)
Bāzār
Dūlāb
Qaşr-e Fīrūzeh
Vasfenārd
Tehran Station
Javādīyeh
Qal'eh Morghī
Tehran South Bus Terminal
Afsarīyeh
Yaftābād
N'ematābād
Dowlatābād
Pārk-e Āzādegān
Shahrak-e Golshahr
Āzādegān Expwy.
Shahr-e Rey (Rey)
Mesgarābād
TO TEHRAN IMAM KHOMEINI INTL. (IKA)
East from Greenwich

CENTRAL TORONTO

TORONTO, CANADA

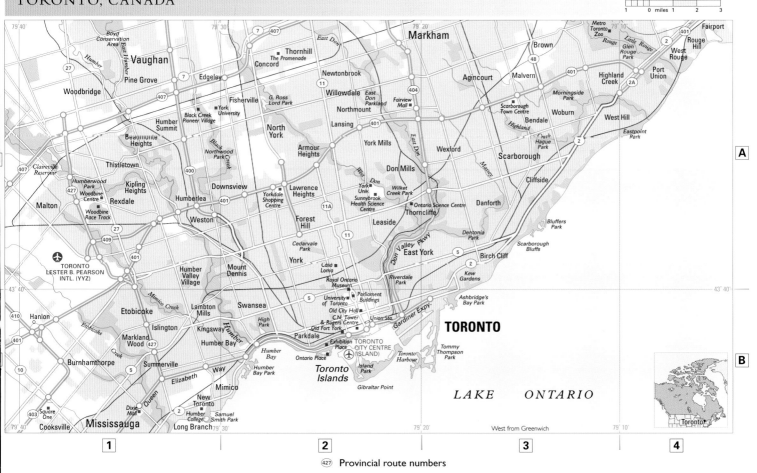

Vaughan
Markham
Thornhill
Concord
Brown
Woodbridge
Pine Grove
Edgeley
Newtonbrook
West Rouge
Fisherville
Willowdale
Port Union
Humber Summit
North York
Northmount
Agincourt
Malvern
Highland Creek
Beaumonde Heights
Lansing
Scarborough Town Centre
Woburn
West Hill
Thistletown
Armour Heights
Wexford
Scarborough
Cliffside
Humberlea
Don Mills
Eastpoint Park
Malton
Downsview
Lawrence Heights
Leaside
Danforth
Weston
Forest Hill
Thorncliffe
Bluffers Park
Cedarvale Park
York
East York
Birch Cliff
TORONTO LESTER B. PEARSON INTL. (YYZ)
Humber Valley Village
Mount Dennis
Kew Gardens
Etobicoke
Lambton Mills
Swansea
TORONTO
Hanlon
Islington
Kingsway
Markland Wood
Humber Bay
Burnhamthorpe
Summerville
Elizabeth
Mimico
New Toronto
LAKE ONTARIO
Cooksville
Mississauga
Long Branch
West from Greenwich

427 Provincial route numbers

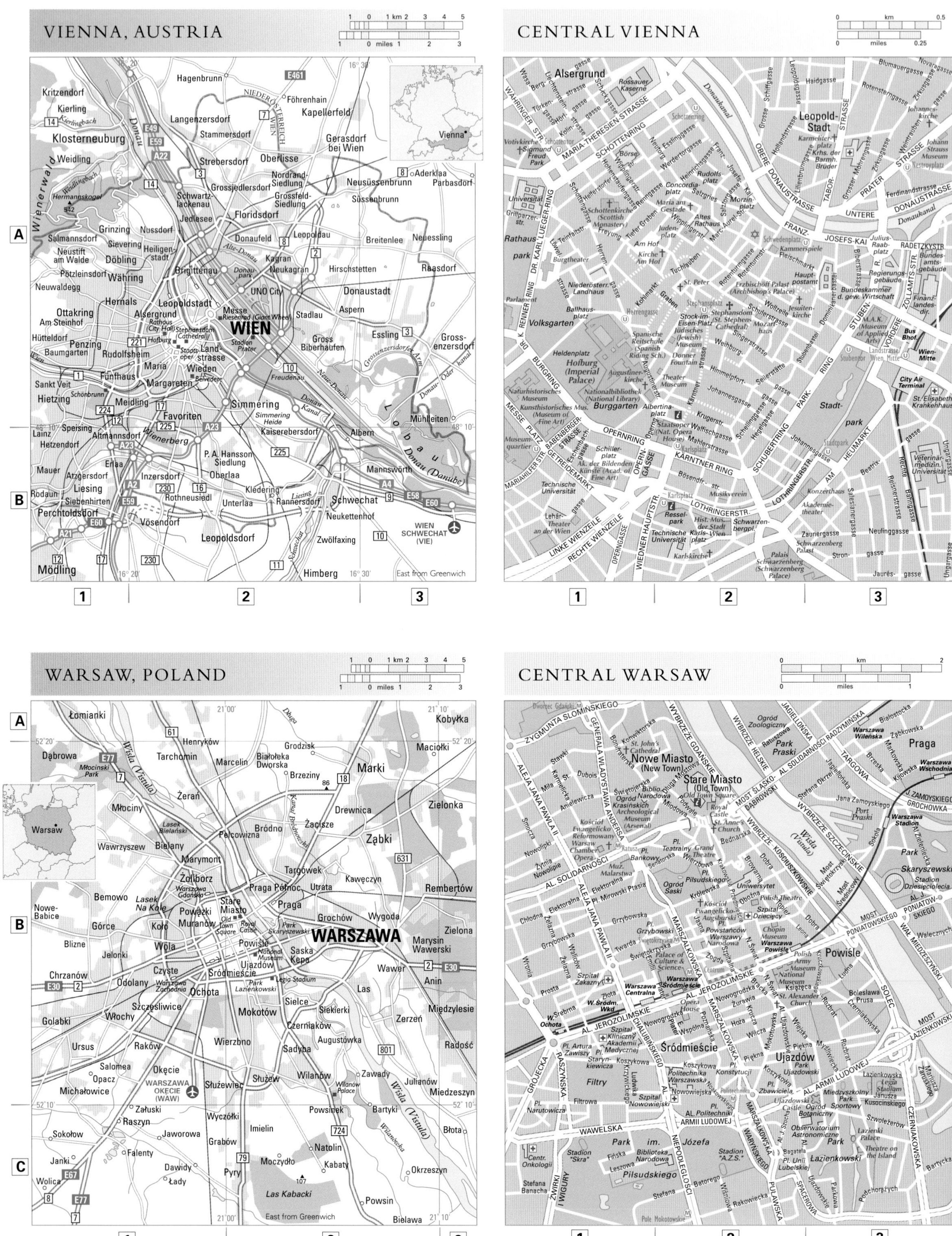
VIENNA, AUSTRIA
CENTRAL VIENNA
WARSAW, POLAND
CENTRAL WARSAW
Klosterneuburg
Hagenbrunn
Langenzersdorf
Stammersdorf
Gerasdorf bei Wien
Kapellerfeld
Föhrenhain
Strebersdorf
Oberlisse
Floridsdorf
Neusüssenbrunn
Süssenbrunn
Aderklaa
Parbasdorf
Kritzendorf
Kierling
Weidling
Grinzing
Nussdorf
Salmannsdorf
Sievering
Neustift am Walde
Döbling
Währing
Pötzleinsdorf
Neuwaldegg
Hernals
Ottakring
Alsergrund
Leopoldstadt
Brigittenau
Donaufeld
Leopoldau
Kagran
Breitenlee
Hirschstetten
Raasdorf
Donaustadt
Aspern
Stadlau
WIEN
Essling
Gross-Enzersdorf
Hütteldorf
Penzing
Baumgarten
Rudolfsheim
Landstrasse
Wieden
Margareten
Fünfhaus
Sankt Veit
Hietzing
Meidling
Simmering
Favoriten
Lainz
Speising
Hetzendorf
Altmannsdorf
Erlaa
Mauer
Atzgersdorf
Liesing
Siebenhirten
Perchtoldsdorf
Rodaun
Inzersdorf
Oberlaa
Unterlaa
Rothneusiedl
Kledering
Rannersdorf
Schwechat
Albern
Mannswörth
Mühlleiten
Vösendorf
Leopoldsdorf
Neukettenhof
Zwölfaxing
Himberg
Mödling
WIEN SCHWECHAT (VIE)
East from Greenwich
Łomianki
Kobyłka
Dąbrowa
Tarchomin
Henryków
Marcelin
Białołęka Dworska
Grodzisk
Marki
Brzeziny
Żerań
Zielonka
Młociny
Drewnica
Bródno
Żacisze
Ząbki
Wawrzyszew
Bielany
Marymont
Targówek
Żoliborz
Bemowo
Praga Północ
Praga
Rembertów
Nowe Babice
Górce
Powązki
Muranów
Koło
Wola
Blizne
Jelonki
Grochów
Wygoda
WARSZAWA
Zielona
Powiśle
Saska Kępa
Marysin Wawerski
Wawer
Czyste
Śródmieście
Chrzanów
Odolany
Ochota
Anin
Szczęśliwice
Las
Golabki
Włochy
Mokotów
Siekierki
Zerzeń
Międzylesie
Ursus
Raków
Wierzbno
Czerniaków
Sadyba
Radość
Salomea
Okęcie
Opacz
Michałowice
WARSZAWA OKĘCIE (WAW)
Służewiec
Służew
Wilanów
Zawady
Julianów
Miedzeszyn
Raszyn
Powsinek
Bartycki
Sokołów
Jaworowa
Wyczółki
Imielin
Błota
Grabów
Natolin
Falenty
Janki
Dawidy
Moczydło
Kabaty
Okrzeszyn
Łady
Pyry
Wolica
Las Kabacki
Powsin
Bielawa
Nowe Miasto
Stare Miasto
Praga
Powiśle
Śródmieście
Ujazdów
Filtry
Park im. Józefa Piłsudskiego
Łazienkowski

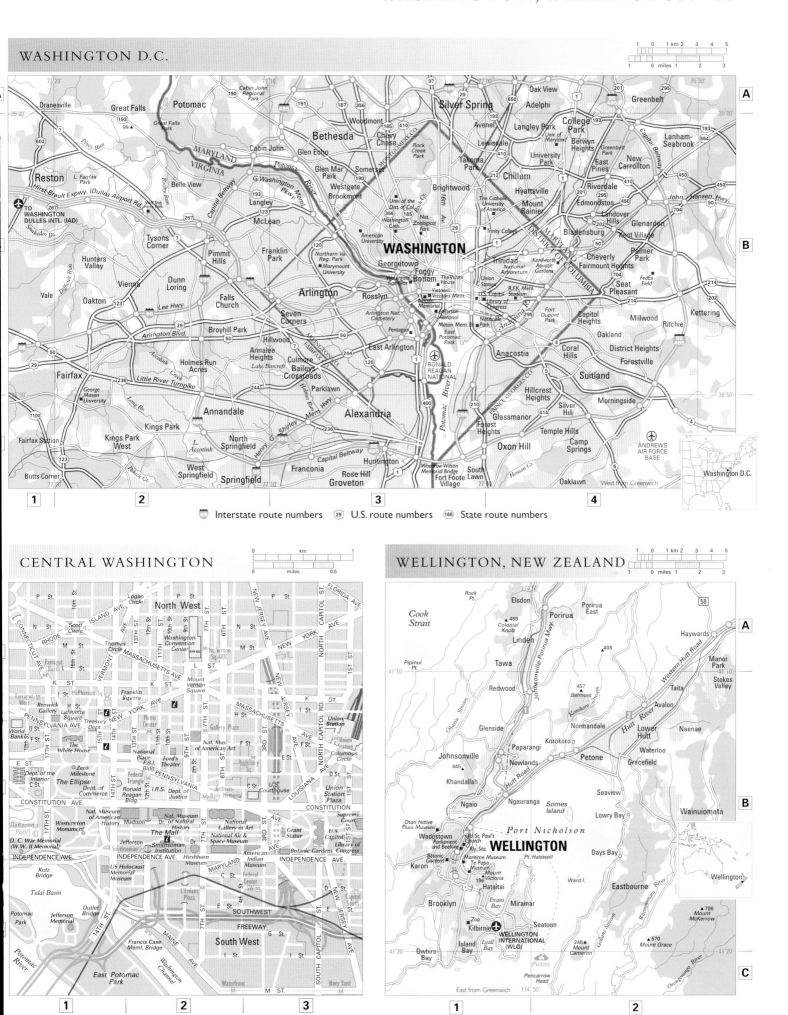

WASHINGTON D.C.

Dranesville · Great Falls · Potomac · Cabin John Regional Park · Oak View · Silver Spring · Adelphi · Greenbelt

Great Falls Park · Woodmont · Chevy Chase · Avenel · Langley Park · College Park · Lanham-Seabrook

Reston · Belle View · Glen Echo · Somerset · Rock Creek Park · Takoma Park · University Park · New Carrollton

Hunters Valley · Tysons Corner · McLean · Langley · Brookmont · Westgate · Brightwood · Hyattsville · Riverdale · Edmonston · Landover · Glenarden

Vale · Vienna · Dunn Loring · Franklin Park · Pimmit Hills · Georgetown · Foggy Bottom · Trinidad · Bladensburg · Kent Village

Oakton · Falls Church · Rosslyn · Arlington · Fort Dupont Park · Cheverly · Fairmount Heights · Palmer Park

Seven Corners · Broyhill Park · Hillwood · East Arlington · Anacostia · Capitol Heights · Oakland · Seat Pleasant · Kettering

Fairfax · Annalee Heights · Culmore · Bailey's Crossroads · Parklawn · Coral Hills · District Heights · Forestville · Ritchie

Annandale · Alexandria · Hillcrest Heights · Suitland · Morningside

Kings Park · Fairfax Station · Kings Park West · North Springfield · Franconia · Huntington · Rose Hill · Groveton · Forest Heights · Glassmanor · Silver Hill · Temple Hills · Camp Springs

Butts Corner · West Springfield · Springfield · Oxon Hill · South Lawn · Oaklawn · Andrews Air Force Base

Washington D.C.

🛡 Interstate route numbers   ⛨ U.S. route numbers   ⬭ State route numbers

CENTRAL WASHINGTON

North West

Logan Circle · Scott Circle · Washington Convention Center · Mount Vernon Square · Thomas Circle · Franklin Square · McPherson Square · Renwick Gallery · Lafayette Square · Metro Center · Gallery Place · Nat. Mus. of American Art · Union Station · Farragut West · Treasury Dept. · National Place · Ford's Theater · F.B.I. Bldg. · Columbus Circle · World Bank · The White House · Judiciary Sq. · Union Station Plaza · Zero Milestone · The Ellipse · Federal Triangle · Ronald Reagan Bldg. · Dept. of Commerce · I.R.S. · Dept. of Justice · Archives-Navy Memorial · U.S. Courthouse · Dept. of the Interior · Reflecting Pool · Washington Monument · Nat. Museum of American History · Nat. Museum of Natural History · National Gallery of Art · Supreme Court · U.S. Capitol · Library of Congress · D.C. War Memorial (W.W. II Memorial) · Madison Dr · The Mall · Smithsonian Institution · National Air & Space Museum · American Indian Museum · Kutz Bridge · Jefferson Dr · Hirshhorn Museum · U.S. Holocaust Memorial Museum · Federal Center · L'Enfant Plaza · Capitol South · Tidal Basin · Potomac Park · Jefferson Memorial · Outlet Bridge · SOUTHWEST · FREEWAY · South West · Francis Case Meml. Bridge · Washington Channel · East Potomac Park · Waterfront · Navy Yard

WELLINGTON, NEW ZEALAND

Cook Strait · Rock Pt. · Elsdon · Porirua · Porirua East · Haywards

Colonial Knob · Linden · Tawa · Manor Park · Stokes Valley · Taita · Avalon

Pipinui Pt. · Redwood · Belmont · Lower Hutt · Naenae

Glenside · Normandale · Korokoro · Waterloo

Johnsonville · Paparangi · Newlands · Petone · Gracefield

Khandallah · Ngaio · Ngauranga · Somes Island · Seaview · Lowry Bay · Wainuiomata

Otari Native Plant Museum · Wadestown · Parliament and Beehive · Old St. Paul's Church · Port Nicholson · Days Bay

Karori · Botanic Gardens · Maritime Museum · Te Papa Museum · WELLINGTON · Pt. Halswell · Ward I. · Eastbourne

Brooklyn · Mount Victoria · Hataitai · Evans Bay · Miramar · Wellington

Zoo · Kilbirnie · Seatoun · Owhiro Bay · Island Bay · Lyall Bay · WELLINGTON INTERNATIONAL (WLG) · Mount Cameron · Mount Grace · Mount McKerrow

Picton · Pencarrow Head

COPYRIGHT PHILIP'S

INDEX TO CITY MAPS

The index contains the names of all the principal places and features shown on the City Maps. Each name is followed by an additional entry in italics giving the name of the City Map within which it is located.

The number in bold type which follows each name refers to the number of the City Map page where that feature or place will be found.

The letter and figure which are immediately after the page number give the grid square on the map within which the feature or place is situated.

The letter represents the latitude and the figure the longitude. The fu geographic reference is provided in the border of the City Maps.

The location given is the centre of the city, suburb or feature and is no necessarily the name. Rivers, canals and roads are indexed to their name Rivers carry the symbol → after their name.

An explanation of the alphabetical order rules and a list of the abbrevia tions used are to be found at the beginning of the World Map Index.

A

Aalām Baghdad 3 B2
Aalsmeer Amsterdam 2 B1
Abbey Wood London 15 B4
Abcoude Amsterdam 2 B2
Ābdīn Cairo 7 A2
Abeno Osaka 23 B2
Aberdeen Hong Kong 12 B1
Aberdour Edinburgh 11 A2
Aberdour Castle Edinburgh 11 A2
Abfanggraben → Munich 21 A3
Ablon-sur-Seine Paris 24 B3
Abramtsevo Moscow 19 B3
Abu Dis Jerusalem 13 B2
Abû en Numrus Cairo 7 B2
Abu Ghosh Jerusalem 13 B1
Acassuso Buenos Aires 7 A2
Accotink, L. Washington 33 C2
Accotink Cr. → Washington 33 B2
Achères Paris 24 A1
Acilia Rome 26 C1
Aclimação São Paulo 27 B2
Acropolis Athens 2 B2
Acton London 15 A2
Açúcar, Pão de Rio de Janeiro 25 B2
Ada Beja Lisbon 16 A1
Adams Park Atlanta 3 B2
Addiscombe London 15 B3
Adelphi Washington 33 A4
Aderklaa Vienna 32 A3
Adler Planetarium Chicago 9 B3
Admiralteyskaya Storona St. Petersburg 27 B2
Áffori Milan 19 A2
Aflandshage Copenhagen 8 B3
Afsarīyeh Tehran 31 B2
Agboyi Cr. → Lagos 14 A2
Ågerup Copenhagen 8 A1
Ägesta Stockholm 29 B2
Aghia Marina Athens 2 C3
Aghia Paraskevi Athens 2 A2
Aghios Dimitrios Athens 2 B2
Aghios Ioannis Rendis Athens 2 B1
Agincourt Toronto 31 A3
Agra Canal Delhi 10 B2
Agricola Oriental Mexico City 18 B2
Água Espraiada → São Paulo 27 B2
Agualva-Cacem Lisbon 16 A1
Ahrensfelde Berlin 5 A4
Ahuntsic Montreal 20 A1
Ai → Osaka 23 A2
Aigremont Paris 24 A1
Air View Park Singapore 28 A2
Airport West Melbourne 18 A1
Ajegunle Lagos 14 B2
Aji Osaka 23 A1
Ajuda Lisbon 16 A1
Akalla Stockholm 29 A1
Akasaka Tokyo 30 A3
Akbarābād Tehran 31 A2
Akershus Slott Oslo 23 A3
Al 'Azamīyah Baghdad 3 A2
Al Quds = Jerusalem Jerusalem 13 B2
Al-Walaja Jerusalem 13 B1
Alaguntan Lagos 14 B2
Alameda San Francisco 26 B3
Alameda Memorial State Beach Park San Francisco 26 B3
Albern Vienna 32 B2
Albert Park Melbourne 18 B1
Alberton Johannesburg 13 B2
Albertslund Copenhagen 8 B2
Alcantara Lisbon 16 A1
Alcatraz I. San Francisco 26 B2
Alcobendas Madrid 17 A2
Alcorcón Madrid 17 B1
Aldershof Berlin 5 B4
Aldo Bonzi Buenos Aires 7 C1
Aleksandrovskoye St. Petersburg 27 B2
Alexander Nevsky Abbey St. Petersburg 27 B2
Alexandra Johannesburg 13 A2
Alexandra Singapore 28 B2
Alexandria Washington 33 C3
Alfortville Paris 24 B3
Algés Lisbon 16 A1
Algiers New Orleans 21 B2
Alhambra Los Angeles 16 B4
Alibey → Istanbul 12 B1
Alibey Barajı Istanbul 12 B1
Alibeyköy Istanbul 12 B1
Alimos Athens 2 B2
Alipur Kolkata 14 B1
Allach Munich 21 A1
Allambie Heights Sydney 29 A2
Allermuir Hill Edinburgh 11 B2
Allstate Arena Chicago 9 A1
Allston Boston 6 A2
Almada Lisbon 16 A2
Almagro Buenos Aires 7 B2
Almargem do Bispo Lisbon 16 A1
Almirante G. Brown, Parque Buenos Aires 7 C2
Almon Jerusalem 13 B2
Almond → Edinburgh 11 B2
Alna Oslo 23 A4
Alnsjøen Oslo 23 A4
Alperton London 15 A2
Alpine New York 22 A2
Alrode Johannesburg 13 B2
Alsemberg Brussels 6 B2
Alsergrund Vienna 32 A2
Alsip Chicago 9 C2
Älsten Stockholm 29 B1
Älta Stockholm 29 B3
Altadena Los Angeles 16 A4
Alte-Donau → Vienna 32 A2
Alter Finkenkrug Berlin 5 A1
Altes Rathaus Munich 21 B2
Altglienicke Berlin 5 B4
Altlandsberg Berlin 5 A5
Altlandsberg Nord Berlin 5 A5
Altmannsdorf Vienna 32 B1
Alto da Boa Vista Rio de Janeiro 25 B1
Alto da Moóca São Paulo 27 B2
Alto do Pina Lisbon 16 A2
Altona Melbourne 18 B1
Alvik Stockholm 29 B1
Alvin Callendar Naval Air Station New Orleans 21 B2
Älvsjo Stockholm 29 B2
Älvvik Stockholm 29 A3
Am Hasenbergl Munich 21 A2
Am Steinhof Vienna 32 A1
Am Wald Munich 21 B2
Ama Keng Singapore 28 A2
Amadora Lisbon 16 A1
Amagasaki Osaka 23 A1
Amager Copenhagen 8 B3
Amāl Qādisiya Baghdad 3 B2
Amalienborg Slot Copenhagen 8 A3
Amata Milan 19 A1
Ambelokipi Athens 2 B2
Ameixoeira Lisbon 16 A2
América São Paulo 27 B1
American Police Hall of Fame Miami 18 B2
American Univ. Washington 33 B3
Amin Baghdad 3 B2
Aminadav Jerusalem 13 B1
Amīrābād Tehran 31 A2
Amora Lisbon 16 B2
Amoreira Lisbon 16 A1
Amper → Munich 21 A1
Amstel-Drecht-Kanaal Amsterdam 2 B2
Amstelveen Amsterdam 2 B2
Amsterdam Amsterdam 2 A2
Amsterdam ✕ (AMS) Amsterdam 2 B1
Amsterdam-Rijnkanaal Amsterdam 2 B3
Amsterdam Zuidoost Amsterdam 2 B2
Amsterdamse Bos Amsterdam 2 B1
Anacosta → Washington 33 B4
Anacostia Washington 33 B4
Anadoluhisarı Istanbul 12 B2
Anadolukavağı Istanbul 12 A2
Anata Jerusalem 13 B2
Ancol Jakarta 12 A1
'Andalus Baghdad 3 B1
Andarai Rio de Janeiro 25 B1
Anderlecht Brussels 6 A1
Anderson Park Atlanta 3 B2
Andingmen Beijing 4 B2
Ang Mo Kio Singapore 28 A3
Ängby Stockholm 29 A1
Angel I. San Francisco 26 A2
Angel Island State Park ⌂ San Francisco 26 A2
Angke, Kali → Jakarta 12 A1
Angyalföld Budapest 7 A2
Anik Mumbai 20 A2
Anin Warsaw 32 B2
Anjou Montreal 20 A2
Annalee Heights Washington 33 B2
Annandale Washington 33 C2
Anne Frankhuis Amsterdam 2 A2
Antony Paris 24 B2
Aoyama Tokyo 30 B3
Ap Lei Chau Hong Kong 12 B1
Apapa Lagos 14 B2
Apapa Quays Lagos 14 B2
Apelação Lisbon 16 A2
Apopka, L. Orlando 23 A1
Apoquindo Santiago 27 B2
Apterkarskiy Ostrov St. Petersburg 27 B2
Ar Kazimiyah Baghdad 3 B1
Ar Ram Jerusalem 13 A2
Ara → Tokyo 30 A4
Arakawa Tokyo 30 A3
Arany-hegyi-patak → Budapest 7 A2
Aravaca Madrid 17 B1
Arbataash Baghdad 3 A1
Arc de Triomphe Paris 24 A2
Arcadia Los Angeles 16 B4
Arcueil Paris 24 B2
Arese Milan 19 A1
Arganzuela Madrid 17 B1
Argenteuil Paris 24 A2
Argiroupoli Athens 2 B2
Argonne Forest Chicago 9 C1
Arima Tokyo 30 B2
Arlanda, Stockholm ✕ (ARN) Stockholm 29 A1
Arlington Boston 6 A1
Arlington Washington 33 B3
Arlington Heights Boston 6 A1
Arlington Nat. Cemetery Washington 33 B3
Armação Rio de Janeiro 25 B2
Armadale Melbourne 18 B2
Armour Heights Toronto 31 A2
Arncliffe Sydney 29 B1
Arnold Arboretum Boston 6 B2
Árpádföld Budapest 7 A3
Arrentela Lisbon 16 B2
Arroyo Seco Park Los Angeles 16 B3
Årsta Stockholm 29 B2
Artane Dublin 10 A2
Artas Jerusalem 13 B2
Arthur's Seat Edinburgh 11 B3
Arts, Place des Montreal 20 A2
As Shawawra Jerusalem 13 B2
Asagaya Tokyo 30 A2
Asahi Osaka 23 A2
Asakusa Tokyo 30 A3
Asati Kolkata 14 C1
Aschheim Munich 21 A3
Ascot Vale Melbourne 18 A1
Ashbridge's Bay Park Toronto 31 B3
Ashburn Chicago 9 C2
Ashburton Melbourne 18 B2
Ashfield Sydney 29 B1
Ashford London 15 B1
Ashtown Dublin 10 A2
Askisto Helsinki 11 B1
Askrikefjärden Stockholm 29 A3
Asnières Paris 24 A2
Aspern Vienna 32 A2
Assago Milan 19 B1
Assendelft Amsterdam 2 A1
Assiano Milan 19 B1
Astoria New York 22 B2
Astrolabe Park Sydney 29 B2
Atarot Jerusalem 13 A2
Atarot ✕ (JRS) Jerusalem 13 A2
Atghara Kolkata 14 B2
Athens = Athina Athens 2 B2
Athina Athens 2 B2
Athina ✕ (ATH) Athens 2 A3
Athinai = Athina Athens 2 B2
Athis-Mons Paris 24 B3
Athlone Cape Town 8 A2
Atholl Johannesburg 13 A2
Atifiya Baghdad 3 A2
Atişalen Istanbul 12 B1
Atlanta Atlanta 3 B2
Atlanta Hartsfield Int. ✕ (ATL) Atlanta 3 C2
Atlanta History Center Atlanta 3 B2
Atlanta Zoo Atlanta 3 B2
Atomium Brussels 6 A2
Attiki Athens 2 A2
Atzgersdorf Vienna 32 B1
Aubervilliers Paris 24 A3
Aubing Munich 21 B1
Auburndale Boston 6 A1
Auchendinny Edinburgh 11 B2
Auckland Park Johannesburg 13 B2
Auderghem Brussels 6 B2
Augustówka Warsaw 32 B2
Aulnay-sous-Bois Paris 24 A3
Aurelio Rome 26 B1
Ausîm Cairo 7 A1
Austerlitz, Gare d' Paris 24 A3
Austin Chicago 9 B2
Avalon Wellington 33 B2
Avedøre Copenhagen 8 B2
Avellaneda Buenos Aires 7 C2
Avenel Washington 33 B4
Avondale Chicago 9 B2
Avondale New Orleans 21 B1
Avondale Heights Melbourne 18 A1
Avtovo St. Petersburg 27 B1
Ayazağa Istanbul 12 B1
Ayer Chawan, Pulau Singapore 28 B2
Ayer Merbau, Pulau Singapore 28 B2
Azabu Tokyo 30 B3
Azcapotzalco Mexico City 18 B1
Azteca, Estadio Mexico City 18 C2
Azucar, Cerro Pan de Santiago 27 A1

B

Baambrugge Amsterdam 2 B2
Baba Channel Karachi 13 B1
Baba I. Karachi 13 B1
Babarpur Delhi 10 A2
Babushkin Moscow 19 A3
Back B. Mumbai 20 B1
Baclaran Manila 17 B2
Bacoor Manila 17 C1
Bacoor B. Manila 17 C1
Badalona Barcelona 4 A2
Badhoevedorp Amsterdam 2 A1
Badli Delhi 10 A1
Bærum Oslo 23 A2
Bağcılar Istanbul 12 B1
Bággio Milan 19 B1
Bāgh-e-Feyz Tehran 31 A1
Baghdād Baghdad 3 A2
Baghdād al Muthana ✕ (BGW) Baghdad 3 B2
Baghdad Int. ✕ (SDA) Baghdad 3 B1
Bagmari Kolkata 14 B2
Bagneux Paris 24 B2
Bagnolet Paris 24 A3
Bagsværd Copenhagen 8 A2
Bagsværd Sø Copenhagen 8 A2
Baguiati Kolkata 14 B2
Bagumbayan Manila 17 C2
Baha'i Temple Chicago 9 A2
Bahçeköy Istanbul 12 A1
Bahçelievler Istanbul 12 B1
Bahtîm Cairo 7 A2
Baile Átha Cliath = Dublin Dublin 10 A2
Baileys Crossroads Washington 33 B3
Bailly Paris 24 A1
Bairro Lopes Lisbon 16 A2
Baisha Guangzhou 11 B2
Baiyun Guangzhou 11 A2
Baiyun Hill Guangzhou 11 B2
Baiyun Mountain Scenic Area Guangzhou 11 A2
Bakırköy Istanbul 12 C1
Bal Harbor Miami 18 A2
Balara Manila 17 B2
Baldia Karachi 13 A1
Baldoyle Dublin 10 A3
Baldwin, L. Orlando 23 A3
Baldwin Hills Los Angeles 16 B2
Baldwin Hills Res. Los Angeles 16 B2
Balgowlah Sydney 29 A2
Balgowlah Heights Sydney 29 A2
Balham London 15 B3
Bali Kolkata 14 B1
Baliganja Kolkata 14 B2
Balingsnäs Stockholm 29 B2
Balingsta Stockholm 29 B2
Balintawak Manila 17 B1
Ballerup Copenhagen 8 A2
Ballinteer Dublin 10 B2
Ballyboden Dublin 10 B2
Ballybrack Dublin 10 B3
Ballyfermot Dublin 10 A1
Ballymorefinn Hill Dublin 10 B1
Ballymun Dublin 10 A2
Balmain Sydney 29 B2
Baluhati Kolkata 14 B1
Balvanera Buenos Aires 7 B2
Balwyn Melbourne 18 A2
Balwyn North Melbourne 18 A2
Banática Lisbon 16 A1
Bandra Mumbai 20 A1
Bandra Pt. Mumbai 20 A1
Bang Kapi Bangkok 3 B2
Bang Na Bangkok 3 B2
Bangbae Seoul 27 C1
Bangkhen Bangkok 3 A2
Bangkok Bangkok 3 B2
Bangkok Don Muang Int. ✕ (BKK) Bangkok 3 A2
Bangkok Noi Bangkok 3 B1
Bangkok Yai Bangkok 3 B1
Banglo Kolkata 14 B1
Bangrak Bangkok 3 B2
Bangsu Bangkok 3 B2
Banks, C. Sydney 29 C2
Banksmeadow Sydney 29 B2
Banstala Kolkata 14 B2
Bantra Kolkata 14 B1
Baoshan Shanghai 28 A1
Bar Giyora Jerusalem 13 B1
Barahanagar Kolkata 14 B2
Barajas Madrid 17 B2
Barajas, Madrid ✕ (MAD) Madrid 17 B2
Barakpur Kolkata 14 A2
Barcarena Lisbon 16 A1
Barcarena, Rib. de → Lisbon 16 A1
Barcelona Barcelona 4 A2
Barcelona-Prat ✕ (BCN) Barcelona 4 B1
Barcroft, L. Washington 33 B3
Barking London 15 A4
Barkingside London 15 A4
Barnes London 15 B2
Barnet London 15 A2
Barra Andai Karachi 13 B2
Barra Funda São Paulo 27 B2
Barracas Buenos Aires 7 B2
Barrackpur = Barakpur Kolkata 14 A2
Barranco Lima 14 B2
Barreiro Lisbon 16 B2
Barreto Rio de Janeiro 25 B2
Bartala Kolkata 14 B1
Barton Park Sydney 29 B1
Bartyki Warsaw 32 C2
Basus Cairo 7 A2
Batanagar Kolkata 14 B1
Bath Beach New York 22 C1
Bath I. Karachi 13 B2
Batir Jerusalem 13 B1
Batok, Bukit Singapore 28 A2
Battersea London 15 B3
Bauman Moscow 19 B3
Baumgarten Vienna 32 A1
Bay, L. Orlando 23 B2
Bay Harbor Islands Miami 18 A2
Bay Hill Orlando 23 B1
Bay Ridge New York 22 C1
Bayit Va-Gan Jerusalem 13 B2
Bayonne New York 22 B1
Bayou Boeuf New Orleans 21 B1
Bayou Segnette State Park ⌂ New Orleans 21 B1
Bayrampaşa Istanbul 12 B1
Bayshore San Francisco 26 B2
Bayt Lahm Jerusalem 13 B2
Bayview San Francisco 26 B2
Bāzār Tehran 31 A2
Beacon Hill Hong Kong 12 A2
Beato Lisbon 16 A2
Beaumont Dublin 10 A2
Beaumonte Heights Toronto 31 A1
Bebek Istanbul 12 B2
Běchovice Prague 25 B3
Beck L. Chicago 9 A1
Beckenham London 15 B3
Beckton London 15 A4
Becontree London 15 A4
Beddington Corner London 15 B3
Bedford Boston 6 A1
Bedford Park Chicago 9 C2
Bedford Park New York 22 A2
Bedford Stuyvesant New York 22 B2
Bedford View Johannesburg 13 B2
Bedok Singapore 28 B3
Bedok, Res. Singapore 28 A3
Beersel Brussels 6 B1
Behala Kolkata 14 B1
Bei Hai Beijing 4 B2
Beicai Shanghai 28 B2
Beijing Beijing 4 B1
Beit Duqu Jerusalem 13 A1
Beit Ghur at-Taht Jerusalem 13 A1
Beit Ghur el-Fawqa Jerusalem 13 A1
Beit Hanina Jerusalem 13 B2
Beit Ij'za Jerusalem 13 A1
Beit Iksa Jerusalem 13 B2
Beit I'nan Jerusalem 13 A1
Beit Jala Jerusalem 13 B2
Beit Lekhem = Bayt Lahm Jerusalem 13 B2
Beit Liqya Jerusalem 13 A1
Beit Nekofa Jerusalem 13 B1
Beit Sahur Jerusalem 13 B2
Beit Sofafa Jerusalem 13 B2
Beit Surik Jerusalem 13 B1
Beit Ur al-Fawqa Jerusalem 13 A1
Beit Zayit Jerusalem 13 B1
Beitaipingzhuan Beijing 4 B1
Beitar Illit Jerusalem 13 B1
Beitin Jerusalem 13 A2
Beitsun Guangzhou 11 B2
Beitunya Jerusalem 13 A1
Beixing Jing Park Shanghai 28 B1
Békásmegyer Budapest 7 A2
Bekkelaget Oslo 23 A3
Bekkestua Oslo 23 A2
Bel Air Los Angeles 16 B2
Bela Vista São Paulo 27 B2
Bélanger Montreal 20 A1
Belas Lisbon 16 A1
Beleghata Kolkata 14 B2
Belém Lisbon 16 A1
Belém, Torre de Lisbon 16 A1
Belénzinho São Paulo 27 B2
Belgachiya Kolkata 14 B2
Belgharia Kolkata 14 B2
Belgrano Buenos Aires 7 B2
Bell Los Angeles 16 C3
Bell Gardens Los Angeles 16 C4
Bellavista Lima 14 B2
Bellavista Santiago 27 C2
Belle Harbor New York 22 C2
Belle Isle Orlando 23 B2
Belle View Washington 33 B2
Bellingham London 15 B3
Bellwood Chicago 9 B1
Belmont Boston 6 A1
Belmont London 15 A2
Belmont, Mt. Wellington 33 B2
Belmont Cragin Chicago 9 B2
Belmont Harbor Chicago 9 B3
Belmore Sydney 29 B1
Belur Kolkata 14 B2
Belvedere Atlanta 3 B3
Belvedere London 15 B4
Belvedere San Francisco 26 A2
Belyayevo Bogorodskoye Moscow 19 C2
Bemowo Warsaw 32 B1
Benaki Museum Athens 2 B2
Bendale Toronto 31 A3
Benefica Rio de Janeiro 25 B1
Benfica Lisbon 16 A1
Benitez Int. ✕ (SCL) Santiago 27 B1
Benito Juárez, Int. ✕ (MEX) Mexico City 18 B2
Bensonhurst New York 22 C2
Berchem-Ste-Agathe Brussels 6 A1
Berg am Laim Munich 21 B2
Bergenfield New York 22 A2
Bergham Munich 21 B2
Bergvliet Cape Town 8 B1
Beri Barcelona 4 A1
Berkeley San Francisco 26 A3
Berlin Berlin 5 A3
Berlin Dom Berlin 5 A3
Berlin Tegel ✕ (TXL) Berlin 5 A2
Bermondsey London 15 B3
Bernabeu, Estadio Madrid 17 B1
Bernal Heights San Francisco 26 B2
Berwyn Chicago 9 B2
Berwyn Heights Washington 33 B4
Besiktas Istanbul 12 B2
Besós → Barcelona 4 A2
Bessie, L. Orlando 23 B1
Bet Horon Jerusalem 13 A1
Bethesda Washington 33 B3
Bethlehem = Bayt Lahm Jerusalem 13 B2
Bethnal Green London 15 A3
Betor Kolkata 14 B1
Beulah Orlando 23 A1
Beulah, L. Orlando 23 A1
Beverley Hills Sydney 29 B1
Beverley Park Sydney 29 B1
Beverly Chicago 9 C3
Beverly Arts Center Chicago 9 C2
Beverly Glen Los Angeles 16 B2
Beverly Hills Los Angeles 16 B2
Beverly Hills-Morgan Park Historic District Chicago 9 C2
Bexley Sydney 29 B1
Bexley □ London 15 B4
Bexleyheath London 15 B4
Beykoz Istanbul 12 B2
Beylerbeyi Istanbul 12 B2
Beyoğlu Istanbul 12 B1
Bezons Paris 24 A2
Bezuidenhout Park Johannesburg 13 B2
Bhadrakali Kolkata 14 A2
Bhalswa Delhi 10 A2
Bhambo Khan Qarmati Karachi 13 B2
Bhatsala Kolkata 14 B1
Bhawanipur Kolkata 14 B2
Bhendkhal Mumbai 20 B2
Bhuleshwar Mumbai 20 B1
Białołeka Dworska Warsaw 32 B2
Bicentennial Park Los Angeles 16 B4
Bicentennial Park Sydney 29 B1
Bickley London 15 B4
Bicutan Manila 17 C2
Bidhan Nagar Kolkata 14 B2
Bidu Jerusalem 13 B1
Bielany Warsaw 32 B1
Bielawa Warsaw 32 C2
Biesdorf Berlin 5 A4
Bièvre → Paris 24 B1
Bièvres Paris 24 B2
Big Sand Lake Orlando 23 B2
Bilston Edinburgh 11 B2
Binacayan Manila 17 C1
Binondo Manila 17 B1
Bintaro Jaya Jakarta 12 B1
Bir Nabala Jerusalem 13 A2
Birak el Kiyam Cairo 7 A1
Birch Cliff Toronto 31 A3
Birkenstein Berlin 5 A5
Birkholz Berlin 5 A4
Birkholzaue Berlin 5 A4
Birrarung Park Melbourne 18 A2
Biscayne Park Miami 18 A2
Bishop Lavis Cape Town 8 A2
Bishopscourt Cape Town 8 A1
Bispebjerg Copenhagen 8 A3
Bissonet Plaza New Orleans 21 A1
Bittsevsky Forest Park Moscow 19 C2
Björknas Stockholm 29 B3
Black Cr. → Toronto 31 A2
Black Creek Pioneer Village Toronto 31 A1
Blackfen London 15 B4
Blackheath London 15 B4
Blackrock Dublin 10 B2
Bladensburg Washington 33 B4
Blair Village Atlanta 3 C2
Blairgowrie Johannesburg 13 A2
Blake House Boston 6 B2
Blakehurst Sydney 29 B1
Blakstad Oslo 23 B1
Blanche, L. Orlando 23 B1
Blankenburg Berlin 5 A3
Blankenfelde Berlin 5 A3
Blizne Warsaw 32 B1
Blota Warsaw 32 C3
Blue Island Chicago 9 D2
Blue Mosque = Sultanahme Camil Istanbul 12 B1
Bluebell Dublin 10 B1
Bluff Hd. Hong Kong 12 B2
Bluffers Park Toronto 31 A3
Blumberg Berlin 5 A4
Blunt Pt. San Francisco 26 A2
Blutenberg Munich 21 B1
Blylaget Oslo 23 B3
Boa Vista, Alto do Rio de Janeiro 25 B1
Boardwalk New York 22 C3
Boavista Lisbon 16 A2
Bobigny Paris 24 A3
Bocanegra Lima 14 B2
Boedo Buenos Aires 7 B2
Bogenhausen Munich 21 B2
Bogorodskoye Moscow 19 B3
Bogota New York 22 A1
Bogstadvatnet Oslo 23 A2
Bohnsdorf Berlin 5 B4
Bois-Colombes Paris 24 A2
Bois-d'Arcy Paris 24 B1
Boissy-St-Léger Paris 24 B4
Boldinasco Milan 19 B1
Bøler Oslo 23 A4
Bollate Milan 19 A1
Bollebeek Brussels 6 A1
Bollensdorf Berlin 5 A5
Bollmora Stockholm 29 B3
Bolshaya Okhta St. Petersburg 27 B2
Bolton Atlanta 3 B2
Bom Retiro São Paulo 27 B2
Bombay = Mumbai Mumbai 20 B2
Bondi Sydney 29 B2
Bondy Paris 24 A3
Bondy, Forêt de Paris 24 A4
Bonifacio Monument Manila 17 B1
Bonnabel Place New Orleans 21 A2
Bonneuil-sur-Marne Paris 24 B4
Bonnington Edinburgh 11 B1
Bonnyrigg and Lasswade Edinburgh 11 B3
Bonsucesso Rio de Janeiro 25 B1
Bonteheuwel Cape Town 8 A2
Boo Stockholm 29 A3
Booterstown Dublin 10 B2
Borisovo Moscow 19 C3
Borle Mumbai 20 A2
Boronia Park Sydney 29 A1
Bosmont Johannesburg 13 B1
Bosön Stockholm 29 A3
Bosporus = Istanbul Boğazı Istanbul 12 B2
Bostancı Istanbul 12 C2
Boston Boston 6 A2
Boston Common Boston 6 A2
Boston Logan Int. ✕ (BOS) Boston 6 A2
Botafogo Rio de Janeiro 25 B1
Botany Sydney 29 B2
Botany B. Sydney 29 B2
Botany Bay △ Sydney 29 B2
Botič → Prague 25 B3
Botica Sete Lisbon 16 A1
Boucherville Montreal 20 A3
Boucherville, Îs. de Montreal 20 A3
Bougival Paris 24 A1
Boulder Pt. Hong Kong 12 B1
Boulogne, Bois de Paris 24 A2
Boulogne-Billancourt Paris 24 A2
Bourg-la-Reine Paris 24 B2
Bouviers Paris 24 B1
Bovenkerk Amsterdam 2 B2
Bovenkerker Polder Amsterdam 2 B2
Bovisa Milan 19 A2
Bow London 15 A3
Boyacıköy Istanbul 12 B2
Boyd Conservation Area Toronto 31 A1
Boyle Heights Los Angeles 16 B3
Braepark Edinburgh 11 B2
Braid Edinburgh 11 B2
Bramley Johannesburg 13 A2
Brandeis Univ. Boston 6 A1
Brandenburger Tor Berlin 5 A3
Brani, Pulau Singapore 28 B3
Branik Prague 25 B2
Brännkyrka Stockholm 29 B2
Brás São Paulo 27 B2
Brasilândia São Paulo 27 A1
Brateyevo Moscow 19 C3
Braybrook Melbourne 18 A1
Brázdim Prague 25 A3
Breakheart Reservation Boston 6 A2
Brede Copenhagen 8 A3
Breezy Point New York 22 C2
Breitenlee Vienna 32 A3
Breña Lima 14 B2
Brent □ London 15 A2
Brent Res. London 15 A2
Brentford London 15 B2
Brentwood Los Angeles 16 B2
Brentwood Park Los Angeles 16 B2
Brera Milan 19 B2
Bresso Milan 19 A2
Brevik Stockholm 29 A3
Břevnov Prague 25 B2
Brickyard, The Chicago 9 B2
Bridge City New Orleans 21 B2
Bridgeport Chicago 9 B3
Bridgetown Cape Town 8 A2
Bridgeview Chicago 9 C2
Brighton Boston 6 A2
Brighton Melbourne 18 B1
Brighton Beach New York 22 C2
Brighton-Le-Sands Sydney 29 B1
Brighton Park Chicago 9 C2
Brightwood Washington 33 B3
Brigittenau Vienna 32 A2
Brimbank Park Melbourne 18 A1
Brisbane San Francisco 26 B2
Britz Berlin 5 B3
Brixton London 15 B3
Broadmeadows Melbourne 18 A1
Broadmoor San Francisco 26 B2
Broadview Chicago 9 B1
Brockley London 15 B3
Bródno Warsaw 32 B2
Bródnowski, Kanal Warsaw 32 B2
Broek Amsterdam 2 A2
Bromley □ London 15 B4
Bromley Common London 15 B4
Bromma Stockholm 29 A1
Bromma ✕ (BMA) Stockholm 29 A1
Brøndby Strand Copenhagen 8 B2
Brøndbyøster Copenhagen 8 B2
Brøndbyvester Copenhagen 8 B2
Brondesbury London 15 A2
Brønnøya Oslo 23 A2
Brønshøj Copenhagen 8 A2
Bronxville New York 22 A3
Brookfield Chicago 9 C1
Brookhaven Atlanta 3 A2
Brookline Boston 6 B2
Brooklyn Cape Town 8 A1
Brooklyn New York 22 C2
Brooklyn Wellington 33 B1
Brooklyn Heights New York 22 B2
Brookmont Washington 33 B3
Brossard Montreal 20 B3
Brou-sur-Chantereine Paris 24 A4
Brown Toronto 31 A3
Broyhill Park Washington 33 B2
Brughério Milan 19 A2
Brunswick Melbourne 18 A1
Brussegem Brussels 6 A1
Brussel Brussels 6 A2
Brussel ✕ (BRU) Brussels 6 A2
Brussels = Brussel Brussels 6 A2
Bruxelles = Brussel Brussels 6 A2
Bruzzano Milan 19 A2
Bry-sur-Marne Paris 24 A4
Bryan, L. Orlando 23 B2
Bryanston Johannesburg 13 A1
Bryn Oslo 23 A1
Brzeziny Warsaw 32 B2
Bubeneč Prague 25 B2
Buc Paris 24 B1
Buchenhain Munich 21 B1
Buchholz Berlin 5 A3
Buckhead Atlanta 3 A2
Buckingham Palace London 15 A
Buckow Berlin 5 B3
Bucktown New Orleans 21 A2
Buda Budapest 7 A2
Buda Castle = Budaváripalota Budapest 7 A2
Budafok Budapest 7 B2
Budaörs Budapest 7 B1
Budapest Budapest 7 B2
Budapest ✕ (BUD) Budapest 7 B3
Budatétény Budapest 7 B2
Budaváripalota Budapest 7 A2
Budding Copenhagen 8 A3
Buenos Aires Buenos Aires 7 B2
Buenos Aires Ezeiza ✕ (EZE) Buenos Aires 7 C1
Bufalotta Rome 26 B2
Bugio Lisbon 16 B1
Buiksloot Amsterdam 2 A2
Buitenveldert Amsterdam 2 B2
Buizingen Brussels 6 B1
Bukhansan Seoul 27 B1
Bukit Panjang Nature Reserve Singapore 28 A2
Bukit Timah Nature Reserve Singapore 28 A2
Bukum, Pulau Singapore 28 B2
Būlāq Cairo 7 A2
Bule Manila 17 C2
Bulim Singapore 28 A2
Bullen Park Melbourne 18 A2
Bund, The Shanghai 28 B1
Bundoora North Melbourne 18 A
Bundoora Park Melbourne 18 A2
Bunker Hill Memorial Boston 6 A2
Bunker I. Karachi 13 B1
Bunkyō Tokyo 30 A3
Bunnefjorden Oslo 23 A3
Buona Vista Park Singapore 28 B
Burbank Chicago 9 C2
Burbank Los Angeles 16 A3
Burden, L. Orlando 23 B1
Burlington Boston 6 A1
Burnham Park Chicago 9 C3
Burnham Park Harbor Chicago 9 B3
Burnhamthorpe Toronto 31 B1
Burnt Oak London 15 A2
Burntisland Edinburgh 11 A2
Burnwynd Edinburgh 11 B1
Burqa Jerusalem 13 A2
Burtus Cairo 7 A1
Burudvatn Oslo 23 A2
Burwood Sydney 29 B1
Bushwick New York 22 B2
Bushy Park London 15 B1
Butantã São Paulo 27 B1
Butcher I. Mumbai 20 B2
Butler, L. Orlando 23 B1
Butts Corner Washington 33 C2
Büyükdere Istanbul 12 B2
Byculla Mumbai 20 B2
Bygdøy Oslo 23 A3
Bywater New Orleans 21 B2

C

C.B.S. Fox Studios Los Angeles 16 B2
C.N.N. Center Atlanta 3 B2
C.N. Tower Toronto 31 B2
Caballito Buenos Aires 7 B2
Cabin John Washington 33 B2
Cabin John Regional Park ⌂ Washington 33 A2
Cabinteely Dublin 10 B3
Cabra Dublin 10 A2
Cabuçú de Baixo → São Paulo 27 A1
Cabuçú de Cima → São Paulo 27 A2
Cachan Paris 24 B2
Cachoeira, Rib. da → São Paulo 27 B1
Cacilhas Lisbon 16 A2
Cahuenga Park Los Angeles 16 B
Cain, L. Orlando 23 B2
Cairo = El Qâhira Cairo 7 A2
Cairo Int. ✕ (CAI) Cairo 7 A3
Caju Rio de Janeiro 25 B1
Čakovice Prague 25 B3
Calcutta = Kolkata Kolkata 14
California Inst. of Tech. Los Angeles 16 B4
California Los Angeles, Univ. of Los Angeles 16 B2
California State Univ. Los Angeles 16 B3
Callao Lima 14 B2
Caloocan Manila 17 B1
Calumet L. Chicago 9 C3
Calumet Park Chicago 9 C3
Calumet Sag Channel → Chicago 9 C2
Calvairate Milan 19 B2
Camarate Lisbon 16 A2
Camaroes Lisbon 16 A1

Camberwell London 15 B3
Camberwell Melbourne 18 B2
Cambridge Boston 6 A2
Cambuci São Paulo 27 B2
Camden □ London 15 A3
Cameron, Mt. Wellington 33 B2
Çamlıca Istanbul 12 B2
Camp Springs Washington 33 C4
Campamento Madrid 17 B1
Campbellfield Melbourne 18 A1
Camperdown Sydney 29 B2
Campo, Casa de Madrid 17 B1
Campo F.C. Barcelona Barcelona 4 A1
Campo Grando Lisbon 16 A2
Campo Pequeño Lisbon 16 A2
Campolide Lisbon 16 A2
Camps Bay Cape Town 8 A1
C'an San Joan Barcelona 4 A2
Cañacao B. Manila 17 C1
Canarsie New York 22 C2
Candiac Montreal 20 B3
Caneças Lisbon 16 A1
Canillas Madrid 17 B2
Canillejas Madrid 17 B2
Canning Town London 15 A4
Canteras de Vallecas Madrid 17 B2
Canterbury Melbourne 18 A2
Canterbury Sydney 29 B1
Canton = Guangzhou Guangzhou 11 B2
Caoheijing Shanghai 28 B1
Capão Redondo São Paulo 27 B1
Caparica Lisbon 16 A2
Caparica, Costa da Lisbon 16 B1
Cape Flats Cape Town 8 B2
Cape Town Cape Town 8 A1
Cape Town Int. ✈ (CPT) Cape Town 8 A2
Capitol Heights Washington 33 B4
Captain Cook Bridge Sydney 29 C1
Captain Cook Landing Place Park Sydney 29 C2
Capuchos Lisbon 16 B1
Carabanchel Alto Madrid 17 B1
Carabanchel Bajo Madrid 17 B1
Carapachay Buenos Aires 7 B1
Caraza Buenos Aires 7 C2
Caridad Manila 17 C1
Carioca, Sa. da Rio de Janeiro 25 B1
Carlstadt New York 22 A1
Carlton Melbourne 18 A1
Carmen de Huechuraba Santiago 27 B1
Carmen de la Legua Lima 14 B2
Carnaxide Lisbon 16 A1
Carnegie Melbourne 18 B2
Carnide Lisbon 16 A1
Carol City Miami 18 A1
Carrascal Santiago 27 B1
Carrickmines Dublin 10 B3
Carrières-sous-Bois Paris 24 A1
Carrières-sous-Poissy Paris 24 A1
Carrières-sur-Seine Paris 24 A2
Carrigeen B. Dublin 10 A3
Cartierville Montreal 20 A1
Casa Loma Toronto 31 A2
Casa Verde São Paulo 27 A1
Casál Morena Rome 26 C2
Casalotti Rome 26 B1
Cascade Heights Atlanta 3 B2
Castèl di Leva Rome 26 C2
Castèl Sant'Angelo Rome 26 B1
Castleknock Dublin 10 A1
Castleton Corners New York 22 C1
Cataouatche, L. New Orleans 21 B1
Catete Rio de Janeiro 25 B1
Catford London 15 B3
Catherine, L. Orlando 23 B2
Caulfield Melbourne 18 B2
Cavite Manila 17 C1
Caxias Lisbon 16 A1
Cebecci Istanbul 12 B1
Cecchignola Rome 26 C2
Cecilienhof, Schloss Berlin 5 B1
Cedar Grove Atlanta 3 C3
Cedarvale Park Toronto 31 A2
Cempaka Putih Jakarta 12 B2
Cengelköy Istanbul 12 B2
Cengkareng Jakarta 12 A1
Centennial Olympic Park Atlanta 3 B2
Centennial Park Sydney 29 B2
Center Hill Atlanta 3 B2
Centocelle Rome 26 B2
Central, Gare Montreal 20 B2
Central Park New York 22 B2
Century City Los Angeles 16 B2
Cerdanyola del Vallès Barcelona 4 A1
Cerillos Santiago 27 B1
Cerro de la Estrella △ Mexico City 18 B2
Cerro de los Angeles Madrid 17 C1
Cerro Navia Santiago 27 B1
Cesano Boscone Milan 19 B1
Cesate Milan 19 A1
Cha Kwo Ling Hong Kong 12 B2
Chacarrita Buenos Aires 7 B2
Chadwell Heath London 15 A4
Chai Chee Singapore 28 B3
Chai Wan Hong Kong 12 B2
Chai Wan Kok Hong Kong 12 A1
Chaidari Athens 2 A1
Chakdaha Kolkata 14 C1
Chalandri Athens 2 A2
Chamartin Madrid 17 B1
Chamberi Madrid 17 B1
Chambourcy Paris 24 A1
Champigny-sur-Marne Paris 24 B4
Champlain, Pont Montreal 20 B2
Champs-sur-Marne Paris 24 A4
Chamrail Kolkata 14 B1
Chanakyapuri Delhi 10 B2
Chanditala Kolkata 14 A1
Changfeng Park Shanghai 28 B1
Changhai = Shanghai Shanghai 28 B2
Changi Singapore 28 A3
Changi, Singapore ✈ (SIN) Singapore 28 A3
Changning Shanghai 28 B1
Chantereine Paris 24 A4
Chantian Guangzhou 11 A2
Chaoyang Beijing 4 B2
Chaoyangmen Beijing 4 B2
Chapelizod Dublin 10 A1
Chapultepec, Bosque de Mexico City 18 B1
Chapultepec, Castillo de Mexico City 18 B1
Charenton-le-Pont Paris 24 B3
Charleroi, Kanal de → Brussels 6 B1
Charles Bridge = Karlův most Prague 25 B2
Charles Gates Dawes House Chicago 9 A2
Charlestown Boston 6 A2
Charlottenburg Berlin 5 A2
Charlottenburg, Schloss Berlin 5 A2
Charlottenlund Copenhagen 8 A3
Charlton London 15 B4
Charneca Lisbon 16 A2
Charneca Lisbon 16 B1
Chase, L. Orlando 23 B1
Châteaufort Paris 24 B1
Châtenay-Malabry Paris 24 B2
Chatham Chicago 9 C3
Châtillon Paris 24 B2
Chatou Paris 24 A1
Chatpur Kolkata 14 B2
Chatswood Sydney 29 A2
Chatuchak Bangkok 3 B2
Chatuchak Park Bangkok 3 B2
Chauki Karachi 13 A1
Chavarria Lima 14 B2
Chaville Paris 24 B2
Chayang Seoul 27 B2
Chelles Paris 24 A4
Chelles, Canal de Paris 24 A4
Chells-le-Pin ✈ Paris 24 A4
Chelsea Boston 6 A2
Chelsea London 15 B2
Chembur Mumbai 20 A2
Chengshousi Beijing 4 C2
Chennevières-sur-Marne Paris 24 B4
Cheongdam Seoul 27 B2
Cheonho Seoul 27 B2
Cheops Cairo 7 B1
Chertanovka → Moscow 19 C2
Chertanovo Moscow 19 C2
Cheryomushki Moscow 19 B2
Chestnut Hill Boston 6 B1
Cheung Sha Wan Hong Kong 12 A1
Cheverly Washington 33 B4
Chevilly-Larue Paris 24 B3
Cheviot Hills Los Angeles 16 B2
Chevy Chase Washington 33 B3
Chhatrapati Shivaji, Mumbai ✈ (BOM) Mumbai 20 A2
Chia Keng Singapore 28 A3
Chiaravalle Milanese Milan 19 B2
Chicago Chicago 9 B3
Chicago, Univ. of Chicago 9 C3
Chicago Harbor Chicago 9 B3
Chicago Lawn Chicago 9 C2
Chicago-Midway ✈ (MDW) Chicago 9 C2
Chicago O'Hare Int. ✈ (ORD) Chicago 9 B1
Chicago Ridge Chicago 9 C2
Chicago River, North Branch → Chicago 9 B2
Chicago Sanitary and Ship Canal Chicago 9 C2
Chicago State Univ. Chicago 9 C3
Chicago Zoo Chicago 9 B1
Chienzui Guangzhou 11 A3
Chik Sha Hong Kong 12 B2
Child's Hill London 15 A2
Chile, Univ. de Santiago 27 B2
Chilla Saroda Delhi 10 B2
Chillum Washington 33 B4
Chilly-Mazarin Paris 24 B2
Chingupota Kolkata 14 C1
Chinna Cr. → Karachi 13 B2
Chiquihuite, Cerro Mexico City 18 A2
Chislehurst London 15 B4
Chiswick London 15 B2
Chiswick House London 15 B2
Chitose Tokyo 30 B2
Chitralada Palace Bangkok 3 B2
Chiyoda Tokyo 30 A3
Choa Chu Kang Singapore 28 A2
Chodov u Prahy Prague 25 B3
Chōfu Tokyo 30 B2
Choisy-le-Roi Paris 24 B3
Cholargos Athens 2 B2
Cholupice Prague 25 C2
Chom Thong Bangkok 3 B1
Chong Pang Singapore 28 A2
Chongwen Beijing 4 B2
Chorrillos Lima 14 C2
Chowpatty Beach Mumbai 20 B1
Christianshavn Copenhagen 8 A3
Chrzanów Warsaw 32 B1
Chuen Lung Hong Kong 12 A1
Chuk Kok Hong Kong 12 A2
Chulalongkom Univ. Bangkok 3 B2
Chuō Tokyo 30 A3
Church End London 15 A2
Church of our Lady = Frauenkirche Munich 21 B2
Churchtown Dublin 10 B2
Ciampino Rome 26 C2
Ciampino ✈ (CIA) Rome 26 C2
Cicero Chicago 9 B2
Cilandak Jakarta 12 B1
Cilincing Jakarta 12 A2
Ciliwung → Jakarta 12 B2
Čimice Prague 25 B2
Cinecittà Rome 26 B2
Cinisello Bálsamo Milan 19 A2
Cinkota Budapest 7 A3
Cipete Jakarta 12 B1
Citta degli Studi Milan 19 B2
Città del Vaticano = Vatican City ■ Rome 26 B1
City, The London 15 A3
City Hall = Rådhus Oslo 23 A3
City Hall = Stadshuset Stockholm 29 B2
City of the Dead Cairo 7 A2
City Park New Orleans 21 A2
City Terrace Los Angeles 16 B3
Ciudad de México Mexico City 18 B2
Ciudad Deportiva Mexico City 18 B2
Ciudad Fin de Semana Madrid 17 B2
Ciudad General Belgrano Buenos Aires 7 C1
Ciudad Lineál Madrid 17 B2
Ciudad Satélite Mexico City 18 A1
Ciudad Universitaria Buenos Aires 7 B2
Ciudad Universitaria Mexico City 18 C1
Claireville Res. Toronto 31 A1
Clamart Paris 24 B2
Clapham London 15 B3
Clapton London 15 A3
Claremont Cape Town 8 A1
Clayhall London 15 A4
Clear, L. Orlando 23 A2
Clermiston Edinburgh 11 B2
Clichy Paris 24 A2
Clichy-sous-Bois Paris 24 A4
Cliffside Toronto 31 A3
Cliffside Park New York 22 B2
Clifton Cape Town 8 A1
Clifton Karachi 13 B2
Clifton New York 22 C1
Clifton Beach Karachi 13 B2
Cliftondale Boston 6 A2
Cloghran Dublin 10 A2
Clonskeagh Dublin 10 B2
Clontarf Dublin 10 A2
Clontarf Sydney 29 A2
Clovelly Sydney 29 B2
Cobras, I. das Rio de Janeiro 25 B2
Coburg Melbourne 18 A1
Cocotá Rio de Janeiro 25 A1
Cœuilly Paris 24 B4
Coina Lisbon 16 B2
Coker Lagos 14 B2
Colaba Mumbai 20 B1
Colaba Pt. Mumbai 20 B1
Colegiales Buenos Aires 7 B2
Colindale London 15 A2
Colinton Edinburgh 11 B2
College Park Atlanta 3 C2
College Park Washington 33 B4
College Point New York 22 B2
Collégien Paris 24 A4
Collier Row London 15 A4
Colliers Wood London 15 B2
Colma San Francisco 26 B2
Colney Hatch London 15 A3
Cologno Monzese Milan 19 A2
Colombes Paris 24 A2
Colonia Güell Barcelona 4 A1
Colonial Knob Wellington 33 A1
Colosseo Rome 26 B1
Colosseum = Colosseo Rome 26 B1
Combault Paris 24 B4
Comércio, Praça do Lisbon 16 A2
Commerce Los Angeles 16 B4
Conceição, I. da Rio de Janeiro 25 B2
Conchalí Santiago 27 B2
Concord Sydney 29 B1
Concord Toronto 31 A2
Concorde, Place de la Paris 24 A2
Concorezzo Milan 19 A2
Condet Jakarta 12 A2
Coney Island New York 22 C2
Congonhas São Paulo ✈ (CGH) São Paulo 27 B2
Conley → Atlanta 3 C3
Connaught Place Delhi 10 B2
Consolação São Paulo 27 B2
Constantia Cape Town 8 B1
Constitución Buenos Aires 7 B2
Constitution Atlanta 3 B2
Convention Center Los Angeles 16 B3
Conway Orlando 23 B2
Conway, L. Orlando 23 B2
Coogee Sydney 29 B2
Cooksville Toronto 31 B1
Coolock Dublin 10 A2
Copacabana Rio de Janeiro 25 B1
Copenhagen = København Copenhagen 8 A2
Coral Gables Miami 18 B2
Coral Hills Washington 33 B4
Corcovado, Cristo Redentar Rio de Janeiro 25 B1
Corduff Dublin 10 A1
Cormano Milan 19 A2
Cornaredo Milan 19 A1
Córsico Milan 19 B1
Corstorphine Edinburgh 11 B2
Corviale Rome 26 B1
Coslada Madrid 17 B2
Cossigny Paris 24 B4
Cotao Lisbon 16 A1
Côte-St-Luc Montreal 20 B2
Cotunduba, I. de Rio de Janeiro 25 B2
Coubron Paris 24 A4
Countryside Chicago 9 C1
County Art Museum Los Angeles 16 B2
Courbevoie Paris 24 A2
Courtry Paris 24 A4
Cowley London 15 A1
Coyoacán Mexico City 18 B2
Craighall Park Johannesburg 13 A2
Craiglockhart Edinburgh 11 B2
Craigmillar Edinburgh 11 B3
Cramond Edinburgh 11 B2
Cramond Bridge Edinburgh 11 B1
Cramond I. Edinburgh 11 B2
Cranford London 15 B1
Crawford Cape Town 8 A2
Crayford London 15 B5
Creekmouth London 15 A4
Crescent, L. Orlando 23 A1
Crescenzago Milan 19 A2
Cressely Paris 24 B1
Cresskill New York 22 A2
Créteil Paris 24 B3
Cricklewood London 15 A2
Cristo Redentor, Corcovado Rio de Janeiro 25 B1
Crockenhill London 15 B4
Croissy-Beaubourg Paris 24 B4
Croissy-sur-Seine Paris 24 A1
Crosby Johannesburg 13 B1
Crosne Paris 24 B3
Cross I. Mumbai 20 B2
Crouch End London 15 A3
Crown Mine Johannesburg 13 B1
Crows Nest Sydney 29 A2
Croydon London 15 B3
Croydon Park Sydney 29 B1
Cruagh Mt. Dublin 10 B2
Crumlin Dublin 10 B2
Cruz de Pau Lisbon 16 B2
Crystal Palace London 15 B3
Csepel Budapest 7 B2
Csepelsziget Budapest 7 B2
Csillaghegy Budapest 7 A2
Csillagtelep Budapest 7 B2
Csömör Budapest 7 A3
Csömöri-patak → Budapest 7 A3
Cuatro Vientos Madrid 17 B1
Cuauhtémoc Mexico City 18 B2
Cuban Museum Miami 18 B2
Cubao Manila 17 B2
Çubuklu Istanbul 12 B2
Cudahy Los Angeles 16 C3
Cuicuilco, Pirámide de Mexico City 18 C1
Culmore Washington 33 B3
Culver City Los Angeles 16 B2
Cupecé → São Paulo 27 B1
Currie Edinburgh 11 B2
Cusago Milan 19 B1
Cusano Milanino Milan 19 A2
Cutler Park Boston 6 B1
Çuvuşabaşi → Istanbul 12 B1
Cuwei Beijing 4 B1
Cypress Hills New York 22 B2
Cypress Park Los Angeles 16 B3
Czernriaków Warsaw 32 B2
Czyste Warsaw 32 B1

D

Da Moóca → São Paulo 27 B2
Ďablice Prague 25 B2
Dąbrowa Warsaw 32 B1
Dachang Shanghai 28 B1
Dachang ✈ Shanghai 28 B1
Dachau Munich 21 A1
Dachau-Ost Munich 21 A1
Dachauer Moos Munich 21 A1
Dadar Mumbai 20 A1
Daebang Seoul 27 B1
Daechi Seoul 27 B2
Dafni Athens 2 B2
Dagenham London 15 A4
Daglfing Munich 21 B2
Dahab, Gezîret el Cairo 7 B2
Daheisha Jerusalem 13 B2
Dahlem Berlin 5 B2
Dahlwitz-Hoppegarten Berlin 5 A5
Dahongmen Beijing 4 C2
Dajiaoting Beijing 4 B2
Dakhnoye St. Petersburg 27 C1
Dalejsky potok → Prague 25 B2
Dalgety Bay Edinburgh 11 A1
Dalkeith Edinburgh 11 B3
Dalkey Dublin 10 B3
Dalkey I. Dublin 10 B3
Dallgow Berlin 5 A1
Dalmeny Edinburgh 11 B1
Dalston London 15 A3
Daly City San Francisco 26 B2
Damaia Lisbon 16 A1
Damarakia Athens 2 B1
Dämeritzsee Berlin 5 B5
Dan Ryan Woods Chicago 9 C2
Danderhall Edinburgh 11 B3
Danderyd Stockholm 29 A2
Danforth Toronto 31 A3
Darakeh Tehran 31 A2
Darband Tehran 31 A2
Darling Point Sydney 29 B2
Darndale Dublin 10 A2
Darrūs Tehran 31 A2
Dartford London 15 B5
Dashi Guangzhou 11 B2
Datansha Guangzhou 11 B2
Datun Beijing 4 B2
Daulatpur Delhi 10 A1
Davidson, Mt. San Francisco 26 B2
Davidson's Mains Edinburgh 11 B2
Dāvūdiyeh Tehran 31 A2
Davydkovo Moscow 19 B1
Dawidy Warsaw 32 C1
Daxing Beijing 4 C2
Days Bay Wellington 33 B2
Decatur Atlanta 3 B3
Dedham Boston 6 B1
Degunino Moscow 19 A2
Deir Dibwan Jerusalem 13 A2
Deir Ibzi'e Jerusalem 13 A1
Dejvice Prague 25 B2
Dekabristov, Ostrov St. Petersburg 27 B1
Del Rey Los Angeles 16 C2
Delhi Delhi 10 B2
Delhi Indira Gandhi Int. ✈ (DEL) Delhi 10 B1
Demarest New York 22 A2
Den Ilp Amsterdam 2 A2
Denistone Heights Sydney 29 A1
Dentonia Park Toronto 31 A3
DePaul Univ. Chicago 9 B3
Deptford London 15 B3
Des Plaines Chicago 9 A1
Desert Wetlands Park Las Vegas 14 B2
Deutsche Oper Berlin 5 A2
Deutsches Museum Munich 21 B2
Devil's Peak Cape Town 8 A1
Dhakuria Kolkata 14 B2
Dharavi Mumbai 20 A2
Diadema São Paulo 27 C2
Diegen Brussels 6 A2
Diemen Amsterdam 2 A2
Diepkloof Johannesburg 13 B1
Dieprivier Cape Town 8 B1
Difficult Run → Washington 33 B2
Dilbeek Brussels 6 A1
Dilli = Delhi Delhi 10 B2
Dirnismaning Munich 21 A2
Disney-M.G.M. Studios Orlando 23 B1
Disney Studios Los Angeles 16 B3
District Heights Washington 33 B4
Djakarta = Jakarta Jakarta 12 A2
Djursholm Stockholm 29 A2
Döberitz Berlin 5 A1
Döbling Vienna 32 A2
Dobong Seoul 27 B2
Dobongsan Seoul 27 A2
Docklands London 15 A3
Doctor Phillips Orlando 23 B2
Dodder → Dublin 10 B1
Dodger Stadium Los Angeles 16 B3
Dogs, Isle of London 15 B3
Dolgoe Ozero St. Petersburg 27 A1
Dollis Hill London 15 A2
Dollymount Dublin 10 A2
Dolni Prague 25 B3
Dolni Chabry Prague 25 B2
Dolni Počernice Prague 25 B3
Dolphins Barn Dublin 10 B2
Dom Pedro II, Parque São Paulo 27 B2
Don Mills Toronto 31 A2
Don Muang Int. ✈ (BKK) Bangkok 3 A2
Donaghmede Dublin 10 A3
Donau-Oder Kanal Vienna 32 A3
Donaufeld Vienna 32 A2
Donaupark Vienna 32 A2
Donaustadt Vienna 32 A2
Dongan Hills New York 22 C1
Dongcheng Beijing 4 B2
Dongdaemung Seoul 27 B2
Donggou Shanghai 28 B2
Dongjak Seoul 27 B1
Dongjiao Guangzhou 11 B2
Dongri Mumbai 20 B2
Dongshanhu Park Guangzhou 11 B2
Dongzhimen Beijing 4 B2
Donnybrook Dublin 10 B2
Doornfontein Johannesburg 13 B2
Dorchester Boston 6 B2
Dorchester B. Boston 6 B2
Dorchester Heights Nat. Historical Site ⌂ Boston 6 B2
Dornach Munich 21 B3
Dorval Int., Montréal ✈ (YUL) Montreal 20 B1
Dos Moninos → São Paulo 27 C2
Douglas Park Chicago 9 B2
Dover Heights Sydney 29 B2
Dowlatābād Tehran 31 B2
Down, L. Orlando 23 A1
Downey Los Angeles 16 C4
Downsview Toronto 31 A1
Dragør Copenhagen 8 B3
Drancy Paris 24 A3
Dranesville Washington 33 A1
Drapetsona Athens 2 B1
Dreilinden Berlin 5 B2
Drewnica Warsaw 32 B2
Drigh Road Karachi 13 A2
Drimnagh Dublin 10 B2
Drogenbos Brussels 6 B1
Druid Hills Atlanta 3 B2
Drumcondra Dublin 10 A2
Drummoyne Sydney 29 B1
Drylaw Edinburgh 11 B2
Dubeč Prague 25 B3
Dublin Dublin 10 A2
Dublin ✈ (DUB) Dublin 10 A2
Dublin B. Dublin 10 B3
Dublin Harbour Dublin 10 A2
Duddingston Edinburgh 11 B3
Dugnano Milan 19 A2
Duivendrecht Amsterdam 2 B2
Dūlāb Tehran 31 B2
Dulles Int. ✈ (IAD) Washington 33 B1
Dulwich London 15 B3
Dum Dum Kolkata 14 B2
Dum Dum Int. ✈ (CCU) Kolkata 14 B2
Dumont New York 22 A2
Dún Laoghaire Dublin 10 B3
Dundrum Dublin 10 B2
Dunearn Singapore 28 B2
Dúnleary = Dún Laoghaire Dublin 10 B3
Dunn Loring Washington 33 B2
Dunning Chicago 9 B2
Dunvegan Johannesburg 13 A2
Duomo Milan 19 B2
Duque de Caxias Rio de Janeiro 25 A1
Duren Sawit Jakarta 12 A2
Dusit Bangkok 3 B2
Dworp Brussels 6 B1
Dyakovo Moscow 19 B2
Dyker Beach Park New York 22 C1
Dzerzhinskiy Moscow 19 B2
Dzerzhinskiy Park Moscow 19 B2

E

E.U.R. = Esposizione Universale di Roma Rome 26 C1
Eagle Rock Los Angeles 16 B3
Ealing □ London 15 A2
Earlsfield London 15 B2
Earlwood Sydney 29 B1
East Arlington Boston 6 A2
East Arlington Washington 33 B3
East Bedfont London 15 B1
East Boston Boston 6 A2
East Don → Toronto 31 A2
East Don Parkland Toronto 31 A2
East Elmhurst New York 22 B2
East Finchley London 15 A2
East Flatbush New York 22 C2
East Ham London 15 A4
East Humber → Toronto 31 A1
East Lamma Channel Hong Kong 12 B1
East Las Vegas = Whitney Las Vegas 14 B2
East Lexington Boston 6 A1
East Los Angeles Los Angeles 16 B3
East Molesey London 15 B1
East New York New York 22 B2
East Pines Washington 33 B4
East Point Atlanta 3 B2
East Potomac Park Washington 33 B3
East River → New York 22 B2
East Rutherford New York 22 A1
East Sheen London 15 B2
East Talpiyot Jerusalem 13 B2
East Wickham London 15 B4
East York Toronto 31 A2
Eastbourne Wellington 33 B2
Eastcote London 15 A1
Easter Howgate Edinburgh 11 B2
Eastpoint Park Toronto 31 A4
Eastwood Sydney 29 A1
Eaton Canyon Park Los Angeles 16 A4
Ebara Tokyo 30 B3
Ebisu Tokyo 30 B3
Ebute-Ikorodu Lagos 14 A2
Ebute-Metta Lagos 14 B2
Eda Tokyo 30 B2
Edendale Johannesburg 13 A2
Edenmore Dublin 10 A2
Edgars Cr. → Melbourne 18 A1
Edgeley Toronto 31 A1
Edgemar San Francisco 26 C2
Edgewater New York 22 B2
Edgewood Orlando 23 B2
Edinburgh Edinburgh 11 B2
Edinburgh ✈ (EDI) Edinburgh 11 B1
Edinburgh Castle Edinburgh 11 B2
Edison Park Chicago 9 A2
Edmondston Washington 33 B4
Edmondstown Dublin 10 B2
Edo → Tokyo 30 B4
Edogawa Tokyo 30 A4
Edsberg Stockholm 29 A1
Edwards L. Melbourne 18 A1
Efzonos Athens 2 B2
Egaleo Athens 2 B2
Egaleo, Oros Athens 2 B1
Eiche Berlin 5 A4
Eiche Süd Berlin 5 A4
Eiffel, Tour Paris 24 A2
Ein Arik Jerusalem 13 A1
Ein Naquba Jerusalem 13 B1
Ein Rafa Jerusalem 13 B1
Eizariya Jerusalem 13 B2
Ejby Copenhagen 8 A2
Ejigbo Lagos 14 A1
Ekeberg Oslo 23 A3
Eknäs Stockholm 29 B3
El 'Abbasiya Cairo 7 A2
El Agustino Lima 14 B2
El Agustino, Cerro Lima 14 B3
El Baragil Cairo 7 A1
El Basâlîn Cairo 7 B2
El Bira Jerusalem 13 A2
El Bosque Santiago 27 C2
El Carmen Santiago 27 B1
El Cortijo Santiago 27 B1
El Cristo, Vaso Regulador Mexico City 18 B1
El Duqqi Cairo 7 A2
El Encinar de los Reyes Madrid 17 A2
El Gezira Cairo 7 A2
El Ghurîya Cairo 7 A2
El Gîza Cairo 7 A2
El Khadr Jerusalem 13 B1
El Khalîfa Cairo 7 A2
El Kôm el Ahmar Cairo 7 A2
El Ma'âdi Cairo 7 B2
El Matarîya Cairo 7 A2
El Mohandessin Cairo 7 A2
El Monte Los Angeles 16 B4
El Muqattam Cairo 7 A2
El Mûskî Cairo 7 A2
El Pardo Madrid 17 A1
El Portal Miami 18 A2
El Prat de Llobregat Barcelona 4 B1
El Qâhira Cairo 7 A2
El Qubba Cairo 7 A2
El Reloj Mexico City 18 C2
El Retiro Madrid 17 B1
El Salto Santiago 27 B2
El Sereno Los Angeles 16 B3
El Talibîya Cairo 7 B2
El Vergel Mexico City 18 C2
El Wâhli Cairo 7 A2
El Zamâlik Cairo 7 A2
El Zeitûn Cairo 7 A2
Eleftherios Venizelos ✈ (ATH) Athens 2 A3
Elephanta Caves Mumbai 20 B2
Elephanta I. Mumbai 20 B2
Ellboda Stockholm 29 A3
Ellenor, L. Orlando 23 B2
Elliniko Olympic Complex Athens 2 B2
Ellis I. New York 22 B1
Elmers End London 15 B3
Elmhurst New York 22 B2
Elmstead London 15 B4
Elmwood New Orleans 21 B1
Elmwood Park Chicago 9 B2
Elmwood Park New York 22 A1
Elsdon Wellington 33 A1
Elsiesrivier Cape Town 8 A2
Elsternwick Melbourne 18 B2
Eltham London 15 B4
Elwood Melbourne 18 B1
Élysée Paris 24 A2
Elysian Park Los Angeles 16 B3
Emāmzādeh Sāleh Tehran 31 A2
Émerainville Paris 24 B4
Emeryville San Francisco 26 A3
Eminönü Istanbul 12 B1
Emirgan Istanbul 12 B2
Emmarentia Johannesburg 13 A2
Empire State Building New York 22 B2
Encantado Rio de Janeiro 25 B1
Encino Los Angeles 16 B2
Encino Res. Los Angeles 16 B1
Enebyberg Stockholm 29 A1
Enfield Sydney 29 B1
Engenho, I. do Rio de Janeiro 25 B2
Englewood Chicago 9 C3
Englewood New York 22 A2
Englewood Cliffs New York 22 A2
Enmore Sydney 29 B2
Enskede Stockholm 29 B2
Enterprise Las Vegas 14 B1
Entrevias Madrid 17 B1
Epcot Orlando 23 B1
Epping Sydney 29 A1
Erenköy Istanbul 12 C2
Erith London 15 B5
Erlaa Vienna 32 B1
Ermington Sydney 29 A1
Ermita Manila 17 B1
Ershatou Guangzhou 11 B2
Erskineville Sydney 29 B2
Erunkan Lagos 14 A2
Erzsébet-Telep Budapest 7 B3
Eschenried Munich 21 A1
Esenler Istanbul 12 B1
Esher London 15 B1
Eskbank Edinburgh 11 B3
Esplugas Barcelona 4 A1
Esposizione Universale di Roma Rome 26 C1
Essendon Melbourne 18 A1
Essendon ✈ (MEB) Melbourne 18 A1
Essingen Stockholm 29 B1
Essling Vienna 32 A3
Est, Gare de l' Paris 24 A3
Estado, Parque do São Paulo 27 B2
Estelle New Orleans 21 B2
Estrela, Basilica da Lisbon 16 A2
Etobicoke Toronto 31 B1
Etobicoke Cr. → Toronto 31 B1
Etterbeek Brussels 6 B2
Eung-am Seoul 27 B1
Eunpyeong Seoul 27 B1
Evanston Chicago 9 A2
Even Sapir Jerusalem 13 B1
Evere Brussels 6 A2
Everett Boston 6 A2
Evergreen Park Chicago 9 C2
Evīn Tehran 31 A2
Ewu Lagos 14 A1
Exhibition Place Toronto 31 B2
Expo 2010 Shanghai 28 B1
Exposições, Palácio das Rio de Janeiro 25 B1
Eyüp Istanbul 12 B1
Ezeiza ✈ (EZE) Buenos Aires 7 C1

F

Fabour, Mt. Singapore 28 B2
Fælledparken Copenhagen 8 A3
Fågelön Stockholm 29 B1
Fagersjö Stockholm 29 B2
Fairfax Washington 33 B2
Fairfax Station Washington 33 C2
Fairland Johannesburg 13 A1
Fairmilehead Edinburgh 11 B2
Fairmount Heights Washington 33 B4
Fairport Toronto 31 A4
Fairview New York 22 B2
Fairview, L. Orlando 23 A2
Falenty Warsaw 32 C1
Faliro Athens 2 B1
Faliro, Ormos Athens 2 B2
Falkenburg Berlin 5 A4
Falkenhagen Berlin 5 A1
Falkensee Berlin 5 A1
Falls Church Washington 33 B2
Falomo Lagos 14 B2
False Bay Cape Town 8 B2
Fangcun Guangzhou 11 B2
Faraḩābād Tehran 31 A2
Farforovskaya St. Petersburg 27 B2
Farningham London 15 B5
Farsta Stockholm 29 B2
Fasanerie-Nord Munich 21 A2
Fasangarten Munich 21 B2
Fatih Istanbul 12 B1
Favoriten Vienna 32 B2
Fawkner Melbourne 18 A1
Fawkner Park Melbourne 18 B1
FedEx Stadium Washington 33 B4
Feldkirchen Munich 21 B3
Feldmoching Munich 21 A2
Feltham London 15 B1
Fener Istanbul 12 B1
Fenerbahçe Istanbul 12 C2
Fengtai Beijing 4 C1
Fengtai Stadium Beijing 4 B1
Ferencváros Budapest 7 B2
Ferihegy, Budapest ✈ (BUD) Budapest 7 B3
Ferndale Johannesburg 13 A2
Fichtenau Berlin 5 B5
Field Museum of Natural History Chicago 9 B3
Fields Corner Boston 6 B2
Fiera Camp Milan 19 B1
Figino Milan 19 B1
Fijir Baghdad 3 A2
Filadelfia Athens 2 A2
Fili-Mazilovo Moscow 19 B1
Filothei Athens 2 A2
Finchley London 15 A2
Fine Arts, Museum of Boston 6 A2
Finglas Dublin 10 A2
Finsbury London 15 A3
Finsbury Park London 15 A3
Fiorito Buenos Aires 7 C2
Firhouse Dublin 10 B2
Fischerhäuser Munich 21 A3
Fisher Island Miami 18 B2
Fishermans Bend Melbourne 18 A1
Fisherville Toronto 31 A2
Fisksätra Stockholm 29 B3
Fitzroy Gardens Melbourne 18 A1
Five Cowrie Cr. → Lagos 14 B2
Five Dock Sydney 29 B1
Fjellstrand Oslo 23 B2
Flamengo Rio de Janeiro 25 B1
Flamingo Orlando 23 B2
Flaminio Rome 26 B1
Flaskebekk Oslo 23 A2
Flatbush New York 22 C2
Flaten Stockholm 29 B2
Flatlands New York 22 C2
Flemington Racecourse Melbourne 18 A1
Flint Pk. Los Angeles 16 B3
Florence Los Angeles 16 C3
Florence Bloom Bird Sanctuary ⌂ Johannesburg 13 A2
Florentia Johannesburg 13 B2
Flores Buenos Aires 7 B2
Flores, Mercado de Mexico City 18 C2
Floresta Buenos Aires 7 B2
Florida Buenos Aires 7 B2
Florida Johannesburg 13 B1
Floridsdorf Vienna 32 A2
Flushing New York 22 B3
Flushing Meadows Corona Park New York 22 B2
Flysta Stockholm 29 A1
Fo Tan Hong Kong 12 A2
Foggy Bottom Washington 33 B3
Föhrenhain Vienna 32 A2
Fontainebleau Johannesburg 13 A1
Fontenay-aux-Roses Paris 24 B2
Fontenay-le-Fleury Paris 24 B1
Fontenay-sous-Bois Paris 24 A3
Foots Cray London 15 B4
Footscray Melbourne 18 A1
Forbidden City = Imperial Palace Museum Beijing 4 B2
Forest Brussels 6 B1
Forest Gate London 15 A4
Forest Heights Washington 33 C3
Forest Hill London 15 B3
Forest Hill Toronto 31 A2
Forest Hills New York 22 B2
Forest Park Chicago 9 B2
Forest Park New York 22 B2
Forest View Chicago 9 C2
Forestville Washington 33 B4
Fornebu Oslo 23 A2
Foro Romano Rome 26 B1
Forstenried Munich 21 B1
Forstenrieder Park Munich 21 B1
Fort Mumbai 20 B1
Fort Dupont Park Washington 33 B4
Fort Foote Village Washington 33 C3
Fort Lee New York 22 A2
Forth, Firth of Edinburgh 11 A3
Forth Rail Bridge Edinburgh 11 A1
Forth Road Bridge Edinburgh 11 A1
Foshan Guangzhou 11 B2
Fót Budapest 7 A3
Fourqueux Paris 24 A1
Fox Studios Sydney 29 B2
Foxrock Dublin 10 B2
Franconia Washington 33 C3
Frank Lloyd Wright Home Chicago 9 B2
Frankel Singapore 28 B3
Franklin Park Boston 6 B2
Franklin Park Chicago 9 B1
Franklin Park Washington 33 B3
Franklin Res. Los Angeles 16 B2
Franz Josef Strauss ✈ (MUC) Munich 21 A2
Frauenkirche Munich 21 B2
Frederiksberg Copenhagen 8 A3
Frederiksdal Copenhagen 8 A2
Fredersdorf Berlin 5 A5
Freguesia Rio de Janeiro 25 A1
Freidrichshain, Volkspark Berlin 5 A3
Freiham Munich 21 B1
Freimann Munich 21 A2
French Quarter New Orleans 21 B2
Fresh Pond Boston 6 A2
Fresnes Paris 24 B2
Freudenau Vienna 32 A2
Friarstown Dublin 10 B1
Friedenau Berlin 5 B3
Friedrichsfelde Berlin 5 B4
Friedrichshagen Berlin 5 B4
Friedrichshain Berlin 5 A3
Friedrichslust Berlin 5 A5
Friherrs Helsinki 11 B1
Frilufts Museum Helsinki 11 B2
Frontón, I. Lima 14 B1
Frunze Moscow 19 B2
Fuchū Tokyo 30 A1
Fuencarral Madrid 17 B1
Fuenlabrada Madrid 17 C1
Fukagawa Tokyo 30 B3
Fukushima Osaka 23 A1
Fulham London 15 B2
Funabori Tokyo 30 A4
Fundão, I. do Rio de Janeiro 25 B1
Fünfhaus Vienna 32 A2
Furesø Copenhagen 8 A2
Fürth Munich 21 B2
Futago-tamagawaen Tokyo 30 B2
Fuxing Dao Shanghai 28 B2
Fuxing Park Shanghai 28 B1
Fuxinglu Beijing 4 B1

G

G. Ross Lord Park Toronto 31 A2
Gaebong Seoul 27 C1
Gage Park Chicago 9 C2
Gagny Paris 24 A4
Galata Istanbul 12 B1
Galata Tower Istanbul 12 B1
Galatsi Athens 2 A2
Galeão, Int. de ✈ (GIG) Rio de Janeiro 25 A1
Galyanovo Moscow 19 B3
Gambir Jakarta 12 A1
Gamboa Rio de Janeiro 25 B1
Gamboloita Milan 19 B2
Gamlebyen Oslo 23 A3
Gangdong Seoul 27 B2
Gangnam Seoul 27 B1
Gangseo Seoul 27 B1
Gangtou Guangzhou 11 A1
Gangwei Guangzhou 11 B2
Ganjiakou Beijing 4 B1
Ganshoren Brussels 6 A1
Gants Hill London 15 A4
Gaoqiao Shanghai 28 A2
Garbagnate Milanese Milan 19 A1
Garbatella Rome 26 B2
Garches Paris 24 A2
Garching Munich 21 A3
Garden City Cairo 7 A2
Garden Reach Kolkata 14 B1
Garder Oslo 23 B2
Garfield New York 22 A1
Garfield Park Chicago 9 B2
Gargareta Athens 2 B2
Garibong Seoul 27 C1
Garvanza Los Angeles 16 B3
Gåshaga Stockholm 29 A3
Gateway ⌂ New York 22 C2
Gateway of India Mumbai 20 B2
Gatow Berlin 5 B1
Gavà Barcelona 4 B1
Gavamar Barcelona 4 B1
Gávea Rio de Janeiro 25 B1
Gávea, Pedra da Rio de Janeiro 25 B1
Gazdagrét Budapest 7 B1
Gaziosmanpaşa Istanbul 12 B1
Gebel el Ahmar Cairo 7 A3
Gebel el Muqattam Cairo 7 A2
Gebel el Tura Cairo 7 B2
Geiselgasteig Munich 21 B2
General San Martin Buenos Aires 7 B1
Gennevilliers Paris 24 A2
Gentilly New Orleans 21 A2
Gentilly Paris 24 B3
Gentofte Copenhagen 8 A3
Genval Brussels 6 B2
George I. Hong Kong 12 B1
Georges River Bridge Sydney 29 C1
Georgetown Washington 33 B3
Georgia Dome Atlanta 3 B2
Gerasdorf bei Wien Vienna 32 A2
Gerberau Munich 21 A1
Gerli Buenos Aires 7 C2
Germiston Johannesburg 13 B2
Gern Munich 21 B2
Getafe Madrid 17 C1
Getty Center, The Los Angeles 16 B2
Geunjeong Seoul 27 C1
Geva Binyamin Jerusalem 13 A2
Geylang Serai Singapore 28 B3
Gharapuri Mumbai 20 B2
Gharb = Shahrak-e Qods Tehran 31 A2
Ghatkopar Mumbai 20 A2
Ghazipur Delhi 10 B2
Ghizri Karachi 13 B2
Ghizri Cr. → Karachi 13 B2
Ghonda Delhi 10 A2
Ghusuri Kolkata 14 B2
Gianicolense Rome 26 B1
Giant Wheel = Riesenrad Vienna 32 A2
Gibraltar Pt. Toronto 31 B2
Gidea Park London 15 A5
Giesing Munich 21 B2
Gilmerton Edinburgh 11 B3
Gilo Jerusalem 13 B2
Gimmersta Stockholm 29 B3
Ginza Tokyo 30 B3
Giv'at Ram Jerusalem 13 B2
Giv'at Ye'arim Jerusalem 13 B1
Giv'at Ze'ev Jerusalem 13 A2
Giv'on Jerusalem 13 A1
Giza = El Giza Cairo 7 A2
Giza Pyramids Cairo 7 B1
Gjersjøen Oslo 23 B3
Gladesville Sydney 29 B1
Gladsakse Copenhagen 8 A2
Glasnevin Dublin 10 A2
Glassmanor Washington 33 C4
Glasthule Dublin 10 B3
Glen Iris Melbourne 18 B2
Glen Mar Park Washington 33 B3
Glen Rouge Park Toronto 31 A4
Glenarden Washington 33 B4
Glenasmole Reservoirs Dublin 10 B1
Glencorse Res. Edinburgh 11 B2
Glencullen Dublin 10 B2
Glendale Los Angeles 16 B3
Glendoo Mt. Dublin 10 B2
Glenhuntly Melbourne 18 B2
Glenside Wellington 33 B1
Glenview Chicago 9 A2
Glenview Countryside Chicago 9 A2
Glenvista Johannesburg 13 B2
Glifada Athens 2 B2
Glostrup Copenhagen 8 B2
Gogar Edinburgh 11 B2
Göktürk Istanbul 12 A1
Golabari Kolkata 14 B2
Golabki Warsaw 32 B1
Gold Coast Chicago 9 B3
Golden Gate San Francisco 26 B2
Golden Gate ⌂ San Francisco 26 A2
Golden Gate Bridge San Francisco 26 B2
Golden Gate Park San Francisco 26 B2
Golden Horn = Haliç Istanbul 12 B1
Golders Green London 15 A2
Golestan Palace Tehran 31 A2
Gollans Stream → Wellington 33 B2
Gonen Jerusalem 13 B2
Gongneung Seoul 27 B2
Goodmayes London 15 A4
Goodwood Cape Town 8 A2
Gopalpur Kolkata 14 B2
Górce Warsaw 32 B1
Gore Hill Sydney 29 A2
Gorelyy → St. Petersburg 27 A3
Gorgie Edinburgh 11 B2

Gorky Park *Moscow* 19 B2
Gosen *Berlin* 5 B5
Gosener kanal *Berlin* 5 B5
Gospel Oak *London* 15 A3
Gotanda *Tokyo* 30 B3
Goth Goli Mar *Karachi* 13 A2
Goth Sher Shah *Karachi* 13 A1
Gotha *Orlando* 23 A1
Gournay-sur-Marne *Paris* 24 A4
Governador, I. do *Rio de Janeiro* 25 A1
Governors I. *New York* 22 B1
Grabów *Warsaw* 32 C1
Grace, Mt. *Wellington* 33 B2
Gracefield *Wellington* 33 B2
Gràcia *Barcelona* 4 A2
Gräfelfing *Munich* 21 B1
Gragoatá *Rio de Janeiro* 25 B2
Grand Bazaar = Kapali Carsi *Istanbul* 12 B1
Grand Central Station *New York* 22 B2
Grand Palace *Bangkok* 3 B1
Grande Place *Brussels* 6 A2
Grankulla = Kauniainen *Helsinki* 11 B1
Grant Park *Chicago* 9 B3
Granton *Edinburgh* 11 B2
Grassy Park *Cape Town* 8 B2
Gratosóglio *Milan* 19 B2
Gratzwalde *Berlin* 5 B5
Gravesend *New York* 22 C2
Grazhdanka *St. Petersburg* 27 B2
Great Falls *Washington* 33 B2
Great Falls Park *Washington* 33 B2
Great Western Forum *Los Angeles* 16 C2
Greco *Milan* 19 A2
Green I. *Hong Kong* 12 B1
Green Point *Cape Town* 8 A1
Green Point Stadium *Cape Town* 8 A1
Greenbelt *Washington* 33 A4
Greenbelt Park *Washington* 33 B4
Greenfield Park *Montreal* 20 B3
Greenford *London* 15 A1
Greenhill *London* 15 A2
Greenhills *Dublin* 10 B1
Greenwich □ *London* 15 B3
Greenwich Observatory *London* 15 B3
Greenwich Village *New York* 22 B2
Greenwood *Boston* 6 A2
Grefsen *Oslo* 23 A3
Gresham Park *Atlanta* 3 B2
Gretna *New Orleans* 21 B2
Gretna City Park *New Orleans* 21 B2
Greve Strand *Copenhagen* 8 B1
Griebnitzsee *Berlin* 5 B1
Griffith Park *Los Angeles* 16 B3
Grimbergen *Brussels* 6 A2
Grinzing *Vienna* 32 A2
Gröbenried *Munich* 21 A1
Grochów *Warsaw* 32 B2
Grodzisk *Warsaw* 32 B2
Groenendaal *Brussels* 6 B2
Grogol Petamburin *Jakarta* 12 A1
Gronsdorf *Munich* 21 B3
Grorud *Oslo* 23 A4
Gross-Glienicke *Berlin* 5 B1
Gross-Hadern *Munich* 21 B1
Gross-Lappen *Munich* 21 A2
Grosse Krampe *Berlin* 5 B5
Grosse Müggelsee *Berlin* 5 B4
Grosse Point Lighthouse *Chicago* 9 A2
Grossenzersdorf *Vienna* 32 A3
Grossenzersdorfer Arm → *Vienna* 32 A3
Grosser Biberhaufen *Vienna* 32 A2
Grosser Wannsee *Berlin* 5 B2
Grossfeld-Siedlung *Vienna* 32 A2
Grosshesselohe *Munich* 21 B2
Grossjedlersdorf *Vienna* 32 A2
Grossziethen *Berlin* 5 B3
Ground Zero *New York* 22 B1
Grove, The *Chicago* 9 A1
Grove Hall *Boston* 6 B2
Grove Park *Atlanta* 3 B2
Grove Park *London* 15 B4
Grove Park *London* 15 B2
Groveton *Washington* 33 C3
Grünau *Berlin* 5 B4
Grunewald *Berlin* 5 B2
Grünwald *Munich* 21 B2
Grünwalder Forst *Munich* 21 B2
Grymes Hill *New York* 22 C1
Guadalupe *Manila* 17 B2
Guadalupe, Basilica de *Mexico City* 18 B2
Guanabara, B. de *Rio de Janeiro* 25 B1
Guanabara, Jardim *Rio de Janeiro* 25 A1
Guanabara, Palácio da *Rio de Janeiro* 25 B1
Guang'anmen *Beijing* 4 B1
Guangqumen *Beijing* 4 B2
Guangzhou *Guangzhou* 11 B2
Guanshuo *Guangzhou* 11 B3
Gudö *Stockholm* 29 B3
Güell, Parque de *Barcelona* 4 A2
Guinardó *Barcelona* 4 A2
Gulbai *Karachi* 13 A1
Güngören *Istanbul* 12 B1
Gunnersbury *London* 15 B2
Guomao *Beijing* 4 B2
Gustavo A. Madero *Mexico City* 18 B2
Guttenberg *New York* 22 B1
Gutuyevskiy, Ostrov *St. Petersburg* 27 B1
Guyancourt *Paris* 24 B1
Gwanak *Seoul* 27 C1
Gwanaksan *Seoul* 27 C1
Gyál *Budapest* 7 B3
Gyáli-patak → *Budapest* 7 B2

H

Haaga *Helsinki* 11 B2
Haar *Munich* 21 B3
Habay *Manila* 17 C1
Hackbridge *London* 15 B3
Hackensack *New York* 22 A1
Hackensack → *New York* 22 B1
Hackney □ *London* 15 A3
Hackney Wick *London* 15 A3
Hadr, Warrâq el *Cairo* 7 A2
Haga *Stockholm* 29 A2
Hagenbrunn *Vienna* 32 A2
Hägersten *Stockholm* 29 B1
Häggvik *Stockholm* 29 A1
Hagonoy *Manila* 17 B2
Hague Park *Toronto* 31 A3
Haidari = Chaidari *Athens* 2 A1
Haidarpur *Delhi* 10 A1
Haidhausen *Munich* 21 B2
Haidian *Beijing* 4 B1
Haight-Ashbury *San Francisco* 26 B2
Hainault *London* 15 A4
Haizhou *Guangzhou* 11 B2
Haizhu Guangchang *Guangzhou* 11 B2
Hakunila *Helsinki* 11 B3
Halandri = Chalandri *Athens* 2 A2
Halásztelek *Budapest* 7 B1
Haliç *Istanbul* 12 B1
Halim Perdana Kusuma Int. ✈ (HLP) *Jakarta* 12 B2
Halle *Brussels* 6 B1
Haltiala *Helsinki* 11 B2
Haltiavuori *Helsinki* 11 B2
Ham *London* 15 B2
Hämeenkylä *Helsinki* 11 B1
Hammarby *Stockholm* 29 B2
Hamme *Brussels* 6 A1
Hammersmith *London* 15 B2
Hampstead *London* 15 A2
Hampstead *Montreal* 20 B2
Hampstead Garden Suburb *London* 15 A2
Hampstead Heath *London* 15 A2
Hampton *London* 15 B1
Hampton Court Palace *London* 15 B1
Hampton Wick *London* 15 B2
Hamrā' *Baghdad* 3 B1
Hanala *Helsinki* 11 A3
Haneda *Tokyo* 30 B3
Haneda, Tōkyō ✈ (HND) *Tokyo* 30 B3
Hang Hau *Hong Kong* 12 B2
Hanging Gardens *Mumbai* 20 B1
Hanlon *Toronto* 31 B1
Hanwell *London* 15 A1
Hanworth *London* 15 B1
Haora *Kolkata* 14 B1
Hapeville *Atlanta* 3 C2
Happy Valley *Hong Kong* 12 B2
Har Adar *Jerusalem* 13 B1
Har Gilo *Jerusalem* 13 B2
Har Homa *Jerusalem* 13 B2
Har Nof *Jerusalem* 13 B2
Harahan *New Orleans* 21 B1
Haren *Brussels* 6 A2
Hareskovby *Copenhagen* 8 A2
Haringey □ *London* 15 A3
Harjusuo *Helsinki* 11 B3
Harlaching *Munich* 21 B2
Harlaw Res. *Edinburgh* 11 B2
Harlem *New York* 22 B2
Harlesden *London* 15 A2
Harlington *London* 15 B1
Harmaja *Helsinki* 11 C2
Harmashatar hegy *Budapest* 7 A2
Harolds Cross *Dublin* 10 B2
Háros *Budapest* 7 B2
Harperrig Res. *Edinburgh* 11 B1
Harrow □ *London* 15 A1
Harrow on the Hill *London* 15 A1
Harrow School *London* 15 A1
Harrow Weald *London* 15 A1
Hartsfield-Atlanta Int. ✈ (ATL) *Atlanta* 3 C2
Harumi *Tokyo* 30 B3
Harvard Univ. *Boston* 6 A2
Harvey *New Orleans* 21 B2
Harwood Heights *Chicago* 9 B2
Ḥasanābād *Tehran* 31 A1
Hasbrouck Heights *New York* 22 A1
Haselhorst *Berlin* 5 A2
Hasköy *Istanbul* 12 B1
Hasle *Oslo* 23 B2
Haslum *Oslo* 23 A2
Hästhagen *Stockholm* 29 B2
Hataitai *Wellington* 33 B1
Hatch End *London* 15 A1
Hatiara *Kolkata* 14 B2
Hauketo *Oslo* 23 A3
Hauz Khas *Delhi* 10 B2
Havel → *Berlin* 5 A2
Havelkanal *Berlin* 5 A2
Havering □ *London* 15 A5
Havering-atte-Bower *London* 15 A5
Haworth *New York* 22 A2
Hawthorne Racecourse *Chicago* 9 C2
Hayes *London* 15 B4
Hayes *London* 15 A1
Hayes End *London* 15 A1
Hayford *Chicago* 9 C2
Haywards *Wellington* 33 A2
Heathfield *Cape Town* 8 B1
Heathrow, London ✈ (LHR) *London* 15 B1
Hebe Haven *Hong Kong* 12 A2
Hedong *Guangzhou* 11 B2
Heidelberg Heights *Melbourne* 18 A2
Heidelberg West *Melbourne* 18 A2
Heidemühle *Berlin* 5 B5
Heideveld *Cape Town* 8 A2
Heiligensee *Berlin* 5 A2
Heiligenstadt *Vienna* 32 A2
Heinersdorf *Berlin* 5 A3
Hélène de Champlain, Parc ⌂ *Montreal* 20 A2
Helenelund *Stockholm* 29 A1
Heliopolis = Masr el Gedida *Cairo* 7 A2
Hellersdorf *Berlin* 5 A4
Hellerup *Copenhagen* 8 A3
Helsingfors = Helsinki *Helsinki* 11 B2
Helsinki *Helsinki* 11 B2
Helsinki-Vantaa ✈ (HEL) *Helsinki* 11 B2
Hen and Chicken B. *Sydney* 29 B1
Henderson *Las Vegas* 14 B2
Hendon *London* 15 A2
Hengsha *Guangzhou* 11 B2
Hennigsdorf *Berlin* 5 A2
Henryków *Warsaw* 32 B1
Henson Cr. → *Washington* 33 C4
Henttaa *Helsinki* 11 B1
Heping Park *Shanghai* 28 B2
Hepingli *Beijing* 4 B2
Herlev *Copenhagen* 8 A2
Herman Eckstein Park *Johannesburg* 13 A2
Hermannskogel *Vienna* 32 A1
Hermiston *Edinburgh* 11 B2
Hermitage and Winter Palace *St. Petersburg* 27 B2
Hermsdorf *Berlin* 5 A2
Hernals *Vienna* 32 A2
Herne Hill *London* 15 B3
Héroes de Churubusco *Mexico City* 18 B2
Herons, Î. aux *Montreal* 20 B2
Herstedøster *Copenhagen* 8 A2
Herttoniemi *Helsinki* 11 B3
Ḥeşārak *Tehran* 31 A1
Heston *London* 15 B1
Hetzendorf *Vienna* 32 B1
Hextable *London* 15 B5
Hialeah *Miami* 18 A1
Hiawassa, L. *Orlando* 23 A2
Hickory Hills *Chicago* 9 C2
Hiekkaharju *Helsinki* 11 B3
Hietaniemi *Helsinki* 11 B2
Hietzing *Vienna* 32 A1
Higashi *Osaka* 23 A2
Higashimurayama *Tokyo* 30 A1
Higashinari *Osaka* 23 A2
Higashisumiyoshi *Osaka* 23 B2
Higashiyodogawa *Osaka* 23 A1
High Park *Toronto* 31 B2
Highbury *London* 15 A3
Highgate *London* 15 A3
Highland Cr. → *Toronto* 31 A3
Highland Creek *Toronto* 31 A3
Highland Park *Los Angeles* 16 B3
Highlands North *Johannesburg* 13 A2
Hillcrest Heights *Washington* 33 C4
Hillend *Edinburgh* 11 A1
Hillingdon □ *London* 15 A1
Hillwood *Washington* 33 B3
Hilmîya *Cairo* 7 A2
Hin Keng *Hong Kong* 12 A2
Hirota *Osaka* 23 A1
Hirschstetten *Vienna* 32 A2
Hither Green *London* 15 B3
Hiyoshi *Tokyo* 30 B2
Hizma *Jerusalem* 13 B2
Hjortekær *Copenhagen* 8 A3
Hjortespring *Copenhagen* 8 A2
Hlubočepy *Prague* 25 B2
Ho Chung *Hong Kong* 12 A2
Hoboken *New York* 22 B1
Hobsons B. *Melbourne* 18 B1
Hochbrück *Munich* 21 A2
Hochelaga *Montreal* 20 A2
Hodgkins *Chicago* 9 C1
Hoegi *Seoul* 27 B2
Hoeilaart *Brussels* 6 B2
Hofburg *Vienna* 32 A2
Hoffman I. *New York* 22 C1
Hohenbrunn *Munich* 21 B3
Hohenschönhausen *Berlin* 5 A4
Holargos = Cholargos *Athens* 2 B2
Holborn *London* 15 A3
Holden, L. *Orlando* 23 B2
Holešovice *Prague* 25 B2
Holland Village *Singapore* 28 B2
Höllriegelskreuth *Munich* 21 B1
Hollywood *Los Angeles* 16 B3
Hollywood Bowl *Los Angeles* 16 B2
Holmenkollen *Oslo* 23 A3
Holmes Run Acres *Washington* 33 B2
Holmgård *Stockholm* 29 B1
Holmlia *Oslo* 23 B3
Holocaust Memorial *Jerusalem* 13 B2
Holyrood House, Palace of *Edinburgh* 11 B3
Holysloot *Amsterdam* 2 A3
Homerton *London* 15 A3
Hometown *Chicago* 9 C2
Hōnanchō *Tokyo* 30 A2
Honcho *Tokyo* 30 A3
Honden *Tokyo* 30 A4
Hondo, Rio → *Los Angeles* 16 B4
Hong Kong *Hong Kong* 12 B1
Hong Kong, Univ. of *Hong Kong* 12 B1
Hong Kong I. *Hong Kong* 12 B2
Hongje *Seoul* 27 B1
Hongjimun Tunnel *Seoul* 27 B1
Hongkou *Shanghai* 28 B1
Hongqiao *Shanghai* 28 B1
Honjyo *Tokyo* 30 A3
Honoré Mercier, Pont *Montreal* 20 B1
Hönow *Berlin* 5 A4
Hook *London* 15 B2
Horikiri *Tokyo* 30 A4
Horn Pond *Boston* 6 A1
Hornchurch *London* 15 A5
Horni *Prague* 25 B3
Horni Počernice *Prague* 25 B3
Hornsey *London* 15 A3
Horoměřice *Prague* 25 B1
Hortaleza *Madrid* 17 B2
Hosoyama *Tokyo* 30 B2
Hostafranchs *Barcelona* 4 A1
Hostivař *Prague* 25 B3
Houbětin *Prague* 25 B3
Houghton *Johannesburg* 13 B2
Houilles *Paris* 24 A2
Hounslow □ *London* 15 B1
Hout Bay *Cape Town* 8 B1
Hove Å → *Copenhagen* 8 B2
Hovedøya *Oslo* 23 A3
Høvik *Oslo* 23 A2
Hovorčovice *Prague* 25 A3
Howard Beach *New York* 22 C2
Howrah = Haora *Kolkata* 14 B1
Howth *Dublin* 10 A3
Howth Hd. *Dublin* 10 A3
Hōya *Tokyo* 30 A2
Hradčany *Prague* 25 B2
Huanghuagang Mausoleum *Guangzhou* 11 B2
Huangpu *Shanghai* 28 B2
Huangpu Jiang → *Shanghai* 28 B1
Huangpu Park *Shanghai* 28 B1
Huangtugang *Beijing* 4 C1
Huascar *Lima* 14 A2
Huay Khwang *Bangkok* 3 B2
Huddinge *Stockholm* 29 B2
Huechuraba *Santiago* 27 B1
Huizingen *Brussels* 6 B1
Humayun's Tomb *Delhi* 10 B2
Humber → *Toronto* 31 A1
Humber B. *Toronto* 31 B2
Humber Bay *Toronto* 31 B2
Humber Bay Park *Toronto* 31 B2
Humber College *Toronto* 31 B1
Humber Summit *Toronto* 31 A1
Humber Valley Village *Toronto* 31 A1
Humberlea *Toronto* 31 A1
Humberwood Park *Toronto* 31 A1
Humboldt Park *Chicago* 9 B2
Humera *Madrid* 17 B1
Hunaydī *Baghdad* 3 B2
Hundige *Copenhagen* 8 B1
Hundige Strand *Copenhagen* 8 B2
Hung Hom *Hong Kong* 12 B2
Hunters Hill *Sydney* 29 B1
Hunters Pt. *San Francisco* 26 B2
Hunters Valley *Washington* 33 B2
Huntington *Washington* 33 C3
Huntington Park *Los Angeles* 16 C3
Hurīya *Baghdad* 3 A1
Hurstville *Sydney* 29 B1
Husan *Jerusalem* 13 B1
Husby *Stockholm* 29 A1
Husum *Copenhagen* 8 A2
Hutt → *Wellington* 33 B2
Hütteldorf *Vienna* 32 A1
Hüvösvölgy *Budapest* 7 A2
Huwon Secret Garden *Seoul* 27 B1
Hvalstad *Oslo* 23 A1
Hvalstrand *Oslo* 23 A2
Hvidovre *Copenhagen* 8 B2
Hwagok *Seoul* 27 B1
Hyattsville *Washington* 33 B4
Hyde Park *Boston* 6 B2
Hyde Park *Chicago* 9 C3
Hyde Park *Johannesburg* 13 A2
Hyde Park *London* 15 A2
Hyde Park *Los Angeles* 16 C3

I

Ibese *Lagos* 14 A2
Ibirapuera *São Paulo* 27 B1
Ibirapuera, Parque do *São Paulo* 27 B2
Icarai *Rio de Janeiro* 25 B2
Içerenköy *Istanbul* 12 C2
Ichgao *Tokyo* 30 B2
Ichigaya *Tokyo* 30 A3
Ichikawa *Tokyo* 30 A4
Ickenham *London* 15 A1
Iddo *Lagos* 14 B2
Idi-Oro *Lagos* 14 A2
Iganmu *Lagos* 14 B2
Igbobi *Lagos* 14 A2
Igbologun *Lagos* 14 B1
Igny *Paris* 24 B2
IJ, Het → *Amsterdam* 2 A2
IJ-meer *Amsterdam* 2 A3
Ijesa-Tedo *Lagos* 14 B1
Ijora *Lagos* 14 B2
Ikebe *Tokyo* 30 B2
Ikebukuro *Tokyo* 30 A3
Ikegami *Tokyo* 30 B3
Ikeja *Lagos* 14 A2
Ikeuchi *Osaka* 23 B2
Ikoyi *Lagos* 14 B2
Ikuata *Lagos* 14 B2
Ikuno *Osaka* 23 A2
Ikuta *Tokyo* 30 B2
Ila *Oslo* 23 A2
Ilford *London* 15 A4
Ilioupoli *Athens* 2 B2
Illinois at Chicago, Univ. of *Chicago* 9 B3
Illinois Institute of Technology *Chicago* 9 B3
Ilpendam *Amsterdam* 2 A2
Ilsos → *Athens* 2 B2
Imbâbah *Cairo* 7 A2
Imielin *Warsaw* 32 C2
Imirim *São Paulo* 27 A2
Imitos *Athens* 2 B2
Imitos, Oros *Athens* 2 B2
Imperial Palace Museum *Beijing* 4 B2
Inagi *Tokyo* 30 B2
Inchcolm *Edinburgh* 11 A2
Inchicore *Dublin* 10 A1
Inchkeith *Edinburgh* 11 A3
Inchmickery *Edinburgh* 11 A2
Incirano *Milan* 19 A1
Independencia *Lima* 14 A2
Independencia *Santiago* 27 B2
India Gate *Delhi* 10 B2
Indian Creek Village *Miami* 18 A2
Indianópolis *São Paulo* 27 B2
Indira Gandhi Int. ✈ (DEL) *Delhi* 10 B1
Industria *Johannesburg* 13 B1
Industrial Canal *New Orleans* 21 A2
Ingierstrand *Oslo* 23 B3
Inglewood *Los Angeles* 16 C3
Ingliston *Edinburgh* 11 B1
Inhaúma *Rio de Janeiro* 25 B1
Inner Port Shelter *Hong Kong* 12 A2
Interlagos *São Paulo* 27 C1
Intramuros *Manila* 17 B1
Invalides *Paris* 24 A2
Inverkeithing *Edinburgh* 11 A1
Inzersdorf *Vienna* 32 B2
Ipanema *Rio de Janeiro* 25 B1
Ipiranga *São Paulo* 27 B2
Ipiranga → *São Paulo* 27 B2
Iponri *Lagos* 14 B2
Ireland's Eye *Dublin* 10 A3
Irving Park *Chicago* 9 B2
Isabel *Rio de Janeiro* 25 B1
Isagatedo *Lagos* 14 A1
Isar → *Munich* 21 A3
Ishbīlīya *Baghdad* 3 A2
Ishøj Strand *Copenhagen* 8 B2
Island Bay *Wellington* 33 B1
Island Park *Toronto* 31 B2
Islev *Copenhagen* 8 A2
Isleworth *London* 15 B2
Islington *Toronto* 31 B1
Islington □ *London* 15 A3
Ismaning *Munich* 21 A3
Ismayloskiy Park *Moscow* 19 B3
Isolo *Lagos* 14 A1
Issy-les-Moulineaux *Paris* 24 B2
İstanbul *Istanbul* 12 C1
İstanbul Boğazı *Istanbul* 12 B2
İstinye *Istanbul* 12 B2
Itä Hakkila *Helsinki* 11 B3
Itaewon *Seoul* 27 B1
Itami *Osaka* 23 A1
Itanhanga *Rio de Janeiro* 25 B1
Ivanhoe *Melbourne* 18 A2
Ivry-sur-Seine *Paris* 24 B3
Ixelles *Brussels* 6 B2
Izmaylovo *Moscow* 19 B3
Iztacalco *Mexico City* 18 B2
Iztapalapa *Mexico City* 18 B2

J

Jaba *Jerusalem* 13 A2
Jacaré *Rio de Janeiro* 25 B1
Jackson Heights *New York* 22 B2
Jackson Park *Chicago* 9 C3
Jacques-Cartier *Montreal* 20 A3
Jacques-Cartier, Pont *Montreal* 20 A2
Jadavpur *Kolkata* 14 C2
Jade Buddha Temple *Shanghai* 28 B1
Jægersborg *Copenhagen* 8 A3
Jægersborg Dyrehave *Copenhagen* 8 A3
Jagadishpur *Kolkata* 14 B1
Jagatpur *Delhi* 10 A2
Jaguaré, Rib. do → *São Paulo* 27 B1
Jahangirpur *Delhi* 10 A2
Jakarta *Jakarta* 12 A2
Jakarta, Teluk *Jakarta* 12 A1
Jakarta Halim Perdana Kusuma Int. ✈ (HLP) *Jakarta* 12 B2
Jalan Kayu *Singapore* 28 A3
Jamaica B. *New York* 22 C3
Jamaica Plain *Boston* 6 B2
Jamakpuri *Delhi* 10 B1
Jamshīdīyeh *Tehran* 31 A2
Jamsil *Seoul* 27 B2
Jamwon *Seoul* 27 B2
Janki *Warsaw* 32 C1
Jaraguá *São Paulo* 27 A1
Jaraguá, Pico de *São Paulo* 27 A1
Jardim Paulista *São Paulo* 27 B1
Järvatältet *Stockholm* 29 A1
Jaskhar *Mumbai* 20 B2
Jatinegara *Jakarta* 12 B2
Javādīyeh *Tehran* 31 B2
Jaworowa *Warsaw* 32 C1
Jean Lafitte Nat. Historical Park ⌂ *New Orleans* 21 B1
Jedlesee *Vienna* 32 A2
Jefferson *New Orleans* 21 B2
Jefferson Memorial *Washington* 33 B3
Jefferson Park *Chicago* 9 B2
Jefferson Park *Los Angeles* 16 B3
Jegi *Seoul* 27 B2
Jelambar *Jakarta* 12 A1
Jelonki *Warsaw* 32 B1
Jerónimos, Mosteiro dos *Lisbon* 16 A1
Jersey City *New York* 22 B1
Jerusalem *Jerusalem* 13 B2
Jessamine, L. *Orlando* 23 B2
Jésus, Î. *Montreal* 20 A1
Jesús Maria *Lima* 14 B2
Jette *Brussels* 6 A1
Jey *Tehran* 31 B2
Jianggao *Guangzhou* 11 A2
Jiangwan *Shanghai* 28 B1
Jib *Jerusalem* 13 A2
Jihād *Baghdad* 3 B1
Jingan *Shanghai* 28 B1
Jinočany *Prague* 25 B1
Jinonice *Prague* 25 B2
Jiulong = Kowloon *Hong Kong* 12 B1
Jiyūgaoka *Tokyo* 30 B3
Jizā'ir *Baghdad* 3 B2
Jizīra *Baghdad* 3 B2
Joglo *Jakarta* 12 B1
Johannesburg *Johannesburg* 13 B2
Johannesburg, Univ. of *Johannesburg* 13 B2
Johanneskirchen *Munich* 21 A2
Johannesstift *Berlin* 5 A2
Johannisthal *Berlin* 5 B4
John F. Kennedy Nat. Historic Site ⌂ *Boston* 6 A2
John McLaren Park *San Francisco* 26 B2
Johnsonville *Wellington* 33 B1
Joinville-le-Pont *Paris* 24 B3
Joli-Bois *Brussels* 6 B2
Jollas *Helsinki* 11 B3
Jongmyo Royal Shrine *Seoul* 27 B1
Jongno *Seoul* 27 B1
Jonstrup *Copenhagen* 8 A2
Joppa *Edinburgh* 11 B3
Jorge Chavez, Int. ✈ (LIM) *Lima* 14 B2
Jorge Newbery ✈ (AEP) *Buenos Aires* 7 B2
Jósefa Piłsudskiego Park *Warsaw* 32 B1
Jōtō *Osaka* 23 A2
Jouy-en-Josas *Paris* 24 B2
Juan Anchorena *Buenos Aires* 7 A2
Juan González Romero *Mexico City* 18 A2
Juárez Int. ✈ (MEX) *Mexico City* 18 B2
Judeira *Jerusalem* 13 A2
Juhdum *Jerusalem* 13 B2
Juhu *Mumbai* 20 A2
Jūjā *Tokyo* 30 A3
Jukskeirivier → *Johannesburg* 13 A2
Julianów *Warsaw* 32 B2
Jung *Seoul* 27 B1
Jungfernheide, Volkspark *Berlin* 5 A2
Jungfernsee *Berlin* 5 B1
Junghwa *Seoul* 27 B2
Jungnang *Seoul* 27 B2
Jungnangcheon → *Seoul* 27 B1
Juniper Green *Edinburgh* 11 B2
Junk B. *Hong Kong* 12 B2
Jurong *Singapore* 28 B2
Jurong, Selat *Singapore* 28 B2
Jurong Industrial Estate *Singapore* 28 B1
Jurujuba, Enseada de *Rio de Janeiro* 25 B2
Jūsō *Osaka* 23 A1
Justice *Chicago* 9 C2
Jwalahari *Delhi* 10 B1

K

Kaapstad = Cape Town *Cape Town* 8 A1
Kabaty *Warsaw* 32 C2
Kadıköy *Istanbul* 12 C2
Kadoma *Osaka* 23 A2
Kafr 'Aqab *Jerusalem* 13 A2
Kâğıthane *Istanbul* 12 B1
Kâğıthane → *Istanbul* 12 B1
Kagran *Vienna* 32 A2
Kahnawake *Montreal* 20 B1
Kaimes *Edinburgh* 11 B2
Kaiserebersdorf *Vienna* 32 B2
Kaivoksela *Helsinki* 11 B2
Kalamaki *Athens* 2 B2
Kalbadevi *Mumbai* 20 B1
Kalipur *Kolkata* 14 A1
Kalithea *Athens* 2 B2
Kalkaji *Delhi* 10 B2
Kalookan = Caloocan *Manila* 17 B1
Kalvebod Fælled *Copenhagen* 8 B3
Kalveboderne *Copenhagen* 8 B3
Kamarhati *Kolkata* 14 A2
Kamata *Tokyo* 30 B3
Kameari *Tokyo* 30 A4
Kameido *Tokyo* 30 A4
Kami-Itabashi *Tokyo* 30 A3
Kamikitazawa *Tokyo* 30 B2
Kamitsuruma *Tokyo* 30 B1
Kamoshida *Tokyo* 30 B2
Kampong Landang *Singapore* 28 A3
Kampong Tanjong Penjuru *Singapore* 28 B2
Kampung Bali *Jakarta* 12 B1
Kanamori *Tokyo* 30 B1
Kanda *Tokyo* 30 A3
Kandilli *Istanbul* 12 B2
Kankurgachi *Kolkata* 14 B2
Kanlıca *Istanbul* 12 B2
Kanonerskiy, Ostrov *St. Petersburg* 27 B1
Kansai Int. ✈ (KIX) *Osaka* 23 B2
Kanzaki → *Osaka* 23 A1
Kapali Carsi *Istanbul* 12 B1
Kapellerfeld *Vienna* 32 A2
Káposztásmegyer *Budapest* 7 A2
Kapotnya *Moscow* 19 C3
Käppala *Stockholm* 29 A3
Kapuk *Jakarta* 12 A1
Käpylä *Helsinki* 11 B2
Karachi *Karachi* 13 A2
Karachi Int. ✈ (KHI) *Karachi* 13 A2
Karkh *Baghdad* 3 A2
Karlin *Prague* 25 B2
Karlsfeld *Munich* 21 A1
Karlshorst *Berlin* 5 B4
Karlův most *Prague* 25 B2
Karol Bagh *Delhi* 10 B2
Karolinenhof *Berlin* 5 B4
Karori *Wellington* 33 B1
Karow *Berlin* 5 A3
Karrādah *Baghdad* 3 B2
Kärsön *Stockholm* 29 B1
Kasai *Tokyo* 30 B4
Kasipur *Kolkata* 14 B2
Kastrup *Copenhagen* 8 B3
Kastrup, København ✈ (CPH) *Copenhagen* 8 B3
Kasuga *Tokyo* 30 A2
Kasuge *Tokyo* 30 A3
Kasumigaseki *Tokyo* 30 B3
Katong *Singapore* 28 B3
Katsushika *Tokyo* 30 A4
Kau Pai Chau *Hong Kong* 12 B2
Kau Yi Chau *Hong Kong* 12 B1
Kaulsdorf *Berlin* 5 B4
Kauniainen *Helsinki* 11 B1
Kawasaki *Tokyo* 30 B3
Kawawa *Tokyo* 30 B2
Kawęczyn *Warsaw* 32 B2
Kayu Putih *Jakarta* 12 B2
Kbely *Prague* 25 B3
Kebayoran Baru *Jakarta* 12 B1
Kebayoran Lama *Jakarta* 12 B1
Kebon Jeruk *Jakarta* 12 B1
Kedar *Jerusalem* 13 B2
Kedoya *Jakarta* 12 B1
Keilor *Melbourne* 18 A1
Keilor North *Melbourne* 18 A1
Keimola *Helsinki* 11 A1
Kelapa Gading *Jakarta* 12 A2
Kelenföld *Budapest* 7 B2
Kelvin *Johannesburg* 13 A2
Kemang *Jakarta* 12 B1
Kemayoran *Jakarta* 12 B2
Kemerburgaz *Istanbul* 12 B1
Kempton Park Races *London* 15 B1
Kenilworth *Cape Town* 8 A1
Kennedy Town *Hong Kong* 12 B1
Kenner *New Orleans* 21 A1
Kensal Green *London* 15 A2
Kensington *Johannesburg* 13 B2
Kensington *London* 15 B2
Kensington *New York* 22 C2
Kensington *Sydney* 29 B2
Kensington Palace *London* 15 A2
Kent Village *Washington* 33 B4
Kentish Town *London* 15 A3
Kenton *London* 15 A2
Kenwood House *London* 15 A3
Kepa *Warsaw* 32 B2
Keppel Harbour *Singapore* 28 B2
Kesariani *Athens* 2 B2
Kettering *Washington* 33 B5
Kew *London* 15 B2
Kew *Melbourne* 18 A2
Kew Gardens *London* 15 B2
Kew Gardens *Toronto* 31 A3
Key Biscayne *Miami* 18 B2
Khalīj *Baghdad* 3 B2
Khandallah *Wellington* 33 B1
Khansā' *Baghdad* 3 A2
Kharavli *Mumbai* 20 B2
Khefren *Cairo* 7 B1
Khichripur *Delhi* 10 B2
Khidirpur *Kolkata* 14 B1
Khimki-Khovrino *Moscow* 19 A2
Khirbet Batin Abu Lihyah *Jerusalem* 13 A1
Khirbet el-Misbah *Jerusalem* 13 A1
Khirbet Jub e-Rum *Jerusalem* 13 B2
Khlong San *Bangkok* 3 B2
Khlong Toey *Bangkok* 3 B2
Khorel *Kolkata* 14 A1
Khorosovo *Moscow* 19 B1
Kiamari *Karachi* 13 B1
Kierling *Vienna* 32 A1
Kierlingbach → *Vienna* 32 A1
Kifisos → *Athens* 2 B2
Kikuna *Tokyo* 30 B2
Kilbarrack *Dublin* 10 A3
Kilbirnie *Wellington* 33 B1
Kilburn *London* 15 A2
Killakee *Dublin* 10 B2
Killester *Dublin* 10 A2
Killiney *Dublin* 10 B3
Killiney B. *Dublin* 10 B3
Kilmacud *Dublin* 10 B2
Kilmainham *Dublin* 10 A2
Kilmainham Gaol *Dublin* 10 A2
Kilmashogue Mt. *Dublin* 10 B2
Kilmore *Dublin* 10 A2
Kilnamanagh *Dublin* 10 B1
Kilo *Helsinki* 11 B1
Kilokri *Delhi* 10 B2
Kiltiernan *Dublin* 10 B2
Kimmage *Dublin* 10 B2
Kindi *Baghdad* 3 B2
Kinghorn *Edinburgh* 11 A2
Kings Domain *Melbourne* 18 A1
Kings Forest = Kongelunden *Copenhagen* 8 B3
Kings Park *Washington* 33 C2
Kings Park West *Washington* 33 C2
Kingsbury *London* 15 A2
Kingsbury *Melbourne* 18 A2
Kingsford *Sydney* 29 B2
Kingsford Smith, Sydney ✈ (SYD) *Sydney* 29 B2
Kingston-upon-Thames □ *London* 15 B2
Kingston Vale *London* 15 B2
Kingsway *Toronto* 31 B1
Kinsaley *Dublin* 10 A2
Kipling Heights *Toronto* 31 A1
Kipseli *Athens* 2 B2
Kirchstockbach *Munich* 21 B3
Kirchtrudering *Munich* 21 B3
Kirikiri *Lagos* 14 B1
Kirke Værløse *Copenhagen* 8 A1
Kirkhill *Edinburgh* 11 B2
Kirkliston *Edinburgh* 11 B1
Kirknewton *Edinburgh* 11 B1
Kirov Palace of Culture *St. Petersburg* 27 B1
Kiryat Anavim *Jerusalem* 13 B1
Kiryat Ha Yovel *Jerusalem* 13 B1
Kısıklı *Istanbul* 12 B2
Kispest *Budapest* 7 B2
Kista *Stockholm* 29 A1
Kita *Osaka* 23 A2
Kita *Tokyo* 30 A3
Kitazawa *Tokyo* 30 B3
Kiu Tsiu *Hong Kong* 12 A2
Kivistö *Helsinki* 11 B2
Kızıltoprak *Istanbul* 12 C2
Kizuri *Osaka* 23 B2
Kjelsås *Oslo* 23 A3
Kladow *Berlin* 5 B1
Klampenborg *Copenhagen* 8 A3
Klečany *Prague* 25 A2
Kledering *Vienna* 32 B2
Klein Jukskei → *Johannesburg* 13 A1
Kleinmachnow *Berlin* 5 B2
Kleinschönebeck *Berlin* 5 B5
Klemetsrud *Oslo* 23 A4
Klender *Jakarta* 12 A2
Kličany *Prague* 25 A2
Klipriviersberg Nature Reserve *Johannesburg* 13 B2
Klosterneuburg *Vienna* 32 A1
Köbánya *Budapest* 7 B2
Kobbegem *Brussels* 6 A1
København *Copenhagen* 8 A2
København ✈ (CPH) *Copenhagen* 8 B3
Kobylisy *Prague* 25 B2
Kobyłka *Warsaw* 32 A3
Kodaira *Tokyo* 30 A1
Kodanaka *Tokyo* 30 B2
Koekelberg *Brussels* 6 A1
Koganei *Tokyo* 30 A2
Kogarah *Sydney* 29 B1
Koivupää *Helsinki* 11 B2
Koja *Jakarta* 12 A2
Koja Utara *Jakarta* 12 A2
Kokhav Ya'akov *Jerusalem* 13 A2
Kokobunji *Tokyo* 30 A1
Kokobunji-Temple *Tokyo* 30 A4
Kolarängen *Stockholm* 29 B3
Kolbotn *Oslo* 23 B3
Kolkata *Kolkata* 14 B2
Kolkata Dum Dum Int. ✈ (CCU) *Kolkata* 14 B2
Kolkata Maidan *Kolkata* 14 B1
Koło *Warsaw* 32 B1
Kolokinthou *Athens* 2 B2
Kolomyagi *St. Petersburg* 27 A1
Kolonos *Athens* 2 B2
Kolsås *Oslo* 23 A2
Komae *Tokyo* 30 B2
Komagome *Tokyo* 30 A3
Komazawa *Tokyo* 30 B3
Kona *Kolkata* 14 B1
Konala *Helsinki* 11 B2
Kondli *Delhi* 10 B2
Kongelunden *Copenhagen* 8 B3
Kongens Lyngby *Copenhagen* 8 A3
Kongo *Helsinki* 11 A1
Konnagar *Kolkata* 14 A2
Konohana *Osaka* 23 A1
Konradshöhe *Berlin* 5 A2
Kopanina *Prague* 25 B1
Koparkhairna *Mumbai* 20 A2
Köpenick *Berlin* 5 B4
Korangi *Karachi* 13 B2
Koremasa *Tokyo* 30 B1
Koridalos *Athens* 2 B1
Korokoro *Wellington* 33 B2
Korokoro Stream → *Wellington* 33 B2
Kosino *Moscow* 19 B4
Kosugi *Tokyo* 30 B2
Kota *Jakarta* 12 A1
Kōtō *Tokyo* 30 A3
Kotrung *Kolkata* 14 A2
Kouponia *Athens* 2 B2
Kowloon *Hong Kong* 12 B1
Kowloon Peak *Hong Kong* 12 A2
Kowloon Res. *Hong Kong* 12 A1
Kowloon Tong *Hong Kong* 12 A2
Kraainem *Brussels* 6 A2
Krailling *Munich* 21 B1
Krampnitz *Berlin* 5 B1
Krampnitzsee *Berlin* 5 B1
Kranji, Sungei → *Singapore* 28 A2
Kranji Industrial Estate *Singapore* 28 A2
Krasno-Presnenskaya *Moscow* 19 B2
Krč *Prague* 25 B2
Kremlin *Moscow* 19 B2
Krestovskiye, Ostrov *St. Petersburg* 27 B1
Kreuzberg *Berlin* 5 A3
Kritzendorf *Vienna* 32 A1
Krumme Lanke *Berlin* 5 B2
Krummensee *Berlin* 5 A5
Krung Thep = Bangkok *Bangkok* 3 B2
Krusboda *Stockholm* 29 B3
Kuangchou = Guangzhou *Guangzhou* 11 B2
Küçükköy *Istanbul* 12 B1
Küçüksu *Istanbul* 12 B2
Kudrovo *St. Petersburg* 27 B3
Kulosaari *Helsinki* 11 B3
Kumla *Stockholm* 29 B3
Kungens kurva *Stockholm* 29 B1
Kungliga Slottet *Stockholm* 29 B2
Kungshatt *Stockholm* 29 B1
Kungsholmen *Stockholm* 29 A2
Kuningan *Jakarta* 12 B1
Kunitachi *Tokyo* 30 A1
Kunming Hu *Beijing* 4 B1
Kunratice *Prague* 25 B2
Kupchino *St. Petersburg* 27 B2
Kuramo Waters *Lagos* 14 B2
Kurbağalı → *Istanbul* 12 C2
Kurihara *Tokyo* 30 A2
Kurla *Mumbai* 20 A2
Kurmuri *Mumbai* 20 A2
Kurume *Tokyo* 30 A2
Kuryanovo *Moscow* 19 C3
Kuskovo *Moscow* 19 B3
Kustia *Kolkata* 14 B2
Kuzguncuk *Istanbul* 12 B2
Kuzyminki *Moscow* 19 B3
Kwai Chung *Hong Kong* 12 A1
Kwangchow = Guangzhou *Guangzhou* 11 B2
Kwun Tong *Hong Kong* 12 B2
Kyje *Prague* 25 B3
Kyūhōji *Osaka* 23 B2

L

La Barceloneta *Barcelona* 4 A2
La Blanca *Santiago* 27 C2
La Boca *Buenos Aires* 7 B2
La Bretèche *Paris* 24 A1
La Campiña *Lima* 14 C2
La Celle-St-Cloud *Paris* 24 A1
La Cisterna *Santiago* 27 C2
La Courneuve *Paris* 24 A3
La Dehesa *Santiago* 27 B2
La Encantada *Lima* 14 C2
La Estación *Madrid* 17 B1
La Floresta *Barcelona* 4 A1
La Florida *Santiago* 27 C2
La Fortuna *Madrid* 17 B1
La França *Barcelona* 4 A1
La Garenne-Colombes *Paris* 24 A
La Giustiniana *Rome* 26 B1
La Grange *Chicago* 9 C1
La Grange Park *Chicago* 9 C1
La Granja *Santiago* 27 C2
La Guardia, New York ✈ (LGA) *New York* 22 B2
La Hulpe *Brussels* 6 B2
La Llacuna *Barcelona* 4 A2
La Loma *Mexico City* 18 A1
La Lucila *Buenos Aires* 7 B2
La Maladrerie *Paris* 24 A1
La Milla, Cerro *Lima* 14 B2
La Monachina *Rome* 26 B1
La Moraleja *Madrid* 17 A2
La Nopalera *Mexico City* 18 C2
La Paternal *Buenos Aires* 7 B2
La Perla *Lima* 14 B2
La Perouse *Sydney* 29 B2
La Pisana *Rome* 26 B1
La Prairie *Montreal* 20 B3
La Punta *Lima* 14 B1
La Puntigala *Barcelona* 4 A2
La Queue-en-Brie *Paris* 24 B4
La Reina *Santiago* 27 B2
La Ribera *Barcelona* 4 A1
La Sagrera *Barcelona* 4 A2
La Salada *Buenos Aires* 7 C2
La Scala *Milan* 19 B2
La Storta *Rome* 26 A1
La Taxonera *Barcelona* 4 A2
La Victoria *Lima* 14 B2
Laajalahti *Helsinki* 11 B1
Laajasalo *Helsinki* 11 B3
Laaksolahti *Helsinki* 11 B1
Labagou National Forest Park *Beijing* 4 A2
Lablâba, W. el → *Cairo* 7 A2
Lachine *Montreal* 20 B1
Lachine, Canal de *Montreal* 20 B2
Lad Phrao *Bangkok* 3 B2
Ladera Heights *Los Angeles* 16 C2
Ládví *Prague* 25 B2
Łady *Warsaw* 32 C1
Lafontaine, Parc *Montreal* 20 A2
Lagoa *Rio de Janeiro* 25 B1
Lagos *Lagos* 14 B2
Lagos Harbour *Lagos* 14 B2
Lagos-Ikeja ✈ (LOS) *Lagos* 14 A1
Lagos Island *Lagos* 14 B2
Lagos Lagoon *Lagos* 14 B2
Laguna de B. *Manila* 17 C2
Laim *Munich* 21 B2
Lainate *Milan* 19 A1
Lainz *Vienna* 32 B1
Lake Buena Vista *Orlando* 23 B1
Lake Cain Hills *Orlando* 23 B2
Lake Fairfax Park *Washington* 33 B2
Lakemba *Sydney* 29 B1
Lakeside *Cape Town* 8 B1
Lakeside *Johannesburg* 13 A2
Lakeview *Chicago* 9 B3
Lakeview *New Orleans* 21 A2
Lakewood Park *Atlanta* 3 B2
Lakhtinskiy *St. Petersburg* 27 B1
Lakhtinskiy Razliv, Oz. *St. Petersburg* 27 B1
Lakshmanpur *Kolkata* 14 B1
Laksi *Bangkok* 3 A2
Lal Qila *Delhi* 10 B2
Lam Tin *Hong Kong* 12 B2
Lambert *Oslo* 23 A3
Lambeth □ *London* 15 B3
Lambrate *Milan* 19 B2
Lambro, Parco *Milan* 19 B2
Lambton Mills *Toronto* 31 B1
Lamma I. *Hong Kong* 12 B1
Landianchang *Beijing* 4 B1
Landover Hills *Washington* 33 B4
Landsmeer *Amsterdam* 2 A2
Landstrasse *Vienna* 32 A2
Landwehr kanal *Berlin* 5 B3
Lane Cove *Sydney* 29 A1
Lane Cove Nat. Park △ *Sydney* 29 A1
Langa *Cape Town* 8 A2
Langenzersdorf *Vienna* 32 A2
Langer See *Berlin* 5 B4
Langley *Washington* 33 B2
Langley Park *Washington* 33 B4
Langwald *Munich* 21 A1
Lanham *Washington* 33 B4
Lankwitz *Berlin* 5 B3
L'Annunziatella *Rome* 26 C2
Lansdowne *Cape Town* 8 A2
Lansing *Toronto* 31 A2
Lanus *Buenos Aires* 7 C2
Lapa *Rio de Janeiro* 25 B1
Laranjeiras *Rio de Janeiro* 25 B1
Las *Warsaw* 32 B2
Las Corts *Barcelona* 4 A1
Las Kabacki *Warsaw* 32 C2
Las Pinas *Manila* 17 C1
Las Rejas *Santiago* 27 B1
Las Vegas *Las Vegas* 14 A2
Las Vegas McCarran Int. ✈ (LAS) *Las Vegas* 14 B2
LaSalle *Montreal* 20 B2
LaSalle Street Station *Chicago* 9 B3
Lasek Bielański *Warsaw* 32 B1
Lasek Na Kole *Warsaw* 32 B1
Latina *Madrid* 17 B1
Lauttasaari *Helsinki* 11 C2
Laval-des-Rapides *Montreal* 20 A
Lavīzān *Tehran* 31 A2
Lavradio *Lisbon* 16 A2
Lawndale *Chicago* 9 B2
Lawne L. *Orlando* 23 A2
Lawrence Heights *Toronto* 31 A2
Layari *Karachi* 13 A2
Layari → *Karachi* 13 A1
Łazienkowski Park *Warsaw* 32
Le Blanc-Mesnil *Paris* 24 A3
Le Bourget *Paris* 24 A3
Le Chenoi *Brussels* 6 B2
Le Chesnay *Paris* 24 B1
Le Christ de Saclay *Paris* 24 B2
Le Kremlin-Bicêtre *Paris* 24 B3
Le Mesnil-le-Roi *Paris* 24 A1
Le Pecq *Paris* 24 A1
Le Perreux *Paris* 24 A3
Le Pin *Paris* 24 A4
Le Plessis-Robinson *Paris* 24 B2
Le Plessis-Trévise *Paris* 24 B4
Le Port-Marly *Paris* 24 A1
Le Pré-St-Gervais *Paris* 24 A3
Le Raincy *Paris* 24 A4
Le Vésinet *Paris* 24 A1
Lea Bridge *London* 15 A3
Leaside *Toronto* 31 A2
Leblon *Rio de Janeiro* 25 B1
Lee *London* 15 B4
Leganés *Madrid* 17 C1
Legazpi *Madrid* 17 B1

Lehtisaari Helsinki 11 B2
Lei Yue Mun Hong Kong 12 B2
Leião Lisbon 16 A1
Leichhardt Sydney 29 B1
Leith Edinburgh 11 B3
Lekki Lagos 14 B2
Leme Rio de Janeiro 25 B1
Lemoyne Montreal 20 B3
Lenin Moscow 19 B2
Lenino Moscow 19 C2
Leninskiye Gory Moscow 19 B2
Lennox Los Angeles 16 C2
Leonia New York 22 A2
Leopardstown Dublin 10 B2
Leopoldau Vienna 32 A2
Leopoldsdorf Vienna 32 B2
Leopoldstadt Vienna 32 A2
Leportovo Moscow 19 B3
Leppävaara Helsinki 11 B1
Les Lilas Paris 24 A3
Les Loges-en-Josas Paris 24 B1
Les Pavillons-sous-Bois Paris 24 A4
Lésigny Paris 24 B4
Lesnozavodskaya St. Petersburg 27 B2
Lester B. Pearson Int., Toronto ✕ (YYZ) Toronto 31 A1
L'Étang-la-Ville Paris 24 A1
Letňany Prague 25 B3
Letopolis = Ausim Cairo 7 A1
Levallois-Perret Paris 24 A2
Levent Istanbul 12 B2
Lewisdale Washington 33 B4
Lewisham ☐ London 15 B3
Lexington Boston 6 A1
Leyton London 15 A3
Leytonstone London 15 A4
L'Haÿ-les-Roses Paris 24 B3
L'Hospitalet de Llobregat Barcelona 4 A1
Lhotka Prague 25 B2
Liangshui He ➔ Beijing 4 C2
Lianhua Chi Beijing 4 B1
Lianhua He ➔ Beijing 4 B1
Libčice nad Vltavou Prague 25 A2
Libeň Prague 25 B2
Liberdade São Paulo 27 B2
Liberton Edinburgh 11 B3
Liberty I. New York 22 B1
Liberty Osaka Museum Osaka 23 B1
Liberty State Park ⌂ New York 22 B1
Libeznice Prague 25 A2
Libuš Prague 25 B2
Lichiao Guangzhou 11 B2
Lichtenburg Berlin 5 A4
Lichterfelde Berlin 5 B2
Lidingö Stockholm 29 A2
Lieshi Lingyuan Guangzhou 11 B2
Liesing Vienna 32 B1
Liesing ➔ Vienna 32 B2
Ligovo St. Petersburg 27 C1
Lijordet Oslo 23 A2
Likhoborka ➔ Moscow 19 A2
Lilla Värtan Stockholm 29 A3
Lille Værløse Copenhagen 8 A2
Liluah Kolkata 14 B1
Lim Chu Kang Singapore 28 A2
Lima Lima 14 B2
Lima Jorge Chavez, Int. ✕ (LIM) Lima 14 B2
Limbiate Milan 19 A1
Limehouse London 15 A3
Limeil-Brévannes Paris 24 B3
Linate, Milano ✕ (LIN) Milan 19 B2
Linbropark Johannesburg 13 A2
Lincoln Center for Performing Arts New York 22 B2
Lincoln Heights Los Angeles 16 B3
Lincoln Memorial Washington 33 B3
Lincoln Park Chicago 9 B3
Lincoln Park New York 22 B1
Lincoln Park San Francisco 26 B1
Lincoln Park Zoo Chicago 9 B3
Lincolnwood Chicago 9 A2
Linda-a-Pastora Lisbon 16 A1
Linden Johannesburg 13 A2
Linden Wellington 33 A1
Lindenberg Berlin 5 A4
Lindøya Oslo 23 A3
Liniers Buenos Aires 7 B1
Linkebeek Brussels 6 B2
Linksfield Johannesburg 13 A2
Linmeyer Johannesburg 13 B2
Linna Helsinki 11 A2
Lintuvaara Helsinki 11 B1
Lion Rock Country Park ⌂ Hong Kong 12 A2
Lion's Head Cape Town 8 A1
Liopetri Athens 2 B2
Lisboa Lisbon 16 A2
Lisboa ✕ (LIS) Lisbon 16 A2
Lisbon = Lisboa Lisbon 16 A2
Lishui Guangzhou 11 A1
Little B. Sydney 29 B2
Little Calumet ➔ Chicago 9 D3
Little Ferry New York 22 A1
Little Lake Conway Orlando 23 B2
Little Red School House Nature Center Chicago 9 C1
Little Rouge ➔ Toronto 31 A4
Liuhang Shanghai 28 A1
Liurong Temple Guangzhou 11 B2
Live Oak Manor New Orleans 21 B1
Livry-Gargan Paris 24 A4
Liwan Guangzhou 11 B2
Ljan Oslo 23 A3
Llano de Can Gineu Barcelona 4 A2
Lo Aranguiz Santiago 27 B2
Lo Boza Santiago 27 B1
Lo Chau Hong Kong 12 B2
Lo Chau Pak Mai Hong Kong 12 B2
Lo Espejo Santiago 27 C1
Lo Hermida Santiago 27 B2
Lo Prado Santiago 27 B1
Lo So Shing Hong Kong 12 B1
Lo Wai Hong Kong 12 A1
Lochanhead Edinburgh 11 B3
Lobau Vienna 32 A3
Lobos, Pt. San Francisco 26 B1
Locham Munich 21 B1
Lochkov Prague 25 B2
Lochhausen Munich 21 A1
Lodi New York 22 A1
Lodi Estate Delhi 10 B2
Logan Int., Boston ✕ (BOS) Boston 6 A2
Logan Square Chicago 9 B2
Lognes-Émerainville ✕ (XLG) Paris 24 B4
Lohme Berlin 5 A5
Lokhet Karachi 13 A2
Lomas Chapultepec Mexico City 18 B1
Lomas de San Angel Inn Mexico City 18 B1
Lombardy East Johannesburg 13 A2
Lomianki Warsaw 32 A1
Lomus Reforma Mexico City 18 B1
London London 15 B3
London Avenue Canal New Orleans 21 A2
London City ✕ (LCY) London 15 A4
London Heathrow ✕ (LHR) London 15 B1
London Zoo London 15 A3
Long B. Sydney 29 B2
Long Branch Toronto 31 B1
Long Brook ➔ Washington 33 C2
Long Ditton London 15 B2
Long Island City New York 22 B2
Longchamp, Hippodrome de Paris 24 A2
Longhua Pagoda Shanghai 28 B1
Longhua Park Shanghai 28 B1
Longjohn Slough Chicago 9 C1
Longtan Hu ➔ Beijing 4 B2
Longue-Pointe Montreal 20 A2
Longueuil Montreal 20 A3
Loni Delhi 10 A2
Loop, The Chicago 9 B3
Lord's Cricket Ground London 15 A3
Loreto Milan 19 B2
Los Angeles Los Angeles 16 B3
Los Angeles Int. ✕ (LAX) Los Angeles 16 C2
Los Cerrillos ✕ (ULC) Santiago 27 B1
Los Nietos Los Angeles 16 C4
Los Olivos Lima 14 A2
Los Remedios ⌂ Mexico City 18 B1
Los Reyes Mexico City 18 B2
Losiny Ostrov ⌂ Moscow 19 A3
Lot Brussels 6 B1
Lotus River Cape Town 8 B2
Lotus Temple Delhi 10 B2
Loughlinstown Dublin 10 B3
Louis Armstrong New Orleans Int. ✕ (MSY) New Orleans 21 A1
Louise, L. Orlando 23 B1
Louisiana Superdome New Orleans 21 B2
Loures Lisbon 16 A1
Louveciennes Paris 24 A1
Louvre, Musée du Paris 24 A3
Lower B. New York 22 C1
Lower Hutt Wellington 33 B2
Lower New York B. = Lower B. New York 22 C1
Lower Shing Mun Res. Hong Kong 12 A1
Lowry Bay Wellington 33 B2
Loyola Univ. Chicago 9 B3
Lu Xun, Tomb of Shanghai 28 B1
Lu Xun Park Shanghai 28 B1
Lübars Berlin 5 A3
Lucy, L. Orlando 23 A2
Ludwigsfeld Munich 21 A1
Luhu Guangzhou 11 B2
Lumiar Lisbon 16 A2
Lundtofte Copenhagen 8 A3
Lung Mei Hong Kong 12 A2
Luojiang Guangzhou 11 B2
Lustheim Munich 21 A2
Luwan Shanghai 28 B1
Luzhniki Sports Centre Moscow 19 B2
Lyndhurst New York 22 B1
Lynn Woods Res. Boston 6 A2
Lyon, Gare de Paris 24 A3
Lyons Chicago 9 C2
Lysaker Oslo 23 A2
Lysakerelva ➔ Oslo 23 A2
Lysolaje Prague 25 B2
Lyublino Moscow 19 B3

M

Ma Nam Wat Hong Kong 12 A2
Ma On Shan Country Park ⌂ Hong Kong 12 A2
Ma'ale Adumim Jerusalem 13 B2
Ma'ale Ha Khamisha Jerusalem 13 B1
Ma'ale Mikhmas Jerusalem 13 A2
Maantiekylä Helsinki 11 A3
Maarifa Baghdad 3 B2
Mabato Pt. Manila 17 C2
Mabel, L. Orlando 23 B1
Macaco, Morro do Rio de Janeiro 25 B2
McCarran Int., Las Vegas ✕ (LAS) Las Vegas 14 B2
McCook Chicago 9 C2
McGill Univ. Montreal 20 A2
Machelen Brussels 6 A2
Machida Tokyo 30 B1
Maciołki Warsaw 32 B2
McKerrow, Mt. Wellington 33 B2
McKinley Park Chicago 9 C2
McLean Washington 33 B2
Macopocho ➔ Santiago 27 B1
MacRitchie Res. Singapore 28 A2
Macul Santiago 27 C2
Madhudaha Kolkata 14 B2
Madhyamgram Kolkata 14 A2
Madin Mexico City 18 A1
Madin, L. Mexico City 18 A1
Madinah Al Mansûr Baghdad 3 B2
Mâdinet Nasr Cairo 7 A2
Madrid Madrid 17 B1
Madrid Barajas ✕ (MAD) Madrid 17 B2
Madrona Barcelona 4 A1
Maesawa Tokyo 30 A2
Magdalena Lima 14 B2
Magdalena Contreras Mexico City 18 C1
Maghreb Baghdad 3 A2
Maginu Tokyo 30 B2
Magliana Rome 26 B1
Magny-les-Hameaux Paris 24 B1
Maheshtala Kolkata 14 C1
Mahim Mumbai 20 A2
Mahim B. Mumbai 20 A1
Mahlsdorf Berlin 5 A4
Mahmoodabad Karachi 13 A2
Mahrauli Delhi 10 B2
Mahul Mumbai 20 A2
Maida Vale London 15 A2
Maidstone Melbourne 18 A1
Maipú Santiago 27 C1
Maisonneuve, Parc Montreal 20 A2
Maisons-Alfort Paris 24 B3
Maisons-Laffitte Paris 24 A1
Maissoneuve Montreal 20 A2
Maitland Cape Town 8 A1
Makasar Jakarta 12 A2
Makati Manila 17 B2
Mäkiniitty Helsinki 11 A2
Malá Strana Prague 25 B2
Malabar Sydney 29 B2
Malabar Hill Mumbai 20 B1
Malabar Pt. Mumbai 20 B1
Malabon Manila 17 B1
Malacañang Palace Manila 17 B1
Malahide Dublin 10 A3
Malakoff Paris 24 B2
Mälarhöjaen Stockholm 29 B1
Malate Manila 17 B1
Malaya Neva St. Petersburg 27 B1
Malaya Okhta St. Petersburg 27 B2
Malchow Berlin 5 A3
Malden Boston 6 A2
Malden London 15 B2
Maleizen Brussels 6 B3
Malešice Prague 25 B3
Malir ➔ Karachi 13 B2
Malleny Mills Edinburgh 11 B2
Malmi Helsinki 11 B2
Malmoya Oslo 23 A3
Måløv Copenhagen 8 A1
Malton Toronto 31 A1
Malvern Johannesburg 13 B2
Malvern Melbourne 18 B2
Malvern Toronto 31 A3
Mampang Prapatan Jakarta 12 B1
Mampukuji Tokyo 30 B2
Man Budrukh Mumbai 20 A2
Man Khurd Mumbai 20 A2
Manakhat Jerusalem 13 B2
Mandaluyong Manila 17 B2
Mandaoli Delhi 10 B2
Mandaqui ➔ São Paulo 27 A2
Mandoli Delhi 10 A2
Mandvi Mumbai 20 B2
Manenberg Cape Town 8 A2
Mang Kung Uk Hong Kong 12 B2
Mang-won Seoul 27 B1
Mangolpuri Delhi 10 A1
Manguinhos ✕ Rio de Janeiro 25 B1
Manhattan New York 22 B2
Manhattan Beach New York 22 C2
Manila Manila 17 B1
Manila Ninoy Aquino Int. ✕ (MNL) Manila 17 B2
Mankkaa Helsinki 11 B1
Manly Sydney 29 A2
Mann, L. Orlando 23 A2
Mann's Chinese Theatre Los Angeles 16 B2
Mannswörth Vienna 32 B3
Manor Park London 15 A4
Manor Park Wellington 33 A2
Manora Karachi 13 B1
Manora Pt. Karachi 13 B1
Manquehue, Cerro Santiago 27 B2
Manzanares, Canal de Madrid 17 C2
Mapo Seoul 27 B1
Mar Vista Los Angeles 16 B2
Maracanã Rio de Janeiro 25 B1
Maraoli Mumbai 20 A2
Marcelin Warsaw 32 B1
Mareil-Marly Paris 24 A1
Margareten Vienna 32 A2
Maria Vienna 32 A2
Maridalen Oslo 23 A3
Maridalsvatnet Oslo 23 A3
Mariendorf Berlin 5 B3
Marienfelde Berlin 5 B3
Marikina Manila 17 B2
Marikina ➔ Manila 17 B2
Marin City San Francisco 26 A1
Marin Pen. San Francisco 26 A1
Marina del Rey Los Angeles 16 C2
Marino Dublin 10 A2
Marisco, Ponta do Rio de Janeiro 25 C1
Markham Toronto 31 A2
Marki Warsaw 32 B2
Markland Wood Toronto 31 B1
Marly, Forêt de Paris 24 A1
Marly-le-Roi Paris 24 A1
Marne-la-Vallée Paris 24 A4
Marolles-en-Brie Paris 24 B4
Maroubra Sydney 29 B2
Marquette Park Chicago 9 C2
Marrero New Orleans 21 B2
Marrickville Sydney 29 B1
Marsfield Sydney 29 A1
Marsha, L. Orlando 23 B2
Marte, Campo de São Paulo 27 B2
Martesana, Naviglio della Milan 19 A2
Martin Luther King Nat. Historic Site ⌂ Atlanta 3 B2
Martinez Buenos Aires 7 A1
Martinkylä Helsinki 11 B2
Martinsried Munich 21 B1
Maruko Tokyo 30 B3
Maryino Moscow 19 B3
Maryland Singapore 28 B2
Marymont Warsaw 32 B1
Marysin Wawerski Warsaw 32 B2
Marzahn Berlin 5 A4
Mascot Sydney 29 B2
Masmo Stockholm 29 B1
Maspeth New York 22 B2
Masr el Gedida Cairo 7 A2
Masr el Qadima Cairo 7 A2
Massachusetts Inst. of Tech. Boston 6 A2
Massamá Lisbon 16 A1
Massey ➔ Toronto 31 A3
Massy Paris 24 B2
Mata Jerusalem 13 B1
Matinha Lisbon 16 A2
Matramam Jakarta 12 B2
Matsubara Osaka 23 B2
Mattapan Boston 6 B2
Mátyásföld Budapest 7 A3
Mau Tso Ngam Hong Kong 12 A2
Mauer Vienna 32 B1
Mauripur Karachi 13 A1
Maxhof Munich 21 B1
Mayfair Johannesburg 13 B2
Maywood Chicago 9 B1
Maywood Los Angeles 16 C3
Maywood New York 22 A1
Maywood Park Race Track Chicago 9 B1
Mazagaon Mumbai 20 B2
Meadowbank Park Sydney 29 A1
Měcholupy Prague 25 B3
Měčice Prague 25 A3
Mecidiyeköy Istanbul 12 B2
Medford Boston 6 A2
Mediodia Madrid 17 B2
Meguro Tokyo 30 B3
Meguro ➔ Tokyo 30 B3
Mehpalpur Delhi 10 B1
Mehrābād ✕ (THR) Tehran 31 A1
Mehram Nagar Delhi 10 B1
Mehrow Berlin 5 A4
Meidling Vienna 32 A2
Méier Rio de Janeiro 25 B1
Meiji Shrine Tokyo 30 A3
Meise Brussels 6 A1
Mejiro Tokyo 30 A3
Melbourne Melbourne 18 A1
Melbourne ✕ (MEL) Melbourne 18 A1
Melkki Helsinki 11 C2
Mellunkylä Helsinki 11 B3
Mellunmäki Helsinki 11 B3
Melrose Boston 6 A2
Melrose New York 22 B2
Melrose Park Chicago 9 B1
Melsbroek Brussels 6 A2
Melville Johannesburg 13 B2
Menteng Jakarta 12 B1
Mérantaise ➔ Paris 24 B1
Mercamadrid Madrid 17 B2
Merced, L. San Francisco 26 B2
Meredale Johannesburg 13 B1
Merlimau, Pulau Singapore 28 B2
Merri Cr. ➔ Melbourne 18 A1
Merrion Dublin 10 B2
Merrionette Park Chicago 9 C2
Merton ☐ London 15 B2
Mesgarābād Tehran 31 B3
Messe Vienna 32 A2
Metairie New Orleans 21 A2
Metanòpoli Milan 19 B2
Metro-Dade Cultural Centre Miami 18 B2
Metro Toronto Zoo Toronto 31 A3
Metropolitan Museum of Art New York 22 B2
Meudon Paris 24 B2
Mevaseret Tsiyon Jerusalem 13 B1
Mevo Beitar Jerusalem 13 B1
México Mexico City 18 B2
México, Ciudad de Mexico City 18 B2
Mexico City Int. ✕ (MEX) Mexico City 18 B2
Meyersdal Johannesburg 13 B2
Mezzate Milan 19 B2
Miadong Seoul 27 B2
Miami Miami 18 B2
Miami Beach Miami 18 B2
Miami Canal Miami 18 A1
Miami Int. ✕ (MIA) Miami 18 B1
Miami Shores Miami 18 A2
Miami Springs Miami 18 B1
Miasto Warsaw 32 B2
Michałowice Warsaw 32 B1
Michle Prague 25 B2
Mid-City Los Angeles 16 B2
Mid City New Orleans 21 B2
Middle Harbour Sydney 29 A2
Middle Hd. Sydney 29 A2
Middle Park Melbourne 18 B1
Middle Village New York 22 B2
Middlesex Fells Reservation Boston 6 A2
Midland Beach New York 22 C1
Midwood New York 22 C2
Miedzeszyn Warsaw 32 B3
Międzylesie Warsaw 32 B3
Miessaari Helsinki 11 C1
Miguel Hidalgo Mexico City 18 B1
Milan = Milano Milan 19 B1
Milanese, Parco Regionale ⌂ Milan 19 A1
Milano Milan 19 B1
Milano Due Milan 19 B2
Milano Linate ✕ (LIN) Milan 19 B2
Milano San Felice Milan 19 B2
Milbertshofen Munich 21 A2
Mill Hill London 15 A2
Millennium Dome London 15 A4
Miller Meadow Chicago 9 B2
Millerhill Edinburgh 11 B3
Milltown Dublin 10 B2
Millwood Washington 33 B4
Milnerton Cape Town 8 A1
Milon-la-Chapelle Paris 24 B1
Milton Boston 6 B2
Milton Bridge Edinburgh 11 B2
Mimico Toronto 31 B2
Mimico Creek ➔ Toronto 31 B1
Minami Osaka 23 A2
Minamitsunashima Tokyo 30 B2
Minato Osaka 23 B1
Minato Tokyo 30 B3
Minshât el Bekkari Cairo 7 A1
Miraflores Lima 14 B2
Miramar Wellington 33 B1
Misericordia, Sa. da Rio de Janeiro 25 B1
Mission San Francisco 26 B2
Mississauga Toronto 31 B1
Mitaka Tokyo 30 A2
Mitcham London 15 B2
Mitcham Common London 15 B3
Mitchell Museum of the American Indian Chicago 9 A2
Mitchell's Plain Cape Town 8 B2
Mitte Berlin 5 A3
Mittel Isarkanal ➔ Munich 21 A3
Mixcoac Mexico City 18 B1
Miyakojima Osaka 23 A2
Mizonokuchi Tokyo 30 B2
Mizue Tokyo 30 A4
Młociński Park Warsaw 32 B1
Mlociny Warsaw 32 B1
Mnevniki Moscow 19 B1
Moba Lagos 14 B2
Moczydło Warsaw 32 C2
Modderfontein Johannesburg 13 A2
Modřany Prague 25 B2
Mogyoród Budapest 7 A3
Moinho Velho, Cor. ➔ São Paulo 27 B2
Mok Seoul 27 B1
Mokotów Warsaw 32 B2
Molenbeek-St-Jean Brussels 6 A1
Molino de Rosas Mexico City 18 B1
Mollem Brussels 6 A1
Mollins de Rey Barcelona 4 A1
Mondeor Johannesburg 13 B2
Moneda, Palacio de la Santiago 27 B2
Monerò Rio de Janeiro 25 A1
Mong Kok Hong Kong 12 B2
Monkstown Dublin 10 B3
Monnickendam Amsterdam 2 A3
Monrovia Los Angeles 16 B4
Monsanto Lisbon 16 A1
Monsanto, Parque Florestal de Lisbon 16 A1
Mont-Royal Montreal 20 A2
Mont-Royal, Parc Montreal 20 A2
Montana de Montjuich Barcelona 4 A1
Montcada i Reixac Barcelona 4 A2
Monte Chingolo Buenos Aires 7 C2
Montebello Los Angeles 16 B4
Montemor Lisbon 16 A1
Monterey Hills Los Angeles 16 B3
Monterey Park Los Angeles 16 B4
Montespaccato Rome 26 B1
Montesson Paris 24 A1
Monteverde Nuovo Rome 26 B1
Montfermeil Paris 24 A4
Montigny-le-Bretonneux Paris 24 B1
Montjay-la-Tour Paris 24 A4
Montjuïc Barcelona 4 A2
Montparnasse, Gare Paris 24 A2
Montréal Montreal 20 A2
Montréal, Î. de Montreal 20 A2
Montréal, Univ. de Montreal 20 B2
Montréal Est Montreal 20 A2
Montréal-Nord Montreal 20 A2
Montréal-Ouest Montreal 20 B1
Montréal Trudeau Int. ✕ (YUL) Montreal 20 B1
Montreuil Paris 24 A3
Montrouge Paris 24 B2
Montserrat Buenos Aires 7 B2
Monza Milan 19 A2
Monzoro Milan 19 B1
Moóca São Paulo 27 B2
Moonachie New York 22 B1
Moonee Ponds Melbourne 18 A1
Moonee Valley Racecourse Melbourne 18 A1
Moosach Munich 21 A2
Mora Mumbai 20 B2
Moratalaz Madrid 17 B2
Mörby Stockholm 29 A2
Morden London 15 B2
Morée ➔ Paris 24 A3
Morgan Park Chicago 9 C3
Moriguchi Osaka 23 A2
Morivione Milan 19 B2
Morningside Edinburgh 11 B2
Morningside Johannesburg 13 A2
Morningside Washington 33 C4
Morningside Park Orlando 23 B2
Morningside Park Toronto 31 A3
Morro Solar, Cerro Lima 14 C2
Mortlake London 15 B2
Mortlake Sydney 29 B1
Morton Grove Chicago 9 A2
Morumbi São Paulo 27 B1
Moscavide Lisbon 16 A2
Moschato Athens 2 B2
Moscow = Moskva Moscow 19 B2
Moskva Moscow 19 B2
Moskvoretskiy Moscow 19 B2
Mosman Sydney 29 B2
Moti Bagh Delhi 10 B2
Motol Prague 25 B2
Motsa Jerusalem 13 B2
Motsa Illit Jerusalem 13 B1
Motspur Park London 15 B2
Mottingham London 15 B4
Mount Dennis Toronto 31 A2
Mount Greenwood Chicago 9 C2
Mount Hood Memorial Park ⌂ Boston 6 A2
Mount Merrion Dublin 10 B2
Mount Olympus Los Angeles 16 B2
Mount Rainier Washington 33 B4
Mount Vernon New York 22 A3
Müggelberge Berlin 5 B4
Müggelheim Berlin 5 B5
Muggiò Milan 19 A2
Mühleiten Vienna 32 A3
Mühlenfliess ➔ Berlin 5 A5
Muiden Amsterdam 2 A3
Muizenberg Cape Town 8 B1
Mujahidpur Delhi 10 B2
Mukandpur Delhi 10 A2
Mukhmas Jerusalem 13 A2
Muko ➔ Osaka 23 A1
Mukojima Tokyo 30 A3
Mulbarton Johannesburg 13 B2
Mullach Íde = Malahide Dublin 10 A3
Mumbai Mumbai 20 B2
Mumbai Chhatrapati Shivaji Int. ✕ (BOM) Mumbai 20 A2
Mumbai Harbour Mumbai 20 B2
Munch-museum Oslo 23 A3
Münchehofe Berlin 5 B5
München Munich 21 B2
München Franz Josef Strauss Int. ✕ (MUC) Munich 21 A2
Munich = München Munich 21 B2
Munkkiniemi Helsinki 11 B2
Munro Buenos Aires 7 B1
Muntinlupa Manila 17 C2
Murai Res. Singapore 28 A2
Muranów Warsaw 32 B1
Murino St. Petersburg 27 B2
Murrayfield Edinburgh 11 B2
Murtala Mohammed International ✕ (LOS) Lagos 14 A1
Musashino Tokyo 30 A2
Museu Nacional Rio de Janeiro 25 B1
Mushin Lagos 14 A2
Musocco Milan 19 B1
Mustansiriya Baghdad 3 A2
Musturud Cairo 7 A2
Muswell Hill London 15 A3
Mutanabi Baghdad 3 B2
Muthana Baghdad 3 B2
Mykerinos Cairo 7 B1
Myllypuro Helsinki 11 B3
Mystic ➔ Boston 6 A2

N

Nacka Stockholm 29 B3
Naenae Wellington 33 B2
Næsrnes Oslo 23 B1
Nagatsuta Tokyo 30 B2
Nagytétény Budapest 7 B1
Nahalin Jerusalem 13 B1
Najafgarh Drain ➔ Delhi 10 B1
Nakahara Tokyo 30 B2
Nakano Tokyo 30 A2
Namsan Park Seoul 27 B1
Namyeong Seoul 27 B1
Nanbiancun Guangzhou 11 B1
Nanchang He ➔ Beijing 4 B1
Nandang Guangzhou 11 B2
Nangal Dewat Delhi 10 B1
Naniwa Osaka 23 B1
Nanole Mumbai 20 A2
Nanpu Bridge Shanghai 28 B2
Nanshi Shanghai 28 B1
Nanterre Paris 24 A2
Napier Mole Karachi 13 B1
Naraina Delhi 10 B1
Nariman Point Mumbai 20 B1
Nariman Pt. Mumbai 20 B1
Nārmak Tehran 31 A2
Naruo Osaka 23 A1
Näsby Stockholm 29 A2
Näsbypark Stockholm 29 A2
National Arboretum Washington 33 B3
National Zoological Park Washington 33 B3
Nationals Park Washington 33 B3
Nativity, Basilica of Jerusalem 13 B2
Natolin Warsaw 32 C2
Naucalpan de Juárez Mexico City 18 B1
Naupada Mumbai 20 A2
Navi Mumbai Mumbai 20 A2
Naviglio di Pavia Milan 19 B1
Naviglio Grande Milan 19 B1
Navotas Manila 17 B1
Navy Pier Chicago 9 B3
Nazal Hikmat Beg Baghdad 3 A2
Nazimabad Karachi 13 A2
Nazlet el Simmân Cairo 7 B1
Nea Alexandria Athens 2 B2
Nea Ionia Athens 2 A2
Nea Liosia Athens 2 A2
Nea Smirni Athens 2 B2
Neapoli Athens 2 B2
Near North Chicago 9 B3
Nebušice Prague 25 B1
Nederhorst Amsterdam 2 B3
Nedlitz Berlin 5 B1
Nee Soon Singapore 28 A2
Needham Boston 6 B1
Needham Heights Boston 6 B1
Nellis Air Force Base ✕ (LSV) Las Vegas 14 A2
N'ematābād Tehran 31 B2
Nerima Tokyo 30 A3
Nesodden Oslo 23 B3
Nesoddtangen Oslo 23 A3
Nesøya Oslo 23 A2
Neu Aubing Munich 21 B1
Neu Buch Berlin 5 A4
Neu Buchhorst Berlin 5 B5
Neu Fahrland Berlin 5 B1
Neu Lindenberg Berlin 5 A4
Neubiberg Munich 21 B3
Neuenhagen Berlin 5 A4
Neuessling Vienna 32 A3
Neuhausen Munich 21 B2
Neuherberg Munich 21 A2
Neuhönow Berlin 5 A5
Neuilly-Plaisance Paris 24 A4
Neuilly-sur-Marne Paris 24 A4
Neuilly-sur-Seine Paris 24 A2
Neukagran Vienna 32 A2
Neukettenhof Vienna 32 B2
Neukölln Berlin 5 B3
Neuperlach Munich 21 B2
Neuried Munich 21 B1
Neustift am Walde Vienna 32 A1
Neusüssenbrunn Vienna 32 A2
Neuwaldegg Vienna 32 A1
Neve Ya'akov Jerusalem 13 A2
Neves Rio de Janeiro 25 B2
New Baghdād Baghdad 3 B2
New Barakpur Kolkata 14 A2
New Brighton New York 22 C1
New Canada Johannesburg 13 B1
New Canada Dam Johannesburg 13 B1
New Carrollton Washington 33 B4
New Cross London 15 B3
New Delhi Delhi 10 B2
New Dorp New York 22 C1
New Dorp Beach New York 22 C1
New Malden London 15 B2
New Milford New York 22 A1
New Mumbai = Navi Mumbai Mumbai 20 A2
New Orleans New Orleans 21 B2
New Orleans Int. ✕ (MSY) New Orleans 21 A1
New Territories Hong Kong 12 A1
New Toronto Toronto 31 B1
New Utrecht New York 22 C2
New York New York 22 B3
New York La Guardia ✕ (LGA) New York 22 B2
Newark B. New York 22 B1
Newbattle Edinburgh 11 B3
Newbury Park London 15 A4
Newcraighall Edinburgh 11 B3
Newham ☐ London 15 A4
Newhaven Edinburgh 11 B2
Newington Edinburgh 11 B2
Newlands Johannesburg 13 B1
Newlands Wellington 33 B1
Newport Melbourne 18 B1
Newton Boston 6 B1
Newtonbrook Toronto 31 A2
Newtongrange Edinburgh 11 B3
Newtonville Boston 6 A1
Newtown Sydney 29 B2
Ngaio Wellington 33 B1
Ngau Chi Wan Hong Kong 12 A2
Ngau Tau Kok Hong Kong 12 B2
Ngauranga Wellington 33 B1
Ngong Shuen Chau Hong Kong 12 B1
Ngua Kok Wan Hong Kong 12 A1
Nhava Sheva Mumbai 20 B2
Nīāvarān Tehran 31 A2
Nibra Kolkata 14 B1
Nidāl Baghdad 3 B2
Niddrie Edinburgh 11 B3
Niddrie Melbourne 18 A1
Nieder Neuendorf Berlin 5 A2
Niederschöneweide Berlin 5 B3
Niederschönhausen Berlin 5 A3
Niemeyer Rio de Janeiro 25 B1
Nieuwendam Amsterdam 2 A2
Nihonbashi Tokyo 30 A3
Niipperi Helsinki 11 B1
Nikea Athens 2 B1
Nikolassee Berlin 5 B2
Nikolskiy Moscow 19 B1
Niles Chicago 9 A2
Nimta Kolkata 14 A2
Ninoy Aquino Int. ✕ (MNL) Manila 17 B2
Nippa Tokyo 30 B2
Nippori Tokyo 30 A3
Nishi Osaka 23 A1
Nishinari Osaka 23 B1
Nishinomiya Osaka 23 A1
Nishiyodogawa Osaka 23 A1
Niterói Rio de Janeiro 25 B2
Nockeby Stockholm 29 B1
Noel Park London 15 A3
Nogatino Moscow 19 B3
Nogent-sur-Marne Paris 24 A3
Noida Delhi 10 B2
Noiseau Paris 24 B4
Noisiel Paris 24 A4
Noisy-le-Grand Paris 24 A4
Noisy-le-Roi Paris 24 A1
Noisy-le-Sec Paris 24 A3
Nokkala Helsinki 11 C1
Nomentano Rome 26 B2
Nonakashinden Tokyo 30 A2
Nongminyundong Jiangxisuo Guangzhou 11 B2
Nonhyeon Seoul 27 B2
Nonthaburi Bangkok 3 A1
Noordgesig Johannesburg 13 B1
Noordzeekanaal Amsterdam 2 A1
Nord, Gare du Paris 24 A3
Nordmarka Oslo 23 A3
Nordrand-Siedlung Vienna 32 A2
Nordstrand Oslo 23 A3
Normandale Wellington 33 B2
Norridge Chicago 9 B2
Norrmalm Stockholm 29 A2
North Arlington New York 22 B1
North Bay Village Miami 18 A2
North Bergen New York 22 B1
North Bull I. Dublin 10 A3
North Cheam London 15 B2
North Cray London 15 B4
North Decatur Atlanta 3 B3
North Druid Hills Atlanta 3 B3
North Esk ➔ Edinburgh 11 B2
North Gyle Edinburgh 11 B2
North Hackensack New York 22 A1
North Harbor Manila 17 B1
North Hd. Sydney 29 A2
North Hollywood Los Angeles 16 B2
North Las Vegas Las Vegas 14 A1
North Las Vegas ✕ (VGT) Las Vegas 14 A1
North Lexington Boston 6 A1
North Miami Miami 18 A2
North Miami Beach Miami 18 A2
North Nazimabad Karachi 13 A2
North Pt. Hong Kong 12 B2
North Queensferry Edinburgh 11 A1
North Quincy Boston 6 B2
North Res. Boston 6 A2
North Riverside Chicago 9 B2
North Saugus Boston 6 A2
North Shore Channel ➔ Chicago 9 B2
North Springfield Washington 33 C2
North Station Boston 6 A2
North Sydney Sydney 29 B2
North Woolwich London 15 A4
North York Toronto 31 A2
Northbridge Sydney 29 A2
Northbridge Park Sydney 29 A2
Northcliff Johannesburg 13 A1
Northcote Melbourne 18 A2
Northeastern Univ. Boston 6 A2
Northern Virginia Regional Park ⌂ Washington 33 B2
Northlake Chicago 9 B1
Northmount Toronto 31 A2
Northolt London 15 A1
Northumberland Heath London 15 B5
Northwestern Station Chicago 9 B3
Northwestern Univ. Chicago 9 A2
Northwood London 15 A1
Northwood Park Toronto 31 A1
Norwood Johannesburg 13 A2
Norwood Park Chicago 9 B2
Noryangjin Seoul 27 B1
Nossa Senhora do Ó São Paulo 27 B1
Nossegem Brussels 6 A3
Notre-Dame Paris 24 A3
Notre-Dame, Basilique Montreal 20 B2
Notre-Dame, Bois Paris 24 B4
Notre-Dame-de-Grace Montreal 20 B2
Notting Hill London 15 A2
Nova Milanese Milan 19 A2
Novate Milanese Milan 19 A1
Novaya Derevnya St. Petersburg 27 A1
Nové Město Prague 25 B2
Novoaleksandrovskoye St. Petersburg 27 B2
Novodevichy Convent Moscow 19 B2
Novogireyevo Moscow 19 B3
Novosaratovka St. Petersburg 27 B3
Nowe-Babice Warsaw 32 B1
Nöykkiö Helsinki 11 B1
Nueva Atzacoalco Mexico City 18 B2
Nueva Pompeya Buenos Aires 7 C2
Nueva Tenochtitlán Mexico City 18 B2
Nuijala Helsinki 11 B1
Nukabukuro Tokyo 30 A2
Nuñez Buenos Aires 7 B2
Nunhead London 15 B3
Ñuñoa Santiago 27 B2
Nusle Prague 25 B2
Nussdorf Vienna 32 A2
Nyanga Cape Town 8 A2
Nymphenburg Munich 21 B2
Nymphenburg, Schloss Munich 21 B2

O

O.R. Tambo Int. ✕ (JNB) Johannesburg 13 A2
Oak Grove Atlanta 3 A3
Oak Hill Boston 6 B1
Oak Lawn Chicago 9 C2
Oak Park Chicago 9 B2
Oak View Washington 33 A4
Oakdale Atlanta 3 A2
Oakland San Francisco 26 B3
Oakland Washington 33 B4
Oaklawn Washington 33 C4
Oakleigh Melbourne 18 B2
Oakton Washington 33 B2
Oakwood Beach New York 22 C1
Oatley Sydney 29 B1
Obalende Lagos 14 B2
Oba's Palace Lagos 14 B2
Oberföhring Munich 21 B2
Oberhaching Munich 21 B2
Oberlaa Vienna 32 B2
Oberlisse Vienna 32 A2
Obermenzing Munich 21 A1
Obermoos Schwaige Munich 21 A1
Oberschleissheim Munich 21 A2
Oberschöneweide Berlin 5 B4
Observatory Johannesburg 13 B2
Óbuda Budapest 7 A2
Obukhovo St. Petersburg 27 B2
Obvodnyy Kanal St. Petersburg 27 B1
Ocean Park Hong Kong 12 B2
Ochota Warsaw 32 B1
Ocoee Orlando 23 A1
Ōdana Tokyo 30 B2
Öden-Stockach Munich 21 B3
Odilampi Helsinki 11 B1
Odivelas Lisbon 16 A1
Odolany Warsaw 32 B1
Oeiras Lisbon 16 A1
Ofin Lagos 14 A3
Ogawa Tokyo 30 A1
Ogden Park Chicago 9 C2
Ogikubo Tokyo 30 A2
Ogogoro Lagos 14 B2
Ogoyo Lagos 14 B2
Ogudu Lagos 14 A2
O'Hare Int., Chicago ✕ (ORD) Chicago 9 B1
Ohariu Stream ➔ Wellington 33 B1
O'Higgins, Parque Santiago 27 B2
Ōimachi Tokyo 30 B3
Ojota Lagos 14 A2
Oke-Ira Lagos 14 B2
Okęcie Warsaw 32 B1
Okęcie, Warszawa ✕ (WAW) Warsaw 32 B1
Okeogbe Lagos 14 B2
Okhla Delhi 10 B2
Okhta ➔ St. Petersburg 27 B2
Okkervil ➔ St. Petersburg 27 B2
Okrzeszyn Warsaw 32 C2
Oksval Oslo 23 A3
Oktyabrskiy Moscow 19 B2
Okubo Tokyo 30 A3
Ōkura Tokyo 30 B1
Olari Helsinki 11 B1
Olaria Rio de Janeiro 25 B1
Old Admiralty St. Petersburg 27 B1
Old City Jerusalem 13 B2
Old City Shanghai 28 B1
Old Fort York Toronto 31 B2
Old Harbor Boston 6 B2
Old Town = Stare Miasto Warsaw 32 B2
Old Town Chicago 9 B3
Old Town Hall = Altes Rathaus Munich 21 B2
Oldbawn Dublin 10 B1
Olgino St. Petersburg 27 A1
Olimpico, Estadio Mexico City 18 C1
Olivais Lisbon 16 A2
Olivar de los Padres Mexico City 18 B1
Olivar del Conde Mexico City 18 B1
Olives, Mt. of Jerusalem 13 B2
Olivia, L. Orlando 23 A1
Olivos Buenos Aires 7 B2
Olona ➔ Milan 19 B1
Olympia, L. Orlando 23 A1
Olympic Green Beijing 4 B2
Olympic Stadium = Turner Field Atlanta 3 B2
Olympic Stadium Beijing 4 B2
Olympic Stadium Helsinki 11 B2
Olympique, Stade Montreal 20 A2
Ōmori Tokyo 30 B3
Onisigun Lagos 14 A2
Ontario Science Centre Toronto 31 A2
Ōokayama Tokyo 30 B3
Oostzaan Tuindorp Amsterdam 2 A2
Opa-Locka Miami 18 A1
Opa-Locka ✕ (OPF) Miami 18 A1
Opacz Warsaw 32 B1
Ophirton Johannesburg 13 B2
Oppegård Oslo 23 B3
Oppem Brussels 6 A1
Oppsal Oslo 23 A4
Ora Jerusalem 13 B1
Oradell New York 22 A1
Oradell Res. New York 22 A1
Orange Bowl Stadium Miami 18 B2
Orangi Karachi 13 A2
Ordrup Copenhagen 8 A3
Orech Prague 25 B1
Oregun Lagos 14 A2
Ørestad Copenhagen 8 B3
Orient Heights Boston 6 A3
Orlando Orlando 23 A2
Orlando Dam Johannesburg 13 B1
Orlando East Johannesburg 13 B1
Orlando Executive ✕ (ORL) Orlando 23 A2
Orlando Int. ✕ (MCO) Orlando 23 B3
Orlovista Orlando 23 A2
Orly Paris 24 B3
Orly, Paris ✕ (ORY) Paris 24 B3
Ormesson-sur-Marne Paris 24 B4
Ormond Melbourne 18 B2
Ormøya Oslo 23 A3
Orpington London 15 B4
Ortaköy Istanbul 12 B2
Ortica Milan 19 B2
Oruba Lagos 14 A2
Ōsaka Osaka 23 B2
Ōsaka Castle Osaka 23 A2
Ōsaka Harbour Osaka 23 B1
Ōsaka Itami Int. ✕ (ITM) Osaka 23 A1
Ōsaka Kansai Int. ✕ (KIX) Osaka 23 B2
Ōsaki Tokyo 30 B3
Osasco São Paulo 27 B1
Osdorf Berlin 5 B3
Osdorp Amsterdam 2 A1
Oshodi Lagos 14 A2
Oslo Oslo 23 A3
Oslo ✕ (OSL) Oslo 23 A4
Ōsone Tokyo 30 B2
Osorun Lagos 14 A2
Ospiate Milan 19 A1
Ostankino Moscow 19 B2
Osterley London 15 B1
Osterley Park London 15 B1
Östermalm Stockholm 29 A2
Österskär Stockholm 29 A3
Östia Malpasso Rome 26 C1
Ostiense Rome 26 B1
Østmarkkapellet Oslo 23 A4
Østøya Oslo 23 A2
Østre Aker Oslo 23 A3
Ōta Tokyo 30 B3
Otaniemi Helsinki 11 B1
Otari Open Air Museum Wellington 33 B1
Otsuka Tokyo 30 A3
Ottakring Vienna 32 A1
Ottavia Rome 26 B1
Ottery Cape Town 8 B2
Ottobrunn Munich 21 B3
Ouderkerk Amsterdam 2 B2
Oulunkylä Helsinki 11 B2
Ourcq, Canal de l' Paris 24 A3
Outer Mission San Francisco 26 B2
Outremont Montreal 20 A2
Overijse Brussels 6 B3
Owhiro Bay Wellington 33 C1
Oworonsoki Lagos 14 A2
Oxgangs Edinburgh 11 B2
Oxon Hill Washington 33 C4
Oyodo Osaka 23 A1
Oyster Rock Mumbai 20 B1
Oyster Rocks Karachi 13 B2
Ozoir-la-Ferrière Paris 24 B4
Ozone Park New York 22 B2

P

Pacific Heights San Francisco 26 B2
Pacific Manor San Francisco 26 C2
Pacific Palisades Los Angeles 16 B1
Pacifica San Francisco 26 C2
Paco Manila 17 B1
Paco de Arcos Lisbon 16 A1
Paddington London 15 A2
Paddington Sydney 29 B2
Paderno Milan 19 A1
Pagewood Sydney 29 B2
Pagote Mumbai 20 B2
Pai, I. do Rio de Janeiro 25 B2
Pak Kok Hong Kong 12 B1
Pak Kong Hong Kong 12 A2
Pakila Helsinki 11 B2
Palacio Real Madrid 17 B1
Palaiseau Paris 24 B2
Palazzolo Milan 19 A1
Paleo Faliro Athens 2 B2
Palermo Buenos Aires 7 B2
Palhais Lisbon 16 B2
Palisades Park New York 22 A1
Palmer Park Washington 33 B4
Palmerston Dublin 10 A1
Palms Los Angeles 16 B2
Paloheinä Helsinki 11 B2
Palomeras Madrid 17 B2
Palos Heights Chicago 9 D2
Palos Hills Chicago 9 C2
Palos Hills Forest Chicago 9 C1
Palos Park Chicago 9 C1
Palpara Kolkata 14 B2
Panchur Kolkata 14 B1
Pandacan Manila 17 B2
Pandan, Selat Singapore 28 B2
Pandan Res. Singapore 28 B2
Pangrati Athens 2 B2
Pangsua, Sungei → Singapore 28 A2
Panihati Kolkata 14 A2
Panjang, Bukit Singapore 28 A2
Panke → Berlin 5 A3
Pankow Berlin 5 A3
Pantheon Rome 26 B1
Panthersville Atlanta 3 B3
Pantin Paris 24 A3
Pantitlán Mexico City 18 B2
Panyu Guangzhou 11 B2
Paparangi Wellington 33 B1
Papiol Barcelona 4 A1
Paradise Las Vegas 14 B2
Paramount Studios Los Angeles 16 B3
Paramus New York 22 A1
Paranaque Manila 17 B1
Paray-Vieille-Poste Paris 24 B3
Pardisān Nature Park Tehran 31 A2
Parel Mumbai 20 B1
Pari São Paulo 27 B2
Parioli Rome 26 B1
Paris Paris 24 A3
Paris Orly ✈ (ORY) Paris 24 B3
Parje Mumbai 20 B2
Pārk-e Āzādegān Tehran 31 B2
Pārk-e Mellat Tehran 31 A2
Park Ridge Chicago 9 A1
Park Royal London 15 A2
Parkchester New York 22 B2
Parkdale Toronto 31 B2
Parkhurst Johannesburg 13 A2
Parklawn Washington 33 B3
Parkmore Johannesburg 13 A2
Parkside San Francisco 26 B2
Parktown Johannesburg 13 B2
Parkview Johannesburg 13 A2
Parkville New York 22 C2
Parkwood Cape Town 8 B1
Parkwood Johannesburg 13 A2
Parow Cape Town 8 A2
Parque Chacabuco Buenos Aires 7 B2
Parque Patricios Buenos Aires 7 B2
Parramatta → Sydney 29 A1
Paşabahçe Istanbul 12 B2
Pasadena Los Angeles 16 B4
Pasar Minggu Jakarta 12 B1
Pasay Manila 17 B1
Pascoe Vale Melbourne 18 A1
Pasig Manila 17 B2
Pasig → Manila 17 B2
Pasila Helsinki 11 B2
Pasing Munich 21 B1
Pasir Panjang Singapore 28 B2
Pasir Ris Singapore 28 A3
Passaic New York 22 A1
Passaic → New York 22 B1
Passirana Milan 19 A1
Patel Nagar Delhi 10 B1
Pateros Manila 17 B2
Pathumwan Bangkok 3 B2
Patipukur Kolkata 14 B2
Patisia Athens 2 A2
Paulo E. Virginia, Gruta Rio de Janeiro 25 B1
Paulshof Berlin 5 A5
Paya Lebar Singapore 28 A3
Peakhurst Sydney 29 B1
Peania Athens 2 B3
Pearson Int. Toronto ✈ (YYZ) Toronto 31 A1
Peckham London 15 B3
Pederstrup Copenhagen 8 A2
Pedralbes Barcelona 4 A1
Pedregal de San Angel, Jardines del Mexico City 18 C1
Peip'ing = Beijing Beijing 4 B1
Peking = Beijing Beijing 4 B1
Pelcowizna Warsaw 32 B2
Peñalolén Santiago 27 B2
Pencarrow Hd. Wellington 33 C2
Peng Siang → Singapore 28 A2
Penge London 15 B3
Penha Rio de Janeiro 25 A1
Penicuik Edinburgh 11 B2
Penjaringan Jakarta 12 A1
Pentagon Washington 33 B3
Penzing Vienna 32 A1
People's Park Shanghai 28 B1
People's Square Shanghai 28 B1
Perales del Rio Madrid 17 C2
Perchtoldsdorf Vienna 32 B1
Perdizes São Paulo 27 B2
Peristeri Athens 2 A2
Perivale London 15 A2
Perk Brussels 6 A2
Perlach Munich 21 B2
Perlacher Forst Munich 21 B2
Pero Milan 19 A1
Peropok, Bukit Singapore 28 B2
Perovo Moscow 19 B3
Pertusella Milan 19 A1
Pesagot Jerusalem 13 A2
Pesanggrahag, Kali → Jakarta 12 B1
Peschiera Borromeo Milan 19 B2
Pesek, Pulau Singapore 28 B2
Pest Budapest 7 B2
Pesterzsébet Budapest 7 B2
Pesthidegkút Budapest 7 A1
Pestimre Budapest 7 B3
Pestlőrinc Budapest 7 B3
Pestújhely Budapest 7 A2
Petas Helsinki 11 B2
Petone Wellington 33 B2
Petrogradskaya Storona St. Petersburg 27 B2
Petroupoli Athens 2 A2
Petrovice Prague 25 B3
Petrovskiy Park Moscow 19 B2
Petrovsko-Razumovskoye Moscow 19 B2
Pettycur Edinburgh 11 A2
Peutie Brussels 6 A2
Pfaueninsel Berlin 5 B1
Phaya Thai Bangkok 3 B2
Phihai Karachi 13 A2
Phillip B. Sydney 29 B2
Phoenix Park Dublin 10 A2
Phra Khanong Bangkok 3 B2
Phra Nakhon Bangkok 3 B2
Phra Pradaeng Bangkok 3 C2
Phranakhon Bangkok 3 B1
Pico Rivera Los Angeles 16 C4
Piedade Lisbon 16 A1
Piedade Rio de Janeiro 25 B1
Piedade, Cova da Lisbon 16 A2
Piedmont Park Atlanta 3 B2
Pierre Elliott Trudeau ✈ (YUL) Montreal 20 B1
Pietralata Rome 26 B2
Pihlajamäki Helsinki 11 B2
Pihlajasaari Helsinki 11 C2
Pilares Rio de Janeiro 25 B1
Pilton Edinburgh 11 B2
Pimmit Hills Washington 33 B2
Pine Castle Orlando 23 B2
Pine Grove Toronto 31 A1
Pine Hills Orlando 23 A2
Pinewood Park Miami 18 A2
Piney Run → Washington 33 B2
Pingzhou Guangzhou 11 B2
Pinheiros → São Paulo 27 B1
Pinjrapur Karachi 13 A2
Pinner London 15 A1
Pinner Green London 15 A1
Pioltello Milan 19 A2
Pipinui Pt. Wellington 33 A1
Piræus = Pireas Athens 2 B1
Piraiévs = Pireas Athens 2 B1
Pirajuçara → São Paulo 27 B1
Pireas Athens 2 B1
Pirinççi Istanbul 12 B1
Pirituba São Paulo 27 A1
Pirkkola Helsinki 11 B2
Pisgat O'mer Jerusalem 13 B2
Pisgat Ze'ev Jerusalem 13 B2
Pisnice Prague 25 C2
Pitampura Delhi 10 A1
Pitkäjärvi Helsinki 11 B1
Planegg Munich 21 B1
Plaza Mayor Madrid 17 B1
Pleasure Island Orlando 23 B1
Plumstead Cape Town 8 B1
Plumstead London 15 B4
Plyushchevo Moscow 19 B3
Po Toi Hong Kong 12 B2
Po Toi O Hong Kong 12 B2
Poasco Milan 19 B2
Podbaba Prague 25 B2
Podoli Prague 25 B2
Pohick Creek → Washington 33 C2
Pointe-Aux-Trembles Montreal 20 A2
Poissy Paris 24 A1
Pok Fu Lam Hong Kong 12 B1
Pokcheong Seoul 27 C2
Polton Edinburgh 11 B3
Polvoranca, Parque de Madrid 17 C1
Polyustrovo St. Petersburg 27 B2
Pomprap Bangkok 3 B2
Pondok Gede Jakarta 12 A2
Pondok Indah Jakarta 12 B1
Pont-Viau Montreal 20 A1
Pontault-Combault Paris 24 B4
Pontinha Lisbon 16 A1
Poplar London 15 A3
Poppintree Dublin 10 A2
Porirua Wellington 33 A2
Porirua East Wellington 33 A2
Port Botany Sydney 29 B2
Port Mearnóg = Portmarnock Dublin 10 A3
Port Melbourne Melbourne 18 B1
Port Nicholson Wellington 33 B2
Port Richmond New York 22 C1
Port Shelter Hong Kong 12 A2
Port Union Toronto 31 A4
Portage Park Chicago 9 B2
Portela, Lisboa ✈ (LIS) Lisbon 16 A2
Porter, L. Orlando 23 A3
Portmarnock Dublin 10 A3
Porto Brandão Lisbon 16 A1
Pôrto Novo Rio de Janeiro 25 A2
Porto Novo Cr. → Lagos 14 B2
Portobello Edinburgh 11 B3
Portrero San Francisco 26 B3
Potomac Washington 33 A2
Potrero Pt. San Francisco 26 B2
Potsdam Berlin 5 B1
Potzham Munich 21 B2
Pötzleinsdorf Vienna 32 A1
Póvoa de Santo Adriao Lisbon 16 A2
Powązki Warsaw 32 B1
Powiśle Warsaw 32 B2
Powsin Warsaw 32 C2
Powsinek Warsaw 32 C2
Poyan Res. Singapore 28 A2
Pozuelo de Alarcón Madrid 17 B1
Prado, Museo del Madrid 17 B1
Prado Churubusco Mexico City 18 B2
Praga Warsaw 32 B2
Prague = Praha Prague 25 B2
Praha Prague 25 B2
Praha ✈ (PRG) Prague 25 B1
Praires, R. des → Montreal 20 A2
Prater Vienna 32 A2
Precotto Milan 19 A2
Prenestino Labicano Rome 26 B2
Prenzlauerberg Berlin 5 A3
Preston Melbourne 18 A1
Pretos Forros, Sa. dos Rio de Janeiro 25 B1
Préville Montreal 20 B3
Přezletice Prague 25 B3
Prima Porta Rome 26 B1
Primavalle Rome 26 B1
Primrose Johannesburg 13 B2
Progreso Nacional Mexico City 18 A2
Prosek Prague 25 B3
Prospect Hill Park Boston 6 A1
Providencia Santiago 27 B2
Průhonice Prague 25 C3
Psichiko Athens 2 A2
Pudong New Area Shanghai 28 B2
Pueblo Libre Lima 14 B2
Pueblo Nuevo Barcelona 4 A2
Pueblo Nuevo Madrid 17 B2
Puerto Madero Buenos Aires 7 B2
Puhuangyu Beijing 4 B2
Puistola Helsinki 11 B3
Pukinmäki Helsinki 11 B2
Pulkovo Int. ✈ (LED) St. Petersburg 27 C1
Pullach Munich 21 B1
Pullman Historic District Chicago 9 C3
Pulo Gadung Jakarta 12 B2
Pūnak Tehran 31 A2
Punchbowl Sydney 29 B1
Punde Mumbai 20 B2
Punggol Singapore 28 A3
Punggol, Sungei → Singapore 28 A3
Punggol Pt. Singapore 28 A3
Punjabi Bagh Delhi 10 A1
Puotila Helsinki 11 B3
Puteaux Paris 24 A2
Putney London 15 B2
Putuo Shanghai 28 B1
Putxet Barcelona 4 A1
Puxi Shanghai 28 B1
Pyramids Cairo 7 B1
Pyry Warsaw 32 C1

Q

Qalandiya Jerusalem 13 A2
Qal'eh Morghī Tehran 31 B2
Qanâ el Ismâ'ilîya Cairo 7 A2
Qāsemābād Tehran 31 A3
Qasr-e Firūzeh Tehran 31 B3
Qatane Jerusalem 13 B1
Qianmen Beijing 4 B2
Qinghuayuan Beijing 4 B1
Qingningsi Shanghai 28 B2
Qolhak Tehran 31 A2
Quadraro Rome 26 B2
Quaid-i-Azam Karachi 13 A1
Quartiere Zingone Milan 19 B1
Qubeiba Jerusalem 13 A1
Quds Baghdad 3 A2
Queen Mary Res. London 15 B1
Queen Victoria Market Melbourne 18 A1
Queensbury London 15 A2
Queenscliffe Sydney 29 A2
Queensferry Edinburgh 11 B1
Queenstown Singapore 28 B2
Quellerina Johannesburg 13 A1
Queluz Lisbon 16 A1
Quezon City Manila 17 B2
Quilicura Santiago 27 B1
Quincy Boston 6 B2
Quincy B. Boston 6 B3
Quinta Normal Santiago 27 B1
Quinto de Stampi Milan 19 B2
Quinto Romano Milan 19 B1
Quirinale Rome 26 B1

R

R.F.K. Memorial Stadium Washington 33 B4
Raasdorf Vienna 32 A3
Radcliffe College Boston 6 A2
Rådhus Oslo 23 A3
Radlice Prague 25 B2
Radość Warsaw 32 B3
Radotin Prague 25 C2
Rafat Jerusalem 13 A2
Raffles Park Singapore 28 B2
Raheny Dublin 10 A3
Rahnsdorf Berlin 5 B5
Rajakylä Helsinki 11 B3
Rajpura Delhi 10 A2
Rákos-patak → Budapest 7 B3
Rákoshegy Budapest 7 B3
Rákoskeresztúr Budapest 7 B3
Rákoskert Budapest 7 B3
Rákosliget Budapest 7 B3
Rákospalota Budapest 7 A2
Rákosszentmihály Budapest 7 A2
Raków Warsaw 32 B1
Ram Jerusalem 13 A2
Rām Allāh Jerusalem 13 A2
Ramadān Baghdad 3 B2
Ramakrishna Puram Delhi 10 B1
Rāmallāh = Rām Allāh Jerusalem 13 A2
Ramanathpur Kolkata 14 A1
Ramat Allon Jerusalem 13 B2
Ramat Eshkol Jerusalem 13 B2
Ramat Razi'el Jerusalem 13 B1
Ramat Shafet Jerusalem 13 B2
Rambler Channel Hong Kong 12 A1
Ramenki Moscow 19 B1
Ramersdorf Munich 21 B2
Rameses Station Cairo 7 A2
Ramos Rio de Janeiro 25 B1
Ramos Mejía Buenos Aires 7 B1
Ramot Jerusalem 13 B2
Rampur Delhi 10 A2
Ramsgate Sydney 29 B1
Rancho Park Los Angeles 16 B2
Rand ✈ (QRA) Johannesburg 13 B2
Randalls I. New York 22 B2
Randburg Johannesburg 13 A1
Randhart Johannesburg 13 B2
Randpark Ridge Johannesburg 13 A1
Randwick Sydney 29 B2
Ranelagh Dublin 10 A2
Rangpuri Delhi 10 B1
Rannersdorf Vienna 32 B2
Ransbèche Brussels 6 B2
Ransdorp Amsterdam 2 A2
Ranvad Mumbai 20 B2
Raposo Lisbon 16 A1
Rastaala Helsinki 11 B1
Rastila Helsinki 11 B3
Raszyn Warsaw 32 C1
Rathfarnham Dublin 10 B2
Ratho Edinburgh 11 B1
Ratho Station Edinburgh 11 B1
Rato Lisbon 16 A2
Ravelston Edinburgh 11 B2
Rawamangun Jakarta 12 B2
Rayners Lane London 15 A1
Raynes Park London 15 B2
Raypur Kolkata 14 C2
Real Felipe, Fuerte Lima 14 B2
Reams, L. Orlando 23 B1
Reclining Buddha = Wat Pho Bangkok 3 B1
Recoleta Buenos Aires 7 B2
Recoleta Santiago 27 B2
Red Fort = Lal Qila Delhi 10 B2
Red Square Moscow 19 B2
Redbridge □ London 15 A4
Redfern Sydney 29 B2
Redwood Wellington 33 B1
Refshaleøen Copenhagen 8 A3
Regents Park Johannesburg 13 B2
Regent's Park London 15 A3
Rego Park New York 22 B2
Reinickendorf Berlin 5 A3
Rekola Helsinki 11 B3
Rembertów Warsaw 32 B2
Rembrandtpark Amsterdam 2 A2
Remedios de Escalada Buenos Aires 7 C2
Rémola, Estany del Barcelona 4 B1
Renca Santiago 27 B1
Rennemoulin Paris 24 A1
Řeporyje Prague 25 B1
Repulse Bay Hong Kong 12 B2
Repy Prague 25 B1
Residenz Munich 21 B2
Reston Washington 33 B2
Retiro Buenos Aires 7 B2
Retiro Madrid 17 B1
Retiro, Puerto Buenos Aires 7 B2
Retreat Cape Town 8 B1
Reutov Moscow 19 B4
Réveillon → Paris 24 B4
Revere Boston 6 A2
Rexdale Toronto 31 A1
Reynosa Tamaulipas Mexico City 18 A1
Rho Milan 19 A1
Rhodes Sydney 29 A1
Rhodon Paris 24 B1
Rhodon → Paris 24 B1
Rialto Towers Melbourne 18 A1
Ribeira Rio de Janeiro 25 A1
Ricarda, Estany de la Barcelona 4 B1
Richmond Melbourne 18 A2
Richmond San Francisco 26 B2
Richmond Hill New York 22 B2
Richmond Park London 15 B2
Richmond-upon-Thames □ London 15 B2
Ridge, The Delhi 10 B2
Ridgefield New York 22 B1
Ridgefield Park New York 22 A1
Ridgewood New York 22 B2
Riem Munich 21 B3
Riesenrad Vienna 32 A2
Rijksmuseum Amsterdam 2 A2
Rikers I. New York 22 B2
Rimac Lima 14 B2
Ringsend Dublin 10 A2
Rinkeby Stockholm 29 A1
Rio Comprido Rio de Janeiro 25 B1
Rio de Janeiro Rio de Janeiro 25 B1
Rio de Janeiro Galeão ✈ (GIG) Rio de Janeiro 25 A1
Rio de Mouro Lisbon 16 A1
Ripollet Barcelona 4 A1
Ris Oslo 23 A3
Risby Copenhagen 8 A1
Rishra Kolkata 14 A2
Ritchie Washington 33 B4
Rithala Delhi 10 A1
Rive Sud, Canal de la Montreal 20 B2
River Edge New York 22 A1
River Forest Chicago 9 B2
River Grove Chicago 9 B1
River Ridge New Orleans 21 B1
Riverdale New York 22 A2
Riverdale Washington 33 B4
Riverdale Park Toronto 31 A2
Riverlea Johannesburg 13 B1
Riverside Chicago 9 C2
Riverwood Sydney 29 B1
Rivière-des-Praires Montreal 20 A2
Rixensart Brussels 6 B3
Riyad Baghdad 3 B2
Rizal Park Manila 17 B1
Rizal Stadium Manila 17 B1
Røa Oslo 23 A2
Robbins Chicago 9 D2
Robertsham Johannesburg 13 B2
Rochelle Park New York 22 A1
Rock Cr. → Washington 33 B3
Rock Creek Park Washington 33 B3
Rock Pt. Wellington 33 A1
Rockaway Beach New York 22 C3
Rockaway Pt. New York 22 C2
Rockdale Sydney 29 B1
Rockefeller Center New York 22 B2
Rocky Run → Washington 33 B2
Roda, Gezîret el Cairo 7 A2
Roda I. = Roda, Gezîret el Cairo 7 A2
Rodaon Vienna 32 B1
Rødovre Copenhagen 8 A2
Rodrigo de Freitas, L. Rio de Janeiro 25 B1
Roehampton London 15 B2
Rogers Park Chicago 9 A3
Roihuvuori Helsinki 11 B3
Roissy-en-Brie Paris 24 B4
Rokytka → Prague 25 B3
Roma Rome 26 B1
Roma Urbe ✈ Rome 26 B2
Római-Fürdö Budapest 7 A2
Romainville Paris 24 A3
Roman Forum = Foro Romano Rome 26 B1
Romano Banco Milan 19 B1
Rome = Roma Rome 26 B1
Romema Jerusalem 13 B2
Romford London 15 A5
Ronald Reagan Nat. Washington ✈ (DCA) Washington 33 B3
Rondebosch Cape Town 8 A1
Roppongi Tokyo 30 B3
Rose, L. Orlando 23 A1
Rose Bowl Los Angeles 16 B4
Rose Hill Washington 33 C3
Rosebank Johannesburg 13 A2
Rosebank New York 22 C1
Rosebery Sydney 29 B2
Rosedal La Candelaria Mexico City 18 B2
Roseland Chicago 9 C3
Rosemead Los Angeles 16 B4
Rosemont Chicago 9 B1
Rosemont Montreal 20 A2
Rosenborg Slot Copenhagen 8 A3
Rosenthal Berlin 5 A3
Rosettenville Johannesburg 13 B2
Rosewell Edinburgh 11 B3
Rosherville Dam Johannesburg 13 B2
Rösjön Stockholm 29 A2
Roslags-Näsby Stockholm 29 A2
Roslin Edinburgh 11 B3
Roslindale Boston 6 B2
Rosny-sous-Bois Paris 24 A4
Rosslyn Washington 33 B3
Rosyth Edinburgh 11 A1
Rotherhithe London 15 B3
Rothneusiedl Vienna 32 B2
Rothschmaige Munich 21 A1
Rouge Hill Toronto 31 A4
Round I. Hong Kong 12 B2
Roxbury Boston 6 B2
Roxeth London 15 A1
Royal Botanic Gardens Edinburgh 11 B2
Royal Observatory Edinburgh 11 B2
Royal Palace = Kungliga Slottet Stockholm 29 B2
Royal Palace = Palacio Real Madrid 17 B1
Royal Park Melbourne 18 A1
Röylä Helsinki 11 B1
Rozas, Portilleros de las Madrid 17 B1
Roztoky Prague 25 B2
Rozzano Milan 19 B1
Rubí → Barcelona 4 A1
Rudolfsheim Vienna 32 A2
Rudolfshöhe Berlin 5 A5
Rudow Berlin 5 B3
Rueil-Malmaison Paris 24 A2
Ruisbroek Brussels 6 B1
Ruislip London 15 A1
Rumelihisarı Istanbul 12 B2
Rungis Paris 24 B3
Rusāfa Baghdad 3 A2
Rush Green London 15 A5
Russa Kolkata 14 C2
Rutherford New York 22 B1
Ruzyně Prague 25 B1
Ruzyne, Praha ✈ (PRG) Prague 25 B1
Rybatskaya St. Petersburg 27 B2
Rydboholm Stockholm 29 A3
Ryde Sydney 29 A1
Ryogoku Tokyo 30 A3
Rzhevka St. Petersburg 27 B3

S

Sa'ādatābād Tehran 31 A2
Saadūn Baghdad 3 B2
Saavedra Buenos Aires 7 B2
Saboli Delhi 10 A2
Sabugo Lisbon 16 A1
Sabzi Mandi Delhi 10 A2
Sacavém Lisbon 16 A2
Saclay Paris 24 B2
Saclay, Étang de Paris 24 B1
Sacomã São Paulo 27 B2
Sacré Cœur Paris 24 A3
Sacrow Berlin 5 B1
Sacrower See Berlin 5 B1
Sadang Seoul 27 C1
Saddle → New York 22 A1
Saddle Brook New York 22 A1
Sadr Karachi 13 A2
Sadr City Baghdad 3 A2
Sadyba Warsaw 32 B2
Safdar Jang's Tomb Delhi 10 B2
Saft el Laban Cairo 7 A2
Saganashkee Slough Chicago 9 C1
Sagene Oslo 23 A3
Sagrada Familia, Templo de Barcelona 4 A2
Sahar, Mumbai ✈ (BOM) Mumbai 20 A2
Sai Kung Hong Kong 12 A2
Sai Wan Ho Hong Kong 12 B2
Sai Ying Pun Hong Kong 12 B1
St-Aubin Paris 24 B1
St-Cloud Paris 24 A2
St-Cyr-l'École Paris 24 B1
St-Cyr-l'École ✈ Paris 24 B1
St-Denis Paris 24 A3
St-Germain, Forêt de Paris 24 A1
St-Germain-en-Laye Paris 24 A1
St-Gilles Brussels 6 B2
St. Helier London 15 B2
St-Hubert Montreal 20 B3
St. Hubert, Galeries Royales Brussels 6 A2
St. Isaac's Cathedral St. Petersburg 27 B1
St-Jacques → Montreal 20 B3
St-Joost-Ten-Node Brussels 6 A2
St. Kilda Melbourne 18 B1
St-Lambert Montreal 20 A3
St-Lambert Paris 24 B1
St-Laurent Montreal 20 A1
St-Léonard Montreal 20 A2
St. Magelungen Stockholm 29 B2
St-Mandé Paris 24 A3
St. Margaret's Dublin 10 A2
St-Martin, Bois Paris 24 B4
St. Mary Cray London 15 B4
St-Maur-des-Fossés Paris 24 B3
St-Maurice Paris 24 B3
St-Michel Montreal 20 A2
St. Nikolaus-Kirken Prague 25 B2
St-Ouen Paris 24 A3
St. Paul's Cray London 15 B4
St. Peters Sydney 29 B2
St. Peter's Square = San Pietro, Piazza Rome 26 B1
St. Petersburg = Sankt-Peterburg St. Petersburg 27 B1
St. Petersburg Pulkovo Int. ✈ (LED) St. Petersburg 27 C1
St-Pierre Montreal 20 B2
St-Pieters Leew Brussels 6 B1
St-Quentin, Étang de Paris 24 B1
St. Stephens Cathedral = Stephansdom Vienna 32 A2
St-Stevens-Woluwe Brussels 6 A2
St-Vincent-de-Paul Montreal 20 A2
St. Xavier Univ. Chicago 9 C2
Ste-Catherine Montreal 20 B2
Ste-Hélène, Î. Montreal 20 A2
Saiwai Tokyo 30 B3
Sakai Osaka 23 B1
Sakai Harbour Osaka 23 B1
Sakra, Pulau Singapore 28 B2
Salam Baghdad 3 A2
Salamanca Madrid 17 B1
Sallynoggin Dublin 10 B3
Salmannsdorf Vienna 32 A1
Salmedina Madrid 17 C2
Salomea Warsaw 32 B1
Salsette I. Mumbai 20 A2
Salt Lake City = Bidhan Nagar Kolkata 14 B2
Salt River Cape Town 8 A1
Salt Water L. Kolkata 14 B2
Saltsjö-Duvnäs Stockholm 29 B3
Samatya Istanbul 12 C1
Sampaloc Manila 17 B1
Samphan Thawong Bangkok 3 B2
Samseon Seoul 27 B2
Samuel Smith Park Toronto 31 B1
Samut Prakan Bangkok 3 B2
San Andrés Barcelona 4 A2
San Angel Mexico City 18 B1
San Basilio Rome 26 B2
San Borja Lima 14 B3
San Bóvio Milan 19 B2
San Bruno, Pt. San Francisco 26 C2
San Bruno Mountain State Park ⌂ San Francisco 26 B2
San Cristóbal Buenos Aires 7 B2
San Cristóbal Madrid 17 B2
San Cristóbal, Cerro Santiago 27 B2
San Cristóbal, Virgen del Santiago 27 B2
San Cristoforo Milan 19 B1
San Donato Milanese Milan 19 B2
San Francisco San Francisco 26 B2
San Francisco B. San Francisco 26 B3
San Francisco Culhuacán Mexico City 18 C2
San Fruttuoso Milan 19 A2
San Gabriel Los Angeles 16 B4
San Gabriel → Los Angeles 16 C4
San Giuliano Milanese Milan 19 B2
San Isidro Lima 14 B2
San Jerónimo Lídice Mexico City 18 C1
San Joaquín Santiago 27 B2
San José Río Hondo Mexico City 18 B1
San Juan → Manila 17 B2
San Juan de Aragón Mexico City 18 B2
San Juan de Aragón, Parque Mexico City 18 B2
San Juan de Lurigancho Lima 14 A2
San Juan del Monte Manila 17 B2
San Juan Ixtacala Mexico City 18 A1
San Just Desvern Barcelona 4 A1
San Justo Buenos Aires 7 C1
San Lorenzo Tezonco Mexico City 18 C2
San Luis Lima 14 B2
San Marino Los Angeles 16 B4
San Martin Barcelona 4 A2
San Martin de Porras Lima 14 B2
San Mateo Tlaltenango Mexico City 18 B1
San Miguel Lima 14 B2
San Miguel Santiago 27 B2
San Nicolás Buenos Aires 7 B2
San Onófrio Rome 26 B1
San Pedro Mártir Barcelona 4 A1
San Pedro Zacatenco Mexico City 18 A2
San Pietro, Piazza Rome 26 B1
San Po Kong Hong Kong 12 A2
San Rafael Chamapa Mexico City 18 B1
San Rafael Hills Los Angeles 16 A3
San Roque Manila 17 B2
San Siro Milan 19 B1
San Souci Sydney 29 B1
San Telmo Buenos Aires 7 B2
San Vicenç dels Horts Barcelona 4 A1
Sandown Johannesburg 13 A2
Sandown Park Races London 15 B1
Sandton Johannesburg 13 A2
Sandvika Oslo 23 A2
Sandyford Dublin 10 B2
Sandymount Dublin 10 B2
Sangenjaya Tokyo 30 B2
Sanggye Seoul 27 B2
Sangley Pt. Manila 17 C1
Sankrail Kolkata 14 B1
Sankt-Peterburg St. Petersburg 27 B1
Sankt Veit Vienna 32 A1
Sanlihe Beijing 4 B1
Sanlintang Shanghai 28 C1
Sans Barcelona 4 A1
Sanssouci Berlin 5 B1
Sant Ambrogio, Basilica di Milan 19 B2
Sant Boi de Llobregat Barcelona 4 A1
Sant Cugat Barcelona 4 A1
Sant Feliu de Llobregat Barcelona 4 A1
Sant Joan Despí Barcelona 4 A1
Santa Ana Manila 17 B2
Santa Anita Park Los Angeles 16 B4
Santa Coloma de Gramenet Barcelona 4 A2
Santa Cruz Manila 17 B1
Santa Cruz Mumbai 20 A1
Santa Cruz, I. de Rio de Janeiro 25 B2
Santa Cruz de Olorde Barcelona 4 A1
Santa Efigênia São Paulo 27 B2
Santa Elena Manila 17 B2
Santa Elena del Gomero Santiago 27 B1
Santa Eulalia Barcelona 4 A2
Santa Fe Springs Los Angeles 16 C4
Santa Iria da Azóia Lisbon 16 A2
Santa Julia Santiago 27 C2
Santa Monica Los Angeles 16 B2
Santa Monica Mts. Los Angeles 16 B2
Santa Rosa de Locobe Santiago 27 B2
Santa Teresa de la Ovalle Santiago 27 B1
Santahamina Helsinki 11 C3
Santana São Paulo 27 A2
Santeny Paris 24 B4
Santiago Santiago 27 B2
Santiago Benitez ✈ (SCL) Santiago 27 B1
Santiago de Surco Lima 14 B2
Santo Amaro Lisbon 16 A1
Santo Amaro São Paulo 27 B1
Santo André Lisbon 16 B2
Santo Antão do Tojal Lisbon 16 A2
Santo Antônio, Qta. de Lisbon 16 B1
Santo Tomas, Univ. of Manila 17 B1
Santos Dumont ✈ (SDU) Rio de Janeiro 25 B2
Santoshpur Kolkata 14 B1
Santragachi Kolkata 14 B1
Santry Dublin 10 A2
Sanyuanli Guangzhou 11 B2
São Caetano do Sul São Paulo 27 B2
São Conrado Rio de Janeiro 25 B1
São Cristóvão Rio de Janeiro 25 B1
São Jorge, Castelo de Lisbon 16 A2
São Julião do Tojal Lisbon 16 A2
São Paulo São Paulo 27 B2
São Paulo Congonhas ✈ (CGH) São Paulo 27 B2
Sapa Kolkata 14 B1
Sapateiro, Cor. do → São Paulo 27 B1
Sarandi Buenos Aires 7 C2
Saraswati → Kolkata 14 A1
Sarecky potok → Prague 25 B2
Sarimbun Singapore 28 A2
Sarimbun Res. Singapore 28 A2
Sarıyer Istanbul 12 A2
Sarriá Barcelona 4 A1
Sarsuna Kolkata 14 C1
Sartrouville Paris 24 A2
Sasad Budapest 7 B2
Sashalom Budapest 7 A3
Saska Warsaw 32 B2
Satalice Prague 25 B3
Satgachi Kolkata 14 B2
Satpukur Kolkata 14 B2
Sätra Stockholm 29 B1
Saúde São Paulo 27 B2
Saugus Boston 6 A2
Saugus → Boston 6 A2
Sault-au-Récollet Montreal 20 A2
Sausalito San Francisco 26 A2
Sawah Besar Jakarta 12 A1
Sawtelle Los Angeles 16 B2
Scald Law Edinburgh 11 B2
Scarborough Toronto 31 A3
Scarborough Bluffs Toronto 31 A3
Sceaux Paris 24 B2
Schaerbeek Brussels 6 A2
Scharfenberg Berlin 5 A2
Schiller Park Chicago 9 B1
Schiller Woods Chicago 9 B1
Schiphol, Amsterdam ✈ (AMS) Amsterdam 2 B1
Schlachtensee Berlin 5 B2
Schlossgarten Berlin 5 A2
Schmargendorf Berlin 5 B2
Schönblick Berlin 5 B5
Schönbrunn Vienna 32 A1
Schöneberg Berlin 5 B3
Schönefeld Berlin 5 B4
Schöneiche Berlin 5 B5
Schönwalde Berlin 5 A1
Schulzendorf Berlin 5 A2
Schwabing Munich 21 B2
Schwanebeck Berlin 5 A4
Schwanenwerder Berlin 5 B2
Schwarzlackenau Vienna 32 A2
Schwechat Vienna 32 B2
Schwechat, Wien ✈ (VIE) Vienna 32 B3
Science and Industry, Museum of Chicago 9 C3
Scitrek Museum Atlanta 3 B2
Scopus, Mt. Jerusalem 13 B2
Scottdale Atlanta 3 B3
Scutari = Üsküdar Istanbul 12 B2
Sea Point Cape Town 8 A1
Seabrook Washington 33 B5
Seacliff San Francisco 26 B2
Seaforth Sydney 29 A2
Seagate New York 22 C1
Seat Pleasant Washington 33 B4
Seaview Wellington 33 B2
SeaWorld Orlando 23 B2
Šeberov Prague 25 B3
Secaucus New York 22 B1
Seddinsee Berlin 5 B5
Seeberg Berlin 5 A5
Seeburg Berlin 5 A1
Seefeld Berlin 5 A5
Seegefeld Berlin 5 A1
Seehof Berlin 5 B2
Segeltorp Stockholm 29 B1
Segrate Milan 19 B2
Seguro Milan 19 B1
Seixal Lisbon 16 B2
Selby Johannesburg 13 B2
Seletar, Pulau Singapore 28 A3
Seletar Hills Singapore 28 A3
Seletar Res. Singapore 28 A2
Selhurst London 15 B3
Sembawang Singapore 28 A2
Senago Milan 19 A1
Sendling Munich 21 B2
Senju Tokyo 30 A3
Senriyama Osaka 23 A2
Sentosa Singapore 28 B2
Seobinggo Seoul 27 B1
Seocho Seoul 27 C1
Seodaemun Seoul 27 B1
Seokkwan Seoul 27 B2
Seongbuk Seoul 27 B2
Seongdong Seoul 27 B2
Seongsu Seoul 27 B2
Seoul Seoul 27 B2
Seoul Nat. Univ. Seoul 27 C1
Seoul Tower Seoul 27 B1
Sepolia Athens 2 A2
Sepulveda Dam Rec. Area ⌂ Los Angeles 16 A2
Serangoon Singapore 28 A3
Serangoon, Pulau Singapore 28 A3
Serangoon, Sungei → Singapore 28 A3
Serangoon Harbour Singapore 28 A3
Seraya, Pulau Singapore 28 B2
Serebryanka → Moscow 19 B3
Serramonte San Francisco 26 C2
Sesto San Giovanni Milan 19 A2
Sesto Ulteriano Milan 19 B2
Setagaya Tokyo 30 B2
Seter Oslo 23 A3
Setia Budi Jakarta 12 B1
Settebagni Rome 26 A2
Settecamini Rome 26 B2
Séttimo Milanese Milan 19 B1
Settsu Osaka 23 A2
Seutula Helsinki 11 A2
Seven Corners Washington 33 B3
Seven Kings London 15 A4
Seventeenth Street Canal New Orleans 21 A2
Sévesco → Milan 19 A1
Sevran Paris 24 A4
Sewri Mumbai 20 B2
Sforzesco, Castello Milan 19 B2
Sha Kok Mei Hong Kong 12 A2
Sha Tin Hong Kong 12 A2
Sha'ar Binyamin Jerusalem 13 A2
Shabrâmant Cairo 7 B2
Shah Mosque Tehran 31 A2
Shahdara Delhi 10 A2
Shahe Guangzhou 11 B2
Shahr-e Rey Tehran 31 B2
Shahrak-e Golshahr Tehran 31 B1
Shahrak-e Qods Tehran 31 A2
Shaikh Aomar Baghdad 3 A2
Shakurbasti Delhi 10 A1
Shalkiya Kolkata 14 B2
Sham Shui Po Hong Kong 12 B1
Shamapur Delhi 10 A1
Shamian Guangzhou 11 B2
Shan Mei Hong Kong 12 A2
Shanghai Shanghai 28 B2
Shanghai Hongqiao ✈ (SHA) Shanghai 28 B1
Shanghai Pudong ✈ (PVG) Shanghai 28 B2
Shankill Dublin 10 B3
Sharp I. Hong Kong 12 A2
Shastrinagar Delhi 10 A2
Shau Kei Wan Hong Kong 12 B2
Shawocun Beijing 4 B1
Shayuan Guangzhou 11 B2
Sheen, L. Orlando 23 B1
Sheepshead Bay New York 22 C2
Shek O Hong Kong 12 B2
Shelter I. Hong Kong 12 B2
Shepherds Bush London 15 A2
Shepperton London 15 B1
Sheremetyevo ✈ (SVO) Moscow 19 A2
Sherman Oaks Los Angeles 16 B2
Sherman Park Chicago 9 C2
Sherwood, L. Orlando 23 A2
Shet Bandar Mumbai 20 B2
Sheung Fa Shan Hong Kong 12 A1
Sheung Lau Wan Hong Kong 12 B2
Sheung Wan Hong Kong 12 B1
Sheva Mumbai 20 B2
Sheva Nhava Mumbai 20 B2
Shiba Tokyo 30 B3
Shibpur Kolkata 14 B1
Shibuya Tokyo 30 B3
Shilihe Beijing 4 B2
Shimogawara Tokyo 30 B1
Shimosalo Tokyo 30 A2
Shimoshakujii Tokyo 30 A2
Shinagawa Tokyo 30 B3
Shing Mun Res. Hong Kong 12 A1
Shinjuku Tokyo 30 A3
Shinjuku Nat. Garden Tokyo 30 A3
Shinkoiwa Tokyo 30 A4
Shinnakano Tokyo 30 A3
Shipai Guangzhou 11 B3
Shirinashi → Osaka 23 B1
Shirogane Tokyo 30 B3
Shitennoji Temple Osaka 23 B2
Shiweitang Guangzhou 11 B2
Shogunle Lagos 14 A2
Shomolu Lagos 14 A2
Shooters Hill London 15 B4
Shoreditch London 15 A3
Shortlands London 15 B4
Shu'afat Jerusalem 13 B2
Shubrâ Cairo 7 A2
Shubrâ el Kheima Cairo 7 A2
Shuikuo Guangzhou 11 A2
Sidcup London 15 B4
Siebenhirten Vienna 32 B1
Siedlung Berlin 5 A1
Siekierki Warsaw 32 B2
Sielce Warsaw 32 B2
Siemensstadt Berlin 5 A2
Sierra Madre Los Angeles 16 B4
Sievering Vienna 32 A2
Sighthill Edinburgh 11 B2
Signal Hill Cape Town 8 A1
Sineung Seoul 27 C1
Sikátorpuszta Budapest 7 A3
Silampur Delhi 10 B2
Sillim Seoul 27 C1
Silver, L. Orlando 23 A2
Silver Hill Washington 33 C4
Silver Lake Los Angeles 16 B3
Silver Spring Washington 33 A3
Silvermine Nature Reserve Cape Town 8 B1
Silvolantekojärvi Helsinki 11 B2
Simei Singapore 28 A3
Simla Kolkata 14 B2
Simmering Vienna 32 A2
Simmeringer Heide Vienna 32 A2
Simonkylä Helsinki 11 B3
Sinchon Seoul 27 B1
Sindang Seoul 27 B2
Singapore ■ Singapore 28 B3
Singapore, Univ. of Singapore 28 B2
Singapore Changi ✈ (SIN) Singapore 28 A3
Sinki, Selat Singapore 28 B2
Sinsa Seoul 27 B2
Sint-Genesius-Rode Brussels 6 B
Sion Mumbai 20 A2
Sipson London 15 B1
Siqeil Cairo 7 A1
Şişli Istanbul 12 B1
Skansen Stockholm 29 B2
Skärholmen Stockholm 29 B1
Skarpäng Stockholm 29 A2
Skarpnäck Stockholm 29 B2
Skaryszewski, Park Warsaw 32 B2
Skokie Chicago 9 A2
Skokie → Chicago 9 A2
Skokie Heritage Museum Chicago 9 A2
Skoklefall Oslo 23 A3
Sköndal Stockholm 29 B2
Skovlunde Copenhagen 8 A2
Skovshoved Copenhagen 8 A3
Skuru Stockholm 29 B3
Sky Lake Orlando 23 B2
Skyland Atlanta 3 A3
Slade Green London 15 B5
Slependen Oslo 23 A2
Slipi Jakarta 12 B1
Slivenec Prague 25 B2
Sloten Amsterdam 2 A1
Sloterdijk Amsterdam 2 A1
Sloterpark Amsterdam 2 A1
Sluhy Prague 25 A3
Służew Warsaw 32 B2
Służewiec Warsaw 32 B2
Smíchov Prague 25 B2
Smith Forest Preserve Chicago 9 B2
Smolny St. Petersburg 27 B2
Smolny Cathedral St. Petersburg 27 B2
Snake Creek Canal Miami 18 A
Snakeden Branch → Washington 33 B1
Snarøya Oslo 23 A2
Snättringe Stockholm 29 B1
Søborg Copenhagen 8 A2
Sobreda Lisbon 16 B1
Söderby Stockholm 29 A3
Södermalm Stockholm 29 B2
Sodpur Kolkata 14 A2
Soeurs, Î. des Montreal 20 B2
Sognsvatn Oslo 23 A3
Soignes, Forêt de Brussels 6 B2
Sok Kwu Wan Hong Kong 12 B
Sokolniki Moscow 19 B3
Sokolniki Park Moscow 19 B3
Sokołów Warsaw 32 C1
Solalinden Munich 21 B3
Sollentuna Stockholm 29 A1
Solln Munich 21 B1
Solna Stockholm 29 A1
Somerset Washington 33 B3
Somerville Boston 6 A2
Somes I. Wellington 33 B2
Sonari Mumbai 20 B2
Søndersø Copenhagen 8 A2
Songpa Seoul 27 B2
Sony Picture Studios Los Angeles 16 B2

Sord = Swords Dublin 10 A2
Soroksár Budapest 7 B2
Soroksari Duna → Budapest 7 B2
Sosenka → Moscow 19 B3
Sosnovka St. Petersburg 27 B2
Sŏul = Seoul Seoul 27 B2
Soundview New York 22 B2
South Beach New York 22 C1
South Bend Park Atlanta 3 B2
South Boston Boston 6 A2
South Decatur Atlanta 3 B3
South Deering Chicago 9 C3
South El Monte Los Angeles 16 B4
South Gate Los Angeles 16 C3
South Harbor Manila 17 B1
South Harrow London 15 A1
South Hd. Sydney 29 B2
South Hills Johannesburg 13 B2
South Lawn Washington 33 C3
South Miami Miami 18 B1
South Norwood London 15 B3
South of Market
San Francisco 26 B2
South Ozone Park
New York 22 B3
South Pasadena Los Angeles 16 B4
South Res. Boston 6 A2
South Ruislip London 15 A1
South San Francisco
San Francisco 26 C2
South San Gabriel
Los Angeles 16 B4
South Shore Chicago 9 C3
Southall London 15 A1
Southborough London 15 B4
Southend London 15 B3
Southern California, Univ.
of Los Angeles 16 B3
Southfields London 15 B2
Southport New Orleans 21 B2
Southwark □ London 15 B3
Southwest Museum
Los Angeles 16 B3
Søvang Copenhagen 8 B3
Soweto Johannesburg 13 B1
Sowhānak Tehran 31 A3
Soya Tokyo 30 A4
Spandau Berlin 5 A1
Spånga Stockholm 29 A1
Spanish Monastery Miami 18 A2
Spectacle I. Boston 6 A3
Speicher-See Munich 21 A3
Speising Vienna 32 B1
Sphinx Cairo 7 B1
Spinaceto Rome 26 C1
Spit Junction Sydney 29 A2
Spořilov Prague 25 B3
Spot Pond Boston 6 A2
Spotswood Melbourne 18 B1
Spring Valley Las Vegas 14 B1
Springeberg Berlin 5 B5
Springfield Washington 33 C2
Squantum Boston 6 B2
Srednaya Rogatka
St. Petersburg 27 C2
Śródmieście Warsaw 32 B2
Staaken Berlin 5 A1
Stabekk Oslo 23 A2
Stadlau Vienna 32 A2
Stadshuset Stockholm 29 B2
Stains Paris 24 A3
Stamford Hill London 15 A3
Stammersdorf Vienna 32 A2
Stanley Hong Kong 12 B2
Stanley Pen. Hong Kong 12 B2
Stanmore London 15 A2
Stapleton New York 22 C1
Staraya Derevnya
St. Petersburg 27 B1
Staré Město Prague 25 B2
Stare Miasto Warsaw 32 B2
Starke, L. Orlando 23 A1
Staten Island Zoo New York 22 C1
Staten Islands Ferry
Terminal New York 22 C1
Statenice Prague 25 B1
Stedelijk Museum
Amsterdam 2 A2
Steele Creek Melbourne 18 A1
Steenokkerzeel Brussels 6 A2
Steer, L. Orlando 23 A2
Steglitz Berlin 5 B2
Stepaside Dublin 10 B2
Stephansdom Vienna 32 A2
Stepney London 15 A3
Sterling Park San Francisco 26 B2
Sticklinge udde Stockholm 29 A2
Stickney Chicago 9 C2
Stillorgan Dublin 10 B2
Stockholm Stockholm 29 B2
Stockholm Arlanda ✕
(ARN) Stockholm 29 A1
Stocksund Stockholm 29 A2
Stodůlky Prague 25 B1
Stoke Newington London 15 A3
Stokes Valley Wellington 33 B2
Stone Canyon Res.
Los Angeles 16 B2
Stone Park Chicago 9 B1
Stonebridge London 15 A2
Stoneham Boston 6 A2
Stony Brook Res. Boston 6 B2
Stony Creek → Chicago 9 C2
Stora Värtan Stockholm 29 A2
Store Hareskov Copenhagen 8 A2
Store Magleby Copenhagen 8 B3
Storholmen Stockholm 29 A2
Stoyka St. Petersburg 27 B2
Straiton Edinburgh 11 B3
Strandfontein Cape Town 8 B2
Strašnice Prague 25 B2
Strasstrudering Munich 21 B3
Stratford London 15 A3
Strathfield Sydney 29 B1
Streatham London 15 B3
Streatham Vale London 15 B3
Strebersdorf Vienna 32 A2
Střešovice Prague 25 B2
Strip, The Las Vegas 14 B2
Střížkov Prague 25 B2
Strombeek-Bever Brussels 6 A2
Stromovka Prague 25 B2
Studio City Los Angeles 16 B2
Stureby Stockholm 29 B2
Stuvsta Stockholm 29 B2
Subhepur Delhi 10 A2
Suburban Canal
New Orleans 21 B1
Sucat Manila 17 C2
Suchdol Prague 25 B2
Sucy-en-Brie Paris 24 B4
Sue, L. Orlando 23 A2
Sugamo Tokyo 30 A3
Sugar Loaf Mt. = Açúcar,
Pão de Rio de Janeiro 25 B2
Suge Tokyo 30 B2
Suginami Tokyo 30 A2
Sugō Tokyo 30 B2
Sui Sai Wan Hong Kong 12 B1
Suita Osaka 23 A2
Suitland Washington 33 B4
Sukchar Kolkata 14 A2
Sultanahme Camil Istanbul 12 B1
Sumida Tokyo 30 A3
Sumida → Tokyo 30 A3
Sumiyoshi Osaka 23 B2
Sumiyoshi Shrine Osaka 23 B1
Summerville Toronto 31 B1
Summit Chicago 9 C2
Sunamachi Tokyo 30 A4
Sunbury-on-Thames
London 15 B1
Sundbyberg Stockholm 29 A1
Sundbyerne Copenhagen 8 B3
Sung Kong Hong Kong 12 B2
Sungei Kadut Industrial
Estate Singapore 28 A2
Sungei Selatar Res.
Singapore 28 A3
Sunrise Manor Las Vegas 14 A2
Sunrise Mountain Natural
Area Las Vegas 14 A2
Sunset Park Las Vegas 14 B2
Sunset Park New York 22 C2
Sunter Jakarta 12 A2
Sunter, Kali → Jakarta 12 B2
Suomenlinna Helsinki 11 C2
Sur Bahr Jerusalem 13 B2
Sura Kolkata 14 B2
Suraksan Seoul 27 A2
Surbiton London 15 B2
Suresnes Paris 24 A2
Surfside Miami 18 A2
Surquillo Lima 14 B2
Surrey Hills Sydney 29 B2
Susaek Seoul 27 B1
Süssenbrunn Vienna 32 A2
Sutton Dublin 10 A3
Sutton London 15 B2
Suyu Seoul 27 B2
Suzukishinden Tokyo 30 A2
Svanemøllen Copenhagen 8 A3
Sveaborg = Suomenlinna
Helsinki 11 C2
Sverdlov Moscow 19 B2
Svestad Oslo 23 B2
Svinö Helsinki 11 C1
Swanley London 15 B4
Swansea Toronto 31 B2
Swinburne I. New York 22 C1
Swords Dublin 10 A2
Sydenham Johannesburg 13 A2
Sydney Sydney 29 B2
Sydney, Univ. of Sydney 29 B2
Sydney Harbour △ Sydney 29 B2
Sydney Harbour Bridge
Sydney 29 B2
Sydney Kingsford Smith ✕
(SYD) Sydney 29 B2
Sydstranden Copenhagen 8 B3
Sylvania Sydney 29 C1
Syon Park London 15 B2
Szczęśliwice Warsaw 32 B1
Széchenyi-hegy Budapest 7 B1
Széphalom Budapest 7 A1

T

Taastrup Copenhagen 8 B1
Tabata Tokyo 30 A3
Tablada Buenos Aires 7 C1
Table B. Cape Town 8 A1
Table Mountain △
Cape Town 8 A1
Table Mt. Cape Town 8 A1
Taboão da Serra São Paulo 27 B1
Täby Stockholm 29 A2
Tacuba Mexico City 18 B1
Tacubaya Mexico City 18 B1
Tafelbaai = Table B.
Cape Town 8 A1
Taft Orlando 23 B2
Tagig → Manila 17 B2
Taguig Manila 17 B2
Tai Lo Shan Hong Kong 12 A2
Tai Po Tsai Hong Kong 12 A2
Tai Seng Singapore 28 A3
Tai Shui Hang Hong Kong 12 A2
Tai Tam B. Hong Kong 12 B2
Tai Tam Tuk Res.
Hong Kong 12 B2
Tai Wai Hong Kong 12 A2
Tai Wan Tau Hong Kong 12 B2
Tai Wo Hau Hong Kong 12 A1
Tainaka Osaka 23 B2
Taishō Osaka 23 B1
Taita Wellington 33 B2
Tajrish Tehran 31 A2
Takaido Tokyo 30 A2
Takaishi Tokyo 30 B2
Takarazuka Osaka 23 A1
Takasago Tokyo 30 A4
Takatsu Tokyo 30 B2
Takeshita Tokyo 30 B2
Takinegawa Tokyo 30 A3
Takoma Park Washington 33 B3
Taksim Istanbul 12 B1
Talaide Lisbon 16 A1
Taliganga Kolkata 14 B2
Talipapa Manila 17 A2
Tallaght Dublin 10 B1
Tallkrogen Stockholm 29 B2
Tama Tokyo 30 B1
Tama → Tokyo 30 B2
Tama Kyūryō Tokyo 30 B2
Tamaden Tokyo 30 B2
Tamagawa-josui → Tokyo 30 A1
Taman Sari Jakarta 12 A1
Tamanduatei → São Paulo 27 B2
Tamboerskloof Cape Town 8 A1
Tambora Jakarta 12 A1
Tamminiemi Helsinki 11 B3
Tammūh Cairo 7 B2
Tampines Singapore 28 A3
Tanah Abang Jakarta 12 B1
Tanah Kusir Jakarta 12 B1
Tangelo Park Orlando 23 B2
Tanjung Duren Jakarta 12 B1
Tanjung Priok Jakarta 12 A2
Tanum Oslo 23 A1
Tapada Lisbon 16 A1
Tapanila Helsinki 11 B3
Tapiales Buenos Aires 7 C1
Tapiola Helsinki 11 B1
Tapsia Kolkata 14 B2
Tara Mumbai 20 A1
Tarabya Istanbul 12 B2
Tårbæk Copenhagen 8 A3
Tarchomin Warsaw 32 B1
Tardeo Mumbai 20 B1
Targówek Warsaw 32 B2
Tårnby Copenhagen 8 B3
Tarqua Bay Lagos 14 B2
Tathong Channel
Hong Kong 12 B2
Tathong Pt. Hong Kong 12 B2
Tatuapé São Paulo 27 B2
Taufkirchen Munich 21 B2
Tavares, I. dos Rio de Janeiro 25 A2
Tavros Athens 2 B2
Tawa Wellington 33 A1
Te Papa Museum
Wellington 33 B1
Teaneck New York 22 A1
Tebet Jakarta 12 B2
Tecamachalco Mexico City 18 B1
Ted Williams Tunnel Boston 6 A2
Teddington London 15 B1
Tegel Berlin 5 A2
Tegel, Berlin ✕ (TXL) Berlin 5 A2
Tegeler See Berlin 5 A2
Tegelort Berlin 5 A2
Teheran = Tehrān Tehran 31 A2
Tehrān Tehran 31 A2
Tehrān Mehrābād ✕
(THR) Tehran 31 A1
Tehrān Pārs Tehran 31 A3
Tei Tong Tsui Hong Kong 12 B2
Tejo, Rio → Lisbon 16 A2
Tekstilyshchik Moscow 19 B3
Telhal Lisbon 16 A1
Telok Blangah Singapore 28 B2
Teltow Berlin 5 B2
Teltow kanal Berlin 5 B3
Tempelhof Berlin 5 B3
Temple City Los Angeles 16 B4
Temple Hills Washington 33 C4
Temple of Heaven Beijing 4 B2
Templeogue Dublin 10 B2
Temppeliaukio Church
Helsinki 11 B2
Tenafly New York 22 A2
Tenayuca, Piramide de
Mexico City 18 A1
Tengah → Singapore 28 A2
Tennoji Osaka 23 B2
Tepalcates Mexico City 18 B2
Tepeyac △ Mexico City 18 B2
Terrazzano Milan 19 A1
Terre des Hommes Montreal 20 A2
Terrugem Lisbon 16 A1
Terrytown New Orleans 21 B2
Tervuren Brussels 6 B3
Tervuren, Park van Brussels 6 B3
Tetuán Madrid 17 B1
Teufelsberg Berlin 5 B2
Thalkirchen Munich 21 B2
Thames Ditton London 15 B1
Thamesmead London 15 A4
Thana Cr. → Mumbai 20 A2
Thiais Paris 24 B3
Thisio Athens 2 B2
Thistletown Toronto 31 A1
Thomastown Melbourne 18 A2
Thompson I. Boston 6 B3
Thon Buri Bangkok 3 B1
Thornbury Melbourne 18 A2
Thorncliffe Toronto 31 A2
Thornhill Toronto 31 A2
Thornton Cape Town 8 A2
Thornton Heath London 15 B3
Threipmuir Res. Edinburgh 11 B2
Throgs Neck New York 22 B3
Tiancun Beijing 4 B1
Tianhe Guangzhou 11 B3
Tibet, L. Orlando 23 B1
Tibidabo Barcelona 4 A1
Tibradden Mt. Dublin 10 B2
Tiburon San Francisco 26 A2
Tiburtino Rome 26 B2
Ticomán Mexico City 18 A2
Tiefersee Berlin 5 B1
Tiergarten Berlin 5 A3
Tijgerhof Cape Town 8 A2
Tijuca Rio de Janeiro 25 B1
Tijuca, Pico da Rio de Janeiro 25 B1
Tijuca △ Rio de Janeiro 25 B1
Tikkurila Helsinki 11 B3
Tilak Nagar Delhi 10 B1
Tilanqiao Shanghai 28 B1
Timah, Bukit Singapore 28 A2
Timiryazev Park Moscow 19 B2
Tin Can I. Lagos 14 B2
Ting Kau Hong Kong 12 A1
Tira Jerusalem 13 A1
Tirsa Cairo 7 B2
Tishrīyaa Baghdad 3 B2
Tiu Keng Leng Hong Kong 12 B2
Tizapán Mexico City 18 C1
Tlalnepantla → Mexico City 18 A1
Tlalpan Mexico City 18 C2
To Kwai Wan Hong Kong 12 B2
Toa Payoh Singapore 28 A3
Točná Prague 25 C2
Toco Hills Atlanta 3 B2
Todt Hill New York 22 C1
Tōkagi Tokyo 30 A4
Tokai Plantation Cape Town 8 B1
Tōkaichiba Tokyo 30 B2
Tōkyō Tokyo 30 B3
Tōkyō B. = Tōkyō-Wan
Tokyo 30 B4
Tōkyō Disneyland Tokyo 30 B4
Tōkyō Haneda ✕ (HND)
Tokyo 30 B3
Tokyo Harbour Tokyo 30 B3
Tōkyō-Wan Tokyo 30 B4
Tolka → Dublin 10 A1
Tollygunge = Taliganga
Kolkata 14 B2
Tolworth London 15 B2
Tomba di Nerone Rome 26 B1
Tommy Thompson Park
Toronto 31 B3
Tondo Manila 17 B1
Tongbinggo Seoul 27 B1
Tongqiao Shanghai 28 A1
Toorak Melbourne 18 B2
Topanga State Park ⌂
Los Angeles 16 B1
Topkapı Istanbul 12 B1
Topkapı Palaca Istanbul 12 B1
Tor di Quinto Rome 26 B1
Tor Pignattara Rome 26 B2
Tor Sapienza Rome 26 B2
Torcy Paris 24 A4
Toronto Toronto 31 B2
Toronto, Univ. of Toronto 31 B2
Toronto City Centre ✕
(YTZ) Toronto 31 B2
Toronto Harbour Toronto 31 B2
Toronto I. Toronto 31 B2
Toronto Lester B. Pearson
Int. ✕ (YYZ) Toronto 31 A1
Torre Lupara Rome 26 B2
Torre Nova Rome 26 B2
Torrellas → Barcelona 4 A1
Torrevécchia Rome 26 B1
Toshima Tokyo 30 A3
Toshimaen Tokyo 30 A2
Tottenham London 15 A3
Tottenham Melbourne 18 A1
Toussus-le-Noble Paris 24 B1
Toussus-le-Noble ✕ (TNF)
Paris 24 B1
Towchāl Cable Car Tehran 31 A2
Tower Hamlets □ London 15 A3
Tower of London London 15 A3
Towra Pt. Sydney 29 C2
Tøyen Oslo 23 A3
Toyonaka Osaka 23 A1
Trafaria Lisbon 16 A1
Traição, Cor. → São Paulo 27 B2
Tranegilde Copenhagen 8 B2
Trångsund Stockholm 29 B2
Trappenfelde Berlin 5 A4
Trastévere Rome 26 B1
Treasure I. San Francisco 26 B2
Třeboradice Prague 25 B3
Třebotov Prague 25 C1
Tremblay-en-France Paris 24 A4
Tremembe → São Paulo 27 A2
Tremont New York 22 A2
Trenno Milan 19 B1
Treptow Berlin 5 B3
Três Rios, Sa. dos
Rio de Janeiro 25 B1
Trevi, Fontana di Rome 26 B1
Trezzano sul Naviglio Milan 19 B1
Trieste Rome 26 B2
Trinidad Washington 33 B4
Trinity Edinburgh 11 B2
Trinity College Dublin 10 A2
Trionfale Rome 26 B1
Triulzo Milan 19 B2
Troja Prague 25 B2
Trollbäcken Stockholm 29 B3
Trombay Mumbai 20 A2
Troparevo Moscow 19 C1
Trudeau, Montréal ✕
(YUL) Montreal 20 B1
Trudyashchikhsya,
Ostrov St. Petersburg 27 B1
Tryvasshøgda Oslo 23 A3
Tseng Lan Shue Hong Kong 12 A2
Tseung Kwan Hong Kong 12 B2
Tsim Sha Tsui Hong Kong 12 B2
Tsing Yi Hong Kong 12 A1
Tsova Jerusalem 13 B1
Tsuen Wan Hong Kong 12 A1
Tsur Hadassa Jerusalem 13 B1
Tsurumi → Tokyo 30 B3
Tsz Wan Shan Hong Kong 12 A2
Tuas Singapore 28 B1
Tuchoměřice Prague 25 B1
Tuckahoe New York 22 A3
Tucuruvi São Paulo 27 A2
Tufello Rome 26 B2
Tufnell Park London 15 A3
Tufts Univ. Boston 6 A2
Tughlakabad Delhi 10 B2
Tullamarine Melbourne 18 A1
Tulse Hill London 15 B3
Tung Lung Chau Hong Kong 12 B2
Tung O Hong Kong 12 B1
Tunis Baghdad 3 A2
Tuomarila Helsinki 11 B1
Tureberg Stockholm 29 A1
Turffontein Johannesburg 13 B2
Turkey L. Orlando 23 A2
Turkey Lake Park Orlando 23 A2
Turner Field Atlanta 3 B2
Turnham Green London 15 B2
Turnhouse Edinburgh 11 B1
Tuscolana, Via Rome 26 B2
Twelve Apostles Cape Town 8 A1
Twentieth Century Fox
Studios Los Angeles 16 B2
Twickenham London 15 B1
Twickenham Rugby
Ground London 15 B1
Twin Peaks San Francisco 26 B2
Two Rock Mt. Dublin 10 B2
Tymon North Dublin 10 B1
Tysons Corner Washington 33 B2

U

U.S. Capitol Washington 33 B3
U.S. Cellular Field Chicago 9 C3
Ubeidiya Jerusalem 13 B2
Uberaba → São Paulo 27 B2
Ubin, Pulau Singapore 28 A3
Uccle Brussels 6 B2
Udelnaya St. Petersburg 27 A2
Udelnoe St. Petersburg 27 B1
Udlding Munich 21 A1
Ueno Tokyo 30 A3
Úholičky Prague 25 B1
Uhříněves Prague 25 B3
Uithoorn Amsterdam 2 B1
Ujazdów Warsaw 32 B2
Újpalota Budapest 7 A2
Újpest Budapest 7 A2
Ukita Tokyo 30 A4
Ullerup Copenhagen 8 B3
Ullevål Oslo 23 A3
Ulriksdal Stockholm 29 A1
Ulyanka St. Petersburg 27 B1
Um Al-Khanazir Island
Baghdad 3 B2
Umeda Osaka 23 A1
Ümraniye Istanbul 12 B2
Underground Atlanta
Atlanta 3 B2
Underhill, L. Orlando 23 A3
Unětický potok → Prague 25 B2
Unhos Lisbon 16 A2
Unidad Santa Fe
Mexico City 18 B1
Union City New York 22 B1
Union Port New York 22 B2
Union Station Toronto 31 B2
United Center Chicago 9 B2
United Nations
Headquarters New York 22 B2
Universal Studios
Los Angeles 16 B2
Universal Studios Orlando 23 B2
Universidad Madrid 17 B1
University Park Washington 33 B4
Unterbiberg Munich 21 B2
Unterföhring Munich 21 A3
Unterhaching Munich 21 B2
Unterlaa Vienna 32 B2
Untermenzing Munich 21 A1
Upper B. New York 22 B1
Upper Elmers End London 15 B3
Upper New York B. =
Upper B. New York 22 C1
Upper Norwood London 15 B3
Upper Peirce Res. Singapore 28 A2
Upper Sydenham London 15 B3
Upper Tooting London 15 B3
Upton London 15 A4
Uptown Chicago 9 B2
Uran Mumbai 20 B2
Urayasu Tokyo 30 B4
Urbe ✕ Rome 26 B2
Urca Rio de Janeiro 25 B2
Uritsk St. Petersburg 27 C1
Úrōm Budapest 7 A2
Ursus Warsaw 32 B1
Ursvik Stockholm 29 A1
Usera Madrid 17 B1
Ushigome Tokyo 30 A3
Usina Rio de Janeiro 25 B1
Üsküdar Istanbul 12 B2
Ust-Slavyanka
St. Petersburg 27 C3
Uteke Stockholm 29 A3
Utrata Warsaw 32 B2
Uttarpara Kolkata 14 B1
Utterslev Mose Copenhagen 8 A2

V

Vaclavské náměstí Prague 25 B2
Vadaul Mumbai 20 A2
Vaires-sur-Marne Paris 24 A4
Valby Copenhagen 8 B2
Valcannuta Rome 26 B1
Valdelatas Madrid 17 A1
Vale Washington 33 B1
Valenton Paris 24 B3
Valera Milan 19 A1
Vallcarca Barcelona 4 A1
Valldoreix Barcelona 4 A1
Vallecas Madrid 17 B2
Vallecas, Cumbres de
Madrid 17 B2
Vallensbæk Copenhagen 8 B2
Vallensbæk Strand
Copenhagen 8 B2
Vallentunasjön Stockholm 29 A2
Valleranello Rome 26 C1
Vallisaari Helsinki 11 C3
Vallvidrera Barcelona 4 A1
Valvidrera → Barcelona 4 A1
Van Gogh Museum
Amsterdam 2 A2
Vanak Tehran 31 A2
Vanda = Vantaa Helsinki 11 B2
Vangede Copenhagen 8 A3
Vaniköy Istanbul 12 B2
Vanlose Copenhagen 8 A2
Vantaa Helsinki 11 B2
Vantaankoski Helsinki 11 B2
Vantaanpuisto Helsinki 11 B2
Vanves Paris 24 B2
Varedo Milan 19 A1
Varkiza Athens 2 C2
Vartiokylä Helsinki 11 B3
Vartiosaari Helsinki 11 B3
Vasamuseet Stockholm 29 B2
Vasco Cape Town 8 A2
Vasfanärd Tehran 31 B2
Vashi Mumbai 20 A2
Vasilyevskiy, Ostrov
St. Petersburg 27 B1
Vatican City ■ Rome 26 B1
Vaucluse Sydney 29 B2
Vaucluse House Sydney 29 B2
Vaucresson Paris 24 A1
Vaughan Toronto 31 A1
Vauhallan Paris 24 B2
Vaujours Paris 24 A4
Vauxhall London 15 B3
Vecsés Budapest 7 B3
Veleň Prague 25 A3
Veleslavin Prague 25 B2
Vélizy-Villacoublay Paris 24 B2
Velka-Chuchle Prague 25 B2
Velké Přílepy Prague 25 B1
Venda Seca Lisbon 16 A1
Venetian Islands Miami 18 B2
Venice Los Angeles 16 C2
Ventas Madrid 17 B1
Ventorro del Cano Madrid 17 B1
Venustiano Carranza
Mexico City 18 B2
Verde → São Paulo 27 A1
Verdi Athens 2 A2
Verdugo Mts. Los Angeles 16 A2
Verdun Montreal 20 B2
Vérhalom Budapest 7 A2
Vermelho → São Paulo 27 B1
Vermont Knolls Los Angeles 16 C3
Vernon Los Angeles 16 B3
Verrières-le-Buisson Paris 24 B2
Versailles Buenos Aires 7 B1
Versailles Paris 24 B1
Veshnyaki Moscow 19 B3
Vesolyy Posolok
St. Petersburg 27 B2
Vestra Helsinki 11 B1
Vestskoven Copenhagen 8 A2
Vicálvaro Madrid 17 B2
Vicente Lopez Buenos Aires 7 B2
Victoria Hong Kong 12 B2
Victoria, Mt. Wellington 33 B1
Victoria, Pont Montreal 20 B2
Victoria and Alfred
Waterfront Cape Town 8 A1
Victoria Gardens Mumbai 20 B2
Victoria Harbour Hong Kong 12 B1
Victoria Island Lagos 14 B2
Victoria L. Johannesburg 13 B2
Victoria Lawn Tennis
Courts Melbourne 18 B2
Victoria Park Singapore 28 B2
Victoria Peak Hong Kong 12 B1
Vienna = Wien Vienna 32 A2
Vienna Washington 33 B2
Vietnam Veterans
Memorial Washington 33 B3
Vietnam War Museum
Chicago 9 B3
Vieux Carré = French
Quarter New Orleans 21 B2
View Park Los Angeles 16 B3
Vigário Geral Rio de Janeiro 25 A1
Vigentino Milan 19 B2
Viggbyholm Stockholm 29 A2
Vighignolo Milan 19 B1
Viikki Helsinki 11 B3
Vikhroli Mumbai 20 A2
Vila Guilherme São Paulo 27 B2
Vila Isabel Rio de Janeiro 25 B1
Vila Jaguára São Paulo 27 B1
Vila Madalena São Paulo 27 B1
Vila Maria São Paulo 27 B2
Vila Mariana São Paulo 27 B2
Vila Prudente São Paulo 27 B2
Viladecans Barcelona 4 B1
Vile Parle Mumbai 20 A2
Villa Adelina Buenos Aires 7 B1
Villa Ballester Buenos Aires 7 B1
Villa Barilari Buenos Aires 7 C2
Villa Bosch Buenos Aires 7 B1
Villa C. Colon Buenos Aires 7 C2
Villa Ciudadela Buenos Aires 7 B1
Villa de Guadalupe
Mexico City 18 B2
Villa Devoto Buenos Aires 7 B1
Villa Diamante Buenos Aires 7 C2
Villa Dominico Buenos Aires 7 C3
Villa Lugano Buenos Aires 7 C2
Villa Lynch Buenos Aires 7 B1
Villa Madero Buenos Aires 7 C1
Villa Sáenz Pena
Buenos Aires 7 B1
Villa Urquiza Buenos Aires 7 B2
Villaverde Madrid 17 B1
Villaverde Bajo Madrid 17 B1
Ville-d'Avray Paris 24 B2
Villecresnes Paris 24 B4
Villejuif Paris 24 B3
Villemomble Paris 24 A4
Villeneuve-la-Garenne
Paris 24 A2
Villeneuve-le-Roi Paris 24 B3
Villeneuve-St-Georges Paris 24 B3
Villeparisis Paris 24 A4
Villevaudé Paris 24 A4
Villiers-le-Bâcle Paris 24 B1
Villiers-sur-Marne Paris 24 B4
Villinki Helsinki 11 C3
Villoresi, Canale Milan 19 A1
Vilvoorde Brussels 6 A2
Vimodrone Milan 19 A2
Vimont Montreal 20 A1
Vincennes Paris 24 A3
Vincennes, Bois de Paris 24 B3
Vineland Orlando 23 B1
Vinings Atlanta 3 A2
Vinohrady Prague 25 B2
Vinoř Prague 25 B3
Violet Hill Hong Kong 12 B2
Virányos Budapest 7 A1
Virginia, L. Orlando 23 A2
Virginia Gardens Miami 18 B1
Virginia Key Miami 18 B2
Viroflay Paris 24 B2
Vironas Athens 2 B2
Virum Copenhagen 8 A2
Visitacion Valley
San Francisco 26 B2
Vista Alegre Lima 14 B3
Vista Alegre Santiago 27 C1
Vista Grove Atlanta 3 A3
Vitacura Santiago 27 B2
Vitinia Rome 26 C1
Vitry-sur-Seine Paris 24 B3
Vizcaya Museum and
Gardens Miami 18 B2
Vladykino Moscow 19 A2
Vokovice Prague 25 B2
Volgelsdorf Berlin 5 B5
Volkhonka-Zil Moscow 19 C2
Vollen Oslo 23 B1
Volodarskoye St. Petersburg 27 B2
Volynkina-Derevnya
St. Petersburg 27 B1
Vondelpark Amsterdam 2 A2
Vösendorf Vienna 32 B2
Voula Athens 2 C2
Vouliagmeni Athens 2 C2
Vredehoek Cape Town 8 A1
Vršovice Prague 25 B2
Vyborgskaya Storona
St. Petersburg 27 B2
Vykhino Moscow 19 B3
Vyšehrad Prague 25 B2

W

Wachterhof Munich 21 B3
Wadala Mumbai 20 A2
Wadestown Wellington 33 B1
Wadi al-Arayis Jerusalem 13 B2
Wadi Fukin Jerusalem 13 B1
Waduk Pluit Jakarta 12 A1
Waggaman New Orleans 21 B1
Wah Fu Hong Kong 12 B1
Wahda Baghdad 3 B2
Währing Vienna 32 A2
Waidmannslust Berlin 5 A3
Wainuiomata Wellington 33 B2
Wainuiomata →
Wellington 33 B2
Wakefield Boston 6 A2
Waldesruh Berlin 5 B4
Waldperlach Munich 21 B3
Waldtrudering Munich 21 B3
Walkinstown Dublin 10 B1
Wall Street New York 22 B1
Walt Disney World Orlando 23 B1
Walter D. Stone Memorial
Zoo Boston 6 A2
Waltham Boston 6 A1
Waltham Forest □ London 15 A3
Walthamstow London 15 A3
Walton-on-Thames London 15 B1
Wambeek Brussels 6 A3
Wan Chai Hong Kong 12 B2
Wandle → London 15 B2
Wandsworth □ London 15 B2
Wangjing Beijing 4 B2
Wannsee Berlin 5 B1
Wansdorf Berlin 5 A1
Wanstead London 15 A4
Wapping London 15 A3
Ward I. Wellington 33 B2
Wards I. New York 22 B2
Warnberg Munich 21 B2
Warner Brothers Studios
Los Angeles 16 B2
Warrâq el 'Arab Cairo 7 A2
Warringal Park Melbourne 18 A2
Warriston Edinburgh 11 B2
Warsaw = Warszawa
Warsaw 32 B2
Warszawa Warsaw 32 B2
Warszawa ✕ (WAW)
Warsaw 32 B1
Wartenberg Berlin 5 A4
Washington Washington 33 B3
Washington Dulles Int. ✕
(IAD) Washington 33 B1
Washington Heights
New York 22 A2
Washington Park Chicago 9 C3
Washington Ronald
Reagan Nat. ✕ (DCA)
Washington 33 B3
Wat Pho Bangkok 3 B1
Water of Leith → Edinburgh 11 B1
Watergraafsmeer Amsterdam 2 A2
Waterland Amsterdam 2 A2
Waterloo Brussels 6 B2
Waterloo Wellington 33 B2
Watermeal-Boitsfort Brussels 6 B2
Watertown Boston 6 A1
Watsonia Melbourne 18 A2
Waverley Boston 6 A1
Waverley Johannesburg 13 A2
Waverley Sydney 29 B2
Wawer Warsaw 32 B2
Wawrzyszew Warsaw 32 B1
Wazirabad Delhi 10 A2
Wazīrīya Baghdad 3 A2
Wazirpur Delhi 10 A2
Wealdstone London 15 A1
Wedding Berlin 5 A3
Weehawken New York 22 B1
Weesp Amsterdam 2 B3
Weidling Vienna 32 A1
Weidlingbach Vienna 32 A1
Weissensee Berlin 5 A3
Wellesley Hills Boston 6 B1
Welling London 15 B4
Wellington Boston 6 A2
Wellington Wellington 33 B1
Wellington Int. ✕ (WLE)
Wellington 33 B1
Weltevreden Park
Johannesburg 13 A1
Wembley London 15 A2
Wemmel Brussels 6 A1
Wemmer Pan Johannesburg 13 B2
Wenceslas Square =
Vaclavské náměstí
Prague 25 B2
Wendenschloss Berlin 5 B4
Wennington London 15 A5
Werneuchen Berlin 5 A5
West Don → Toronto 31 A2
West Drayton London 15 A1
West Ham London 15 A4
West Harrow London 15 A1
West Heath London 15 B4
West Hill Toronto 31 A3
West Hollywood Los Angeles 16 B2
West Lamma Channel
Hong Kong 12 B1
West Los Angeles, Univ. of
Los Angeles 16 C2
West Medford Boston 6 A2
West Miami Miami 18 B1
West Molesey London 15 B1
West New York New York 22 B1
West of Twin Peaks
San Francisco 26 B2
West Park Johannesburg 13 A1
West Rouge Toronto 31 A4
West Roxbury Boston 6 B2
West Springfield
Washington 33 C2
West Town Chicago 9 B2
West Wharf Karachi 13 B1
Westchester Chicago 9 B1
Westchester Los Angeles 16 C2
Westchester New York 22 A2
Westcliff Johannesburg 13 B2
Westdene Johannesburg 13 B1
Westend Helsinki 11 C1
Wester Hailes Edinburgh 11 B2
Westerham Munich 21 B2
Western Addition
San Francisco 26 B2
Westgate New Orleans 21 A1
Westgate Washington 33 B3
Westlake Cape Town 8 B1
Westlake Los Angeles 16 B3
Westlake San Francisco 26 B2
Westminster London 15 A3
Westmount Montreal 20 B2
Weston Toronto 31 A1
Westwego New Orleans 21 B2
Westwood Village
Los Angeles 16 B2
Westzaan Amsterdam 2 A1
Wetton Cape Town 8 B2
Wexford Toronto 31 A3
Weybridge London 15 B1
Wezembeek-Oppem Brussels 6 A2
White Cloud Hill = Baiyun
Hill Guangzhou 11 B2
White House, The
Washington 33 B3
Whitechapel London 15 A3
Whitehall Dublin 10 A2
Whitney Las Vegas 14 B2
Whittier Los Angeles 16 C4
Whitton London 15 B1
Wieden Vienna 32 A2
Wien Vienna 32 A2
Wien-Schwechat ✕ (VIE)
Vienna 32 B3
Wienerberg Vienna 32 B2
Wierzbno Warsaw 32 B2
Wijde Wormer Amsterdam 2 A2
Wilanów Warsaw 32 B2
Wilanówka → Warsaw 32 C2
Wilds, The Johannesburg 13 B2
Wilhelmshagen Berlin 5 B5
Wilket Creek Park Toronto 31 A2
Wilkieston Edinburgh 11 B1
Will Rogers State
Historical Park ⌂
Los Angeles 16 B1
Willbrook Dublin 10 B2
Willesden London 15 A2
Willesden Green London 15 A2
Williamsbridge New York 22 A2
Williamsburg New York 22 B2
Williamsburg Orlando 23 B2
Williamstown Melbourne 18 B1
Willis, L. Orlando 23 B2
Willoughby Sydney 29 A2
Willow Springs Chicago 9 C1
Willowdale New Orleans 21 A1
Willowdale Toronto 31 A2
Wilmersdorf Berlin 5 B2
Wilmette Chicago 9 A2
Wilmington London 15 B5
Wimbledon London 15 B2
Wimbledon Common
London 15 B2
Wimbledon Park London 15 B2
Wimbledon Tennis
Ground London 15 B2
Winchester Boston 6 A2
Winchester Las Vegas 14 B2
Windermere Cape Town 8 A2
Windermere Orlando 23 B1
Windsor Johannesburg 13 A1
Windsor, Gare Montreal 20 B2
Windsor Hills Los Angeles 16 C2
Windy Arbour Dublin 10 B2
Winning Munich 21 B2
Wissous Paris 24 B2
Wittenau Berlin 5 A2
Witwatersrand, Univ. of
Johannesburg 13 B2
Wlochy Warsaw 32 B1
Wo Mei Hong Kong 12 A2
Wo Yi Hop Hong Kong 12 A1
Woburn Boston 6 A2
Woburn Toronto 31 A3
Wola Warsaw 32 B1
Wolf Trap Farm Park
Washington 33 B2
Wolgok Seoul 27 B2
Wolica Warsaw 32 C2
Wolica Warsaw 32 C1
Wollaston Boston 6 B2
Woltersdorf Berlin 5 B5
Woluwe-St-Lambert Brussels 6 A2
Woluwe-St-Pierre Brussels 6 B2
Wong Chuk Hang
Hong Kong 12 B2
Wong Chuk Wan
Hong Kong 12 A2
Wong Chuk Yeung
Hong Kong 12 A2
Wood Green London 15 A3
Wood Ridge New York 22 A1
Woodbine Race Track
Toronto 31 A1
Woodbridge Toronto 31 A1
Woodford London 15 A4
Woodford Bridge London 15 A4
Woodford Green London 15 A4
Woodhaven New York 22 B2
Woodhouselee Edinburgh 11 B2
Woodlands Singapore 28 A2
Woodmont Washington 33 B3
Woodside London 15 B3
Woodside New York 22 B2
Woodstock Cape Town 8 A1
Woollahra Sydney 29 B2
Woolooware B. Sydney 29 C1
Woolwich London 15 B4
Worker's Stadium Beijing 4 B2
World Financial Centre
Shanghai 28 B2
World Trade Center, Site
of former = Ground
Zero New York 22 B1
Worli Mumbai 20 A1
Worth Chicago 9 C2
Wren's Nest Atlanta 3 B2
Wrigley Field Chicago 9 B3
Wuhlgarten Berlin 5 A4
Wujiaochang Shanghai 28 B2
Würm → Munich 21 B1
Würm-kanal Munich 21 A1
Wusong Shanghai 28 A1
Wyczółki Warsaw 32 C1
Wygoda Warsaw 32 B2
Wynberg Cape Town 8 B1

X

Xabregas Lisbon 16 A2
Xiaogang Park Guangzhou 11 B2
Xiaoping Guangzhou 11 A2
Xiasha chong Guangzhou 11 B1
Xichang Guangzhou 11 B2
Xicheng Beijing 4 B1
Xidan Beijing 4 B2
Xizhimen Beijing 4 B1
Xochimilco, Parque
Ecológico Mexico City 18 C2
Xuanwu Beijing 4 B2
Xuhui Shanghai 28 B1

Y

Yaba Lagos 14 A2
Yaftābād Tehran 31 B1
Yahara Tokyo 30 A2
Yaho Tokyo 30 A1
Yakire Tokyo 30 A4
Yaksu Seoul 27 B2
Yamada Osaka 23 A2
Yamada Tokyo 30 B2
Yamato → Osaka 23 B1
Yan Kit Singapore 28 A3
Yanbu Guangzhou 11 B1
Yangcheon Seoul 27 B1
Yanghuayuan Beijing 4 C1
Yangjae Seoul 27 C2
Yangjiazhuang Shanghai 28 A1
Yangjing Shanghai 28 B2
Yangpu Shanghai 28 B2
Yangpu Park Shanghai 28 B2
Yannawa Bangkok 3 B2
Yarmūk Baghdad 3 B1
Yarra Bend Park Melbourne 18 A2
Yarraville Melbourne 18 A1
Yauza → Moscow 19 B4
Yeading London 15 A1
Yedikule Istanbul 12 C1
Yenikapı Istanbul 12 B1
Yeniköy Istanbul 12 B2
Yeongdeungpo Seoul 27 B1
Yeongdong Seoul 27 B2
Yeouido Seoul 27 B1
Yerba Buena I.
San Francisco 26 B2
Yerres Paris 24 B4
Yerushalayim =
Jerusalem Jerusalem 13 B2
Yiexiu Guangzhou 11 B2
Yiheyuan Beijing 4 B1
Yıldız Park Istanbul 12 B2
Yinhangzhen Shanghai 28 A2
Yishun Singapore 28 A3
Ylästö Helsinki 11 B2
Yodo → Osaka 23 A2
Yongdingman Beijing 4 B2
Yongfucun Guangzhou 11 B2
Yongsan Seoul 27 B1
Yonkers New York 22 A2
York Toronto 31 A2
York Mills Toronto 31 A2
York Univ. Toronto 31 A1
You'anmen Beijing 4 B1
Youngsfield Cape Town 8 B1
Yuanmingyuan Park Beijing 4 A1
Yuanxiatian Guangzhou 11 A2
Yuexiu Park Guangzhou 11 B2
Yugo-Zarad Moscow 19 B2
Yuhalixqui, Volcan
Mexico City 18 C2
Yung Shue Wan Hong Kong 12 B1
Yūsofābād Tehran 31 A2

Z

Zaandam Amsterdam 2 A1
Zaandijk Amsterdam 2 A1
Zaanstad Amsterdam 2 A1
Záběhlice Prague 25 B2
Zabki Warsaw 32 B2
Zacisze Warsaw 32 B2
Zahrā Baghdad 3 A1
Zalov Prague 25 A2
Zaluski Warsaw 32 C1
Zamdorf Munich 21 B2
Zanevka St. Petersburg 27 B3
Zapote Manila 17 C1
Zaventem Brussels 6 A2
Zawady Warsaw 32 B2
Zāwiyet Abû Musallam
Cairo 7 B1
Zawrā' Park Baghdad 3 B2
Zbraslav Prague 25 C2
Zbuzany Prague 25 B1
Zdiby Prague 25 B2
Zeekoevlei Cape Town 8 B2
Zehlendorf Berlin 5 B2
Zenne → Brussels 6 A2
Zephyr Field New Orleans 21 B1
Żerań Warsaw 32 B1
Żerzeń Warsaw 32 B2
Zeytinburnu Istanbul 12 C1
Zhabei Shanghai 28 B1
Zhdanov Moscow 19 B3
Zhenru Shanghai 28 B1
Zhernovka St. Petersburg 27 B2
Zhicun Guangzhou 11 B2
Zhongguancun Beijing 4 B1
Zhongshan Park Shanghai 28 B1
Zhoucun Guangzhou 11 A2
Zhoujiadu Shanghai 28 B1
Zhoujiazhen Shanghai 28 B2
Zhouzhuangzi Beijing 4 B1
Zhulebino Moscow 19 B4
Zhushadi Guangzhou 11 A3
Zielona Warsaw 32 B3
Zielonka Warsaw 32 B2
Zizhuyuan Park Beijing 4 B1
Žižkov Prague 25 B2
Zličín Prague 25 B1
Zografou Athens 2 B2
Żoliborz Warsaw 32 B2
Zonnebloem Cape Town 8 A1
Zugliget Budapest 7 B1
Zugló Budapest 7 A2
Zuiderwoude Amsterdam 2 A3
Zumbi Rio de Janeiro 25 A1
Zunderdorp Amsterdam 2 A2
Zuvuvu → São Paulo 27 C1
Zwanenburg Amsterdam 2 A1
Zwölfaxing Vienna 32 B2

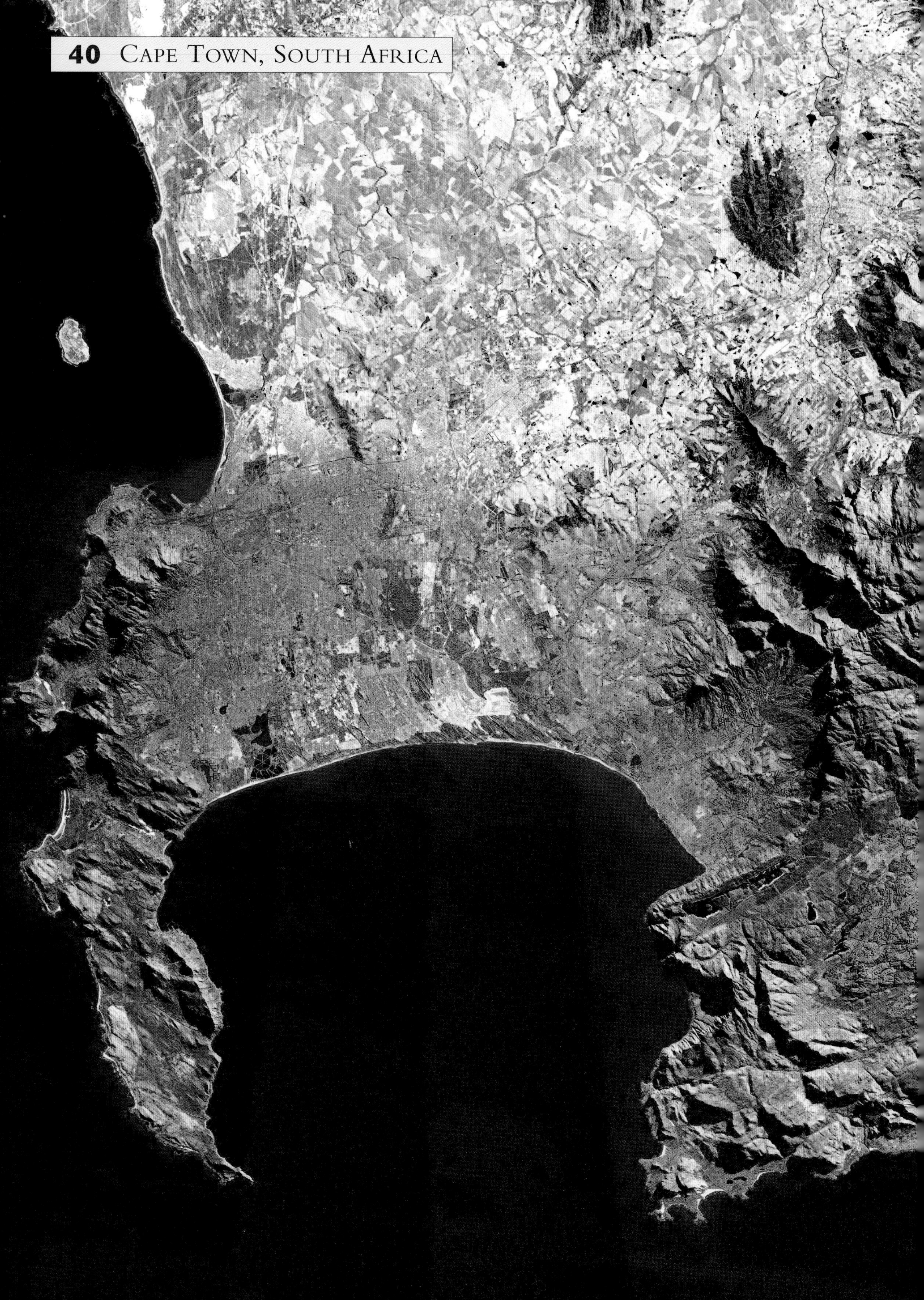

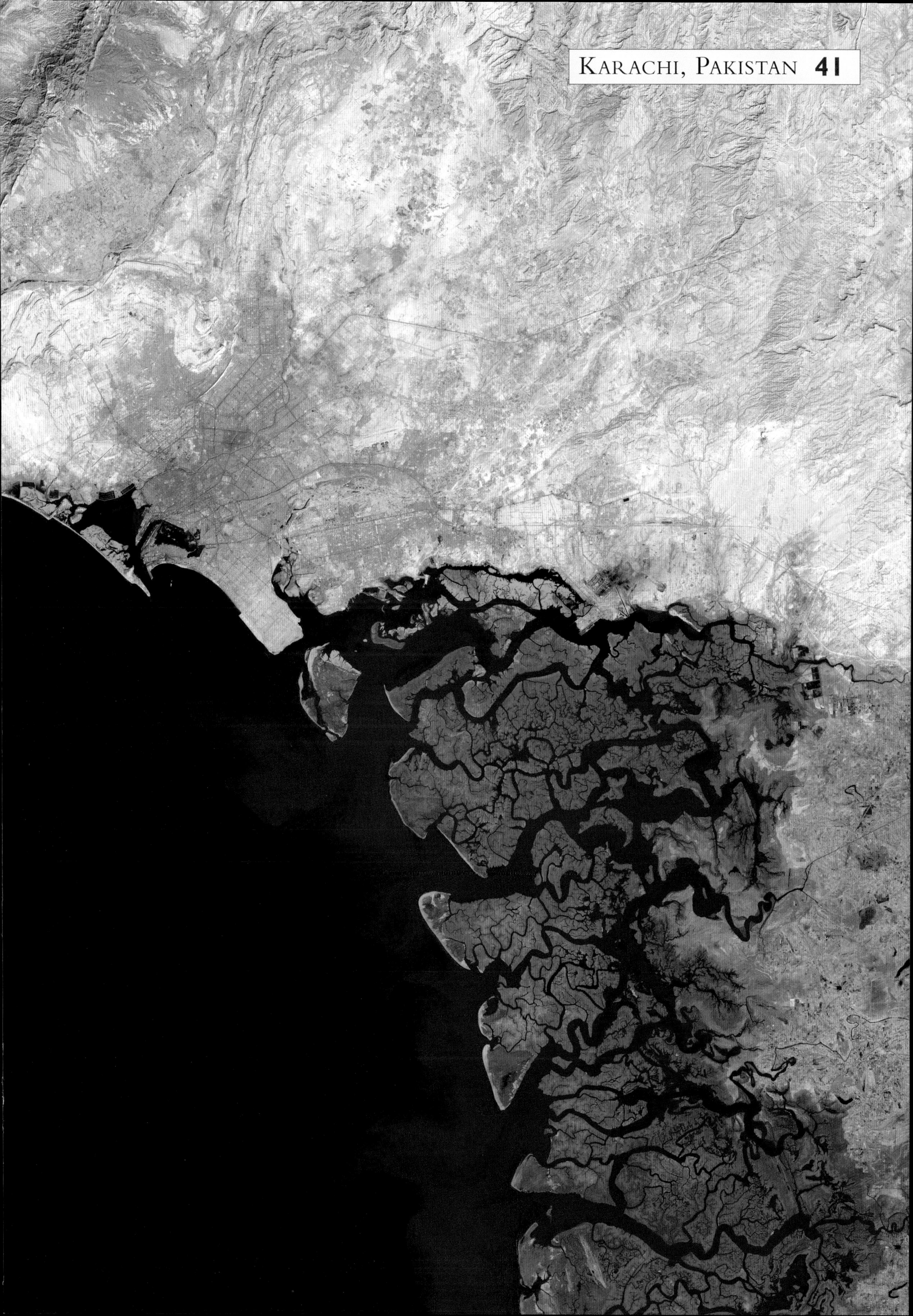

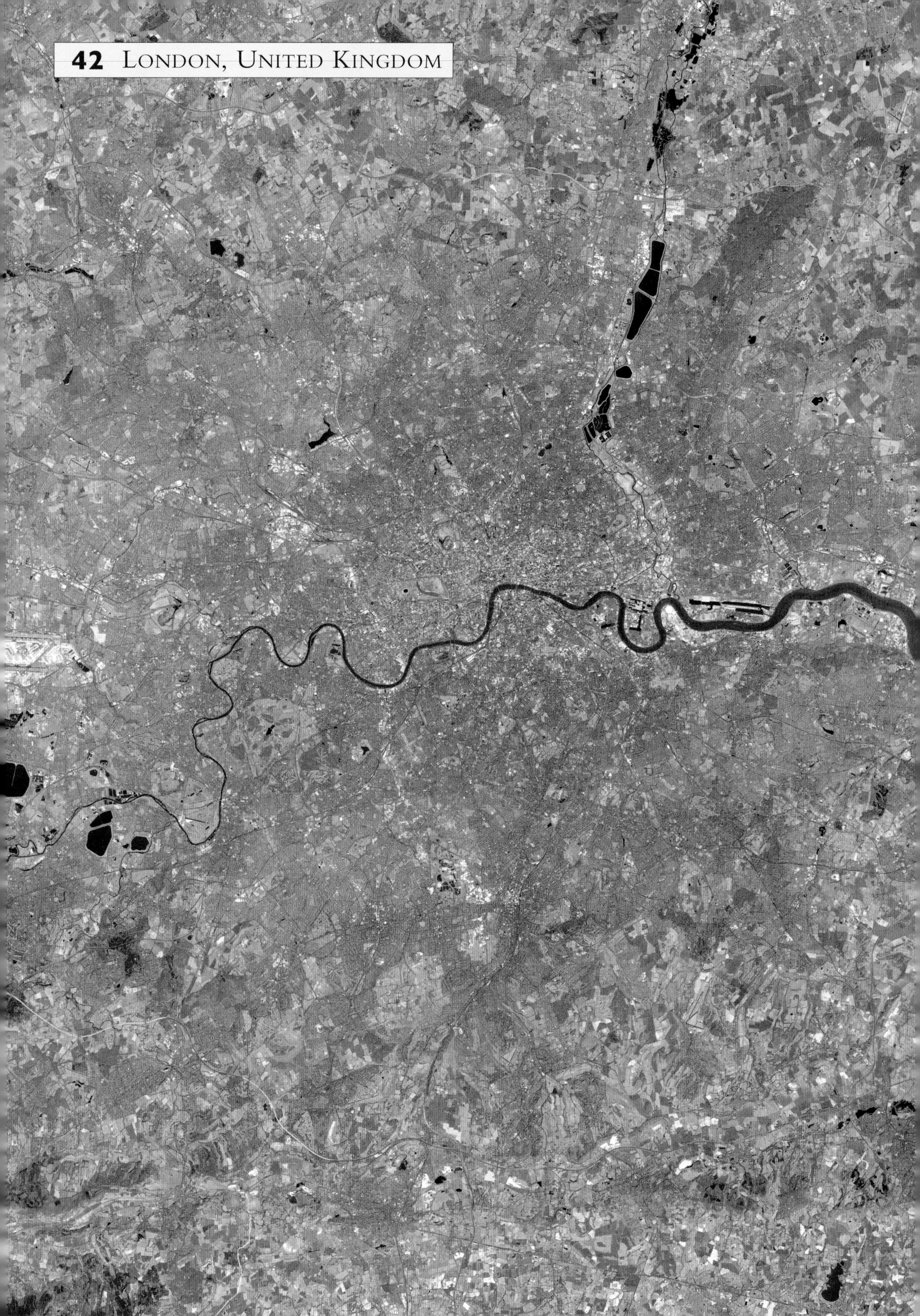

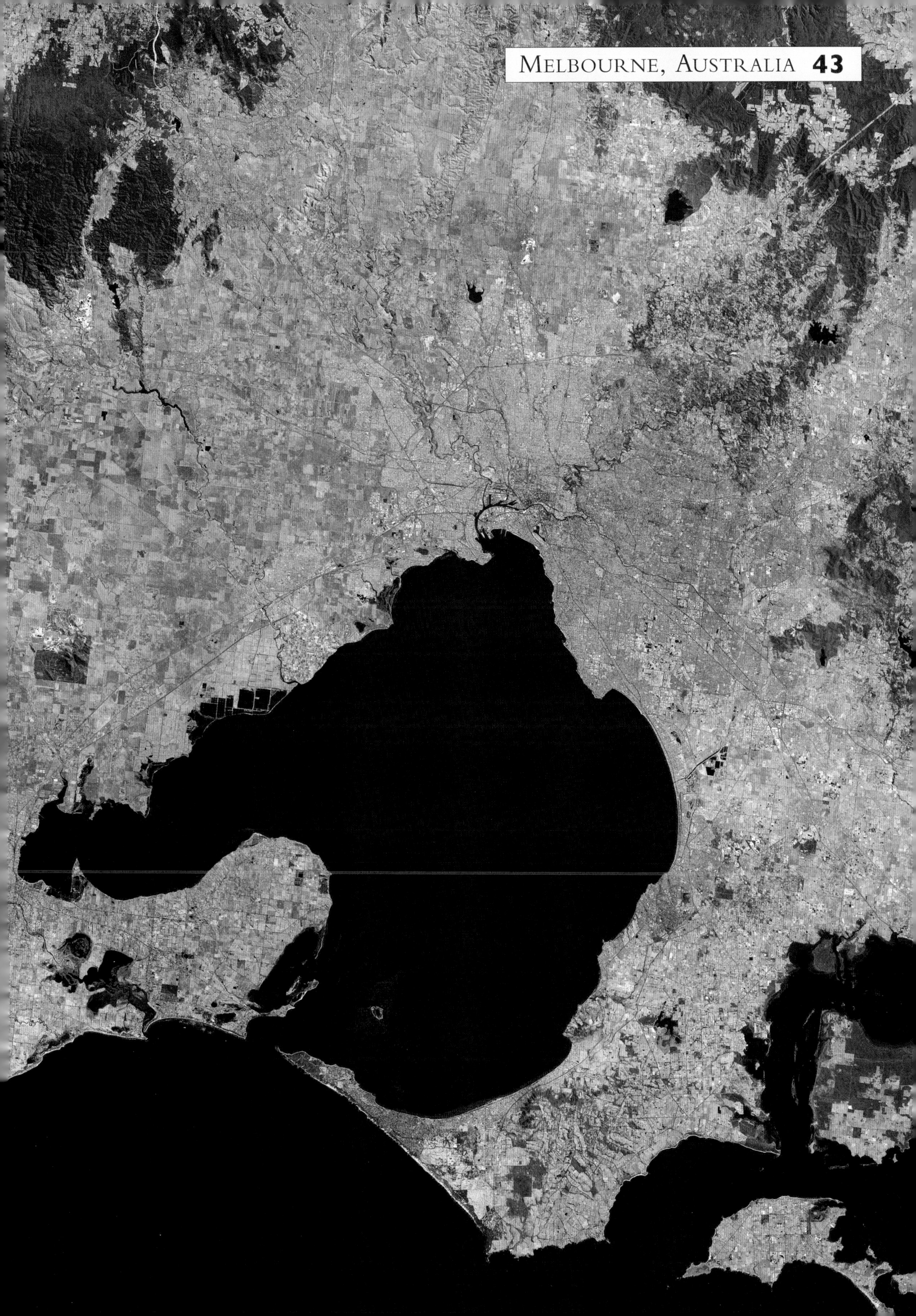

MELBOURNE, AUSTRALIA 43

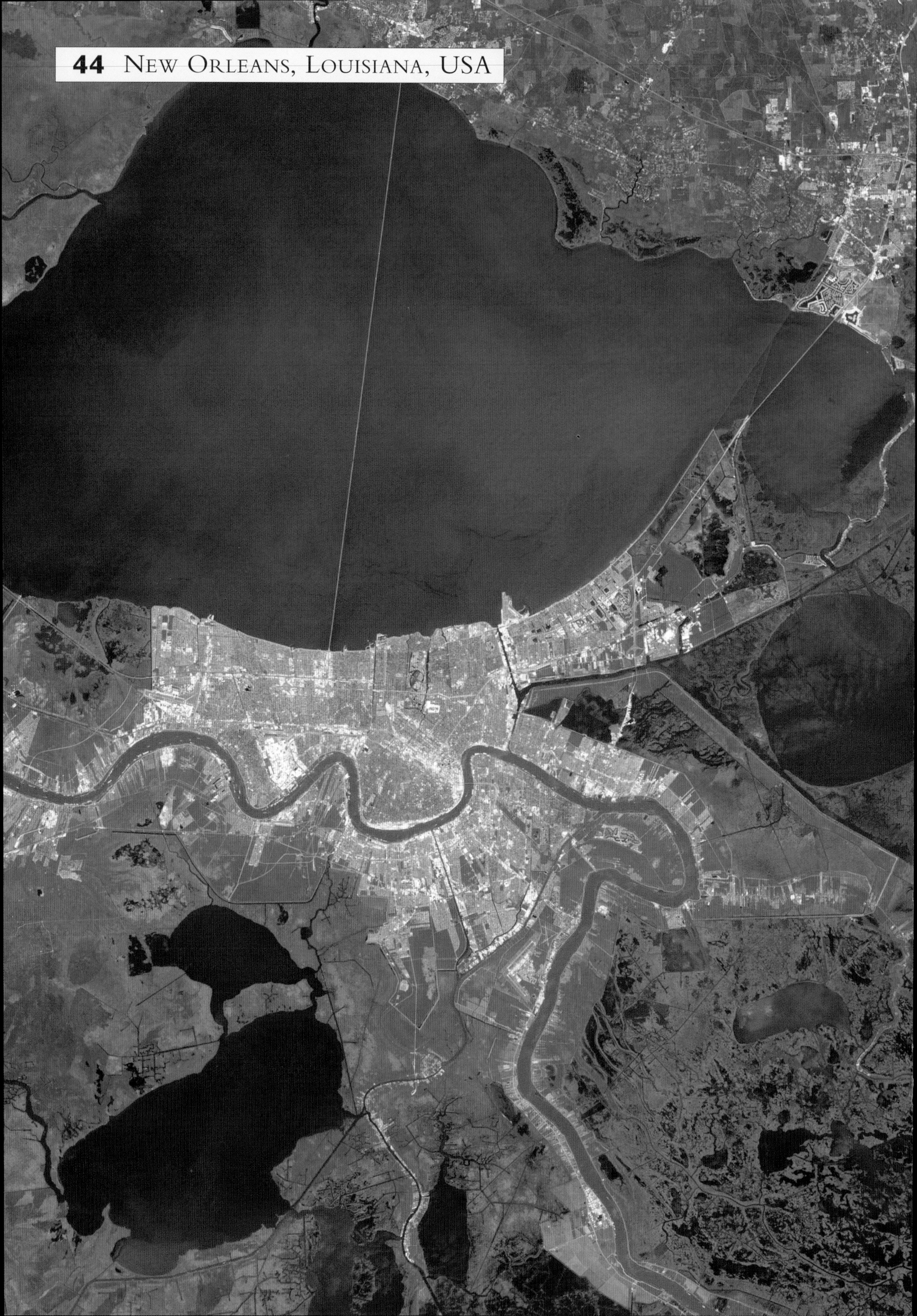

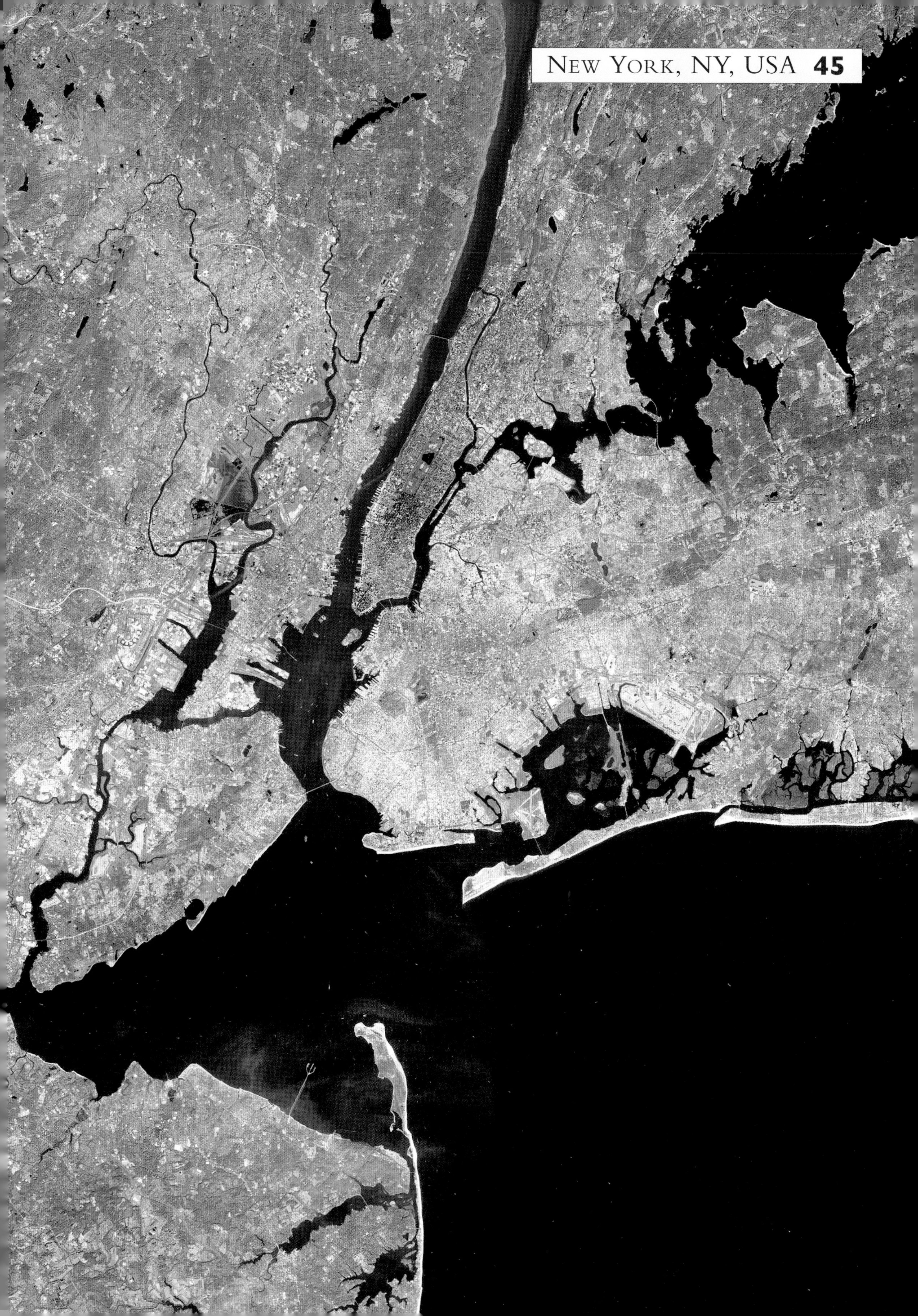

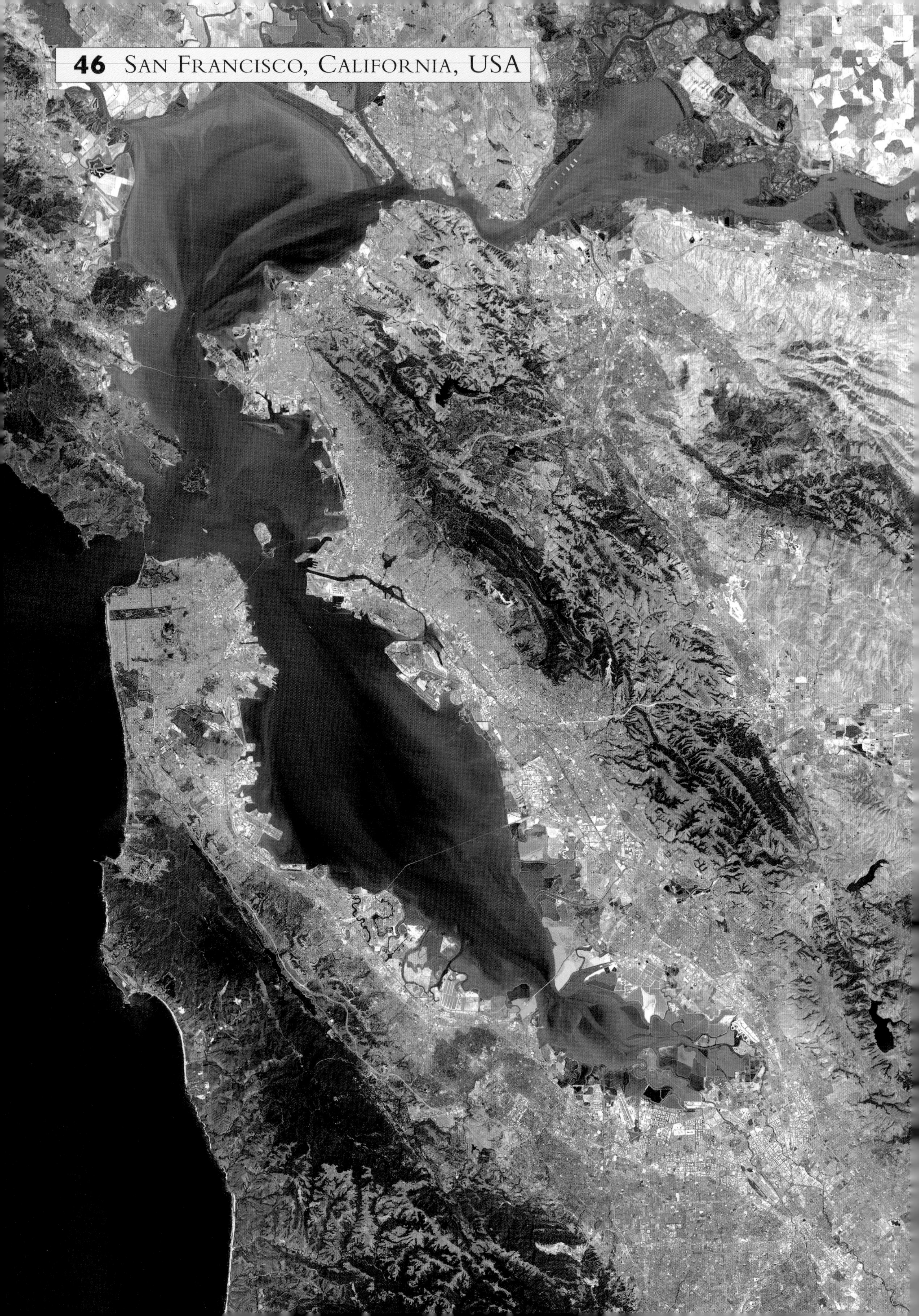

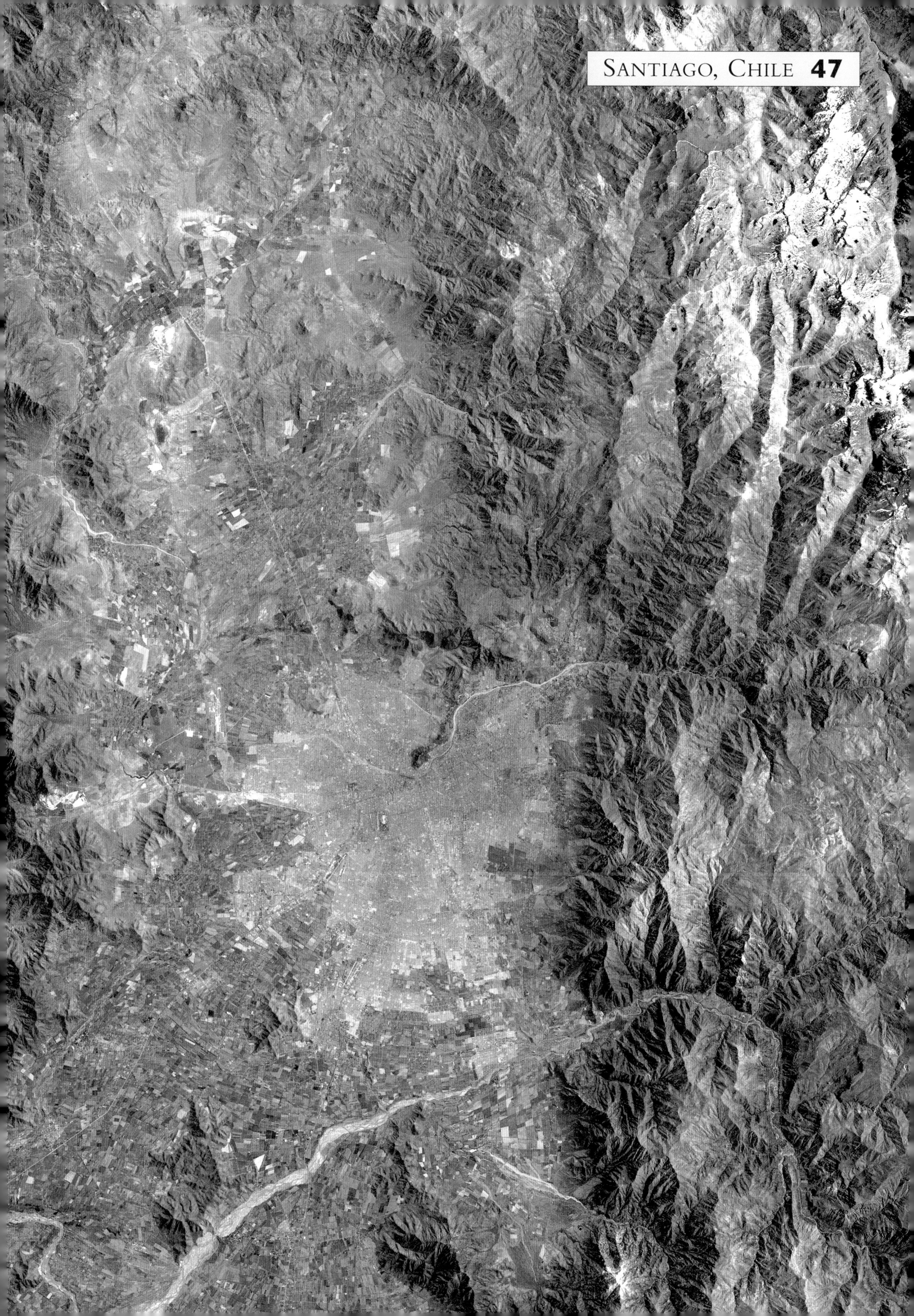

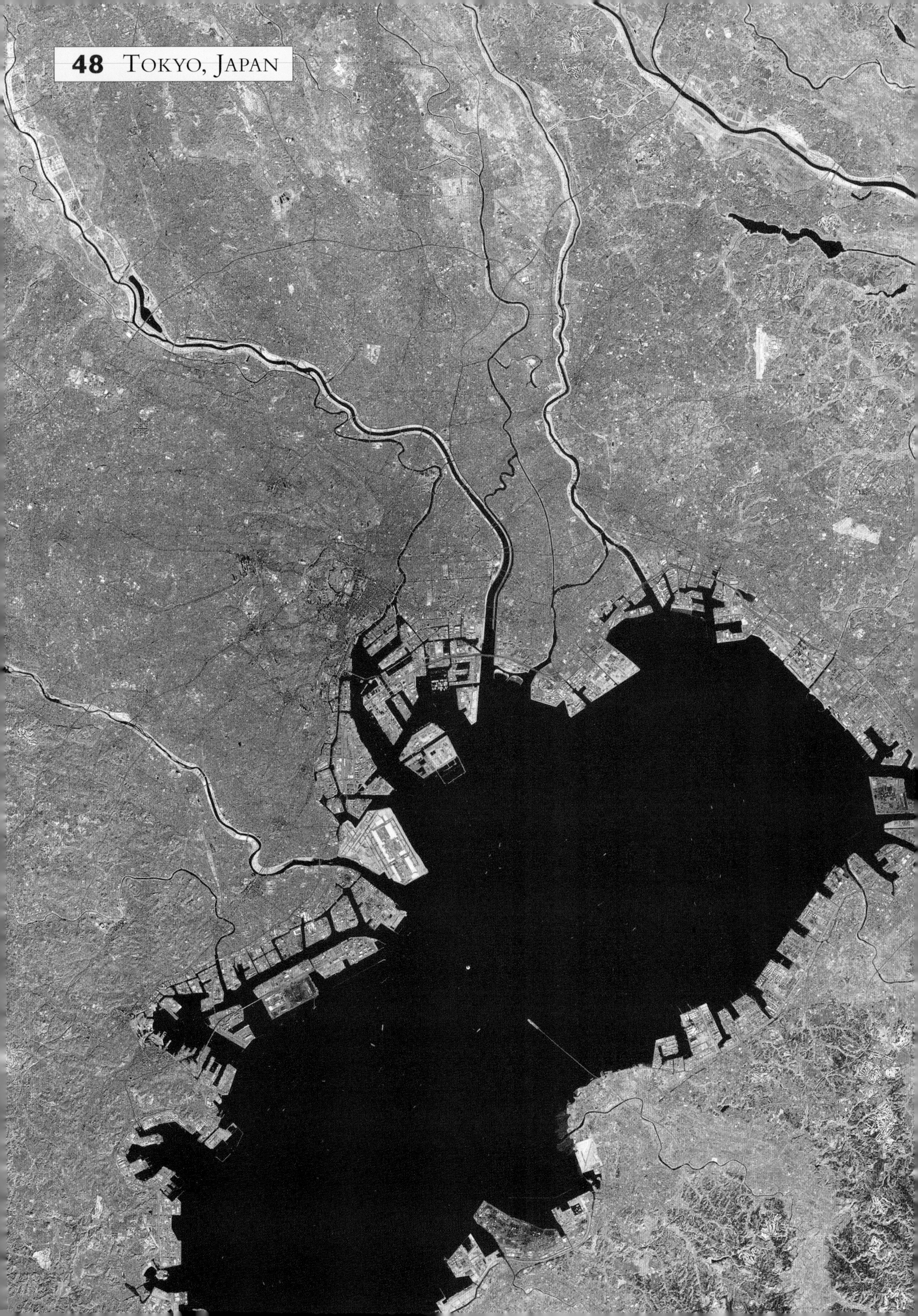

WORLD
MAPS

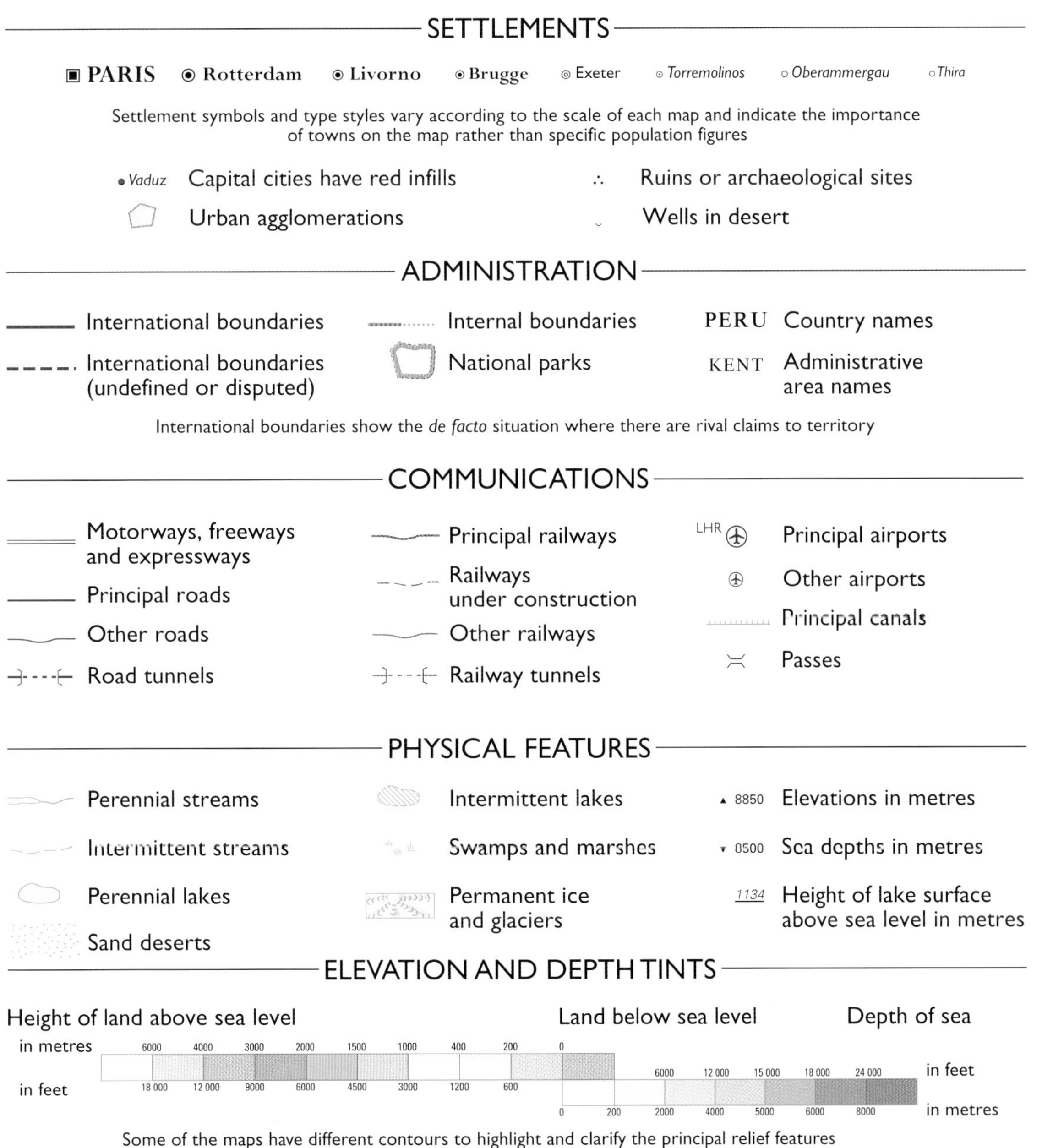

SETTLEMENTS

■ **PARIS**   ◉ Rotterdam   ◎ Livorno   ◉ Brugge   ◎ Exeter   ○ Torremolinos   ○ Oberammergau   ○ Thira

Settlement symbols and type styles vary according to the scale of each map and indicate the importance
of towns on the map rather than specific population figures

● *Vaduz*   Capital cities have red infills   ∴ Ruins or archaeological sites

⬠   Urban agglomerations   ⌣ Wells in desert

ADMINISTRATION

—————— International boundaries   ··········· Internal boundaries   **PERU** Country names

– – – –. International boundaries   ⬡ National parks   KENT Administrative
            (undefined or disputed)   area names

International boundaries show the *de facto* situation where there are rival claims to territory

COMMUNICATIONS

══════ Motorways, freeways
            and expressways   ——— Principal railways   ^{LHR} ✈ Principal airports

—————— Principal roads   – –‿– – Railways
                                                                            under construction   ✈ Other airports

———— Other roads   ——— Other railways   ············ Principal canals

⊣ - - ⊢ Road tunnels   ⊣ - - ⊢ Railway tunnels   ≍ Passes

PHYSICAL FEATURES

～～ Perennial streams   ◌ Intermittent lakes   ▲ 8850 Elevations in metres

– –～– Intermittent streams   Swamps and marshes   ▼ 0500 Sea depths in metres

◯ Perennial lakes   Permanent ice   *1134* Height of lake surface
                                                    and glaciers   above sea level in metres

    Sand deserts

ELEVATION AND DEPTH TINTS

Height of land above sea level   Land below sea level   Depth of sea

in metres   6000 4000 3000 2000 1500 1000 400 200   0

in feet   18 000 12 000 9000 6000 4500 3000 1200 600

                                    6000 12 000 15 000 18 000 24 000   in feet

                                    0   200   2000   4000   5000   6000   8000   in metres

Some of the maps have different contours to highlight and clarify the principal relief features

Equatorial Scale 1:95 000 000

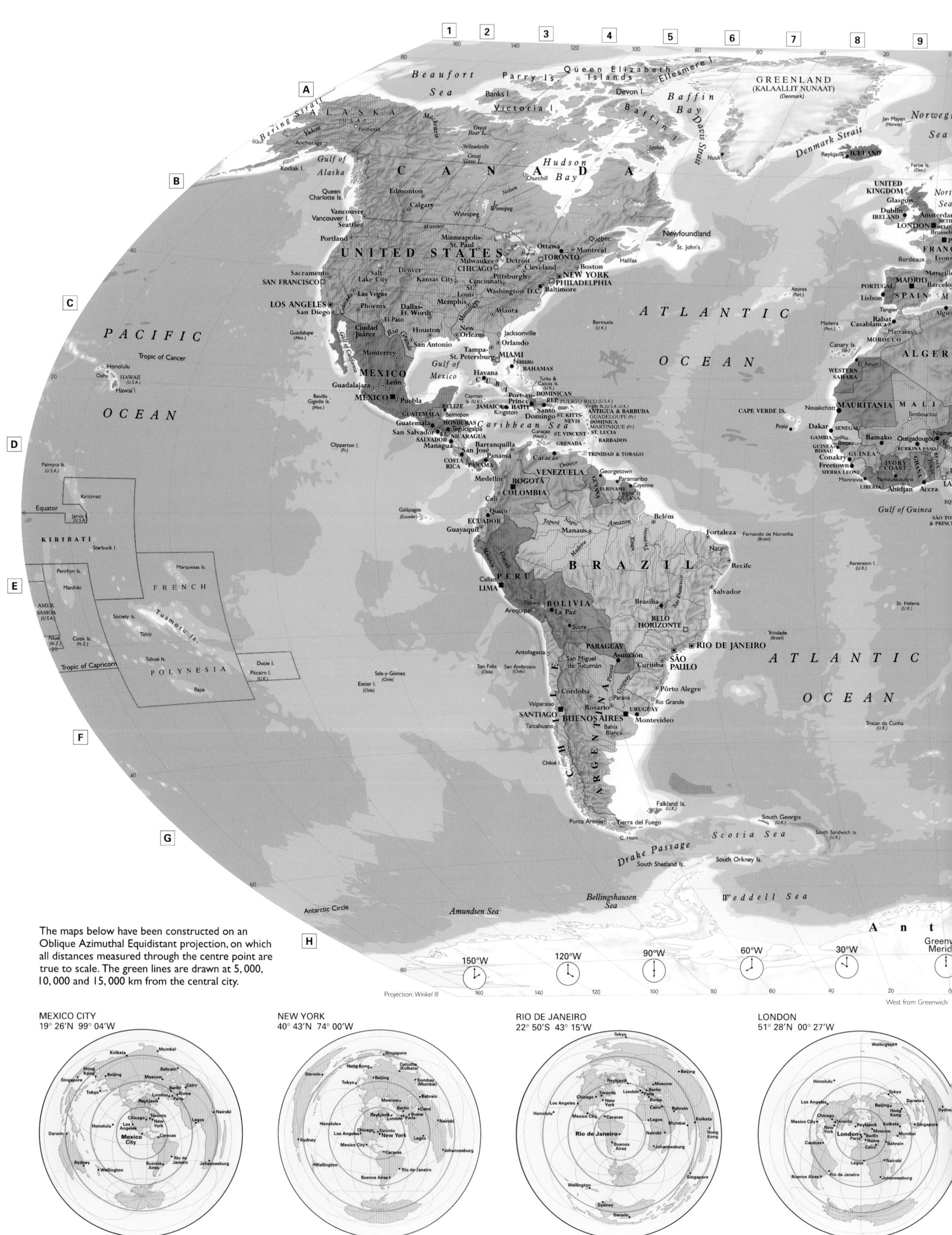

The maps below have been constructed on an Oblique Azimuthal Equidistant projection, on which all distances measured through the centre point are true to scale. The green lines are drawn at 5,000, 10,000 and 15,000 km from the central city.

Projection: Winkel III

West from Greenwich

MEXICO CITY
19° 26'N 99° 04'W

NEW YORK
40° 43'N 74° 00'W

RIO DE JANEIRO
22° 50'S 43° 15'W

LONDON
51° 28'N 00° 27'W

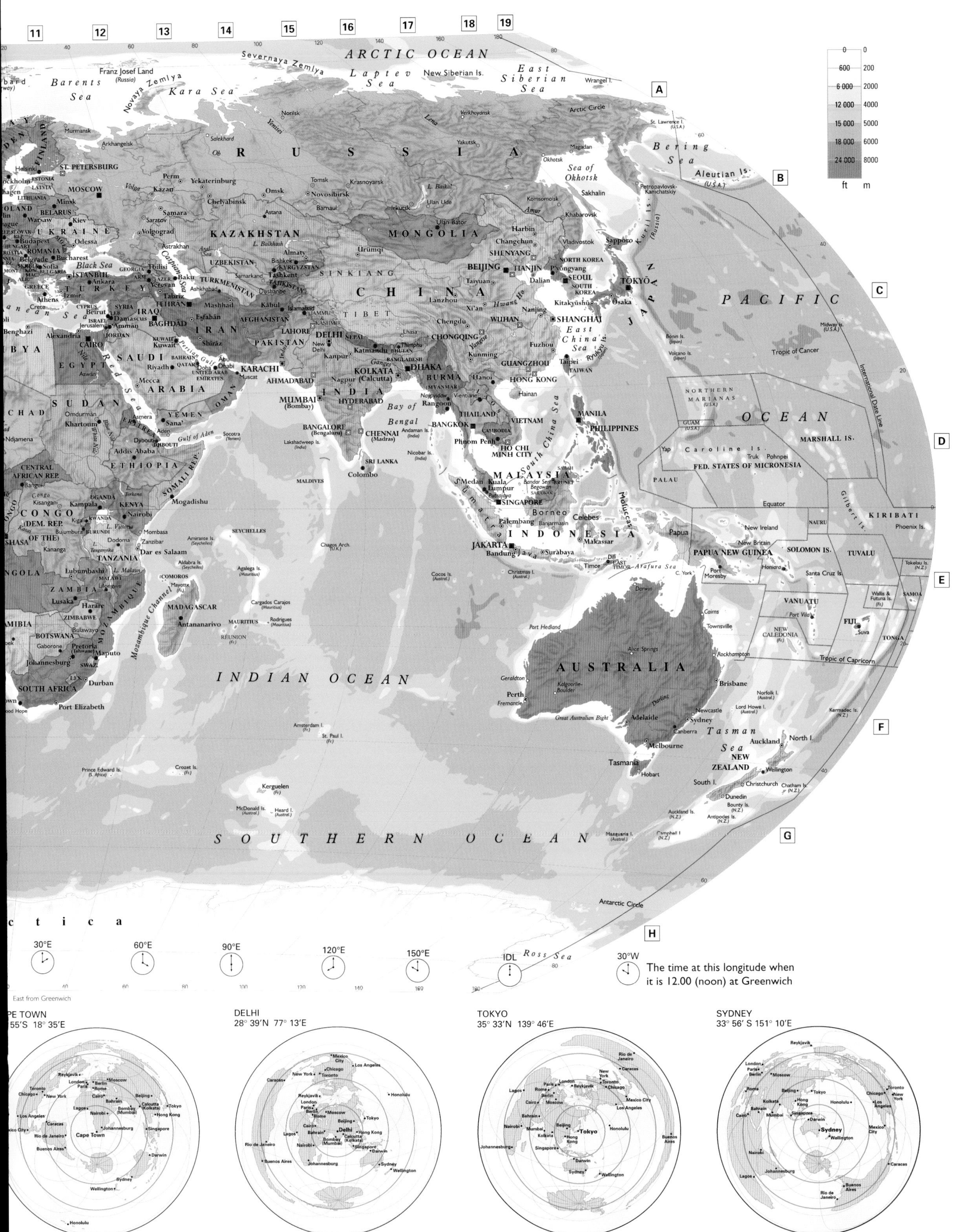

11   12   13   14   15   16   17   18   19

ARCTIC OCEAN

Severnaya Zemlya
Laptev Sea   New Siberian Is.   *East Siberian Sea*   Wrangel I.

Barents Sea   Franz Josef Land *(Russia)*   Novaya Zemlya
Kara Sea

Arctic Circle

A

Murmansk   Norilsk   Yerkhoyansk   St. Lawrence I. *(U.S.A.)*

Arkhangelsk   Salekhard   Yakutsk   Magadan

Bering Sea

ST. PETERSBURG   Ob   Yenisey   Lena   Okhotsk   Petropavlovsk-Kamchatskiy

MOSCOW   Perm   Yekaterinburg   Tomsk   Krasnoyarsk   *Sea of Okhotsk*   Sakhalin   Aleutian Is. *(U.S.A.)*

B

R U S S I A

Kazan   Omsk   Novosibirsk   L. Baikal   Irkutsk   Ulan Ude   Amur   Khabarovsk

Samara   **KAZAKHSTAN**   Barnaul   Ulan-Bator   Harbin   Vladivostok   Sapporo

Saratov   Volgograd   Astana   Changchun   SHENYANG   Komsomolsk

Astrakhan   L. Balkhash   Almaty   **MONGOLIA**   **BEIJING**   TIANJIN   Dalian   C

Black Sea   **UZBEKISTAN**   Bishkek   **NORTH KOREA**   PYONGYANG

ISTANBUL   **TURKMENISTAN**   Tashkent   **KYRGYZSTAN**   Ürümqi   Taiyuan   **SEOUL**   **TÔKYÔ**   *PACIFIC*

Baku   Samarkand   **SOUTH KOREA**   Ôsaka

Ashkhabad   Dushanbe   **S I N K I A N G**   **C H I N A**   Hwang He   Xi'an   Kitakyūshū

TEHRAN   Mashhad   **AFGHANISTAN**   Lanzhou   Nanjing   *East China Sea*

KÂBUL   Islamabad   **JAMMU**   Lhasa   Chengdu   **WUHAN**   **SHANGHAI**

I R A N   **LAHORE**   **T I B E T**   **CHONGQING**   Yangtze   Fuzhou   D

Esfahan   **PAKISTAN**   DELHI   NEPAL   Kunming   **GUANGZHOU**   Taipei   **TAIWAN**

Shiraz   New Delhi   Katmandu   BHUTAN   **HONG KONG**   Hainan

KARACHI   Kanpur   Ganga   **BANGLADESH**   **DHAKA**   Hanoi

AHMADABAD   Nagpur   **I N D I A**   KOLKATA (Calcutta)   **BURMA**   Naypyidaw

MUMBAI (Bombay)   HYDERABAD   *Bay of Bengal*   Rangoon   **THAILAND**   **VIETNAM**   **MANILA**

BANGALORE (Bengaluru)   CHENNAI (Madras)   Andaman Is. *(India)*   **BANGKOK**   **CAMBODIA**   **PHILIPPINES**

Lakshadweep Is. *(India)*   Nicobar Is. *(India)*   Phnom Penh   **HO CHI MINH CITY**

SRI LANKA   Colombo   **MALDIVES**   *South China Sea*

M A L A Y S I A   Medan   Kuala Lumpur   **SINGAPORE**   **BRUNEI**   Bandar Seri Begawan   SARAWAK

Palembang   Banjarmasin   Borneo   Celebes   *Moluccas*   Papua   **NORTHERN MARIANAS** *(U.S.A.)*

JAKARTA   Bandung   Java   Surabaya   **I N D O N E S I A**   Makassar

INDIAN OCEAN   *SOUTHERN OCEAN*

A U S T R A L I A

(lower map body reproduced faithfully where legible)

30°E   60°E   90°E   120°E   150°E   IDL   *Ross Sea*   30°W
East from Greenwich

The time at this longitude when it is 12.00 (noon) at Greenwich

CAPE TOWN
33° 55'S 18° 35'E

DELHI
28° 39'N 77° 13'E

TOKYO
35° 33'N 139° 46'E

SYDNEY
33° 56'S 151° 10'E

4 ARCTIC OCEAN

100 0 200 400 600 800 1000 1200 1400 km
100 0 200 400 600 800 1000 miles

1:35 000 000

18 17 16 15

JAPAN

PACIFIC OCEAN

Tufts Abyssal Plain
Gilbert Seamounts
Aleutian Trench
Aleutian Islands (U.S.A.)
Aleutian Basin
Bowers Basin
Near Is. (U.S.A.)
7822
Bowers Ridge
Dutch Harbor
D
Komandorskiye Ostrova
Mys Lopatka
Kurilskiye Ostrova (Russia)
Kuril Basin
Hokkaido
SAPPORO
La Perouse Str.
Yuzhno-Sakhalinsk
Unimak I. 2857
Bering Sea
Petropavlovsk-Kamchatskiy
Gora 4750
Klyuchevskaya
Ust-Kamchatsk
Ostrov Karaginskiy
Poluostrov Kamchatka
Sakhalin (Russia) 1609
Sakhalinskiy Zaliv
Sea of Okhotsk
Vanino
Amur
Khabarovsk
Nikolayevsk
Udskaya Guba
Komsomolsk-na-Amur
Pribilof Is. (U.S.A.)
42
Kodiak I. 1362
Bristol Bay
Alaska Pen.
St. Matthew (U.S.A.)
Mys Navarin
Anadyrskiy Zaliv
2453
C
Penzhinskaya G.
Gizhiginskaya Guba
Tauiskaya Guba
Magadan
Kolymskoye Nagorye
Stanovoy Khrebet
Okhotsk
Nunivak
St. Lawrence I. (U.S.A.)
Nome
Bering Str.
Provid-eniya
Mys Dezhneva
Anadyr
Omolon
Aldan
Yakutsk
1
G. of Alaska
Seward
Prince William Sd.
Cordova
Mt. McKinley 6194
Anchorage
Yukon
Kuskokwim
Norton Sd.
Prince of Wales
Kotzebue Sd.
Chukotskoye Nagorye
Kolyma
3147
Verkhoyansk
Lena
2295
Zhigansk
Olekma
Queen Charlotte Is.
Prince Rupert
44 Alexander Arch.
Mt. St. Elias 5489
Skagway Mt. Logan 5959
Fairbanks
Juneau
4019
4949
Pt. Hope
C. Lisburne
Chukchi Sea
Proliv Longa
Pevek
Nizhne-Kolymsk
Srednekolymsk
Indigirka
Yana
Kazachye
Bulun
Tiksi
Olenek
4949
ALASKA (U.S.A.)
Koyukuk
C. Halkett
Harrison Bay
Ostrov Vrangelya (Russia)
1096
Chaunskaya
46
East Siberian Sea
Novosibirskiye Ostrova
Lyakhovskiye Ostrova
Kotelnyy
Anabar
Nordvik
Rocky Mountains
Whitehorse
Dawson
Stewart
2782
Peel
Porcupine
Fort Yukon
Prudhoe Bay 2761
Fort McPherson
C. Pt. Barrow
Beaufort Sea
Mackenzie Bay
Tuktoyaktuk 2882
C. Bathurst
Canada Abyssal Plain
Canada Basin
Chukchi Plateau
3327
Mendeleyev Ridge
B
Lyakhovskiye Ostrova
374
Laptev Sea
Ostrova Petra
Poluostrov Taymyr
Khatanga
Ozero Taymyr
2
Dawson Creek
Fort Nelson
Liard
Fort Simpson
Mackenzie
Fort Good Hope
Tulita
Great Bear Lake
North
C. Kellett
Banks I.
C. Prince Alfred
371
Prince Patrick I.
Borden I.
A
3700
3546
3849
North
4007
Amundsen Basin
4100
4484
Severnaya Zemlya
Oktyabrskoy Revolyutsii
965
Mys Chelyuskin
Gory Byrranga
Putorana
Nizhnyaya Tunguska
Fort Vermilion
Athabasca
Yellowknife
Great Slave Lake
Coppermine
Kugluktuk
Dolphin & Union Sd.
Prince Albert Pen.
M'Clure Str.
Queen Melville I.
Parry Is.
North Magnetic Pole 2007
Alpha Ridge
Makarov Basin
Lomonosov Ridge
Nansen Basin
O. Ushakova
O. Vise
Dikson
Norilsk
Igarka
Yenisey
Taz
America
NUNAVUT
Athabasca Lake
King William I.
Victoria Island
Viscount Melville Sd.
M'Clintock Chan.
Elizabeth
Prince of Bathurst
Melville I.
Ellef Ringnes I.
Sverdrup Is.
2104
POLE 4346
NORTH
Nansiv
Amundsen Basin
Arctic Mid-Ocean Ridge
3741
3910
Zemlya Frantsa Iosifa (Russia)
O. Greem-Bell
Z. Vilcheka
O. Belyy
Novaya Zemlya
1547
Kara Sea
Gydanskiy Poluostrov
Dudinka
Gydanskiy Poluostrov
Novyy Urengoy
3
Churchill
Hudson Bay
Southampton I.
Coats I.
Mansel I.
Chesterfield Inlet
Roes Welcome Sd.
Back
Boothia Pen.
Somerset I.
Resolute
Axel Heiberg I.
Islands
Nansen Sd.
Eureka
2616
Alert
C. Columbia
Lincoln Sea
Norway Basin
Independence Fjord
90
O. Uedineniya
Z. Aleksandry
Poluostrov Yamal
Baydaratskaya Guba
Novyy Port
Nadym
Nizhnevartovsk
Surgut
ft m
12 000 4000
6000 2000
4500 1500
3000 1000
1200 400
600 200
0 0
500 1500
1000 3000
2000 6000
3000 9000
4000 12 000
5000 15 000
m ft
Foxe Chan.
Foxe Basin
Prince Charles I.
Hudson Str.
Baffin I.
Iqaluit
2147
C. Dyer
K. York
2469
Knud Rasmussen Land
Uummannaq
Kong Frederik VIII.s Land
Peary Land
McKinley Sea
Nansen Basin
Morris Jesup
Nordkapp
Zemlya Frantsa Iosifa
A
Novaya Zemlya
Nordaustlandet
1342
Vorkuta
Amderma
1894
Narodnaya
Uralskie Gory
Berezovo
Ob
Tobolsk
4
C. Wolstenholme
Nettilling L.
Baffin Bay
Upernavik
Qeqertarsuaq
Uummannaq
Qeqertarsuaq
Resolution I.
Chidley
Ungava Bay
Davis Str.
Kronprins Frederik Land
2170
Kong Frederik VIII.s Land
Independence Fjord
3238
A
1717
2571
Longyearbyen
Edgeøya
Vestspitsbergen
Svalbard (Norway)
Barents Sea
Belushya Guba
O. Kolguyev
Mys Kanin Nos
YEKATERINBURG
PERM
UFA
Labrador
Nuuk
Paamiut
2850
Kong Frederik IX.s Kyst
3360
Mt. Forel 3360
Kong Christian IX.s Land
GREENLAND (KALAALLIT NUNAAT) (Denmark)
Kejser Franz Joseph Fd.
Kong Oscar Fjord
Ittoqqortoormiit
2277
Jan Mayen (Norway)
Greenland Sea
B
Bjørnøya (Norway)
480
Vardø
Kirkenes
Varangerfjorden
Mezen
Mezen
Arkhangelsk
Syktyvkar
Ukhta
Pechora
Naryan-Mar
SAMARA
5
Hamilton Inlet
Labrador Sea
Qaqortoq
Alluitsup Paa
Nunap Isua (Kap Farvel)
Breiðafjörður
Gunnbjørn Fjeld 3700
Tasiilaq
Kangikajik
Denmark Str.
Icelandic Plateau
Mohns Ridge
Norwegian Sea
Arctic Circle
70
Lofoten
Narvik
Tromsø
Hammerfest
Nordkapp
Murmansk
Kola Poluostrov
Kandalaksha
Beloye More
Severodvinsk
Onega
Onezhskoye Ozero
Volga
NIZHNIY NOVGOROD
Saratov
Horn
Fontur
Reykjavík
ICELAND
Öræfajökull 2119
Norwegian Basin
C
3800
Trondheim
2469
NORWAY
SWEDEN
Oulu
FINLAND
Helsinki
Ladozhskoye Ozero
ST. PETERBURG
Chudskoye Ozero
MOSKVA
VOLGOGRAD
Charlie Gibbs Fracture Zone
4563
Iceland Basin
ATLANTIC
Mid-Atlantic Ridge
OCEAN
Føroyar (Den.)
Shetland Is. (U.K.)
Bergen
STOCKHOLM
Gulf of Bothnia
OSLO
Tornio
Tallinn
EST.
Riga
LAT.
LITH.
RUSSIA
Saratov
KHARKIV
ROSTOV
DONETSK
King's Trough
Rockall (U.K.)
Hebrides (U.K.)
Orkney Is. (U.K.)
North Sea
Oslo
KØBENHAVN
DENMARK
Kaliningrad (Russia)
Vilnius
BELARUS
KYYIV
Sea of Azov
Rockall Trough
UNITED KINGDOM
SCOTLAND
Edinburgh
GLASGOW
HAMBURG
BERLIN
POLAND
WARSZAWA
Wisła
UKRAINE
MOLDOVA
ODESA
Black Sea
6 7 8 9

Maximum extent of sea ice
Minimum extent of sea ice (September 2007)
Ice caps and permanent ice shelf

DUBLIN
IRELAND
WALES
ENGLAND
LONDON
NETH.
AMSTERDAM
GERMANY
PRAHA
Elbe
Kraków
Lviv
ROMANIA
C. Clear

Projection : Zenithal Equidistant
West from Greenwich   East from Greenwich

COPYRIGHT PHILIP'S

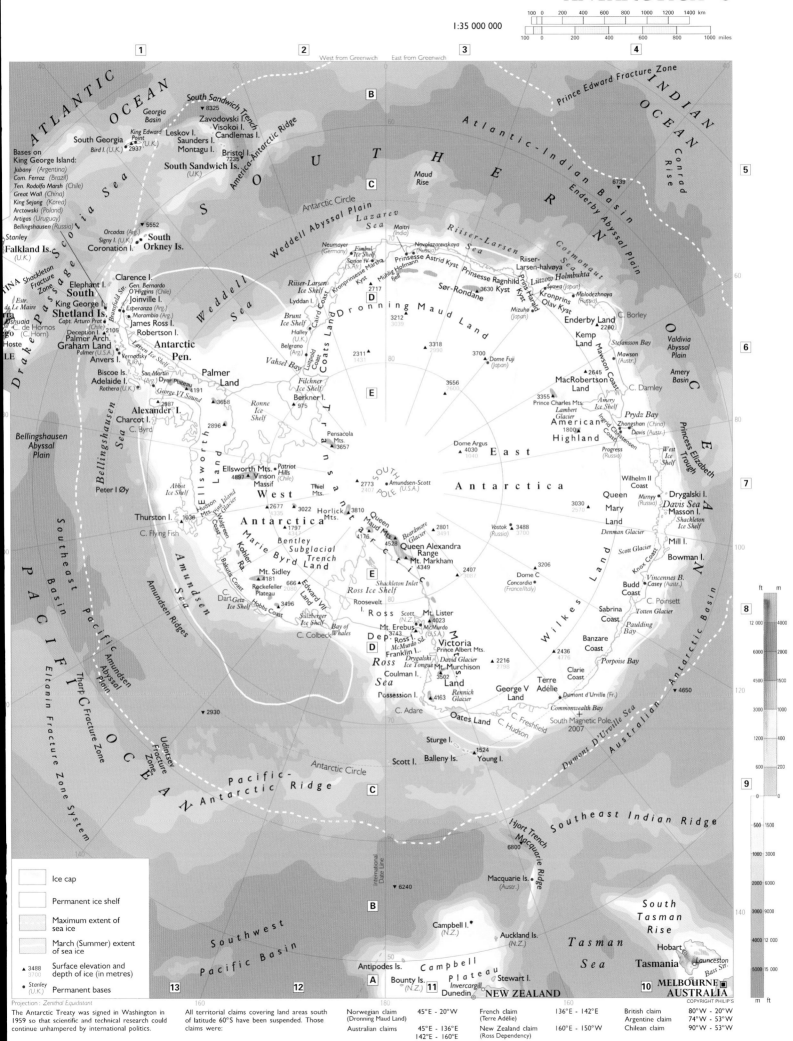

1:35 000 000

100 0 200 400 600 800 1000 1200 1400 km
100 0 200 400 600 800 1000 miles

West from Greenwich   East from Greenwich

Legend

| Ice cap |
| Permanent ice shelf |
| Maximum extent of sea ice |
| March (Summer) extent of sea ice |

▲ 3488 / 3700   Surface elevation and depth of ice (in metres)
● Stanley (U.K.)   Permanent bases

Projection : Zenithal Equidistant

The Antarctic Treaty was signed in Washington in 1959 so that scientific and technical research could continue unhampered by international politics.

All territorial claims covering land areas south of latitude 60°S have been suspended. Those claims were:

Norwegian claim (Dronning Maud Land)	45°E – 20°W
Australian claims	45°E – 136°E / 142°E – 160°E
French claim (Terre Adélie)	136°E – 142°E
New Zealand claim (Ross Dependency)	160°E – 150°W
British claim	80°W – 20°W
Argentine claim	74°W – 53°W
Chilean claim	90°W – 53°W

Bases on King George Island:
Jubany (Argentina)
Com. Ferraz (Brazil)
Ten. Rodolfo Marsh (Chile)
Great Wall (China)
King Sejong (Korea)
Arctowski (Poland)
Artigas (Uruguay)
Bellingshausen (Russia)

COPYRIGHT PHILIP'S

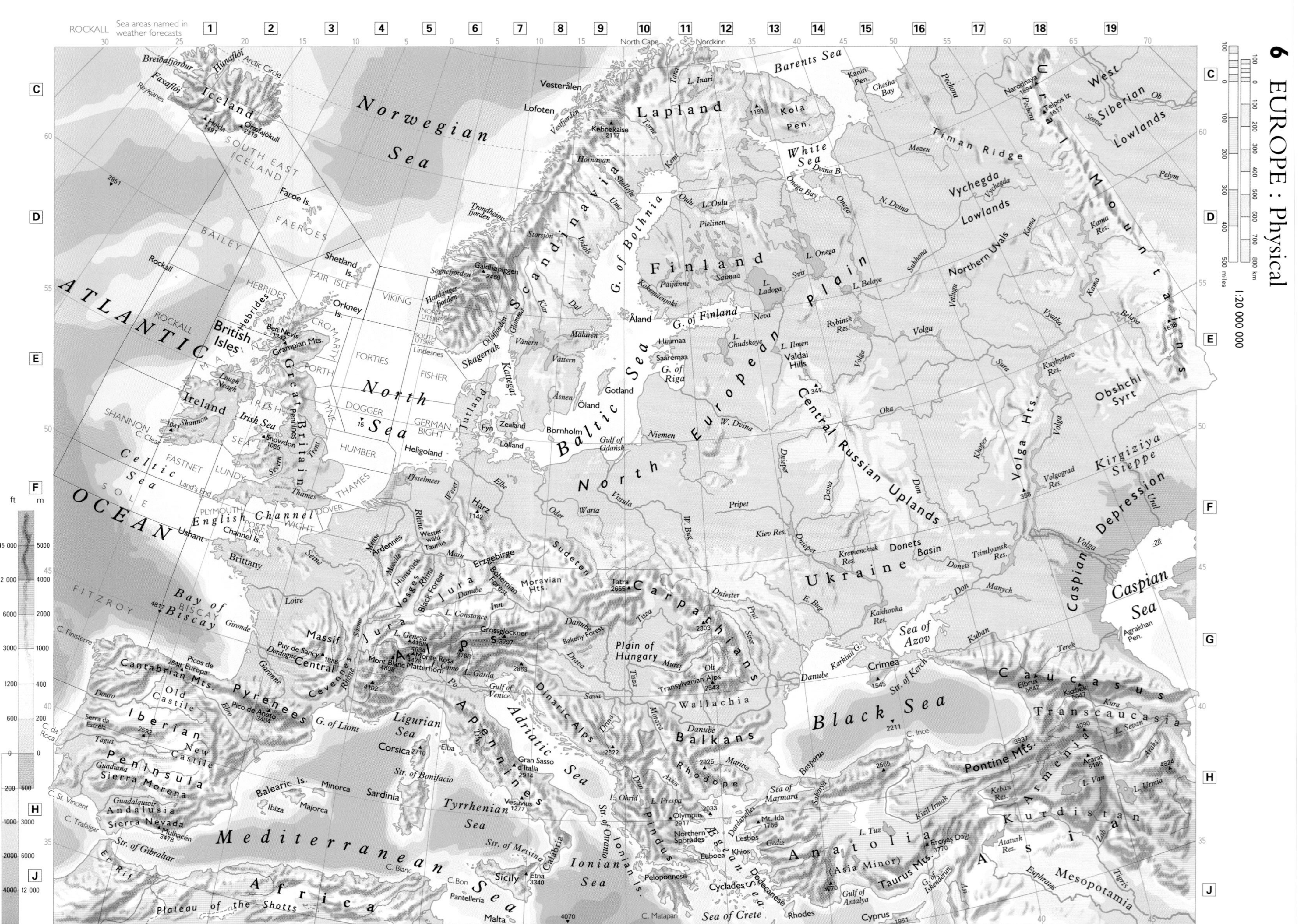

1:20 000 000

1:20 000 000

100 0 100 200 300 400 500 600 700 800 km
100 0 100 200 300 400 500 miles
Norwegian Sea
North Sea
ATLANTIC OCEAN
Bay of Biscay
English Channel
Baltic Sea
G. of Bothnia
White Sea
Black Sea
Caspian Sea
Adriatic Sea
Ionian Sea
Tyrrhenian Sea
Ægean Sea
Mediterranean Sea
Africa
Asia
ICELAND
UNITED KINGDOM
IRELAND
NORWAY
SWEDEN
FINLAND
DENMARK
ESTONIA
LATVIA
LITHUANIA
BELARUS
UKRAINE
RUSSIA
POLAND
GERMANY
NETHERLANDS
BELGIUM
LUX.
FRANCE
SWITZERLAND
AUSTRIA
CZECH REP.
SLOVAK REP.
HUNGARY
ROMANIA
MOLDOVA
BULGARIA
SERBIA
CROATIA
SLOVENIA
BOSNIA-HERZ.
MONTENEGRO
ALBANIA
MACEDONIA
KOSOVO
GREECE
ITALY
SPAIN
PORTUGAL
TURKEY
GEORGIA
ARMENIA
AZERBAIJAN
KAZAKHSTAN
IRAN
IRAQ
SYRIA
CYPRUS
MALTA
TUNISIA
ALGERIA
MOROCCO
■ LONDON Capital Cities
Projection: Bonne
West from Greenwich
East from Greenwich
COPYRIGHT PHILIP'S

SCANDINAVIA

ICELAND
on same scale

FÆROE ISLANDS
on same scale

50 0 25 50 75 100 125 150 175 km

50 0 25 50 75 100 125 miles

1:6 0

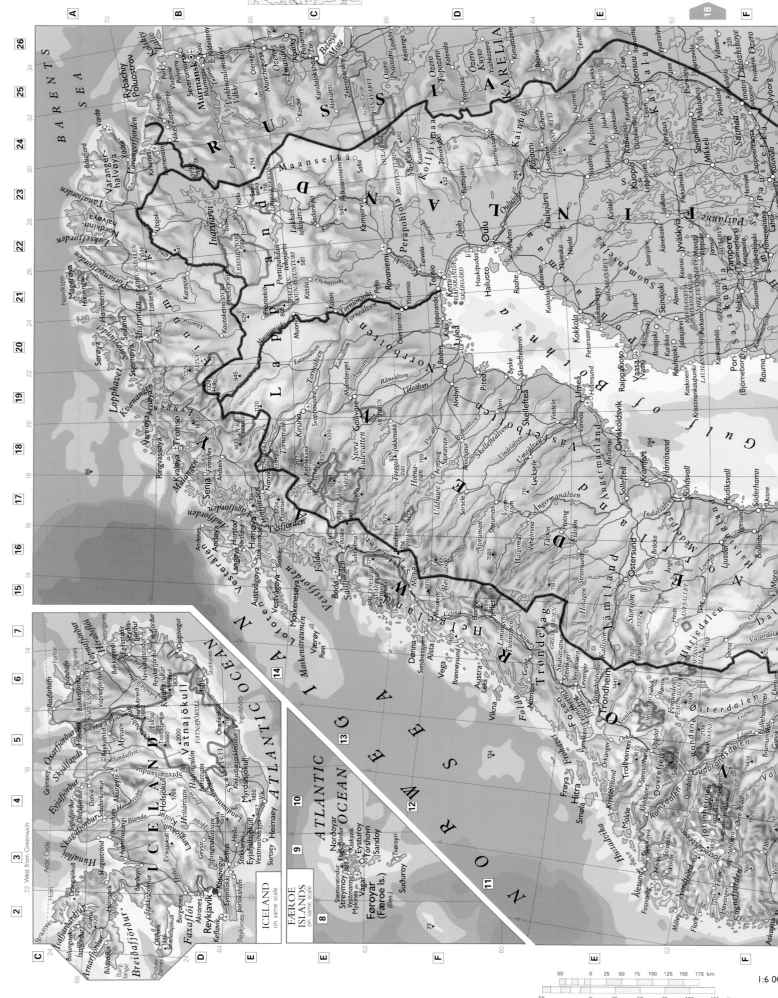

STOCKHOLM
Uppsala
Gotland (Sweden)
Öland
Bornholm (Denmark)
KØBENHAVN (Copenhagen)
Malmö
Göteborg (Gothenburg)
Kattegat
Skagerrak
NORTH SEA
BALTIC SEA
Gulf of Riga
Gulf of Finland
ESTONIA
Tallinn
Tartu
LATVIA
Riga
LITHUANIA
Vilnius
Kaunas
KALININGRAD
Kaliningrad (Russia)
BELARUS
MINSK
POLAND
WARSZAWA (Warsaw)
Kraków
Gdańsk
DENMARK
GERMANY
BERLIN
HAMBURG
Bremen
Hannover
CZECH REP.
PRAHA (Prague)
UKRAINE
Lviv (Lvov)
COPYRIGHT PHILIP'S
Projection: Conical with two standard parallels
East from Greenwich

1:2 000 000

Projection : Lambert's Conformal Conic

West from Greenwich

COPYRIGHT PHILIP'S

ATLANTIC OCEAN

SCOTLAND
Kintyre
Mull of Oa
Brodick
Arran
Campbeltown
Mull of Kintyre
Firth of Clyde
Ailsa Craig
Cairnryan
Stranraer
Portpatrick
L. Ryan

North Channel
Mts. of Antrim

NORTHERN IRELAND

Malin Hd.
Inishtrahull
Trawbreaga B.
Malin
Carndonagh
Lough Swilly
Fanad Hd.
Mulroy B.
Giants Causeway
Portstewart
Portrush
Coleraine
Ballymoney
Ballycastle
Fair Hd.
Cushendall
Garron Pt.
GLENARIFF
Carnlough
Larne
Carrickfergus
Donaghadee
Newtownards
Bangor
Belfast
Belfast L.
Strangford L.
Ards Pen.
Portaferry
Ballyquintin Pt.
Strangford
Saltfield
DOWN
Downpatrick
Ardglass
St. John's Pt.
Dundrum B.
Newcastle
Slieve Donard 852
Mourne Mts.
Kilkeel
Carlingford L.
Greenore
Warrenpoint

Tory I.
Horn Hd.
Sheep Haven
Dunfanaghy
Bloody Foreland
Gweedore
Errigal 752
Derryveagh Mts.
GLENVEAGH
Rathmelton
Letterkenny
Lifford
Strabane
Sion Mills
Newtownstewart
Londonderry
LONDONDERRY
Limavady
Ballymena
Antrim
Lough Neagh
Randalstown
Ballyclare
Newtownabbey
Lisburn
Craigavon
Lurgan
Portadown
Armagh
ARMAGH
Middletown
Keady
Newry

The Rosses
Crohy Hd.
Dungloe
Inishfree B.
Aran I.
The Rosses
Glenties
Glencolumbkille
Rossan Pt.
Dawros Hd.
Loughros More B.
Ardara
Lavagh More 676
DONEGAL
Killybegs
Slieve League 601
St. John's Pt.
Donegal
Ballyshannon
Bundoran
Donegal Bay

Magherafelt
Moneymore
Cookstown
Coalisland
Dungannon
TYRONE
Omagh
Sperrin Mts.
Sawel Mt. 683
Mountfield
Castlederg
Drumquin
Derg
ULSTER
FERMANAGH
Lower L. Erne
Enniskillen
Lisnaskea
Upper Erne
Clones
MONAGHAN
Monaghan
Castleblaney
Crossmaglen 577
Slieve Gullion
Dundalk
Dundalk Bay
LOUTH
Ardee
Dunleer
Clogher Hd.
Drogheda
Balbriggan
Skerries
Rush
Lambay I.
Malahide
Howth Hd.
DUBLIN
Dun Laoghaire
Killiney
Bray
Greystones

Irvinestown
Dromore
Ballygawley
Aughnacloy
Belturbet
Cavan
CAVAN
L. Gowna
Kingscourt
Ceanannus Mor (Kells)
Oldcastle
Blackwater
MEATH
An Uaimh (Navan)
Trim
Dunshaughlin
Swords
Cloncurry
Maynooth
Clondalkin

ATLANTIC OCEAN

Broad Haven
Erris Hd.
Portacloy
Downpatrick Hd.
Belmullet
Mullet Pen.
Inishkea North
Inishkea South
Blacksod Bay
Achill Hd. 672
Achill I.
Clare I.
Clew Bay
Louisburgh
Croagh Patrick 765
Westport
Newport
Corraun Pen.
Inishturk
Killary Harbour
Inishbofin
Inishshark
Mweelrea 819
CONNEMARA
Slyne Hd.
Clifden
Roundstone
Bertraghboy B.
Kilkieran B.
Spiddle
Aran Is.
Inishmore
Inishmaan
Inisheer
Black Hd.
Burren
Cliffs of Moher
Hags Hd.
Liscannor Bay
Lahinch
Mal Bay
Milltown Malbay
Mutton I.
Kilrush
Loop Hd.
Kilkee
Mouth of the Shannon
Kerry Hd.
Ballybunion
Smerwick Harbour
Brandon B.
Tralee B.
Brandon Mt. 953
Dingle
Dunmore Hd.
Great Blasket I.
Inishvickillane
Dingle Bay
Valencia I.
Puffin I.
Great Skellig
Ballinskelligs B.
Dursey I.
Crow Hd.
Dunmanus B.
Mizen Hd.
Long I.
Sherkin I.
Clear I.
C. Clear
Fastnet Rock

Belmullet
Crossmolina
Ballina
Killala B.
Killala
Nephin Beg Range 627
Nephin 806
L. Conn
Foxford
L. Cullin
Swinford
MAYO
Castlebar
Ballinrobe
Lough Mask
CONNACHT
Partry Mts. 683
Tuam
Lough Corrib
Oughterard
GALWAY
Galway
Galway Bay
Clarinbridge
Athenry
Loughrea
Gort 368
Kinvarra
CLARE
Lisdoonvarna 345
Ennistimon
Crusheen
Feakle
Tulla
Ennis
Sixmilebridge
Shannon Airport
Limerick
Silvermine Mts. 694
Keeper Hill

Ballina
Sligo Bay
Sligo
Dromore West
Coolaney
Collooney
SLIGO
Ballymote
Bricklieve Mts.
Boyle
Carrick-on-Shannon
LEITRIM
Leitrim
Lough Allen
L. Key
L. Arrow
Ballaghaderreen
ROSCOMMON
Castlerea
Frenchpark
Ballyhaunis
Knock
Claremorris
Glennamaddy
Mount Bellew Bridge
Ballinasloe
Aughrim
Shannonbridge
OFFALY
Ferbane
Clara
Tullamore
Portarlington
Monasterevin
KILDARE
Kildare
Droichead Nua
Naas
Rathangan
Edenderry
Allen
Bog of Allen
Grand Canal
Clane

Lough Melvin
Manorhamilton
Glenfarne
Belcoo
Florencecourt
Swanlinbar
Dowra
Drumshanbo
Drumkeeran
Carrigallen
Mohill
Longford
LONGFORD
Granard
Castlepollard
Edgeworthstown
WESTMEATH
Mullingar
Castlepollard
L. Owel
L. Ennell
L. Derravaragh
L. Sheelin
Finea
Ballyjamesduff
Cootehill
Carrickmacross
Annalee
Clones

IRELAND
Leinster

Strokestown
Roscommon
Lanesborough
Lough Ree
Athlone
Moate
Kilbeggan
Ballymore
Horseleap
Ballycumber
Clara
Birr
Banagher
Shannonbridge
Shannonharbour
Cloghan
Ferbane

Roscommon
Castlerea
Strokestown
Elphin
Knockcroghery

Connaught
Moycullen
Spiddal
Barna
Salthill
Oranmore

Tuam
Athenry
Loughrea
Kilconnell
Mount Bellew
Ballygar
Creggs

Portumna
Lough Derg
Borrisokane
Nenagh
Dromineer
Puckane
Terryglass
Borrisokane

Roscrea
Shinrone
Moneygall
Cloughjordan
Toomevara

Slieve Bloom Mts.
Arderin 528
Mountrath
Mountmellick
Port Laoise
LAOIS
Abbeyleix
Durrow
Rathdowney
Borris-in-Ossory
Ballacolla

Birr
Kinnitty
Roscrea
Templemore
Thurles
Holycross
Littleton
Two-Mile-Borris

TIPPERARY
Tipperary
Golden Vale
Golden
Cashel
Cahir
Clonmel
Carrick-on-Suir
Slievenamon 722
Fethard
Mullinahone
Killenaule
Ballingarry

Galtymore 920
Galty Mts.
Mitchelstown
Fermoy
Rath Luirc (Charleville) 519
Buttevant
Kilfinnane
Kilmallock
Bruff
LIMERICK
Limerick
Rathkeale
Askeaton
Foynes
Glin
Newcastle West
Abbeyfeale
Listowel
Ardfert
Tralee
Castleisland
Killorglin
KERRY
Slieve Mish 853
Castlemaine
Killarney
Macgillycuddy's Reeks
Carrauntoohill 1041
KILLARNEY
L. Leane
Kenmare River
Kenmare
Sneem
Caherciveen
Glenbeigh
L. Currane
Waterville
Caha Mts. 686
Castletown Bearhaven
Bear I.
Bantry Bay
Bantry
Whiddy I.
Glengarriff
Ballydehob
Skull
Schull
Baltimore
Galley Hd.
Clonakilty B.
Clonakilty
Timoleague
Dunmanway
Bandon
Kinsale
Old Head of Kinsale

Munster
CORK
Boggeragh Mts. 646
Macroom
Blarney
Cork
Cork Harbour
Crosshaven
Cobh
Carrigaline
Passage West
Midleton
Youghal
Youghal B.
WATERFORD
Dungarvan
Dungarvan Harbour
Lismore
Cappoquin
Comeragh Mts. 792
Knockmealdown Mts. 795
Tallow
Blackwater
Nagles Mts. 429
Mallow
Millstreet
Kanturk
Newmarket
Nad
Mushera

Waterford
Waterford Harbour
Passage East
Dunmore East
Tramore
Tramore B.
Hook Hd.
Carnsore Pt.
Saltee Is.
Kilmore Quay
Rosslare Harbour
Rosslare
Greenore Pt.
Wexford Harbour
Wexford
WEXFORD
New Ross
Enniscorthy
Blackstairs Mt. 796
Mt. Leinster 794
KILKENNY
Kilkenny
Callan
Thomastown
Graiguenamanagh
Bagenalstown
Muine Bheag
Carlow
CARLOW
Tullow
Hacketstown
Shillelagh
Tinahely
Gorey
Ballycanew
Arklow
Mizen Hd.
Wicklow Hd.
WICKLOW
WICKLOW MTS.
Wicklow
Rathdrum
Lugnaquilla 926
Aughrim
Baltinglass
Rathnew
Ashford
Roundwood
Blessington
Poulaphouca Res.
Kippure 754
Sallins
Kilcullen
Athy
Carlow

IRISH SEA

Celtic Sea

St. George's Channel

WALES
St. David's Hd.
St. David's
St. Brides Bay

1:2 000 000

Key to Scottish unitary
authorities on map
1 CITY OF ABERDEEN   8 EAST RENFREWSHIRE
2 DUNDEE CITY   9 NORTH LANARKSHIRE
3 WEST DUNBARTONSHIRE   10 FALKIRK
4 EAST DUNBARTONSHIRE   11 CLACKMANNANSHIRE
5 CITY OF GLASGOW   12 WEST LOTHIAN
6 INVERCLYDE   13 CITY OF EDINBURGH
7 RENFREWSHIRE   14 MIDLOTHIAN

ORKNEY IS.
on same scale

SHETLAND IS.
on same scale

Projection : Lambert's Conformal Conic

West from Greenwich

COPYRIGHT PHILIP'S

1:2 000 000

Key to English unitary
authorities on map

25 HARTLEPOOL
26 DARLINGTON
27 STOCKTON-ON-TEES
28 MIDDLESBROUGH
29 REDCAR AND CLEVELAND
30 BLACKPOOL
31 BLACKBURN WITH DARWEN
32 HALTON
33 WARRINGTON
34 KINGSTON UPON HULL
35 NORTH EAST LINCOLNSHIRE
36 STOKE-ON-TRENT
37 TELFORD AND WREKIN
38 DERBY CITY
39 CITY OF NOTTINGHAM
40 LEICESTER CITY
41 RUTLAND
42 PETERBOROUGH
43 MILTON KEYNES
44 LUTON
45 NORTH SOMERSET
46 CITY OF BRISTOL
47 BATH AND NORTH EAST SOMERSET
48 SWINDON
49 READING
50 WOKINGHAM
51 WINDSOR AND MAIDENHEAD
52 SLOUGH
53 BRACKNELL FOREST
54 THURROCK
55 SOUTHEND-ON-SEA
56 MEDWAY
57 PLYMOUTH
58 TORBAY
59 POOLE
60 BOURNEMOUTH
61 SOUTHAMPTON
62 PORTSMOUTH
63 BRIGHTON AND HOVE
64 BEDFORD
65 CENTRAL BEDFORDSHIRE

Key to Welsh unitary
authorities on map

15 SWANSEA
16 NEATH PORT TALBOT
17 BRIDGEND
18 RHONDDA CYNON TAFF
19 MERTHYR TYDFIL
20 CAERPHILLY
21 BLAENAU GWENT
22 TORFAEN
23 CARDIFF
24 NEWPORT

N O R T H   S E A

I R I S H   S E A

North Channel

NORTHERN IRELAND

SCOTLAND

ENGLAND

WALES

ISLE OF MAN

Firth of Clyde

The Wash

Edinburgh
Glasgow
Newcastle-upon-Tyne
Sunderland
Middlesbrough
Leeds
Bradford
Manchester
Liverpool
Sheffield
Derby
Nottingham
Kingston upon Hull
Lincoln
Stoke-on-Trent
Chester
Belfast
Carlisle
Lancaster
Blackpool
Preston
Bolton
Warrington
York
Scarborough
Whitby
Harrogate
Doncaster
Grimsby
Skegness
Boston

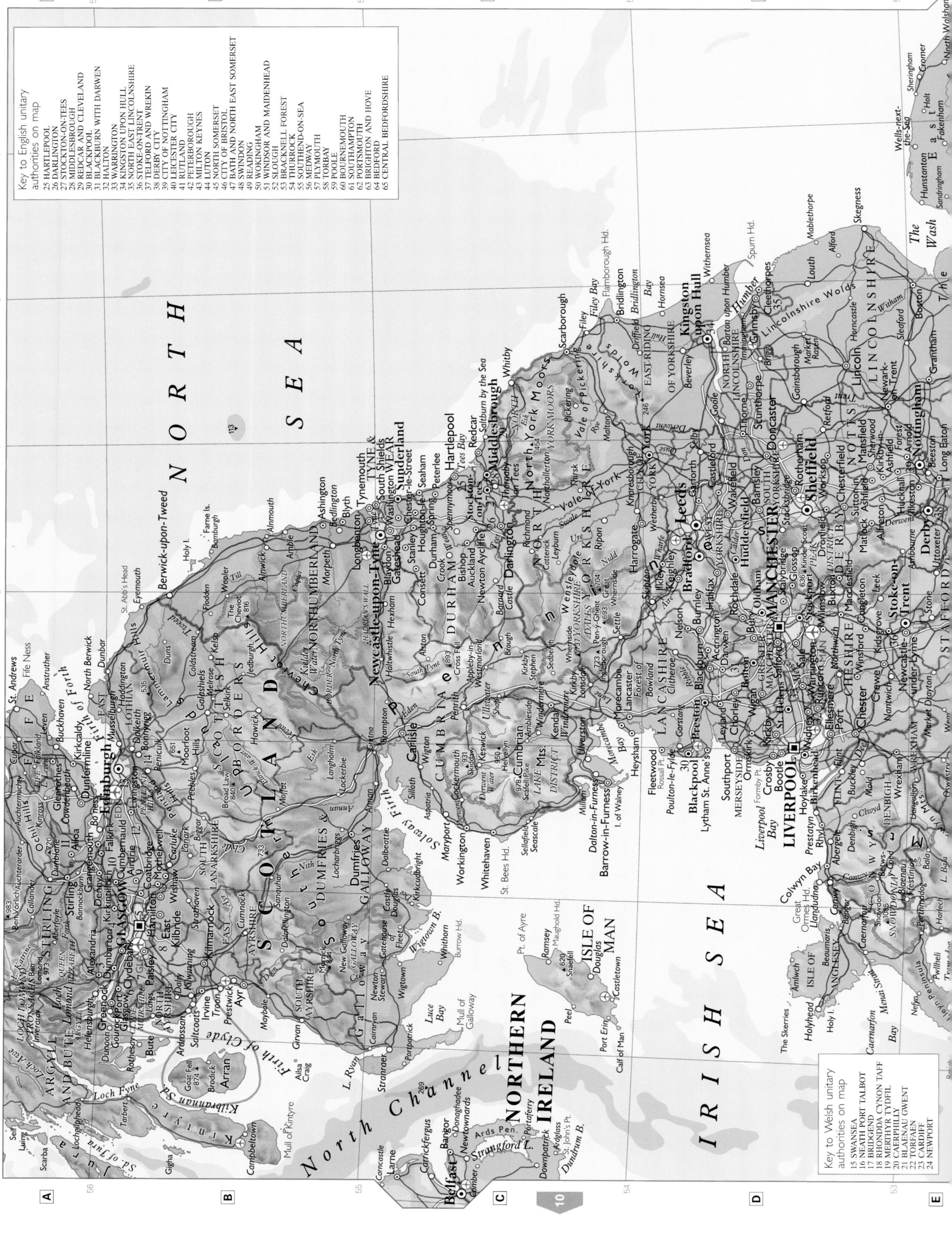

W A L E S
E N G L A N D
F R A N C E
E N G L I S H C H A N N E L
Bristol Channel
Strait of Dover
CHANNEL ISLANDS (U.K.)
ISLES OF SCILLY on same scale
LONDON
BIRMINGHAM
Cardiff
Bristol
Southampton
Portsmouth
Bournemouth
Plymouth
Brighton
Swansea
Newport
Oxford
Cambridge
Reading
Gloucester
Exeter
Le Havre
Rouen
Caen
Cherbourg
Calais
Dieppe
Guernsey
Jersey
Alderney
Sark
ISLE OF WIGHT
Lyme Bay
Cardigan Bay
Baie de la Seine
Baie de la Somme
NORMANDIE
HAUTE
Projection : Lambert's Conformal Conic
COPYRIGHT PHILIP'S

1:5 000 000
km
miles
ATLANTIC OCEAN
NORTH SEA
IRISH SEA
CELTIC SEA
English Channel
Bristol Channel
St. George's Channel
North Channel
Shetland Is.
(U.K.)
Yell
Unst
Fetlar
Mainland
Lerwick
Foula
Fair Isle
Orkney Is.
Westray
Sanday
Stronsay
Kirkwall
Hoy
South Ronaldsay
Pentland Firth
Thurso
Wick
C. Wrath
Helmsdale
Outer Hebrides
Lewis
Stornoway
Harris
St. Kilda
(U.K.)
North Uist
Benbecula
South Uist
Barra
Inner Hebrides
Sea of the Hebrides
North Minch
Skye
Portree
Rhum
Eigg
Coll
Tiree
Mull
Colonsay
Jura
Islay
North West Highlands
Ullapool
Lairg
Golspie
Tain
Invergordon
Dingwall
Moray Firth
Buckie
Banff
Fraserburgh
Peterhead
Nairn
Elgin
Inverness
Huntly
Inverurie
Aberdeen
CAIRNGORMS
Glen More
L. Ness
Aviemore
SCOTLAND
Grampian Mts.
Mallaig
Fort William
Ben Nevis
1342
Tobermory
Oban
Ballater
Stonehaven
Montrose
Arbroath
Forfar
Dundee
St. Andrews
Perth
Glenrothes
Kirkcaldy
Stirling
Dunfermline
Dunbar
L. Lomond
L. Awe
L. Fyne
Dumbarton
Greenock
GLASGOW
Motherwell
Edinburgh
Paisley
Hamilton
East Kilbride
Irvine
Arran
Campbeltown
Kilmarnock
Ayr
Firth of Clyde
Southern Uplands
Galashiels
Berwick-upon-Tweed
Jedburgh
Hawick
Cheviot Hills
Alnwick
Girvan
Dumfries
Kirkcudbright
Stranraer
Annan
Carlisle
NORTHUMBERLAND
Newcastle-upon-Tyne
South Shields
Sunderland
Hexham
Gateshead
Durham
Hartlepool
Redcar
Middlesbrough
Stockton-on-Tees
Darlington
N. YORK MOORS
Scarborough
Pennines
Workington
Whitehaven
Cumbrian Mts.
LAKE DISTRICT
Barrow-in-Furness
Mull of Galloway
Douglas
I. of Man
UNITED KINGDOM
YORKSHIRE DALES
Lancaster
Harrogate
York
Bridlington
Beverley
Kingston upon Hull
Blackpool
Preston
Blackburn
Burnley
Keighley
Leeds
Bradford
Halifax
Huddersfield
Barnsley
Bolton
Doncaster
Scunthorpe
Grimsby
Humber
MANCHESTER
Oldham
LIVERPOOL
Rotherham
Sheffield
Louth
Warrington
Stockport
Chesterfield
Lincoln
Skegness
Anglesey
Holyhead
Bangor
Colwyn Bay
Chester
Crewe
PEAK DISTRICT
Mansfield
Nottingham
Boston
The Wash
Cromer
Wrexham
Snowdon
1085
Stoke on Trent
Derby
Stafford
Grantham
King's Lynn
Norwich
THE BROADS
Great Yarmouth
Lowestoft
Pwllheli
SNOWDONIA
Cambrian Mts.
Shrewsbury
Telford
Welshpool
Wolverhampton
Nuneaton
Leicester
ENGLAND
Peterborough
Thetford
Corby
Cardigan Bay
Aberystwyth
BIRMINGHAM
Coventry
Rugby
Redditch
Ely
Bury St. Edmunds
Worcester
Royal Leamington Spa
Northampton
Cambridge
Ipswich
Felixstowe
Hereford
Bedford
Milton Keynes
Stevenage
Harwich
Colchester
WALES
Brecon
BRECON BEACONS
Cheltenham
Gloucester
Cotswold Hills
Oxford
Luton
Harlow
Hemel Hempstead
Chelmsford
Fishguard
Carmarthen
Haverfordwest
Milford Haven
Pembroke
PEMBROKESHIRE COAST
Merthyr Tydfil
Neath
Llanelli
Swansea
Port Talbot
Rhondda
Cwmbran
Newport
Cardiff
Barry
Bristol
Bath
Swindon
High Wycombe
Watford
Slough
Basildon
Southend-on-Sea
LONDON
Thames
Reading
Newbury
Basingstoke
Reigate
Chatham
Margate
Canterbury
Maidstone
Dover
Guildford
Crawley
Ashford
Folkestone
Hastings
Str. of Dover
EXMOOR
Exmoor
Barnstaple
Weston-super-Mare
Taunton
Salisbury
Winchester
Southampton
Fareham
Havant
Brighton
Worthing
Eastbourne
Bude
Yeovil
NEW FOREST
Bournemouth
Poole
Portsmouth
Isle of Wight
Newport
Weymouth
Exeter
Exmouth
Torbay
Dartmoor
DARTMOOR
Plymouth
Newquay
Truro
St. Austell
Falmouth
Penzance
Land's End
Isles of Scilly
IRELAND
NORTHERN IRELAND
Malin Hd.
Buncrana
Letterkenny
Aran I.
GLENVEAGH
Lifford
Londonderry
Coleraine
Ballymena
Larne
Antrim
Lough Neagh
Belfast
Bangor
Lisburn
Lurgan
Portadown
Armagh
Newry
Donegal
Ulster
Omagh
Bundoran
Lower L. Erne
Enniskillen
Clones
Sligo
Ballina
Achill I.
L. Conn
Castlebar
Westport
Leitrim
Cavan
Castleblaney
Dundalk
Drogheda
Connacht
Lough Mask
Connemara
Roscommon
Longford
Lough Ree
Ceanannus Mor
Mullingar
Boyne
Lough Corrib
Athlone
Tullamore
Liffey
DUBLIN
Dun Laoghaire
Bray
Ballinasloe
Galway
Galway B.
Aran Is.
BURREN
Ennis
Lough Derg
Birr
Port Laoise
Leinster
Athy
Carlow
Wicklow Mts.
926
Arklow
Kilrush
Limerick
Nenagh
Thurles
Kilkenny
Shannon
Listowel
Tipperary
Munster
Tralee
Clonmel
Carrick-on-Suir
Wexford
Rosslare
Dingle
Carrauntoohill
1041
Killarney
Mallow
Blackwater
Waterford
Valencia I.
Macgillycuddy's Reeks
Dungarvan
Cork
Youghal
Cobh
Bantry
Bandon
Kinsale
C. Clear
Channel Is.
(U.K.)
Alderney
C. de la Hague
Guernsey
St. Peter Port
Sark
St. Helier
Jersey
Cotentin
Pte. de Barfleur
Cherbourg
Valognes
Bayeux
Le Havre
Trouville-sur-Mer
Caen
Lisieux
Elbeuf
Seine
Fécamp
Bolbec
Pays de Caux
Rouen
Dieppe
Le Tréport
Abbeville
Amiens
Picardie
FRANCE
Boulogne-sur-Mer
Le Touquet-Paris-Plage
Calais
Gris-Nez
Dunkerque
St-Omer
Béthune
LILLE
Bruay-la-Buissière
Lens
Valenciennes
St-Quentin
Cambrai
Laon
Tourcoing
Villeneuve d'Ascq
Oostende
Zeebrugge
Brugge
Gent
Vlissingen
NETHERLA
's-Gravenha
(Den Haag
Hoek van Holla
ROTTERD
Askøyna
Bergen
Osøyro
Stord
Bømlo
NORW
Haugesund
Kopervik
Åkrahamn
Stava
Projection: Conical with two standard parallels
East from Greenwich
West from Greenwich
COPYRIGHT PHILIP'S
20

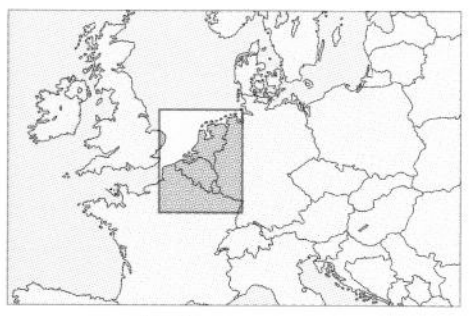

1:2 500 000

10 0 10 20 30 40 50 60 70 80 90 km
10 0 10 20 30 40 50 60 miles

NORTH SEA

UNITED KINGDOM

NETHERLANDS

AMSTERDAM
's-Gravenhage (Den Haag)
ROTTERDAM
Utrecht
Groningen

BELGIUM

BRUSSEL (Bruxelles)
Antwerpen
Gent (Gand)
Brugge
Namur
Charleroi
Liège

LUXEMBOURG

Luxembourg

GERMANY

Münster
Dortmund
Essen
Düsseldorf
Köln
Bonn
Koblenz
Wiesbaden
Mainz
Saarbrücken
Kaiserslautern
Strasbourg

FRANCE

PARIS
Reims
Amiens
Nancy
Lille

Projection: Lambert's Conformal Conic

COPYRIGHT PHILIP'S

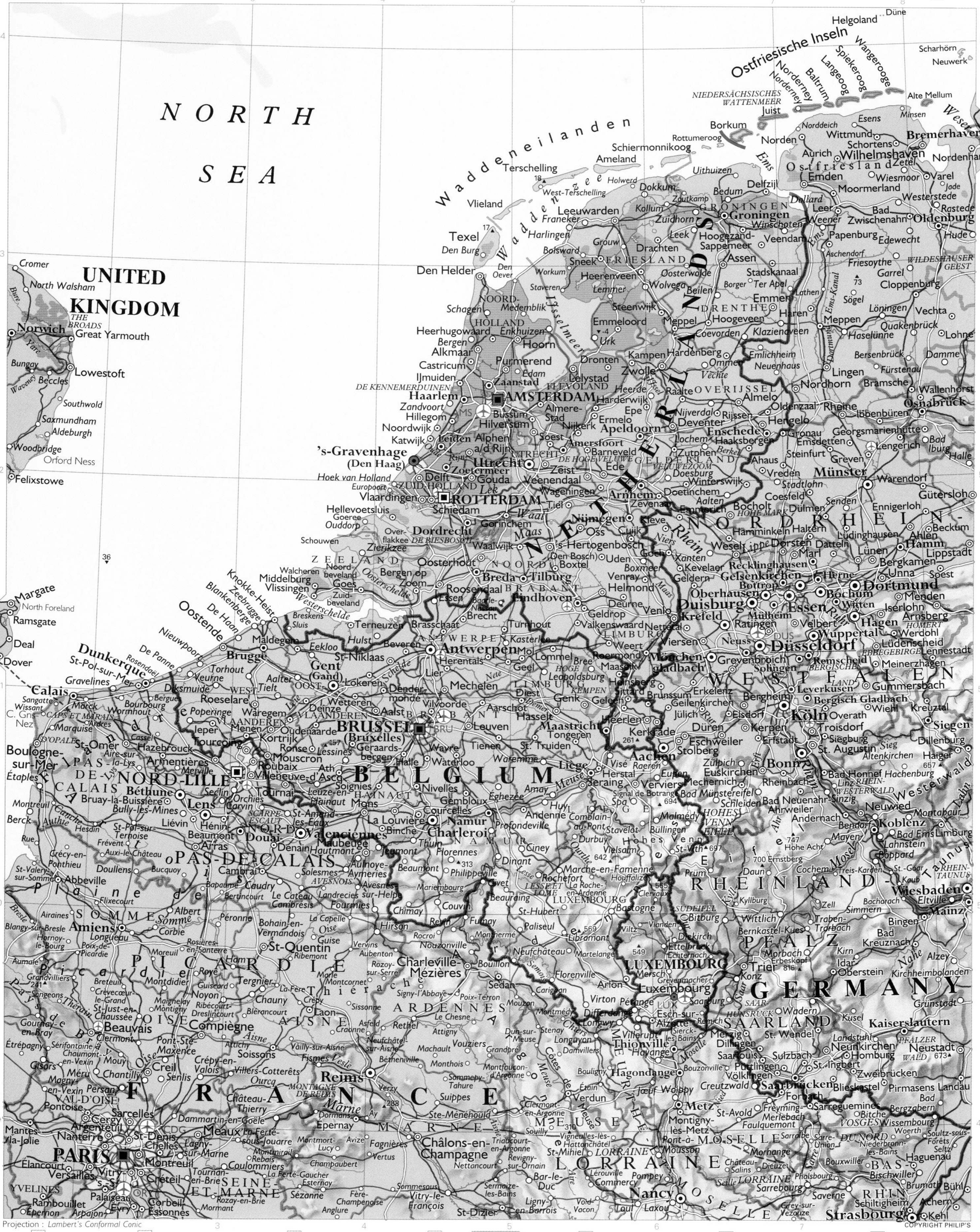

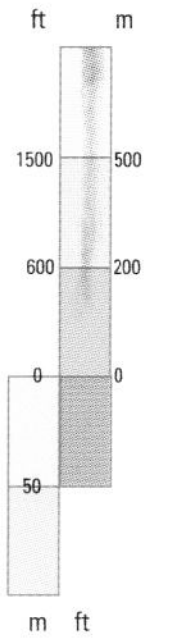

Underlined towns give their name to the administrative area in which they stand.

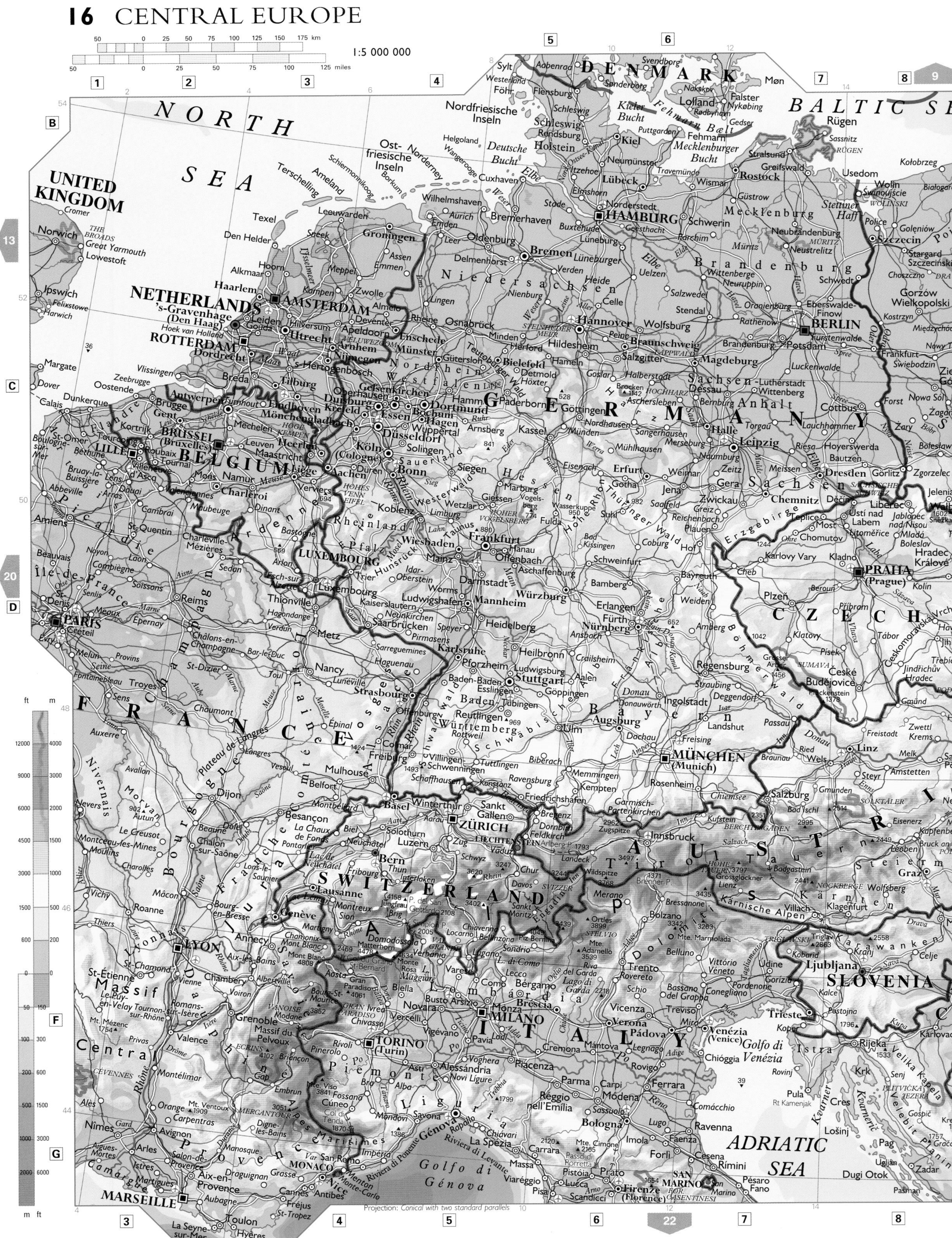

1:5 000 000

50   0   25   50   75   100   125   150   175 km
50   0   25   50   75   100   125 miles

NORTH SEA

BALTIC SEA

DENMARK

UNITED KINGDOM

NETHERLANDS

BELGIUM

LUXEMBOURG

GERMANY

CZECH

FRANCE

SWITZERLAND

AUSTRIA

SLOVENIA

ITALY

ADRIATIC SEA

Golfo di Génova

Projection: Conical with two standard parallels

LITHUANIA
BELARUS
POLAND
UKRAINE
SLOVAK REP.
HUNGARY
ROMANIA
MOLDOVA
SERBIA
BOSNIA-HERZEGOVINA
BULGARIA
Zatoka Gdańska
Kaliningrad (Russia)
Vilnius
MINSK
Gdynia
Gdańsk
Elbląg
Olsztyn
Hrodna
Białystok
Bydgoszcz
Toruń
Włocławek
Płock
WARSZAWA (Warsaw)
Łódź
Radom
Lublin
Kielce
Częstochowa
Wrocław
Opole
Katowice
Kraków
Tarnów
Rzeszów
Brest
Pinsk
Mazyr
Homyel
Babruysk
Baranavichy
Mahilyow
Lutsk
Rivne
Lviv (Lvov)
Ternopil
Khmelnytskyy
Vinnytsya
Zhytomyr
KYYIV (Kiev)
Bila Tserkva
Chernivtsi
Ivano-Frankivsk
Uzhhorod
Košice
Prešov
Žilina
Bratislava
Ostrava
Miskolc
Debrecen
Nyíregyháza
BUDAPEST
Győr
Szeged
Pécs
Kecskemét
Oradea
Cluj-Napoca
Satu Mare
Baia Mare
Suceava
Iași
Chișinău (Kishinev)
Tiraspol
Bacău
Brașov
Sibiu
Arad
Timișoara
Craiova
Pitești
Ploiești
BUCUREȘTI (Bucharest)
Constanța
Galați
Brăila
Tulcea
Novi Sad
BEOGRAD (Belgrade)
Sarajevo
Ruse
Varna
Dobrich
Carpații Meridionali
Transilvania
Vychodné Beskydy
Vojvodina
Moldova
Podilska Vysochyna
Dunărea (Danube)
Dnister
Vistula
Wisła
Pripet Marshes
Dobruja
Walachia
East from Greenwich
COPYRIGHT PHILIP'S

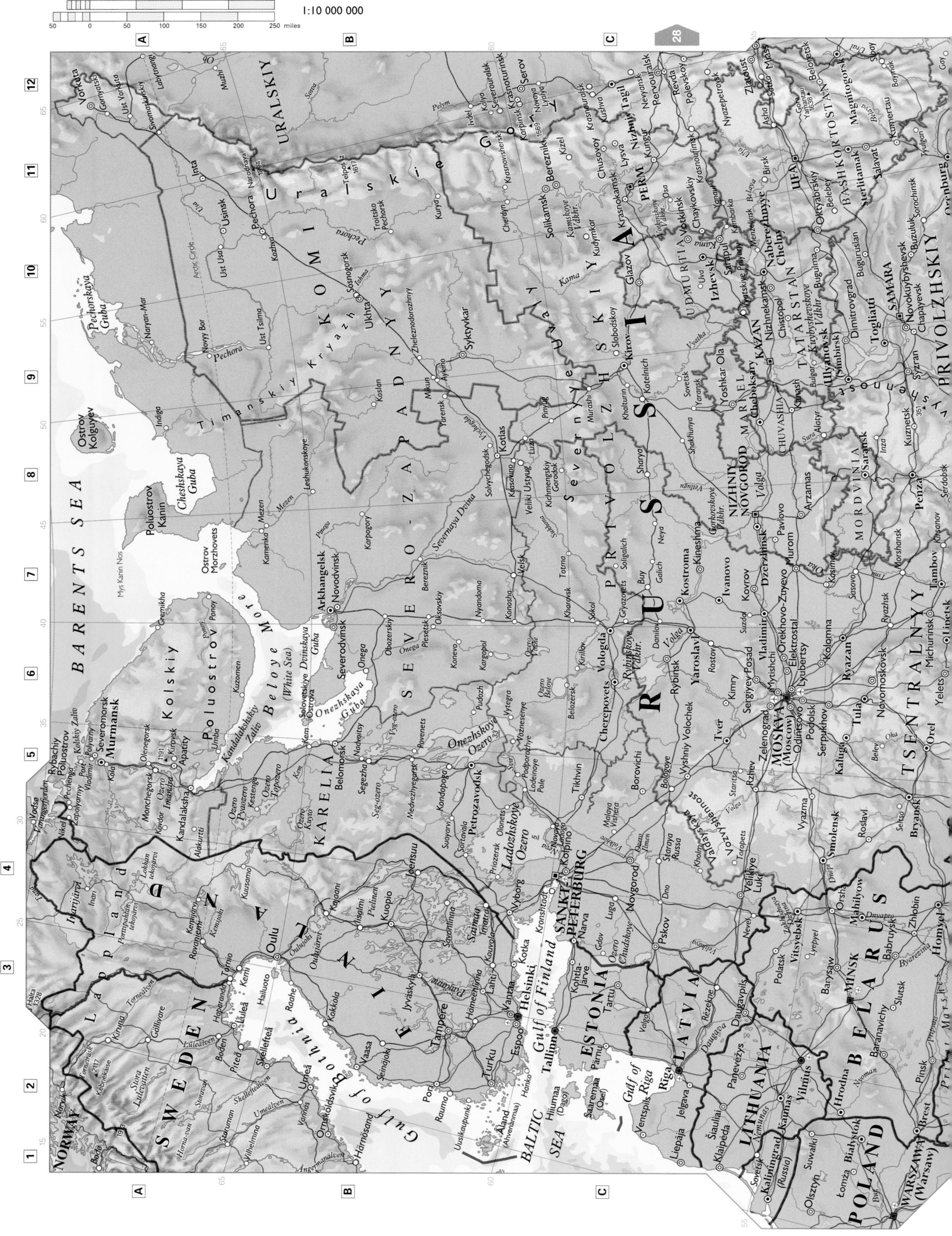

UKRAINE
KAZAKHSTAN
Kirgiz Steppe
Caspian Depression
ROMANIA
MOLDOVA
BULGARIA
Carpathians
Sea of Azov
CRIMEA
YUZHNYY
KALMYKIA
BLACK SEA
CASPIAN SEA
Caucasus Mountains
Ustyurt Plateau
Garabogazköl Aylagy (Kara Bogaz Gol)
TURKMENISTAN
GEORGIA
ARMENIA
AZERBAIJAN
TURKEY
Anadolu
Kuzey Anadolu Dağları
Toros Dağları
Kurdistan
IRAN
IRAQ
SYRIA
CYPRUS
LEBANON
MEDITERRANEAN SEA
Bādiyat ash Shām
Dasht-e Kavīr
KYYIV (Kiev)
KHARKIV
DNIPROPETROVSK
DONETSK
VOLGOGRAD
ROSTOV
ODESA
CHIȘINĂU
BUCUREȘTI (Bucharest)
İSTANBUL
ANKARA
İZMIR (Smyrna)
TBILISI
YEREVAN
BAKI (Baku)
TABRĪZ
TEHRĀN
QOM
AL MAWṢIL (Mosul)
HALAB (Aleppo)
BAYRŪT (Beirut)
DIMASHQ
Projection: Conical with two standard parallels
East from Greenwich
COPYRIGHT PHILIP'S

1:5 000 000

50 25 0 25 50 75 100 125 150 175 km
50 0 25 50 75 100 125 miles

Major labels:

UNITED KINGDOM

GERMANY

LUXEMBOURG

BELGIUM

SWITZERLAND

AUSTRIA

ITALY

ANDORRA

SPAIN

FRANCE

English Channel

Str. of Dover

Bay of Biscay

MEDITERRANEAN SEA

Golfe du Lion

Golfe de Gascogne

Corse (Corsica)

Selected cities and features:
PARIS, BRUSSELS (Bruxelles), LILLE, ZÜRICH, MILANO, TORINO (Turin), MONACO, MARSEILLE, LYON, Bordeaux, Nantes, Rennes, Brest, Cherbourg, Le Havre, Rouen, Caen, Le Mans, Tours, Orléans, Dijon, Besançon, Nancy, Metz, Strasbourg, Mulhouse, Reims, Amiens, Limoges, Clermont-Ferrand, Toulouse, Montpellier, Nîmes, Nice, Cannes, Grenoble, Genève, Lausanne, Bern, Basel, Luzern, Frankfurt, Mannheim, Karlsruhe, Stuttgart, Nürnberg, Augsburg, München, Genova (Genoa), La Spezia, Bayonne, Biarritz, Perpignan, Narbonne, Béziers, Carcassonne, Pau, Lourdes, Pamplona-Iruña, Donosti-San Sebastián, Bilbao, Santander

Massif Central, Pyrénées, Alpes, Jura, Vosges, Ardennes, Cotentin, Bretagne, Normandie, Aquitaine, Gascogne, Provence

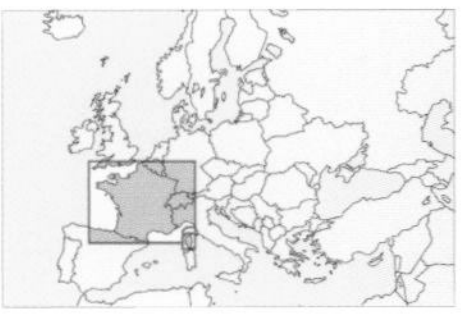

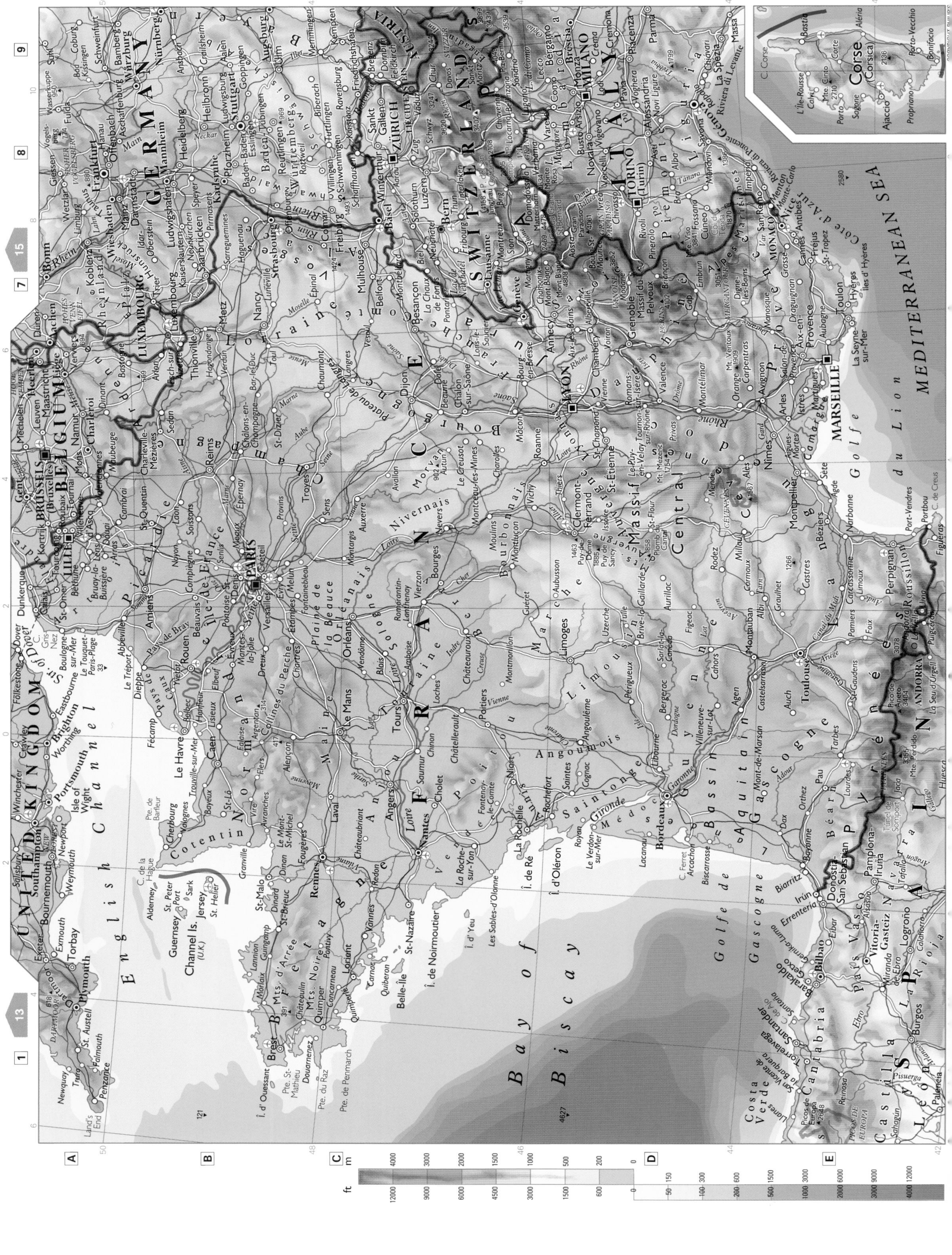

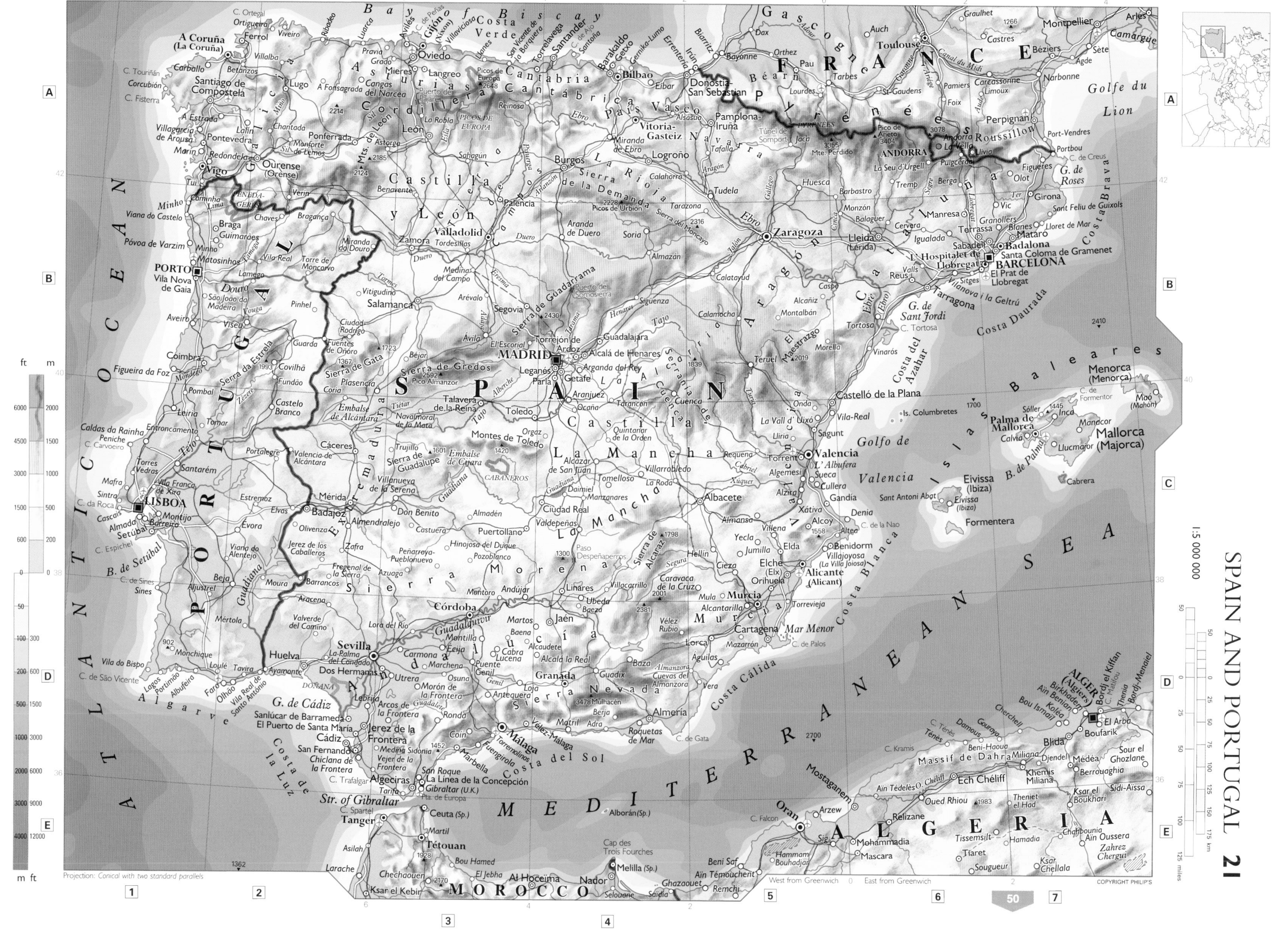
1:5 000 000
km
miles
MADRID
LISBOA
BARCELONA
Valencia
Sevilla
Zaragoza
Bilbao
Málaga
Granada
Córdoba
Murcia
Alicante (Alicant)
PORTO
ALGER (Algers)
Oran
Tanger
Mallorca (Majorca)
Menorca (Menorca)
Eivissa (Ibiza)
Formentera
ANDORRA
Str. of Gibraltar
G. de Cádiz
Golfo de Valencia
Golfe du Lion
Costa Brava
Costa del Sol
Costa de la Luz
Algarve
Sierra Nevada
Sierra Morena
Montes de Toledo
Sierra de Guadarrama
Sierra de Gredos
Projection: Conical with two standard parallels
West from Greenwich
East from Greenwich
COPYRIGHT PHILIP'S
ft
m

SWITZERLAND
AUSTRIA
SLOVENIA
CROATIA
FRANCE
ALGERIA
TUNISIA
MALTA
SAN MARINO
VATICAN CITY
MONACO
LYON
MARSEILLE
TORINO (Turin)
MILANO
ZAGREB
Ljubljana
Venézia (Venice)
Firenze (Florence)
ROMA
NÁPOLI
Palermo
Messina
Catánia
Cágliari
Corse (Corsica)
Sardegna (Sardinia)
Sicilia
TUNIS
Valletta
Trieste
Génova
Bologna
Ancona
Pescara
Bari
LIGURIAN SEA
TYRRHENIAN SEA
ADRIATIC SEA
MEDITERRANEAN
Golfo di Génova
Golfo di Venézia
Str. of Sicily
Str. di Messina
Bouches de Bonifacio
Ísole Eólie
Ísole Égadi
Ísole Pelagie (Italy)
Lampedusa
Pantelleria (Italy)
Elba
Ústica (Italy)
Golfe de Tunis
Golfe de Hammamet
Gozo
ft m
12000 4000
9000 3000
6000 2000
4500 1500
3000 1000
1500 500
600 200
0 0
50 150
100 300
200 600
500 1500
1000 3000
2000 6000
3000 9000
4000 12000
m ft

17
HUNGARY
ROMANIA
UKRAINE
SERBIA
BOSNIA-HERZEGOVINA
MONTENEGRO
KOSOVO
ALBANIA
MACEDONIA
BULGARIA
GREECE
TURKEY
BLACK SEA
Æ G E A N S E A
Sea of Thrace
Mirtoo Sea
Sea of Crete
IONIAN SEA
MEDITERRANEAN SEA
Marmara Denizi
Str. of Otranto
Ionioi Nisoi (Ionian Is.)
Transilvania
Carpații Meridionali
Wallachia
Dobruja
Vojvodina
Stara Planina
Rhodopi Planina
Makedonia
Thrace
Peloponnese
Cyclades
Dodecanese
Kriti (Crete)
BEOGRAD (Belgrade)
BUCUREȘTI (Bucharest)
SOFIYA
Skopje
Tiranë
Sarajevo
Podgorica
Priština
ATHINA (Athens)
Thessaloniki (Salonica)
ISTANBUL
IZMIR (Smyrna)
BURSA
Constanța
Varna
Burgas
Plovdiv
Edirne
Lesbos
Chios
Samos
Rhodes
Karpathos
Kythira
Iraklio
Chania
Limnos
Thasos
Kerkyra (Corfu)
Kefalonia (Cephalonia)
Zakynthos (Zante)
Evia
Northern Sporades
East from Greenwich
COPYRIGHT PHILIP'S
19
51

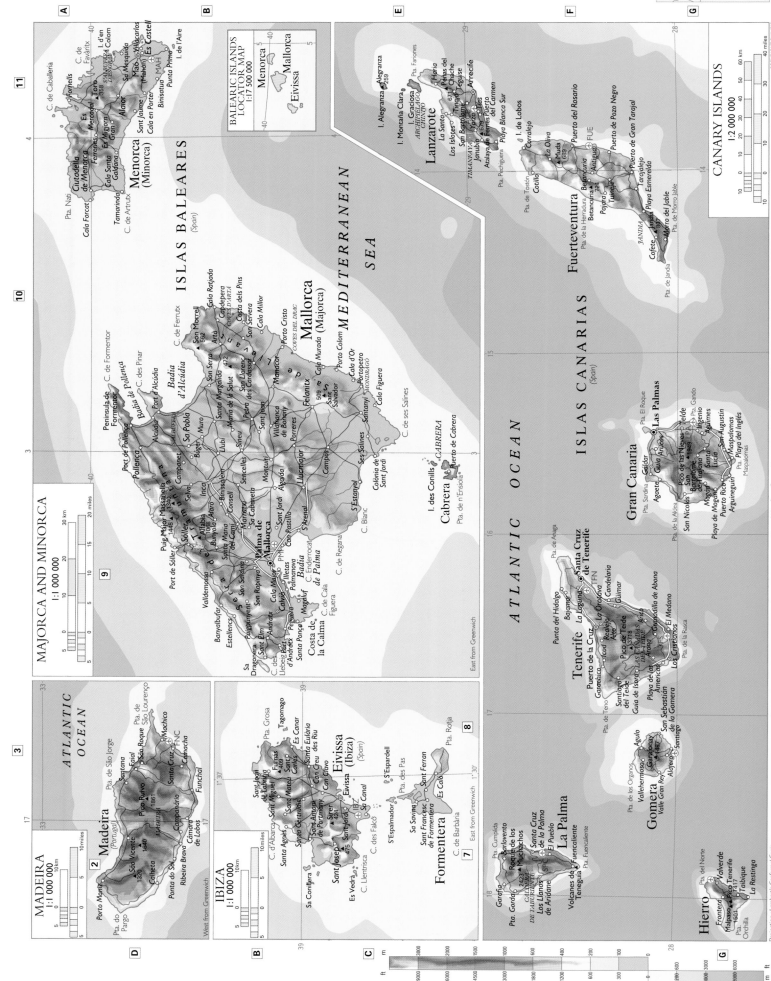

ISLAS BALEARES
(Spain)

MEDITERRANEAN SEA

Menorca
(Minorca)

BALEARIC ISLANDS
LOCATOR MAP
1:7 500 000

Menorca

Mallorca

Eivissa

Mallorca
(Majorca)

Cabrera

MAJORCA AND MINORCA
1:1 000 000

Badia
de Palma

Palma
de Mallorca

MADEIRA
1:1 000 000

Madeira
(Portugal)

ATLANTIC
OCEAN

IBIZA
1:1 000 000

Eivissa
(Ibiza)

Formentera

ATLANTIC OCEAN

ISLAS CANARIAS
(Spain)

Lanzarote

Fuerteventura

Gran Canaria
Las Palmas

Tenerife
Santa Cruz
de Tenerife

Gomera

La Palma

Hierro

CANARY ISLANDS
1:2 000 000

CRETE
1:1 300 000

Kriti (Crete) (Greece)

CYPRUS
1:1 300 000

MALTA
1:1 000 000

CORFU
1:1 000 000

RHODES
1:1 000 000

Projection: Lambert's Conformal Conic

East from Greenwich

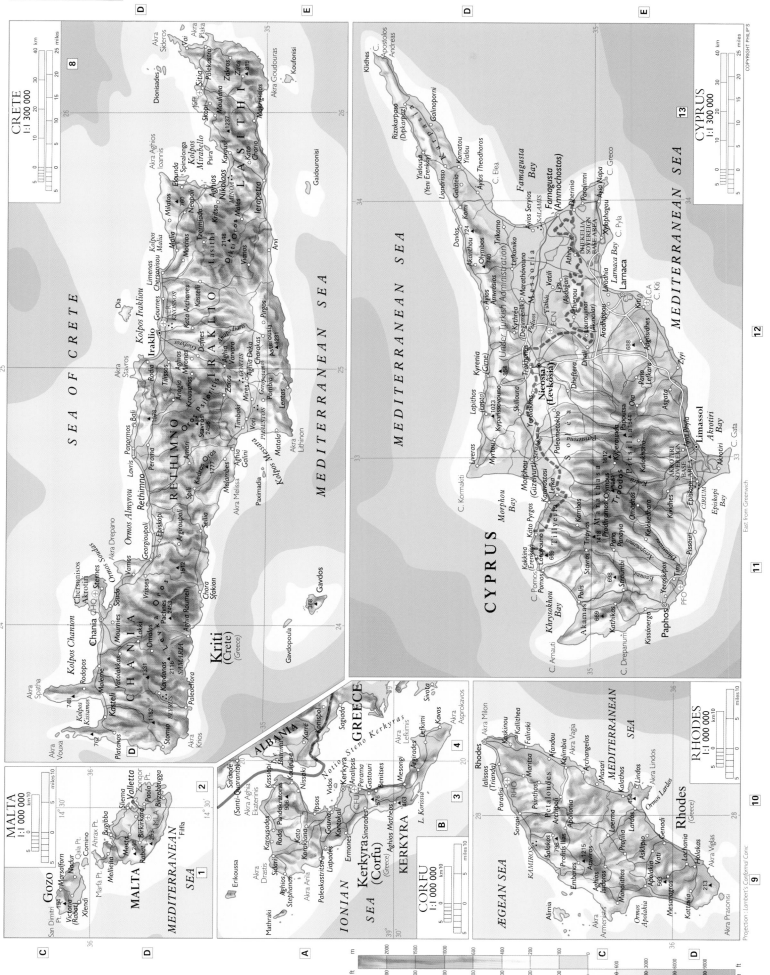

100 0 200 400 600 800 1000 1200 1400 km

100 0 200 400 600 800 1000 miles

1:47 000 000

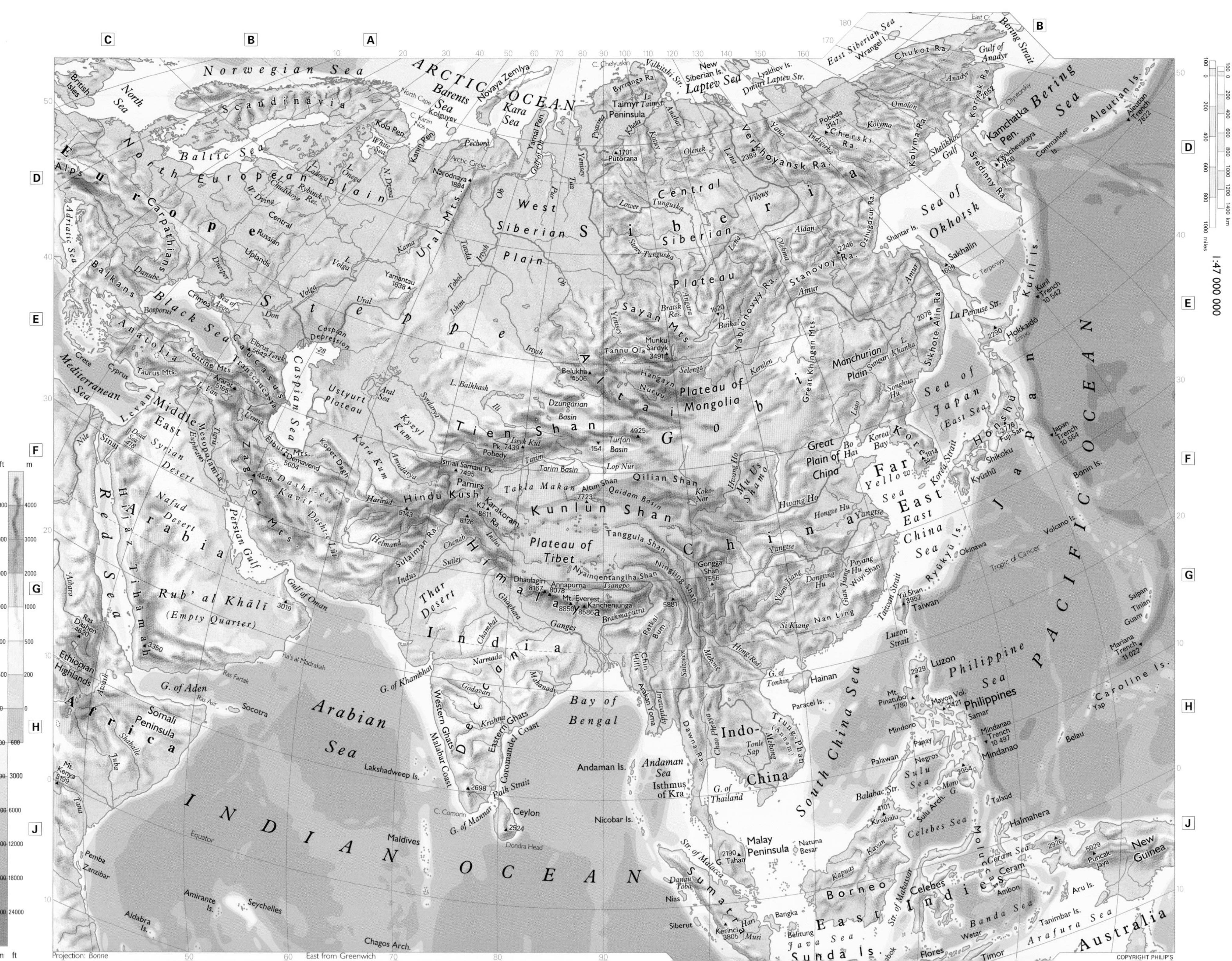

ft m
12 000 4000
9000 3000
6000 2000
3000 1000
1500 500
600 200
0 0
200 600
1000 3000
2000 6000
4000 12000
6000 18000
8000 24000
m ft

1:47 000 000

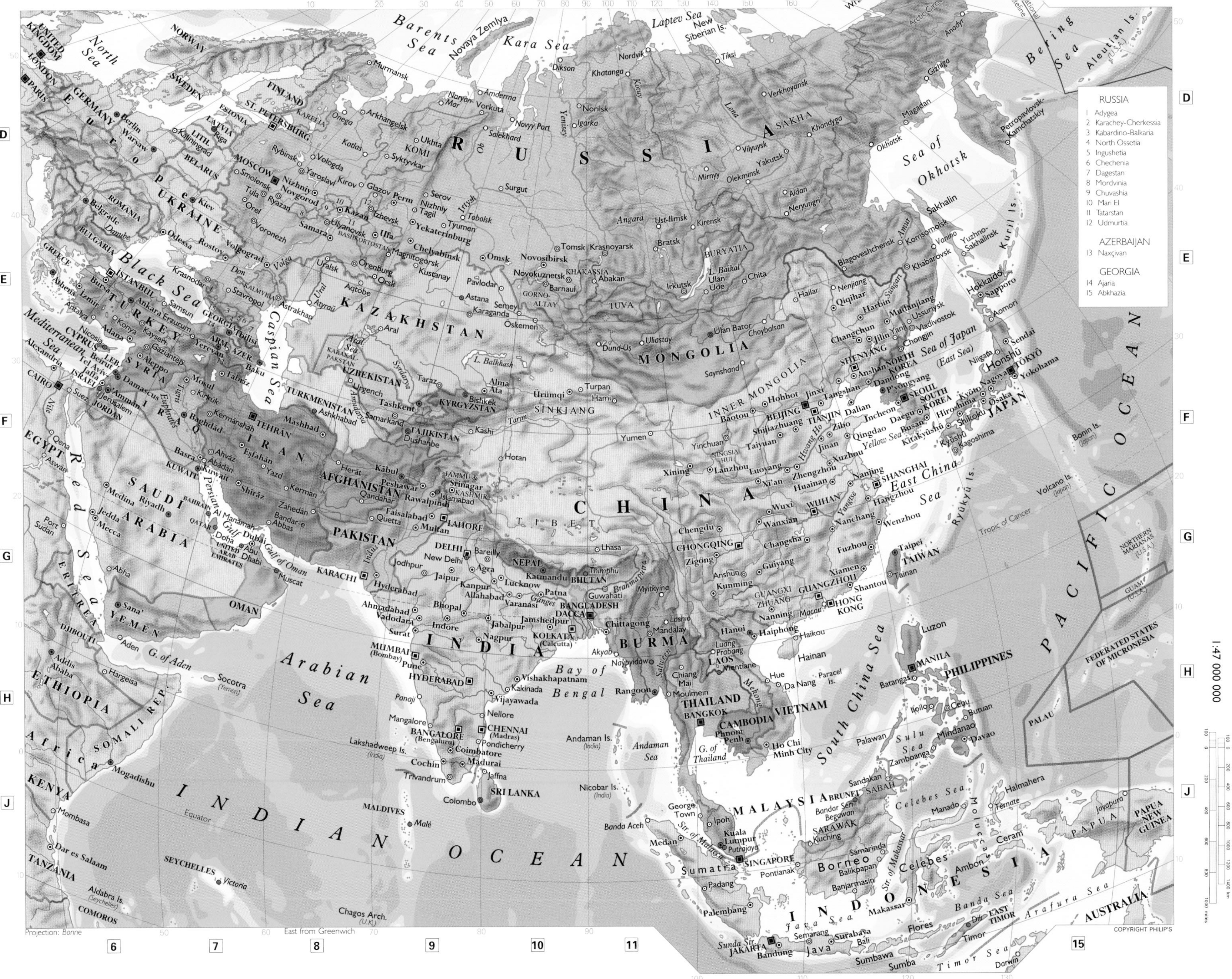
RUSSIA
1 Adygea
2 Karachey-Cherkessia
3 Kabardino-Balkaria
4 North Ossetia
5 Ingushetia
6 Chechenia
7 Dagestan
8 Mordvinia
9 Chuvashia
10 Mari El
11 Tatarstan
12 Udmurtia
AZERBAIJAN
13 Naxçivan
GEORGIA
14 Ajaria
15 Abkhazia
RUSSIA
CHINA
MONGOLIA
KAZAKHSTAN
INDIA
IRAN
SAUDI ARABIA
TURKEY
PAKISTAN
AFGHANISTAN
BURMA
THAILAND
VIETNAM
LAOS
CAMBODIA
MALAYSIA
INDONESIA
PHILIPPINES
JAPAN
NORTH KOREA
SOUTH KOREA
TAIWAN
NEPAL
BHUTAN
BANGLADESH
SRI LANKA
MALDIVES
OMAN
YEMEN
ETHIOPIA
SOMALI REP.
KENYA
TANZANIA
EGYPT
ERITREA
DJIBOUTI
SYRIA
JORDAN
ISRAEL
LEB.
CYPRUS
GREECE
BULGARIA
ROMANIA
UKRAINE
BELARUS
GERMANY
SWEDEN
NORWAY
FINLAND
ESTONIA
LATVIA
LITH.
UNITED KINGDOM
UZBEKISTAN
TURKMENISTAN
KYRGYZSTAN
TAJIKISTAN
GEORGIA
ARM.
AZER.
KUWAIT
BAHRAIN
QATAR
SINGAPORE
BRUNEI
EAST TIMOR
PAPUA NEW GUINEA
AUSTRALIA
PACIFIC OCEAN
INDIAN OCEAN
Caspian Sea
Black Sea
Red Sea
Arabian Sea
Bay of Bengal
South China Sea
East China Sea
Sea of Japan (East Sea)
Sea of Okhotsk
Bering Sea
Barents Sea
Kara Sea
Laptev Sea
Mediterranean Sea
North Sea
Andaman Sea
Projection: Bonne
East from Greenwich
COPYRIGHT PHILIP'S

100 0 100 200 300 400 500 600 700 800 km
100 0 100 200 300 400 500 miles

1:20 000 000

RUSSIA	
1	Adygea
2	Karachey-Cherkessia
3	Kabardino-Balkaria
4	North Ossetia
5	Ingushetia
6	Chechenia
7	Dagestan
8	Mordvinia
9	Chuvashia
10	Mari El
11	Tatarstan
12	Udmurtia
13	Khakassia
AZERBAIJAN	
14	Naxçivan
GEORGIA	UKRAINE
15	Ajaria 17 Crimea
16	Abkhazia

Projection: Conical Orthomorphic with two standard parallels

East from Greenwich

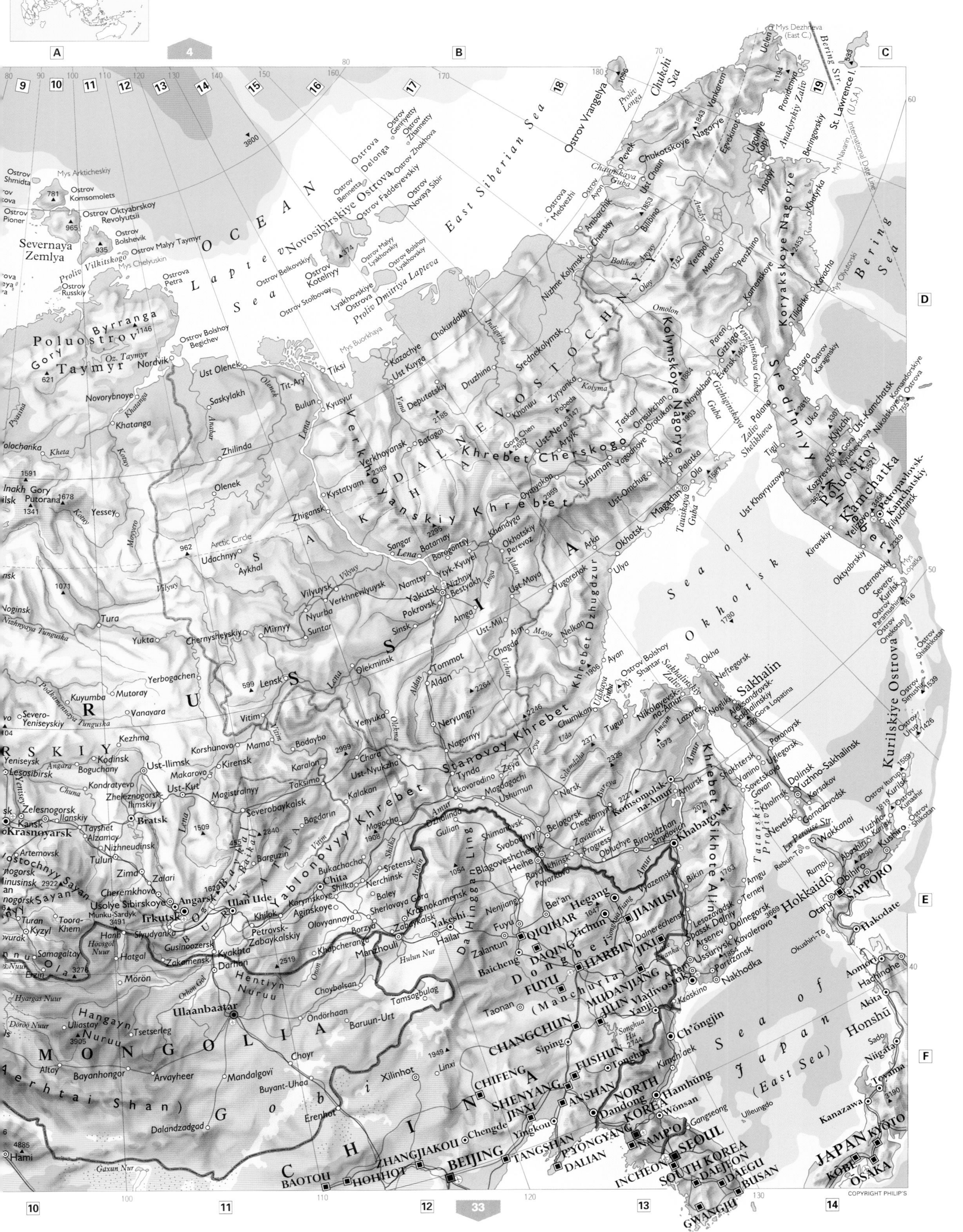

East Siberian Sea
Laptev Sea
OCEAN
Chukchi Sea
Bering Sea
Sea of Okhotsk
Sea of Japan (East Sea)
Bering Str.
Ostrov Vrangelya
Novosibirskiye Ostrova
Severnaya Zemlya
Poluostrov Taymyr
Byrranga Gory
Verkhoyanskiy Khrebet
Khrebet Cherskogo
Kolymskoye Nagorye
Chukotskoye Nagorye
Koryakskoye Nagorye
Sredinnyy Khrebet
Poluostrov Kamchatka
Stanovoy Khrebet
Yablonovyy Khrebet
Khrebet Dzhugdzur
Khrebet Sikhote Alin
Kurilskiye Ostrova
Sakhalin
Hokkaidō
Honshū
DALNEVOSTOCHNY
SAKHA
BURYATIYA
RUSSIA
MONGOLIA
CHINA
NORTH KOREA
SOUTH KOREA
JAPAN
Gobi
Dongbei (Manchuria)
Da Hinggan Ling
Hentiyn Nuruu
Hangayn Nuruu
Oz. Baykal
Lena
Yakutsk
Magadan
Petropavlovsk-Kamchatskiy
Khabarovsk
Vladivostok
Irkutsk
Ulan Ude
Chita
Bratsk
Krasnoyarsk
Ulaanbaatar
HARBIN
QIQIHAR
CHANGCHUN
SHENYANG
BEIJING
P'YŎNGYANG
SEOUL
BUSAN
SAPPORO
KYOTO
ŌSAKA
Arctic Circle
International Date Line
COPYRIGHT PHILIP'S

1:5 000 000
km
miles
SEA OF OKHOTSK
SEA OF JAPAN
(EAST SEA)
Yamato Rise
Sakhalin
(Russia)
La Perouse Strait
(Sōya-Kaikyō)
HOKKAIDŌ
Hokkaidō
SAPPORO
Asahikawa
Otaru
Hakodate
Kushiro
Obihiro
Abashiri
Wakkanai
Rishiri-Tō
Rebun-Tō
Okushiri-Tō
Ishikari-Wan
(Otaru-Wan)
Uchiura-Wan
Tsugaru-Kaikyō
TŌHOKU
Honshū
SENDAI
Akita
Niigata
Sado
Aomori
Morioka
Hachinohe
Sendai-Wan
Tsuruoka
Sakata
Yamagata
Hirosaki
Noshiro
Oga-Hantō
Tobi-Shima
Awa-Shima
Ostrov Moneron
(Russia)
Ostrov Kunashir
RUSSIA
PRIMORSKIY KRAY
Sikhote Alin
CHINA
HEILONGJIANG
JILIN
NORTH KOREA
Dongbei
(Manchuria)
Vladivostok
Ussuriysk
Nakhodka
Artem
Lake Khanka
Zaliv Petra Velikogo
Dalnerechensk
Lesozavodsk
Arsenev
Spassk Dalniy
Partizansk
Olga
Plastun
Terney
Svetlaya
Bikin
JIXI
Mishan
Hulin
Suifenhe
Hunchun
Tumen
Ch'ŏngjin
Najin
JAMUSI
Hegang
Fujin
Shuangyashan
Baoqing
Qitaihe
Boli
Linkou
Muling
Dongning
Suiyang

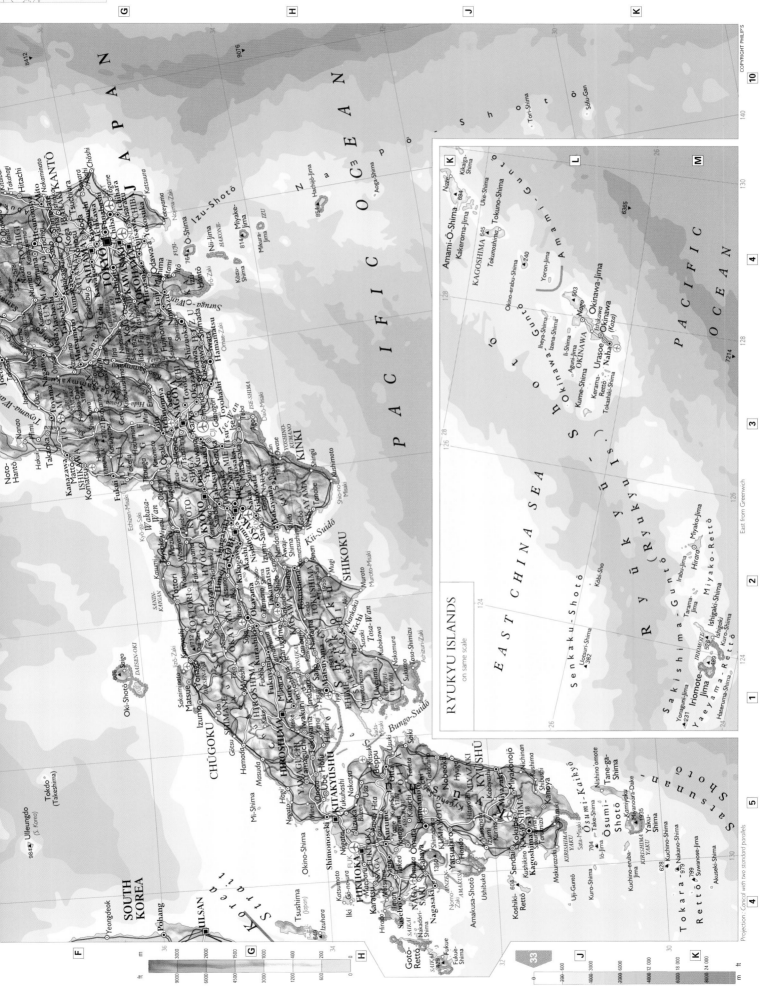

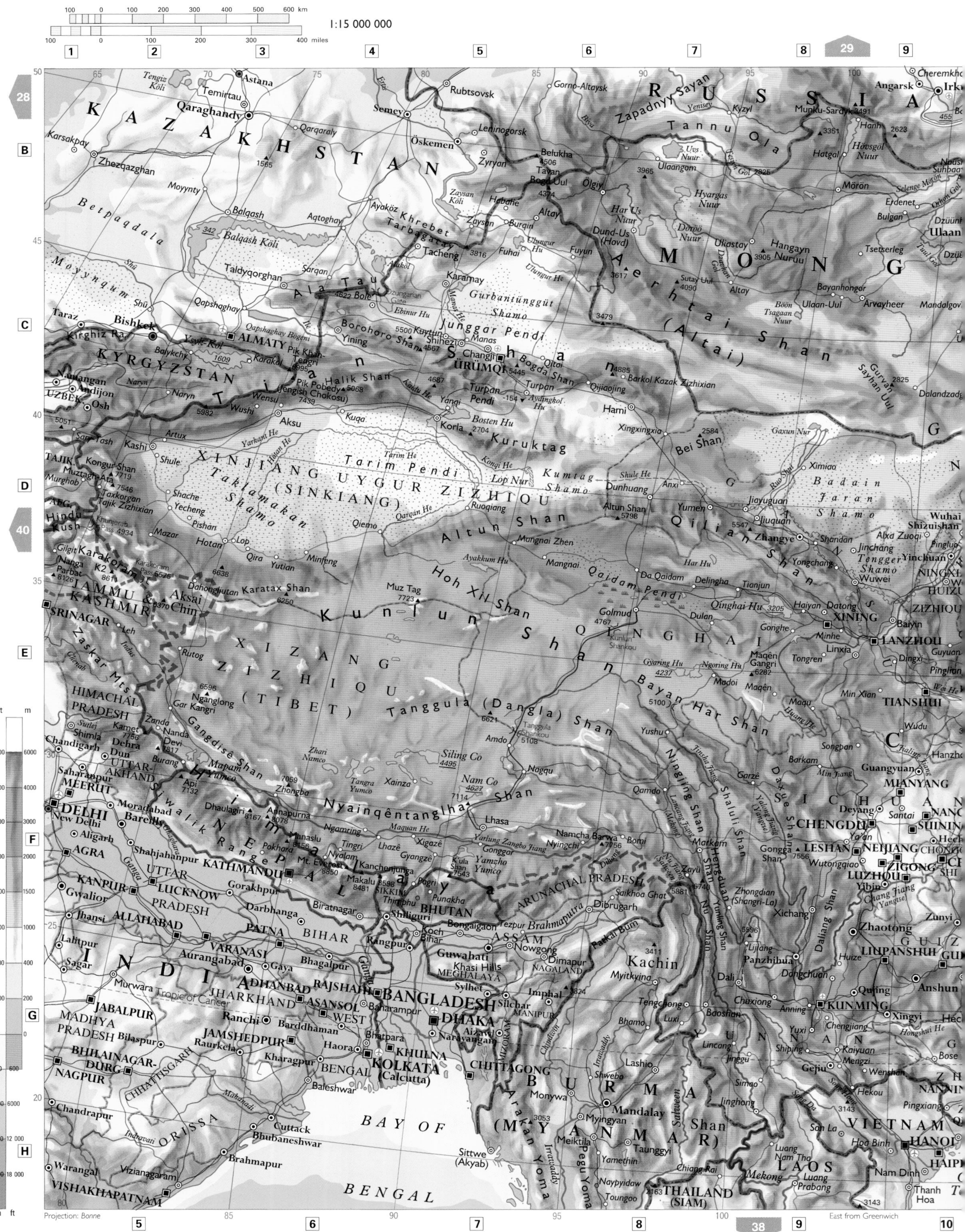
1:15 000 000
km
miles
Projection: Bonne
East from Greenwich
KAZAKHSTAN
RUSSIA
MONGOLIA
KYRGYZSTAN
XINJIANG UYGUR ZIZHIQU (SINKIANG)
XIZANG ZIZHIQU (TIBET)
QINGHAI
SICHUAN
JAMMU & KASHMIR
HIMACHAL PRADESH
UTTARAKHAND
UTTAR PRADESH
MADHYA PRADESH
CHHATTISGARH
ORISSA
JHARKHAND
BIHAR
WEST BENGAL
NEPAL
BHUTAN
SIKKIM
ARUNACHAL PRADESH
ASSAM
NAGALAND
MANIPUR
MEGHALAYA
MIZORAM
INDIA
BANGLADESH
BURMA (MYANMAR)
THAILAND (SIAM)
LAOS
VIETNAM
BAY OF BENGAL
Tropic of Cancer
Astana
Qaraghandy
Temirtau
Zhezqazghan
Balqash
Balqash Köli
Betpaqdala
Moyynqum
Taldyqorghan
Bishkek
ALMATY
Taraz
Ysyk-Köl
Semey
Öskemen
Rubtsovsk
Leninogorsk
Zaysan Köli
Ala Tau
Tian Shan
Pik Pobedy (Jengish Chokosu) 7439
Pik Khan-Tengri 6995
Kashi
Aksu
Kuqa
Yining
ÜRÜMQI
Turpan
Turpan Pendi
Hami
Junggar Pendi
Gurbantünggüt Shamo
Bogda Shan
Karamay
Tacheng
Altay
Tarim Pendi
Taklamakan Shamo
Lop Nur
Kuruktag
Kumtag Shamo
Dunhuang
Altun Shan
Hotan
Yecheng
Kunlun Shan
Hoh Xil Shan
Qaidam Pendi
Golmud
XINING
LANZHOU
Qinghai Hu
Bayan Har Shan
Tanggula (Dangla) Shan
Nyainqêntanglha Shan
Gangdisê Shan
Lhasa
Siling Co
Nam Co
Himalaya
Mt. Everest 8850
Kanchenjunga
Makalu 8481
Annapurna 8078
Dhaulagiri 8167
K2 8611
Karakoram
Nanga Parbat 8126
Hindu Kush
Aksai Chin
SRINAGAR
Leh
Zaskar Mts.
Siwalik Range
KATHMANDU
Thimphu
DELHI
New Delhi
AGRA
KANPUR
LUCKNOW
ALLAHABAD
VARANASI
PATNA
JABALPUR
NAGPUR
BHILAINAGAR-DURG
KOLKATA (Calcutta)
DHAKA
CHITTAGONG
Brahmaputra
Ganga
Guwahati
Khasi Hills
Imphal
Kachin
Mandalay
Sittwe (Akyab)
Arakan Yoma
Pegu Yoma
Irrawaddy
Salween
Shan
Naypyidaw
Chiang Rai
Mekong
Luang Prabang
HANOI
KUNMING
CHENGDU
CHONGQING
ZIGONG
LESHAN
NEIJIANG
LUZHOU
MIANYANG
Hengduan Shan
Daxue Shan
Shaluli Shan
Ningjing Shan
Daliang Shan
Tarbagatay
Tannu Ola
Zapadnyy Sayan
Aerhtai Shan (Altai)
Hangayn Nuruu
Uliastay
Ulaangom
Badain Jaran Shamo
Tengger Shamo
Gurvan Sayhan Uul
Qilian Shan
Bei Shan
Jiayuguan
Zhangye
Wuwei
TIANSHUI
Yushu
Qamdo
Nagqu
Xichang
Zhaotong
LIUPANSHUI
Dali
Lijiang
Baoshan
Lashio
Myitkyina
Bhamo
Jinghong
Simao
Gejiu
Hekou
Nam Dinh
Thanh Hoa
Cuttack
Bhubaneshwar
Brahmapur
VISHAKHAPATNAM
Warangal
Chandrapur
Baleshwar
Kharagpur
JAMSHEDPUR
Ranchi
DHANBAD
ASANSOL
RAJSHAHI
Gaya
Aurangabad
Bhagalpur
Darbhanga
Gorakhpur
Biratnagar
Jhansi
Gwalior
Meerut
Saharanpur
Chandigarh
Shimla
Dehra Dun
Moradabad
Bareilly
Aligarh
Shahjahanpur

HONG KONG, MACAU AND SHENZHEN

1:1 000 000

a

37

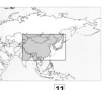

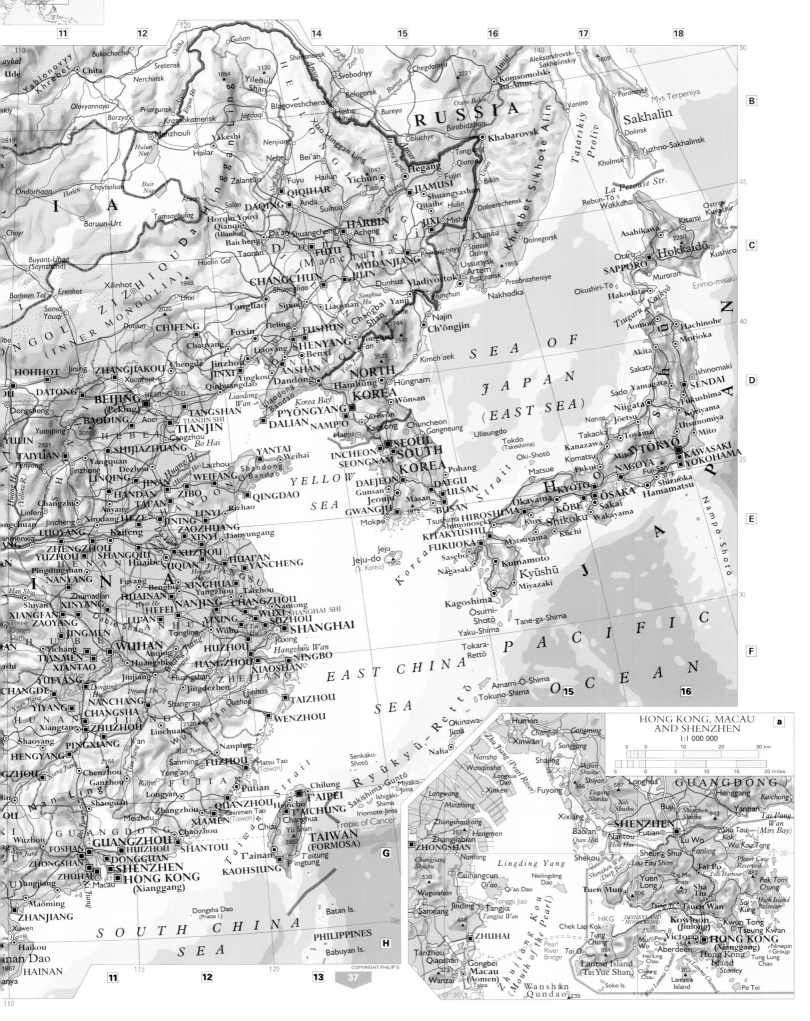

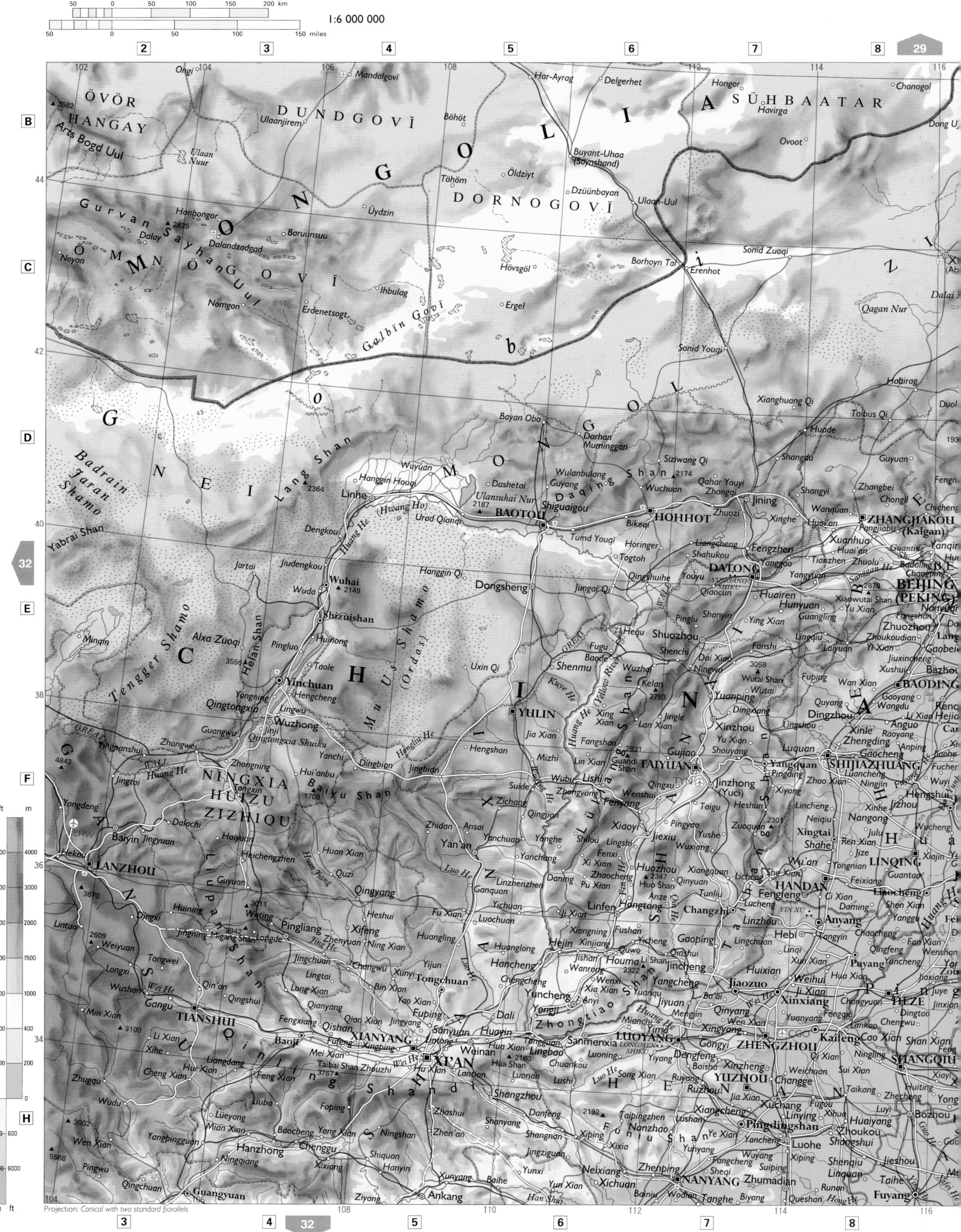

1:6 000 000

Projection: Conical with two standard parallels

HARBIN
CHANGCHUN
SHENYANG
DALIAN
P'YŎNGYANG
SEOUL
BUSAN
NORTH KOREA
SOUTH KOREA
SEA OF JAPAN (EAST SEA)
YELLOW SEA (Huang Hai)
Korea Bay
Bo Hai
Liaodong Wan
Korea Strait
Vladivostok
QINGDAO
INCHEON
DAEGU
DAEJEON
GWANGJU
ULSAN
JAPAN
RUSSIA
East from Greenwich
COPYRIGHT PHILIP'S

1:12 500 000

100 0 100 200 300 400 500 km
100 0 50 100 150 200 250 300 350 miles
BURMA (MYANMAR)
RANGOON (Yangon)
Moulmein
G. of Martaban
THAILAND
BANGKOK (Krung Thep)
Vientiane (Viangchan)
Nong Khai
Udon Thani
Khon Kaen
Nakhon Ratchasima
LAOS
Ba Don
Dong Hoi
Hue
Da Nang
VIETNAM
Quang Ngai
Qui Nhon
Nha Trang
Cam Ranh
Phan Rang
Bien Hoa
THANH PHO HO CHI MINH (Saigon)
Vung Tau
Can Tho
Soc Trang
Bac Lieu
Ca Mau
Con Son
Paracel Is.
CAMBODIA
PHNOM PENH
Siemreap
Tonlé Sap
Battambang
Gulf of Thailand
ANDAMAN SEA
Mergui (Myeik)
Tavoy (Dawei)
Chumphon
Surat Thani
Nakhon Si Thammarat
Phuket
Hat Yai
Songkhla
Malay Peninsula
SOUTH CHINA SEA
Spratly Is.
Spratly I.
Amboyna Cay
Kota Bharu
Kuala Terengganu
George Town
Butterworth
Ipoh
PENINSULAR MALAYSIA
KUALA LUMPUR
Kuantan
Klang
Seremban
Melaka
Johor Bahru
SINGAPORE
Straits of Malacca
MALAYSIA
Kota Kinabalu
Gunung Kinabalu 4101
SABAH
LABUAN
Bandar Seri Begawan
BRUNEI
Miri
Bintulu
Sibu
SARAWAK
Kuching
Pontianak
Kepulauan Natuna Besar (Indonesia)
Kepulauan Anambas (Indonesia)
Banda Aceh
Lhokseumawe
ACEH
MEDAN
Tebingtinggi
Pematangsiantar
Simeulue
Nias
Padangsidempuan
SUMATERA UTARA
Pekanbaru
RIAU
Padang
SUMATERA BARAT
Siberut
Kepulauan Mentawai
Jambi
JAMBI
PALEMBANG
SUMATERA SELATAN
Bengkulu
BENGKULU
LAMPUNG
Bandar Lampung
Enggano
Bangka
Pangkalpinang
Belitung
Selat Karimata
KALIMANTAN BARAT
KALIMANTAN TENGAH
Banjarmasin
KALIMANTAN SELATAN
Palangkaraya
INDONESIA
Greater Sunda Islands
JAVA SEA
JAKARTA
Bogor
BANDUNG
Cirebon
Semarang
JAWA TENGAH
Yogyakarta
Surakarta
Madiun
Kediri
Malang
SURABAYA
JAWA TIMUR
Madura
BALI
Denpasar
Lombok
Mataram
NUSA TENGGARA BARAT
Jawa (Java)
INDIAN OCEAN
Java Trench
6073
6650
Projection: Mercator
East from Greenwich

JAVA AND MADURA
1:7 500 000
BALI
1:2 000 000
JAKARTA
Bogor
BANDUNG
Tasikmalaya
Cirebon
Semarang
Yogyakarta
Surakarta
SURABAYA
Malang
Madura
Selat Madura
Bali
Lombok
Denpasar
Singaraja
Mataram
Jawa
BALI SEA
INDIAN OCEAN
Luzon
MANILA
Quezon City
Mindanao
DAVAO
Zamboanga
Cebu
Iloilo
Bacolod
Samar
Leyte
Mindoro
Sibuyan Sea
Visayan Sea
Bohol Sea
Sulu Arch.
Mindanao Trench
CELEBES SEA
PACIFIC OCEAN
Halmahera
Ternate
Tidore
Manado
Gorontalo
Sulawesi (Celebes)
MOLUCCA SEA
Seram
Ambon
Buru
MALUKU
BANDA SEA
Kepulauan Aru
Kepulauan Tanimbar
IRIAN JAYA
PAPUA
Pegunungan Maoke
Jayapura
Merauke
PAPUA NEW GUINEA
Flores
Sumba
Sawu Sea
Timor
Dili
EAST TIMOR
Kupang
NUSA TENGGARA TIMUR
ARAFURA SEA
Equator
COPYRIGHT PHILIP'S

1:6 000 000

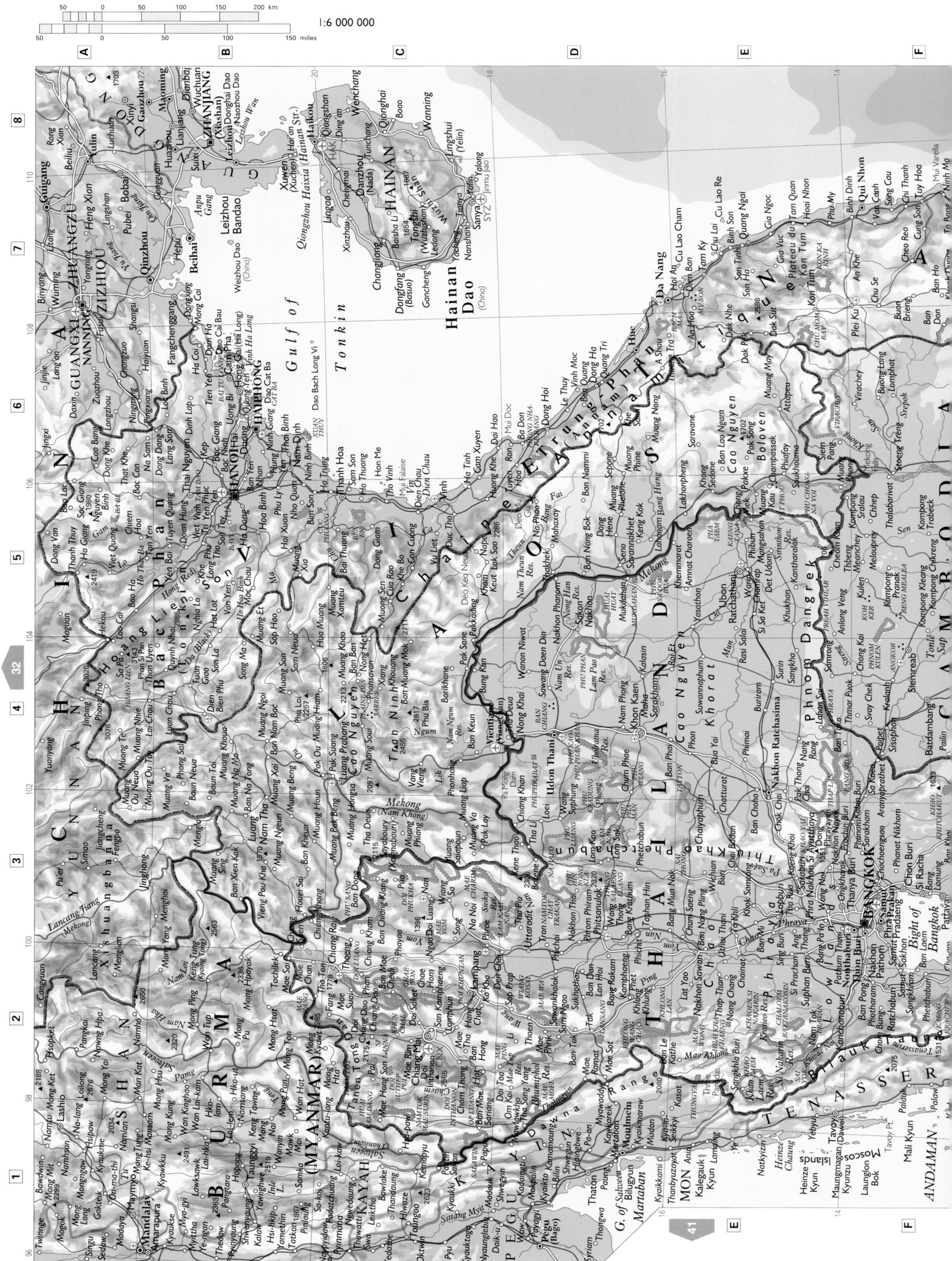

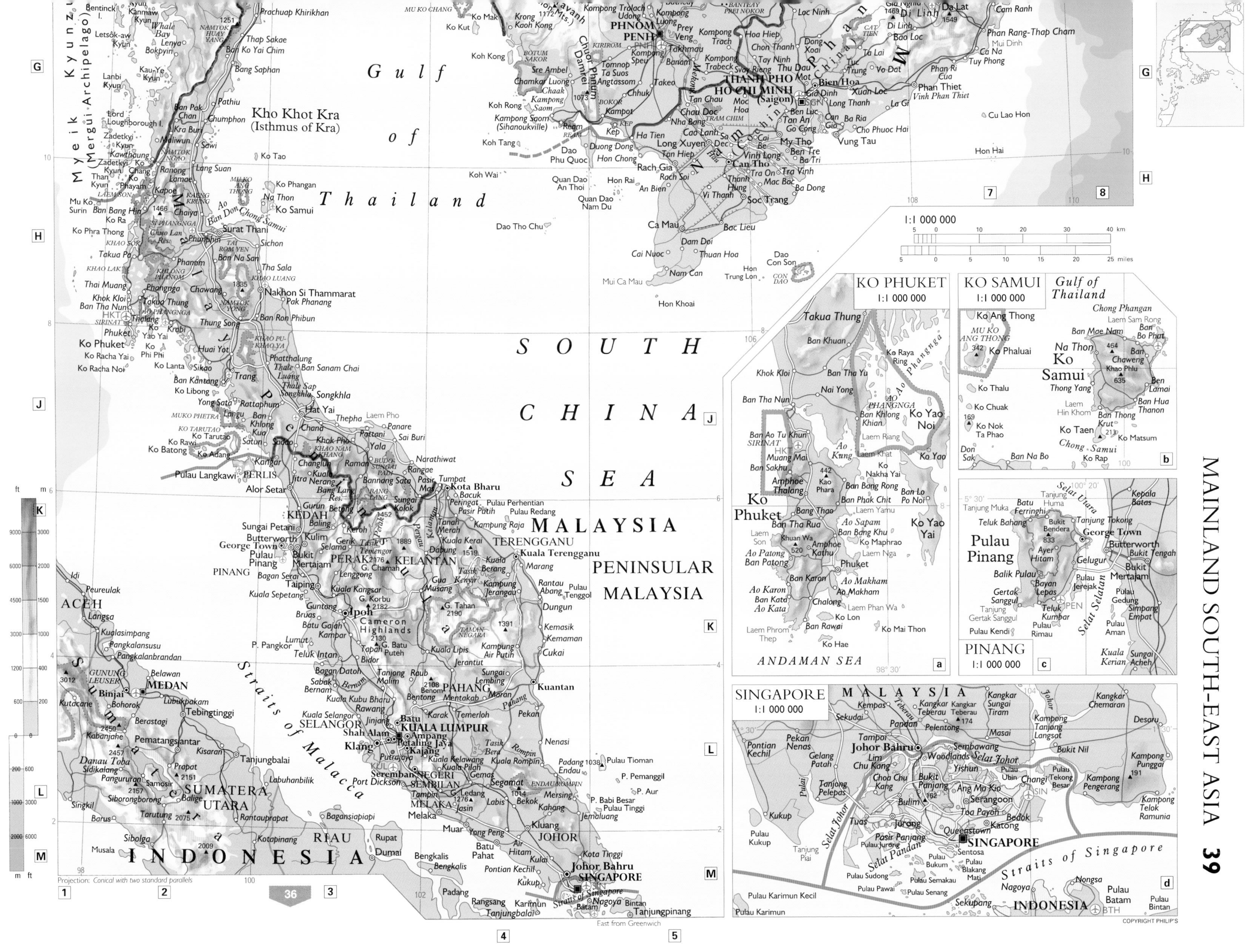

Gulf of Thailand

KO SAMUI
1:1 000 000

b

d

c

Pulau
Pinang

PINANG
1:1 000 000

KO PHUKET
1:1 000 000

a

ANDAMAN SEA

MALAYSIA

SINGAPORE
1:1 000 000

Strait of Singapore

INDONESIA

SOUTH

CHINA

SEA

MALAYSIA

PENINSULAR
MALAYSIA

TERENGGANU

PAHANG

KELANTAN

PERAK

KEDAH

PERLIS

PINANG

SELANGOR

SEMBILAN

MELAKA

JOHOR

KUALA LUMPUR

SINGAPORE

Straits of Malacca

Straits of Malacca

INDONESIA

SUMATERA
UTARA

ACEH

RIAU

Gulf

of

Thailand

Kho Khot Kra
(Isthmus of Kra)

Mergui Archipelago

Kyunzu

MU KO CHANG

Projection: Conical with two standard parallels

East from Greenwich

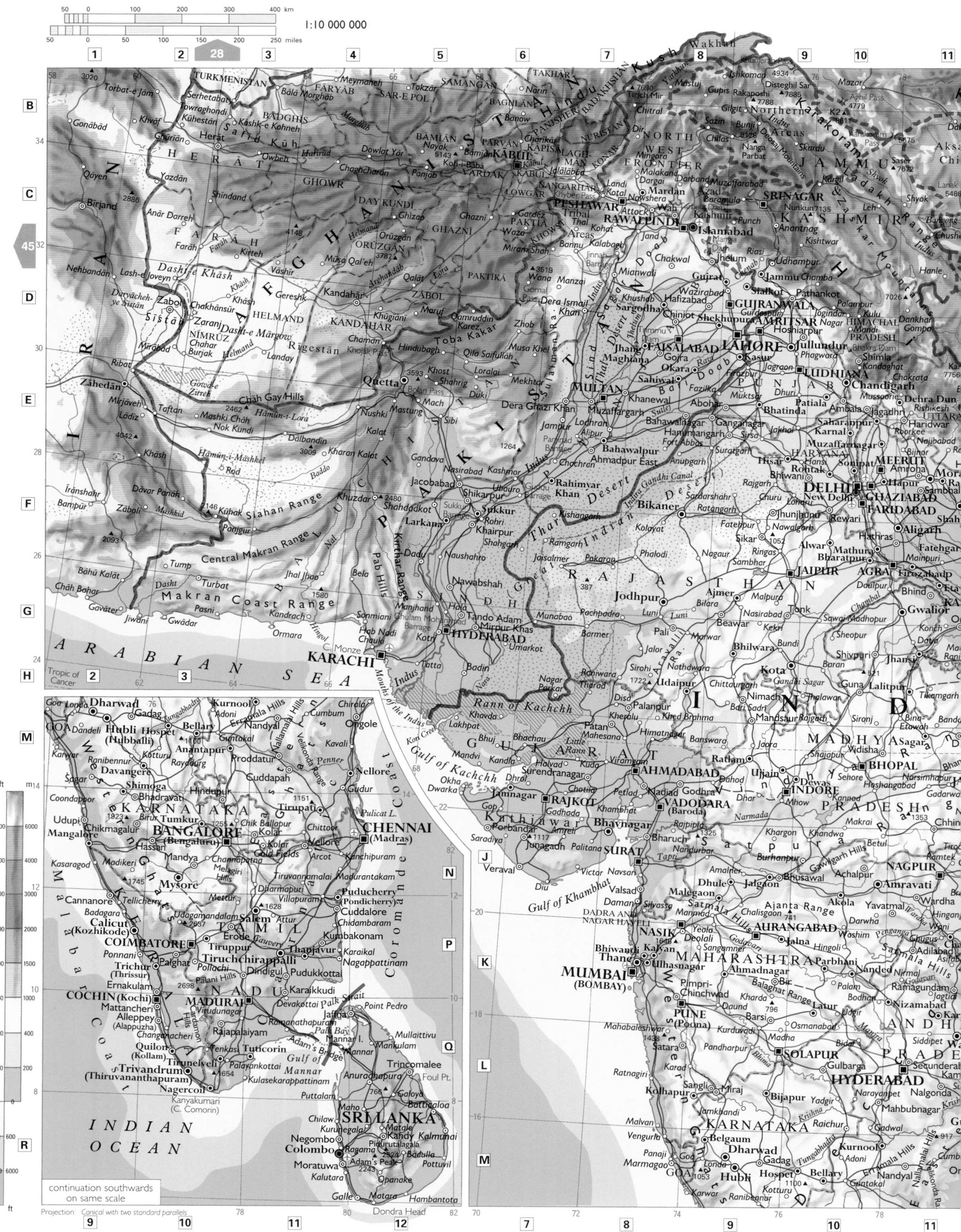

continuation southwards
on same scale

Projection: Conical with two standard parallels

XIZANG ZIZHIQU (TIBET)

CHINA

QINGHAI

SICHUAN

YUNNAN

NEPAL

BHUTAN

BANGLADESH

BURMA (MYANMAR)

THAILAND

BAY OF BENGAL

INDIAN OCEAN

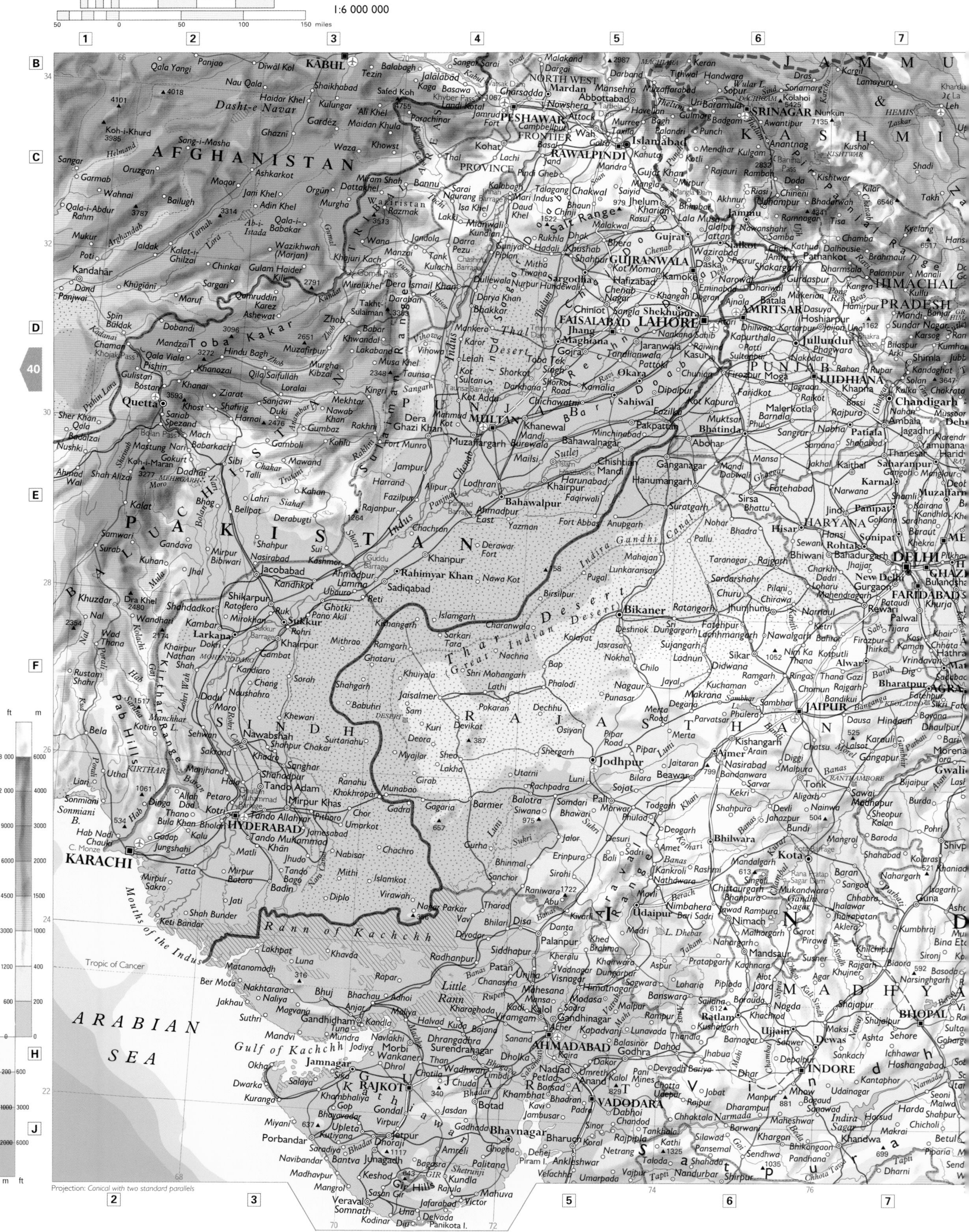

1:6 000 000

Projection: Conical with two standard parallels

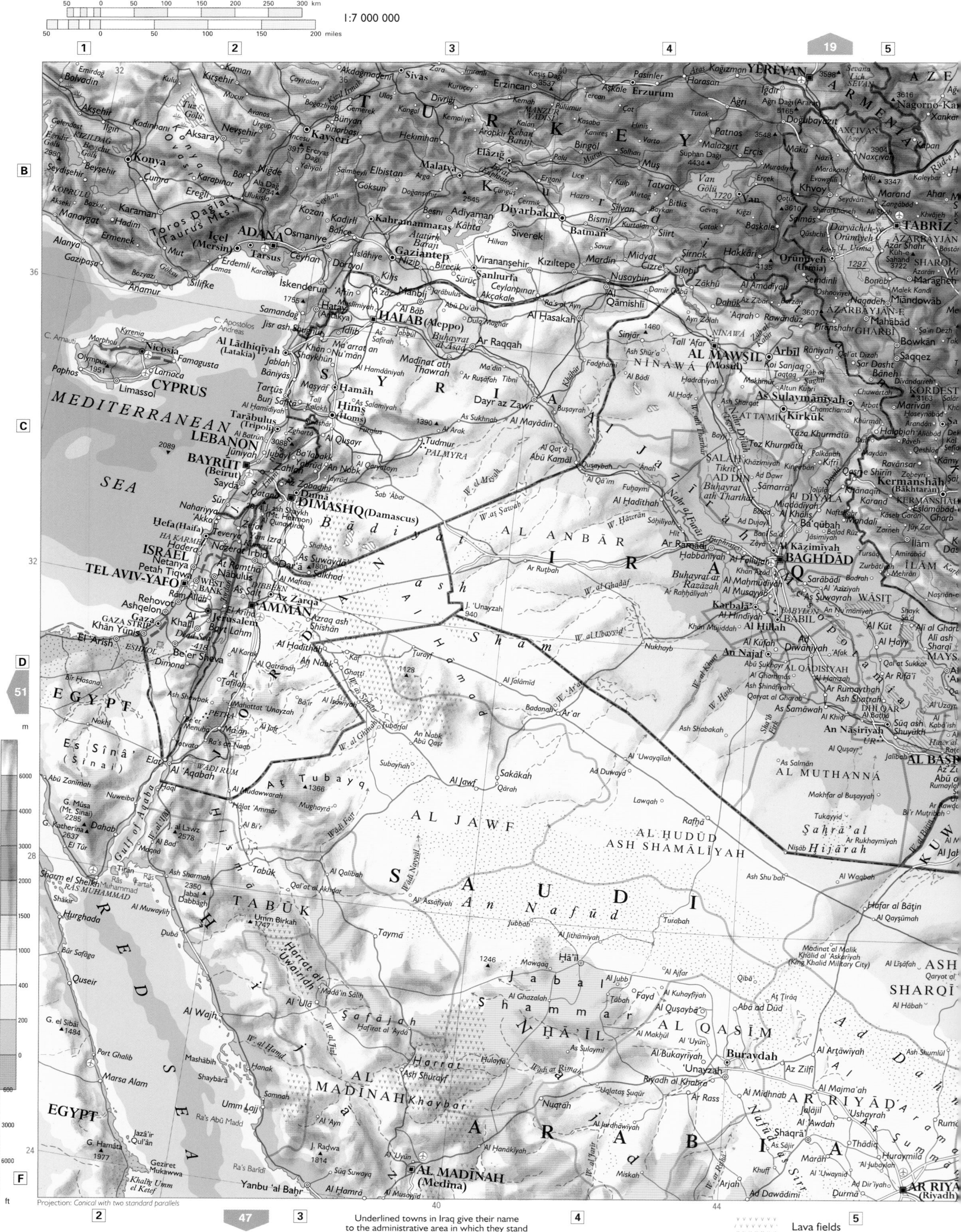

1:7 000 000

Projection: Conical with two standard parallels

Underlined towns in Iraq give their name
to the administrative area in which they stand

Lava fields

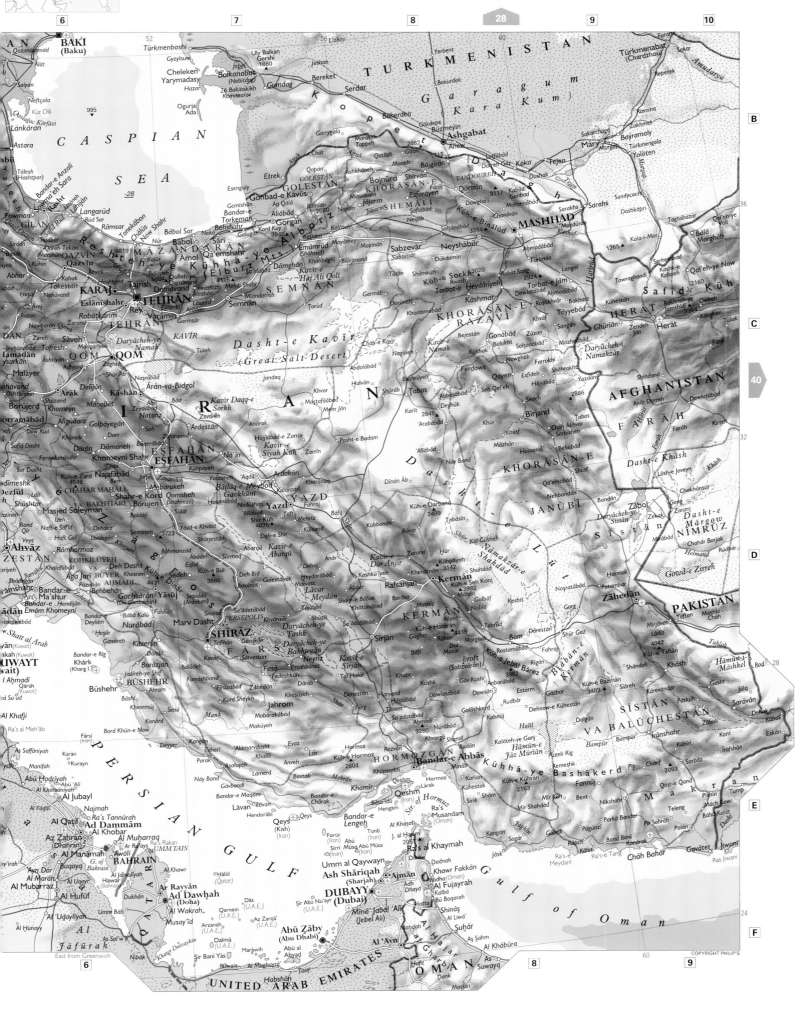

TURKMENISTAN

CASPIAN SEA

GOLESTĀN

KHORĀSĀN-E SHEMĀLI

MASHHAD

HERĀT

MAZANDARĀN

Elburz Mts
Kūhhā-ye Alborz

SEMNĀN

KHORĀSĀN-E RAZAVI

AFGHANISTAN

KARAJ

TEHRĀN

TEHRĀN

Dasht-e Kavīr
(Great Salt Desert)

QOM

QOM

I R A N

FARĀH

ESFAHĀN

ESFAHĀN

YAZD

Yazd

KHORĀSĀN-E JANŪBĪ

PAKISTAN

Zāhedān

CHAHĀR MAHĀLL
VA BAKHTIĀRĪ

ZESTĀN

Ahvāz

KOHKILŪYEH VA
BŪYER AHMADĪ

FĀRS

SHĪRĀZ

Kavīr-e Sīrjān

KERMĀN

Kermān

SĪSTĀN VA
BALŪCHESTĀN

BŪSHEHR

Būshehr

PERSIAN GULF

HORMOZGĀN

Bandar-e Abbās

Kūhhā-ye Bashākerd

MAKRĀN

Al Jubayl

Ad Dammām
Al Khobar

BAHRAIN

QATAR

Ad Dawḩah
(Doha)

Qeshm

Ra's
Musandam
(Oman)

Str. of Hormuz

Ras al Khaymah

Gulf of Oman

Umm al Qaywayn
Ash Shāriqah
(Sharjah)

Ajman

DUBAYY
(Dubai)

Al Fujayrah

Abū Ẕāby
(Abu Dhabi)

UNITED ARAB EMIRATES

OMAN

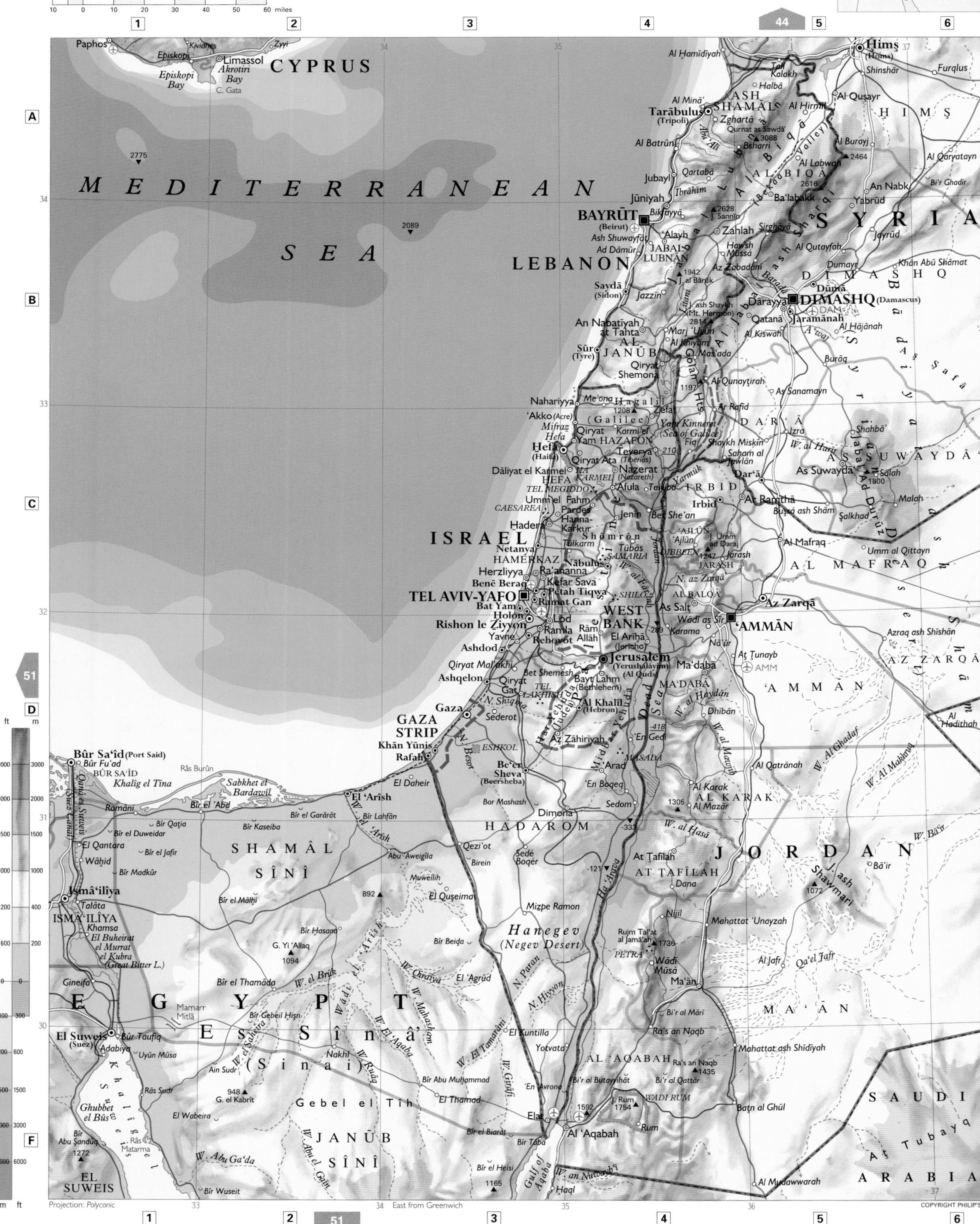

1:2 500 000

=== 1974 Cease Fire Lines

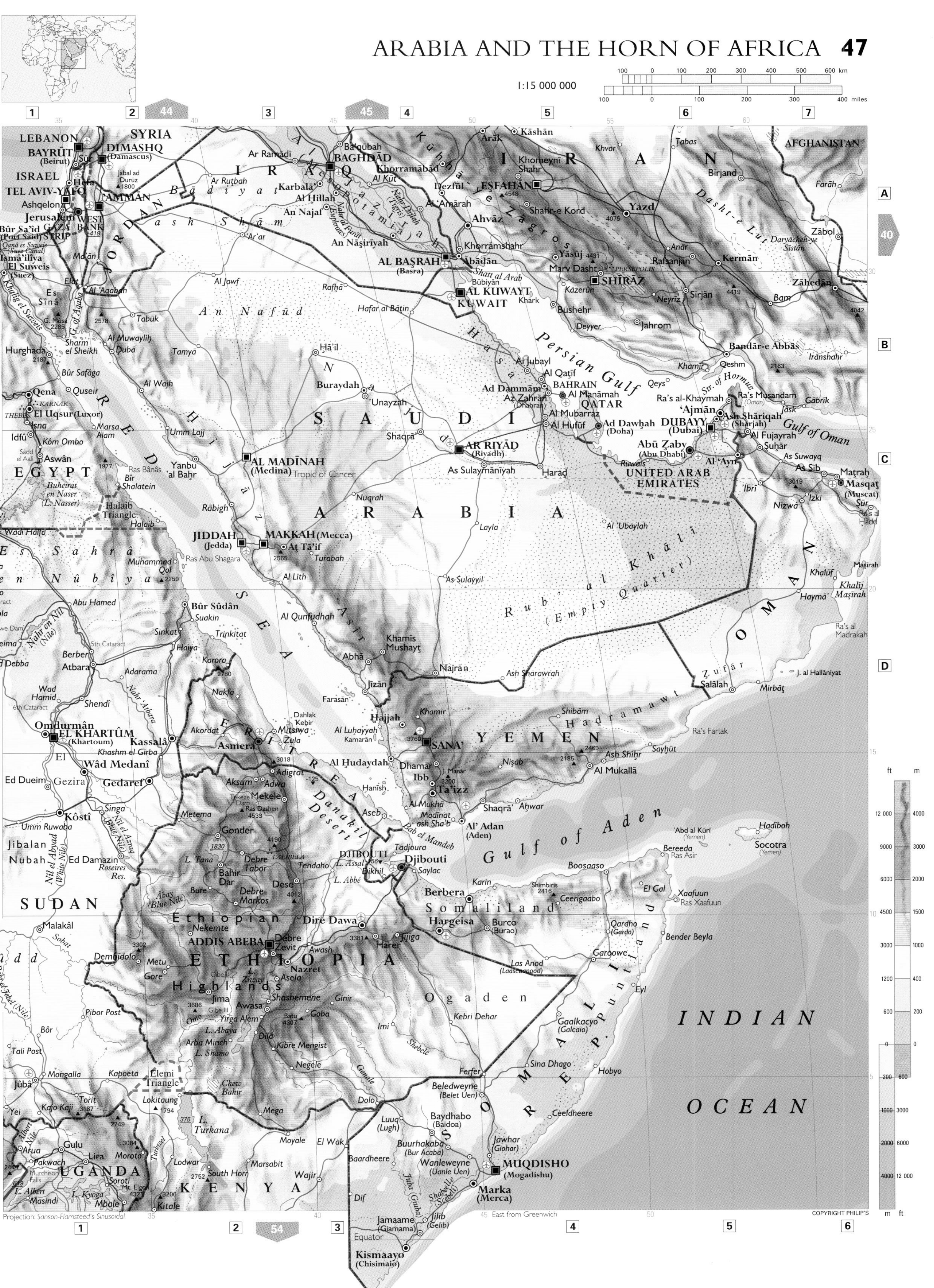
1:15 000 000
100 0 100 200 300 400 500 600 km
100 0 100 200 300 400 miles
SYRIA
DIMASHQ (Damascus)
LEBANON
BAYRŪT (Beirut)
ISRAEL
TEL AVIV-YAFO
Jerusalem
WEST BANK
GAZA STRIP
'AMMĀN
JORDAN
IRAQ
BAGHDĀD
Ba'qūbah
Ar Ramādī
Ar Ruṭbah
Karbalā'
Al Ḥillah
An Najaf
An Nāṣirīyah
Al Kūt
Al 'Amārah
AL BAṢRAH (Basra)
Bādiyat ash Shām
Ar'ar
Al Jawf
Rafḥā'
Hafar al Bāṭin
AL KUWAYT
KUWAIT
IRAN
Kāshān
Arāk
Khomeynī Shahr
EṢFAHĀN
Shahr-e Kord
Yazd
Dezfūl
Ahvāz
Khorramshahr
Ābādān
Khorramābād
Kūhhā-ye Zāgros
Yāsūj
Marv Dasht
PERSEPOLIS
SHĪRĀZ
Kāzerūn
Būshehr
Khark
Deyyer
Jahrom
Neyrīz
Sīrjān
Rafsanjān
Anār
Kermān
Bam
Zāhedān
Zābol
Daryācheh-ye Sīstān
Dasht-e Lut
Tabas
Khvor
Bīrjand
Farāh
AFGHANISTAN
Bandar-e 'Abbās
Qeshm
Khamir
Īrānshahr
Persian Gulf
Str. of Hormuz
Gulf of Oman
Ra's Musandam
Jāsk
Gābrīk
SAUDI ARABIA
An Nafūd
Ḥā'il
Buraydah
'Unayzah
Shaqrā'
AR RIYĀD (Riyadh)
As Sulaymānīyah
Al Jubayl
Al Qaṭīf
Ad Dammām
Az Zahrān (Dhahran)
Al Mubarraz
Al Hufūf
Harad
BAHRAIN
Al Manāmah
QATAR
Ad Dawḥah (Doha)
UNITED ARAB EMIRATES
Abū Ẓaby (Abu Dhabi)
DUBAYY (Dubai)
'Ajmān
Ash Shāriqah (Sharjah)
Ra's al-Khaymah
Al Fujayrah
Ṣuḥār
Al 'Ayn
Ruwais
OMAN
Masqaṭ (Muscat)
Maṭraḥ
As Sīb
As Suwayq
Nizwá
'Ibrī
Izkī
Ṣūr
Ra's al Hadd
Maṣīrah
Khalīj Maṣīrah
Khalūf
Haymā'
Ra's al Madrakah
Ẓufār
Ṣalālah
Mirbāṭ
J. al Hallāniyat
Rub' al Khālī (Empty Quarter)
Al 'Ubaylah
Layla
As Sulayyil
Nuqrah
Tabūk
Al Muwayliḥ
Ḍubā
Tamyā
Al Wajh
Umm Lajj
Yanbu 'al Baḥr
AL MADĪNAH (Medina)
Tropic of Cancer
Rābigh
JIDDAH (Jedda)
MAKKAH (Mecca)
Aṭ Ṭā'if
Turabah
Al Līth
Al Qunfudhah
Abhā
Khamīs Mushayṭ
Jīzān
Najrān
Sharawrah
Ḥijāz
'Asīr
YEMEN
ṢANA'
Ḥajjah
Khamir
Al Ḥudaydah
Dhamār
Ibb
Ta'izz
Al Mukhā
Aden
Al' Adan (Aden)
Shaqrā'
Aḥwar
Niṣāb
Shibām
Ḥadramawt
Ash Shiḥr
Al Mukallā
Sayḥūt
Ra's Fartak
Bab el Mandeb
Gulf of Aden
'Abd al Kūrī (Yemen)
Socotra (Yemen)
Hadiboh
EGYPT
El Suweis (Suez)
Ismâ'ilîya
Bûr Sa'îd (Port Said)
Qanâ es Suweis (Suez Canal)
Es Sînâ'
G. Mûsa 2285
Sharm el Sheikh
G. of Aqaba
Khalig el Suweis
Hurghada
Bûr Safâga
Qena
Quseir
El Uqsur (Luxor)
Isna
Idfû
Kôm Ombo
Aswân
Marsa Alam
Ras Bânâs
Bîr Shalatein
Buheirat en Naser (L. Nasser)
Halaib Triangle
Halaib
Wadi Halfa
RED SEA
Es Sahrâ en Nûbîya
Muhammad Qol
Bûr Sûdân
Suakin
Sinkat
Trinkitat
Abu Hamed
Nahr en Nîl (Nile)
Berber
Atbara
Adarama
Nahr 'Atbara
Shendî
Wad Hamid
Omdurmân
EL KHARTÛM (Khartoum)
Kassalâ
Khashm el Girba
Wâd Medanî
Gedaref
Gezira
Ed Dueim
Kôstî
Singa
Umm Ruwaba
Jibalan Nubah
Ed Damazin
Roseires Res.
SUDAN
Malakâl
Sobat
Bahr el Jebel (Nile)
Pibor Post
Bôr
Tali Post
Mongalla
Jûba
Kapoeta
Elemi Triangle
Torit
Kajo Kaji
Yei
ERITREA
Asmera
Akordat
Mitsiwa
Zula
Karora
Nakfa
Dahlak Kebir
Danakil Desert
Aseb
Adigrat
Aksum
Adwa
Mekele
Ras Dashen 4533
Gonder
Metema
L. Tana
Bahir Dar
Debre Tabor
Dese
Debre Markos
Abay (Blue Nile)
Bure
LALIBELA
DJIBOUTI
Djibouti
Tadjoura
Dikhil
L. Assal
L. Abbé
Tendaho
ETHIOPIA
Ethiopian Highlands
ADDIS ABEBA
Debre Zevit
Nazret
Asela
Awash
Dire Dawa
Harer
Jijiga
Nekemte
Dembidolo
Metu
Gore
Jima
Shashemene
Awasa
Yirga Alem
Goba
Ginir
L. Abaya
Arba Minch
L. Shamo
Dila
Kibre Mengist
Negele
Mega
Ogaden
Kebri Dehar
Imi
Shebele
Genale
Chew Bahir
L. Turkana
SOMALIA
Somaliland
Saylac
Berbera
Hargeisa
Burco (Burao)
Boosaaso
Ceerigaabo
Shimbiris 2416
El Gal
Xaafuun
Ras Xaafuun
Qardho (Gardo)
Bender Beyla
Garoowe
Las Anod (Lascaanood)
Eyl
Puntland
Gaalkacyo (Galcaio)
Hobyo
Sina Dhago
Ceeldheere
Beledweyne (Belet Uen)
Ferfer
Dolo
Luuq (Lugh)
Baydhabo (Baidoa)
Buurhakaba (Bur Acaba)
Wanleweyne (Uanle Uen)
Jawhar (Giohar)
MUQDISHO (Mogadishu)
Marka (Merca)
Baardheere
Jamaame (Giamama)
Jilib (Gelib)
Kismaayo (Chisimaio)
Equator
INDIAN OCEAN
KENYA
Moyale
El Wak
Lodwar
Marsabit
Wajir
South Horr
Lokitaung
Dif
UGANDA
Gulu
Lira
Moroto
Arua
Pakwach
Murchison Falls
Soroti
Mt. Elgon 4321
L. Albert
L. Kyoga
Masindi
Mbale
Kitale
Projection: Sanson-Flamsteed's Sinusoidal
East from Greenwich
COPYRIGHT PHILIP'S

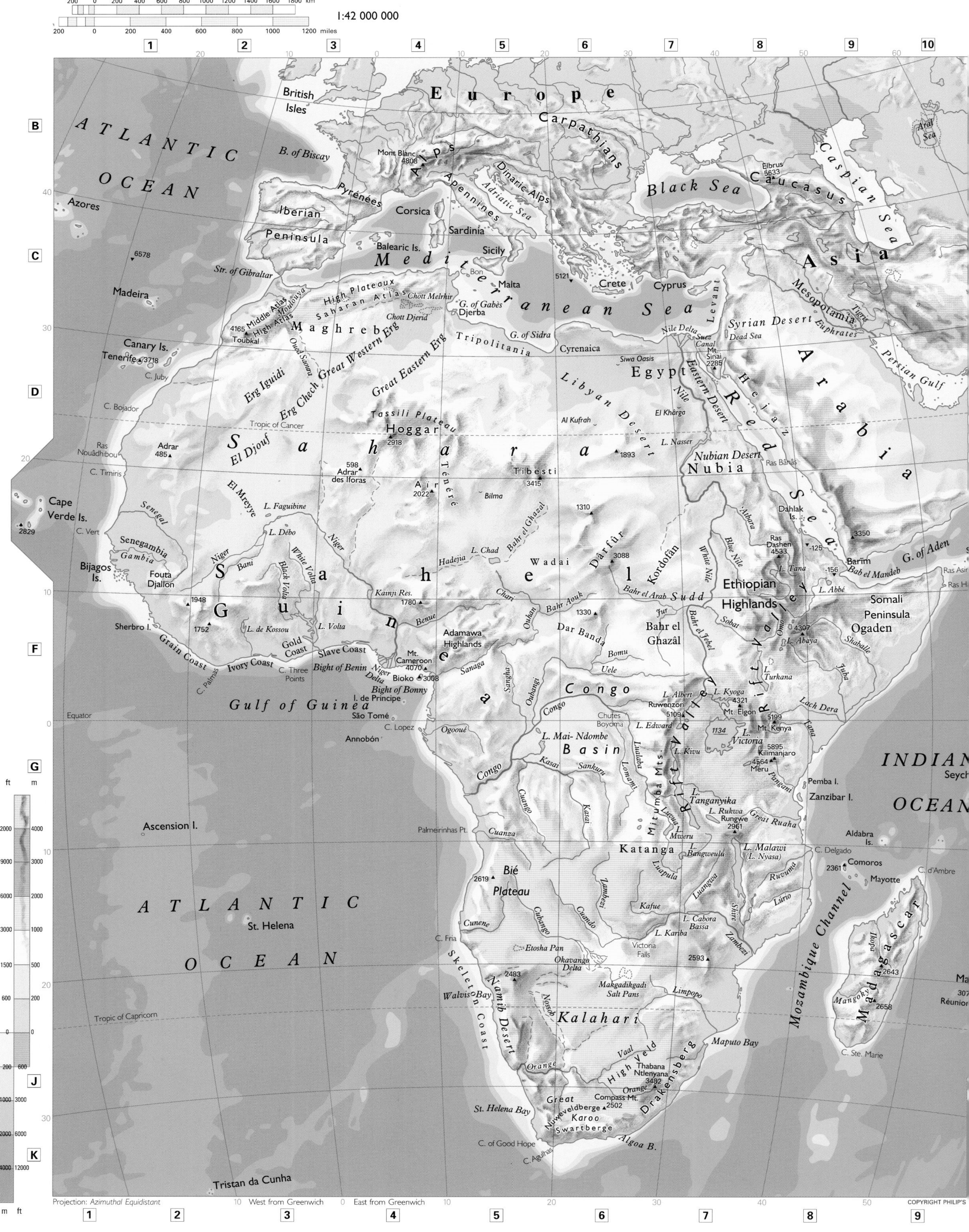

1:42 000 000

Projection: Azimuthal Equidistant

COPYRIGHT PHILIP'S

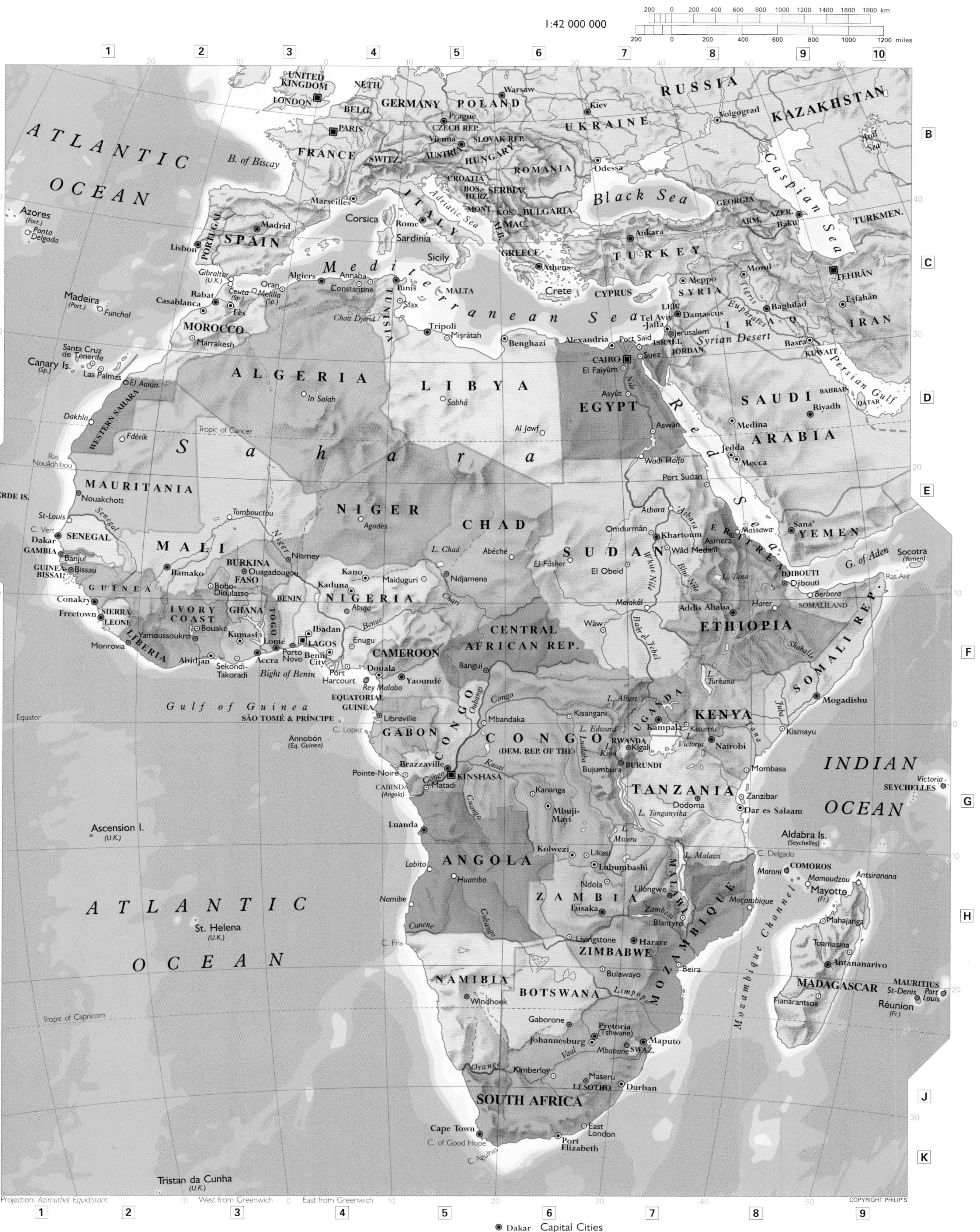
1:42 000 000
200 0 200 400 600 800 1000 1200 1400 1600 1800 km
200 0 200 400 600 800 1000 1200 miles
ATLANTIC OCEAN
B. of Biscay
Azores (Port.)
Ponta Delgada
UNITED KINGDOM
LONDON
NETH.
BELG.
GERMANY
POLAND
Warsaw
PARIS
FRANCE
Prague
CZECH REP.
Vienna
SLOVAK REP.
AUSTRIA
HUNGARY
SWITZ.
ROMANIA
CROATIA
BOS.-HERZ.
SERBIA
MONT.
KOS.
BULGARIA
MAC.
ALB.
GREECE
Athens
RUSSIA
Kiev
UKRAINE
Volgograd
Odessa
KAZAKHSTAN
Aral Sea
Caspian Sea
Black Sea
GEORGIA
ARM.
AZER.
Baku
TURKMEN.
Marseilles
Corsica
ITALY
Adriatic Sea
Rome
Sardinia
Sicily
Madrid
SPAIN
PORTUGAL
Lisbon
Gibraltar (U.K.)
Ceuta (Sp.)
Melilla (Sp.)
Oran
Algiers
Annaba
Constantine
Tunis
TUNISIA
Sfax
MALTA
Mediterranean Sea
Crete
CYPRUS
Ankara
TURKEY
Aleppo
SYRIA
Mosul
TEHRAN
Esfahan
IRAN
LEB.
Damascus
Tel Aviv -Jaffa
Jerusalem
ISRAEL
JORDAN
IRAQ
Baghdad
Tigris
Euphrates
Syrian Desert
Basra
KUWAIT
Persian Gulf
Madeira (Port.)
Funchal
Rabat
Casablanca
Fès
MOROCCO
Marrakesh
Chott Djerid
Tripoli
Misratah
Benghazi
Alexandria
Port Said
CAIRO
Suez
El Faiyûm
Asyût
Nile
Santa Cruz de Tenerife
Canary Is. (Sp.)
Las Palmas
El Aaiún
ALGERIA
In Salah
LIBYA
Sabhâ
EGYPT
Red Sea
SAUDI ARABIA
BAHRAIN
QATAR
Riyadh
Medina
Jedda
Mecca
Dakhla
WESTERN SAHARA
Fdérik
Tropic of Cancer
Sahara
Al Jawf
Aswân
Wadi Halfa
Port Sudan
Ras Nouâdhibou
MAURITANIA
Nouakchott
Tombouctou
NIGER
Agadès
CHAD
Atbara
Omdurmân
Khartoum
Wad Medani
Asmera
ERITREA
Massawa
Sana'
YEMEN
G. of Aden
Socotra (Yemen)
St-Louis
C. Vert
Dakar
SENEGAL
GAMBIA
Banjul
GUINEA-BISSAU
Bissau
Senegal
MALI
Niger
Niamey
BURKINA FASO
Ouagadougou
Bamako
L. Chad
Abéché
Ndjamena
SUDAN
El Fasher
El Obeid
White Nile
Blue Nile
L. Tana
DJIBOUTI
Djibouti
Ras Asir
Berbera
SOMALILAND
GUINEA
Conakry
Freetown
SIERRA LEONE
Bobo-Dioulasso
BENIN
Kano
Maiduguri
Kaduna
NIGERIA
Abuja
Chari
Malakâl
Addis Ababa
Harer
ETHIOPIA
IVORY COAST
GHANA
TOGO
Yamoussoukro
Bouaké
Kumasi
LIBERIA
Monrovia
Abidjan
Sekondi-Takoradi
Accra
Lomé
Porto Novo
Ibadan
LAGOS
Benin City
Enugu
Benue
CENTRAL AFRICAN REP.
Wâw
Bahr el Jebel
Shabelle
SOMALI REP.
Bight of Benin
Port Harcourt
CAMEROON
Douala
Yaoundé
Rey Malabo
Bangui
Gulf of Guinea
EQUATORIAL GUINEA
SÃO TOMÉ & PRÍNCIPE
Libreville
Equator
Congo
Oubangui
Mbandaka
Kisangani
L. Albert
UGANDA
Kampala
KENYA
L. Turkana
Juba
Mogadishu
C. Lopez
GABON
CONGO
Annobón (Eq. Guinea)
CONGO (DEM. REP. OF THE)
L. Edward
Lualaba
RWANDA
Kigali
L. Kivu
BURUNDI
Bujumbura
Kisumu
L. Victoria
Nairobi
Tana
Kismayu
INDIAN OCEAN
Victoria
SEYCHELLES
Brazzaville
KINSHASA
Kasai
Pointe-Noire
CABINDA (Angola)
Matadi
Kwango
Kananga
TANZANIA
Dodoma
Zanzibar
Dar es Salaam
Mombasa
L. Tanganyika
Mbuji-Mayi
Luanda
Ascension I. (U.K.)
L. Mweru
Aldabra Is. (Seychelles)
C. Delgado
Kolwezi
Likasi
Lubumbashi
L. Malawi
MALAWI
Lobito
ANGOLA
Huambo
Moroni
COMOROS
Mamoudzou
Mayotte (Fr.)
Antsiranana
Ndola
ZAMBIA
Lusaka
Lilongwe
Zambezi
Blantyre
MOZAMBIQUE
Moçambique
Mozambique Channel
Namibe
Cunene
Cubango
ATLANTIC OCEAN
St. Helena (U.K.)
C. Fria
Livingstone
Harare
ZIMBABWE
Mahajanga
Toamasina
Antananarivo
MADAGASCAR
Beira
Bulawayo
NAMIBIA
BOTSWANA
Limpopo
Windhoek
MAURITIUS
Port Louis
St-Denis
Réunion (Fr.)
Fianarantsoa
Tropic of Capricorn
Gaborone
Pretoria (Tshwane)
Johannesburg
Maputo
Mbabane
SWAZ.
Vaal
Orange
Kimberley
Maseru
LESOTHO
Durban
SOUTH AFRICA
Cape Town
C. of Good Hope
C. Agulhas
East London
Port Elizabeth
Tristan da Cunha (U.K.)
Projection: Azimuthal Equidistant
West from Greenwich
East from Greenwich
COPYRIGHT PHILIP'S
Dakar Capital Cities

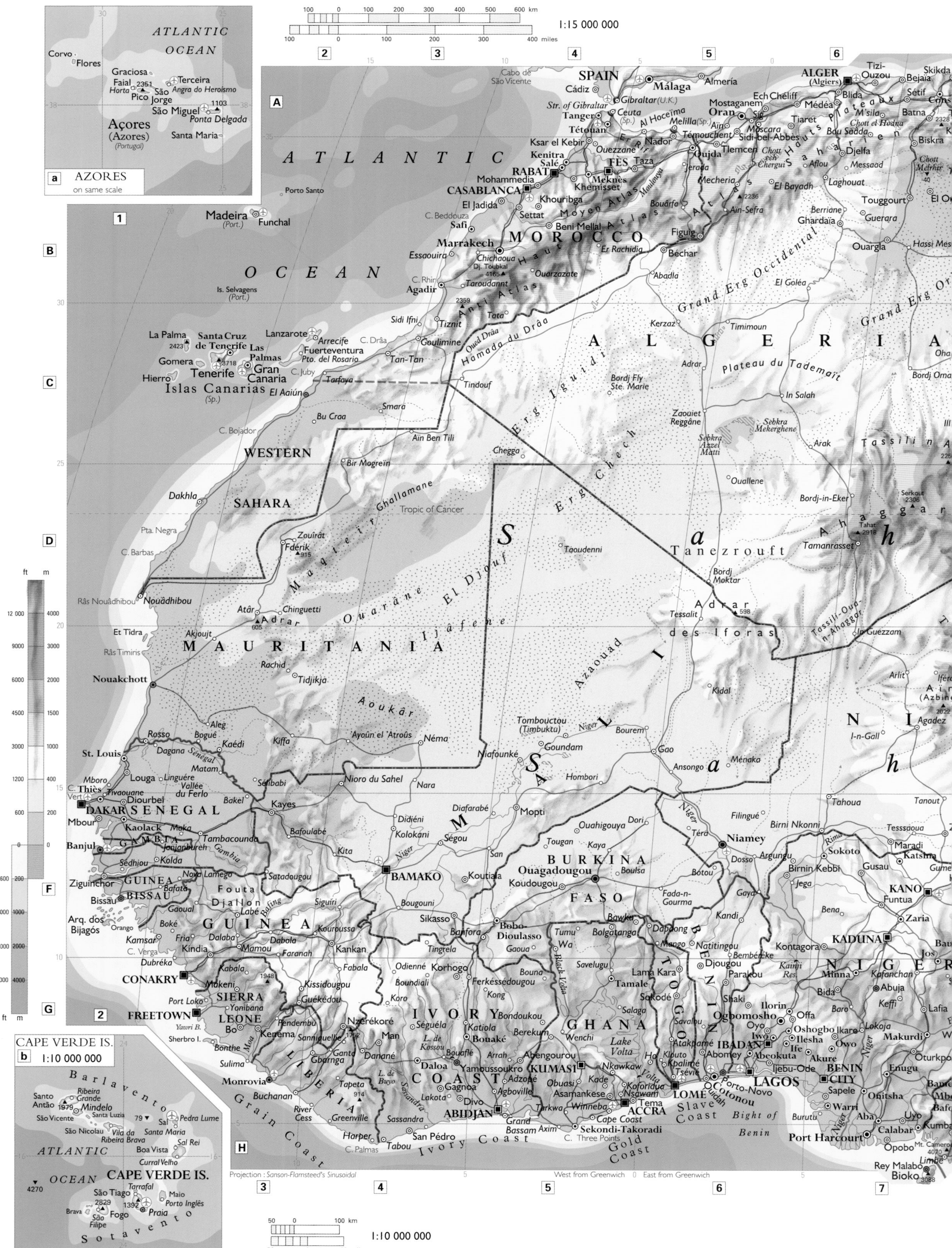

100 0   100   200   300   400   500   600 km
100 0   100   200   300   400 miles
1:15 000 000

a AZORES on same scale

ATLANTIC OCEAN

Corvo · Flores
Graciosa
Faial 2351 Terceira
Horta ▲ São Jorge Angra do Heroísmo
Pico
São Miguel ▲ 1103
Ponta Delgada
Santa Maria

Açores (Azores) (Portugal)

ATLANTIC OCEAN

Madeira (Port.) · Funchal
Porto Santo

Is. Selvagens (Port.)

La Palma · Santa Cruz de Tenerife
2423 Lanzarote
Gomera Arrecife C. Dráa
Hierro Tenerife Fuerteventura
Gran Pto. del Rosario
Canaria C. Juby
Islas Canarias (Sp.) El Aaiún

SPAIN
Cabo de São Vicente
Cádiz · Málaga · Almería
Str. of Gibraltar (U.K.)
Tanger Gibraltar (Sp.) Melilla (Sp.)
Tétouan Al Hoceima Nador
Ksar el Kebir Oujda
Kenitra Taza
Salé RABAT FES
Mohammedia Meknes Khemisset
CASABLANCA
El Jadida Khouriga Settat
Safi Beni Mellal
C. Beddouza MOROCCO
Essaouira Chichaoua
Marrakech Er Rachidia
Dj. Toubkal 4165 Bouárfa
C. Rhir Ouarzazate
Agadir Taroudannt
Sidi Ifni 2359 Abadla
Tiznit Béchar
Tata Goulimine
Hamada du Drâa
Tan-Tan Kerzaz
Tarfaya Tindouf

ALGER (Algiers) Tizi-Ouzou Skikda
Mostaganem Ech Cheliff Médéa Blida Sétif Bejaia
Oran Mascara Tiaret Bou Saâda M'sila 2328
Tlemcen Aflou Djelfa Biskra
Jerada Messaad
Mecheria El Bayadh Laghouat
Ghardaïa Ouargla
El Goléa Guerara Touggourt
Grand Erg Occidental Hassi Messaoud
Adrar Timimoun In Salah
Plateau du Tademaït Bordj Oma
Erg Chech Reggâne
Ouallene Arak Tassili n A
Bordj-in-Eker Serkout 2306
Tamanrasset 2918 A h a g g a r Tahat
Adrar des Iforas 598 In-Guezzam
Tessalit Arlit
Kidal Aïr (Azbine) 2022
I-n-Gall Agadez

WESTERN SAHARA

Dakhla
Pta. Negra
C. Barbas

Tropic of Cancer
Zouîrat Fdérik 915
Taoudenni

Rás Nouâdhibou · Nouâdhibou
Atâr Chinguetti Adrar 605
Et Tidra Akjoujt
Rás Timiris MAURITANIA
Rachid Ijâfene
Nouakchott Tidjikja
Aoukâr

Tombouctou (Timbuktu)
Bourem
Niafounké Gao
Goundam Ansongo Ménaka
Hombori Tahoua
Tessaoua

SENEGAL
St. Louis
Aleg Kaédi Kiffa Néma Nara
Rosso Bogué 'Ayoûn el 'Atroûs
Dagana Sénégal
Louga Matam Sélibabi
C. Vert Tivaouane Diourbel
DAKAR Mbour Linguère Vallée du Ferlo
Kaolack Moka Bakel Nioro du Sahel
GAMBIA Tambacounda Kayes
Banjul Joujanbureh Diamou
Sédhiou Kolda BAMAKO
Ziguinchor GUINEA Satadougou Kita
BISSAU Bafatá Koutiala Ségou
GUINEA-BISSAU Fouta Djallon Siguiri Bougouni Sikasso
Arq. dos Bijagós Labé Kouroussa
Boké Dalaba Kankan
Kamsar Fria Dabola Tingrela Banfora
C. Verga Kindia Mamou Faranah Odienné Korhogo
Dubréka Kabala 1948 Fabala Boundiali
CONAKRY Makeni Kissidougou Bouna
SIERRA Port Loko Guéckédou Kong
Yonibana Karo Ferkéssédougou
FREETOWN LEONE Kenema Nzérékoré IVORY
Bo Sanniquellie Séguéla Katiola
Sherbro I. Pendembu Man L. de Kossou Bouaké
Sulima Ganta Danané Arrah COAST
Monrovia LIBERIA Tapeta 914 Daloa Gagnoa Yamoussoukro
Buchanan River Cess Greenville Divo Lakota
Harper San Pédro Sassandra
C. Palmas Tabou Grain Coast Ivory Coast

BURKINA FASO
Ouahigouya Dori
Tougan Kaya Filingué
Ouagadougou Boulsa Niamey
Koudougou Fada-N'Gourma Birni Nkonni
Dosso Téra Argungu
Kandi Gaya Jega Birnin Kebbi Sokoto Maradi Katsina
Bawku Bobo-Dioulasso Bimini Gusau Gumel
Tumu Bolgatanga Dapaong KANO Funtua
Wa Mango Natitingou Kontagora ZARIA Kaduna
Savelugu Bembéréké Kainji Res. Minna Zaria
Tamale Djougou Parakou Bida KADUNA
GHANA Salaga Lama Kara Shaki Abuja
Wenchi Sokodé Kafanchan
Bouna TOGO BENIN Ilorin NIGER
Lake Volta Atakpamé Savalou Ogbomosho Offa Jos
Ho Abomey Oyo Iwo Lafia
Black Volta KUMASI Kpalimé Iseyin Ikare Keffi
Obuasi Asamankese Nkawkaw Porto-Novo Oshogbo Ilesha Makurdi
Koforidua LOME Cotonou IBADAN Ife Akure BENIN
Nsawam Tsévié Abeokuta Ijebu-Ode CITY Enugu
ACCRA Tema LAGOS Sapele Onitsha
Winneba Cape Coast Slave Coast Warri Aba Calabar
Sekondi-Takoradi Bight of Benin Opobo Uyo Kumba
Gold Coast Port Harcourt
C. Three Points Rey Malabo Mt. Cameroun 4070
Bioko Limbe 3008

Projection : Sanson-Flamsteed's Sinusoidal
West from Greenwich | East from Greenwich

S a h a r a
S a h e l
M A L I
NIGER
BENIN

ft m
12 000 4000
9000 3000
6000 2000
4500 1500
3000 1000
1200 400
600 200
0
600 200
3000 1000
6000 2000
12 000 4000
ft m

CAPE VERDE IS.
b 1:10 000 000

Barlavento
Santo Antão 1979 Ribeira Grande
Mindelo Santa Luzia
São Vicente 79
São Nicolau Sal
Ribeira Brava Pedra Lume
Vila da Santa Maria
Maio
Boa Vista Sal Rei
ATLANTIC OCEAN
CAPE VERDE IS.
4270 Tarrafal Porto Inglês
São Tiago
2829 1392 Praia
Brava Fogo São Filipe
Sotavento

50 0 50 100 km
50 0 50 miles
1:10 000 000

MEDITERRANEAN SEA
LIBYA
EGYPT
CHAD
SUDAN
CENTRAL AFRICAN REPUBLIC
ETHIOPIA
ERITREA
SAUDI ARABIA
SYRIA
IRAQ
JORDAN
LEBANON
ISRAEL
CYPRUS
GREECE
TURKEY
ITALY
MALTA
CONGO
RED SEA
Sahara
TUNIS
TARĀBULUS (Tripoli)
BANGHĀZĪ
EL ISKANDARÎYA (Alexandria)
EL QÂHIRA (Cairo)
EL GÎZA
EL KHARTÛM (Khartoum)
Omdurmân
Ndjamena
Bangui
DIMASHQ (Damascus)
BAYRŪT (Beirut)
TEL AVIV-YAFO
'AMMAN
Jerusalem
HALAB (Aleppo)
Tibesti
Emi Koussi 3415
Lac Tchad
COPYRIGHT PHILIPS

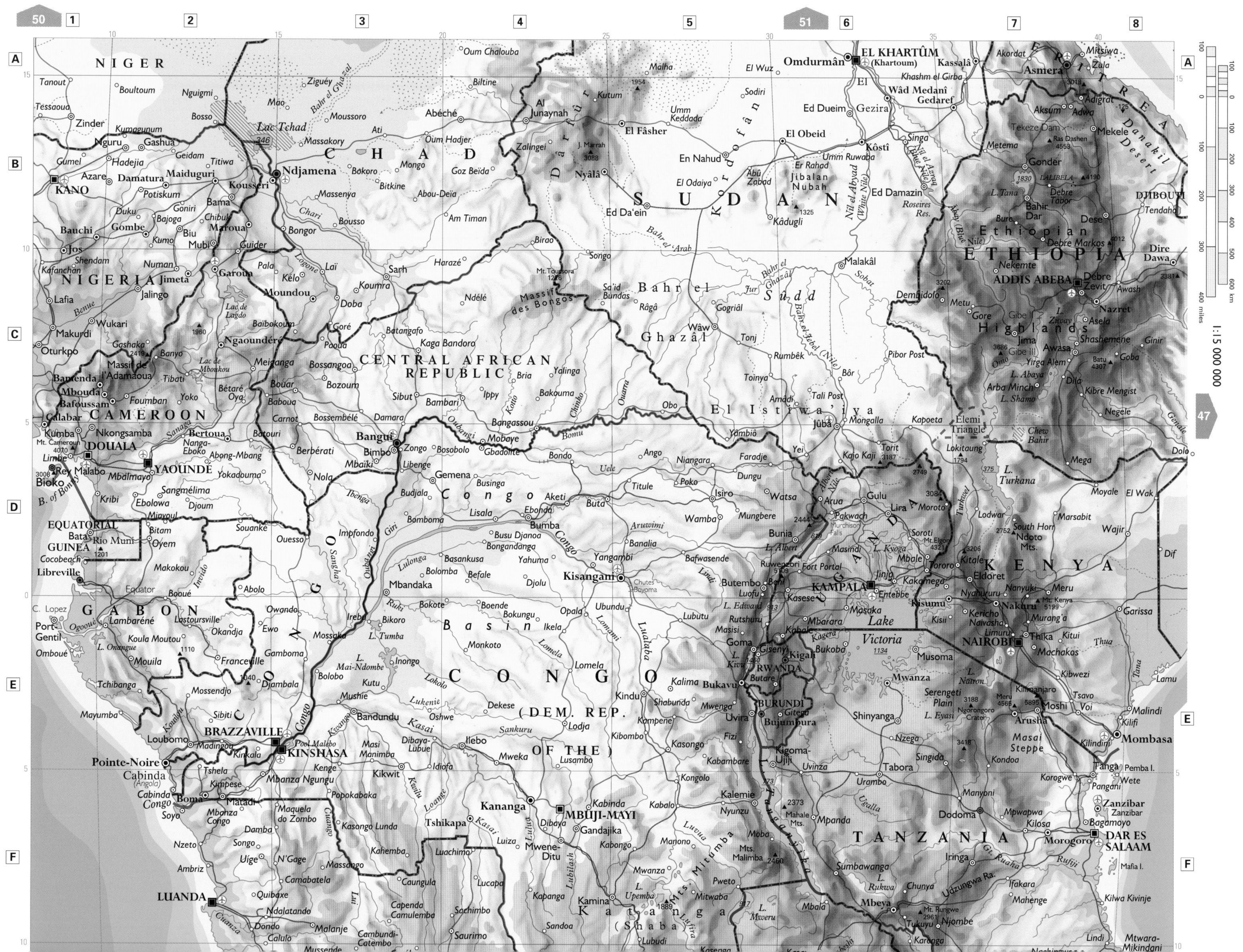

1:15 000 000

ATLANTIC OCEAN
INDIAN OCEAN
SOUTH AFRICA
NAMIBIA
BOTSWANA
ZIMBABWE
ZAMBIA
MOZAMBIQUE
MADAGASCAR
LESOTHO
SWAZILAND
LUSAKA
HARARE
MAPUTO
PRETORIA
JOHANNESBURG
CAPE TOWN
PORT ELIZABETH
DURBAN
ANTANANARIVO
Windhoek
Gaborone
Bloemfontein
Bulawayo
Beira
Skeleton Coast
Namib Desert
Kalahari
Mozambique Channel
Tropic of Capricorn
COMOROS 1:8 000 000 a
SEYCHELLES 1:2 500 000 b
RÉUNION 1:2 500 000 c
MAURITIUS 1:2 500 000 d
MADAGASCAR on same scale as main map
1:2 500 000
1:8 000 000
East from Greenwich
Projection: Sanson-Flamsteed's Sinusoidal
COPYRIGHT PHILIPS

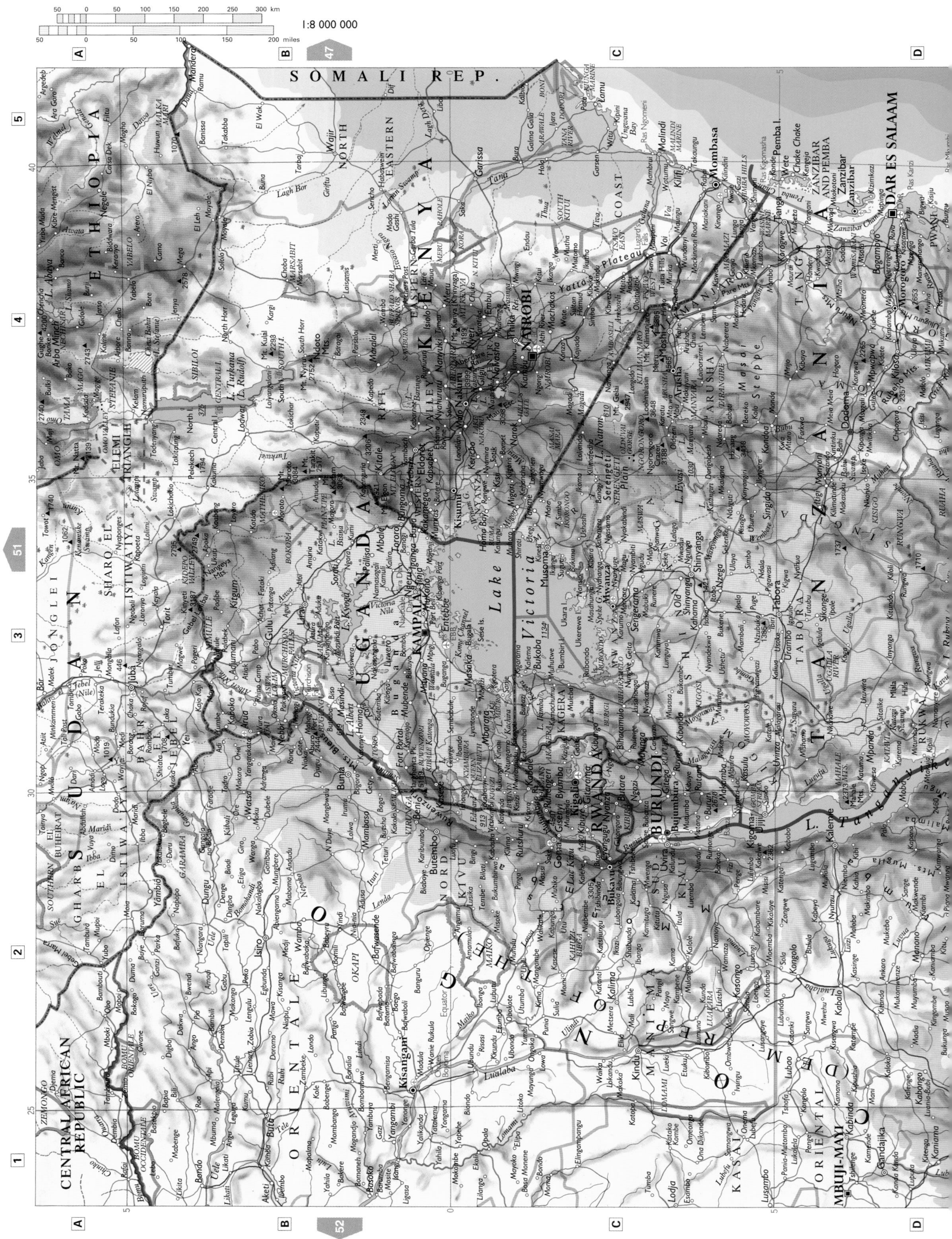

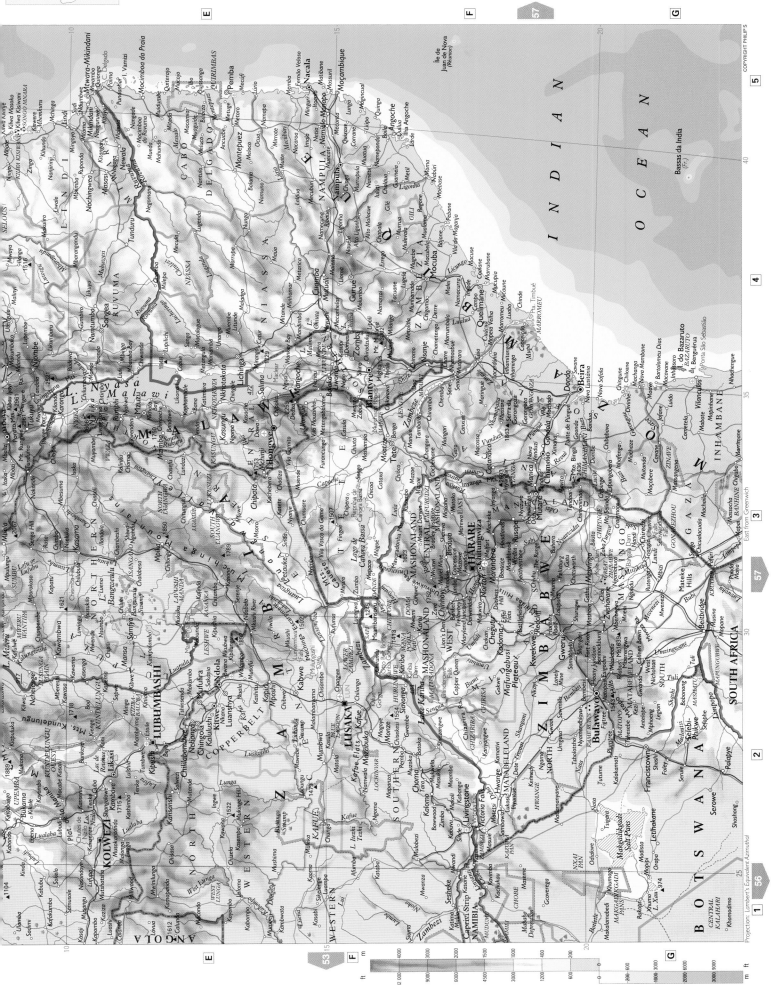

Projection: Lambert's Equivalent Azimuthal

East from Greenwich

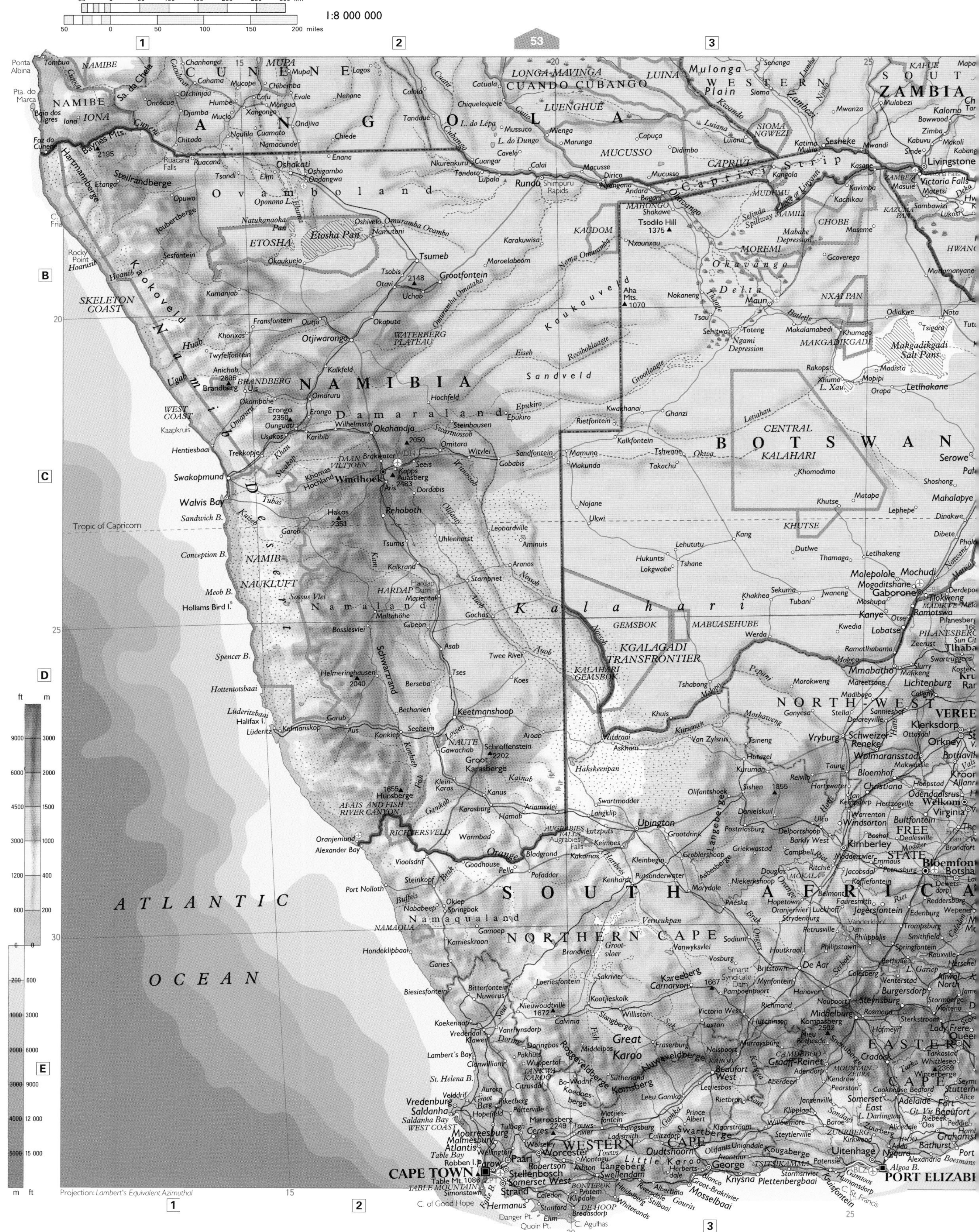

Projection: Lambert's Equivalent Azimuthal

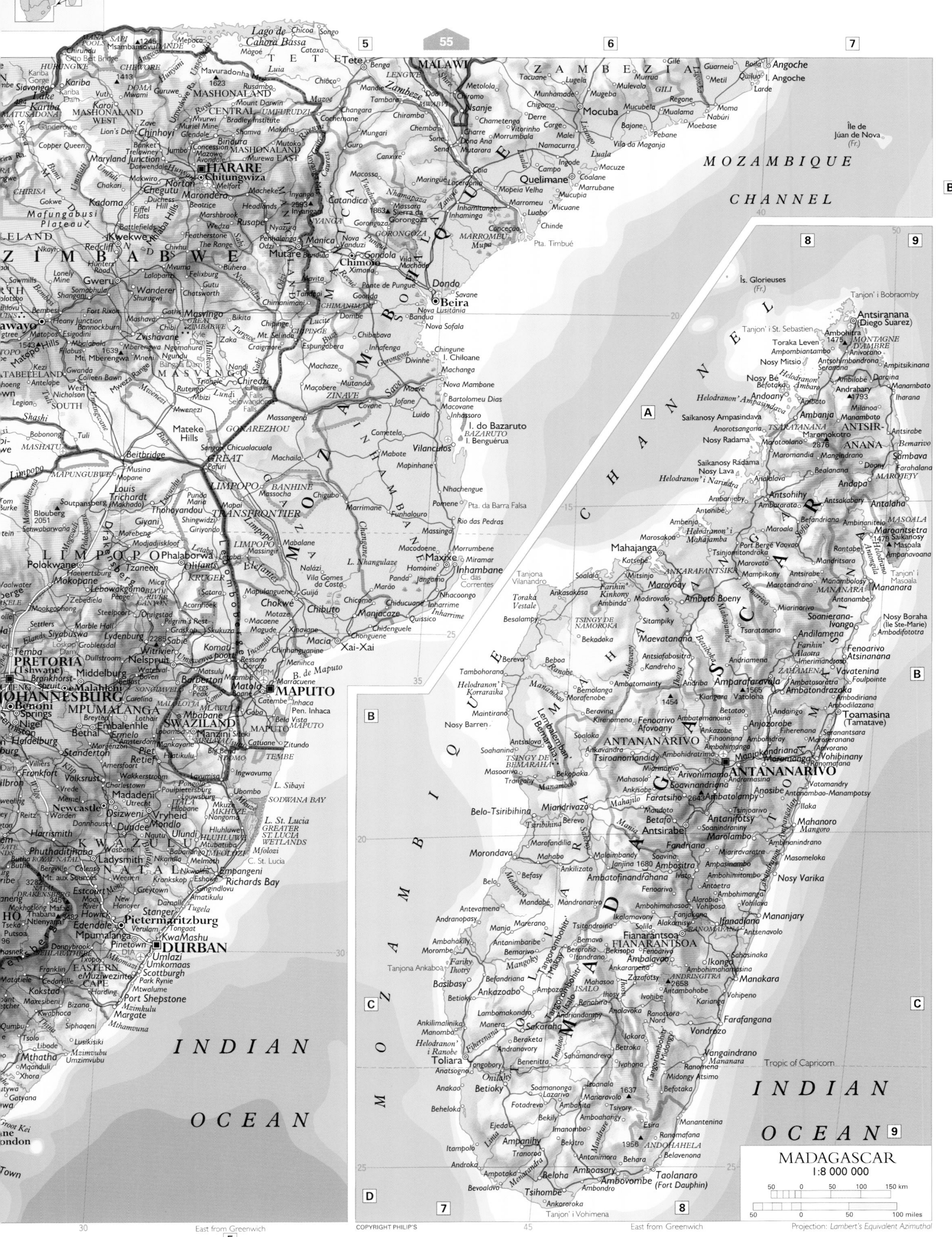
ZIMBABWE
MOZAMBIQUE
MALAWI
SWAZILAND
HARARE
MAPUTO
Mbabane
Beira
Tete
Quelimane
Mocuba
Angoche
Chimoio
Mutare
Gweru
Masvingo
Kariba
Chinhoyi
Chitungwiza
Marondera
Kadoma
Kwekwe
Polokwane
Phalaborwa
Nelspruit
PRETORIA (Tshwane)
JOHANNESBURG
Middelburg
MPUMALANGA
LIMPOPO
KWAZULU NATAL
Pietermaritzburg
DURBAN
Richards Bay
Newcastle
Ladysmith
Port Shepstone
Inhambane
Maxixe
Xai-Xai
Chibuto
Vilanculos
I. do Bazaruto
Lago de Cahora Bassa
MOZAMBIQUE CHANNEL
INDIAN OCEAN
Île de Júan de Nova (Fr.)
Îs. Glorieuses (Fr.)
MADAGASCAR
ANTANANARIVO
Antsiranana (Diego Suarez)
Mahajanga
Toamasina (Tamatave)
Antsirabe
Fianarantsoa
Toliara
Morondava
Taolanaro (Fort Dauphin)
Ambovombe
Manakara
Farafangana
Mananjary
Nosy Bé
Nosy Boraha (Île Ste-Marie)
Sambava
Antalaha
Tropic of Capricorn
MADAGASCAR
1:8 000 000
50 0 50 100 150 km
50 0 50 100 miles
East from Greenwich
Projection: Lambert's Equivalent Azimuthal

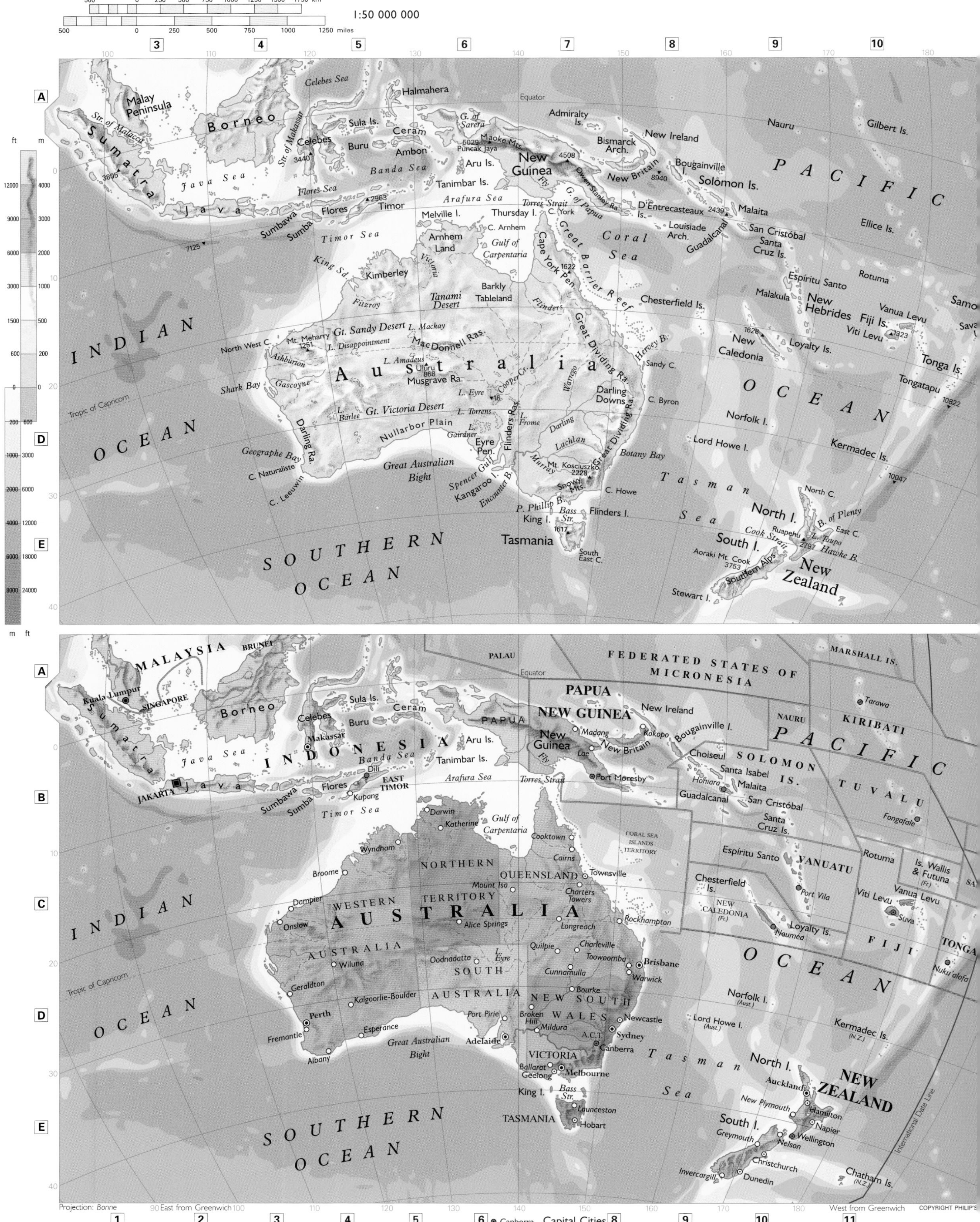

500 0 250 500 750 1000 1250 1500 1750 km
500 0 250 500 750 1000 1250 miles
1:50 000 000

3   4   5   6   7   8   9   10

Physical (upper map)

ft m
12000 4000
9000 3000
6000 2000
3000 1000
1500 500
600 200
0 0
200 600
1000 3000
2000 6000
4000 12000
6000 18000
8000 24000
m ft

Celebes Sea
Malay Peninsula
Halmahera
Equator
Admiralty Is.
Nauru
Gilbert Is.
Borneo
Sula Is.
Ceram
G. of Sarera
New Ireland
New Britain
Bismarck Arch.
Bougainville
Solomon Is.
PACIFIC
Sumatra
Str. of Makassar
Buru
Ambon
6029 Puncak Jaya
Maoke Mts.
New Guinea 4508
8940
Malaita
Ellice Is.
3805
Java Sea
Banda Sea
Aru Is.
Tanimbar Is.
Arafura Sea
Fly
Owen Stanley Ra.
D'Entrecasteaux 2439
Louisiade Arch.
Guadalcanal
San Cristóbal Santa Cruz Is.
Rotuma
3440
Flores Sea
Timor
Torres Strait
G. of Papua
Espíritu Santo
New Hebrides
Vanua Levu
Samo
7125
Sumbawa Sumba Flores
2963
Melville I.
Thursday I.
C. York
Coral Sea
Chesterfield Is.
Malakula
Fiji Is.
Viti Levu
1923
Timor Sea
C. Arnhem
Cape York Pen.
1622
Great Barrier Reef
1628
Tonga Is.
INDIAN
Arnhem Land
Gulf of Carpentaria
New Caledonia
Tongatapu
North West C.
Mt. Meharry 1251
L. Mackay
L. Disappointment
MacDonnell Ras.
Loyalty Is.
10822
Ashburton
Kimberley
Barkly Tableland
Flinders Ra.
Hervey B.
Shark Bay
Gascoyne
Fitzroy
Tanami Desert
Great Dividing Ra.
Sandy C.
Norfolk I.
OCEAN
King Sd.
Gt. Sandy Desert
Australia
Cooper Cr.
Warrego
Darling Downs
C. Byron
OCEAN
Victoria
L. Amadeus
Uluru 868
Musgrave Ra.
L. Eyre 16
Darling
Lord Howe I.
Tropic of Capricorn
L. Barlee
Gt. Victoria Desert
L. Torrens
L. Frome
Lachlan
Botany Bay
Darling Ra.
Nullarbor Plain
L. Gairdner
Eyre Pen.
Murray
Mt. Kosciuszko 2228
C. Howe
Tasman
North C.
Geographe Bay
Spencer Gulf
Snowy Mts.
Great Dividing Ra.
B. of Plenty
East C.
C. Naturaliste
Kangaroo I.
Encounter B.
Ruapehu 2797
Hawke B.
Great Australian Bight
P. Phillip B.
Bass Str.
Flinders I.
Sea
North I.
C. Leeuwin
King I.
1617
Cook Strait
South I.
Tasmania
South East C.
Aoraki Mt. Cook 3753
New Zealand
SOUTHERN
OCEAN
Southern Alps
Stewart I.

Political (lower map)

m ft

MALAYSIA
BRUNEI
PALAU
FEDERATED STATES OF MICRONESIA
MARSHALL IS.
Equator
Kuala Lumpur
SINGAPORE
Borneo
Sula Is.
Ceram
PAPUA
New Ireland
NAURU
Tarawa
KIRIBATI
Sumatra
Celebes
Buru
PAPUA NEW GUINEA
Madang
Kokopo
Bougainville I.
PACIFIC
INDONESIA
Aru Is.
New Guinea
Lae
New Britain
Choiseul
SOLOMON IS.
TUVALU
Java Sea
Makassar
Banda Sea
Fly
Santa Isabel
JAKARTA
Java
Dili
EAST TIMOR
Tanimbar Is.
Port Moresby
Honiara
Malaita
Fongafale
Sumbawa Sumba Flores
Kupang
Arafura Sea
Torres Strait
Guadalcanal
San Cristóbal
Santa Cruz Is.
Timor Sea
Darwin
Espíritu Santo
VANUATU
Rotuma
Is. Wallis & Futuna (Fr.)
Katherine
Gulf of Carpentaria
CORAL SEA ISLANDS TERRITORY
Vanua Levu
Wyndham
Cooktown
Suva
INDIAN
Broome
NORTHERN
Cairns
Townsville
Chesterfield Is.
Port Vila
Viti Levu
FIJI
Dampier
WESTERN
TERRITORY
Mount Isa
QUEENSLAND
NEW CALEDONIA (Fr.)
TONGA
Onslow
AUSTRALIA
Charters Towers
Nouméa
Loyalty Is.
OCEAN
AUSTRALIA
Alice Springs
Longreach
Rockhampton
Norfolk I. (Aust.)
Wiluna
Oodnadatta
Quilpie
Charleville
Nuku'alofa
Geraldton
L. Eyre
SOUTH-
Toowoomba
Brisbane
Tropic of Capricorn
Kalgoorlie-Boulder
AUSTRALIA
Cunnamulla
Warwick
Perth
Port Pirie
NEW SOUTH
Bourke
Lord Howe I. (Aust.)
Fremantle
Esperance
Broken Hill
WALES
Newcastle
Kermadec Is. (N.Z.)
Albany
Adelaide
Mildura
A.C.T.
Sydney
Great Australian Bight
Ballarat
VICTORIA
Canberra
Tasman
North I.
Auckland
NEW ZEALAND
Geelong
Melbourne
New Plymouth
Hamilton
King I.
Bass Str.
Sea
Napier
TASMANIA
Launceston
South I.
Wellington
Hobart
Greymouth
Nelson
SOUTHERN
Christchurch
OCEAN
Invercargill
Dunedin
Chatham Is. (N.Z.)

1:6 000 000

50 0 50 100 150 200 km
50 0 50 100 150 miles

FIJI
on same scale a

178 E   180

Great Sea Reef   Kia   Ringgold Is.   Udu Pt.

PACIFIC OCEAN

Yaqaga   Labasa   Natewa Bay   Rabi
Yasawa Group   Yadua   Bua   Buca   Somosomo   Qamea
Vanua Levu ▲1031   Savusavu   Taveuni
Savusavu Bay   Kanacea   Naitaba   Vanua Balavu
Nabouwalu   Namenalala   Vacata   Lomaloma
Viwa   Nacula   Koro   Northern Lau Group
Naviti   Tavua   Makogai   Vatu   Cicia   Tuvuca
Waya   Vomo   Rakiraki   Nairai   Vara   Mago
Mamanuca Group   Tomanivi ▲1323   Levuka   Wakaya   Batiki   Sawaleke   Lakeba   Oneata
Lautoka   KORO INITI   Ovalau   Nausori   Lakeba Passage   Moce
Nadi   Keiyasi   Viti Levu   Yunidewa   Nayau   Southern Lau Group
Sigatoka   Navua   Suva   Gau   Namuka-i-Lau   Yagasa Cluster
Korolevu   Yanuca   Beqa   Totoya   Fulaga   Ogea Levu
Vatulele   Ono   Vunisea   Moala   Matuku   Kabara   Ogea Driki

KORO SEA   Northern Lau Group   Southern Lau Group

Nothern Lau Group

East from Greenwich   180   West from Greenwich

Kadavu   Tavuki   Vunisea   Kadavu Passage

SAMOA   1858 ▲   Pu'apu'a
Falelima   Asau   Safune   Salelologa
Savai'i   Satupa'itea   Taga   Mulifanua   Apia   Falefa
Manono   Apolima   Afega   Amaile
OLE PUPU PU'E   Safata Bay   'Upolu

PACIFIC OCEAN

AMERICAN SAMOA (U.S.A.)
Ofu   Olosega   Ta'ū
Leone   Tutuila   Pago Pago   Luma   Manu'a Is.
Vaitogi   AMERICAN SAMOA

172 W   West from Greenwich

SAMOAN ISLANDS
on same scale b

2   3

TONGA
on same scale c

174 W   170

PACIFIC OCEAN

Fonualei   Toku
Vava'u   Neiafu
Late   Vava'u Group
Home Reef
Disney Reef
Kao   Ha'ano   Foa   Lifuka
Tofua   Kotu Group   Uiha   Ha'apai Group
Fonuafo'ou   Nomuka   Mango   Oto Tolu Group
Hunga Ha'apai   Nomuka Group   Tonumea

TONGA
Nuku'alofa   Tongatapu
Tongatapu Group   Eua

West from Greenwich

1

TASMAN SEA

PACIFIC OCEAN

North Island

4   64

174   176   178   34

C. Reinga   North C.
C. Maria van Diemen
Houhora Heads
Ahipara B.   Rangaunu B.
Kaitaia   Mangonui   Doubtless B.
Tauroa Pt.   Okaihau   B. of Islands   C. Brett
Rawene   Waitangi   Opua
Hokianga Harbour   Kaikohe   Hikurangi
Whangarei   Whangarei Harb.
Waipoua Forest   Bream Hd.   Bream B.
Dargaville   Waipu   Little Barrier I.
Great Barrier I.
Kaipara Harbour   C. Rodney
Helensville   Warkworth   Cuvier I.
Hauraki Gulf   Coromandel   Whitianga
Takapuna   Whangamata
AUCKLAND   Thames   Mayor I.
Manukau   Papakura   Waihi   Tauranga Harb.
Waiuku   Pukekohe   Mount Maunganui   Whakaari (White I.)
Mercer   Paeroa   Te Aroha   Waihi   Runaway
Waikato   Huntly   Morrinsville   Tauranga   Bay of Plenty
Raglan   Ngaruawahia   Cambridge   Te Puke   Opotiki   East C.
Hamilton   Matamata   Whakatane   Raukumara Ra.
Kawhia   Te Awamutu   Rotorua   Taneatua   ▲1753
Kawhia Harbour   Otorohanga   Kinleith   L. Tarawera   Motu
Te Kuiti   Mangakino   L. Rotorua   Murupara   UREWERA
North Taranaki Bight   Waitomo Caves   Mokai   Waikaremoana   Tolaga Bay
Mokau   Wairakei   Taupo   Ruatahuna   Ormond
New Plymouth   Ongarue   L. Taupo   Gisborne
Waitara   Taumarunui   Turangi   Poverty Bay
Inglewood   WHANGANUI   Whangamomona   Ruapehu ▲2797   Nuhaka
Mt. Taranaki or Mt. Egmont ▲2518   EGMONT   TONGARIRO   Wairoa
Opunake   Stratford   Ohakune   Waiouru   Mohaka   Hawke Bay
Kaponga   Eltham   Raetihi   Bay View   C. Kidnappers
Hawera   Waverley   Taihape   Napier
South Taranaki Bight   Patea   Mangaweka   Ruahine Ra.   Hastings
Wanganui   Hunterville   Waipawa
Marton   Halcombe   Waipukurau
Bulls   Feilding   Dannevirke
Palmerston North   Woodville
Foxton   Shannon   Pahiatua
Paraparaumu   Levin   Eketahuna
Otaki   Tararua Ra.   C. Turnagain
Kapiti I.   Featherston   Masterton
Upper Hutt   Carterton   Greytown
Petone   Lower Hutt   Martinborough
Wellington   L. Wairarapa
Cook Strait   Palliser B.

PACIFIC OCEAN

40   42

C. Farewell
Collingwood   Golden B.   D'Urville I.
Takaka   ABEL TASMAN   Tasman B.
KAHURANGI   Tasman Mts.   Motueka   Pelorus
Karamea   Takaka   Nelson   Havelock   Picton
Karamea Bight   Tadmor   Richmond   Wakefield   Waitohi
Seddonville   Granity   Richmond   Blenheim   Ward
Westport   Lyell   Murchison   NELSON LAKES   Seddon
Matiri Ra.   Inangahua   Rotoroa   ▲2885 Tapuae-o-Uenuku
Buller Gr.   Reefton   Mt. Travers ▲2337   Clarence
PAPAROA   Spenser Mts.   Kaikoura
Punakaiki   Blackball   Hanmer Springs
Runanga   Stillwater   Waiau
Greymouth   Kumara   L. Brunner   Culverden
Hokitika   Jacksons   Waikari   Hurunui
Ross   ARTHUR'S PASS   Waipara   Pegasus Bay
Amberley   Rangiora   New Brighton
Otira   Oxford   Kaiapoi   Christchurch
Whitecliffs   Springfield   Lyttelton
Aoraki/Mt. Cook ▲3753   Coalgate   Darfield   Riccarton   Banks Pen.
MT. COOK   Methven   Lincoln   Little River   Akaroa
Mount Cook   Staveley   L. Ellesmere
Jackson B.   Haast   Fairlie   C. Saunders
MOUNT ASPIRING   Mt. Aspiring ▲3033   Geraldine   Ashburton
Milford Sd.   Earnslaw ▲2819   L. Wanaka   Temuka   Canterbury Bight
Sutherland Falls   Milford Sound   Wanaka   Timaru
Bligh Sound   Arrowtown   St. Andrews
George Sound   Queenstown   Cromwell   Waimate
Secretary I.   Anau   Kingston   Alexandra   Oamaru
Doubtful Sd.   Mossburn   Clyde   Maheno
FIORDLAND   L. Manapouri   Roxburgh   Hampden
Breaksea Sd.   Lumsden   Waikouaiti   Danback
Resolution I.   Ohai   Gore   Lawrence   Palmerston
Dusky Sd.   Nightcaps   Tapanui   Milton   Port Chalmers
Chalky Inlet   Winton   Clinton   Balclutha   DUNEDIN
Preservation Inlet   Edievale   Mataura   Kaitangata   Otago Harbour
Te Waewae B.   Riverton   Wyndham   Owaka   C. Saunders
Orepuki   Invercargill   Tokanui   Nugget Pt.
Solander I.   Bluff   Ruapuke I.   Tautuku   Tahakopa
Halfmoon Bay   Foveaux Str.
Stewart I. (Rakiura)
RAKIURA   South West C.   Port Pegasus

Projection : Conical with two standard parallels

166   168   170   172   East from Greenwich

South Island
Westland Bight   WESTLAND
Southern Alps (Tiritiri o te Moana)
Canterbury Plains
Southland

TAHITI & MOOREA
1:1 000 000 d

17°30'S

Papetoai   Pte. Aroa   B. de Matavai   Pte. Vénus
Paopao   Mahina
Mt. Tohiea ▲1207   Papeete   Arue   Pirae   Papenoo   Tiarei
Haapiti   Afareaitu   Faaa   Pirae   Hitiaa
Moorea (France)   Mt. Aorai ▲2060   Mt. Orohena ▲2241
Pte. Nuupere   Punaauia   Faaone
Paea   Mt. Tetufera ▲1798   Lac Vaihiria
PACIFIC OCEAN   Papara   Taravao   Afaahiti   Isthme de Taravao
Maraa   Papeari   Mataiea   Pueu   Pte. Tatutua
Atimaono   Vairao   ▲Mt. Roonu 1332
Teahupoo   Presqu'île de Taiarapu   Tautira

149°45'W   149°30'W   West from Greenwich   149°15'W

COPYRIGHT PHILIP'S

10   0   10 km
10   0   10 miles
1:1 000 000

ft   m
9000   3000
6000   2000
3000   1000
1200   400
600   200
0
200   600
2000   6000
4000   12 000
6000   18 000
m   ft

A   B   C   D   E

64

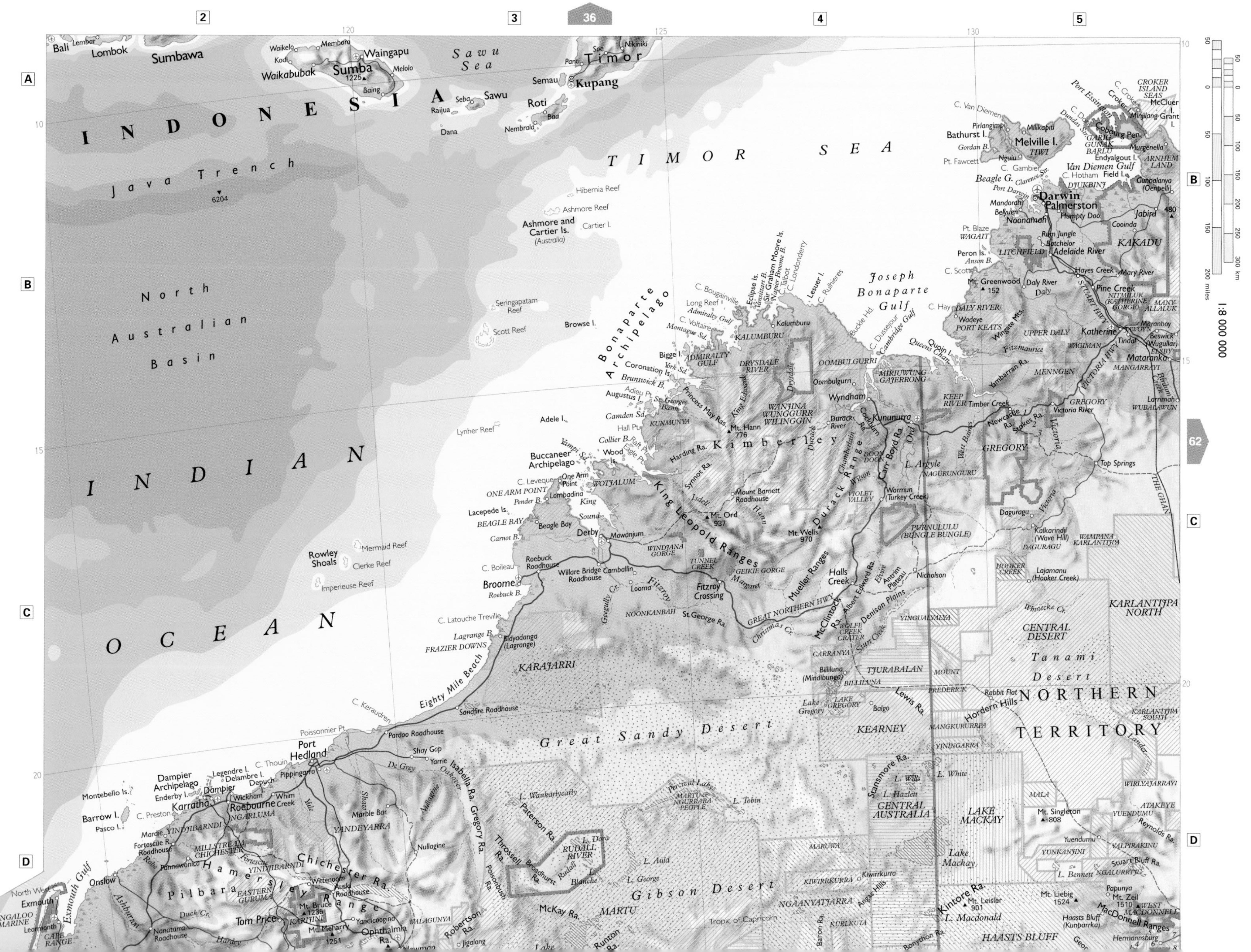

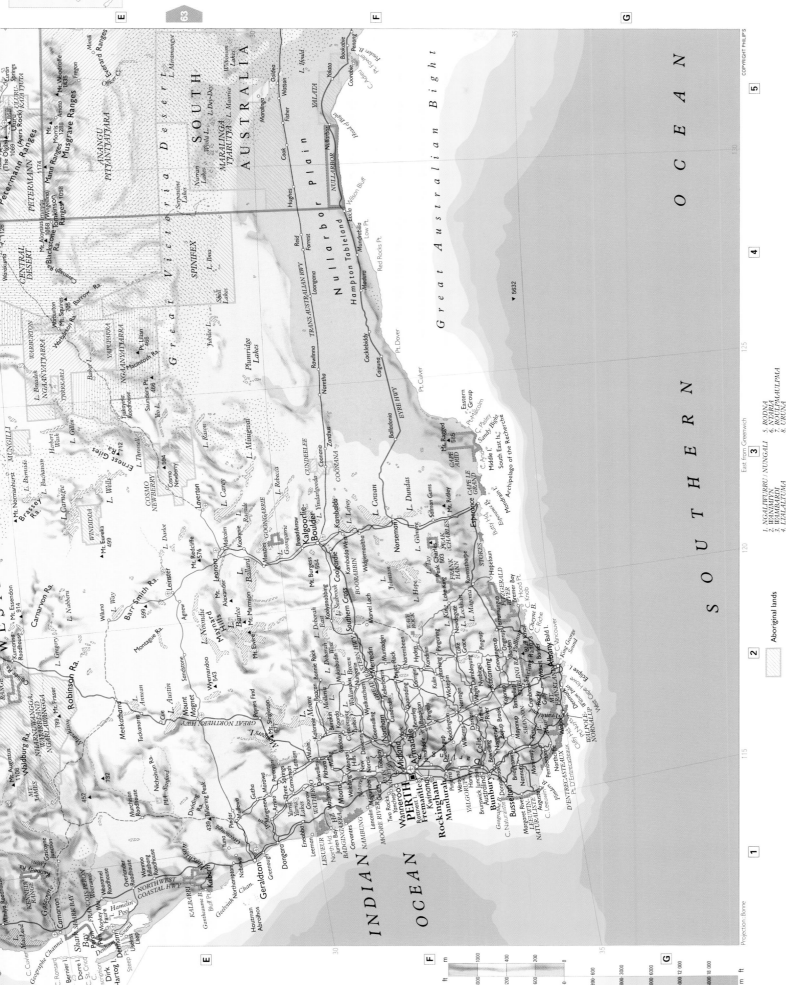

Projection: Bonne

East from Greenwich

Aboriginal lands

*1. NGALIWURRU / NUNGALI
2. WANJINA
3. WAMBARDI
4. LTIALALTUMA
5. RODNA
6. NYARDA
7. ROOLPMAULPMA
8. URUNA*

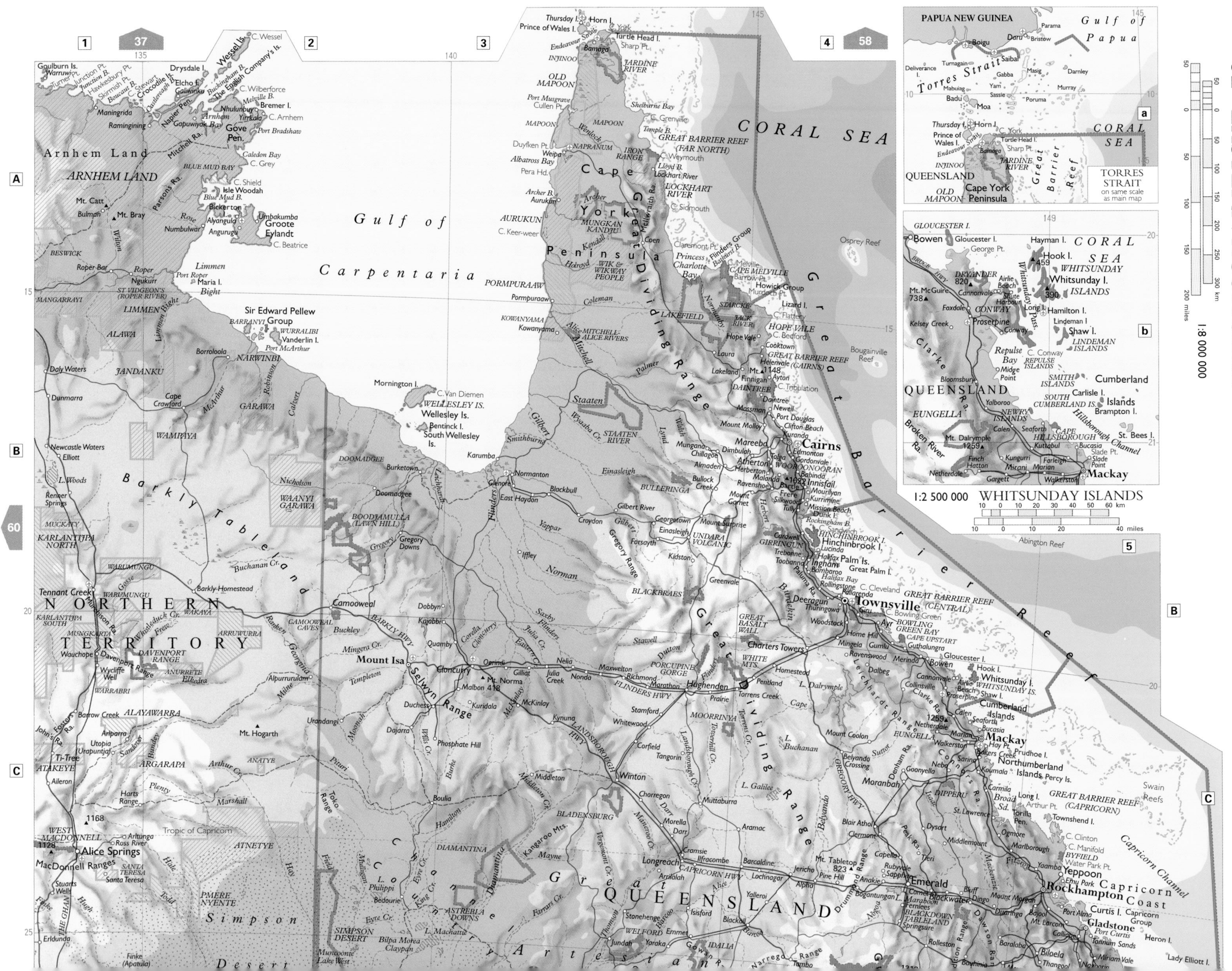

1:8 000 000

a

PAPUA NEW GUINEA

Gulf of Papua

TORRES STRAIT
on same scale
as main map

CORAL SEA

Great Barrier Reef

QUEENSLAND

OLD MAPOON
Cape York Peninsula

b

CORAL SEA

WHITSUNDAY ISLANDS

QUEENSLAND

EUNGELLA

Mackay

GLOUCESTER I.
Bowen
Proserpine
Airlie Beach
CONWAY

SMITH ISLANDS
SOUTH CUMBERLAND IS.
LINDEMAN ISLANDS
Cumberland Islands
Brampton I.

WHITSUNDAY ISLANDS

1:2 500 000

CORAL SEA

Great Barrier Reef

GREAT BARRIER REEF (CENTRAL)

Townsville

Cairns

Great Dividing Range

Cape York Peninsula

Gulf of Carpentaria

ARNHEM LAND

Barkly Tableland

NORTHERN TERRITORY

Tennant Creek

Alice Springs

MacDonnell Ranges

QUEENSLAND

Mount Isa

Great Artesian Basin

Mackay

Rockhampton

GREAT BARRIER REEF (CAPRICORN)

Gladstone

Emerald

Capricorn Channel

Tropic of Capricorn

Simpson Desert

Aboriginal lands

on same scale

Projection: Bonne

COPYRIGHT PHILIP'S

East from Greenwich

7 8 9
6
1 2 3 4 5

B | Moskva Yekaterinburg Tomsk Okhotsk Poluostrov Kamchatka Aleutian Basin
Volga Novosibirsk Irkutsk Chita Sea of Okhotsk Komandorskiye Ostrova (Russia) Near Is. (U.S.A.)
Astana (Aqmola) Semey Oz. Baykal Amur Petropavlovsk-Kamchatskiy Aleutian Tr. 7822
KAZAKHSTAN Altai Blagoveshchensk Sakhalin Kuril-Kamchatka Trench Aleutian Tr.
C | Aral Sea Balqash Kol Ulaanbaatar Khabarovsk La Perouse Str. Kurilskiye Ostrova (Russia) 10,542 Northwest Emperor Trough Chinook
Almaty Ürümqi MONGOLIA Changchun Harbin Vladivostok Sapporo Shatsky Rise Pacific Seamount Chain
Toshkent KYRGYZSTAN Altai Shenyang Hakodate Sea of Japan
D | TAJIKISTAN CHINA Beijing NORTH KOREA Sendai 10,554 Basin Midway (U.S.A.)
AFGHANISTAN Kabul Srinagar Kunlun Shan Lanzhou Taiyuan Dalian Seoul SOUTH KOREA Nagoya Kyōto Tōkyō Japan Trench
PAKISTAN Himalaya XIZANG Xi'an Qingdao Kitakyūshū Osaka JAPAN Yokohama
Lahore Lhasa Chongqing Nanjing Wuhan Yellow Sea Shikoku Kyūshū
Delhi 8850 Mt Everest NEPAL Changsha Shanghai East China Sea Okinawa Minami-Tori-Shima (Japan) Lisianski (U.S.A.)
E | Kanpur Ganga Brahmaputra Kunming Fuzhou Ryūkyū-rettō (Japan) Iwo-Jima (Japan) Ogasawara Gunto (Japan) Mid - Pacific Wake I. (U.S.A.)
Kolkata (Calcutta) Dhaka Guangzhou Hong Kong TAIWAN Kazan-Rettō (Japan) Mountains
INDIA BANGLADESH BURMA Mandalay Macau Taipei Philippine Sea West Mariana Basin NORTHERN MARIANAS (U.S.A.) P
Hyderabad LAOS Hanoi Hainan C. Engano Philippine Basin Tinian Saipan East Mariana Basin MARSHALL IS.
F | Bay of Rangoon Luzon GUAM (U.S.A.) Enewetak Atoll Bikini Atoll Ralik Chain
Bengal THAILAND Paracel Is. Manila Challenger 11,022 Deep Yap Kwajalein Majuro
Chennai (Madras) Bangkok South China Sea Mindoro PHILIPPINES Samar Mariana Trench Caroline Is. Micronesia Jaluit I.
CAMBODIA Phnom Penh Palawan 10,497 Chuuk Pohnpei Palikir
SRI LANKA Nicobar Is. (India) Thanh Pho Ho Chi Minh G. of Thailand Sulu Sea Mindanao Melekeok FED. STATES OF MICRONESIA Butaritari
G | Colombo Andaman Is. (India) MALAYSIA Davao Mindanao Trench PALAU East Caroline Basin Pohnpei Melanesian Basin Tarawa Gilbert Is.
Kuala Lumpur PEN. MALAYSIA BRUNEI SABAH Celebes Sea 4101 West Caroline Basin Eauripik Rise Mela Solomon Rise Howland Baker
Sumatera Singapore SARAWAK Maluku Halmahera Phoenix Is. Abari Ende
H | Palembang Borneo Sulawesi Seram Puncak Jaya 5029 PAPUA Admiralty Is. New Ireland NAURU Banaba
INDONESIA Makassar Buru Banda Sea New Guinea Bismarck Arch. Kokopo Bougainville SOLOMON IS. Fongafale TUVALU
Jakarta Jawa Java Sea Flores Sea 7440 Lae New Britain Honiara Guadalcanal Santa Cruz Is. Rotuma Îs. Wallis & Futuna (Fr.)
Surabaya Bali Flores Dili EAST TIMOR Timor Arafura Sea Port Moresby 9165 Espiritu Santo Vanua Levu
Christmas I. (Austral.) Sumbawa Sumba Torres Strait C. York Louisiade Arch. Coral Sea Basin VANUATU Port Vila West Fiji Basin Viti Levu FIJI
Cocos Is. (Austral.) North Australian Basin C. Arnhem Gulf of Carpentaria Cairns Coral Sea Îs. Chesterfield 7570 Suva Nuku'alofa
INDIAN Exmouth Plateau Darwin NEW CALEDONIA (Fr.) Noumea Îs. Loyauté 10,822
Broome North West C. Townsville Rockhampton Great Barrier Reef Middleton Basin South Fiji Basin
Wharton Basin Mount Isa AUSTRALIA Great Dividing Ra. Brisbane Lord Howe Rise Norfolk Ridge Kermadec Is. (N.Z.) Kermadec Tr. 10,047
OCEAN Alice Springs Darling Lord Howe I. (Austral.) Norfolk I. (Austral.)
Geraldton Perth Basin L. Eyre Sydney New Caledonia Trough NEW ZEALAND
Perth Naturaliste Plateau Great Australian Bight Murray Canberra Mt Kosciuszko 2228 Tasman Sea Auckland
Albany Adelaide Melbourne Bass Str. Tasman Aoraki Mt. Cook 3753 Cook Strait Wellington
Broken Ridge South Australian Basin Tasmania Hobart East Tasman Plateau Tasman Basin Christchurch Chatham Rise
Mid SOUTHERN South Tasman Rise Dunedin Bounty Trough Bounty Is. (N.Z.)
Nouvelle Amsterdam (Fr.) I. St. Paul (Fr.) OCEAN Invercargill Antipodes Is. (N.Z.)
Îs. Crozet (Fr.) Indian Ridge Auckland Is. (N.Z.) Campbell Plateau
Kerguelen (Fr.) Campbell I. (N.Z.)
Heard I. (Austral.) Macquarie I. (Austral.)

ft m
12 000 4000
9000 3000
6000 2000
3000 1000
1500 500
600 200
0 0
200 600
1000 3000
2000 6000
4000 12 000
6000 18 000
8000 24 000
m ft

40 60 80 100 110 120 130 140 150 160 170 180

1 2 3 4 5 6 7 8 9

Arctic Circle
ALASKA
(U.S.A.)
Anchorage
15
6959
16 **17** **18** **19** **20**
l Bay
Gulf of Alaska
Juneau
C A N A D A
B
Prince of Wales I.
(U.S.A.) Prince Rupert
Edmonton
L. Winnipeg
Newfoundland
(U.S.A.)
Queen Charlotte Is.
(Canada)
R o c k y
Calgary
Winnipeg
Tufts
Vancouver
Regina
St. Lawrence
50
Abyssal
Vancouver I.
Victoria
L. Superior
Québec
St. John's
C
Plain
Seattle
Boise
Minneapolis
L. Huron
Montréal
Portland
Spake
Missouri
Toronto
Ottawa
Boston
orthe ast
Detroit
L. Ontario
Buffalo
northeast
Salt Lake
Chicago
L. Erie
New York
C. Mendocino
City
Pittsburgh
Philadelphia
40
Mendocino Fracture Zone
Denver
Colorado
Kansas City
Cincinnati
Baltimore
D
Sacramento
St. Louis
Washington D.C.
San Francisco
4418
UNITED STATES
A T L A N T I C
6741
Oklahoma City
Memphis
C. Hatteras
Los Angeles
Phoenix
Bermuda
P a c i f i c
San Diego
Dallas
(U.K.)
Murray Fracture Zone
Houston
Jacksonville
30
Ciudad
San Antonio
New
Sargasso Sea
E
Juárez
Orleans
Guadalupe
Tampa
OCEAN
(Mex.)
Golfo de California
Monterrey
Miami
BAHAMAS
Molokai Fracture Zone
Gulf of Mexico
Ridge
Tropic of Cancer
Baja California
La Habana
West Indies
B a s i n
C. San Lucas
CUBA
Kauai
Honolulu
Canal de Yucatán
I.
Oahu Maui
HAWAIIAN IS.
Guadalajara
9200
Hilo
4205
(U.S.A.)
5610
Mérida
HAITI DOMINICAN REP.
20
Hawaii
Mexico
7680
JAMAICA
Puebla
BELIZE
Kingston
PUERTO
Leeward
I F I C
Clarion Fracture Zone
Is. Revilla Gigedo
Acapulco
GUATEMALA
HONDURAS
Caribbean Sea
RICO
Is.
(Mex.)
Middle America Trench
Guatemala
(U.S.A.)
F
6662
BARBADOS
San Salvador
NICARAGUA
Windward Is.
Î. Clipperton
EL SALVADOR Managua
Barranquilla
Maracaibo
(Fr.)
Guatemala
COSTA San José
Colón
Panamá
Caracas
Clipperton Fracture Zone
Basin
RICA
PANAMA
10
Cooper Ridge
I. del Coco
Bogotá
G
(Costa Rica)
Medellín
Cocos Ridge
Cali
Teraina
I. de Malpelo
VENEZUELA
EA N
Tabuaeran
(Colombia)
COLOMBIA
Kiritimati
Galápagos Fracture Zone
Orinoco
Jarvis I.
Galápagos
Quito
0
myra Is.
(U.S.A.)
(Ecuador)
Carnegie Ridge
ECUADOR
(U.S.A.)
Equator
Guayaquil
Iquitos
H
A T I
Malden I.
C. Paliñas
Amazonas
Starbuck I.
BRAZIL
manihiki
Penrhyn
Vostok I.
ukapuka
(Tongareva)
Caroline I.
Trujillo
Manihiki
(Millennium I.)
lateau
Flint I.
Marquesas Fracture Zone
10
Suwarrow Is.
Îs. Marquises
6369
PERU
J
Nuku Hiva
Yupanqui
Îs. de la
Hiva Oa
Lima
Société
Basin
Cusco
Bora Bora
Mendaña
Huahine
Rangiroa
Fracture Zone
Arequipa
L. Titicaca
Nevado Ancohuma
Cook Is.
Raiatea Tahiti
6866
7550
(N.Z.)
Papeete
Îs. Tuamotu
P e r u
B a s i n
La Paz
Aitutaki
FRENCH POLYNESIA
Peru-
BOLIVIA
Atiu
Arica
Rarotonga
Îs. Gambier
20
Mangaia
Muroroa
Iquique
Îs. Tubuai
Rapa
Tropic of Capricorn
Chile
PARAGUAY
Oeno I.
Antofagasta
K
Henderson I.
8050
Asunción
Pitcairn I. Ducie I.
Easter Fracture Zone
Sala y Gómez Ridge
Trench
San Miguel
(U.K.)
Sala-y-Gómez
San Felix
de Tucumán
(Chile)
San Ambrosio
(Chile)
I. de Pascua
(Chile)
Porto
(Chile)
Alegre
Challenger Fracture Zone
Roggeveen
Arch. de
30
Córdoba
Basin
Juan Fernández
Aconcagua
L
S o u t h w e s t
(Chile)
Valparaíso
6962
Rosario
URUGUAY
Santiago
Buenos
Montevideo
Aires
Río de la Plata
40
P a c i f i c
Concepción
ARGENTINA
B a s i n
ATLANTIC
M
Menard Fracture Zone
OCEAN
6212
50
Pacific-Antarctic Ridge
East
Punta Arenas
Falkland Is.
South Georgia
N
S o u t h e a s t
C. de Hornos
Tierra del Fuego
(U.K.)
(U.K.)
Pacific Basin
Drake Passage
West from Greenwich
COPYRIGHT PHILIP'S

1:35 000 000

100 0 200 400 600 800 1000 1200 1400 km
100 0 200 400 600 800 1000 miles
Asia
ARCTIC OCEAN
Lincoln Sea
Greenland
Iceland
Denmark Strait
Petermann Peak 2940
Mt. Forel 3360
Chukchi Sea
Wrangel I.
Bering Strait
Bering Sea
St. Lawrence I.
Nunivak I.
Unimak I.
Kotzebue Sound
Norton Sound
Pt. Hope
Barrow Pt.
Icy C.
Beaufort Sea
Brooks Ra.
2761
Colville
Koyukuk
Yukon
Kuskokwim
Porcupine
Mackenzie Bay
Alaska
Alaska Range
Mt. McKinley 6194
Iliamna L.
Bristol Bay
Alaska Peninsula
Cook Inlet
Kodiak I.
Gulf of Alaska
Mt. St. Elias 5489
Mt. Logan 5959
Sverdrup Is.
Parry Is.
Queen Elizabeth Is.
Axel Heiberg I.
Ellesmere I.
Kane Basin
Melville I.
Bathurst I.
Devon I.
Banks I.
M'Clure Strait
Viscount Melville Sd.
Lancaster Sd.
Bylot I.
Somerset I.
Prince of Wales I.
M'Clintock Channel
Victoria I.
Amundsen Gulf
C. Bathurst
Baffin Bay
Baffin Island
Gulf of Boothia
Boothia Pen.
King William I.
Melville Pen.
Foxe Basin
Foxe Channel
Disko I.
Davis Strait
Cumberland Sd.
Frobisher B.
2147
Cape Farewell
Labrador Sea
Hudson Strait
Southampton I.
C. Chidley
Wolstenholme
Coats I.
Mansel I.
Ungava Peninsula
Ungava B.
1652
Hudson Bay
Chesterfield Inlet
Thelon
Back
Great Bear L.
Great Slave L.
Mackenzie
Mackenzie Mts.
2762
Stewart
Pelly
Teslin
Liard
Stikine
Skeena
Alexander Archipelago
Queen Charlotte Islands
Hecate Str.
Queen Charlotte Str.
Mt. Waddington 3994
Coast Mountains
Rocky Mountains
Mt. Robson 3954
Fraser
Vancouver I.
Str. of Juan de Fuca
C. Flattery
Canadian Shield
Great Plains
Athabasca
L. Athabasca
Peace
Reindeer L.
Churchill
Nelson
C. Churchill
Severn
Attawapiskat
Albany
James Bay
Rupert
Eastmain
La Grande
Belcher Is.
C. Henrietta Maria
Grande Baleine
L. Caniapiscau
Smallwood Res.
Labrador
Churchill
C. Harrison
Hamilton Inlet
Str. of Belle Isle
Newfoundland
C. Race
Gulf of St. Lawrence
I. de Anticosti
Pr. Edward I.
Cape Breton I.
Nova Scotia
Sable I.
C. Sable
B. of Fundy
St. John
St. Lawrence
Laurentian Plateau
L. Winnipegosis
L. Winnipeg
L. Manitoba
Saskatchewan
Selkirk Mts.
L. Seul
L. of the Woods
Red L.
L. Nipigon
L. Superior
Great Lakes
Ottawa
Mt. Washington 1917
C. Cod
Nantucket I.
Long I.
Hudson
Ontario
Niagara Falls
L. Huron
L. Michigan
L. Erie
Souris
Missouri
Red
Yellowstone
Black Hills
Powder
4202
Mt. Rainier 4392
Cascade Range
Columbia
Columbia Plateau
Snake
C. Blanco
C. Mendocino
Coast Ranges
Mt. Shasta 4317
Sacramento
Sierra Nevada
San Joaquin
Mt. Whitney 4418
Death Valley -86
Great Salt Lake
Great Basin
Wasatch Ra.
Green
Mt. Elbert 4399
Blanca Peak 4378
Platte
James
Des Moines
Mississippi
Illinois
Wabash
Ohio
Cumberland Plateau
Tennessee
Allegheny Mts.
Appalachian Mts.
Mt. Mitchell 2037
Blue Ridge
Piedmont
Savannah
C. Charles
Chesapeake B.
C. Hatteras
C. Fear
Bermuda
Pt. Conception
Channel Is.
3506
Colorado
Grand Canyon
Colorado Plateau
Arkansas
Canadian
Ozark Plateau
Ouachita Mts.
Rio Grande
3659
Gila
Llano Estacado
Red
Brazos
Pecos
Edwards Plateau
Alabama
Chattahoochee
Apalachee B.
Florida
C. Canaveral
L. Okeechobee
The Everglades
Florida Strait
ATLANTIC OCEAN
Sargasso Sea
Bahamas
PACIFIC OCEAN
Baja California
3078
Guadalupe
Pta. Baja
I. de Cedros
Pta. Falsa
Gulf of California
Yaqui
Conchos
Western Sierra Madre
3050
Mexican Plateau
Eastern Sierra Madre
Rio Grande
Mississippi River Delta
Laguna Madre
Gulf of Mexico
Tropic of Cancer
C. San Lucas
B. de Banderas
Santiago
C. Rojo
C. Corrientes
Clarion Fracture Zone
Revilla Gigedo Is.
Popocatepetl 5452
5610 Pico de Orizaba
Balsas
Sa. Madre del Sur
Isthmus of Tehuantepec
Gulf of Campeche
Yucatán Peninsula
Yucatán Channel
Yucatán Basin
Cuba
Greater Antilles
Cayman Is.
Cayman Trench
Jamaica
Windward Passage
Hispaniola
3175
West Indies
Mona Passage
Caribbean Sea
Colombian Basin
G. of Honduras
C. Gracias a Dios
Coco
Pta. de Perlas
Sierra Madre
4093
G. of Tehuantepec
Guatemala Trench
G. of Fonseca
L. Nicaragua
C. Blanco
Pta. Burica
3819
Central America
Panama Canal
Isthmus of Panama
G. of Panamá
G. of Darién
Sierra Nevada de Santa Marta 5775
Maracaibo
G. of Venezuela
Andes
Magdalena
Cord. de Mérida
ft m
9000 3000
6000 2000
3000 1000
1500 500
600 200
0 0
200 600
1000 3000
2000 6000
4000 12000
6000 18000
8000 24000
m ft
Projection: Bonne
West from Greenwich
COPYRIGHT PHILIP'S

1:35 000 000

100 0 200 400 600 800 1000 1200 1400 km

100 0 200 400 600 800 1000 miles

Projection: Bonne

7 ■ MÉXICO Capital Cities **8** **9** West from Greenwich **10** **11** **12**

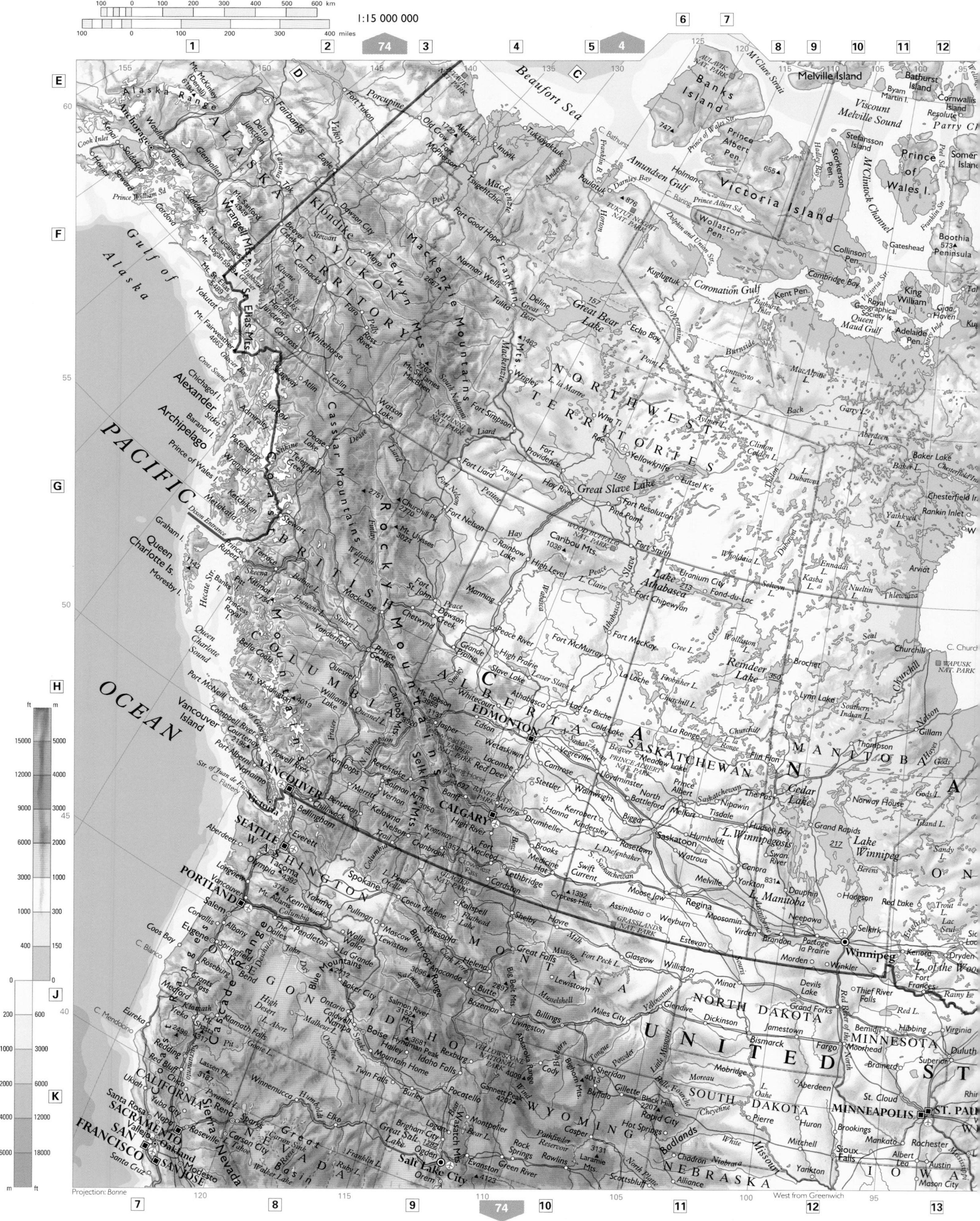

1:15 000 000

Projection: Bonne

West from Greenwich

NORTHERN CANADA
continuation northwards on same scale as main map
ARCTIC OCEAN
GREENLAND (KALAALLIT NUNAAT)
(Denmark)
Kronprins Frederik Land
Lincoln Sea
Alert
Ellesmere Island
Axel Heiberg Island
Eureka
Sverdrup Islands
Ellef Ringnes Island
Amund Ringnes I.
Queen Elizabeth Islands
N.W.T.
NUNAVUT
Parry Islands
Melville Island
Prince Patrick Island
Borden Island
Brock I.
Mackenzie King I.
King Christian I.
Lougheed I.
Emerald I.
Eglinton I.
Meighen I.
C. Isachsen
C. Thomas Hubbard
C. Columbia
North Magnetic Pole 2007
Grinnell Pen.
Graham I.
Cornwall I.
Norwegian Bay
Grise Fiord
Coburg I.
Jones Sound
Devon Island
Bathurst Island
Cornwallis Island
Resolute
Byam Martin I.
Viscount Melville Sound
Lowther I.
Parry Channel
Stefansson Island
Prince of Wales I.
Somerset Island
Lancaster Sound
Baffin Bay
Kane Basin
Knud Rasmussen Land
Qaanaaq (Thule)
Qeqertarsuaq
Uummannaq (Dundas)
Kap York
Melville Bugt
Lauge Koch Kyst
Petermann Gletscher
Sermersuaq (Humboldt Gletscher)
Nares Strait
Robeson Channel
Kennedy Channel
Hans I.
Lake Hazen
Barbeau Pk.
Greely Fiord
Agassiz Icecap
Prince of Wales Icefield
Nansen Sd.
Wellington Channel
Peel Sd.
Prince Regent Inlet
Brodeur Pen.
Arctic Bay
Nanisivik
Borden Pen.
Bylot I.
Pond Inlet
Eclipse Sd.
Baffin Island
Davis Strait
Nunavik
Clyde River
C. Raper
C. Adair
Home B.
Qikiqtarjuaq
Cumberland Peninsula
Pangnirtung
Cumberland Sd.
Hoare B.
C. Mercy
C. Dyer
Hall Peninsula
Iqaluit
Frobisher Bay
Meta Incognita Peninsula
Kimmirut
Resolution I.
Hudson Strait
Igloolik
Hall Beach
Rowley I.
Melville Peninsula
Foxe Basin
Prince Charles I.
Air Force I.
Spicer Is.
Taverner Bay
Koukdjuak
Nettilling L.
Amadjuak L.
Repulse Bay
Vansittart I.
C. Dorchester
Foxe Channel
Foxe Pen.
Kinngait
Salisbury I.
Mill I.
Southampton I.
Coral Harbour
Bell Pen.
Nottingham I.
Charles I.
Digges Is.
Coats I.
Mansel I.
Salluit
Ivujivik
Kangiqsujuaq
Cratère du Nouveau-Québec
Quaqtaq
Akpatok I.
C. Chidley
Péninsule d'Ungava
Ungava Bay
Kangirsuk
Torngat Mts.
Mt. d'Iberville
Mt. Caubvick
Kangiqsualujjuaq
Hebron
Smith I.
Puvirnituq
Ottawa Is.
Kuujjuaq
Inukjuak
Nain
Hopedale
C. Harrison
Labrador Sea
ATLANTIC OCEAN
Rigolet
Cartwright
NEWFOUNDLAND & LABRADOR
Port Hope Simpson
Belle Isle
C. Bauld
St. Anthony
Grey Is.
Str. of Belle Isle
Labrador
Smallwood Res.
North West River
Happy Valley-Goose Bay
Churchill
Churchill Falls
Kawawachikamach
Schefferville
Petitsikapau L.
Esker
Labrador City
Fermont
L. Ashuanipi
Sleeper Is.
King George Is.
Bakers Dozen Is.
Sanikiluaq
Belcher Is.
L. à l'Eau Claire
L. Minto
Kuujjuarapik
Grande Baleine
L. Bienville
L. Caniapiscau
La Grande
C. Henrietta Maria
Pte. Louis XIV
James Bay
Chisasibi
Wemindji
Twin Is.
Akimiski I.
Eastmain
Attawapiskat
Fort Albany
Charlton I.
Waskaganish
L. Mistassini
Moosonee
QUÉBEC
ONTARIO
Mts. Otish
Mts. Groulx
Gagnon
Rés. Manicouagan
St-Augustin
Natashquan
Havre-St-Pierre
Romaine
Long Range Mts.
Baie Verte
Notre Dame B.
Lewisporte
Gander
Bonavista
Grand Falls-Windsor
Deer Lake
Corner Brook
Stephenville
Newfoundland
Carbonear
St. John's
Placentia
Avalon Pen.
C. Race
Marystown
Channel-Port aux Basques
ST-PIERRE et MIQUELON (Fr.)
Cabot Strait
Gulf of St. Lawrence
Î. d'Anticosti
Dét. de Jacques-Cartier
Dét. d'Honguedo
Sept-Îles
Port-Cartier
Baie-Comeau
Gaspé
Pén. de la Gaspésie
Matane
Rimouski
Îs. de la Madeleine
PRINCE EDWARD I.
Charlottetown
Summerside
Cape Breton I.
Sydney
Glace Bay
Port Hawkesbury
NOVA SCOTIA
Halifax
Dartmouth
Truro
New Glasgow
Antigonish
Kentville
Bridgewater
Liverpool
Yarmouth
Digby
C. Sable
Sable I. (Nova Scotia)
B. of Fundy
NEW BRUNSWICK
Saint John
Fredericton
Moncton
Amherst
Miramichi
Bathurst
Campbellton
Edmundston
Rivière-du-Loup
Chaleur B.
Northumberland Str.
Chibougamau
Dolbeau-Mistassini
Alma
Chicoutimi
Saguenay
Jonquière
Roberval
L. St-Jean
La Tuque
Québec
Lévis
St-Georges
Shawinigan
Trois-Rivières
Thetford Mines
Drummondville
St-Hyacinthe
Sherbrooke
Granby
MONTRÉAL
Hull
OTTAWA
Rés. Gouin
Rés. Cabonga
Val-d'Or
Amos
Rouyn-Noranda
Matagami
L. Matagami
Cochrane
L. Abitibi
Timmins
Kirkland Lake
New Liskeard
Kapuskasing
Hearst
Nakina
Greenstone
Marathon
Wawa
Chapleau
Sudbury
Elliot Lake
Sault Ste. Marie
North Bay
L. Nipissing
Pembroke
Parry Sound
Huntsville
Georgian Bay
Manitoulin I.
Orillia
Barrie
Owen Sound
Lake Huron
Peterborough
Belleville
Kingston
Brockville
Cornwall
Oshawa
TORONTO
Hamilton
Kitchener
London
Sarnia
Windsor
DETROIT
Toledo
CLEVELAND
Erie
L. Erie
L. Ontario
Niagara Falls
Buffalo
Rochester
Syracuse
Albany
NEW YORK
Binghamton
Elmira
Jamestown
PENNSYLVANIA
Adirondack Mts.
Burlington
Montpelier
VERMONT
NEW HAMPSHIRE
Concord
Manchester
Lowell
MAINE
Augusta
Bangor
Lewiston
Portland
BOSTON
MASS.
Springfield
Hartford
CONN.
New Haven
R.I.
PROVIDENCE
C. Cod
Green Bay
L. Michigan
Sheboygan
Grand Rapids
Lansing
Flint
Saginaw
Petoskey
Traverse City
Cadillac
Manistique
L. Champlain
Ottawa
St. Lawrence (St-Laurent)

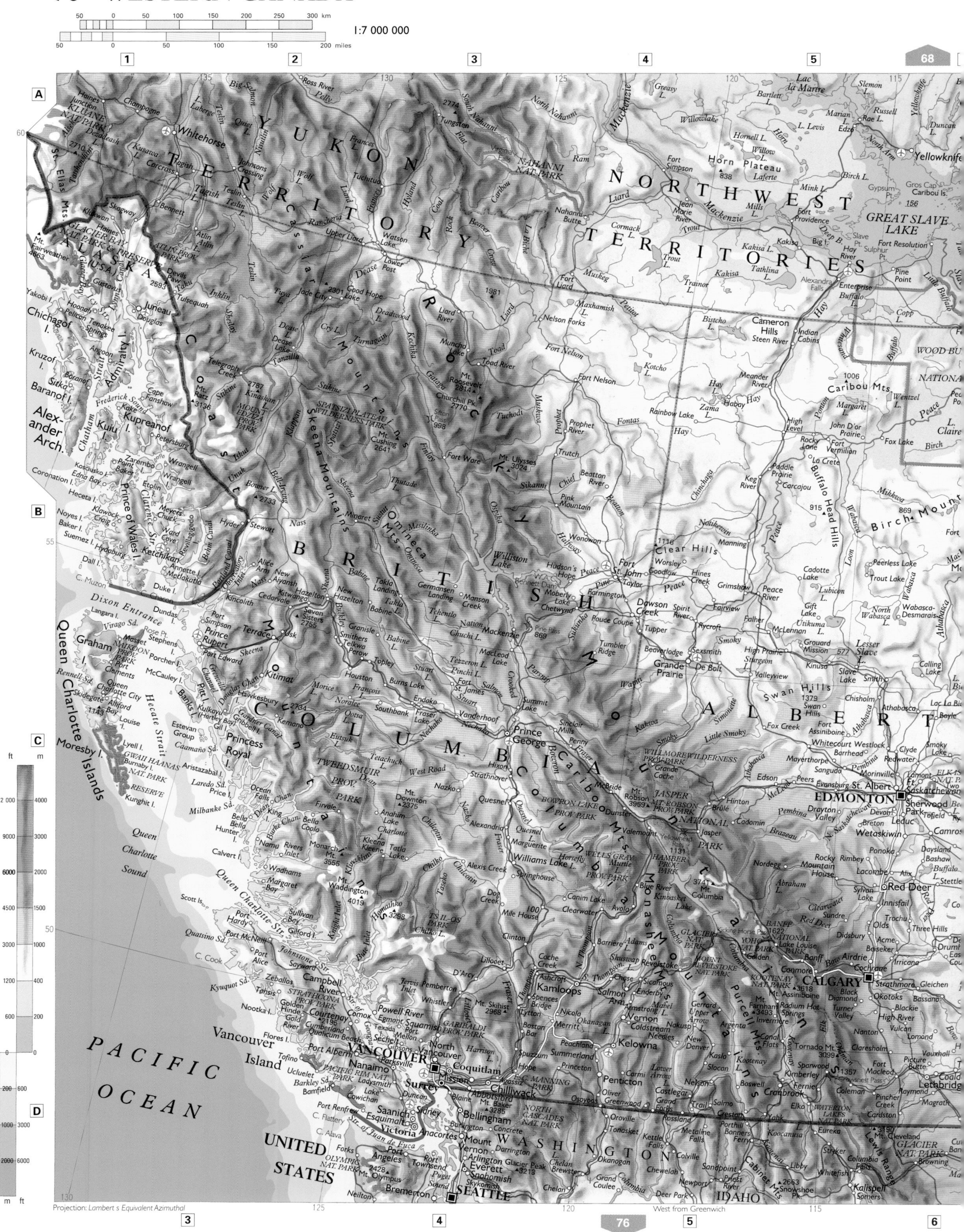

1:7 000 000

Projection: Lambert s Equivalent Azimuthal

West from Greenwich

HUDSON BAY
MANITOBA
SASKATCHEWAN
NORTH DAKOTA
MINNESOTA
MONTANA
LAKE WINNIPEG
Lake Winnipegosis
Lake Manitoba
Reindeer Lake
Wollaston L.
Lake Athabasca
Lake of the Woods
Cedar Lake
Southern Indian L.
Nueltin L.
Churchill
Thompson
The Pas
Flin Flon
Prince Albert
Saskatoon
Regina
Moose Jaw
Swift Current
Brandon
Winnipeg
Portage la Prairie
Yorkton
Dauphin
Kenora
Dryden
Minot
Grand Forks
Duluth
WAPUSK NATIONAL PARK
PRINCE ALBERT NAT. PARK
RIDING MOUNTAIN NAT. PARK
GRASSLANDS NAT. PARK
Cypress Hills
72
82
7 8 9 10 11
A B C D

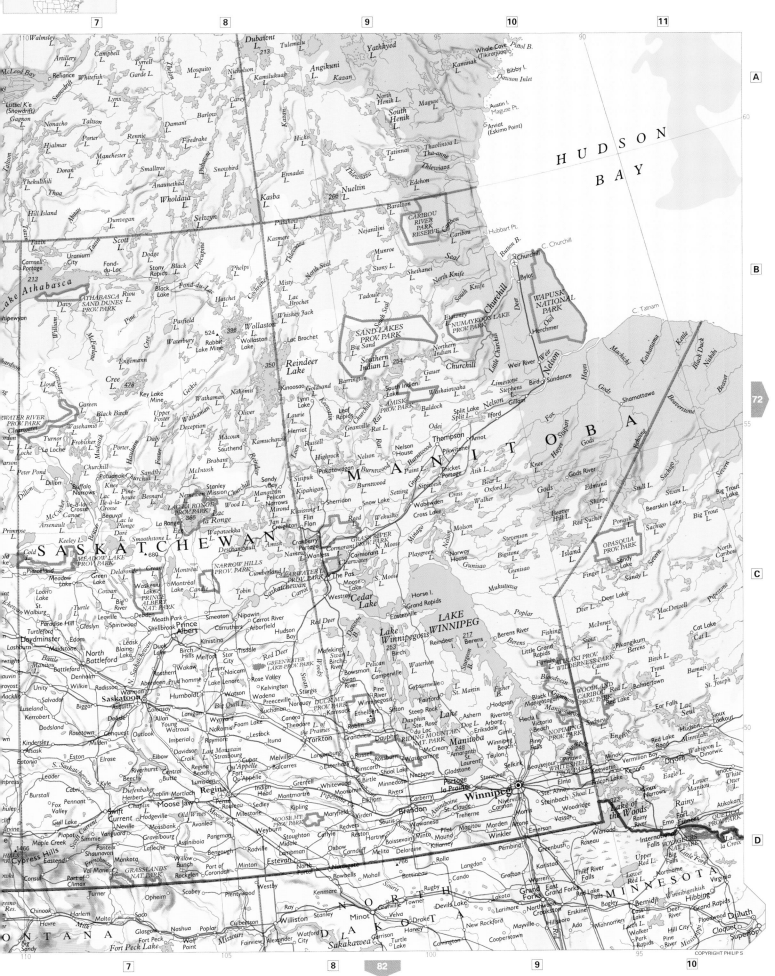

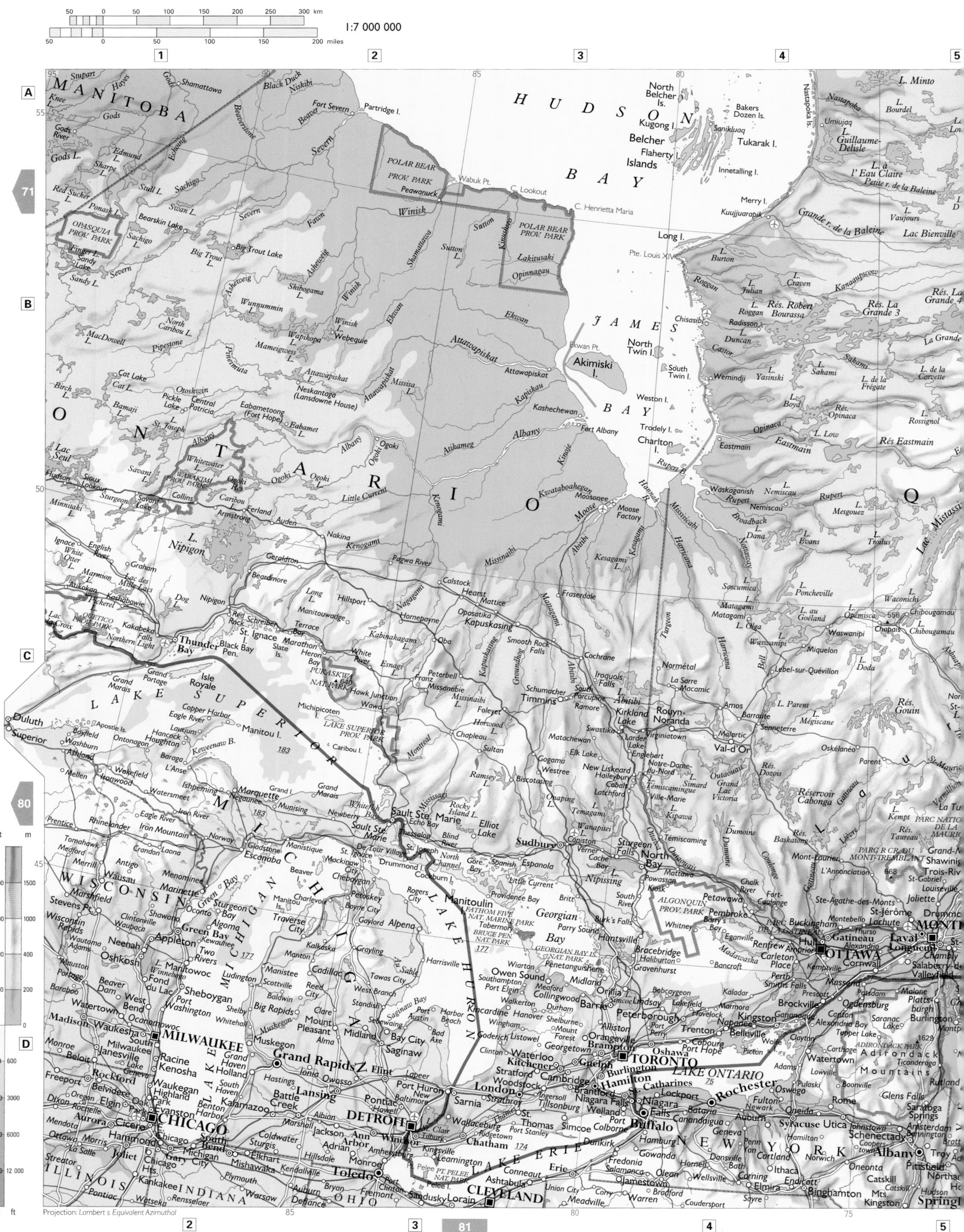

1:7 000 000

Projection: Lambert's Equivalent Azimuthal

6 69 7 8 9

A B C D

LABRADOR SEA

NEWFOUNDLAND & LABRADOR

Labrador

Newfoundland

Q u é b e c

Smallwood Reservoir

Mealy Mts.

Long Range Mts.

Str. of Belle Isle

Île d'Anticosti

Dét. de Jacques-Cartier

Dét. d'Honguedo

GULF OF ST. LAWRENCE

St. Lawrence

Pén. de la Gaspésie

Mts. Chic-Chocs

NEW BRUNSWICK

PRINCE EDWARD ISLAND

NOVA SCOTIA

Cape Breton Island

Cabot Strait

Îs. de la Madeleine (Québec)

Bay of Fundy

Sable I. (Nova Scotia)

ST-PIERRE-ET-MIQUELON (France)

MAINE

UNITED STATES

ATLANTIC OCEAN

Nain, Hopedale, Postville, Makkovik, Cartwright, Happy Valley-Goose Bay, North West River, Churchill Falls, Labrador City, Wabush, Fermont, Schefferville, Sept-Îles, Port-Cartier, Havre-St-Pierre, Natashquan, Kegaska, St-Augustin, Blanc-Sablon, L'Anse au Loup, Red Bay, St. Anthony, Roddickton, Corner Brook, Stephenville, Deer Lake, Grand Falls-Windsor, Gander, Lewisporte, Bonavista, Clarenville, St. John's, Mt Pearl, Argentia, Placentia, Marystown, Burin, Channel-Port aux Basques, Sydney, Glace Bay, Louisbourg, Halifax, Dartmouth, Truro, Amherst, New Glasgow, Yarmouth, Charlottetown, Summerside, Moncton, Saint John, Fredericton, Bathurst, Campbellton, Edmundston, Gaspé, Rimouski, Matane, Rivière-du-Loup, Chicoutimi, Baie-Comeau, Lévis, Bangor, Augusta, Portland, BOSTON

Mts. Otish 1128

1:15 000 000

PACIFIC OCEAN
Anchorage
Washington D.C.
San Francisco
Honolulu
3363 miles
5412 km
2010 miles
3234 km
2438 miles
3923 km
2785 miles
4482 km
2395 miles
3854 km
Tropic of Cancer
BRITISH COLUMBIA
VANCOUVER
WASHINGTON
SEATTLE
Spokane
PORTLAND
OREGON
IDAHO
Boise
NEVADA
SACRAMENTO
SAN FRANCISCO
SAN JOSE
Fresno
LAS VEGAS
LOS ANGELES
SAN BERNARDINO
SAN DIEGO
TIJUANA
Mexicali
ARIZONA
PHOENIX
Tucson
UTAH
Salt Lake City
Colorado Plateau
SONORA
ALBERTA
West from Greenwich
ALASKA
on same scale
RUSSIA
ARCTIC OCEAN
CHUKCHI SEA
North Slope
Brooks Range
Bering Strait
Seward Peninsula
ALASKA
Fairbanks
Alaska Range
Mt. McKinley (Denali) 6194
Anchorage
Kodiak I.
Gulf of Alaska
BERING SEA
Aleutian Islands
Pribilof Is.
Alaska Peninsula
International Date Line
Projection: Albers' Equal Area with two standard parallels

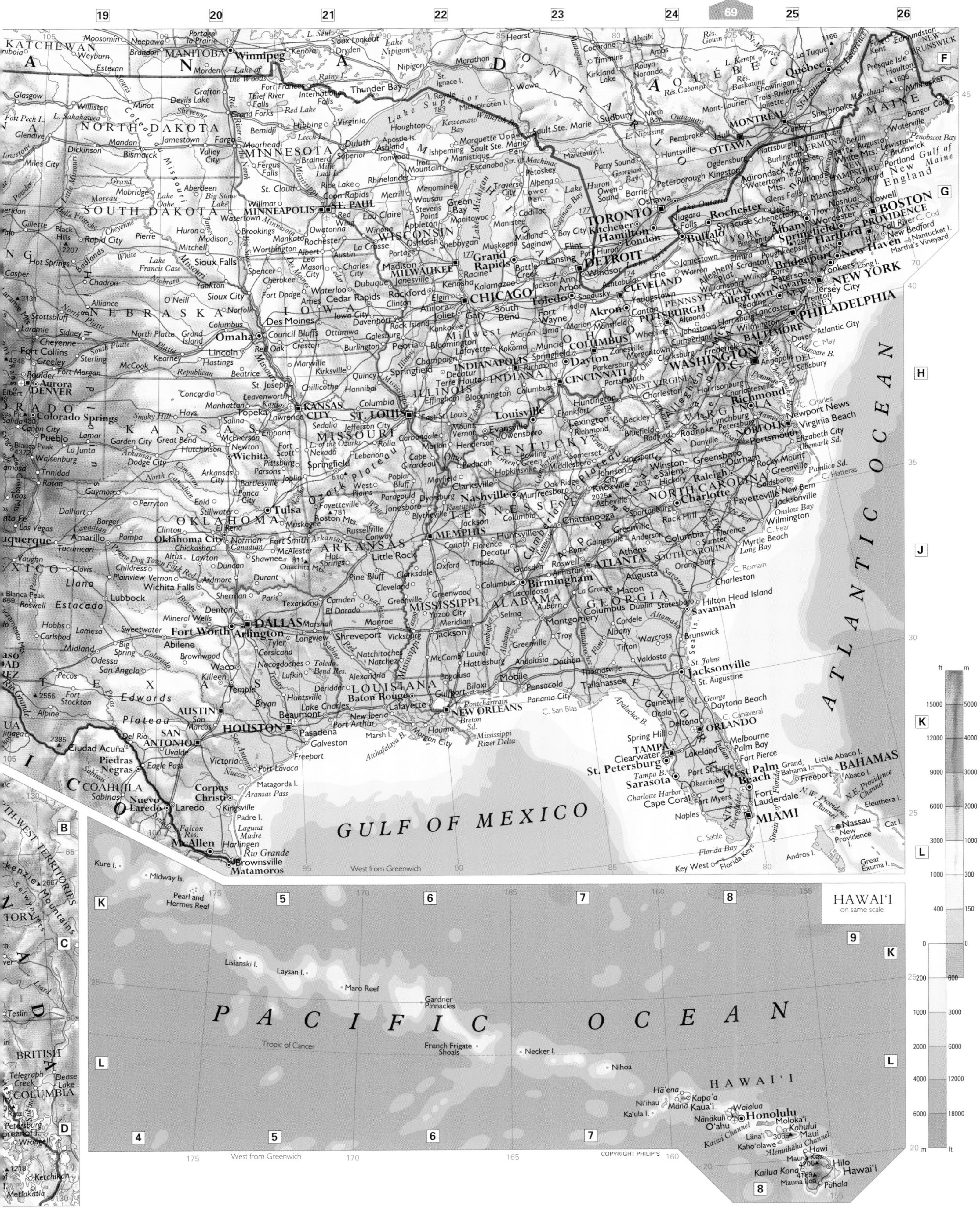

GULF OF MEXICO
ATLANTIC OCEAN
PACIFIC OCEAN
HAWAI'I
on same scale
NORTH DAKOTA
SOUTH DAKOTA
NEBRASKA
KANSAS
OKLAHOMA
TEXAS
MINNESOTA
IOWA
MISSOURI
ARKANSAS
LOUISIANA
WISCONSIN
ILLINOIS
INDIANA
OHIO
KENTUCKY
TENNESSEE
MISSISSIPPI
ALABAMA
GEORGIA
SOUTH CAROLINA
NORTH CAROLINA
VIRGINIA
WEST VIRGINIA
NEW YORK
PENNSYLVANIA
MAINE
ONTARIO
QUEBEC
MANITOBA
CHICAGO
DETROIT
TORONTO
BOSTON
NEW YORK
PHILADELPHIA
WASHINGTON D.C.
ATLANTA
DALLAS
HOUSTON
NEW ORLEANS
MIAMI
ORLANDO
TAMPA
MEMPHIS
ST. LOUIS
KANSAS CITY
MINNEAPOLIS
ST. PAUL
MILWAUKEE
CLEVELAND
PITTSBURGH
BALTIMORE
INDIANAPOLIS
CINCINNATI
COLUMBUS
AUSTIN
SAN ANTONIO
BAHAMAS
Honolulu
West from Greenwich
COPYRIGHT PHILIP'S

1:6 700 000

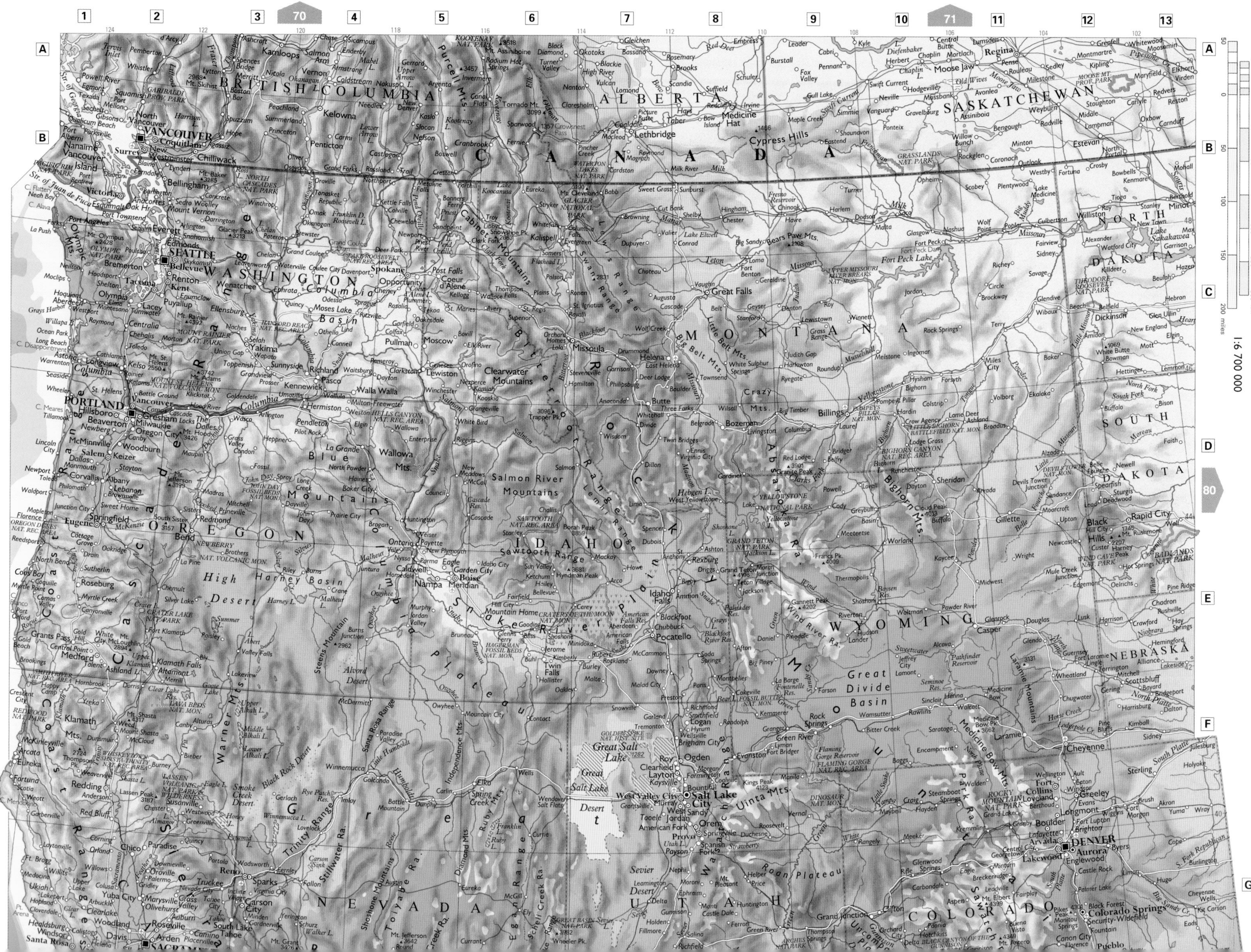
50 0 50 100 150 200 250 300 km
50 0 50 100 150 200 miles
CANADA
BRITISH COLUMBIA
ALBERTA
SASKATCHEWAN
NORTH DAKOTA
SOUTH DAKOTA
NEBRASKA
MONTANA
WYOMING
IDAHO
WASHINGTON
OREGON
UTAH
NEVADA
COLORADO
VANCOUVER
SEATTLE
PORTLAND
DENVER
Salt Lake City
Great Salt Lake
Great Divide Basin
Bighorn Mts.
Salmon River Mountains
Clearwater Mountains
Sawtooth Range
Snake River Plain
Wind River Ra.
Uinta Mts.
Laramie Mountains
Medicine Bow Mts.
Cypress Hills
Bears Paw Mts.
Little Belt Mts.
Crazy Mts.
Blue Mountains
Wallowa Mts.
High Desert
Harney Basin
Great Salt Lake Desert
Coast Ranges
Klamath Mts.
Warner Mts.
Trinity Range
Stillwater Ra.
Independence Mts.
Ruby Mts.
Roan Plateau
Fort Peck Lake
Great Falls
Billings
Boise
Pocatello
Idaho Falls
Spokane
Missoula
Helena
Cheyenne
Casper
Rapid City
Colorado Springs

PACIFIC OCEAN
Golfo de California
BAJA CALIFORNIA
BAJA CALIFORNIA SUR
ARIZONA
NEW MEXICO
TEXAS
CHIHUAHUA
SONORA
MEXICO
Colorado Plateau
Sonoran Desert
Mojave Desert
Painted Desert
Llano Estacado
Cedros Trench
Isla Guadalupe (Mexico)
LOS ANGELES
SAN DIEGO
TIJUANA
PHOENIX
Tucson
Albuquerque
El Paso
CIUDAD JUÁREZ
Chihuahua
Hermosillo
LAS VEGAS
Lava fields
West from Greenwich
Projection: Albers Equal Area with two standard parallels
COPYRIGHT PHILIP'S

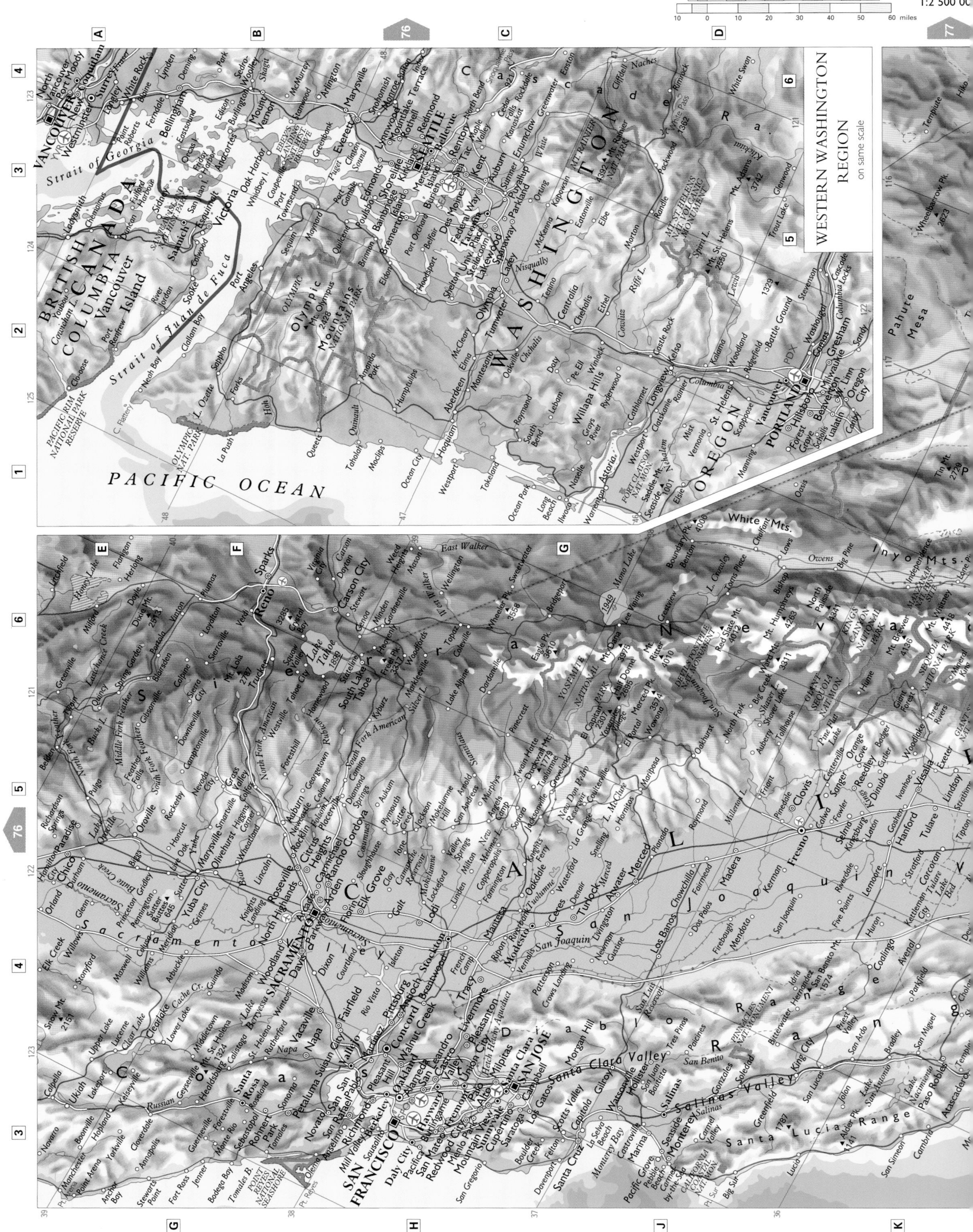

1:2 500 000

WESTERN WASHINGTON
REGION
on same scale

PACIFIC OCEAN
ARIZONA
MEXICO
BAJA CALIFORNIA
LOS ANGELES
SAN DIEGO
TIJUANA
LAS VEGAS
SAN BERNARDINO
Mojave Desert
Salton Sea
Chocolate Mts.
Channel Islands
San Pedro Channel
Santa Barbara Channel
Gulf of Santa Catalina
Santa Catalina I.
San Clemente I.
San Nicolas I.
Santa Cruz I.
Santa Rosa I.
San Miguel I.
Lake Mead
Death Valley
Amargosa Range
Bakersfield
Santa Barbara
Ventura
Oxnard
Lancaster
Palmdale
Victorville
Barstow
Riverside
Indio
El Centro
Mexicali
Yuma
Needles
Kingman
Blythe
Oceanside
Escondido
Chula Vista
Lompoc
Santa Maria
86
West from Greenwich
Lava fields
Projection: Bonne
COPYRIGHT PHILIP'S
m 3000 2000 1500 1000 400 200 0
ft 9000 6000 4500 3000 1200 600 0
200 600 1500 3000 6000 m
600 1500 3000 6000 ft

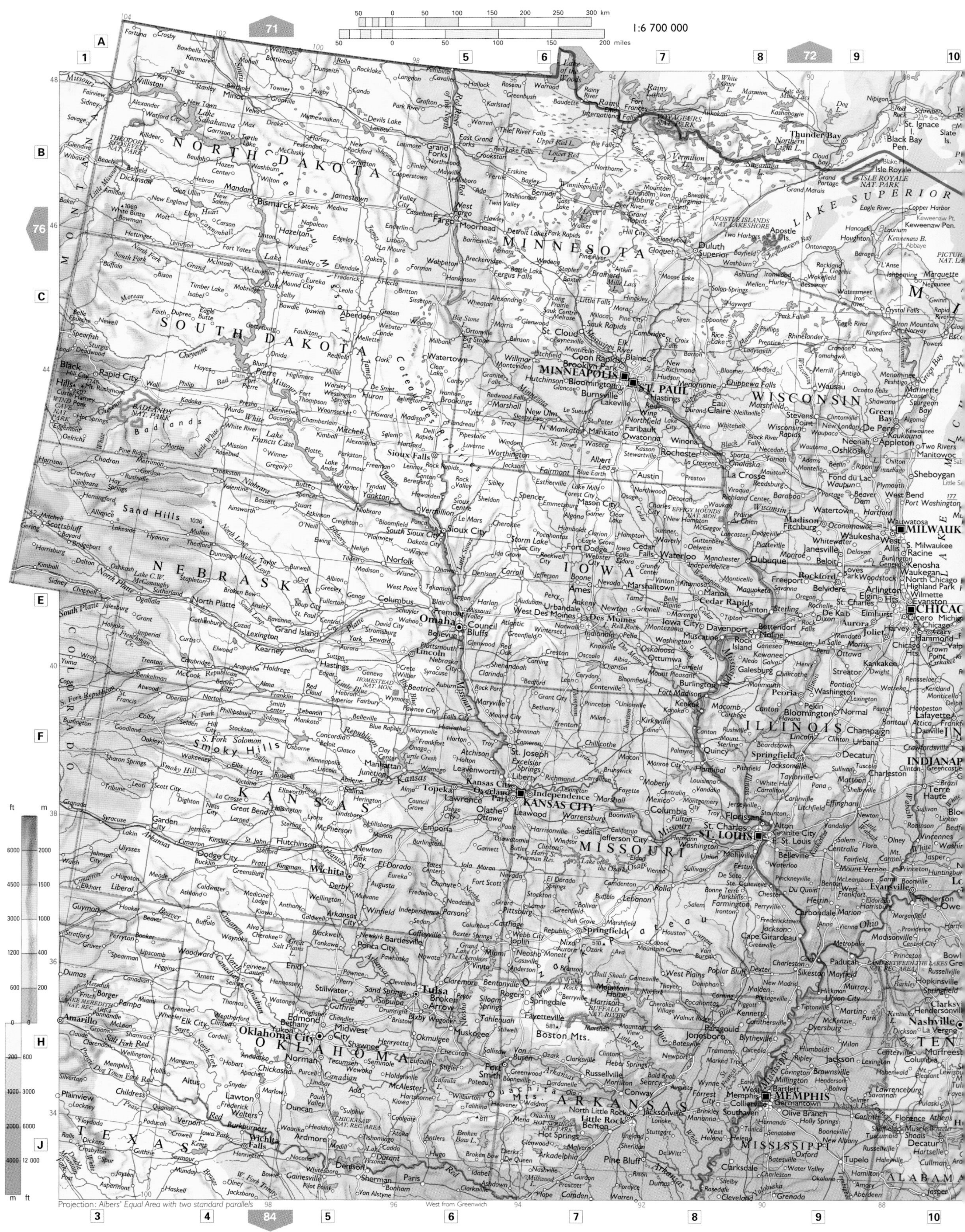
1:6 700 000
50 0 50 100 150 200 250 300 km
50 0 50 100 150 200 miles
71
72
76
84
NORTH DAKOTA
SOUTH DAKOTA
NEBRASKA
KANSAS
OKLAHOMA
TEXAS
MINNESOTA
IOWA
MISSOURI
ARKANSAS
WISCONSIN
ILLINOIS
MISSISSIPPI
LAKE SUPERIOR
MINNEAPOLIS
ST. PAUL
KANSAS CITY
ST. LOUIS
MEMPHIS
MILWAUKEE
Bismarck
Pierre
Lincoln
Omaha
Topeka
Des Moines
Oklahoma City
Tulsa
Wichita
Little Rock
Springfield
Sioux Falls
Fargo
Duluth
Madison
Sand Hills
Badlands
Smoky Hills
Boston Mts.
Ouachita Mts
Projection: Albers' Equal Area with two standard parallels
West from Greenwich

11
12
13
72
14
15
16
17
18
19
20
B
D
E
F
G
H
J
ONTARIO
QUÉBEC
CANADA
LAKE HURON
LAKE ERIE
LAKE ONTARIO
Georgian Bay
Gulf of Maine
ATLANTIC
OCEAN
NEW YORK
PENNSYLVANIA
OHIO
WEST VIRGINIA
VIRGINIA
NORTH CAROLINA
SOUTH CAROLINA
GEORGIA
MARYLAND
DELAWARE
NEW JERSEY
VERMONT
NEW HAMPSHIRE
MAINE
MA.
CT.
R.I.
NEW BRUNS.
Appalachian Mountains
Allegheny Plateau
Blue Ridge
Adirondack Mts.
TORONTO
MONTRÉAL
OTTAWA
Québec
DETROIT
CLEVELAND
PITTSBURGH
COLUMBUS
CINCINNATI
PHILADELPHIA
BALTIMORE
WASHINGTON D.C.
NEW YORK
BOSTON
PROVIDENCE
NORFOLK
Richmond
Charlotte
Raleigh
Buffalo
Rochester
Albany
Hartford
Toledo
Akron
Columbia
Atlanta
Chesapeake Bay
Delaware Bay
Pamlico Sound
Albemarle Sd.
Long Island Sound
Cape Cod
Nantucket I.
C. Hatteras
C. Lookout
C. Fear
11
85
12
13
14
15
16
17
18
COPYRIGHT PHILIP'S

1:2 500 000

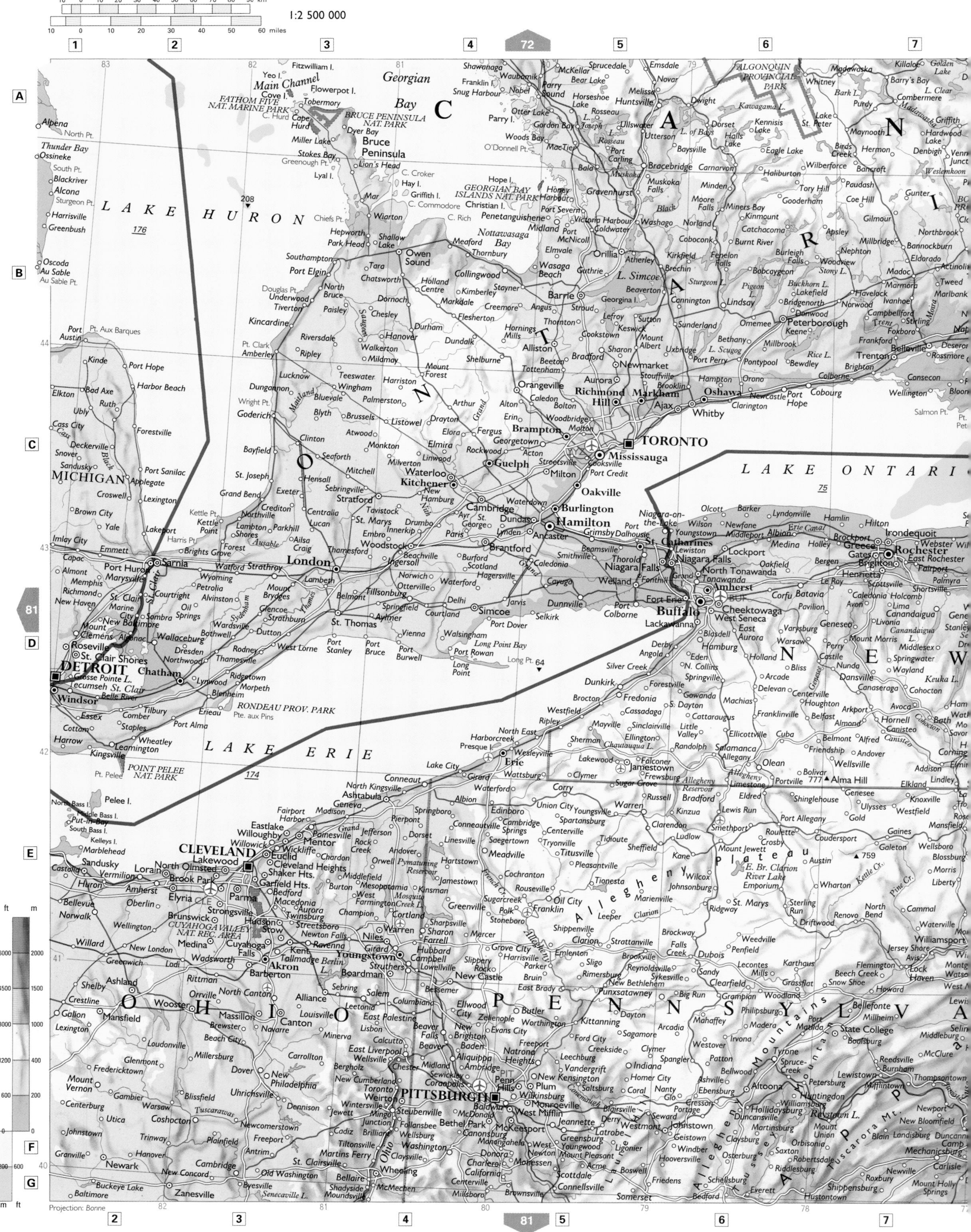
72
81
LAKE HURON
176
LAKE ONTARIO
75
LAKE ERIE
174
Georgian Bay
MICHIGAN
ONTARIO
CANADA
NEW YORK
OHIO
PENNSYLVANIA
TORONTO
Mississauga
Hamilton
Burlington
Oakville
Brampton
Kitchener
Waterloo
Guelph
Cambridge
London
Sarnia
Port Huron
DETROIT
Windsor
Chatham
St. Thomas
Woodstock
Brantford
St. Catharines
Niagara Falls
Buffalo
Amherst
Cheektowaga
Rochester
Irondequoit
Oshawa
Whitby
Ajax
Markham
Richmond Hill
Barrie
Orillia
Owen Sound
Peterborough
Belleville
Erie
Jamestown
Olean
Bradford
CLEVELAND
Lakewood
Euclid
Parma
Elyria
Lorain
Akron
Canton
Youngstown
Warren
Mansfield
PITTSBURGH
Altoona
State College
Allegheny Plateau
Allegheny Mountains
ALGONQUIN PROVINCIAL PARK
BRUCE PENINSULA NAT. PARK
FATHOM FIVE NAT. MARINE PARK
GEORGIAN BAY ISLANDS NAT. PARK
POINT PELEE NAT. PARK
RONDEAU PROV. PARK
CUYAHOGA VALLEY NAT. REC. AREA
Projection: Bonne

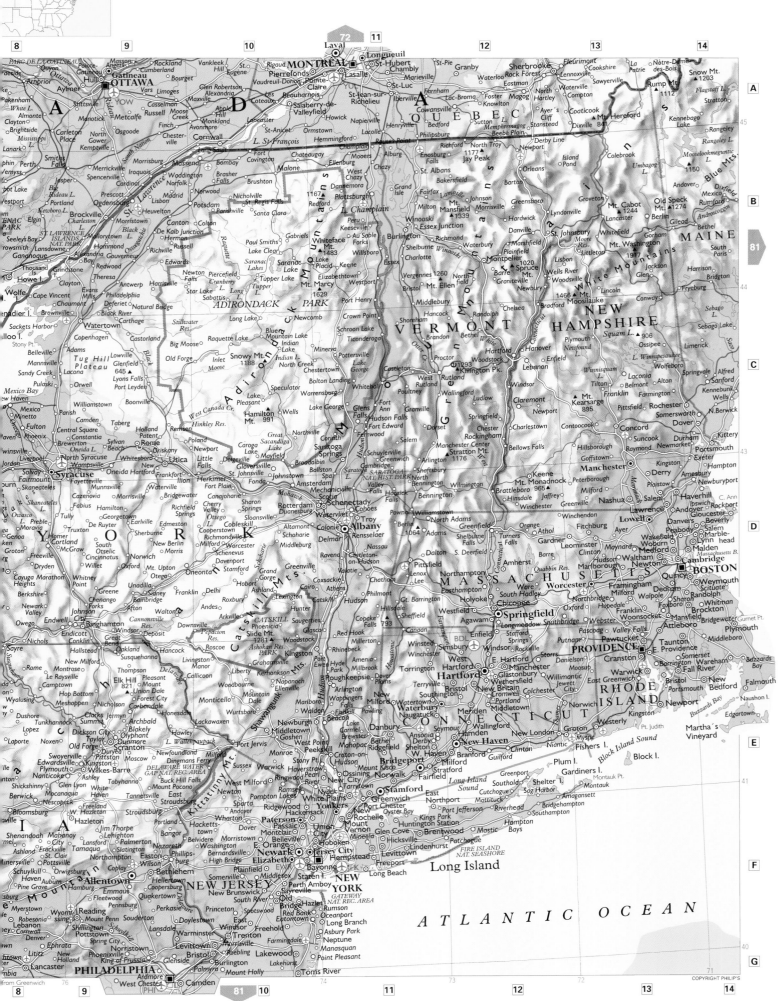

QUÉBEC

ONTARIO

VERMONT

NEW HAMPSHIRE

MAINE

MASSACHUSETTS

RHODE ISLAND

CONNECTICUT

NEW YORK

NEW JERSEY

ADIRONDACK PARK

Adirondack Mountains

Green Mountains

White Mountains

Catskill Mts.

Long Island

Block Island Sound

ATLANTIC OCEAN

72

81

81

1:6 700 000

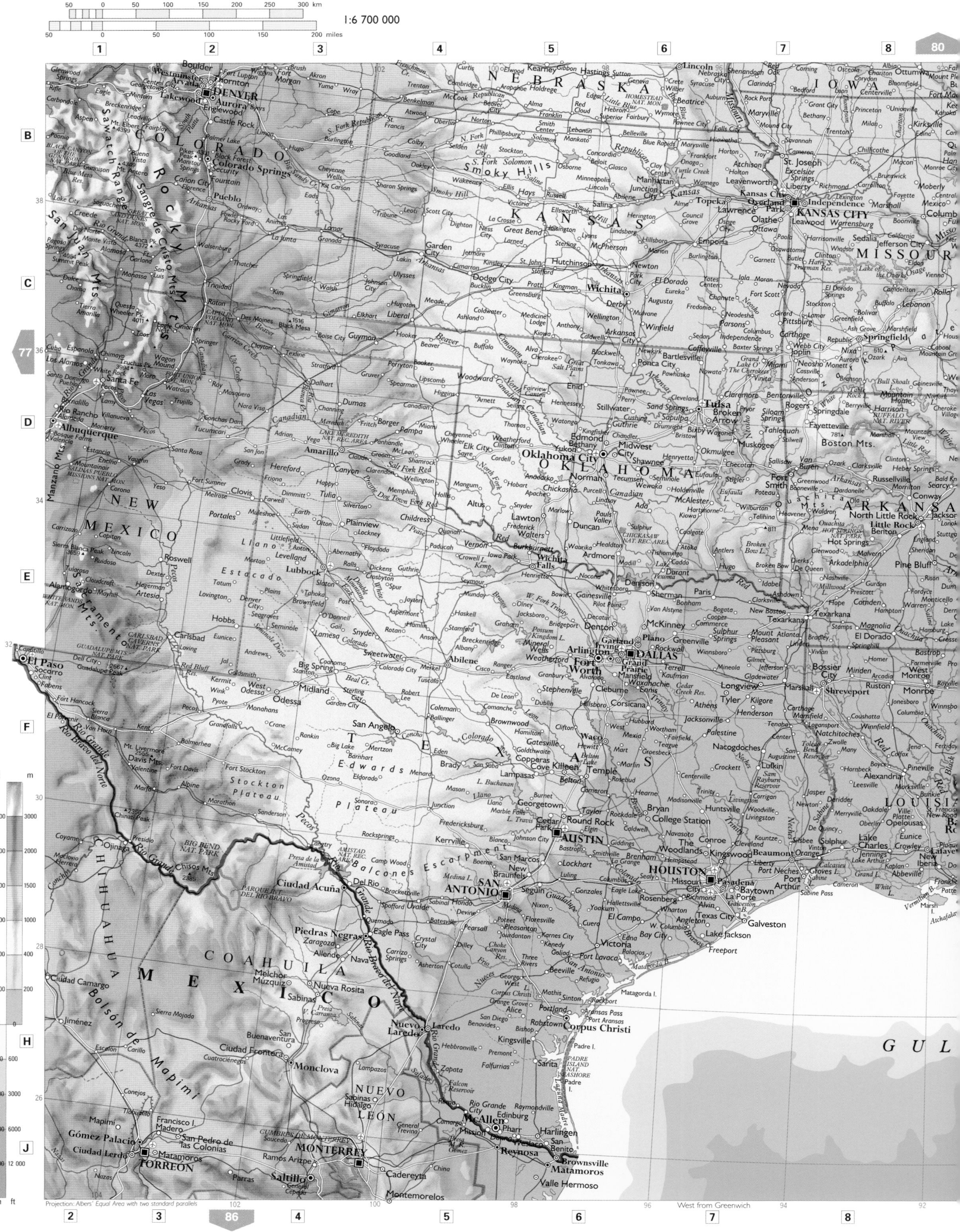

10 11 12 81 13 14 15 16 17
B
C
D
E
F
G
H
J
10 11 12 15 88 16
ILLINOIS
INDIANA
OHIO
PENNSYLVANIA
MARYLAND
DELAWARE
WEST VIRGINIA
VIRGINIA
KENTUCKY
TENNESSEE
NORTH CAROLINA
SOUTH CAROLINA
GEORGIA
ALABAMA
MISSISSIPPI
FLORIDA
BAHAMAS
PITTSBURGH
PHILADELPHIA
COLUMBUS
INDIANAPOLIS
CINCINNATI
BALTIMORE
WASHINGTON D.C.
NORFOLK
MEMPHIS
ATLANTA
NEW ORLEANS
JACKSONVILLE
ORLANDO
TAMPA
MIAMI
Nashville
Louisville
Charlotte
Richmond
Montgomery
Tallahassee
Columbia
Raleigh
Birmingham
Appalachian Mountains
Cumberland Plateau
Blue Ridge
ATLANTIC OCEAN
GULF OF MEXICO
Florida Keys
Str. of Florida
Key West
Dry Tortugas
Lake Okeechobee
Chesapeake Bay
Delaware Bay
Pamlico Sound
Albemarle Sd.
Sea Islands
COPYRIGHT PHILIP'S

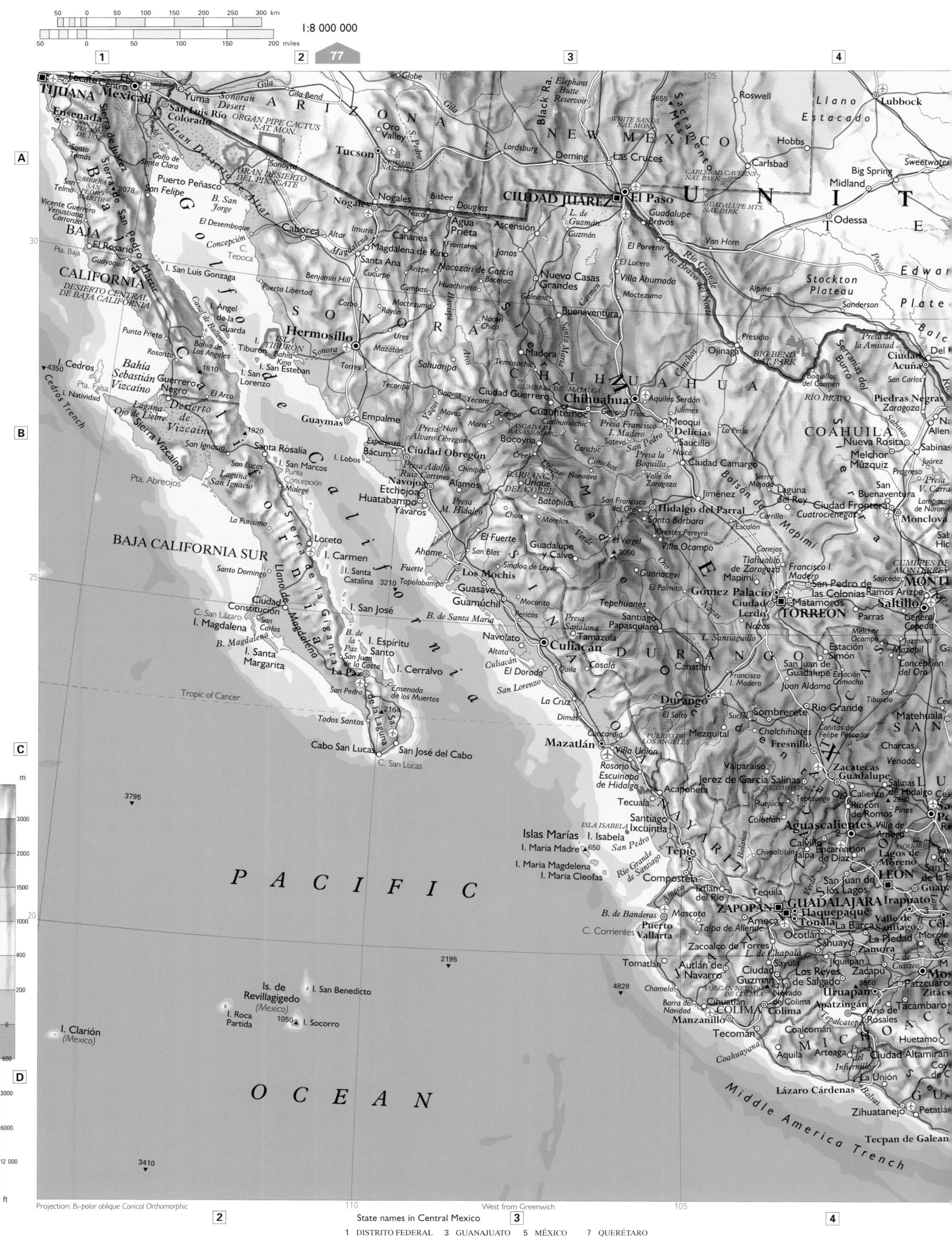

50 0   50 100 150 200 250 300 km

1:8 000 000

50 0   50 100 150 200 miles

77

PACIFIC

OCEAN

State names in Central Mexico

1 DISTRITO FEDERAL 3 GUANAJUATO 5 MÉXICO 7 QUERÉTARO
2 AGUASCALIENTES   4 HIDALGO   6 MORELOS 8 TLAXCALA

Projection: Bi-polar oblique Conical Orthomorphic

West from Greenwich

5 6 85 7 8

A B C D E

GULF OF MEXICO

Sigsbee Deep

Golfo de Campeche

Golfo de Tehuantepec

Canal de Yucatán

Yucatan Basin

Banco Campeche

Tropic of Cancer

UNITED STATES

TEXAS

ARKANSAS

LOUISIANA

MISSISSIPPI

ALABAMA

GEORGIA

FLORIDA

Dallas
Fort Worth
Austin
Houston
San Antonio
Corpus Christi
Shreveport
Jackson
Montgomery
Tallahassee
Mobile
Pensacola
New Orleans
Baton Rouge
Galveston
Beaumont
Port Arthur
Lake Charles
Lafayette
Biloxi
Gulfport
Clearwater
Apalachee Bay
Mississippi River Delta
Padre Island Nat. Seashore
Laguna Madre

MEXICO

McAllen
Reynosa
Matamoros
Brownsville
Nuevo Laredo
Tampico
Ciudad Madero
Tuxpan
Poza Rica
Xalapa
Veracruz
Puebla
Oaxaca
Acapulco
Coatzacoalcos
Villahermosa
Campeche
Mérida
Progreso
Cancún
Cozumel
Chetumal
Ciudad del Carmen
Tuxtla Gutiérrez
Tapachula
Salina Cruz
Istmo de Tehuantepec

TAMAULIPAS
VERACRUZ
PUEBLA
OAXACA
CHIAPAS
TABASCO
CAMPECHE
YUCATÁN
QUINTANA ROO

CUBA

BELIZE
Belmopan
Belize City

GUATEMALA
Guatemala
Flores

HONDURAS
Tegucigalpa
San Pedro Sula
La Ceiba

Golfo de Honduras

Barrier Reef

Ambergris Cay

Banco Chinchorro

Isla Mujeres

I. Desterrada
I. Pérez (Mexico)

Sierra Madre de Chiapas

88
88

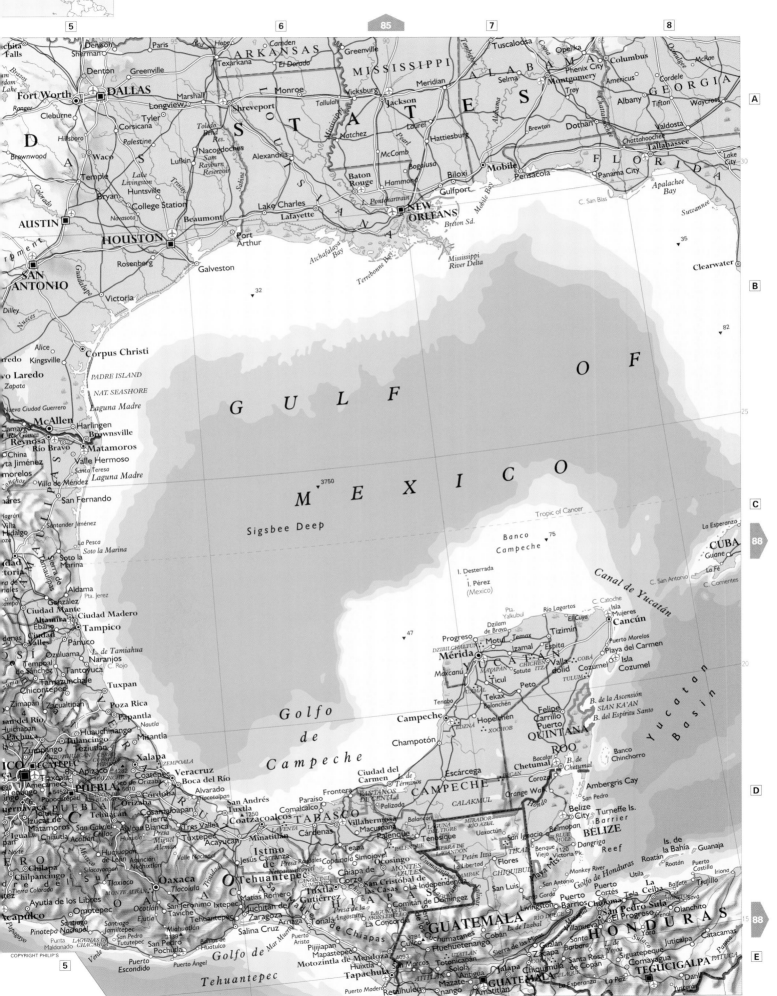

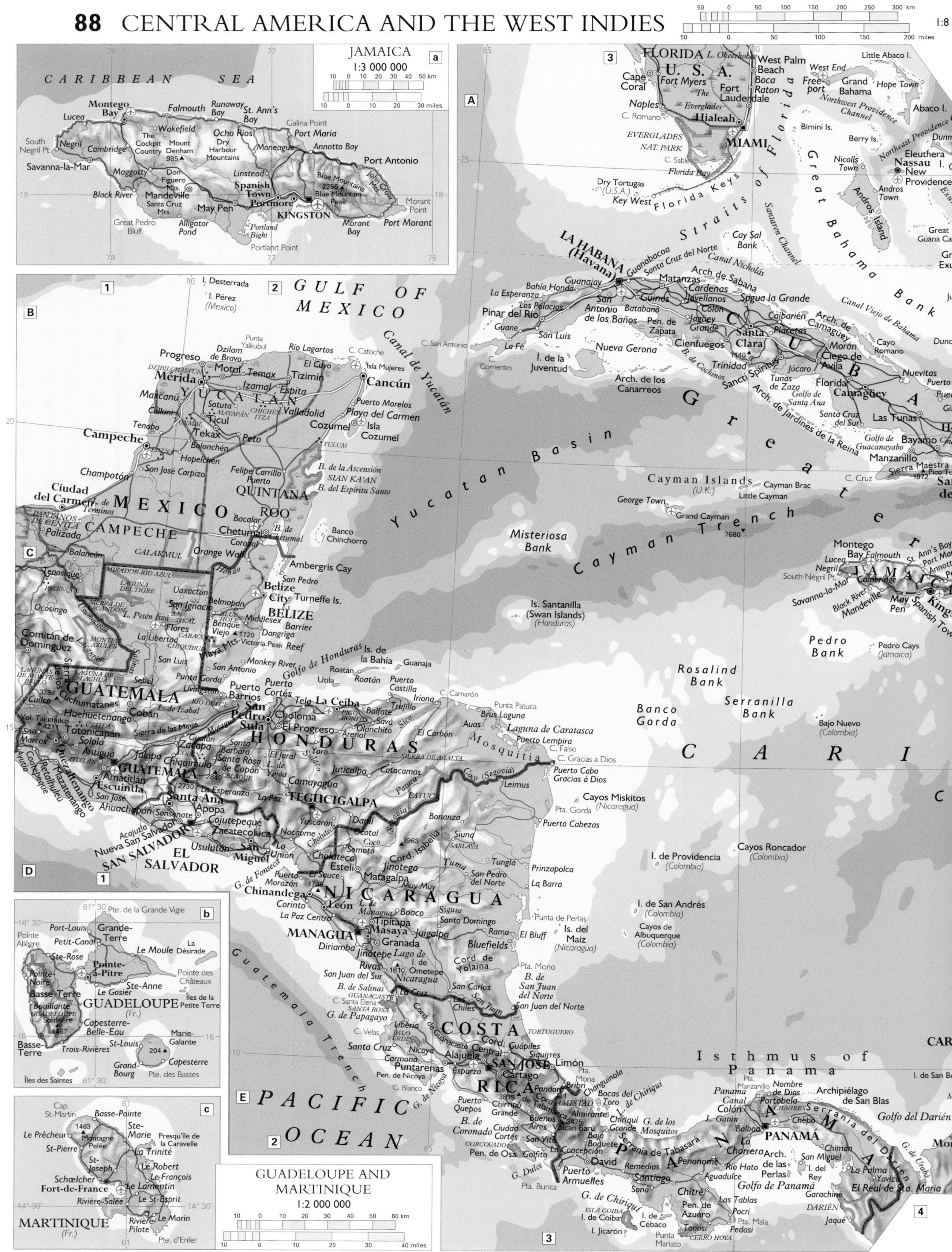

PUERTO RICO d
1:3 000 000
10 0 10 20 30 40 50 km
10 0 10 20 30 miles

VIRGIN ISLANDS e
1:2 000 000
10 0 10 20 30 km
10 0 10 20 miles

ATLANTIC OCEAN

PUERTO RICO
(U.S.A.)

Pta. Agujereada
Isabela
Aguadilla
Arecibo
Barceloneta
Manati
Vega Baja
Bayamón
SAN JUAN
Rio Grande
Dewey
Mayagüez
San Sebastián
Utuado
Carolina
Sierra de Luquillo
Fajardo
Pta. Puerca
Culebra
Adjuntas
Cordillera Central
Caguas
Naguabo
Vieques
San German
Mts. de Uroyan
Cerro 1338 de Punta
Yauco
Cayey
Coamo
Humacao
Yabucoa
Esperanza
Ponce
Guanica
Guayama
Pta. Aguila
I. Caja de Muertos

Rufling Pt.
The Settlement
Anegada
East Pt.
Virgin Islands
(U.K.)
Jost Van Dyke I.
Great Camanoe
Haps
Guana I.
Lollik I.
Tortola
Beef I.
Virgin Gorda
Virgin Is.
(U.S.A.)
Cruz Bay
Road Town
Spanish Town
Charlotte Amalie
St. John I.
Peter I.
VIRGIN IS.
St. Thomas I.

ST. LUCIA f
1:1 000 000
5 0 10 km
5 0 10 miles

Cap Point
Gros Islet
Pte. Hardy
Esperance Bay
Castries
Marquis
Girard
Anse la Raye
Dennery
Canaries
Millet
Soufrière
Mt. Gimie 950
750 Petit Piton
Trou Gras Pt.
Micoud
Soufrière Bay
796 Gros Piton
Vierge Pt.
Gros Piton Pt.
Choiseul
Laborie
Vieux Fort
C. Moule à Chique

ATLANTIC OCEAN
Crab Hill
North Point
Spring Hall
Fustic
Boscobelle
Portland
245 Belleplaine
Speightstown
Westmoreland
Bathsheba
Hillcrest
Alleynes Bay
Mt. Hillaby 840
Martin's Bay
Holetown
Bridgefield
Massiah Street
Jackson
BARBADOS
Black Rock
Ellerton
Six Cross Roads
Bridgetown
Ivy
Edey
The Crane
Carlisle Bay
Oistins
St. Martins
Worthing
Oistins Bay
Chancery Lane
South Point
BGI

BARBADOS g
1:1 000 000
5 0 10 km
5 0 10 miles

ATLANTIC OCEAN

Town
Bight
at I.
vador I.
Conception I.
Rum Cay
Long I.
Tropic of Cancer
Clarence Town
Samana Cay
Crooked I. Passage
Crooked I.
Plana Cays
Albert Town
Snug Corner
Mayaguana I.
Acklins I.
Mira por vos Cay
Turks & Caicos Is.
(U.K.)
Hogsty Reef
Caicos Is.
Little Inagua I.
Cockburn Town
INAGUA
Turks Is.
Lake Rose
Great Inagua I.
Matthew Town
Mouchoir Bank
Silver Bank
Navidad Bank
Moa
Baracoa
Pta. de Maisi
Monte Cristi
LA ISABELA
Santiago de los Caballeros
San Francisco de Macoris
Milwaukee Deep 9200
Puerto Rico Trench
ANAMO
Maisi
Cap-Haitien
Puerto Plata
La Vega
Nagua
Samana
7
Paso de los Vientos (Windward Passage)
Jean Rabel
Port-de-Paix
Fort Liberté
Cord.
3175
Sánchez
Los
Sabana de la Mar
Bayamón
SAN JUAN
Cap-à-Foux
Gonaïves
Hinche
Central
San Juan
Hato Mayor
Carolina
St. Thomas
Virgin Gorda
Anegada
Virgin Is.
Sombrero (U.K.)
G. de la Gonâve
Higuey
C. Engaño
Aguadilla
Arecibo
Fajardo
Road Town
Anguilla (U.K.)
HAITI
DOMINICAN
San Pedro de Macoris
1338
Culebra
Charlotte Amalie
St.-Martin (Fr.)
Jérémie
Î. de la Gonâve
PORT-AU-PRINCE
REP.
Mayagüez
Ponce
Vieques
St.-Barthélemy (Fr.)
Petit Goâve
2680
SANTO
Isla Mona
Guayama
St. Maarten (Neth.)
Massif de la Hotte
Jacmel
SIERRA DE BAORUCO
DOMINGO
ESTE
B. de Yuma
(U.S.A.)
Frederiksted
Saba (Neth.)
Barbuda
Les Cayes
Aquin
Goâve
San Cristóbal
PUERTO
St. Croix
St. Eustatius
Mt. Liamuiga 1156
Pointe-à-Gravois
î. à Vache
Pedernales
Bani
Barahona
RICO
Christiansted
(U.S.A.)
(Neth.)
Basseterre
ST. KITTS & NEVIS
ANTIGUA & BARBUDA
L. Enriquillo
(U.S.A.)
Nevis
St. John's
I. Beata
C. Beata
Redonda
Soufrière Hills
Antigua
Hispaniola
Brades 914
Montserrat (U.K.)
Guadeloupe Passage
Ste.-Rose
Le Moule
La Désirade
A n t i l l e s
Beata Ridge
Ste.-Rose 1467
La Désirade
GUADELOUPE
Pointe-à-Pitre
Marie-Galante (Fr.)
(Fr.)
Basse-Terre
Grand-Bourg
I. des Saintes (Fr.)
1419
Dominica Passage
Portsmouth
Morne
DOMINICA
Diablotin
MORNE TROIS PITONS
Roseau
Martinique Passage
I. de Aves
(Venezuela)
Mt. Pelée 1397
Ste.-Marie
V e n e z u e l a n
Le François
S E A
Rivière-Pilote
Fort-de-France
MARTINIQUE
Basin
St. Lucia Channel (Fr.)
BEAN
Castries 950
ST. LUCIA
Soufrière
St. Vincent Passage
Soufrière 1234
St. Vincent
Speightstown 340
Kingstown
BARBADOS
ombian
Bequia
Bridgetown
Basin
Canouan
ST. VINCENT & THE GRENADINES
Lesser
ABC
Carriacou
The Grenadines
Islands
840
St. George's
GRENADA
Oranjestad
Aruba
Curaçao
Bonaire
I. Blanquilla (Ven.)
Tobago
(Neth.)
(Neth.)
(Neth.)
Is. Los Hermanos (Ven.)
Scarborough
Pta. Gallinas
ARC. LOS ROQUES
I. Orchila
NUEVA
Is. Los Testigos (Ven.)
Port of Spain 940
COLOMBIA
C. San Román
Is. Las Aves
Is. Los Roques
(Ven.)
ESPARTA
Trinidad
Pen. de la Guajira
Paraguaná
(Ven.)
(Ven.)
I. de Margarita
Arima
Santa Marta
Ríohacha
Uribia
Punta Cardón
I. La Tortuga (Ven.)
La Asunción
Rio Claro
TAYRONA
GUAJIRA
Golfo de Venezuela
Puerto Cumarebo
CERRO EL COPEY
Porlamar
Guiria
SA. NEVADA DE STA. MARTA
Punta Espada
MÉDANOS DE CORO
LAGUNA LA RESTINGA
TRINIDAD & TOBAGO
Cienaga
Santa Marta 5775
San Rafael
Coro
La Vela
MAIQUETIA
Carúpano
G. de Paria
San Fernando
Soledad
Altagracia
FALCÓN
CUEVA DE LA QUEBRADA
CARACAS
Cariaco
Sabanalarga
Mene de Mauroa
Tucacas
HENRI PITTIER
La Guaira
I. La Tortuga
SUCRE
Fundación
La Concepción
Santa Rita
Baragua
Puerto Cabello
MARACAY
MIRANDA
Cumaná
TURIMIQUIRE 2640
EL GUACHARO
Magdalena
MARACAIBO
Cabimas
LARA
CARABOBO
Higuerote
Puerto La Cruz
Caripito
Valledupar
CESAR
Ciudad Ojeda
CERRO SAROARE
YARACUY
VALENCIA
ARAGUA
Los Teques
Rio Chico
Barcelona
Maturín
CESAR
ZULIA
BARQUISIMETO
Villa de Cura
San Juan de los Morros
Anaco
MONAGAS
Machiques
Lago de Maracaibo
Carora
Yaritagua
El Sombrero
Valle de la Pascua
Cantaura
MARIUSA DELTA
SIERRA DE PERIJÁ
TRUJILLO
DESPENA
San Carlos
COJEDES
Aragua de Orituco
El Tigre
Tucupita
Perijá
Betijoque
Valera
El Guache
Calabozo
G U A R I C O
Pariaguán
AMACURO
El Banco
San Carlos del Zulia
PORTUGUESA
El Baúl
Santa Maria de Ipire
Los Barrancos
COLOMBIA
NORTE DE
Trujillo
Guanare
Soledad
Ciudad Guayana
Ocaña
Mérida
Barinas
Libertad
ANZOÁTEGUI
El Pao
Sierra Imataca
SANTANDER
MÉRIDA
AGUARO-GUARIQUITO
CORD. DE MÉRIDA
BARINAS
San Fernando de Apure
Ciudad Bolivia
Ciudad Bolívar
Upata
BOLÍVAR
Simití
Cúcuta
TÁCHIRA
V E N E Z U E L A
Bruzual
Mapire
Guasipati
El Callao
Caicara
Embalse de Guri
Tumeremo

West from Greenwich
COPYRIGHT PHILIP'S
92
5 6 7
3000 2000 1500 1000 400 200 0
6000 12 000 18 000 24 000 ft
6000 4500 3000 1200 600 0 200 2000 4000 6000 8000 m

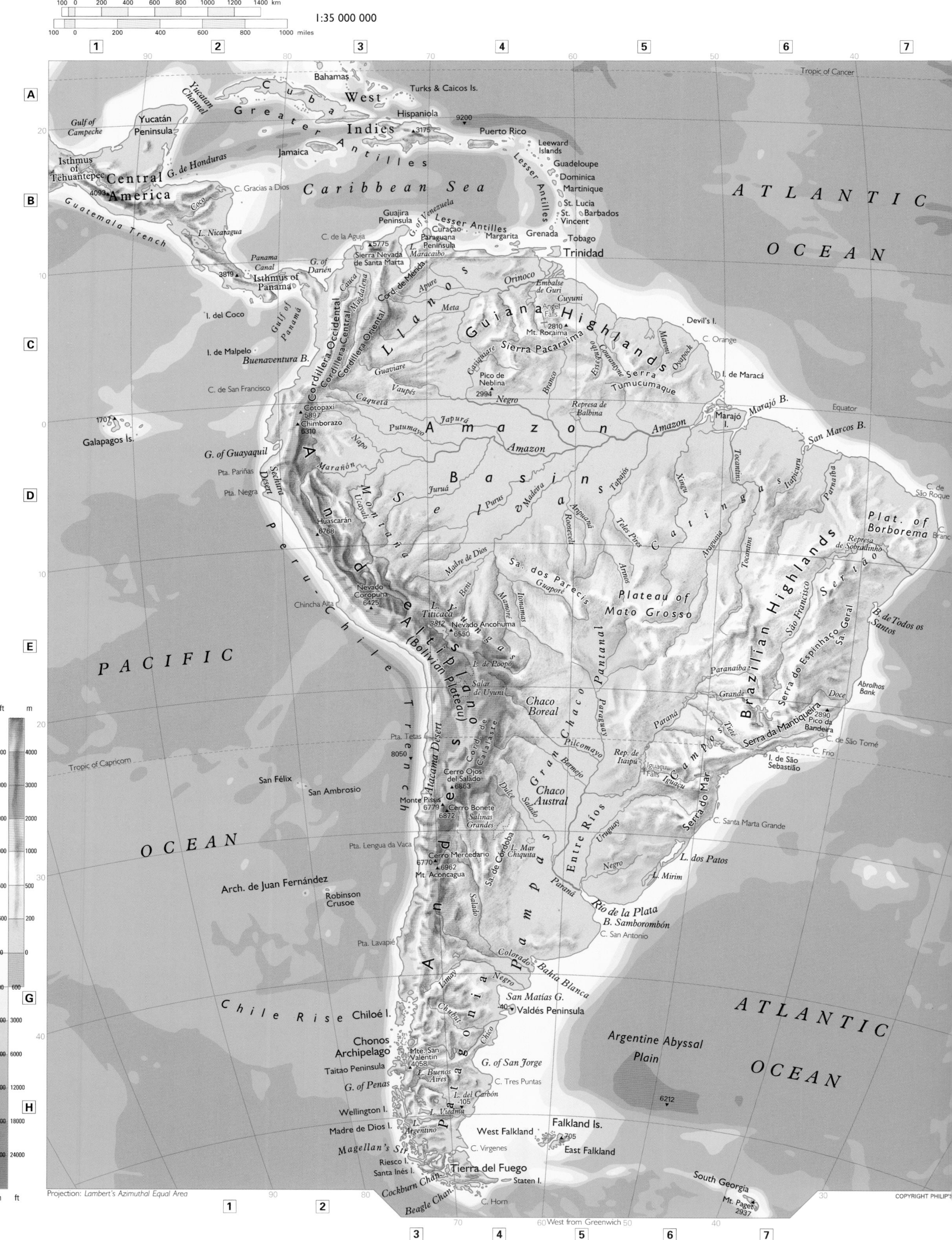

1:35 000 000

ATLANTIC OCEAN
PACIFIC OCEAN
Caribbean Sea
MEXICO
CUBA
BAHAMAS
HAITI
DOMINICAN REP.
JAMAICA
BELIZE
GUATEMALA
HONDURAS
EL SALVADOR
NICARAGUA
COSTA RICA
PANAMA
VENEZUELA
COLOMBIA
GUYANA
SURINAME
FRENCH GUIANA
ECUADOR
PERU
BOLIVIA
BRAZIL
PARAGUAY
URUGUAY
CHILE
ARGENTINA
FALKLAND IS. (U.K.)
Galapagos Is. (Ecuador)
Arch. de Juan Fernández (Chile)
South Georgia (U.K.)
Tropic of Cancer
Equator
Tropic of Capricorn
Projection: Lambert's Azimuthal Equal Area
COPYRIGHT PHILIP'S
LIMA Capital Cities

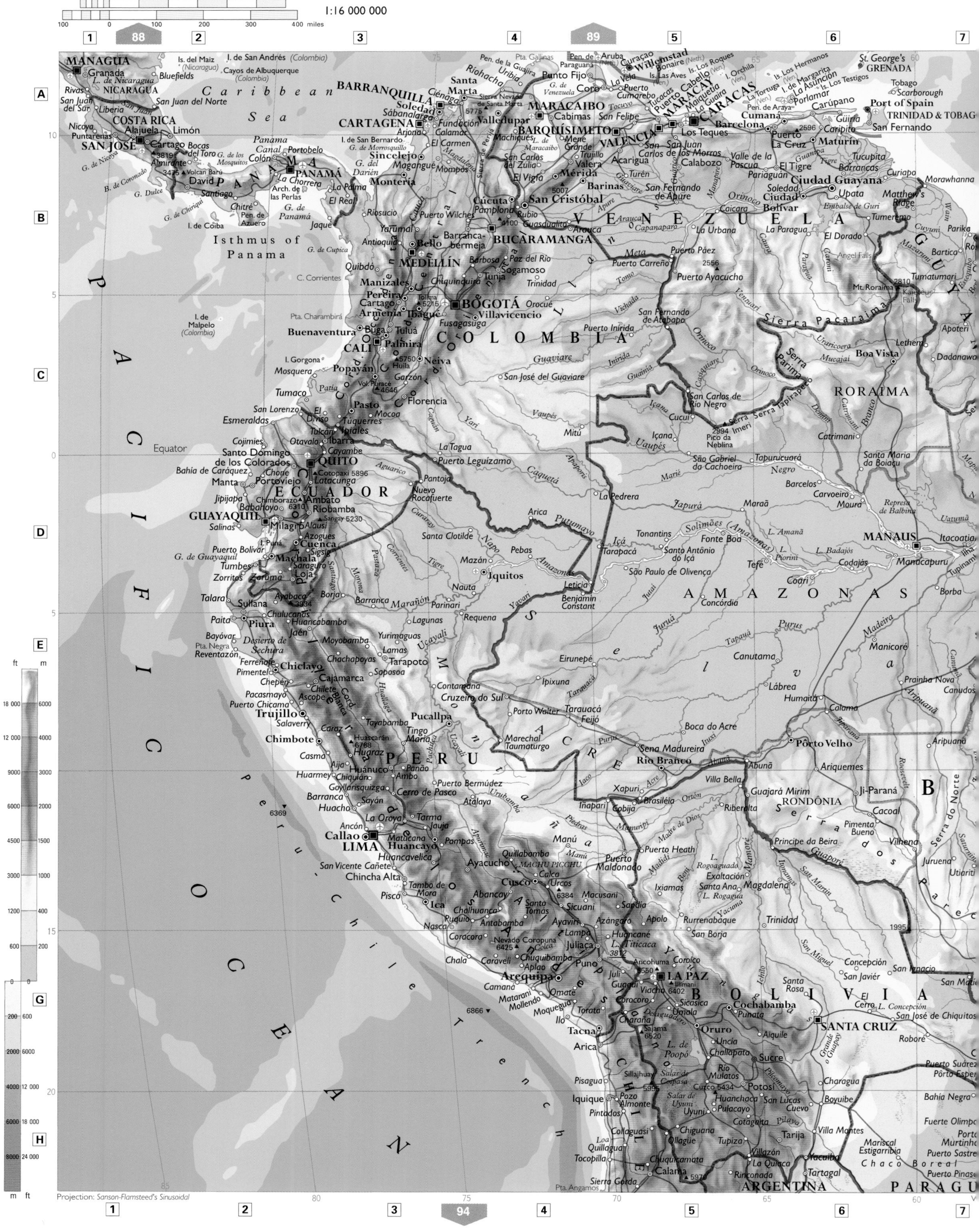

1:16 000 000

Projection: Sanson-Flamsteed's Sinusoidal

TRINIDAD AND TOBAGO
1:2 500 000
ATLANTIC OCEAN
Tobago
Trinidad
Port of Spain
San Fernando
Golfo de Paria
VENEZUELA
Paramaribo
Cayenne
FRENCH GUIANA
AMAPÁ
Macapá
Equator
BELÉM
São Luís
FORTALEZA
Teresina
MARANHÃO
PARÁ
CEARÁ
PIAUÍ
NATAL
João Pessoa
RECIFE
MACEIÓ
Aracaju
TOCANTINS
BAHIA
SALVADOR
BRASÍLIA
GOIÂNIA
GOIÁS
MINAS GERAIS
BELO HORIZONTE
VITÓRIA
RIO DE JANEIRO
CAMPINAS
Fernando de Noronha (Braz.)
Atol das Rocas (Braz.)
São Pedro & São Paulo (Braz.)
Trindade (Braz.)
Martin Vaz (Braz.)
Banco dos Abrolhos
COPYRIGHT PHILIP'S

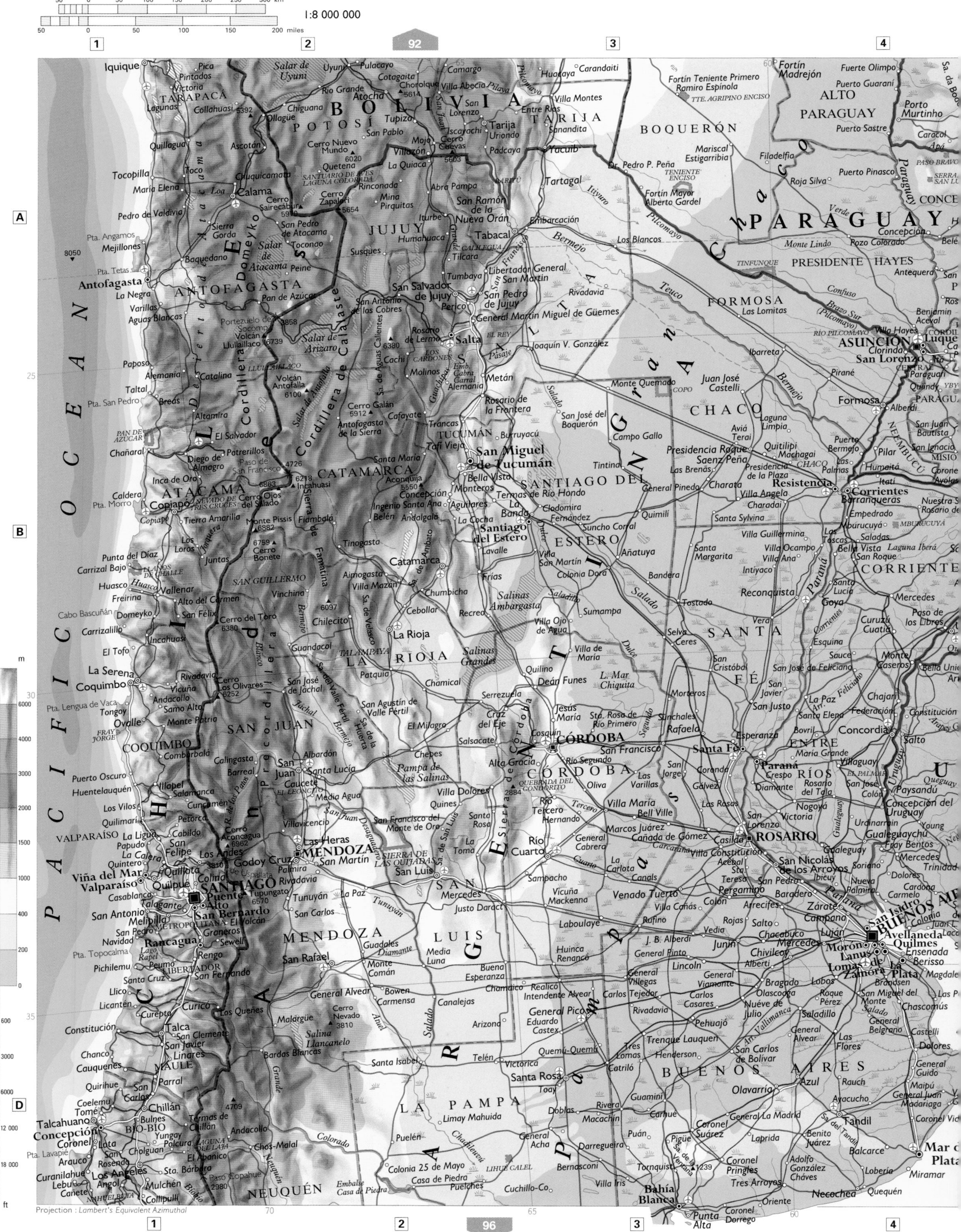

Projection : Lambert's Equivalent Azimuthal

1:8 000 000

93

ATLANTIC OCEAN
BRAZIL
SÃO PAULO
RIO DE JANEIRO
BELO HORIZONTE
CURITIBA
PORTO ALEGRE
Florianópolis
Tropic of Capricorn
5
6
7
A
B
C
D
West from Greenwich
55
50
45
40
25
30
35

1:16 000 000

Projection: Sanson-Flamsteed's Sinusoidal

West from Greenwich

COPYRIGHT PHILIP

Countries / Regions

PARAGUAY
ASUNCIÓN
URUGUAY
MONTEVIDEO
BUENOS AIRES
SANTIAGO
CÓRDOBA
ROSARIO
MENDOZA
SÃO PAULO
CURITIBA
PORTO ALEGRE
RIO DE JANEIRO

Oceans

PACIFIC OCEAN
ATLANTIC OCEAN
Argentine Abyssal Plain

Selected place names

Antofagasta, Mejillones, Taltal, Chañaral, Caldera, Copiapó, Carrizal Bajo, Huasco, Vallenar, La Serena, Coquimbo, Ovalle, Illapel, Los Vilos, Viña del Mar, Valparaíso, San Antonio, San Bernardo, Rancagua, Pichilemu, San Fernando, Curicó, Talca, Linares, Cauquenes, San Carlos, Chillán, Talcahuano, Concepción, Coronel, Lota, Los Ángeles, Lebu, Cañete, Temuco, Valdivia, La Unión, Osorno, Puerto Varas, Puerto Montt, Ancud, Castro, Quellón, I. de Chiloé, Coihaique, Balmaceda, Puerto Aisén

Calama, Chuquicamata, Salta, Jujuy, San Salvador de Jujuy, San Miguel de Tucumán, Catamarca, Santiago del Estero, La Rioja, San Juan, Mendoza, Godoy Cruz, San Luis, Río Cuarto, San Rafael, Neuquén, General Roca, Bahía Blanca, Mar del Plata, Necochea, Tandil, Azul, Olavarría, Santa Rosa, Río Gallegos, Puerto Santa Cruz, Puerto San Julián, Comodoro Rivadavia, San Jorge, Sarmiento, Perito Moreno, Esquel, Trelew, Rawson, Puerto Madryn, Pen. Valdés, Viedma, Carmen de Patagones

ASUNCIÓN, Concepción, Pedro Juan Caballero, Villa Hayes, Formosa, Resistencia, Corrientes, Posadas, Paraná, Santa Fe, Rafaela, Gualeguay, Concordia, Salto, Paysandú, Mercedes, Durazno, Melo, Rivera, Treinta y Tres, Rocha, Maldonado, La Plata, Avellaneda

São Paulo, Santos, Curitiba, Paranaguá, Joinville, Blumenau, Florianópolis, Lages, Caxias do Sul, Novo Hamburgo, Canoas, Porto Alegre, Pelotas, Rio Grande, Ribeirão Preto, Araraquara, Campinas, Sorocaba, Jundiaí, Marília, Bauru, Londrina, Maringá, Foz do Iguaçu, Cascavel, Ponta Grossa, Rio de Janeiro

FALKLAND ISLANDS (ISLAS MALVINAS) (U.K.)
West Falkland, East Falkland, Stanley, Port Darwin, Weddell I.
South Georgia (U.K.), Grytviken, Mt. Paget 2934, Bird I., King Edward Pt.

Tierra del Fuego, Río Grande, Ushuaia, Punta Arenas, Porvenir, San Sebastián, Isla Grande de Tierra del Fuego, I. de los Estados (Staten I.), C. Horn (Hornos)

Estrecho de Magallanes (Magellan's Str.), Canal Beagle, Cerro Fitz Roy 3405, Cerro Murallón 2831, Cerro Paine 2360, El Calafate, Puerto Natales

Cerro Aconcagua 6960, Cerro Ojos del Salado 6863, Cerro Mercedario

Tropic of Capricorn

INDEX TO WORLD MAPS

The index contains the names of all the principal places and features shown on the World Maps. Each name is followed by an additional entry in italics giving the country or region within which it is located. The alphabetical order of names composed of two or more words is governed primarily by the first word, then by the second, and then by the country or region name that follows. This is an example of the rule:

Mīr Kūh *Iran*	26°22N 58°55E	**45** E8
Mīr Shahdād *Iran*	26°15N 58°29E	**45** E8
Mira *Italy*	45°26N 12°8E	**22** B5
Mira por vos Cay *Bahamas*	22°9N 74°30W	**89** B5

Physical features composed of a proper name (Erie) and a description (Lake) are positioned alphabetically by the proper name. The description is positioned after the proper name and is usually abbreviated:

Erie, L. *N. Amer.*	42°15N 81°0W	**82** D4

Where a description forms part of a settlement or administrative name, however, it is always written in full and put in its true alphabetical position:

Mount Morris *U.S.A.*	42°44N 77°52W	**82** D7

Names beginning with M' and Mc are indexed as if they were spelled Mac. Names beginning St. are alphabetized under Saint, but Sankt, Sint, Sant', Santa and San are all spelt in full and are alphabetized accordingly. If the same place name occurs two or more times in the index and all are in the same country, each is followed by the name of the administrative subdivision in which it is located.

The geographical co-ordinates which follow each name in the index give the latitude and longitude of each place. The first co-ordinate indicates latitude – the distance north or south of the Equator. The second co-ordinate indicates longitude – the distance east or west of the Greenwich Meridian. Both latitude and longitude are measured in degrees and minutes (there are 60 minutes in a degree).

The latitude is followed by N(orth) or S(outh) and the longitude by E(ast) or W(est).

The number in bold type which follows the geographical co-ordinates refers to the number of the map page where that feature or place will be found. This is usually the largest scale at which the place or feature appears.

The letter and figure that are immediately after the page number give the grid square on the map page, within which the feature is situated. The letter represents the latitude and the figure the longitude. A lower-case letter immediately after the page number refers to an inset map on that page.

In some cases the feature itself may fall within the specified square, while the name is outside. This is usually the case only with features that are larger than a grid square.

Rivers are indexed to their mouths or confluences, and carry the symbol ➜ after their names. The following symbols are also used in the index: ■ country, ☑ overseas territory or dependency, □ first-order administrative area, △ national park, ⌂ other park (provincial park, nature reserve or game reserve), ✈ (LHR) principal airport (and location identifier), ◎ Australian aboriginal land.

Abbreviations used in the index

A.C.T. – Australian Capital Territory
A.R. – Autonomous Region
Afghan. – Afghanistan
Afr. – Africa
Ala. – Alabama
Alta. – Alberta
Amer. – America(n)
Ant. – Antilles
Arch. – Archipelago
Ariz. – Arizona
Ark. – Arkansas
Atl. Oc. – Atlantic Ocean
B. – Baie, Bahía, Bay, Bucht, Bugt
B.C. – British Columbia
Bangla. – Bangladesh
Barr. – Barrage
Bos.-H. – Bosnia-Herzegovina
C. – Cabo, Cap, Cape, Coast
C.A.R. – Central African Republic
C. Prov. – Cape Province
Calif. – California
Cat. – Catarata
Cent. – Central
Chan. – Channel
Colo. – Colorado
Conn. – Connecticut
Cord. – Cordillera
Cr. – Creek
Czech. – Czech Republic
D.C. – District of Columbia
Del. – Delaware
Dem. – Democratic
Dep. – Dependency
Des. – Desert
Dét. – Détroit
Dist. – District
Dj. – Djebel
Dom. Rep. – Dominican Republic

E. – East
El Salv. – El Salvador
Eq. Guin. – Equatorial Guinea
Est. – Estrecho
Falk. Is. – Falkland Is.
Fd. – Fjord
Fla. – Florida
Fr. – French
G. – Golfe, Golfo, Gulf, Guba, Gebel
Ga. – Georgia
Gt. – Great, Greater
Guinea-Biss. – Guinea-Bissau
H.K. – Hong Kong
H.P. – Himachal Pradesh
Hants. – Hampshire
Harb. – Harbor, Harbour
Hd. – Head
Hts. – Heights
I.(s). – Île, Ilha, Insel, Isla, Island, Isle
Ill. – Illinois
Ind. – Indiana
Ind. Oc. – Indian Ocean
Ivory C. – Ivory Coast
J. – Jabal, Jebel
Jaz. – Jazīrah
Junc. – Junction
K. – Kap, Kapp
Kans. – Kansas
Kep. – Kepulauan
Ky. – Kentucky
L. – Lac, Lacul, Lago, Lagoa, Lake, Limni, Loch, Lough
La. – Louisiana
Ld. – Land
Liech. – Liechtenstein
Lux. – Luxembourg
Mad. P. – Madhya Pradesh
Madag. – Madagascar
Man. – Manitoba
Mass. – Massachusetts

Md. – Maryland
Me. – Maine
Medit. S. – Mediterranean Sea
Mich. – Michigan
Minn. – Minnesota
Miss. – Mississippi
Mo. – Missouri
Mont. – Montana
Mozam. – Mozambique
Mt.(s) – Mont, Montaña, Mountain
Mte. – Monte
Mti. – Monti
N. – Nord, Norte, North, Northern, Nouveau, Nahal, Nahr
N.B. – New Brunswick
N.C. – North Carolina
N. Cal. – New Caledonia
N. Dak. – North Dakota
N.H. – New Hampshire
N.I. – North Island
N.J. – New Jersey
N. Mex. – New Mexico
N.S. – Nova Scotia
N.S.W. – New South Wales
N.W.T. – North West Territory
N.Y. – New York
N.Z. – New Zealand
Nac. – Nacional
Nat. – National
Nebr. – Nebraska
Neths. – Netherlands
Nev. – Nevada
Nfld & L. – Newfoundland and Labrador
Nic. – Nicaragua
O. – Oued, Ouadi
Occ. – Occidentale
Okla. – Oklahoma
Ont. – Ontario
Or. – Orientale

Oreg. – Oregon
Os. – Ostrov
Oz. – Ozero
P. – Pass, Passo, Pasul, Pulau
P.E.I. – Prince Edward Island
Pa. – Pennsylvania
Pac. Oc. – Pacific Ocean
Papua N.G. – Papua New Guinea
Pass. – Passage
Peg. – Pegunungan
Pen. – Peninsula, Péninsule
Phil. – Philippines
Pk. – Peak
Plat. – Plateau
Prov. – Province, Provincial
Pt. – Point
Pta. – Ponta, Punta
Pte. – Pointe
Qué. – Québec
Queens. – Queensland
R. – Rio, River
R.I. – Rhode Island
Ra. – Range
Raj. – Rajasthan
Recr. – Recreational, Récréatif
Reg. – Region
Rep. – Republic
Res. – Reserve, Reservoir
Rhld-Pfz. – Rheinland-Pfalz
S. – South, Southern, Sur
Si. Arabia – Saudi Arabia
S.C. – South Carolina
S. Dak. – South Dakota
S.I. – South Island
S. Leone – Sierra Leone
Sa. – Serra, Sierra
Sask. – Saskatchewan
Scot. – Scotland
Sd. – Sound
Sev. – Severnaya
Sib. – Siberia

Sprs. – Springs
St. – Saint
Sta. – Santa
Ste. – Sainte
Sto. – Santo
Str. – Strait, Stretto
Switz. – Switzerland
Tas. – Tasmania
Tenn. – Tennessee
Terr. – Territory, Territoire
Tex. – Texas
Tg. – Tanjung
Trin. & Tob. – Trinidad & Tobago
U.A.E. – United Arab Emirates
U.K. – United Kingdom
U.S.A. – United States of America
Ut. P. – Uttar Pradesh
Va. – Virginia
Vdkhr. – Vodokhranilishche
Vdskh. – Vodoskhovyshche
Vf. – Vírful
Vic. – Victoria
Vol. – Volcano
Vt. – Vermont
W. – Wadi, West
W. Va. – West Virginia
Wall. & F. Is. – Wallis and Futuna Is.
Wash. – Washington
Wis. – Wisconsin
Wlkp. – Wielkopolski
Wyo. – Wyoming
Yorks. – Yorkshire

A

A Coruña *Spain* 43°20N 8°25W **21** A1
A Estrada *Spain* 42°43N 8°27W **21** A1
A Fonsagrada *Spain* 43°8N 7°4W **21** A2
A Shau *Vietnam* 16°6N 107°22E **38** D6
Aabenraa *Denmark* 55°3N 9°25E **9** J13
Aachen *Germany* 50°45N 6°6E **16** C4
Aalborg *Denmark* 57°2N 9°54E **9** H13
Aalen *Germany* 48°51N 10°6E **16** D6
Aalst *Belgium* 50°56N 4°2E **15** D4
Aalten *Neths.* 51°56N 6°35E **15** C6
Aalter *Belgium* 51°5N 3°28E **15** C3
Äänekoski *Finland* 62°36N 25°44E **8** E21
Aarau *Switz.* 47°23N 8°4E **20** C8
Aare → *Switz.* 47°33N 8°14E **20** C8
Aarhus = Århus *Denmark* 56°8N 10°11E **9** H14
Aarschot *Belgium* 50°59N 4°49E **15** D4
Aba *Dem. Rep. of the Congo* 3°58N 30°17E **54** B3
Aba *Nigeria* 5°10N 7°19E **50** G7
Abaco I. *Bahamas* 26°25N 77°10W **88** A4
Ābādān *Iran* 30°22N 48°20E **45** D6
Ābādeh *Iran* 31°8N 52°40E **45** D7
Abadla *Algeria* 31°2N 2°45W **50** B5
Abaetetuba *Brazil* 1°40S 48°50W **93** D9
Abagnar Qi = Xilinhot *China* 43°52N 116°2E **34** C9
Abah, Tanjung *Indonesia* 8°46S 115°38E **37** K18
Abai *Paraguay* 25°58S 55°54W **95** B4
Abakan *Russia* 53°40N 91°10E **29** D10
Abancay *Peru* 13°35S 72°55W **92** F4
Abang, Gunung *Indonesia* 8°16S 115°25E **37** J18
Abariringa *Kiribati* 2°50S 171°40W **64** H10
Abarqū *Iran* 31°10N 53°20E **45** D7
Abashiri *Japan* 44°0N 144°15E **30** B12
Abashiri-Wan *Japan* 44°0N 144°30E **30** C12
Ābay = Nîl el Azraq → *Sudan* 15°38N 32°31E **51** E12
Abay *Kazakhstan* 49°38N 72°53E **28** E8
Abaya, L. *Ethiopia* 6°30N 37°50E **47** F2
Abaza *Russia* 52°39N 90°6E **28** D10
'Abbāsābād *Iran* 33°34N 58°23E **45** C8
Abbay = Nîl el Azraq → *Sudan* 15°38N 32°31E **51** E12
Abbaye, Pt. *U.S.A.* 46°58N 88°8W **80** B9
Abbé, L. *Ethiopia* 11°8N 41°47E **47** E3
Abbeville *France* 50°6N 1°49E **20** A4
Abbeville *Ala., U.S.A.* 31°34N 85°15W **85** F12
Abbeville *La., U.S.A.* 29°58N 92°8W **84** G8
Abbeville *S.C., U.S.A.* 34°11N 82°23W **85** D13
Abbeyfeale *Ireland* 52°23N 9°18W **10** D2
Abbot Ice Shelf *Antarctica* 73°0S 92°0W **5** D16
Abbotsford *Canada* 49°5N 122°20W **70** D4
Abbottabad *Pakistan* 34°10N 73°15E **42** B5
ABC Islands *W. Indies* 12°15N 69°0W **89** D6
Abd al Kūrī *Yemen* 12°5N 52°20E **47** E5
Ābdar *Iran* 30°16N 55°19E **45** D7
'Abdolābād *Iran* 34°12N 56°30E **45** C8
Abdulpur *Bangla.* 24°15N 88°59E **43** G13
Abéché *Chad* 13°50N 20°35E **51** F10
Abel Tasman △ *N.Z.* 40°59S 173°3E **59** D4
Abengourou *Ivory C.* 6°42N 3°27W **50** G5
Åbenrå = Aabenraa *Denmark* 55°3N 9°25E **9** J13
Abeokuta *Nigeria* 7°3N 3°19E **50** G6
Aberaeron *U.K.* 52°15N 4°15W **13** E3
Aberayron = Aberaeron *U.K.* 52°15N 4°15W **13** E3
Aberchirder *U.K.* 57°34N 2°37W **11** D6
Abercorn *Australia* 25°12S 151°5E **63** D5
Aberdare *U.K.* 51°43N 3°27W **13** F4
Aberdare △ *Kenya* 0°22S 36°44E **54** C4
Aberdare Ra. *Kenya* 0°15S 36°50E **54** C4
Aberdeen *Australia* 32°9S 150°56E **63** E5
Aberdeen *Canada* 52°20N 106°8W **71** C7
Aberdeen *China* 22°14N 114°8E **33** a
Aberdeen *S. Africa* 32°28S 24°2E **56** E3
Aberdeen *U.K.* 57°9N 2°5W **11** D6
Aberdeen *Idaho, U.S.A.* 42°57N 112°50W **76** E7
Aberdeen *Md., U.S.A.* 39°31N 76°10W **81** F15
Aberdeen *Miss., U.S.A.* 33°49N 88°33W **85** E10
Aberdeen *S. Dak., U.S.A.* 45°28N 98°29W **80** C4
Aberdeen *Wash., U.S.A.* 46°59N 123°50W **78** D3
Aberdeen, City of □ *U.K.* 57°10N 2°10W **11** D6
Aberdeen L. *Canada* 64°30N 99°0W **68** E12
Aberdeenshire □ *U.K.* 57°17N 2°36W **11** D6
Aberdovey = Aberdyfi *U.K.* 52°33N 4°3W **13** E3
Aberdyfi *U.K.* 52°33N 4°3W **13** E3
Aberfeldy *U.K.* 56°37N 3°51W **11** E5
Aberfoyle *U.K.* 56°11N 4°23W **11** E4
Abergavenny *U.K.* 51°49N 3°1W **13** F4
Abergele *U.K.* 53°17N 3°35W **12** D4
Abernathy *U.S.A.* 33°50N 101°51W **84** E4
Abert, L. *U.S.A.* 42°38N 120°14W **76** E3
Aberystwyth *U.K.* 52°25N 4°5W **13** E3
Abhā *Si. Arabia* 18°0N 42°34E **47** D3
Abhar *Iran* 36°9N 49°13E **45** B6
Abhayapuri *India* 26°24N 90°38E **43** F14
Abidjan *Ivory C.* 5°26N 3°58W **50** G5
Abilene *Kans., U.S.A.* 38°55N 97°13W **80** F5
Abilene *Tex., U.S.A.* 32°28N 99°43W **84** E5
Abingdon *U.K.* 51°40N 1°17W **13** F6
Abingdon *U.S.A.* 36°43N 81°59W **81** G13
Abington Reef *Australia* 18°0S 149°35E **62** B4
Abitau → *Canada* 59°53N 109°3W **71** B7
Abitibi → *Canada* 51°3N 80°55W **72** B3
Abitibi, L. *Canada* 48°40N 79°40W **72** C4
Abkhaz Republic = Abkhazia □ *Georgia* 43°12N 41°5E **19** F7
Abkhazia □ *Georgia* 43°12N 41°5E **19** F7
Abminga *Australia* 26°8S 134°51E **63** D1
Åbo = Turku *Finland* 60°30N 22°19E **9** F20
Abohar *India* 30°10N 74°10E **42** D6
Abomey *Benin* 7°10N 2°5E **50** G6
Abong-Mbang *Cameroon* 4°0N 13°8E **52** D2
Abou-Deïa *Chad* 11°20N 19°20E **51** F9
Aboyne *U.K.* 57°4N 2°47W **11** D6
Abra Pampa *Argentina* 22°43S 65°42W **94** A2
Abraham L. *Canada* 52°15N 116°35W **70** C5
Abreojos, Pta. *Mexico* 26°50N 113°40W **86** B2
Abrolhos, Banco dos *Brazil* 18°0S 38°0W **90** E7
Abrud *Romania* 46°19N 23°5E **17** E12
Absaroka Range *U.S.A.* 44°45N 109°50W **76** D9
Abu *India* 24°41N 72°50E **42** G5
Abū al Abyad *U.A.E.* 24°11N 53°50E **45** E7
Abū al Khaşīb *Iraq* 30°25N 48°0E **44** D5
Abū 'Alī *Si. Arabia* 27°20N 49°27E **45** E6
Abū 'Alī → *Lebanon* 34°25N 35°50E **46** A4
Abu Dhabi = Abū Ẓāby *U.A.E.* 24°28N 54°22E **45** E7
Abū Du'ān *Syria* 36°25N 38°15E **44** B3
Abu el Gaĭn, W. → *Egypt* 29°35N 33°30E **46** F2
Abu Ga'da, W. → *Egypt* 29°15N 32°53E **46** F1
Abū Ḩadrīyah *Si. Arabia* 27°20N 48°58E **45** E6
Abu Hamed *Sudan* 19°32N 33°13E **51** E12
Abū Kamāl *Syria* 34°30N 41°0E **44** C4
Abū Madd, Ra's *Si. Arabia* 24°50N 37°7E **44** E3
Abū Mūsā *U.A.E.* 25°52N 55°3E **45** E7
Abū Qaşr *Si. Arabia* 30°21N 38°34E **44** D3
Abu Shagara, Ras *Sudan* 21°4N 37°19E **51** D13
Abu Simbel *Egypt* 22°18N 31°40E **51** D12
Abū Şukhayr *Iraq* 31°54N 44°30E **44** D5
Abu Zabad *Sudan* 12°25N 29°10E **51** F11
Abū Ẓāby *U.A.E.* 24°28N 54°22E **45** E7
Abū Zeydābād *Iran* 33°54N 51°45E **45** C6
Abuja *Nigeria* 9°5N 7°32E **50** G7
Abukuma-Gawa → *Japan* 38°6N 140°52E **30** E10
Abukuma-Sammyaku *Japan* 37°30N 140°45E **30** F10
Abunã *Brazil* 9°40S 65°20W **92** E5
Abunã → *Brazil* 9°41S 65°20W **92** E5
Aburo *Dem. Rep. of the Congo* 2°4N 30°53E **54** B3
Abut Hd. *N.Z.* 43°7S 170°15E **59** E3
Ābyek *Iran* 36°4N 50°33E **45** B6
Acadia △ *U.S.A.* 44°20N 68°13W **81** C19
Açailândia *Brazil* 4°57S 47°0W **93** D9
Acajutla *El Salv.* 13°36N 89°50W **88** D2
Acámbaro *Mexico* 20°2N 100°44W **86** D4
Acaponeta *Mexico* 22°30N 105°22W **86** C3
Acapulco *Mexico* 16°51N 99°55W **87** D5
Acaraí, Serra *Brazil* 1°50N 57°50W **92** C7
Acarigua *Venezuela* 9°33N 69°12W **92** B5
Acatlán *Mexico* 18°12N 98°3W **87** D5
Acayucán *Mexico* 17°57N 94°55W **87** D6
Accomac *U.S.A.* 37°43N 75°40W **81** G16
Accra *Ghana* 5°35N 0°6W **50** G5
Accrington *U.K.* 53°45N 2°22W **12** D5
Acebal *Argentina* 33°20S 60°50W **94** C3
Aceh □ *Indonesia* 4°15N 97°30E **36** D1
Achalpur *India* 21°22N 77°32E **40** J10
Acharnes *Greece* 38°5N 23°44E **23** E10
Acheloos → *Greece* 38°19N 21°7E **23** E9
Acheng *China* 45°30N 126°58E **35** B14
Acher *India* 23°10N 72°32E **42** H5
Achill Hd. *Ireland* 53°58N 10°15W **10** C1
Achill I. *Ireland* 53°58N 10°1W **10** C1
Achinsk *Russia* 56°20N 90°20E **29** D10
Acireale *Italy* 37°37N 15°10E **22** F6
Ackerman *U.S.A.* 33°19N 89°11W **85** E10
Acklins I. *Bahamas* 22°30N 74°0W **89** B5
Acme *Canada* 51°33N 113°30W **70** C6
Acme *U.S.A.* 40°8N 79°26W **82** F5
Aconcagua, Cerro *Argentina* 32°39S 70°0W **94** C2
Aconquija, Mt. *Argentina* 27°0S 66°0W **94** B2
Açores, Is. dos *Atl. Oc.* 38°0N 27°0W **50** a
Acornhoek *S. Africa* 24°37S 31°2E **57** C5
Acraman, L. *Australia* 32°2S 135°23E **63** E2
Acre = 'Akko *Israel* 32°55N 35°4E **46** C4
Acre □ *Brazil* 9°1S 71°0W **92** E4
Acre → *Brazil* 8°45S 67°22W **92** E5
Actinolite *Canada* 44°32N 77°19W **82** B7
Acton *Canada* 43°38N 80°3W **82** C4
Ad Dammām *Si. Arabia* 26°20N 50°5E **45** E6
Ad Dāmūr *Lebanon* 33°43N 35°27E **46** B4
Ad Dawādimī *Si. Arabia* 24°35N 44°15E **44** E5
Ad Dawḩah *Qatar* 25°15N 51°35E **45** E6
Ad Dawr *Iraq* 34°27N 43°47E **44** C4
Ad Dir'īyah *Si. Arabia* 24°44N 46°35E **44** E5
Ad Dīwānīyah *Iraq* 32°0N 45°0E **44** D5
Ad Dujayl *Iraq* 33°51N 44°14E **44** C5
Ad Duwayd *Si. Arabia* 30°15N 42°17E **44** D4
Ada *Minn., U.S.A.* 47°18N 96°31W **80** B5
Ada *Okla., U.S.A.* 34°46N 96°41W **84** D6
Adabiya *Egypt* 29°53N 32°28E **46** F1
Adair, C. *Canada* 71°30N 71°34W **69** C17
Adaja → *Spain* 41°32N 4°52W **21** B3
Adak *U.S.A.* 51°45N 176°45W **74** E4
Adak I. *U.S.A.* 51°45N 176°45W **74** E4
Adamaoua, Massif de l' *Cameroon* 7°20N 12°20E **51** G8
Adamawa Highlands = Adamaoua, Massif de l' *Cameroon* 7°20N 12°20E **51** G8
Adamello, Mte. *Italy* 46°9N 10°30E **20** C9
Adaminaby *Australia* 36°0S 148°45E **63** F4
Adams *Mass., U.S.A.* 42°38N 73°7W **83** D11
Adams *N.Y., U.S.A.* 43°49N 76°1W **83** C8
Adams *Wis., U.S.A.* 43°57N 89°49W **80** D9
Adams, Mt. *U.S.A.* 46°12N 121°30W **78** D5
Adam's Bridge *Sri Lanka* 9°15N 79°40E **40** Q11
Adams L. *Canada* 51°10N 119°40W **70** C5
Adam's Peak *Sri Lanka* 6°48N 80°30E **40** R12
Adana *Turkey* 37°0N 35°16E **44** B2
Adang, Ko *Thailand* 6°33N 99°18E **39** J2
Adapazarı = Sakarya *Turkey* 40°48N 30°25E **19** F5
Adarama *Sudan* 17°10N 34°52E **51** E12
Adare, C. *Antarctica* 71°0S 171°0E **5** D11
Adaut *Indonesia* 8°8S 131°7E **37** F8
Adavale *Australia* 25°52S 144°32E **63** D3
Adda → *Italy* 45°8N 9°53E **20** D8
Addis Ababa = Addis Abeba *Ethiopia* 9°2N 38°42E **47** F2
Addis Abeba *Ethiopia* 9°2N 38°42E **47** F2
Addison *U.S.A.* 42°1N 77°14W **82** D7
Addo *S. Africa* 33°32S 25°45E **56** E4
Addo △ *S. Africa* 33°30S 25°50E **56** E4
Ādeh *Iran* 37°42N 45°11E **44** B5
Adel *U.S.A.* 31°8N 83°25W **85** F13
Adelaide *Australia* 34°52S 138°30E **63** E2
Adelaide *S. Africa* 32°42S 26°20E **56** E4
Adelaide I. *Antarctica* 67°15S 68°30W **5** C17
Adelaide Pen. *Canada* 68°15N 97°30W **68** D12
Adelaide River *Australia* 13°15S 131°7E **60** B5
Adelaide Village *Bahamas* 25°0N 77°31W **88** A4
Adelanto *U.S.A.* 34°35N 117°22W **79** L9
Adele I. *Australia* 15°32S 123°9E **60** C3
Adélie, Terre *Antarctica* 68°0S 140°0E **5** C10
Adelie Land = Adélie, Terre *Antarctica* 68°0S 140°0E **5** C10
Aden = Al 'Adan *Yemen* 12°45N 45°0E **47** E4
Aden, G. of *Ind. Oc.* 12°30N 47°30E **47** E4
Adendorp *S. Africa* 32°15S 24°30E **56** E3
Adh Dhayd *U.A.E.* 25°17N 55°53E **45** E7
Adhoi *India* 23°26N 70°32E **42** H4
Adi *Indonesia* 4°15S 133°30E **37** E8
Adieu, C. *Australia* 32°0S 132°10E **61** F5
Adieu Pt. *Australia* 15°14S 124°35E **60** C3
Adige → *Italy* 45°9N 12°20E **22** B5
Adigrat *Ethiopia* 14°20N 39°26E **47** E2
Adilabad *India* 19°33N 78°20E **40** K11
Adirondack ⌓ *U.S.A.* 44°0N 74°20W **83** C10
Adirondack Mts. *U.S.A.* 44°0N 74°0W **83** C10
Adis Abeba = Addis Abeba *Ethiopia* 9°2N 38°42E **47** F2
Adjumani *Uganda* 3°20N 31°50E **54** B3
Adjuntas *Puerto Rico* 18°10N 66°43W **89** d
Adlavik Is. *Canada* 55°0N 58°40W **73** B8
Admiralty G. *Australia* 14°20S 125°55E **60** B4
Admiralty Gulf ✪ *Australia* 14°16S 125°52E **60** B4
Admiralty I. *U.S.A.* 57°30N 134°30W **70** B2
Admiralty Inlet *Canada* 72°30N 86°0W **69** C14
Admiralty Is. *Papua N. G.* 2°0S 147°0E **58** B7
Adolfo González Chaves *Argentina* 38°2S 60°5W **94** D3
Adolfo Ruiz Cortines, Presa *Mexico* 27°15N 109°6W **86** B3
Adonara *Indonesia* 8°15S 123°5E **37** F6
Adoni *India* 15°33N 77°18E **40** M10
Adour → *France* 43°32N 1°32W **20** E3
Adra *India* 23°30N 86°42E **43** H12
Adra *Spain* 36°43N 3°3W **21** D4
Adrano *Italy* 37°40N 14°50E **22** F6
Adrar *Algeria* 27°51N 0°11E **50** C6
Adrar *Mauritania* 20°30N 7°30W **50** D3
Adrar des Iforas *Africa* 19°40N 1°40E **50** E6
Adrian *Mich., U.S.A.* 41°54N 84°2W **81** E11
Adrian *Tex., U.S.A.* 35°16N 102°40W **84** D3
Adriatic Sea *Medit. S.* 43°0N 16°0E **22** C6
Adua *Indonesia* 1°45S 129°50E **37** E7
Adwa *Ethiopia* 14°15N 38°52E **47** E2
Adygea □ *Russia* 45°0N 40°0E **19** F7
Adzhar Republic = Ajaria □ *Georgia* 41°30N 42°0E **19** F7
Adzopé *Ivory C.* 6°7N 3°49W **50** G5
Ægean Sea *Medit. S.* 38°30N 25°0E **23** E11
Aerhtai Shan *Mongolia* 46°40N 92°45E **32** B7
Afaahiti *Tahiti* 17°45S 149°17W **59** d
'Afak *Iraq* 32°4N 45°15E **44** C5
Afandou *Greece* 36°18N 28°12E **25** C10
Afareaitu *Moorea* 17°33S 149°47W **59** d
Afghanistan ■ *Asia* 33°0N 65°0E **40** C4
Aflou *Algeria* 34°7N 2°3E **50** B6
Afognak I. *U.S.A.* 58°15N 152°30W **74** D9
Africa 10°0N 20°0E **48** E6
'Afrīn *Syria* 36°32N 36°50E **44** B3
Afton *N.Y., U.S.A.* 42°14N 75°32W **83** D9
Afton *Wyo., U.S.A.* 42°44N 110°56W **76** E8
Afuá *Brazil* 0°15S 50°20W **93** D8
'Afula *Israel* 32°37N 35°17E **46** C4
Afyon *Turkey* 38°45N 30°33E **19** G5
Afyonkarahisar = Afyon *Turkey* 38°45N 30°33E **19** G5
Āgā Jarī *Iran* 30°42N 49°50E **45** D6
Agadés = Agadez *Niger* 16°58N 7°59E **50** E7
Agadez *Niger* 16°58N 7°59E **50** E7
Agadir *Morocco* 30°28N 9°55W **50** B4
Agaete *Canary Is.* 28°6N 15°43W **24** F4
Agalega Is. *Mauritius* 11°0S 57°0E **3** E12
Agar *India* 23°40N 76°2E **42** H7
Agartala *India* 23°50N 91°23E **41** H17
Agassiz *Canada* 49°14N 121°46W **70** D4
Agassiz Icecap *Canada* 80°15N 76°0W **69** A16
Agats *Indonesia* 5°33S 138°0E **37** F9
Agattu I. *U.S.A.* 52°25N 173°35E **74** E2
Agawam *U.S.A.* 42°5N 72°37W **83** D12
Agboville *Ivory C.* 5°55N 4°15W **50** G5
Ağdam *Azerbaijan* 40°0N 46°58E **44** B5
Agde *France* 43°19N 3°28E **20** E5
Agen *France* 44°12N 0°38E **20** D4
Āgh Kand *Iran* 37°15N 48°4E **45** B6
Aghia Deka *Greece* 35°3N 24°58E **25** D6
Aghia Ekaterinis, Akra *Greece* 39°50N 19°50E **25** A3
Aghia Galini *Greece* 35°6N 24°41E **25** D6
Aghia Varvara *Greece* 35°8N 25°1E **25** D7
Aghios Efstratios *Greece* 39°34N 24°58E **23** E11
Aghios Ioannis, Akra *Greece* 35°20N 25°40E **25** D7
Aghios Isidoros *Greece* 36°9N 27°51E **25** C9
Aghios Matheos *Greece* 39°30N 19°47E **25** B3
Aghios Nikolaos *Greece* 35°11N 25°41E **25** D7
Aghios Stephanos *Greece* 39°46N 19°39E **25** A3
Aghiou Orous, Kolpos *Greece* 40°6N 24°0E **23** D11
Aginskoye *Russia* 51°6N 114°32E **29** D12
Agnew *Australia* 28°1S 120°31E **61** E3
Agori *India* 24°33N 82°57E **43** G10
Agra *India* 27°17N 77°58E **42** F7
Ağri *Turkey* 39°44N 43°3E **19** G7
Agri → *Italy* 40°13N 16°44E **22** D7
Ağrı Dağı *Turkey* 39°50N 44°15E **44** B5
Ağrı Karakose = Ağri *Turkey* 39°44N 43°3E **19** G7
Agrigento *Italy* 37°19N 13°34E **22** F5
Agrinio *Greece* 38°37N 21°27E **23** E9
Agua Caliente *Mexico* 32°29N 116°59W **79** N10
Agua Caliente Springs *U.S.A.* 32°56N 116°19W **79** N10
Água Clara *Brazil* 20°25S 52°45W **93** H8
Agua Fria ⌓ *U.S.A.* 34°14N 112°0W **77** J8
Agua Hechicera *Mexico* 32°28N 116°15W **79** N10
Agua Prieta *Mexico* 31°18N 109°34W **86** A3
Aguadilla *Puerto Rico* 18°26N 67°10W **89** d
Aguadulce *Panama* 8°15N 80°32W **88** E3
Aguanga *U.S.A.* 33°27N 116°51W **79** M10
Aguanish *Canada* 50°14N 62°2W **73** B7
Aguanish → *Canada* 50°13N 62°5W **73** B7
Aguapey → *Argentina* 29°7S 56°36W **94** B4
Aguaray Guazú → *Paraguay* 24°47S 57°19W **94** A4
Aguarico → *Ecuador* 0°59S 75°11W **92** D3
Aguaro-Guariquito △ *Venezuela* 8°20N 66°35W **89** E6
Aguas Blancas *Chile* 24°15S 69°55W **94** A2
Aguas Calientes, Sierra de *Argentina* 25°26S 66°40W **94** B2
Aguascalientes *Mexico* 21°53N 102°18W **86** C4
Aguascalientes □ *Mexico* 22°0N 102°20W **86** C4
Aguila, Punta *Puerto Rico* 17°57N 67°13W **89** d
Aguilares *Argentina* 27°26S 65°35W **94** B2
Águilas *Spain* 37°23N 1°35W **21** D5
Agüimes *Canary Is.* 27°58N 15°27W **24** G4
Aguja, C. de la *Colombia* 11°18N 74°12W **90** B3
Agujereada, Pta. *Puerto Rico* 18°30N 67°8W **89** d
Agulhas, C. *S. Africa* 34°52S 20°0E **56** E3
Agulo *Canary Is.* 28°11N 17°12W **24** F2
Agung, Gunung *Indonesia* 8°20S 115°28E **37** J18
Aguni-Jima *Japan* 26°30N 127°10E **31** L3
Agur *Uganda* 2°28N 32°55E **54** B3
Agusan → *Phil.* 9°0N 125°30E **37** C7
Aha Mts. *Botswana* 19°45S 21°0E **56** B3
Ahaggar *Algeria* 23°0N 6°30E **50** D7
Ahar *Iran* 38°35N 47°0E **44** B5
Ahipara B. *N.Z.* 35°5S 173°5E **59** A4
Ahiri *India* 19°30N 80°0E **40** K12
Ahmad Wal *Pakistan* 29°18N 65°58E **42** E1
Ahmadabad *India* 23°0N 72°40E **42** H5
Aḩmadābād *Khorāsān, Iran* 35°3N 60°50E **45** C9
Aḩmadābād *Khorāsān, Iran* 35°49N 59°42E **45** C8
Aḩmadī *Iran* 27°56N 56°42E **45** E8
Ahmadnagar *India* 19°7N 74°46E **40** K9
Ahmadpur East *Pakistan* 29°12N 71°10E **42** E4
Ahmadpur Lamma *Pakistan* 28°19N 70°3E **42** E4
Ahmedabad = Ahmadabad *India* 23°0N 72°40E **42** H5
Ahmednagar = Ahmadnagar *India* 19°7N 74°46E **40** K9
Ahome *Mexico* 25°55N 109°11W **86** B3
Ahoskie *U.S.A.* 36°17N 76°59W **85** C16
Ahram *Iran* 28°52N 51°16E **45** D6
Ahrax Pt. *Malta* 36°0N 14°22E **25** D1
Ahuachapán *El Salv.* 13°54N 89°52W **88** D2
Ahvāz *Iran* 31°20N 48°40E **45** D6
Ahvenanmaa = Åland *Finland* 60°15N 20°0E **9** F19
Aḩwar *Yemen* 13°30N 46°40E **47** E4
Ai → *India* 26°26N 90°44E **43** F14
Ai-Ais *Namibia* 27°54S 17°59E **56** D2
Ai-Ais and Fish River Canyon △ *Namibia* 24°45S 17°15E **56** C2
Aichi □ *Japan* 35°0N 137°15E **31** G8
Aigrettes, Pte. des *Réunion* 21°3S 55°13E **53** c
Aiguá *Uruguay* 34°13S 54°46W **95** C5
Aigues-Mortes *France* 43°35N 4°12E **20** E6
Aihui = Heihe *China* 50°10N 127°30E **33** A14
Aija *Peru* 9°50S 77°45W **92** E3
Aikawa *Japan* 38°2N 138°15E **30** E9
Aiken *U.S.A.* 33°34N 81°43W **85** E14
Aileron *Australia* 22°39S 133°20E **62** C1
Aillik *Canada* 55°11N 59°18W **73** A8
Ailsa Craig *Canada* 43°8N 81°33W **82** C3
Ailsa Craig *U.K.* 55°15N 5°6W **11** F3
Aim *Russia* 59°0N 133°55E **29** D14
Aimere *Indonesia* 8°45S 121°3E **37** F6
Aimogasta *Argentina* 28°33S 66°50W **94** B2
Aïn Ben Tili *Mauritania* 25°59N 9°27W **50** C4
Aïn-Sefra *Algeria* 32°47N 0°37W **50** B5
Ain Sudr *Egypt* 29°50N 33°6E **46** F2
Aïn Témouchent *Algeria* 35°16N 1°8W **50** A5
Ainaži *Latvia* 57°50N 24°24E **9** H21
Ainsworth *U.S.A.* 42°33N 99°52W **80** D4
Aiquile *Bolivia* 18°10S 65°10W **92** G5
Aïr *Niger* 18°30N 8°0E **50** E7
Air Force I. *Canada* 67°58N 74°5W **69** D17
Air Hitam *Malaysia* 1°55N 103°11E **39** M4
Airdrie *Canada* 51°18N 114°2W **70** C6
Airdrie *U.K.* 55°52N 3°57W **11** F5
Aire → *U.K.* 53°43N 0°55W **12** D7
Aire, I. de l' *Spain* 39°48N 4°16E **24** B11
Airlie Beach *Australia* 20°16S 148°43E **62** b
Aisne → *France* 49°26N 2°50E **20** B5
Ait *India* 25°54N 79°14E **43** G8
Aitkin *U.S.A.* 46°32N 93°42W **80** B7
Aitutaki *Cook Is.* 18°52S 159°45W **65** J12
Aiud *Romania* 46°19N 23°44E **17** E12
Aix-en-Provence *France* 43°32N 5°27E **20** E6
Aix-la-Chapelle = Aachen *Germany* 50°45N 6°6E **16** C4
Aix-les-Bains *France* 45°41N 5°53E **20** D6
Aizawl *India* 23°40N 92°44E **41** H18
Aizkraukle *Latvia* 56°36N 25°11E **9**
Aizpute *Latvia* 56°43N 21°40E **9**
Aizuwakamatsu *Japan* 37°30N 139°56E **30**
Ajaccio *France* 41°55N 8°40E **20**
Ajai ⌓ *Uganda* 2°52N 31°16E **54**
Ajaigarh *India* 24°52N 80°16E **43**
Ajalpan *Mexico* 18°22N 97°15W **87**
Ajanta Ra. *India* 20°28N 75°50E **40**
Ajari Rep. = Ajaria □ *Georgia* 41°30N 42°0E **19**
Ajaria □ *Georgia* 41°30N 42°0E **19**
Ajax *Canada* 43°50N 79°1W **82**
Ajdābiyā *Libya* 30°54N 20°4E **51**
Ajka *Hungary* 47°4N 17°31E **17**
'Ajlūn *Jordan* 32°18N 35°47E **46**
'Ajlūn □ *Jordan* 32°18N 35°47E **46**
'Ajmān *U.A.E.* 25°25N 55°30E **45**
Ajmer *India* 26°28N 74°37E **42**
Ajnala *India* 31°50N 74°48E **42**
Ajo *U.S.A.* 32°22N 112°52W **77**
Ajo, C. de *Spain* 43°31N 3°35W **21**
Akabira *Japan* 43°33N 142°5E **30**
Akagera △ *Rwanda* 1°31S 30°33E **54**
Akamas *Cyprus* 35°3N 32°18E **25**
Akan △ *Japan* 43°20N 144°20E **30**
Akanthou *Cyprus* 35°22N 33°45E **25**
Akaroa *N.Z.* 43°49S 172°59E **59**
Akashi *Japan* 34°45N 134°58E **31**
Akbarpur *Bihar, India* 24°39N 83°58E **43**
Akbarpur *Ut. P., India* 26°25N 82°32E **43**
Akçakale *Turkey* 36°41N 38°56E **44**
Akdoğan = Lysi *Cyprus* 35°6N 33°41E **25**
Akelamo *Indonesia* 1°35N 129°40E **37**
Aketi *Dem. Rep. of the Congo* 2°38N 23°47E **52**
Akhisar *Turkey* 38°56N 27°48E **23**
Akhnur *India* 32°52N 74°45E **43**
Akhtyrka = Okhtyrka *Ukraine* 50°25N 35°0E **19**
Aki *Japan* 33°30N 133°54E **31**
Akimiski I. *Canada* 52°50N 81°30W **72**
Akincilar = Louroujina *Cyprus* 35°0N 33°28E **25**
Akita *Japan* 39°45N 140°7E **30**
Akita □ *Japan* 39°40N 140°30E **30**
Akjoujt *Mauritania* 19°45N 14°15W **50**
Akkeshi *Japan* 43°2N 144°51E **30**
'Akko *Israel* 32°55N 35°4E **46**
Aklavik *Canada* 68°12N 135°0W **68**
Aklera *India* 24°26N 76°32E **42**
Akō *Japan* 34°45N 134°24E **31**
Akola *India* 20°42N 77°2E **40**
Akordat *Eritrea* 15°30N 37°40E **47**
Akpatok I. *Canada* 60°25N 68°8W **69**
Åkrahamn *Norway* 59°15N 5°10E **9**
Akranes *Iceland* 64°19N 22°5W **8**
Akron *Colo., U.S.A.* 40°10N 103°13W **76**
Akron *Ohio, U.S.A.* 41°5N 81°31W **82**
Akrotiri *Cyprus* 34°36N 32°57E **25**
Akrotiri Bay *Cyprus* 34°35N 33°10E **25**
Aksai Chin *China* 35°15N 79°55E **43**
Aksaray *Turkey* 38°25N 34°2E **44**
Aksay = Aqsay *Kazakhstan* 51°11N 53°0E **19**
Akşehir *Turkey* 38°18N 31°30E **44**
Akşehir Gölü *Turkey* 38°30N 31°25E **19**
Aksu *China* 41°5N 80°10E **32**
Aksum *Ethiopia* 14°5N 38°40E **47**
Aktsyabrski *Belarus* 52°38N 28°53E **17**
Aktyubinsk = Aqtöbe *Kazakhstan* 50°17N 57°10E **19**
Akure *Nigeria* 7°15N 5°5E **50**
Akureyri *Iceland* 65°40N 18°6W **8**
Akuseki-Shima *Japan* 29°27N 129°37E **31**
Akyab = Sittwe *Burma* 20°18N 92°45E **41**
Al 'Adan *Yemen* 12°45N 45°0E **47**
Al Aḩsā = Hasa *Si. Arabia* 25°50N 49°0E **45**
Al Ajfar *Si. Arabia* 27°26N 43°0E **44**
Al Amādīyah *Iraq* 37°5N 43°30E **44**
Al 'Amārah *Iraq* 31°55N 47°15E **44**
Al Anbār □ *Iraq* 33°25N 42°0E **44**
Al 'Aqabah *Jordan* 29°31N 35°0E **46**
Al 'Aqabah □ *Jordan* 29°30N 35°0E **46**
Al Arak *Syria* 34°38N 38°35E **44**
Al 'Aramah *Si. Arabia* 25°30N 46°0E **44**
Al 'Arţāwīyah *Si. Arabia* 26°31N 45°20E **44**
Al 'Āşimah = 'Ammān □ *Jordan* 31°40N 36°30E **46**
Al 'Assāfīyah *Si. Arabia* 28°17N 38°59E **44**
Al 'Awdah *Si. Arabia* 25°32N 45°41E **44**
Al 'Ayn *Si. Arabia* 25°4N 38°6E **44**
Al 'Ayn *U.A.E.* 24°15N 55°45E **45**
Al 'Azīzīyah *Iraq* 32°54N 45°4E **44**
Al Bāb *Syria* 36°23N 37°29E **44**
Al Bad' *Si. Arabia* 28°28N 35°1E **44**
Al Bādī *Iraq* 35°56N 41°32E **44**
Al Baḩrah *Kuwait* 29°40N 47°52E **44**
Al Baḩral Mayyit = Dead Sea *Asia* 31°30N 35°30E **46**
Al Balqā' □ *Jordan* 32°5N 35°45E **46**
Al Bārūk, J. *Lebanon* 33°39N 35°40E **46**
Al Başrah *Iraq* 30°30N 47°50E **44**
Al Baţḩā *Iraq* 31°6N 45°53E **44**
Al Batrūn *Lebanon* 34°15N 35°40E **46**
Al Bayḑā *Libya* 32°50N 21°44E **51**
Al Biqā *Lebanon* 34°10N 36°10E **46**
Al Bi'r *Si. Arabia* 28°51N 36°16E **44**
Al Bukayrīyah *Si. Arabia* 26°9N 43°40E **44**
Al Burayj *Syria* 34°15N 36°46E **46**
Al Faḑilī *Si. Arabia* 26°58N 49°10E **45**
Al Fallūjah *Iraq* 33°20N 43°55E **44**
Al Fāw *Iraq* 30°0N 48°30E **45**
Al Fujayrah *U.A.E.* 25°7N 56°18E **45**
Al Ghadaf, W. → *Jordan* 31°26N 36°43E **46**
Al Ghammās *Iraq* 31°45N 44°37E **44**
Al Ghazālah *Si. Arabia* 26°48N 41°19E **44**
Al Ḩadīthah *Iraq* 34°0N 41°13E **44**
Al Ḩadīthah *Si. Arabia* 31°28N 37°8E **46**
Al Ḩaḑr *Iraq* 35°35N 42°44E **44**
Al Hājānah *Syria* 33°20N 36°33E **46**

Ḥajar al Gharbī Oman 24°10N 56°15E 45 E8
Ḥāmad Si. Arabia 31°30N 39°30E 44 D3
Ḥamdānīyah Syria 35°25N 36°50E 44 C3
Ḥamīdīyah Syria 34°42N 35°57E 46 A4
Ḥammār Iraq 30°57N 46°51E 44 D5
Ḥamrā' Si. Arabia 24°2N 38°55E 44 E3
Ḥamzah Iraq 31°43N 44°58E 44 D5
Ḥanākīyah Si. Arabia 24°51N 40°31E 44 E4
Harūj al Aswad Libya 27°0N 17°10E 51 C9
Ḥasakah Syria 36°35N 40°45E 44 B4
Ḥayy Iraq 32°5N 46°5E 44 C5
Ḥijarah Asia 30°0N 44°0E 44 D4
Ḥillah Iraq 32°30N 44°25E 44 C5
Hindīyah Iraq 32°30N 44°10E 44 C5
Hirmil Lebanon 34°26N 36°24E 46 A5
Hoceïma Morocco 35°8N 3°58W 50 A5
Ḥudaydah Yemen 14°50N 43°0E 47 E3
Ḥudūd ash Shamālīyah □ Si. Arabia 29°10N 42°30E 44 D4
Hufūf Si. Arabia 25°25N 49°45E 45 E6
Ḥumayḍah Si. Arabia 29°14N 34°56E 44 D2
Ḥunayy Si. Arabia 25°58N 48°45E 45 E6
Īsāwīyah Si. Arabia 30°43N 37°59E 44 D3
Jafr Jordan 30°18N 36°14E 46 E5
Jāfūrah Si. Arabia 25°0N 50°15E 45 E7
Jaghbūb Libya 29°42N 24°38E 51 C10
Jahrah Kuwait 29°25N 47°40E 44 D5
Jalāmīd Si. Arabia 31°20N 40°6E 44 D3
Jamalīyah Qatar 25°37N 51°5E 45 E6
Janūb □ Lebanon 33°20N 35°20E 46 B4
Jawf Libya 24°10N 23°24E 51 D10
Jawf Si. Arabia 29°55N 39°40E 44 D3
Jawf □ Si. Arabia 29°30N 39°30E 44 D3
Jazair = Algeria ■ Africa 28°30N 2°0E 50 C6
Jazirah Iraq 33°30N 44°0E 44 C5
Jithāmīyah Si. Arabia 27°41N 41°43E 44 E4
Jubayl Si. Arabia 27°0N 49°50E 45 E6
Jubaylah Si. Arabia 24°55N 46°25E 44 E5
Jubb Si. Arabia 27°11N 42°17E 44 E4
Junaynah Sudan 13°27N 22°45E 51 F10
Kabā'ish Iraq 30°58N 47°0E 44 D5
Karak Jordan 31°11N 35°42E 46 D4
Karak □ Jordan 31°0N 36°0E 46 E5
Kāẓimīyah Iraq 33°22N 44°18E 44 C5
Khābūrah Oman 23°57N 57°5E 45 F8
Khafji Si. Arabia 28°24N 48°29E 45 E6
Khalīl West Bank 31°32N 35°6E 46 D4
Khāliş Iraq 33°49N 44°32E 44 C5
Kharsānīyah Si. Arabia 27°13N 49°18E 45 E6
Khaşab Oman 26°14N 56°15E 45 E8
Khawr Qatar 25°41N 51°30E 45 E6
Khiḍr Iraq 31°12N 45°33E 44 D5
Khiyām Lebanon 33°20N 35°36E 46 B4
Khobar Si. Arabia 26°17N 50°12E 45 E6
Khums Libya 32°40N 14°17E 51 B8
Kiswah Syria 33°23N 36°14E 46 B5
Kūfah Iraq 32°2N 44°24E 44 C5
Kufrah Libya 24°17N 23°15E 51 D10
Kuhayfīyah Si. Arabia 27°12N 43°3E 44 E4
Kūt Iraq 32°30N 46°0E 44 C5
Kuwayt Kuwait 29°30N 48°0E 44 D5
Labwah Lebanon 34°11N 36°20E 46 A5
Lādhiqīyah Syria 35°30N 35°45E 44 C2
Līth Si. Arabia 20°9N 40°15E 47 C3
Liwā' Oman 24°31N 56°36E 45 E8
Luḩayyah Yemen 15°45N 42°40E 47 D3
Madīnah Iraq 30°57N 47°16E 44 D5
Madīnah Si. Arabia 24°35N 39°52E 44 E3
Mafraq Jordan 32°17N 36°14E 46 C5
Mafraq □ Jordan 32°17N 36°15E 46 C5
Maghreb = Morocco ■ N. Afr. 32°0N 5°50W 50 B4
Maḩmūdīyah Iraq 33°3N 44°21E 44 C5
Majma'ah Si. Arabia 25°57N 45°22E 44 E5
Makhruq, W. → Jordan 31°28N 37°0E 46 D6
Makḩūl Si. Arabia 26°37N 42°39E 44 E4
Manāmah Bahrain 26°10N 50°30E 45 E6
Maqwa' Kuwait 29°10N 47°59E 44 D5
Marāḩ Si. Arabia 25°35N 49°35E 45 E6
Marj Libya 32°25N 20°30E 51 B10
Maţlā Kuwait 29°24N 47°40E 44 D5
Mawşil Iraq 36°15N 43°5E 44 B4
Mayādin Syria 35°1N 40°27E 44 C4
Mazār Jordan 31°4N 35°41E 46 D4
Midhnab Si. Arabia 25°50N 44°18E 44 E5
Minā' Lebanon 34°24N 35°49E 46 A4
Miqdādīyah Iraq 34°0N 45°0E 44 C5
Mubarraz Si. Arabia 25°30N 49°40E 45 E6
Mudawwarah Jordan 29°19N 36°0E 46 F5
Mughayrā' U.A.E. 24°5N 53°32E 45 E7
Muḩarraq Bahrain 26°15N 50°40E 45 E6
Mukallā Yemen 14°33N 49°2E 47 E4
Mukhā Yemen 13°18N 43°15E 47 E3
Musayjīd Si. Arabia 24°5N 39°5E 44 E3
Musayyib Iraq 32°49N 44°20E 44 C5
Muthanná □ Iraq 30°30N 45°15E 44 D5
Muwayliḩ Si. Arabia 27°40N 35°30E 44 E2
Qādisīyah □ Iraq 32°0N 45°0E 44 D5
Qā'im Iraq 34°21N 41°7E 44 C4
Qalībah Si. Arabia 28°24N 37°42E 44 D3
Qāmishlī Syria 37°2N 41°14E 44 B4
Qaryatayn Syria 34°12N 37°13E 46 A6
Qaşīm □ Si. Arabia 26°0N 43°0E 44 E4
Qaţ'ā Syria 34°40N 40°48E 44 C4
Qaţīf Si. Arabia 26°35N 50°0E 45 E6
Qatrānah Jordan 31°12N 36°6E 46 D5
Qaţrūn Libya 24°56N 15°3E 51 D9
Qayşūmah Si. Arabia 28°20N 46°7E 44 D5
Quds = Jerusalem Israel/West Bank 31°47N 35°10E 46 D4
Qunayţirah Syria 33°5N 35°45E 46 B4
Qunfudhah Si. Arabia 19°3N 41°4E 47 D3
Qurnah Iraq 31°1N 47°25E 44 D5
Quşayr Iraq 30°39N 45°50E 44 D5
Quşayr Syria 34°31N 36°34E 46 A5
Qutayfah Syria 33°44N 36°36E 46 B5
'Ubaylah Si. Arabia 21°59N 50°57E 47 C5
Al 'Uḍaylīyah Si. Arabia 25°8N 49°18E 45 E6
Al 'Ulā Si. Arabia 26°35N 38°0E 44 E3
Al 'Uqayr Si. Arabia 25°40N 50°15E 45 E6
Al 'Uwaynid Si. Arabia 24°50N 46°0E 44 E5
Al 'Uwayqīlah Si. Arabia 30°30N 42°10E 44 D4
Al 'Uyūn Ḥijāz, Si. Arabia 24°33N 39°35E 44 E3
Al 'Uyūn Najd, Si. Arabia 26°30N 43°50E 44 E4
Al 'Uzayr Iraq 31°19N 47°25E 44 D5
Al Wajh Si. Arabia 26°10N 36°30E 44 E3
Al Wakrah Qatar 25°10N 51°40E 45 E6
Al Waqbah Si. Arabia 28°48N 45°33E 44 D5
Al Wari'āh Si. Arabia 27°51N 47°25E 44 E5
Al Yaman = Yemen ■ Asia 15°0N 44°0E 47 E3
Ala Dağ Turkey 37°44N 35°9E 44 B2
Ala Tau Asia 45°30N 80°40E 32 B5
Ala Tau Shankou = Dzungarian Gate Asia 45°10N 82°0E 32 B5
Alabama □ U.S.A. 33°0N 87°0W 85 E11
Alabama → U.S.A. 31°8N 87°57W 85 F11
Alabaster U.S.A. 33°15N 86°49W 85 E11
Alaçam Dağları Turkey 39°18N 28°49E 23 E13
Alacant = Alicante Spain 38°23N 0°30W 21 C5
Alachua U.S.A. 29°47N 82°30W 85 G13
Alagoa Grande Brazil 7°3S 35°35W 93 E11
Alagoas □ Brazil 9°0S 36°0W 93 E11
Alagoinhas Brazil 12°7S 38°20W 93 F11
Alaheaieatnu = Altaelva → Norway 69°54N 23°17E 8 B20
Alaior Spain 39°57N 4°8E 24 B11
Alajero Canary Is. 28°3N 17°13W 24 F2
Alajuela Costa Rica 10°2N 84°8W 88 D3
Alakamisy Madag. 21°19S 47°14E 57 C8
Alaknanda → India 30°8N 78°36E 43 D8
Alakurtti Russia 66°58N 30°25E 8 C24
Alamarvdasht Iran 27°37N 52°59E 45 E7
Alameda Calif., U.S.A. 37°46N 122°15W 78 H4
Alameda N. Mex., U.S.A. 35°11N 106°37W 77 J10
Alamo U.S.A. 37°22N 115°10W 79 H11
Alamogordo U.S.A. 32°54N 105°57W 77 K11
Alamos Mexico 27°1N 108°56W 86 B3
Alamosa U.S.A. 37°28N 105°52W 77 H11
Åland Finland 60°15N 20°0E 9 F19
Ålands hav Europe 60°0N 19°30E 9 G18
Alania = North Ossetia □ Russia 43°30N 44°30E 19 F7
Alanya Turkey 36°38N 32°0E 44 B1
Alaotra, Farihin' Madag. 17°30S 48°30E 57 B8
Alapayevsk Russia 57°52N 61°42E 28 D7
Alappuzha = Alleppey India 9°30N 76°28E 40 Q10
Alaró Spain 39°42N 2°47E 24 B9
Alarobia-Vohiposa Madag. 20°59S 47°9E 57 C8
Alaşehir Turkey 38°23N 28°30E 23 E13
Alaska □ U.S.A. 64°0N 154°0W 74 C9
Alaska, G. of Pac. Oc. 58°0N 145°0W 68 F3
Alaska Peninsula U.S.A. 56°0N 159°0W 74 D8
Alaska Range U.S.A. 62°50N 151°0W 68 E1
Älät Azerbaijan 39°58N 49°25E 45 B6
Alatau Shan = Ala Tau Asia 45°30N 80°40E 32 B5
Alatyr Russia 54°55N 46°35E 18 D8
Alausi Ecuador 2°0S 78°50W 92 D3
Alava, C. U.S.A. 48°10N 124°44W 76 B1
Alavo = Alavus Finland 62°35N 23°36E 8 E20
Alavus Finland 62°35N 23°36E 8 E20
Alawa ☼ Australia 15°42S 134°39E 62 B1
Alawoona Australia 34°45S 140°30E 63 E3
Alayawarra ☼ Australia 22°0S 134°30E 62 C1
'Alayh Lebanon 33°46N 35°33E 46 B4
Alba Italy 44°42N 8°2E 20 D8
Alba-Iulia Romania 46°8N 23°39E 17 E12
Albacete Spain 39°0N 1°50W 21 C5
Albacutya, L. Australia 35°45S 141°58E 63 F3
Albanel, L. Canada 50°55N 73°12W 72 B5
Albania ■ Europe 41°0N 20°0E 23 D9
Albany Australia 35°1S 117°58E 61 G2
Albany Ga., U.S.A. 31°35N 84°10W 85 F12
Albany N.Y., U.S.A. 42°39N 73°45W 83 D11
Albany Oreg., U.S.A. 44°38N 123°6W 76 D2
Albany Tex., U.S.A. 32°44N 99°18W 84 E5
Albany → Canada 52°17N 81°31W 72 B3
Albarca, C. d' Spain 39°4N 1°22E 24 B7
Albardón Argentina 31°20S 68°30W 94 C2
Albatross B. Australia 12°45S 141°30E 62 A3
Albemarle U.S.A. 35°21N 80°12W 85 D14
Albemarle Sd. U.S.A. 36°5N 76°0W 85 C16
Alberche → Spain 39°58N 4°46W 21 C3
Alberdi Paraguay 26°14S 58°20W 94 B4
Alberga → Australia 27°6S 135°33E 63 D2
Albert, L. Africa 1°30N 31°0E 54 B3
Albert, L. Australia 35°30S 139°10E 63 F2
Albert Edward Ra. Australia 18°17S 127°57E 60 C4
Albert Lea U.S.A. 43°39N 93°22W 80 D7
Albert Nile → Uganda 3°36N 32°2E 54 B3
Albert Town Bahamas 22°37N 74°33W 89 B5
Alberta □ Canada 54°40N 115°0W 70 C6
Alberti Argentina 35°1S 60°16W 94 D3
Albertinia S. Africa 34°11S 21°34E 56 E3
Alberton Canada 46°50N 64°0W 73 C7
Albertville France 45°40N 6°22E 20 D7
Albertville U.S.A. 34°16N 86°13W 85 D11
Albi France 43°56N 2°9E 20 E5
Albia U.S.A. 41°2N 92°48W 80 E7
Albina Suriname 5°37N 54°15W 93 B8
Albina, Ponta Angola 15°52S 11°44E 56 B1
Albion Mich., U.S.A. 42°15N 84°45W 81 D11
Albion Nebr., U.S.A. 41°42N 98°0W 80 E4
Albion Pa., U.S.A. 41°53N 80°22W 82 E4
Alborán Medit. S. 35°57N 3°0W 21 E4
Ålborg = Aalborg Denmark 57°2N 9°54E 9 H13
Alborz, Reshteh-ye Kūhhā-ye Iran 36°0N 52°0E 45 C7
Albufeira Portugal 37°5N 8°15W 21 D1
Albuquerque U.S.A. 35°5N 106°39W 77 J10
Albuquerque, Cayos de Caribbean 12°10N 81°50W 88 D3
Alburg U.S.A. 44°59N 73°18W 83 B11
Albury Australia 36°3S 146°56E 63 F4
Alcalá de Henares Spain 40°28N 3°22W 21 B4
Alcalá la Real Spain 37°27N 3°57W 21 D4
Álcamo Italy 37°59N 12°55E 22 F5
Alcañiz Spain 41°2N 0°8W 21 B5
Alcântara Brazil 2°20S 44°30W 93 D10
Alcántara, Embalse de Spain 39°44N 6°50W 21 C2
Alcantarilla Spain 37°59N 1°12W 21 D5
Alcaraz, Sierra de Spain 38°40N 2°20W 21 C4
Alcaudete Spain 37°35N 4°5W 21 D3
Alcázar de San Juan Spain 39°24N 3°12W 21 C4
Alchevsk Ukraine 48°30N 38°45E 19 E6
Alcira = Alzira Spain 39°9N 0°30W 21 C5
Alcova U.S.A. 42°34N 106°43W 76 E10
Alcoy Spain 38°43N 0°30W 21 C5
Alcúdia Spain 39°51N 3°7E 24 B10
Alcúdia, B. d' Spain 39°47N 3°15E 24 B10
Aldabra Is. Seychelles 9°22S 46°28E 49 G8
Aldama Mexico 22°55N 98°4W 87 C5
Aldan Russia 58°40N 125°30E 29 D13
Aldan → Russia 63°28N 129°35E 29 C13
Aldea, Pta. de la Canary Is. 28°0N 15°50W 24 G4
Aldeburgh U.K. 52°10N 1°37E 13 E9
Alder Pk. U.S.A. 35°53N 121°22W 78 K5
Alderney U.K. 49°42N 2°11W 13 H5
Aldershot U.K. 51°15N 0°44W 13 F7
Aledo U.S.A. 41°12N 90°45W 80 E8
Aleg Mauritania 17°3N 13°55W 50 E3
Alegranza Canary Is. 29°23N 13°32W 24 E6
Alegranza, I. Canary Is. 29°23N 13°32W 24 E6
Alegre Brazil 20°50S 41°30W 95 A7
Alegrete Brazil 29°40S 56°0W 95 B4
Aleksandriya = Oleksandriya Ukraine 50°37N 26°19E 17 C14
Aleksandrov Gay Russia 50°9N 48°34E 19 D8
Aleksandrovsk-Sakhalinskiy Russia 50°50N 142°20E 33 A17
Aleksandry, Zemlya Russia 80°25N 48°0E 28 A5
Além Paraíba Brazil 21°52S 42°41W 95 A7
Alemania Argentina 25°40S 65°30W 94 B2
Alemania Chile 25°10S 69°55W 94 B2
Alençon France 48°27N 0°4E 20 B4
Alenquer Brazil 1°56S 54°46W 93 D8
'Alenuihähä Channel U.S.A. 20°30N 156°0W 75 L8
Aleppo = Ḥalab Syria 36°10N 37°15E 44 B3
Aléria France 42°5N 9°26E 20 E9
Alert Canada 83°2N 60°0W 69 A20
Alès France 44°9N 4°5E 20 D6
Alessándria Italy 44°54N 8°37E 20 D8
Ålesund Norway 62°28N 6°12E 8 E12
Aleutian Basin Pac. Oc. 57°0N 177°0E 64 B9
Aleutian Is. Pac. Oc. 52°0N 175°0W 64 B10
Aleutian Range U.S.A. 60°0N 154°0W 74 D9
Aleutian Trench Pac. Oc. 48°0N 180°0E 4 D17
Alexander U.S.A. 47°51N 103°39W 80 B2
Alexander, Mt. Australia 28°58S 120°16E 61 E3
Alexander Arch. U.S.A. 56°0N 136°0W 68 F4
Alexander Bay S. Africa 28°40S 16°30E 56 D2
Alexander City U.S.A. 32°56N 85°58W 85 E12
Alexander I. Antarctica 69°0S 70°0W 5 C17
Alexandra Australia 37°8S 145°40E 63 F4
Alexandra N.Z. 45°14S 169°25E 59 F2
Alexandra Falls Canada 60°29N 116°18W 70 A5
Alexandria = El Iskandarîya Egypt 31°13N 29°58E 51 B11
Alexandria B.C., Canada 52°35N 122°27W 70 C4
Alexandria Ont., Canada 45°19N 74°38W 83 A10
Alexandria Romania 43°57N 25°24E 17 G13
Alexandria S. Africa 33°38S 26°28E 56 E4
Alexandria U.K. 55°59N 4°35W 11 F4
Alexandria La., U.S.A. 31°18N 92°27W 84 F8
Alexandria Minn., U.S.A. 45°53N 95°22W 80 C6
Alexandria S. Dak., U.S.A. 43°39N 97°47W 80 D5
Alexandria Va., U.S.A. 38°49N 77°5W 81 F15
Alexandria Bay U.S.A. 44°20N 75°55W 83 B9
Alexandrina, L. Australia 35°25S 139°10E 63 F2
Alexandroupoli Greece 40°50N 25°54E 23 D11
Alexis → Canada 52°33N 56°8W 73 B8
Alexis Creek Canada 52°10N 123°20W 70 C4
Aleysk Russia 52°40N 83°0E 28 D9
Alfabia Spain 39°44N 2°44E 24 B9
Alfenas Brazil 21°20S 46°10W 95 A6
Alford Aberds., U.K. 57°14N 2°41W 11 D6
Alford Lincs., U.K. 53°15N 0°10E 12 D8
Alfred Maine, U.S.A. 43°29N 70°43W 83 C14
Alfred N.Y., U.S.A. 42°16N 77°48W 82 D7
Alfreton U.K. 53°6N 1°24W 12 D6
Algaida Spain 39°33N 2°53E 24 B9
Ålgård Norway 58°46N 5°53E 9 G11
Algarve Portugal 36°58N 8°20W 21 D1
Algeciras Spain 36°9N 5°28W 21 D3
Algemesí Spain 39°11N 0°27W 21 C5
Alger Algeria 36°42N 3°8E 50 A6
Algeria ■ Africa 28°30N 2°0E 50 C6
Algha Kazakhstan 49°53N 57°20E 19 E10
Alghero Italy 40°33N 8°19E 22 D3
Algiers = Alger Algeria 36°42N 3°8E 50 A6
Algoa B. S. Africa 33°50S 25°45E 56 E4
Algoma U.S.A. 44°36N 87°26W 80 C10
Algona U.S.A. 43°4N 94°14W 80 D6
Algonac U.S.A. 42°37N 82°32W 82 D2
Algonquin ⌂ Canada 45°50N 78°30W 72 C4
Algorta Uruguay 32°25S 57°23W 96 C5
Alhambra U.S.A. 34°5N 118°7W 79 L8
Alhucemas = Al Hoceïma Morocco 35°8N 3°58W 50 A5
'Alī al Gharbī Iraq 32°30N 46°45E 44 C5
'Alī ash Sharqī Iraq 32°7N 46°44E 44 C5
Äli Bayramlı Azerbaijan 39°59N 48°52E 45 B6
'Alī Khēl Afghan. 33°57N 69°43E 42 C3
Alī Shāh Iran 38°9N 45°50E 44 B5
'Alīābād Golestān, Iran 36°40N 54°33E 45 B7
'Alīābād Khorāsān, Iran 32°30N 57°30E 45 C8
'Alīābād Kordestān, Iran 35°4N 46°58E 44 C5
'Alīābād Yazd, Iran 31°41N 53°49E 45 D7
Aliağa Turkey 38°47N 26°59E 23 E12
Aliakmonas → Greece 40°30N 22°36E 23 D10
Alicante Spain 38°23N 0°30W 21 C5
Alice S. Africa 32°48S 26°55E 56 E4
Alice U.S.A. 27°45N 98°5W 84 H5
Alice → Queens., Australia 24°2S 144°50E 62 C3
Alice → Queens., Australia 15°35S 142°20E 62 B3
Alice Arm Canada 55°29N 129°31W 70 B3
Alice Springs Australia 23°40S 133°50E 62 C1
Alicedale S. Africa 33°15S 26°4E 56 E4
Aliceville U.S.A. 33°8N 88°9W 85 E10
Aliganj India 27°30N 79°10E 43 F8
Aligarh Raj., India 25°55N 76°15E 42 G7
Aligarh Ut. P., India 27°55N 78°10E 42 F8
Alīgūdarz Iran 33°25N 49°45E 45 C6
Alimia Greece 36°16N 27°43E 25 C9
Alingsås Sweden 57°56N 12°31E 9 H15
Alipur Pakistan 29°25N 70°55E 42 E4
Alipur Duar India 26°30N 89°35E 41 F16
Aliquippa U.S.A. 40°37N 80°15W 82 F4
Alitus = Alytus Lithuania 54°24N 24°3E 9 J21
Aliwal North S. Africa 30°45S 26°45E 56 E4
Alix Canada 52°24N 113°11W 70 C6
Aljustrel Portugal 37°55N 8°10W 21 D1
Alkmaar Neths. 52°37N 4°45E 15 B4
All American Canal U.S.A. 32°45N 115°15W 77 K6
Allagash → U.S.A. 47°5N 69°3W 81 B19
Allah Dad Pakistan 25°38N 67°34E 42 G2
Allahabad India 25°25N 81°58E 43 G9
Allan Canada 51°53N 106°4W 71 C7
Allanridge S. Africa 27°45S 26°40E 56 D4
Allegany U.S.A. 42°6N 78°30W 82 D6
Allegheny → U.S.A. 40°27N 80°1W 82 F5
Allegheny Mts. U.S.A. 38°15N 80°10W 81 F13
Allegheny Plateau U.S.A. 41°30N 78°30W 81 E14
Allegheny Res. U.S.A. 41°50N 79°0W 82 E6
Allègre, Pte. Guadeloupe 16°22N 61°46W 88 b
Allen, Bog of Ireland 53°15N 7°0W 10 C5
Allen, L. Ireland 54°8N 8°4W 10 B3
Allendale U.S.A. 33°1N 81°18W 85 E14
Allende Mexico 28°20N 100°51W 86 B4
Allentown U.S.A. 40°37N 75°29W 83 F9
Alleppey India 9°30N 76°28E 40 Q10
Aller → Germany 52°56N 9°12E 16 B5
Alleynes B. Barbados 13°13N 59°39W 89 g
Alliance Nebr., U.S.A. 42°6N 102°52W 80 D2
Alliance Ohio, U.S.A. 40°55N 81°6W 82 F3
Allier → France 46°57N 3°4E 20 C5
Alliford Bay Canada 53°12N 131°58W 70 C2
Alligator Pond Jamaica 17°52N 77°34W 88 a
Alliston Canada 44°9N 79°52W 82 B5
Alloa U.K. 56°7N 3°47W 11 E5
Allora Australia 28°2S 152°0E 63 D5
Alluitsup Paa Greenland 60°30N 45°35W 4 C5
Alma Canada 48°35N 71°40W 73 C5
Alma Ga., U.S.A. 31°33N 82°28W 85 F13
Alma Kans., U.S.A. 39°1N 96°17W 80 F5
Alma Mich., U.S.A. 43°23N 84°39W 81 D11
Alma Nebr., U.S.A. 40°6N 99°22W 80 E4
Alma Wis., U.S.A. 44°20N 91°55W 80 C8
Alma Ata = Almaty Kazakhstan 43°15N 76°57E 32 C4
Alma Hill U.S.A. 42°2N 78°0W 82 D7
Almada Portugal 38°41N 9°8W 21 C1
Almaden Australia 17°22S 144°40E 62 B3
Almadén Spain 38°49N 4°52W 21 C3
Almalyk = Olmaliq Uzbekistan 40°50N 69°35E 28 E7
Almanor, L. U.S.A. 40°14N 121°9W 76 F3
Almansa Spain 38°51N 1°5W 21 C5
Almanzor, Pico Spain 40°15N 5°18W 21 B3
Almanzora → Spain 37°14N 1°46W 21 D5
Almaty Kazakhstan 43°15N 76°57E 32 C4
Almazán Spain 41°30N 2°30W 21 B4
Almeirim Brazil 1°30S 52°34W 93 D8
Almelo Neths. 52°22N 6°42E 15 B6
Almendralejo Spain 38°41N 6°26W 21 C2
Almere-Stad Neths. 52°20N 5°15E 15 B5
Almería Spain 36°52N 2°27W 21 D4
Almirante Panama 9°10N 82°30W 88 E3
Almond U.S.A. 42°19N 77°44W 82 D7
Almont U.S.A. 42°55N 83°3W 82 D1
Almonte Canada 45°14N 76°12W 83 A8
Almora India 29°38N 79°40E 43 E8
Almyrou, Ormos Greece 35°23N 24°20E 25 D6
Alness U.K. 57°41N 4°16W 11 D4
Alnmouth U.K. 55°24N 1°37W 12 B6
Alnwick U.K. 55°24N 1°42W 12 B6
Aloi Uganda 2°16N 33°10E 54 B3
Alon Burma 22°12N 95°5E 41 H19
Alor Indonesia 8°15S 124°30E 37 F6
Alor Setar Malaysia 6°7N 100°22E 39 J3
Alot India 23°56N 75°40E 42 H6
Aloysius, Mt. Australia 26°0S 128°38E 61 E4
Alpaugh U.S.A. 35°53N 119°29W 78 K7
Alpena U.S.A. 45°4N 83°27W 81 C12
Alpha Australia 23°39S 146°37E 62 C4
Alpha Ridge Arctic 84°0N 118°0W 4 A2
Alphen aan den Rijn Neths. 52°7N 4°40E 15 B4
Alpine Ariz., U.S.A. 33°51N 109°9W 77 K9
Alpine Calif., U.S.A. 32°50N 116°46W 79 N10
Alpine Tex., U.S.A. 30°22N 103°40W 84 F3
Alps Europe 46°30N 9°30E 20 C8
Alpurrurulam Australia 20°59S 137°50E 62 C2
Alsace □ France 48°15N 7°25E 20 B7
Alsask Canada 51°21N 109°59W 71 C7
Alsasua Spain 42°54N 2°10W 21 A4
Alsek → U.S.A. 59°10N 138°12W 70 B1
Alsta Norway 65°58N 12°40E 8 D15
Alston U.K. 54°49N 2°25W 12 C5
Alta Norway 69°57N 23°10E 8 B20
Alta Gracia Argentina 31°40S 64°30W 94 C3
Alta Sierra U.S.A. 35°42N 118°33W 79 K8
Altaelva → Norway 69°54N 23°17E 8 B20
Altafjorden Norway 70°5N 23°5E 8 A20
Altai = Aerhtai Shan Mongolia 46°40N 92°45E 32 B7
Altai = Gorno-Altay □ Russia 51°0N 86°0E 28 D9
Altamaha → U.S.A. 31°20N 81°20W 85 F14
Altamira Brazil 3°12S 52°10W 93 D8
Altamira Chile 25°47S 69°51W 94 B2
Altamira Mexico 22°24N 97°55W 87 C5
Altamont U.S.A. 42°42N 74°2W 83 D10
Altamura Italy 40°49N 16°33E 22 D7
Altanbulag Mongolia 50°16N 106°30E 32 A10
Altar Mexico 30°43N 111°44W 86 A2
Altar, Gran Desierto de Mexico 31°50N 114°10W 86 B2
Altata Mexico 24°40N 107°55W 86 C3
Altavista U.S.A. 37°6N 79°17W 81 G14
Altay China 47°48N 88°10E 32 B6
Altay Mongolia 46°22N 96°15E 32 B8
Altea Spain 38°38N 0°2W 21 C5
Altiplano Bolivia 17°0S 68°0W 92 G5
Alto Araguaia Brazil 17°15S 53°20W 93 G8
Alto Cuchumatanes = Cuchumatanes, Sierra de los Guatemala 15°35N 91°25W 88 C1
Alto del Carmen Chile 28°46S 70°30W 94 B1
Alto Ligonha Mozam. 15°30S 38°11E 55 F4
Alto Molocue Mozam. 15°50S 37°35E 55 F4
Alto Paraguay □ Paraguay 21°0S 58°30W 94 A4
Alto Paraná □ Paraguay 25°30S 54°50W 95 B5
Alton Canada 43°54N 80°5W 82 C4
Alton U.K. 51°9N 0°59W 13 F7
Alton Ill., U.S.A. 38°53N 90°11W 80 F8
Alton N.H., U.S.A. 43°27N 71°13W 83 C13
Altona Canada 49°6N 97°33W 71 D9
Altoona U.S.A. 40°31N 78°24W 82 F6
Altun Kupri Iraq 35°45N 44°9E 44 C5
Altun Shan Xinjiang Uygur, China 39°18N 93°42E 32 D7
Altun Shan Xinjiang Uygur, China 38°30N 88°0E 32 D6
Alturas U.S.A. 41°29N 120°32W 76 F3
Altus U.S.A. 34°38N 99°20W 84 D5
Alucra Turkey 40°22N 38°47E 19 F6
Alūksne Latvia 57°24N 27°3E 9 H22
Alunite U.S.A. 35°59N 114°55W 79 K12
Alusi Indonesia 7°35S 131°40E 37 F8
Alva U.S.A. 36°48N 98°40W 84 C5
Alvarado Mexico 18°46N 95°46W 87 D5
Alvarado U.S.A. 32°24N 97°13W 84 E6
Alvaro Obregón, Presa Mexico 27°52N 109°52W 86 B3
Alvear Argentina 29°5S 56°30W 94 B4
Alvesta Sweden 56°54N 14°35E 9 H16
Alvin U.S.A. 29°26N 95°15W 84 G7
Alvinston Canada 42°49N 81°52W 82 D3
Älvkarleby Sweden 60°34N 17°26E 9 F17
Alvord Desert U.S.A. 42°30N 118°25W 76 E4
Älvsbyn Sweden 65°40N 21°0E 8 D19
Alwar India 27°38N 76°34E 42 F7
Alxa Zuoqi China 38°50N 105°40E 34 E3
Alyangula Australia 13°55S 136°30E 62 A2
Alyata = Älät Azerbaijan 39°58N 49°25E 45 B6
Alyth U.K. 56°38N 3°13W 11 E5
Alytus Lithuania 54°24N 24°3E 9 J21
Alzada U.S.A. 45°2N 104°25W 76 D11
Alzamay Russia 55°33N 98°39E 29 D10
Alzira Spain 39°9N 0°30W 21 C5
Am Timan Chad 11°0N 20°10E 51 F10
Amadeus, L. Australia 24°54S 131°0E 61 D5
Amadi Dem. Rep. of the Congo 3°40N 26°40E 54 B2
Amâdi Sudan 5°29N 30°25E 51 G12
Amadjuak L. Canada 65°0N 71°8W 69 E17
Amagansett U.S.A. 40°59N 72°9W 83 F12
Amagasaki Japan 34°42N 135°23E 31 G7
Amagi Japan 33°25N 130°39E 31 H5
Amahai Indonesia 3°20S 128°55E 37 E7
Amaile Samoa 13°59S 171°22W 59 b
Amakusa-Shotō Japan 32°15N 130°10E 31 H5
Åmål Sweden 59°3N 12°42E 9 G15
Amaliada Greece 37°47N 21°22E 23 F9
Amalner India 21°5N 75°5E 40 J9
Amamapare Indonesia 4°53S 136°38E 37 E9
Amambaí Brazil 23°5S 55°13W 95 A4
Amambaí → Brazil 23°22S 53°56W 95 A5
Amambay □ Paraguay 23°0S 56°0W 95 A4
Amambay, Cordillera de S. Amer. 23°0S 55°45W 95 A4
Amami-Guntō Japan 27°16N 129°21E 31 L4
Amami-Ō-Shima Japan 28°16N 129°21E 31 K4
Aman, Pulau Malaysia 5°16N 100°24E 39 c
Amaná, L. Brazil 2°35S 64°40W 92 D6
Amanat → India 24°7N 84°4E 43 G11
Amanda Park U.S.A. 47°28N 123°55W 78 C3
Amankeldi Kazakhstan 50°10N 65°10E 28 D7
Amapá Brazil 2°5N 50°50W 93 C8
Amapá □ Brazil 1°40N 52°0W 93 C8
Amarante Brazil 6°14S 42°50W 93 E10
Amaranth Canada 50°36N 98°43W 71 C9
Amargosa → U.S.A. 36°14N 116°51W 79 J10
Amargosa Desert U.S.A. 36°40N 116°30W 79 J10
Amargosa Range U.S.A. 36°20N 116°45W 79 J10
Amari Greece 35°13N 24°40E 25 D6
Amarillo U.S.A. 35°13N 101°50W 84 D4
Amarkantak India 22°40N 81°45E 43 H9
Amaro, Mte. Italy 42°5N 14°5E 22 C6
Amarpur India 25°5N 87°0E 43 G12
Amarwara India 22°18N 79°10E 43 H8
Amasya Turkey 40°40N 35°50E 19 F6
Amata Australia 26°9S 131°9E 61 E5

Amatikulu S. Africa 29°3S 31°33E 57 D5
Amatitlán Guatemala 14°29N 90°38W 88 D1
Amay Belgium 50°33N 5°19E 15 D5
Amazon = Amazonas → S. Amer. 0°5S 50°0W 93 D8
Amazonas □ Brazil 5°0S 65°0W 92 E6
Amazonas → S. Amer. 0°5S 50°0W 93 D8
Ambah India 26°43N 78°13E 42 F8
Ambahakily Madag. 21°36S 43°41E 57 C7
Ambahita Madag. 24°1S 45°16E 57 C8
Ambala India 30°23N 76°56E 42 D7
Ambalavao Madag. 21°50S 46°56E 57 C8
Ambanja Madag. 13°40S 48°27E 57 A8
Ambararata Madag. 15°3S 48°33E 57 B8
Ambarchik Russia 69°40N 162°20E 29 C17
Ambarijeby Madag. 14°56S 47°41E 57 A8
Ambaro, Helodranon' Madag. 13°23S 48°38E 57 A8
Ambato Ecuador 1°5S 78°42W 92 D3
Ambato Madag. 13°24S 48°29E 57 A8
Ambato, Sierra de Argentina 28°25S 66°10W 94 B2
Ambato Boeny Madag. 16°28S 46°43E 57 B8
Ambatofinandrahana Madag. 20°33S 46°48E 57 C8
Ambatolampy Madag. 19°20S 47°35E 57 B8
Ambatomainty Madag. 17°41S 45°40E 57 B8
Ambatomanoina Madag. 18°18S 47°37E 57 B8
Ambatondrazaka Madag. 17°55S 48°28E 57 B8
Ambatosoratra Madag. 17°37S 48°31E 57 B8
Ambenja Madag. 15°17S 46°58E 57 B8
Amberg Germany 49°26N 11°52E 16 D6
Ambergris Cay Belize 18°0N 87°55W 87 D7
Amberley Canada 44°2N 81°42W 82 B3
Amberley N.Z. 43°9S 172°44E 59 E4
Ambikapur India 23°15N 83°15E 43 H10
Ambilobé Madag. 13°10S 49°3E 57 A8
Ambinanindrano Madag. 20°5S 48°23E 57 C8
Ambinanitelo Madag. 15°21S 49°35E 57 B8
Ambinda Madag. 16°25S 45°52E 57 B8
Amble U.K. 55°20N 1°36W 12 B6
Ambleside U.K. 54°26N 2°58W 12 C5
Ambo Peru 10°5S 76°10W 92 F3
Amboahangy Madag. 24°15S 46°22E 57 C8
Ambodifototra Madag. 16°59S 49°52E 57 B8
Ambodilazana Madag. 18°6S 49°10E 57 B8
Ambodiriana Madag. 17°55S 49°18E 57 B8
Ambohidratrimo Madag. 18°50S 47°26E 57 B8
Ambohidray Madag. 18°36S 48°18E 57 B8
Ambohimahamasina Madag. 21°56S 47°11E 57 C8
Ambohimahasoa Madag. 21°7S 47°13E 57 C8
Ambohimanga Madag. 20°52S 47°36E 57 C8
Ambohimitombo Madag. 20°43S 47°26E 57 C8
Ambohitra Madag. 12°30S 49°10E 57 A8
Amboise France 47°24N 1°2E 20 C4
Ambon Indonesia 3°43S 128°12E 37 E7
Ambondro Madag. 25°13S 45°44E 57 D8
Amboseli, L. Kenya 2°40S 37°10E 54 C4
Amboseli △ Kenya 2°37S 37°13E 54 C4
Ambositra Madag. 20°31S 47°25E 57 C8
Ambovombe Madag. 25°11S 46°5E 57 D8
Amboy U.S.A. 34°33N 115°45W 79 L11
Amboyna Cay S. China Sea 7°50N 112°50E 36 C4
Ambridge U.S.A. 40°36N 80°14W 82 F4
Ambriz Angola 7°48S 13°8E 52 F2
Amchitka I. U.S.A. 51°32N 179°0E 74 E3
Amderma Russia 69°45N 61°30E 28 C7
Amdhi India 23°51N 81°27E 43 H9
Amdo China 32°20N 91°40E 32 E7
Ameca Mexico 20°33N 104°2W 86 C4
Ameca → Mexico 20°41N 105°18W 86 C3
Amecameca de Juárez Mexico 19°8N 98°46W 87 D5
Amed Indonesia 8°19S 115°39E 37 J18
Ameland Neths. 53°27N 5°45E 15 A5
Amenia U.S.A. 41°51N 73°33W 83 E11
America-Antarctica Ridge S. Ocean 59°0S 16°0W 5 B2
American Falls U.S.A. 42°47N 112°51W 76 E7
American Falls Res. U.S.A. 42°47N 112°52W 76 E7
American Fork U.S.A. 40°23N 111°48W 76 F8
American Highland Antarctica 73°0S 75°0E 5 D6
American Samoa ☑ Pac. Oc. 14°20S 170°0W 59 b
American Samoa △ Amer. Samoa 14°15S 170°28W 59 b
Americana Brazil 22°45S 47°20W 95 A6
Americus U.S.A. 32°4N 84°14W 85 E12
Amersfoort Neths. 52°9N 5°23E 15 B5
Amersfoort S. Africa 26°59S 29°53E 57 D4
Amery Basin S. Ocean 68°15S 74°30E 5 C6
Amery Ice Shelf Antarctica 69°30S 72°0E 5 C6
Ames U.S.A. 42°2N 93°37W 80 D7
Amesbury U.S.A. 42°51N 70°56W 83 D14
Amet India 25°18N 73°56E 42 G5
Amga Russia 60°50N 132°0E 29 C14
Amga → Russia 62°38N 134°32E 29 C14
Amgu Russia 45°45N 137°15E 30 B8
Amgun → Russia 52°56N 139°38E 29 D14
Amherst Canada 45°48N 64°8W 73 C7
Amherst Mass., U.S.A. 42°23N 72°31W 83 D12
Amherst N.Y., U.S.A. 42°59N 78°48W 82 D6
Amherst Ohio, U.S.A. 41°24N 82°14W 82 E2
Amherst I. Canada 44°8N 76°43W 83 B8
Amherstburg Canada 42°6N 83°6W 72 D3
Amiata, Mte. Italy 42°53N 11°37E 22 C4
Amidon U.S.A. 46°29N 103°19W 80 B2
Amiens France 49°54N 2°16E 20 B5
Aminuis Namibia 23°43S 19°21E 56 C2
Amīrābād Iran 33°20N 46°16E 44 C5
Amirante Is. Seychelles 6°0S 53°0E 26 J7
Amisk △ Canada 56°43N 98°0W 71 B9
Amisk L. Canada 54°35N 102°15W 71 C8
Amistad, Presa de la Mexico 29°26N 101°3W 86 B4
Amistad △ U.S.A. 29°32N 101°12W 84 G4
Amite U.S.A. 30°44N 90°30W 85 F9
Amla India 21°56N 78°7E 42 J8
Amlapura Indonesia 8°27S 115°37E 37 J18
Amlia I. U.S.A. 52°4N 173°30W 74 E5
Amlwch U.K. 53°24N 4°20W 12 D3
'Ammān Jordan 31°57N 35°52E 46 D4
'Ammān □ Jordan 31°40N 36°30E 46 D5
'Ammān ✕ (AMM) Jordan 31°45N 36°2E 46 D5
Ammanford U.K. 51°48N 3°59W 13 F4
Ammassalik = Tasiilaq Greenland 65°40N 37°20W 4 C6
Ammochostos = Famagusta Cyprus 35°8N 33°55E 25 D12
Ammon U.S.A. 43°28N 111°58W 76 E8
Amnat Charoen Thailand 15°51N 104°38E 38 E5
Amnura Bangla. 24°37N 88°25E 43 G13
Āmol Iran 36°23N 52°20E 45 B7
Amorgos Greece 36°50N 25°57E 23 F11
Amory U.S.A. 33°59N 88°29W 85 E10
Amos Canada 48°35N 78°5W 72 C4
Åmot Norway 59°57N 9°54E 9 G13
Amoy = Xiamen China 24°25N 118°4E 33 G12
Ampanavoana Madag. 15°41S 50°22E 57 B9
Ampang Malaysia 3°8N 101°45E 39 L3
Ampangalana, Lakandranon' Madag. 22°48S 47°50E 57 C8
Ampanihy Madag. 24°40S 44°45E 57 C7
Amparafaravola Madag. 17°35S 48°13E 57 B8
Ampasinambo Madag. 20°31S 48°0E 57 C8
Ampasindava, Helodranon' Madag. 13°40S 48°15E 57 A8
Ampasindava, Saikanosy Madag. 13°42S 47°55E 57 A8
Ampenan Indonesia 8°34S 116°4E 37 K18
Amper → Germany 48°29N 11°55E 16 D6
Amphoe Kathu Thailand 7°55N 98°21E 39 a
Amphoe Thalang Thailand 8°1N 98°20E 39 a
Ampitsikinana Madag. 12°57S 49°49E 57 A8
Ampombiantambo Madag. 12°42S 48°57E 57 A8
Ampotaka Madag. 25°3S 44°41E 57 D7
Ampoza Madag. 22°20S 44°44E 57 C7
Amqui Canada 48°28N 67°27W 73 C6
Amravati India 20°55N 77°45E 40 J10
Amreli India 21°35N 71°17E 42 J4
Amritsar India 31°35N 74°57E 42 D6
Amroha India 28°53N 78°30E 43 E8
Amsterdam Neths. 52°23N 4°54E 15 B4
Amsterdam U.S.A. 42°56N 74°11W 83 D10
Amsterdam ✕ (AMS) Neths. 52°18N 4°45E 15 B4
Amsterdam, I. = Nouvelle Amsterdam, Î. Ind. Oc. 38°30S 77°30E 3 F13
Amstetten Austria 48°7N 14°51E 16 D8
Amudarya → Uzbekistan 43°58N 59°34E 28 E6
Amund Ringnes I. Canada 78°20N 96°25W 69 B12
Amundsen Abyssal Plain S. Ocean 65°0S 125°0W 5 C14
Amundsen Basin Arctic 87°30N 80°0E 4 A
Amundsen Gulf Canada 71°0N 124°0W 68 C7
Amundsen Ridges S. Ocean 69°15S 123°0W 5 C14
Amundsen-Scott Antarctica 90°0S 166°0E 5 E
Amundsen Sea Antarctica 72°0S 115°0W 5 D15
Amuntai Indonesia 2°28S 115°25E 36 E5
Amur → Russia 52°56N 141°10E 29 D15
Amurang Indonesia 1°5N 124°40E 37 D6
Amursk Russia 50°14N 136°54E 29 D14
Amyderya = Amudarya → Uzbekistan 43°58N 59°34E 28 E6
An Bien Vietnam 9°45N 105°0E 39 H5
An Cabhán = Cavan Ireland 54°0N 7°22W 10 B4
An Cóbh = Cóbh Ireland 51°51N 8°17W 10 E3
An Daingean = Dingle Ireland 52°9N 10°17W 10 D1
An Hoa Vietnam 15°40N 108°5E 38 E7
An Khe Vietnam 13°57N 108°51E 38 F7
An Longfort = Longford Ireland 53°43N 7°49W 10 C4
An Muileann gCearr = Mullingar Ireland 53°31N 7°21W 10 C4
An Nabatīyah at Tahta Lebanon 33°23N 35°27E 46 B4
An Nabk Si. Arabia 31°20N 37°20E 44 D3
An Nabk Syria 34°2N 36°44E 46 A5
An Nafūd Si. Arabia 28°15N 41°0E 44 D4
An Najaf Iraq 32°3N 44°15E 44 C5
An Nás = Naas Ireland 53°12N 6°40W 10 C5
An Nāşirīyah Iraq 31°0N 46°15E 44 D5
An Nhon = Binh Dinh Vietnam 13°55N 109°7E 38 F7
An Nu'ayrīyah Si. Arabia 27°30N 48°30E 45 E6
An Nu'mānīyah Iraq 32°32N 45°25E 44 C5
An Ros = Rush Ireland 53°31N 6°6W 10 C5
An tAonach = Nenagh Ireland 52°52N 8°11W 10 D3
An Thoi, Quan Dao Vietnam 9°58N 104°0E 39 H5
An tInbhear Mór = Arklow Ireland 52°48N 6°10W 10 D5
An Uaimh Ireland 53°39N 6°41W 10 C5
Anabar → Russia 73°8N 113°36E 29 B12
Anaconda U.S.A. 46°8N 112°57W 76 C7
Anacortes U.S.A. 48°30N 122°37W 78 B4
Anadarko U.S.A. 35°4N 98°15W 84 D5
Anadolu Turkey 39°0N 30°0E 19 G5
Anadyr Russia 64°35N 177°20E 29 C18
Anadyr → Russia 64°55N 176°5E 29 C18
Anadyrskiy Zaliv Russia 64°0N 180°0E 29 C19
Anaga, Pta. de Canary Is. 28°34N 16°9W 24 F3
'Ānah Iraq 34°25N 42°0E 44 C4
Anaheim U.S.A. 33°50N 117°55W 79 M9
Anahim Lake Canada 52°28N 125°18W 70 C3
Anakapalle India 17°42N 83°6E 41 L13
Anakie Australia 23°32S 147°45E 62 C4
Analalava Madag. 14°35S 48°0E 57 A8
Analavoka Madag. 22°23S 46°30E 57 C8
Analipsis Greece 39°36N 19°55E 25 A3
Anambar → Pakistan 30°15N 68°50E 42 D3
Anambas, Kepulauan Indonesia 3°20N 106°30E 36 D3
Anambas Is. = Anambas, Kepulauan Indonesia 3°20N 106°30E 36 D3
Anamosa U.S.A. 42°7N 91°17W 80 D8
Anamur Turkey 36°8N 32°58E 44 B2
Anan Japan 33°54N 134°40E 31 H7
Anand India 22°32N 72°59E 42 H5
Anangu Pitjantjatjara ◎ Australia 27°0S 132°0E 61 E5
Anantapur India 14°39N 77°42E 40 M10
Anantnag India 33°45N 75°10E 43 C6
Ananyiv Ukraine 47°44N 29°58E 17 E15
Anapodiaris → Greece 34°59N 25°20E 25 E7
Anápolis Brazil 16°15S 48°50W 93 G9
Anapu → Brazil 1°53S 50°53W 93 D8
Anār Iran 30°55N 55°13E 45 D7
Anārak Iran 33°25N 53°40E 45 C7
Anas → India 23°26N 74°0E 42 H5
Anatolia = Anadolu Turkey 39°0N 30°0E 19 G5
Anatsogno Madag. 23°33S 43°46E 57 C7
Añatuya Argentina 28°20S 62°50W 94 B3
Anatye ◎ Australia 22°29S 137°3E 62 C2
Anaunethad L. Canada 60°55N 104°25W 71 A8
Anbyŏn N. Korea 39°1N 127°35E 35 E14
Ancaster Canada 43°13N 79°59W 82 C5
Anchor Bay U.S.A. 38°48N 123°34W 78 G3
Anchorage U.S.A. 61°13N 149°54W 68 E2
Anci China 39°20N 116°40E 34 E9
Ancohuma, Nevado Bolivia 16°0S 68°50W 92 G5
Ancón Peru 11°50S 77°10W 92 F3
Ancona Italy 43°38N 13°30E 22 C5
Ancud Chile 42°0S 73°50W 96 E2
Ancud, G. de Chile 42°0S 73°0W 96 E2
Anda China 46°24N 125°19E 33 B14
Andacollo Argentina 37°10S 70°42W 94 D1
Andacollo Chile 30°14S 71°6W 94 C1
Andaingo Madag. 18°12S 48°17E 57 B8
Andalgalá Argentina 27°40S 66°30W 94 B2
Åndalsnes Norway 62°35N 7°43E 8 E12
Andalucía □ Spain 37°35N 5°0W 21 D3
Andalusia = Andalucía □ Spain 37°35N 5°0W 21 D3
Andalusia U.S.A. 31°18N 86°29W 85 F11
Andaman Is. Ind. Oc. 12°30N 92°45E 27 G11
Andaman Sea Ind. Oc. 13°0N 96°0E 36 B1
Andamooka Australia 30°27S 137°9E 63 E2
Andapa Madag. 14°39S 49°39E 57 A8
Andara Namibia 18°2S 21°9E 56 B3
Andenes Norway 69°19N 16°18E 8 B17
Andenne Belgium 50°28N 5°5E 15 D5
Anderson Alaska, U.S.A. 64°25N 149°15W 74 C10
Anderson Calif., U.S.A. 40°27N 122°18W 76 F2
Anderson Ind., U.S.A. 40°10N 85°41W 81 E11
Anderson Mo., U.S.A. 36°39N 94°27W 80 G6
Anderson S.C., U.S.A. 34°31N 82°39W 85 D13
Anderson → Canada 69°42N 129°0W 68 D6
Andes U.S.A. 42°12N 74°47W 83 D10
Andes, Cord. de los S. Amer. 20°0S 68°0W 92 H5
Andfjorden Norway 69°10N 16°20E 8 B17
Andhra Pradesh □ India 18°0N 79°0E 40 L11
Andijon Uzbekistan 41°10N 72°15E 32 C3
Andikíthira = Antikythira Greece 35°52N 23°15E 23 G10
Andilamena Madag. 17°1S 48°35E 57 B8
Andīmeshk Iran 32°27N 48°21E 45 C6
Andizhan = Andijon Uzbekistan 41°10N 72°15E 32 C3
Andoany Madag. 13°25S 48°16E 57 A8
Andohahela △ Madag. 24°4S 46°44E 57 C8
Andong S. Korea 36°40N 128°43E 35 F15
Andorra ■ Europe 42°30N 1°30E 20 E4
Andorra La Vella Andorra 42°31N 1°32E 20 E4
Andover U.K. 51°12N 1°29W 13 F6
Andover Maine, U.S.A. 44°38N 70°45W 83 B14
Andover Mass., U.S.A. 42°40N 71°8W 83 D13
Andover N.J., U.S.A. 40°59N 74°45W 83 F10
Andover N.Y., U.S.A. 42°10N 77°48W 82 D7
Andover Ohio, U.S.A. 41°36N 80°34W 82 E4
Andøya Norway 69°10N 15°50E 8 B16
Andradina Brazil 20°54S 51°23W 93 H8
Andrahary Madag. 13°37S 49°17E 57 A8
Andramasina Madag. 19°11S 47°35E 57 B8
Andranopasy Madag. 21°17S 43°44E 57 C7
Andranovory Madag. 23°8S 44°10E 57 C7
Andratx Spain 39°39N 2°25E 24 B9
Andreanof Is. U.S.A. 51°30N 176°0W 74 E4
Andrews S.C., U.S.A. 33°27N 79°34W 85 E15
Andrews Tex., U.S.A. 32°19N 102°33W 84 E3
Ándria Italy 41°13N 16°17E 22 D7
Andriamena Madag. 17°26S 47°30E 57 B8
Andriandampy Madag. 22°45S 45°41E 57 C8
Andriba Madag. 17°30S 46°58E 57 B8
Andringitra △ Madag. 22°13S 46°5E 57 C8
Androka Madag. 24°58S 44°2E 57 C7
Andros Greece 37°50N 24°57E 23 F11
Andros I. Bahamas 24°30N 78°0W 88 B4
Andros Town Bahamas 24°43N 77°47W 88 B4
Androscoggin → U.S.A. 43°58N 69°52W 83 C14
Andselv Norway 69°4N 18°34E 8 B18
Andújar Spain 38°3N 4°5W 21 C3
Andulo Angola 11°25S 16°45E 52 G3
Anegada Br. Virgin Is. 18°45N 64°20W 89 e
Anegada Passage W. Indies 18°15N 63°45W 89 C7
Aneto, Pico de Spain 42°37N 0°40E 21 A6
Ang Mo Kio Singapore 1°23N 103°50E 39 d
Ang Thong Thailand 14°35N 100°31E 38 E3
Ang Thong, Ko Thailand 9°37N 99°41E 39 b
Ang Thong, Mu Ko △ Thailand 9°40N 99°43E 39 H2
Angamos, Punta Chile 23°1S 70°32W 94 A1
Angara → Russia 58°5N 94°20E 29 D10
Angarsk Russia 52°30N 104°0E 32 A9
Angas Hills Australia 23°0S 127°50E 60 D4
Angaston Australia 34°30S 139°8E 63 E2
Ånge Sweden 62°31N 15°35E 8 E16
Ángel, Salto = Angel Falls Venezuela 5°57N 62°30W 92 B6
Ángel de la Guarda, I. Mexico 29°20N 113°25W 86 B2
Angel Falls Venezuela 5°57N 62°30W 92 B6
Angeles Phil. 15°9N 120°33E 37 A6
Ängelholm Sweden 56°15N 12°50E 9 H15
Angels Camp U.S.A. 38°4N 120°32W 78 G6
Ångermanälven → Sweden 64°0N 17°20E 8 E17
Ångermanland Sweden 63°36N 17°45E 8 E17
Angers Canada 45°31N 75°29W 83 A9
Angers France 47°30N 0°35W 20 C3
Ängesån → Sweden 66°16N 22°47E 8 C20
Angikuni L. Canada 62°12N 99°59W 71 A9
Angkor Cambodia 13°22N 103°50E 38 F4
Angledool Australia 29°5S 147°55E 63 D4
Anglesey U.K. 53°17N 4°20W 12 D3
Anglesey, Isle of □ U.K. 53°16N 4°18W 12 D3
Angleton U.S.A. 29°10N 95°26W 84 G7
Anglisidhes Cyprus 34°51N 33°27E 25 E12
Angmagssalik = Tasiilaq Greenland 65°40N 37°20W 4 C6
Ango Dem. Rep. of the Congo 4°10N 26°5E 54 B2
Angoche Mozam. 16°8S 39°55E 55 F4
Angoche, I. Mozam. 16°20S 39°50E 55 F4
Angol Chile 37°56S 72°45W 94 D1
Angola Ind., U.S.A. 41°38N 85°0W 81 E11
Angola N.Y., U.S.A. 42°38N 79°2W 82 D5
Angola ■ Africa 12°0S 18°0E 53 G3
Angoulême France 45°39N 0°10E 20 D4
Angoumois France 45°50N 0°25E 20 D3
Angra do Heroismo Azores 38°39N 27°13W 50 a
Angra dos Reis Brazil 23°0S 44°10W 95 A7
Angtassom Cambodia 11°1N 104°41E 39 G5
Angu Dem. Rep. of the Congo 3°23N 24°30E 54 B1
Anguang China 45°15N 123°45E 35 B12
Anguilla ☑ W. Indies 18°14N 63°5W 89 C7
Anguo China 38°28N 115°15E 34 E8
Angurugu Australia 14°0S 136°25E 62 A2
Angus Canada 44°19N 79°53W 82 B5
Angus □ U.K. 56°46N 2°56W 11 E6
Angwa → Zimbabwe 16°0S 30°23E 57 B5
Anhanduí → Brazil 21°46S 52°9W 95 A5
Anholt Denmark 56°42N 11°33E 9 H14
Anhui □ China 32°0N 117°0E 33 E12
Anhwei = Anhui □ China 32°0N 117°0E 33 E12
Anichab Namibia 21°0S 14°46E 56 C1
Animas → U.S.A. 36°43N 108°13W 77 H9
Anin Burma 15°36N 97°50E 38 E1
Anivorano Madag. 18°44S 48°58E 57 B8
Anjalankoski Finland 60°45N 26°51E 8 F22
Anjar India 23°6N 70°10E 42 H4
Anjou France 47°20N 0°15W 20 C3
Anjouan Comoros Is. 12°15S 44°20E 53 a
Anjozorobe Madag. 18°22S 47°52E 57 B8
Anju N. Korea 39°36N 125°40E 35 E13
Ankaboa, Tanjona Madag. 21°58S 43°20E 57 C7
Ankang China 32°40N 109°1E 34 H5
Ankara Turkey 39°57N 32°54E 19 G5
Ankarafantsika △ Madag. 16°8S 47°5E 57 B8
Ankaramena Madag. 21°57S 46°39E 57 C8
Ankaratra Madag. 19°25S 47°12E 53 H9
Ankasakasa Madag. 16°21S 44°52E 57 B7
Ankavandra Madag. 18°46S 45°18E 57 B8
Ankazoabo Madag. 22°18S 44°31E 57 C7
Ankazobe Madag. 18°20S 47°10E 57 B8
Ankeny U.S.A. 41°44N 93°36W 80 E7
Ankilimalinika Madag. 22°58S 43°45E 57 C7
Ankilizato Madag. 20°25S 45°1E 57 C8
Ankisabe Madag. 19°17S 46°29E 57 B8
Ankoro Dem. Rep. of the Congo 6°45S 26°55E 54 D2
Ankororoka Madag. 25°30S 45°11E 57 D8
Anlong Veng Cambodia 14°14N 104°5E 38 E5
Anmyeondo S. Korea 36°25N 126°25E 35 F14
Ann, C. U.S.A. 42°38N 70°35W 83 D14
Ann Arbor U.S.A. 42°17N 83°45W 81 D12
Anna U.S.A. 37°28N 89°15W 80 G9
Annaba Algeria 36°50N 7°46E 50 A7
Annalee → Ireland 54°2N 7°24W 10 B4
Annam = Trung-Phan Vietnam 17°0N 109°0E 38 D6
Annamitique, Chaîne Asia 17°0N 106°0E 38 D6
Annan U.K. 54°59N 3°16W 11 G5
Annan → U.K. 54°58N 3°16W 11 G5
Annapolis U.S.A. 38°59N 76°30W 81 F15
Annapolis Royal Canada 44°44N 65°32W 73 D6
Annapurna Nepal 28°34N 83°50E 43 E10
Annean, L. Australia 26°54S 118°14E 61 E2
Annecy France 45°55N 6°8E 20 D7
Annette I. U.S.A. 55°9N 131°28W 70 B2
Anning China 24°55N 102°26E 32 G9
Anniston U.S.A. 33°39N 85°50W 85 E12
Annobón Atl. Oc. 1°25S 5°36E 49 G4
Annotto B. Jamaica 18°17N 76°45W 88 a
Annville U.S.A. 40°20N 76°31W 83 F8
Anogia Greece 35°16N 24°52E 25 D6
Anorotsangana Madag. 13°56S 47°55E 57 A8
Anosibe Madag. 19°26S 48°13E 57 B8
Anping Hebei, China 38°15N 115°30E 34 E8
Anping Liaoning, China 41°5N 123°30E 35 D12
Anqing China 30°30N 117°3E 33 E12
Anqiu China 36°25N 119°10E 35 F10
Ansai China 36°50N 109°20E 34 F5
Ansan S. Korea 37°21N 126°52E 35 F14
Ansbach Germany 49°28N 10°34E 16 D6
Anse Boileau Seychelles 4°43S 55°29E 53 b
Anse Royale Seychelles 4°44S 55°31E 53 b
Anshan China 41°5N 122°58E 35 D1
Anshun China 26°18N 105°57E 32 F1
Ansley U.S.A. 41°18N 99°23W 80 E
Anson U.S.A. 32°45N 99°54W 84 E
Anson B. Australia 13°20S 130°6E 60 B
Ansongo Mali 15°25N 0°35E 50 E
Ansonia U.S.A. 41°21N 73°5W 83 E1
Anstruther U.K. 56°14N 2°41W 11 E
Ansudu Indonesia 2°11S 139°22E 37 E
Antabamba Peru 14°40S 73°0W 92 F
Antakya = Hatay Turkey 36°14N 36°10E 44 B
Antalaha Madag. 14°57S 50°20E 57 A
Antalya Turkey 36°52N 30°45E 19 G
Antalya Körfezi Turkey 36°15N 31°30E 19 G
Antambohobe Madag. 22°20S 46°47E 57 C
Antanambao-Manampotsy Madag. 19°29S 48°34E 57 B
Antanambe Madag. 16°26S 49°52E 57 B
Antananarivo Madag. 18°55S 47°31E 57 B
Antananarivo □ Madag. 19°0S 47°0E 57 B
Antanifotsy Madag. 19°39S 47°19E 57 B
Antanimbaribe Madag. 21°30S 44°48E 57 C
Antanimora Madag. 24°49S 45°40E 57 C
Antarctic Pen. Antarctica 67°0S 60°0W 5 C1
Antarctica 90°0S 0°0 5 E
Antelope Zimbabwe 21°2S 28°31E 55 G
Antequera Paraguay 24°8S 57°7W 94 A
Antequera Spain 37°5N 4°33W 21 D
Antero, Mt. U.S.A. 38°41N 106°15W 76 G1
Antevamena Madag. 21°2S 44°8E 57 C
Anthony Kans., U.S.A. 37°9N 98°2W 80 G
Anthony N. Mex., U.S.A. 32°0N 106°36W 77 K1
Anti Atlas Morocco 30°0N 8°30W 50 C
Anti-Lebanon = Sharqi, Al Jabal ash Lebanon 33°40N 36°10E 46 B
Antibes France 43°34N 7°6E 20 E
Anticosti, Î. d' Canada 49°30N 63°0W 73 C
Antigo U.S.A. 45°9N 89°9W 80 C
Antigonish Canada 45°38N 61°58W 73 C
Antigua Canary Is. 28°24N 14°1W 24 F
Antigua Guatemala 14°34N 90°41W 88 D
Antigua W. Indies 17°0N 61°50W 89 C
Antigua & Barbuda ■ W. Indies 17°20N 61°48W 89 C
Antikythira Greece 35°52N 23°15E 23 G
Antilla Cuba 20°40N 75°50W 88 B
Antilles = West Indies Cent. Amer. 15°0N 65°0W 89 D
Antioch U.S.A. 38°1N 121°48W 78 G
Antioquia Colombia 6°40N 75°55W 92 B
Antipodes Is. Pac. Oc. 49°45S 178°40E 64 M
Antlers U.S.A. 34°14N 95°37W 84 D
Antoetra Madag. 20°46S 47°20E 57 C
Antofagasta Chile 23°50S 70°30W 94 A
Antofagasta □ Chile 24°0S 69°0W 94 A
Antofagasta de la Sierra Argentina 26°5S 67°20W 94 B
Antofalla Argentina 25°30S 68°5W 94 B
Antofalla, Salar de Argentina 25°40S 67°45W 94 B
Anton U.S.A. 33°49N 102°10W 84 E
Antongila, Helodrano Madag. 15°30S 49°50E 57 B
Antonibé Madag. 15°7S 47°24E 57 B
Antonibé, Presqu'île d' Madag. 14°55S 47°20E 57 A
Antonina Brazil 25°26S 48°42W 95 B
Antrim U.K. 54°43N 6°14W 10 B
Antrim U.S.A. 40°7N 81°21W 82 F
Antrim □ U.K. 54°56N 6°25W 10 B
Antrim, Mts. of U.K. 55°3N 6°14W 10 A
Antrim Plateau Australia 18°8S 128°20E 60 C
Antsakabary Madag. 15°3S 48°56E 57 B
Antsalova Madag. 18°40S 44°37E 57 B
Antsenavolo Madag. 21°24S 48°3E 57 C
Antsiafabositra Madag. 17°18S 46°57E 57 B
Antsirabe Antananarivo, Madag. 19°55S 47°2E 57 B
Antsirabe Antsiranana, Madag. 14°0S 49°59E 57 A
Antsirabe Mahajanga, Madag. 15°57S 48°58E 57 B
Antsiranana Madag. 12°25S 49°20E 57 A
Antsiranana □ Madag. 12°16S 49°17E 57 A
Antsohihy Madag. 14°50S 47°59E 57 A
Antsohimbondrona Seranana Madag. 13°7S 48°48E 57 A
Antu China 42°30N 128°20E 35 C1
Antwerp = Antwerpen Belgium 51°13N 4°25E 15 C
Antwerp U.S.A. 44°12N 75°37W 83 B
Antwerpen Belgium 51°13N 4°25E 15 C
Antwerpen □ Belgium 51°15N 4°40E 15 C
Anupgarh India 29°10N 73°10E 42 E
Anuppur India 23°6N 81°41E 43 H
Anuradhapura Sri Lanka 8°22N 80°28E 40 Q1
Anurrete ◎ Australia 20°50S 135°38E 62 C
Anveh Iran 27°23N 54°11E 45 E
Anvers = Antwerpen Belgium 51°13N 4°25E 15 C
Anvers I. Antarctica 64°30S 63°40W 5 C1
Anxi China 40°30N 95°43E 32 C
Anxious B. Australia 33°24S 134°45E 63 E
Anyang China 36°5N 114°21E 34 F
Anyang S. Korea 37°23N 126°55E 35 F1
Anyer Indonesia 6°4S 105°53E 37 G1
Anyi China 35°2N 111°2E 34 G
Anza U.S.A. 33°35N 116°39W 79 M1
Anze China 36°10N 112°12E 34 F
Anzhero-Sudzhensk Russia 56°10N 86°0E 28 D
Ánzio Italy 41°27N 12°37E 22 D
Ao Makham Thailand 7°50N 98°24E 39
Ao Phangnga △ Thailand 8°10N 98°32E 39
Aoga-Shima Japan 32°28N 139°46E 31 H
Aohan Qi China 43°18N 119°43E 35 C1
Aoji N. Korea 42°31N 130°23E 35 C1
Aomen = Macau China 22°12N 113°33E 33
Aomori Japan 40°45N 140°45E 30 D1

NDEX TO WORLD MAPS
Aspatria 101

omori □ Japan 40°45N 140°40E 30 D10
onla India 28°16N 79°11E 43 E8
orai, Mt. Tahiti 17°34S 149°30W 59 d
oraki Mount Cook
N.Z. 43°36S 170°9E 59 E3
oral, Phnum Cambodia 12°0N 104°15E 39 G5
osta Italy 45°45N 7°20E 20 D7
otearoa = New Zealand ■
Oceania 40°0S 176°0E 59 D6
oukâr Mauritania 17°40N 10°0W 50 E4
ozou, Couloir d' Chad 22°0N 19°0E 51 D9
pá → S. Amer. 22°6S 58°2W 94 A4
pache U.S.A. 34°54N 98°22W 84 D5
pache Junction
U.S.A. 33°25N 111°33W 77 K8
palachee B. U.S.A. 30°0N 84°0W 85 G13
palachicola U.S.A. 29°43N 84°59W 85 G12
palachicola →
U.S.A. 29°43N 84°58W 85 G12
paporis → Colombia 1°23S 69°25W 92 D5
parados da Serra △
Brazil 29°10S 50°8W 95 B5
parri Phil. 18°22N 121°38E 37 A6
patity Russia 67°34N 33°22E 8 C25
patula = Finke
Australia 25°34S 134°35E 62 D1
patzingán Mexico 19°5N 102°21W 86 D4
peldoorn Neths. 52°13N 5°57E 15 B5
pennines = Appennini
Italy 44°30N 10°0E 22 B4
pi Nepal 30°0N 80°57E 32 F5
pia Samoa 13°50S 171°50W 59 b
piacás, Serra dos Brazil 9°50S 57°0W 92 E7
pies → S. Africa 25°15S 28°8E 57 D4
pizaco Mexico 19°25N 98°8W 87 D5
plao Peru 16°0S 72°40W 92 G4
po, Mt. Phil. 6°53N 125°14E 37 C7
polakia Greece 36°5N 27°48E 25 C9
polakia, Ormos Greece 36°5N 27°45E 25 C9
pollonia = Marsá Súsah
Libya 32°52N 21°59E 51 B10
pollonia Greece 36°15N 27°58E 25 C9
polo Bolivia 14°30S 68°30W 92 F5
popa El Salv. 13°48N 89°10W 88 D2
poré → Brazil 19°27S 50°57W 93 G8
postle Is. U.S.A. 47°0N 90°40W 80 B8
postle Islands ⌂ U.S.A. 46°55N 91°0W 80 B8
póstoles Argentina 28°0S 56°0W 95 B4
postolos Andreas, C.
Cyprus 35°42N 34°35E 25 D13
poteri Guyana 4°2N 58°32W 92 C7
ppalachian Mts. U.S.A. 38°0N 80°0W 81 G14
ppennini Italy 44°30N 10°0E 22 B4
pple Hill Canada 45°13N 74°46W 83 A10
pple Valley U.S.A. 34°32N 117°14W 79 L9
ppleby-in-Westmorland
U.K. 54°35N 2°29W 12 C5
ppledore U.K. 51°3N 4°13W 13 F3
ppleton U.S.A. 44°16N 88°25W 80 C9
pprouague →
Fr. Guiana 4°30N 51°57W 93 C8
prilia Italy 41°36N 12°39E 22 D5
psley Canada 44°45N 78°6W 82 B6
pucarana Brazil 23°55S 51°33W 95 A5
pure → Venezuela 7°37N 66°25W 92 B5
purimac → Peru 12°17S 73°56W 92 F4
q Qālā Iran 37°10N 54°30E 45 B7
qaba = Al 'Aqabah
Jordan 29°31N 35°0E 46 F4
qaba, G. of Red Sea 29°0N 34°40E 44 D2
qabah, Khalīj al = Aqaba, G. of
Red Sea 29°0N 34°40E 44 D2
qdā Iran 32°26N 53°37E 45 C7
qrah Iraq 36°46N 43°45E 44 B4
qsay Kazakhstan 51°11N 53°0E 19 D9
qtaū Kazakhstan 43°39N 51°12E 19 F9
qtöbe Kazakhstan 50°17N 57°10E 19 D10
qtoghay Kazakhstan 46°57N 79°40E 28 E8
quidauana Brazil 20°30S 55°50W 93 H7
quila Mexico 18°36N 103°30W 86 D4
quiles Serdán Mexico 28°36N 105°53W 86 B3
quin Haiti 18°16N 73°24W 89 C5
quitain, Bassin France 44°0N 0°30W 20 D3
r Horqin Qi China 43°45N 120°0E 35 C11
r Rafid Syria 32°57N 35°52E 46 C4
r Raḥḥālīyah Iraq 32°44N 43°23E 44 C4
r Ramādī Iraq 33°25N 43°20E 44 C4
r Ramthā Jordan 32°34N 36°0E 46 C5
r Raqqah Syria 35°59N 39°8E 44 C3
r Rashidiya = Er Rachidia
Morocco 31°58N 4°20W 50 B5
r Rass Si. Arabia 25°50N 43°40E 44 E4
r Rayyan Qatar 25°17N 51°25E 45 E6
r Rifā'ī Iraq 31°50N 46°10E 44 D5
r Riyāḍ Si. Arabia 24°41N 46°42E 44 E5
r Ru'ays Qatar 26°8N 51°12E 45 E6
r Rukhaymīyah Iraq 29°22N 45°38E 44 D5
r Rumaythah Iraq 31°31N 45°12E 44 D5
r Ruṣāfah Syria 35°45N 38°49E 44 C3
r Ruṭbah Iraq 33°0N 40°15E 44 C4
ra India 25°35N 84°32E 43 G11
rab U.S.A. 34°19N 86°30W 85 D11
rab, Bahr el → Sudan 9°0N 29°30E 51 G11
rab, Shatt al → Asia 29°57N 48°34E 45 D6
rabābād Iran 33°2N 57°41E 45 C8
rabia Asia 25°0N 45°0E 26 F6
rabian Desert = Es Sahrā' Esh
Sharqîya Egypt 27°30N 32°30E 51 C12
rabian Gulf = Persian Gulf
Asia 27°0N 50°0E 45 E6
rabian Sea Ind. Oc. 16°0N 65°0E 26 G8
racaju Brazil 10°55S 37°4W 93 F11
racati Brazil 4°30S 37°44W 93 D11
raçatuba Brazil 21°10S 50°30W 95 A5
racena Spain 37°53N 6°38W 21 D2
raçuaí Brazil 16°52S 42°4W 93 G10
rad Israel 31°15N 35°12E 46 D4
rad Romania 46°10N 21°20E 17 E11
rādān Iran 35°21N 52°30E 45 C7
radhippou Cyprus 34°57N 33°36E 25 E12
Arafura Sea E. Indies 9°0S 135°0E 58 B6
Aragón □ Spain 41°25N 0°40W 21 B5
Aragón → Spain 42°13N 1°44W 21 A5
Araguacema Brazil 8°50S 49°20W 93 E9
Araguaia → Brazil 5°21S 48°41W 93 E9
Araguaína Brazil 7°12S 48°12W 93 E9
Araguari Brazil 18°38S 48°11W 93 G9
Araguari → Brazil 1°15N 49°55W 93 C9
Arain India 26°27N 75°2E 42 F6
Arak Algeria 25°20N 3°45E 50 C6
Arāk Iran 34°0N 49°40E 45 C6
Arakan Coast Burma 19°0N 94°0E 41 K19
Arakan Yoma Burma 20°0N 94°40E 41 K19
Araks = Aras, Rūd-e →
Asia 40°5N 48°29E 44 B5
Aral Kazakhstan 46°41N 61°45E 28 E7
Aral Sea Asia 44°30N 60°0E 28 E7
Aral Tengizi = Aral Sea
Asia 44°30N 60°0E 28 E7
Aralsk = Aral Kazakhstan 46°41N 61°45E 28 E7
Aralskoye More = Aral Sea
Asia 44°30N 60°0E 28 E7
Aramac Australia 22°58S 145°14E 62 C4
Aran I. Ireland 55°0N 8°30W 10 A3
Aran Is. Ireland 53°6N 9°38W 10 C2
Aranda de Duero Spain 41°39N 3°42W 21 B4
Arandān Iran 35°23N 46°55E 44 C5
Aranjuez Spain 40°1N 3°40W 21 B4
Aranos Namibia 24°9S 19°7E 56 C2
Aransas Pass U.S.A. 27°55N 97°9W 84 H6
Aranyaprathet Thailand 13°41N 102°30E 38 F4
Arapahoe U.S.A. 40°18N 99°54W 80 E4
Arapey Grande →
Uruguay 30°55S 57°49W 94 C4
Arapgir Turkey 39°5N 38°30E 44 B3
Arapiraca Brazil 9°45S 36°39W 93 E11
Arapongas Brazil 23°29S 51°28W 95 A5
Ar'ar Si. Arabia 30°59N 41°2E 44 D4
Araranguá Brazil 29°0S 49°30W 95 B6
Araraquara Brazil 21°50S 48°0W 93 H9
Ararás, Serra das Brazil 25°0S 53°10W 95 B5
Ararat Australia 37°16S 142°58E 63 F3
Ararat, Mt. = Ağrı Dağı
Turkey 39°50N 44°15E 44 B5
Araria India 26°9N 87°33E 43 F12
Araripe, Chapada do
Brazil 7°20S 40°0W 93 E11
Araruama, L. de Brazil 22°53S 42°12W 95 A7
Aras, Rūd-e → Asia 40°5N 48°29E 44 B5
Arauca Colombia 7°0N 70°40W 92 B4
Arauca → Venezuela 7°24N 66°35W 92 B5
Arauco Chile 37°16S 73°25W 94 D1
Aravalli Range India 25°0N 73°30E 42 G5
Arawale ⌂ Kenya 1°24S 40°9E 54 C5
Araxá Brazil 19°35S 46°55W 93 G9
Araya, Pen. de Venezuela 10°40N 64°0W 92 A6
Arba Minch Ethiopia 6°0N 37°30E 47 F2
Arbat Iraq 35°25N 45°35E 44 C5
Árbatax Italy 39°56N 9°42E 22 E3
Arbil Iraq 36°15N 44°5E 44 B5
Arborfield Canada 53°6N 103°39W 71 C8
Arborg Canada 50°54N 97°13W 71 C9
Arbroath U.K. 56°34N 2°35W 11 E6
Arbuckle U.S.A. 39°1N 122°3W 78 F4
Arcachon France 44°40N 1°10W 20 D3
Arcade U.S.A. 42°32N 78°25W 82 D6
Arcadia Fla., U.S.A. 27°13N 81°52W 85 H14
Arcadia La., U.S.A. 32°33N 92°55W 84 E8
Arcadia Pa., U.S.A. 40°47N 78°51W 82 F6
Arcata U.S.A. 40°52N 124°5W 76 F1
Archangel = Arkhangelsk
Russia 64°38N 40°36E 18 B7
Archangelos Greece 36°13N 28°7E 25 C10
Archbald U.S.A. 41°30N 75°32W 83 E9
Archer → Australia 13°28S 141°41E 62 A3
Archer B. Australia 13°20S 141°30E 62 A3
Archer Bend = Mungkan
Kandju △ Australia 13°35S 142°52E 62 A3
Archers Post Kenya 0°35N 37°35E 54 B4
Arches △ U.S.A. 38°45N 109°25W 76 G9
Archipel-de-Mingan △
Canada 50°13N 63°10W 73 B7
Archipiélago Chinijo △
Canary Is. 29°20N 13°30W 24 E6
Archipiélago Los Roques △
Venezuela 11°50N 66°44W 89 D6
Arckaringa Cr. →
Australia 28°10S 135°22E 63 D2
Arco U.S.A. 43°38N 113°18W 76 E7
Arcos de la Frontera
Spain 36°45N 5°49W 21 D3
Arcot India 12°53N 79°20E 40 N11
Arctic Bay Canada 73°1N 85°7W 69 C14
Arctic Mid-Ocean Ridge
Arctic 87°0N 90°0E 4 A
Arctic Ocean Arctic 78°0N 160°0W 4 B18
Arctic Red River = Tsiigehtchic
Canada 67°15N 134°0W 68 D5
Arctowski Antarctica 62°30S 58°0W 5 C18
Arda → Bulgaria 41°40N 26°30E 23 D12
Ardabīl Iran 38°15N 48°18E 45 B6
Ardabīl □ Iran 38°15N 48°20E 45 B6
Ardakān = Sepīdān Iran 30°20N 52°5E 45 D7
Ardakān Iran 32°19N 53°59E 45 C7
Ardara Ireland 54°46N 8°25W 10 B3
Ardee Ireland 53°52N 6°33W 10 C5
Arden Canada 44°43N 76°56W 82 B8
Arden Calif., U.S.A. 38°36N 121°33W 78 G5
Arden Nev., U.S.A. 36°1N 115°14W 79 J11
Ardenne Belgium 49°50N 5°5E 15 E5
Ardennes = Ardenne
Belgium 49°50N 5°5E 15 E5
Arderin Ireland 53°2N 7°39W 10 C4
Ardestān Iran 33°20N 52°25E 45 C7
Ardfert Ireland 52°20N 9°47W 10 D2
Ardglass U.K. 54°17N 5°36W 10 B6
Ardivachar Pt. U.K. 57°23N 7°26W 11 D1
Ardlethan Australia 34°22S 146°53E 63 E4
Ardmore Okla., U.S.A. 34°10N 97°8W 84 D6
Ardmore Pa., U.S.A. 40°2N 75°17W 83 F9
Ardnamurchan, Pt. of
U.K. 56°43N 6°14W 11 E2
Ardnave Pt. U.K. 55°53N 6°20W 11 F2
Ardrossan Australia 34°26S 137°53E 63 E2
Ardrossan U.K. 55°39N 4°49W 11 F4
Ards Pen. U.K. 54°33N 5°34W 10 B6
Arecibo Puerto Rico 18°29N 66°43W 89 d
Areia Branca Brazil 5°0S 37°0W 93 E11
Arena, Pt. U.S.A. 38°57N 123°44W 78 G3
Arenal Honduras 15°21N 86°50W 88 C2
Arendal Norway 58°28N 8°46E 9 G13
Arequipa Peru 16°20S 71°30W 92 G4
Arévalo Spain 41°3N 4°43W 21 B3
Arezzo Italy 43°25N 11°53E 22 C4
Arga Turkey 38°21N 37°59E 44 B3
Arganda del Rey Spain 40°19N 3°26W 21 B4
Argarapa ⚙ Australia 22°20S 134°58E 62 C1
Argenta Canada 50°11N 116°56W 70 C5
Argentan France 48°45N 0°1W 20 B3
Argentário, Mte. Italy 42°24N 11°9E 22 C4
Argentia Canada 47°18N 53°58W 73 C9
Argentina ■ S. Amer. 35°0S 66°0W 96 D3
Argentine Basin Atl. Oc. 45°0S 45°0W 96 F6
Argentino, L. Argentina 50°10S 73°0W 96 G2
Argeș → Romania 44°5N 26°38E 17 F14
Arghandab → Afghan. 31°30N 64°15E 42 D1
Argirades Greece 39°27N 19°58E 25 B3
Argiroupoli Greece 35°17N 24°20E 25 D6
Argolikos Kolpos Greece 37°20N 22°52E 23 F10
Argos Greece 37°40N 22°43E 23 F10
Argostoli Greece 38°11N 20°29E 23 E9
Arguello, Pt. U.S.A. 34°35N 120°39W 79 L6
Arguineguín Canary Is. 27°46N 15°41W 24 G4
Argun → Russia 53°20N 121°28E 33 A13
Argungu Nigeria 12°40N 4°31E 50 F6
Argus Pk. U.S.A. 35°52N 117°26W 79 K9
Argyle, L. Australia 16°20S 128°40E 60 C4
Argyll ⌂ U.K. 56°6N 5°0W 11 E4
Argyll & Bute □ U.K. 56°13N 5°28W 11 E3
Århus Denmark 56°8N 10°11E 9 H14
Ariadnoye Russia 45°8N 134°25E 30 B7
Ariamsvlei Namibia 28°9S 19°51E 56 D2
Ariana Tunisia 36°52N 10°12E 51 A8
Arica Chile 18°32S 70°20W 92 G4
Arica Colombia 2°0S 71°50W 92 D4
Arico Canary Is. 28°9N 16°29W 24 F3
Arid, C. Australia 34°1S 123°10E 61 F3
Arida Japan 34°5N 135°8E 31 G7
Aride Seychelles 4°13S 55°40E 53 b
Ariège → France 43°30N 1°25E 20 E4
Arila, Akra Greece 39°43N 19°39E 25 A3
Arima Trin. & Tob. 10°38N 61°17W 93 K15
Arinos → Brazil 10°25S 58°20W 92 F7
Ario de Rosales Mexico 19°12N 101°43W 86 D4
Aripo, Mt. Trin. & Tob. 10°45N 61°15W 93 K15
Aripuanã Brazil 9°25S 60°30W 92 E6
Aripuanã → Brazil 5°7S 60°25W 92 E6
Ariquemes Brazil 9°55S 63°6W 92 E6
Arisaig U.K. 56°55N 5°51W 11 E3
Aristazabal I. Canada 52°40N 129°10W 70 C3
Arivonimamo Madag. 19°1S 47°11E 57 B8
Arizaro, Salar de
Argentina 24°40S 67°50W 94 A2
Arizona Argentina 35°45S 65°25W 94 D2
Arizona □ U.S.A. 34°0N 112°0W 77 J8
Arizpe Mexico 30°20N 110°10W 86 A2
'Arjah Si. Arabia 24°43N 44°17E 44 E5
Arjeplog Sweden 66°3N 17°54E 8 C17
Arjepluovve = Arjeplog
Sweden 66°3N 17°54E 8 C17
Arjona Colombia 10°14N 75°22W 92 A3
Arjuna Indonesia 7°49S 112°34E 37 G15
Arka Russia 60°15N 142°0E 29 C15
Arkadelphia U.S.A. 34°7N 93°4W 84 D8
Arkaig, L. U.K. 56°59N 5°10W 11 E3
Arkalyk = Arqalyk
Kazakhstan 50°13N 66°50E 28 D7
Arkansas □ U.S.A. 35°0N 92°30W 84 D8
Arkansas → U.S.A. 33°47N 91°4W 84 E9
Arkansas City U.S.A. 37°4N 97°2W 80 G5
Arkaroola Australia 30°20S 139°22E 63 E2
Arkhangelsk Russia 64°38N 40°36E 18 B7
Arki India 31°9N 76°58E 42 D7
Arklow Ireland 52°48N 6°10W 10 D5
Arkport U.S.A. 42°24N 77°42W 82 D7
Arkticheskiy, Mys
Russia 81°10N 95°0E 29 A10
Arkville U.S.A. 42°9N 74°37W 83 D10
Arlanzón → Spain 42°3N 4°17W 21 A3
Arlbergpass Austria 47°9N 10°12E 16 E6
Arlbergtunnel Austria 47°9N 10°12E 16 E6
Arles France 43°41N 4°40E 20 E6
Arlington S. Africa 28°1S 27°53E 57 D4
Arlington N.Y., U.S.A. 41°42N 73°54W 83 E11
Arlington Oreg., U.S.A. 45°43N 120°12W 76 D3
Arlington S. Dak., U.S.A. 44°22N 97°8W 80 C5
Arlington Tex., U.S.A. 32°44N 97°6W 84 E6
Arlington Va., U.S.A. 38°53N 77°7W 81 F15
Arlington Vt., U.S.A. 43°5N 73°9W 83 C11
Arlington Wash., U.S.A. 48°12N 122°8W 78 B4
Arlington Heights
U.S.A. 42°5N 87°59W 80 D10
Arlit Niger 19°0N 7°38E 50 E7
Arlon Belgium 49°42N 5°49E 15 E5
Arlparra Australia 22°11S 134°30E 62 C1
Arltunga Australia 23°26S 134°41E 62 C1
Armadale Australia 32°9S 116°0E 61 F2
Armagh U.K. 54°21N 6°39W 10 B5
Armagh □ U.K. 54°18N 6°37W 10 B5
Armando Bermudez △
Dom. Rep. 19°3N 71°0W 89 C5
Armavir Russia 45°2N 41°7E 19 E7
Armenia Colombia 4°35N 75°45W 92 C3
Armenia ■ Asia 40°20N 45°0E 19 F7
Armenistis, Akra Greece 36°8N 27°42E 25 C9
Armidale Australia 30°30S 151°40E 63 E5
Armour U.S.A. 43°19N 98°21W 80 D4
Armstrong B.C.,
Canada 50°25N 119°10W 70 C5
Armstrong Ont., Canada 50°18N 89°4W 72 B2
Arnarfjörður Iceland 65°48N 23°40W 8 D2
Arnaud → Canada 59°59N 69°46W 69 F18
Arnauti, C. Cyprus 35°6N 32°17E 25 D11
Arnett U.S.A. 36°8N 99°46W 84 C5
Arnhem Neths. 51°58N 5°55E 15 C5
Arnhem, C. Australia 12°20S 137°30E 62 A2
Arnhem B. Australia 12°20S 136°10E 62 A2
Arnhem Land Australia 13°10S 134°30E 62 A1
Arnhem Land ⚙
Australia 12°50S 134°50E 62 A1
Arno → Italy 43°41N 10°17E 22 C4
Arno Bay Australia 33°54S 136°34E 63 E2
Arnold U.K. 53°1N 1°7W 12 D6
Arnold U.S.A. 38°15N 120°21W 78 G6
Arnot Canada 55°56N 96°41W 71 B9
Arnøya Norway 70°9N 20°40E 8 A19
Arnprior Canada 45°26N 76°21W 83 A8
Arnsberg Germany 51°24N 8°5E 16 C5
Aroa, Pte. Moorea 17°28S 149°46W 59 d
Aroab Namibia 26°41S 19°39E 56 D2
Aron India 25°57N 77°56E 42 G6
Arona Canary Is. 28°6N 16°40W 24 F3
Aros → Mexico 29°9N 107°57W 86 B3
Arqalyk Kazakhstan 50°13N 66°50E 28 D7
Arrah = Ara India 25°35N 84°32E 43 G11
Arrah Ivory C. 6°40N 3°58W 50 G5
Arran U.K. 55°34N 5°12W 11 F3
Arras France 50°17N 2°46E 20 A5
Arrecife Canary Is. 28°57N 13°37W 24 F6
Arrecifes Argentina 34°6S 60°9W 94 C3
Arrée, Mts. d' France 48°26N 3°55W 20 B2
Arriaga Mexico 16°14N 93°54W 87 D6
Arrilalah Australia 23°43S 143°54E 62 C3
Arrino Australia 29°30S 115°40E 61 E2
Arrow, L. Ireland 54°3N 8°19W 10 B3
Arrowtown N.Z. 44°57S 168°50E 59 F2
Arroyo Grande U.S.A. 35°7N 120°35W 79 K6
Ars Iran 37°9N 47°46E 44 B5
Arsenault L. Canada 55°6N 108°32W 71 B7
Arsenev Russia 44°10N 133°15E 30 B6
Arta Greece 39°8N 21°2E 23 E9
Artà Spain 39°41N 3°21E 24 B10
Artà, Coves d' Spain 39°40N 3°24E 24 B10
Arteaga Mexico 18°28N 102°25W 86 D4
Artem Russia 43°22N 132°13E 30 C6
Artemovsk Russia 54°45N 93°35E 29 D10
Artemovsk Ukraine 48°35N 38°0E 19 E6
Artesia U.S.A. 32°51N 104°24W 77 K11
Arthur Canada 43°50N 80°32W 82 C4
Arthur → Australia 41°2S 144°40E 63 G3
Arthur Cr. → Australia 22°30S 136°25E 62 C2
Arthur Pt. Australia 22°7S 150°3E 62 C5
Arthur River Australia 33°20S 117°2E 61 F2
Arthur's Pass N.Z. 42°54S 171°35E 59 E3
Arthur's Pass △ N.Z. 42°53S 171°42E 59 E3
Arthur's Town Bahamas 24°38N 75°42W 89 B4
Artigas Antarctica 62°30S 58°0W 5 C18
Artigas Uruguay 30°20S 56°30W 94 C4
Artillery L. Canada 63°9N 107°52W 71 A7
Artois France 50°20N 2°30E 20 A5
Artrutx, C. de Spain 39°55N 3°49E 24 B10
Arts Bogd Uul Mongolia 44°40N 102°20E 34 B2
Artsyz Ukraine 46°4N 29°26E 17 E15
Artux China 39°40N 76°10E 32 D4
Artvin Turkey 41°14N 41°44E 19 F7
Artyk Russia 64°12N 145°6E 29 C15
Aru, Kepulauan Indonesia 6°0S 134°30E 37 F8
Aru Is. = Aru, Kepulauan
Indonesia 6°0S 134°30E 37 F8
Arua Uganda 3°1N 30°58E 54 B3
Aruanã Brazil 14°54S 51°10W 93 F8
Aruba ☒ W. Indies 12°30N 70°0W 89 D6
Arucas Canary Is. 28°7N 15°32W 24 F4
Arué Tahiti 17°31S 149°30W 59 d
Arun → Nepal 26°55N 87°10E 43 F12
Arun → U.K. 50°49N 0°33W 13 G7
Arunachal Pradesh □
India 28°0N 95°0E 41 F19
Arusha Tanzania 3°20S 36°40E 54 C4
Arusha □ Tanzania 4°0S 36°30E 54 C4
Arusha △ Tanzania 3°16S 36°47E 54 C4
Arusha Chini Tanzania 3°32S 37°20E 54 C4
Aruwimi →
Dem. Rep. of the Congo 1°13N 23°36E 54 B1
Arvada Colo., U.S.A. 39°48N 105°5W 76 G11
Arvada Wyo., U.S.A. 44°39N 106°8W 76 D10
Arvayheer Mongolia 46°15N 102°48E 32 B9
Arvi Greece 34°59N 25°28E 25 E7
Arviat Canada 61°6N 93°59W 71 A10
Arvidsjaur Sweden 65°35N 19°10E 8 D18
Arvika Sweden 59°40N 12°36E 9 G15
Arvin U.S.A. 35°12N 118°50W 79 K8
Arwal India 25°15N 84°41E 43 G11
Arxan China 47°11N 119°57E 33 B12
Arys Kazakhstan 42°26N 68°48E 28 E7
Arzamas Russia 55°27N 43°55E 18 C7
Arzanah U.A.E. 24°47N 52°34E 45 E7
Aş Şafā Syria 33°10N 37°0E 46 B6
As Saffānīyah Si. Arabia 27°55N 48°50E 45 E6
As Safīrah Syria 36°5N 37°21E 44 B3
Aş Şahm Oman 24°10N 56°53E 45 E8
As Sājir Si. Arabia 25°11N 44°36E 44 E5
As Salamīyah Syria 35°1N 37°2E 44 C3
As Salmān Iraq 30°30N 44°32E 44 D5
As Salṭ Jordan 32°2N 35°43E 46 C4
As Sal'w'a Qatar 24°23N 50°50E 45 E6
As Samāwah Iraq 31°15N 45°15E 44 D5
As Sanamayn Syria 33°3N 36°10E 46 B5
As Sohar = Şuḥār Oman 24°20N 56°40E 45 E8
As Sukhnah Syria 34°52N 38°52E 44 C3
As Sulaymānīyah Iraq 35°35N 45°29E 44 C5
As Sulaymī Si. Arabia 26°17N 41°21E 44 E4
As Sulayyil Si. Arabia 20°27N 45°34E 47 C4
As Summān Si. Arabia 25°0N 47°0E 44 E5
As Suwaydā' Syria 32°40N 36°30E 46 C5
As Suwaydā' □ Syria 32°45N 36°45E 46 C5
As Suwayq Oman 23°51N 57°26E 45 F8
Aş Şuwayrah Iraq 32°55N 45°0E 44 C5
Asab Namibia 25°30S 18°0E 56 D2
Asad, Buḥayrat al Syria 36°0N 38°15E 44 C3
Asahi Japan 35°43N 140°39E 31 E10
Asahi-Gawa → Japan 34°36N 133°58E 31 G6
Asahigawa = Asahikawa
Japan 43°46N 142°22E 30 C11
Asahikawa Japan 43°46N 142°22E 30 C11
Asaluyeh Iran 27°29N 52°37E 45 E7
Asamankese Ghana 5°50N 0°40W 50 G5
Asan → India 26°37N 78°24E 43 F8
Asansol India 23°40N 87°1E 43 H12
Asau Samoa 13°27S 172°33W 59 b
Asbesberge S. Africa 29°0S 23°0E 56 D3
Asbestos Canada 45°47N 71°58W 73 C5
Asbury Park U.S.A. 40°13N 74°1W 83 F10
Ascensión Mexico 31°6N 107°59W 86 A3
Ascensión, B. de la
Mexico 19°40N 87°30W 87 D7
Ascension I. Atl. Oc. 7°57S 14°23W 49 G2
Aschaffenburg Germany 49°58N 9°6E 16 D5
Aschersleben Germany 51°45N 11°29E 16 C6
Áscoli Piceno Italy 42°51N 13°34E 22 C5
Ascope Peru 7°46S 79°8W 92 E3
Ascotán Chile 21°45S 68°17W 94 A2
Aseb Eritrea 13°0N 42°40E 47 E3
Asela Ethiopia 8°0N 39°0E 47 F2
Asenovgrad Bulgaria 42°1N 24°51E 23 C11
Aşgabat = Ashgabat
Turkmenistan 38°0N 57°50E 45 B8
Asgata Cyprus 34°46N 33°15E 25 E12
Ash Fork U.S.A. 35°13N 112°29W 77 J7
Ash Grove U.S.A. 37°19N 93°35W 80 G7
Ash Shabakah Iraq 30°49N 43°39E 44 D4
Ash Shamāl □ Lebanon 34°25N 36°0E 46 A5
Ash Shāmīyah Iraq 31°55N 44°35E 44 D5
Ash Shāriqah U.A.E. 25°23N 55°26E 45 E7
Ash Sharmah Si. Arabia 28°1N 35°16E 44 D2
Ash Sharqāt Iraq 35°27N 43°16E 44 C4
Ash Shaṭrah Iraq 31°30N 46°10E 44 D5
Ash Shawbak Jordan 30°32N 35°34E 44 D2
Ash Shiḥr Yemen 14°45N 49°36E 47 E4
Ash Shināfiyah Iraq 31°35N 44°39E 44 D5
Ash Shu'bah Si. Arabia 28°54N 44°44E 44 D5
Ash Shumlūl Si. Arabia 26°31N 47°20E 44 E5
Ash Shūr'a Iraq 35°58N 43°13E 44 C4
Ash Shurayf Si. Arabia 25°43N 39°14E 44 E3
Ash Shuwayfāt Lebanon 33°45N 35°30E 46 B4
Asha Russia 55°0N 57°16E 18 D10
Ashbourne U.K. 53°2N 1°43W 12 D6
Ashburn U.S.A. 31°43N 83°39W 85 F13
Ashburton N.Z. 43°53S 171°48E 59 E3
Ashburton → Australia 21°40S 114°56E 60 D1
Ashcroft Canada 50°40N 121°20W 70 C4
Ashdod Israel 31°49N 34°35E 46 D3
Ashdown U.S.A. 33°40N 94°8W 84 E7
Asheboro U.S.A. 35°43N 79°49W 85 D15
Ashern Canada 51°11N 98°21W 71 C9
Asherton U.S.A. 28°27N 99°46W 84 G5
Asheville U.S.A. 35°36N 82°33W 85 D13
Ashewat Pakistan 31°22N 68°32E 42 D3
Asheweig → Canada 54°17N 87°12W 72 B2
Ashford Australia 29°15S 151°3E 63 D5
Ashford U.K. 51°8N 0°53E 13 F8
Ashgabat Turkmenistan 38°0N 57°50E 45 B8
Ashibetsu Japan 43°31N 142°11E 30 C11
Ashikaga Japan 36°28N 139°29E 31 F9
Ashington U.K. 55°11N 1°33W 12 B6
Ashizuri-Uwakai △
Japan 32°56N 132°32E 31 H6
Ashizuri-Zaki Japan 32°44N 133°0E 31 H6
Ashkarkot Afghan. 33°3N 67°58E 42 C2
Ashkhabad = Ashgabat
Turkmenistan 38°0N 57°50E 45 B8
Āshkhāneh Iran 37°26N 56°55E 45 B8
Ashland Kans., U.S.A. 37°11N 99°46W 80 G4
Ashland Ky., U.S.A. 38°28N 82°38W 81 F12
Ashland Maine, U.S.A. 46°38N 68°24W 81 B19
Ashland Mont., U.S.A. 45°36N 106°16W 76 D10
Ashland Ohio, U.S.A. 40°52N 82°19W 82 F2
Ashland Oreg., U.S.A. 42°12N 122°43W 76 E2
Ashland Pa., U.S.A. 40°45N 76°22W 83 F8
Ashland Va., U.S.A. 37°46N 77°29W 81 G15
Ashland Wis., U.S.A. 46°35N 90°53W 80 B8
Ashley N. Dak., U.S.A. 46°2N 99°22W 80 B4
Ashley Pa., U.S.A. 41°12N 75°55W 83 E9
Ashmore and Cartier Is.
Ind. Oc. 12°15S 123°0E 60 B3
Ashmore Reef Australia 12°14S 123°5E 60 B3
Ashmyany Belarus 54°26N 25°52E 17 A13
Ashokan Res. U.S.A. 41°56N 74°13W 83 E10
Ashoknagar India 24°34N 77°43E 42 G7
Ashqelon Israel 31°42N 34°35E 46 D3
Ashta India 23°1N 76°43E 42 H7
Ashtabula U.S.A. 41°52N 80°47W 82 E4
Ashtiyān Iran 34°31N 50°0E 45 C6
Ashton S. Africa 33°50S 20°5E 56 E3
Ashton U.S.A. 44°4N 111°27W 76 D8
Ashuanipi, L. Canada 52°45N 66°15W 73 B6
Ashuapmushuan →
Canada 48°37N 72°20W 72 C5
Ashville U.S.A. 40°34N 78°33W 82 F6
Asia 45°0N 75°0E 26 E9
Asia, Kepulauan
Indonesia 1°0N 131°13E 37 D8
Asifabad India 19°20N 79°24E 40 K11
Asinara Italy 41°4N 8°16E 22 D3
Asinara, G. dell' Italy 41°0N 8°30E 22 D3
Asino Russia 57°0N 86°0E 28 D9
Asipovichy Belarus 53°19N 28°33E 17 B15
'Asir Si. Arabia 18°40N 42°30E 47 D3
Asir, Ras Somali Rep. 11°55N 51°10E 47 E5
Askham S. Africa 26°59S 20°47E 56 D3
Askim Norway 59°35N 11°10E 9 G14
Askja Iceland 65°3N 16°48W 8 D5
Asklipio Greece 36°4N 27°56E 25 C9
Askøyna Norway 60°29N 5°10E 8 F11
Asmara = Asmera
Eritrea 15°19N 38°55E 47 D2
Asmera Eritrea 15°19N 38°55E 47 D2
Åsnen Sweden 56°37N 14°45E 9 H16
Aso Kujū △ Japan 32°53N 131°6E 31 H5
Aspatria U.K. 54°47N 3°19W 12 C4

Aspen *U.S.A.* 39°11N 106°49W **76** G10
Aspermont *U.S.A.* 33°8N 100°14W **84** E4
Aspiring, Mt. *N.Z.* 44°23S 168°46E **59** F2
Asprokavos, Akra *Greece* 39°21N 20°6E **25** B4
Aspur *India* 23°58N 74°7E **42** H6
Asquith *Canada* 52°8N 107°13W **71** C7
Assab = Aseb *Eritrea* 13°0N 42°40E **47** E3
Assal, L. *Djibouti* 11°40N 42°26E **47** E3
Assam □ *India* 26°0N 93°0E **41** G18
Assateague Island △ *U.S.A.* 38°15N 75°10W **81** F16
Asse *Belgium* 50°24N 4°10E **15** D4
Assen *Neths.* 53°0N 6°35E **15** A6
Assiniboia *Canada* 49°40N 105°59W **71** D7
Assiniboine → *Canada* 49°53N 97°8W **71** D9
Assiniboine, Mt. *Canada* 50°52N 115°39W **70** C5
Assis *Brazil* 22°40S 50°20W **95** A5
Assisi *Italy* 43°4N 12°37E **22** C5
Assynt, L. *U.K.* 58°10N 5°3W **11** C3
Astana *Kazakhstan* 51°10N 71°30E **32** A3
Āstāneh *Iran* 37°17N 49°59E **45** B6
Astara *Azerbaijan* 38°30N 48°50E **45** B6
Astarabad = Gorgān *Iran* 36°55N 54°30E **45** B7
Asterousia *Greece* 34°59N 25°3E **25** E7
Asti *Italy* 44°54N 8°12E **20** D8
Astipalea *Greece* 36°32N 26°22E **23** F12
Astorga *Spain* 42°29N 6°8W **21** A2
Astoria *U.S.A.* 46°11N 123°50W **78** D3
Astrakhan *Russia* 46°25N 48°5E **19** E8
Astrebla Downs *Australia* 24°12S 140°34E **62** C3
Asturias □ *Spain* 43°15N 6°0W **21** A3
Asunción *Paraguay* 25°10S 57°30W **94** B4
Asunción Nochixtlán *Mexico* 17°28N 97°14W **87** D5
Aswa → *Uganda* 3°43N 31°55E **54** B3
Aswa-Lolim △ *Uganda* 2°43N 31°35E **54** B3
Aswân *Egypt* 24°4N 32°57E **51** D12
Aswan High Dam = Sadd el Aali *Egypt* 23°54N 32°54E **51** D12
Asyût *Egypt* 27°11N 31°4E **51** C12
At Ṭafilah *Jordan* 30°45N 35°30E **46** E4
At Ṭafilah □ *Jordan* 30°45N 35°30E **46** E4
Aṭ Ṭā'if *Si. Arabia* 21°5N 40°27E **47** C3
At Ta'mīm □ *Iraq* 35°30N 44°20E **44** C5
Aṭ Ṭirāq *Si. Arabia* 27°19N 44°33E **44** E5
Aṭ Ṭubayq *Si. Arabia* 29°30N 37°0E **44** D3
Aṭ Ṭunayb *Jordan* 31°48N 35°57E **46** D4
Atacama □ *Chile* 27°30S 70°0W **94** B2
Atacama, Desierto de *Chile* 24°0S 69°20W **94** A2
Atacama, Salar de *Chile* 23°30S 68°20W **94** A2
Atakeye ✪ *Australia* 22°30S 133°45E **60** D5
Atakpamé *Togo* 7°31N 1°13E **50** G6
Atalaya *Peru* 10°45S 73°50W **92** F4
Atalaya de Femes *Canary Is.* 28°56N 13°47W **24** F6
Atami *Japan* 35°5N 139°4E **31** G9
Atamyrat *Turkmenistan* 37°50N 65°12E **28** F7
Atapupu *Indonesia* 9°0S 124°51E **37** F6
Atâr *Mauritania* 20°30N 13°5W **50** D3
Atari *Pakistan* 30°56N 74°2E **42** D6
Atascadero *U.S.A.* 35°29N 120°40W **78** K6
Atasū *Kazakhstan* 48°30N 71°0E **28** E8
Atatürk Barajı *Turkey* 37°28N 38°30E **19** G6
Atauro *E. Timor* 8°10S 125°30E **37** F7
Ataviros *Greece* 36°12N 27°50E **25** C9
Atbara *Sudan* 17°42N 33°59E **51** E12
'Atbara, Nahr → *Sudan* 17°40N 33°56E **51** E12
Atbasar *Kazakhstan* 51°48N 68°20E **28** D7
Atchafalaya B. *U.S.A.* 29°25N 91°25W **84** G9
Atchison *U.S.A.* 39°34N 95°7W **80** F6
Āteshān *Iran* 35°35N 52°37E **45** C7
Ath *Belgium* 50°38N 3°47E **15** D3
Athabasca *Canada* 54°45N 113°20W **70** C6
Athabasca → *Canada* 58°40N 110°50W **71** B6
Athabasca, L. *Canada* 59°15N 109°15W **71** B7
Athabasca Sand Dunes △ *Canada* 59°4N 108°43W **71** B7
Athboy *Ireland* 53°37N 6°56W **10** C5
Athenry *Ireland* 53°18N 8°44W **10** C3
Athens = Athina *Greece* 37°58N 23°43E **23** F10
Athens *Ala., U.S.A.* 34°48N 86°58W **85** D11
Athens *Ga., U.S.A.* 33°57N 83°23W **85** E13
Athens *N.Y., U.S.A.* 42°16N 73°49W **83** D11
Athens *Ohio, U.S.A.* 39°20N 82°6W **81** F12
Athens *Pa., U.S.A.* 41°57N 76°31W **83** E8
Athens *Tenn., U.S.A.* 35°27N 84°36W **85** D12
Athens *Tex., U.S.A.* 32°12N 95°51W **84** E7
Atherley *Canada* 44°37N 79°20W **82** B5
Atherton *Australia* 17°17S 145°30E **62** B4
Athi River *Kenya* 1°28S 36°58E **54** C4
Athienou *Cyprus* 35°3N 33°32E **25** D12
Athina *Greece* 37°58N 23°43E **23** F10
Athinai = Athina *Greece* 37°58N 23°43E **23** F10
Athlone *Ireland* 53°25N 7°56W **10** C4
Athna *Cyprus* 35°3N 33°47E **25** D12
Athol *U.S.A.* 42°36N 72°14W **83** D12
Atholl, Forest of *U.K.* 56°51N 3°50W **11** E5
Atholville *Canada* 47°59N 66°43W **73** C6
Athos *Greece* 40°9N 24°22E **23** D11
Athy *Ireland* 53°0N 7°0W **10** C5
Ati *Chad* 13°13N 18°20E **51** F9
Atiak *Uganda* 3°12N 32°2E **54** B3
Atik L. *Canada* 55°15N 96°0W **71** B9
Atikaki △ *Canada* 51°30N 95°31W **71** C9
Atikameg → *Canada* 52°30N 82°46W **72** B3
Atikokan *Canada* 48°45N 91°37W **72** C1
Atikonak L. *Canada* 52°40N 64°32W **73** B7
Atimaono *Tahiti* 17°46S 149°28W **59** d
Atitlán △ *Cent. Amer.* 14°38N 91°10W **87** E6
Atiu *Cook Is.* 20°0S 158°10W **65** J12
Atka *Russia* 60°50N 151°48E **29** C16
Atka I. *U.S.A.* 52°7N 174°30W **74** E5
Atkinson *U.S.A.* 42°32N 98°59W **80** D4
Atlanta *Ga., U.S.A.* 33°45N 84°23W **85** E12
Atlanta *Tex., U.S.A.* 33°7N 94°10W **84** E7
Atlantic *U.S.A.* 41°24N 95°1W **80** E6
Atlantic City *U.S.A.* 39°21N 74°27W **81** F16
Atlantic-Indian Basin *Antarctica* 60°0S 30°0E **5** B4
Atlantic Ocean 0°0 20°0W **2** D8
Atlas Mts. = Haut Atlas *Morocco* 32°30N 5°0W **50** B4
Atlin *Canada* 59°31N 133°41W **70** B2
Atlin, L. *Canada* 59°26N 133°45W **70** B2
Atlin △ *Canada* 59°10N 134°30W **70** B2
Atmore *U.S.A.* 31°2N 87°29W **85** F11
Atoka *U.S.A.* 34°23N 96°8W **84** D6
Atolia *U.S.A.* 35°19N 117°37W **79** K9
Atqasuk *U.S.A.* 70°28N 157°24W **74** A8
Atrai → *Bangla.* 24°7N 89°22E **43** G13
Atrak = Atrek → *Turkmenistan* 37°35N 53°58E **45** B8
Atrauli *India* 28°2N 78°20E **42** E8
Atrek → *Turkmenistan* 37°35N 53°58E **45** B8
Atsuta *Japan* 43°24N 141°26E **30** C10
Attalla *U.S.A.* 34°1N 86°6W **85** D11
Attapeu *Laos* 14°48N 106°50E **38** E6
Attawapiskat *Canada* 52°56N 82°24W **72** B3
Attawapiskat → *Canada* 52°57N 82°18W **72** B3
Attawapiskat L. *Canada* 52°18N 87°54W **72** B2
Attica *Ind., U.S.A.* 40°18N 87°15W **80** E10
Attica *Ohio, U.S.A.* 41°4N 82°53W **82** E2
Attikamagen L. *Canada* 55°0N 66°30W **73** B6
Attleboro *U.S.A.* 41°57N 71°17W **83** E13
Attock *Pakistan* 33°52N 72°20E **42** C5
Attopu = Attapeu *Laos* 14°48N 106°50E **38** E6
Attu I. *U.S.A.* 52°55N 172°55E **74** E2
Attur *India* 11°35N 78°30E **40** P11
Atuel → *Argentina* 36°17S 66°50W **94** D2
Atura *Uganda* 2°7N 32°20E **54** B3
Åtvidaberg *Sweden* 58°12N 16°0E **9** G17
Atwater *U.S.A.* 37°21N 120°37W **78** H6
Atwood *Canada* 43°40N 81°1W **82** C3
Atwood *U.S.A.* 39°48N 101°3W **80** F3
Atyraū *Kazakhstan* 47°5N 52°0E **19** E9
Au Sable *U.S.A.* 44°25N 83°20W **82** B1
Au Sable → *U.S.A.* 44°25N 83°20W **81** C12
Au Sable Forks *U.S.A.* 44°27N 73°41W **83** B11
Au Sable Pt. *U.S.A.* 44°20N 83°20W **82** B1
Auas *Honduras* 15°29N 84°20W **88** C3
Auasberg *Namibia* 22°37S 17°13E **56** C2
Aubagne *France* 43°17N 5°37E **20** E6
Aube → *France* 48°34N 3°43E **20** B5
Auberry *U.S.A.* 37°7N 119°29W **78** H7
Auburn *Ala., U.S.A.* 32°36N 85°29W **85** E12
Auburn *Calif., U.S.A.* 38°54N 121°4W **78** G5
Auburn *Ind., U.S.A.* 41°22N 85°4W **81** E11
Auburn *Maine, U.S.A.* 44°6N 70°14W **81** C18
Auburn *N.Y., U.S.A.* 42°56N 76°34W **83** D8
Auburn *Nebr., U.S.A.* 40°23N 95°51W **80** E6
Auburn *Pa., U.S.A.* 40°36N 76°6W **83** F8
Auburn *Wash., U.S.A.* 47°18N 122°14W **78** C4
Auburn Ra. *Australia* 25°15S 150°30E **63** D5
Auburndale *U.S.A.* 28°4N 81°48W **85** G14
Aubusson *France* 45°57N 2°11E **20** D5
Auch *France* 43°39N 0°36E **20** E4
Auchterarder *U.K.* 56°18N 3°41W **11** E5
Auchtermuchty *U.K.* 56°18N 3°13W **11** E5
Auckland *N.Z.* 36°52S 174°46E **59** B5
Auckland Is. *Pac. Oc.* 50°40S 166°5E **64** N8
Aude → *France* 43°13N 3°14E **20** E5
Auden *Canada* 50°14N 87°53W **72** B2
Audubon *U.S.A.* 41°43N 94°56W **80** E6
Augathella *Australia* 25°48S 146°35E **63** D4
Aughnacloy *U.K.* 54°25N 6°59W **10** B5
Aughrim *Ireland* 53°18N 8°19W **10** C3
Augrabies Falls *S. Africa* 28°35S 20°20E **56** D3
Augrabies Falls △ *S. Africa* 28°40S 20°22E **56** D3
Augsburg *Germany* 48°25N 10°52E **16** D6
Augusta *Australia* 34°19S 115°9E **61** F2
Augusta *Italy* 37°13N 15°13E **22** F6
Augusta *Ark., U.S.A.* 35°17N 91°22W **84** D9
Augusta *Ga., U.S.A.* 33°28N 81°58W **85** E14
Augusta *Kans., U.S.A.* 37°41N 96°59W **80** G5
Augusta *Maine, U.S.A.* 44°19N 69°47W **81** C19
Augusta *Mont., U.S.A.* 47°30N 112°24W **76** C7
Augustów *Poland* 53°51N 23°0E **17** B12
Augustus, Mt. *Australia* 24°20S 116°50E **61** D2
Augustus I. *Australia* 15°20S 124°30E **60** C3
Aujuittuq = Grise Fiord *Canada* 76°25N 82°57W **69** B15
Aukštaitija △ *Lithuania* 55°15N 26°0E **9** J22
Aukum *U.S.A.* 38°34N 120°43W **78** G6
Aulavik △ *Canada* 73°42N 119°55W **68** C8
Auld, L. *Australia* 22°25S 123°50E **60** D3
Ault *U.S.A.* 40°35N 104°44W **76** F11
Aunis *France* 46°5N 0°50W **20** C3
Aunu'u *Amer. Samoa* 14°20S 170°31W **59** b
Auponhia *Indonesia* 1°58S 125°27E **37** E7
Aur, Pulau *Malaysia* 2°35N 104°10E **39** L5
Auraiya *India* 26°28N 79°33E **43** F8
Aurangabad *Bihar, India* 24°45N 84°18E **43** G11
Aurangabad *Maharashtra, India* 19°50N 75°23E **40** K9
Aurich *Germany* 53°28N 7°28E **16** B4
Aurillac *France* 44°55N 2°26E **20** D5
Aurora *Canada* 44°0N 79°28W **82** C5
Aurora *S. Africa* 32°40S 18°29E **56** E2
Aurora *Colo., U.S.A.* 39°43N 104°49W **76** G11
Aurora *Ill., U.S.A.* 41°45N 88°19W **80** E9
Aurora *Mo., U.S.A.* 36°58N 93°43W **80** G7
Aurora *N.Y., U.S.A.* 42°45N 76°42W **83** D8
Aurora *Nebr., U.S.A.* 40°52N 98°0W **80** E5
Aurora *Ohio, U.S.A.* 41°21N 81°20W **82** E3
Aurukun *Australia* 13°20S 141°45E **62** A3
Aurukun ✪ *Australia* 13°36S 141°48E **62** A3
Aus *Namibia* 26°35S 16°12E **56** D2
Ausable → *Canada* 43°19N 81°46W **82** C3
Auschwitz = Oświęcim *Poland* 50°2N 19°11E **17** C10
Auski Roadhouse *Australia* 22°22S 118°41E **60** D2
Austin *Minn., U.S.A.* 43°40N 92°58W **80** D7
Austin *Nev., U.S.A.* 39°30N 117°4W **76** G5
Austin *Pa., U.S.A.* 41°38N 78°6W **82** E6
Austin *Tex., U.S.A.* 30°17N 97°45W **84** F6
Austin, L. *Australia* 27°40S 118°0E **61** E2
Austin I. *Canada* 61°10N 94°0W **71** A10
Austra *Norway* 65°8N 11°55E **8** D14
Austral Is. = Tubuaï, Îs. *French Polynesia* 25°0S 150°0W **65** K13
Austral Seamount Chain *Pac. Oc.* 24°0S 150°0W **65** K13
Australia ■ *Oceania* 23°0S 135°0E **58** D6
Australian-Antarctic Basin *S. Ocean* 60°0S 120°0E **5** C9
Australian Capital Territory □ *Australia* 35°30S 149°0E **63** F4
Australind *Australia* 33°17S 115°42E **61** F2
Austria ■ *Europe* 47°0N 14°0E **16** E8
Austvågøya *Norway* 68°20N 14°40E **8** B16
Autlán de Navarro *Mexico* 19°46N 104°22W **86** D4
Autun *France* 46°58N 4°17E **20** C6
Auvergne □ *France* 45°20N 3°15E **20** D5
Auvergne, Mts. d' *France* 45°20N 2°55E **20** D5
Auxerre *France* 47°48N 3°32E **20** C5
Auyuittuq △ *Canada* 67°30N 66°0W **69** D18
Av-Dovurak *Russia* 51°17N 91°35E **29** D10
Ava *U.S.A.* 36°57N 92°40W **80** G7
Avallon *France* 47°30N 3°53E **20** C5
Avalon *U.S.A.* 33°21N 118°20W **79** M8
Avalon Pen. *Canada* 47°30N 53°20W **73** C9
Avanos *Turkey* 38°43N 34°51E **44** B2
Avaré *Brazil* 23°4S 48°58W **95** A6
Avawatz Mts. *U.S.A.* 35°40N 116°30W **79** K10
Aveiro *Brazil* 3°10S 55°5W **93** D7
Aveiro *Portugal* 40°37N 8°38W **21** B1
Āvej *Iran* 35°40N 49°15E **45** C6
Avellaneda *Argentina* 34°40S 58°22W **94** C4
Avellino *Italy* 40°54N 14°47E **22** D6
Avenal *U.S.A.* 36°0N 120°8W **78** K6
Aversa *Italy* 40°58N 14°12E **22** D6
Avery *U.S.A.* 47°15N 115°49W **76** C6
Aves, I. de *W. Indies* 15°45N 63°55W **89** C7
Aves, Is. las *Venezuela* 12°0N 67°30W **89** D6
Avesta *Sweden* 60°9N 16°10E **9** F17
Aveyron → *France* 44°5N 1°16E **20** D4
Avezzano *Italy* 42°2N 13°25E **22** C5
Aviá Terai *Argentina* 26°45S 60°50W **94** B3
Aviemore *U.K.* 57°12N 3°50W **11** D5
Avignon *France* 43°57N 4°50E **20** E6
Ávila *Spain* 40°39N 4°43W **21** B3
Avila Beach *U.S.A.* 35°11N 120°44W **79** K6
Avilés *Spain* 43°35N 5°57W **21** A3
Avis *U.S.A.* 41°11N 77°19W **82** E7
Avoca *U.S.A.* 42°25N 77°25W **82** D7
Avoca → *Australia* 35°40S 143°43E **63** F3
Avoca → *Ireland* 52°48N 6°10W **10** D5
Avola *Canada* 51°45N 119°19W **70** C5
Avola *Italy* 36°56N 15°7E **22** F6
Avon *U.S.A.* 42°55N 77°45W **82** D7
Avon → *Australia* 31°40S 116°7E **61** F2
Avon → *Bristol, U.K.* 51°29N 2°41W **13** F5
Avon → *Dorset, U.K.* 50°44N 1°46W **13** G6
Avon → *Warks., U.K.* 52°0N 2°8W **13** E5
Avon Park *U.S.A.* 27°36N 81°31W **85** H14
Avondale *Zimbabwe* 17°43S 30°58E **55** F3
Avonlea *Canada* 50°0N 105°0W **71** D8
Avonmore *Canada* 45°10N 74°58W **83** A10
Avonmouth *U.K.* 51°30N 2°42W **13** F5
Avranches *France* 48°40N 1°20W **20** B3
Awa-Shima *Japan* 38°27N 139°14E **30** E9
A'waj → *Syria* 33°23N 36°20E **46** B5
Awaji-Shima *Japan* 34°30N 134°50E **31** G7
'Awālī *Bahrain* 26°0N 50°30E **45** E6
Awantipur *India* 33°55N 75°3E **43** C6
Awasa *Ethiopia* 7°2N 38°28E **47** F2
Awash *Ethiopia* 9°1N 40°10E **47** F3
Awatere → *N.Z.* 41°37S 174°10E **59** D5
Awbārī *Libya* 26°46N 12°57E **51** C8
Awbārī, Idehan *Libya* 27°10N 11°30E **51** C8
Awe, L. *U.K.* 56°17N 5°16W **11** E3
Awjilah *Libya* 29°8N 21°7E **51** C10
Axe → *U.K.* 50°42N 3°4W **13** F5
Axel Heiberg I. *Canada* 80°0N 90°0W **69** B14
Axim *Ghana* 4°51N 2°15W **50** H5
Axios → *Greece* 40°57N 22°35E **23** D10
Axminster *U.K.* 50°46N 3°0W **13** G4
Ayabaca *Peru* 4°40S 79°53W **92** D3
Ayabe *Japan* 35°20N 135°20E **31** G7
Ayacucho *Argentina* 37°5S 58°20W **94** D4
Ayacucho *Peru* 13°0S 74°0W **92** F4
Ayaguz = Ayaköz *Kazakhstan* 48°10N 80°10E **32** B5
Ayakkum Hu *China* 37°30N 89°20E **32** D6
Ayaköz *Kazakhstan* 48°10N 80°10E **32** B5
Ayamonte *Spain* 37°12N 7°24W **21** D2
Ayan *Russia* 56°30N 138°16E **29** D14
Ayaviri *Peru* 14°50S 70°35W **92** F4
Aydın *Turkey* 37°51N 27°51E **23** F12
Aydingkol Hu *China* 42°40N 89°15E **32** C6
Ayer *U.S.A.* 42°34N 71°35W **83** D13
Ayer Hitam *Malaysia* 5°24N 100°16E **39** c
Ayer's Cliff *Canada* 45°10N 72°3W **83** A12
Ayers Rock = Uluru *Australia* 25°23S 131°5E **61** E5
Ayeyarwady = Irrawaddy → *Burma* 15°50N 95°6E **41** M19
Áyia Napa *Cyprus* 34°59N 34°0E **25** E13
Áyia Phyla *Cyprus* 34°43N 33°1E **25** E12
Áyios Amvrósios *Cyprus* 35°20N 33°35E **25** D12
Áyios Seryios *Cyprus* 35°12N 33°53E **25** D12
Áyios Theodhoros *Cyprus* 35°22N 34°1E **25** D13
Aykhal *Russia* 66°0N 111°30E **29** C12
Aykino *Russia* 62°15N 49°56E **18** B8
Aylesbury *U.K.* 51°49N 0°49W **13** F7
Aylmer *Canada* 42°46N 80°59W **82** D4
Aylmer, L. *Canada* 64°5N 108°30W **68** E10
'Ayn, Wādī al *Oman* 22°15N 55°28E **45** F7
Ayn Dār *Si. Arabia* 25°55N 49°10E **45** E7
Ayn Zālah *Iraq* 36°45N 42°35E **44** B4
Ayolas *Paraguay* 27°10S 56°59W **94** B4
Ayon, Ostrov *Russia* 69°50N 169°0E **29** C17
'Ayoûn el 'Atroûs *Mauritania* 16°38N 9°37W **50** E4
Ayr *Australia* 19°35S 147°25E **62** B4
Ayr *Canada* 43°17N 80°27W **82** C4
Ayr *U.K.* 55°28N 4°38W **11** F4
Ayr → *U.K.* 55°28N 4°38W **11** F4
Ayre, Pt. of *I. of Man* 54°25N 4°21W **12** C3
Ayton *Australia* 15°56S 145°22E **62** B4
Aytos *Bulgaria* 42°42N 27°16E **23** C12
Ayu, Kepulauan *Indonesia* 0°35N 131°5E **37** D8
Ayutla *Guatemala* 14°40N 92°10W **88** D1
Ayutla de los Libres *Mexico* 16°54N 99°13W **87** D5
Ayvacık *Turkey* 39°36N 26°24E **23** E12
Ayvalık *Turkey* 39°20N 26°46E **23** E12
Az Zabadānī *Syria* 33°43N 36°5E **46** B5
Az Ẓāhirīyah *West Bank* 31°25N 34°58E **46** D3
Aẓ Ẓahrān *Si. Arabia* 26°10N 50°7E **45** E6
Az Zarqā *Jordan* 32°5N 36°4E **46** C5
Az Zarqā' *U.A.E.* 24°53N 53°4E **45** E7
Az Zarqā □ *Jordan* 32°5N 36°4E **46** C5
Az Zāwiyah *Libya* 32°52N 12°56E **51** B8
Az Zībār *Iraq* 36°52N 44°4E **44** B5
Az Zilfī *Si. Arabia* 26°12N 44°52E **44** E5
Az Zubayr *Iraq* 30°26N 47°40E **44** D5
Azad Kashmir □ *Pakistan* 33°50N 73°50E **43** C5
Azamgarh *India* 26°5N 83°13E **43** F10
Azangaro *Peru* 14°55S 70°13W **92** F4
Azaouad *Mali* 19°0N 3°0W **50** E5
Āzār Shahr *Iran* 37°45N 45°59E **44** B5
Azarān *Iran* 37°25N 47°16E **44** B5
Azārbayjan = Azerbaijan ■ *Asia* 40°20N 48°0E **19** F8
Āzarbāyjān-e Gharbī □ *Iran* 37°0N 44°30E **44** B5
Āzarbāyjān-e Sharqī □ *Iran* 37°20N 47°0E **44** B5
Azare *Nigeria* 11°55N 10°10E **50** F8
A'zāz *Syria* 36°36N 37°4E **44** B3
Azbine = Aïr *Niger* 18°30N 8°0E **50** E7
Azerbaijan ■ *Asia* 40°20N 48°0E **19** F8
Azimganj *India* 24°14N 88°16E **43** G13
Azogues *Ecuador* 2°35S 78°0W **92** D3
Azores = Açores, Is. dos *Atl. Oc.* 38°0N 27°0W **50** a
Azov *Russia* 47°3N 39°25E **19** E6
Azov, Sea of *Europe* 46°0N 36°30E **19** E6
Azovskoye More = Azov, Sea of *Europe* 46°0N 36°30E **19** E6
Azraq ash Shīshān *Jordan* 31°50N 36°49E **46** D5
Aztec *U.S.A.* 36°49N 107°59W **77** H10
Azúa de Compostela *Dom. Rep.* 18°25N 70°44W **89** C5
Azuaga *Spain* 38°16N 5°39W **21** C3
Azuero, Pen. de *Panama* 7°30N 80°30W **88** E3
Azul *Argentina* 36°42S 59°43W **94** D4
Azumino *Japan* 36°20N 137°50E **31** F8
Azusa *U.S.A.* 34°8N 117°52W **79** L9
Azzel Matti, Sebkra *Algeria* 26°10N 0°43E **50** C6

B

Ba Be △ *Vietnam* 22°25N 105°37E **38** A5
Ba Don *Vietnam* 17°45N 106°26E **38** D6
Ba Dong *Vietnam* 9°40N 106°33E **39** H6
Ba Ngoi *Vietnam* 11°54N 109°10E **39** G7
Ba Ria *Vietnam* 10°30N 107°10E **39** G6
Ba Tri *Vietnam* 10°2N 106°36E **39** G6
Ba Vi △ *Vietnam* 21°1N 105°22E **38** B5
Ba Xian = Bazhou *China* 39°8N 116°22E **34** E9
Baa *Indonesia* 10°50S 123°0E **60** B3
Baardheere *Somali Rep.* 2°20N 42°27E **47** G3
Baarle-Nassau *Belgium* 51°27N 4°56E **15** C4
Bab el Mandeb *Red Sea* 12°35N 43°25E **47** E3
Bābā, Koh-i- *Afghan.* 34°30N 67°0E **40** B5
Baba Burnu *Turkey* 39°29N 26°2E **23** E12
Bābā Kalū *Iran* 30°7N 50°49E **45** D6
Babadag *Romania* 44°53N 28°44E **17** F15
Babaeski *Turkey* 41°26N 27°6E **23** D12
Babahoyo *Ecuador* 1°40S 79°30W **92** D3
Babai = Sarju → *India* 27°21N 81°23E **43** F9
Babar *Indonesia* 8°0S 129°30E **37** F7
Babar *Pakistan* 31°7N 69°32E **42** D3
Babarkach *Pakistan* 29°45N 68°0E **42** E3
Babb *U.S.A.* 48°51N 113°27W **76** B7
Baberu *India* 25°33N 80°43E **43** G9
Babi Besar, Pulau *Malaysia* 2°25N 103°59E **39** L4
Bābil □ *Iraq* 32°30N 44°30E **44** C5
Babinda *Australia* 17°20S 145°56E **62** B4
Babine *Canada* 55°22N 126°37W **70** B3
Babine → *Canada* 55°45N 127°44W **70** B3
Babine L. *Canada* 54°48N 126°0W **70** C3
Babo *Indonesia* 2°30S 133°30E **37** E8
Bābol *Iran* 36°40N 52°50E **45** B7
Bābol Sar *Iran* 36°45N 52°45E **45** B7
Baboua *C.A.R.* 5°49N 14°58E **52** C2
Babruysk *Belarus* 53°10N 29°15E **17** B15
Babuhri *India* 26°49N 69°43E **42** F3
Babusar Pass *Pakistan* 35°12N 73°59E **43** B5
Babuyan Chan. *Phil.* 18°40N 121°30E **37** A6
Babylon *Iraq* 32°34N 44°22E **44** C5
Bac Can *Vietnam* 22°8N 105°49E **38** A5
Bac Giang *Vietnam* 21°16N 106°11E **38** B6
Bac Lieu *Vietnam* 9°17N 105°43E **39** H5
Bac Ninh *Vietnam* 21°13N 106°4E **38** B6
Bac Phan *Vietnam* 22°0N 105°0E **38** B5
Bacabal *Brazil* 4°15S 44°45W **93** D10
Bacalar *Mexico* 18°43N 88°27W **87** D7
Bacan, Kepulauan *Indonesia* 0°35S 127°30E **37** E7
Bacarra *Phil.* 18°15N 120°37E **37** A
Bacău *Romania* 46°35N 26°55E **17** E1
Bacerac *Mexico* 30°18N 108°50W **86** A
Bach Long Vi, Dao *Vietnam* 20°10N 107°40E **38** B
Bach Ma △ *Vietnam* 16°11N 107°49E **38** D
Bachhwara *India* 25°35N 85°54E **43** G
Back → *Canada* 65°10N 104°0W **68** D
Bacolod *Phil.* 10°40N 122°57E **37** B
Bacuk *Malaysia* 6°4N 102°25E **39** J
Bácum *Mexico* 27°33N 110°5W **86** B
Bād *Iran* 33°41N 52°1E **45** C
Bad → *U.S.A.* 44°21N 100°22W **80** C
Bad Axe *U.S.A.* 43°48N 83°0W **82** C
Bad Ischl *Austria* 47°44N 13°38E **16** E
Bad Kissingen *Germany* 50°11N 10°4E **16** C
Bada Barabil *India* 22°7N 85°24E **43** H
Badagara *India* 11°35N 75°40E **40** P
Badain Jaran Shamo *China* 40°23N 102°0E **32** C
Badajós, L. *Brazil* 3°15S 62°50W **92** D
Badajoz *Spain* 38°50N 6°59W **21** C
Badakhshān □ *Afghan.* 36°30N 71°0E **40** A
Badaling *China* 40°20N 116°0E **34** D
Badalona *Spain* 41°26N 2°15E **21** B
Badalzai *Afghan.* 29°50N 65°35E **42** E
Badampahar *India* 22°10N 86°10E **41** H
Badanah *Si. Arabia* 30°58N 41°30E **44** D
Badarinath *India* 30°45N 79°30E **43** D
Badas, Kepulauan *Indonesia* 0°45N 107°5E **36** D
Baddo → *Pakistan* 28°0N 64°20E **40** F
Bade *Indonesia* 7°10S 139°35E **37** F
Baden *Austria* 48°1N 16°13E **16** D
Baden *U.S.A.* 40°38N 80°14W **82** F
Baden-Baden *Germany* 48°44N 8°13E **16** D
Baden-Württemberg □ *Germany* 48°20N 8°40E **16** D
Badgam *India* 34°1N 74°45E **43** B
Badgastein *Austria* 47°7N 13°9E **16** E
Badger *Canada* 49°0N 56°4W **73** C
Badger *U.S.A.* 36°38N 119°1W **78** J
Bādghīs □ *Afghan.* 35°0N 63°0E **40** B
Badgingarra △ *Australia* 30°23S 115°22E **61** F
Badin *Pakistan* 24°38N 68°54E **42** G
Badlands *U.S.A.* 43°55N 102°30W **80** D
Badlands △ *U.S.A.* 43°38N 102°56W **80** D
Badrah *Iraq* 33°6N 45°58E **44** C
Badrain Jaran Shamo *China* 40°40N 103°20E **34** D
Badrinath *India* 30°44N 79°29E **43** D
Badu *Australia* 10°7S 142°11E **62**
Badulla *Sri Lanka* 7°1N 81°7E **40** R
Badung, Bukit *Indonesia* 8°49S 115°10E **37** K
Badung, Selat *Indonesia* 8°40S 115°22E **37** K
Baena *Spain* 37°37N 4°20W **21** D
Baengnyeongdo *S. Korea* 37°57N 124°40E **35** F
Baeza *Spain* 37°57N 3°25W **21** D
Bafatá *Guinea-Biss.* 12°8N 14°40W **50** F
Baffin B. *N. Amer.* 72°0N 64°0W **66** B
Baffin I. *Canada* 68°0N 75°0W **69** D
Bafing → *Mali* 13°49N 10°50W **50** F
Bafliyūn *Syria* 36°37N 36°59E **44** B
Bafoulabé *Mali* 13°50N 10°55W **50** F
Bafoussam *Cameroon* 5°28N 10°25E **52** C
Bāfq *Iran* 31°40N 55°25E **45** D
Bafra *Turkey* 41°34N 35°54E **19** F
Bāft *Iran* 29°15N 56°38E **45** D
Bafwasende *Dem. Rep. of the Congo* 1°3N 27°5E **54** B
Bagaha *India* 27°6N 84°5E **43** F
Bagamoyo *Tanzania* 6°28S 38°55E **54** D
Bagan Datoh *Malaysia* 3°59N 100°47E **39**
Bagan Serai *Malaysia* 5°1N 100°32E **39**
Baganga *Phil.* 7°34N 126°33E **37** C
Bagani *Namibia* 18°7S 21°41E **56** B
Bagansiapiapi *Indonesia* 2°12N 100°50E **36** D
Bagasra *India* 21°30N 71°0E **42** J
Bagaud *India* 22°19N 75°53E **42** H
Bagdad *U.S.A.* 34°35N 115°53W **79** L
Bagdarin *Russia* 54°26N 113°36E **29** D
Bagé *Brazil* 31°20S 54°15W **95** C
Bagenalstown = Muine Bheag *Ireland* 52°42N 6°58W **10** D
Baggs *U.S.A.* 41°2N 107°39W **76** F
Bagh *Pakistan* 33°59N 73°45E **43** C
Baghain → *India* 25°32N 81°1E **43** G
Baghdād *Iraq* 33°20N 44°23E **44** C
Bagheria *Italy* 38°5N 13°30E **22** E
Baghlān *Afghan.* 32°12N 68°46E **40** A
Baghlān □ *Afghan.* 36°0N 68°30E **40** B
Bagley *U.S.A.* 47°32N 95°24W **80** B
Bago = Pegu *Burma* 17°20N 96°29E **41** L
Bagodar *India* 24°5N 85°52E **43** G
Bagrationovsk *Russia* 54°23N 20°39E **9** J
Baguio *Phil.* 16°26N 120°34E **37** A
Bah *India* 26°53N 78°36E **43** F
Bahadurganj *India* 26°16N 87°49E **43** F
Bahadurgarh *India* 28°40N 76°57E **42** E
Bahama, Canal Viejo de *W. Indies* 22°10N 77°30W **88** B
Bahamas ■ *N. Amer.* 24°0N 75°0W **89** B
Bahār *Iran* 34°54N 48°26E **45** C
Baharampur *India* 24°2N 88°27E **43** G
Baharu Pandan = Pandan *Malaysia* 1°32N 103°46E **39**
Bahawalnagar *Pakistan* 30°0N 73°15E **42** E
Bahawalpur *Pakistan* 29°24N 71°40E **42** E
Bäherden *Turkmenistan* 38°25N 57°26E **45** B
Baheri *India* 28°45N 79°34E **43** E
Bahgul → *India* 27°45N 79°36E **43** F
Bahi *Tanzania* 5°58S 35°21E **54** D
Bahi Swamp *Tanzania* 6°10S 35°0E **54** D
Bahia = Salvador *Brazil* 13°0S 38°30W **93** F
Bahia □ *Brazil* 12°0S 42°0W **93** F
Bahía, Is. de la *Honduras* 16°45N 86°15W **88** C
Bahía Blanca *Argentina* 38°35S 62°13W **94** D
Bahía de Caráquez *Ecuador* 0°40S 80°27W **92** D

hia Kino Mexico 28°47N 111°58W 86 B2
hía Laura Argentina 48°10S 66°30W 96 F3
hia Negra Paraguay 20°5S 58°5W 92 H7
hir Dar Ethiopia 11°37N 37°10E 47 E2
hmanzād Iran 31°15N 51°47E 45 D6
hraich India 27°38N 81°37E 43 F9
hrain ■ Asia 26°0N 50°35E 45 E6
hror India 27°51N 76°20E 42 F7
hū Kalāt Iran 25°43N 61°25E 45 E9
i Bung, Mui = Ca Mau, Mui
Vietnam 8°38N 104°44E 39 H5
i Thuong Vietnam 19°54N 105°23E 38 C5
ia Mare Romania 47°40N 23°35E 17 E12
ião Brazil 2°40S 49°40W 93 D9
ïbokoum Chad 7°46N 15°43E 51 G9
icheng China 45°38N 122°42E 35 B12
idoa = Baydhabo
Somali Rep. 3°8N 43°30E 47 G3
ie-Comeau Canada 49°12N 68°10W 73 C6
ie-St-Paul Canada 47°28N 70°32W 73 C5
ie Ste-Anne Seychelles 4°18S 55°45E 53 b
ie-Trinité Canada 49°25N 67°20W 73 C6
ie Verte Canada 49°55N 56°12W 73 C8
ihar India 22°6N 80°33E 43 H9
ihe Hubei, China 32°50N 110°5E 34 H6
ihe Jilin, China 42°27N 128°9E 35 C15
'ījī Iraq 35°0N 43°30E 44 C4
ijnath India 29°55N 79°37E 43 E8
ikal, L. = Baykal, Oz.
Russia 53°0N 108°0E 29 D11
ikonur = Bayqongyr
Kazakhstan 45°40N 63°20E 28 E7
ikunthpur India 23°15N 82°33E 43 H10
ile Átha Cliath = Dublin
Ireland 53°21N 6°15W 10 C5
ile Átha Fhirdhia = Ardee
Ireland 53°52N 6°33W 10 C5
ile Átha I = Athy Ireland 53°0N 7°0W 10 C5
ile Átha Luain = Athlone
Ireland 53°25N 7°56W 10 C4
ile Átha Troim = Trim
Ireland 53°33N 6°48W 10 C5
ile Brigín = Balbriggan
Ireland 53°37N 6°11W 10 C5
ileşti Romania 44°1N 23°20E 17 F12
inbridge Ga., U.S.A. 30°55N 84°35W 85 F12
inbridge N.Y., U.S.A. 42°18N 75°29W 83 D9
inbridge Island
U.S.A. 47°38N 122°32W 78 C4
ine China 42°0N 128°0E 33 C14
ing Indonesia 10°14S 120°34E 37 F6
iniu China 32°50N 112°15E 34 H7
'ir Jordan 30°45N 36°55E 46 E5
ird Mts. U.S.A. 67°0N 160°0W 74 B8
iriki = Tarawa Kiribati 1°30N 173°0E 64 G9
irin Youqi China 43°30N 118°35E 35 C10
irin Zuoqi China 43°58N 119°15E 35 C10
irnsdale Australia 37°48S 147°36E 63 F4
isha China 34°20N 112°32E 34 G7
isha Li China 19°12N 109°20E 38 C7
ishan China 42°43N 127°14E 35 C14
itadi Nepal 29°35N 80°25E 43 E9
iyin China 36°45N 104°14E 34 F3
iyu Shan China 37°15N 107°30E 34 F4
j Baj India 22°30N 88°5E 43 H13
ja Hungary 46°12N 18°59E 17 E10
ja, Pta. Mexico 29°58N 115°49W 86 B1
ja California Mexico 31°10N 115°12W 86 A1
ja California □ Mexico 30°0N 115°0W 86 B2
ja California Sur □
Mexico 25°50N 111°50W 86 B2
jag India 22°40N 81°21E 43 H9
jamar Canary Is. 28°33N 16°20W 24 F3
jana India 23°7N 71°49E 42 H4
jatrejo Indonesia 8°29S 114°19E 37 J17
jera Indonesia 8°31S 115°2E 37 J18
jgīrān Iran 37°36N 58°24E 45 B8
jimba, Mt. Australia 29°17S 152°6E 63 D5
jo Boquete Panama 8°46N 82°27W 88 E3
jo Nuevo Caribbean 15°40N 78°50W 88 C4
joga Nigeria 10°57N 11°20E 51 F8
jool Australia 23°40S 150°35E 62 C5
kel Senegal 14°56N 12°20W 50 F3
ker Calif., U.S.A. 35°16N 116°4W 79 K10
ker Mont., U.S.A. 46°22N 104°17W 76 C11
ker, L. Canada 64°0N 96°0W 68 E12
ker, Mt. U.S.A. 48°50N 121°49W 76 B3
ker City U.S.A. 44°47N 117°50W 76 D5
ker I. Pac. Oc. 0°10N 176°35W 64 G10
ker I. U.S.A. 55°20N 133°40W 70 B2
ker L. Australia 26°54S 126°5E 61 E4
ker Lake Canada 64°20N 96°3W 68 E12
kers Creek Australia 21°13S 149°7E 62 C4
kers Dozen Is. Canada 56°45N 78°45W 72 A4
kersfield Calif., U.S.A. 35°23N 119°1W 79 K8
kersfield Vt., U.S.A. 44°45N 72°48W 83 B12
kharden = Bäherden
Turkmenistan 38°25N 57°26E 45 B8
khtarān = Kermānshāh
Iran 34°23N 47°0E 44 C5
khtarān = Kermānshāh □
Iran 34°0N 46°30E 44 C5
kı Azerbaijan 40°29N 49°56E 45 A6
kkafjörður Iceland 66°2N 14°48W 8 C6
kkagerði Iceland 65°31N 13°49W 8 D7
kony Hungary 47°10N 17°30E 17 E9
kony Forest = Bakony
Hungary 47°10N 17°30E 17 E9
kouma C.A.R. 5°40N 22°56E 52 C4
kswaho India 24°15N 79°18E 43 G8
ku = Bakı Azerbaijan 40°29N 49°56E 45 A6
kutis Coast Antarctica 74°0S 120°0W 5 D15
ky = Bakı Azerbaijan 40°29N 49°56E 45 A6
la Canada 45°1N 79°37W 82 A5
la U.K. 52°54N 3°36W 12 E4
la, L. U.K. 52°53N 3°37W 12 E4
labac I. Phil. 8°0N 117°0E 36 C5
labac Str. E. Indies 7°53N 117°5E 36 C5
labagh Afghan. 34°25N 70°12E 42 B4
'labakk Lebanon 34°0N 36°10E 46 B5
Balabalangan, Kepulauan
Indonesia 2°20S 117°30E 36 E5
Balad Iraq 34°0N 44°9E 44 C5
Balad Rūz Iraq 33°42N 45°5E 44 C5
Bālādeh Fārs, Iran 29°17N 51°56E 45 D6
Bālādeh Māzandaran, Iran 36°12N 51°48E 45 B6
Balaghat India 21°49N 80°12E 40 J12
Balaghat Ra. India 18°50N 76°30E 40 K10
Balaguer Spain 41°50N 0°50E 21 B6
Balaklava Ukraine 44°30N 33°30E 19 F5
Balakovo Russia 52°4N 47°55E 18 D8
Balamau India 27°10N 80°21E 43 F9
Balancán Mexico 17°48N 91°32W 87 D6
Balashov Russia 51°30N 43°10E 19 D7
Balasinor India 22°57N 73°23E 42 H5
Balasore = Baleshwar
India 21°35N 87°3E 41 J15
Balaton Hungary 46°50N 17°40E 17 E9
Balbina, Represa de Brazil 2°0S 59°30W 92 D7
Balboa Panama 8°57N 79°34W 88 E4
Balbriggan Ireland 53°37N 6°11W 10 C5
Balcarce Argentina 38°0S 58°10W 94 D4
Balcarres Canada 50°50N 103°35W 71 C8
Balchik Bulgaria 43°28N 28°11E 23 C13
Balclutha N.Z. 46°15S 169°45E 59 G2
Balcones Escarpment
U.S.A. 29°30N 99°15W 84 G5
Bald I. Australia 34°57S 118°27E 61 F2
Bald Knob U.S.A. 35°19N 91°34W 84 D9
Baldock L. Canada 56°33N 97°57W 71 B9
Baldwin Mich., U.S.A. 43°54N 85°51W 81 D11
Baldwin Pa., U.S.A. 40°21N 79°58W 82 F5
Baldwinsville U.S.A. 43°10N 76°20W 83 C8
Baldy Peak U.S.A. 33°54N 109°34W 77 K9
Baleares, Is. Spain 39°30N 3°0E 24 B10
Balearic Is. = Baleares, Is.
Spain 39°30N 3°0E 24 B10
Baleine ↝ Canada 58°15N 67°40W 73 A6
Baleine, Petite R. de la ↝
Canada 56°0N 76°45W 72 A4
Baler Phil. 15°46N 121°34E 37 A6
Baleshare U.K. 57°31N 7°22W 11 D1
Baleshwar India 21°35N 87°3E 41 J15
Baley Russia 51°36N 116°37E 29 D12
Balfate Honduras 15°48N 86°25W 88 C2
Balgo Australia 20°9S 127°58E 60 D4
Bali Greece 35°25N 24°47E 25 D6
Bali India 25°11N 73°17E 42 G5
Bali Indonesia 8°20S 115°0E 37 J18
Bali □ Indonesia 8°20S 115°0E 37 J18
Bali, Selat Indonesia 8°18S 114°25E 37 J17
Bali Barat △ Indonesia 8°12S 114°35E 37 J17
Bali Sea Indonesia 8°0S 115°0E 37 J17
Baliapal India 21°40N 87°17E 43 J12
Balik Pulau Malaysia 5°21N 100°14E 39 c
Balıkeşir Turkey 39°39N 27°53E 23 E12
Balikpapan Indonesia 1°10S 116°55E 36 E5
Balimbing Phil. 5°5N 119°58E 37 C5
Baling Malaysia 5°41N 100°55E 39 K3
Balkan Mts. = Stara Planina
Bulgaria 43°15N 23°0E 23 C10
Balkanabat Turkmenistan 39°30N 54°22E 45 B7
Balkhash = Balqash
Kazakhstan 46°50N 74°50E 28 E8
Balkhash, Ozero = Balqash Köli
Kazakhstan 46°0N 74°50E 32 B3
Ballachulish U.K. 56°41N 5°8W 11 E3
Balladonia Australia 32°27S 123°51E 61 F3
Ballaghaderreen Ireland 53°55N 8°34W 10 C3
Ballarat Australia 37°33S 143°50E 63 F3
Ballard, L. Australia 29°20S 120°40E 61 E3
Ballater U.K. 57°3N 3°3W 11 D5
Ballenas, Canal de
Mexico 29°10N 113°29W 86 B2
Balleny Is. Antarctica 66°30S 163°0E 5 C11
Ballia India 25°46N 84°12E 43 G11
Ballina Australia 28°50S 153°31E 63 D5
Ballina Ireland 54°7N 9°9W 10 B2
Ballinasloe Ireland 53°20N 8°13W 10 C3
Ballinger U.S.A. 31°45N 99°57W 84 F5
Ballinrobe Ireland 53°38N 9°13W 10 C2
Ballinskelligs B. Ireland 51°48N 10°13W 10 E1
Ballston Spa U.S.A. 43°0N 73°51W 83 D11
Ballyboghil Ireland 53°32N 6°16W 10 C5
Ballybunion Ireland 52°31N 9°40W 10 D2
Ballycanew Ireland 52°37N 6°19W 10 D5
Ballycastle U.K. 55°12N 6°15W 10 A5
Ballyclare U.K. 54°46N 6°0W 10 B5
Ballydehob Ireland 51°34N 9°28W 10 E2
Ballygawley U.K. 54°27N 7°2W 10 B4
Ballyhaunis Ireland 53°46N 8°46W 10 C3
Ballyheige Ireland 52°23N 9°49W 10 D2
Ballymena U.K. 54°52N 6°17W 10 B5
Ballymoney U.K. 55°5N 6°31W 10 A5
Ballymote Ireland 54°5N 8°31W 10 B3
Ballynahinch U.K. 54°24N 5°54W 10 B6
Ballyquintin Pt. U.K. 54°20N 5°30W 10 B6
Ballyshannon Ireland 54°30N 8°11W 10 B3
Balmaceda Chile 46°0S 71°50W 96 F2
Balmertown Canada 51°4N 93°41W 71 C10
Balmoral Australia 37°15S 141°48E 63 F3
Balmorhea U.S.A. 30°59N 103°45W 84 F3
Balochistan = Baluchistan □
Pakistan 27°30N 65°0E 40 F4
Balonne ↝ Australia 28°47S 147°56E 63 D4
Balotra India 25°50N 72°14E 42 G5
Balqash Kazakhstan 46°50N 74°50E 28 E8
Balqash Köli Kazakhstan 46°0N 74°50E 32 B3
Balrampur India 27°30N 82°20E 43 F10
Balranald Australia 34°38S 143°33E 63 E3
Balsas ↝ Brazil 7°15S 44°35W 93 E9
Balsas ↝ Mexico 17°55N 102°10W 86 D4
Balsas del Norte Mexico 18°0N 99°46W 87 D5
Balta Ukraine 47°56N 29°45E 17 D15
Bălţi Moldova 47°48N 27°58E 17 E14
Baltic Sea Europe 57°0N 19°0E 9 H18
Baltimore Ireland 51°29N 9°22W 10 E2
Baltimore Md., U.S.A. 39°17N 76°36W 81 F15
Baltimore Ohio, U.S.A. 39°51N 82°36W 82 G2
Baltinglass Ireland 52°56N 6°43W 10 D5
Baltit Pakistan 36°15N 74°40E 43 A6
Baltiysk Russia 54°41N 19°58E 9 J18
Baluchistan □ Pakistan 27°30N 65°0E 40 F4
Balurghat India 25°15N 88°44E 43 G13
Balvi Latvia 57°8N 27°15E 9 H22
Balya Turkey 39°44N 27°35E 23 E12
Balykchy Kyrgyzstan 42°26N 76°12E 32 C4
Balyqshy Kazakhstan 47°4N 51°52E 19 E9
Bam Iran 29°7N 58°14E 45 D8
Bama Nigeria 11°33N 13°41E 51 F8
Bamaga Australia 10°50S 142°25E 62 A3
Bamaji L. Canada 51°9N 91°25W 72 B1
Bamako Mali 12°34N 7°55W 50 F4
Bambari C.A.R. 5°40N 20°35E 52 C4
Bambaroo Australia 18°50S 146°10E 62 B4
Bamberg Germany 49°54N 10°54E 16 D6
Bamberg U.S.A. 33°18N 81°2W 85 E14
Bambili
Dem. Rep. of the Congo 3°40N 26°0E 54 B2
Bamburgh U.K. 55°37N 1°43W 12 B6
Bamenda Cameroon 5°57N 10°11E 52 C2
Bamfield Canada 48°45N 125°10W 70 D3
Bāmīān □ Afghan. 35°0N 67°0E 40 B5
Bamiancheng China 43°15N 124°2E 35 C13
Bampūr Iran 27°15N 60°21E 45 E9
Bampūr ↝ Iran 27°24N 59°0E 45 E8
Ban Ao Tu Khun Thailand 8°9N 98°20E 39 a
Ban Ban Laos 19°31N 103°30E 38 C4
Ban Bang Hin Thailand 9°32N 98°35E 39 H2
Ban Bang Khu Thailand 7°57N 98°23E 39 a
Ban Bang Rong Thailand 8°3N 98°25E 39 a
Ban Bo Phut Thailand 9°33N 100°2E 39 b
Ban Chaweng Thailand 9°32N 100°3E 39 b
Ban Chiang Thailand 17°30N 103°10E 38 D4
Ban Chiang Klang
Thailand 19°25N 100°55E 38 C3
Ban Choho Thailand 15°2N 102°9E 38 E4
Ban Dan Lan Hoi Thailand 17°0N 99°35E 38 D2
Ban Don = Surat Thani
Thailand 9°6N 99°20E 39 H2
Ban Don Vietnam 12°53N 107°48E 38 F6
Ban Don, Ao ↝ Thailand 9°20N 99°25E 39 H2
Ban Dong Thailand 19°30N 100°59E 38 C3
Ban Hong Thailand 18°18N 98°50E 38 C2
Ban Hua Thanon Thailand 9°26N 100°1E 39 b
Ban Kantang Thailand 7°25N 99°31E 39 J2
Ban Karon Thailand 7°51N 98°18E 39 a
Ban Kata Thailand 7°50N 98°18E 39 a
Ban Keun Laos 18°22N 102°35E 38 C4
Ban Khai Thailand 12°46N 101°18E 38 F3
Ban Kheun Laos 20°13N 101°7E 38 B3
Ban Khlong Khian
Thailand 8°10N 98°26E 39 a
Ban Khlong Kua Thailand 6°57N 100°8E 39 J3
Ban Khuan Thailand 8°20N 98°25E 39 a
Ban Ko Yai Chim
Thailand 11°17N 99°26E 39 G2
Ban Laem Thailand 13°13N 99°59E 38 F2
Ban Lamai Thailand 9°28N 100°3E 39 b
Ban Lao Ngam Laos 15°28N 106°10E 38 E6
Ban Le Kathe Thailand 15°49N 98°53E 38 E2
Ban Lo Po Noi Thailand 8°1N 98°34E 39 a
Ban Mae Chedi Thailand 19°11N 99°31E 38 C2
Ban Mae Nam Thailand 9°34N 100°0E 39 b
Ban Mae Sariang
Thailand 18°10N 97°56E 38 C1
Ban Mê Thuôt = Buon Ma Thuot
Vietnam 12°40N 108°3E 38 F7
Ban Mi Thailand 15°3N 100°32E 38 E3
Ban Muang Mo Laos 19°4N 103°58E 38 C4
Ban Na Bo Thailand 9°19N 99°41E 39 b
Ban Na San Thailand 8°53N 99°52E 39 H2
Ban Na Tong Laos 20°56N 101°47E 38 B3
Ban Nam Bac Laos 20°38N 102°20E 38 B4
Ban Nammi Laos 17°7N 105°40E 38 D5
Ban Nong Bok Laos 17°5N 104°48E 38 D5
Ban Nong Pling
Thailand 15°40N 100°10E 38 E3
Ban Pak Chan Thailand 10°32N 98°51E 39 G2
Ban Patong Thailand 7°54N 98°18E 39 a
Ban Phai Thailand 16°4N 102°44E 38 D4
Ban Phak Chit Thailand 8°0N 98°24E 39 a
Ban Pong Thailand 13°50N 99°55E 38 F2
Ban Rawai Thailand 7°47N 98°20E 39 a
Ban Ron Phibun Thailand 8°9N 99°51E 39 H2
Ban Sakhu Thailand 8°4N 98°18E 39 a
Ban Sanam Chai
Thailand 7°33N 100°25E 39 J3
Ban Tak Thailand 17°2N 99°4E 38 D2
Ban Tako Thailand 14°5N 102°40E 38 E4
Ban Tha Nun Thailand 8°12N 98°18E 39 a
Ban Tha Rua Thailand 7°59N 98°22E 39 a
Ban Tha Yu Thailand 8°17N 98°22E 39 a
Ban Thong Krut Thailand 9°25N 99°57E 39 b
Ban Xien Kok Laos 20°54N 100°39E 38 B3
Ban Yen Nhan Vietnam 20°57N 106°2E 38 B6
Banaba Kiribati 0°45S 169°50E 64 H8
Banalia
Dem. Rep. of the Congo 1°32N 25°5E 54 B2
Banam Cambodia 11°20N 105°17E 39 G5
Bananal, I. do Brazil 11°30S 50°30W 93 F8
Banaras = Varanasi
India 25°22N 83°0E 43 G10
Banas ↝ Gujarat, India 23°45N 71°25E 42 H4
Banas ↝ Mad. P., India 24°15N 81°30E 43 G9
Bânâs, Ras Egypt 23°57N 35°59E 51 D13
Banbridge U.K. 54°22N 6°16W 10 B5
Banbury U.K. 52°4N 1°20W 13 E6
Banchory U.K. 57°3N 2°29W 11 D6
Bancroft Canada 45°3N 77°51W 82 A7
Band Bonī Iran 25°30N 59°33E 45 E8
Band Qīr Iran 31°39N 48°53E 45 D6
Banda Mad. P., India 24°3N 78°57E 43 G8
Banda Ut. P., India 25°30N 80°26E 43 G9
Banda, Kepulauan
Indonesia 4°37S 129°50E 37 E7
Banda Aceh Indonesia 5°35N 95°20E 36 C1
Banda Banda, Mt.
Australia 31°10S 152°28E 63 E5
Banda Elat Indonesia 5°40S 133°5E 37 F8
Banda Is. = Banda, Kepulauan
Indonesia 4°37S 129°50E 37 E7
Banda Sea Indonesia 6°0S 130°0E 37 F8
Bandai-Asahi △ Japan 37°38N 140°5E 30 F10
Bandai-San Japan 37°36N 140°4E 30 F10
Bandān Iran 31°23N 60°44E 45 D9
Bandanaira Indonesia 4°32S 129°54E 37 E7
Bandanwara India 26°9N 74°38E 42 F6
Bandar = Machilipatnam
India 16°12N 81°8E 41 L12
Bandar-e Abbās Iran 27°15N 56°15E 45 E8
Bandar-e Anzalī Iran 37°30N 49°30E 45 B6
Bandar-e Bushehr = Būshehr
Iran 28°55N 50°55E 45 D6
Bandar-e Chārak Iran 26°45N 54°20E 45 E7
Bandar-e Deylam Iran 30°5N 50°10E 45 D6
Bandar-e Emām Khomeynī
Iran 30°30N 49°5E 45 D6
Bandar-e Lengeh Iran 26°35N 54°58E 45 E7
Bandar-e Maqām Iran 26°56N 53°29E 45 E7
Bandar-e Ma'shur Iran 30°35N 49°10E 45 D6
Bandar-e Rīg Iran 29°29N 50°38E 45 D6
Bandar-e Torkeman Iran 37°0N 54°10E 45 B7
Bandar Labuan Malaysia 5°20N 115°14E 36 C5
Bandar Lampung
Indonesia 5°20S 105°10E 36 F3
Bandar Maharani = Muar
Malaysia 2°3N 102°34E 39 L4
Bandar Penggaram = Batu Pahat
Malaysia 1°50N 102°56E 39 M4
Bandar Seri Begawan
Brunei 4°52N 115°0E 36 D5
Bandar Sri Aman
Malaysia 1°15N 111°32E 36 D4
Bandawe Malawi 11°58S 34°5E 55 E3
Bandeira, Pico da Brazil 20°26S 41°47W 95 A7
Bandera Argentina 28°55S 62°20W 94 B3
Banderas, B. de Mexico 20°40N 105°25W 86 C3
Bandhavgarh India 23°40N 81°2E 43 H9
Bandi ↝ India 26°12N 75°47E 42 F6
Bandikui India 27°3N 76°34E 42 F7
Bandırma Turkey 40°20N 28°0E 23 D13
Bandjarmasin = Banjarmasin
Indonesia 3°20S 114°35E 36 E4
Bandon Ireland 51°44N 8°44W 10 E3
Bandon ↝ Ireland 51°43N 8°37W 10 E3
Bandula Mozam. 19°0S 33°7E 55 F3
Bandundu
Dem. Rep. of the Congo 3°15S 17°22E 52 E3
Bandung Indonesia 6°54S 107°36E 37 G12
Bāneh Iran 35°59N 45°53E 44 C5
Banes Cuba 21°0N 75°42W 89 B4
Banff Canada 51°10N 115°34W 70 C5
Banff U.K. 57°40N 2°33W 11 D6
Banff △ Canada 51°30N 116°15W 70 C5
Bang Fai ↝ Laos 16°57N 104°45E 38 D5
Bang Hieng ↝ Laos 16°10N 105°10E 38 D5
Bang Krathum Thailand 16°34N 100°18E 38 D3
Bang Lamung Thailand 13°3N 100°56E 38 F3
Bang Lang △ Thailand 5°58N 101°19E 39 K3
Bang Lang Res. Thailand 6°6N 101°17E 39 J3
Bang Mun Nak Thailand 16°2N 100°23E 38 D3
Bang Pa In Thailand 14°14N 100°35E 38 E3
Bang Rakam Thailand 16°45N 100°7E 38 D3
Bang Saphan Thailand 11°14N 99°28E 39 G2
Bang Thao Thailand 7°59N 98°18E 39 a
Bangaduni I. India 21°34N 88°52E 43 J13
Bangala Dam Zimbabwe 21°7S 31°25E 55 G3
Bangalore India 12°59N 77°40E 40 N10
Banganga ↝ India 27°6N 77°25E 42 F6
Bangaon India 23°0N 88°47E 43 H13
Bangassou C.A.R. 4°55N 23°7E 52 D4
Banggai Indonesia 1°34S 123°30E 37 E6
Banggai, Kepulauan
Indonesia 1°40S 123°30E 37 E6
Banggai Arch. = Banggai,
Kepulauan Indonesia 1°40S 123°30E 37 E6
Banggi, Pulau Malaysia 7°17N 117°12E 36 C5
Banghāzī Libya 32°11N 20°3E 51 B10
Bangka Sulawesi, Indonesia 1°50N 125°5E 37 D7
Bangka Sumatera,
Indonesia 2°0S 105°50E 36 E3
Bangka, Selat Indonesia 2°30S 105°30E 36 E3
Bangka-Belitung □
Indonesia 2°30S 107°0E 36 E3
Bangkalan Indonesia 7°2S 112°46E 37 G15
Bangkinang Indonesia 0°18N 101°5E 36 D2
Bangko Indonesia 2°5S 102°9E 36 E2
Bangkok Thailand 13°45N 100°35E 38 F3
Bangkok, Bight of
Thailand 12°55N 100°30E 38 F3
Bangla = West Bengal □
India 23°0N 88°0E 43 H13
Bangladesh ■ Asia 24°0N 90°0E 41 H17
Bangli Indonesia 8°27S 115°21E 37 J18
Bangong Co China 33°45N 78°43E 43 C8
Bangor Down, U.K. 54°40N 5°40W 10 B6
Bangor Gwynedd, U.K. 53°14N 4°8W 12 D3
Bangor Maine, U.S.A. 44°48N 68°46W 81 C19
Bangor Pa., U.S.A. 40°52N 75°13W 83 F9
Bangued Phil. 17°40N 120°37E 37 A6
Bangui C.A.R. 4°23N 18°35E 52 D3
Banguru
Dem. Rep. of the Congo 0°30N 27°10E 54 B2
Bangweulu, L. Zambia 11°0S 30°0E 55 E3
Bangweulu Swamp
Zambia 11°20S 30°15E 55 E3
Banhine △ Mozam. 22°49S 32°55E 57 C5
Baní Dom. Rep. 18°16N 70°22W 89 C5
Banī Sa'd Iraq 33°34N 44°32E 44 C5
Banihal Pass India 33°30N 75°12E 43 C6
Banissa Kenya 3°55N 40°19E 54 B4
Bāniyās Syria 35°10N 36°0E 44 C3
Banja Luka Bos.-H. 44°49N 17°11E 22 B7
Banjar India 31°38N 77°21E 42 D7
Banjar ↝ India 22°36N 80°22E 43 H9
Banjarmasin Indonesia 3°20S 114°35E 36 E4
Banjul Gambia 13°28N 16°40W 50 F2
Banka India 24°53N 86°55E 43 G12
Banket Zimbabwe 17°27S 30°19E 55 F3
Bankipore India 25°35N 85°10E 41 G14
Banks I. = Moa Australia 10°11S 142°16E 62 a
Banks I. B.C., Canada 53°20N 130°0W 70 C3
Banks I. N.W.T., Canada 73°15N 121°30W 68 C7
Banks Pen. N.Z. 43°45S 173°15E 59 E4
Banks Str. Australia 40°40S 148°10E 63 G4
Bankura India 23°11N 87°18E 43 H12
Banmankhi India 25°53N 87°11E 43 G12
Bann ↝ Armagh, U.K. 54°30N 6°31W 10 B5
Bann ↝ L'derry., U.K. 55°8N 6°41W 10 A5
Bannang Sata Thailand 6°16N 101°16E 39 J3
Banning U.S.A. 33°56N 116°53W 79 M10
Bannockburn Canada 44°39N 77°33W 82 B7
Bannockburn U.K. 56°5N 3°55W 11 E5
Bannockburn Zimbabwe 20°17S 29°48E 55 G2
Bannu Pakistan 33°0N 70°18E 40 C7
Bano India 22°40N 84°55E 43 H11
Bansgaon India 26°33N 83°21E 43 F10
Banská Bystrica
Slovak Rep. 48°46N 19°14E 17 D10
Banswara India 23°32N 74°24E 42 H6
Bantaeng Indonesia 5°32S 119°56E 37 F5
Banteay Prei Nokor
Cambodia 11°56N 105°40E 39 G5
Banten □ Indonesia 6°30S 106°0E 37 G11
Bantry Ireland 51°41N 9°27W 10 E2
Bantry B. Ireland 51°37N 9°44W 10 E2
Bantva India 21°29N 70°12E 42 J4
Banyak, Kepulauan
Indonesia 2°10N 97°10E 36 D1
Banyalbufar Spain 39°42N 2°31E 24 B9
Banyo Cameroon 6°52N 11°45E 52 C2
Banyuwangi Indonesia 8°13S 114°21E 37 J17
Banzare Coast Antarctica 68°0S 125°0E 5 C9
Bao Ha Vietnam 22°11N 104°21E 38 A5
Bao Lac Vietnam 22°57N 105°40E 38 A5
Bao Loc Vietnam 11°32N 107°48E 39 G6
Bao'an China 22°34N 113°52E 33 a
Baocheng China 33°12N 106°56E 34 H4
Baode China 39°1N 111°5E 34 E6
Baodi China 39°38N 117°20E 35 E9
Baoding China 38°50N 115°28E 34 E8
Baoji China 34°20N 107°5E 34 G4
Baoshan China 25°10N 99°5E 32 F8
Baotou China 40°32N 110°2E 34 D6
Baoying China 33°17N 119°20E 35 H10
Baoyou = Ledong China 18°41N 109°5E 38 C7
Bap India 27°23N 72°18E 42 F5
Bapatla India 15°55N 80°30E 41 M12
Bāqerābād Iran 33°2N 51°58E 45 C6
Ba'qūbah Iraq 33°45N 44°50E 44 C5
Baquedano Chile 23°20S 69°52W 94 A2
Bar Montenegro 42°8N 19°6E 23 C8
Bar Ukraine 49°4N 27°40E 17 D14
Bar Bigha India 25°21N 85°47E 43 G11
Bar Harbor U.S.A. 44°23N 68°13W 81 C19
Bar-le-Duc France 48°47N 5°10E 20 B6
Bara India 25°16N 81°43E 43 G9
Bara Banki India 26°55N 81°12E 43 F9
Barabai Indonesia 2°32S 115°34E 36 E5
Baraboo U.S.A. 43°28N 89°45W 80 D9
Baracoa Cuba 20°20N 74°30W 89 B5
Baradā ↝ Syria 33°33N 36°34E 46 B5
Baradero Argentina 33°52S 59°29W 94 C4
Baraga U.S.A. 46°47N 88°30W 80 B9
Baragoi Kenya 1°47N 36°47E 54 B4
Barah ↝ India 27°42N 77°5E 42 F6
Barahona Dom. Rep. 18°13N 71°7W 89 C5
Barail Range India 25°15N 93°20E 41 G18
Barakaldo Spain 43°18N 2°59W 21 A4
Barakar ↝ India 24°7N 86°14E 43 G12
Barakot India 21°33N 84°59E 43 J11
Barakpur India 22°47N 88°21E 43 H13
Baralaba Australia 24°13S 149°50E 62 C4
Baralzon L. Canada 60°0N 98°3W 71 B9
Baramula India 34°15N 74°20E 43 B6
Baran India 25°9N 76°40E 42 G7
Baran ↝ Pakistan 25°13N 68°17E 42 G3
Baranavichy Belarus 53°10N 26°0E 17 B14
Baranof U.S.A. 57°5N 134°50W 70 B2
Baranof I. U.S.A. 57°0N 135°0W 68 F4
Barapasi Indonesia 2°15S 137°5E 37 E9
Barasat India 22°46N 88°31E 43 H13
Barat Daya, Kepulauan
Indonesia 7°30S 128°0E 37 F7
Barataria B. U.S.A. 29°20N 89°55W 85 G10
Barauda India 23°33N 75°15E 42 H6
Baraut India 29°13N 77°7E 42 E7
Barbacena Brazil 21°15S 43°56W 95 A7
Barbados ■ W. Indies 13°10N 59°30W 89 g
Barbària, C. de Spain 38°39N 1°24E 24 C7
Barbas, C. W. Sahara 22°20N 16°42W 50 D2
Barbastro Spain 42°2N 0°5E 21 A6
Barbeau Pk. Canada 81°54N 75°1W 69 A16
Barberton S. Africa 25°42S 31°2E 57 D5
Barberton U.S.A. 41°1N 81°39W 82 E3
Barbosa Colombia 5°57N 73°37W 92 B4
Barbourville U.S.A. 36°52N 83°53W 81 G12
Barbuda W. Indies 17°30N 61°40W 89 C7
Barcaldine Australia 23°43S 145°6E 62 C4
Barcellona Pozzo di Gotto
Italy 38°9N 15°13E 22 E6
Barcelona Spain 41°22N 2°10E 21 B7
Barcelona Venezuela 10°10N 64°40W 92 A6
Barceloneta Puerto Rico 18°27N 66°32W 89 d
Barcelos Brazil 1°0S 63°0W 92 D6
Barcoo ↝ Australia 25°30S 142°50E 62 D3
Bardaï Chad 21°25N 17°0E 51 D9
Bardas Blancas
Argentina 35°49S 69°45W 94 D2
Bardawîl, Sabkhet el
Egypt 31°10N 33°15E 46 D2
Barddhaman India 23°14N 87°39E 43 H12
Bardejov Slovak Rep. 49°18N 21°15E 17 D11
Bardera = Baardheere
Somali Rep. 2°20N 42°27E 47 G3
Bardīyah Libya 31°45N 25°5E 51 B10
Bardsey I. U.K. 52°45N 4°47W 12 E3
Bardstown U.S.A. 37°49N 85°28W 81 G11
Bareilly India 28°22N 79°27E 43 E8

Barela India 23°6N 80°3E 43 H9
Barents Sea Arctic 73°0N 39°0E 4 B9
Barfleur, Pte. de France 49°42N 1°16W 20 B3
Bargara Australia 24°50S 152°25E 62 C5
Bargi Dam India 22°59N 80°0E 43 H9
Barguzin Russia 53°37N 109°37E 29 D11
Barh India 25°29N 85°46E 43 G11
Barhaj India 26°18N 83°44E 43 F10
Barharwa India 24°52N 87°47E 43 G12
Barhi India 24°15N 85°25E 43 G11
Bari India 26°39N 77°39E 42 F7
Bari Italy 41°8N 16°51E 22 D7
Bari Doab Pakistan 30°20N 73°0E 42 D5
Bari Sadri India 24°28N 74°30E 42 G6
Barīdī, Ra's Si. Arabia 24°17N 37°31E 44 E3
Barīm Yemen 12°39N 43°25E 48 E8
Barinas Venezuela 8°36N 70°15W 92 B4
Baring, C. Canada 70°0N 117°30W 68 D8
Baringo, L. Kenya 0°47N 36°16E 54 B4
Barisal Bangla. 22°45N 90°20E 41 H17
Barisal □ Bangla. 22°45N 90°20E 41 H17
Barisan, Pegunungan Indonesia 3°30S 102°15E 36 E2
Barito → Indonesia 4°0S 114°50E 36 E4
Baritú △ Argentina 23°43S 64°40W 94 A3
Barjūj, Wadi → Libya 25°26N 12°12E 51 C8
Bark L. Canada 45°27N 77°51W 82 A7
Barkakana India 23°37N 85°29E 43 H11
Barkam China 31°51N 102°28E 32 E9
Barker U.S.A. 43°20N 78°33W 82 C6
Barkley, L. U.S.A. 37°1N 88°14W 85 C10
Barkley Sound Canada 48°50N 125°10W 70 D3
Barkly East S. Africa 30°58S 27°33E 56 E4
Barkly Homestead Australia 19°52S 135°50E 62 B2
Barkly Tableland Australia 17°50S 136°40E 62 B2
Barkly West S. Africa 28°5S 24°31E 56 D3
Barkol Kazak Zizhixian China 43°37N 93°2E 32 C7
Bârlad Romania 46°15N 27°38E 17 E14
Bârlad → Romania 45°38N 27°32E 17 F14
Barlee, L. Australia 29°15S 119°30E 61 E2
Barlee, Mt. Australia 24°38S 128°13E 61 D4
Barletta Italy 41°19N 16°17E 22 D7
Barlovento Canary Is. 28°48N 17°48W 24 F2
Barlovento C. Verde Is. 17°0N 25°0W 50 b
Barlow L. Canada 62°0N 103°0W 71 A8
Barmedman Australia 34°9S 147°21E 63 E4
Barmer India 25°45N 71°20E 42 G4
Barmera Australia 34°15S 140°28E 63 E3
Barmouth U.K. 52°44N 4°4W 12 E3
Barna → India 25°21N 83°3E 43 G10
Barnagar India 23°7N 75°19E 42 H6
Barnala India 30°23N 75°33E 42 D6
Barnard Castle U.K. 54°33N 1°55W 12 C6
Barnaul Russia 53°20N 83°40E 28 D9
Barnesville Ga., U.S.A. 33°3N 84°9W 85 E12
Barnesville Minn., U.S.A. 46°43N 96°28W 80 B5
Barnet □ U.K. 51°38N 0°9W 13 F7
Barneveld Neths. 52°7N 5°36E 15 B5
Barnhart U.S.A. 31°8N 101°10W 84 F4
Barnsley U.K. 53°34N 1°27W 12 D6
Barnstable U.S.A. 41°42N 70°18W 81 E18
Barnstaple U.K. 51°5N 4°4W 13 F3
Barnstaple Bay = Bideford Bay U.K. 51°5N 4°20W 13 F3
Barnwell U.S.A. 33°15N 81°23W 85 E14
Baro Nigeria 8°35N 6°18E 50 G7
Baroda = Vadodara India 22°20N 73°10E 42 H5
Baroda India 25°29N 76°35E 42 G7
Baroe S. Africa 33°13S 24°33E 56 E3
Baron Ra. Australia 23°30S 127°45E 60 D4
Barotseland Zambia 15°0S 24°0E 53 H4
Barpeta India 26°20N 91°10E 41 F17
Barqa Libya 27°0N 23°0E 51 C10
Barques, Pt. Aux U.S.A. 44°4N 82°58W 82 B2
Barquísimeto Venezuela 10°4N 69°19W 92 A5
Barr Smith Range Australia 27°4S 120°20E 61 E3
Barra Brazil 11°5S 43°10W 93 F10
Barra U.K. 57°0N 7°29W 11 E1
Barra, Sd. of U.K. 57°4N 7°25W 11 D1
Barra de Navidad Mexico 19°12N 104°41W 86 D4
Barra do Corda Brazil 5°30S 45°10W 93 E9
Barra do Garças Brazil 15°54S 52°16W 93 G8
Barra do Piraí Brazil 22°30S 43°50W 95 A7
Barra Falsa, Pta. da Mozam. 22°58S 35°37E 57 C6
Barra Hd. U.K. 56°47N 7°40W 11 E1
Barra Mansa Brazil 22°35S 44°12W 95 A7
Barraba Australia 30°21S 150°35E 63 E5
Barrackpur = Barakpur India 22°47N 88°21E 43 H13
Barraigh = Barra U.K. 57°0N 7°29W 11 E1
Barranca Lima, Peru 10°45S 77°50W 92 F3
Barranca Loreto, Peru 4°50S 76°50W 92 D3
Barranca del Cobre △ Mexico 27°18N 107°40W 86 B3
Barrancabermeja Colombia 7°0N 73°50W 92 B4
Barrancas Venezuela 8°55N 62°5W 92 B6
Barrancos Portugal 38°10N 6°58W 21 C2
Barranqueras Argentina 27°30S 59°0W 94 B4
Barranquilla Colombia 11°0N 74°50W 92 A4
Barraute Canada 48°26N 77°38W 72 C4
Barre Mass., U.S.A. 42°25N 72°6W 83 D12
Barre Vt., U.S.A. 44°12N 72°30W 83 B12
Barreal Argentina 31°33S 69°28W 94 C2
Barreiras Brazil 12°8S 45°0W 93 F10
Barreirinhas Brazil 2°30S 42°50W 93 D10
Barreiro Portugal 38°39N 9°5W 21 C1
Barren, Nosy Madag. 18°25S 43°40E 57 B7
Barretos Brazil 20°30S 48°35W 93 H9
Barrhead Canada 54°10N 114°24W 70 C6
Barrie Canada 44°24N 79°40W 82 B5
Barrier Ra. Australia 31°0S 141°30E 63 E3
Barrier Reef Belize 17°9N 88°3W 87 D7
Barrière Canada 51°12N 120°7W 70 C4
Barrington U.S.A. 41°44N 71°18W 83 E13
Barrington L. Canada 56°55N 100°15W 71 B8
Barrington Tops Australia 32°6S 151°28E 63 E5
Barringun Australia 29°1S 145°41E 63 D4
Barron U.S.A. 45°24N 91°51W 80 C8
Barrow U.S.A. 71°18N 156°47W 74 A8
Barrow → Ireland 52°25N 6°58W 10 D5
Barrow, Pt. U.S.A. 71°23N 156°29W 74 A8
Barrow Creek Australia 21°30S 133°55E 62 C1
Barrow I. Australia 20°45S 115°20E 60 D2
Barrow-in-Furness U.K. 54°7N 3°14W 12 C4
Barrow Pt. Australia 14°20S 144°40E 62 A3
Barrow Ra. Australia 26°0S 127°40E 61 E4
Barrow Str. Canada 74°20N 95°0W 4 B3
Barry U.K. 51°24N 3°16W 13 F4
Barry's Bay Canada 45°29N 77°41W 82 A7
Barsat Pakistan 36°10N 72°45E 43 A5
Barsi India 18°10N 75°50E 40 K9
Barsoi India 25°48N 87°57E 41 G15
Barstow U.S.A. 34°54N 117°1W 79 L9
Bartica Guyana 6°25N 58°40W 92 B7
Bartle Frere Australia 17°27S 145°50E 62 B4
Bartlesville U.S.A. 36°45N 95°59W 84 C7
Bartlett Calif., U.S.A. 36°29N 118°2W 78 J8
Bartlett Tenn., U.S.A. 35°12N 89°52W 85 D10
Bartlett, L. Canada 63°5N 118°20W 70 A5
Bartolomeu Dias Mozam. 21°10S 35°8E 55 G4
Barton U.S.A. 44°45N 72°11W 83 B12
Barton upon Humber U.K. 53°41N 0°25W 12 D7
Bartow U.S.A. 27°54N 81°50W 85 H14
Barú, Volcan Panama 8°55N 82°35W 88 E3
Barumba Dem. Rep. of the Congo 1°3N 23°37E 54 B1
Barung, Nusa Indonesia 8°30S 113°30E 37 H15
Baruun Urt Mongolia 46°46N 113°15E 33 B11
Baruunsuu Mongolia 43°43N 105°35E 34 C3
Barwani India 22°2N 74°57E 42 H6
Barysaw Belarus 54°17N 28°28E 17 A15
Barzān Iraq 36°55N 44°3E 44 B5
Bāsa'idū Iran 26°35N 55°20E 45 E7
Basal Pakistan 33°33N 72°13E 42 C5
Basankusa Dem. Rep. of the Congo 1°5N 19°50E 52 D3
Basarabeasca Moldova 46°21N 28°58E 17 E15
Basarabia = Bessarabiya Moldova 47°0N 28°10E 17 E15
Basawa Afghan. 34°15N 70°50E 42 B4
Bascuñán, C. Chile 28°52S 71°35W 94 B1
Basel Switz. 47°35N 7°35E 20 C7
Bashākerd, Kūhhā-ye Iran 26°42N 58°35E 45 E8
Bashaw Canada 52°35N 112°58W 70 C6
Bāshī Iran 28°41N 51°4E 45 D6
Bashkir Republic = Bashkortostan □ Russia 54°0N 57°0E 18 D10
Bashkortostan □ Russia 54°0N 57°0E 18 D10
Basibasy Madag. 22°10S 43°40E 57 C7
Basilan I. Phil. 6°35N 122°0E 37 C6
Basilan Str. Phil. 6°50N 122°0E 37 C6
Basildon U.K. 51°34N 0°28E 13 F8
Basim = Washim India 20°3N 77°0E 40 J10
Basin U.S.A. 44°23N 108°2W 76 D9
Basingstoke U.K. 51°15N 1°5W 13 F6
Baskatong, Rés. Canada 46°46N 75°50W 72 C4
Basle = Basel Switz. 47°35N 7°35E 20 C7
Basoda India 23°52N 77°54E 42 H7
Basoko Dem. Rep. of the Congo 1°16N 23°40E 54 B1
Basque Provinces = País Vasco □ Spain 42°50N 2°45W 21 A4
Basra = Al Başrah Iraq 30°30N 47°50E 44 D5
Bass Str. Australia 39°15S 146°30E 63 F4
Bassano Canada 50°48N 112°20W 70 C6
Bassano del Grappa Italy 45°46N 11°44E 22 B4
Bassas da India Ind. Oc. 22°0S 39°0E 55 G4
Basse-Pointe Martinique 14°52N 61°8W 88 c
Basse-Terre Guadeloupe 16°0N 61°44W 88 b
Basse Terre Trin. & Tob. 10°7N 61°19W 93 K15
Bassein Burma 16°45N 94°30E 41 L19
Basses, Pte. des Guadeloupe 15°52N 61°17W 88 b
Basseterre St. Kitts & Nevis 17°17N 62°43W 89 C7
Bassett U.S.A. 42°35N 99°32W 80 D4
Bassi India 30°44N 76°21E 42 D7
Bastak Iran 27°15N 54°25E 45 E7
Baştām Iran 36°29N 55°4E 45 B7
Bastar India 19°15N 81°40E 41 K12
Basti India 26°52N 82°55E 43 F10
Bastia France 42°40N 9°30E 20 E8
Bastogne Belgium 50°1N 5°43E 15 D5
Bastrop La., U.S.A. 32°47N 91°55W 84 E9
Bastrop Tex., U.S.A. 30°7N 97°19W 84 F6
Basuo = Dongfang China 18°50N 108°33E 38 C7
Bat Yam Israel 32°2N 34°44E 46 C3
Bata Eq. Guin. 1°57N 9°50E 52 D1
Bataan □ Phil. 14°40N 120°25E 37 B6
Batabanó Cuba 22°41N 82°18W 88 B3
Batabanó, G. de Cuba 22°30N 82°30W 88 B3
Batac Phil. 18°3N 120°34E 37 A6
Batagai Russia 67°38N 134°38E 29 C14
Batala India 31°48N 75°12E 42 D6
Batama Dem. Rep. of the Congo 0°58N 26°33E 54 B2
Batamay Russia 63°30N 129°15E 29 C13
Batang Indonesia 6°55S 109°45E 37 G13
Batangafo C.A.R. 7°25N 18°20E 52 C3
Batangas Phil. 13°35N 121°10E 37 B6
Batanta Indonesia 0°55S 130°40E 37 E8
Batatais Brazil 20°54S 47°37W 95 A6
Batavia U.S.A. 43°0N 78°11W 82 D6
Batchelor Australia 13°4S 131°1E 60 B5
Batdambang Cambodia 13°7N 103°12E 38 F4
Batemans B. Australia 35°40S 150°12E 63 F5
Batemans Bay Australia 35°44S 150°11E 63 F5
Batesburg-Leesville U.S.A. 33°54N 81°33W 85 E14
Batesville Ark., U.S.A. 35°46N 91°39W 84 D9
Batesville Miss., U.S.A. 34°19N 89°57W 85 D10
Batesville Tex., U.S.A. 28°58N 99°37W 84 G5
Bath Canada 44°11N 76°47W 83 B8
Bath U.K. 51°23N 2°22W 13 F5
Bath Maine, U.S.A. 43°55N 69°49W 81 D19
Bath N.Y., U.S.A. 42°20N 77°19W 82 D7
Bath & North East Somerset □ U.K. 51°21N 2°27W 13 F5
Batheay Cambodia 11°59N 104°57E 39 G5
Bathsheba Barbados 13°13N 59°32W 89 g
Bathurst Australia 33°25S 149°31E 63 E4
Bathurst Canada 47°37N 65°43W 73 C6
Bathurst S. Africa 33°30S 26°50E 56 E4
Bathurst, C. Canada 70°34N 128°0W 68 C6
Bathurst B. Australia 14°16S 144°25E 62 A3
Bathurst Harb. Australia 43°15S 146°10E 63 G4
Bathurst I. Australia 11°30S 130°10E 60 B5
Bathurst I. Canada 76°0N 100°30W 69 B11
Bathurst Inlet Canada 66°50N 108°1W 68 D10
Batiki Fiji 17°48S 179°10E 59 a
Batlow Australia 35°31S 148°9E 63 F4
Batman Turkey 37°55N 41°5E 44 B4
Baţn al Ghūl Jordan 29°36N 35°56E 46 F4
Batna Algeria 35°34N 6°15E 50 A7
Batoka Zambia 16°45S 27°15E 55 F2
Baton Rouge U.S.A. 30°27N 91°11W 84 F9
Batong, Ko Thailand 6°32N 99°12E 39 J2
Batopilas Mexico 27°1N 107°44W 86 B3
Batouri Cameroon 4°30N 14°25E 52 D2
Båtsfjord Norway 70°38N 29°39E 8 A23
Battambang = Batdambang Cambodia 13°7N 103°12E 38 F4
Batticaloa Sri Lanka 7°43N 81°45E 40 R12
Battipáglia Italy 40°37N 14°58E 22 D6
Battle U.K. 50°55N 0°30E 13 G8
Battle → Canada 52°43N 108°15W 71 C7
Battle Creek U.S.A. 42°19N 85°11W 81 D11
Battle Ground U.S.A. 45°47N 122°32W 78 E4
Battle Harbour Canada 52°16N 55°35W 73 B8
Battle Lake U.S.A. 46°17N 95°43W 80 B6
Battle Mountain U.S.A. 40°38N 116°56W 76 F5
Battlefields Zimbabwe 18°37S 29°47E 55 F2
Battleford Canada 52°45N 108°15W 71 C7
Batu Ethiopia 6°55N 39°45E 47 F2
Batu Malaysia 3°15N 101°40E 39 L3
Batu, Kepulauan Indonesia 0°30S 98°25E 36 E1
Batu Ferringhi Malaysia 5°28N 100°15E 39 c
Batu Gajah Malaysia 4°28N 101°3E 39 K3
Batu Is. = Batu, Kepulauan Indonesia 0°30S 98°25E 36 E1
Batu Pahat Malaysia 1°50N 102°56E 39 M4
Batu Puteh, Gunung Malaysia 4°15N 101°31E 39 K3
Batuata Indonesia 6°12S 122°42E 37 F6
Batugondang, Tanjung Indonesia 8°6S 114°29E 37 J17
Batukaru, Gunung Indonesia 8°20S 115°5E 37 J18
Batumi Georgia 41°39N 41°44E 19 F7
Batur, Gunung Indonesia 8°14S 115°23E 37 J18
Batura Sar Pakistan 36°30N 74°31E 43 A6
Baturaja Indonesia 4°11S 104°15E 36 E2
Baturité Brazil 4°28S 38°45W 93 D11
Baturiti Indonesia 8°19S 115°11E 37 J18
Bau Malaysia 1°25N 110°9E 36 D4
Baubau Indonesia 5°25S 122°38E 37 F6
Baucau E. Timor 8°27S 126°27E 37 F7
Bauchi Nigeria 10°22N 9°48E 50 F7
Baudette U.S.A. 48°43N 94°36W 80 A6
Bauer, C. Australia 32°44S 134°4E 63 E1
Bauhinia Australia 24°35S 149°18E 62 C4
Baukau = Baucau E. Timor 8°27S 126°27E 37 F7
Bauld, C. Canada 51°38N 55°26W 69 G20
Bauru Brazil 22°10S 49°0W 95 A6
Bausi India 24°48N 87°1E 43 G12
Bauska Latvia 56°24N 24°15E 9 H21
Bautzen Germany 51°10N 14°26E 16 C8
Bavānāt Iran 30°28N 53°27E 45 D7
Bavaria = Bayern □ Germany 48°50N 12°0E 16 D6
Bavispe → Mexico 29°15N 109°11W 86 B3
Bawdwin Burma 23°5N 97°20E 41 H20
Bawean Indonesia 5°46S 112°35E 36 F4
Bawku Ghana 11°3N 0°19W 50 F5
Bawlake Burma 19°11N 97°21E 41 K20
Baxley U.S.A. 31°47N 82°21W 85 F13
Baxter U.S.A. 46°21N 94°17W 80 B6
Baxter Springs U.S.A. 37°2N 94°44W 80 G6
Baxter State △ U.S.A. 46°5N 68°57W 81 B19
Bay City Mich., U.S.A. 43°36N 83°54W 81 D12
Bay City Tex., U.S.A. 28°59N 95°58W 84 G7
Bay Minette U.S.A. 30°53N 87°46W 85 F11
Bay Roberts Canada 47°36N 53°16W 73 C9
Bay St. Louis U.S.A. 30°19N 89°20W 85 F10
Bay Springs U.S.A. 31°59N 89°17W 85 F10
Bay View N.Z. 39°25S 176°50E 59 C6
Baya Dem. Rep. of the Congo 11°53S 27°25E 55 E2
Bayamo Cuba 20°20N 76°40W 88 B4
Bayamón Puerto Rico 18°24N 66°10W 89 d
Bayan Har Shan China 34°0N 98°0E 32 E8
Bayan Hot = Alxa Zuoqi China 38°50N 105°40E 34 E3
Bayan Lepas Malaysia 5°17N 100°16E 39 c
Bayan Obo China 41°52N 109°59E 34 D5
Bayan-Ovoo = Erdenetsogt Mongolia 42°55N 106°5E 34 C4
Bayana India 26°55N 77°18E 42 F7
Bayanaūyl Kazakhstan 50°45N 75°45E 28 D8
Bayanhongor Mongolia 46°8N 102°43E 32 B9
Bayard N. Mex., U.S.A. 32°46N 108°8W 77 K9
Bayard Nebr., U.S.A. 41°45N 103°20W 80 E2
Baybay Phil. 10°40N 124°55E 37 B6
Baydaratskaya Guba Russia 69°0N 67°30E 28 C7
Baydhabo Somali Rep. 3°8N 43°30E 47 G3
Bayern □ Germany 48°50N 12°0E 16 D6
Bayeux France 49°17N 0°42W 20 B3
Bayfield Canada 43°34N 81°42W 82 C3
Bayfield U.S.A. 46°49N 90°49W 80 B8
Bayındır Turkey 38°13N 27°39E 23 E12
Baykal, Oz. Russia 53°0N 108°0E 29 D11
Baykan Turkey 38°7N 41°44E 44 B4
Baymak Russia 52°36N 58°19E 18 D10
Baynes Mts. Namibia 17°15S 13°0E 56 B1
Bayombong Phil. 16°30N 121°10E 37 A6
Bayonne France 43°30N 1°28W 20 E3
Bayonne U.S.A. 40°40N 74°6W 83 F10
Bayovar Peru 5°50S 81°0W 92 E2
Bayqongyr Kazakhstan 45°40N 63°20E 28 E7
Bayram-Ali = Baýramaly Turkmenistan 37°37N 62°10E 45 B9
Baýramaly Turkmenistan 37°37N 62°10E 45 B9
Bayramiç Turkey 39°48N 26°36E 23 E12
Bayreuth Germany 49°56N 11°35E 16 D6
Bayrūt Lebanon 33°53N 35°31E 46 B4
Bays, L. of Canada 45°15N 79°4W 82 A5
Baysville Canada 45°9N 79°7W 82 A5
Bayt Lahm West Bank 31°43N 35°12E 46 D4
Baytown U.S.A. 29°43N 94°59W 84 G7
Baza Spain 37°30N 2°47W 21 D4
Bazaruto, I. do Mozam. 21°40S 35°28E 57 C6
Bazaruto △ Mozam. 21°42S 35°26E 57 C6
Bazhou China 39°8N 116°22E 34 E9
Bazmān, Kūh-e Iran 28°4N 60°1E 45 D9
Beach U.S.A. 46°58N 104°0W 80 B2
Beach City U.S.A. 40°39N 81°35W 82 F3
Beachport Australia 37°29S 140°0E 63 F3
Beachville Canada 43°5N 80°49W 82 C4
Beachy Hd. U.K. 50°44N 0°15E 13 G8
Beacon Australia 30°26S 117°52E 61 F2
Beacon U.S.A. 41°30N 73°58W 83 E11
Beaconsfield Australia 41°11S 146°48E 63 G4
Beagle, Canal S. Amer. 55°0S 68°30W 96 H3
Beagle Bay Australia 16°58S 122°40E 60 C3
Beagle Bay ✪ Australia 16°53S 122°40E 60 C3
Beagle G. Australia 12°15S 130°25E 60 B5
Béal an Átha = Ballina Ireland 54°7N 9°9W 10 B2
Béal Átha na Sluaighe = Ballinasloe Ireland 53°20N 8°13W 10 C3
Bealanana Madag. 14°33S 48°44E 57 A8
Beals Cr. → U.S.A. 32°10N 100°51W 84 E4
Beamsville Canada 43°12N 79°28W 82 C5
Bear → Calif., U.S.A. 38°56N 121°36W 78 G5
Bear → Utah, U.S.A. 41°30N 112°8W 74 G17
Bear I. Ireland 51°38N 9°50W 10 E2
Bear L. Canada 55°8N 96°0W 71 B9
Bear L. U.S.A. 41°59N 111°21W 76 F8
Bear Lake Canada 45°27N 79°35W 82 A5
Beardmore Canada 49°36N 87°57W 72 C2
Beardmore Glacier Antarctica 84°30S 170°0E 5 E11
Beardstown U.S.A. 40°1N 90°26W 80 E8
Bearma → India 24°20N 79°51E 43 G8
Béarn France 43°20N 0°30W 20 E3
Bearpaw Mts. U.S.A. 48°12N 109°30W 76 B9
Bearskin Lake Canada 53°58N 91°2W 72 B1
Beas → India 31°10N 74°59E 42 D6
Beata, C. Dom. Rep. 17°40N 71°30W 89 C5
Beata, I. Dom. Rep. 17°34N 71°31W 89 C5
Beatrice U.S.A. 40°16N 96°45W 80 E5
Beatrice Zimbabwe 18°15S 30°55E 55 F3
Beatrice, C. Australia 14°20S 136°55E 62 A2
Beatton → Canada 56°15N 120°45W 70 B4
Beatton River Canada 57°26N 121°20W 70 B4
Beatty U.S.A. 36°54N 116°46W 78 J10
Beau Bassin Mauritius 20°13S 57°27E 53 d
Beauce, Plaine de la France 48°10N 1°45E 20 B4
Beauceville Canada 46°13N 70°46W 73 C5
Beaudesert Australia 27°59S 153°0E 63 D5
Beaufort Malaysia 5°30N 115°40E 36 C5
Beaufort N.C., U.S.A. 34°43N 76°40W 85 D16
Beaufort S.C., U.S.A. 32°26N 80°40W 85 E14
Beaufort Sea Arctic 72°0N 140°0W 66 B5
Beaufort West S. Africa 32°18S 22°36E 56 E3
Beauharnois Canada 45°20N 73°52W 83 A11
Beaulieu → Canada 62°3N 113°11W 70 A6
Beauly U.K. 57°30N 4°28W 11 D4
Beauly → U.K. 57°29N 4°27W 11 D4
Beaumaris U.K. 53°16N 4°6W 12 D3
Beaumont Belgium 50°15N 4°14E 15 D4
Beaumont U.S.A. 30°5N 94°6W 84 F7
Beaune France 47°2N 4°50E 20 C6
Beaupré Canada 47°3N 70°54W 73 C5
Beauraing Belgium 50°7N 4°57E 15 D4
Beausejour Canada 50°5N 96°35W 71 C9
Beauvais France 49°25N 2°8E 20 B5
Beauval Canada 55°9N 107°37W 71 B7
Beaver Okla., U.S.A. 36°49N 100°31W 84 C4
Beaver Pa., U.S.A. 40°42N 80°19W 82 F4
Beaver Utah, U.S.A. 38°17N 112°38W 76 G7
Beaver → B.C., Canada 59°52N 124°20W 70 B4
Beaver → Ont., Canada 55°55N 87°48W 72 A2
Beaver → Sask., Canada 55°26N 107°45W 71 B7
Beaver → U.S.A. 36°35N 99°30W 84 C5
Beaver City U.S.A. 40°8N 99°50W 80 E4
Beaver Creek Canada 63°0N 141°0W 68 E3
Beaver Dam U.S.A. 43°28N 88°50W 80 D9
Beaver Falls U.S.A. 40°46N 80°20W 82 F4
Beaver Hill L. Canada 54°5N 94°50W 71 C10
Beaver I. U.S.A. 45°40N 85°33W 81 C11
Beavercreek U.S.A. 39°43N 84°11W 81 F11
Beaverhill L. Canada 53°27N 112°32W 70 C6
Beaverlodge Canada 55°11N 119°29W 70 B5
Beaverstone → Canada 54°59N 89°25W 72 B2
Beaverton Canada 44°26N 79°9W 82 B5
Beaverton U.S.A. 45°29N 122°48W 78 E4
Beawar India 26°3N 74°18E 42 F6
Bebedouro Brazil 21°0S 48°25W 95 A6
Bebera, Tanjung Indonesia 8°44S 115°51E 37
Beboa Madag. 17°22S 44°33E 57
Becán Mexico 18°34N 89°31W 87
Bécancour Canada 46°20N 72°26W 81
Beccles U.K. 52°27N 1°35E 13
Bečej Serbia 45°36N 20°3E 23
Béchar Algeria 31°38N 2°18W 50
Becharof L. U.S.A. 57°56N 156°23W 74
Beckley U.S.A. 37°47N 81°11W 81
Beddouza, C. Morocco 32°33N 9°9W 50
Bedford Canada 45°7N 72°59W 83
Bedford S. Africa 32°40S 26°10E 56
Bedford U.K. 52°8N 0°28W 13
Bedford Ind., U.S.A. 38°52N 86°29W 80
Bedford Iowa, U.S.A. 40°40N 94°44W 80
Bedford Ohio, U.S.A. 41°23N 81°32W 82
Bedford Pa., U.S.A. 40°1N 78°30W 82
Bedford Va., U.S.A. 37°20N 79°31W 81
Bedford □ U.K. 52°4N 0°28W 13
Bedford, C. Australia 15°14S 145°21E 62
Bedok Singapore 1°19N 103°56E 3
Bedourie Australia 24°30S 139°30E 62
Bedugul Indonesia 8°17S 115°10E 37
Bedum Neths. 53°18N 6°36E 15
Beebe Plain Canada 45°1N 72°9W 83
Beech Creek U.S.A. 41°5N 77°36W 82
Beechy Canada 50°53N 107°24W 71
Beed = Bir India 19°4N 75°46E 40
Beef I. Br. Virgin Is. 18°26N 64°30W 8
Beenleigh Australia 27°43S 153°10E 63
Be'er Menuha Israel 30°19N 35°8E 44
Be'er Sheva Israel 31°15N 34°48E 46
Beersheba = Be'er Sheva Israel 31°15N 34°48E 46
Beestekraal S. Africa 25°23S 27°38E 57
Beeston U.K. 52°56N 1°14W 12
Beeton Canada 44°5N 79°47W 82
Beeville U.S.A. 28°24N 97°45W 84
Befale Dem. Rep. of the Congo 0°25N 20°45E 52
Befandriana Mahajanga, Madag. 15°16S 48°32E 57
Befandriana Toliara, Madag. 21°55S 44°0E 57
Befasy Madag. 20°33S 44°23E 57
Befotaka Antsiranana, Madag. 13°15S 48°16E 57
Befotaka Fianarantsoa, Madag. 23°49S 47°0E 57
Bega Australia 36°41S 149°51E 63
Begusarai India 25°24N 86°9E 43
Behābād Iran 32°24N 59°47E 45
Behala India 22°30N 88°18E 43
Behara Madag. 24°55S 46°20E 57
Behbehān Iran 30°30N 50°15E 45
Behm Canal U.S.A. 55°10N 131°0W 70
Behshahr Iran 36°45N 53°35E 45
Bei Jiang → China 23°2N 112°58E 33
Bei Shan China 41°30N 96°0E 32
Bei'an China 48°10N 126°20E 33
Beihai China 21°28N 109°6E 33
Beijing China 39°53N 116°21E 34
Beijing □ China 39°55N 116°20E 34
Beilen Neths. 52°52N 6°27E 15
Beilpajah Australia 32°54S 143°52E 63
Beinn na Faoghla = Benbecula U.K. 57°26N 7°21W 11
Beipiao China 41°52N 120°32E 35
Beira Mozam. 19°50S 34°52E 55
Beirut = Bayrūt Lebanon 33°53N 35°31E 46
Beiseker Canada 51°23N 113°32W 70
Beit Lekhem = Bayt Lahm West Bank 31°43N 35°12E 46
Beitaolaizhao China 44°58N 125°58E 35
Beitbridge Zimbabwe 22°12S 30°0E 55
Beizhen = Binzhou China 37°20N 118°2E 35
Beizhen China 41°38N 121°54E 35
Beizhengzhen China 44°31N 123°30E 35
Beja Portugal 38°2N 7°53W 21
Béja Tunisia 36°43N 9°12E 51
Bejaïa Algeria 36°42N 5°2E 50
Béjar Spain 40°23N 5°46W 21
Bejestān Iran 34°30N 58°5E 45
Bekaa Valley = Al Biqā Lebanon 34°10N 36°10E 46
Bekasi Indonesia 6°14S 106°59E 37
Békéscsaba Hungary 46°40N 21°5E 17
Bekily Madag. 24°13S 45°19E 57
Bekisopa Madag. 21°40S 45°54E 57
Bekitro Madag. 24°33S 45°18E 57
Bekodoka Madag. 16°58S 45°7E 57
Bekok Malaysia 2°20N 103°7E 39
Bekopaka Madag. 19°9S 44°48E 57
Bela India 25°50N 82°0E 43
Bela Pakistan 26°12N 66°20E 42
Bela Bela S. Africa 24°51S 28°19E 57
Bela Crkva Serbia 44°55N 21°27E 23
Bela Vista Brazil 22°12S 56°20W 94
Bela Vista Mozam. 26°10S 32°44E 57
Belan → India 24°2N 81°45E 43
Belarus ■ Europe 53°30N 27°0E 17
Belau = Palau ■ Palau 7°30N 134°30E 58
Belavenona Madag. 24°50S 47°4E 57
Belawan Indonesia 3°33N 98°32E 36
Belaya → Russia 54°40N 56°0E 18
Belaya Tserkov = Bila Tserkva Ukraine 49°45N 30°10E 17
Belaya Zemlya, Ostrova Russia 81°36N 62°18E 28
Belcher Is. Canada 56°15N 78°45W 72
Belden U.S.A. 40°2N 121°17W 78
Belebey Russia 54°7N 54°7E 18
Beledweyne Somali Rep. 4°30N 45°5E 47
Belém Brazil 1°20S 48°30W 93
Belén Argentina 27°40S 67°5W 94
Belén Paraguay 23°30S 57°6W 94
Belen U.S.A. 34°40N 106°46W 77
Belet Uen = Beledweyne Somali Rep. 4°30N 45°5E 47

lev Russia 53°50N 36°5E 18 D6
lfair U.S.A. 47°27N 122°50W 78 C4
lfast S. Africa 25°42S 30°2E 57 D5
lfast U.K. 54°37N 5°56W 10 B6
lfast Maine, U.S.A. 44°26N 69°1W 81 C19
lfast N.Y., U.S.A. 42°21N 78°7W 82 D6
lfast L. U.K. 54°40N 5°50W 10 B6
lfield U.S.A. 46°53N 103°12W 80 B2
lfort France 47°38N 6°50E 20 C7
lfry U.S.A. 45°9N 109°1W 76 D9
lgaum India 15°55N 74°35E 40 M9
lgavi = Belgaum India 15°55N 74°35E 40 M9
lgium ■ Europe 50°30N 5°0E 15 D4
lgorod Russia 50°35N 36°35E 19 D6
lgorod-Dnestrovskiy = Bilhorod-Dnistrovskyy Ukraine 46°11N 30°23E 19 E5
lgrade = Beograd Serbia 44°50N 20°37E 23 B9
lgrade U.S.A. 45°47N 111°11W 76 D8
lgrano Antarctica 77°52S 34°37W 5 D1
lhaven U.S.A. 35°33N 76°37W 85 D16
li Drim → Europe 42°6N 20°25E 23 C9
limbing Indonesia 8°24S 115°2E 37 J18
linyu Indonesia 1°35S 105°50E 36 E3
liton Is. = Belitung Indonesia 3°10S 107°50E 36 E3
litung Indonesia 3°10S 107°50E 36 E3
lize ■ Cent. Amer. 17°0N 88°30W 87 D7
lize City Belize 17°25N 88°10W 87 D7
lkovskiy, Ostrov Russia 75°32N 135°44E 29 B14
ll → Canada 49°48N 77°38W 72 C4
ll I. Canada 50°46N 55°35W 73 B8
ll-Irving → Canada 56°12N 129°5W 70 B3
ll Peninsula Canada 63°50N 82°0W 69 E15
ll Ville Argentina 32°40S 62°40W 94 C3
lla Bella Canada 52°10N 128°10W 70 C3
lla Coola Canada 52°25N 126°40W 70 C3
lla Unión Uruguay 30°15S 57°40W 94 C4
lla Vista Corrientes, Argentina 28°33S 59°0W 94 B4
lla Vista Tucuman, Argentina 27°10S 65°25W 94 B2
llaire U.S.A. 40°1N 80°45W 82 F4
llary India 15°10N 76°56E 40 M10
llata Australia 29°53S 149°46E 63 D4
lle Fourche U.S.A. 44°40N 103°51W 80 C2
lle Fourche → U.S.A. 44°26N 102°18W 75 G19
lle Glade U.S.A. 26°41N 80°40W 85 H14
lle-Île France 47°20N 3°10W 20 C2
lle Isle Canada 51°57N 55°25W 73 B8
lle Isle, Str. of Canada 51°30N 56°30W 73 B8
lle Plaine U.S.A. 41°54N 92°17W 80 E7
lle River Canada 42°18N 82°43W 82 D2
llefontaine U.S.A. 40°22N 83°46W 81 E12
llefonte U.S.A. 40°55N 77°47W 82 F7
lleoram Canada 47°31N 55°25W 73 C8
lleplaine Barbados 13°15N 59°34W 89 g
lleville Canada 44°10N 77°23W 82 B7
lleville Ill., U.S.A. 38°31N 89°59W 80 F9
lleville Kans., U.S.A. 39°50N 97°38W 80 F5
lleville N.J., U.S.A. 40°47N 74°9W 83 F10
lleville N.Y., U.S.A. 43°46N 76°10W 83 C8
llevue Idaho, U.S.A. 43°28N 114°16W 76 E6
llevue Nebr., U.S.A. 41°9N 95°54W 80 E6
llevue Ohio, U.S.A. 41°17N 82°51W 82 E2
llevue Wash., U.S.A. 47°37N 122°12W 78 C4
llin = Kangirsuk Canada 60°0N 70°0W 69 F18
llingen Australia 30°25S 152°50E 63 E5
llingham U.S.A. 48°46N 122°29W 78 B4
llingshausen Abyssal Plain S. Ocean 64°0S 90°0W 5 C16
llingshausen Sea Antarctica 66°0S 80°0W 5 C17
llinzona Switz. 46°11N 9°1E 20 C8
llo Colombia 6°20N 75°33W 92 B3
llows Falls U.S.A. 43°8N 72°27W 83 C12
llpat Pakistan 29°0N 68°5E 42 E3
lluno Italy 46°9N 12°13E 22 A5
llwood U.S.A. 40°36N 78°20W 82 F6
lmont Canada 42°53N 81°5W 82 D3
lmont S. Africa 29°28S 24°22E 56 D3
lmont U.S.A. 42°14N 78°2W 82 D6
lmonte Brazil 16°0S 39°0W 93 G11
lmopan Belize 17°18N 88°30W 87 D7
lmullet Ireland 54°14N 9°58W 10 B2
lo Horizonte Brazil 19°55S 43°56W 93 G10
lo-Tsiribihina Madag. 19°40S 44°30E 57 B7
logorsk Russia 51°0N 128°20E 29 D13
loha Madag. 25°10S 45°3E 57 D8
loit Kans., U.S.A. 39°28N 98°6W 80 F4
loit Wis., U.S.A. 42°31N 89°2W 80 D9
lokorovichi Ukraine 51°7N 28°2E 17 C15
lomorsk Russia 64°35N 34°54E 18 B5
lonia India 23°15N 91°30E 41 H17
loretsk Russia 53°58N 58°24E 18 D10
lorussia = Belarus ■ Europe 53°30N 27°0E 17 B14
lovo Russia 54°30N 86°0E 28 D9
loyarskiy Russia 63°42N 66°40E 28 C7
loye, Ozero Russia 60°10N 37°35E 18 B6
loye More Russia 66°30N 38°0E 18 A6
lozersk Russia 60°1N 37°45E 18 B6
lpre U.S.A. 39°17N 81°34W 81 F13
lrain India 28°23N 80°55E 43 E9
lt U.S.A. 47°23N 110°55W 76 C8
ltana Australia 30°48S 138°25E 63 E2
lterra Brazil 2°45S 55°0W 93 D8
lton U.S.A. 31°3N 97°28W 84 F6
lton L. U.S.A. 31°6N 97°28W 84 F6
ltsy = Bălţi Moldova 47°48N 27°58E 17 E14
lturbet Ireland 54°6N 7°26W 10 B4
lukha Russia 49°50N 86°50E 32 B6
luran Malaysia 5°48N 117°35E 36 C5
lushya Guba Russia 71°32N 52°19E 28 B6
lvidere Ill., U.S.A. 42°15N 88°50W 80 D9
lvidere N.J., U.S.A. 40°50N 75°5W 83 F9
Belyando → Australia 21°38S 146°50E 62 C4
Belyando Crossing Australia 21°32S 146°51E 62 C4
Belyuen Australia 12°34S 130°42E 60 B5
Belyy, Ostrov Russia 73°30N 71°0E 28 B8
Belyy Yar Russia 58°26N 84°39E 28 D9
Belzoni U.S.A. 33°11N 90°29W 85 E9
Bemaraha, Lembalemban' i Madag. 18°40S 44°45E 57 B7
Bemarivo Madag. 21°45S 44°45E 57 C7
Bemarivo → Antsiranana, Madag. 14°9S 50°9E 57 A9
Bemarivo → Mahajanga, Madag. 15°27S 47°40E 57 B8
Bemavo Madag. 21°33S 45°25E 57 C8
Bembéréke Benin 10°11N 2°43E 50 F6
Bembesi Zimbabwe 20°0S 28°58E 55 G2
Bembesi → Zimbabwe 18°57S 27°47E 55 F2
Bemetara India 21°42N 81°32E 43 J9
Bemidji U.S.A. 47°28N 94°53W 80 B6
Bemolanga Madag. 17°44S 45°6E 57 B8
Ben Iran 32°32N 50°45E 45 C6
Ben Cruachan U.K. 56°26N 5°8W 11 E3
Ben Dearg U.K. 57°47N 4°56W 11 D4
Ben En △ Vietnam 19°37N 105°30E 38 C5
Ben Gardane Tunisia 33°11N 11°11E 51 B8
Ben Hope U.K. 58°25N 4°36W 11 C4
Ben Lawers U.K. 56°32N 4°14W 11 E4
Ben Lomond N.S.W., Australia 30°1S 151°43E 63 E5
Ben Lomond Tas., Australia 41°38S 147°42E 63 G4
Ben Lomond U.K. 56°11N 4°38W 11 E4
Ben Lomond △ Australia 41°33S 147°39E 63 G4
Ben Luc Vietnam 10°39N 106°29E 39 G6
Ben Macdhui U.K. 57°4N 3°40W 11 D5
Ben Mhor U.K. 57°15N 7°18W 11 D1
Ben More Argyll & Bute, U.K. 56°26N 6°1W 11 E2
Ben More Stirling, U.K. 56°23N 4°32W 11 E4
Ben More Assynt U.K. 58°8N 4°52W 11 C4
Ben Nevis U.K. 56°48N 5°1W 11 E3
Ben Quang Vietnam 17°3N 106°55E 38 D6
Ben Tre Vietnam 10°14N 106°23E 39 G6
Ben Vorlich U.K. 56°21N 4°14W 11 E4
Ben Wyvis U.K. 57°40N 4°35W 11 D4
Bena Nigeria 11°20N 5°50E 50 F7
Benalla Australia 36°30S 146°0E 63 F4
Benares = Varanasi India 25°22N 83°0E 43 G10
Benavente Spain 42°2N 5°43W 21 A3
Benavides U.S.A. 27°36N 98°25W 84 H5
Benbecula U.K. 57°26N 7°21W 11 D1
Benbonyathe Hill Australia 30°25S 139°11E 63 E2
Bend U.S.A. 44°4N 121°19W 76 D3
Bender Beyla Somali Rep. 9°30N 50°48E 47 F5
Bendery = Tighina Moldova 46°50N 29°30E 17 E15
Bendigo Australia 36°40S 144°15E 63 F3
Benē Beraq Israel 32°6N 34°51E 46 C3
Benenitra Madag. 23°27S 45°5E 57 C8
Benevento Italy 41°8N 14°45E 22 D6
Beng Mealea Cambodia 13°28N 104°14E 38 F5
Benga Mozam. 16°11S 33°40E 55 F3
Bengal, Bay of Ind. Oc. 15°0N 90°0E 41 M17
Bengaluru = Bangalore India 12°59N 77°40E 40 N10
Bengbu China 32°58N 117°20E 35 H9
Benghazi = Banghāzī Libya 32°11N 20°3E 51 B10
Bengkalis Indonesia 1°30N 102°10E 39 M4
Bengkulu Indonesia 3°50S 102°12E 36 E2
Bengkulu □ Indonesia 3°48S 102°16E 36 E2
Bengough Canada 49°25N 105°10W 71 D7
Benguela Angola 12°37S 13°25E 53 G2
Benguérua, I. Mozam. 21°58S 35°28E 57 C6
Beni Dem. Rep. of the Congo 0°30N 29°27E 54 B2
Beni → Bolivia 10°23S 65°24W 92 F5
Beni Mellal Morocco 32°21N 6°21W 50 B4
Beni Suef Egypt 29°5N 31°6E 51 C12
Beniah L. Canada 63°23N 112°17W 70 A6
Benidorm Spain 38°33N 0°9W 21 C5
Benin ■ Africa 10°0N 2°0E 50 G6
Benin, Bight of W. Afr. 5°0N 3°0E 50 H6
Benin City Nigeria 6°20N 5°31E 50 G7
Benito Juárez Argentina 37°40S 59°43W 94 D4
Benitses Greece 39°32N 19°55E 25 A3
Benjamin Aceval Paraguay 24°58S 57°34W 94 A4
Benjamin Constant Brazil 4°40S 70°15W 92 D4
Benjamin Hill Mexico 30°9N 111°7W 86 A2
Benkelman U.S.A. 40°3N 101°32W 80 E3
Bennett Canada 59°51N 135°0W 70 B2
Bennett, L. Australia 22°50S 131°2E 60 D5
Bennetta, Ostrov Russia 76°21N 148°56E 29 B15
Bennettsville U.S.A. 34°37N 79°41W 85 D15
Bennington N.H., U.S.A. 43°0N 71°55W 83 D11
Bennington Vt., U.S.A. 42°53N 73°12W 83 D11
Benom Malaysia 3°50N 102°6E 39 L4
Benoni S. Africa 26°11S 28°18E 57 D4
Benque Viejo Belize 17°5N 89°8W 87 D7
Benson Ariz., U.S.A. 31°58N 110°18W 77 L8
Benson Minn., U.S.A. 45°19N 95°36W 80 C6
Bent Iran 26°20N 59°31E 45 E8
Benteng Indonesia 6°10S 120°30E 37 F6
Bentinck I. Australia 17°3S 139°35E 62 B2
Bentinck I. Burma 11°45N 98°3E 39 G2
Bentley Subglacial Trench Antarctica 80°0S 115°0W 5 E15
Bento Gonçalves Brazil 29°10S 51°31W 95 B5
Benton Ark., U.S.A. 34°34N 92°35W 84 D8
Benton Calif., U.S.A. 37°48N 118°32W 78 H8
Benton Ill., U.S.A. 38°0N 88°55W 80 F9
Benton Pa., U.S.A. 41°12N 76°23W 83 E8
Benton Harbor U.S.A. 42°6N 86°27W 80 D10
Bentong Malaysia 3°31N 101°55E 39 L3
Bentonville U.S.A. 36°22N 94°13W 84 C7
Benue → Nigeria 7°48N 6°46E 50 G7
Benxi China 41°20N 123°48E 35 D12
Beo Indonesia 4°25N 126°50E 37 D7
Beograd Serbia 44°50N 20°37E 23 B9
Beolgyo S. Korea 34°51N 127°21E 35 G14
Beppu Japan 33°15N 131°30E 31 H5
Beqa Fiji 18°23S 178°8E 59 a
Beqaa Valley = Al Biqā Lebanon 34°10N 36°10E 46 A5
Ber Mota India 23°27N 68°34E 42 H3
Bera, Tasik Malaysia 3°5N 102°38E 39 L4
Berach → India 25°15N 75°2E 42 G6
Beraketa Madag. 23°7S 44°25E 57 C7
Berastagi Indonesia 3°11N 98°31E 39 L2
Berat Albania 40°43N 19°59E 23 D8
Berau = Tanjungredeb Indonesia 2°9N 117°29E 36 D5
Berau, Teluk Indonesia 2°30S 132°30E 37 E8
Beravina Madag. 18°10S 45°14E 57 B8
Berber Sudan 18°0N 34°0E 51 E12
Berbera Somali Rep. 10°30N 45°2E 47 E4
Berbérati C.A.R. 4°15N 15°40E 52 D3
Berbice → Guyana 6°20N 57°32W 92 B7
Berdichev = Berdychiv Ukraine 49°57N 28°30E 17 D15
Berdsk Russia 54°47N 83°2E 28 D9
Berdyansk Ukraine 46°45N 36°50E 19 E6
Berdychiv Ukraine 49°57N 28°30E 17 D15
Berea U.S.A. 37°34N 84°17W 81 G11
Berebere Indonesia 2°25N 128°45E 37 D7
Bereeda Somali Rep. 11°45N 51°0E 47 E5
Berehove Ukraine 48°15N 22°35E 17 D12
Bereket Turkmenistan 39°16N 55°32E 45 B7
Berekum Ghana 7°29N 2°34W 50 G5
Berens → Canada 52°25N 97°2W 71 C9
Berens I. Canada 52°18N 97°18W 71 C9
Berens River Canada 52°25N 97°0W 71 C9
Beresford U.S.A. 43°5N 96°47W 80 D5
Berestechko Ukraine 50°22N 25°5E 17 C13
Berevo Mahajanga, Madag. 17°14S 44°17E 57 B7
Berevo Toliara, Madag. 19°44S 44°58E 57 B7
Bereza = Byaroza Belarus 52°31N 24°51E 17 B13
Berezhany Ukraine 49°26N 24°58E 17 D13
Berezina = Byarezina → Belarus 52°33N 30°14E 17 B16
Bereznik Russia 62°51N 42°40E 18 B7
Berezniki Russia 59°24N 56°46E 18 C10
Berezovo Russia 64°0N 65°0E 28 C7
Berga Spain 42°6N 1°48E 21 A6
Bergama Turkey 39°8N 27°11E 23 E12
Bérgamo Italy 45°41N 9°43E 20 D8
Bergen Neths. 52°40N 4°43E 15 B4
Bergen Norway 60°20N 5°20E 8 F11
Bergen U.S.A. 43°5N 77°57W 82 C7
Bergen op Zoom Neths. 51°28N 4°18E 15 C4
Bergerac France 44°51N 0°30E 20 D4
Bergholz U.S.A. 40°31N 80°53W 82 F4
Bergisch Gladbach Germany 50°59N 7°8E 15 D7
Bergville S. Africa 28°52S 29°18E 57 D4
Berhala, Selat Indonesia 1°0S 104°15E 36 E2
Berhampore = Baharampur India 24°2N 88°27E 43 G13
Berhampur = Brahmapur India 19°15N 84°54E 41 K14
Bering Glacier U.S.A. 60°20N 143°30W 74 C11
Bering Sea Pac. Oc. 58°0N 171°0W 66 D2
Bering Strait Pac. Oc. 65°30N 169°0W 74 B6
Beringovskiy Russia 63°3N 179°19E 29 C18
Berisso Argentina 34°56S 57°50W 94 C4
Berja Spain 36°50N 2°56W 21 D4
Berkeley U.S.A. 37°51N 122°16W 78 H4
Berkner I. Antarctica 79°30S 50°0W 5 D18
Berkshire U.S.A. 42°19N 76°11W 83 D8
Berkshire Downs U.K. 51°33N 1°29W 13 F6
Berlin Germany 52°31N 13°23E 16 B7
Berlin Md., U.S.A. 38°20N 75°13W 81 F16
Berlin N.H., U.S.A. 44°28N 71°11W 83 B13
Berlin N.Y., U.S.A. 42°42N 73°23W 83 D11
Berlin Wis., U.S.A. 43°58N 88°57W 80 D9
Berlin L. U.S.A. 41°3N 81°0W 82 E4
Bermejo → Formosa, Argentina 26°51S 58°23W 94 B4
Bermejo → San Juan, Argentina 32°30S 67°30W 94 C2
Bermejo, Paso = Uspallata, P. de Argentina 32°37S 69°22W 94 C2
Bermen, L. Canada 53°35N 68°55W 73 B6
Bermuda ☐ Atl. Oc. 32°18N 64°45W 67 F13
Bern Switz. 46°57N 7°28E 20 C7
Bernalillo U.S.A. 35°18N 106°33W 77 J10
Bernardo de Irigoyen Argentina 26°15S 53°40W 95 B5
Bernardsville U.S.A. 40°43N 74°34W 83 F10
Bernasconi Argentina 37°55S 63°44W 94 D3
Bernburg Germany 51°47N 11°44E 16 C6
Berne = Bern Switz. 46°57N 7°28E 20 C7
Berneray U.K. 57°43N 7°11W 11 D1
Bernier I. Australia 24°50S 113°12E 61 D1
Bernina, Piz Switz. 46°20N 9°54E 20 C8
Beroroha Madag. 21°40S 45°10E 57 C8
Beroun Czech Rep. 49°57N 14°5E 16 D8
Berri Australia 34°14S 140°35E 63 E3
Berriane Algeria 32°50N 3°46E 50 B6
Berry Australia 34°46S 150°43E 63 E5
Berry France 46°50N 2°0E 20 C5
Berry Is. Bahamas 25°40N 77°50W 88 A4
Berryessa, L. U.S.A. 38°31N 122°6W 78 G4
Berryville U.S.A. 36°22N 93°34W 84 C8
Berseba Namibia 26°0S 17°46E 56 D2
Bershad Ukraine 48°22N 29°31E 17 D15
Berthold U.S.A. 48°19N 101°44W 80 A3
Berthoud U.S.A. 40°19N 105°5W 76 F11
Bertoua Cameroon 4°30N 13°45E 52 D2
Bertraghboy B. Ireland 53°22N 9°54W 10 C2
Berwick U.S.A. 41°3N 76°14W 83 E8
Berwick-upon-Tweed U.K. 55°46N 2°0W 12 B6
Berwyn Mts. U.K. 52°54N 3°26W 12 E4
Besal Pakistan 35°4N 73°56E 43 B5
Besalampy Madag. 16°43S 44°29E 57 B7
Besançon France 47°15N 6°2E 20 C7
Besar Indonesia 2°40S 116°0E 36 E5
Besnard L. Canada 55°25N 106°0W 71 B7
Besni Turkey 37°41N 37°52E 44 B3
Besor, N. → Egypt 31°28N 34°22E 46 D3
Bessarabiya Moldova 47°0N 28°10E 17 E15
Bessarabka = Basarabeasca Moldova 46°21N 28°58E 17 E15
Bessemer Ala., U.S.A. 33°24N 86°58W 85 E11
Bessemer Mich., U.S.A. 46°29N 90°3W 80 B8
Bessemer Pa., U.S.A. 40°59N 80°30W 82 F4
Beswick Australia 14°34S 132°53E 60 B5
Beswick ✪ Australia 14°31S 132°51E 60 B5
Bet She'an Israel 32°30N 35°30E 46 C4
Bet Shemesh Israel 31°44N 35°0E 46 D4
Betafo Madag. 19°50S 46°51E 57 B8
Betancuria Canary Is. 28°25N 14°3W 24 F5
Betanzos Spain 43°15N 8°12W 21 A1
Bétaré Oya Cameroon 5°40N 14°5E 52 C2
Betatao Madag. 18°11S 47°52E 57 B8
Bethal S. Africa 26°27S 29°28E 57 D4
Bethanien Namibia 26°31S 17°8E 56 D2
Bethany Canada 44°11N 78°34W 82 B6
Bethany Mo., U.S.A. 40°16N 94°2W 80 E6
Bethany Okla., U.S.A. 35°31N 97°38W 84 D6
Bethel Alaska, U.S.A. 60°48N 161°45W 74 C7
Bethel Conn., U.S.A. 41°22N 73°25W 83 E11
Bethel Maine, U.S.A. 44°25N 70°47W 83 B14
Bethel Vt., U.S.A. 43°50N 72°38W 83 C12
Bethel Park U.S.A. 40°19N 80°2W 82 F4
Bethlehem = Bayt Lahm West Bank 31°43N 35°12E 46 D4
Bethlehem S. Africa 28°14S 28°18E 57 D4
Bethlehem U.S.A. 40°37N 75°23W 83 F9
Bethulie S. Africa 30°30S 25°59E 56 E4
Béthune France 50°30N 2°38E 20 A5
Betioky Madag. 23°48S 44°20E 57 C7
Betong Malaysia 1°24N 111°31E 36 D4
Betong Thailand 5°45N 101°5E 39 K3
Betoota Australia 25°45S 140°42E 62 D3
Betpaqdala Kazakhstan 45°45N 70°30E 32 B3
Betroka Madag. 23°16S 46°0E 57 C8
Betsiamites Canada 48°56N 68°40W 73 C6
Betsiamites → Canada 48°56N 68°38W 73 C6
Betsiboka → Madag. 16°3S 46°36E 57 B8
Bettendorf U.S.A. 41°32N 90°30W 80 E8
Bettiah India 26°48N 84°33E 43 F11
Betul India 21°58N 77°59E 40 J10
Betws-y-Coed U.K. 53°5N 3°48W 12 D4
Beulah Mich., U.S.A. 44°38N 86°6W 80 C10
Beulah N. Dak., U.S.A. 47°16N 101°47W 80 B3
Beveren Belgium 51°12N 4°16E 15 C4
Beverley Australia 32°9S 116°56E 61 F2
Beverley U.K. 53°51N 0°26W 12 D7
Beverly U.S.A. 42°33N 70°53W 83 D14
Beverly Hills Calif., U.S.A. 34°5N 118°24W 79 L8
Beverly Hills Fla., U.S.A. 28°55N 82°28W 85 G13
Bevoalavo Madag. 25°13S 45°26E 57 D8
Bewas → India 23°59N 79°21E 43 H8
Bewdley Canada 44°5N 78°19W 82 B6
Bexhill U.K. 50°51N 0°29E 13 G8
Beyānlū Iran 36°0N 47°51E 44 C5
Beyneu Kazakhstan 45°18N 55°9E 19 E10
Beypazarı Turkey 40°10N 31°56E 19 F5
Beyşehir Gölü Turkey 37°41N 31°33E 44 B1
Béziers France 43°20N 3°12E 20 E5
Bezwada = Vijayawada India 16°31N 80°39E 41 L12
Bhabua India 25°3N 83°37E 43 G10
Bhachau India 23°20N 70°16E 42 H4
Bhadar → Gujarat, India 22°17N 72°20E 42 H5
Bhadar → Gujarat, India 21°27N 69°47E 42 J3
Bhadarwah India 32°58N 75°46E 43 C6
Bhadgaon = Bhaktapur Nepal 27°38N 85°24E 43 F11
Bhadohi India 25°25N 82°34E 43 G10
Bhadra India 29°8N 75°14E 42 E6
Bhadrakh India 21°10N 86°30E 41 J15
Bhadran India 22°19N 72°6E 42 H5
Bhadravati India 13°49N 75°40E 40 N9
Bhag Pakistan 29°2N 67°49E 42 E2
Bhagalpur India 25°10N 87°0E 43 G12
Bhagirathi → Uttarakhand, India 30°8N 78°35E 43 D8
Bhagirathi → W. Bengal, India 23°25N 88°23E 43 H13
Bhakkar Pakistan 31°40N 71°5E 42 D4
Bhakra Dam India 31°30N 76°45E 42 D7
Bhaktapur Nepal 27°38N 85°24E 43 F11
Bhamo Burma 24°15N 97°15E 41 G20
Bhandara India 21°5N 79°42E 40 J11
Bhanpura India 24°31N 75°44E 42 G6
Bhanrer Ra. India 23°40N 79°45E 43 H8
Bhaptiahi India 26°19N 86°44E 43 F12
Bharat = India ■ Asia 20°0N 78°0E 40 K11
Bharatpur Chhattisgarh, India 23°44N 81°46E 43 H9
Bharatpur Raj., India 27°15N 77°30E 42 F7
Bharatpur Nepal 27°34N 84°10E 43 F11
Bharno India 23°14N 84°53E 43 H11
Bharuch India 21°47N 73°0E 40 J8
Bhatinda India 30°15N 74°57E 42 D6
Bhatpara India 22°50N 88°25E 43 H13
Bhattu India 29°36N 75°19E 42 E6
Bhaun Pakistan 32°55N 72°40E 42 C5
Bhaunagar = Bhavnagar India 21°45N 72°10E 40 J8
Bhavnagar India 21°45N 72°10E 40 J8
Bhawari India 25°42N 73°4E 42 G5
Bhayavadar India 21°51N 70°15E 42 J4
Bhera Pakistan 32°29N 72°57E 42 C5
Bhikangaon India 21°52N 75°57E 42 J6
Bhilai = Bhilainagar-Durg India 21°13N 81°26E 41 J12
Bhilainagar-Durg India 21°13N 81°26E 41 J12
Bhilsa = Vidisha India 23°28N 77°53E 42 H7
Bhilwara India 25°25N 74°38E 42 G6
Bhima → India 16°25N 77°17E 40 L10
Bhimbar Pakistan 32°59N 74°3E 43 C6
Bhind India 26°30N 78°46E 43 F8
Bhinga India 27°43N 81°56E 43 F9
Bhinmal India 25°0N 72°15E 42 G5
Bhisho S. Africa 32°50S 27°23E 56 E4
Bhiwandi India 19°20N 73°0E 40 K8
Bhiwani India 28°50N 76°9E 42 E7
Bhogava → India 22°26N 72°20E 42 H5
Bhola Bangla. 22°45N 90°35E 41 H17
Bholari Pakistan 25°19N 68°13E 42 G3
Bhopal India 23°20N 77°30E 42 H7
Bhubaneshwar India 20°15N 85°50E 41 J14
Bhuj India 23°15N 69°49E 42 H3
Bhumiphol Res. Thailand 17°20N 98°40E 38 D2
Bhusawal India 21°3N 75°46E 40 J9
Bhutan ■ Asia 27°25N 90°30E 41 F17
Biafra, B. of = Bonny, Bight of Africa 3°30N 9°20E 52 D1
Biak Indonesia 1°10S 136°6E 37 E9
Biała Podlaska Poland 52°4N 23°6E 17 B12
Białogard Poland 54°2N 15°58E 16 A8
Białystok Poland 53°10N 23°10E 17 B12
Biaora India 23°56N 76°56E 42 H7
Bīārjmand Iran 36°6N 55°53E 45 B7
Biaro Indonesia 2°5N 125°26E 37 D7
Biarritz France 43°29N 1°33W 20 E3
Bibai Japan 43°19N 141°52E 30 C10
Bibby I. Canada 61°55N 93°0W 71 A10
Biberach Germany 48°5N 9°47E 16 D5
Bibungwa Dem. Rep. of the Congo 2°40S 28°15E 54 C2
Bicester U.K. 51°54N 1°9W 13 F6
Bicheno Australia 41°52S 148°18E 63 G4
Bichia India 22°27N 80°42E 43 H9
Bickerton I. Australia 13°45S 136°10E 62 A2
Bid = Bir India 19°4N 75°46E 40 K9
Bida Nigeria 9°3N 5°58E 50 G7
Bidar India 17°55N 77°35E 40 L10
Biddeford U.S.A. 43°30N 70°28W 81 D18
Bideford U.K. 51°1N 4°13W 13 F3
Bideford Bay U.K. 51°5N 4°20W 13 F3
Bidhuna India 26°49N 79°31E 43 F8
Bīdokht Iran 34°20N 58°46E 45 C8
Bidor Malaysia 4°6N 101°15E 39 K3
Bidyadanga Australia 18°45S 121°43E 60 C3
Bié, Planalto de Angola 12°0S 16°0E 53 G3
Bieber U.S.A. 41°7N 121°8W 76 F3
Biel Switz. 47°8N 7°14E 20 C7
Bielefeld Germany 52°1N 8°33E 16 B5
Biella Italy 45°34N 8°3E 20 D8
Bielsk Podlaski Poland 52°47N 23°12E 17 B12
Bielsko-Biała Poland 49°50N 19°2E 17 D10
Bien Hoa Vietnam 10°57N 106°49E 39 G6
Bienne = Biel Switz. 47°8N 7°14E 20 C7
Bienville, L. Canada 55°5N 72°40W 72 A5
Biesiesfontein S. Africa 30°57S 17°58E 56 E2
Big → Canada 54°50N 58°55W 73 B8
Big B. Canada 55°43N 60°35W 73 A7
Big Bear City U.S.A. 34°16N 116°51W 79 L10
Big Bear Lake U.S.A. 34°15N 116°56W 79 L10
Big Belt Mts. U.S.A. 46°30N 111°25W 76 C8
Big Bend Swaziland 26°50S 31°58E 57 D5
Big Bend △ U.S.A. 29°20N 103°5W 84 G3
Big Black → U.S.A. 32°3N 91°4W 85 E9
Big Blue → U.S.A. 39°35N 96°34W 80 F5
Big Creek U.S.A. 37°11N 119°14W 78 H7
Big Cypress ⌂ U.S.A. 26°0N 81°10W 85 H14
Big Desert Australia 35°45S 141°10E 63 F3
Big Falls U.S.A. 48°12N 93°48W 80 A7
Big Fork → U.S.A. 48°31N 93°43W 80 A7
Big Horn Mts. = Bighorn Mts. U.S.A. 44°25N 107°0W 76 D10
Big I. Canada 61°7N 116°45W 70 A5
Big Lake U.S.A. 31°12N 101°28W 84 F4
Big Moose U.S.A. 43°49N 74°58W 83 C10
Big Muddy Cr. → U.S.A. 48°8N 104°36W 76 B11
Big Pine U.S.A. 37°10N 118°17W 78 H8
Big Piney U.S.A. 42°32N 110°7W 76 E8
Big Quill L. Canada 51°55N 104°30W 71 C8
Big Rapids U.S.A. 43°42N 85°29W 81 D11
Big Rideau L. Canada 44°40N 76°15W 83 B8
Big River Canada 53°50N 107°0W 71 C7
Big Run U.S.A. 40°57N 78°55W 82 F6
Big Sable Pt. U.S.A. 44°3N 86°1W 80 C10
Big Salmon → Canada 61°52N 134°55W 70 A2
Big Sand L. Canada 57°45N 99°45W 71 B9
Big Sandy U.S.A. 48°11N 110°7W 76 B8
Big Sandy → U.S.A. 38°25N 82°36W 81 F12
Big Sandy Cr. = Sandy Cr. → U.S.A. 41°51N 109°47W 76 F9
Big Sandy Cr. → U.S.A. 38°7N 102°29W 76 G12
Big Sioux → U.S.A. 42°29N 96°27W 80 D5
Big South Fork ⌂ U.S.A. 36°27N 84°47W 85 C12
Big Spring U.S.A. 32°15N 101°28W 84 E4
Big Stone City U.S.A. 45°18N 96°28W 80 C5
Big Stone Gap U.S.A. 36°52N 82°47W 81 G12
Big Stone L. U.S.A. 45°18N 96°27W 80 C5
Big Sur U.S.A. 36°15N 121°48W 78 J5
Big Timber U.S.A. 45°50N 109°57W 76 D9
Big Trout L. Canada 53°40N 90°0W 72 B2
Big Trout Lake Canada 53°45N 90°0W 72 B2
Biğa Turkey 40°13N 27°14E 23 D12
Bigadiç Turkey 39°22N 28°7E 23 E13
Biggar Canada 52°4N 108°0W 71 C7
Biggar U.K. 55°38N 3°32W 11 F5
Bigge I. Australia 14°35S 125°10E 60 B4
Biggenden Australia 25°31S 152°4E 63 D5
Biggleswade U.K. 52°5N 0°14W 13 E7
Biggs U.S.A. 39°25N 121°43W 78 F5
Bighorn U.S.A. 46°10N 107°27W 76 C10
Bighorn → U.S.A. 46°10N 107°28W 76 C10
Bighorn Canyon ⌂ U.S.A. 45°10N 108°0W 76 D10
Bighorn L. U.S.A. 44°55N 108°15W 76 D9

Bighorn Mts. U.S.A. 44°25N 107°0W 76 D10
Bigstone L. Canada 53°42N 95°44W 71 C9
Bigwa Tanzania 7°10S 39°10E 54 D4
Bihać Bos.-H. 44°49N 15°57E 16 F8
Bihar India 25°5N 85°40E 43 G11
Bihar □ India 25°0N 86°0E 43 G12
Biharamulo Tanzania 2°25S 31°25E 54 C3
Biharamulo ⌂ Tanzania 2°24S 31°26E 54 C3
Bihariganj India 25°44N 86°59E 43 G12
Bihor, Munţii Romania 46°29N 22°47E 17 E12
Bijagós, Arquipélago dos
Guinea-Biss. 11°15N 16°10W 50 F2
Bijaipur India 26°2N 77°20E 42 F7
Bijapur Chhattisgarh, India 18°50N 80°50E 41 K12
Bijapur Karnataka, India 16°50N 75°55E 40 L9
Bījār Iran 35°52N 47°35E 44 C5
Bijawar India 24°38N 79°30E 43 G8
Bijeljina Bos.-H. 44°46N 19°14E 23 B8
Bijnor India 29°27N 78°11E 42 E8
Bikaner India 28°2N 73°18E 42 E5
Bikapur India 26°30N 82°7E 43 F10
Bikeqi China 40°43N 111°20E 34 D6
Bikfayyā Lebanon 33°55N 35°41E 46 B4
Bikin Russia 46°50N 134°20E 30 A7
Bikin → Russia 46°51N 134°2E 30 A7
Bikini Atoll Marshall Is. 12°0N 167°30E 64 F8
Bikita Zimbabwe 20°6S 31°41E 57 C5
Bīkkū Bīttī Libya 22°0N 19°12E 51 D9
Bila Tserkva Ukraine 49°45N 30°10E 17 D16
Bilara India 26°14N 73°53E 42 F5
Bilaspur Chhattisgarh, India 22°2N 82°15E 43 H10
Bilaspur Punjab, India 31°19N 76°50E 42 D7
Bilauk Taungdan Thailand 13°0N 99°0E 38 F2
Bilbao Spain 43°16N 2°56W 21 A4
Bilbo = Bilbao Spain 43°16N 2°56W 21 A4
Bildudalur Iceland 65°41N 23°36W 8 D2
Bílé Karpaty Europe 49°5N 18°0E 17 D9
Bilecik Turkey 40°5N 30°5E 19 F5
Bilgram India 27°11N 80°2E 43 F9
Bilhaur India 26°51N 80°5E 43 F9
Bilhorod-Dnistrovskyy Ukraine 46°11N 30°23E 19 E5
Bilibino Russia 68°3N 166°20E 29 C17
Bilibiza Mozam. 12°30S 40°20E 55 E5
Billiluna Australia 19°37S 127°41E 60 C4
Billings U.S.A. 45°47N 108°30W 76 D9
Billiton Is. = Belitung Indonesia 3°10S 107°50E 36 E3
Bilma Niger 18°50N 13°30E 51 E8
Biloela Australia 24°24S 150°31E 62 C5
Biloxi U.S.A. 30°24N 88°53W 85 F10
Bilpa Morea Claypan Australia 25°0S 140°0E 62 D3
Biltine Chad 14°40N 20°50E 51 F10
Bim Son Vietnam 20°4N 105°51E 38 B5
Bima Indonesia 8°22S 118°49E 37 F5
Bimbo C.A.R. 4°15N 18°33E 52 D3
Bimini Is. Bahamas 25°42N 79°25W 88 A4
Bin Xian Heilongjiang, China 45°42N 127°32E 35 B14
Bin Xian Shaanxi, China 35°2N 108°4E 34 G5
Bina-Etawah India 24°13N 78°14E 42 G8
Binalong Australia 34°40S 148°39E 63 E4
Bīnālūd, Kūh-e Iran 36°30N 58°30E 45 B8
Binatang = Bintangau Malaysia 2°10N 111°40E 36 D4
Binche Belgium 50°26N 4°10E 15 D4
Bindki India 26°2N 80°36E 43 F9
Bindura Zimbabwe 17°18S 31°18E 55 F3
Bingara Australia 29°52S 150°36E 63 D5
Bingham U.S.A. 45°3N 69°53W 81 C19
Binghamton U.S.A. 42°6N 75°55W 83 D9
Bingöl Turkey 38°53N 40°29E 44 B4
Binh Dinh Vietnam 13°55N 109°7E 38 F7
Binh Son Vietnam 15°20N 108°40E 38 E7
Binhai China 34°2N 119°49E 35 G10
Binisatua Spain 39°50N 4°11E 24 B11
Binissalem Spain 39°41N 2°50E 24 B9
Binjai Indonesia 3°20N 98°30E 36 D3
Binnaway Australia 31°28S 149°24E 63 E4
Binongko Indonesia 5°57S 124°2E 37 F6
Binscarth Canada 50°37N 101°17W 71 C8
Bintan Indonesia 1°0N 104°0E 36 D2
Bintangau Malaysia 2°10N 111°40E 36 D4
Bintulu Malaysia 3°10N 113°0E 36 D4
Bintuni Indonesia 2°7S 133°32E 37 E8
Binzert = Bizerte Tunisia 37°15N 9°50E 51 A7
Binzhou China 37°20N 118°2E 35 F10
Bío Bío □ Chile 37°35S 72°0W 94 D1
Biobío → Chile 36°49S 73°10W 94 D1
Bioko Eq. Guin. 3°30N 8°40E 52 D1
Bir India 19°4N 75°46E 40 K9
Bîr Abu Muḩammad Egypt 29°44N 34°14E 46 F3
Bi'r al Butayyihāt Jordan 29°47N 35°20E 46 F4
Bi'r al Mārī Jordan 30°4N 35°33E 46 E4
Bi'r al Qattār Jordan 29°47N 35°32E 46 F4
Bîr Atrun Sudan 18°15N 26°40E 51 E11
Bîr Beiḑa Egypt 30°25N 34°29E 46 E3
Bîr el 'Abd Egypt 31°2N 33°0E 46 D2
Bîr el Biarât Egypt 29°30N 34°43E 46 F3
Bîr el Duweidar Egypt 30°56N 32°32E 46 E1
Bîr el Garârât Egypt 31°3N 33°34E 46 D2
Bîr el Heisi Egypt 29°22N 34°36E 46 F3
Bîr el Jafir Egypt 30°50N 32°41E 46 E1
Bîr el Mâlḩi Egypt 30°38N 33°19E 46 E2
Bîr el Thamâda Egypt 30°12N 33°27E 46 E2
Bîr Gebeil Ḩişn Egypt 30°2N 33°18E 46 E2
Bi'r Ghadīr Syria 34°6N 37°3E 46 A6
Bîr Ḩasana Egypt 30°29N 33°46E 46 E2
Bîr Kaseiba Egypt 31°0N 33°17E 46 E2
Bîr Lahfân Egypt 31°0N 33°51E 46 E2
Bîr Madkûr Egypt 30°44N 32°33E 46 E1
Bîr Mogreïn Mauritania 25°10N 11°25W 50 C3
Bi'r Muţribah Kuwait 29°54N 47°17E 44 D5
Bîr Qaţia Egypt 30°58N 32°45E 46 E1
Bîr Shalatein Egypt 23°5N 35°25E 51 D13
Birāk Libya 27°31N 14°20E 51 C8
Biratnagar Nepal 26°27N 87°17E 43 F12
Birawa Dem. Rep. of the Congo 2°20S 28°48E 54 C2
Birch → Canada 58°28N 112°17W 70 B6
Birch Hills Canada 52°59N 105°25W 71 C7
Birch I. Canada 52°26N 99°54W 71 C9
Birch L. N.W.T., Canada 62°4N 116°33W 70 A5
Birch L. Ont., Canada 51°23N 92°18W 72 B1
Birch Mts. Canada 57°30N 113°10W 70 B6
Birch River Canada 52°24N 101°6W 71 C8
Birchip Australia 35°56S 142°55E 63 F3
Bird Canada 56°30N 94°13W 71 B10
Bird I. = Aves, I. de W. Indies 15°45N 63°55W 89 C7
Bird I. S. Georgia 54°0S 38°3W 96 G9
Birds Creek Canada 45°6N 77°52W 82 A7
Birdsville Australia 25°51S 139°20E 62 D2
Birdum Cr. → Australia 15°14S 133°0E 60 C5
Birecik Turkey 37°2N 38°0E 44 B3
Birein Israel 30°50N 34°28E 46 E3
Bireuen Indonesia 5°14N 96°39E 36 C1
Birganj Nepal 27°1N 84°52E 43 F11
Birigüi Brazil 21°18S 50°16W 95 A5
Bīrjand Iran 32°53N 59°13E 45 C8
Birkenhead U.K. 53°23N 3°2W 12 D4
Bîrlad = Bârlad Romania 46°15N 27°38E 17 E14
Birmingham U.K. 52°29N 1°52W 13 E6
Birmingham U.S.A. 33°31N 86°48W 85 E11
Birmingham Int. ✈ (BHX) U.K. 52°26N 1°45W 13 E6
Birmitrapur India 22°24N 84°46E 41 H14
Birni Nkonni Niger 13°55N 5°15E 50 F7
Birnin Kebbi Nigeria 12°32N 4°12E 50 F6
Birobidzhan Russia 48°50N 132°50E 33 B15
Birr Ireland 53°6N 7°54W 10 C4
Birrie → Australia 29°43S 146°37E 63 D4
Birsilpur India 28°11N 72°15E 42 E5
Birsk Russia 55°25N 55°30E 18 C10
Birtle Canada 50°30N 101°5W 71 C8
Birur India 13°30N 75°55E 40 N9
Biržai Lithuania 56°11N 24°45E 9 H21
Birzebbugga Malta 35°50N 14°32E 25 D2
Bisa Indonesia 1°15S 127°28E 37 E7
Bisalpur India 28°14N 79°48E 43 E8
Bisbee U.S.A. 31°27N 109°55W 77 L9
Biscarrosse France 44°22N 1°20W 20 D3
Biscay, B. of Atl. Oc. 45°0N 2°0W 20 D1
Biscayne B. U.S.A. 25°40N 80°12W 85 J14
Biscoe Is. Antarctica 66°0S 67°0W 5 C17
Biscotasing Canada 47°18N 82°9W 72 C3
Bishkek Kyrgyzstan 42°54N 74°46E 32 C3
Bishnupur India 23°8N 87°20E 43 H12
Bisho = Bhisho S. Africa 32°50S 27°23E 56 E4
Bishop Calif., U.S.A. 37°22N 118°24W 78 H8
Bishop Tex., U.S.A. 27°35N 97°48W 84 H6
Bishop Auckland U.K. 54°39N 1°40W 12 C6
Bishop's Falls Canada 49°2N 55°30W 73 C8
Bishop's Stortford U.K. 51°52N 0°10E 13 F8
Bisina, L. Uganda 1°38N 33°56E 54 B3
Biskra Algeria 34°50N 5°44E 50 B7
Bismarck U.S.A. 46°48N 100°47W 80 B3
Bismarck Arch. Papua N. G. 2°30S 150°0E 58 B7
Biso Uganda 1°44N 31°26E 54 B3
Bison U.S.A. 45°31N 102°28W 80 C2
Bīsotūn Iran 34°23N 47°26E 44 C5
Bissagos = Bijagós, Arquipélago dos Guinea-Biss. 11°15N 16°10W 50 F2
Bissau Guinea-Biss. 11°45N 15°45W 50 F2
Bistcho L. Canada 59°45N 118°50W 70 B5
Bistriţa Romania 47°9N 24°35E 17 E13
Bistriţa → Romania 46°30N 26°57E 17 E14
Biswan India 27°29N 81°2E 43 F9
Bitam Gabon 2°5N 11°25E 52 D2
Bitkine Chad 11°59N 18°13E 51 F9
Bitlis Turkey 38°20N 42°3E 44 B4
Bitola Macedonia 41°1N 21°20E 23 D9
Bitolj = Bitola Macedonia 41°1N 21°20E 23 D9
Bitter Creek U.S.A. 41°33N 108°33W 76 F9
Bitterfontein S. Africa 31°1S 18°32E 56 E2
Bitterroot → U.S.A. 46°52N 114°7W 76 C6
Bitterroot Range U.S.A. 46°0N 114°20W 76 C6
Bitterwater U.S.A. 36°23N 121°0W 78 J6
Biu Nigeria 10°40N 12°3E 51 F8
Biwa-Ko Japan 35°15N 136°10E 31 G8
Biwabik U.S.A. 47°32N 92°21W 80 B7
Bixby U.S.A. 35°57N 95°53W 84 D7
Biyang China 32°38N 113°21E 34 H7
Biysk Russia 52°40N 85°0E 28 D9
Bizana S. Africa 30°50S 29°52E 57 E4
Bizen Japan 34°43N 134°8E 31 G7
Bizerte Tunisia 37°15N 9°50E 51 A7
Bjargtangar Iceland 65°30N 24°30W 8 D1
Bjelovar Croatia 45°56N 16°49E 22 B7
Björneborg = Pori Finland 61°29N 21°48E 8 F19
Bjørnevatn Norway 69°40N 30°0E 8 B24
Bjørnøya Arctic 74°30N 19°0E 4 B8
Black = Da → Vietnam 21°15N 105°20E 38 B5
Black → Canada 44°42N 79°19W 82 B5
Black → Ariz., U.S.A. 33°44N 110°13W 77 K8
Black → Ark., U.S.A. 35°38N 91°20W 84 D9
Black → La., U.S.A. 31°16N 91°50W 84 F9
Black → Mich., U.S.A. 42°59N 82°27W 82 D2
Black → N.Y., U.S.A. 43°59N 76°4W 83 C8
Black → Wis., U.S.A. 43°57N 91°22W 80 D8
Black Bay Pen. Canada 48°38N 88°21W 72 C2
Black Birch L. Canada 56°53N 107°45W 71 B7
Black Canyon of the Gunnison ⌂ U.S.A. 38°40N 107°35W 76 G10
Black Diamond Canada 50°45N 114°14W 70 C6
Black Duck → Canada 56°51N 89°2W 72 A2
Black Forest = Schwarzwald Germany 48°30N 8°20E 16 D5
Black Forest U.S.A. 39°0N 104°43W 76 G11
Black Hd. Ireland 53°9N 9°16W 10 C2
Black Hills U.S.A. 44°0N 103°45W 80 D2
Black I. Canada 51°12N 96°30W 71 C9
Black L. Canada 59°12N 105°15W 71 B7
Black L. Mich., U.S.A. 45°28N 84°16W 81 C11
Black L. N.Y., U.S.A. 44°31N 75°36W 83 B9
Black Lake Canada 59°11N 105°20W 71 B7
Black Mesa U.S.A. 36°58N 102°58W 84 C3
Black Mt. = Mynydd Du U.K. 51°52N 3°50W 13 F4
Black Mts. U.K. 51°55N 3°7W 13 F4
Black Range U.S.A. 33°15N 107°50W 77 K10
Black River Jamaica 18°0N 77°50W 88 a
Black River U.S.A. 44°0N 75°47W 83 C9
Black River Falls U.S.A. 44°18N 90°51W 80 C8
Black Rock Barbados 13°7N 59°37W 89 g
Black Rock Desert U.S.A. 41°10N 118°50W 76 F4
Black Sea Eurasia 43°30N 35°0E 19 F6
Black Tickle Canada 53°28N 55°45W 73 B8
Black Volta → Africa 8°41N 1°33W 50 G5
Black Warrior → U.S.A. 32°32N 87°51W 85 E11
Blackall Australia 24°25S 145°45E 62 C4
Blackball N.Z. 42°22S 171°26E 59 E3
Blackbraes △ Australia 19°10S 144°10E 62 B3
Blackbull Australia 17°55S 141°45E 62 B3
Blackburn U.K. 53°45N 2°29W 12 D5
Blackburn, Mt. U.S.A. 61°44N 143°26W 74 C11
Blackburn with Darwen □ U.K. 53°45N 2°29W 12 D5
Blackdown Tableland △ Australia 23°52S 149°8E 62 C4
Blackfoot U.S.A. 43°11N 112°21W 76 E7
Blackfoot → U.S.A. 46°52N 113°53W 76 C7
Blackfoot Res. U.S.A. 42°55N 111°39W 76 E8
Blackpool U.K. 53°49N 3°3W 12 D4
Blackpool □ U.K. 53°49N 3°3W 12 D4
Blackriver U.S.A. 44°46N 83°17W 82 B1
Blacks Harbour Canada 45°3N 66°49W 73 C6
Blacksburg U.S.A. 37°14N 80°25W 81 G13
Blacksod B. Ireland 54°6N 10°0W 10 B1
Blackstairs Mt. Ireland 52°33N 6°48W 10 D5
Blackstone U.S.A. 37°5N 78°0W 81 G14
Blackstone Ra. Australia 26°0S 128°30E 61 E4
Blackwater = West Road → Canada 53°18N 122°53W 70 C4
Blackwater Australia 23°35S 148°53E 62 C4
Blackwater → Meath, Ireland 53°39N 6°41W 10 C4
Blackwater → Waterford, Ireland 52°4N 7°52W 10 D4
Blackwater → U.K. 54°31N 6°35W 10 B5
Blackwell U.S.A. 36°48N 97°17W 84 C6
Blackwells Corner U.S.A. 35°37N 119°47W 79 K7
Bladensburg △ Australia 22°30S 142°59E 62 C3
Blaenau Ffestiniog U.K. 53°0N 3°56W 12 E4
Blaenau Gwent □ U.K. 51°48N 3°12W 13 F4
Blagodarnoye = Blagodarnyy Russia 45°7N 43°37E 19 E7
Blagodarnyy Russia 45°7N 43°37E 19 E7
Blagoevgrad Bulgaria 42°2N 23°5E 23 C10
Blagoveshchensk Russia 50°20N 127°30E 33 A14
Blahkiuh Indonesia 8°31S 115°12E 37 J18
Blain U.S.A. 40°20N 77°31W 82 F7
Blaine Minn., U.S.A. 45°10N 93°13W 80 C7
Blaine Wash., U.S.A. 48°59N 122°45W 78 B4
Blaine Lake Canada 52°51N 106°52W 71 C7
Blair U.S.A. 41°33N 96°8W 80 E5
Blair Athol Australia 22°42S 147°31E 62 C4
Blair Atholl U.K. 56°46N 3°50W 11 E5
Blairgowrie U.K. 56°35N 3°21W 11 E5
Blairsden U.S.A. 39°47N 120°37W 78 F6
Blairsville U.S.A. 40°26N 79°16W 82 F5
Blakang Mati, Pulau Singapore 1°15N 103°50E 39 d
Blake Pt. U.S.A. 48°11N 88°25W 80 A9
Blakely Ga., U.S.A. 31°23N 84°56W 85 F12
Blakely Pa., U.S.A. 41°28N 75°37W 83 E9
Blambangan, Semenanjung Indonesia 8°42S 114°29E 37 K17
Blanc, C. Spain 39°21N 2°51E 24 B9
Blanc, Mont Europe 45°48N 6°50E 20 D7
Blanca, B. Argentina 39°10S 61°30W 96 D4
Blanca, Cord. Peru 9°10S 77°35W 92 E3
Blanca Peak U.S.A. 37°35N 105°29W 77 H11
Blanche, C. Australia 33°1S 134°9E 63 E1
Blanche, L. S. Austral., Australia 29°15S 139°40E 63 D2
Blanche, L. W. Austral., Australia 22°25S 123°17E 60 D3
Blanchisseuse Trin. & Tob. 10°48N 61°18W 93 K15
Blanco S. Africa 33°55S 22°23E 56 E3
Blanco U.S.A. 30°6N 98°25W 84 F5
Blanco → Argentina 30°20S 68°42W 94 C2
Blanco, C. Costa Rica 9°34N 85°8W 88 E2
Blanco, C. U.S.A. 42°51N 124°34W 76 E1
Blanda → Iceland 65°37N 20°9W 8 D3
Blandford Forum U.K. 50°51N 2°9W 13 G5
Blanding U.S.A. 37°37N 109°29W 77 H9
Blanes Spain 41°40N 2°48E 21 B7
Blankenberge Belgium 51°20N 3°9E 15 C3
Blanquilla Venezuela 11°51N 64°37W 89 D7
Blanquillo Uruguay 32°53S 55°37W 95 C4
Blantyre Malawi 15°45S 35°0E 55 F4
Blarney Ireland 51°56N 8°33W 10 E3
Blasdell U.S.A. 42°48N 78°50W 82 D6
Blåvands Huk Denmark 55°33N 8°4E 9 J13
Blaydon U.K. 54°58N 1°42W 12 C6
Blayney Australia 33°32S 149°14E 63 E4
Blaze, Pt. Australia 12°56S 130°11E 60 B5
Blekinge Sweden 56°25N 15°20E 9 H16
Blenheim Canada 42°20N 82°0W 82 D3
Blenheim N.Z. 41°38S 173°57E 59 D4
Bletchley U.K. 51°59N 0°44W 13 F7
Bleus, Monts Dem. Rep. of the Congo 1°30N 30°30E 54 B3
Blida Algeria 36°30N 2°49E 50 A6
Bligh Sound N.Z. 44°47S 167°32E 59 F1
Bligh Water Fiji 17°0S 178°0E 59 a
Blind River Canada 46°10N 82°58W 72 C3
Bliss Idaho, U.S.A. 42°56N 114°57W 76 E6
Bliss N.Y., U.S.A. 42°34N 78°15W 82 D6
Blissfield U.S.A. 40°24N 81°58W 82 F3
Blitar Indonesia 8°5S 112°11E 37 H15
Block I. U.S.A. 41°11N 71°35W 83 E13
Block Island Sd. U.S.A. 41°15N 71°40W 83 E13
Bloemfontein S. Africa 29°6S 26°7E 56 D4
Bloemhof S. Africa 27°38S 25°32E 56 D4
Blois France 47°35N 1°20E 20 C4
Blönduós Iceland 65°40N 20°12W 8 D3
Blongas Indonesia 8°53S 116°2E 37 K19
Bloodvein → Canada 51°47N 96°43W 71 C9
Bloody Foreland Ireland 55°10N 8°17W 10 A3
Bloomer U.S.A. 45°6N 91°29W 80 C8
Bloomfield Canada 43°59N 77°14W 82 C7
Bloomfield Iowa, U.S.A. 40°45N 92°25W 80 E7
Bloomfield N. Mex., U.S.A. 36°43N 107°59W 77 H10
Bloomfield Nebr., U.S.A. 42°36N 97°39W 80 D5
Bloomington Ill., U.S.A. 40°28N 89°0W 80 E9
Bloomington Ind., U.S.A. 39°10N 86°32W 80 F10
Bloomington Minn., U.S.A. 44°50N 93°17W 80 C7
Bloomsburg U.S.A. 41°0N 76°27W 83 F8
Bloomsbury Australia 20°48S 148°38E 62 b
Blora Indonesia 6°57S 111°25E 37 G14
Blossburg U.S.A. 41°41N 77°4W 82 E7
Blouberg S. Africa 23°8S 28°59E 57 C4
Blountstown U.S.A. 30°27N 85°3W 85 F12
Blue Earth U.S.A. 43°38N 94°6W 80 D6
Blue Hole △ Belize 17°24N 88°30W 88 C2
Blue Lagoon △ Zambia 15°28S 27°26E 55 F2
Blue Mesa Res. U.S.A. 38°28N 107°20W 76 G10
Blue Mountain Lake U.S.A. 43°51N 74°27W 83 C10
Blue Mountain Pk. Jamaica 18°3N 76°36W 88 a
Blue Mt. U.S.A. 40°30N 76°30W 83 F8
Blue Mts. Jamaica 18°3N 76°36W 88 a
Blue Mts. Maine, U.S.A. 44°50N 70°35W 83 B14
Blue Mts. Oreg., U.S.A. 45°0N 118°20W 76 D4
Blue Mud B. Australia 13°30S 136°0E 62 A2
Blue Mud Bay ✪ Australia 13°25S 136°2E 62 A2
Blue Nile = Nîl el Azraq → Sudan 15°38N 32°31E 51 E12
Blue Rapids U.S.A. 39°41N 96°39W 80 F5
Blue Ridge U.S.A. 36°40N 80°50W 81 G13
Blue River Canada 52°6N 119°18W 70 C5
Bluefield U.S.A. 37°15N 81°17W 81 G13
Bluefields Nic. 12°20N 83°50W 88 D3
Bluevale Canada 43°51N 81°15W 82 C3
Bluff Australia 23°35S 149°4E 62 C4
Bluff N.Z. 46°37S 168°20E 59 G2
Bluff U.S.A. 37°17N 109°33W 77 H9
Bluff Knoll Australia 34°24S 118°15E 61 F2
Bluff Pt. Australia 27°50S 114°5E 61 E1
Bluffton U.S.A. 40°44N 85°11W 81 E11
Blumenau Brazil 27°0S 49°0W 95 B6
Blunt U.S.A. 44°31N 99°59W 80 C4
Bly U.S.A. 42°24N 121°3W 76 E3
Blyde River Canyon ⌂ S. Africa 24°37S 31°2E 57 C5
Blyth Canada 43°44N 81°26W 82 C3
Blyth U.K. 55°8N 1°31W 12 B6
Blythe U.S.A. 33°37N 114°36W 79 M12
Blytheville U.S.A. 35°56N 89°55W 85 D10
Bo S. Leone 7°55N 11°50W 50 G3
Bo Duc Vietnam 11°58N 106°50E 39 G6
Bo Hai China 39°0N 119°0E 35 E10
Bo Hai Haixia Asia 38°25N 121°10E 35 E11
Bo Xian = Bozhou China 33°55N 115°41E 34 H8
Boa Vista Brazil 2°48N 60°30W 92 C6
Boa Vista C. Verde Is. 16°0N 22°49W 50 b
Boaco Nic. 12°29N 85°35W 88 D2
Bo'ai China 35°10N 113°3E 34 G7
Boalsburg U.S.A. 40°47N 77°49W 82 F7
Boane Mozam. 26°6S 32°19E 57 D5
Boao China 19°8N 110°34E 38 C8
Boardman U.S.A. 41°2N 80°40W 82 E4
Bobadah Australia 32°19S 146°41E 63 E4
Bobbili India 18°35N 83°30E 41 K13
Bobcaygeon Canada 44°33N 78°33W 82 B6
Bobo-Dioulasso Burkina Faso 11°8N 4°13W 50 F5
Bobonong Botswana 21°58S 28°20E 55 G2
Bóbr → Poland 52°4N 15°4E 16 B8
Bobraomby, Tanjon' i Madag. 12°40S 49°10E 57 A8
Bobruysk = Babruysk Belarus 53°10N 29°15E 17 B15
Boby, Pic Madag. 22°12S 46°55E 53 J9
Boca del Río Mexico 19°5N 96°4W 87 D5
Boca do Acre Brazil 8°50S 67°27W 92 E5
Boca Raton U.S.A. 26°21N 80°5W 85 H14
Bocas del Dragón = Dragon's Mouths Venezuela 11°0N 61°50W 93 K15
Bocas del Toro Panama 9°15N 82°20W 88 E3
Bochnia Poland 49°58N 20°27E 17 D11
Bochum Germany 51°28N 7°13E 16 C4
Bocoyna Mexico 27°52N 107°35W 86 B3
Bodaybo Russia 57°50N 114°0E 29 D12
Boddam U.K. 59°56N 1°17W 11 B7
Boddington Australia 32°50S 116°30E 61 F2
Bodega Bay U.S.A. 38°20N 123°3W 78 G3
Boden Sweden 65°50N 21°42E 8 D19
Bodensee Europe 47°35N 9°25E 20 C8
Bodhan India 18°40N 77°44E 40 K10
Bodmin U.K. 50°28N 4°43W 13 G3
Bodmin Moor U.K. 50°33N 4°36W 13 G3
Bodø Norway 67°17N 14°24E 8 C16
Bodrog → Hungary 48°11N 21°22E 17 D11
Bodrum Turkey 37°3N 27°30E 23 F12
Boende Dem. Rep. of the Congo 0°24S 21°12E 52 E4
Boerne U.S.A. 29°47N 98°44W 84 G5
Boesmans → S. Africa 33°42S 26°39E 56 E4
Bogalusa U.S.A. 30°47N 89°52W 85 F
Bogan → Australia 30°20S 146°55E 63 D
Bogan Gate Australia 33°7S 147°49E 63 E
Bogantungan Australia 23°41S 147°17E 62 C
Bogata U.S.A. 33°28N 95°13W 84 E
Bogda Shan China 43°35N 89°40E 32 C
Boggabilla Australia 28°36S 150°24E 63 D
Boggabri Australia 30°45S 150°5E 63 E
Boggeragh Mts. Ireland 52°2N 8°55W 10 D
Bognor Regis U.K. 50°47N 0°40W 13 G
Bogo Phil. 11°3N 124°0E 37 B
Bogong, Mt. Australia 36°47S 147°17E 63 F
Bogor Indonesia 6°36S 106°48E 37 G
Bogotá Colombia 4°34N 74°0W 92 C
Bogotol Russia 56°15N 89°50E 28 D
Bogra Bangla. 24°51N 89°22E 41 G
Boguchany Russia 58°40N 97°30E 29 D
Bogué Mauritania 16°45N 14°10W 50 E
Bohemian Forest = Böhmerwald Germany 49°8N 13°14E 16 D
Böhmerwald Germany 49°8N 13°14E 16 D
Bohol □ Phil. 9°50N 124°10E 37 C
Bohol Sea Phil. 9°0N 124°0E 37 C
Bohorok Indonesia 3°30N 98°12E 39 L
Böhöt Mongolia 45°13N 108°16E 34 B
Bohuslän Sweden 58°25N 12°0E 9 G
Boi, Pta. do Brazil 23°55S 45°15W 95 A
Boiaçu Brazil 0°27S 61°46W 92 D
Boigu Australia 9°16S 142°13E 62
Boileau, C. Australia 17°40S 122°7E 60 C
Boise U.S.A. 43°37N 116°13W 76 E
Boise City U.S.A. 36°44N 102°31W 84 C
Boissevain Canada 49°15N 100°5W 71 D
Bojador, C. W. Sahara 26°0N 14°30W 50 C
Bojana → Albania 41°52N 19°22E 23 D
Bojnūrd Iran 37°30N 57°20E 45 B
Bojonegoro Indonesia 7°11S 111°54E 37 G
Bokaro India 23°46N 85°55E 43 H
Boké Guinea 10°56N 14°17W 50 F
Bokhara → Australia 29°55S 146°42E 63 D
Boknafjorden Norway 59°14N 5°40E 9 G
Bokor △ Cambodia 10°50N 104°1E 39 G
Bokora ⌂ Uganda 2°12N 31°32E 54 B
Bokoro Chad 12°25N 17°14E 51 F
Bokpyin Burma 11°18N 98°42E 39 G
Bokungu Dem. Rep. of the Congo 0°35S 22°50E 52 E
Bolan → Pakistan 28°38N 67°42E 42 E
Bolan Pass Pakistan 29°50N 67°20E 40 E
Bolaños → Mexico 21°12N 104°5W 86 C
Bolbec France 49°30N 0°30E 20 B
Boldājī Iran 31°56N 51°3E 45 D
Bole China 44°55N 81°37E 32 C
Bolekhiv Ukraine 49°0N 23°57E 17 D
Bolesławiec Poland 51°17N 15°37E 16 C
Bolgatanga Ghana 10°44N 0°53W 50 F
Bolgrad = Bolhrad Ukraine 45°40N 28°32E 17 F
Bolhrad Ukraine 45°40N 28°32E 17 F
Bolinao Phil. 16°23N 119°54E 37 A
Bolivar Mo., U.S.A. 37°37N 93°25W 80 G
Bolivar N.Y., U.S.A. 42°4N 78°10W 82 D
Bolivar Tenn., U.S.A. 35°12N 89°0W 85 D
Bolivia ■ S. Amer. 17°6S 64°0W 92 G
Bolivian Plateau = Altiplano Bolivia 17°0S 68°0W 92 G
Bollnäs Sweden 61°21N 16°24E 8 F
Bollon Australia 28°2S 147°29E 63 D
Bolmen Sweden 56°55N 13°40E 9 H
Bolobo Dem. Rep. of the Congo 2°6S 16°20E 52 E
Bologna Italy 44°29N 11°20E 22 B
Bologoye Russia 57°55N 34°5E 18 C
Bolomba Dem. Rep. of the Congo 0°35N 19°0E 52 D
Bolonchén Mexico 20°1N 89°45W 87 D
Boloven, Cao Nguyen Laos 15°10N 106°30E 38 E
Bolpur India 23°40N 87°45E 43 H
Bolsena, L. di Italy 42°36N 11°56E 22 C
Bolshevik, Ostrov Russia 78°30N 102°0E 29 B
Bolshoy Anyuy → Russia 68°30N 160°49E 29 C
Bolshoy Begichev, Ostrov Russia 74°20N 112°30E 29 B
Bolshoy Kamen Russia 43°7N 132°19E 30 C
Bolshoy Kavkas = Caucasus Mountains Eurasia 42°50N 44°0E 19 F
Bolshoy Lyakhovskiy, Ostrov Russia 73°35N 142°0E 29 B
Bolshoy Tyuters, Ostrov Russia 59°51N 27°13E 9 G
Bolsward Neths. 53°3N 5°32E 15 A
Bolt Head U.K. 50°12N 3°48W 13 G
Bolton Canada 43°54N 79°45W 82 C
Bolton U.K. 53°35N 2°26W 12 D
Bolton Landing U.S.A. 43°32N 73°35W 83 C
Bolu Turkey 40°45N 31°35E 19 F
Bolungavík Iceland 66°9N 23°15W 8 C
Bolvadin Turkey 38°45N 31°4E 44 B
Bolzano Italy 46°31N 11°22E 22 A
Bom Jesus da Lapa Brazil 13°15S 43°25W 93 F
Boma Dem. Rep. of the Congo 5°50S 13°4E 52 F
Bombala Australia 36°56S 149°15E 63 F
Bombay = Mumbai India 18°56N 72°50E 40 K
Bombay U.S.A. 44°56N 74°34W 83 B
Bombedor, Pta. Venezuela 9°53N 61°37W 93 L1
Bomberai, Semenanjung Indonesia 3°0S 133°0E 37 E
Bomboma Dem. Rep. of the Congo 2°25N 18°55E 52 D
Bombombwa Dem. Rep. of the Congo 1°40N 25°40E 54 B
Bomi China 29°50N 95°45E 32 F
Bomili Dem. Rep. of the Congo 1°45N 27°5E 54 B

mlo Norway 59°37N 5°13E 9 G11
mokandi ➜ Dem. Rep. of the Congo 3°39N 26°8E 54 B2
mu ➜ C.A.R. 4°40N 22°30E 52 D4
n, C. = Ra's aț Tib Tunisia 37°1N 11°2E 22 F4
n Acceuil Mauritius 20°10S 57°39E 53 d
n Echo △ Canada 44°55N 77°16W 82 B7
n Sar Pa Vietnam 12°24N 107°35E 38 F6
nāb Iran 36°35N 48°41E 45 B6
naigarh India 21°50N 84°57E 43 J11
naire W. Indies 12°10N 68°15W 89 D6
nampak Mexico 16°44N 91°5W 87 D6
nang Australia 37°11S 148°41E 63 F4
nanza Nic. 13°54N 84°35W 88 D3
naparte Arch. Australia 14°0S 124°30E 60 B3
nar Bridge U.K. 57°54N 4°20W 11 D4
nasse Trin. & Tob. 10°5N 61°54W 93 K15
naventure Canada 48°5N 65°32W 73 C6
navista Canada 48°40N 53°5W 73 C9
navista, C. Canada 48°42N 53°5W 73 C9
navista B. Canada 48°45N 53°25W 73 C9
ndo Dem. Rep. of the Congo 3°55N 23°53E 54 B1
ndoukou Ivory C. 8°2N 2°47W 50 G5
ndowoso Indonesia 7°55S 113°49E 37 G15
ne, Teluk Indonesia 4°10S 120°50E 37 E6
nerate Indonesia 7°25S 121°5E 37 F6
nerate, Kepulauan Indonesia 6°30S 121°10E 37 F6
'ness U.K. 56°1N 3°37W 11 E5
nete, Cerro Argentina 27°55S 68°40W 94 B2
ng Son = Hoai Nhon Vietnam 14°28N 109°1E 38 E7
ngaigaon India 26°28N 90°34E 32 F7
ngandanga Dem. Rep. of the Congo 1°24N 21°3E 52 D4
ngor Chad 10°35N 15°20E 51 F9
ngos, Massif des C.A.R. 8°40N 22°25E 52 C4
nham U.S.A. 33°35N 96°11W 84 E6
ni △ Kenya 1°35S 41°18E 54 C5
nifacio France 41°24N 9°10E 20 F8
nifacio, Bouches de Medit. S. 41°12N 9°15E 22 D3
nin Is. = Ogasawara Gunto Pac. Oc. 27°0N 142°0E 27 F16
nn Germany 50°46N 7°6E 16 C4
nne Terre U.S.A. 37°55N 90°33W 80 G8
nners Ferry U.S.A. 48°42N 116°19W 76 B5
nney, L. Australia 37°50S 140°20E 63 F3
nnie Rock Australia 30°29S 118°22E 61 F2
nny, Bight of Africa 3°30N 9°20E 52 D1
nnyrigg U.K. 55°53N 3°6W 11 F5
nnyville Canada 54°20N 110°45W 71 C6
noi Indonesia 1°45S 137°41E 37 E9
nsall U.S.A. 33°16N 117°14W 79 M9
ntang Indonesia 0°10N 117°30E 36 D5
ntebok △ S. Africa 34°5S 20°28E 56 E3
nthe S. Leone 7°30N 12°33W 50 G3
ntoc Phil. 17°7N 120°58E 37 A6
nython Ra. Australia 23°40S 128°45E 60 D4
odjamulla △ Australia 18°15S 138°6E 62 B2
okabie Australia 31°50S 132°41E 61 F5
oker U.S.A. 36°27N 100°32W 84 C4
oligal Australia 33°58S 144°53E 63 E3
ön Tsagaan Nuur Mongolia 45°35N 99°9E 32 B8
onah Australia 27°58S 152°41E 63 D5
one Iowa, U.S.A. 42°4N 93°53W 80 D7
one N.C., U.S.A. 36°13N 81°41W 85 C14
oneville Ark., U.S.A. 35°8N 93°55W 84 D8
oneville Miss., U.S.A. 34°39N 88°34W 85 D10
onville Calif., U.S.A. 39°1N 123°22W 78 F3
onville Ind., U.S.A. 38°3N 87°16W 80 F10
onville Mo., U.S.A. 38°58N 92°44W 80 F7
onville N.Y., U.S.A. 43°29N 75°20W 83 C9
orabbin △ Australia 31°30S 120°10E 61 F3
orindal Australia 30°22S 146°11E 63 E4
orowa Australia 34°28S 148°44E 63 E4
osaaso Somali Rep. 11°12N 49°18E 47 E4
othia, Gulf of Canada 71°0N 90°0W 68 C14
othia Pen. Canada 71°0N 94°0W 68 C13
otle U.K. 53°28N 3°1W 12 D4
oué Gabon 0°5S 11°55E 52 E2
quilla, Presa de la Mexico 27°31N 105°30W 86 B3
quillas del Carmen Mexico 29°11N 102°58W 86 B4
r Serbia 44°5N 22°7E 23 B10
r Sudan 6°10N 31°40E 51 G12
r Mashash Israel 31°7N 34°50E 46 D3
ra Bora French Polynesia 16°30S 151°45W 65 J12
rah Peak U.S.A. 44°8N 113°47W 76 D7
rås Sweden 57°43N 12°56E 9 H15
rāzjān Iran 29°22N 51°10E 45 D6
rba Brazil 4°12S 59°34W 92 D7
rborema, Planalto da Brazil 7°0S 37°0W 90 D7
rd Khūn-e Now Iran 28°3N 51°28E 45 D6
rda, C. Australia 35°45S 136°34E 63 F2
rdeaux France 44°50N 0°36W 20 D3
rden Australia 34°3S 118°12E 61 F2
rden-Carleton Canada 46°18N 63°47W 73 C7
rden I. Canada 78°30N 111°30W 69 B9
rden Pen. Canada 73°0N 83°0W 69 C15
rder Ranges △ Australia 28°24S 152°56E 63 D5
rders = Scottish Borders □ U.K. 55°35N 2°50W 11 F6
rdertown Australia 36°19S 140°45E 63 F3
rðeyri Iceland 65°12N 21°6W 8 D3
rdj Fly Ste. Marie Algeria 27°19N 2°32W 50 C5
rdj-in-Eker Algeria 24°9N 5°3E 50 D7
rdj Mokhtar Algeria 21°20N 0°56E 50 D6
rdj Omar Driss Algeria 28°10N 6°40E 50 C7
rehamwood U.K. 51°40N 0°15W 13 F7
rgarnes Iceland 64°32N 21°55W 8 D3
Børgefjellet Norway 65°20N 13°45E 8 D15
Borger Neths. 52°54N 6°44E 15 B6
Borger U.S.A. 35°39N 101°24W 84 D4
Borgholm Sweden 56°52N 16°39E 9 H17
Borhoyn Tal Mongolia 43°50N 111°58E 34 C6
Borikhane Laos 18°33N 103°43E 38 C4
Borisoglebsk Russia 51°27N 42°5E 19 D7
Borisov = Barysaw Belarus 54°17N 28°28E 17 A15
Borja Peru 4°20S 77°40W 92 D3
Borkou Chad 18°15N 18°50E 51 E9
Borkum Germany 53°34N 6°40E 16 B4
Borlänge Sweden 60°29N 15°26E 9 F16
Borley, C. Antarctica 66°15S 52°30E 5 C5
Borneo E. Indies 1°0N 115°0E 36 D5
Bornholm Denmark 55°10N 15°0E 9 J16
Borobudur △ Indonesia 7°36S 110°12E 37 G14
Borogontsy Russia 62°42N 131°8E 29 C14
Borohoro Shan China 44°6N 83°10E 32 C5
Boron U.S.A. 35°0N 117°39W 79 L9
Borongan Phil. 11°37N 125°26E 37 B7
Borovichi Russia 58°25N 33°55E 18 C5
Borrego Springs U.S.A. 33°15N 116°23W 79 M10
Borrisokane Ireland 53°0N 8°7W 10 D3
Borroloola Australia 16°4S 136°17E 62 B2
Borşa Romania 47°41N 24°50E 17 E13
Borsad India 22°25N 72°54E 42 H5
Borth U.K. 52°29N 4°2W 13 E3
Borūjerd Iran 33°55N 48°50E 45 C6
Boryeong S. Korea 36°21N 126°36E 35 F14
Boryslav Ukraine 49°18N 23°28E 17 D12
Borzya Russia 50°24N 116°31E 33 A12
Bosa Italy 40°18N 8°30E 22 D3
Bosanska Gradiška Bos.-H. 45°10N 17°15E 22 B7
Boscastle U.K. 50°41N 4°42W 13 G3
Boscobelle Barbados 13°17N 59°35W 89 g
Bose China 23°53N 106°35E 32 G10
Boseong S. Korea 34°46N 127°5E 35 G14
Boshan China 36°28N 117°49E 35 F9
Boshof S. Africa 28°31S 25°13E 56 D4
Boshrūyeh Iran 33°50N 57°30E 45 C8
Bosna ➜ Bos.-H. 45°4N 18°29E 23 B8
Bosna i Hercegovina = Bosnia-Herzegovina ■ Europe 44°0N 18°0E 22 B7
Bosnia-Herzegovina ■ Europe 44°0N 18°0E 22 B7
Bosnik Indonesia 1°5S 136°10E 37 E9
Bosobolo Dem. Rep. of the Congo 4°15N 19°50E 52 D3
Bosporus = İstanbul Boğazı Turkey 41°5N 29°3E 23 D13
Bosque Farms U.S.A. 35°51N 106°42W 77 J10
Bossangoa C.A.R. 6°35N 17°30E 52 C3
Bossier City U.S.A. 32°31N 93°44W 84 E8
Bosso Niger 13°43N 13°19E 51 F8
Bostan Pakistan 30°26N 67°2E 42 D2
Bostānābād Iran 37°50N 46°50E 44 B5
Bosten Hu China 41°55N 87°40E 32 C6
Boston U.K. 52°59N 0°2W 12 E7
Boston U.S.A. 42°22N 71°3W 83 D13
Boston Bar Canada 49°52N 121°30W 70 D4
Boston Mts. U.S.A. 35°42N 93°15W 84 D8
Boswell Canada 49°28N 116°45W 70 D5
Boswell U.S.A. 40°10N 79°2W 82 F5
Botad India 22°15N 71°40E 42 H4
Botany B. Australia 33°58S 151°11E 58 E8
Botene Laos 17°35N 101°12E 38 D3
Bothaville S. Africa 27°23S 26°34E 56 D4
Bothnia, G. of Europe 62°0N 20°0E 8 F19
Bothwell Australia 42°20S 147°1E 63 G4
Bothwell Canada 42°38N 81°52W 82 D3
Botletle ➜ Botswana 20°10S 23°15E 56 C3
Botoşani Romania 47°42N 26°41E 17 E14
Botou Burkina Faso 12°42N 1°59E 50 F6
Botou China 38°4N 116°34E 34 E9
Botshabelo S. Africa 29°14S 26°44E 56 D4
Botswana ■ Africa 22°0S 24°0E 56 C3
Bottineau U.S.A. 48°50N 100°27W 80 A3
Bottrop Germany 51°31N 6°58E 15 C6
Botucatu Brazil 22°55S 48°30W 95 A6
Botum Sakor △ Cambodia 11°5N 103°15E 39 G4
Botwood Canada 49°6N 55°23W 73 C8
Bou Saâda Algeria 35°11N 4°9E 50 A6
Bouaflé Ivory C. 7°1N 5°47W 50 G4
Bouaké Ivory C. 7°40N 5°2W 50 G4
Bouar C.A.R. 6°0N 15°40E 52 C3
Bouârfa Morocco 32°32N 1°58W 50 B5
Boucaut B. Australia 12°0S 134°25E 62 A1
Bouctouche Canada 46°30N 64°45W 73 C7
Bougainville, C. Australia 13°57S 126°4E 60 B4
Bougainville I. Papua N. G. 6°0S 155°0E 58 B8
Bougainville Reef Australia 15°30S 147°5E 62 B4
Bougie = Bejaïa Algeria 36°42N 5°2E 50 A7
Bougouni Mali 11°30N 7°20W 50 F4
Bouillon Belgium 49°44N 5°3E 15 E5
Boulder Colo., U.S.A. 40°1N 105°17W 76 F11
Boulder Mont., U.S.A. 46°14N 112°7W 76 C7
Boulder City U.S.A. 35°58N 114°49W 79 K12
Boulder Creek U.S.A. 37°7N 122°7W 78 H4
Boulder Dam = Hoover Dam U.S.A. 36°1N 114°44W 79 K12
Boulia Australia 22°52S 139°51E 62 C2
Boulogne-sur-Mer France 50°42N 1°36E 20 A4
Boulsa Burkina Faso 12°39N 0°34W 50 F5
Boultoum Niger 14°45N 10°25E 51 F8
Bouma △ Fiji 16°50S 179°52E 59 a
Boun Neua Laos 21°38N 101°54E 38 B3
Boun Tai Laos 21°23N 101°58E 38 B3
Bouna Ivory C. 9°10N 3°0W 50 G5
Boundary Peak U.S.A. 37°51N 118°21W 78 H8
Boundiali Ivory C. 9°30N 6°20W 50 G4
Bountiful U.S.A. 40°53N 111°52W 76 F8
Bounty Is. Pac. Oc. 48°0S 178°30E 64 M9
Bounty Trough Pac. Oc. 46°0S 178°0E 64 M9
Bourbonnais France 46°28N 3°0E 20 C5
Bourdel L. Canada 56°43N 74°10W 72 A5
Bourem Mali 17°0N 0°24W 50 E5
Bourg-en-Bresse France 46°13N 5°12E 20 C6
Bourg-St-Maurice France 45°35N 6°46E 20 D7
Bourgas = Burgas Bulgaria 42°33N 27°29E 23 C12
Bourges France 47°9N 2°25E 20 C5
Bourget Canada 45°26N 75°9W 83 A9
Bourgogne □ France 47°0N 4°50E 20 C6
Bourke Australia 30°8S 145°55E 63 E4
Bourne U.K. 52°47N 0°22W 12 E7
Bournemouth U.K. 50°43N 1°52W 13 G6
Bournemouth □ U.K. 50°43N 1°52W 13 G6
Bouse U.S.A. 33°56N 114°0W 79 M13
Bousso Chad 10°34N 16°52E 51 F9
Bouvet I. = Bouvetøya Antarctica 54°26S 3°24E 2 G10
Bouvetøya Antarctica 54°26S 3°24E 2 G10
Bovill U.S.A. 46°51N 116°24W 76 C5
Bovril Argentina 31°21S 59°26W 94 C4
Bow ➜ Canada 49°57N 111°41W 70 C6
Bow Island Canada 49°50N 111°23W 76 B8
Bowbells U.S.A. 48°48N 102°15W 80 A2
Bowdle U.S.A. 45°27N 99°39W 80 C4
Bowelling Australia 33°25S 116°30E 61 F2
Bowen Argentina 35°0S 67°31W 94 D2
Bowen Australia 20°0S 148°16E 62 J6
Bowen Mts. Australia 37°0S 147°50E 63 F4
Bowers Basin Pac. Oc. 53°45N 176°0E 4 D16
Bowers Ridge Pac. Oc. 54°0N 180°0E 4 D17
Bowie Ariz., U.S.A. 32°19N 109°29W 77 K9
Bowie Tex., U.S.A. 33°34N 97°51W 84 E6
Bowkān Iran 36°31N 46°12E 44 B5
Bowland, Forest of U.K. 54°0N 2°30W 12 D5
Bowling Green Ky., U.S.A. 36°59N 86°27W 80 G10
Bowling Green Ohio, U.S.A. 41°23N 83°39W 81 E12
Bowling Green, C. Australia 19°19S 147°25E 62 B4
Bowling Green Bay △ Australia 19°26S 146°57E 62 B4
Bowman U.S.A. 46°11N 103°24W 80 B2
Bowman I. Antarctica 65°0S 104°0E 5 C8
Bowmanville = Clarington Canada 43°55N 78°41W 82 C6
Bowmore U.K. 55°45N 6°17W 11 F2
Bowral Australia 34°26S 150°27E 63 E5
Bowraville Australia 30°37S 152°52E 63 E5
Bowron ➜ Canada 54°3N 121°50W 70 C4
Bowron Lake △ Canada 53°10N 121°5W 70 C4
Bowser L. Canada 56°30N 129°30W 70 B3
Bowsman Canada 52°14N 101°12W 71 C8
Bowwood Zambia 17°5S 26°20E 55 F2
Box Cr. ➜ Australia 34°10S 143°50E 63 E3
Boxmeer Neths. 51°38N 5°56E 15 C5
Boxtel Neths. 51°36N 5°20E 15 C5
Boyce U.S.A. 31°23N 92°40W 84 F8
Boyd L. Canada 52°46N 76°42W 72 B4
Boyle Canada 54°35N 112°49W 70 C6
Boyle Ireland 53°59N 8°18W 10 C3
Boyne ➜ Ireland 53°43N 6°15W 10 C5
Boyne City U.S.A. 45°13N 85°1W 81 C11
Boynton Beach U.S.A. 26°32N 80°4W 85 H14
Boyoma, Chutes Dem. Rep. of the Congo 0°35N 25°23E 54 B2
Boysen Res. U.S.A. 43°25N 108°11W 76 E9
Boyuibe Bolivia 20°25S 63°17W 92 G6
Boyup Brook Australia 33°50S 116°23E 61 F2
Boz Dağları Turkey 38°20N 28°0E 23 E13
Bozburun Turkey 36°43N 28°4E 23 F13
Bozcaada Turkey 39°49N 26°3E 23 E12
Bozdoğan Turkey 37°40N 28°17E 23 F13
Bozeman U.S.A. 45°41N 111°2W 76 D8
Bozen = Bolzano Italy 46°31N 11°22E 22 A4
Bozhou China 33°55N 115°41E 34 H8
Bozoum C.A.R. 6°25N 16°35E 52 C3
Bozyazı Turkey 36°6N 33°0E 44 B2
Bra Italy 44°42N 7°51E 20 D7
Brabant □ Belgium 50°46N 4°30E 15 D4
Brabant L. Canada 55°58N 103°43W 71 B8
Brač Croatia 43°20N 16°40E 22 C7
Bracadale, L. U.K. 57°20N 6°30W 11 D2
Bracciano, L. di Italy 42°7N 12°14E 22 C5
Bracebridge Canada 45°2N 79°19W 82 A5
Bräcke Sweden 62°45N 15°26E 8 E16
Brackettville U.S.A. 29°19N 100°25W 84 G4
Bracknell U.K. 51°25N 0°43W 13 F7
Bracknell Forest □ U.K. 51°25N 0°44W 13 F7
Brad Romania 46°10N 22°50E 17 E12
Bradenton U.S.A. 27°30N 82°34W 85 H13
Bradford Canada 44°7N 79°34W 82 B5
Bradford U.K. 53°47N 1°45W 12 D6
Bradford Pa., U.S.A. 41°58N 78°38W 82 E6
Bradford Vt., U.S.A. 43°59N 72°9W 83 C12
Bradley Ark., U.S.A. 33°6N 93°39W 84 E8
Bradley Calif., U.S.A. 35°52N 120°48W 78 K6
Bradley Institute Zimbabwe 17°7S 31°25E 55 F3
Brady U.S.A. 31°9N 99°20W 84 F5
Braeside Canada 45°28N 76°24W 83 A8
Braga Portugal 41°35N 8°25W 21 B1
Bragado Argentina 35°2S 60°27W 94 D3
Bragança Brazil 1°0S 47°2W 93 D9
Bragança Portugal 41°48N 6°50W 21 B2
Bragança Paulista Brazil 22°55S 46°32W 95 A6
Brahestad = Raahe Finland 64°40N 24°28E 8 D21
Brahmanbaria Bangla. 23°58N 91°15E 41 H17
Brahmani ➜ India 20°39N 86°46E 41 J15
Brahmapur India 19°15N 84°54E 41 K14
Brahmaputra ➜ Asia 23°40N 90°35E 43 H13
Braich-y-pwll U.K. 52°47N 4°46W 12 E3
Braidwood Australia 35°27S 149°49E 63 F4
Brăila Romania 45°19N 27°59E 17 F14
Brainerd U.S.A. 46°22N 94°12W 80 B6
Braintree U.K. 51°53N 0°34E 13 F8
Braintree U.S.A. 42°13N 71°0W 83 D14
Brak ➜ S. Africa 29°35S 22°55E 56 D3
Brakwater Namibia 22°28S 17°3E 56 C2
Brampton Canada 43°45N 79°45W 82 C5
Brampton U.K. 54°57N 2°44W 12 C5
Brampton I. Australia 20°49S 149°16E 62 b
Branco ➜ Brazil 1°20S 61°50W 92 D6
Branco, C. Brazil 7°9S 34°47W 90 D7
Brandberg Namibia 21°10S 14°33E 56 C1
Brandberg △ Namibia 21°10S 14°30E 56 C1
Brandenburg = Neubrandenburg Germany 53°33N 13°15E 16 B7
Brandenburg Germany 52°25N 12°33E 16 B7
Brandenburg □ Germany 52°50N 13°0E 16 B6
Brandfort S. Africa 28°40S 26°30E 56 D4
Brandon Canada 49°50N 99°57W 71 D9
Brandon U.S.A. 43°48N 73°6W 83 C11
Brandon B. Ireland 52°17N 10°8W 10 D1
Brandon Mt. Ireland 52°15N 10°15W 10 D1
Brandsen Argentina 35°10S 58°15W 94 D4
Brandvlei S. Africa 30°25S 20°30E 56 E3
Branford U.S.A. 41°17N 72°49W 83 E12
Braniewo Poland 54°25N 19°50E 17 A10
Bransfield Str. Antarctica 63°0S 59°0W 5 C18
Branson U.S.A. 36°39N 93°13W 80 G7
Brantford Canada 43°10N 80°15W 82 C4
Bras d'Or L. Canada 45°50N 60°50W 73 C7
Brasher Falls U.S.A. 44°49N 74°47W 83 B10
Brasil = Brazil ■ S. Amer. 12°0S 50°0W 93 F9
Brasil, Planalto Brazil 18°0S 46°30W 90 E6
Brasiléia Brazil 11°0S 68°45W 92 F5
Brasília Brazil 15°47S 47°55W 93 G9
Brasília Legal Brazil 3°49S 55°36W 93 D7
Braslaw Belarus 55°38N 27°0E 9 J22
Braşov Romania 45°38N 25°35E 17 F13
Brasschaat Belgium 51°19N 4°27E 15 C4
Brassey, Banjaran Malaysia 5°0N 117°15E 36 D5
Brassey Ra. Australia 25°8S 122°15E 61 E3
Brasstown Bald U.S.A. 34°53N 83°49W 85 D13
Brastad Sweden 58°23N 11°30E 9 G14
Bratislava Slovak Rep. 48°10N 17°7E 17 D9
Bratsk Russia 56°10N 101°30E 29 D11
Brattleboro U.S.A. 42°51N 72°34W 83 D12
Braunau Austria 48°15N 13°3E 16 D7
Braunschweig Germany 52°15N 10°31E 16 B6
Braunton U.K. 51°7N 4°10W 13 F3
Brava C. Verde Is. 15°0N 24°40W 50 b
Bravo del Norte, Rio = Grande, Rio ➜ N. Amer. 25°58N 97°9W 84 J6
Brawley U.S.A. 32°59N 115°31W 79 N11
Bray Ireland 53°13N 6°7W 10 C5
Bray, Mt. Australia 14°0S 134°30E 62 A1
Bray, Pays de France 49°46N 1°26E 20 B4
Brazeau ➜ Canada 52°55N 115°14W 70 C5
Brazil U.S.A. 39°32N 87°8W 80 F10
Brazil ■ S. Amer. 12°0S 50°0W 93 F9
Brazilian Highlands = Brasil, Planalto Brazil 18°0S 46°30W 90 E6
Brazo Sur ➜ S. Amer. 25°21S 57°42W 94 B4
Brazos ➜ U.S.A. 28°53N 95°23W 84 G7
Brazzaville Congo 4°9S 15°12E 52 E3
Brčko Bos.-H. 44°54N 18°46E 23 B8
Bré = Bray Ireland 53°13N 6°7W 10 C5
Breaden, L. Australia 25°51S 125°28E 61 E4
Breaksea Sd. N.Z. 45°35S 166°35E 59 F1
Bream B. N.Z. 35°56S 174°28E 59 A5
Bream Hd. N.Z. 35°51S 174°36E 59 A5
Breas Chile 25°29S 70°24W 94 B1
Brebes Indonesia 6°52S 109°3E 37 G13
Brechin Canada 44°32N 79°10W 82 B5
Brechin U.K. 56°44N 2°39W 11 E6
Brecht Belgium 51°21N 4°38E 15 C4
Breckenridge Colo., U.S.A. 39°29N 106°3W 76 G10
Breckenridge Minn., U.S.A. 46°16N 96°35W 80 B5
Breckenridge Tex., U.S.A. 32°45N 98°54W 84 E5
Breckland U.K. 52°30N 0°40E 13 E8
Brecon U.K. 51°57N 3°23W 13 F4
Brecon Beacons U.K. 51°53N 3°26W 13 F4
Brecon Beacons △ U.K. 51°50N 3°30W 13 F4
Breda Neths. 51°35N 4°45E 15 C4
Bredasdorp S. Africa 34°33S 20°2E 56 E3
Bree Belgium 51°8N 5°35E 15 C5
Bregenz Austria 47°30N 9°45E 16 E5
Breiðafjörður Iceland 65°15N 23°15W 8 D2
Brejo Brazil 3°41S 42°47W 93 D10
Bremen Germany 53°4N 8°47E 16 B5
Bremer Bay Australia 34°21S 119°20E 61 F2
Bremer I. Australia 12°5S 136°45E 62 A2
Bremerhaven Germany 53°33N 8°36E 16 B5
Bremerton U.S.A. 47°34N 122°37W 78 C4
Brenham U.S.A. 30°10N 96°24W 84 F6
Brennerpass Austria 47°2N 11°30E 16 E6
Brent U.S.A. 32°56N 87°10W 85 E11
Brentwood U.K. 51°37N 0°19E 13 F8
Brentwood Calif., U.S.A. 37°56N 121°42W 78 H5
Brentwood N.Y., U.S.A. 40°47N 73°15W 83 F11
Bréscia Italy 45°33N 10°15E 20 D9
Breskens Neths. 51°23N 3°33E 15 C3
Breslau = Wrocław Poland 51°5N 17°5E 17 C9
Bressanone Italy 46°43N 11°39E 22 A4
Bressay U.K. 60°9N 1°6W 11 A7
Brest Belarus 52°10N 23°40E 17 B12
Brest France 48°24N 4°31W 20 B1
Brest □ Belarus 52°30N 26°10E 17 B13
Brest-Litovsk = Brest Belarus 52°10N 23°40E 17 B12
Bretagne □ France 48°10N 3°0W 20 B2
Breton Canada 53°7N 114°28W 70 C6
Breton Sd. U.S.A. 29°35N 89°15W 85 G10
Brett, C. N.Z. 35°10S 174°20E 59 A5
Brevard U.S.A. 35°14N 82°44W 85 D13
Breves Brazil 1°40S 50°29W 93 D8
Brewarrina Australia 30°0S 146°51E 63 E4
Brewer U.S.A. 44°48N 68°46W 81 C19
Brewer, Mt. U.S.A. 36°44N 118°28W 78 J8
Brewster N.Y., U.S.A. 41°24N 73°36W 83 E11
Brewster Ohio, U.S.A. 40°43N 81°36W 82 F3
Brewster Wash., U.S.A. 48°6N 119°47W 76 B4
Brewster, Kap = Kangikajik Greenland 70°7N 22°0W 4 B6
Brewton U.S.A. 31°7N 87°4W 85 F11
Breyten S. Africa 26°16S 30°0E 57 D5
Bria C.A.R. 6°30N 21°58E 52 C4
Briançon France 44°54N 6°39E 20 D7
Bribie I. Australia 27°0S 153°10E 63 D5
Bribri Costa Rica 9°38N 82°50W 88 E3
Bridgefield Barbados 13°9N 59°36W 89 g
Bridgehampton U.S.A. 40°56N 72°19W 83 F12
Bridgend U.K. 51°30N 3°34W 13 F4
Bridgend □ U.K. 51°36N 3°36W 13 F4
Bridgenorth Canada 44°23N 78°23W 82 B6
Bridgeport Calif., U.S.A. 38°15N 119°14W 78 G7
Bridgeport Conn., U.S.A. 41°11N 73°12W 83 E11
Bridgeport N.Y., U.S.A. 43°9N 75°58W 83 C9
Bridgeport Nebr., U.S.A. 41°40N 103°6W 80 E2
Bridgeport Tex., U.S.A. 33°13N 97°45W 84 E6
Bridger U.S.A. 45°18N 108°55W 76 D9
Bridgeton U.S.A. 39°26N 75°14W 81 F16
Bridgetown Australia 33°58S 116°7E 61 F2
Bridgetown Barbados 13°6N 59°37W 89 g
Bridgetown Canada 44°55N 65°18W 73 D6
Bridgewater Australia 42°44S 147°14E 63 G4
Bridgewater Canada 44°25N 64°31W 73 D7
Bridgewater Mass., U.S.A. 41°59N 70°58W 83 E14
Bridgewater N.Y., U.S.A. 42°53N 75°15W 83 D9
Bridgewater, C. Australia 38°23S 141°23E 63 F3
Bridgnorth U.K. 52°32N 2°25W 13 E5
Bridgton U.S.A. 44°3N 70°42W 83 B14
Bridgwater U.K. 51°8N 2°59W 13 F5
Bridgwater B. U.K. 51°15N 3°15W 13 F4
Bridlington U.K. 54°5N 0°12W 12 C7
Bridlington B. U.K. 54°4N 0°10W 12 C7
Bridport Australia 40°59S 147°23E 63 G4
Bridport U.K. 50°44N 2°45W 13 G5
Brig Switz. 46°18N 7°59E 20 C7
Brigg U.K. 53°34N 0°28W 12 D7
Brigham City U.S.A. 41°31N 112°1W 76 F7
Bright Australia 36°42S 146°56E 63 F4
Brighton Australia 35°5S 138°30E 63 F2
Brighton Canada 44°2N 77°44W 82 B7
Brighton Trin. & Tob. 10°13N 61°39W 93 K15
Brighton U.K. 50°49N 0°7W 13 G7
Brighton U.S.A. 43°8N 77°34W 82 C7
Brightside Canada 45°7N 76°29W 83 A8
Brilliant U.S.A. 40°15N 80°39W 82 F4
Brindisi Italy 40°39N 17°55E 23 D7
Brinkley U.S.A. 34°53N 91°12W 84 D9
Brinnon U.S.A. 47°41N 122°54W 78 C4
Brion, Î. Canada 47°46N 61°26W 73 C7
Brisay Canada 54°26N 70°31W 73 B5
Brisbane Australia 27°25S 153°2E 63 D5
Brisbane ➜ Australia 27°24S 153°9E 63 D5
Bristol U.K. 51°26N 2°35W 13 F5
Bristol Conn., U.S.A. 41°40N 72°57W 83 E12
Bristol Pa., U.S.A. 40°6N 74°51W 83 F10
Bristol R.I., U.S.A. 41°40N 71°16W 83 E13
Bristol Tenn., U.S.A. 36°36N 82°11W 85 C13
Bristol Vt., U.S.A. 44°8N 73°4W 83 B11
Bristol, City of □ U.K. 51°27N 2°36W 13 F5
Bristol B. U.S.A. 58°0N 160°0W 74 D8
Bristol Channel U.K. 51°18N 4°30W 13 F3
Bristol I. Antarctica 58°45S 28°0W 5 B1
Bristol L. U.S.A. 34°28N 115°41W 77 J6
Bristow U.S.A. 35°50N 96°23W 84 D6
Britain = Great Britain Europe 54°0N 2°15W 6 E5
British Columbia □ Canada 55°0N 125°15W 70 C3
British Indian Ocean Terr. = Chagos Arch. ☒ Ind. Oc. 6°0S 72°0E 27 J9
British Isles Europe 54°0N 4°0W 14 D5
British Mts. N. Amer. 68°50N 140°0W 74 B12
British Virgin Is. ☒ W. Indies 18°30N 64°30W 89 e
Brits S. Africa 25°37S 27°48E 57 D4
Britstown S. Africa 30°37S 23°30E 56 E3
Britt Canada 45°46N 80°34W 72 C3
Brittany = Bretagne □ France 48°10N 3°0W 20 B2
Britton U.S.A. 45°48N 97°45W 80 C5
Brive-la-Gaillarde France 45°10N 1°32E 20 D4
Brixen = Bressanone Italy 46°43N 11°39E 22 A4
Brixham U.K. 50°23N 3°31W 13 G4
Brno Czech Rep. 49°10N 16°35E 17 D9
Broach = Bharuch India 21°47N 73°0E 40 J8
Broad ➜ U.S.A. 34°1N 81°4W 85 D14
Broad Arrow Australia 30°23S 121°15E 61 F3
Broad B. U.K. 58°14N 6°18W 11 C2
Broad Haven Ireland 54°20N 9°55W 10 B2
Broad Law U.K. 55°30N 3°21W 11 F5
Broad Pk. = Faichan Kangri India 35°48N 76°34E 43 B7
Broad Sd. Australia 22°0S 149°45E 62 C4
Broadalbin U.S.A. 43°4N 74°12W 83 C10
Broadback ➜ Canada 51°21N 78°52W 72 B4
Broadhurst Ra. Australia 22°30S 122°30E 60 D3
Broads, The U.K. 52°45N 1°30E 12 E9
Broadus U.S.A. 45°27N 105°25W 76 D11
Brochet Canada 57°53N 101°40W 71 B8
Brochet, L. Canada 58°36N 101°35W 71 B8
Brock I. Canada 77°52N 114°19W 69 B9
Brocken Germany 51°47N 10°37E 16 C6
Brockport U.S.A. 43°13N 77°56W 82 C7
Brockton U.S.A. 42°5N 71°1W 83 D13
Brockville Canada 44°35N 75°41W 83 B9
Brockway Mont., U.S.A. 47°18N 105°45W 76 C11
Brockway Pa., U.S.A. 41°15N 78°47W 82 E6
Brocton U.S.A. 42°23N 79°26W 82 D5
Brodeur Pen. Canada 72°30N 88°10W 69 C14
Brodick U.K. 55°35N 5°9W 11 F3
Brodnica Poland 53°15N 19°25E 17 B10

Brody *Ukraine* 50°5N 25°10E **17** C13
Brogan *U.S.A.* 44°15N 117°31W **76** D5
Broken Arrow *U.S.A.* 36°3N 95°48W **84** C7
Broken Bow *Nebr., U.S.A.* 41°24N 99°38W **80** E4
Broken Bow *Okla., U.S.A.* 34°2N 94°44W **84** D7
Broken Bow Lake *U.S.A.* 34°9N 94°40W **84** D7
Broken Hill *Australia* 31°58S 141°29E **63** E3
Broken Ridge *Ind. Oc.* 30°0S 94°0E **64** L1
Broken River Ra. *Australia* 21°0S 148°22E **62** b
Bromley □ *U.K.* 51°24N 0°2E **13** F8
Bromo Tengger Semeru △ *Indonesia* 7°56S 112°57E **37** H15
Bromsgrove *U.K.* 52°21N 2°2W **13** E5
Brønderslev *Denmark* 57°16N 9°57E **9** H13
Bronkhorstspruit *S. Africa* 25°46S 28°45E **57** D4
Brønnøysund *Norway* 65°28N 12°14E **8** D15
Brook Park *U.S.A.* 41°23N 81°48W **82** E4
Brookhaven *U.S.A.* 31°35N 90°26W **85** F9
Brookings *Oreg., U.S.A.* 42°3N 124°17W **76** E1
Brookings *S. Dak., U.S.A.* 44°19N 96°48W **80** C5
Brooklin *Canada* 43°55N 78°55W **82** C6
Brooklyn Park *U.S.A.* 45°6N 93°23W **80** C7
Brooks *Canada* 50°35N 111°55W **70** C6
Brooks Range *U.S.A.* 68°0N 152°0W **74** B9
Brooksville *U.S.A.* 28°33N 82°23W **85** G13
Brookton *Australia* 32°22S 117°0E **61** F2
Brookville *U.S.A.* 41°10N 79°5W **82** E5
Broom, L. *U.K.* 57°55N 5°15W **11** D3
Broome *Australia* 18°0S 122°15E **60** C3
Brora *U.K.* 58°0N 3°52W **11** C5
Brora → *U.K.* 58°0N 3°51W **11** C5
Brosna → *Ireland* 53°14N 7°58W **10** C4
Brothers *U.S.A.* 43°49N 120°36W **76** E3
Brough *U.K.* 54°32N 2°18W **12** C5
Brough Hd. *U.K.* 59°8N 3°20W **11** B5
Broughton Island = Qikiqtarjuaq *Canada* 67°33N 63°0W **69** D19
Brown, L. *Australia* 31°5S 118°15E **61** F2
Brown, Pt. *Australia* 32°32S 133°50E **63** E1
Brown City *U.S.A.* 43°13N 82°59W **82** C2
Brown Willy *U.K.* 50°35N 4°37W **13** G3
Brownfield *U.S.A.* 33°11N 102°17W **84** E3
Browning *U.S.A.* 48°34N 113°1W **76** B7
Brownsville *Oreg., U.S.A.* 44°24N 122°59W **76** D2
Brownsville *Pa., U.S.A.* 40°1N 79°53W **82** F5
Brownsville *Tenn., U.S.A.* 35°36N 89°16W **85** D10
Brownsville *Tex., U.S.A.* 25°54N 97°30W **84** J6
Brownville *U.S.A.* 44°0N 75°59W **83** C9
Brownwood *U.S.A.* 31°43N 98°59W **84** F5
Browse I. *Australia* 14°7S 123°33E **60** B3
Bruas *Malaysia* 4°30N 100°47E **39** K3
Bruay-la-Buissière *France* 50°29N 2°33E **20** A5
Bruce, Mt. *Australia* 22°37S 118°8E **60** D2
Bruce Pen. *Canada* 45°0N 81°30W **82** B3
Bruce Peninsula △ *Canada* 45°14N 81°36W **82** A3
Bruce Rock *Australia* 31°52S 118°8E **61** F2
Bruck an der Leitha *Austria* 48°1N 16°47E **17** D9
Bruck an der Mur *Austria* 47°24N 15°16E **16** E8
Brue → *U.K.* 51°13N 2°59W **13** F5
Bruges = Brugge *Belgium* 51°13N 3°13E **15** C3
Brugge *Belgium* 51°13N 3°13E **15** C3
Bruin *U.S.A.* 41°3N 79°43W **82** E5
Brûk, W. el → *Egypt* 30°15N 33°50E **46** E2
Brûlé *Canada* 53°15N 117°58W **70** C5
Brûlé, L. *Canada* 53°35N 64°4W **73** B7
Brumado *Brazil* 14°14S 41°40W **93** F10
Brumunddal *Norway* 60°53N 10°56E **8** F14
Bruneau *U.S.A.* 42°53N 115°48W **76** E6
Bruneau → *U.S.A.* 42°56N 115°57W **76** E6
Brunei = Bandar Seri Begawan *Brunei* 4°52N 115°0E **36** D5
Brunei ■ *Asia* 4°50N 115°0E **36** D5
Brunner, L. *N.Z.* 42°37S 171°27E **59** E3
Brunssum *Neths.* 50°57N 5°59E **15** D5
Brunswick = Braunschweig *Germany* 52°15N 10°31E **16** B6
Brunswick *Ga., U.S.A.* 31°10N 81°30W **85** F14
Brunswick *Maine, U.S.A.* 43°55N 69°58W **81** D19
Brunswick *Md., U.S.A.* 39°19N 77°38W **81** F15
Brunswick *Mo., U.S.A.* 39°26N 93°8W **80** F7
Brunswick *Ohio, U.S.A.* 41°14N 81°51W **82** E3
Brunswick, Pen. de *Chile* 53°30S 71°30W **96** G2
Brunswick B. *Australia* 15°15S 124°50E **60** C3
Brunswick Junction *Australia* 33°15S 115°50E **61** F2
Brunt Ice Shelf *Antarctica* 75°30S 25°0W **5** D2
Brus Laguna *Honduras* 15°47N 84°35W **88** C3
Brush *U.S.A.* 40°15N 103°37W **76** F12
Brushton *U.S.A.* 44°50N 74°31W **83** B10
Brusque *Brazil* 27°5S 49°0W **95** B6
Brussel *Belgium* 50°51N 4°21E **15** D4
Brussel ✕ (BRU) *Belgium* 50°54N 4°29E **15** D5
Brussels = Brussel *Belgium* 50°51N 4°21E **15** D4
Brussels *Canada* 43°44N 81°15W **82** C3
Bruthen *Australia* 37°42S 147°50E **63** F4
Bruxelles = Brussel *Belgium* 50°51N 4°21E **15** D4
Bryan *Ohio, U.S.A.* 41°28N 84°33W **81** E11
Bryan *Tex., U.S.A.* 30°40N 96°22W **84** F6
Bryan, Mt. *Australia* 33°30S 139°5E **63** E2
Bryansk *Russia* 53°13N 34°25E **18** D4
Bryce Canyon △ *U.S.A.* 37°30N 112°10W **77** H7
Bryne *Norway* 58°44N 5°38E **9** G11
Bryson City *U.S.A.* 35°26N 83°27W **85** D13
Bsharri *Lebanon* 34°15N 36°0E **46** A5
Bū Baqarah *U.A.E.* 25°35N 56°25E **45** E8
Bu Craa *W. Sahara* 26°45N 12°50W **50** C3
Bū Ḩasā *U.A.E.* 23°30N 53°20E **45** F7
Bua *Fiji* 16°48S 178°37E **59** a
Bua → *Malaysia* 12°45S 34°16E **55** E3
Bua Yai *Thailand* 15°33N 102°26E **38** E4
Buan *S. Korea* 35°44N 126°44E **35** G14
Buapinang *Indonesia* 4°40S 121°30E **37** E6
Bubanza *Burundi* 3°6S 29°23E **54** C2
Bubi → *Zimbabwe* 22°20S 31°7E **55** G3
Būbiyān *Kuwait* 29°45N 48°15E **45** D6
Buca *Fiji* 16°38S 179°52E **59** a
Bucaramanga *Colombia* 7°0N 73°0W **92** B4
Bucasia *Australia* 21°2S 149°10E **62** b
Buccaneer Arch. *Australia* 16°7S 123°20E **60** C3
Buccoo Reef *Trin. & Tob.* 11°10N 60°51W **93** J16
Buchach *Ukraine* 49°5N 25°25E **17** D13
Buchan *U.K.* 57°32N 2°21W **11** D6
Buchan Ness *U.K.* 57°29N 1°46W **11** D7
Buchanan *Canada* 51°40N 102°45W **71** C8
Buchanan *Liberia* 5°57N 10°2W **50** G3
Buchanan, L. *Queens., Australia* 21°35S 145°52E **62** C4
Buchanan, L. *W. Austral., Australia* 25°33S 123°2E **61** E3
Buchanan, L. *U.S.A.* 30°45N 98°25W **84** F5
Buchanan Cr. → *Australia* 19°13S 136°33E **62** B2
Buchans *Canada* 48°50N 56°52W **73** C8
Bucharest = Bucureşti *Romania* 44°27N 26°10E **17** F14
Bucheon *S. Korea* 37°28N 126°45E **35** F14
Buchon, Pt. *U.S.A.* 35°15N 120°54W **78** K6
Buck Hill Falls *U.S.A.* 41°11N 75°16W **83** E9
Buckeye Lake *U.S.A.* 39°55N 82°29W **82** G2
Buckhannon *U.S.A.* 39°0N 80°8W **81** F13
Buckhaven *U.K.* 56°11N 3°3W **11** E5
Buckhorn L. *Canada* 44°29N 78°23W **82** B6
Buckie *U.K.* 57°41N 2°58W **11** D6
Buckingham *Canada* 45°37N 75°24W **72** C4
Buckingham *U.K.* 51°59N 0°57W **13** F7
Buckingham B. *Australia* 12°10S 135°40E **62** A2
Buckinghamshire □ *U.K.* 51°53N 0°55W **13** F7
Buckle Hd. *Australia* 14°26S 127°52E **60** B4
Buckleboo *Australia* 32°54S 136°12E **63** E2
Buckley *U.K.* 53°10N 3°5W **12** D4
Buckley → *Australia* 20°10S 138°49E **62** C2
Bucklin *U.S.A.* 37°33N 99°38W **80** G4
Bucks L. *U.S.A.* 39°54N 121°12W **78** F5
Bucureşti *Romania* 44°27N 26°10E **17** F14
Bucyrus *U.S.A.* 40°48N 82°59W **81** E12
Budalin *Burma* 22°20N 95°10E **41** H19
Budapest *Hungary* 47°29N 19°3E **17** E10
Budaun *India* 28°5N 79°10E **43** E8
Budd Coast *Antarctica* 68°0S 112°0E **5** C8
Bude *U.K.* 50°49N 4°34W **13** G3
Budennovsk *Russia* 44°50N 44°10E **19** F7
Budge Budge = Baj Baj *India* 22°30N 88°5E **43** H13
Budgewoi *Australia* 33°13S 151°34E **63** E5
Budjala *Dem. Rep. of the Congo* 2°50N 19°40E **52** D3
Budo-Sungai Padi △ *Thailand* 6°19N 101°42E **39** J3
Buellton *U.S.A.* 34°37N 120°12W **79** L6
Buena Esperanza *Argentina* 34°45S 65°15W **94** C2
Buena Park *U.S.A.* 33°52N 117°59W **79** M9
Buena Vista *Colo., U.S.A.* 38°51N 106°8W **76** G10
Buena Vista *Va., U.S.A.* 37°44N 79°21W **81** G14
Buena Vista Lake Bed *U.S.A.* 35°12N 119°18W **79** K7
Buenaventura *Colombia* 3°53N 77°4W **92** C3
Buenaventura *Mexico* 29°51N 107°29W **86** B3
Buenos Aires *Argentina* 34°36S 58°22W **94** C4
Buenos Aires *Costa Rica* 9°10N 83°20W **88** E3
Buenos Aires □ *Argentina* 36°30S 60°0W **94** D4
Buenos Aires, L. *Argentina* 46°35S 72°30W **96** F2
Buffalo *Mo., U.S.A.* 37°39N 93°6W **80** G7
Buffalo *N.Y., U.S.A.* 42°53N 78°53W **82** D6
Buffalo *Okla., U.S.A.* 36°50N 99°38W **84** C5
Buffalo *S. Dak., U.S.A.* 45°35N 103°33W **80** C2
Buffalo *Wyo., U.S.A.* 44°21N 106°42W **76** D10
Buffalo → *Canada* 60°5N 115°5W **70** A5
Buffalo → *S. Africa* 28°43S 30°37E **57** D5
Buffalo ⌓ *U.S.A.* 36°14N 92°36W **84** C8
Buffalo Head Hills *Canada* 57°25N 115°55W **70** B5
Buffalo L. *Alta., Canada* 52°27N 112°54W **70** C6
Buffalo L. *N.W.T., Canada* 60°12N 115°25W **70** A5
Buffalo Narrows *Canada* 55°51N 108°29W **71** B7
Buffalo Springs ⌓ *Kenya* 0°32N 37°35E **54** B4
Buffels → *S. Africa* 29°36S 17°3E **56** D2
Buford *U.S.A.* 34°10N 84°0W **85** D12
Bug = Buh → *Ukraine* 46°59N 31°58E **19** E5
Bug → *Poland* 52°31N 21°5E **17** B11
Buga *Colombia* 4°0N 76°15W **92** C3
Bugala I. *Uganda* 0°40S 32°20E **54** C3
Buganda *Uganda* 0°0 31°30E **54** C3
Buganga *Uganda* 0°3S 32°0E **54** C3
Bugel, Tanjung *Indonesia* 6°26S 111°3E **37** G14
Búger *Spain* 39°45N 2°59E **24** B9
Bugibba *Malta* 35°57N 14°25E **25** D1
Bugsuk I. *Phil.* 8°12N 117°18E **36** C5
Bugulma *Russia* 54°33N 52°48E **18** D9
Bugungu ⌓ *Uganda* 2°17N 31°50E **54** B3
Buguruslan *Russia* 53°39N 52°26E **18** D9
Buh → *Ukraine* 46°59N 31°58E **19** E5
Buhera *Zimbabwe* 19°18S 31°29E **57** B5
Buhl *U.S.A.* 42°36N 114°46W **76** E6
Builth Wells *U.K.* 52°9N 3°25W **13** E4
Buir Nur *Mongolia* 47°50N 117°42E **33** B12
Buji *China* 22°37N 114°5E **33** a
Bujumbura *Burundi* 3°16S 29°18E **54** C2
Bukachacha *Russia* 52°55N 116°50E **29** D12
Bukama *Dem. Rep. of the Congo* 9°10S 25°50E **55** D2
Bukavu *Dem. Rep. of the Congo* 2°20S 28°52E **54** C2
Bukene *Tanzania* 4°15S 32°48E **54** C3
Bukhara = Buxoro *Uzbekistan* 39°48N 64°25E **28** F7
Bukhoro = Buxoro *Uzbekistan* 39°48N 64°25E **28** F7
Bukhtarma Res. = Zaysan Köli *Kazakhstan* 48°0N 83°0E **28** E9
Bukima *Tanzania* 1°50S 33°25E **54** C3
Bukit Bendera *Malaysia* 5°25N 100°15E **39** c
Bukit Mertajam *Malaysia* 5°22N 100°28E **39** c
Bukit Nil *Malaysia* 1°22N 104°12E **39** d
Bukit Panjang *Singapore* 1°23N 103°46E **39** d
Bukit Tengah *Malaysia* 5°22N 100°25E **39** c
Bukittinggi *Indonesia* 0°20S 100°20E **36** E2
Bukoba *Tanzania* 1°20S 31°49E **54** C3
Bukum, Pulau *Singapore* 1°14N 103°46E **39** d
Bukuya *Uganda* 0°40N 31°52E **54** B3
Būl, Kuh-e *Iran* 30°48N 52°45E **45** D7
Bula *Indonesia* 3°6S 130°30E **37** E8
Bulahdelah *Australia* 32°23S 152°13E **63** E5
Bulan *Phil.* 12°40N 123°52E **37** B6
Bulandshahr *India* 28°28N 77°51E **42** E7
Bulawayo *Zimbabwe* 20°7S 28°32E **55** G2
Buldan *Turkey* 38°2N 28°50E **23** E13
Buldir I. *U.S.A.* 52°21N 175°56E **74** E3
Bulgan *Mongolia* 48°45N 103°34E **32** B9
Bulgar *Russia* 54°57N 49°4E **18** D8
Bulgaria ■ *Europe* 42°35N 25°30E **23** C11
Buli, Teluk *Indonesia* 0°48N 128°25E **37** D7
Buliluyan, C. *Phil.* 8°20N 117°15E **36** C5
Bulim *Singapore* 1°22N 103°43E **39** d
Bulkley → *Canada* 55°15N 127°40W **70** B3
Bull Shoals L. *U.S.A.* 36°22N 92°35W **84** C8
Bulleringa △ *Australia* 17°39S 143°56E **62** B3
Bullhead City *U.S.A.* 35°8N 114°32W **79** K12
Büllingen *Belgium* 50°25N 6°16E **15** D6
Bullock Creek *Australia* 17°43S 144°31E **62** B3
Bulloo → *Australia* 28°43S 142°30E **63** D3
Bulloo L. *Australia* 28°43S 142°25E **63** D3
Bulls *N.Z.* 40°10S 175°24E **59** D5
Bulman *Australia* 13°39S 134°20E **62** A1
Bulnes *Chile* 36°42S 72°19W **94** D1
Bulsar = Valsad *India* 20°40N 72°58E **40** J8
Bultfontein *S. Africa* 28°18S 26°10E **56** D4
Bulukumba *Indonesia* 5°33S 120°11E **37** F6
Bulun *Russia* 70°37N 127°30E **29** B13
Bumba *Dem. Rep. of the Congo* 2°13N 22°30E **52** D4
Bumbah, Khalīj *Libya* 32°20N 23°15E **51** B10
Bumbiri I. *Tanzania* 1°40S 31°55E **54** C3
Bumhpa Bum *Burma* 26°51N 97°14E **41** F20
Bumi → *Zimbabwe* 17°0S 28°20E **55** F2
Buna *Kenya* 2°58N 39°30E **54** B4
Bunaken *Indonesia* 1°37N 124°46E **37** D6
Bunazi *Tanzania* 1°3S 31°23E **54** C3
Bunbury *Australia* 33°20S 115°35E **61** F2
Bunclody *Ireland* 52°39N 6°40W **10** D5
Buncrana *Ireland* 55°8N 7°27W **10** A4
Bundaberg *Australia* 24°54S 152°22E **63** C5
Bundey → *Australia* 21°46S 135°37E **62** C2
Bundi *India* 25°30N 75°35E **42** G6
Bundjalung △ *Australia* 29°16S 153°21E **63** D5
Bundoran *Ireland* 54°28N 8°16W **10** B3
Bung Kan *Thailand* 18°23N 103°37E **38** C4
Bungay *U.K.* 52°27N 1°28E **13** E9
Bungil Cr. → *Australia* 27°5S 149°5E **63** D4
Bungle Bungle = Purnululu △ *Australia* 17°20S 128°20E **60** C4
Bungo-Suidō *Japan* 33°0N 132°15E **31** H6
Bungoma *Kenya* 0°34N 34°34E **54** B3
Bungotakada *Japan* 33°35N 131°25E **31** H5
Bungu *Tanzania* 7°35S 39°0E **54** D4
Bunia *Dem. Rep. of the Congo* 1°35N 30°20E **54** B3
Bunji *Pakistan* 35°45N 74°40E **43** B6
Bunkie *U.S.A.* 30°57N 92°11W **84** F8
Bunnell *U.S.A.* 29°28N 81°16W **85** G14
Buntok *Indonesia* 1°40S 114°58E **36** E4
Bunya Mts. △ *Australia* 26°51S 151°34E **63** D5
Bunyola *Spain* 39°41N 2°42E **24** B9
Bunyu *Indonesia* 3°35N 117°50E **36** D5
Buol *Indonesia* 1°15N 121°32E **37** D6
Buon Brieng *Vietnam* 13°9N 108°12E **38** F7
Buon Ho *Vietnam* 12°57N 108°18E **38** F7
Buon Ma Thuot *Vietnam* 12°40N 108°3E **38** F7
Buong Long *Cambodia* 13°44N 106°59E **38** F6
Buorkhaya, Mys *Russia* 71°50N 132°40E **29** B14
Buqayq *Si. Arabia* 26°0N 49°45E **45** E6
Bur Acaba = Buurhakaba *Somali Rep.* 3°12N 44°20E **47** G3
Bûr Safâga *Egypt* 26°43N 33°57E **44** E2
Bûr Sa'îd *Egypt* 31°16N 32°18E **51** B12
Bûr Sûdân *Sudan* 19°32N 37°9E **51** E13
Bura *Kenya* 1°4S 39°58E **54** C4
Burakin *Australia* 30°31S 117°10E **61** F2
Burang *China* 30°15N 81°10E **32** E5
Burao = Burco *Somali Rep.* 9°32N 45°32E **47** F4
Burāq *Syria* 33°11N 36°29E **46** B5
Buraydah *Si. Arabia* 26°20N 43°59E **44** E4
Burbank *U.S.A.* 34°12N 118°18W **79** L8
Burco *Somali Rep.* 9°32N 45°32E **47** F4
Burda *India* 25°50N 77°35E **42** G7
Burdekin → *Australia* 19°38S 147°25E **62** B4
Burdur *Turkey* 37°45N 30°17E **19** G5
Burdwan = Barddhaman *India* 23°14N 87°39E **43** H12
Bure *Ethiopia* 10°40N 37°4E **47** E2
Bure → *U.K.* 52°38N 1°43E **12** E9
Bureya → *Russia* 49°27N 129°30E **29** E13
Burford *Canada* 43°7N 80°27W **82** C4
Burgas *Bulgaria* 42°33N 27°29E **23** C12
Burgeo *Canada* 47°37N 57°38W **73** C8
Burgersdorp *S. Africa* 31°0S 26°20E **56** E4
Burgess, Mt. *Australia* 30°50S 121°5E **61** F3
Burghead *U.K.* 57°43N 3°30W **11** D5
Burgos *Spain* 42°21N 3°41W **21** A4
Burgsvik *Sweden* 57°3N 18°19E **9** H18
Burgundy = Bourgogne □ *France* 47°0N 4°50E **20** C6
Burhaniye *Turkey* 39°30N 26°58E **23** E12
Burhanpur *India* 21°18N 76°14E **40** J10
Burhi Gandak → *India* 25°20N 86°37E **43** G12
Burhner → *India* 22°43N 80°31E **43** H9
Burias I. *Phil.* 12°55N 123°5E **37** B6
Burica, Pta. *Costa Rica* 8°3N 82°51W **88** E3
Burien *U.S.A.* 47°28N 122°20W **78** C4
Burigi, L. *Tanzania* 2°2S 31°22E **54** C3
Burigi ⌓ *Tanzania* 2°20S 31°6E **54** C3
Burin *Canada* 47°1N 55°14W **73** C8
Buriram *Thailand* 15°0N 103°0E **38** E4
Burkburnett *U.S.A.* 34°6N 98°34W **84** D5
Burke → *Australia* 23°12S 139°33E **62** C2
Burke Chan. *Canada* 52°10N 127°30W **70** C3
Burketown *Australia* 17°45S 139°33E **62** B2
Burkina Faso ■ *Africa* 12°0N 1°0W **50** F5
Burk's Falls *Canada* 45°37N 79°24W **72** C4
Burleigh Falls *Canada* 44°33N 78°12W **82** B6
Burley *U.S.A.* 42°32N 113°48W **76** E7
Burlingame *U.S.A.* 37°35N 122°21W **78** H4
Burlington *Canada* 43°18N 79°45W **82** C5
Burlington *Colo., U.S.A.* 39°18N 102°16W **76** G12
Burlington *Iowa, U.S.A.* 40°49N 91°14W **80** E8
Burlington *Kans., U.S.A.* 38°12N 95°45W **80** F6
Burlington *N.C., U.S.A.* 36°6N 79°26W **85** C15
Burlington *N.J., U.S.A.* 40°4N 74°51W **83** F10
Burlington *Vt., U.S.A.* 44°29N 73°12W **83** B11
Burlington *Wash., U.S.A.* 48°28N 122°20W **78** B4
Burlington *Wis., U.S.A.* 42°41N 88°17W **80** D9
Burma ■ *Asia* 21°0N 96°30E **41** J20
Burnaby I. *Canada* 52°25N 131°19W **70** C2
Burnet *U.S.A.* 30°45N 98°14W **84** F5
Burney *U.S.A.* 40°53N 121°40W **76** F3
Burnham *U.S.A.* 40°38N 77°34W **82** F7
Burnham-on-Sea *U.K.* 51°14N 3°0W **13** F5
Burnie *Australia* 41°4S 145°56E **63** G4
Burnley *U.K.* 53°47N 2°14W **12** D5
Burns *U.S.A.* 43°35N 119°3W **76** E4
Burns Junction *U.S.A.* 42°47N 117°51W **76** E5
Burns Lake *Canada* 54°14N 125°45W **70** C3
Burnside → *Canada* 66°51N 108°4W **68** D10
Burnside, L. *Australia* 25°22S 123°0E **61** E3
Burnsville *U.S.A.* 44°47N 93°17W **80** C7
Burnt River *Canada* 44°41N 78°42W **82** B6
Burntwood → *Canada* 56°8N 96°34W **71** B9
Burntwood L. *Canada* 55°22N 100°26W **71** B8
Burqān *Kuwait* 29°0N 47°57E **44** D5
Burqin *China* 47°43N 87°0E **32** B6
Burra *Australia* 33°40S 138°55E **63** E2
Burray *U.K.* 58°51N 2°54W **11** C6
Burren *Ireland* 53°9N 9°5W **10** C2
Burren △ *Ireland* 53°1N 8°58W **10** C3
Burren Junction *Australia* 30°7S 148°58E **63** E4
Burrinjuck, L. *Australia* 35°0S 148°36E **63** F4
Burro, Serranías del *Mexico* 28°56N 102°5W **86** B4
Burrow Hd. *U.K.* 54°41N 4°24W **11** G4
Burrum Coast △ *Australia* 25°13S 152°36E **63** D5
Burruyacú *Argentina* 26°30S 64°40W **94** B3
Burry Port *U.K.* 51°41N 4°15W **13** F3
Bursa *Turkey* 40°15N 29°5E **23** D13
Burstall *Canada* 50°39N 109°54W **71** C7
Burton *Ohio, U.S.A.* 41°28N 81°8W **82** E3
Burton *S.C., U.S.A.* 32°26N 80°43W **85** E14
Burton, L. *Canada* 54°45N 78°20W **72** B4
Burton upon Trent *U.K.* 52°48N 1°38W **12** E6
Buru *Indonesia* 3°30S 126°30E **37** E7
Burûn, Râs *Egypt* 31°14N 33°7E **46** D2
Burundi ■ *Africa* 3°15S 30°0E **54** C3
Bururi *Burundi* 3°57S 29°37E **54** C2
Burutu *Nigeria* 5°20N 5°29E **50** G7
Burwell *U.S.A.* 41°47N 99°8W **80** E4
Burwick *U.K.* 58°45N 2°58W **11** C5
Bury *U.K.* 53°35N 2°17W **12** D5
Bury St. Edmunds *U.K.* 52°15N 0°43E **13** E8
Buryatia □ *Russia* 53°0N 110°0E **29** D12
Busan *S. Korea* 35°5N 129°0E **35** G15
Busango Swamp *Zambia* 14°15S 25°45E **55** E2
Busaso = Boosaaso *Somali Rep.* 11°12N 49°18E **47** E4
Buşayrah *Syria* 35°9N 40°26E **44** C4
Būshehr *Iran* 28°55N 50°55E **45** D6
Būshehr □ *Iran* 28°20N 51°45E **45** D6
Bushenyi *Uganda* 0°35S 30°10E **54** C3
Bushire = Būshehr *Iran* 28°55N 50°55E **45** D6
Businga *Dem. Rep. of the Congo* 3°16N 20°59E **52** D4
Buşra ash Shām *Syria* 32°30N 36°25E **46** C5
Busselton *Australia* 33°42S 115°15E **61** F2
Bussum *Neths.* 52°16N 5°10E **15** B5
Busto Arsízio *Italy* 45°37N 8°51E **20** D8
Busu Djanoa *Dem. Rep. of the Congo* 1°43N 21°23E **52** D4
Busuanga I. *Phil.* 12°10N 120°0E **37** B6
Busungbiu *Indonesia* 8°16S 114°58E **37** J17
Buta *Dem. Rep. of the Congo* 2°50N 24°53E **54** B1
Butare *Rwanda* 2°31S 29°52E **54** C2
Butaritari *Kiribati* 3°30N 174°0E **64** G9
Bute *U.K.* 55°48N 5°2W **11** F3
Bute Inlet *Canada* 50°40N 124°53W **70** C4
Butemba *Uganda* 1°9N 31°37E **54** B3
Butembo *Dem. Rep. of the Congo* 0°9N 29°18E **54** B2
Butere *Kenya* 0°13N 34°30E **54** B3
Butha Qi *China* 48°0N 122°32E **33** B13
Butiaba *Uganda* 1°50N 31°20E **54** B3
Butler *Mo., U.S.A.* 38°16N 94°20W **80** F6
Butler *Pa., U.S.A.* 40°52N 79°54W **82** F5
Buton *Indonesia* 5°0S 122°45E **37** E6
Butte *Mont., U.S.A.* 46°0N 112°32W **76** C7
Butte *Nebr., U.S.A.* 42°58N 98°51W **80** D4
Butte Creek → *U.S.A.* 39°12N 121°56W **78** F
Butterworth = Gcuwa *S. Africa* 32°20S 28°11E **57** E
Butterworth *Malaysia* 5°24N 100°23E **39**
Buttevant *Ireland* 52°14N 8°40W **10** D
Buttfield, Mt. *Australia* 24°45S 128°9E **61** D
Button B. *Canada* 58°45N 94°23W **71** B1
Buttonwillow *U.S.A.* 35°24N 119°28W **79** K
Butty Hd. *Australia* 33°54S 121°39E **61** F
Butuan *Phil.* 8°57N 125°33E **37** C
Butung = Buton *Indonesia* 5°0S 122°45E **37** E
Buturlinovka *Russia* 50°50N 40°35E **19** D
Butwal *Nepal* 27°33N 83°31E **43** F1
Buurhakaba *Somali Rep.* 3°12N 44°20E **47** G
Buxa Duar *India* 27°45N 89°35E **43** F1
Buxar *India* 25°34N 83°58E **43** G1
Buxoro *Uzbekistan* 39°48N 64°25E **28** F
Buxtehude *Germany* 53°28N 9°39E **16** B
Buxton *U.K.* 53°16N 1°54W **12** D
Buy *Russia* 58°28N 41°28E **18** C
Buyant-Uhaa *Mongolia* 44°55N 110°11E **34** B
Buyo, L. de *Ivory C.* 6°16N 7°10W **50** G
Büyük Menderes → *Turkey* 37°28N 27°11E **23** F1
Büyükçekmece *Turkey* 41°2N 28°35E **23** D1
Buyun Shan *China* 40°4N 122°43E **35** D1
Buzău *Romania* 45°10N 26°50E **17** F1
Buzău → *Romania* 45°26N 27°44E **17** F1
Buzen *Japan* 33°35N 131°5E **31** H
Buzi → *Mozam.* 19°50S 34°43E **55** F
Büzmeýin *Turkmenistan* 38°3N 58°12E **45** B
Buzuluk *Russia* 52°48N 52°12E **18** D
Buzzards Bay *U.S.A.* 41°45N 70°37W **83** E1
Bwana Mkubwa *Zambia* 13°8S 28°38E **55** E
Bwindi △ *Uganda* 1°2S 29°42E **54** C
Byam Martin I. *Canada* 75°15N 104°15W **69** B1
Byarezina → *Belarus* 52°33N 30°14E **17** B1
Byaroza *Belarus* 52°31N 24°51E **17** B1
Bydgoszcz *Poland* 53°10N 18°0E **17** B
Byelorussia = Belarus ■ *Europe* 53°30N 27°0E **17** B1
Byers *U.S.A.* 39°43N 104°14W **76** G1
Byesville *U.S.A.* 39°58N 81°32W **82** G
Byfield △ *Australia* 22°52S 150°45E **62** C
Bykhaw *Belarus* 53°31N 30°14E **17** B1
Bykhov = Bykhaw *Belarus* 53°31N 30°14E **17** B1
Bylas *U.S.A.* 33°8N 110°7W **77** K
Bylot *Canada* 58°25N 94°8W **71** B1
Bylot I. *Canada* 73°13N 78°34W **69** C1
Byrd, C. *Antarctica* 69°38S 76°7W **5** C1
Byrock *Australia* 30°40S 146°27E **63** E
Byron, C. *Australia* 28°43S 153°38E **63** D
Byron Bay *Australia* 28°43S 153°37E **63** D
Byrranga, Gory *Russia* 75°0N 100°0E **29** B1
Byrranga Mts. = Byrranga, Gory *Russia* 75°0N 100°0E **29** B1
Byske *Sweden* 64°57N 21°11E **8** D1
Byskeälven → *Sweden* 64°57N 21°13E **8** D1
Bytom *Poland* 50°25N 18°54E **17** C1
Bytów *Poland* 54°10N 17°30E **17** A
Byumba *Rwanda* 1°35S 30°4E **54** C

C

C.W. McConaughy, L. *U.S.A.* 41°14N 101°40W **80** E
Ca → *Vietnam* 18°45N 105°45E **38** C
Ca Mau *Vietnam* 9°7N 105°8E **39** H
Ca Mau, Mui *Vietnam* 8°38N 104°44E **39** H
Ca Na *Vietnam* 11°20N 108°54E **39** G
Caacupé *Paraguay* 25°23S 57°5W **94** B
Caaguazú △ *Paraguay* 26°5S 55°31W **95** B
Caála *Angola* 12°46S 15°30E **53** G
Caamaño Sd. *Canada* 52°55N 129°25W **70** C
Caazapá *Paraguay* 26°8S 56°19W **94** B
Caazapá □ *Paraguay* 26°10S 56°0W **95** B
Caballeria, C. de *Spain* 40°5N 4°5E **24** A
Cabanatuan *Phil.* 15°30N 120°58E **37**
Cabano *Canada* 47°40N 68°56W **73** C
Cabazon *U.S.A.* 33°55N 116°47W **79** M1
Cabedelo *Brazil* 7°0S 34°50W **93** E1
Cabildo *Chile* 32°30S 71°5W **94** C
Cabimas *Venezuela* 10°23N 71°25W **92** A
Cabinda *Angola* 5°33S 12°11E **52** F
Cabinda □ *Angola* 5°0S 12°30E **52** F
Cabinet Mts. *U.S.A.* 48°10N 115°50W **76** B
Cabo Blanco *Argentina* 47°15S 65°47W **96** F
Cabo Frio *Brazil* 22°51S 42°3W **95** A
Cabo Pantoja *Peru* 1°0S 75°10W **92** D
Cabo San Lucas *Mexico* 22°53N 109°54W **86** C
Cabo Verde = Cape Verde Is. ■ *Atl. Oc.* 16°0N 24°0W **50**
Cabonga, Réservoir *Canada* 47°20N 76°40W **72** C
Cabool *U.S.A.* 37°7N 92°6W **80** G
Caboolture *Australia* 27°5S 152°58E **63**
Cabora Bassa Dam = Cahora Bassa, Lago de *Mozam.* 15°20S 32°50E **55**
Caborca *Mexico* 30°37N 112°6W **86** A
Cabot, Mt. *U.S.A.* 44°30N 71°25W **83** B
Cabot Hd. *Canada* 45°14N 81°17W **82** A
Cabot Str. *Canada* 47°15N 59°40W **73**
Cabra *Spain* 37°30N 4°28W **21** D
Cabrera *Spain* 39°8N 2°57E **24** B
Cabri *Canada* 50°35N 108°25W **71** C
Cabriel → *Spain* 39°14N 1°3W **21** C
Caçador *Brazil* 26°47S 51°0W **95** B
Čačak *Serbia* 43°54N 20°20E **23** C
Caçapava do Sul *Brazil* 30°30S 53°30W **95**
Cáceres *Brazil* 16°5S 57°40W **92** G
Cáceres *Spain* 39°26N 6°23W **21** C
Cache Bay *Canada* 46°22N 80°0W **72** C
Cache Cr. → *U.S.A.* 38°42N 121°42W **78** G
Cache Creek *Canada* 50°48N 121°19W **70** C
Cachi *Argentina* 25°5S 66°10W **94** B
Cachimbo, Serra do *Brazil* 9°30S 55°30W **93** E

chinal de la Sierra
'hile 24°58S 69°32W 94 A2
choeira Brazil 12°30S 39°0W 93 F11
choeira do Sul Brazil 30°3S 52°53W 95 C5
choeiro de Itapemirim
Brazil 20°51S 41°7W 95 A7
coal Brazil 11°32S 61°18W 92 F6
cólo Angola 10°9S 19°21E 52 G3
conda Angola 13°48S 15°8E 53 G3
ddo U.S.A. 34°7N 96°16W 84 D6
der Idris U.K. 52°42N 3°53W 13 E4
dereyta de Jiménez
Mexico 25°36N 100°0W 86 B5
dibarrawirracanna, L.
Australia 28°52S 135°27E 63 D2
dillac U.S.A. 44°15N 85°24W 81 C11
diz Phil. 10°57N 123°15E 37 B6
diz Spain 36°30N 6°20W 21 D2
diz Calif., U.S.A. 34°30N 115°28W 79 L11
diz Ohio, U.S.A. 40°22N 81°0W 82 F4
diz, G. de Spain 36°40N 7°0W 21 D2
diz L. U.S.A. 34°18N 115°24W 77 J6
dney Park Australia 27°55S 134°3E 63 D1
domin Canada 53°2N 117°20W 70 C5
dotte Lake Canada 56°26N 116°23W 70 B5
doux Australia 30°46S 117°7E 61 F2
en France 49°10N 0°22W 20 B3
ernarfon U.K. 53°8N 4°16W 12 D3
ernarfon B. U.K. 53°4N 4°40W 12 D3
ernarvon = Caernarfon
U.K. 53°8N 4°16W 12 D3
erphilly U.K. 51°35N 3°13W 13 F4
erphilly □ U.K. 51°37N 3°12W 13 F4
esarea Israel 32°30N 34°53E 46 C3
etité Brazil 13°50S 42°32W 93 F10
fayate Argentina 26°2S 66°0W 94 B2
fu Angola 16°30S 15°8E 56 B2
gayan de Oro Phil. 8°30N 124°40E 37 C6
gliari Italy 39°13N 9°7E 22 E3
gliari, G. di Italy 39°8N 9°11E 22 E3
guán → Colombia 0°8S 74°18W 92 D4
guas Puerto Rico 18°14N 66°2W 89 d
ha Mts. Ireland 51°45N 9°40W 10 E2
hama Angola 16°17S 14°19E 56 B1
her Ireland 52°22N 7°56W 10 D4
hersiveen Ireland 51°56N 10°14W 10 E1
hora Bassa, Lago de
Mozam. 15°20S 32°50E 55 F3
hore Pt. Ireland 52°33N 6°12W 10 D5
hors France 44°27N 1°27E 20 D4
hul Moldova 45°50N 28°15E 17 F15
Bau, Dao Vietnam 21°10N 107°27E 38 B6
Be Vietnam 10°20N 106°2E 39 G6
Nuoc Vietnam 8°56N 105°1E 39 H5
a Mozam. 17°51S 35°24E 55 F4
anda Angola 11°2S 23°31E 55 E1
barién Cuba 22°30N 79°30W 88 B4
cara Venezuela 7°38N 66°10W 92 B5
có Brazil 6°20S 37°0W 93 E11
cos Is. Turks & Caicos 21°40N 71°40W 89 B5
cos Passage W. Indies 22°45N 72°45W 89 B5
guna Australia 32°16S 125°29E 61 F4
rd Coast Antarctica 75°0S 25°0W 5 D1
rn Gorm U.K. 57°7N 3°39W 11 D5
rngorm Mts. U.K. 57°6N 3°42W 11 D5
rngorms △ U.K. 57°10N 3°50W 11 D5
rnryan U.K. 54°59N 5°1W 11 G3
rns Australia 16°57S 145°45E 62 B4
rns L. Canada 51°42N 94°30W 71 C10
ro = El Qâhira Egypt 30°2N 31°13E 51 B12
ro Ga., U.S.A. 30°52N 84°13W 85 F12
ro Ill., U.S.A. 37°0N 89°11W 80 G9
ro N.Y., U.S.A. 42°18N 74°0W 83 D10
seal = Cashel Ireland 52°30N 7°53W 10 D4
sleán an Bharraigh = Castlebar
reland 53°52N 9°18W 10 C2
thness U.K. 58°25N 3°35W 11 C5
thness, Ord of U.K. 58°8N 3°36W 11 C5
a de Muertos, I.
Puerto Rico 17°54N 66°32W 89 d
amarca Peru 7°5S 78°28W 92 E3
azeiras Brazil 6°52S 38°30W 93 E11
a d'Or Spain 39°23N 3°14E 24 B10
a en Porter Spain 39°52N 4°8E 24 B11
a Figuera Spain 39°20N 3°10E 24 B10
a Figuera, C. de Spain 39°27N 2°31E 24 B9
a Forcat Spain 40°0N 3°47E 24 B10
a Major Spain 39°33N 2°37E 24 B9
a Mezquida = Sa Mesquida
pain 39°55N 4°16E 24 B11
a Millor Spain 39°35N 3°22E 24 B10
a Murada Spain 39°27N 3°17E 24 B10
a Ratjada Spain 39°43N 3°27E 24 B10
a Santa Galdana
pain 39°56N 3°58E 24 B10
abar Nigeria 4°57N 8°20E 50 H7
abogie Canada 45°18N 76°43W 83 A8
abozo Venezuela 9°0N 67°28W 92 B5
ábria □ Italy 39°0N 16°30E 22 E7
ahorra Spain 42°18N 1°59W 21 A5
ais France 50°57N 1°56E 20 A4
ais U.S.A. 45°11N 67°17W 81 C20
akmul ⌓ Mexico 18°14N 89°48W 87 D7
alaste, Cord. de
Argentina 25°0S 67°0W 94 B2
ama Brazil 8°0S 62°50W 92 E6
ama Chile 22°30S 68°55W 94 A2
amar Colombia 10°15N 74°55W 92 A4
amian Group Phil. 11°50N 119°55E 37 B5
amocha Spain 40°50N 1°17W 21 B5
ang Indonesia 4°37N 95°37E 36 D1
anscio, Sarīr Libya 27°30N 21°30E 51 C10
apan Phil. 13°25N 121°7E 37 B6
ărași Romania 44°12N 27°20E 17 F14
atayud Spain 41°20N 1°40W 21 B5
auag Phil. 13°55N 122°15E 37 B6
avite, C. Phil. 13°26N 120°20E 37 B6
bayog Phil. 12°4N 124°38E 37 B6
ca Peru 13°22S 72°0W 92 F4
casieu L. U.S.A. 29°55N 93°18W 84 G8
Calcium U.S.A. 44°1N 75°50W 83 B9
Calcutta = Kolkata
India 22°34N 88°21E 43 H13
Calcutta U.S.A. 40°40N 80°34W 82 F4
Caldas da Rainha Portugal 39°24N 9°8W 21 C1
Calder → U.K. 53°44N 1°22W 12 D6
Caldera Chile 27°5S 70°55W 94 B1
Caldera de Taburiente △
Canary Is. 28°43N 17°52W 24 F2
Caldwell Idaho, U.S.A. 43°40N 116°41W 76 E5
Caldwell Kans., U.S.A. 37°2N 97°37W 80 G5
Caldwell Tex., U.S.A. 30°32N 96°42W 84 F6
Caledon Canada 43°51N 79°51W 82 C5
Caledon S. Africa 34°14S 19°26E 56 E2
Caledon → S. Africa 30°31S 26°5E 56 E4
Caledon B. Australia 12°45S 137°0E 62 A2
Caledonia Canada 43°7N 79°58W 82 C5
Caledonia U.S.A. 42°58N 77°51W 82 D7
Calemba Angola 16°0S 15°44E 56 B2
Calen Australia 20°56S 148°48E 62 b
Caletones Chile 34°6S 70°27W 94 C1
Calexico U.S.A. 32°40N 115°30W 79 N11
Calf of Man I. of Man 54°3N 4°48W 12 C3
Calgary Canada 51°0N 114°10W 70 C6
Calheta Madeira 32°44N 17°11W 24 D2
Calhoun U.S.A. 34°30N 84°57W 85 D12
Cali Colombia 3°25N 76°35W 92 C3
Calicut India 11°15N 75°43E 40 P9
Caliente U.S.A. 37°37N 114°31W 77 H6
California Mo., U.S.A. 38°38N 92°34W 80 F7
California Pa., U.S.A. 40°4N 79°54W 82 F5
California □ U.S.A. 37°30N 119°30W 78 H7
California, Baja, T.N. = Baja
California □ Mexico 30°0N 115°0W 86 B2
California, Baja, T.S. = Baja
California Sur □
Mexico 25°50N 111°50W 86 B2
California, G. de Mexico 27°0N 111°0W 86 B2
California City U.S.A. 35°10N 117°55W 79 K9
California Hot Springs
U.S.A. 35°51N 118°41W 79 K8
Calilegua △ Argentina 25°36S 64°50W 94 B3
Calingasta Argentina 31°15S 69°30W 94 C2
Calipatria U.S.A. 33°8N 115°31W 79 M11
Calistoga U.S.A. 38°35N 122°35W 78 G4
Calitzdorp S. Africa 33°33S 21°42E 56 E3
Callabonna, L. Australia 29°40S 140°5E 63 D3
Callan Ireland 52°32N 7°24W 10 D4
Callander U.K. 56°15N 4°13W 11 E4
Callao Peru 12°3S 77°8W 92 F3
Callicoon U.S.A. 41°46N 75°3W 83 E9
Calling Lake Canada 55°15N 113°12W 70 B6
Calliope Australia 24°0S 151°16E 62 C5
Calma, Costa de la Spain 39°28N 2°26E 24 B9
Calne U.K. 51°26N 2°0W 13 F6
Calola Angola 16°25S 17°48E 56 B2
Caloundra Australia 26°45S 153°10E 63 D5
Calpella U.S.A. 39°14N 123°12W 78 F3
Calpine U.S.A. 39°40N 120°27W 78 F6
Calstock Canada 49°47N 84°9W 72 C3
Caltagirone Italy 37°14N 14°31E 22 F6
Caltanissetta Italy 37°29N 14°4E 22 F6
Calulo Angola 10°1S 14°56E 52 G2
Calvert → Australia 16°17S 137°44E 62 B2
Calvert I. Canada 51°30N 128°0W 70 C3
Calvert Ra. Australia 24°0S 122°30E 60 D3
Calvi France 42°34N 8°45E 20 E8
Calvià Spain 39°34N 2°31E 24 B9
Calvillo Mexico 21°51N 102°43W 86 C4
Calvinia S. Africa 31°28S 19°45E 56 E2
Calwa U.S.A. 36°42N 119°46W 78 J7
Cam → U.K. 52°21N 0°16E 13 E8
Cam Lam = Ba Ngoi
Vietnam 11°54N 109°10E 39 G7
Cam Pha Vietnam 21°7N 107°18E 38 B6
Cam Ranh Vietnam 11°54N 109°12E 39 G7
Cam Xuyen Vietnam 18°15N 106°0E 38 C6
Camabatela Angola 8°20S 15°26E 52 F3
Camacha Madeira 32°41N 16°49W 24 D3
Camacupa Angola 11°58S 17°22E 53 G3
Camagüey Cuba 21°20N 77°55W 88 B4
Camaná Peru 16°30S 72°50W 92 G4
Camanche Res. U.S.A. 38°14N 121°1W 78 G6
Camaquã Brazil 30°51S 51°49W 95 C5
Camaquã → Brazil 31°17S 51°47W 95 C5
Câmara de Lobos
Madeira 32°39N 16°59W 24 D3
Camargo Mexico 26°19N 98°50W 87 B5
Camarillo U.S.A. 34°13N 119°2W 79 L7
Camarón, C. Honduras 16°0N 85°5W 88 C2
Camarones Argentina 44°50S 65°40W 96 E3
Camas U.S.A. 45°35N 122°24W 78 E4
Camas Valley U.S.A. 43°2N 123°40W 76 E2
Camballin Australia 17°59S 124°12E 60 C3
Cambará Brazil 23°2S 50°5W 95 A5
Cambay = Khambhat
India 22°23N 72°33E 42 H5
Cambay, G. of = Khambhat, G. of
India 20°45N 72°30E 40 J8
Cambodia ■ Asia 12°15N 105°0E 38 F5
Camborne U.K. 50°12N 5°19W 13 G2
Cambrai France 50°11N 3°14E 20 A5
Cambria U.S.A. 35°34N 121°5W 78 K5
Cambrian Mts. U.K. 52°3N 3°57W 13 E4
Cambridge Canada 43°23N 80°15W 82 C4
Cambridge Jamaica 18°18N 77°54W 88 a
Cambridge N.Z. 37°54S 175°29E 59 B5
Cambridge U.K. 52°12N 0°8E 13 E8
Cambridge Mass., U.S.A. 42°23N 71°7W 83 D13
Cambridge Minn., U.S.A. 45°34N 93°13W 80 C7
Cambridge N.Y., U.S.A. 43°2N 73°22W 83 C11
Cambridge Nebr.,
U.S.A. 40°17N 100°10W 80 E3
Cambridge Ohio, U.S.A. 40°2N 81°35W 82 F3
Cambridge Bay Canada 69°10N 105°0W 68 D11
Cambridge G. Australia 14°55S 128°15E 60 B4
Cambridge Springs
U.S.A. 41°48N 80°4W 82 E4
Cambridgeshire □ U.K. 52°25N 0°7W 13 E7
Cambuci Brazil 21°35S 41°55W 95 A7
Cambundi-Catembo
Angola 10°10S 17°35E 52 G3
Camden Australia 34°1S 150°43E 63 B5
Camden Ala., U.S.A. 31°59N 87°17W 85 F11
Camden Ark., U.S.A. 33°35N 92°50W 84 E8
Camden Maine, U.S.A. 44°13N 69°4W 81 C19
Camden N.J., U.S.A. 39°55N 75°7W 83 G9
Camden N.Y., U.S.A. 43°20N 75°45W 83 C9
Camden S.C., U.S.A. 34°16N 80°36W 85 D14
Camden Sd. Australia 15°27S 124°25E 60 C3
Camdenton U.S.A. 38°1N 92°45W 80 F7
Camelford U.K. 50°37N 4°42W 13 G3
Cameron Ariz., U.S.A. 35°53N 111°25W 77 J8
Cameron La., U.S.A. 29°48N 93°20W 84 G8
Cameron Mo., U.S.A. 39°44N 94°14W 80 F6
Cameron Tex., U.S.A. 30°51N 96°59W 84 F6
Cameron Highlands
Malaysia 4°27N 101°22E 39 K3
Cameron Hills Canada 59°48N 118°0W 70 B5
Cameroon ■ Africa 6°0N 12°30E 51 G7
Cameroun, Mt. Cameroon 4°13N 9°10E 52 D1
Cametá Brazil 2°12S 49°30W 93 D9
Camilla U.S.A. 31°14N 84°12W 85 F12
Caminha Portugal 41°50N 8°50W 21 B1
Camino U.S.A. 38°44N 120°41W 78 G6
Cammal U.S.A. 41°24N 77°28W 82 E7
Camocim Brazil 2°55S 40°50W 93 D10
Camooweal Australia 19°56S 138°7E 62 B2
Camooweal Caves △
Australia 20°1S 138°11E 62 C2
Camopi Fr. Guiana 3°12N 52°17W 93 C8
Camp Hill U.S.A. 40°14N 76°55W 82 F8
Camp Nelson U.S.A. 36°8N 118°39W 79 J8
Camp Pendleton
U.S.A. 33°13N 117°24W 79 M9
Camp Verde U.S.A. 34°34N 111°51W 77 J8
Camp Wood U.S.A. 29°40N 100°1W 84 G4
Campana Argentina 34°10S 58°55W 94 C4
Campana, I. Chile 48°20S 75°20W 96 F1
Campanário Madeira 32°39N 17°2W 24 D2
Campanet Spain 39°46N 2°58E 24 B9
Campánia □ Italy 41°0N 14°30E 22 D6
Campbell S. Africa 28°48S 23°44E 56 D3
Campbell Calif., U.S.A. 37°17N 121°57W 78 H5
Campbell Ohio, U.S.A. 41°5N 80°37W 82 E4
Campbell I. Pac. Oc. 52°30S 169°0E 64 N8
Campbell L. Canada 63°14N 106°55W 71 A7
Campbell Plateau S. Ocean 50°0S 170°0E 5 A11
Campbell River Canada 50°5N 125°20W 70 C3
Campbellford Canada 44°18N 77°48W 82 B7
Campbellpur Pakistan 33°46N 72°26E 42 C5
Campbellsville U.S.A. 37°21N 85°20W 81 G11
Campbellton Canada 47°57N 66°43W 73 C6
Campbelltown Australia 34°4S 150°49E 63 E5
Campbeltown U.K. 55°26N 5°36W 11 F3
Campeche Mexico 19°51N 90°32W 87 D6
Campeche □ Mexico 19°0N 90°30W 87 D6
Campeche, Golfo de
Mexico 19°30N 93°0W 87 D6
Camperdown Australia 38°14S 143°9E 63 F3
Camperville Canada 51°59N 100°9W 71 C8
Câmpina Romania 45°10N 25°45E 17 F13
Campina Grande Brazil 7°20S 35°47W 93 E11
Campinas Brazil 22°50S 47°0W 95 A6
Campo Grande Brazil 20°25S 54°40W 93 H8
Campo Maior Brazil 4°50S 42°12W 93 D10
Campo Mourão Brazil 24°3S 52°22W 95 A5
Campobasso Italy 41°34N 14°39E 22 D6
Campos Brazil 21°50S 41°20W 95 A7
Campos Spain 39°26N 3°1E 24 B10
Campos Belos Brazil 13°10S 47°3W 93 F9
Campos Novos Brazil 27°21S 51°50W 95 B5
Camptonville U.S.A. 39°27N 121°3W 78 F5
Camptown U.S.A. 41°44N 76°14W 83 E8
Câmpulung Romania 45°17N 25°3E 17 F13
Camrose Canada 53°0N 112°50W 70 C6
Camsell Portage
Canada 59°37N 109°15W 71 B7
Çan Turkey 40°2N 27°3E 23 D12
Can Clavo Spain 38°57N 1°27E 24 C7
Can Creu Spain 38°58N 1°28E 24 C7
Can Gio Vietnam 10°25N 106°58E 39 G6
Can Pastilla Spain 39°32N 2°42E 24 B9
Can Tho Vietnam 10°2N 105°46E 39 G5
Canaan U.S.A. 42°2N 73°20W 83 D11
Canada ■ N. Amer. 60°0N 100°0W 68 G11
Canada Abyssal Plain
Arctic 80°0N 140°0W 4 B18
Canada Basin Arctic 80°0N 145°0W 4 B18
Cañada de Gómez
Argentina 32°40S 61°30W 94 C3
Canadian U.S.A. 35°55N 100°23W 84 D4
Canadian → U.S.A. 35°28N 95°3W 84 D7
Canadian Shield Canada 53°0N 75°0W 66 D12
Canajoharie U.S.A. 42°54N 74°35W 83 D10
Çanakkale Turkey 40°8N 26°24E 23 D12
Çanakkale Boğazı
Turkey 40°17N 26°32E 23 D12
Canal Flats Canada 50°10N 115°48W 70 C5
Canalejas Argentina 35°15S 66°34W 94 D2
Canals Argentina 33°35S 62°53W 94 C3
Canandaigua U.S.A. 42°54N 77°17W 82 D7
Canandaigua L. U.S.A. 42°47N 77°19W 82 D7
Cananea Mexico 31°0N 110°18W 86 A2
Canarias, Is. Atl. Oc. 28°30N 16°0W 24 F4
Canaries St. Lucia 13°55N 61°4W 89 f
Canarreos, Arch. de los
Cuba 21°35N 81°40W 88 B3
Canary Is. = Canarias, Is.
Atl. Oc. 28°30N 16°0W 24 F4
Canaseraga U.S.A. 42°27N 77°45W 82 D7
Canatlán Mexico 24°31N 104°47W 86 C4
Canaveral, C. U.S.A. 28°27N 80°32W 85 G14
Canaveral ⌓ U.S.A. 28°28N 80°34W 85 G14
Canavieiras Brazil 15°39S 39°0W 93 G11
Canberra Australia 35°15S 149°8E 63 F4
Canby Calif., U.S.A. 41°27N 120°52W 76 F3
Canby Minn., U.S.A. 44°43N 96°16W 80 C5
Canby Oreg., U.S.A. 45°16N 122°42W 78 E4
Cancún Mexico 21°8N 86°44W 87 C7
Candelaria Argentina 27°29S 55°44W 95 B4
Candelaria Canary Is. 28°22N 16°22W 24 F3
Candelo Australia 36°47S 149°43E 63 F4
Candi Dasa Indonesia 8°30S 115°34E 37 J18
Candia = Iraklio Greece 35°20N 25°12E 25 D7
Candle L. Canada 53°50N 105°18W 71 C7
Candlemas I. Antarctica 57°3S 26°40W 5 B1
Cando U.S.A. 48°32N 99°12W 80 A4
Canea = Chania Greece 35°30N 24°4E 25 D6
Canelones Uruguay 34°32S 56°17W 95 C4
Cañete Chile 37°50S 73°30W 94 D1
Cangas del Narcea Spain 43°10N 6°32W 21 A2
Cangshan China 34°50N 118°2E 35 G10
Canguaretama Brazil 6°20S 35°5W 93 E11
Canguçu Brazil 31°22S 52°43W 95 C5
Canguçu, Serra do
Brazil 31°20S 52°40W 95 C5
Cangzhou China 38°19N 116°52E 34 E9
Caniapiscau → Canada 56°40N 69°30W 73 A6
Caniapiscau, L. Canada 54°10N 69°55W 73 B6
Canicattì Italy 37°21N 13°51E 22 F5
Canim Lake Canada 51°47N 120°54W 70 C4
Canindeyu □ Paraguay 24°10S 55°0W 95 A5
Canisteo U.S.A. 42°16N 77°36W 82 D7
Canisteo → U.S.A. 42°7N 77°8W 82 D7
Cañitas de Felipe Pescador
Mexico 23°36N 102°43W 86 C4
Çankırı Turkey 40°40N 33°37E 19 F5
Cankuzo Burundi 3°10S 30°31E 54 C3
Canmore Canada 51°7N 115°18W 70 C5
Cann River Australia 37°35S 149°7E 63 F4
Canna U.K. 57°3N 6°33W 11 D2
Cannanore India 11°53N 75°27E 40 P9
Cannes France 43°32N 7°1E 20 E7
Canning Town = Port Canning
India 22°23N 88°40E 43 H13
Cannington Canada 44°20N 79°2W 82 B5
Cannock U.K. 52°41N 2°1W 13 E5
Cannonball → U.S.A. 46°26N 100°35W 80 B3
Cannondale Mt.
Australia 25°13S 148°57E 62 D4
Cannonsville Res. U.S.A. 42°4N 75°22W 83 D9
Cannonvale Australia 20°17S 148°43E 62 b
Canoas Brazil 29°56S 51°11W 95 B5
Canoe L. Canada 55°10N 108°15W 71 B7
Cañon City U.S.A. 38°27N 105°14W 76 G11
Cañón de Río Blanco △
Mexico 18°43N 97°15W 87 D5
Cañón del Sumidero △
Mexico 19°22N 96°24W 87 D5
Canonniers Pt. Mauritius 20°2S 57°32E 53 d
Canora Canada 51°40N 102°30W 71 C8
Canowindra Australia 33°35S 148°38E 63 E4
Canso Canada 45°20N 61°0W 73 C7
Cantabria □ Spain 43°10N 4°0W 21 A4
Cantabrian Mts. = Cantábrica,
Cordillera Spain 43°0N 5°10W 21 A3
Cantábrica, Cordillera
Spain 43°0N 5°10W 21 A3
Cantal, Plomb du France 45°3N 2°45E 20 D5
Canterbury Australia 25°23S 141°53E 62 D3
Canterbury U.K. 51°16N 1°6E 13 F9
Canterbury Bight N.Z. 44°16S 171°55E 59 F3
Canterbury Plains N.Z. 43°55S 171°22E 59 E3
Cantil U.S.A. 35°18N 117°58W 79 K9
Canton = Guangzhou
China 23°6N 113°13E 33 G11
Canton Ga., U.S.A. 34°14N 84°29W 85 D12
Canton Ill., U.S.A. 40°33N 90°2W 80 E8
Canton Miss., U.S.A. 32°37N 90°2W 85 E9
Canton Mo., U.S.A. 40°8N 91°32W 80 E8
Canton N.Y., U.S.A. 44°36N 75°10W 83 B9
Canton Ohio, U.S.A. 40°48N 81°23W 82 F3
Canton Pa., U.S.A. 41°39N 76°51W 82 E8
Canton S. Dak., U.S.A. 43°18N 96°35W 80 D5
Canton L. U.S.A. 36°6N 98°35W 84 C5
Canudos Brazil 7°13S 58°5W 92 E7
Canumã → Brazil 3°55S 59°10W 92 D7
Canutama Brazil 6°30S 64°20W 92 E6
Canutillo U.S.A. 31°55N 106°36W 84 F1
Canvey U.K. 51°31N 0°37E 13 F8
Canyon U.S.A. 34°59N 101°55W 84 D4
Canyon De Chelly ⌓
U.S.A. 36°10N 109°20W 77 H9
Canyonlands △ U.S.A. 38°15N 110°0W 77 G9
Canyons of the Ancients ⌓
U.S.A. 37°30N 108°55W 77 H9
Canyonville U.S.A. 42°56N 123°17W 76 E2
Cao Bang Vietnam 22°40N 106°15E 38 A6
Cao He → China 40°10N 124°32E 35 D13
Cao Lanh Vietnam 10°27N 105°38E 39 G5
Cao Xian China 34°50N 115°35E 34 G8
Cap-aux-Meules Canada 47°23N 61°52W 73 C7
Cap-Chat Canada 49°6N 66°40W 73 C6
Cap-de-la-Madeleine
Canada 46°22N 72°31W 72 C5
Cap-Haïtien Haiti 19°40N 72°20W 89 C5
Cap Pt. St. Lucia 14°7N 60°57W 89 f
Capac U.S.A. 43°1N 82°56W 82 C2
Capanaparo → Venezuela 7°1N 67°7W 92 B5
Cape → Australia 20°59S 146°51E 62 C4
Cape Arid △ Australia 33°58S 123°13E 61 F3
Cape Barren I. Australia 40°25S 148°15E 63 G4
Cape Breton Highlands △
Canada 46°50N 60°40W 73 C7
Cape Breton I. Canada 46°0N 60°30W 73 C7
Cape Charles U.S.A. 37°16N 76°1W 81 G15
Cape Coast Ghana 5°5N 1°15W 50 G5
Cape Cod ⌓ U.S.A. 41°56N 70°6W 81 E18
Cape Coral U.S.A. 26°33N 81°57W 85 H14
Cape Crawford Australia 16°41S 135°43E 62 B2
Cape Dorset = Kinngait
Canada 64°14N 76°32W 69 E16
Cape Fear → U.S.A. 33°53N 78°1W 85 E15
Cape Girardeau U.S.A. 37°19N 89°32W 80 G9
Cape Hatteras ⌓
U.S.A. 35°30N 75°28W 85 D17
Cape Hillsborough △
Australia 20°54S 149°2E 62 b
Cape Le Grand △
Australia 33°54S 122°26E 61 F3
Cape Lookout ⌓
U.S.A. 35°45N 76°25W 85 D16
Cape May U.S.A. 38°56N 74°56W 81 F16
Cape May Point U.S.A. 38°56N 74°58W 81 F16
Cape Melville △
Australia 14°26S 144°28E 62 A3
Cape Range △ Australia 22°3S 114°0E 60 D1
Cape St. George Canada 48°28N 59°14W 73 C8
Cape Tormentine Canada 46°8N 63°47W 73 C7
Cape Town S. Africa 33°55S 18°22E 56 E2
Cape Upstart △
Australia 19°45S 147°48E 62 B4
Cape Verde Is. ■ Atl. Oc. 16°0N 24°0W 50 b
Cape Vincent U.S.A. 44°8N 76°20W 83 B8
Cape York Peninsula
Australia 12°0S 142°30E 62 A3
Capela Brazil 10°30S 37°0W 93 F11
Capella Australia 23°2S 148°1E 62 C4
Capesterre-Belle-Eau
Guadeloupe 16°4N 61°36W 88 b
Capesterre-de-Marie-Galante
Guadeloupe 15°53N 61°14W 88 b
Capim → Brazil 1°40S 47°47W 93 D9
Capitan U.S.A. 33°35N 105°35W 77 K11
Capitán Arturo Prat
Antarctica 63°0S 61°0W 5 C17
Capitol Reef △ U.S.A. 38°15N 111°10W 77 G8
Capitola U.S.A. 36°59N 121°57W 78 J5
Capoche → Mozam. 15°35S 33°0E 55 F3
Capraia Italy 43°2N 9°50E 20 E8
Capri Italy 40°33N 14°14E 22 D6
Capricorn Coast
Australia 23°16S 150°49E 62 C5
Capricorn Group
Australia 23°30S 151°55E 62 C5
Capricorn Ra. Australia 23°20S 116°50E 60 D2
Caprivi ⌓ Namibia 17°55S 22°37E 56 B3
Caprivi Strip Namibia 18°0S 23°0E 56 B3
Captains Flat Australia 35°35S 149°27E 63 F4
Capulin Volcano ⌓
U.S.A. 36°47N 103°58W 84 C3
Caquetá → Colombia 1°15S 69°15W 92 D5
Caracal Romania 44°8N 24°22E 17 F13
Caracas Venezuela 10°30N 66°55W 92 A5
Caracol Belize 16°45N 89°6W 88 C2
Caracol Mato Grosso do Sul,
Brazil 22°18S 57°1W 94 A4
Caracol Piauí, Brazil 9°15S 43°22W 93 E10
Carajas Brazil 6°5S 50°23W 93 E8
Carajás, Serra dos Brazil 6°0S 51°30W 93 E8
Carangola Brazil 20°44S 42°5W 95 A7
Caransebeş Romania 45°28N 22°18E 17 F12
Caraquet Canada 47°48N 64°57W 73 C6
Caratasca, L. de
Honduras 15°20N 83°40W 88 C3
Caratinga Brazil 19°50S 42°10W 93 G10
Caraúbas Brazil 5°43S 37°33W 93 E11
Caravaca de la Cruz Spain 38°8N 1°52W 21 C5
Caravelas Brazil 17°45S 39°15W 93 G11
Caraveli Peru 15°45S 73°25W 92 G4
Caravelle, Presqu'île de la
Martinique 14°46N 60°48W 88 c
Caraz Peru 9°3S 77°47W 92 E3
Carazinho Brazil 28°16S 52°46W 95 B5
Carballo Spain 43°13N 8°41W 21 A1
Carberry Canada 49°50N 99°25W 71 D9
Carbó Mexico 29°42N 110°58W 86 B2
Carbón, L. del Argentina 49°35S 68°21W 90 G3
Carbonara, C. Italy 39°6N 9°31E 22 E3
Carbondale Colo.,
U.S.A. 39°24N 107°13W 76 G10
Carbondale Ill., U.S.A. 37°44N 89°13W 80 G9
Carbondale Pa., U.S.A. 41°35N 75°30W 83 E9
Carbonear Canada 47°42N 53°13W 73 C9
Carbónia Italy 39°10N 8°30E 22 E3
Carcajou Canada 57°47N 117°6W 70 B5
Carcarana → Argentina 32°27S 60°48W 94 C3
Carcasse, C. Haiti 18°30N 74°28W 89 C5
Carcassonne France 43°13N 2°20E 20 E5
Carcross Canada 60°13N 134°45W 70 A2
Cardamon Hills India 9°30N 77°15E 40 Q10
Cardamon Mts. = Kravanh,
Chuor Phnum
Cambodia 12°0N 103°32E 39 G4
Cárdenas Cuba 23°0N 81°30W 88 B3
Cárdenas San Luis Potosí,
Mexico 22°0N 99°38W 87 C5
Cárdenas Tabasco,
Mexico 17°59N 93°22W 87 D6
Cardiff U.K. 51°29N 3°10W 13 F4
Cardiff □ U.K. 51°31N 3°12W 13 F4
Cardiff-by-the-Sea
U.S.A. 33°1N 117°17W 79 M9
Cardigan U.K. 52°5N 4°40W 13 E3
Cardigan B. U.K. 52°30N 4°30W 13 E3
Cardinal Canada 44°47N 75°23W 83 B9
Cardona Uruguay 33°53S 57°18W 94 C4
Cardoso, Ilha do Brazil 25°8S 47°58W 95 B5
Cardston Canada 49°15N 113°20W 70 D6
Cardwell Australia 18°14S 146°2E 62 B4
Careen L. Canada 57°0N 108°11W 71 B7
Carei Romania 47°40N 22°29E 17 E12
Carey U.S.A. 43°19N 113°57W 76 E7
Carey, L. Australia 29°0S 122°15E 61 E3
Carey L. Canada 62°12N 102°55W 71 A8
Carhué Argentina 37°10S 62°50W 94 D3
Caria Turkey 37°20N 28°10E 23 F13
Cariacica Brazil 20°16S 40°25W 93 H10
Caribbean Sea W. Indies 15°0N 75°0W 89 D5
Cariboo Mts. Canada 53°0N 121°0W 70 C4
Caribou U.S.A. 46°52N 68°1W 81 B19
Caribou → Man.,
Canada 59°20N 94°44W 71 B10
Caribou → N.W.T.,
Canada 61°27N 125°45W 70 A3
Caribou I. Canada 47°22N 85°49W 72 C2
Caribou Is. Canada 61°55N 113°15W 70 A6
Caribou L. Man., Canada 59°21N 96°10W 71 B9
Caribou L. Ont., Canada 50°25N 89°5W 72 B2
Caribou Mts. Canada 59°12N 115°40W 70 B5

Caribou River ⌓ *Canada* 59°35N 96°35W **71** B9
Carichic *Mexico* 27°56N 107°3W **86** B3
Carinda *Australia* 30°28S 147°41E **63** E4
Carinhanha *Brazil* 14°15S 44°46W **93** F10
Carinhanha ➜ *Brazil* 14°20S 43°47W **93** F10
Carinthia = Kärnten □ *Austria* 46°52N 13°30E **16** E8
Caripito *Venezuela* 10°8N 63°6W **92** A6
Carleton, Mt. *Canada* 47°23N 66°53W **73** C6
Carleton Place *Canada* 45°8N 76°9W **83** A8
Carlin *U.S.A.* 40°43N 116°7W **76** F5
Carlingford L. *U.K.* 54°3N 6°9W **10** B5
Carlinville *U.S.A.* 39°17N 89°53W **80** F9
Carlisle *U.K.* 54°54N 2°56W **12** C5
Carlisle *U.S.A.* 40°12N 77°12W **82** F7
Carlisle B. *Barbados* 13°5N 59°37W **89** g
Carlisle I. *Australia* 20°49S 149°18E **62** b
Carlos Casares *Argentina* 35°32S 61°20W **94** D3
Carlos Tejedor *Argentina* 35°25S 62°25W **94** D3
Carlow *Ireland* 52°50N 6°56W **10** D5
Carlow □ *Ireland* 52°43N 6°50W **10** D5
Carlsbad *Calif., U.S.A.* 33°10N 117°21W **79** M9
Carlsbad *N. Mex., U.S.A.* 32°25N 104°14W **77** K11
Carlsbad Caverns △ *U.S.A.* 32°10N 104°35W **77** K11
Carluke *U.K.* 55°45N 3°50W **11** F5
Carlyle *Canada* 49°40N 102°20W **71** D8
Carmacks *Canada* 62°5N 136°16W **68** E4
Carman *Canada* 49°30N 98°0W **71** D9
Carmarthen *U.K.* 51°52N 4°19W **13** F3
Carmarthen B. *U.K.* 51°40N 4°30W **13** F3
Carmarthenshire □ *U.K.* 51°55N 4°13W **13** F3
Carmaux *France* 44°3N 2°10E **20** D5
Carmel *U.S.A.* 41°26N 73°41W **83** E11
Carmel-by-the-Sea *U.S.A.* 36°33N 121°55W **78** J5
Carmel Valley *U.S.A.* 36°29N 121°43W **78** J5
Carmelo *Uruguay* 34°0S 58°20W **94** C4
Carmen *Paraguay* 27°13S 56°12W **95** B4
Carmen ➜ *Mexico* 30°42N 106°29W **86** A3
Carmen, I. *Mexico* 25°57N 111°12W **86** B2
Carmen de Patagones *Argentina* 40°50S 63°0W **96** E4
Carmensa *Argentina* 35°15S 67°40W **94** D2
Carmi *Canada* 49°36N 119°8W **70** D5
Carmi *U.S.A.* 38°5N 88°10W **80** F9
Carmichael *U.S.A.* 38°38N 121°19W **78** G5
Carmila *Australia* 21°55S 149°24E **62** C4
Carmona *Costa Rica* 10°0N 85°15W **88** E2
Carmona *Spain* 37°28N 5°42W **21** D3
Carn Ban *U.K.* 57°7N 4°15W **11** D4
Carn Eige *U.K.* 57°17N 5°8W **11** D3
Carnamah *Australia* 29°41S 115°53E **61** E2
Carnarvon *Australia* 24°51S 113°42E **61** D1
Carnarvon *Canada* 45°3N 78°41W **82** A6
Carnarvon *S. Africa* 30°56S 22°8E **56** E3
Carnarvon △ *Australia* 24°54S 148°2E **62** C4
Carnarvon Ra. *Queens., Australia* 25°15S 148°30E **62** D4
Carnarvon Ra. *W. Austral., Australia* 25°20S 120°45E **61** E3
Carnation *U.S.A.* 47°39N 121°55W **78** C5
Carncastle *U.K.* 54°54N 5°53W **10** B6
Carndonagh *Ireland* 55°16N 7°15W **10** A4
Carnduff *Canada* 49°10N 101°50W **71** D8
Carnegie *U.S.A.* 40°24N 80°5W **82** F4
Carnegie, L. *Australia* 26°5S 122°30E **61** E3
Carnegie Ridge *Pac. Oc.* 1°0S 87°0W **65** H19
Carnic Alps = Karnische Alpen *Europe* 46°36N 13°0E **16** E7
Carniche Alpi = Karnische Alpen *Europe* 46°36N 13°0E **16** E7
Carnot *C.A.R.* 4°59N 15°56E **52** D3
Carnot, C. *Australia* 34°57S 135°38E **63** E2
Carnot B. *Australia* 17°20S 122°15E **60** C3
Carnoustie *U.K.* 56°30N 2°42W **11** E6
Carnsore Pt. *Ireland* 52°10N 6°22W **10** D5
Caro *U.S.A.* 43°29N 83°24W **81** D12
Caroga Lake *U.S.A.* 43°8N 74°28W **83** C10
Carol City *U.S.A.* 25°56N 80°14W **85** J14
Carolina *Brazil* 7°10S 47°30W **93** E9
Carolina *Puerto Rico* 18°23N 65°58W **89** d
Carolina *S. Africa* 26°5S 30°6E **57** D5
Caroline I. *Kiribati* 9°58S 150°13W **65** H12
Caroline Is. *Micronesia* 8°0N 150°0E **64** G6
Caroni *Trin. & Tob.* 10°34N 61°23W **93** K15
Caroní ➜ *Venezuela* 8°21N 62°43W **92** B6
Caronie = Nébrodi, Monti *Italy* 37°54N 14°35E **22** F6
Caroona *Australia* 31°24S 150°26E **63** E5
Carpathians *Europe* 49°30N 21°0E **17** D11
Carpații Meridionali *Romania* 45°30N 25°0E **17** F13
Carpentaria, G. of *Australia* 14°0S 139°0E **62** A2
Carpentras *France* 44°3N 5°2E **20** D6
Carpi *Italy* 44°47N 10°53E **22** B4
Carpinteria *U.S.A.* 34°24N 119°31W **79** L7
Carr Boyd Ra. *Australia* 16°15S 128°35E **60** C4
Carra, L. *Ireland* 53°41N 9°14W **10** C2
Carrabelle *U.S.A.* 29°51N 84°40W **85** G12
Carraig Mhachdire Rois = Carrickmacross *Ireland* 53°59N 6°43W **10** C5
Carraig na Siúire = Carrick-on-Suir *Ireland* 52°21N 7°24W **10** D4
Carranya ☼ *Australia* 19°13S 127°46E **60** C4
Carranza, Presa V. *Mexico* 27°20N 100°50W **86** B4
Carrara *Italy* 44°5N 10°6E **20** D9
Carrauntoohill *Ireland* 52°0N 9°45W **10** D2
Carrick-on-Shannon *Ireland* 53°57N 8°5W **10** C3
Carrick-on-Suir *Ireland* 52°21N 7°24W **10** D4
Carrickfergus *U.K.* 54°43N 5°49W **10** B6
Carrickmacross *Ireland* 53°59N 6°43W **10** C5
Carrieton *Australia* 32°25S 138°31E **63** E2
Carrillo *Mexico* 26°54N 103°55W **86** B4
Carrington *U.S.A.* 47°27N 99°8W **80** B4
Carrizal Bajo *Chile* 28°5S 71°20W **94** B1
Carrizalillo *Chile* 29°5S 71°30W **94** B1
Carrizo Cr. ➜ *U.S.A.* 36°55N 103°55W **77** H12
Carrizo Plain ⌓ *U.S.A.* 35°11N 119°47W **78** K7
Carrizo Springs *U.S.A.* 28°31N 99°52W **84** G5
Carrizozo *U.S.A.* 33°38N 105°53W **77** K11
Carroll *U.S.A.* 42°4N 94°52W **80** D6
Carrollton *Ga., U.S.A.* 33°35N 85°5W **85** E12
Carrollton *Ill., U.S.A.* 39°18N 90°24W **80** F8
Carrollton *Ky., U.S.A.* 38°41N 85°11W **81** F11
Carrollton *Mo., U.S.A.* 39°22N 93°30W **80** F7
Carrollton *Ohio, U.S.A.* 40°34N 81°5W **82** F3
Carron ➜ *U.K.* 57°53N 4°22W **11** D4
Carron, L. *U.K.* 57°22N 5°35W **11** D3
Carrot ➜ *Canada* 53°50N 101°17W **71** C8
Carrot River *Canada* 53°17N 103°35W **71** C8
Carruthers *Canada* 52°52N 109°16W **71** C7
Carson *Calif., U.S.A.* 33°49N 118°16W **79** M8
Carson *N. Dak., U.S.A.* 46°25N 101°34W **80** B3
Carson ➜ *U.S.A.* 39°45N 118°40W **78** F8
Carson City *U.S.A.* 39°10N 119°46W **78** F7
Carson Sink *U.S.A.* 39°50N 118°25W **76** G4
Cartagena *Colombia* 10°25N 75°33W **92** A3
Cartagena *Spain* 37°38N 0°59W **21** D5
Cartago *Colombia* 4°45N 75°55W **92** C3
Cartago *Costa Rica* 9°50N 83°55W **88** E3
Cartersville *U.S.A.* 34°10N 84°48W **85** D12
Carterton *N.Z.* 41°2S 175°31E **59** D5
Carthage *Tunisia* 36°52N 10°20E **22** F4
Carthage *Ill., U.S.A.* 40°25N 91°8W **80** E8
Carthage *Mo., U.S.A.* 37°11N 94°19W **80** G6
Carthage *N.Y., U.S.A.* 43°59N 75°37W **81** D16
Carthage *Tex., U.S.A.* 32°9N 94°20W **84** E7
Cartier I. *Australia* 12°31S 123°29E **60** B3
Cartwright *Canada* 53°41N 56°58W **73** B8
Caruaru *Brazil* 8°15S 35°55W **93** E11
Carúpano *Venezuela* 10°39N 63°15W **92** A6
Caruthersville *U.S.A.* 36°11N 89°39W **80** G9
Carvoeiro *Brazil* 1°30S 61°59W **92** D6
Carvoeiro, C. *Portugal* 39°21N 9°24W **21** C1
Cary *U.S.A.* 35°47N 78°46W **85** D15
Casa de Piedra *Argentina* 38°5S 67°28W **94** D2
Casa de Piedra, Embalse *Argentina* 38°5S 67°32W **94** D2
Casa Grande *U.S.A.* 32°53N 111°45W **77** K8
Casa Nova *Brazil* 9°25S 41°5W **93** E10
Casablanca *Chile* 33°20S 71°25W **94** C1
Casablanca *Morocco* 33°36N 7°36W **50** B4
Cascada de Basaseachic △ *Mexico* 28°9N 108°15W **86** B3
Cascade *Seychelles* 4°39S 55°29E **53** b
Cascade *Idaho, U.S.A.* 44°31N 116°2W **76** D5
Cascade *Mont., U.S.A.* 47°16N 111°42W **76** C8
Cascade Locks *U.S.A.* 45°40N 121°54W **78** E5
Cascade Ra. *U.S.A.* 47°0N 121°30W **78** D5
Cascade Res. *U.S.A.* 44°32N 116°3W **76** D5
Cascades, Pte. des *Réunion* 21°9S 55°51E **53** c
Cascais *Portugal* 38°41N 9°25W **21** C1
Cascavel *Brazil* 24°57S 53°28W **95** A5
Cáscina *Italy* 43°41N 10°33E **22** C4
Casco B. *U.S.A.* 43°45N 70°0W **81** D19
Caserta *Italy* 41°4N 14°20E **22** D6
Casey *Antarctica* 66°0S 76°0E **5** C8
Caseyr, Raas = Asir, Ras *Somali Rep.* 11°55N 51°10E **47** E5
Cashel *Ireland* 52°30N 7°53W **10** D4
Casiguran *Phil.* 16°22N 122°7E **37** A6
Casilda *Argentina* 33°10S 61°10W **94** C3
Casino *Australia* 28°52S 153°3E **63** D5
Casiquiare ➜ *Venezuela* 2°1N 67°7W **92** C5
Casma *Peru* 9°30S 78°20W **92** E3
Casmalia *U.S.A.* 34°50N 120°32W **79** L6
Caspe *Spain* 41°14N 0°1W **21** B5
Casper *U.S.A.* 42°51N 106°19W **76** E10
Caspian Depression *Eurasia* 47°0N 48°0E **19** E8
Caspian Sea *Eurasia* 43°0N 50°0E **19** F9
Cass City *U.S.A.* 43°36N 83°10W **82** C1
Cass Lake *U.S.A.* 47°23N 94°37W **80** B6
Cassadaga *U.S.A.* 42°20N 79°19W **82** D5
Casselman *Canada* 45°19N 75°5W **83** A9
Casselton *U.S.A.* 46°54N 97°13W **80** B5
Cassiar Mts. *Canada* 59°30N 130°30W **70** B2
Cassino *Italy* 41°30N 13°49E **22** D5
Cassville *U.S.A.* 36°41N 93°52W **80** G7
Castaic *U.S.A.* 34°30N 118°38W **79** L8
Castalia *U.S.A.* 41°24N 82°49W **82** E2
Castanhal *Brazil* 1°18S 47°55W **93** D9
Castara *Trin. & Tob.* 11°17N 60°42W **93** J16
Castellammare di Stábia *Italy* 40°42N 14°29E **22** D6
Castelli *Argentina* 36°7S 57°47W **94** D4
Castelló de la Plana *Spain* 39°58N 0°3W **21** C5
Castelo *Brazil* 20°33S 41°14W **95** A7
Castelo Branco *Portugal* 39°50N 7°31W **21** C2
Castelsarrasin *France* 44°2N 1°7E **20** E4
Castelvetrano *Italy* 37°41N 12°47E **22** F5
Casterton *Australia* 37°30S 141°30E **63** F3
Castile *U.S.A.* 42°38N 78°3W **82** D6
Castilla-La Mancha □ *Spain* 39°30N 3°30W **21** C4
Castilla y Leon □ *Spain* 42°0N 5°0W **21** B3
Castillos *Uruguay* 34°12S 53°52W **95** C5
Castle Dale *U.S.A.* 39°13N 111°1W **76** G8
Castle Douglas *U.K.* 54°56N 3°56W **11** G5
Castle Rock *Colo., U.S.A.* 39°22N 104°51W **76** G11
Castle Rock *Wash., U.S.A.* 46°17N 122°54W **78** D4
Castlebar *Ireland* 53°52N 9°18W **10** C2
Castlebay *U.K.* 56°57N 7°31W **11** E1
Castleblaney *Ireland* 54°7N 6°44W **10** B5
Castlederg *U.K.* 54°42N 7°35W **10** B4
Castleford *U.K.* 53°43N 1°21W **12** D6
Castlegar *Canada* 49°20N 117°40W **70** D5
Castlemaine *Australia* 37°2S 144°12E **63** F3
Castlemaine *Ireland* 52°10N 9°42W **10** D2
Castlepollard *Ireland* 53°41N 7°19W **10** C4
Castlerea *Ireland* 53°46N 8°29W **10** C3
Castlereagh ➜ *Australia* 30°12S 147°32E **63** E4
Castlereagh B. *Australia* 12°10S 135°10E **62** A2
Castleton *U.S.A.* 43°37N 73°11W **83** C11
Castleton-on-Hudson *U.S.A.* 42°31N 73°45W **83** D11
Castletown *I. of Man* 54°5N 4°38W **12** C3
Castletown Bearhaven *Ireland* 51°39N 9°55W **10** E2
Castor *Canada* 52°15N 111°50W **70** C6
Castor ➜ *Canada* 53°24N 78°58W **72** B4
Castorland *U.S.A.* 43°53N 75°31W **83** C9
Castres *France* 43°37N 2°13E **20** E5
Castricum *Neths.* 52°33N 4°40E **15** B4
Castries *St. Lucia* 14°2N 60°58W **89** f
Castro *Brazil* 24°45S 50°0W **95** A6
Castro *Chile* 42°30S 73°50W **96** E2
Castro Alves *Brazil* 12°46S 39°33W **93** F11
Castro Valley *U.S.A.* 37°41N 122°5W **78** H4
Castroville *U.S.A.* 36°46N 121°45W **78** J5
Castuera *Spain* 38°43N 5°37W **21** C3
Cat Ba, Dao *Vietnam* 20°50N 107°0E **38** B6
Cat Ba △ *Vietnam* 20°47N 107°3E **38** B6
Cat I. *Bahamas* 24°30N 75°30W **89** B4
Cat L. *Canada* 51°40N 91°50W **72** B1
Cat Lake *Canada* 51°40N 91°50W **72** B1
Cat Tien △ *Vietnam* 11°25N 107°17E **39** G6
Catacamas *Honduras* 14°54N 85°56W **88** D2
Cataguases *Brazil* 21°23S 42°39W **95** A7
Catalão *Brazil* 18°10S 47°57W **93** G9
Çatalca *Turkey* 41°8N 28°27E **23** D13
Catalina *Canada* 48°31N 53°4W **73** C9
Catalina *Chile* 25°13S 69°43W **94** B2
Catalina *U.S.A.* 32°30N 110°50W **77** K8
Catalonia = Cataluña □ *Spain* 41°40N 1°15E **21** B6
Cataluña □ *Spain* 41°40N 1°15E **21** B6
Catamarca *Argentina* 28°30S 65°50W **94** B2
Catamarca □ *Argentina* 27°0S 65°50W **94** B2
Catanduanes □ *Phil.* 13°50N 124°20E **37** B6
Catanduva *Brazil* 21°5S 48°58W **95** A6
Catánia *Italy* 37°30N 15°6E **22** F6
Catanzaro *Italy* 38°54N 16°35E **22** E7
Catarman *Phil.* 12°28N 124°35E **37** B6
Catatumbo-Bari △ *Colombia* 9°3N 73°12W **89** E5
Catchacoma *Canada* 44°44N 78°19W **82** B6
Cateel *Phil.* 7°47N 126°24E **37** C7
Catembe *Mozam.* 26°0S 32°33E **57** D5
Caterham *U.K.* 51°15N 0°4W **13** F7
Cathair na Mart = Westport *Ireland* 53°48N 9°31W **10** C2
Cathair Saidhbhín = Cahersiveen *Ireland* 51°56N 10°14W **10** E1
Cathcart *S. Africa* 32°18S 27°10E **56** E4
Cathedral City *U.S.A.* 33°47N 116°28W **79** M10
Cathlamet *U.S.A.* 46°12N 123°23W **78** D3
Catlettsburg *U.S.A.* 38°25N 82°36W **81** F12
Catoche, C. *Mexico* 21°35N 87°5W **87** C7
Catriló *Argentina* 36°26S 63°24W **94** D3
Catrimani *Brazil* 0°27N 61°41W **92** C6
Catrimani ➜ *Brazil* 0°28N 61°44W **92** C6
Catskill *U.S.A.* 42°14N 73°52W **83** D11
Catskill ⌓ *U.S.A.* 42°8N 74°39W **83** D10
Catskill Mts. *U.S.A.* 42°10N 74°25W **83** D10
Catt, Mt. *Australia* 13°49S 134°23E **62** A1
Cattaraugus *U.S.A.* 42°20N 78°52W **82** D6
Catterick *U.K.* 54°23N 1°37W **12** C6
Catuala *Angola* 16°25S 19°2E **56** B2
Catuane *Mozam.* 26°48S 32°18E **57** D5
Catur *Mozam.* 13°45S 35°30E **55** E4
Caubvick, Mt. = Iberville, Mt. d' *Canada* 58°50N 63°50W **69** F19
Cauca ➜ *Colombia* 8°54N 74°28W **92** B4
Caucaia *Brazil* 3°40S 38°35W **93** D11
Caucasus Mountains *Eurasia* 42°50N 44°0E **19** F7
Caucete *Argentina* 31°38S 68°20W **94** C2
Cauese Mts. *Mozam.* 15°14S 30°33E **55** F3
Caungula *Angola* 8°26S 18°38E **52** F3
Cauquenes *Chile* 36°0S 72°22W **94** D1
Caura ➜ *Venezuela* 7°38N 64°53W **92** B6
Cauresi ➜ *Mozam.* 17°8S 33°0E **55** F3
Causapscal *Canada* 48°19N 67°12W **73** C6
Cauvery ➜ *India* 11°9N 78°52E **40** P11
Caux, Pays de *France* 49°38N 0°35E **20** B4
Cavalier *U.S.A.* 48°48N 97°37W **80** A5
Cavan *Ireland* 54°0N 7°22W **10** B4
Cavan □ *Ireland* 54°1N 7°16W **10** C4
Cave Creek *U.S.A.* 33°50N 111°57W **77** K8
Cavenagh Ra. *Australia* 26°12S 127°55E **61** E4
Cavendish *Australia* 37°31S 142°2E **63** F3
Caviana, I. *Brazil* 0°10N 50°10W **93** C8
Cavite *Phil.* 14°29N 120°54E **37** B6
Cawndilla L. *Australia* 32°30S 142°15E **63** E3
Cawnpore = Kanpur *India* 26°28N 80°20E **43** F9
Caxias *Brazil* 4°55S 43°20W **93** D10
Caxias do Sul *Brazil* 29°10S 51°10W **95** B5
Cay Sal Bank *Bahamas* 23°45N 80°0W **88** B4
Cayambe *Ecuador* 0°3N 78°8W **92** C3
Cayenne *Fr. Guiana* 5°5N 52°18W **93** B8
Cayey *Puerto Rico* 18°7N 66°10W **89** d
Cayman Brac *Cayman Is.* 19°43N 79°49W **88** C4
Cayman Is. ☒ *W. Indies* 19°40N 80°30W **88** C3
Cayman Trench *Caribbean* 17°0N 83°0W **66** H11
Cayuga *Canada* 42°59N 79°50W **82** D5
Cayuga *U.S.A.* 42°54N 76°44W **83** D8
Cayuga Heights *U.S.A.* 42°27N 76°29W **83** D8
Cayuga L. *U.S.A.* 42°41N 76°41W **83** D8
Cazenovia *U.S.A.* 42°56N 75°51W **83** D9
Cazombo *Angola* 11°54S 22°56E **53** G4
Ceanannus Mor *Ireland* 53°44N 6°53W **10** C5
Ceará = Fortaleza *Brazil* 3°45S 38°35W **93** D11
Ceará □ *Brazil* 5°0S 40°0W **93** E11
Ceará-Mirim *Brazil* 5°38S 35°25W **93** E11
Ceatharlach = Carlow *Ireland* 52°50N 6°56W **10** D5
Cébaco, I. de *Panama* 7°33N 81°9W **88** E3
Cebollar *Argentina* 29°10S 66°35W **94** B2
Cebu *Phil.* 10°18N 123°54E **37** B6
Cecil Plains *Australia* 27°30S 151°11E **63** D5
Cedar ➜ *U.S.A.* 41°17N 91°21W **80** E8
Cedar City *U.S.A.* 37°41N 113°4W **77** H7
Cedar Creek Res. *U.S.A.* 32°11N 96°4W **84** E6
Cedar Falls *Iowa, U.S.A.* 42°32N 92°27W **80** D7
Cedar Falls *Wash., U.S.A.* 47°25N 121°45W **78** C5
Cedar Key *U.S.A.* 29°8N 83°2W **85** G13
Cedar L. *Canada* 53°10N 100°0W **71** C9
Cedar Park *U.S.A.* 30°30N 97°49W **84** F6
Cedar Rapids *U.S.A.* 41°59N 91°40W **80** E8
Cedartown *U.S.A.* 34°1N 85°15W **85** D12
Cedarvale *Canada* 55°1N 128°22W **70** B3
Cedarville *S. Africa* 30°23S 29°3E **57** E4
Cedral *Mexico* 23°50N 100°45W **86** C4
Cedro *Brazil* 6°34S 39°3W **93** E11
Cedros, I. *Mexico* 28°12N 115°15W **86** B1
Cedros B. *Trin. & Tob.* 10°16N 61°54W **93** K15
Ceduna *Australia* 32°7S 133°46E **63** E1
Ceel Gaal *Somali Rep.* 10°58N 50°20E **47** E5
Ceel Waaq *Kenya* 2°49N 40°56E **52** D8
Ceeldheere *Somali Rep.* 3°50N 47°8E **47** G4
Ceerigaabo *Somali Rep.* 10°35N 47°20E **47** E4
Cefalù *Italy* 38°2N 14°1E **22** E6
Cegléd *Hungary* 47°11N 19°47E **17** E10
Cekik *Indonesia* 8°12S 114°27E **37** J17
Celaque △ *Honduras* 14°30N 88°43W **88** D2
Celaya *Mexico* 20°31N 100°37W **86** C4
Celebes = Sulawesi *Indonesia* 2°0S 120°0E **37** E6
Celebes Sea *Indonesia* 3°0N 123°0E **37** D6
Celina *U.S.A.* 40°33N 84°35W **81** E11
Celje *Slovenia* 46°16N 15°18E **16** E8
Celle *Germany* 52°37N 10°4E **16** B6
Celtic Sea *Atl. Oc.* 50°9N 9°34W **14** F2
Cenderwasih, Teluk *Indonesia* 3°0S 135°20E **37** E9
Center *N. Dak., U.S.A.* 47°7N 101°18W **80** B3
Center *Tex., U.S.A.* 31°48N 94°11W **84** F7
Centerburg *U.S.A.* 40°18N 82°42W **82** F2
Centerville *Calif., U.S.A.* 36°44N 119°30W **78** J7
Centerville *Iowa, U.S.A.* 40°44N 92°52W **80** E7
Centerville *Pa., U.S.A.* 40°3N 79°59W **82** F5
Centerville *Tenn., U.S.A.* 35°47N 87°28W **85** D11
Centerville *Tex., U.S.A.* 31°16N 95°59W **84** F7
Central = Tsentralnyy □ *Russia* 52°0N 40°0E **28** D4
Central □ *Kenya* 0°30S 37°30E **54** C4
Central □ *Malawi* 13°30S 33°30E **55** E3
Central □ *Zambia* 14°25S 28°50E **55** E2
Central, Cordillera *Colombia* 5°0N 75°0W **92** C4
Central, Cordillera *Costa Rica* 10°10N 84°5W **88** D3
Central, Cordillera *Dom. Rep.* 19°15N 71°0W **89** C5
Central, Cordillera *Puerto Rico* 18°8N 66°35W **89** d
Central African Rep. ■ *Africa* 7°0N 20°0E **51** G9
Central America *America* 12°0N 85°0W **66** H11
Central Australia ☼ *Australia* 22°30S 128°30E **60** D4
Central Bedfordshire □ *U.K.* 52°5N 0°20W **13** E7
Central Butte *Canada* 50°48N 106°31W **71** C7
Central City *Colo., U.S.A.* 39°48N 105°31W **76** G11
Central City *Ky., U.S.A.* 37°18N 87°7W **80** G10
Central City *Nebr., U.S.A.* 41°7N 98°0W **80** E4
Central Desert ☼ *Australia* 19°56S 130°46E **60** C5
Central I. *Kenya* 3°30N 36°0E **54** B4
Central Island △ *Kenya* 2°33N 36°1E **54** B4
Central Kalahari ⌓ *Botswana* 22°36S 23°58E **56** C3
Central Makran Range *Pakistan* 26°30N 64°15E **40** F4
Central Pacific Basin *Pac. Oc.* 8°0N 175°0W **64** G10
Central Patricia *Canada* 51°30N 90°9W **72** B1
Central Point *U.S.A.* 42°23N 122°55W **76** E2
Central Russian Uplands *Europe* 54°0N 36°0E **6** E13
Central Siberian Plateau *Russia* 65°0N 105°0E **26** B12
Central Square *U.S.A.* 43°17N 76°9W **83** C8
Centralia *Canada* 43°17N 81°28W **82** C3
Centralia *Ill., U.S.A.* 38°32N 89°8W **80** F9
Centralia *Mo., U.S.A.* 39°13N 92°8W **80** F7
Centralia *Wash., U.S.A.* 46°43N 122°58W **78** D4
Centre de Flacq *Mauritius* 20°12S 57°43E **53** d
Centreville *N.Y., U.S.A.* 42°28N 78°14W **82** D6
Centreville *Pa., U.S.A.* 41°44N 79°45W **82** E5
Cephalonia = Kefalonia *Greece* 38°15N 20°30E **23** E9
Cepu *Indonesia* 7°9S 111°35E **37** G14
Ceram = Seram *Indonesia* 3°10S 129°0E **37** E7
Ceram Sea = Seram Sea *Indonesia* 2°30S 128°30E **37** E7
Ceredigion □ *U.K.* 52°16N 4°15W **13** E3
Ceres *Argentina* 29°55S 61°55W **94** B3
Ceres *S. Africa* 33°21S 19°18E **56** E2
Ceres *U.S.A.* 37°35N 120°57W **78** H6
Cerf *Seychelles* 4°38S 55°40E **53** b
Cerignola *Italy* 41°17N 15°53E **22** D6
Cerigo = Kythira *Greece* 36°8N 23°0E **23** F10
Çerkezköy *Turkey* 41°17N 28°0E **23** D12
Cerralvo, I. *Mexico* 24°15N 109°55W **86** C3
Cerritos *Mexico* 22°25N 100°16W **86** C4
Cerro Chato *Uruguay* 33°6S 55°8W **95** C4
Cerro Cofre de Perote △ *Mexico* 19°29N 97°8W **87** D5
Cerro Corá △ *Paraguay* 22°35S 56°2W **95** A4
Cerro de Pasco *Peru* 10°45S 76°10W **92** F3
Cerro el Copey △ *Venezuela* 10°59N 63°53W **89** D7
Cerro Hoya △ *Panama* 7°17N 80°45W **88** E3
Cerro Saroche △ *Venezuela* 10°8N 69°38W **89** D
Cerventes *Australia* 30°31S 115°3E **61** F
Cervera *Spain* 41°40N 1°16E **21** B
Cesena *Italy* 44°8N 12°15E **22** B
Cēsis *Latvia* 57°18N 25°15E **9** H2
Česká Rep. = Czech Rep. ■ *Europe* 50°0N 15°0E **16** D
České Budějovice *Czech Rep.* 48°55N 14°25E **16** D
Českomoravská Vrchovina *Czech Rep.* 49°30N 15°40E **16** D
Çeşme *Turkey* 38°20N 26°23E **23** E1
Cessnock *Australia* 32°50S 151°21E **63** E
Cetinje *Montenegro* 42°23N 18°59E **23** C
Cetraro *Italy* 39°31N 15°55E **22** E
Ceuta *N. Afr.* 35°52N 5°18W **21** E
Cévennes *France* 44°10N 3°50E **20** D
Ceyhan *Turkey* 37°4N 35°47E **44** B
Ceylanpınar *Turkey* 36°50N 40°2E **44** B
Ceylon = Sri Lanka ■ *Asia* 7°30N 80°50E **40** R1
Cha-am *Thailand* 12°48N 99°58E **38** F
Chacabuco *Argentina* 34°40S 60°27W **94** C
Chachapoyas *Peru* 6°15S 77°50W **92** E
Chachoengsao *Thailand* 13°42N 101°5E **38** F
Chachran *Pakistan* 28°55N 70°30E **42** E
Chachro *Pakistan* 25°5N 70°15E **42** G
Chaco □ *Argentina* 26°30S 61°0W **94** B
Chaco □ *Paraguay* 26°0S 60°0W **94** B
Chaco ➜ *U.S.A.* 36°46N 108°39W **77** H
Chaco △ *Argentina* 27°0S 59°30W **94** B
Chaco Austral *S. Amer.* 27°0S 61°30W **96** B
Chaco Boreal *S. Amer.* 22°0S 60°0W **92** H
Chaco Central *S. Amer.* 24°0S 61°0W **96** A
Chaco Culture ⌓ *U.S.A.* 36°3N 107°58W **77** H
Chacon, C. *U.S.A.* 54°42N 132°0W **70** C
Chad ■ *Africa* 15°0N 17°15E **51** F
Chad, L. = Tchad, L. *Chad* 13°30N 14°30E **51** F
Chadileuvú ➜ *Argentina* 37°46S 66°0W **94** D
Chadiza *Zambia* 14°45S 32°27E **55** E
Chadron *U.S.A.* 42°50N 103°0W **80** D
Chadyr-Lunga = Ciadâr-Lunga *Moldova* 46°3N 28°51E **17** E
Chae Hom *Thailand* 18°43N 99°35E **38** C
Chae Son △ *Thailand* 18°42N 99°20E **38** C
Chaem ➜ *Thailand* 18°11N 98°38E **38** C
Chaeryŏng *N. Korea* 38°24N 125°36E **35** E
Chagai Hills = Chāh Gay Hills *Afghan.* 29°30N 64°0E **40** E
Chagda *Russia* 58°45N 130°38E **29** D
Chaghcharān *Afghan.* 34°31N 65°15E **40** E
Chagos Arch. ☒ *Ind. Oc.* 6°0S 72°0E **27**
Chagres △ *Panama* 9°33N 79°37W **88**
Chaguanas *Trin. & Tob.* 10°30N 61°26W **93** K
Chāh Ākhvor *Iran* 32°41N 59°40E **45**
Chāh Bahar *Iran* 25°20N 60°40E **45**
Chāh-e Kavīr *Iran* 34°29N 56°52E **45**
Chāh Gay Hills *Afghan.* 29°30N 64°0E **40**
Chahār Borjak *Afghan.* 30°17N 62°3E **40**
Chahār Mahāll va Bakhtīarī □ *Iran* 32°0N 49°0E **45**
Chai Badan *Thailand* 15°12N 101°8E **38**
Chaibasa *India* 22°42N 85°49E **43** H
Chainat *Thailand* 15°11N 100°8E **38**
Chaiya *Thailand* 9°23N 99°14E **39**
Chaiyaphum *Thailand* 15°48N 102°2E **38**
Chaj Doab *Pakistan* 32°15N 73°0E **42**
Chajari *Argentina* 30°42S 58°0W **94**
Chak Amru *Pakistan* 32°22N 75°11E **42**
Chakar ➜ *Pakistan* 29°29N 68°2E **42**
Chakari *Zimbabwe* 18°5S 29°51E **57**
Chake Chake *Tanzania* 5°15S 39°45E **54**
Chakhānsūr *Afghan.* 31°10N 62°0E **40**
Chakonipau, L. *Canada* 56°18N 68°30W **73**
Chakradharpur *India* 22°45N 85°40E **43** H
Chakrata *India* 30°42N 77°51E **42**
Chakwal *Pakistan* 32°56N 72°53E **42**
Chala *Peru* 15°48S 74°20W **92**
Chalchihuites *Mexico* 23°29N 103°53W **86**
Chalcis = Chalkida *Greece* 38°27N 23°42E **23** E
Chaleur B. *Canada* 47°55N 65°30W **73**
Chalfant *U.S.A.* 37°32N 118°21W **78**
Chalhuanca *Peru* 14°15S 73°15W **92**
Chalisgaon *India* 20°30N 75°10E **40**
Chalk River *Canada* 46°1N 77°27W **72**
Chalkida *Greece* 38°27N 23°42E **23** E
Chalky Inlet *N.Z.* 46°3S 166°31E **59**
Challapata *Bolivia* 18°53S 66°50W **92**
Challenger Deep *Pac. Oc.* 11°30N 142°0E **64**
Challenger Fracture Zone *Pac. Oc.* 35°0S 105°0W **65** L
Challis *U.S.A.* 44°30N 114°14W **76**
Chalmette *U.S.A.* 29°56N 89°57W **85** G
Chaloem Rattanakosin △ *Thailand* 14°40N 99°17E **38**
Chalon-sur-Saône *France* 46°48N 4°50E **20**
Chalong *Thailand* 7°50N 98°22E **3**
Châlons-en-Champagne *France* 48°58N 4°20E **20**
Chálten, Cerro = Fitz Roy, Cerro *Argentina* 49°17S 73°5W **96**
Chālūs *Iran* 36°38N 51°26E **45**
Cham, Cu Lao *Vietnam* 15°57N 108°30E **38**
Chama *U.S.A.* 36°54N 106°35W **77** H
Chamaicó *Argentina* 35°3S 64°58W **94**
Chaman *Pakistan* 30°58N 66°25E **40**
Chamba *India* 32°35N 76°10E **42**
Chamba *Tanzania* 11°37S 37°0E **55**
Chambal ➜ *India* 26°29N 79°15E **43**
Chamberlain *U.S.A.* 43°49N 99°20W **80**
Chamberlain ➜ *Australia* 15°30S 127°54E **60**
Chamberlain L. *U.S.A.* 46°14N 69°19W **81** E
Chambers *U.S.A.* 35°11N 109°26W **77**
Chambersburg *U.S.A.* 39°56N 77°40W **81** E
Chambéry *France* 45°34N 5°55E **20**

ambeshi → Zambia 11°53S 29°48E 52 G6
ambly Canada 45°27N 73°17W 83 A11
ambord Canada 48°25N 72°6W 73 C5
amchamal Iraq 35°32N 44°50E 44 C5
amela Mexico 19°32N 105°5W 86 D3
amical Argentina 30°22S 66°27W 94 C2
amkar Luong Cambodia 11°0N 103°45E 39 G4
amoli India 30°24N 79°21E 43 D8
amonix-Mont Blanc France 45°55N 6°51E 20 D7
amouchuane = Ashuapmushuan → Canada 48°37N 72°20W 72 C5
ampa India 22°2N 82°43E 43 H10
ampagne Canada 60°49N 136°30W 70 A1
ampagne France 48°40N 4°20E 20 B6
ampaign U.S.A. 40°7N 88°15W 80 E9
ampassak Laos 14°53N 105°52E 38 E5
ampawat India 29°20N 80°6E 43 E9
ampdoré, L. Canada 55°55N 65°49W 73 A6
ampion U.S.A. 41°19N 80°51W 82 E4
amplain U.S.A. 44°59N 73°27W 83 B11
amplain, L. U.S.A. 44°40N 73°20W 83 B11
ampotón Mexico 19°21N 90°43W 87 D6
ampua India 22°5N 85°40E 43 H11
ana Thailand 6°55N 100°44E 39 J3
añaral Chile 26°23S 70°40W 94 B1
anārān Iran 36°39N 59°6E 45 B8
anasma India 23°44N 72°5E 42 H5
ancery Lane Barbados 13°3N 59°30W 89 g
anco Chile 35°44S 72°32W 94 D1
and India 21°57N 79°7E 43 J8
andan India 24°38N 86°40E 43 G12
andan Chauki India 28°33N 80°47E 43 E9
andannagar India 22°52N 88°24E 43 H13
andausi India 28°27N 78°49E 43 E8
andeleur Is. U.S.A. 29°55N 88°57W 85 G10
andeleur Sd. U.S.A. 29°55N 89°0W 85 G10
andigarh India 30°43N 76°47E 42 D7
andil India 22°58N 86°3E 43 H12
andler Australia 27°0S 133°19E 63 D1
andler Canada 48°18N 64°46W 73 C7
andler Ariz., U.S.A. 33°18N 111°50W 77 K8
andler Okla., U.S.A. 35°42N 96°53W 84 D6
andod India 21°59N 73°28E 42 J5
andpur Bangla. 23°8N 90°45E 41 H17
andrapur India 19°57N 79°25E 40 K11
ānf Iran 26°38N 60°29E 45 E9
ang Pakistan 26°59N 68°30E 42 F3
ang, Ko Thailand 12°0N 102°23E 39 G4
ang, Ko Thailand 9°50N 98°27E 39 H2
'ang Chiang = Chang Jiang → China 31°48N 121°10E 33 E13
ang Jiang → China 31°48N 121°10E 33 E13
ang-won S. Korea 35°16N 128°37E 35 G15
angan China 22°48N 113°48E 33 a
anganacheri India 9°25N 76°31E 40 Q10
anganassery = Changanacheri India 9°25N 76°31E 40 Q10
angane → Mozam. 24°30S 33°30E 57 C5
angbai China 41°25N 128°5E 35 D15
angbai Shan China 42°20N 129°0E 35 C15
angchiak'ou = Zhangjiakou China 40°48N 114°55E 34 D8
'angchou = Changzhou China 31°47N 119°58E 33 E12
angchun China 43°57N 125°17E 35 C13
angchunling China 45°18N 125°27E 35 B13
angde China 29°4N 111°35E 33 F11
angdo-ri N. Korea 38°30N 127°40E 35 E14
angge China 34°13N 113°46E 34 G7
anghai = Shanghai China 31°15N 121°26E 33 E13
anghua Taiwan 24°2N 120°30E 33 G13
anghŭngni N. Korea 40°24N 128°19E 35 D15
angi Singapore 1°23N 103°59E 39 d
angi, Singapore ✕ (SIN) Singapore 1°23N 103°59E 39 M4
angji China 44°1N 87°19E 32 C6
angjiang China 19°20N 108°55E 38 C7
angjiang Shuiku China 22°29N 113°27E 33 a
angjin N. Korea 40°23N 127°15E 35 D14
angjin-ho N. Korea 40°30N 127°15E 35 D14
angli China 39°40N 119°13E 35 E10
angling China 44°20N 123°58E 35 B12
anglun Malaysia 6°25N 100°26E 39 J3
angping China 40°14N 116°12E 34 D9
angsha China 28°12N 113°0E 33 F11
angshan Qundao China 39°11N 122°32E 35 E12
angtu China 42°46N 124°6E 35 C13
anguinola Panama 9°26N 82°31W 88 E3
angwu China 35°10N 107°45E 34 G4
angyi China 36°40N 119°30E 35 F10
angyŏn N. Korea 38°15N 125°6E 35 E13
angyuan China 35°15N 114°42E 34 G8
angzhi China 36°10N 113°6E 34 F7
angzhou China 31°47N 119°58E 33 E12
anhanga Angola 16°0S 14°8E 56 B1
ania Greece 35°30N 24°4E 25 D6
ania ☐ Greece 35°30N 24°0E 25 D6
anion, Kolpos Greece 35°33N 23°55E 25 D5
annapatna India 12°40N 77°15E 40 N10
annel Is. U.K. 49°19N 2°24W 13 H5
annel Is. U.S.A. 33°40N 119°15W 79 M7
annel Islands △ U.S.A. 34°0N 119°24W 79 L7
annel-Port aux Basques Canada 47°30N 59°9W 73 C8
annel Tunnel Europe 51°0N 1°30E 13 F9
anning U.S.A. 35°41N 102°20W 84 D3
antada Spain 42°36N 7°46W 21 A2
anthaburi Thailand 12°38N 102°12E 38 F4
antrey Inlet Canada 67°48N 96°20W 68 D12
anute U.S.A. 37°41N 95°27W 80 G6
ao Phraya → Thailand 13°40N 100°31E 38 F3
Chao Phraya Lowlands Thailand 15°30N 100°0E 38 E3
Chaocheng China 36°4N 115°37E 34 F8
Chaoyang China 41°35N 120°22E 35 D11
Chaozhou China 23°42N 116°32E 33 G12
Chapaev Kazakhstan 50°25N 51°10E 19 D9
Chapais Canada 49°47N 74°51W 72 C5
Chapala Mozam. 15°50S 37°35E 55 F4
Chapala, L. de Mexico 20°15N 103°0W 86 C4
Chapayev = Chapaev Kazakhstan 50°25N 51°10E 19 D9
Chapayevsk Russia 53°0N 49°40E 18 D8
Chapecó Brazil 27°14S 52°41W 95 B5
Chapel Hill U.S.A. 35°55N 79°4W 85 D15
Chapleau Canada 47°50N 83°24W 72 C3
Chaplin Canada 50°28N 106°40W 71 C7
Chaplin L. Canada 50°22N 106°36W 71 C7
Chappell U.S.A. 41°6N 102°28W 80 E2
Chapra = Chhapra India 25°48N 84°44E 43 G11
Chara Russia 56°54N 118°20E 29 D12
Charadai Argentina 27°35S 59°55W 94 B4
Charagua Bolivia 19°45S 63°10W 92 G6
Charakas Greece 35°1N 25°7E 25 D7
Charambirá, Punta Colombia 4°16N 77°32W 92 C3
Charaña Bolivia 17°30S 69°25W 92 G5
Charanwala India 27°51N 72°10E 42 F5
Charata Argentina 27°13S 61°14W 94 B3
Charcas Mexico 23°8N 101°7W 86 C4
Charcot I. Antarctica 70°0S 70°0W 5 C17
Chard U.K. 50°52N 2°58W 13 G5
Chardon U.S.A. 41°35N 81°12W 82 E3
Chardzhou = Türkmenabat Turkmenistan 39°6N 63°34E 45 B9
Charente → France 45°57N 1°5W 20 D3
Chari → Chad 12°58N 14°31E 51 F8
Chārīkār Afghan. 35°0N 69°10E 40 B6
Chariton U.S.A. 41°1N 93°19W 80 E7
Chariton → U.S.A. 39°19N 92°58W 80 F7
Chärjew = Türkmenabat Turkmenistan 39°6N 63°34E 45 B9
Charkhari India 25°24N 79°45E 43 G8
Charkhi Dadri India 28°37N 76°17E 42 E7
Charleroi Belgium 50°24N 4°27E 15 D4
Charleroi U.S.A. 40°9N 79°57W 82 F5
Charles, C. U.S.A. 37°7N 75°58W 81 G16
Charles, Peak Australia 32°52S 121°11E 61 F3
Charles City U.S.A. 43°4N 92°41W 80 D7
Charles I. Canada 62°39N 74°15W 69 E17
Charles L. Canada 59°50N 110°33W 71 B6
Charles Town U.S.A. 39°17N 77°52W 81 F15
Charlesbourg Canada 46°51N 71°16W 81 B18
Charleston Ill., U.S.A. 39°30N 88°10W 80 F9
Charleston Miss., U.S.A. 34°1N 90°4W 85 D9
Charleston Mo., U.S.A. 36°55N 89°21W 80 G9
Charleston S.C., U.S.A. 32°46N 79°56W 85 E15
Charleston W. Va., U.S.A. 38°21N 81°38W 81 F13
Charleston L. Canada 44°32N 76°0W 83 B9
Charleston Peak U.S.A. 36°16N 115°42W 79 J11
Charlestown Ireland 53°58N 8°48W 10 C3
Charlestown S. Africa 27°26S 29°53E 57 D4
Charlestown Ind., U.S.A. 38°27N 85°40W 81 F11
Charlestown N.H., U.S.A. 43°14N 72°25W 83 C12
Charlestown of Aberlour U.K. 57°28N 3°14W 11 D5
Charleville = Rath Luirc Ireland 52°21N 8°40W 10 D3
Charleville Australia 26°24S 146°15E 63 D4
Charleville-Mézières France 49°44N 4°40E 20 B6
Charlevoix U.S.A. 45°19N 85°16W 81 C11
Charlotte Mich., U.S.A. 42°34N 84°50W 81 D11
Charlotte N.C., U.S.A. 35°13N 80°50W 85 D14
Charlotte Vt., U.S.A. 44°19N 73°16W 83 B11
Charlotte Amalie U.S. Virgin Is. 18°21N 64°56W 89 e
Charlotte Harbor U.S.A. 26°57N 82°4W 85 H13
Charlotte L. Canada 52°12N 125°19W 70 C3
Charlottesville U.S.A. 38°2N 78°30W 81 F14
Charlottetown Nfld. & L., Canada 52°46N 56°7W 73 B8
Charlottetown P.E.I., Canada 46°14N 63°8W 73 C7
Charlotteville Trin. & Tob. 11°20N 60°33W 93 J16
Charlton Australia 36°16S 143°24E 63 F3
Charlton I. Canada 52°0N 79°20W 72 B4
Charny Canada 46°43N 71°15W 73 C5
Charolles France 46°27N 4°16E 20 C6
Charre Mozam. 17°13S 35°10E 55 F4
Charsadda Pakistan 34°7N 71°45E 42 B4
Charters Towers Australia 20°5S 146°13E 62 C4
Chartres France 48°29N 1°30E 20 B4
Chascomús Argentina 35°30S 58°0W 94 D4
Chase Canada 50°50N 119°41W 70 C5
Chasefu Zambia 11°55S 33°8E 55 E3
Chashma Barrage Pakistan 32°27N 71°20E 42 C4
Chāt Iran 37°59N 55°16E 45 B7
Châteaubriant France 47°43N 1°23W 20 C3
Chateaugay U.S.A. 44°56N 74°5W 83 B10
Châteauguay, L. Canada 56°26N 70°3W 73 A5
Châteaulin France 48°11N 4°8W 20 B1
Châteauroux France 46°50N 1°40E 20 C4
Châteaux, Pte. des Guadeloupe 16°15N 61°10W 88 b
Châtellerault France 46°50N 0°30E 20 C4
Chatham = Miramichi Canada 47°2N 65°28W 73 C6
Chatham Canada 42°24N 82°11W 82 D2
Chatham U.K. 51°22N 0°32E 13 F8
Chatham U.S.A. 42°21N 73°36W 83 D11
Chatham Is. Pac. Oc. 44°0S 176°40W 64 M10
Chatham Rise Pac. Oc. 43°30S 180°0E 64 M10
Chatmohar Bangla. 24°15N 89°15E 43 G13
Chatra India 24°12N 84°56E 43 G11
Chatrapur India 19°22N 85°2E 41 K14
Chats, L. des Canada 45°30N 76°20W 83 A8
Chatsu India 26°36N 75°57E 42 F6
Chatsworth Canada 44°27N 80°54W 82 B4
Chatsworth Zimbabwe 19°38S 31°13E 55 F3
Châttagâm = Chittagong Bangla. 22°19N 91°48E 41 H17
Chattahoochee U.S.A. 30°42N 84°51W 85 F12
Chattahoochee → U.S.A. 30°54N 84°57W 85 F12
Chattanooga U.S.A. 35°3N 85°19W 85 D12
Chatteris U.K. 52°28N 0°2E 13 E8
Chatturat Thailand 15°40N 101°51E 38 E3
Chau Doc Vietnam 10°42N 105°7E 39 G5
Chaukan Pass Burma 27°8N 97°10E 41 F20
Chaumont France 48°7N 5°8E 20 B6
Chaumont U.S.A. 44°4N 76°8W 83 B8
Chaunskaya G. Russia 69°0N 169°0E 29 C17
Chautauqua L. U.S.A. 42°10N 79°24W 82 D5
Chauvin Canada 52°45N 110°10W 71 C6
Chaves Brazil 0°15S 49°55W 93 D9
Chaves Portugal 41°45N 7°32W 21 B2
Chawang Thailand 8°25N 99°30E 39 H2
Chaykovskiy Russia 56°47N 54°9E 18 C9
Chazy U.S.A. 44°53N 73°26W 83 B11
Cheb Czech Rep. 50°9N 12°28E 16 C7
Cheboksary Russia 56°8N 47°12E 18 C8
Cheboygan U.S.A. 45°39N 84°29W 81 C11
Chech, Erg Africa 25°0N 2°15W 50 D5
Chechenia ☐ Russia 43°30N 45°29E 19 F8
Checheno-Ingush Republic = Chechenia ☐ Russia 43°30N 45°29E 19 F8
Chechnya = Chechenia ☐ Russia 43°30N 45°29E 19 F8
Checotah U.S.A. 35°28N 95°31W 84 D7
Chedabucto B. Canada 45°25N 61°8W 73 C7
Cheduba I. Burma 18°45N 93°40E 41 K18
Cheektowaga U.S.A. 42°54N 78°45W 82 D6
Cheepie Australia 26°33S 145°0E 63 D4
Chegdomyn Russia 51°7N 133°1E 29 D14
Chegga Mauritania 25°27N 5°40W 50 C4
Chegutu Zimbabwe 18°10S 30°14E 55 F3
Chehalis U.S.A. 46°40N 122°58W 78 D4
Chehalis → U.S.A. 46°57N 123°50W 78 D3
Cheju = Jeju S. Korea 33°31N 126°32E 35 H14
Cheju-do = Jeju-do S. Korea 33°29N 126°34E 35 H14
Chek Lap Kok China 22°18N 113°56E 33 a
Chekiang = Zhejiang ☐ China 29°0N 120°0E 33 F13
Chela, Sa. da Angola 16°20S 13°20E 56 B1
Chelan U.S.A. 47°51N 120°1W 76 C3
Chelan, L. U.S.A. 48°11N 120°30W 76 B3
Cheleken = Hazar Turkmenistan 39°34N 53°16E 19 G9
Cheleken Yarymadasy Turkmenistan 39°30N 53°15E 45 B7
Chelforó Argentina 39°0S 66°33W 96 D3
Chelkar = Shalqar Kazakhstan 47°48N 59°39E 28 E6
Chelm Poland 51°8N 23°30E 17 C12
Chelmno Poland 53°20N 18°30E 17 B10
Chelmsford U.K. 51°44N 0°29E 13 F8
Chelsea Australia 38°5S 145°8E 63 F4
Chelsea U.S.A. 43°59N 72°27W 83 C12
Cheltenham U.K. 51°54N 2°4W 13 F5
Chelyabinsk Russia 55°10N 61°24E 28 D7
Chelyuskin, C. = Chelyuskin, Mys Russia 77°30N 103°0E 29 B11
Chelyuskin, Mys Russia 77°30N 103°0E 29 B11
Chemainus Canada 48°55N 123°42W 78 B3
Chemba Mozam. 17°9S 34°53E 53 H6
Chemin Grenier Mauritius 20°29S 57°28E 53 d
Chemnitz Germany 50°51N 12°54E 16 C7
Chemult U.S.A. 43°14N 121°47W 76 E3
Chen, Gora Russia 65°16N 141°50E 29 C15
Chenab → Pakistan 30°23N 71°2E 42 D4
Chenab Nagar Pakistan 31°45N 72°55E 42 D5
Chenango Forks U.S.A. 42°15N 75°51W 83 D9
Cheney U.S.A. 47°30N 117°35W 76 C5
Cheng Xian China 33°43N 105°42E 34 H3
Chengcheng China 35°8N 109°56E 34 G5
Chengchou = Zhengzhou China 34°45N 113°34E 34 G7
Chengde China 40°59N 117°58E 35 D9
Chengdu China 30°38N 104°2E 32 E9
Chenggu China 33°10N 107°21E 34 H4
Chengjiang China 24°39N 103°0E 32 G9
Chengmai China 19°50N 109°58E 38 C7
Chengshan Jiao China 37°25N 122°44E 35 F12
Ch'engtu = Chengdu China 30°38N 104°2E 32 E9
Chengwu China 34°58N 115°50E 34 G8
Chengyang China 36°18N 120°21E 35 F11
Chenjiagang China 34°23N 119°47E 35 G10
Chennai India 13°8N 80°19E 40 N12
Chenzhou China 25°47N 113°1E 33 F11
Cheò, Eilean a' = Skye U.K. 57°15N 6°10W 11 D2
Cheo Reo Vietnam 13°20N 108°25E 38 F7
Cheom Ksan Cambodia 14°13N 104°56E 38 E5
Cheonan S. Korea 36°48N 127°9E 35 F14
Cheongdo S. Korea 35°38N 128°42E 35 G15
Cheongju S. Korea 36°39N 127°27E 35 F14
Cheorwon S. Korea 38°15N 127°10E 35 E14
Chepén Peru 7°15S 79°23W 92 E3
Chepes Argentina 31°20S 66°35W 94 C2
Chepo Panama 9°10N 79°6W 88 E4
Chepstow U.K. 51°38N 2°41W 13 F5
Chequamegon B. U.S.A. 46°39N 90°51W 80 B8
Cher → France 47°21N 0°29E 20 C4
Cheraw U.S.A. 34°42N 79°53W 85 D15
Cherbourg France 49°39N 1°40W 20 B3
Cherdyn Russia 60°24N 56°29E 18 B10
Cheremkhovo Russia 53°8N 103°1E 32 A9
Cherepovets Russia 59°5N 37°55E 18 C6
Chergui, Chott ech Algeria 34°21N 0°25E 50 B6
Cherikov = Cherykaw Belarus 53°32N 31°20E 17 B16
Cherkasy Ukraine 49°27N 32°4E 19 E5
Cherkessk Russia 44°15N 42°5E 19 F7
Cherlak Russia 54°15N 74°55E 28 D8
Chernaya Russia 70°30N 89°10E 29 B9
Chernigov = Chernihiv Ukraine 51°28N 31°20E 18 D5
Chernigovka Russia 44°19N 132°34E 30 B6
Chernihiv Ukraine 51°28N 31°20E 18 D5
Chernivtsi Ukraine 48°15N 25°52E 17 D13
Chernobyl = Chornobyl Ukraine 51°20N 30°15E 17 C16
Chernogorsk Russia 53°49N 91°18E 29 D10
Chernovtsy = Chernivtsi Ukraine 48°15N 25°52E 17 D13
Chernyakhovsk Russia 54°36N 21°48E 9 J19
Chernysheyskiy Russia 63°0N 112°30E 29 C12
Cherokee Iowa, U.S.A. 42°45N 95°33W 80 D6
Cherokee Okla., U.S.A. 36°45N 98°21W 84 C5
Cherokee Village U.S.A. 36°18N 91°31W 84 C9
Cherokees, Grand Lake O' The U.S.A. 36°28N 94°55W 84 C7
Cherrapunji India 25°17N 91°47E 41 G17
Cherry Valley U.S.A. 42°48N 74°45W 83 D10
Cherskiy Russia 68°45N 161°18E 29 C17
Cherskogo Khrebet Russia 65°0N 143°0E 29 C15
Chersonisos Greece 35°18N 25°22E 25 D7
Chersonisos Akrotiri Greece 35°30N 24°10E 25 D6
Cherven Belarus 53°45N 28°28E 17 B15
Chervonohrad Ukraine 50°25N 24°10E 17 C13
Cherwell → U.K. 51°44N 1°14W 13 F6
Cherykaw Belarus 53°32N 31°20E 17 B16
Chesapeake U.S.A. 36°49N 76°16W 81 G15
Chesapeake B. U.S.A. 38°0N 76°10W 81 F14
Cheshire ☐ U.K. 53°14N 2°30W 12 D5
Cheshskaya Guba Russia 67°20N 47°0E 18 A8
Cheshunt U.K. 51°43N 0°1W 13 F7
Chesil Beach U.K. 50°37N 2°33W 13 G5
Chesley Canada 44°17N 81°5W 82 B3
Chester U.K. 53°12N 2°53W 12 D5
Chester Calif., U.S.A. 40°19N 121°14W 76 F3
Chester Ill., U.S.A. 37°55N 89°49W 80 G9
Chester Mont., U.S.A. 48°31N 110°58W 76 B8
Chester Pa., U.S.A. 39°51N 75°22W 81 F16
Chester S.C., U.S.A. 34°43N 81°12W 85 D14
Chester Vt., U.S.A. 43°16N 72°36W 83 C12
Chester W. Va., U.S.A. 40°37N 80°34W 82 F4
Chester-le-Street U.K. 54°51N 1°34W 12 C6
Chesterfield U.K. 53°15N 1°25W 12 D6
Chesterfield, Îs. N. Cal. 19°52S 158°15E 58 C8
Chesterfield Inlet Canada 63°30N 90°45W 68 E13
Chesterton Ra. Australia 25°30S 147°27E 63 D4
Chesterton Range △ Australia 26°16S 147°22E 63 D4
Chestertown U.S.A. 43°40N 73°48W 83 C11
Chesterville Canada 45°6N 75°14W 83 A9
Chesuncook L. U.S.A. 46°0N 69°21W 81 C19
Chéticamp Canada 46°37N 60°59W 73 C7
Chetumal Mexico 18°30N 88°20W 87 D7
Chetumal, B. de Cent. Amer. 18°40N 88°10W 87 D7
Chetwynd Canada 55°45N 121°36W 70 B4
Cheung Chau China 22°13N 114°1E 33 a
Cheviot, The U.K. 55°29N 2°9W 12 B5
Cheviot Hills U.K. 55°20N 2°30W 12 B5
Cheviot Ra. Australia 25°20S 143°45E 62 D3
Chew Bahir Ethiopia 4°40N 36°50E 47 G2
Chewelah U.S.A. 48°17N 117°43W 76 B5
Chewore △ Zimbabwe 16°0S 29°52E 55 F2
Cheyenne Okla., U.S.A. 35°37N 99°40W 84 D5
Cheyenne Wyo., U.S.A. 41°8N 104°49W 76 F11
Cheyenne → U.S.A. 44°41N 101°18W 80 C3
Cheyenne Wells U.S.A. 38°49N 102°21W 76 G12
Cheyne B. Australia 34°35S 118°50E 61 F2
Chhabra India 24°40N 76°54E 42 G7
Chhaktala India 22°6N 74°11E 42 H6
Chhapra India 25°48N 84°44E 43 G11
Chhata India 27°42N 77°30E 42 F7
Chhatarpur Jharkhand, India 24°23N 84°11E 43 G11
Chhatarpur Mad. P., India 24°55N 79°35E 43 G8
Chhattisgarh ☐ India 22°0N 82°0E 43 J10
Chhep Cambodia 13°45N 105°24E 38 F5
Chhindwara Mad. P., India 23°3N 79°29E 43 H8
Chhindwara Mad. P., India 22°2N 78°59E 43 H8
Chhlong Cambodia 12°15N 105°58E 39 F5
Chhota Tawa → India 22°14N 76°36E 42 H7
Chhoti Kali Sindh → India 24°2N 75°31E 42 G6
Chhuikhadan India 21°32N 80°59E 43 J9
Chhuk Cambodia 10°46N 104°28E 39 G5
Chi → Thailand 15°11N 104°43E 38 E5
Chi Thanh Vietnam 13°17N 109°16E 38 F7
Chiai Taiwan 23°29N 120°25E 33 G13
Chiang Dao Thailand 19°22N 98°58E 38 C2
Chiang Dao, Doi Thailand 19°23N 98°54E 38 C2
Chiang Kham Thailand 19°32N 100°18E 38 C3
Chiang Khan Thailand 17°52N 101°36E 38 D3
Chiang Khong Thailand 20°17N 100°24E 38 B3
Chiang Mai Thailand 18°47N 98°59E 38 C2
Chiang Rai Thailand 19°52N 99°50E 38 C2
Chiang Saen Thailand 20°16N 100°5E 38 B3
Chiapa → Mexico 16°42N 93°0W 87 D6
Chiapa de Corzo Mexico 16°42N 93°0W 87 D6
Chiapas ☐ Mexico 16°30N 92°30W 87 D6
Chiapas, Sa. Madre de Mexico 15°40N 93°0W 87 D6
Chiautla de Tapia Mexico 18°18N 98°36W 87 D5
Chiávari Italy 44°19N 9°19E 20 D8
Chiavenna Italy 46°19N 9°24E 20 C8
Chiba Japan 35°30N 140°7E 31 G10
Chiba ☐ Japan 35°30N 140°20E 31 G10
Chibabava Mozam. 20°17S 33°35E 57 C5
Chibemba Cunene, Angola 15°48S 14°8E 53 H2
Chibemba Huila, Angola 16°20S 15°20E 56 B2
Chibi Zimbabwe 20°18S 30°25E 57 C5
Chibia Angola 15°10S 13°42E 53 H2
Chibougamau Canada 49°56N 74°24W 72 C5
Chibougamau, L. Canada 49°50N 74°20W 72 C5
Chibuk Nigeria 10°52N 12°50E 51 F8
Chibuto Mozam. 24°40S 33°33E 57 C5
Chic-Chocs, Mts. Canada 48°55N 66°0W 73 C6
Chicacole = Srikakulam India 18°14N 83°58E 41 K13
Chicago U.S.A. 41°52N 87°38W 80 E10
Chicago Heights U.S.A. 41°30N 87°38W 80 E10
Chichagof I. U.S.A. 57°30N 135°30W 68 F4
Chichaoua Morocco 31°32N 8°44W 50 B4
Chichawatni Pakistan 30°32N 72°42E 42 D5
Chichén-Itzá Mexico 20°37N 88°35W 87 C7
Chicheng China 40°55N 115°55E 34 D8
Chichester U.K. 50°50N 0°47W 13 G7
Chichester Ra. Australia 22°12S 119°15E 60 D2
Chichibu Japan 35°59N 139°10E 31 F9
Chichibu-Tama-Kai △ Japan 35°52N 138°42E 31 G9
Ch'ich'ihaerh = Qiqihar China 47°26N 124°0E 33 B13
Chicholi India 22°1N 77°40E 42 H8
Chickasaw ⌂ U.S.A. 34°26N 97°0W 84 D6
Chickasha U.S.A. 35°3N 97°58W 84 D6
Chiclana de la Frontera Spain 36°26N 6°9W 21 D2
Chiclayo Peru 6°42S 79°50W 92 E3
Chico U.S.A. 39°44N 121°50W 78 F5
Chico → Chubut, Argentina 44°0S 67°0W 96 E3
Chico → Santa Cruz, Argentina 50°0S 68°30W 96 G3
Chicoa Mozam. 15°36S 32°20E 55 F3
Chicomo Mozam. 24°31S 34°6E 57 C5
Chicomostoc Mexico 22°28N 102°46W 86 C4
Chicontepec Mexico 20°58N 98°10W 87 C5
Chicopee U.S.A. 42°9N 72°37W 83 D12
Chicoutimi Canada 48°28N 71°5W 73 C5
Chicualacuala Mozam. 22°6S 31°42E 57 C5
Chidambaram India 11°20N 79°45E 40 P11
Chidenguele Mozam. 24°55S 34°11E 57 C5
Chidley, C. Canada 60°23N 64°26W 69 E19
Chiducuane Mozam. 24°35S 34°25E 57 C5
Chiede Angola 17°15S 16°22E 56 B2
Chiefs Pt. Canada 44°41N 81°18W 82 B3
Chiem Hoa Vietnam 22°12N 105°17E 38 A5
Chiemsee Germany 47°53N 12°28E 16 E7
Chiengi Zambia 8°45S 29°10E 55 D2
Chiengmai = Chiang Mai Thailand 18°47N 98°59E 38 C2
Chiese → Italy 45°8N 10°25E 20 D9
Chieti Italy 42°21N 14°10E 22 C6
Chifeng China 42°18N 118°58E 35 C10
Chignecto B. Canada 45°30N 64°40W 73 C7
Chiguana Bolivia 21°0S 67°58W 94 A2
Chigwell U.K. 51°37N 0°6E 13 F8
Chihli, G. of = Bo Hai China 39°0N 119°0E 35 E10
Chihuahua Mexico 28°38N 106°5W 86 B3
Chihuahua ☐ Mexico 28°30N 106°0W 86 B3
Chiili = Shīeli Kazakhstan 44°20N 66°15E 28 E7
Chik Bollapur India 13°25N 77°45E 40 N10
Chikmagalur India 13°15N 75°45E 40 N9
Chikwawa Malawi 16°2S 34°50E 55 F3
Chilanga Zambia 15°33S 28°16E 55 F2
Chilapa Mexico 17°36N 99°10W 87 D5
Chilas Pakistan 35°25N 74°5E 43 B6
Chilaw Sri Lanka 7°30N 79°50E 40 R11
Chilcotin → Canada 51°44N 122°23W 70 C4
Childers Australia 25°15S 152°17E 63 D5
Childress U.S.A. 34°25N 100°13W 84 D4
Chile ■ S. Amer. 35°0S 72°0W 96 D2
Chile Rise Pac. Oc. 38°0S 92°0W 65 L18
Chilecito Argentina 29°10S 67°30W 94 B2
Chilete Peru 7°10S 78°50W 92 E3
Chililabombwe Zambia 12°18S 27°43E 55 E2
Chilim Pakistan 35°5N 75°5E 43 B6
Chilin = Jilin China 43°44N 126°30E 35 C14
Chilka L. India 19°40N 85°25E 41 K14
Chilko → Canada 52°0N 123°40W 70 C4
Chilko L. Canada 51°20N 124°10W 70 C4
Chillagoe Australia 17°7S 144°33E 62 B3
Chillán Chile 36°40S 72°10W 94 D1
Chillicothe Ill., U.S.A. 40°55N 89°29W 80 E9
Chillicothe Mo., U.S.A. 39°48N 93°33W 80 F7
Chillicothe Ohio, U.S.A. 39°20N 82°59W 81 F12
Chilliwack Canada 49°10N 121°54W 70 D4
Chilo India 27°25N 73°32E 42 F5
Chiloane, I. Mozam. 20°40S 34°55E 57 C5
Chiloé, I. de Chile 42°30S 73°50W 96 E2
Chilpancingo Mexico 17°33N 99°30W 87 D5
Chiltern Hills U.K. 51°40N 0°53W 13 F7
Chilton U.S.A. 44°2N 88°10W 80 C9
Chilubi Zambia 11°5S 29°58E 55 E2
Chilubula Zambia 10°14S 30°51E 55 E3
Chilumba Malawi 10°28S 34°12E 55 E3
Chilung Taiwan 25°3N 121°45E 33 F13
Chilwa, L. Malawi 15°15S 35°40E 55 F4
Chimaltitán Mexico 21°35N 103°50W 86 C4
Chimán Panama 8°45N 78°40W 88 E4
Chimanimani Zimbabwe 19°48S 32°52E 57 B5
Chimay Belgium 50°3N 4°20E 15 D4
Chimayo U.S.A. 36°0N 105°56W 77 J11
Chimba Zambia 9°54S 31°28E 55 D3
Chimborazo Ecuador 1°29S 78°55W 92 D3
Chimbote Peru 9°0S 78°35W 92 E3
Chimboy Uzbekistan 42°57N 59°47E 28 E6
Chimkent = Shymkent Kazakhstan 42°18N 69°36E 28 E7
Chimoio Mozam. 19°4S 33°30E 55 F3
Chimpembe Zambia 9°31S 29°33E 55 D2
Chin ☐ Burma 22°0N 93°0E 41 J18

Chin Hills Burma 22°30N 93°30E 41 H18
Chin Ling Shan = Qinling Shandi China 33°50N 108°10E 34 H5
China Mexico 25°42N 99°14W 87 B5
China ■ Asia 30°0N 110°0E 33 E11
China, Great Plain of Asia 35°0N 115°0E 26 E13
China Lake U.S.A. 35°44N 117°37W 79 K9
Chinan = Jinan China 36°38N 117°1E 34 F9
Chinandega Nic. 12°35N 87°12W 88 D2
Chinati Peak U.S.A. 29°57N 104°29W 84 G2
Chincha Alta Peru 13°25S 76°7W 92 F3
Chinchaga → Canada 58°53N 118°20W 70 B5
Chinchilla Australia 26°45S 150°38E 63 D5
Chinchorro, Banco Mexico 18°35N 87°22W 87 D7
Chinchou = Jinzhou China 41°5N 121°3E 35 D11
Chincoteague U.S.A. 37°56N 75°23W 81 G16
Chinde Mozam. 18°35S 36°30E 55 F4
Chindwin → Burma 21°26N 95°15E 41 J19
Chineni India 33°2N 75°15E 43 C6
Chinga Mozam. 15°13S 38°35E 55 F4
Chingola Zambia 12°31S 27°53E 55 E2
Chingole Malawi 13°4S 34°17E 55 E3
Ch'ingtao = Qingdao China 36°5N 120°20E 35 F11
Chinguetti Mauritania 20°25N 12°24W 50 D3
Chingune Mozam. 20°33S 34°58E 57 C5
Chinhanguanine Mozam. 25°21S 32°30E 57 D5
Chinhoyi Zimbabwe 17°20S 30°8E 55 F3
Chini India 31°32N 78°15E 42 D8
Chiniot Pakistan 31°45N 73°0E 42 D5
Chínipas Mexico 27°23N 108°32W 86 B3
Chinji Pakistan 32°42N 72°22E 42 C5
Chinju = Jinju S. Korea 35°12N 128°2E 35 G15
Chinko → C.A.R. 4°50N 23°53E 52 D4
Chinle U.S.A. 36°9N 109°33W 77 H9
Chinmen Tao Taiwan 24°26N 118°22E 33 G12
Chinnampo = Namp'o N. Korea 38°52N 125°10E 35 E13
Chino Japan 35°59N 138°9E 31 G9
Chino U.S.A. 34°1N 117°41W 79 L9
Chino Valley U.S.A. 34°45N 112°27W 77 J7
Chinon France 47°10N 0°15E 20 C4
Chinook U.S.A. 48°35N 109°14W 76 B9
Chinook Trough Pac. Oc. 44°0N 175°0W 64 C10
Chinsali Zambia 10°30S 32°2E 55 E3
Chinsura = Chunchura India 22°53N 88°27E 43 H13
Chióggia Italy 45°13N 12°17E 22 B5
Chios Greece 38°27N 26°9E 23 E12
Chipata Zambia 13°38S 32°28E 55 E3
Chipindo Angola 13°49S 15°48E 53 G3
Chipinge Zimbabwe 20°13S 32°28E 55 G3
Chipinge ⌓ Zimbabwe 20°14S 33°0E 55 G3
Chipley U.S.A. 30°47N 85°32W 85 F12
Chipman Canada 46°6N 65°53W 73 C6
Chipoka Malawi 13°57S 34°28E 55 E3
Chippenham U.K. 51°27N 2°6W 13 F5
Chippewa → U.S.A. 44°25N 92°5W 80 C7
Chippewa Falls U.S.A. 44°56N 91°24W 80 C8
Chipping Norton U.K. 51°56N 1°32W 13 F6
Chiputneticook Lakes N. Amer. 45°35N 67°35W 81 C20
Chiquián Peru 10°10S 77°0W 92 F3
Chiquibul △ Belize 16°49N 88°52W 88 C2
Chiquimula Guatemala 14°51N 89°37W 88 D2
Chiquinquira Colombia 5°37N 73°50W 92 B4
Chirala India 15°50N 80°26E 40 M12
Chiramba Mozam. 16°55S 34°39E 55 F3
Chirawa India 28°14N 75°42E 42 E6
Chirchiq Uzbekistan 41°29N 69°35E 28 E7
Chiredzi Zimbabwe 21°0S 31°38E 57 C5
Chiricahua ⌓ U.S.A. 32°0N 109°20W 77 K9
Chiricahua Peak U.S.A. 31°51N 109°18W 77 L9
Chirikof I. U.S.A. 55°50N 155°40W 74 D8
Chiriquí, G. de Panama 8°0N 82°10W 88 E3
Chiriquí, L. de Panama 9°10N 82°0W 88 E3
Chirisa ⌓ Zimbabwe 17°53S 28°15E 55 F2
Chirivira Falls Zimbabwe 21°10S 32°12E 55 G3
Chirmiri India 23°15N 82°20E 43 H10
Chirripó Grande, Cerro Costa Rica 9°29N 83°29W 88 E3
Chirundu Zimbabwe 16°3S 28°50E 57 B4
Chisamba Zambia 14°55S 28°20E 55 E2
Chisapani Nepal 28°37N 81°16E 43 E9
Chisasibi Canada 53°50N 79°0W 72 B4
Chisholm Canada 54°55N 114°10W 70 C6
Chisholm U.S.A. 47°29N 92°53W 80 B7
Chishtian Mandi Pakistan 29°50N 72°55E 42 E5
Chisimaio = Kismaayo Somali Rep. 0°22S 42°32E 47 H3
Chisimba Falls Zambia 10°12S 30°56E 55 E3
Chişinău Moldova 47°2N 28°50E 17 E15
Chisos Mts. U.S.A. 29°5N 103°15W 84 G3
Chistopol Russia 55°25N 50°38E 18 C9
Chita Russia 52°0N 113°35E 33 A11
Chitipa Malawi 9°41S 33°19E 55 D3
Chitose Japan 42°49N 141°39E 30 C10
Chitral Pakistan 35°50N 71°56E 40 B7
Chitré Panama 7°59N 80°27W 88 E3
Chittagong Bangla. 22°19N 91°48E 41 H17
Chittagong □ Bangla. 24°5N 91°0E 41 G17
Chittaurgarh India 24°52N 74°38E 42 G6
Chittoor India 13°15N 79°5E 40 N11
Chitungwiza Zimbabwe 18°0S 31°6E 55 F3
Chiusi Italy 43°1N 11°57E 22 C4
Chivasso Italy 45°11N 7°53E 20 D7
Chivhu Zimbabwe 19°2S 30°52E 55 F3
Chivilcoy Argentina 34°55S 60°0W 94 C4
Chiwanda Tanzania 11°23S 34°55E 55 E3
Chizarira Zimbabwe 17°36S 27°45E 55 F2
Chizarira △ Zimbabwe 17°44S 27°52E 55 F2
Chizela Zambia 13°8S 25°0E 55 E2
Chkalov = Orenburg Russia 51°45N 55°6E 18 D10
Chloride U.S.A. 35°25N 114°12W 79 K12
Cho-do N. Korea 38°30N 124°40E 35 E13
Cho Phuoc Hai Vietnam 10°26N 107°18E 39 G6
Choa Chu Kang Singapore 1°22N 103°41E 39 d
Chobe △ Botswana 18°37S 24°23E 56 B4
Chocolate Mts. U.S.A. 33°15N 115°15W 79 M11
Choctawhatchee → U.S.A. 30°25N 86°8W 85 F11
Choele Choel Argentina 39°11S 65°40W 96 D3
Choiseul St. Lucia 13°47N 61°3W 89 f
Choiseul Solomon Is. 7°0S 156°40E 58 B8
Choix Mexico 26°43N 108°17W 86 B3
Chojnice Poland 53°42N 17°32E 17 B9
Chok Chai Thailand 14°44N 102°10E 38 E4
Chōkai-San Japan 39°6N 140°3E 30 E10
Choke Canyon Res. U.S.A. 28°30N 98°20W 84 G5
Chokurdakh Russia 70°38N 147°55E 29 B15
Cholame U.S.A. 35°44N 120°18W 78 K6
Cholet France 47°4N 0°52W 20 C3
Cholguan Chile 37°10S 72°3W 94 D1
Choloma Honduras 15°37N 87°57W 88 C2
Choluteca Honduras 13°20N 87°14W 88 D2
Choluteca → Honduras 13°0N 87°20W 88 D2
Chom Bung Thailand 13°37N 99°36E 38 F2
Chom Thong Thailand 18°25N 98°41E 38 C2
Choma Zambia 16°48S 26°59E 55 F2
Chomolungma = Everest, Mt. Nepal 28°5N 86°58E 43 E12
Chomun India 27°15N 75°40E 42 F6
Chomutov Czech Rep. 50°28N 13°23E 16 C7
Chon Buri Thailand 13°21N 101°1E 38 F3
Chon Thanh Vietnam 11°24N 106°36E 39 G6
Chone Ecuador 0°40S 80°0W 92 D3
Chong Kai Cambodia 13°57N 103°35E 38 F4
Chong Mek Thailand 15°10N 105°27E 38 E5
Chong Phangan Thailand 9°39N 100°0E 39 b
Chong Samui Thailand 9°21N 99°50E 39 b
Ch'ŏngjin N. Korea 41°47N 129°50E 35 D15
Chŏngju N. Korea 39°40N 125°5E 35 E13
Chongli China 40°58N 115°15E 34 D8
Chŏngp'yŏng N. Korea 39°47N 127°23E 35 E14
Chongqing China 29°35N 106°25E 32 F10
Chongqing Shi □ China 30°0N 108°0E 32 F10
Chonguene Mozam. 25°3S 33°49E 57 C5
Chongwe Zambia 15°21S 28°42E 55 F2
Chonogol Mongolia 45°54N 115°22E 34 B8
Chonos, Arch. de los Chile 45°0S 75°0W 96 F2
Chop Ukraine 48°26N 22°12E 17 D12
Chopim → Brazil 25°35S 53°5W 95 B5
Chor Pakistan 25°31N 69°46E 42 G3
Chora Sfakion Greece 35°15N 24°9E 25 D6
Chorbat La India 34°42N 76°37E 43 B7
Chorley U.K. 53°39N 2°38W 12 D5
Chornobyl Ukraine 51°20N 30°15E 17 C16
Chorolque, Cerro Bolivia 20°59S 66°5W 94 A2
Chorregon Australia 22°40S 143°32E 62 C3
Chorro el Indio △ Venezuela 7°43N 72°9W 89 E5
Chortkiv Ukraine 49°2N 25°46E 17 D13
Chorzów Poland 50°18N 18°57E 17 C10
Chos-Malal Argentina 37°20S 70°15W 94 D1
Ch'osan N. Korea 40°50N 125°47E 35 D13
Choszczno Poland 53°7N 15°25E 16 B8
Choteau U.S.A. 47°49N 112°11W 76 C7
Chotila India 22°23N 71°15E 42 H4
Chotta Udepur India 22°19N 74°1E 42 H6
Chowchilla U.S.A. 37°7N 120°16W 78 H6
Choybalsan Mongolia 48°4N 114°30E 33 B11
Choyr Mongolia 46°24N 108°30E 33 B10
Christchurch N.Z. 43°33S 172°47E 59 E4
Christchurch U.K. 50°44N 1°47W 13 G6
Christian I. Canada 44°50N 80°12W 82 B4
Christian Sd. U.S.A. 55°56N 134°40W 75 D13
Christiana S. Africa 27°52S 25°8E 56 D4
Christiansted U.S. Virgin Is. 17°45N 64°42W 89 C7
Christie B. Canada 62°32N 111°10W 71 A6
Christina → Canada 56°40N 111°3W 71 B6
Christmas Cr. → Australia 18°29S 125°23E 60 C4
Christmas I. = Kiritimati Kiribati 1°58N 157°27W 65 G12
Christmas I. Ind. Oc. 10°30S 105°40E 64 J2
Christopher, L. Australia 24°49S 127°42E 61 D4
Chtimba Malawi 10°35S 34°13E 55 E3
Chū = Shū Kazakhstan 43°36N 73°42E 28 E8
Chu → Vietnam 19°53N 105°45E 38 C5
Chu Lai Vietnam 15°28N 108°45E 38 E7
Chu Mom Ray △ Vietnam 14°29N 107°38E 38 E6
Chu Se Vietnam 13°40N 108°8E 38 F7
Chuadanga Bangla. 23°38N 88°51E 43 H13
Chuak, Ko Thailand 9°28N 99°41E 39 b
Ch'uanchou = Quanzhou China 24°55N 118°34E 33 G12
Chuankou China 34°20N 110°59E 34 G6
Chubbuck U.S.A. 42°55N 112°28W 76 E7
Chūbu □ Japan 36°45N 137°30E 31 F8
Chūbu-Sangaku △ Japan 36°30N 137°40E 31 F8
Chubut → Argentina 43°20S 65°5W 96 E3
Chuchi L. Canada 55°12N 124°30W 70 B4
Chuda India 22°29N 71°41E 42 H4
Chudskoye, Ozero Russia 58°13N 27°30E 9 G22
Chugach Mts. U.S.A. 60°45N 147°0W 74 C10
Chūgoku □ Japan 35°0N 133°0E 31 G6
Chūgoku-Sanchi Japan 35°0N 133°0E 31 G6
Chugwater U.S.A. 41°46N 104°50W 76 F11
Chuka Kenya 0°20S 37°39E 54 C4
Chukchi Plateau Arctic 78°0N 165°0W 4 B17
Chukchi Sea Russia 68°0N 175°0W 29 C19
Chukotskiy Mys Russia 64°14N 173°8W 74 C5
Chukotskiy Poluostrov Russia 66°0N 174°0W 74 B5
Chukotskoye Nagorye Russia 68°0N 175°0E 29 C18
Chula Vista U.S.A. 32°38N 117°5W 79 N9
Chulucanas Peru 5°8S 80°10W 92 E2
Chulym → Russia 57°43N 83°51E 28 D9
Chum Phae Thailand 16°40N 102°6E 38 D4
Chum Saeng Thailand 15°55N 100°15E 38 E3
Chumar India 32°40N 78°35E 43 C8
Chumbicha Argentina 29°0S 66°10W 94 B2
Chumikan Russia 54°40N 135°10E 29 D14
Chumphon Thailand 10°35N 99°14E 39 G2
Chumuare Mozam. 14°31S 31°50E 55 E3
Chuna → Russia 57°47N 94°37E 29 D10
Chuncheon S. Korea 37°58N 127°44E 35 F14
Chunchura India 22°53N 88°27E 43 H13
Chunga Zambia 15°0S 26°2E 55 F2
Chunggang-ŭp N. Korea 41°48N 126°48E 35 D14
Chunghwa N. Korea 38°52N 125°47E 35 E13
Chungju S. Korea 36°58N 127°58E 35 F14
Chungking = Chongqing China 29°35N 106°25E 32 F10
Chungt'iaoshan = Zhongtiao Shan China 35°0N 111°10E 34 G6
Chunian Pakistan 30°57N 74°0E 42 D6
Chunya Tanzania 8°30S 33°27E 55 D3
Chunyang China 43°38N 129°23E 35 C15
Chupara Pt. Trin. & Tob. 10°49N 61°22W 93 K15
Chuquibamba Peru 15°47S 72°44W 92 G4
Chuquicamata Chile 22°15S 69°0W 94 A2
Chur Switz. 46°52N 9°32E 20 C8
Churachandpur India 24°20N 93°40E 41 G18
Church Stretton U.K. 52°32N 2°48W 13 E5
Churchill Canada 58°47N 94°11W 71 B10
Churchill → Man., Canada 58°47N 94°12W 71 B10
Churchill → Nfld. & L., Canada 53°19N 60°10W 73 B7
Churchill, C. Canada 58°46N 93°12W 71 B10
Churchill Falls Canada 53°36N 64°19W 73 B7
Churchill L. Canada 55°55N 108°20W 71 B7
Churchill Pk. Canada 58°10N 125°10W 70 B3
Churki India 23°50N 83°12E 43 H10
Churu India 28°20N 74°50E 42 E6
Churún Merú = Angel Falls Venezuela 5°57N 62°30W 92 B6
Chushal India 33°40N 78°40E 43 C8
Chuska Mts. U.S.A. 36°15N 108°50W 77 H9
Chusovoy Russia 58°22N 57°50E 18 C10
Chute-aux-Outardes Canada 49°7N 68°24W 73 C6
Chuuk = Truk Micronesia 7°25N 151°46E 64 G7
Chuvash Republic = Chuvashia □ Russia 55°30N 47°0E 18 C8
Chuvashia □ Russia 55°30N 47°0E 18 C8
Chuwārtah Iraq 35°43N 45°34E 44 C5
Chuxiong China 25°2N 101°28E 32 F9
Chuy Uruguay 33°41S 53°27W 95 C5
Ci Xian China 36°20N 114°25E 34 F8
Ciadâr-Lunga Moldova 46°3N 28°51E 17 E15
Ciamis Indonesia 7°20S 108°21E 37 G13
Cianjur Indonesia 6°49S 107°8E 37 G12
Cianorte Brazil 23°37S 52°37W 95 A5
Cibola U.S.A. 33°17N 114°42W 79 M12
Cicero U.S.A. 41°51N 87°44W 80 E10
Cicia Fiji 17°45S 179°18W 59 a
Ciéagas del Cataumbo △ Venezuela 9°25N 71°54W 89 E5
Ciechanów Poland 52°52N 20°38E 17 B11
Ciego de Ávila Cuba 21°50N 78°50W 88 B4
Ciénaga Colombia 11°1N 74°15W 92 A4
Cienfuegos Cuba 22°10N 80°30W 88 B3
Cieszyn Poland 49°45N 18°35E 17 D10
Cieza Spain 38°17N 1°23W 21 C5
Cihuatlán Mexico 19°14N 104°35W 86 D4
Cijara, Embalse de Spain 39°18N 4°52W 21 C3
Cijulang Indonesia 7°42S 108°27E 37 G13
Cilacap Indonesia 7°43S 109°0E 37 G13
Cill Airne = Killarney Ireland 52°4N 9°30W 10 D2
Cill Chainnigh = Kilkenny Ireland 52°39N 7°15W 10 D4
Cill Dara = Kildare Ireland 53°9N 6°55W 10 C5
Cill Mhantáin = Wicklow Ireland 52°59N 6°3W 10 D5
Cilo Dağı Turkey 37°28N 43°55E 19 G7
Cima U.S.A. 35°14N 115°30W 79 K11
Cimarron Kans., U.S.A. 37°48N 100°21W 80 G3
Cimarron N. Mex., U.S.A. 36°31N 104°55W 77 H11
Cimarron → U.S.A. 36°10N 96°16W 84 C6
Cimişlia Moldova 46°34N 28°44E 17 E15
Cimone, Mte. Italy 44°12N 10°42E 22 B4
Cinca → Spain 41°26N 0°21E 21 B6
Cincar Bos.-H. 43°55N 17°5E 22 C7
Cincinnati U.S.A. 39°9N 84°27W 81 F11
Cincinnatus U.S.A. 42°33N 75°54W 83 D9
Çine Turkey 37°37N 28°2E 23 F13
Ciney Belgium 50°18N 5°5E 15 D5
Cinto, Mte. France 42°24N 8°54E 20 E8
Circle Alaska, U.S.A. 65°50N 144°4W 74 B11
Circle Mont., U.S.A. 47°25N 105°35W 76 C11
Circleville U.S.A. 39°36N 82°57W 81 F12
Cirebon Indonesia 6°45S 108°32E 37 G13
Cirencester U.K. 51°43N 1°57W 13 F6
Cirium Cyprus 34°40N 32°53E 25 E11
Cisco U.S.A. 32°23N 98°59W 84 E5
Citlaltépetl = Orizaba, Pico de Mexico 18°58N 97°15W 87 D5
Citrus Heights U.S.A. 38°42N 121°17W 78 G5
Citrusdal S. Africa 32°35S 19°0E 56 E2
Città del Vaticano = Vatican City ■ Europe 41°54N 12°27E 22 D5
Città di Castello Italy 43°27N 12°14E 22 C5
Ciudad Acuña Mexico 29°18N 100°55W 86 B4
Ciudad Altamirano Mexico 18°20N 100°40W 86 D4
Ciudad Anáhuac Mexico 27°14N 100°7W 86 B4
Ciudad Bolívar Venezuela 8°5N 63°36W 92 B6
Ciudad Camargo Mexico 27°40N 105°10W 86 B3
Ciudad Cortés Costa Rica 8°55N 84°0W 88 E3
Ciudad de México Mexico 19°24N 99°9W 87 D5
Ciudad de Valles Mexico 22°0N 99°0W 87 C5
Ciudad del Carmen Mexico 18°38N 91°50W 87 D6
Ciudad del Este Paraguay 25°30S 54°50W 95 B5
Ciudad Delicias = Delicias Mexico 28°13N 105°28W 86 B3
Ciudad Frontera Mexico 26°56N 101°27W 86 B4
Ciudad Guayana Venezuela 8°0N 62°30W 92 B6
Ciudad Guerrero Mexico 28°33N 107°30W 86 B3
Ciudad Guzmán Mexico 19°41N 103°29W 86 D4
Ciudad Juárez Mexico 31°44N 106°29W 86 A3
Ciudad Lerdo Mexico 25°32N 103°32W 86 B4
Ciudad Madero Mexico 22°19N 97°50W 87 C5
Ciudad Mante Mexico 22°44N 98°59W 87 C5
Ciudad Obregón Mexico 27°29N 109°56W 86 B3
Ciudad Real Spain 38°59N 3°55W 21 C4
Ciudad Rodrigo Spain 40°35N 6°32W 21 B2
Ciudad Victoria Mexico 23°44N 99°8W 87 C5
Ciudadela Spain 40°0N 3°50E 24 B10
Civitanova Marche Italy 43°18N 13°44E 22 C5
Civitavécchia Italy 42°6N 11°48E 22 C4
Cizre Turkey 37°19N 42°10E 44 B4
Clackmannanshire □ U.K. 56°10N 3°43W 11 E5
Clacton-on-Sea U.K. 51°47N 1°11E 13 F9
Claire, L. Canada 58°35N 112°5W 70 B6
Clairton U.S.A. 40°18N 79°53W 82 F5
Clallam Bay U.S.A. 48°15N 124°16W 78 B2
Clanton U.S.A. 32°51N 86°38W 85 E11
Clanwilliam S. Africa 32°11S 18°52E 56 E2
Clara Ireland 53°21N 7°37W 10 C4
Claraville U.S.A. 35°24N 118°20W 79 K8
Clare Australia 33°50S 138°37E 63 E2
Clare U.S.A. 43°49N 84°46W 81 D11
Clare □ Ireland 52°45N 9°0W 10 D3
Clare → Ireland 53°20N 9°2W 10 C2
Clare I. Ireland 53°49N 10°0W 10 C1
Claremont Calif., U.S.A. 34°6N 117°43W 79 L9
Claremont N.H., U.S.A. 43°23N 72°20W 83 C12
Claremont Pt. Australia 14°1S 143°41E 62 A3
Claremore U.S.A. 36°19N 95°36W 84 C7
Claremorris Ireland 53°45N 9°0W 10 C3
Clarence → Australia 29°25S 153°22E 63 D5
Clarence → N.Z. 42°10S 173°56E 59 E4
Clarence, I. Chile 54°0S 72°0W 96 G2
Clarence I. Antarctica 61°10S 54°0W 5 C18
Clarence Str. Australia 12°0S 131°0E 60 B5
Clarence Town Bahamas 23°6N 74°59W 89 B5
Clarendon Pa., U.S.A. 41°47N 79°6W 82 E5
Clarendon Tex., U.S.A. 34°56N 100°53W 84 D4
Clarenville Canada 48°10N 54°1W 73 C9
Claresholm Canada 50°2N 113°33W 70 D6
Clarie Coast Antarctica 68°0S 135°0E 5 C9
Clarinda U.S.A. 40°44N 95°2W 80 E6
Clarington Canada 43°55N 78°41W 82 C6
Clarion Iowa, U.S.A. 42°44N 93°44W 80 D7
Clarion Pa., U.S.A. 41°13N 79°23W 82 E5
Clarion → U.S.A. 41°7N 79°41W 82 E5
Clarion Fracture Zone Pac. Oc. 20°0N 120°0W 66 H7
Clark U.S.A. 44°53N 97°44W 80 C5
Clark, Pt. Canada 44°4N 81°45W 82 B3
Clark Fork U.S.A. 48°9N 116°11W 76 B5
Clark Fork → U.S.A. 48°9N 116°15W 76 B5
Clarkdale U.S.A. 34°46N 112°3W 77 J7
Clarke City Canada 50°12N 66°38W 73 B6
Clarke I. Australia 40°32S 148°10E 63 G4
Clarke Ra. Australia 20°40S 148°30E 62 b
Clarks Fork Yellowstone → U.S.A. 45°32N 108°50W 76 D9
Clark's Harbour Canada 43°25N 65°38W 73 D6
Clarks Hill L. = J. Strom Thurmond L. U.S.A. 33°40N 82°12W 85 E13
Clarks Summit U.S.A. 41°30N 75°42W 83 E9
Clarksburg U.S.A. 39°17N 80°30W 81 F13
Clarksdale U.S.A. 34°12N 90°35W 85 D9
Clarkston U.S.A. 46°25N 117°3W 76 C5
Clarksville Ark., U.S.A. 35°28N 93°28W 84 D8
Clarksville Tenn., U.S.A. 36°32N 87°21W 85 C11
Clarksville Tex., U.S.A. 33°37N 95°3W 84 E7
Clatskanie U.S.A. 46°6N 123°12W 78 D3
Claude U.S.A. 35°7N 101°22W 84 D4
Claveria Phil. 18°37N 121°4E 37 A6
Clay U.S.A. 38°17N 121°10W 78 G5
Clay Center U.S.A. 39°23N 97°8W 80 F5
Claypool U.S.A. 33°25N 110°51W 77 K8
Claysburg U.S.A. 40°17N 78°27W 82 F6
Claysville U.S.A. 40°7N 80°25W 82 F4
Clayton Canada 45°11N 76°19W 83 A8
Clayton N. Mex., U.S.A. 36°27N 103°11W 77 H12
Clayton N.Y., U.S.A. 44°14N 76°5W 83 B8
Clear, C. Ireland 51°25N 9°32W 10 E2
Clear, L. Canada 45°26N 77°12W 82 A7
Clear Hills Canada 56°40N 119°30W 70 B5
Clear I. Ireland 51°26N 9°30W 10 E2
Clear L. U.S.A. 39°2N 122°47W 78 F4
Clear Lake Iowa, U.S.A. 43°8N 93°23W 80 D7
Clear Lake S. Dak., U.S.A. 44°45N 96°41W 80 C5
Clear Lake Res. U.S.A. 41°56N 121°5W 76 F3
Clearfield Pa., U.S.A. 41°2N 78°27W 82 E6
Clearfield Utah, U.S.A. 41°7N 112°2W 76 F7
Clearlake U.S.A. 38°57N 122°38W 78 G4
Clearwater Canada 51°38N 120°2W 70 C4
Clearwater U.S.A. 27°59N 82°48W 85 H13
Clearwater → Alta., Canada 52°22N 114°57W 70 C6
Clearwater → Alta., Canada 56°44N 111°23W 71 B6
Clearwater L. Canada 53°34N 99°49W 71 C9
Clearwater Lake ⌓ Canada 54°0N 101°0W 71 C8
Clearwater Mts. U.S.A. 46°5N 115°20W 76 C6
Clearwater River ⌓ Canada 56°55N 109°10W 71 B7
Cleburne U.S.A. 32°21N 97°23W 84 E6
Clee Hills U.K. 52°26N 2°35W 13 E5
Cleethorpes U.K. 53°33N 0°3W 12 D7
Cleeve Cloud U.K. 51°56N 2°0W 13 F6
Clemson U.S.A. 34°41N 82°50W 85 D13
Clerke Reef Australia 17°22S 119°20E 60 C2
Clermont Australia 22°49S 147°39E 62 C4
Clermont U.S.A. 28°33N 81°46W 85 G14
Clermont-Ferrand France 45°46N 3°4E 20 D5
Clervaux Lux. 50°4N 6°2E 15 D6
Clevedon U.K. 51°26N 2°52W 13 F5
Cleveland Miss., U.S.A. 33°45N 90°43W 85 E9
Cleveland Ohio, U.S.A. 41°29N 81°41W 82 E3
Cleveland Okla., U.S.A. 36°19N 96°28W 84 C6
Cleveland Tenn., U.S.A. 35°10N 84°53W 85 D12
Cleveland Tex., U.S.A. 30°21N 95°5W 84 F7
Cleveland, C. Australia 19°11S 147°1E 62 B4
Cleveland, Mt. U.S.A. 48°56N 113°51W 76 B7
Cleveland Heights U.S.A. 41°31N 81°33W 82 E3
Clevelândia Brazil 26°24S 52°23W 95 B5
Clew B. Ireland 53°50N 9°49W 10 C2
Clewiston U.S.A. 26°45N 80°56W 85 H1
Clifden Ireland 53°29N 10°1W 10 C
Clifden N.Z. 46°1S 167°42E 59 G
Cliffdell U.S.A. 46°56N 121°5W 78 D
Cliffy Hd. Australia 35°1S 116°29E 61 G
Clifton Australia 27°59S 151°53E 63 D5
Clifton Ariz., U.S.A. 33°3N 109°18W 77 K
Clifton Colo., U.S.A. 39°7N 108°25W 76 G
Clifton Tex., U.S.A. 31°47N 97°35W 84 F
Clifton Beach Australia 16°46S 145°39E 62 B
Clinch → U.S.A. 35°53N 84°29W 85 D1
Clingmans Dome U.S.A. 35°34N 83°30W 85 D1
Clint U.S.A. 31°35N 106°14W 84 F
Clinton B.C., Canada 51°6N 121°35W 70 C
Clinton Ont., Canada 43°37N 81°32W 82 C
Clinton N.Z. 46°12S 169°23E 59 G
Clinton Ark., U.S.A. 35°36N 92°28W 84 D
Clinton Conn., U.S.A. 41°17N 72°32W 83 E1
Clinton Ill., U.S.A. 40°9N 88°57W 80 E
Clinton Ind., U.S.A. 39°40N 87°24W 80 F1
Clinton Iowa, U.S.A. 41°51N 90°12W 80 E
Clinton Mass., U.S.A. 42°25N 71°41W 83 D1
Clinton Miss., U.S.A. 32°20N 90°20W 85 E
Clinton Mo., U.S.A. 38°22N 93°46W 80 F
Clinton N.C., U.S.A. 35°0N 78°22W 85 D1
Clinton Okla., U.S.A. 35°31N 98°58W 84 D
Clinton S.C., U.S.A. 34°29N 81°53W 85 D1
Clinton Tenn., U.S.A. 36°6N 84°8W 85 C1
Clinton Wash., U.S.A. 47°59N 122°21W 78 C
Clinton, C. Australia 22°30S 150°45E 62 C
Clinton Colden L. Canada 63°58N 107°27W 68 E1
Clintonville U.S.A. 44°37N 88°46W 80 C
Clipperton, Î. Pac. Oc. 10°18N 109°13W 65 F1
Clipperton Fracture Zone Pac. Oc. 19°0N 122°0W 65 G1
Clisham U.K. 57°58N 6°49W 11 D
Clitheroe U.K. 53°53N 2°22W 12 D
Clo-oose Canada 48°39N 124°49W 78 B
Cloates, Pt. Australia 22°43S 113°40E 60 D
Clocolan S. Africa 28°55S 27°34E 57 D
Clodomira Argentina 27°35S 64°14W 94 B
Clogher Hd. Ireland 53°48N 6°14W 10 C
Clonakilty Ireland 51°37N 8°53W 10 E
Clonakilty B. Ireland 51°35N 8°51W 10 E
Cloncurry Australia 20°40S 140°28E 62 C
Cloncurry Ireland 53°25N 6°48W 10 C
Cloncurry → Australia 18°37S 140°40E 62 B
Clondalkin Ireland 53°19N 6°25W 10 C
Clones Ireland 54°11N 7°15W 10 B
Clonmel Ireland 52°21N 7°42W 10 D
Cloquet U.S.A. 46°43N 92°28W 80 B
Clorinda Argentina 25°16S 57°45W 94 B
Cloud Peak U.S.A. 44°23N 107°11W 76 D1
Cloudcroft U.S.A. 32°58N 105°45W 77 K1
Cloverdale U.S.A. 38°48N 123°1W 78 C
Clovis Calif., U.S.A. 36°49N 119°42W 78 J
Clovis N. Mex., U.S.A. 34°24N 103°12W 77 J1
Cloyne Canada 44°49N 77°11W 82 B
Cluain Meala = Clonmel Ireland 52°21N 7°42W 10 D
Cluj-Napoca Romania 46°47N 23°38E 17 E
Clunes Australia 37°20S 143°45E 63
Clutha → N.Z. 46°20S 169°49E 59
Clwyd → U.K. 53°19N 3°31W 12
Clyde Canada 54°9N 113°39W 70
Clyde N.Z. 45°12S 169°20E 59
Clyde U.S.A. 43°5N 76°52W 82
Clyde → U.K. 55°55N 4°30W 11
Clyde, Firth of U.K. 55°22N 5°1W 11
Clyde Muirshiel ⌓ U.K. 55°50N 4°40W 11
Clyde River Canada 70°30N 68°30W 69 C
Clydebank U.K. 55°54N 4°23W 11
Clymer N.Y., U.S.A. 42°1N 79°37W 82
Clymer Pa., U.S.A. 40°40N 79°1W 82
Coachella U.S.A. 33°41N 116°10W 79 M
Coachella Canal U.S.A. 32°43N 114°57W 79 N
Coahoma U.S.A. 32°18N 101°18W 84
Coahuayana → Mexico 18°41N 103°45W 86
Coahuila □ Mexico 27°20N 102°0W 86
Coal → Canada 59°39N 126°57W 70
Coalane Mozam. 17°48S 37°2E 55
Coalcomán Mexico 18°47N 103°9W 86
Coaldale Canada 49°45N 112°35W 70
Coalgate U.S.A. 34°32N 96°13W 84
Coalinga U.S.A. 36°9N 120°21W 78
Coalisland U.K. 54°33N 6°42W 10
Coalville U.K. 52°44N 1°23W 12
Coalville U.S.A. 40°55N 111°24W 76

mo Puerto Rico 18°5N 66°22W 89 d
ari Brazil 4°8S 63°7W 92 D6
ast □ Kenya 2°40S 39°45E 54 C4
ast Mts. Canada 55°0N 129°20W 70 C3
ast Ranges U.S.A. 39°0N 123°0W 78 G4
atbridge U.K. 55°52N 4°6W 11 F4
atepec Mexico 19°27N 96°58W 87 D5
atepeque Guatemala 14°46N 91°55W 88 D1
atesville U.S.A. 39°59N 75°50W 81 F16
aticook Canada 45°10N 71°46W 83 A13
ats I. Canada 62°30N 83°0W 69 E15
ats Land Antarctica 77°0S 25°0W 5 D1
atzacoalcos Mexico 18°7N 94°25W 87 D6
bá Mexico 20°31N 87°45W 87 C7
balt Canada 47°25N 79°42W 72 C4
bán Guatemala 15°30N 90°21W 88 C1
bar Australia 31°27S 145°48E 63 E4
bh Ireland 51°51N 8°17W 10 E3
bija Bolivia 11°0S 68°50W 92 F5
bleskill U.S.A. 42°41N 74°29W 83 D10
boconk Canada 44°39N 78°48W 82 B6
bourg Canada 43°58N 78°10W 82 C6
bourg Pen. Australia 11°20S 132°15E 60 B5
bram Australia 35°54S 145°40E 63 F4
bué Mozam. 12°0S 34°58E 55 E3
burg Germany 50°15N 10°58E 16 C6
burg I. Canada 75°57N 79°26W 69 B16
canada = Kakinada
India 16°57N 82°11E 41 L13
chabamba Bolivia 17°26S 66°10W 92 G5
chemane Mozam. 17°0S 32°54E 55 F3
chin India 9°58N 76°20E 40 Q10
chin China = Nam-Phan
Vietnam 10°30N 106°0E 39 G6
chran U.S.A. 32°23N 83°21W 85 E13
chrane Alta., Canada 51°11N 114°30W 70 C6
chrane Ont., Canada 49°0N 81°0W 72 C3
chrane Chile 47°15S 72°33W 96 F2
chrane → Canada 59°0N 103°40W 71 B8
chrane, L. Chile 47°10S 72°0W 96 F2
chranton U.S.A. 41°31N 80°3W 82 E4
ckburn Australia 32°5S 141°0E 63 E3
ckburn, Canal Chile 54°30S 72°0W 96 G2
ckburn I. Canada 45°55N 83°22W 72 C3
ckburn Ra. Australia 15°46S 128°0E 60 C4
ckermouth U.K. 54°40N 3°22W 12 C4
cklebiddy Australia 32°0S 126°3E 61 F4
ckpit Country, The
Jamaica 18°15N 77°45W 88 a
co → Cent. Amer. 15°0N 83°8W 88 D3
co, I. del Pac. Oc. 5°25N 87°55W 65 G19
coa U.S.A. 28°21N 80°44W 85 G14
cobeach Gabon 0°59N 9°34E 52 D1
cos B. Trin. & Tob. 10°25N 61°2W 93 K15
cos Is. Ind. Oc. 12°10S 96°55E 64 J1
cos Ridge Pac. Oc. 4°0N 88°0W 65 G19
d, C. U.S.A. 42°5N 70°10W 81 D18
dajás Brazil 3°55S 62°0W 92 D6
dó Brazil 4°30S 43°55W 93 D10
dy U.S.A. 44°32N 109°3W 76 D9
e Hill Canada 44°52N 77°50W 82 B7
elemu Chile 36°30S 72°48W 94 D1
en Australia 13°52S 143°12E 62 A3
eur d'Alene U.S.A. 47°41N 116°46W 76 C5
eur d'Alene L.
U.S.A. 47°32N 116°49W 76 C5
evorden Neths. 52°40N 6°44E 15 B6
fete Canary Is. 28°6N 14°23W 24 F5
feyville U.S.A. 37°2N 95°37W 80 G6
ffin B. Australia 34°38S 135°28E 63 E2
ffin Bay Australia 34°37S 135°29E 63 E2
ffin Bay △ Australia 34°34S 135°19E 63 E2
ffin Bay Peninsula
Australia 34°32S 135°15E 63 E2
ffs Harbour Australia 30°16S 153°5E 63 E5
gnac France 45°41N 0°20W 20 D3
hocton U.S.A. 42°30N 77°30W 82 D7
hocton → U.S.A. 42°9N 77°6W 82 D7
hoes U.S.A. 42°46N 73°42W 83 D11
huna Australia 35°45S 144°15E 63 F3
iba, I. de Panama 7°30N 81°40W 88 E3
ig → Argentina 51°0S 69°10W 96 G3
igeach, Rubha U.K. 58°6N 5°26W 11 C3
ihaique Chile 45°30S 71°45W 96 F2
imbatore India 11°2N 76°59E 40 P10
imbra Brazil 19°55S 57°48W 92 G7
imbra Portugal 40°15N 8°27W 21 B1
ín Spain 36°40N 4°48W 21 D3
ipasa, Salar de Bolivia 19°26S 68°9W 92 G5
jimies Ecuador 0°20N 80°0W 92 C3
jutepeque El Salv. 13°41N 88°54W 88 D2
keville U.S.A. 42°5N 110°57W 76 E8
lac Australia 38°21S 143°35E 63 F3
latina Brazil 19°32S 40°37W 93 G10
lbeck, C. Antarctica 77°6S 157°48W 5 D13
lborne Canada 44°0N 77°53W 82 C7
lby U.S.A. 39°24N 101°3W 80 F3
lca → Peru 15°55S 72°43W 92 G4
lchester U.K. 51°54N 0°55E 13 F8
lchester U.S.A. 41°35N 72°20W 83 E12
ld L. Canada 54°33N 110°5W 71 C7
ld Lake Canada 54°27N 110°10W 71 C6
ldstream Canada 50°13N 119°11W 70 C5
ldstream U.K. 55°39N 2°15W 11 F6
ldwater Canada 44°42N 79°40W 82 B5
ldwater Kans., U.S.A. 37°16N 99°20W 80 G4
ldwater Mich., U.S.A. 41°57N 85°0W 81 E11
lebrook U.S.A. 44°54N 71°30W 83 B13
leman Canada 49°40N 114°30W 70 D6
leman U.S.A. 31°50N 99°26W 84 F5
leman → Australia 15°6S 141°38E 62 B3
lenso S. Africa 28°44S 29°50E 57 D4
leraine Australia 37°36S 141°40E 63 F3
leraine U.K. 55°8N 6°41W 10 A5
leridge, L. N.Z. 43°17S 171°30E 59 E3
lesberg S. Africa 30°45S 25°5E 56 E4
leville U.S.A. 38°34N 119°30W 78 G7
lfax Calif., U.S.A. 39°6N 120°57W 78 F6
lfax La., U.S.A. 31°31N 92°42W 84 F8
lfax Wash., U.S.A. 46°53N 117°22W 76 C5

Colhué Huapi, L.
Argentina 45°30S 69°0W 96 F3
Coligny S. Africa 26°17S 26°15E 56 D4
Colima Mexico 19°14N 103°43W 86 D4
Colima □ Mexico 19°10N 104°0W 86 D4
Colima, Nevado de
Mexico 19°33N 103°38W 86 D4
Colina Chile 33°13S 70°45W 94 C1
Colinas Brazil 6°0S 44°10W 93 E10
Coll U.K. 56°39N 6°34W 11 E2
Collahuasi Chile 21°5S 68°45W 94 A2
Collarenebri Australia 29°33S 148°34E 63 D4
Colleen Bawn Zimbabwe 21°0S 29°12E 55 G2
College U.S.A. 64°52N 147°49W 74 C10
College Park U.S.A. 33°39N 84°27W 85 E12
College Station U.S.A. 30°37N 96°21W 84 F6
Collie Australia 33°22S 116°8E 61 F2
Collier B. Australia 16°10S 124°15E 60 C3
Collier Ra. Australia 24°45S 119°10E 61 D2
Collier Range △
Australia 24°39S 119°7E 61 D2
Collierville U.S.A. 35°3N 89°40W 85 D10
Collina, Passo di Italy 44°2N 10°56E 22 B4
Collingwood Canada 44°29N 80°13W 82 B4
Collingwood N.Z. 40°41S 172°40E 59 D4
Collins Canada 50°17N 89°27W 72 B2
Collins Bay Canada 44°14N 76°36W 83 B8
Collinson Pen. Canada 69°58N 101°24W 68 D11
Collinsville Australia 20°30S 147°56E 62 C4
Collipulli Chile 37°55S 72°30W 94 D1
Collooney Ireland 54°11N 8°29W 10 B3
Colmar France 48°5N 7°20E 20 B7
Colo → Australia 33°25S 150°52E 63 E5
Cologne = Köln Germany 50°56N 6°57E 16 C4
Colom, I. d'en Spain 39°58N 4°16E 24 B11
Coloma U.S.A. 38°48N 120°53W 78 G6
Colomb-Béchar = Béchar
Algeria 31°38N 2°18W 50 B5
Colombia ■ S. Amer. 3°45N 73°0W 92 C4
Colombian Basin
Caribbean 14°0N 76°0W 88 D4
Colombo Sri Lanka 6°56N 79°58E 40 R11
Colón B. Aires, Argentina 33°53S 61°7W 94 C3
Colón Entre Ríos,
Argentina 32°12S 58°10W 94 C4
Colón Cuba 22°42N 80°54W 88 B3
Colón Panama 9°20N 79°54W 88 E4
Colón, Arch. de Ecuador 0°0 91°0W 90 D1
Colònia de Sant Jordi
Spain 39°19N 2°59E 24 B9
Colonia del Sacramento
Uruguay 34°25S 57°50W 94 C4
Colonia Dora Argentina 28°34S 62°59W 94 B3
Colonial Beach U.S.A. 38°15N 76°58W 81 F15
Colonie U.S.A. 42°43N 73°50W 83 D11
Colonsay Canada 51°59N 105°52W 71 C7
Colonsay U.K. 56°5N 6°12W 11 E2
Colorado □ U.S.A. 39°30N 105°30W 76 G11
Colorado → Argentina 39°50S 62°8W 96 D4
Colorado → N. Amer. 31°45N 114°40W 77 L6
Colorado → U.S.A. 28°36N 95°59W 84 G7
Colorado City U.S.A. 32°24N 100°52W 84 E4
Colorado Plateau U.S.A. 37°0N 111°0W 77 H8
Colorado River Aqueduct
U.S.A. 33°50N 117°23W 79 L12
Colorado Springs
U.S.A. 38°50N 104°49W 76 G11
Colotlán Mexico 22°6N 103°16W 86 C4
Colstrip U.S.A. 45°53N 106°38W 76 D10
Colton U.S.A. 44°33N 74°56W 83 B10
Columbia Ky., U.S.A. 37°6N 85°18W 81 G11
Columbia La., U.S.A. 32°6N 92°5W 84 E8
Columbia Miss., U.S.A. 31°15N 89°50W 85 F10
Columbia Mo., U.S.A. 38°57N 92°20W 80 F7
Columbia Pa., U.S.A. 40°2N 76°30W 83 F8
Columbia S.C., U.S.A. 34°0N 81°2W 85 D14
Columbia Tenn., U.S.A. 35°37N 87°2W 85 D11
Columbia → N. Amer. 46°15N 124°5W 78 D2
Columbia, C. Canada 83°6N 69°57W 69 A18
Columbia, District of □
U.S.A. 38°55N 77°0W 81 F15
Columbia, Mt. Canada 52°8N 117°20W 70 C5
Columbia Basin U.S.A. 46°45N 119°5W 76 C4
Columbia Falls U.S.A. 48°23N 114°11W 76 B6
Columbia Mts. Canada 52°0N 119°0W 70 C5
Columbia Plateau
U.S.A. 44°0N 117°30W 76 E5
Columbiana U.S.A. 40°53N 80°42W 82 F4
Columbretes, Is. Spain 39°50N 0°50E 21 C6
Columbus Ga., U.S.A. 32°28N 84°59W 85 E12
Columbus Ind., U.S.A. 39°13N 85°55W 81 F11
Columbus Kans., U.S.A. 37°10N 94°50W 80 G6
Columbus Miss., U.S.A. 33°30N 88°25W 85 E10
Columbus Mont.,
U.S.A. 45°38N 109°15W 76 D9
Columbus N. Mex.,
U.S.A. 31°50N 107°38W 77 L10
Columbus Nebr., U.S.A. 41°26N 97°22W 80 E5
Columbus Ohio, U.S.A. 39°58N 83°0W 81 F12
Columbus Tex., U.S.A. 29°42N 96°33W 84 G6
Colusa U.S.A. 39°13N 122°1W 78 F4
Colville U.S.A. 48°33N 117°54W 76 B5
Colville → U.S.A. 70°25N 150°30W 74 A9
Colville, C. N.Z. 36°29S 175°21E 59 B5
Colwood Canada 48°26N 123°29W 78 B3
Colwyn Bay U.K. 53°18N 3°44W 12 D4
Comácchio Italy 44°42N 12°11E 22 B5
Comalcalco Mexico 18°16N 93°13W 87 D6
Comallo Argentina 41°0S 70°5W 96 E2
Comanche U.S.A. 31°54N 98°36W 84 F5
Comandante Ferraz
Antarctica 62°30S 58°0W 5 C18
Comayagua Honduras 14°25N 87°37W 88 D2
Combahee → U.S.A. 32°31N 80°31W 85 E14
Combarbalá Chile 31°11S 71°2W 94 C1
Combe Martin U.K. 51°12N 4°3W 13 F3
Comber Canada 42°14N 82°33W 82 D2
Comber U.K. 54°33N 5°45W 10 B6
Combermere Canada 45°22N 77°37W 82 A7
Comblain-au-Pont
Belgium 50°29N 5°35E 15 D5

Comeragh Mts. Ireland 52°18N 7°34W 10 D4
Comet Australia 23°36S 148°38E 62 C4
Comilla Bangla. 23°28N 91°10E 41 H17
Comino Malta 36°1N 14°20E 25 C1
Comino, C. Italy 40°32N 9°49E 22 D3
Comitán de Domínguez
Mexico 16°15N 92°8W 87 D6
Commerce Ga., U.S.A. 34°12N 83°28W 85 D13
Commerce Tex., U.S.A. 33°15N 95°54W 84 E7
Committee B. Canada 68°30N 86°30W 69 D14
Commodore, C. Canada 44°47N 80°54W 82 B4
Commonwealth B.
Antarctica 67°0S 144°0E 5 C10
Commoron Cr. →
Australia 28°22S 150°8E 63 D5
Communism Pk. = imeni Ismail
Samani, Pik Tajikistan 39°0N 72°2E 28 F8
Como Italy 45°47N 9°5E 20 D8
Como, Lago di Italy 46°0N 9°11E 20 D8
Comodoro Rivadavia
Argentina 45°50S 67°40W 96 F3
Comorin, C. = Kannyakumari
India 8°3N 77°40E 40 Q10
Comoros ■ Ind. Oc. 12°10S 44°15E 53 a
Comox Canada 49°42N 124°55W 70 D4
Compiègne France 49°24N 2°50E 20 B5
Compostela Mexico 21°14N 104°55W 86 C4
Comprida, I. Brazil 24°50S 47°42W 95 A6
Compton Canada 45°14N 71°49W 83 A13
Compton U.S.A. 33°53N 118°13W 79 M8
Comrat Moldova 46°18N 28°40E 17 E15
Con Cuong Vietnam 19°2N 104°54E 38 C5
Con Dao △ Vietnam 8°42N 106°35E 39 H6
Con Son, Dao Vietnam 8°41N 106°37E 39 H6
Conakry Guinea 9°29N 13°49W 50 G3
Conara Australia 41°50S 147°26E 63 G4
Concarneau France 47°52N 3°56W 20 C2
Conceição Mozam. 18°47S 36°7E 55 F4
Conceição da Barra
Brazil 18°35S 39°45W 93 G11
Conceição do Araguaia
Brazil 8°0S 49°2W 93 E9
Concepción Argentina 27°20S 65°35W 94 B2
Concepción Bolivia 16°15S 62°8W 92 G6
Concepción Chile 36°50S 73°0W 94 D1
Concepción Mexico 18°15N 90°5W 87 D6
Concepción Paraguay 23°22S 57°26W 94 A4
Concepción □ Chile 37°0S 72°30W 94 D1
Concepción → Mexico 30°32N 113°2W 86 A2
Concepción, Est. de
Chile 50°30S 74°55W 96 G2
Concepción, L. Bolivia 17°20S 61°20W 92 G6
Concepción, Pta.
Mexico 26°53N 111°50W 86 B2
Concepción del Oro
Mexico 24°38N 101°25W 86 C4
Concepción del Uruguay
Argentina 32°35S 58°20W 94 C4
Conception, Pt. U.S.A. 34°27N 120°28W 79 L6
Conception B. Canada 47°45N 53°0W 73 C9
Conception B. Namibia 23°55S 14°22E 56 C1
Conception I. Bahamas 23°52N 75°9W 89 B4
Concession Zimbabwe 17°27S 30°56E 55 F3
Conchas Dam U.S.A. 35°22N 104°11W 77 J11
Concho U.S.A. 34°28N 109°36W 77 J9
Concho → U.S.A. 31°34N 99°43W 84 F5
Conchos → Chihuahua,
Mexico 29°35N 104°25W 86 B4
Conchos → Chihuahua,
Mexico 27°29N 106°45W 86 B3
Conchos → Tamaulipas,
Mexico 24°55N 97°38W 87 B5
Concord Calif., U.S.A. 37°59N 122°2W 78 H4
Concord N.C., U.S.A. 35°25N 80°35W 85 D14
Concord N.H., U.S.A. 43°12N 71°32W 83 C13
Concordia Antarctica 75°6S 123°23E 5 D9
Concordia Argentina 31°20S 58°2W 94 C4
Concórdia Amazonas,
Brazil 4°36S 66°36W 92 D5
Concórdia Sta. Catarina,
Brazil 27°14S 52°1W 95 B5
Concordia Mexico 23°17N 106°4W 86 C3
Concordia U.S.A. 39°34N 97°40W 80 F5
Concrete U.S.A. 48°32N 121°45W 76 B3
Condamine Australia 26°56S 150°9E 63 D5
Conde U.S.A. 45°9N 98°6W 80 C4
Condeúba Brazil 14°52S 42°0W 93 F10
Condobolin Australia 33°4S 147°6E 63 E4
Condon U.S.A. 45°14N 120°11W 76 D3
Conegliano Italy 45°53N 12°18E 22 B5
Conejera, I. = Conills, I. des
Spain 39°11N 2°58E 24 B9
Conejos Mexico 26°14N 103°53W 86 B4
Conemaugh → U.S.A. 40°28N 79°19W 82 F5
Confuso → Paraguay 25°9S 57°34W 94 B4
Congleton U.K. 53°10N 2°13W 12 D5
Congo (Brazzaville) = Congo ■
Africa 1°0S 16°0E 52 E3
Congo (Kinshasa) = Congo, Dem.
Rep. of the ■ Africa 3°0S 23°0E 52 E4
Congo ■ Africa 1°0S 16°0E 52 E3
Congo → Africa 6°4S 12°24E 52 F2
Congo, Dem. Rep. of the ■
Africa 3°0S 23°0E 52 E4
Congo Basin Africa 0°10S 24°30E 52 E4
Congonhas Brazil 20°30S 43°52W 95 A7
Congress U.S.A. 34°9N 112°51W 77 J7
Conills, I. des Spain 39°11N 2°58E 24 B9
Coniston Canada 46°29N 80°51W 72 C3
Conjeeveram = Kanchipuram
India 12°52N 79°45E 40 N11
Conklin Canada 55°38N 111°5W 71 B6
Conklin U.S.A. 42°2N 75°49W 83 D9
Conn, L. Ireland 54°3N 9°15W 10 B2
Connacht □ Ireland 53°43N 9°12W 10 C2
Conneaut U.S.A. 41°57N 80°34W 82 E4
Conneautville U.S.A. 41°45N 80°22W 82 E4
Connecticut □ U.S.A. 41°30N 72°45W 83 E12
Connecticut → U.S.A. 41°16N 72°20W 83 E12
Connell U.S.A. 46°40N 118°52W 76 C4

Connellsville U.S.A. 40°1N 79°35W 82 F5
Connemara Ireland 53°29N 9°45W 10 C2
Connemara △ Ireland 53°32N 9°52W 10 C2
Connersville U.S.A. 39°39N 85°8W 81 F11
Connors Ra. Australia 21°40S 149°10E 62 C4
Conquest Canada 51°32N 107°14W 71 C7
Conrad U.S.A. 48°10N 111°57W 76 B8
Conrad Rise S. Ocean 53°0S 41°0E 5 B5
Conran, C. Australia 37°49S 148°44E 63 F4
Conroe U.S.A. 30°19N 95°27W 84 F7
Consecon Canada 44°0N 77°31W 82 C7
Conselheiro Lafaiete
Brazil 20°40S 43°48W 95 A7
Consell Spain 39°40N 2°49E 24 B9
Consett U.K. 54°51N 1°50W 12 C6
Consort Canada 52°1N 110°46W 71 C6
Constance = Konstanz
Germany 47°40N 9°10E 16 E5
Constance, L. = Bodensee
Europe 47°35N 9°25E 20 C8
Constanța Romania 44°14N 28°38E 17 F15
Constantine Algeria 36°25N 6°42E 50 A7
Constitución Chile 35°20S 72°30W 94 D1
Constitución Uruguay 31°0S 57°50W 94 C4
Constitución de 1857 △
Mexico 32°4N 115°55W 86 A1
Consul Canada 49°20N 109°30W 71 D7
Contact U.S.A. 41°46N 114°45W 76 F6
Contai India 21°54N 87°46E 43 J12
Contamana Peru 7°19S 74°55W 92 E4
Contas → Brazil 14°17S 39°1W 93 F11
Contoocook U.S.A. 43°13N 71°45W 83 C13
Contwoyto L. Canada 65°42N 110°50W 68 D9
Conway = Conwy U.K. 53°17N 3°50W 12 D4
Conway = Conwy →
U.K. 53°17N 3°50W 12 D4
Conway Australia 20°24S 148°41E 62 b
Conway Canada 44°6N 76°54W 82 B8
Conway Ark., U.S.A. 35°5N 92°26W 84 D8
Conway N.H., U.S.A. 43°59N 71°7W 83 C13
Conway S.C., U.S.A. 33°51N 79°3W 85 E15
Conway, C. Australia 20°34S 148°46E 62 b
Conway, L. Australia 28°17S 135°35E 63 D2
Conwy U.K. 53°17N 3°50W 12 D4
Conwy □ U.K. 53°10N 3°44W 12 D4
Conwy → U.K. 53°17N 3°50W 12 D4
Coober Pedy Australia 29°1S 134°43E 63 D1
Cooch Behar = Koch Bihar
India 26°22N 89°29E 41 F16
Cooinda Australia 13°15S 130°5E 60 B5
Cook Australia 30°37S 130°25E 61 F5
Cook U.S.A. 47°51N 92°41W 80 B7
Cook, B. Chile 55°10S 70°0W 96 H3
Cook, C. Canada 50°8N 127°55W 70 C3
Cook, Mt. = Aoraki Mount Cook
N.Z. 43°36S 170°9E 59 E3
Cook Inlet U.S.A. 60°0N 152°0W 68 F1
Cook Is. ⊠ Pac. Oc. 17°0S 160°0W 65 J12
Cook Strait N.Z. 41°15S 174°29E 59 D5
Cookeville U.S.A. 36°10N 85°30W 85 C12
Cookhouse S. Africa 32°44S 25°47E 56 E4
Cooks Harbour Canada 51°36N 55°52W 73 B8
Cookshire Canada 45°25N 71°38W 83 A13
Cookstown Canada 44°11N 79°42W 82 B5
Cookstown U.K. 54°39N 6°45W 10 B5
Cooksville Canada 43°35N 79°38W 82 C5
Cooktown Australia 15°30S 145°16E 62 B4
Coolabah Australia 31°1S 146°43E 63 E4
Cooladdi Australia 26°37S 145°23E 63 D4
Coolah Australia 31°48S 149°41E 63 E4
Coolamon Australia 34°46S 147°8E 63 E4
Coolgardie Australia 30°55S 121°8E 61 F3
Coolidge U.S.A. 32°59N 111°31W 77 K8
Coolidge Dam U.S.A. 33°10N 110°32W 77 K8
Cooma Australia 36°12S 149°8E 63 F4
Coon Rapids U.S.A. 45°9N 93°19W 80 C7
Coonabarabran
Australia 31°14S 149°18E 63 E4
Coonamble Australia 30°56S 148°27E 63 E4
Coonana Australia 31°0S 123°0E 61 F3
Coonana ✪ Australia 30°51S 122°53E 61 F3
Coondapoor India 13°42N 74°40E 40 N9
Cooninnie, L. Australia 26°4S 139°59E 63 D2
Cooper U.S.A. 33°23N 95°42W 84 E7
Cooper Cr. → Australia 28°29S 137°46E 63 D2
Cooper Ridge Pac. Oc. 10°0N 150°30W 65 G12
Cooperstown N. Dak.,
U.S.A. 47°27N 98°8W 80 B4
Cooperstown N.Y.,
U.S.A. 42°42N 74°56W 83 D10
Coorabie Australia 31°54S 132°18E 61 F5
Coorong, The Australia 35°50S 139°20E 63 F2
Coorow Australia 29°53S 116°2E 61 E2
Cooroy Australia 26°22S 152°54E 63 D5
Coos Bay U.S.A. 43°22N 124°13W 76 E1
Coosa → U.S.A. 32°30N 86°16W 85 E11
Cootamundra Australia 34°36S 148°1E 63 E4
Cootehill Ireland 54°4N 7°5W 10 B4
Copahue Paso Argentina 37°49S 71°8W 94 D1
Copainalá Mexico 17°4N 93°18W 87 D6
Copake U.S.A. 42°7N 73°31W 83 D11
Copán Honduras 14°50N 89°9W 88 D2
Cope U.S.A. 39°40N 102°51W 76 G12
Copenhagen = København
Denmark 55°40N 12°26E 9 J15
Copenhagen U.S.A. 43°54N 75°41W 83 C9
Copiapó Chile 27°30S 70°20W 94 B1
Copiapó → Chile 27°19S 70°56W 94 B1
Coplay U.S.A. 40°44N 75°29W 83 F9
Copo △ Argentina 25°53S 61°54W 94 B3
Copp L. Canada 60°14N 114°40W 70 A6
Coppename → Suriname 5°48N 55°55W 93 B7
Copper Canyon = Barranca del
Cobre ⌂ Mexico 27°18N 107°40W 86 B3
Copper Center U.S.A. 61°58N 145°18W 74 C10
Copper Harbor U.S.A. 47°28N 87°53W 80 B10
Copper Queen Zimbabwe 17°29S 29°18E 55 F2
Copperas Cove U.S.A. 31°8N 97°54W 84 F6
Copperbelt □ Zambia 13°15S 27°30E 55 E2
Coppermine = Kugluktuk
Canada 67°50N 115°5W 68 D8

Coppermine → Canada 67°49N 116°4W 68 D8
Copperopolis U.S.A. 37°58N 120°38W 78 H6
Coquet → U.K. 55°20N 1°32W 12 B6
Coquille U.S.A. 43°11N 124°11W 76 E1
Coquimbo Chile 30°0S 71°20W 94 C1
Coquimbo □ Chile 31°0S 71°0W 94 C1
Coquitlam Canada 49°17N 122°45W 70 D4
Corabia Romania 43°48N 24°30E 17 G13
Coracora Peru 15°5S 73°45W 92 G4
Coraki Australia 28°59S 153°17E 63 D5
Coral U.S.A. 40°29N 79°10W 82 F5
Coral Bay Australia 23°8S 113°46E 60 D1
Coral Gables U.S.A. 25°43N 80°16W 85 J14
Coral Harbour Canada 64°8N 83°10W 69 E15
Coral Sea Pac. Oc. 15°0S 150°0E 58 C8
Coral Sea Basin Pac. Oc. 14°0S 152°0E 64 J7
Coral Sea Islands Terr. □
Australia 20°0S 155°0E 58 C8
Coral Springs U.S.A. 26°16N 80°16W 85 H14
Coraopolis U.S.A. 40°31N 80°10W 82 F4
Corato Italy 41°9N 16°25E 22 D7
Corbett △ India 29°20N 79°0E 43 E8
Corbin U.S.A. 36°57N 84°6W 81 G11
Corby U.K. 52°30N 0°41W 13 E7
Corcaigh = Cork Ireland 51°54N 8°29W 10 E3
Corcoran U.S.A. 36°6N 119°33W 78 J7
Corcovado △ Costa Rica 8°33N 83°35W 88 E3
Corcubión Spain 42°56N 9°12W 21 A1
Cordele U.S.A. 31°58N 83°47W 85 F13
Cordell U.S.A. 35°17N 98°59W 84 D5
Córdoba Argentina 31°20S 64°10W 94 C3
Córdoba Mexico 18°53N 96°56W 87 D5
Córdoba Spain 37°50N 4°50W 21 D3
Córdoba □ Argentina 31°22S 64°15W 94 C3
Córdoba, Sierra de
Argentina 31°10S 64°25W 94 C3
Cordova U.S.A. 60°33N 145°45W 68 E2
Corella → Australia 19°34S 140°47E 62 B3
Corfield Australia 21°40S 143°21E 62 C3
Corfu = Kerkyra Greece 39°38N 19°50E 25 A3
Corfu U.S.A. 42°57N 78°24W 82 D6
Corfu, Str. of = Kerkyras, Notio
Steno Greece 39°34N 20°0E 25 A4
Coria Spain 39°58N 6°33W 21 C2
Corigliano Cálabro Italy 39°36N 16°31E 22 E7
Coringa Is. Australia 16°58S 149°58E 62 B4
Corinth = Korinthos
Greece 37°56N 22°55E 23 F10
Corinth Miss., U.S.A. 34°56N 88°31W 85 D10
Corinth N.Y., U.S.A. 43°15N 73°49W 83 C11
Corinth, G. of = Korinthiakos
Kolpos Greece 38°16N 22°30E 23 E10
Corinto Brazil 18°20S 44°30W 93 G10
Corinto Nic. 12°30N 87°10W 88 D2
Cork Ireland 51°54N 8°29W 10 E3
Cork □ Ireland 51°57N 8°40W 10 E3
Cork Harbour Ireland 51°47N 8°16W 10 E3
Çorlu Turkey 41°11N 27°49E 23 D12
Cormack L. Canada 60°56N 121°37W 70 A4
Cormorant Canada 54°14N 100°35W 71 C8
Cormorant L. Canada 54°15N 100°50W 71 C8
Corn Is. = Maíz, Is. del
Nic. 12°15N 83°4W 88 D3
Cornélio Procópio Brazil 23°7S 50°40W 95 A5
Corner Brook Canada 48°57N 57°58W 73 C8
Cornești Moldova 47°21N 28°1E 17 E15
Corning Ark., U.S.A. 36°25N 90°35W 85 C9
Corning Calif., U.S.A. 39°56N 122°11W 76 G2
Corning Iowa, U.S.A. 40°59N 94°44W 80 E6
Corning N.Y., U.S.A. 42°9N 77°3W 82 D7
Cornwall Canada 45°2N 74°44W 83 A10
Cornwall U.S.A. 40°17N 76°25W 83 F8
Cornwall □ U.K. 50°26N 4°40W 13 G3
Cornwall I. Canada 77°37N 94°38W 69 B13
Cornwallis I. Canada 75°8N 95°0W 69 B13
Corny Pt. Australia 34°55S 137°0E 63 E2
Coro Venezuela 11°25N 69°41W 92 A5
Coroatá Brazil 4°8S 44°0W 93 D10
Corocoro Bolivia 17°15S 68°28W 92 G5
Coroico Bolivia 16°0S 67°50W 92 G5
Coromandel N.Z. 36°45S 175°31E 59 B5
Coromandel Coast India 12°30N 81°0E 40 N12
Corona Calif., U.S.A. 33°53N 117°34W 79 M9
Corona N. Mex., U.S.A. 34°15N 105°36W 77 J11
Coronado U.S.A. 32°41N 117°10W 79 N9
Coronado, B. de Costa Rica 9°0N 83°40W 88 E3
Coronados, Is. Los
Mexico 32°26N 117°19W 79 N9
Coronation Canada 52°5N 111°27W 70 C6
Coronation Gulf
Canada 68°25N 110°0W 68 D10
Coronation I. Antarctica 60°45S 46°0W 5 C18
Coronation Is. Australia 14°57S 124°55E 60 B3
Coronda Argentina 31°58S 60°56W 94 C3
Coronel Chile 37°0S 73°10W 94 D1
Coronel Bogado
Paraguay 27°11S 56°18W 94 B4
Coronel Dorrego
Argentina 38°40S 61°10W 94 D3
Coronel Oviedo
Paraguay 25°24S 56°30W 94 B4
Coronel Pringles
Argentina 38°0S 61°30W 94 D3
Coronel Suárez
Argentina 37°30S 61°52W 94 D3
Coronel Vidal Argentina 37°28S 57°45W 94 D4
Coropuna, Nevado Peru 15°30S 72°41W 92 G4
Corowa Australia 35°58S 146°21E 63 F4
Corozal □ Belize 18°23N 88°23W 87 D7
Corozal Pt. Trin. & Tob. 10°45N 61°37W 93 K15
Corpus Argentina 27°10S 55°30W 95 B4
Corpus Christi U.S.A. 27°47N 97°24W 84 H6
Corpus Christi, L. U.S.A. 28°2N 97°52W 84 G6
Corralejo Canary Is. 28°43N 13°53W 24 F6
Corraun Pen. Ireland 53°54N 9°54W 10 C2
Correntes, C. das Mozam. 24°6S 35°34E 57 C6
Corrib, L. Ireland 53°27N 9°16W 10 C2
Corrientes Argentina 27°30S 58°45W 94 B4
Corrientes □ Argentina 28°0S 57°0W 94 B4
Corrientes → Argentina 30°42S 59°38W 94 C4

Corrientes ➝ Peru 3°43S 74°35W 92 D4
Corrientes, C. Colombia 5°30N 77°34W 92 B3
Corrientes, C. Cuba 21°43N 84°30W 88 B3
Corrientes, C. Mexico 20°25N 105°42W 86 C3
Corrigan U.S.A. 31°0N 94°52W 84 F7
Corrigin Australia 32°20S 117°53E 61 F2
Corriverton Guyana 5°55N 57°20W 92 B7
Corry U.S.A. 41°55N 79°39W 82 E5
Corse □ France 42°0N 9°0E 20 F8
Corse, C. France 43°1N 9°25E 20 E8
Corsica = Corse □ France 42°0N 9°0E 20 F8
Corsicana U.S.A. 32°6N 96°28W 84 E6
Corte France 42°19N 9°11E 20 E8
Cortés, Mar de = California, G. de Mexico 27°0N 111°0W 86 B2
Cortez U.S.A. 37°21N 108°35W 77 H9
Cortland N.Y., U.S.A. 42°36N 76°11W 83 D8
Cortland Ohio, U.S.A. 41°20N 80°44W 82 E4
Çorum Turkey 40°30N 34°57E 19 F5
Corumbá Brazil 19°0S 57°30W 92 G7
Corunna = A Coruña Spain 43°20N 8°25W 21 A1
Corvallis U.S.A. 44°34N 123°16W 76 D2
Corvette, L. de la Canada 53°25N 74°3W 72 B5
Corvo Azores 39°43N 31°8W 50 a
Corydon U.S.A. 40°46N 93°19W 80 E7
Cosalá Mexico 24°23N 106°41W 86 C3
Cosamaloapan de Carpio Mexico 18°22N 95°48W 87 D5
Cosenza Italy 39°18N 16°15E 22 E7
Coshocton U.S.A. 40°16N 81°51W 82 F3
Cosmo Newberry Australia 28°0S 122°54E 61 E3
Cosmo Newberry ✪ Australia 27°59S 122°53E 61 E3
Cosmonaut Sea S. Ocean 66°30S 40°0E 5 C5
Coso Junction U.S.A. 36°3N 117°57W 79 J9
Coso Pk. U.S.A. 36°13N 117°44W 79 J9
Cosquín Argentina 31°15S 64°30W 94 C3
Costa Blanca Spain 38°25N 0°10W 21 C5
Costa Brava Spain 41°30N 3°0E 21 B7
Costa Cálida Spain 37°30N 1°30W 21 D5
Costa Daurada Spain 41°12N 1°15E 21 B6
Costa de la Luz Spain 36°15N 5°58W 21 D2
Costa del Azahar Spain 39°39N 0°13W 21 B6
Costa del Sol Spain 36°30N 4°30W 21 D3
Costa dels Pins Spain 39°38N 3°26E 24 B10
Costa Mesa U.S.A. 33°38N 117°55W 79 M9
Costa Rica ■ Cent. Amer. 10°0N 84°0W 88 E3
Costa Smeralda Italy 41°5N 9°35E 22 D3
Costa Verde Spain 43°22N 4°14W 21 A3
Cosumnes ➝ U.S.A. 38°16N 121°26W 78 G5
Cotabato Phil. 7°14N 124°15E 37 C6
Cotagaita Bolivia 20°45S 65°40W 94 A2
Côte d'Azur France 43°25N 7°10E 20 E7
Côte-d'Ivoire = Ivory Coast ■ Africa 7°30N 5°0W 50 G4
Coteau des Prairies U.S.A. 45°20N 97°50W 80 C5
Coteau du Missouri U.S.A. 47°0N 100°0W 80 B4
Cotentin France 49°15N 1°30W 20 B3
Cotillo Canary Is. 28°41N 14°1W 24 F5
Cotonou Benin 6°20N 2°25E 50 G6
Cotopaxi Ecuador 0°40S 78°30W 92 D3
Cotswold Hills U.K. 51°42N 2°10W 13 F5
Cottage Grove U.S.A. 43°48N 123°3W 76 E2
Cottam Canada 42°8N 82°45W 82 D2
Cottbus Germany 51°45N 14°20E 16 C8
Cottonwood U.S.A. 34°45N 112°1W 77 J7
Cotulla U.S.A. 28°26N 99°14W 84 G5
Coudersport U.S.A. 41°46N 78°1W 82 E6
Couedic, C. du Australia 36°5S 136°40E 63 F2
Coulee City U.S.A. 47°37N 119°17W 76 C4
Coulee Dam Nat. Recr. Area = Lake Roosevelt ⌓ U.S.A. 48°5N 118°14W 76 B4
Coulman I. Antarctica 73°35S 170°0E 5 D11
Coulonge ➝ Canada 45°52N 76°46W 72 C4
Coulterville U.S.A. 37°43N 120°12W 78 H6
Council Alaska, U.S.A. 64°53N 163°41W 74 C7
Council Idaho, U.S.A. 44°44N 116°26W 76 D5
Council Bluffs U.S.A. 41°16N 95°52W 80 E6
Council Grove U.S.A. 38°40N 96°29W 80 F5
Coupeville U.S.A. 48°13N 122°41W 78 B4
Courantyne ➝ S. Amer. 5°55N 57°5W 92 B7
Courcelles Belgium 50°28N 4°22E 15 D4
Courtenay Canada 49°45N 125°0W 70 D4
Courtland Canada 42°51N 80°38W 82 D4
Courtland U.S.A. 38°20N 121°34W 78 G5
Courtrai = Kortrijk Belgium 50°50N 3°17E 15 D3
Courtright Canada 42°49N 82°28W 82 D2
Coushatta U.S.A. 32°1N 93°21W 84 E8
Coutts Crossing Australia 29°49S 152°55E 63 D5
Couva Trin. & Tob. 10°25N 61°27W 93 K15
Couvin Belgium 50°3N 4°29E 15 D4
Cove I. Canada 45°17N 81°44W 82 A3
Coventry U.K. 52°25N 1°28W 13 E6
Covilhã Portugal 40°17N 7°31W 21 B2
Covington Ga., U.S.A. 33°36N 83°51W 85 E13
Covington Ky., U.S.A. 39°5N 84°30W 81 F11
Covington Tenn., U.S.A. 35°34N 89°39W 85 D10
Covington Va., U.S.A. 37°47N 79°59W 81 G14
Cowal, L. Australia 33°40S 147°25E 63 E4
Cowan, L. Australia 31°45S 121°45E 61 F3
Cowan L. Canada 54°0N 107°15W 71 C7
Cowangie Australia 35°12S 141°26E 63 F3
Cowansville Canada 45°14N 72°46W 83 A12
Coward Springs Australia 29°24S 136°49E 63 D2
Cowcowing Lakes Australia 30°55S 117°20E 61 F2
Cowdenbeath U.K. 56°7N 3°21W 11 E5
Cowell Australia 33°39S 136°56E 63 E2
Cowes U.K. 50°45N 1°18W 13 G6
Cowichan L. Canada 48°53N 124°17W 78 B2
Cowlitz ➝ U.S.A. 46°6N 122°55W 78 D4
Cowra Australia 33°49S 148°42E 63 E4
Coxilha Grande Brazil 28°18S 51°30W 95 B5
Coxim Brazil 18°30S 54°55W 93 G8
Cox's Bazar Bangla. 21°26N 91°59E 41 J17
Coyote Wells U.S.A. 32°44N 115°58W 79 N11
Coyuca de Benítez Mexico 17°2N 100°4W 87 D4
Coyuca de Catalán Mexico 18°20N 100°39W 86 D4
Cozad U.S.A. 40°52N 99°59W 80 E4
Cozumel Mexico 20°31N 86°55W 87 C7
Cozumel, Isla Mexico 20°30N 86°40W 87 C7
Crab Hill Barbados 13°19N 59°38W 89 g
Cracow = Kraków Poland 50°4N 19°57E 17 C10
Cracow Australia 25°17S 150°17E 63 D5
Cradle Mt.-Lake St. Clair △ Australia 41°49S 147°56E 63 G4
Cradock Australia 32°6S 138°31E 63 E2
Cradock S. Africa 32°8S 25°36E 56 E4
Craig U.S.A. 40°31N 107°33W 76 F10
Craigavon U.K. 54°27N 6°23W 10 B5
Craigmore Zimbabwe 20°28S 32°50E 55 G3
Craik Canada 51°3N 105°49W 71 C7
Crailsheim Germany 49°8N 10°5E 16 D6
Craiova Romania 44°21N 23°48E 17 F12
Cramsie Australia 23°20S 144°15E 62 C3
Cranberry L. U.S.A. 44°11N 74°50W 83 B10
Cranberry Portage Canada 54°35N 101°23W 71 C8
Cranbrook Australia 34°18S 117°33E 61 F2
Cranbrook Canada 49°30N 115°46W 70 D5
Crandon U.S.A. 45°34N 88°54W 80 C9
Crane Oreg., U.S.A. 43°25N 118°35W 76 E4
Crane Tex., U.S.A. 31°24N 102°21W 84 F3
Crane, The Barbados 13°6N 59°27W 89 g
Cranston U.S.A. 41°47N 71°26W 83 E13
Crater L. U.S.A. 42°56N 122°6W 76 E2
Crater Lake △ U.S.A. 42°55N 122°10W 76 E2
Craters of the Moon ⌓ U.S.A. 43°25N 113°30W 76 E7
Crateús Brazil 5°10S 40°39W 93 E10
Crato Brazil 7°10S 39°25W 93 E11
Craven, L. Canada 54°20N 76°56W 72 B4
Crawford U.S.A. 42°41N 103°25W 80 D2
Crawfordsville U.S.A. 40°2N 86°54W 80 E10
Crawley U.K. 51°7N 0°11W 13 F7
Crazy Mts. U.S.A. 46°12N 110°20W 76 C8
Crean L. Canada 54°5N 106°9W 71 C7
Crediton Canada 43°17N 81°33W 82 C3
Crediton U.K. 50°47N 3°40W 13 G4
Cree ➝ Canada 58°57N 105°47W 71 B7
Cree ➝ U.K. 54°55N 4°25W 11 G4
Cree L. Canada 57°30N 106°30W 71 B7
Creede U.S.A. 37°51N 106°56W 77 H10
Creekside U.S.A. 40°40N 79°11W 82 F5
Creel Mexico 27°45N 107°38W 86 B3
Creemore Canada 44°19N 80°6W 82 B4
Creighton Canada 54°45N 101°54W 71 C8
Creighton U.S.A. 42°28N 97°54W 80 D5
Crema Italy 45°22N 9°41E 20 D8
Cremona Italy 45°7N 10°2E 20 D9
Cres Croatia 44°58N 14°25E 16 F8
Crescent City U.S.A. 41°45N 124°12W 76 F1
Crespo Argentina 32°2S 60°19W 94 C3
Cresson U.S.A. 40°28N 78°36W 82 F6
Crestline Calif., U.S.A. 34°14N 117°18W 79 L9
Crestline Ohio, U.S.A. 40°47N 82°44W 82 F2
Creston Canada 49°10N 116°31W 70 D5
Creston Calif., U.S.A. 35°32N 120°33W 78 K6
Creston Iowa, U.S.A. 41°4N 94°22W 80 E6
Crestview Calif., U.S.A. 37°46N 118°58W 78 H8
Crestview Fla., U.S.A. 30°46N 86°34W 85 F11
Crete = Kriti Greece 35°15N 25°0E 25 D7
Crete U.S.A. 40°38N 96°58W 80 E5
Créteil France 48°47N 2°27E 20 B5
Creus, C. de Spain 42°20N 3°19E 21 A7
Creuse ➝ France 47°0N 0°34E 20 C4
Crewe U.K. 53°6N 2°26W 12 D5
Crewkerne U.K. 50°53N 2°48W 13 G5
Crianlarich U.K. 56°24N 4°37W 11 E4
Criciúma Brazil 28°40S 49°23W 95 B6
Crieff U.K. 56°22N 3°50W 11 E5
Crimea □ Ukraine 45°30N 33°10E 19 E5
Crimean Pen. = Krymskyy Pivostriv Ukraine 45°0N 34°0E 19 F5
Crişul Alb ➝ Romania 46°42N 21°17E 17 E11
Crişul Negru ➝ Romania 46°42N 21°16E 17 E11
Crna ➝ Macedonia 41°33N 21°59E 23 D9
Crna Gora = Montenegro ■ Europe 42°40N 19°20E 23 C8
Crna Gora Macedonia 42°10N 21°30E 23 C9
Crna Reka = Crna ➝ Macedonia 41°33N 21°59E 23 D9
Croagh Patrick Ireland 53°46N 9°40W 10 C2
Croatia ■ Europe 45°20N 16°0E 16 F9
Crocker, Banjaran Malaysia 5°40N 116°30E 36 C5
Crockett U.S.A. 31°19N 95°27W 84 F7
Crocodile = Umgwenya ➝ Mozam. 25°14S 32°18E 57 D5
Crocodile Is. Australia 12°3S 134°58E 62 A1
Crohy Hd. Ireland 54°55N 8°26W 10 B3
Croix, L. la Canada 48°20N 92°15W 72 C1
Croker, C. Australia 10°58S 132°35E 60 B5
Croker, C. Canada 44°58N 80°59W 82 B4
Croker I. Australia 11°12S 132°32E 60 B5
Croker Islands Seas ✪ Australia 11°20S 132°30E 60 B5
Cromarty U.K. 57°40N 4°2W 11 D4
Cromer U.K. 52°56N 1°17E 12 E9
Cromwell N.Z. 45°3S 169°14E 59 F2
Cromwell U.S.A. 41°36N 72°39W 83 E12
Crook U.K. 54°43N 1°45W 12 C6
Crooked ➝ Canada 54°50N 122°54W 70 C4
Crooked ➝ U.S.A. 44°32N 121°16W 76 D3
Crooked I. Bahamas 22°50N 74°10W 89 B5
Crooked Island Passage Bahamas 22°55N 74°35W 89 B5
Crookston Minn., U.S.A. 47°47N 96°37W 80 B5
Crookston Nebr., U.S.A. 42°56N 100°45W 80 D3
Crookwell Australia 34°28S 149°24E 63 E4
Crosby U.K. 53°30N 3°3W 12 D4
Crosby N. Dak., U.S.A. 48°55N 103°18W 80 A2
Crosby Pa., U.S.A. 41°45N 78°23W 82 E6
Crosbyton U.S.A. 33°40N 101°14W 84 E4
Cross City U.S.A. 29°38N 83°7W 85 G13
Cross Fell U.K. 54°43N 2°28W 12 C5
Cross L. Canada 54°45N 97°30W 71 C9
Cross Lake Canada 54°37N 97°47W 71 C9
Cross Sound U.S.A. 58°0N 135°0W 68 F4
Crossett U.S.A. 33°8N 91°58W 84 E9
Crosshaven Ireland 51°47N 8°17W 10 E3
Crossmaglen U.K. 54°5N 6°36W 10 B5
Crossmolina Ireland 54°6N 9°20W 10 B2
Crossville U.S.A. 35°57N 85°2W 85 D12
Croswell U.S.A. 43°16N 82°37W 82 C2
Croton-on-Hudson U.S.A. 41°12N 73°55W 83 E11
Crotone Italy 39°5N 17°8E 22 E7
Crow ➝ Canada 59°41N 124°20W 70 B4
Crow Agency U.S.A. 45°36N 107°28W 76 D10
Crow Hd. Ireland 51°35N 10°9W 10 E1
Crowell U.S.A. 33°59N 99°43W 84 E5
Crowley U.S.A. 30°13N 92°22W 84 F8
Crowley, L. U.S.A. 37°35N 118°42W 78 H8
Crown Point Ind., U.S.A. 41°25N 87°22W 80 E10
Crown Point N.Y., U.S.A. 43°57N 73°26W 83 C11
Crown Pt. Trin. & Tob. 11°18N 60°51W 93 J16
Crownpoint U.S.A. 35°41N 108°9W 77 J9
Crows Landing U.S.A. 37°23N 121°6W 78 H5
Crows Nest Australia 27°16S 152°4E 63 D5
Crowsnest Pass Canada 49°40N 114°40W 70 D6
Croydon Australia 18°13S 142°14E 62 B3
Croydon □ U.K. 51°22N 0°5W 13 F7
Crozet, Îs. Ind. Oc. 46°27S 52°0E 3 G12
Crusheen Ireland 52°57N 8°53W 10 D3
Cruz, C. Cuba 19°50N 77°50W 88 C4
Cruz Alta Brazil 28°45S 53°40W 95 B5
Cruz Bay U.S. Virgin Is. 18°20N 64°48W 89 e
Cruz del Eje Argentina 30°45S 64°50W 94 C3
Cruzeiro Brazil 22°33S 45°0W 95 A7
Cruzeiro do Oeste Brazil 23°46S 53°4W 95 A5
Cruzeiro do Sul Brazil 7°35S 72°35W 92 E4
Cry L. Canada 58°45N 129°0W 70 B3
Crystal Bay U.S.A. 39°15N 120°0W 78 F7
Crystal Brook Australia 33°21S 138°12E 63 E2
Crystal City U.S.A. 28°41N 99°50W 84 G5
Crystal Falls U.S.A. 46°5N 88°20W 80 B9
Crystal River U.S.A. 28°54N 82°35W 85 G13
Crystal Springs U.S.A. 31°59N 90°21W 85 F9
Csongrád Hungary 46°43N 20°12E 17 E11
Cu Lao Hon Vietnam 10°54N 108°18E 39 G7
Cua Rao Vietnam 19°16N 104°27E 38 C5
Cuácua ➝ Mozam. 17°54S 37°0E 55 F4
Cuamato Angola 17°2S 15°7E 56 B2
Cuamba Mozam. 14°45S 36°22E 55 E4
Cuando ➝ Angola 17°30S 23°15E 53 H4
Cuando Cubango □ Angola 16°25S 20°0E 56 B3
Cuangar Angola 17°36S 18°39E 56 B2
Cuango = Kwango ➝ Dem. Rep. of the Congo 3°14S 17°22E 52 E3
Cuanza ➝ Angola 9°21S 13°9E 52 F2
Cuarto ➝ Argentina 33°25S 63°2W 94 C3
Cuatrociénegas Mexico 26°59N 102°5W 86 B4
Cuauhtémoc Mexico 28°25N 106°52W 86 B3
Cuba N. Mex., U.S.A. 36°1N 107°4W 77 H10
Cuba N.Y., U.S.A. 42°13N 78°17W 82 D6
Cuba ■ W. Indies 22°0N 79°0W 88 B4
Cubango ➝ Africa 18°50S 22°25E 56 B3
Cuc Phuong △ Vietnam 20°17N 105°38E 38 B5
Cuchumatanes, Sierra de los Guatemala 15°35N 91°25W 88 C1
Cuckfield U.K. 51°1N 0°8W 13 F7
Cucuí Brazil 1°12N 66°50W 92 C5
Cucurpé Mexico 30°20N 110°43W 86 A2
Cúcuta Colombia 7°54N 72°31W 92 B4
Cuddalore India 11°46N 79°45E 40 P11
Cuddapah India 14°30N 78°47E 40 M11
Cuddapan, L. Australia 25°45S 141°26E 62 D3
Cue Australia 27°25S 117°54E 61 E2
Cuenca Ecuador 2°50S 79°9W 92 D3
Cuenca Spain 40°5N 2°10W 21 B4
Cuenca, Serranía de Spain 39°55N 1°50W 21 C5
Cuernavaca Mexico 18°55N 99°15W 87 D5
Cuero U.S.A. 29°6N 97°17W 84 G6
Cueva de la Quebrada del Toro △ Venezuela 10°46N 69°3W 89 D6
Cuevas del Almanzora Spain 37°18N 1°58W 21 D5
Cuevo Bolivia 20°15S 63°30W 92 H6
Cuiabá Brazil 15°30S 56°0W 93 G7
Cuiabá ➝ Brazil 17°5S 56°36W 93 G7
Cuihangcun China 22°27N 113°32E 33 a
Cuijk Neths. 51°44N 5°50E 15 C5
Cuilco Guatemala 15°24N 91°58W 88 C1
Cuillin Hills U.K. 57°13N 6°15W 11 D2
Cuillin Sd. U.K. 57°4N 6°20W 11 D2
Cuilo = Kwilu ➝ Dem. Rep. of the Congo 3°22S 17°22E 52 E3
Cuito ➝ Angola 18°1S 20°48E 56 B3
Cuitzeo, L. de Mexico 19°55N 101°5W 86 D4
Cukai Malaysia 4°13N 103°25E 39 K4
Culbertson U.S.A. 48°9N 104°31W 76 B11
Culcairn Australia 35°41S 147°3E 63 F4
Culebra Puerto Rico 18°19N 65°18W 89 d
Culgoa ➝ Australia 29°56S 146°20E 63 D4
Culgoa Flood Plain △ Australia 28°58S 147°5E 63 D4
Culiacán Mexico 24°50N 107°23W 86 C3
Culiacán ➝ Mexico 24°30N 107°42W 86 C3
Culik Indonesia 8°21S 115°37E 37 J18
Culion Phil. 11°54N 119°58E 37 B6
Cullarin Ra. Australia 34°30S 149°30E 63 E4
Cullen U.K. 57°42N 2°49W 11 D6
Cullen Pt. Australia 11°57S 141°54E 62 A3
Cullera Spain 39°9N 0°17W 21 C5
Cullman U.S.A. 34°11N 86°51W 85 D11
Cullompton U.K. 50°51N 3°24W 13 G4
Culpeper U.S.A. 38°30N 78°0W 81 F14
Culuene ➝ Brazil 12°56S 52°51W 93 F8
Culver, Pt. Australia 32°54S 124°43E 61 F3
Culverden N.Z. 42°47S 172°49E 59 E4
Cumaná Venezuela 10°30N 64°5W 92 A6
Cumberland B.C., Canada 49°40N 125°0W 70 D4
Cumberland Ont., Canada 45°29N 75°24W 83 A9
Cumberland U.S.A. 39°39N 78°46W 81 F14
Cumberland ➝ U.S.A. 37°9N 88°25W 80 G9
Cumberland, L. U.S.A. 36°52N 85°9W 81 G11
Cumberland Gap ⌓ U.S.A. 36°36N 83°40W 81 G12
Cumberland I. U.S.A. 30°50N 81°25W 85 F14
Cumberland Is. Australia 20°35S 149°10E 62 b
Cumberland Island ⌓ U.S.A. 30°12N 81°24W 85 F14
Cumberland L. Canada 54°3N 102°18W 71 C8
Cumberland Pen. Canada 67°0N 64°0W 69 D19
Cumberland Plateau U.S.A. 36°0N 85°0W 85 D12
Cumberland Sd. Canada 65°30N 66°0W 69 D18
Cumbernauld U.K. 55°57N 3°58W 11 F5
Cumborah Australia 29°40S 147°45E 63 D4
Cumbres de Majalca △ Mexico 28°48N 106°30W 86 B3
Cumbres de Monterrey △ Mexico 25°26N 100°25W 86 B4
Cumbria □ U.K. 54°42N 2°52W 12 C5
Cumbrian Mts. U.K. 54°30N 3°0W 12 C5
Cumbum India 15°40N 79°10E 40 M11
Cuminá ➝ Brazil 1°30S 56°0W 93 D7
Cummings Mt. U.S.A. 35°2N 118°34W 79 K8
Cummins Australia 34°16S 135°43E 63 E2
Cumnock Australia 32°59S 148°46E 63 E4
Cumnock U.K. 55°28N 4°17W 11 F4
Cumpas Mexico 30°2N 109°48W 86 B3
Cumplida, Pta. Canary Is. 28°50N 17°48W 24 F2
Cunco Chile 38°55S 72°2W 96 D2
Cuncumén Chile 31°53S 70°38W 94 C1
Cundeelee ✪ Australia 30°43S 123°25E 61 F3
Cunderdin Australia 31°37S 117°12E 61 F2
Cunene ➝ Angola 17°20S 11°50E 56 B1
Cúneo Italy 44°23N 7°32E 20 D7
Cung Son Vietnam 13°2N 108°58E 38 F7
Çüngüş Turkey 38°13N 39°17E 44 B3
Cunillera, I. = Sa Conillera Spain 38°59N 1°13E 24 C7
Cunnamulla Australia 28°2S 145°38E 63 D4
Cupar Canada 50°57N 104°10W 71 C8
Cupar U.K. 56°19N 3°1W 11 E5
Cupertino U.S.A. 37°19N 122°2W 78 H4
Cupica, G. de Colombia 6°25N 77°30W 92 B3
Curaçao W. Indies 12°10N 69°0W 89 D6
Curanilahue Chile 37°29S 73°28W 94 D1
Curaray ➝ Peru 2°20S 74°5W 92 D4
Curepipe Mauritius 20°19S 57°31E 53 d
Curepto Chile 35°8S 72°1W 94 D1
Curiapo Venezuela 8°33N 61°5W 92 B6
Curicó Chile 34°55S 71°20W 94 C1
Curieuse Seychelles 4°15S 55°44E 53 b
Curitiba Brazil 25°20S 49°10W 95 B6
Curitibanos Brazil 27°18S 50°36W 95 B5
Currabubula Australia 31°16S 150°44E 63 E5
Currais Novos Brazil 6°13S 36°30W 93 E11
Curral Velho C. Verde Is. 16°8N 22°48W 50 b
Curralinho Brazil 1°45S 49°46W 93 D9
Currane, L. Ireland 51°49N 10°4W 10 E1
Currant U.S.A. 38°44N 115°28W 76 G6
Currawinya △ Australia 28°55S 144°27E 63 D3
Current ➝ U.S.A. 36°15N 90°55W 84 C9
Currie Australia 39°56S 143°53E 63 F3
Currie U.S.A. 40°16N 114°45W 76 F6
Curtea de Argeş Romania 45°12N 24°42E 17 F13
Curtin Springs Australia 25°20S 131°45E 61 E5
Curtis U.S.A. 40°38N 100°31W 80 E3
Curtis Group Australia 39°30S 146°37E 63 F4
Curtis I. Australia 23°35S 151°10E 62 C5
Curuápanema ➝ Brazil 2°25S 55°2W 93 D7
Curuçá Brazil 0°43S 47°50W 93 D9
Curuguaty Paraguay 24°31S 55°42W 95 A4
Çürüksu Çayi = Büyük Menderes ➝ Turkey 37°28N 27°11E 23 F12
Curup Indonesia 3°26S 102°13E 36 E2
Cururupu Brazil 1°50S 44°50W 93 D10
Curuzú Cuatiá Argentina 29°50S 58°5W 94 B4
Curvelo Brazil 18°45S 44°27W 93 G10
Cusco Peru 13°32S 72°0W 92 F4
Cushendall U.K. 55°5N 6°4W 10 A5
Cushing U.S.A. 35°59N 96°46W 84 D6
Cushing, Mt. Canada 57°35N 126°57W 70 B3
Cusihuiriáchic Mexico 28°14N 106°50W 86 B3
Custer U.S.A. 43°46N 103°36W 80 D2
Cut Bank U.S.A. 48°38N 112°20W 76 B7
Cutchogue U.S.A. 41°1N 72°30W 83 E12
Cuthbert U.S.A. 31°46N 84°48W 85 F12
Cutler U.S.A. 36°31N 119°17W 78 J7
Cuttaburra ➝ Australia 29°43S 144°22E 63 D3
Cuttack India 20°25N 85°57E 41 J14
Cuvier, C. Australia 23°14S 113°22E 61 D1
Cuvier I. N.Z. 36°27S 175°50E 59 B5
Cuxhaven Germany 53°51N 8°41E 16 B5
Cuyahoga Falls U.S.A. 41°8N 81°29W 82 E3
Cuyahoga Valley △ U.S.A. 41°24N 81°33W 82 E3
Cuyo Phil. 10°51N 121°2E 37 B6
Cuyuni ➝ Guyana 6°23N 58°41W 92 B7
Cuzco = Cusco Peru 13°32S 72°0W 92 F4
Cuzco Bolivia 20°0S 66°50W 92 H5
Cwmbran U.K. 51°39N 3°2W 13 F4
Cyangugu Rwanda 2°29S 28°54E 54 C2
Cyclades Greece 37°0N 24°30E 23 F11
Cygnet Australia 43°8S 147°1E 63 G4
Cynthiana U.S.A. 38°23N 84°18W 81 F11
Cypress Hills Canada 49°40N 109°30W 71
Cypress Hills ⌓ Canada 49°40N 109°30W 71
Cyprus ■ Asia 35°0N 33°0E 25
Cyrenaica = Barqa Libya 27°0N 23°0E 51
Cyrene Libya 32°53N 21°52E 51
Czar Canada 52°27N 110°50W 71
Czech Rep. ■ Europe 50°0N 15°0E 16
Częstochowa Poland 50°49N 19°7E 17

D

Da ➝ Vietnam 21°15N 105°20E 38
Da Hinggan Ling China 48°0N 121°0E 33
Da Lat Vietnam 11°56N 108°25E 39
Da Nang Vietnam 16°4N 108°13E 38
Da Qaidam China 37°50N 95°15E 32
Da Yunhe ➝ Hopei, China 39°10N 117°10E 35
Da Yunhe ➝ Jiangsu, China 34°25N 120°5E 35
Da'an China 45°30N 124°7E 35
Daan Viljoen ⌓ Namibia 22°2S 16°45E 56
Daba Shan China 32°0N 109°0E 33
Dabbagh, Jabal Si. Arabia 27°52N 35°45E 44
Dabhoi India 22°10N 73°20E 42
Dabie Shan China 31°20N 115°20E 33
Dabola Guinea 10°50N 11°5W 50
Dabung Malaysia 5°23N 102°1E 39
Dacca = Dhaka Bangla. 23°43N 90°26E 43
Dacca = Dhaka □ Bangla. 24°25N 90°25E 43
Dachau Germany 48°15N 11°26E 16
Dachigam △ India 34°10N 75°0E 42
Dacre Canada 45°22N 76°57W 82
Dadanawa Guyana 2°50N 59°30W 92
Dade City U.S.A. 28°22N 82°11W 85
Dadhar Pakistan 29°28N 67°39E 42
Ḑadnah U.A.E. 25°32N 56°22E 45
Dadra & Nagar Haveli □ India 20°5N 73°0E 40
Dadri = Charkhi Dadri India 28°37N 76°17E 42
Dadu Pakistan 26°45N 67°45E 42
Daegu S. Korea 35°50N 128°37E 35
Daejeon S. Korea 36°20N 127°28E 35
Daejeong S. Korea 33°8N 126°17E 35
Daet Phil. 14°2N 122°55E 37
Dafeng China 33°3N 120°45E 35
Dafnes Greece 35°13N 25°3E 25
Dagana Senegal 16°30N 15°35W 50
Dagestan □ Russia 42°30N 47°0E 19
Daggett U.S.A. 34°52N 116°52W 79
Daghestan Republic = Dagestan □ Russia 42°30N 47°0E 19
Dağlıq Qarabağ = Nagorno-Karabakh □ Azerbaijan 39°55N 46°45E 44
Dagö = Hiiumaa Estonia 58°50N 22°45E 9
Dagu China 38°59N 117°40E 35
Dagupan Phil. 16°3N 120°20E 37
Daguragu Australia 17°33S 130°30E 60
Daguragu ✪ Australia 17°24S 130°48E 60
Dahab Egypt 28°31N 34°31E 44
Dahlak Kebir Eritrea 15°50N 40°10E 47
Dahlonega U.S.A. 34°32N 83°59W 85
Dahod India 22°50N 74°15E 42
Dahongliutan China 35°45N 79°20E 43
Dahra Libya 29°30N 17°50E 51
Dahūk Iraq 36°50N 43°1E 44
Dai Hao Vietnam 18°1N 106°25E 38
Dai-Sen Japan 35°22N 133°32E 31
Dai Xian China 39°4N 112°58E 34
Daicheng China 38°42N 116°38E 34
Daikondi = Day Kundī □ Afghan. 34°0N 66°0E 40
Daingean Ireland 53°18N 7°17W 10
Daintree Australia 16°20S 145°20E 62
Daintree △ Australia 16°8S 145°2E 62
Daiō-Misaki Japan 34°15N 136°45E 31
Daisen-Oki △ Japan 35°23N 133°34E 31
Daisetsu-Zan Japan 43°30N 142°57E 30
Daisetsu-Zan △ Japan 43°30N 142°55E 30
Dajarra Australia 21°42S 139°30E 62
Dajiawa China 37°9N 119°0E 35
Dak Dam Cambodia 12°20N 107°21E 38
Dak Nhe Vietnam 15°28N 107°48E 38
Dak Pek Vietnam 15°4N 107°44E 38
Dak Song Vietnam 12°19N 107°35E 39
Dak Sui Vietnam 14°55N 107°43E 38
Dakar Senegal 14°34N 17°29W 50
Dakhla W. Sahara 23°50N 15°53W 50
Dakhla, El Wâhât el Egypt 25°30N 28°50E 51
Dakor India 22°45N 73°11E 42
Dakota City U.S.A. 42°25N 96°25W 80
Đakovica Kosovo 42°22N 20°26E 23
Dalachi China 36°48N 105°0E 34
Dalai Nur China 43°20N 116°45E 34
Dālakī Iran 29°26N 51°17E 45
Dalälven ➝ Sweden 60°12N 16°43E 9
Dalaman ➝ Turkey 36°41N 28°43E 23
Dalandzadgad Mongolia 43°27N 104°30E 34
Dalap-Uliga-Darrit = Majuro Marshall Is. 7°9N 171°12E 64
Dalarna Sweden 61°0N 14°0E 8
Dalay Mongolia 43°28N 103°30E 34
Dālbandīn Pakistan 29°0N 64°23E 40
Dalbeattie U.K. 54°56N 3°50W 11
Dalbeg Australia 20°16S 147°18E 62
Dalby Australia 27°10S 151°17E 63
Dale City U.S.A. 38°38N 77°19W 81
Dale Hollow L. U.S.A. 36°32N 85°27W 85
Dalgān Iran 27°31N 59°19E 45
Dalhart U.S.A. 36°4N 102°31W 84
Dalhousie Canada 48°5N 66°26W 73
Dalhousie India 32°38N 75°58E 42
Dali Shaanxi, China 34°48N 109°58E 34
Dali Yunnan, China 25°40N 100°10E 41
Dalian China 38°50N 121°40E 35

aliang Shan China 28°0N 102°45E 32 F9
aling He → China 40°55N 121°40E 35 D11
iliyat el Karmel Israel 32°43N 35°2E 46 C4
alkeith U.K. 55°53N 3°4W 11 F5
allas Oreg., U.S.A. 44°55N 123°19W 76 D2
allas Tex., U.S.A. 32°47N 96°48W 84 E6
alles, The U.S.A. 45°36N 121°10W 76 D3
almā U.A.E. 24°30N 52°20E 45 E7
almacija Croatia 43°20N 17°0E 22 C7
almas, L. Canada 53°30N 71°50W 73 B5
almatia = Dalmacija
Croatia 43°20N 17°0E 22 C7
almau India 26°4N 81°2E 43 F9
almellington U.K. 55°19N 4°23W 11 F4
alnegorsk Russia 44°32N 135°33E 30 B7
alnerechensk Russia 45°50N 133°40E 30 B6
alnevostochnyy □
Russia 67°0N 140°0E 29 C14
aloa Ivory C. 7°0N 6°30W 50 G4
alry U.K. 55°42N 4°43W 11 F4
alrymple, L. Australia 20°40S 147°0E 62 C4
alrymple, Mt. Australia 21°1S 148°39E 62 b
alsland Sweden 58°50N 12°15E 9 G15
altenganj India 24°0N 84°4E 43 H11
alton Ga., U.S.A. 34°46N 84°58W 85 D12
alton Mass., U.S.A. 42°28N 73°11W 83 D11
alton Nebr., U.S.A. 41°25N 102°58W 80 E2
alton-in-Furness U.K. 54°10N 3°11W 12 C4
alvík Iceland 65°58N 18°32W 8 D4
alvvadis = Jokkmokk
Sweden 66°35N 19°50E 8 C18
alwallinu Australia 30°17S 116°40E 61 F2
aly → Australia 13°35S 130°19E 60 B5
aly City U.S.A. 37°42N 122°27W 78 H4
aly L. Canada 56°32N 105°39W 71 B7
aly River Australia 13°46S 130°42E 60 B5
aly River-Port Keats ☼
Australia 14°13S 129°36E 60 B4
aly Waters Australia 16°15S 133°24E 62 B1
am Doi Vietnam 8°50N 105°12E 39 H5
am Ha Vietnam 21°21N 107°36E 38 B6
aman India 20°25N 72°57E 40 J8
āmaneh Iran 33°1N 50°29E 45 C6
amanhûr Egypt 31°0N 30°30E 51 B12
amant L. Canada 61°45N 105°5W 71 A7
amar Indonesia 7°7S 128°40E 37 F7
amara C.A.R. 4°58N 18°42E 52 D3
amaraland Namibia 20°0S 15°0E 56 C2
amascus = Dimashq
Syria 33°30N 36°18E 46 B5
amatura Nigeria 11°45N 11°55E 52 B2
amāvand Iran 35°47N 52°0E 45 C7
amāvand, Qolleh-ye
Iran 35°56N 52°10E 45 C7
amba Angola 6°44S 15°20E 52 F3
ămbovița → Romania 44°12N 26°26E 17 F14
ame Marie Haiti 18°36N 74°26W 89 C5
āmghān Iran 36°10N 54°17E 45 B7
amiel Spain 39°4N 3°37W 21 C4
amietta = Dumyât
Egypt 31°24N 31°48E 51 B12
aming China 36°15N 115°6E 34 F8
amīr Qābū Syria 36°58N 41°51E 44 B4
ammam = Ad Dammām
Si. Arabia 26°20N 50°5E 45 E6
amodar → India 23°17N 87°35E 43 H12
amoh India 23°50N 79°28E 43 H8
ampier Australia 20°41S 116°42E 60 D2
ampier, Selat Indonesia 0°40S 131°0E 37 E8
ampier Arch.
Australia 20°38S 116°32E 60 D2
amrei, Chuor Phnum
Cambodia 11°30N 103°0E 39 G4
amyang S. Korea 35°19N 126°59E 35 G14
ana Indonesia 11°0S 121°15E 37 F6
ana Jordan 30°41N 35°37E 46 E4
ana, L. Canada 50°53N 77°20W 72 B4
ana, Mt. U.S.A. 37°54N 119°12W 78 H7
anakil Desert Ethiopia 12°45N 41°0E 47 E3
anané Ivory C. 7°16N 8°9W 50 G4
anbury U.S.A. 41°24N 73°28W 83 E11
anby L. U.S.A. 34°13N 115°5W 77 J6
and Afghan. 31°28N 65°32E 42 D1
ande ⌂ Zimbabwe 15°56S 30°16E 55 F3
andeldhura Nepal 29°20N 80°35E 43 E9
andeli India 15°5N 74°30E 40 M9
andenong Australia 38°0S 145°15E 63 F4
andong China 40°10N 124°20E 35 D13
anfeng China 33°45N 110°25E 34 H6
anger Is. = Pukapuka
Cook Is. 10°53S 165°49W 65 J11
anger Pt. S. Africa 34°40S 19°17E 56 E2
angla Shan = Tanggula Shan
China 32°40N 92°10E 32 E7
angrek, Mts. = Dangrek, Phnom
Thailand 14°20N 104°0E 38 E5
angrek, Phnom
Thailand 14°20N 104°0E 38 E5
angriga Belize 17°0N 88°13W 87 D7
angshan China 34°27N 116°22E 34 G9
aniel U.S.A. 42°52N 110°4W 76 E8
aniel's Harbour
Canada 50°13N 57°35W 73 B8
anielskuil S. Africa 28°11S 23°33E 56 D3
anielson U.S.A. 41°48N 71°53W 83 E13
anilov Russia 58°16N 40°13E 18 C7
aning China 36°28N 110°45E 34 F6
ank Oman 23°33N 56°16E 45 F8
ankhar Gompa India 32°10N 78°10E 42 C8
anlí Honduras 14°4N 86°35W 88 D2
anmark = Denmark ■
Europe 55°45N 10°0E 9 J14
annemora U.S.A. 44°43N 73°44W 83 B11
annevirke N.Z. 40°12S 176°8E 59 D6
annhauser S. Africa 28°0S 30°3E 57 D5
ansville U.S.A. 42°34N 77°42W 82 D7
anta India 24°11N 72°46E 42 G5
antan India 21°57N 87°20E 43 J12
ante = Xaafuun
Somali Rep. 10°25N 51°16E 47 E5
Danube = Dunărea →
Europe 45°20N 29°40E 17 F15
Danvers U.S.A. 42°34N 70°56W 83 D14
Danville Ill., U.S.A. 40°8N 87°37W 80 E10
Danville Ky., U.S.A. 37°39N 84°46W 81 G11
Danville Pa., U.S.A. 40°58N 76°37W 83 F8
Danville Va., U.S.A. 36°36N 79°23W 81 G14
Danville Vt., U.S.A. 44°25N 72°9W 83 B12
Danzhou China 19°31N 109°33E 38 C7
Danzig = Gdańsk
Poland 54°22N 18°40E 17 A10
Dapaong Togo 10°55N 0°16E 50 F6
Daqing China 46°35N 125°0E 33 B13
Daqing Shan China 40°40N 111°0E 34 D6
Daqq-e Sorkh, Kavīr
Iran 33°45N 52°50E 45 C7
Dar Banda Africa 8°0N 23°0E 48 F6
Dar el Beida = Casablanca
Morocco 33°36N 7°36W 50 B4
Dar es Salaam Tanzania 6°50S 39°12E 54 D4
Dar Mazār Iran 29°14N 57°20E 45 D8
Dar'ā Syria 32°36N 36°7E 46 C5
Dar'ā □ Syria 32°55N 36°10E 46 C5
Dārāb Iran 28°50N 54°30E 45 D7
Daraban Pakistan 31°44N 70°20E 42 D4
Daraina Madag. 13°12S 49°40E 57 A8
Daraj Libya 30°10N 10°28E 51 B8
Dārān Iran 32°59N 50°24E 45 C6
Dārayyā Syria 33°28N 36°15E 46 B5
Darband Pakistan 34°20N 72°50E 42 B5
Darband, Kūh-e Iran 31°34N 57°8E 45 D8
Darbhanga India 26°15N 85°55E 43 F11
D'Arcy Canada 50°33N 122°29W 70 C4
Dardanelle Ark., U.S.A. 35°13N 93°9W 84 D8
Dardanelle Calif.,
U.S.A. 38°20N 119°50W 78 G7
Dardanelles = Çanakkale Boğazı
Turkey 40°17N 26°32E 23 D12
Dārestān Iran 29°9N 58°42E 45 D8
Dârfûr Sudan 13°40N 24°0E 51 F10
Dargai Pakistan 34°25N 71°55E 42 B4
Dargaville N.Z. 35°57S 173°52E 59 A4
Darhan Mongolia 49°37N 106°21E 32 B10
Darhan Muminggan
China 41°40N 110°28E 34 D6
Darıca Turkey 40°45N 29°23E 23 D13
Darién, G. del Colombia 9°0N 77°0W 92 B3
Darién △ Panama 7°36N 77°57W 88 E4
Dariganga = Ovoot
Mongolia 45°21N 113°45E 34 B7
Darjeeling = Darjiling
India 27°3N 88°18E 43 F13
Darjiling India 27°3N 88°18E 43 F13
Darkan Australia 33°20S 116°43E 61 F2
Darkhana Pakistan 30°39N 72°11E 42 D5
Darkhazīneh Iran 31°54N 48°39E 45 D6
Darkot Pass Pakistan 36°45N 73°26E 43 A5
Darling → Australia 34°4S 141°54E 63 E3
Darling Downs
Australia 27°30S 150°30E 63 D5
Darling Ra. Australia 32°30S 116°20E 61 F2
Darlington U.K. 54°32N 1°33W 12 C6
Darlington U.S.A. 34°18N 79°52W 85 D15
Darlington □ U.K. 54°32N 1°33W 12 C6
Darlington, L. S. Africa 33°10S 25°9E 56 E4
Darlot, L. Australia 27°48S 121°35E 61 E3
Darłowo Poland 54°25N 16°25E 16 A9
Darmstadt Germany 49°51N 8°39E 16 D5
Darnah Libya 32°45N 22°45E 51 B10
Darnall S. Africa 29°23S 31°18E 57 D5
Darnley Australia 9°35S 143°46E 62 a
Darnley, C. Antarctica 68°0S 69°0E 5 C6
Darnley B. Canada 69°30N 123°30W 68 D7
Darr → Australia 23°39S 143°50E 62 C3
Darra Pezu Pakistan 32°19N 70°44E 42 C4
Darrequeira Argentina 37°42S 63°10W 94 D3
Darrington U.S.A. 48°15N 121°36W 76 B3
Dart → U.K. 50°24N 3°39W 13 G4
Dart, C. Antarctica 73°6S 126°20W 5 D14
Dartford U.K. 51°26N 0°13E 13 F8
Dartmoor U.K. 50°38N 3°57W 13 G4
Dartmoor △ U.K. 50°37N 3°59W 13 G4
Dartmouth Canada 44°40N 63°30W 73 D7
Dartmouth U.K. 50°21N 3°36W 13 G4
Dartmouth Res.
Australia 26°4S 145°18E 63 D4
Dartuch, C. = Artrutx, C. de
Spain 39°55N 3°49E 24 B10
Darvel, Teluk = Lahad Datu,
Telok Malaysia 4°50N 118°20E 37 D5
Darwen U.K. 53°42N 2°29W 12 D5
Darwendale Zimbabwe 17°41S 30°33E 57 B5
Darwha India 20°15N 77°45E 40 J10
Darwin Australia 12°25S 130°51E 60 B5
Darwin U.S.A. 36°15N 117°35W 79 J9
Darya Khan Pakistan 31°48N 71°6E 42 D4
Daryācheh-ye Bakhtegān
Iran 29°40N 53°50E 45 D7
Daryoi Amu = Amudarya →
Uzbekistan 43°58N 59°34E 28 E6
Dās U.A.E. 25°20N 53°30E 45 E7
Dashen, Ras Ethiopia 13°8N 38°26E 47 E2
Dashetai China 41°0N 109°5E 34 D5
Dashiqiao China 40°38N 122°30E 35 D12
Dashköpri Turkmenistan 36°16N 62°8E 45 B9
Dashoguz Turkmenistan 41°49N 59°58E 28 E6
Dasht → Pakistan 25°10N 61°40E 40 G2
Daska Pakistan 32°20N 74°20E 42 C6
Dasuya India 31°49N 75°38E 42 D6
Datça Turkey 36°46N 27°40E 23 F12
Datia India 25°39N 78°27E 43 G8
Datong Qinghai, China 36°20N 102°50E 32 D9
Datong Shanxi, China 40°6N 113°18E 34 D7
Dattakhel Pakistan 32°54N 69°46E 42 C3
Datu, Tanjung Indonesia 2°5N 109°39E 36 D3
Datu Piang Phil. 7°2N 124°30E 37 C6
Datuk, Tanjong = Datu, Tanjung
Indonesia 2°5N 109°39E 36 D3
Daud Khel Pakistan 32°53N 71°34E 42 C4
Daudnagar India 25°2N 84°24E 43 G11
Daugava → Latvia 57°4N 24°3E 9 H21
Daugavpils Latvia 55°53N 26°32E 9 J22
Daulpur India 26°45N 77°59E 42 F7
Daung Kyun Burma 12°13N 98°4E 39 F1
Dauphin Canada 51°9N 100°5W 71 C8
Dauphin U.S.A. 40°22N 76°56W 82 F8
Dauphin L. Canada 51°20N 99°45W 71 C9
Dauphiné France 45°15N 5°25E 20 D6
Dausa India 26°52N 76°20E 42 F7
Davangere India 14°25N 75°55E 40 M9
Davao Phil. 7°0N 125°40E 37 C7
Davao G. Phil. 6°30N 125°48E 37 C7
Dāvar Panāh = Sarāvān
Iran 27°25N 62°15E 45 E9
Davenport Calif., U.S.A. 37°1N 122°12W 78 H4
Davenport Iowa, U.S.A. 41°32N 90°35W 80 E8
Davenport Wash., U.S.A. 47°39N 118°9W 76 C4
Davenport Ra. Australia 20°28S 134°0E 62 C1
Davenport Range △
Australia 20°36S 134°22E 62 C1
Daventry U.K. 52°16N 1°10W 13 E6
David Panama 8°30N 82°30W 88 E3
David City U.S.A. 41°15N 97°8W 80 E5
David Glacier Antarctica 75°20S 162°0E 5 D11
David Gorodok = Davyd Haradok
Belarus 52°4N 27°8E 17 B14
Davidson Canada 51°16N 105°59W 71 C7
Davidson Mts. U.S.A. 68°41N 142°22W 74 B11
Davis Antarctica 68°34S 77°55E 5 C6
Davis U.S.A. 38°33N 121°44W 78 G5
Davis Dam U.S.A. 35°12N 114°34W 79 K12
Davis Inlet Canada 55°50N 60°59W 73 A7
Davis Mts. U.S.A. 30°50N 103°55W 84 F3
Davis Sea Antarctica 66°0S 92°0E 5 C7
Davis Str. N. Amer. 65°0N 58°0W 69 D19
Davlos Cyprus 35°25N 33°54E 25 D12
Davos Switz. 46°48N 9°49E 20 C8
Davy L. Canada 58°53N 108°18W 71 B7
Davyd Haradok Belarus 52°4N 27°8E 17 B14
Dawei = Tavoy Burma 14°2N 98°12E 38 E2
Dawes Ra. Australia 24°40S 150°40E 62 C5
Dawlish U.K. 50°35N 3°28W 13 G4
Dawna Ra. Burma 16°30N 98°30E 38 D2
Dawros Hd. Ireland 54°50N 8°33W 10 B3
Dawson U.S.A. 31°46N 84°27W 85 F12
Dawson, I. Chile 53°50S 70°50W 96 G2
Dawson B. Canada 52°53N 100°49W 71 C8
Dawson City Canada 64°10N 139°30W 68 E4
Dawson Creek Canada 55°45N 120°15W 70 B4
Dawson Inlet Canada 61°50N 93°25W 71 A10
Dawson Ra. Australia 24°30S 149°48E 62 C4
Dax France 43°44N 1°3W 20 E3
Daxian China 31°15N 107°23E 32 E10
Daxindian China 37°30N 120°50E 35 F11
Daxing China 39°47N 116°24E 34 E9
Daxinggou China 43°25N 129°40E 35 C15
Daxue Shan China 30°30N 101°30E 32 E9
Day Kundī □ Afghan. 34°0N 66°0E 40 C5
Daylesford Australia 37°21S 144°9E 63 F3
Dayr az Zawr Syria 35°20N 40°5E 44 C4
Daysland Canada 52°50N 112°20W 70 C6
Dayton Nev., U.S.A. 39°14N 119°36W 78 F7
Dayton Ohio, U.S.A. 39°45N 84°12W 81 F11
Dayton Pa., U.S.A. 40°53N 79°15W 82 F5
Dayton Tenn., U.S.A. 35°30N 85°1W 85 D12
Dayton Wash., U.S.A. 46°19N 117°59W 76 C5
Dayton Wyo., U.S.A. 44°53N 107°16W 76 D10
Daytona Beach U.S.A. 29°13N 81°1W 85 G14
Dayville U.S.A. 44°28N 119°32W 76 D4
De Aar S. Africa 30°39S 24°0E 56 E3
De Biesbosch △ Neths. 51°45N 4°48E 15 C4
De Funiak Springs
U.S.A. 30°43N 86°7W 85 F11
De Grey → Australia 20°12S 119°13E 60 D2
De Haan Belgium 51°16N 3°2E 15 C3
De Hoge Veluwe △ Neths. 52°5N 5°46E 15 B5
De Hoop ⌂ S. Africa 34°30S 20°28E 56 E3
De Kalb Junction U.S.A. 44°30N 75°16W 83 B9
De Kennemerduinen △
Neths. 52°27N 4°33E 15 B4
De Land U.S.A. 29°2N 81°18W 85 G14
De Leon U.S.A. 32°7N 98°32W 84 E5
De Long Mts. U.S.A. 68°10N 163°30W 74 B7
De Panne Belgium 51°6N 2°34E 15 C2
De Pere U.S.A. 44°27N 88°4W 80 C9
De Queen U.S.A. 34°2N 94°21W 84 D7
De Quincy U.S.A. 30°27N 93°26W 84 F8
De Ruyters U.S.A. 42°45N 75°53W 83 D9
De Smet U.S.A. 44°23N 97°33W 80 C5
De Soto U.S.A. 38°8N 90°34W 80 F8
De Tour Village U.S.A. 46°0N 83°56W 81 B12
De Witt U.S.A. 34°18N 91°20W 84 D9
Dead Sea Asia 31°30N 35°30E 46 D4
Deadwood U.S.A. 44°23N 103°44W 80 C2
Deadwood L. Canada 59°10N 128°30W 70 B3
Deal U.K. 51°13N 1°25E 13 F9
Deal I. Australia 39°30S 147°20E 63 F4
Dealesville S. Africa 28°41S 25°44E 56 D4
Dean → Canada 52°49N 126°58W 70 C3
Dean, Forest of U.K. 51°45N 2°33W 13 F5
Dean Chan. Canada 52°30N 127°15W 70 C3
Deán Funes Argentina 30°20S 64°20W 94 C3
Dease → Canada 59°56N 128°32W 70 B3
Dease L. Canada 58°40N 130°5W 70 B2
Dease Lake Canada 58°25N 130°6W 70 B2
Death Valley U.S.A. 36°15N 116°50W 79 J10
Death Valley △ U.S.A. 36°29N 117°6W 79 J9
Death Valley Junction
U.S.A. 36°20N 116°25W 79 J10
Deatnu = Tana →
Norway 70°30N 28°14E 8 A23
Debagram India 23°51N 90°33E 43 H14
Debar Macedonia 41°31N 20°30E 23 D9
Debden Canada 53°30N 106°50W 71 C7
Dębica Poland 50°2N 21°25E 17 C11
DeBolt Canada 55°12N 118°1W 70 B5
Deborah East, L.
Australia 30°45S 119°30E 61 F2
Deborah West, L.
Australia 30°45S 119°5E 61 F2
Debre Markos Ethiopia 10°20N 37°40E 47 E2
Debre Tabor Ethiopia 11°50N 38°26E 47 E2
Debre Zebit Ethiopia 11°48N 38°30E 47 F2
Debrecen Hungary 47°33N 21°42E 17 E11
Decatur Ala., U.S.A. 34°36N 86°59W 85 D11
Decatur Ga., U.S.A. 33°46N 84°16W 85 E12
Decatur Ill., U.S.A. 39°51N 88°57W 80 F9
Decatur Ind., U.S.A. 40°50N 84°56W 81 E11
Decatur Tex., U.S.A. 33°14N 97°35W 84 E6
Deccan India 18°0N 79°0E 40 L11
Deception Bay Australia 27°10S 153°5E 63 D5
Deception I. Antarctica 63°0S 60°15W 5 C17
Deception L. Canada 56°33N 104°13W 71 B8
Dechhu India 26°46N 72°20E 42 F5
Děčín Czech Rep. 50°47N 14°12E 16 C8
Deckerville U.S.A. 43°32N 82°44W 82 C2
Decorah U.S.A. 43°18N 91°48W 80 D8
Dedéagach = Alexandroupoli
Greece 40°50N 25°54E 23 D11
Dedham U.S.A. 42°15N 71°10W 83 D13
Dedza Malawi 14°20S 34°20E 55 E3
Dee → Aberds., U.K. 57°9N 2°5W 11 D6
Dee → Dumf. & Gall., U.K. 54°51N 4°3W 11 G4
Dee → Wales, U.K. 53°22N 3°17W 12 D4
Deep B. Canada 61°15N 116°35W 70 A5
Deep Bay = Shenzhen Wan
China 22°27N 113°55E 33 a
Deepwater Australia 29°25S 151°51E 63 D5
Deer → Canada 58°23N 94°13W 71 B10
Deer L. Canada 52°40N 94°20W 71 C10
Deer Lake Nfld. & L.,
Canada 49°11N 57°27W 73 C8
Deer Lake Ont., Canada 52°36N 94°20W 71 C10
Deer Lodge U.S.A. 46°24N 112°44W 76 C7
Deer Park U.S.A. 47°57N 117°28W 76 C5
Deer River U.S.A. 47°20N 93°48W 80 B7
Deeragun Australia 19°16S 146°33E 62 B4
Defiance U.S.A. 41°17N 84°22W 81 E11
Degana India 26°50N 74°20E 42 F6
Dégelis Canada 47°30N 68°35W 73 C6
Deggendorf Germany 48°50N 12°57E 16 D7
Degh → Pakistan 31°3N 73°21E 42 D5
Degirmenlik = Kythréa
Cyprus 35°15N 33°29E 25 D12
Deh Bīd Iran 30°39N 53°11E 45 D7
Deh Dasht Iran 30°47N 50°33E 45 D6
Deh-e Shīr Iran 31°29N 53°45E 45 D7
Dehaj Iran 30°42N 54°53E 45 D7
Dehak Iran 27°11N 62°37E 45 E9
Dehdez Iran 31°43N 50°17E 45 D6
Dehej India 21°44N 72°40E 42 J5
Dehestān Iran 28°30N 55°35E 45 D7
Dehgolān Iran 35°17N 47°25E 44 C5
Dehibat Tunisia 32°0N 10°47E 51 B8
Dehlorān Iran 32°41N 47°16E 44 C5
Dehnow-e Kūhestān
Iran 27°58N 58°32E 45 E8
Dehra Dun India 30°20N 78°4E 42 D8
Dehri India 24°50N 84°15E 43 G11
Dehui China 44°30N 125°40E 35 B13
Deinze Belgium 50°59N 3°32E 15 D3
Dej Romania 47°10N 23°52E 17 E12
Deka → Zimbabwe 18°4S 26°42E 56 B4
DeKalb U.S.A. 41°56N 88°46W 80 E9
Dekese
Dem. Rep. of the Congo 3°24S 21°24E 52 E4
Del Mar U.S.A. 32°58N 117°16W 79 N9
Del Norte U.S.A. 37°41N 106°21W 77 H10
Del Rio U.S.A. 29°22N 100°54W 84 G4
Delambre I. Australia 20°26S 117°5E 60 D2
Delano U.S.A. 35°46N 119°15W 79 K7
Delano Peak U.S.A. 38°22N 112°22W 76 G7
Delareyville S. Africa 26°41S 25°26E 56 D4
Delaronde L. Canada 54°3N 107°3W 71 C7
Delavan U.S.A. 42°38N 88°39W 80 D9
Delaware U.S.A. 40°18N 83°4W 81 E12
Delaware □ U.S.A. 39°0N 75°20W 81 F16
Delaware → U.S.A. 39°15N 75°20W 83 G9
Delaware B. U.S.A. 39°0N 75°10W 81 F16
Delaware Water Gap ⌂
U.S.A. 41°10N 74°55W 83 E10
Delay → Canada 56°56N 71°28W 73 A5
Delegate Australia 37°4S 148°56E 63 F4
Delevan U.S.A. 42°29N 78°29W 82 D6
Delft Neths. 52°1N 4°22E 15 B4
Delfzijl Neths. 53°20N 6°55E 15 A6
Delgado, C. Mozam. 10°45S 40°40E 55 E5
Delgerhet Mongolia 45°50N 110°30E 34 B6
Delgo Sudan 20°6N 30°40E 51 D12
Delhi Canada 42°51N 80°30W 82 D4
Delhi India 28°39N 77°13E 42 E7
Delhi La., U.S.A. 32°28N 91°30W 84 E9
Delhi N.Y., U.S.A. 42°17N 74°55W 83 D10
Delia Canada 51°38N 112°23W 70 C6
Delice Turkey 39°54N 34°2E 19 G5
Delicias Mexico 28°13N 105°28W 86 B3
Delījān Iran 33°59N 50°40E 45 C6
Déline Canada 65°11N 123°25W 68 D7
Delingha China 37°23N 97°23E 32 D8
Delisle Canada 51°55N 107°8W 71 C7
Deliverance I. Australia 9°31S 141°34E 62 a
Dell City U.S.A. 31°56N 105°12W 84 F2
Dell Rapids U.S.A. 43°50N 96°43W 80 D5
Delmar U.S.A. 42°37N 73°47W 83 D11
Delmenhorst Germany 53°3N 8°37E 16 B5
Delonga, Ostrova
Russia 76°40N 149°20E 29 B15
Deloraine Australia 41°30S 146°40E 63 G4
Deloraine Canada 49°15N 100°29W 71 D8
Delphi U.S.A. 40°36N 86°41W 80 E10
Delphos U.S.A. 40°51N 84°21W 81 E11
Delportshoop S. Africa 28°22S 24°20E 56 D3
Delray Beach U.S.A. 26°28N 80°4W 85 H14
Delta Colo., U.S.A. 38°44N 108°4W 76 G9
Delta Utah, U.S.A. 39°21N 112°35W 76 G7
Delta Dunării ⌂
Romania 45°15N 29°25E 17 F15
Delta Junction U.S.A. 64°2N 145°44W 68 E2
Deltona U.S.A. 28°54N 81°16W 85 G14
Delungra Australia 29°39S 150°51E 63 D5
Delvada India 20°46N 71°2E 42 J4
Delvinë Albania 39°59N 20°6E 23 E9
Demak Indonesia 6°53S 110°38E 37 G14
Demanda, Sierra de la
Spain 42°15N 3°0W 21 A4
Demavend = Damāvand,
Qolleh-ye Iran 35°56N 52°10E 45 C7
Dembia C.A.R. 4°56N 23°53E 54 B1
Dembia
Dem. Rep. of the Congo 3°33N 25°48E 54 B2
Dembidolo Ethiopia 8°34N 34°50E 47 F1
Demchok India 32°42N 79°29E 43 C8
Demer → Belgium 50°57N 4°42E 15 D4
Deming N. Mex.,
U.S.A. 32°16N 107°46W 77 K10
Deming Wash., U.S.A. 48°50N 122°13W 78 B4
Demini → Brazil 0°46S 62°56W 92 D6
Demirci Turkey 39°2N 28°38E 23 E13
Demirköy Turkey 41°49N 27°45E 23 D12
Demopolis U.S.A. 32°31N 87°50W 85 E11
Dempo Indonesia 4°2S 103°15E 36 E2
Den Bosch = 's-Hertogenbosch
Neths. 51°42N 5°17E 15 C5
Den Burg Neths. 53°3N 4°47E 15 A4
Den Chai Thailand 17°59N 100°4E 38 D3
Den Haag = 's-Gravenhage
Neths. 52°7N 4°17E 15 B4
Den Helder Neths. 52°57N 4°45E 15 B4
Den Oever Neths. 52°56N 5°2E 15 B5
Denair U.S.A. 37°32N 120°48W 78 H6
Denali = McKinley, Mt.
U.S.A. 63°4N 151°0W 68 E1
Denau Uzbekistan 38°16N 67°54E 28 F7
Denbigh Canada 45°8N 77°15W 82 A7
Denbigh U.K. 53°12N 3°25W 12 D4
Denbighshire □ U.K. 53°8N 3°22W 12 D4
Dendang Indonesia 3°7S 107°56E 36 E3
Dendermonde Belgium 51°2N 4°5E 15 C4
Dengfeng China 34°25N 113°2E 34 G7
Dengkou China 40°18N 106°55E 34 D4
Denham Australia 25°56S 113°31E 61 E1
Denham, Mt. Jamaica 18°13N 77°32W 88 a
Denham Ra. Australia 21°55S 147°46E 62 C4
Denham Sd. Australia 25°45S 113°15E 61 E1
Denholm Canada 52°39N 108°1W 71 C7
Denia Spain 38°49N 0°8E 21 C6
Denial B. Australia 32°14S 133°32E 63 E1
Deniliquin Australia 35°30S 144°58E 63 F3
Denimoo = Fort Resolution
Canada 61°10N 113°40W 70 A6
Denison Iowa, U.S.A. 42°1N 95°21W 80 D6
Denison Tex., U.S.A. 33°45N 96°33W 84 E6
Denison Plains Australia 18°35S 128°0E 60 C4
Denizli Turkey 37°42N 29°2E 19 G4
Denman Glacier
Antarctica 66°45S 100°0E 5 C8
Denmark Australia 34°59S 117°25E 61 F2
Denmark ■ Europe 55°45N 10°0E 9 J14
Denmark Str. Atl. Oc. 66°0N 30°0W 66 C17
Dennery St. Lucia 13°55N 60°54W 89 f
Dennison U.S.A. 40°24N 81°19W 82 F3
Denny U.K. 56°1N 3°55W 11 E5
Denpasar Indonesia 8°39S 115°13E 37 K18
Denpasar ✈ (DPS)
Indonesia 8°44S 115°10E 37 K18
Denton Mont., U.S.A. 47°19N 109°57W 76 C9
Denton Tex., U.S.A. 33°13N 97°8W 84 E6
D'Entrecasteaux, Pt.
Australia 34°50S 115°57E 61 F2
D'Entrecasteaux △
Australia 34°20S 115°33E 61 F2
D'Entrecasteaux Is.
Papua N. G. 9°0S 151°0E 58 B8
Denver Colo., U.S.A. 39°42N 104°59W 76 G11
Denver Pa., U.S.A. 40°14N 76°8W 83 F8
Denver City U.S.A. 32°58N 102°50W 84 E3
Deoband India 29°42N 77°43E 42 E7
Deogarh India 25°32N 73°54E 42 G5
Deoghar India 24°30N 86°42E 43 G12
Deolali India 19°58N 73°50E 40 K8
Deoli = Devli India 25°50N 75°20E 42 G6
Deora India 26°22N 70°55E 42 F4
Deori India 23°24N 79°1E 43 H8
Deoria India 26°31N 83°48E 43 F10
Deosai Mts. Pakistan 35°40N 75°0E 43 B6
Deosri India 26°46N 90°29E 43 F14
Depalpur India 22°51N 75°33E 42 H6
Deposit U.S.A. 42°4N 75°25W 83 D9
Depuch I. Australia 20°37S 117°44E 60 D2
Deputatskiy Russia 69°18N 139°54E 29 C14
Dera Ghazi Khan
Pakistan 30°5N 70°43E 42 D4
Dera Ismail Khan
Pakistan 31°50N 70°50E 42 D4
Derabugti Pakistan 29°2N 69°9E 42 E3
Derawar Fort Pakistan 28°46N 71°20E 42 E4
Derbent Russia 42°5N 48°15E 19 F8
Derby Australia 17°18S 123°38E 60 C3
Derby U.K. 52°56N 1°28W 12 E6
Derby Conn., U.S.A. 41°19N 73°5W 83 E11
Derby Kans., U.S.A. 37°33N 97°16W 80 G5
Derby N.Y., U.S.A. 42°41N 78°58W 82 D6
Derby City □ U.K. 52°56N 1°28W 12 E6
Derby Line U.S.A. 45°0N 72°6W 83 B12
Derbyshire □ U.K. 53°11N 1°38W 12 D6
Derdepoort S. Africa 24°38S 26°24E 56 C4
Dereham U.K. 52°41N 0°57E 13 E8
Derg → U.K. 54°44N 7°26W 10 B4
Derg, L. Ireland 53°0N 8°20W 10 D3
Deridder U.S.A. 30°51N 93°17W 84 F8
Dermott U.S.A. 33°32N 91°26W 84 E9
Derry = Londonderry
U.K. 55°0N 7°20W 10 B4
Derry = Londonderry □
U.K. 55°0N 7°20W 10 B4
Derry N.H., U.S.A. 42°53N 71°19W 83 D13
Derry Pa., U.S.A. 40°20N 79°18W 82 F5
Derryveagh Mts. Ireland 54°56N 8°11W 10 B3
Derwent → Cumb., U.K. 54°39N 3°33W 12 C4
Derwent → Derby, U.K. 52°57N 1°28W 12 E6

Derwent → N. Yorks.,
U.K. 53°45N 0°58W **12** D7
Derwent Water U.K. 54°35N 3°9W **12** C4
Des Moines Iowa, U.S.A. 41°35N 93°37W **80** E7
Des Moines N. Mex.,
U.S.A. 36°46N 103°50W **77** H12
Des Moines Wash.,
U.S.A. 47°24N 122°19W **78** C4
Des Moines → U.S.A. 40°23N 91°25W **80** E8
Desaguadero →
Argentina 34°30S 66°46W **94** C2
Desaguadero → Bolivia 16°35S 69°5W **92** G5
Desaru Malaysia 1°31N 104°17E **39** d
Descanso, Pta. Mexico 32°21N 117°3W **79** N9
Deschaillons-sur-St-Laurent
Canada 46°32N 72°7W **73** C5
Deschambault L.
Canada 54°50N 103°30W **71** C8
Deschutes → U.S.A. 45°38N 120°55W **76** D3
Dese Ethiopia 11°5N 39°40E **47** E2
Deseado → Argentina 47°45S 65°54W **96** F3
Deseronto Canada 44°12N 77°3W **82** B7
Desert Center U.S.A. 33°43N 115°24W **79** M11
Desert Hot Springs
U.S.A. 33°58N 116°30W **79** M10
Deshnok India 27°48N 73°21E **42** F5
Desierto Central de Baja
California △ Mexico 29°40N 114°50W **86** B2
Desna → Ukraine 50°33N 30°32E **17** C16
Desolación, I. Chile 53°0S 74°0W **96** G2
Despeñaperros, Paso
Spain 38°24N 3°30W **21** C4
Dessau Germany 51°51N 12°14E **16** C7
Dessye = Dese Ethiopia 11°5N 39°40E **47** E2
D'Estrees B. Australia 35°55S 137°45E **63** F2
Desuri India 25°18N 73°35E **42** G5
Det Udom Thailand 14°54N 105°5E **38** E5
Dete Zimbabwe 18°38S 26°50E **56** B4
Detmold Germany 51°56N 8°52E **16** C5
Detour, Pt. U.S.A. 45°40N 86°40W **80** C10
Detroit U.S.A. 42°19N 83°12W **82** D1
Detroit Lakes U.S.A. 46°49N 95°51W **80** B6
Deua △ Australia 35°32S 149°46E **63** F4
Deurne Neths. 51°27N 5°49E **15** C5
Deutsche Bucht Germany 54°15N 8°0E **16** A5
Deutschland = Germany ■
Europe 51°0N 10°0E **16** C6
Deva Romania 45°53N 22°55E **17** F12
Devakottai India 9°55N 78°45E **40** Q11
Devaprayag India 30°13N 78°35E **43** D8
Deventer Neths. 52°15N 6°10E **15** B6
Deveron → U.K. 57°41N 2°32W **11** D6
Devgadh Bariya India 22°40N 73°55E **42** H5
Devikot India 26°42N 71°12E **42** F4
Devils Den U.S.A. 35°46N 119°58W **78** K7
Devils Hole = Death Valley △
U.S.A. 36°29N 117°6W **79** J9
Devils Lake U.S.A. 48°7N 98°52W **80** A4
Devils Paw Canada 58°47N 134°0W **70** B2
Devils Postpile △ U.S.A. 37°37N 119°5W **78** H7
Devils Tower U.S.A. 44°35N 104°42W **76** D11
Devils Tower △
U.S.A. 44°48N 104°55W **76** D11
Devine U.S.A. 29°8N 98°54W **84** G5
Devizes U.K. 51°22N 1°58W **13** F6
Devli India 25°50N 75°20E **42** G6
Devon Canada 53°24N 113°44W **70** C6
Devon □ U.K. 50°50N 3°40W **13** G4
Devon I. Canada 75°10N 85°0W **69** B15
Devonport Australia 41°10S 146°22E **63** G4
Devonport N.Z. 36°49S 174°49E **59** B5
Dewas India 22°59N 76°3E **42** H7
Dewetsdorp S. Africa 29°33S 26°39E **56** D4
Dewey Puerto Rico 18°18N 65°18W **89** d
Dexter Maine, U.S.A. 45°1N 69°18W **81** C19
Dexter Mo., U.S.A. 36°48N 89°57W **80** G9
Dexter N. Mex., U.S.A. 33°12N 104°22W **77** K11
Dey-Dey, L. Australia 29°12S 131°4E **61** E5
Deyang China 31°3N 104°27E **32** B5
Deyhūk Iran 33°15N 57°30E **45** C8
Deyyer Iran 27°55N 51°55E **45** E6
Dez → Iran 31°39N 48°52E **45** D6
Dezadeash L. Canada 60°28N 136°58W **70** A1
Dezful Iran 32°20N 48°30E **45** C6
Dezhneva, Mys Russia 66°5N 169°40W **29** C19
Dezhou China 37°26N 116°18E **34** F9
Dhadhar → India 24°56N 85°24E **43** G11
Dhahiriya = Az Ẓāhirīyah
West Bank 31°25N 34°58E **46** D3
Dhahran = Az Ẓahrān
Si. Arabia 26°10N 50°7E **45** E6
Dhak Pakistan 32°25N 72°33E **42** C5
Dhaka Bangla. 23°43N 90°26E **43** H14
Dhaka □ Bangla. 24°25N 90°25E **43** G14
Dhali Cyprus 35°1N 33°25E **25** D12
Dhamār Yemen 14°30N 44°20E **47** E3
Dhampur India 29°19N 78°33E **43** E8
Dhamtari India 20°42N 81°35E **41** J12
Dhanbad India 23°50N 86°30E **43** H12
Dhangarhi Nepal 28°55N 80°40E **43** E9
Dhankuta Nepal 26°55N 87°40E **43** F12
Dhanpuri India 23°13N 81°30E **43** H9
Dhar India 22°13N 75°18E **42** H6
Dharampur India 22°13N 75°18E **42** H6
Dharamsala = Dharmsala
India 32°16N 76°23E **42** C7
Dharan Nepal 26°49N 87°17E **43** F12
Dhariwal India 31°57N 75°19E **42** D6
Dharla → Bangla. 25°46N 89°42E **43** G13
Dharmapuri India 12°10N 78°10E **40** N11
Dharmjaygarh India 22°28N 83°13E **43** H10
Dharmsala India 32°16N 76°23E **42** C7
Dharni India 21°33N 76°53E **42** J7
Dharwad India 15°30N 75°4E **40** M9
Dhasan → India 25°48N 79°24E **43** G8
Dhaulagiri Nepal 28°39N 83°28E **43** E10
Dhebar, L. India 24°10N 74°0E **42** G6
Dheftera Cyprus 35°5N 33°16E **25** D12
Dhenkanal India 20°45N 85°35E **41** J14
Dherinia Cyprus 35°3N 33°57E **25** D12

Dhī Qār □ Iraq 31°0N 46°15E **44** D5
Dhiarrizos → Cyprus 34°41N 32°34E **25** E11
Dhībān Jordan 31°30N 35°46E **46** D4
Dhilwan India 31°31N 75°21E **42** D6
Dhimarkhera India 23°28N 80°22E **43** H9
Dhodhekánisos = Dodecanese
Greece 36°35N 27°0E **23** F12
Dholka India 22°44N 72°29E **42** H5
Dhoraji India 21°45N 70°37E **42** J4
Dhrangadhra India 22°59N 71°31E **42** H4
Dhrol India 22°33N 70°25E **42** H4
Dhuburi India 26°2N 89°59E **41** F16
Dhule India 20°58N 74°50E **40** J9
Di Linh Vietnam 11°35N 108°4E **39** G7
Di Linh, Cao Nguyen
Vietnam 11°30N 108°0E **39** G7
Dia Greece 35°28N 25°14E **25** D7
Diablo Range U.S.A. 37°20N 121°25W **78** J5
Diafarabé Mali 14°9N 4°57W **50** F5
Diamante Argentina 32°5S 60°40W **94** C3
Diamante → Argentina 34°30S 66°46W **94** C2
Diamantina Brazil 18°17S 43°40W **93** G10
Diamantina →
Australia 26°45S 139°10E **63** D2
Diamantina △ Australia 23°33S 141°23E **62** C3
Diamantino Brazil 14°30S 56°30W **93** F7
Diamond Bar U.S.A. 34°1N 117°48W **79** L9
Diamond Harbour
India 22°11N 88°14E **43** H13
Diamond Is. Australia 17°25S 151°5E **62** B5
Diamond Mts. U.S.A. 39°50N 115°50W **76** G6
Diamond Springs
U.S.A. 38°42N 120°49W **78** G6
Diaoyu Dao = Senkaku-Shotō
E. China Sea 25°45N 123°30E **31** M1
Diaoyu Tai = Senkaku-Shotō
E. China Sea 25°45N 123°30E **31** M1
Dībā U.A.E. 25°45N 56°16E **45** E8
Dibai India 28°13N 78°15E **42** E8
Dibaya
Dem. Rep. of the Congo 6°30S 22°57E **52** F4
Dibaya-Lubue
Dem. Rep. of the Congo 4°12S 19°54E **52** E3
Dibbeen △ Jordan 32°20N 35°45E **46** C4
D'Iberville, Lac Canada 55°55N 73°15W **72** A5
Dibete Botswana 23°45S 26°32E **56** C4
Dibrugarh India 27°29N 94°55E **41** F19
Dickens U.S.A. 33°37N 100°50W **84** E4
Dickinson U.S.A. 46°53N 102°47W **80** B2
Dickson U.S.A. 36°5N 87°23W **85** C11
Dickson City U.S.A. 41°28N 75°36W **83** E9
Didiéni Mali 13°53N 8°6W **50** F4
Didsbury Canada 51°35N 114°10W **70** C6
Didwana India 27°23N 74°36E **42** F6
Diefenbaker, L. Canada 51°0N 106°55W **71** C7
Diego de Almagro Chile 26°22S 70°3W **94** B1
Diego Suarez = Antsiranana
Madag. 12°25S 49°20E **57** A8
Diekirch Lux. 49°52N 6°10E **15** E6
Dien Ban Vietnam 15°53N 108°16E **38** E7
Dien Bien Phu Vietnam 21°20N 103°0E **38** B4
Dien Chau, Vinh
Vietnam 19°0N 105°59E **38** C5
Dien Khanh Vietnam 12°15N 109°6E **39** F7
Dieppe France 49°54N 1°4E **20** B4
Dierks U.S.A. 34°7N 94°1W **84** D7
Diest Belgium 50°58N 5°4E **15** D5
Differdange Lux. 49°31N 5°54E **15** E5
Dig India 27°28N 77°20E **42** F7
Digba
Dem. Rep. of the Congo 4°25N 25°48E **54** B2
Digby Canada 44°38N 65°50W **73** D6
Digges Is. Canada 62°40N 77°50W **69** E16
Diggi India 26°22N 75°26E **42** F6
Dighinala Bangla. 23°15N 92°5E **41** H18
Dighton U.S.A. 38°29N 100°28W **80** F3
Digne-les-Bains France 44°5N 6°12E **20** D7
Digos Phil. 6°45N 125°20E **37** C7
Digranes Iceland 66°4N 14°44W **8** C6
Digul → Indonesia 7°7S 138°42E **37** F9
Dihang = Brahmaputra →
Asia 23°40N 90°35E **43** H13
Dijlah, Nahr → Asia 31°0N 47°25E **44** D5
Dijon France 47°20N 5°3E **20** C6
Dikhil Djibouti 11°8N 42°20E **47** E3
Dikkil = Dikhil Djibouti 11°8N 42°20E **47** E3
Diksmuide Belgium 51°2N 2°52E **15** C2
Dikson Russia 73°40N 80°5E **28** B9
Dikti Oros Greece 35°8N 25°30E **25** D7
Dila Ethiopia 6°21N 38°22E **47** F2
Dili E. Timor 8°39S 125°34E **37** F7
Dilley U.S.A. 28°40N 99°10W **84** G5
Dilli = Delhi India 28°39N 77°13E **42** E7
Dillingham U.S.A. 59°3N 158°28W **74** D8
Dillon Canada 55°56N 108°35W **71** B7
Dillon Mont., U.S.A. 45°13N 112°38W **76** D7
Dillon S.C., U.S.A. 34°25N 79°22W **85** D15
Dillon → Canada 55°56N 108°56W **71** B7
Dillsburg U.S.A. 40°7N 77°2W **82** F7
Dilolo
Dem. Rep. of the Congo 10°28S 22°18E **52** G4
Dimapur India 25°54N 93°45E **41** G18
Dimas Mexico 23°43N 106°47W **86** C3
Dimashq Syria 33°30N 36°18E **46** B5
Dimashq □ Syria 33°30N 36°30E **46** B5
Dimbaza S. Africa 32°50S 27°14E **56** E4
Dimboola Australia 36°28S 142°7E **63** F3
Dîmbovița = Dâmbovița →
Romania 44°12N 26°26E **17** F14
Dimbulah Australia 17°8S 145°4E **62** B4
Dimitrovgrad Bulgaria 42°5N 25°35E **23** C11
Dimitrovgrad Russia 54°14N 49°39E **18** D8
Dimitrovo = Pernik
Bulgaria 42°35N 23°2E **23** C10
Dimmitt U.S.A. 34°33N 102°19W **84** D3
Dimona Israel 31°2N 35°1E **46** D4
Dinagat I. Phil. 10°10N 125°40E **37** B7
Dinajpur Bangla. 25°33N 88°43E **41** G16
Dinan France 48°28N 2°2W **20** B2

Dīnān Āb Iran 32°4N 56°49E **45** C8
Dinant Belgium 50°16N 4°55E **15** D4
Dinapur India 25°38N 85°5E **43** G11
Dīnār, Kūh-e Iran 30°42N 51°46E **45** D6
Dinara Planina Croatia 44°0N 16°30E **22** C7
Dinard France 48°38N 2°6W **20** B2
Dinaric Alps = Dinara Planina
Croatia 44°0N 16°30E **22** C7
Dindigul India 10°25N 78°0E **40** P11
Dindori India 22°57N 81°5E **43** H9
Ding Xian = Dingzhou
China 38°30N 114°59E **34** E8
Dinga Pakistan 25°26N 67°10E **42** G2
Ding'an China 19°42N 110°19E **38** C8
Dingbian China 37°35N 107°32E **34** F4
Dingle Ireland 52°9N 10°17W **10** D1
Dingle B. Ireland 52°3N 10°20W **10** D1
Dingmans Ferry U.S.A. 41°13N 74°55W **83** E10
Dingo Australia 23°38S 149°19E **62** C4
Dingtao China 35°5N 115°35E **34** G8
Dingwall U.K. 57°36N 4°26W **11** D4
Dingxi China 35°30N 104°33E **34** G3
Dingxiang China 38°30N 112°58E **34** E7
Dingzhou China 38°30N 114°59E **34** E8
Dinh, Mui Vietnam 11°22N 109°1E **39** G7
Dinh Lap Vietnam 21°33N 107°6E **38** B6
Dinin → Ireland 52°43N 7°18N **10** D4
Dinira △ Venezuela 9°57N 70°6W **89** E6
Dinokwe Botswana 23°29S 26°37E **56** C4
Dinorwic Canada 49°41N 92°30W **71** D10
Dinosaur △ Canada 50°47N 111°30W **70** C6
Dinosaur △ U.S.A. 40°30N 108°45W **76** F9
Dinuba U.S.A. 36°32N 119°23W **78** J7
Dionisades Greece 35°20N 26°10E **25** D8
Diourbel Senegal 14°39N 16°12W **50** F2
Dipalpur Pakistan 30°40N 73°39E **42** D5
Dipkarpaz = Rizokarpaso
Cyprus 35°36N 34°23E **25** D13
Diplo Pakistan 24°35N 69°35E **42** G3
Dipolog Phil. 8°36N 123°20E **37** C6
Dipperu △ Australia 21°56S 148°42E **62** C4
Dir Pakistan 35°8N 71°59E **40** B7
Dire Dawa Ethiopia 9°35N 41°45E **47** F3
Dirfis Oros Greece 38°40N 23°54E **23** E10
Diriamba Nic. 11°51N 86°19W **88** D2
Dirk Hartog I. Australia 25°50S 113°5E **61** E1
Dirranbandi Australia 28°33S 148°17E **63** D4
Disa India 24°18N 72°10E **42** G5
Disappointment, C.
U.S.A. 46°18N 124°5W **76** C1
Disappointment, L.
Australia 23°20S 122°40E **60** D3
Disaster B. Australia 37°15S 149°58E **63** F4
Discovery B. Australia 38°10S 140°40E **63** F3
Discovery B. China 22°18N 114°1E **33** a
Disko = Qeqertarsuaq
Greenland 69°45N 53°30W **4** C5
Disney Reef Tonga 19°17S 174°7W **59** c
Disneyland Hong Kong
China 22°18N 114°2E **33** a
Diss U.K. 52°23N 1°7E **13** E9
Disteghil Sar Pakistan 36°20N 75°12E **43** A6
District of Columbia □
U.S.A. 38°55N 77°0W **81** F15
Distrito Federal □ Brazil 15°45S 47°45W **93** G9
Distrito Federal □
Mexico 19°15N 99°10W **87** D5
Diu India 20°45N 70°58E **42** J4
Dīvāndarreh Iran 35°55N 47°2E **44** C5
Divide U.S.A. 45°45N 112°45W **76** D7
Dividing Ra. Australia 24°7S 116°0E **61** E2
Divinópolis Brazil 20°10S 44°54W **93** H10
Divnoye Russia 45°55N 43°21E **19** E7
Divo Ivory C. 5°48N 5°15W **50** G4
Dīwāl Kol Afghan. 34°23N 67°52E **42** B2
Dixie Mt. U.S.A. 39°55N 120°16W **78** F6
Dixon Calif., U.S.A. 38°27N 121°49W **78** G5
Dixon Ill., U.S.A. 41°50N 89°29W **80** E9
Dixon Entrance U.S.A. 54°30N 132°0W **68** F6
Dixville Canada 45°4N 71°46W **83** A13
Diyālā □ Iraq 33°45N 44°50E **44** C5
Diyālā → Iraq 33°13N 44°30E **44** C5
Diyarbakır Turkey 37°55N 40°18E **44** B4
Diyodar India 24°8N 71°50E **42** G4
Djakarta = Jakarta
Indonesia 6°9S 106°52E **37** G12
Djamba Angola 16°45S 13°58E **56** B1
Djambala Congo 2°32S 14°30E **52** E2
Djanet Algeria 24°35N 9°32E **50** D7
Djawa = Jawa Indonesia 7°0S 110°0E **36** F3
Djelfa Algeria 34°40N 3°15E **50** B6
Djema C.A.R. 6°3N 25°15E **54** A2
Djerba, Î. de Tunisia 33°50N 10°48E **51** B8
Djerid, Chott Tunisia 33°42N 8°30E **50** B7
Djibouti Djibouti 11°30N 43°5E **47** E3
Djibouti ■ Africa 12°0N 43°0E **47** E3
Djolu Dem. Rep. of the Congo 0°35N 22°5E **52** D4
Djougou Benin 9°40N 1°45E **50** G6
Djoum Cameroon 2°41N 12°35E **52** D2
Djourab, Erg du Chad 16°40N 18°50E **51** E9
Djugu
Dem. Rep. of the Congo 1°55N 30°35E **54** B3
Djukbinj △ Australia 12°11S 131°2E **60** B5
Djúpivogur Iceland 64°39N 14°17W **8** D6
Dmitriya Lapteva, Proliv
Russia 73°0N 140°0E **29** B15
Dnepr = Dnipro →
Ukraine 46°30N 32°18E **19** E5
Dneprodzerzhinsk =
Dniprodzerzhynsk
Ukraine 48°32N 34°37E **19** E5
Dnepropetrovsk =
Dnipropetrovsk
Ukraine 48°30N 35°0E **19** E6
Dnestr = Dnister →
Europe 46°18N 30°17E **17** E16
Dnieper = Dnipro →
Ukraine 46°30N 32°18E **19** E5
Dniester = Dnister →
Europe 46°18N 30°17E **17** E16

Dnipro → Ukraine 46°30N 32°18E **19** E5
Dniprodzerzhynsk
Ukraine 48°32N 34°37E **19** E5
Dnipropetrovsk Ukraine 48°30N 35°0E **19** E6
Dnister → Europe 46°18N 30°17E **17** E16
Dno Russia 57°50N 29°58E **9** H23
Dnyapro = Dnipro →
Ukraine 46°30N 32°18E **19** E5
Do Gonbadān = Gachsārān
Iran 30°15N 50°45E **45** D6
Doaktown Canada 46°33N 66°8W **73** C6
Doan Hung Vietnam 21°30N 105°10E **38** B5
Doany Madag. 14°21S 49°30E **57** A8
Doba Chad 8°40N 16°50E **51** G9
Dobandi Pakistan 31°13N 66°50E **42** D2
Dobbyn Australia 19°44S 140°2E **62** B3
Dobele Latvia 56°37N 23°16E **9** H20
Doberai, Jazirah Indonesia 1°25S 133°0E **37** E8
Doblas Argentina 37°5S 64°0W **94** D3
Dobo Indonesia 5°45S 134°15E **37** F8
Doboj Bos.-H. 44°46N 18°4E **23** B8
Dobrich Bulgaria 43°37N 27°49E **23** C12
Dobruja Europe 44°30N 28°15E **17** F15
Dobrush Belarus 52°25N 31°22E **17** B16
Doc, Mui Vietnam 17°58N 106°30E **38** D6
Docker River = Kaltukatjara
Australia 24°52S 129°5E **61** D4
Doctor Arroyo Mexico 23°40N 100°11W **86** C4
Doctor Pedro P. Peña
Paraguay 22°27S 62°21W **94** A3
Doda India 33°10N 75°34E **43** C6
Doda, L. Canada 49°25N 75°13W **72** C4
Dodecanese Greece 36°35N 27°0E **23** F12
Dodekanisa = Dodecanese
Greece 36°35N 27°0E **23** F12
Dodge City U.S.A. 37°45N 100°1W **80** G3
Dodge L. Canada 59°50N 105°36W **71** B7
Dodgeville U.S.A. 42°58N 90°8W **80** D8
Dodoma Tanzania 6°8S 35°45E **54** D4
Dodoma □ Tanzania 6°0S 36°0E **54** D4
Dodori △ Kenya 1°55S 41°7E **54** C5
Dodsland Canada 51°50N 108°45W **71** C7
Dodson U.S.A. 48°24N 108°15W **76** B9
Doesburg Neths. 52°1N 6°9E **15** B6
Doetinchem Neths. 51°59N 6°18E **15** C6
Dog Creek Canada 51°35N 122°14W **70** C4
Dog L. Man., Canada 51°2N 98°31W **71** C9
Dog L. Ont., Canada 48°48N 89°30W **72** C2
Dogran Pakistan 31°48N 73°35E **42** D5
Doğubayazıt Turkey 39°31N 44°5E **44** B5
Doha = Ad Dawḥah
Qatar 25°15N 51°35E **45** E6
Dohazari Bangla. 22°10N 92°5E **41** H18
Dohrighat India 26°16N 83°31E **43** F10
Doi Indonesia 2°14N 127°49E **37** D7
Doi Inthanon Thailand 18°35N 98°29E **38** C2
Doi Inthanon △ Thailand 18°33N 98°34E **38** C2
Doi Khuntan △ Thailand 18°33N 99°14E **38** C2
Doi Luang Thailand 18°30N 101°0E **38** C3
Doi Luang △ Thailand 19°22N 99°35E **38** C2
Doi Phukha △ Thailand 19°8N 101°9E **38** C3
Doi Saket Thailand 18°52N 99°9E **38** C2
Doi Suthep Pui △
Thailand 18°49N 98°53E **38** C2
Doi Toa Thailand 17°55N 98°30E **38** D2
Dois Irmãos, Sa. Brazil 9°0S 42°30W **93** E10
Dokkum Neths. 53°20N 5°59E **15** A5
Dokri Pakistan 27°25N 68°7E **42** F3
Dolak, Pulau Indonesia 8°0S 138°30E **37** F9
Dolbeau-Mistassini
Canada 48°53N 72°14W **73** C5
Dole France 47°7N 5°31E **20** C6
Dolgellau U.K. 52°45N 3°53W **12** E4
Dolgelley = Dolgellau
U.K. 52°45N 3°53W **12** E4
Dolinsk Russia 47°21N 142°48E **29** E15
Dollard Neths. 53°20N 7°10E **15** A7
Dolo Ethiopia 4°11N 42°3E **47** G3
Dolomites = Dolomiti
Italy 46°23N 11°51E **22** A4
Dolomiti Italy 46°23N 11°51E **22** A4
Dolores Argentina 36°20S 57°40W **94** D4
Dolores Uruguay 33°34S 58°15W **94** C4
Dolores U.S.A. 37°28N 108°30W **77** H9
Dolores → U.S.A. 38°49N 109°17W **76** G9
Dolphin, C. Falk. Is. 51°10S 59°0W **96** G5
Dolphin and Union Str.
Canada 69°5N 114°45W **68** D8
Dom Pedrito Brazil 31°0S 54°40W **95** C5
Doma → Zimbabwe 16°28S 30°12E **55** F3
Domariaganj → India 26°17N 83°44E **43** F10
Domasi Malawi 15°15S 35°22E **55** F4
Dombarovskiy Russia 50°46N 59°32E **28** D6
Dombås Norway 62°4N 9°8E **8** E13
Dome Argus Antarctica 80°50S 76°30E **5** E6
Dome C. Antarctica 75°12S 123°37E **5** D9
Dome Fuji Antarctica 77°20S 39°45E **5** D5
Domel I. = Letsôk-aw Kyun
Burma 11°30N 98°25E **39** G2
Domeyko Chile 29°0S 71°0W **94** B1
Domeyko, Cordillera
Chile 24°30S 69°0W **94** A2
Dominador Chile 24°21S 69°20W **94** A2
Dominica ■ W. Indies 15°20N 61°20W **89** C7
Dominica Passage
W. Indies 15°10N 61°20W **89** C7
Dominican Rep. ■
W. Indies 19°0N 70°30W **89** C5
Domodóssola Italy 46°7N 8°17E **20** C8
Domville, Mt. Australia 28°1S 151°15E **63** D5
Don → Russia 47°4N 39°18E **19** E6
Don → Aberds., U.K. 57°11N 2°5W **11** D6
Don → S. Yorks., U.K. 53°41N 0°52W **12** D7
Don, C. Australia 11°18S 131°46E **60** B5
Don Benito Spain 38°53N 5°51W **21** C3
Don Figuerero Mts.
Jamaica 18°5N 77°36W **88** a
Don Sak Thailand 9°18N 99°41E **39** b
Dona Ana = Nhamaabué
Mozam. 17°25S 35°5E **55** F4

Donaghadee U.K. 54°39N 5°33W **10** ■
Donaghmore Ireland 52°52N 7°36W **10** ■
Donald Australia 36°23S 143°0E **63** ■
Donaldsonville U.S.A. 30°6N 90°59W **84** ■
Donalsonville U.S.A. 31°3N 84°53W **85** ■
Donau = Dunărea →
Europe 45°20N 29°40E **17** ■
Donauwörth Germany 48°43N 10°47E **16** ■
Doncaster U.K. 53°32N 1°6W **12** ■
Dondo Angola 9°45S 14°25E **52** ■
Dondo Mozam. 19°33S 34°46E **55** ■
Dondo, Teluk Indonesia 0°50N 120°30E **37** ■
Dondra Head Sri Lanka 5°55N 80°40E **40** ■
Donegal Ireland 54°39N 8°5W **10** ■
Donegal □ Ireland 54°53N 8°0W **10** ■
Donegal B. Ireland 54°31N 8°49W **10** ■
Donets → Russia 47°33N 40°55E **19** ■
Donets Basin Ukraine 49°0N 38°0E **6** ■
Donetsk Ukraine 48°0N 37°45E **19** ■
Dong Ba Thin Vietnam 12°8N 109°13E **38** ■
Dong Dang Vietnam 21°54N 106°42E **38** ■
Dong Giam Vietnam 19°25N 105°31E **38** ■
Dong Ha Vietnam 16°55N 107°8E **38** ■
Dong Hene Laos 16°40N 105°18E **38** ■
Dong Hoi Vietnam 17°29N 106°36E **38** ■
Dong Khe Vietnam 22°26N 106°27E **38** ■
Dong Phrayayen
Thailand 14°20N 101°22E **38** ■
Dong Ujimqin Qi China 45°32N 116°55E **34** ■
Dong Van Vietnam 23°16N 105°22E **38** ■
Dong Xoai Vietnam 11°32N 106°55E **39** ■
Dongara Australia 29°14S 114°57E **61** ■
Dongbei China 45°0N 125°0E **35** ■
Dongchuan China 26°8N 103°1E **32** ■
Dongfang China 18°50N 108°33E **38** ■
Dongfeng China 42°40N 125°34E **35** ■
Donggala Indonesia 0°30S 119°40E **37** ■
Donggang China 39°52N 124°10E **35** ■
Dongguan China 22°58N 113°44E **33** ■
Dongguang China 37°50N 116°30E **34** ■
Donghae S. Korea 37°29N 129°7E **35** ■
Dongjingcheng China 44°5N 129°10E **35** ■
Dongliao He → China 42°58N 123°32E **35** ■
Dongnae S. Korea 35°12N 129°5E **35** ■
Dongning China 44°2N 131°5E **35** ■
Dongola Sudan 19°9N 30°22E **51** ■
Dongping China 35°55N 116°20E **34** ■
Dongsha Dao
S. China Sea 20°45N 116°43E **33** ■
Dongsheng China 39°50N 110°0E **34** ■
Dongtai China 32°51N 120°21E **35** ■
Dongting Hu China 29°18N 112°45E **33** ■
Dongying China 37°37N 118°37E **35** ■
Dongyinggang China 37°50N 118°58E **35** ■
Donington, C. Australia 34°45S 136°0E **63** ■
Doniphan U.S.A. 36°37N 90°50W **80** ■
Dønna Norway 66°6N 12°30E **8** ■
Donna U.S.A. 26°9N 98°4W **84** ■
Donnacona Canada 46°41N 71°41W **73** ■
Donnelly's Crossing
N.Z. 35°42S 173°38E **59** ■
Donnybrook Australia 33°34S 115°48E **61** ■
Donnybrook S. Africa 29°59S 29°48E **57** ■
Donora U.S.A. 40°11N 79°52W **82** ■
Donostia = Donostia-San
Sebastián Spain 43°17N 1°58W **21** ■
Donostia-San Sebastián
Spain 43°17N 1°58W **21** ■
Donwood Canada 44°19N 78°16W **82** ■
Doomadgee Australia 17°56S 138°49E **62** ■
Doomadgee ◎ Australia 17°56S 138°49E **62** ■
Doon → U.K. 55°27N 4°39W **11** ■
Doon Doon ◎ Australia 16°18S 128°14E **60** ■
Dora, L. Australia 22°0S 123°0E **60** ■
Dora Báltea → Italy 45°11N 8°3E **20** ■
Doran L. Canada 61°13N 108°6W **71** ■
Dorchester U.K. 50°42N 2°27W **13** ■
Dorchester, C. Canada 65°27N 77°27W **69** ■
Dordabis Namibia 22°52S 17°38E **56** ■
Dordogne → France 45°2N 0°36W **20** ■
Dordrecht Neths. 51°48N 4°39E **15** ■
Dordrecht S. Africa 31°20S 27°3E **56** ■
Doré L. Canada 54°46N 107°17W **71** ■
Dori Burkina Faso 14°3N 0°2W **50** ■
Doring → S. Africa 31°54S 18°39E **56** ■
Doringbos S. Africa 31°59S 19°16E **56** ■
Dorking U.K. 51°14N 0°19W **13** ■
Dornbirn Austria 47°25N 9°45E **16** ■
Dornie U.K. 57°17N 5°31W **11** ■
Dornoch Canada 44°18N 80°51W **82** ■
Dornoch U.K. 57°53N 4°2W **11** ■
Dornoch Firth U.K. 57°51N 4°4W **11** ■
Dornogovĭ □ Mongolia 44°0N 110°0E **34** ■
Doro, Kavo Greece 38°9N 24°38E **23** ■
Dorohoi Romania 47°56N 26°23E **17** ■
Döröö Nuur Mongolia 48°0N 93°0E **32** ■
Dorr Iran 33°17N 50°38E **45** ■
Dorre I. Australia 25°13S 113°12E **61** ■
Dorrigo Australia 30°20S 152°44E **63** ■
Dorris U.S.A. 41°58N 121°55W **76** ■
Dorset Canada 45°14N 78°54W **82** ■
Dorset Ohio, U.S.A. 41°40N 80°40W **82** ■
Dorset □ U.K. 50°45N 2°26W **13** ■
Dortmund Germany 51°30N 7°28E **16** ■
Dörtyol Turkey 36°50N 36°13E **44** ■
Doruma
Dem. Rep. of the Congo 4°42N 27°33E **54** ■
Dorüneh Iran 35°10N 57°18E **45** ■
Dos Bahías, C. Argentina 44°58S 65°32W **96** ■
Dos Hermanas Spain 37°16N 5°55W **21** ■
Dos Palos U.S.A. 36°59N 120°37W **78** ■
Dosso Niger 13°0N 3°13E **50** ■
Dothan U.S.A. 31°13N 85°24W **85** ■
Doty U.S.A. 46°38N 123°17W **78** ■
Douai France 50°21N 3°4E **20** ■
Douala Cameroon 4°0N 9°45E **52** ■
Douarnenez France 48°6N 4°21W **20** ■
Double Island Pt.
Australia 25°56S 153°11E **63** ■

ouble Mountain Fork →
U.S.A. 33°16N 100°0W 84 E4
oubs → France 46°53N 5°1E 20 C6
oubtful Sd. N.Z. 45°20S 166°49E 59 F1
oubtless B. N.Z. 34°55S 173°26E 59 A4
ouglas Canada 45°31N 76°56W 82 A8
ouglas I. of Man 54°10N 4°28W 12 C3
ouglas S. Africa 29°4S 23°46E 56 D3
ouglas Ariz., U.S.A. 31°21N 109°33W 77 L9
ouglas Ga., U.S.A. 31°31N 82°51W 85 F13
ouglas Wyo., U.S.A. 42°45N 105°24W 76 E11
ouglas Apsley △
Australia 41°45S 148°11E 63 G4
ouglas Chan. Canada 53°40N 129°20W 70 C3
ouglas Pt. Canada 44°19N 81°37W 82 B3
ouglasville U.S.A. 33°45N 84°45W 85 E12
ounreay U.K. 58°35N 3°44W 11 C5
ourada, Serra Brazil 13°10S 48°45W 93 F9
ourados Brazil 22°9S 54°50W 95 A5
ourados → Brazil 21°58S 54°18W 95 A5
ourados, Serra dos
Brazil 23°30S 53°30W 95 A5
ouro → Europe 41°8N 8°40W 21 B1
ove → U.K. 52°51N 1°36W 12 E6
ove Creek U.S.A. 37°46N 108°54W 77 H9
over Australia 43°18S 147°2E 63 G4
over U.K. 51°7N 1°19E 13 F9
over Del., U.S.A. 39°10N 75°32W 81 F16
over N.H., U.S.A. 43°12N 70°56W 83 C14
over N.J., U.S.A. 40°53N 74°34W 83 F10
over Ohio, U.S.A. 40°32N 81°29W 82 F3
over, Pt. Australia 32°32S 125°32E 61 F4
over, Str. of Europe 51°0N 1°30E 13 G9
over-Foxcroft U.S.A. 45°11N 69°13W 81 C19
over Plains U.S.A. 41°43N 73°35W 83 E11
ovey = Dyfi → U.K. 52°32N 4°3W 13 E3
ovrefjell Norway 62°15N 9°33E 8 E13
ow Rūd Iran 33°28N 49°4E 45 C6
owa Malawi 13°38S 33°58E 55 E3
owagiac U.S.A. 41°59N 86°6W 80 E10
owerin Australia 31°12S 117°2E 61 F2
owgha'i Iran 36°54N 58°32E 45 B8
owlatābād Kermān,
Iran 28°20N 56°40E 45 D8
owlatābād Khorāsān,
Iran 35°16N 59°29E 45 C8
own □ U.K. 54°23N 6°2W 10 B5
owney Calif., U.S.A. 33°56N 118°9W 79 M8
owney Idaho, U.S.A. 42°26N 112°7W 76 E7
ownham Market U.K. 52°37N 0°23E 13 E8
ownieville U.S.A. 39°34N 120°50W 78 F6
ownpatrick U.K. 54°20N 5°43W 10 B6
ownpatrick Hd. Ireland 54°20N 9°21W 10 B2
ownsville U.S.A. 42°5N 75°0W 83 D10
ownton, Mt. Canada 52°42N 124°52W 70 C4
owsārī Iran 28°25N 57°59E 45 D8
oyle U.S.A. 40°2N 120°6W 78 E6
oylestown U.S.A. 40°21N 75°10W 83 F9
ozois, Rés. Canada 47°30N 77°5W 72 C4
ra Khel Pakistan 27°58N 66°45E 42 F2
râa, C. Morocco 28°47N 11°0W 50 C3
râa, Hamada du Algeria 28°0N 7°0W 50 C4
râa, Oued → Morocco 28°40N 11°10W 50 C3
rac, Coves del Spain 39°31N 3°19E 24 B10
rachten Neths. 53°7N 6°5E 15 A6
rāgăşani Romania 44°39N 24°17E 17 F13
ragichyn Belarus 52°15N 25°8E 17 B13
ragoman, Prokhod
Bulgaria 42°58N 22°53E 23 C10
ragon's Mouths
Venezuela 11°0N 61°50W 93 K15
raguignan France 43°32N 6°27E 20 E7
rain U.S.A. 43°40N 123°19W 76 E2
rake U.S.A. 47°55N 100°23W 80 B3
rake Passage S. Ocean 58°0S 68°0W 5 B17
rakensberg S. Africa 31°0S 28°0E 57 D4
rama Greece 41°9N 24°10E 23 D11
rammen Norway 59°42N 10°12E 9 G14
rangajökull Iceland 66°9N 22°15W 8 C2
ras India 34°25N 75°48E 43 B6
rastis, Akra Greece 39°48N 19°40E 25 A3
rau = Drava →
Croatia 45°33N 18°55E 23 B8
rava → Croatia 45°33N 18°55E 23 B8
rayton Canada 43°46N 80°40W 82 C4
rayton Valley Canada 53°12N 114°58W 70 C6
renthe □ Neths. 52°52N 6°40E 15 B6
repano, Akra Greece 35°28N 24°14E 25 D6
repanum, C. Cyprus 34°54N 32°19E 25 E11
resden Canada 42°35N 82°11W 82 D2
resden Germany 51°3N 13°44E 16 C7
resden U.S.A. 42°41N 76°57W 82 D8
reux France 48°44N 1°23E 20 B4
riffield U.K. 54°0N 0°26W 12 C7
riftwood U.S.A. 41°20N 78°8W 82 E6
riggs U.S.A. 43°44N 111°6W 76 E8
rin → Albania 42°1N 19°38E 23 C8
rina → Bos.-H. 44°53N 19°21E 23 B8
røbak Norway 59°39N 10°39E 9 G14
robeta-Turnu Severin
Romania 44°39N 22°41E 17 F12
rochia Moldova 48°2N 27°48E 17 D14
rogheda Ireland 53°43N 6°22W 10 C5
rogichin = Dragichyn
Belarus 52°15N 25°8E 17 B13
rogobych = Drohobych
Ukraine 49°20N 23°30E 17 D12
rohobych Ukraine 49°20N 23°30E 17 D12
roichead Atha = Drogheda
Ireland 53°43N 6°22W 10 C5
roichead na Bandan = Bandon
Ireland 51°44N 8°44W 10 E3
roichead Nua Ireland 53°11N 6°48W 10 C5
roitwich U.K. 52°16N 2°8W 13 E5
rôme → France 44°46N 4°46E 20 D6
romedary, C. Australia 36°17S 150°10E 63 F5
romore U.K. 54°31N 7°28W 10 B4
romore West Ireland 54°15N 8°52W 10 B3
ronfield U.K. 53°19N 1°27W 12 D6
ronning Maud Land
Antarctica 72°30S 12°0E 5 D3
Dronten Neths. 52°32N 5°43E 15 B5
Druk Yul = Bhutan ■
Asia 27°25N 90°30E 41 F17
Drumbo Canada 43°16N 80°35W 82 C4
Drumcliff Ireland 54°20N 8°29W 10 B3
Drumheller Canada 51°25N 112°40W 70 C6
Drummond U.S.A. 46°40N 113°9W 76 C7
Drummond I. U.S.A. 46°1N 83°39W 81 B12
Drummond Pt. Australia 34°9S 135°16E 63 E2
Drummond Ra.
Australia 23°45S 147°10E 62 C4
Drummondville Canada 45°55N 72°25W 72 C5
Drumright U.S.A. 35°59N 96°36W 84 D6
Druskininkai Lithuania 54°3N 23°58E 9 J20
Drut → Belarus 53°8N 30°5E 17 B16
Druzhina Russia 68°14N 145°18E 29 C15
Dry Harbour Mts.
Jamaica 18°19N 77°24W 88 a
Dry Tortugas U.S.A. 24°38N 82°55W 85 F13
Dryander △ Australia 20°13S 148°34E 62 b
Dryden Canada 49°47N 92°50W 71 D10
Dryden U.S.A. 42°30N 76°18W 83 D8
Drygalski I. Antarctica 66°0S 92°0E 5 C7
Drygalski Ice Tongue
Antarctica 75°24S 163°30E 5 D11
Drysdale → Australia 13°59S 126°51E 60 B4
Drysdale I. Australia 11°41S 136°0E 62 A2
Drysdale River △
Australia 14°56S 127°2E 60 B4
Du Gué → Canada 57°21N 70°45W 72 A5
Du Quoin U.S.A. 38°1N 89°14W 80 F9
Duanesburg U.S.A. 42°45N 74°11W 83 D10
Duaringa Australia 23°42S 149°42E 62 C4
Duarte, Pico Dom. Rep. 19°2N 70°59W 89 D5
Dubai = Dubayy U.A.E. 25°18N 55°20E 45 E7
Dubăsari Moldova 47°15N 29°10E 17 E15
Dubăsari Vdkhr.
Moldova 47°30N 29°0E 17 E15
Dubawnt → Canada 64°33N 100°6W 71 A8
Dubawnt L. Canada 63°8N 101°28W 71 A8
Dubayy U.A.E. 25°18N 55°20E 45 E7
Dubbo Australia 32°11S 148°35E 63 E4
Dubele
Dem. Rep. of the Congo 2°56N 29°35E 54 B2
Dublin Ireland 53°21N 6°15W 10 C5
Dublin Ga., U.S.A. 32°32N 82°54W 85 E13
Dublin Tex., U.S.A. 32°5N 98°21W 84 E5
Dublin □ Ireland 53°24N 6°20W 10 C5
Dublin ✕ (DUB) Ireland 53°26N 6°15W 10 C5
Dubno Ukraine 50°25N 25°45E 17 C13
Dubois Idaho, U.S.A. 44°10N 112°14W 76 D7
DuBois Pa., U.S.A. 41°7N 78°46W 82 E6
Dubossary = Dubăsari
Moldova 47°15N 29°10E 17 E15
Dubossary Vdkhr. = Dubăsari
Vdkhr. Moldova 47°30N 29°0E 17 E15
Dubovka Russia 49°5N 44°50E 19 E7
Dubrajpur India 23°48N 87°25E 43 H12
Dubréka Guinea 9°46N 13°31W 50 G3
Dubrovitsa = Dubrovytsya
Ukraine 51°31N 26°35E 17 C14
Dubrovnik Croatia 42°39N 18°6E 23 C8
Dubrovytsya Ukraine 51°31N 26°35E 17 C14
Dubuque U.S.A. 42°30N 90°41W 80 D8
Duc Tho Vietnam 18°32N 105°35E 38 C5
Duchesne U.S.A. 40°10N 110°24W 76 F8
Duchess Australia 21°20S 139°50E 62 C2
Ducie I. Pac. Oc. 24°40S 124°48W 65 K15
Duck → U.S.A. 36°2N 87°52W 85 C11
Duck Cr. → Australia 22°37S 116°53E 60 D2
Duck Lake Canada 52°50N 106°16W 71 C7
Duck Mountain ⌂
Canada 51°45N 101°0W 71 C8
Duckwall, Mt. U.S.A. 37°58N 120°7W 78 H6
Dudhi India 24°15N 83°10E 43 G10
Dudinka Russia 69°30N 86°13E 29 C9
Dudley U.K. 52°31N 2°5W 13 E5
Dudwa India 28°30N 80°41E 43 E9
Dudwa △ India 28°30N 80°40E 43 E9
Duero = Douro → Europe 41°8N 8°40W 21 B1
Dufftown U.K. 57°27N 3°8W 11 D5
Dugi Otok Croatia 44°0N 15°3E 16 G8
Duisburg Germany 51°26N 6°45E 16 C4
Duiwelskloof = Modjadjiskloof
S. Africa 23°42S 30°10E 57 C5
Dūkdamīn Iran 35°59N 57°43E 45 C8
Dukelský Průsmyk
Slovak Rep. 49°25N 21°42E 17 D11
Dukhān Qatar 25°25N 50°50E 45 E6
Duki Pakistan 30°14N 68°25E 40 D6
Duku Nigeria 10°43N 10°43E 51 F8
Dulan China 36°0N 97°11E 32 D8
Dulce U.S.A. 36°56N 107°0W 77 H10
Dulce → Argentina 30°32S 62°33W 94 C3
Dulce, G. Costa Rica 8°40N 83°20W 88 E3
Dulf Iraq 35°7N 45°51E 44 C5
Duliu China 39°2N 116°55E 34 E9
Dullewala Pakistan 31°50N 71°25E 42 D4
Dullstroom S. Africa 25°27S 30°7E 57 D5
Dulq Maghār Syria 36°22N 38°39E 44 B3
Duluth U.S.A. 46°47N 92°6W 80 B7
Dum Dum India 22°39N 88°26E 43 H13
Dum Duma India 27°40N 95°40E 41 F19
Dūmā Syria 33°34N 36°24E 46 B5
Dumaguete Phil. 9°17N 123°15E 37 C6
Dumai Indonesia 1°35N 101°28E 36 D2
Dumaran Phil. 10°33N 119°50E 37 B5
Dumas Ark., U.S.A. 33°53N 91°29W 84 E9
Dumas Tex., U.S.A. 35°52N 101°58W 84 D4
Dumayr Syria 33°39N 36°42E 46 B5
Dumbarton U.K. 55°57N 4°33W 11 F4
Dumbleyung Australia 33°17S 117°42E 61 F2
Dumfries U.K. 55°4N 3°37W 11 F5
Dumfries & Galloway □
U.K. 55°9N 3°58W 11 F5
Dumka India 24°12N 87°15E 43 G12
Dumoine → Canada 46°13N 77°51W 72 C4
Dumoine, L. Canada 46°55N 77°55W 72 C4
Dumont d'Urville
Antarctica 66°40S 140°0E 5 C10
Dumont d'Urville Sea
S. Ocean 63°30S 138°0E 5 C9
Dumraon India 25°33N 84°8E 43 G11
Dumyât Egypt 31°24N 31°48E 51 B12
Dún Dealgan = Dundalk
Ireland 54°1N 6°24W 10 B5
Dún Garbhán = Dungarvan
Ireland 52°5N 7°37W 10 D4
Dún Laoghaire Ireland 53°17N 6°8W 10 C5
Duna = Dunărea →
Europe 45°20N 29°40E 17 F15
Dunagiri India 30°31N 79°52E 43 D8
Dunaj = Dunărea →
Europe 45°20N 29°40E 17 F15
Dunakeszi Hungary 47°37N 19°8E 17 E10
Dunărea → Europe 45°20N 29°40E 17 F15
Dunaújváros Hungary 46°58N 18°57E 17 E10
Dunav = Dunărea →
Europe 45°20N 29°40E 17 F15
Dunay Russia 42°52N 132°22E 30 C6
Dunback N.Z. 45°23S 170°36E 59 F3
Dunbar U.K. 56°0N 2°31W 11 E6
Dunblane U.K. 56°11N 3°58W 11 E5
Duncan Canada 48°45N 123°40W 78 B3
Duncan Ariz., U.S.A. 32°43N 109°6W 77 K9
Duncan Okla., U.S.A. 34°30N 97°57W 84 D6
Duncan, L. Canada 53°29N 77°58W 72 B4
Duncan L. Canada 62°51N 113°58W 70 A6
Duncan Town Bahamas 22°15N 75°45W 88 B4
Duncannon U.S.A. 40°23N 77°2W 82 F7
Duncansby Head U.K. 58°38N 3°1W 11 C5
Duncansville U.S.A. 40°25N 78°26W 82 F6
Dund-Us Mongolia 48°1N 91°38E 32 B7
Dundalk Canada 44°10N 80°24W 82 B4
Dundalk Ireland 54°1N 6°24W 10 B5
Dundalk U.S.A. 39°15N 76°31W 81 F15
Dundalk Bay Ireland 53°55N 6°15W 10 C5
Dundas = Uummannaq
Greenland 77°33N 68°52W 69 B18
Dundas Canada 43°17N 79°59W 82 C5
Dundas, L. Australia 32°35S 121°50E 61 F3
Dundas I. Canada 54°30N 130°50W 70 C2
Dundas Str. Australia 11°15S 131°35E 60 B5
Dundee S. Africa 28°11S 30°15E 57 D5
Dundee U.K. 56°28N 2°59W 11 E6
Dundee U.S.A. 42°32N 76°59W 82 D8
Dundee City □ U.K. 56°30N 2°58W 11 E6
Dundgovĭ □ Mongolia 45°10N 106°0E 34 B4
Dundrum U.K. 54°16N 5°52W 10 B6
Dundrum B. U.K. 54°13N 5°47W 10 B6
Dunedin N.Z. 45°50S 170°33E 59 F3
Dunedin U.S.A. 28°1N 82°46W 85 G13
Dunfanaghy Ireland 55°11N 7°58W 10 A4
Dunfermline U.K. 56°5N 3°27W 11 E5
Dungannon Canada 43°51N 81°36W 82 C3
Dungannon U.K. 54°31N 6°46W 10 B5
Dungarpur India 23°52N 73°45E 42 H5
Dungarvan Ireland 52°5N 7°37W 10 D4
Dungarvan Harbour
Ireland 52°4N 7°35W 10 D4
Dungeness U.K. 50°54N 0°59E 13 G8
Dunglow Ireland 54°57N 8°21W 10 B3
Dungo, L. do Angola 17°15S 19°0E 56 B2
Dungog Australia 32°22S 151°46E 63 E5
Dungu
Dem. Rep. of the Congo 3°40N 28°32E 54 B2
Dungun Malaysia 4°45N 103°25E 39 K4
Dunhua China 43°20N 128°14E 35 C15
Dunhuang China 40°8N 94°36E 32 C7
Dunk I. Australia 17°59S 146°29E 62 B4
Dunkeld Australia 33°25S 149°29E 63 E4
Dunkeld U.K. 56°34N 3°35W 11 E5
Dunkerque France 51°2N 2°20E 20 A5
Dunkery Beacon U.K. 51°9N 3°36W 13 F4
Dunkirk = Dunkerque
France 51°2N 2°20E 20 A5
Dunkirk U.S.A. 42°29N 79°20W 82 D5
Dúnleary = Dún Laoghaire
Ireland 53°17N 6°8W 10 C5
Dunleer Ireland 53°50N 6°24W 10 C5
Dunmanus B. Ireland 51°31N 9°50W 10 E2
Dunmanway Ireland 51°43N 9°6W 10 E2
Dunmarra Australia 16°42S 133°25E 62 B1
Dunmore U.S.A. 41°25N 75°38W 83 E9
Dunmore East Ireland 52°9N 7°0W 10 D5
Dunmore Hd. Ireland 52°10N 10°35W 10 D1
Dunmore Town
Bahamas 25°30N 76°39W 88 A4
Dunn U.S.A. 35°19N 78°37W 85 D15
Dunnellon U.S.A. 29°3N 82°28W 85 G13
Dunnet Hd. U.K. 58°40N 3°21W 11 C5
Dunning U.S.A. 41°50N 100°6W 80 E3
Dunnville Canada 42°54N 79°36W 82 D5
Dunolly Australia 36°51S 143°44E 63 F3
Dunoon U.K. 55°57N 4°56W 11 F4
Duns U.K. 55°47N 2°20W 11 F6
Dunseith U.S.A. 48°50N 100°3W 80 A3
Dunshaughlin Ireland 53°31N 6°33W 10 C5
Dunsmuir U.S.A. 41°13N 122°16W 76 F2
Dunstable U.K. 51°53N 0°32W 13 F7
Dunstan Mts. N.Z. 44°53S 169°35E 59 F2
Dunster Canada 53°8N 119°50W 70 C5
Dunvegan U.K. 57°27N 6°35W 11 D2
Dunvegan L. Canada 60°8N 107°10W 71 A7
Duolun China 42°12N 116°28E 34 C9
Duong Dong Vietnam 10°13N 103°58E 39 G4
Dupree U.S.A. 45°4N 101°35W 80 C3
Dupuyer U.S.A. 48°13N 112°30W 76 B7
Duque de Caxias Brazil 22°46S 43°18W 95 A7
Durack → Australia 15°33S 127°52E 60 C4
Durack Ra. Australia 16°50S 127°40E 60 C4
Durack River Australia 15°32S 127°39E 60 C4
Durance → France 43°55N 4°45E 20 E6
Durand U.S.A. 44°38N 91°58W 80 C8
Durango Mexico 24°3N 104°39W 86 C4
Durango U.S.A. 37°16N 107°53W 77 H10
Durango □ Mexico 24°50N 105°20W 86 C4
Durant Miss., U.S.A. 33°4N 89°51W 85 E10
Durant Okla., U.S.A. 33°59N 96°25W 84 E6
Durazno Uruguay 33°25S 56°31W 94 C4
Durazzo = Durrës
Albania 41°19N 19°28E 23 D8
Durban S. Africa 29°49S 31°1E 57 D5
Durbuy Belgium 50°21N 5°28E 15 D5
Düren Germany 50°48N 6°29E 16 C4
Durg = Bhilainagar-Durg
India 21°13N 81°26E 41 J12
Durgapur India 23°30N 87°20E 43 H12
Durham Canada 44°10N 80°49W 82 B4
Durham U.K. 54°47N 1°34W 12 C6
Durham Calif., U.S.A. 39°39N 121°48W 78 F5
Durham N.C., U.S.A. 35°59N 78°54W 85 D15
Durham N.H., U.S.A. 43°8N 70°56W 83 C14
Durham □ U.K. 54°42N 1°45W 12 C6
Durlas = Thurles Ireland 52°41N 7°49W 10 D4
Ḑurmā Si. Arabia 24°37N 46°8E 44 E5
Durmitor Montenegro 43°10N 19°0E 23 C8
Durness U.K. 58°34N 4°45W 11 C4
Durrës Albania 41°19N 19°28E 23 D8
Durrow Ireland 52°51N 7°24W 10 D4
Dursey I. Ireland 51°36N 10°12W 10 E1
Dursley U.K. 51°40N 2°21W 13 F5
Dursunbey Turkey 39°35N 28°37E 23 E13
Duru
Dem. Rep. of the Congo 4°14N 28°50E 54 B2
Durūz, Jabal ad Jordan 32°35N 36°40E 46 C5
D'Urville, Tanjung
Indonesia 1°28S 137°54E 37 E9
D'Urville I. N.Z. 40°50S 173°55E 59 D4
Duryea U.S.A. 41°20N 75°45W 83 E9
Dushak Turkmenistan 37°13N 60°1E 45 B9
Dushanbe Tajikistan 38°33N 68°48E 28 F7
Dushore U.S.A. 41°31N 76°24W 83 E8
Dusky Sd. N.Z. 45°47S 166°30E 59 F1
Dussejour, C. Australia 14°45S 128°13E 60 B4
Düsseldorf Germany 51°14N 6°47E 16 C4
Dutch Harbor U.S.A. 53°53N 166°32W 74 E6
Dutlwe Botswana 23°58S 23°46E 56 C3
Dutton Canada 42°39N 81°30W 82 D3
Dutton → Australia 20°44S 143°10E 62 C3
Dutywa S. Africa 32°8S 28°18E 57 E4
Duwayhin, Khawr
U.A.E. 24°20N 51°25E 45 E6
Duyfken Pt. Australia 12°33S 141°38E 62 A3
Duyun China 26°18N 107°29E 32 F10
Dvina, Severnaya →
Russia 64°32N 40°30E 18 B7
Dvinsk = Daugavpils
Latvia 55°53N 26°32E 9 J22
Dvinskaya Guba Russia 65°0N 39°0E 18 B6
Dwarka India 22°18N 69°8E 42 H3
Dwellingup Australia 32°43S 116°4E 61 F2
Dwight Canada 45°20N 79°1W 82 A5
Dwight U.S.A. 41°5N 88°26W 80 E9
Dyatlovo = Dzyatlava
Belarus 53°28N 25°28E 17 B13
Dyce U.K. 57°13N 2°12W 11 D6
Dyer, C. Canada 66°37N 61°16W 69 D19
Dyer Bay Canada 45°10N 81°20W 82 A3
Dyer Plateau Antarctica 70°45S 65°30W 5 D17
Dyersburg U.S.A. 36°3N 89°23W 85 C10
Dyfi → U.K. 52°32N 4°3W 13 E3
Dymer Ukraine 50°47N 30°18E 17 C16
Dysart Australia 22°32S 148°23E 62 C4
Dzaoudzi Mayotte 12°47S 45°16E 53 a
Dzavhan Gol →
Mongolia 48°54N 93°23E 32 B7
Dzerzhinsk Russia 56°14N 43°30E 18 C7
Dzhalinda Russia 53°26N 124°0E 29 D13
Dzhambul = Taraz
Kazakhstan 42°54N 71°22E 32 C3
Dzhankoy Ukraine 45°40N 34°20E 19 E5
Dzhezkazgan = Zhezqazghan
Kazakhstan 47°44N 67°40E 28 E7
Dzhizak = Jizzax
Uzbekistan 40°6N 67°50E 28 E7
Dzhugdzur, Khrebet
Russia 57°30N 138°0E 29 D14
Dzhungarskiye Vorota =
Dzungarian Gate Asia 45°10N 82°0E 32 B5
Działdowo Poland 53°15N 20°15E 17 B11
Dzibilchaltún Mexico 21°10N 89°35W 87 C7
Dzierżoniów Poland 50°45N 16°39E 17 C9
Dzilam de Bravo Mexico 21°24N 88°53W 87 C7
Dzūkija △ Lithuania 54°10N 24°30E 9 J21
Dzungarian Basin = Junggar
Pendi China 44°30N 86°0E 32 C6
Dzungarian Gate Asia 45°10N 82°0E 32 B5
Dzüünbayan Mongolia 44°29N 110°2E 34 B6
Dzüünharaa Mongolia 48°52N 106°28E 32 B10
Dzüünmod Mongolia 47°45N 106°58E 32 B10
Dzyarzhynsk Belarus 53°40N 27°1E 17 B14
Dzyatlava Belarus 53°28N 25°28E 17 B13

E

E.C. Manning ⌂ Canada 49°5N 120°45W 70 D4
Eabamet L. Canada 51°30N 87°46W 72 B2
Eabametoong Canada 51°30N 88°0W 72 B2
Éadan Doire = Edenderry
Ireland 53°21N 7°4W 10 C4
Eads U.S.A. 38°29N 102°47W 76 G12
Eagar U.S.A. 34°6N 109°17W 77 J9
Eagle Alaska, U.S.A. 64°47N 141°12W 68 E3
Eagle Colo., U.S.A. 39°39N 106°50W 76 G10
Eagle → Canada 53°36N 57°26W 73 B8
Eagle Butte U.S.A. 45°0N 101°10W 80 C3
Eagle Grove U.S.A. 42°40N 93°54W 80 D7
Eagle L. Canada 49°42N 93°13W 71 D10
Eagle L. Calif., U.S.A. 40°39N 120°45W 76 F3
Eagle L. Maine, U.S.A. 46°20N 69°22W 81 B19
Eagle Lake Canada 45°8N 78°29W 82 A6
Eagle Lake Maine, U.S.A. 47°3N 68°36W 81 B19
Eagle Lake Tex., U.S.A. 29°35N 96°20W 84 G6
Eagle Mountain
U.S.A. 33°49N 115°27W 79 M11
Eagle Nest U.S.A. 36°33N 105°16W 77 H11
Eagle Pass U.S.A. 28°43N 100°30W 84 G4
Eagle Pk. U.S.A. 38°10N 119°25W 78 G7
Eagle Pt. Australia 16°11S 124°23E 60 C3
Eagle River Mich., U.S.A. 47°25N 88°18W 80 B9
Eagle River Wis., U.S.A. 45°55N 89°15W 80 C9
Eaglehawk Australia 36°44S 144°15E 63 F3
Eagles Mere U.S.A. 41°25N 76°33W 83 E8
Ealing □ U.K. 51°31N 0°20W 13 F7
Ear Falls Canada 50°38N 93°13W 71 C10
Earle U.S.A. 35°16N 90°28W 85 D9
Earlimart U.S.A. 35°53N 119°16W 79 K7
Earlville U.S.A. 42°44N 75°32W 83 D9
Earn → U.K. 56°21N 3°18W 11 E5
Earn, L. U.K. 56°23N 4°13W 11 E4
Earnslaw, Mt. N.Z. 44°32S 168°27E 59 F2
Earth U.S.A. 34°14N 102°24W 84 D3
Easley U.S.A. 34°50N 82°36W 85 D13
East Anglia U.K. 52°30N 1°0E 12 E9
East Angus Canada 45°30N 71°40W 73 C5
East Antarctica Antarctica 80°0S 90°0E 5 D7
East Aurora U.S.A. 42°46N 78°37W 82 D6
East Ayrshire □ U.K. 55°26N 4°11W 11 F4
East Bengal Bangla. 24°0N 90°0E 41 G17
East Beskids = Vychodné Beskydy
Europe 49°20N 22°0E 17 D11
East Brady U.S.A. 40°59N 79°37W 82 F5
East Branch Clarion River L.
U.S.A. 41°35N 78°35W 82 E6
East C. = Dezhneva, Mys
Russia 66°5N 169°40W 29 C19
East C. N.Z. 37°42S 178°35E 59 B7
East Caroline Basin
Pac. Oc. 4°0N 146°45E 64 G6
East Chicago U.S.A. 41°38N 87°27W 80 E10
East China Sea Asia 30°0N 126°0E 33 F14
East Coulee Canada 51°23N 112°27W 70 C6
East Dereham = Dereham
U.K. 52°41N 0°57E 13 E8
East Dunbartonshire □
U.K. 55°57N 4°13W 11 F4
East Falkland Falk. Is. 51°30S 58°30W 96 G5
East Grand Forks U.S.A. 47°56N 97°1W 80 B5
East Greenwich U.S.A. 41°40N 71°27W 83 E13
East Grinstead U.K. 51°7N 0°0 13 F8
East Hartford U.S.A. 41°46N 72°39W 83 E12
East Haydon Australia 18°0S 141°30E 62 B3
East Helena U.S.A. 46°35N 111°56W 76 C8
East Indies Asia 0°0 120°0E 26 J14
East Kilbride U.K. 55°47N 4°11W 11 F4
East Lamma Channel
China 22°14N 114°9E 33 a
East Lansing U.S.A. 42°44N 84°29W 81 D11
East Liverpool U.S.A. 40°37N 80°35W 82 F4
East London S. Africa 33°0S 27°55E 57 E4
East Lothian □ U.K. 55°58N 2°44W 11 F6
East Main = Eastmain
Canada 52°10N 78°30W 72 B4
East Mariana Basin
Pac. Oc. 12°0N 153°0E 64 F7
East Northport U.S.A. 40°53N 73°20W 83 F11
East Orange U.S.A. 40°46N 74°12W 83 F10
East Pacific Rise Pac. Oc. 15°0S 110°0W 65 J17
East Palestine U.S.A. 40°50N 80°33W 82 F4
East Pine Canada 55°48N 120°12W 70 B4
East Point U.S.A. 33°41N 84°25W 85 E12
East Providence U.S.A. 41°49N 71°23W 83 E13
East Pt. Br. Virgin Is. 18°40N 64°18W 89 e
East Pt. Canada 46°27N 61°58W 73 C7
East Renfrewshire □
U.K. 55°46N 4°21W 11 F4
East Retford = Retford
U.K. 53°19N 0°56W 12 D7
East Riding of Yorkshire □
U.K. 53°55N 0°30W 12 D7
East Rochester U.S.A. 43°7N 77°29W 82 C7
East St. Louis U.S.A. 38°37N 90°9W 80 F8
East Schelde = Oosterschelde →
Neths. 51°33N 4°0E 15 C4
East Sea = Japan, Sea of
Asia 40°0N 135°0E 30 E7
East Siberian Sea Russia 73°0N 160°0E 29 B17
East Stroudsburg U.S.A. 41°1N 75°11W 83 E9
East Sussex □ U.K. 50°56N 0°19E 13 G8
East Tasman Plateau
Pac. Oc. 43°30S 152°0E 64 M7
East Tawas U.S.A. 44°17N 83°29W 81 C12
East Timor ■ Asia 8°50S 126°0E 37 F7
East Toorale Australia 30°27S 145°28E 63 E4
East Walker → U.S.A. 38°52N 119°10W 78 G7
East Windsor U.S.A. 40°17N 74°34W 83 F10
Eastbourne N.Z. 41°19S 174°55E 59 D5
Eastbourne U.K. 50°46N 0°18E 13 G8
Eastend Canada 49°32N 108°50W 71 D7
Easter Fracture Zone
Pac. Oc. 25°0S 115°0W 65 K16
Easter I. = Pascua, I. de
Chile 27°7S 109°23W 65 K17
Eastern □ Kenya 0°0 38°30E 54 C4
Eastern Cape □ S. Africa 32°0S 26°0E 56 E4
Eastern Cr. → Australia 20°40S 141°35E 62 C3
Eastern Desert = Es Sahrâ' Esh
Sharqîya Egypt 27°30N 32°30E 51 C12
Eastern Ghats India 14°0N 78°50E 40 N11
Eastern Group = Lau Group
Fiji 17°0S 178°30W 59 a
Eastern Group Australia 33°30S 124°30E 61 F3
Eastern Guruma ☼
Australia 22°0S 117°30E 60 D2
Eastern Transvaal =
Mpumalanga □
S. Africa 26°0S 30°0E 57 D5
Easterville Canada 53°8N 99°49W 71 C9
Easthampton U.S.A. 42°16N 72°40W 83 D12
Eastlake U.S.A. 41°40N 81°26W 82 E3
Eastland U.S.A. 32°24N 98°49W 84 E5
Eastleigh U.K. 50°58N 1°21W 13 G6
Eastmain Canada 52°10N 78°30W 72 B4
Eastmain → Canada 52°27N 78°26W 72 B4
Eastman Canada 45°18N 72°19W 83 A12
Eastman U.S.A. 32°12N 83°11W 85 E13
Easton Md., U.S.A. 38°47N 76°5W 81 F15
Easton Pa., U.S.A. 40°41N 75°13W 83 F9

Easton Wash., U.S.A. 47°14N 121°11W 78 C5
Eastport U.S.A. 44°56N 67°0W 81 C20
Eastsound U.S.A. 48°42N 122°55W 78 B4
Eaton U.S.A. 40°32N 104°42W 76 F11
Eatonia Canada 51°13N 109°25W 71 C7
Eatonton U.S.A. 33°20N 83°23W 85 E13
Eatontown U.S.A. 40°19N 74°4W 83 F10
Eatonville U.S.A. 46°52N 122°16W 78 D4
Eau Claire U.S.A. 44°49N 91°30W 80 C8
Eau Claire, L. à l' Canada 56°10N 74°25W 72 A5
Eauripik Rise Pac. Oc. 2°0N 142°0E 64 G6
Ebano Mexico 22°13N 98°24W 87 C5
Ebbw Vale U.K. 51°46N 3°12W 13 F4
Ebeltoft Denmark 56°12N 10°41E 9 H14
Ebensburg U.S.A. 40°29N 78°44W 82 F6
Eberswalde-Finow Germany 52°50N 13°49E 16 B7
Ebetsu Japan 43°7N 141°34E 30 C10
Ebey's Landing ⌂ U.S.A. 48°12N 122°41W 78 B4
Ebinur Hu China 44°55N 82°55E 32 C5
Ebolowa Cameroon 2°55N 11°10E 52 D2
Ebonda Dem. Rep. of the Congo 2°12N 22°21E 52 D4
Ebre = Ebro → Spain 40°43N 0°54E 21 B6
Ebro → Spain 40°43N 0°54E 21 B6
Ecatepec de Morelos Mexico 19°36N 99°3W 87 D5
Ecbatana = Hamadān Iran 34°52N 48°32E 45 C6
Eceabat Turkey 40°11N 26°21E 23 D12
Ech Chéliff Algeria 36°10N 1°20E 50 A6
Echigo-Sammyaku Japan 36°50N 139°50E 31 F9
Echizen-Misaki Japan 35°59N 135°57E 31 G7
Echo Bay N.W.T., Canada 66°5N 117°55W 68 D8
Echo Bay Ont., Canada 46°29N 84°4W 72 C3
Echoing → Canada 55°51N 92°5W 72 B1
Echternach Lux. 49°49N 6°25E 15 E6
Echuca Australia 36°10S 144°45E 63 F3
Écija Spain 37°30N 5°10W 21 D3
Eclipse I. Australia 35°5S 117°58E 61 G2
Eclipse Is. Australia 13°54S 126°19E 60 B4
Eclipse Sd. Canada 72°38N 79°0W 69 C16
Ecuador ■ S. Amer. 2°0S 78°0W 92 D3
Ed Da'ein Sudan 11°26N 26°9E 52 B5
Ed Damazin Sudan 11°46N 34°21E 51 F12
Ed Dâmer Sudan 17°27N 34°0E 51 E12
Ed Dar el Beida = Casablanca Morocco 33°36N 7°36W 50 B4
Ed Debba Sudan 18°0N 30°51E 51 E12
Ed Déffa Egypt 30°40N 26°30E 51 B11
Ed Dueim Sudan 14°0N 32°10E 51 F12
Edam Canada 53°11N 108°46W 71 C7
Edam Neths. 52°31N 5°3E 15 B5
Eday U.K. 59°11N 2°47W 11 B6
Eddrachillis B. U.K. 58°17N 5°14W 11 C3
Eddystone U.K. 50°11N 4°16W 13 G3
Eddystone Pt. Australia 40°59S 148°20E 63 G4
Ede Neths. 52°4N 5°40E 15 B5
Edehon L. Canada 60°25N 97°15W 71 A9
Eden Australia 37°3S 149°55E 63 F4
Eden N.C., U.S.A. 36°29N 79°53W 85 C15
Eden N.Y., U.S.A. 42°39N 78°55W 82 D6
Eden Tex., U.S.A. 31°13N 99°51W 84 F5
Eden → U.K. 54°57N 3°1W 12 C4
Edenburg S. Africa 29°43S 25°58E 56 D4
Edendale S. Africa 29°39S 30°18E 57 D5
Edenderry Ireland 53°21N 7°4W 10 C4
Edenton U.S.A. 36°4N 76°39W 85 C16
Edenville S. Africa 27°37S 27°34E 57 D4
Eder → Germany 51°12N 9°28E 16 C5
Edessa Greece 40°48N 22°5E 23 D10
Edfu = Idfû Egypt 24°55N 32°49E 51 D12
Edgar U.S.A. 40°22N 97°58W 80 E5
Edgartown U.S.A. 41°23N 70°31W 83 E14
Edge Hill U.K. 52°8N 1°26W 13 E6
Edgefield U.S.A. 33°47N 81°56W 85 E14
Edgeley U.S.A. 46°22N 98°43W 80 B4
Edgemont U.S.A. 43°18N 103°50W 80 D2
Edgeøya Svalbard 77°45N 22°30E 4 B9
Édhessa = Edessa Greece 40°48N 22°5E 23 D10
Edievale N.Z. 45°49S 169°22E 59 F2
Edina U.S.A. 40°10N 92°11W 80 E7
Edinboro U.S.A. 41°52N 80°8W 82 E4
Edinburg U.S.A. 26°18N 98°10W 84 H5
Edinburgh U.K. 55°57N 3°13W 11 F5
Edinburgh ✕ (EDI) U.K. 55°54N 3°22W 11 F5
Edinburgh, City of □ U.K. 55°57N 3°17W 11 F5
Edineț Moldova 48°9N 27°18E 17 D14
Edirne Turkey 41°40N 26°34E 23 D12
Edison U.S.A. 48°33N 122°27W 78 B4
Edithburgh Australia 35°5S 137°43E 63 F2
Edmeston U.S.A. 42°42N 75°15W 83 D9
Edmond U.S.A. 35°39N 97°29W 84 D6
Edmonds U.S.A. 47°48N 122°22W 78 C4
Edmonton Australia 17°2S 145°46E 62 B4
Edmonton Canada 53°30N 113°30W 70 C6
Edmund L. Canada 54°45N 93°17W 72 B1
Edmundston Canada 47°23N 68°20W 73 C6
Edna U.S.A. 28°59N 96°39W 84 G6
Edremit Turkey 39°34N 27°0E 23 E12
Edremit Körfezi Turkey 39°30N 26°45E 23 E12
Edson Canada 53°35N 116°28W 70 C5
Eduardo Castex Argentina 35°50S 64°18W 94 D3
Edward → Australia 35°5S 143°30E 63 F3
Edward, L. Africa 0°25S 29°40E 54 C2
Edward VII Land Antarctica 80°0S 150°0W 5 E13
Edwards Calif., U.S.A. 34°50N 117°40W 79 L9
Edwards N.Y., U.S.A. 44°20N 75°15W 83 B9
Edwards Plateau U.S.A. 30°45N 101°20W 84 F4
Edwardsville U.S.A. 41°15N 75°56W 83 E9
Edzná Mexico 19°39N 90°19W 87 D6
Edzo Canada 62°49N 116°4W 70 A5
Eeklo Belgium 51°11N 3°33E 15 C3
Eesti = Estonia ■ Europe 58°30N 25°30E 9 G21
Effigy Mounds ⌂ U.S.A. 43°5N 91°11W 80 D8
Effingham U.S.A. 39°7N 88°33W 80 F9
Égadi, Ísole Italy 37°55N 12°16E 22 F5
Egan Range U.S.A. 39°35N 114°55W 76 G6
Eganville Canada 45°32N 77°5W 82 A7
Eger = Cheb Czech Rep. 50°9N 12°28E 16 C7
Eger Hungary 47°53N 20°27E 17 E11
Egersund Norway 58°26N 6°1E 9 G12
Egg L. Canada 55°5N 105°30W 71 B7
Éghezée Belgium 50°35N 4°55E 15 D4
Egilsstaðir Iceland 65°16N 14°25W 8 D6
Egio Greece 38°15N 22°5E 23 E10
Eglinton I. Canada 75°48N 118°30W 69 B8
Egmont Canada 49°45N 123°56W 70 D4
Egmont, C. N.Z. 39°16S 173°45E 59 C4
Egmont, Mt. = Taranaki, Mt. N.Z. 39°17S 174°5E 59 C5
Egmont △ N.Z. 39°17S 174°4E 59 C5
Egra India 21°54N 87°32E 43 J12
Eğridir Turkey 37°52N 30°51E 19 G5
Eğridir Gölü Turkey 37°53N 30°50E 44 B1
Egvekinot Russia 66°19N 179°50W 29 C19
Egypt ■ Africa 28°0N 31°0E 51 C12
Ehime □ Japan 33°30N 132°40E 31 H6
Ehrenberg U.S.A. 33°36N 114°31W 79 M12
Eibar Spain 43°11N 2°28W 21 A4
Eidsvold Australia 25°25S 151°12E 63 D5
Eidsvoll Norway 60°19N 11°14E 9 F14
Eielson U.S.A. 64°40N 147°4W 74 C10
Eifel Germany 50°15N 6°50E 16 C4
Eiffel Flats Zimbabwe 18°20S 30°0E 55 F3
Eigg U.K. 56°54N 6°10W 11 E2
Eighty Mile Beach Australia 19°30S 120°40E 60 C3
Eil, L. U.K. 56°51N 5°16W 11 E3
Eildon, L. Australia 37°10S 146°0E 63 F4
Eilean Sar = Western Isles □ U.K. 57°30N 7°10W 11 D1
Einasleigh Australia 18°32S 144°5E 62 B3
Einasleigh → Australia 17°30S 142°17E 62 B3
Eindhoven Neths. 51°26N 5°28E 15 C5
Eire = Ireland ■ Europe 53°50N 7°52W 10 C4
Eiríksjökull Iceland 64°46N 20°24W 8 D3
Eirunepé Brazil 6°35S 69°53W 92 E5
Eiseb → Namibia 20°33S 20°59E 56 C2
Eisenach Germany 50°58N 10°19E 16 C6
Eisenerz Austria 47°32N 14°54E 16 E8
Eivissa Spain 38°54N 1°26E 24 C7
Ejeda Madag. 24°20S 44°31E 57 C7
Ejutla Mexico 16°34N 96°44W 87 D5
Ekalaka U.S.A. 45°53N 104°33W 76 D11
Ekaterinburg = Yekaterinburg Russia 56°50N 60°30E 28 D7
Ekenäs = Raasepori Finland 60°0N 23°26E 9 G20
Eketahuna N.Z. 40°38S 175°43E 59 D5
Ekibastuz Kazakhstan 51°50N 75°10E 28 D8
Ekoli Dem. Rep. of the Congo 0°23S 24°13E 54 C1
Eksjö Sweden 57°40N 14°58E 9 H16
Ekuma → Namibia 18°40S 16°2E 56 B2
Ekwan → Canada 53°12N 82°15W 72 B3
Ekwan Pt. Canada 53°16N 82°7W 72 B3
El Aaiún W. Sahara 27°9N 13°12W 50 C3
El Abanico Chile 37°20S 71°31W 94 D1
El 'Agrûd Egypt 30°14N 34°24E 46 E3
El Alamein Egypt 30°48N 28°58E 51 B11
El 'Aqaba, W. → Egypt 30°7N 33°54E 46 E2
El Arīḥā West Bank 31°52N 35°27E 46 D4
El 'Arîsh Egypt 31°8N 33°50E 46 D2
El 'Arîsh, W. → Egypt 31°8N 33°47E 46 D2
El Asnam = Ech Chéliff Algeria 36°10N 1°20E 50 A6
El Bayadh Algeria 33°40N 1°1E 50 B6
El Bluff Nic. 11°59N 83°40W 88 D3
El Cajon U.S.A. 32°48N 116°58W 79 N10
El Calafate Argentina 50°19S 72°15W 96 G2
El Campo U.S.A. 29°12N 96°16W 84 G6
El Capitan U.S.A. 37°44N 119°38E 78 H7
El Carbón Honduras 15°25N 85°32W 88 C2
El Carmen Colombia 9°43N 75°8W 92 B3
El Centro U.S.A. 32°48N 115°34W 79 N11
El Cerro Bolivia 17°30S 61°40W 92 G6
El Compadre Mexico 32°20N 116°14W 79 N10
El Cuy Argentina 39°55S 68°25W 96 D3
El Cuyo Mexico 21°31N 87°41W 87 C7
El Daheir Egypt 31°13N 34°10E 46 D3
El Descanso Mexico 32°12N 116°58W 79 N10
El Desemboque Mexico 30°33N 113°1W 86 A2
El Diviso Colombia 1°22N 78°14W 92 C3
El Djouf Mauritania 20°0N 9°0W 50 D4
El Dorado Mexico 24°17N 107°21W 86 C3
El Dorado Ark., U.S.A. 33°12N 92°40W 84 E8
El Dorado Kans., U.S.A. 37°49N 96°52W 80 G5
El Dorado Venezuela 6°55N 61°37W 92 B6
El Dorado Springs U.S.A. 37°52N 94°1W 80 G6
El Escorial Spain 40°35N 4°7W 21 B3
El Faiyûm Egypt 29°19N 30°50E 51 C12
El Fâsher Sudan 13°33N 25°26E 51 F11
El Ferrol = Ferrol Spain 43°29N 8°15W 21 A1
El Fuerte Mexico 26°25N 108°39W 86 B3
El Geneina = Al Junaynah Sudan 13°27N 22°45E 51 F10
El Gezira □ Sudan 15°0N 33°0E 51 F12
El Gîza Egypt 30°0N 31°12E 51 C12
El Gogorrón △ Mexico 21°49N 100°57W 86 C4
El Goléa Algeria 30°30N 2°50E 50 B6
El Golfo de Santa Clara Mexico 31°42N 114°30W 86 A2
El Guácharo △ Venezuela 10°8N 63°21W 89 D7
El Guache △ Venezuela 9°45N 69°30W 89 E6
El Iskandarîya Egypt 31°13N 29°58E 51 B11
El Istiwa'iya Sudan 5°0N 28°0E 51 G11
El Jadida Morocco 33°11N 8°17W 50 B4
El Jardal Honduras 14°54N 88°50W 88 D2
El Kef □ Tunisia 36°0N 9°0E 51 A7
El Khârga Egypt 25°30N 30°33E 51 C12
El Khartûm Sudan 15°31N 32°35E 51 E12
El Khartûm Bahrî Sudan 15°40N 32°31E 51 E12
El Kuntilla Egypt 30°1N 34°45E 46 E3
El Leoncito △ Argentina 31°58S 69°10W 94 C2
El Lucero Mexico 30°37N 106°31W 86 A3
El Maestrazgo Spain 40°30N 0°25W 21 B5
El Mahalla el Kubra Egypt 31°0N 31°0E 51 B12
El Malpais ⌂ U.S.A. 34°53N 108°0W 77 J10
El Mansûra Egypt 31°0N 31°19E 51 B12
El Medano Canary Is. 28°3N 16°32W 24 F3
El Ménia = El Goléa Algeria 30°30N 2°50E 50 B6
El Milagro Argentina 30°59S 65°59W 94 C2
El Minyâ Egypt 28°7N 30°33E 51 C12
El Monte U.S.A. 34°4N 118°1W 79 L8
El Obeid Sudan 13°8N 30°10E 51 F12
El Odaiya Sudan 12°8N 28°12E 51 F11
El Oro Mexico 19°51N 100°7W 87 D4
El Oued Algeria 33°20N 6°58E 50 B7
El Palmar △ Argentina 32°10S 58°31W 94 C4
El Palmito, Presa Mexico 25°40N 105°30W 86 B3
El Paso U.S.A. 31°45N 106°29W 84 F1
El Pinacate y Gran Desierto de Altar = Gran Desierto del Pinacate ⌂ Mexico 31°51N 113°32W 86 A2
El Portal U.S.A. 37°41N 119°47W 78 H7
El Porvenir Mexico 31°15N 105°51W 86 A3
El Prat de Llobregat Spain 41°19N 2°5E 21 B7
El Progreso Honduras 15°26N 87°51W 88 C2
El Pueblito Mexico 29°6N 105°7W 86 B3
El Pueblo Canary Is. 28°36N 17°47W 24 F2
El Puerto de Santa María Spain 36°36N 6°13W 21 D2
El Qâhira Egypt 30°2N 31°13E 51 B12
El Qantara Egypt 30°51N 32°20E 46 E1
El Quseima Egypt 30°40N 34°15E 46 E3
El Real de Santa María Panama 8°0N 77°40W 92 B3
El Reno U.S.A. 35°32N 97°57W 75 H20
El Rey △ Argentina 24°40S 64°34W 94 A3
El Rio U.S.A. 34°14N 119°10W 79 L7
El Roque, Pta. Canary Is. 28°10N 15°25W 24 F4
El Rosarito Mexico 28°38N 114°4W 86 B2
El Salto Mexico 23°47N 105°22W 86 C3
El Salvador ■ Cent. Amer. 13°50N 89°0W 88 D2
El Sauce Nic. 13°0N 86°40W 88 D2
El Sueco Mexico 29°54N 106°24W 86 B3
El Suweis Egypt 29°58N 32°31E 51 C12
El Tamarâni, W. → Egypt 30°7N 34°43E 46 E3
El Thamad Egypt 29°40N 34°28E 46 F3
El Tigre Venezuela 8°44N 64°15W 92 B6
El Tîh, Gebel Egypt 29°40N 33°50E 46 F2
El Tofo Chile 29°22S 71°18W 94 B1
El Tránsito Chile 28°52S 70°17W 94 B1
El Tûr Egypt 28°14N 33°36E 44 D2
El Turbio Argentina 51°45S 72°5W 96 G2
El Uqsur Egypt 25°41N 32°38E 51 C12
El Vergel Mexico 26°28N 106°22W 86 B3
El Vigía Venezuela 8°38N 71°39W 92 B4
El Wabeira Egypt 29°34N 33°6E 46 F2
El Wuz Sudan 15°5N 30°7E 51 E12
Elat Israel 29°30N 34°56E 46 F3
Elâzığ Turkey 38°37N 39°14E 44 B3
Elba Italy 42°46N 10°17E 22 C4
Elba U.S.A. 31°25N 86°4W 85 F11
Elbasan Albania 41°9N 20°9E 23 D9
Elbe U.S.A. 46°45N 122°10W 78 D4
Elbe → Europe 53°50N 9°0E 16 B5
Elbert, Mt. U.S.A. 39°7N 106°27W 76 G10
Elberton U.S.A. 34°7N 82°52W 85 D13
Elbeuf France 49°17N 1°2E 20 B4
Elbing = Elbląg Poland 54°10N 19°25E 17 A10
Elbistan Turkey 38°13N 37°15E 44 B3
Elbląg Poland 54°10N 19°25E 17 A10
Elbow Canada 51°7N 106°35W 71 C7
Elbrus Russia 43°21N 42°30E 19 F7
Elburz Mts. = Alborz, Reshteh-ye Kūhhā-ye Iran 36°0N 52°0E 45 C7
Elche Spain 38°15N 0°42W 21 C5
Elcho I. Australia 11°55S 135°45E 62 A2
Elda Spain 38°29N 0°47W 21 C5
Eldama Ravine Kenya 0°3N 35°43E 54 B4
Elde → Germany 53°7N 11°15E 16 B6
Eldon Mo., U.S.A. 38°21N 92°35W 80 F7
Eldon Wash., U.S.A. 47°33N 123°3W 78 C3
Eldora U.S.A. 42°22N 93°5W 80 D7
Eldorado Argentina 26°28S 54°43W 95 B5
Eldorado Canada 44°35N 77°31W 82 B7
Eldorado Ill., U.S.A. 37°49N 88°26W 80 G9
Eldorado Tex., U.S.A. 30°52N 100°36W 84 F4
Eldoret Kenya 0°30N 35°17E 54 B4
Eldred U.S.A. 41°58N 78°23W 82 E6
Elea, C. Cyprus 35°19N 34°4E 25 D13
Eleanora, Pk. Australia 32°57S 121°9E 61 F3
Elefantes → Africa 24°10S 32°40E 57 C5
Elektrostal Russia 55°41N 38°32E 18 C6
Elemi Triangle Africa 5°0N 35°20E 54 B4
Elephant Butte Res. U.S.A. 33°9N 107°11W 77 K10
Elephant I. Antarctica 61°0S 55°0W 5 C18
Eleuthera Bahamas 25°0N 76°20W 88 B4
Elgin Canada 44°36N 76°13W 83 B8
Elgin U.K. 57°39N 3°19W 11 D5
Elgin Ill., U.S.A. 42°2N 88°17W 80 D9
Elgin N. Dak., U.S.A. 46°24N 101°51W 80 B3
Elgin Oreg., U.S.A. 45°34N 117°55W 76 D5
Elgin Tex., U.S.A. 30°21N 97°22W 84 F6
Elgon, Mt. Africa 1°10N 34°30E 54 B3
Eliase Indonesia 8°21S 130°48E 37 F8
Elim Namibia 17°48S 15°31E 56 B2
Elim S. Africa 34°35S 19°45E 56 E2
Elista Russia 46°16N 44°14E 19 E7
Elizabeth Australia 34°42S 138°41E 63 E2
Elizabeth U.S.A. 40°39N 74°12W 83 F10
Elizabeth City U.S.A. 36°18N 76°14W 85 C16
Elizabethton U.S.A. 36°21N 82°13W 85 C13
Elizabethtown Ky., U.S.A. 37°42N 85°52W 81 G11
Elizabethtown N.Y., U.S.A. 44°13N 73°36W 83 B11
Elizabethtown Pa., U.S.A. 40°9N 76°36W 83 F8
Ełk Poland 53°50N 22°21E 17 B12
Elk → Canada 49°11N 115°14W 70 C5
Elk → U.S.A. 34°46N 87°16W 85 D11
Elk City U.S.A. 35°25N 99°25W 84 D5
Elk Creek U.S.A. 39°36N 122°32W 78 F4
Elk Grove U.S.A. 38°25N 121°22W 78 G5
Elk Island △ Canada 53°35N 112°59W 70 C6
Elk Lake Canada 47°40N 80°25W 72 C3
Elk Point Canada 53°54N 110°55W 71 C6
Elk River Idaho, U.S.A. 46°47N 116°11W 76 C5
Elk River Minn., U.S.A. 45°18N 93°35W 80 C7
Elkedra → Australia 21°8S 136°22E 62 C2
Elkhart Ind., U.S.A. 41°41N 85°58W 81 E11
Elkhart Kans., U.S.A. 37°0N 101°54W 80 G3
Elkhorn Canada 49°59N 101°14W 71 D8
Elkhorn → U.S.A. 41°8N 96°19W 80 E5
Elkhovo Bulgaria 42°10N 26°35E 23 C12
Elkin U.S.A. 36°15N 80°51W 85 C14
Elkins U.S.A. 38°55N 79°51W 81 F14
Elkland U.S.A. 41°59N 77°19W 82 E7
Elko Canada 49°20N 115°10W 70 D5
Elko U.S.A. 40°50N 115°46W 76 F6
Elkton U.S.A. 43°49N 83°11W 82 C1
Ellas = Greece ■ Europe 40°0N 23°0E 23 E9
Ellef Ringnes I. Canada 78°30N 102°2W 69 B11
Ellen, Mt. U.S.A. 44°9N 72°56W 83 B12
Ellenburg U.S.A. 44°54N 73°48W 83 B11
Ellendale U.S.A. 46°0N 98°32W 80 B4
Ellensburg U.S.A. 46°59N 120°34W 76 C3
Ellenville U.S.A. 41°43N 74°24W 83 E10
Ellerton Barbados 13°7N 59°33W 89 g
Ellery, Mt. Australia 37°28S 148°47E 63 F4
Ellesmere, L. N.Z. 43°47S 172°28E 59 G4
Ellesmere I. Canada 79°30N 80°0W 69 B16
Ellesmere Island = Quttinirpaaq △ Canada 82°13N 72°13W 69 A17
Ellesmere Port U.K. 53°17N 2°54W 12 D5
Ellice Is. = Tuvalu ■ Pac. Oc. 8°0S 178°0E 58 B10
Ellicottville U.S.A. 42°17N 78°40W 82 D6
Ellington U.S.A. 42°13N 79°6W 82 D5
Elliot Australia 17°33S 133°32E 62 B1
Elliot S. Africa 31°22S 27°48E 57 E4
Elliot Lake Canada 46°25N 82°35W 72 C3
Elliotdale = Xhora S. Africa 31°55S 28°38E 57 E4
Ellis U.S.A. 38°56N 99°34W 80 F4
Elliston Australia 33°39S 134°53E 63 E1
Ellisville U.S.A. 31°36N 89°12W 85 F10
Ellon U.K. 57°22N 2°4W 11 D6
Ellore = Eluru India 16°48N 81°8E 41 L12
Ellsworth Kans., U.S.A. 38°44N 98°14W 80 F4
Ellsworth Maine, U.S.A. 44°33N 68°25W 81 C19
Ellsworth Land Antarctica 76°0S 89°0W 5 D16
Ellsworth Mts. Antarctica 78°30S 85°0W 5 D16
Ellwood City U.S.A. 40°52N 80°17W 82 F4
Elma Canada 49°52N 95°55W 71 D9
Elma U.S.A. 47°0N 123°25W 78 D3
Elmalı Turkey 36°44N 29°56E 19 G4
Elmhurst U.S.A. 41°53N 87°56W 80 E10
Elmira Canada 43°36N 80°33W 82 C4
Elmira U.S.A. 42°6N 76°48W 82 D8
Elmira Heights U.S.A. 42°8N 76°50W 82 D8
Elmore Australia 36°30S 144°37E 63 F3
Elmshorn Germany 53°43N 9°40E 16 B5
Elmvale Canada 44°35N 79°52W 82 B5
Elora Canada 43°41N 80°26W 82 C4
Elounda Greece 35°16N 25°42E 25 D7
Eloy U.S.A. 32°45N 111°33W 77 K8
Elphin Canada 44°55N 76°37W 83 B8
Elrose Canada 51°12N 108°0W 71 C7
Elsby △ Australia 14°55S 133°10E 60 B5
Elsie U.S.A. 45°52N 123°36W 78 E3
Elsinore = Helsingør Denmark 56°2N 12°35E 9 H15
Eltanin Fracture Zone System S. Ocean 54°0S 130°0W 5 B14
Eltham N.Z. 39°26S 174°19E 59 C5
Eluru India 16°48N 81°8E 41 L12
Elvas Portugal 38°50N 7°10W 21 C2
Elverum Norway 60°53N 11°34E 8 F14
Elvire → Australia 17°51S 128°11E 60 C4
Elvire, Mt. Australia 29°22S 119°36E 61 E2
Elwell, L. = Tiber Res. U.S.A. 48°19N 111°6W 76 B8
Elwood Ind., U.S.A. 40°17N 85°50W 81 E11
Elwood Nebr., U.S.A. 40°36N 99°52W 80 E4
Elx = Elche Spain 38°15N 0°42W 21 C5
Ely U.K. 52°24N 0°16E 13 E8
Ely Minn., U.S.A. 47°55N 91°51W 80 B8
Ely Nev., U.S.A. 39°15N 114°54W 76 G6
Elyria U.S.A. 41°22N 82°7W 82 E2
eMalahleni S. Africa 25°51S 29°14E 57 D4
Emāmrūd Iran 36°30N 55°0E 45 B7
Embarcación Argentina 23°10S 64°0W 94 A3
Embetsu Japan 44°44N 141°47E 30 B10
Embi Kazakhstan 48°50N 58°8E 28 E6
Embi → Kazakhstan 46°55N 53°28E 19 E9
Embonas Greece 36°13N 27°51E 25 C9
Embro Canada 43°9N 80°54W 82 C4
Embrun France 44°34N 6°30E 20 D7
Embu Kenya 0°32S 37°38E 54 C4
Emden Germany 53°21N 7°12E 16 B4
Emerald Australia 23°32S 148°10E 62 C4
Emerald I. Canada 76°48N 114°10W 69 B9
Emerson Canada 49°0N 97°10W 71 D9
Emet Turkey 39°20N 29°15E 23 E13
Emi Koussi Chad 19°45N 18°55E 51 E9
Eminabad Pakistan 32°2N 74°8E 42 C6
Emine, Nos Bulgaria 42°40N 27°56E 23 C12
Emissi, Tarso Chad 21°27N 18°36E 51 D9
Emlenton U.S.A. 41°11N 79°43W 82 E5
Emmaus S. Africa 29°2S 25°15E 56 D4
Emmaus U.S.A. 40°32N 75°30W 83 F9
Emmeloord Neths. 52°44N 5°46E 15 B5
Emmen Neths. 52°48N 6°57E 15 B6
Emmet Australia 24°45S 144°30E 62 C3
Emmetsburg U.S.A. 43°7N 94°41W 80 D6
Emmett Idaho, U.S.A. 43°52N 116°30W 76 E5
Emmett Mich., U.S.A. 42°59N 82°46W 82 D2
Emmonak U.S.A. 62°47N 164°31W 74
Emo Canada 48°38N 93°50W 71 D
Empalme Mexico 27°58N 110°51W 86
Empangeni S. Africa 28°50S 31°52E 57
Empedrado Argentina 28°0S 58°46W 94
Emperor Seamount Chain Pac. Oc. 40°0N 170°0E 64
Emperor Trough Pac. Oc. 43°0N 175°30E 64
Emporia Kans., U.S.A. 38°25N 96°11W 80
Emporia Va., U.S.A. 36°42N 77°32W 81 G
Emporium U.S.A. 41°31N 78°14W 82
Empress Canada 50°57N 110°0W 71
Empty Quarter = Rub' al Khālī Si. Arabia 19°0N 48°0E 47
Ems → Germany 53°20N 7°12E 16
Emsdale Canada 45°32N 79°19W 82
Emu China 43°40N 128°6E 35 C
Emu Park Australia 23°13S 150°50E 62
eMuziwezinto S. Africa 30°15S 30°45E 57
'En 'Avrona Israel 29°43N 35°0E 46
'En Boqeq Israel 31°12N 35°21E 45
'En Gedi Israel 31°28N 35°25E 45
En Nahud Sudan 12°45N 28°25E 51 F
Ena Japan 35°25N 137°25E 31
Enana Namibia 17°30S 16°23E 56
Enard B. U.K. 58°5N 5°20W 11
Enare träsk = Inarijärvi Finland 69°0N 28°0E 8 B
Enarotali Indonesia 3°55S 136°21E 37
Encampment U.S.A. 41°12N 106°47W 76 F
Encantadas, Serra Brazil 30°40S 53°0W 95
Encarnación Paraguay 27°15S 55°50W 95
Encarnación de Díaz Mexico 21°31N 102°14W 86
Encinitas U.S.A. 33°3N 117°17W 79
Encino U.S.A. 34°39N 105°28W 77 J
Encounter B. Australia 35°45S 138°45E 63
Endako Canada 54°6N 125°2W 70
Endau Kenya 1°18S 38°31E 54
Endau Rompin △ Malaysia 2°27N 103°45E 39
Ende Indonesia 8°45S 121°40E 37
Endeavour Str. Australia 10°45S 142°0E 62
Enderbury Kiribati 3°8S 171°5W 64 H
Enderby Canada 50°35N 119°10W 70
Enderby Abyssal Plain S. Ocean 60°0S 40°0E 5
Enderby I. Australia 20°35S 116°30E 60
Enderby Land Antarctica 66°0S 53°0E 5
Enderlin U.S.A. 46°38N 97°36W 80
Enderrocat, C. Spain 39°28N 2°43E 24
Endicott U.S.A. 42°6N 76°4W 83
Endicott Mts. U.S.A. 68°0N 152°0W 74
Endwell U.S.A. 42°6N 76°2W 83
Endyalgout I. Australia 11°40S 132°35E 60
Eneabba Australia 29°49S 115°16E 61
Enewetak Atoll Marshall Is. 11°30N 162°15E 64
Enez Turkey 40°45N 26°5E 23 D
Enfer, Pte. d' Martinique 14°22N 60°54W 8
Enfield Canada 44°56N 63°32W 73
Enfield Conn., U.S.A. 41°58N 72°36W 83 E
Enfield N.C., U.S.A. 36°11N 77°41W 85 C
Enfield N.H., U.S.A. 43°39N 72°9W 83 C
Engadin Switz. 46°45N 10°10E 20
Engaño, C. Dom. Rep. 18°30N 68°20W 89
Engaño, C. Phil. 18°35N 122°23E 37
Engaru Japan 44°3N 143°31E 30 B
Engcobo = Ngcobo S. Africa 31°37S 28°0E 57
Engels Russia 51°28N 46°6E 19
Engemann L. Canada 58°0N 106°55W 71
Enggano Indonesia 5°20S 102°40E 36
England U.S.A. 34°33N 91°58W 84
England □ U.K. 53°0N 2°0W 13
Englee Canada 50°45N 56°5W 73
Englehart Canada 47°49N 79°52W 72
Englewood U.S.A. 39°38N 104°59W 76 G
English → Canada 49°12N 91°5W 71 C
English Bazar = Ingraj Bazar India 24°58N 88°10E 43 G
English Channel Europe 50°0N 2°0W 13
English Company's Is., The Australia 11°50S 136°32E 62
English River Canada 49°14N 91°0W 72
Enid U.S.A. 36°24N 97°53W 84
Eniwa Japan 43°0N 141°30E 30 C
Enkhuizen Neths. 52°42N 5°17E 15
Enna Italy 37°34N 14°16E 22
Ennadai L. Canada 60°58N 101°20W 71
Ennedi Chad 17°15N 22°0E 51 E
Enngonia Australia 29°21S 145°50E 63
Ennis Ireland 52°51N 8°59W 10
Ennis Mont., U.S.A. 45°21N 111°44W 76
Ennis Tex., U.S.A. 32°20N 96°38W 84
Enniscorthy Ireland 52°30N 6°34W 10
Enniskillen U.K. 54°21N 7°39W 10
Ennistimon Ireland 52°57N 9°17W 10
Enns → Austria 48°14N 14°32E 16
Eno Finland 62°47N 30°10E 8 E
Enontekiö Finland 68°23N 23°37E 8 B
Enosburg Falls U.S.A. 44°55N 72°48W 83 B
Enriquillo, L. Dom. Rep. 18°20N 72°5W 89
Enschede Neths. 52°13N 6°53E 15
Ensenada Argentina 34°55S 57°55W 94
Ensenada Mexico 31°52N 116°37W 86
Ensenada de los Muertos Mexico 23°59N 109°51W 86
Enshi China 30°18N 109°29E 33 E
Ensiola, Pta. de n' Spain 39°7N 2°55E 24
Entebbe Uganda 0°4N 32°28E 54
Enterprise Canada 60°47N 115°45W 70
Enterprise Ala., U.S.A. 31°19N 85°51W 85 F
Enterprise Oreg., U.S.A. 45°25N 117°17W 76
Entre Ríos Bolivia 21°30S 64°25W 94
Entre Ríos □ Argentina 30°30S 58°30W 94
Entroncamento Portugal 39°28N 8°28W 21
Enugu Nigeria 6°30N 7°30E 50
Enumclaw U.S.A. 47°12N 121°59W 78
Eochail = Youghal Ireland 51°56N 7°52W 10

lie, Ís. Italy 38°30N 14°57E 22 E6
e Neths. 52°21N 5°59E 15 B5
ernay France 49°3N 3°56E 20 B5
hesus Turkey 37°55N 27°22E 23 F12
hraim U.S.A. 39°22N 111°35W 76 G8
hrata Pa., U.S.A. 40°11N 76°11W 83 F8
hrata Wash., U.S.A. 47°19N 119°33W 76 C4
nal France 48°10N 6°27E 20 B7
iskopi Cyprus 34°40N 32°54E 25 E11
iskopi Greece 35°20N 24°20E 25 D6
iskopi Bay Cyprus 34°35N 32°50E 25 E11
som U.K. 51°19N 0°16W 13 F7
ukiro Namibia 21°40S 19°9E 56 C2
uatoria = El Istiwa'iya Sudan 5°0N 28°0E 51 G11
uatorial Guinea ■ Africa 2°0N 8°0E 52 D1
Hai China 25°48N 100°11E 41 G22
Rachidia Morocco 31°58N 4°20W 50 B5
Rahad Sudan 12°45N 30°32E 51 F12
Rif Morocco 35°1N 4°1W 50 A5
iwadī Myit = Irrawaddy → Burma 15°50N 95°6E 41 M19
iwadī Myitwanya = Irrawaddy, Mouths of the Burma 15°30N 95°0E 41 M19
awan △ Thailand 14°25N 98°58E 38 E2
bil = Arbīl Iraq 36°15N 44°5E 44 B5
çek Turkey 38°39N 43°36E 44 B4
ciyaş Dağı Turkey 38°30N 35°30E 44 B2
d Hungary 47°22N 18°56E 17 E10
dao Jiang → China 42°37N 128°0E 35 C14
dek Turkey 40°23N 27°47E 23 D12
dene = Ulaan-Uul Mongolia 46°4N 100°49E 34 B6
denet Mongolia 49°2N 104°5E 32 B9
denetsogt Mongolia 42°55N 106°5E 34 C4
ebus, Mt. Antarctica 77°35S 167°0E 5 D11
echim Brazil 27°35S 52°15W 95 B5
eğli Konya, Turkey 37°31N 34°4E 44 B2
eğli Zonguldak, Turkey 41°15N 31°24E 19 F5
enhot China 43°48N 112°2E 34 C7
esma → Spain 41°26N 4°45W 21 B3
enisdam S. Africa 28°30S 26°50E 56 D4
furt Germany 50°58N 11°2E 16 C6
gani Turkey 38°17N 39°49E 44 B3
gel Mongolia 43°8N 109°5E 34 C5
geni Vozvyshennost Russia 47°0N 44°0E 19 E7
gli Latvia 56°54N 25°38E 9 H21
boll, L. U.K. 58°30N 4°42W 11 C4
ce Italy 38°2N 12°35E 22 E5
e U.S.A. 42°8N 80°5W 82 D4
e, L. N. Amer. 42°15N 81°0W 82 D4
e Canal U.S.A. 43°5N 78°43W 82 C7
eau Canada 42°16N 81°57W 82 D3
gavo = Ceerigaabo Somali Rep. 10°35N 47°20E 47 E4
koussa Greece 39°53N 19°34E 25 A3
ksdale Canada 50°52N 98°7W 71 C9
manthos Greece 37°57N 21°50E 23 F9
mo-misaki Japan 41°50N 143°15E 30 D11
n Canada 43°45N 80°7W 82 C4
n Pt. Trin. & Tob. 10°3N 61°39W 93 K15
npura India 25°9N 73°3E 42 G5
skay U.K. 57°4N 7°18W 11 D1
trea ■ Africa 14°0N 38°30E 47 D2
angen Germany 49°36N 11°0E 16 D6
dunda Australia 25°14S 133°12E 62 D1
nelo Neths. 52°18N 5°35E 15 B5
nelo S. Africa 26°31S 29°59E 57 D4
menek Turkey 36°38N 33°0E 44 B2
mones Greece 39°37N 19°46E 25 A3
noupoli Greece 37°28N 24°57E 23 F11
nakulam India 9°59N 76°22E 40 Q10
ne → Ireland 54°30N 8°16W 10 B3
ne, Lower L. U.K. 54°28N 7°47W 10 B4
ne, Upper L. U.K. 54°14N 7°32W 10 B4
nest Giles Ra. Australia 27°0S 123°45E 61 E3
ode India 11°24N 77°45E 40 P10
omanga Australia 26°40S 143°11E 63 D3
ongo Namibia 21°39S 15°58E 56 C2
ramala Hills India 15°30N 78°15E 40 M11
rentería Spain 43°19N 1°54W 21 A5
rigal Ireland 55°2N 8°6W 10 A3
rinundra △ Australia 37°20S 148°47E 63 F4
ris Hd. Ireland 54°19N 10°0W 10 B1
skine U.S.A. 47°40N 96°0W 80 B6
tis = Irtysh → Russia 61°4N 68°52E 28 C7
win U.S.A. 36°9N 82°25W 85 C13
zgebirge Germany 50°27N 12°55E 16 C7
zin Russia 50°15N 95°10E 29 D10
zincan Turkey 39°46N 39°30E 44 B3
zurum Turkey 39°57N 41°15E 44 B4
Caló Spain 38°40N 1°30E 24 C8
Canar Spain 39°2N 1°36E 24 B8
Mercadal Spain 39°59N 4°5E 24 B11
Migjorn Gran Spain 39°57N 4°3E 24 B11
Sahrâ' Esh Sharqîya Egypt 27°30N 32°30E 51 C12
Sînâ' Egypt 29°0N 34°0E 46 F2
Vedrà Spain 38°52N 1°12E 24 C7
ambo Dem. Rep. of the Congo 3°48S 23°30E 54 C1
an-Misaki Japan 41°40N 141°10E 30 D10
ashi Hokkaidō, Japan 44°56N 142°35E 30 B11
ashi Hokkaidō, Japan 41°52N 140°7E 30 D10
bjerg Denmark 55°29N 8°29E 9 J13
bo = Espoo Finland 60°12N 24°40E 9 F21
calante U.S.A. 37°47N 111°36W 77 H8
calante → U.S.A. 37°24N 110°57W 77 H8
calón Mexico 26°45N 104°20W 86 B4
cambia → U.S.A. 30°32N 87°11W 85 F11
canaba U.S.A. 45°45N 87°4W 80 C10
ch-sur-Alzette Lux. 49°32N 6°0E 15 E6
condido U.S.A. 33°7N 117°5W 79 M9
cuinapa de Hidalgo Mexico 22°50N 105°50W 86 C3
cuintla Guatemala 14°20N 90°48W 88 D1
enguly Turkmenistan 37°37N 53°59E 28 F6
fahān Iran 32°39N 51°43E 45 C6
fahān □ Iran 32°50N 51°50E 45 C6
Esfarāyen Iran 37°4N 57°30E 45 B8
Esfīdeh Iran 33°39N 59°46E 45 C8
Esh Sham = Dimashq Syria 33°30N 36°18E 46 B5
Esha Ness U.K. 60°29N 1°38W 11 A7
Esher U.K. 51°21N 0°20W 13 F7
Eshkol △ Israel 31°20N 34°30E 46 D3
Eshowe S. Africa 28°50S 31°30E 57 D5
Esigodini Zimbabwe 20°18S 28°56E 57 C4
Esil = Ishim → Russia 57°45N 71°10E 28 D8
Esira Madag. 24°20S 46°42E 57 C8
Esk → Dumf. & Gall., U.K. 54°58N 3°2W 11 G5
Esk → N. Yorks., U.K. 54°30N 0°37W 12 C7
Eskān Iran 26°48N 63°9E 45 E9
Esker Siding Canada 53°53N 66°25W 73 B6
Eskifjörður Iceland 65°3N 13°55W 8 D7
Eskilstuna Sweden 59°22N 16°32E 9 G17
Eskimo Point = Arviat Canada 61°6N 93°59W 71 A10
Eskişehir Turkey 39°50N 30°30E 19 G5
Esla → Spain 41°29N 6°3W 21 B2
Eslāmābād-e Gharb Iran 34°10N 46°30E 44 C5
Eslāmshahr Iran 35°40N 51°10E 45 C6
Eşme Turkey 38°23N 28°58E 23 E13
Esmeraldas Ecuador 1°0N 79°40W 92 C3
Esna = Isna Egypt 25°17N 32°30E 51 C12
Esnagi L. Canada 48°36N 84°33W 72 C3
España = Spain ■ Europe 39°0N 4°0W 21 B4
Espanola Canada 46°15N 81°46W 72 C3
Espanola U.S.A. 35°59N 106°5W 77 J10
Esparza Costa Rica 9°59N 84°40W 88 E3
Esperance Australia 33°45S 121°55E 61 F3
Esperance B. Australia 33°48S 121°55E 61 F3
Esperance Harbour St. Lucia 14°4N 60°55W 89 f
Esperanza Antarctica 65°0S 55°0W 5 C18
Esperanza Argentina 31°29S 61°3W 94 C3
Esperanza Puerto Rico 18°6N 65°28W 89 d
Espichel, C. Portugal 38°22N 9°16W 21 C1
Espigão, Serra do Brazil 26°35S 50°30W 95 B5
Espinazo, Sierra del = Espinhaço, Serra do Brazil 17°30S 43°30W 93 G10
Espinhaço, Serra do Brazil 17°30S 43°30W 93 G10
Espinilho, Serra do Brazil 28°30S 55°0W 95 B5
Espírito Santo □ Brazil 20°0S 40°45W 93 H10
Espírito Santo do Pinhal Brazil 22°10S 46°46W 95 A6
Espíritu Santo Vanuatu 15°15S 166°50E 58 C9
Espíritu Santo, B. del Mexico 19°20N 87°35W 87 D7
Espíritu Santo, I. Mexico 24°30N 110°22W 86 C2
Espita Mexico 21°1N 88°19W 87 C7
Espoo Finland 60°12N 24°40E 9 F21
Espungabera Mozam. 20°29S 32°45E 57 C5
Esquel Argentina 42°55S 71°20W 96 E2
Esquimalt Canada 48°26N 123°25W 78 B3
Esquina Argentina 30°0S 59°30W 94 C4
Essaouira Morocco 31°32N 9°42W 50 B4
Essen Belgium 51°28N 4°28E 15 C4
Essen Germany 51°28N 7°2E 16 C4
Essendon, Mt. Australia 25°0S 120°29E 61 E3
Essequibo → Guyana 6°50N 58°30W 92 B7
Essex Canada 42°10N 82°49W 82 D2
Essex Calif., U.S.A. 34°44N 115°15W 79 L11
Essex N.Y., U.S.A. 44°19N 73°21W 83 B11
Essex □ U.K. 51°54N 0°27E 13 F8
Essex Junction U.S.A. 44°29N 73°7W 83 B11
Esslingen Germany 48°44N 9°18E 16 D5
Estación Camacho Mexico 24°25N 102°18W 86 C4
Estación Simón Mexico 24°42N 102°35W 86 C4
Estados, I. de Los Argentina 54°40S 64°30W 96 G4
Eşţahbānāt Iran 29°8N 54°4E 45 D7
Estância Brazil 11°16S 37°26W 93 F11
Estancia U.S.A. 34°46N 106°4W 77 J10
Estārm Iran 28°21N 58°21E 45 D8
Estcourt S. Africa 29°0S 29°53E 57 D4
Este △ Dom. Rep. 18°14N 68°42W 89 C6
Estelí Nic. 13°9N 86°22W 88 D2
Estellencs Spain 39°39N 2°29E 24 B9
Esterhazy Canada 50°37N 102°5W 71 C8
Estevan Canada 49°10N 102°59W 71 D8
Estevan Group Canada 53°3N 129°38W 70 C3
Estherville U.S.A. 43°24N 94°50W 80 D6
Eston Canada 51°8N 108°40W 71 C7
Estonia ■ Europe 58°30N 25°30E 9 G21
Estreito Brazil 6°32S 47°25W 93 E9
Estrela, Serra da Portugal 40°10N 7°45W 21 B2
Estremoz Portugal 38°51N 7°39W 21 C2
Estrondo, Serra do Brazil 7°20S 48°0W 93 E9
Esztergom Hungary 47°47N 18°44E 17 E10
Et Tîdra Mauritania 19°45N 16°20W 50 E2
Etah India 27°35N 78°40E 43 F8
Étampes France 48°26N 2°10E 20 B5
Etanga Namibia 17°55S 13°0E 56 B1
Etawah India 26°48N 79°6E 43 F8
Etawney L. Canada 57°50N 96°50W 71 B9
Etchojoa Mexico 26°55N 109°38W 86 B3
eThekwini = Durban S. Africa 29°49S 31°1E 57 D5
Ethel U.S.A. 46°32N 122°46W 78 D4
Ethelbert Canada 51°32N 100°25W 71 C8
Ethiopia ■ Africa 8°0N 40°0E 47 F3
Ethiopian Highlands Ethiopia 10°0N 37°0E 47 F2
Etive, L. U.K. 56°29N 5°10W 11 E3
Etna Italy 37°50N 14°55E 22 F6
Etoile Dem. Rep. of the Congo 11°33S 27°30E 55 E2
Etolin Strait U.S.A. 60°20N 165°15W 74 C6
Etosha △ Namibia 19°0S 16°0E 56 B2
Etosha Pan Namibia 18°40S 16°30E 56 B2
Etowah U.S.A. 35°20N 84°32W 85 D12
Etrek Turkmenistan 37°36N 54°46E 45 B7
Ettelbruck Lux. 49°51N 6°5E 15 E6
Ettrick Water → U.K. 55°31N 2°55W 11 F6
Etuku Dem. Rep. of the Congo 3°42S 25°45E 54 C2
Etzná-Tixmucuy = Edzná Mexico 19°39N 90°19W 87 D6
Eua Tonga 21°22S 174°56W 59 c
Euboea = Evia Greece 38°30N 24°0E 23 E11
Eucla Australia 31°41S 128°52E 61 F4
Euclid U.S.A. 41°34N 81°32W 82 E3
Eucumbene, L. Australia 36°2S 148°40E 63 F4
Eudora U.S.A. 33°7N 91°16W 84 E9
Eufaula Ala., U.S.A. 31°54N 85°9W 85 F12
Eufaula Okla., U.S.A. 35°17N 95°35W 84 D7
Eufaula L. U.S.A. 35°18N 95°21W 84 D7
Eugene U.S.A. 44°5N 123°4W 76 D2
Eugowra Australia 33°22S 148°24E 63 E4
Eulo Australia 28°10S 145°3E 63 D4
Eungella △ Australia 20°57S 148°40E 62 b
Eunice La., U.S.A. 30°30N 92°25W 84 F8
Eunice N. Mex., U.S.A. 32°26N 103°10W 77 K12
Eupen Belgium 50°37N 6°3E 15 D6
Euphrates = Furāt, Nahr al → Asia 31°0N 47°25E 44 D5
Eureka Canada 80°0N 85°56W 69 B14
Eureka Calif., U.S.A. 40°47N 124°9W 76 F1
Eureka Kans., U.S.A. 37°49N 96°17W 80 G5
Eureka Mont., U.S.A. 48°53N 115°3W 76 B6
Eureka Nev., U.S.A. 39°31N 115°58W 76 G6
Eureka S. Dak., U.S.A. 45°46N 99°38W 80 C4
Eureka, Mt. Australia 26°35S 121°35E 61 E3
Eureka Sd. Canada 79°0N 85°0W 69 B15
Euroa Australia 36°44S 145°35E 63 F4
Europa, Picos de Spain 43°10N 4°49W 21 A3
Europa, Pt. Gib. 36°3N 5°21W 21 D3
Europe 50°0N 20°0E 6 E10
Europoort Neths. 51°57N 4°10E 15 C4
Eustis U.S.A. 28°51N 81°41W 85 G14
Eutsuk L. Canada 53°20N 126°45W 70 C3
Evale Angola 16°33S 15°44E 56 B2
Evans U.S.A. 40°23N 104°41W 76 F11
Evans, L. Canada 50°50N 77°0W 72 B4
Evans City U.S.A. 40°46N 80°4W 82 F4
Evans Head Australia 29°7S 153°27E 63 D5
Evansburg Canada 53°36N 114°59W 70 C5
Evanston Ill., U.S.A. 42°3N 87°40W 80 D10
Evanston Wyo., U.S.A. 41°16N 110°58W 76 F8
Evansville U.S.A. 37°58N 87°35W 80 G10
Evaz Iran 27°46N 53°59E 45 E7
Eveleth U.S.A. 47°28N 92°32W 80 B7
Evensk Russia 62°12N 159°30E 29 C16
Everard, L. Australia 31°30S 135°0E 63 E2
Everard Ranges Australia 27°5S 132°28E 61 E5
Everest, Mt. Nepal 28°5N 86°58E 43 E12
Everett Pa., U.S.A. 40°1N 78°23W 82 F6
Everett Wash., U.S.A. 47°59N 122°12W 78 C4
Everglades, The U.S.A. 25°50N 81°0W 85 J14
Everglades △ U.S.A. 25°30N 81°0W 85 J14
Everglades City U.S.A. 25°52N 81°23W 85 J14
Evergreen Ala., U.S.A. 31°26N 86°57W 85 F11
Evergreen Mont., U.S.A. 48°14N 114°17W 76 B6
Evesham U.K. 52°6N 1°56W 13 E6
Evia Greece 38°30N 24°0E 23 E11
Evje Norway 58°36N 7°51E 9 G12
Évora Portugal 38°33N 7°57W 21 C2
Evowghlī Iran 38°43N 45°13E 44 B5
Évreux France 49°3N 1°8E 20 B4
Evros → Greece 41°40N 26°34E 23 D12
Évry France 48°38N 2°27E 20 B5
Évvoia = Evia Greece 38°30N 24°0E 23 E11
Ewe, L. U.K. 57°49N 5°38W 11 D3
Ewing U.S.A. 42°16N 98°21W 80 D4
Ewo Congo 0°48S 14°45E 52 E2
Exaltación Bolivia 13°10S 65°20W 92 F5
Excelsior Springs U.S.A. 39°20N 94°13W 80 F6
Exe → U.K. 50°41N 3°29W 13 G4
Exeter Canada 43°21N 81°29W 82 C3
Exeter U.K. 50°43N 3°31W 13 G4
Exeter Calif., U.S.A. 36°18N 119°9W 78 J7
Exeter N.H., U.S.A. 42°59N 70°57W 83 D14
Exmoor U.K. 51°12N 3°45W 13 F4
Exmoor △ U.K. 51°8N 3°42W 13 F4
Exmouth Australia 21°54S 114°10E 60 D1
Exmouth U.K. 50°37N 3°25W 13 G4
Exmouth G. Australia 22°15S 114°15E 60 D1
Exmouth Plateau Ind. Oc. 19°0S 114°0E 64 J3
Expedition △ Australia 25°41S 149°7E 63 D4
Expedition Ra. Australia 24°30S 149°12E 62 C4
Extremadura □ Spain 39°30N 6°5W 21 C2
Exuma Sound Bahamas 24°30N 76°20W 88 B4
Eyasi, L. Tanzania 3°30S 35°0E 54 C4
Eye Pen. U.K. 58°13N 6°10W 11 C2
Eyemouth U.K. 55°52N 2°5W 11 F6
Eyjafjallajökull Iceland 63°38N 19°36W 8 E4
Eyjafjörður Iceland 66°15N 18°30W 8 C4
Eyl Somali Rep. 8°0N 49°50E 47 F4
Eyre (North), L. Australia 28°30S 137°20E 63 D2
Eyre (South), L. Australia 29°18S 137°25E 63 D2
Eyre, L. Australia 29°30S 137°26E 58 D6
Eyre Mts. N.Z. 45°25S 168°25E 59 F2
Eyre Pen. Australia 33°30S 136°17E 63 E2
Eysturoy Færoe Is. 62°13N 6°54W 8 E9
Eyvān = Jūy Zar Iran 33°50N 46°18E 44 C5
Eyvānkī Iran 35°24N 51°56E 45 C6
Ezine Turkey 39°48N 26°20E 23 E12
Ezouza → Cyprus 34°44N 32°27E 25 E11

F

F.Y.R.O.M. = Macedonia ■ Europe 41°53N 21°40E 23 D9
Faaa Tahiti 17°34S 149°35W 59 d
Faaone Tahiti 17°40S 149°21W 59 d
Fabala Guinea 9°44N 9°5W 50 G4
Fabens U.S.A. 31°30N 106°10W 84 F1
Fabius U.S.A. 42°50N 75°59W 83 D9
Fabriano Italy 43°20N 12°54E 22 C5
Fachi Niger 18°6N 11°34E 51 E8
Fada Chad 17°13N 21°34E 51 E10
Fada-n-Gourma Burkina Faso 12°10N 0°30E 50 F6
Faddeyevskiy, Ostrov Russia 76°0N 144°0E 29 B15
Fadghāmī Syria 35°53N 40°52E 44 C4
Faenza Italy 44°17N 11°53E 22 B4
Færoe Is. = Føroyar ☑ Atl. Oc. 62°0N 7°0W 8 F9
Făgăras Romania 45°48N 24°58E 17 F13
Fagersta Sweden 60°1N 15°46E 9 F16
Fagnano, L. Argentina 54°30S 68°0W 96 G3
Fahlīān Iran 30°11N 51°28E 45 D6
Fahraj Kermān, Iran 29°0N 59°0E 45 D8
Fahraj Yazd, Iran 31°46N 54°36E 45 D7
Faial Azores 38°34N 28°42W 50 a
Faial Madeira 32°47N 16°53W 24 D3
Faichan Kangri India 35°48N 76°34E 43 B7
Fair Haven N.Y., U.S.A. 43°18N 76°42W 83 C8
Fair Haven Vt., U.S.A. 43°36N 73°16W 81 D17
Fair Hd. U.K. 55°14N 6°9W 10 A5
Fair Isle U.K. 59°32N 1°38W 14 B6
Fair Oaks U.S.A. 38°39N 121°16W 78 G5
Fairbanks U.S.A. 64°51N 147°43W 68 E2
Fairbury U.S.A. 40°8N 97°11W 80 E5
Fairfax U.S.A. 44°40N 73°1W 83 B11
Fairfield Ala., U.S.A. 33°29N 86°55W 85 E11
Fairfield Calif., U.S.A. 38°15N 122°3W 78 G4
Fairfield Conn., U.S.A. 41°9N 73°16W 83 E11
Fairfield Idaho, U.S.A. 43°21N 114°44W 76 E6
Fairfield Ill., U.S.A. 38°23N 88°22W 80 F9
Fairfield Iowa, U.S.A. 40°56N 91°57W 80 E8
Fairfield Tex., U.S.A. 31°44N 96°10W 84 F6
Fairford Canada 51°37N 98°38W 71 C9
Fairhope U.S.A. 30°31N 87°54W 85 F11
Fairlie N.Z. 44°5S 170°49E 59 F3
Fairmead U.S.A. 37°5N 120°10W 78 H6
Fairmont Minn., U.S.A. 43°39N 94°28W 80 D6
Fairmont W. Va., U.S.A. 39°29N 80°9W 81 F13
Fairmount Calif., U.S.A. 34°45N 118°26W 79 L8
Fairmount N.Y., U.S.A. 43°5N 76°12W 83 C8
Fairplay U.S.A. 39°15N 106°2W 76 G10
Fairport U.S.A. 43°6N 77°27W 82 C7
Fairport Harbor U.S.A. 41°45N 81°17W 82 E3
Fairview Canada 56°5N 118°25W 70 B5
Fairview Mont., U.S.A. 47°51N 104°3W 76 C11
Fairview Okla., U.S.A. 36°16N 98°29W 84 C5
Fairweather, Mt. U.S.A. 58°55N 137°32W 70 B1
Faisalabad Pakistan 31°30N 73°5E 42 D5
Faith U.S.A. 45°2N 102°2W 80 C2
Faizabad India 26°45N 82°10E 43 F10
Fajardo Puerto Rico 18°20N 65°39W 89 d
Fajr, W. → Si. Arabia 29°10N 38°10E 44 D3
Fakenham U.K. 52°51N 0°51E 12 E8
Fakfak Indonesia 2°55S 132°18E 37 E8
Faku China 42°32N 123°21E 35 C12
Falaise France 48°54N 0°12W 20 B3
Falam Burma 23°0N 93°45E 41 H18
Falcó, C. des Spain 38°50N 1°23E 24 C7
Falcón, Presa Mexico 26°35N 99°10W 87 B5
Falcon Lake Canada 49°42N 95°15W 71 D9
Falcon Res. U.S.A. 26°34N 99°10W 84 H5
Falconara Maríttima Italy 43°37N 13°24E 22 C5
Falcone, C. del Italy 40°58N 8°12E 22 D3
Falconer U.S.A. 42°7N 79°12W 82 D5
Falefa Samoa 13°54S 171°31W 59 b
Falelatai Samoa 13°55S 171°59W 59 b
Falelima Samoa 13°32S 172°41W 59 b
Faleshty = Fălești Moldova 47°32N 27°44E 17 E14
Fălești Moldova 47°32N 27°44E 17 E14
Falfurrias U.S.A. 27°14N 98°9W 84 H5
Falher Canada 55°44N 117°15W 70 B5
Faliraki Greece 36°22N 28°12E 25 C10
Falkenberg Sweden 56°54N 12°30E 9 H15
Falkirk U.K. 56°0N 3°47W 11 F5
Falkirk □ U.K. 55°58N 3°49W 11 F5
Falkland U.K. 56°16N 3°12W 11 E5
Falkland Is. ☑ Atl. Oc. 51°30S 59°0W 96 G5
Falkland Sd. Falk. Is. 52°0S 60°0W 96 G5
Fall River U.S.A. 41°43N 71°10W 83 E13
Fallbrook U.S.A. 33°23N 117°15W 79 M9
Fallon U.S.A. 39°28N 118°47W 76 G4
Falls City U.S.A. 40°3N 95°36W 80 E6
Falls Creek U.S.A. 41°9N 78°48W 82 E6
Falmouth Jamaica 18°30N 77°40W 88 a
Falmouth U.K. 50°9N 5°5W 13 G2
Falmouth U.S.A. 41°33N 70°37W 83 E14
Falsa, Pta. Mexico 27°51N 115°3W 86 B1
False B. S. Africa 34°15S 18°40E 56 E2
Falso, C. Honduras 15°12N 83°21W 88 C3
Falster Denmark 54°45N 11°55E 9 J14
Fălticeni Romania 47°21N 26°20E 17 E14
Falun Sweden 60°37N 15°37E 8 F16
Famagusta Cyprus 35°8N 33°55E 25 D12
Famagusta Bay Cyprus 35°15N 34°0E 25 D13
Famatina, Sierra de Argentina 27°30S 68°0W 94 B2
Family L. Canada 51°54N 95°27W 71 C9
Famoso U.S.A. 35°37N 119°12W 79 K7
Fan Xian China 35°55N 115°38E 34 G8
Fanad Hd. Ireland 55°17N 7°38W 10 A4
Fandriana Madag. 20°14S 47°21E 57 C8
Fang Thailand 19°55N 99°13E 38 C2
Fangcheng China 33°18N 112°59E 34 H7
Fangshan Beijing, China 39°41N 116°0E 34 E9
Fangshan Shanxi, China 38°3N 111°25E 34 E6
Fangzi China 36°33N 119°10E 35 F10
Fanjakana Madag. 21°10S 46°53E 57 C8
Fanjiatun China 43°40N 125°15E 35 C13
Fanling China 22°30N 114°8E 33 a
Fannich, L. U.K. 57°38N 4°59W 11 D4
Fannūj Iran 26°35N 59°38E 45 E8
Fanø Denmark 55°25N 8°25E 9 J13
Fano Italy 43°50N 13°1E 22 C5
Fanshi China 39°12N 113°20E 34 E7
Fao = Al Fāw Iraq 30°0N 48°30E 45 D6
Faqirwali Pakistan 29°27N 73°0E 42 E5
Far East = Dalnevostochnyy □ Russia 67°0N 140°0E 29 C14
Far East Asia 40°0N 130°0E 26 E14
Faradje Dem. Rep. of the Congo 3°50N 29°45E 54 B2
Farafangana Madag. 22°49S 47°50E 57 C8
Farāh Afghan. 32°20N 62°7E 40 C3
Farāh □ Afghan. 32°25N 62°10E 40 C3
Farahalana Madag. 14°26S 50°10E 57 A9
Faranah Guinea 10°3N 10°45W 50 F3
Farasān, Jazā'ir Si. Arabia 16°45N 41°55E 47 D3
Farasan Is. = Farasān, Jazā'ir Si. Arabia 16°45N 41°55E 47 D3
Faratsiho Madag. 19°24S 46°57E 57 B8
Fareham U.K. 50°51N 1°11W 13 G6
Farewell, C. N.Z. 40°29S 172°43E 59 D4
Farewell C. = Nunap Isua Greenland 59°48N 43°55W 66 D15
Farghona Uzbekistan 40°23N 71°19E 28 E8
Fargo U.S.A. 46°53N 96°48W 80 B5
Fār'iah, W. al → West Bank 32°12N 35°27E 46 C4
Faribault U.S.A. 44°18N 93°16W 80 C7
Faridabad India 28°26N 77°19E 42 E6
Faridkot India 30°44N 74°45E 42 D6
Faridpur Bangla. 23°15N 89°55E 43 H13
Faridpur India 28°13N 79°33E 43 E8
Farīmān Iran 35°40N 59°49E 45 C8
Fariones, Pta. Canary Is. 29°13N 13°28W 24 E6
Farleigh Australia 21°4S 149°8E 62 b
Farmerville U.S.A. 32°47N 92°24W 84 E8
Farmingdale U.S.A. 40°12N 74°10W 83 F10
Farmington Canada 55°54N 120°30W 70 B4
Farmington Calif., U.S.A. 37°55N 120°59W 78 H6
Farmington Maine, U.S.A. 44°40N 70°9W 81 C18
Farmington Mo., U.S.A. 37°47N 90°25W 80 G8
Farmington N.H., U.S.A. 43°24N 71°4W 83 C13
Farmington N. Mex., U.S.A. 36°44N 108°12W 77 H9
Farmington Utah, U.S.A. 40°59N 111°53W 76 F8
Farmington → U.S.A. 41°51N 72°38W 83 E12
Farmville U.S.A. 37°18N 78°24W 81 G14
Farne Is. U.K. 55°38N 1°37W 12 B6
Farnham Canada 45°17N 72°59W 83 A12
Farnham, Mt. Canada 50°29N 116°30W 70 C5
Faro Brazil 2°10S 56°39W 93 D7
Faro Canada 62°11N 133°22W 68 E5
Faro Portugal 37°2N 7°55W 21 D2
Fårö Sweden 57°55N 19°5E 9 H18
Farquhar, C. Australia 23°50S 113°36E 61 D1
Farrars Cr. → Australia 25°35S 140°43E 62 D3
Farrāshband Iran 28°57N 52°5E 45 D7
Farrell U.S.A. 41°13N 80°30W 82 E4
Farrokhī Iran 33°50N 59°31E 45 C8
Farruch, C. = Ferrutx, C. de Spain 39°47N 3°21E 24 B10
Farrukhabad India 27°24N 79°34E 43 F8
Fārs □ Iran 29°30N 55°0E 45 D7
Farsala Greece 39°17N 22°23E 23 E10
Fārsī Iran 27°58N 50°11E 45 E6
Farson U.S.A. 42°7N 109°26W 76 E9
Farsund Norway 58°5N 6°55E 9 G12
Fartak, Râs Si. Arabia 28°5N 34°34E 44 D2
Fartak, Ra's Yemen 15°38N 52°15E 47 D5
Fartura, Serra da Brazil 26°21S 52°52W 95 B5
Fārūj Iran 37°14N 58°14E 45 B8
Farvel, Kap = Nunap Isua Greenland 59°48N 43°55W 66 D15
Farwell U.S.A. 34°23N 103°2W 84 D3
Fāryāb □ Afghan. 36°0N 65°0E 40 B4
Fasā Iran 29°0N 53°39E 45 D7
Fasano Italy 40°50N 17°22E 22 D7
Fastiv Ukraine 50°7N 29°57E 17 C15
Fastnet Rock Ireland 51°22N 9°37W 10 E2
Fastov = Fastiv Ukraine 50°7N 29°57E 17 C15
Fatagar, Tanjung Indonesia 2°46S 131°57E 37 E8
Fatehabad Haryana, India 29°31N 75°27E 42 E6
Fatehabad Ut. P., India 27°1N 78°19E 42 F8
Fatehgarh India 27°25N 79°35E 43 F8
Fatehpur Bihar, India 24°38N 85°14E 43 G11
Fatehpur Raj., India 28°0N 74°40E 42 F6
Fatehpur Ut. P., India 25°56N 81°13E 43 G9
Fatehpur Ut. P., India 27°10N 81°13E 43 F9
Fatehpur Sikri India 27°6N 77°40E 42 F6
Fathom Five ⌂ Canada 45°17N 81°40W 82 A3
Fatima Canada 47°24N 61°53W 73 C7
Faulkton U.S.A. 45°2N 99°8W 80 C4
Faure I. Australia 25°52S 113°50E 61 E1
Fauresmith S. Africa 29°44S 25°17E 56 D4
Fauske Norway 67°17N 15°25E 8 C16
Favara Italy 37°19N 13°39E 22 F5
Faváritx, C. de Spain 40°0N 4°15E 24 B11
Favignana Italy 37°56N 12°20E 22 F5
Fawcett, Pt. Australia 11°46S 130°2E 60 B5
Fawn → Canada 55°20N 87°35W 72 A2
Fawnskin U.S.A. 34°16N 116°56W 79 L10
Faxaflói Iceland 64°29N 23°0W 8 D2
Faya-Largeau Chad 17°58N 19°6E 51 E9
Fayd Si. Arabia 27°1N 42°52E 44 E4
Fayette Ala., U.S.A. 33°41N 87°50W 85 E11
Fayette Mo., U.S.A. 39°9N 92°41W 80 F7
Fayette N.Y., U.S.A. 42°48N 76°48W 83 D8
Fayetteville Ark., U.S.A. 36°4N 94°10W 84 C7
Fayetteville N.C., U.S.A. 35°3N 78°53W 85 D15
Fayetteville N.Y., U.S.A. 43°1N 76°0W 83 C9
Fayetteville Tenn., U.S.A. 35°9N 86°34W 85 D11
Faylakah Kuwait 29°27N 48°20E 45 D6
Fazilka India 30°27N 74°2E 42 D6
Fazilpur Pakistan 29°18N 70°29E 42 E4
Fdérik Mauritania 22°40N 12°45W 50 D3
Feakle Ireland 52°56N 8°40W 10 D3
Feale → Ireland 52°27N 9°37W 10 D2
Fear, C. U.S.A. 33°50N 77°58W 85 E16
Feather → U.S.A. 38°47N 121°36W 76 G3
Feather Falls U.S.A. 39°36N 121°16W 78 F5
Featherston N.Z. 41°6S 175°20E 59 D5
Featherstone Zimbabwe 18°42S 30°55E 55 F3
Fécamp France 49°45N 0°22E 20 B4

Fedala = Mohammedia
Morocco 33°44N 7°21W 50 B4
Federación Argentina 31°0S 57°55W 94 C4
Féderal Argentina 30°57S 58°48W 96 C5
Federal Way U.S.A. 47°18N 122°19W 78 C4
Fedeshkūh Iran 28°49N 53°50E 45 D7
Fehmarn Germany 54°27N 11°7E 16 A6
Fehmarn Bælt Europe 54°35N 11°20E 9 J14
Fehmarn Belt = Fehmarn Bælt
Europe 54°35N 11°20E 9 J14
Fei Xian China 35°18N 117°59E 35 G9
Feicheng China 36°14N 116°45E 34 F9
Feijó Brazil 8°9S 70°21W 92 E4
Feilding N.Z. 40°13S 175°35E 59 D5
Feira de Santana Brazil 12°15S 38°57W 93 F11
Feixiang China 36°30N 114°45E 34 F8
Felanitx Spain 39°28N 3°9E 24 B10
Feldkirch Austria 47°15N 9°37E 16 E5
Félicité Seychelles 4°19S 55°52E 53 b
Felipe Carrillo Puerto
Mexico 19°38N 88°3W 87 D7
Felixburg Zimbabwe 19°29S 30°51E 57 B5
Felixstowe U.K. 51°58N 1°23E 13 F9
Felton U.S.A. 37°3N 122°4W 78 H4
Femer Bælt = Fehmarn Bælt
Europe 54°35N 11°20E 9 J14
Femunden Norway 62°10N 11°53E 8 E14
Fen He → China 35°36N 110°42E 34 G6
Fenelon Falls Canada 44°32N 78°45W 82 B6
Feng Xian Jiangsu, China 34°43N 116°35E 34 G9
Feng Xian Shaanxi,
China 33°54N 106°40E 34 H4
Fengcheng China 40°28N 124°5E 35 D13
Fengfeng China 36°28N 114°8E 34 F8
Fengning China 41°10N 116°33E 34 D9
Fengqiu China 35°2N 114°25E 34 G8
Fengrun China 39°48N 118°8E 35 E10
Fengxiang China 34°29N 107°25E 34 G4
Fengyang China 32°51N 117°29E 35 H9
Fengzhen China 40°25N 113°2E 34 D7
Fenoarivo Fianarantsoa,
Madag. 21°43S 46°24E 57 C8
Fenoarivo Fianarantsoa,
Madag. 20°52S 46°53E 57 C8
Fenoarivo Afovoany
Madag. 18°26S 46°34E 57 B8
Fenoarivo Atsinanana
Madag. 17°22S 49°25E 57 B8
Fens, The U.K. 52°38N 0°2W 12 E7
Fenton U.S.A. 42°48N 83°42W 81 D12
Fenxi China 36°40N 111°31E 34 F6
Fenyang China 37°18N 111°48E 34 F6
Feodosiya Ukraine 45°2N 35°16E 19 E6
Ferbane Ireland 53°16N 7°50W 10 C4
Ferdows Iran 33°58N 58°2E 45 C8
Ferfer Somali Rep. 5°4N 45°9E 47 F4
Fergana = Farghona
Uzbekistan 40°23N 71°19E 28 E8
Fergus Canada 43°43N 80°24W 82 C4
Fergus Falls U.S.A. 46°17N 96°4W 80 B5
Ferizaj = Uroševac
Kosovo 42°23N 21°10E 23 C9
Ferkéssédougou Ivory C. 9°35N 5°6W 50 G4
Ferland Canada 50°19N 88°27W 72 B2
Ferlo, Vallée du Senegal 15°15N 14°15W 50 E3
Fermanagh □ U.K. 54°21N 7°40W 10 B4
Fermo Italy 43°9N 13°43E 22 C5
Fermont Canada 52°47N 67°5W 73 B6
Fermoy Ireland 52°9N 8°16W 10 D3
Fernández Argentina 27°55S 63°50W 94 B3
Fernandina Beach
U.S.A. 30°40N 81°27W 85 F14
Fernando de Noronha
Brazil 4°0S 33°10W 93 D12
Fernando Póo = Bioko
Eq. Guin. 3°30N 8°40E 52 D1
Ferndale Canada 44°58N 81°17W 82 B3
Ferndale U.S.A. 48°51N 122°36W 78 B4
Fernie Canada 49°30N 115°5W 70 D5
Fernlees Australia 23°51S 148°7E 62 C4
Fernley U.S.A. 39°36N 119°15W 76 G4
Fernwood U.S.A. 43°16N 73°38W 83 C11
Ferozepore = Firozpur
India 30°55N 74°40E 42 D6
Ferrara Italy 44°50N 11°35E 22 B4
Ferreñafe Peru 6°42S 79°50W 92 E3
Ferrerías Spain 39°59N 4°1E 24 B11
Ferret, C. France 44°38N 1°15W 20 D3
Ferriday U.S.A. 31°38N 91°33W 84 F9
Ferrol Spain 43°29N 8°15W 21 A1
Ferron U.S.A. 39°5N 111°8W 76 G8
Ferrutx, C. de Spain 39°47N 3°21E 24 B10
Ferryland Canada 47°2N 52°53W 73 C9
Fertile U.S.A. 47°32N 96°17W 80 B5
Fès Morocco 34°0N 5°0W 50 B5
Fessenden U.S.A. 47°39N 99°38W 80 B4
Festus U.S.A. 38°13N 90°24W 80 F8
Feteşti Romania 44°22N 27°51E 17 F14
Fethiye Turkey 36°36N 29°6E 19 G4
Fetlar U.K. 60°36N 0°52W 11 A8
Feuilles → Canada 58°47N 70°4W 69 F17
Fez = Fès Morocco 34°0N 5°0W 50 B5
Fezzan Libya 27°0N 13°0E 51 C8
Fiambalá Argentina 27°45S 67°37W 94 B2
Fianarantsoa Madag. 21°26S 47°5E 57 C8
Fianarantsoa □ Madag. 22°30S 47°0E 57 C8
Ficksburg S. Africa 28°51S 27°53E 57 D4
Field → Australia 23°48S 138°0E 62 C2
Field I. Australia 12°5S 132°23E 60 B5
Fier Albania 40°43N 19°33E 23 D8
Fife □ U.K. 56°16N 3°1W 11 E5
Fife ⌂ U.K. 56°15N 3°15W 11 E5
Fife Ness U.K. 56°17N 2°35W 11 E6
Fifth Cataract Sudan 18°22N 33°50E 51 E12
Figeac France 44°37N 2°2E 20 D5
Figtree Zimbabwe 20°22S 28°20E 55 G2
Figueira da Foz Portugal 40°7N 8°54W 21 B1
Figueres Spain 42°18N 2°58E 21 A7
Figuig Morocco 32°5N 1°11W 50 B5
Fihaonana Madag. 18°36S 47°12E 57 B8
Fiherenana Madag. 18°29S 48°24E 57 B8
Fiherenana → Madag. 23°19S 43°37E 57 C7
Fiji ■ Pac. Oc. 17°20S 179°0E 59 a
Filabus Zimbabwe 20°34S 29°20E 57 C4
Filadelfia Paraguay 22°21S 60°2W 94 A3
Filchner Ice Shelf Antarctica 79°0S 40°0W 5 D1
Filey U.K. 54°12N 0°18W 12 C7
Filey B. U.K. 54°12N 0°15W 12 C7
Filfla Malta 35°47N 14°24E 25 D1
Filiatra Greece 37°9N 21°35E 23 F9
Filingué Niger 14°21N 3°22E 50 F6
Filipstad Sweden 59°43N 14°9E 9 G16
Fillmore Calif., U.S.A. 34°24N 118°55W 79 L8
Fillmore Utah, U.S.A. 38°58N 112°20W 76 G7
Fimbul Ice Shelf S. Ocean 69°30S 1°0W 5 C2
Finch Canada 45°11N 75°7W 83 A9
Finch Hatton Australia 20°25S 148°39E 62 b
Findhorn → U.K. 57°38N 3°38W 11 D5
Findlay U.S.A. 41°2N 83°39W 81 E12
Fine U.S.A. 44°14N 75°8W 83 B9
Finger L. Canada 53°33N 93°30W 72 B1
Finger Lakes U.S.A. 42°40N 76°30W 83 D8
Fingoè Mozam. 15°55S 31°50E 55 F3
Finisterre, C. = Fisterra, C.
Spain 42°50N 9°19W 21 A1
Finke Australia 25°34S 134°35E 62 D1
Finke Gorge △ Australia 24°8S 132°49E 60 D5
Finland ■ Europe 63°0N 27°0E 8 E22
Finland, G. of Europe 60°0N 26°0E 9 G22
Finlay → Canada 57°0N 125°10W 70 B3
Finley Australia 35°38S 145°35E 63 F4
Finley U.S.A. 47°31N 97°50W 80 B5
Finn → Ireland 54°51N 7°28W 10 B4
Finnigan, Mt. Australia 15°49S 145°17E 62 B4
Finniss, C. Australia 33°8S 134°51E 63 E1
Finnmark Norway 69°37N 23°57E 8 B20
Finnsnes Norway 69°14N 18°0E 8 B18
Finspång Sweden 58°43N 15°47E 9 G16
Fiora → Italy 42°20N 11°34E 22 C4
Fiordland △ N.Z. 45°46S 167°0E 59 F1
Fīq Syria 32°46N 35°41E 46 C4
Firat = Furāt, Nahr al →
Asia 31°0N 47°25E 44 D5
Fire Island ⌂ U.S.A. 40°38N 73°8W 83 F11
Firebag → Canada 57°45N 111°21W 71 B6
Firebaugh U.S.A. 36°52N 120°27W 78 J6
Firedrake L. Canada 61°25N 104°30W 71 A8
Firenze Italy 43°46N 11°15E 22 C4
Firk, Sha'ib → Iraq 30°59N 44°34E 44 D5
Firozabad India 27°10N 78°25E 43 F8
Firozpur India 30°55N 74°40E 42 D6
Firozpur-Jhirka India 27°48N 76°57E 42 F7
Fīrūzābād Iran 28°52N 52°35E 45 D7
Fīrūzkūh Iran 35°50N 52°50E 45 C7
Firvale Canada 52°27N 126°13W 70 C3
Fish → Namibia 28°7S 17°10E 56 D2
Fish → S. Africa 31°30S 20°16E 56 E3
Fish River Canyon
Namibia 27°40S 17°35E 56 D2
Fisher B. Canada 51°35N 97°13W 71 C9
Fishers I. U.S.A. 41°15N 72°0W 83 E13
Fishguard U.K. 52°0N 4°58W 13 E3
Fishing L. Canada 52°10N 95°24W 71 C9
Fishkill U.S.A. 41°32N 73°54W 83 E11
Fisterra, C. Spain 42°50N 9°19W 21 A1
Fitchburg Mass., U.S.A. 42°35N 71°48W 83 D13
Fitchburg Wis., U.S.A. 42°58N 89°28W 80 D9
Fitz Roy Argentina 47°0S 67°0W 96 F3
Fitz Roy, Cerro Argentina 49°17S 73°5W 96 F2
Fitzgerald Canada 59°51N 111°36W 70 B6
Fitzgerald U.S.A. 31°43N 83°15W 85 F13
Fitzgerald River △
Australia 33°53S 119°55E 61 F3
Fitzmaurice → Australia 14°45S 130°5E 60 B5
Fitzroy → Queens.,
Australia 23°32S 150°52E 62 C5
Fitzroy → W. Austral.,
Australia 17°31S 123°35E 60 C3
Fitzroy Crossing
Australia 18°9S 125°38E 60 C4
Fitzwilliam I. Canada 45°30N 81°45W 82 A3
Fiume = Rijeka Croatia 45°20N 14°21E 16 F8
Five Points U.S.A. 36°26N 120°6W 78 J6
Fizi Dem. Rep. of the Congo 4°17S 28°55E 54 C2
Flagstaff U.S.A. 35°12N 111°39W 77 J8
Flagstaff L. U.S.A. 45°12N 70°18W 83 A14
Flaherty I. Canada 56°15N 79°15W 72 A4
Flåm Norway 60°50N 7°7E 8 F12
Flambeau → U.S.A. 45°18N 91°14W 80 C8
Flamborough Hd. U.K. 54°7N 0°5W 12 C7
Flaming Gorge ⌂
U.S.A. 41°10N 109°25W 76 F9
Flaming Gorge Res.
U.S.A. 41°10N 109°25W 76 F9
Flamingo, Teluk Indonesia 5°30S 138°0E 37 F9
Flanders = Flandre
Europe 50°50N 2°30E 15 B9
Flandre Europe 50°50N 2°30E 15 B9
Flandre-Occidentale = West-
Vlaanderen □ Belgium 51°0N 3°0E 15 D2
Flandre-Orientale = Oost-
Vlaanderen □ Belgium 51°5N 3°50E 15 C3
Flandreau U.S.A. 44°3N 96°36W 80 C5
Flanigan U.S.A. 40°10N 119°53W 78 E7
Flannan Is. U.K. 58°9N 7°52W 11 C1
Flåsjön Sweden 64°5N 15°40E 8 D16
Flat → Canada 61°33N 125°18W 70 A3
Flat I. Mauritius 19°53S 57°35E 53 d
Flathead L. U.S.A. 47°51N 114°8W 76 C6
Flattery, C. Australia 14°58S 145°21E 62 A4
Flattery, C. U.S.A. 48°23N 124°29W 78 B2
Flatwoods U.S.A. 38°31N 82°43W 81 F12
Fleetwood U.K. 53°55N 3°1W 12 D4
Fleetwood U.S.A. 40°27N 75°49W 83 F9
Flekkefjord Norway 58°18N 6°39E 9 G12
Flemington U.S.A. 41°7N 77°28W 82 E7
Flensburg Germany 54°47N 9°27E 16 A5
Flers France 48°47N 0°33W 20 B3
Flesherton Canada 44°16N 80°33W 82 B4
Flesko, Tanjung
Indonesia 0°29N 124°30E 37 D6
Fleurieu Pen. Australia 35°40S 138°5E 63 F2
Flevoland □ Neths. 52°30N 5°30E 15 B5
Flin Flon Canada 54°46N 101°53W 71 C8
Flinders → Australia 17°36S 140°36E 62 B3
Flinders B. Australia 34°19S 115°19E 61 F2
Flinders Group Australia 14°11S 144°15E 62 A3
Flinders I. S. Austral.,
Australia 33°44S 134°41E 63 E1
Flinders I. Tas., Australia 40°0S 148°0E 63 G4
Flinders Ranges
Australia 31°30S 138°30E 63 E2
Flinders Reefs Australia 17°37S 148°31E 62 B4
Flint U.K. 53°15N 3°8W 12 D4
Flint U.S.A. 43°1N 83°41W 81 D12
Flint → U.S.A. 30°57N 84°34W 85 F12
Flint I. Kiribati 11°26S 151°48W 65 J12
Flintshire □ U.K. 53°17N 3°17W 12 D4
Flodden U.K. 55°37N 2°8W 12 B5
Floodwood U.S.A. 46°56N 92°55W 80 B7
Flora U.S.A. 38°40N 88°29W 80 F9
Florala U.S.A. 31°0N 86°20W 85 F11
Florence = Firenze Italy 43°46N 11°15E 22 C4
Florence Ala., U.S.A. 34°48N 87°41W 85 D11
Florence Ariz., U.S.A. 33°2N 111°23W 77 K8
Florence Colo., U.S.A. 38°23N 105°8W 76 G11
Florence Oreg., U.S.A. 43°58N 124°7W 76 D1
Florence S.C., U.S.A. 34°12N 79°46W 85 D15
Florence, L. Australia 28°53S 138°9E 63 D2
Florencia Colombia 1°36N 75°36W 92 C3
Florennes Belgium 50°15N 4°35E 15 D4
Florenville Belgium 49°40N 5°19E 15 E5
Flores Azores 39°26N 31°13W 50 a
Flores Guatemala 16°59N 89°50W 88 C2
Flores Indonesia 8°35S 121°0E 37 F6
Flores I. Canada 49°20N 126°10W 70 D3
Flores Sea Indonesia 6°30S 120°0E 37 F6
Floreşti Moldova 47°53N 28°17E 17 E15
Floresville U.S.A. 29°8N 98°10W 84 G5
Floriano Brazil 6°50S 43°0W 93 E10
Florianópolis Brazil 27°30S 48°30W 95 B6
Florida Cuba 21°32N 78°14W 88 B4
Florida Uruguay 34°7S 56°10W 95 C4
Florida □ U.S.A. 28°0N 82°0W 85 G14
Florida, Straits of U.S.A. 25°0N 80°0W 88 B4
Florida B. U.S.A. 25°0N 80°45W 88 B3
Florida Keys U.S.A. 24°40N 81°0W 85 J14
Florin U.S.A. 38°30N 121°24W 78 G5
Florina Greece 40°48N 21°26E 23 D9
Florissant U.S.A. 38°47N 90°19W 80 F8
Florø Norway 61°35N 5°1E 8 F11
Flower Station Canada 45°10N 76°41W 83 A8
Flowerpot I. Canada 45°18N 81°38W 82 A3
Floydada U.S.A. 33°59N 101°20W 84 E4
Fluk Indonesia 1°42S 127°44E 37 E7
Flushing = Vlissingen
Neths. 51°26N 3°34E 15 C3
Fly → Papua N. G. 8°25S 143°0E 58 B7
Flying Fish, C. Antarctica 72°6S 102°29W 5 D15
Foa Tonga 19°45S 174°18W 59 c
Foam Lake Canada 51°40N 103°32W 71 C8
Foça Turkey 38°39N 26°46E 23 E12
Fochabers U.K. 57°37N 3°6W 11 D5
Focşani Romania 45°41N 27°15E 17 F14
Fóggia Italy 41°27N 15°34E 22 D6
Fogo Canada 49°43N 54°17W 73 C9
Fogo C. Verde Is. 15°5N 24°20W 50 b
Fogo I. Canada 49°40N 54°5W 73 C9
Föhr Germany 54°43N 8°30E 16 A5
Foix France 42°58N 1°38E 20 E4
Folda Nord-Trøndelag,
Norway 64°32N 10°30E 8 D14
Folda Nordland, Norway 67°38N 14°50E 8 C16
Foley Botswana 21°34S 27°21E 56 C4
Foley U.S.A. 30°24N 87°41W 85 F11
Foleyet Canada 48°15N 82°25W 72 C3
Folgefonna Norway 60°3N 6°23E 9 F12
Foligno Italy 42°57N 12°42E 22 C5
Folkestone U.K. 51°5N 1°12E 13 F9
Folkston U.S.A. 30°50N 82°0W 85 F13
Follansbee U.S.A. 40°19N 80°35W 82 F4
Folsom L. U.S.A. 38°42N 121°9W 78 G5
Fomboni Comoros Is. 12°18S 43°46E 53 a
Fond-du-Lac Canada 59°19N 107°12W 71 B7
Fond du Lac U.S.A. 43°47N 88°27W 80 D9
Fond-du-Lac → Canada 59°17N 106°0W 71 B7
Fonda U.S.A. 42°57N 74°22W 83 D10
Fondi Italy 41°21N 13°25E 22 D5
Fongafale Tuvalu 8°31S 179°13E 58 B10
Fonsagrada = A Fonsagrada
Spain 43°8N 7°4W 21 A2
Fonseca, G. de
Cent. Amer. 13°10N 87°40W 88 D2
Fontainebleau France 48°24N 2°40E 20 B5
Fontana U.S.A. 34°6N 117°26W 79 L9
Fontas → Canada 58°14N 121°48W 70 B4
Fonte Boa Brazil 2°33S 66°0W 92 D5
Fontenay-le-Comte
France 46°28N 0°48W 20 C3
Fontenelle Res. U.S.A. 42°1N 110°3W 76 E8
Fontur Iceland 66°23N 14°32W 8 C6
Fonuafo'ou Tonga 20°19S 175°25W 59 c
Fonualei Tonga 18°1S 174°19W 59 c
Foochow = Fuzhou
China 26°5N 119°16E 33 F12
Foping China 33°41N 108°0E 34 H5
Foraker, Mt. U.S.A. 62°58N 151°24W 74 C9
Forbes Australia 33°22S 148°5E 63 E4
Forbesganj India 26°17N 87°18E 43 F12
Ford City Calif., U.S.A. 35°9N 119°27W 79 K7
Ford City Pa., U.S.A. 40°46N 79°32W 82 F5
Førde Norway 61°27N 5°53E 8 F11
Fords Bridge Australia 29°41S 145°29E 63 D4
Fordyce U.S.A. 33°49N 92°25W 84 E8
Forel, Mt. Greenland 66°52N 36°55W 4 C6
Foremost Canada 49°26N 111°34W 70 D6
Forest Canada 43°6N 82°0W 82 C3
Forest U.S.A. 32°22N 89°29W 85 E10
Forest City Iowa, U.S.A. 43°16N 93°39W 80 D7
Forest City N.C., U.S.A. 35°20N 81°52W 85 D14
Forest City Pa., U.S.A. 41°39N 75°28W 83 E9
Forest Grove U.S.A. 45°31N 123°7W 78 E3
Forestburg Canada 52°35N 112°1W 70 C6
Foresthill U.S.A. 39°1N 120°49W 78 F6
Forestier Pen. Australia 43°0S 148°0E 63 G4
Forestville Canada 48°48N 69°2W 73 C6
Forestville Calif., U.S.A. 38°28N 122°54W 78 G4
Forestville N.Y., U.S.A. 42°28N 79°10W 82 D5
Forfar U.K. 56°39N 2°53W 11 E6
Forillon △ Canada 48°46N 64°12W 73 C7
Forks U.S.A. 47°57N 124°23W 78 C2
Forksville U.S.A. 41°29N 76°35W 83 E8
Forlì Italy 44°13N 12°3E 22 B5
Forman U.S.A. 46°7N 97°38W 80 B5
Formby Pt. U.K. 53°33N 3°6W 12 D4
Formentera Spain 38°43N 1°27E 24 C7
Formentor, C. de Spain 39°58N 3°13E 24 B10
Formentor, Pen. de
Spain 39°56N 3°11E 24 B10
Former Yugoslav Republic of
Macedonia = Macedonia ■
Europe 41°53N 21°40E 23 D9
Fórmia Italy 41°15N 13°37E 22 D5
Formosa = Taiwan ■
Asia 23°30N 121°0E 33 G13
Formosa Argentina 26°15S 58°10W 94 B4
Formosa Brazil 15°32S 47°20W 93 G9
Formosa □ Argentina 25°0S 60°0W 94 B4
Formosa, Serra Brazil 12°0S 55°0W 93 F8
Formosa B. = Ungwana B.
Kenya 2°45S 40°20E 54 C5
Fornells Spain 40°3N 4°7E 24 A11
Føroyar ☑ Atl. Oc. 62°0N 7°0W 8 F9
Forres U.K. 57°37N 3°37W 11 D5
Forrest Australia 30°51S 128°6E 61 F4
Forrest, Mt. Australia 24°48S 127°45E 61 D4
Forrest City U.S.A. 35°1N 90°47W 85 D9
Forsayth Australia 18°33S 143°34E 62 B3
Forssa Finland 60°49N 23°38E 8 F20
Forst Germany 51°45N 14°37E 16 C8
Forster Australia 32°12S 152°31E 63 E5
Forsyth U.S.A. 46°16N 106°41W 76 C10
Fort Abbas Pakistan 29°12N 72°52E 42 E5
Fort Albany Canada 52°15N 81°35W 72 B3
Fort Assiniboine
Canada 54°20N 114°45W 70 C6
Fort Augustus U.K. 57°9N 4°42W 11 D4
Fort Beaufort S. Africa 32°46S 26°40E 56 E4
Fort Benton U.S.A. 47°49N 110°40W 76 C8
Fort Bragg U.S.A. 39°26N 123°48W 76 G2
Fort Bridger U.S.A. 41°19N 110°23W 76 F8
Fort Chipewyan Canada 58°42N 111°8W 71 B6
Fort Clatsop = Lewis and Clark ⌂
U.S.A. 46°8N 123°53W 78 D3
Fort Collins U.S.A. 40°35N 105°5W 76 F11
Fort-Coulonge Canada 45°50N 76°45W 72 C4
Fort Covington U.S.A. 44°59N 74°29W 83 B10
Fort Dauphin = Taolanaro
Madag. 25°2S 47°0E 57 D8
Fort Davis U.S.A. 30°35N 103°54W 84 F3
Fort-de-France Martinique 14°36N 61°2W 88 c
Fort Defiance U.S.A. 35°45N 109°5W 77 J9
Fort Dodge U.S.A. 42°30N 94°11W 80 D6
Fort Edward U.S.A. 43°16N 73°35W 83 C11
Fort Erie Canada 42°54N 78°56W 82 D6
Fort Fairfield U.S.A. 46°46N 67°50W 81 B20
Fort Frances Canada 48°36N 93°24W 71 D10
Fort Franklin = Déline
Canada 65°11N 123°25W 68 D7
Fort Garland U.S.A. 37°26N 105°26W 77 H11
Fort George = Chisasibi
Canada 53°50N 79°0W 72 B4
Fort Good Hope Canada 66°14N 128°40W 68 D6
Fort Hancock U.S.A. 31°18N 105°51W 84 F2
Fort Hope = Eabametoong
Canada 51°30N 88°0W 72 B2
Fort Irwin U.S.A. 35°16N 116°41W 79 K10
Fort Kent U.S.A. 47°15N 68°36W 81 B19
Fort Klamath U.S.A. 42°42N 122°0W 76 D3
Fort Laramie U.S.A. 42°13N 104°31W 76 E11
Fort Lauderdale U.S.A. 26°7N 80°8W 85 H14
Fort Liard Canada 60°14N 123°30W 70 A4
Fort Liberté Haiti 19°42N 71°51W 89 C5
Fort Lupton U.S.A. 40°5N 104°49W 76 F11
Fort MacKay Canada 57°12N 111°41W 70 B6
Fort Macleod Canada 49°45N 113°30W 70 D6
Fort McMurray Canada 56°44N 111°7W 70 B6
Fort McPherson
Canada 67°30N 134°55W 68 D5
Fort Madison U.S.A. 40°38N 91°27W 80 E8
Fort Meade U.S.A. 27°45N 81°48W 85 H14
Fort Morgan U.S.A. 40°15N 103°48W 76 F12
Fort Myers U.S.A. 26°39N 81°52W 85 H14
Fort Nelson Canada 58°50N 122°44W 70 B4
Fort Nelson → Canada 59°32N 124°0W 70 B4
Fort Norman = Tulita
Canada 64°57N 125°30W 68 E6
Fort Payne U.S.A. 34°26N 85°43W 85 D12
Fort Peck U.S.A. 48°1N 106°27W 76 B10
Fort Peck Dam U.S.A. 48°0N 106°26W 76 C10
Fort Peck L. U.S.A. 48°0N 106°26W 76 C10
Fort Pierce U.S.A. 27°27N 80°20W 85 H14
Fort Pierre U.S.A. 44°21N 100°22W 80 C3
Fort Plain U.S.A. 42°56N 74°37W 83 D10
Fort Portal Uganda 0°40N 30°20E 54 B3
Fort Providence Canada 61°3N 117°40W 70 A5
Fort Qu'Appelle
Canada 50°45N 103°50W 71 C8
Fort Resolution Canada 61°10N 113°40W 70 A6
Fort Rixon Zimbabwe 20°2S 29°17E 55 G2
Fort Ross U.S.A. 38°32N 123°13W 78 G3
Fort Rupert = Waskaganish
Canada 51°30N 78°40W 72 B4
Fort St. James Canada 54°30N 124°10W 70 C4
Fort St. John Canada 56°15N 120°50W 70 B4
Fort Saskatchewan
Canada 53°40N 113°15W 70 C6
Fort Scott U.S.A. 37°50N 94°42W 80 G6
Fort Severn Canada 56°0N 87°40W 72 A2
Fort Shevchenko
Kazakhstan 44°35N 50°23E 19 F9
Fort Simpson Canada 61°45N 121°15W 7
Fort Smith Canada 60°0N 111°51W 7
Fort Smith U.S.A. 35°23N 94°25W 8
Fort Stockton U.S.A. 30°53N 102°53W 8
Fort Sumner U.S.A. 34°28N 104°15W 7
Fort Thompson U.S.A. 44°3N 99°26W 8
Fort Union ⌂ U.S.A. 35°54N 105°1W 77
Fort Valley U.S.A. 32°33N 83°53W 85
Fort Vermilion Canada 58°24N 116°0W 7
Fort Walton Beach
U.S.A. 30°25N 86°36W 85
Fort Ware Canada 57°26N 125°41W 7
Fort Wayne U.S.A. 41°4N 85°9W 81
Fort William U.K. 56°49N 5°7W 1
Fort Worth U.S.A. 32°43N 97°19W 8
Fort Yates U.S.A. 46°5N 100°38W 8
Fort Yukon U.S.A. 66°34N 145°16W 6
Fortaleza Brazil 3°45S 38°35W 93
Forteau Canada 51°28N 56°58W 7
Fortescue → Australia 21°0S 116°4E 6
Fortescue River Roadhouse
Australia 21°17S 116°8E 6
Forth → U.K. 56°9N 3°50W 1
Forth, Firth of U.K. 56°5N 2°55W 1
Fortrose U.K. 57°35N 4°9W 1
Fortuna Calif., U.S.A. 40°36N 124°9W 7
Fortuna N. Dak., U.S.A. 48°55N 103°47W 8
Fortune Canada 47°4N 55°50W 7
Fortune B. Canada 47°30N 55°22W 7
Forūr Iran 26°17N 54°32E 4
Fosen Norway 63°50N 10°20E 8
Foshan China 23°6N 113°9E 33
Fosnavåg Norway 62°22N 5°38E 8
Fossano Italy 44°33N 7°43E 2
Fossil U.S.A. 45°0N 120°9W 7
Fossil Butte ⌂ U.S.A. 41°50N 110°27W 7
Foster Canada 45°17N 72°30W 83
Foster → Canada 55°47N 105°49W 7
Fosters Ra. Australia 21°35S 133°48E 6
Fostoria U.S.A. 41°10N 83°25W 81
Fotadrevo Madag. 24°3S 45°1E 5
Fougères France 48°21N 1°14W 2
Foul Pt. Sri Lanka 8°35N 81°18E 4
Foula U.K. 60°10N 2°5W 1
Foulness I. U.K. 51°36N 0°55E 1
Foulpointe Madag. 17°41S 49°31E 5
Foumban Cameroon 5°45N 10°50E 5
Foumbouni Comoros Is. 11°51S 43°29E
Fountain U.S.A. 38°41N 104°42W 7
Fountain Hills U.S.A. 33°37N 111°43W 7
Fountain Springs
U.S.A. 35°54N 118°51W 7
Four Mountains, Is. of
U.S.A. 53°0N 170°0W 7
Fouriesburg S. Africa 28°38S 28°14E 5
Fourni Greece 37°36N 26°32E 2
Fouta Djallon Guinea 11°20N 12°10W 5
Foux, Cap-à- Haiti 19°43N 73°27W 8
Foveaux Str. N.Z. 46°42S 168°10E 5
Fowey U.K. 50°20N 4°39W 1
Fowler Calif., U.S.A. 36°38N 119°41W 7
Fowler Colo., U.S.A. 38°8N 104°2W 7
Fowler, Pt. Australia 32°2S 132°3E 6
Fowlers B. Australia 31°59S 132°34E 6
Fowman Iran 37°13N 49°19E 4
Fox → Canada 56°3N 93°18W 7
Fox Creek Canada 54°24N 116°48W 7
Fox Is. U.S.A. 53°0N 168°0W 7
Fox Lake Canada 58°28N 114°31W 7
Fox Valley Canada 50°30N 109°25W 7
Foxboro Canada 44°14N 77°26W 8
Foxboro U.S.A. 42°4N 71°16W 8
Foxdale Australia 20°22S 148°35E
Foxe Basin Canada 66°0N 77°0W 6
Foxe Chan. Canada 65°0N 80°0W 6
Foxe Pen. Canada 65°0N 76°0W 6
Foxford Ireland 53°59N 9°7W 1
Foxton N.Z. 40°29S 175°18E 5
Foyle, Lough U.K. 55°7N 7°4W 1
Foynes Ireland 52°37N 9°7W 1
Foz do Cunene Angola 17°15S 11°48E 5
Foz do Iguaçu Brazil 25°30S 54°30W 9
Frackville U.S.A. 40°47N 76°14W 8
Fraile Muerto Uruguay 32°31S 54°32W 9
Framingham U.S.A. 42°18N 71°24W 8
Franca Brazil 20°33S 47°30W 9
Francavilla Fontana
Italy 40°32N 17°35E 2
France ■ Europe 47°0N 3°0E 2
Frances Australia 36°41S 140°55E 6
Frances → Canada 60°16N 129°10W 7
Frances L. Canada 61°23N 129°30W 7
Franceville Gabon 1°40S 13°32E 5
Franche-Comté □ France 46°50N 5°55E 2
Francis Case, L. U.S.A. 43°4N 98°34W 8
Francisco Beltrão Brazil 26°5S 53°4W 9
Francisco Ignacio Madero
Coahuila, Mexico 25°48N 103°18W 86
Francisco Ignacio Madero Durango,
Mexico 24°26N 104°18W 86
Francisco Ignacio Madero, Presa
Mexico 28°10N 105°37W 86
Francistown Botswana 21°7S 27°33E 57
François Canada 47°35N 56°45W 73
François L. Canada 54°0N 125°30W 70
Francois Peron △
Australia 25°42S 113°33E 61
Francs Pk. U.S.A. 43°58N 109°20W 76
Franeker Neths. 53°12N 5°33E 15
Frank Hann △ Australia 32°52S 120°19E 61
Frankford Canada 44°12N 77°36W 82
Frankfort S. Africa 27°17S 28°30E 57
Frankfort Ind., U.S.A. 40°17N 86°31W 80
Frankfort Kans., U.S.A. 39°42N 96°25W 80
Frankfort Ky., U.S.A. 38°12N 84°52W 81
Frankfort N.Y., U.S.A. 43°2N 75°4W 83
Frankfurt Brandenburg,
Germany 52°20N 14°32E 16
Frankfurt Hessen, Germany 50°7N 8°41E 16
Fränkische Alb Germany 49°10N 11°23E 16

Frankland → Australia 35°0S 116°48E 61 G2
Franklin Ky., U.S.A. 36°43N 86°35W 80 G10
Franklin La., U.S.A. 29°48N 91°30W 84 G9
Franklin Mass., U.S.A. 42°5N 71°24W 83 D13
Franklin N.H., U.S.A. 43°27N 71°39W 83 C13
Franklin N.Y., U.S.A. 42°20N 75°9W 83 D9
Franklin Nebr., U.S.A. 40°6N 98°57W 80 E4
Franklin Pa., U.S.A. 41°24N 79°50W 82 E5
Franklin Va., U.S.A. 36°41N 76°56W 81 G15
Franklin W. Va., U.S.A. 38°39N 79°20W 81 F14
Franklin B. Canada 69°45N 126°0W 68 D6
Franklin D. Roosevelt L. U.S.A. 48°18N 118°9W 76 B4
Franklin-Gordon Wild Rivers △ Australia 42°19S 145°51E 63 G4
Franklin I. Antarctica 76°10S 168°30E 5 D11
Franklin I. Canada 45°24N 80°20W 82 A4
Franklin L. U.S.A. 40°25N 115°22W 76 F6
Franklin Mts. Canada 65°0N 125°0W 68 D6
Franklin Str. Canada 72°0N 96°0W 68 C12
Franklinton U.S.A. 30°51N 90°9W 85 F9
Franklinville U.S.A. 42°20N 78°27W 82 D6
Frankston Australia 38°8S 145°8E 63 F4
ransfontein Namibia 20°12S 15°1E 56 C2
rantsa Iosifa, Zemlya Russia 82°0N 55°0E 28 A6
ranz Canada 48°25N 84°30W 72 C3
ranz Josef Land = Frantsa Iosifa, Zemlya Russia 82°0N 55°0E 28 A6
raser U.S.A. 42°32N 82°57W 82 D2
raser → B.C., Canada 49°7N 123°11W 78 A3
raser → Nfld. & L., Canada 56°39N 62°10W 73 A7
raser, Mt. Australia 25°35S 118°20E 61 E2
raser I. Australia 25°15S 153°10E 63 D5
raser Lake Canada 54°0N 124°50W 70 C4
raserburg S. Africa 31°55S 21°30E 56 E3
raserburgh U.K. 57°42N 2°1W 11 D6
raserdale Canada 49°55N 81°37W 72 C3
ray Bentos Uruguay 33°10S 58°15W 94 C4
ray Jorge △ Chile 30°42S 71°40W 94 C1
razier Downs ❂ Australia 18°48S 121°42E 60 C3
redericia Denmark 55°34N 9°45E 9 J13
rederick Md., U.S.A. 39°25N 77°25W 81 F15
rederick Okla., U.S.A. 34°23N 99°1W 84 D5
rederick S. Dak., U.S.A. 45°50N 98°31W 80 C4
redericksburg Pa., U.S.A. 40°27N 76°26W 83 F8
redericksburg Tex., U.S.A. 30°16N 98°52W 84 F5
redericksburg Va., U.S.A. 38°18N 77°28W 81 F15
redericktown Mo., U.S.A. 37°34N 90°18W 80 G8
redericktown Ohio, U.S.A. 40°29N 82°33W 82 F2
rederico Westphalen Brazil 27°22S 53°24W 95 B5
redericton Canada 45°57N 66°40W 73 C6
redericton Junction Canada 45°41N 66°40W 73 C6
rederikshåb = Paamiut Greenland 62°0N 49°43W 4 C5
rederikshamn = Hamina Finland 60°34N 27°12E 8 F22
rederikshavn Denmark 57°28N 10°31E 9 H14
rederiksted U.S. Virgin Is. 17°43N 64°53W 89 C7
redonia Ariz., U.S.A. 36°57N 112°32W 77 H7
redonia Kans., U.S.A. 37°32N 95°49W 80 G6
redonia N.Y., U.S.A. 42°26N 79°20W 82 D5
redrikstad Norway 59°13N 10°57E 9 G14
ree State ☐ S. Africa 28°30S 27°0E 56 D4
reehold U.S.A. 40°16N 74°17W 83 F10
reeland U.S.A. 41°1N 75°54W 83 E9
reels, C. Nfld. & L., Canada 49°15N 53°30W 73 C9
reels, C. Nfld. & L., Canada 46°37N 53°32W 73 C9
reeman Calif., U.S.A. 35°35N 117°53W 79 K9
reeman S. Dak., U.S.A. 43°21N 97°26W 80 D5
reeport Bahamas 26°30N 78°47W 88 A4
reeport Ill., U.S.A. 42°17N 89°36W 80 D9
reeport N.Y., U.S.A. 40°39N 73°35W 83 F11
reeport Ohio, U.S.A. 40°12N 81°15W 82 F3
reeport Pa., U.S.A. 40°41N 79°41W 82 F5
reeport Tex., U.S.A. 28°57N 95°21W 84 G7
reetown S. Leone 8°30N 13°17W 50 G3
reeville U.S.A. 42°30N 76°20W 83 D8
régate, L. de la Canada 53°15N 74°45W 72 B5
regenal de la Sierra Spain 38°10N 6°39W 21 C2
regon Australia 26°45S 132°1E 61 E5
reibourg = Fribourg Switz. 46°49N 7°9E 20 C7
reiburg Germany 47°59N 7°51E 16 E4
reire Chile 38°54S 72°38W 96 D2
reirina Chile 28°30S 71°10W 94 B1
reising Germany 48°24N 11°45E 16 D6
reistadt Austria 48°30N 14°30E 16 D8
réjus France 43°25N 6°44E 20 E7
remantle Australia 32°7S 115°47E 61 F2
remont Calif., U.S.A. 37°32N 121°57W 78 H4
remont Mich., U.S.A. 43°28N 85°57W 81 D11
remont Nebr., U.S.A. 41°26N 96°30W 80 E5
remont Ohio, U.S.A. 41°21N 83°7W 81 E12
remont → U.S.A. 38°24N 110°42W 76 G8
rench Camp U.S.A. 37°53N 121°16W 78 H5
rench Cays = Plana Cays Bahamas 22°38N 73°30W 89 B5
rench Creek → U.S.A. 41°24N 79°50W 82 E5
rench Frigate Shoals U.S.A. 23°45N 166°10W 75 L6
rench Guiana ⊠ S. Amer. 4°0N 53°0W 93 C8
rench Polynesia ⊠ Pac. Oc. 20°0S 145°0W 65 J13
rench Riviera = Côte d'Azur France 43°25N 7°10E 20 E7
renchman Cr. → N. Amer. 48°31N 107°10W 76 B10
Frenchman Cr. → U.S.A. 40°14N 100°50W 80 E3
Fresco → Brazil 7°15S 51°30W 93 E8
Freshfield, C. Antarctica 68°25S 151°10E 5 C10
Fresnillo Mexico 23°10N 102°53W 86 C4
Fresno U.S.A. 36°44N 119°47W 78 J7
Fresno Res. U.S.A. 48°36N 109°57W 76 B9
Frew → Australia 20°0S 135°38E 62 C2
Frewsburg U.S.A. 42°3N 79°10W 82 D5
Freycinet △ Australia 42°11S 148°19E 63 G4
Freycinet Pen. Australia 42°10S 148°25E 63 G4
Fria Guinea 10°27N 13°38W 50 F3
Fria, C. Namibia 18°0S 12°0E 56 B1
Friant U.S.A. 36°59N 119°43W 78 J7
Frias Argentina 28°40S 65°5W 94 B2
Fribourg Switz. 46°49N 7°9E 20 C7
Friday Harbor U.S.A. 48°32N 123°1W 78 B3
Friedens U.S.A. 40°3N 78°59W 82 F6
Friedrichshafen Germany 47°39N 9°30E 16 E5
Friendship U.S.A. 42°12N 78°8W 82 D6
Friesland ☐ Neths. 53°5N 5°50E 15 A5
Frigate Seychelles 4°35S 55°56E 53 b
Frio → U.S.A. 28°26N 98°11W 84 G5
Frio, C. Brazil 22°50S 41°50W 90 F6
Friona U.S.A. 34°38N 102°43W 84 D3
Fritch U.S.A. 35°38N 101°36W 84 D4
Frobisher B. Canada 62°30N 66°0W 69 E18
Frobisher Bay = Iqaluit Canada 63°44N 68°31W 69 E18
Frobisher L. Canada 56°20N 108°15W 71 B7
Frohavet Norway 64°0N 9°30E 8 E13
Frome U.K. 51°14N 2°19W 13 F5
Frome → U.K. 50°41N 2°6W 13 G5
Frome, L. Australia 30°45S 139°45E 63 E2
Front Range U.S.A. 40°25N 105°45W 75 G18
Front Royal U.S.A. 38°55N 78°12W 81 F14
Frontenac △ Canada 44°32N 76°30W 83 B8
Frontera Canary Is. 27°47N 17°59W 24 G2
Frontera Mexico 18°32N 92°38W 87 D6
Fronteras Mexico 30°56N 109°31W 86 A3
Frosinone Italy 41°38N 13°19E 22 D5
Frostburg U.S.A. 39°39N 78°56W 81 F14
Frostisen Norway 68°14N 17°10E 8 B17
Frøya Norway 63°43N 8°40E 8 E13
Frunze = Bishkek Kyrgyzstan 42°54N 74°46E 32 C3
Frutal Brazil 20°0S 49°0W 93 H9
Frýdek-Místek Czech Rep. 49°40N 18°20E 17 D10
Fryeburg U.S.A. 44°1N 70°59W 83 B14
Fu Xian = Wafangdian China 39°38N 121°58E 35 E11
Fu Xian China 36°0N 109°20E 34 G5
Fucheng China 37°50N 116°10E 34 F9
Fuchou = Fuzhou China 26°5N 119°16E 33 F12
Fuchū Japan 34°34N 133°14E 31 G6
Fuencaliente Canary Is. 28°28N 17°50W 24 F2
Fuencaliente, Pta. Canary Is. 28°27N 17°51W 24 F2
Fuengirola Spain 36°32N 4°41W 21 D3
Fuentes de Oñoro Spain 40°33N 6°52W 21 B2
Fuerte → Mexico 25°54N 109°22W 86 B3
Fuerte Olimpo Paraguay 21°0S 57°51W 94 A4
Fuerteventura Canary Is. 28°30N 14°0W 24 F6
Fuerteventura ✈ (FUE) Canary Is. 28°24N 13°52W 24 F6
Fufeng China 34°22N 108°0E 34 G5
Fugou China 34°3N 114°25E 34 G8
Fugu China 39°2N 111°3E 34 E6
Fuhai China 47°2N 87°25E 32 B6
Fuḩaymī Iraq 34°16N 42°10E 44 C4
Fuji Japan 35°9N 138°39E 31 G9
Fuji-Hakone-Izu △ Japan 35°15N 138°45E 31 G9
Fuji-San Japan 35°22N 138°44E 31 G9
Fuji-Yoshida Japan 35°30N 138°46E 31 G9
Fujian ☐ China 26°0N 118°0E 33 F12
Fujin China 47°16N 132°1E 33 B15
Fujinomiya Japan 35°10N 138°40E 31 G9
Fujisawa Japan 35°22N 139°29E 31 G9
Fujiyama, Mt. = Fuji-San Japan 35°22N 138°44E 31 G9
Fukagawa Japan 43°43N 142°2E 30 C11
Fukien = Fujian ☐ China 26°0N 118°0E 33 F12
Fukuchiyama Japan 35°19N 135°9E 31 G7
Fukue-Shima Japan 32°40N 128°45E 31 H4
Fukui Japan 36°5N 136°10E 31 F8
Fukui ☐ Japan 36°0N 136°12E 31 G8
Fukuoka Japan 33°39N 130°21E 31 H5
Fukuoka ☐ Japan 33°30N 131°0E 31 H5
Fukushima Japan 37°44N 140°28E 30 F10
Fukushima ☐ Japan 37°30N 140°15E 30 F10
Fukuyama Japan 34°35N 133°20E 31 G6
Fulaga Fiji 19°8S 178°33W 59 a
Fulda Germany 50°32N 9°40E 16 C5
Fulda → Germany 51°25N 9°39E 16 C5
Fulford Harbour Canada 48°47N 123°27W 78 B3
Fullerton Calif., U.S.A. 33°53N 117°56W 79 M9
Fullerton Nebr., U.S.A. 41°22N 97°58W 80 E5
Fulongquan China 44°20N 124°42E 35 B13
Fulton Mo., U.S.A. 38°52N 91°57W 80 F8
Fulton N.Y., U.S.A. 43°19N 76°25W 83 C8
Funabashi Japan 35°45N 140°0E 31 G10
Funafuti = Fongafale Tuvalu 8°31S 179°13E 58 B10
Funchal Madeira 32°38N 16°54W 24 D3
Funchal ✈ (FNC) Madeira 32°42N 16°45W 24 D3
Fundación Colombia 10°31N 74°11W 92 A4
Fundão Portugal 40°8N 7°30W 21 B2
Fundy, B. of Canada 45°0N 66°0W 73 D6
Fundy △ Canada 45°35N 65°10W 73 C6
Funhalouro Mozam. 23°3S 34°25E 57 C5
Funing Hebei, China 39°53N 119°12E 35 E10
Funing Jiangsu, China 33°45N 119°50E 35 H10
Funiu Shan China 33°30N 112°20E 34 H7
Funtua Nigeria 11°30N 7°18E 50 F7
Fuping Hebei, China 38°48N 114°12E 34 E8
Fuping Shaanxi, China 34°42N 109°10E 34 G5
Furano Japan 43°21N 142°23E 30 C11
Furāt, Nahr al → Asia 31°0N 47°25E 44 D5
Fūrg Iran 28°18N 55°13E 45 D7
Furnás Spain 39°3N 1°32E 24 B8
Furnas, Represa de Brazil 20°50S 45°30W 95 A6
Furneaux Group Australia 40°10S 147°50E 63 G4
Furqlus Syria 34°36N 37°8E 46 A6
Fürstenwalde Germany 52°22N 14°3E 16 B8
Fürth Germany 49°28N 10°59E 16 D6
Furukawa Japan 38°34N 140°58E 30 E10
Fury and Hecla Str. Canada 69°56N 84°0W 69 D15
Fusagasugá Colombia 4°21N 74°22W 92 C4
Fushan Shandong, China 37°30N 121°15E 35 F11
Fushan Shanxi, China 35°58N 111°51E 34 G6
Fushun China 41°50N 123°56E 35 D12
Fusong China 42°20N 127°15E 35 C14
Fustic Barbados 13°16N 59°38W 89 g
Futian China 22°32N 114°4E 33 a
Fuxin China 42°5N 121°48E 35 C11
Fuyang China 33°0N 115°48E 34 H8
Fuyang He → China 38°12N 117°0E 34 E9
Fuyong China 22°40N 113°49E 33 a
Fuyu Heilongjiang, China 47°49N 124°27E 33 B13
Fuyu Jilin, China 45°12N 124°43E 35 B13
Fuyun China 47°0N 89°28E 32 B6
Fuzhou = Linchuan China 27°57N 116°15E 33 F12
Fuzhou China 26°5N 119°16E 33 F12
Fyn Denmark 55°20N 10°30E 9 J14
Fyne, L. U.K. 55°59N 5°23W 11 F3

G

Gaalkacyo Somali Rep. 6°30N 47°30E 47 F4
Gabba Australia 9°45S 142°38E 62 a
Gabela Angola 11°0S 14°24E 52 G2
Gabès Tunisia 33°53N 10°2E 51 B8
Gabès, G. de Tunisia 34°0N 10°30E 51 B8
Gabon ■ Africa 0°10S 10°0E 52 E2
Gaborone Botswana 24°45S 25°57E 56 C4
Gabriels U.S.A. 44°26N 74°12W 83 B10
Gābrīk Iran 25°44N 58°28E 45 E8
Gabrovo Bulgaria 42°52N 25°19E 23 C11
Gāch Sār Iran 36°7N 51°19E 45 B6
Gachsārān Iran 30°15N 50°45E 45 D6
Gadag India 15°30N 75°45E 40 M9
Gadap Pakistan 25°5N 67°28E 42 G2
Gadarwara India 22°50N 78°50E 43 H8
Gadhada India 22°0N 71°35E 42 J4
Gadra Pakistan 25°40N 70°38E 42 G4
Gadsden U.S.A. 34°1N 86°1W 85 D11
Gadwal India 16°10N 77°50E 40 L10
Gaffney U.S.A. 35°5N 81°39W 85 D14
Gafsa Tunisia 34°24N 8°43E 51 B7
Gagaria India 25°43N 70°46E 42 G4
Găgăuzia ☐ Moldova 46°10N 28°40E 17 E15
Gagnoa Ivory C. 6°56N 5°16W 50 G4
Gagnon Canada 51°50N 68°5W 73 B6
Gagnon, L. Canada 62°3N 110°27W 71 A6
Gahini Rwanda 1°50S 30°30E 54 C3
Gahmar India 25°27N 83°49E 43 G10
Gai Xian = Gaizhou China 40°22N 122°20E 35 D12
Gaidouronisi Greece 34°53N 25°41E 25 E7
Gail U.S.A. 32°46N 101°27W 84 E4
Gaillimh = Galway Ireland 53°17N 9°3W 10 C2
Gaines U.S.A. 41°46N 77°35W 82 E7
Gainesville Fla., U.S.A. 29°40N 82°20W 85 G13
Gainesville Ga., U.S.A. 34°18N 83°50W 85 D13
Gainesville Mo., U.S.A. 36°36N 92°26W 80 G7
Gainesville Tex., U.S.A. 33°38N 97°8W 84 E6
Gainsborough U.K. 53°24N 0°46W 12 D7
Gairdner, L. Australia 31°30S 136°0E 63 E2
Gairloch U.K. 57°43N 5°41W 11 D3
Gairloch, L. U.K. 57°43N 5°45W 11 D3
Gaizhou China 40°22N 122°20E 35 D12
Gaj → Pakistan 26°26N 67°21E 42 F2
Gakuch Pakistan 36°7N 73°45E 43 A5
Galán, Cerro Argentina 25°55S 66°52W 94 B2
Galana → Kenya 3°9S 40°8E 54 C5
Galápagos = Colón, Arch. de Ecuador 0°0 91°0W 90 D1
Galapagos Fracture Zone Pac. Oc. 3°0N 110°0W 65 G17
Galapagos Rise Pac. Oc. 15°0S 95°0W 65 J18
Galashiels U.K. 55°37N 2°49W 11 F6
Galateia Cyprus 35°25N 34°4E 25 D13
Galați Romania 45°27N 28°2E 17 F15
Galatina Italy 40°10N 18°10E 23 D8
Galax U.S.A. 36°40N 80°56W 81 G13
Galbīn Govĭ Mongolia 43°0N 107°0E 34 C4
Galcaio = Gaalkacyo Somali Rep. 6°30N 47°30E 47 F4
Galdhøpiggen Norway 61°38N 8°18E 8 F13
Galeana Chihuahua, Mexico 30°7N 107°38W 86 A3
Galeana Nuevo León, Mexico 24°50N 100°4W 86 A3
Galela Indonesia 1°50N 127°49E 37 D7
Galeota Pt. Trin. & Tob. 10°8N 60°59W 93 K16
Galera Pt. Trin. & Tob. 10°49N 60°54W 93 K16
Galesburg U.S.A. 40°57N 90°22W 80 E8
Galestan △ Iran 37°30N 56°0E 45 B8
Galeton U.S.A. 41°44N 77°39W 82 E7
Galich Russia 58°22N 42°24E 18 C7
Galicia ☐ Spain 42°43N 7°45W 21 A2
Galilee = Hagalil Israel 32°53N 35°18E 46 C4
Galilee, L. Australia 22°20S 145°50E 62 C4
Galilee, Sea of = Yam Kinneret Israel 32°45N 35°35E 46 C4
Galina Pt. Jamaica 18°24N 76°58W 88 a
Galinoporni Cyprus 35°31N 34°18E 25 D13
Galion U.S.A. 40°44N 82°47W 82 F2
Galiuro Mts. U.S.A. 32°30N 110°20W 77 K8
Galiwinku Australia 12°2S 135°34E 62 A2
Gallan Hd. U.K. 58°15N 7°2W 11 C1
Gallatin U.S.A. 36°24N 86°27W 85 C11
Galle Sri Lanka 6°5N 80°10E 40 R12
Gállego → Spain 41°39N 0°51W 21 B5
Gallegos → Argentina 51°35S 69°0W 96 G3
Galley Hd. Ireland 51°32N 8°55W 10 E3
Gallinas, Pta. Colombia 12°28N 71°40W 92 A4
Gallipoli = Gelibolu Turkey 40°28N 26°43E 23 D12
Gallipoli Italy 40°3N 17°58E 23 D8
Gallipolis U.S.A. 38°49N 82°12W 81 F12
Gällivare Sweden 67°9N 20°40E 8 C19
Galloo I. U.S.A. 43°55N 76°25W 83 C8
Galloway U.K. 55°1N 4°29W 11 F4
Galloway, Mull of U.K. 54°39N 4°52W 11 G4
Galloway ⌓ U.K. 55°3N 4°20W 11 F4
Gallup U.S.A. 35°32N 108°45W 77 J9
Galoya Sri Lanka 8°10N 80°55E 40 Q12
Galt U.S.A. 38°15N 121°18W 78 G5
Galty Mts. Ireland 52°22N 8°10W 10 D3
Galtymore Ireland 52°21N 8°11W 10 D3
Galugāh Iran 36°43N 53°48E 45 B7
Galva U.S.A. 41°10N 90°3W 80 E8
Galveston U.S.A. 29°18N 94°48W 84 G7
Galveston B. U.S.A. 29°36N 94°50W 84 G7
Gálvez Argentina 32°0S 61°14W 94 C3
Galway Ireland 53°17N 9°3W 10 C2
Galway ☐ Ireland 53°22N 9°1W 10 C2
Galway B. Ireland 53°13N 9°10W 10 C2
Gam → Vietnam 21°55N 105°12E 38 B5
Gamagōri Japan 34°50N 137°14E 31 G8
Gambat Pakistan 27°17N 68°26E 42 F3
Gambhir → India 26°58N 77°27E 42 F6
Gambia ■ W. Afr. 13°25N 16°0W 50 F2
Gambia → W. Afr. 13°28N 16°34W 50 F2
Gambier U.S.A. 40°22N 82°23W 82 F2
Gambier, C. Australia 11°56S 130°57E 60 B5
Gambier, Îs. French Polynesia 23°8S 134°58W 65 K14
Gambier Is. Australia 35°3S 136°30E 63 F2
Gambo Canada 48°47N 54°13W 73 C9
Gamboli Pakistan 29°53N 68°24E 42 E3
Gamboma Congo 1°55S 15°52E 52 E3
Gamka → S. Africa 33°18S 21°39E 56 E3
Gamkab → Namibia 28°4S 17°54E 56 D2
Gamlakarleby = Kokkola Finland 63°50N 23°8E 8 E20
Gammon → Canada 51°24N 95°44W 71 C9
Gammon Ranges △ Australia 30°38S 139°8E 63 E2
Gamtoos → S. Africa 33°58S 25°1E 56 E4
Gan Jiang → China 29°15N 116°0E 33 F12
Ganado U.S.A. 35°43N 109°33W 77 J9
Gananoque Canada 44°20N 76°10W 83 B8
Ganāveh Iran 29°35N 50°35E 45 D6
Gäncä Azerbaijan 40°45N 46°20E 19 F8
Gancheng China 18°51N 108°37E 38 C7
Gand = Gent Belgium 51°2N 3°42E 15 C3
Ganda Angola 13°3S 14°35E 53 G2
Gandajika Dem. Rep. of the Congo 6°46S 23°58E 52 F4
Gandak → India 25°39N 85°13E 43 G11
Gandava Pakistan 28°32N 67°32E 42 E2
Gander Canada 48°58N 54°35W 73 C9
Gander L. Canada 48°58N 54°35W 73 C9
Ganderowe Falls Zimbabwe 17°20S 29°10E 55 F2
Gandhi Sagar India 24°40N 75°40E 42 G6
Gandhinagar India 23°15N 72°45E 42 H5
Gandia Spain 38°58N 0°9W 21 C5
Gando, Pta. Canary Is. 27°55N 15°22W 24 G4
Ganedidalem = Gani Indonesia 0°48S 128°14E 37 E7
Ganga → India 23°20N 90°30E 43 H14
Ganga Sagar India 21°38N 88°5E 43 J13
Gangan → India 28°38N 78°58E 43 E8
Ganganagar India 29°56N 73°56E 42 E5
Gangapur India 26°32N 76°49E 42 F7
Gangaw Burma 22°5N 94°5E 41 H19
Gangdisê Shan China 31°20N 81°0E 43 D9
Ganges = Ganga → India 23°20N 90°30E 43 H14
Ganges Canada 48°51N 123°31W 70 D4
Ganges, Mouths of the India 21°30N 90°0E 43 J14
Ganggyeong S. Korea 36°10N 127°0E 35 F14
Gangneung S. Korea 37°45N 128°54E 35 F15
Gangoh India 29°46N 77°18E 42 E7
Gangotri India 30°50N 79°10E 43 D8
Gangotri △ India 30°50N 79°10E 43 D8
Gangseong S. Korea 38°24N 128°30E 35 E15
Gangtok India 27°20N 88°37E 41 F16
Gangu China 34°40N 105°15E 34 G3
Gangyao China 44°12N 126°37E 35 B14
Gani Indonesia 0°48S 128°14E 37 E7
Ganj India 27°45N 78°57E 43 F8
Gannett Peak U.S.A. 43°11N 109°39W 76 E9
Ganquan China 36°20N 109°20E 34 F5
Gansu ☐ China 36°0N 104°0E 34 G3
Ganta Liberia 7°15N 8°59W 50 G4
Gantheaume, C. Australia 36°4S 137°32E 63 F2
Gantheaume B. Australia 27°40S 114°10E 61 E1
Gantsevichi = Hantsavichy Belarus 52°49N 26°30E 17 B14
Ganyem = Genyem Indonesia 2°46S 140°12E 37 E10
Ganyu China 34°50N 119°8E 35 G10
Ganzhou China 25°51N 114°56E 33 F11
Gao Mali 16°15N 0°5W 50 E5
Gaobeidian China 39°19N 115°51E 34 E8
Gaocheng China 38°2N 114°49E 34 E8
Gaomi China 36°20N 119°42E 35 F10
Gaoping China 35°45N 112°55E 34 G7
Gaotang China 36°50N 116°15E 34 F9
Gaoua Burkina Faso 10°20N 3°8W 50 F5
Gaoual Guinea 11°45N 13°25W 50 F3
Gaoxiong = Kaohsiung Taiwan 22°35N 120°16E 33 G13
Gaoyang China 38°40N 115°45E 34 E8
Gaoyou Hu China 32°45N 119°20E 35 H10
Gaoyuan China 37°8N 117°58E 35 F9
Gap France 44°33N 6°5E 20 D7
Gapat → India 24°30N 82°28E 43 G10
Gapuwiyak Australia 12°25S 135°43E 62 A2
Gar China 32°10N 79°58E 32 E4
Gara, L. Ireland 53°57N 8°26W 10 C3
Garabogazköl Aylagy Turkmenistan 41°0N 53°30E 19 F9
Garachico Canary Is. 28°22N 16°46W 24 F3
Garachiné Panama 8°0N 78°12W 88 E4
Garafia Canary Is. 28°48N 17°57W 24 F2
Garagum Turkmenistan 39°30N 60°0E 45 B8
Garah Australia 29°5S 149°38E 63 D4
Garajonay Canary Is. 28°7N 17°14W 24 F2
Garamba △ Dem. Rep. of the Congo 4°10N 29°40E 54 B2
Garanhuns Brazil 8°50S 36°30W 93 E11
Garautha India 25°34N 79°18E 43 G8
Garawa ❂ Australia 16°43S 136°51E 62 B2
Garba Tula Kenya 0°30N 38°32E 54 B4
Garberville U.S.A. 40°6N 123°48W 76 F2
Garbiyang India 30°8N 80°54E 43 D9
Gard → France 43°51N 4°37E 20 E6
Garda, L. di Italy 45°40N 10°41E 22 B4
Garde L. Canada 62°50N 106°13W 71 A7
Garden City Ga., U.S.A. 32°6N 81°9W 85 E14
Garden City Kans., U.S.A. 37°58N 100°53W 80 G3
Garden City Tex., U.S.A. 31°52N 101°29W 84 F4
Garden Grove U.S.A. 33°47N 117°55W 79 M9
Gardēz Afghan. 33°37N 69°9E 42 C3
Gardiner Maine, U.S.A. 44°14N 69°47W 81 C19
Gardiner Mont., U.S.A. 45°2N 110°22W 76 D8
Gardiners I. U.S.A. 41°6N 72°6W 83 E12
Gardner U.S.A. 42°34N 71°59W 83 D13
Gardner Canal Canada 53°27N 128°8W 70 C3
Gardner Pinnacles U.S.A. 25°0N 167°55W 75 L6
Gardnerville U.S.A. 38°56N 119°45W 78 G7
Gardo = Qardho Somali Rep. 9°30N 49°6E 47 F4
Garey U.S.A. 34°53N 120°19W 79 L6
Garfield U.S.A. 47°1N 117°9W 76 C5
Garforth U.K. 53°47N 1°24W 12 D6
Gargantua, C. Canada 47°36N 85°2W 81 B11
Gargett Australia 21°9S 148°46E 62 b
Garibaldi ⌓ Canada 49°50N 122°40W 70 D4
Gariep, L. S. Africa 30°40S 25°40E 56 E4
Garies S. Africa 30°32S 17°59E 56 E2
Garig Gunak Barlu △ Australia 11°26S 131°58E 60 B5
Garigliano → Italy 41°13N 13°45E 22 D5
Garissa Kenya 0°25S 39°40E 54 C4
Garland Tex., U.S.A. 32°54N 96°38W 84 E6
Garland Utah, U.S.A. 41°45N 112°10W 76 F7
Garmāb Iran 35°25N 56°45E 45 C8
Garmisch-Partenkirchen Germany 47°30N 11°6E 16 E6
Garmsār Iran 35°20N 52°25E 45 C7
Garner U.S.A. 43°6N 93°36W 80 D7
Garnett U.S.A. 38°17N 95°14W 80 F6
Garo Hills India 25°30N 90°30E 43 G14
Garoe = Garoowe Somali Rep. 8°25N 48°33E 47 F4
Garonne → France 45°2N 0°36W 20 D3
Garoowe Somali Rep. 8°25N 48°33E 47 F4
Garot India 24°19N 75°41E 42 G6
Garoua Cameroon 9°19N 13°21E 51 G8
Garrauli India 25°5N 79°22E 43 G8
Garrison Mont., U.S.A. 46°31N 112°49W 76 C7
Garrison N. Dak., U.S.A. 47°40N 101°25W 80 B3
Garrison Res. = Sakakawea, L. U.S.A. 47°30N 101°25W 80 B3
Garron Pt. U.K. 55°3N 5°59W 10 A6
Garry → U.K. 56°44N 3°47W 11 E5
Garry, L. Canada 65°58N 100°18W 68 D11
Garrygala Turkmenistan 38°31N 56°29E 45 B8
Garsen Kenya 2°20S 40°5E 54 C5
Garson L. Canada 56°19N 110°2W 71 B6
Garstang U.K. 53°55N 2°46W 12 D5
Garu India 23°40N 84°14E 43 H11
Garub Namibia 26°37S 16°0E 56 D2
Garut Indonesia 7°14S 107°53E 37 G12
Garvie Mts. N.Z. 45°30S 168°50E 59 F2
Garwa = Garoua Cameroon 9°19N 13°21E 51 G8
Garwa India 24°11N 83°47E 43 G10
Gary U.S.A. 41°36N 87°20W 80 E10
Garzê China 31°38N 100°1E 32 E9
Garzón Colombia 2°10N 75°40W 92 C3
Gas-San Japan 38°32N 140°1E 30 E10
Gasan Kuli = Esenguly Turkmenistan 37°37N 53°59E 28 F6
Gascogne France 43°45N 0°20E 20 E4
Gascogne, G. de Europe 44°0N 2°0W 20 D2
Gascony = Gascogne France 43°45N 0°20E 20 E4
Gascoyne → Australia 24°52S 113°37E 61 D1
Gascoyne Junction Australia 25°2S 115°17E 61 E2
Gashaka Nigeria 7°20N 11°29E 51 G8
Gasherbrum Pakistan 35°40N 76°40E 43 B7
Gashua Nigeria 12°54N 11°0E 51 F8
Gasparillo Trin. & Tob. 10°18N 61°26W 93 K15
Gaspé Canada 48°52N 64°30W 73 C7
Gaspé, C. de Canada 48°48N 64°7W 73 C7
Gaspé Pen. = Gaspésie, Pén. de la Canada 48°45N 65°40W 73 C6
Gaspésie, Pén. de la Canada 48°45N 65°40W 73 C6
Gaspésie ⌓ Canada 48°55N 66°10W 73 C6
Gasteiz = Vitoria-Gasteiz Spain 42°50N 2°41W 21 A4
Gastonia U.S.A. 35°16N 81°11W 85 D14
Gastre Argentina 42°20S 69°15W 96 E3

Gata, C. *Cyprus* 34°34N 33°2E **25** E12
Gata, C. de *Spain* 36°41N 2°13W **21** D4
Gata, Sierra de *Spain* 40°20N 6°45W **21** B2
Gataga → *Canada* 58°35N 126°59W **70** B3
Gatchina *Russia* 59°35N 30°9E **9** G24
Gatehouse of Fleet *U.K.* 54°53N 4°12W **11** G4
Gates *U.S.A.* 43°9N 77°42W **82** C7
Gateshead *U.K.* 54°57N 1°35W **12** C6
Gateshead I. *Canada* 70°36N 100°26W **68** C11
Gatesville *U.S.A.* 31°26N 97°45W **84** F6
Gateway ⌂ *U.S.A.* 40°38N 73°51W **83** F11
Gaths *Zimbabwe* 20°2S 30°32E **55** G3
Gatineau *Canada* 45°29N 75°39W **83** A9
Gatineau → *Canada* 45°27N 75°42W **72** C4
Gatineau △ *Canada* 45°40N 76°0W **72** C4
Gatton *Australia* 27°32S 152°17E **63** D5
Gatún, L. *Panama* 9°7N 79°56W **88** E4
Gatwick, London ✈ (LGW) *U.K.* 51°10N 0°11W **13** F7
Gatyana *S. Africa* 32°16S 28°31E **57** E4
Gau *Fiji* 18°2S 179°18E **59** a
Gauer L. *Canada* 57°0N 97°50W **71** B9
Gauhati = Guwahati *India* 26°10N 91°45E **41** F17
Gauja → *Latvia* 57°10N 24°16E **9** H21
Gaujas △ *Latvia* 57°10N 24°50E **9** H21
Gaula → *Norway* 63°21N 10°14E **8** E14
Gauri Phanta *India* 28°41N 80°36E **43** E9
Gaustatoppen *Norway* 59°48N 8°40E **9** G13
Gauteng □ *S. Africa* 26°0S 28°0E **57** D4
Gāv Koshī *Iran* 28°38N 57°12E **45** D8
Gāvakān *Iran* 29°37N 53°10E **45** D7
Gavāter *Iran* 25°10N 61°31E **45** E9
Gāvbandī *Iran* 27°12N 53°4E **45** E7
Gavdopoula *Greece* 34°56N 24°0E **25** E6
Gavdos *Greece* 34°50N 24°5E **25** E6
Gaviota *U.S.A.* 34°29N 120°13W **79** L6
Gāvkhūnī, Bātlāq-e *Iran* 32°6N 52°52E **45** C7
Gävle *Sweden* 60°40N 17°9E **8** F17
Gawachab *Namibia* 27°4S 17°55E **56** D2
Gawilgarh Hills *India* 21°15N 76°45E **40** J10
Gawler *Australia* 34°30S 138°42E **63** E2
Gawler Ranges *Australia* 32°30S 135°45E **63** E2
Gaxun Nur *China* 42°22N 100°30E **32** D9
Gay *Russia* 51°27N 58°27E **18** D10
Gaya *India* 24°47N 85°4E **43** G11
Gaya *Niger* 11°52N 3°28E **50** F6
Gaylord *U.S.A.* 45°2N 84°41W **81** C11
Gayndah *Australia* 25°35S 151°32E **63** D5
Gaysin = Haysyn *Ukraine* 48°57N 29°25E **17** D15
Gayvoron = Hayvoron *Ukraine* 48°22N 29°52E **17** D15
Gaza *Gaza Strip* 31°30N 34°28E **46** D3
Gaza □ *Mozam.* 23°10S 32°45E **57** C5
Gaza Strip ■ *Asia* 31°29N 34°25E **46** D3
Gazanjyk = Bereket *Turkmenistan* 39°16N 55°32E **45** B7
Gāzbor *Iran* 28°5N 58°51E **45** D8
Gazi *Dem. Rep. of the Congo* 1°3N 24°30E **54** B1
Gaziantep *Turkey* 37°6N 37°23E **44** B3
Gazimağusa = Famagusta *Cyprus* 35°8N 33°55E **25** D12
Gbadolite *Dem. Rep. of the Congo* 4°17N 21°1E **52** D4
Gbarnga *Liberia* 7°19N 9°13W **50** G4
Gcoverega *Botswana* 19°8S 24°18E **56** B3
Gcuwa *S. Africa* 32°20S 28°11E **57** E4
Gdańsk *Poland* 54°22N 18°40E **17** A10
Gdańska, Zatoka *Poland* 54°30N 19°20E **17** A10
Gdov *Russia* 58°48N 27°55E **9** G22
Gdynia *Poland* 54°35N 18°33E **17** A10
Gebe *Indonesia* 0°5N 129°25E **37** D7
Gebze *Turkey* 40°47N 29°25E **23** D13
Gedaref *Sudan* 14°2N 35°28E **51** F13
Gede, Tanjung *Indonesia* 6°46S 105°12E **37** G11
Gediz → *Turkey* 38°35N 26°48E **23** E12
Gedser *Denmark* 54°35N 11°55E **9** J14
Gedung, Pulau *Malaysia* 5°17N 100°23E **39** c
Geegully Cr. → *Australia* 18°32S 123°41E **60** C3
Geel *Belgium* 51°10N 4°59E **15** C4
Geelong *Australia* 38°10S 144°22E **63** F3
Geelvink B. = Cenderwasih, Teluk *Indonesia* 3°0S 135°20E **37** E9
Geelvink Chan. *Australia* 28°30S 114°0E **61** E1
Geesthacht *Germany* 53°26N 10°22E **16** B6
Geidam *Nigeria* 12°57N 11°57E **51** F8
Geikie → *Canada* 57°45N 103°52W **71** B8
Geikie Gorge △ *Australia* 18°3S 125°41E **60** C4
Geistown *U.S.A.* 40°18N 78°52W **82** F6
Geita *Tanzania* 2°48S 32°12E **54** C3
Gejiu *China* 23°20N 103°10E **32** G9
Gel, Meydān-e *Iran* 29°4N 54°50E **45** D7
Gela *Italy* 37°4N 14°15E **22** F6
Gelang Patah *Malaysia* 1°27N 103°35E **39** d
Gelderland □ *Neths.* 52°5N 6°10E **15** B6
Geldrop *Neths.* 51°25N 5°32E **15** C5
Geleen *Neths.* 50°57N 5°49E **15** D5
Gelib = Jilib *Somali Rep.* 0°29N 42°46E **47** G3
Gelibolu *Turkey* 40°28N 26°43E **23** D12
Gelsenkirchen *Germany* 51°32N 7°6E **16** C4
Gelugur *Malaysia* 5°22N 100°18E **39** c
Gemas *Malaysia* 2°37N 102°36E **39** L4
Gembloux *Belgium* 50°34N 4°43E **15** D4
Gemena *Dem. Rep. of the Congo* 3°13N 19°48E **52** D3
Gemerek *Turkey* 39°15N 36°10E **44** B3
Gemlik *Turkey* 40°26N 29°9E **23** D13
Gemsbok △ *Botswana* 25°5S 21°1E **56** D3
Genadi *Greece* 36°2N 27°56E **25** C9
Genale → *Ethiopia* 6°2N 39°1E **47** F2
General Acha *Argentina* 37°20S 64°38W **94** D3
General Alvear *B. Aires, Argentina* 36°0S 60°0W **94** D4
General Alvear *Mendoza, Argentina* 35°0S 67°40W **94** D2
General Artigas *Paraguay* 26°52S 56°16W **94** B4
General Belgrano *Argentina* 36°35S 58°47W **94** D4
General Bernardo O'Higgins *Antarctica* 63°0S 58°3W **5** C18
General Cabrera *Argentina* 32°53S 63°52W **94** C3
General Cepeda *Mexico* 25°21N 101°22W **86** B4
General Guido *Argentina* 36°40S 57°50W **94** D4
General Juan Madariaga *Argentina* 37°0S 57°0W **94** D4
General La Madrid *Argentina* 37°17S 61°20W **94** D3
General MacArthur *Phil.* 11°18N 125°28E **37** B7
General Martin Miguel de Güemes *Argentina* 24°50S 65°0W **94** A3
General Pico *Argentina* 35°45S 63°50W **94** D3
General Pinedo *Argentina* 27°15S 61°20W **94** B3
General Pinto *Argentina* 34°45S 61°50W **94** C3
General Roca *Argentina* 39°2S 67°35W **96** D3
General Santos *Phil.* 6°5N 125°14E **37** C7
General Treviño *Mexico* 26°14N 99°29W **87** B5
General Trías *Mexico* 28°21N 106°22W **86** B3
General Viamonte *Argentina* 35°1S 61°3W **94** D3
General Villegas *Argentina* 35°5S 63°0W **94** D3
Genesee *Idaho, U.S.A.* 46°33N 116°56W **76** C5
Genesee *Pa., U.S.A.* 41°59N 77°54W **82** E7
Genesee → *U.S.A.* 43°16N 77°36W **82** C7
Geneseo *Ill., U.S.A.* 41°27N 90°9W **80** E8
Geneseo *N.Y., U.S.A.* 42°48N 77°49W **82** D7
Geneva = Genève *Switz.* 46°12N 6°9E **20** C7
Geneva *Ala., U.S.A.* 31°2N 85°52W **85** F12
Geneva *N.Y., U.S.A.* 42°52N 76°59W **82** D8
Geneva *Nebr., U.S.A.* 40°32N 97°36W **80** E5
Geneva *Ohio, U.S.A.* 41°48N 80°57W **82** E4
Geneva, L. = Léman, L. *Europe* 46°26N 6°30E **20** C7
Genève *Switz.* 46°12N 6°9E **20** C7
Genil → *Spain* 37°42N 5°19W **21** D3
Genk *Belgium* 50°58N 5°32E **15** D5
Gennargentu, Mti. del *Italy* 40°1N 9°19E **22** D3
Genoa = Génova *Italy* 44°25N 8°57E **20** D8
Genoa *Australia* 37°29S 149°35E **63** F4
Genoa *N.Y., U.S.A.* 42°40N 76°32W **83** D8
Genoa *Nebr., U.S.A.* 41°27N 97°44W **80** E5
Genoa *Nev., U.S.A.* 39°2N 119°50W **78** F7
Génova *Italy* 44°25N 8°57E **20** D8
Génova, G. di *Italy* 44°0N 9°0E **22** C3
Genriyetty, Ostrov *Russia* 77°6N 156°30E **29** B16
Gent *Belgium* 51°2N 3°42E **15** C3
Genteng *Jawa Barat, Indonesia* 7°22S 106°24E **37** G12
Genteng *Jawa Timur, Indonesia* 8°22S 114°9E **37** J17
Genyem *Indonesia* 2°46S 140°12E **37** E10
Geochang *S. Korea* 35°41N 127°55E **35** G14
Geographe B. *Australia* 33°30S 115°15E **61** F2
Geographe Chan. *Australia* 24°30S 113°0E **61** D1
Georga, Zemlya *Russia* 80°30N 49°0E **28** A5
George *S. Africa* 33°58S 22°29E **56** E3
George → *Canada* 58°49N 66°10W **73** A6
George, L. *N.S.W., Australia* 35°10S 149°25E **63** F4
George, L. *S. Austral., Australia* 37°25S 140°0E **63** F3
George, L. *W. Austral., Australia* 22°45S 123°40E **60** D3
George, L. *Uganda* 0°5N 30°10E **54** B3
George, L. *Fla., U.S.A.* 29°17N 81°36W **85** G14
George, L. *N.Y., U.S.A.* 43°37N 73°33W **83** C11
George Gill Ra. *Australia* 24°22S 131°45E **60** D5
George Pt. *Australia* 20°6S 148°36E **62** b
George River = Kangiqsualujjuaq *Canada* 58°30N 65°59W **69** F18
George Sound *N.Z.* 44°52S 167°25E **59** F1
George Town *Australia* 41°6S 146°49E **63** G4
George Town *Bahamas* 23°33N 75°47W **88** B4
George Town *Cayman Is.* 19°20N 81°24W **88** C3
George Town *Malaysia* 5°25N 100°20E **39** c
George V Land *Antarctica* 69°0S 148°0E **5** C10
George VI Sound *Antarctica* 71°0S 68°0W **5** D17
George West *U.S.A.* 28°20N 98°7W **84** G5
Georgetown = Janjanbureh *Gambia* 13°30N 14°47W **50** F3
Georgetown *Australia* 18°17S 143°33E **62** B3
Georgetown *Ont., Canada* 43°40N 79°56W **82** C5
Georgetown *P.E.I., Canada* 46°13N 62°24W **73** C7
Georgetown *Guyana* 6°50N 58°12W **92** B7
Georgetown *Calif., U.S.A.* 38°54N 120°50W **78** G6
Georgetown *Colo., U.S.A.* 39°42N 105°42W **76** G11
Georgetown *Ky., U.S.A.* 38°13N 84°33W **81** F11
Georgetown *N.Y., U.S.A.* 42°46N 75°44W **83** D9
Georgetown *Ohio, U.S.A.* 38°52N 83°54W **81** F12
Georgetown *S.C., U.S.A.* 33°23N 79°17W **85** E15
Georgetown *Tex., U.S.A.* 30°38N 97°41W **84** F6
Georgia □ *U.S.A.* 32°50N 83°15W **85** E13
Georgia ■ *Asia* 42°0N 43°0E **19** F7
Georgia, Str. of *N. Amer.* 49°25N 124°0W **78** A3
Georgia Basin *S. Ocean* 50°45S 35°30W **5** B1
Georgian B. *Canada* 45°15N 81°0W **82** A4
Georgian Bay Islands △ *Canada* 44°53N 79°52W **82** B5
Georgievka *Kazakhstan* 49°19N 84°34E **28** E9
Georgina → *Australia* 23°30S 139°47E **62** C2
Georgina I. *Canada* 44°22N 79°17W **82** B5
Georgioupoli *Greece* 35°20N 24°15E **25** D6
Georgiyevsk *Russia* 44°12N 43°28E **19** F7
Gera *Germany* 50°53N 12°4E **16** C7
Geraardsbergen *Belgium* 50°45N 3°53E **15** D3
Geral, Serra *Brazil* 26°25S 50°0W **95** B6
Geral de Goiás, Serra *Brazil* 12°0S 46°0W **93** F9
Geraldine *U.S.A.* 47°36N 110°16W **76** C8
Geraldton *Australia* 28°48S 114°32E **61** E1
Geraldton *Canada* 49°44N 87°10W **72** C2
Gerdine, Mt. *U.S.A.* 61°35N 152°27W **74** C9
Gereshk *Afghan.* 31°47N 64°35E **40** D4
Gerik *Malaysia* 5°50N 101°15E **39** K3
Gering *U.S.A.* 41°50N 103°40W **80** E2
Gerlach *U.S.A.* 40°39N 119°21W **76** F4
German Bight = Deutsche Bucht *Germany* 54°15N 8°0E **16** A5
Germansen Landing *Canada* 55°43N 124°40W **70** B4
Germantown *U.S.A.* 35°5N 89°49W **85** D10
Germany ■ *Europe* 51°0N 10°0E **16** C6
Germī *Iran* 39°1N 48°3E **45** B6
Germiston *S. Africa* 26°13S 28°10E **57** D4
Gernika-Lumo *Spain* 43°19N 2°40W **21** A4
Gero *Japan* 35°48N 137°14E **31** G8
Gerokgak *Indonesia* 8°11S 114°27E **37** J17
Gerona = Girona *Spain* 41°58N 2°46E **21** B7
Geropotamos → *Greece* 35°3N 24°50E **25** D6
Gerrard *Canada* 50°30N 117°17W **70** C5
Gertak Sanggul *Malaysia* 5°17N 100°12E **39** c
Gertak Sanggul, Tanjung *Malaysia* 5°16N 100°11E **39** c
Gerung *Indonesia* 8°43S 116°7E **37** K19
Geser *Indonesia* 3°50S 130°54E **37** E8
Getafe *Spain* 40°18N 3°43W **21** B4
Gettysburg *Pa., U.S.A.* 39°50N 77°14W **81** F15
Gettysburg *S. Dak., U.S.A.* 45°1N 99°57W **80** C4
Getxo *Spain* 43°21N 2°59W **21** A4
Getz Ice Shelf *Antarctica* 75°0S 130°0W **5** D14
Geyser *U.S.A.* 47°16N 110°30W **76** C8
Geyserville *U.S.A.* 38°42N 122°54W **78** G4
Ghadāf, W. al → *Iraq* 32°56N 43°30E **44** C4
Ghadāmis *Libya* 30°11N 9°29E **51** B8
Ghaggar → *India* 29°30N 74°53E **42** E6
Ghaghara → *India* 25°45N 84°40E **43** G11
Ghaghat → *Bangla.* 25°19N 89°38E **43** G13
Ghagra *India* 23°17N 84°33E **43** H11
Ghagra → *India* 27°29N 81°9E **43** F9
Ghallamane *Mauritania* 23°15N 10°0W **50** D4
Ghana ■ *W. Afr.* 8°0N 1°0W **50** G5
Ghansor *India* 22°39N 80°1E **43** H9
Ghanzi *Botswana* 21°50S 21°34E **56** C3
Gharb Bahr el Ghazâl □ *Sudan* 7°30N 27°30E **48** F6
Ghardaïa *Algeria* 32°20N 3°37E **50** B6
Gharm *Tajikistan* 39°0N 70°20E **28** F8
Gharyān *Libya* 32°10N 13°0E **51** B8
Ghāt *Libya* 24°59N 10°11E **51** D8
Ghatal *India* 22°40N 87°46E **43** H12
Ghatampur *India* 26°8N 80°13E **43** F9
Ghats, Eastern *India* 14°0N 78°50E **40** N11
Ghats, Western *India* 14°0N 75°0E **40** N9
Ghatsila *India* 22°36N 86°29E **43** H12
Ghaṭṭī *Si. Arabia* 31°16N 37°31E **44** D3
Ghawdex = Gozo *Malta* 36°3N 14°15E **25** C1
Ghazal, Bahr el → *Chad* 13°0N 15°47E **51** F9
Ghazâl, Bahr el → *Sudan* 9°31N 30°25E **51** G12
Ghaziabad *India* 28°42N 77°26E **42** E7
Ghazipur *India* 25°38N 83°35E **43** G10
Ghaznī *Afghan.* 33°30N 68°28E **42** C3
Ghaznī □ *Afghan.* 32°10N 68°20E **40** C6
Ghent = Gent *Belgium* 51°2N 3°42E **15** C3
Gheorghe Gheorghiu-Dej = Onești *Romania* 46°17N 26°47E **17** E14
Ghīnah, Wādī al → *Si. Arabia* 30°27N 38°14E **44** D3
Ghizar → *Pakistan* 36°15N 73°43E **43** A5
Ghotaru *India* 27°20N 70°1E **42** F4
Ghotki *Pakistan* 28°5N 69°21E **42** E3
Ghowr □ *Afghan.* 34°0N 64°20E **40** C4
Ghughri *India* 22°39N 80°41E **43** H9
Ghugus *India* 19°58N 79°12E **40** K11
Ghulam Mohammad Barrage *Pakistan* 25°30N 68°20E **42** G3
Ghūrīān *Afghan.* 34°17N 61°25E **40** B2
Gia Dinh *Vietnam* 10°49N 106°42E **39** G6
Gia Lai = Plei Ku *Vietnam* 13°57N 108°0E **38** F7
Gia Nghia *Vietnam* 11°58N 107°42E **39** G6
Gia Ngoc *Vietnam* 14°50N 108°58E **38** E7
Gia Vuc *Vietnam* 14°42N 108°34E **38** E7
Giamama = Jamaame *Somali Rep.* 0°4N 42°44E **47** G3
Gianitsa *Greece* 40°46N 22°24E **23** D10
Giant Forest *U.S.A.* 36°36N 118°43W **78** J8
Giant Sequoia ⌂ *U.S.A.* 36°10N 118°35W **78** K8
Giants Causeway *U.K.* 55°16N 6°29W **10** A5
Gianyar *Indonesia* 8°32S 115°20E **37** K18
Giarabub = Al Jaghbūb *Libya* 29°42N 24°38E **51** C10
Giarre *Italy* 37°43N 15°11E **22** F6
Gibara *Cuba* 21°9N 76°11W **88** B4
Gibbon *U.S.A.* 40°45N 98°51W **80** E4
Gibe II *Ethiopia* 8°0N 37°30E **47** F2
Gibe III *Ethiopia* 7°0N 37°30E **47** F2
Gibeon *Namibia* 25°9S 17°43E **56** D2
Gibraltar ⊘ *Europe* 36°7N 5°22W **21** D3
Gibraltar, Str. of *Medit. S.* 35°55N 5°40W **21** E3
Gibraltar Range △ *Australia* 29°31S 152°19E **63** D5
Gibson Desert *Australia* 24°0S 126°0E **60** D4
Gibsons *Canada* 49°24N 123°32W **70** D4
Gibsonville *U.S.A.* 39°46N 120°54W **78** F6
Giddings *U.S.A.* 30°11N 96°56W **84** F6
Giebnegáisi = Kebnekaise *Sweden* 67°53N 18°33E **8** C18
Giessen *Germany* 50°34N 8°41E **16** C5
Gīfān *Iran* 37°54N 57°28E **45** B8
Gift Lake *Canada* 55°53N 115°49W **70** B5
Gifu *Japan* 35°30N 136°45E **31** G8
Gifu □ *Japan* 35°40N 137°0E **31** G8
Giganta, Sa. de la *Mexico* 26°0N 111°39W **86** B2
Gigha *U.K.* 55°42N 5°44W **11** F3
Giglio *Italy* 42°20N 10°52E **22** C4
Gijón *Spain* 43°32N 5°42W **21** A3
Gikongoro *Rwanda* 2°28S 29°34E **54** C2
Gil I. *Canada* 53°12N 129°15W **70** C3
Gila → *U.S.A.* 32°43N 114°33W **77** K6
Gila Bend *U.S.A.* 32°57N 112°43W **77** K7
Gila Bend Mts. *U.S.A.* 33°10N 113°0W **77** K7
Gila Cliff Dwellings ⌂ *U.S.A.* 33°12N 108°16W **77** K9
Gīlān □ *Iran* 37°0N 50°0E **45** B6
Gilbert → *Australia* 16°35S 141°15E **62** B3
Gilbert Is. *Kiribati* 1°0N 172°0E **58** A10
Gilbert River *Australia* 18°9S 142°52E **62** B3
Gilbert Seamounts *Pac. Oc.* 52°50N 150°10W **4** D18
Gilead *U.S.A.* 44°24N 70°59W **83** B14
Gilf el Kebîr, Hadabat el *Egypt* 23°50N 25°50E **51** D11
Gilford I. *Canada* 50°40N 126°30W **70** C3
Gilgandra *Australia* 31°43S 148°39E **63** E4
Gilgil *Kenya* 0°30S 36°20E **54** C4
Gilgit *India* 35°50N 74°15E **43** B6
Gilgit → *Pakistan* 35°44N 74°37E **43** B6
Gili ⌂ *Mozam.* 16°39S 38°27E **55** F4
Gilimanuk *Indonesia* 8°10S 114°26E **37** J17
Gillam *Canada* 56°20N 94°40W **71** B10
Gillen, L. *Australia* 26°11S 124°38E **61** E3
Gilles, L. *Australia* 32°50S 136°45E **63** E2
Gillette *U.S.A.* 44°18N 105°30W **76** D11
Gilliat *Australia* 20°40S 141°28E **62** C3
Gillingham *U.K.* 51°23N 0°33E **13** F8
Gilmer *U.S.A.* 32°44N 94°57W **84** E7
Gilmore, L. *Australia* 32°29S 121°37E **61** F3
Gilmour *Canada* 44°48N 77°37W **82** B7
Gilroy *U.S.A.* 37°1N 121°34W **78** H5
Gimcheon *S. Korea* 36°11N 128°4E **35** F15
Gimhae *S. Korea* 35°14N 128°53E **35** G15
Gimhwa *S. Korea* 38°17N 127°28E **35** E14
Gimie, Mt. *St. Lucia* 13°54N 61°0W **89** f
Gimje *S. Korea* 35°48N 126°45E **35** G14
Gimli *Canada* 50°40N 97°0W **71** C9
Gin Gin *Australia* 25°0S 151°58E **63** D5
Gingin *Australia* 31°22S 115°54E **61** F2
Gingindlovu *S. Africa* 29°2S 31°30E **57** D5
Ginir *Ethiopia* 7°6N 40°40E **47** F3
Giofyros → *Greece* 35°20N 25°6E **25** D7
Giohar = Jawhar *Somali Rep.* 2°48N 45°30E **47** G4
Giona, Oros *Greece* 38°38N 22°14E **23** E10
Gir △ *India* 21°0N 71°0E **42** J4
Gir Hills *India* 21°0N 71°0E **42** J4
Girab *India* 26°2N 70°38E **42** F4
Girâfi, W. → *Egypt* 29°58N 34°39E **46** F3
Girard *St. Lucia* 13°59N 60°57W **89** D7
Girard *Kans., U.S.A.* 37°31N 94°51W **80** G6
Girard *Ohio, U.S.A.* 41°9N 80°42W **82** E4
Girard *Pa., U.S.A.* 42°0N 80°19W **82** E4
Girdle Ness *U.K.* 57°9N 2°3W **11** D6
Giresun *Turkey* 40°55N 38°30E **19** F6
Girga *Egypt* 26°17N 31°55E **51** C12
Giri → *India* 30°28N 77°41E **42** D7
Giridih *India* 24°10N 86°21E **43** G12
Giriftu *Kenya* 1°59N 39°46E **54** B4
Giriyondo *S. Africa* 23°15S 31°40E **57** C5
Girne = Kyrenia *Cyprus* 35°20N 33°20E **25** D12
Giron = Kiruna *Sweden* 67°52N 20°15E **8** C19
Girona *Spain* 41°58N 2°46E **21** B7
Gironde → *France* 45°32N 1°7W **20** D3
Girraween △ *Australia* 28°46S 151°54E **63** D5
Girringun △ *Australia* 18°15S 145°32E **62** B4
Giru *Australia* 19°30S 147°5E **62** B4
Girvan *U.K.* 55°14N 4°51W **11** F4
Gisborne *N.Z.* 38°39S 178°5E **59** C7
Gisenyi *Rwanda* 1°41S 29°15E **54** C2
Gislaved *Sweden* 57°19N 13°32E **9** H15
Gitarama *Rwanda* 2°4S 29°45E **54** C2
Gitega *Burundi* 3°26S 29°56E **54** C2
Githio *Greece* 36°46N 22°34E **23** F10
Giuba = Juba → *Somali Rep.* 1°30N 42°35E **47** G3
Giurgiu *Romania* 43°52N 25°57E **17** G13
Giza = El Gîza *Egypt* 30°0N 31°12E **51** C12
Giza Pyramids *Egypt* 29°58N 31°9E **51** C12
Gizab *Afghan.* 33°22N 66°17E **42** C1
Gizhiga *Russia* 62°3N 160°30E **29** C17
Gizhiginskaya Guba *Russia* 61°0N 158°0E **29** C16
Giżycko *Poland* 54°2N 21°48E **17** A11
Gjakovë = Đakovica *Kosovo* 42°22N 20°26E **23** C9
Gjirokastër *Albania* 40°7N 20°10E **23** D9
Gjoa Haven *Canada* 68°38N 95°53W **68** D12
Gjøvik *Norway* 60°47N 10°43E **8** F14
Glace Bay *Canada* 46°11N 59°58W **73** C8
Glacier △ *Canada* 51°15N 117°30W **70** C5
Glacier △ *U.S.A.* 48°42N 113°48W **76** B7
Glacier Bay *U.S.A.* 58°40N 136°0W **68** F4
Glacier Bay ⌂ *U.S.A.* 58°45N 136°30W **70** B1
Glacier Peak *U.S.A.* 48°7N 121°7W **76** B3
Gladewater *U.S.A.* 32°33N 94°56W **84** E7
Gladstone *Queens., Australia* 23°52S 151°16E **62** C5
Gladstone *S. Austral., Australia* 33°15S 138°22E **63** E2
Gladstone *Canada* 50°13N 98°57W **71** C9
Gladstone *U.S.A.* 45°51N 87°1W **80** C10
Gladwin *U.S.A.* 43°59N 84°29W **81** D11
Glagah *Indonesia* 8°13S 114°18E **37** J17
Glåma = Glomma → *Norway* 59°12N 10°57E **9** G14
Gláma *Iceland* 65°48N 23°0W **8** D2
Glamis *U.S.A.* 32°55N 115°5W **79** N11
Glamorgan, Vale of □ *U.K.* 51°28N 3°25W **13** F4
Glasco *Kans., U.S.A.* 39°22N 97°50W **80** F5
Glasco *N.Y., U.S.A.* 42°3N 73°57W **83** D11
Glasgow *U.K.* 55°51N 4°15W **11** F4
Glasgow *Ky., U.S.A.* 37°0N 85°55W **81** G11
Glasgow *Mont., U.S.A.* 48°12N 106°38W **76** B10
Glasgow, City of □ *U.K.* 55°51N 4°12W **11** F4
Glasgow Int. ✈ (GLA) *U.K.* 55°51N 4°21W **11** F4
Glaslyn *Canada* 53°22N 108°21W **71** C7
Glastonbury *U.K.* 51°9N 2°43W **13** F5
Glastonbury *U.S.A.* 41°43N 72°37W **83** E12
Glazov *Russia* 58°9N 52°40E **18** C9
Gleichen *Canada* 50°52N 113°3W **70** C6
Gleiwitz = Gliwice *Poland* 50°22N 18°41E **17** C10
Glen *U.S.A.* 44°7N 71°11W **83** B13
Glen Affric *U.K.* 57°17N 5°1W **11** D3
Glen Canyon *U.S.A.* 37°30N 110°40W **77** H8
Glen Canyon ⌂ *U.S.A.* 37°15N 111°0W **77** H8
Glen Canyon Dam *U.S.A.* 36°57N 111°29W **77** H8
Glen Coe *U.K.* 56°40N 5°0W **11** E3
Glen Cove *U.S.A.* 40°51N 73°38W **83** F11
Glen Garry *U.K.* 57°3N 5°7W **11** D3
Glen Innes *Australia* 29°44S 151°44E **63** D5
Glen Lyon *U.S.A.* 41°10N 76°5W **83** E8
Glen Mor *U.K.* 57°9N 4°37W **11** D4
Glen More ⌂ *U.K.* 57°8N 3°40W **11** D5
Glen Moriston *U.K.* 57°11N 4°52W **11** D4
Glen Robertson *Canada* 45°22N 74°30W **83** A1
Glen Spean *U.K.* 56°53N 4°40W **11** E4
Glen Ullin *U.S.A.* 46°49N 101°50W **80** B3
Glenariff △ *Ireland* 55°2N 6°10W **10** A5
Glenbeigh *Ireland* 52°3N 9°58W **10** D2
Glencoe *Canada* 42°45N 81°43W **82** D3
Glencoe *U.S.A.* 44°46N 94°9W **80** C6
Glencolumbkille *Ireland* 54°43N 8°42W **10** B3
Glendale *Ariz., U.S.A.* 33°32N 112°11W **77** K7
Glendale *Calif., U.S.A.* 34°9N 118°15W **79** L8
Glendale *Zimbabwe* 17°22S 31°5E **55** F3
Glendive *U.S.A.* 47°7N 104°43W **76** C11
Glendo *U.S.A.* 42°30N 105°2W **76** E1
Glenelg → *Australia* 38°4S 140°59E **63** F3
Glenfield *U.S.A.* 43°43N 75°24W **83** C9
Glengad Hd. *Ireland* 55°20N 7°10W **10** A4
Glengarriff *Ireland* 51°45N 9°34W **10** E2
Glenmont *U.S.A.* 40°31N 82°6W **82** F2
Glenmorgan *Australia* 27°14S 149°42E **63** D4
Glenn *U.S.A.* 39°31N 122°1W **78** F4
Glennallen *U.S.A.* 62°7N 145°33W **68** E2
Glennamaddy *Ireland* 53°37N 8°33W **10** C
Glenns Ferry *U.S.A.* 42°57N 115°18W **76** E
Glenora *Canada* 44°2N 77°3W **82** B
Glenorchy *Australia* 42°49S 147°18E **63** G
Glenore *Australia* 17°50S 141°12E **62** B
Glenreagh *Australia* 30°2S 153°1E **63** E
Glenrock *U.S.A.* 42°52N 105°52W **76** E1
Glenrothes *U.K.* 56°12N 3°10W **11** E
Glens Falls *U.S.A.* 43°19N 73°39W **83** C1
Glenside *U.S.A.* 40°6N 75°9W **83** F
Glenties *Ireland* 54°49N 8°16W **10** B
Glenveagh △ *Ireland* 55°3N 8°1W **10** A
Glenville *U.S.A.* 38°56N 80°50W **81** F1
Glenwood *Canada* 49°0N 54°58W **73** C
Glenwood *Ark., U.S.A.* 34°20N 93°33W **84** D
Glenwood *Iowa, U.S.A.* 41°3N 95°45W **80** E
Glenwood *Minn., U.S.A.* 45°39N 95°23W **80** C
Glenwood *Wash., U.S.A.* 46°1N 121°17W **78** D
Glenwood Springs *U.S.A.* 39°33N 107°19W **76** G1
Glettinganes *Iceland* 65°30N 13°37W **8** D
Glin *Ireland* 52°34N 9°17W **10** D
Gliwice *Poland* 50°22N 18°41E **17** C1
Globe *U.S.A.* 33°24N 110°47W **77** K
Głogów *Poland* 51°37N 16°5E **16** C
Glomma → *Norway* 59°12N 10°57E **9** G1
Glorieuses, Îs. *Ind. Oc.* 11°30S 47°20E **57** A
Glossop *U.K.* 53°27N 1°56W **12** D
Gloucester *Australia* 32°0S 151°59E **63** E
Gloucester *U.K.* 51°53N 2°15W **13** F
Gloucester *U.S.A.* 42°37N 70°40W **83** D1
Gloucester I. *Australia* 20°0S 148°30E **62**
Gloucester Island △ *Australia* 20°2S 148°30E **62**
Gloucester Point *U.S.A.* 37°15N 76°30W **81** G1
Gloucestershire □ *U.K.* 51°46N 2°15W **13**
Gloversville *U.S.A.* 43°3N 74°21W **83** C1
Glovertown *Canada* 48°40N 54°3W **73**
Glusk *Belarus* 52°53N 28°41E **17** B1
Gmünd *Austria* 48°45N 15°0E **16** D
Gmunden *Austria* 47°55N 13°48E **16** E
Gniezno *Poland* 52°30N 17°35E **17** B
Gnowangerup *Australia* 33°58S 117°59E **61**
Go Cong *Vietnam* 10°22N 106°40E **39**
Gō-no-ura *Japan* 33°44N 129°40E **31**
Goa *India* 15°33N 73°59E **40** M
Goa □ *India* 15°33N 73°59E **40** M
Goalen Hd. *Australia* 36°33S 150°4E **63**
Goalpara *India* 26°10N 90°40E **41** F1
Goaltor *India* 22°43N 87°10E **43** H1
Goalundo Ghat *Bangla.* 23°50N 89°47E **43** H
Goat Fell *U.K.* 55°38N 5°11W **11**
Goba *Ethiopia* 7°1N 39°59E **47**
Goba *Mozam.* 26°15S 32°13E **57** D
Gobabis *Namibia* 22°30S 19°0E **56**
Gobi *Asia* 44°0N 110°0E **34**
Gobō *Japan* 33°53N 135°10E **31**
Gochas *Namibia* 24°59S 18°55E **56**
Godalming *U.K.* 51°11N 0°36W **13**
Godavari → *India* 16°25N 82°18E **41** L
Godavari Pt. *India* 17°0N 82°20E **41** L
Godbout *Canada* 49°20N 67°38W **73**
Godda *India* 24°50N 87°13E **43** G
Goderich *Canada* 43°45N 81°41W **82**
Godfrey Ra. *Australia* 24°0S 117°0E **61**
Godhavn = Qeqertarsuaq *Greenland* 69°15N 53°38W **4**
Godhra *India* 22°49N 73°40E **42**
Godoy Cruz *Argentina* 32°56S 68°52W **94**
Gods → *Canada* 56°22N 92°51W **72**

ls L. Canada 54°40N 94°15W 72 B1
ls River Canada 54°50N 94°5W 71 C10
lthåb = Nuuk
reenland 64°10N 51°35W 67 C14
ie Hoop, Kaap die = Good
Hope, C. of S. Africa 34°24S 18°30E 56 E2
land, L. au Canada 49°50N 76°48W 72 C4
lands, L. aux Canada 55°27N 64°17W 73 A7
ree Neths. 51°50N 4°0E 15 C3
s Neths. 51°30N 3°55E 15 C3
fstown U.S.A. 43°1N 71°36W 83 C13
gama Canada 47°35N 81°43W 72 C3
gebic, L. U.S.A. 46°30N 89°35W 80 B9
gra = Ghaghara →
ndia 25°45N 84°40E 43 G11
griâl Sudan 8°30N 28°8E 51 G11
hana India 29°8N 76°42E 42 E7
harganj India 23°1N 77°41E 42 H7
→ India 22°4N 74°46E 42 H6
ânia Brazil 16°43S 49°20W 93 G9
ás Brazil 15°55S 50°10W 93 G8
ás □ Brazil 12°10S 48°0W 93 F9
o-Erê Brazil 24°12S 53°1W 95 A5
5 Japan 34°21N 135°42E 31 G7
ra Pakistan 31°10N 72°40E 42 D5
kçeada Turkey 40°10N 25°50E 23 D11
kova Körfezi Turkey 36°55N 27°50E 23 F12
ksu → Turkey 36°19N 34°5E 44 B2
kteik Burma 22°26N 97°0E 41 H20
kurt Pakistan 29°40N 67°26E 42 E2
kwe Zimbabwe 18°7S 28°58E 57 B4
a India 28°3N 80°32E 43 E9
akganj India 26°8N 89°52E 43 F13
an Heights = Hagolan
yria 33°0N 35°45E 46 C4
āshkerd Iran 27°59N 57°16E 45 E8
conda U.S.A. 40°58N 117°30W 76 F5
d U.S.A. 41°52N 77°50W 82 E7
d Beach U.S.A. 42°25N 124°25W 76 E1
d Coast W. Afr. 4°0N 1°40W 50 H5
d Hill U.S.A. 42°26N 123°3W 76 E2
d River Canada 49°46N 126°3W 70 D3
den Canada 51°20N 116°59W 70 C5
den B. N.Z. 40°40S 172°50E 59 D4
den Gate U.S.A. 37°48N 122°29W 77 H2
den Gate Highlands △
Africa 28°40S 28°40E 57 D4
den Hinde Canada 49°40N 125°44W 70 D3
den Lake Canada 45°34N 77°21W 82 A7
den Spike ⌂ U.S.A. 41°37N 112°33W 76 F7
den Vale Ireland 52°33N 8°17W 10 D3
dendale U.S.A. 45°49N 120°50W 76 D3
dfield U.S.A. 37°42N 117°14W 77 H5
dsand L. Canada 57°2N 101°8W 71 B8
dsboro U.S.A. 35°23N 77°59W 85 D16
dsmith U.S.A. 31°59N 102°37W 84 F3
dthwaite U.S.A. 31°27N 98°34W 84 F5
eniów Poland 53°35N 14°50E 16 B8
estān □ Iran 37°20N 55°25E 45 B7
estānak Iran 30°36N 54°14E 45 D7
eta U.S.A. 34°27N 119°50W 79 L7
ito Costa Rica 8°41N 83°5W 88 E3
fo Aranci Italy 40°59N 9°38E 22 D3
ad U.S.A. 28°40N 97°23W 84 G6
mud China 36°25N 94°53E 32 D7
pāyegān Iran 33°27N 50°18E 45 C6
ra Pakistan 33°37N 72°56E 42 C5
spie U.K. 57°58N 3°59W 11 D5
na
em. Rep. of the Congo 1°37S 29°10E 54 C2
nal Pass Pakistan 31°56N 69°20E 42 D3
nati → India 25°32N 83°11E 43 G10
nbari
em. Rep. of the Congo 2°45N 29°3E 54 B2
nbe Nigeria 10°19N 11°2E 51 F8
nbe → Tanzania 4°38S 31°40E 54 C3
nbe Stream △
anzania 4°42S 29°37E 54 C2
nel = Homyel Belarus 52°28N 31°0E 17 B16
nera Canary Is. 28°7N 17°14W 24 F2
nez Palacio Mexico 25°34N 103°30W 86 B4
nīshān Iran 37°4N 54°6E 45 B7
nogomo Indonesia 6°39S 134°43E 37 F8
noh India 23°52N 86°10E 43 H12
npa = Ganta Liberia 7°15N 8°59W 50 G4
nābād Iran 34°15N 58°45E 45 C8
naïves Haiti 19°20N 72°42W 89 C5
narezhou △ Zimbabwe 21°32S 31°55E 55 G3
nâve, G. de la Haiti 19°29N 72°42W 89 C5
nâve, Île de la Haiti 18°51N 73°3W 89 C5
nbad-e Kāvūs Iran 37°20N 55°25E 45 B7
nda India 27°9N 81°58E 43 F9
ndal India 21°58N 70°52E 42 J4
nder Ethiopia 12°39N 37°30E 47 E2
ndia India 21°23N 80°10E 40 J12
ndola Mozam. 19°10S 33°37E 55 F3
nen Turkey 40°6N 27°39E 23 D12
ngbei China 22°12N 113°32E 33 a
ngchangling China 41°7N 123°27E 35 D12
ngga Shan China 29°40N 101°55E 32 F9
nggar China 29°23N 91°7E 32 F7
nghe China 36°18N 100°32E 32 D9
ngju S. Korea 36°27N 127°7E 35 F14
ngming China 22°47N 113°53E 33 a
ngolgon Australia 30°21S 146°54E 63 E4
ngyi China 34°45N 112°58E 34 G7
ngzhuling China 43°30N 124°40E 35 C13
niri Nigeria 11°30N 12°15E 51 F8
nzales Calif., U.S.A. 36°30N 121°26W 78 J5
nzales Tex., U.S.A. 29°30N 97°27W 84 G6
nzález Mexico 22°48N 98°25W 87 C5
od Hope, C. of S. Africa 34°24S 18°30E 56 E2
od Hope Lake
Canada 59°16N 129°18W 70 B3
oderham Canada 44°54N 78°21W 82 B6
odhouse S. Africa 28°57S 18°13E 56 D2
oding U.S.A. 42°56N 114°43W 76 E6
odland U.S.A. 39°21N 101°43W 80 F3
odlands Mauritius 20°2S 57°39E 53 d
odlow Canada 56°20N 120°8W 70 B4
Goodooga Australia 29°3S 147°28E 63 D4
Goodsprings U.S.A. 35°49N 115°27W 79 K11
Goole U.K. 53°42N 0°53W 12 D7
Goolgowi Australia 33°58S 145°41E 63 E4
Goomalling Australia 31°15S 116°49E 61 F2
Goomeri Australia 26°12S 152°6E 63 D5
Goonda Mozam. 19°48S 33°57E 55 F3
Goondiwindi Australia 28°30S 150°21E 63 D5
Goongarrie, L. Australia 30°3S 121°9E 61 F3
Goongarrie △ Australia 30°7S 121°30E 61 F3
Goonyella Australia 21°47S 147°58E 62 C4
Goose → Canada 53°20N 60°35W 73 B7
Goose Creek U.S.A. 32°59N 80°2W 85 E14
Goose L. U.S.A. 41°56N 120°26W 76 F3
Gop India 22°5N 69°50E 42 H3
Gopalganj India 26°28N 84°30E 43 F11
Göppingen Germany 48°42N 9°39E 16 D5
Gorakhpur India 26°47N 83°23E 43 F10
Goražde Bos.-H. 43°38N 18°58E 23 C8
Gorda, Pta. Canary Is. 28°45N 18°0W 24 F2
Gorda, Pta. Nic. 14°20N 83°10W 88 D3
Gordan B. Australia 11°35S 130°10E 60 B5
Gordon U.S.A. 42°48N 102°12W 80 D2
Gordon → Australia 42°27S 145°30E 63 G4
Gordon Bay Canada 45°12N 79°47W 82 A5
Gordon L. Alta., Canada 56°30N 110°25W 71 B6
Gordon L. N.W.T.,
Canada 63°5N 113°11W 70 A6
Gordonvale Australia 17°5S 145°50E 62 B4
Goré Chad 7°59N 16°31E 51 G9
Gore Ethiopia 8°12N 35°32E 47 F2
Gore N.Z. 46°5S 168°58E 59 G2
Gore Bay Canada 45°57N 82°28W 72 C3
Gorey Ireland 52°41N 6°18W 10 D5
Gorg Iran 29°29N 59°43E 45 D8
Gorgān Iran 36°55N 54°30E 45 B7
Gorgona, I. Colombia 3°0N 78°10W 92 C3
Gorham U.S.A. 44°23N 71°10W 83 B13
Goriganga → India 29°45N 80°23E 43 E9
Gorinchem Neths. 51°50N 4°59E 15 C4
Goris Armenia 39°31N 46°22E 19 G8
Gorízia Italy 45°56N 13°37E 22 B5
Gorkiy = Nizhniy Novgorod
Russia 56°20N 44°0E 18 C7
Gorkovskoye Vdkhr.
Russia 57°2N 43°4E 18 C7
Gorleston U.K. 52°35N 1°44E 13 E9
Görlitz Germany 51°9N 14°58E 16 C8
Gorlovka = Horlivka
Ukraine 48°19N 38°5E 19 E6
Gorman U.S.A. 34°47N 118°51W 79 L8
Gorna Dzhumayo = Blagoevgrad
Bulgaria 42°2N 23°5E 23 C10
Gorna Oryakhovitsa
Bulgaria 43°7N 25°40E 23 C11
Gorno-Altay □ Russia 51°0N 86°0E 28 D9
Gorno-Altaysk Russia 51°50N 86°5E 28 D9
Gornozavodsk Russia 46°33N 141°50E 29 E15
Gornyatski Russia 67°32N 64°3E 18 A11
Gornyy Russia 44°57N 133°59E 30 B6
Gorodenka = Horodenka
Ukraine 48°41N 25°29E 17 D13
Gorodok = Horodok
Ukraine 49°46N 23°32E 17 D12
Gorokhov = Horokhiv
Ukraine 50°30N 24°45E 17 C13
Goromonzi Zimbabwe 17°52S 31°22E 55 F3
Gorong, Kepulauan
Indonesia 3°59S 131°25E 37 E8
Gorongose → Mozam. 20°30S 34°40E 57 C5
Gorongoza Mozam. 18°44S 34°2E 55 F3
Gorongoza, Sa. da Mozam. 18°27S 34°2E 55 F3
Gorongoza △ Mozam. 18°50S 34°29E 57 B5
Gorontalo Indonesia 0°35N 123°5E 37 D6
Gorontalo □ Indonesia 0°50N 122°20E 37 D6
Gort Ireland 53°3N 8°49W 10 C3
Gortis Greece 35°4N 24°58E 25 D6
Goryeong S. Korea 35°44N 128°15E 35 G15
Gorzów Wielkopolski
Poland 52°43N 15°15E 16 B8
Gosford Australia 33°23S 151°18E 63 E5
Goshen Calif., U.S.A. 36°21N 119°25W 78 J7
Goshen Ind., U.S.A. 41°35N 85°50W 81 E11
Goshen N.Y., U.S.A. 41°24N 74°20W 83 E10
Goshogawara Japan 40°48N 140°27E 30 D10
Goslar Germany 51°54N 10°25E 16 C6
Gospič Croatia 44°35N 15°23E 16 F8
Gosport U.K. 50°48N 1°9W 13 G6
Gosse → Australia 19°32S 134°37E 62 B1
Göta älv → Sweden 57°42N 11°54E 9 H14
Göta kanal Sweden 58°30N 15°58E 9 G16
Götaland Sweden 57°30N 14°30E 9 H16
Göteborg Sweden 57°43N 11°59E 9 H14
Gotha Germany 50°56N 10°42E 16 C6
Gothenburg = Göteborg
Sweden 57°43N 11°59E 9 H14
Gothenburg U.S.A. 40°56N 100°10W 80 E3
Gotland Sweden 57°30N 18°33E 9 H18
Gotō-Rettō Japan 32°55N 129°5E 31 H4
Gotska Sandön Sweden 58°24N 19°15E 9 G18
Gōtsu Japan 35°0N 132°14E 31 G6
Göttingen Germany 51°31N 9°55E 16 C5
Gottwaldov = Zlín
Czech Rep. 49°14N 17°40E 17 D9
Goubangzi China 41°20N 121°52E 35 D11
Gouda Neths. 52°1N 4°42E 15 B4
Goudouras, Akra Greece 34°59N 26°6E 25 E8
Gouin, Rés. Canada 48°35N 74°40W 72 C5
Goulburn Australia 34°44S 149°44E 63 E4
Goulburn Is. Australia 11°40S 133°20E 62 A1
Goulimine Morocco 28°56N 10°0W 50 C3
Goundam Mali 16°27N 3°40W 50 E5
Gourits → S. Africa 34°21S 21°52E 56 E3
Gournes Greece 35°19N 25°16E 25 D7
Gourock U.K. 55°57N 4°49W 11 F4
Gouverneur U.S.A. 44°20N 75°28W 83 B9
Gouvia Greece 39°39N 19°50E 25 A3
Gove Peninsula
Australia 12°17S 136°49E 62 A2
Governador Valadares
Brazil 18°15S 41°57W 93 G10
Governor's Harbour
Bahamas 25°10N 76°14W 88 A4
Govindgarh India 24°23N 81°18E 43 G9
Gowan Ra. Australia 25°0S 145°0E 62 D4
Gowanda U.S.A. 42°28N 78°56W 82 D6
Gower U.K. 51°35N 4°10W 13 F3
Gowna, L. Ireland 53°51N 7°34W 10 C4
Goya Argentina 29°10S 59°10W 94 B4
Goyang S. Korea 37°39N 126°50E 35 F14
Goyder Lagoon Australia 27°3S 138°58E 63 D2
Goyllarisquizga Peru 10°31S 76°24W 92 F3
Goz Beïda Chad 12°10N 21°20E 51 F10
Gozo Malta 36°3N 14°15E 25 C1
Graaff-Reinet S. Africa 32°13S 24°32E 56 E3
Gračac Croatia 44°18N 15°57E 16 F8
Gracias a Dios, C.
Honduras 15°0N 83°10W 88 D3
Graciosa Azores 39°4N 28°0W 50 a
Graciosa, I. Canary Is. 29°15N 13°32W 24 E6
Grado Spain 43°23N 6°4W 21 A2
Grady U.S.A. 34°49N 103°19W 77 J12
Grafham Water U.K. 52°19N 0°18W 13 E7
Grafton Australia 29°38S 152°58E 63 D5
Grafton N. Dak., U.S.A. 48°25N 97°25W 80 A5
Grafton W. Va., U.S.A. 39°21N 80°2W 81 F13
Graham Canada 49°20N 90°30W 72 C1
Graham U.S.A. 33°6N 98°35W 84 E5
Graham, Mt. U.S.A. 32°42N 109°52W 77 K9
Graham Bell, Ostrov = Greem-
Bell, Ostrov Russia 81°0N 62°0E 28 A7
Graham I. B.C., Canada 53°40N 132°30W 70 C2
Graham I. Nunavut,
Canada 77°25N 90°30W 69 B13
Graham Land Antarctica 65°0S 64°0W 5 C17
Grahamstown S. Africa 33°19S 26°31E 56 E4
Grahamsville U.S.A. 41°51N 74°33W 83 E10
Grain Coast W. Afr. 4°20N 10°0W 50 H3
Grajagan Indonesia 8°35S 114°13E 37 K17
Grajaú Brazil 5°50S 46°4W 93 E9
Grajaú → Brazil 3°41S 44°48W 93 D10
Grampian U.S.A. 40°58N 78°37W 82 F6
Grampian Highlands = Grampian
Mts. U.K. 56°50N 4°0W 11 E5
Grampian Mts. U.K. 56°50N 4°0W 11 E5
Grampians, The
Australia 37°15S 142°20E 63 F3
Gran Canaria Canary Is. 27°55N 15°35W 24 G4
Gran Chaco S. Amer. 25°0S 61°0W 94 B3
Gran Desierto del Pinacate ⌂
Mexico 31°51N 113°32W 86 A2
Gran Paradiso Italy 45°33N 7°17E 20 D7
Gran Sasso d'Itália Italy 42°27N 13°42E 22 C5
Granada Nic. 11°58N 86°0W 88 D2
Granada Spain 37°10N 3°35W 21 D4
Granada U.S.A. 38°4N 102°19W 76 G12
Granadilla de Abona
Canary Is. 28°7N 16°33W 24 F3
Granard Ireland 53°47N 7°30W 10 C4
Granbury U.S.A. 32°27N 97°47W 84 E6
Granby Canada 45°25N 72°45W 83 A12
Granby U.S.A. 40°5N 105°56W 76 F11
Grand → Canada 42°51N 79°34W 82 D5
Grand → Mo., U.S.A. 39°23N 93°7W 80 F7
Grand → S. Dak.,
U.S.A. 45°40N 100°45W 80 C3
Grand Bahama I.
Bahamas 26°40N 78°30W 88 A4
Grand Baie Mauritius 20°0S 57°35E 53 d
Grand Bank Canada 47°6N 55°48W 73 C8
Grand Bassam Ivory C. 5°10N 3°49W 50 G5
Grand Bend Canada 43°18N 81°45W 82 C3
Grand-Bourg Guadeloupe 15°53N 61°19W 88 b
Grand Canal = Da Yunhe →
China 39°10N 117°10E 35 E9
Grand Canyon U.S.A. 36°3N 112°9W 77 H7
Grand Canyon △
U.S.A. 36°15N 112°30W 77 H7
Grand Canyon-Parashant ⌂
U.S.A. 36°30N 113°45W 77 H7
Grand Cayman
Cayman Is. 19°20N 81°20W 88 C3
Grand Coulee U.S.A. 47°57N 119°0W 76 C4
Grand Coulee Dam
U.S.A. 47°57N 118°59W 76 C4
Grand Erg de Bilma Niger 18°30N 14°0E 51 E8
Grand Falls Canada 47°3N 67°44W 73 C6
Grand Falls-Windsor
Canada 48°56N 55°40W 73 C8
Grand Forks Canada 49°0N 118°30W 70 D5
Grand Forks U.S.A. 47°55N 97°3W 80 B5
Grand Gorge U.S.A. 42°21N 74°29W 83 D10
Grand Haven U.S.A. 43°4N 86°13W 80 D10
Grand I. Mich., U.S.A. 46°31N 86°40W 80 B10
Grand I. N.Y., U.S.A. 43°0N 78°58W 82 D6
Grand Island U.S.A. 40°55N 98°21W 80 E4
Grand Isle La., U.S.A. 29°14N 90°0W 85 G9
Grand Isle Vt., U.S.A. 44°43N 73°18W 83 B11
Grand Junction U.S.A. 39°4N 108°33W 76 G9
Grand L. N.B., Canada 45°57N 66°7W 73 C6
Grand L. Nfld. & L.,
Canada 49°0N 57°30W 73 C8
Grand L. Nfld. & L.,
Canada 53°40N 60°30W 73 B7
Grand L. U.S.A. 29°55N 92°47W 84 G8
Grand Lake U.S.A. 40°15N 105°49W 76 F11
Grand Manan I. Canada 44°45N 66°52W 73 D6
Grand Marais Mich.,
U.S.A. 46°40N 85°59W 81 B11
Grand Marais Minn.,
U.S.A. 47°45N 90°25W 72 C1
Grand-Mère Canada 46°36N 72°40W 72 C5
Grand Portage U.S.A. 47°58N 89°41W 80 B9
Grand Prairie U.S.A. 32°44N 96°59W 84 E6
Grand Rapids Canada 53°12N 99°19W 71 C9
Grand Rapids Mich.,
U.S.A. 42°58N 85°40W 81 D11
Grand Rapids Minn.,
U.S.A. 47°14N 93°31W 80 B7
Grand St-Bernard, Col du
Europe 45°50N 7°10E 20 D7
Grand Staircase-Escalante ⌂
U.S.A. 37°25N 111°50W 77 H8
Grand Teton U.S.A. 43°54N 110°50W 76 E8
Grand Teton △ U.S.A. 43°50N 110°50W 76 E8
Grand Union Canal U.K. 52°7N 0°53W 13 E7
Grande → Jujuy,
Argentina 24°20S 65°2W 94 A2
Grande → Mendoza,
Argentina 36°52S 69°45W 94 D2
Grande → Bolivia 15°51S 64°39W 92 G6
Grande → Bahia, Brazil 11°30S 44°30W 93 F10
Grande → Minas Gerais,
Brazil 20°6S 51°4W 93 H8
Grande, B. Argentina 50°30S 68°20W 96 G3
Grande, Rio → N. Amer. 25°58N 97°9W 84 J6
Grande Anse Seychelles 4°18S 55°45E 53 b
Grande Baleine →
Canada 55°16N 77°47W 72 A4
Grande Cache Canada 53°53N 119°8W 70 C5
Grande Comore
Comoros Is. 11°35S 43°20E 53 a
Grande-Entrée Canada 47°30N 61°40W 73 C7
Grande Prairie Canada 55°10N 118°50W 70 B5
Grande-Rivière Canada 48°26N 64°30W 73 C7
Grande-Terre Guadeloupe 16°20N 61°25W 88 b
Grande-Vallée Canada 49°14N 65°8W 73 C6
Grande Vigie, Pte. de la
Guadeloupe 16°32N 61°27W 88 b
Grandfalls U.S.A. 31°20N 102°51W 84 F3
Grands-Jardins ⌂
Canada 47°41N 70°51W 73 C5
Grandview Canada 51°10N 100°42W 71 C8
Grandview U.S.A. 46°15N 119°54W 76 C4
Graneros Chile 34°5S 70°45W 94 C1
Grangemouth U.K. 56°1N 3°42W 11 E5
Granger U.S.A. 41°35N 109°58W 76 F9
Grangeville U.S.A. 45°56N 116°7W 76 D5
Granisle Canada 54°53N 126°13W 70 C3
Granite City U.S.A. 38°42N 90°8W 80 F8
Granite Falls U.S.A. 44°49N 95°33W 80 C6
Granite L. Canada 48°8N 57°5W 73 C8
Granite Mt. U.S.A. 33°5N 116°28W 79 M10
Granite Pk. U.S.A. 45°10N 109°48W 76 D9
Graniteville U.S.A. 44°8N 72°29W 83 B12
Granity N.Z. 41°39S 171°51E 59 D3
Granja Brazil 3°7S 40°50W 93 D10
Granollers Spain 41°39N 2°18E 21 B7
Grant U.S.A. 40°50N 101°43W 80 E3
Grant, Mt. U.S.A. 38°34N 118°48W 76 G4
Grant City U.S.A. 40°29N 94°25W 80 E6
Grant I. Australia 11°10S 132°52E 60 B5
Grant Range U.S.A. 38°30N 115°25W 76 G6
Grantham U.K. 52°55N 0°38W 12 E7
Grantown-on-Spey U.K. 57°20N 3°36W 11 D5
Grants U.S.A. 35°9N 107°52W 77 J10
Grants Pass U.S.A. 42°26N 123°19W 76 E2
Grantsville U.S.A. 40°36N 112°28W 76 F7
Granville France 48°50N 1°35W 20 B3
Granville N. Dak.,
U.S.A. 48°16N 100°47W 80 A3
Granville N.Y., U.S.A. 43°24N 73°16W 83 C11
Granville Ohio, U.S.A. 40°4N 82°31W 82 F2
Granville L. Canada 56°18N 100°30W 71 B8
Graskop S. Africa 24°56S 30°49E 57 C5
Grass → Canada 56°3N 96°33W 71 B9
Grass Range U.S.A. 47°2N 108°48W 76 C9
Grass River ⌂ Canada 54°40N 100°50W 71 C8
Grass Valley Calif.,
U.S.A. 39°13N 121°4W 78 F6
Grass Valley Oreg.,
U.S.A. 45°22N 120°47W 76 D3
Grasse France 43°38N 6°56E 20 E7
Grassflat U.S.A. 41°0N 78°6W 82 E6
Grasslands △ Canada 49°11N 107°38W 71 D7
Grassy Australia 40°3S 144°5E 63 G3
Graulhet France 43°45N 1°59E 20 E4
Gravelbourg Canada 49°50N 106°35W 71 D7
's-Gravenhage Neths. 52°7N 4°17E 15 B4
Gravenhurst Canada 44°52N 79°20W 82 B5
Gravesend Australia 29°35S 150°20E 63 D5
Gravesend U.K. 51°26N 0°22E 13 F8
Gravois, Pointe-à- Haiti 18°15N 73°56W 89 C5
Grayling U.S.A. 44°40N 84°43W 81 C11
Grays U.K. 51°28N 0°21E 13 F8
Grays Harbor U.S.A. 46°59N 124°1W 76 C1
Grays L. U.S.A. 43°4N 111°26W 76 E8
Grays River U.S.A. 46°21N 123°37W 78 D3
Graz Austria 47°4N 15°27E 16 E8
Greasy L. Canada 62°55N 122°12W 70 A4
Great Abaco I. = Abaco I.
Bahamas 26°25N 77°10W 88 A4
Great Artesian Basin
Australia 23°0S 144°0E 62 C3
Great Australian Bight
Australia 33°30S 130°0E 61 F5
Great Bahama Bank
Bahamas 23°15N 78°0W 88 B4
Great Barrier I. N.Z. 36°11S 175°25E 59 B5
Great Barrier Reef
Australia 18°0S 146°50E 62 B4
Great Barrier Reef ⌂
Australia 20°0S 150°0E 62 B4
Great Barrington
U.S.A. 42°12N 73°22W 83 D11
Great Basalt Wall △
Australia 19°52S 145°43E 62 B4
Great Basin U.S.A. 40°0N 117°0W 76 G5
Great Basin △ U.S.A. 38°56N 114°15W 76 G6
Great Bear → Canada 65°0N 126°0W 68 E7
Great Bear L. Canada 65°30N 120°0W 68 D8
Great Belt = Store Bælt
Denmark 55°20N 11°0E 9 J14
Great Bend Kans., U.S.A. 38°22N 98°46W 80 F4
Great Bend Pa., U.S.A. 41°58N 75°45W 83 E9
Great Blasket I. Ireland 52°6N 10°32W 10 D1
Great Britain Europe 54°0N 2°15W 6 E5
Great Camanoe
Br. Virgin Is. 18°30N 64°35W 89 e
Great Coco I. = Koko Kyunzu
Burma 14°7N 93°22E 41 M18
Great Codroy Canada 47°51N 59°16W 73 C8
Great Divide, The = Great
Dividing Ra. Australia 23°0S 146°0E 62 C4
Great Divide Basin
U.S.A. 42°0N 108°0W 76 E9
Great Dividing Ra.
Australia 23°0S 146°0E 62 C4
Great Driffield = Driffield
U.K. 54°0N 0°26W 12 C7
Great Exuma I. Bahamas 23°30N 75°50W 88 B4
Great Falls U.S.A. 47°30N 111°17W 76 C8
Great Fish = Groot-Vis →
S. Africa 33°28S 27°5E 56 E4
Great Guana Cay
Bahamas 24°0N 76°20W 88 B4
Great Himalayan △
India 31°30N 77°30E 42 D7
Great Inagua I. Bahamas 21°0N 73°20W 89 B5
Great Indian Desert = Thar Desert
India 28°0N 72°0E 42 F5
Great Karoo S. Africa 31°55S 21°0E 56 E3
Great Khingan Mts. = Da
Hinggan Ling China 48°0N 121°0E 33 B13
Great Lake Australia 41°50S 146°40E 63 G4
Great Lakes N. Amer. 46°0N 84°0W 66 E11
Great Limpopo Transfrontier △
Africa 23°0S 31°45E 57 C5
Great Malvern U.K. 52°7N 2°18W 13 E5
Great Miami → U.S.A. 39°7N 84°49W 81 F11
Great Ormes Head U.K. 53°20N 3°52W 12 D4
Great Ouse → U.K. 52°48N 0°21E 12 E8
Great Palm I. Australia 18°45S 146°40E 62 B4
Great Pedro Bluff Jamaica 17°51N 77°44W 88 a
Great Pee Dee →
U.S.A. 33°21N 79°10W 85 E15
Great Plains N. Amer. 47°0N 105°0W 66 E9
Great Ruaha → Tanzania 7°56S 37°52E 54 D4
Great Sacandaga L.
U.S.A. 43°6N 74°16W 83 C10
Great Saint Bernard Pass = Grand
St-Bernard, Col du
Europe 45°50N 7°10E 20 D7
Great Salt Desert = Kavīr, Dasht-e
Iran 34°30N 55°0E 45 C7
Great Salt L. U.S.A. 41°15N 112°40W 76 F7
Great Salt Lake Desert
U.S.A. 40°50N 113°30W 76 F7
Great Salt Plains L.
U.S.A. 36°45N 98°8W 84 C5
Great Sand Dunes ⌂
U.S.A. 37°48N 105°45W 77 H11
Great Sandy △ Australia 26°13S 153°2E 63 D5
Great Sandy Desert
Australia 21°0S 124°0E 60 D3
Great Sangi = Sangihe, Pulau
Indonesia 3°35N 125°30E 37 D7
Great Sea Reef Fiji 16°15S 179°0E 59 a
Great Sitkin I. U.S.A. 52°3N 176°6W 74 E4
Great Skellig Ireland 51°47N 10°33W 10 E1
Great Slave L. Canada 61°23N 115°38W 70 A5
Great Smoky Mts. △
U.S.A. 35°40N 83°40W 85 D13
Great Stour = Stour →
U.K. 51°18N 1°22E 13 F9
Great Victoria Desert
Australia 29°30S 126°30E 61 E4
Great Wall Antarctica 62°30S 58°0W 5 C18
Great Wall China 38°30N 109°30E 34 E5
Great Whernside U.K. 54°10N 1°58W 12 C6
Great Yarmouth U.K. 52°37N 1°44E 13 E9
Great Zab = Zāb al Kabīr →
Iraq 36°1N 43°24E 44 B4
Great Zimbabwe
Zimbabwe 20°16S 30°54E 55 G3
Greater Antilles W. Indies 17°40N 74°0W 89 C5
Greater London □ U.K. 51°31N 0°6W 13 F7
Greater Manchester □
U.K. 53°30N 2°15W 12 D5
Greater St. Lucia Wetlands ⌂
S. Africa 28°6S 32°27E 57 D5
Greater Sudbury = Sudbury
Canada 46°30N 81°0W 72 C3
Greater Sunda Is. Indonesia 7°0S 112°0E 36 F4
Greco, C. Cyprus 34°57N 34°5E 25 E13
Gredos, Sierra de Spain 40°20N 5°0W 21 B3
Greece U.S.A. 43°13N 77°41W 82 C7
Greece ■ Europe 40°0N 23°0E 23 E9
Greeley Colo., U.S.A. 40°25N 104°42W 76 F11
Greeley Nebr., U.S.A. 41°33N 98°32W 80 E4
Greely Fd. Canada 80°30N 85°0W 69 A15
Greem-Bell, Ostrov Russia 81°0N 62°0E 28 A7
Green → Ky., U.S.A. 37°54N 87°30W 80 G10
Green → Utah, U.S.A. 38°11N 109°53W 76 G9
Green B. U.S.A. 45°0N 87°30W 80 C10
Green Bay U.S.A. 44°31N 88°0W 80 C9
Green C. Australia 37°13S 150°1E 63 F5
Green Cove Springs
U.S.A. 29°59N 81°42W 85 G14
Green Lake Canada 54°17N 107°47W 71 C7
Green Mts. U.S.A. 43°45N 72°45W 83 C12
Green River Utah,
U.S.A. 38°59N 110°10W 76 G8
Green River Wyo.,
U.S.A. 41°32N 109°28W 76 F9
Green Valley U.S.A. 31°52N 110°56W 77 L8
Greenbank U.S.A. 48°6N 122°34W 78 B4
Greenbush Mich., U.S.A. 44°35N 83°19W 82 B1
Greenbush Minn., U.S.A. 48°42N 96°11W 80 A5
Greencastle U.S.A. 39°38N 86°52W 80 F10
Greene U.S.A. 42°20N 75°46W 83 D9
Greeneville U.S.A. 36°10N 82°50W 85 C13
Greenfield Calif., U.S.A. 36°19N 121°15W 78 J5
Greenfield Calif., U.S.A. 35°15N 119°0W 79 K8
Greenfield Ind., U.S.A. 39°47N 85°46W 81 F11
Greenfield Iowa, U.S.A. 41°18N 94°28W 80 E6
Greenfield Mass.,
U.S.A. 42°35N 72°36W 83 D12
Greenfield Mo., U.S.A. 37°25N 93°51W 80 G7
Greenfield Park Canada 45°29N 73°28W 83 A11
Greenland ☑ N. Amer. 66°0N 45°0W 67 C15

124 Greenland Sea INDEX TO WORLD MAP

Greenland Sea Arctic 73°0N 10°0W 4 B7
Greenock U.K. 55°57N 4°46W 11 F4
Greenore Ireland 54°2N 6°8W 10 B5
Greenore Pt. Ireland 52°14N 6°19W 10 D5
Greenough Australia 28°58S 114°43E 61 E1
Greenough → Australia 28°51S 114°38E 61 E1
Greenough Pt. Canada 44°58N 81°26W 82 B3
Greenport U.S.A. 41°6N 72°22W 83 E12
Greensboro Ga., U.S.A. 33°35N 83°11W 85 E13
Greensboro N.C., U.S.A. 36°4N 79°48W 85 C15
Greensboro Vt., U.S.A. 44°36N 72°18W 83 B12
Greensburg Ind., U.S.A. 39°20N 85°29W 81 F11
Greensburg Kans., U.S.A. 37°36N 99°18W 80 G4
Greensburg Pa., U.S.A. 40°18N 79°33W 82 F5
Greenstone = Geraldton Canada 49°44N 87°10W 72 C2
Greenstone Pt. U.K. 57°55N 5°37W 11 D3
Greenvale Australia 18°59S 145°7E 62 B4
Greenville Liberia 5°1N 9°6W 50 G4
Greenville Ala., U.S.A. 31°50N 86°38W 85 F11
Greenville Calif., U.S.A. 40°8N 120°57W 78 E6
Greenville Maine, U.S.A. 45°28N 69°35W 81 C19
Greenville Mich., U.S.A. 43°11N 85°15W 81 D11
Greenville Miss., U.S.A. 33°24N 91°4W 85 E9
Greenville Mo., U.S.A. 37°8N 90°27W 80 G8
Greenville N.C., U.S.A. 35°37N 77°23W 85 D16
Greenville N.H., U.S.A. 42°46N 71°49W 83 D13
Greenville N.Y., U.S.A. 42°25N 74°1W 83 D10
Greenville Ohio, U.S.A. 40°6N 84°38W 81 E11
Greenville Pa., U.S.A. 41°24N 80°23W 82 E4
Greenville S.C., U.S.A. 34°51N 82°24W 85 D13
Greenville Tex., U.S.A. 33°8N 96°7W 84 E6
Greenwater Lake ⌓ Canada 52°32N 103°30W 71 C8
Greenwich Conn., U.S.A. 41°2N 73°38W 83 E11
Greenwich N.Y., U.S.A. 43°5N 73°30W 83 C11
Greenwich Ohio, U.S.A. 41°2N 82°31W 82 E2
Greenwich □ U.K. 51°29N 0°1E 13 F8
Greenwood Canada 49°10N 118°40W 70 D5
Greenwood Ark., U.S.A. 35°13N 94°16W 84 D7
Greenwood Ind., U.S.A. 39°37N 86°7W 80 F10
Greenwood Miss., U.S.A. 33°31N 90°11W 85 E9
Greenwood S.C., U.S.A. 34°12N 82°10W 85 D13
Greenwood, Mt. Australia 13°48S 130°4E 60 B5
Gregory U.S.A. 43°14N 99°26W 80 D4
Gregory → Australia 17°53S 139°17E 62 B2
Gregory, L. S. Austral., Australia 28°55S 139°0E 63 D2
Gregory, L. W. Austral., Australia 20°0S 127°40E 60 D4
Gregory, L. W. Austral., Australia 25°38S 119°58E 61 E2
Gregory △ Australia 15°38S 131°15E 60 C5
Gregory Downs Australia 18°35S 138°45E 62 B2
Gregory Ra. Queens., Australia 19°30S 143°40E 62 B3
Gregory Ra. W. Austral., Australia 21°20S 121°12E 60 D3
Greifswald Germany 54°5N 13°23E 16 A7
Greiz Germany 50°39N 12°10E 16 C7
Gremikha Russia 67°59N 39°47E 18 A6
Grenaa Denmark 56°25N 10°53E 9 H14
Grenada U.S.A. 33°47N 89°49W 85 E10
Grenada ■ W. Indies 12°10N 61°40W 89 D7
Grenadier I. U.S.A. 44°3N 76°22W 83 B8
Grenadines, The St. Vincent 12°40N 61°20W 89 D7
Grenen Denmark 57°44N 10°40E 9 H14
Grenfell Australia 33°52S 148°8E 63 E4
Grenfell Canada 50°30N 102°56W 71 C8
Grenoble France 45°12N 5°42E 20 D6
Grenville, C. Australia 12°0S 143°13E 62 A3
Grenville Chan. Canada 53°40N 129°46W 70 C3
Gresham U.S.A. 45°30N 122°25W 78 E4
Gresik Indonesia 7°13S 112°38E 37 G15
Gretna U.K. 55°0N 3°3W 11 F5
Gretna U.S.A. 29°54N 90°3W 85 G9
Grevenmacher Lux. 49°41N 6°26E 15 E6
Grey → Canada 47°34N 57°6W 73 C8
Grey → N.Z. 42°27S 171°12E 59 E3
Grey, C. Australia 13°0S 136°35E 62 A2
Grey Is. Canada 50°50N 55°35W 69 G20
Grey Ra. Australia 27°0S 143°30E 63 D3
Greybull U.S.A. 44°30N 108°3W 76 D9
Greymouth N.Z. 42°29S 171°13E 59 E3
Greystones Ireland 53°9N 6°5W 10 C5
Greytown N.Z. 41°5S 175°29E 59 D5
Greytown S. Africa 29°1S 30°36E 57 D5
Gribbell I. Canada 53°23N 129°0W 70 C3
Gridley U.S.A. 39°22N 121°42W 78 F5
Griekwastad S. Africa 28°49S 23°15E 56 D3
Griffin U.S.A. 33°15N 84°16W 85 E12
Griffith Australia 34°18S 146°2E 63 E4
Griffith Canada 45°15N 77°10W 82 A7
Griffith I. Canada 44°50N 80°55W 82 B4
Grimaylov = Hrymayliv Ukraine 49°20N 26°5E 17 D14
Grimes U.S.A. 39°4N 121°54W 78 F5
Grimsay U.K. 57°29N 7°14W 11 D1
Grimsby Canada 43°12N 79°34W 82 C5
Grimsby U.K. 53°34N 0°5W 12 D7
Grímsey Iceland 66°33N 17°58W 8 C5
Grimshaw Canada 56°10N 117°40W 70 B5
Grimstad Norway 58°20N 8°35E 9 G13
Grindstone I. U.S.A. 44°17N 76°27W 83 B8
Grinnell U.S.A. 41°45N 92°43W 80 E7
Grinnell Pen. Canada 76°40N 95°0W 69 B13
Gris-Nez, C. France 50°52N 1°35E 20 A4
Grise Fiord Canada 76°25N 82°57W 69 B15
Groais I. Canada 50°55N 55°35W 73 B8
Groblersdal S. Africa 25°15S 29°25E 57 D4
Grodno = Hrodna Belarus 53°42N 23°52E 17 B12
Grodzyanka = Hrodzyanka Belarus 53°31N 28°42E 17 B15
Groesbeck U.S.A. 31°31N 96°32W 84 F6
Grójec Poland 51°50N 20°58E 17 C11
Grong Norway 64°25N 12°8E 8 D15
Groningen Neths. 53°15N 6°35E 15 A6
Groningen □ Neths. 53°16N 6°40E 15 A6
Groom U.S.A. 35°12N 101°6W 84 D4
Groot → S. Africa 33°45S 24°36E 56 E3
Groot-Berg → S. Africa 32°47S 18°8E 56 E2
Groot-Brakrivier S. Africa 34°2S 22°18E 56 E3
Groot Karasberge Namibia 27°20S 18°40E 56 D2
Groot Kei → S. Africa 32°41S 28°22E 57 E4
Groot-Vis → S. Africa 33°28S 27°5E 56 E4
Grootdrink S. Africa 28°33S 21°42E 56 D3
Groote Eylandt Australia 14°0S 136°40E 62 A2
Grootfontein Namibia 19°31S 18°6E 56 B2
Grootlaagte → Africa 20°55S 21°27E 56 C3
Grootvloer → S. Africa 30°0S 20°40E 56 E3
Gros C. Canada 61°59N 113°32W 70 A6
Gros Islet St. Lucia 14°5N 60°58W 89 f
Gros Morne △ Canada 49°40N 57°50W 73 C8
Gros Piton St. Lucia 13°49N 61°5W 89 f
Gros Piton Pt. St. Lucia 13°49N 61°5W 89 f
Grossa, Pta. Spain 39°6N 1°36E 24 B8
Grosse Point U.S.A. 42°23N 82°54W 82 D2
Grosser Arber Germany 49°6N 13°8E 16 D7
Grosseto Italy 42°46N 11°8E 22 C4
Grossglockner Austria 47°5N 12°40E 16 E7
Groswater B. Canada 54°20N 57°40W 73 B8
Groton Conn., U.S.A. 41°21N 72°5W 83 E12
Groton N.Y., U.S.A. 42°36N 76°22W 83 D8
Groton S. Dak., U.S.A. 45°27N 98°6W 80 C4
Grouard Mission Canada 55°33N 116°9W 70 B5
Groundhog → Canada 48°45N 82°58W 72 C3
Grouw Neths. 53°5N 5°51E 15 A5
Grove City U.S.A. 41°10N 80°5W 82 E4
Grove Hill U.S.A. 31°42N 87°47W 85 F11
Groveland U.S.A. 37°50N 120°14W 78 H6
Grover Beach U.S.A. 35°7N 120°37W 79 K6
Groves U.S.A. 29°57N 93°54W 84 G8
Groveton U.S.A. 44°36N 71°31W 83 B13
Groznyy Russia 43°20N 45°45E 19 F8
Grudziądz Poland 53°30N 18°47E 17 B10
Gruinard B. U.K. 57°56N 5°35W 11 D3
Grundy Center U.S.A. 42°22N 92°47W 80 D7
Gruver U.S.A. 36°16N 101°24W 84 C4
Gryazi Russia 52°30N 39°58E 18 D6
Gryazovets Russia 58°50N 40°10E 18 C7
Grytviken S. Georgia 54°19S 36°33W 96 G9
Gua India 22°18N 85°20E 43 H11
Gua Musang Malaysia 4°53N 101°58E 39 K3
Guacanayabo, G. de Cuba 20°40N 77°20W 88 B4
Guachípas → Argentina 25°40S 65°30W 94 B2
Guadalajara Mexico 20°40N 103°20W 86 C4
Guadalajara Spain 40°37N 3°12W 21 B4
Guadalcanal Solomon Is. 9°32S 160°12E 58 B9
Guadales Argentina 34°30S 67°55W 94 C2
Guadalete → Spain 36°35N 6°13W 21 D2
Guadalquivir → Spain 36°47N 6°22W 21 D2
Guadalupe = Guadeloupe ☒ W. Indies 16°20N 61°40W 88 b
Guadalupe Mexico 22°45N 102°31W 86 C4
Guadalupe U.S.A. 34°58N 120°34W 79 L6
Guadalupe → U.S.A. 28°27N 96°47W 84 G6
Guadalupe, Sierra de Spain 39°28N 5°30W 21 C3
Guadalupe Bravos Mexico 31°20N 106°10W 86 A3
Guadalupe I. Pac. Oc. 29°0N 118°50W 66 G8
Guadalupe Mts. △ U.S.A. 31°40N 104°30W 84 F2
Guadalupe Peak U.S.A. 31°50N 104°52W 84 F2
Guadalupe y Calvo Mexico 26°6N 106°58W 86 B3
Guadarrama, Sierra de Spain 41°0N 4°0W 21 B4
Guadeloupe ☒ W. Indies 16°20N 61°40W 88 b
Guadeloupe △ Guadeloupe 16°10N 61°40W 88 b
Guadeloupe Passage W. Indies 16°50N 62°15W 89 C7
Guadiana → Portugal 37°14N 7°22W 21 D2
Guadix Spain 37°18N 3°11W 21 D4
Guafo, Boca del Chile 43°35S 74°0W 96 E2
Guaico Trin. & Tob. 10°35N 61°9W 93 K15
Guainía → Colombia 2°1N 67°7W 92 C5
Guaíra Brazil 24°5S 54°10W 95 A5
Guaíra □ Paraguay 25°45S 56°30W 94 B4
Guaire = Gorey Ireland 52°41N 6°18W 10 D5
Guaitecas, Is. Chile 44°0S 74°30W 96 E2
Guajará-Mirim Brazil 10°50S 65°20W 92 F5
Guajira, Pen. de la Colombia 12°0N 72°0W 92 A4
Gualán Guatemala 15°8N 89°22W 88 C2
Gualeguay Argentina 33°10S 59°14W 94 C4
Gualeguaychú Argentina 33°3S 59°31W 94 C4
Gualequay → Argentina 33°19S 59°39W 94 C4
Guam ☒ Pac. Oc. 13°27N 144°45E 64 F6
Guaminí Argentina 37°1S 62°28W 94 D3
Guamúchil Mexico 25°28N 108°6W 86 B3
Guana I. Br. Virgin Is. 18°30N 64°30W 89 e
Guanabacoa Cuba 23°8N 82°18W 88 B3
Guanacaste, Cordillera de Costa Rica 10°40N 85°4W 88 D2
Guanacaste △ Costa Rica 10°57N 85°30W 88 D2
Guanaceví Mexico 25°56N 105°57W 86 B3
Guanahani = San Salvador I. Bahamas 24°0N 74°40W 89 B5
Guanaja Honduras 16°30N 85°55W 88 C2
Guanajay Cuba 22°56N 82°42W 88 B3
Guanajuato Mexico 21°1N 101°15W 86 C4
Guanajuato □ Mexico 21°0N 101°0W 86 C4
Guandacol Argentina 29°30S 68°40W 94 B2
Guandi Shan China 37°53N 111°29E 34 F6
Guane Cuba 22°10N 84°7W 88 B3
Guangdong □ China 23°0N 113°0E 33 G11
Guangling China 39°47N 114°22E 34 E8
Guangrao China 37°5N 118°25E 35 F10
Guangwu China 37°48N 105°57E 34 F3
Guangxi Zhuangzu Zizhiqu □ China 24°0N 109°0E 33 G10
Guangyuan China 32°26N 105°51E 34 H3
Guangzhou China 23°6N 113°13E 33 G11
Guanica Puerto Rico 17°58N 66°55W 89 d
Guanipa → Venezuela 9°56N 62°26W 92 B6
Guannan China 34°8N 119°21E 35 G10
Guantánamo Cuba 20°10N 75°14W 89 B4
Guantánamo B. Cuba 19°59N 75°10W 89 C4
Guantao China 36°42N 115°25E 34 F8
Guanting Shuiku China 40°14N 115°35E 34 D8
Guanyun China 34°20N 119°18E 35 G10
Guapay = Grande → Bolivia 15°51S 64°39W 92 G6
Guápiles Costa Rica 10°10N 83°46W 88 D3
Guapo B. Trin. & Tob. 10°12N 61°41W 93 K15
Guaporé Brazil 28°51S 51°54W 95 B5
Guaporé → Brazil 11°55S 65°4W 92 F5
Guaqui Bolivia 16°41S 68°54W 92 G5
Guaramacal △ Venezuela 9°13N 70°12W 89 E5
Guarapari Brazil 20°40S 40°30W 95 A7
Guarapuava Brazil 25°20S 51°30W 95 B5
Guaratinguetá Brazil 22°49S 45°9W 95 A6
Guaratuba Brazil 25°53S 48°38W 95 B6
Guarda Portugal 40°32N 7°20W 21 B2
Guardafui, C. = Asir, Ras Somali Rep. 11°55N 51°10E 47 E5
Guárico □ Venezuela 8°40N 66°35W 92 B5
Guarujá Brazil 24°2S 46°25W 95 A6
Guarulhos Brazil 23°29S 46°33W 95 A6
Guasave Mexico 25°34N 108°27W 86 B3
Guasdualito Venezuela 7°15N 70°44W 92 B4
Guatemala Guatemala 14°40N 90°22W 88 D1
Guatemala ■ Cent. Amer. 15°40N 90°30W 88 C1
Guatemala Basin Pac. Oc. 11°0N 95°0W 65 F18
Guatemala Trench Pac. Oc. 14°0N 95°0W 66 H10
Guatopo △ Venezuela 10°5N 66°30W 89 D6
Guatuaro Pt. Trin. & Tob. 10°19N 60°59W 93 K16
Guaviare → Colombia 4°3N 67°44W 92 C5
Guaxupé Brazil 21°10S 47°5W 95 A6
Guayaguayare Trin. & Tob. 10°8N 61°2W 93 K15
Guayama Puerto Rico 17°59N 66°7W 89 d
Guayaquil Ecuador 2°15S 79°52W 92 D3
Guayaquil Mexico 29°59N 115°4W 86 B1
Guayaquil, G. de Ecuador 3°10S 81°0W 92 D2
Guaymas Mexico 27°56N 110°54W 86 B2
Guba Dem. Rep. of the Congo 10°38S 26°27E 55 E2
Gubkin Russia 51°17N 37°32E 19 D6
Gubkinskiy Russia 64°27N 76°36E 28 C8
Gudbrandsdalen Norway 61°33N 10°10E 8 F14
Guddu Barrage Pakistan 28°30N 69°50E 42 E3
Gudur India 14°12N 79°55E 40 M11
Guecho = Getxo Spain 43°21N 2°59W 21 A4
Guékédou Guinea 8°40N 10°5W 50 G3
Guelmine = Goulimine Morocco 28°56N 10°0W 50 C3
Guelph Canada 43°35N 80°20W 82 C4
Guerara Algeria 32°51N 4°22E 50 B6
Guéret France 46°11N 1°51E 20 C4
Guerneville U.S.A. 38°30N 123°0W 78 G4
Guernica = Gernika-Lumo Spain 43°19N 2°40W 21 A4
Guernsey U.K. 49°26N 2°35W 13 H5
Guernsey U.S.A. 42°16N 104°45W 76 E11
Guerrero □ Mexico 17°40N 100°0W 87 D5
Gügher Iran 29°28N 56°27E 45 D8
Guia Canary Is. 28°8N 15°38W 24 F4
Guia de Isora Canary Is. 28°12N 16°46W 24 F3
Guia Lopes da Laguna Brazil 21°26S 56°7W 95 A4
Guiana Highlands S. Amer. 5°10N 60°40W 90 C4
Guidónia-Montecélio Italy 42°1N 12°45E 22 C5
Guigang China 23°8N 109°35E 33 G10
Guijá Mozam. 24°27S 33°0E 57 C5
Guildford U.K. 51°14N 0°34W 13 F7
Guilford U.S.A. 41°17N 72°41W 83 E12
Guilin China 25°18N 110°15E 33 F11
Guillaume-Delisle, L. Canada 56°15N 76°17W 72 A4
Güímar Canary Is. 28°18N 16°24W 24 F3
Guimarães Portugal 41°28N 8°24W 21 B1
Guimaras □ Phil. 10°35N 122°37E 37 B6
Guinda U.S.A. 38°50N 122°12W 78 G4
Guinea Africa 8°0N 8°0E 48 F4
Guinea ■ W. Afr. 10°20N 11°30W 50 F3
Guinea, Gulf of Atl. Oc. 3°0N 2°30E 49 F4
Guinea-Bissau ■ Africa 12°0N 15°0W 50 F3
Güines Cuba 22°50N 82°0W 88 B3
Guingamp France 48°34N 3°10W 20 B2
Güiria Venezuela 10°32N 62°18W 93 K14
Guiuan Phil. 11°5N 125°55E 37 B7
Guiyang China 26°32N 106°40E 32 F10
Guizhou □ China 27°0N 107°0E 32 F10
Gujar Khan Pakistan 33°16N 73°19E 42 C5
Gujarat □ India 23°20N 71°0E 42 H4
Gujiao China 37°54N 112°8E 34 F7
Gujranwala Pakistan 32°10N 74°12E 42 C6
Gujrat Pakistan 32°40N 74°2E 42 C6
Gulbarga India 17°20N 76°50E 40 L10
Gulbene Latvia 57°8N 26°52E 9 H22
Gulf, The = Persian Gulf Asia 27°0N 50°0E 45 E6
Gulf Islands ⌓ U.S.A. 30°10N 87°10W 85 F11
Gulfport U.S.A. 30°22N 89°6W 85 F10
Gulgong Australia 32°20S 149°49E 63 E4
Gulian China 52°56N 122°21E 33 A13
Gulistan Pakistan 30°30N 66°35E 42 D2
Gulja = Yining China 43°58N 81°10E 32 C5
Gull Lake Canada 50°10N 108°29W 71 C7
Güllük Turkey 37°14N 27°35E 23 F12
Gulmarg India 34°3N 74°25E 43 B6
Gülshat Kazakhstan 46°38N 74°21E 28 C8
Gulu Uganda 2°48N 32°17E 54 B3
Gulwe Tanzania 6°30S 36°25E 54 D4
Gumal → Pakistan 31°40N 71°50E 42 D4
Gumbaz Pakistan 30°2N 69°0E 42 D3
Gumel Nigeria 12°39N 9°22E 50 F7
Gumi S. Korea 36°10N 128°12E 35 F15
Gumla India 23°3N 84°33E 43 H11
Gumlu Australia 19°53S 147°41E 62 B4
Gumma □ Japan 36°30N 138°20E 31 F9
Gumzai Indonesia 5°28S 134°42E 37 F8
Guna India 24°40N 77°19E 42 G7
Gunbalanya Australia 12°20S 133°4E 60 B5
Gundabooka △ Australia 30°30S 145°20E 63 E4
Gunisao → Canada 53°56N 97°53W 71 C9
Gunisao L. Canada 53°33N 96°15W 71 C9
Gunjyal Pakistan 32°20N 71°55E 42 C4
Gunnbjørn Fjeld Greenland 68°55N 29°47W 4 C6
Gunnedah Australia 30°59S 150°15E 63 E5
Gunnewin Australia 25°59S 148°33E 63 D4
Gunningbar Cr. → Australia 31°14S 147°6E 63 E4
Gunnison Colo., U.S.A. 38°33N 106°56W 76 G10
Gunnison Utah, U.S.A. 39°9N 111°49W 76 G8
Gunnison → U.S.A. 39°4N 108°35W 76 G9
Gunsan S. Korea 35°59N 126°45E 35 G14
Guntakal India 15°11N 77°27E 40 M10
Gunter Canada 44°52N 77°32W 82 B7
Guntersville U.S.A. 34°21N 86°18W 85 D11
Guntong Malaysia 4°36N 101°3E 39 K3
Guntur India 16°23N 80°30E 41 L12
Gunung Ciremay △ Indonesia 6°53S 108°24E 37 G13
Gunungapi Indonesia 6°45S 126°30E 37 F7
Gunungsitoli Indonesia 1°15N 97°30E 36 D1
Gunza Angola 10°50S 13°50E 52 G2
Guo He → China 32°59N 117°10E 35 H9
Guoyang China 33°32N 116°12E 34 H9
Gupis Pakistan 36°15N 73°20E 43 A5
Gurbantünggüt Shamo China 45°8N 87°20E 32 B6
Gurdaspur India 32°5N 75°31E 42 C6
Gurdon U.S.A. 33°55N 93°9W 84 E8
Gurgaon India 28°27N 77°1E 42 E7
Gurgueia → Brazil 6°50S 43°24W 93 E10
Gurha India 25°12N 71°39E 42 G4
Guri, Embalse de Venezuela 7°50N 62°52W 92 B6
Gurkha Nepal 28°5N 84°40E 43 E11
Gurla Mandhata = Naimona'nyi Feng Nepal 30°26N 81°18E 43 D9
Gurley Australia 29°45S 149°48E 63 D4
Gurnet Point U.S.A. 42°1N 70°34W 83 D14
Guro Mozam. 17°26S 32°30E 55 F3
Gurué Mozam. 15°25S 36°58E 55 F4
Gurun Malaysia 5°49N 100°27E 39 K3
Gürün Turkey 38°43N 37°15E 19 G6
Gurupá Brazil 1°25S 51°35W 93 D8
Gurupá, I. Grande de Brazil 1°25S 51°45W 93 D8
Gurupi Brazil 11°43S 49°4W 93 F9
Gurupi → Brazil 1°13S 46°6W 93 D9
Guruwe Zimbabwe 16°40S 30°42E 57 B5
Gurvan Sayhan Uul Mongolia 43°50N 104°0E 34 C3
Guryev = Atyraū Kazakhstan 47°5N 52°0E 19 E9
Gusau Nigeria 12°12N 6°40E 50 F7
Gushan China 39°50N 123°35E 35 E12
Gushgy = Serhetabat Turkmenistan 35°20N 62°18E 45 C9
Gusinoozersk Russia 51°16N 106°27E 29 D11
Gustavus U.S.A. 58°25N 135°44W 70 B1
Gustine U.S.A. 37°16N 121°0W 78 H6
Güstrow Germany 53°47N 12°10E 16 B7
Gütersloh Germany 51°54N 8°24E 16 C5
Gutha Australia 28°58S 115°55E 61 E2
Guthalungra Australia 19°52S 147°50E 62 B4
Guthrie Canada 44°28N 79°32W 82 B5
Guthrie Okla., U.S.A. 35°53N 97°25W 84 D6
Guthrie Tex., U.S.A. 33°37N 100°19W 84 E4
Guttenberg U.S.A. 42°47N 91°6W 80 D8
Gutu Zimbabwe 19°41S 31°9E 57 B5
Guwahati India 26°10N 91°45E 41 F17
Guy Fawkes River △ Australia 30°0S 152°20E 63 D5
Guyana ■ S. Amer. 5°0N 59°0W 92 C7
Guyane française = French Guiana ☒ S. Amer. 4°0N 53°0W 93 C8
Guyang China 41°0N 110°5E 34 D6
Guyenne France 44°30N 0°40E 20 D4
Guymon U.S.A. 36°41N 101°29W 84 C4
Guyra Australia 30°15S 151°40E 63 E5
Guyuan Hebei, China 41°37N 115°40E 34 D8
Guyuan Ningxia Huizu, China 36°0N 106°20E 34 F4
Güzelyurt = Morphou Cyprus 35°12N 32°59E 25 D11
Guzhen China 33°22N 117°18E 35 H9
Guzmán, L. de Mexico 31°20N 107°30W 86 A3
Gwa Burma 17°36N 94°34E 41 L19
Gwaai Zimbabwe 19°15S 27°45E 55 F2
Gwaai → Zimbabwe 17°59S 26°52E 55 F2
Gwabegar Australia 30°37S 148°59E 63 E4
Gwādar Pakistan 25°10N 62°18E 40 G3
Gwaii Haanas △ Canada 52°21N 131°26W 70 C2
Gwalior India 26°12N 78°10E 42 F8
Gwanda Zimbabwe 20°55S 29°0E 55 G2
Gwane Dem. Rep. of the Congo 4°45N 25°48E 54 B2
Gwangju S. Korea 35°9N 126°54E 35 G14
Gwanju = Gwangju S. Korea 35°9N 126°54E 35 G14
Gweebarra B. Ireland 54°51N 8°23W 10 B3
Gweedore Ireland 55°3N 8°13W 10 A3
Gweru Zimbabwe 19°28S 29°45E 55 F2
Gwinn U.S.A. 46°19N 87°27W 80 B10
Gwydir → Australia 29°27S 149°48E 63 D4
Gwynedd □ U.K. 52°52N 4°10W 12 E3
Gyandzha = Gäncä Azerbaijan 40°45N 46°20E 19 F8
Gyangzê China 29°5N 89°47E 32
Gyaring Hu China 34°50N 97°40E 32
Gydanskiy Poluostrov Russia 70°0N 78°0E 28
Gyeongju S. Korea 35°51N 129°14E 35
Gympie Australia 26°11S 152°38E 63
Gyöngyös Hungary 47°48N 19°56E 17
Győr Hungary 47°41N 17°40E 17
Gypsum Pt. Canada 61°53N 114°35W 70
Gypsumville Canada 51°45N 98°40W 71
Gyula Hungary 46°38N 21°17E 17
Gyumri Armenia 40°47N 43°50E 19
Gyzylarbat = Serdar Turkmenistan 39°4N 56°23E 45
Gyzyletrek = Etrek Turkmenistan 37°36N 54°46E 45

H

Ha 'Arava → Israel 30°50N 35°20E 46
Ha Coi Vietnam 21°26N 107°46E 38
Ha Dong Vietnam 20°58N 105°46E 38
Ha Giang Vietnam 22°50N 104°59E 38
Ha Karmel △ Israel 32°45N 35°5E 46
Ha Long = Hong Gai Vietnam 20°57N 107°5E 38
Ha Long, Vinh Vietnam 20°56N 107°3E 38
Ha Tien Vietnam 10°23N 104°29E 39
Ha Tinh Vietnam 18°20N 105°54E 38
Ha Trung Vietnam 19°58N 105°50E 38
Haaksbergen Neths. 52°9N 6°45E 15
Ha'ano Tonga 19°41S 174°18W 5
Ha'apai Group Tonga 19°47S 174°27W 5
Haapiti Moorea 17°34S 149°52W 5
Haapsalu Estonia 58°56N 23°30E 9
Haarlem Neths. 52°23N 4°39E 15
Haast → N.Z. 43°50S 169°2E 59
Haasts Bluff Australia 23°22S 132°0E 60
Haasts Bluff ✪ Australia 23°39S 130°34E 60
Hab → Pakistan 24°53N 66°41E 42
Hab Nadi Chauki Pakistan 25°0N 66°50E 42
Habahe China 48°3N 86°23E 32
Habaswein Kenya 1°2N 39°30E 54
Habay Canada 58°50N 118°44W 70
Ḩabbānīyah Iraq 33°17N 43°29E 44
Habirag China 42°17N 115°42E 34
Haboro Japan 44°22N 141°42E 30
Ḩabshān U.A.E. 23°50N 53°37E 45
Hachijō-Jima Japan 33°5N 139°45E 31
Hachiman Japan 35°45N 136°57E 31
Hachinohe Japan 40°30N 141°29E 30
Hachiōji Japan 35°40N 139°20E 31
Hackensack U.S.A. 40°52N 74°4W 83
Hackettstown U.S.A. 40°51N 74°50W 83
Hadali Pakistan 32°16N 72°11E 42
Hadarba, Ras Sudan 22°4N 36°51E 51
Hadarom □ Israel 31°0N 35°0E 46
Ḩadd, Ra's al Oman 22°35N 59°50E 47
Haddington U.K. 55°57N 2°47W 11
Hadejia Nigeria 12°30N 10°5E 50
Hadera Israel 32°27N 34°55E 46
Hadera, N. → Israel 32°28N 34°52E 46
Haderslev Denmark 55°15N 9°30E 9
Hadhramaut = Ḩaḑramawt Yemen 15°30N 49°30E 47
Hadībоh Yemen 12°39N 54°2E 47
Hadley B. Canada 72°31N 108°12W 68
Hadong S. Korea 35°5N 127°44E 35
Ḩaḑramawt Yemen 15°30N 49°30E 47
Ḩadrānīyah Iraq 35°38N 43°14E 44
Hadrian's Wall U.K. 55°0N 2°30W 12
Hae, Ko Thailand 7°44N 98°22E 3
Haeju N. Korea 38°3N 125°45E 35
Häʻena U.S.A. 22°14N 159°34W 75
Haenam S. Korea 34°34N 126°35E 35
Haenertsburg S. Africa 24°0S 29°50E 57
Haerhpin = Harbin China 45°48N 126°40E 35
Hafar al Bāţin Si. Arabia 28°32N 45°52E 44
Ḩafīrat al 'Aydā Si. Arabia 26°26N 39°12E 44
Ḩafit Oman 23°59N 55°49E 45
Hafizabad Pakistan 32°5N 73°40E 42
Haflong India 25°10N 93°5E 41
Haft Gel Iran 31°30N 49°32E 45
Hagalil Israel 32°53N 35°18E 46
Hagemeister I. U.S.A. 58°39N 160°54W 74
Hagen Germany 51°21N 7°27E 16
Hagerman U.S.A. 33°7N 104°20W 77
Hagerman Fossil Beds ⌓ U.S.A. 42°48N 114°57W 76
Hagerstown U.S.A. 39°39N 77°43W 81
Hagersville Canada 42°58N 80°3W 82
Hagfors Sweden 60°3N 13°45E 9
Hagi Japan 34°30N 131°22E 31
Hagolan Syria 33°0N 35°45E 46
Hagondange France 49°16N 6°11E 20
Hags Hd. Ireland 52°57N 9°28W 10
Hague, C. de la France 49°44N 1°56W 20
Hague, The = 's-Gravenhage Neths. 52°7N 4°17E 15
Haguenau France 48°49N 7°47E 20
Hai Duong Vietnam 20°56N 106°19E 38
Haicheng China 40°50N 122°45E 35
Haidar Khel Afghan. 33°58N 68°38E 42
Haidarâbâd = Hyderabad India 17°22N 78°29E 40
Haidargarh India 26°37N 81°22E 43
Haifa = Hefa Israel 32°46N 35°0E 46
Haikou China 20°1N 110°16E 38
Ḩā'il Si. Arabia 27°28N 41°45E 44
Ḩā'il □ Si. Arabia 26°40N 41°40E 44
Hailar China 49°10N 119°38E 33
Hailey U.S.A. 43°31N 114°19W 76
Haileybury Canada 47°30N 79°38W 72
Hailin China 44°37N 129°30E 35
Hailun China 47°28N 126°50E 33
Hailuoto Finland 65°3N 24°45E 8

nan □ China 19°0N 109°30E 38 C7
nan Dao China 19°0N 109°30E 38 C7
nan Str. = Qiongzhou Haixia
'hina 20°10N 110°15E 38 B8
naut □ Belgium 50°30N 4°0E 15 D4
nes Alaska, U.S.A. 59°14N 135°26W 70 B1
nes Oreg., U.S.A. 44°55N 117°56W 76 D5
nes City U.S.A. 28°7N 81°38W 85 G14
nes Junction
'anada 60°45N 137°30W 70 A1
phong Vietnam 20°47N 106°41E 38 B6
ti ■ W. Indies 19°0N 72°30W 89 C5
ya Sudan 18°20N 36°21E 51 E13
yan China 36°53N 100°59E 32 D9
yang China 36°47N 121°9E 35 F11
yuan China 36°35N 105°52E 34 F3
zhou China 34°37N 119°7E 35 G10
zhou Wan China 34°50N 119°20E 35 G10
Ali Qoli, Kavīr Iran 35°55N 54°50E 45 C7
dúböszörmény
lungary 47°40N 21°30E 17 E11
i Ibrahim Iraq 36°40N 44°30E 44 B5
īābād Iran 33°37N 60°0E 45 C9
ipur India 25°45N 85°13E 43 G11
jah Yemen 15°42N 43°36E 47 D3
jīābād Iran 28°19N 55°55E 45 D7
jīābād-e Zarrīn Iran 33°9N 54°51E 45 C7
nówka Poland 52°47N 23°35E 17 B12
kansson, Mts.
Dem. Rep. of the Congo 8°40S 25°45E 55 D2
kkâri Turkey 37°34N 43°44E 44 B4
kken-Zan Japan 34°10N 135°54E 31 G7
kkōda San Japan 40°50N 141°0E 30 D10
kodate Japan 41°45N 140°44E 30 D10
kos Namibia 23°13S 16°21E 56 C2
ku-San Japan 36°9N 136°46E 31 F8
ku-San △ Japan 36°15N 136°45E 31 F8
kui Japan 36°53N 136°47E 31 F8
a Pakistan 25°43N 68°20E 40 G6
ab Syria 36°10N 37°15E 44 B3
abjah Iraq 35°10N 45°58E 44 C5
aib Sudan 22°12N 36°30E 51 D13
aib Triangle Africa 22°30N 35°20E 51 D13
at 'Ammār Si. Arabia 29°10N 36°4E 44 D3
bā Lebanon 34°34N 36°6E 46 A5
berstadt Germany 51°54N 11°3E 16 C6
combe N.Z. 40°8S 175°30E 59 D5
con, Mt. Phil. 13°16N 121°0E 37 B6
de Fjäll = Haltiatunturi
inland 69°17N 21°18E 8 B19
den Norway 59°9N 11°23E 9 G14
dia Bangla. 22°1N 88°3E 43 H13
dwani India 29°31N 79°30E 43 E8
e → Australia 24°56S 135°53E 62 C2
esowen U.K. 52°27N 2°3W 13 E5
esworth U.K. 52°20N 1°31E 13 E9
eyville U.S.A. 34°14N 87°37W 85 D11
f Dome U.S.A. 37°44N 119°32E 78 H7
fmoon Bay N.Z. 46°50S 168°5E 59 G2
fway → Canada 56°12N 121°32W 70 B4
ia India 24°50N 82°19E 43 G10
iburton Canada 45°3N 78°30W 82 A6
ifax Australia 18°32S 146°22E 62 B4
ifax Canada 44°38N 63°35W 73 D7
ifax U.K. 53°43N 1°52W 12 D6
ifax U.S.A. 40°25N 76°55W 82 F8
ifax B. Australia 18°50S 147°0E 62 B4
ifax I. Namibia 26°38S 15°4E 56 D2
k Shan China 42°20N 81°22E 32 C5
l → Iran 27°40N 58°30E 45 E8
mun △ Indonesia 6°42S 106°26E 37 G11
kida = Chalkida
reece 38°27N 23°42E 23 E10
kirk U.K. 58°30N 3°29W 11 C5
Beach Canada 68°46N 81°12W 69 D15
Pen. Canada 63°30N 66°0W 69 E18
Pt. Australia 15°40S 124°23E 60 C3
and Sweden 57°8N 12°47E 9 H15
āniyat, Jazā'ir al
man 17°30N 55°58E 47 D6
asan S. Korea 33°22N 126°32E 35 H14
e Belgium 50°44N 4°13E 15 D4
e Germany 51°30N 11°56E 16 C6
efors Sweden 59°47N 14°31E 9 G16
lett Australia 33°25S 138°55E 63 E2
lettsville U.S.A. 29°27N 96°57W 84 G6
ey Antarctica 75°35S 26°39W 5 D1
im S. Korea 33°24N 126°15E 35 H14
ingdalselva →
'orway 60°23N 9°35E 8 F13
lock U.S.A. 48°47N 96°57W 80 A5
ls Creek Australia 18°16S 127°38E 60 C4
s Gap Australia 37°8S 142°34E 63 F3
s Lake Canada 45°7N 78°45W 82 A6
sberg Sweden 59°5N 15°7E 9 G16
lstead U.S.A. 41°58N 75°45W 83 E9
mahera Indonesia 0°40N 128°0E 37 D7
mstad Sweden 56°41N 12°52E 9 H15
singborg = Helsingborg
weden 56°3N 12°42E 9 H15
singland Sweden 61°40N 16°5E 8 F17
stead U.K. 51°57N 0°40E 13 F8
tiatunturi Finland 69°17N 21°18E 8 B19
ton □ U.K. 53°22N 2°45W 12 D5
twhistle U.K. 54°58N 2°26W 12 C5
ūl Qatar 25°40N 52°40E 45 E7
vad India 23°1N 71°11E 42 H4
vān Iran 33°57N 56°15E 45 C8
mab Namibia 28°7S 19°16E 56 D2
mada Japan 34°56N 132°4E 31 G6
madān Iran 34°52N 48°32E 45 C6
madān □ Iran 35°0N 49°0E 45 C6
māh Syria 35°5N 36°40E 44 C3
mamatsu Japan 34°45N 137°45E 31 G8
mar Norway 60°48N 11°7E 8 F14
mâta, Gebel Egypt 24°17N 35°0E 44 E2
matonbetsu Japan 45°10N 142°20E 30 B11
mbantota Sri Lanka 6°10N 81°10E 40 R12
mber ⌂ Canada 52°20N 118°0W 70 C5
mburg Germany 53°33N 9°59E 16 B5
Hamburg Ark., U.S.A. 33°14N 91°48W 84 E9
Hamburg N.Y., U.S.A. 42°43N 78°50W 82 D6
Hamburg Pa., U.S.A. 40°33N 75°59W 83 F9
Ḥamḍ, W. al → Si. Arabia 24°55N 36°20E 44 E3
Hamden U.S.A. 41°23N 72°54W 83 E12
Häme Finland 61°38N 25°10E 8 F21
Hämeenlinna Finland 61°0N 24°28E 8 F21
Hamelin Pool Australia 26°22S 114°20E 61 E1
Hameln Germany 52°6N 9°21E 16 B5
Hamerkaz □ Israel 32°15N 34°55E 46 C3
Hamersley Ra. Australia 22°0S 117°45E 60 D2
Hamhŭng N. Korea 39°54N 127°30E 35 E14
Hami China 42°55N 93°25E 32 C7
Hamilton Australia 37°45S 142°2E 63 F3
Hamilton Canada 43°15N 79°50W 82 C5
Hamilton N.Z. 37°47S 175°19E 59 B5
Hamilton U.K. 55°46N 4°2W 11 F4
Hamilton Ala., U.S.A. 34°9N 87°59W 85 D11
Hamilton Mont., U.S.A. 46°15N 114°10W 76 C6
Hamilton N.Y., U.S.A. 42°50N 75°33W 83 D9
Hamilton Ohio, U.S.A. 39°24N 84°34W 81 F11
Hamilton Tex., U.S.A. 31°42N 98°7W 84 F5
Hamilton → Queens., Australia 23°30S 139°47E 62 C2
Hamilton → S. Austral., Australia 26°40S 135°19E 63 D2
Hamilton City U.S.A. 39°45N 122°1W 78 F4
Hamilton I. Australia 20°21S 148°56E 62 b
Hamilton Inlet Canada 54°0N 57°30W 73 B8
Hamilton Mt. U.S.A. 43°25N 74°22W 83 C10
Hamina Finland 60°34N 27°12E 8 F22
Hamirpur H.P., India 31°41N 76°31E 42 D7
Hamirpur Ut. P., India 25°57N 80°9E 43 G9
Hamlet U.S.A. 34°53N 79°42W 85 D15
Hamley Bridge Australia 34°17S 138°35E 63 E2
Hamlin = Hameln Germany 52°6N 9°21E 16 B5
Hamlin N.Y., U.S.A. 43°17N 77°55W 82 C7
Hamlin Tex., U.S.A. 32°53N 100°8W 84 E4
Hamm Germany 51°40N 7°50E 16 C4
Ḥammār, Hawr al Iraq 30°50N 47°10E 44 D5
Hammerfest Norway 70°39N 23°41E 8 A20
Hammond Ind., U.S.A. 41°38N 87°30W 80 E10
Hammond La., U.S.A. 30°30N 90°28W 85 F9
Hammond N.Y., U.S.A. 44°27N 75°42W 83 B9
Hammondsport U.S.A. 42°25N 77°13W 82 D7
Hammonton U.S.A. 39°39N 74°48W 81 F16
Hampden N.Z. 45°18S 170°50E 59 F3
Hampshire □ U.K. 51°7N 1°23W 13 F6
Hampshire Downs U.K. 51°15N 1°10W 13 F6
Hampton N.B., Canada 45°32N 65°51W 73 C6
Hampton Ont., Canada 43°58N 78°45W 82 C6
Hampton Ark., U.S.A. 33°32N 92°28W 84 E8
Hampton Iowa, U.S.A. 42°45N 93°13W 80 D7
Hampton N.H., U.S.A. 42°57N 70°50W 83 D14
Hampton S.C., U.S.A. 32°52N 81°7W 85 E14
Hampton Va., U.S.A. 37°2N 76°21W 81 G15
Hampton Bays U.S.A. 40°53N 72°30W 83 F12
Hampton Tableland Australia 32°0S 127°0E 61 F4
Hamyang S. Korea 35°32N 127°42E 35 G14
Han Pijesak Bos.-H. 44°5N 18°57E 23 B8
Han Shui → China 30°34N 114°17E 33 E11
Hanak Si. Arabia 25°32N 37°0E 44 E3
Hanamaki Japan 39°23N 141°7E 30 E10
Hanang Tanzania 4°30S 35°25E 54 C4
Hanau Germany 50°7N 8°56E 16 C5
Hanbogd = Ihbulag Mongolia 43°11N 107°10E 34 C4
Hancheng China 35°31N 110°25E 34 G6
Hancock Mich., U.S.A. 47°8N 88°35W 80 B9
Hancock N.Y., U.S.A. 41°57N 75°17W 83 E9
Hancock Vt., U.S.A. 43°55N 72°50W 83 C12
Handa Japan 34°53N 136°55E 31 G8
Handa I. U.K. 58°23N 5°11W 11 C3
Handan China 36°35N 114°28E 34 F8
Handeni Tanzania 5°25S 38°2E 54 D4
Handwara India 34°21N 74°20E 43 B6
Hanegev Israel 30°50N 35°0E 46 E4
Hanford U.S.A. 36°20N 119°39W 78 J7
Hanford Reach ⌂ U.S.A. 46°40N 119°30W 76 C4
Hang Chat Thailand 18°20N 99°21E 38 C2
Hangang → S. Korea 37°50N 126°30E 35 F14
Hangayn Nuruu Mongolia 47°30N 99°0E 32 B8
Hangchou = Hangzhou China 30°18N 120°11E 33 E13
Hanggin Houqi China 40°58N 107°4E 34 D4
Hanggin Qi China 39°52N 108°50E 34 E5
Hangu China 39°18N 117°53E 35 E9
Hangzhou China 30°18N 120°11E 33 E13
Hangzhou Wan China 30°15N 120°45E 33 E13
Hanh Mongolia 51°32N 100°35E 32 A9
Hanhongor Mongolia 43°55N 104°28E 34 C3
Hania = Chania Greece 35°30N 24°4E 25 D6
Ḥanīdh Si. Arabia 26°35N 48°38E 45 E6
Ḥanīsh Yemen 13°45N 42°46E 47 E3
Hankinson U.S.A. 46°4N 96°54W 80 B5
Hankö Finland 59°50N 22°57E 9 G20
Hanksville U.S.A. 38°22N 110°43W 76 G8
Hanle India 32°42N 79°4E 43 C8
Hanmer Springs N.Z. 42°32S 172°50E 59 E4
Hann → Australia 17°26S 126°17E 60 C4
Hann, Mt. Australia 15°45S 126°0E 60 C4
Hanna Canada 51°40N 111°54W 70 C6
Hanna U.S.A. 41°52N 106°34W 76 F10
Hannah B. Canada 51°40N 80°0W 72 B4
Hannibal Mo., U.S.A. 39°42N 91°22W 80 F8
Hannibal N.Y., U.S.A. 43°19N 76°35W 83 C8
Hannover Germany 52°22N 9°46E 16 B5
Hanoi Vietnam 21°5N 105°55E 38 B5
Hanover = Hannover Germany 52°22N 9°46E 16 B5
Hanover Canada 44°9N 81°2W 82 B3
Hanover S. Africa 31°4S 24°29E 56 E3
Hanover N.H., U.S.A. 43°42N 72°17W 83 C12
Hanover Ohio, U.S.A. 40°4N 82°16W 82 F2
Hanover Pa., U.S.A. 39°48N 76°59W 81 F15
Hanover, I. Chile 51°0S 74°50W 96 G2
Hans Lollik I. U.S. Virgin Is. 18°24N 64°53W 89 e
Hansdiha India 24°36N 87°5E 43 G12
Hansi H.P., India 32°27N 77°50E 42 C7
Hansi Haryana, India 29°10N 75°57E 42 E6
Hanson, L. Australia 31°0S 136°15E 63 E2
Hanting China 36°46N 119°12E 35 F10
Hantsavichy Belarus 52°49N 26°30E 17 B14
Hanumangarh India 29°35N 74°19E 42 E6
Hanyin China 32°54N 108°28E 34 H5
Hanzhong China 33°10N 107°1E 34 H4
Hanzhuang China 34°33N 117°23E 35 G9
Haora India 22°34N 88°18E 43 H13
Haparanda Sweden 65°52N 24°8E 8 D21
Happy U.S.A. 34°45N 101°52W 84 D4
Happy Camp U.S.A. 41°48N 123°23W 76 F2
Happy Valley-Goose Bay Canada 53°15N 60°20W 73 B7
Hapsu N. Korea 41°13N 128°51E 35 D15
Hapur India 28°45N 77°45E 42 E7
Ḥaql Si. Arabia 29°10N 34°58E 46 F3
Har Indonesia 5°16S 133°14E 37 F8
Har-Ayrag Mongolia 45°47N 109°16E 34 B5
Har Hu China 38°20N 97°38E 32 D8
Har Us Nuur Mongolia 48°0N 92°0E 32 B7
Har Yehuda Israel 31°35N 34°57E 46 D3
Ḥaraḍ Si. Arabia 24°22N 49°0E 47 C4
Haramosh Pakistan 35°50N 74°54E 43 B6
Haranomachi Japan 37°38N 140°58E 30 F10
Harare Zimbabwe 17°43S 31°2E 55 F3
Harazé Chad 9°57N 20°48E 51 G10
Harbin China 45°48N 126°40E 35 B14
Harbor Beach U.S.A. 43°51N 82°39W 82 C2
Harborcreek U.S.A. 42°9N 79°57W 82 D5
Harbour Breton Canada 47°29N 55°50W 73 C8
Harbour Deep Canada 50°25N 56°32W 73 B8
Harda India 22°27N 77°5E 42 H7
Hardangerfjorden Norway 60°5N 6°0E 9 F12
Hardangervidda Norway 60°7N 7°20E 8 F12
Hardap ⌂ Namibia 24°29S 17°45E 56 C2
Hardap Dam Namibia 24°32S 17°50E 56 C2
Hardenberg Neths. 52°34N 6°37E 15 B6
Harderwijk Neths. 52°21N 5°38E 15 B5
Hardey → Australia 22°45S 116°8E 60 D2
Hardin U.S.A. 45°44N 107°37W 76 D10
Harding S. Africa 30°35S 29°55E 57 E4
Harding Ra. Australia 16°17S 124°55E 60 C3
Hardisty Canada 52°40N 111°18W 70 C6
Hardoi India 27°26N 80°6E 43 F9
Hardwar = Haridwar India 29°58N 78°9E 42 E8
Hardwick U.S.A. 44°30N 72°22W 83 B12
Hardwood Lake Canada 45°12N 77°26W 82 A7
Hardy, Pen. Chile 55°30S 68°20W 96 H3
Hardy, Pte. St. Lucia 14°6N 60°56W 89 f
Hare B. Canada 51°15N 55°45W 73 B8
Hareid Norway 62°22N 6°1E 8 E12
Harer Ethiopia 9°20N 42°8E 47 F3
Hargeisa Somali Rep. 9°30N 44°2E 47 F3
Hari → Indonesia 1°16S 104°5E 36 E2
Haria Canary Is. 29°8N 13°32W 24 E6
Haridwar India 29°58N 78°9E 42 E8
Harim, Jabal al Oman 25°58N 56°14E 45 E8
Haringhata → Bangla. 22°0N 89°58E 41 J16
Harīr, W. al → Syria 32°44N 35°59E 46 C4
Harīrūd → Asia 37°24N 60°38E 45 B9
Härjedalen Sweden 62°22N 13°5E 8 E15
Harlan Iowa, U.S.A. 41°39N 95°19W 80 E6
Harlan Ky., U.S.A. 36°51N 83°19W 81 G12
Harlech U.K. 52°52N 4°6W 12 E3
Harlem U.S.A. 48°32N 108°47W 76 B9
Harlingen Neths. 53°11N 5°25E 15 A5
Harlingen U.S.A. 26°12N 97°42W 84 H6
Harlow U.K. 51°46N 0°8E 13 F8
Harlowton U.S.A. 46°26N 109°50W 76 C9
Harnai Pakistan 30°6N 67°56E 42 D2
Harney Basin U.S.A. 43°0N 119°30W 76 E4
Harney L. U.S.A. 43°14N 119°8W 76 E4
Harney Peak U.S.A. 43°52N 103°32W 80 D2
Härnösand Sweden 62°38N 17°55E 8 E17
Haroldswick U.K. 60°48N 0°50W 11 A8
Harp L. Canada 55°5N 61°50W 73 A7
Harper Liberia 4°25N 7°43W 50 H4
Harper, Mt. U.S.A. 64°14N 143°51W 74 C11
Harrai India 22°37N 79°13E 43 H8
Harrand Pakistan 29°28N 70°3E 42 E4
Harricana → Canada 50°56N 79°32W 72 B4
Harriman U.S.A. 35°56N 84°33W 85 D12
Harrington Harbour Canada 50°31N 59°30W 73 B8
Harris U.K. 57°50N 6°55W 11 D2
Harris, L. Australia 31°10S 135°10E 63 E2
Harris, Sd. of U.K. 57°44N 7°6W 11 D1
Harris Pt. Canada 43°6N 82°9W 82 C2
Harrisburg Ill., U.S.A. 37°44N 88°32W 80 G9
Harrisburg Nebr., U.S.A. 41°33N 103°44W 80 E2
Harrisburg Pa., U.S.A. 40°16N 76°53W 82 F8
Harrismith S. Africa 28°15S 29°8E 57 D4
Harrison Ark., U.S.A. 36°14N 93°7W 84 C8
Harrison Maine, U.S.A. 44°7N 70°39W 83 B14
Harrison Nebr., U.S.A. 42°41N 103°53W 80 D2
Harrison, C. Canada 54°55N 57°55W 73 B8
Harrison Bay U.S.A. 70°40N 151°0W 74 A9
Harrison L. Canada 49°33N 121°50W 70 D4
Harrisonburg U.S.A. 38°27N 78°52W 81 F14
Harrisonville U.S.A. 38°39N 94°21W 80 F6
Harriston Canada 43°57N 80°53W 82 C4
Harrisville Mich., U.S.A. 44°39N 83°17W 82 B1
Harrisville N.Y., U.S.A. 44°9N 75°19W 83 B9
Harrisville Pa., U.S.A. 41°8N 80°0W 82 E5
Harrodsburg U.S.A. 37°46N 84°51W 81 G11
Harrogate U.K. 54°0N 1°33W 12 C6
Harrow Canada 42°2N 82°55W 82 D2
Harrow □ U.K. 51°35N 0°21W 13 F7
Harrowsmith Canada 44°24N 76°40W 83 B8
Harry S. Truman Res. U.S.A. 38°16N 93°24W 80 F7
Harsīn Iran 34°18N 47°33E 44 C5
Harstad Norway 68°48N 16°30E 8 B17
Harsud India 22°6N 76°44E 42 H7
Hart U.S.A. 43°42N 86°22W 80 D10
Hart, L. Australia 31°10S 136°25E 63 E2
Hartbees → S. Africa 28°45S 20°32E 56 D3
Hartford Conn., U.S.A. 41°46N 72°41W 83 E12
Hartford Ky., U.S.A. 37°27N 86°55W 80 G10
Hartford S. Dak., U.S.A. 43°38N 96°57W 80 D5
Hartford Vt., U.S.A. 43°40N 72°20W 83 C12
Hartford Wis., U.S.A. 43°19N 88°22W 80 D9
Hartford City U.S.A. 40°27N 85°22W 81 E11
Hartland Canada 46°20N 67°32W 73 C6
Hartland Pt. U.K. 51°1N 4°32W 13 F3
Hartlepool U.K. 54°42N 1°13W 12 C6
Hartlepool □ U.K. 54°42N 1°17W 12 C6
Hartley Bay Canada 53°25N 129°15W 70 C3
Hartmannberge Namibia 17°0S 13°0E 56 B1
Hartney Canada 49°30N 100°35W 71 D8
Harts → S. Africa 28°24S 24°17E 56 D3
Harts Range Australia 23°6S 134°55E 62 C1
Hartselle U.S.A. 34°27N 86°56W 85 D11
Hartshorne U.S.A. 34°51N 95°34W 84 D7
Hartstown U.S.A. 41°33N 80°23W 82 E4
Hartsville U.S.A. 34°23N 80°4W 85 D14
Hartswater S. Africa 27°34S 24°43E 56 D3
Hartwell U.S.A. 34°21N 82°56W 85 D13
Harunabad Pakistan 29°35N 73°8E 42 E5
Harvand Iran 28°25N 55°43E 45 D7
Harvey Australia 33°5S 115°54E 61 F2
Harvey Ill., U.S.A. 41°36N 87°50W 80 E10
Harvey N. Dak., U.S.A. 47°47N 99°56W 80 B4
Harwich U.K. 51°56N 1°17E 13 F9
Haryana □ India 29°0N 76°10E 42 E7
Haryn → Belarus 52°7N 27°17E 17 B14
Harz Germany 51°38N 10°44E 16 C6
Hasa Si. Arabia 25°50N 49°0E 45 E6
Ḥasā, W. al → Jordan 31°4N 35°29E 46 D4
Ḥasanābād Iran 32°8N 52°44E 45 C7
Ḥasb, W. → Iraq 31°45N 44°17E 44 D5
Hasdo → India 21°44N 82°44E 43 J10
Hashimoto Japan 34°19N 135°37E 31 G7
Hashtjerd Iran 35°52N 50°40E 45 C6
Hashtpur = Tālesh Iran 37°58N 48°58E 45 B6
Haskell U.S.A. 33°10N 99°44W 84 E5
Haskovo = Khaskovo Bulgaria 41°56N 25°30E 23 D11
Haslemere U.K. 51°5N 0°43W 13 F7
Hasselt Belgium 50°56N 5°21E 15 D5
Hassi Messaoud Algeria 31°51N 6°1E 50 B7
Hässleholm Sweden 56°10N 13°46E 9 H15
Hastings N.Z. 39°39S 176°52E 59 C6
Hastings U.K. 50°51N 0°35E 13 G8
Hastings Mich., U.S.A. 42°39N 85°17W 81 D11
Hastings Minn., U.S.A. 44°44N 92°51W 80 C7
Hastings Nebr., U.S.A. 40°35N 98°23W 80 E4
Hastings Ra. Australia 31°15S 152°14E 63 E5
Hat Lot Vietnam 21°15N 104°7E 38 B5
Hat Yai Thailand 7°1N 100°27E 39 J3
Hatanbulag = Ergel Mongolia 43°8N 109°5E 34 C5
Hatay Turkey 36°14N 36°10E 44 B3
Hatch U.S.A. 32°40N 107°9W 77 K10
Hatchet L. Canada 58°36N 103°40W 71 B8
Hateruma-Shima Japan 24°3N 123°47E 31 M1
Hatfield Australia 33°54S 143°49E 63 E3
Hatgal Mongolia 50°26N 100°9E 32 A9
Hathras India 27°36N 78°6E 42 F8
Hatia Bangla. 22°30N 91°5E 41 H17
Hato Mayor Dom. Rep. 18°46N 69°15W 89 C6
Hatta India 24°7N 79°36E 43 G8
Hatta U.A.E. 24°45N 56°4E 45 E8
Hattah Australia 34°48S 142°17E 63 E3
Hattah Kulkyne △ Australia 34°16S 142°33E 63 E3
Hatteras, C. U.S.A. 35°14N 75°32W 85 D17
Hattiesburg U.S.A. 31°20N 89°17W 85 F10
Hatvan Hungary 47°40N 19°45E 17 E10
Hau → Vietnam 9°30N 106°13E 39 H6
Haugesund Norway 59°23N 5°13E 9 G11
Haukipudas Finland 65°12N 25°20E 8 D21
Haultain → Canada 55°51N 106°46W 71 B7
Hauraki G. N.Z. 36°35S 175°5E 59 B5
Haut Atlas Morocco 32°30N 5°0W 50 B4
Hautes Fagnes = Hohes Venn Belgium 50°30N 6°5E 15 D6
Hauts Plateaux Algeria 35°0N 1°0E 50 B6
Havana = La Habana Cuba 23°8N 82°22W 88 B3
Havana U.S.A. 40°18N 90°4W 80 E8
Havant U.K. 50°51N 0°58W 13 G7
Havasor = Kığzı Turkey 38°18N 43°25E 44 B4
Havasu, L. U.S.A. 34°18N 114°28W 79 L12
Havel → Germany 52°50N 12°3E 16 B7
Havelian Pakistan 34°2N 73°10E 42 B5
Havelock Canada 44°26N 77°53W 82 B7
Havelock N.Z. 41°17S 173°48E 59 D4
Havelock U.S.A. 34°53N 76°54W 85 D16
Haverfordwest U.K. 51°48N 4°58W 13 F3
Haverhill U.K. 52°5N 0°28E 13 E8
Haverhill U.S.A. 42°47N 71°5W 83 D13
Haverstraw U.S.A. 41°12N 73°58W 83 E11
Havirga Mongolia 45°41N 113°5E 34 B7
Havířov Czech Rep. 49°46N 18°20E 17 D10
Havlíčkův Brod Czech Rep. 49°36N 15°33E 16 D8
Havre U.S.A. 48°33N 109°41W 76 B9
Havre-Aubert Canada 47°12N 61°56W 73 C7
Havre-St.-Pierre Canada 50°18N 63°33W 73 B7
Haw → U.S.A. 35°36N 79°3W 85 D15
Hawai'i U.S.A. 19°30N 155°30W 75 M8
Hawai'i □ U.S.A. 19°30N 156°30W 75 M8
Hawaiian Is. Pac. Oc. 20°30N 156°0W 65 E12
Hawaiian Ridge Pac. Oc. 24°0N 165°0W 65 E11
Hawarden U.S.A. 43°0N 96°29W 80 D5
Hawea, L. N.Z. 44°28S 169°19E 59 F2
Hawera N.Z. 39°35S 174°19E 59 C5
Hawi U.S.A. 20°14N 155°50W 75 L8
Hawick U.K. 55°26N 2°47W 11 F6
Hawk Junction Canada 48°5N 84°38W 72 C3
Hawke B. N.Z. 39°25S 177°20E 59 C6
Hawker Australia 31°59S 138°22E 63 E2
Hawke's Bay Canada 50°36N 57°10W 73 B8
Hawkesbury Canada 45°37N 74°37W 72 C5
Hawkesbury I. Canada 53°37N 129°3W 70 C3
Hawkesbury Pt. Australia 11°55S 134°5E 62 A1
Hawkinsville U.S.A. 32°17N 83°28W 85 E13
Hawley Minn., U.S.A. 46°53N 96°19W 80 B5
Hawley Pa., U.S.A. 41°28N 75°11W 83 E9
Ḥawrān, W. → Iraq 33°58N 42°34E 44 C4
Hawsh Mūssá Lebanon 33°45N 35°55E 46 B4
Hawthorne U.S.A. 38°32N 118°38W 76 G4
Hay Australia 34°30S 144°51E 63 E3
Hay → Australia 24°50S 138°0E 62 C2
Hay → Canada 60°50N 116°26W 70 A5
Hay, C. Australia 14°5S 129°29E 60 B4
Hay I. Canada 44°53N 80°58W 82 B4
Hay L. Canada 58°50N 118°50W 70 B5
Hay-on-Wye U.K. 52°5N 3°8W 13 E4
Hay Point Australia 21°18S 149°17E 62 C4
Hay River Canada 60°51N 115°44W 70 A5
Hay Springs U.S.A. 42°41N 102°41W 80 D2
Haya = Tehoru Indonesia 3°23S 129°30E 37 E7
Hayachine-San Japan 39°34N 141°29E 30 E10
Hayastan = Armenia ■ Asia 40°20N 45°0E 19 F7
Haydān, W. al → Jordan 31°29N 35°34E 46 D4
Hayden U.S.A. 40°30N 107°16W 76 F10
Hayes U.S.A. 44°23N 101°1W 80 C3
Hayes → Canada 57°3N 92°12W 72 A1
Hayes, Mt. U.S.A. 63°37N 146°43W 74 C10
Hayes Creek Australia 13°43S 131°22E 60 B5
Hayle U.K. 50°11N 5°26W 13 G2
Hayling I. U.K. 50°48N 0°59W 13 G7
Hayman I. Australia 20°4S 148°53E 62 b
Hayrabolu Turkey 41°12N 27°5E 23 D12
Hays Canada 50°6N 111°48W 70 C6
Hays U.S.A. 38°53N 99°20W 80 F4
Haysyn Ukraine 48°57N 29°25E 17 D15
Hayvoron Ukraine 48°22N 29°52E 17 D15
Hayward Calif., U.S.A. 37°40N 122°4W 78 H4
Hayward Wis., U.S.A. 46°1N 91°29W 80 B8
Haywards Heath U.K. 51°0N 0°5W 13 G7
Hazafon □ Israel 32°40N 35°20E 46 C4
Hazar Turkmenistan 39°34N 53°16E 19 G9
Hazārān, Kūh-e Iran 29°35N 57°20E 45 D8
Hazard U.S.A. 37°15N 83°12W 81 G12
Hazaribag India 23°58N 85°26E 43 H11
Hazaribag Road India 24°12N 85°57E 43 G11
Hazelton Canada 55°20N 127°42W 70 B3
Hazelton U.S.A. 46°29N 100°17W 80 B3
Hazen U.S.A. 47°18N 101°38W 80 B3
Hazen, L. Canada 81°47N 71°1W 69 A17
Hazlehurst Ga., U.S.A. 31°52N 82°36W 85 F13
Hazlehurst Miss., U.S.A. 31°52N 90°24W 85 F9
Hazlet U.S.A. 40°25N 74°12W 83 F10
Hazleton U.S.A. 40°57N 75°59W 83 F9
Hazlett, L. Australia 21°30S 128°48E 60 D4
Hazro Turkey 38°15N 40°47E 44 B4
Head of Bight Australia 31°30S 131°25E 61 F5
Headlands Zimbabwe 18°15S 32°2E 55 F3
Healdsburg U.S.A. 38°37N 122°52W 78 G4
Healdton U.S.A. 34°14N 97°29W 84 D6
Healesville Australia 37°35S 145°30E 63 F4
Healy U.S.A. 63°52N 148°58W 74 C10
Heany Junction Zimbabwe 20°6S 28°54E 57 C4
Heard I. Ind. Oc. 53°6S 72°36E 3 G13
Hearne U.S.A. 30°53N 96°36W 84 F6
Hearst Canada 49°40N 83°41W 72 C3
Heart → U.S.A. 46°46N 100°50W 80 B3
Heart's Content Canada 47°54N 53°27W 73 C9
Heath, Pte. Canada 49°8N 61°40W 73 C7
Heathrow, London ✈ (LHR) U.K. 51°28N 0°27W 13 F7
Heavener U.S.A. 34°53N 94°36W 84 D7
Hebbronville U.S.A. 27°18N 98°41W 84 H5
Hebei □ China 39°0N 116°0E 34 E9
Hebel Australia 28°58S 147°47E 63 D4
Heber U.S.A. 32°44N 115°32W 79 N11
Heber Springs U.S.A. 35°30N 92°2W 84 D8
Hebgen L. U.S.A. 44°52N 111°20W 76 D8
Hebi China 35°57N 114°7E 34 G8
Hebrides U.K. 57°30N 7°0W 6 D4
Hebrides, Sea of the U.K. 57°5N 7°0W 11 D2
Hebron = Al Khalīl West Bank 31°32N 35°6E 46 D4
Hebron Canada 58°5N 62°30W 69 F19
Hebron N. Dak., U.S.A. 46°54N 102°3W 80 B2
Hebron Nebr., U.S.A. 40°10N 97°35W 80 E5
Hecate Str. Canada 53°10N 130°30W 70 C2
Heceta I. U.S.A. 55°46N 133°40W 70 B2
Hechi China 24°40N 108°2E 32 G10
Hechuan China 30°2N 106°12E 32 E10
Hecla U.S.A. 45°53N 98°9W 80 C4
Hecla I. Canada 51°10N 96°43W 71 C9
Hede Sweden 62°23N 13°30E 8 E15
Hedemora Sweden 60°18N 15°58E 9 F16
Heerde Neths. 52°24N 6°2E 15 B6
Heerenveen Neths. 52°57N 5°55E 15 B5
Heerhugowaard Neths. 52°40N 4°51E 15 B4
Heerlen Neths. 50°55N 5°58E 15 D5
Ḫefa Israel 32°46N 35°0E 46 C4
Ḫefa □ Israel 32°40N 35°0E 46 C4
Hefei China 31°52N 117°18E 33 E12
Hegang China 47°20N 130°19E 33 B15
Hei Ling Chau China 22°15N 114°2E 33 a
Heichengzhen China 36°24N 106°3E 34 F4
Heidelberg Germany 49°24N 8°42E 16 D5
Heidelberg S. Africa 34°6S 20°59E 56 E3
Heihe China 50°10N 127°30E 33 A14
Heilbron S. Africa 27°16S 27°59E 57 D4
Heilbronn Germany 49°9N 9°13E 16 D5
Heilongjiang □ China 48°0N 126°0E 33 B14
Heilunkiang = Heilongjiang □ China 48°0N 126°0E 33 B14
Heimaey Iceland 63°26N 20°17W 8 E3
Heinola Finland 61°13N 26°2E 8 F22
Heinze Chaung Burma 14°42N 97°52E 38 E1
Heinze Kyun Burma 14°25N 97°45E 38 E1
Heishan China 41°40N 122°5E 35 D12
Heishui China 42°8N 119°30E 35 C10

Hejaz = Ḩijāz Si. Arabia 24°0N 40°0E 44 E3
Hejian China 38°25N 116°5E 34 E9
Hejin China 35°35N 110°42E 34 G6
Hekımhan Turkey 38°50N 37°55E 44 B3
Hekla Iceland 63°56N 19°35W 8 E4
Hekou Gansu, China 36°10N 103°26E 34 F2
Hekou Yunnan, China 22°30N 103°59E 32 G9
Helan Shan China 38°30N 105°55E 34 E3
Helen Atoll Palau 2°40N 132°0E 37 D8
Helena Ark., U.S.A. 34°32N 90°36W 85 D9
Helena Mont., U.S.A. 46°36N 112°2W 76 C7
Helendale U.S.A. 34°44N 117°19W 79 L9
Helensburgh U.K. 56°1N 4°43W 11 E4
Helensville N.Z. 36°41S 174°29E 59 B5
Helenvale Australia 15°43S 145°14E 62 B4
Helgeland Norway 66°7N 13°29E 8 C15
Helgoland Germany 54°10N 7°53E 16 A4
Heligoland = Helgoland Germany 54°10N 7°53E 16 A4
Heligoland B. = Deutsche Bucht Germany 54°15N 8°0E 16 A5
Hell Hole Gorge △ Australia 25°31S 144°12E 62 D3
Hella Iceland 63°50N 20°24W 8 E3
Hellas = Greece ■ Europe 40°0N 23°0E 23 E9
Hellertown U.S.A. 40°35N 75°21W 83 F9
Hellespont = Çanakkale Boğazı Turkey 40°17N 26°32E 23 D12
Hellevoetsluis Neths. 51°50N 4°8E 15 C4
Hellín Spain 38°31N 1°40W 21 C5
Hells Canyon ⌂ U.S.A. 45°30N 117°45W 76 D5
Hell's Gate △ Kenya 0°54S 36°19E 54 C4
Helmand □ Afghan. 31°20N 64°0E 40 D4
Helmand → Afghan. 31°12N 61°34E 40 D2
Helmeringhausen Namibia 25°54S 16°57E 56 D2
Helmond Neths. 51°29N 5°41E 15 C5
Helmsdale U.K. 58°7N 3°39W 11 C5
Helmsdale → U.K. 58°8N 3°43W 11 C5
Helong China 42°40N 129°0E 35 C15
Helper U.S.A. 39°41N 110°51W 76 G8
Helsingborg Sweden 56°3N 12°42E 9 H15
Helsingfors = Helsinki Finland 60°10N 24°55E 9 F21
Helsingør Denmark 56°2N 12°35E 9 H15
Helsinki Finland 60°10N 24°55E 9 F21
Helston U.K. 50°6N 5°17W 13 G2
Helvellyn U.K. 54°32N 3°1W 12 C4
Helwân Egypt 29°50N 31°20E 51 C12
Hemel Hempstead U.K. 51°44N 0°28W 13 F7
Hemet U.S.A. 33°45N 116°58W 79 M10
Hemingford U.S.A. 42°19N 103°4W 80 D2
Hemis △ India 34°10N 77°15E 42 B7
Hemmingford Canada 45°3N 73°35W 83 A11
Hempstead N.Y., U.S.A. 40°42N 73°37W 83 F11
Hempstead Tex., U.S.A. 30°6N 96°5W 84 F6
Hemse Sweden 57°15N 18°22E 9 H18
Henan □ China 34°0N 114°0E 34 H8
Henares → Spain 40°24N 3°30W 21 B4
Henashi-Misaki Japan 40°37N 139°51E 30 D9
Henderson Argentina 36°18S 61°43W 94 D3
Henderson Ky., U.S.A. 37°50N 87°35W 80 G10
Henderson N.C., U.S.A. 36°20N 78°25W 85 C15
Henderson N.Y., U.S.A. 43°50N 76°10W 83 C8
Henderson Nev., U.S.A. 36°2N 114°58W 79 J12
Henderson Tenn., U.S.A. 35°26N 88°38W 85 D10
Henderson Tex., U.S.A. 32°9N 94°48W 84 E7
Henderson I. Pac. Oc. 24°22S 128°19W 65 K15
Hendersonville N.C., U.S.A. 35°19N 82°28W 85 D13
Hendersonville Tenn., U.S.A. 36°18N 86°37W 85 C11
Hendījān Iran 30°14N 49°43E 45 D6
Hendorābī Iran 26°40N 53°37E 45 E7
Hengcheng China 38°18N 106°28E 34 E4
Hengdaohezi China 44°52N 129°0E 35 B15
Hengduan Shan China 28°30N 98°50E 32 F8
Hengelo Neths. 52°16N 6°48E 15 B6
Henggang China 22°39N 114°12E 33 a
Hengmen China 22°33N 113°35E 33 a
Hengshan China 37°58N 109°5E 34 F5
Hengshui China 37°41N 115°40E 34 F8
Hengyang China 26°59N 112°22E 33 F11
Henley-on-Thames U.K. 51°32N 0°54W 13 F7
Henlopen, C. U.S.A. 38°48N 75°6W 81 F16
Hennenman S. Africa 27°59S 27°1E 56 D4
Hennessey U.S.A. 36°6N 97°54W 84 C6
Henri Pittier △ Venezuela 10°26N 67°37W 89 D6
Henrietta N.Y., U.S.A. 43°4N 77°37W 82 C7
Henrietta Tex., U.S.A. 33°49N 98°12W 84 E5
Henrietta, Ostrov = Genriyetty, Ostrov Russia 77°6N 156°30E 29 B16
Henrietta Maria, C. Canada 55°9N 82°20W 72 A3
Henry U.S.A. 41°7N 89°22W 80 E9
Henryetta U.S.A. 35°27N 95°59W 84 D7
Henryville Canada 45°8N 73°11W 83 A11
Hensall Canada 43°26N 81°30W 82 C3
Hentiesbaai Namibia 22°8S 14°18E 56 C1
Hentiyn Nuruu Mongolia 48°30N 108°30E 33 B10
Henty Australia 35°30S 147°3E 63 F4
Henzada Burma 17°38N 95°26E 41 L19
Heppner U.S.A. 45°21N 119°33W 76 D4
Hepworth Canada 44°37N 81°9W 82 B3
Hequ China 39°20N 111°15E 34 E6
Héraðsflói Iceland 65°42N 14°12W 8 D6
Héraðsvötn → Iceland 65°45N 19°25W 8 D4
Heraklion = Iraklio Greece 35°20N 25°12E 25 D7
Herald Cays Australia 16°58S 149°9E 62 B4
Herāt Afghan. 34°20N 62°7E 40 B3
Herāt □ Afghan. 35°0N 62°0E 40 B3
Herbert Canada 50°30N 107°10W 71 C7
Herbert → Australia 18°31S 146°17E 62 B4
Herberton Australia 17°20S 145°25E 62 B4
Herbertsdale S. Africa 34°1S 21°46E 56 E3
Herceg-Novi Montenegro 42°30N 18°33E 23 C8
Herchmer Canada 57°22N 94°10W 71 B10
Herðubreið Iceland 65°11N 16°21W 8 D5
Hereford U.K. 52°4N 2°43W 13 E5
Hereford U.S.A. 34°49N 102°24W 84 D3
Herefordshire □ U.K. 52°8N 2°40W 13 E5
Herentals Belgium 51°12N 4°51E 15 C4
Herford Germany 52°7N 8°39E 16 B5
Herington U.S.A. 38°40N 96°57W 80 F5
Herkimer U.S.A. 43°2N 74°59W 83 D10
Herlen → Asia 48°48N 117°0E 33 B12
Herlong U.S.A. 40°8N 120°8W 78 E6
Herm U.K. 49°30N 2°28W 13 H5
Hermann U.S.A. 38°42N 91°27W 80 F8
Hermannsburg Australia 23°57S 132°45E 60 D5
Hermanus S. Africa 34°27S 19°12E 56 E2
Hermidale Australia 31°30S 146°42E 63 E4
Hermiston U.S.A. 45°51N 119°17W 76 D4
Hermon Canada 45°6N 77°37W 82 A7
Hermon U.S.A. 44°28N 75°14W 83 B9
Hermon, Mt. = Shaykh, J. ash Lebanon 33°25N 35°50E 46 B4
Hermosillo Mexico 29°10N 111°0W 86 B2
Hernád → Hungary 47°56N 21°8E 17 D11
Hernandarias Paraguay 25°20S 54°40W 95 B5
Hernandez U.S.A. 36°24N 120°46W 78 J6
Hernando Argentina 32°28S 63°40W 94 C3
Hernando U.S.A. 34°50N 90°0W 85 D10
Herndon U.S.A. 40°43N 76°51W 82 F8
Herne Germany 51°32N 7°14E 15 C7
Herne Bay U.K. 51°21N 1°8E 13 F9
Herning Denmark 56°8N 8°58E 9 H13
Heroica Caborca = Caborca Mexico 30°37N 112°6W 86 A2
Heroica Nogales = Nogales Mexico 31°19N 110°56W 86 A2
Heron Bay Canada 48°40N 86°25W 72 C2
Heron I. Australia 23°27S 151°55E 62 C5
Herradura, Pta. de la Canary Is. 28°26N 14°8W 24 F5
Herreid U.S.A. 45°50N 100°4W 80 C3
Herrin U.S.A. 37°48N 89°2W 80 G9
Herriot Canada 56°22N 101°16W 71 B8
Herschel I. Canada 69°35N 139°5W 4 C1
Hershey U.S.A. 40°17N 76°39W 83 F8
Herstal Belgium 50°40N 5°38E 15 D5
Hertford U.K. 51°48N 0°4W 13 F7
Hertfordshire □ U.K. 51°51N 0°5W 13 F7
's-Hertogenbosch Neths. 51°42N 5°17E 15 C5
Hertzogville S. Africa 28°9S 25°30E 56 D4
Hervey B. Australia 25°0S 152°52E 62 C5
Herzliyya Israel 32°10N 34°50E 46 C3
Ḩeşār Fārs, Iran 29°52N 50°16E 45 D6
Ḩeşār Markazī, Iran 35°50N 49°12E 45 C6
Heshui China 35°48N 108°0E 34 G5
Heshun China 37°22N 113°32E 34 F7
Hesperia U.S.A. 34°25N 117°18W 79 L9
Hesse = Hessen □ Germany 50°30N 9°0E 16 C5
Hessen □ Germany 50°30N 9°0E 16 C5
Hetauda Nepal 27°25N 85°2E 43 F11
Hetch Hetchy Aqueduct U.S.A. 37°29N 122°19W 78 H5
Hettinger U.S.A. 46°0N 102°42W 80 B2
Heuksando S. Korea 34°40N 125°30E 35 G13
Heunghae S. Korea 36°12N 129°21E 35 F15
Heuvelton U.S.A. 44°37N 75°25W 83 B9
Hewitt U.S.A. 31°28N 97°12W 84 F6
Hexham U.K. 54°58N 2°4W 12 C5
Hexigten Qi China 43°18N 117°30E 35 C9
Ḩeydarābād Iran 30°33N 55°38E 45 D7
Heysham U.K. 54°3N 2°53W 12 C5
Heywood Australia 38°8S 141°37E 63 F3
Heze China 35°14N 115°20E 34 G8
Hi Vista U.S.A. 34°45N 117°46W 79 L9
Hialeah U.S.A. 25°51N 80°16W 85 J14
Hiawatha U.S.A. 39°51N 95°32W 80 F6
Hibbing U.S.A. 47°25N 92°56W 80 B7
Hibernia Reef Australia 12°0S 123°23E 60 B3
Hickman U.S.A. 36°34N 89°11W 80 G9
Hickory U.S.A. 35°44N 81°21W 85 D14
Hicks, Pt. Australia 37°49S 149°17E 63 F4
Hicks L. Canada 61°25N 100°0W 71 A9
Hicksville U.S.A. 40°46N 73°32W 83 F11
Hida-Gawa → Japan 35°26N 137°3E 31 G8
Hida-Sammyaku Japan 36°30N 137°40E 31 F8
Hidaka-Sammyaku Japan 42°35N 142°45E 30 C11
Hidalgo □ Mexico 20°30N 99°0W 87 C5
Hidalgo, Presa M. Mexico 26°30N 108°35W 86 B3
Hidalgo del Parral Mexico 26°56N 105°40W 86 B3
Hierro Canary Is. 27°44N 18°0W 24 G1
Higashiajima-San Japan 37°40N 140°10E 30 F10
Higashiōsaka Japan 34°39N 135°37E 31 G7
Higgins U.S.A. 36°7N 100°2W 84 C4
Higgins Corner U.S.A. 39°2N 121°5W 78 F5
High Bridge U.S.A. 40°40N 74°54W 83 F10
High Desert U.S.A. 43°40N 120°20W 76 E3
High Island Res. China 22°22N 114°21E 33 a
High Level Canada 58°31N 117°8W 70 B5
High Point U.S.A. 35°57N 80°0W 85 D15
High Prairie Canada 55°30N 116°30W 70 B5
High River Canada 50°30N 113°50W 70 C6
High Tatra = Tatry Slovak Rep. 49°20N 20°0E 17 D11
High Veld Africa 27°0S 27°0E 48 J6
High Wycombe U.K. 51°37N 0°45W 13 F7
Highland □ U.K. 57°17N 4°21W 11 D4
Highland Park U.S.A. 42°11N 87°48W 80 D10
Highmore U.S.A. 44°31N 99°27W 80 C4
Highrock L. Canada 55°45N 100°30W 71 B8
Higüey Dom. Rep. 18°37N 68°42W 89 C6
Hiiumaa Estonia 58°50N 22°45E 9 G20
Ḩijārah, Şaḩrā' al Iraq 30°25N 44°30E 44 D5
Ḩijāz Si. Arabia 24°0N 40°0E 44 E3
Hijo = Tagum Phil. 7°33N 125°53E 37 C7
Hikari Japan 33°58N 131°58E 31 H5
Hiko U.S.A. 37°32N 115°14W 78 H11
Hikone Japan 35°15N 136°10E 31 G8
Hikurangi Gisborne, N.Z. 37°55S 178°4E 59 C6
Hikurangi Northland, N.Z. 35°36S 174°17E 59 A5
Hildesheim Germany 52°9N 9°56E 16 B5
Hill → Australia 30°23S 115°3E 61 F2
Hill City Idaho, U.S.A. 43°18N 115°3W 76 E6
Hill City Kans., U.S.A. 39°22N 99°51W 80 F4
Hill City Minn., U.S.A. 46°59N 93°36W 80 B7
Hill City S. Dak., U.S.A. 43°56N 103°35W 80 D2
Hill Island L. Canada 60°30N 109°50W 71 A7
Hillaby, Mt. Barbados 13°12N 59°35W 89 g
Hillcrest Barbados 13°13N 59°31W 89 g
Hillcrest Center U.S.A. 35°23N 118°57W 79 K8
Hillegom Neths. 52°18N 4°35E 15 B4
Hillsboro Kans., U.S.A. 38°21N 97°12W 80 F5
Hillsboro N. Dak., U.S.A. 47°26N 97°3W 80 B5
Hillsboro Ohio, U.S.A. 39°12N 83°37W 81 F12
Hillsboro Oreg., U.S.A. 45°31N 122°59W 78 E4
Hillsboro Tex., U.S.A. 32°1N 97°8W 84 E6
Hillsborough Grenada 12°28N 61°28W 89 D7
Hillsborough U.S.A. 43°7N 71°54W 83 C13
Hillsborough Channel Australia 20°56S 149°15E 62 b
Hillsdale Mich., U.S.A. 41°56N 84°38W 81 E11
Hillsdale N.Y., U.S.A. 42°11N 73°32W 83 D11
Hillsport Canada 49°27N 85°34W 72 C2
Hillston Australia 33°30S 145°31E 63 E4
Hilo U.S.A. 19°44N 155°5W 75 M8
Hilton U.S.A. 43°17N 77°48W 82 C7
Hilton Head Island U.S.A. 32°13N 80°45W 85 E14
Hilversum Neths. 52°14N 5°10E 15 B5
Himachal Pradesh □ India 31°30N 77°0E 42 D7
Himalaya Asia 29°0N 84°0E 43 E11
Himalchuli Nepal 28°27N 84°38E 43 E11
Himatnagar India 23°37N 72°57E 42 H5
Himeji Japan 34°50N 134°40E 31 G7
Himi Japan 36°50N 136°55E 31 F8
Ḩimş Syria 34°40N 36°45E 46 A5
Ḩimş □ Syria 34°30N 37°0E 46 A6
Hinche Haiti 19°9N 72°1W 89 C5
Hinchinbrook I. Australia 18°20S 146°15E 62 B4
Hinchinbrook Island △ Australia 18°14S 146°6E 62 B4
Hinckley U.K. 52°33N 1°22W 13 E6
Hinckley U.S.A. 46°1N 92°56W 80 B7
Hindaun India 26°44N 77°5E 42 F7
Hindmarsh, L. Australia 36°5S 141°55E 63 F3
Hindu Bagh Pakistan 30°56N 67°50E 42 D2
Hindu Kush Asia 36°0N 71°0E 40 B7
Hindupur India 13°49N 77°32E 40 N10
Hines Creek Canada 56°20N 118°40W 70 B5
Hinesville U.S.A. 31°51N 81°36W 85 F14
Hinganghat India 20°30N 78°52E 40 J11
Hingham U.S.A. 48°33N 110°25W 76 B8
Hingir India 21°57N 83°41E 43 J10
Hingoli India 19°41N 77°15E 40 K10
Hinna = Imi Ethiopia 6°28N 42°10E 47 F3
Hinnøya Norway 68°35N 15°50E 8 B16
Hinojosa del Duque Spain 38°30N 5°9W 21 C3
Hinsdale U.S.A. 42°47N 72°29W 83 D12
Hinthada = Henzada Burma 17°38N 95°26E 41 L19
Hinton Canada 53°26N 117°34W 70 C5
Hinton U.S.A. 37°40N 80°54W 81 G13
Hios = Chios Greece 38°27N 26°9E 23 E12
Hirado Japan 33°22N 129°33E 31 H4
Hirakud Dam India 21°32N 83°45E 41 J13
Hiran → India 23°6N 79°21E 43 H8
Hirapur India 24°22N 79°13E 43 G8
Hirara Japan 24°48N 125°17E 31 M2
Hiratsuka Japan 35°19N 139°21E 31 G9
Hiroo Japan 42°17N 143°19E 30 C11
Hirosaki Japan 40°34N 140°28E 30 D10
Hiroshima Japan 34°24N 132°30E 31 G6
Hiroshima □ Japan 34°50N 133°0E 31 G6
Hisar India 29°12N 75°45E 42 E6
Hisb, Sha'ib = Ḩasb, W. → Iraq 31°45N 44°17E 44 D5
Ḩismá Si. Arabia 28°30N 36°0E 44 D3
Hispaniola W. Indies 19°0N 71°0W 89 C5
Hīt Iraq 33°38N 42°49E 44 C4
Hita Japan 33°20N 130°58E 31 H5
Hitachi Japan 36°36N 140°39E 31 F10
Hitchin U.K. 51°58N 0°16W 13 F7
Hitiaa Tahiti 17°36S 149°18W 59 d
Hitoyoshi Japan 32°13N 130°45E 31 H5
Hitra Norway 63°30N 8°45E 8 E13
Hiva Oa French Polynesia 9°45S 139°0W 65 H14
Hixon Canada 53°25N 122°35W 70 C4
Ḩiyyon, N. → Israel 30°25N 35°10E 46 E4
Hjalmar L. Canada 61°33N 109°25W 71 A7
Hjälmaren Sweden 59°18N 15°40E 9 G16
Hjørring Denmark 57°29N 9°59E 9 H13
Hjort Trench S. Ocean 58°0S 157°30E 5 B10
Hkakabo Razi Burma 28°25N 97°23E 41 E20
Hkamti = Singkaling Hkamti Burma 26°0N 95°39E 41 G19
Hlobane S. Africa 27°42S 31°0E 57 D5
Hluhluwe S. Africa 28°1S 32°15E 57 D5
Hluhluwe ⌂ S. Africa 22°10S 32°5E 57 C5
Hlyboka Ukraine 48°5N 25°56E 17 D13
Ho Ghana 6°37N 0°27E 50 G6
Ho Chi Minh City = Thanh Pho Ho Chi Minh Vietnam 10°58N 106°40E 39 G6
Ho Hoa Binh Vietnam 20°50N 105°0E 38 B5
Ho Thac Ba Vietnam 21°42N 105°1E 38 A5
Ho Thuong Vietnam 19°32N 105°48E 38 C5
Hoa Binh Vietnam 20°50N 105°20E 38 B5
Hoa Hiep Vietnam 11°34N 105°51E 39 G5
Hoai Nhon Vietnam 14°28N 109°1E 38 E7
Hoang Lien △ Vietnam 21°30N 105°32E 38 B5
Hoang Lien Son Vietnam 22°0N 104°0E 38 A4
Hoanib → Namibia 19°27S 12°46E 56 B2
Hoare B. Canada 65°17N 62°30W 69 D19
Hoarusib → Namibia 19°3S 12°36E 56 B2
Hobart Australia 42°50S 147°21E 63 G4
Hobart U.S.A. 35°1N 99°6W 84 D5
Hobbs U.S.A. 32°42N 103°8W 77 K12
Hobbs Coast Antarctica 74°50S 131°0W 5 D14
Hobe Sound U.S.A. 27°4N 80°8W 85 H14
Hoboken U.S.A. 40°44N 74°3W 83 F10
Hobro Denmark 56°39N 9°46E 9 H13
Hoburgen Sweden 56°55N 18°7E 9 H18
Hobyo Somali Rep. 5°25N 48°30E 47 F4
Hochfeld Namibia 21°28S 17°58E 56 C2
Hodaka-Dake Japan 36°17N 137°39E 31 F8
Hodeida = Al Ḩudaydah Yemen 14°50N 43°0E 47 E3
Hodgeville Canada 50°7N 106°58W 71 C7
Hodgson Canada 51°13N 97°36W 71 C9
Hódmezővásárhely Hungary 46°28N 20°22E 17 E11
Hodna, Chott el Algeria 35°26N 4°43E 50 A6
Hodonín Czech Rep. 48°50N 17°10E 17 D9
Hoek van Holland Neths. 52°0N 4°7E 15 C4
Hoengseong S. Korea 37°29N 127°59E 35 F14
Hoeryong N. Korea 42°30N 129°45E 35 C15
Hoeyang N. Korea 38°43N 127°36E 35 E14
Hof Germany 50°19N 11°55E 16 C6
Hofmeyr S. Africa 31°39S 25°50E 56 E4
Höfn Iceland 64°15N 15°13W 8 D6
Hofors Sweden 60°31N 16°15E 9 F17
Hofsjökull Iceland 64°49N 18°48W 8 D4
Hōfu Japan 34°3N 131°34E 31 G5
Hogan Group Australia 39°13S 147°1E 63 F4
Hogarth, Mt. Australia 21°48S 136°58E 62 C2
Hoge Kempen △ Belgium 51°6N 5°35E 15 C5
Hoggar = Ahaggar Algeria 23°0N 6°30E 50 D7
Hogsty Reef Bahamas 21°41N 73°48W 89 B5
Hoh → U.S.A. 47°45N 124°29W 78 C2
Hoh Xil Shan China 36°30N 89°0E 32 D6
Hohenwald U.S.A. 35°33N 87°33W 85 D11
Hoher Rhön = Rhön Germany 50°24N 9°58E 16 C5
Hohes Venn Belgium 50°30N 6°5E 15 D6
Hohhot China 40°52N 111°40E 34 D6
Hoi An Vietnam 15°30N 108°19E 38 E7
Hoi Xuan Vietnam 20°25N 105°9E 38 B5
Hoisington U.S.A. 38°31N 98°47W 80 F4
Hōjō Japan 33°58N 132°46E 31 H6
Hokianga Harbour N.Z. 35°31S 173°22E 59 A4
Hokitika N.Z. 42°42S 171°0E 59 E3
Hokkaidō □ Japan 43°30N 143°0E 30 C11
Hola Kenya 1°29S 40°2E 54 B4
Holakas Greece 35°57N 27°53E 25 D9
Holbrook Australia 35°42S 147°18E 63 F4
Holbrook U.S.A. 34°54N 110°10W 77 J8
Holcomb U.S.A. 42°54N 77°25W 82 D7
Holden U.S.A. 39°6N 112°16W 76 G7
Holdenville U.S.A. 35°5N 96°24W 84 D6
Holdrege U.S.A. 40°26N 99°23W 80 E4
Holetown Barbados 13°11N 59°38W 89 g
Holguín Cuba 20°50N 76°20W 88 B4
Hollams Bird I. Namibia 24°40S 14°30E 56 C1
Holland = Netherlands ■ Europe 52°0N 5°30E 15 C5
Holland Mich., U.S.A. 42°47N 86°7W 80 D10
Holland N.Y., U.S.A. 42°38N 78°32W 82 D6
Holland Centre Canada 44°23N 80°47W 82 B4
Holland Patent U.S.A. 43°14N 75°15W 83 C9
Hollandale U.S.A. 33°10N 90°51W 85 E9
Holley U.S.A. 43°14N 78°2W 82 C6
Hollidaysburg U.S.A. 40°26N 78°24W 82 F6
Hollis U.S.A. 34°41N 99°55W 84 D5
Hollister Calif., U.S.A. 36°51N 121°24W 78 J5
Hollister Idaho, U.S.A. 42°21N 114°35W 76 E6
Holly Hill U.S.A. 29°16N 81°3W 85 G14
Holly Springs U.S.A. 34°46N 89°27W 85 D10
Hollywood U.S.A. 26°0N 80°8W 85 J14
Holman Canada 70°44N 117°44W 68 C8
Hólmavík Iceland 65°42N 21°40W 8 D3
Holmen U.S.A. 43°58N 91°15W 80 D8
Holmes Reefs Australia 16°27S 148°0E 62 B4
Holmsund Sweden 63°41N 20°20E 8 E19
Holon Israel 32°2N 34°47E 46 C3
Holroyd → Australia 14°10S 141°36E 62 A3
Holstebro Denmark 56°22N 8°37E 9 H13
Holsworthy U.K. 50°48N 4°22W 13 G3
Holton Canada 54°31N 57°12W 73 B8
Holton U.S.A. 39°28N 95°44W 80 F6
Holtville U.S.A. 32°49N 115°23W 79 N11
Holwerd Neths. 53°22N 5°54E 15 A5
Holy Cross U.S.A. 62°12N 159°46W 74 C8
Holy I. Anglesey, U.K. 53°17N 4°37W 12 D3
Holy I. Northumberland, U.K. 55°40N 1°47W 12 B6
Holyhead U.K. 53°18N 4°38W 12 D3
Holyoke Colo., U.S.A. 40°35N 102°18W 76 F12
Holyoke Mass., U.S.A. 42°12N 72°37W 83 D12
Holyrood Canada 47°27N 53°8W 73 C9
Homa Bay Kenya 0°36S 34°30E 54 C3
Homalin Burma 24°55N 95°0E 41 G19
Homand Iran 32°28N 59°37E 45 C8
Homathko → Canada 51°0N 124°56W 70 C4
Hombori Mali 15°20N 1°38W 50 E5
Home B. Canada 68°40N 67°10W 69 D18
Home Hill Australia 19°43S 147°25E 62 B4
Home Reef Tonga 18°59S 174°47W 59 c
Homedale U.S.A. 43°37N 116°56W 76 E5
Homer Alaska, U.S.A. 59°39N 151°33W 68 F1
Homer La., U.S.A. 32°48N 93°4W 84 E8
Homer N.Y., U.S.A. 42°38N 76°10W 83 D8
Homer City U.S.A. 40°32N 79°10W 82 F5
Homestead Australia 20°20S 145°40E 62 C4
Homestead U.S.A. 25°28N 80°29W 85 J14
Homestead ⌂ U.S.A. 40°17N 96°50W 80 E5
Homoine Mozam. 23°55S 35°8E 57 C6
Homs = Ḩimş Syria 34°40N 36°45E 46 A5
Homyel Belarus 52°28N 31°0E 17 B16
Hon Chong Vietnam 10°25N 104°30E 39 G5
Hon Hai Vietnam 10°0N 109°0E 39 H7
Hon Me Vietnam 19°23N 105°56E 38 C5
Honan = Henan □ China 34°0N 114°0E 34 H8
Honbetsu Japan 43°7N 143°37E 30 C11
Honcut U.S.A. 39°20N 121°32W 78
Honda, Bahía Cuba 22°54N 83°10W 88
Hondeklipbaai S. Africa 30°19S 17°17E 56
Hondo Japan 32°27N 130°12E 31
Hondo U.S.A. 29°21N 99°9W 84
Hondo, Río → Belize 18°25N 88°21W 87
Honduras ■ Cent. Amer. 14°40N 86°30W 88
Honduras, G. de Caribbean 16°50N 87°0W 88
Hønefoss Norway 60°10N 10°18E 9
Honesdale U.S.A. 41°34N 75°16W 83
Honey Harbour Canada 44°52N 79°49W 82
Honey L. U.S.A. 40°15N 120°19W 78
Honfleur France 49°25N 0°13E 20
Hong → Vietnam 20°16N 106°34E 38
Hong Gai Vietnam 20°57N 107°5E 38
Hong He → China 32°25N 115°35E 34
Hong Kong □ China 22°11N 114°14E 3
Hong Kong I. China 22°16N 114°12E 3
Hong Kong Int. ✕ (HKG) China 22°19N 113°57E 3
Hongcheon S. Korea 37°44N 127°53E 35
Hongjiang China 27°7N 109°59E 33
Hongliu He → China 38°0N 109°50E 34
Hongor Mongolia 45°45N 112°50E 34
Hongseong S. Korea 36°37N 126°38E 35
Hongshan China 36°38N 117°58E 35
Hongshui He → China 23°48N 109°30E 33
Hongtong China 36°16N 111°40E 34
Honguedo, Détroit d' Canada 49°15N 64°0W 73
Hongwon N. Korea 40°0N 127°56E 35
Hongze Hu China 33°15N 118°35E 35
Honiara Solomon Is. 9°27S 159°57E 58
Honiton U.K. 50°47N 3°11W 13
Honjō Japan 39°23N 140°3E 30
Honningsvåg Norway 70°59N 25°59E 8
Honolulu U.S.A. 21°19N 157°52W 75
Honshū Japan 36°0N 138°0E 30
Hood, Mt. U.S.A. 45°23N 121°42W 76
Hood, Pt. Australia 34°23S 119°34E 61
Hood River U.S.A. 45°43N 121°31W 76
Hoodsport U.S.A. 47°24N 123°9W 78
Hoogeveen Neths. 52°44N 6°28E 15
Hoogezand-Sappemeer Neths. 53°9N 6°45E 15
Hooghly = Hugli → India 21°56N 88°4E 43
Hooghly-Chinsura = Chunchura India 22°53N 88°27E 43
Hook Hd. Ireland 52°7N 6°56W 10
Hook I. Australia 20°4S 149°0E 6
Hook of Holland = Hoek van Holland Neths. 52°0N 4°7E 15
Hooker U.S.A. 36°52N 101°13W 84
Hooker Creek = Lajamanu Australia 18°23S 130°38E 60
Hooker Creek ✪ Australia 18°6S 130°23E 60
Hoonah U.S.A. 58°7N 135°27W 70
Hooper Bay U.S.A. 61°32N 166°6W 74
Hoopeston U.S.A. 40°28N 87°40W 80
Hoopstad S. Africa 27°50S 25°55E 56
Hoorn Neths. 52°38N 5°4E 15
Hoover U.S.A. 33°24N 86°49W 85
Hoover Dam U.S.A. 36°1N 114°44W 79
Hooversville U.S.A. 40°9N 78°55W 82
Hop Bottom U.S.A. 41°42N 75°46W 83
Hope Canada 49°25N 121°25W 70
Hope Ariz., U.S.A. 33°43N 113°42W 79 M
Hope Ark., U.S.A. 33°40N 93°36W 84
Hope, L. S. Austral., Australia 28°24S 139°18E 63
Hope, L. W. Austral., Australia 32°35S 120°15E 61
Hope, Pt. U.S.A. 68°21N 166°47W 66
Hope I. Canada 44°55N 80°11W 82
Hope Town Bahamas 26°35N 76°57W 88
Hope Vale Australia 15°16S 145°20E 62
Hope Vale ✪ Australia 15°8S 145°15E 62
Hopedale Canada 55°28N 60°13W 73
Hopedale U.S.A. 42°8N 71°33W 83
Hopefield S. Africa 33°3S 18°22E 56
Hopei = Hebei □ China 39°0N 116°0E 34
Hopelchén Mexico 19°46N 89°51W 87
Hopetoun Vic., Australia 35°42S 142°22E 63
Hopetoun W. Austral., Australia 33°57S 120°7E 61
Hopetown S. Africa 29°34S 24°3E 56
Hopewell U.S.A. 37°18N 77°17W 81
Hopkins, L. Australia 24°15S 128°35E 60
Hopkinsville U.S.A. 36°52N 87°29W 80
Hopland U.S.A. 38°58N 123°7W 78
Hoquiam U.S.A. 46°59N 123°53W 78
Hordern Hills Australia 20°15S 130°0E 60
Horinger China 40°28N 111°48E 34
Horizontina Brazil 27°37S 54°19W 95
Horlick Mts. Antarctica 84°0S 102°0W 5
Horlivka Ukraine 48°19N 38°5E 19
Hormak Iran 29°58N 60°51E 45
Hormoz Iran 27°35N 55°0E 45
Hormoz, Jaz.-ye Iran 27°8N 56°28E 45
Hormozgān □ Iran 27°30N 56°0E 45
Hormuz, Kūh-e Iran 27°27N 55°10E 45
Hormuz, Str. of The Gulf 26°30N 56°30E 45
Horn Austria 48°39N 15°40E 16
Horn → Canada 61°30N 118°1W 70
Horn, Cape = Hornos, C. de Chile 55°50S 67°30W 96
Horn Head Ireland 55°14N 8°0W 10
Horn I. Australia 10°37S 142°17E 62
Horn Plateau Canada 62°15N 119°15W 70
Hornavan Sweden 66°15N 17°30E 8
Hornbeck U.S.A. 31°20N 93°24W 84
Hornbrook U.S.A. 41°55N 122°33W 76
Horncastle U.K. 53°13N 0°7W 12
Hornell U.S.A. 42°20N 77°40W 82
Hornell L. Canada 62°20N 119°25W 70
Hornepayne Canada 49°14N 84°48W 72
Hornings Mills Canada 44°9N 80°12W 82

rnitos U.S.A. 37°30N 120°14W 78 H6
rnos, C. de Chile 55°50S 67°30W 96 H3
rnsby Australia 33°42S 151°2E 63 B5
rnsea U.K. 53°55N 0°11W 12 D7
robetsu = Noboribetsu Japan 42°24N 141°6E 30 C10
rodenka Ukraine 48°41N 25°29E 17 D13
rodok Khmelnytskyy, Ukraine 49°10N 26°34E 17 D14
rodok Lviv, Ukraine 49°46N 23°32E 17 D12
rokhiv Ukraine 50°30N 24°45E 17 C13
rqin Youyi Qianqi China 46°5N 122°3E 35 A12
rqin Zuoyi Zhongqi China 44°8N 123°18E 35 B12
rqueta Paraguay 23°15S 56°55W 94 A4
rse Cr. → U.S.A. 41°57N 103°58W 76 F12
rse I. Canada 53°20N 99°6W 71 C9
rse Is. Canada 50°15N 55°50W 73 B8
rsefly L. Canada 52°25N 121°0W 70 C4
rseheads U.S.A. 42°10N 76°49W 82 D8
rsens Denmark 55°52N 9°51E 9 J13
rseshoe Lake Canada 45°17N 79°51W 82 A5
rsham Australia 36°44S 142°13E 63 F3
rsham U.K. 51°4N 0°20W 13 F7
rta Azores 38°32N 28°38W 50 a
rten Norway 59°25N 10°32E 9 G14
rton U.S.A. 39°40N 95°32W 80 F6
rton → Canada 69°56N 126°52W 68 D6
rwood L. Canada 48°5N 82°20W 72 C3
sapete = Hospet India 15°15N 76°20E 40 M10
seynābād Khuzestān, Iran 32°45N 48°20E 45 C6
seynābād Kordestān, Iran 35°33N 47°8E 44 C5
shangabad India 22°45N 77°45E 42 H7
shiarpur India 31°30N 75°58E 42 D6
spet India 15°15N 76°20E 40 M10
ste, I. Chile 55°0S 69°0W 96 H3
t Thailand 18°8N 98°29E 38 C2
t Creek Range U.S.A. 38°40N 116°20W 76 G5
t Springs Ark., U.S.A. 34°31N 93°3W 84 D8
t Springs S. Dak., U.S.A. 43°26N 103°29W 80 D2
t Springs △ U.S.A. 34°31N 93°3W 84 D8
tagen Sweden 63°59N 14°12E 8 E16
tan China 37°25N 79°55E 32 D4
tan He → China 40°22N 80°56E 32 C5
tazel S. Africa 27°17S 22°58E 56 D3
tchkiss U.S.A. 38°48N 107°43W 76 G10
tham, C. Australia 12°2S 131°18E 60 B5
ting Sweden 64°8N 16°15E 8 D17
tte, Massif de la Haiti 18°30N 73°45W 89 C5
ttentotsbaai Namibia 26°8S 14°59E 56 D1
u Hai China 22°32N 113°56E 33 a
uei Sai Laos 20°18N 100°26E 38 B3
uffalize Belgium 50°8N 5°48E 15 D5
ughton Mich., U.S.A. 47°7N 88°34W 80 B9
ughton N.Y., U.S.A. 42°25N 78°10W 82 D6
ughton L. U.S.A. 44°21N 84°44W 81 C11
ughton-le-Spring U.K. 54°51N 1°28W 12 C6
uhora Heads N.Z. 34°49S 173°9E 59 A4
ulton U.S.A. 46°8N 67°51W 81 B20
uma China 35°36N 111°20E 34 G6
uma U.S.A. 29°36N 90°43W 85 G9
usatonic → U.S.A. 41°10N 73°7W 83 E11
uston Canada 54°25N 126°39W 70 C3
uston Mo., U.S.A. 37°22N 91°58W 80 G8
uston Tex., U.S.A. 29°45N 95°21W 84 G7
ut = Mogwadi → S. Africa 23°4S 29°36E 57 C4
utkraal S. Africa 30°23S 24°5E 56 E3
utman Abrolhos Australia 28°43S 113°48E 61 E1
vd = Dund-Us Mongolia 48°1N 91°38E 32 B7
ve U.K. 50°50N 0°10W 13 G7
venweep △ U.S.A. 37°20N 109°0W 77 H9
veyzeh Iran 31°27N 48°4E 45 D6
vsgöl Mongolia 43°37N 109°39E 34 C5
vsgöl Nuur Mongolia 51°0N 100°30E 32 A9
war, Wadi → Sudan 17°30N 27°8E 51 E11
ward Australia 25°16S 152°32E 63 D5
ward Pa., U.S.A. 41°1N 77°40W 82 F7
ward S. Dak., U.S.A. 44°1N 97°32W 80 C5
we U.S.A. 43°48N 113°0W 76 E7
we, C. Australia 37°30S 150°0E 63 F5
we, West Cape Australia 35°8S 117°36E 61 G2
we I. Canada 44°16N 76°17W 83 B8
well U.S.A. 42°36N 83°56W 81 D12
wick Canada 45°11N 73°51W 83 A11
wick S. Africa 29°28S 30°14E 57 D5
wick Group Australia 14°20S 145°30E 62 A4
witt, L. Australia 27°40S 138°40E 63 D2
wland I. Pac. Oc. 0°48N 176°38W 64 G10
wrah = Haora India 22°34N 88°18E 43 H13
wth Ireland 53°23N 6°6W 10 C5
wth Hd. Ireland 53°22N 6°4W 10 C5
xter Germany 51°46N 9°22E 16 C5
y U.K. 58°50N 3°15W 11 C5
yanger Norway 61°13N 6°4E 8 F12
yerswerda Germany 51°26N 14°14E 16 C8
ylake U.K. 53°24N 3°10W 12 D4
a-an = Pa-an Burma 16°51N 97°40E 41 L20
unan Pass Burma 27°30N 96°55E 41 F20
adec Králové Czech Rep. 50°15N 15°50E 16 C8
odna Belarus 53°42N 23°52E 17 B12
odzyanka Belarus 53°31N 28°42E 17 B15
on → Slovak Rep. 47°49N 18°45E 17 E10
vatska = Croatia ■ Europe 45°20N 16°0E 16 F9
ymayliv Ukraine 49°20N 26°5E 17 D14
enwi Burma 23°22N 97°55E 41 H20
iamen = Xiamen China 24°25N 118°4E 33 G12
Hsian = Xi'an China 34°15N 109°0E 34 G5
Hsinchu Taiwan 24°48N 120°58E 33 G13
Hsinhailien = Lianyungang China 34°40N 119°11E 35 G10
Hsüchou = Xuzhou China 34°18N 117°10E 35 G9
Hu Xian China 34°8N 108°42E 34 G5
Hua Hin Thailand 12°34N 99°58E 38 F2
Hua Muang Laos 20°13N 103°52E 38 B4
Hua Shan China 34°28N 110°4E 34 G6
Hua Xian Henan, China 35°30N 114°30E 34 G8
Hua Xian Shaanxi, China 34°30N 109°48E 34 G5
Huab → Namibia 20°52S 13°25E 56 B2
Huachinera Mexico 30°9N 108°55W 86 A3
Huacho Peru 11°10S 77°35W 92 F3
Huade China 41°55N 113°59E 34 D7
Huadian China 43°0N 126°40E 35 C14
Huahine, Î. French Polynesia 16°46S 150°58W 65 J12
Huai Hat △ Thailand 16°52N 104°17E 38 D5
Huai He → China 33°0N 118°30E 33 E12
Huai Kha Khaeng △ Thailand 15°48N 99°7E 38 E2
Huai Nam Dang △ Thailand 19°30N 98°30E 38 C2
Huai Yot Thailand 7°45N 99°37E 39 J2
Huai'an Hebei, China 40°30N 114°20E 34 D8
Huai'an Jiangsu, China 33°30N 119°10E 35 H10
Huaibei China 34°0N 116°48E 34 G9
Huaide = Gongzhuling China 43°30N 124°40E 35 C13
Huaidezhen China 43°48N 124°50E 35 C13
Huaihua China 27°32N 109°57E 33 F10
Huainan China 32°38N 116°58E 33 E12
Huairen China 39°48N 113°20E 34 E7
Huairou China 40°20N 116°35E 34 D9
Huaiyang China 33°40N 114°52E 34 H8
Huaiyin China 33°30N 119°2E 35 H10
Huaiyuan China 32°55N 117°10E 35 H9
Huajuapán de León Mexico 17°48N 97°46W 87 D5
Hualapai Peak U.S.A. 35°5N 113°54W 77 J7
Huallaga → Peru 5°15S 75°30W 92 E3
Huambo Angola 12°42S 15°54E 53 G3
Huan Jiang → China 34°28N 109°0E 34 G5
Huan Xian China 36°33N 107°7E 34 F4
Huancabamba Peru 5°10S 79°15W 92 E3
Huancane Peru 15°10S 69°44W 92 G5
Huancavelica Peru 12°50S 75°5W 92 F3
Huancayo Peru 12°5S 75°12W 92 F3
Huanchaca Bolivia 20°15S 66°40W 92 H5
Huang Hai = Yellow Sea China 35°0N 123°0E 35 G12
Huang He → China 37°55N 118°50E 35 F10
Huang Xian China 37°38N 120°30E 35 F11
Huangdao China 36°0N 120°7E 35 G11
Huanghua China 38°22N 117°20E 35 E9
Huanghuagang China 38°20N 117°54E 35 E9
Huangling China 35°34N 109°15E 34 G5
Huanglong China 35°30N 109°59E 34 G5
Huangshan China 29°42N 118°25E 33 F12
Huangshi China 30°10N 115°3E 33 E12
Huangsongdian China 43°45N 127°25E 35 C14
Huanren China 41°23N 125°20E 35 D13
Huantai China 36°58N 117°56E 35 F9
Huánuco Peru 9°55S 76°15W 92 E3
Huaraz Peru 9°30S 77°32W 92 E3
Huarmey Peru 10°5S 78°5W 92 F3
Huascarán, Nevado Peru 9°7S 77°37W 92 E3
Huasco Chile 28°30S 71°15W 94 B1
Huasco → Chile 28°27S 71°13W 94 B1
Huasna U.S.A. 35°6N 120°24W 79 K6
Huatabampo Mexico 26°50N 109°38W 86 B3
Huauchinango Mexico 20°12N 98°3W 87 C5
Huautla de Jiménez Mexico 18°8N 96°51W 87 D5
Huayin China 34°35N 110°5E 34 G6
Hubballi = Hubli India 15°22N 75°15E 40 M9
Hubbard Ohio, U.S.A. 41°9N 80°34W 82 E4
Hubbard Tex., U.S.A. 31°51N 96°48W 84 F6
Hubbard Glacier Canada 60°18N 139°22W 68 E4
Hubbart Pt. Canada 59°21N 94°41W 71 B10
Hubei □ China 31°0N 112°0E 33 E11
Hubli India 15°22N 75°15E 40 M9
Huch'ang N. Korea 41°25N 127°2E 35 D14
Hucknall U.K. 53°3N 1°13W 12 D6
Huddersfield U.K. 53°39N 1°47W 12 D6
Hudiksvall Sweden 61°43N 17°10E 8 F17
Hudson Canada 50°6N 92°9W 72 B1
Hudson Mass., U.S.A. 42°23N 71°34W 83 D13
Hudson N.Y., U.S.A. 42°15N 73°46W 83 D11
Hudson Wis., U.S.A. 44°58N 92°45W 80 C7
Hudson Wyo., U.S.A. 42°54N 108°35W 76 E9
Hudson → U.S.A. 40°42N 74°2W 83 F10
Hudson, C. Antarctica 68°21S 153°45E 5 C10
Hudson Bay Nunavut, Canada 60°0N 86°0W 69 F14
Hudson Bay Sask., Canada 52°51N 102°23W 71 C8
Hudson Falls U.S.A. 43°18N 73°35W 83 C11
Hudson Mts. Antarctica 74°32S 99°20W 5 D16
Hudson Str. Canada 62°0N 70°0W 69 E18
Hudson's Hope Canada 56°0N 121°54W 70 B4
Hue Vietnam 16°30N 107°35E 38 D6
Huehuetenango Guatemala 15°20N 91°28W 88 C1
Huejúcar Mexico 22°21N 103°13W 86 C4
Huelva Spain 37°18N 6°57W 21 D2
Huentelauquén Chile 31°38S 71°33W 94 C1
Huerta, Sa. de la Argentina 31°10S 67°30W 94 C2
Huesca Spain 42°8N 0°25W 21 A5
Huetamo Mexico 18°35N 100°53W 86 D4
Hugh → Australia 25°1S 134°1E 62 D1
Hughenden Australia 20°52S 144°10E 62 C3
Hughesville U.S.A. 41°14N 76°44W 83 E8
Hugli → India 21°56N 88°4E 43 J13
Hugo Colo., U.S.A. 39°8N 103°28W 76 G12
Hugo Okla., U.S.A. 34°1N 95°31W 84 D7
Hugoton U.S.A. 37°11N 101°21W 80 G3
Hui Xian = Huixian China 35°27N 113°12E 34 G7
Hui Xian China 33°50N 106°4E 34 H4
Hui'anbu China 37°28N 106°38E 34 F4
Huichapan Mexico 20°23N 99°39W 87 C5
Huichon N. Korea 40°10N 126°16E 35 D14
Huifa He → China 43°0N 127°50E 35 C14
Huila, Nevado del Colombia 3°0N 76°0W 92 C3
Huimin China 37°27N 117°28E 35 F9
Huinan China 42°40N 126°2E 35 C14
Huinca Renancó Argentina 34°51S 64°22W 94 C3
Huining China 35°38N 105°0E 34 G3
Huinong China 39°5N 106°35E 34 E4
Huiting China 34°5N 116°5E 34 G9
Huixian China 35°27N 113°12E 34 G7
Huixtla Mexico 15°9N 92°28W 87 D6
Huize China 26°24N 103°15E 32 F9
Huizhou China 23°0N 114°23E 33 G11
Hukawng Valley Burma 26°30N 96°30E 41 F20
Hukuntsi Botswana 23°58S 21°45E 56 C3
Ḩulayfā' Si. Arabia 25°58N 40°45E 44 E4
Hulin China 45°48N 132°59E 33 B15
Hulin He → China 45°0N 122°10E 35 B12
Hull = Kingston upon Hull U.K. 53°45N 0°21W 12 D7
Hull Canada 45°26N 75°43W 83 A9
Hull → U.K. 53°44N 0°20W 12 D7
Hulst Neths. 51°17N 4°2E 15 C4
Huludao China 40°45N 120°50E 35 D11
Hulun Nur China 49°0N 117°30E 33 B12
Huma, Tanjung Malaysia 5°29N 100°16E 39 c
Humacao Puerto Rico 18°9N 65°50W 89 d
Humahuaca Argentina 23°10S 65°25W 94 A2
Humaitá Brazil 7°35S 63°1W 92 E6
Humaitá Paraguay 27°2S 58°31W 94 B4
Humansdorp S. Africa 34°2S 24°46E 56 E3
Humbe Angola 16°40S 14°55E 56 B1
Humber → U.K. 53°42N 0°27W 12 D7
Humboldt Canada 52°15N 105°9W 71 C7
Humboldt Iowa, U.S.A. 42°44N 94°13W 80 D6
Humboldt Tenn., U.S.A. 35°50N 88°55W 85 D10
Humboldt → U.S.A. 39°59N 118°36W 76 G4
Humboldt Gletscher = Sermersuaq Greenland 79°30N 62°0W 69 B19
Hume U.S.A. 36°48N 118°54W 78 J8
Hume, L. Australia 36°0S 147°5E 63 F4
Humen China 22°50N 113°40E 33 a
Humenné Slovak Rep. 48°55N 21°50E 17 D11
Humphreys, Mt. U.S.A. 37°17N 118°40W 78 H8
Humphreys Peak U.S.A. 35°21N 111°41W 77 J8
Humptulips U.S.A. 47°14N 123°57W 78 C3
Hūn Libya 29°2N 16°0E 51 C9
Hun Jiang → China 40°50N 125°38E 35 D13
Húnaflói Iceland 65°50N 20°50W 8 D3
Hunan □ China 27°30N 112°0E 33 F11
Hunchun China 42°52N 130°28E 35 C16
Hundewali Pakistan 31°55N 72°38E 42 D5
Hundred Mile House Canada 51°38N 121°18W 70 C4
Hunedoara Romania 45°40N 22°50E 17 F12
Hung Yen Vietnam 20°39N 106°4E 38 B6
Hunga Ha'apai Tonga 20°41S 175°7W 59 c
Hungary ■ Europe 47°20N 19°20E 17 E10
Hungary, Plain of Europe 47°0N 20°0E 6 F10
Hungerford Australia 28°58S 144°24E 63 D3
Hŭngnam N. Korea 39°49N 127°45E 35 E14
Hunjiang China 41°54N 126°26E 35 D14
Hunsberge Namibia 27°45S 17°12E 56 D2
Hunsrück Germany 49°56N 7°27E 16 D4
Hunstanton U.K. 52°56N 0°29E 12 E8
Hunter U.S.A. 42°13N 74°13W 83 D10
Hunter I. Australia 40°30S 144°45E 63 G3
Hunter I. Canada 51°55N 128°0W 70 C3
Hunter Ra. Australia 32°45S 150°15E 63 E5
Hunters Road Zimbabwe 19°9S 29°49E 55 F2
Hunterville N.Z. 39°56S 175°35E 59 C5
Huntingburg U.S.A. 38°18N 86°57W 80 F10
Huntingdon U.K. 52°20N 0°11W 13 E7
Huntingdon U.S.A. 40°30N 78°1W 82 F6
Huntington Ind., U.S.A. 40°53N 85°30W 81 E11
Huntington N.Y., U.S.A. 40°52N 73°26W 83 F11
Huntington Oreg., U.S.A. 44°21N 117°16W 76 D5
Huntington Utah, U.S.A. 39°20N 110°58W 76 G8
Huntington W. Va., U.S.A. 38°25N 82°27W 81 F12
Huntington Beach U.S.A. 33°40N 118°5W 79 M9
Huntly N.Z. 37°34S 175°11E 59 B5
Huntly U.K. 57°27N 2°47W 11 D6
Huntsville Canada 45°20N 79°14W 82 A5
Huntsville Ala., U.S.A. 34°44N 86°35W 85 D11
Huntsville Tex., U.S.A. 30°43N 95°33W 84 F7
Hunyani → Zimbabwe 15°57S 30°39E 55 F3
Hunyuan China 39°42N 113°42E 34 E7
Hunza → India 35°54N 74°20E 43 B6
Huo Shan China 36°26N 111°52E 34 F6
Huo Xian = Huozhou China 36°36N 111°42E 34 F6
Huolin Gol China 45°32N 119°38E 35 B10
Huong Khe Vietnam 18°13N 105°41E 38 C5
Huonville Australia 43°0S 147°5E 63 G4
Huozhou China 36°36N 111°42E 34 F6
Hupeh = Hubei □ China 31°0N 112°0E 33 E11
Ḩūr Iran 30°50N 57°7E 45 D8
Hurānd Iran 38°51N 47°22E 44 B5
Ḩuraymīla Si. Arabia 25°8N 46°8E 44 E5
Hurd, C. Canada 45°13N 81°44W 82 A3
Hure Qi China 42°45N 121°45E 35 C11
Hurghada Egypt 27°15N 33°50E 51 C12
Hurley N. Mex., U.S.A. 32°42N 108°8W 77 K9
Hurley Wis., U.S.A. 46°27N 90°11W 80 B8
Huron Calif., U.S.A. 36°12N 120°6W 78 J6
Huron Ohio, U.S.A. 41°24N 82°33W 82 E2
Huron S. Dak., U.S.A. 44°22N 98°13W 80 C4
Huron, L. U.S.A. 44°30N 82°40W 82 B2
Huron East Canada 43°38N 81°18W 82 C3
Hurricane U.S.A. 37°11N 113°17W 77 H7
Hurungwe △ Zimbabwe 16°7S 29°5E 55 F2
Hurunui → N.Z. 42°54S 173°18E 59 E4
Húsavík Iceland 66°3N 17°21W 8 C5
Huşi Romania 46°41N 28°7E 17 E15
Hustadvika Norway 63°0N 7°0E 8 E12
Hustontown U.S.A. 40°3N 78°2W 82 F6
Hutchinson Kans., U.S.A. 38°5N 97°56W 80 F5
Hutchinson Minn., U.S.A. 44°54N 94°22W 80 C6
Hutte Sauvage, L. de la Canada 56°15N 64°45W 73 A7
Hutton, Mt. Australia 25°51S 148°20E 63 D4
Huy Belgium 50°31N 5°15E 15 D5
Huzhou China 30°51N 120°8E 33 E13
Hvammstangi Iceland 65°24N 20°57W 8 D3
Hvar Croatia 43°11N 16°28E 22 C7
Hvítá → Iceland 64°30N 21°58W 8 D3
Hwacheon-Cheosuji S. Korea 38°5N 127°50E 35 E14
Hwang Ho = Huang He → China 37°55N 118°50E 35 F10
Hwange Zimbabwe 18°18S 26°30E 55 F2
Hwange △ Zimbabwe 19°0S 26°30E 56 B4
Hyannis Mass., U.S.A. 41°39N 70°17W 81 E18
Hyannis Nebr., U.S.A. 42°0N 101°46W 80 E3
Hyargas Nuur Mongolia 49°0N 93°0E 32 B7
Hydaburg U.S.A. 55°15N 132°50W 70 B2
Hyde Park U.S.A. 41°47N 73°56W 83 E11
Hyden Australia 32°24S 118°53E 61 F2
Hyder U.S.A. 55°55N 130°5W 70 B2
Hyderabad India 17°22N 78°29E 40 L11
Hyderabad Pakistan 25°23N 68°24E 42 G3
Hydra Greece 37°20N 23°28E 23 F10
Hyères France 43°8N 6°9E 20 E7
Hyères, Îs. d' France 43°0N 6°20E 20 E7
Hyesan N. Korea 41°20N 128°10E 35 D15
Hyland → Canada 59°52N 128°12W 70 B3
Hyndman Peak U.S.A. 43°45N 114°8W 76 E6
Hyōgo □ Japan 35°15N 134°50E 31 G7
Hyrum U.S.A. 41°38N 111°51W 76 F8
Hysham U.S.A. 46°18N 107°14W 76 C10
Hythe U.K. 51°4N 1°5E 13 F9
Hyūga Japan 32°25N 131°35E 31 H5
Hyvinge = Hyvinkää Finland 60°38N 24°50E 8 F21
Hyvinkää Finland 60°38N 24°50E 8 F21

I

I-n-Gall Niger 16°51N 7°1E 50 E7
Iaco → Brazil 9°3S 68°34W 92 E5
Iakora Madag. 23°6S 46°40E 57 C8
Ialissos Greece 36°24N 28°10E 25 C10
Ialomiţa → Romania 44°42N 27°51E 17 F14
Iaşi Romania 47°10N 27°40E 17 E14
Ib → India 21°34N 83°48E 43 J10
Iba Phil. 15°22N 120°0E 37 A6
Ibadan Nigeria 7°22N 3°58E 50 G6
Ibagué Colombia 4°20N 75°20W 92 C3
Ibar → Serbia 43°43N 20°45E 23 C9
Ibaraki □ Japan 36°10N 140°10E 31 F10
Ibarra Ecuador 0°21N 78°7W 92 C3
Ibb Yemen 14°2N 44°10E 47 E3
Ibembo Dem. Rep. of the Congo 2°35N 23°35E 54 B1
Ibenga → Congo 2°19N 18°9E 52 D3
Ibera, L. Argentina 28°30S 57°9W 94 B4
Iberian Peninsula Europe 40°0N 5°0W 6 H5
Iberville Canada 45°19N 73°17W 83 A11
Iberville, Mt. d' Canada 58°50N 63°50W 69 F19
Ibiá Brazil 19°30S 46°30W 93 G9
Ibiapaba, Sa. da Brazil 4°0S 41°30W 93 D10
Ibicuí → Brazil 29°25S 56°47W 95 B4
Ibicuy Argentina 33°55S 59°10W 94 C4
Ibiza = Eivissa Spain 38°54N 1°26E 24 C7
Ibo Mozam. 12°22S 40°40E 55 E5
Ibonma Indonesia 3°29S 133°31E 37 E8
Ibotirama Brazil 12°13S 43°12W 93 F10
Ibrāhīm → Lebanon 34°4N 35°38E 46 A4
'Ibrī Oman 23°14N 56°30E 45 F8
Ibu Indonesia 1°35N 127°33E 37 D7
Ibusuki Japan 31°12N 130°40E 31 J5
Ica Peru 14°0S 75°48W 92 F3
Iça → Brazil 2°55S 67°58W 92 D5
Icacos Pt. Trin. & Tob. 10°3N 61°57W 93 K15
Içana Brazil 0°21N 67°19W 92 C5
Içana → Brazil 0°26N 67°19W 92 C5
İçel Turkey 36°51N 34°36E 44 B2
Iceland ■ Europe 64°45N 19°0W 8 D4
Iceland Basin Atl. Oc. 61°0N 19°0W 4 C7
Icelandic Plateau Arctic 64°0N 10°0W 4 C7
Ich'ang = Yichang China 30°40N 111°20E 33 E11
Ichchapuram India 19°10N 84°40E 41 K14
Ichhawar India 23°1N 77°1E 42 H7
Ichihara Japan 35°28N 140°5E 31 G10
Ichikawa Japan 35°43N 139°54E 31 G9
Ichilo → Bolivia 15°57S 64°50W 92 G6
Ichinohe Japan 40°13N 141°17E 30 D10
Ichinomiya Japan 35°18N 136°48E 31 G8
Ichinoseki Japan 38°55N 141°8E 30 E10
Icod Canary Is. 28°22N 16°43W 24 F3
Içoguz Turkmenistan 40°11N 58°24E 28 E6
Icy C. U.S.A. 70°20N 161°52W 66 B3
Ida Grove U.S.A. 42°21N 95°28W 80 D6
Idabel U.S.A. 33°54N 94°50W 84 E7
Idaho □ U.S.A. 45°0N 115°0W 76 D6
Idaho City U.S.A. 43°50N 115°50W 76 E6
Idaho Falls U.S.A. 43°30N 112°2W 76 E7
Idalia △ Australia 24°49S 144°36E 62 C3
Idar-Oberstein Germany 49°43N 7°16E 16 D4
Idensalmi = Iisalmi Finland 63°32N 27°10E 8 E22
Idfû Egypt 24°55N 32°49E 51 D12
Ídhra = Hydra Greece 37°20N 23°28E 23 F10
Idi Indonesia 5°2N 97°37E 36 C1
Idi, Oros = Psiloritis, Oros Greece 35°15N 24°45E 25 D6
Idiofa Dem. Rep. of the Congo 4°55S 19°42E 52 E3
Idlib Syria 35°55N 36°36E 44 C3
Idria U.S.A. 36°25N 120°41W 78 J6
Idutywa = Dutywa S. Africa 32°8S 28°18E 57 E4
Ieper Belgium 50°51N 2°53E 15 D2
Ierapetra Greece 35°1N 25°44E 25 E7
Iesi Italy 43°31N 13°14E 22 C5
Ifakara Tanzania 8°8S 36°41E 52 F7
'Ifāl, W. al → Si. Arabia 28°7N 35°3E 44 D2
Ifanadiana Madag. 21°19S 47°39E 57 C8
Ife Nigeria 7°30N 4°31E 50 G6
Iffley Australia 18°53S 141°12E 62 B3
Iforas, Adrar des Africa 19°40N 1°40E 50 E6
Ifould, L. Australia 30°52S 132°6E 61 F5
Iganga Uganda 0°37N 33°28E 54 B3
Igarapava Brazil 20°3S 47°47W 93 H9
Igarka Russia 67°30N 86°33E 28 C9
Iggesund Sweden 61°39N 17°10E 8 F17
Iglésias Italy 39°19N 8°32E 22 E3
Igloolik = Iglulik Canada 69°20N 81°49W 69 D15
Igluligaarjuk = Chesterfield Inlet Canada 63°30N 90°45W 68 E13
Iglulik Canada 69°20N 81°49W 69 D15
Ignace Canada 49°30N 91°40W 72 C1
İğneada Burnu Turkey 41°53N 28°2E 23 D13
Igoumenitsa Greece 39°32N 20°18E 23 E9
Igrim Russia 63°12N 64°30E 28 C7
Iguaçu → Brazil 25°36S 54°36W 95 B5
Iguaçu, Cat. del Brazil 25°41S 54°26W 95 B5
Iguaçu △ Brazil 25°30S 54°0W 95 B5
Iguaçu Falls = Iguaçu, Cat. del Brazil 25°41S 54°26W 95 B5
Iguala Mexico 18°21N 99°32W 87 D5
Igualada Spain 41°37N 1°37E 21 B6
Iguassu = Iguaçu → Brazil 25°36S 54°36W 95 B5
Iguatu Brazil 6°20S 39°18W 93 E11
Iguazú △ Argentina 25°42S 54°22W 95 B5
Iguidi, Erg Africa 27°0N 7°0W 50 C4
Iharana Madag. 13°25S 50°0E 57 A9
Ihbulag Mongolia 43°11N 107°10E 34 C4
Iheya-Shima Japan 27°4N 127°58E 31 L3
Ihosy Madag. 22°24S 46°8E 57 C8
Ihotry, Farihy Madag. 21°56S 43°41E 57 C7
Ii Finland 65°19N 25°22E 8 D21
Ii-Shima Japan 26°43N 127°47E 31 L3
Iida Japan 35°35N 137°50E 31 G8
Iijoki → Finland 65°20N 25°20E 8 D21
Iisalmi Finland 63°32N 27°10E 8 E22
Iiyama Japan 36°51N 138°22E 31 F9
Iizuka Japan 33°38N 130°42E 31 H5
Ijâfene Mauritania 20°40N 8°0W 50 D4
Ijebu-Ode Nigeria 6°47N 3°58E 50 G6
IJmuiden Neths. 52°28N 4°35E 15 B4
Ijo älv = Iijoki → Finland 65°20N 25°20E 8 D21
IJssel → Neths. 52°35N 5°50E 15 B5
IJsselmeer Neths. 52°45N 5°20E 15 B5
Ijuí Brazil 28°23S 53°55W 95 B5
Ijuí → Brazil 27°58S 55°20W 95 B4
Ikalamavony Madag. 21°9S 46°35E 57 C8
Ikanga Kenya 1°42S 38°4E 54 C4
Ikare Nigeria 7°32N 5°40E 50 G7
Ikaria Greece 37°35N 26°10E 23 F12
Ikeda Japan 34°1N 133°48E 31 G6
Ikela Dem. Rep. of the Congo 1°6S 23°6E 52 E4
Iki Japan 33°45N 129°42E 31 H4
Ikimba L. Tanzania 1°30S 31°20E 54 C3
Ikongo Madag. 21°52S 47°27E 57 C8
Ikopa → Madag. 16°45S 46°40E 57 B8
Ikorongo △ Tanzania 1°50S 34°53E 54 C3
Ikparjuk = Arctic Bay Canada 73°1N 85°7W 69 C14
Iksan S. Korea 35°59N 127°0E 35 G14
Ikungu Tanzania 1°33S 33°42E 54 C3
Ikuntji = Haasts Bluff Australia 23°22S 132°0E 60 D5
Ilagan Phil. 17°7N 121°53E 37 A6
Ilaka Madag. 19°33S 48°52E 57 B8
Īlām Iran 33°36N 46°36E 44 C5
Ilam Nepal 26°58N 87°58E 43 F12
Īlām □ Iran 33°0N 47°0E 44 C5
Ilanskiy Russia 56°14N 96°3E 29 D10
Iława Poland 53°36N 19°34E 17 B10
Ile → Kazakhstan 45°53N 77°10E 28 E8
Île-à-la-Crosse Canada 55°27N 107°53W 71 B7
Île-à-la-Crosse, Lac Canada 55°40N 107°45W 71 B7
Île-de-France □ France 49°0N 2°20E 20 B5
Ilebo Dem. Rep. of the Congo 4°17S 20°55E 52 E4
Ilek Russia 51°32N 53°21E 28 D6
Ilek → Russia 51°30N 53°22E 18 D9
Ilesha Nigeria 7°37N 4°40E 50 G6
Ilford Canada 56°4N 95°35W 71 B9
Ilfracombe Australia 23°30S 144°30E 62 C3
Ilfracombe U.K. 51°12N 4°8W 13 F3
Ilha Grande, Represa Brazil 23°10S 53°5W 95 A5
Ilha Grande △ Brazil 23°10S 53°20W 95 A5
Ilhéus Brazil 14°49S 39°2W 93 F11
Ili = Ile → Kazakhstan 45°53N 77°10E 28 E8
Iliamna L. U.S.A. 59°30N 155°0W 74 D8
Ilias, Profitis Greece 36°17N 27°56E 25 C9
Iligan Phil. 8°12N 124°13E 37 C6
Ilion U.S.A. 43°1N 75°2W 83 D9
Ilkeston U.K. 52°58N 1°19W 12 E6
Ilkley U.K. 53°56N 1°48W 12 D6
Illampu = Ancohuma, Nevado Bolivia 16°0S 68°50W 92 G5
Illana B. Phil. 7°35N 123°45E 37 C6
Illapel Chile 32°0S 71°10W 94 C1
Iller → Germany 48°23N 9°58E 16 D6
Illetas Spain 39°32N 2°35E 24 B9

Illimani, Nevado Bolivia 16°30S 67°50W 92 G5
Illinois □ U.S.A. 40°15N 89°30W 80 E9
Illinois → U.S.A. 38°58N 90°28W 80 F8
Illizi Algeria 26°31N 8°32E 50 C7
Ilma, L. Australia 29°13S 127°46E 61 E4
Ilmajoki Finland 62°44N 22°34E 8 E20
Ilmen, Ozero Russia 58°15N 31°10E 18 C5
Ilo Peru 17°40S 71°20W 92 G4
Iloilo Phil. 10°45N 122°33E 37 B6
Ilomantsi Finland 62°38N 30°57E 8 E24
Ilorin Nigeria 8°30N 4°35E 50 G6
Ilwaco U.S.A. 46°19N 124°3W 78 D2
Ilwaki Indonesia 7°55S 126°30E 37 F7
Imabari Japan 34°4N 133°0E 31 G6
Imaloto → Madag. 23°27S 45°13E 57 C8
Imandra, Ozero Russia 67°30N 33°0E 8 C25
Imanombo Madag. 24°26S 45°49E 57 C8
Imari Japan 33°15N 129°52E 31 H4
Imatra Finland 61°12N 28°48E 8 F23
imeni 26 Bakinskikh Komissarov = Neftçala Azerbaijan 39°19N 49°12E 45 B6
imeni 26 Bakinskikh Komissarov Turkmenistan 39°22N 54°10E 45 B7
imeni Ismail Samani, Pik Tajikistan 39°0N 72°2E 28 F8
Imeri, Serra Brazil 0°50N 65°25W 92 C5
Imerimandroso Madag. 17°26S 48°35E 57 B8
Imfolozi ⌂ S. Africa 28°18S 31°50E 57 D5
Imi Ethiopia 6°28N 42°10E 47 F3
Imlay U.S.A. 40°40N 118°9W 76 F4
Imlay City U.S.A. 43°2N 83°5W 82 D1
Immingham U.K. 53°37N 0°13W 12 D7
Immokalee U.S.A. 26°25N 81°25W 85 H14
Ímola Italy 44°20N 11°42E 22 B4
Imperatriz Brazil 5°30S 47°29W 93 E9
Impéria Italy 43°53N 8°3E 20 E8
Imperial Canada 51°21N 105°28W 71 C7
Imperial Calif., U.S.A. 32°51N 115°34W 79 N11
Imperial Nebr., U.S.A. 40°31N 101°39W 80 E3
Imperial Beach U.S.A. 32°35N 117°6W 79 N9
Imperial Dam U.S.A. 32°55N 114°25W 79 N12
Imperial Res. U.S.A. 32°53N 114°28W 79 N12
Imperial Valley U.S.A. 33°0N 115°30W 79 N11
Imperieuse Reef Australia 17°36S 118°50E 60 C2
Impfondo Congo 1°40N 18°0E 52 D3
Imphal India 24°48N 93°56E 41 G18
İmroz = Gökçeada Turkey 40°10N 25°50E 23 D11
Imuris Mexico 30°47N 110°52W 86 A2
Imuruan B. Phil. 10°40N 119°10E 37 B5
In Guezzam Algeria 19°37N 5°52E 50 E7
In Salah Algeria 27°10N 2°32E 50 C6
Ina Japan 35°50N 137°55E 31 G8
Inagua △ Bahamas 21°5N 73°24W 89 B5
Inangahua N.Z. 41°52S 171°59E 59 D3
Inanwatan Indonesia 2°8S 132°10E 37 E8
Iñapari Peru 11°0S 69°40W 92 F5
Inari Finland 68°54N 27°1E 8 B22
Inarijärvi Finland 69°0N 28°0E 8 B23
Inawashiro-Ko Japan 37°29N 140°6E 30 F10
Inca Spain 39°43N 2°54E 24 B9
Inca de Oro Chile 26°45S 69°54W 94 B2
Incahuasi Argentina 27°2S 68°18W 94 B2
Incahuasi Chile 29°12S 71°5W 94 B1
İnce Burun Turkey 42°7N 34°56E 19 F5
İncesu Turkey 38°38N 35°11E 44 B2
Incheon S. Korea 37°27N 126°40E 35 F14
İncirliova Turkey 37°50N 27°41E 23 F12
Incline Village U.S.A. 39°10N 119°58W 76 G4
Incomáti → Mozam. 25°46S 32°43E 57 D5
Indalsälven → Sweden 62°36N 17°30E 8 E17
Indaw Burma 24°15N 96°5E 41 G20
Independence Calif., U.S.A. 36°48N 118°12W 78 J8
Independence Iowa, U.S.A. 42°28N 91°54W 80 D8
Independence Kans., U.S.A. 37°14N 95°42W 80 G6
Independence Ky., U.S.A. 38°57N 84°33W 81 F11
Independence Mo., U.S.A. 39°6N 94°25W 80 F6
Independence Fjord Greenland 82°10N 29°0W 4 A6
Independence Mts. U.S.A. 41°20N 116°0W 76 F5
Index U.S.A. 47°50N 121°33W 78 C5
India ■ Asia 20°0N 78°0E 40 K11
Indian → U.S.A. 27°59N 80°34W 85 H14
Indian Cabins Canada 59°52N 117°40W 70 B5
Indian Harbour Canada 54°27N 57°13W 73 B8
Indian Head Canada 50°30N 103°41W 71 C8
Indian L. U.S.A. 43°46N 74°16W 83 C10
Indian Lake U.S.A. 43°47N 74°16W 83 C10
Indian Springs U.S.A. 36°35N 115°40W 79 J11
Indiana U.S.A. 40°37N 79°9W 82 F5
Indiana □ U.S.A. 40°0N 86°0W 81 F11
Indianapolis U.S.A. 39°46N 86°9W 80 F10
Indianola Iowa, U.S.A. 41°22N 93°34W 80 E7
Indianola Miss., U.S.A. 33°27N 90°39W 85 E9
Indiga Russia 67°38N 49°9E 18 A8
Indigirka → Russia 70°48N 148°54E 29 B15
Indio U.S.A. 33°43N 116°13W 79 M10
Indira Gandhi Canal India 28°0N 72°0E 42 F5
Indira Sagar India 22°15N 76°40E 42 H7
Indo-China Asia 15°0N 102°0E 26 G12
Indonesia ■ Asia 5°0S 115°0E 36 F5
Indore India 22°42N 75°53E 42 H6
Indramayu Indonesia 6°20S 108°19E 37 G13
Indravati → India 19°20N 80°20E 41 K12
Indre → France 47°16N 0°11E 20 C4
Indulkana Australia 26°58S 133°5E 63 D1
Indus → Pakistan 24°20N 67°47E 42 G2
Indus, Mouths of the Pakistan 24°0N 68°0E 42 H3
İnebolu Turkey 41°55N 33°40E 19 F5
Infiernillo, Presa del Mexico 18°35N 101°50W 86 D4
Ingenio Canary Is. 27°55N 15°26W 24 G4
Ingenio Santa Ana Argentina 27°25S 65°40W 94 B2
Ingersoll Canada 43°4N 80°55W 82 C4
Ingham Australia 18°43S 146°10E 62 B4
Ingleborough U.K. 54°10N 2°22W 12 C5
Inglewood Queens., Australia 28°25S 151°2E 63 D5
Inglewood Vic., Australia 36°29S 143°53E 63 F3
Inglewood N.Z. 39°9S 174°14E 59 C5
Inglewood U.S.A. 33°58N 118°21W 79 M8
Ingólfshöfði Iceland 63°48N 16°39W 8 E5
Ingolstadt Germany 48°46N 11°26E 16 D6
Ingomar U.S.A. 46°35N 107°23W 76 C10
Ingonish Canada 46°42N 60°18W 73 C7
Ingraj Bazar India 24°58N 88°10E 43 G13
Ingrid Christensen Coast Antarctica 69°30S 76°0E 5 C6
Ingulec = Inhulec Ukraine 47°42N 33°14E 19 E5
Ingushetia □ Russia 43°20N 44°50E 19 F8
Ingwavuma S. Africa 27°9S 31°59E 57 D5
Inhaca Mozam. 26°1S 32°57E 57 D5
Inhafenga Mozam. 20°36S 33°53E 57 C5
Inhambane Mozam. 23°54S 35°30E 57 C6
Inhambane □ Mozam. 22°30S 34°20E 57 C5
Inhaminga Mozam. 18°26S 35°0E 55 F4
Inharrime Mozam. 24°30S 35°0E 57 C6
Inharrime → Mozam. 24°30S 35°0E 57 C6
Inhulec Ukraine 47°42N 33°14E 19 E5
Ining = Yining China 43°58N 81°10E 32 C5
Inírida → Colombia 3°55N 67°52W 92 C5
Inis = Ennis Ireland 52°51N 8°59W 10 D3
Inis Córthaidh = Enniscorthy Ireland 52°30N 6°34W 10 D5
Inishbofin Ireland 53°37N 10°13W 10 C1
Inisheer Ireland 53°3N 9°32W 10 C2
Inishfree B. Ireland 55°4N 8°23W 10 A3
Inishkea North Ireland 54°9N 10°11W 10 B1
Inishkea South Ireland 54°7N 10°12W 10 B1
Inishmaan Ireland 53°5N 9°35W 10 C2
Inishmore Ireland 53°8N 9°45W 10 C2
Inishmurray I. Ireland 54°26N 8°39W 10 B3
Inishowen Pen. Ireland 55°14N 7°15W 10 A4
Inishshark Ireland 53°37N 10°16W 10 C1
Inishturk Ireland 53°42N 10°7W 10 C1
Inishvickillane Ireland 52°3N 10°37W 10 D1
Injinoo ✪ Australia 10°56S 142°15E 62 A3
Injune Australia 25°53S 148°32E 63 D4
Inklin → N. Amer. 58°50N 133°10W 70 B2
Inland Kaikoura Ra. N.Z. 41°59S 173°41E 59 D4
Inland Sea = Setonaikai Japan 34°20N 133°30E 31 G6
Inle L. Burma 20°30N 96°58E 41 J20
Inlet U.S.A. 43°45N 74°48W 83 C10
Inn → Austria 48°35N 13°28E 16 D7
Innamincka Australia 27°44S 140°46E 63 D3
Inner Hebrides U.K. 57°0N 6°30W 11 E2
Inner Mongolia = Nei Mongol Zizhiqu □ China 42°0N 112°0E 34 D7
Inner Sound U.K. 57°30N 5°55W 11 D3
Innerkip Canada 43°13N 80°42W 82 C4
Innetalling I. Canada 56°0N 79°0W 72 A4
Innisfail Australia 17°33S 146°5E 62 B4
Innisfail Canada 52°2N 113°57W 70 C6
In'noshima Japan 34°19N 133°10E 31 G6
Innsbruck Austria 47°16N 11°23E 16 E6
Inny → Ireland 53°30N 7°50W 10 C4
Inongo Dem. Rep. of the Congo 1°55S 18°30E 52 E3
Inoucdjouac = Inukjuak Canada 58°25N 78°15W 69 F16
Inowrocław Poland 52°50N 18°12E 17 B10
Inscription, C. Australia 25°29S 112°59E 61 E1
Insein Burma 16°50N 96°5E 41 L20
Inta Russia 66°5N 60°8E 18 A11
Intendente Alvear Argentina 35°12S 63°32W 94 D3
Interlaken Switz. 46°41N 7°50E 20 C7
Interlaken U.S.A. 42°37N 76°44W 83 D8
International Falls U.S.A. 48°36N 93°25W 80 A7
Intiyaco Argentina 28°43S 60°5W 94 B3
Inukjuak Canada 58°25N 78°15W 69 F16
Inútil, B. Chile 53°30S 70°15W 96 G2
Inuvik Canada 68°16N 133°40W 68 D5
Inveraray U.K. 56°14N 5°5W 11 E3
Inverbervie U.K. 56°51N 2°17W 11 E6
Invercargill N.Z. 46°24S 168°24E 59 G2
Inverclyde □ U.K. 55°55N 4°49W 11 F4
Inverell Australia 29°45S 151°8E 63 D5
Invergordon U.K. 57°41N 4°10W 11 D4
Inverloch Australia 38°38S 145°45E 63 F4
Invermere Canada 50°30N 116°2W 70 C5
Inverness Canada 46°15N 61°19W 73 C7
Inverness U.K. 57°29N 4°13W 11 D4
Inverness U.S.A. 28°50N 82°20W 85 G13
Inverurie U.K. 57°17N 2°23W 11 D6
Investigator Group Australia 34°45S 134°20E 63 E1
Investigator Str. Australia 35°30S 137°0E 63 F2
Inya Russia 50°28N 86°37E 28 D9
Inyanga Zimbabwe 18°12S 32°40E 55 F3
Inyangani Zimbabwe 18°5S 32°50E 55 F3
Inyantue Zimbabwe 18°33S 26°39E 56 B4
Inyo Mts. U.S.A. 36°40N 118°0W 78 J9
Inyokern U.S.A. 35°39N 117°49W 79 K9
Inyonga Tanzania 6°45S 32°5E 54 D3
Inza Russia 53°55N 46°25E 18 D8
Iō-Jima Japan 30°48N 130°18E 31 J5
Ioannina Greece 39°42N 20°47E 23 E9
Iola U.S.A. 37°55N 95°24W 80 G6
Iona U.K. 56°20N 6°25W 11 E2
Ione U.S.A. 38°21N 120°56W 78 G6
Ionia U.S.A. 42°59N 85°4W 81 D11
Ionian Is. = Ionioi Nisoi Greece 38°40N 20°0E 23 E9
Ionian Sea Medit. S. 37°30N 17°30E 23 E7
Ionioi Nisoi Greece 38°40N 20°0E 23 E9
Ios Greece 36°41N 25°20E 23 F11
Iowa □ U.S.A. 42°18N 93°30W 80 D7
Iowa → U.S.A. 41°10N 91°1W 80 E8
Iowa City U.S.A. 41°40N 91°32W 80 E8
Iowa Falls U.S.A. 42°31N 93°16W 80 D7
Iowa Park U.S.A. 33°57N 98°40W 84 E5
Ipala Tanzania 4°30S 32°52E 54 C3
Ipameri Brazil 17°44S 48°9W 93 G9
Ipatinga Brazil 19°32S 42°30W 93 G10
Ipiales Colombia 0°50N 77°37W 92 C3
Ipin = Yibin China 28°45N 104°32E 32 F9
Ipixuna Brazil 7°0S 71°40W 92 E4
Ipoh Malaysia 4°35N 101°5E 39 K3
Ippy C.A.R. 6°5N 21°7E 52 C4
İpsala Turkey 40°55N 26°23E 23 D12
Ipswich Australia 27°35S 152°40E 63 D5
Ipswich U.K. 52°4N 1°10E 13 E9
Ipswich Mass., U.S.A. 42°41N 70°50W 83 D14
Ipswich S. Dak., U.S.A. 45°27N 99°2W 80 C4
Ipu Brazil 4°23S 40°44W 93 D10
Iqaluit Canada 63°44N 68°31W 69 E18
Iqaluktuutiaq = Cambridge Bay Canada 69°10N 105°0W 68 D11
Iquique Chile 20°19S 70°5W 92 H4
Iquitos Peru 3°45S 73°10W 92 D4
Irabu-Jima Japan 24°50N 125°10E 31 M2
Iracoubo Fr. Guiana 5°30N 53°10W 93 B8
Īrafshān Iran 26°42N 61°56E 45 E9
Iraklio Greece 35°20N 25°12E 25 D7
Iraklio □ Greece 35°10N 25°10E 25 D7
Iráklion = Iraklio Greece 35°20N 25°12E 25 D7
Irakliou, Kolpos Greece 35°23N 25°8E 25 D7
Irala Paraguay 25°55S 54°35W 95 B5
Iran ■ Asia 33°0N 53°0E 45 C7
Iran, Pegunungan Malaysia 2°20N 114°50E 36 D4
Iran Ra. = Iran, Pegunungan Malaysia 2°20N 114°50E 36 D4
Īrānshahr Iran 27°15N 60°40E 45 E9
Irapuato Mexico 20°41N 101°28W 86 C4
Iraq ■ Asia 33°0N 44°0E 44 C5
Irati Brazil 25°25S 50°38W 95 B5
Irbid Jordan 32°35N 35°48E 46 C4
Irbid □ Jordan 32°30N 35°50E 46 C5
Irebu Dem. Rep. of the Congo 0°40S 17°46E 52 E3
Ireland ■ Europe 53°50N 7°52W 10 C4
Iri = Iksan S. Korea 35°59N 127°0E 35 G14
Irian Jaya = Papua □ Indonesia 4°0S 137°0E 37 E9
Irian Jaya Barat □ Indonesia 2°5S 132°50E 37 E8
Iringa Tanzania 7°48S 35°43E 54 D4
Iringa □ Tanzania 7°48S 35°43E 54 D4
Iriomote △ Japan 24°29N 123°53E 31 M1
Iriomote-Jima Japan 24°19N 123°48E 31 M1
Iriona Honduras 15°57N 85°11W 88 C2
Iriri → Brazil 3°52S 52°37W 93 D8
Irish Republic ■ Europe 53°50N 7°52W 10 C4
Irish Sea Europe 53°38N 4°48W 12 D3
Irkutsk Russia 52°18N 104°20E 32 A9
Irma Canada 52°55N 111°14W 71 C6
Irō-Zaki Japan 34°36N 138°51E 31 G9
Iron Baron Australia 32°58S 137°11E 63 E2
Iron Gate = Portile de Fier Europe 44°44N 22°30E 17 F12
Iron Knob Australia 32°46S 137°8E 63 E2
Iron Mountain U.S.A. 45°49N 88°4W 80 C9
Iron Range △ Australia 12°34S 143°18E 62 A3
Iron River U.S.A. 46°6N 88°39W 80 B9
Irondequoit U.S.A. 43°13N 77°35W 82 C7
Ironton Mo., U.S.A. 37°36N 90°38W 80 G8
Ironton Ohio, U.S.A. 38°32N 82°41W 81 F12
Ironwood U.S.A. 46°27N 90°9W 80 B8
Ironwood Forest ⌂ U.S.A. 32°32N 111°28W 77 K8
Iroquois Canada 44°51N 75°19W 83 B9
Iroquois Falls Canada 48°46N 80°41W 72 C3
Irpin Ukraine 50°30N 30°15E 17 C16
Irrara Cr. → Australia 29°35S 145°31E 63 D4
Irrawaddy □ Burma 17°0N 95°0E 41 L19
Irrawaddy → Burma 15°50N 95°6E 41 M19
Irrawaddy, Mouths of the Burma 15°30N 95°0E 41 M19
Irricana Canada 51°19N 113°37W 70 C6
Irrunytju Australia 26°3S 128°56E 61 E4
Irtysh → Russia 61°4N 68°52E 28 C7
Irumu Dem. Rep. of the Congo 1°32N 29°53E 54 B2
Irún Spain 43°20N 1°52W 21 A5
Irunea = Pamplona-Iruña Spain 42°48N 1°38W 21 A5
Irvine Canada 49°57N 110°16W 71 D6
Irvine U.K. 55°37N 4°41W 11 F4
Irvine Calif., U.S.A. 33°41N 117°46W 79 M9
Irvine Ky., U.S.A. 37°42N 83°58W 81 G12
Irvinestown U.K. 54°28N 7°39W 10 B4
Irving U.S.A. 32°48N 96°56W 84 E6
Irvona U.S.A. 40°46N 78°33W 82 F6
Irwin → Australia 29°15S 114°54E 61 E1
Irymple Australia 34°14S 142°8E 63 E3
Isa Khel Pakistan 32°41N 71°17E 42 C4
Isaac → Australia 22°55S 149°20E 62 C4
Isabel U.S.A. 45°24N 101°26W 80 C3
Isabel, I. Mexico 21°51N 105°55W 86 C3
Isabela Phil. 6°40N 121°59E 37 C6
Isabela Puerto Rico 18°30N 67°2W 89 d
Isabella, Cord. Nic. 13°30N 85°25W 88 D2
Isabella Ra. Australia 21°0S 121°4E 60 D3
Isachsen, C. Canada 79°20N 105°28W 69 B10
Ísafjarðardjúp Iceland 66°10N 23°0W 8 C2
Ísafjörður Iceland 66°5N 23°9W 8 C2
Isagarh India 24°48N 77°51E 42 G7
Isahaya Japan 32°52N 130°2E 31 H5
Isaka Tanzania 3°56S 32°59E 54 C3
Isalo △ Madag. 22°23S 45°16E 57 C8
Isan → India 26°51N 80°7E 43 F9
Isana = Içana → Brazil 0°26N 67°19W 92 C5
Isangano △ Zambia 11°8S 30°35E 55 E3
Isar → Germany 48°48N 12°57E 16 D7
Íschia Italy 40°44N 13°57E 22 D5
Isdell → Australia 16°27S 124°51E 60 C3
Ise Japan 34°25N 136°45E 31 G8
Ise-Shima △ Japan 34°25N 136°48E 31 G8
Ise-Wan Japan 34°43N 136°43E 31 G8
Iseramagazi Tanzania 4°37S 32°10E 54 C3
Isère → France 44°59N 4°51E 20 D6
Isérnia Italy 41°36N 14°14E 22 D6
Isfahan = Eşfahān Iran 32°39N 51°43E 45 C6
Ishigaki Japan 24°26N 124°10E 31 M2
Ishigaki-Shima Japan 24°20N 124°10E 31 M2
Ishikari Japan 43°20N 141°15E 30 C10
Ishikari-Gawa → Japan 43°15N 141°23E 30 C10
Ishikari-Sammyaku Japan 43°30N 143°0E 30 C11
Ishikari-Wan Japan 43°25N 141°1E 30 C10
Ishikawa Japan 26°25N 127°49E 31 L3
Ishikawa □ Japan 36°30N 136°30E 31 F8
Ishim Russia 56°10N 69°30E 28 D7
Ishim → Russia 57°45N 71°10E 28 D8
Ishinomaki Japan 38°32N 141°20E 30 E10
Ishioka Japan 36°11N 140°16E 31 F10
Ishkoman Pakistan 36°30N 73°50E 43 A5
Ishpeming U.S.A. 46°29N 87°40W 80 B10
Isil Kul Russia 54°55N 71°16E 28 D8
Isiolo Kenya 0°24N 37°33E 54 B4
Isiro Dem. Rep. of the Congo 2°53N 27°40E 54 B2
Isisford Australia 24°15S 144°21E 62 C3
İskenderun Turkey 36°32N 36°10E 44 B3
İskenderun Körfezi Turkey 36°40N 35°50E 19 G6
Iskŭr → Bulgaria 43°45N 24°25E 23 C11
Iskut → Canada 56°45N 131°49W 70 B2
Isla → U.K. 56°32N 3°20W 11 E5
Isla Coiba △ Panama 7°33N 81°36W 88 E3
Isla de Salamanca △ Colombia 10°59N 74°40W 89 D5
Isla Gorge △ Australia 25°10S 149°57E 62 D4
Isla Isabel △ Mexico 21°54N 105°58W 86 C3
Isla Tiburón y San Esteban ⌂ Mexico 29°0N 112°27W 86 B2
Isla Vista U.S.A. 34°25N 119°53W 79 L7
Islam Headworks Pakistan 29°49N 72°33E 42 E5
Islamabad Pakistan 33°40N 73°10E 42 C5
Islamgarh Pakistan 27°51N 70°48E 42 F4
Islamkot Pakistan 24°42N 70°13E 42 G4
Islampur Bihar, India 25°9N 85°12E 43 G11
Islampur W. Bengal, India 26°16N 88°12E 43 F13
Island = Iceland ■ Europe 64°45N 19°0W 8 D4
Island L. Canada 53°47N 94°25W 71 C10
Island Lagoon Australia 31°30S 136°40E 63 E2
Island Pond U.S.A. 44°49N 71°53W 83 B13
Islands, B. of Canada 49°11N 58°15W 73 C8
Islands, B. of N.Z. 35°15S 174°6E 59 A5
Islay U.K. 55°46N 6°10W 11 F2
Isle → France 44°55N 0°15W 20 D3
Isle aux Morts Canada 47°35N 59°0W 73 C8
Isle of Wight □ U.K. 50°41N 1°17W 13 G6
Isle Royale △ U.S.A. 48°0N 88°55W 80 B9
Isleton U.S.A. 38°10N 121°37W 78 G5
Ismail = Izmayil Ukraine 45°22N 28°46E 17 F15
Ismâ'ilîya Egypt 30°37N 32°18E 51 B12
Isna Egypt 25°17N 32°30E 51 C12
Isoanala Madag. 23°50S 45°44E 57 C8
Isparta Turkey 37°47N 30°30E 19 G5
Íspica Italy 36°47N 14°55E 22 F6
Israel ■ Asia 32°0N 34°50E 46 D3
Issoire France 45°32N 3°15E 20 D5
Issyk-Kul = Balykchy Kyrgyzstan 42°26N 76°12E 32 C4
Issyk-Kul, Ozero = Ysyk-Köl Kyrgyzstan 42°25N 77°15E 28 E8
İstanbul Turkey 41°0N 28°58E 23 D13
İstanbul Boğazı Turkey 41°5N 29°3E 23 D13
Istiea Greece 38°57N 23°9E 23 E10
Isto, Mt. U.S.A. 69°12N 143°48W 74 B11
Istokpoga, L. U.S.A. 27°23N 81°17W 85 H14
Istra Croatia 45°10N 14°0E 16 F7
Istres France 43°31N 4°59E 20 E6
Istria = Istra Croatia 45°10N 14°0E 16 F7
Itá Paraguay 25°29S 57°21W 94 B4
Itaberaba Brazil 12°32S 40°18W 93 F10
Itabira Brazil 19°37S 43°13W 93 G10
Itabirito Brazil 20°15S 43°48W 95 A7
Itabuna Brazil 14°48S 39°16W 93 F11
Itacaunas → Brazil 5°21S 49°8W 93 E9
Itacoatiara Brazil 3°8S 58°25W 92 D7
Itaipú, Represa de Brazil 25°30S 54°30W 95 B5
Itaituba Brazil 4°10S 55°50W 93 D7
Itajaí Brazil 27°50S 48°39W 95 B6
Itajubá Brazil 22°24S 45°30W 95 A6
Itaka Tanzania 8°50S 32°49E 55 D3
Itala ⌂ S. Africa 27°30S 31°7E 57 D5
Italy ■ Europe 42°0N 13°0E 22 C5
Itamaraju Brazil 17°4S 39°32W 93 G11
Itampolo Madag. 24°41S 43°57E 57 C7
Itandrano Madag. 21°47S 45°17E 57 C8
Itapecuru Mirim Brazil 3°24S 44°20W 93 D10
Itaperuna Brazil 21°10S 41°54W 95 A7
Itapetininga Brazil 23°36S 48°7W 95 A6
Itapeva Brazil 23°59S 48°59W 95 A6
Itapicuru → Bahia, Brazil 11°47S 37°32W 93 F11
Itapicuru → Maranhão, Brazil 2°52S 44°12W 93 D10
Itapipoca Brazil 3°30S 39°35W 93 D11
Itapuá □ Paraguay 26°40S 55°40W 95 B4
Itaquí Brazil 29°8S 56°30W 94 B4
Itararé Brazil 24°6S 49°23W 95 A6
Itarsi India 22°36N 77°51E 42 H7
Itatí Argentina 27°16S 58°15W 94 B4
Itatiaia △ Brazil 22°22S 44°38W 95 A7
Itchen → U.K. 50°55N 1°22W 13 G6
Itezhi Tezhi, L. Zambia 15°30S 25°30E 55 F2
Ithaca = Ithaki Greece 38°25N 20°40E 23 E9
Ithaca U.S.A. 42°27N 76°30W 83 D8
Ithaki Greece 38°25N 20°40E 23 E9
Itiquira → Brazil 17°18S 56°44W 93 G7
Itiyuro → Argentina 22°40S 63°50W 94 A3
Itō Japan 34°58N 139°5E 31
Itoigawa Japan 37°2N 137°51E 31
Itonamas → Bolivia 12°28S 64°24W 92
Ittoqqortoormiit Greenland 70°20N 23°0W 4
Itu Brazil 23°17S 47°15W 95
Itu Aba I. S. China Sea 10°23N 114°21E 36
Ituiutaba Brazil 19°0S 49°25W 93
Itumbiara Brazil 18°20S 49°10W 93
Ituna Canada 51°10N 103°24W 71
Itunge Port Tanzania 9°40S 33°55E 55
Iturbe Argentina 23°0S 65°25W 94
Ituri → Dem. Rep. of the Congo 1°40N 27°1E 54
Iturup, Ostrov Russia 45°0N 148°0E 29
Ituxi → Brazil 7°18S 64°51W 92
Itzehoe Germany 53°55N 9°31E 16
Ivahona Madag. 23°27S 46°10E 57
Ivaí → Brazil 23°18S 53°42W 95
Ivalo Finland 68°38N 27°35E 8
Ivalojoki → Finland 68°40N 27°40E 8
Ivanava Belarus 52°7N 25°29E 17
Ivanhoe Australia 32°56S 144°20E 63
Ivanhoe Canada 44°23N 77°28W 82
Ivanhoe Calif., U.S.A. 36°23N 119°13W 78
Ivanhoe Minn., U.S.A. 44°28N 96°15W 80
Ivano-Frankivsk Ukraine 48°40N 24°40E 17 D
Ivanovo = Ivanava Belarus 52°7N 25°29E 17
Ivanovo Russia 57°5N 41°0E 18
Ivato Madag. 20°37S 47°10E 57
Ivatsevichy Belarus 52°43N 25°21E 17
Ivdel Russia 60°42N 60°24E 18
Ivinheima → Brazil 23°14S 53°42W 95
Ivinhema Brazil 22°10S 53°37W 95
Ivohibe Madag. 22°31S 46°57E 57
Ivory Coast W. Afr. 4°20N 5°0W 50
Ivory Coast ■ Africa 7°30N 5°0W 50
Ivrea Italy 45°28N 7°52E 20
Ivujivik Canada 62°24N 77°55W 69
Ivvavik △ Canada 69°6N 139°30W 68
Ivybridge U.K. 50°23N 3°56W 13
Iwaizumi Japan 39°50N 141°45E 30
Iwaki Japan 37°3N 140°55E 31
Iwakuni Japan 34°15N 132°8E 31
Iwamizawa Japan 43°12N 141°46E 30
Iwanai Japan 42°58N 140°30E 30
Iwata Japan 34°42N 137°51E 31
Iwate □ Japan 39°30N 141°30E 30
Iwate-San Japan 39°51N 141°0E 30
Iwo Nigeria 7°39N 4°9E 50
Iwŏn N. Korea 40°19N 128°39E 35
Ixiamas Bolivia 13°50S 68°5W 92
Ixopo S. Africa 30°11S 30°5E 57
Ixtepec Mexico 16°34N 95°6W 87
Ixtlán del Río Mexico 21°2N 104°22W 86
Iyo Japan 33°45N 132°45E 31
Izabal, L. de Guatemala 15°30N 89°10W 88
Izamal Mexico 20°56N 89°1W 87
Izena-Shima Japan 26°56N 127°56E 31
Izhevsk Russia 56°51N 53°14E 18
Izhma → Russia 65°19N 52°54E 18
Izmayil Ukraine 45°22N 28°46E 17
İzmir Turkey 38°25N 27°8E 23
İzmit = Kocaeli Turkey 40°45N 29°50E 19
İznik Gölü Turkey 40°27N 29°30E 23
Izra Syria 32°51N 36°15E 46
Izu-Hantō Japan 34°45N 139°0E 31
Izu-Shotō Japan 34°30N 140°0E 31
Izúcar de Matamoros Mexico 18°36N 98°28W 87
Izumi Japan 32°5N 130°22E 31
Izumi-Sano Japan 34°23N 135°18E 31
Izumo Japan 35°20N 132°46E 31
Izyaslav Ukraine 50°5N 26°50E 17

J

J.F.K. Int. ✈ (JFK) U.S.A. 40°38N 73°47W 83
J. Strom Thurmond L. U.S.A. 33°40N 82°12W 85
Jabalpur India 23°9N 79°58E 43
Jabbūl Syria 36°4N 37°30E 44
Jabiru Australia 12°40S 132°53E 60
Jablah Syria 35°20N 36°0E 44
Jablonec nad Nisou Czech Rep. 50°43N 15°10E 16
Jaboatão Brazil 8°7S 35°1W 93
Jaboticabal Brazil 21°15S 48°17W 95
Jaca Spain 42°35N 0°33W 21
Jacareí Brazil 23°20S 46°0W 95
Jacarèzinho Brazil 23°5S 49°58W 95
Jack River △ Australia 14°58S 144°19E 62
Jackman U.S.A. 45°37N 70°15W 81
Jacksboro U.S.A. 33°13N 98°10W 84
Jackson Barbados 13°7N 59°36W 8
Jackson Ala., U.S.A. 31°31N 87°53W 85
Jackson Calif., U.S.A. 38°21N 120°46W 78
Jackson Ky., U.S.A. 37°33N 83°23W 81
Jackson Mich., U.S.A. 42°15N 84°24W 81
Jackson Minn., U.S.A. 43°37N 95°1W 80
Jackson Miss., U.S.A. 32°18N 90°12W 85
Jackson Mo., U.S.A. 37°23N 89°40W 80
Jackson N.H., U.S.A. 44°10N 71°11W 83
Jackson Ohio, U.S.A. 39°3N 82°39W 81
Jackson Tenn., U.S.A. 35°37N 88°49W 85
Jackson Wyo., U.S.A. 43°29N 110°46W 76
Jackson B. N.Z. 43°58S 168°42E 59
Jackson L. U.S.A. 43°52N 110°36W 76
Jacksons N.Z. 42°46S 171°32E 59
Jackson's Arm Canada 49°52N 56°47W 73
Jacksonville Ala., U.S.A. 33°49N 85°46W 85
Jacksonville Ark., U.S.A. 34°52N 92°7W 84
Jacksonville Calif., U.S.A. 37°52N 120°24W 78
Jacksonville Fla., U.S.A. 30°20N 81°39W 85

cksonville Ill., U.S.A. 39°44N 90°14W 80 F8
cksonville N.C., U.S.A. 34°45N 77°26W 85 D16
cksonville Tex., U.S.A. 31°58N 95°17W 84 F7
cksonville Beach U.S.A. 30°17N 81°24W 85 F14
cmel Haiti 18°14N 72°32W 89 C5
cob Lake U.S.A. 36°43N 112°13W 77 H7
cobabad Pakistan 28°20N 68°29E 42 E3
cobina Brazil 11°11S 40°30W 93 F10
cques-Cartier, Dét. de Canada 50°0N 63°30W 73 C7
cques-Cartier, Mt. Canada 48°57N 66°0W 73 C6
cques-Cartier △ Canada 47°15N 71°33W 73 C5
cuí → Brazil 30°2S 51°15W 95 C5
cumba U.S.A. 32°37N 116°11W 79 N10
cundá → Brazil 1°57S 50°26W 93 D8
de City Canada 59°15N 129°37W 70 B3
de Mt. = Yü Shan Taiwan 23°25N 120°52E 33 G13
én Peru 5°25S 78°40W 92 E3
én Spain 37°44N 3°43W 21 D4
farabad India 20°52N 71°22E 42 J4
ffa = Tel Aviv-Yafo Israel 32°4N 34°48E 46 C3
ffa, C. Australia 36°58S 139°40E 63 F2
ffna Sri Lanka 9°45N 80°2E 40 Q12
ffrey U.S.A. 42°49N 72°2W 83 D12
gadhri India 30°10N 77°20E 42 D7
gadishpur India 25°30N 84°21E 43 G11
gdalpur India 19°3N 82°0E 41 K13
gdaqi China 50°25N 124°7E 33 A13
gersfontein S. Africa 29°44S 25°27E 56 D4
ghīn → Iran 27°17N 57°13E 45 E8
godina Serbia 44°5N 21°15E 23 C9
graon India 30°50N 75°25E 42 D6
gtial India 18°50N 79°0E 40 K11
guariaíva Brazil 24°10S 49°50W 95 A6
guaribe → Brazil 4°25S 37°45W 93 D11
güey Grande Cuba 22°35N 81°7W 88 B3
hanabad India 25°13N 84°59E 43 G11
hazpur India 25°37N 75°17E 42 G6
hrom Iran 28°30N 53°31E 45 D7
jon India 31°21N 76°9E 42 D7
lolo Indonesia 1°5N 127°30E 37 D7
lolo, Selat Indonesia 0°5N 129°5E 37 D7
pur India 27°0N 75°50E 42 F6
s India 26°15N 81°32E 43 F9
salmer India 26°55N 70°54E 42 F4
singhnagar India 23°38N 78°34E 43 H8
taran India 26°12N 73°56E 42 F5
thari India 23°14N 78°37E 43 H8
arm Iran 36°58N 56°27E 45 B8
kam → India 23°54N 74°13E 42 H6
karta Indonesia 6°9S 106°52E 37 G12
khal India 29°48N 75°50E 42 E6
khau India 23°13N 68°43E 42 H3
kobstad = Pietarsaari Finland 63°40N 22°43E 8 E20
U.S.A. 32°7N 103°12W 77 K12
ājil Si. Arabia 25°40N 45°27E 44 E5
ālābād Afghan. 34°30N 70°29E 42 B4
alabad India 27°41N 79°42E 43 F8
alpur Jattan Pakistan 32°38N 74°11E 42 C6
ama U.S.A. 34°29N 120°29W 79 L6
andhar = Jullundur India 31°20N 75°40E 42 D6
apa Guatemala 14°39N 89°59W 88 D2
apa Enriquez = Xalapa Mexico 19°32N 96°55W 87 D5
asjärvi Finland 62°29N 22°47E 8 E20
aun India 26°8N 79°25E 43 F8
dhaka → Bangla. 26°16N 89°16E 43 F13
esar India 27°29N 78°19E 42 F8
eswar Nepal 26°38N 85°48E 43 F11
gaon India 21°0N 75°42E 40 J9
bah Iraq 30°35N 46°32E 44 D5
ingo Nigeria 8°55N 11°25E 51 G8
sco □ Mexico 20°20N 103°40W 86 D4
kot Pakistan 35°14N 73°24E 43 B5
na India 19°48N 75°38E 40 K9
ón → Spain 41°47N 1°4W 21 B5
or India 25°21N 72°37E 42 G5
oa Mexico 21°38N 102°58W 86 C4
paiguri India 26°32N 88°46E 41 F16
oan Mexico 21°14N 99°29W 87 C5
uit I. Marshall Is. 6°0N 169°30E 64 G8
ūlā Iraq 34°16N 45°10E 44 C5
naame Somali Rep. 0°4N 42°44E 47 G3
naica ■ W. Indies 18°10N 77°30W 88 a
nalpur Bangla. 24°52N 89°56E 41 G16
nalpur India 25°18N 86°28E 43 G12
nalpurganj India 23°2N 87°59E 43 H13
nanxim → Brazil 4°43S 56°18W 93 D7
nbewangi Indonesia 8°17S 114°7E 37 J17
nbi Indonesia 1°38S 103°30E 36 E2
nbi □ Indonesia 1°30S 102°30E 36 E2
nbongan, Pulau Malaysia 6°45N 117°20E 36 C5
nbusar India 22°3N 72°51E 42 H5
nes → S. Dak., U.S.A. 42°52N 97°18W 80 D5
nes → Va., U.S.A. 36°56N 76°27W 81 G15
nes B. Canada 54°0N 80°0W 72 B3
nes Ranges Australia 24°10S 132°30E 60 D5
nes Ross I. Antarctica 63°58S 57°50W 5 C18
nesabad Pakistan 25°17N 69°15E 42 G3
nestown Australia 33°10S 138°32E 63 E2
nestown S. Africa 31°6S 26°45E 56 E4
nestown N. Dak., U.S.A. 46°54N 98°42W 80 B4
nestown N.Y., U.S.A. 42°6N 79°14W 82 D5
nestown Pa., U.S.A. 41°29N 80°27W 82 E4
nīlābād Iran 34°24N 48°28E 45 C6
nira → India 21°35N 88°28E 43 J13
nkhandi India 16°30N 75°15E 40 L9
nmu India 32°43N 74°54E 42 C6
nmu & Kashmir □ India 34°25N 77°0E 43 B7
Jamnagar India 22°30N 70°6E 42 H4
Jamni → India 25°13N 78°35E 43 G8
Jampur Pakistan 29°39N 70°40E 42 E4
Jamrud Pakistan 33°59N 71°24E 42 C4
Jämsä Finland 61°53N 25°10E 8 F21
Jamshedpur India 22°44N 86°12E 43 H12
Jamtara India 23°59N 86°49E 43 H12
Jämtland Sweden 63°31N 14°0E 8 E16
Jan L. Canada 54°56N 102°55W 71 C8
Jan Mayen Arctic 71°0N 9°0W 4 B7
Janakpur Nepal 26°42N 85°55E 43 F11
Janaúba Brazil 15°48S 43°19W 93 G10
Jand Pakistan 33°30N 72°6E 42 C5
Jandanku ✪ Australia 16°20S 135°45E 62 B2
Jandaq Iran 34°3N 54°22E 45 C7
Jandia Canary Is. 28°6N 14°21W 24 F5
Jandia, Pta. de Canary Is. 28°3N 14°31W 24 F5
Jandía △ Canary Is. 28°4N 14°19W 24 F5
Jandola Pakistan 32°20N 70°9E 42 C4
Jandowae Australia 26°45S 151°7E 63 D5
Janesville U.S.A. 42°41N 89°1W 80 D9
Jangamo Mozam. 24°6S 35°21E 57 C6
Janghai India 25°33N 82°19E 43 G10
Jangheung S. Korea 34°41N 126°52E 35 G14
Janjanbureh Gambia 13°30N 14°47W 50 F3
Janjgir India 22°1N 82°34E 43 J10
Janjina Madag. 20°30S 45°50E 57 C8
Janos Mexico 30°54N 108°10W 86 A3
Januária Brazil 15°25S 44°25W 93 G10
Janûb Sînî □ Egypt 29°30N 33°50E 46 F2
Janubio Canary Is. 28°56N 13°50W 24 F6
Jaora India 23°40N 75°10E 42 H6
Japan ■ Asia 36°0N 136°0E 31 G8
Japan, Sea of Asia 40°0N 135°0E 30 E7
Japan Trench Pac. Oc. 32°0N 142°0E 64 D6
Japen = Yapen Indonesia 1°50S 136°0E 37 E9
Japla India 24°33N 84°1E 43 G11
Japurá → Brazil 3°8S 65°46W 92 D5
Jaquarão Brazil 32°34S 53°23W 95 C5
Jaqué Panama 7°27N 78°8W 88 E4
Jarābulus Syria 36°49N 38°1E 44 B3
Jarama → Spain 40°24N 3°32W 21 B4
Jaranwala Pakistan 31°15N 73°26E 42 D5
Jarash Jordan 32°17N 35°54E 46 C4
Jarash □ Jordan 32°17N 35°54E 46 C4
Jardim Brazil 21°28S 56°2W 94 A4
Jardín América Argentina 27°3S 55°14W 95 B4
Jardine River △ Australia 11°9S 142°21E 62 A3
Jardines de la Reina, Arch. de los Cuba 20°50N 78°50W 88 B4
Jargalang China 43°5N 122°55E 35 C12
Jari → Brazil 1°9S 51°54W 93 D8
Jarīr, W. al → Si. Arabia 25°38N 42°30E 44 E4
Jarosław Poland 50°2N 22°42E 17 C12
Jarrahdale Australia 32°24S 116°5E 61 F2
Jarrahi → Iran 30°49N 48°48E 45 D6
Jartai China 39°45N 105°48E 34 E3
Jarud Qi China 44°28N 120°50E 35 B11
Järvenpää Finland 60°29N 25°5E 8 F21
Jarvis Canada 42°53N 80°6W 82 D4
Jarvis I. Pac. Oc. 0°15S 160°5W 65 H11
Jarwa India 27°38N 82°30E 43 F10
Jasdan India 22°2N 71°12E 42 H4
Jashpurnagar India 22°54N 84°9E 43 H11
Jasidih India 24°31N 86°39E 43 G12
Jāsimīyah Iraq 33°45N 44°41E 44 C5
Jasin Malaysia 2°20N 102°26E 39 L4
Jāsk Iran 25°38N 57°45E 45 E8
Jasło Poland 49°45N 21°30E 17 D11
Jaso India 24°30N 80°29E 43 G9
Jasper Alta., Canada 52°55N 118°5W 70 C5
Jasper Ont., Canada 44°52N 75°57W 83 B9
Jasper Ala., U.S.A. 33°50N 87°17W 85 E11
Jasper Fla., U.S.A. 30°31N 82°57W 85 F13
Jasper Ind., U.S.A. 38°24N 86°56W 80 F10
Jasper Tex., U.S.A. 30°56N 94°1W 84 F7
Jasper △ Canada 52°50N 118°8W 70 C5
Jasrasar India 27°43N 73°49E 42 F5
Jászberény Hungary 47°30N 19°55E 17 E10
Jataí Brazil 17°58S 51°48W 93 G8
Jati Pakistan 24°20N 68°19E 42 G3
Jatibarang Indonesia 6°28S 108°18E 37 G13
Jatiluwih Indonesia 8°23S 115°8E 37 J18
Jatinegara Indonesia 6°13S 106°52E 37 G12
Játiva = Xátiva Spain 38°59N 0°32W 21 C5
Jaú Brazil 22°10S 48°30W 95 A6
Jauja Peru 11°45S 75°15W 92 F3
Jaunpur India 25°46N 82°44E 43 G10
Java = Jawa Indonesia 7°0S 110°0E 36 F3
Java Sea Indonesia 4°35S 107°15E 36 E3
Java Trench Ind. Oc. 9°0S 105°0E 36 F3
Jawa Indonesia 7°0S 110°0E 36 F3
Jawa Barat □ Indonesia 7°0S 107°0E 37 G12
Jawa Tengah □ Indonesia 7°0S 110°0E 37 G14
Jawa Timur □ Indonesia 8°0S 113°0E 37 G15
Jawad India 24°36N 74°51E 42 G6
Jawhar Somali Rep. 2°48N 45°30E 47 G4
Jawoyn ✪ Australia 14°16S 132°28E 60 B5
Jay Peak U.S.A. 44°55N 72°32W 83 B12
Jaya, Puncak Indonesia 3°57S 137°17E 37 E9
Jayanti India 26°45N 89°40E 41 F16
Jayapura Indonesia 2°28S 140°38E 37 E10
Jayawijaya, Pegunungan Indonesia 5°0S 139°0E 37 F9
Jaynagar India 26°43N 86°9E 43 F12
Jaypur = Jeypore India 18°50N 82°38E 41 K13
Jayrūd Syria 33°49N 36°44E 44 C3
Jayton U.S.A. 33°15N 100°34W 84 E4
Jāz Mūrīān, Hāmūn-e Iran 27°20N 58°55E 45 E8
Jazīreh-ye Shīf Iran 29°4N 50°54E 45 D6
Jazminal Mexico 24°52N 101°24W 86 C4
Jazzīn Lebanon 33°31N 35°35E 46 B4
Jean U.S.A. 35°47N 115°20W 79 K11
Jean Marie River Canada 61°32N 120°38W 70 A4
Jean-Rabel Haiti 19°50N 73°5W 89 C5
Jeanerette U.S.A. 29°55N 91°40W 84 G9
Jeanette, Ostrov = Zhannetty, Ostrov Russia 76°43N 158°0E 29 B16
Jeannette U.S.A. 40°20N 79°36W 82 F5
Jebāl Bārez, Kūh-e Iran 28°30N 58°20E 45 D8
Jebel, Bahr el → Sudan 9°30N 30°25E 51 G12
Jebel Ali = Minā' Jabal 'Alī U.A.E. 25°2N 55°8E 45 E7
Jecheon S. Korea 37°8N 128°12E 35 F15
Jedburgh U.K. 55°29N 2°33W 11 F6
Jedda = Jiddah Si. Arabia 21°29N 39°10E 47 C2
Jeddore L. Canada 48°3N 55°55W 73 C8
Jędrzejów Poland 50°35N 20°15E 17 C11
Jefferson Iowa, U.S.A. 42°1N 94°23W 80 D6
Jefferson Ohio, U.S.A. 41°44N 80°46W 82 E4
Jefferson Tex., U.S.A. 32°46N 94°21W 84 E7
Jefferson, Mt. Nev., U.S.A. 38°47N 116°56W 76 G5
Jefferson, Mt. Oreg., U.S.A. 44°41N 121°48W 76 D3
Jefferson City Mo., U.S.A. 38°34N 92°10W 80 F7
Jefferson City Tenn., U.S.A. 36°7N 83°30W 85 C13
Jeffersontown U.S.A. 38°12N 85°35W 81 F11
Jeffersonville U.S.A. 38°17N 85°44W 81 F11
Jeffrey City U.S.A. 42°30N 107°49W 76 E10
Jega Nigeria 12°15N 4°23E 50 F6
Jeju S. Korea 33°31N 126°32E 35 H14
Jeju-do S. Korea 33°29N 126°34E 35 H14
Jēkabpils Latvia 56°29N 25°57E 9 H21
Jekyll I. U.S.A. 31°4N 81°25W 85 F14
Jelenia Góra Poland 50°50N 15°45E 16 C8
Jelgava Latvia 56°41N 23°49E 9 H20
Jemaluang Malaysia 2°16N 103°52E 39 L4
Jember Indonesia 8°11S 113°41E 37 H15
Jena Germany 50°54N 11°35E 16 C6
Jena U.S.A. 31°41N 92°8W 84 F8
Jengish Chokoso = Pobedy, Pik Kyrgyzstan 42°0N 79°58E 32 C4
Jenīn West Bank 32°28N 35°18E 46 C4
Jenkins U.S.A. 37°10N 82°38W 81 G12
Jenner U.S.A. 38°27N 123°7W 78 G3
Jennings U.S.A. 30°13N 92°40W 84 F8
Jeong-eup S. Korea 35°35N 126°50E 35 G14
Jeongseon S. Korea 37°20N 128°45E 35 F15
Jeonju S. Korea 35°50N 127°4E 35 G14
Jepara Indonesia 7°40S 109°14E 37 G14
Jeparit Australia 36°8S 142°1E 63 F3
Jequié Brazil 13°51S 40°5W 93 F10
Jequitinhonha Brazil 16°30S 41°0W 93 G10
Jequitinhonha → Brazil 15°51S 38°53W 93 G11
Jerada Morocco 34°17N 2°10W 50 B5
Jerantut Malaysia 3°56N 102°22E 39 L4
Jerejak, Pulau Malaysia 5°19N 100°19E 39 c
Jérémie Haiti 18°40N 74°10W 89 C5
Jerez, Pta. Mexico 22°58N 97°40W 87 C5
Jerez de García Salinas Mexico 22°39N 103°0W 86 C4
Jerez de la Frontera Spain 36°41N 6°7W 21 D2
Jerez de los Caballeros Spain 38°20N 6°45W 21 C2
Jericho = El Arīḩā West Bank 31°52N 35°27E 46 D4
Jericho Australia 23°38S 146°6E 62 C4
Jerid, Chott el = Djerid, Chott Tunisia 33°42N 8°30E 50 B7
Jerilderie Australia 35°20S 145°41E 63 F4
Jermyn U.S.A. 41°32N 75°33W 83 E9
Jerome U.S.A. 42°44N 114°31W 76 E6
Jerramungup Australia 33°55S 118°55E 61 F2
Jersey U.K. 49°11N 2°7W 13 H5
Jersey City U.S.A. 40°42N 74°4W 83 F10
Jersey Shore U.S.A. 41°12N 77°15W 82 E7
Jerseyville U.S.A. 39°7N 90°20W 80 F8
Jerusalem Israel/West Bank 31°47N 35°10E 46 D4
Jervis B. Australia 35°8S 150°46E 63 F5
Jervis Inlet Canada 50°0N 123°57W 70 C4
Jesi = Iesi Italy 43°31N 13°14E 22 C5
Jessore Bangla. 23°10N 89°10E 41 H16
Jesup U.S.A. 31°36N 81°53W 85 F14
Jesús Carranza Mexico 17°26N 95°2W 87 D5
Jesús María Argentina 30°59S 64°5W 94 C3
Jetmore U.S.A. 38°4N 99°54W 80 F4
Jetpur India 21°45N 70°10E 42 J4
Jevnaker Norway 60°15N 10°26E 9 F14
Jewett U.S.A. 40°22N 81°2W 82 F3
Jewett City U.S.A. 41°36N 71°59W 83 E13
Jeyḩūnābād Iran 34°58N 48°59E 45 C6
Jeypore India 18°50N 82°38E 41 K13
Jha Jha India 24°46N 86°22E 43 G12
Jhaarkand = Jharkhand □ India 24°0N 85°50E 43 H11
Jhabua India 22°46N 74°36E 42 H6
Jhajjar India 28°37N 76°42E 42 E7
Jhal Pakistan 28°17N 67°27E 42 E2
Jhal Jhao Pakistan 26°20N 65°35E 40 F4
Jhalawar India 24°40N 76°10E 42 G7
Jhalida India 23°22N 85°58E 43 H11
Jhalrapatan India 24°33N 76°10E 42 G7
Jhang Maghiana Pakistan 31°15N 72°22E 42 D5
Jhansi India 25°30N 78°36E 43 G8
Jhargram India 22°27N 86°59E 43 H12
Jharia India 23°45N 86°26E 43 H12
Jharkhand □ India 24°0N 85°50E 43 H11
Jharsuguda India 21°56N 84°5E 41 J14
Jhelum Pakistan 33°0N 73°45E 42 C5
Jhelum → Pakistan 31°20N 72°10E 42 D5
Jhilmilli India 23°24N 82°51E 43 H10
Jhudo Pakistan 24°58N 69°18E 42 G3
Jhunjhunu India 28°10N 75°30E 42 E6
Ji-Paraná Brazil 10°52S 62°57W 92 F6
Ji Xian China 36°7N 110°40E 34 F6
Jia Xian Henan, China 33°59N 113°12E 34 H7
Jia Xian Shaanxi, China 38°12N 110°28E 34 E6
Jiaji = Qionghai China 19°15N 110°26E 38 C8
Jiamusi China 46°40N 130°26E 33 B15
Ji'an Jiangxi, China 27°6N 114°59E 33 F11
Ji'an Jilin, China 41°5N 126°10E 35 D14
Jianchang China 40°55N 120°35E 35 D11
Jiangsu □ China 33°0N 120°0E 35 H11
Jiangxi □ China 27°30N 116°0E 33 F12
Jiangyuan China 42°2N 126°34E 35 C14
Jianping China 41°53N 119°42E 35 D10
Jiao Xian = Jiaozhou China 36°18N 120°1E 35 F11
Jiaohe Hebei, China 38°2N 116°20E 34 E9
Jiaohe Jilin, China 43°40N 127°22E 35 C14
Jiaonan China 35°52N 119°58E 35 G10
Jiaozhou China 36°18N 120°1E 35 F11
Jiaozhou Wan China 36°5N 120°10E 35 F11
Jiaozuo China 35°16N 113°12E 34 G7
Jiawang China 34°28N 117°26E 35 G9
Jiaxiang China 35°25N 116°20E 34 G9
Jiaxing China 30°49N 120°45E 33 E13
Jiayi = Chiai Taiwan 23°29N 120°25E 33 G13
Jiayuguan China 39°49N 98°18E 32 D8
Jibuti = Djibouti ■ Africa 12°0N 43°0E 47 E3
Jicarón, I. Panama 7°10N 81°50W 88 E3
Jiddah Si. Arabia 21°29N 39°10E 47 C2
Jido India 29°2N 94°58E 41 E19
Jieshou China 33°18N 115°22E 34 H8
Jiexiu China 37°2N 111°55E 34 F6
Jigalong Australia 23°21S 120°47E 60 D3
Jigalong ✪ Australia 23°21S 120°46E 60 D3
Jigni India 25°45N 79°25E 43 G8
Jihlava Czech Rep. 49°28N 15°35E 16 D8
Jihlava → Czech Rep. 48°55N 16°36E 17 D9
Jijiga Ethiopia 9°20N 42°50E 47 F3
Jilib Somali Rep. 0°29N 42°46E 47 G3
Jilin China 43°44N 126°30E 35 C14
Jilin □ China 44°0N 127°0E 35 C14
Jilong = Chilung Taiwan 25°3N 121°45E 33 F13
Jim Thorpe U.S.A. 40°52N 75°44W 83 F9
Jima Ethiopia 7°40N 36°47E 47 F2
Jimbaran, Teluk Indonesia 8°46S 115°9E 37 K18
Jiménez Mexico 27°8N 104°54W 86 B4
Jimeta Nigeria 9°17N 12°28E 52 C2
Jimo China 36°23N 120°30E 35 F11
Jin Xian = Jinzhou China 38°55N 121°42E 35 E11
Jinan China 36°38N 117°1E 34 F9
Jinchang China 38°30N 102°10E 32 D9
Jincheng China 35°29N 112°50E 34 G7
Jind India 29°19N 76°22E 42 E7
Jindabyne Australia 36°25S 148°35E 63 F4
Jinding China 22°22N 113°33E 33 a
Jindo S. Korea 34°28N 126°15E 35 G14
Jindřichův Hradec Czech Rep. 49°10N 15°2E 16 D8
Jing He → China 34°27N 109°4E 34 G5
Jingbian China 37°20N 108°30E 34 F5
Jingchuan China 35°20N 107°20E 34 G4
Jingdezhen China 29°20N 117°11E 33 F12
Jinggu China 23°35N 100°41E 32 G9
Jinghai China 38°55N 116°55E 34 E9
Jinghong China 22°0N 100°45E 32 G9
Jingle China 38°20N 111°55E 34 E6
Jingmen China 31°0N 112°10E 33 E11
Jingning China 35°30N 105°43E 34 G3
Jingpo Hu China 43°55N 128°55E 35 C15
Jingtai China 37°10N 104°6E 34 F3
Jingtanggang China 39°10N 119°5E 35 E10
Jingyang China 34°30N 108°50E 34 G5
Jingyu China 42°25N 126°45E 35 C14
Jingyuan China 36°30N 104°40E 34 F3
Jingziguan China 33°15N 111°0E 34 H6
Jinhua China 29°8N 119°38E 33 F12
Jining Nei Monggol Zizhiqu, China 41°5N 113°0E 34 D7
Jining Shandong, China 35°22N 116°34E 34 G9
Jinja Uganda 0°25N 33°12E 54 B3
Jinjang Malaysia 3°13N 101°39E 39 L3
Jinji China 37°58N 106°8E 34 F4
Jinju S. Korea 35°12N 128°2E 35 G15
Jinmu Jiao China 18°9N 109°34E 38 C7
Jinnah Barrage Pakistan 32°58N 71°33E 42 C4
Jinotega Nic. 13°6N 85°59W 88 D2
Jinotepe Nic. 11°50N 86°10W 88 D2
Jinsha Jiang → China 28°50N 104°36E 32 F9
Jinxi China 40°52N 120°50E 35 D11
Jinxiang China 35°5N 116°22E 34 G9
Jinzhong China 37°42N 112°46E 34 F7
Jinzhou Liaoning, China 38°55N 121°42E 35 E11
Jinzhou Liaoning, China 41°5N 121°3E 35 D11
Jiparaná → Brazil 8°3S 62°52W 92 E6
Jipijapa Ecuador 1°0S 80°40W 92 D2
Jiquilpan Mexico 19°59N 102°43W 86 D4
Jirisan S. Korea 35°20N 127°44E 35 G14
Jiroft Iran 28°45N 57°50E 45 D8
Jishan China 35°34N 110°58E 34 G6
Jishou China 28°21N 109°43E 33 F10
Jisr ash Shughūr Syria 35°49N 36°18E 44 C3
Jitarning Australia 32°48S 117°57E 61 F2
Jitra Malaysia 6°16N 100°25E 39 J3
Jiu → Romania 43°47N 23°48E 17 F12
Jiudengkou China 39°56N 106°40E 34 E4
Jiujiang China 29°42N 115°58E 33 F12
Jiulong = Kowloon China 22°19N 114°11E 33 G11
Jiuquan China 39°50N 98°20E 32 D8
Jiutai China 44°10N 125°50E 35 B13
Jiuxincheng China 39°17N 115°59E 34 E8
Jiwani Pakistan 25°1N 61°44E 40 G2
Jixi China 45°20N 130°50E 35 B16
Jiyang China 37°0N 117°12E 35 F9
Jiyuan China 35°7N 112°57E 34 G7
Jīzān Si. Arabia 17°0N 42°20E 47 D3
Jize China 36°54N 114°56E 34 F8
Jizhou China 37°35N 115°30E 34 F8
Jizl, Wādī al → Si. Arabia 25°39N 38°25E 44 E3
Jizō-Zaki Japan 35°34N 133°20E 31 G6
Jizzax Uzbekistan 40°6N 67°50E 28 E7
Joaçaba Brazil 27°5S 51°31W 95 B5
João Pessoa Brazil 7°10S 34°52W 93 E12
Joaquín V. González Argentina 25°10S 64°0W 94 B3
Jobat India 22°25N 74°34E 42 H6
Jodhpur India 26°23N 73°8E 42 F5
Jodiya India 22°42N 70°18E 42 H4
Joensuu Finland 62°37N 29°49E 8 E23
Jōetsu Japan 37°12N 138°10E 31 F9
Jofane Mozam. 21°15S 34°18E 57 C5
Jogbani India 26°25N 87°15E 43 F12
Jõgeva Estonia 58°45N 26°24E 9 G22
Jogjakarta = Yogyakarta Indonesia 7°49S 110°22E 37 G14
Johannesburg S. Africa 26°11S 28°2E 57 D4
Johannesburg U.S.A. 35°22N 117°38W 79 K9
Johilla → India 23°37N 81°14E 43 H9
John Crow Mts. Jamaica 18°5N 76°25W 88 a
John Day U.S.A. 44°25N 118°57W 76 D4
John Day → U.S.A. 45°44N 120°39W 76 D3
John Day Fossil Beds △ U.S.A. 44°33N 119°38W 76 D4
John D'or Prairie Canada 58°30N 115°8W 70 B5
John H. Kerr Res. U.S.A. 36°36N 78°18W 85 C15
John o' Groats U.K. 58°38N 3°4W 11 C5
Johnnie U.S.A. 36°25N 116°5W 79 J10
John's Ra. Australia 21°55S 133°23E 62 C1
Johnson U.S.A. 44°38N 72°41W 83 B12
Johnson City Kans., U.S.A. 37°34N 101°45W 80 G3
Johnson City N.Y., U.S.A. 42°7N 75°58W 83 D9
Johnson City Tenn., U.S.A. 36°19N 82°21W 85 C13
Johnson City Tex., U.S.A. 30°17N 98°25W 84 F5
Johnsonburg U.S.A. 41°29N 78°41W 82 E6
Johnsondale U.S.A. 35°58N 118°32W 79 K8
Johnsons Crossing Canada 60°29N 133°18W 70 A2
Johnston, L. Australia 32°25S 120°30E 61 F3
Johnston Falls = Mambilima Falls Zambia 10°31S 28°45E 55 E2
Johnston I. Pac. Oc. 17°10N 169°8W 65 F11
Johnstone Str. Canada 50°28N 126°0W 70 C3
Johnstown Ireland 52°45N 7°33W 10 D4
Johnstown N.Y., U.S.A. 43°0N 74°22W 83 C10
Johnstown Ohio, U.S.A. 40°9N 82°41W 82 F2
Johnstown Pa., U.S.A. 40°20N 78°55W 82 F6
Johor □ Malaysia 2°5N 103°20E 39 M4
Johor, Selat Asia 1°28N 103°47E 39 d
Johor Bahru Malaysia 1°28N 103°46E 39 d
Jõhvi Estonia 59°22N 27°27E 9 G22
Joinville Brazil 26°15S 48°55W 95 B6
Joinville I. Antarctica 65°0S 55°30W 5 C18
Jojutla Mexico 18°37N 99°11W 87 D5
Jokkmokk Sweden 66°35N 19°50E 8 C18
Jökulsá á Bru → Iceland 65°40N 14°16W 8 D6
Jökulsá á Fjöllum → Iceland 66°10N 16°30W 8 C5
Jökulsárgljúfur △ Iceland 65°56N 16°31W 8 D5
Jolfā Āzarbājān-e Sharqī, Iran 38°57N 45°38E 44 B5
Jolfā Eşfahan, Iran 32°58N 51°37E 45 C6
Joliet U.S.A. 41°32N 88°5W 80 E9
Joliette Canada 46°3N 73°24W 72 C5
Jolo Phil. 6°0N 121°0E 37 C6
Jolon U.S.A. 35°58N 121°9W 78 K5
Jombang Indonesia 7°33S 112°14E 37 G15
Jonava Lithuania 55°8N 24°12E 9 J21
Jones Sound Canada 76°0N 85°0W 69 B15
Jonesboro Ark., U.S.A. 35°50N 90°42W 85 D9
Jonesboro La., U.S.A. 32°15N 92°43W 84 E8
Joniškis Lithuania 56°13N 23°35E 9 H20
Jönköping Sweden 57°45N 14°8E 9 H16
Jonquière Canada 48°27N 71°14W 73 C5
Joplin U.S.A. 37°6N 94°31W 80 G6
Jora India 26°20N 77°49E 42 F6
Jordan Mont., U.S.A. 47°19N 106°55W 76 C10
Jordan N.Y., U.S.A. 43°4N 76°29W 83 C8
Jordan ■ Asia 31°0N 36°0E 46 E5
Jordan → Asia 31°48N 35°32E 46 D4
Jordan Valley U.S.A. 42°59N 117°3W 76 E5
Jorhat India 26°45N 94°12E 41 F19
Jörn Sweden 65°4N 20°1E 8 D19
Jorong Indonesia 3°58S 114°56E 36 E4
Jørpeland Norway 59°3N 6°1E 9 G12
Jorquera → Chile 28°3S 69°58W 94 B2
Jos Nigeria 9°53N 8°51E 50 G7
José Batlle y Ordóñez Uruguay 33°20S 55°10W 95 C4
Joseph, L. Canada 45°10N 79°44W 82 A5
Joseph Bonaparte G. Australia 14°35S 128°50E 60 B4
Joseph L. Canada 52°45N 65°18W 73 B6
Joshinath India 30°34N 79°34E 43 D8
Joshinetsu-Kōgen △ Japan 36°42N 138°32E 31 F9
Joshua Tree U.S.A. 34°8N 116°19W 79 L10
Joshua Tree △ U.S.A. 33°55N 116°0W 79 M10
Jost Van Dyke I. Br. Virgin Is. 18°29N 64°47W 89 e
Jostedalsbreen Norway 61°40N 6°59E 8 F12
Jotunheimen Norway 61°35N 8°25E 8 F13
Joubertberge Namibia 18°30S 14°0E 56 B1
Jourdanton U.S.A. 28°55N 98°33W 84 G5
Jovellanos Cuba 22°40N 81°10W 88 B3
Ju Xian China 36°35N 118°20E 35 F10
Juan Aldama Mexico 24°19N 103°21W 86 C4
Juan Bautista Alberdi Argentina 34°26S 61°48W 94 C3
Juan de Fuca, Str of. N. Amer. 48°15N 124°0W 78 B3
Juán de Nova Ind. Oc. 17°3S 43°45E 57 B7
Juan Fernández, Arch. de Pac. Oc. 33°50S 80°0W 90 G2
Juan José Castelli Argentina 25°27S 60°57W 94 B3
Juan L. Lacaze Uruguay 34°26S 57°25W 94 C4
Juankoski Finland 63°3N 28°19E 8 E23
Juárez Mexico 27°37N 100°44W 86 B4
Juárez, Sierra de Mexico 32°0N 116°0W 86 A1
Juàzeiro Brazil 9°30S 40°30W 93 E10
Juàzeiro do Norte Brazil 7°10S 39°18W 93 E11
Jûbâ Sudan 4°50N 31°35E 51 H12
Juba → Somali Rep. 1°30N 42°35E 47 G3

Jubany Antarctica 62°30S 58°0W 5 C18
Jubayl Lebanon 34°5N 35°39E 46 A4
Jubbah Si. Arabia 28°2N 40°56E 44 D4
Jubbal India 31°5N 77°40E 42 D7
Jubbulpore = Jabalpur India 23°9N 79°58E 43 H8
Jubilee L. Australia 29°0S 126°50E 61 E4
Juby, C. Morocco 28°0N 12°59W 50 C3
Júcar = Xúquer → Spain 39°5N 0°10W 21 C5
Júcaro Cuba 21°37N 78°51W 88 B4
Juchitán de Zaragoza Mexico 16°26N 95°1W 87 D5
Judea = Har Yehuda Israel 31°35N 34°57E 46 D3
Judith → U.S.A. 47°44N 109°39W 76 C9
Judith, Pt. U.S.A. 41°22N 71°29W 83 E13
Judith Gap U.S.A. 46°41N 109°45W 76 C9
Juigalpa Nic. 12°6N 85°26W 88 D2
Juiz de Fora Brazil 21°43S 43°19W 95 A7
Jujuy ☐ Argentina 23°20S 65°40W 94 A2
Julesburg U.S.A. 40°59N 102°16W 76 F12
Juli Peru 16°10S 69°25W 92 G5
Julia Cr. → Australia 20°0S 141°11E 62 C3
Julia Creek Australia 20°39S 141°44E 62 C3
Juliaca Peru 15°25S 70°10W 92 G4
Julian U.S.A. 33°4N 116°38W 79 M10
Julian, L. Canada 54°25N 77°57W 72 B4
Julianatop Suriname 3°40N 56°30W 93 C7
Julianehåb = Qaqortoq Greenland 60°43N 46°0W 4 C5
Julimes Mexico 28°25N 105°27W 86 B3
Jullundur India 31°20N 75°40E 42 D6
Julu China 37°15N 115°2E 34 F8
Jumbo Zimbabwe 17°30S 30°58E 55 F3
Jumbo Pk. U.S.A. 36°12N 114°11W 79 J12
Jumentos Cays Bahamas 23°0N 75°40W 88 B4
Jumilla Spain 38°28N 1°19W 21 C5
Jumla Nepal 29°15N 82°13E 43 E10
Jumna = Yamuna → India 25°30N 81°53E 43 G9
Jumunjin S. Korea 37°55N 128°54E 35 F15
Junagadh India 21°30N 70°30E 42 J4
Junan China 35°12N 118°53E 35 G10
Junction Tex., U.S.A. 30°29N 99°46W 84 F5
Junction Utah, U.S.A. 38°14N 112°13W 77 G7
Junction B. Australia 11°52S 133°55E 62 A1
Junction City Kans., U.S.A. 39°2N 96°50W 80 F5
Junction City Oreg., U.S.A. 44°13N 123°12W 76 D2
Junction Pt. Australia 11°45S 133°50E 62 A1
Jundah Australia 24°46S 143°2E 62 C3
Jundiaí Brazil 24°30S 47°0W 95 A6
Juneau U.S.A. 58°18N 134°25W 70 B2
Junee Australia 34°53S 147°35E 63 E4
Jungar Qi China 39°49N 110°57E 34 E6
Jungfrau Switz. 46°32N 7°58E 20 C7
Junggar Pendi China 44°30N 86°0E 32 C6
Jungshahi Pakistan 24°52N 67°44E 42 G2
Juniata → U.S.A. 40°24N 77°1W 82 F7
Junín Argentina 34°33S 60°57W 94 C3
Junín de los Andes Argentina 39°45S 71°0W 96 D2
Jūniyah Lebanon 33°59N 35°38E 46 B4
Juntas Chile 28°24S 69°58W 94 B2
Juntura U.S.A. 43°45N 118°5W 76 E4
Jur, Nahr el → Sudan 8°45N 29°15E 51 G11
Jura = Jura, Mts. du Europe 46°40N 6°5E 20 C7
Jura = Schwäbische Alb Germany 48°20N 9°30E 16 D5
Jura U.K. 56°0N 5°50W 11 F3
Jura, Mts. du Europe 46°40N 6°5E 20 C7
Jura, Sd. of U.K. 55°57N 5°45W 11 F3
Jurbarkas Lithuania 55°4N 22°46E 9 J20
Jurien Bay Australia 30°18S 115°2E 61 F2
Jūrmala Latvia 56°58N 23°34E 9 H20
Jurong Singapore 1°19N 103°42E 39 d
Juruá → Brazil 2°37S 65°44W 92 D5
Juruena Brazil 13°0S 58°10W 92 F7
Juruena → Brazil 7°20S 58°3W 92 E7
Juruti Brazil 2°9S 56°4W 93 D7
Justo Daract Argentina 33°52S 65°12W 94 C2
Jutaí → Brazil 2°43S 66°57W 92 D5
Juticalpa Honduras 14°40N 86°12W 88 D2
Jutland = Jylland Denmark 56°25N 9°30E 9 H13
Juuka Finland 63°13N 29°17E 8 E23
Juventud, I. de la Cuba 21°40N 82°40W 88 B3
Jūy Zar Iran 33°50N 46°18E 44 C5
Juye China 35°22N 116°5E 34 G9
Jwaneng Botswana 24°45S 24°50E 53 J4
Jylland Denmark 56°25N 9°30E 9 H13
Jyväskylä Finland 62°14N 25°50E 8 E21

K

K2 Pakistan 35°58N 76°32E 43 B7
Kaakha = Kaka Turkmenistan 37°21N 59°36E 45 B8
Kaapkruis Namibia 21°55S 13°57E 56 C1
Kaapstad = Cape Town S. Africa 33°55S 18°22E 56 E2
Kabaena Indonesia 5°15S 122°0E 37 F6
Kabala S. Leone 9°38N 11°37W 50 G3
Kabale Uganda 1°15S 30°0E 54 C3
Kabalo Dem. Rep. of the Congo 6°0S 27°0E 54 D2
Kabambare Dem. Rep. of the Congo 4°41S 27°39E 54 C2
Kabango Dem. Rep. of the Congo 8°35S 28°30E 55 D2
Kabanjahe Indonesia 3°6N 98°30E 36 D1
Kabankalan Phil. 9°59N 122°49E 37 C6
Kabara Fiji 18°59S 178°56W 59 a
Kabardino-Balkaria ☐ Russia 43°30N 43°30E 19 F7
Kabarega Falls = Murchison Falls Uganda 2°15N 31°30E 54 B3
Kabarnet Kenya 0°30N 35°45E 54 B4
Kabasalan Phil. 7°47N 122°44E 37 C6
Kabat Indonesia 8°16S 114°19E 37 J17
Kabin Buri Thailand 13°57N 101°43E 38 F3
Kabinakagami L. Canada 48°54N 84°25W 72 C3
Kabinda Dem. Rep. of the Congo 6°19S 24°20E 52 F4
Kabompo Zambia 13°36S 24°14E 55 E1
Kabompo → Zambia 14°11S 23°11E 53 G4
Kabondo Dem. Rep. of the Congo 8°58S 25°40E 55 D2
Kabongo Dem. Rep. of the Congo 7°22S 25°33E 54 D2
Kabrīt, G. el Egypt 29°42N 33°16E 46 F2
Kabūd Gonbad Iran 37°5N 59°45E 45 B8
Kābul Afghan. 34°28N 69°11E 42 B3
Kābul ☐ Afghan. 34°30N 69°0E 40 B6
Kābul → Pakistan 33°55N 72°14E 42 C5
Kabunga Dem. Rep. of the Congo 1°38S 28°3E 54 C2
Kaburuang Indonesia 3°50N 126°30E 37 D7
Kabwe Zambia 14°30S 28°29E 55 E2
Kachchh, Gulf of India 22°50N 69°15E 42 H3
Kachchh, Rann of India 24°0N 70°0E 42 H4
Kachchhidhana India 21°44N 78°46E 43 J8
Kachebera Zambia 13°50S 32°50E 55 E3
Kachikau Botswana 18°8S 24°26E 56 B3
Kachin ☐ Burma 26°0N 97°30E 41 G20
Kachira, L. Uganda 0°40S 31°7E 54 C3
Kachīry Kazakhstan 53°10N 75°50E 28 D8
Kachnara India 23°50N 75°6E 42 H6
Kachot Cambodia 11°30N 103°3E 39 G4
Kaçkar Turkey 40°45N 41°10E 19 F7
Kadam, Mt. Uganda 1°45N 34°45E 54 B3
Kadan Kyun Burma 12°30N 98°20E 38 F2
Kadanai → Afghan. 31°22N 65°45E 42 D1
Kadavu Fiji 19°0S 178°15E 59 a
Kadavu Passage Fiji 18°45S 178°0E 59 a
Kade Ghana 6°7N 0°56W 50 G5
Kadhimain = Al Kāẓimīyah Iraq 33°22N 44°18E 44 C5
Kadi India 23°18N 72°23E 42 H5
Kadina Australia 33°55S 137°43E 63 E2
Kadipur India 26°10N 82°23E 43 F10
Kadirli Turkey 37°23N 36°5E 44 B3
Kadiyevka = Stakhanov Ukraine 48°35N 38°40E 19 E6
Kadoka U.S.A. 43°50N 101°31W 80 D3
Kadoma Zimbabwe 18°20S 29°52E 55 F2
Kâdugli Sudan 11°0N 29°45E 51 F11
Kaduna Nigeria 10°30N 7°21E 50 F7
Kaédi Mauritania 16°9N 13°28W 50 E3
Kaeng Khoi Thailand 14°35N 101°0E 38 E3
Kaeng Krachan △ Thailand 12°57N 99°23E 38 F2
Kaeng Krung △ Thailand 9°30N 98°50E 39 H2
Kaeng Tana △ Thailand 15°25N 105°32E 38 E5
Kaesŏng N. Korea 37°58N 126°35E 35 F14
Kāf Si. Arabia 31°25N 37°29E 44 D3
Kafan = Kapan Armenia 39°18N 46°27E 44 B5
Kafanchan Nigeria 9°40N 8°20E 50 G7
Kafinda Zambia 12°32S 30°20E 55 E3
Kafue Zambia 15°46S 28°9E 55 F2
Kafue → Zambia 15°30S 29°0E 53 H5
Kafue △ Zambia 15°12S 25°38E 55 F2
Kafue Flats Zambia 15°40S 27°25E 55 F2
Kafulwe Zambia 9°0S 29°1E 55 D2
Kaga Afghan. 34°14N 70°10E 42 B4
Kaga Bandoro C.A.R. 7°0N 19°10E 52 C3
Kagawa ☐ Japan 34°15N 134°0E 31 G7
Kagera ☐ Tanzania 2°0S 31°30E 54 C3
Kagera → Uganda 0°57S 31°47E 54 C3
Kağızman Turkey 40°5N 43°10E 44 B4
Kagoshima Japan 31°35N 130°33E 31 J5
Kagoshima ☐ Japan 31°30N 130°30E 31 J5
Kagul = Cahul Moldova 45°50N 28°15E 17 F15
Kahak Iran 36°6N 49°46E 45 B6
Kahama Tanzania 4°8S 32°30E 54 C3
Kahan Pakistan 29°18N 68°54E 42 E3
Kahang Malaysia 2°12N 103°32E 39 L4
Kahayan → Indonesia 3°40S 114°0E 36 E4
Kahe Tanzania 3°30S 37°25E 54 C4
Kahemba Dem. Rep. of the Congo 7°18S 18°55E 52 F3
Kahnūj Iran 27°55N 57°40E 45 E8
Kahoka U.S.A. 40°25N 91°44W 80 E8
Kaho'olawe U.S.A. 20°33N 156°37W 75 L8
Kahramanmaraş Turkey 37°37N 36°53E 44 B3
Kâhta Turkey 37°46N 38°36E 44 B3
Kahului U.S.A. 20°54N 156°28W 75 L8
Kahurangi △ N.Z. 41°10S 172°32E 59 D4
Kahuta Pakistan 33°35N 73°24E 42 C5
Kahuzi-Biega △ Dem. Rep. of the Congo 1°50S 27°55E 54 C2
Kai, Kepulauan Indonesia 5°55S 132°45E 37 F8
Kai Besar Indonesia 5°35S 133°0E 37 F8
Kai Is. = Kai, Kepulauan Indonesia 5°55S 132°45E 37 F8
Kai Kecil Indonesia 5°45S 132°40E 37 F8
Kaiapoi N.Z. 43°24S 172°40E 59 E4
Kaidu He → China 41°46N 86°31E 32 C6
Kaieteur Falls Guyana 5°1N 59°10W 92 B7
Kaifeng China 34°48N 114°21E 34 G8
Kaikohe N.Z. 35°25S 173°49E 59 A4
Kaikoura N.Z. 42°25S 173°43E 59 E4
Kailash = Kangrinboqe Feng China 31°0N 81°25E 43 D9
Kailu China 43°38N 121°18E 35 C11
Kailua Kona U.S.A. 19°39N 155°59W 75 M8
Kaimana Indonesia 3°39S 133°45E 37 E8
Kaimanawa Mts. N.Z. 39°15S 175°56E 59 C5
Kaimganj India 27°33N 79°24E 43 F8
Kaimur Hills India 24°30N 82°0E 43 G10
Kainab → Namibia 28°32S 19°34E 56 D2
Kainji Res. Nigeria 10°1N 4°40E 50 F6
Kainuu Finland 64°30N 29°7E 8 D23
Kaipara Harbour N.Z. 36°25S 174°14E 59 B5
Kaipokok B. Canada 54°54N 59°47W 73 B8
Kaira India 22°45N 72°50E 42 H5
Kairana India 29°24N 77°15E 42 E7
Kaironi Indonesia 0°47S 133°40E 37 E8
Kairouan Tunisia 35°45N 10°5E 51 A8
Kaiserslautern Germany 49°26N 7°45E 16 D4
Kaitaia N.Z. 35°8S 173°17E 59 A4
Kaitangata N.Z. 46°17S 169°51E 59 G2
Kaithal India 29°48N 76°26E 42 E7
Kaitu → Pakistan 33°10N 70°30E 42 C4
Kaiwi Channel U.S.A. 21°15N 157°30W 75 L8
Kaiyuan Liaoning, China 42°28N 124°1E 35 C13
Kaiyuan Yunnan, China 23°40N 103°12E 32 G9
Kaiyuh Mts. U.S.A. 64°30N 158°0W 74 C8
Kajaani Finland 64°17N 27°46E 8 D22
Kajabbi Australia 20°0S 140°1E 62 C3
Kajana = Kajaani Finland 64°17N 27°46E 8 D22
Kajang Malaysia 2°59N 101°48E 39 L3
Kajiado Kenya 1°53S 36°48E 54 C4
Kajo Kaji Sudan 3°58N 31°40E 51 H12
Kaka Turkmenistan 37°21N 59°36E 45 B8
Kakabeka Falls Canada 48°24N 89°37W 72 C2
Kakadu △ Australia 12°0S 132°3E 60 B5
Kakamas S. Africa 28°45S 20°33E 56 D3
Kakamega Kenya 0°20N 34°46E 54 B3
Kakanui Mts. N.Z. 45°10S 170°30E 59 F3
Kakdwip India 21°53N 88°11E 43 J13
Kake Japan 34°36N 132°19E 31 G6
Kake U.S.A. 56°59N 133°57W 70 B2
Kakegawa Japan 34°45N 138°1E 31 G9
Kakeroma-Jima Japan 28°8N 129°14E 31 K4
Kākhak Iran 34°9N 58°38E 45 C8
Kakhovka Ukraine 46°45N 33°30E 19 E5
Kakhovske Vdskh. Ukraine 47°5N 34°0E 19 E5
Kakinada India 16°57N 82°11E 41 L13
Kakisa Canada 60°56N 117°25W 70 A5
Kakisa → Canada 61°3N 118°10W 70 A5
Kakisa L. Canada 60°56N 117°43W 70 A5
Kakogawa Japan 34°46N 134°51E 31 G7
Kakuma Kenya 3°43N 34°52E 54 B3
Kakwa → Canada 54°37N 118°28W 70 C5
Kāl Gūsheh Iran 30°59N 58°12E 45 D8
Kal Sefīd Iran 34°52N 47°23E 44 C5
Kalaallit Nunaat = Greenland ☒ N. Amer. 66°0N 45°0W 67 C15
Kalabagh Pakistan 33°0N 71°28E 42 C4
Kalabahi Indonesia 8°13S 124°31E 37 F6
Kalach Russia 50°22N 41°0E 19 D7
Kaladan → Burma 20°20N 93°5E 41 J18
Kaladar Canada 44°37N 77°5W 82 B7
Kalahari Africa 24°0S 21°30E 56 C3
Kalahari Gemsbok △ S. Africa 25°30S 20°30E 56 D3
Kalajoki Finland 64°12N 24°10E 8 D21
Kalakamati Botswana 20°40S 27°25E 57 C4
Kalakan Russia 55°15N 116°45E 29 D12
K'alak'unlun Shank'ou = Karakoram Pass Asia 35°33N 77°50E 43 B7
Kalam Pakistan 35°34N 72°30E 43 B5
Kalama Dem. Rep. of the Congo 2°52S 28°35E 54 C2
Kalama U.S.A. 46°1N 122°51W 78 E4
Kalamata Greece 37°3N 22°10E 23 F10
Kalamazoo U.S.A. 42°17N 85°35W 81 D11
Kalamazoo → U.S.A. 42°40N 86°10W 80 D10
Kalambo Falls Tanzania 8°37S 31°35E 55 D3
Kalan Turkey 39°7N 39°32E 44 B3
Kalannie Australia 30°22S 117°5E 61 F2
Kalāntarī Iran 32°10N 54°8E 45 C7
Kalao Indonesia 7°21S 121°0E 37 F6
Kalaotoa Indonesia 7°20S 121°50E 37 F6
Kalasin Thailand 16°26N 103°30E 38 D4
Kalāt Iran 25°29N 59°22E 45 E8
Kalat Pakistan 29°8N 66°31E 40 E5
Kalāteh Iran 36°33N 55°41E 45 B7
Kalāteh-ye Ganj Iran 27°31N 57°55E 45 E8
Kalbā U.A.E. 25°5N 56°22E 45 E8
Kalbarri Australia 27°40S 114°10E 61 E1
Kalbarri △ Australia 27°51S 114°30E 61 E1
Kalburgi = Gulbarga India 17°20N 76°50E 40 L10
Kalce Slovenia 45°54N 14°13E 16 F8
Kale Turkey 37°27N 28°49E 23 F13
Kalegauk Kyun Burma 15°33N 97°35E 38 E1
Kalehe Dem. Rep. of the Congo 2°6S 28°50E 54 C2
Kalema Tanzania 1°12S 31°55E 54 C3
Kalemie Dem. Rep. of the Congo 5°55S 29°9E 54 D2
Kalewa Burma 23°10N 94°15E 41 H19
Kaleybar Iran 38°47N 47°2E 44 B5
Kalgoorlie-Boulder Australia 30°40S 121°22E 61 F3
Kali → India 27°6N 79°55E 43 F8
Kali Sindh → India 25°32N 76°17E 42 G6
Kaliakra, Nos Bulgaria 43°21N 28°30E 23 C13
Kalianda Indonesia 5°50S 105°45E 36 F3
Kalibo Phil. 11°43N 122°22E 37 B6
Kalima Dem. Rep. of the Congo 2°33S 26°32E 54 C2
Kalimantan Indonesia 0°0 114°0E 36 E4
Kalimantan Barat ☐ Indonesia 0°0 110°30E 36 E4
Kalimantan Selatan ☐ Indonesia 2°30S 115°30E 36 E5
Kalimantan Tengah ☐ Indonesia 2°0S 113°30E 36 E4
Kalimantan Timur ☐ Indonesia 1°30N 116°30E 36 D5
Kálimnos Greece 37°0N 27°0E 23 F12
Kalimpong India 27°4N 88°35E 43 F13
Kaliningrad Russia 54°42N 20°32E 9 J19
Kalinkavichy Belarus 52°12N 29°20E 17 B15
Kalinkovichi = Kalinkavichy Belarus 52°12N 29°20E 17 B15
Kaliro Uganda 0°56N 33°30E 54 B3
Kalispell U.S.A. 48°12N 114°19W 76 B6
Kalisz Poland 51°45N 18°8E 17 C10
Kaliua Tanzania 5°5S 31°48E 54 D3
Kalix = Kalixälven → Sweden 65°50N 23°11E 8 D20
Kalix Sweden 65°53N 23°12E 8 D20
Kalixälven → Sweden 65°50N 23°11E 8 D20
Kalka India 30°46N 76°57E 42 D7
Kalkarindji Australia 17°30S 130°47E 60 C5
Kalkaska U.S.A. 44°44N 85°11W 81 C11
Kalkfeld Namibia 20°57S 16°14E 56 C2
Kalkfontein Botswana 22°4S 20°57E 56 C3
Kalkrand Namibia 24°1S 17°35E 56 C2
Kallavesi Finland 62°58N 27°30E 8 E22
Kallsjön Sweden 63°38N 13°0E 8 E15
Kalmar Sweden 56°40N 16°20E 9 H17
Kalmykia ☐ Russia 46°5N 46°1E 19 E8
Kalna India 23°13N 88°25E 43 H13
Kalnai India 22°46N 83°30E 43 H10
Kalocsa Hungary 46°32N 19°0E 17 E10
Kalokhorio Cyprus 34°51N 33°2E 25 E12
Kaloko Dem. Rep. of the Congo 6°47S 25°48E 54 D2
Kalol Gujarat, India 22°37N 73°31E 42 H5
Kalol Gujarat, India 23°15N 72°33E 42 H5
Kalomo Zambia 17°0S 26°30E 55 F2
Kalpi India 26°8N 79°47E 43 F8
Kaltag U.S.A. 64°20N 158°43W 74 C8
Kaltukatjara Australia 24°52S 129°5E 61 D4
Kalu Pakistan 25°5N 67°39E 42 G2
Kaluga Russia 54°35N 36°10E 18 D6
Kalulushi Zambia 12°50S 28°3E 55 E2
Kalumburu Australia 13°55S 126°35E 60 B4
Kalumburu ✪ Australia 14°17S 126°38E 60 B4
Kalush Ukraine 49°3N 24°23E 17 D13
Kalutara Sri Lanka 6°35N 80°0E 40 R12
Kalya Russia 60°15N 59°59E 18 B10
Kalyan India 19°15N 73°9E 40 K8
Kama Dem. Rep. of the Congo 3°30S 27°5E 54 C2
Kama → Russia 55°45N 52°0E 18 C9
Kamachumu Tanzania 1°37S 31°37E 54 C3
Kamaishi Japan 39°16N 141°53E 30 E10
Kamalia Pakistan 30°44N 72°42E 42 D5
Kaman India 27°39N 77°16E 42 F6
Kamanjab Namibia 19°35S 14°51E 56 B2
Kamapanda Zambia 12°5S 24°0E 55 E1
Kamarān Yemen 15°21N 42°35E 47 D3
Kamativi Zimbabwe 18°20S 27°6E 56 B4
Kambalda West Australia 31°10S 121°37E 61 F3
Kambangan, Nusa Indonesia 7°40S 108°10E 37 G13
Kambar Pakistan 27°37N 68°1E 42 F3
Kambarka Russia 56°15N 54°11E 18 C9
Kambolé Zambia 8°47S 30°48E 55 D3
Kambos Cyprus 35°2N 32°44E 25 D11
Kambove Dem. Rep. of the Congo 10°51S 26°33E 55 E2
Kamchatka, Poluostrov Russia 57°0N 160°0E 29 D17
Kamchatka Pen. = Kamchatka, Poluostrov Russia 57°0N 160°0E 29 D17
Kamchiya → Bulgaria 43°4N 27°44E 23 C12
Kame Ruins Zimbabwe 20°7S 28°25E 55 G2
Kameda-Hantō Japan 41°50N 141°0E 30 D10
Kamen Russia 53°50N 81°30E 28 D9
Kamen-Rybolov Russia 44°46N 132°2E 30 B5
Kamenjak, Rt Croatia 44°47N 13°55E 16 F7
Kamenka Russia 65°58N 44°0E 18 A7
Kamenka Bugskaya = Kamyanka-Buzka Ukraine 50°8N 24°16E 17 C13
Kamensk Uralskiy Russia 56°25N 62°2E 28 D7
Kamenskoye Russia 62°45N 165°30E 29 C17
Kameoka Japan 35°0N 135°35E 31 G7
Kamet India 30°55N 79°35E 43 D8
Kamiah U.S.A. 46°14N 116°2W 76 C5
Kamieskroon S. Africa 30°9S 17°56E 56 E2
Kamilukuak L. Canada 62°22N 101°40W 71 A8
Kamin-Kashyrskyy Ukraine 51°39N 24°56E 17 C13
Kamina Dem. Rep. of the Congo 8°45S 25°0E 55 D2
Kaminak L. Canada 62°10N 95°0W 71 A10
Kaminoyama Japan 38°9N 140°17E 30 E10
Kamiros Greece 36°20N 27°56E 25 C9
Kamituga Dem. Rep. of the Congo 3°2S 28°10E 54 C2
Kamla → India 25°35N 86°36E 43 G12
Kamloops Canada 50°40N 120°20W 70 C4
Kamo Japan 37°39N 139°3E 30 F9
Kamoke Pakistan 32°4N 74°4E 42 C6
Kampala Uganda 0°20N 32°30E 54 B3
Kampar Malaysia 4°18N 101°9E 39 K3
Kampar → Indonesia 0°30N 103°8E 36 D2
Kampen Neths. 52°33N 5°53E 15 B5
Kampene Dem. Rep. of the Congo 3°36S 26°40E 52 E5
Kamphaeng Phet Thailand 16°28N 99°30E 38 D2
Kampolombo, L. Zambia 11°37S 29°42E 55 E2
Kampong Cham Cambodia 12°0N 105°30E 39 G5
Kampong Chhnang Cambodia 12°20N 104°35E 39 F5
Kampong Pengerang Malaysia 1°22N 104°7E 39 d
Kampong Punggai Malaysia 1°27N 104°18E 39 d
Kampong Saom Cambodia 10°38N 103°30E 39 G4
Kampong Saom, Chaak Cambodia 10°50N 103°32E 39 G4
Kampong Tanjong Langsat Malaysia 1°28N 104°1E 39 d
Kampong Telok Ramunia Malaysia 1°22N 104°15E 39 d
Kampot Cambodia 10°36N 104°10E 39 G5
Kampuchea = Cambodia ■ Asia 12°15N 105°0E 38 F5
Kampung Air Putih Malaysia 4°15N 103°10E 39 K4
Kampung Jerangau Malaysia 4°50N 103°10E 39 K4
Kampung Raja Malaysia 5°45N 102°35E 39 K4
Kampungbaru = Tolitoli Indonesia 1°5N 120°50E 37 D
Kamrau, Teluk Indonesia 3°30S 133°36E 37 E
Kamsack Canada 51°34N 101°54W 71 C
Kamsar Guinea 10°40N 14°36W 50 F
Kamskoye Vdkhr. Russia 58°41N 56°7E 18 C1
Kamuchawie L. Canada 56°18N 101°59W 71 B
Kamui-Misaki Japan 43°20N 140°21E 30 C1
Kamyanets-Podilskyy Ukraine 48°45N 26°40E 17 D1
Kamyanka-Buzka Ukraine 50°8N 24°16E 17 C1
Kāmyārān Iran 34°47N 46°56E 44 C
Kamyshin Russia 50°10N 45°24E 19 D
Kanaaupscow → Canada 54°2N 76°30W 72 B
Kanab U.S.A. 37°3N 112°32W 77 H
Kanab Cr. → U.S.A. 36°24N 112°38W 77 H
Kanacea Lau Group, Fiji 17°15S 179°6W 59
Kanacea Taveuni, Fiji 16°59S 179°56E 59
Kanaga I. U.S.A. 51°45N 177°22W 74 E
Kanagi Japan 40°54N 140°27E 30 D1
Kanairiktok → Canada 55°2N 60°18W 73 A
Kananga Dem. Rep. of the Congo 5°55S 22°18E 52 F
Kanash Russia 55°30N 47°32E 18 C
Kanaskat U.S.A. 47°19N 121°54W 78 C
Kanastraíon, Ákra = Paliouri, Akra Greece 39°57N 23°45E 23 E1
Kanawha → U.S.A. 38°50N 82°9W 81 F
Kanazawa Japan 36°30N 136°38E 31 F
Kanchanaburi Thailand 14°2N 99°31E 38
Kanchenjunga Nepal 27°50N 88°10E 43 F
Kanchipuram India 12°52N 79°45E 40 N
Kandaghat India 30°59N 77°7E 42 D
Kandahār Afghan. 31°32N 65°43E 40 D
Kandahār ☐ Afghan. 31°0N 65°0E 40
Kandalaksha Russia 67°9N 32°30E 8 C
Kandalakshskiy Zaliv Russia 66°0N 35°0E 18 A
Kandangan Indonesia 2°50S 115°20E 36
Kandanos Greece 35°19N 23°44E 25 D
Kandavu = Kadavu Fiji 19°0S 178°15E 59
Kandavu Passage = Kadavu Passage Fiji 18°45S 178°0E 59
Kandhkot Pakistan 28°16N 69°8E 42
Kandhla India 29°18N 77°19E 42
Kandi Benin 11°7N 2°55E 50
Kandi India 23°58N 88°5E 43 H
Kandiaro Pakistan 27°4N 68°13E 42
Kandla India 23°0N 70°10E 42
Kandos Australia 32°45S 149°58E 63
Kandreho Madag. 17°29S 46°6E 57
Kandy Sri Lanka 7°18N 80°43E 40 R
Kane U.S.A. 41°40N 78°49W 82
Kane Basin Greenland 79°1N 70°0W 69 B
Kang Botswana 23°41S 22°50E 56
Kangān Fārs, Iran 27°50N 52°3E 45
Kangān Hormozgān, Iran 25°48N 57°28E 45
Kangar Malaysia 6°27N 100°12E 39
Kangaroo I. Australia 35°45S 137°0E 63
Kangaroo Mts. Australia 23°29S 141°51E 62
Kangasala Finland 61°28N 24°4E 8 F
Kangāvar Iran 34°40N 48°0E 45
Kangdong N. Korea 39°9N 126°5E 35 E
Kangean, Kepulauan Indonesia 6°55S 115°23E 36
Kangean Is. = Kangean, Kepulauan Indonesia 6°55S 115°23E 36
Kanggye N. Korea 41°0N 126°35E 35 D
Kangikajik Greenland 70°7N 22°0W 4
Kangiqliniq = Rankin Inlet Canada 62°30N 93°0W 68 E
Kangiqsualujjuaq Canada 58°30N 65°59W 69 F
Kangiqsujuaq Canada 61°30N 72°0W 69 E
Kangiqtugaapik = Clyde River Canada 70°30N 68°30W 69 C
Kangirsuk Canada 60°0N 70°0W 69 F
Kangkar Chemaran Malaysia 1°34N 104°12E 39
Kangkar Sungai Tiram Malaysia 1°35N 103°55E 39
Kangkar Teberau Malaysia 1°32N 103°51E 39
Kangping China 42°43N 123°18E 35 C
Kangra India 32°6N 76°16E 42
Kangrinboqe Feng China 31°0N 81°25E 43
Kangto China 27°50N 92°35E 41 F
Kanha △ India 22°15N 80°40E 43
Kanhar → India 24°28N 83°8E 43 G
Kaniama Dem. Rep. of the Congo 7°30S 24°12E 54
Kaniapiskau = Caniapiscau → Canada 56°40N 69°30W 73
Kaniapiskau, L. = Caniapiscau, L. Canada 54°10N 69°55W 73
Kanin, Poluostrov Russia 68°0N 45°0E 18
Kanin Nos, Mys Russia 68°39N 43°32E 18
Kanin Pen. = Kanin, Poluostrov Russia 68°0N 45°0E 18
Kaniva Australia 36°22S 141°18E 63
Kanjut Sar Pakistan 36°7N 75°25E 43
Kankaanpää Finland 61°44N 22°50E 8
Kankakee U.S.A. 41°7N 87°52W 80 E
Kankakee → U.S.A. 41°23N 88°15W 80
Kankan Guinea 10°23N 9°15W 50
Kankendy = Xankändi Azerbaijan 39°52N 46°49E 44
Kanker India 20°10N 81°40E 41 J
Kankroli India 25°4N 73°53E 42
Kanmaw Kyun Burma 11°40N 98°28E 39
Kannapolis U.S.A. 35°30N 80°37W 85 D
Kannauj India 27°3N 79°56E 43
Kannod India 22°45N 76°40E 40 H
Kannur = Cannanore India 11°53N 75°27E 40
Kannyakumari India 8°3N 77°40E 40 C

no Nigeria 12°2N 8°30E 50 F7
n'onji Japan 34°7N 133°39E 31 G6
nowit Malaysia 2°14N 112°20E 36 D4
noya Japan 31°25N 130°50E 31 J5
npetlet Burma 21°10N 93°59E 41 J18
npur India 26°28N 80°20E 43 F9
nsas □ U.S.A. 38°30N 99°0W 80 F4
nsas → U.S.A. 39°7N 94°37W 80 F6
nsas City Kans., U.S.A. 39°7N 94°38W 80 F6
nsas City Mo., U.S.A. 39°6N 94°35W 80 F6
nsenia Dem. Rep. of the Congo 10°20S 26°0E 55 E2
nsk Russia 56°20N 95°37E 29 D10
nsu = Gansu □ China 36°0N 104°0E 34 G3
ntaphor India 22°35N 76°34E 42 H7
ntharalak Thailand 14°39N 104°39E 38 E5
nthi = Contai India 21°54N 87°46E 43 J12
ntli → India 28°20N 75°30E 42 E6
ntō □ Japan 36°15N 139°30E 31 F9
ntō-Sanchi Japan 35°59N 138°50E 31 G9
nturk Ireland 52°11N 8°54W 10 D3
numa Japan 36°34N 139°42E 31 F9
nus Namibia 27°50S 18°39E 56 D2
nye Botswana 24°55S 25°28E 56 C4
nzenze Dem. Rep. of the Congo 10°30S 25°12E 55 E2
nzi, Ras Tanzania 7°1S 39°33E 54 D4
o Tonga 19°40S 175°1W 59 c
o Phara Thailand 8°3N 98°22E 39 a
ohsiung Taiwan 22°35N 120°16E 33 G13
okoveld Namibia 19°15S 14°30E 56 B1
olack Senegal 14°5N 16°8W 50 F2
oshan China 44°38N 124°50E 35 B13
pa'a U.S.A. 22°5N 159°19W 75 L8
padvanj India 23°5N 73°0E 42 H5
pan Armenia 39°18N 46°27E 44 B5
panga Dem. Rep. of the Congo 8°30S 22°40E 52 F4
pchagai = Qapshaghay Kazakhstan 43°51N 77°14E 28 E8
pedo Kenya 1°10N 36°6E 54 B4
pela = Velika Kapela Croatia 45°10N 15°5E 16 F8
pema Dem. Rep. of the Congo 10°45S 28°22E 55 E2
penguria Kenya 1°14N 35°7E 54 B4
pfenberg Austria 47°26N 15°18E 16 E8
piri Mposhi Zambia 13°59S 28°43E 55 E2
pīsā □ Afghan. 35°0N 69°20E 40 B6
piskau → Canada 52°47N 81°55W 72 B3
pit Malaysia 2°0N 112°55E 36 D4
piti I. N.Z. 40°50S 174°56E 59 D5
plan U.S.A. 30°0N 92°17W 84 F8
poe Thailand 9°34N 98°32E 39 H2
poeta Sudan 4°50N 33°35E 51 H12
posvár Hungary 46°25N 17°47E 17 E9
powsin U.S.A. 46°59N 122°13W 78 D4
pps Namibia 22°32S 17°18E 56 C2
psabet Kenya 0°12N 35°6E 54 B4
psan N. Korea 41°4N 128°19E 35 D15
psukas = Marijampolė Lithuania 54°33N 23°19E 9 J20
ptai L. Bangla. 22°40N 92°20E 41 H18
puas → Indonesia 0°25S 109°20E 36 E3
puas Hulu, Pegunungan Malaysia 1°30N 113°30E 36 D4
puas Hulu Ra. = Kapuas Hulu, Pegunungan Malaysia 1°30N 113°30E 36 D4
pulo Dem. Rep. of the Congo 8°18S 29°15E 55 D2
punda Australia 34°20S 138°56E 63 E2
puni N.Z. 39°29S 174°8E 59 C5
purthala India 31°23N 75°25E 42 D6
puskasing Canada 49°25N 82°30W 72 C3
puskasing → Canada 49°49N 82°0W 72 C3
putar, Mt. Australia 30°15S 150°10E 63 E5
putir Kenya 2°5N 35°28E 54 B4
ra Russia 69°10N 65°0E 28 C7
ra Bogaz Gol, Zaliv = Garabogazköl Aylagy Turkmenistan 41°0N 53°30E 19 F9
ra-Kala = Garrygala Turkmenistan 38°31N 56°29E 45 B8
ra Kalpak Republic = Qoraqalpoghistan □ Uzbekistan 43°0N 58°0E 28 E6
ra Kum = Garagum Turkmenistan 39°30N 60°0E 45 B8
ra Sea Russia 75°0N 70°0E 28 B8
rabiğa Turkey 40°23N 27°17E 23 D12
rabük Turkey 41°12N 32°37E 19 F5
raburun Turkey 38°41N 26°28E 23 E12
rabutak = Qarabutaq Kazakhstan 50°0N 60°14E 28 E7
racabey Turkey 40°12N 28°21E 23 D13
racasu Turkey 37°43N 28°35E 23 F13
rachey-Cherkessia □ Russia 43°40N 41°30E 19 F7
rachi Pakistan 24°50N 67°0E 42 G2
rad India 17°15N 74°10E 40 L9
raganda = Qaraghandy Kazakhstan 49°50N 73°10E 32 B3
ragayly = Qaraghayly Kazakhstan 49°26N 76°0E 28 E8
raginskiy, Ostrov Russia 58°45N 164°0E 29 D17
ragiye, Vpadina Kazakhstan 43°27N 51°45E 19 F9
ragiye Depression = Karagiye, Vpadina Kazakhstan 43°27N 51°45E 19 F9
ragola Road India 25°29N 87°23E 43 G12
raikal India 10°59N 79°50E 40 P11
raikkudi India 10°5N 78°45E 40 P11
raj Iran 35°48N 51°0E 45 C6
rajarri ☼ Australia 19°0S 122°30E 60 C3
rak Malaysia 3°25N 102°2E 39 L4
rakalpakstan = Qoraqalpoghistan □ Uzbekistan 43°0N 58°0E 28 E6
rakelong Indonesia 4°35N 126°50E 37 D7
Karakitang Indonesia 3°14N 125°28E 37 D7
Karakol Kyrgyzstan 42°30N 78°20E 32 C4
Karakoram Pass Asia 35°33N 77°50E 43 B7
Karakoram Ra. Pakistan 35°30N 77°0E 43 B7
Karakuwisa Namibia 18°56S 19°40E 56 B2
Karalon Russia 57°5N 115°50E 29 D12
Karama Jordan 31°57N 35°35E 46 D4
Karaman Turkey 37°14N 33°13E 44 B2
Karamay China 45°30N 84°58E 32 B5
Karambu Indonesia 3°53S 116°6E 36 E5
Karamea Bight N.Z. 41°22S 171°40E 59 D3
Karamnasa → India 25°31N 83°52E 43 G10
Karān Si. Arabia 27°43N 49°49E 45 E6
Karand Iran 34°16N 46°15E 44 C5
Karanganyar Indonesia 7°38S 109°37E 37 G13
Karangasem = Amlapura Indonesia 8°27S 115°37E 37 J18
Karanjia India 21°47N 85°58E 43 J11
Karasburg Namibia 28°0S 18°44E 56 D2
Karasino Russia 66°50N 86°50E 28 C9
Karasjok Norway 69°27N 25°30E 8 B21
Karasuk Russia 53°44N 78°2E 28 D8
Karasuyama Japan 36°39N 140°9E 31 F10
Karatau, Khrebet = Qarataū Kazakhstan 43°30N 69°30E 28 E7
Karatax Shan China 35°57N 81°0E 32 D5
Karatsu Japan 33°26N 129°58E 31 H5
Karaul Russia 70°6N 82°15E 28 B9
Karauli India 26°30N 77°4E 42 F7
Karavostasi Cyprus 35°8N 32°50E 25 D11
Karawang Indonesia 6°30S 107°15E 37 G12
Karawanken Europe 46°30N 14°40E 16 E8
Karayazı Turkey 39°41N 42°9E 19 G7
Karazhal = Qarazhal Kazakhstan 48°2N 70°49E 28 E8
Karbalā' Iraq 32°36N 44°3E 44 C5
Karcag Hungary 47°19N 20°57E 17 E11
Karcha → Pakistan 34°45N 76°10E 43 B7
Karchana India 25°17N 81°56E 43 G9
Karditsa Greece 39°23N 21°54E 23 E9
Kärdla Estonia 59°0N 22°45E 9 G20
Kardzhali = Kŭrdzhali Bulgaria 41°38N 25°21E 23 D11
Kareeberge S. Africa 30°59S 21°50E 56 E3
Kareha → India 25°44N 86°21E 43 G12
Kareima Sudan 18°30N 31°49E 51 E12
Karelia □ Russia 65°30N 32°30E 18 A5
Karelian Republic = Karelia □ Russia 65°30N 32°30E 18 A5
Karera India 25°32N 78°9E 42 G8
Kārevāndar Iran 27°53N 60°44E 45 E9
Kargasok Russia 59°3N 80°53E 28 D9
Kargat Russia 55°10N 80°15E 28 D9
Kargi Kenya 2°31N 37°34E 54 B4
Kargil India 34°32N 76°12E 43 B7
Kargopol Russia 61°30N 38°58E 18 B6
Karhal India 27°1N 78°57E 43 F8
Karīān Iran 26°57N 57°14E 45 E8
Karianga Madag. 22°25S 47°22E 57 C8
Kariba Zimbabwe 16°28S 28°50E 55 F2
Kariba, L. Zimbabwe 16°40S 28°25E 55 F2
Kariba Dam Zimbabwe 16°30S 28°35E 55 F2
Kariba Gorge Zambia 16°30S 28°50E 55 F2
Karibib Namibia 22°0S 15°56E 56 C2
Karijini △ Australia 23°8S 118°15E 60 D2
Karimata, Kepulauan Indonesia 1°25S 109°0E 36 E3
Karimata, Selat Indonesia 2°0S 108°40E 36 E3
Karimata Is. = Karimata, Kepulauan Indonesia 1°25S 109°0E 36 E3
Karimnagar India 18°26N 79°10E 40 K11
Karimun, Pulau Indonesia 1°3N 103°22E 39 d
Karimun Kecil, Pulau Indonesia 1°8N 103°22E 39 d
Karimunjawa, Kepulauan Indonesia 5°50S 110°30E 36 F4
Karin Somali Rep. 10°50N 45°52E 47 E4
Karīt Iran 33°29N 56°55E 45 C8
Kariya Japan 34°58N 137°1E 31 G8
Kariyangwe Zimbabwe 18°0S 27°38E 57 B4
Karjala Finland 62°0N 30°25E 8 F24
Karkaralinsk = Qarqaraly Kazakhstan 49°26N 75°30E 32 B4
Karkheh → Iran 31°2N 47°29E 44 D5
Karkinitska Zatoka Ukraine 45°56N 33°0E 19 E5
Karkinitskiy Zaliv = Karkinitska Zatoka Ukraine 45°56N 33°0E 19 E5
Karkuk = Kirkūk Iraq 35°30N 44°21E 44 C5
Karlantijpa North ☼ Australia 19°27S 133°33E 60 C5
Karlantijpa South ☼ Australia 20°31S 133°9E 60 D5
Karleby = Kokkola Finland 63°50N 23°8E 8 E20
Karlovac Croatia 45°31N 15°36E 16 F8
Karlovo Bulgaria 42°38N 24°47E 23 C11
Karlovy Vary Czech Rep. 50°13N 12°51E 16 C7
Karlsbad = Karlovy Vary Czech Rep. 50°13N 12°51E 16 C7
Karlsena, Mys Russia 77°0N 67°42E 28 B7
Karlshamn Sweden 56°10N 14°51E 9 H16
Karlskoga Sweden 59°28N 14°33E 9 G16
Karlskrona Sweden 56°10N 15°35E 9 H16
Karlsruhe Germany 49°0N 8°23E 16 D5
Karlstad Sweden 59°23N 13°30E 9 G15
Karlstad U.S.A. 48°35N 96°31W 80 A5
Karmi'el Israel 32°55N 35°18E 46 C4
Karnak Egypt 25°43N 32°39E 51 C12
Karnal India 29°42N 77°2E 42 E7
Karnali → Nepal 28°45N 81°16E 43 E9
Karnaphuli Res. = Kaptai L. Bangla. 22°40N 92°20E 41 H18
Karnaprayag India 30°16N 79°15E 43 D8
Karnataka □ India 13°15N 77°0E 40 N10
Karnes City U.S.A. 28°53N 97°54W 84 G6
Karnische Alpen Europe 46°36N 13°0E 16 E7
Kärnten □ Austria 46°52N 13°30E 16 E8
Karoi Zimbabwe 16°48S 29°45E 55 F2
Karon, Ao Thailand 7°51N 98°17E 39 a
Karonga Malawi 9°57S 33°55E 55 D3
Karoo △ S. Africa 32°18S 22°27E 56 E3
Karoonda Australia 35°1S 139°59E 63 F2
Karor Pakistan 31°15N 70°59E 42 D4
Karora Sudan 17°44N 38°15E 51 E13
Karpasia Cyprus 35°32N 34°15E 25 D13
Karpathos Greece 35°37N 27°10E 23 G12
Karpinsk Russia 59°45N 60°1E 18 C11
Karpogory Russia 64°0N 44°27E 18 B7
Karpuz Burnu = Apostolos Andreas, C. Cyprus 35°42N 34°35E 25 D13
Karratha Australia 20°44S 116°52E 60 D2
Kars Turkey 40°40N 43°5E 19 F7
Karsakpay Kazakhstan 47°55N 66°40E 28 E7
Karshi = Qarshi Uzbekistan 38°53N 65°48E 28 F7
Karsiyang India 26°56N 88°18E 43 F13
Karsog India 31°23N 77°12E 42 D7
Kartala Comoros Is. 11°45S 43°21E 53 a
Kartaly Russia 53°3N 60°40E 28 D7
Kartapur India 31°27N 75°32E 42 D6
Karthaus U.S.A. 41°8N 78°9W 82 E6
Karufa Indonesia 3°50S 133°20E 37 E8
Karuma ⌂ Uganda 2°5N 32°15E 54 B3
Karumba Australia 17°31S 140°50E 62 B3
Karumo Tanzania 2°25S 32°50E 54 C3
Karumwa Tanzania 3°12S 32°38E 54 C3
Kārūn → Iran 30°26N 48°10E 45 D6
Karungu Kenya 0°50S 34°10E 54 C3
Karviná Czech Rep. 49°53N 18°31E 17 D10
Karwan → India 27°26N 78°4E 42 F8
Karwar India 14°55N 74°13E 40 M9
Karwi India 25°12N 80°57E 43 G9
Karymskoye Russia 51°36N 114°21E 29 D12
Kasache Malawi 13°25S 34°20E 55 E3
Kasai → Dem. Rep. of the Congo 3°30S 16°10E 52 E3
Kasai-Oriental □ Dem. Rep. of the Congo 5°0S 24°30E 54 D1
Kasaji Dem. Rep. of the Congo 10°25S 23°27E 55 E1
Kasama Zambia 10°16S 31°9E 55 E3
Kasan-dong N. Korea 41°18N 126°55E 35 D14
Kasandra Kolpos Greece 40°5N 23°30E 23 D10
Kasane Namibia 17°34S 24°50E 56 B3
Kasanga Tanzania 8°30S 31°10E 55 D3
Kasanka △ Zambia 11°34S 30°15E 55 E3
Kasaragod India 12°30N 74°58E 40 N9
Kasba L. Canada 60°20N 102°10W 71 A8
Kāseh Garān Iran 34°5N 46°2E 44 C5
Kasempa Zambia 13°30S 25°44E 55 E2
Kasenga Dem. Rep. of the Congo 10°20S 28°45E 55 E2
Kasese Uganda 0°13N 30°3E 54 B3
Kasganj India 27°48N 78°42E 43 F8
Kashabowie Canada 48°40N 90°26W 72 C1
Kashaf Iran 35°58N 61°7E 45 C9
Kāshān Iran 34°5N 51°30E 45 C6
Kashechewan Canada 52°18N 81°37W 72 B3
Kashgar = Kashi China 39°30N 76°2E 32 D4
Kashi China 39°30N 76°2E 32 D4
Kashimbo Dem. Rep. of the Congo 11°12S 26°19E 55 E2
Kashipur India 29°15N 79°0E 43 E8
Kashiwazaki Japan 37°22N 138°33E 31 F9
Kashk-e Kohneh Afghan. 34°55N 62°30E 40 B3
Kashkū'iyeh Iran 30°31N 55°40E 45 D7
Kāshmar Iran 35°16N 58°26E 45 C8
Kashmir Asia 34°0N 76°0E 43 C7
Kashmor Pakistan 28°28N 69°32E 42 E3
Kashun Noerh = Gaxun Nur China 42°22N 100°30E 32 C9
Kasiari India 22°8N 87°14E 43 H12
Kasimov Russia 54°55N 41°20E 18 D7
Kasinge Dem. Rep. of the Congo 6°15S 26°58E 54 D2
Kasiruta Indonesia 0°25S 127°12E 37 E7
Kaskaskia → U.S.A. 37°58N 89°57W 80 G9
Kaskattama → Canada 57°3N 90°4W 71 B10
Kaskinen Finland 62°22N 21°15E 8 E19
Kaskö = Kaskinen Finland 62°22N 21°15E 8 E19
Kaslo Canada 49°55N 116°55W 70 D5
Kasmere L. Canada 59°34N 101°10W 71 B8
Kasongo Dem. Rep. of the Congo 4°30S 26°33E 54 C2
Kasongo Lunda Dem. Rep. of the Congo 6°35S 16°49E 52 F3
Kasos Greece 35°20N 26°55E 23 G12
Kassalâ Sudan 15°30N 36°0E 51 E13
Kassel Germany 51°18N 9°26E 16 C5
Kassiopi Greece 39°48N 19°53E 25 A3
Kasson U.S.A. 44°2N 92°45W 80 C7
Kastamonu Turkey 41°25N 33°43E 19 F5
Kasteli Greece 35°29N 23°38E 25 D5
Kastelli Greece 35°12N 25°20E 25 D7
Kasterlee Belgium 51°15N 4°59E 15 C4
Kastoria Greece 40°30N 21°19E 23 D9
Kasulu Tanzania 4°37S 30°5E 54 C3
Kasumi Japan 35°38N 134°38E 31 G7
Kasungu Malawi 13°0S 33°29E 55 E3
Kasungu △ Malawi 12°53S 33°9E 55 E3
Kasur Pakistan 31°5N 74°25E 42 D6
Kata, Ao Thailand 7°48N 98°18E 39 J2
Kata Archanes Greece 35°15N 25°10E 25 D7
Kata Tjuta Australia 25°20S 130°50E 61 E5
Kataba Zambia 16°5S 25°10E 55 F2
Katahdin, Mt. U.S.A. 45°54N 68°56W 81 C19
Kataka = Cuttack India 20°25N 85°57E 41 J14
Katako Kombe Dem. Rep. of the Congo 3°25S 24°20E 54 C1
Katale Tanzania 4°52S 31°7E 54 C3
Katanda Katanga, Dem. Rep. of the Congo 7°52S 24°13E 54 D1
Katanda Nord-Kivu, Dem. Rep. of the Congo 0°55S 29°21E 54 C2
Katanga □ Dem. Rep. of the Congo 8°0S 25°0E 54 D2
Katangi India 21°56N 79°50E 40 J11
Katanning Australia 33°40S 117°33E 61 F2
Katavi △ Tanzania 6°51S 31°10E 54 D3
Katavi Swamp Tanzania 6°50S 31°10E 54 D3
Katerini Greece 40°18N 22°37E 23 D10
Katghora India 22°30N 82°33E 43 H10
Katha Burma 24°10N 96°30E 41 G20
Katherîna, Gebel Egypt 28°30N 33°57E 44 D2
Katherine Australia 14°27S 132°20E 60 B5
Katherine Gorge = Nitmiluk △ Australia 14°6S 132°15E 60 B5
Kathi India 21°47N 74°3E 42 J6
Kathiawar India 22°20N 71°0E 42 H4
Kathikas Cyprus 34°55N 32°25E 25 E11
Kathmandu Nepal 27°45N 85°20E 43 F11
Kathua India 32°23N 75°34E 42 C6
Katihar India 25°34N 87°36E 43 G12
Katima Mulilo Namibia 17°28S 24°13E 56 B3
Katimbira Malawi 12°40S 34°0E 55 E3
Katingan = Mendawai → Indonesia 3°30S 113°0E 36 E4
Katiola Ivory C. 8°10N 5°10W 50 G4
Katiti ☼ Australia 25°1S 131°11E 61 E5
Katmandu = Kathmandu Nepal 27°45N 85°20E 43 F11
Katni = Murwara India 23°46N 80°28E 43 H9
Katni India 23°51N 80°24E 43 H9
Kato Chorio Greece 35°3N 25°47E 25 D7
Kato Korakiana Greece 39°42N 19°45E 25 A3
Káto Pyrgos Cyprus 35°11N 32°41E 25 D11
Katompe Dem. Rep. of the Congo 6°2S 26°23E 54 D2
Katong Singapore 1°18N 103°53E 39 d
Katonga → Uganda 0°34N 31°50E 54 B3
Katoomba Australia 33°41S 150°19E 63 E5
Katowice Poland 50°17N 19°5E 17 C10
Katrine, L. U.K. 56°15N 4°30W 11 E4
Katrineholm Sweden 59°0N 16°12E 9 G17
Katsepe Madag. 15°45S 46°15E 57 B8
Katsina Nigeria 13°0N 7°32E 50 F7
Katsumoto Japan 33°51N 129°42E 31 H4
Katsuura Japan 35°10N 140°20E 31 G10
Katsuyama Japan 36°3N 136°30E 31 F8
Kattavia Greece 35°57N 27°46E 25 D9
Kattegat Denmark 56°40N 11°20E 9 H14
Katumba Dem. Rep. of the Congo 7°40S 25°17E 54 D2
Katwa India 23°30N 88°5E 43 H13
Katwijk Neths. 52°12N 4°24E 15 B4
Kau-Ye Kyun Burma 11°1N 98°32E 39 G2
Kaua'i U.S.A. 22°3N 159°30W 75 L8
Kaudom ⌂ Namibia 18°45S 20°51E 56 B3
Kaufman U.S.A. 32°35N 96°19W 84 E6
Kauhajoki Finland 62°25N 22°10E 8 E20
Kaukauna U.S.A. 44°17N 88°17W 80 C9
Kaukauveld Namibia 20°0S 20°15E 56 C3
Ka'ula I. U.S.A. 21°40N 160°33W 75 L7
Kaunas Lithuania 54°54N 23°54E 9 J20
Kaunia Bangla. 25°46N 89°26E 43 G13
Kautokeino Norway 69°0N 23°4E 8 B20
Kauwapur India 27°31N 82°18E 43 F10
Kavacha Russia 60°16N 169°51E 29 C17
Kavala Greece 40°57N 24°28E 23 D11
Kavalerovo Russia 44°15N 135°4E 30 B7
Kavali India 14°55N 80°1E 40 M12
Kavār Iran 29°11N 52°44E 45 D7
Kavi India 22°12N 72°38E 42 H5
Kavimba Botswana 18°2S 24°38E 56 B3
Kavīr, Dasht-e Iran 34°30N 55°0E 45 C7
Kavīr △ Iran 34°40N 52°0E 45 C7
Kavos Greece 39°23N 20°3E 25 B4
Kaw Fr. Guiana 4°30N 52°15W 93 C8
Kawagama L. Canada 45°18N 78°45W 82 A6
Kawagoe Japan 35°55N 139°29E 31 G9
Kawaguchi Japan 35°52N 139°45E 31 G9
Kawah Ijen △ Indonesia 8°3S 114°14E 37 H15
Kawambwa Zambia 9°48S 29°3E 55 D2
Kawanoe Japan 34°1N 133°34E 31 G6
Kawardha India 22°0N 81°17E 43 J9
Kawasaki Japan 35°31N 139°43E 31 G9
Kawasi Indonesia 1°38S 127°28E 37 E7
Kawawachikamach Canada 54°48N 66°50W 73 B6
Kawerau N.Z. 38°7S 176°42E 59 C6
Kawhia N.Z. 38°4S 174°49E 59 C5
Kawhia Harbour N.Z. 38°5S 174°51E 59 C5
Kawio, Kepulauan Indonesia 4°30N 125°30E 37 D7
Kawthaung Burma 10°5N 98°36E 39 H2
Kawthoolei = Kayin □ Burma 18°0N 97°30E 41 L20
Kawthule = Kayin □ Burma 18°0N 97°30E 41 L20
Kaya Burkina Faso 13°4N 1°10W 50 F5
Kayah □ Burma 19°15N 97°15E 41 K20
Kayak I. U.S.A. 59°56N 144°23W 74 D11
Kayan → Indonesia 2°55N 117°35E 36 D5
Kaycee U.S.A. 43°43N 106°38W 76 E10
Kayeli Indonesia 3°20S 127°10E 37 E7
Kayenta U.S.A. 36°44N 110°15W 77 H8
Kayes Mali 14°25N 11°30W 50 F3
Kayin □ Burma 18°0N 97°30E 41 L20
Kayoa Indonesia 0°1N 127°28E 37 D7
Kayomba Zambia 13°11S 24°2E 55 E1
Kayseri Turkey 38°45N 35°30E 44 B2
Kaysville U.S.A. 41°2N 111°56W 76 F8
Kazachye Russia 70°52N 135°58E 29 B14
Kazakhstan ■ Asia 50°0N 70°0E 28 E8
Kazan Russia 55°50N 49°10E 18 C8
Kazan → Canada 64°2N 95°29W 71 A9
Kazan-Rettō Pac. Oc. 25°0N 141°0E 64 E6
Kazanlŭk Bulgaria 42°38N 25°20E 23 C11
Kazatin = Kozyatyn Ukraine 49°45N 28°50E 17 D15
Kāzerūn Iran 29°38N 51°40E 45 D6
Kazi Magomed = Qazımämmäd Azerbaijan 40°3N 49°0E 45 A6
Kazuma Pan △ Zimbabwe 18°20S 25°48E 55 F2
Kazuno Japan 40°10N 140°45E 30 D10
Kazym → Russia 63°54N 65°50E 28 C7
Kea Greece 37°35N 24°22E 23 F11
Keady U.K. 54°15N 6°42W 10 B5
Kearney U.S.A. 40°42N 99°5W 80 E4
Kearney ☼ Australia 20°10S 128°4E 60 D4
Kearny U.S.A. 33°3N 110°55W 77 K8
Kearsarge, Mt. U.S.A. 43°22N 71°50W 83 C13
Keban Turkey 38°50N 38°50E 19 G6
Keban Barajı Turkey 38°41N 38°33E 44 B3
Kebnekaise Sweden 67°53N 18°33E 8 C18
Kebri Dehar Ethiopia 6°45N 44°17E 47 F3
Kebumen Indonesia 7°42S 109°40E 37 G13
Kechika → Canada 59°41N 127°12W 70 B3
Kecskemét Hungary 46°57N 19°42E 17 E10
Kedah □ Malaysia 5°50N 100°40E 39 K3
Kėdainiai Lithuania 55°15N 24°2E 9 J21
Kedarnath India 30°44N 79°4E 43 D8
Kedgwick Canada 47°40N 67°20W 73 C6
Kediri Indonesia 7°51S 112°1E 37 G15
Kedros Oros Greece 35°11N 24°37E 25 D6
Keeler U.S.A. 36°29N 117°52W 78 J9
Keeley L. Canada 54°54N 108°8W 71 C7
Keeling Is. = Cocos Is. Ind. Oc. 12°10S 96°55E 64 J1
Keelung = Chilung Taiwan 25°3N 121°45E 33 F13
Keene Canada 44°15N 78°10W 82 B6
Keene Calif., U.S.A. 35°13N 118°33W 79 K8
Keene N.H., U.S.A. 42°56N 72°17W 83 D12
Keene N.Y., U.S.A. 44°16N 73°46W 83 B11
Keep River △ Australia 15°49S 129°8E 60 C4
Keeper Hill Ireland 52°45N 8°16W 10 D3
Keerweer, C. Australia 14°0S 141°32E 62 A3
Keeseville U.S.A. 44°29N 73°30W 83 B11
Keetmanshoop Namibia 26°35S 18°8E 56 D2
Keewatin Canada 49°46N 94°34W 71 D10
Keewatin → Canada 56°29N 100°46W 71 B8
Kefalonia Greece 38°15N 20°30E 23 E9
Kefamenanu Indonesia 9°28S 124°29E 37 F6
Kefar Sava Israel 32°11N 34°54E 46 C3
Keffi Nigeria 8°55N 7°43E 50 G7
Keflavík Iceland 64°2N 22°35W 8 D2
Keg River Canada 57°54N 117°55W 70 B5
Kegaska Canada 50°9N 61°18W 73 B7
Kehancha Kenya 1°11S 34°37E 54 C3
Keighley U.K. 53°52N 1°54W 12 D6
Keila Estonia 59°18N 24°25E 9 G21
Keimoes S. Africa 28°41S 20°59E 56 D3
Keitele Finland 63°10N 26°20E 8 E22
Keith Australia 36°6S 140°20E 63 F3
Keith U.K. 57°32N 2°57W 11 D6
Keiyasi Fiji 17°53S 177°46E 59 a
Keizer U.S.A. 44°57N 123°1W 76 D2
Kejimkujik △ Canada 44°25N 65°25W 73 D6
Kejserr Franz Joseph Fd. Greenland 73°30N 24°30W 4 B6
Kekri India 26°0N 75°10E 42 G6
Kelan China 38°43N 111°31E 34 E6
Kelang = Klang Malaysia 3°2N 101°26E 39 L3
Kelantan □ Malaysia 5°10N 102°0E 39 K3
Kelantan → Malaysia 6°13N 102°14E 39 J4
Kelkit → Turkey 40°45N 36°32E 19 F6
Kellerberrin Australia 31°36S 117°38E 61 F2
Kellett, C. Canada 72°0N 126°0W 4 B1
Kelleys I. U.S.A. 41°36N 82°42W 82 E2
Kellogg U.S.A. 47°32N 116°7W 76 C5
Kells = Ceanannus Mor Ireland 53°44N 6°53W 10 C5
Kélo Chad 9°10N 15°45E 51 G9
Kelokedhara Cyprus 34°48N 32°39E 25 E11
Kelowna Canada 49°50N 119°25W 70 D5
Kelsey Creek Australia 20°26S 148°31E 62 b
Kelseyville U.S.A. 38°59N 122°50W 78 G4
Kelso N.Z. 45°54S 169°15E 59 F2
Kelso U.K. 55°36N 2°26W 11 F6
Kelso U.S.A. 46°9N 122°54W 78 D4
Keluang = Kluang Malaysia 2°3N 103°18E 39 L4
Kelvington Canada 52°10N 103°30W 71 C8
Kem Russia 65°0N 34°38E 18 B5
Kem → Russia 64°57N 34°41E 18 B5
Kema Indonesia 1°22N 125°8E 37 D7
Kemah Turkey 39°32N 39°5E 44 B3
Kemaman Malaysia 4°12N 103°18E 39 K4
Kemano Canada 53°35N 128°0W 70 C3
Kemasik Malaysia 4°25N 103°27E 39 K4
Kemerovo Russia 55°20N 86°5E 28 D9
Kemi Finland 65°44N 24°34E 8 D21
Kemi älv = Kemijoki → Finland 65°47N 24°32E 8 D21
Kemi träsk = Kemijärvi Finland 66°43N 27°22E 8 C22
Kemijärvi Finland 66°43N 27°22E 8 C22
Kemijoki → Finland 65°47N 24°32E 8 D21
Kemmerer U.S.A. 41°48N 110°32W 76 F8
Kemmuna = Comino Malta 36°1N 14°20E 25 C1
Kemp, L. U.S.A. 33°46N 99°9W 84 E5
Kemp Land Antarctica 69°0S 55°0E 5 C5
Kempas Malaysia 1°33N 103°42E 39 d
Kempsey Australia 31°1S 152°50E 63 E5
Kempt, L. Canada 47°25N 74°22W 72 C5
Kempten Germany 47°45N 10°17E 16 E6
Kempton Australia 42°31S 147°12E 63 G4
Kemptville Canada 45°0N 75°38W 83 B9
Ken → India 25°13N 80°27E 43 G9
Kenai U.S.A. 60°33N 151°16W 68 E1
Kenai Mts. U.S.A. 60°0N 150°0W 74 D9
Kenamuke Swamp Sudan 5°55N 33°48E 54 A3
Kendai India 22°45N 82°37E 43 H10
Kendal Indonesia 6°56S 110°14E 37 G14
Kendal U.K. 54°20N 2°44W 12 C5
Kendall Australia 31°35S 152°44E 63 E5
Kendall U.S.A. 25°40N 80°19W 85 J14
Kendall → Australia 14°4S 141°35E 62 A3
Kendallville U.S.A. 41°27N 85°16W 81 E11
Kendari Indonesia 3°50S 122°30E 37 E6
Kendawangan Indonesia 2°32S 110°17E 36 E4
Kendi, Pulau Malaysia 5°13N 100°11E 39 c
Kendrapara India 20°35N 86°30E 41 J15
Kendrew S. Africa 32°32S 24°30E 56 E3

Kene Thao *Laos* 17°44N 101°10E **38** D3
Kenedy *U.S.A.* 28°49N 97°51W **84** G6
Kenema *S. Leone* 7°50N 11°14W **50** G3
Keng Kok *Laos* 16°26N 105°12E **38** D5
Keng Tawng *Burma* 20°45N 98°18E **41** J21
Keng Tung *Burma* 21°18N 99°39E **38** B2
Kengeja *Tanzania* 5°26S 39°45E **54** D4
Kenhardt *S. Africa* 29°19S 21°12E **56** D3
Kenitra *Morocco* 34°15N 6°40W **50** B4
Kenli *China* 37°30N 118°20E **35** F10
Kenmare *Ireland* 51°53N 9°36W **10** E2
Kenmare *U.S.A.* 48°41N 102°5W **80** A2
Kenmare River *Ireland* 51°48N 9°51W **10** E2
Kennebago Lake *U.S.A.* 45°4N 70°40W **83** A14
Kennebec *U.S.A.* 43°54N 99°52W **80** D4
Kennebec → *U.S.A.* 43°45N 69°46W **81** D19
Kennebunk *U.S.A.* 43°23N 70°33W **83** C14
Kennedy *Zimbabwe* 18°52S 27°10E **56** B4
Kennedy Channel *Arctic* 80°50N 66°0W **69** A18
Kennedy Ra. *Australia* 24°45S 115°10E **61** D2
Kennedy Range △ *Australia* 24°34S 115°2E **61** D2
Kenner *U.S.A.* 29°59N 90°14W **85** G9
Kennet → *U.K.* 51°27N 0°57W **13** F7
Kenneth Ra. *Australia* 23°50S 117°8E **61** D2
Kennett *U.S.A.* 36°14N 90°3W **80** G8
Kennewick *U.S.A.* 46°12N 119°7W **76** C4
Kennisis Lake *Canada* 45°13N 78°36W **82** A6
Kenogami → *Canada* 51°6N 84°28W **72** B3
Kenora *Canada* 49°47N 94°29W **71** D10
Kenosha *U.S.A.* 42°35N 87°49W **80** D10
Kensington *Canada* 46°28N 63°34W **73** C7
Kent *Ohio, U.S.A.* 41°9N 81°22W **82** E3
Kent *Tex., U.S.A.* 31°4N 104°13W **84** F2
Kent *Wash., U.S.A.* 47°22N 122°14W **78** C4
Kent □ *U.K.* 51°12N 0°40E **13** F8
Kent Group *Australia* 39°30S 147°20E **63** F4
Kent Pen. *Canada* 68°30N 107°0W **68** D10
Kentaū *Kazakhstan* 43°32N 68°36E **28** E7
Kentland *U.S.A.* 40°46N 87°27W **80** E10
Kenton *U.S.A.* 40°39N 83°37W **81** E12
Kentucky □ *U.S.A.* 37°0N 84°0W **81** G11
Kentucky → *U.S.A.* 38°41N 85°11W **81** F11
Kentucky L. *U.S.A.* 37°1N 88°16W **80** G9
Kentville *Canada* 45°6N 64°29W **73** C7
Kentwood *U.S.A.* 30°56N 90°31W **85** F9
Kenya ■ *Africa* 1°0N 38°0E **54** B4
Kenya, Mt. *Kenya* 0°10S 37°18E **54** C4
Kenyir, Tasik *Malaysia* 5°1N 102°54E **39** K4
Keo Neua, Deo *Vietnam* 18°23N 105°10E **38** C5
Keokuk *U.S.A.* 40°24N 91°24W **80** E8
Keoladeo △ *India* 27°0N 77°20E **42** F7
Keonjhargarh *India* 21°28N 85°35E **43** J11
Kep *Cambodia* 10°29N 104°19E **39** G5
Kep *Vietnam* 21°24N 106°16E **38** B6
Kep △ *Cambodia* 10°30N 104°18E **39** G5
Kepala Batas *Malaysia* 5°31N 100°26E **39** c
Kepi *Indonesia* 6°32S 139°19E **37** F9
Kerala □ *India* 11°0N 76°15E **40** P10
Kerama-Rettō *Japan* 26°5N 127°15E **31** L3
Keran *Pakistan* 34°35N 73°59E **43** B5
Kerang *Australia* 35°40S 143°55E **63** F3
Keraudren, C. *Australia* 19°58S 119°45E **60** C2
Kerch *Ukraine* 45°20N 36°20E **19** E6
Kerguelen *Ind. Oc.* 49°15S 69°10E **3** G13
Kerian, Kuala *Malaysia* 5°10N 100°25E **39** c
Kericho *Kenya* 0°22S 35°15E **54** C4
Kerinci *Indonesia* 1°40S 101°15E **36** E2
Kerki = Atamyrat *Turkmenistan* 37°50N 65°12E **28** F7
Kerkrade *Neths.* 50°53N 6°4E **15** D6
Kerkyra *Greece* 39°38N 19°50E **25** A3
Kerkyras, Notio Steno *Greece* 39°34N 20°0E **25** A4
Kermadec Is. *Pac. Oc.* 30°0S 178°15W **58** E11
Kermadec Trench *Pac. Oc.* 30°30S 176°0W **64** L10
Kermān *Iran* 30°15N 57°1E **45** D8
Kerman *U.S.A.* 36°43N 120°4W **78** J6
Kermān □ *Iran* 30°0N 57°0E **45** D8
Kermān, Bīābān-e *Iran* 28°45N 59°45E **45** D8
Kermānshāh *Iran* 34°23N 47°0E **44** C5
Kermānshāh □ *Iran* 34°0N 46°30E **44** C5
Kermit *U.S.A.* 31°52N 103°6W **84** F3
Kern → *U.S.A.* 35°16N 119°18W **79** K7
Kernow = Cornwall □ *U.K.* 50°26N 4°40W **13** G3
Kernville *U.S.A.* 35°45N 118°26W **79** K8
Keroh *Malaysia* 5°43N 101°1E **39** K3
Kerrera *U.K.* 56°24N 5°33W **11** E3
Kerrobert *Canada* 51°56N 109°8W **71** C7
Kerrville *U.S.A.* 30°3N 99°8W **84** F5
Kerry □ *Ireland* 52°7N 9°35W **10** D2
Kerry Hd. *Ireland* 52°25N 9°56W **10** D2
Kerulen = Herlen → *Asia* 48°48N 117°0E **33** B12
Kerzaz *Algeria* 29°29N 1°37W **50** C5
Kesagami → *Canada* 51°40N 79°45W **72** B4
Kesagami L. *Canada* 50°23N 80°15W **72** B3
Keşan *Turkey* 40°49N 26°38E **23** D12
Kesennuma *Japan* 38°54N 141°35E **30** E10
Keshit *Iran* 29°43N 58°17E **45** D8
Kestell *S. Africa* 28°17S 28°42E **57** D4
Kestenga *Russia* 65°50N 31°45E **8** D24
Keswick *U.K.* 54°36N 3°8W **12** C4
Ket → *Russia* 58°55N 81°32E **28** D9
Ketapang *Jawa Timur, Indonesia* 8°9S 114°23E **37** J17
Ketapang *Kalimantan Barat, Indonesia* 1°55S 110°0E **36** E4
Ketchikan *U.S.A.* 55°21N 131°39W **70** B2
Ketchum *U.S.A.* 43°41N 114°22W **76** E6
Ketef, Khalîg Umm el *Egypt* 23°40N 35°35E **44** F2
Keti Bandar *Pakistan* 24°8N 67°27E **42** G2
Ketri *India* 28°1N 75°50E **42** E6
Kętrzyn *Poland* 54°7N 21°22E **17** A11
Kettering *U.K.* 52°24N 0°43W **13** E7
Kettering *U.S.A.* 39°41N 84°10W **81** F11
Kettle → *Canada* 56°40N 89°34W **71** B11
Kettle Falls *U.S.A.* 48°37N 118°3W **76** B4
Kettle Point *Canada* 43°10N 82°1W **82** C2
Kettle Pt. *Canada* 43°13N 82°1W **82** C2
Kettleman City *U.S.A.* 36°1N 119°58W **78** J7
Keuka L. *U.S.A.* 42°30N 77°9W **82** D7
Keuruu *Finland* 62°16N 24°41E **8** E21
Kewanee *U.S.A.* 41°14N 89°56W **80** E9
Kewaunee *U.S.A.* 44°27N 87°31W **80** C10
Keweenaw B. *U.S.A.* 47°0N 88°15W **80** B9
Keweenaw Pen. *U.S.A.* 47°15N 88°15W **80** B9
Keweenaw Pt. *U.S.A.* 47°25N 87°43W **80** B10
Key, L. *Ireland* 54°0N 8°15W **10** C3
Key Lake Mine *Canada* 57°5N 105°32W **71** B7
Key Largo *U.S.A.* 25°5N 80°27W **85** J14
Key West *U.S.A.* 24°33N 81°48W **88** B3
Keynsham *U.K.* 51°24N 2°29W **13** F5
Keyser *U.S.A.* 39°26N 78°59W **81** F14
Kezhma *Russia* 58°59N 101°9E **29** D11
Kezi *Zimbabwe* 20°58S 28°32E **57** C4
Kgalagadi Transfrontier ⌓ *Africa* 25°10S 21°0E **56** D3
Khabarovsk *Russia* 48°30N 135°5E **33** B16
Khabr *Iran* 28°51N 56°22E **45** D8
Khābūr → *Syria* 35°17N 40°35E **44** C4
Khachmas = Xaçmaz *Azerbaijan* 41°31N 48°42E **19** F8
Khachrod *India* 23°25N 75°20E **42** H6
Khadro *Pakistan* 26°11N 68°50E **42** F3
Khadzhilyangar = Dahongliutan *China* 35°45N 79°20E **43** B8
Khaga *India* 25°47N 81°7E **43** G9
Khagaria *India* 25°30N 86°32E **43** G12
Khaipur *Pakistan* 29°34N 72°17E **42** E5
Khair *India* 27°57N 77°46E **42** F7
Khairabad *India* 27°33N 80°47E **43** F9
Khairagarh *India* 21°27N 81°2E **43** J9
Khairpur *Pakistan* 27°32N 68°49E **42** F3
Khairpur Nathan Shah *Pakistan* 27°6N 67°44E **42** F2
Khairwara *India* 23°58N 73°38E **42** H5
Khaisor → *Pakistan* 31°17N 68°59E **42** D3
Khajuri Kach *Pakistan* 32°4N 69°51E **42** C3
Khakassia □ *Russia* 53°0N 90°0E **28** D9
Khakhea *Botswana* 24°48S 23°22E **56** C3
Khalafābād *Iran* 30°54N 49°24E **45** D6
Khalilabad *India* 26°48N 83°5E **43** F10
Khalīlī *Iran* 27°38N 53°17E **45** E7
Khalkhāl *Iran* 37°37N 48°32E **45** B6
Khalkís = Chalkida *Greece* 38°27N 23°42E **23** E10
Khalmer Yu *Russia* 67°58N 65°1E **28** C7
Khalturin *Russia* 58°40N 48°50E **18** C8
Khalūf *Oman* 20°30N 58°13E **47** C6
Kham Keut *Laos* 18°15N 104°43E **38** C5
Khamaria *India* 23°5N 80°48E **43** H9
Khambhaliya *India* 22°14N 69°41E **42** H3
Khambhat *India* 22°23N 72°33E **42** H5
Khambhat, G. of *India* 20°45N 72°30E **40** J8
Khamīr *Iran* 26°57N 55°36E **45** E7
Khamir *Yemen* 16°2N 44°0E **47** D3
Khamīs Mushayṭ *Si. Arabia* 18°18N 42°44E **47** D3
Khamsa *Egypt* 30°27N 32°23E **46** E1
Khan → *Namibia* 22°37S 14°56E **56** C2
Khān Abū Shāmat *Syria* 33°39N 36°53E **46** B5
Khān Azād *Iraq* 33°7N 44°22E **44** C5
Khān Mujiddah *Iraq* 32°21N 43°48E **44** C4
Khān Shaykhūn *Syria* 35°26N 36°38E **44** C3
Khan Tengri, Pik *Asia* 42°12N 80°10E **32** C5
Khān Yūnis *Gaza Strip* 31°21N 34°18E **46** D3
Khanai *Pakistan* 30°30N 67°8E **42** D2
Khānaqīn *Iraq* 34°23N 45°25E **44** C5
Khānbāghī *Iran* 36°10N 55°25E **45** B7
Khandwa *India* 21°49N 76°22E **40** J10
Khandyga *Russia* 62°42N 135°35E **29** C14
Khanewal *Pakistan* 30°20N 71°55E **42** D4
Khangah Dogran *Pakistan* 31°50N 73°37E **42** D5
Khanh Duong *Vietnam* 12°44N 108°44E **38** F7
Khaniá = Chania *Greece* 35°30N 24°4E **25** D6
Khaniadhana *India* 25°1N 78°8E **42** G8
Khanka, L. *Asia* 45°0N 132°24E **30** B6
Khankendy = Xankändi *Azerbaijan* 39°52N 46°49E **44** B5
Khanna *India* 30°42N 76°16E **42** D7
Khanozai *Pakistan* 30°37N 67°19E **42** D2
Khanpur *Pakistan* 28°42N 70°35E **42** E4
Khanty-Mansiysk *Russia* 61°0N 69°0E **28** C7
Khao Khitchakut △ *Thailand* 12°50N 102°10E **38** F4
Khao Laem △ *Thailand* 14°56N 98°31E **38** E2
Khao Laem Res. *Thailand* 14°50N 98°30E **38** E2
Khao Lak △ *Malaysia* 8°38N 98°18E **39** H2
Khao Luang △ *Thailand* 8°34N 99°42E **39** H2
Khao Nam Khang △ *Thailand* 6°32N 100°35E **39** J3
Khao Phlu *Thailand* 9°29N 99°59E **39** b
Khao Pu-Khao Ya △ *Thailand* 7°26N 99°57E **39** J2
Khao Sam Roi Yot △ *Thailand* 12°13N 99°57E **39** F2
Khao Sok △ *Thailand* 8°55N 98°38E **39** H2
Khao Yai △ *Thailand* 14°21N 101°29E **38** E3
Khaoen Si Nakarin △ *Thailand* 14°47N 99°0E **38** E2
Khapalu *Pakistan* 35°10N 76°20E **43** B7
Khapcheranga *Russia* 49°42N 112°24E **29** E12
Khaptao △ *Nepal* 29°20N 81°10E **43** E9
Kharaghoda *India* 23°11N 71°46E **42** H4
Kharagpur *India* 22°20N 87°25E **43** H12
Kharan Kalat *Pakistan* 28°34N 65°21E **40** E4
Kharānaq *Iran* 32°20N 54°45E **45** C7
Kharda *India* 18°40N 75°34E **40** K9
Khardung La *India* 34°20N 77°43E **43** B7
Kharg = Khārk, Jazīreh-ye *Iran* 29°15N 50°28E **45** D6
Khârga, El Wâhât el *Egypt* 25°10N 30°35E **51** C12
Khargon *India* 21°45N 75°40E **40** J9
Khari → *India* 25°54N 74°31E **42** G6
Kharian *Pakistan* 32°49N 73°52E **42** C5
Khārk, Jazīreh-ye *Iran* 29°15N 50°28E **45** D6
Kharkiv *Ukraine* 49°58N 36°20E **19** E6
Kharkov = Kharkiv *Ukraine* 49°58N 36°20E **19** E6
Kharovsk *Russia* 59°56N 40°13E **18** C7
Kharsawangarh *India* 22°48N 85°50E **43** H11
Kharta *Turkey* 40°55N 29°7E **23** D13
Khartoum = El Khartûm *Sudan* 15°31N 32°35E **51** E12
Khartum *Canada* 45°15N 77°55W **82** A7
Khasan *Russia* 42°25N 130°40E **30** C5
Khāsh *Iran* 28°15N 61°15E **45** D9
Khashm el Girba *Sudan* 14°59N 35°58E **51** F13
Khaskovo *Bulgaria* 41°56N 25°30E **23** D11
Khatanga *Russia* 72°0N 102°20E **29** B11
Khatanga → *Russia* 72°55N 106°0E **29** B11
Khatauli *India* 29°17N 77°43E **42** E7
Khatra *India* 22°59N 86°51E **43** H12
Khātūnābād *Iran* 30°1N 55°25E **45** D7
Khatyrka *Russia* 62°3N 175°15E **29** C18
Khavda *India* 23°51N 69°43E **42** H3
Khawr Fakkān *U.A.E.* 25°21N 56°22E **45** E8
Khaybar, Ḥarrat *Si. Arabia* 25°30N 39°45E **44** E4
Khāzimiyah *Iraq* 34°46N 43°37E **44** C4
Khe Bo *Vietnam* 19°8N 104°41E **38** C5
Khe Long *Vietnam* 21°29N 104°46E **38** B5
Khe Sanh *Vietnam* 16°37N 106°45E **38** D6
Khed Brahma *India* 24°7N 73°5E **42** G5
Khekra *India* 28°52N 77°20E **42** E7
Khemarak Phouminville = Krong Kaoh Kong *Cambodia* 11°37N 102°59E **39** G4
Khemisset *Morocco* 33°50N 6°1W **50** B4
Khemmarat *Thailand* 16°10N 105°15E **38** D5
Khenāmān *Iran* 30°27N 56°29E **45** D8
Khenchela *Algeria* 35°28N 7°11E **50** A7
Khersān → *Iran* 31°33N 50°22E **45** D6
Kherson *Ukraine* 46°35N 32°35E **19** E5
Kheta → *Russia* 71°54N 102°6E **29** B11
Khewari *Pakistan* 26°36N 68°52E **42** F3
Khilchipur *India* 24°2N 76°34E **42** G7
Khilok *Russia* 51°30N 110°45E **29** D12
Khios = Chios *Greece* 38°27N 26°9E **23** E12
Khirsadoh *India* 22°11N 78°47E **43** H8
Khiuma = Hiiumaa *Estonia* 58°50N 22°45E **9** G20
Khiva *Uzbekistan* 41°30N 60°18E **28** E7
Khlong Khlung *Thailand* 16°12N 99°43E **38** D2
Khlong Lan △ *Thailand* 16°12N 99°13E **38** D2
Khlong Phanom △ *Thailand* 8°45N 98°38E **39** H2
Khmelnik *Ukraine* 49°33N 27°58E **17** D14
Khmelnytskyy *Ukraine* 49°23N 27°0E **17** D14
Khmelnytskyy □ *Ukraine* 49°50N 26°40E **17** D14
Khmer Rep. = Cambodia ■ *Asia* 12°15N 105°0E **38** F5
Khoai, Hon *Vietnam* 8°26N 104°50E **39** H5
Khodoriv *Ukraine* 49°24N 24°19E **17** D13
Khodzent = Khŭjand *Tajikistan* 40°17N 69°37E **28** E7
Khojak Pass *Afghan.* 30°51N 66°34E **42** D2
Khok Kloi *Thailand* 8°17N 98°19E **39** a
Khok Pho *Thailand* 6°43N 101°6E **39** J3
Khok Samrong *Thailand* 15°3N 100°43E **38** E3
Kholm *Russia* 57°10N 31°15E **18** C5
Kholmsk *Russia* 47°40N 142°5E **29** E15
Khomas Hochland *Namibia* 22°40S 16°0E **56** C2
Khomeyn *Iran* 33°40N 50°7E **45** C6
Khomeynī Shahr *Iran* 32°41N 51°31E **45** C6
Khomodino *Botswana* 22°46S 23°52E **56** C3
Khon Kaen *Thailand* 16°30N 102°47E **38** D4
Khong → *Cambodia* 13°32N 105°58E **38** F5
Khong Sedone *Laos* 15°34N 105°49E **38** E5
Khonuu *Russia* 66°30N 143°12E **29** C15
Khoper → *Russia* 49°30N 42°20E **19** D6
Khorāsān-e Janūb □ *Iran* 32°0N 59°0E **45** C8
Khorāsān-e Razavī □ *Iran* 35°0N 59°0E **45** C8
Khorāsān-e Shemālī □ *Iran* 37°0N 57°0E **45** B8
Khorat = Nakhon Ratchasima *Thailand* 14°59N 102°12E **38** E4
Khorat, Cao Nguyen *Thailand* 15°30N 102°50E **38** E4
Khorixas *Namibia* 20°16S 14°59E **56** C1
Khorramābād *Khorāsān, Iran* 35°6N 57°57E **45** C8
Khorramābād *Lorestān, Iran* 33°30N 48°25E **45** C6
Khorrāmshahr *Iran* 30°29N 48°15E **45** D6
Khorugh *Tajikistan* 37°30N 71°36E **28** F8
Khosravī *Iran* 30°48N 51°28E **45** D6
Khosrowābād *Khuzestān, Iran* 30°10N 48°25E **45** D6
Khosrowābād *Kordestān, Iran* 35°31N 47°38E **44** C5
Khost *Pakistan* 30°13N 67°35E **42** D2
Khosūyeh *Iran* 28°32N 54°26E **45** D7
Khotyn *Ukraine* 48°31N 26°27E **17** D14
Khouribga *Morocco* 32°58N 6°57W **50** B4
Khowst *Afghan.* 33°22N 69°58E **42** C3
Khowst □ *Afghan.* 33°20N 70°0E **40** C6
Khoyniki *Belarus* 51°54N 29°55E **17** C15
Khram, Ko *Thailand* 12°42N 100°47E **38** F3
Khrysokhou B. *Cyprus* 35°6N 32°25E **25** D11
Khuan Wa *Thailand* 7°53N 98°17E **39** a
Khudzhand = Khŭjand *Tajikistan* 40°17N 69°37E **28** E7
Khuff *Si. Arabia* 24°55N 44°53E **44** E5
Khūgīānī *Afghan.* 31°34N 66°32E **42** D2
Khuis *Botswana* 26°40S 21°49E **56** D3
Khuiyala *India* 27°9N 70°25E **42** F4
Khŭjand *Tajikistan* 40°17N 69°37E **28** E7
Khujner *India* 23°47N 76°36E **42** H7
Khukhan *Thailand* 14°42N 104°12E **38** E5
Khulna *Bangla.* 22°45N 89°34E **41** H16
Khulna □ *Bangla.* 22°25N 89°35E **41** H16
Khumago *Botswana* 20°26S 24°32E **56** C3
Khunjerab △ *Pakistan* 36°40N 75°30E **43** A6
Khunjerab Pass = Kinjirap Daban *Asia* 36°40N 75°25E **40** A9
Khūnsorkh *Iran* 27°9N 56°7E **45** E8
Khunti *India* 23°5N 85°17E **43** H11
Khūr *Iran* 32°55N 58°18E **45** C8
Khurai *India* 24°3N 78°23E **42** G8
Khurayṣ *Si. Arabia* 25°6N 48°2E **45** E6
Khurja *India* 28°15N 77°58E **42** E7
Khūrmāl *Iraq* 35°18N 46°2E **44** C5
Khurr, Wādī al *Iraq* 32°3N 43°52E **44** C4
Khūsf *Iran* 32°46N 58°53E **45** C8
Khushab *Pakistan* 32°20N 72°20E **42** C5
Khust *Ukraine* 48°10N 23°18E **17** D12
Khutse ⌓ *Botswana* 23°31S 24°12E **56** C3
Khuzdar *Pakistan* 27°52N 66°30E **42** F2
Khūzestān □ *Iran* 31°0N 49°0E **45** D6
Khvāf *Iran* 34°33N 60°8E **45** C9
Khvājeh *Iran* 38°9N 46°35E **44** B5
Khvānsār *Iran* 29°56N 54°8E **45** D7
Khvor *Iran* 33°45N 55°0E **45** C7
Khvorgū *Iran* 27°34N 56°27E **45** E8
Khvormūj *Iran* 28°40N 51°30E **45** D6
Khvoy *Iran* 38°35N 45°0E **44** B5
Khwae Noi → *Thailand* 14°1N 99°32E **38** E2
Khyber Pass *Afghan.* 34°10N 71°8E **42** B4
Kia *Fiji* 16°16S 179°8E **59** a
Kiabukwa *Dem. Rep. of the Congo* 8°40S 24°48E **55** D1
Kiama *Australia* 34°40S 150°50E **63** E5
Kiamba *Phil.* 6°2N 124°46E **37** C6
Kiambi *Dem. Rep. of the Congo* 7°15S 28°0E **54** D2
Kiambu *Kenya* 1°8S 36°50E **54** C4
Kiangara *Madag.* 17°58S 47°2E **57** B8
Kiangsi = Jiangxi □ *China* 27°30N 116°0E **33** F12
Kiangsu = Jiangsu □ *China* 33°0N 120°0E **35** H11
Kibale △ *Uganda* 0°16N 30°18E **54** B3
Kibanga Port *Uganda* 0°10N 32°58E **54** B3
Kibara *Tanzania* 2°8S 33°30E **54** C3
Kibare, Mts. *Dem. Rep. of the Congo* 8°25S 27°10E **54** D2
Kibira △ *Burundi* 3°6S 30°24E **54** C3
Kiboga *Uganda* 0°58N 31°45E **54** B3
Kibombo *Dem. Rep. of the Congo* 3°57S 25°53E **54** C2
Kibondo *Tanzania* 3°35S 30°45E **54** C3
Kibre Mengist *Ethiopia* 5°54N 38°59E **47** F2
Kıbrıs = Cyprus ■ *Asia* 35°0N 33°0E **25** E12
Kibumbu *Burundi* 3°32S 29°45E **54** C2
Kibungo *Rwanda* 2°10S 30°32E **54** C3
Kibuye *Burundi* 3°39S 29°59E **54** C2
Kibuye *Rwanda* 2°3S 29°21E **54** C2
Kibwesa *Tanzania* 6°30S 29°58E **54** D2
Kibwezi *Kenya* 2°27S 37°57E **54** C4
Kichha *India* 28°53N 79°30E **43** E8
Kichha → *India* 28°41N 79°18E **43** E8
Kichmengskiy Gorodok *Russia* 59°59N 45°48E **18** B8
Kicking Horse Pass *Canada* 51°28N 116°16W **70** C5
Kidal *Mali* 18°26N 1°22E **50** E6
Kidderminster *U.K.* 52°24N 2°15W **13** E5
Kidepo Valley △ *Uganda* 3°52N 33°50E **54** B3
Kidete *Tanzania* 6°25S 37°17E **54** D4
Kidnappers, C. *N.Z.* 39°38S 177°5E **59** C6
Kidsgrove *U.K.* 53°5N 2°14W **12** D5
Kidston *Australia* 18°52S 144°8E **62** B3
Kidugallo *Tanzania* 6°49S 38°15E **54** D4
Kiel *Germany* 54°19N 10°8E **16** A6
Kiel Canal = Nord-Ostsee-Kanal *Germany* 54°12N 9°32E **16** A5
Kielce *Poland* 50°52N 20°42E **17** C11
Kielder Water *U.K.* 55°11N 2°31W **12** B5
Kieler Bucht *Germany* 54°35N 10°25E **16** A6
Kienge *Dem. Rep. of the Congo* 10°30S 27°30E **55** E2
Kiev = Kyyiv *Ukraine* 50°30N 30°28E **17** C16
Kiffa *Mauritania* 16°37N 11°24W **50** E3
Kifrī *Iraq* 34°45N 45°0E **44** C5
Kigali *Rwanda* 1°59S 30°4E **54** C3
Kigarama *Tanzania* 1°1S 31°50E **54** C3
Kigezi ⌓ *Uganda* 0°34S 29°55E **54** C2
Kigoma □ *Tanzania* 5°0S 30°0E **54** D3
Kigoma-Ujiji *Tanzania* 4°55S 29°36E **54** C2
Kigomasha, Ras *Tanzania* 4°58S 38°58E **54** C4
Kigosi ⌓ *Tanzania* 4°10S 31°45E **54** C3
Kığzı *Turkey* 38°18N 43°25E **44** B4
Kihnu *Estonia* 58°9N 24°1E **9** G21
Kii-Sanchi *Japan* 34°20N 136°0E **31** G8
Kii-Suidō *Japan* 33°40N 134°45E **31** H7
Kiira Dam *Uganda* 0°25N 33°0E **54** B3
Kikaiga-Shima *Japan* 28°19N 129°59E **31** K4
Kikiak = Rigolet *Canada* 54°10N 58°23W **73** B8
Kikinda *Serbia* 45°50N 20°30E **23** B9
Kikládhes = Cyclades *Greece* 37°0N 24°30E **23** F11
Kikuchi *Japan* 32°59N 130°47E **31** H5
Kikwit *Dem. Rep. of the Congo* 5°0S 18°45E **52** F3
Kilar *India* 33°6N 76°25E **42** C7
Kilbeggan *Ireland* 53°22N 7°30W **10** C4
Kilbrannan Sd. *U.K.* 55°37N 5°26W **11** F3
Kilbuck Mts. *U.S.A.* 60°36N 159°53W **74** C8
Kilchu *N. Korea* 40°57N 129°25E **35** D15
Kilcoy *Australia* 26°59S 152°30E **63** D5
Kildare *Ireland* 53°9N 6°55W **10** C5
Kildare □ *Ireland* 53°10N 6°50W **10** C5
Kildinstroy *Russia* 68°48N 33°6E **8** B25
Kilfinnane *Ireland* 52°21N 8°28W **10** D3
Kilgarvan *Ireland* 51°54N 9°27W **10** E2
Kilgore *U.S.A.* 32°23N 94°53W **84** E7
Kilgoris *Kenya* 1°0S 34°53E **54** C3
Kilifi *Kenya* 3°40S 39°48E **54** C4
Kilimanjaro *Tanzania* 3°7S 37°20E **54** C4
Kilimanjaro □ *Tanzania* 4°0S 38°0E **54** C4
Kilimanjaro △ *Tanzania* 3°2S 37°19E **54** C4
Kilindini *Kenya* 4°4S 39°40E **54** C4
Kilis *Turkey* 36°42N 37°6E **44** B3
Kiliya *Ukraine* 45°28N 29°16E **17** F15
Kilkee *Ireland* 52°41N 9°39W **10** D2
Kilkeel *U.K.* 54°4N 6°0W **10** B5
Kilkenny *Ireland* 52°39N 7°15W **10** D4
Kilkenny □ *Ireland* 52°35N 7°15W **10** D4
Kilkieran B. *Ireland* 53°20N 9°41W **10** C2
Kilkis *Greece* 40°58N 22°57E **23** D10
Killala *Ireland* 54°13N 9°12W **10** B2
Killala B. *Ireland* 54°16N 9°8W **10** B2
Killaloe *Canada* 45°33N 77°25W **82** A7
Killaloe *Ireland* 52°48N 8°28W **10** D3
Killarney *Australia* 28°20S 152°18E **63** D5
Killarney *Canada* 49°10N 99°40W **71** D9
Killarney *Ireland* 52°4N 9°30W **10** D2
Killarney △ *Ireland* 52°0N 9°33W **10** D2
Killary Harbour *Ireland* 53°38N 9°52W **10** C2
Killdeer *U.S.A.* 47°22N 102°45W **80** B2
Killeen *U.S.A.* 31°7N 97°44W **84** F6
Killin *U.K.* 56°28N 4°19W **11** E4
Killiney *Ireland* 53°15N 6°7W **10** C5
Killington Pk. *U.S.A.* 43°36N 72°49W **83** C12
Killini *Greece* 37°54N 22°25E **23** F10
Killorglin *Ireland* 52°6N 9°47W **10** D2
Killybegs *Ireland* 54°38N 8°26W **10** B3
Kilmarnock *U.K.* 55°37N 4°29W **11** F4
Kilmore *Australia* 37°25S 144°53E **63** F3
Kilmore Quay *Ireland* 52°10N 6°36W **10** D5
Kilondo *Tanzania* 9°45S 34°20E **55** D3
Kilosa *Tanzania* 6°48S 37°0E **54** D4
Kilrush *Ireland* 52°38N 9°29W **10** D2
Kilwa Kisiwani *Tanzania* 8°58S 39°32E **55** D4
Kilwa Kivinje *Tanzania* 8°45S 39°25E **55** D4
Kilwa Masoko *Tanzania* 8°55S 39°30E **55** D4
Kilwinning *U.K.* 55°39N 4°43W **11** F4
Kim *U.S.A.* 37°15N 103°21W **77** H12
Kimaam *Indonesia* 7°58S 138°53E **37** F9
Kimamba *Tanzania* 6°45S 37°10E **54** D4
Kimba *Australia* 33°8S 136°23E **63** E2
Kimball *Nebr., U.S.A.* 41°14N 103°40W **80** E2
Kimball *S. Dak., U.S.A.* 43°45N 98°57W **80** D4
Kimberley *Australia* 16°20S 127°0E **60** C
Kimberley *B.C., Canada* 49°40N 115°59W **70** D
Kimberley *Ont., Canada* 44°23N 80°32W **82** B
Kimberley *S. Africa* 28°43S 24°46E **56** D
Kimberly *U.S.A.* 42°32N 114°22W **76** E
Kimch'aek *N. Korea* 40°40N 129°10E **35** D1
Kimhae = Gimhae *S. Korea* 35°14N 128°53E **35** G1
Kimmirut *Canada* 62°50N 69°50W **69** E1
Kimpese *Dem. Rep. of the Congo* 5°35S 14°26E **52** F
Kimry *Russia* 56°55N 37°15E **18** C
Kinabalu, Gunung *Malaysia* 6°3N 116°14E **36** C
Kinango *Kenya* 4°8S 39°19E **54** C
Kinaskan L. *Canada* 57°38N 130°8W **70** B
Kinbasket L. *Canada* 52°0N 118°10W **70** C
Kincardine *Canada* 44°10N 81°40W **82** B
Kincolith *Canada* 55°0N 129°57W **70** C
Kinda *Dem. Rep. of the Congo* 9°18S 25°4E **55** D
Kinde *U.S.A.* 43°56N 83°0W **82** C
Kinder Scout *U.K.* 53°24N 1°52W **12** D
Kindersley *Canada* 51°30N 109°10W **71** C
Kindia *Guinea* 10°0N 12°52W **50** F
Kindu *Dem. Rep. of the Congo* 2°55S 25°50E **54** C
Kineshma *Russia* 57°30N 42°5E **18** C
Kinesi *Tanzania* 1°25S 33°50E **54** C
King, L. *Australia* 33°10S 119°35E **61** F
King, Mt. *Australia* 25°10S 147°30E **62** D
King Christian I. *Canada* 77°48N 101°40W **69** B1
King City *U.S.A.* 36°13N 121°8W **78** J
King Cr. → *Australia* 24°35S 139°30E **62** C
King Edward → *Australia* 14°14S 126°35E **60** B
King Edward Point *S. Georgia* 54°17S 36°30W **96** G
King Frederick VI Land = Kong Frederik VI.s Kyst *Greenland* 63°0N 43°0W **4** C
King George B. *Falk. Is.* 51°30S 60°30W **96** G
King George I. *Antarctica* 60°0S 60°0W **5** C1
King George Is. *Canada* 57°20N 80°30W **69** F1
King George Sd. *Australia* 35°5S 118°0E **61** G
King I. = Kadan Kyun *Burma* 12°30N 98°20E **38** F
King I. *Australia* 39°50S 144°0E **63** F
King I. *Canada* 52°10N 127°40W **70** C
King Khalid Military City = Madinat al Malik Khālid al Askarīyah *Si. Arabia* 27°54N 45°31E **44** E
King Leopold Ranges *Australia* 17°30S 125°45E **60** C
King of Prussia *U.S.A.* 40°5N 75°23W **83** F
King Sd. *Australia* 16°50S 123°20E **60** C
King Sejong *Antarctica* 62°30S 58°0W **5** C1
King William I. *Canada* 69°10N 97°25W **68** D1
King William's Town *S. Africa* 32°51S 27°22E **56** E
Kingaok = Bathurst Inlet *Canada* 66°50N 108°1W **68** D1
Kingaroy *Australia* 26°32S 151°51E **63** D
Kingfisher *U.S.A.* 35°52N 97°56W **84**
Kingirbān *Iraq* 34°40N 44°54E **44** C
Kingisepp = Kuressaare *Estonia* 58°15N 22°30E **9** G2
Kingisepp *Russia* 59°25N 28°40E **9** G2
Kingman *Ariz., U.S.A.* 35°12N 114°4W **79** K1
Kingman *Kans., U.S.A.* 37°39N 98°7W **80**
Kingoonya *Australia* 30°55S 135°19E **63**
Kingri *Pakistan* 30°27N 69°49E **42** D
Kings → *U.S.A.* 36°3N 119°50W **78**
Kings Canyon *Australia* 24°15S 131°34E **60** D
Kings Canyon △ *U.S.A.* 36°50N 118°40W **78**
King's Lynn *U.K.* 52°45N 0°24E **12**
Kings Park *U.S.A.* 40°53N 73°16W **83** F

ngs Peak U.S.A. 40°46N 110°23W 76 F8
ngsbridge U.K. 50°17N 3°47W 13 G4
ngsburg U.S.A. 36°31N 119°33W 78 J7
ngscote Australia 35°40S 137°38E 63 F2
ngscourt Ireland 53°55N 6°48W 10 C5
ngsford U.S.A. 45°48N 88°4W 80 C9
ngsland U.S.A. 30°48N 81°41W 85 F14
ngsport U.S.A. 36°33N 82°33W 85 C13
ngston Canada 44°14N 76°30W 83 B8
ngston Jamaica 18°0N 76°50W 88 a
ngston N.Z. 45°20S 168°43E 59 F2
ngston N.H., U.S.A. 42°56N 71°3W 83 D13
ngston N.Y., U.S.A. 41°56N 73°59W 83 E11
ngston Pa., U.S.A. 41°16N 75°54W 83 E9
ngston R.I., U.S.A. 41°29N 71°30W 83 E13
ngston Pk. U.S.A. 35°44N 115°55W 79 K11
ngston South East Australia 36°51S 139°55E 63 F2
ngston upon Hull U.K. 53°45N 0°21W 12 D7
ngston upon Hull □ U.K. 53°45N 0°21W 12 D7
ngston-upon-Thames □ U.K. 51°24N 0°17W 13 F7
ngstown St. Vincent 13°10N 61°10W 89 D7
ngstree U.S.A. 33°40N 79°50W 85 E15
ngsville Canada 42°2N 82°45W 82 D2
ngsville U.S.A. 27°31N 97°52W 84 H6
ngussie U.K. 57°6N 4°2W 11 D4
ngwood U.S.A. 30°2N 95°16W 84 F7
nık Turkey 39°6N 27°24E 23 E12
nistino Canada 52°57N 105°2W 71 C7
njirap Daban Asia 36°40N 75°25E 40 A9
nkala Congo 4°18S 14°49E 52 E2
nki □ Japan 33°45N 136°0E 31 H8
nleith N.Z. 38°20S 175°56E 59 C5
nlochleven U.K. 56°43N 5°0W 11 E4
nmount Canada 44°48N 78°45W 82 B6
nna Sweden 57°32N 12°42E 9 H15
nnairds Hd. U.K. 57°43N 2°1W 11 D6
nngait Canada 64°14N 76°32W 69 E16
no Nuevo Mexico 28°52N 112°3W 86 B2
noje → Canada 52°8N 81°25W 72 B3
nomoto Japan 35°30N 136°13E 31 G8
noni Uganda 0°41S 30°28E 54 C3
noosao Canada 57°5N 102°1W 71 B8
nross U.K. 56°13N 3°25W 11 E5
nsale Ireland 51°42N 8°31W 10 E3
nsale, Old Hd. of Ireland 51°37N 8°33W 10 E3
nsha = Chang Jiang → China 31°48N 121°10E 33 E13
nshasa Dem. Rep. of the Congo 4°20S 15°15E 52 E3
nsley U.S.A. 37°55N 99°25W 80 G4
nsman U.S.A. 41°26N 80°35W 82 E4
nston U.S.A. 35°16N 77°35W 85 D16
ntamani Indonesia 8°14S 115°19E 37 J18
ntore U.K. 57°14N 2°20W 11 D6
ntore Ra. Australia 23°15S 128°47E 60 D4
ntyre U.K. 55°30N 5°35W 11 F3
ntyre, Mull of U.K. 55°17N 5°47W 11 F3
nushseo → Canada 55°15N 83°45W 72 A3
nuso Canada 55°20N 115°25W 70 B5
nvarra Ireland 53°8N 8°56W 10 C3
nyangiri Tanzania 4°25S 34°37E 54 C3
nyeti Sudan 3°57N 32°54E 54 B3
nzua U.S.A. 41°52N 78°58W 82 E6
osk Canada 46°6N 78°53W 72 C4
owa Kans., U.S.A. 37°1N 98°29W 80 G4
owa Okla., U.S.A. 34°43N 95°54W 84 D7
pahigan L. Canada 55°20N 101°55W 71 B8
panga Tanzania 6°15S 35°20E 54 D4
parissia Greece 37°15N 21°40E 23 F9
parissiakos Kolpos Greece 37°25N 21°25E 23 F9
pawa, L. Canada 46°50N 79°0W 72 C4
pembawe Tanzania 7°38S 33°27E 54 D3
pengere Ra. Tanzania 9°12S 34°15E 55 D3
pili Tanzania 7°28S 30°32E 54 D3
pini Kenya 2°30S 40°32E 54 C5
pling Canada 50°6N 102°38W 71 C8
ppure Ireland 53°11N 6°21W 10 C5
pushi Dem. Rep. of the Congo 11°48S 27°12E 55 E2
ranomena Madag. 18°17S 46°2E 57 B8
rensk Russia 57°50N 107°55E 29 D11
rghiz Range Asia 42°40N 73°40E 32 C3
rghizia = Kyrgyzstan ■ Asia 42°0N 75°0E 28 E8
rgiziya Steppe Eurasia 50°0N 55°0E 19 E10
ribati ■ Pac. Oc. 3°0S 180°0E 58 B10
rikkale Turkey 39°51N 33°32E 19 G5
rillov Russia 59°49N 38°24E 18 C6
rin = Jilin China 43°44N 126°30E 35 C14
rinyaga = Kenya, Mt. Kenya 0°10S 37°18E 54 C4
rirom △ Cambodia 11°18N 104°5E 39 G5
rishima Yaku △ Japan 31°24N 130°50E 31 J5
ritimati Kiribati 1°58N 157°27W 65 G12
rkby U.K. 53°30N 2°54W 12 D5
rkby-in-Ashfield U.K. 53°6N 1°14W 12 D6
rkby Lonsdale U.K. 54°12N 2°36W 12 C5
rkby Stephen U.K. 54°29N 2°21W 12 C5
rkcaldy U.K. 56°7N 3°9W 11 E5
rkcudbright U.K. 54°50N 4°2W 11 G4
rkenes Norway 69°40N 30°5E 8 B24
rkfield Canada 44°34N 78°59W 82 B6
rkintilloch U.K. 55°56N 4°8W 11 F4
rkjubæjarklaustur Iceland 63°47N 18°4W 8 E4
rkkonummi Finland 60°8N 24°26E 9 F21
rkland U.S.A. 47°40N 122°12W 78 C4
rkland Lake Canada 48°9N 80°2W 72 C3
rklareli Turkey 41°44N 27°15E 23 D12
rksville U.S.A. 40°12N 92°35W 80 E7
rkūk Iraq 35°30N 44°21E 44 C5
rkwall U.K. 58°59N 2°58W 11 C6
rkwood S. Africa 33°22S 25°15E 56 E4
rov Russia 58°35N 49°40E 18 C8
Kirovabad = Gäncä Azerbaijan 40°45N 46°20E 19 F8
Kirovograd = Kirovohrad Ukraine 48°35N 32°20E 19 E5
Kirovohrad Ukraine 48°35N 32°20E 19 E5
Kirovsk Russia 67°32N 33°41E 18 A5
Kirovskiy Kamchatka, Russia 54°27N 155°42E 29 D16
Kirovskiy Primorsk, Russia 45°7N 133°30E 30 B6
Kirriemuir U.K. 56°41N 3°1W 11 E5
Kirsanov Russia 52°35N 42°40E 18 D7
Kırşehir Turkey 39°14N 34°5E 44 B2
Kirthar △ Pakistan 25°45N 67°30E 42 F2
Kirthar Range Pakistan 27°0N 67°0E 42 F2
Kirtland U.S.A. 36°44N 108°21W 77 H9
Kiruna Sweden 67°52N 20°15E 8 C19
Kirundu Dem. Rep. of the Congo 0°50S 25°35E 54 C2
Kiryū Japan 36°24N 139°20E 31 F9
Kisaga Tanzania 4°30S 34°23E 54 C3
Kisanga Dem. Rep. of the Congo 2°30N 26°35E 54 B2
Kisangani Dem. Rep. of the Congo 0°35N 25°15E 54 B2
Kisar Indonesia 8°5S 127°10E 37 F7
Kisarawe Tanzania 6°53S 39°0E 54 D4
Kisarazu Japan 35°23N 139°55E 31 G9
Kiselevsk Russia 54°5N 86°39E 28 D9
Kish = Qeys Iran 26°32N 53°58E 45 E7
Kishanganga → Pakistan 34°18N 73°28E 43 B5
Kishanganj India 26°3N 88°14E 43 F13
Kishangarh Raj., India 26°34N 74°52E 42 F6
Kishangarh Raj., India 27°50N 70°30E 42 F4
Kishinev = Chişinău Moldova 47°2N 28°50E 17 E15
Kishiwada Japan 34°28N 135°22E 31 G7
Kishtwar India 33°20N 75°48E 42 C6
Kisigo ⌂ Tanzania 6°27S 34°17E 54 D3
Kisii Kenya 0°40S 34°45E 54 C3
Kisiju Tanzania 7°23S 39°19E 54 D4
Kisizi Uganda 1°0S 29°58E 54 C2
Kiska I. U.S.A. 51°59N 177°30E 74 E3
Kiskőrös Hungary 46°37N 19°20E 17 E10
Kiskunfélegyháza Hungary 46°42N 19°53E 17 E10
Kiskunhalas Hungary 46°28N 19°37E 17 E10
Kislovodsk Russia 43°50N 42°45E 19 F7
Kismaayo Somali Rep. 0°22S 42°32E 47 H3
Kiso-Gawa → Japan 35°20N 136°45E 31 G8
Kiso-Sammyaku Japan 35°45N 137°45E 31 G8
Kisofukushima Japan 35°52N 137°43E 31 G8
Kisoro Uganda 1°17S 29°48E 54 C2
Kissamos = Kasteli Greece 35°29N 23°38E 25 D5
Kissamos, Kolpos Greece 35°30N 23°38E 25 D5
Kissidougou Guinea 9°5N 10°5W 50 G3
Kissimmee U.S.A. 28°18N 81°24W 85 G14
Kissimmee → U.S.A. 27°9N 80°52W 85 H14
Kississing L. Canada 55°10N 101°20W 71 B8
Kissónerga Cyprus 34°49N 32°24E 25 E11
Kisumu Kenya 0°3S 34°45E 54 C3
Kiswani Tanzania 4°5S 37°57E 54 C4
Kiswere Tanzania 9°27S 39°30E 55 D4
Kit Carson U.S.A. 38°46N 102°48W 76 G12
Kita Mali 13°5N 9°25W 50 F4
Kitaibaraki Japan 36°50N 140°45E 31 F10
Kitakami Japan 39°20N 141°10E 30 E10
Kitakami-Gawa → Japan 38°25N 141°19E 30 E10
Kitakami-Sammyaku Japan 39°30N 141°30E 30 E10
Kitakata Japan 37°39N 139°52E 30 F9
Kitakyūshū Japan 33°50N 130°50E 31 H5
Kitale Kenya 1°0N 35°0E 54 B4
Kitami Japan 43°48N 143°54E 30 C11
Kitami-Sammyaku Japan 44°22N 142°43E 30 B11
Kitangiri, L. Tanzania 4°5S 34°20E 54 C3
Kitaya Tanzania 10°38S 40°8E 55 E5
Kitchener Canada 43°27N 80°29W 82 C4
Kitee Finland 62°5N 30°8E 8 E24
Kitega = Gitega Burundi 3°26S 29°56E 54 C2
Kitengo Dem. Rep. of the Congo 7°26S 24°8E 54 D1
Kitgum Uganda 3°17N 32°52E 54 B3
Kíthira = Kythira Greece 36°8N 23°0E 23 F10
Kithnos Greece 37°26N 24°27E 23 F11
Kiti Cyprus 34°50N 33°34E 25 E12
Kiti, C. Cyprus 34°48N 33°36E 25 E12
Kitimat Canada 54°3N 128°38W 70 C3
Kitinen → Finland 67°14N 27°27E 8 C22
Kitsuki Japan 33°25N 131°37E 31 H5
Kittakittaooloo, L. Australia 28°3S 138°14E 63 D2
Kittanning U.S.A. 40°49N 79°31W 82 F5
Kittatinny Mt. U.S.A. 41°19N 74°39W 83 F10
Kittery U.S.A. 43°5N 70°45W 83 C14
Kittilä Finland 67°40N 24°51E 8 C21
Kitui Kenya 1°17S 38°0E 54 C4
Kitwanga Canada 55°6N 128°4W 70 B3
Kitwe Zambia 12°54S 28°13E 55 E2
Kiunga Marine ⌂ Kenya 2°1S 41°14E 54 C5
Kivarli India 24°33N 72°46E 42 G5
Kivertsi Ukraine 50°50N 25°28E 17 C13
Kividhes Cyprus 34°46N 32°51E 25 E11
Kivu, L. Dem. Rep. of the Congo 1°48S 29°0E 54 C2
Kiwirrkurra Australia 22°49S 127°45E 60 D4
Kiwirrkurra ✪ Australia 22°49S 127°45E 60 D4
Kiyev = Kyyiv Ukraine 50°30N 30°28E 17 C16
Kiyevskoye Vdkhr. = Kyyivske Vdskh. Ukraine 51°0N 30°25E 17 C16
Kizel Russia 59°3N 57°40E 18 C10
Kiziba-Baluba ⌂ Congo 11°50S 28°0E 55 E2
Kiziguru Rwanda 1°46S 30°23E 54 C3
Kızıl Irmak → Turkey 41°44N 35°58E 19 F6
Kizil Jilga China 35°26N 78°50E 43 B8
Kızıldağ △ Turkey 38°5N 31°20E 44 B1
Kızıltepe Turkey 37°12N 40°35E 44 B4
Kizimkazi Tanzania 6°28S 39°30E 54 D4
Kizlyar Russia 43°51N 46°40E 19 F8
Kizyl-Arvat = Serdar Turkmenistan 39°4N 56°23E 45 B8
Kjölur Iceland 64°50N 19°25W 8 D4
Kladno Czech Rep. 50°10N 14°7E 16 C8
Klaeng Thailand 12°47N 101°39E 38 F3
Klagenfurt Austria 46°38N 14°20E 16 E8
Klaipėda Lithuania 55°43N 21°10E 9 J19
Klaksvík Færoe Is. 62°14N 6°35W 8 E9
Klamath → U.S.A. 41°33N 124°5W 76 F1
Klamath Falls U.S.A. 42°13N 121°46W 76 E3
Klamath Mts. U.S.A. 41°50N 123°20W 76 F2
Klamono Indonesia 1°8S 131°30E 37 E8
Klang Malaysia 3°2N 101°26E 39 L3
Klappan → Canada 58°0N 129°43W 70 B3
Klarälven → Sweden 59°23N 13°32E 9 G15
Klatovy Czech Rep. 49°23N 13°18E 16 D7
Klawer S. Africa 31°44S 18°36E 56 E2
Klazienaveen Neths. 52°44N 7°0E 15 B6
Kleena Kleene Canada 52°0N 124°59W 70 C4
Klein-Karas Namibia 27°33S 18°7E 56 D2
Klerksdorp S. Africa 26°53S 26°38E 56 D4
Kletsk = Klyetsk Belarus 53°5N 26°45E 17 B14
Kletskiy Russia 49°16N 43°11E 19 E7
Klickitat U.S.A. 45°49N 121°9W 76 D3
Klickitat → U.S.A. 45°42N 121°17W 78 E5
Klidhes Cyprus 35°42N 34°36E 25 D13
Klinaklini → Canada 51°21N 125°40W 70 C3
Klip → S. Africa 27°3S 29°3E 57 D4
Klipdale S. Africa 34°19S 19°57E 56 E2
Klipplaat S. Africa 33°1S 24°22E 56 E3
Kłodzko Poland 50°28N 16°38E 17 C9
Klondike Canada 64°0N 139°26W 68 E4
Klong Wang Chao △ Thailand 16°20N 99°9E 38 D2
Klouto Togo 6°57N 0°44E 50 G6
Kluane △ Canada 60°45N 139°30W 70 A1
Kluane L. Canada 61°15N 138°40W 68 E4
Kluang Malaysia 2°3N 103°18E 39 L4
Kluczbork Poland 50°58N 18°12E 17 C10
Klukwan U.S.A. 59°24N 135°54W 70 B1
Klungkung Indonesia 8°32S 115°24E 37 K18
Klyetsk Belarus 53°5N 26°45E 17 B14
Klyuchevskaya, Gora Russia 55°50N 160°30E 29 D17
Klyuchi Russia 56°18N 160°51E 29 D17
Knaresborough U.K. 54°1N 1°28W 12 C6
Knee L. Man., Canada 55°3N 94°45W 72 A1
Knee L. Sask., Canada 55°51N 107°0W 71 B7
Knight Inlet Canada 50°45N 125°40W 70 C3
Knighton U.K. 52°21N 3°3W 13 E4
Knights Ferry U.S.A. 37°50N 120°40W 78 H6
Knights Landing U.S.A. 38°48N 121°43W 78 G5
Knob, C. Australia 34°32S 119°16E 61 F2
Knock Ireland 53°48N 8°55W 10 C3
Knockmealdown Mts. Ireland 52°14N 7°56W 10 D4
Knokke-Heist Belgium 51°21N 3°17E 15 C3
Knossos Greece 35°16N 25°10E 25 D7
Knowlton Canada 45°13N 72°31W 83 A12
Knox U.S.A. 41°18N 86°37W 80 E10
Knox Coast Antarctica 66°30S 108°0E 5 C8
Knoxville Iowa, U.S.A. 41°19N 93°6W 80 E7
Knoxville Pa., U.S.A. 41°57N 77°27W 82 E7
Knoxville Tenn., U.S.A. 35°58N 83°55W 85 D13
Knud Rasmussen Land Greenland 78°0N 60°0W 69 B20
Knysna S. Africa 34°2S 23°2E 56 E3
Ko Kha Thailand 18°11N 99°24E 38 C2
Ko Tarutao △ Thailand 6°31N 99°26E 39 J2
Ko Yao Thailand 8°7N 98°35E 39 a
Koartac = Quaqtaq Canada 60°55N 69°40W 69 E18
Koba Indonesia 6°37S 134°37E 37 F8
Kobarid Slovenia 46°15N 13°30E 16 E7
Kobayashi Japan 31°56N 130°59E 31 J5
Kōbe Japan 34°41N 135°13E 31 G7
København Denmark 55°40N 12°26E 9 J15
Kōbi-Sho Japan 25°56N 123°41E 31 M1
Koblenz Germany 50°21N 7°36E 16 C4
Kobryn Belarus 52°15N 24°22E 17 B13
Kobuk → U.S.A. 66°54N 160°38W 74 B7
Kocaeli Turkey 40°45N 29°50E 19 F4
Kočani Macedonia 41°55N 22°25E 23 D10
Koch Bihar India 26°22N 89°29E 41 F16
Kochas India 25°15N 83°56E 43 G10
Kochi = Cochin India 9°58N 76°20E 40 Q10
Kōchi Japan 33°30N 133°35E 31 H6
Kōchi □ Japan 33°40N 133°30E 31 H6
Kochiu = Gejiu China 23°20N 103°10E 32 G9
Kodarma India 24°28N 85°36E 43 G11
Kodi Indonesia 9°33S 119°0E 60 A2
Kodiak U.S.A. 57°47N 152°24W 74 D9
Kodiak I. U.S.A. 57°30N 152°45W 74 D9
Kodinar India 20°46N 70°46E 42 J4
Kodinsk Russia 58°40N 99°10E 29 D10
Koedoesberge S. Africa 32°40S 20°11E 56 E3
Koes Namibia 26°0S 19°15E 56 D2
Koffiefontein S. Africa 29°30S 25°0E 56 D4
Kofiau Indonesia 1°11S 129°50E 37 E7
Koforidua Ghana 6°3N 0°17W 50 G5
Kōfu Japan 35°40N 138°30E 31 G9
Koga Japan 36°11N 139°43E 31 F9
Kogaluc → Canada 56°12N 61°44W 73 A7
Køge Denmark 55°27N 12°11E 9 J15
Kogon Uzbekistan 39°43N 64°33E 28 F7
Koh-i-Khurd Afghan. 33°30N 65°59E 42 C1
Koh-i-Maran Pakistan 29°18N 66°50E 42 E2
Koh Ker Cambodia 13°47N 104°32E 38 F5
Koh Wai Thailand 11°54N 102°25E 39 H4
Kohat Pakistan 33°40N 71°29E 42 C4
Kohima India 25°35N 94°10E 41 G19
Kohkīlūyeh va Būyer Aḥmadi □ Iran 31°30N 50°30E 45 D6
Kohler Ra. Antarctica 77°0S 110°0W 5 D15
Kohlu Pakistan 29°54N 69°15E 42 E3
Kohtla-Järve Estonia 59°20N 27°20E 9 G22
Koi Sanjaq Iraq 36°5N 44°38E 44 B5
Koillismaa Finland 65°44N 28°36E 8 D23
Koin N. Korea 40°28N 126°18E 35 D14
Kojō N. Korea 38°58N 127°58E 35 E14
Kojonup Australia 33°48S 117°10E 61 F2
Kojūr Iran 36°23N 51°43E 45 B6
Kokand = Qŭqon Uzbekistan 40°31N 70°56E 28 E8
Kokas Indonesia 2°42S 132°26E 37 E8
Kokchetav = Kökshetaū Kazakhstan 53°20N 69°25E 28 D7
Kokemäenjoki → Finland 61°32N 21°44E 8 F19
Kokkola Finland 63°50N 23°8E 8 E20
Koko Kyunzu Burma 14°7N 93°22E 41 M18
Kokomo U.S.A. 40°29N 86°8W 80 E10
Kokopo Papua N. G. 4°22S 152°19E 64 H7
Koksan N. Korea 38°46N 126°40E 35 E14
Kökshetaū Kazakhstan 53°20N 69°25E 28 D7
Koksoak → Canada 58°30N 68°10W 69 F18
Kokstad S. Africa 30°32S 29°29E 57 E4
Kokubu Japan 31°44N 130°46E 31 J5
Kola Indonesia 5°35S 134°30E 37 F8
Kola Russia 68°45N 33°8E 8 B25
Kola Pen. = Kolskiy Poluostrov Russia 67°30N 38°0E 18 A6
Kolachi → Pakistan 27°8N 67°2E 42 F2
Kolahoi India 34°12N 75°22E 43 B6
Kolaka Indonesia 4°3S 121°46E 37 E6
Kolar India 13°12N 78°15E 40 N11
Kolar Gold Fields India 12°58N 78°16E 40 N11
Kolaras India 25°14N 77°36E 42 G6
Kolari Finland 67°20N 23°48E 8 C20
Kolayat India 27°50N 72°50E 42 F5
Kolda Senegal 12°55N 14°57W 50 F3
Kolding Denmark 55°30N 9°29E 9 J13
Kolepom = Dolak, Pulau Indonesia 8°0S 138°30E 37 F9
Kolguyev, Ostrov Russia 69°20N 48°30E 18 A8
Kolhapur India 16°43N 74°15E 40 L9
Kolimbia Greece 36°15N 28°10E 25 C10
Kolín Czech Rep. 50°2N 15°9E 16 C8
Kolkas rags Latvia 57°46N 22°37E 9 H20
Kolkata India 22°34N 88°21E 43 H13
Kollam = Quilon India 8°50N 76°38E 40 Q10
Kollum Neths. 53°17N 6°10E 15 A6
Kolmanskop Namibia 26°45S 15°14E 56 D2
Köln Germany 50°56N 6°57E 16 C4
Koło Poland 52°14N 18°40E 17 B10
Kołobrzeg Poland 54°10N 15°35E 16 A8
Kolokani Mali 13°35N 7°45W 50 F4
Kolomna Russia 55°8N 38°45E 18 C6
Kolomyya Ukraine 48°31N 25°2E 17 D13
Kolonodale Indonesia 2°0S 121°19E 37 E6
Kolosib India 24°15N 92°45E 41 G18
Kolpashevo Russia 58°20N 83°5E 28 D9
Kolpino Russia 59°44N 30°39E 18 C5
Kolskiy Poluostrov Russia 67°30N 38°0E 18 A6
Kolskiy Zaliv Russia 69°23N 34°0E 8 B26
Kolwezi Dem. Rep. of the Congo 10°40S 25°25E 55 E2
Kolyma → Russia 69°30N 161°0E 29 C17
Kolymskoye Nagorye Russia 63°0N 157°0E 29 C16
Kôm Ombo Egypt 24°25N 32°52E 51 D12
Komandorskiye Is. = Komandorskiye Ostrova Russia 55°0N 167°0E 29 D17
Komandorskiye Ostrova Russia 55°0N 167°0E 29 D17
Komárno Slovak Rep. 47°49N 18°5E 17 E10
Komatipoort S. Africa 25°25S 31°55E 57 D5
Komatou Yialou Cyprus 35°25N 34°8E 25 D13
Komatsu Japan 36°25N 136°30E 31 F8
Komatsushima Japan 34°0N 134°35E 31 H7
Kome Channel Uganda 0°10S 32°35E 54 C3
Komi Cyprus 35°24N 34°0E 25 D13
Komi □ Russia 64°0N 55°0E 18 B10
Kommunarsk = Alchevsk Ukraine 48°30N 38°45E 19 E6
Kommunizma, Pik = imeni Ismail Samani, Pik Tajikistan 39°0N 72°2E 28 F8
Komodo Indonesia 8°37S 119°20E 37 F5
Komoran, Pulau Indonesia 8°18S 138°45E 37 F9
Komoro Japan 36°19N 138°26E 31 F9
Komotini Greece 41°9N 25°26E 23 D11
Kompasberg S. Africa 31°45S 24°32E 56 E3
Kompong Chhnang = Kampong Chhnang Cambodia 12°20N 104°35E 39 F5
Kompong Chikreng Cambodia 13°5N 104°18E 38 F5
Kompong Kleang Cambodia 13°6N 104°8E 38 F5
Kompong Luong Cambodia 11°49N 104°48E 39 G5
Kompong Pranak Cambodia 13°35N 104°55E 38 F5
Kompong Som = Kampong Saom Cambodia 10°38N 103°30E 39 G4
Kompong Som, Chhung = Kampong Saom, Chaak Cambodia 10°50N 103°32E 39 G4
Kompong Speu Cambodia 11°26N 104°32E 39 G5
Kompong Sralau Cambodia 14°5N 105°46E 38 E5
Kompong Thom Cambodia 12°35N 104°51E 38 F5
Kompong Trabeck Cambodia 13°6N 105°14E 38 F5
Kompong Trach Cambodia 11°25N 105°48E 39 G5
Kompong Tralach Cambodia 11°54N 104°47E 39 G5
Komrat = Comrat Moldova 46°18N 28°40E 17 E15
Komsberg S. Africa 32°40S 20°45E 56 E3
Komsomolets, Ostrov Russia 80°30N 95°0E 29 A10
Komsomolsk-na-Amur Russia 50°30N 137°0E 33 A16
Kon Ka Kinh △ Vietnam 14°19N 107°23E 38 E6
Kon Tum Vietnam 14°24N 108°0E 38 E7
Kon Tum, Plateau du Vietnam 14°30N 108°30E 38 E7
Konar □ Afghan. 34°30N 71°3E 40 B7
Konārī Iran 28°13N 51°36E 45 D6
Konch India 26°0N 79°10E 43 G8
Konde Tanzania 4°57S 39°45E 54 C4
Kondinin Australia 32°34S 118°8E 61 F2
Kondoa Tanzania 4°55S 35°50E 54 C4
Kondopoga Russia 62°12N 34°17E 18 B5
Kondratyevo Russia 57°22N 98°15E 29 D10
Köneürgench Turkmenistan 42°19N 59°10E 28 E6
Konevo Russia 62°8N 39°20E 18 B6
Kong = Khong → Cambodia 13°32N 105°58E 38 F5
Kong Ivory C. 8°54N 4°36W 50 G5
Kong, Koh Cambodia 11°20N 103°0E 39 G4
Kong Christian IX.s Land Greenland 68°0N 36°0W 4 C6
Kong Christian X.s Land Greenland 74°0N 29°0W 4 B6
Kong Frederik IX.s Land Greenland 67°0N 52°0W 4 C5
Kong Frederik VI.s Kyst Greenland 63°0N 43°0W 4 C5
Kong Frederik VIII.s Land Greenland 78°30N 26°0W 4 B6
Kong Oscar Fjord Greenland 72°20N 24°0W 4 B6
Konglu Burma 27°13N 97°57E 41 F20
Kongola Namibia 17°45S 23°20E 56 B3
Kongolo Kasai-Or., Dem. Rep. of the Congo 5°26S 24°49E 54 D1
Kongolo Katanga, Dem. Rep. of the Congo 5°22S 27°0E 54 D2
Kongsberg Norway 59°39N 9°39E 9 G13
Kongsvinger Norway 60°12N 12°2E 9 F15
Kongur Shan China 38°34N 75°18E 32 D4
Kongwa Tanzania 6°11S 36°26E 54 D4
Koni Dem. Rep. of the Congo 10°40S 27°11E 55 E2
Konin Poland 52°12N 18°15E 17 B10
Konjic Bos.-H. 43°42N 17°58E 23 C7
Konkiep Namibia 26°49S 17°15E 56 D2
Konosha Russia 61°0N 40°5E 18 B7
Kōnosu Japan 36°3N 139°31E 31 F9
Konotop Ukraine 51°12N 33°7E 19 D5
Konqi He → China 41°0N 86°0E 32 C6
Końskie Poland 51°15N 20°23E 17 C11
Konstanz Germany 47°40N 9°10E 16 E5
Kont Iran 26°55N 61°50E 45 E9
Kontagora Nigeria 10°23N 5°27E 50 F7
Kontiolahti Finland 62°46N 29°50E 8 E23
Kontokali Greece 39°38N 19°51E 25 A3
Konya Turkey 37°52N 32°35E 44 B2
Konza Kenya 1°45S 37°7E 54 C4
Koocanusa, L. Canada 49°20N 115°15W 76 B6
Kookynie Australia 29°17S 121°22E 61 E3
Koolyanobbing Australia 30°48S 119°36E 61 F2
Koonibba Australia 31°54S 133°25E 63 E1
Koorawatha Australia 34°2S 148°33E 63 E4
Koorda Australia 30°48S 117°35E 61 F2
Kooskia U.S.A. 46°9N 115°59W 76 C6
Kootenay → U.S.A. 49°19N 117°39W 70 D5
Kootenay △ Canada 51°0N 116°0W 70 C5
Kootenay L. Canada 49°45N 116°50W 76 B5
Kootjieskolk S. Africa 31°15S 20°21E 56 E3
Kopaonik Serbia 43°10N 20°50E 23 C9
Kópavogur Iceland 64°6N 21°55W 8 D3
Koper Slovenia 45°31N 13°44E 16 F7
Kopervik Norway 59°17N 5°17E 9 G11
Kopet Dagh Asia 38°0N 58°0E 45 B8
Kopi Australia 33°24S 135°40E 63 E2
Köping Sweden 59°31N 16°3E 9 G17
Koppeh Dāgh = Kopet Dagh Asia 38°0N 58°0E 45 B8
Koppies S. Africa 27°20S 27°30E 57 D4
Koprivnica Croatia 46°12N 16°45E 22 A7
Köprülü △ Turkey 37°20N 31°5E 44 B1
Kopychyntsi Ukraine 49°7N 25°58E 17 D13
Kora △ Kenya 0°14S 38°44E 54 C4
Korab Macedonia 41°44N 20°40E 23 D9
Koral India 21°50N 73°12E 42 J5
Korba India 22°20N 82°45E 43 H10
Korbu, Gunung Malaysia 4°41N 101°18E 39 K3
Korçë Albania 40°37N 20°50E 23 D9
Korčula Croatia 42°56N 16°57E 22 C7
Kord Kūy Iran 36°48N 54°7E 45 B7
Kord Sheykh Iran 28°31N 52°53E 45 D7
Kordestān □ Iran 36°0N 47°0E 44 C5
Kordofân Sudan 13°0N 29°0E 51 F11
Korea, North ■ Asia 40°0N 127°0E 35 E14
Korea, South ■ Asia 36°0N 128°0E 35 G15
Korea Bay Korea 39°0N 124°0E 35 E13
Korea Strait Asia 34°0N 129°30E 35 H15
Korets Ukraine 50°40N 27°5E 17 C14
Korhogo Ivory C. 9°29N 5°28W 50 G4
Korinthiakos Kolpos Greece 38°16N 22°30E 23 E10
Korinthos Greece 37°56N 22°55E 23 F10
Korissia, L. Greece 39°27N 19°53E 25 B3
Kōriyama Japan 37°24N 140°23E 30 F10
Korla China 41°45N 86°4E 32 C6
Kormakiti, C. Cyprus 35°23N 32°56E 25 D11
Korneshty = Corneşti Moldova 47°21N 28°1E 17 E15
Koro Fiji 17°19S 179°23E 59 a
Koro Ivory C. 8°32N 7°30W 50 G4
Koro Sea Fiji 17°30S 179°45W 59 a
Korogwe Tanzania 5°5S 38°25E 54 D4
Korolevu Fiji 18°12S 177°46E 59 a

Koronadal *Phil.* 6°12N 124°51E **37** C6
Körös → *Hungary* 46°43N 20°12E **17** E11
Korosten *Ukraine* 50°54N 28°36E **17** C15
Korostyshev *Ukraine* 50°19N 29°4E **17** C15
Korovou *Fiji* 17°47S 178°32E **59** a
Koroyanitu △ *Fiji* 17°40S 177°35E **59** a
Korraraika, Helodranon' i *Madag.* 17°45S 43°57E **57** B7
Korsakov *Russia* 46°36N 142°42E **29** E15
Korshunovo *Russia* 58°37N 110°10E **29** D12
Korsør *Denmark* 55°20N 11°9E **9** J14
Kortrijk *Belgium* 50°50N 3°17E **15** D3
Korwai *India* 24°7N 78°5E **42** G8
Koryakskoye Nagorye *Russia* 61°0N 171°0E **29** C18
Kos *Greece* 36°50N 27°15E **23** F12
Kosan *N. Korea* 38°52N 127°25E **35** E14
Kościan *Poland* 52°5N 16°40E **17** B9
Kosciusko *U.S.A.* 33°4N 89°35W **85** E10
Kosciuszko, Mt. *Australia* 36°27S 148°16E **63** F4
Kosha *Sudan* 20°50N 30°30E **51** D12
K'oshih = Kashi *China* 39°30N 76°2E **32** D4
Koshiki-Rettō *Japan* 31°45N 129°49E **31** J4
Kosi *India* 27°48N 77°29E **42** F7
Kosi → *India* 28°41N 78°57E **43** E8
Košice *Slovak Rep.* 48°42N 21°15E **17** D11
Koskinou *Greece* 36°23N 28°13E **25** C10
Koslan *Russia* 63°34N 49°14E **18** B8
Kosŏng *N. Korea* 38°40N 128°22E **35** E15
Kosovo ■ *Europe* 42°30N 21°0E **23** C9
Kosovska Mitrovica *Kosovo* 42°54N 20°52E **23** C9
Kossou, L. de *Ivory C.* 6°59N 5°31W **50** G4
Koster *S. Africa* 25°52S 26°54E **56** D4
Kôstî *Sudan* 13°8N 32°43E **51** F12
Kostopil *Ukraine* 50°51N 26°22E **17** C14
Kostroma *Russia* 57°50N 40°58E **18** C7
Kostrzyn *Poland* 52°35N 14°39E **16** B8
Koszalin *Poland* 54°11N 16°8E **16** A9
Kot Addu *Pakistan* 30°30N 71°0E **42** D4
Kot Kapura *India* 30°35N 74°50E **42** D6
Kot Moman *Pakistan* 32°13N 73°0E **42** C5
Kot Sultan *Pakistan* 30°46N 70°56E **42** D4
Kota *India* 25°14N 75°49E **42** G6
Kota Barrage *India* 25°6N 75°51E **42** G6
Kota Belud *Malaysia* 6°21N 116°26E **36** C5
Kota Bharu *Malaysia* 6°7N 102°14E **39** J4
Kota Kinabalu *Malaysia* 6°0N 116°4E **36** C5
Kota Tinggi *Malaysia* 1°44N 103°53E **39** M4
Kotaagung *Indonesia* 5°38S 104°29E **36** F2
Kotabaru *Indonesia* 3°20S 116°20E **36** E5
Kotabumi *Indonesia* 4°49S 104°54E **36** E2
Kotamobagu *Indonesia* 0°57N 124°31E **37** D6
Kotapinang *Indonesia* 1°53N 100°5E **39** M3
Kotcho L. *Canada* 59°7N 121°12W **70** B4
Kotdwara *India* 29°45N 78°32E **43** E8
Kotelnich *Russia* 58°22N 48°24E **18** C8
Kotelnikovo *Russia* 47°38N 43°8E **19** E7
Kotelnyy, Ostrov *Russia* 75°10N 139°0E **29** B14
Kothari → *India* 25°20N 75°4E **42** G6
Kothi *Chhattisgarh, India* 23°21N 82°3E **43** H10
Kothi *Mad. P., India* 24°45N 80°40E **43** G9
Kotiro *Pakistan* 26°17N 67°13E **42** F2
Kotka *Finland* 60°28N 26°58E **8** F22
Kotlas *Russia* 61°17N 46°43E **18** B8
Kotli *Pakistan* 33°30N 73°55E **42** C5
Kotlik *U.S.A.* 63°2N 163°33W **74** C7
Kotma *India* 23°12N 81°58E **43** H9
Kotor *Montenegro* 42°25N 18°47E **23** C8
Kotovsk *Ukraine* 47°45N 29°35E **17** E15
Kotputli *India* 27°43N 76°12E **42** F7
Kotri *Pakistan* 25°22N 68°22E **42** G3
Kotto → *C.A.R.* 4°14N 22°2E **52** D4
Kotturu *India* 14°45N 76°10E **40** M10
Kotu Group *Tonga* 20°0S 174°45W **59** c
Kotuy → *Russia* 71°54N 102°6E **29** B11
Kotzebue *U.S.A.* 66°53N 162°39W **74** B7
Kotzebue Sound *U.S.A.* 66°20N 163°0W **74** B7
Kouchibouguac △ *Canada* 46°50N 65°0W **73** C6
Koudougou *Burkina Faso* 12°10N 2°20W **50** F5
Koufonisi *Greece* 34°56N 26°8E **25** E8
Kougaberge *S. Africa* 33°48S 23°50E **56** E3
Kouilou → *Congo* 4°10S 12°5E **52** E2
Koukdjuak → *Canada* 66°43N 73°0W **69** D17
Koula Moutou *Gabon* 1°15S 12°25E **52** E2
Koulen = Kulen *Cambodia* 13°50N 104°40E **38** F5
Kouloura *Greece* 39°42N 19°54E **25** A3
Koumala *Australia* 21°38S 149°15E **62** C4
Koumra *Chad* 8°50N 17°35E **51** G9
Kountze *U.S.A.* 30°22N 94°19W **84** F7
Kouris → *Cyprus* 34°38N 32°54E **25** E11
Kourou *Fr. Guiana* 5°9N 52°39W **93** B8
Kouroussa *Guinea* 10°45N 9°45W **50** F4
Kousseri *Cameroon* 12°0N 14°55E **51** F8
Koutiala *Mali* 12°25N 5°23W **50** F4
Kouvola *Finland* 60°52N 26°43E **8** F22
Kovdor *Russia* 67°34N 30°24E **8** C24
Kovel *Ukraine* 51°11N 24°38E **17** C13
Kovrov *Russia* 56°25N 41°25E **18** C7
Kowanyama *Australia* 15°29S 141°44E **62** B3
Kowanyama ☼ *Australia* 15°20S 141°47E **62** B3
Kowloon *China* 22°19N 114°11E **33** G11
Kowŏn *N. Korea* 39°26N 127°14E **35** E14
Koyampattur = Coimbatore *India* 11°2N 76°59E **40** P10
Köyceğiz *Turkey* 36°57N 28°40E **23** F13
Koyukuk → *U.S.A.* 64°55N 157°32W **74** C8
Koza = Okinawa *Japan* 26°19N 127°46E **31** L3
Kozan *Turkey* 37°26N 35°50E **44** B2
Kozani *Greece* 40°19N 21°47E **23** D9
Kozhikode = Calicut *India* 11°15N 75°43E **40** P9
Kozhva *Russia* 65°10N 57°0E **18** A10
Kōzu-Shima *Japan* 34°13N 139°10E **31** G9
Kozyatyn *Ukraine* 49°45N 28°50E **17** D15
Kozyrevsk *Russia* 56°3N 159°51E **29** D16
Kpalimé *Togo* 6°57N 0°44E **50** G6
Kra, Isthmus of = Kra, Kho Khot *Thailand* 10°15N 99°30E **39** G2
Kra, Kho Khot *Thailand* 10°15N 99°30E **39** G2
Kra Buri *Thailand* 10°22N 98°46E **39** G2
Kraai → *S. Africa* 30°40S 26°45E **56** E4
Krabi *Thailand* 8°4N 98°55E **39** H2
Kracheh = Kratie *Cambodia* 12°32N 106°10E **38** F6
Kragan *Indonesia* 6°43S 111°38E **37** G14
Kragerø *Norway* 58°52N 9°25E **9** G13
Kragujevac *Serbia* 44°2N 20°56E **23** B9
Krakatau = Rakata, Pulau *Indonesia* 6°10S 105°20E **37** G11
Krakatoa = Rakata, Pulau *Indonesia* 6°10S 105°20E **37** G11
Krakor *Cambodia* 12°32N 104°12E **38** F5
Kraków *Poland* 50°4N 19°57E **17** C10
Kraksaan *Indonesia* 7°43S 113°23E **37** G16
Kralanh *Cambodia* 13°35N 103°25E **38** F4
Kraljevo *Serbia* 43°44N 20°41E **23** C9
Kramatorsk *Ukraine* 48°50N 37°30E **19** E6
Kramfors *Sweden* 62°55N 17°48E **8** E17
Kranj *Slovenia* 46°16N 14°22E **16** E8
Krankskop *S. Africa* 28°0S 30°47E **57** D5
Krasavino *Russia* 60°58N 46°29E **18** B8
Krasieo Res. *Thailand* 14°49N 99°30E **38** E2
Kraskino *Russia* 42°44N 130°48E **30** C5
Kraśnik *Poland* 50°55N 22°15E **17** C12
Krasnoarmeysk *Russia* 51°0N 45°42E **28** D5
Krasnodar *Russia* 45°5N 39°0E **19** E6
Krasnokamensk *Russia* 50°3N 118°0E **29** D12
Krasnokamsk *Russia* 58°4N 55°48E **18** C10
Krasnoperekopsk *Ukraine* 46°0N 33°54E **19** E5
Krasnorechenskiy *Russia* 44°41N 135°14E **30** B7
Krasnoselkup *Russia* 65°20N 82°10E **28** C9
Krasnoturinsk *Russia* 59°46N 60°12E **18** C11
Krasnoufimsk *Russia* 56°36N 57°38E **18** C10
Krasnouralsk *Russia* 58°21N 60°3E **18** C11
Krasnovishersk *Russia* 60°23N 57°3E **18** B10
Krasnoyarsk *Russia* 56°8N 93°0E **29** D10
Krasnyy Kut *Russia* 50°50N 47°0E **19** D8
Krasnyy Luch *Ukraine* 48°13N 39°0E **19** E6
Krasnyy Yar *Russia* 46°43N 48°23E **19** E8
Kratie *Cambodia* 12°32N 106°10E **38** F6
Krau *Indonesia* 3°19S 140°5E **37** E10
Kravanh, Chuor Phnum *Cambodia* 12°0N 103°32E **39** G4
Krefeld *Germany* 51°20N 6°33E **16** C4
Kremen *Croatia* 44°28N 15°53E **16** F8
Kremenchuk *Ukraine* 49°5N 33°25E **19** E5
Kremenchuksk Vdskh. *Ukraine* 49°20N 32°30E **19** E5
Kremenets *Ukraine* 50°8N 25°43E **17** C13
Kremmling *U.S.A.* 40°4N 106°24W **76** F10
Krems *Austria* 48°25N 15°36E **16** D8
Kretinga *Lithuania* 55°53N 21°15E **9** J19
Kribi *Cameroon* 2°57N 9°56E **52** D1
Krichev = Krychaw *Belarus* 53°40N 31°41E **17** B16
Kril'on, Mys *Russia* 45°53N 142°5E **30** B11
Krios, Akra *Greece* 35°13N 23°34E **25** D5
Krishna → *India* 15°57N 80°59E **41** M12
Krishnanagar *India* 23°24N 88°33E **43** H13
Kristiansand *Norway* 58°8N 8°1E **9** G13
Kristianstad *Sweden* 56°2N 14°9E **9** H16
Kristiansund *Norway* 63°7N 7°45E **8** E12
Kristiinankaupunki *Finland* 62°16N 21°21E **8** E19
Kristinehamn *Sweden* 59°18N 14°7E **9** G16
Kristinestad = Kristiinankaupunki *Finland* 62°16N 21°21E **8** E19
Kriti *Greece* 35°15N 25°0E **25** D7
Kritsa *Greece* 35°10N 25°41E **25** D7
Krivoy Rog = Kryvyy Rih *Ukraine* 47°51N 33°20E **19** E5
Krk *Croatia* 45°8N 14°40E **16** F8
Krokodil = Umgwenya → *Mozam.* 25°14S 32°18E **57** D5
Krong Kaoh Kong *Cambodia* 11°37N 102°59E **39** G4
Kronprins Frederik Land *Greenland* 81°0N 45°0W **4** B5
Kronprins Olav Kyst *Antarctica* 69°0S 42°0E **5** C5
Kronprinsesse Märtha Kyst *Antarctica* 73°30S 10°0W **5** D2
Kronshtadt *Russia* 59°57N 29°51E **18** B4
Kroonstad *S. Africa* 27°43S 27°19E **56** D4
Kropotkin *Russia* 45°28N 40°28E **19** E7
Krosno *Poland* 49°42N 21°46E **17** D11
Krotoszyn *Poland* 51°42N 17°23E **17** C9
Krousonas *Greece* 35°13N 24°59E **25** D6
Kruger △ *S. Africa* 24°50S 26°10E **57** C5
Krugersdorp *S. Africa* 26°5S 27°46E **57** D4
Kruisfontein *S. Africa* 33°59S 24°43E **56** E3
Krung Thep = Bangkok *Thailand* 13°45N 100°35E **38** F3
Krupki *Belarus* 54°19N 29°8E **17** A15
Kruševac *Serbia* 43°35N 21°28E **23** C9
Krychaw *Belarus* 53°40N 31°41E **17** B16
Krymskiy Poluostrov = Krymskyy Pivostriv *Ukraine* 45°0N 34°0E **19** F5
Krymskyy Pivostriv *Ukraine* 45°0N 34°0E **19** F5
Kryvyy Rih *Ukraine* 47°51N 33°20E **19** E5
Ksar el Kebir *Morocco* 35°0N 6°0W **50** B4
Ksar es Souk = Er Rachidia *Morocco* 31°58N 4°20W **50** B5
Kuah *Malaysia* 6°19N 99°51E **39** J2
Kuala Belait *Malaysia* 4°35N 114°11E **36** D4
Kuala Berang *Malaysia* 5°5N 103°1E **39** K4
Kuala Dungun = Dungun *Malaysia* 4°45N 103°25E **39** K4
Kuala Kangsar *Malaysia* 4°46N 100°56E **39** K3
Kuala Kelawang *Malaysia* 2°56N 102°5E **39** L4
Kuala Kerai *Malaysia* 5°30N 102°12E **39** K4
Kuala Kubu Bharu *Malaysia* 3°34N 101°39E **39** L3
Kuala Lipis *Malaysia* 4°10N 102°3E **39** K4
Kuala Lumpur *Malaysia* 3°9N 101°41E **39** L3
Kuala Nerang *Malaysia* 6°16N 100°37E **39** J3
Kuala Pilah *Malaysia* 2°45N 102°15E **39** L4
Kuala Rompin *Malaysia* 2°49N 103°29E **39** L4
Kuala Selangor *Malaysia* 3°20N 101°15E **39** L3
Kuala Sepetang *Malaysia* 4°49N 100°28E **39** K3
Kuala Terengganu *Malaysia* 5°20N 103°8E **39** K4
Kualajelai *Indonesia* 2°58S 110°46E **36** E4
Kualakapuas *Indonesia* 2°55S 114°20E **36** E4
Kualakurun *Indonesia* 1°10S 113°50E **36** E4
Kualapembuang *Indonesia* 3°14S 112°38E **36** E4
Kualasimpang *Indonesia* 4°17N 98°3E **36** D1
Kuancheng *China* 40°37N 118°30E **35** D10
Kuandang *Indonesia* 0°56N 123°1E **37** D6
Kuandian *China* 40°45N 124°45E **35** D13
Kuangchou = Guangzhou *China* 23°6N 113°13E **33** G11
Kuantan *Malaysia* 3°49N 103°20E **39** L4
Kuba = Quba *Azerbaijan* 41°21N 48°32E **19** F8
Kuban → *Russia* 45°20N 37°30E **19** E6
Kubokawa *Japan* 33°12N 133°8E **31** H6
Kubu *Indonesia* 8°16S 115°35E **37** J18
Kubutambahan *Indonesia* 8°5S 115°10E **37** J18
Kucar, Tanjung *Indonesia* 8°39S 114°34E **37** K18
Kuchaman *India* 27°13N 74°47E **42** F6
Kuchinda *India* 21°44N 84°21E **43** J11
Kuching *Malaysia* 1°33N 110°25E **36** D4
Kuchino-eruba-Jima *Japan* 30°28N 130°12E **31** J5
Kuchino-Shima *Japan* 29°57N 129°55E **31** K4
Kuchinotsu *Japan* 32°36N 130°11E **31** H5
Kucing = Kuching *Malaysia* 1°33N 110°25E **36** D4
Kud → *Pakistan* 26°5N 66°20E **42** F2
Kuda *India* 23°10N 71°15E **42** H4
Kudat *Malaysia* 6°55N 116°55E **36** C5
Kudus *Indonesia* 6°48S 110°51E **37** G14
Kudymkar *Russia* 59°1N 54°39E **18** C9
Kueiyang = Guiyang *China* 26°32N 106°40E **32** F10
Kufra Oasis = Al Kufrah *Libya* 24°17N 23°15E **51** D10
Kufstein *Austria* 47°35N 12°11E **16** E7
Kugaaruk = Pelly Bay *Canada* 68°38N 89°50W **69** D14
Kugluktuk *Canada* 67°50N 115°5W **68** D8
Kugong I. *Canada* 56°18N 79°50W **72** A4
Kūh Dasht *Iran* 33°32N 47°36E **44** C5
Kūh-e-Jebāl Bārez *Iran* 29°0N 58°0E **45** D8
Kūhak *Iran* 27°12N 63°10E **45** E9
Kuhan *Pakistan* 28°19N 67°14E **42** E2
Kūhbonān *Iran* 31°23N 56°19E **45** D8
Kūhestak *Iran* 26°47N 57°2E **45** E8
Kuhin *Iran* 36°22N 49°40E **45** B6
Kūhīrī *Iran* 26°55N 61°2E **45** E9
Kuhmo *Finland* 64°7N 29°31E **8** D23
Kuhn Chae △ *Thailand* 19°8N 99°24E **38** C2
Kūhpāyeh *Eşfahan, Iran* 32°44N 52°20E **45** C7
Kūhpāyeh *Kermān, Iran* 30°35N 57°15E **45** D8
Kūhrān, Kūh-e *Iran* 26°46N 58°12E **45** E8
Kui Buri *Thailand* 12°3N 99°52E **39** F2
Kuiburi △ *Thailand* 12°10N 99°37E **39** F2
Kuichong *China* 22°38N 114°25E **33** a
Kuiseb → *Namibia* 22°59S 14°31E **56** C1
Kuito *Angola* 12°22S 16°55E **53** G3
Kuiu I. *U.S.A.* 57°45N 134°10W **70** B2
Kujang *N. Korea* 39°57N 126°1E **35** E14
Kuji *Japan* 40°11N 141°46E **30** D10
Kujū-San *Japan* 33°5N 131°15E **31** H5
Kukës *Albania* 42°5N 20°27E **23** C9
Kukup *Malaysia* 1°20N 103°27E **39** d
Kukup, Pulau *Malaysia* 1°18N 103°25E **39** d
Kula *Turkey* 38°32N 28°40E **23** E13
K'ula Shan *Bhutan* 28°14N 90°36E **32** F7
Kulachi *Pakistan* 31°56N 70°27E **42** D4
Kulai *Malaysia* 1°44N 103°35E **39** M4
Kulasekarappattinam *India* 8°20N 78°5E **40** Q11
Kuldīga *Latvia* 56°58N 21°59E **9** H19
Kulen *Cambodia* 13°50N 104°40E **38** F5
Kulgam *India* 33°36N 75°2E **43** C6
Kulgera *Australia* 25°50S 133°18E **62** D1
Kulim *Malaysia* 5°22N 100°34E **39** K3
Kulin *Australia* 32°40S 118°2E **61** F2
Kulkayu = Hartley Bay *Canada* 53°25N 129°15W **70** C3
Kullu *India* 31°58N 77°6E **42** D7
Kŭlob *Tajikistan* 37°55N 69°50E **28** F7
Kulsary = Qulsary *Kazakhstan* 46°59N 54°1E **19** E9
Kulti *India* 23°43N 86°50E **43** H12
Kulunda *Russia* 52°35N 78°57E **28** D8
Kulungar *Afghan.* 34°0N 69°2E **42** C3
Kūlvand *Iran* 31°21N 54°35E **45** D7
Kulwin *Australia* 35°2S 142°42E **63** F3
Kulyab = Kŭlob *Tajikistan* 37°55N 69°50E **28** F7
Kuma → *Russia* 44°55N 47°0E **19** F8
Kumaganum *Nigeria* 13°8N 10°38E **51** F8
Kumagaya *Japan* 36°9N 139°22E **31** F9
Kumai *Indonesia* 2°44S 111°43E **36** E4
Kumamba, Kepulauan *Indonesia* 1°36S 138°45E **37** E9
Kumamoto *Japan* 32°45N 130°45E **31** H5
Kumamoto ☐ *Japan* 32°55N 130°55E **31** H5
Kumanovo *Macedonia* 42°9N 21°42E **23** C9
Kumara *N.Z.* 42°37S 171°12E **59** E3
Kumarina Roadhouse *Australia* 24°41S 119°32E **61** D2
Kumasi *Ghana* 6°41N 1°38W **50** G5
Kumba *Cameroon* 4°36N 9°24E **52** D1
Kumbakonam *India* 10°58N 79°25E **40** P11
Kumbarilla *Australia* 27°15S 150°55E **63** D5
Kumbhraj *India* 24°22N 77°3E **42** G7
Kumbia *Australia* 26°41S 151°39E **63** D5
Kŭmch'on *N. Korea* 38°10N 126°29E **35** E14
Kume-Shima *Japan* 26°20N 126°47E **31** L3
Kumertau *Russia* 52°45N 55°57E **18** D10
Kumharsain *India* 31°19N 77°27E **42** D7
Kumi *Uganda* 1°30N 33°58E **54** B3
Kumo *Nigeria* 10°1N 11°12E **51** F8
Kumo älv = Kokemäenjoki → *Finland* 61°32N 21°44E **8** F19
Kumon Bum *Burma* 26°30N 97°15E **41** F20
Kumtag Shamo *China* 39°40N 92°0E **32** D7
Kunashir, Ostrov *Russia* 44°0N 146°0E **29** E15
Kunda *Estonia* 59°30N 26°34E **9** G22
Kunda *India* 25°43N 81°31E **43** G9
Kundar → *Pakistan* 31°56N 69°19E **42** D3
Kundelungu △ *Dem. Rep. of the Congo* 10°30S 27°40E **55** E2
Kundelungu Ouest △ *Dem. Rep. of the Congo* 9°55S 27°17E **55** D2
Kundian *Pakistan* 32°27N 71°28E **42** C4
Kundla *India* 21°21N 71°25E **42** J4
Kung, Ao *Thailand* 8°5N 98°24E **39** a
Kunga → *Bangla.* 21°46N 89°30E **43** J13
Kunghit I. *Canada* 52°6N 131°3W **70** C2
Kungrad = Qŭnghirot *Uzbekistan* 43°2N 58°50E **28** E6
Kungsbacka *Sweden* 57°30N 12°5E **9** H15
Kungur *Russia* 57°25N 56°57E **18** C10
Kungurri *Australia* 21°4S 148°45E **62** b
Kunhar → *Pakistan* 34°20N 73°30E **43** B5
Kuningan *Indonesia* 6°59S 108°29E **37** G13
Kunlong *Burma* 23°20N 98°50E **41** H21
Kunlun Shan *Asia* 36°0N 86°30E **32** D6
Kunlun Shankou *China* 35°38N 94°4E **32** D7
Kunming *China* 25°1N 102°41E **32** F9
Kunmunya ☼ *Australia* 15°26S 124°42E **60** C3
Kununurra *Australia* 15°40S 128°50E **60** C4
Kunwari → *India* 26°26N 79°11E **43** F8
Kunya-Urgench = Köneürgench *Turkmenistan* 42°19N 59°10E **28** E6
Kuopio *Finland* 62°53N 27°35E **8** E22
Kupa → *Croatia* 45°28N 16°24E **16** F9
Kupang *Indonesia* 10°19S 123°39E **37** F6
Kupreanof I. *U.S.A.* 56°50N 133°30W **70** B2
Kupyansk-Uzlovoi *Ukraine* 49°40N 37°43E **19** E6
Kuqa *China* 41°35N 82°30E **32** C5
Kür → *Azerbaijan* 39°29N 49°15E **19** G8
Kür Dili *Azerbaijan* 39°3N 49°13E **45** B6
Kura = Kür → *Azerbaijan* 39°29N 49°15E **19** G8
Kuranda *Australia* 16°48S 145°35E **62** B4
Kuranga *India* 22°4N 69°10E **42** H3
Kurashiki *Japan* 34°40N 133°50E **31** G6
Kurayn *Si. Arabia* 27°39N 49°50E **45** E6
Kurayoshi *Japan* 35°26N 133°50E **31** G6
Kūrchatov *Kazakhstan* 50°45N 78°32E **28** D8
Kŭrdzhali *Bulgaria* 41°38N 25°21E **23** D11
Kure *Japan* 34°14N 132°32E **31** G6
Kure I. *U.S.A.* 28°25N 178°25W **75** K4
Kuressaare *Estonia* 58°15N 22°30E **9** G20
Kurgan *Russia* 55°26N 65°18E **28** D7
Kuri *India* 26°37N 70°43E **42** F4
Kuria Maria Is. = Hallāniyat, Jazā'ir al *Oman* 17°30N 55°58E **47** D6
Kuridala *Australia* 21°16S 140°29E **62** C3
Kurigram *Bangla.* 25°49N 89°39E **41** G16
Kurikka *Finland* 62°36N 22°24E **8** E20
Kuril Basin *Pac. Oc.* 47°0N 150°0E **4** E15
Kuril Is. = Kurilskiye Ostrova *Russia* 45°0N 150°0E **29** E16
Kuril-Kamchatka Trench *Pac. Oc.* 44°0N 153°0E **64** C7
Kurilsk *Russia* 45°14N 147°53E **29** E15
Kurilskiye Ostrova *Russia* 45°0N 150°0E **29** E16
Kurino *Japan* 31°57N 130°43E **31** J5
Kurinskaya Kosa = Kür Dili *Azerbaijan* 39°3N 49°13E **45** B6
Kurlkuta ☼ *Australia* 24°0S 127°5E **60** D4
Kurnool *India* 15°45N 78°0E **40** M11
Kuro-Shima *Kagoshima, Japan* 30°50N 129°57E **31** J4
Kuro-Shima *Okinawa, Japan* 24°14N 124°1E **31** M2
Kurow *N.Z.* 44°44S 170°29E **59** F3
Kurram → *Pakistan* 32°36N 71°20E **42** C4
Kurri Kurri *Australia* 32°50S 151°28E **63** E5
Kurrimine *Australia* 17°47S 146°6E **62** B4
Kurseong = Karsiyang *India* 26°56N 88°18E **43** F13
Kurshskiy Zaliv *Russia* 55°9N 21°6E **9** J19
Kursk *Russia* 51°42N 36°11E **18** D6
Kuruçay *Turkey* 39°39N 38°29E **44** B3
Kurukshetra = Thanesar *India* 30°1N 76°52E **42** D7
Kuruktag *China* 41°0N 89°0E **32** C6
Kuruman *S. Africa* 27°28S 23°28E **56** D3
Kuruman → *S. Africa* 26°56S 20°39E **56** D3
Kurume *Japan* 33°15N 130°30E **31** H5
Kurunegala *Sri Lanka* 7°30N 80°23E **40** R12
Kurya *Russia* 61°42N 57°9E **18** B10
Kus Gölü *Turkey* 40°10N 27°55E **23** D12
Kuşadası *Turkey* 37°52N 27°15E **23** F12
Kusamba *Indonesia* 8°34S 115°27E **37** K18
Kusatsu *Japan* 36°37N 138°36E **31** F9
Kusawa L. *Canada* 60°20N 136°13W **70** A1
Kushalgarh *India* 23°10N 74°27E **42** H6
Kushikino *Japan* 31°44N 130°16E **31** J5
Kushima *Japan* 31°29N 131°14E **31** J5
Kushimoto *Japan* 33°28N 135°47E **31** H7
Kushiro *Japan* 43°0N 144°25E **30** C12
Kushiro-Gawa → *Japan* 42°59N 144°23E **30** C12
Kushiro Shitsugen △ *Japan* 43°9N 144°26E **30** C12
Kūshk *Iran* 28°46N 56°51E **45** D8
Kushka = Serhetabat *Turkmenistan* 35°20N 62°18E **45** C9
Kushol *India* 33°40N 76°36E **43** C7
Kushtia *Bangla.* 23°55N 89°5E **41** H16
Kushva *Russia* 58°18N 59°45E **18** C10
Kuskokwim → *U.S.A.* 60°5N 162°25W **74** C7
Kuskokwim B. *U.S.A.* 59°45N 162°25W **74** D7
Kuskokwim Mts. *U.S.A.* 62°30N 156°0W **74** C8
Kusmi *India* 23°17N 83°55E **43** H10
Kusŏng *N. Korea* 39°59N 125°15E **35** E13
Kussharo-Ko *Japan* 43°38N 144°21E **30** C12
Kustanay = Qostanay *Kazakhstan* 53°10N 63°35E **28** D7
Kut, Ko *Thailand* 11°40N 102°35E **39** G4
Kuta *Indonesia* 8°43S 115°11E **37** K18
Kütahya *Turkey* 39°30N 30°2E **19** G5
Kutaisi *Georgia* 42°19N 42°40E **19** F7
Kutaraja = Banda Aceh *Indonesia* 5°35N 95°20E **36** C1
Kutch, Gulf of = Kachchh, Gulf of *India* 22°50N 69°15E **42** H3
Kutch, Rann of = Kachchh, Rann of *India* 24°0N 70°0E **42** H4
Kutiyana *India* 21°36N 70°2E **42** J4
Kutno *Poland* 52°15N 19°23E **17** B10
Kuttabul *Australia* 21°1S 148°54E **62**
Kutu *Dem. Rep. of the Congo* 2°40S 18°11E **52** E3
Kutum *Sudan* 14°10N 24°40E **51** F10
Kuujjuaq *Canada* 58°6N 68°15W **69** F18
Kuujjuarapik *Canada* 55°20N 77°35W **72** A
Kuusamo *Finland* 65°57N 29°8E **8** D2
Kuusankoski *Finland* 60°55N 26°38E **8** F2
Kuwait = Al Kuwayt *Kuwait* 29°30N 48°0E **44** D5
Kuwait ■ *Asia* 29°30N 47°30E **44** D
Kuwana *Japan* 35°5N 136°43E **31** G
Kuwana → *India* 26°25N 83°15E **43** F1
Kuybyshev = Samara *Russia* 53°8N 50°6E **18** D
Kuybyshev *Russia* 55°27N 78°19E **28** D
Kuybyshevskoye Vdkhr. *Russia* 55°2N 49°30E **18** C
Kuye He → *China* 38°23N 110°46E **34** E
Kūyeh *Iran* 38°45N 47°57E **44** B
Kuyto, Ozero *Russia* 65°6N 31°20E **8** D2
Kuytun *China* 44°25N 85°0E **32** C
Kuyumba *Russia* 60°58N 96°59E **29** C1
Kuzey Anadolu Dağları *Turkey* 41°0N 36°45E **19** F
Kuznetsk *Russia* 53°12N 46°40E **18** D
Kuzomen *Russia* 66°22N 36°50E **18** A
Kvænangen *Norway* 70°5N 21°15E **8** A1
Kvaløya *Norway* 69°40N 18°30E **8** B1
Kvarner *Croatia* 44°50N 14°10E **16** F
Kvarnerič *Croatia* 44°43N 14°37E **16** F
Kwabhaca *S. Africa* 30°51S 29°0E **57** E
Kwai = Khwae Noi → *Thailand* 14°1N 99°32E **38** E
Kwajalein *Marshall Is.* 9°5N 167°20E **64** G
Kwakhanai *Botswana* 21°39S 21°16E **56** C
Kwakoegron *Suriname* 5°12N 55°25W **93** B
Kwale *Kenya* 4°15S 39°31E **54** C
KwaMashu *S. Africa* 29°45S 30°58E **57** D
Kwando → *Africa* 18°27S 23°32E **56** B
Kwangchow = Guangzhou *China* 23°6N 113°13E **33** G1
Kwangdaeri *N. Korea* 40°34N 127°33E **35** D1
Kwango → *Dem. Rep. of the Congo* 3°14S 17°22E **52** E
Kwangsi-Chuang = Guangxi Zhuangzu Zizhiqu ☐ *China* 24°0N 109°0E **33** G1
Kwangtung = Guangdong ☐ *China* 23°0N 113°0E **33** G1
Kwataboahegan → *Canada* 51°9N 80°50W **72** B
Kwatisore *Indonesia* 3°18S 134°50E **37**
KwaZulu Natal ☐ *S. Africa* 29°0S 30°0E **57**
Kweichow = Guizhou ☐ *China* 27°0N 107°0E **32** F1
Kwekwe *Zimbabwe* 18°58S 29°48E **55**
Kwidzyn *Poland* 53°44N 18°55E **17** B
Kwilu → *Dem. Rep. of the Congo* 3°22S 17°22E **52**
Kwinana *Australia* 32°15S 115°47E **61**
Kwoka *Indonesia* 0°31S 132°27E **37**
Kwun Tong *China* 22°19N 114°13E **33**
Kyabra Cr. → *Australia* 25°36S 142°55E **63**
Kyabram *Australia* 36°19S 145°4E **63**
Kyaikto *Burma* 17°20N 97°3E **38**
Kyaing Tong = Keng Tung *Burma* 21°18N 99°39E **38**
Kyakhta *Russia* 50°30N 106°25E **29** D
Kyambura ⌓ *Uganda* 0°7S 30°9E **54**
Kyancutta *Australia* 33°8S 135°33E **63**
Kyaukpadaung *Burma* 20°52N 95°8E **41** J
Kyaukpyu *Burma* 19°28N 93°30E **41** K
Kyaukse *Burma* 21°36N 96°10E **41** J
Kyburz *U.S.A.* 38°47N 120°18W **78**
Kyelang *India* 32°35N 77°2E **42**
Kyenjojo *Uganda* 0°40N 30°37E **54**
Kyle *Canada* 50°50N 108°2W **71**
Kyle Dam *Zimbabwe* 20°15S 31°0E **55**
Kyle of Lochalsh *U.K.* 57°17N 5°44W **11**
Kymijoki → *Finland* 60°30N 26°55E **8** F
Kymmene älv = Kymijoki → *Finland* 60°30N 26°55E **8** F
Kyneton *Australia* 37°10S 144°29E **63**
Kynuna *Australia* 21°37S 141°55E **62**
Kyō-ga-Saki *Japan* 35°45N 135°15E **31**
Kyoga, L. *Uganda* 1°35N 33°0E **54**
Kyogle *Australia* 28°40S 153°0E **63**
Kyŏngju = Gyeongju *S. Korea* 35°51N 129°14E **35** G
Kyongpyaw *Burma* 17°12N 95°10E **41** L
Kyŏngsŏng *N. Korea* 41°35N 129°36E **35** D
Kyōto *Japan* 35°0N 135°45E **31**
Kyōto ☐ *Japan* 35°15N 135°45E **31**
Kyparissovouno *Cyprus* 35°19N 33°10E **25** D
Kyperounda *Cyprus* 34°56N 32°58E **25** E
Kypros = Cyprus ■ *Asia* 35°0N 33°0E **25** E

renia *Cyprus* 35°20N 33°20E **25** D12
rgyzstan ■ *Asia* 42°0N 75°0E **28** E8
ro älv = Kyrönjoki ➤
Finland 63°14N 21°45E **8** E19
rönjoki ➤ *Finland* 63°14N 21°45E **8** E19
statyam *Russia* 67°20N 123°10E **29** C13
thira *Greece* 36°8N 23°0E **23** F10
thréa *Cyprus* 35°15N 33°29E **25** D12
unhla *Burma* 23°25N 95°15E **41** H19
uquot Sound *Canada* 50°2N 127°22W **70** D3
ūshū *Japan* 33°0N 131°0E **31** H5
ūshū *Japan* 33°0N 131°0E **31** H5
ushu-Palau Ridge
Pac. Oc. 20°0N 136°0E **64** E5
ūshū-Sanchi *Japan* 32°35N 131°17E **31** H5
ustendil *Bulgaria* 42°16N 22°41E **23** C10
usyur *Russia* 70°19N 127°30E **29** B13
yiv *Ukraine* 50°30N 30°28E **17** C16
yiv Vdskh.
Ukraine 51°0N 30°25E **17** C16
zyl *Russia* 51°50N 94°30E **32** A7
zyl Kum *Uzbekistan* 42°30N 65°0E **28** E7
zyl Kyya *Kyrgyzstan* 40°16N 72°8E **28** E8
zyl Orda = Qyzylorda
Kazakhstan 44°48N 65°28E **28** E7

Alcarria *Spain* 40°31N 2°45W **21** B4
Amistad △ *Cent. Amer.* 8°18W **88** E3
Asunción *Venezuela* 11°2N 63°53W **92** A6
Baie *Canada* 48°19N 70°53W **73** C5
Banda *Argentina* 27°45S 64°10W **94** B3
Barca *Mexico* 20°17N 102°34W **86** C4
Barge *U.S.A.* 42°16N 110°12W **76** E8
Barra *Nic.* 12°54N 83°33W **88** D3
Belle *U.S.A.* 26°46N 81°26W **85** H14
Biche ➤ *Canada* 59°57N 123°50W **70** B4
Biche, L. *Canada* 54°50N 112°3W **70** C6
Brea *Trin. & Tob.* 10°15N 61°37W **93** K15
Calera *Chile* 32°58S 71°14W **94** C1
Campana △ *Chile* 32°58S 71°14W **94** C1
Canal = Sa Canal
Spain 38°51N 1°23E **24** C7
Carlota *Argentina* 33°30S 63°20W **94** C3
Ceiba *Honduras* 15°40N 86°50W **88** C2
Chaux-de-Fonds *Switz.* 47°7N 6°50E **20** C7
Chorrera *Panama* 8°53N 79°47W **88** E4
Cocha *Argentina* 27°50S 65°40W **94** B2
Concepción *Panama* 8°31N 82°37W **88** E3
Concordia *Mexico* 16°5N 92°38W **87** D6
Coruña = A Coruña
Spain 43°20N 8°25W **21** A1
Crescent *U.S.A.* 43°50N 91°18W **80** D8
Crete *Canada* 58°11N 116°24W **70** B5
Crosse *Kans., U.S.A.* 38°32N 99°18W **80** F4
Crosse *Wis., U.S.A.* 43°48N 91°15W **80** D8
Cruz *Costa Rica* 11°4N 85°39W **88** D2
Cruz *Mexico* 23°55N 106°54W **86** C3
Désirade *Guadeloupe* 16°18N 61°3W **88** b
Digue *Seychelles* 4°20S 55°51E **53** b
Esperanza *Argentina* 31°29S 61°3W **94** C3
Esperanza *Honduras* 14°15N 88°10W **88** D2
Estrada = A Estrada
Spain 42°43N 8°27W **21** A1
Fayette *U.S.A.* 34°42N 85°17W **85** D12
Fé *Cuba* 22°2N 84°15W **88** B3
Follette *U.S.A.* 36°23N 84°7W **85** C12
Gi *Vietnam* 10°40N 107°45E **39** G6
Grande *U.S.A.* 45°20N 118°5W **76** D4
Grande ➤ *Canada* 53°50N 79°0W **72** B5
Grande 3, Rés.
Canada 53°40N 75°10W **72** B4
Grande 4, Rés. *Canada* 54°0N 73°15W **72** B5
Grange *Calif., U.S.A.* 37°42N 120°27W **78** H6
Grange *Ga., U.S.A.* 33°2N 85°2W **85** E12
Grange *Ky., U.S.A.* 38°24N 85°22W **81** F11
Grange *Tex., U.S.A.* 29°54N 96°52W **84** G6
Guaira *Venezuela* 10°36N 66°56W **92** A5
Habana *Cuba* 23°8N 82°22W **88** B3
Independencia *Mexico* 16°15N 92°1W **87** D6
Isabela *Dom. Rep.* 19°58N 71°2W **89** C5
Junta *U.S.A.* 37°59N 103°33W **76** H12
Laguna *Canary Is.* 28°28N 16°18W **24** F3
Libertad = Puerto Libertad
Mexico 29°55N 112°43W **86** B2
Libertad *Guatemala* 16°47N 90°7W **88** C1
Ligua *Chile* 32°30S 71°16W **94** C1
Línea de la Concepción
Spain 36°15N 5°23W **21** D3
Loche *Canada* 56°29N 109°26W **71** B7
Louvière *Belgium* 50°27N 4°10E **15** D4
Lune *Trin. & Tob.* 10°3N 61°22W **93** K15
Malbaie *Canada* 47°40N 70°10W **73** C5
Malinche △ *Mexico* 19°15N 98°3W **87** D5
Mancha *Spain* 39°10N 2°54W **21** C4
Martre, L. *Canada* 63°15N 117°55W **70** A5
Mesa *Mexico* 32°30N 116°57W **79** N10
Mesa *U.S.A.* 32°46N 117°1W **79** N9
Mesilla *U.S.A.* 32°16N 106°48W **77** K10
Misión *Mexico* 32°6N 116°53W **86** A1
Moure *U.S.A.* 48°21N 98°18W **80** B4
Negra *Chile* 23°46S 70°18W **94** A1
Oliva *Canary Is.* 28°36N 13°57W **24** F6
Orotava *Canary Is.* 28°22N 16°31W **24** F3
Oroya *Peru* 11°32S 75°54W **92** F3
Palma *Canary Is.* 28°40N 17°50W **24** F2
Palma *U.S.A.* 8°15N 78°0W **88** E4
Palma del Condado
Spain 37°21N 6°38W **21** D2
Paloma *Chile* 30°35S 71°0W **94** C1
Pampa □ *Argentina* 36°50S 66°0W **94** D2
Paragua *Venezuela* 6°50N 63°20W **92** B6
Paz *Entre Rios,*
Argentina 30°50S 59°45W **94** C4
Paz *San Luis,*
Argentina 33°30S 67°20W **94** C2
Paz *Bolivia* 16°20S 68°10W **92** G5
Paz *Honduras* 13°44N 87°47W **88** D2

La Paz *Mexico* 24°10N 110°18W **86** C2
La Paz Centro *Nic.* 12°20N 86°41W **88** D2
La Pedrera *Colombia* 1°18S 69°43W **92** D5
La Pérade *Canada* 46°35N 72°12W **73** C5
La Perla *Mexico* 28°18N 104°32W **86** B4
La Perouse Str. *Asia* 45°40N 142°0E **30** B11
La Pesca *Mexico* 23°46N 97°47W **87** C5
La Piedad *Mexico* 20°21N 102°0W **86** C4
La Pine *U.S.A.* 43°40N 121°30W **76** E3
La Plata *Argentina* 35°0S 57°55W **94** C4
La Pocatière *Canada* 47°22N 70°2W **73** C5
La Porte *U.S.A.* 29°40N 95°1W **84** G7
La Purisima *Mexico* 26°10N 112°4W **86** B2
La Push *U.S.A.* 47°55N 124°38W **78** C2
La Quiaca *Argentina* 22°5S 65°35W **94** A2
La Restinga *Canary Is.* 27°38N 17°59W **24** G2
La Rioja *Argentina* 29°20S 67°0W **94** B2
La Rioja □ *Argentina* 29°30S 67°0W **94** B2
La Rioja □ *Spain* 42°20N 2°20W **21** A4
La Robla *Spain* 42°50N 5°41W **21** A3
La Roche-en-Ardenne
Belgium 50°11N 5°35E **15** D5
La Roche-sur-Yon *France* 46°40N 1°25W **20** C3
La Rochelle *France* 46°10N 1°9W **20** C3
La Roda *Spain* 39°13N 2°15W **21** C4
La Romaine *Canada* 50°13N 60°40W **73** B7
La Romana *Dom. Rep.* 18°27N 68°57W **89** C6
La Ronge *Canada* 55°5N 105°20W **71** B7
La Rumorosa *Mexico* 32°34N 116°6W **79** N10
La Sabina = Sa Savina
Spain 38°44N 1°25E **24** C7
La Salle *U.S.A.* 41°20N 89°6W **80** E9
La Santa *Canary Is.* 29°5N 13°40W **24** E6
La Sarre *Canada* 48°45N 79°15W **72** C4
La Scie *Canada* 49°57N 55°36W **73** C8
La Selva Beach *U.S.A.* 36°56N 121°51W **78** J5
La Serena *Chile* 29°55S 71°10W **94** B1
La Seu d'Urgell *Spain* 42°22N 1°23E **21** A6
La Seyne-sur-Mer *France* 43°7N 5°52E **20** E6
La Soufrière *St. Vincent* 13°20N 61°11W **89** D7
La Spézia *Italy* 44°7N 9°50E **20** D8
La Tagua *Colombia* 0°3N 74°40W **92** C4
La Tortuga *Venezuela* 11°0N 65°22W **89** D6
La Trinité *Martinique* 14°45N 60°58W **88** c
La Tuque *Canada* 47°30N 72°50W **73** C5
La Unión *Chile* 40°10S 73°0W **96** E2
La Unión *El Salv.* 13°20N 87°50W **88** D2
La Unión *Mexico* 17°58N 101°49W **86** D4
La Urbana *Venezuela* 7°8N 66°56W **92** B5
La Vache Pt.
Trin. & Tob. 10°47N 61°28W **93** K15
La Vall d'Uixó *Spain* 39°49N 0°15W **21** C5
La Vega *Dom. Rep.* 19°20N 70°30W **89** C5
La Vela de Coro
Venezuela 11°27N 69°34W **92** A5
La Venta *Mexico* 18°5N 94°3W **87** D6
La Vergne *U.S.A.* 36°1N 86°35W **85** C11
La Villa Joiosa = Villajoyosa
Spain 38°30N 0°12W **21** C5
Laascaanood = Las Anod
Somali Rep. 8°26N 47°19E **47** F4
Labasa *Fiji* 16°30S 179°27E **59** a
Labdah = Leptis Magna
Libya 32°40N 14°12E **51** B8
Labe = Elbe ➤ *Europe* 53°50N 9°0E **16** B5
Labé *Guinea* 11°24N 12°16W **50** F3
Laberge, L. *Canada* 61°11N 135°12W **70** A1
Labis *Malaysia* 2°22N 103°2E **39** L4
Laborie *St. Lucia* 13°45N 61°2W **89** f
Laboulaye *Argentina* 34°10S 63°30W **94** C3
Labrador *Canada* 53°20N 61°0W **73** B7
Labrador City *Canada* 52°57N 66°55W **73** B6
Labrador Sea *Atl. Oc.* 57°0N 54°0W **69** F21
Lábrea *Brazil* 7°15S 64°51W **92** E6
Labuan □ *Malaysia* 5°20N 115°12E **36** C4
Labuha *Indonesia* 0°30S 127°30E **37** E7
Labuhan *Indonesia* 6°22S 105°50E **37** G11
Labuhanbajo *Indonesia* 8°28S 119°54E **37** F6
Labuhanbilik *Indonesia* 2°31N 100°10E **39** L3
Labuk, Telok *Malaysia* 6°10N 117°50E **36** C5
Labyrinth, L. *Australia* 30°40S 135°11E **63** E2
Labytnangi *Russia* 66°39N 66°21E **18** A12
Lac-Bouchette *Canada* 48°16N 72°11W **73** C5
Lac Édouard *Canada* 47°40N 72°16W **73** C5
Lac La Biche *Canada* 54°45N 111°58W **70** C6
Lac la Martre = Wha Ti
Canada 63°8N 117°16W **68** E8
Lac La Ronge △ *Canada* 55°9N 104°41W **71** B7
Lac-Mégantic *Canada* 45°35N 70°53W **73** C5
Lac Thien *Vietnam* 12°25N 108°11E **38** F7
Lacanau *France* 44°58N 1°5W **20** D3
Lacantún ➤ *Mexico* 16°36N 90°39W **87** D6
Laccadive Is. = Lakshadweep Is.
India 10°0N 72°30E **3** D13
Lacepede B. *Australia* 36°40S 139°40E **63** F2
Lacepede Is. *Australia* 16°55S 122°0E **60** C3
Lacerdónia *Mozam.* 18°3S 35°35E **55** F4
Lacey *U.S.A.* 47°7N 122°49W **78** C4
Lachania *Greece* 35°58N 27°54E **25** D9
Lachhmangarh *India* 27°50N 75°4E **42** F6
Lachi *Pakistan* 33°25N 71°20E **42** C4
Lachine *Canada* 45°30N 73°40W **72** C5
Lachlan ➤ *Australia* 34°22S 143°55E **63** E3
Lachute *Canada* 45°39N 74°21W **72** C5
Lackagh Hills *Ireland* 54°16N 8°10W **10** B3
Lackawanna *U.S.A.* 42°50N 78°50W **82** D6
Lackawaxen *U.S.A.* 41°29N 74°59W **83** E10
Lacolle *Canada* 45°5N 73°22W **83** A11
Lacombe *Canada* 52°30N 113°44W **70** C6
Lacona *U.S.A.* 43°39N 76°10W **83** C8
Laconia *Greece* 36°53N 22°35E **23** F10
Laconia □ *U.S.A.* 43°32N 71°28W **83** C13
Ladakh Ra. *India* 34°0N 78°0E **43** C8
Ladismith *S. Africa* 33°28S 21°15E **56** E3
Lādīz *Iran* 28°55N 61°15E **45** D9
Ladnun *India* 27°38N 74°25E **42** F6
Ladoga, L. = Ladozhskoye Ozero
Russia 61°15N 30°30E **8** F24
Ladozhskoye Ozero
Russia 61°15N 30°30E **8** F24

Lady Elliott I. *Australia* 24°7S 152°42E **62** C5
Lady Grey *S. Africa* 30°43S 27°13E **56** E4
Ladybrand *S. Africa* 29°9S 27°29E **56** D4
Ladysmith *Canada* 49°0N 123°49W **78** B3
Ladysmith *S. Africa* 28°32S 29°46E **57** D4
Ladysmith *U.S.A.* 45°28N 91°12W **80** C8
Lae *Papua N. G.* 6°40S 147°2E **58** B7
Laem Hin Khom *Thailand* 9°25N 99°56E **39** b
Laem Khat *Thailand* 8°6N 98°20E **39** a
Laem Nga *Thailand* 7°55N 98°27E **39** a
Laem Ngop *Thailand* 12°10N 102°26E **39** F4
Laem Phan Wa *Thailand* 7°47N 98°25E **39** a
Laem Pho *Thailand* 6°55N 101°19E **39** J3
Laem Phrom Thep *Thailand* 7°45N 98°19E **39** a
Laem Riang *Thailand* 8°7N 98°27E **39** a
Laem Sam Rong *Thailand* 9°35N 100°4E **39** b
Laem Son *Thailand* 7°59N 98°16E **39** a
Laem Son △ *Thailand* 9°29N 98°24E **39** H2
Laem Yamu *Thailand* 7°59N 98°26E **39** a
Laerma *Greece* 36°9N 27°57E **25** C9
Læsø *Denmark* 57°15N 11°5E **9** H14
Lafayette *Ind., U.S.A.* 40°25N 86°54W **80** E10
Lafayette *La., U.S.A.* 30°14N 92°1W **84** F8
Lafayette *Tenn., U.S.A.* 36°31N 86°2W **85** C11
Laferte ➤ *Canada* 61°53N 117°44W **70** A5
Lafia *Nigeria* 8°30N 8°34E **50** G7
Laflèche *Canada* 49°45N 106°40W **71** D7
Lagan ➤ *U.K.* 54°36N 5°55W **10** B6
Lagarfljót ➤ *Iceland* 65°40N 14°18W **8** D6
Lagdo, L. de *Cameroon* 8°40N 14°0E **51** G8
Lågen ➤ *Oppland, Norway* 61°8N 10°25E **8** F14
Lågen ➤ *Vestfold, Norway* 59°3N 10°3E **9** G14
Lages *Brazil* 27°48S 50°20W **95** B5
Laghouat *Algeria* 33°50N 2°59E **50** B6
Lagoa do Peixe △ *Brazil* 31°12S 50°55W **95** C5
Lagoa Vermelha *Brazil* 28°13S 51°32W **95** B5
Lagonoy G. *Phil.* 13°35N 123°50E **37** B6
Lagos *Nigeria* 6°25N 3°27E **50** G6
Lagos *Portugal* 37°5N 8°41W **21** D1
Lagos de Moreno
Mexico 21°21N 101°55W **86** C4
Lagrange = Bidyadanga
Australia 18°45S 121°43E **60** C3
Lagrange B. *Australia* 18°38S 121°42E **60** C3
Laguna *Brazil* 28°30S 48°50W **95** B6
Laguna *U.S.A.* 35°2N 107°25W **77** J10
Laguna, Sa. de la
Mexico 23°35N 109°55W **86** C3
Laguna Beach *U.S.A.* 33°33N 117°47W **79** M9
Laguna de la Restinga △
Venezuela 10°58N 64°0W **89** D6
Laguna de Lachuá △
Guatemala 15°55N 90°40W **88** C1
Laguna del Laja △ *Chile* 37°27S 71°20W **94** D1
Laguna del Tigre △
Guatemala 17°32N 90°56W **88** C1
Laguna Limpia
Argentina 26°32S 59°45W **94** B4
Lagunas *Chile* 21°0S 69°45W **94** A2
Lagunas *Peru* 5°10S 75°35W **92** E3
Lagunas de Chacahua △
Mexico 16°0N 97°43W **87** D5
Lagunas de Montebello △
Mexico 16°4N 91°42W **87** D6
Lahad Datu *Malaysia* 5°0N 118°20E **37** D5
Lahad Datu, Telok
Malaysia 4°50N 118°20E **37** D5
Lahan Sai *Thailand* 14°25N 102°52E **38** E4
Lahanam *Laos* 16°16N 105°16E **38** D5
Lahar *India* 26°12N 78°57E **43** F8
Laharpur *India* 27°43N 80°56E **43** F9
Lahat *Indonesia* 3°45S 103°30E **36** E2
Lahewa *Indonesia* 1°22N 97°12E **36** D1
Lāhījān *Iran* 37°10N 50°6E **45** B6
Lahn ➤ *Germany* 50°19N 7°37E **16** C4
Laholm *Sweden* 56°30N 13°2E **9** H15
Lahore *Pakistan* 31°32N 74°22E **42** D6
Lahri *Pakistan* 29°11N 68°13E **42** E3
Lāhrūd *Iran* 38°30N 47°52E **44** B5
Lahti *Finland* 60°58N 25°40E **8** F21
Lahtis = Lahti *Finland* 60°58N 25°40E **8** F21
Laï *Chad* 9°25N 16°18E **51** G9
Lai Chau *Vietnam* 22°5N 103°3E **38** A4
Laila = Layla *Si. Arabia* 22°10N 46°40E **47** C4
Laingsburg *S. Africa* 33°9S 20°52E **56** E3
Lainioälven ➤ *Sweden* 67°35N 22°40E **8** C20
Lairg *U.K.* 58°2N 4°24W **11** C4
Laisamis *Kenya* 1°36N 37°48E **54** B4
Laiwu *China* 36°15N 117°40E **35** F9
Laixi *China* 36°50N 120°31E **35** F11
Laiyang *China* 36°59N 120°45E **35** F11
Laiyuan *China* 39°20N 114°40E **34** E8
Laizhou *China* 37°8N 119°57E **35** F10
Laizhou Wan *China* 37°30N 119°30E **35** F10
Laja ➤ *Mexico* 20°55N 100°46W **86** C4
Lajamanu *Australia* 18°23S 130°38E **60** C5
Lak Sao *Laos* 18°11N 104°59E **38** C5
Lakaband *Pakistan* 31°2N 69°15E **42** D3
Lake *U.S.A.* 44°33N 110°24W **76** D8
Lake Alpine *U.S.A.* 38°29N 120°0W **78** G7
Lake Andes *U.S.A.* 43°9N 98°32W **80** D4
Lake Arthur *U.S.A.* 30°5N 92°41W **84** F8
Lake Bindegolly △
Australia 28°0S 144°12E **63** D3
Lake Cargelligo
Australia 33°15S 146°22E **63** E4
Lake Charles *U.S.A.* 30°14N 93°13W **84** F8
Lake City *Colo., U.S.A.* 38°2N 107°19W **76** G10
Lake City *Fla., U.S.A.* 30°11N 82°38W **85** F13
Lake City *Mich., U.S.A.* 44°20N 85°13W **81** C11
Lake City *Minn., U.S.A.* 44°27N 92°16W **80** C7
Lake City *Pa., U.S.A.* 42°1N 80°21W **82** D4
Lake City *S.C., U.S.A.* 33°52N 79°45W **85** E15
Lake Cowichan *Canada* 48°49N 124°3W **78** D4
Lake District △ *U.K.* 54°30N 3°21W **12** C4
Lake Elsinore *U.S.A.* 33°38N 117°20W **79** M9
Lake Eyre △ *Australia* 28°40S 137°31E **63** D2
Lake Gairdner △
Australia 31°41S 135°51E **63** E2
Lake George *U.S.A.* 43°26N 73°43W **83** C11

Lake Grace *Australia* 33°7S 118°28E **61** F2
Lake Gregory ○
Australia 20°12S 127°27E **60** D4
Lake Harbour = Kimmirut
Canada 62°50N 69°50W **69** E18
Lake Havasu City
U.S.A. 34°27N 114°22W **79** L12
Lake Hughes *U.S.A.* 34°41N 118°26W **79** L8
Lake Isabella *U.S.A.* 35°38N 118°28W **79** K8
Lake Jackson *U.S.A.* 29°3N 95°27W **84** G7
Lake King *Australia* 33°5S 119°45E **61** F2
Lake Lenore *Canada* 52°24N 104°59W **71** C8
Lake Louise *Canada* 51°30N 116°10W **70** C5
Lake Mackay ○
Australia 21°48S 129°46E **60** D4
Lake Mburo △ *Uganda* 0°33S 30°56E **54** C3
Lake Mead △ *U.S.A.* 36°30N 114°22W **79** K12
Lake Meredith △
U.S.A. 35°50N 101°50W **84** D4
Lake Mills *U.S.A.* 43°25N 93°32W **80** D7
Lake Nakura △ *Kenya* 0°21S 36°5E **54** C4
Lake Placid *U.S.A.* 44°17N 73°59W **83** B11
Lake Pleasant *U.S.A.* 43°28N 74°25W **83** C10
Lake Providence *U.S.A.* 32°48N 91°10W **84** E9
Lake Roosevelt △
U.S.A. 47°51N 118°14W **76** B4
Lake St. Peter *Canada* 45°18N 78°2W **82** A6
Lake Superior △ *Canada* 47°45N 84°45W **72** C3
Lake Torrens ○
Australia 30°55S 137°40E **63** E2
Lake Village *U.S.A.* 33°20N 91°17W **84** E9
Lake Wales *U.S.A.* 27°54N 81°35W **85** H14
Lake Worth *U.S.A.* 26°37N 80°3W **85** H14
Lakeba *Fiji* 18°13S 178°47W **59** a
Lakeba Passage *Fiji* 18°0S 178°45W **59** a
Lakefield *Canada* 44°25N 78°16W **82** B6
Lakefield △ *Australia* 15°24S 144°26E **62** B3
Lakehurst *U.S.A.* 40°1N 74°19W **83** F10
Lakeland *Australia* 15°49S 144°57E **62** B3
Lakeland *U.S.A.* 28°3N 81°57W **85** G14
Lakemba = Lakeba *Fiji* 18°13S 178°47W **59** a
Lakeport *Calif., U.S.A.* 39°3N 122°55W **78** F4
Lakeport *Mich., U.S.A.* 43°7N 82°30W **82** C2
Lakes Entrance *Australia* 37°50S 148°0E **63** F4
Lakeside *Calif., U.S.A.* 32°52N 116°55W **79** N10
Lakeside *Nebr., U.S.A.* 42°3N 102°26W **80** D2
Lakeview *U.S.A.* 42°11N 120°21W **76** E3
Lakeville *U.S.A.* 44°39N 93°14W **80** C7
Lakewood *Colo., U.S.A.* 39°42N 105°4W **76** G11
Lakewood *N.J., U.S.A.* 40°6N 74°13W **83** F10
Lakewood *N.Y., U.S.A.* 42°6N 79°19W **82** D5
Lakewood *Ohio, U.S.A.* 41°28N 81°47W **82** E3
Lakewood *Wash.,*
U.S.A. 47°11N 122°32W **78** C4
Lakha *India* 26°9N 70°54E **42** F4
Lakhimpur *India* 27°57N 80°46E **43** F9
Lakhisarai *India* 25°11N 86°5E **43** G12
Lakhnadon *India* 22°36N 79°36E **43** H8
Lakhnau = Lucknow
India 26°50N 81°0E **43** F9
Lakhonpheng *Laos* 15°54N 105°34E **38** E5
Lakhpat *India* 23°48N 68°47E **42** H3
Lakin *U.S.A.* 37°57N 101°15W **80** G3
Lakitusaki ➤ *Canada* 54°21N 82°25W **72** B3
Lakki *Greece* 35°24N 23°57E **25** D5
Lakki *Pakistan* 32°36N 70°55E **42** C4
Lakonikos Kolpos
Greece 36°40N 22°40E **23** F10
Lakor *Indonesia* 8°15S 128°17E **37** F7
Lakota *Ivory C.* 5°50N 5°30W **50** G4
Lakota *U.S.A.* 48°2N 98°21W **80** A4
Laksar *India* 29°46N 78°3E **42** E8
Laksefjorden *Norway* 70°45N 26°50E **8** A22
Lakselv *Norway* 70°2N 25°0E **8** A21
Lakshadweep Is. *India* 10°0N 72°30E **3** D13
Lakshmanpur *India* 22°58N 83°3E **43** H10
Lakshmikantapur *India* 22°5N 88°19E **43** H13
Lala Musa *Pakistan* 32°40N 73°57E **42** C5
Lalaghat *India* 24°5N 92°40E **41** E18
Lalago *Tanzania* 3°28S 33°58E **54** C3
Lalapanzi *Zimbabwe* 19°20S 30°15E **55** F3
L'Albufera *Spain* 39°20N 0°27W **21** C5
Lalganj *India* 25°52N 85°13E **43** G11
Lalgola *India* 24°25N 88°15E **43** G13
Lālī *Iran* 32°21N 49°6E **45** C6
Lalibela *Ethiopia* 12°3N 39°0E **47** E2
Lalín *Spain* 42°40N 8°5W **21** A1
Lalin He ➤ *China* 45°32N 125°40E **35** B13
Lalitapur *Nepal* 27°40N 85°20E **43** F11
Lalitpur *India* 24°42N 78°28E **43** G8
Lalkua *India* 29°5N 79°31E **43** E8
Lalsot *India* 26°34N 76°20E **42** F7
Lam Nam Nan △
Thailand 17°55N 100°28E **38** D3
Lam Pao Res. *Thailand* 16°50N 103°15E **38** D4
Lama Kara *Togo* 9°30N 1°15E **50** G6
Lamae *Thailand* 9°54N 99°5E **39** H2
Lamaing *Burma* 15°25N 97°53E **38** E1
Lamar *Colo., U.S.A.* 38°5N 102°37W **76** G12
Lamar *Mo., U.S.A.* 37°30N 94°16W **80** G6
Lamas *Peru* 6°28S 76°31W **92** E3
Lamayuru *India* 34°10N 76°30E **43** B7
Lambaréné *Gabon* 0°41S 10°12E **52** E2
Lambasa = Labasa *Fiji* 16°30S 179°27E **59** a
Lambay I. *Ireland* 53°29N 6°1W **10** C5
Lambert's Bay *S. Africa* 32°5S 18°17E **56** E2
Lambeth *Canada* 42°54N 81°18W **82** D3
Lambomakondro
Madag. 22°41S 44°44E **57** C7
Lame Deer *U.S.A.* 45°37N 106°40W **76** D10
Lamego *Portugal* 41°5N 7°52W **21** B2
Lamèque *Canada* 47°45N 64°38W **73** C7
Lameroo *Australia* 35°19S 140°33E **63** F3
Lamesa *U.S.A.* 32°44N 101°58W **84** E4
Lamia *Greece* 38°55N 22°26E **23** E10
Lamington △ *Australia* 28°13S 153°12E **63** D5
Lamma I. *China* 22°12N 114°7E **33** a
Lammermuir Hills *U.K.* 55°50N 2°40W **11** F6
Lamoille ➤ *U.S.A.* 44°38N 73°13W **83** B11

Lamon B. *Phil.* 14°30N 122°20E **37** B6
Lamont *Canada* 53°46N 112°50W **70** C6
Lamont *Calif., U.S.A.* 35°15N 118°55W **79** K8
Lamont *Wyo., U.S.A.* 42°13N 107°29W **76** E10
Lampa *Peru* 15°22S 70°22W **92** G4
Lampang *Thailand* 18°16N 99°32E **38** C2
Lampasas *U.S.A.* 31°4N 98°11W **84** F5
Lampazos de Naranjo
Mexico 27°1N 100°31W **86** B4
Lampedusa *Medit. S.* 35°36N 12°40E **22** G5
Lampeter *U.K.* 52°7N 4°4W **13** E3
Lamphun *Thailand* 18°40N 99°2E **38** C2
Lampione *Medit. S.* 35°33N 12°20E **22** G5
Lampman *Canada* 49°25N 102°50W **71** D8
Lampung □ *Indonesia* 5°30S 104°30E **36** F2
Lamta *India* 22°8N 80°7E **43** H9
Lamu *Kenya* 2°16S 40°55E **54** C5
Lamy *U.S.A.* 35°29N 105°53W **77** J11
Lan Xian *China* 38°15N 111°35E **34** E6
Lāna'i *U.S.A.* 20°50N 156°55W **75** L8
Lanak La *China* 34°27N 79°32E **43** B8
Lanak'o Shank'ou = Lanak La
China 34°27N 79°32E **43** B8
Lanark *Canada* 45°1N 76°22W **83** A8
Lanark *U.K.* 55°40N 3°47W **11** F5
Lanbi Kyun *Burma* 10°50N 98°20E **39** G2
Lancang Jiang ➤
China 21°40N 101°10E **32** G9
Lancashire □ *U.K.* 53°50N 2°48W **12** D5
Lancaster *Canada* 45°10N 74°30W **83** A10
Lancaster *U.K.* 54°3N 2°48W **12** C5
Lancaster *Calif., U.S.A.* 34°42N 118°8W **79** L8
Lancaster *Ky., U.S.A.* 37°37N 84°35W **81** G11
Lancaster *N.H., U.S.A.* 44°29N 71°34W **83** B13
Lancaster *Ohio, U.S.A.* 39°43N 82°36W **81** F12
Lancaster *Pa., U.S.A.* 40°2N 76°19W **83** F8
Lancaster *S.C., U.S.A.* 34°43N 80°46W **85** D14
Lancaster *Wis., U.S.A.* 42°51N 90°43W **80** D8
Lancaster Sd. *Canada* 74°13N 84°0W **69** C15
Lancelin *Australia* 31°0S 115°18E **61** F2
Lanchow = Lanzhou
China 36°1N 103°52E **34** F2
Lanciano *Italy* 42°14N 14°23E **22** C6
Lancun *China* 36°25N 120°10E **35** F11
Land Between the Lakes △
U.S.A. 36°25N 88°0W **85** C11
Landeck *Austria* 47°9N 10°34E **16** E6
Lander *U.S.A.* 42°50N 108°44W **76** E9
Lander ➤ *Australia* 22°0S 132°0E **60** D5
Landes *France* 44°0N 1°0W **20** D3
Landi Kotal *Pakistan* 34°7N 71°6E **42** B4
Landisburg *U.S.A.* 40°21N 77°19W **82** F7
Landmannalaugar *Iceland* 63°59N 19°4W **8** E4
Land's End *U.K.* 50°4N 5°44W **13** G2
Landsborough Cr. ➤
Australia 22°28S 144°35E **62** C3
Landshut *Germany* 48°34N 12°8E **16** D7
Lanesboro *U.S.A.* 41°57N 75°34W **83** E9
Lanett *U.S.A.* 32°52N 85°12W **85** E12
Lang Shan *China* 41°0N 106°30E **34** D4
Lang Son *Vietnam* 21°52N 106°42E **38** B6
Lang Suan *Thailand* 9°57N 99°4E **39** H2
La'nga Co *China* 30°45N 81°15E **43** D9
Langar *Iran* 35°23N 60°25E **45** C9
Langara I. *Canada* 54°14N 133°1W **70** C2
Langarūd *Iran* 37°11N 50°8E **45** B6
Langdon *U.S.A.* 48°45N 98°22W **80** A4
Langeberg *S. Africa* 33°55S 21°0E **56** E3
Langeberge *S. Africa* 28°15S 22°33E **56** D3
Langeland *Denmark* 54°56N 10°48E **9** J14
Langenburg *Canada* 50°51N 101°43W **71** C8
Langfang *China* 39°30N 116°41E **34** E9
Langholm *U.K.* 55°9N 3°0W **11** F5
Langjökull *Iceland* 64°39N 20°12W **8** D3
Langkawi, Pulau
Malaysia 6°25N 99°45E **39** J2
Langklip *S. Africa* 28°12S 20°20E **56** D3
Langkon *Malaysia* 6°30N 116°40E **36** C5
Langley *Canada* 49°7N 122°39W **78** A4
Langøya *Norway* 68°45N 14°50E **8** B16
Langreo *Spain* 43°18N 5°40W **21** A3
Langres *France* 47°52N 5°20E **20** C6
Langres, Plateau de *France* 47°45N 5°3E **20** C6
Langsa *Indonesia* 4°30N 97°57E **36** D1
Langtang □ *Nepal* 28°10N 85°30E **43** E11
Langtou *China* 40°1N 124°19E **35** D13
Langtry *U.S.A.* 29°49N 101°34W **84** G4
Langu *Thailand* 6°53N 99°47E **39** J2
Languedoc *France* 43°58N 3°55E **20** E5
Langwang *China* **33** a
Lanigan *Canada* 51°51N 105°2W **71** C7
Lankao *China* 34°48N 114°50E **34** G8
Länkäran *Azerbaijan* 38°48N 48°52E **45** B6
Lannion *France* 48°46N 3°29W **20** B2
L'Annonciation *Canada* 46°25N 74°55W **72** C5
Lansang △ *Thailand* 16°45N 99°0E **38** D2
Lansdale *U.S.A.* 40°14N 75°17W **83** F9
Lansdowne *Australia* 31°48S 152°30E **63** E5
Lansdowne *Canada* 44°24N 76°1W **83** B8
Lansdowne *India* 29°50N 78°41E **43** E8
Lansdowne House = Neskantaga
Canada 52°14N 87°53W **72** B2
L'Anse *U.S.A.* 46°45N 88°27W **80** B9
L'Anse au Loup *Canada* 51°32N 56°50W **73** B8
L'Anse aux Meadows
Canada 51°36N 55°32W **73** B8
L'Anse la Raye *St. Lucia* 13°55N 61°3W **89** f
Lansford *U.S.A.* 40°50N 75°53W **83** F9
Lanshantou *China* 35°5N 119°20E **35** G10
Länsi-Turunmaa *Finland* 60°18N 22°18E **9** F20
Lansing *U.S.A.* 42°44N 84°33W **81** D11
Lanta, Ko *Thailand* 7°35N 99°3E **39** J2
Lantau I. *China* 22°15N 113°56E **33** a
Lantian *China* 34°11N 109°20E **34** G5
Lanus *Argentina* 34°42S 58°23W **94** C4
Lanusei *Italy* 39°52N 9°34E **22** E3
Lanzarote *Canary Is.* 29°0N 13°40W **24** F6
Lanzarote ✈ (ACE)
Canary Is. 28°57N 13°40W **24** F6
Lanzhou *China* 36°1N 103°52E **34** F2

Lao Cai *Vietnam* 22°30N 103°57E **38** A4
Laoag *Phil.* 18°7N 120°34E **37** A6
Laoang *Phil.* 12°32N 125°8E **37** B7
Laoha He → *China* 43°25N 120°35E **35** C11
Laois □ *Ireland* 52°57N 7°36W **10** D4
Laon *France* 49°33N 3°35E **20** B5
Laona *U.S.A.* 45°34N 88°40W **80** C9
Laos ■ *Asia* 17°45N 105°0E **38** D5
Lapa *Brazil* 25°46S 49°44W **95** B6
Lapeer *U.S.A.* 43°3N 83°19W **81** D12
Lapithos *Cyprus* 35°21N 33°11E **25** D12
Lapland = Lappland *Europe* 68°7N 24°0E **8** B21
LaPorte *Ind., U.S.A.* 41°36N 86°43W **80** E10
Laporte *Pa., U.S.A.* 41°25N 76°30W **83** E8
Lappeenranta *Finland* 61°3N 28°12E **8** F23
Lappland *Europe* 68°7N 24°0E **8** B21
Lappo = Lapua *Finland* 62°58N 23°0E **8** E20
Laprida *Argentina* 37°34S 60°45W **94** D3
Lâpsekı *Turkey* 40°20N 26°41E **23** D12
Laptev Sea *Russia* 76°0N 125°0E **29** B13
Lapua *Finland* 62°58N 23°0E **8** E20
L'Àquila *Italy* 42°22N 13°22E **22** C5
Lār *Iran* 27°40N 54°14E **45** E7
Laramie *U.S.A.* 41°19N 105°35W **76** F11
Laramie Mts. *U.S.A.* 42°0N 105°30W **76** F11
Laranjeiras do Sul *Brazil* 25°23S 52°23W **95** B5
Larantuka *Indonesia* 8°21S 122°55E **37** F6
Larat *Indonesia* 7°0S 132°0E **37** F8
Larde *Mozam.* 16°28S 39°43E **55** F4
Larder Lake *Canada* 48°5N 79°40W **72** C4
Lardos *Greece* 36°6N 28°1E **25** C10
Lardos, Akra = Lindos, Akra *Greece* 36°4N 28°10E **25** C10
Lardos, Ormos *Greece* 36°4N 28°2E **25** C10
Lare *Kenya* 0°20N 37°56E **54** C4
Laredo *U.S.A.* 27°30N 99°30W **84** H5
Laredo Sd. *Canada* 52°30N 128°53W **70** C3
Largo *U.S.A.* 27°54N 82°47W **85** H13
Largs *U.K.* 55°47N 4°52W **11** F4
Lariang *Indonesia* 1°26S 119°17E **37** E5
Larimore *U.S.A.* 47°54N 97°38W **80** B5
Larisa *Greece* 39°36N 22°27E **23** E10
Larkana *Pakistan* 27°32N 68°18E **42** F3
Larnaca *Cyprus* 34°55N 33°38E **25** E12
Larnaca Bay *Cyprus* 34°53N 33°45E **25** E12
Larne *U.K.* 54°51N 5°51W **10** B6
Larned *U.S.A.* 38°11N 99°6W **80** F4
Larose *U.S.A.* 29°34N 90°23W **85** G9
Larrimah *Australia* 15°35S 133°12E **60** C5
Larsen Ice Shelf *Antarctica* 67°0S 62°0W **5** C17
Larvik *Norway* 59°4N 10°2E **9** G14
Las Animas *U.S.A.* 38°4N 103°13W **76** G12
Las Anod *Somali Rep.* 8°26N 47°19E **47** F4
Las Brenãs *Argentina* 27°5S 61°7W **94** B3
Las Cañadas del Teide △ *Canary Is.* 28°15N 16°37W **24** F3
Las Cejas *Argentina* 26°53S 64°44W **96** B4
Las Chimeneas *Mexico* 32°8N 116°5W **79** N10
Las Cruces *U.S.A.* 32°19N 106°47W **77** K10
Las Flores *Argentina* 36°10S 59°7W **94** D4
Las Heras *Argentina* 32°51S 68°49W **94** C2
Las Lajas *Argentina* 38°30S 70°25W **96** D2
Las Lomitas *Argentina* 24°43S 60°35W **94** A3
Las Palmas *Argentina* 27°8S 58°45W **94** B4
Las Palmas *Canary Is.* 28°7N 15°26W **24** F4
Las Palmas → *Mexico* 32°31N 116°58W **79** N10
Las Palmas ✈ (LPA) *Canary Is.* 27°55N 15°25W **24** G4
Las Piedras *Uruguay* 34°44S 56°14W **95** C4
Las Pipinas *Argentina* 35°30S 57°19W **94** D4
Las Plumas *Argentina* 43°40S 67°15W **96** E3
Las Rosas *Argentina* 32°30S 61°35W **94** C3
Las Tablas *Panama* 7°49N 80°14W **88** E3
Las Toscas *Argentina* 28°21S 59°18W **94** B4
Las Truchas *Mexico* 17°57N 102°13W **86** D4
Las Tunas *Cuba* 20°58N 76°59W **88** B4
Las Varillas *Argentina* 31°50S 62°50W **94** C3
Las Vegas *N. Mex., U.S.A.* 35°36N 105°13W **77** J11
Las Vegas *Nev., U.S.A.* 36°10N 115°8W **79** J11
Las Vegas McCarran Int. ✈ (LAS) *U.S.A.* 36°5N 115°9W **79** J11
Lascano *Uruguay* 33°35S 54°12W **95** C5
Lash-e Joveyn *Afghan.* 31°45N 61°30E **40** D2
Lashburn *Canada* 53°10N 109°40W **71** C7
Lashio *Burma* 22°56N 97°45E **41** H20
Lashkar *India* 26°10N 78°10E **42** F8
Lasithi *Greece* 35°11N 25°31E **25** D7
Lasithi □ *Greece* 35°5N 25°50E **25** D7
Lāsjerd *Iran* 35°24N 53°4E **45** C7
Lassen Pk. *U.S.A.* 40°29N 121°30W **76** F3
Lassen Volcanic △ *U.S.A.* 40°30N 121°20W **76** F3
Last Mountain L. *Canada* 51°5N 105°14W **71** C7
Lastchance Cr. → *U.S.A.* 40°2N 121°15W **78** E5
Lastoursville *Gabon* 0°55S 12°38E **52** E2
Lastovo *Croatia* 42°46N 16°55E **22** C7
Lat Yao *Thailand* 15°45N 99°48E **38** E2
Latacunga *Ecuador* 0°50S 78°35W **92** D3
Latakia = Al Lādhiqīyah *Syria* 35°30N 35°45E **44** C2
Latchford *Canada* 47°20N 79°50W **72** C4
Late *Tonga* 18°48S 174°39W **59** c
Latehar *India* 23°45N 84°30E **43** H11
Latham *Australia* 29°44S 116°20E **61** E2
Lathi *India* 27°43N 71°23E **42** F4
Lathrop Wells *U.S.A.* 36°39N 116°24W **79** J10
Latina *Italy* 41°28N 12°52E **22** D5
Latium = Lazio □ *Italy* 42°10N 12°30E **22** C5
Laton *U.S.A.* 36°26N 119°41W **78** J7
Latouche Treville, C. *Australia* 18°27S 121°49E **60** C3
Latrobe *Australia* 41°14S 146°30E **63** G4
Latrobe *U.S.A.* 40°19N 79°23W **82** F5
Latvia ■ *Europe* 56°50N 24°0E **9** H21
Lau Fau Shan *China* 22°28N 113°59E **33** a
Lau Group *Fiji* 17°0S 178°30W **59** a
Lauchhammer *Germany* 51°29N 13°47E **16** C7
Lauge Koch Kyst *Greenland* 75°45N 57°45W **69** B20
Laughlin *U.S.A.* 35°10N 114°34W **77** J6
Laukaa *Finland* 62°24N 25°56E **8** E21
Launceston *Australia* 41°24S 147°8E **63** G4
Launceston *U.K.* 50°38N 4°22W **13** G3
Laune → *Ireland* 52°7N 9°47W **10** D2
Launglon Bok *Burma* 13°50N 97°54E **38** F1
Laura *Australia* 15°32S 144°32E **62** B3
Laurel *Miss., U.S.A.* 31°41N 89°8W **85** F10
Laurel *Mont., U.S.A.* 45°40N 108°46W **76** D9
Laurel Hill *U.S.A.* 40°14N 79°6W **82** F5
Laurencekirk *U.K.* 56°50N 2°28W **11** E6
Laurens *U.S.A.* 34°30N 82°1W **85** D13
Laurentian Plateau *Canada* 52°0N 70°0W **73** B6
Lauria *Italy* 40°2N 15°50E **22** E6
Laurie L. *Canada* 56°35N 101°57W **71** B8
Laurinburg *U.S.A.* 34°47N 79°28W **85** D15
Laurium *U.S.A.* 47°14N 88°27W **80** B9
Lausanne *Switz.* 46°32N 6°38E **20** C7
Laut *Indonesia* 4°45N 108°0E **36** D3
Laut, Pulau *Indonesia* 3°40S 116°10E **36** E5
Laut Kecil, Kepulauan *Indonesia* 4°45S 115°40E **36** E5
Lautoka *Fiji* 17°37S 177°27E **59** a
Lava Beds ⌓ *U.S.A.* 41°40N 121°30W **76** F3
Lavagh More *Ireland* 54°46N 8°6W **10** B3
Laval *France* 48°4N 0°48W **20** B3
Laval-des-Rapides *Canada* 45°33N 73°42W **72** C5
Lavalle *Argentina* 28°15S 65°15W **94** B2
Lāvān *Iran* 26°48N 53°22E **45** E7
Lavant *Canada* 45°3N 76°42W **83** A8
Lāvar Meydān *Iran* 30°20N 54°30E **45** D7
Laverton *Australia* 28°44S 122°29E **61** E3
Lavras *Brazil* 21°20S 45°0W **95** A7
Lavrio *Greece* 37°40N 24°4E **23** F11
Lavris *Greece* 35°25N 24°40E **25** D6
Lavumisa *Swaziland* 27°20S 31°55E **57** D5
Lavushi Manda △ *Zambia* 12°46S 31°0E **55** E3
Lawaki *Fiji* 17°40S 178°35E **59** a
Lawas *Malaysia* 4°55N 115°25E **36** D5
Lawele *Indonesia* 5°13S 122°57E **37** F6
Lawn Hill = Boodjamulla △ *Australia* 18°15S 138°6E **62** B2
Lawqah *Si. Arabia* 29°49N 42°45E **44** D4
Lawrence *N.Z.* 45°55S 169°41E **59** F2
Lawrence *Ind., U.S.A.* 39°50N 86°2W **80** F10
Lawrence *Kans., U.S.A.* 38°58N 95°14W **80** F6
Lawrence *Mass., U.S.A.* 42°43N 71°10W **83** D13
Lawrenceburg *Ind., U.S.A.* 39°6N 84°52W **81** F11
Lawrenceburg *Tenn., U.S.A.* 35°14N 87°20W **85** D11
Lawrenceville *Ga., U.S.A.* 33°57N 83°59W **85** E13
Lawrenceville *Pa., U.S.A.* 41°59N 77°8W **82** E7
Laws *U.S.A.* 37°24N 118°20W **78** H8
Lawton *U.S.A.* 34°37N 98°25W **84** D5
Lawu *Indonesia* 7°40S 111°13E **37** G14
Laxford, L. *U.K.* 58°24N 5°6W **11** C3
Laylá *Si. Arabia* 22°10N 46°40E **47** C4
Laylān *Iraq* 35°18N 44°31E **44** C5
Laysan I. *U.S.A.* 25°50N 171°50W **75** K5
Layton *U.S.A.* 41°4N 111°58W **76** F8
Laytonville *U.S.A.* 39°41N 123°29W **76** G2
Lazarev *Russia* 52°13N 141°30E **29** D15
Lazarev Sea *S. Ocean* 67°30S 3°0W **5** C2
Lazarivo *Madag.* 23°54S 44°59E **57** C8
Lázaro Cárdenas *Mexico* 17°55N 102°11W **86** D4
Lazio □ *Italy* 42°10N 12°30E **22** C5
Lazo *Russia* 43°25N 133°55E **30** C6
Le Bic *Canada* 48°20N 68°41W **73** C6
Le Creusot *France* 46°48N 4°24E **20** C6
Le François *Martinique* 14°38N 60°57W **88** c
Le Gosier *Guadeloupe* 16°14N 61°29W **88** b
Le Gris Gris *Mauritius* 20°31S 57°32E **53** d
Le Havre *France* 49°30N 0°5E **20** B4
Le Lamentin *Martinique* 14°35N 61°2W **88** c
Le Mans *France* 48°0N 0°10E **20** C4
Le Marin *Martinique* 14°27N 60°55W **88** c
Le Mars *U.S.A.* 42°47N 96°10W **80** D5
Le Mont-St-Michel *France* 48°40N 1°30W **20** B3
Le Moule *Guadeloupe* 16°20N 61°22W **88** b
Le Moyne, L. *Canada* 56°45N 68°47W **73** A6
Le Port *Réunion* 20°56S 55°18E **53** c
Le Prêcheur *Martinique* 14°50N 61°12W **88** c
Le Puy-en-Velay *France* 45°3N 3°52E **20** D5
Le Raysville *U.S.A.* 41°50N 76°10W **83** E8
Le Robert *Martinique* 14°40N 60°56W **88** c
Le Roy *U.S.A.* 42°58N 77°59W **82** D7
Le St-Esprit *Martinique* 14°34N 60°56W **88** c
Le Sueur *U.S.A.* 44°28N 93°55W **80** C7
Le Tampon *Réunion* 21°16S 55°32E **53** c
Le Thuy *Vietnam* 17°14N 106°49E **38** D6
Le Touquet-Paris-Plage *France* 50°30N 1°36E **20** A4
Le Tréport *France* 50°3N 1°20E **20** A4
Le Verdon-sur-Mer *France* 45°33N 1°4W **20** D3
Lea → *U.K.* 51°31N 0°1E **13** F8
Leach *Cambodia* 12°21N 103°46E **39** F4
Lead *U.S.A.* 44°21N 103°46W **80** C2
Leader *Canada* 50°50N 109°30W **71** C7
Leadville *U.S.A.* 39°15N 106°18W **76** G10
Leaf → *U.S.A.* 30°59N 88°44W **85** F10
Leaf Rapids *Canada* 56°30N 99°59W **71** B9
Leamington *Canada* 42°3N 82°36W **82** D2
Leamington *U.S.A.* 39°32N 112°17W **76** G7
Leamington Spa = Royal Leamington Spa *U.K.* 52°18N 1°31W **13** E6
Leandro Norte Alem *Argentina* 27°34S 55°15W **95** B4
Leane, L. *Ireland* 52°2N 9°32W **10** D2
Learmonth *Australia* 22°13S 114°10E **60** D1
Leask *Canada* 53°5N 106°45W **71** C7
Leatherhead *U.K.* 51°18N 0°20W **13** F7
Leavdnja = Lakselv *Norway* 70°2N 25°0E **8** A21
Leavenworth *Kans., U.S.A.* 39°19N 94°55W **80** F6
Leavenworth *Wash., U.S.A.* 47°36N 120°40W **76** C3
Leawood *U.S.A.* 38°58N 94°37W **80** F6
Lebak *Phil.* 6°32N 124°5E **37** C6
Lebam *U.S.A.* 46°34N 123°33W **78** D3
Lebanon *Ind., U.S.A.* 40°3N 86°28W **80** E10
Lebanon *Kans., U.S.A.* 39°49N 98°33W **80** F4
Lebanon *Ky., U.S.A.* 37°34N 85°15W **81** G11
Lebanon *Mo., U.S.A.* 37°41N 92°40W **80** G7
Lebanon *N.H., U.S.A.* 43°39N 72°15W **83** C12
Lebanon *Oreg., U.S.A.* 44°32N 122°55W **76** D2
Lebanon *Pa., U.S.A.* 40°20N 76°26W **83** F8
Lebanon *Tenn., U.S.A.* 36°12N 86°18W **85** C11
Lebanon ■ *Asia* 34°0N 36°0E **46** B5
Lebec *U.S.A.* 34°51N 118°52W **79** L8
Lebel-sur-Quévillon *Canada* 49°3N 76°59W **72** C4
Lebomboberge *S. Africa* 24°30S 32°0E **57** C5
Lębork *Poland* 54°33N 17°46E **17** A9
Lebrija *Spain* 36°53N 6°5W **21** D2
Lebu *Chile* 37°40S 73°47W **94** D1
Lecce *Italy* 40°23N 18°11E **23** D8
Lecco *Italy* 45°51N 9°23E **20** D8
Lech → *Germany* 48°43N 10°56E **16** D6
Lecontes Mills *U.S.A.* 41°5N 78°17W **82** E6
Łęczyca *Poland* 52°5N 19°15E **17** B10
Ledang, Gunung *Malaysia* 2°22N 102°37E **39** L4
Ledong *China* 18°41N 109°5E **38** C7
Leduc *Canada* 53°15N 113°30W **70** C6
Lee *U.S.A.* 42°19N 73°15W **83** D11
Lee → *Ireland* 51°53N 8°56W **10** E3
Lee Vining *U.S.A.* 37°58N 119°7W **78** H7
Leech L. *U.S.A.* 47°10N 94°24W **80** B6
Leechburg *U.S.A.* 40°37N 79°36W **82** F5
Leeds *U.K.* 53°48N 1°33W **12** D6
Leeds *U.S.A.* 33°33N 86°33W **85** E11
Leek *Neths.* 53°10N 6°24E **15** A6
Leek *U.K.* 53°7N 2°1W **12** D5
Leeman *Australia* 29°57S 114°58E **61** E1
Leeper *U.S.A.* 41°22N 79°18W **82** E5
Leer *Germany* 53°13N 7°26E **16** B4
Leesburg *U.S.A.* 28°49N 81°53W **85** G14
Leesville *U.S.A.* 31°8N 93°16W **84** F8
Leeton *Australia* 34°33S 146°23E **63** E4
Leetonia *U.S.A.* 40°53N 80°45W **82** F4
Leeu Gamka *S. Africa* 32°47S 21°59E **56** E3
Leeuwarden *Neths.* 53°15N 5°48E **15** A5
Leeuwin, C. *Australia* 34°20S 115°9E **61** F2
Leeuwin Naturaliste △ *Australia* 34°6S 115°3E **61** F2
Leeward Is. *Atl. Oc.* 16°30N 63°30W **89** C7
Lefka *Cyprus* 35°6N 32°51E **25** D11
Lefkada *Greece* 38°40N 20°43E **23** E9
Lefkimi *Greece* 39°25N 20°3E **25** B4
Lefkimis, Akra *Greece* 39°29N 20°4E **25** B4
Lefkoniko *Cyprus* 35°18N 33°44E **25** D12
Lefroy *Canada* 44°16N 79°34W **82** B5
Lefroy, L. *Australia* 31°21S 121°40E **61** F3
Leganés *Spain* 40°19N 3°45W **21** B4
Legazpi *Phil.* 13°10N 123°45E **37** B6
Legendre I. *Australia* 20°22S 116°55E **60** D2
Leghorn = Livorno *Italy* 43°33N 10°19E **22** C4
Legian *Indonesia* 8°42S 115°10E **37** K18
Legionowo *Poland* 52°25N 20°50E **17** B11
Legnago *Italy* 45°11N 11°18E **22** B4
Legnica *Poland* 51°12N 16°10E **16** C9
Leh *India* 34°9N 77°35E **43** B7
Lehigh Acres *U.S.A.* 26°36N 81°39W **85** H14
Lehighton *U.S.A.* 40°50N 75°43W **83** F9
Lehututu *Botswana* 23°54S 21°55E **56** C3
Leiah *Pakistan* 30°58N 70°58E **42** D4
Leicester *U.K.* 52°38N 1°8W **13** E6
Leicester City □ *U.K.* 52°38N 1°8W **13** E6
Leicestershire □ *U.K.* 52°41N 1°17W **13** E6
Leichhardt → *Australia* 17°35S 139°48E **62** B2
Leichhardt Ra. *Australia* 20°46S 147°40E **62** C4
Leiden *Neths.* 52°9N 4°30E **15** B4
Leie → *Belgium* 51°2N 3°45E **15** C3
Leimus *Nic.* 14°40N 84°3W **88** D3
Leine → *Germany* 52°43N 9°36E **16** B5
Leinster *Australia* 27°51S 120°36E **61** E3
Leinster □ *Ireland* 53°3N 7°8W **10** C4
Leinster, Mt. *Ireland* 52°37N 6°46W **10** D5
Leipzig *Germany* 51°18N 12°22E **16** C7
Leiria *Portugal* 39°46N 8°53W **21** C1
Leirvik *Norway* 59°47N 5°28E **9** G11
Leisler, Mt. *Australia* 23°23S 129°20E **60** D4
Leith *U.K.* 55°59N 3°11W **11** F5
Leith Hill *U.K.* 51°11N 0°22W **13** F7
Leitir Ceanainn = Letterkenny *Ireland* 54°57N 7°45W **10** B4
Leitrim *Ireland* 54°0N 8°5W **10** B3
Leitrim □ *Ireland* 54°8N 8°0W **10** B4
Leizhou Bandao *China* 21°0N 110°0E **33** G11
Lek → *Neths.* 51°54N 4°35E **15** C4
Leka *Norway* 65°5N 11°35E **8** D14
Leland *Mich., U.S.A.* 45°1N 85°45W **81** C11
Leland *Miss., U.S.A.* 33°24N 90°54W **85** E9
Leleque *Argentina* 42°28S 71°0W **96** E2
Leling *China* 37°44N 117°13E **35** F9
Lelystad *Neths.* 52°30N 5°25E **15** B5
Léman, L. *Europe* 46°26N 6°30E **20** C7
Lembar *Indonesia* 8°45S 116°4E **37** K19
Lembongan, Nusa *Indonesia* 8°40S 115°27E **37** K18
Lembuak *Indonesia* 8°36S 116°11E **37** K19
Lemera *Dem. Rep. of the Congo* 3°0S 28°55E **54** C2
Lemhi Ra. *U.S.A.* 44°0N 113°0W **76** D7
Lemmer *Neths.* 52°51N 5°43E **15** B5
Lemmon *U.S.A.* 45°57N 102°10W **80** C2
Lemon Grove *U.S.A.* 32°44N 117°1W **79** N9
Lemoore *U.S.A.* 36°18N 119°46W **78** J7
Lemvig *Denmark* 56°33N 8°20E **9** H13
Lena → *Russia* 72°52N 126°40E **29** B13
Lenadoon Pt. *Ireland* 54°18N 9°3W **10** B2
Lenggong *Malaysia* 5°6N 100°58E **39** K3
Lengua de Vaca, Pta. *Chile* 30°14S 71°38W **94** C1
Lengwe △ *Malawi* 16°14S 34°45E **55** F3
Leninogorsk *Kazakhstan* 50°20N 83°30E **28** D9
Leninsk *Russia* 48°40N 45°15E **19** E8
Leninsk-Kuznetskiy *Russia* 54°44N 86°10E **28** D9
Lenkoran = Länkäran *Azerbaijan* 38°48N 48°52E **45** B6
Lenmalu *Indonesia* 1°45S 130°15E **37** E8
Lennox *U.S.A.* 43°21N 96°53W **80** D5
Lennoxville *Canada* 45°22N 71°51W **83** A13
Lenoir *U.S.A.* 35°55N 81°32W **85** D14
Lenoir City *U.S.A.* 35°48N 84°16W **85** D12
Lenore L. *Canada* 52°30N 104°59W **71** C8
Lenox *U.S.A.* 42°22N 73°17W **83** D11
Lens *France* 50°26N 2°50E **20** A5
Lensk *Russia* 60°48N 114°55E **29** C12
Lentas *Greece* 34°56N 24°56E **25** E6
Lentini *Italy* 37°17N 15°0E **22** F6
Lenwood *U.S.A.* 34°53N 117°7W **79** L9
Lenya *Burma* 11°33N 98°57E **39** G2
Leoben *Austria* 47°22N 15°5E **16** E8
Leodhas = Lewis *U.K.* 58°9N 6°40W **11** C2
Leola *U.S.A.* 45°43N 98°56W **80** C4
Leominster *U.K.* 52°14N 2°43W **13** E5
Leominster *U.S.A.* 42°32N 71°46W **83** D13
León *Mexico* 21°6N 101°41W **86** C4
León *Nic.* 12°27N 86°51W **88** D2
León *Spain* 42°38N 5°34W **21** A3
Leon *U.S.A.* 40°44N 93°45W **80** E7
Leon → *U.S.A.* 31°14N 97°28W **84** F6
León, Montes de *Spain* 42°30N 6°18W **21** A2
Leonardtown *U.S.A.* 38°17N 76°38W **81** F15
Leonardville *Namibia* 23°29S 18°49E **56** C2
Leone *Amer. Samoa* 14°23S 170°48W **59** b
Leongatha *Australia* 38°30S 145°58E **63** F4
Leonora *Australia* 28°49S 121°19E **61** E3
Leopoldina *Brazil* 21°28S 42°40W **95** A7
Leopoldsburg *Belgium* 51°7N 5°13E **15** C5
Leoti *U.S.A.* 38°29N 101°21W **80** F3
Leova *Moldova* 46°28N 28°15E **17** E15
Leoville *Canada* 53°39N 107°33W **71** C7
Lepel = Lyepyel *Belarus* 54°50N 28°40E **9** J23
Lépo, L. do *Angola* 17°0S 19°0E **56** B2
Leppävirta *Finland* 62°29N 27°46E **8** E22
Leptis Magna *Libya* 32°40N 14°12E **51** B8
Leribe *Lesotho* 28°51S 28°3E **57** D4
Lérida = Lleida *Spain* 41°37N 0°39E **21** B6
Lerwick *U.K.* 60°9N 1°9W **11** A7
Les Cayes *Haiti* 18°15N 73°46W **89** C5
Les Coteaux *Canada* 45°15N 74°13W **83** A10
Les Escoumins *Canada* 48°21N 69°24W **73** C6
Les Sables-d'Olonne *France* 46°30N 1°45W **20** C3
Lesbos *Greece* 39°10N 26°20E **23** E12
Leshan *China* 29°33N 103°41E **32** F9
Leshukonskoye *Russia* 64°54N 45°46E **18** B8
Leshwe ⌓ *Dem. Rep. of the Congo* 12°45S 29°30E **55** E2
Leskov I. *Antarctica* 56°0S 28°0W **5** B1
Leskovac *Serbia* 43°0N 21°58E **23** C9
Lesopilnoye *Russia* 46°44N 134°20E **30** A7
Lesotho ■ *Africa* 29°40S 28°0E **57** D4
Lesozavodsk *Russia* 45°30N 133°29E **30** B6
Lesse → *Belgium* 50°15N 4°54E **15** D4
Lesse et Lomme ⌓ *Belgium* 50°8N 5°9E **15** D5
Lesser Antilles *W. Indies* 15°0N 61°0W **89** D7
Lesser Slave L. *Canada* 55°30N 115°25W **70** B5
Lesser Sunda Is. *Indonesia* 8°0S 120°0E **37** F6
Lessines *Belgium* 50°42N 3°50E **15** D3
Lester B. Pearson Int., Toronto ✈ (YYZ) *Canada* 43°46N 79°35W **82** C5
Lestock *Canada* 51°19N 103°59W **71** C8
Lesuer I. *Australia* 13°50S 127°17E **60** B4
Lesueur △ *Australia* 30°11S 115°10E **61** F2
Lésvos = Lesbos *Greece* 39°10N 26°20E **23** E12
Leszno *Poland* 51°50N 16°30E **17** C9
Letaba *S. Africa* 23°59S 31°50E **57** C5
Letchworth *U.K.* 51°59N 0°13W **13** F7
Lethbridge *Canada* 49°45N 112°45W **70** D6
Lethem *Guyana* 3°20N 59°50W **92** C7
Leti, Kepulauan *Indonesia* 8°10S 128°0E **37** F7
Leti Is. = Leti, Kepulauan *Indonesia* 8°10S 128°0E **37** F7
Letiahau → *Botswana* 21°16S 24°0E **56** C3
Leticia *Colombia* 4°9S 70°0W **92** D5
Leting *China* 39°23N 118°55E **35** E10
Letjiesbos *S. Africa* 32°34S 22°16E **56** E3
Letlhakane *Botswana* 21°27S 25°30E **56** C4
Letlhakeng *Botswana* 24°0S 24°59E **56** C3
Letpadan *Burma* 17°45N 95°45E **41** L19
Letpan *Burma* 19°28N 94°10E **41** K19
Letsôk-aw Kyun *Burma* 11°30N 98°25E **39** G2
Letterkenny *Ireland* 54°57N 7°45W **10** B4
Leucadia *U.S.A.* 33°4N 117°18W **79** M9
Leuchars *U.K.* 56°24N 2°53W **11** E6
Leuser, Gunung *Indonesia* 3°46N 97°12E **36** D1
Leuven *Belgium* 50°52N 4°42E **15** D4
Leuze-en-Hainaut *Belgium* 50°36N 3°37E **15** D3
Levanger *Norway* 63°45N 11°19E **8** E14
Levelland *U.S.A.* 33°35N 102°23W **84** E3
Leven *U.K.* 56°12N 3°0W **11** E6
Leven, L. *U.K.* 56°12N 3°22W **11** E5
Leven, Toraka *Madag.* 12°30S 47°45E **57** A8
Leveque C. *Australia* 16°20S 123°0E **60** C3
Levice *Slovak Rep.* 48°13N 18°35E **17** D10
Levin *N.Z.* 40°37S 175°18E **59** D5
Lévis *Canada* 46°48N 71°9W **73** C5
Levis, L. *Canada* 62°37N 117°58W **70** A5
Levittown *N.Y., U.S.A.* 40°44N 73°31W **83** F11
Levittown *Pa., U.S.A.* 40°9N 74°51W **83** F10
Levka Oros *Greece* 35°18N 24°3E **25** D6
Levkás = Lefkada *Greece* 38°40N 20°43E **23** E9
Levkôsia = Nicosia *Cyprus* 35°10N 33°25E **25** D12
Levskigrad = Karlovo *Bulgaria* 42°38N 24°47E **23** C11
Levuka *Fiji* 17°34S 179°0E **5**
Lewes *U.K.* 50°52N 0°1E **13**
Lewes *U.S.A.* 38°46N 75°9W **81**
Lewis *U.K.* 58°9N 6°40W **11**
Lewis → *U.S.A.* 45°51N 122°48W **78**
Lewis, Butt of *U.K.* 58°31N 6°16W **11**
Lewis and Clark ⌓ *U.S.A.* 46°8N 123°53W **78**
Lewis Ra. *Australia* 20°3S 128°50E **60**
Lewis Range *U.S.A.* 48°5N 113°5W **76**
Lewis Run *U.S.A.* 41°52N 78°40W **82**
Lewisburg *Pa., U.S.A.* 40°58N 76°54W **8**
Lewisburg *Tenn., U.S.A.* 35°27N 86°48W **85**
Lewisburg *W. Va., U.S.A.* 37°48N 80°27W **81**
Lewisporte *Canada* 49°15N 55°3W **73**
Lewiston *Idaho, U.S.A.* 46°25N 117°1W **76**
Lewiston *Maine, U.S.A.* 44°6N 70°13W **81**
Lewiston *N.Y., U.S.A.* 43°11N 79°3W **82**
Lewistown *Mont., U.S.A.* 47°4N 109°26W **76**
Lewistown *Pa., U.S.A.* 40°36N 77°34W **82**
Lexington *Ill., U.S.A.* 40°39N 88°47W **8**
Lexington *Ky., U.S.A.* 38°3N 84°30W **81**
Lexington *Mich., U.S.A.* 43°16N 82°32W **82**
Lexington *Mo., U.S.A.* 39°11N 93°52W **80**
Lexington *N.C., U.S.A.* 35°49N 80°15W **85**
Lexington *N.Y., U.S.A.* 42°15N 74°22W **83**
Lexington *Nebr., U.S.A.* 40°47N 99°45W **8**
Lexington *Ohio, U.S.A.* 40°41N 82°35W **82**
Lexington *Tenn., U.S.A.* 35°39N 88°24W **85**
Lexington *Va., U.S.A.* 37°47N 79°27W **81**
Lexington Park *U.S.A.* 38°16N 76°27W **81**
Leyburn *U.K.* 54°19N 1°48W **12**
Leyland *U.K.* 53°42N 2°43W **12**
Leyte □ *Phil.* 11°0N 125°0E **37**
Lezhë *Albania* 41°47N 19°39E **23**
Lhasa *China* 29°25N 90°58E **32**
Lhazê *China* 29°5N 87°38E **32**
L'Hermite, I. *Chile* 55°50S 68°0W **96**
Lhokkruet *Indonesia* 4°55N 95°24E **36**
Lhokseumawe *Indonesia* 5°10N 97°10E **36**
L'Hospitalet de Llobregat *Spain* 41°21N 2°6E **21**
Li *Thailand* 17°48N 98°57E **38**
Li Shan *China* 35°30N 111°56E **34**
Li Xian *Gansu, China* 34°10N 105°5E **34**
Li Xian *Hebei, China* 38°30N 115°35E **34**
Liancourt Rocks = Tokdo *Asia* 37°15N 131°52E **31**
Lianga *Phil.* 8°38N 126°6E **37**
Liangcheng *Nei Monggol Zizhiqu, China* 40°28N 112°25E **34**
Liangcheng *Shandong, China* 35°32N 119°37E **35**
Liangdang *China* 33°56N 106°18E **34**
Liangpran *Indonesia* 1°4N 114°23E **36**
Lianshanguan *China* 40°53N 123°43E **35**
Lianshui *China* 33°42N 119°20E **35**
Lianyungang *China* 34°40N 119°11E **35**
Liao He → *China* 41°0N 121°50E **35**
Liaocheng *China* 36°28N 115°58E **34**
Liaodong Bandao *China* 40°0N 122°30E **35**
Liaodong Wan *China* 40°20N 121°10E **35**
Liaoning □ *China* 41°40N 122°30E **35**
Liaotung, G. of = Liaodong Wan *China* 40°20N 121°10E **35**
Liaoyang *China* 41°15N 122°58E **35**
Liaoyuan *China* 42°58N 125°2E **35**
Liaozhong *China* 41°23N 122°50E **35**
Liapades *Greece* 39°42N 19°40E **25**
Liard → *Canada* 61°51N 121°18W **70**
Liard River *Canada* 59°25N 126°5W **70**
Liari *Pakistan* 25°37N 66°30E **42**
Libau = Liepāja *Latvia* 56°30N 21°0E **9**
Libby *U.S.A.* 48°23N 115°33W **76**
Libenge *Dem. Rep. of the Congo* 3°40N 18°55E **52**
Liberal *U.S.A.* 37°3N 100°55W **80**
Liberec *Czech Rep.* 50°47N 15°7E **16**
Liberia *Costa Rica* 10°40N 85°30W **88**
Liberia ■ *W. Afr.* 6°30N 9°30W **50**
Libertador □ *Chile* 34°15S 70°45W **94**
Liberty *Mo., U.S.A.* 39°15N 94°25W **80**
Liberty *N.Y., U.S.A.* 41°48N 74°45W **83**
Liberty *Pa., U.S.A.* 41°34N 77°6W **82**
Liberty *Tex., U.S.A.* 30°3N 94°48W **84**
Liberty-Newark Int. ✈ (EWR) *U.S.A.* 40°42N 74°10W **83**
Lîbîya, Sahrâ' *Africa* 25°0N 25°0E **51**
Libobo, Tanjung *Indonesia* 0°54S 128°28E **37**
Libode *S. Africa* 31°33S 29°2E **57**
Libong, Ko *Thailand* 7°15N 99°23E **39**
Libourne *France* 44°55N 0°14W **20**
Libramont *Belgium* 49°55N 5°23E **15**
Libreville *Gabon* 0°25N 9°26E **52**
Libya ■ *N. Afr.* 27°0N 17°0E **51**
Libyan Desert = Lîbîya, Sahrâ' *Africa* 25°0N 25°0E **51**
Libyan Plateau = Ed Déffa *Egypt* 30°40N 26°30E **51**
Licantén *Chile* 35°55S 72°0W **94**
Licata *Italy* 37°6N 13°56E **22**
Licheng *China* 36°28N 113°20E **34**
Lichfield *U.K.* 52°41N 1°49W **13**
Lichinga *Mozam.* 13°13S 35°11E **55**
Lichtenburg *S. Africa* 26°8S 26°8E **56**
Licking → *U.S.A.* 39°6N 84°30W **81**
Licungo → *Mozam.* 17°40S 37°15E **55**
Lida *Belarus* 53°53N 25°15E **17**
Lidköping *Sweden* 58°31N 13°7E **9**
Liebig, Mt. *Australia* 23°18S 131°22E **60**
Liechtenstein ■ *Europe* 47°8N 9°35E **20**
Liège *Belgium* 50°38N 5°35E **15**
Liège □ *Belgium* 50°32N 5°35E **15**
Liegnitz = Legnica *Poland* 51°12N 16°10E **16**
Lieksa *Finland* 63°18N 30°2E **8**
Lienart *Dem. Rep. of the Congo* 3°3N 25°31E **54**

enyünchiangshih = Lianyungang China 34°40N 119°11E 35 G10
enz Austria 46°50N 12°46E 16 E7
epāja Latvia 56°30N 21°0E 9 H19
er Belgium 51°7N 4°34E 15 C4
etuva = Lithuania ■ Europe 55°30N 24°0E 9 J21
èvre → Canada 45°31N 75°26W 72 C4
ffey → Ireland 53°21N 6°13W 10 C5
fford Ireland 54°51N 7°29W 10 B4
fudzin Russia 44°21N 134°58E 30 B7
ghtning Ridge Australia 29°22S 148°0E 63 D4
gonha → Mozam. 16°54S 39°9E 55 F4
gonier U.S.A. 40°15N 79°14W 82 F5
guria □ Italy 44°30N 8°50E 20 D8
gurian Sea Medit. S. 43°20N 9°0E 22 C3
hou Reefs and Cays Australia 17°25S 151°40E 62 B5
hué Calel △ Argentina 38°0S 65°10W 94 D2
iang China 26°55N 100°20E 32 F9
k → Laos 18°31N 102°30E 38 C4
kasi Dem. Rep. of the Congo 10°55S 26°48E 55 E2
koma I. Malawi 12°3S 34°45E 55 E3
kumburu Tanzania 9°43S 35°8E 55 D4
Île-Rousse France 42°38N 8°57E 20 E8
lle France 50°38N 3°3E 20 A5
lle Bælt Denmark 55°20N 9°45E 9 J13
llehammer Norway 61°8N 10°30E 8 F14
llesand Norway 58°15N 8°23E 9 G13
llian Pt. Australia 27°40S 126°6E 61 E4
llooet Canada 50°44N 121°57W 70 C4
llooet → Canada 49°15N 121°57W 70 D4
longwe Malawi 14°0S 33°48E 55 E3
loy Phil. 8°4N 122°39E 37 C6
m → Europe 43°45N 19°15E 23 C8
m Chu Kang Singapore 1°26N 103°43E 39 d
ma Indonesia 3°39S 127°58E 37 E7
ma Peru 12°3S 77°2W 92 F3
ma Mont., U.S.A. 44°38N 112°36W 76 D7
ma N.Y., U.S.A. 42°54N 77°36W 82 D7
ma Ohio, U.S.A. 40°44N 84°6W 81 E11
ma → Portugal 41°41N 8°50W 21 B1
massol Cyprus 34°42N 33°1E 25 E12
mavady U.K. 55°3N 6°56W 10 A5
may → Argentina 39°0S 68°0W 96 D3
may Mahuida Argentina 37°10S 66°45W 94 D2
mbang Malaysia 4°42N 115°6E 36 D5
mbaži Latvia 57°31N 24°42E 9 H21
mbdi India 22°34N 71°51E 42 H4
mbe Cameroon 4°1N 9°10E 52 D1
mburg Germany 50°22N 8°4E 16 C5
mburg □ Belgium 51°2N 5°25E 15 C5
mburg □ Neths. 51°20N 5°55E 15 C5
meira Brazil 22°35S 47°28W 95 A6
merick Ireland 52°40N 8°37W 10 D3
merick U.S.A. 43°41N 70°48W 83 C14
merick □ Ireland 52°30N 8°50W 10 D3
mestone U.S.A. 42°2N 78°38W 82 D6
mestone → Canada 56°31N 94°7W 71 B10
mfjorden Denmark 56°55N 9°0E 9 H13
mia = Lima → Portugal 41°41N 8°50W 21 B1
mingen Norway 64°48N 13°35E 8 D15
mmen ✪ Australia 15°15S 135°30E 62 B2
mmen Bight Australia 14°40S 135°35E 62 A2
mmen Bight → Australia 15°7S 135°44E 62 B2
mnos Greece 39°50N 25°5E 23 E11
moges Canada 45°20N 75°16W 83 A9
moges France 45°50N 1°15E 20 D4
món Costa Rica 10°0N 83°2W 88 E3
mon U.S.A. 39°16N 103°41W 76 G12
mousin □ France 45°30N 1°30E 20 D4
noux France 43°4N 2°12E 20 E5
mpopo □ S. Africa 24°5S 29°0E 57 C4
mpopo → Africa 25°5S 33°30E 57 D5
muru Kenya 1°2S 36°35E 54 C4
n Xian China 37°57N 110°58E 34 F6
nares Chile 35°50S 71°40W 94 D1
nares Mexico 24°52N 99°34W 87 C5
nares Spain 38°10N 3°40W 21 C4
ncang China 23°58N 100°1E 32 G9
ncheng China 37°25N 114°30E 34 F8
nchuan China 27°57N 116°15E 33 F12
ncoln = Beamsville Canada 43°12N 79°28W 82 C5
ncoln Argentina 34°55S 61°30W 94 C3
ncoln N.Z. 43°38S 172°30E 59 E4
ncoln U.K. 53°14N 0°32W 12 D7
ncoln Calif., U.S.A. 38°54N 121°17W 78 G5
ncoln Ill., U.S.A. 40°9N 89°22W 80 E9
ncoln Kans., U.S.A. 39°3N 98°9W 80 F4
ncoln Maine, U.S.A. 45°22N 68°30W 81 C19
ncoln N.H., U.S.A. 44°3N 71°40W 83 B13
ncoln N. Mex., U.S.A. 33°30N 105°23W 77 K11
ncoln Nebr., U.S.A. 40°49N 96°41W 80 E5
ncoln City U.S.A. 44°57N 124°1W 76 D1
ncoln Hav = Lincoln Sea Arctic 84°0N 55°0W 69 A21
ncoln Sea Arctic 84°0N 55°0W 69 A21
ncolnshire □ U.K. 53°14N 0°32W 12 D7
ncolnshire Wolds U.K. 53°26N 0°13W 12 D7
ncolnton U.S.A. 35°29N 81°16W 85 D14
nd U.S.A. 46°58N 118°37W 76 C4
nda U.S.A. 39°8N 121°34W 78 F5
ndeman I. Australia 20°27S 149°3E 62 b
ndeman Islands △ Australia 20°28S 149°5E 62 b
nden Guyana 6°0N 58°10W 92 B7
nden Ala., U.S.A. 32°18N 87°48W 85 E11
nden Calif., U.S.A. 38°1N 121°5W 78 G5
nden Tex., U.S.A. 33°1N 94°22W 84 E7
ndenhurst U.S.A. 40°41N 73°23W 83 F11
ndesnes Norway 57°58N 7°3E 9 H12
ndi Tanzania 9°58S 39°38E 55 D4
ndi □ Tanzania 9°40S 38°30E 55 D4
ndi → Dem. Rep. of the Congo 0°33N 25°5E 54 B2
Lindley U.S.A. 42°1N 77°8W 82 D7
Lindos Greece 36°6N 28°4E 25 C10
Lindos, Akra Greece 36°4N 28°10E 25 C10
Lindsay Canada 44°22N 78°43W 82 B6
Lindsay Calif., U.S.A. 36°12N 119°5W 78 J7
Lindsay Okla., U.S.A. 34°50N 97°38W 84 D6
Lindsborg U.S.A. 38°35N 97°40W 80 F5
Line Islands Pac. Oc. 7°0N 160°0W 65 H12
Linesville U.S.A. 41°39N 80°26W 82 E4
Linfen China 36°3N 111°30E 34 F6
Ling Xian China 37°22N 116°30E 34 F9
Lingao China 19°56N 109°42E 38 C7
Lingayen Phil. 16°1N 120°14E 37 A6
Lingayen G. Phil. 16°10N 120°15E 37 A6
Lingbao China 34°31N 110°51E 34 G6
Lingbi China 33°33N 117°33E 35 H9
Lingchuan China 35°45N 113°12E 34 G7
Lingding Yang China 22°25N 113°44E 33 a
Lingen Germany 52°31N 7°19E 16 B4
Lingga Indonesia 0°12S 104°37E 36 E2
Lingga, Kepulauan Indonesia 0°10S 104°30E 36 E2
Lingga Arch. = Lingga, Kepulauan Indonesia 0°10S 104°30E 36 E2
Lingle U.S.A. 42°8N 104°21W 76 E11
Lingqiu China 39°28N 114°22E 34 E8
Lingshi China 36°48N 111°48E 34 F6
Lingshou China 38°20N 114°20E 34 E8
Lingshui China 18°27N 110°0E 38 C8
Lingtai China 35°0N 107°40E 34 G4
Linguère Senegal 15°25N 15°5W 50 E2
Lingwu China 38°6N 106°20E 34 E4
Lingyuan China 41°10N 119°15E 35 D10
Linhares Brazil 19°25S 40°4W 93 G10
Linhe China 40°48N 107°20E 34 D4
Linjiang China 41°50N 127°0E 35 D14
Linköping Sweden 58°28N 15°36E 9 G16
Linkou China 45°15N 130°18E 35 B16
Linnhe, L. U.K. 56°36N 5°25W 11 E3
Linosa Medit. S. 35°51N 12°50E 22 G5
Linqi China 35°45N 113°52E 34 G7
Linqing China 36°50N 115°42E 34 F8
Linqu China 36°25N 118°30E 35 F10
Linquan China 33°5N 115°17E 34 H8
Lins Brazil 21°40S 49°44W 95 A6
Linshu China 34°53N 118°38E 35 G10
Linstead Jamaica 18°8N 77°2W 88 a
Linta → Madag. 25°2S 44°5E 57 D7
Lintao China 35°26N 103°58E 34 G2
Linton Ind., U.S.A. 39°2N 87°10W 80 F10
Linton N. Dak., U.S.A. 46°16N 100°14W 80 B3
Lintong China 34°20N 109°10E 34 G5
Linwood Canada 43°35N 80°43W 82 C4
Linxi Hebei, China 39°42N 118°26E 35 E10
Linxi Liaoning, China 43°36N 118°2E 35 C10
Linxia China 35°36N 103°10E 32 D9
Linyanti → Africa 17°50S 25°5E 56 B4
Linyi Shandong, China 37°13N 116°54E 34 F9
Linyi Shandong, China 35°5N 118°21E 35 G10
Linying China 33°48N 113°56E 34 H7
Linz Austria 48°18N 14°18E 16 D8
Linzhenzhen China 36°30N 109°59E 34 F5
Linzhou China 36°4N 113°49E 34 F7
Linzi China 36°50N 118°20E 35 F10
Lion, G. du France 43°10N 4°0E 20 E6
Lionárisso Cyprus 35°28N 34°8E 25 D13
Lions, G. of = Lion, G. du France 43°10N 4°0E 20 E6
Lion's Den Zimbabwe 17°15S 30°5E 55 F3
Lion's Head Canada 44°58N 81°15W 82 B3
Lipa Phil. 13°57N 121°10E 37 B6
Lipali Mozam. 15°50S 35°50E 55 F4
Lípari Italy 38°26N 14°58E 22 E6
Lípari, Is. = Eólie, Ís. Italy 38°30N 14°57E 22 E6
Lipcani Moldova 48°14N 26°48E 17 D14
Liperi Finland 62°31N 29°24E 8 E23
Lipetsk Russia 52°37N 39°35E 18 D6
Lipkany = Lipcani Moldova 48°14N 26°48E 17 D14
Lipovets Ukraine 49°12N 29°1E 17 D15
Lippe → Germany 51°39N 6°36E 16 C4
Lipscomb U.S.A. 36°14N 100°16W 84 C4
Liptrap, C. Australia 38°50S 145°55E 63 F4
Lira Uganda 2°17N 32°57E 54 B3
Liria = Lliria Spain 39°37N 0°35W 21 C5
Lisala Dem. Rep. of the Congo 2°12N 21°38E 52 D4
Lisboa Portugal 38°42N 9°8W 21 C1
Lisbon = Lisboa Portugal 38°42N 9°8W 21 C1
Lisbon N. Dak., U.S.A. 46°27N 97°41W 80 B5
Lisbon N.H., U.S.A. 44°13N 71°55W 83 B13
Lisbon N.Y., U.S.A. 44°43N 75°19W 83 B9
Lisbon Ohio, U.S.A. 40°46N 80°46W 82 F4
Lisbon Falls U.S.A. 44°0N 70°4W 81 D18
Lisburn U.K. 54°31N 6°3W 10 B5
Lisburne, C. U.S.A. 68°53N 166°13W 74 B6
Liscannor B. Ireland 52°55N 9°24W 10 D2
Lisdoonvarna Ireland 53°2N 9°18W 10 C2
Lishi China 37°31N 111°8E 34 F6
Lishu China 43°20N 124°18E 35 C13
Lisianski I. U.S.A. 26°2N 174°0W 75 K5
Lisichansk = Lysychansk Ukraine 48°55N 38°30E 19 E6
Lisieux France 49°10N 0°12E 20 B4
Liski Russia 51°3N 39°30E 19 D6
Lismore Australia 28°44S 153°21E 63 D5
Lismore Ireland 52°8N 7°55W 10 D4
Lista Norway 58°7N 6°39E 9 G12
Lister, Mt. Antarctica 78°0S 162°0E 5 D11
Liston Australia 28°39S 152°6E 63 D5
Listowel Canada 43°44N 80°58W 82 C4
Listowel Ireland 52°27N 9°29W 10 D2
Litani → Lebanon 33°20N 35°15E 46 B4
Litchfield Calif., U.S.A. 40°24N 120°23W 78 E6
Litchfield Conn., U.S.A. 41°45N 73°11W 83 E11
Litchfield Ill., U.S.A. 39°11N 89°39W 80 F9
Litchfield Minn., U.S.A. 45°8N 94°32W 80 C6
Litchfield △ Australia 13°14S 131°1E 60 B5
Lithgow Australia 33°25S 150°8E 63 E5
Lithino, Akra Greece 34°55N 24°44E 25 E6
Lithuania ■ Europe 55°30N 24°0E 9 J21
Lititz U.S.A. 40°9N 76°18W 83 F8
Litoměřice Czech Rep. 50°33N 14°10E 16 C8
Little Abaco Bahamas 26°50N 77°30W 88 A4
Little Barrier I. N.Z. 36°12S 175°8E 59 B5
Little Belt Mts. U.S.A. 46°40N 110°45W 76 C8
Little Bighorn Battlefield ⌂ U.S.A. 45°34N 107°25W 76 D10
Little Blue → U.S.A. 39°42N 96°41W 80 F5
Little Buffalo → Canada 61°0N 113°46W 70 A6
Little Cayman Cayman Is. 19°41N 80°3W 88 C3
Little Churchill → Canada 57°30N 95°22W 71 B9
Little Colorado → U.S.A. 36°12N 111°48W 77 H8
Little Current Canada 45°55N 82°0W 72 C3
Little Current → Canada 50°57N 84°36W 72 B3
Little Diomede I. U.S.A. 65°45N 168°56W 74 B6
Little Falls Minn., U.S.A. 45°59N 94°22W 80 C6
Little Falls N.Y., U.S.A. 43°3N 74°51W 83 C10
Little Fork → U.S.A. 48°31N 93°35W 80 A7
Little Grand Rapids Canada 52°0N 95°29W 71 C9
Little Humboldt → U.S.A. 41°1N 117°43W 76 F5
Little Inagua I. Bahamas 21°40N 73°50W 89 B5
Little Karoo S. Africa 33°45S 21°0E 56 E3
Little Khingan Mts. = Xiao Hinggan Ling China 49°0N 127°0E 33 B14
Little Lake U.S.A. 35°56N 117°55W 79 K9
Little Laut Is. = Laut Kecil, Kepulauan Indonesia 4°45S 115°40E 36 E5
Little Mecatina = Petit-Mécatina → Canada 50°40N 59°30W 73 B8
Little Minch U.K. 57°35N 6°45W 11 D2
Little Missouri → U.S.A. 47°36N 102°25W 80 B2
Little Ouse → U.K. 52°22N 1°12E 13 E9
Little Rann India 23°25N 71°25E 42 H4
Little Red → U.S.A. 35°11N 91°27W 84 D9
Little River N.Z. 43°45S 172°49E 59 E4
Little Rock U.S.A. 34°45N 92°17W 84 D8
Little Ruaha → Tanzania 7°57S 37°53E 54 D4
Little Sable Pt. U.S.A. 43°38N 86°33W 80 D10
Little Sioux → U.S.A. 41°48N 96°4W 80 E5
Little Smoky → Canada 54°44N 117°11W 70 C5
Little Snake → U.S.A. 40°27N 108°26W 76 F9
Little Tobago Trin. & Tob. 11°18N 60°30W 93 J16
Little Valley U.S.A. 42°15N 78°48W 82 D6
Little Wabash → U.S.A. 37°55N 88°5W 80 G9
Little White → U.S.A. 43°40N 100°40W 80 D3
Little Zab = Zāb aş Şaghīr → Iraq 35°17N 43°29E 44 C4
Littlefield U.S.A. 33°55N 102°20W 84 E3
Littlehampton U.K. 50°49N 0°32W 13 G7
Littleton U.S.A. 44°18N 71°46W 83 B13
Liu He → China 40°55N 121°35E 35 D11
Liuba China 33°38N 106°55E 34 H4
Liugou China 40°57N 118°15E 35 D10
Liuhe China 42°17N 125°43E 35 C13
Liukang Tenggaja = Sabalana, Kepulauan Indonesia 6°45S 118°50E 37 F5
Liuli Tanzania 11°3S 34°38E 55 E3
Liupan Shan China 35°39N 106°12E 34 G4
Liupanshui China 26°38N 104°48E 32 F9
Liuwa Plain Zambia 14°20S 22°30E 53 G4
Liuzhou China 24°22N 109°22E 33 G10
Liuzhuang China 33°12N 120°18E 35 H11
Livadhia Cyprus 34°57N 33°38E 25 E12
Livadia Greece 38°27N 22°54E 23 E10
Live Oak Calif., U.S.A. 39°17N 121°40W 78 F5
Live Oak Fla., U.S.A. 30°18N 82°59W 85 F13
Lively Canada 46°26N 81°9W 81 B13
Liveras Cyprus 35°23N 32°57E 25 D11
Livermore U.S.A. 37°41N 121°47W 78 H5
Livermore, Mt. U.S.A. 30°38N 104°11W 84 F2
Livermore Falls U.S.A. 44°29N 70°11W 81 C18
Liverpool Australia 33°54S 150°58E 63 B5
Liverpool Canada 44°5N 64°41W 73 D7
Liverpool U.K. 53°25N 3°0W 12 D4
Liverpool U.S.A. 43°6N 76°13W 83 C8
Liverpool Bay U.K. 53°30N 3°20W 12 D4
Liverpool Ra. Australia 31°50S 150°30E 63 E5
Livingston Guatemala 15°50N 88°50W 88 C2
Livingston U.K. 55°54N 3°30W 11 F5
Livingston Ala., U.S.A. 32°35N 88°11W 85 E10
Livingston Calif., U.S.A. 37°23N 120°43W 78 H6
Livingston Mont., U.S.A. 45°40N 110°34W 76 D8
Livingston S.C., U.S.A. 33°38N 81°7W 85 E14
Livingston Tenn., U.S.A. 36°23N 85°19W 85 C12
Livingston Tex., U.S.A. 30°43N 94°56W 84 F7
Livingston, L. U.S.A. 30°50N 95°10W 84 F7
Livingston Manor U.S.A. 41°54N 74°50W 83 E10
Livingstone Zambia 17°46S 25°52E 55 F2
Livingstone Mts. Tanzania 9°40S 34°20E 55 D3
Livingstonia Malawi 10°38S 34°5E 55 E3
Livny Russia 52°30N 37°30E 18 D6
Livonia Mich., U.S.A. 42°23N 83°23W 81 D12
Livonia N.Y., U.S.A. 42°49N 77°40W 82 D7
Livorno Italy 43°33N 10°19E 22 C4
Livramento Brazil 30°55S 55°30W 95 C4
Liwale Tanzania 9°48S 37°58E 55 D4
Liwonde Malaysia 15°4S 35°13E 55 F4
Liwonde △ Malawi 14°48S 35°20E 55 E4
Lizard I. Australia 14°42S 145°30E 62 A4
Lizard Pt. U.K. 49°57N 5°13W 13 H2
Ljubljana Slovenia 46°4N 14°33E 16 E8
Ljungan → Sweden 62°18N 17°23E 8 E17
Ljungby Sweden 56°49N 13°55E 9 H15
Ljusdal Sweden 61°46N 16°3E 8 F17
Ljusnan → Sweden 61°12N 17°8E 8 F17
Ljusne Sweden 61°13N 17°7E 8 F17
Llancanelo, Salina Argentina 35°40S 69°8W 94 D2
Llandeilo U.K. 51°53N 3°59W 13 F4
Llandovery U.K. 51°59N 3°48W 13 F4
Llandrindod Wells U.K. 52°14N 3°22W 13 E4
Llandudno U.K. 53°19N 3°50W 12 D4
Llanelli U.K. 51°41N 4°10W 13 F3
Llanes Spain 43°25N 4°50W 21 A3
Llangollen U.K. 52°58N 3°11W 12 E4
Llanidloes U.K. 52°27N 3°31W 13 E4
Llano U.S.A. 30°45N 98°41W 84 F5
Llano → U.S.A. 30°39N 98°26W 84 F5
Llano Estacado U.S.A. 33°30N 103°0W 84 E3
Llanos S. Amer. 5°0N 71°35W 92 C4
Llanos de Challe △ Chile 28°8S 71°1W 94 B1
Llanquihue, L. Chile 41°10S 72°50W 96 E1
Llanwrtyd Wells U.K. 52°7N 3°38W 13 E4
Llebeig, C. des Spain 39°33N 2°18E 24 B9
Lleida Spain 41°37N 0°39E 21 B6
Llentrisca, C. Spain 38°52N 1°15E 24 C7
Llera de Canales Mexico 23°19N 99°1W 87 C5
Lleyn Peninsula U.K. 52°51N 4°36W 12 E3
Llico Chile 34°46S 72°5W 94 C1
Lliria Spain 39°37N 0°35W 21 C5
Llobregat → Spain 41°19N 2°5E 21 B7
Lloret de Mar Spain 41°41N 2°53E 21 B7
Lloseta Spain 39°43N 2°52E 24 B9
Lloyd B. Australia 12°45S 143°27E 62 A3
Lloyd L. Canada 57°22N 108°57W 71 B7
Lloydminster Canada 53°17N 110°0W 71 C7
Llubí Spain 39°42N 3°0E 24 B10
Llucmajor Spain 39°29N 2°53E 24 B9
Llullaillaco, Volcán S. Amer. 24°43S 68°30W 94 A2
Llullaillaco △ Chile 24°50S 68°51W 94 A2
Lo → Vietnam 21°18N 105°25E 38 B5
Loa U.S.A. 38°24N 111°39W 76 G8
Loa → Chile 21°26S 70°41W 94 A1
Loaita I. S. China Sea 10°41N 114°25E 36 B4
Loange → Dem. Rep. of the Congo 4°17S 20°2E 52 E4
Lobatse Botswana 25°12S 25°40E 56 D4
Lobería Argentina 38°10S 58°40W 94 D4
Lobito Angola 12°18S 13°35E 53 G2
Lobos Argentina 35°10S 59°0W 94 D4
Lobos, I. Mexico 27°20N 110°36W 86 B2
Lobos, I. de Canary Is. 28°45N 13°50W 24 F6
Loc Binh Vietnam 21°46N 106°54E 38 B6
Loc Ninh Vietnam 11°50N 106°34E 39 G6
Locarno Switz. 46°10N 8°47E 20 C8
Loch Baghasdail = Lochboisdale U.K. 57°9N 7°20W 11 D1
Loch Garman = Wexford Ireland 52°20N 6°28W 10 D5
Loch Lomond and the Trossachs △ U.K. 56°10N 4°40W 11 E4
Loch Nam Madadh = Lochmaddy U.K. 57°36N 7°10W 11 D1
Lochaber U.K. 56°59N 5°1W 11 E3
Locharbriggs U.K. 55°7N 3°35W 11 F5
Lochboisdale U.K. 57°9N 7°20W 11 D1
Loche, L. La Canada 56°30N 109°30W 71 B7
Lochem Neths. 52°9N 6°26E 15 B6
Loches France 47°7N 1°0E 20 C4
Lochgilphead U.K. 56°2N 5°26W 11 E3
Lochinvar △ Zambia 15°55S 27°15E 55 F2
Lochinver U.K. 58°9N 5°14W 11 C3
Lochmaddy U.K. 57°36N 7°10W 11 D1
Lochnagar Australia 23°33S 145°38E 62 C4
Lochnagar U.K. 56°57N 3°15W 11 E5
Lochy, L. U.K. 57°0N 4°53W 11 E4
Lock Australia 33°34S 135°46E 63 E2
Lock Haven U.S.A. 41°8N 77°28W 82 E7
Lockeford U.S.A. 38°10N 121°9W 78 G5
Lockeport Canada 43°47N 65°4W 73 D6
Lockerbie U.K. 55°7N 3°21W 11 F5
Lockhart U.S.A. 29°53N 97°40W 84 G6
Lockhart, L. Australia 33°15S 119°3E 61 F2
Lockhart River Australia 12°58S 143°30E 62 A3
Lockhart River ✪ Australia 12°47S 143°20E 62 A3
Lockney U.S.A. 34°7N 101°27W 84 D4
Lockport U.S.A. 43°10N 78°42W 82 C6
Lod Israel 31°57N 34°54E 46 D3
Lodeinoye Pole Russia 60°44N 33°33E 18 B5
Lodge Bay Canada 52°14N 55°51W 73 B8
Lodge Grass U.S.A. 45°19N 107°22W 76 D10
Lodhran Pakistan 29°32N 71°30E 42 E4
Lodi Italy 45°19N 9°30E 20 D8
Lodi Calif., U.S.A. 38°8N 121°16W 78 G5
Lodi Ohio, U.S.A. 41°2N 82°1W 82 E3
Lodja Dem. Rep. of the Congo 3°30S 23°23E 54 C1
Lodwar Kenya 3°7N 35°36E 54 B4
Łódź Poland 51°45N 19°27E 17 C10
Loei Thailand 17°29N 101°35E 38 D3
Loengo Dem. Rep. of the Congo 4°48S 26°30E 54 C2
Loeriesfontein S. Africa 31°0S 19°26E 56 E2
Lofoten Norway 68°30N 13°59E 8 B16
Logan Iowa, U.S.A. 41°39N 95°47W 80 E6
Logan Ohio, U.S.A. 39°32N 82°25W 81 F12
Logan Utah, U.S.A. 41°44N 111°50W 76 F8
Logan W. Va., U.S.A. 37°51N 81°59W 81 G13
Logan, Mt. Canada 60°34N 140°23W 68 E3
Logandale U.S.A. 36°36N 114°29W 79 J12
Logansport Ind., U.S.A. 40°45N 86°22W 80 E10
Logansport La., U.S.A. 31°58N 94°0W 84 F8
Logone → Chad 12°6N 15°2E 51 F9
Logroño Spain 42°28N 2°27W 21 A4
Lohardaga India 23°27N 84°45E 43 H11
Loharia India 23°45N 74°14E 42 H6
Loharu India 28°27N 75°49E 42 E6
Lohri Wah → Pakistan 27°27N 67°37E 42 F2
Loi-kaw Burma 19°40N 97°17E 41 K20
Loimaa Finland 60°50N 23°5E 8 F20
Loir → France 47°33N 0°32W 20 C3
Loire → France 47°16N 2°10W 20 C2
Loja Ecuador 3°59S 79°16W 92 D3
Loja Spain 37°10N 4°10W 21 D3
Loji = Kawasi Indonesia 1°38S 127°28E 37 E7
Lokandu Dem. Rep. of the Congo 2°30S 25°45E 54 C2
Lokeren Belgium 51°6N 3°59E 15 C3
Lokgwabe Botswana 24°10S 21°50E 56 C3
Lokichar Kenya 2°23N 35°39E 54 B4
Lokichokio Kenya 4°19N 34°13E 54 B3
Lokitaung Kenya 4°12N 35°48E 54 B4
Lokkan tekojärvi Finland 67°55N 27°35E 8 C22
Lokoja Nigeria 7°47N 6°45E 50 G7
Lokolo → Dem. Rep. of the Congo 1°43S 18°23E 52 E3
Lola, Mt. U.S.A. 39°26N 120°22W 78 F6
Lolgorien Kenya 1°14S 34°48E 54 C3
Loliondo Tanzania 2°2S 35°39E 54 C4
Lolland Denmark 54°45N 11°30E 9 J14
Lolo U.S.A. 46°45N 114°5W 76 C6
Lom Bulgaria 43°48N 23°12E 23 C10
Lom Kao Thailand 16°53N 101°14E 38 D3
Lom Sak Thailand 16°47N 101°15E 38 D3
Loma U.S.A. 47°56N 110°30W 76 C8
Loma Linda U.S.A. 34°3N 117°16W 79 L9
Lomaloma Fiji 17°17S 178°59W 59 a
Lomami → Dem. Rep. of the Congo 0°46N 24°16E 54 B1
Lomas de Zamora Argentina 34°45S 58°24W 94 C4
Lombadina Australia 16°31S 122°54E 60 C3
Lombárdia □ Italy 45°40N 9°30E 20 D8
Lombardy = Lombárdia □ Italy 45°40N 9°30E 20 D8
Lomblen Indonesia 8°30S 123°32E 37 F6
Lombok Indonesia 8°45S 116°30E 36 F5
Lombok, Selat Indonesia 8°30S 115°50E 37 K18
Lomé Togo 6°9N 1°20E 50 G6
Lomela Dem. Rep. of the Congo 2°19S 23°15E 52 E4
Lomela → Dem. Rep. of the Congo 0°15S 20°40E 52 E4
Lommel Belgium 51°14N 5°19E 15 C5
Lomond Canada 50°24N 112°36W 70 C6
Lomond, L. U.K. 56°8N 4°38W 11 E4
Lomonosov Ridge Arctic 88°0N 140°0E 4 A
Lomphat Cambodia 13°30N 106°59E 38 F6
Lompobatang Indonesia 5°24S 119°56E 37 F5
Lompoc U.S.A. 34°38N 120°28W 79 L6
Łomża Poland 53°10N 22°2E 17 B12
Lon, Ko Thailand 7°47N 98°23E 39 a
Loncoche Chile 39°20S 72°50W 96 D2
Londa India 15°30N 74°30E 40 M9
Londiani Kenya 0°10S 35°33E 54 C4
London Canada 42°59N 81°15W 82 D3
London U.K. 51°30N 0°3W 13 F7
London Ky., U.S.A. 37°8N 84°5W 81 G11
London Ohio, U.S.A. 39°53N 83°27W 81 F12
London, Greater □ U.K. 51°36N 0°5W 13 F7
London Gatwick ✈ (LGW) U.K. 51°10N 0°11W 13 F7
London Heathrow ✈ (LHR) U.K. 51°28N 0°27W 13 F7
London Stansted ✈ (STN) U.K. 51°54N 0°14E 13 F8
Londonderry U.K. 55°0N 7°20W 10 B4
Londonderry □ U.K. 55°0N 7°20W 10 B4
Londonderry, C. Australia 13°45S 126°55E 60 B4
Londonderry, I. Chile 55°0S 71°0W 96 H2
Londres Argentina 27°43S 67°7W 96 B3
Londrina Brazil 23°18S 51°10W 95 A5
Lone Pine U.S.A. 36°36N 118°4W 78 J8
Lonely Mine Zimbabwe 19°30S 28°49E 57 B4
Long B. U.S.A. 33°35N 78°45W 85 E15
Long Beach Calif., U.S.A. 33°46N 118°11W 79 M8
Long Beach N.Y., U.S.A. 40°35N 73°39W 83 F11
Long Beach Wash., U.S.A. 46°21N 124°3W 78 D2
Long Branch U.S.A. 40°18N 74°0W 83 F11
Long Creek U.S.A. 44°43N 119°6W 76 D4
Long Eaton U.K. 52°53N 1°15W 12 E6
Long I. Australia 20°22S 148°51E 62 b
Long I. Bahamas 23°20N 75°10W 89 B4
Long I. Canada 54°50N 79°20W 72 B4
Long I. Ireland 51°30N 9°34W 10 E2
Long I. U.S.A. 40°45N 73°30W 83 F11
Long Island Sd. U.S.A. 41°10N 73°0W 83 E12
Long L. Canada 49°30N 86°50W 72 C2
Long L. U.S.A. 44°0N 74°23W 83 C10
Long Lake U.S.A. 43°58N 74°25W 83 C10
Long Point Canada 42°34N 80°25W 82 D4
Long Point B. Canada 42°40N 80°10W 82 D4
Long Prairie U.S.A. 45°59N 94°52W 80 C6
Long Prairie → U.S.A. 46°20N 94°36W 80 B6
Long Pt. Ont., Canada 42°35N 80°2W 82 D4
Long Pt. Ont., Canada 43°56N 76°53W 83 C8
Long Range Mts. Canada 49°30N 57°30W 73 C8
Long Reef Australia 14°1S 125°48E 60 B4
Long Sault Canada 45°2N 74°53W 83 A10
Long Str. = Longa, Proliv Russia 70°0N 175°0E 29 C18
Long Thanh Vietnam 10°47N 106°57E 39 G6
Long Xian China 34°55N 106°55E 34 G4
Long Xuyen Vietnam 10°19N 105°28E 39 G5
Longa, Proliv Russia 70°0N 175°0E 29 C18
Longbenton U.K. 55°1N 1°31W 12 B6
Longboat Key U.S.A. 27°23N 82°39W 85 H13
Longde China 35°30N 106°20E 34 G4
Longford Australia 41°32S 147°3E 63 G4
Longford Ireland 53°43N 7°49W 10 C4
Longford □ Ireland 53°42N 7°45W 10 C4
Longhua Guangdong, China 22°39N 114°0E 33 a
Longhua Hebei, China 41°18N 117°45E 35 D9
Longido Tanzania 2°43S 36°42E 54 C4
Longiram Indonesia 0°5S 115°45E 36 E5
Longjing China 42°46N 129°25E 35 C15
Longkou China 37°40N 120°18E 35 F11
Longmeadow U.S.A. 42°3N 72°34W 83 D12
Longmen Shiku China 34°34N 112°10E 34 G7

Longmont U.S.A. 40°10N 105°6W 76 F11
Longnawan Indonesia 1°51N 114°55E 36 D4
Longreach Australia 23°28S 144°14E 62 C3
Longueuil Canada 45°31N 73°29W 83 A11
Longview Tex., U.S.A. 32°30N 94°44W 84 E7
Longview Wash., U.S.A. 46°8N 122°57W 78 D4
Longxi China 34°53N 104°40E 34 G3
Longxue Dao China 22°41N 113°38E 33 a
Longyan China 25°10N 117°0E 33 F12
Longyearbyen Svalbard 78°13N 15°40E 4 B8
Lonoke U.S.A. 34°47N 91°54W 84 D9
Lonquimay Chile 38°26S 71°14W 96 D2
Lons-le-Saunier France 46°40N 5°31E 20 C6
Looe U.K. 50°22N 4°28W 13 G3
Lookout, C. Canada 55°18N 83°56W 72 A3
Lookout, C. U.S.A. 34°35N 76°32W 85 D16
Loolmalasin Tanzania 3°0S 35°53E 54 C4
Loon → Alta., Canada 57°8N 115°3W 70 B5
Loon → Man., Canada 55°53N 101°59W 71 B8
Loon Lake Canada 54°2N 109°10W 71 C7
Loongana Australia 30°52S 127°5E 61 F4
Loop Hd. Ireland 52°34N 9°56W 10 D2
Lop China 37°3N 80°11E 32 D5
Lop Nor = Lop Nur China 40°20N 90°10E 32 C7
Lop Nur China 40°20N 90°10E 32 C7
Lopatina, Gora Russia 50°47N 143°10E 29 D15
Lopatka, Mys Russia 50°52N 156°40E 29 D16
Lopburi Thailand 14°48N 100°37E 38 E3
Lopez U.S.A. 41°27N 76°20W 83 E8
Lopez, C. Gabon 0°47S 8°40E 52 E1
Lopphavet Norway 70°27N 21°15E 8 A19
Lora → Afghan. 31°35N 66°32E 40 D4
Lora, Hāmūn-i- Pakistan 29°38N 64°58E 40 E4
Lora Cr. → Australia 28°10S 135°22E 63 D2
Lora del Río Spain 37°39N 5°33W 21 D3
Lorain U.S.A. 41°28N 82°11W 82 E2
Loralai Pakistan 30°20N 68°41E 42 D3
Lorca Spain 37°41N 1°42W 21 D5
Lord Howe I. Pac. Oc. 31°33S 159°6E 58 E8
Lord Howe Rise Pac. Oc. 30°0S 162°30E 64 L8
Lord Loughborough I. Burma 10°25N 97°54E 39 G1
Lordsburg U.S.A. 32°21N 108°43W 77 K9
Lorestān □ Iran 33°30N 48°40E 45 C6
Loreto Brazil 7°5S 45°10W 93 E9
Loreto Mexico 26°0N 111°21W 86 B2
Lorient France 47°45N 3°23W 20 C2
Lormi India 22°17N 81°41E 43 H9
Lorn U.K. 56°26N 5°10W 11 E3
Lorn, Firth of U.K. 56°20N 5°40W 11 E3
Lorne Australia 38°33S 143°59E 63 F3
Lorovouno Cyprus 35°8N 32°36E 25 D11
Lorraine □ France 48°53N 6°0E 20 B7
Los Alamos Calif., U.S.A. 34°44N 120°17W 79 L6
Los Alamos N. Mex., U.S.A. 35°53N 106°19W 77 J10
Los Altos U.S.A. 37°23N 122°7W 78 H4
Los Andes Chile 32°50S 70°40W 94 C1
Los Angeles Chile 37°28S 72°23W 94 D1
Los Angeles U.S.A. 34°4N 118°15W 79 M8
Los Angeles, Bahia de Mexico 28°56N 113°34W 86 B2
Los Angeles Aqueduct U.S.A. 35°22N 118°5W 79 K9
Los Angeles Int. ✈ (LAX) U.S.A. 33°57N 118°25W 79 M8
Los Banos U.S.A. 37°4N 120°51W 78 H6
Los Blancos Argentina 23°40S 62°30W 94 A3
Los Cardones △ Argentina 25°8S 65°55W 94 B2
Los Chiles Costa Rica 11°2N 84°43W 88 D3
Los Cristianos Canary Is. 28°3N 16°42W 24 F3
Los Gatos U.S.A. 37°14N 121°59W 78 H5
Los Haïtises △ Dom. Rep. 19°4N 69°36W 89 C6
Los Hermanos Is. Venezuela 11°45N 64°25W 89 D7
Los Islotes Canary Is. 29°4N 13°44W 24 E6
Los Llanos de Aridane Canary Is. 28°38N 17°54W 24 F2
Los Loros Chile 27°50S 70°6W 94 B1
Los Lunas U.S.A. 34°48N 106°44W 77 J10
Los Mochis Mexico 25°45N 108°57W 86 B3
Los Olivos U.S.A. 34°40N 120°7W 79 L6
Los Palacios Cuba 22°35N 83°15W 88 B3
Los Queñes Chile 35°1S 70°48W 94 D1
Los Reyes de Salgado Mexico 19°35N 102°29W 86 D4
Los Roques Is. Venezuela 11°50N 66°45W 89 D6
Los Teques Venezuela 10°21N 67°2W 92 A5
Los Testigos, Is. Venezuela 11°23N 63°6W 92 A6
Los Vilos Chile 32°10S 71°30W 94 C1
Lošinj Croatia 44°30N 14°30E 16 F8
Loskop Dam S. Africa 25°23S 29°20E 57 D4
Lossiemouth U.K. 57°42N 3°17W 11 D5
Lostwithiel U.K. 50°24N 4°41W 13 G3
Lot → France 44°18N 0°20E 20 D4
Lota Chile 37°5S 73°10W 94 D1
Loṭfābād Iran 37°32N 59°20E 45 B8
Lothair S. Africa 26°22S 30°27E 57 D5
Lotta → Europe 68°42N 31°6E 8 B24
Loubomo Congo 4°9S 12°47E 52 E2
Loudonville U.S.A. 40°38N 82°14W 82 F2
Louga Senegal 15°45N 16°5W 50 E2
Loughborough U.K. 52°47N 1°11W 12 E6
Lougheed I. Canada 77°26N 105°6W 69 B10
Loughrea Ireland 53°12N 8°33W 10 C3
Loughros More B. Ireland 54°48N 8°32W 10 B3
Louis Trichardt S. Africa 23°1S 29°43E 57 C4
Louis XIV, Pte. Canada 54°37N 79°45W 72 B4
Louisa U.S.A. 38°7N 82°36W 81 F12
Louisbourg Canada 45°55N 60°0W 73 C8
Louisburgh Ireland 53°46N 9°49W 10 C2
Louise I. Canada 52°55N 131°50W 70 C2
Louiseville Canada 46°20N 72°56W 72 C5
Louisiade Arch. Papua N. G. 11°10S 153°0E 58 C8
Louisiana U.S.A. 39°27N 91°3W 80 F8
Louisiana □ U.S.A. 30°50N 92°0W 84 F9
Louisville Ky., U.S.A. 38°15N 85°46W 81 F11
Louisville Miss., U.S.A. 33°7N 89°3W 85 E10
Louisville Ohio, U.S.A. 40°50N 81°16W 82 F3
Louisville Ridge Pac. Oc. 31°0S 172°30W 64 L10
Loulé Portugal 37°9N 8°0W 21 D1
Loup City U.S.A. 41°17N 98°58W 80 E4
Loups Marins, Lacs des Canada 56°30N 73°45W 72 A5
Lourdes France 43°6N 0°3W 20 E3
Lourdes-de-Blanc-Sablon Canada 51°24N 57°12W 73 B8
Louroujina Cyprus 35°0N 33°28E 25 E12
Louth Australia 30°30S 145°8E 63 E4
Louth Ireland 53°58N 6°32W 10 C5
Louth U.K. 53°22N 0°1W 12 D7
Louth □ Ireland 53°56N 6°34W 10 C5
Louvain = Leuven Belgium 50°52N 4°42E 15 D4
Louwsburg S. Africa 27°37S 31°7E 57 D5
Lovech Bulgaria 43°8N 24°42E 23 C11
Loveland U.S.A. 40°24N 105°5W 76 F11
Lovell U.S.A. 44°50N 108°24W 76 D9
Lovelock U.S.A. 40°11N 118°28W 76 F4
Loviisa Finland 60°28N 26°12E 8 F22
Lovina Indonesia 8°9S 115°1E 37 J18
Loving U.S.A. 32°17N 104°6W 77 K11
Lovington U.S.A. 32°57N 103°21W 77 K12
Lovisa = Loviisa Finland 60°28N 26°12E 8 F22
Low, L. Canada 52°29N 76°17W 72 B4
Low Pt. Australia 32°25S 127°25E 61 F4
Low Tatra = Nízké Tatry Slovak Rep. 48°55N 19°30E 17 D10
Lowa Dem. Rep. of the Congo 1°25S 25°47E 54 C2
Lowa → Dem. Rep. of the Congo 1°24S 25°51E 54 C2
Lowell U.S.A. 42°38N 71°19W 83 D13
Lowellville U.S.A. 41°2N 80°32W 82 E4
Löwen → Namibia 26°51S 18°17E 56 D2
Lower Alkali L. U.S.A. 41°16N 120°2W 76 F3
Lower Arrow L. Canada 49°40N 118°5W 70 D5
Lower California = Baja California Mexico 31°10N 115°12W 86 A1
Lower Hutt N.Z. 41°10S 174°55E 59 D5
Lower Lake U.S.A. 38°55N 122°37W 78 G4
Lower Manitou L. Canada 49°15N 93°0W 71 D10
Lower Post Canada 59°58N 128°30W 70 B3
Lower Red L. U.S.A. 47°58N 95°0W 80 B6
Lower Saxony = Niedersachsen □ Germany 52°50N 9°0E 16 B5
Lower Tunguska = Tunguska, Nizhnyaya → Russia 65°48N 88°4E 29 C9
Lower Zambezi △ Zambia 15°25S 29°40E 55 F2
Lowestoft U.K. 52°29N 1°45E 13 E9
Lowgar □ Afghan. 34°0N 69°0E 40 B6
Łowicz Poland 52°6N 19°55E 17 B10
Lowther I. Canada 74°33N 97°30W 69 C12
Lowville U.S.A. 43°47N 75°29W 83 C9
Loxton Australia 34°28S 140°31E 63 E3
Loxton S. Africa 31°30S 22°22E 56 E3
Loyalton U.S.A. 39°41N 120°14W 78 F6
Loyalty Is. = Loyauté, Îs. N. Cal. 20°50S 166°30E 58 D9
Loyang = Luoyang China 34°40N 112°26E 34 G7
Loyauté, Îs. N. Cal. 20°50S 166°30E 58 D9
Loyev = Loyew Belarus 51°56N 30°46E 17 C16
Loyew Belarus 51°56N 30°46E 17 C16
Loyoro Uganda 3°22N 34°14E 54 B3
Ltalaltuma ⚙ Australia 23°57S 132°25E 60 D5
Lu Wo China 22°33N 114°6E 33 a
Luachimo Angola 7°23S 20°48E 52 F4
Luajan → India 24°44N 85°1E 43 G11
Lualaba → Dem. Rep. of the Congo 0°26N 25°20E 54 B2
Luambe △ Zambia 12°30S 32°15E 55 E3
Luampa Zambia 15°4S 24°20E 55 F1
Lu'an China 31°45N 116°29E 33 E12
Luan Chau Vietnam 21°38N 103°24E 38 B4
Luan He → China 39°20N 119°5E 35 E10
Luan Xian China 39°40N 118°40E 35 E10
Luancheng China 37°53N 114°40E 34 F8
Luanda Angola 8°50S 13°15E 52 F2
Luang, Doi Thailand 18°30N 101°15E 38 C3
Luang, Thale Thailand 7°30N 100°15E 39 J3
Luang Nam Tha Laos 20°58N 101°30E 38 B3
Luang Prabang Laos 19°52N 102°10E 38 C4
Luangwa Zambia 15°35S 30°16E 55 F3
Luangwa → Zambia 14°25S 30°25E 55 E3
Luangwa Valley Zambia 13°30S 31°30E 55 E3
Luangwe = Loange → Dem. Rep. of the Congo 4°17S 20°2E 52 E4
Luanne China 40°55N 117°40E 35 D9
Luanping China 40°53N 117°23E 35 D9
Luanshya Zambia 13°3S 28°28E 55 E2
Luapula □ Zambia 11°0S 29°0E 55 E2
Luapula → Africa 9°26S 28°33E 55 D2
Luarca Spain 43°32N 6°32W 21 A2
Luashi Dem. Rep. of the Congo 10°50S 23°36E 55 E1
Luau Angola 10°40S 22°10E 52 G4
Lubana, Ozero = Lubānas Ezers Latvia 56°45N 27°0E 9 H22
Lubānas Ezers Latvia 56°45N 27°0E 9 H22
Lubang Is. Phil. 13°50N 120°12E 37 B6
Lubango Angola 14°55S 13°30E 53 G2
Lubao Dem. Rep. of the Congo 5°17S 25°42E 54 D2
Lubbock U.S.A. 33°35N 101°51W 84 E4
Lübeck Germany 53°52N 10°40E 16 B6
Lubefu Dem. Rep. of the Congo 4°47S 24°27E 54 C1
Lubefu → Dem. Rep. of the Congo 4°10S 23°0E 54 C1
Lubero = Luofu Dem. Rep. of the Congo 0°10S 29°15E 54 C2
Lubicon L. Canada 56°23N 115°56W 70 B5
Lubilash → Dem. Rep. of the Congo 6°2S 23°45E 52 F4
Lubin Poland 51°24N 16°11E 16 C9
Lublin Poland 51°12N 22°38E 17 C12
Lubnān = Lebanon ■ Asia 34°0N 36°0E 46 B5
Lubnān, Jabal Lebanon 33°45N 35°40E 46 B4
Lubny Ukraine 50°3N 32°58E 28 D4
Lubongola Dem. Rep. of the Congo 2°35S 27°50E 54 C2
Lubudi Dem. Rep. of the Congo 9°57S 25°58E 52 F5
Lubudi → Dem. Rep. of the Congo 9°0S 25°35E 55 D2
Lubuklinggau Indonesia 3°15S 102°55E 36 E2
Lubuksikaping Indonesia 0°10N 100°15E 36 D2
Lubumbashi Dem. Rep. of the Congo 11°40S 27°28E 55 E2
Lubunda Dem. Rep. of the Congo 5°12S 26°41E 54 D2
Lubungu Zambia 14°35S 26°24E 55 E2
Lubutu Dem. Rep. of the Congo 0°45S 26°30E 54 C2
Lucan Canada 43°11N 81°24W 82 C3
Lucania, Mt. Canada 61°1N 140°27W 68 E3
Lucas Channel = Main Channel Canada 45°21N 81°45W 82 A3
Lucca Italy 43°50N 10°29E 22 C4
Luce Bay U.K. 54°45N 4°48W 11 G4
Lucea Jamaica 18°27N 78°10W 88 a
Lucedale U.S.A. 30°56N 88°35W 85 F10
Lucena Phil. 13°56N 121°37E 37 B6
Lucena Spain 37°27N 4°31W 21 D3
Lučenec Slovak Rep. 48°18N 19°42E 17 D10
Lucerne = Luzern Switz. 47°3N 8°18E 20 C8
Lucerne U.S.A. 39°6N 122°48W 78 F4
Lucerne Valley U.S.A. 34°27N 116°57W 79 L10
Lucero Mexico 30°49N 106°30W 86 A3
Lucheng China 36°20N 113°11E 34 F7
Lucheringo → Mozam. 11°43S 36°17E 55 E4
Lucia U.S.A. 36°2N 121°33W 78 J5
Lucinda Australia 18°32S 146°20E 62 B4
Luckenwalde Germany 52°5N 13°10E 16 B7
Luckhoff S. Africa 29°44S 24°43E 56 D3
Lucknow Canada 43°57N 81°31W 82 C3
Lucknow India 26°50N 81°0E 43 F9
Lüda = Dalian China 38°50N 121°40E 35 E11
Lüderitz Namibia 26°41S 15°8E 56 D2
Lüderitzbaai Namibia 26°36S 15°8E 56 D2
Ludhiana India 30°57N 75°56E 42 D6
Ludington U.S.A. 43°57N 86°27W 80 D10
Ludlow U.K. 52°22N 2°42W 13 E5
Ludlow Calif., U.S.A. 34°43N 116°10W 79 L10
Ludlow Pa., U.S.A. 41°43N 78°56W 82 E6
Ludlow Vt., U.S.A. 43°24N 72°42W 83 C12
Ludvika Sweden 60°8N 15°14E 9 F16
Ludwigsburg Germany 48°53N 9°11E 16 D5
Ludwigshafen Germany 49°29N 8°26E 16 D5
Lueki Dem. Rep. of the Congo 3°20S 25°48E 54 C2
Luena Dem. Rep. of the Congo 9°28S 25°43E 55 D2
Luena Zambia 10°40S 30°25E 55 E3
Luena Flats Zambia 14°47S 23°17E 53 G4
Luenha = Ruenya → Africa 16°24S 33°48E 55 F3
Lüeyang China 33°22N 106°10E 34 H4
Lufira → Dem. Rep. of the Congo 9°30S 27°0E 55 D2
Lufkin U.S.A. 31°21N 94°44W 84 F7
Lufupa Dem. Rep. of the Congo 10°37S 24°56E 55 E1
Luga Russia 58°40N 29°55E 9 G23
Lugano Switz. 46°1N 8°57E 20 C8
Lugansk = Luhansk Ukraine 48°38N 39°15E 19 E6
Lugard's Falls Kenya 3°6S 38°41E 54 C4
Lugela Mozam. 16°25S 36°43E 55 F4
Lugenda → Mozam. 11°25S 38°33E 55 E4
Lugh = Luuq Somali Rep. 3°48N 42°34E 47 G3
Lugnaquillia Ireland 52°58N 6°28W 10 D5
Lugo Italy 44°25N 11°54E 22 B4
Lugo Spain 43°2N 7°35W 21 A2
Lugoj Romania 45°42N 21°57E 17 F11
Lugovoy = Qulan Kazakhstan 42°55N 72°43E 28 E8
Luhansk Ukraine 48°38N 39°15E 19 E6
Lui → Angola 8°21S 17°33E 52 F3
Luiana Angola 17°25S 22°59E 56 B3
Luiana → Angola 17°24S 23°3E 53 H4
Luichow Pen. = Leizhou Bandao China 21°0N 110°0E 33 G11
Luimneach = Limerick Ireland 52°40N 8°37W 10 D3
Luing U.K. 56°14N 5°39W 11 E3
Luís Correia Brazil 3°0S 41°35W 93 D10
Luitpold Coast Antarctica 78°30S 32°0W 5 D1
Luiza Dem. Rep. of the Congo 7°40S 22°30E 52 F4
Luizi Dem. Rep. of the Congo 6°0S 27°25E 54 D2
Luján Argentina 34°45S 59°5W 94 C4
Lukanga Swamp Zambia 14°30S 27°40E 55 E2
Lukenie → Dem. Rep. of the Congo 3°0S 18°50E 52 E3
Lukolela Dem. Rep. of the Congo 5°23S 24°32E 54 D1
Lukosi Zimbabwe 18°30S 26°30E 55 F2
Łuków Poland 51°55N 22°23E 17 C12
Lukusuzi △ Zambia 12°43S 32°36E 55 E3
Lülang Shan China 38°0N 111°15E 34 F6
Luleå Sweden 65°35N 22°10E 8 D20
Luleälven → Sweden 65°35N 22°10E 8 D20
Lüleburgaz Turkey 41°23N 27°22E 23 D12
Lulima Dem. Rep. of the Congo 4°12S 25°36E 54 C2
Luling U.S.A. 29°41N 97°39W 84 G6
Lulong China 39°53N 118°51E 35 E10
Lulonga → Dem. Rep. of the Congo 1°0N 18°10E 52 D3
Lulua → Dem. Rep. of the Congo 4°30S 20°30E 52 E4
Luma Amer. Samoa 14°16S 169°33W 59 b
Lumajang Indonesia 8°8S 113°13E 37 H15
Lumbala N'guimbo Angola 14°18S 21°18E 53 G4
Lumberton U.S.A. 34°37N 79°0W 85 D15
Lumsden Canada 50°39N 104°52W 71 C8
Lumsden N.Z. 45°44S 168°27E 59 F2
Lumut Malaysia 4°13N 100°37E 39 K3
Lumut, Tanjung Indonesia 3°50S 105°58E 36 E3
Luna India 23°43N 69°16E 42 H3
Lunavada India 23°8N 73°37E 42 H5
Lund Sweden 55°44N 13°12E 9 J15
Lundazi Zambia 12°20S 33°7E 55 E3
Lundi → Zimbabwe 21°43S 32°34E 55 G3
Lundu Malaysia 1°40N 109°50E 36 D3
Lundy U.K. 51°10N 4°41W 13 F3
Lune → U.K. 54°0N 2°51W 12 C5
Lüneburg Germany 53°15N 10°24E 16 B6
Lüneburg Heath = Lüneburger Heide Germany 53°10N 10°12E 16 B6
Lüneburger Heide Germany 53°10N 10°12E 16 B6
Lunenburg Canada 44°22N 64°18W 73 D7
Lunéville France 48°36N 6°30E 20 B7
Lunga → Zambia 14°34S 26°25E 55 E2
Lunga Lunga Kenya 4°33S 39°7E 54 C4
Lunglei India 22°55N 92°45E 41 H18
Luni India 26°0N 73°6E 42 G5
Luni → India 24°41N 71°14E 42 G4
Luninets = Luninyets Belarus 52°15N 26°50E 17 B14
Luning U.S.A. 38°30N 118°11W 76 G4
Luninyets Belarus 52°15N 26°50E 17 B14
Lunkaransar India 28°29N 73°44E 42 E5
Lunsemfwa → Zambia 14°54S 30°12E 55 E3
Lunsemfwa Falls Zambia 14°30S 29°6E 55 E2
Luo He → China 34°35N 110°20E 34 G6
Luochuan China 35°45N 109°26E 34 G5
Luofu Dem. Rep. of the Congo 0°10S 29°15E 54 C2
Luohe China 33°32N 114°2E 34 H8
Luonan China 34°5N 110°10E 34 G6
Luoning China 34°35N 111°40E 34 G6
Luoyang China 34°40N 112°26E 34 G7
Luozigou China 43°42N 130°18E 35 C16
Lupilichi Mozam. 11°47S 35°13E 55 E4
Luquan China 38°4N 114°17E 34 E8
Luque Paraguay 25°19S 57°25W 94 B4
Luquillo, Sierra de Puerto Rico 18°20N 65°47W 89 d
Luray U.S.A. 38°40N 78°28W 81 F14
Lurgan U.K. 54°28N 6°19W 10 B5
Lúrio Mozam. 13°32S 40°30E 55 E5
Lusaka Zambia 15°28S 28°16E 55 F2
Lusaka □ Zambia 15°30S 29°0E 55 F2
Lusambo Dem. Rep. of the Congo 4°58S 23°28E 54 C1
Lusangaye Dem. Rep. of the Congo 4°54S 26°0E 54 C2
Luseland Canada 52°5N 109°24W 71 C7
Lusenga Plain △ Zambia 9°22S 29°14E 55 D2
Lushan China 33°45N 112°55E 34 H7
Lushi China 34°3N 111°3E 34 G6
Lushnjë Albania 40°55N 19°41E 23 D8
Lushoto Tanzania 4°47S 38°20E 54 C4
Lüshun China 38°45N 121°15E 35 E11
Lusk U.S.A. 42°46N 104°27W 76 E11
Lūt, Dasht-e Iran 31°30N 58°0E 45 D8
Luta = Dalian China 38°50N 121°40E 35 E11
Lutherstadt Wittenberg Germany 51°53N 12°39E 16 C7
Luton U.K. 51°53N 0°24W 13 F7
Luton □ U.K. 51°53N 0°24W 13 F7
Łutselk'e Canada 62°24N 110°44W 71 A6
Lutsk Ukraine 50°50N 25°15E 17 C13
Lutto = Lotta → Europe 68°42N 31°6E 8 B24
Lützow Holmbukta Antarctica 69°10S 37°30E 5 C4
Lutzputs S. Africa 28°3S 20°40E 56 D3
Luuq Somali Rep. 3°48N 42°34E 47 G3
Luverne Ala., U.S.A. 31°43N 86°16W 85 F11
Luverne Minn., U.S.A. 43°39N 96°13W 80 D5
Luvua Dem. Rep. of the Congo 8°48S 25°17E 55 D2
Luvua → Dem. Rep. of the Congo 6°50S 27°30E 54 D2
Luvuvhu → S. Africa 22°25S 31°18E 57 C5
Luwegu → Tanzania 8°31S 37°23E 55 D4
Luwero Uganda 0°50N 32°28E 54 B3
Luwuk Indonesia 0°56S 122°47E 37 E6
Luxembourg Lux. 49°37N 6°9E 15 E6
Luxembourg □ Belgium 49°58N 5°30E 15 E5
Luxembourg ■ Europe 49°45N 6°0E 15 E5
Luxembourg ✈ (LUX) Lux. 49°37N 6°10E 15 E6
Luxi China 24°27N 98°36E 32 G8
Luxor = El Uqsur Egypt 25°41N 32°38E 51 C12
Luyi China 33°50N 115°35E 34 H8
Luza Russia 60°39N 47°10E 18 B8
Luzern Switz. 47°3N 8°18E 20 C8
Luzhou China 28°52N 105°20E 32 F10
Luziânia Brazil 16°20S 48°0W 93 G9
Luzon Phil. 16°0N 121°0E 37 A6
Lviv Ukraine 49°50N 24°0E 17 D13
Lviv □ Ukraine 49°30N 23°45E 17 D12
Lvov = Lviv Ukraine 49°50N 24°0E 17 D13
Lyakhavichy Belarus 53°2N 26°32E 17 B14
Lyakhovskiye, Ostrova Russia 73°40N 141°0E 29 B15
Lyal I. Canada 44°57N 81°24W 82 B3
Lybster U.K. 58°18N 3°15W 11 C5
Lycksele Sweden 64°38N 18°40E 8 D18
Lydda = Lod Israel 31°57N 34°54E 46 D3
Lyddan I. Antarctica 74°0S 21°0W 5 D2
Lydenburg S. Africa 25°10S 30°29E 57 D5
Lydia Turkey 38°48N 28°19E 23 E13
Lyell N.Z. 41°48S 172°4E 59 D4
Lyell I. Canada 52°40N 131°35W 70 C2
Lyepyel Belarus 54°50N 28°40E 9 J23
Lykens U.S.A. 40°34N 76°42W 83 F8
Lyman U.S.A. 41°20N 110°18W 76 F8
Lyme B. U.K. 50°42N 2°53W 13 G4
Lyme Regis U.K. 50°43N 2°57W 13 G5
Lymington U.K. 50°45N 1°32W 13 G6
Lyna → Poland 54°37N 21°14E 17 A11
Lynchburg U.S.A. 37°25N 79°9W 81 G14
Lynd → Australia 16°28S 143°18E 62 B3
Lynd Ra. Australia 25°30S 149°20E 63 D4
Lynden Canada 43°14N 80°9W 82 C4
Lynden U.S.A. 48°57N 122°27W 78 B4
Lyndhurst Australia 30°15S 138°18E 63 E2
Lyndon → Australia 23°29S 114°6E 61 D1
Lyndonville N.Y., U.S.A. 43°20N 78°23W 82 C6
Lyndonville Vt., U.S.A. 44°31N 72°1W 83 B12
Lyngen Norway 69°45N 20°30E 8 B19
Lynher Reef Australia 15°27S 121°55E 60 C3
Lynn U.S.A. 42°28N 70°57W 83 D14
Lynn Canal U.S.A. 58°50N 135°15W 74 D12
Lynn Haven U.S.A. 30°15N 85°39W 85 F12
Lynn Lake Canada 56°51N 101°3W 71 B8
Lynnwood U.S.A. 47°49N 122°18W 78 C4
Lynton U.K. 51°13N 3°50W 13 F4
Lyntupy Belarus 55°4N 26°23E 9 J22
Lynx L. Canada 62°25N 106°15W 71 A7
Lyon France 45°46N 4°50E 20 D6
Lyonnais France 45°45N 4°15E 20 D6
Lyons = Lyon France 45°46N 4°50E 20 D6
Lyons Ga., U.S.A. 32°12N 82°19W 85 E13
Lyons Kans., U.S.A. 38°21N 98°12W 80 F4
Lyons N.Y., U.S.A. 43°5N 77°0W 82 C8
Lyons → Australia 25°2S 115°9E 61 E2
Lyons Falls U.S.A. 43°37N 75°22W 83 C9
Lys = Leie → Belgium 51°2N 3°45E 15 C3
Lysi Cyprus 35°6N 33°41E 25 D12
Lysva Russia 58°7N 57°49E 18 C10
Lysychansk Ukraine 48°55N 38°30E 19 E6
Lytham St. Anne's U.K. 53°45N 3°0W 12 D4
Lyttelton N.Z. 43°35S 172°44E 59 E4
Lytton Canada 50°13N 121°31W 70 C4
Lyubertsy Russia 55°40N 37°51E 18 C6
Lyuboml Ukraine 51°11N 24°4E 17 C13

M

Ma → Vietnam 19°47N 105°56E 38 C5
Ma'adaba Jordan 31°43N 35°47E 46 D4
Maamba Zambia 17°17S 26°28E 56 B4
Ma'ān Jordan 30°12N 35°44E 46 E4
Ma'ān □ Jordan 30°0N 36°0E 46 F
Maanselkä Finland 63°52N 28°32E 8 E2
Maarianhamina = Mariehamn Finland 60°5N 19°55E 9 F1
Ma'arrat an Nu'mān Syria 35°43N 36°43E 44 C3
Maas → Neths. 51°45N 4°32E 15 C
Maaseik Belgium 51°6N 5°45E 15 C
Maasin Phil. 10°8N 124°50E 37 B
Maastricht Neths. 50°50N 5°40E 15 D
Maave Mozam. 21°4S 34°47E 57 C
Mababe Depression Botswana 18°50S 24°15E 56 B
Mabalane Mozam. 23°37S 32°31E 57 C
Mabel L. Canada 50°35N 118°43W 70 C
Mabenge Dem. Rep. of the Congo 4°15N 24°12E 54 B
Maberly Canada 44°50N 76°32W 83 B
Mablethorpe U.K. 53°20N 0°15E 12 D
Maboma Dem. Rep. of the Congo 2°30N 28°10E 54 B
Mabuasehube △ Botswana 25°5S 21°10E 56 D
Mabuiag Australia 9°57S 142°11E 62
Mac Bac Vietnam 9°46N 106°7E 39 H
Macachín Argentina 37°10S 63°43W 94 D
Macaé Brazil 22°20S 41°43W 95 A
McAlester U.S.A. 34°56N 95°46W 84 D
McAllen U.S.A. 26°12N 98°14W 84 H
MacAlpine L. Canada 66°32N 102°45W 68 D1
Macamic Canada 48°45N 79°0W 72 C
Macao = Macau China 22°12N 113°33E 33
Macapá Brazil 0°5N 51°4W 93 C
Macarao △ Venezuela 10°22N 67°7W 89 D
McArthur → Australia 15°54S 136°40E 62 B
McArthur, Port Australia 16°4S 136°23E 62 B
Macau Brazil 5°15S 36°40W 93 E1
Macau China 22°12N 113°33E 33
McBride Canada 53°20N 120°19W 70 C
McCall U.S.A. 44°55N 116°6W 76 D
McCamey U.S.A. 31°8N 102°14W 84 F
McCammon U.S.A. 42°39N 112°12W 76 E
McCarran Int., Las Vegas ✈ (LAS) U.S.A. 36°5N 115°9W 79 J1
McCauley I. Canada 53°40N 130°15W 70 C
McCleary U.S.A. 47°3N 123°16W 78 C
Macclenny U.S.A. 30°17N 82°7W 85 F1
Macclesfield U.K. 53°15N 2°8W 12 D
M'Clintock Chan. Canada 72°0N 102°0W 68 C1
McClintock Ra. Australia 18°44S 127°38E 60 C
McCloud U.S.A. 41°15N 122°8W 76 F
McCluer I. Australia 11°5S 133°0E 60 B
McClure U.S.A. 40°42N 77°19W 82 F
McClure, L. U.S.A. 37°35N 120°16W 78 H
M'Clure Str. Canada 75°0N 119°0W 69 C
McClusky U.S.A. 47°29N 100°27W 80 B
McComb U.S.A. 31°15N 90°27W 85 F
McCook U.S.A. 40°12N 100°38W 80 E
McCreary Canada 50°47N 99°29W 71 C
McCullough Mt. U.S.A. 35°35N 115°13W 79 K1
McCusker → Canada 55°32N 108°39W 71 B
McDermitt U.S.A. 41°59N 117°43W 76 F
McDonald U.S.A. 40°22N 80°14W 82 F
Macdonald, L. Australia 23°30S 129°0E 60 D
McDonald Is. Ind. Oc. 53°0S 73°0E 3 G
MacDonnell Ranges Australia 23°40S 133°0E 60 D
MacDowell L. Canada 52°15N 92°45W 72 B
Macduff U.K. 57°40N 2°31W 11 D
Macedonia U.S.A. 41°19N 81°31W 82 E
Macedonia □ Greece 40°39N 22°0E 23 D

cedonia ■ Europe 41°53N 21°40E 23 D9
ceió Brazil 9°40S 35°41W 93 E11
cerata Italy 43°18N 13°27E 22 C5
Farland U.S.A. 35°41N 119°14W 79 K7
Farlane → Canada 59°12N 107°58W 71 B7
cfarlane, L. Australia 32°0S 136°40E 63 E2
Gehee U.S.A. 33°38N 91°24W 84 E9
Gill U.S.A. 39°23N 114°47W 76 G6
cgillycuddy's Reeks reland 51°58N 9°45W 10 E2
Graw U.S.A. 42°36N 76°8W 83 D8
Gregor U.S.A. 43°1N 91°11W 80 D8
Gregor Ra. Australia 27°0S 142°45E 63 D3
Guire, Mt. Australia 20°18S 148°23E 62 b
ch Pakistan 29°50N 67°20E 42 E2
ch Kowr Iran 25°48N 61°28E 45 E9
chado = Jiparaná → 3razil 8°3S 62°52W 92 E6
chagai Argentina 26°56S 60°2W 94 B3
chakos Kenya 1°30S 37°15E 54 C4
chala Ecuador 3°20S 79°57W 92 D3
changa Mozam. 20°59S 35°0E 57 C6
chattie, L. Australia 24°50S 139°48E 62 C2
cheke Zimbabwe 18°5S 31°51E 57 B5
chhu → India 23°6N 70°46E 42 H4
chiara △ Pakistan 34°40N 73°30E 42 B5
chias Maine, U.S.A. 44°43N 67°28W 81 C20
chias N.Y., U.S.A. 42°25N 78°29W 82 D6
chichi → Canada 57°3N 92°6W 71 B10
chico Madeira 32°43N 16°44W 24 D3
chilipatnam India 16°12N 81°8E 41 L12
chiques Venezuela 10°4N 72°34W 92 A4
chu Picchu Peru 13°8S 72°30W 92 F4
chynlleth U.K. 52°35N 3°50W 13 E4
cia Mozam. 25°2S 33°8E 57 D5
Ilwraith Ra. Australia 13°50S 143°20E 62 A3
Innes L. Canada 52°13N 93°45W 71 C10
Intosh U.S.A. 45°55N 101°21W 80 C3
Intosh L. Canada 55°45N 105°0W 71 B8
cintosh Ra. Australia 27°39S 125°32E 61 E4
cintyre → Australia 28°37S 150°47E 63 D5
ckay Australia 21°8S 149°11E 62 K7
ckay U.S.A. 43°55N 113°37W 76 E7
cKay → Canada 57°10N 111°38W 70 B6
ckay, L. Australia 22°30S 129°0E 60 D4
Kay Ra. Australia 23°0S 122°30E 60 D3
Keesport U.S.A. 40°20N 79°51W 82 F5
Kellar Canada 45°30N 79°55W 82 A5
Kenna U.S.A. 46°56N 122°33W 78 D4
ckenzie Canada 55°20N 123°5W 70 B4
Kenzie U.S.A. 36°8N 88°31W 85 C10
ckenzie → Australia 23°38S 149°46E 62 C4
ckenzie → Canada 69°10N 134°20W 68 D5
Kenzie → U.S.A. 44°7N 123°6W 76 D2
ckenzie Bay Canada 69°0N 137°30W 66 C6
ckenzie City = Linden Guyana 6°0N 58°10W 92 B7
ckenzie King I. Canada 77°45N 111°0W 69 B9
ckenzie Mts. Canada 64°0N 130°0W 66 C6
ckinac, Straits of U.S.A. 45°50N 84°40W 75 F23
ckinaw City U.S.A. 45°47N 84°44W 81 C11
Kinlay Australia 21°16S 141°18E 62 C3
Kinlay → Australia 20°50S 141°28E 62 C3
Kinley, Mt. U.S.A. 63°4N 151°0W 68 E1
Kinley Sea Arctic 82°0N 0°0 4 A7
Kinney U.S.A. 33°12N 96°37W 84 E6
ckinnon Road Kenya 3°40S 39°1E 54 C4
Kittrick U.S.A. 35°18N 119°37W 79 K7
cklin Canada 52°20N 109°56W 71 C7
cksville Australia 30°40S 152°56E 63 E5
Laughlin U.S.A. 45°49N 100°49W 80 C3
clean Australia 29°26S 153°16E 63 D5
Lean U.S.A. 35°14N 100°36W 84 D4
Leansboro U.S.A. 38°6N 88°32W 80 F9
clear S. Africa 31°2S 28°23E 57 E4
clear, C. Malawi 13°58S 34°49E 55 E3
cleay → Australia 30°56S 153°0E 63 E5
Lennan Canada 55°42N 116°50W 70 B5
Leod → Canada 54°9N 115°44W 70 C5
cLeod, L. Australia 24°9S 113°47E 61 D1
cLeod B. Canada 62°53N 110°0W 71 A7
cLeod Lake Canada 54°58N 123°0W 70 C4
Loughlin, Mt. U.S.A. 42°27N 122°19W 76 E2
Mechen U.S.A. 39°57N 80°44W 82 G4
Minnville Oreg., U.S.A. 45°13N 123°12W 76 D2
Minnville Tenn., U.S.A. 35°41N 85°46W 85 D12
Murdo Antarctica 77°51S 166°37E 5 D11
Murdo Sd. Antarctica 77°0S 170°0E 5 D11
Murray = Fort McMurray Canada 56°44N 111°7W 70 B6
Murray U.S.A. 48°19N 122°14W 78 B4
çobere Mozam. 21°13S 32°47E 55 G3
codoene Mozam. 23°32S 35°5E 57 C6
comb U.S.A. 40°27N 90°40W 80 E8
âcon France 46°19N 4°50E 20 C6
con Ga., U.S.A. 32°51N 83°38W 85 E13
con Miss., U.S.A. 33°7N 88°34W 85 E10
con Mo., U.S.A. 39°44N 92°28W 80 F7
cossa Mozam. 17°55S 33°56E 55 F3
coun L. Canada 56°32N 103°40W 71 B8
covane Mozam. 21°30S 35°2E 57 C6
Pherson U.S.A. 38°22N 97°40W 80 F5
Pherson Pk. U.S.A. 34°53N 119°53W 79 L7
Pherson Ra. Australia 28°15S 153°15E 63 D5
cquarie → Australia 30°7S 147°24E 63 E4
cquarie Harbour Australia 42°15S 145°23E 63 G4
cquarie I. Pac. Oc. 54°36S 158°55E 64 N7
cquarie Ridge S. Ocean 57°0S 159°0E 5 B10
acRobertson Land Antarctica 71°0S 64°0E 5 D6
acroom Ireland 51°54N 8°57W 10 E3
acTier Canada 45°8N 79°47W 82 A5
Macubela Mozam. 16°53S 37°49E 55 F4
Macuira △ Colombia 12°9N 71°21W 89 D5
Macumba → Australia 27°52S 137°12E 63 D2
Macuro Venezuela 10°42N 61°55W 93 K15
Macusani Peru 14°4S 70°29W 92 F4
Macuse Mozam. 17°45S 37°10E 55 F4
Macuspana Mexico 17°46N 92°36W 87 D6
Macusse Angola 17°48S 20°23E 56 B3
Ma'dabā □ Jordan 31°43N 35°47E 46 D4
Madadeni S. Africa 27°43S 30°3E 57 D5
Madagascar ■ Africa 20°0S 47°0E 57 C8
Madā'in Sālih Si. Arabia 26°46N 37°57E 44 E3
Madama Niger 22°0N 13°40E 51 D8
Madame, I. Canada 45°30N 60°58W 73 C7
Madang Papua N. G. 5°12S 145°49E 58 B7
Madaripur Bangla. 23°19N 90°15E 41 H17
Madauk Burma 17°56N 96°52E 41 L20
Madawaska Canada 45°30N 78°0W 82 A7
Madawaska → Canada 45°27N 76°21W 82 A7
Madaya Burma 22°12N 96°10E 41 H20
Maddalena Italy 41°16N 9°23E 22 D3
Madeira Atl. Oc. 32°50N 17°0W 24 D3
Madeira → Brazil 3°22S 58°45W 92 D7
Madeleine, Îs. de la Canada 47°30N 61°40W 73 C7
Madera Mexico 29°12N 108°7W 86 B3
Madera Calif., U.S.A. 36°57N 120°3W 78 J6
Madera Pa., U.S.A. 40°49N 78°26W 82 F6
Madha India 18°0N 75°30E 40 L9
Madhavpur India 21°15N 69°58E 42 J3
Madhepura India 26°11N 86°23E 43 F12
Madhubani India 26°21N 86°7E 43 F12
Madhupur India 24°16N 86°39E 43 G12
Madhya Pradesh □ India 22°50N 78°0E 42 J8
Madidi → Bolivia 12°32S 66°52W 92 F5
Madikeri India 12°30N 75°45E 40 N9
Madikwe ⌓ S. Africa 27°38S 32°15E 57 D5
Madill U.S.A. 34°6N 96°46W 84 D6
Madimba Dem. Rep. of the Congo 4°58S 15°5E 52 E3
Ma'din Syria 35°45N 39°36E 44 C3
Madinat al Malik Khālid al Askarīyah Si. Arabia 27°54N 45°31E 44 E5
Madingou Congo 4°10S 13°33E 52 E2
Madirovalo Madag. 16°26S 46°32E 57 B8
Madison Calif., U.S.A. 38°41N 121°59W 78 G5
Madison Fla., U.S.A. 30°28N 83°25W 85 F13
Madison Ind., U.S.A. 38°44N 85°23W 81 F11
Madison Nebr., U.S.A. 41°50N 97°27W 80 E5
Madison Ohio, U.S.A. 41°46N 81°3W 82 E3
Madison S. Dak., U.S.A. 44°0N 97°7W 80 C5
Madison Wis., U.S.A. 43°4N 89°24W 80 D9
Madison → U.S.A. 45°56N 111°31W 76 D8
Madison Heights U.S.A. 37°25N 79°8W 81 G14
Madisonville Ky., U.S.A. 37°20N 87°30W 80 G10
Madisonville Tex., U.S.A. 30°57N 95°55W 84 F7
Madista Botswana 21°15S 25°6E 56 C4
Madiun Indonesia 7°38S 111°32E 37 G14
Mado Gashi Kenya 0°44N 39°10E 54 B4
Madoc Canada 44°30N 77°28W 82 B7
Madoi China 34°46N 98°18E 32 E8
Madona Latvia 56°53N 26°5E 9 H22
Madrakah, Ra's al Oman 19°0N 57°50E 47 D6
Madras = Chennai India 13°8N 80°19E 40 N12
Madras = Tamil Nadu □ India 11°0N 77°0E 40 P10
Madras U.S.A. 44°38N 121°8W 76 D3
Madre, L. U.S.A. 25°15N 97°30W 84 J6
Madre, Sierra Phil. 17°0N 122°0E 37 A6
Madre de Dios → Bolivia 10°59S 66°8W 92 F5
Madre de Dios, I. Chile 50°20S 75°10W 96 G1
Madre del Sur, Sierra Mexico 17°30N 100°0W 87 D5
Madre Occidental, Sierra Mexico 27°0N 107°0W 86 B3
Madre Oriental, Sierra Mexico 25°0N 100°0W 86 C5
Madri India 24°16N 73°32E 42 G5
Madrid Spain 40°24N 3°42W 21 B4
Madrid U.S.A. 44°45N 75°8W 83 B9
Madula Dem. Rep. of the Congo 0°27N 25°22E 54 B2
Madura Australia 31°55S 127°0E 61 F4
Madura Indonesia 7°30S 114°0E 37 G15
Madura, Selat Indonesia 7°30S 113°20E 37 G15
Madurai India 9°55N 78°10E 40 Q11
Madurantakam India 12°30N 79°50E 40 N11
Mae Chan Thailand 20°9N 99°52E 38 B2
Mae Charim △ Thailand 18°17N 100°59E 38 C3
Mae Hong Son Thailand 19°16N 97°56E 38 C2
Mae Khlong → Thailand 13°24N 100°0E 38 F3
Mae Moei △ Thailand 17°26N 98°7E 38 D2
Mae Phang △ Thailand 19°7N 99°13E 38 C2
Mae Phrik Thailand 17°27N 99°7E 38 D2
Mae Ping △ Thailand 17°37N 98°51E 38 D2
Mae Ramat Thailand 16°58N 98°31E 38 D2
Mae Rim Thailand 18°54N 98°57E 38 C2
Mae Sai Thailand 20°20N 99°55E 38 B2
Mae Sot Thailand 16°43N 98°34E 38 D2
Mae Suai Thailand 19°39N 99°33E 38 C2
Mae Tha Thailand 18°28N 99°8E 38 C2
Mae Tup Res. Thailand 17°52N 98°45E 38 D2
Mae Wa △ Thailand 17°23N 99°16E 38 D2
Mae Wong △ Thailand 15°54N 99°12E 38 E2
Mae Yom △ Thailand 18°43N 100°15E 38 C3
Maebaru Japan 33°33N 130°12E 31 H5
Maebashi Japan 36°24N 139°4E 31 F9
Maelpaeg L. Canada 48°20N 56°30W 73 C8
Maesteg U.K. 51°36N 3°40W 13 F4
Maestra, Sierra Cuba 20°15N 77°0W 88 B4
Maevatanana Madag. 16°56S 46°49E 57 B8
Mafeking = Mafikeng S. Africa 25°50S 25°38E 56 D4
Mafeking Canada 52°40N 101°10W 71 C8
Mafeteng Lesotho 29°51S 27°15E 56 D4
Maffra Australia 37°53S 146°58E 63 F4
Mafia I. Tanzania 7°45S 39°50E 54 D4
Mafikeng S. Africa 25°50S 25°38E 56 D4
Mafra Brazil 26°10S 49°55W 95 B6
Mafra Portugal 38°55N 9°20W 21 C1
Mafungabusi Plateau Zimbabwe 18°30S 29°8E 55 F2
Magadan Russia 59°38N 150°50E 29 D16
Magadi Kenya 1°54S 36°19E 54 C4
Magadi, L. Kenya 1°54S 36°19E 54 C4
Magallanes, Estrecho de Chile 52°30S 75°0W 96 G2
Magangué Colombia 9°14N 74°45W 92 B4
Magdagachi Russia 53°27N 125°48E 29 D13
Magdalen Is. = Madeleine, Îs. de la Canada 47°30N 61°40W 73 C7
Magdalena Argentina 35°5S 57°30W 94 D4
Magdalena Bolivia 13°13S 63°57W 92 F6
Magdalena U.S.A. 34°7N 107°15W 77 J10
Magdalena → Colombia 11°6N 74°51W 92 A4
Magdalena → Mexico 30°40N 112°25W 86 A2
Magdalena, B. Mexico 24°35N 112°0W 86 C2
Magdalena, I. Mexico 24°40N 112°15W 86 C2
Magdalena, Llano de Mexico 25°0N 111°25W 86 C2
Magdalena de Kino Mexico 30°38N 110°57W 86 A2
Magdeburg Germany 52°7N 11°38E 16 B6
Magdelaine Cays Australia 16°33S 150°18E 62 B5
Magee U.S.A. 31°52N 89°44W 85 F10
Magelang Indonesia 7°29S 110°13E 37 G14
Magellan's Str. = Magallanes, Estrecho de Chile 52°30S 75°0W 96 G2
Magenta, L. Australia 33°30S 119°2E 61 F2
Magerøya Norway 71°3N 25°40E 8 A21
Maggiore, Lago Italy 45°57N 8°39E 20 D8
Maggotty Jamaica 18°9N 77°46W 88 a
Maghâgha Egypt 28°38N 30°50E 51 C12
Magherafelt U.K. 54°45N 6°37W 10 B5
Maghreb N. Afr. 32°0N 4°0W 48 C3
Magistralnyy Russia 56°16N 107°36E 29 D11
Magnetic Pole (North) Arctic 82°42N 114°24W 4 A1
Magnetic Pole (South) Antarctica 64°8S 138°8E 5 C9
Magnitogorsk Russia 53°27N 59°4E 18 D10
Magnolia Ark., U.S.A. 33°16N 93°14W 84 E8
Magnolia Miss., U.S.A. 31°9N 90°28W 85 F9
Mago Fiji 17°26S 179°8W 59 a
Magog Canada 45°18N 72°9W 83 A12
Magoro Uganda 1°45N 34°12E 54 B3
Mağosa = Famagusta Cyprus 35°8N 33°55E 25 D12
Magoulades Greece 39°45N 19°42E 25 A3
Magoye Zambia 16°1S 27°30E 55 F2
Magpie, L. Canada 51°0N 64°41W 73 B7
Magrath Canada 49°25N 112°50W 70 D6
Magu Tanzania 2°30S 33°30E 54 C3
Maguarinho, C. Brazil 0°15S 48°30W 93 D9
Magude Mozam. 25°2S 32°40E 57 D5
Mağusa = Famagusta Cyprus 35°8N 33°55E 25 D12
Maguse L. Canada 61°37N 95°10W 71 A9
Maguse Pt. Canada 61°20N 93°50W 71 A10
Magvana India 23°13N 69°22E 42 H3
Magwe Burma 20°10N 95°0E 41 J19
Magyarorszag = Hungary ■ Europe 47°20N 19°20E 17 E10
Maha Sarakham Thailand 16°12N 103°16E 38 D4
Mahābād Iran 36°50N 45°45E 44 B5
Mahabharat Lekh Nepal 28°30N 82°0E 43 E10
Mahabo Madag. 20°23S 44°40E 57 C7
Mahadeo Hills India 22°20N 78°30E 43 H8
Mahaffey U.S.A. 40°53N 78°44W 82 F6
Mahagi Dem. Rep. of the Congo 2°20N 31°0E 54 B3
Mahajamba → Madag. 15°33S 47°8E 57 B8
Mahajamba, Helodranon' i Madag. 15°24S 47°5E 57 B8
Mahajan India 28°48N 73°56E 42 E5
Mahajanga Madag. 15°40S 46°25E 57 B8
Mahajanga □ Madag. 17°0S 47°0E 57 B8
Mahajilo → Madag. 19°42S 45°22E 57 B8
Mahakam → Indonesia 0°35S 117°17E 36 E5
Mahalapye Botswana 23°1S 26°51E 56 C4
Mahale Mts. Tanzania 6°20S 30°0E 54 D3
Mahale Mts. △ Tanzania 6°10S 29°50E 54 D2
Maḩallāt Iran 33°55N 50°30E 45 C6
Māhān Iran 30°5N 57°18E 45 D8
Mahan → India 23°30N 82°50E 43 H10
Mahanadi → India 20°20N 86°25E 41 J15
Mahananda → India 25°12N 87°52E 43 G12
Mahanoro Madag. 19°54S 48°48E 57 B8
Mahanoy City U.S.A. 40°49N 76°9W 83 F8
Maharashtra □ India 20°30N 75°30E 40 J9
Mahasham, W. → Egypt 30°15N 34°10E 46 E3
Mahasoa Madag. 22°12S 46°6E 57 C8
Mahasolo Madag. 19°7S 46°22E 57 B8
Mahattat ash Shīdīyah Jordan 29°55N 35°55E 46 F4
Mahattat 'Unayzah Jordan 30°30N 35°47E 46 E4
Mahavavy → Madag. 15°57S 45°54E 57 B8
Mahaxay Laos 17°22N 105°12E 38 D5
Mahbubnagar India 16°45N 77°59E 40 L10
Mahda U.A.E. 25°20N 56°15E 45 E8
Maḩdah Oman 24°24N 55°59E 45 E7
Mahdia Tunisia 35°28N 11°0E 51 A8
Mahe India 33°10N 78°32E 43 C8
Mahé Seychelles 5°0S 55°30E 53 b
Mahé ✈ (SEZ) Seychelles 4°40S 55°31E 53 b
Mahébourg Mauritius 20°24S 57°42E 53 d
Mahendragarh India 28°17N 76°14E 42 E7
Mahendranagar Nepal 28°55N 80°20E 43 E9
Mahenge Tanzania 8°45S 36°41E 55 D4
Maheno N.Z. 45°10S 170°50E 59 F3
Mahesana India 23°39N 72°26E 42 H5
Maheshwar India 22°11N 75°35E 42 H6
Mahgawan India 26°29N 78°37E 43 F8
Mahi → India 22°15N 72°55E 42 H5
Mahia Pen. N.Z. 39°9S 177°55E 59 C6
Mahilyow Belarus 53°55N 30°18E 17 B16
Mahina Tahiti 17°30S 149°27W 59 d
Mahinerangi, L. N.Z. 45°50S 169°56E 59 F2
Mahmud Kot Pakistan 30°16N 71°0E 42 D4
Mahnomen U.S.A. 47°19N 95°58W 80 B6
Mahoba India 25°15N 79°55E 43 G8
Mahón = Maó Spain 39°53N 4°16E 24 B11
Mahone Bay Canada 44°27N 64°23W 73 D7
Mahopac U.S.A. 41°22N 73°45W 83 E11
Mahuva India 21°5N 71°48E 42 J4
Mai-Ndombe, L. Dem. Rep. of the Congo 2°0S 18°20E 52 E3
Mai Thon, Ko Thailand 7°40N 98°28E 39 a
Maicuru → Brazil 2°14S 54°17W 93 D8
Maidan Khula Afghan. 33°36N 69°50E 42 C3
Maidenhead U.K. 51°31N 0°42W 13 F7
Maidstone Canada 53°5N 109°20W 71 C7
Maidstone U.K. 51°16N 0°32E 13 F8
Maiduguri Nigeria 12°0N 13°20E 51 F8
Maigh Nuad = Maynooth Ireland 53°23N 6°34W 10 C5
Maihar India 24°16N 80°45E 43 G9
Maikala Ra. India 22°0N 81°0E 41 J12
Maiko △ Dem. Rep. of the Congo 0°30S 27°50E 54 C2
Mailani India 28°17N 80°21E 43 E9
Mailsi Pakistan 29°48N 72°15E 42 E5
Main → Germany 50°0N 8°18E 16 C5
Main → U.K. 54°48N 6°18W 10 B5
Main Channel Canada 45°21N 81°45W 82 A3
Main Range △ Australia 28°11S 152°27E 63 D5
Main Ridge Trin. & Tob. 11°16N 60°40W 93 J16
Maine France 48°20N 0°15W 20 C3
Maine □ U.S.A. 45°20N 69°0W 81 C19
Maine → Ireland 52°9N 9°45W 10 D2
Maine, G. of U.S.A. 43°0N 68°30W 81 D19
Maingkwan Burma 26°15N 96°37E 41 F20
Mainistir an Corann = Midleton Ireland 51°55N 8°10W 10 E3
Mainit, L. Phil. 9°31N 125°30E 37 C7
Mainland Orkney, U.K. 58°59N 3°8W 11 C5
Mainland Shet., U.K. 60°15N 1°22W 11 A7
Mainpuri India 27°18N 79°4E 43 F8
Maintirano Madag. 18°3S 44°1E 57 B7
Mainz Germany 50°1N 8°14E 16 C5
Maio C. Verde Is. 15°10N 23°10W 50 b
Maipú Argentina 36°52S 57°50W 94 D4
Maiquetía Venezuela 10°36N 66°57W 92 A5
Mairabari India 26°30N 92°22E 41 F18
Maisi Cuba 20°17N 74°9W 89 B5
Maisi, Pta. de Cuba 20°10N 74°10W 89 B5
Maitland N.S.W., Australia 32°33S 151°36E 63 E5
Maitland S. Austral., Australia 34°23S 137°40E 63 E2
Maitland → Canada 43°45N 81°43W 82 C3
Maitri Antarctica 70°0S 3°0W 5 D3
Maíz, Is. del Nic. 12°15N 83°4W 88 D3
Maizuru Japan 35°25N 135°22E 31 G7
Majalengka Indonesia 6°50S 108°13E 37 G13
Majanji Uganda 0°16N 34°0E 54 B3
Majene Indonesia 3°38S 118°57E 37 E5
Majete ⌓ Malawi 15°54S 34°34E 55 F3
Majorca = Mallorca Spain 39°30N 3°0E 24 B10
Majuro Marshall Is. 7°9N 171°12E 64 G9
Mak, Ko Thailand 11°49N 102°29E 39 G4
Maka Senegal 13°40N 14°10W 50 F3
Makaha Zimbabwe 17°20S 32°39E 57 B5
Makalamabedi Botswana 20°19S 23°51E 56 C3
Makale Indonesia 3°6S 119°51E 37 E5
Makalu Asia 27°55N 87°8E 43 F12
Makalu-Barun △ Nepal 27°45N 87°10E 43 F12
Makamba Burundi 4°8S 29°49E 54 C2
Makarikari = Makgadikgadi Salt Pans Botswana 20°40S 25°45E 56 C4
Makarov Basin Arctic 87°0N 150°0W 4 A
Makarovo Russia 57°40N 107°45E 29 D11
Makassar Indonesia 5°10S 119°20E 37 F5
Makassar, Selat Indonesia 1°0S 118°20E 37 E5
Makassar, Str. of = Makassar, Selat Indonesia 1°0S 118°20E 37 E5
Makat = Maqat Kazakhstan 47°39N 53°19E 19 E9
Makedonija = Macedonia ■ Europe 41°53N 21°40E 23 D9
Makeni S. Leone 8°55N 12°5W 50 G3
Makeyevka = Makiyivka Ukraine 48°0N 38°0E 19 E6
Makgadikgadi △ Botswana 20°27S 24°47E 56 C3
Makgadikgadi Salt Pans Botswana 20°40S 25°45E 56 C4
Makhachkala Russia 43°0N 47°30E 19 F8
Makhado = Louis Trichardt S. Africa 23°1S 29°43E 57 C4
Makham, Ao Thailand 7°51N 98°25E 39 a
Makhfar al Buşayyah Iraq 30°0N 46°10E 44 D5
Makhmūr Iraq 35°46N 43°35E 44 C4
Makian Indonesia 0°20N 127°20E 37 D7
Makindu Kenya 2°18S 37°50E 54 C4
Makīnsk Kazakhstan 52°37N 70°26E 28 D8
Makira = San Cristóbal Solomon Is. 10°30S 161°0E 58 C9
Makiyivka Ukraine 48°0N 38°0E 19 E6
Makkah Si. Arabia 21°30N 39°54E 47 C2
Makkovik Canada 55°10N 59°10W 73 A8
Makó Hungary 46°14N 20°33E 17 E11
Makogai Fiji 17°28S 179°0E 59 a
Makokou Gabon 0°40N 12°50E 52 D2
Makongo Dem. Rep. of the Congo 3°25N 26°17E 54 B2
Makoro Dem. Rep. of the Congo 3°10N 29°59E 54 B2
Makrai India 22°2N 77°0E 42 H7
Makran Coast Range Pakistan 25°40N 64°0E 40 G4
Makrana India 27°2N 74°46E 42 F6
Makrigialos Greece 35°2N 25°59E 25 D7
Mākū Iran 39°15N 44°31E 44 B5
Makunda Botswana 22°30S 20°7E 56 C3
Makurazaki Japan 31°15N 130°20E 31 J5
Makurdi Nigeria 7°43N 8°35E 50 G7
Makushin Volcano U.S.A. 53°53N 166°55W 74 E6
Makūyeh Iran 28°7N 53°9E 45 D7
Makwassie S. Africa 27°17S 26°0E 56 D4
Makwiro Zimbabwe 17°58S 30°25E 57 B5
Mal B. Ireland 52°50N 9°30W 10 D2
Mala = Mallow Ireland 52°8N 8°39W 10 D3
Mala ☼ Australia 21°39S 130°45E 60 D5
Mala, Pta. Panama 7°28N 80°2W 88 E3
Malabar Coast India 11°0N 75°0E 40 P9
Malacca, Straits of Indonesia 3°0N 101°0E 39 L3
Malad City U.S.A. 42°12N 112°15W 76 E7
Maladzyechna Belarus 54°20N 26°50E 17 A14
Málaga Spain 36°43N 4°23W 21 D3
Malagarasi Tanzania 5°5S 30°50E 54 D3
Malagarasi → Tanzania 5°12S 29°47E 54 D2
Malagasy Rep. = Madagascar ■ Africa 20°0S 47°0E 57 C8
Malahide Ireland 53°26N 6°9W 10 C5
Malaimbandy Madag. 20°20S 45°36E 57 C8
Malaita Solomon Is. 9°0S 161°0E 58 B9
Malakâl Sudan 9°33N 31°40E 51 G12
Malakand Pakistan 34°40N 71°55E 42 B4
Malakula Vanuatu 16°15S 167°30E 58 C9
Malakwal Pakistan 32°34N 73°13E 42 C5
Malamala Indonesia 3°21S 120°55E 37 E6
Malanda Australia 17°22S 145°35E 62 B4
Malang Indonesia 7°59S 112°45E 37 G15
Malanga Mozam. 13°28S 36°7E 55 E4
Malangen Norway 69°24N 18°37E 8 B18
Malanje Angola 9°36S 16°17E 52 F3
Mälaren Sweden 59°30N 17°10E 9 G17
Malargüe Argentina 35°32S 69°30W 94 D2
Malartic Canada 48°9N 78°9W 72 C4
Malaryta Belarus 51°50N 24°3E 17 C13
Malaspina Glacier U.S.A. 59°50N 140°30W 74 D11
Malatya Turkey 38°25N 38°20E 44 B3
Malawi ■ Africa 11°55S 34°0E 55 E3
Malawi, L. Africa 12°30S 34°30E 55 E3
Malay Pen. Asia 7°25N 100°0E 39 J3
Malaya Vishera Russia 58°55N 32°25E 18 C5
Malaybalay Phil. 8°5N 125°7E 37 C7
Malāyer Iran 34°19N 48°51E 45 C6
Malaysia ■ Asia 5°0N 110°0E 39 K4
Malazgirt Turkey 39°10N 42°33E 44 B4
Malbon Australia 21°5S 140°17E 62 C3
Malbooma Australia 30°41S 134°11E 63 E1
Malbork Poland 54°3N 19°1E 17 B10
Malcolm Australia 28°51S 121°25E 61 E3
Malcolm, Pt. Australia 33°48S 123°45E 61 F3
Maldah India 25°2N 88°9E 43 G13
Maldegem Belgium 51°14N 3°26E 15 C3
Malden Mass., U.S.A. 42°26N 71°3W 83 D13
Malden Mo., U.S.A. 36°34N 89°57W 80 G9
Malden I. Kiribati 4°3S 155°1W 65 H12
Maldives ■ Ind. Oc. 5°0N 73°0E 27 H9
Maldon U.K. 51°44N 0°42E 13 F8
Maldonado Uruguay 34°59S 55°0W 95 C5
Maldonado, Pta. Mexico 16°20N 98°33W 87 D5
Malé Karpaty Slovak Rep. 48°30N 17°20E 17 D9
Maleas, Akra Greece 36°28N 23°7E 23 F10
Malebo, Pool Africa 4°17S 15°20E 52 E3
Malegaon India 20°30N 74°38E 40 J9
Malei Mozam. 17°12S 36°58E 55 F4
Malek Kandī Iran 37°9N 46°6E 44 B5
Malela Dem. Rep. of the Congo 4°22S 26°8E 54 C2
Malema Mozam. 14°57S 37°20E 55 E4
Maleme Greece 35°31N 23°49E 25 D5
Malerkotla India 30°32N 75°58E 42 D6
Males Greece 35°6N 25°35E 25 D7
Malgomaj Sweden 64°40N 16°30E 8 D17
Malha Sudan 15°8N 25°10E 51 E11
Malhargarh India 24°17N 74°59E 42 G6
Malheur → U.S.A. 44°4N 116°59W 76 D5
Malheur L. U.S.A. 43°20N 118°48W 76 E4
Mali ■ Africa 17°0N 3°0W 50 E5
Mali → Burma 25°42N 97°30E 41 G20
Mali Kyun Burma 13°0N 98°20E 38 F2
Malia Greece 35°17N 25°32E 25 D7
Malia, Kolpos Greece 35°19N 25°27E 25 D7
Malibu U.S.A. 34°2N 118°41W 79 L8
Maliku Indonesia 0°39S 123°16E 37 E6
Malili Indonesia 2°42S 121°6E 37 E6
Malimba, Mts. Dem. Rep. of the Congo 7°30S 29°30E 54 D2
Malin Hd. Ireland 55°23N 7°23W 10 A4
Malin Pen. Ireland 55°20N 7°17W 10 A4
Malindi Kenya 3°12S 40°5E 54 C5
Malindi Marine ⌓ Kenya 3°15S 40°7E 54 C5
Malines = Mechelen Belgium 51°2N 4°29E 15 C4
Malino Indonesia 1°0N 121°0E 37 D6
Malinyi Tanzania 8°56S 36°0E 55 D4
Malita Phil. 6°19N 125°39E 37 C7
Maliwun Burma 10°17N 98°40E 39 G2
Maliya India 23°5N 70°46E 42 H4
Malka Mari △ Kenya 4°11N 40°46E 54 B5
Malkara Turkey 40°53N 26°53E 23 D12
Mallacoota Inlet Australia 37°34S 149°40E 63 F4
Mallaig U.K. 57°0N 5°50W 11 D3
Mallawan India 27°4N 80°12E 43 F9
Mallawi Egypt 27°44N 30°44E 51 C12
Mallicolo = Malakula Vanuatu 16°15S 167°30E 58 C9
Mallorca Spain 39°30N 3°0E 24 B10
Mallorytown Canada 44°29N 75°53W 83 B9
Mallow Ireland 52°8N 8°39W 10 D3
Malmberget Sweden 67°11N 20°40E 8 C19
Malmédy Belgium 50°25N 6°2E 15 D6
Malmesbury S. Africa 33°28S 18°41E 56 E2

Malmivaara = Malmberget *Sweden* 67°11N 20°40E **8** C19
Malmö *Sweden* 55°36N 12°59E **9** J15
Malolo *Fiji* 17°45S 177°11E **59** a
Malolos *Phil.* 14°50N 120°49E **37** B6
Malolotja ⌓ *Swaziland* 26°4S 31°6E **57** D5
Malombe L. *Malawi* 14°40S 35°15E **55** E4
Malone *U.S.A.* 44°51N 74°18W **83** B10
Måløy *Norway* 61°57N 5°6E **8** F11
Malpaso *Canary Is.* 27°43N 18°3W **24** G1
Malpaso, Presa = Netzahualcóyotl, Presa *Mexico* 17°8N 93°35W **87** D6
Malpelo, I. de *Colombia* 4°3N 81°35W **92** C2
Malpur *India* 23°21N 73°27E **42** H5
Malpura *India* 26°17N 75°23E **42** F6
Malta *Idaho, U.S.A.* 42°18N 113°22W **76** E7
Malta *Mont., U.S.A.* 48°21N 107°52W **76** B10
Malta ■ *Europe* 35°55N 14°26E **25** D2
Maltahöhe *Namibia* 24°55S 17°0E **56** C2
Malton *Canada* 43°42N 79°38W **82** C5
Malton *U.K.* 54°8N 0°49W **12** C7
Maluku *Indonesia* 1°0S 127°0E **37** E7
Maluku □ *Indonesia* 3°0S 128°0E **37** E7
Maluku Sea = Molucca Sea *Indonesia* 0°0 125°0E **37** E6
Malvan *India* 16°2N 73°30E **40** L8
Malvern *U.S.A.* 34°22N 92°49W **84** D8
Malvern Hills *U.K.* 52°0N 2°19W **13** E5
Malvinas, Is. = Falkland Is. ☑ *Atl. Oc.* 51°30S 59°0W **96** G5
Malya *Tanzania* 3°5S 33°38E **54** C3
Malyn *Ukraine* 50°46N 29°3E **17** C15
Malyy Lyakhovskiy, Ostrov *Russia* 74°7N 140°36E **29** B15
Malyy Taymyr, Ostrov *Russia* 78°6N 107°15E **29** B11
Mama *Russia* 58°18N 112°54E **29** D12
Mamanguape *Brazil* 6°50S 35°4W **93** E11
Mamanuca Group *Fiji* 17°35S 177°5E **59** a
Mamarr Mitlā *Egypt* 30°2N 32°54E **46** E1
Mamasa *Indonesia* 2°55S 119°20E **37** E5
Mambasa *Dem. Rep. of the Congo* 1°22N 29°3E **54** B2
Mamberamo → *Indonesia* 2°0S 137°50E **37** E9
Mambilima Falls *Zambia* 10°31S 28°45E **55** E2
Mambirima *Dem. Rep. of the Congo* 11°25S 27°33E **55** E2
Mambo *Tanzania* 4°52S 38°22E **54** C4
Mambrui *Kenya* 3°5S 40°5E **54** C5
Mamburao *Phil.* 13°13N 120°39E **37** B6
Mameigwess L. *Canada* 52°35N 87°50W **72** B2
Mammoth *U.S.A.* 32°43N 110°39W **77** K8
Mammoth Cave △ *U.S.A.* 37°8N 86°13W **80** G10
Mamoré → *Bolivia* 10°23S 65°53W **92** F5
Mamou *Guinea* 10°15N 12°0W **50** F3
Mamoudzou *Mayotte* 12°48S 45°14E **53** a
Mampikony *Madag.* 16°6S 47°38E **57** B8
Mamuju *Indonesia* 2°41S 118°50E **37** E5
Mamuno *Botswana* 22°16S 20°1E **56** C3
Man *Ivory C.* 7°30N 7°40W **50** G4
Man, I. of *U.K.* 54°15N 4°30W **12** C3
Man-Bazar *India* 23°4N 86°39E **43** H12
Man Na *Burma* 23°27N 97°19E **41** H20
Mänä *U.S.A.* 22°2N 159°47W **75** L8
Mana → *Fr. Guiana* 5°45N 53°55W **93** B8
Mana Pools △ *Zimbabwe* 15°56S 29°25E **55** F2
Manaar, G. of = Mannar, G. of *Asia* 8°30N 79°0E **40** Q11
Manacapuru *Brazil* 3°16S 60°37W **92** D6
Manacor *Spain* 39°34N 3°13E **24** B10
Manado *Indonesia* 1°29N 124°51E **37** D6
Managua *Nic.* 12°6N 86°20W **88** D2
Managua, L. de *Nic.* 12°20N 86°30W **88** D2
Manakara *Madag.* 22°8S 48°1E **57** C8
Manali *India* 32°16N 77°10E **42** C7
Manama = Al Manāmah *Bahrain* 26°10N 50°30E **45** E6
Manambao → *Madag.* 17°35S 44°0E **57** B7
Manambato *Madag.* 13°43S 49°7E **57** A8
Manambolo → *Madag.* 19°18S 44°22E **57** B7
Manambolosy *Madag.* 16°2S 49°40E **57** B8
Mananara *Madag.* 16°10S 49°46E **57** B8
Mananara → *Madag.* 23°21S 47°42E **57** C8
Mananara ⌓ *Madag.* 16°14S 49°45E **57** B8
Mananjary *Madag.* 21°13S 48°20E **57** C8
Manantenina *Madag.* 24°17S 47°19E **57** C8
Manaos = Manaus *Brazil* 3°0S 60°0W **92** D7
Manapire → *Venezuela* 7°42N 66°7W **92** B5
Manapouri *N.Z.* 45°34S 167°39E **59** F1
Manapouri, L. *N.Z.* 45°32S 167°32E **59** F1
Manār, Jabal *Yemen* 14°2N 44°17E **47** E3
Manaravolo *Madag.* 23°59S 45°39E **57** C8
Manas *China* 44°17N 86°10E **32** C6
Manas → *India* 26°12N 90°40E **41** F17
Manas He → *China* 45°38N 85°12E **32** B6
Manaslu *Nepal* 28°33N 84°33E **43** E11
Manasquan *U.S.A.* 40°8N 74°3W **83** F10
Manassa *U.S.A.* 37°11N 105°56W **77** H11
Manatí *Puerto Rico* 18°26N 66°29W **89** d
Manaung *Burma* 18°45N 93°40E **41** K18
Manaus *Brazil* 3°0S 60°0W **92** D7
Manawan L. *Canada* 55°24N 103°14W **71** B8
Manbij *Syria* 36°31N 37°57E **44** B3
Manchegorsk *Russia* 67°54N 32°58E **28** C4
Manchester *U.K.* 53°29N 2°12W **12** D5
Manchester *Calif., U.S.A.* 38°58N 123°41W **78** G3
Manchester *Conn., U.S.A.* 41°47N 72°31W **83** E12
Manchester *Ga., U.S.A.* 32°51N 84°37W **85** E12
Manchester *Iowa, U.S.A.* 42°29N 91°27W **80** D8
Manchester *Ky., U.S.A.* 37°9N 83°46W **81** G12
Manchester *N.H., U.S.A.* 42°59N 71°28W **83** D13
Manchester *N.Y., U.S.A.* 42°56N 77°16W **82** D7
Manchester *Pa., U.S.A.* 40°4N 76°43W **83** F8
Manchester *Tenn., U.S.A.* 35°29N 86°5W **85** D11
Manchester *Vt., U.S.A.* 43°10N 73°5W **83** C11
Manchester Int. ✈ (MAN) *U.K.* 53°21N 2°17W **12** D5
Manchester L. *Canada* 61°28N 107°29W **71** A7
Manchhar L. *Pakistan* 26°25N 67°39E **42** F2
Manchuria = Dongbei *China* 45°0N 125°0E **35** D13
Manchurian Plain *China* 47°0N 124°0E **26** D14
Mand → *India* 21°42N 83°15E **43** J10
Mand → *Iran* 28°20N 52°30E **45** D7
Manda *Ludewe, Tanzania* 10°30S 34°40E **55** E3
Manda *Mbeya, Tanzania* 7°58S 32°29E **54** D3
Manda *Mbeya, Tanzania* 8°30S 32°49E **55** D3
Mandabé *Madag.* 21°0S 44°55E **57** C7
Mandaguari *Brazil* 23°32S 51°42W **95** A5
Mandah = Töhöm *Mongolia* 44°27N 108°2E **34** B5
Mandal *Norway* 58°2N 7°25E **9** G12
Mandala, Puncak *Indonesia* 4°44S 140°20E **37** E10
Mandalay *Burma* 22°0N 96°4E **41** J20
Mandale = Mandalay *Burma* 22°0N 96°4E **41** J20
Mandalgarh *India* 25°12N 75°6E **42** G6
Mandalgovĭ *Mongolia* 45°45N 106°10E **34** B4
Mandalī *Iraq* 33°43N 45°28E **44** C5
Mandan *U.S.A.* 46°50N 100°54W **80** B3
Mandar, Teluk *Indonesia* 3°32S 119°21E **37** E5
Mandaue *Phil.* 10°20N 123°56E **37** B6
Mandera *Kenya* 3°55N 41°53E **54** B5
Mandeville *Jamaica* 18°2N 77°31W **88** a
Mandi *India* 31°39N 76°58E **42** D7
Mandi Burewala *Pakistan* 30°9N 72°41E **42** D5
Mandi Dabwali *India* 29°58N 74°42E **42** E6
Mandimba *Mozam.* 14°20S 35°40E **55** E4
Mandioli *Indonesia* 0°40S 127°20E **37** E7
Mandla *India* 22°39N 80°30E **43** H9
Mandorah *Australia* 12°32S 130°42E **60** B5
Mandoto *Madag.* 19°34S 46°17E **57** B8
Mandra *Pakistan* 33°23N 73°12E **42** C5
Mandrare → *Madag.* 25°10S 46°30E **57** D8
Mandritsara *Madag.* 15°50S 48°49E **57** B8
Mandronarivo *Madag.* 21°7S 45°38E **57** C8
Mandsaur *India* 24°3N 75°8E **42** G6
Mandurah *Australia* 32°36S 115°48E **61** F2
Mandvi *India* 22°51N 69°22E **42** H3
Mandya *India* 12°30N 77°0E **40** N10
Mandzai *Pakistan* 30°55N 67°6E **42** D2
Maneh *Iran* 37°39N 57°7E **45** B8
Manera *Madag.* 22°55S 44°20E **57** C7
Maneroo Cr. → *Australia* 23°21S 143°53E **62** C3
Manfalût *Egypt* 27°20N 30°52E **51** C12
Manfredónia *Italy* 41°38N 15°55E **22** D6
Mangabeiras, Chapada das *Brazil* 10°0S 46°30W **93** F9
Mangaia *Cook Is.* 21°55S 157°55W **65** K12
Mangalia *Romania* 43°50N 28°35E **17** G15
Mangalore *India* 12°55N 74°47E **40** N9
Mangaluru = Mangalore *India* 12°55N 74°47E **40** N9
Mangan *India* 27°31N 88°32E **43** F13
Mangawan *India* 24°41N 81°33E **43** G9
Mangaweka *N.Z.* 39°48S 175°47E **59** C5
Manggar *Indonesia* 2°50S 108°10E **36** E3
Manggawitu *Indonesia* 4°8S 133°32E **37** E8
Mangghystaū Tübegi *Kazakhstan* 44°30N 52°30E **28** E6
Manggis *Indonesia* 8°29S 115°31E **37** J18
Mangindrano *Madag.* 14°17S 48°58E **57** A8
Mangkalihat, Tanjung *Indonesia* 1°2N 118°59E **37** D5
Mangkururrpa ☼ *Australia* 20°35S 129°43E **60** D4
Mangla *Pakistan* 33°7N 73°39E **42** C5
Mangla Dam *Pakistan* 33°9N 73°44E **43** C5
Manglaur *India* 29°44N 77°49E **42** E7
Mangnai *China* 37°52N 91°43E **32** D7
Mangnai Zhen *China* 38°24N 90°14E **32** D7
Mango *Togo* 10°20N 0°30E **50** F6
Mango *Tonga* 20°17S 174°29W **59** c
Mangoche *Malawi* 14°25S 35°16E **55** E4
Mangoky → *Madag.* 21°29S 43°41E **57** C7
Mangole *Indonesia* 1°50S 125°55E **37** E6
Mangombe *Dem. Rep. of the Congo* 1°20S 26°48E **54** C2
Mangonui *N.Z.* 35°1S 173°32E **59** A4
Mangoro → *Madag.* 20°0S 48°45E **57** B8
Mangrol *Mad. P., India* 21°7N 70°7E **42** J4
Mangrol *Raj., India* 25°20N 76°31E **42** G6
Mangueira, L. da *Brazil* 33°0S 52°50W **95** C5
Mangum *U.S.A.* 34°53N 99°30W **84** D5
Mangurri *Australia* 28°58S 134°22E **63** A1
Mangyshlak, Poluostrov = Mangghystaū Tübegi *Kazakhstan* 44°30N 52°30E **28** E6
Manhattan *U.S.A.* 39°11N 96°35W **80** F5
Manhiça *Mozam.* 25°23S 32°49E **57** D5
Mania → *Madag.* 19°42S 45°22E **57** B8
Manica *Mozam.* 18°58S 32°59E **57** B5
Manica □ *Mozam.* 19°10S 33°45E **57** B5
Manicaland □ *Zimbabwe* 19°0S 32°30E **55** F3
Manicoré *Brazil* 5°48S 61°16W **92** E6
Manicouagan → *Canada* 49°30N 68°30W **73** C6
Manicouagan, Rés. *Canada* 51°5N 68°40W **73** B6
Maniema □ *Dem. Rep. of the Congo* 3°0S 26°0E **54** C2
Manīfah *Si. Arabia* 27°44N 49°0E **45** E6
Manifold, C. *Australia* 22°41S 150°50E **62** C5
Manigotagan *Canada* 51°6N 96°18W **71** C9
Manigotagan → *Canada* 51°7N 96°20W **71** C9
Manihari *India* 25°21N 87°38E **43** G12
Manihiki *Cook Is.* 10°24S 161°1W **65** J11
Manihiki Plateau *Pac. Oc.* 11°0S 164°0W **65** J11
Manika, Plateau de la *Dem. Rep. of the Congo* 10°0S 25°5E **55** E2
Manikpur *India* 25°4N 81°7E **43** G9
Manila *Phil.* 14°35N 120°58E **37** B6
Manila *U.S.A.* 40°59N 109°43W **76** F9
Manila B. *Phil.* 14°40N 120°35E **37** B6
Manilla *Australia* 30°45S 150°43E **63** E5
Maningrida *Australia* 12°3S 134°13E **62** A1
Manipur □ *India* 25°0N 94°0E **41** G19
Manipur → *Burma* 23°45N 94°20E **41** H19
Manisa *Turkey* 38°38N 27°30E **23** E12
Manistee *U.S.A.* 44°15N 86°19W **80** C10
Manistee → *U.S.A.* 44°15N 86°21W **80** C10
Manistique *U.S.A.* 45°57N 86°15W **80** C10
Manitoba □ *Canada* 53°30N 97°0W **71** B9
Manitoba, L. *Canada* 51°0N 98°45W **71** C9
Manitou *Canada* 49°15N 98°32W **71** D9
Manitou, L. *Canada* 50°55N 65°17W **73** B6
Manitou Is. *U.S.A.* 45°8N 86°0W **80** C10
Manitou L. *Canada* 52°43N 109°43W **71** C7
Manitou Springs *U.S.A.* 38°52N 104°55W **76** G11
Manitoulin I. *Canada* 45°40N 82°30W **72** C3
Manitouwadge *Canada* 49°8N 85°48W **72** C2
Manitowoc *U.S.A.* 44°5N 87°40W **80** C10
Manizales *Colombia* 5°5N 75°32W **92** B3
Manja *Madag.* 21°26S 44°20E **57** C7
Manjacaze *Mozam.* 24°45S 34°0E **57** C5
Manjakandriana *Madag.* 18°55S 47°47E **57** B8
Manjhand *Pakistan* 25°50N 68°10E **42** G3
Manjimup *Australia* 34°15S 116°6E **61** F2
Manjra → *India* 18°49N 77°52E **40** K10
Mankato *Kans., U.S.A.* 39°47N 98°13W **80** F4
Mankato *Minn., U.S.A.* 44°10N 94°0W **80** C6
Mankayane *Swaziland* 26°40S 31°4E **57** D5
Mankera *Pakistan* 31°23N 71°26E **42** D4
Mankota *Canada* 49°25N 107°5W **71** D7
Manlay = Üydzin *Mongolia* 44°9N 107°0E **34** B4
Manmad *India* 20°18N 74°28E **40** J9
Mann Ranges *Australia* 26°6S 130°5E **61** E5
Manna *Indonesia* 4°25S 102°55E **36** E2
Mannahill *Australia* 32°25S 140°0E **63** E3
Mannar *Sri Lanka* 9°1N 79°54E **40** Q11
Mannar, G. of *Asia* 8°30N 79°0E **40** Q11
Mannar I. *Sri Lanka* 9°5N 79°45E **40** Q11
Mannheim *Germany* 49°29N 8°29E **16** D5
Manning *Canada* 56°53N 117°39W **70** B5
Manning *Oreg., U.S.A.* 45°45N 123°13W **78** E3
Manning *S.C., U.S.A.* 33°42N 80°13W **85** E14
Mannum *Australia* 34°50S 139°20E **63** E2
Manoharpur *India* 22°23N 85°12E **43** H11
Manokwari *Indonesia* 0°54S 134°0E **37** E8
Manombo *Madag.* 22°57S 43°28E **57** C7
Manono *Dem. Rep. of the Congo* 7°15S 27°25E **54** D2
Manono *Samoa* 13°50S 172°5W **59** b
Manorhamilton *Ireland* 54°18N 8°9W **10** B3
Manosque *France* 43°49N 5°47E **20** E6
Manotick *Canada* 45°13N 75°41W **83** A9
Manouane → *Canada* 49°30N 71°10W **73** C5
Manouane, L. *Canada* 50°45N 70°45W **73** B5
Manp'o *N. Korea* 41°6N 126°24E **35** D14
Manpojin = Manp'o *N. Korea* 41°6N 126°24E **35** D14
Manpur *Chhattisgarh, India* 23°17N 83°35E **43** H10
Manpur *Mad. P., India* 22°26N 75°37E **42** H6
Manresa *Spain* 41°48N 1°50E **21** B6
Mansa *Gujarat, India* 23°27N 72°45E **42** H5
Mansa *Punjab, India* 30°0N 75°27E **42** E6
Mansa *Zambia* 11°13S 28°55E **55** E2
Mansehra *Pakistan* 34°20N 73°15E **42** B5
Mansel I. *Canada* 62°0N 80°0W **69** E15
Mansfield *Australia* 37°4S 146°6E **63** F4
Mansfield *U.K.* 53°9N 1°11W **12** D6
Mansfield *La., U.S.A.* 32°2N 93°43W **84** E8
Mansfield *Mass., U.S.A.* 42°2N 71°13W **83** D13
Mansfield *Ohio, U.S.A.* 40°45N 82°31W **82** F2
Mansfield *Pa., U.S.A.* 41°48N 77°5W **82** E7
Mansfield *Tex., U.S.A.* 32°33N 97°8W **84** E6
Mansfield, Mt. *U.S.A.* 44°33N 72°49W **83** B12
Manson Creek *Canada* 55°37N 124°32W **70** B4
Manta *Ecuador* 1°0S 80°40W **92** D2
Mantadia △ *Madag.* 18°54S 48°21E **57** B8
Mantalingajan, Mt. *Phil.* 8°55N 117°45E **36** C5
Mantare *Tanzania* 2°42S 33°13E **54** C3
Manteca *U.S.A.* 37°48N 121°13W **78** H5
Manteo *U.S.A.* 35°55N 75°40W **85** D17
Mantes-la-Jolie *France* 48°58N 1°41E **20** B4
Manthani *India* 18°40N 79°35E **40** K11
Manti *U.S.A.* 39°16N 111°38W **76** G8
Mantiqueira, Serra da *Brazil* 22°0S 44°0W **95** A7
Manton *U.S.A.* 44°25N 85°24W **81** C11
Mántova *Italy* 45°9N 10°48E **22** B4
Mänttä Vilppula *Finland* 62°3N 24°40E **8** E21
Mantua = Mántova *Italy* 45°9N 10°48E **22** B4
Manú *Peru* 12°10S 70°51W **92** F4
Manú → *Peru* 12°16S 70°55W **92** F4
Manu'a Is. *Amer. Samoa* 14°13S 169°35W **59** b
Manuel Alves → *Brazil* 11°19S 48°28W **93** F9
Manui *Indonesia* 3°35S 123°5E **37** E6
Manukau *N.Z.* 37°0S 174°52E **59** B5
Manuripi → *Bolivia* 11°6S 67°36W **92** F5
Many *U.S.A.* 31°34N 93°29W **84** F8
Manyallaluk ☼ *Australia* 14°16S 132°49E **60** B5
Manyani *Kenya* 3°5S 38°30E **54** C4
Manyara, L. *Tanzania* 3°40S 35°50E **54** C4
Manyara △ *Tanzania* 3°35S 35°50E **54** C4
Manych-Gudilo, Ozero *Russia* 46°24N 42°38E **19** E7
Manyonga → *Tanzania* 4°10S 34°15E **54** C3
Manyoni *Tanzania* 5°45S 34°55E **54** D3
Manzai *Pakistan* 32°12N 70°15E **42** C4
Manzanar ⌓ *U.S.A.* 36°44N 118°9W **78** J7
Manzanares *Spain* 39°2N 3°22W **21** C4
Manzanillo *Cuba* 20°20N 77°31W **88** B4
Manzanillo *Mexico* 19°3N 104°20W **86** D4
Manzanillo, Pta. *Panama* 9°30N 79°40W **88** E4
Manzano Mts. *U.S.A.* 34°40N 106°20W **77** J10
Manzarīyeh *Iran* 34°53N 50°50E **45** C6
Manzhouli *China* 49°35N 117°25E **33** B12
Manzini *Swaziland* 26°30S 31°25E **57** D5
Manzouli = Manzhouli *China* 49°35N 117°25E **33** B12
Manzur Vadisi △ *Turkey* 39°10N 39°30E **44** B3
Mao *Chad* 14°4N 15°19E **51** F9
Maó *Spain* 39°53N 4°16E **24** B11
Maoke, Pegunungan *Indonesia* 3°40S 137°30E **37** E9
Maolin *China* 43°58N 123°30E **35** C12
Maoming *China* 21°50N 110°54E **33** G11
Maoxing *China* 45°28N 124°40E **35** B13
Mapam Yumco *China* 30°45N 81°28E **43** D9
Mapastepec *Mexico* 15°26N 92°54W **87** D6
Maphrao, Ko *Thailand* 7°56N 98°26E **39** a
Mapia, Kepulauan *Indonesia* 0°50N 134°20E **37** D8
Mapimí *Mexico* 25°49N 103°51W **86** B4
Mapimí, Bolsón de *Mexico* 27°0N 104°15W **86** B4
Mapinga *Tanzania* 6°40S 39°12E **54** D4
Mapinhane *Mozam.* 22°20S 35°0E **57** C6
Maple Creek *Canada* 49°55N 109°29W **71** D7
Maple Valley *U.S.A.* 47°25N 122°3W **78** C4
Mapleton *U.S.A.* 44°2N 123°52W **76** D2
Mapoon ☼ *Australia* 11°44S 142°8E **62** A3
Mapuera → *Brazil* 1°5S 57°2W **92** D7
Mapulanguene *Mozam.* 24°29S 32°6E **57** C5
Mapungubwe △ *S. Africa* 22°12S 29°22E **55** G2
Maputo *Mozam.* 25°58S 32°32E **57** D5
Maputo □ *Mozam.* 26°0S 32°25E **57** D5
Maputo, B. de *Mozam.* 25°50S 32°45E **48** J7
Maputo ⌓ *Mozam.* 26°23S 32°48E **57** D5
Maqat *Kazakhstan* 47°39N 53°19E **19** E9
Maqên *China* 34°24N 100°6E **32** E9
Maqên Gangri *China* 34°55N 99°18E **32** E8
Maqiaohe *China* 44°40N 130°30E **35** B16
Maqnā *Si. Arabia* 28°25N 34°50E **44** D2
Maqteïr *Mauritania* 21°50N 11°40W **50** D3
Maqu *China* 33°52N 101°42E **32** E9
Maquan He = Brahmaputra → *Asia* 23°40N 90°35E **43** H13
Maquela do Zombo *Angola* 6°0S 15°15E **52** F3
Maquinchao *Argentina* 41°15S 68°50W **96** E3
Maquoketa *U.S.A.* 42°4N 90°40W **80** D8
Mar *Canada* 44°49N 81°12W **82** B3
Mar, Serra do *Brazil* 25°30S 49°0W **95** B6
Mar Chiquita, L. *Argentina* 30°40S 62°50W **94** C3
Mar del Plata *Argentina* 38°0S 57°30W **94** D4
Mar Menor *Spain* 37°40N 0°45W **21** D5
Mara *Tanzania* 1°30S 34°32E **54** C3
Mara □ *Tanzania* 1°45S 34°20E **54** C3
Maraã *Brazil* 1°52S 65°25W **92** D5
Maraa *Tahiti* 17°46S 149°34W **59** d
Marabá *Brazil* 5°20S 49°5W **93** E9
Maraboon, L. *Australia* 23°41S 148°0E **62** C4
Maracá, I. de *Brazil* 2°10N 50°30W **93** C8
Maracaibo *Venezuela* 10°40N 71°37W **92** A4
Maracaibo, L. de *Venezuela* 9°40N 71°30W **92** B4
Maracaju *Brazil* 21°38S 55°9W **95** A4
Maracas Bay Village *Trin. & Tob.* 10°46N 61°28W **93** K15
Maracay *Venezuela* 10°15N 67°28W **92** A5
Marādah *Libya* 29°15N 19°15E **51** C9
Maradi *Niger* 13°29N 7°20E **50** F7
Marāgheh *Iran* 37°30N 46°12E **44** B5
Marāh *Si. Arabia* 25°0N 45°35E **44** E5
Marajó, I. de *Brazil* 1°0S 49°30W **93** D9
Marākand *Iran* 38°51N 45°16E **44** B5
Marakele △ *S. Africa* 24°30S 25°30E **57** C4
Maralal *Kenya* 1°0N 36°38E **54** B4
Maralinga *Australia* 30°13S 131°32E **61** F5
Maralinga Tjarutja ☼ *Australia* 29°30S 131°0E **61** E5
Marambio *Antarctica* 64°0S 56°0W **5** C18
Maran *Malaysia* 3°35N 102°45E **39** L4
Marana *U.S.A.* 32°27N 111°13W **77** K8
Maranboy *Australia* 14°40S 132°39E **60** B5
Marand *Iran* 38°30N 45°45E **44** B5
Marang *Malaysia* 5°12N 103°13E **39** K4
Maranguape *Brazil* 3°55S 38°50W **93** D11
Maranhão = São Luís *Brazil* 2°39S 44°15W **93** D10
Maranhão □ *Brazil* 5°0S 46°0W **93** E9
Maranoa → *Australia* 27°50S 148°37E **63** D4
Marañón → *Peru* 4°30S 73°35W **92** D4
Marão *Mozam.* 24°18S 34°2E **57** C5
Maraş = Kahramanmaraş *Turkey* 37°37N 36°53E **44** B3
Marathasa *Cyprus* 34°59N 32°51E **25** E11
Marathon *Australia* 20°51S 143°32E **62** C3
Marathon *Canada* 48°44N 86°23W **72** C2
Marathon *N.Y., U.S.A.* 42°27N 76°2W **83** D8
Marathon *Tex., U.S.A.* 30°12N 103°15W **84** F3
Marathóvouno *Cyprus* 35°13N 33°37E **25** D12
Maratua *Indonesia* 2°10N 118°35E **37** D5
Maraval *Trin. & Tob.* 10°42N 61°31W **93** K15
Maravatío *Mexico* 19°54N 100°27W **86** D4
Marāwih *U.A.E.* 24°18N 53°18E **45** E7
Marbella *Spain* 36°30N 4°57W **21** D3
Marble Bar *Australia* 21°9S 119°44E **60** D2
Marble Falls *U.S.A.* 30°35N 98°16W **84** F5
Marblehead *Mass., U.S.A.* 42°29N 70°51W **83** D14
Marblehead *Ohio, U.S.A.* 41°32N 82°44W **82** E2
Marburg *Germany* 50°47N 8°46E **16** C5
Marca, Pta. do *Angola* 16°31S 11°43E **53** H2
March *U.K.* 52°33N 0°5E **13** E8
Marche *France* 46°5N 1°20E **20** C4
Marche-en-Famenne *Belgium* 50°14N 5°19E **15** D5
Marchena *Spain* 37°18N 5°23W **21** D3
Marco Island *U.S.A.* 25°58N 81°44W **85** J14
Marcos Juárez *Argentina* 32°42S 62°5W **94** C3
Marcus Baker, Mt. *U.S.A.* 61°26N 147°45W **74** C10
Marcus I. = Minami-Tori-Shima *Pac. Oc.* 24°20N 153°58E **6**
Marcy, Mt. *U.S.A.* 44°7N 73°56W **83**
Mardan *Pakistan* 34°20N 72°0E **4**
Mardie *Australia* 21°12S 115°59E **6**
Mardin *Turkey* 37°20N 40°43E **4**
Maree, L. *U.K.* 57°40N 5°26W **1**
Mareeba *Australia* 16°59S 145°28E **6**
Mareetsane *S. Africa* 26°9S 25°25E **5**
Marek = Stanke Dimitrov *Bulgaria* 42°17N 23°9E **23**
Marengo *U.S.A.* 41°48N 92°4W **8**
Marerano *Madag.* 21°23S 44°52E **5**
Marfa *U.S.A.* 30°19N 104°1W **8**
Marfa Pt. *Malta* 35°59N 14°19E **2**
Margaret → *Australia* 18°9S 125°41E **6**
Margaret Bay *Canada* 51°20N 127°35W **7**
Margaret L. *Canada* 58°56N 115°25W **7**
Margaret River *Australia* 33°57S 115°4E **6**
Margarita, I. de *Venezuela* 11°0N 64°0W **9**
Margaritovo *Russia* 43°25N 134°45E **3**
Margate *S. Africa* 30°50S 30°20E **5**
Margate *U.K.* 51°23N 1°23E **1**
Margherita Pk. *Uganda* 0°22N 29°51E **5**
Marghilon *Uzbekistan* 40°27N 71°42E **2**
Mārgow, Dasht-e *Afghan.* 30°40N 62°30E **4**
Marguerite *Canada* 52°30N 122°25W **7**
Mari El □ *Russia* 56°30N 48°0E **1**
Mari Indus *Pakistan* 32°57N 71°34E **4**
Mari Republic = Mari El □ *Russia* 56°30N 48°0E **1**
María de la Salut *Spain* 39°40N 3°5E **24**
María Elena *Chile* 22°18S 69°40W **9**
María Grande *Argentina* 31°45S 59°55W **9**
Maria I. *N. Terr., Australia* 14°52S 135°45E **6**
Maria I. *Tas., Australia* 42°35S 148°0E **6**
Maria Island △ *Australia* 42°38S 148°5E **6**
Maria van Diemen, C. *N.Z.* 34°29S 172°40E **5**
Mariakani *Kenya* 3°50S 39°27E **5**
Mariala △ *Australia* 25°57S 145°2E **6**
Marian *Australia* 21°9S 148°57E **6**
Marian L. *Canada* 63°0N 116°15W **7**
Mariana Trench *Pac. Oc.* 13°0N 145°0E **6**
Marianna *Ark., U.S.A.* 34°46N 90°46W **8**
Marianna *Fla., U.S.A.* 30°46N 85°14W **85**
Marias → *U.S.A.* 47°56N 110°30W **7**
Marías, Is. *Mexico* 21°25N 106°28W **8**
Mariato, Punta *Panama* 7°12N 80°52W **8**
Maribor *Slovenia* 46°36N 15°40E **1**
Marico → *Africa* 23°35S 26°57E **5**
Maricopa *Ariz., U.S.A.* 33°4N 112°3W **7**
Maricopa *Calif., U.S.A.* 35°4N 119°24W **7**
Marié → *Brazil* 0°27S 66°26W **9**
Marie Byrd Land *Antarctica* 79°30S 125°0W **5**
Marie-Galante *Guadeloupe* 15°56N 61°16W **8**
Mariecourt = Kangiqsujuaq *Canada* 61°30N 72°0W **69**
Mariehamn *Finland* 60°5N 19°55E **9**
Mariembourg *Belgium* 50°6N 4°31E **15**
Mariental *Namibia* 24°36S 18°0E **5**
Marienville *U.S.A.* 41°28N 79°8W **8**
Mariestad *Sweden* 58°43N 13°50E **9**
Marietta *Ga., U.S.A.* 33°57N 84°33W **85**
Marietta *Ohio, U.S.A.* 39°25N 81°27W **81**
Marieville *Canada* 45°26N 73°10W **83**
Mariinsk *Russia* 56°10N 87°20E **28**
Marijampolė *Lithuania* 54°33N 23°19E **9**
Marília *Brazil* 22°13S 50°0W **9**
Marín *Spain* 42°23N 8°42W **21**
Marina *U.S.A.* 36°41N 121°48W **7**
Marinduque *Phil.* 13°25N 122°0E **37**
Marine City *U.S.A.* 42°43N 82°30W **82**
Marinette *U.S.A.* 45°6N 87°38W **80**
Maringá *Brazil* 23°26S 52°2W **9**
Marion *Ala., U.S.A.* 32°38N 87°19W **85**
Marion *Ill., U.S.A.* 37°44N 88°56W **80**
Marion *Ind., U.S.A.* 40°32N 85°40W **81**
Marion *Iowa, U.S.A.* 42°2N 91°36W **80**
Marion *Kans., U.S.A.* 38°21N 97°1W **80**
Marion *N.C., U.S.A.* 35°41N 82°1W **85**
Marion *Ohio, U.S.A.* 40°35N 83°8W **81**
Marion *S.C., U.S.A.* 34°11N 79°24W **85**
Marion *Va., U.S.A.* 36°50N 81°31W **81**
Marion, L. *U.S.A.* 33°28N 80°10W **85**
Mariposa *U.S.A.* 37°29N 119°58W **78**
Mariscal Estigarribia *Paraguay* 22°3S 60°40W **94**
Maritime Alps = Maritimes, Alpes *Europe* 44°10N 7°10E **20**
Maritimes, Alpes *Europe* 44°10N 7°10E **20**
Maritsa = Evros → *Greece* 41°40N 26°34E **23**
Maritsa *Greece* 36°22N 28°8E **25**
Mariupol *Ukraine* 47°5N 37°31E **19**
Mariusa △ *Venezuela* 9°24N 61°27W **8**
Marīvān *Iran* 35°30N 46°25E **44**
Marj 'Uyūn *Lebanon* 33°21N 35°34E **46**
Marka *Somali Rep.* 1°48N 44°50E **47**
Markam *China* 29°42N 98°38E **41**
Markdale *Canada* 44°19N 80°39W **82**
Marked Tree *U.S.A.* 35°32N 90°25W **85**
Market Drayton *U.K.* 52°54N 2°29W **12**
Market Harborough *U.K.* 52°29N 0°55W **13**
Market Rasen *U.K.* 53°24N 0°20W **12**
Markham *Canada* 43°52N 79°16W **82**
Markham, Mt. *Antarctica* 83°0S 164°0E **5**
Markleeville *U.S.A.* 38°42N 119°47W **78**
Markovo *Russia* 64°40N 170°24E **29**
Marks *Russia* 51°45N 46°50E **18**
Marksville *U.S.A.* 31°8N 92°4W **84**
Marla *Australia* 27°19S 133°33E **63**
Marlbank *Canada* 44°26N 77°6W **82**
Marlboro *U.S.A.* 41°36N 73°59W **83**
Marlborough *Australia* 22°46S 149°52E **62**

ˈlborough U.K. 51°25N 1°43W 13 F6
ˈlborough U.S.A. 42°21N 71°33W 83 D13
ˈlborough Downs
.K. 51°27N 1°53W 13 F6
ˈlin U.S.A. 31°18N 96°54W 84 F6
ˈlow U.K. 51°34N 0°46W 13 F7
ˈlow U.S.A. 34°39N 97°58W 84 D6
ˈmagao India 15°25N 73°56E 40 M8
ˈmara Turkey 40°35N 27°34E 23 D12
ˈmara, Sea of = Marmara
enizi Turkey 40°45N 28°15E 23 D13
ˈmara Denizi Turkey 40°45N 28°15E 23 D13
ˈmaris Turkey 36°50N 28°14E 23 F13
ˈmion, Mt. Australia 29°16S 119°50E 61 E2
ˈmion L. Canada 48°55N 91°20W 72 C1
ˈmolada, Mte. Italy 46°26N 11°51E 22 A4
ˈmora Canada 44°28N 77°41W 82 B7
ˈmugao = Marmagao
dia 15°25N 73°56E 40 M8
ne → France 48°47N 2°29E 20 B5
o Reef U.S.A. 25°25N 170°35W 75 K5
oala Madag. 15°23S 47°59E 57 B8
oantsetra Madag. 15°26S 49°44E 57 B8
oelaboom Namibia 19°15S 18°53E 56 B2
ofandilia Madag. 20°7S 44°34E 57 C7
ojejy △ Madag. 14°26S 49°21E 57 A8
olambo Madag. 20°2S 48°7E 57 C8
omandia Madag. 14°13S 48°5E 57 A8
omokotro Madag. 14°0S 49°0E 57 A8
ondera Zimbabwe 18°5S 31°42E 55 F3
oni → Fr. Guiana 5°30N 54°0W 93 B8
oochydore Australia 26°29S 153°5E 63 D5
oona Australia 37°27S 142°54E 63 F3
osakoa Madag. 15°26S 46°38E 57 B8
oseranana Madag. 18°32S 48°51E 57 B8
otandrano Madag. 16°10S 48°50E 57 B8
otaolano Madag. 12°47S 49°15E 57 A8
oua Cameroon 10°40N 14°20E 51 F8
ovato Madag. 15°48S 48°5E 57 B8
ovoay Madag. 16°6S 46°39E 57 B8
quard S. Africa 28°40S 27°28E 56 D4
quesas Fracture Zone
c. Oc. 9°0S 125°0W 65 H15
quesas Is. = Marquises, Îs.
ench Polynesia 9°30S 140°0W 65 H14
quette U.S.A. 46°33N 87°24W 80 B10
quis St. Lucia 14°2N 60°54W 89 f
quises, Îs.
ench Polynesia 9°30S 140°0W 65 H14
ra, Djebel Sudan 13°10N 24°22E 51 F10
racuene Mozam. 25°45S 32°35E 57 D5
rakech Morocco 31°9N 8°0W 50 B4
rawah Australia 40°55S 144°42E 63 G3
ree Australia 29°39S 138°1E 63 D2
rero U.S.A. 29°53N 90°6W 85 G9
rimane Mozam. 22°58S 33°34E 57 C5
romeu Mozam. 18°15S 36°25E 57 B6
romeu △ Mozam. 19°0S 36°0E 57 B6
rowie Cr. →
ustralia 33°23S 145°40E 63 E4
rubane Mozam. 18°0S 37°0E 55 F4
rupa Mozam. 13°8S 37°30E 55 E4
s Hill U.S.A. 46°31N 67°52W 81 B20
sá 'Alam Egypt 25°5N 34°54E 47 B1
sā Matrûh Egypt 31°19N 27°9E 51 B11
sā Sūsah Libya 32°52N 21°59E 51 B10
sabit Kenya 2°18N 38°0E 54 B4
sabit △ Kenya 2°23N 37°56E 54 B4
sala Italy 37°48N 12°26E 22 F5
salforn Malta 36°4N 14°16E 25 C1
sden Australia 33°47S 147°32E 63 E4
seille France 43°18N 5°23E 20 E6
seilles = Marseille
ance 43°18N 5°23E 20 E6
sh I. U.S.A. 29°34N 91°53W 84 G9
shall Ark., U.S.A. 35°55N 92°38W 84 D8
shall Mich., U.S.A. 42°16N 84°58W 81 D11
shall Minn., U.S.A. 44°27N 95°47W 80 C6
shall Mo., U.S.A. 39°7N 93°12W 80 F7
shall Tex., U.S.A. 32°33N 94°23W 84 E7
shall → Australia 22°59S 136°59E 62 C2
shall Is. ■ Pac. Oc. 9°0N 171°0E 58 A10
shalltown U.S.A. 42°3N 92°55W 80 D7
shbrook Zimbabwe 18°33S 31°9E 57 B5
shfield Mo., U.S.A. 37°15N 92°54W 80 G7
shfield Vt., U.S.A. 44°20N 72°20W 83 B12
shfield Wis., U.S.A. 44°40N 90°10W 80 C8
shūn Iran 36°19N 49°23E 45 B6
sta Sweden 59°37N 17°52E 9 G17
t U.S.A. 31°33N 96°50W 84 F6
taban Burma 16°30N 97°35E 41 L20
taban, G. of Burma 16°5N 96°30E 41 L20
tapura Kalimantan Selatan,
donesia 3°22S 114°47E 36 E4
tapura Sumatera Selatan,
donesia 4°19S 104°22E 36 E2
te R. Gómez, Presa
exico 26°10N 99°0W 87 B5
telange Belgium 49°49N 5°43E 15 E5
tha's Vineyard
S.A. 41°25N 70°38W 83 E14
tigny Switz. 46°6N 7°3E 20 C7
tigues France 43°24N 5°4E 20 E6
tin Slovak Rep. 49°6N 18°58E 17 D10
tin S. Dak., U.S.A. 43°11N 101°44W 80 D3
tin Tenn., U.S.A. 36°21N 88°51W 85 C10
tin L. U.S.A. 32°41N 85°55W 85 E12
tina Franca Italy 40°42N 17°20E 22 D7
tinborough N.Z. 41°14S 175°29E 59 D5
tinez Calif., U.S.A. 38°1N 122°8W 78 G4
tinez Ga., U.S.A. 33°31N 82°5W 85 E13
tinique ☑ W. Indies 14°40N 61°0W 88 c
tinique Passage
'. Indies 15°15N 61°0W 89 C7
tinópolis Brazil 22°11S 51°12W 95 A5
tin's Bay Barbados 13°12N 59°29W 89 g
tins Ferry U.S.A. 40°6N 80°44W 82 F4
tinsburg Pa., U.S.A. 40°19N 78°20W 82 F6
tinsburg W. Va.,
S.A. 39°27N 77°58W 81 F15
Martinsville Ind.,
U.S.A. 39°26N 86°25W 80 F10
Martinsville Va., U.S.A. 36°41N 79°52W 81 G14
Marton N.Z. 40°4S 175°23E 59 D5
Martos Spain 37°44N 3°58W 21 D4
Martu ✪ Australia 22°30S 122°30E 60 D3
Marudi Malaysia 4°11N 114°19E 36 D4
Maruf Afghan. 31°30N 67°6E 40 D5
Marugame Japan 34°15N 133°40E 31 G6
Marunga Angola 17°28S 20°2E 56 B3
Marungu, Mts.
Dem. Rep. of the Congo 7°30S 30°0E 54 D3
Maruwa ✪ Australia 22°30S 127°30E 60 D4
Marv Dasht Iran 29°50N 52°40E 45 D7
Marvast Iran 30°30N 54°15E 45 D7
Marvel Loch Australia 31°28S 119°29E 61 F2
Marwar India 25°43N 73°45E 42 G5
Mary Turkmenistan 37°40N 61°50E 45 B9
Maryborough = Port Laoise
Ireland 53°2N 7°18W 10 C4
Maryborough Queens.,
Australia 25°31S 152°37E 63 D5
Maryborough Vic.,
Australia 37°3S 143°44E 63 F3
Maryfield Canada 49°50N 101°35W 71 D8
Maryland ☐ U.S.A. 39°0N 76°30W 81 F15
Maryland Junction
Zimbabwe 17°45S 30°31E 55 F3
Maryport U.K. 54°44N 3°28W 12 C4
Mary's Harbour Canada 52°18N 55°51W 73 B8
Marystown Canada 47°10N 55°10W 73 C8
Marysville Calif., U.S.A. 39°9N 121°35W 78 F5
Marysville Kans., U.S.A. 39°51N 96°39W 80 F5
Marysville Mich., U.S.A. 42°54N 82°29W 82 D2
Marysville Ohio, U.S.A. 40°14N 83°22W 81 E12
Marysville Wash., U.S.A. 48°3N 122°11W 78 B4
Maryville Mo., U.S.A. 40°21N 94°52W 80 E6
Maryville Tenn., U.S.A. 35°46N 83°58W 85 D13
Marzūq Libya 25°53N 13°57E 51 C8
Marzūq, Idehān Libya 24°50N 13°51E 51 D8
Masada Israel 31°15N 35°20E 46 D4
Masahunga Tanzania 2°6S 33°18E 54 C3
Masai Malaysia 1°29N 103°55E 39 d
Masai Mara △ Kenya 1°25S 35°5E 54 C4
Masai Steppe Tanzania 4°30S 36°30E 54 C4
Masaka Uganda 0°21S 31°45E 54 C3
Masalembo, Kepulauan
Indonesia 5°35S 114°30E 36 F4
Masalima, Kepulauan
Indonesia 5°4S 117°5E 36 F5
Masamba Indonesia 2°30S 120°15E 37 E6
Masan S. Korea 35°11N 128°32E 35 G15
Masandam, Ra's Oman 26°30N 56°30E 45 E8
Masasi Tanzania 10°45S 38°52E 55 E4
Masaya Nic. 12°0N 86°7W 88 D2
Masbate Phil. 12°21N 123°36E 37 B6
Mascara Algeria 35°26N 0°6E 50 A6
Mascarene Is. Ind. Oc. 22°0S 55°0E 48 J9
Mascota Mexico 20°32N 104°49W 86 C4
Masela Indonesia 8°9S 129°51E 37 F7
Maseru Lesotho 29°18S 27°30E 56 D4
Mashaba Zimbabwe 20°2S 30°29E 55 G3
Mashābih Si. Arabia 25°35N 36°30E 44 E3
Mashang China 36°48N 117°57E 35 F9
Mashatu △ Botswana 22°45S 29°5E 57 C4
Masherbrum Pakistan 35°38N 76°18E 43 B7
Mashhad Iran 36°20N 59°35E 45 B8
Mashīz Iran 29°56N 56°37E 45 D8
Māshkel, Hāmūn-i-
Pakistan 28°20N 62°56E 45 D9
Mashki Chāh Pakistan 29°5N 62°30E 40 E3
Mashonaland Zimbabwe 16°30S 31°0E 53 H6
Mashonaland Central ☐
Zimbabwe 17°30S 31°0E 57 B5
Mashonaland East ☐
Zimbabwe 18°0S 32°0E 57 B5
Mashonaland West ☐
Zimbabwe 17°30S 29°30E 57 B4
Mashrakh India 26°7N 84°48E 43 F11
Masi Manimba
Dem. Rep. of the Congo 4°40S 17°54E 52 E3
Masig Australia 9°45S 143°24E 62 a
Masindi Uganda 1°40N 31°43E 54 B3
Masindi Port Uganda 1°43N 32°2E 54 B3
Masinga Res. Kenya 0°58S 37°38E 54 C4
Maşīrah, Jazīrat Oman 21°0N 58°50E 47 C6
Maşīrah, Khalīj Oman 20°10N 58°10E 47 C6
Masisi
Dem. Rep. of the Congo 1°23S 28°49E 54 C2
Masjed Soleyman Iran 31°55N 49°18E 45 D6
Mask, L. Ireland 53°36N 9°22W 10 C2
Maskin Oman 23°44N 56°52E 45 F8
Masoala, Tanjon' i
Madag. 15°59S 50°13E 57 B9
Masoala △ Madag. 15°30S 50°12E 57 B9
Masoarivo Madag. 19°3S 44°19E 57 B7
Masohi = Amahai
Indonesia 3°20S 128°55E 37 E7
Masomeloka Madag. 20°17S 48°37E 57 C8
Mason Nev., U.S.A. 38°56N 119°8W 78 G7
Mason Tex., U.S.A. 30°45N 99°14W 84 F5
Mason City U.S.A. 43°9N 93°12W 80 D7
Maspalomas Canary Is. 27°46N 15°35W 24 G4
Maspalomas, Pta.
Canary Is. 27°43N 15°36W 24 G4
Masqat Oman 23°37N 58°36E 47 C6
Massa Italy 44°1N 10°9E 20 D9
Massachusetts ☐ U.S.A. 42°30N 72°0W 83 D13
Massachusetts B.
U.S.A. 42°25N 70°50W 83 D14
Massakory Chad 13°0N 15°49E 51 F9
Massamba Mozam. 15°58S 33°31E 55 F3
Massanella Spain 39°48N 2°51E 24 B9
Massangena Mozam. 21°34S 33°0E 57 C5
Massango Angola 8°2S 16°21E 52 F3
Massawa = Mitsiwa
Eritrea 15°35N 39°25E 47 D2
Massena U.S.A. 44°56N 74°54W 83 B10
Massenya Chad 11°21N 16°9E 51 F9
Masset Canada 54°2N 132°10W 70 C2
Massiah Street Barbados 13°9N 59°29W 89 g
Massif Central France 44°55N 3°0E 20 D5
Massillon U.S.A. 40°48N 81°32W 82 F3
Massinga Mozam. 23°15S 35°22E 57 C6
Massingir Mozam. 23°51S 32°4E 57 C5
Masson-Angers Canada 45°32N 75°25W 83 A9
Masson I. Antarctica 66°10S 93°20E 5 C7
Mastanli = Momchilgrad
Bulgaria 41°33N 25°23E 23 D11
Masterton N.Z. 40°56S 175°39E 59 D5
Mastic U.S.A. 40°47N 72°54W 83 F12
Mastuj Pakistan 36°20N 72°36E 43 A5
Mastung Pakistan 29°50N 66°56E 40 E5
Masty Belarus 53°27N 24°38E 17 B13
Masuda Japan 34°40N 131°51E 31 G5
Masuku = Franceville
Gabon 1°40S 13°32E 52 E2
Masurian Lakes = Mazurski,
Pojezierze Poland 53°50N 21°0E 17 B11
Masvingo Zimbabwe 20°8S 30°49E 55 G3
Masvingo ☐ Zimbabwe 21°0S 31°30E 55 G3
Maswa △ Tanzania 2°58S 34°19E 54 C3
Maşyāf Syria 35°4N 36°20E 44 C3
Mata-au = Clutha →
N.Z. 46°20S 169°49E 59 G2
Matabeleland Zimbabwe 18°0S 27°0E 53 H5
Matabeleland North ☐
Zimbabwe 19°0S 28°0E 55 F2
Matabeleland South ☐
Zimbabwe 21°0S 29°0E 55 G2
Matachewan Canada 47°56N 80°39W 72 C3
Matadi
Dem. Rep. of the Congo 5°52S 13°31E 52 F2
Matagalpa Nic. 13°0N 85°58W 88 D2
Matagami Canada 49°45N 77°34W 72 C4
Matagami, L. Canada 49°50N 77°40W 72 C4
Matagorda B. U.S.A. 28°40N 96°12W 84 G6
Matagorda I. U.S.A. 28°15N 96°30W 84 G6
Mataiea Tahiti 17°46S 149°25W 59 d
Matak Indonesia 3°18N 106°16E 36 D3
Matala Greece 34°59N 24°45E 25 E6
Matam Senegal 15°34N 13°17W 50 E3
Matamoros Campeche,
Mexico 18°50N 90°50W 87 D6
Matamoros Coahuila,
Mexico 25°32N 103°15W 86 B4
Matamoros Tamaulipas,
Mexico 25°53N 97°30W 87 B5
Ma'ṭan as Sarra Libya 21°45N 22°0E 51 D10
Matandu → Tanzania 8°45S 34°19E 55 D3
Matane Canada 48°50N 67°33W 73 C6
Matanomadh India 23°33N 68°57E 42 H3
Matanzas Cuba 23°0N 81°40W 88 B3
Matapa Botswana 23°11S 24°39E 56 C3
Matapan, C. = Tenaro, Akra
Greece 36°22N 22°27E 23 F10
Matapédia Canada 48°0N 66°59W 73 C6
Matapo △ Zimbabwe 20°30S 29°40E 55 G2
Matara Sri Lanka 5°58N 80°30E 40 S12
Mataram Indonesia 8°35S 116°7E 37 K19
Matarani Peru 17°0S 72°10W 92 G4
Mataranka Australia 14°55S 133°4E 60 B5
Matarma, Râs Egypt 30°27N 32°44E 46 E1
Mataró Spain 41°32N 2°29E 21 B7
Matatiele S. Africa 30°20S 28°49E 57 E4
Mataura N.Z. 46°11S 168°51E 59 G2
Matavai, B. de Tahiti 17°30S 149°23W 59 d
Matehuala Mexico 23°39N 100°39W 86 C4
Mateke Hills Zimbabwe 21°48S 31°0E 55 G3
Matelot Trin. & Tob. 10°50N 61°7W 93 K15
Matera Italy 40°40N 16°36E 22 D7
Matetsi Zimbabwe 18°12S 26°0E 55 F2
Matheniko △ Uganda 2°49N 34°27E 54 B3
Mathis U.S.A. 28°6N 97°50W 84 G6
Mathraki Greece 39°48N 19°31E 25 A3
Mathura India 27°30N 77°40E 42 F7
Mati Phil. 6°55N 126°15E 37 C7
Matiali India 26°56N 88°49E 43 F13
Matías Romero Mexico 16°53N 95°2W 87 D5
Matibane Mozam. 14°49S 40°45E 55 E5
Matiri Ra. N.Z. 41°38S 172°20E 59 D4
Matjiesfontein S. Africa 33°14S 20°35E 56 E3
Matla → India 21°40N 88°40E 43 J13
Matlamanyane
Botswana 19°33S 25°57E 56 B4
Matli Pakistan 25°2N 68°39E 42 G3
Matlock U.K. 53°9N 1°33W 12 D6
Mato Grosso ☐ Brazil 14°0S 55°0W 93 F8
Mato Grosso, Planalto do
Brazil 15°0S 55°0W 93 G8
Mato Grosso do Sul ☐
Brazil 18°0S 55°0W 93 G8
Matobo = Matapo △
Zimbabwe 20°30S 29°40E 55 G2
Matochkin Shar, Proliv
Russia 73°23N 55°12E 28 B6
Matopo Hills Zimbabwe 20°36S 28°20E 55 G2
Matopos Zimbabwe 20°20S 28°29E 55 G2
Matosinhos Portugal 41°11N 8°42W 21 B1
Matroosberg S. Africa 33°23S 19°40E 56 E2
Maţruḩ Oman 23°37N 58°30E 47 C6
Matsu Tao Taiwan 26°8N 119°56E 33 F12
Matsue Japan 35°25N 133°10E 31 G6
Matsum, Ko Thailand 9°22N 99°59E 39 b
Matsumae Japan 41°26N 140°7E 30 D10
Matsumae-Hantō
Japan 41°30N 140°15E 30 D10
Matsumoto Japan 36°15N 138°0E 31 F9
Matsusaka Japan 34°34N 136°32E 31 G8
Matsushima Japan 38°20N 141°10E 30 E10
Matsuura Japan 33°20N 129°49E 31 H4
Matsuyama Japan 33°45N 132°45E 31 H6
Mattagami → Canada 50°43N 81°29W 72 B3
Mattancheri India 9°50N 76°15E 40 Q10
Mattawa Canada 46°20N 78°45W 72 C4
Matterhorn Switz. 45°58N 7°39E 20 D7
Matthew Town
Bahamas 20°57N 73°40W 89 B5
Matthews Ridge Guyana 7°37N 60°10W 92 B6
Mattice Canada 49°40N 83°20W 72 C3
Mattituck U.S.A. 40°59N 72°32W 83 F12
Mattō Japan 36°31N 136°34E 31 F8
Mattoon U.S.A. 39°29N 88°23W 80 F9
Matucana Peru 11°55S 76°25W 92 F3
Matuku Fiji 19°10S 179°44E 59 a
Matūn = Khowst
Afghan. 33°22N 69°58E 42 C3
Matura B. Trin. & Tob. 10°39N 61°1W 93 K15
Maturín Venezuela 9°45N 63°11W 92 B6
Matusadona △ Zimbabwe 16°58S 28°42E 55 F2
Mau Mad. P., India 26°17N 78°41E 43 F8
Mau Ut. P., India 25°56N 83°33E 43 G10
Mau Ut. P., India 25°17N 81°23E 43 G9
Mau Escarpment Kenya 0°40S 36°0E 54 C4
Mau Ranipur India 25°16N 79°8E 43 G8
Maua Kenya 0°14N 37°56E 54 C4
Maua Mozam. 13°53S 37°10E 55 E4
Maubeuge France 50°17N 3°57E 20 A6
Maubin Burma 16°44N 95°39E 41 L19
Maud, Pt. Australia 23°6S 113°45E 60 D1
Maud Rise S. Ocean 66°0S 3°0E 5 C3
Maude Australia 34°29S 144°18E 63 E3
Maudin Sun Burma 16°0N 94°30E 41 M19
Maués Brazil 3°20S 57°45W 92 D7
Mauganj India 24°50N 81°55E 43 G9
Maughold Hd. I. of Man 54°18N 4°18W 12 C3
Maui U.S.A. 20°48N 156°20W 75 L8
Maulamyaing = Moulmein
Burma 16°30N 97°40E 41 L20
Maule ☐ Chile 36°5S 72°30W 94 D1
Maumee U.S.A. 41°34N 83°39W 81 E12
Maumee → U.S.A. 41°42N 83°28W 81 E12
Maumere Indonesia 8°38S 122°13E 37 F6
Maun Botswana 20°0S 23°26E 56 C3
Mauna Kea U.S.A. 19°50N 155°28W 75 M8
Mauna Loa U.S.A. 19°30N 155°35W 75 M8
Maunath Bhanjan = Mau
India 25°56N 83°33E 43 G10
Maungmagan Kyunzu
Burma 14°0N 97°48E 38 E1
Maungu Kenya 3°33S 38°45E 54 C4
Maupin U.S.A. 45°11N 121°5W 76 D3
Maurepas, L. U.S.A. 30°15N 90°30W 85 F9
Maurice, L. Australia 29°30S 131°0E 61 E5
Mauricie △ Canada 46°45N 73°0W 72 C5
Mauritania ■ Africa 20°50N 10°0W 50 E3
Mauritius ■ Ind. Oc. 20°0S 57°0E 53 d
Mauston U.S.A. 43°48N 90°5W 80 D8
Mavli India 24°45N 73°55E 42 G5
Mavuradonha Mts.
Zimbabwe 16°30S 31°30E 55 F3
Mawa
Dem. Rep. of the Congo 2°45N 26°40E 54 B2
Mawai India 22°30N 81°4E 43 H9
Mawana India 29°6N 77°58E 42 E7
Mawand Pakistan 29°33N 68°38E 42 E3
Mawjib, W. al → Jordan 31°28N 35°36E 46 D4
Mawk Mai Burma 20°14N 97°37E 41 J20
Mawlaik Burma 23°40N 94°26E 41 H19
Mawlamyine = Moulmein
Burma 16°30N 97°40E 41 L20
Mawqaq Si. Arabia 27°25N 41°8E 44 E4
Mawson Antarctica 67°30S 62°53E 5 C6
Mawson Coast Antarctica 68°30S 63°0E 5 C6
Max U.S.A. 47°49N 101°18W 80 B3
Maxcanú Mexico 20°35N 90°0W 87 C6
Maxesibeni S. Africa 30°49S 29°23E 57 E4
Maxhamish L. Canada 59°50N 123°17W 70 B4
Maxixe Mozam. 23°54S 35°17E 57 C6
Maxville Canada 45°17N 74°51W 83 A10
Maxwell U.S.A. 39°17N 122°11W 78 F4
Maxwelton Australia 20°43S 142°41E 62 C3
May, C. U.S.A. 38°56N 74°58W 81 F16
May Pen Jamaica 17°58N 77°15W 88 a
Maya → Russia 60°28N 134°28E 29 D14
Maya Mts. Belize 16°30N 89°0W 87 D7
Mayaguana Bahamas 22°30N 72°44W 89 B5
Mayagüez Puerto Rico 18°12N 67°9W 89 d
Mayāmey Iran 36°24N 55°42E 45 B7
Mayanup Australia 33°57S 116°27E 61 F2
Mayapán Mexico 20°29N 89°11W 87 C7
Mayarí Cuba 20°40N 75°41W 89 B4
Mayaro B. Trin. & Tob. 10°14N 60°59W 93 K16
Maybell U.S.A. 40°31N 108°5W 76 F9
Maybole U.K. 55°21N 4°42W 11 F4
Maydān Iraq 34°55N 45°37E 44 C5
Mayenne → France 47°30N 0°32W 20 C3
Mayer U.S.A. 34°24N 112°14W 77 J7
Mayerthorpe Canada 53°57N 115°8W 70 C5
Mayfield Ky., U.S.A. 36°44N 88°38W 80 G9
Mayfield N.Y., U.S.A. 43°6N 74°16W 83 C10
Mayhill U.S.A. 32°53N 105°29W 77 K11
Maykop Russia 44°35N 40°10E 19 F7
Maymyo Burma 22°2N 96°28E 38 A1
Maynard Mass., U.S.A. 42°26N 71°27W 83 D13
Maynard Wash., U.S.A. 47°59N 122°55W 78 C4
Maynard Hills Australia 28°28S 119°49E 61 E2
Mayne → Australia 23°40S 141°55E 62 C3
Maynooth Canada 45°13N 77°56W 82 A7
Maynooth Ireland 53°23N 6°34W 10 C5
Mayo Canada 63°38N 135°57W 68 E4
Mayo ☐ Ireland 53°53N 9°3W 10 C2
Mayon Volcano Phil. 13°15N 123°41E 37 B6
Mayor I. N.Z. 37°16S 176°17E 59 B6
Mayotte ☑ Ind. Oc. 12°50S 45°10E 53 a
Maysān ☐ Iraq 31°55N 47°15E 44 D5
Maysville U.S.A. 38°39N 83°46W 81 F12
Mayu Indonesia 1°30N 126°30E 37 D7
Mayville N. Dak., U.S.A. 47°30N 97°20W 80 B5
Mayville N.Y., U.S.A. 42°15N 79°30W 82 D5
Mazabuka Zambia 15°52S 27°44E 55 F2
Mazagān = El Jadida
Morocco 33°11N 8°17W 50 B4
Mazagão Brazil 0°7S 51°16W 93 D8
Mazán Peru 3°30S 73°0W 92 D4
Māzandarān ☐ Iran 36°30N 52°0E 45 B7
Mazapil Mexico 24°39N 101°34W 86 C4
Mazar China 36°32N 77°1E 32 D4
Mazara del Vallo Italy 37°39N 12°35E 22 F5
Mazarrón Spain 37°38N 1°19W 21 D5
Mazaruni → Guyana 6°25N 58°35W 92 B7
Mazatán Mexico 29°0N 110°8W 86 B2
Mazatenango Guatemala 14°35N 91°30W 88 D1
Mazatlán Mexico 23°13N 106°25W 86 C3
Mažeikiai Lithuania 56°20N 22°20E 9 H20
Māzhān Iran 32°30N 59°0E 45 C8
Mazīnān Iran 36°19N 56°56E 45 B8
Mazoe Mozam. 16°42S 33°7E 55 F3
Mazoe → Mozam. 16°20S 33°30E 55 F3
Mazowe Zimbabwe 17°28S 30°58E 55 F3
Mazurski, Pojezierze
Poland 53°50N 21°0E 17 B11
Mazyr Belarus 51°59N 29°15E 17 B15
Mba Fiji 17°33S 177°41E 59 a
Mbabane Swaziland 26°18S 31°6E 57 D5
Mbaïki C.A.R. 3°53N 18°1E 52 D3
Mbala Zambia 8°46S 31°24E 55 D3
Mbalabala Zimbabwe 20°27S 29°3E 57 C4
Mbale Uganda 1°8N 34°12E 54 B3
Mbalmayo Cameroon 3°33N 11°33E 52 D2
Mbamba Bay Tanzania 11°13S 34°49E 55 E3
Mbandaka
Dem. Rep. of the Congo 0°1N 18°18E 52 D3
Mbanza Congo Angola 6°18S 14°16E 52 F2
Mbanza Ngungu
Dem. Rep. of the Congo 5°12S 14°53E 52 F2
Mbarangandu Tanzania 10°11S 36°48E 55 D4
Mbarara Uganda 0°35S 30°40E 54 C3
Mbengga = Beqa Fiji 18°23S 178°8E 59 a
Mbenkuru → Tanzania 9°25S 39°50E 55 D4
Mberengwa Zimbabwe 20°29S 29°57E 55 G2
Mberengwa, Mt.
Zimbabwe 20°37S 29°55E 55 G2
Mbesuma Zambia 10°0S 32°2E 55 E3
Mbeya Tanzania 8°54S 33°29E 55 D3
Mbeya ☐ Tanzania 8°15S 33°30E 54 D3
Mbhashe → S. Africa 32°15S 28°54E 57 E4
Mbinga Tanzania 10°50S 35°0E 55 E4
Mbini = Río Muni ☐
Eq. Guin. 1°30N 10°0E 52 D2
Mbizi Zimbabwe 21°23S 31°1E 55 G3
Mbouda Cameroon 5°38N 10°15E 50 G8
M'boukou, L. de
Cameroon 6°23N 12°50E 52 C2
Mbour Senegal 14°22N 16°54W 50 F2
Mbuji-Mayi
Dem. Rep. of the Congo 6°9S 23°40E 54 D1
Mbulu Tanzania 3°45S 35°30E 54 C4
Mburucuyá Argentina 28°1S 58°14W 94 B4
Mburucuyá △ Argentina 28°1S 58°12W 94 B4
Mchinga Tanzania 9°44S 39°45E 55 D4
Mchinji Malawi 13°47S 32°58E 55 E3
Mdantsane S. Africa 32°56S 27°46E 57 E4
Mead, L. U.S.A. 36°0N 114°44W 79 J12
Meade U.S.A. 37°17N 100°20W 80 G3
Meade River = Atqasuk
U.S.A. 70°28N 157°24W 74 A8
Meadow Lake Canada 54°10N 108°26W 71 C7
Meadow Lake △ Canada 54°27N 109°0W 71 C7
Meadow Valley Wash →
U.S.A. 36°40N 114°34W 79 J12
Meadville U.S.A. 41°39N 80°9W 82 E4
Meaford Canada 44°36N 80°35W 82 B4
Meakan Dake Japan 45°15N 144°0E 30 C11
Mealy Mts. Canada 53°10N 58°0W 73 B8
Meander River Canada 59°2N 117°42W 70 B5
Meares, C. U.S.A. 45°37N 124°0W 76 D1
Mearim → Brazil 3°4S 44°35W 93 D10
Meath ☐ Ireland 53°40N 6°57W 10 C5
Meath Park Canada 53°27N 105°22W 71 C7
Meaux France 48°58N 2°50E 20 B5
Mebechi-Gawa →
Japan 40°31N 141°31E 30 D10
Mebulu, Tanjung
Indonesia 8°50S 115°5E 37 K18
Mecanhelas Mozam. 15°12S 35°54E 55 F4
Mecca = Makkah
Si. Arabia 21°30N 39°54E 47 C2
Mecca U.S.A. 33°34N 116°5W 79 M10
Mechanicsburg U.S.A. 40°13N 77°1W 82 F8
Mechanicville U.S.A. 42°54N 73°41W 83 D11
Mechelen Belgium 51°2N 4°29E 15 C4
Mecheria Algeria 33°35N 0°18W 50 B5
Mecklenburg Germany 53°33N 11°40E 16 B7
Mecklenburger Bucht
Germany 54°20N 11°40E 16 A6
Meconta Mozam. 14°59S 39°50E 55 E4
Mecubúri Mozam. 14°39S 38°54E 55 E4
Mecubúri → Mozam. 14°10S 40°30E 55 E5
Mecúfi Mozam. 13°20S 40°32E 55 E5
Medan Indonesia 3°40N 98°38E 36 D1
Médanos de Coro △
Venezuela 11°35N 69°44W 89 D6
Medanosa, Pta. Argentina 48°8S 66°0W 96 F3
Médéa Algeria 36°12N 2°50E 50 A6
Medellín Colombia 6°15N 75°35W 92 B3
Medelpad Sweden 62°33N 16°30E 8 E17
Medemblik Neths. 52°46N 5°8E 15 B5
Medford Mass., U.S.A. 42°25N 71°7W 83 D13
Medford Oreg., U.S.A. 42°19N 122°52W 76 E2
Medford Wis., U.S.A. 45°9N 90°20W 80 C8
Medgidia Romania 44°15N 28°19E 17 F15
Media Agua Argentina 31°58S 68°25W 94 C2
Media Luna Argentina 34°45S 66°44W 94 C2
Medianeira Brazil 25°17S 54°5W 95 B5
Mediaş Romania 46°9N 24°22E 17 E13
Medicine Bow U.S.A. 41°54N 106°12W 76 F10
Medicine Bow Mts.
U.S.A. 40°40N 106°0W 76 F10
Medicine Bow Pk.
U.S.A. 41°21N 106°19W 76 F10
Medicine Hat Canada 50°0N 110°45W 71 D6
Medicine Lake U.S.A. 48°30N 104°30W 76 B11
Medicine Lodge U.S.A. 37°17N 98°35W 80 G4
Medina = Al Madīnah
Si. Arabia 24°35N 39°52E 44 E3
Medina N. Dak., U.S.A. 46°54N 99°18W 80 B4
Medina N.Y., U.S.A. 43°13N 78°23W 82 C6
Medina Ohio, U.S.A. 41°8N 81°52W 82 E3
Medina → U.S.A. 29°16N 98°29W 84 G5

Medina del Campo *Spain* 41°18N 4°55W **21** B3
Medina L. *U.S.A.* 29°32N 98°56W **84** G5
Medina Sidonia *Spain* 36°28N 5°57W **21** D3
Medinipur *India* 22°25N 87°21E **43** H12
Mediterranean Sea *Europe* 35°0N 15°0E **6** H7
Médoc *France* 45°10N 0°50W **20** D3
Medveditsa → *Russia* 49°35N 42°41E **19** E7
Medvezhi, Ostrava *Russia* 71°0N 161°0E **29** B17
Medvezhyegorsk *Russia* 63°0N 34°25E **18** B5
Medway □ *U.K.* 51°25N 0°32E **13** F8
Medway → *U.K.* 51°27N 0°46E **13** F8
Meekatharra *Australia* 26°32S 118°29E **61** E2
Meeker *U.S.A.* 40°2N 107°55W **76** F10
Meerut *India* 29°1N 77°42E **42** E7
Meeteetse *U.S.A.* 44°9N 108°52W **76** D9
Mega *Ethiopia* 3°57N 38°19E **47** G2
Megara *Greece* 37°58N 23°22E **23** F10
Megasini *India* 21°38N 86°21E **43** J12
Meghalaya □ *India* 25°50N 91°0E **41** G17
Meghna → *Bangla.* 22°50N 90°50E **41** H17
Megion *Russia* 61°3N 76°6E **28** C8
Mégiscane, L. *Canada* 48°35N 75°55W **72** C4
Meharry, Mt. *Australia* 22°59S 118°35E **60** D2
Mehlville *U.S.A.* 38°31N 90°19W **80** F8
Mehndawal *India* 26°58N 83°5E **43** F10
Mehr Jān *Iran* 33°50N 55°6E **45** C7
Mehrābād *Iran* 36°53N 47°55E **44** B5
Mehrān *Iran* 33°7N 46°10E **44** C5
Mehrān → *Iran* 26°45N 55°26E **45** E7
Mehrgarh *Pakistan* 29°30N 67°30E **42** E2
Mehrīz *Iran* 31°35N 54°28E **45** D7
Mei Xian *China* 34°18N 107°55E **34** G4
Meighen I. *Canada* 80°0N 99°30W **69** B12
Meihekou *China* 42°32N 125°40E **35** C13
Meiktila *Burma* 20°53N 95°54E **41** J19
Meissen *Germany* 51°9N 13°29E **16** C7
Meitan *China* 27°45N 107°29E **32** F10
Meizhou *China* 24°16N 116°6E **33** G12
Meja *India* 25°9N 82°7E **43** G10
Mejillones *Chile* 23°10S 70°30W **94** A1
Mekele *Ethiopia* 13°33N 39°30E **47** E2
Mekerghene, Sebkra *Algeria* 26°21N 1°30E **50** C6
Mekhtar *Pakistan* 30°30N 69°15E **40** D6
Meknès *Morocco* 33°57N 5°33W **50** B4
Mekong → *Asia* 9°30N 106°15E **39** H6
Mekongga *Indonesia* 3°39S 121°15E **37** E6
Mekvari = Kür → *Azerbaijan* 39°29N 49°15E **19** G8
Melagiri Hills *India* 12°20N 77°30E **40** N10
Melaka *Malaysia* 2°15N 102°15E **39** L4
Melaka □ *Malaysia* 2°15N 102°15E **39** L4
Melambes *Greece* 35°8N 24°40E **25** D6
Melanesia *Pac. Oc.* 4°0S 155°0E **64** H7
Melanesian Basin *Pac. Oc.* 0°5N 160°35E **64** G8
Melaya *Indonesia* 8°17S 114°30E **37** J17
Melbourne *Australia* 37°48S 144°58E **63** F4
Melbourne *U.S.A.* 28°5N 80°37W **85** G14
Melchor Múzquiz *Mexico* 27°53N 101°31W **86** B4
Melchor Ocampo *Mexico* 24°51N 101°39W **86** C4
Melekeok *Palau* 7°27N 134°38E **64** G5
Mélèzes → *Canada* 57°40N 69°29W **72** A5
Melfort *Canada* 52°50N 104°37W **71** C8
Melfort *Zimbabwe* 18°0S 31°25E **55** F3
Melhus *Norway* 63°17N 10°18E **8** E14
Melilla *N. Afr.* 35°21N 2°57W **21** E4
Melipilla *Chile* 33°42S 71°15W **94** C1
Melissa *Canada* 45°24N 79°14W **82** A5
Melissa, Akra *Greece* 35°6N 24°33E **25** D6
Melita *Canada* 49°15N 101°0W **71** D8
Melitopol *Ukraine* 46°50N 35°22E **19** E6
Melk *Austria* 48°13N 15°20E **16** D8
Mellansel *Sweden* 63°26N 18°19E **8** E18
Mellen *U.S.A.* 46°20N 90°40W **80** B8
Mellerud *Sweden* 58°41N 12°28E **9** G15
Mellette *U.S.A.* 45°9N 98°30W **80** C4
Mellieha *Malta* 35°57N 14°22E **25** D1
Melo *Uruguay* 32°20S 54°10W **95** C5
Melolo *Indonesia* 9°53S 120°40E **37** F6
Melouprey *Cambodia* 13°48N 105°16E **38** F5
Melrhir, Chott *Algeria* 34°13N 6°30E **50** B7
Melrose *Australia* 32°42S 146°57E **63** E4
Melrose *U.K.* 55°36N 2°43W **11** F6
Melrose *Minn., U.S.A.* 45°40N 94°49W **80** C6
Melrose *N. Mex., U.S.A.* 34°26N 103°38W **77** J12
Melstone *U.S.A.* 46°36N 107°52W **76** C10
Melton Mowbray *U.K.* 52°47N 0°54W **12** E7
Melun *France* 48°32N 2°39E **20** B5
Melville *Canada* 50°55N 102°50W **71** C8
Melville, C. *Australia* 14°11S 144°30E **62** A3
Melville, L. *Canada* 53°30N 60°0W **73** B8
Melville B. *Australia* 12°0S 136°45E **62** A2
Melville B. *Greenland* 75°30N 63°0W **69** B19
Melville I. *Australia* 11°30S 131°0E **60** B5
Melville I. *Canada* 75°30N 112°0W **69** B9
Melville Pen. *Canada* 68°0N 84°0W **69** D15
Melvin, Lough *Ireland* 54°26N 8°10W **10** B3
Memba *Mozam.* 14°11S 40°30E **55** E5
Memboro *Indonesia* 9°30S 119°30E **60** A2
Memel = Klaipėda *Lithuania* 55°43N 21°10E **9** J19
Memel *S. Africa* 27°38S 29°36E **57** D4
Memmingen *Germany* 47°58N 10°10E **16** E6
Mempawah *Indonesia* 0°30N 109°5E **36** D3
Memphis *Mich., U.S.A.* 42°54N 82°46W **82** D2
Memphis *Tenn., U.S.A.* 35°8N 90°2W **85** D9
Memphis *Tex., U.S.A.* 34°44N 100°33W **84** D4
Memphrémagog, L. *N. Amer.* 45°8N 72°17W **83** B12
Mena *U.S.A.* 34°35N 94°15W **84** D7
Menado = Manado *Indonesia* 1°29N 124°51E **37** D6
Menai Strait *U.K.* 53°11N 4°13W **12** D3
Ménaka *Mali* 15°59N 2°18E **50** E6
Menan = Chao Phraya → *Thailand* 13°40N 100°31E **38** F3
Menarandra → *Madag.* 25°17S 44°30E **57** D7
Menard *U.S.A.* 30°55N 99°47W **84** F5
Menard Fracture Zone *Pac. Oc.* 43°0S 97°0W **65** M18
Mendaña Fracture Zone *Pac. Oc.* 16°0S 91°0W **65** J18
Mendawai → *Indonesia* 3°30S 113°0E **36** E4
Mende *France* 44°31N 3°30E **20** D5
Mendeleyev Ridge *Arctic* 80°0N 178°0W **4** B17
Mendhar *India* 33°35N 74°10E **43** C6
Mendip Hills *U.K.* 51°17N 2°40W **13** F5
Mendocino *U.S.A.* 39°19N 123°48W **76** G2
Mendocino, C. *U.S.A.* 40°26N 124°25W **76** F1
Mendocino Fracture Zone *Pac. Oc.* 40°0N 142°0W **65** D13
Mendota *Calif., U.S.A.* 36°45N 120°23W **78** J6
Mendota *Ill., U.S.A.* 41°33N 89°7W **80** E9
Mendoyo *Indonesia* 8°23S 114°42E **37** J17
Mendoza *Argentina* 32°50S 68°52W **94** C2
Mendoza □ *Argentina* 33°0S 69°0W **94** C2
Mene Grande *Venezuela* 9°49N 70°56W **92** B4
Menemen *Turkey* 38°34N 27°3E **23** E12
Menen *Belgium* 50°47N 3°7E **15** D3
Mengcheng *China* 33°12N 116°36E **34** H9
Menggala *Indonesia* 4°30S 105°15E **36** E3
Mengjin *China* 34°55N 112°45E **34** G7
Mengyin *China* 35°40N 117°58E **35** G9
Mengzi *China* 23°20N 103°22E **32** G9
Menihek *Canada* 54°28N 56°36W **73** B6
Menihek L. *Canada* 54°0N 67°0W **73** B6
Menin = Menen *Belgium* 50°47N 3°7E **15** D3
Menindee *Australia* 32°20S 142°25E **63** E3
Menindee L. *Australia* 32°20S 142°25E **63** E3
Meningie *Australia* 35°50S 139°18E **63** F2
Menjangan, Pulau *Indonesia* 8°7S 114°31E **37** J17
Menlo Park *U.S.A.* 37°27N 122°12W **78** H4
Menngen ☼ *Australia* 15°21S 131°16E **60** C5
Menominee *U.S.A.* 45°6N 87°37W **80** C10
Menominee → *U.S.A.* 45°6N 87°35W **80** C10
Menomonie *U.S.A.* 44°53N 91°55W **80** C8
Menongue *Angola* 14°48S 17°52E **53** G3
Menorca *Spain* 40°0N 4°0E **24** B11
Mentakab *Malaysia* 3°29N 102°21E **39** L4
Mentawai, Kepulauan *Indonesia* 2°0S 99°0E **36** E1
Menton *France* 43°50N 7°29E **20** E7
Mentor *U.S.A.* 41°40N 81°21W **82** E3
Menzelinsk *Russia* 55°47N 53°11E **18** C9
Menzies *Australia* 29°40S 121°2E **61** E3
Meob B. *Namibia* 24°25S 14°34E **56** C1
Me'ona *Israel* 33°1N 35°15E **46** B4
Meoqui *Mexico* 28°17N 105°29W **86** B3
Mepaco *Mozam.* 15°57S 30°48E **55** F3
Meppel *Neths.* 52°42N 6°12E **15** B6
Merak *Indonesia* 6°10N 106°26E **37** F12
Meramangye, L. *Australia* 28°25S 132°13E **61** E5
Meran = Merano *Italy* 46°40N 11°9E **22** A4
Merano *Italy* 46°40N 11°9E **22** A4
Merauke *Indonesia* 8°29S 140°24E **37** F10
Merbein *Australia* 34°10S 142°2E **63** E3
Merca = Marka *Somali Rep.* 1°48N 44°50E **47** G3
Merced *U.S.A.* 37°18N 120°29W **78** H6
Merced → *U.S.A.* 37°21N 120°59W **78** H6
Merced Pk. *U.S.A.* 37°36N 119°24W **78** H7
Mercedes *B. Aires, Argentina* 34°40S 59°30W **94** C4
Mercedes *Corrientes, Argentina* 29°10S 58°5W **94** B4
Mercedes *San Luis, Argentina* 33°40S 65°21W **94** C2
Mercedes *Uruguay* 33°12S 58°0W **94** C4
Merceditas *Chile* 28°20S 70°35W **94** B1
Mercer *N.Z.* 37°16S 175°5E **59** B5
Mercer *U.S.A.* 41°14N 80°15W **82** E4
Mercer Island *U.S.A.* 47°34N 122°13W **78** C4
Mercury *U.S.A.* 36°40N 115°59W **79** J11
Mercy, C. *Canada* 65°0N 63°30W **69** E19
Mere *U.K.* 51°6N 2°16W **13** F5
Merebuk, Gunung *Indonesia* 8°13S 114°39E **37** J17
Meredith, C. *Falk. Is.* 52°15S 60°40W **96** G4
Meredith, L. *U.S.A.* 35°43N 101°33W **84** D4
Mergui *Burma* 12°26N 98°34E **38** F2
Mergui Arch. = Myeik Kyunzu *Burma* 11°30N 97°30E **39** G1
Mérida *Mexico* 20°58N 89°37W **87** C7
Mérida *Spain* 38°55N 6°25W **21** C2
Mérida *Venezuela* 8°24N 71°8W **92** B4
Mérida, Cord. de *Venezuela* 9°0N 71°0W **92** B4
Meriden *U.K.* 52°26N 1°38W **13** E6
Meriden *U.S.A.* 41°32N 72°48W **83** E12
Meridian *Calif., U.S.A.* 39°9N 121°55W **78** F5
Meridian *Idaho, U.S.A.* 43°37N 116°24W **76** E5
Meridian *Miss., U.S.A.* 32°22N 88°42W **85** E10
Merinda *Australia* 20°2S 148°11E **62** C4
Merir *Palau* 4°10N 132°30E **37** D8
Merirumã *Brazil* 1°15N 54°50W **93** C8
Merkel *U.S.A.* 32°28N 100°1W **84** E4
Mermaid Reef *Australia* 17°6S 119°36E **60** C2
Merowe Dam *Sudan* 18°35N 31°56E **51** E12
Merredin *Australia* 31°28S 118°18E **61** F2
Merrick *U.K.* 55°8N 4°28W **11** F4
Merrickville *Canada* 44°55N 75°50W **83** B9
Merrill *Oreg., U.S.A.* 42°1N 121°36W **76** E3
Merrill *Wis., U.S.A.* 45°11N 89°41W **80** C9
Merrimack → *U.S.A.* 42°49N 70°49W **83** D14
Merriman *U.S.A.* 42°55N 101°42W **80** D3
Merritt *Canada* 50°10N 120°45W **70** C4
Merritt Island *U.S.A.* 28°21N 80°42W **85** G14
Merriwa *Australia* 32°6S 150°22E **63** E5
Merry I. *Canada* 55°29N 77°31W **72** A4
Merryville *U.S.A.* 30°45N 93°33W **84** F8
Mersch *Lux.* 49°44N 6°7E **15** E6
Mersea I. *U.K.* 51°47N 0°58E **13** F8
Merseburg *Germany* 51°22N 11°59E **16** C6
Mersey → *U.K.* 53°25N 3°1W **12** D4
Merseyside □ *U.K.* 53°31N 3°2W **12** D4
Mersin = İçel *Turkey* 36°51N 34°36E **44** B2
Mersing *Malaysia* 2°25N 103°50E **39** L4
Merta *India* 26°39N 74°4E **42** F6
Merta Road *India* 26°43N 73°55E **42** F5
Merthyr Tydfil *U.K.* 51°45N 3°22W **13** F4
Merthyr Tydfil □ *U.K.* 51°46N 3°21W **13** F4
Merti *Kenya* 1°4N 38°40E **54** B4
Mértola *Portugal* 37°40N 7°40W **21** D2
Mertzon *U.S.A.* 31°16N 100°49W **84** F4
Meru *Kenya* 0°3N 37°40E **54** B4
Meru *Tanzania* 3°15S 36°46E **54** C4
Meru △ *Kenya* 0°5N 38°10E **54** B4
Meru Betiri △ *Indonesia* 8°27S 113°51E **37** H15
Mesa *U.S.A.* 33°25N 111°50W **77** K8
Mesa Verde △ *U.S.A.* 37°11N 108°29W **77** H9
Mesanagros *Greece* 36°1N 27°49E **25** C9
Mesaoría *Cyprus* 35°12N 33°14E **25** D12
Mesgouez, L. *Canada* 51°20N 75°0W **72** B5
Meshed = Mashhad *Iran* 36°20N 59°35E **45** B8
Meshgīn Shahr *Iran* 38°30N 47°45E **44** B5
Meshoppen *U.S.A.* 41°36N 76°3W **83** E8
Mesilinka → *Canada* 56°6N 124°30W **70** B4
Mesologi *Greece* 38°21N 21°28E **23** E9
Mesongi *Greece* 39°29N 19°56E **25** B3
Mesopotamia = Al Jazirah *Iraq* 33°30N 44°0E **44** C5
Mesopotamia *U.S.A.* 41°27N 80°57W **82** E4
Mesquite *U.S.A.* 36°48N 114°4W **77** H6
Messaad *Algeria* 34°8N 3°30E **50** B6
Messalo → *Mozam.* 12°25S 39°15E **55** E4
Messara, Kolpos *Greece* 35°6N 24°47E **25** D6
Messina = Musina *S. Africa* 22°20S 30°5E **57** C5
Messina *Italy* 38°11N 15°34E **22** E6
Messina, Str. di *Italy* 38°15N 15°35E **22** F6
Messini *Greece* 37°4N 22°1E **23** F10
Messiniakos Kolpos *Greece* 36°45N 22°5E **23** F10
Mesta = Nestos → *Europe* 40°54N 24°49E **23** D11
Meta → *S. Amer.* 6°12N 67°28W **92** B5
Meta Incognita Pen. *Canada* 62°45N 68°30W **69** E18
Metabetchouan *Canada* 48°26N 71°52W **73** C5
Metairie *U.S.A.* 29°59N 90°9W **85** G9
Metaline Falls *U.S.A.* 48°52N 117°22W **76** B5
Metán *Argentina* 25°30S 65°0W **94** B3
Metangula *Mozam.* 12°40S 34°50E **55** E3
Metcalfe *Canada* 45°14N 75°28W **83** A9
Metema *Ethiopia* 12°58N 36°12E **47** E2
Metengobalame *Mozam.* 14°49S 34°30E **55** E3
Methven *N.Z.* 43°38S 171°40E **59** E3
Metil *Mozam.* 16°24S 39°0E **55** F4
Metlakatla *U.S.A.* 55°8N 131°35W **68** F5
Metropolis *U.S.A.* 37°9N 88°44W **80** G9
Metropolitana □ *Chile* 33°30S 70°50W **94** C1
Metu *Ethiopia* 8°18N 35°35E **47** F2
Metz *France* 49°8N 6°10E **20** B7
Meulaboh *Indonesia* 4°11N 96°3E **36** D1
Meureudu *Indonesia* 5°19N 96°10E **36** C1
Meuse → *Europe* 50°45N 5°41E **15** D5
Mexia *U.S.A.* 31°41N 96°29W **84** F6
Mexiana, I. *Brazil* 0°0 49°30W **93** D9
Mexicali *Mexico* 32°40N 115°30W **79** N11
Mexican Plateau *Mexico* 25°0N 104°0W **66** G9
Mexican Water *U.S.A.* 36°57N 109°32W **77** H9
México *Mexico* 19°24N 99°9W **87** D5
Mexico *Maine, U.S.A.* 44°34N 70°33W **83** B14
Mexico *Mo., U.S.A.* 39°10N 91°53W **80** F8
Mexico *N.Y., U.S.A.* 43°28N 76°14W **83** C8
México □ *Mexico* 19°20N 99°30W **87** D5
Mexico ■ *Cent. Amer.* 25°0N 105°0W **86** C4
México, Ciudad de *Mexico* 19°24N 99°9W **87** D5
Mexico, G. of *Cent. Amer.* 25°0N 90°0W **87** C7
Mexico B. *U.S.A.* 43°35N 76°20W **83** C8
Meydān-e Naftūn *Iran* 31°56N 49°18E **45** D6
Meydani, Ra's-e *Iran* 25°24N 59°6E **45** E8
Meyers Chuck *U.S.A.* 55°45N 132°15W **70** B2
Meymaneh *Afghan.* 35°53N 64°38E **40** B4
Mezen *Russia* 65°50N 44°20E **18** A7
Mezen → *Russia* 65°44N 44°22E **18** A7
Mézenc, Mt. *France* 44°54N 4°11E **20** D6
Mezhdurechensk *Russia* 53°41N 88°3E **28** D9
Mezhdurechenskiy *Russia* 59°36N 65°56E **28** D7
Mezőkövesd *Hungary* 47°49N 20°35E **17** E11
Mezőtúr *Hungary* 47°1N 20°41E **17** E11
Mezquital *Mexico* 23°29N 104°23W **86** C4
Mfolozi → *S. Africa* 28°25S 32°26E **57** D5
Mgeta *Tanzania* 8°22S 36°6E **55** D4
Mhlaba Hills *Zimbabwe* 18°30S 30°30E **55** F3
Mhow *India* 22°33N 75°50E **42** H6
Miahuatlán *Mexico* 16°20N 96°36W **87** D5
Miami *Fla., U.S.A.* 25°46N 80°11W **85** J14
Miami *Okla., U.S.A.* 36°53N 94°53W **84** C7
Miami *Tex., U.S.A.* 35°42N 100°38W **84** D4
Miami Beach *U.S.A.* 25°47N 80°7W **85** J14
Mian Xian *China* 33°10N 106°32E **34** H4
Mianchi *China* 34°48N 111°48E **34** G6
Mīāndarreh *Iran* 35°37N 53°39E **45** C7
Mīāndowāb *Iran* 37°0N 46°5E **44** B5
Miandrivazo *Madag.* 19°31S 45°29E **57** B8
Mīāneh *Iran* 37°30N 47°40E **44** B5
Mianwali *Pakistan* 32°38N 71°28E **42** C4
Mianyang *China* 31°22N 104°47E **32** E9
Miaodao Qundao *China* 38°10N 120°45E **35** E11
Miarinarivo *Antananarivo, Madag.* 18°57S 46°55E **57** B8
Miarinarivo *Toamasina, Madag.* 16°38S 48°15E **57** B8
Miariravaratra *Madag.* 20°13S 47°31E **57** C8
Miass *Russia* 54°59N 60°6E **18** D11
Mica *S. Africa* 24°10S 30°48E **57** C5
Michalovce *Slovak Rep.* 48°47N 21°58E **17** D11
Michigan □ *U.S.A.* 44°0N 85°0W **81** C11
Michigan, L. *U.S.A.* 44°0N 87°0W **80** D10
Michigan City *U.S.A.* 41°43N 86°54W **80** E10
Michipicoten I. *Canada* 47°40N 85°40W **72** C2
Michoacán □ *Mexico* 19°10N 101°50W **86** D4
Michurin *Bulgaria* 42°9N 27°51E **23** C12
Michurinsk *Russia* 52°58N 40°27E **18** D7
Micoud *St. Lucia* 13°49N 60°54W **89** f
Micronesia *Pac. Oc.* 11°0N 160°0E **64** G7
Micronesia, Federated States of ■ *Pac. Oc.* 9°0N 150°0E **58** A8
Mid-Indian Ridge *Ind. Oc.* 30°0S 75°0E **64** M1
Mid-Oceanic Ridge *Ind. Oc.* 42°0S 90°0E **64** M1
Mid-Pacific Seamounts *Pac. Oc.* 18°0N 177°0W **64** F10
Midai *Indonesia* 3°0N 107°47E **36** D3
Midale *Canada* 49°25N 103°20W **71** D8
Middelburg *Neths.* 51°30N 3°36E **15** C3
Middelburg *Eastern Cape, S. Africa* 31°30S 25°0E **56** E4
Middelburg *Mpumalanga, S. Africa* 25°49S 29°28E **57** D4
Middelpos *S. Africa* 31°55S 20°13E **56** E3
Middelwit *S. Africa* 24°51S 27°3E **56** C4
Middle Alkali L. *U.S.A.* 41°27N 120°5W **76** F3
Middle America Trench = Guatemala Trench *Pac. Oc.* 14°0N 95°0W **66** H10
Middle Bass I. *U.S.A.* 41°41N 82°48W **82** E2
Middle East *Asia* 35°0N 40°0E **26** E5
Middle Fork Feather → *U.S.A.* 38°33N 121°30W **78** F5
Middle I. *Australia* 34°6S 123°11E **61** F3
Middle Loup → *U.S.A.* 41°17N 98°24W **80** E4
Middleboro *U.S.A.* 41°54N 70°55W **83** E14
Middleburg *Fla., U.S.A.* 30°4N 81°52W **85** F14
Middleburg *Pa., U.S.A.* 40°47N 77°3W **82** F7
Middleburgh *U.S.A.* 42°36N 74°20W **83** D10
Middlebury *U.S.A.* 44°1N 73°10W **83** B11
Middlefield *U.S.A.* 41°27N 81°4W **82** E3
Middlemount *Australia* 22°50S 148°40E **62** C4
Middleport *N.Y., U.S.A.* 43°13N 78°29W **82** C6
Middleport *Ohio, U.S.A.* 39°0N 82°3W **81** F12
Middlesboro *U.S.A.* 36°36N 83°43W **81** G12
Middlesbrough *U.K.* 54°35N 1°13W **12** C6
Middlesbrough □ *U.K.* 54°28N 1°13W **12** C6
Middlesex *Belize* 17°2N 88°31W **88** C2
Middlesex *N.J., U.S.A.* 40°36N 74°30W **83** F10
Middlesex *N.Y., U.S.A.* 42°42N 77°16W **82** D7
Middleton *Australia* 22°22S 141°32E **62** C3
Middleton *Canada* 44°57N 65°4W **73** D6
Middleton Cr. → *Australia* 22°35S 141°51E **62** C3
Middleton I. *U.S.A.* 59°26N 146°20W **74** D10
Middletown *U.K.* 54°17N 6°51W **10** B5
Middletown *Calif., U.S.A.* 38°45N 122°37W **78** G4
Middletown *Conn., U.S.A.* 41°34N 72°39W **83** E12
Middletown *N.Y., U.S.A.* 41°27N 74°25W **83** E10
Middletown *Ohio, U.S.A.* 39°31N 84°24W **81** F11
Middletown *Pa., U.S.A.* 40°12N 76°44W **83** F8
Midge Point *Australia* 20°39S 148°43E **62** b
Midhurst *Canada* 44°26N 79°43W **82** B5
Midhurst *U.K.* 50°59N 0°44W **13** G7
Midi, Canal du → *France* 43°45N 1°21E **20** E4
Midland *Australia* 31°54S 116°1E **61** F2
Midland *Canada* 44°45N 79°50W **82** B5
Midland *Calif., U.S.A.* 33°52N 114°48W **79** M12
Midland *Mich., U.S.A.* 43°37N 84°14W **81** D11
Midland *Pa., U.S.A.* 40°39N 80°27W **82** F4
Midland *Tex., U.S.A.* 32°0N 102°3W **84** F3
Midlands □ *Zimbabwe* 19°40S 29°0E **55** F2
Midleton *Ireland* 51°55N 8°10W **10** E3
Midlothian □ *U.K.* 55°51N 3°5W **11** F5
Midnapore = Medinipur *India* 22°25N 87°21E **43** H12
Midongy, Tangorombohitr' i *Madag.* 23°30S 47°0E **57** C8
Midongy Atsimo *Madag.* 23°35S 47°1E **57** C8
Midway Is. *Pac. Oc.* 28°13N 177°22W **75** K4
Midway Wells *U.S.A.* 32°41N 115°7W **79** N11
Midwest *U.S.A.* 42°0N 90°0W **75** G22
Midwest *Wyo., U.S.A.* 43°25N 106°16W **76** E10
Midwest City *U.S.A.* 35°27N 97°24W **84** D6
Midyat *Turkey* 37°25N 41°23E **44** B4
Midžor *Bulgaria* 43°24N 22°40E **23** C10
Mie □ *Japan* 34°30N 136°10E **31** G8
Międzychód *Poland* 52°35N 15°53E **16** B8
Międzyrzec Podlaski *Poland* 51°58N 22°45E **17** C12
Mielec *Poland* 50°15N 21°25E **17** C11
Mienga *Angola* 17°12S 19°48E **56** B2
Miercurea-Ciuc *Romania* 46°21N 25°48E **17** E13
Mieres *Spain* 43°18N 5°48W **21** A3
Mifflintown *U.S.A.* 40°34N 77°24W **82** F7
Mifraz Hefa *Israel* 32°52N 35°0E **46** C4
Migang Shan *China* 35°32N 106°13E **34** G4
Migori *Kenya* 1°4S 34°28E **54** C3
Miguasha, Parc de ⌂ *Canada* 48°5N 66°19W **73** C6
Miguel Alemán, Presa *Mexico* 18°15N 96°32W **87** D5
Mihara *Japan* 34°24N 133°5E **31** G6
Mikese *Tanzania* 6°48S 37°55E **54** D4
Mikhaylovgrad = Montana *Bulgaria* 43°27N 23°16E **23** C10
Mikhaylovka *Russia* 50°3N 43°5E **19** D7
Mikines *Greece* 37°43N 22°46E **23** F10
Mikkeli *Finland* 61°43N 27°15E **8** F22
Mikkwa → *Canada* 58°25N 114°46W **70** B6
Mikonos = Mykonos *Greece* 37°30N 25°25E **23** F11
Mikumi *Tanzania* 7°26S 37°0E **54** D4
Mikumi △ *Tanzania* 7°35S 37°15E **54** D4
Mikun *Russia* 62°20N 50°0E **18** B9
Milaca *U.S.A.* 45°45N 93°39W **80** C7
Milagro *Ecuador* 2°11S 79°36W **92** D3
Milan = Milano *Italy* 45°28N 9°10E **20** D8
Milan *Mo., U.S.A.* 40°12N 93°7W **80** E7
Milan *Tenn., U.S.A.* 35°55N 88°46W **85**
Milange *Mozam.* 16°3S 35°45E **55**
Milano *Italy* 45°28N 9°10E **20**
Milanoa *Madag.* 13°35S 49°47E **57**
Milâs *Turkey* 37°20N 27°50E **23**
Milatos *Greece* 35°18N 25°34E **25**
Milazzo *Italy* 38°13N 15°15E **22**
Milbank *U.S.A.* 45°13N 96°38W **80**
Milbanke Sd. *Canada* 52°19N 128°33W **70**
Milden *Canada* 51°29N 107°32W **71**
Mildenhall *U.K.* 52°21N 0°32E **13**
Mildmay *Canada* 44°3N 81°7W **82**
Mildura *Australia* 34°13S 142°9E **63**
Miles *Australia* 26°40S 150°9E **63**
Miles City *U.S.A.* 46°25N 105°51W **76**
Milestone *Canada* 49°59N 104°31W **71**
Miletus *Turkey* 37°30N 27°18E **23**
Milford *Calif., U.S.A.* 40°10N 120°22W **78**
Milford *Conn., U.S.A.* 41°14N 73°3W **83**
Milford *Del., U.S.A.* 38°55N 75°26W **81**
Milford *Mass., U.S.A.* 42°8N 71°31W **83**
Milford *N.H., U.S.A.* 42°50N 71°39W **83**
Milford *N.Y., U.S.A.* 42°35N 74°56W **83**
Milford *Pa., U.S.A.* 41°19N 74°48W **83**
Milford *Utah, U.S.A.* 38°24N 113°1W **76**
Milford Haven *U.K.* 51°42N 5°7W **13**
Milford Sd. *N.Z.* 44°41S 167°47E **59**
Milford Sound *N.Z.* 44°41S 167°55E **59**
Milḥ, Baḥr al = Razāzah, Buḥayrat ar *Iraq* 32°40N 43°35E **44**
Milikapiti *Australia* 11°26S 130°40E **60**
Miling *Australia* 30°30S 116°17E **61**
Milk River *Canada* 49°10N 112°5W **70**
Mill → *U.S.A.* 42°57N 83°23W **82**
Mill I. *Antarctica* 66°0S 101°30E **5**
Mill I. *Canada* 63°58N 77°47W **69**
Mill Valley *U.S.A.* 37°54N 122°32W **78**
Millau *France* 44°8N 3°4E **20**
Millbridge *Canada* 44°41N 77°36W **82**
Millbrook *Canada* 44°10N 78°29W **82**
Millbrook *Ala., U.S.A.* 32°29N 86°22W **85**
Millbrook *N.Y., U.S.A.* 41°47N 73°42W **83**
Mille Lacs, L. des *Canada* 48°45N 90°35W **72**
Mille Lacs L. *U.S.A.* 46°15N 93°39W **80**
Milledgeville *U.S.A.* 33°5N 83°14W **85**
Millen *U.S.A.* 32°48N 81°57W **85**
Millennium I. = Caroline I. *Kiribati* 9°58S 150°13W **65**
Miller *U.S.A.* 44°31N 98°59W **80**
Miller Lake *Canada* 45°6N 81°26W **82**
Millersburg *Ohio, U.S.A.* 40°33N 81°55W **82**
Millersburg *Pa., U.S.A.* 40°32N 76°58W **82**
Millerton *U.S.A.* 41°57N 73°31W **83**
Millerton L. *U.S.A.* 37°1N 119°41W **7**
Millet *St. Lucia* 13°55N 60°59W
Millheim *U.S.A.* 40°54N 77°29W **82**
Millicent *Australia* 37°34S 140°21E **63**
Millington *U.S.A.* 35°20N 89°53W **85**
Millinocket *U.S.A.* 45°39N 68°43W **81**
Millmerran *Australia* 27°53S 151°16E **63**
Millom *U.K.* 54°13N 3°16W **12**
Mills L. *Canada* 61°30N 118°20W **70**
Millsboro *U.S.A.* 40°0N 80°0W **82**
Millstream Chichester △ *Australia* 21°35S 117°6E **60**
Millstreet *Ireland* 52°4N 9°4W **10**
Milltown Malbay *Ireland* 52°52N 9°24W **10**
Millville *N.J., U.S.A.* 39°24N 75°2W **81**
Millville *Pa., U.S.A.* 41°7N 76°32W **83**
Millwood L. *U.S.A.* 33°42N 93°58W **84**
Milne → *Australia* 21°10S 137°33E **62**
Milo *U.S.A.* 45°15N 68°59W **81**
Milon, Akra *Greece* 36°15N 28°11E **25**
Milos *Greece* 36°44N 24°25E **23**
Milparinka *Australia* 29°46S 141°57E **63**
Milpitas *U.S.A.* 37°26N 121°55W **78**
Milton *N.S., Canada* 44°4N 64°45W **73**
Milton *Ont., Canada* 43°31N 79°53W **82**
Milton *N.Z.* 46°7S 169°59E **59**
Milton *Calif., U.S.A.* 38°3N 120°51W **78**
Milton *Fla., U.S.A.* 30°38N 87°3W **85**
Milton *Pa., U.S.A.* 41°1N 76°51W **82**
Milton *Vt., U.S.A.* 44°38N 73°7W **83**
Milton-Freewater *U.S.A.* 45°56N 118°23W **76**
Milton Keynes *U.K.* 52°1N 0°44W **13**
Milton Keynes □ *U.K.* 52°1N 0°44W **1**
Milverton *Canada* 43°34N 80°55W **82**
Milwaukee *U.S.A.* 43°2N 87°54W **80**
Milwaukee Deep *Atl. Oc.* 19°50N 68°0W **8**
Milwaukie *U.S.A.* 45°26N 122°38W **7**
Mimili *Australia* 27°0S 132°42E **61**
Min Jiang → *Fujian, China* 26°0N 119°35E **33**
Min Jiang → *Sichuan, China* 28°45N 104°40E **32**
Min Xian *China* 34°25N 104°5E **34**
Minā' Jabal 'Alī *U.A.E.* 25°2N 55°8E **4**
Mina Pirquitas *Argentina* 22°40S 66°30W **94**
Mīnā Su'ud *Si. Arabia* 28°45N 48°28E **45**
Mīnā'al Aḥmadī *Kuwait* 29°5N 48°10E **45**
Minago → *Canada* 54°33N 98°59W **71**
Minaki *Canada* 49°59N 94°40W **71**
Minamata *Japan* 32°10N 130°30E **31**
Minami-Arapusa △ *Japan* 35°30N 138°9E **31**
Minami-Tori-Shima *Pac. Oc.* 24°20N 153°58E **64**
Minas *Uruguay* 34°20S 55°10W **95**
Minas, Sierra de las *Guatemala* 15°9N 89°31W **88**
Minas Basin *Canada* 45°20N 64°12W **73**
Minas Gerais □ *Brazil* 18°50S 46°0W **93**
Minatitlán *Mexico* 17°59N 94°31W **87**
Minbu *Burma* 20°10N 94°52E **41**
Minchinabad *Pakistan* 30°10N 73°34E **42**
Mindanao *Phil.* 8°0N 125°0E **37**
Mindanao Sea = Bohol Sea *Phil.* 9°0N 124°0E **37**

ndanao Trench
Pac. Oc. 12°0N 126°6E 37 B7
ndelo C. Verde Is. 16°24N 25°0W 50 b
nden Canada 44°55N 78°43W 82 B6
nden Germany 52°17N 8°55E 16 B5
nden La., U.S.A. 32°37N 93°17W 84 E8
nden Nev., U.S.A. 38°57N 119°46W 78 G7
ndibungu = Billiluna
ustralia 19°37S 127°41E 60 C4
ndiptana Indonesia 5°55S 140°22E 37 F10
ndoro Phil. 13°0N 121°0E 37 B6
ndoro Str. Phil. 12°30N 120°30E 37 B6
ne Japan 34°12N 131°7E 31 G5
nehead U.K. 51°12N 3°29W 13 F4
neola N.Y., U.S.A. 40°44N 73°38W 83 F11
neola Tex., U.S.A. 32°40N 95°29W 84 E7
neral King U.S.A. 36°27N 118°36W 78 J8
neral Wells U.S.A. 32°48N 98°7W 84 E5
ners Bay Canada 44°49N 78°46W 82 B6
nersville U.S.A. 40°41N 76°16W 83 F8
nerva N.Y., U.S.A. 43°47N 73°59W 83 C11
nerva Ohio, U.S.A. 40°44N 81°6W 82 F3
netto U.S.A. 43°24N 76°28W 83 C8
nfeng China 37°4N 82°46E 32 D5
ngäçevir Su Anbarı
zerbaijan 40°57N 46°50E 19 F8
ngan Canada 50°20N 64°0W 73 B7
ngechaurskoye Vdkhr. =
lingäçevir Su Anbarı
zerbaijan 40°57N 46°50E 19 F8
ngela Australia 19°52S 146°38E 62 B4
ngenew Australia 29°12S 115°21E 61 E2
ngera Cr. →
ustralia 20°38S 137°45E 62 C2
ngin Burma 22°50N 94°30E 41 H19
ngo Junction U.S.A. 40°19N 80°37W 82 F4
ngora Pakistan 34°48N 72°22E 43 B5
ngteke Daban = Mintaka Pass
akistan 37°0N 74°58E 43 A6
ngyuegue China 43°2N 128°50E 35 C15
nhe China 36°9N 102°45E 32 D9
nho = Miño → Spain 41°52N 8°40W 21 A2
nho Portugal 41°25N 8°20W 21 B1
nidoka U.S.A. 42°45N 113°29W 76 E7
nigwal, L. Australia 29°31S 123°14E 61 E3
nilya → Australia 23°45S 114°0E 61 D1
nilya Roadhouse
ustralia 23°55S 114°0E 61 D1
nipi L. Canada 52°25N 60°45W 73 B7
njilang Australia 11°8S 132°33E 60 B5
nk L. Canada 61°54N 117°40W 70 A5
nna Nigeria 9°37N 6°30E 50 G7
nneapolis Kans., U.S.A. 39°8N 97°42W 80 F5
nneapolis Minn.,
.S.A. 44°57N 93°16W 80 C7
nnedosa Canada 50°14N 99°50W 71 C9
nnesota □ U.S.A. 46°0N 94°15W 80 B6
nnesota → U.S.A. 44°54N 93°9W 80 C7
nnewaukan U.S.A. 48°4N 99°15W 80 A4
nnipa Australia 32°51S 135°9E 63 E2
nnitaki L. Canada 49°57N 92°10W 72 C1
no Japan 35°32N 136°55E 31 G8
no → Spain 41°52N 8°40W 21 A2
norca = Menorca Spain 40°0N 4°0E 24 B11
not U.S.A. 48°14N 101°18W 80 A3
nqin China 38°38N 103°20E 34 E2
nsk Belarus 53°52N 27°30E 17 B14
nsk Mazowiecki
oland 52°10N 21°33E 17 B11
ntabie Australia 27°15S 133°7E 63 D1
ntaka Pass Pakistan 37°0N 74°58E 43 A6
nto Canada 46°5N 66°5W 73 C6
nto, L. Canada 57°13N 75°0W 72 A5
nton Canada 49°10N 104°35W 71 D8
nturn U.S.A. 39°35N 106°26W 76 G10
nudasht Iran 37°17N 56°7E 45 B8
nusinsk Russia 53°43N 91°20E 29 D10
nutang India 28°15N 96°30E 41 E20
nvoul Gabon 2°9N 12°8E 52 D2
nya Konka = Gongga Shan
hina 29°40N 101°55E 32 F9
nzhong China 22°37N 113°30E 33 a
quelon Canada 49°25N 76°27W 72 C4
quelon St-P. & M. 47°8N 56°22W 73 C8
Kūh Iran 26°22N 58°55E 45 E8
Shahdād Iran 26°15N 58°29E 45 E8
a Italy 45°26N 12°8E 22 B5
a por vos Cay
ahamas 22°9N 74°30W 89 B5
abello, Kolpos Greece 35°10N 25°50E 25 D7
ador-Rio Azul △
uatemala 17°45N 89°50W 88 C2
aj India 16°50N 74°45E 40 L9
am Shah Pakistan 33°0N 70°2E 42 C4
amar Argentina 38°15S 57°50W 94 D4
amar Mozam. 23°50S 35°35E 57 C6
amichi Canada 47°2N 65°28W 73 C6
amichi B. Canada 47°15N 65°0W 73 C7
anda Brazil 20°10S 56°15W 93 H7
anda → Brazil 19°25S 57°20W 92 G7
anda de Ebro Spain 42°41N 2°57W 21 A4
anda do Douro
ortugal 41°30N 6°16W 21 B2
andópolis Brazil 21°9S 51°6W 95 A5
ango Malawi 13°32S 34°58E 55 E3
ani Australia 21°8S 148°53E 62 b
assol Brazil 20°46S 49°28W 95 A6
bāṭ Oman 17°0N 54°45E 47 D5
es Greece 35°4N 24°56E 25 D6
i Malaysia 4°23N 113°59E 36 D4
iam Vale Australia 24°20S 151°33E 62 C5
im, L. S. Amer. 32°45S 52°50W 95 C5
iuwung Gajerrong ❂
ustralia 15°0S 128°45E 60 C4
jāveh Iran 29°1N 61°30E 45 D9
nyy Antarctica 66°50S 92°30E 5 C14
nyy Russia 62°33N 113°53E 29 C12
okhan Pakistan 27°46N 68°6E 42 F3
ond L. Canada 55°6N 102°47W 71 B8
pur Pakistan 33°32N 73°56E 43 C5
Mirpur Batoro Pakistan 24°44N 68°16E 42 G3
Mirpur Bibiwari Pakistan 28°33N 67°44E 42 E2
Mirpur Khas Pakistan 25°30N 69°0E 42 G3
Mirpur Sakro Pakistan 24°33N 67°41E 42 G2
Mirs Bay = Tai Pang Wan
China 22°33N 114°24E 33 a
Mirtağ Turkey 38°23N 41°56E 44 B4
Mirtoo Sea Greece 37°0N 23°20E 23 F10
Miryang S. Korea 35°31N 128°44E 35 G15
Mirzapur India 25°10N 82°34E 43 G10
Mirzapur-cum-Vindhyachal =
Mirzapur India 25°10N 82°34E 43 G10
Misantla Mexico 19°56N 96°50W 87 D5
Misawa Japan 40°41N 141°24E 30 D10
Miscou I. Canada 47°57N 64°31W 73 C7
Mish'āb, Ra's al
Si. Arabia 28°15N 48°43E 45 D6
Mishamo Tanzania 5°41S 30°41E 54 D3
Mishan China 45°37N 131°48E 33 B5
Mishawaka U.S.A. 41°40N 86°11W 80 E10
Mishima Japan 35°10N 138°52E 31 G9
Misión Mexico 32°6N 116°53W 79 N10
Misiones □ Argentina 27°0S 55°0W 95 B5
Misiones □ Paraguay 27°0S 56°0W 94 B4
Miskah Si. Arabia 24°49N 42°56E 44 E4
Miskitos, Cayos Nic. 14°26N 82°50W 88 D3
Miskolc Hungary 48°7N 20°50E 17 D11
Misoke
Dem. Rep. of the Congo 0°42S 28°2E 54 C2
Misool Indonesia 1°52S 130°10E 37 E8
Mişr = Egypt ■ Africa 28°0N 31°0E 51 C12
Mişrātah Libya 32°24N 15°3E 51 B9
Missanabie Canada 48°20N 84°6W 72 C3
Missinaibi → Canada 50°43N 81°29W 72 B3
Missinaibi L. Canada 48°23N 83°40W 72 C3
Mission Canada 49°10N 122°15W 70 D4
Mission S. Dak., U.S.A. 43°18N 100°39W 80 D3
Mission Tex., U.S.A. 26°13N 98°20W 84 H5
Mission Beach Australia 17°53S 146°6E 62 B4
Mission Viejo U.S.A. 33°36N 117°40W 79 M9
Missisa L. Canada 52°20N 85°7W 72 B2
Missisicabi → Canada 51°14N 79°31W 72 B4
Mississagi → Canada 46°15N 83°9W 72 C3
Mississauga Canada 43°32N 79°35W 82 C5
Mississippi □ U.S.A. 33°0N 90°0W 85 E10
Mississippi → U.S.A. 29°9N 89°15W 85 G10
Mississippi L. Canada 45°5N 76°10W 83 A8
Mississippi River Delta
U.S.A. 29°10N 89°15W 85 G10
Mississippi Sd. U.S.A. 30°20N 89°0W 85 F10
Missoula U.S.A. 46°52N 114°1W 76 C6
Missouri □ U.S.A. 38°25N 92°30W 80 F7
Missouri → U.S.A. 38°49N 90°7W 80 F8
Missouri City U.S.A. 29°37N 95°32W 84 G7
Missouri Valley U.S.A. 41°34N 95°53W 80 E6
Mist U.S.A. 45°59N 123°15W 78 E3
Mistassibi → Canada 48°53N 72°13W 73 B5
Mistassini Canada 48°53N 72°12W 73 C5
Mistassini → Canada 48°42N 72°20W 73 C5
Mistassini, L. Canada 51°0N 73°30W 72 B5
Mistastin L. Canada 55°57N 63°20W 73 A7
Mistinibi, L. Canada 55°56N 64°17W 73 A7
Mistissini Canada 48°53N 72°12W 73 C5
Misty L. Canada 58°53N 101°40W 71 B8
Misurata = Mişrātah
Libya 32°24N 15°3E 51 B9
Mitande Mozam. 14°6S 35°58E 55 E4
Mitchell Australia 26°29S 147°58E 63 D4
Mitchell Canada 43°28N 81°12W 82 C3
Mitchell Nebr., U.S.A. 41°57N 103°49W 80 E2
Mitchell Oreg., U.S.A. 44°34N 120°9W 76 D3
Mitchell S. Dak., U.S.A. 43°43N 98°2W 80 D4
Mitchell → Australia 15°12S 141°35E 62 B3
Mitchell, Mt. U.S.A. 35°46N 82°16W 85 D13
Mitchell-Alice Rivers △
Australia 15°28S 142°5E 62 B3
Mitchell Ra. Australia 12°49S 135°36E 62 A2
Mitchelstown Ireland 52°15N 8°16W 10 D3
Mitha Tiwana Pakistan 32°13N 72°6E 42 C5
Mithi Pakistan 24°44N 69°48E 42 G3
Mithrao Pakistan 27°28N 69°40E 42 F3
Mitilini Greece 39°6N 26°35E 23 E12
Mitla Pass = Mamarr Mitlā
Egypt 30°2N 32°54E 46 E1
Mito Japan 36°20N 140°30E 31 F10
Mitrovica = Kosovska Mitrovica
Kosovo 42°54N 20°52E 23 C9
Mitsamiouli Comoros Is. 11°20S 43°16E 53 a
Mitsinjo Madag. 16°1S 45°52E 57 B8
Mitsiwa Eritrea 15°35N 39°25E 47 D2
Mitsukaidō Japan 36°1N 139°59E 31 F9
Mittagong Australia 34°28S 150°29E 63 E5
Mittimatalik = Pond Inlet
Canada 72°40N 77°0W 69 C16
Mitú Colombia 1°15N 70°13W 92 C4
Mitumba Tanzania 7°8S 31°2E 54 D3
Mitumba, Mts.
Dem. Rep. of the Congo 7°0S 27°30E 54 D2
Mitwaba
Dem. Rep. of the Congo 8°2S 27°17E 55 D2
Mityana Uganda 0°23N 32°2E 54 B3
Mixteco → Mexico 18°11N 98°30W 87 D5
Miyagi □ Japan 38°15N 140°45E 30 E10
Miyāh, W. el → Syria 34°44N 39°57E 44 C3
Miyake-Jima Japan 34°5N 139°30E 31 G9
Miyako Japan 39°40N 141°59E 30 E10
Miyako-Jima Japan 24°45N 125°20E 31 M2
Miyako-Rettō Japan 24°24N 125°0E 31 M2
Miyakonojō Japan 31°40N 131°5E 31 J5
Miyani India 21°50N 69°26E 42 J3
Miyanoura-Dake Japan 30°20N 130°31E 31 J5
Miyazaki Japan 31°56N 131°30E 31 J5
Miyazaki □ Japan 32°30N 131°30E 31 H5
Miyazu Japan 35°35N 135°10E 31 G7
Miyet, Bahr el = Dead Sea
Asia 31°30N 35°30E 46 D4
Miyoshi Japan 34°48N 132°51E 31 G6
Miyun China 40°28N 116°50E 34 D9
Miyun Shuiku China 40°30N 117°0E 35 D9
Mizdah Libya 31°30N 13°0E 51 B8
Mizen Hd. Cork, Ireland 51°27N 9°50W 10 E2
Mizen Hd. Wicklow, Ireland 52°51N 6°4W 10 D5
Mizhi China 37°47N 110°12E 34 F6
Mizoram □ India 23°30N 92°40E 41 H18
Mizpe Ramon Israel 30°34N 34°49E 46 E3
Mizuho Antarctica 70°30S 41°0E 5 D5
Mizusawa Japan 39°8N 141°8E 30 E10
Mjölby Sweden 58°20N 15°10E 9 G16
Mjøsa Norway 60°40N 11°0E 8 F14
Mkata Tanzania 5°45S 38°20E 54 D4
Mkhaya △ Swaziland 26°34S 31°45E 57 D5
Mkhuze △ S. Africa 29°27S 29°30E 57 D4
Mkokotoni Tanzania 5°55S 39°15E 54 D4
Mkomazi Tanzania 4°40S 38°7E 54 C4
Mkomazi → S. Africa 30°12S 30°50E 57 E5
Mkomazi △ Tanzania 4°4S 30°2E 54 C3
Mkulwe Tanzania 8°37S 32°20E 55 D3
Mkumbi, Ras Tanzania 7°38S 39°55E 54 D4
Mkushi Zambia 14°25S 29°15E 55 E2
Mkushi River Zambia 13°32S 29°45E 55 E2
Mkuze S. Africa 27°10S 32°0E 57 D5
Mladá Boleslav
Czech Rep. 50°27N 14°53E 16 C8
Mlala Hills Tanzania 6°50S 31°40E 54 D3
Mlange = Mulanje, Mt.
Malawi 16°2S 35°33E 55 F4
Mława Poland 53°9N 20°25E 17 B11
Mlawula △ Swaziland 26°12S 32°2E 57 D5
Mljet Croatia 42°43N 17°30E 22 C7
Mmabatho S. Africa 25°49S 25°30E 56 D4
Mo i Rana Norway 66°20N 14°7E 8 C16
Moa Australia 10°11S 142°16E 62 a
Moa Cuba 20°40N 74°56W 89 B4
Moa Indonesia 8°0S 128°0E 37 F7
Moa → S. Leone 6°59N 11°36W 50 G3
Moab U.S.A. 38°35N 109°33W 76 G9
Moala Fiji 18°36S 179°53E 59 a
Moama Australia 36°7S 144°46E 63 F3
Moamba Mozam. 25°36S 32°15E 57 D5
Moapa U.S.A. 36°40N 114°37W 79 J12
Moate Ireland 53°24N 7°44W 10 C4
Moba Dem. Rep. of the Congo 7°0S 29°48E 54 D2
Mobārakābād Iran 28°24N 53°20E 45 D7
Mobaye C.A.R. 4°25N 21°5E 52 D4
Moberly U.S.A. 39°25N 92°26W 80 F7
Moberly Lake Canada 55°50N 121°44W 70 B4
Mobile U.S.A. 30°41N 88°3W 85 F10
Mobile B. U.S.A. 30°30N 88°0W 85 F11
Mobridge U.S.A. 45°32N 100°26W 80 C3
Moc Chau Vietnam 20°50N 104°38E 38 B5
Moc Hoa Vietnam 10°46N 105°56E 39 G5
Mocabe Kasari
Dem. Rep. of the Congo 9°58S 26°12E 55 D2
Moçambique = Mozambique ■
Africa 19°0S 35°0E 55 F4
Moçambique Mozam. 15°3S 40°42E 55 F5
Mocanaqua U.S.A. 41°9N 76°8W 83 E8
Moce Fiji 18°40S 178°29W 59 a
Mochima △ Venezuela 10°30N 64°5W 89 D7
Mochos Greece 35°16N 25°27E 25 D7
Mochudi Botswana 24°27S 26°7E 56 C4
Mocimboa da Praia
Mozam. 11°25S 40°20E 55 E5
Moclips U.S.A. 47°14N 124°13W 78 C2
Mocoa Colombia 1°7N 76°35W 92 C3
Mococa Brazil 21°28S 47°0W 95 A6
Mocorito Mexico 25°29N 107°55W 86 B3
Moctezuma Mexico 29°48N 109°42W 86 B3
Moctezuma → Mexico 21°59N 98°34W 87 C5
Mocuba Mozam. 16°54S 36°57E 55 F4
Modane France 45°12N 6°40E 20 D7
Modasa India 23°30N 73°21E 42 H5
Modder → S. Africa 29°2S 24°37E 56 D3
Modderrivier S. Africa 29°2S 24°38E 56 D3
Módena Italy 44°40N 10°55E 22 B4
Modena U.S.A. 37°48N 113°56W 77 H7
Modesto U.S.A. 37°39N 121°0W 78 H6
Módica Italy 36°52N 14°46E 22 F6
Modimolle S. Africa 24°42S 28°22E 57 C4
Modjadjiskloof S. Africa 23°42S 30°10E 57 C5
Moe Australia 38°12S 146°19E 63 F4
Moebase Mozam. 17°3S 38°41E 55 F4
Moengo Suriname 5°45N 54°20W 93 B8
Moffat U.K. 55°21N 3°27W 11 F5
Moga India 30°48N 75°8E 42 D6
Mogadishu = Muqdisho
Somali Rep. 2°2N 45°25E 47 G4
Mogador = Essaouira
Morocco 31°32N 9°42W 50 B4
Mogalakwena →
S. Africa 22°38S 28°40E 57 C4
Mogami-Gawa →
Japan 38°45N 140°0E 30 E10
Mogán Canary Is. 27°53N 15°43W 24 G4
Mogaung Burma 25°20N 97°0E 41 G20
Mogi-Mirim Brazil 22°29S 47°0W 93 H9
Mogilev = Mahilyow
Belarus 53°55N 30°18E 17 B16
Mogilev-Podolskiy = Mohyliv-
Podilskyy Ukraine 48°26N 27°48E 17 D14
Mogincual Mozam. 15°35S 40°25E 55 F5
Mogocha Russia 53°40N 119°50E 29 D12
Mogok Burma 23°0N 96°40E 41 H20
Mogollon Rim U.S.A. 34°10N 110°50W 77 J8
Mogumber Australia 31°2S 116°3E 61 F2
Mogwadi → S. Africa 23°4S 29°36E 57 C4
Mohács Hungary 45°58N 18°41E 17 F10
Mohales Hoek Lesotho 30°7S 27°26E 56 E4
Mohall U.S.A. 48°46N 101°31W 80 A3
Moḥammadābād Iran 37°52N 59°5E 45 B8
Mohammedia Morocco 33°44N 7°21W 50 B4
Mohana → India 24°43N 85°0E 43 G11
Mohanlalganj India 26°41N 80°58E 43 F9
Mohave, L. U.S.A. 35°12N 114°34W 79 K12
Mohawk → U.S.A. 42°47N 73°41W 83 D11
Mohéli Comoros Is. 12°20S 43°40E 53 a
Mohenjodaro Pakistan 27°19N 68°7E 42 F3
Moher, Cliffs of Ireland 52°58N 9°27W 10 D2
Mohicanville Res. U.S.A. 40°45N 82°9W 82 F2
Mohns Ridge Arctic 72°30N 5°0W 4 B7
Mohoro Tanzania 8°6S 39°8E 54 D4
Mohsenābād Iran 36°40N 59°35E 45 B8
Mohyliv-Podilskyy
Ukraine 48°26N 27°48E 17 D14
Moidart, L. U.K. 56°47N 5°52W 11 E3
Moira → Canada 44°21N 77°24W 82 B7
Moisaküla Estonia 58°3N 25°12E 9 G21
Moisie Canada 50°12N 66°1W 73 B6
Moisie → Canada 50°14N 66°5W 73 B6
Mojave U.S.A. 35°3N 118°10W 79 K8
Mojave △ U.S.A. 35°7N 115°32W 79 K11
Mojave Desert U.S.A. 35°0N 116°30W 79 L10
Moji das Cruzes Brazil 23°31S 46°11W 95 A6
Moji-Guaçu → Brazil 20°53S 48°10W 95 A6
Mojo Bolivia 21°48S 65°33W 94 A2
Mojokerto Indonesia 7°28S 112°26E 37 G15
Mokai N.Z. 38°32S 175°56E 59 C5
Mokambo
Dem. Rep. of the Congo 12°25S 28°20E 55 E2
Mokameh India 25°24N 85°55E 43 G11
Mokau N.Z. 38°42S 174°39E 59 C5
Mokelumne → U.S.A. 38°13N 121°28W 78 G5
Mokelumne Hill
U.S.A. 38°18N 120°43W 78 G6
Mokhotlong Lesotho 29°22S 29°2E 57 D4
Mokoan, L. Australia 36°27S 146°5E 63 C4
Mokokchung India 26°15N 94°30E 41 F19
Mokolo → S. Africa 23°14S 27°43E 57 C4
Mokopane S. Africa 24°10S 28°55E 57 C4
Mokpo S. Korea 34°50N 126°25E 35 G14
Mokra Gora Europe 42°50N 20°30E 23 C9
Mol Belgium 51°11N 5°5E 15 C5
Molchanovo Russia 57°40N 83°50E 28 D9
Mold U.K. 53°9N 3°8W 12 D4
Moldavia = Moldova ■
Europe 47°0N 28°0E 17 E15
Moldavia Romania 46°30N 27°0E 17 E14
Molde Norway 62°45N 7°9E 8 E12
Moldova ■ Europe 47°0N 28°0E 17 E15
Moldoveanu, Vf.
Romania 45°36N 24°45E 17 F13
Mole → U.K. 51°24N 0°21W 13 F7
Molepolole Botswana 24°28S 25°28E 56 C4
Molfetta Italy 41°12N 16°36E 22 D7
Moline U.S.A. 41°30N 90°31W 80 E8
Molinos Argentina 25°28S 66°15W 94 B2
Moliro
Dem. Rep. of the Congo 8°12S 30°30E 54 D3
Mollendo Peru 17°0S 72°0W 92 G4
Mollerin, L. Australia 30°30S 117°35E 61 F2
Molo Kenya 0°15S 35°44E 54 C4
Molodechno = Maladzyechna
Belarus 54°20N 26°50E 17 A14
Molodezhnaya Antarctica 67°40S 45°51E 5 C9
Moloka'i U.S.A. 21°8N 157°0W 75 L8
Molokai Fracture Zone
Pac. Oc. 28°0N 125°0W 65 E15
Molong Australia 33°5S 148°54E 63 E4
Molopo → Africa 28°30S 20°12E 56 D3
Molson L. Canada 54°22N 96°40W 71 C9
Molteno S. Africa 31°22S 26°22E 56 E4
Molu Indonesia 6°45S 131°40E 37 F8
Molucca Sea Indonesia 0°0 125°0E 37 E6
Moluccas = Maluku
Indonesia 1°0S 127°0E 37 E7
Moma
Dem. Rep. of the Congo 1°35S 23°52E 54 C1
Moma Mozam. 16°47S 39°4E 55 F4
Mombasa Kenya 4°3S 39°40E 54 C4
Mombetsu Japan 44°21N 143°22E 30 B11
Momchilgrad Bulgaria 41°33N 25°23E 23 D11
Momi Dem. Rep. of the Congo 1°42S 27°0E 54 C2
Mompós Colombia 9°14N 74°26W 92 B4
Møn Denmark 54°57N 12°20E 9 J15
Mon □ Burma 16°0N 97°30E 41 L20
Mona, Canal de la = Mona
Passage W. Indies 18°30N 67°45W 89 C6
Mona, Isla Puerto Rico 18°5N 67°54W 89 C6
Mona, Pta. Costa Rica 9°37N 82°36W 88 E3
Mona Passage W. Indies 18°30N 67°45W 89 C6
Monaca U.S.A. 40°41N 80°17W 82 F4
Monaco ■ Europe 43°46N 7°23E 20 E7
Monadhliath Mts. U.K. 57°10N 4°4W 11 D4
Monadnock, Mt. U.S.A. 42°52N 72°7W 83 D12
Monaghan Ireland 54°15N 6°57W 10 B5
Monaghan □ Ireland 54°11N 6°56W 10 B5
Monahans U.S.A. 31°36N 102°54W 84 F3
Monapo Mozam. 14°56S 40°19E 55 E5
Monar, L. U.K. 57°26N 5°8W 11 D3
Monarch Mt. Canada 51°55N 125°57W 70 C3
Monashee Mts. Canada 51°0N 118°43W 70 C5
Monasterevin Ireland 53°8N 7°4W 10 C4
Monastir = Bitola
Macedonia 41°1N 21°20E 23 D9
Monastir Tunisia 35°50N 10°49E 51 A8
Monbetsu Japan 42°30N 142°10E 30 C11
Moncayo, Sierra del
Spain 41°48N 1°50W 21 B5
Monchegorsk Russia 67°54N 32°58E 8 C25
Mönchengladbach
Germany 51°11N 6°27E 16 C4
Monchique Portugal 37°19N 8°38W 21 D1
Moncks Corner U.S.A. 33°12N 80°1W 85 E14
Monclova Mexico 26°54N 101°25W 86 B4
Moncton Canada 46°7N 64°51W 73 C7
Mondego → Portugal 40°9N 8°52W 21 B1
Mondeodo Indonesia 3°34S 122°9E 37 E6
Mondovì Italy 44°23N 7°49E 20 D7
Mondrain I. Australia 34°9S 122°14E 61 F3
Moneague Jamaica 18°16N 77°7W 88 a
Moneron, Ostrov
Russia 46°15N 141°16E 30 A10
Monessen U.S.A. 40°9N 79°54W 82 F5
Monett U.S.A. 36°55N 93°55W 80 G7
Moneymore U.K. 54°41N 6°40W 10 B5
Monforte de Lemos Spain 42°31N 7°33W 21 A2
Mong Hpayak Burma 20°52N 99°55E 38 B2
Mong Hsat Burma 20°31N 99°15E 38 B2
Mong Hsu Burma 21°54N 98°30E 41 J21
Mong Kung Burma 21°35N 97°35E 41 J20
Mong Nai Burma 20°32N 97°46E 41 J20
Mong Ton Burma 20°17N 98°45E 41 J21
Mong Yai Burma 22°21N 98°3E 41 H21
Mong Yang Burma 21°50N 99°41E 38 B2
Mongalla Sudan 5°8N 31°42E 51 G12
Mongers, L. Australia 29°25S 117°5E 61 E2
Monghyr = Munger
India 25°23N 86°30E 43 G12
Mongibello = Etna Italy 37°50N 14°55E 22 F6
Mongo Chad 12°14N 18°43E 51 F9
Mongolia ■ Asia 47°0N 103°0E 32 B9
Mongolia, Plateau of
Asia 45°0N 105°0E 26 D12
Mongu Zambia 15°16S 23°12E 53 H4
Mônguà Angola 16°43S 15°20E 56 B2
Monifieth U.K. 56°30N 2°48W 11 E6
Monkey Bay Malawi 14°7S 35°1E 55 E4
Monkey Mia Australia 25°48S 113°43E 61 E1
Monkey River Belize 16°22N 88°29W 87 D7
Monkland Canada 45°11N 74°52W 83 A10
Monkoto
Dem. Rep. of the Congo 1°38S 20°35E 52 E4
Monkton Canada 43°35N 81°5W 82 C3
Monmouth U.K. 51°48N 2°42W 13 F5
Monmouth Ill., U.S.A. 40°55N 90°39W 80 E8
Monmouth Oreg.,
U.S.A. 44°51N 123°14W 76 D2
Monmouthshire □ U.K. 51°48N 2°54W 13 F5
Mono, Pta. Nic. 12°0N 83°30W 88 D3
Mono L. U.S.A. 38°1N 119°1W 78 H7
Monolith U.S.A. 35°7N 118°22W 79 K8
Monolithos Greece 36°7N 27°45E 25 C9
Monongahela U.S.A. 40°12N 79°56W 82 F5
Monópoli Italy 40°57N 17°18E 22 D7
Monos I. Trin. & Tob. 10°42N 61°44W 93 K15
Monroe Ga., U.S.A. 33°47N 83°43W 85 E13
Monroe La., U.S.A. 32°30N 92°7W 84 E8
Monroe Mich., U.S.A. 41°55N 83°24W 81 E12
Monroe N.C., U.S.A. 34°59N 80°33W 85 D14
Monroe N.Y., U.S.A. 41°20N 74°11W 83 E10
Monroe Utah, U.S.A. 38°38N 112°7W 76 G7
Monroe Wash., U.S.A. 47°51N 121°58W 78 C5
Monroe Wis., U.S.A. 42°36N 89°38W 80 D9
Monroe City U.S.A. 39°39N 91°44W 80 F8
Monroeton U.S.A. 41°43N 76°29W 83 E8
Monroeville Ala.,
U.S.A. 31°31N 87°20W 85 F11
Monroeville Pa., U.S.A. 40°26N 79°45W 82 F5
Monrovia Liberia 6°18N 10°47W 50 G3
Mons Belgium 50°27N 3°58E 15 D3
Monse Indonesia 4°7S 123°15E 37 E6
Mont-de-Marsan France 43°54N 0°31W 20 E3
Mont-Joli Canada 48°37N 68°10W 73 C6
Mont-Laurier Canada 46°35N 75°30W 72 C4
Mont-Louis Canada 49°15N 65°44W 73 C6
Mont-St-Michel, Le
France 48°40N 1°30W 20 B3
Mont-Tremblant △
Canada 46°30N 74°30W 72 C5
Montagne d'Ambre △
Madag. 12°37S 49°8E 57 A8
Montagu S. Africa 33°45S 20°8E 56 E3
Montagu I. Antarctica 58°25S 26°20W 5 B1
Montague Canada 46°10N 62°39W 73 C7
Montague, I. Mexico 31°45N 114°48W 86 A2
Montague I. U.S.A. 60°0N 147°30W 74 D10
Montague Ra. Australia 27°15S 119°30E 61 E2
Montague Sd. Australia 14°28S 125°20E 60 B4
Montalbán Spain 40°50N 0°45W 21 B5
Montalvo U.S.A. 34°15N 119°12W 79 L7
Montana Bulgaria 43°27N 23°16E 23 C10
Montaña Peru 6°0S 73°0W 92 E4
Montana □ U.S.A. 47°0N 110°0W 76 C9
Montaña Clara, I.
Canary Is. 29°17N 13°33W 24 E6
Montargis France 47°59N 2°43E 20 C5
Montauban France 44°2N 1°21E 20 D4
Montauk U.S.A. 41°3N 71°57W 83 E13
Montauk Pt. U.S.A. 41°4N 71°51W 83 E13
Montbéliard France 47°31N 6°48E 20 C7
Montceau-les-Mines
France 46°40N 4°23E 20 C6
Montclair U.S.A. 40°49N 74°12W 83 F10
Monte Albán Mexico 17°2N 96°46W 87 D5
Monte Alegre Brazil 2°0S 54°0W 93 D8
Monte Azul Brazil 15°9S 42°53W 93 G10
Monte-Carlo Monaco 43°44N 7°25E 20 E7
Monte Caseros
Argentina 30°10S 57°50W 94 C4
Monte Comán Argentina 34°40S 67°53W 94 C2
Monte Cristi Dom. Rep. 19°52N 71°39W 89 C5
Monte Lindo →
Paraguay 23°56S 57°12W 94 A4
Monte Patria Chile 30°42S 70°58W 94 C1
Monte Quemado
Argentina 25°53S 62°41W 94 B3
Monte Rio U.S.A. 38°28N 123°0W 78 G4
Monte Santu, C. di Italy 40°5N 9°44E 22 D3
Monte Vista U.S.A. 37°35N 106°9W 77 H10
Monteagudo Argentina 27°14S 54°8W 95 B5
Montebello Canada 45°40N 74°55W 72 C5
Montebello Is. Australia 20°30S 115°45E 60 D2
Montecarlo Argentina 26°34S 54°47W 95 B5
Montecito U.S.A. 34°26N 119°40W 79 L7
Montecristo Italy 42°20N 10°19E 22 C4
Montego Bay Jamaica 18°28N 77°55W 88 a
Montélimar France 44°33N 4°45E 20 D6
Montello U.S.A. 43°48N 89°20W 80 D9
Montemorelos Mexico 25°12N 99°49W 87 B5
Montenegro Brazil 29°39S 51°29W 95 B5
Montenegro ■ Europe 42°40N 19°20E 23 C8
Montepuez Mozam. 13°8S 38°59E 55 E4
Montepuez → Mozam. 12°32S 40°27E 55 E5
Monterey U.S.A. 36°37N 121°55W 78 J5
Monterey B. U.S.A. 36°45N 122°0W 78 J5
Montería Colombia 8°46N 75°53W 92 B3
Monteros Argentina 27°11S 65°30W 94 B2
Monterrey Mexico 25°40N 100°19W 86 B4
Montes Azules △ Mexico 16°21N 91°3W 87 D6
Montes Claros Brazil 16°30S 43°50W 93 G10

Montesano U.S.A. 46°59N 123°36W **78** D3
Montesilvano Italy 42°29N 14°8E **22** C6
Montevideo Uruguay 34°50S 56°11W **95** C4
Montevideo U.S.A. 44°57N 95°43W **80** C6
Montezuma U.S.A. 41°35N 92°32W **80** E7
Montezuma Castle ○ U.S.A. 34°39N 111°45W **77** J8
Montgomery U.K. 52°34N 3°8W **13** E4
Montgomery Ala., U.S.A. 32°23N 86°19W **85** E11
Montgomery Pa., U.S.A. 41°10N 76°53W **82** E8
Montgomery W. Va., U.S.A. 38°11N 81°19W **81** F13
Montgomery City U.S.A. 38°59N 91°30W **80** F8
Monticello Ark., U.S.A. 33°38N 91°47W **84** E9
Monticello Fla., U.S.A. 30°33N 83°52W **85** F13
Monticello Ind., U.S.A. 40°45N 86°46W **80** E10
Monticello Iowa, U.S.A. 42°15N 91°12W **80** D8
Monticello Ky., U.S.A. 36°50N 84°51W **81** G11
Monticello Minn., U.S.A. 45°18N 93°48W **80** C7
Monticello Miss., U.S.A. 31°33N 90°7W **85** F9
Monticello N.Y., U.S.A. 41°39N 74°42W **83** E10
Monticello Utah, U.S.A. 37°52N 109°21W **77** H9
Montijo Portugal 38°41N 8°54W **21** C1
Montilla Spain 37°36N 4°40W **21** D3
Montluçon France 46°22N 2°36E **20** C5
Montmagny Canada 46°58N 70°34W **73** C5
Montmartre Canada 50°14N 103°27W **71** C8
Montmorillon France 46°26N 0°50E **20** C4
Monto Australia 24°52S 151°6E **62** C5
Montongbuwoh Indonesia 8°33S 116°4E **37** K19
Montoro Spain 38°1N 4°27W **21** C3
Montour Falls U.S.A. 42°21N 76°51W **82** D8
Montoursville U.S.A. 41°15N 76°55W **82** E8
Montpelier Idaho, U.S.A. 42°19N 111°18W **76** E8
Montpelier Vt., U.S.A. 44°16N 72°35W **83** B12
Montpellier France 43°37N 3°52E **20** E5
Montréal Canada 45°30N 73°33W **83** A11
Montreal → Canada 47°14N 84°39W **72** C3
Montreal L. Canada 54°20N 105°45W **71** C7
Montreal Lake Canada 54°3N 105°46W **71** C7
Montreux Switz. 46°26N 6°55E **20** C7
Montrose U.K. 56°44N 2°27W **11** E6
Montrose Colo., U.S.A. 38°29N 107°53W **76** G10
Montrose Pa., U.S.A. 41°50N 75°53W **83** E9
Monts, Pte. des Canada 49°20N 67°12W **73** C6
Montserrat ☑ W. Indies 16°40N 62°10W **89** C7
Montuiri Spain 39°34N 2°59E **24** B9
Monywa Burma 22°7N 95°11E **41** H19
Monza Italy 45°35N 9°16E **20** B8
Monze Zambia 16°17S 27°29E **55** F2
Monze, C. Pakistan 24°47N 66°37E **42** G2
Monzón Spain 41°52N 0°10E **21** B6
Mooers U.S.A. 44°58N 73°35W **83** B11
Mooi → S. Africa 28°45S 30°34E **57** D5
Mooi River S. Africa 29°13S 29°50E **57** D4
Moomba Australia 28°6S 140°12E **63** A3
Moonah → Australia 22°3S 138°33E **62** C2
Moonda, L. Australia 25°52S 140°25E **62** D3
Moonie Australia 27°46S 150°20E **63** D5
Moonie → Australia 29°19S 148°43E **63** D4
Moonta Australia 34°6S 137°32E **63** E2
Moora Australia 30°37S 115°58E **61** F2
Moorcroft U.S.A. 44°16N 104°57W **76** D11
Moore → Australia 31°22S 115°30E **61** F2
Moore, L. Australia 29°50S 117°35E **61** E2
Moore Falls Canada 44°48N 78°48W **82** B6
Moore Park Australia 24°43S 152°17E **62** C5
Moore Res. U.S.A. 44°20N 71°53W **83** B13
Moore River △ Australia 31°7S 115°39E **61** F2
Moorea French Polynesia 17°30S 149°50W **59** d
Moorefield U.S.A. 39°4N 78°58W **81** F14
Moorfoot Hills U.K. 55°44N 3°8W **11** F5
Moorhead U.S.A. 46°53N 96°45W **80** B5
Moorpark U.S.A. 34°17N 118°53W **79** L8
Mooreesburg S. Africa 33°6S 18°38E **56** E2
Moorrinya △ Australia 21°42S 144°58E **62** C3
Moose → Canada 51°20N 80°25W **72** B3
Moose → U.S.A. 43°38N 75°24W **83** C9
Moose Creek Canada 45°15N 74°58W **83** A10
Moose Factory Canada 51°16N 80°32W **72** B3
Moose Jaw Canada 50°24N 105°30W **71** C7
Moose Jaw → Canada 50°34N 105°18W **71** C7
Moose Lake Canada 53°46N 100°8W **71** C8
Moose Lake U.S.A. 46°27N 92°46W **80** B7
Moose Mountain △ Canada 49°48N 102°25W **71** D8
Moosehead L. U.S.A. 45°38N 69°40W **81** C19
Mooselookmeguntic L. U.S.A. 44°55N 70°49W **83** B14
Moosilauke, Mt. U.S.A. 44°3N 71°40W **83** B13
Moosomin Canada 50°9N 101°40W **71** C8
Moosonee Canada 51°17N 80°39W **72** B3
Moosup U.S.A. 41°43N 71°53W **83** E13
Mopane S. Africa 22°37S 29°52E **57** C4
Mopeia Velha Mozam. 17°30S 35°40E **55** F4
Mopipi Botswana 21°6S 24°55E **56** C3
Mopoi C.A.R. 5°6N 26°54E **54** A2
Mopti Mali 14°30N 4°0W **50** F5
Moqor Afghan. 32°50N 67°42E **42** C2
Moquegua Peru 17°15S 70°46W **92** G4
Mora Sweden 61°2N 14°38E **8** F16
Mora Minn., U.S.A. 45°53N 93°18W **80** C7
Mora N. Mex., U.S.A. 35°58N 105°20W **77** J11
Moradabad India 28°50N 78°50E **43** E8
Morafenobe Madag. 17°50S 44°53E **57** B7
Moramanga Madag. 18°56S 48°12E **57** B8
Moran Kans., U.S.A. 37°55N 95°10W **80** G6
Moran Wyo., U.S.A. 43°50N 110°31W **76** E8
Moranbah Australia 22°1S 148°6E **62** C4
Morang = Biratnagar Nepal 26°27N 87°17E **43** F12
Morant Bay Jamaica 17°53N 76°25W **88** a
Morant Cays Jamaica 17°22N 76°0W **88** D4
Morant Pt. Jamaica 17°55N 76°12W **88** a
Morar India 26°14N 78°14E **42** F8
Morar, L. U.K. 56°57N 5°40W **11** E3

Moratuwa Sri Lanka 6°45N 79°55E **40** R11
Morava → Serbia 44°36N 21°4E **23** B9
Morava → Slovak Rep. 48°10N 16°59E **17** D9
Moravia U.S.A. 42°43N 76°25W **83** D8
Moravian Hts. = Českomoravská Vrchovina Czech Rep. 49°30N 15°40E **16** D8
Morawa Australia 29°13S 116°0E **61** E2
Morawhanna Guyana 8°30N 59°40W **92** B7
Moray □ U.K. 57°31N 3°18W **11** D5
Moray Firth U.K. 57°40N 3°52W **11** D5
Morbi India 22°50N 70°42E **42** H4
Morden Canada 49°15N 98°10W **71** D9
Mordovian Republic = Mordvinia □ Russia 54°20N 44°30E **18** D7
Mordvinia □ Russia 54°20N 44°30E **18** D7
Moreau → U.S.A. 45°18N 100°43W **80** C3
Morebeng S. Africa 23°30S 29°55E **57** C4
Morecambe U.K. 54°5N 2°52W **12** C5
Morecambe B. U.K. 54°7N 3°0W **12** C5
Moree Australia 29°28S 149°54E **63** D4
Morehead U.S.A. 38°11N 83°26W **81** F12
Morehead City U.S.A. 34°43N 76°43W **85** D16
Morel → India 26°13N 76°36E **42** F7
Morelia Mexico 19°42N 101°7W **86** D4
Morella Australia 23°0S 143°52E **62** C3
Morella Spain 40°35N 0°5W **21** B5
Morelos Mexico 26°42N 107°40W **86** B3
Morelos □ Mexico 18°45N 99°0W **87** D5
Moremi △ Botswana 19°18S 23°10E **56** B3
Morena India 26°30N 78°4E **42** F8
Morena, Sierra Spain 38°20N 4°0W **21** C3
Moreno Valley U.S.A. 33°56N 117°14W **79** M10
Moresby I. Canada 52°30N 131°40W **70** C2
Moreton I. Australia 27°10S 153°25E **63** D5
Moreton Island △ Australia 27°2S 153°24E **63** D5
Morgan U.S.A. 41°2N 111°41W **76** F8
Morgan City U.S.A. 29°42N 91°12W **84** G9
Morgan Hill U.S.A. 37°8N 121°39W **78** H5
Morganfield U.S.A. 37°41N 87°55W **80** G10
Morganton U.S.A. 35°45N 81°41W **85** D14
Morgantown U.S.A. 39°38N 79°57W **81** F14
Morgenzon S. Africa 26°45S 29°36E **57** D4
Morghak Iran 29°7N 57°54E **45** D8
Morhar → India 25°29N 85°11E **43** G11
Mori Japan 42°6N 140°35E **30** C10
Moriarty U.S.A. 34°59N 106°3W **77** J10
Morice L. Canada 53°50N 127°40W **70** C3
Morinville Canada 53°49N 113°41W **70** C6
Morioka Japan 39°45N 141°8E **30** E10
Moris Mexico 28°10N 108°32W **86** B3
Morlaix France 48°36N 3°52W **20** B2
Mornington Australia 38°15S 145°5E **63** F4
Mornington, I. Chile 49°50S 75°30W **96** F1
Mornington I. Australia 16°30S 139°30E **62** B2
Moro Pakistan 26°40N 68°0E **42** F2
Moro → Pakistan 29°42N 67°22E **42** E2
Moro G. Phil. 6°30N 123°0E **37** C6
Morocco ■ N. Afr. 32°0N 5°50W **50** B4
Morogoro Tanzania 6°50S 37°40E **54** D4
Morogoro □ Tanzania 8°0S 37°0E **54** D4
Moroleón Mexico 20°8N 101°12W **86** C4
Morombe Madag. 21°45S 43°22E **57** C7
Moron Argentina 34°39S 58°37W **94** C4
Morón Cuba 22°8N 78°39W **88** B4
Mörön Mongolia 49°38N 100°9E **32** B9
Morón de la Frontera Spain 37°6N 5°28W **21** D3
Morona → Peru 4°40S 77°10W **92** D3
Morondava Madag. 20°17S 44°17E **57** C7
Morongo Valley U.S.A. 34°3N 116°37W **79** L10
Moroni Comoros Is. 11°40S 43°16E **53** a
Moroni U.S.A. 39°32N 111°35W **76** G8
Morotai Indonesia 2°10N 128°30E **37** D7
Moroto Uganda 2°28N 34°42E **54** B3
Moroto, Mt. Uganda 2°30N 34°43E **54** B3
Morpeth Canada 42°23N 81°50W **82** D3
Morpeth U.K. 55°10N 1°41W **12** B6
Morphou Cyprus 35°12N 32°59E **25** D11
Morphou Bay Cyprus 35°15N 32°50E **25** D11
Morrilton U.S.A. 35°9N 92°44W **84** D8
Morrinhos Brazil 17°45S 49°10W **93** G9
Morrinsville N.Z. 37°40S 175°32E **59** B5
Morris Canada 49°25N 97°22W **71** D9
Morris Ill., U.S.A. 41°22N 88°26W **80** E9
Morris Minn., U.S.A. 45°35N 95°55W **80** C6
Morris N.Y., U.S.A. 42°33N 75°15W **83** D9
Morris Pa., U.S.A. 41°35N 77°17W **82** E7
Morris, Mt. Australia 26°9S 131°4E **61** E5
Morris Jesup, Kap Greenland 83°40N 34°0W **66** A16
Morrisburg Canada 44°55N 75°7W **83** B9
Morristown Ariz., U.S.A. 33°51N 112°37W **77** K7
Morristown N.J., U.S.A. 40°48N 74°29W **83** F10
Morristown N.Y., U.S.A. 44°35N 75°39W **83** B9
Morristown Tenn., U.S.A. 36°13N 83°18W **85** C13
Morrisville N.Y., U.S.A. 42°53N 75°35W **83** D9
Morrisville Pa., U.S.A. 40°13N 74°47W **83** F10
Morrisville Vt., U.S.A. 44°34N 72°36W **83** B12
Morro, Pta. Chile 27°6S 71°0W **94** B1
Morro Bay U.S.A. 35°22N 120°51W **78** K6
Morro del Jable Canary Is. 28°3N 14°23W **24** F5
Morro Jable, Pta. de Canary Is. 28°2N 14°20W **24** F5
Morrocoy △ Venezuela 10°48N 68°13W **89** D6
Morrosquillo, G. de Colombia 9°35N 75°40W **88** E4
Morrumbene Mozam. 23°31S 35°16E **57** C6
Morshansk Russia 53°28N 41°50E **18** D7
Morteros Argentina 30°50S 62°0W **94** C3
Mortlach Canada 50°27N 106°4W **71** C7
Mortlake Australia 38°5S 142°50E **63** F3
Morton Tex., U.S.A. 33°44N 102°46W **84** E3
Morton Wash., U.S.A. 46°34N 122°17W **78** D4
Moruga Trin. & Tob. 10°4N 61°16W **93** K15
Morundah Australia 34°57S 146°19E **63** E4
Moruya Australia 35°58S 150°3E **63** F5
Morvan France 47°5N 4°3E **20** C5

Morven Australia 26°22S 147°5E **63** D4
Morvern U.K. 56°38N 5°44W **11** E3
Morwell Australia 38°10S 146°22E **63** F4
Morzhovets, Ostrov Russia 66°44N 42°35E **18** A7
Mosakahiken = Moose Lake Canada 53°46N 100°8W **71** C8
Moscos Is. Burma 14°0N 97°30E **38** F1
Moscow = Moskva Russia 55°45N 37°37E **18** C6
Moscow Idaho, U.S.A. 46°44N 117°0W **76** C5
Moscow Pa., U.S.A. 41°20N 75°31W **83** E9
Mosel → Europe 50°22N 7°36E **20** A7
Moselle = Mosel → Europe 50°22N 7°36E **20** A7
Moses Lake U.S.A. 47°8N 119°17W **76** C4
Mosgiel N.Z. 45°53S 170°21E **59** F3
Moshaweng → S. Africa 26°35S 22°50E **56** D3
Moshchnyy, Ostrov Russia 60°1N 27°50E **9** F22
Moshi Tanzania 3°22S 37°18E **54** C4
Moshupa Botswana 24°46S 25°29E **56** C4
Mosjøen Norway 65°51N 13°12E **8** D15
Moskenesøya Norway 67°58N 13°0E **8** C15
Moskenstraumen Norway 67°47N 12°45E **8** C15
Moskva Russia 55°45N 37°37E **18** C6
Mosonmagyaróvár Hungary 47°52N 17°18E **17** E9
Mosquera Colombia 2°35N 78°24W **92** C3
Mosquero U.S.A. 35°47N 103°58W **77** J12
Mosquitia Honduras 15°20N 84°10W **88** C3
Mosquito Coast = Mosquitia Honduras 15°20N 84°10W **88** C3
Mosquito Creek L. U.S.A. 41°18N 80°46W **82** E4
Mosquito L. Canada 62°35N 103°20W **71** A8
Mosquitos, G. de los Panama 9°15N 81°10W **88** E3
Moss Norway 59°27N 10°40E **9** G14
Moss Vale Australia 34°32S 150°25E **63** E5
Mossaka Congo 1°15S 16°45E **52** E3
Mossbank Canada 49°56N 105°56W **71** D7
Mossburn N.Z. 45°41S 168°15E **59** F2
Mosselbaai S. Africa 34°11S 22°8E **56** E3
Mossendjo Congo 2°55S 12°42E **52** E2
Mossgiel Australia 33°15S 144°5E **63** E3
Mossman Australia 16°21S 145°15E **62** B4
Mossoró Brazil 5°10S 37°15W **93** E11
Mossuril Mozam. 14°58S 40°42E **55** E5
Most Czech Rep. 50°31N 13°38E **16** C7
Mosta Malta 35°55N 14°26E **25** D1
Mostaganem Algeria 35°54N 0°5E **50** A6
Mostar Bos.-H. 43°22N 17°50E **23** C7
Mostardas Brazil 31°2S 50°51W **95** C5
Mostiska = Mostyska Ukraine 49°48N 23°4E **17** D12
Mosty = Masty Belarus 53°27N 24°38E **17** B13
Mostyska Ukraine 49°48N 23°4E **17** D12
Mosul = Al Mawşil Iraq 36°15N 43°5E **44** B4
Motagua → Guatemala 15°44N 88°14W **88** C2
Motala Sweden 58°32N 15°1E **9** G16
Motaze Mozam. 24°48S 32°52E **57** C5
Moth India 25°43N 78°57E **43** G8
Motherwell U.K. 55°47N 3°58W **11** F5
Motihari India 26°30N 84°55E **43** F11
Motozintla de Mendoza Mexico 15°22N 92°14W **87** D6
Motril Spain 36°31N 3°37W **21** D4
Mott U.S.A. 46°23N 102°20W **80** B2
Motueka N.Z. 41°7S 173°1E **59** D4
Motueka → N.Z. 41°5S 173°1E **59** D4
Motul Mexico 21°6N 89°17W **87** C7
Mouchalagane → Canada 50°56N 68°41W **73** B6
Moudros Greece 39°50N 25°18E **23** E11
Mouhoun = Black Volta → Africa 8°41N 1°33W **50** G5
Mouila Gabon 1°50S 11°0E **52** E2
Moulamein Australia 35°3S 144°1E **63** F3
Moule à Chique, C. St. Lucia 13°43N 60°57W **89** f
Mouliana Greece 35°10N 25°59E **25** D7
Moulins France 46°35N 3°19E **20** C5
Moulmein Burma 16°30N 97°40E **41** L20
Moulouya, O. → Morocco 35°5N 2°25W **50** B5
Moultrie U.S.A. 31°11N 83°47W **85** F13
Moultrie, L. U.S.A. 33°20N 80°5W **85** E14
Mound City Mo., U.S.A. 40°7N 95°14W **80** E6
Mound City S. Dak., U.S.A. 45°44N 100°4W **80** C3
Moundou Chad 8°40N 16°10E **51** G9
Moundsville U.S.A. 39°55N 80°44W **82** G4
Moung Cambodia 12°46N 103°27E **38** F4
Mount Airy U.S.A. 36°31N 80°37W **85** C14
Mount Albert Canada 44°8N 79°19W **82** B5
Mount Aspiring △ N.Z. 44°19S 168°47E **59** F2
Mount Barker S. Austral., Australia 35°5S 138°52E **63** F2
Mount Barker W. Austral., Australia 34°38S 117°40E **61** F2
Mount Barnett Roadhouse Australia 16°39S 125°57E **60** C4
Mount Bellew Bridge Ireland 53°28N 8°31W **10** C3
Mount Brydges Canada 42°54N 81°29W **82** D3
Mount Burr Australia 37°34S 140°26E **63** F3
Mount Carmel = Ha Karmel △ Israel 32°45N 35°5E **46** C4
Mount Carmel Ill., U.S.A. 38°25N 87°46W **80** F10
Mount Carmel Pa., U.S.A. 40°47N 76°26W **83** F8
Mount Clemens U.S.A. 42°35N 82°53W **82** D2
Mount Coolon Australia 21°25S 147°25E **62** C4
Mount Darwin Zimbabwe 16°47S 31°38E **55** F3
Mount Desert I. U.S.A. 44°21N 68°20W **81** C19
Mount Dora U.S.A. 28°48N 81°38W **85** G14
Mount Ebenezer Australia 25°6S 132°34E **61** E5

Mount Edziza △ Canada 57°30N 130°45W **70** B2
Mount Elgon △ E. Afr. 1°4N 34°42E **54** B3
Mount Field △ Australia 42°39S 146°35E **63** G4
Mount Fletcher S. Africa 30°40S 28°30E **57** E4
Mount Forest Canada 43°59N 80°43W **82** C4
Mount Frankland △ Australia 31°47S 116°37E **61** F2
Mount Frederick ◎ Australia 19°39S 129°18E **60** C4
Mount Gambier Australia 37°50S 140°46E **63** F3
Mount Garnet Australia 17°37S 145°6E **62** B4
Mount Holly U.S.A. 39°59N 74°47W **83** G10
Mount Holly Springs U.S.A. 40°7N 77°12W **82** F7
Mount Hope N.S.W., Australia 32°51S 145°51E **63** E4
Mount Hope S. Austral., Australia 34°7S 135°23E **63** E2
Mount Isa Australia 20°42S 139°26E **62** C2
Mount James ◎ Australia 24°51S 116°54E **61** D2
Mount Jewett U.S.A. 41°44N 78°39W **82** E6
Mount Kaputar △ Australia 30°16S 150°10E **63** E5
Mount Kenya △ Kenya 0°7S 37°21E **54** C4
Mount Kisco U.S.A. 41°12N 73°44W **83** E11
Mount Laguna U.S.A. 32°52N 116°25W **79** N10
Mount Larcom Australia 23°48S 150°59E **62** C5
Mount Lofty Ranges Australia 34°35S 139°5E **63** E2
Mount Magnet Australia 28°2S 117°47E **61** E2
Mount Maunganui N.Z. 37°40S 176°14E **59** B6
Mount Molloy Australia 16°42S 145°20E **62** B4
Mount Morgan Australia 23°40S 150°25E **62** C5
Mount Morris U.S.A. 42°44N 77°52W **82** D7
Mount Pearl Canada 47°31N 52°47W **73** C9
Mount Penn U.S.A. 40°20N 75°54W **83** F9
Mount Perry Australia 25°13S 151°42E **63** D5
Mount Pleasant Iowa, U.S.A. 40°58N 91°33W **80** E8
Mount Pleasant Mich., U.S.A. 43°36N 84°46W **81** D11
Mount Pleasant Pa., U.S.A. 40°9N 79°33W **82** F5
Mount Pleasant S.C., U.S.A. 32°47N 79°52W **85** E15
Mount Pleasant Tenn., U.S.A. 35°32N 87°12W **85** D11
Mount Pleasant Tex., U.S.A. 33°9N 94°58W **84** E7
Mount Pleasant Utah, U.S.A. 39°33N 111°27W **76** G8
Mount Pocono U.S.A. 41°7N 75°22W **83** E9
Mount Rainier △ U.S.A. 46°55N 121°50W **78** D5
Mount Revelstoke △ Canada 51°5N 118°30W **70** C5
Mount Robson △ Canada 53°0N 119°0W **70** C5
Mount St. Helens △ U.S.A. 46°14N 122°11W **78** D4
Mount Selinda Zimbabwe 20°24S 32°43E **57** C5
Mount Shasta U.S.A. 41°19N 122°19W **76** F2
Mount Signal U.S.A. 32°39N 115°37W **79** N11
Mount Sterling Ill., U.S.A. 39°59N 90°45W **80** F8
Mount Sterling Ky., U.S.A. 38°4N 83°56W **81** F12
Mount Surprise Australia 18°10S 144°17E **62** B3
Mount Union U.S.A. 40°23N 77°53W **82** F7
Mount Upton U.S.A. 42°26N 75°23W **83** D9
Mount Vernon Ill., U.S.A. 38°19N 88°55W **80** F9
Mount Vernon Ind., U.S.A. 37°56N 87°54W **80** G10
Mount Vernon N.Y., U.S.A. 40°54N 73°49W **83** F11
Mount Vernon Ohio, U.S.A. 40°23N 82°29W **82** F2
Mount Vernon Wash., U.S.A. 48°25N 122°20W **78** B4
Mount William △ Australia 40°56S 148°14E **63** G4
Mountain Ash U.K. 51°40N 3°23W **13** F4
Mountain Center U.S.A. 33°42N 116°44W **79** M10
Mountain City Nev., U.S.A. 41°50N 115°58W **76** F6
Mountain City Tenn., U.S.A. 36°29N 81°48W **85** C14
Mountain Dale U.S.A. 41°41N 74°32W **83** E10
Mountain Grove U.S.A. 37°8N 92°16W **80** G7
Mountain Home Ark., U.S.A. 36°20N 92°23W **84** C8
Mountain Home Idaho, U.S.A. 43°8N 115°41W **76** E6
Mountain Iron U.S.A. 47°32N 92°37W **80** B7
Mountain Pass U.S.A. 35°29N 115°35W **79** K11
Mountain View Ark., U.S.A. 35°52N 92°7W **84** D8
Mountain View Calif., U.S.A. 37°23N 122°5W **78** H4
Mountain Zebra △ S. Africa 32°14S 25°27E **56** E4
Mountainair U.S.A. 34°31N 106°15W **77** J10
Mountlake Terrace U.S.A. 47°47N 122°18W **78** C4
Mountmellick Ireland 53°7N 7°20W **10** C4
Mountrath Ireland 53°0N 7°28W **10** C4
Moura Australia 24°35S 149°58E **62** C4
Moura Brazil 1°32S 61°38W **92** D6
Moura Portugal 38°7N 7°30W **21** C2
Mourdi, Dépression du Chad 18°10N 23°0E **51** E10
Mourilyan Australia 17°35S 146°3E **62** B4

Mourne → U.K. 54°52N 7°26W **10** B4
Mourne Mts. U.K. 54°10N 6°0W **10** B5
Mournies Greece 35°29N 24°1E **25** D6
Mouscron Belgium 50°45N 3°12E **15** D3
Moussoro Chad 13°41N 16°35E **51** F9
Moutong Indonesia 0°28N 121°13E **37** D6
Movas Mexico 28°10N 109°25W **86** B3
Moville Ireland 55°11N 7°3W **10** A4
Mowandjum Australia 17°22S 123°40E **60** C3
Moy → Ireland 54°8N 9°8W **10** B2
Moya Comoros Is. 12°18S 44°18E **53** a
Moyale Ethiopia 3°34N 39°5E **54** G2
Moyale Kenya 3°30N 39°4E **54** B4
Moyen Atlas Morocco 33°0N 5°0W **50** B4
Moyo Indonesia 8°10S 117°40E **36** F5
Moyobamba Peru 6°0S 77°0W **92** E3
Moyowosi → Tanzania 3°50S 31°10E **54** C3
Moyyero → Russia 68°44N 103°42E **29** C11
Moyynqum Kazakhstan 44°12N 71°0E **32** C2
Moyynty Kazakhstan 47°10N 73°18E **28** E8
Mozambique = Moçambique Mozam. 15°3S 40°42E **55** F5
Mozambique ■ Africa 19°0S 35°0E **55** F4
Mozambique Chan. Africa 17°30S 42°30E **57** B7
Mozdok Russia 43°45N 44°48E **19** F7
Mozdūrān Iran 36°9N 60°35E **45** B9
Mozhnābād Iran 34°7N 60°6E **45** C9
Mozyr = Mazyr Belarus 51°59N 29°15E **17** B15
Mpanda Tanzania 6°23S 31°1E **54** D3
Mphoeng Zimbabwe 21°10S 27°51E **55** G2
Mpika Zambia 11°51S 31°25E **55** E3
Mpulungu Zambia 8°51S 31°5E **55** D3
Mpumalanga S. Africa 29°50S 30°33E **57** D5
Mpumalanga □ S. Africa 26°0S 30°0E **57** D5
Mpwapwa Tanzania 6°23S 36°30E **54** D4
Mqanduli S. Africa 31°49S 28°45E **57** E4
Msaken Tunisia 35°49N 10°33E **51** A8
Msambansovu Zimbabwe 15°50S 30°3E **55** F3
M'sila Algeria 35°46N 4°30E **50** A6
Msoro Zambia 13°35S 31°50E **55** E3
Mstislavl = Mstsislaw Belarus 54°0N 31°50E **17** A16
Mstsislaw Belarus 54°0N 31°50E **17** A16
Mtama Tanzania 10°17S 39°21E **55** E4
Mtamvuna = Mthamvuna → S. Africa 31°6S 30°12E **57** E5
Mthamvuna → S. Africa 31°6S 30°12E **57** E5
Mthatha S. Africa 31°36S 28°49E **57** E4
Mtilikwe → Zimbabwe 21°9S 31°30E **55** G3
Mtito Andei Kenya 2°41S 38°10E **54** C4
Mtubatuba S. Africa 28°30S 32°8E **57** D5
Mtwalume S. Africa 30°30S 30°38E **57** E5
Mtwara-Mikindani Tanzania 10°20S 40°20E **55** E5
Mu Gia, Deo Vietnam 17°40N 105°47E **38** D5
Mu Ko Chang △ Thailand 11°59N 102°22E **39** b
Mu Ko Surin Thailand 9°25N 97°55E **39** b
Mu Us Shamo China 39°0N 109°0E **34** E5
Muang Beng Laos 20°23N 101°46E **38** B3
Muang Chiang Rai = Chiang Rai Thailand 19°52N 99°50E **38** C2
Muang Et Laos 20°49N 104°1E **38** B5
Muang Hiam Laos 20°5N 103°22E **38** B4
Muang Hongsa Laos 19°43N 101°20E **38** B3
Muang Houn Laos 20°8N 101°23E **38** B3
Muang Kau Laos 15°6N 105°47E **38** E5
Muang Khao Laos 19°38N 103°32E **38** C4
Muang Khong Laos 14°7N 105°51E **38** E5
Muang Khoua Laos 21°5N 102°31E **38** B4
Muang Liap Laos 18°29N 101°40E **38** C3
Muang Mai Thailand 8°5N 98°21E **39** a
Muang May Laos 14°49N 106°56E **38** E6
Muang Ngeun Laos 20°36N 101°3E **38** B3
Muang Ngoi Laos 20°43N 102°41E **38** B4
Muang Nong Laos 16°22N 106°30E **38** D6
Muang Ou Neua Laos 22°18N 101°48E **38** A3
Muang Ou Tay Laos 22°7N 101°48E **38** A3
Muang Pak Beng Laos 19°54N 101°8E **38** C3
Muang Phalane Laos 16°39N 105°34E **38** D5
Muang Phiang Laos 19°6N 101°32E **38** C3
Muang Phine Laos 16°32N 106°2E **38** D6
Muang Saiapoun Laos 18°24N 101°31E **38** C3
Muang Sing Laos 21°11N 101°9E **38** A3
Muang Son Laos 20°27N 103°19E **38** B4
Muang Soui Laos 19°33N 102°52E **38** C4
Muang Va Laos 21°53N 102°19E **38** B4
Muang Xai Laos 20°42N 101°59E **38** B3
Muang Xamteu Laos 19°59N 104°38E **38** C5
Muar Malaysia 2°3N 102°34E **39** L4
Muarabungo Indonesia 1°28S 102°52E **36** E2
Muaraenim Indonesia 3°40S 103°50E **36** E2
Muarajuloi Indonesia 0°12S 114°3E **36** E4
Muarakaman Indonesia 0°2S 116°45E **36** E5
Muaratebo Indonesia 1°30S 102°26E **36** E2
Muaratembesi Indonesia 1°42S 103°8E **36** E2
Muarateweh Indonesia 0°58S 114°52E **36** E4
Mubarakpur India 26°6N 83°18E **43** F10
Mubarraz = Al Mubarraz Si. Arabia 25°30N 49°40E **45** E6
Mubende Uganda 0°33N 31°22E **54** B3
Mubi Nigeria 10°18N 13°16E **51** F8
Mucajaí → Brazil 2°25N 60°52W **92** C6
Muchachos, Roque de los Canary Is. 28°44N 17°52W **24** F2
Muchinga Mts. Zambia 11°30S 31°30E **55** E3
Muck U.K. 56°50N 6°15W **11** E2
Muckadilla Australia 26°35S 148°23E **63** D4
Muckaty ◎ Australia 18°37S 133°52E **62** B1
Muckle Flugga U.K. 60°51N 0°54W **11** A8
Mucuri Brazil 18°0S 39°36W **93** G11
Mucusso Angola 18°1S 21°25E **56** B3
Muda Canary Is. 28°34N 13°57W **24** F6
Mudanjiang China 44°38N 129°30E **35** B15
Mudanya Turkey 40°25N 28°50E **23** D13
Muddy Cr. → U.S.A. 38°24N 110°42W **76** G8
Mudgee Australia 32°32S 149°31E **63** E4
Mudjatik → Canada 56°1N 107°36W **71** B7

ecate *Mozam.* 14°55S 39°40E **55** E4
eda *Mozam.* 11°36S 39°28E **55** E4
eller Ranges
ustralia 18°18S 126°46E **60** C4
ende *Mozam.* 14°28S 33°0E **55** E3
erto, Mar *Mexico* 16°10N 94°10W **87** D6
fulira *Zambia* 12°32S 28°15E **55** E2
fumbiro Range *Africa* 1°25S 29°30E **54** C2
ghal Sarai *India* 25°18N 83°7E **43** G10
ghayrā' *Si. Arabia* 29°17N 37°41E **44** D3
gi *Japan* 33°40N 134°25E **31** H7
gila, Mts.
em. Rep. of the Congo 7°0S 28°50E **54** D2
ğla *Turkey* 37°15N 28°22E **23** F13
gu *Nepal* 29°45N 82°30E **43** E10
hala *Congo* 5°40S 28°43E **54** D2
hammad, Râs *Egypt* 27°44N 34°16E **44** E2
hammad Qol *Sudan* 20°53N 37°9E **51** D13
hammadabad *India* 26°4N 83°25E **43** F10
hesi → *Tanzania* 7°0S 35°20E **54** D4
hlhausen *Germany* 51°12N 10°27E **16** C6
hlig Hofmann fjell
ntarctica 72°30S 5°0E **5** D3
hos *Finland* 64°47N 25°59E **8** D21
hu *Estonia* 58°36N 23°11E **9** G20
hutwe *Tanzania* 1°35S 31°45E **54** C3
i Wo *China* 22°16N 113°59E **33** a
ika *Japan* 37°5N 138°50E **31** F9
ine Bheag *Ireland* 52°42N 6°58W **10** D5
ineachán = Monaghan
eland 54°15N 6°57W **10** B5
ir, L. *Australia* 34°30S 116°40E **61** F2
ir of Ord *U.K.* 57°32N 4°28W **11** D4
jeres, I. *Mexico* 21°13N 86°43W **88** B2
jnak = Muynak
zbekistan 43°44N 59°10E **28** E6
ka, Tanjung *Malaysia* 5°28N 100°11E **39** c
kacheve *Ukraine* 48°27N 22°45E **17** D12
kachevo = Mukacheve
kraine 48°27N 22°45E **17** D12
kah *Malaysia* 2°55N 112°5E **36** D4
kandwara *India* 24°49N 75°59E **42** G6
kawwa, Geziret
gypt 23°55N 35°53E **44** F2
kdahan *Thailand* 16°32N 104°43E **38** D5
kdahan △ *Thailand* 16°26N 104°45E **38** D5
kden = Shenyang
hina 41°48N 123°27E **35** D12
kerian *India* 31°57N 75°37E **42** D6
kinbudin *Australia* 30°55S 118°5E **61** F2
kishi
em. Rep. of the Congo 8°30S 24°44E **55** D1
ko Phetra △ *Thailand* 6°57N 99°33E **39** J2
komuko *Indonesia* 2°30S 101°10E **36** E2
komwenze
em. Rep. of the Congo 6°49S 27°15E **54** D2
kono *Uganda* 0°21N 32°45E **54** B3
ktsar *India* 30°30N 74°30E **42** D6
kur = Moqor *Afghan.* 32°50N 67°42E **42** C2
kutuwa → *Canada* 53°10N 97°24W **71** C9
kwela *Zambia* 17°0S 26°40E **55** F2
la *Spain* 38°3N 1°33W **21** C5
la → *Pakistan* 27°57N 67°36E **42** F2
lange
em. Rep. of the Congo 3°40S 27°10E **54** C2
lanje, Mt. *Malawi* 16°2S 35°33E **55** F4
lchatna → *U.S.A.* 59°40N 157°7W **74** D8
lchén *Chile* 37°45S 72°20W **94** D1
lde → *Germany* 51°53N 12°15E **16** C7
le Creek Junction
.S.A. 43°23N 104°13W **76** E11
leba *Tanzania* 1°50S 31°37E **54** C3
legé *Mexico* 26°53N 111°59W **86** B2
leshoe *U.S.A.* 34°13N 102°43W **84** D3
lgrave *Canada* 45°38N 61°31W **73** C7
lgrave I. = Badu
ustralia 10°7S 142°11E **62** a
lhacén *Spain* 37°4N 3°20W **21** D4
lhouse *France* 47°40N 7°20E **20** C7
lifanua *Samoa* 13°50S 171°59W **59** b
ling *China* 44°35N 130°10E **35** B16
ll *U.K.* 56°25N 5°56W **11** E3
ll, Sound of *U.K.* 56°30N 5°50W **11** E3
llach Íde = Malahide
eland 53°26N 6°9W **10** C5
llaittivu *Sri Lanka* 9°15N 80°49E **40** Q12
llen *U.S.A.* 42°3N 101°1W **80** D3
llengudgery
ustralia 31°43S 147°23E **63** B4
llens *U.S.A.* 37°35N 81°23W **81** G13
ller, Pegunungan
ndonesia 0°30N 113°30E **36** D4
llet Pen. *Ireland* 54°13N 10°2W **10** B1
llewa *Australia* 28°29S 115°30E **61** E2
lligan → *Australia* 25°0S 139°0E **62** D2
llingar *Ireland* 53°31N 7°21W **10** C4
llins *U.S.A.* 34°12N 79°15W **85** D15
llumbimby *Australia* 28°30S 153°30E **63** D5
lobezi *Zambia* 16°45S 25°7E **55** F2
longa Plain *Zambia* 16°20S 22°40E **53** H4
lroy B. *Ireland* 55°15N 7°46W **10** A4
ltan *Pakistan* 30°15N 71°36E **42** D4
lumbe, Mts.
em. Rep. of the Congo 8°40S 27°30E **55** D2
lungushi Dam
ambia 14°48S 28°48E **55** E2
lvane *U.S.A.* 37°29N 97°15W **80** G5
lwala, L. *Australia* 35°59S 146°1E **63** C4
mbai *India* 18°56N 72°50E **40** K8
mbwa *Zambia* 15°0S 27°0E **55** F2
mias *Kenya* 0°20N 34°29E **54** C4
n → *Thailand* 15°19N 105°30E **38** E5
na *Indonesia* 5°0S 122°30E **37** F6
nabao *India* 25°45N 70°17E **42** G4
namagi *Estonia* 57°43N 27°4E **9** H22
ncan *Indonesia* 8°34S 115°11E **37** K18
ncar *Indonesia* 8°26S 114°20E **37** J17
nchen *Germany* 48°8N 11°34E **16** D6
nchen-Gladbach =
lönchengladbach
ermany 51°11N 6°27E **16** C4

Muncho Lake *Canada* 59°0N 125°50W **70** B3
Munch'ŏn *N. Korea* 39°14N 127°19E **35** E14
Muncie *U.S.A.* 40°12N 85°23W **81** E11
Muncoonie L. West
Australia 25°12S 138°40E **62** D2
Mundabbera *Australia* 25°36S 151°18E **63** D5
Munday *U.S.A.* 33°27N 99°38W **84** E5
Münden *Germany* 51°25N 9°38E **16** C5
Mundiwindi *Australia* 23°47S 120°9E **60** D3
Mundo Novo *Brazil* 11°50S 40°29W **93** F10
Mundra *India* 22°54N 69°48E **42** H3
Mundrabilla *Australia* 31°52S 127°51E **61** F4
Mungallala *Australia* 26°28S 147°34E **63** D4
Mungallala Cr. →
Australia 28°53S 147°5E **63** D4
Mungana *Australia* 17°8S 144°27E **62** B3
Mungaoli *India* 24°24N 78°7E **42** G8
Mungari *Mozam.* 17°12S 33°30E **55** F3
Mungbere
Dem. Rep. of the Congo 2°36N 28°28E **54** B2
Mungeli *India* 22°4N 81°41E **43** H9
Munger *India* 25°23N 86°30E **43** G12
Mungerannie *Australia* 28°1S 138°39E **63** A2
Mungilli ✪ *Australia* 25°14S 124°17E **61** E3
Mungkan Kandju △
Australia 13°35S 142°52E **62** A3
Mungkarta ✪ *Australia* 20°22S 134°2E **62** C1
Munich = München
Germany 48°8N 11°34E **16** D6
Munising *U.S.A.* 46°25N 86°40W **80** B10
Munku-Sardyk *Russia* 51°45N 100°20E **29** D11
Munnsville *U.S.A.* 42°58N 75°35W **83** D9
Muñoz Gamero, Pen.
Chile 52°30S 73°5W **96** G2
Munroe L. *Canada* 59°13N 98°35W **71** B9
Munsan *S. Korea* 37°51N 126°48E **35** F14
Münster *Germany* 51°58N 7°37E **16** C4
Munster □ *Ireland* 52°18N 8°44W **10** D3
Muntadgin *Australia* 31°45S 118°33E **61** F2
Muntok *Indonesia* 2°5S 105°10E **36** E3
Muong Nhie *Vietnam* 22°12N 102°28E **38** A4
Muong Sen *Vietnam* 19°24N 104°8E **38** C5
Muong Te *Vietnam* 22°24N 102°49E **38** A4
Muong Xia *Vietnam* 20°19N 104°50E **38** B5
Muonio *Finland* 67°57N 23°40E **8** C20
Muonio älv = Muonionjoki →
Finland 67°11N 23°34E **8** C20
Muonioälven = Muonionjoki →
Finland 67°11N 23°34E **8** C20
Muonionjoki → *Finland* 67°11N 23°34E **8** C20
Mupa → *Mozam.* 18°58S 35°54E **55** F4
Muping *China* 37°22N 121°36E **35** F11
Muqdisho *Somali Rep.* 2°2N 45°25E **47** G4
Mur → *Austria* 46°18N 16°52E **17** E9
Murakami *Japan* 38°14N 139°29E **30** E9
Muralag = Prince of Wales I.
Australia 10°40S 142°10E **62** A3
Murallón, Cerro *Chile* 49°48S 73°30W **96** F2
Muranda *Rwanda* 1°52S 29°20E **54** C2
Murang'a *Kenya* 0°45S 37°9E **54** C4
Murashi *Russia* 59°30N 49°0E **18** C8
Murat → *Turkey* 38°46N 40°0E **19** G7
Muratlı *Turkey* 41°10N 27°29E **23** D12
Murayama *Japan* 38°30N 140°25E **30** E10
Murchison → *Australia* 27°45S 114°0E **61** E1
Murchison, Mt.
Antarctica 73°25S 166°20E **5** D11
Murchison Falls *Uganda* 2°15N 31°30E **54** B3
Murchison Falls △
Uganda 2°17N 31°48E **54** B3
Murchison Ra. *Australia* 20°0S 134°10E **62** C1
Murchison Rapids
Malawi 15°55S 34°35E **55** F3
Murchison Roadhouse
Australia 27°39S 116°14E **61** E2
Murcia *Spain* 38°5N 1°10W **21** D5
Murcia □ *Spain* 37°50N 1°30W **21** D5
Murdo *U.S.A.* 43°53N 100°43W **80** D3
Murdoch Pt. *Australia* 14°37S 144°55E **62** A3
Mureş → *Romania* 46°15N 20°13E **17** E11
Mureşul = Mureş →
Romania 46°15N 20°13E **17** E11
Murewa *Zimbabwe* 17°39S 31°47E **57** B5
Murfreesboro *N.C.,*
U.S.A. 36°27N 77°6W **85** C16
Murfreesboro *Tenn.,*
U.S.A. 35°51N 86°24W **85** D11
Murgap *Tajikistan* 38°10N 74°2E **28** F8
Murgap → *Turkmenistan* 38°18N 61°12E **45** B9
Murgenella *Australia* 11°34S 132°56E **60** B5
Murgha Kibzai *Pakistan* 30°44N 69°25E **42** D3
Murghob = Murgap
Tajikistan 38°10N 74°2E **28** F8
Murgon *Australia* 26°15S 151°54E **63** D5
Muri *India* 23°22N 85°52E **43** H11
Muria *Indonesia* 6°36S 110°53E **37** G14
Muriaé *Brazil* 21°8S 42°23W **95** A7
Muriel Mine *Zimbabwe* 17°14S 30°40E **55** F3
Müritz *Germany* 53°25N 12°42E **16** B7
Murliganj *India* 25°54N 86°59E **43** G12
Murmansk *Russia* 68°57N 33°10E **8** B25
Murmashi *Russia* 68°47N 32°42E **8** B25
Muro *Spain* 39°44N 3°3E **24** B10
Murom *Russia* 55°35N 42°3E **18** C7
Muroran *Japan* 42°25N 141°0E **30** C10
Muroto *Japan* 33°18N 134°9E **31** H7
Muroto-Misaki *Japan* 33°15N 134°10E **31** H7
Murphy *U.S.A.* 43°13N 116°33W **76** E5
Murphys *U.S.A.* 38°8N 120°28W **78** G6
Murray *Australia* 9°56S 144°2E **62** a
Murray *Ky., U.S.A.* 36°37N 88°19W **80** G9
Murray *Utah, U.S.A.* 40°40N 111°53W **76** F8
Murray → *Australia* 35°20S 139°22E **63** F2
Murray, L. *U.S.A.* 34°3N 81°13W **85** D14
Murray Bridge *Australia* 35°6S 139°14E **63** F2
Murray Fracture Zone
Pac. Oc. 35°0N 130°0W **65** D14
Murray Harbour *Canada* 46°0N 62°28W **73** C7
Murray River △
Australia 34°23S 140°32E **63** E3

Murraysburg *S. Africa* 31°58S 23°47E **56** E3
Murree *Pakistan* 33°56N 73°28E **42** C5
Murrieta *U.S.A.* 33°33N 117°13W **79** M9
Murrumbidgee →
Australia 34°43S 143°12E **63** E3
Murrumburrah
Australia 34°32S 148°22E **63** E4
Murrurundi *Australia* 31°42S 150°51E **63** E5
Murshidabad *India* 24°11N 88°19E **43** G13
Murtle L. *Canada* 52°8N 119°38W **70** C5
Murtoa *Australia* 36°35S 142°28E **63** F3
Murungu *Tanzania* 4°12S 31°10E **54** C3
Mururoa
French Polynesia 21°52S 138°55W **65** K14
Murwara *India* 23°46N 80°28E **43** H9
Murwillumbah
Australia 28°18S 153°27E **63** D5
Mürzzuschlag *Austria* 47°36N 15°41E **16** E8
Muş *Turkey* 38°45N 41°30E **44** B4
Mûsa, Gebel *Egypt* 28°33N 33°59E **44** D2
Musa Khel *Pakistan* 30°59N 69°52E **42** D3
Mūsa Qal'eh *Afghan.* 32°20N 64°50E **40** C4
Musafirkhana *India* 26°22N 81°48E **43** F9
Musala *Bulgaria* 42°13N 23°37E **23** C10
Musala *Indonesia* 1°41N 98°28E **36** D1
Musan *N. Korea* 42°12N 129°12E **35** C15
Musangu
Dem. Rep. of the Congo 10°28S 23°55E **55** E1
Musasa *Tanzania* 3°25S 31°30E **54** C3
Musay'īd *Qatar* 25°0N 51°33E **45** E6
Muscat = Masqat *Oman* 23°37N 58°36E **47** C6
Muscatine *U.S.A.* 41°25N 91°3W **80** E8
Muscle Shoals *U.S.A.* 34°45N 87°40W **85** D11
Musengezi = Unsengedsi →
Zimbabwe 15°43S 31°14E **55** F3
Musgrave Harbour
Canada 49°27N 53°58W **73** C9
Musgrave Ranges
Australia 26°0S 132°0E **61** E5
Mushie
Dem. Rep. of the Congo 2°56S 16°55E **52** E3
Musi → *Indonesia* 2°20S 104°56E **36** E2
Musina *S. Africa* 22°20S 30°5E **57** C5
Muskeg → *Canada* 60°20N 123°20W **70** A4
Muskegon *U.S.A.* 43°14N 86°16W **80** D10
Muskegon → *U.S.A.* 43°14N 86°21W **80** D10
Muskegon Heights
U.S.A. 43°12N 86°16W **80** D10
Muskogee *U.S.A.* 35°45N 95°22W **84** D7
Muskoka, L. *Canada* 45°0N 79°25W **82** B5
Muskoka Falls *Canada* 44°59N 79°17W **82** B5
Muskwa → *Canada* 58°47N 122°48W **70** B4
Muslīmiyah *Syria* 36°19N 37°12E **44** B3
Musofu *Zambia* 13°30S 29°0E **55** E2
Musoma *Tanzania* 1°30S 33°48E **54** C3
Musquaro, L. *Canada* 50°38N 61°5W **73** B7
Musquodoboit Harbour
Canada 44°50N 63°9W **73** D7
Musselburgh *U.K.* 55°57N 3°2W **11** F5
Musselshell → *U.S.A.* 47°21N 107°57W **76** C10
Mussende *Angola* 10°32S 16°5E **52** G3
Mussoorie *India* 30°27N 78°6E **42** D8
Mussuco *Angola* 17°2S 19°3E **56** B2
Mustafakemalpaşa
Turkey 40°2N 28°24E **23** D13
Mustang *Nepal* 29°10N 83°55E **43** E10
Musters, L. *Argentina* 45°20S 69°25W **96** F3
Musudan *N. Korea* 40°50N 129°43E **35** D15
Muswellbrook *Australia* 32°16S 150°56E **63** E5
Mût *Egypt* 25°28N 28°58E **51** C11
Mut *Turkey* 36°40N 33°28E **44** B2
Mutanda *Mozam.* 21°0S 33°34E **57** C5
Mutanda *Zambia* 12°24S 26°13E **55** E2
Mutare *Zimbabwe* 18°58S 32°38E **55** F3
Mutawintji △ *Australia* 31°10S 142°30E **63** E3
Mutha *Kenya* 1°48S 38°26E **54** C4
Muting *Indonesia* 7°23S 140°20E **37** F10
Mutki = Mirtağ *Turkey* 38°23N 41°56E **44** B4
Mutoko *Zimbabwe* 17°24S 32°13E **57** B5
Mutomo *Kenya* 1°51S 38°12E **54** C4
Mutoray *Russia* 60°56N 101°0E **29** C11
Mutsamudu *Comoros Is.* 12°10S 44°25E **53** a
Mutshatsha
Dem. Rep. of the Congo 10°35S 24°20E **55** E1
Mutsu *Japan* 41°5N 140°55E **30** D10
Mutsu-Wan *Japan* 41°5N 140°55E **30** D10
Muttaburra *Australia* 22°38S 144°29E **62** C3
Mutton I. *Ireland* 52°49N 9°32W **10** D2
Mutuáli *Mozam.* 14°55S 37°0E **55** E4
Mutumba *Burundi* 3°25S 29°21E **54** C2
Muweilih *Egypt* 30°42N 34°19E **46** E3
Muy Muy *Nic.* 12°39N 85°36W **88** D2
Muyinga *Burundi* 3°14S 30°33E **54** C3
Muynak *Uzbekistan* 43°44N 59°10E **28** E6
Muyunkum, Peski = Moyynqum
Kazakhstan 44°12N 71°0E **32** C3
Muz Tag *China* 36°25N 87°25E **32** D6
Muzaffarabad *Pakistan* 34°25N 73°30E **43** B5
Muzaffargarh *Pakistan* 30°5N 71°14E **42** D4
Muzaffarnagar *India* 29°26N 77°40E **42** E7
Muzaffarpur *India* 26°7N 85°23E **43** F11
Muzafirpur *Pakistan* 30°58N 69°9E **42** D3
Muzhi *Russia* 65°25N 64°40E **18** A11
Muztagh-Ata *China* 38°17N 75°7E **32** D4
Mvuma *Zimbabwe* 19°16S 30°30E **55** F3
Mvurwi *Zimbabwe* 17°0S 30°57E **55** F3
Mwabvi ⌂ *Malawi* 16°42S 35°0E **55** F3
Mwadui *Tanzania* 3°26S 33°32E **54** C3
Mwali = Mohéli
Comoros Is. 12°20S 43°40E **53** a
Mwambo *Tanzania* 10°30S 40°22E **55** E5
Mwandi *Zambia* 17°30S 24°51E **55** F1
Mwanza
Dem. Rep. of the Congo 7°55S 26°43E **54** D2
Mwanza *Tanzania* 2°30S 32°58E **54** C3
Mwanza *Zambia* 16°58S 24°28E **55** F1
Mwanza □ *Tanzania* 2°0S 33°0E **54** C3
Mwaya *Tanzania* 9°32S 33°55E **55** D3
Mweelrea *Ireland* 53°39N 9°49W **10** C2
Mweka
Dem. Rep. of the Congo 4°50S 21°34E **52** E4

Mwene-Ditu
Dem. Rep. of the Congo 6°35S 22°27E **52** F4
Mwenezi *Zimbabwe* 21°15S 30°48E **55** G3
Mwenezi → *Mozam.* 22°40S 31°50E **55** G3
Mwenga
Dem. Rep. of the Congo 3°1S 28°28E **54** C2
Mweru, L. *Zambia* 9°0S 28°40E **55** D2
Mweru Wantipa △
Zambia 8°39S 29°25E **55** D2
Mweza Range *Zimbabwe* 21°0S 30°0E **55** G3
Mwilambwe
Dem. Rep. of the Congo 8°7S 25°5E **54** D2
Mwimbi *Tanzania* 8°38S 31°39E **55** D3
Mwingi *Kenya* 0°56S 38°4E **54** C4
Mwinilunga *Zambia* 11°43S 24°25E **55** E1
My Son *Vietnam* 15°48N 108°7E **38** E7
My Tho *Vietnam* 10°29N 106°23E **39** G6
Myajlar *India* 26°15N 70°20E **42** F4
Myanaung *Burma* 18°18N 95°22E **41** K19
Myanmar = Burma ■
Asia 21°0N 96°30E **41** J20
Myaungmya *Burma* 16°30N 94°40E **41** L19
Mycenæ = Mikines
Greece 37°43N 22°46E **23** F10
Myeik = Mergui *Burma* 12°26N 98°34E **38** F2
Myeik Kyunzu *Burma* 11°30N 97°30E **39** G1
Myerstown *U.S.A.* 40°22N 76°19W **83** F8
Myingyan *Burma* 21°30N 95°20E **41** J19
Myitkyina *Burma* 25°24N 97°26E **41** G20
Mykines *Færoe Is.* 62°7N 7°35W **8** E9
Mykolayiv *Ukraine* 46°58N 32°0E **19** E5
Mykonos *Greece* 37°30N 25°25E **23** F11
Mymensingh *Bangla.* 24°45N 90°24E **41** G17
Mynydd Du *U.K.* 51°52N 3°50W **13** F4
Mýrdalsjökull *Iceland* 63°40N 19°6W **8** E4
Myrtle Beach *U.S.A.* 33°42N 78°53W **85** E15
Myrtle Creek *U.S.A.* 43°1N 123°17W **76** E2
Myrtle Point *U.S.A.* 43°4N 124°8W **76** E1
Myrtou *Cyprus* 35°18N 33°4E **25** D12
Mysia *Turkey* 39°50N 27°0E **23** E12
Mysore = Karnataka □
India 13°15N 77°0E **40** N10
Mysore *India* 12°17N 76°41E **40** N10
Mystic *U.S.A.* 41°21N 71°58W **83** E13
Mysuru = Mysore *India* 12°17N 76°41E **40** N10
Myszków *Poland* 50°45N 19°22E **17** C10
Mytishchi *Russia* 55°50N 37°50E **18** C6
Mývatn *Iceland* 65°36N 17°0W **8** D5
Mzimba *Malawi* 11°55S 33°39E **55** E3
Mzimkulu → *S. Africa* 30°44S 30°28E **57** E5
Mzimvubu → *S. Africa* 31°38S 29°33E **57** E4
Mzuzu *Malawi* 11°30S 33°55E **55** E3

N

Na Clocha Liatha = Greystones
Ireland 53°9N 6°5W **10** C5
Na Haeo △ *Thailand* 17°31N 100°58E **38** D3
Na Hearadh = Harris
U.K. 57°50N 6°55W **11** D2
Na Noi *Thailand* 18°19N 100°43E **38** C3
Na Phao *Laos* 17°35N 105°44E **38** D5
Na Sam *Vietnam* 22°3N 106°37E **38** A6
Na Thon *Thailand* 9°32N 99°56E **39** b
Naab → *Germany* 49°1N 12°2E **16** D6
Naantali *Finland* 60°29N 22°2E **9** F20
Naas *Ireland* 53°12N 6°40W **10** C5
Nababeep *S. Africa* 29°36S 17°46E **56** D2
Nabadwip = Navadwip
India 23°34N 88°20E **43** H13
Nabawa *Australia* 28°30S 114°48E **61** E1
Nabberu, L. *Australia* 25°50S 120°30E **61** E3
Naberezhnyye Chelny
Russia 55°42N 52°19E **18** C9
Nabeul *Tunisia* 36°30N 10°44E **51** A8
Nabha *India* 30°26N 76°14E **42** D7
Nabīd *Iran* 29°40N 57°38E **45** D8
Nabire *Indonesia* 3°15S 135°26E **37** E9
Nabisar *Pakistan* 25°8N 69°40E **42** G3
Nabisipi → *Canada* 50°14N 62°13W **73** B7
Nabiswera *Uganda* 1°27N 32°15E **54** B3
Nāblus = Nābulus
West Bank 32°14N 35°15E **46** C4
Nabouwalu *Fiji* 17°0S 178°45E **59** a
Nābulus *West Bank* 32°14N 35°15E **46** C4
Nacala *Mozam.* 14°32S 40°34E **55** E5
Nacaome *Honduras* 13°31N 87°30W **88** D2
Nacaroa *Mozam.* 14°22S 39°56E **55** E4
Naches *U.S.A.* 46°44N 120°42W **76** C3
Naches → *U.S.A.* 46°38N 120°31W **78** D6
Nachicapau, L. *Canada* 56°40N 68°5W **73** A6
Nachingwea *Tanzania* 10°23S 38°49E **55** E4
Nachna *India* 27°34N 71°41E **42** F4
Nacimiento, L. *U.S.A.* 35°46N 120°53W **78** K6
Naco *Mexico* 31°19N 109°56W **86** A3
Nacogdoches *U.S.A.* 31°36N 94°39W **84** F7
Nácori Chico *Mexico* 29°40N 108°57W **86** B3
Nacozari de García
Mexico 30°25N 109°38W **86** A3
Nacula *Fiji* 16°54S 177°27E **59** a
Nada = Danzhou *China* 19°31N 109°33E **38** C7
Nådendal = Naantali
Finland 60°29N 22°2E **9** F20
Nadi *Fiji* 17°42S 177°20E **59** a
Nadiad *India* 22°41N 72°56E **42** H5
Nador *Morocco* 35°14N 2°58W **50** B5
Nadur *Malta* 36°2N 14°18E **25** C1
Nadūshan *Iran* 32°2N 53°35E **45** C7
Nadvirna *Ukraine* 48°37N 24°30E **17** D13
Nadvoitsy *Russia* 63°52N 34°14E **18** B5
Nadvornaya = Nadvirna
Ukraine 48°37N 24°30E **17** D13
Nadym *Russia* 65°35N 72°42E **28** C8
Nadym → *Russia* 66°12N 72°0E **28** C8
Nærbø *Norway* 58°40N 5°39E **9** G11
Næstved *Denmark* 55°13N 11°44E **9** J14
Nafpaktos *Greece* 38°24N 21°50E **23** E9
Nafplio *Greece* 37°33N 22°50E **23** F10
Naft-e Safid *Iran* 31°40N 49°17E **45** D6

Naftshahr *Iran* 34°0N 45°30E **44** C5
Nafud Desert = An Nafūd
Si. Arabia 28°15N 41°0E **44** D4
Naga *Phil.* 13°38N 123°15E **37** B6
Nagagami *Canada* 50°23N 84°20W **72** B3
Nagahama *Japan* 35°23N 136°16E **31** G8
Nagai *Japan* 38°6N 140°2E **30** E10
Nagaland □ *India* 26°0N 94°30E **41** G19
Nagano *Japan* 36°40N 138°10E **31** F9
Nagano □ *Japan* 36°15N 138°0E **31** F9
Nagaoka *Japan* 37°27N 138°51E **31** F9
Nagaon = Nowgong
India 26°20N 92°50E **41** F18
Nagappattinam *India* 10°46N 79°51E **40** P11
Nagar → *Bangla.* 24°27N 89°12E **43** G13
Nagar Parkar *Pakistan* 24°28N 70°46E **42** G4
Nagasaki *Japan* 32°47N 129°50E **31** H4
Nagasaki □ *Japan* 32°50N 129°40E **31** H4
Nagato *Japan* 34°19N 131°5E **31** G5
Nagaur *India* 27°15N 73°45E **42** F5
Nagda *India* 23°27N 75°25E **42** H6
Nagercoil *India* 8°12N 77°26E **40** Q10
Nagina *India* 29°30N 78°30E **43** E8
Nagīneh *Iran* 34°20N 57°15E **45** C8
Nagir *Pakistan* 36°12N 74°42E **43** A6
Nagles Mts. *Ireland* 52°8N 8°30W **10** D3
Nagod *India* 24°34N 80°36E **43** G9
Nagoorin *Australia* 24°17S 151°15E **62** C5
Nagorno-Karabakh □
Azerbaijan 39°55N 46°45E **44** B5
Nagornyy *Russia* 55°58N 124°57E **29** D13
Nagoya *Japan* 35°10N 136°50E **31** G8
Nagpur *India* 21°8N 79°10E **40** J11
Nagqu *China* 31°29N 92°3E **32** E7
Nagua *Dom. Rep.* 19°23N 69°50W **89** C6
Naguabo *Puerto Rico* 18°13N 65°44W **89** d
Nagurunguru ✪
Australia 16°45S 129°45E **60** C4
Nagykanizsa *Hungary* 46°28N 17°0E **17** E9
Nagykőrös *Hungary* 47°5N 19°48E **17** E10
Naha *Japan* 26°13N 127°42E **31** L3
Nahan *India* 30°33N 77°18E **42** D7
Nahanni △ *Canada* 61°36N 125°41W **70** A4
Nahanni Butte *Canada* 61°2N 123°31W **70** A4
Nahargarh *Mad. P., India* 24°10N 75°14E **42** G6
Nahargarh *Raj., India* 24°55N 76°50E **42** G7
Nahariyya *Israel* 33°1N 35°5E **44** C2
Nahāvand *Iran* 34°10N 48°22E **45** C6
Nahuel Huapi, L.
Argentina 41°0S 71°32W **96** E2
Nahuelbuta △ *Chile* 37°44S 72°57W **94** D1
Nai Yong *Thailand* 8°14N 98°22E **39** a
Naicá *Mexico* 27°53N 105°31W **86** B3
Naicam *Canada* 52°30N 104°30W **71** C8
Naikoon ⌂ *Canada* 53°55N 131°55W **70** C2
Naimisharanya *India* 27°21N 80°30E **43** F9
Naimona'nyi Feng *Nepal* 30°26N 81°18E **43** D9
Nain *Canada* 56°34N 61°40W **73** A7
Nā'īn *Iran* 32°54N 53°0E **45** C7
Naini Tal *India* 29°30N 79°30E **43** E8
Nainpur *India* 22°30N 80°10E **43** H9
Nainwa *India* 25°46N 75°51E **42** G6
Nairai *Fiji* 17°49S 179°15E **59** a
Nairn *U.K.* 57°35N 3°53W **11** D5
Nairobi *Kenya* 1°17S 36°48E **54** C4
Nairobi △ *Kenya* 1°22S 36°50E **54** C4
Naissaar *Estonia* 59°34N 24°29E **9** G21
Naitaba *Fiji* 17°0S 179°16W **59** a
Naivasha *Kenya* 0°40S 36°30E **54** C4
Naivasha, L. *Kenya* 0°48S 36°20E **54** C4
Najaf = An Najaf *Iraq* 32°3N 44°15E **44** C5
Najafābād *Iran* 32°40N 51°15E **45** C6
Najd *Si. Arabia* 26°30N 42°0E **47** B3
Najibabad *India* 29°40N 78°20E **42** E8
Najin *N. Korea* 42°12N 130°15E **35** C16
Najmah *Si. Arabia* 26°42N 50°6E **45** E6
Najrān *Si. Arabia* 17°34N 44°18E **47** D3
Naju *S. Korea* 35°3N 126°43E **35** G14
Nakadōri-Shima *Japan* 32°57N 129°4E **31** H4
Nakalagba
Dem. Rep. of the Congo 2°50N 27°58E **54** B2
Nakaminato *Japan* 36°21N 140°36E **31** F10
Nakamura *Japan* 32°59N 132°56E **31** H6
Nakano *Japan* 36°45N 138°22E **31** F9
Nakano-Shima *Japan* 29°51N 129°52E **31** K4
Nakashibetsu *Japan* 43°33N 144°59E **30** C12
Nakfa *Eritrea* 16°40N 38°32E **47** D2
Nakha Yai, Ko *Thailand* 8°3N 98°28E **39** a
Nakhichevan = Naxçıvan
Azerbaijan 39°12N 45°15E **44** B5
Nakhichevan Rep. = Naxçıvan □
Azerbaijan 39°25N 45°26E **44** B5
Nakhl *Egypt* 29°55N 33°43E **46** F2
Nakhl-e Taqī *Iran* 27°28N 52°36E **45** E7
Nakhodka *Russia* 42°53N 132°54E **30** C6
Nakhon Nayok
Thailand 14°12N 101°13E **38** E3
Nakhon Pathom
Thailand 13°49N 100°3E **38** F3
Nakhon Phanom
Thailand 17°23N 104°43E **38** D5
Nakhon Ratchasima
Thailand 14°59N 102°12E **38** E4
Nakhon Sawan
Thailand 15°35N 100°10E **38** E3
Nakhon Si Thammarat
Thailand 8°29N 100°0E **39** H3
Nakhon Thai *Thailand* 17°5N 100°44E **38** D3
Nakhtarana *India* 23°20N 69°15E **42** H3
Nakina *Canada* 50°10N 86°40W **72** B2
Nakodar *India* 31°8N 75°31E **42** D6
Naktong → *S. Korea* 35°7N 128°57E **35** G15
Nakuru *Kenya* 0°15S 36°4E **54** C4
Nakuru, L. *Kenya* 0°23S 36°5E **54** C4
Nakusp *Canada* 50°20N 117°45W **70** C5
Nal *Pakistan* 27°40N 66°12E **42** F2
Nal → *Pakistan* 25°20N 65°30E **42** G1
Nalázi *Mozam.* 24°3S 33°20E **57** C5
Nalchik *Russia* 43°30N 43°33E **19** F7
Nalgonda *India* 17°6N 79°15E **40** L11

Nalhati *India* 24°17N 87°52E **43** G12
Naliya *India* 23°16N 68°50E **42** H3
Nallamalai Hills *India* 15°30N 78°50E **40** M11
Nalubaale Dam *Uganda* 0°30N 33°5E **54** B3
Nam Can *Vietnam* 8°46N 104°59E **39** H5
Nam-ch'on *N. Korea* 38°15N 126°26E **35** E14
Nam Co *China* 30°30N 90°45E **32** E7
Nam Dinh *Vietnam* 20°25N 106°5E **38** B6
Nam Du, Quan Dao *Vietnam* 9°41N 104°21E **39** H5
Nam Loi → *Burma* 21°3N 98°27E **38** B2
Nam Nao △ *Thailand* 16°44N 101°32E **38** D3
Nam Ngum Res. *Laos* 18°35N 102°34E **38** C4
Nam-Phan *Vietnam* 10°30N 106°0E **39** G6
Nam Phong *Thailand* 16°42N 102°52E **38** D4
Nam Theun Res. *Laos* 17°51N 105°3E **38** D5
Nam Tok *Thailand* 14°21N 99°4E **38** E2
Nam Un Res. *Thailand* 17°13N 103°44E **38** D4
Namacunde *Angola* 17°18S 15°50E **56** B2
Namacurra *Mozam.* 17°30S 36°50E **57** B6
Namak, Daryācheh-ye *Iran* 34°30N 52°0E **45** C7
Namak, Kavir-e *Iran* 34°30N 57°30E **45** C8
Namakzār, Daryācheh-ye *Iran* 34°0N 60°30E **45** C9
Namaland *Namibia* 26°0S 17°0E **56** C2
Namanga *Kenya* 2°33S 36°47E **54** C4
Namangan *Uzbekistan* 41°0N 71°40E **32** C3
Namapa *Mozam.* 13°43S 39°50E **55** E4
Namaqualand *S. Africa* 30°0S 17°25E **56** E2
Namasagali *Uganda* 1°2N 33°0E **54** B3
Namber *Indonesia* 1°2S 134°49E **37** E8
Nambour *Australia* 26°32S 152°58E **63** D5
Nambouwalu = Nabouwalu *Fiji* 17°0S 178°45E **59** a
Nambucca Heads *Australia* 30°37S 153°0E **63** E5
Nambung △ *Australia* 30°30S 115°5E **61** F2
Namcha Barwa *China* 29°40N 95°10E **32** F8
Namche Bazar *Nepal* 27°51N 86°47E **43** F12
Namchonjŏm = Nam-ch'on *N. Korea* 38°15N 126°26E **35** E14
Namecunda *Mozam.* 14°54S 37°37E **55** E4
Namenalala *Fiji* 17°8S 179°9E **59** a
Nameponda *Mozam.* 15°50S 39°50E **55** F4
Nametil *Mozam.* 15°40S 39°21E **55** F4
Namew L. *Canada* 54°14N 101°56W **71** C8
Namgia *India* 31°48N 78°40E **43** D8
Namialo *Mozam.* 14°55S 39°59E **55** E4
Namib Desert *Namibia* 22°30S 15°0E **56** C2
Namib-Naukluft ⌓ *Namibia* 24°40S 15°16E **56** C2
Namibe *Angola* 15°7S 12°11E **53** H2
Namibe □ *Angola* 16°35S 12°30E **56** B1
Namibia ■ *Africa* 22°0S 18°9E **56** C2
Namibwoestyn = Namib Desert *Namibia* 22°30S 15°0E **56** C2
Namjeju *S. Korea* 33°14N 126°33E **35** H14
Namlea *Indonesia* 3°18S 127°5E **37** E7
Namoi → *Australia* 30°12S 149°30E **63** E4
Nampa *U.S.A.* 43°34N 116°34W **76** E5
Namp'o *N. Korea* 38°52N 125°10E **35** E13
Nampō-Shotō *Japan* 32°0N 140°0E **31** J10
Nampula *Mozam.* 15°6S 39°15E **55** F4
Namrole *Indonesia* 3°46S 126°46E **37** E7
Namse Shankou *China* 30°0N 82°25E **43** E10
Namsen → *Norway* 64°28N 11°37E **8** D14
Namsos *Norway* 64°29N 11°30E **8** D14
Namtok Chat Trakan ⌓ *Thailand* 17°17N 100°40E **38** D3
Namtok Huay Yang △ *Thailand* 11°35N 99°30E **39** G2
Namtok Mae Surin △ *Thailand* 18°55N 98°2E **38** C2
Namtok Ngao △ *Thailand* 9°58N 98°46E **39** H2
Namtok Phlew △ *Thailand* 12°32N 102°13E **38** F4
Namtok Yong △ *Thailand* 8°15N 98°45E **39** H2
Namtsy *Russia* 62°43N 129°37E **29** C13
Namtu *Burma* 23°5N 97°28E **41** H20
Namtumbo *Tanzania* 10°30S 36°4E **55** E4
Namu *Canada* 51°52N 127°50W **70** C3
Namuka-i-Lau *Fiji* 18°53S 178°37W **59** a
Namur *Belgium* 50°27N 4°52E **15** D4
Namur □ *Belgium* 50°17N 5°0E **15** D4
Namuruputh *Kenya* 4°34N 35°57E **54** B4
Namutoni *Namibia* 18°49S 16°55E **56** B2
Namwala *Zambia* 15°44S 26°30E **55** F2
Namwon *S. Korea* 35°23N 127°23E **35** G14
Namyang *N. Korea* 42°57N 129°52E **35** C15
Nan *Thailand* 18°48N 100°46E **38** C3
Nan → *Thailand* 15°42N 100°9E **38** E3
Nan-ch'ang = Nanchang *China* 28°42N 115°55E **33** F12
Nan Ling *China* 25°0N 112°30E **33** F11
Nanaimo *Canada* 49°10N 124°0W **70** D4
Nänäkuli *U.S.A.* 21°24N 158°9W **75** L8
Nanam *N. Korea* 41°44N 129°40E **35** D15
Nanango *Australia* 26°40S 152°0E **63** D5
Nanao *Japan* 37°0N 137°0E **31** F8
Nanchang *China* 28°42N 115°55E **33** F12
Nanching = Nanjing *China* 32°2N 118°47E **33** E12
Nanchong *China* 30°43N 106°2E **32** E10
Nancy *France* 48°42N 6°12E **20** B7
Nanda Devi *India* 30°23N 79°59E **43** D8
Nanda Devi △ *India* 30°30N 79°50E **43** D8
Nanda Kot *India* 30°17N 80°5E **43** D9
Nandan *Japan* 34°10N 134°42E **31** G7
Nanded *India* 19°10N 77°20E **40** K10
Nandewar Ra. *Australia* 30°15S 150°35E **63** E5
Nandi = Nadi *Fiji* 17°42S 177°20E **59** a
Nandi *Zimbabwe* 20°58S 31°44E **55** G3
Nandigram *India* 22°1N 87°58E **43** H12
Nandurbar *India* 21°20N 74°15E **40** J9
Nandyal *India* 15°30N 78°30E **40** M11
Nang Rong *Thailand* 14°38N 102°48E **38** E4
Nanga-Eboko *Cameroon* 4°41N 12°22E **52** D2
Nanga Parbat *Pakistan* 35°10N 74°35E **43** B6
Nangade *Mozam.* 11°5S 39°36E **55** E4
Nangapinoh *Indonesia* 0°20S 111°44E **36** E4
Nangarhār □ *Afghan.* 34°20N 70°0E **40** B7
Nangatayap *Indonesia* 1°32S 110°34E **36** E4
Nangeya Mts. *Uganda* 3°30N 33°30E **54** B3
Nangong *China* 37°23N 115°22E **34** F8
Nanhuang *China* 36°58N 121°48E **35** F11
Nanisivik *Canada* 73°2N 84°33W **69** C15
Nanjing *China* 32°2N 118°47E **33** E12
Nanjirinji *Tanzania* 9°41S 39°5E **55** D4
Nankana Sahib *Pakistan* 31°27N 73°38E **42** D5
Nanking = Nanjing *China* 32°2N 118°47E **33** E12
Nankoku *Japan* 33°39N 133°44E **31** H6
Nanlang *China* 22°30N 113°32E **33** a
Nanning *China* 22°48N 108°20E **32** G10
Nannup *Australia* 33°59S 115°48E **61** F2
Nanpara *India* 27°52N 81°33E **43** F9
Nanping *China* 26°38N 118°10E **33** F12
Nansei-Shotō = Ryūkyū-Rettō *Japan* 26°0N 126°0E **31** M3
Nansen Basin *Arctic* 84°0N 50°0E **4** A10
Nansen Sd. *Canada* 81°0N 91°0W **69** A13
Nansha *China* 22°45N 113°34E **33** a
Nanshan *China* 18°19N 109°10E **38** C7
Nanshan I. *S. China Sea* 10°45N 115°49E **36** B5
Nansio *Tanzania* 2°3S 33°4E **54** C3
Nantawarrinna ❂ *Australia* 30°49S 138°58E **63** B2
Nantes *France* 47°12N 1°33W **20** C3
Nanticoke *U.S.A.* 41°12N 76°0W **83** E8
Nanton *Canada* 50°21N 113°46W **70** C6
Nantong *China* 32°1N 120°52E **33** E13
Nantou *China* 22°32N 113°55E **33** a
Nantucket *U.S.A.* 41°17N 70°6W **81** E18
Nantucket I. *U.S.A.* 41°16N 70°5W **81** E18
Nantulo *Mozam.* 12°32S 38°45E **55** E4
Nantwich *U.K.* 53°4N 2°31W **12** D5
Nanty Glo *U.S.A.* 40°28N 78°50W **82** F6
Nanuku Passage *Fiji* 16°45S 179°15W **59** a
Nanuque *Brazil* 17°50S 40°21W **93** G10
Nanusa, Kepulauan *Indonesia* 4°45N 127°1E **37** D7
Nanutarra Roadhouse *Australia* 22°32S 115°30E **60** D2
Nanyang *China* 33°11N 112°30E **34** H7
Nanyuan *China* 39°47N 116°24E **34** E9
Nanyuki *Kenya* 0°2N 37°4E **54** B4
Nanzhao *China* 33°30N 112°20E **34** H7
Nao, C. de la *Spain* 38°44N 0°14E **21** C6
Naococane, L. *Canada* 52°50N 70°45W **73** B5
Napa *U.S.A.* 38°18N 122°17W **78** G4
Napa → *U.S.A.* 38°10N 122°19W **78** G4
Napanee *Canada* 44°15N 77°0W **82** B8
Napanoch *U.S.A.* 41°44N 74°22W **83** E10
Nape *Laos* 18°18N 105°6E **38** C5
Nape Pass = Keo Neua, Deo *Vietnam* 18°23N 105°10E **38** C5
Napier *N.Z.* 39°30S 176°56E **59** C6
Napier Broome B. *Australia* 14°2S 126°37E **60** B4
Napier Pen. *Australia* 12°4S 135°43E **62** A2
Napierville *Canada* 45°11N 73°25W **83** A11
Naples = Nápoli *Italy* 40°50N 14°15E **22** D6
Naples *U.S.A.* 26°8N 81°48W **85** H14
Napo → *Peru* 3°20S 72°40W **92** D4
Napoleon *N. Dak., U.S.A.* 46°30N 99°46W **80** B4
Napoleon *Ohio, U.S.A.* 41°23N 84°8W **81** E11
Nápoli *Italy* 40°50N 14°15E **22** D6
Napopo *Dem. Rep. of the Congo* 4°15N 28°0E **54** B2
Naqadeh *Iran* 36°57N 45°23E **44** B5
Naqb, Ra's an *Jordan* 29°48N 35°44E **46** F4
Naqqāsh *Iran* 35°40N 49°6E **45** C6
Nara *Japan* 34°40N 135°49E **31** G7
Nara *Mali* 15°10N 7°20W **50** E4
Nara □ *Japan* 34°30N 136°0E **31** G8
Nara Canal *Pakistan* 24°30N 69°20E **42** G3
Nara Visa *U.S.A.* 35°37N 103°6W **77** J12
Naracoorte *Australia* 36°58S 140°45E **63** F3
Naradhan *Australia* 33°34S 146°17E **63** E4
Naraini *India* 25°11N 80°29E **43** G9
Naranjos *Mexico* 21°21N 97°41W **87** C5
Narasapur *India* 16°26N 81°40E **41** L12
Narathiwat *Thailand* 6°30N 101°48E **39** J3
Narayangadh = Bharatpur *Nepal* 27°34N 84°10E **43** F11
Narayanganj *Bangla.* 23°40N 90°33E **41** H17
Narayanpet *India* 16°45N 77°30E **40** L10
Narberth *U.K.* 51°47N 4°44W **13** F3
Narbonne *France* 43°11N 3°0E **20** E5
Nardīn *Iran* 37°3N 55°59E **45** B7
Nardò *Italy* 40°11N 18°2E **23** D8
Narembeen *Australia* 32°7S 118°24E **61** F2
Narendranagar *India* 30°10N 78°18E **42** D8
Nares Str. *Arctic* 80°0N 70°0W **69** B18
Naretha *Australia* 31°0S 124°45E **61** F3
Narew → *Poland* 52°26N 20°41E **17** B11
Nari → *Pakistan* 28°0N 67°40E **42** F2
Narin *Afghan.* 36°5N 69°0E **40** A6
Narindra, Helodranon' i *Madag.* 14°55S 47°30E **57** A8
Narita *Japan* 35°47N 140°19E **31** G10
Nariva Swamp *Trin. & Tob.* 10°26N 61°4W **93** K15
Narmada → *India* 21°38N 72°36E **42** J5
Narnaul *India* 28°5N 76°11E **42** E7
Narodnaya *Russia* 65°5N 59°58E **18** A10
Narok *Kenya* 1°55S 35°52E **54** C4
Narooma *Australia* 36°14S 150°4E **63** F5
Narowal *Pakistan* 32°6N 74°52E **42** C6
Narrabri *Australia* 30°19S 149°46E **63** E4
Narran → *Australia* 28°37S 148°12E **63** D4
Narrandera *Australia* 34°42S 146°31E **63** E4
Narrogin *Australia* 32°58S 117°14E **61** F2
Narromine *Australia* 32°12S 148°12E **63** E4
Narrow Hills ⌓ *Canada* 54°0N 104°37W **71** C8
Narsimhapur *India* 22°54N 79°14E **43** H8
Narsinghgarh *India* 23°45N 76°40E **42** H7
Naruto *Japan* 34°11N 134°37E **31** G7
Narva *Estonia* 59°23N 28°12E **18** C4
Narva → *Russia* 59°27N 28°2E **9** G23
Narva Bay = Narva Laht *Estonia* 59°35N 27°35E **9** G22
Narva Laht *Estonia* 59°35N 27°35E **9** G22
Narvik *Norway* 68°28N 17°26E **8** B17
Narwana *India* 29°39N 76°6E **42** E7
Narwinbi ❂ *Australia* 16°7S 136°17E **62** B2
Naryan-Mar *Russia* 67°42N 53°12E **18** A9
Narym *Russia* 59°0N 81°30E **28** D9
Naryn *Kyrgyzstan* 41°26N 75°58E **32** C4
Naryn Qum *Kazakhstan* 47°30N 49°0E **28** E5
Nasa *Norway* 66°29N 15°23E **8** C16
Nasau *Fiji* 17°19S 179°27E **59** a
Nasca *Peru* 14°50S 74°57W **92** F4
Nasca Ridge *Pac. Oc.* 20°0S 80°0W **65** K19
Naseby *N.Z.* 45°1S 170°10E **59** F3
Naselle *U.S.A.* 46°22N 123°49W **78** D3
Naser, Buheirat en *Egypt* 23°0N 32°30E **51** D12
Nashik = Nasik *India* 19°58N 73°50E **40** K8
Nashua *Mont., U.S.A.* 48°8N 106°22W **76** B10
Nashua *N.H., U.S.A.* 42°45N 71°28W **83** D13
Nashville *Ark., U.S.A.* 33°57N 93°51W **84** E8
Nashville *Ga., U.S.A.* 31°12N 83°15W **85** F13
Nashville *Tenn., U.S.A.* 36°10N 86°47W **85** C11
Nasik *India* 19°58N 73°50E **40** K8
Nasirabad *India* 26°15N 74°45E **42** F6
Nasirabad *Pakistan* 28°23N 68°24E **42** E3
Nasiri = Ahvāz *Iran* 31°20N 48°40E **45** D6
Nasiriyah = An Nāṣirīyah *Iraq* 31°0N 46°15E **44** D5
Naskaupi → *Canada* 53°47N 60°51W **73** B7
Naṣrābād *Iran* 34°8N 51°26E **45** C6
Naṣrīān-e Pā'īn *Iran* 32°52N 46°52E **44** C5
Nass → *Canada* 55°0N 129°40W **70** C3
Nassau *Bahamas* 25°5N 77°20W **88** A4
Nassau *U.S.A.* 42°31N 73°37W **83** D11
Nassau, B. *Chile* 55°20S 68°0W **96** H3
Nasser, L. = Naser, Buheirat en *Egypt* 23°0N 32°30E **51** D12
Nässjö *Sweden* 57°39N 14°42E **9** H16
Nastapoka → *Canada* 56°55N 76°33W **72** A4
Nastapoka, Is. *Canada* 56°55N 76°50W **72** A4
Nata *Botswana* 20°12S 26°12E **56** C4
Nata → *Botswana* 20°14S 26°10E **56** C4
Natal *Brazil* 5°47S 35°13W **93** E11
Natal *Indonesia* 0°35N 99°7E **36** D1
Natal Drakensberg ⌓ *S. Africa* 29°27S 29°30E **57** D4
Naṭanz *Iran* 33°30N 51°55E **45** C6
Natashquan *Canada* 50°14N 61°46W **73** B7
Natashquan → *Canada* 50°7N 61°50W **73** B7
Natchez *U.S.A.* 31°34N 91°24W **84** F9
Natchitoches *U.S.A.* 31°46N 93°5W **84** F8
Natewa B. *Fiji* 16°35S 179°40E **59** a
Nathalia *Australia* 36°1S 145°13E **63** F4
Nathdwara *India* 24°55N 73°50E **42** G5
Nati, Pta. *Spain* 40°3N 3°50E **24** A10
Natimuk *Australia* 36°42S 142°0E **63** F3
Nation → *Canada* 55°30N 123°32W **70** B4
National City *U.S.A.* 32°40N 117°5W **79** N9
Natitingou *Benin* 10°20N 1°26E **50** F6
Natividad, I. *Mexico* 27°52N 115°11W **86** B1
Natkyizin *Burma* 14°57N 97°59E **38** E1
Natron, L. *Tanzania* 2°20S 36°0E **54** C4
Natrona Heights *U.S.A.* 40°37N 79°44W **82** F5
Natukanaoka Pan *Namibia* 18°40S 15°45E **56** B2
Natuna Besar, Kepulauan *Indonesia* 4°0N 108°15E **36** D3
Natuna Is. = Natuna Besar, Kepulauan *Indonesia* 4°0N 108°15E **36** D3
Natuna Selatan, Kepulauan *Indonesia* 2°45N 109°0E **36** D3
Natural Bridge *U.S.A.* 44°5N 75°30W **83** B9
Natural Bridges ⌓ *U.S.A.* 37°36N 110°0W **77** H9
Naturaliste, C. *Tas., Australia* 40°50S 148°15E **63** G4
Naturaliste, C. *W. Austral., Australia* 33°32S 115°0E **61** F2
Naturaliste Plateau *Ind. Oc.* 34°0S 112°0E **64** L3
Nau Qala *Afghan.* 34°5N 68°5E **42** B3
Naugatuck *U.S.A.* 41°30N 73°3W **83** E11
Naujaat = Repulse Bay *Canada* 66°30N 86°30W **69** D14
Naumburg *Germany* 51°9N 11°47E **16** C6
Nauru ■ *Pac. Oc.* 1°0S 166°0E **58** B9
Naushahra = Nowshera *Pakistan* 34°0N 72°0E **40** C8
Naushahro *Pakistan* 26°50N 68°7E **42** F3
Naushon I. *U.S.A.* 41°29N 70°45W **83** E14
Nausori *Fiji* 18°2S 178°32E **59** a
Nauta *Peru* 4°31S 73°35W **92** D4
Nautanwa *India* 27°20N 83°25E **43** F10
Naute ⌓ *Namibia* 26°55S 17°57E **56** D2
Nautla *Mexico* 20°13N 96°47W **87** C5
Nava *Mexico* 28°25N 100°45W **86** B4
Navadwip *India* 23°34N 88°20E **43** H13
Navahrudak *Belarus* 53°40N 25°50E **17** B13
Navajo Res. *U.S.A.* 36°48N 107°36W **77** H10
Navalmoral de la Mata *Spain* 39°52N 5°33W **21** C3
Navan = An Uaimh *Ireland* 53°39N 6°41W **10** C5
Navarin, Mys *Russia* 62°15N 179°5E **29** C18
Navarino, I. *Chile* 55°0S 67°40W **96** H3
Navarra □ *Spain* 42°40N 1°40W **21** A5
Navarre *U.S.A.* 40°43N 81°31W **82** F3
Navarro → *U.S.A.* 39°11N 123°45W **78** F3
Navasota *U.S.A.* 30°23N 96°5W **84** F6
Navassa I. *W. Indies* 18°30N 75°0W **89** C5
Naver → *U.K.* 58°32N 4°14W **11** C4
Navibandar *India* 21°26N 69°48E **42** J3
Navidad *Chile* 33°57S 71°50W **94** C1
Naviraí *Brazil* 23°8S 54°13W **95** A5
Naviti *Fiji* 17°7S 177°15E **59** a
Navlakhi *India* 22°58N 70°28E **42** H4
Năvodari *Romania* 44°19N 28°36E **17** F15
Navoi *Uzbekistan* 40°9N 65°22E **28** E7
Navojoa *Mexico* 27°6N 109°26W **86** B3
Navolato *Mexico* 24°47N 107°42W **86** C3
Navsari *India* 20°57N 72°59E **40** J8
Navua *Fiji* 18°12S 178°11E **59** a
Nawa Kot *Pakistan* 28°21N 71°24E **42** E4
Nawab Khan *Pakistan* 30°17N 69°12E **42** D3
Nawabganj *Ut. P., India* 26°56N 81°14E **43** F9
Nawabganj *Ut. P., India* 28°32N 79°40E **43** E8
Nawabshah *Pakistan* 26°15N 68°25E **42** F3
Nawada *India* 24°50N 85°33E **43** G11
Nawakot *Nepal* 27°55N 85°10E **43** F11
Nawalgarh *India* 27°50N 75°15E **42** F6
Nawanshahr *India* 32°33N 74°48E **43** C6
Nawar, Dasht-i- *Afghan.* 33°52N 68°0E **42** C3
Nawoiy = Navoi *Uzbekistan* 40°9N 65°22E **28** E7
Naxçıvan *Azerbaijan* 39°12N 45°15E **44** B5
Naxçıvan □ *Azerbaijan* 39°25N 45°26E **44** B5
Naxos *Greece* 37°8N 25°25E **23** F11
Nay, Mui *Vietnam* 12°54N 109°26E **38** F7
Nāy Band *Būshehr, Iran* 27°20N 52°40E **45** E7
Nāy Band *Khorāsān, Iran* 32°20N 57°34E **45** C8
Nayakhan *Russia* 61°56N 159°0E **29** C16
Nayarit □ *Mexico* 22°0N 105°0W **86** C4
Nayau *Fiji* 18°6S 178°10E **59** a
Nayoro *Japan* 44°21N 142°28E **30** B11
Naypyidaw *Burma* 19°44N 96°12E **38** C1
Nayyāl, W. → *Si. Arabia* 28°35N 39°4E **44** D3
Nazaré *Brazil* 13°2S 39°0W **93** F11
Nazareth = Naẓerat *Israel* 32°42N 35°17E **46** C4
Nazareth *U.S.A.* 40°44N 75°19W **83** F9
Nazarovo *Russia* 57°2N 90°40E **29** D10
Nazas *Mexico* 25°14N 104°8W **86** B4
Nazas → *Mexico* 25°12N 104°12W **86** B4
Nazca = Nasca *Peru* 14°50S 74°57W **92** F4
Naze *Japan* 28°22N 129°27E **31** K4
Naze, The *U.K.* 51°53N 1°18E **13** F9
Naẓerat *Israel* 32°42N 35°17E **46** C4
Nāzīk *Iran* 39°1N 45°4E **44** B5
Nazilli *Turkey* 37°55N 28°15E **23** F13
Nazko *Canada* 53°1N 123°37W **70** C4
Nazko → *Canada* 53°7N 123°34W **70** C4
Nazret *Ethiopia* 8°32N 39°22E **47** F2
Nchanga *Zambia* 12°30S 27°49E **55** E2
Ncheu *Malawi* 14°50S 34°47E **55** E3
Ndala *Tanzania* 4°45S 33°15E **54** C3
Ndalatando *Angola* 9°12S 14°48E **52** F2
Ndareda *Tanzania* 4°12S 35°30E **54** C4
Ndélé *C.A.R.* 8°25N 20°36E **52** C4
Ndjamena *Chad* 12°10N 15°0E **51** F8
Ndola *Zambia* 13°0S 28°34E **55** E2
Ndomo ⌓ *S. Africa* 26°52S 32°15E **57** D5
Ndoto Mts. *Kenya* 2°0N 37°0E **54** B4
Nduguti *Tanzania* 4°18S 34°41E **54** C3
Neagh, Lough *U.K.* 54°37N 6°25W **10** B5
Neah Bay *U.S.A.* 48°22N 124°37W **78** B2
Neale, L. *Australia* 24°15S 130°0E **60** D5
Neales → *Australia* 28°8S 136°47E **63** D2
Neapoli *Greece* 35°15N 25°37E **25** D7
Near Is. *U.S.A.* 52°30N 174°0E **74** E2
Neath *U.K.* 51°39N 3°48W **13** F4
Neath Port Talbot □ *U.K.* 51°42N 3°45W **13** F4
Nebbi *Uganda* 2°28N 31°6E **54** B3
Nebine Cr. → *Australia* 29°27S 146°56E **63** D4
Nebitdag = Balkanabat *Turkmenistan* 39°30N 54°22E **45** B7
Nebo *Australia* 21°42S 148°42E **62** C4
Nebraska □ *U.S.A.* 41°30N 99°30W **80** E4
Nebraska City *U.S.A.* 40°41N 95°52W **80** E6
Nébrodi, Monti *Italy* 37°54N 14°35E **22** F6
Necedah *U.S.A.* 44°2N 90°4W **80** C8
Nechako → *Canada* 53°55N 122°42W **70** C4
Neches → *U.S.A.* 29°58N 93°51W **84** G8
Neckar → *Germany* 49°27N 8°29E **16** D5
Necker I. *U.S.A.* 23°35N 164°42W **75** L7
Necochea *Argentina* 38°30S 58°50W **94** D4
Nederland = Netherlands ■ *Europe* 52°0N 5°30E **15** C5
Needles *Canada* 49°53N 118°7W **70** D5
Needles *U.S.A.* 34°51N 114°37W **79** L12
Needles, The *U.K.* 50°39N 1°35W **13** G6
Neembucú □ *Paraguay* 27°0S 58°0W **94** B4
Neemuch = Nimach *India* 24°30N 74°56E **42** G6
Neenah *U.S.A.* 44°11N 88°28W **80** C9
Neepawa *Canada* 50°15N 99°30W **71** C9
Neftçala *Azerbaijan* 39°19N 49°12E **45** B6
Neftegorsk *Russia* 53°1N 142°58E **29** D15
Neftekumsk *Russia* 44°46N 44°50E **19** F7
Nefteyugansk *Russia* 61°5N 72°42E **28** C8
Nefyn *U.K.* 52°56N 4°31W **12** E3
Negapatam = Nagappattinam *India* 10°46N 79°51E **40** P11
Negara *Indonesia* 8°22S 114°37E **37** J17
Negaunee *U.S.A.* 46°30N 87°36W **80** B10
Negele *Ethiopia* 5°20N 39°36E **47** F2
Negeri Sembilan □ *Malaysia* 2°45N 102°10E **39** L4
Negev Desert = Hanegev *Israel* 30°50N 35°0E **46** E4
Negombo *Sri Lanka* 7°12N 79°50E **40** R11
Negotin *Serbia* 44°16N 22°37E **23** B10
Negra, Pta. *Mauritania* 22°54N 16°18W **50** D2
Negra, Pta. *Peru* 6°6S 81°10W **92** E2
Negrais, C. = Maudin Sun *Burma* 16°0N 94°30E **41** M19
Negril *Jamaica* 18°22N 78°20W **88** a
Negro → *Argentina* 41°2S 62°47W **96** E4
Negro → *Brazil* 3°0S 60°0W **92** D7
Negro → *Uruguay* 33°24S 58°22W **94** C4
Negros *Phil.* 9°30N 122°40E **37** C6
Neguac *Canada* 47°15N 65°5W **73** C6
Nehalem → *U.S.A.* 45°40N 123°56W **78** E3
Nehāvand *Iran* 35°56N 49°31E **45** C6
Nehbandān *Iran* 31°35N 60°5E **45** D9
Nehe *China* 48°29N 124°50E **33** B13
Nei Mongol Zizhiqu □ *China* 42°0N 112°0E **34** D7
Neiafu *Tonga* 18°39S 173°59W
Neiges, Piton des *Réunion* 21°5S 55°29E **3**
Neijiang *China* 29°35N 104°55E **3**
Neilingding Dao *China* 22°25N 113°48E
Neillsville *U.S.A.* 44°34N 90°36W **8**
Neilton *U.S.A.* 47°25N 123°53W **7**
Neiqiu *China* 37°15N 114°30E **3**
Neiva *Colombia* 2°56N 75°18W **9**
Neixiang *China* 33°10N 111°52E **3**
Nejanilini L. *Canada* 59°33N 97°48W **7**
Nejd = Najd *Si. Arabia* 26°30N 42°0E **4**
Nekā *Iran* 36°39N 53°19E **4**
Nekemte *Ethiopia* 9°4N 36°30E **4**
Neksø *Denmark* 55°4N 15°8E **9**
Nelia *Australia* 20°39S 142°12E **6**
Neligh *U.S.A.* 42°8N 98°2W **8**
Nelkan *Russia* 57°40N 136°4E **29**
Nellore *India* 14°27N 79°59E **40**
Nelson *Canada* 49°30N 117°20W **7**
Nelson *N.Z.* 41°18S 173°16E **5**
Nelson *U.K.* 53°50N 2°13W **1**
Nelson *Ariz., U.S.A.* 35°31N 113°19W **7**
Nelson *Nev., U.S.A.* 35°42N 114°49W **79**
Nelson → *Canada* 54°33N 98°2W **7**
Nelson, C. *Australia* 38°26S 141°32E **6**
Nelson, Estrecho *Chile* 51°30S 75°0W **9**
Nelson Forks *Canada* 59°30N 124°0W **7**
Nelson House *Canada* 55°47N 98°51W **7**
Nelson L. *Canada* 55°48N 100°7W **7**
Nelson Lakes △ *N.Z.* 41°55S 172°44E **5**
Nelspoort *S. Africa* 32°7S 23°0E **5**
Nelspruit *S. Africa* 25°29S 30°59E **5**
Néma *Mauritania* 16°40N 7°15W **5**
Neman = Nemunas → *Lithuania* 55°25N 21°10E **9**
Nembrala *Indonesia* 10°53S 122°50E **6**
Nemeiben L. *Canada* 55°20N 105°20W **7**
Nemiscau *Canada* 51°18N 76°54W **7**
Nemiscau, L. *Canada* 51°25N 76°40W **7**
Nemunas → *Lithuania* 55°25N 21°10E **9**
Nemuro *Japan* 43°20N 145°35E **30**
Nemuro-Kaikyō *Japan* 43°30N 145°30E **30**
Nen Jiang → *China* 45°28N 124°30E **35**
Nenagh *Ireland* 52°52N 8°11W **1**
Nenana *U.S.A.* 64°34N 149°5W **74**
Nenasi *Malaysia* 3°9N 103°23E **3**
Nene → *U.K.* 52°49N 0°11E **1**
Nenjiang *China* 49°10N 125°10E **33**
Neno *Malawi* 15°25S 34°40E **5**
Neodesha *U.S.A.* 37°25N 95°41W **8**
Neora Valley △ *India* 27°0N 88°45E **4**
Neosho *U.S.A.* 36°52N 94°22W **8**
Neosho → *U.S.A.* 36°48N 95°18W **8**
Nepal ■ *Asia* 28°0N 84°30E **43**
Nepalganj *Nepal* 28°5N 81°40E **4**
Nepalganj Road *India* 28°1N 81°41E **4**
Nephi *U.S.A.* 39°43N 111°50W **7**
Nephin *Ireland* 54°1N 9°22W **1**
Nephin Beg Range *Ireland* 54°0N 9°40W **1**
Neptune *U.S.A.* 40°13N 74°2W **83**
Neqāb *Iran* 36°42N 57°25E **4**
Nerang *Australia* 27°58S 153°20E **6**
Nerchinsk *Russia* 52°0N 116°39E **29**
Néret, L. *Canada* 54°45N 70°44W **7**
Neretva → *Croatia* 43°1N 17°27E **2**
Neringa *Lithuania* 55°20N 21°5E **9**
Neris → *Lithuania* 55°8N 24°16E **9**
Neryungri *Russia* 57°38N 124°28E **29**
Nescopeck *U.S.A.* 41°3N 76°12W **8**
Neskantaga *Canada* 52°14N 87°53W **7**
Neskaupstaður *Iceland* 65°9N 13°42W **8**
Ness, L. *U.K.* 57°15N 4°32W **11**
Ness City *U.S.A.* 38°27N 99°54W **8**
Nesterov = Zhovkva *Ukraine* 50°4N 23°58E **17**
Nestos → *Europe* 40°54N 24°49E **23**
Nesvizh = Nyasvizh *Belarus* 53°14N 26°38E **17**
Netanya *Israel* 32°20N 34°51E **4**
Netarhat *India* 23°29N 84°16E **43**
Nete → *Belgium* 51°7N 4°14E **15**
Netherdale *Australia* 21°10S 148°33E **6**
Netherlands ■ *Europe* 52°0N 5°30E **15**
Netherlands Antilles = ABC Islands *W. Indies* 12°15N 69°0W **8**
Netrang *India* 21°39N 73°21E **4**
Nettilling L. *Canada* 66°30N 71°0W **69**
Netzahualcóyotl, Presa *Mexico* 17°8N 93°35W **87**
Neubrandenburg *Germany* 53°33N 13°15E **16**
Neuchâtel *Switz.* 47°0N 6°55E **20**
Neuchâtel, Lac de *Switz.* 46°53N 6°50E **20**
Neufchâteau *Belgium* 49°50N 5°25E **15**
Neumayer *Antarctica* 71°0S 68°30W **5**
Neumünster *Germany* 54°4N 9°58E **16**
Neunkirchen *Germany* 49°20N 7°9E **16**
Neuquén *Argentina* 38°55S 68°0W **96**
Neuquén □ *Argentina* 38°0S 69°50W **94**
Neuruppin *Germany* 52°55N 12°48E **16**
Neuse → *U.S.A.* 35°6N 76°29W **85**
Neusiedler See *Austria* 47°50N 16°47E **17**
Neustrelitz *Germany* 53°21N 13°4E **16**
Neva → *Russia* 59°56N 30°20E **18**
Nevada *Iowa, U.S.A.* 42°1N 93°27W **80**
Nevada *Mo., U.S.A.* 37°51N 94°22W **80**
Nevada □ *U.S.A.* 39°0N 117°0W **76**
Nevada City *U.S.A.* 39°16N 121°1W **78**
Nevado, Cerro *Argentina* 35°30S 68°32W **94**
Nevado de Colima = Volcán de Colima △ *Mexico* 19°30N 103°40W **86**
Nevado de Tres Cruces △ *Chile* 27°13S 69°5W **94**
Nevel *Russia* 56°0N 29°55E **18**
Nevelsk *Russia* 46°40N 141°51E **29**
Nevers *France* 47°0N 3°9E **20**
Nevertire *Australia* 31°50S 147°44E **63**
Neville *Canada* 49°58N 107°39W **71**
Nevinnomyssk *Russia* 44°40N 42°0E **19**
Nevis *St. Kitts & Nevis* 17°0N 62°30W **89**

›vşehir Turkey 38°33N 34°40E 44 B2
vyansk Russia 57°30N 60°13E 18 C11
›w – U.S.A. 38°10N 81°12W 81 F13
›w Aiyansh Canada 55°12N 129°4W 70 B3
›w Albany Ind., U.S.A. 38°18N 85°49W 81 F11
›w Albany Miss.. U.S.A. 34°29N 89°0W 85 D10
›w Albany Pa., U.S.A. 41°36N 76°27W 83 E8
›w Amsterdam Guyana 6°15N 57°36W 92 B7
›w Baltimore U.S.A. 42°41N 82°44W 82 D2
›w Bedford U.S.A. 41°38N 70°56W 83 E14
›w Berlin N.Y., U.S.A. 42°37N 75°20W 83 D9
›w Berlin U.S.A. 40°50N 76°57W 82 F8
›w Bern U.S.A. 35°7N 77°3W 85 D16
›w Bethlehem U.S.A. 41°0N 79°20W 82 F5
›w Bight Bahamas 24°19N 75°24W 89 B4
›w Bloomfield U.S.A. 40°25N 77°11W 82 F7
›w Boston U.S.A. 33°28N 94°25W 84 E7
›w Braunfels U.S.A. 29°42N 98°8W 84 G5
›w Brighton N.Z. 43°29S 172°43E 59 E4
›w Brighton U.S.A. 40°42N 80°19W 82 F4
›w Britain Papua N. G. 5°50S 150°20E 58 B8
›w Britain U.S.A. 41°40N 72°47W 83 E12
›w Brunswick U.S.A. 40°30N 74°27W 83 F10
›w Brunswick ☐ ·Canada 46°50N 66°30W 73 C6
›w Caledonia ☑ Pac. Oc. 21°0S 165°0E 58 D9
›w Caledonia Trough Pac. Oc. 30°0S 165°0E 64 L8
›w Castile = Castilla-La Mancha ☐ Spain 39°30N 3°30W 21 C4
›w Castle Ind., U.S.A. 39°55N 85°22W 81 F11
›w Castle Pa., U.S.A. 41°0N 80°21W 82 F4
›w City U.S.A. 41°9N 73°59W 83 E11
›w Concord U.S.A. 39°59N 81°54W 82 G3
›w Cumberland U.S.A. 40°30N 80°36W 82 F4
›w Cuyama U.S.A. 34°57N 119°38W 79 L7
›w Delhi India 28°36N 77°11E 42 E7
›w Denver Canada 50°0N 117°25W 70 D5
›w Don Pedro Res. U.S.A. 37°43N 120°24W 78 H6
›w England U.S.A. 43°0N 71°0W 75 G25
›w England N. Dak., U.S.A. 46°32N 102°52W 80 B2
›w England Ra. Australia 30°20S 151°45E 63 E5
›w Forest △ U.K. 50°53N 1°34W 13 G6
›w Galloway U.K. 55°N 4°9W 11 F4
›w Glasgow Canada 45°35N 62°36W 73 C7
›w Guinea Oceania 4°0S 136°0E 58 B6
›w Hamburg Canada 43°23N 80°42W 82 C4
›w Hampshire ☐ U.S.A. 44°0N 71°30W 83 C13
›w Hampton U.S.A. 43°3N 92°19W 80 D7
›w Hanover S. Africa 29°22S 30°31E 57 D5
›w Hartford U.S.A. 43°4N 75°18W 83 C9
›w Haven Conn., U.S.A. 41°18N 72°55W 83 E12
›w Haven Mich., U.S.A. 42°44N 82°48W 82 D2
›w Haven N.Y., U.S.A. 43°28N 76°52W 83 C8
›w Hazelton Canada 55°20N 127°30W 70 B3
›w Hebrides = Vanuatu ■ Pac. Oc. 15°0S 168°0E 58 C9
›w Holland U.S.A. 40°6N 76°5W 83 F8
›w Iberia U.S.A. 30°1N 91°49W 84 F9
›w Ireland Papua N. G. 3°20S 151°50E 58 B8
›w Jersey ☐ U.S.A. 40°0N 74°30W 81 F16
›w Kensington U.S.A. 39°45N 79°46W 82 F5
›w Lexington U.S.A. 39°43N 82°13W 81 F12
›w Liskeard Canada 47°31N 79°41W 72 C4
›w London U.S.A. 41°22N 72°6W 83 E12
›w London Ohio, U.S.A. 41°5N 82°24W 82 E2
›w London Wis., U.S.A. 44°23N 88°45W 80 C9
›w Madrid U.S.A. 39°39N 80°52W 81 F13
›w Martinsville U.S.A. 39°39N 80°52W 81 F13
›w Meadows U.S.A. 44°58N 116°18W 76 D5
›w Melones L. U.S.A. 37°57N 120°31W 78 H6
›w Mexico ☐ U.S.A. 34°30N 106°0W 77 J11
›w Milford Conn.. U.S.A. 41°35N 73°25W 83 E11
›w Milford Pa., U.S.A. 41°19N 74°48W 83 E9
›w Norcia Australia 30°57S 116°13E 61 F2
›w Norfolk Australia 42°46S 147°2E 63 G4
›w Orleans U.S.A. 29°57N 90°4W 85 G9
›w Philadelphia U.S.A. 40°30N 81°27W 82 F3
›w Plymouth N.Z. 39°4S 174°5E 59 C5
›w Plymouth U.S.A. 43°58N 116°49W 76 E5
›w Port Richey U.S.A. 28°16N 82°43W 85 G13
›w Providence Bahamas 25°25N 78°35W 88 A4
›w Quay U.K. 52°13N 4°21W 13 E3
›w Radnor U.K. 52°15N 3°10W 13 E4
›w Richmond Canada 48°15N 65°45W 73 C6
›w Richmond U.S.A. 45°7N 92°32W 80 C7
›w River Gorge △ U.S.A. 37°53N 81°5W 81 G13
›w Roads U.S.A. 30°42N 91°26W 84 F9
›w Rochelle U.S.A. 40°55N 73°46W 83 F11
›w Rockford U.S.A. 47°41N 99°8W 80 B4
›w Romney U.K. 50°59N 0°57E 13 G8
›w Ross Ireland 52°23N 6°57W 10 D5
›w Salem U.S.A. 46°51N 101°25W 80 B3
›w Scone = Scone U.K. 53°24S 3°24W 11 E5
·Ostrov Russia 75°10N 150°0E 29 B16
›w Siberian Is. = Novosibirskiye
Ostrova Russia 75°0N 142°0E 29 B15
›w Smyrna Beach U.S.A. 29°1N 80°56W 85 G14
›w South Wales ☐ Australia 33°0S 146°0E 63 E4
›w Tecumseth = Alliston Canada 44°9N 79°52W 82 B5

New Town U.S.A. 47°59N 102°30W 80 B2
New Tredegar U.K. 51°44N 3°16W 13 F4
New Ulm U.S.A. 44°19N 94°28W 80 C6
New Waterford Canada 46°13N 60°4W 73 C7
New Westminster Canada 49°13N 122°55W 78 A4
New York U.S.A. 40°43N 74°0W 83 F11
New York ☐ U.S.A. 43°0N 75°0W 83 D9
New York J.F. Kennedy Int. ✈ (JFK) U.S.A. 40°38N 73°47W 83 F11
New Zealand ■ Oceania 40°0S 176°0E 59 D6
Newaj → India 24°24N 76°49E 42 G7
Newala Tanzania 10°58S 39°18E 55 E4
Newark Del., U.S.A. 39°41N 75°46W 81 F16
Newark N.J., U.S.A. 40°44N 74°10W 83 F10
Newark N.Y., U.S.A. 43°3N 77°6W 82 C7
Newark Ohio, U.S.A. 40°3N 82°24W 82 F2
Newark Liberty Int. ✈ (EWR) U.S.A. 40°42N 74°10W 83 F10
Newark-on-Trent U.K. 53°5N 0°48W 12 D7
Newark Valley U.S.A. 42°14N 76°11W 83 D8
Newberg U.S.A. 45°18N 122°58W 76 D2
Newberry Mich., U.S.A. 46°21N 85°30W 81 B11
Newberry S.C., U.S.A. 34°17N 81°37W 85 D14
Newberry Springs U.S.A. 34°50N 116°41W 79 L10
Newboro L. Canada 44°38N 76°20W 83 B8
Newbridge = Droichead Nua Ireland 53°11N 6°48W 10 C5
Newburgh Canada 44°19N 76°52W 82 B8
Newburgh U.S.A. 41°30N 74°1W 83 E10
Newbury U.K. 51°24N 1°20W 13 F6
Newbury N.H., U.S.A. 43°19N 72°3W 83 D12
Newbury Vt., U.S.A. 44°5N 72°4W 83 B12
Newburyport U.S.A. 42°49N 70°53W 83 D14
Newcastle Australia 33°0S 151°46E 63 E5
Newcastle N.B., Canada 47°1N 65°38W 73 C6
Newcastle Ont., Canada 43°55N 78°35W 72 D4
Newcastle S. Africa 27°45S 29°58E 57 D4
Newcastle U.K. 54°13N 5°54W 10 B6
Newcastle Calif., U.S.A. 38°53N 121°8W 78 G5
Newcastle Wyo., U.S.A. 43°50N 104°11W 76 E11
Newcastle Emlyn U.K. 52°2N 4°28W 13 E3
Newcastle Ra. Australia 15°45S 130°15E 60 C5
Newcastle-under-Lyme U.K. 53°1N 2°14W 12 D5
Newcastle-upon-Tyne U.K. 54°58N 1°36W 12 C6
Newcastle Waters Australia 17°30S 133°28E 62 B1
Newcastle West Ireland 52°27N 9°3W 10 D2
Newcomb U.S.A. 43°58N 74°10W 83 C10
Newcomerstown U.S.A. 40°16N 81°36W 82 F3
Newdegate Australia 33°6S 119°0E 61 F2
Newell Australia 16°20S 145°16E 62 B4
Newell U.S.A. 44°43N 103°25W 80 C2
Newenham, C. U.S.A. 58°39N 162°11W 74 D7
Newfane U.S.A. 43°17N 78°43W 82 C6
Newfield U.S.A. 42°18N 76°33W 83 D8
Newfound L. U.S.A. 43°40N 71°47W 83 C13
Newfoundland Canada 49°0N 55°0W 73 C8
Newfoundland U.S.A. 41°18N 75°19W 83 E9
Newfoundland & Labrador ☐ Canada 53°0N 58°0W 73 B8
Newhaven U.K. 50°47N 0°3E 13 G8
Newkirk U.S.A. 36°53N 97°3W 84 C6
Newlyn U.K. 50°6N 5°34W 13 G2
Newman Australia 23°18S 119°45E 60 D2
Newman U.S.A. 37°19N 121°1W 78 H5
Newmarket Canada 44°3N 79°28W 82 B5
Newmarket Ireland 52°13N 9°0W 10 D2
Newmarket U.K. 52°15N 0°25E 13 E8
Newmarket U.S.A. 43°5N 70°56W 83 C14
Newnan U.S.A. 33°23N 84°48W 85 E12
Newport Ireland 53°53N 9°33W 10 C2
Newport I. of W., U.K. 50°42N 1°17W 13 G6
Newport Newport, U.K. 51°35N 3°0W 13 F5
Newport Ark., U.S.A. 35°37N 91°16W 84 D9
Newport Ky., U.S.A. 39°5N 84°29W 81 F11
Newport N.H., U.S.A. 43°22N 72°10W 83 C12
Newport N.Y., U.S.A. 43°11N 75°1W 83 C9
Newport Oreg., U.S.A. 44°39N 124°3W 76 D1
Newport Pa., U.S.A. 40°29N 77°8W 82 F7
Newport R.I., U.S.A. 41°29N 71°19W 83 E13
Newport Tenn., U.S.A. 35°58N 83°11W 85 D13
Newport Vt., U.S.A. 44°56N 72°13W 83 B12
Newport Wash., U.S.A. 48°11N 117°3W 76 B5
Newport ☐ U.K. 51°33N 3°1W 13 F4
Newport Beach U.S.A. 33°37N 117°56W 79 M9
Newport News U.S.A. 36°58N 76°25W 81 G15
Newquay U.K. 50°25N 5°6W 13 G2
Newry U.K. 54°11N 6°21W 10 B5
Newry Islands △ Australia 20°51S 148°56E 62 b
Newton Ill., U.S.A. 38°59N 88°10W 80 F9
Newton Iowa, U.S.A. 41°42N 93°3W 80 E7
Newton Kans., U.S.A. 38°3N 97°21W 80 F5
Newton Mass., U.S.A. 42°21N 71°12W 83 D13
Newton Miss., U.S.A. 32°19N 89°10W 85 E10
Newton N.C., U.S.A. 35°40N 81°13W 85 D14
Newton N.J., U.S.A. 41°3N 74°45W 83 E10
Newton Tex., U.S.A. 30°51N 93°46W 84 F8
Newton Abbot U.K. 50°32N 3°37W 13 G4
Newton Aycliffe U.K. 54°37N 1°34W 12 C6
Newton Falls N.Y., U.S.A. 44°12N 74°59W 83 B10
Newton Falls Ohio, U.S.A. 41°11N 80°59W 82 E4
Newton Stewart U.K. 54°57N 4°30W 11 G4
Newtonmore U.K. 57°4N 4°8W 11 D4
Newtown U.K. 52°31N 3°19W 13 E4
Newtownabbey U.K. 54°40N 5°56W 10 B6
Newtownards U.K. 54°36N 5°42W 10 B6
Newtownbarry = Bunclody Ireland 52°39N 6°40W 10 D5
Newtownstewart U.K. 54°43N 7°23W 10 B4
Newville U.S.A. 40°10N 77°24W 82 F7
Neya Russia 58°21N 43°49E 18 C7

Neyrīz Iran 29°15N 54°19E 45 D7
Neyshābūr Iran 36°10N 58°50E 45 B8
Nezhin = Nizhyn Ukraine 51°5N 31°55E 19 D5
Nezperce U.S.A. 46°14N 116°14W 76 C5
Ngaanyatjarra ☉ Australia 22°40S 127°0E 60 D4
Ngabang Indonesia 0°23N 109°55E 36 D3
Ngabordamlu, Tanjung Indonesia 6°56S 134°11E 37 F8
N'Gage Angola 7°46S 15°15E 52 F3
Ngaliwurru Nungali ☉ Australia 15°42S 130°4E 60 C5
Ngalurrtju ☉ Australia 22°44S 132°5E 60 D5
Ngami Depression Botswana 20°30S 22°46E 56 C3
Ngamo Zimbabwe 19°3S 27°32E 55 F2
Ngamring China 29°14N 87°10E 32 F6
Nganglong Kangri China 33°0N 81°0E 43 C9
Ngao Thailand 18°46N 99°59E 38 C2
Ngaoundéré Cameroon 7°15N 13°35E 52 C2
Ngapara N.Z. 44°57S 170°46E 59 F3
Ngara Tanzania 2°29S 30°40E 54 C3
Ngarluma ☉ Australia 21°30S 118°30E 60 D2
Ngawi Indonesia 7°24S 111°26E 37 G14
Ngcobo S. Africa 31°37S 28°0E 57 E4
Ngeru Uganda 0°27N 33°10E 54 B3
Nghia Lo Vietnam 21°33N 104°28E 38 B5
Ngoma Malawi 13°8S 33°45E 55 E3
Ngomahura Zimbabwe 20°26S 30°43E 55 G3
Ngomba Tanzania 8°20S 32°53E 55 D3
Ngomeni, Ras Kenya 2°59S 40°14E 54 C5
Ngong Kenya 1°22S 36°39E 54 C4
Ngoring Hu China 34°55N 97°5E 32 E8
Ngorongoro Tanzania 3°11S 35°32E 54 C4
Ngorongoro ☉ Tanzania 2°40S 35°30E 54 C4
Ngozi Burundi 2°54S 29°50E 54 C2
Ngudu Tanzania 2°58S 33°25E 54 C3
Nguigmi Niger 14°20N 13°20E 51 F8
Nguiu Australia 11°46S 130°38E 60 B5
Ngukurr Australia 14°44S 134°44E 62 A1
Ngum → Laos 18°31N 102°32E 38 C4
Ngundu Zimbabwe 18°11S 31°15E 55 F3
Ngunga Tanzania 3°37S 33°37E 54 C3
Nguru Nigeria 12°56N 10°29E 51 F8
Nguru Mts. Tanzania 6°0S 37°30E 54 D4
Ngusi Malawi 14°0S 34°50E 55 E3
Nguyen Binh Vietnam 22°39N 105°56E 38 A5
Nha Bang Vietnam 10°37N 105°0E 39 G5
Nha Trang Vietnam 12°16N 109°10E 39 F7
Nhacoongo Mozam. 24°18S 35°14E 57 C6
Nhamaabué Mozam. 17°25S 35°5E 55 F4
Nhamundá → Brazil 2°12S 56°41W 93 D7
Nhangulaze, L. Mozam. 24°0S 34°30E 57 C5
Nharnuwangga, Wajarri and Ngarlawangga ☉ Australia 24°50S 118°0E 61 D2
Nhill Australia 36°18S 141°40E 63 F3
Nho Quan Vietnam 20°18N 105°45E 38 B5
Nhulunbuy Australia 12°10S 137°20E 62 A2
Nia-nia Dem. Rep. of the Congo 1°30N 27°40E 54 B2
Niafounké Mali 16°0N 4°5W 50 E5
Niagara Falls Canada 43°7N 79°5W 82 C5
Niagara Falls U.S.A. 43°5N 79°4W 82 C5
Niagara-on-the-Lake Canada 43°15N 79°4W 82 C5
Niah Malaysia 3°58N 113°46E 36 D4
Niamey Niger 13°27N 2°6E 50 F6
Niangara Dem. Rep. of the Congo 3°42N 27°50E 54 B2
Niantic U.S.A. 41°20N 72°11W 83 E12
Nias Indonesia 1°0N 97°30E 36 D1
Niassa ☐ Mozam. 13°30S 36°0E 55 E4
Niassa → Mozam. 12°4S 36°57E 55 E4
Nibāk Si. Arabia 24°25N 50°50E 45 E7
Nicaragua ■ Cent. Amer. 11°40N 85°30W 88 D2
Nicaragua, L. de Nic. 12°0N 85°30W 88 D2
Nicastro Italy 38°59N 16°19E 22 E7
Nice France 43°42N 7°14E 20 E7
Niceville U.S.A. 30°31N 86°30W 85 F11
Nichicun, L. Canada 53°5N 71°0W 73 B5
Nichinan Japan 31°38N 131°23E 31 J5
Nicholás, Canal W. Indies 23°30N 80°5W 88 B3
Nicholasville U.S.A. 37°53N 84°34W 81 G11
Nichols U.S.A. 42°1N 76°22W 83 D8
Nicholson Australia 18°2S 128°54E 60 C4
Nicholson U.S.A. 41°37N 75°47W 83 E9
Nicholson → Australia 17°31S 139°36E 62 B2
Nicholson L. Canada 62°40N 102°40W 71 A8
Nicholson Ra. Australia 27°15S 116°45E 61 E2
Nicholville U.S.A. 44°41N 74°39W 83 B10
Nicobar Is. Ind. Oc. 8°0N 93°30E 27 H11
Nicola Canada 50°12N 120°40W 70 C4
Nicolls Town Bahamas 25°8N 78°0W 88 A4
Nicosia Cyprus 35°10N 33°25E 25 D12
Nicoya Costa Rica 10°9N 85°27W 88 D2
Nicoya, G. de Costa Rica 10°0N 85°0W 88 E3
Nicoya, Pen. de Costa Rica 9°45N 85°40W 88 E2
Nidd → U.K. 53°59N 1°23W 12 D6
Nidderdale U.K. 54°5N 1°46W 12 C6
Niedersachsen ☐ Germany 52°50N 9°0E 16 B5
Niekerkshoop S. Africa 29°19S 22°51E 56 D3
Niemba Dem. Rep. of the Congo 5°58S 28°24E 54 D2
Niemen = Nemunas → Lithuania 55°25N 21°10E 9 J19
Nienburg Germany 52°39N 9°13E 16 B5
Nieu Bethesda S. Africa 31°51S 24°34E 56 E3
Nieuw Amsterdam Suriname 5°53N 55°5W 93 B7
Nieuw Nickerie Suriname 6°0N 56°59W 93 B7
Nieuwoudtville S. Africa 31°23S 19°7E 56 E2
Nieuwpoort Belgium 51°8N 2°45E 15 C2
Nieves, Pico de las Canary Is. 27°57N 15°35W 24 G4
Niğde Turkey 37°58N 34°40E 44 B2
Nigel S. Africa 26°27S 28°25E 57 D4
Niger ■ W. Afr. 17°30N 10°0E 50 E7
Niger → W. Afr. 5°33N 6°33E 50 G7

Nigeria ■ W. Afr. 8°30N 8°0E 50 G7
Nighasin India 28°14N 80°52E 43 E9
Nightcaps N.Z. 45°57S 168°2E 59 F2
Nihoa U.S.A. 23°6N 161°58W 75 L7
Nihon = Japan ■ Asia 36°0N 136°0E 31 G8
Nii-Jima Japan 34°20N 139°15E 31 G9
Niigata Japan 37°58N 139°0E 30 F9
Niigata ☐ Japan 37°15N 138°45E 31 F9
Niihama Japan 33°55N 133°16E 31 H6
Ni'ihau U.S.A. 21°54N 160°9W 75 L7
Niimi Japan 34°59N 133°28E 31 G6
Niitsu Japan 37°48N 139°7E 30 F9
Nijil Jordan 30°32N 35°33E 46 E4
Nijkerk Neths. 52°13N 5°30E 15 B5
Nijmegen Neths. 51°50N 5°52E 15 C5
Nijverdal Neths. 52°22N 6°28E 15 B6
Nīk Pey Iran 36°50N 48°10E 45 B6
Nikel Russia 69°24N 30°13E 8 B24
Nikiniki Indonesia 9°49S 124°30E 60 A3
Nikkō Japan 36°45N 139°35E 31 F9
Nikkō △ Japan 36°56N 139°37E 31 F9
Nikolayev = Mykolayiv Ukraine 46°58N 32°0E 19 E5
Nikolayevsk Russia 50°0N 45°35E 19 E8
Nikolayevsk-na-Amur Russia 53°8N 140°44E 29 D15
Nikolskoye Russia 55°12N 166°0E 29 D17
Nikopol Ukraine 47°35N 34°25E 19 E5
Nīkshahr Iran 26°15N 60°10E 45 E9
Nikšić Montenegro 42°50N 18°57E 23 C8
Nīl, Nahr en → Africa 30°10N 31°6E 51 B12
Nīl el Abyad → Sudan 15°38N 32°31E 51 E12
Nīl el Azraq → Sudan 15°38N 32°31E 51 E12
Nila Indonesia 6°44S 129°31E 37 F7
Niland U.S.A. 33°14N 115°31W 79 M11
Nile = Nīl, Nahr en → Africa 30°10N 31°6E 51 B12
Niles Mich., U.S.A. 41°50N 86°15W 80 E10
Niles Ohio, U.S.A. 41°11N 80°46W 82 E4
Nim Ka Thana India 27°44N 75°48E 42 F6
Nimach India 24°37N 74°45E 42 G6
Nimbahera India 24°37N 74°45E 42 G6
Nîmes France 43°50N 4°23E 20 E6
Nimfaíon, Ákra = Pines, Akra Greece 40°5N 24°20E 23 D11
Nimmitabel Australia 36°29S 149°15E 63 F4
Nīnawā Iraq 36°25N 43°10E 44 B4
Nīnawā ☐ Iraq 36°15N 43°0E 44 B4
Nindigully Australia 28°21S 148°50E 63 D4
Ninepin Group China 22°16N 114°21E 33 a
Nineveh = Nīnawā Iraq 36°25N 43°10E 44 B4
Ning Xian China 35°30N 107°58E 34 G4
Ningaloo Marine △ Australia 22°23S 113°32E 60 D1
Ning'an China 44°22N 129°20E 35 B15
Ningbo China 29°51N 121°28E 33 F13
Ningcheng China 41°32N 119°53E 35 D10
Ningjin China 37°35N 114°57E 34 F8
Ningjing Shan China 31°0N 98°20E 32 D8
Ningling China 34°25N 115°22E 34 G8
Ningpo = Ningbo China 29°51N 121°28E 33 F13
Ningqiang China 32°47N 106°15E 34 H4
Ningshan China 33°21N 108°21E 34 H5
Ningsia Hui A.R. = Ningxia Huizu Zizhiqu ☐ China 38°0N 106°0E 34 F4
Ningwu China 39°0N 112°18E 34 E7
Ningxia Huizu Zizhiqu ☐ China 38°0N 106°0E 34 F4
Ningyang China 35°47N 116°45E 34 G9
Ninh Binh Vietnam 20°15N 105°55E 38 B5
Ninh Giang Vietnam 20°44N 106°24E 38 B6
Ninh Hoa Vietnam 12°30N 109°7E 38 F7
Ninh Ma Vietnam 12°48N 109°21E 38 F7
Ninove Belgium 50°51N 4°2E 15 D4
Nioaque Brazil 21°5S 55°50W 95 A4
Niobrara U.S.A. 42°45N 98°2W 80 D4
Niobrara → U.S.A. 42°46N 98°3W 80 D4
Nioro du Sahel Mali 15°15N 9°30W 50 E4
Niort France 46°19N 0°29W 20 C3
Nipawin Canada 53°20N 104°0W 71 C8
Nipigon Canada 49°0N 88°17W 72 C2
Nipigon, L. Canada 49°50N 88°30W 72 C2
Nipishish L. Canada 54°12N 60°45W 73 B7
Nipissing, L. Canada 46°20N 80°0W 72 C4
Nipomo U.S.A. 35°3N 120°29W 79 K6
Nippon = Japan ■ Asia 36°0N 136°0E 31 G8
Nipton U.S.A. 35°28N 115°16W 79 K11
Niquelândia Brazil 14°33S 48°23W 93 F9
Nīr Iran 38°2N 47°59E 44 B5
Nirasaki Japan 35°42N 138°27E 31 G9
Nirmal India 19°3N 78°20E 40 K11
Nirmali India 26°20N 86°35E 43 F12
Niš Serbia 43°19N 21°58E 23 C9
Nişāb Si. Arabia 29°11N 44°43E 44 D5
Nişāb Yemen 14°25N 46°29E 47 E4
Nishinomiya Japan 34°45N 135°20E 31 G7
Nishino'omote Japan 30°43N 130°59E 31 J5
Nishiwaki Japan 34°59N 134°58E 31 G7
Niskibi → Canada 56°29N 88°9W 72 A2
Nisqually → U.S.A. 47°7N 122°42W 78 C4
Nissaki Greece 39°43N 19°52E 25 A3
Nissum Bredning Denmark 56°40N 8°20E 9 H13
Nistru = Dnister → Europe 46°18N 30°17E 17 E16
Nisutlin → Canada 60°14N 132°34W 70 A2
Nitchequon Canada 53°10N 70°58W 73 B5
Niterói Brazil 22°53S 43°7W 95 A7
Nith → Canada 43°12N 80°23W 82 C4
Nith → U.K. 55°14N 3°33W 11 F5
Nitmiluk △ Australia 14°6S 132°15E 60 B5
Nitra Slovak Rep. 48°19N 18°4E 17 D10
Nitra → Slovak Rep. 47°46N 18°10E 17 E10
Niue ☑ Pac. Oc. 19°2S 169°54W 65 J11
Niut Indonesia 0°55N 109°54E 36 D4
Niuzhuang China 40°58N 122°28E 35 D12
Nivala Finland 63°56N 24°57E 8 E21
Nivelles Belgium 50°35N 4°20E 15 D4
Nivernais France 47°15N 3°30E 20 C5

Niverville Canada 49°36N 97°3W 71 D9
Niwas India 23°3N 80°26E 43 H9
Nixa U.S.A. 37°3N 93°18W 80 G7
Nixon U.S.A. 29°16N 97°46W 84 G6
Nizamabad India 18°45N 78°7E 40 K11
Nizamghat India 28°20N 95°45E 41 E19
Nizhne Kolymsk Russia 68°34N 160°55E 29 C17
Nizhnekamsk Russia 55°38N 51°49E 18 C9
Nizhneudinsk Russia 54°54N 99°3E 29 D10
Nizhnevartovsk Russia 60°56N 76°38E 28 C8
Nizhniy Bestyakh Russia 61°57N 129°54E 29 C13
Nizhniy Novgorod Russia 56°20N 44°0E 18 C7
Nizhniy Tagil Russia 57°55N 59°57E 18 C10
Nizhyn Ukraine 51°5N 31°55E 19 D5
Nizip Turkey 37°5N 37°50E 44 B3
Nízké Tatry Slovak Rep. 48°55N 19°30E 17 D10
Nizwá Oman 22°56N 57°32E 47 C6
Njakwa Malawi 11°1S 33°56E 55 E3
Njazidja = Grande Comore Comoros Is. 11°35S 43°20E 53 a
Njinjo Tanzania 8°48S 38°54E 55 D4
Njombe Tanzania 9°20S 34°50E 55 D3
Njombe → Tanzania 6°56S 35°6E 54 D4
Nkana Zambia 12°50S 28°8E 55 E2
Nkandla S. Africa 28°37S 31°5E 57 D5
Nkawkaw Ghana 6°36N 0°49W 50 G5
Nkayi Zimbabwe 19°41S 29°20E 55 F2
Nkhotakota Malawi 12°56S 34°15E 55 E3
Nkhotakota △ Malawi 12°50S 34°0E 55 E3
Nkondwe Tanzania 5°51S 30°51E 54 D3
Nkongsamba Cameroon 4°55N 9°55E 52 D1
Nkurenkuru Namibia 17°42S 18°32E 56 B2
Nmai → Burma 25°30N 97°25E 41 G20
Noakhali = Maijdi Bangla. 22°48N 91°10E 41 H17
Nobel Canada 45°25N 80°6W 82 A4
Nobeoka Japan 32°36N 131°41E 31 H5
Noblesville U.S.A. 40°3N 86°1W 80 E10
Noboribetsu Japan 42°24N 141°6E 30 C10
Noccundra Australia 27°50S 142°36E 63 D3
Nocera Inferiore Italy 40°44N 14°38E 22 D6
Nocona U.S.A. 33°47N 97°44W 84 E6
Noda Japan 35°56N 139°52E 31 G9
Nogales Mexico 31°19N 110°56W 86 A2
Nogales U.S.A. 31°21N 110°56W 77 L8
Nōgata Japan 33°48N 130°44E 31 H5
Noggerup Australia 33°32S 116°5E 61 F2
Noginsk Russia 64°30N 90°50E 29 C10
Nogliki Russia 51°50N 143°10E 29 D15
Nogoa → Australia 23°40S 147°55E 62 C4
Nogoyá Argentina 32°24S 59°48W 94 C4
Nohar India 29°11N 74°49E 42 E6
Nohta India 23°40N 79°34E 43 H8
Noires, Mts. France 48°11N 3°40W 20 B2
Noirmoutier, Î. de France 46°58N 2°10W 20 C2
Nojane Botswana 23°15S 20°14E 56 C3
Nojima-Zaki Japan 34°54N 139°53E 31 G9
Nok Kundi Pakistan 28°50N 62°45E 40 E3
Nok Ta Phao, Ko Thailand 9°23N 99°40E 39 b
Nokaneng Botswana 19°40S 22°17E 56 B3
Nokia Finland 61°30N 23°30E 8 F20
Nokomis Canada 51°35N 105°0W 71 C8
Nokomis L. Canada 57°0N 103°0W 71 B8
Nola C.A.R. 3°35N 16°4E 52 D3
Noma Omuramba → Namibia 18°52S 20°53E 56 B3
Nombre de Dios Panama 9°34N 79°28W 88 E4
Nome U.S.A. 64°30N 165°25W 74 C6
Nomgon Mongolia 42°50N 105°8E 34 C3
Nomo-Zaki Japan 32°35N 129°44E 31 H4
Nomuka Tonga 20°17S 174°48W 59 c
Nomuka Group Tonga 20°20S 174°48W 59 c
Nonacho L. Canada 61°42N 109°40W 71 A7
Nonda Australia 20°40S 142°28E 62 C3
Nong Chang Thailand 15°23N 99°51E 38 E2
Nong Han Res. Thailand 17°11N 104°9E 38 D5
Nong Het Laos 19°29N 103°59E 38 C4
Nong Khai Thailand 17°50N 102°46E 38 D4
Nong'an China 44°25N 125°5E 35 B13
Nongoma S. Africa 27°58S 31°35E 57 D5
Nongsa Indonesia 1°11N 104°8E 39 d
Nonoava Mexico 27°28N 106°44W 86 B3
Nonsan S. Korea 36°12N 127°5E 35 F14
Nonthaburi Thailand 13°50N 100°29E 38 F3
Noonamah Australia 12°40S 131°4E 60 B5
Noondie, L. Australia 28°30S 119°30E 61 E2
Noonkanbah ☉ Australia 18°30S 124°50E 60 C3
Noord Brabant ☐ Neths. 51°40N 5°0E 15 C5
Noord Holland ☐ Neths. 52°30N 4°45E 15 B4
Noordbeveland Neths. 51°35N 3°50E 15 C3
Noordoostpolder Neths. 52°45N 5°45E 15 B5
Noordwijk Neths. 52°14N 4°26E 15 B4
Noosa Heads Australia 26°25S 153°6E 63 D5
Nootka I. Canada 49°32N 126°42W 70 D3
Nopiming ☑ Canada 50°30N 95°37W 71 C9
Noralee Canada 53°59N 126°26W 70 C3
Noranda = Rouyn-Noranda Canada 48°20N 79°0W 72 C4
Norco U.S.A. 33°56N 117°33W 79 M9
Nord-Kivu ☐ Dem. Rep. of the Congo 1°0S 29°0E 54 C2
Nord-Ostsee-Kanal → Germany 54°12N 9°32E 16 A5
Nordaustlandet Svalbard 79°14N 23°0E 4 B9
Nordegg Canada 52°29N 116°5W 70 C5
Norderney Germany 53°42N 7°9E 16 B4
Norderstedt Germany 53°42N 10°1E 16 B5
Nordfjord Norway 61°55N 5°30E 8 F11
Nordfriesische Inseln Germany 54°40N 8°20E 16 A5
Nordhausen Germany 51°30N 10°47E 16 C6
Norðoyar Færoe Is. 62°17N 6°35W 8 E9
Nordkapp Norway 71°10N 25°50E 8 A21
Nordkapp Svalbard 80°31N 20°0E 4 A9
Nordkinnhalvøya Norway 70°55N 27°40E 8 A22
Nordrhein-Westfalen ☐ Germany 51°45N 7°30E 16 C4

Nordvik Russia 74°2N 111°32E 29 B12
Nore → Ireland 52°25N 6°58W 10 D4
Noreland Canada 44°43N 78°48W 82 B6
Norfolk = Simcoe Canada 42°50N 80°23W 82 D4
Norfolk N.Y., U.S.A. 44°48N 74°59W 83 B10
Norfolk Nebr., U.S.A. 42°2N 97°25W 80 D5
Norfolk Va., U.S.A. 36°50N 76°17W 81 G15
Norfolk □ U.K. 52°39N 0°54E 13 E8
Norfolk Broads △ U.K. 52°45N 1°30E 13 E9
Norfolk I. Pac. Oc. 28°58S 168°3E 58 D9
Norfolk Ridge Pac. Oc. 29°0S 168°0E 64 K8
Norfork L. U.S.A. 36°15N 92°14W 84 C8
Norge = Norway ■ Europe 63°0N 11°0E 8 E14
Norilsk Russia 69°20N 88°6E 29 C9
Norma, Mt. Australia 20°55S 140°42E 62 C3
Normal U.S.A. 40°31N 88°59W 80 E9
Norman U.S.A. 35°13N 97°26W 84 D6
Norman → Australia 19°18S 141°51E 62 B3
Norman Wells Canada 65°17N 126°51W 68 D6
Normanby → Australia 14°23S 144°10E 62 A3
Normandie France 48°45N 0°10E 20 B4
Normandin Canada 48°49N 72°31W 72 C5
Normandy = Normandie France 48°45N 0°10E 20 B4
Normanhurst, Mt. Australia 25°4S 122°30E 61 E3
Normanton Australia 17°40S 141°10E 62 B3
Normétal Canada 49°0N 79°22W 72 C4
Norquay Canada 51°53N 102°5W 71 C8
Norquinco Argentina 41°51S 70°55W 96 E2
Norrbottens län □ Sweden 66°50N 20°0E 8 C19
Norris Point Canada 49°31N 57°53W 73 C8
Norristown U.S.A. 40°7N 75°21W 83 F9
Norrköping Sweden 58°37N 16°11E 9 G17
Norrland Sweden 62°15N 15°45E 8 E16
Norrtälje Sweden 59°46N 18°42E 9 G18
Norseman Australia 32°8S 121°43E 61 F3
Norsk Russia 52°30N 130°5E 29 D14
Norte, Pta. del Canary Is. 27°51N 17°57W 24 G2
Norte, Serra do Brazil 11°20S 59°0W 92 F7
North, C. Canada 47°2N 60°20W 73 C7
North Adams U.S.A. 42°42N 73°7W 83 D11
North America 40°0N 100°0W 66 F10
North Arm Canada 62°0N 114°30W 70 A5
North Augusta U.S.A. 33°30N 81°59W 85 E14
North Australian Basin Ind. Oc. 14°30S 116°30E 64 J3
North Ayrshire □ U.K. 55°45N 4°44W 11 F4
North Bass I. U.S.A. 41°40N 82°56W 82 E2
North Battleford Canada 52°50N 108°17W 71 C7
North Bay Canada 46°20N 79°30W 72 C4
North Belcher Is. Canada 56°50N 79°50W 72 A4
North Bend Oreg., U.S.A. 43°24N 124°14W 76 E1
North Bend Pa., U.S.A. 41°20N 77°42W 82 E7
North Bend Wash., U.S.A. 47°30N 121°47W 78 C5
North Bennington U.S.A. 42°56N 73°15W 83 D11
North Berwick U.K. 56°4N 2°42W 11 E6
North Berwick U.S.A. 43°18N 70°44W 83 C14
North Bruce Canada 44°22N 81°26W 82 B3
North C. Canada 47°5N 64°0W 73 C7
North C. N.Z. 34°23S 173°4E 59 A4
North Canadian → U.S.A. 35°22N 95°37W 84 D7
North Canton U.S.A. 40°53N 81°24W 82 F3
North Cape = Nordkapp Norway 71°10N 25°50E 8 A21
North Caribou L. Canada 52°50N 90°40W 72 B1
North Carolina □ U.S.A. 35°30N 80°0W 85 D15
North Cascades △ U.S.A. 48°45N 121°10W 76 B3
North Channel Canada 46°0N 83°0W 72 C3
North Channel U.K. 55°13N 5°52W 11 F3
North Charleston U.S.A. 32°53N 79°58W 85 E15
North Chicago U.S.A. 42°19N 87°51W 80 D10
North Collins U.S.A. 42°35N 78°56W 82 D6
North Creek U.S.A. 43°42N 73°59W 83 C11
North Dakota □ U.S.A. 47°30N 100°15W 80 B3
North Downs U.K. 51°19N 0°21E 13 F8
North East U.S.A. 42°13N 79°50W 82 D5
North East Frontier Agency = Arunachal Pradesh □ India 28°0N 95°0E 41 F19
North East Lincolnshire □ U.K. 53°34N 0°2W 12 D7
North Eastern □ Kenya 1°30N 40°0E 54 B5
North Esk → U.K. 56°46N 2°24W 11 E6
North European Plain Europe 55°0N 25°0E 6 E10
North Foreland U.K. 51°22N 1°28E 13 F9
North Fork U.S.A. 37°14N 119°21W 78 H7
North Fork American → U.S.A. 38°57N 120°59W 78 G5
North Fork Feather → U.S.A. 38°33N 121°30W 78 F5
North Fork Grand → U.S.A. 45°47N 102°16W 80 C2
North Fork Red → U.S.A. 34°24N 99°14W 84 D5
North Frisian Is. = Nordfriesische Inseln Germany 54°40N 8°20E 16 A5
North Gower Canada 45°8N 75°43W 83 A9
North Hd. Australia 30°14S 114°59E 61 F1
North Henik L. Canada 61°45N 97°40W 71 A9
North Highlands U.S.A. 38°40N 121°23W 78 G5
North Horr Kenya 3°20N 37°8E 54 B4
North I. Kenya 4°5N 36°5E 54 B4
North I. N.Z. 38°0S 175°0E 59 C5
North I. Seychelles 4°25S 55°13E 53 b
North Kingsville U.S.A. 41°54N 80°42W 82 E4
North Kitui ⌂ Kenya 0°15S 38°29E 54 C4
North Knife → Canada 58°53N 94°45W 71 B10
North Koel → India 24°45N 83°50E 43 G10
North Korea ■ Asia 40°0N 127°0E 35 E14
North Lakhimpur India 27°14N 94°7E 41 F19
North Lanarkshire □ U.K. 55°52N 3°56W 11 F5
North Las Vegas U.S.A. 36°11N 115°7W 79 J11
North Lincolnshire □ U.K. 53°36N 0°30W 12 D7
North Little Rock U.S.A. 34°45N 92°16W 84 D8
North Loup → U.S.A. 41°17N 98°24W 80 E4
North Luangwa △ Zambia 11°49S 32°9E 55 E3
North Magnetic Pole Arctic 82°42N 114°24W 4 A1
North Mankato U.S.A. 44°10N 94°2W 80 C6
North Minch U.K. 58°5N 5°55W 11 C3
North Moose L. Canada 54°4N 100°12W 71 C8
North Myrtle Beach U.S.A. 33°48N 78°42W 85 E15
North Nahanni → Canada 62°15N 123°20W 70 A4
North Olmsted U.S.A. 41°25N 81°56W 82 E3
North Ossetia □ Russia 43°30N 44°30E 19 F7
North Pagai, I. = Pagai Utara, Pulau Indonesia 2°35S 100°0E 36 E2
North Palisade U.S.A. 37°6N 118°31W 78 H8
North Platte U.S.A. 41°8N 100°46W 80 E3
North Platte → U.S.A. 41°7N 100°42W 80 E3
North Pole Arctic 90°0N 0°0 4 A
North Portal Canada 49°0N 102°33W 71 D8
North Powder U.S.A. 45°2N 117°55W 76 D5
North Pt. Barbados 13°20N 59°37W 89 g
North Pt. Trin. & Tob. 11°21N 60°31W 93 J16
North Pt. U.S.A. 45°2N 83°16W 82 A1
North Rhine Westphalia = Nordrhein-Westfalen □ Germany 51°45N 7°30E 16 C4
North River Canada 53°49N 57°6W 73 B8
North Ronaldsay U.K. 59°22N 2°26W 11 B6
North Saskatchewan → Canada 53°15N 105°5W 71 C7
North Sea Europe 56°0N 4°0E 6 D6
North Seal → Canada 58°50N 98°7W 71 B9
North Slope ☆ U.S.A. 69°15N 152°0W 74 B9
North Somerset □ U.K. 51°24N 2°45W 13 F5
North Sydney Canada 46°12N 60°15W 73 C7
North Syracuse U.S.A. 43°8N 76°7W 83 C8
North Taranaki Bight N.Z. 38°50S 174°15E 59 C5
North Thompson → Canada 50°40N 120°20W 70 C4
North Tonawanda U.S.A. 43°2N 78°53W 82 C6
North Troy U.S.A. 45°0N 72°24W 83 B12
North Twin I. Canada 53°20N 80°0W 72 B4
North Tyne → U.K. 55°0N 2°8W 12 B5
North Uist U.K. 57°40N 7°15W 11 D1
North Vancouver Canada 49°19N 123°4W 78 A3
North Vernon U.S.A. 39°0N 85°38W 81 F11
North Wabasca L. Canada 56°0N 113°55W 70 B6
North Walsham U.K. 52°50N 1°22E 12 E9
North West = Severo-Zapadnyy □ Russia 65°0N 40°0E 28 C4
North-West □ S. Africa 27°0S 25°0E 56 D4
North West C. Australia 21°45S 114°9E 60 D1
North West Christmas I. Ridge Pac. Oc. 6°30N 165°0W 65 F11
North West Frontier □ Pakistan 34°0N 72°0E 42 C4
North West Highlands U.K. 57°33N 4°58W 11 D4
North West River Canada 53°30N 60°10W 73 B7
North Western □ Zambia 13°30S 25°30E 55 E2
North Wildwood U.S.A. 39°0N 74°48W 81 F16
North York Moors U.K. 54°23N 0°53W 12 C7
North York Moors △ U.K. 54°27N 0°51W 12 C7
North Yorkshire □ U.K. 54°15N 1°25W 12 C6
Northallerton U.K. 54°20N 1°26W 12 C6
Northam Australia 31°35S 116°42E 61 F2
Northam S. Africa 24°56S 27°18E 56 C4
Northampton Australia 28°27S 114°33E 61 E1
Northampton U.K. 52°15N 0°53W 13 E7
Northampton Mass., U.S.A. 42°19N 72°38W 83 D12
Northampton Pa., U.S.A. 40°41N 75°30W 83 F9
Northamptonshire □ U.K. 52°16N 0°55W 13 E7
Northbridge U.S.A. 42°9N 71°39W 83 D13
Northbrook Canada 44°44N 77°9W 82 B7
Northcliffe Australia 34°39S 116°7E 61 F2
Northeast Pacific Basin Pac. Oc. 32°0N 145°0W 65 D13
Northeast Providence Chan. W. Indies 26°0N 76°0W 88 A4
Northern = Limpopo □ S. Africa 24°5S 29°0E 57 C4
Northern □ Malawi 11°0S 34°0E 55 E3
Northern □ Zambia 10°30S 31°0E 55 E3
Northern Areas □ Pakistan 36°30N 73°0E 43 A5
Northern Cape □ S. Africa 30°0S 20°0E 56 D3
Northern Circars India 17°30N 82°30E 41 L13
Northern Indian L. Canada 57°20N 97°20W 71 B9
Northern Ireland □ U.K. 54°45N 7°0W 10 B5
Northern Lau Group Fiji 17°30S 178°59W 59 a
Northern Light L. Canada 48°15N 90°39W 72 C1
Northern Marianas ☒ Pac. Oc. 17°0N 145°0E 64 F6
Northern Province □ S. Africa 24°0S 29°0E 57 C4
Northern Range Trin. & Tob. 10°46N 61°15W 93 K15
Northern Sporades Greece 39°15N 23°30E 23 E10
Northern Territory □ Australia 20°0S 133°0E 60 D5
Northfield Minn., U.S.A. 44°27N 93°9W 80 C7
Northfield Vt., U.S.A. 44°9N 72°40W 83 B12
Northgate Canada 49°0N 102°16W 71 D8
Northland □ N.Z. 35°30S 173°30E 59 A4
Northome U.S.A. 47°52N 94°17W 80 B6
Northport Ala., U.S.A. 33°14N 87°35W 85 E11
Northport Wash., U.S.A. 48°55N 117°48W 76 B5
Northumberland □ U.K. 55°12N 2°0W 12 B6
Northumberland, C. Australia 38°5S 140°40E 63 F3
Northumberland Is. Australia 21°30S 149°50E 62 C4
Northumberland Str. Canada 46°20N 64°0W 73 C7
Northville Canada 43°10N 81°56W 82 C3
Northville U.S.A. 43°13N 74°11W 83 C10
Northwest Pacific Basin Pac. Oc. 32°0N 165°0E 64 D8
Northwest Providence Channel W. Indies 26°0N 78°0W 88 A4
Northwest Territories □ Canada 63°0N 118°0W 68 E8
Northwich U.K. 53°15N 2°31W 12 D5
Northwood Iowa, U.S.A. 43°27N 93°13W 80 D7
Northwood N. Dak., U.S.A. 47°44N 97°34W 80 B5
Norton U.S.A. 39°50N 99°53W 80 F4
Norton Zimbabwe 17°52S 30°40E 55 F3
Norton Sd. U.S.A. 63°50N 164°0W 74 C7
Norwalk Calif., U.S.A. 33°54N 118°4W 79 M8
Norwalk Conn., U.S.A. 41°7N 73°22W 83 E11
Norwalk Iowa, U.S.A. 41°29N 93°41W 80 E7
Norwalk Ohio, U.S.A. 41°15N 82°37W 82 E2
Norway Maine, U.S.A. 44°13N 70°32W 81 C18
Norway Mich., U.S.A. 45°47N 87°55W 80 C10
Norway ■ Europe 63°0N 11°0E 8 E14
Norway House Canada 53°59N 97°50W 71 C9
Norwegian B. Canada 77°30N 90°0W 69 B14
Norwegian Basin Atl. Oc. 68°0N 2°0W 4 C7
Norwegian Sea Atl. Oc. 66°0N 1°0E 6 B6
Norwich Canada 42°59N 80°36W 82 D4
Norwich U.K. 52°38N 1°18E 13 E9
Norwich Conn., U.S.A. 41°31N 72°5W 83 E12
Norwich N.Y., U.S.A. 42°32N 75°32W 83 D9
Norwood Canada 44°23N 77°59W 82 B7
Norwood U.S.A. 44°45N 75°0W 83 B10
Nosappu-Misaki Japan 45°26N 141°39E 30 C12
Noshiro Japan 40°12N 140°0E 30 D10
Noşratābād Iran 29°55N 60°0E 45 D8
Noss Hd. U.K. 58°28N 3°3W 11 C5
Nossob → S. Africa 26°55S 20°45E 56 D3
Nosy Barren Madag. 18°25S 43°40E 53 H8
Nosy Bé Madag. 13°25S 48°15E 53 G9
Nosy Boraha Madag. 16°50S 49°55E 57 B8
Nosy Lava Madag. 14°33S 47°36E 57 A8
Nosy Varika Madag. 20°35S 48°32E 57 C8
Noteć → Poland 52°44N 15°26E 16 B8
Notikewin → Canada 57°2N 117°38W 70 B5
Noto-Hantō Japan 37°15N 136°40E 31 F8
Notodden Norway 59°35N 9°17E 9 G13
Notre Dame B. Canada 49°45N 55°30W 73 C8
Notre-Dame-de-Koartac = Quaqtaq Canada 60°55N 69°40W 69 E18
Notre-Dame-des-Bois Canada 45°24N 71°4W 83 A13
Notre-Dame-d'Ivugivic = Ivujivik Canada 62°24N 77°55W 69 E16
Notre-Dame-du-Nord Canada 47°36N 79°30W 72 C4
Nottawasaga B. Canada 44°35N 80°15W 82 B4
Nottaway → Canada 51°22N 78°55W 72 B4
Nottingham U.K. 52°58N 1°10W 12 E6
Nottingham, City of □ U.K. 52°58N 1°10W 12 E6
Nottingham I. Canada 63°20N 77°55W 69 E16
Nottinghamshire □ U.K. 53°10N 1°3W 12 D6
Nottoway → U.S.A. 36°33N 76°55W 81 G15
Notwane → Botswana 23°35S 26°58E 56 C4
Nouâdhibou Mauritania 20°54N 17°0W 50 D2
Nouâdhibou, Râs Mauritania 20°50N 17°0W 50 D2
Nouakchott Mauritania 18°9N 15°58W 50 E2
Nouméa N. Cal. 22°17S 166°30E 58 D9
Noupoort S. Africa 31°10S 24°57E 56 E3
Nouveau Comptoir = Wemindji Canada 53°0N 78°49W 72 B4
Nouvelle Amsterdam, Î. Ind. Oc. 38°30S 77°30E 3 F13
Nouvelle-Calédonie = New Caledonia ☒ Pac. Oc. 21°0S 165°0E 58 D9
Nova Esperança Brazil 23°8S 52°24W 95 A5
Nova Friburgo Brazil 22°16S 42°30W 95 A7
Nova Iguaçu Brazil 22°45S 43°28W 95 A7
Nova Iorque Brazil 7°0S 44°5W 93 E10
Nova Lamego Guinea-Biss. 12°19N 14°11W 50 F3
Nova Lusitânia Mozam. 19°50S 34°34E 55 F3
Nova Mambone Mozam. 21°0S 35°3E 57 C6
Nova Scotia □ Canada 45°10N 63°0W 73 C7
Nova Sofala Mozam. 20°7S 34°42E 57 C5
Nova Venécia Brazil 18°45S 40°24W 93 G10
Nova Zagora Bulgaria 42°32N 26°1E 23 C11
Novar Canada 45°27N 79°15W 82 A5
Novara Italy 45°28N 8°38E 20 D8
Novato U.S.A. 38°6N 122°35W 78 G4
Novaya Ladoga Russia 60°7N 32°16E 18 B5
Novaya Lyalya Russia 59°4N 60°45E 18 C11
Novaya Sibir, Ostrov Russia 75°10N 150°0E 29 B16
Novaya Zemlya Russia 75°0N 56°0E 28 B6
Nové Zámky Slovak Rep. 48°2N 18°8E 17 D10
Novgorod Russia 58°30N 31°25E 18 C5
Novgorod-Severskiy = Novhorod-Siverskyy Ukraine 52°2N 33°10E 18 D5
Novhorod-Siverskyy Ukraine 52°2N 33°10E 18 D5
Novi Lígure Italy 44°46N 8°47E 20 D8
Novi Pazar Serbia 43°12N 20°28E 23 C9
Novi Sad Serbia 45°18N 19°52E 23 B8
Novo Hamburgo Brazil 29°37S 51°7W 95 B5
Novo Mesto Slovenia 45°47N 15°12E 22 B6
Novoaltaysk Russia 53°30N 84°0E 28 D9
Novocherkassk Russia 47°27N 40°15E 19 E7
Novodvinsk Russia 64°25N 40°42E 18 B7
Novogrudok = Navahrudak Belarus 53°40N 25°50E 17 B13
Novohrad-Volynskyy Ukraine 50°34N 27°35E 17 C14
Novokachalinsk Russia 45°5N 132°0E 30 B5
Novokuybyshevsk Russia 53°7N 49°58E 18 D8
Novokuznetsk Russia 53°45N 87°10E 28 D9
Novolazarevskaya Antarctica 71°0S 12°0E 5 D3
Novomoskovsk Russia 54°5N 38°15E 18 D6
Novorossiysk Russia 44°43N 37°46E 19 F6
Novorybnoye Russia 72°50N 105°50E 29 B11
Novoselytsya Ukraine 48°14N 26°15E 17 D14
Novoshakhtinsk Russia 47°46N 39°58E 19 E6
Novosibirsk Russia 55°0N 83°5E 28 D9
Novosibirskiye Ostrova Russia 75°0N 142°0E 29 B15
Novotroitsk Russia 51°10N 58°15E 18 D10
Novouzensk Russia 50°32N 48°17E 19 D8
Novovolynsk Ukraine 50°45N 24°4E 17 C13
Novska Croatia 45°19N 17°0E 22 B7
Novyy Bor Russia 66°43N 52°19E 18 A9
Novyy Port Russia 67°40N 72°30E 28 C8
Novyy Urengoy Russia 65°48N 76°52E 28 C8
Nowa Sól Poland 51°48N 15°44E 16 C8
Nowata U.S.A. 36°42N 95°38W 84 C7
Nowbarān Iran 35°8N 49°42E 45 C6
Nowghāb Iran 33°53N 59°4E 45 C8
Nowgong Assam, India 26°20N 92°50E 41 F18
Nowgong Mad. P., India 25°4N 79°27E 43 G8
Nowra Australia 34°53S 150°35E 63 E5
Nowshera Pakistan 34°0N 72°0E 40 C8
Nowy Sącz Poland 49°40N 20°41E 17 D11
Nowy Targ Poland 49°29N 20°2E 17 D11
Nowy Tomyśl Poland 52°19N 16°10E 16 B9
Noxen U.S.A. 41°25N 76°4W 83 E8
Noyabr'sk Russia 64°34N 76°21E 28 C8
Noyon France 49°34N 2°59E 20 B5
Noyon Mongolia 43°2N 102°4E 34 C2
Nqutu S. Africa 28°13S 30°32E 57 D5
Nsanje Malawi 16°55S 35°12E 55 F4
Nsawam Ghana 5°50N 0°24W 50 G5
Nseluka Zambia 9°58S 31°16E 55 D3
Nsomba Zambia 10°45S 29°51E 55 E2
Ntaria ✪ Australia 24°0S 132°41E 60 D5
Ntungamo Uganda 0°56S 30°17E 54 C3
Nu Jiang → China 29°58N 97°25E 32 F8
Nu Shan China 26°0N 99°20E 41 G21
Nuba Mts. = Nubah, Jibalan Sudan 12°0N 31°0E 51 F12
Nubah, Jibalan Sudan 12°0N 31°0E 51 F12
Nubia Africa 21°0N 32°0E 48 D7
Nubian Desert = Nûbîya, Es Sahrâ en Sudan 21°30N 33°30E 51 D12
Nûbîya, Es Sahrâ en Sudan 21°30N 33°30E 51 D12
Nuboai Indonesia 2°10S 136°30E 37 E9
Nubra → India 34°35N 77°35E 43 B7
Nueces → U.S.A. 27°51N 97°30W 84 H6
Nueltin L. Canada 60°30N 99°30W 71 A9
Nuestra Señora del Rosario de Caá-Catí Argentina 27°45S 57°36W 94 B4
Nueva Ciudad Guerrero Mexico 26°34N 99°12W 87 B5
Nueva Gerona Cuba 21°53N 82°49W 88 B3
Nueva Palmira Uruguay 33°52S 58°20W 94 C4
Nueva Rosita Mexico 27°57N 101°13W 86 B4
Nueva San Salvador El Salv. 13°40N 89°18W 88 D2
Nuéve de Julio Argentina 35°30S 61°0W 94 D3
Nuevitas Cuba 21°30N 77°20W 88 B4
Nuevo, G. Argentina 43°0S 64°30W 96 E4
Nuevo Casas Grandes Mexico 30°25N 107°55W 86 A3
Nuevo Laredo Mexico 27°30N 99°31W 87 B5
Nuevo León □ Mexico 25°20N 100°0W 86 C5
Nuevo Rocafuerte Ecuador 0°55S 75°27W 92 D3
Nugget Pt. N.Z. 46°27S 169°50E 59 G2
Nuhaka N.Z. 39°3S 177°45E 59 C6
Nukey Bluff Australia 32°26S 135°29E 63 E2
Nukhayb Iraq 32°4N 42°3E 44 C4
Nuku Hiva French Polynesia 8°54S 140°6W 65 H13
Nuku'alofa Tonga 21°10S 175°12W 59 c
Nukus Uzbekistan 42°27N 59°41E 28 E6
Nulato U.S.A. 64°43N 158°6W 74 C8
Nullagine Australia 21°53S 120°7E 60 D3
Nullagine → Australia 21°20S 120°20E 60 D3
Nullarbor Australia 31°28S 130°55E 61 F5
Nullarbor △ Australia 32°39S 130°0E 61 F2
Nullarbor Plain Australia 31°10S 129°0E 61 F4
Numalla, L. Australia 28°43S 144°20E 63 D3
Numan Nigeria 9°29N 12°3E 51 G8
Numata Japan 36°45N 139°4E 31 F9
Numaykoos Lake ⌂ Canada 57°52N 95°58W 71 B9
Numazu Japan 35°7N 138°51E 31 G9
Numbulwar Australia 14°15S 135°45E 62 A2
Numfoor Indonesia 1°0S 134°50E 37 E8
Numurkah Australia 36°5S 145°26E 63 F4
Nunaksaluk I. Canada 55°49N 60°20W 73 A7
Nunap Isua Greenland 59°48N 43°55W 66 D15
Nunavik Greenland 71°50N 54°25W 69 C21
Nunavut □ Canada 66°0N 85°0W 69 D15
Nunda U.S.A. 42°35N 77°56W 82 D7
Nuneaton U.K. 52°32N 1°27W 13 E6
Nungarin Australia 31°12S 118°6E 61 F2
Nungo Mozam. 13°23S 37°43E 55 E4
Nungwe Tanzania 2°48S 32°2E 54 C3
Nunivak I. U.S.A. 60°10N 166°30W 74 C6
Nunkun India 33°57N 76°2E 43
Núoro Italy 40°20N 9°20E 22
Nūr Iran 36°33N 52°1E 45
Nūrābād Hormozgān, Iran 27°47N 57°12E 45
Nūrābād Lorestān, Iran 34°4N 47°58E 44
Nuremberg = Nürnberg Germany 49°27N 11°3E 16
Nuri Mexico 28°5N 109°22W 86
Nuriootpa Australia 34°27S 139°0E 63
Nuristān □ Afghan. 35°20N 71°0E 40
Nurmes Finland 63°33N 29°10E 8 E
Nürnberg Germany 49°27N 11°3E 16
Nurpur Pakistan 31°53N 71°54E 42
Nurran, L. = Terewah, L. Australia 29°52S 147°35E 63
Nurrari Lakes Australia 29°1S 130°5E 61
Nusa Dua Indonesia 8°48S 115°14E 37 K
Nusa Tenggara Barat □ Indonesia 8°50S 117°30E 36
Nusa Tenggara Timur □ Indonesia 9°30S 122°0E 37
Nusaybin Turkey 37°3N 41°10E 19
Nushki Pakistan 29°35N 66°0E 42
Nuuk Greenland 64°10N 51°35W 67 C
Nuupere, Pte. Moorea 17°36S 149°47W 5
Nuwakot Nepal 28°10N 83°55E 43 E
Nuwayb'ī, W. an → Si. Arabia 29°18N 34°57E 46
Nuweiba' Egypt 28°59N 34°39E 44
Nuwerus S. Africa 31°8S 18°24E 56
Nuweveldberge S. Africa 32°10S 21°45E 56
Nuyts, Pt. Australia 35°4S 116°38E 61
Nuyts Arch. Australia 32°35S 133°20E 63
Nxai Pan △ Botswana 19°50S 24°46E 56
Nxaunxau Botswana 18°57S 21°4E 56
Nyabing Australia 33°33S 118°9E 61
Nyack U.S.A. 41°5N 73°55W 83 E
Nyagan Russia 62°30N 65°38E 28
Nyahanga Tanzania 2°20S 33°37E 54
Nyahua Tanzania 5°25S 33°23E 54
Nyahururu Kenya 0°2N 36°27E 54
Nyainqêntanglha Shan China 30°0N 90°0E 32
Nyakanazi Tanzania 3°2S 31°10E 54
Nyâlâ Sudan 12°2N 24°58E 51 F
Nyalam China 28°32N 86°4E 32
Nyamandhlovu Zimbabwe 19°55S 28°16E 55
Nyambiti Tanzania 2°48S 33°27E 54
Nyamira Kenya 0°36S 34°52E 54
Nyamwaga Tanzania 1°27S 34°33E 54
Nyandekwa Tanzania 3°57S 32°32E 54
Nyandoma Russia 61°40N 40°12E 18
Nyanga △ Zimbabwe 18°17S 32°46E 55
Nyangana Namibia 18°0S 20°40E 56
Nyanguge Tanzania 2°30S 33°12E 54
Nyanji Zambia 14°25S 31°46E 55
Nyanza Rwanda 2°20S 29°42E 54
Nyanza □ Kenya 0°10S 34°15E 54
Nyanza-Lac Burundi 4°21S 29°36E 54
Nyasa, L. = Malawi, L. Africa 12°30S 34°30E 55
Nyasvizh Belarus 53°14N 26°38E 17 B
Nyazepetrovsk Russia 56°3N 59°36E 18 C
Nyazura Zimbabwe 18°40S 32°16E 55
Nyazwidzi → Zimbabwe 20°0S 31°17E 55
Nybro Sweden 56°44N 15°55E 9 H
Nyda Russia 66°40N 72°58E 28
Nyeboe Land Greenland 82°0N 57°0W 69 A
Nyeri Kenya 0°23S 36°56E 54
Nyika △ Malawi 10°30S 33°53E 55
Nyimba Zambia 14°33S 30°50E 55
Nyingchi China 29°32N 94°25E 32
Nyíregyháza Hungary 47°58N 21°47E 17 E
Nyiru, Mt. Kenya 2°8N 36°50E 54
Nykarleby = Uusikaarlepyy Finland 63°32N 22°31E 8 E
Nykøbing Nordjylland, Denmark 56°48N 8°51E 9 H
Nykøbing Sjælland, Denmark 54°56N 11°52E 9 J
Nykøbing Sjælland, Denmark 55°55N 11°40E 9 J
Nyköping Sweden 58°45N 17°1E 9 G
Nylstroom = Modimolle S. Africa 24°42S 28°22E 57
Nymagee Australia 32°7S 146°20E 63
Nymboida △ Australia 29°38S 152°26E 63
Nynäshamn Sweden 58°54N 17°57E 9 G
Nyngan Australia 31°30S 147°8E 63
Nyoma Rap India 33°10N 78°40E 43
Nyoman = Nemunas → Lithuania 55°25N 21°10E 9 J
Nysa Poland 50°30N 17°22E 17
Nysa → Europe 52°4N 14°46E 16
Nyslott = Savonlinna Finland 61°52N 28°53E 8 F
Nyssa U.S.A. 43°53N 117°0W 76
Nystad = Uusikaupunki Finland 60°47N 21°25E 8 F
Nyunzu Dem. Rep. of the Congo 5°57S 27°58E 54
Nyurba Russia 63°17N 118°28E 29 C
Nzega Tanzania 4°10S 33°12E 54
Nzérékoré Guinea 7°49N 8°48W 50
Nzeto Angola 7°10S 12°52E 52
Nzilo, Chutes de Dem. Rep. of the Congo 10°18S 25°27E 55
Nzubuka Tanzania 4°45S 32°50E 54
Nzwani = Anjouan Comoros Is. 12°15S 44°20E 53

O

O Le Pupū Pu'e △ Samoa 13°59S 171°43W 59
Ō-Shima Hokkaidō, Japan 41°30N 139°22E 30 D
Ō-Shima Shizuoka, Japan 34°44N 139°24E 31 G
Oa, Mull of U.K. 55°35N 6°20W 11 F

acoma U.S.A. 43°48N 99°24W 80 D4
ahe, L. U.S.A. 44°27N 100°24W 80 C3
ahe Dam U.S.A. 44°27N 100°24W 80 C3
ahu U.S.A. 21°28N 157°58W 75 L8
ak Harbor U.S.A. 48°18N 122°39W 78 B4
ak Hill U.S.A. 37°59N 81°9W 81 G13
ak Island U.S.A. 33°55N 78°10W 85 E15
ak Ridge U.S.A. 36°1N 84°16W 85 C12
ak View U.S.A. 34°24N 119°18W 79 L7
akan-Dake Japan 43°27N 144°10E 30 C12
akdale Calif., U.S.A. 37°46N 120°51W 78 H6
akdale La., U.S.A. 30°49N 92°40W 84 F8
akes U.S.A. 46°8N 98°6W 80 B4
akesdale U.S.A. 47°8N 117°15W 76 C5
akey Australia 27°25S 151°43E 63 D5
akfield U.S.A. 43°4N 78°16W 82 C6
akham U.K. 52°40N 0°43W 13 E7
akhurst U.S.A. 37°19N 119°40W 78 H7
akland Calif., U.S.A. 37°48N 122°18W 78 H4
akland Pa., U.S.A. 41°57N 75°36W 83 E9
akley Idaho, U.S.A. 42°15N 113°53W 76 E7
akley Kans., U.S.A. 39°8N 100°51W 80 F3
akover → Australia 21°0S 120°40E 60 D3
akridge U.S.A. 43°45N 122°28W 76 E2
akville Canada 43°27N 79°41W 82 C5
akville U.S.A. 46°51N 123°14W 78 D3
amaru N.Z. 45°5S 170°59E 59 F3
asis Calif., U.S.A. 37°29N 117°55W 78 H9
asis Calif., U.S.A. 33°28N 116°6W 79 M10
ates Land Antarctica 69°0S 160°0E 5 C11
atman U.S.A. 35°1N 114°19W 79 K12
axaca Mexico 17°3N 96°43W 87 D5
axaca □ Mexico 17°0N 96°30W 87 D5
→ Russia 66°45N 69°30E 28 C7
a Canada 49°4N 84°7W 72 C3
ama Japan 35°30N 135°45E 31 G7
an U.K. 56°25N 5°29W 11 E3
bia = Hobyo Somali Rep. 5°25N 48°30E 47 F4
erá Argentina 27°21S 55°2W 95 B4
erhausen Germany 51°28N 6°51E 16 C4
erlin Kans., U.S.A. 39°49N 100°32W 80 F3
erlin La., U.S.A. 30°37N 92°46W 84 F8
erlin Ohio, U.S.A. 41°18N 82°13W 82 E2
eron Australia 33°45S 149°52E 63 E4
i, Kepulauan Indonesia 1°23S 127°45E 37 E7
idos Brazil 1°50S 55°30W 93 D7
ihiro Japan 42°56N 143°12E 30 C11
ilatu Indonesia 1°25S 127°20E 37 E7
luchye Russia 49°1N 131°4E 29 E14
o C.A.R. 5°20N 26°32E 54 A2
oyan Russia 51°15N 36°21E 28 D4
ozerskaya = Obozerskiy Russia 63°34N 40°21E 18 B7
ozerskiy Russia 63°34N 40°21E 18 B7
servatory Inlet Canada 55°10N 129°54W 70 B3
shchi Syrt Russia 52°0N 53°0E 6 E16
skaya Guba Russia 69°0N 73°0E 28 C8
uasi Ghana 6°17N 1°40W 50 G5
ala U.S.A. 29°11N 82°8W 85 G13
ampo Chihuahua, Mexico 28°11N 108°23W 86 B3
ampo Tamaulipas, Mexico 22°50N 99°20W 87 C5
aña Spain 39°55N 3°30W 21 C4
cidental, Cordillera Colombia 5°0N 76°0W 92 C3
cidental, Grand Erg Algeria 30°20N 1°0E 50 B6
ean City Md., U.S.A. 38°20N 75°5W 81 F16
ean City N.J., U.S.A. 39°17N 74°35W 81 F16
ean City Wash., U.S.A. 47°4N 124°10W 78 C2
ean Falls Canada 52°18N 127°48W 70 C3
ean I. = Banaba Kiribati 0°45S 169°50E 64 H8
ean Park U.S.A. 46°30N 124°3W 78 D2
eano U.S.A. 35°6N 120°37W 79 K6
eanport U.S.A. 40°19N 74°3W 83 F10
eanside U.S.A. 33°12N 117°23W 79 M9
hil Hills U.K. 56°14N 3°40W 11 E5
ho Rios Jamaica 18°24N 77°6W 88 a
illa U.S.A. 31°36N 83°15W 85 F13
mulgee → U.S.A. 31°58N 82°33W 85 F13
niţa Moldova 48°25N 27°30E 17 D14
onee → U.S.A. 31°58N 82°33W 85 F13
onomowoc U.S.A. 43°7N 88°30W 80 D9
onto U.S.A. 44°53N 87°52W 80 C10
onto Falls U.S.A. 44°52N 88°9W 80 C9
osingo Mexico 16°53N 92°6W 87 D6
otal Nic. 13°41N 86°31W 88 D2
otlán Jalisco, Mexico 20°21N 102°46W 86 C4
otlán Oaxaca, Mexico 16°48N 96°40W 87 D5
ussi = Pante Macassar E. Timor 9°30S 123°58E 37 F6
a Japan 35°11N 132°30E 31 G6
áðahraun Iceland 65°5N 17°0W 8 D5
aejin N. Korea 41°34N 129°40E 30 D4
ate Japan 40°16N 140°34E 30 D10
awara Japan 35°20N 139°6E 31 G9
da Norway 60°3N 6°35E 9 F12
ei → Canada 56°6N 96°54W 71 B9
emiş Turkey 38°15N 28°0E 23 E13
endaalsrus S. Africa 27°48S 26°45E 56 D4
ense Denmark 55°22N 10°23E 9 J14
er → Europe 53°33N 14°38E 16 B8
esa Ukraine 46°30N 30°45E 19 E5
essa = Odesa Ukraine 46°30N 30°45E 19 E5
essa Canada 44°17N 76°43W 83 B8
essa Tex., U.S.A. 31°52N 102°23W 84 F3
essa Wash., U.S.A. 47°20N 118°41W 76 C4
iakwe Botswana 20°12S 25°17E 56 C4
ienné Ivory C. 9°30N 7°34W 50 G4
intsovo Russia 55°40N 37°16E 18 C6
Donnell U.S.A. 32°58N 101°50W 84 E4
Donnell Pt. Canada 45°5N 80°5W 82 A4
orheiu Secuiesc Romania 46°21N 25°21E 17 E13
ra = Oder → Europe 53°33N 14°38E 16 B8
Odzi Zimbabwe 19°0S 32°20E 57 B5
Odzi → Zimbabwe 19°45S 32°23E 57 B5
Oeiras Brazil 7°0S 42°8W 93 E10
Oelrichs U.S.A. 43°11N 103°14W 80 D2
Oelwein U.S.A. 42°41N 91°55W 80 D8
Oeno I. Pac. Oc. 24°0S 131°0W 65 K14
Oenpelli = Gunbalanya Australia 12°20S 133°4E 60 B5
Ofanto → Italy 41°22N 16°13E 22 D7
Offa Nigeria 8°13N 4°42E 50 G6
Offaly □ Ireland 53°15N 7°30W 10 C4
Offenbach Germany 50°6N 8°44E 16 C5
Offenburg Germany 48°28N 7°56E 16 D4
Officer Cr. → Australia 27°46S 132°30E 61 E5
Ofolanga Tonga 19°38S 174°27W 59 c
Ofotfjorden Norway 68°27N 17°0E 8 B17
Ofu Amer. Samoa 14°11S 169°41W 59 b
Ōfunato Japan 39°4N 141°43E 30 E10
Oga Japan 39°55N 139°50E 30 E9
Oga-Hantō Japan 39°58N 139°47E 30 E9
Ogaden Ethiopia 7°30N 45°30E 47 F3
Ōgaki Japan 35°21N 136°37E 31 G8
Ogallala U.S.A. 41°8N 101°43W 80 E3
Ogasawara Gunto Pac. Oc. 27°0N 142°0E 27 F16
Ogbomosho Nigeria 8°1N 4°11E 50 G6
Ogden U.S.A. 41°13N 111°58W 76 F8
Ogdensburg U.S.A. 44°42N 75°30W 83 B9
Ogea Driki Fiji 19°12S 178°27W 59 a
Ogea Levu Fiji 19°8S 178°24W 59 a
Ogeechee → U.S.A. 31°50N 81°3W 85 F14
Ogilby U.S.A. 32°49N 114°50W 79 N12
Óglio → Italy 45°2N 10°39E 22 B4
Ogmore Australia 22°37S 149°35E 62 C4
Ogoki Canada 51°38N 85°58W 72 B2
Ogoki → Canada 51°38N 85°57W 72 B2
Ogoki L. Canada 50°50N 87°10W 72 B2
Ogoki Res. Canada 50°45N 88°15W 72 B2
Ogooué → Gabon 1°0S 9°0E 52 E1
Ogowe = Ogooué → Gabon 1°0S 9°0E 52 E1
Ogre Latvia 56°49N 24°36E 9 H21
Ogurja Ada Turkmenistan 38°55N 53°2E 45 B7
Ohai N.Z. 45°55S 168°0E 59 F2
Ohakune N.Z. 39°24S 175°24E 59 C5
Ohata Japan 41°24N 141°10E 30 D10
Ohau, L. N.Z. 44°15S 169°53E 59 F2
Ohio □ U.S.A. 40°15N 82°45W 82 F2
Ohio → U.S.A. 36°59N 89°8W 80 G9
Ohře → Czech Rep. 50°30N 14°10E 16 C8
Ohrid Macedonia 41°8N 20°52E 23 D9
Ohridsko Jezero Macedonia 41°8N 20°52E 23 D9
Ohrigstad S. Africa 24°39S 30°36E 57 C5
Oiapoque Brazil 3°50N 51°50W 93
Oikou China 38°35N 117°42E 35 E9
Oil City U.S.A. 41°26N 79°42W 82 E5
Oil Springs Canada 42°47N 82°7W 82 D2
Oildale U.S.A. 35°25N 119°1W 79 K7
Oise → France 49°0N 2°4E 20 B5
Oistins Barbados 13°4N 59°33W 89 g
Oistins B. Barbados 13°4N 59°33W 89 g
Ōita Japan 33°14N 131°36E 31 H5
Ōita □ Japan 33°15N 131°30E 31 H5
Oiticica Brazil 5°3S 41°5W 93 E10
Ojai U.S.A. 34°27N 119°15W 79 L7
Ojinaga Mexico 29°34N 104°25W 86 B4
Ojiya Japan 37°18N 138°48E 31 F9
Ojo Caliente Mexico 21°53N 102°15W 86 C4
Ojo de Liebre, L. Mexico 27°45N 114°15W 86 B2
Ojos del Salado, Cerro Argentina 27°0S 68°40W 94 B2
Oka → Russia 56°20N 43°59E 18 C7
Okaba Indonesia 8°6S 139°42E 37 F9
Okahandja Namibia 22°0S 16°59E 56 C2
Okanagan L. Canada 50°0N 119°30W 70 D5
Okandja Gabon 0°35S 13°45E 52 E2
Okanogan U.S.A. 48°22N 119°35W 76 B4
Okanogan → U.S.A. 48°6N 119°44W 76 B4
Okanogan Range N. Amer. 49°0N 119°55W 70 D5
Okapi △ Dem. Rep. of the Congo 2°30N 27°20E 54 B2
Okaputa Namibia 20°5S 17°0E 56 C2
Okara Pakistan 30°50N 73°31E 42 D5
Okaukuejo Namibia 19°10S 16°0E 56 B2
Okavango Delta Botswana 18°45S 22°45E 56 B3
Okavango Swamp = Okavango Delta Botswana 18°45S 22°45E 56 B3
Okaya Japan 36°5N 138°10E 31 F9
Okayama Japan 34°40N 133°54E 31 G6
Okayama □ Japan 35°0N 133°50E 31 G6
Okazaki Japan 34°57N 137°10E 31 G8
Okeechobee U.S.A. 27°15N 80°50W 85 H14
Okeechobee, L. U.S.A. 27°0N 80°50W 85 H14
Okefenokee ⌓ U.S.A. 30°45N 82°18W 85 F13
Okefenokee Swamp U.S.A. 30°40N 82°20W 85 F13
Okehampton U.K. 50°44N 4°0W 13 G4
Okha India 22°27N 69°4E 42 H3
Okha Russia 53°40N 143°0E 29 D15
Okhotsk Russia 59°20N 143°10E 29 D15
Okhotsk, Sea of Asia 55°0N 145°0E 29 D15
Okhotskiy Perevoz Russia 61°52N 135°35E 29 C14
Okhtyrka Ukraine 50°25N 35°0E 19 D5
Oki-Shotō Japan 36°5N 133°15E 31 F6
Okiep S. Africa 29°39S 17°53E 56 D2
Okinawa Japan 26°19N 127°46E 31 L3
Okinawa □ Japan 26°40N 128°0E 31 L4
Okinawa-Guntō Japan 26°40N 128°0E 31 L4
Okinawa-Jima Japan 26°32N 128°0E 31 L4
Okino-erabu-Shima Japan 27°21N 128°33E 31 L4
Oklahoma □ U.S.A. 35°20N 97°30W 84 D6
Oklahoma City U.S.A. 35°30N 97°30W 84 D6
Okmulgee U.S.A. 35°37N 95°58W 84 D7
Oknitsa = Ocniţa Moldova 48°25N 27°30E 17 D14
Okolo Uganda 2°37N 31°8E 54 B3
Okolona U.S.A. 34°0N 88°45W 85 E10
Okombahe Namibia 21°23S 15°22E 56 C2
Okotoks Canada 50°43N 113°58W 70 C6
Oksovskiy Russia 62°33N 39°57E 18 B6
Oktyabrsk = Qandyaghash Kazakhstan 49°28N 57°25E 19 E10
Oktyabrskiy = Aktsyabrski Belarus 52°38N 28°53E 17 B15
Oktyabrskiy Bashkortostan, Russia 54°28N 53°28E 18 D9
Oktyabrskiy Kamchatka, Russia 52°39N 156°14E 29 D16
Oktyabrskoy Revolyutsii, Ostrov Russia 79°30N 97°0E 29 B10
Okuru N.Z. 43°55S 168°55E 59 E2
Okushiri-Tō Japan 42°15N 139°30E 30 C9
Okwa → Botswana 22°30S 23°0E 56 C3
Ola Russia 59°35N 151°17E 29 D16
Ola U.S.A. 35°2N 93°13W 84 D8
Ólafsfjörður Iceland 66°4N 18°39W 8 C4
Ólafsvík Iceland 64°53N 23°43W 8 D2
Olancha U.S.A. 36°17N 118°1W 79 J8
Olancha Pk. U.S.A. 36°16N 118°7W 79 J8
Olanchito Honduras 15°30N 86°30W 88 C2
Öland Sweden 56°45N 16°38E 9 H17
Olary Australia 32°18S 140°19E 63 E3
Olascoaga Argentina 35°15S 60°39W 94 D3
Olathe U.S.A. 38°53N 94°49W 80 F6
Olavarría Argentina 36°55S 60°20W 94 D3
Oława Poland 50°57N 17°20E 17 C9
Ólbia Italy 40°55N 9°31E 22 D3
Olcott U.S.A. 43°20N 78°42W 82 C6
Old Bahama Chan. = Bahama, Canal Viejo de W. Indies 22°10N 77°30W 88 B4
Old Baldy Pk. = San Antonio, Mt. U.S.A. 34°17N 117°38W 79 L9
Old Bridge U.S.A. 40°25N 74°22W 83 F10
Old Castile = Castilla y Leon □ Spain 42°0N 5°0W 21 B3
Old Crow Canada 67°30N 139°55W 68 D4
Old Dale U.S.A. 34°8N 115°47W 79 L11
Old Forge N.Y., U.S.A. 43°43N 74°58W 83 C10
Old Forge Pa., U.S.A. 41°22N 75°45W 83 E9
Old Mapoon ☼ Australia 11°5S 142°22E 62 A3
Old Perlican Canada 48°5N 53°1W 73 C9
Old Shinyanga Tanzania 3°33S 33°27E 54 C3
Old Speck Mt. U.S.A. 44°34N 70°57W 83 B14
Old Town U.S.A. 44°56N 68°39W 81 C19
Old Washington U.S.A. 40°2N 81°27W 82 F3
Old Wives L. Canada 50°5N 106°0W 71 C7
Oldbury U.K. 51°38N 2°33W 13 F5
Oldcastle Ireland 53°46N 7°10W 10 C4
Oldeani Tanzania 3°22S 35°35E 54 C4
Oldenburg Germany 53°9N 8°13E 16 B5
Oldenzaal Neths. 52°19N 6°53E 15 B6
Oldham U.K. 53°33N 2°7W 12 D5
Oldman → Canada 49°57N 111°42W 70 D6
Oldmeldrum U.K. 57°20N 2°19W 11 D6
Olds Canada 51°50N 114°10W 70 C6
Olduvai Gorge Tanzania 2°57S 35°23E 54 C4
Öldziyt Mongolia 44°40N 109°1E 34 B5
Olean U.S.A. 42°5N 78°26W 82 D6
Olekma → Russia 60°22N 120°42E 29 C13
Olekminsk Russia 60°25N 120°30E 29 C13
Oleksandriya Ukraine 50°37N 26°19E 17 C14
Olema U.S.A. 38°3N 122°47W 78 G4
Olenegorsk Russia 68°9N 33°18E 8 B25
Olenek Russia 68°28N 112°18E 29 C12
Olenek → Russia 73°0N 120°10E 29 B13
Oléron, Î. d' France 45°55N 1°15W 20 D3
Oleśnica Poland 51°13N 17°22E 17 C9
Olevsk Ukraine 51°12N 27°39E 17 C14
Olga, L. Canada 49°47N 77°15W 72 C4
Olgas, The = Kata Tjuta Australia 25°20S 130°50E 61 E5
Ölgiy Mongolia 48°56N 89°57E 32 B6
Olhão Portugal 37°3N 7°48W 21 D2
Olifants = Elefantes → Africa 24°10S 32°40E 57 C5
Olifants → Namibia 25°30S 19°30E 56 C2
Olifantshoek S. Africa 27°57S 22°42E 56 D3
Ólimbos, Óros = Olymbos Oros Greece 40°6N 22°23E 23 D10
Olímpia Brazil 20°44S 48°54W 95 A6
Olinda Brazil 8°1S 34°51W 93 E12
Oliva Argentina 32°0S 63°38W 94 C3
Olivares, Cerro los Argentina 30°18S 69°55W 94 C2
Olive Branch U.S.A. 34°57N 89°49W 85 D10
Olivehurst U.S.A. 39°6N 121°34W 78 F5
Olivenza Spain 38°41N 7°9W 21 C2
Oliver Canada 49°13N 119°37W 70 D5
Oliver L. Canada 56°56N 103°22W 71 B8
Ollagüe Chile 21°15S 68°10W 94 A2
Olmaliq Uzbekistan 40°50N 69°35E 28 E7
Olney Ill., U.S.A. 38°44N 88°5W 80 F9
Olney Tex., U.S.A. 33°22N 98°45W 84 E5
Oloitokitok Kenya 2°56S 37°30E 54 C4
Olomane → Canada 50°14N 60°37W 73 B7
Olomouc Czech Rep. 49°38N 17°12E 17 D9
Olonets Russia 61°0N 32°54E 18 B5
Olongapo Phil. 14°50N 120°18E 37 B6
Olosega Amer. Samoa 14°10S 169°37W 59 b
Olosenga = Swains I. Amer. Samoa 11°11S 171°4W 65 J11
Olot Spain 42°11N 2°30E 21 A7
Olovyannaya Russia 50°58N 115°35E 29 D12
Oloy → Russia 66°29N 159°29E 29 C16
Olsztyn Poland 53°48N 20°29E 17 B11
Olt → Romania 43°43N 24°51E 17 G13
Olteniţa Romania 44°7N 26°42E 17 F14
Olton U.S.A. 34°11N 102°8W 84 D3
Olymbos Cyprus 35°21N 33°45E 25 D12
Olymbos Oros Greece 40°6N 22°23E 23 D10
Olympia Greece 37°39N 21°39E 23 F9
Olympia U.S.A. 47°3N 122°53W 78 D4
Olympic △ U.S.A. 47°45N 123°43W 78 C3
Olympic Dam Australia 30°30S 136°55E 63 E2
Olympic Mts. U.S.A. 47°55N 123°45W 78 C3
Olympus Cyprus 34°56N 32°52E 25 E11
Olympus, Mt. = Olymbos Oros Greece 40°6N 22°23E 23 D10
Olympus, Mt. = Uludağ Turkey 40°4N 29°13E 23 D13
Olympus, Mt. U.S.A. 47°48N 123°43W 78 C3
Olyphant U.S.A. 41°27N 75°36W 83 E9
Olyutorskiy, Mys Russia 59°55N 170°27E 29 D18
Om → Russia 54°59N 73°22E 28 D8
Om Koi Thailand 17°48N 98°22E 38 D2
Ōma Japan 41°45N 141°5E 30 D10
Ōmachi Japan 36°30N 137°50E 31 F8
Omae-Zaki Japan 34°36N 138°14E 31 G9
Ōmagari Japan 39°27N 140°29E 30 E10
Omagh U.K. 54°36N 7°19W 10 B4
Omagh □ U.K. 54°35N 7°15W 10 B4
Omaha U.S.A. 41°17N 95°58W 80 E6
Omak U.S.A. 48°25N 119°31W 76 B4
Omalos Greece 35°19N 23°55E 25 D5
Oman ■ Asia 23°0N 58°0E 47 C6
Oman, G. of Asia 24°30N 58°30E 45 E8
Omaruru Namibia 21°26S 16°0E 56 C2
Omaruru → Namibia 22°7S 14°15E 56 C1
Omate Peru 16°45S 71°0W 92 G4
Ombai, Selat Indonesia 8°30S 124°50E 37 F6
Omboué Gabon 1°35S 9°15E 52 E1
Ombrone → Italy 42°42N 11°5E 22 C4
Omdurmân Sudan 15°40N 32°28E 51 E12
Omemee Canada 44°18N 78°33W 82 B6
Omeonga Dem. Rep. of the Congo 3°40S 24°22E 54 C1
Ometepe, I. de Nic. 11°32N 85°35W 88 D2
Ometepec Mexico 16°41N 98°25W 87 D5
Ominato Japan 41°17N 141°10E 30 D10
Omineca → Canada 56°3N 124°16W 70 B4
Omineca Mts. Canada 56°30N 125°30W 70 B3
Omitara Namibia 22°16S 18°2E 56 C2
Ōmiya = Saitama Japan 35°54N 139°38E 31 G9
Ommen Neths. 52°31N 6°26E 15 B6
Ömnögovĭ □ Mongolia 43°15N 104°0E 34 C3
Omo → Ethiopia 6°25N 36°10E 47 F2
Omodhos Cyprus 34°51N 32°48E 25 E11
Omolon → Russia 68°42N 158°36E 29 C16
Omono-Gawa → Japan 39°46N 140°3E 30 E10
Ompha Canada 45°0N 76°50W 83 B8
Omsk Russia 55°0N 73°12E 28 D8
Omsukchan Russia 62°32N 155°48E 29 C16
Ōmu Japan 44°34N 142°58E 30 B11
Omul, Vf. Romania 45°27N 25°29E 17 F13
Ōmura Japan 32°56N 129°57E 31 H4
Omuramba Omatako → Namibia 17°45S 20°25E 56 B2
Omuramba Ovambo → Namibia 18°45S 16°59E 56 B2
Ōmuta Japan 33°5N 130°26E 31 H5
Onaga U.S.A. 39°29N 96°10W 80 F5
Onalaska U.S.A. 43°53N 91°14W 80 D8
Onancock U.S.A. 37°43N 75°45W 81 G16
Onangue, L. Gabon 0°57S 10°4E 52 E2
Onaping L. Canada 47°3N 81°30W 72 C3
Onavas Mexico 28°31N 109°35W 86 B3
Onawa U.S.A. 42°2N 96°6W 80 D5
Oncócua Angola 16°30S 13°25E 56 B1
Onda Spain 39°55N 0°17W 21 C5
Ondangwa Namibia 17°57S 16°4E 56 B2
Ondjiva Angola 16°48S 15°50E 56 B2
Öndörhaan Mongolia 47°19N 110°39E 33 B11
Öndverðarnes Iceland 64°52N 24°0W 8 D2
One Arm Point Australia 16°26S 123°3E 60 C3
One Arm Point ☼ Australia 16°31S 122°53E 60 C3
One Tree Australia 34°11S 144°43E 63 E3
Oneata Fiji 18°26S 178°25W 59 a
Onega Russia 64°0N 38°10E 18 B6
Onega → Russia 63°58N 38°2E 18 B6
Onega, G. of = Onezhskaya Guba Russia 64°24N 36°38E 18 B6
Onega, L. = Onezhskoye Ozero Russia 61°44N 35°22E 18 B6
Oneida U.S.A. 43°6N 75°39W 83 C9
Oneida L. U.S.A. 43°12N 75°54W 83 C9
O'Neill U.S.A. 42°27N 98°39W 80 D4
Onekotan, Ostrov Russia 49°25N 154°45E 29 E16
Onema Dem. Rep. of the Congo 4°35S 24°30E 54 C1
Oneonta U.S.A. 42°27N 75°4W 83 D9
Oneşti Romania 46°17N 26°47E 17 E14
Onezhskaya Guba Russia 64°24N 36°38E 18 B6
Onezhskoye Ozero Russia 61°44N 35°22E 18 B6
Ongarue N.Z. 38°42S 175°19E 59 C5
Ongea Levu = Ogea Levu Fiji 19°8S 178°24W 59 a
Ongers → S. Africa 31°4S 23°13E 56 E3
Ongerup Australia 33°58S 118°28E 61 F2
Ongi Mongolia 45°27N 103°54E 34 B2
Ongjin N. Korea 37°56N 125°21E 35 F13
Ongkharak Thailand 14°8N 101°1E 38 E3
Ongniud Qi China 43°0N 118°38E 35 C10
Ongoka Dem. Rep. of the Congo 1°20S 26°0E 54 C2
Ongole India 15°33N 80°2E 40 M12
Ongon = Havirga Mongolia 45°41N 113°5E 34 B7
Onida U.S.A. 44°42N 100°4W 80 C3
Onilahy → Madag. 23°34S 43°45E 57 C7
Onitsha Nigeria 6°6N 6°42E 50 G7
Ono Fiji 18°55S 178°29E 59 a
Onoda Japan 33°59N 131°11E 31 G5
Onslow Australia 21°40S 115°12E 60 D2
Onslow B. U.S.A. 34°20N 77°15W 85 D16
Ontake-San Japan 35°53N 137°29E 31 G8
Ontario Calif., U.S.A. 34°4N 117°39W 79 L9
Ontario Oreg., U.S.A. 44°2N 116°58W 76 D5
Ontario □ Canada 48°0N 83°0W 72 B2
Ontario, L. N. Amer. 43°20N 78°0W 82 C7
Ontonagon U.S.A. 46°52N 89°19W 80 B9
Onyx U.S.A. 35°41N 118°14W 79 K8
Oodnadatta Australia 27°33S 135°30E 63 D2
Ooldea Australia 30°27S 131°50E 61 F5
Oombulgurri Australia 15°15S 127°45E 60 C4
Oombulgurri ☼ Australia 15°10S 127°50E 60 C4
Oorindi Australia 20°40S 141°1E 62 C3
Oost-Vlaanderen □ Belgium 51°5N 3°50E 15 C3
Oostende Belgium 51°15N 2°54E 15 C2
Oosterhout Neths. 51°39N 4°47E 15 C4
Oosterschelde → Neths. 51°33N 4°0E 15 C4
Oosterwolde Neths. 53°0N 6°17E 15 B6
Ootacamund = Udagamandalam India 11°30N 76°44E 40 P10
Ootsa L. Canada 53°50N 126°2W 70 C3
Ooty = Udagamandalam India 11°30N 76°44E 40 P10
Op Luang △ Thailand 18°12N 98°32E 38 C2
Opala Dem. Rep. of the Congo 0°40S 24°20E 54 C1
Opanake Sri Lanka 6°35N 80°40E 40 R12
Opasatika Canada 49°30N 82°50W 72 C3
Opasquia ⌓ Canada 53°33N 93°5W 72 B1
Opava Czech Rep. 49°57N 17°58E 17 D9
Opelika U.S.A. 32°39N 85°23W 85 E12
Opelousas U.S.A. 30°32N 92°5W 84 F8
Opémisca, L. Canada 49°56N 74°52W 72 C5
Opheim U.S.A. 48°51N 106°24W 76 B10
Ophthalmia Ra. Australia 23°15S 119°30E 60 D2
Opinaca → Canada 52°15N 78°2W 72 B4
Opinaca, Rés. Canada 52°39N 76°20W 72 B4
Opinnagau → Canada 54°12N 82°25W 72 B3
Opiscotéo, L. Canada 53°10N 68°10W 73 B6
Opobo Nigeria 4°35N 7°34E 50 H7
Opole Poland 50°42N 17°58E 17 C9
Oponono L. Namibia 18°8S 15°45E 56 B2
Oporto = Porto Portugal 41°8N 8°40W 21 B1
Opotiki N.Z. 38°1S 177°19E 59 C6
Opp U.S.A. 31°17N 86°16W 85 F11
Oppdal Norway 62°35N 9°41E 8 E13
Opportunity U.S.A. 47°39N 117°15W 76 C5
Opua N.Z. 35°19S 174°9E 59 A5
Opunake N.Z. 39°26S 173°52E 59 C4
Opuwo Namibia 18°3S 13°45E 56 B1
Ora Cyprus 34°51N 33°12E 25 E12
Oracle U.S.A. 32°37N 110°46W 77 K8
Oradea Romania 47°2N 21°58E 17 E11
Öræfajökull Iceland 64°2N 16°39W 8 D5
Orai India 25°58N 79°30E 43 G8
Oral = Zhayyq → Kazakhstan 47°0N 51°48E 19 E9
Oral Kazakhstan 51°20N 51°20E 19 D9
Oran Algeria 35°45N 0°39W 50 A5
Orange Australia 33°15S 149°7E 63 E4
Orange France 44°8N 4°47E 20 D6
Orange Calif., U.S.A. 33°47N 117°51W 79 M9
Orange Mass., U.S.A. 42°35N 72°19W 83 D12
Orange Tex., U.S.A. 30°6N 93°44W 84 F8
Orange Va., U.S.A. 38°15N 78°7W 81 F14
Orange → S. Africa 28°41S 16°28E 56 D2
Orange, C. Brazil 4°20N 51°30W 93 C8
Orange Cove U.S.A. 36°38N 119°19W 78 J7
Orange Free State = Free State □ S. Africa 28°30S 27°0E 56 D4
Orange Grove U.S.A. 27°58N 97°56W 84 H6
Orange Walk Belize 18°6N 88°33W 87 D7
Orangeburg U.S.A. 33°30N 80°52W 85 E14
Orangeville Canada 43°55N 80°5W 82 C4
Orango Guinea-Biss. 11°5N 16°0W 50 F2
Oranienburg Germany 52°45N 13°14E 16 B7
Oranje = Orange → S. Africa 28°41S 16°28E 56 D2
Oranjemund Namibia 28°38S 16°29E 56 D2
Oranjerivier S. Africa 29°40S 24°12E 56 D3
Oranjestad Aruba 12°32N 70°2W 89 D5
Orapa Botswana 21°15S 25°30E 53 J5
Oras Phil. 12°9N 125°28E 37 B7
Orbetello Italy 42°27N 11°13E 22 C4
Orbisonia U.S.A. 40°15N 77°54W 82 F7
Orbost Australia 37°40S 148°29E 63 F4
Orcadas Antarctica 60°44S 44°37W 5 C18
Orcas I. U.S.A. 48°42N 122°56W 78 B4
Orchard City U.S.A. 38°50N 107°58W 76 G10
Orchard Homes U.S.A. 46°55N 114°4W 76 C6
Orchila, I. Venezuela 11°48N 66°10W 89 D6
Orchilla, Pta. Canary Is. 27°42N 18°10W 24 G1
Orcutt U.S.A. 34°52N 120°27W 79 L6
Ord U.S.A. 41°36N 98°56W 80 E4
Ord → Australia 15°33S 128°15E 60 C4
Ord, Mt. Australia 17°20S 125°34E 60 C4
Ord Mts. U.S.A. 34°39N 116°45W 79 L10
Orderville U.S.A. 37°17N 112°38W 77 H7
Ordos = Mu Us Shamo China 39°0N 109°0E 34 E5
Ordu Turkey 40°55N 37°53E 19 F6
Ordway U.S.A. 38°13N 103°46W 76 G12
Ore Dem. Rep. of the Congo 3°17N 29°30E 54 B2
Ore Mts. = Erzgebirge Germany 50°27N 12°55E 16 C7
Örebro Sweden 59°20N 15°18E 9 G16
Oregon U.S.A. 42°1N 89°20W 80 D9
Oregon □ U.S.A. 44°0N 121°0W 76 E3
Oregon City U.S.A. 45°21N 122°36W 78 E4
Oregon Dunes ⌓ U.S.A. 43°40N 124°10W 76 E1
Orekhovo-Zuyevo Russia 55°50N 38°55E 18 C6
Orel Russia 52°57N 36°3E 18 D6
Orem U.S.A. 40°19N 111°42W 76 F8
Ören Turkey 37°3N 27°57E 23 F12
Orenburg Russia 51°45N 55°6E 18 D10
Orense = Ourense Spain 42°19N 7°55W 21 A2
Orepuki N.Z. 46°19S 167°46E 59 G1
Orestes Pereyra Mexico 26°30N 105°39W 86 B3
Orestiada Greece 41°30N 26°33E 23 D12
Orfanos Kolpos Greece 40°33N 24°0E 23 D11
Orford Ness U.K. 52°5N 1°35E 13 E9

Organ Pipe Cactus ⌂ U.S.A. 32°0N 112°52W 77 K7
Organos, Pta. de los Canary Is. 28°12N 17°17W 24 F2
Orgaz Spain 39°39N 3°53W 21 C4
Orgeyev = Orhei Moldova 47°24N 28°50E 17 E15
Orhaneli Turkey 39°54N 28°59E 23 E13
Orhangazi Turkey 40°29N 29°18E 23 D13
Orhei Moldova 47°24N 28°50E 17 E15
Orhon Gol → Mongolia 50°21N 106°0E 32 A10
Oriental, Cordillera Colombia 6°0N 73°0W 92 B4
Oriental, Grand Erg Algeria 30°0N 6°30E 50 B7
Orientale □ Dem. Rep. of the Congo 2°20N 26°0E 54 B2
Oriente Argentina 38°44S 60°37W 94 D3
Orihuela Spain 38°7N 0°55W 21 C5
Orillia Canada 44°40N 79°24W 82 B5
Orinoco → Venezuela 9°15N 61°30W 92 B6
Orion Canada 49°27N 110°49W 71 D6
Oriskany U.S.A. 43°10N 75°20W 83 C9
Orissa □ India 20°0N 84°0E 41 K14
Orissaare Estonia 58°34N 23°5E 9 G20
Oristano Italy 39°54N 8°36E 22 E3
Oristano, G. di Italy 39°50N 8°29E 22 E3
Orizaba Mexico 18°51N 97°6W 87 D5
Orizaba, Pico de Mexico 18°58N 97°15W 87 D5
Orkanger Norway 63°18N 9°52E 8 E13
Orkla → Norway 63°18N 9°51E 8 E13
Orkney S. Africa 26°58S 26°40E 56 D4
Orkney □ U.K. 59°2N 3°13W 11 B5
Orkney Is. U.K. 59°0N 3°0W 11 B6
Orland U.S.A. 39°45N 122°12W 78 F4
Orlando U.S.A. 28°32N 81°22W 85 G14
Orléanais France 48°0N 2°0E 20 C5
Orléans France 47°54N 1°52E 20 C4
Orleans U.S.A. 44°49N 72°12W 83 B12
Orléans, Î. d' Canada 46°54N 70°58W 73 C5
Ormara Pakistan 25°16N 64°33E 40 G4
Ormoc Phil. 11°0N 124°37E 37 B6
Ormond N.Z. 38°33S 177°56E 59 C6
Ormond Beach U.S.A. 29°17N 81°3W 85 G14
Ormskirk U.K. 53°35N 2°54W 12 D5
Ormstown Canada 45°8N 74°0W 83 A11
Örnsköldsvik Sweden 63°17N 18°40E 8 E18
Oro N. Korea 40°1N 127°27E 35 D14
Oro → Mexico 25°35N 105°2W 86 B3
Oro Grande U.S.A. 34°36N 117°20W 79 L9
Oro Valley U.S.A. 32°26N 110°58W 77 K8
Orocué Colombia 4°48N 71°20W 92 C4
Orofino U.S.A. 46°29N 116°15W 76 C5
Orohena, Mt. Tahiti 17°37S 149°28W 59 d
Orol Dengizi = Aral Sea Asia 44°30N 60°0E 28 E7
Oromocto Canada 45°54N 66°29W 73 C6
Orono Canada 43°59N 78°37W 82 C6
Orono U.S.A. 44°53N 68°40W 81 C19
Oronsay U.K. 56°1N 6°15W 11 E2
Oroquieta Phil. 8°32N 123°44E 37 C6
Orosei Italy 40°23N 9°42E 22 D3
Orosháza Hungary 46°32N 20°42E 17 E11
Orotukan Russia 62°16N 151°42E 29 C16
Oroville Calif., U.S.A. 39°31N 121°33W 78 F5
Oroville Wash., U.S.A. 48°56N 119°26W 76 B4
Oroville, L. U.S.A. 39°33N 121°29W 78 F5
Orroroo Australia 32°43S 138°38E 63 E2
Orrville U.S.A. 40°50N 81°46W 82 F3
Orsha Belarus 54°30N 30°25E 18 D5
Orsk Russia 51°12N 58°34E 28 D6
Orşova Romania 44°41N 22°25E 17 F12
Ortaca Turkey 36°49N 28°45E 23 F13
Ortegal, C. Spain 43°43N 7°52W 21 A2
Orthez France 43°29N 0°48W 20 E3
Ortigueira Spain 43°40N 7°50W 21 A2
Orting U.S.A. 47°6N 122°12W 78 C4
Ortles Italy 46°31N 10°33E 20 C9
Ortón → Bolivia 10°50S 67°0W 92 F5
Ortonville U.S.A. 45°19N 96°27W 80 C5
Orūmīyeh Iran 37°40N 45°0E 44 B5
Orūmīyeh, Daryācheh-ye Iran 37°50N 45°30E 44 B5
Oruro Bolivia 18°0S 67°9W 92 G5
Orust Sweden 58°10N 11°40E 9 G14
Oruzgān □ Afghan. 33°0N 66°0E 40 C5
Orvieto Italy 42°43N 12°7E 22 C5
Orwell N.Y., U.S.A. 43°35N 75°50W 83 C9
Orwell Ohio, U.S.A. 41°32N 80°52W 82 E4
Orwell → U.K. 51°59N 1°18E 13 F9
Orwigsburg U.S.A. 40°38N 76°6W 83 F8
Oryakhovo Bulgaria 43°40N 23°57E 23 C10
Oryol = Orel Russia 52°57N 36°3E 18 D6
Osa Russia 57°17N 55°26E 18 C10
Osa, Pen. de Costa Rica 8°0N 84°0W 88 E3
Osage U.S.A. 43°17N 92°49W 80 D7
Osage → U.S.A. 38°36N 91°57W 80 F8
Osage City U.S.A. 38°38N 95°50W 80 F6
Ōsaka Japan 34°42N 135°30E 31 G7
Osawatomie U.S.A. 38°31N 94°57W 80 F6
Osborne U.S.A. 39°26N 98°42W 80 F4
Osceola Ark., U.S.A. 35°42N 89°58W 85 D10
Osceola Iowa, U.S.A. 41°2N 93°46W 80 E7
Oscoda U.S.A. 44°26N 83°20W 82 B1
Ösel = Saaremaa Estonia 58°30N 22°30E 9 G20
Osgoode Canada 45°8N 75°36W 83 A9
Osh Kyrgyzstan 40°37N 72°49E 32 C3
Oshakati Namibia 17°45S 15°40E 53 H3
Oshawa Canada 43°50N 78°50W 82 C6
Oshigambo Namibia 17°45S 16°5E 56 B2
Oshika-Hantō Japan 38°20N 141°30E 30 E10
Oshkosh Nebr., U.S.A. 41°24N 102°21W 80 E2
Oshkosh Wis., U.S.A. 44°1N 88°33W 80 C9
Oshmyany = Ashmyany Belarus 54°26N 25°52E 17 A13
Oshnovīyeh Iran 37°2N 45°6E 44 B5
Oshogbo Nigeria 7°48N 4°37E 50 G6
Oshtorīnān Iran 34°1N 48°38E 45 C6
Oshwe Dem. Rep. of the Congo 3°25S 19°28E 52 E3
Osijek Croatia 45°34N 18°41E 23 B8
Osipovichi = Asipovichy Belarus 53°19N 28°33E 17 B15
Osiyan India 26°43N 72°55E 42 F5
Osizweni S. Africa 27°49S 30°7E 57 D5
Oskaloosa U.S.A. 41°18N 92°39W 80 E7
Oskarshamn Sweden 57°15N 16°27E 9 H17
Oskélanéo Canada 48°5N 75°15W 72 C4
Öskemen Kazakhstan 50°0N 82°36E 28 D9
Oslo Norway 59°54N 10°43E 9 G14
Oslofjorden Norway 59°20N 10°35E 9 G14
Osmanabad India 18°5N 76°10E 40 K10
Osmaniye Turkey 37°5N 36°10E 44 B3
Osnabrück Germany 52°17N 8°3E 16 B5
Osório Brazil 29°53S 50°17W 95 B5
Osorno Chile 40°25S 73°0W 96 E2
Osorno, Vol. Chile 41°0S 72°30W 96 D2
Osoyoos Canada 49°0N 119°30W 70 D5
Osøyro Norway 60°9N 5°30E 8 F11
Ospika → Canada 56°20N 124°0W 70 B4
Osprey Reef Australia 13°52S 146°36E 62 A4
Oss Neths. 51°46N 5°32E 15 C5
Ossa, Mt. Australia 41°52S 146°3E 63 G4
Ossa, Oros Greece 39°47N 22°42E 23 E10
Ossabaw I. U.S.A. 31°50N 81°5W 85 F14
Ossining U.S.A. 41°10N 73°55W 83 E11
Ossipee U.S.A. 43°41N 71°7W 83 C13
Ossokmanuan L. Canada 53°25N 65°0W 73 B7
Ossora Russia 59°20N 163°13E 29 D17
Ostend = Oostende Belgium 51°15N 2°54E 15 C2
Oster Ukraine 50°57N 30°53E 17 C16
Österbotten = Pohjanmaa Finland 62°58N 22°50E 8 E20
Osterburg U.S.A. 40°16N 78°31W 82 F6
Österdalälven → Sweden 60°30N 15°7E 8 F15
Østerdalen Norway 61°40N 10°50E 8 F14
Östermyra = Seinäjoki Finland 62°40N 22°51E 8 E20
Österreich = Austria ■ Europe 47°0N 14°0E 16 E8
Östersund Sweden 63°10N 14°38E 8 E16
Ostfriesische Inseln Germany 53°42N 7°0E 16 B4
Ostrava Czech Rep. 49°51N 18°18E 17 D10
Ostróda Poland 53°42N 19°58E 17 B10
Ostroh Ukraine 50°20N 26°30E 17 C14
Ostrołęka Poland 53°4N 21°32E 17 B11
Ostrów Mazowiecka Poland 52°50N 21°51E 17 B11
Ostrów Wielkopolski Poland 51°36N 17°44E 17 C9
Ostrowiec-Świętokrzyski Poland 50°55N 21°22E 17 C11
Ostuni Italy 40°44N 17°35E 23 D7
Ōsumi-Kaikyō Japan 30°55N 131°0E 31 J5
Ōsumi-Shotō Japan 30°30N 130°0E 31 J5
Osuna Spain 37°14N 5°8W 21 D3
Oswegatchie → U.S.A. 44°42N 75°30W 83 B9
Oswego U.S.A. 43°27N 76°31W 83 C8
Oswego → U.S.A. 43°27N 76°30W 83 C8
Oswestry U.K. 52°52N 3°3W 12 E4
Oświęcim Poland 50°2N 19°11E 17 C10
Otago □ N.Z. 45°15S 170°0E 59 F2
Otago Harbour N.Z. 45°47S 170°42E 59 F3
Otaheite B. Trin. & Tob. 10°15N 61°30W 93 K15
Ōtake Japan 34°12N 132°13E 31 G6
Otaki N.Z. 40°45S 175°10E 59 D5
Otaru Japan 43°10N 141°0E 30 C10
Otaru-Wan = Ishikari-Wan Japan 43°25N 141°1E 30 C10
Otavalo Ecuador 0°13N 78°20W 92 C3
Otavi Namibia 19°40S 17°24E 56 B2
Otawara Japan 36°50N 140°5E 31 F10
Otchinjau Angola 16°30S 13°56E 56 B1
Otego U.S.A. 42°23N 75°10W 83 D9
Otelnuk, L. Canada 56°9N 68°12W 73 A6
Othello U.S.A. 46°50N 119°10W 76 C4
Otish, Mts. Canada 52°22N 70°30W 73 B5
Otjiwarongo Namibia 20°30S 16°33E 56 C2
Oto Tolu Group Tonga 20°21S 174°32W 59 c
Otoineppu Japan 44°44N 142°16E 30 B11
Otorohanga N.Z. 38°12S 175°14E 59 C5
Otoskwin → Canada 52°13N 88°6W 72 B2
Otra → Norway 58°9N 8°1E 9 G13
Otranto Italy 40°9N 18°28E 23 D8
Otranto, C. d' Italy 40°7N 18°30E 23 D8
Otranto, Str. of Italy 40°15N 18°40E 23 D8
Otse S. Africa 25°2S 25°45E 56 D4
Otsego L. U.S.A. 42°45N 74°53W 83 D10
Ōtsu Japan 35°0N 135°50E 31 G7
Ōtsuki Japan 35°36N 138°57E 31 G9
Ottawa = Outaouais → Canada 45°27N 74°8W 72 C5
Ottawa Canada 45°26N 75°42W 83 A9
Ottawa Ill., U.S.A. 41°21N 88°51W 80 E9
Ottawa Kans., U.S.A. 38°37N 95°16W 80 F6
Ottawa Is. Canada 59°35N 80°10W 69 F15
Otter Cr. → U.S.A. 44°13N 73°17W 83 B11
Otter Lake Canada 45°17N 79°56W 82 A5
Otterville Canada 42°55N 80°36W 82 D4
Ottery St. Mary U.K. 50°44N 3°17W 13 G4
Otto Beit Bridge Zimbabwe 15°59S 28°56E 55 F2
Ottosdal S. Africa 26°46S 25°59E 56 D4
Ottumwa U.S.A. 41°1N 92°25W 80 E7
Oturkpo Nigeria 7°16N 8°8E 50 G7
Otway, B. Chile 53°30S 74°0W 96 G2
Otway, C. Australia 38°52S 143°30E 63 F3
Otwock Poland 52°5N 21°20E 17 B11
Ou → Laos 20°4N 102°13E 38 B4
Ou-Sammyaku Japan 39°20N 140°35E 30 E10
Ouachita → U.S.A. 31°38N 91°49W 84 F9
Ouachita, L. U.S.A. 34°34N 93°12W 84 D8
Ouachita Mts. U.S.A. 34°30N 94°30W 84 D7
Ouagadougou Burkina Faso 12°25N 1°30W 50 F5
Ouahigouya Burkina Faso 13°31N 2°25W 50 F5
Ouahran = Oran Algeria 35°45N 0°39W 50 A5
Ouallene Algeria 24°41N 1°11E 50 D6
Ouarâne Mauritania 21°0N 10°30W 50 D3
Ouargla Algeria 31°59N 5°16E 50 B7
Ouarra → C.A.R. 5°5N 24°26E 52 C4
Ouarzazate Morocco 30°55N 6°50W 50 B4
Oubangi → Dem. Rep. of the Congo 0°30S 17°50E 52 E3
Ouddorp Neths. 51°50N 3°57E 15 C3
Oude Rijn → Neths. 52°12N 4°24E 15 B4
Oudenaarde Belgium 50°50N 3°37E 15 D3
Oudtshoorn S. Africa 33°35S 22°14E 56 E3
Ouessant, Î. d' France 48°28N 5°6W 20 B1
Ouesso Congo 1°37N 16°5E 52 D3
Ouest, Pte. de l' Canada 49°52N 64°40W 73 C7
Ouezzane Morocco 34°51N 5°35W 50 B4
Oughter, L. Ireland 54°1N 7°28W 10 B4
Oughterard Ireland 53°26N 9°18W 10 C2
Ouidah Benin 6°25N 2°0E 50 G6
Oujda Morocco 34°41N 1°55W 50 B5
Oulainen Finland 64°17N 24°47E 8 D21
Oulu Finland 65°1N 25°29E 8 D21
Oulujärvi Finland 64°25N 27°15E 8 D22
Oulujoki → Finland 65°1N 25°30E 8 D21
Oum Chalouba Chad 15°48N 20°46E 51 E10
Oum Hadjer Chad 13°18N 19°41E 51 F9
Ounasjoki → Finland 66°31N 25°40E 8 C21
Ounguati Namibia 22°0S 15°46E 56 C2
Ounianga Kébir Chad 19°4N 20°29E 51 E10
Our → Lux. 49°55N 6°5E 15 E6
Ouray U.S.A. 38°1N 107°40W 77 G10
Ourense Spain 42°19N 7°55W 21 A2
Ouricuri Brazil 7°53S 40°5W 93 E10
Ourinhos Brazil 23°0S 49°54W 95 A6
Ouro Fino Brazil 22°16S 46°25W 95 A6
Ouro Prêto Brazil 20°20S 43°30W 95 A7
Ourthe → Belgium 50°29N 5°35E 15 D5
Ouse → E. Sussex, U.K. 50°47N 0°4E 13 G8
Ouse → N. Yorks., U.K. 53°44N 0°55W 12 D7
Outaouais → Canada 45°27N 74°8W 72 C5
Outardes → Canada 49°24N 69°30W 73 C6
Outer Hebrides U.K. 57°30N 7°40W 11 D1
Outjo Namibia 20°5S 16°7E 56 C2
Outlook Canada 51°30N 107°0W 71 C7
Outokumpu Finland 62°43N 29°1E 8 E23
Ouyen Australia 35°1S 142°22E 63 F3
Ovalau Fiji 17°40S 178°48E 59 a
Ovalle Chile 30°33S 71°18W 94 C1
Ovamboland Namibia 18°30S 16°0E 56 B2
Overflakkee Neths. 51°44N 4°10E 15 C4
Overijssel □ Neths. 52°25N 6°35E 15 B6
Overland Park U.S.A. 38°58N 94°40W 80 F6
Overlander Roadhouse Australia 26°19S 114°28E 61 E1
Overton U.S.A. 36°33N 114°27W 79 J12
Övertorneå Sweden 66°23N 23°38E 8 C20
Ovid U.S.A. 42°41N 76°49W 83 D8
Oviedo Spain 43°25N 5°50W 21 A3
Ovišrags Latvia 57°33N 21°44E 9 H19
Ovoot Mongolia 45°21N 113°45E 34 B7
Övör Hangay □ Mongolia 45°0N 102°30E 34 B2
Øvre Ardal Norway 61°19N 7°48E 8 F12
Ovruch Ukraine 51°25N 28°45E 17 C15
Owaka N.Z. 46°27S 169°40E 59 G2
Owambo = Ovamboland Namibia 18°30S 16°0E 56 B2
Owasco L. U.S.A. 42°50N 76°31W 83 D8
Owase Japan 34°7N 136°12E 31 G8
Owatonna U.S.A. 44°5N 93°14W 80 C7
Owbeh Afghan. 34°28N 63°10E 40 B3
Owego U.S.A. 42°6N 76°16W 83 D8
Owen Falls Dam = Nalubaale Dam Uganda 0°30N 33°5E 54 B3
Owen Sound Canada 44°35N 80°55W 82 B4
Owen Stanley Ra. Papua N. G. 8°30S 147°0E 58 B7
Oweniny → Ireland 54°8N 9°34W 10 B2
Owens → U.S.A. 36°32N 117°59W 78 J9
Owens L. U.S.A. 36°26N 117°57W 79 J9
Owensboro U.S.A. 37°46N 87°7W 80 G10
Owl → Canada 57°51N 92°44W 71 B10
Owo Nigeria 7°10N 5°39E 50 G7
Owosso U.S.A. 43°0N 84°10W 81 D11
Owyhee U.S.A. 41°57N 116°6W 76 F5
Owyhee → U.S.A. 43°49N 117°2W 76 E5
Owyhee, L. U.S.A. 43°38N 117°14W 76 E5
Ox Mts. = Slieve Gamph Ireland 54°6N 9°0W 10 B3
Öxarfjörður Iceland 66°15N 16°45W 8 C5
Oxbow Canada 49°14N 102°10W 71 D8
Oxelösund Sweden 58°43N 17°5E 9 G17
Oxford N.Z. 43°18S 172°11E 59 E4
Oxford U.K. 51°46N 1°15W 13 F6
Oxford Mass., U.S.A. 42°7N 71°52W 83 D13
Oxford Miss., U.S.A. 34°22N 89°31W 85 D10
Oxford N.C., U.S.A. 36°19N 78°35W 85 C15
Oxford N.Y., U.S.A. 42°27N 75°36W 83 D9
Oxford Ohio, U.S.A. 39°31N 84°45W 81 F11
Oxford L. Canada 54°51N 95°37W 71 C9
Oxfordshire □ U.K. 51°48N 1°16W 13 F6
Oxnard U.S.A. 34°12N 119°11W 79 L7
Oxus = Amudarya → Uzbekistan 43°58N 59°34E 28 E6
Oya Malaysia 2°55N 111°55E 36 D4
Oyama Japan 36°18N 139°48E 31 F9
Oyem Gabon 1°34N 11°31E 52 D2
Oyen Canada 51°22N 110°28W 71 C6
Oykel → U.K. 57°56N 4°26W 11 D4
Oymyakon Russia 63°25N 142°44E 29 C15
Oyo Nigeria 7°46N 3°56E 50 G6
Oyster Bay U.S.A. 40°52N 73°32W 83 F11
Ōyūbari Japan 43°1N 142°5E 30 C11
Ozamiz Phil. 8°15N 123°50E 37 C6
Ozark Ala., U.S.A. 31°28N 85°39W 85 F12
Ozark Ark., U.S.A. 35°29N 93°50W 84 D8
Ozark Mo., U.S.A. 37°1N 93°12W 80 G7
Ozark Plateau U.S.A. 37°20N 91°40W 80 G8
Ozarks, L. of the U.S.A. 38°12N 92°38W 80 F7
Ózd Hungary 48°14N 20°15E 17 D11
Ozernovskiy Russia 51°30N 156°31E 29 D16
Ozette, L. U.S.A. 48°6N 124°38W 78 B2
Ozieri Italy 40°35N 9°0E 22 D3
Ozona U.S.A. 30°43N 101°12W 84 F4
Ozuluama Mexico 21°40N 97°51W 87 C5

P

Pa-an Burma 16°51N 97°40E 41 L20
Pa Mong Dam Thailand 18°0N 102°22E 38 D4
Pa Sak → Thailand 15°30N 101°0E 38 E3
Paamiut Greenland 62°0N 49°43W 4 C5
Paarl S. Africa 33°45S 18°56E 56 E2
Pab Hills Pakistan 26°30N 66°45E 42 F2
Pabbay U.K. 57°46N 7°14W 11 D1
Pabianice Poland 51°40N 19°20E 17 C10
Pabna Bangla. 24°1N 89°18E 41 G16
Pabo Uganda 3°1N 32°10E 54 B3
Pacaja → Brazil 1°56S 50°50W 93 D8
Pacaraima, Sa. S. Amer. 4°0N 62°30W 92 C6
Pacasmayo Peru 7°20S 79°35W 92 E3
Pachitea → Peru 8°46S 74°33W 92 E4
Pachmarhi India 22°28N 78°26E 43 H8
Pachnes Greece 35°16N 24°4E 25 D6
Pachpadra India 25°58N 72°10E 42 G5
Pachuca Mexico 20°7N 98°44W 87 C5
Pacific Antarctic Ridge Pac. Oc. 43°0S 115°0W 5 B13
Pacific Grove U.S.A. 36°38N 121°56W 78 J5
Pacific Ocean 10°0N 140°0W 65 G14
Pacific Rim △ Canada 48°40N 124°45W 78 B2
Pacifica U.S.A. 37°37N 122°27W 78 H4
Pacitan Indonesia 8°12S 111°7E 37 H14
Packwood U.S.A. 46°36N 121°40W 78 D5
Padaido, Kepulauan Indonesia 1°15S 136°30E 37 E9
Padang Riau, Indonesia 1°30N 102°30E 39 M4
Padang Sumatera Barat, Indonesia 1°0S 100°20E 36 E2
Padang Endau Malaysia 2°40N 103°38E 39 L4
Padangpanjang Indonesia 0°40S 100°20E 36 E2
Padangsidempuan Indonesia 1°30N 99°15E 36 D1
Paddle Prairie Canada 57°57N 117°29W 70 B5
Paderborn Germany 51°42N 8°45E 16 C5
Padma India 24°12N 85°22E 43 G11
Pádova Italy 45°25N 11°53E 22 B4
Padra India 22°15N 73°7E 42 H5
Padrauna India 26°54N 83°59E 43 F10
Padre I. U.S.A. 27°10N 97°25W 84 H6
Padre Island ⌂ U.S.A. 27°0N 97°25W 84 H6
Padstow U.K. 50°33N 4°58W 13 G3
Padua = Pádova Italy 45°25N 11°53E 22 B4
Paducah Ky., U.S.A. 37°5N 88°37W 80 G9
Paducah Tex., U.S.A. 34°1N 100°18W 84 D4
Paea Tahiti 17°41S 149°35W 59 d
Paektu-san N. Korea 41°59N 128°4E 35 D15
Paeroa N.Z. 37°23S 175°41E 59 B5
Pafúri Mozam. 22°28S 31°17E 57 C5
Pag Croatia 44°25N 15°3E 16 F8
Pagadian Phil. 7°55N 123°30E 37 C6
Pagai Selatan, Pulau Indonesia 3°0S 100°15E 36 E2
Pagai Utara, Pulau Indonesia 2°35S 100°0E 36 E2
Pagalu = Annobón Atl. Oc. 1°25S 5°36E 49 G4
Pagara India 24°22N 80°1E 43 G9
Pagastikos Kolpos Greece 39°15N 23°0E 23 E10
Pagatan Indonesia 3°33S 115°59E 36 E5
Page U.S.A. 36°57N 111°27W 77 H8
Paget, Mt. S. Georgia 54°26S 36°31W 96 G9
Pago Pago Amer. Samoa 14°16S 170°43W 59 b
Pagosa Springs U.S.A. 37°16N 107°1W 77 H10
Pagri China 27°45N 89°10E 32 F6
Pagwa River Canada 50°2N 85°14W 72 B2
Pāhala U.S.A. 19°12N 155°29W 75 M8
Pahang □ Malaysia 3°30N 102°45E 39 L4
Pahang → Malaysia 3°30N 103°9E 39 L4
Pahiatua N.Z. 40°27S 175°50E 59 D5
Pahokee U.S.A. 26°50N 80°40W 85 H14
Pahrump U.S.A. 36°12N 115°59W 79 J11
Pahute Mesa U.S.A. 37°20N 116°45W 78 H10
Pai Thailand 19°19N 98°27E 38 C2
Paicines U.S.A. 36°44N 121°17W 78 J5
Paide Estonia 58°53N 25°33E 9 G21
Paignton U.K. 50°26N 3°35W 13 G4
Päijänne Finland 61°30N 25°30E 8 F21
Pailani India 25°45N 80°26E 43 G9
Pailin Cambodia 12°46N 102°36E 38 F4
Painan Indonesia 1°21S 100°34E 36 E2
Painesville U.S.A. 41°43N 81°15W 82 E3
Paint Hills = Wemindji Canada 53°0N 78°49W 72 B4
Paint L. Canada 55°28N 97°57W 71 B9
Painted Desert U.S.A. 36°0N 111°0W 77 H8
Paintsville U.S.A. 37°49N 82°48W 81 G12
País Vasco □ Spain 42°50N 2°45W 21 A4
Paisley Canada 44°18N 81°16W 82 B3
Paisley U.K. 55°50N 4°25W 11 F4
Paisley U.S.A. 42°42N 120°32W 76 E3
Paita Peru 5°11S 81°9W 92 E2
Pajares, Puerto de Spain 42°58N 5°46W 21 A3
Pak Lay Laos 18°15N 101°27E 38 C3
Pak Ou Laos 20°3N 102°12E 38 B4
Pak Phanang Thailand 8°21N 100°12E 39 H3
Pak Sane Laos 18°22N 103°39E 38 C4
Pak Song Laos 15°11N 106°14E 38 E6
Pak Suang Laos 19°58N 102°15E 38 C4
Pak Tam Chung China 22°24N 114°19E 33 a
Pak Thong Chai Thailand 14°43N 102°1E 38 E4
Pakaur India 24°38N 87°51E 43 G12
Pakch'ŏn N. Korea 39°44N 125°35E 35 E13
Pakenham Canada 45°18N 76°18W 83 A8
Pakhuis S. Africa 32°9S 19°5E 56 E2
Pakistan ■ Asia 30°0N 70°0E 42 E4
Pakkading Laos 18°19N 103°59E 38 C4
Pakokku Burma 21°20N 95°0E 41 J19
Pakowki L. Canada 49°20N 111°0W 71 D6
Pakpattan Pakistan 30°25N 73°27E 42 D5
Paktīā □ Afghan. 33°30N 69°15E 40 C6
Paktīkā □ Afghan. 32°30N 69°0E 40 C6
Pakwach Uganda 2°28N 31°27E 54 B3
Pakxe Laos 15°5N 105°52E 38 E5
Pal Lahara India 21°27N 85°11E 43 J
Pala Chad 9°25N 15°5E 51
Pala Dem. Rep. of the Congo 6°45S 29°30E 54
Pala U.S.A. 33°22N 117°5W 79
Palabek Uganda 3°22N 32°33E 54
Palacios U.S.A. 28°42N 96°13W 84
Palagruža Croatia 42°24N 16°15E 22
Palakkad = Palghat India 10°46N 76°42E 40 P
Palam India 19°0N 77°0E 40 K
Palampur India 32°10N 76°30E 42
Palana Australia 39°45S 147°55E 63
Palana Russia 59°10N 159°59E 29 D
Palanan Phil. 17°8N 122°29E 37
Palanan Pt. Phil. 17°17N 122°30E 37
Palandri Pakistan 33°42N 73°40E 43
Palanga Lithuania 55°58N 21°3E 9 J
Palangkaraya Indonesia 2°16S 113°56E 36
Palani Hills India 10°14N 77°33E 40 P
Palanpur India 24°10N 72°25E 42
Palapye Botswana 22°30S 27°7E 56
Palas Pakistan 35°4N 73°14E 43
Palasponga India 21°47N 85°34E 43 J
Palatka Russia 60°6N 150°54E 29 C
Palatka U.S.A. 29°39N 81°38W 85 G
Palau ■ Palau 7°30N 134°30E 58
Palauk Burma 13°10N 98°40E 38
Palaw Burma 12°58N 98°39E 38
Palawan Phil. 9°30N 118°30E 36
Palayankottai India 8°45N 77°45E 40 Q
Paldiski Estonia 59°23N 24°9E 9 G
Palekastro Greece 35°12N 26°15E 25
Paleleh Indonesia 1°10N 121°50E 37
Palembang Indonesia 3°0S 104°50E 36
Palencia Spain 42°1N 4°34W 21
Palenque Mexico 17°29N 92°1W 87
Paleochora Greece 35°16N 23°39E 25
Paleokastritsa Greece 39°40N 19°41E 25
Paleometokho Cyprus 35°7N 33°11E 25 D
Palermo Italy 38°7N 13°22E 22
Palermo U.S.A. 39°26N 121°33W 76
Palestina Chile 23°50S 69°47W 96
Palestine Asia 32°0N 35°0E 46
Palestine U.S.A. 31°46N 95°38W 84
Paletwa Burma 21°10N 92°50E 41 J
Palghat India 10°46N 76°42E 40 P
Palgrave, Mt. Australia 23°22S 115°58E 60
Pali India 25°50N 73°20E 42
Palikir Micronesia 6°55N 158°9E 64
Paliouri, Akra Greece 39°57N 23°45E 23 F
Palisades Res. U.S.A. 43°20N 111°12W 76
Paliseul Belgium 49°54N 5°8E 15
Palitana India 21°32N 71°49E 42
Palizada Mexico 18°15N 92°5W 87
Palk Bay Asia 9°30N 79°15E 40 Q
Palk Strait Asia 10°0N 79°45E 40 Q
Palkānah Iraq 35°49N 44°26E 44
Palkot India 22°53N 84°39E 43 H
Pallanza = Verbánia Italy 45°56N 8°33E 20
Pallarenda Australia 19°12S 146°46E 62
Pallinup → Australia 34°27S 118°50E 61
Pallisa Uganda 1°12N 33°43E 54
Pallu India 28°59N 74°14E 42
Palm Bay U.S.A. 28°2N 80°35W 85 G
Palm Beach U.S.A. 26°43N 80°2W 85 H
Palm Coast U.S.A. 29°35N 81°12W 85 G
Palm Desert U.S.A. 33°43N 116°22W 79 M
Palm-Grove △ Australia 24°57S 149°21E 62
Palm Is. Australia 18°40S 146°35E 62
Palm Springs U.S.A. 33°50N 116°33W 79 M
Palma Mozam. 10°46S 40°29E 55
Palma, B. de Spain 39°30N 2°39E 24
Palma de Mallorca Spain 39°35N 2°39E 24
Palma Nova = Palmanova Spain 39°32N 2°34E 24
Palma Soriano Cuba 20°15N 76°0W 88
Palmanova Spain 39°32N 2°34E 24
Palmares Brazil 8°41S 35°28W 93 E
Palmas Brazil 26°29S 52°0W 95
Palmas, C. Liberia 4°27N 7°46W 50
Pálmas, G. di Italy 39°0N 8°30E 22
Palmdale U.S.A. 34°35N 118°7W 79
Palmeira das Missões Brazil 27°55S 53°17W 95
Palmeira dos Índios Brazil 9°25S 36°37W 93 E
Palmer Antarctica 64°35S 65°0W 5 C
Palmer U.S.A. 61°36N 149°7W 68
Palmer → Australia 16°0S 142°26E 62
Palmer Arch. Antarctica 64°15S 65°0W 5 C
Palmer Lake U.S.A. 39°7N 104°55W 76 G
Palmer Land Antarctica 73°0S 63°0W 5 D
Palmerston Australia 12°31S 130°59E 60
Palmerston Canada 43°50N 80°51W 82
Palmerston N.Z. 45°29S 170°43E 59
Palmerston North N.Z. 40°21S 175°39E 59
Palmerton U.S.A. 40°48N 75°37W 83
Palmetto U.S.A. 27°31N 82°34W 85 H
Palmi Italy 38°21N 15°51E 22
Palmira Argentina 32°59S 68°34W 94
Palmira Colombia 3°32N 76°16W 92
Palmyra = Tudmur Syria 34°36N 38°15E 44
Palmyra Mo., U.S.A. 39°48N 91°32W 80
Palmyra N.J., U.S.A. 40°0N 75°1W 83
Palmyra N.Y., U.S.A. 43°5N 77°18W 82
Palmyra Pa., U.S.A. 40°18N 76°36W 83
Palmyra Is. Pac. Oc. 5°52N 162°5W 65 G
Palo Alto U.S.A. 37°27N 122°10W 78
Palo Seco Trin. & Tob. 10°4N 61°36W 93 K
Palo Verde U.S.A. 33°26N 114°44W 79 M
Palo Verde △ Costa Rica 10°21N 85°21W 88
Palomar Mt. U.S.A. 33°22N 116°50W 79 M
Palopo Indonesia 3°0S 120°16E 37
Palos, C. de Spain 37°38N 0°40W 21
Palos Verdes, Pt. U.S.A. 33°46N 118°25W 79
Palos Verdes Estates U.S.A. 33°48N 118°23W 79
Palu Indonesia 1°0S 119°52E 37

lu Turkey 38°45N 40°0E 44 B3
luma Ra. Australia 19°9S 146°22E 62 B4
lwal India 28°8N 77°19E 42 E7
manukan Indonesia 6°16S 107°49E 37 G12
mekasan Indonesia 7°10S 113°28E 37 G15
menang Indonesia 8°24S 116°6E 37 J19
miers France 43°7N 1°39E 20 E4
mir Tajikistan 37°40N 73°0E 28 F8
mlico → U.S.A. 35°20N 76°28W 85 D16
mlico Sd. U.S.A. 35°20N 76°0W 85 D17
mpa U.S.A. 35°32N 100°58W 84 D4
mpa de las Salinas Argentina 32°1S 66°58W 94 C2
mpanua Indonesia 4°16S 120°8E 37 E6
mpas Argentina 35°0S 63°0W 94 D3
mpas Peru 12°20S 74°50W 92 F4
mplona Colombia 7°23N 72°39W 92 B4
mplona-Iruña Spain 42°48N 1°38W 21 A5
mpoenpoort S. Africa 31°3S 22°40E 56 E3
n de Azúcar Chile 24°5S 68°10W 94 A2
n de Azúcar △ Chile 26°0S 70°40W 94 B1
na U.S.A. 39°23N 89°5W 80 F9
naca U.S.A. 37°47N 114°23W 77 H6
naitan Indonesia 6°36S 105°12E 37 G11
naji India 15°25N 73°50E 40 M8
namá Panama 9°0N 79°25W 88 E4
nama ■ Cent. Amer. 8°48N 79°55W 88 E4
namá, G. de Panama 8°4N 79°20W 88 E4
nama, Isthmus of Cent. Amer. 9°0N 79°0W 88 E4
nama Basin Pac. Oc. 5°0N 83°30W 65 G19
nama Canal Panama 9°10N 79°37W 88 E4
nama City U.S.A. 30°10N 85°40W 85 F12
namint Range U.S.A. 36°20N 117°20W 79 J9
namint Springs U.S.A. 36°20N 117°28W 79 J9
não Peru 9°55S 75°55W 92 E3
nare Thailand 6°51N 101°30E 39 J3
nay Phil. 11°10N 122°30E 37 B6
nay G. Phil. 11°0N 122°30E 37 B6
nčevo Serbia 44°52N 20°41E 23 B9
nda Mozam. 24°2S 34°45E 57 C5
ndan Malaysia 1°32N 103°46E 39 d
ndan Phil. 11°45N 122°10E 37 B6
ndan, Selat Singapore 1°15N 103°44E 39 d
ndan Tampoi = Tampoi Malaysia 1°30N 103°39E 39 d
ndegelang Indonesia 6°25S 106°5E 37 G12
ndhana India 21°42N 76°13E 42 J7
ndharpur India 17°41N 75°20E 40 L9
ndo Uruguay 34°44S 56°0W 95 C4
ndo, L. = Hope, L. Australia 28°24S 139°18E 63 D2
ndokratoras Greece 39°45N 19°50E 25 A3
ndora Costa Rica 9°43N 83°3W 88 E3
nevėžys Lithuania 55°42N 24°25E 9 J21
ng-Long Burma 23°11N 98°45E 41 H21
ng Sida △ Thailand 14°5N 102°17E 38 E4
ng-Yang Burma 22°7N 98°48E 41 H21
nga Dem. Rep. of the Congo 1°52N 26°18E 54 B2
ngalanes, Canal des = Ampangalana, Lakandranon' Madag. 22°48S 47°50E 57 C8
ngandaran Indonesia 7°40S 108°39E 37 G13
ngandaran △ Indonesia 7°43S 108°42E 37 G13
ngani Tanzania 5°25S 38°58E 54 D4
ngani → Tanzania 5°26S 38°58E 54 D4
ngfou = Bengbu China 32°58N 117°20E 35 H9
ngga, Tanjung Indonesia 8°54S 116°2E 37 K19
ngil Dem. Rep. of the Congo 3°10S 26°35E 54 C2
ngjiabu China 40°38N 115°26E 34 D8
ngkah, Tanjung Indonesia 6°51S 112°33E 37 G15
ngkajene Indonesia 4°46S 119°34E 37 E5
ngkalanbrandan Indonesia 4°1N 98°20E 36 D1
ngkalanbuun Indonesia 2°41S 111°37E 36 E4
ngkalpinang Indonesia 2°10S 106°10E 36 E3
ngkor, Pulau Malaysia 4°13N 100°34E 39 K3
ngniqtuuq = Pangnirtung Canada 66°8N 65°43W 69 D18
ngnirtung Canada 66°8N 65°43W 69 D18
ngody Russia 65°52N 74°27E 28 C8
ngong Tso = Bangong Co China 33°45N 78°43E 43 C8
nguitch U.S.A. 37°50N 112°26W 77 H7
ngutaran Group Phil. 6°18N 120°34E 37 C6
nhandle U.S.A. 35°21N 101°23W 84 D4
ni Mines India 22°29N 73°50E 42 H5
nia-Mutombo Dem. Rep. of the Congo 5°11S 23°51E 54 D1
nikota I. India 20°46N 71°21E 42 J4
nipat India 29°25N 77°2E 42 E7
njal Range = Pir Panjal Range India 32°30N 76°50E 42 C7
njgur Pakistan 27°0N 64°5E 40 F4
njim = Panaji India 15°25N 73°50E 40 M8
njin China 41°3N 122°2E 35 D12
njnad → Pakistan 28°57N 70°30E 42 E4
njnad Barrage Pakistan 29°22N 71°15E 42 E4
njwai Afghan. 31°26N 65°27E 42 D1
nmunjŏm N. Korea 37°59N 126°38E 35 F14
nna India 24°40N 80°15E 43 G9
nna △ India 24°40N 80°0E 43 G8
nna Hills India 24°40N 81°15E 43 G9
nnawonica Australia 21°39S 116°19E 60 D2
no Akil Pakistan 27°51N 69°7E 42 F3
no Lefkara Cyprus 34°53N 33°20E 25 E12
no Panayia Cyprus 34°55N 32°38E 25 E11
norama Brazil 21°21S 51°51W 95 A5
normos Greece 35°25N 24°41E 25 D6
Pansemal India 21°39N 74°42E 42 J6
Panshan = Panjin China 41°3N 122°2E 35 D12
Panshi China 42°58N 126°5E 35 C14
Pantanal Brazil 17°30S 57°40W 92 H7
Pantanos de Centla △ Mexico 18°25N 92°35W 87 D6
Pantar Indonesia 8°28S 124°10E 37 F6
Pante Macassar E. Timor 9°30S 123°58E 37 F6
Pantelleria Italy 36°50N 11°57E 22 F4
Pánuco Mexico 22°3N 98°10W 87 C5
Panzhihua China 26°33N 101°44E 32 F9
Paola Malta 35°52N 14°30E 25 D2
Paola U.S.A. 38°35N 94°53W 80 F6
Paonia U.S.A. 38°52N 107°36W 76 G10
Paopao Moorea 17°30S 149°49W 59 d
Paoting = Baoding China 38°50N 115°28E 34 E8
Paot'ou = Baotou China 40°32N 110°2E 34 D6
Paoua C.A.R. 7°9N 16°20E 52 C3
Pápa Hungary 47°22N 17°30E 17 E9
Papa Stour U.K. 60°20N 1°42W 11 A7
Papa Westray U.K. 59°20N 2°55W 11 B6
Papagayo → Mexico 16°46N 99°43W 87 D5
Papagayo, G. de Costa Rica 10°30N 85°50W 88 D2
Papakura N.Z. 37°4S 174°59E 59 B5
Papantla Mexico 20°27N 97°19W 87 C5
Papar Malaysia 5°45N 116°0E 36 C5
Papara Tahiti 17°43S 149°31W 59 d
Paparoa △ N.Z. 42°7S 171°26E 59 E3
Papeete Tahiti 17°32S 149°34W 59 d
Papenoo Tahiti 17°30S 149°25W 59 d
Papenoo → Tahiti 17°30S 149°25W 59 d
Papetoai Moorea 17°29S 149°52W 59 d
Paphos Cyprus 34°46N 32°25E 25 E11
Paposo Chile 25°0S 70°30W 94 B1
Papoutsa Cyprus 34°54N 33°4E 25 E12
Papua □ Indonesia 4°0S 137°0E 37 E9
Papua, G. of Papua N. G. 9°0S 144°50E 58 B7
Papua New Guinea ■ Oceania 8°0S 145°0E 58 B7
Papudo Chile 32°29S 71°27W 94 C1
Papun Burma 18°2N 97°30E 41 K20
Papunya Australia 23°15S 131°54E 60 D5
Pará = Belém Brazil 1°20S 48°30W 93 D9
Pará □ Brazil 3°20S 52°0W 93 D8
Paraburdoo Australia 23°14S 117°32E 60 D2
Paracatu Brazil 17°10S 46°50W 93 G9
Paracel Is. S. China Sea 15°50N 112°0E 36 A4
Parachilna Australia 31°10S 138°21E 63 E2
Parachinar Pakistan 33°55N 70°5E 42 C4
Paradip India 20°15N 86°35E 41 J15
Paradise Calif., U.S.A. 39°46N 121°37W 78 F5
Paradise Nev., U.S.A. 36°5N 115°8W 79 J11
Paradise → Canada 53°27N 57°19W 73 B8
Paradise Hill Canada 53°32N 109°28W 71 C7
Paradise River Canada 53°27N 57°17W 73 B8
Paradise Valley U.S.A. 41°30N 117°32W 76 F5
Paradisi Greece 36°18N 28°7E 25 C10
Parado Indonesia 8°42S 118°30E 37 F5
Paragould U.S.A. 36°3N 90°29W 85 C9
Paragua → Venezuela 6°55N 62°55W 92 B6
Paraguaçú → Brazil 12°45S 38°54W 93 F11
Paraguaçu Paulista Brazil 22°22S 50°35W 95 A5
Paraguaná, Pen. de Venezuela 12°0N 70°0W 92 A5
Paraguarí Paraguay 25°36S 57°0W 94 B4
Paraguarí □ Paraguay 26°0S 57°10W 94 B4
Paraguay ■ S. Amer. 23°0S 57°0W 94 A4
Paraguay → Paraguay 27°18S 58°38W 94 B4
Paraíba = João Pessoa Brazil 7°10S 34°52W 93 E12
Paraíba □ Brazil 7°0S 36°0W 93 E11
Paraíba do Sul → Brazil 21°37S 41°3W 95 A7
Paraíso Mexico 18°24N 93°14W 87 D6
Parak Iran 27°38N 52°25E 45 E7
Parakou Benin 9°25N 2°40E 50 G6
Paralimni Cyprus 35°2N 33°58E 25 D12
Paramaribo Suriname 5°50N 55°10W 93 B7
Páramos del Batallón y La Negra △ Venezuela 8°2N 71°55W 89 E5
Paramushir, Ostrov Russia 50°24N 156°0E 29 D16
Paran → Israel 30°20N 35°10E 46 E4
Paraná Argentina 31°45S 60°30W 94 C3
Paraná Brazil 12°30S 47°48W 93 F9
Paraná □ Brazil 24°30S 51°0W 95 A5
Paraná → Argentina 33°43S 59°15W 94 C4
Paranaguá Brazil 25°30S 48°30W 95 B6
Paranaíba Brazil 19°40S 51°11W 93 G8
Paranaíba → Brazil 20°6S 51°4W 93 H8
Paranapanema → Brazil 22°40S 53°9W 95 A5
Paranapiacaba, Serra do Brazil 24°31S 48°35W 95 A6
Paranavaí Brazil 23°4S 52°56W 95 A5
Parang Maguindanao, Phil. 7°23N 124°16E 37 C6
Parang Sulu, Phil. 5°55N 120°54E 37 C6
Parângul Mare, Vf. Romania 45°20N 23°37E 17 F12
Paraparaumu N.Z. 40°57S 175°3E 59 D5
Parbati → Mad. P., India 25°50N 76°30E 42 G7
Parbati → Raj., India 26°54N 77°53E 42 F7
Parbhani India 19°8N 76°52E 40 K10
Parchim Germany 53°26N 11°52E 16 B6
Pardes Hanna-Karkur Israel 32°28N 34°57E 46 C3
Pardo → Bahia, Brazil 15°40S 39°0W 93 G11
Pardo → Mato Grosso, Brazil 21°46S 52°9W 95 A5
Pardoo Roadhouse Australia 20°6S 119°3E 60 D2
Pardubice Czech Rep. 50°3N 15°45E 16 C8
Pare Indonesia 7°43S 112°12E 37 G15
Pare Mts. Tanzania 4°0S 37°45E 54 C4
Parecis, Serra dos Brazil 13°0S 60°0W 92 F7
Paren Russia 62°30N 163°15E 29 C17
Parent Canada 47°55N 74°35W 72 C5
Parent, L. Canada 48°31N 77°1W 72 C4
Parepare Indonesia 4°0S 119°40E 37 E5
Parga Greece 39°15N 20°29E 23 E9
Pargas = Länsi-Turunmaa Finland 60°18N 22°18E 9 F20
Pargo, Pta. do Madeira 32°49N 17°17W 24 D2
Parham Canada 44°39N 76°43W 83 B8
Paria → U.S.A. 36°52N 111°36W 77 H8
Paria, G. de Venezuela 10°20N 62°0W 93 K14
Pariaguán Venezuela 8°51N 64°34W 92 B6
Paricutín, Cerro Mexico 19°28N 102°15W 86 D4
Parigi Indonesia 0°50S 120°5E 37 E6
Parika Guyana 6°50N 58°20W 92 B7
Parikkala Finland 61°33N 29°31E 8 F23
Parima, Serra Brazil 2°30N 64°0W 92 C6
Parinari Peru 4°35S 74°25W 92 D4
Pariñas, Pta. Peru 4°30S 82°0W 90 D2
Parintins Brazil 2°40S 56°50W 93 D7
Pariparit Kyun Burma 14°52N 93°41E 41 M18
Paris Canada 43°12N 80°25W 82 C4
Paris France 48°53N 2°20E 20 B5
Paris Idaho, U.S.A. 42°14N 111°24W 76 E8
Paris Ky., U.S.A. 38°13N 84°15W 81 F11
Paris Tenn., U.S.A. 36°18N 88°19W 85 C10
Paris Tex., U.S.A. 33°40N 95°33W 84 E7
Parish U.S.A. 43°25N 76°8W 83 C8
Parishville U.S.A. 44°38N 74°49W 83 B10
Pariti Indonesia 10°1S 123°45E 60 B3
Park U.S.A. 48°45N 122°18W 78 B4
Park City U.S.A. 37°48N 97°20W 80 G5
Park Falls U.S.A. 45°56N 90°27W 80 C8
Park Head Canada 44°36N 81°9W 82 B3
Park Hills U.S.A. 37°51N 90°51W 80 G8
Park Range U.S.A. 40°41N 106°41W 76 F10
Park Rapids U.S.A. 46°55N 95°4W 80 B6
Park River U.S.A. 48°24N 97°45W 80 A5
Park Rynie S. Africa 30°25S 30°45E 57 E5
Parkā Bandar Iran 25°55N 59°35E 45 E8
Parkano Finland 62°1N 23°0E 8 E20
Parker Ariz., U.S.A. 34°9N 114°17W 79 L12
Parker Pa., U.S.A. 41°5N 79°41W 82 E5
Parker Dam U.S.A. 34°18N 114°8W 79 L12
Parkersburg U.S.A. 39°16N 81°34W 81 F13
Parkes Australia 33°9S 148°11E 63 E4
Parkfield U.S.A. 35°54N 120°26W 78 K6
Parkhill Canada 43°15N 81°38W 82 C3
Parkland U.S.A. 47°9N 122°26W 78 C4
Parkston U.S.A. 43°24N 97°59W 80 D5
Parksville Canada 49°20N 124°21W 70 D4
Parkway U.S.A. 38°32N 121°26W 78 G5
Parla Spain 40°14N 3°46W 21 B4
Parma Italy 44°48N 10°20E 20 D9
Parma Idaho, U.S.A. 43°47N 116°57W 76 E5
Parma Ohio, U.S.A. 41°24N 81°43W 82 E3
Parnaguá Brazil 10°10S 44°38W 93 F10
Parnaíba Brazil 2°54S 41°47W 93 D10
Parnaíba → Brazil 3°0S 41°50W 93 D10
Parnassos Greece 38°35N 22°30E 23 E10
Pärnu Estonia 58°28N 24°33E 9 G21
Paro Dzong Bhutan 27°32N 89°53E 43 F13
Paroo → Australia 31°28S 143°32E 63 E3
Paros Greece 37°5N 25°12E 23 F11
Parowan U.S.A. 37°51N 112°50W 77 H7
Parral Chile 36°10S 71°52W 94 D1
Parramatta Australia 33°48S 151°1E 63 B5
Parras Mexico 25°25N 102°11W 86 B4
Parrett → U.K. 51°12N 3°1W 13 F4
Parris I. U.S.A. 32°20N 80°41W 85 E14
Parrsboro Canada 45°30N 64°25W 73 C7
Parry Channel Canada 74°15N 94°0W 69 C13
Parry I. Canada 45°18N 80°10W 82 A4
Parry Is. Canada 77°0N 110°0W 69 B10
Parry Sound Canada 45°20N 80°0W 82 A5
Parsaloi Kenya 1°16N 36°51E 54 B4
Parsnip → Canada 55°10N 123°2W 70 B4
Parsons U.S.A. 37°20N 95°16W 80 G6
Parsons Ra. Australia 13°30S 135°15E 62 A2
Partinico Italy 38°3N 13°7E 22 E5
Partizansk Russia 43°8N 133°9E 29 E14
Partridge I. Canada 55°59N 87°37W 72 A2
Partry Mts. Ireland 53°40N 9°28W 10 C2
Paru → Brazil 1°33S 52°38W 93 D8
Parvān □ Afghan. 35°0N 69°0E 40 B6
Parvatipuram India 18°50N 83°25E 41 K13
Parvatsar India 26°52N 74°49E 42 F6
Parys S. Africa 26°52S 27°29E 57 D4
Pas, Pta. des Spain 38°46N 1°26E 24 C7
Pas, The Canada 53°45N 101°15W 71 C8
Pasadena Canada 49°1N 57°36W 73 C8
Pasadena Calif., U.S.A. 34°9N 118°8W 79 L8
Pasadena Tex., U.S.A. 29°43N 95°13W 84 G7
Pasaje → Argentina 25°39S 63°56W 94 B3
Pasar Indonesia 8°27S 114°54E 37 J17
Pascagoula U.S.A. 30°21N 88°33W 85 F10
Pascagoula → U.S.A. 30°23N 88°37W 85 F10
Pașcani Romania 47°14N 26°45E 17 E14
Pasco U.S.A. 46°14N 119°6W 76 C4
Pasco I. Australia 20°57S 115°20E 60 D2
Pascoag U.S.A. 41°57N 71°42W 83 E13
Pascua, I. de Chile 27°7S 109°23W 65 K17
Pasfield L. Canada 58°24N 105°20W 71 B7
Pasir Mas Malaysia 6°2N 102°8E 39 J4
Pasir Panjang Singapore 1°18N 103°46E 39 d
Pasir Putih Malaysia 5°50N 102°24E 39 K4
Pasirian Indonesia 8°13S 113°8E 37 H15
Paskūh Iran 27°34N 61°39E 45 E9
Pasley, C. Australia 33°52S 123°35E 61 F3
Pašman Croatia 43°58N 15°20E 16 G8
Pasni Pakistan 25°15N 63°27E 40 G3
Paso Bravo △ Paraguay 22°32S 57°5W 94 A4
Paso Cantinela Mexico 32°33N 115°47W 79 N11
Paso de Indios Argentina 43°55S 69°0W 96 E3
Paso de los Libres Argentina 29°44S 57°10W 94 B4
Paso de los Toros Uruguay 32°45S 56°30W 94 C4
Paso Robles U.S.A. 35°38N 120°41W 78 K6
Paspébiac Canada 48°3N 65°17W 73 C6
Pasrur Pakistan 32°16N 74°43E 42 C6
Passage East Ireland 52°14N 7°0W 10 D5
Passage West Ireland 51°52N 8°21W 10 E3
Passaic U.S.A. 40°51N 74°7W 83 F10
Passau Germany 48°34N 13°28E 16 D7
Passero, C. Italy 36°41N 15°10E 22 F6
Passo Fundo Brazil 28°10S 52°20W 95 B5
Passos Brazil 20°45S 46°37W 93 H9
Pastavy Belarus 55°4N 26°50E 9 J22
Pastaza → Peru 4°50S 76°52W 92 D3
Pasto Colombia 1°13N 77°17W 92 C3
Pasuruan Indonesia 7°40S 112°44E 37 G15
Patagonia Argentina 45°0S 69°0W 96 F3
Patagonia U.S.A. 31°33N 110°45W 77 L8
Patambar Iran 29°45N 60°17E 45 D9
Patan = Lalitapur Nepal 27°40N 85°20E 43 F11
Patan = Somnath India 20°53N 70°22E 42 J4
Patan India 23°54N 72°14E 42 H5
Patani Indonesia 0°20N 128°50E 37 D7
Pataudi India 28°18N 76°48E 42 E7
Patchewollock Australia 35°22S 142°12E 63 F3
Patchogue U.S.A. 40°46N 73°1W 83 F11
Pate Kenya 2°10S 41°0E 54 C5
Patea N.Z. 39°45S 174°30E 59 C5
Patensie S. Africa 33°46S 24°49E 56 E3
Paternò Italy 37°34N 14°54E 22 F6
Pateros U.S.A. 48°3N 119°54W 76 B4
Paterson U.S.A. 40°54N 74°9W 83 F10
Paterson Ra. Australia 21°45S 122°10E 60 D3
Pathankot India 32°18N 75°45E 42 C6
Pathein = Bassein Burma 16°45N 94°30E 41 L19
Pathfinder Res. U.S.A. 42°28N 106°51W 76 E10
Pathiu Thailand 10°42N 99°19E 39 G2
Pathum Thani Thailand 14°1N 100°32E 38 E3
Pati Indonesia 6°45S 111°1E 37 G14
Patía → Colombia 2°13N 78°40W 92 C3
Patiala Punjab, India 30°23N 76°26E 42 D7
Patiala Ut. P., India 27°43N 79°1E 43 F8
Patkai Bum India 27°0N 95°30E 41 F19
Patmos Greece 37°21N 26°36E 23 F12
Patna India 25°35N 85°12E 43 G11
Pato Branco Brazil 26°13S 52°40W 95 B5
Patong, Ao Thailand 7°54N 98°17E 39 a
Patonga Uganda 2°45N 33°15E 54 B3
Patos Brazil 6°55S 37°16W 93 E11
Patos, L. dos Brazil 31°20S 51°0W 95 C5
Patos, Río de los → Argentina 31°18S 69°25W 94 C2
Patos de Minas Brazil 18°35S 46°32W 93 G9
Patquía Argentina 30°2S 66°55W 94 C2
Patra Greece 38°14N 21°47E 23 E9
Patraikos Kolpos Greece 38°17N 21°30E 23 E9
Patras = Patra Greece 38°14N 21°47E 23 E9
Patriot Hills Antarctica 82°20S 81°25W 5 E16
Patrocínio Brazil 18°57S 47°0W 93 G9
Pattani Thailand 6°48N 101°15E 39 J3
Pattaya Thailand 12°52N 100°55E 38 F3
Patten U.S.A. 46°0N 68°38W 81 B19
Patterson Calif., U.S.A. 37°28N 121°8W 78 H5
Patterson La., U.S.A. 29°42N 91°18W 84 G9
Patterson, Mt. U.S.A. 38°29N 119°20W 78 G7
Patti Punjab, India 31°17N 74°54E 42 D6
Patti Ut. P., India 25°55N 82°12E 43 G10
Pattoki Pakistan 31°5N 73°52E 42 D5
Patton U.S.A. 40°38N 78°39W 82 F6
Patuakhali Bangla. 22°20N 90°25E 41 H17
Patuanak Canada 55°55N 107°43W 71 B7
Patuca → Honduras 15°50N 84°18W 88 C3
Patuca, Punta Honduras 15°49N 84°14W 88 C3
Patuca △ Honduras 14°30N 85°30W 88 D2
Patvinsuo △ Finland 63°7N 30°45E 8 E24
Pátzcuaro Mexico 19°31N 101°38W 86 D4
Pau France 43°19N 0°25W 20 E3
Paudash Canada 44°59N 77°58W 82 B7
Pauk Burma 21°27N 94°30E 41 J19
Paul I. Canada 56°30N 61°20W 73 A7
Paul Smiths U.S.A. 44°26N 74°15W 83 B10
Paulatuk Canada 69°25N 124°0W 68 D7
Paulding Bay S. Ocean 66°0S 118°0E 5 C8
Paulistana Brazil 8°9S 41°9W 93 E10
Paulo Afonso Brazil 9°21S 38°15W 93 E11
Paulpietersburg S. Africa 27°23S 30°50E 57 D5
Pauls Valley U.S.A. 34°44N 97°13W 84 D6
Pauma Valley U.S.A. 33°16N 116°58W 79 M10
Pauri India 30°9N 78°47E 43 D8
Pāveh Iran 35°3N 46°22E 44 C5
Pavia Italy 45°7N 9°8E 20 D8
Pavilion U.S.A. 42°52N 78°1W 82 D6
Pāvilosta Latvia 56°53N 21°14E 9 H19
Pavlodar Kazakhstan 52°33N 77°0E 28 D8
Pavlograd = Pavlohrad Ukraine 48°30N 35°52E 19 E6
Pavlohrad Ukraine 48°30N 35°52E 19 E6
Pavlovo Russia 55°58N 43°5E 18 C7
Pavlovsk Russia 50°26N 40°5E 19 D7
Pavlovskaya Russia 46°17N 39°47E 19 E6
Pawai, Pulau Singapore 1°11N 103°44E 39 d
Pawayan India 28°4N 80°6E 43 E9
Pawhuska U.S.A. 36°40N 96°20W 84 C6
Pawling U.S.A. 41°34N 73°36W 83 E11
Pawnee U.S.A. 36°20N 96°48W 84 C6
Pawnee City U.S.A. 40°7N 96°9W 80 E5
Pawtucket U.S.A. 41°53N 71°23W 83 E13
Paxi Greece 39°14N 20°12E 23 E9
Paximadia Greece 35°0N 24°35E 25 E6
Paxton U.S.A. 40°27N 88°6W 80 E9
Payakumbuh Indonesia 0°20S 100°35E 36 E2
Payette U.S.A. 44°5N 116°56W 76 D5
Payne Bay = Kangirsuk Canada 60°0N 70°0W 69 F18
Payne L. Canada 59°30N 74°30W 69 F17
Paynes Find Australia 29°15S 117°42E 61 E2
Paynesville U.S.A. 45°23N 94°43W 80 C6
Paysandú Uruguay 32°19S 58°8W 94 C4
Payson Ariz., U.S.A. 34°14N 111°20W 77 J8
Payson Utah, U.S.A. 40°3N 111°44W 76 F8
Paz → Guatemala 13°44N 90°10W 88 D1
Paz, B. de la Mexico 24°9N 110°25W 86 C2
Pāzanān Iran 30°35N 49°59E 45 D6
Pazardzhik Bulgaria 42°12N 24°20E 23 C11
Pe Ell U.S.A. 46°34N 123°18W 78 D3
Peabody U.S.A. 42°31N 70°56W 83 D14
Peace → Canada 59°0N 111°25W 70 B6
Peace Point Canada 59°7N 112°27W 70 B6
Peace River Canada 56°15N 117°18W 70 B5
Peach Springs U.S.A. 35°32N 113°25W 77 J7
Peachland Canada 49°47N 119°45W 70 D5
Peachtree City U.S.A. 33°25N 84°35W 85 E12
Peak, The = Kinder Scout U.K. 53°24N 1°52W 12 D6
Peak Charles △ Australia 32°42S 121°10E 61 F3
Peak District △ U.K. 53°24N 1°46W 12 D6
Peak Hill Australia 32°47S 148°11E 63 E4
Peak Ra. Australia 22°50S 148°20E 62 C4
Peake Cr. → Australia 28°2S 136°7E 63 D2
Peale, Mt. U.S.A. 38°26N 109°14W 76 G9
Pearblossom U.S.A. 34°30N 117°55W 79 L9
Pearl → U.S.A. 30°11N 89°32W 85 F10
Pearl and Hermes Reef U.S.A. 27°55N 175°45W 75 K4
Pearl River U.S.A. 41°4N 74°2W 83 E10
Pearl River Bridge China 22°15N 113°48E 33 a
Pearsall U.S.A. 28°54N 99°6W 84 G5
Pearson Int. Toronto ✕ (YYZ) Canada 43°46N 79°35W 82 C5
Peary Land Greenland 82°40N 33°0W 4 A6
Pease → U.S.A. 34°12N 99°2W 84 D5
Peawanuck Canada 55°15N 85°12W 72 A2
Pebane Mozam. 17°10S 38°8E 55 F4
Pebas Peru 3°10S 71°46W 92 D4
Pebble Beach U.S.A. 36°34N 121°57W 78 J5
Peć Kosovo 42°40N 20°17E 23 C9
Pechenga Russia 69°29N 31°4E 8 B24
Pechenizhyn Ukraine 48°30N 24°48E 17 D13
Pechiguera, Pta. Canary Is. 28°51N 13°53W 24 F6
Pechora Russia 65°10N 57°11E 18 A10
Pechora → Russia 68°13N 54°15E 18 A9
Pechorskaya Guba Russia 68°40N 54°0E 18 A9
Pechory Russia 57°48N 27°40E 9 H22
Pecos N. Mex., U.S.A. 35°35N 105°41W 77 J11
Pecos Tex., U.S.A. 31°26N 103°30W 84 F3
Pecos → U.S.A. 29°42N 101°22W 84 G4
Pécs Hungary 46°5N 18°15E 17 E10
Pedasí Panama 7°32N 80°2W 88 E3
Pedder, L. Australia 42°55S 146°10E 63 G4
Peddie S. Africa 33°14S 27°7E 56 E4
Pedernales Dom. Rep. 18°2N 71°44W 89 C5
Pedieos → Cyprus 35°10N 33°54E 25 D12
Pedirka Desert Australia 26°47S 134°11E 63 A1
Pedra Azul Brazil 16°2S 41°17W 93 G10
Pedra Lume C. Verde Is. 16°40N 22°52W 50 b
Pedreiras Brazil 4°32S 44°40W 93 D10
Pedro Afonso Brazil 9°0S 48°10W 93 E9
Pedro Cays Jamaica 17°5N 77°48W 88 C4
Pedro de Valdivia Chile 22°55S 69°38W 94 A2
Pedro Juan Caballero Paraguay 22°30S 55°40W 95 A4
Pee Dee = Great Pee Dee → U.S.A. 33°21N 79°10W 85 E15
Peebinga Australia 34°52S 140°57E 63 E3
Peebles U.K. 55°40N 3°11W 11 F5
Peekskill U.S.A. 41°17N 73°55W 83 E11
Peel I. of Man 54°13N 4°40W 12 C3
Peel → Australia 30°50S 150°29E 63 E5
Peel → Canada 67°0N 135°0W 68 D5
Peel Sd. Canada 73°0N 96°0W 68 C12
Peera Peera Poolanna L. Australia 26°30S 138°0E 63 D2
Peerless Lake Canada 56°37N 114°40W 70 B6
Peers Canada 53°40N 116°0W 70 C5
Pegasus Bay N.Z. 43°20S 173°10E 59 E4
Pegu Burma 17°20N 96°29E 41 L20
Pegu Yoma Burma 19°0N 96°0E 41 K20
Pehuajó Argentina 35°45S 62°0W 94 D3
Pei Xian China 34°44N 116°55E 34 G9
Peine Chile 23°45S 68°8W 94 A2
Peine Germany 52°19N 10°14E 16 B6
Peip'ing = Beijing China 39°53N 116°21E 34 E9
Peipus, L. = Chudskoye, Ozero Russia 58°13N 27°30E 9 G22
Peixe Brazil 12°0S 48°40W 93 F9
Peixe → Brazil 21°31S 51°58W 93 H8
Peje = Peć Kosovo 42°40N 20°17E 23 C9
Pekalongan Indonesia 6°53S 109°40E 37 G13
Pekan Malaysia 3°30N 103°25E 39 L4
Pekan Nenas Malaysia 1°31N 103°31E 39 d
Pekanbaru Indonesia 0°30N 101°15E 36 D2
Pekin U.S.A. 40°35N 89°40W 80 E9
Peking = Beijing China 39°53N 116°21E 34 E9
Pekutatan Indonesia 8°25S 114°49E 37 J17
Pelabuhan Ratu, Teluk Indonesia 7°5S 106°30E 37 G12
Pelagie, Is. Italy 35°39N 12°33E 22 G5
Pelaihari Indonesia 3°55S 114°45E 36 E4
Peleaga, Vf. Romania 45°22N 22°55E 17 F12
Pelée, Mt. Martinique 14°48N 61°10W 88 c
Pelee, Pt. Canada 41°54N 82°31W 72 D3
Pelee I. Canada 41°47N 82°40W 82 E2
Pelekech Kenya 3°52N 35°8E 54 B4
Peleng Indonesia 1°20S 123°30E 37 E6
Pelentong Malaysia 1°32N 103°49E 39 d
Pelican U.S.A. 57°58N 136°14W 70 B1
Pelican L. Canada 52°28N 100°20W 71 C8
Pelican Narrows Canada 55°10N 102°56W 71 B8
Pelješac Croatia 42°55N 17°25E 22 C7
Pelkosenniemi Finland 67°6N 27°28E 8 C22
Pella S. Africa 29°1S 19°6E 56 D2
Pella U.S.A. 41°25N 92°55W 80 E7
Pello Finland 66°47N 23°59E 8 C20
Pelly → Canada 62°47N 137°19W 70 A1
Pelly Bay Canada 68°38N 89°50W 69 D14
Peloponnesos □ Greece 37°10N 22°0E 23 F10
Pelorus Sd. N.Z. 40°59S 173°59E 59 D4
Pelotas Brazil 31°42S 52°23W 95 C5
Pelotas → Brazil 27°28S 51°55W 95 B5
Pelvoux, Massif du France 44°52N 6°20E 20 D7

Pemalang Indonesia 6°53S 109°23E 37 G13
Pemanggil, Pulau Malaysia 2°37N 104°21E 39 L5
Pematangsiantar Indonesia 2°57N 99°5E 36 D1
Pemba Mozam. 12°58S 40°30E 55 E5
Pemba Zambia 16°30S 27°28E 55 F2
Pemba Channel Tanzania 5°0S 39°37E 54 D4
Pemba I. Tanzania 5°0S 39°45E 54 D4
Pemberton Australia 34°30S 116°0E 61 F2
Pemberton Canada 50°25N 122°50W 70 C4
Pembina → Canada 54°45N 114°17W 70 C6
Pembroke Canada 45°50N 77°7W 72 C4
Pembroke U.K. 51°41N 4°55W 13 F3
Pembrokeshire □ U.K. 51°52N 4°56W 13 F3
Pembrokeshire Coast △ U.K. 51°50N 5°2W 13 F2
Pen-y-Ghent U.K. 54°10N 2°14W 12 C5
Penal Trin. & Tob. 10°9N 61°29W 93 K15
Penang = Pinang Malaysia 5°25N 100°15E 39 c
Penápolis Brazil 21°30S 50°0W 95 A6
Peñarroya-Pueblonuevo Spain 38°19N 5°16W 21 C3
Penarth U.K. 51°26N 3°11W 13 F4
Peñas, C. de Spain 43°42N 5°52W 21 A3
Peñas, G. de Chile 47°0S 75°0W 96 F2
Peñas del Chache Canary Is. 29°6N 13°33W 24 E6
Pench △ India 21°45N 79°20E 43 J8
Pench'i = Benxi China 41°20N 123°48E 35 D12
Pend Oreille → U.S.A. 49°4N 117°37W 76 B5
Pend Oreille, L. U.S.A. 48°10N 116°21W 76 B5
Pendembu S. Leone 9°7N 11°14W 50 G3
Pender B. Australia 16°45S 122°42E 60 C3
Pendleton U.S.A. 45°40N 118°47W 76 D4
Pendra India 22°46N 81°57E 43 H9
Penedo Brazil 10°15S 36°36W 93 F11
Penelokan Indonesia 8°17S 115°22E 37 J18
Penetanguishene Canada 44°50N 79°55W 82 B5
Penfield U.S.A. 41°13N 78°35W 82 E6
Peng Chau China 22°17N 114°2E 33 a
Penge Kasai-Or., Dem. Rep. of the Congo 5°30S 24°33E 54 D1
Penge Sud-Kivu, Dem. Rep. of the Congo 4°27S 28°25E 54 C2
Penglai China 37°48N 120°42E 35 F11
Penguin Australia 41°8S 146°6E 63 G4
Penhalonga Zimbabwe 18°52S 32°40E 55 F3
Peniche Portugal 39°19N 9°22W 21 C1
Penicuik U.K. 55°50N 3°13W 11 F5
Penida, Nusa Indonesia 8°45S 115°30E 37 K18
Peninsular Malaysia □ Malaysia 4°0N 102°0E 39 L4
Penitente, Serra do Brazil 8°45S 46°20W 93 E9
Penkridge U.K. 52°44N 2°6W 12 E5
Penmarch, Pte. de France 47°48N 4°22W 20 C1
Penn Hills U.S.A. 40°28N 79°52W 82 F5
Penn Yan U.S.A. 42°40N 77°3W 82 D7
Pennant Canada 50°32N 108°14W 71 C7
Penner → India 14°35N 80°10E 40 M12
Pennines U.K. 54°45N 2°27W 12 C5
Pennington U.S.A. 39°15N 121°47W 78 F5
Pennsburg U.S.A. 40°23N 75°29W 83 F9
Pennsylvania □ U.S.A. 40°45N 77°30W 81 E15
Penny Canada 53°51N 121°20W 70 C4
Penobscot → U.S.A. 44°30N 68°48W 81 C19
Penobscot B. U.S.A. 44°35N 68°50W 81 C19
Penola Australia 37°25S 140°48E 63 F3
Penong Australia 31°56S 133°1E 61 F5
Penonomé Panama 8°31N 80°21W 88 E3
Penrhyn Cook Is. 9°0S 158°0W 65 H12
Penrith Australia 33°43S 150°38E 63 E5
Penrith U.K. 54°40N 2°45W 12 C5
Penryn U.K. 50°9N 5°7W 13 G2
Pensacola U.S.A. 30°25N 87°13W 85 F11
Pensacola Mts. Antarctica 84°0S 40°0W 5 E1
Pense Canada 50°25N 104°59W 71 C8
Penshurst Australia 37°49S 142°20E 63 F3
Penticton Canada 49°30N 119°38W 70 D5
Pentland Australia 20°32S 145°25E 62 C4
Pentland Firth U.K. 58°43N 3°10W 11 C5
Pentland Hills U.K. 55°48N 3°25W 11 F5
Pentland Hills ⌓ U.K. 55°50N 3°20W 11 F5
Penza Russia 53°15N 45°5E 18 D8
Penzance U.K. 50°7N 5°33W 13 G2
Penzhino Russia 63°30N 167°55E 29 C17
Penzhinskaya Guba Russia 61°30N 163°0E 29 C17
Peoria Ariz., U.S.A. 33°43N 112°14W 77 K7
Peoria Ill., U.S.A. 40°42N 89°36W 80 E9
Pepacton Res. U.S.A. 42°5N 74°58W 83 D10
Pepani → S. Africa 25°49S 22°47E 56 D3
Pera Hd. Australia 12°55S 141°37E 62 A3
Perabumulih Indonesia 3°27S 104°15E 36 E2
Perak □ Malaysia 5°0N 101°0E 39 K3
Perak → Malaysia 4°0N 100°50E 39 K3
Perama Kerkyra, Greece 39°34N 19°54E 25 A3
Perama Kriti, Greece 35°20N 24°40E 25 D6
Perancak Indonesia 8°24S 114°37E 37 J17
Peräpohjola Finland 66°16N 26°10E 8 C22
Percé Canada 48°31N 64°13W 73 C7
Perche, Collines du France 48°30N 0°40E 20 B4
Percival Lakes Australia 21°25S 125°0E 60 D4
Percy Is. Australia 21°39S 150°16E 62 C5
Perdido, Mte. Spain 42°40N 0°5E 21 A6
Perdu, Mt. = Perdido, Mte. Spain 42°40N 0°5E 21 A6
Pereira Colombia 4°49N 75°43W 92 C3
Perenjori Australia 29°26S 116°16E 61 E2
Pereyaslav-Khmelnytskyy Ukraine 50°3N 31°28E 19 D5
Pérez, I. Mexico 22°24N 89°42W 87 C7
Pergamino Argentina 33°52S 60°30W 94 C3
Pergau → Malaysia 5°23N 102°2E 39 K3
Perham U.S.A. 46°36N 95°34W 80 B6
Perhentian, Kepulauan Malaysia 5°54N 102°42E 39 K4
Péribonka → Canada 48°45N 72°5W 73 C5
Péribonka, L. Canada 50°1N 71°10W 73 B5
Péribonka, La Centrale de Canada 49°30N 71°10W 73 C5
Perico Argentina 24°20S 65°5W 94 A2
Pericos Mexico 25°3N 107°42W 86 B3
Périgueux France 45°10N 0°42E 20 D4
Perijá, Sierra de Colombia 9°30N 73°3W 92 B4
Perijá △ Venezuela 9°30N 72°55W 89 E5
Peringat Malaysia 6°2N 102°17E 39 J4
Peristerona → Cyprus 35°8N 33°5E 25 D12
Perito Moreno Argentina 46°36S 70°56W 96 F2
Perkasie U.S.A. 40°22N 75°18W 83 F9
Perlas, Arch. de las Panama 8°41N 79°7W 88 E4
Perlas, Punta de Nic. 12°30N 83°30W 88 D3
Perlis □ Malaysia 6°30N 100°15E 39 J2
Perm Russia 58°0N 56°10E 18 C10
Pernambuco = Recife Brazil 8°0S 35°0W 93 E12
Pernambuco □ Brazil 8°0S 37°0W 93 E11
Pernatty Lagoon Australia 31°30S 137°12E 63 E2
Pernik Bulgaria 42°35N 23°2E 23 C10
Peron Is. Australia 13°9S 130°4E 60 B5
Peron Pen. Australia 26°0S 113°10E 61 E1
Perow Canada 54°35N 126°10W 70 C3
Perpignan France 42°42N 2°53E 20 E5
Perris U.S.A. 33°47N 117°14W 79 M9
Perry Fla., U.S.A. 30°7N 83°35W 85 F13
Perry Ga., U.S.A. 32°28N 83°44W 85 E13
Perry Iowa, U.S.A. 41°51N 94°6W 80 E6
Perry N.Y., U.S.A. 42°42N 78°0W 82 D7
Perry Okla., U.S.A. 36°17N 97°14W 84 C6
Perryton U.S.A. 36°24N 100°48W 84 C4
Perryville U.S.A. 37°43N 89°52W 80 G9
Persepolis Iran 29°55N 52°50E 45 D7
Pershotravensk Ukraine 50°13N 27°40E 17 C14
Persia = Iran ■ Asia 33°0N 53°0E 45 C7
Persian Gulf Asia 27°0N 50°0E 45 E6
Perth Australia 31°57S 115°52E 61 F2
Perth Canada 44°55N 76°15W 83 B8
Perth U.K. 56°24N 3°26W 11 E5
Perth & Kinross □ U.K. 56°45N 3°55W 11 E5
Perth Amboy U.S.A. 40°30N 74°15W 83 F10
Perth-Andover Canada 46°44N 67°42W 73 C6
Perth Basin Ind. Oc. 30°0S 108°0E 64 L2
Peru Ind., U.S.A. 40°45N 86°4W 80 E10
Peru N.Y., U.S.A. 44°35N 73°32W 83 B11
Peru ■ S. Amer. 4°0S 75°0W 92 D4
Peru Basin Pac. Oc. 20°0S 95°0W 65 J18
Peru-Chile Trench Pac. Oc. 20°0S 72°0W 92 G3
Perúgia Italy 43°7N 12°23E 22 C5
Pervomaysk Ukraine 48°10N 30°46E 19 E5
Pervouralsk Russia 56°59N 59°59E 18 C10
Pésaro Italy 43°54N 12°55E 22 C5
Pescara Italy 42°28N 14°13E 22 C6
Peshawar Pakistan 34°2N 71°37E 42 B4
Peshkopi Albania 41°41N 20°25E 23 D9
Peshtigo U.S.A. 45°3N 87°45W 80 C10
Pesqueira Brazil 8°20S 36°42W 93 E11
Petah Tiqwa Israel 32°6N 34°53E 46 C3
Petaling Jaya Malaysia 3°4N 101°42E 39 L3
Petaloudes Greece 36°18N 28°5E 25 C10
Petaluma U.S.A. 38°14N 122°39W 78 G4
Pétange Lux. 49°33N 5°55E 15 E5
Petaro Pakistan 25°31N 68°18E 42 G3
Petatlán Mexico 17°31N 101°16W 86 D4
Petauke Zambia 14°14S 31°20E 55 E3
Petawawa Canada 45°54N 77°17W 72 C4
Petén Itzá, L. Guatemala 16°58N 89°50W 88 C2
Peter I. Br. Virgin Is. 18°22N 64°35W 89 e
Peter I. Øy Antarctica 69°0S 91°0W 5 C16
Peter Pond L. Canada 55°55N 108°44W 71 B7
Peterbell Canada 48°36N 83°21W 72 C3
Peterborough Australia 32°58S 138°51E 63 E2
Peterborough Canada 44°20N 78°20W 82 B6
Peterborough U.K. 52°35N 0°15W 13 E7
Peterborough U.S.A. 42°53N 71°57W 83 D13
Peterborough □ U.K. 52°35N 0°15W 13 E7
Peterculter U.K. 57°6N 2°16W 11 D6
Peterhead U.K. 57°31N 1°48W 11 D7
Peterlee U.K. 54°47N 1°20W 12 C6
Petermann ☼ Australia 24°58S 130°0E 61 E4
Petermann Gletscher Greenland 80°30N 60°0W 69 A20
Petermann Ranges Australia 26°0S 130°30E 61 E5
Petersburg Alaska, U.S.A. 56°48N 132°58W 68 F5
Petersburg Pa., U.S.A. 40°34N 78°3W 82 F6
Petersburg Va., U.S.A. 37°14N 77°24W 81 G15
Petersburg W. Va., U.S.A. 39°1N 79°5W 81 F14
Petersfield U.K. 51°1N 0°56W 13 F7
Petit-Canal Guadeloupe 16°25N 61°31W 88 b
Petit Goâve Haiti 18°27N 72°51W 89 C5
Petit Lac Manicouagan Canada 51°25N 67°40W 73 B6
Petit-Mécatina → Canada 50°40N 59°30W 73 B8
Petit-Mécatina, Î. du Canada 50°30N 59°25W 73 B8
Petit Piton St. Lucia 13°51N 61°5W 89 f
Petit-Saguenay Canada 48°15N 70°4W 73 C5
Petitcodiac Canada 45°57N 65°11W 73 C6
Petite Terre, Îles de la Guadeloupe 16°13N 61°9W 88 b
Petitot → Canada 60°14N 123°29W 70 A4
Petitsikapau L. Canada 54°37N 66°25W 73 B6
Petlad India 22°30N 72°45E 42 H5
Peto Mexico 20°8N 88°55W 87 C7
Petone N.Z. 41°13S 174°53E 59 D5
Petorca Chile 32°15S 70°56W 94 C1
Petoskey U.S.A. 45°22N 84°57W 81 C11
Petra Jordan 30°20N 35°22E 46 E4
Petra Spain 39°37N 3°6E 24 B10
Petra, Ostrov Russia 76°15N 118°30E 29 B12
Petra Velikogo, Zaliv Russia 42°40N 132°0E 30 C6
Petre, Pt. Canada 43°50N 77°9W 82 C7
Petrich Bulgaria 41°24N 23°13E 23 D10
Petrified Forest △ U.S.A. 35°0N 109°30W 77 J9
Petrikov = Pyetrikaw Belarus 52°11N 28°29E 17 B15
Petrolândia Brazil 9°5S 38°20W 93 E11
Petrolia Canada 42°54N 82°9W 82 D2
Petrolina Brazil 9°24S 40°30W 93 E10
Petropavl Kazakhstan 54°53N 69°13E 28 D7
Petropavlovsk = Petropavl Kazakhstan 54°53N 69°13E 28 D7
Petropavlovsk-Kamchatskiy Russia 53°3N 158°43E 29 D16
Petrópolis Brazil 22°33S 43°9W 95 A7
Petroşani Romania 45°28N 23°20E 17 F12
Petrovaradin Serbia 45°16N 19°55E 23 B8
Petrovsk Russia 52°22N 45°19E 18 D8
Petrovsk-Zabaykalskiy Russia 51°20N 108°55E 29 D11
Petrozavodsk Russia 61°41N 34°20E 18 B5
Petrus Steyn S. Africa 27°38S 28°8E 57 D4
Petrusburg S. Africa 29°4S 25°26E 56 D4
Peumo Chile 34°21S 71°12W 94 C1
Peureulak Indonesia 4°48N 97°45E 36 D1
Pevek Russia 69°41N 171°19E 29 C18
Pforzheim Germany 48°52N 8°41E 16 D5
Pha Taem △ Thailand 15°32N 105°30E 38 E5
Phaestos Greece 35°2N 24°50E 25 D6
Phagwara India 31°10N 75°40E 42 D6
Phala Botswana 23°45S 26°50E 56 C4
Phalera = Phulera India 26°52N 75°16E 42 F6
Phalodi India 27°12N 72°24E 42 F5
Phaluai, Ko Thailand 9°32N 99°41E 39 b
Phan Thailand 19°28N 99°43E 38 C2
Phan Rang-Thap Cham Vietnam 11°34N 109°0E 39 G7
Phan Ri Cua Vietnam 11°16N 108°40E 39 G7
Phan Si Pan Vietnam 22°18N 103°46E 38 A4
Phan Thiet Vietnam 11°1N 108°9E 39 G7
Phan Thiet, Vinh Vietnam 10°53N 108°5E 39 G7
Phanat Nikhom Thailand 13°27N 101°11E 38 F3
Phangan, Ko Thailand 9°45N 100°0E 39 H3
Phangnga Thailand 8°28N 98°30E 39 H2
Phangnga, Ao Thailand 8°16N 98°33E 39 a
Phanom Sarakham Thailand 13°45N 101°21E 38 F3
Phaphund India 26°36N 79°28E 43 F8
Pharenda India 27°5N 83°17E 43 F10
Pharr U.S.A. 26°12N 98°11W 84 H5
Phatthalung Thailand 7°39N 100°6E 39 J3
Phayam, Ko Thailand 9°44N 98°25E 39 H2
Phayao Thailand 19°11N 99°55E 38 C2
Phelps U.S.A. 42°58N 77°3W 82 D7
Phelps L. Canada 59°15N 103°15W 71 B8
Phenix City U.S.A. 32°28N 85°0W 85 E12
Phet Buri = Phetchaburi Thailand 13°1N 99°55E 38 F2
Phetchabun Thailand 16°25N 101°8E 38 D3
Phetchabun, Thiu Khao Thailand 16°0N 101°20E 38 E3
Phetchaburi Thailand 13°1N 99°55E 38 F2
Phi Phi, Ko Thailand 7°45N 98°46E 39 J2
Phiafay Laos 14°48N 106°0E 38 E6
Phibun Mangsahan Thailand 15°14N 105°14E 38 E5
Phichai Thailand 17°22N 100°10E 38 D3
Phichit Thailand 16°26N 100°22E 38 D3
Philadelphia Miss., U.S.A. 32°46N 89°7W 85 E10
Philadelphia N.Y., U.S.A. 44°9N 75°43W 83 B9
Philadelphia Pa., U.S.A. 39°57N 75°9W 83 G9
Philip U.S.A. 44°2N 101°40W 80 C3
Philip Smith Mts. U.S.A. 68°0N 148°0W 74 B10
Philippeville Belgium 50°12N 4°33E 15 D4
Philippi U.S.A. 39°9N 80°3W 81 F13
Philippi L. Australia 24°20S 138°55E 62 C2
Philippine Basin Pac. Oc. 17°0N 132°0E 64 F5
Philippine Sea Pac. Oc. 18°0N 125°0E 64 F4
Philippine Trench = Mindanao Trench Pac. Oc. 12°0N 126°6E 37 B7
Philippines ■ Asia 12°0N 123°0E 37 B6
Philippolis S. Africa 30°15S 25°16E 56 E4
Philippopolis = Plovdiv Bulgaria 42°8N 24°44E 23 C11
Philipsburg Canada 45°2N 73°5W 83 A11
Philipsburg Mont., U.S.A. 46°20N 113°18W 76 C7
Philipsburg Pa., U.S.A. 40°54N 78°13W 82 F6
Philipstown = Daingean Ireland 53°18N 7°17W 10 C4
Philipstown S. Africa 30°28S 24°30E 56 E3
Phillip I. Australia 38°30S 145°12E 63 F4
Phillips U.S.A. 45°42N 90°24W 80 C8
Phillipsburg Kans., U.S.A. 39°45N 99°19W 80 F4
Phillipsburg N.J., U.S.A. 40°42N 75°12W 83 F9
Philmont U.S.A. 42°15N 73°39W 83 D11
Philomath U.S.A. 44°32N 123°22W 76 D2
Phimai Thailand 15°13N 102°30E 38 E4
Phitsanulok Thailand 16°50N 100°12E 38 D3
Phnom Kulen △ Cambodia 13°38N 104°15E 38 F5
Phnom Penh Cambodia 11°33N 104°55E 39 G5
Phnum Penh = Phnom Penh Cambodia 11°33N 104°55E 39 G5
Phoenicia U.S.A. 42°5N 74°14W 83 D10
Phoenix Mauritius 20°17S 57°30E 53 d
Phoenix Ariz., U.S.A. 33°26N 112°4W 77 K7
Phoenix N.Y., U.S.A. 43°14N 76°18W 83 C8
Phoenix Is. Kiribati 3°30S 172°0W 64 H10
Phoenixville U.S.A. 40°8N 75°31W 83 F9
Phon Thailand 15°49N 102°36E 38 E4
Phong → Thailand 16°23N 102°56E 38 D4
Phong Nha -Ke Bang △ Vietnam 17°30N 106°10E 38 D6
Phong Saly Laos 21°42N 102°9E 38 B4
Phong Tho Vietnam 22°32N 103°21E 38 A4
Phonhong Laos 18°30N 102°25E 38 C4
Phonsavan Laos 19°27N 103°10E 38 C4
Phonum Thailand 8°49N 98°48E 39 H2
Phosphate Hill Australia 21°53S 139°58E 62 C2
Photharam Thailand 13°41N 99°51E 38 F2
Phra Nakhon Si Ayutthaya Thailand 14°25N 100°30E 38 E3
Phra Pradaeng Thailand 13°39N 100°33E 38 F3
Phra Thong, Ko Thailand 9°5N 98°17E 39 H2
Phrae Thailand 18°7N 100°9E 38 C3
Phrom Phiram Thailand 17°2N 100°12E 38 D3
Phu Bia Laos 19°10N 103°0E 38 C4
Phu Chong-Na Yoi △ Thailand 14°25N 105°30E 38 E5
Phu Hin Rong Kla △ Thailand 17°0N 100°59E 38 D3
Phu Kao -Phu Phan Kham △ Thailand 16°53N 102°31E 38 D4
Phu Kradung △ Thailand 17°2N 101°44E 38 D3
Phu Loi Laos 20°14N 103°14E 38 B4
Phu Luang △ Thailand 17°15N 101°29E 38 D3
Phu Ly Vietnam 20°35N 105°50E 38 B5
Phu My Vietnam 14°15N 109°5E 38 E7
Phu Pha Man △ Thailand 16°44N 101°50E 38 D3
Phu Phan △ Thailand 17°0N 103°56E 38 D4
Phu Quoc, Dao Vietnam 10°20N 104°0E 39 G4
Phu Sa Dok Bua △ Thailand 16°11N 104°45E 38 D5
Phu Sang △ Thailand 19°35N 100°19E 38 C3
Phu Tho Vietnam 21°24N 105°13E 38 B5
Phu Wiang △ Thailand 16°39N 102°23E 38 D4
Phuc Yen Vietnam 21°16N 105°45E 38 B5
Phuket Thailand 7°53N 98°24E 39 a
Phuket, Ko Thailand 8°0N 98°22E 39 a
Phul India 30°19N 75°14E 42 D6
Phulad India 25°38N 73°49E 42 G5
Phulchari Bangla. 25°11N 89°37E 43 G13
Phulera India 26°52N 75°16E 42 F6
Phulpur India 25°31N 82°49E 43 G10
Phunphin Thailand 9°7N 99°12E 39 H2
Phuntsholing Bhutan 26°51N 89°23E 43 F13
Phuphrabat Thailand 17°40N 102°45E 38 D4
Pia ☼ Australia 27°7S 116°23E 61 E2
Piacenza Italy 45°1N 9°40E 20 D8
Piai, Tanjung Malaysia 1°17N 103°30E 39 d
Pian Cr. → Australia 30°2S 148°12E 63 E4
Pian-Upe ⌓ Uganda 1°44N 34°20E 54 B3
Pianosa Italy 42°35N 10°5E 22 C4
Piao'ertun China 41°50N 123°47E 35 D12
Piapot Canada 49°59N 109°8W 71 D7
Piatra Neamţ Romania 46°56N 26°21E 17 E14
Piauí □ Brazil 7°0S 43°0W 93 E10
Piauí → Brazil 6°38S 42°42W 93 E10
Piave → Italy 45°32N 12°44E 22 B5
Pibor Post Sudan 6°47N 33°3E 51 G12
Picardie □ France 49°50N 3°0E 20 B5
Picardy = Picardie □ France 49°50N 3°0E 20 B5
Picayune U.S.A. 30°32N 89°41W 85 F10
Pichhor India 25°58N 78°20E 43 G8
Pichilemu Chile 34°22S 72°0W 94 C1
Pichor India 25°11N 78°11E 42 G8
Pickerel L. Canada 48°40N 91°25W 72 C1
Pickering U.K. 54°15N 0°46W 12 C7
Pickering, Vale of U.K. 54°14N 0°45W 12 C7
Pickle Lake Canada 51°30N 90°12W 72 B1
Pickwick L. U.S.A. 35°4N 88°15W 85 D10
Pico Azores 38°28N 28°20W 50 a
Pico Bonito △ Honduras 15°34N 86°48W 88 C2
Pico Truncado Argentina 46°40S 68°0W 96 F3
Picos Brazil 7°5S 41°28W 93 E10
Picton Canada 44°1N 77°9W 82 B7
Picton N.Z. 41°18S 174°3E 59 D5
Pictou Canada 45°41N 62°42W 73 C7
Picture Butte Canada 49°55N 112°45W 70 D6
Pictured Rocks ⌓ U.S.A. 46°30N 86°30W 80 B10
Picún Leufú Argentina 39°30S 69°5W 96 D3
Pidurutalagala Sri Lanka 7°10N 80°50E 40 R12
Piedmont = Piemonte □ Italy 45°0N 8°0E 20 D7
Piedmont Ala., U.S.A. 33°55N 85°37W 85 E12
Piedmont S.C., U.S.A. 34°42N 82°28W 66 F11
Piedras Negras Mexico 28°42N 100°31W 86 B4
Pieksämäki Finland 62°18N 27°10E 8 E22
Pielinen Finland 63°15N 29°40E 8 E23
Piemonte □ Italy 45°0N 8°0E 20 D7
Pienaarsrivier S. Africa 25°15S 28°18E 57 D4
Piercefield U.S.A. 44°13N 74°35W 83 B10
Pierceland Canada 54°20N 109°46W 71 C7
Pierpont U.S.A. 41°45N 80°34W 82 E4
Pierre U.S.A. 44°22N 100°21W 80 C3
Pierrefonds Canada 45°29N 73°51W 83 A11
Pierreville Trin. & Tob. 10°16N 61°0W 93 K16
Piet Retief S. Africa 27°1S 30°50E 57 D5
Pietarsaari Finland 63°40N 22°43E 8 E20
Pietermaritzburg S. Africa 29°35S 30°25E 57 D5
Pietersburg = Polokwane S. Africa 23°54S 29°25E 57 C4
Pietrosul, Vf. Maramureş, Romania 47°35N 24°43E 17 E13
Pietrosul, Vf. Suceava, Romania 47°12N 25°18E 17 E13
Pigeon L. Canada 44°27N 78°30W 82 B6
Piggott U.S.A. 36°23N 90°11W 85 C9
Pigüe Argentina 37°36S 62°25W 94 D3
Pihani India 27°36N 80°15E 43 F9
Pihlajavesi Finland 61°45N 28°45E 8 F23
Pijijiapan Mexico 15°42N 93°14W 87 D6
Pikangikum Canada 51°49N 94°0W 71 C10
Pikes Peak U.S.A. 38°50N 105°3W 76 G11
Piketberg S. Africa 32°55S 18°40E 56 E2
Pikeville U.S.A. 37°29N 82°31W 81 G12
Pikou China 39°18N 122°22E 35 E12
Pikwitonei Canada 55°35N 97°9W 71 B9
Piła Poland 53°10N 16°48E 17 B9
Pilanesberg △ S. Africa 25°15S 27°4E 56 D4
Pilani India 28°22N 75°33E 42 E6
Pilar Paraguay 26°50S 58°20W 94 B4
Pilaya → Bolivia 20°55S 64°4W 92 H6
Pilbara Australia 21°15S 118°16E 60
Pilcomaya → Paraguay 25°21S 57°42W 94
Pilgrim's Rest S. Africa 24°55S 30°44E 57
Pilibhit India 28°40N 79°50E 43
Pilica → Poland 51°52N 21°17E 17
Pilipinas = Philippines ■ Asia 12°0N 123°0E 37
Pilkhawa India 28°43N 77°42E 42
Pilliga Australia 30°21S 148°54E 63
Pilos Greece 36°55N 21°42E 23
Pilot Mound Canada 49°15N 98°54W 71
Pilot Point U.S.A. 33°24N 96°58W 84
Pilot Rock U.S.A. 45°29N 118°50W 76
Pima U.S.A. 32°54N 109°50W 77
Pimba Australia 31°18S 136°46E 63
Pimenta Bueno Brazil 11°35S 61°10W 92
Pimentel Peru 6°45S 79°55W 92
Pin Valley △ India 31°50N 77°50E 42
Pinang Malaysia 5°25N 100°15E
Pinang □ Malaysia 5°20N 100°20E 39
Pinar, C. des Spain 39°53N 3°12E 24
Pinar del Río Cuba 22°26N 83°40W 88
Pınarhisar Turkey 41°37N 27°30E 23
Pinatubo, Mt. Phil. 15°8N 120°21E 37
Pinawa Canada 50°9N 95°50W 71
Pincher Creek Canada 49°30N 113°57W 70
Pinchi L. Canada 54°38N 124°30W 70
Pinckneyville U.S.A. 38°5N 89°23W 80
Pińczów Poland 50°32N 20°32E 17
Pindar Australia 28°30S 115°47E 61
Pindera Downs ☼ Australia 29°24S 142°37E 63
Pindi Gheb Pakistan 33°14N 72°21E 42
Pindos Oros Greece 40°0N 21°0E 23
Pindus Mts. = Pindos Oros Greece 40°0N 21°0E 23
Pine → B.C., Canada 56°8N 120°43W 70
Pine → Sask., Canada 58°50N 105°38W 71
Pine Bluff U.S.A. 34°13N 92°1W 84
Pine Bluffs U.S.A. 41°11N 104°4W 76
Pine City U.S.A. 45°50N 92°59W 80
Pine Cr. → U.S.A. 41°10N 77°16W 82
Pine Creek Australia 13°50S 131°50E 60
Pine Falls Canada 50°34N 96°11W 71
Pine Flat L. U.S.A. 36°50N 119°20W 78
Pine Grove U.S.A. 40°33N 76°23W 83
Pine Hill Australia 23°38S 146°57E 62
Pine Island Glacier Antarctica 76°0S 100°0W 5
Pine Pass Canada 55°25N 122°42W 70
Pine Point Canada 60°50N 114°28W 70
Pine Ridge U.S.A. 43°2N 102°33W 80
Pine River Canada 51°45N 100°30W 71
Pine River U.S.A. 46°43N 94°24W 80
Pine Valley U.S.A. 32°50N 116°32W 79
Pinecrest U.S.A. 38°12N 120°1W 78
Pinedale Calif., U.S.A. 36°50N 119°48W 78
Pinedale Wyo., U.S.A. 42°52N 109°52W 76
Pinega → Russia 64°30N 44°19E 18
Pinehouse L. Canada 55°32N 106°35W 71
Pineimuta → Canada 52°8N 88°33W 72
Pinerolo Italy 44°53N 7°21E 20
Pines, Akra Greece 40°5N 24°20E 23
Pinetop-Lakeside U.S.A. 34°9N 109°58W 77
Pinetown S. Africa 29°48S 30°54E 57
Pineville U.S.A. 31°19N 92°26W 84
Ping → Thailand 15°42N 100°9E 38
Pingaring Australia 32°40S 118°32E 61
Pingding China 37°47N 113°38E 34
Pingdingshan China 33°43N 113°27E 34
Pingdu China 36°42N 119°59E 35
Pingelly Australia 32°32S 117°5E 61
Pingliang China 35°35N 106°31E 34
Pinglu China 39°31N 112°30E 34
Pingluo China 38°52N 106°30E 34
Pingquan China 41°1N 118°37E 35
Pingrup Australia 33°32S 118°29E 61
P'ingtung Taiwan 22°38N 120°30E 33
Pingwu China 32°25N 104°30E 34
Pingxiang Guangxi Zhuangzu, China 22°6N 106°46E 32
Pingxiang Jiangxi, China 27°43N 113°48E 33
Pingyao China 37°12N 112°10E 34
Pingyi China 35°30N 117°35E 35
Pingyin China 36°20N 116°25E 34
Pingyuan China 37°10N 116°22E 34
Pinheiro Brazil 2°31S 45°5W 93
Pinheiro Machado Brazil 31°34S 53°23W 95
Pinhel Portugal 40°50N 7°1W 21
Pini Indonesia 0°10N 98°40E 36
Pinios → Greece 39°55N 22°41E 23
Pinjarra Australia 32°37S 115°52E 61
Pink Mountain Canada 57°3N 122°52W 70
Pinnacle Is. = Senkaku-Shotō E. China Sea 25°45N 123°30E 31
Pinnacles ⌓ U.S.A. 36°25N 121°12W 78
Pinnaroo Australia 35°17S 140°53E 63
Pinon Hills U.S.A. 34°26N 117°39W 79
Pinos Mexico 22°18N 101°34W 86
Pinos, Mt. U.S.A. 34°49N 119°8W 79
Pinos Pt. U.S.A. 36°38N 121°57W 77
Pinrang Indonesia 3°46S 119°41E 37
Pins, Pte. aux Canada 42°15N 81°51W 82
Pinsk Belarus 52°10N 26°1E 17
Pintados Chile 20°35S 69°40W 92
Pinyug Russia 60°5N 48°0E 18
Pioche U.S.A. 37°56N 114°27W 77
Piombino Italy 42°55N 10°32E 22
Pioner, Ostrov Russia 79°50N 92°0E 29
Piopiotaki = Milford Sd. N.Z. 44°41S 167°47E 59
Piorini, L. Brazil 3°15S 62°35W 92
Piotrków Trybunalski Poland 51°23N 19°43E 17
Pīp Iran 26°45N 60°10E 45
Pipar India 26°25N 73°31E 42
Pipar Road India 26°27N 73°27E 42
Piparia Mad. P., India 22°45N 78°23E 42

paria *Mad. P., India* 21°49N 77°37E **42** J7
pestone *U.S.A.* 44°0N 96°19W **80** D5
pestone → *Canada* 52°53N 89°23W **72** B2
pestone Cr. → *Canada* 49°38N 100°15W **71** D8
plan *Pakistan* 32°17N 71°21E **42** C4
ploda *India* 23°37N 74°56E **42** H6
pmuacan, Rés. *Canada* 49°45N 70°30W **73** C5
ppingarra *Australia* 20°27S 118°42E **60** D2
qua *U.S.A.* 40°9N 84°15W **81** E11
quiri → *Brazil* 24°3S 54°14W **95** A5
r Panjal Range *India* 32°30N 76°50E **42** C7
r Sohrāb *Iran* 25°44N 60°54E **45** E9
racicaba *Brazil* 22°45S 47°40W **95** A6
racuruca *Brazil* 3°50S 41°50W **93** D10
rae *Tahiti* 17°31S 149°32W **59** d
ræus = Pireas *Greece* 37°57N 23°42E **23** F10
raiévs = Pireas *Greece* 37°57N 23°42E **23** F10
rajuí *Brazil* 21°59S 49°29W **95** A6
ram I. *India* 21°36N 72°21E **42** J5
rané *Argentina* 25°42S 59°6W **94** B4
rānshahr *Iran* 36°41N 45°8E **44** B5
rapora *Brazil* 17°20S 44°56W **93** G10
rawa *India* 24°10N 76°2E **42** G7
reas *Greece* 37°57N 23°42E **23** F10
rgos *Greece* 37°40N 21°27E **23** F9
ribebuy *Paraguay* 25°26S 57°2W **94** B4
rimapun *Indonesia* 6°20S 138°24E **37** F9
rin Planina *Bulgaria* 41°40N 23°30E **23** D10
rineos = Pyrénées *Europe* 42°45N 0°18E **20** E4
ripiri *Brazil* 4°15S 41°46W **93** D10
rlangimpi *Australia* 11°24S 130°26E **60** B5
rmasens *Germany* 49°12N 7°36E **16** D4
rot *Serbia* 43°9N 22°33E **23** C10
ru *Indonesia* 3°4S 128°12E **37** E7
ru *U.S.A.* 34°25N 118°48W **79** L8
sa *Italy* 43°43N 10°23E **22** C4
sagua *Chile* 19°40S 70°15W **92** G4
sco *Peru* 13°50S 76°12W **92** F3
sek *Czech Rep.* 49°19N 14°10E **16** D8
shan *China* 37°30N 78°33E **32** D4
shīn *Iran* 26°6N 61°47E **45** E9
shin *Pakistan* 30°35N 67°0E **42** D2
shin Lora → *Pakistan* 29°9N 64°5E **42** E1
sing *Indonesia* 5°8S 121°53E **37** F6
smo Beach *U.S.A.* 35°9N 120°38W **79** K6
ssis, Cerro *Argentina* 27°45S 68°48W **94** B2
ssouri *Cyprus* 34°40N 32°42E **25** E11
stóia *Italy* 43°55N 10°54E **22** C4
stol B. *Canada* 62°25N 92°37W **71** A10
suerga → *Spain* 41°33N 4°52W **21** B3
t → *U.S.A.* 40°47N 122°6W **76** F2
tarpunga L. *Australia* 34°24S 143°30E **63** E3
tcairn I. *Pac. Oc.* 25°5S 130°5W **65** K14
tch L. *Trin. & Tob.* 10°12N 61°39W **93** K15
teå *Sweden* 65°20N 21°25E **8** D19
teälven → *Sweden* 65°20N 21°25E **8** D19
teşti *Romania* 44°52N 24°54E **17** F13
thapuram *India* 17°10N 82°15E **41** L13
thara *Australia* 30°20S 116°35E **61** F2
thoragarh *India* 29°35N 80°13E **43** E9
thoro *Pakistan* 25°31N 69°23E **42** G3
tlochry *U.K.* 56°42N 3°44W **11** E5
tsilia *Cyprus* 34°55N 33°0E **25** E12
tt I. *Canada* 53°30N 129°50W **70** C3
ttsburg *Calif., U.S.A.* 38°2N 121°53W **78** G5
ttsburg *Kans., U.S.A.* 37°25N 94°42W **80** G6
ttsburg *Tex., U.S.A.* 33°0N 94°59W **84** E7
ttsburgh *U.S.A.* 40°26N 79°58W **82** F5
ttsfield *Ill., U.S.A.* 39°36N 90°49W **80** F8
ttsfield *Maine, U.S.A.* 44°47N 69°23W **81** C19
ttsfield *Mass., U.S.A.* 42°27N 73°15W **83** D11
ttsfield *N.H., U.S.A.* 43°18N 71°20W **83** C13
ttston *U.S.A.* 41°19N 75°47W **83** E9
ttsworth *Australia* 27°41S 151°37E **63** D5
turi → *Australia* 22°35S 138°30E **62** C2
ura *Peru* 5°15S 80°38W **92** E2
xley *U.S.A.* 35°58N 119°18W **78** K7
zhou = Pei Xian *China* 34°44N 116°55E **34** G9
zhou *China* 34°18N 117°57E **35** G9
acentia *Canada* 47°20N 54°0W **73** C9
acentia B. *Canada* 47°0N 54°40W **73** C9
acerville *U.S.A.* 38°44N 120°48W **78** G6
acetas *Cuba* 22°15N 79°44W **88** B4
aine des Jarres *Laos* 19°27N 103°10E **38** C4
ainfield *N.J., U.S.A.* 40°37N 74°25W **83** F10
ainfield *Ohio, U.S.A.* 40°13N 81°43W **82** F3
ainfield *Vt., U.S.A.* 44°17N 72°26W **83** B12
ains *Mont., U.S.A.* 47°28N 114°53W **76** C6
ains *Tex., U.S.A.* 33°11N 102°50W **84** E3
ainview *Nebr., U.S.A.* 42°21N 97°47W **80** D5
ainview *Tex., U.S.A.* 34°11N 101°43W **84** D4
ainwell *U.S.A.* 42°27N 85°38W **81** D11
aistow *U.S.A.* 42°50N 71°6W **83** D13
aka, Akra *Greece* 35°11N 26°19E **25** D8
ana Cays *Bahamas* 22°38N 73°30W **89** B5
anada *U.S.A.* 37°16N 120°19W **78** H6
ano *U.S.A.* 33°1N 96°42W **84** E6
ant City *U.S.A.* 28°1N 82°7W **85** G13
aquemine *U.S.A.* 30°17N 91°14W **84** F9
asencia *Spain* 40°3N 6°8W **21** B2
aster City *U.S.A.* 32°47N 115°51W **79** N11
aster Rock *Canada* 46°53N 67°22W **73** C6
astun *Russia* 44°45N 136°19E **30** B8
ata, Río de la → *S. Amer.* 34°45S 57°30W **94** C4
átani → *Italy* 37°23N 13°16E **22** F5
atanos *Greece* 35°28N 23°33E **25** D5
atte *U.S.A.* 43°23N 98°51W **80** D4
atte → *Mo., U.S.A.* 39°16N 94°50W **80** F6
atte → *Nebr., U.S.A.* 41°4N 95°53W **80** E6
atteville *U.S.A.* 42°44N 90°29W **80** D8
attsburgh *U.S.A.* 44°42N 73°28W **83** B11
attsmouth *U.S.A.* 41°1N 95°53W **80** E6
auen *Germany* 50°30N 12°8E **16** C7
avinas *Latvia* 56°35N 25°46E **9** H21
aya Blanca *Canary Is.* 28°55N 13°37W **24** F6
aya Blanca Sur *Canary Is.* 28°51N 13°50W **24** F6
Playa de las Americas *Canary Is.* 28°5N 16°43W **24** F3
Playa de Mogán *Canary Is.* 27°48N 15°47W **24** G4
Playa del Carmen *Mexico* 20°37N 87°4W **87** C7
Playa del Inglés *Canary Is.* 27°45N 15°33W **24** G4
Playa Esmerelda *Canary Is.* 28°8N 14°16W **24** F5
Playgreen L. *Canada* 54°0N 98°15W **71** C9
Pleasant Bay *Canada* 46°51N 60°48W **73** C7
Pleasant Hill *U.S.A.* 37°57N 122°4W **78** H4
Pleasant Mount *U.S.A.* 41°44N 75°26W **83** E9
Pleasanton *Calif., U.S.A.* 37°39N 121°52W **78** H5
Pleasanton *Tex., U.S.A.* 28°58N 98°29W **84** G5
Pleasantville *N.J., U.S.A.* 39°24N 74°32W **81** F16
Pleasantville *Pa., U.S.A.* 41°35N 79°34W **82** E5
Plei Ku *Vietnam* 13°57N 108°0E **38** F7
Plenty → *Australia* 23°25S 136°31E **62** C2
Plenty, B. of *N.Z.* 37°45S 177°0E **59** B6
Plentywood *U.S.A.* 48°47N 104°34W **76** B11
Plesetsk *Russia* 62°43N 40°20E **18** B7
Plessisville *Canada* 46°14N 71°47W **73** C5
Plétipi, L. *Canada* 51°44N 70°6W **73** B5
Pleven *Bulgaria* 43°26N 24°37E **23** C11
Plevlja *Montenegro* 43°21N 19°21E **23** C8
Plevna *Canada* 44°58N 76°59W **82** B8
Płock *Poland* 52°32N 19°40E **17** B10
Plöckenstein *Germany* 48°46N 13°51E **16** D7
Ploieşti *Romania* 44°57N 26°5E **17** F14
Plonge, Lac la *Canada* 55°8N 107°20W **71** B7
Plovdiv *Bulgaria* 42°8N 24°44E **23** C11
Plover Cove Res. *China* 22°28N 114°15E **33** a
Plum *U.S.A.* 40°29N 79°47W **82** F5
Plum I. *U.S.A.* 41°11N 72°12W **83** E12
Plumas *U.S.A.* 39°45N 120°4W **78** F6
Plummer *U.S.A.* 47°20N 116°53W **76** C5
Plumridge Lakes *Australia* 29°26S 125°16E **61** E4
Plumtree *Zimbabwe* 20°27S 27°55E **55** G2
Plungė *Lithuania* 55°53N 21°59E **9** J19
Plymouth *Trin. & Tob.* 11°14N 60°48W **93** J16
Plymouth *U.K.* 50°22N 4°10W **13** G3
Plymouth *Calif., U.S.A.* 38°29N 120°51W **78** G6
Plymouth *Ind., U.S.A.* 41°21N 86°19W **80** E10
Plymouth *Mass., U.S.A.* 41°57N 70°40W **83** E14
Plymouth *N.C., U.S.A.* 35°52N 76°43W **85** D16
Plymouth *N.H., U.S.A.* 43°46N 71°41W **83** C13
Plymouth *Pa., U.S.A.* 41°14N 75°57W **83** E9
Plymouth *Wis., U.S.A.* 43°45N 87°59W **80** D10
Plympton-Wyoming = Wyoming *Canada* 42°57N 82°7W **82** D2
Plynlimon = Pumlumon Fawr *U.K.* 52°28N 3°46W **13** E4
Plyusa *Russia* 58°28N 29°27E **9** G23
Plzeň *Czech Rep.* 49°45N 13°22E **16** D7
Pmere Nyente ☼ *Australia* 24°15S 136°10E **62** C2
Po → *Italy* 44°57N 12°4E **22** B5
Po Hai = Bo Hai *China* 39°0N 119°0E **35** E10
Po Toi *China* 22°10N 114°16E **33** a
Pobeda *Russia* 65°12N 146°12E **29** C15
Pobedy, Pik *Kyrgyzstan* 42°0N 79°58E **32** C4
Pocahontas *Ark., U.S.A.* 36°16N 90°58W **85** C9
Pocahontas *Iowa, U.S.A.* 42°44N 94°40W **80** D6
Pocatello *U.S.A.* 42°52N 112°27W **76** E7
Pocomoke City *U.S.A.* 38°5N 75°34W **81** F16
Poços de Caldas *Brazil* 21°50S 46°33W **95** A6
Podgorica *Montenegro* 42°30N 19°19E **23** C8
Podilska Vysochyna *Ukraine* 49°0N 28°0E **17** D14
Podolsk *Russia* 55°25N 37°30E **18** C6
Podporozhye *Russia* 60°55N 34°2E **18** B5
Pofadder *S. Africa* 29°10S 19°22E **56** D2
Pogranichnyy *Russia* 44°25N 131°24E **30** B5
Poh *Indonesia* 0°46S 122°51E **37** E6
Pohang *S. Korea* 36°1N 129°23E **35** F15
Pohjanmaa *Finland* 62°58N 22°50E **8** E20
Pohnpei *Micronesia* 6°55N 158°10E **64** G7
Pohri *India* 25°32N 77°22E **42** G6
Poinsett, C. *Antarctica* 65°42S 113°18E **5** C8
Point Arena *U.S.A.* 38°55N 123°41W **78** G3
Point Baker *U.S.A.* 56°21N 133°37W **70** B2
Point Fortin *Trin. & Tob.* 10°10N 61°42W **93** K15
Point Hope *U.S.A.* 68°21N 166°47W **74** B6
Point L. *Canada* 65°15N 113°4W **68** D9
Point Lisas *Trin. & Tob.* 10°24N 61°29W **93** K15
Point Pedro *Sri Lanka* 9°50N 80°15E **40** Q12
Point Pelee △ *Canada* 41°57N 82°31W **82** E2
Point Pleasant *N.J., U.S.A.* 40°5N 74°4W **83** F10
Point Pleasant *W. Va., U.S.A.* 38°51N 82°8W **81** F12
Point Reyes ⌓ *U.S.A.* 38°10N 122°55W **78** G2
Pointe-à-Pitre *Guadeloupe* 16°10N 61°32W **88** b
Pointe-au-Pic = La Malbaie *Canada* 47°40N 70°10W **73** C5
Pointe-Claire *Canada* 45°26N 73°50W **83** A11
Pointe-Gatineau *Canada* 45°27N 75°41W **83** A9
Pointe-Noire *Congo* 4°48S 11°53E **52** E2
Pointe-Noire *Guadeloupe* 16°14N 61°47W **88** b
Poipet *Cambodia* 13°39N 102°33E **38** F4
Poisonbush Ra. *Australia* 22°30S 121°30E **60** D3
Poissonnier Pt. *Australia* 19°57S 119°10E **60** C2
Poitiers *France* 46°35N 0°20E **20** C4
Poitou *France* 46°40N 0°10W **20** C3
Pojoaque *U.S.A.* 35°54N 106°1W **77** J10
Pokaran *India* 27°0N 71°50E **42** F4
Pokataroo *Australia* 29°30S 148°36E **63** D4
Pokhara *Nepal* 28°14N 83°58E **43** E10
Poko *Dem. Rep. of the Congo* 3°7N 26°52E **54** B2
Pokrovsk *Russia* 61°29N 129°0E **29** C13
Pola = Pula *Croatia* 44°54N 13°57E **16** F7
Polacca *U.S.A.* 35°50N 110°23W **77** J8
Polan *Iran* 25°30N 61°10E **45** E9
Poland *U.S.A.* 43°13N 75°3W **83** C9
Poland ■ *Europe* 52°0N 20°0E **17** C10
Polar Bear ⌓ *Canada* 55°0N 83°45W **72** A2
Polatsk *Belarus* 55°30N 28°50E **9** J23
Polcura *Chile* 37°17S 71°43W **94** D1
Polesye = Pripet Marshes *Europe* 52°10N 27°10E **17** B15
Polevskoy *Russia* 56°26N 60°11E **18** C11
Police *Poland* 53°33N 14°33E **16** B8
Police, Pte. *Seychelles* 4°51S 55°32E **53** b
Poligiros *Greece* 40°23N 23°25E **23** D10
Polillo Is. *Phil.* 14°56N 122°0E **37** B6
Polis *Cyprus* 35°2N 32°26E **25** D11
Polk *U.S.A.* 41°22N 79°56W **82** E5
Pollachi *India* 10°35N 77°0E **40** P10
Pollença *Spain* 39°54N 3°1E **24** B10
Pollença, B. de *Spain* 39°53N 3°8E **24** B10
Polnovat *Russia* 63°50N 65°54E **28** C7
Polokwane *S. Africa* 23°54S 29°25E **57** C4
Polokwane → *S. Africa* 22°25S 30°5E **57** C5
Polonne *Ukraine* 50°6N 27°30E **17** C14
Polonnoye = Polonne *Ukraine* 50°6N 27°30E **17** C14
Polson *U.S.A.* 47°41N 114°9W **76** C6
Poltava *Ukraine* 49°35N 34°35E **19** E5
Põltsamaa *Estonia* 58°39N 25°58E **9** G21
Polunochnoye *Russia* 60°52N 60°25E **28** C7
Põlva *Estonia* 58°3N 27°3E **9** G22
Polyarny *Russia* 69°8N 33°20E **8** B25
Polyarnyye Zori *Russia* 67°22N 32°30E **8** C25
Polynesia *Pac. Oc.* 10°0S 162°0W **65** F11
Polynésie française = French Polynesia ⊠ *Pac. Oc.* 20°0S 145°0W **65** J13
Pombal *Portugal* 39°55N 8°40W **21** C1
Pombia *Greece* 35°0N 24°51E **25** E6
Pomene *Mozam.* 22°53S 35°33E **57** C6
Pomeroy *Ohio, U.S.A.* 39°2N 82°2W **81** F12
Pomeroy *Wash., U.S.A.* 46°28N 117°36W **76** C5
Pomézia *Italy* 41°40N 12°30E **22** D5
Pomona *Australia* 26°22S 152°52E **63** D5
Pomona *U.S.A.* 34°4N 117°45W **79** L9
Pomorskie, Pojezierze *Poland* 53°40N 16°37E **17** B9
Pomos *Cyprus* 35°9N 32°33E **25** D11
Pomos, C. *Cyprus* 35°10N 32°33E **25** D11
Pompano Beach *U.S.A.* 26°14N 80°7W **85** H14
Pompeys Pillar *U.S.A.* 45°59N 107°57W **76** D10
Pompeys Pillar ⌓ *U.S.A.* 46°0N 108°0W **76** D10
Pompton Lakes *U.S.A.* 41°0N 74°17W **83** F10
Ponape = Pohnpei *Micronesia* 6°55N 158°10E **64** G7
Ponask L. *Canada* 54°0N 92°41W **72** B1
Ponca *U.S.A.* 42°34N 96°43W **80** D5
Ponca City *U.S.A.* 36°42N 97°5W **84** C6
Ponce *Puerto Rico* 18°1N 66°37W **89** d
Ponchatoula *U.S.A.* 30°26N 90°26W **85** F9
Poncheville, L. *Canada* 50°10N 76°55W **72** B4
Pond *U.S.A.* 35°43N 119°20W **79** K7
Pond Inlet *Canada* 72°40N 77°0W **69** C16
Pondicherry = Puducherry *India* 11°59N 79°50E **40** P11
Ponds, I. of *Canada* 53°27N 55°52W **73** B8
Ponferrada *Spain* 42°32N 6°35W **21** A2
Ponnani *India* 10°45N 75°59E **40** P9
Ponoka *Canada* 52°42N 113°40W **70** C6
Ponorogo *Indonesia* 7°52S 111°27E **37** G14
Ponoy *Russia* 67°0N 41°13E **18** A7
Ponoy → *Russia* 66°59N 41°17E **18** A7
Ponta Delgada *Azores* 37°44N 25°40W **50** a
Ponta do Sol *Madeira* 32°42N 17°7W **24** D2
Ponta Grossa *Brazil* 25°7S 50°10W **95** B5
Ponta Porã *Brazil* 22°20S 55°35W **95** A4
Pontarlier *France* 46°54N 6°20E **20** C7
Pontchartrain, L. *U.S.A.* 30°5N 90°5W **85** F9
Ponte do Pungué *Mozam.* 19°30S 34°33E **55** F3
Ponte Nova *Brazil* 20°25S 42°54W **95** A7
Ponteix *Canada* 49°46N 107°29W **71** D7
Pontevedra *Spain* 42°26N 8°40W **21** A1
Pontiac *Ill., U.S.A.* 40°53N 88°38W **80** E9
Pontiac *Mich., U.S.A.* 42°38N 83°18W **81** D12
Pontian Kechil *Malaysia* 1°29N 103°23E **39** d
Pontianak *Indonesia* 0°3S 109°15E **36** E3
Pontine Is. = Ponziane, Ísole *Italy* 40°55N 12°57E **22** D5
Pontine Mts. = Kuzey Anadolu Dağları *Turkey* 41°0N 36°45E **19** F6
Pontivy *France* 48°5N 2°58W **20** B2
Pontoise *France* 49°3N 2°5E **20** B5
Ponton → *Canada* 58°27N 116°11W **70** B5
Pontypool *Canada* 44°6N 78°38W **82** B6
Pontypool *U.K.* 51°42N 3°2W **13** F4
Ponziane, Ísole *Italy* 40°55N 12°57E **22** D5
Poochera *Australia* 32°43S 134°51E **63** E1
Poole *U.K.* 50°43N 1°59W **13** G6
Poole ☐ *U.K.* 50°43N 1°59W **13** G6
Poona = Pune *India* 18°29N 73°57E **40** K8
Pooncarie *Australia* 33°22S 142°31E **63** E3
Poopelloe L. *Australia* 31°40S 144°0E **63** E3
Poopó, L. de *Bolivia* 18°30S 67°35W **92** G5
Popayán *Colombia* 2°27N 76°36W **92** C3
Poperinge *Belgium* 50°51N 2°42E **15** D2
Popiltah L. *Australia* 33°10S 141°42E **63** E3
Popio L. *Australia* 33°10S 141°52E **63** E3
Poplar *U.S.A.* 48°7N 105°12W **76** B11
Poplar → *Canada* 53°0N 97°19W **71** C9
Poplar Bluff *U.S.A.* 36°46N 90°24W **80** G8
Poplarville *U.S.A.* 30°51N 89°32W **85** F10
Popocatépetl, Volcán *Mexico* 19°2N 98°38W **87** D5
Popokabaka *Dem. Rep. of the Congo* 5°41S 16°40E **52** F3
Poprad *Slovak Rep.* 49°3N 20°18E **17** D11
Porali → *Pakistan* 25°58N 66°26E **42** G2
Porbandar *India* 21°44N 69°43E **42** J3
Porcher I. *Canada* 53°50N 130°30W **70** C2
Porcupine → *Canada* 59°11N 104°46W **71** B8
Porcupine → *U.S.A.* 66°34N 145°19W **68** D2
Porcupine Gorge △ *Australia* 20°22S 144°26E **62** C3
Pordenone *Italy* 45°57N 12°39E **22** B5
Pori *Finland* 61°29N 21°48E **8** F19
Porkhov *Russia* 57°45N 29°38E **9** H23
Porlamar *Venezuela* 10°57N 63°51W **92** A6
Pormpuraaw *Australia* 14°59S 141°26E **62** A3
Pormpuraaw ☼ *Australia* 14°55S 141°47E **62** A3
Poronaysk *Russia* 49°13N 143°0E **33** B17
Poroshiri-Dake *Japan* 42°41N 142°52E **30** C11
Poroto Mts. *Tanzania* 9°0S 33°30E **55** D3
Porpoise B. *Antarctica* 66°0S 127°0E **5** C9
Porreres *Spain* 39°31N 3°2E **24** B10
Porsangerfjorden *Norway* 70°40N 25°40E **8** A21
Porsgrunn *Norway* 59°10N 9°40E **9** G13
Port Alberni *Canada* 49°14N 124°50W **70** D4
Port Alfred *S. Africa* 33°36S 26°55E **56** E4
Port Alice *Canada* 50°20N 127°25W **70** C3
Port Allegany *U.S.A.* 41°48N 78°17W **82** E6
Port Allen *U.S.A.* 30°27N 91°12W **84** F9
Port Alma *Australia* 23°38S 150°53E **62** C5
Port Alma *Canada* 42°10N 82°14W **82** D2
Port Angeles *U.S.A.* 48°7N 123°27W **78** B3
Port Antonio *Jamaica* 18°10N 76°26W **88** a
Port Aransas *U.S.A.* 27°50N 97°4W **84** H6
Port Arthur *Australia* 43°7S 147°50E **63** G4
Port Arthur *U.S.A.* 29°54N 93°56W **84** G8
Port au Choix *Canada* 50°43N 57°22W **73** B8
Port au Port B. *Canada* 48°40N 58°50W **73** C8
Port-au-Prince *Haiti* 18°40N 72°20W **89** C5
Port Augusta *Australia* 32°30S 137°50E **63** E2
Port Austin *U.S.A.* 44°3N 83°1W **82** B2
Port Bell *Uganda* 0°18N 32°35E **54** B3
Port Bergé Vaovao *Madag.* 15°33S 47°40E **57** B8
Port Blandford *Canada* 48°20N 54°10W **73** C9
Port Bradshaw *Australia* 12°30S 137°20E **62** A2
Port Broughton *Australia* 33°37S 137°56E **63** E2
Port Bruce *Canada* 42°39N 81°0W **82** D4
Port Burwell *Canada* 42°40N 80°48W **82** D4
Port Canning *India* 22°23N 88°40E **43** H13
Port Carling *Canada* 45°7N 79°35W **82** A5
Port-Cartier *Canada* 50°2N 66°50W **73** B6
Port Chalmers *N.Z.* 45°49S 170°30E **59** F3
Port Charlotte *U.S.A.* 26°59N 82°6W **85** H13
Port Chester *U.S.A.* 41°0N 73°40W **83** F11
Port Clements *Canada* 53°40N 132°10W **70** C2
Port Clinton *U.S.A.* 41°31N 82°56W **81** E12
Port Colborne *Canada* 42°50N 79°10W **82** D5
Port Coquitlam *Canada* 49°15N 122°45W **78** A4
Port Credit *Canada* 43°33N 79°35W **82** C5
Port Curtis *Australia* 23°57S 151°20E **62** C5
Port d'Alcúdia *Spain* 39°50N 3°7E **24** B10
Port Dalhousie *Canada* 43°13N 79°16W **82** C5
Port d'Andratx *Spain* 39°32N 2°23E **24** B9
Port Darwin *Australia* 12°24S 130°45E **60** B5
Port Darwin *Falk. Is.* 51°50S 59°0W **96** G5
Port Davey *Australia* 43°16S 145°55E **63** G4
Port-de-Paix *Haiti* 19°50N 72°50W **89** C5
Port de Pollença *Spain* 39°54N 3°4E **24** B10
Port de Sóller *Spain* 39°48N 2°42E **24** B9
Port Dickson *Malaysia* 2°30N 101°49E **39** L3
Port Douglas *Australia* 16°30S 145°30E **62** B4
Port Dover *Canada* 42°47N 80°12W **82** D4
Port Edward *Canada* 54°12N 130°10W **70** C2
Port Elgin *Canada* 44°25N 81°25W **82** B3
Port Elizabeth *S. Africa* 33°58S 25°40E **56** E4
Port Ellen *U.K.* 55°38N 6°11W **11** F2
Port Erin *I. of Man* 54°5N 4°45W **12** C3
Port Essington *Australia* 11°15S 132°10E **60** B5
Port Ewen *U.S.A.* 41°54N 73°59W **83** E11
Port Fairy *Australia* 38°22S 142°12E **63** F3
Port Gamble *U.S.A.* 47°51N 122°34W **78** C4
Port-Gentil *Gabon* 0°40S 8°50E **52** E1
Port Germein *Australia* 33°1S 138°1E **63** E2
Port Ghalib *Egypt* 25°20N 34°50E **44** E2
Port Gibson *U.S.A.* 31°58N 90°59W **84** F9
Port Glasgow *U.K.* 55°56N 4°41W **11** F4
Port Harcourt *Nigeria* 4°40N 7°10E **50** H7
Port Hardy *Canada* 50°41N 127°30W **70** C3
Port Harrison = Inukjuak *Canada* 58°25N 78°15W **69** F16
Port Hawkesbury *Canada* 45°36N 61°22W **73** C7
Port Hedland *Australia* 20°25S 118°35E **60** D2
Port Henry *U.S.A.* 44°3N 73°28W **83** B11
Port Hood *Canada* 46°0N 61°32W **73** C7
Port Hope *Canada* 43°56N 78°20W **82** C6
Port Hope *U.S.A.* 43°57N 82°43W **82** C2
Port Hope Simpson *Canada* 52°33N 56°18W **73** B8
Port Hueneme *U.S.A.* 34°7N 119°12W **79** L7
Port Huron *U.S.A.* 42°58N 82°26W **82** D2
Port Jefferson *U.S.A.* 40°57N 73°3W **83** F11
Port Jervis *U.S.A.* 41°22N 74°41W **83** E10
Port Kenny *Australia* 33°10S 134°41E **63** E1
Port Lairge = Waterford *Ireland* 52°15N 7°8W **10** D4
Port Laoise *Ireland* 53°2N 7°18W **10** C4
Port Lavaca *U.S.A.* 28°37N 96°38W **84** G6
Port Leyden *U.S.A.* 43°35N 75°21W **83** C9
Port Lincoln *Australia* 34°42S 135°52E **63** E2
Port Loko *S. Leone* 8°48N 12°46W **50** G3
Port-Louis *Guadeloupe* 16°28N 61°32W **88** b
Port Louis *Mauritius* 20°10S 57°30E **53** d
Port MacDonnell *Australia* 38°5S 140°48E **63** F3
Port McNeill *Canada* 50°35N 127°6W **70** C3
Port Macquarie *Australia* 31°25S 152°25E **63** E5
Port Maria *Jamaica* 18°22N 76°54W **88** a
Port Matilda *U.S.A.* 40°48N 78°3W **82** F6
Port McNicoll *Canada* 44°44N 79°48W **82** B5
Port Mellon *Canada* 49°32N 123°31W **70** D4
Port-Menier *Canada* 49°51N 64°15W **73** C7
Port Moody *Canada* 49°17N 122°51W **78** A4
Port Morant *Jamaica* 17°54N 76°19W **88** a
Port Moresby *Papua N. G.* 9°24S 147°8E **58** B7
Port Musgrave *Australia* 11°55S 141°50E **62** A3
Port Neches *U.S.A.* 30°0N 93°59W **84** G8
Port Nolloth *S. Africa* 29°17S 16°52E **56** D2
Port Nouveau-Québec = Kangiqsualujjuaq *Canada* 58°30N 65°59W **69** F18
Port of Climax *Canada* 49°10N 108°20W **71** D7
Port of Coronach *Canada* 49°7N 105°31W **71** D7
Port of Spain *Trin. & Tob.* 10°40N 61°31W **93** K15
Port Orange *U.S.A.* 29°9N 80°59W **85** G14
Port Orchard *U.S.A.* 47°32N 122°38W **78** C4
Port Orford *U.S.A.* 42°45N 124°30W **76** E1
Port Pegasus *N.Z.* 47°12S 167°41E **59** G1
Port Perry *Canada* 44°6N 78°56W **82** B6
Port Phillip B. *Australia* 38°10S 144°50E **63** F3
Port Pirie *Australia* 33°10S 138°1E **63** E2
Port Renfrew *Canada* 48°30N 124°20W **78** B2
Port Roper *Australia* 14°45S 135°25E **62** A2
Port Rowan *Canada* 42°40N 80°30W **82** D4
Port Safaga = Bûr Safâga *Egypt* 26°43N 33°57E **44** E2
Port Said = Bûr Sa'îd *Egypt* 31°16N 32°18E **51** B12
Port St. Joe *U.S.A.* 29°49N 85°18W **85** G12
Port St. Johns = Umzimvubu *S. Africa* 31°38S 29°33E **57** E4
Port St. Lucie *U.S.A.* 27°18N 80°21W **85** H14
Port Sanilac *U.S.A.* 43°26N 82°33W **82** C2
Port Severn *Canada* 44°48N 79°43W **82** B5
Port Shepstone *S. Africa* 30°44S 30°28E **57** E5
Port Simpson *Canada* 54°30N 130°20W **70** C2
Port Stanley = Stanley *Falk. Is.* 51°40S 59°51W **96** G5
Port Stanley *Canada* 42°40N 81°10W **82** D3
Port Sudan = Bûr Sûdân *Sudan* 19°32N 37°9E **51** E13
Port Sulphur *U.S.A.* 29°29N 89°42W **85** G10
Port Talbot *U.K.* 51°35N 3°47W **13** F4
Port Townsend *U.S.A.* 48°7N 122°45W **78** B4
Port-Vendres *France* 42°32N 3°8E **20** E5
Port Vila *Vanuatu* 17°45S 168°18E **58** C9
Port Vladimir *Russia* 69°25N 33°6E **8** B25
Port Wakefield *Australia* 34°12S 138°10E **63** E2
Port Washington *U.S.A.* 43°23N 87°53W **80** D10
Porta Orientalis *Romania* 45°6N 22°18E **17** F12
Portacloy *Ireland* 54°20N 9°46W **10** B2
Portadown *U.K.* 54°25N 6°27W **10** B5
Portaferry *U.K.* 54°23N 5°33W **10** B6
Portage *Pa., U.S.A.* 40°23N 78°41W **82** F6
Portage *Wis., U.S.A.* 43°33N 89°28W **80** D9
Portage la Prairie *Canada* 49°58N 98°18W **71** D9
Portageville *U.S.A.* 36°26N 89°42W **80** G9
Portalegre *Portugal* 39°19N 7°25W **21** C2
Portales *U.S.A.* 34°11N 103°20W **77** J12
Portarlington *Ireland* 53°9N 7°14W **10** C4
Portbou *Spain* 42°25N 3°9E **21** A7
Porter L. *N.W.T., Canada* 61°41N 108°5W **71** A7
Porter L. *Sask., Canada* 56°20N 107°20W **71** B7
Porterville *S. Africa* 33°0S 19°0E **56** E2
Porterville *U.S.A.* 36°4N 119°1W **78** J8
Porthcawl *U.K.* 51°29N 3°42W **13** F4
Porthill *U.S.A.* 48°59N 116°30W **76** B5
Porthmadog *U.K.* 52°55N 4°8W **12** E3
Portile de Fier *Europe* 44°44N 22°30E **17** F12
Portimão *Portugal* 37°8N 8°32W **21** D1
Portishead *U.K.* 51°29N 2°46W **13** F5
Portknockie *U.K.* 57°42N 2°51W **11** D6
Portland *N.S.W., Australia* 33°20S 150°0E **63** E5
Portland *Vic., Australia* 38°20S 141°35E **63** F3
Portland *Canada* 44°42N 76°12W **83** B8
Portland *Conn., U.S.A.* 41°34N 72°38W **83** E12
Portland *Maine, U.S.A.* 43°39N 70°16W **81** D18
Portland *Mich., U.S.A.* 42°52N 84°54W **81** D11
Portland *Oreg., U.S.A.* 45°32N 122°37W **78** E4
Portland *Pa., U.S.A.* 40°55N 75°6W **83** F9
Portland *Tex., U.S.A.* 27°53N 97°20W **84** H6
Portland, I. of *U.K.* 50°33N 2°26W **13** G5
Portland B. *Australia* 38°15S 141°45E **63** F3
Portland Bight *Jamaica* 17°52N 77°5W **88** a
Portland Bill *U.K.* 50°31N 2°28W **13** G5
Portland Canal *U.S.A.* 55°56N 130°0W **70** B2
Portland Int. ✈ (PDX) *U.S.A.* 45°35N 122°36W **78** E4
Portland Pt. *Jamaica* 17°42N 77°11W **88** a
Portmadoc = Porthmadog *U.K.* 52°55N 4°8W **12** E3
Portmore *Jamaica* 17°53N 76°53W **88** a
Porto *France* 42°16N 8°42E **20** E8
Porto *Portugal* 41°8N 8°40W **21** B1
Pôrto Alegre *Brazil* 30°5S 51°10W **95** C5
Porto Colom *Spain* 39°26N 3°15E **24** B10
Porto Cristo *Spain* 39°33N 3°20E **24** B10
Porto de Moz *Brazil* 1°41S 52°13W **93** D8
Porto Empédocle *Italy* 37°17N 13°32E **22** F5
Porto Esperança *Brazil* 19°37S 57°29W **92** G7
Porto Franco *Brazil* 6°20S 47°24W **93** E9
Porto Inglês *C. Verde Is.* 15°21N 23°10W **50** b
Porto Mendes *Brazil* 24°30S 54°15W **95** A5
Porto Moniz *Madeira* 32°52N 17°11W **24** D2
Porto Murtinho *Brazil* 21°45S 57°55W **92** H7
Porto Nacional *Brazil* 10°40S 48°30W **93** F9
Porto-Novo *Benin* 6°23N 2°42E **50** G6
Porto Primavera, Represa *Brazil* 22°10S 52°45W **95** A5
Porto Santo, I. de *Madeira* 33°45N 16°25W **50** B2
Porto São José *Brazil* 22°43S 53°10W **95** A5
Porto Seguro *Brazil* 16°26S 39°5W **93** G11
Porto Tórres *Italy* 40°50N 8°24E **22** D3
Porto União *Brazil* 26°10S 51°10W **95** B5
Porto-Vecchio *France* 41°35N 9°16E **20** F8
Pôrto Velho *Brazil* 8°46S 63°54W **92** E6
Porto Walter *Brazil* 8°15S 72°40W **92** E4
Portobelo *Panama* 9°35N 79°42W **88** E4
Portoferráio *Italy* 42°48N 10°20E **22** C4
Portola *U.S.A.* 39°49N 120°28W **78** F6
Portopetro *Spain* 39°22N 3°13E **24** B10

154 Portoscuso
INDEX TO WORLD MAP

Portoscuso Italy 39°12N 8°24E 22 E3
Portoviejo Ecuador 1°7S 80°28W 92 D2
Portpatrick U.K. 54°51N 5°7W 11 G3
Portree U.K. 57°25N 6°12W 11 D2
Portrush U.K. 55°12N 6°40W 10 A5
Portsmouth Dominica 15°34N 61°27W 89 C7
Portsmouth U.K. 50°48N 1°6W 13 G6
Portsmouth N.H., U.S.A. 43°5N 70°45W 83 C14
Portsmouth Ohio, U.S.A. 38°44N 82°57W 81 F12
Portsmouth R.I., U.S.A. 41°36N 71°15W 83 E13
Portsmouth Va., U.S.A. 36°58N 76°23W 81 G15
Portsmouth □ U.K. 50°48N 1°6W 13 G6
Portsoy U.K. 57°41N 2°41W 11 D6
Portstewart U.K. 55°11N 6°43W 10 A5
Porttipahdan tekojärvi Finland 68°5N 26°40E 8 B22
Portugal ■ Europe 40°0N 8°0W 21 C1
Portumna Ireland 53°6N 8°14W 10 C3
Portville U.S.A. 42°3N 78°20W 82 D6
Poruma Australia 10°2S 143°4E 62 a
Porvenir Chile 53°10S 70°16W 96 G2
Posadas Argentina 27°30S 55°50W 95 B4
Posht-e Badam Iran 33°2N 55°23E 45 C7
Poso Indonesia 1°20S 120°55E 37 E6
Poso, Danau Indonesia 1°52S 120°35E 37 E6
Posse Brazil 14°4S 46°18W 93 F9
Possession I. Antarctica 72°4S 172°0E 5 D11
Possum Kingdom L. U.S.A. 32°52N 98°26W 84 E5
Post U.S.A. 33°12N 101°23W 84 E4
Post Falls U.S.A. 47°43N 116°57W 76 C5
Postavy = Pastavy Belarus 55°4N 26°50E 9 J22
Poste-de-la-Baleine = Kuujjuarapik Canada 55°20N 77°35W 72 A4
Postmasburg S. Africa 28°18S 23°5E 56 D3
Postojna Slovenia 45°46N 14°12E 16 F8
Poston U.S.A. 34°0N 114°24W 79 M12
Postville Canada 54°54N 59°47W 73 B8
Posyet Russia 42°39N 130°48E 30 C5
Potchefstroom S. Africa 26°41S 27°7E 56 D4
Poteau U.S.A. 35°3N 94°37W 84 D7
Poteet U.S.A. 29°2N 98°35W 84 G5
Potenza Italy 40°38N 15°48E 22 D6
Poteriteri, L. N.Z. 46°5S 167°10E 59 G1
Potgietersrus = Mokopane S. Africa 24°10S 28°55E 57 C4
Poti Georgia 42°10N 41°38E 19 F7
Potiskum Nigeria 11°39N 11°2E 51 F8
Potomac → U.S.A. 38°0N 76°23W 81 F14
Potosí Bolivia 19°38S 65°50W 92 G5
Potosi Mt. U.S.A. 35°57N 115°29W 79 K11
Pototan Phil. 10°54N 122°38E 37 B6
Potrerillos Chile 26°30S 69°30W 94 B2
Potsdam Germany 52°23N 13°3E 16 B7
Potsdam U.S.A. 44°40N 74°59W 83 B10
Pottersville U.S.A. 43°43N 73°50W 83 C11
Pottstown U.S.A. 40°15N 75°39W 83 F9
Pottsville U.S.A. 40°41N 76°12W 83 F8
Pottuvil Sri Lanka 6°55N 81°50E 40 R12
Pouce Coupé Canada 55°40N 120°10W 70 B4
Poughkeepsie U.S.A. 41°42N 73°56W 83 E11
Poulaphouca Res. Ireland 53°8N 6°30W 10 C5
Poulsbo U.S.A. 47°44N 122°38W 78 C4
Poultney U.S.A. 43°31N 73°14W 83 C11
Poulton-le-Fylde U.K. 53°51N 2°58W 12 D5
Pouso Alegre Brazil 22°14S 45°57W 95 A6
Pouthisat Cambodia 12°34N 103°50E 38 F4
Považská Bystrica Slovak Rep. 49°8N 18°27E 17 D10
Povenets Russia 62°50N 34°50E 18 B5
Poverty B. N.Z. 38°43S 178°2E 59 C7
Póvoa de Varzim Portugal 41°25N 8°46W 21 B1
Povorotnyy, Mys Russia 42°40N 133°2E 30 C6
Povungnituk = Puvirnituq Canada 60°2N 77°10W 69 E16
Powassan Canada 46°5N 79°25W 72 C4
Poway U.S.A. 32°58N 117°2W 79 N9
Powder → U.S.A. 46°45N 105°26W 76 C11
Powder River U.S.A. 43°2N 106°59W 76 E10
Powell U.S.A. 44°45N 108°46W 76 D9
Powell, L. U.S.A. 36°57N 111°29W 77 H8
Powell River Canada 49°50N 124°35W 70 D4
Powers U.S.A. 45°41N 87°32W 80 C10
Pownal U.S.A. 42°45N 73°14W 83 D11
Powys □ U.K. 52°20N 3°20W 13 E4
Poyang Hu China 29°5N 116°20E 33 F12
Poyarkovo Russia 49°36N 128°41E 29 E13
Poza Rica Mexico 20°33N 97°27W 87 C5
Požarevac Serbia 44°35N 21°18E 23 B9
Poznań Poland 52°25N 16°55E 17 B9
Pozo U.S.A. 35°20N 120°24W 79 K6
Pozo Almonte Chile 20°10S 69°50W 92 H5
Pozo Colorado Paraguay 23°30S 58°45W 94 A4
Pozoblanco Spain 38°23N 4°51W 21 C3
Pozzuoli Italy 40°49N 14°7E 22 D6
Prachin Buri Thailand 14°0N 101°25E 38 F3
Prachuap Khirikhan Thailand 11°49N 99°48E 39 G2
Prado Brazil 17°20S 39°13W 93 G11
Prague = Praha Czech Rep. 50°4N 14°25E 16 C8
Praha Czech Rep. 50°4N 14°25E 16 C8
Praia C. Verde Is. 15°2N 23°34W 50 b
Prainha Brazil 1°45S 53°30W 93 D8
Prainha Nova Brazil 7°10S 60°30W 92 E6
Prairie Australia 20°50S 144°35E 62 C3
Prairie City U.S.A. 44°28N 118°43W 76 D4
Prairie Dog Town Fork Red → U.S.A. 34°34N 99°58W 84 D5
Prairie du Chien U.S.A. 43°3N 91°9W 80 D8
Prairies, L. of the Canada 51°16N 101°32W 71 C8
Prambanan △ Indonesia 7°45S 110°28E 37 G14
Pran Buri Thailand 12°23N 99°55E 38 F2
Prapat Indonesia 2°41N 98°58E 36 D1
Praslin Seychelles 4°18S 55°45E 53 b
Prasonisi, Akra Greece 35°42N 27°46E 25 D9
Prata Brazil 19°25S 48°54W 93 G9
Pratabpur India 23°28N 83°15E 43 H10
Pratapgarh Raj., India 24°2N 74°40E 42 G6
Pratapgarh Ut. P., India 25°56N 81°59E 43 G9
Pratas I. = Dongsha Dao S. China Sea 20°45N 116°43E 33 G12
Prato Italy 43°53N 11°6E 22 C4
Pratt U.S.A. 37°39N 98°44W 80 G4
Prattville U.S.A. 32°28N 86°29W 85 E11
Pravia Spain 43°30N 6°12W 21 A2
Praya Indonesia 8°39S 116°17E 36 F5
Prayag = Allahabad India 25°25N 81°58E 43 G9
Preah Vihear Cambodia 14°23N 104°41E 38 E5
Preble U.S.A. 42°44N 76°8W 83 D8
Precipice △ Australia 25°18S 150°5E 63 D5
Precordillera Argentina 30°0S 69°1W 94 C2
Preeceville Canada 51°57N 102°40W 71 C8
Preiļi Latvia 56°18N 26°43E 9 H22
Premont U.S.A. 27°22N 98°7W 84 H5
Prentice U.S.A. 45°33N 90°17W 80 C8
Preobrazheniye Russia 42°54N 133°54E 30 C6
Preparis I. = Pariparit Kyun Burma 14°52N 93°41E 41 M18
Preparis North Channel Ind. Oc. 15°27N 94°5E 41 M18
Preparis South Channel Ind. Oc. 14°33N 93°30E 41 M18
Přerov Czech Rep. 49°28N 17°27E 17 D9
Prescott Canada 44°45N 75°30W 83 B9
Prescott Ariz., U.S.A. 34°33N 112°28W 77 J7
Prescott Ark., U.S.A. 33°48N 93°23W 84 E8
Prescott Valley U.S.A. 34°40N 112°18W 77 J7
Preservation Inlet N.Z. 46°8S 166°35E 59 G1
Presho U.S.A. 43°54N 100°3W 80 D3
Presidencia de la Plaza Argentina 27°0S 59°50W 94 B4
Presidencia Roque Saenz Peña Argentina 26°45S 60°30W 94 B3
Presidente Epitácio Brazil 21°56S 52°6W 93 H8
Presidente Hayes □ Paraguay 24°0S 59°0W 94 A4
Presidente Prudente Brazil 22°5S 51°25W 95 A5
Presidio Mexico 29°29N 104°23W 86 B4
Presidio U.S.A. 29°34N 104°22W 84 G2
Prešov Slovak Rep. 49°0N 21°15E 17 D11
Prespa, L. = Prespansko Jezero Macedonia 40°55N 21°0E 23 D9
Prespansko Jezero Macedonia 40°55N 21°0E 23 D9
Presque I. U.S.A. 42°10N 80°6W 82 D4
Presque Isle U.S.A. 46°41N 68°1W 81 B19
Prestatyn U.K. 53°20N 3°24W 12 D4
Presteigne U.K. 52°17N 3°0W 13 E5
Preston Canada 43°23N 80°21W 82 C4
Preston U.K. 53°46N 2°42W 12 D5
Preston Idaho, U.S.A. 42°6N 111°53W 76 E8
Preston Minn., U.S.A. 43°40N 92°5W 80 D7
Preston, C. Australia 20°51S 116°12E 60 D2
Prestonsburg U.S.A. 37°40N 82°47W 81 G12
Prestwick U.K. 55°29N 4°37W 11 F4
Pretoria S. Africa 25°44S 28°12E 57 D4
Preveza Greece 38°57N 20°45E 23 E9
Prey Veng Cambodia 11°35N 105°29E 39 G5
Pribilof Is. U.S.A. 57°0N 170°0W 74 D6
Příbram Czech Rep. 49°41N 14°2E 16 D8
Price U.S.A. 39°36N 110°49W 76 G8
Price I. Canada 52°23N 128°41W 70 C3
Prichard U.S.A. 30°44N 88°5W 85 F10
Priekule Latvia 56°26N 21°35E 9 H19
Prienai Lithuania 54°38N 23°57E 9 J20
Prieska S. Africa 29°40S 22°42E 56 D3
Priest L. U.S.A. 48°35N 116°52W 76 B5
Priest River U.S.A. 48°11N 116°55W 76 B5
Priest Valley U.S.A. 36°10N 120°39W 78 J6
Prievidza Slovak Rep. 48°46N 18°36E 17 D10
Prikaspiyskaya Nizmennost = Caspian Depression Eurasia 47°0N 48°0E 19 E8
Prilep Macedonia 41°21N 21°32E 23 D9
Priluki = Pryluky Ukraine 50°30N 32°24E 19 D5
Prime Seal I. Australia 40°3S 147°43E 63 G4
Primo Tapia Mexico 32°16N 116°54W 79 N10
Primorskiy Kray □ Russia 45°0N 135°0E 30 B7
Primrose L. Canada 54°55N 109°45W 71 C7
Prince Albert Canada 53°15N 105°50W 71 C7
Prince Albert S. Africa 33°12S 22°2E 56 E3
Prince Albert △ Canada 54°0N 106°25W 71 C7
Prince Albert Mts. Antarctica 76°0S 161°30E 5 D11
Prince Albert Pen. Canada 72°30N 116°0W 68 C8
Prince Albert Sd. Canada 70°25N 115°0W 68 C9
Prince Alfred, C. Canada 74°20N 124°40W 4 B1
Prince Charles I. Canada 67°47N 76°12W 69 D16
Prince Charles Mts. Antarctica 72°0S 67°0E 5 D6
Prince Edward Fracture Zone Ind. Oc. 46°0S 35°0E 5 A4
Prince Edward I. □ Canada 46°20N 63°20W 73 C7
Prince Edward Is. Ind. Oc. 46°35S 38°0E 3 G11
Prince George Canada 53°55N 122°50W 70 C4
Prince of Wales, C. U.S.A. 65°36N 168°5W 74 B6
Prince of Wales I. Australia 10°40S 142°10E 62 A3
Prince of Wales I. Canada 73°0N 99°0W 68 C12
Prince of Wales I. U.S.A. 55°47N 132°50W 68 F5
Prince of Wales Icefield Canada 78°15N 79°0W 69 B16
Prince of Wales Str. Canada 73°0N 117°0W 68 C8
Prince Patrick I. Canada 77°0N 120°0W 69 B8
Prince Regent Inlet Canada 73°0N 90°0W 69 C14
Prince Rupert Canada 54°20N 130°20W 70 C2
Prince William Sd. U.S.A. 60°40N 147°0W 68 E2
Princes Town Trin. & Tob. 10°16N 61°23W 93 K15
Princess Charlotte B. Australia 14°25S 144°0E 62 A3
Princess Elizabeth Trough S. Ocean 64°10S 83°0E 5 C7
Princess May Ranges Australia 15°30S 125°30E 60 C4
Princess Royal I. Canada 53°0N 128°40W 70 C3
Princeton Canada 49°27N 120°30W 70 D4
Princeton Calif., U.S.A. 39°24N 122°1W 78 F4
Princeton Ill., U.S.A. 41°23N 89°28W 80 E9
Princeton Ind., U.S.A. 38°21N 87°34W 80 F10
Princeton Ky., U.S.A. 37°7N 87°53W 80 G10
Princeton Mo., U.S.A. 40°24N 93°35W 80 E7
Princeton N.J., U.S.A. 40°21N 74°39W 83 F10
Princeton W. Va., U.S.A. 37°22N 81°6W 81 G13
Príncipe São Tomé & Príncipe 1°37N 7°25E 48 F4
Príncipe da Beira Brazil 12°20S 64°30W 92 F6
Prineville U.S.A. 44°18N 120°51W 76 D3
Prins Harald Kyst Antarctica 70°0S 35°1E 5 D4
Prinsesse Astrid Kyst Antarctica 70°45S 12°30E 5 D3
Prinsesse Ragnhild Kyst Antarctica 70°15S 27°30E 5 D4
Prinzapolca Nic. 13°20N 83°35W 88 D3
Priozersk Russia 61°2N 30°7E 8 F24
Pripet = Prypyat → Europe 51°20N 30°15E 17 C16
Pripet Marshes Europe 52°10N 27°10E 17 B15
Pripyat Marshes = Pripet Marshes Europe 52°10N 27°10E 17 B15
Pripyats = Prypyat → Europe 51°20N 30°15E 17 C16
Prishtine = Priština Kosovo 42°40N 21°13E 23 C9
Priština Kosovo 42°40N 21°13E 23 C9
Privas France 44°45N 4°37E 20 D6
Privolzhskaya Vozvyshennost Russia 51°0N 46°0E 19 D8
Privolzhskiy □ Russia 56°0N 50°0E 28 D6
Prizren Kosovo 42°13N 20°45E 23 C9
Probolinggo Indonesia 7°46S 113°13E 37 G15
Proctor U.S.A. 43°40N 73°2W 83 C11
Proddatur India 14°45N 78°30E 40 M11
Prodhromos Cyprus 34°57N 32°50E 25 E11
Profilia Greece 36°5N 27°51E 25 C9
Profondeville Belgium 50°23N 4°52E 15 D4
Progreso Coahuila, Mexico 27°28N 100°59W 86 B4
Progreso Yucatán, Mexico 21°20N 89°40W 87 C7
Progress Antarctica 66°22S 76°22E 5 C12
Progress Russia 49°45N 129°37E 29 E13
Prokopyevsk Russia 54°0N 86°45E 28 D9
Prokuplje Serbia 43°16N 21°36E 23 C9
Prome Burma 18°49N 95°13E 41 K19
Prophet → Canada 58°48N 122°40W 70 B4
Prophet River Canada 58°6N 122°43W 70 B4
Propriá Brazil 10°13S 36°51W 93 F11
Propriano France 41°41N 8°52E 20 F8
Proserpine Australia 20°21S 148°36E 62 b
Prosna → Poland 52°6N 17°44E 17 B9
Prospect U.S.A. 43°18N 75°9W 83 C9
Prosser U.S.A. 46°12N 119°46W 76 C4
Prostějov Czech Rep. 49°30N 17°9E 17 D9
Proston Australia 26°8S 151°32E 63 D5
Provence France 43°40N 5°46E 20 E6
Providence Ky., U.S.A. 37°24N 87°46W 80 G10
Providence R.I., U.S.A. 41°49N 71°24W 83 E13
Providence Bay Canada 45°41N 82°15W 72 C3
Providence Mts. U.S.A. 35°10N 115°15W 79 K11
Providencia, I. de Colombia 13°25N 81°26W 88 D3
Provideniya Russia 64°23N 173°18W 29 C19
Provincetown U.S.A. 42°3N 70°11W 81 D18
Provins France 48°33N 3°15E 20 B5
Provo U.S.A. 40°14N 111°39W 76 F8
Provost Canada 52°25N 110°20W 71 C6
Prudhoe Bay U.S.A. 70°18N 148°22W 74 A10
Prudhoe I. Australia 21°19S 149°41E 62 C4
Prud'homme Canada 52°20N 105°54W 71 C7
Pruszków Poland 52°9N 20°49E 17 B11
Prut → Romania 45°28N 28°10E 17 F15
Pruzhany Belarus 52°33N 24°28E 17 B13
Prydz B. Antarctica 69°0S 74°0E 5 C6
Pryluky Ukraine 50°30N 32°24E 19 D5
Pryor U.S.A. 36°19N 95°19W 84 C7
Prypyat → Europe 51°20N 30°15E 17 C16
Przemyśl Poland 49°50N 22°45E 17 D12
Przhevalsk = Karakol Kyrgyzstan 42°30N 78°20E 32 C4
Psara Greece 38°37N 25°38E 23 E11
Psiloritis, Oros Greece 35°15N 24°45E 25 D6
Psira Greece 35°12N 25°52E 25 D7
Pskov Russia 57°50N 28°25E 9 H23
Ptich = Ptsich → Belarus 52°9N 28°52E 17 B15
Ptichia = Vidos Greece 39°38N 19°55E 25 A3
Ptolemaida Greece 40°30N 21°43E 23 D9
Ptsich → Belarus 52°9N 28°52E 17 B15
Pu Xian China 36°24N 111°6E 34 F6
Pua Thailand 19°11N 100°55E 38 C3
Puán Argentina 37°30S 62°45W 94 D3
Pu'apu'a Samoa 13°34S 172°9W 59 b
Pucallpa Peru 8°25S 74°30W 92 E4
Puch'on = Bucheon S. Korea 37°28N 126°45E 35 F14
Pudasjärvi Finland 65°23N 26°53E 8 D22
Pudozh Russia 61°48N 36°32E 18 B6
Puducherry India 11°59N 79°50E 40 P11
Pudukkottai India 10°28N 78°47E 40 P11
Puebla Mexico 19°3N 98°12W 87 D5
Puebla □ Mexico 18°50N 98°0W 87 D5
Pueblo U.S.A. 38°16N 104°37W 76 G11
Puelches Argentina 38°5S 65°51W 94 D2
Puelén Argentina 37°32S 67°38W 94 D2
Puente Alto Chile 33°32S 70°35W 94 C1
Puente-Genil Spain 37°22N 4°47W 21 D3
Puerca, Pta. Puerto Rico 18°13N 65°36W 89 d
Puerco → U.S.A. 34°22N 107°50W 77 J10
Puerto Aisén Chile 45°27S 73°0W 96 F2
Puerto Ángel Mexico 15°40N 96°29W 87 D5
Puerto Arista Mexico 15°56N 93°48W 87 D6
Puerto Armuelles Panama 8°20N 82°51W 88 E3
Puerto Ayacucho Venezuela 5°40N 67°35W 92 B5
Puerto Barrios Guatemala 15°40N 88°32W 88 C2
Puerto Bermejo Argentina 26°55S 58°34W 94 B4
Puerto Bermúdez Peru 10°20S 74°58W 92 F4
Puerto Bolívar Ecuador 3°19S 79°55W 92 D3
Puerto Cabello Venezuela 10°28N 68°1W 92 A5
Puerto Cabezas Nic. 14°0N 83°30W 88 D3
Puerto Cabo Gracias á Dios Nic. 15°0N 83°10W 88 D3
Puerto Carreño Colombia 6°12N 67°22W 92 B5
Puerto Castilla Honduras 16°0N 86°0W 88 C2
Puerto Chicama Peru 7°45S 79°20W 92 E3
Puerto Coig Argentina 50°54S 69°15W 96 G3
Puerto Cortés Honduras 15°51N 88°0W 88 C2
Puerto Cumarebo Venezuela 11°29N 69°30W 92 A5
Puerto de Alcudia = Port d'Alcúdia Spain 39°50N 3°7E 24 B10
Puerto de Cabrera Spain 39°8N 2°56E 24 B9
Puerto de Gran Tarajal Canary Is. 28°13N 14°1W 24 F5
Puerto de la Cruz Canary Is. 28°24N 16°32W 24 F3
Puerto de los Angeles △ Mexico 23°39N 105°45W 86 C3
Puerto de Pozo Negro Canary Is. 28°19N 13°55W 24 F6
Puerto de Sóller = Port de Sóller Spain 39°48N 2°42E 24 B9
Puerto del Carmen Canary Is. 28°55N 13°38W 24 F6
Puerto del Rosario Canary Is. 28°30N 13°52W 24 F6
Puerto Deseado Argentina 47°55S 66°0W 96 F3
Puerto Escondido Mexico 15°50N 97°3W 87 D5
Puerto Heath Bolivia 12°34S 68°39W 92 F5
Puerto Inírida Colombia 3°53N 67°52W 92 C5
Puerto Juárez Mexico 21°11N 86°49W 87 C7
Puerto La Cruz Venezuela 10°13N 64°38W 92 A6
Puerto Leguízamo Colombia 0°12S 74°46W 92 D4
Puerto Lempira Honduras 15°16N 83°46W 88 C3
Puerto Libertad Mexico 29°55N 112°43W 86 B2
Puerto Limón Colombia 3°23N 73°30W 92 C4
Puerto Lobos Argentina 42°0S 65°3W 96 E3
Puerto Madryn Argentina 42°48S 65°4W 96 E3
Puerto Maldonado Peru 12°30S 69°10W 92 F5
Puerto Manatí Cuba 21°22N 76°50W 88 B4
Puerto Montt Chile 41°28S 73°0W 96 E2
Puerto Morazán Nic. 12°51N 87°11W 88 D2
Puerto Morelos Mexico 20°50N 86°52W 87 C7
Puerto Natales Chile 51°45S 72°15W 96 G2
Puerto Oscuro Chile 31°24S 71°35W 94 C1
Puerto Padre Cuba 21°13N 76°35W 88 B4
Puerto Páez Venezuela 6°13N 67°28W 92 B5
Puerto Peñasco Mexico 31°20N 113°33W 86 A2
Puerto Pinasco Paraguay 22°36S 57°50W 94 A4
Puerto Plata Dom. Rep. 19°48N 70°45W 89 C5
Puerto Pollensa = Port de Pollença Spain 39°54N 3°4E 24 B10
Puerto Princesa Phil. 9°46N 118°45E 37 C5
Puerto Quepos Costa Rica 9°29N 84°6W 88 E3
Puerto Rico Canary Is. 27°47N 15°42W 24 G4
Puerto Rico ☒ W. Indies 18°15N 66°45W 89 d
Puerto Rico Trench Atl. Oc. 19°50N 66°0W 89 C6
Puerto San Julián Argentina 49°18S 67°43W 96 F3
Puerto Santa Cruz Argentina 50°0S 68°32W 96 G3
Puerto Sastre Paraguay 22°2S 57°55W 94 A4
Puerto Suárez Bolivia 18°58S 57°52W 92 G7
Puerto Vallarta Mexico 20°37N 105°15W 86 C3
Puerto Varas Chile 41°19S 72°59W 96 E2
Puerto Wilches Colombia 7°21N 73°54W 92 B4
Puertollano Spain 38°43N 4°7W 21 C3
Pueu Tahiti 17°44S 149°13W 59 d
Pueyrredón, L. Argentina 47°20S 72°0W 96 F2
Puffin I. Ireland 51°50N 10°24W 10 E1
Pugachev Russia 52°0N 48°49E 18 D8
Pugal India 28°30N 72°48E 42 E5
Puge Tanzania 4°45S 33°11E 54 C3
Puget Sound U.S.A. 47°50N 122°30W 76 C2
Pugŏdong N. Korea 42°5N 130°0E 35 C16
Pugu Tanzania 6°55S 39°4E 54 D4
Pūgūnzī Iran 25°49N 59°10E 45 E8
Puig Major Spain 39°48N 2°47E 24 B9
Puigcerdà Spain 42°24N 1°50E 21 A6
Puigpunyent Spain 39°38N 2°32E 24 B9
Pujon-ho N. Korea 40°35N 127°35E 35 D14
Pukaki, L. N.Z. 44°4S 170°1E 59 F3
Pukapuka Cook Is. 10°53S 165°49W 65 J11
Pukaskwa △ Canada 48°20N 86°0W 72 C2
Pukatawagan Canada 55°45N 101°20W 71 B8
Pukchin N. Korea 40°12N 125°45E 35 D13
Pukch'ŏng N. Korea 40°14N 128°10E 35 D15
Pukekohe N.Z. 37°12S 174°55E 59 B5
Pukhrayan India 26°14N 79°51E 43 F8
Puksubaek-san N. Korea 40°42N 127°45E 35 D14
Pula Croatia 44°54N 13°57E 16 F7
Pulacayo Bolivia 20°25S 66°41W 92 H5
Pulandian China 39°25N 121°58E 35 E11
Pulaski N.Y., U.S.A. 43°34N 76°8W 83 C8
Pulaski Tenn., U.S.A. 35°12N 87°2W 85 D11
Pulaski Va., U.S.A. 37°3N 80°47W 81 G
Pulau → Indonesia 5°50S 138°15E 37
Pulau Gili Indonesia 8°21S 116°1E 37 J
Pulawy Poland 51°23N 21°59E 17 C
Pulga U.S.A. 39°48N 121°29W 78
Pulicat L. India 13°40N 80°15E 40 N
Pullman U.S.A. 46°44N 117°10W 76
Pulog, Mt. Phil. 16°40N 120°50E 37
Pułtusk Poland 52°43N 21°6E 17 B
Pumlumon Fawr U.K. 52°28N 3°46W 13
Puná, I. Ecuador 2°55S 80°5W 92
Punaauia Tahiti 17°37S 149°34W 5
Punakaiki N.Z. 42°7S 171°20E 59
Punakha Dzong Bhutan 27°42N 89°52E 41 F
Punasar India 27°6N 73°6E 42
Punata Bolivia 17°32S 65°50W 92
Punch India 33°48N 74°4E 43
Punch → Pakistan 33°12N 73°40E 42
Punda Maria S. Africa 22°40S 31°5E 57
Pune India 18°29N 73°57E 40
P'ungsan N. Korea 40°50N 128°9E 35 D
Pungue, Ponte de Mozam. 19°0S 34°0E 55
Punjab □ India 31°0N 76°0E 42
Punjab □ Pakistan 32°0N 72°30E 42
Puno Peru 15°55S 70°3W 92
Punpun → India 25°31N 85°18E 43 G
Punta, Cerro de Puerto Rico 18°10N 66°37W 89
Punta Alta Argentina 38°53S 62°4W 96
Punta Arenas Chile 53°10S 71°0W 96
Punta del Díaz Chile 28°0S 70°45W 94
Punta del Hidalgo Canary Is. 28°33N 16°19W 24
Punta Gorda Belize 16°10N 88°45W 87
Punta Gorda U.S.A. 26°56N 82°3W 85 H
Punta Prieta Mexico 28°58N 114°17W 86
Punta Prima Spain 39°48N 4°16E 24 B
Puntarenas Costa Rica 10°0N 84°50W 88
Puntland Somali Rep. 8°0N 49°0E 47
Punto Fijo Venezuela 11°50N 70°13W 92
Punxsutawney U.S.A. 40°57N 78°59W 82
Pupuan Indonesia 8°19S 115°0E 37 J
Puquio Peru 14°45S 74°10W 92
Pur → Russia 67°31N 77°55E 28
Puracé, Vol. Colombia 2°21N 76°23W 92
Puralia = Puruliya India 23°17N 86°24E 43 H
Puranpur India 28°31N 80°9E 43
Purbalingga Indonesia 7°23S 109°21E 37 G
Purbeck, Isle of U.K. 50°39N 1°59W 13
Purcell U.S.A. 35°1N 97°22W 84
Purcell Mts. Canada 49°55N 116°15W 70
Purdy Canada 45°19N 77°44W 82
Puri India 19°50N 85°58E 41 K
Purmerend Neths. 52°32N 4°58E 15
Purnia India 25°45N 87°31E 43 G
Purnululu △ Australia 17°20S 128°20E 60
Pursat = Pouthisat Cambodia 12°34N 103°50E 38
Purukcahu Indonesia 0°35S 114°35E 36
Puruliya India 23°17N 86°24E 43 H
Purus → Brazil 3°42S 61°28W 92
Puruvesi Finland 61°50N 29°30E 8 F
Purvis U.S.A. 31°9N 89°25W 85 F
Purwa India 26°28N 80°47E 43
Purwakarta Indonesia 6°35S 107°29E 37 G
Purwo, Tanjung Indonesia 8°44S 114°21E 37 K
Purwodadi Indonesia 7°7S 110°55E 37 G
Purwokerto Indonesia 7°25S 109°14E 37 G
Puryŏng N. Korea 42°5N 129°43E 35 C
Pusa India 25°59N 85°41E 43 G
Pusan = Busan S. Korea 35°5N 129°0E 35 G
Pushkin Russia 59°45N 30°25E 9 G
Pushkino Russia 51°16N 47°0E 19
Put-in-Bay U.S.A. 41°39N 82°49W 82
Putahow L. Canada 59°54N 100°40W 71
Putao Burma 27°28N 97°30E 41 F
Putaruru N.Z. 38°2S 175°50E 59
Puteran Indonesia 7°5S 114°0E 37 G
Putian China 25°23N 119°0E 33 F
Putignano Italy 40°51N 17°7E 22
Puting, Tanjung Indonesia 3°31S 111°46E 36
Putnam U.S.A. 41°55N 71°55W 83 E
Putorana, Gory Russia 69°0N 95°0E 29 C
Putrajaya Malaysia 2°55N 101°40E 39
Puttalam Sri Lanka 8°1N 79°55E 40 Q
Puttgarden Germany 54°30N 11°10E 16
Putumayo → S. Amer. 3°7S 67°58W 92
Putussibau Indonesia 0°50N 112°56E 36 D
Puvirnituq Canada 60°2N 77°10W 69 E
Puy-de-Dôme France 45°46N 2°57E 20 D
Puyallup U.S.A. 47°12N 122°18W 78
Puyang China 35°40N 115°1E 34 G
Pūzeh Rīg Iran 27°20N 58°40E 45 E
Pwani □ Tanzania 7°0S 39°0E 54 D
Pweto Dem. Rep. of the Congo 8°25S 28°51E 55 D
Pwllheli U.K. 52°53N 4°25W 12 E
Pyaozero, Ozero Russia 66°5N 30°58E 8 C
Pyapon Burma 16°20N 95°40E 41 L
Pyasina → Russia 73°30N 87°0E 29 B
Pyatigorsk Russia 44°2N 43°6E 19 F
Pyè = Prome Burma 18°49N 95°13E 41 K
Pyeongtaek S. Korea 37°1N 127°4E 35 F
Pyetrikaw Belarus 52°11N 28°29E 17 B
Pyhäjoki → Finland 64°28N 24°14E 8 D
Pyinmana Burma 19°45N 96°12E 41 K
Pyla, C. Cyprus 34°56N 33°51E 25 E
Pymatuning Res. U.S.A. 41°30N 80°28W 82 E
Pyŏktong N. Korea 40°50N 125°50E 35 D
Pyŏnggang N. Korea 38°24N 127°17E 35 E
P'yŏngsong N. Korea 39°14N 125°52E 35 E
P'yŏngyang N. Korea 39°0N 125°30E 35 E
Pyote U.S.A. 31°32N 103°8W 84 F
Pyramid L. U.S.A. 40°1N 119°35W 76 F
Pyramid Pk. U.S.A. 36°25N 116°37W 79 J
Pyramids Egypt 29°58N 31°9E 51 C
Pyrénées Europe 42°45N 0°18E 20 E
Pyu Burma 18°30N 96°28E 41 K

Q

anaaq Greenland	77°30N 69°10W	69 B18
chasnek S. Africa	30°6S 28°42E	57 E4
el Jafr Jordan	30°20N 36°25E	46 E5
emābād Iran	31°44N 60°2E	45 D9
emshahr Iran	36°30N 52°53E	45 B7
gan Nur Jilin, China	45°15N 124°18E	35 B13
gan Nur Nei Monggol Zizhiqu,		
China	43°30N 114°55E	34 C8
har Youyi Zhongqi		
China	41°12N 112°40E	34 D7
ran	34°23N 47°0E	44 C5
dam Pendi China	37°0N 95°0E	32 D8
ariyeh Iran	31°1N 48°22E	45 D6
la, Ras il Malta	36°2N 14°20E	25 C1
la-i-Jadid = Spīn Būldak		
Afghan.	31°1N 66°25E	42 D2
la Point = Qala, Ras il		
Malta	36°2N 14°20E	25 C1
la Viala Pakistan	30°49N 67°17E	42 D2
la Yangi Afghan.	34°20N 66°30E	42 B2
'at al Akhḍar Si. Arabia	28°4N 37°9E	44 E3
'at Dīzah Iraq	36°11N 45°7E	44 B5
'at Ṣāliḥ Iraq	31°31N 47°16E	44 D5
'at Sukkar Iraq	31°51N 46°5E	44 D5
mani'tuaq = Baker Lake		
Canada	64°20N 96°3W	68 E12
mdo China	31°15N 97°6E	32 E8
mea Fiji	16°45S 179°45W	59 a
mruddin Karez		
Pakistan	31°45N 68°20E	42 D3
adahār = Kandahār		
Afghan.	31°32N 65°43E	40 D4
adahār = Kandahār □		
Afghan.	31°0N 65°0E	40 D4
ndyaghash		
Kazakhstan	49°28N 57°25E	19 E10
pān Iran	37°40N 55°47E	45 B7
pshaghay Kazakhstan	43°51N 77°14E	28 E8
qortoq Greenland	60°43N 46°0W	4 C5
ra Qash → China	35°0N 78°30E	43 B8
rabutau Kazakhstan	50°0N 60°0E	28 E7
raghandy Kazakhstan	49°50N 73°10E	32 B3
raghayly Kazakhstan	49°26N 76°0E	28 E8
rah Si. Arabia	29°55N 40°3E	44 D4
rataū Ongtüstik Qazaqstan,		
Kazakhstan	43°30N 69°30E	28 E7
rataū Zhambyl,		
Kazakhstan	43°10N 70°28E	28 E8
razhal Kazakhstan	48°2N 70°49E	28 E8
rdho Somali Rep.	9°30N 49°6E	47 F4
reh → Iran	39°25N 47°22E	44 B5
reh Tekān Iran	36°38N 49°29E	45 B6
rnein U.A.E.	24°56N 52°52E	45 E7
rqan W.→ China	39°30N 88°30E	32 D6
qraraly Kazakhstan	49°26N 75°30E	32 B4
rshi Uzbekistan	38°53N 65°48E	28 F7
rtaba Lebanon	34°4N 35°50E	46 A4
rūh Kuwait	28°49N 48°46E	45 D6
ryat al Gharab Iraq	31°27N 44°48E	44 D5
ryat al 'Ulyā Si. Arabia	27°33N 47°42E	44 E5
sr 'Amra Jordan	31°48N 36°35E	44 D3
sr-e Qand Iran	26°15N 60°45E	45 E9
sr-e Shīrīn Iran	34°31N 45°35E	44 C5
sr Farāfra Egypt	27°0N 28°1E	51 C11
suittuq = Resolute		
Canada	74°42N 94°54W	69 C13
tanā Syria	33°26N 36°4E	46 B5
tar ■ Asia	25°30N 51°15E	45 E6
tlīsh Iran	37°50N 57°19E	45 B8
ttāra, Munkhafed el		
Egypt	29°30N 27°30E	51 C11
ttāra Depression = Qattâra,		
Munkhafed el Egypt	29°30N 27°30E	51 C11
wām al Ḥamzah = Al Ḥamzah		
Iraq	31°43N 44°58E	44 D5
yen Iran	33°40N 59°10E	45 C8
zaqstan = Kazakhstan ■		
azamämmäd Azerbaijan	40°3N 49°0E	45 A6
zvīn □ Iran	36°15N 50°0E	45 B6
zvīn Iran	36°20N 50°0E	45 B6
na Egypt	26°10N 32°43E	51 C12
qertarsuaq Greenland	69°45N 53°30W	4 C5
qortaq Greenland	69°15N 53°38W	4 C5
shlāq Iran	34°55N 46°28E	44 C5
shm Iran	26°55N 56°10E	45 E8
ys Iran	26°32N 53°58E	45 E7
zel Owzen → Iran	36°45N 49°22E	45 B6
zi'ot Israel	30°52N 34°26E	46 E3
Xian China	34°40N 114°48E	34 G8
an Gorlos China	45°5N 124°42E	35 B13
an Hai China	22°32N 113°54E	33 a
an Xian China	34°31N 108°15E	34 G5
an'an China	40°0N 118°41E	35 E10
anjin China	47°34N 133°4E	33 B15
anshan China	22°15N 113°31E	33 a
anyang China	22°40N 107°58E	34 G4
ao China	22°50N 113°39E	33 a
ao Dao China	22°25N 113°38E	33 a
emo China	39°56N 112°55E	34 E7
iaojing China	38°8N 85°32E	32 D6
la Saifullah Pakistan	30°45N 68°17E	42 D3
lian Shan China	38°30N 98°0E	32 C8
n He → China	35°1N 113°22E	34 G7
n Ling = Qinling Shandi		
China	33°50N 108°10E	34 H5
an'an China	34°48N 105°40E	34 G3
ngcheng China	33°35N 116°45E	34 H9
ngchuan China	32°35N 105°9E	34 H3
ngfeng China	35°52N 115°8E	34 G8
nghai □ China	36°0N 98°0E	32 C8
nghai Hu China	36°40N 100°10E	32 C9

Qinghecheng China	41°28N 124°15E	35 D13
Qinghemen China	41°48N 121°25E	35 D11
Qingjian China	37°8N 110°8E	34 F6
Qingjiang = Huaiyin		
China	33°30N 119°2E	35 H10
Qingshui China	34°48N 106°8E	34 G4
Qingshuihe China	39°55N 111°35E	34 E6
Qingtongxia China	38°2N 106°3E	34 E4
Qingtongxia Shuiku		
China	37°50N 105°58E	34 F3
Qingxu China	37°34N 112°22E	34 F7
Qingyang China	36°2N 107°55E	34 F4
Qingyuan China	42°10N 124°55E	35 C13
Qingyun China	37°45N 117°20E	35 F9
Qinhuangdao China	39°56N 119°30E	35 E10
Qinling Shandi China	33°50N 108°10E	34 H5
Qinshui China	35°40N 112°8E	34 G7
Qinyang = Jiyuan China	35°7N 112°57E	34 G7
Qinyang China	35°5N 112°56E	34 G7
Qinyuan China	36°29N 112°20E	34 F7
Qinzhou China	21°58N 108°38E	32 G10
Qionghai China	19°15N 110°26E	38 C8
Qiongshan China	19°51N 110°26E	38 C8
Qiongzhou Haixia		
China	20°10N 110°15E	38 B8
Qiqihar China	47°26N 124°0E	33 B13
Qira China	37°0N 80°48E	32 D5
Qiraïya, W. → Egypt	30°27N 34°0E	46 E3
Qiryat Ata Israel	32°47N 35°6E	46 C4
Qiryat Gat Israel	31°32N 34°46E	46 D3
Qiryat Mal'akhi Israel	31°44N 34°44E	46 D3
Qiryat Shemona Israel	33°13N 35°35E	46 B4
Qiryat Yam Israel	32°51N 35°4E	46 C4
Qishan China	34°25N 107°38E	34 G4
Qitai China	44°2N 89°35E	32 C6
Qitaihe China	45°48N 130°51E	30 B5
Qixia China	37°17N 120°52E	35 F11
Qızlağaç Körfäzi		
Azerbaijan	39°9N 49°0E	45 B6
Qojūr Iran	36°12N 47°55E	44 B5
Qom Iran	34°40N 51°0E	45 C6
Qom □ Iran	34°40N 51°0E	45 C6
Qomolangma Feng = Everest, Mt.		
Nepal	28°5N 86°58E	43 E12
Qomsheh Iran	32°0N 51°55E	45 D6
Qoqek = Tacheng China	46°40N 82°58E	32 B5
Qoqon = Qŭqon		
Uzbekistan	40°31N 70°56E	28 E8
Qoraqalpoghistan □		
Uzbekistan	43°0N 58°0E	28 E6
Qorveh Iran	35°10N 47°48E	44 C5
Qosshaghyl Kazakhstan	46°40N 54°0E	19 E9
Qostanay Kazakhstan	53°10N 63°35E	28 D7
Quabbin Res. U.S.A.	42°20N 72°20W	83 D12
Quairading Australia	32°0S 117°21E	61 F2
Quakertown U.S.A.	40°26N 75°21W	83 F9
Qualicum Beach		
Canada	49°22N 124°26W	70 D4
Quambatook Australia	35°49S 143°34E	63 F3
Quambone Australia	30°57S 147°53E	63 E4
Quamby Australia	20°22S 140°17E	62 C3
Quan Long = Ca Mau		
Vietnam	9°7N 105°8E	39 H5
Quanah U.S.A.	34°18N 99°44W	84 D5
Quang Ngai Vietnam	15°13N 108°58E	38 E7
Quang Tri Vietnam	16°45N 107°13E	38 D6
Quang Yen Vietnam	20°56N 106°52E	38 B6
Quantock Hills U.K.	51°8N 3°10W	13 F4
Quanyang China	42°21N 127°32E	35 C14
Quanzhou China	24°55N 118°34E	33 G12
Qu'Appelle → Canada	50°33N 103°53W	71 C8
Quaqtaq Canada	60°55N 69°40W	69 E18
Quarai Brazil	30°15S 56°20W	94 C4
Quartu Sant'Elena Italy	39°15N 9°10E	22 E3
Quartzsite U.S.A.	33°40N 114°13W	79 M12
Quatre Bornes Mauritius	20°15S 57°28E	53 d
Quatsino Sd. Canada	50°25N 127°58W	70 C3
Quba Azerbaijan	41°21N 48°32E	19 F8
Qūchān Iran	37°10N 58°27E	45 B8
Queanbeyan Australia	35°17S 149°14E	63 F4
Québec Canada	46°52N 71°13W	73 C5
Québec □ Canada	48°0N 74°0W	73 B6
Quebrada del Condorito △		
Argentina	31°49S 64°40W	94 C3
Queen Alexandra Ra.		
Antarctica	85°0S 170°0E	5 E11
Queen Charlotte City		
Canada	53°15N 132°2W	70 C2
Queen Charlotte Is.		
Canada	53°20N 132°10W	70 C2
Queen Charlotte Sd.		
Canada	51°0N 128°0W	70 C3
Queen Charlotte Strait		
Canada	50°45N 127°10W	70 C3
Queen Elizabeth △ Uganda	0°0 30°0E	54 C3
Queen Elizabeth □ U.K.	56°7N 4°30W	11 E4
Queen Elizabeth Is.		
Canada	76°0N 95°0W	69 B13
Queen Mary Land		
Antarctica	70°0S 95°0E	5 D7
Queen Maud G.		
Canada	68°15N 102°30W	68 D11
Queen Maud Land = Dronning		
Maud Land Antarctica	72°30S 12°0E	5 D3
Queen Maud Mts.		
Antarctica	86°0S 160°0W	5 E12
Queens Channel		
Australia	15°0S 129°30E	60 C4
Queenscliff Australia	38°16S 144°39E	63 F3
Queensland □ Australia	22°0S 142°0E	62 C3
Queenstown Australia	42°4S 145°35E	63 G4
Queenstown N.Z.	45°1S 168°40E	59 F2
Queenstown Singapore	1°18N 103°48E	39 d
Queenstown S. Africa	31°52S 26°52E	56 E4
Queets U.S.A.	47°32N 124°19W	78 C2
Queguay Grande →		
Uruguay	32°9S 58°9W	94 C4
Queimadas Brazil	11°0S 39°38W	93 F11
Quelimane Mozam.	17°53S 36°58E	55 F4
Quellón Chile	43°7S 73°37W	96 E2

Quelpart = Jeju-do		
S. Korea	33°29N 126°34E	35 H14
Quemado N. Mex.,		
U.S.A.	34°20N 108°30W	77 J9
Quemado Tex., U.S.A.	28°56N 100°37W	84 G4
Quemú-Quemú		
Argentina	36°3S 63°36W	94 D3
Quequén Argentina	38°30S 58°30W	94 D4
Querétaro Mexico	20°36N 100°23W	86 C4
Querétaro □ Mexico	21°0N 99°55W	86 C5
Quesnel Canada	53°0N 122°30W	70 C4
Quesnel → Canada	52°58N 122°29W	70 C4
Quesnel L. Canada	52°30N 121°20W	70 C4
Questa U.S.A.	36°42N 105°36W	77 H11
Quetico △ Canada	48°30N 91°45W	72 C1
Quetta Pakistan	30°15N 66°55E	42 D2
Quezaltenango		
Guatemala	14°50N 91°30W	88 D1
Quezon City Phil.	14°37N 121°2E	37 B6
Qufar Si. Arabia	27°26N 41°37E	44 E4
Qufu China	35°35N 116°59E	34 G9
Qui Nhon Vietnam	13°40N 109°13E	38 F7
Quibala Angola	10°46S 14°59E	52 G2
Quibaxe Angola	8°24S 14°27E	52 F2
Quibdó Colombia	5°42N 76°40W	92 B3
Quiberon France	47°29N 3°9W	20 C2
Quiet L. Canada	61°5N 133°5W	70 A2
Quiindy Paraguay	25°58S 57°14W	94 B4
Quilá Mexico	24°23N 107°13W	86 C3
Quilán, C. Chile	43°15S 74°30W	96 E2
Quilcene U.S.A.	47°49N 122°53W	78 C4
Quilimarí Chile	32°5S 71°30W	94 C1
Quilino Argentina	30°14S 64°29W	94 C3
Quillabamba Peru	12°50S 72°50W	92 F4
Quillagua Chile	21°40S 69°40W	94 A2
Quillota Chile	32°54S 71°16W	94 C1
Quilmes Argentina	34°43S 58°15W	94 C4
Quilon India	8°50N 76°38E	40 Q10
Quilpie Australia	26°35S 144°11E	63 D3
Quilpué Chile	33°5S 71°33W	94 C1
Quilua Mozam.	16°17S 39°54E	55 F4
Quimili Argentina	27°40S 62°30W	94 B3
Quimper France	48°0N 4°9W	20 B1
Quimperlé France	47°53N 3°33W	20 C2
Quinault → U.S.A.	47°21N 124°18W	78 C2
Quincy Calif., U.S.A.	39°56N 120°57W	78 F6
Quincy Fla., U.S.A.	30°35N 84°34W	85 F12
Quincy Ill., U.S.A.	39°56N 91°23W	80 F8
Quincy Mass., U.S.A.	42°14N 71°0W	83 D14
Quincy Wash., U.S.A.	47°14N 119°51W	76 C4
Quines Argentina	32°13S 65°48W	94 C2
Quinga Mozam.	15°49S 40°15E	55 F5
Quintana Roo □ Mexico	19°40N 88°30W	87 D7
Quintanar de la Orden		
Spain	39°36N 3°5W	21 C4
Quintero Chile	32°45S 71°30W	94 C1
Quirihue Chile	36°15S 72°35W	94 D1
Quirimbas △ Mozam.	12°30S 40°15E	55 E5
Quirindi Australia	31°28S 150°40E	63 E5
Quirinópolis Brazil	18°32S 50°30W	93 G8
Quissanga Mozam.	12°24S 40°28E	55 E5
Quissico Mozam.	24°42S 34°44E	57 C5
Quitilipi Argentina	26°50S 60°13W	94 B3
Quitman U.S.A.	30°47N 83°34W	85 F13
Quito Ecuador	0°15S 78°35W	92 D3
Quixadá Brazil	4°55S 39°0W	93 D11
Quixaxe Mozam.	15°17S 40°4E	55 F5
Qujing China	25°32N 103°41E	32 F9
Qulan Kazakhstan	42°55N 72°43E	28 E8
Qul'ân, Jazâ'ir Egypt	24°22N 35°31E	44 E2
Qulsary Kazakhstan	46°59N 54°1E	19 E9
Qumbu S. Africa	31°10S 28°48E	57 E4
Quneitra = Al Qunayţirah		
Syria	33°5N 35°45E	46 B4
Qŭnghirot Uzbekistan	43°2N 58°50E	28 E6
Quoin I. Australia	14°54S 129°32E	60 B4
Quoin Pt. S. Africa	34°46S 19°37E	56 E2
Quorn Australia	32°25S 138°5E	63 E2
Qŭqon Uzbekistan	40°31N 70°56E	28 E8
Qurnat as Sawdā'		
Lebanon	34°18N 36°6E	46 A5
Quşaybā' Si. Arabia	26°53N 43°35E	44 E4
Qusaybah Iraq	34°24N 40°59E	44 C4
Quseir Egypt	26°7N 34°16E	44 E2
Qūshchī Iran	37°59N 45°3E	44 B5
Quthing Lesotho	30°25S 27°36E	57 E4
Qūṭīābād Iran	35°47N 48°30E	45 C6
Quttinirpaaq △ Canada	82°13N 72°13W	69 A17
Quwo China	35°38N 111°25E	34 G6
Quyang China	38°35N 114°40E	34 E8
Quynh Nhai Vietnam	21°49N 103°33E	38 B4
Quyon Canada	45°31N 76°14W	83 A8
Quzhou China	28°57N 118°54E	33 F12
Quzi China	36°20N 107°20E	34 F4
Qyzylorda Kazakhstan	44°48N 65°28E	28 E7

R

Ra, Ko Thailand	9°13N 98°16E	39 H2
Raahe Finland	64°40N 24°28E	8 D21
Raalte Neths.	52°23N 6°16E	15 B6
Raas Indonesia	7°8S 114°33E	37 G16
Raasay Scotland	57°25N 6°4W	11 D2
Raasay, Sd. of U.K.	57°30N 6°8W	11 D2
Raasepori Finland	60°0N 23°32E	9 G20
Rába → Hungary	47°38N 17°38E	17 E9
Rabai Kenya	3°50S 39°31E	54 C4
Rabat = Victoria Malta	36°3N 14°14E	25 C1
Rabat Malta	35°53N 14°24E	25 D1
Rabat Morocco	34°2N 6°48W	50 B4
Rabaul Papua N. G.	4°24S 152°18E	58 B8
Rabbit Flat Australia	20°11S 130°1E	60 D5
Rabbit Lake Mine Canada	58°4N 103°40W	71 B8
Rabi Fiji	16°30S 179°59W	59 a
Rābigh Si. Arabia	22°50N 39°5E	47 C2
Rabwah = Chenab Nagar		
Pakistan	31°45N 72°55E	42 D5

Race, C. Canada	46°40N 53°5W	73 C9
Rach Gia Vietnam	10°5N 105°5E	39 G5
Rach Soi Vietnam	9°57N 105°7E	39 H5
Rachid Mauritania	18°45N 11°35W	50 E3
Racha Noi, Ko Thailand	7°30N 98°19E	39 J2
Racha Yai, Ko Thailand	7°36N 98°21E	39 J2
Racibórz Poland	50°7N 18°18E	17 C10
Racine U.S.A.	42°44N 87°47W	80 D10
Rackerby U.S.A.	39°26N 121°22W	78 F5
Radama, Nosy Madag.	14°0S 47°47E	57 A8
Radama, Saikanosy		
Madag.	14°16S 47°53E	57 A8
Rădăuţi Romania	47°50N 25°59E	17 E13
Radcliff U.S.A.	37°51N 85°57W	81 G11
Radekhiv Ukraine	50°25N 24°32E	17 C13
Radekhov = Radekhiv		
Ukraine	50°25N 24°32E	17 C13
Radford U.S.A.	37°8N 80°34W	81 G13
Radhanpur India	23°50N 71°38E	42 H4
Radhwa, Jabal Si. Arabia	24°34N 38°18E	44 E3
Radisson Qué., Canada	53°47N 77°37W	72 B4
Radisson Sask., Canada	52°30N 107°20W	71 C7
Radium Hot Springs		
Canada	50°35N 116°2W	70 C5
Radnor Forest U.K.	52°17N 3°10W	13 E4
Radom Poland	51°23N 21°12E	17 C11
Radomsko Poland	51°5N 19°28E	17 C10
Radomyshl Ukraine	50°30N 29°12E	17 C15
Radstock, C. Australia	33°12S 134°20E	63 E1
Raduzhnyy Russia	62°5N 77°28E	28 C8
Radviliškis Lithuania	55°49N 23°33E	9 J20
Radville Canada	49°30N 104°15W	71 D8
Rae Canada	62°50N 116°3W	70 A5
Rae Bareli India	26°18N 81°20E	43 F9
Rae Isthmus Canada	66°40N 87°30W	69 D14
Raeren Belgium	50°41N 6°7E	15 D6
Raeside, L. Australia	29°20S 122°0E	61 E3
Raetihi N.Z.	39°25S 175°17E	59 C5
Rafaela Argentina	31°10S 61°30W	94 C3
Rafah Gaza Strip	31°18N 34°14E	46 D3
Rafai C.A.R.	4°59N 23°58E	54 B1
Raffaḍ Si. Arabia	29°35N 43°35E	44 D4
Rafsanjān Iran	30°30N 56°5E	45 D8
Raft Pt. Australia	16°4S 124°26E	60 C3
Râgâ Sudan	8°28N 25°41E	51 G11
Ragachow Belarus	53°8N 30°5E	17 B16
Ragama Sri Lanka	7°0N 79°50E	40 R11
Ragged, Mt. Australia	33°27S 123°25E	61 F3
Ragged Pt. Barbados	13°10N 59°26W	89 g
Raghunathpalli India	22°14N 84°48E	43 H11
Raghunathpur India	23°33N 86°40E	43 H12
Raglan N.Z.	37°55S 174°55E	59 B5
Ragusa Italy	36°55N 14°44E	22 F6
Raha Indonesia	4°55S 123°0E	37 E6
Rahaeng = Tak Thailand	16°52N 99°8E	38 D2
Rahatgarh India	23°47N 78°22E	43 H8
Rahimyar Khan Pakistan	28°30N 70°25E	42 E4
Rāhjerd Iran	34°22N 50°22E	45 C6
Rahole △ Kenya	0°5N 38°57E	54 B4
Rahon India	31°3N 76°7E	42 D7
Rai, Hon Vietnam	9°48N 104°38E	39 H5
Raiatéa, Î.		
French Polynesia	16°50S 151°25W	65 J12
Raichur India	16°10N 77°20E	40 L10
Raiganj India	25°37N 88°10E	43 G13
Raigarh India	21°56N 83°25E	41 J13
Raijua Indonesia	10°37S 121°36E	37 F6
Raikot India	30°41N 75°42E	42 D6
Railton Australia	41°25S 146°28E	63 G4
Rainbow Bridge △		
U.S.A.	37°5N 110°58W	77 H8
Rainbow Lake Canada	58°30N 119°23W	70 B5
Rainier U.S.A.	46°53N 122°41W	78 D4
Rainier, Mt. U.S.A.	46°52N 121°46W	78 D5
Rainy L. Canada	48°42N 93°10W	71 D10
Rainy River Canada	48°43N 94°29W	71 D10
Raippaluoto Finland	63°13N 21°14E	8 E19
Raipur India	21°17N 81°45E	41 J12
Raisen India	23°20N 77°48E	42 H8
Raisio Finland	60°28N 22°11E	9 F20
Raj Nandgaon India	21°5N 81°5E	41 J12
Raj Nilgiri India	21°28N 86°46E	43 J12
Raja, Ujung Indonesia	3°40N 96°25E	36 D1
Raja Ampat, Kepulauan		
Indonesia	0°30S 130°0E	37 E8
Rajahmundry India	17°1N 81°48E	41 L12
Rajaji △ India	30°10N 78°20E	42 D8
Rajang → Malaysia	2°30N 112°0E	36 D4
Rajanpur Pakistan	29°6N 70°19E	42 E4
Rajapalaiyam India	9°25N 77°35E	40 Q10
Rajasthan □ India	26°45N 73°30E	42 F5
Rajasthan Canal = Indira Gandhi		
Canal India	28°0N 72°0E	42 F5
Rajauri India	33°25N 74°21E	43 C6
Rajgarh Mad. P., India	24°2N 76°45E	42 G7
Rajgarh Raj., India	27°14N 76°38E	42 F7
Rajgarh Raj., India	28°40N 75°25E	42 E6
Rajgir India	25°2N 85°25E	43 G11
Rajkot India	22°15N 70°56E	42 H4
Rajmahal Hills India	24°30N 87°30E	43 G12
Rajpipla India	21°50N 73°30E	40 J8
Rajpur India	22°18N 74°21E	42 H6
Rajpura India	30°25N 76°32E	42 D7
Rajsamand = Kankroli		
India	25°4N 73°53E	42 G5
Rajshahi Bangla.	24°22N 88°39E	41 G16
Rajshahi □ Bangla.	25°0N 89°0E	43 G13
Rajula India	21°3N 71°26E	42 J4
Rakaia N.Z.	43°45S 172°1E	59 E4
Rakaia → N.Z.	43°36S 172°15E	59 E4
Rakan, Ra's Qatar	26°10N 51°20E	45 E6
Rakaposhi Pakistan	36°10N 74°25E	43 A6
Rakata, Pulau Indonesia	6°10S 105°20E	37 G11
Rakhiv Ukraine	48°3N 24°12E	17 D13
Rakhni Pakistan	30°4N 69°56E	42 D3
Rakhni → Pakistan	29°31N 69°36E	42 E3
Rakiraki Fiji	17°22S 178°11E	59 a
Rakitnoye Russia	45°36N 134°17E	30 B7
Rakiura = Stewart I.		
N.Z.	46°58S 167°54E	59 G1

Rakiura △ N.Z.	47°0S 167°50E	59 G1
Rakops Botswana	21°1S 24°28E	56 C3
Rakvere Estonia	59°20N 26°25E	9 G22
Raleigh U.S.A.	35°47N 78°39W	85 D15
Ralik Chain Pac. Oc.	8°0N 168°0E	64 G8
Ralls U.S.A.	33°41N 101°24W	84 E4
Ralston U.S.A.	41°30N 76°57W	83 E8
Ram → Canada	62°1N 123°41W	70 A4
Rām Allāh West Bank	31°55N 35°10E	46 D4
Rama Nic.	12°9N 84°15W	88 D3
Ramakona India	21°43N 78°50E	43 J8
Rāmallāh = Rām Allāh		
West Bank	31°55N 35°10E	46 D4
Raman Thailand	6°29N 101°18E	39 J3
Ramanathapuram India	9°25N 78°55E	40 Q11
Ramanujganj India	23°48N 83°42E	43 H10
Ramat Gan Israel	32°4N 34°48E	46 C3
Ramatlhabama S. Africa	25°37S 25°33E	56 D4
Ramban India	33°14N 75°12E	43 C6
Rambi = Rabi Fiji	16°30S 179°59W	59 a
Rambipuji Indonesia	8°12S 113°37E	37 H15
Rame Hd. Australia	37°47S 149°30E	63 F4
Ramechhap Nepal	27°25N 86°10E	43 F12
Ramganga → India	27°5N 79°58E	43 F8
Ramgarh Jharkhand,		
India	23°40N 85°35E	43 H11
Ramgarh Raj., India	27°16N 75°14E	42 F6
Ramgarh Raj., India	27°30N 70°36E	42 F4
Rāmhormoz Iran	31°15N 49°35E	45 D6
Ramiān Iran	37°3N 55°16E	45 B7
Ramingining Australia	12°19S 135°3E	62 A2
Ramla Israel	31°55N 34°52E	46 D3
Ramm = Rum Jordan	29°39N 35°26E	46 F4
Ramm, Jabal Jordan	29°35N 35°24E	46 F4
Ramnad = Ramanathapuram		
India	9°25N 78°55E	40 Q11
Ramnagar Jammu & Kashmir,		
India	32°47N 75°18E	43 C6
Ramnagar Uttarakhand,		
India	29°24N 79°7E	43 E8
Râmnicu Sărat Romania	45°26N 27°3E	17 F14
Râmnicu Vâlcea		
Romania	45°9N 24°21E	17 F13
Ramona U.S.A.	33°2N 116°52W	79 M10
Ramore Canada	48°30N 80°25W	72 C3
Ramotswa Botswana	24°50S 25°52E	56 C4
Rampur H.P., India	31°26N 77°43E	42 D7
Rampur Mad. P., India	23°25N 73°53E	42 H5
Rampur Ut. P., India	28°50N 79°5E	43 E8
Rampur Hat India	24°10N 87°50E	43 G12
Rampura India	24°30N 75°27E	42 G6
Ramrama Tola India	21°52N 79°55E	43 J8
Ramree I. Burma	19°0N 93°40E	41 K19
Rāmsar Iran	36°53N 50°41E	45 B6
Ramsey I. of Man	54°20N 4°22W	12 C3
Ramsey U.S.A.	41°4N 74°9W	83 E10
Ramsey L. Canada	47°13N 82°15W	72 C3
Ramsgate U.K.	51°20N 1°25E	13 F9
Ramtek India	21°20N 79°15E	40 J11
Ramu Kenya	3°55N 41°10E	54 B5
Rana Pratap Sagar Dam		
India	24°58N 75°38E	42 G6
Ranaghat India	23°15N 88°35E	43 H13
Ranahu Pakistan	25°55N 69°45E	42 G3
Ranau Malaysia	6°2N 116°40E	36 C5
Rancagua Chile	34°10S 70°50W	94 C1
Rancheria → Canada	60°13N 129°7W	70 A3
Ranchester U.S.A.	44°54N 107°10W	76 D10
Ranchi India	23°19N 85°27E	43 H11
Rancho Cordova		
U.S.A.	38°36N 121°18W	78 G5
Rancho Cucamonga		
U.S.A.	34°10N 117°30W	79 L9
Randalstown U.K.	54°45N 6°19W	10 B5
Randers Denmark	56°29N 10°1E	9 H14
Randfontein S. Africa	26°8S 27°45E	57 D4
Randle U.S.A.	46°32N 121°57W	78 D5
Randolph Mass., U.S.A.	42°10N 71°2W	83 D14
Randolph N.Y., U.S.A.	42°10N 78°59W	82 D6
Randolph Utah, U.S.A.	41°40N 111°11W	76 F8
Randolph Vt., U.S.A.	43°55N 72°40W	83 C12
Randsburg U.S.A.	35°22N 117°39W	79 K9
Råneälven → Sweden	65°50N 22°20E	8 D20
Rangae Thailand	6°19N 101°44E	39 J3
Rangaunu B. N.Z.	34°51S 173°15E	59 A4
Range, The Zimbabwe	19°2S 31°2E	55 F3
Rangeley U.S.A.	44°58N 70°39W	83 B14
Rangeley L. U.S.A.	44°55N 70°43W	83 B14
Rangely U.S.A.	40°5N 108°48W	76 F9
Ranger U.S.A.	32°28N 98°41W	84 E5
Rangia India	26°28N 91°38E	41 F17
Rangiora N.Z.	43°19S 172°36E	59 E4
Rangitaiki → N.Z.	37°54S 176°49E	59 B6
Rangitata → N.Z.	43°45S 171°15E	59 E3
Rangitoto ke te tonga = D'Urville		
I. N.Z.	40°50S 173°55E	59 D4
Rangkasbitung		
Indonesia	6°21S 106°15E	37 G12
Rangoon Burma	16°45N 96°20E	41 L20
Rangoon → Burma	16°28N 96°40E	41 L20
Rangpur Bangla.	25°42N 89°22E	41 G16
Rangsang Indonesia	1°20N 103°9E	39 M4
Ranibennur India	14°35N 75°30E	40 M9
Raniganj Ut. P., India	27°3N 82°13E	43 F10
Raniganj W. Bengal, India	23°40N 87°5E	41 H15
Ranikhet India	29°39N 79°25E	43 E8
Raniwara India	24°50N 72°10E	42 G5
Rāniya Iraq	36°15N 44°53E	44 B5
Ranka India	23°59N 83°47E	43 H10
Ranken → Australia	20°31S 137°36E	62 C2
Rankin U.S.A.	31°13N 101°56W	84 F4
Rankin Inlet Canada	62°30N 93°0W	68 E13
Rankins Springs		
Australia	33°49S 146°14E	63 E4
Rannoch, L. U.K.	56°41N 4°20W	11 E4
Rannoch Moor U.K.	56°38N 4°48W	11 E4
Ranobe, Helodranon' i		
Madag.	23°3S 43°33E	57 C7
Ranohira Madag.	22°29S 45°24E	57 C8
Ranomafana Toamasina,		
Madag.	18°57S 48°50E	57 B8

Ranomafana Toliara, Madag. 24°34S 47°0E 57 C8
Ranomafana △ Madag. 21°16S 47°25E 57 C8
Ranomena Madag. 23°25S 47°17E 57 C8
Ranong Thailand 9°56N 98°40E 39 H2
Ranotsara Nord Madag. 22°48S 46°36E 57 C8
Rānsa Iran 33°39N 48°18E 45 C6
Ransiki Indonesia 1°30S 134°10E 37 E8
Rantabe Madag. 15°42S 49°39E 57 B8
Rantau Abang Malaysia 4°52N 103°24E 39 K4
Rantauprapat Indonesia 2°15N 99°50E 36 D1
Rantemario Indonesia 3°15S 119°57E 37 E5
Ranthambore △ India 26°10N 76°30E 42 F7
Rantoul U.S.A. 40°19N 88°9W 80 E9
Raohe China 46°47N 134°0E 30 A7
Raoyang China 38°15N 115°45E 34 E8
Rap, Ko Thailand 9°19N 99°58E 39 b
Rapa French Polynesia 27°35S 144°20W 65 K13
Rapa Nui = Pascua, I. de Chile 27°7S 109°23W 65 K17
Rapallo Italy 44°21N 9°14E 20 D8
Rapar India 23°34N 70°38E 42 H4
Rāpch Iran 25°40N 59°15E 45 E8
Rapel, Lago Chile 34°20S 71°14W 94 C1
Raper, C. Canada 69°44N 67°6W 69 D18
Rapid City U.S.A. 44°5N 103°14W 80 C2
Rapid River U.S.A. 45°55N 86°58W 80 C10
Rapla Estonia 59°1N 24°52E 9 G21
Rapti → India 26°18N 83°41E 43 F10
Raquette → U.S.A. 45°0N 74°42W 83 B10
Raquette Lake U.S.A. 43°49N 74°40W 83 C10
Rara △ Nepal 29°30N 82°10E 43 E10
Rarotonga Cook Is. 21°30S 160°0W 65 K12
Ra's al 'Ayn Syria 36°45N 40°12E 44 B4
Ra's al Khaymah U.A.E. 25°50N 55°59E 45 E7
Ra's aţ Tib Tunisia 37°1N 11°2E 22 F4
Rasa, Pta. Argentina 36°20S 56°41W 94 D4
Rasca, Pta. de la Canary Is. 27°59N 16°41W 24 G3
Raseiniai Lithuania 55°25N 23°5E 9 J20
Rashmi India 25°4N 74°22E 42 G6
Rasht Iran 37°20N 49°40E 45 B6
Rasi Salai Thailand 15°20N 104°9E 38 E5
Rason L. Australia 28°45S 124°25E 61 E3
Rasra India 25°50N 83°50E 43 G10
Rasul Pakistan 32°42N 73°34E 42 C5
Rat → Canada 49°35N 97°10W 71 D9
Rat Buri = Ratchaburi Thailand 13°30N 99°54E 38 F2
Rat Islands U.S.A. 52°0N 178°0E 74 E3
Rat L. Canada 56°10N 99°40W 71 B9
Ratak Chain Pac. Oc. 1°0N 170°0E 64 G8
Ratangarh India 28°5N 74°35E 42 E6
Raţāwī Iraq 30°38N 47°13E 44 D5
Ratchaburi Thailand 13°30N 99°54E 38 F2
Rath India 25°36N 79°37E 43 G8
Rath Luirc Ireland 52°21N 8°40W 10 D3
Rathangan Ireland 53°13N 7°1W 10 C4
Rathdrum Ireland 52°56N 6°14W 10 D5
Rathenow Germany 52°37N 12°19E 16 B7
Rathkeale Ireland 52°32N 8°56W 10 D3
Rathlin I. U.K. 55°18N 6°14W 10 A5
Rathmelton Ireland 55°2N 7°38W 10 A4
Ratibor = Racibórz Poland 50°7N 18°18E 17 C10
Ratlam India 23°20N 75°0E 42 H6
Ratmanova Ostrov Russia 65°46N 169°6W 74 B6
Ratnagiri India 16°57N 73°18E 40 L8
Ratodero Pakistan 27°48N 68°18E 42 F3
Raton U.S.A. 36°54N 104°24W 77 H11
Rattaphum Thailand 7°8N 100°16E 39 J3
Rattray Hd. U.K. 57°38N 1°50W 11 D7
Ratz, Mt. Canada 57°23N 132°12W 70 B2
Raub Malaysia 3°47N 101°52E 39 L3
Rauch Argentina 36°45S 59°5W 94 D4
Raudales Mexico 17°27N 93°39W 87 D6
Raufarhöfn Iceland 66°27N 15°57W 8 C6
Raufoss Norway 60°44N 10°37E 8 F14
Raukumara Ra. N.Z. 38°5S 177°55E 59 C6
Rauma Finland 61°10N 21°30E 8 F19
Raumo = Rauma Finland 61°10N 21°30E 8 F19
Raung, Gunung Indonesia 8°8S 114°3E 37 J17
Raurkela India 22°14N 84°50E 43 H11
Rausu-Dake Japan 44°4N 145°7E 30 B12
Rava-Ruska Ukraine 50°15N 23°42E 17 C12
Rava Russkaya = Rava-Ruska Ukraine 50°15N 23°42E 17 C12
Ravalli U.S.A. 47°17N 114°11W 76 C6
Ravānsar Iran 34°43N 46°40E 44 C5
Rāvar Iran 31°20N 56°51E 45 D8
Ravena U.S.A. 42°28N 73°49W 83 D11
Ravenna Italy 44°25N 12°12E 22 B5
Ravenna Nebr., U.S.A. 41°1N 98°55W 80 E4
Ravenna Ohio, U.S.A. 41°9N 81°15W 82 E3
Ravensburg Germany 47°46N 9°36E 16 E5
Ravenshoe Australia 17°37S 145°29E 62 B4
Ravensthorpe Australia 33°35S 120°2E 61 F3
Ravenswood Australia 20°6S 146°54E 62 C4
Ravenswood U.S.A. 38°57N 81°46W 81 F13
Ravi → Pakistan 30°35N 71°49E 42 D4
Rawalpindi Pakistan 33°38N 73°8E 42 C5
Rawang Malaysia 3°20N 101°35E 39 L3
Rawene N.Z. 35°25S 173°32E 59 A4
Rawi, Ko Thailand 6°33N 99°14E 39 J2
Rawlinna Australia 30°58S 125°28E 61 F4
Rawlins U.S.A. 41°47N 107°14W 76 F10
Rawlinson Ra. Australia 24°40S 128°30E 61 D4
Rawson Argentina 43°15S 65°5W 96 E3
Raxaul India 26°59N 84°51E 43 F11
Ray U.S.A. 48°21N 103°10W 80 A2
Ray, C. Canada 47°33N 59°15W 73 C8
Ray Mts. U.S.A. 66°0N 152°0W 74 B9
Raya Ring, Ko Thailand 8°18N 98°29E 39 a
Rayadurg India 14°40N 76°50E 40 M10
Rayagada India 19°15N 83°20E 41 K13
Raychikhinsk Russia 49°46N 129°25E 29 E13
Rāyen Iran 29°34N 57°26E 45 D8
Rayleigh U.K. 51°36N 0°37E 13 F8
Raymond Canada 49°30N 112°35W 70 D6
Raymond Calif., U.S.A. 37°13N 119°54W 78 H7
Raymond N.H., U.S.A. 43°2N 71°11W 83 C13
Raymond Wash., U.S.A. 46°41N 123°44W 78 D3
Raymondville U.S.A. 26°29N 97°47W 84 H6
Raymore Canada 51°25N 104°31W 71 C8
Rayón Mexico 29°43N 110°35W 86 B2
Rayong Thailand 12°40N 101°20E 38 F3
Raystown L. U.S.A. 40°25N 78°5W 82 F6
Rayville U.S.A. 32°29N 91°46W 84 E9
Raz, Pte. du France 48°2N 4°47W 20 C1
Razan Iran 35°23N 49°2E 45 C6
Razāzah, Buhayrat ar Iraq 32°40N 43°35E 44 C4
Razazah, L. = Razāzah, Buhayrat ar Iraq 32°40N 43°35E 44 C4
Razdel'naya = Rozdilna Ukraine 46°50N 30°2E 17 E16
Razdolnoye Russia 43°30N 131°52E 30 C5
Razeh Iran 32°47N 48°9E 45 C6
Razgrad Bulgaria 43°33N 26°34E 23 C12
Razim, Lacul Romania 44°50N 29°0E 17 F15
Razmak Pakistan 32°45N 69°50E 42 C3
Re, Cu Lao Vietnam 15°22N 109°8E 38 E7
Ré, Î. de France 46°12N 1°30W 20 C3
Reading U.K. 51°27N 0°58W 13 F7
Reading U.S.A. 40°20N 75°56W 83 F9
Reading □ U.K. 51°27N 0°58W 13 F7
Realicó Argentina 35°0S 64°15W 94 D3
Ream Cambodia 10°34N 103°39E 39 G4
Ream △ Cambodia 10°30N 103°45E 39 G4
Reay Forest U.K. 58°22N 4°55W 11 C4
Rebecca, L. Australia 30°0S 122°15E 61 F3
Rebi Indonesia 6°23S 134°7E 37 F8
Rebiana, Sahrâ' Libya 24°30N 21°0E 51 D10
Reboly Russia 63°49N 30°47E 8 E24
Rebun-Tō Japan 45°23N 141°2E 30 B10
Recherche, Arch. of the Australia 34°15S 122°50E 61 F3
Rechna Doab Pakistan 31°35N 73°30E 42 D5
Rechytsa Belarus 52°21N 30°24E 17 B16
Recife Brazil 8°0S 35°0W 93 E12
Recife Seychelles 4°36S 55°42E 53 b
Recklinghausen Germany 51°37N 7°12E 15 C7
Reconquista Argentina 29°10S 59°45W 94 B4
Recreo Argentina 29°25S 65°10W 94 B2
Red = Hong → Vietnam 20°16N 106°34E 38 B5
Red → U.S.A. 31°1N 91°45W 84 F9
Red Bank U.S.A. 40°21N 74°5W 83 F10
Red Bay Canada 51°44N 56°25W 73 B8
Red Bluff U.S.A. 40°11N 122°15W 76 F2
Red Bluff Res. U.S.A. 31°54N 103°55W 77 L12
Red Cliffs Australia 34°19S 142°11E 63 E3
Red Cloud U.S.A. 40°5N 98°32W 80 E4
Red Creek U.S.A. 43°14N 76°45W 83 C8
Red Deer Canada 52°20N 113°50W 70 C6
Red Deer → Alta., Canada 50°58N 110°0W 71 C7
Red Deer → Man., Canada 52°53N 101°1W 71 C8
Red Deer L. Canada 52°55N 101°20W 71 C8
Red Hook U.S.A. 41°55N 73°53W 83 E11
Red Indian L. Canada 48°35N 57°0W 73 C8
Red L. Canada 51°3N 93°49W 71 C10
Red Lake Canada 51°3N 93°49W 71 C10
Red Lake Falls U.S.A. 47°53N 96°16W 80 B5
Red Lake Road Canada 49°59N 93°25W 71 C10
Red Lodge U.S.A. 45°11N 109°15W 76 D9
Red Mountain U.S.A. 35°37N 117°38W 79 K9
Red Oak U.S.A. 41°1N 95°14W 80 E6
Red River of the North → N. Amer. 49°0N 97°15W 80 A5
Red Rock Canada 48°55N 88°15W 72 C2
Red Rock, L. U.S.A. 41°22N 92°59W 80 E7
Red Rocks Pt. Australia 32°13S 127°32E 61 F4
Red Sea Asia 25°0N 36°0E 47 C2
Red Slate Mt. U.S.A. 37°31N 118°52W 78 H8
Red Sucker L. Canada 54°9N 93°40W 72 B1
Red Tower Pass = Turnu Roşu, P. Romania 45°33N 24°17E 17 F13
Red Wing U.S.A. 44°34N 92°31W 80 C7
Redang, Pulau Malaysia 5°49N 103°2E 39 K4
Redange Lux. 49°46N 5°52E 15 E5
Redcar U.K. 54°37N 1°4W 12 C6
Redcar & Cleveland □ U.K. 54°29N 1°0W 12 C7
Redcliff Canada 50°10N 110°50W 76 A8
Redcliffe Australia 27°12S 153°6E 63 D5
Redcliffe, Mt. Australia 28°30S 121°30E 61 E3
Reddersburg S. Africa 29°41S 26°10E 56 D4
Redding U.S.A. 40°35N 122°24W 76 F2
Redditch U.K. 52°18N 1°55W 13 E6
Redfield U.S.A. 44°53N 98°31W 80 C4
Redford U.S.A. 44°38N 73°48W 83 B11
Redhead Trin. & Tob. 10°44N 60°58W 93 K16
Redlands U.S.A. 34°4N 117°11W 79 M9
Redmond Oreg., U.S.A. 44°17N 121°11W 76 D3
Redmond Wash., U.S.A. 47°40N 122°7W 78 C4
Redon France 47°40N 2°6W 20 C2
Redonda Antigua & B. 16°58N 62°19W 89 C7
Redondela Spain 42°15N 8°38W 21 A1
Redondo Beach U.S.A. 33°50N 118°23W 79 M8
Redoubt Volcano U.S.A. 60°29N 152°45W 74 C9
Redruth U.K. 50°14N 5°14W 13 G2
Redvers Canada 49°35N 101°40W 71 D8
Redwater Canada 53°55N 113°6W 70 C6
Redwood U.S.A. 44°18N 75°48W 83 B9
Redwood △ U.S.A. 41°40N 124°5W 76 F1
Redwood City U.S.A. 37°30N 122°15W 78 H4
Redwood Falls U.S.A. 44°32N 95°7W 80 C6
Ree, L. Ireland 53°35N 8°0W 10 C3
Reed City U.S.A. 43°53N 85°31W 81 D11
Reed L. Canada 54°38N 100°30W 71 C8
Reedley U.S.A. 36°36N 119°27W 78 J7
Reedsburg U.S.A. 43°32N 90°0W 80 D8
Reedsport U.S.A. 43°42N 124°6W 76 E1
Reedsville U.S.A. 40°39N 77°35W 82 F7
Reefton N.Z. 42°6S 171°51E 59 E3
Reese → U.S.A. 40°48N 117°4W 76 F5
Refugio U.S.A. 28°18N 97°17W 84 G6
Regana, C. de Spain 39°25N 2°43E 24 B9
Regensburg Germany 49°1N 12°6E 16 D7
Reggâne = Zaouiet Reggâne Algeria 26°32N 0°3E 50 C6
Réggio di Calábria Italy 38°6N 15°39E 22 E6
Réggio nell'Emília Italy 44°43N 10°36E 22 B4
Reghin Romania 46°46N 24°42E 17 E13
Regina Canada 50°27N 104°35W 71 C8
Regina Beach Canada 50°47N 105°0W 71 C8
Registro Brazil 24°29S 47°49W 95 A6
Rehar → India 23°55N 82°40E 43 H10
Rehli India 23°38N 79°5E 43 H8
Rehoboth Namibia 23°15S 17°4E 56 C2
Rehovot Israel 31°54N 34°48E 46 D3
Reichenbach Germany 50°37N 12°17E 16 C7
Reid Australia 30°49S 128°26E 61 F4
Reidsville U.S.A. 36°21N 79°40W 85 C15
Reigate U.K. 51°14N 0°12W 13 F7
Reims France 49°15N 4°1E 20 B6
Reina Adelaida, Arch. Chile 52°20S 74°0W 96 G2
Reina Sofia, Tenerife ✈ (TFS) Canary Is. 28°3N 16°33W 24 F3
Reindeer → Canada 55°36N 103°11W 71 B8
Reindeer I. Canada 52°30N 98°0W 71 C9
Reindeer L. Canada 57°15N 102°15W 71 B8
Reinga, C. N.Z. 34°25S 172°43E 59 A4
Reinosa Spain 43°2N 4°15W 21 A3
Reitz S. Africa 27°48S 28°29E 57 D4
Reivilo S. Africa 27°36S 24°8E 56 D3
Reliance Canada 63°0N 109°20W 71 A7
Remanso Brazil 9°41S 42°4W 93 E10
Remarkable, Mt. Australia 32°48S 138°10E 63 E2
Rembang Indonesia 6°42S 111°21E 37 G14
Remedios Panama 8°15N 81°50W 88 E3
Remeshk Iran 26°55N 58°50E 45 E8
Remich Lux. 49°32N 6°22E 15 E6
Remscheid Germany 51°11N 7°12E 15 C7
Ren Xian China 37°8N 114°40E 34 F8
Rendang Indonesia 8°26S 115°25E 37 J18
Rendsburg Germany 54°17N 9°39E 16 A5
Renfrew Canada 45°30N 76°40W 83 A8
Renfrewshire □ U.K. 55°49N 4°38W 11 F4
Rengat Indonesia 0°30S 102°45E 36 E2
Rengo Chile 34°24S 70°50W 94 C1
Reni = Taranagar India 28°43N 74°50E 42 E6
Reni Ukraine 45°28N 28°15E 17 F15
Renmark Australia 34°11S 140°43E 63 E3
Rennell Sd. Canada 53°23N 132°35W 70 C2
Renner Springs Australia 18°20S 133°47E 62 B1
Rennes France 48°7N 1°41W 20 B3
Rennick Glacier Antarctica 70°30S 161°45E 5 D11
Rennie L. Canada 61°32N 105°35W 71 A7
Reno U.S.A. 39°31N 119°48W 78 F7
Reno → Italy 44°38N 12°16E 22 B5
Renovo U.S.A. 41°20N 77°45W 82 E7
Renqiu China 38°43N 116°5E 34 E9
Rensselaer Ind., U.S.A. 40°57N 87°9W 80 E10
Rensselaer N.Y., U.S.A. 42°38N 73°45W 83 D11
Renton U.S.A. 47°28N 122°12W 78 C4
Renukoot India 24°12N 83°2E 43 G10
Reotipur India 25°33N 83°45E 43 G10
Republic Mo., U.S.A. 37°7N 93°29W 80 G7
Republic Wash., U.S.A. 48°39N 118°44W 76 B4
Republican → U.S.A. 39°4N 96°48W 80 F5
Repulse B. Australia 20°35S 148°46E 62 b
Repulse Bay Canada 66°30N 86°30W 69 D14
Requena Peru 5°5S 73°52W 92 E4
Requena Spain 39°30N 1°4W 21 C5
Reserve U.S.A. 33°43N 108°45W 77 K9
Resht = Rasht Iran 37°20N 49°40E 45 B6
Resistencia Argentina 27°30S 59°0W 94 B4
Reşiţa Romania 45°18N 21°53E 17 F11
Reso = Raisio Finland 60°28N 22°11E 9 F20
Resolute Canada 74°42N 94°54W 69 C13
Resolution I. Canada 61°30N 65°0W 69 E19
Resolution I. N.Z. 45°40S 166°40E 59 F1
Ressano Garcia Mozam. 25°25S 32°0E 57 D5
Reston Canada 49°33N 101°6W 71 D8
Retalhuleu Guatemala 14°33N 91°46W 88 D1
Retenue, L. de Dem. Rep. of the Congo 11°0S 27°0E 55 E2
Retford U.K. 53°19N 0°56W 12 D7
Rethimno Greece 35°18N 24°30E 25 D6
Rethimno □ Greece 35°23N 24°28E 25 D6
Reti Pakistan 28°5N 69°48E 42 E3
Réunion ⊠ Ind. Oc. 21°0S 56°0E 53 c
Reus Spain 41°10N 1°5E 21 B6
Reutlingen Germany 48°29N 9°12E 16 D5
Reval = Tallinn Estonia 59°22N 24°48E 9 G21
Revda Russia 56°48N 59°57E 18 C10
Revelganj India 25°50N 84°40E 43 G11
Revelstoke Canada 51°0N 118°10W 70 C5
Reventazón Peru 6°10S 80°58W 92 E2
Revillagigedo, Is. de Pac. Oc. 18°40N 112°0W 86 D2
Revuè → Mozam. 19°50S 34°0E 55 F3
Rewa India 24°33N 81°25E 43 G9
Rewari India 28°15N 76°40E 42 E7
Rexburg U.S.A. 43°49N 111°47W 76 E8
Rey Iran 35°35N 51°25E 45 C6
Rey, I. del Panama 8°20N 78°30W 88 E4
Rey, L. del Mexico 27°1N 103°26W 86 B4
Rey Malabo Eq. Guin. 3°45N 8°50E 52 D1
Reyðarfjörður Iceland 65°2N 14°13W 8 D6
Reyes, Pt. U.S.A. 38°0N 123°0W 78 H3
Reykjahlið Iceland 65°40N 16°55W 8 D5
Reykjanes Iceland 63°48N 22°40W 8 E2
Reykjavík Iceland 64°10N 21°57W 8 D3
Reynolds Ra. Australia 22°30S 133°0E 60 D5
Reynoldsville U.S.A. 41°6N 78°53W 82 E6
Reynosa Mexico 26°7N 98°18W 87 B5
Rēzekne Latvia 56°30N 27°17E 9 H22
Rezvān Iran 27°34N 56°6E 45 E8
Rhayader U.K. 52°18N 3°29W 13 E4
Rhein → Europe 51°52N 6°2E 15 C6
Rhein-Main-Donau-Kanal Germany 49°1N 11°27E 16 D6
Rheine Germany 52°17N 7°26E 16 B4
Rheinland-Pfalz □ Germany 50°0N 7°0E 16 C4
Rhin = Rhein → Europe 51°52N 6°2E 15 C6
Rhine = Rhein → Europe 51°52N 6°2E 15 C6
Rhinebeck U.S.A. 41°56N 73°55W 83 E11
Rhineland-Palatinate = Rheinland-Pfalz □ Germany 50°0N 7°0E 16 C4
Rhinelander U.S.A. 45°38N 89°25W 80 C9
Rhinns Pt. U.K. 55°40N 6°29W 11 F2
Rhino Camp Uganda 3°0N 31°22E 54 B3
Rhir, Cap Morocco 30°38N 9°54W 50 B4
Rhode Island □ U.S.A. 41°40N 71°30W 83 E13
Rhodes Greece 36°15N 28°10E 25 C10
Rhodope Mts. = Rhodopi Planina Bulgaria 41°40N 24°20E 23 D11
Rhodopi Planina Bulgaria 41°40N 24°20E 23 D11
Rhön Germany 50°24N 9°58E 16 C5
Rhondda U.K. 51°39N 3°31W 13 F4
Rhondda Cynon Taff □ U.K. 51°42N 3°27W 13 F4
Rhône → France 43°28N 4°42E 20 E6
Rhum U.K. 57°0N 6°20W 11 E2
Rhyl U.K. 53°20N 3°29W 12 D4
Riachão Brazil 7°20S 46°37W 93 E9
Riasi India 33°10N 74°50E 43 C6
Riau □ Indonesia 0°0 102°35E 36 E2
Riau, Kepulauan □ Indonesia 0°30N 104°20E 36 D2
Riau Arch. = Riau, Kepulauan □ Indonesia 0°30N 104°20E 36 D2
Ribadeo Spain 43°35N 7°5W 21 A2
Ribas do Rio Pardo Brazil 20°27S 53°46W 93 H8
Ribauè Mozam. 14°57S 38°17E 55 E4
Ribble → U.K. 53°52N 2°25W 12 D5
Ribe Denmark 55°19N 8°44E 9 J13
Ribeira Brava Madeira 32°41N 17°4W 24 D2
Ribeira Grande C. Verde Is. 17°0N 25°4W 50 b
Ribeirão Prêto Brazil 21°10S 47°50W 95 A6
Riberalta Bolivia 11°0S 66°0W 92 F5
Riccarton N.Z. 43°32S 172°37E 59 E4
Rice U.S.A. 34°5N 114°51W 79 L12
Rice L. Canada 44°12N 78°10W 82 B6
Rice Lake U.S.A. 45°30N 91°44W 80 C8
Rich, C. Canada 44°43N 80°38W 82 B4
Richards Bay S. Africa 28°48S 32°6E 57 D5
Richardson → Canada 58°25N 111°14W 71 B6
Richardson Lakes U.S.A. 44°46N 70°58W 81 C18
Richardson Springs U.S.A. 39°51N 121°46W 78 F5
Riche, C. Australia 34°36S 118°47E 61 F2
Richey U.S.A. 47°39N 105°4W 76 C11
Richfield U.S.A. 38°46N 112°5W 76 G7
Richfield Springs U.S.A. 42°51N 74°59W 83 D10
Richford U.S.A. 45°0N 72°40W 83 B12
Richibucto Canada 46°42N 64°54W 73 C7
Richland Ga., U.S.A. 32°5N 84°40W 85 E12
Richland Wash., U.S.A. 46°17N 119°18W 76 C4
Richland Center U.S.A. 43°21N 90°23W 80 D8
Richlands U.S.A. 37°6N 81°48W 81 G13
Richmond Australia 20°43S 143°8E 62 C3
Richmond N.Z. 41°20S 173°12E 59 D4
Richmond U.K. 54°25N 1°43W 12 C6
Richmond Calif., U.S.A. 37°56N 122°21W 78 H4
Richmond Ind., U.S.A. 39°50N 84°53W 81 F11
Richmond Ky., U.S.A. 37°45N 84°18W 81 G11
Richmond Mich., U.S.A. 42°49N 82°45W 82 D2
Richmond Mo., U.S.A. 39°17N 93°58W 80 F7
Richmond Tex., U.S.A. 29°35N 95°46W 84 G7
Richmond Utah, U.S.A. 41°56N 111°48W 76 F8
Richmond Va., U.S.A. 37°33N 77°27W 81 G15
Richmond Vt., U.S.A. 44°24N 72°59W 83 B12
Richmond Hill Canada 43°52N 79°27W 82 C5
Richmond Ra. Australia 29°0S 152°45E 63 D5
Richmondville U.S.A. 42°38N 74°33W 83 D10
Richtersveld △ S. Africa 28°15S 17°10E 56 D2
Richville U.S.A. 44°25N 75°23W 83 B9
Richwood U.S.A. 38°14N 80°32W 81 F13
Ridder = Leninogorsk Kazakhstan 50°20N 83°30E 28 D9
Riddlesburg U.S.A. 40°9N 78°15W 82 F6
Ridgecrest U.S.A. 35°38N 117°40W 79 K9
Ridgefield Conn., U.S.A. 41°17N 73°30W 83 E11
Ridgefield Wash., U.S.A. 45°49N 122°45W 78 E4
Ridgeland Miss., U.S.A. 32°26N 90°8W 85 E9
Ridgeland S.C., U.S.A. 32°29N 80°59W 85 E14
Ridgetown Canada 42°26N 81°52W 82 D3
Ridgewood U.S.A. 40°59N 74°7W 83 F10
Ridgway U.S.A. 41°25N 78°44W 82 E6
Riding Mountain △ Canada 50°50N 100°0W 71 C9
Ridley, Mt. Australia 33°12S 122°7E 61 F3
Riebeek-Oos S. Africa 33°10S 26°10E 56 E4
Ried Austria 48°14N 13°30E 16 D7
Riesa Germany 51°17N 13°17E 16 C7
Riet → S. Africa 29°0S 23°54E 56 D3
Rietbron S. Africa 32°54S 23°10E 56 E3
Rietfontein Namibia 21°58S 20°58E 56 C3
Rieti Italy 42°24N 12°51E 22 C5
Rif = Er Rif Morocco 35°1N 4°1W 50 A5
Riffe L. U.S.A. 46°32N 122°26W 78 D4
Rifle U.S.A. 39°32N 107°47W 76 G10
Rift Valley Africa 7°0N 30°0E 48 G7
Rift Valley □ Kenya 0°20N 36°0E 54 B4
Riga Latvia 56°53N 24°8E 9 H21
Riga, G. of Latvia 57°40N 23°45E 9 H20
Rīgān Iran 28°37N 58°58E 45 D8
Rigas Jūras Līcis = Riga, G. of Latvia 57°40N 23°45E 9 H20
Rigaud Canada 45°29N 74°18W 83 A10
Rigby U.S.A. 43°40N 111°55W 76 E8
Rīgestān Afghan. 30°15N 65°0E 40 D4
Riggins U.S.A. 45°25N 116°19W 76
Rigolet Canada 54°10N 58°23W 73
Rihand Dam India 24°9N 83°2E 43
Riihimäki Finland 60°45N 24°48E 8
Riiser-Larsen-halvøya Antarctica 68°0S 35°0E 5
Riiser-Larsen Ice Shelf S. Ocean 74°0S 19°0W 5
Riiser-Larsen Sea S. Ocean 67°30S 22°0E 5
Rijeka Croatia 45°20N 14°21E 16
Rijssen Neths. 52°19N 6°31E 15
Rikuchū-Kaigan △ Japan 39°20N 142°0E 30
Rikuzentakada Japan 39°0N 141°40E 30
Riley U.S.A. 43°32N 119°28W 76
Rima → Nigeria 13°4N 5°10E 50
Rimah, Wadi ar → Si. Arabia 26°5N 41°30E 44
Rimau, Pulau Malaysia 5°15N 100°16E 3
Rimbey Canada 52°35N 114°15W 70
Rimersburg U.S.A. 41°3N 79°30W 82
Rímini Italy 44°3N 12°33E 22
Rimouski Canada 48°27N 68°30W 73
Rimrock U.S.A. 46°40N 121°7W 78
Rinca Indonesia 8°45S 119°35E 37
Rincón de Romos Mexico 22°14N 102°18W 86
Rinconada Argentina 22°26S 66°10W 94
Rind → India 25°53N 80°33E 43
Ringas India 27°21N 75°34E 42
Ringgold Is. Fiji 16°15S 179°25W 5
Ringkøbing Denmark 56°5N 8°15E 9
Ringvassøya Norway 69°56N 19°15E 8
Ringwood U.S.A. 41°7N 74°15W 83
Rinjani, Gunung Indonesia 8°24S 116°28E 36
Rio Branco Brazil 9°58S 67°49W 92
Río Branco Uruguay 32°40S 53°40W 95
Río Bravo Mexico 25°59N 98°6W 87
Rio Bravo ⌓ N. Amer. 29°2N 102°45W 86
Río Bravo del Norte → Mexico 25°57N 97°9W 87
Rio Brilhante Brazil 21°48S 54°33W 95
Rio Claro Brazil 22°19S 47°35W 95
Rio Claro Trin. & Tob. 10°20N 61°25W 93
Río Colorado Argentina 39°0S 64°0W 96
Río Cuarto Argentina 33°10S 64°25W 94
Rio das Pedras Mozam. 23°8S 35°28E 57
Rio de Janeiro Brazil 22°54S 43°12W 95
Rio de Janeiro □ Brazil 22°50S 43°0W 95
Rio do Sul Brazil 27°13S 49°37W 95
Río Dulce △ Guatemala 15°43N 88°50W 88
Río Gallegos Argentina 51°35S 69°15W 96
Río Grande Argentina 53°50S 67°45W 96
Rio Grande Brazil 32°0S 52°20W 95
Río Grande Mexico 23°50N 103°2W 86
Río Grande Puerto Rico 18°23N 65°50W 8
Rio Grande → N. Amer. 25°58N 97°9W 84
Rio Grande City U.S.A. 26°23N 98°49W 84
Río Grande de Santiago → Mexico 21°36N 105°26W 86
Rio Grande do Norte □ Brazil 5°40S 36°0W 93
Rio Grande do Sul □ Brazil 30°0S 53°0W 95
Río Hato Panama 8°22N 80°10W 88
Río Lagartos Mexico 21°36N 88°10W 87
Rio Largo Brazil 9°28S 35°50W 93
Río Mulatos Bolivia 19°40S 66°50W 92
Río Muni □ Eq. Guin. 1°30N 10°0E 52
Rio Negro Brazil 26°0S 49°55W 95
Rio Pardo Brazil 30°0S 52°30W 95
Río Pilcomayo △ Argentina 25°5S 58°5W 94
Río Platano ⌓ Honduras 15°45N 85°0W 88
Rio Rancho U.S.A. 35°14N 106°41W 77
Río Segundo Argentina 31°40S 63°59W 94
Río Tercero Argentina 32°15S 64°8W 94
Rio Verde Brazil 17°50S 51°0W 93
Río Verde Mexico 21°56N 99°59W 87
Rio Vista U.S.A. 38°10N 121°42W 78
Riobamba Ecuador 1°50S 78°45W 92
Riohacha Colombia 11°33N 72°55W 92
Riosucio Colombia 7°27N 77°7W 92
Riou L. Canada 59°7N 106°25W 71
Ripley Canada 44°4N 81°35W 82
Ripley Calif., U.S.A. 33°32N 114°39W 79
Ripley N.Y., U.S.A. 42°16N 79°43W 82
Ripley Tenn., U.S.A. 35°45N 89°32W 85
Ripley W. Va., U.S.A. 38°49N 81°43W 81
Ripon U.K. 54°9N 1°31W 12
Ripon Calif., U.S.A. 37°44N 121°7W 78
Ripon Wis., U.S.A. 43°51N 88°50W 80
Rishā', W. ar → Si. Arabia 25°33N 44°5E 44
Rishiri-Rebun-Sarobetsu △ Japan 45°26N 141°30E 30
Rishiri-Tō Japan 45°11N 141°15E 30
Rishon le Ziyyon Israel 31°58N 34°48E 46
Rison U.S.A. 33°58N 92°11W 84
Risør Norway 58°43N 9°13E 9
Rita Blanca Cr. → U.S.A. 35°40N 102°29W 84
Ritter, Mt. U.S.A. 37°41N 119°12W 78
Rittman U.S.A. 40°58N 81°47W 82
Ritzville U.S.A. 47°8N 118°23W 76
Riva del Garda Italy 45°53N 10°50E 22
Rivadavia B. Aires, Argentina 35°29S 62°59W 94
Rivadavia Mendoza, Argentina 33°13S 68°30W 94
Rivadavia Salta, Argentina 24°5S 62°54W 94
Rivadavia Chile 29°57S 70°35W 94
Rivas Nic. 11°30N 85°50W 88
Rivash Iran 35°28N 58°26E 45
River Cess Liberia 5°30N 9°32W 50
River Jordan Canada 48°26N 124°3W 78
Rivera Argentina 37°12S 63°14W 94
Rivera Uruguay 31°0S 55°50W 95
Riverbank U.S.A. 37°44N 120°56W 78
Riverdale U.S.A. 36°26N 119°52W 78

erhead U.S.A. 40°55N 72°40W 83 F12
erhurst Canada 50°55N 106°50W 71 C7
ers Canada 50°2N 100°14W 71 C8
ers Inlet Canada 51°42N 127°15W 70 C3
ersdale Canada 44°55N 81°20W 82 B3
ersdale S. Africa 34°7S 21°15E 56 E3
erside U.S.A. 33°59N 117°22W 79 M9
erton Australia 34°10S 138°46E 63 E2
erton Canada 51°1N 97°0W 71 C9
erton N.Z. 46°21S 168°0E 59 G2
erton U.S.A. 43°2N 108°23W 76 E9
era = Côte d'Azur
rance 43°25N 7°10E 20 E7
era U.S.A. 35°4N 114°35W 79 K12
era di Levante Italy 44°15N 9°30E 20 D8
era di Ponente Italy 44°10N 8°20E 20 D8
ère-au-Renard
anada 48°59N 64°23W 73 C7
ère-du-Loup Canada 47°50N 69°30W 73 C6
ère-Pentecôte Canada 49°57N 67°1W 73 C6
ère-Pilote Martinique 14°26N 60°53W 88 c
ère St-Paul Canada 51°28N 57°45W 73 B8
ère-Salée Martinique 14°31N 61°0W 88 c
ne Ukraine 50°40N 26°10E 17 C14
ne □ Ukraine 51°15N 26°30E 17 C14
oli Italy 45°3N 7°31E 20 D7
oli B. Australia 37°32S 140°3E 63 F3
adh = Ar Riyāḑ
. Arabia 24°41N 46°42E 44 E5
adh al Khabrā'
. Arabia 26°2N 43°33E 44 E4
Turkey 41°0N 40°30E 19 F7
nao China 35°25N 119°30E 35 G10
okarpaso Cyprus 35°36N 34°23E 25 D13
uto, C. Italy 38°53N 17°5E 22 E7
kan Norway 59°54N 8°33E 9 G13
d Town Br. Virgin Is. 18°27N 64°37W 89 e
n Plateau U.S.A. 39°20N 109°20W 76 G9
nne France 46°3N 4°4E 20 C6
noke Ala., U.S.A. 33°9N 85°22W 85 E12
noke Va., U.S.A. 37°16N 79°56W 81 G14
noke → U.S.A. 35°57N 76°42W 85 D16
noke I. U.S.A. 35°53N 75°39W 85 D17
noke Rapids U.S.A. 36°28N 77°40W 85 C16
tán Honduras 16°18N 86°35W 88 C2
āt Sang Iran 35°35N 59°10E 45 C8
āṭkarīm Iran 35°25N 50°59E 45 C6
bins I. Australia 40°42S 145°0E 63 G4
e → Australia 21°42S 116°15E 60 D2
ert Bourassa, Rés.
anada 53°40N 76°55W 72 B4
ert Lee U.S.A. 31°54N 100°29W 84 F4
ertsdale U.S.A. 40°11N 78°6W 82 F6
ertsganj India 24°44N 83°4E 43 G10
ertson S. Africa 33°46S 19°50E 56 E2
ertson I. Antarctica 65°15S 59°30W 5 C18
ertson Ra. Australia 23°15S 121°0E 60 D3
ertstown Australia 33°58S 139°5E 63 E2
erval Canada 48°32N 72°15W 73 C5
eson Chan. N. Amer. 82°0N 61°30W 69 A19
esonia U.S.A. 40°21N 76°8W 83 F8
inson U.S.A. 39°0N 87°44W 80 F10
inson → Australia 16°3S 137°16E 62 B2
inson Crusoe I.
ac. Oc. 33°38S 78°52W 90 F2
inson Ra. Australia 25°40S 119°0E 61 E2
invale Australia 34°40S 142°45E 63 E3
lin Canada 51°14N 101°21W 71 C8
oré Bolivia 18°10S 59°45W 92 G7
son, Mt. Canada 53°10N 119°10W 70 C5
stown U.S.A. 27°47N 97°40W 84 H6
a, C. da Portugal 38°40N 9°31W 21 C1
a Partida, I. Mexico 19°1N 112°2W 86 D2
as, Atol das Brazil 4°0S 34°1W 93 D12
ha Uruguay 34°30S 54°25W 95 C5
dale U.K. 53°38N 2°9W 12 D5
hefort Belgium 50°9N 5°12E 15 D5
hefort France 45°56N 0°57W 20 D3
helle U.S.A. 41°56N 89°4W 80 E9
her River Canada 61°23N 112°44W 70 A6
hester U.K. 51°23N 0°31E 13 F8
hester Ind., U.S.A. 41°4N 86°13W 80 E10
hester Minn., U.S.A. 44°1N 92°28W 80 C7
hester N.H., U.S.A. 43°18N 70°59W 83 C14
hester N.Y., U.S.A. 43°10N 77°37W 82 C7
k → Canada 60°7N 127°7W 70 A3
k, The Australia 35°15S 147°2E 63 F4
k Creek U.S.A. 41°40N 80°52W 82 E4
k Falls U.S.A. 41°47N 89°41W 80 E9
k Hill U.S.A. 34°56N 81°1W 85 D14
k Island U.S.A. 41°30N 90°34W 80 E8
k Port U.S.A. 40°25N 95°31W 80 E6
k Rapids U.S.A. 43°26N 96°10W 80 D5
k Sound Bahamas 24°54N 76°12W 88 B4
k Springs Mont.,
.S.A. 46°49N 106°15W 76 C10
k Springs Wyo.,
.S.A. 41°35N 109°14W 76 F9
k Valley U.S.A. 43°12N 96°18W 80 D5
kall Atl. Oc. 57°37N 13°42W 6 D3
kdale Tex., U.S.A. 30°39N 97°0W 84 F6
kdale Wash., U.S.A. 47°22N 121°28W 78 C5
keby = Mungkan Kandju △
ustralia 13°35S 142°52E 62 A3
kefeller Plateau
ntarctica 76°0S 130°0W 5 E14
kford U.S.A. 42°16N 89°6W 80 D9
kglen Canada 49°11N 105°57W 71 D7
khampton Australia 23°22S 150°32E 62 C5
kingham Australia 32°15S 115°38E 61 F2
kingham N.C.,
.S.A. 34°57N 79°46W 85 D15
kingham Vt.,
.S.A. 43°11N 72°29W 83 C12
kingham B. Australia 18°5S 146°10E 62 B4
klake U.S.A. 48°47N 99°15W 80 A4
kland Canada 45°33N 75°17W 83 A9
kland Idaho, U.S.A. 42°34N 112°53W 76 E7
kland Maine, U.S.A. 44°6N 69°7W 81 C19
kland Mich., U.S.A. 46°44N 89°11W 80 B9

Rocklin U.S.A. 38°48N 121°14W 78 G5
Rockly B. Trin. & Tob. 11°9N 60°46W 93 J16
Rockmart U.S.A. 34°0N 85°3W 85 D12
Rockport Mass., U.S.A. 42°39N 70°37W 83 D14
Rockport Tex., U.S.A. 28°2N 97°3W 84 G6
Rocksprings U.S.A. 30°1N 100°13W 84 F4
Rockville Conn., U.S.A. 41°52N 72°28W 83 E12
Rockville Md., U.S.A. 39°5N 77°9W 81 F15
Rockwall U.S.A. 32°56N 96°28W 84 E6
Rockwell City U.S.A. 42°24N 94°38W 80 D6
Rockwood Canada 43°37N 80°8W 82 C4
Rockwood Maine, U.S.A. 45°41N 69°45W 81 C19
Rockwood Tenn., U.S.A. 35°52N 84°41W 85 D12
Rocky Ford U.S.A. 38°3N 103°43W 76 G12
Rocky Gully Australia 34°30S 116°57E 61 F2
Rocky Harbour Canada 49°36N 57°55W 73 C8
Rocky Island L. Canada 46°55N 83°0W 72 C3
Rocky Lane Canada 58°31N 116°22W 70 B5
Rocky Mount U.S.A. 35°57N 77°48W 85 D16
Rocky Mountain △ U.S.A. 40°25N 105°45W 76 F11
Rocky Mountain House Canada 52°22N 114°55W 70 C6
Rocky Mts. N. Amer. 49°0N 115°0W 76 B6
Rocky Point Namibia 19°3S 12°30E 56 B2
Rod Pakistan 28°10N 63°5E 40 E3
Roda Greece 39°48N 19°46E 25 A3
Rødbyhavn Denmark 54°39N 11°22E 9 J14
Roddickton Canada 50°51N 56°8W 73 B8
Rodez France 44°21N 2°33E 20 D5
Ródhos = Rhodes Greece 36°15N 28°10E 25 C10
Rodia Greece 35°22N 25°1E 25 D7
Rodna ✿ Australia 23°45S 132°4E 60 D5
Rodney Canada 42°34N 81°41W 82 D3
Rodney, C. N.Z. 36°17S 174°50E 59 B5
Rodopos Greece 35°34N 23°45E 25 D5
Rodrigues Ind. Oc. 19°45S 63°20E 3 E13
Roe → U.K. 55°6N 6°59W 10 A5
Roebling U.S.A. 40°7N 74°47W 83 F10
Roebourne Australia 20°44S 117°9E 60 D2
Roebuck B. Australia 18°5S 122°20E 60 C3
Roebuck Roadhouse Australia 17°59S 122°36E 60 C3
Roermond Neths. 51°12N 6°0E 15 C6
Roes Welcome Sd. Canada 65°0N 87°0W 69 E14
Roeselare Belgium 50°57N 3°7E 15 D3
Rogachev = Ragachow Belarus 53°8N 30°5E 17 B16
Rogagua, L. Bolivia 13°43S 66°50W 92 F5
Rogatyn Ukraine 49°24N 24°36E 17 D13
Rogers U.S.A. 36°20N 94°7W 84 C7
Rogers City U.S.A. 45°25N 83°49W 81 C12
Rogersville Canada 46°44N 65°26W 73 C6
Roggan → Canada 54°24N 79°25W 72 B4
Roggan L. Canada 54°8N 77°50W 72 B4
Roggeveen Basin Pac. Oc. 31°30S 95°30W 65 L18
Roggeveldberge S. Africa 32°10S 20°10E 56 E3
Rogoaguado, L. Bolivia 13°0S 65°30W 92 F5
Rogojampi Indonesia 8°19S 114°17E 37 J17
Rogue → U.S.A. 42°26N 124°26W 76 E1
Rohnert Park U.S.A. 38°16N 122°40W 78 G4
Rohri Pakistan 27°45N 68°51E 42 F3
Rohri Canal Pakistan 26°15N 68°27E 42 F3
Rohtak India 28°55N 76°43E 42 E7
Roi Et Thailand 16°4N 103°40E 38 D4
Roja Latvia 57°29N 22°43E 9 H20
Rojas Argentina 34°10S 60°45W 94 C3
Rojo, C. Mexico 21°33N 97°20W 87 C5
Rokan → Indonesia 2°0N 100°50E 36 D2
Rokiškis Lithuania 55°55N 25°35E 9 J21
Rolândia Brazil 23°18S 51°23W 95 A5
Rolla Mo., U.S.A. 37°57N 91°46W 80 G8
Rolla N. Dak., U.S.A. 48°52N 99°37W 80 A4
Rolleston Australia 24°28S 148°35E 62 C4
Rollingstone Australia 19°2S 146°24E 62 B4
Roma Australia 26°32S 148°49E 63 D4
Roma Italy 41°54N 12°28E 22 D5
Roma Sweden 57°32N 18°26E 9 H18
Roma-Los Saenz U.S.A. 26°24N 99°1W 84 H5
Romain, C. U.S.A. 33°0N 79°22W 85 E15
Romaine → Canada 50°18N 63°47W 73 B7
Roman Romania 46°57N 26°55E 17 E14
Romang Indonesia 7°30S 127°20E 37 F7
Români Egypt 30°59N 32°38E 46 E1
Romania ■ Europe 46°0N 25°0E 17 F12
Romano, Cayo Cuba 22°0N 77°30W 88 B4
Romans-sur-Isère France 45°3N 5°3E 20 D6
Romanzof C. U.S.A. 61°49N 166°6W 74 C6
Romblon Phil. 12°33N 122°17E 37 B6
Rome = Roma Italy 41°54N 12°28E 22 D5
Rome Ga., U.S.A. 34°15N 85°10W 85 D12
Rome N.Y., U.S.A. 43°13N 75°27W 83 C9
Rome Pa., U.S.A. 41°51N 76°21W 83 E8
Romney U.S.A. 39°21N 78°45W 81 F14
Romney Marsh U.K. 51°2N 0°54E 13 F8
Rømø Denmark 55°10N 8°30E 9 J13
Romorantin-Lanthenay France 47°21N 1°45E 20 C4
Rompin → Malaysia 2°49N 103°29E 39 L4
Romsdalen Norway 62°25N 7°52E 8 E12
Romsey U.K. 51°0N 1°29W 13 G6
Ron Vietnam 17°53N 106°27E 38 D6
Rona U.K. 57°34N 5°59W 11 D3
Ronan U.S.A. 47°32N 114°6W 76 C6
Roncador, Cayos Colombia 13°32N 80°4W 88 D3
Roncador, Serra do Brazil 12°30S 52°30W 93 F8
Ronda Spain 36°46N 5°12W 21 D3
Rondane Norway 61°57N 9°50E 8 F13
Rondônia □ Brazil 11°0S 63°0W 92 F6
Rondonópolis Brazil 16°28S 54°38W 93 G8
Rondu Pakistan 35°32N 75°10E 43 B6
Rong, Koh Cambodia 10°45N 103°15E 39 G4
Rongcheng China 37°9N 122°23E 35 F12

Ronge, L. la Canada 55°6N 105°17W 71 B7
Rønne Denmark 55°6N 14°43E 9 J16
Ronne Ice Shelf Antarctica 77°30S 60°0W 5 D18
Ronsard, C. Australia 24°46S 113°10E 61 D1
Ronse Belgium 50°45N 3°35E 15 D3
Roodepoort S. Africa 26°11S 27°54E 57 D4
Roof Butte U.S.A. 36°28N 109°5W 77 H9
Rooiboklaagte → Namibia 20°50S 21°0E 56 C3
Rooniu, Mt. Tahiti 17°49S 149°12W 59 d
Roorkee India 29°52N 77°59E 42 E7
Roosendaal Neths. 51°32N 4°29E 15 C4
Roosevelt U.S.A. 40°18N 109°59W 76 F9
Roosevelt → Brazil 7°35S 60°20W 92 E6
Roosevelt, Mt. Canada 58°26N 125°20W 70 B3
Roosevelt I. Antarctica 79°30S 162°0W 5 D12
Roper → Australia 14°43S 135°27E 62 A2
Roper Bar Australia 14°44S 134°44E 62 A1
Roper River = St. Vidgeon's ✿ Australia 14°47S 134°53E 62 A1
Roque Pérez Argentina 35°25S 59°24W 94 D4
Roquetas de Mar Spain 36°46N 2°36W 21 D4
Roraima □ Brazil 2°0N 61°30W 92 C6
Roraima, Mt. Venezuela 5°10N 60°40W 92 B6
Røros Norway 62°35N 11°23E 8 E14
Ros Comáin = Roscommon Ireland 53°38N 8°11W 10 C3
Rosa Zambia 9°33S 31°15E 55 D3
Rosa, Monte Europe 45°57N 7°53E 20 D7
Rosalia U.S.A. 47°14N 117°22W 76 C5
Rosamond U.S.A. 34°52N 118°10W 79 L8
Rosario Argentina 33°0S 60°40W 94 C3
Rosário Brazil 3°0S 44°15W 93 D10
Rosario Baja Calif., Mexico 30°0N 115°50W 86 B1
Rosario Sinaloa, Mexico 22°58N 105°53W 86 C3
Rosario Paraguay 24°30S 57°35W 94 A4
Rosario de la Frontera Argentina 25°50S 65°0W 94 B3
Rosario de Lerma Argentina 24°59S 65°35W 94 A2
Rosario del Tala Argentina 32°20S 59°10W 94 C4
Rosário do Sul Brazil 30°15S 54°55W 95 C5
Rosarito Mexico 32°20N 117°2W 79 N9
Roscoe U.S.A. 41°56N 74°55W 83 E10
Roscommon Ireland 53°38N 8°11W 10 C3
Roscommon □ Ireland 53°49N 8°23W 10 C3
Roscrea Ireland 52°57N 7°49W 10 D4
Rose → Australia 14°16S 135°45E 62 A2
Rose, L. Bahamas 21°0N 73°30W 89 B5
Rose Belle Mauritius 20°24S 57°36E 53 d
Rose Blanche Canada 47°38N 58°45W 73 C8
Rose Hill Mauritius 20°14S 57°27E 53 d
Rose Pt. Canada 54°11N 131°39W 70 C2
Rose Valley Canada 52°19N 103°49W 71 C8
Roseau Dominica 15°17N 61°24W 89 C7
Roseau U.S.A. 48°51N 95°46W 80 A6
Rosebery Australia 41°46S 145°33E 63 G4
Rosebud S. Dak., U.S.A. 43°14N 100°51W 80 D3
Rosebud Tex., U.S.A. 31°4N 96°59W 84 F6
Roseburg U.S.A. 43°13N 123°20W 76 E2
Rosedale U.S.A. 33°51N 91°2W 85 E9
Rosehearty U.K. 57°42N 2°7W 11 D6
Roseires Res. Sudan 11°51N 34°23E 51 F12
Roseland U.S.A. 38°25N 122°43W 78 G4
Rosemary Canada 50°46N 112°5W 70 C6
Rosenberg U.S.A. 29°34N 95°49W 84 G7
Rosenheim Germany 47°51N 12°7E 16 E7
Roses, G. de Spain 42°10N 3°15E 21 A7
Rosetown Canada 51°35N 107°59W 71 C7
Roseville Calif., U.S.A. 38°45N 121°17W 78 G5
Roseville Mich., U.S.A. 42°30N 82°56W 82 D2
Roseville Pa., U.S.A. 41°51N 76°57W 83 E8
Rosewood Australia 27°38S 152°36E 63 D5
Roshkhvār Iran 34°58N 59°37E 45 C8
Rosignano Marittimo Italy 43°24N 10°28E 22 C4
Rosignol Guyana 6°15N 57°30W 92 B7
Roşiori de Vede Romania 44°9N 25°0E 17 F13
Roskilde Denmark 55°38N 12°3E 9 J15
Roslavl Russia 53°57N 32°55E 18 D5
Rosmead S. Africa 31°29S 25°8E 56 E4
Ross Australia 42°2S 147°30E 63 G4
Ross N.Z. 42°53S 170°49E 59 E3
Ross Dependency Antarctica 76°0S 170°0W 5 D12
Ross I. Antarctica 77°30S 168°0E 5 D11
Ross Ice Shelf Antarctica 80°0S 180°0E 5 E12
Ross L. U.S.A. 48°44N 121°4W 76 B3
Ross Mhic Thriúin = New Ross Ireland 52°23N 6°57W 10 D5
Ross-on-Wye U.K. 51°54N 2°34W 13 F5
Ross River Australia 23°44S 134°30E 62 C1
Ross River Canada 62°30N 131°30W 70 A2
Ross Sea Antarctica 74°0S 178°0E 5 D11
Rossall Pt. U.K. 53°55N 3°3W 12 D4
Rossan Pt. Ireland 54°42N 8°47W 10 B3
Rossano Italy 39°36N 16°39E 22 E7
Rossburn Canada 50°40N 100°49W 71 C8
Rosseau Canada 45°16N 79°39W 82 A5
Rosseau, L. Canada 45°10N 79°35W 82 A5
Rosses, The Ireland 55°2N 8°20W 10 A3
Rossignol, L. Canada 52°43N 73°40W 72 B5
Rossignol L. Canada 44°12N 65°10W 73 D6
Rossiya = Russia ■ Eurasia 62°0N 105°0E 29 C11
Rossland Canada 49°6N 117°50W 70 D5
Rosslare Ireland 52°17N 6°24W 10 D5
Rosslare Harbour Ireland 52°15N 6°20W 10 D5
Rossmore Canada 44°8N 77°23W 82 B7
Rosso Mauritania 16°40N 15°45W 50 E2
Rossosh Russia 50°15N 39°28E 19 D6
Røssvatnet Norway 65°45N 14°5E 8 D16
Røst Norway 67°32N 12°10E 8 C15
Rosthern Canada 52°40N 106°20W 71 C7
Rostock Germany 54°5N 12°8E 16 A7
Rostov Don, Russia 47°15N 39°45E 19 E6
Rostov Yaroslavl, Russia 57°14N 39°25E 18 C6

Roswell Ga., U.S.A. 34°2N 84°22W 85 D12
Roswell N. Mex., U.S.A. 33°24N 104°32W 77 K11
Rotan U.S.A. 32°51N 100°28W 84 E4
Rother → U.K. 50°59N 0°45E 13 G8
Rothera Antarctica 67°34S 68°8W 5 C17
Rotherham U.K. 53°26N 1°20W 12 D6
Rothes U.K. 57°32N 3°13W 11 D5
Rothesay Canada 45°23N 66°0W 73 C6
Rothesay U.K. 55°50N 5°3W 11 F3
Roti Indonesia 10°50S 123°0E 37 F6
Roto Australia 33°0S 145°30E 63 E4
Rotondo, Mte. France 42°14N 9°8E 20 E8
Rotoroa, L. N.Z. 41°55S 172°39E 59 D4
Rotorua N.Z. 38°9S 176°16E 59 C6
Rotorua, L. N.Z. 38°5S 176°18E 59 C6
Rotterdam Neths. 51°55N 4°30E 15 C4
Rotterdam U.S.A. 42°48N 74°1W 83 D10
Rottnest I. Australia 32°0S 115°27E 61 F2
Rottumeroog Neths. 53°33N 6°34E 15 A6
Rottweil Germany 48°9N 8°37E 16 D5
Rotuma Fiji 12°25S 177°5E 58 C10
Roubaix France 50°40N 3°10E 20 A5
Rouen France 49°27N 1°4E 20 B4
Rouleau Canada 50°10N 104°56W 71 C8
Roulpmaulpma ✿ Australia 23°52S 133°0E 60 D5
Round I. Mauritius 19°51S 57°45E 53 d
Round Mountain U.S.A. 38°43N 117°4W 76 G5
Round Mt. Australia 30°26S 152°16E 63 E5
Round Rock U.S.A. 30°31N 97°41W 84 F6
Roundup U.S.A. 46°27N 108°33W 76 C9
Rousay U.K. 59°10N 3°2W 11 B5
Rouses Point U.S.A. 44°59N 73°22W 83 B11
Rouseville U.S.A. 41°28N 79°42W 82 E5
Rousse = Ruse Bulgaria 43°48N 25°59E 23 C12
Roussillon France 42°30N 2°35E 20 E5
Rouxville S. Africa 30°25S 26°50E 56 E4
Rouyn-Noranda Canada 48°20N 79°0W 72 C4
Rovaniemi Finland 66°29N 25°41E 8 C21
Rovereto Italy 45°53N 11°3E 22 B4
Rovigo Italy 45°4N 11°47E 22 B4
Rovinj Croatia 45°5N 13°40E 16 F7
Rovno = Rivne Ukraine 50°40N 26°10E 17 C14
Rovuma = Ruvuma → Tanzania 10°29S 40°28E 55 E5
Row'ān Iran 35°8N 48°51E 45 C6
Rowena Australia 29°48S 148°55E 63 D4
Rowley I. Canada 69°6N 77°52W 69 D16
Rowley Shoals Australia 17°30S 119°0E 60 C2
Roxas Phil. 11°36N 122°49E 37 B6
Roxboro U.S.A. 36°24N 78°59W 85 C15
Roxborough Trin. & Tob. 11°15N 60°35W 93 J16
Roxburgh N.Z. 45°33S 169°19E 59 F2
Roxbury N.Y., U.S.A. 42°17N 74°33W 83 D10
Roxbury Pa., U.S.A. 40°6N 77°39W 82 F7
Roxby Downs Australia 30°43S 136°46E 63 E2
Roy Mont., U.S.A. 47°20N 108°58W 76 C9
Roy N. Mex., U.S.A. 35°57N 104°12W 77 J11
Roy Utah, U.S.A. 41°10N 112°2W 76 F7
Royal Bardia △ Nepal 28°20N 81°20E 43 E9
Royal Canal Ireland 53°30N 7°13W 10 C4
Royal Chitawan △ Nepal 26°30N 84°30E 43 F11
Royal Geographical Society Is. Canada 68°56N 100°15W 68 D11
Royal Leamington Spa U.K. 52°18N 1°31W 13 E6
Royal Natal △ S. Africa 28°43S 28°51E 57 D4
Royal Tunbridge Wells U.K. 51°7N 0°16E 13 F8
Royale, Isle U.S.A. 48°0N 88°54W 80 B9
Royan France 45°37N 1°2W 20 D3
Royston U.K. 52°3N 0°0 13 E7
Rozdilna Ukraine 46°50N 30°2E 17 E16
Rozhyshche Ukraine 50°54N 25°15E 17 C13
Rtishchevo Russia 52°18N 43°46E 18 D7
Ruacaná Namibia 17°27S 14°21E 56 B1
Ruaha △ Tanzania 7°41S 34°30E 54 D3
Ruahine Ra. N.Z. 39°55S 176°2E 59 C6
Ruapehu N.Z. 39°17S 175°35E 59 C5
Ruapuke I. N.Z. 46°46S 168°31E 59 G2
Ruâq, W. → Egypt 30°0N 33°49E 46 F2
Rub' al Khālī Si. Arabia 19°0N 48°0E 47 D4
Rubeho Mts. Tanzania 6°50S 36°25E 54 D4
Rubh a' Mhail U.K. 55°56N 6°8W 11 F2
Rubha Hunish U.K. 57°42N 6°20W 11 D2
Rubha Robhanais = Lewis, Butt of U.K. 58°31N 6°16W 11 C2
Rubicon → U.S.A. 38°53N 121°4W 78 G5
Rubio Venezuela 7°43N 72°22W 92 B4
Rubondo △ Tanzania 2°18S 31°58E 54 C3
Rubtsovsk Russia 51°30N 81°10E 32 A5
Ruby L. U.S.A. 40°10N 115°28W 76 F6
Ruby Mts. U.S.A. 40°30N 115°20W 76 F6
Rubyvale Australia 23°25S 147°42E 62 C4
Rūd Sar Iran 37°8N 50°18E 45 B6
Rudall Australia 33°43S 136°17E 63 E2
Rudall → Australia 22°34S 122°13E 60 D3
Rudall River △ Australia 22°38S 122°30E 60 D3
Rudbar Iran 36°48N 49°23E 45 B6
Rudewa Tanzania 10°7S 34°40E 55 E3
Rudnyy Kazakhstan 52°57N 63°7E 28 D7
Rudolfa, Ostrov Russia 81°45N 58°30E 28 A6
Rudyard U.S.A. 46°14N 84°36W 81 B11
Ruenya → Africa 16°24S 33°48E 55 F3
Ruffling Pt. Br. Virgin Is. 18°44N 64°27W 89 e
Rufiji → Tanzania 7°50S 39°15E 54 D4
Rufino Argentina 34°20S 62°50W 94 C3
Rufunsa Zambia 15°4S 29°34E 55 F2
Rugby U.K. 52°23N 1°16W 13 E6
Rugby U.S.A. 48°22N 100°0W 80 A4
Rügen Germany 54°22N 13°24E 16 A7
Ruhengeri Rwanda 1°30S 29°36E 54 C2
Ruhnu Estonia 57°48N 23°15E 9 H20
Ruhr → Germany 51°27N 6°43E 16 C4
Ruhuhu → Tanzania 10°31S 34°34E 55 E3

Ruidoso U.S.A. 33°20N 105°41W 77 K11
Ruijin China 25°48N 116°0E 33 F12
Ruivo, Pico Madeira 32°45N 16°56W 24 D3
Rujm Tal'at al Jamā'ah Jordan 30°24N 35°30E 46 E4
Ruk Pakistan 27°50N 68°42E 42 F3
Rukhla Pakistan 32°27N 71°57E 42 C4
Ruki → Dem. Rep. of the Congo 0°5N 18°17E 52 E3
Rukwa □ Tanzania 7°0S 31°30E 54 D3
Rukwa, L. Tanzania 8°0S 32°20E 54 D3
Rulenge Burundi 2°43S 30°37E 54 C3
Rulhieres, C. Australia 13°56S 127°22E 60 B4
Rum = Rhum U.K. 57°0N 6°20W 11 E2
Rum Jordan 29°39N 35°26E 46 F4
Rum Cay Bahamas 23°40N 74°58W 89 B5
Rum Jungle Australia 13°0S 130°59E 60 B5
Ruma △ Kenya 0°39S 34°18E 54 C3
Rumāḩ Si. Arabia 25°29N 47°10E 44 E5
Rumania = Romania ■ Europe 46°0N 25°0E 17 F12
Rumaylah Iraq 30°47N 47°37E 44 D5
Rumbêk Sudan 6°54N 29°37E 51 G11
Rumford U.S.A. 44°33N 70°33W 83 B14
Rumia Poland 54°37N 18°25E 17 A10
Rumoi Japan 43°56N 141°39E 30 C10
Rumonge Burundi 3°59S 29°26E 54 C2
Rumphi Malaysia 11°1S 33°52E 55 E3
Rumson U.S.A. 40°23N 74°0W 83 F11
Rumuruti Kenya 0°17N 36°32E 54 B4
Runan China 33°0N 114°30E 34 H8
Runanga N.Z. 42°25S 171°15E 59 E3
Runaway, C. N.Z. 37°32S 177°59E 59 B6
Runaway Bay Jamaica 18°27N 77°20W 88 a
Runcorn U.K. 53°21N 2°44W 12 D5
Rundu Namibia 17°52S 19°43E 56 B2
Rungwa Tanzania 6°55S 33°32E 54 D3
Rungwa → Tanzania 7°36S 31°50E 54 D3
Rungwa ⌓ Tanzania 6°53S 34°2E 54 D3
Rungwe Tanzania 9°11S 33°32E 55 D3
Rungwe, Mt. Tanzania 9°8S 33°40E 52 F6
Runton Ra. Australia 23°31S 123°6E 60 D3
Ruo Shui → China 41°0N 100°16E 32 C9
Ruokolahti Finland 61°17N 28°50E 8 F23
Ruoqiang China 38°55N 88°10E 32 D6
Rupa India 27°15N 92°21E 41 F18
Rupar India 31°2N 76°38E 42 D7
Rupat Indonesia 1°45N 101°40E 36 D2
Rupen → India 23°28N 71°31E 42 H4
Rupert U.S.A. 42°37N 113°41W 76 E7
Rupert → Canada 51°29N 78°45W 72 B4
Rupert B. Canada 51°35N 79°0W 72 B4
Rupert House = Waskaganish Canada 51°30N 78°40W 72 B4
Rupsa India 21°37N 87°1E 43 J12
Rurrenabaque Bolivia 14°30S 67°32W 92 F5
Rusambo Zimbabwe 16°30S 32°4E 55 F3
Rusape Zimbabwe 18°35S 32°8E 55 F3
Ruschuk = Ruse Bulgaria 43°48N 25°59E 23 C12
Ruse Bulgaria 43°48N 25°59E 23 C12
Rush Ireland 53°31N 6°6W 10 C5
Rushan China 36°56N 121°30E 35 F11
Rushden U.K. 52°18N 0°35W 13 E7
Rushmore, Mt. U.S.A. 43°53N 103°28W 80 D2
Rushville Ill., U.S.A. 40°7N 90°34W 80 E8
Rushville Ind., U.S.A. 39°37N 85°27W 81 F11
Rushville Nebr., U.S.A. 42°43N 102°28W 80 D2
Russas Brazil 4°55S 37°50W 93 D11
Russell Canada 50°50N 101°20W 71 C8
Russell Kans., U.S.A. 38°54N 98°52W 80 F4
Russell N.Y., U.S.A. 44°27N 75°9W 83 B9
Russell Pa., U.S.A. 41°56N 79°8W 82 E5
Russell Cave ⌓ U.S.A. 34°59N 85°49W 85 D12
Russell L. Man., Canada 56°15N 101°30W 71 B8
Russell L. N.W.T., Canada 63°5N 115°44W 70 A5
Russellkonda India 19°57N 84°42E 41 K14
Russellville Ala., U.S.A. 34°30N 87°44W 85 D11
Russellville Ark., U.S.A. 35°17N 93°8W 84 D8
Russellville Ky., U.S.A. 36°51N 86°53W 80 G10
Russia ■ Eurasia 62°0N 105°0E 29 C11
Russian → U.S.A. 38°27N 123°8W 78 G3
Rustam Shahr Pakistan 26°58N 66°6E 42 F2
Rustavi Georgia 41°30N 45°0E 19 F8
Rustenburg S. Africa 25°41S 27°14E 56 D4
Ruston U.S.A. 32°32N 92°38W 84 E8
Rutana Burundi 3°55S 30°0E 54 C3
Ruteng Indonesia 8°35S 120°30E 37 F6
Rutenga Zimbabwe 21°15S 30°44E 55 G3
Ruth U.S.A. 43°42N 82°45W 82 C2
Rutherford U.S.A. 38°26N 122°24W 78 G4
Rutland U.S.A. 43°37N 72°58W 83 C12
Rutland □ U.K. 52°38N 0°40W 13 E7
Rutland Water U.K. 52°39N 0°38W 13 E7
Rutledge → Canada 61°4N 112°0W 71 A6
Rutledge L. Canada 61°33N 110°47W 71 A6
Rutog China 33°27N 79°42E 32 E4
Rutshuru Dem. Rep. of the Congo 1°13S 29°25E 54 C2
Ruvu Tanzania 6°49S 38°43E 54 D4
Ruvu → Tanzania 6°23S 38°52E 54 D4
Ruvuba △ Burundi 3°3S 29°33E 54 C2
Ruvuma □ Tanzania 10°20S 36°0E 55 E4
Ruvuma → Tanzania 10°29S 40°28E 55 E5
Ruwais U.A.E. 24°5N 52°50E 45 E7
Ruwenzori Africa 0°30N 29°55E 54 B2
Ruwenzori △ Uganda 0°20N 30°0E 54 B2
Ruya → Zimbabwe 16°27S 32°5E 57 B5
Ruyang China 34°9N 112°27E 34 G7
Ruyigi Burundi 3°29S 30°15E 54 C3
Ruzhou China 34°11N 112°52E 34 G7
Ružomberok Slovak Rep. 49°3N 19°17E 17 D10
Rwanda ■ Africa 2°0S 30°0E 54 C3
Ryan, L. U.K. 55°0N 5°2W 11 G3
Ryazan Russia 54°40N 39°40E 18 D6
Ryazhsk Russia 53°45N 40°3E 18 D7
Rybachiy Poluostrov Russia 69°43N 32°0E 8 B25
Rybachye = Balykchy Kyrgyzstan 42°26N 76°12E 32 C4

Rybinsk *Russia* 58°5N 38°50E 18 C6
Rybinskoye Vdkhr. *Russia* 58°30N 38°25E 18 C6
Rybnitsa = Râbnița *Moldova* 47°45N 29°0E 17 E15
Rycroft *Canada* 55°45N 118°40W 70 B5
Ryde *U.K.* 50°43N 1°9W 13 G6
Ryderwood *U.S.A.* 46°23N 123°3W 78 D3
Rye *U.K.* 50°57N 0°45E 13 G8
Rye → *U.K.* 54°11N 0°44W 12 C7
Rye Bay *U.K.* 50°52N 0°49E 13 G8
Rye Patch Res. *U.S.A.* 40°28N 118°19W 76 F4
Ryegate *U.S.A.* 46°18N 109°15W 76 C9
Ryley *Canada* 53°17N 112°26W 70 C6
Rylstone *Australia* 32°46S 149°58E 63 E4
Ryn Peski = Naryn Qum *Kazakhstan* 47°30N 49°0E 28 E5
Ryōtsu *Japan* 38°5N 138°26E 30 E9
Rypin *Poland* 53°3N 19°25E 17 B10
Ryūgasaki *Japan* 35°54N 140°11E 31 G10
Ryukyu Is. = Ryūkyū-Rettō *Japan* 26°0N 126°0E 31 M3
Ryūkyū-Rettō *Japan* 26°0N 126°0E 31 M3
Rzeszów *Poland* 50°5N 21°58E 17 C11
Rzhev *Russia* 56°20N 34°20E 18 C5

S

Sa Cabaneta *Spain* 39°37N 2°45E 24 B9
Sa Canal *Spain* 38°51N 1°23E 24 C7
Sa Conillera *Spain* 38°59N 1°13E 24 C7
Sa Dec *Vietnam* 10°20N 105°46E 39 G5
Sa Dragonera *Spain* 39°35N 2°19E 24 B9
Sa Kaeo *Thailand* 13°49N 102°4E 38 F4
Sa Mesquida *Spain* 39°55N 4°16E 24 B11
Sa Pa *Vietnam* 22°20N 103°47E 38 A4
Sa Savina *Spain* 38°44N 1°25E 24 C7
Sa'ādatābād *Fārs, Iran* 30°10N 53°5E 45 D7
Sa'ādatābād *Hormozgān, Iran* 28°3N 55°53E 45 D7
Sa'ādatābād *Kermān, Iran* 29°40N 55°51E 45 D7
Saale → *Germany* 51°56N 11°54E 16 C6
Saalfeld *Germany* 50°38N 11°21E 16 C6
Saanich *Canada* 48°29N 123°26W 78 B3
Saar → *Europe* 49°41N 6°32E 15 E6
Saarbrücken *Germany* 49°14N 6°59E 16 D4
Saaremaa *Estonia* 58°30N 22°30E 9 G20
Saarijärvi *Finland* 62°43N 25°16E 8 E21
Sab 'Ābar *Syria* 33°46N 37°41E 44 C3
Saba *W. Indies* 17°38N 63°14W 89 C7
Šabac *Serbia* 44°48N 19°42E 23 B8
Sabadell *Spain* 41°28N 2°7E 21 B7
Sabah □ *Malaysia* 6°0N 117°0E 36 C5
Sabak Bernam *Malaysia* 3°46N 100°58E 39 L3
Sabalān, Kūhhā-ye *Iran* 38°15N 47°45E 44 B5
Sabalana, Kepulauan *Indonesia* 6°45S 118°50E 37 F5
Sábana de la Mar *Dom. Rep.* 19°7N 69°24W 89 C6
Sábanalarga *Colombia* 10°38N 74°55W 92 A4
Sabang *Indonesia* 5°50N 95°15E 36 C1
Sabarmati → *India* 22°18N 72°22E 42 H5
Sabattis *U.S.A.* 44°6N 74°40W 83 B10
Saberania *Indonesia* 2°5S 138°18E 37 E9
Sabhā *Libya* 27°9N 14°29E 51 C8
Sabi → *India* 28°29N 76°44E 42 E7
Sabie *S. Africa* 25°10S 30°48E 57 D5
Sabinal *Mexico* 30°57N 107°30W 86 A3
Sabinal *U.S.A.* 29°19N 99°28W 84 G5
Sabinas *Mexico* 27°51N 101°7W 86 B4
Sabinas → *Mexico* 27°37N 100°42W 86 B4
Sabinas Hidalgo *Mexico* 26°30N 100°10W 86 B4
Sabine → *U.S.A.* 29°59N 93°47W 84 G8
Sabine L. *U.S.A.* 29°53N 93°51W 84 G8
Sabine Pass *U.S.A.* 29°44N 93°54W 84 G8
Sablayan *Phil.* 12°50N 120°50E 37 B6
Sable *Canada* 55°30N 68°21W 73 A6
Sable, C. *Canada* 43°29N 65°38W 73 D6
Sable, C. *U.S.A.* 25°9N 81°8W 88 A3
Sable I. *Canada* 44°0N 60°0W 73 D8
Sabrina Coast *Antarctica* 68°0S 120°0E 5 C9
Sabulubek *Indonesia* 1°36S 98°40E 36 E1
Sabzevār *Iran* 36°15N 57°40E 45 B8
Sabzvārān = Jiroft *Iran* 28°45N 57°50E 45 D8
Sac City *U.S.A.* 42°25N 95°0W 80 D6
Săcele *Romania* 45°37N 25°41E 17 F13
Sacheon *S. Korea* 35°0N 128°6E 35 G15
Sachigo → *Canada* 55°6N 88°58W 72 A2
Sachigo, L. *Canada* 53°50N 92°12W 72 B1
Sachimbo *Angola* 9°14S 20°16E 52 F4
Sachsen □ *Germany* 50°55N 13°10E 16 C7
Sachsen-Anhalt □ *Germany* 52°0N 12°0E 16 C7
Sackets Harbor *U.S.A.* 43°57N 76°7W 83 C8
Sackville *Canada* 45°54N 64°22W 73 C7
Saco *Maine, U.S.A.* 43°30N 70°27W 83 C14
Saco *Mont., U.S.A.* 48°28N 107°21W 76 B10
Sacramento *U.S.A.* 38°35N 121°29W 78 G5
Sacramento → *U.S.A.* 38°3N 121°56W 78 G5
Sacramento Mts. *U.S.A.* 32°30N 105°30W 77 K11
Sacramento Valley *U.S.A.* 39°30N 122°0W 78 G5
Sada-Misaki *Japan* 33°20N 132°5E 31 H6
Sadabad *India* 27°27N 78°3E 42 F8
Sadani *Tanzania* 5°58S 38°35E 54 D4
Sadani △ *Tanzania* 6°3S 38°47E 54 D4
Sadao *Thailand* 6°38N 100°26E 39 J3
Sadd el Aali *Egypt* 23°54N 32°54E 51 D12
Saddle Mt. *U.S.A.* 45°58N 123°41W 78 E3
Sadimi *Dem. Rep. of the Congo* 9°25S 23°32E 55 D1
Sado *Japan* 38°0N 138°25E 30 F9
Sadra *India* 23°21N 72°43E 42 H5
Sadri *India* 25°11N 73°26E 42 G5
Sæby *Denmark* 57°21N 10°30E 9 H14
Saegertown *U.S.A.* 41°43N 80°9W 82 E4
Şafājah *Si. Arabia* 26°25N 39°0E 44 E3
Safata B. *Samoa* 14°0S 171°50W 59 b
Säffle *Sweden* 59°8N 12°55E 9 G15
Safford *U.S.A.* 32°50N 109°43W 77 K9
Saffron Walden *U.K.* 52°1N 0°16E 13 E8
Safi *Morocco* 32°18N 9°20W 50 B4
Şafīābād *Iran* 36°45N 57°58E 45 B8
Safid Dasht *Iran* 33°27N 48°11E 45 C6
Safid Kūh *Afghan.* 34°45N 63°0E 40 B3
Safid Rūd → *Iran* 37°23N 50°11E 45 B6
Safipur *India* 26°44N 80°21E 43 F9
Şāfitā *Syria* 34°48N 36°7E 44 C3
Safune *Samoa* 13°25S 172°21W 59 b
Şafwān *Iraq* 30°7N 47°43E 44 D5
Sag Harbor *U.S.A.* 41°0N 72°18W 83 F12
Saga *Japan* 33°15N 130°16E 31 H5
Saga □ *Japan* 33°15N 130°20E 31 H5
Sagae *Japan* 38°22N 140°17E 30 E10
Sagaing *Burma* 21°52N 95°59E 41 J19
Sagamore *U.S.A.* 40°46N 79°14W 82 F5
Saganaga L. *Canada* 48°14N 90°52W 80 A8
Saganthit Kyun *Burma* 11°56N 98°29E 39 G2
Sagar *Karnataka, India* 14°14N 75°6E 40 M9
Sagar *Mad. P., India* 23°50N 78°44E 43 H8
Sagara, L. *Tanzania* 5°20S 31°0E 54 D3
Sagarmatha = Everest, Mt. *Nepal* 28°5N 86°58E 43 E12
Sagarmatha △ *Nepal* 27°55N 86°45E 43 F12
Saginaw *U.S.A.* 43°26N 83°56W 81 D12
Saginaw B. *U.S.A.* 43°50N 83°40W 81 D12
Saglouc = Salluit *Canada* 62°14N 75°38W 69 E16
Sagone *France* 42°7N 8°42E 20 E8
Sagua la Grande *Cuba* 22°50N 80°10W 88 B3
Saguache *U.S.A.* 38°5N 106°8W 76 G10
Saguaro △ *U.S.A.* 32°12N 110°38W 77 K8
Saguenay → *Canada* 48°22N 71°0W 73 C5
Sagunt *Spain* 39°42N 0°18W 21 C5
Sagunto = Sagunt *Spain* 39°42N 0°18W 21 C5
Sagwara *India* 23°41N 74°1E 42 H6
Sahagún *Spain* 42°18N 5°2W 21 A3
Saḩam al Jawlān *Syria* 32°45N 35°55E 46 C4
Sahamandrevo *Madag.* 23°15S 45°35E 57 C8
Sahand, Kūh-e *Iran* 37°44N 46°27E 44 B5
Sahara *Africa* 23°0N 5°0E 50 D6
Saharan Atlas = Saharien, Atlas *Algeria* 33°30N 1°0E 50 B6
Saharanpur *India* 29°58N 77°33E 42 E7
Saharien, Atlas *Algeria* 33°30N 1°0E 50 B6
Saharsa *India* 25°53N 86°36E 43 G12
Sahasinaka *Madag.* 21°49S 47°49E 57 C8
Sahaswan *India* 28°5N 78°45E 43 E8
Saheira, W. el → *Egypt* 30°5N 33°25E 46 E2
Sahel *Africa* 16°0N 5°0E 50 E5
Sahibganj *India* 25°12N 87°40E 43 G12
Sāḩilīyah *Iraq* 33°43N 42°42E 44 C4
Sahiwal *Pakistan* 30°45N 73°8E 42 D5
Şahneh *Iran* 34°29N 47°41E 44 C5
Sahrawi = Western Sahara ■ *Africa* 25°0N 13°0W 50 D3
Sahuaripa *Mexico* 29°3N 109°14W 86 B3
Sahuarita *U.S.A.* 31°57N 110°58W 77 L8
Sahuayo de Díaz *Mexico* 20°4N 102°43W 86 C4
Sai → *India* 25°39N 82°47E 43 G10
Sai Buri *Thailand* 6°43N 101°45E 39 J3
Sai Kung *China* 22°23N 114°16E 33 a
Sai Thong △ *Thailand* 15°56N 101°10E 38 E3
Sai Yok △ *Thailand* 14°25N 98°40E 38 E2
Saibai I. *Australia* 9°25S 142°40E 62 a
Sa'id Bundâs *Sudan* 8°24N 24°48E 51 G10
Sa'īdābād = Sīrjān *Iran* 29°30N 55°45E 45 D7
Sa'īdābād *Iran* 36°8N 54°11E 45 B7
Sa'īdīyeh *Iran* 36°20N 48°55E 45 B6
Saidpur *Bangla.* 25°48N 89°0E 41 G16
Saidpur *India* 25°33N 83°11E 43 G10
Saidu Sharif *Pakistan* 34°43N 72°24E 43 B5
Saigō *Japan* 36°12N 133°20E 31 F6
Saigon = Thanh Pho Ho Chi Minh *Vietnam* 10°58N 106°40E 39 G6
Saijō *Japan* 33°55N 133°11E 31 H6
Saikai △ *Japan* 33°12N 129°36E 31 H4
Saikanosy Masoala *Madag.* 15°45S 50°10E 57 B9
Saikhoa Ghat *India* 27°50N 95°40E 41 F19
Saiki *Japan* 32°58N 131°51E 31 H5
Sailana *India* 23°28N 74°55E 42 H6
Sailolof *Indonesia* 1°15S 130°46E 37 E8
Saimaa *Finland* 61°15N 28°15E 8 F23
Saimen = Saimaa *Finland* 61°15N 28°15E 8 F23
Şa'in Dezh *Iran* 36°40N 46°25E 44 B5
St. Abb's Head *U.K.* 55°55N 2°8W 11 F6
St. Alban's *Canada* 47°51N 55°50W 73 C8
St. Albans *U.K.* 51°45N 0°19W 13 F7
St. Albans *Vt., U.S.A.* 44°49N 73°5W 83 B11
St. Albans *W. Va., U.S.A.* 38°23N 81°50W 81 F13
St. Alban's Head *U.K.* 50°34N 2°4W 13 G5
St. Albert *Canada* 53°37N 113°32W 70 C6
St-André *Réunion* 20°57S 55°39E 53 c
St. Andrew's *Canada* 47°45N 59°15W 73 C8
St. Andrews *U.K.* 56°20N 2°47W 11 E6
St-Anicet *Canada* 45°8N 74°22W 83 A10
St. Annes *Canada* 49°40N 96°39W 71 D9
St. Anns B. *Canada* 46°22N 60°25W 73 C7
St. Ann's Bay *Jamaica* 18°26N 77°12W 88 a
St. Anthony *Canada* 51°22N 55°35W 73 B8
St. Anthony *U.S.A.* 43°58N 111°41W 76 E8
St-Antoine *Canada* 46°22N 64°45W 73 C7
St. Arnaud *Australia* 36°40S 143°16E 63 F3
St-Augustin *Canada* 51°13N 58°38W 73 B8
St-Augustin → *Canada* 51°16N 58°40W 73 B8
St. Augustine *U.S.A.* 29°54N 81°19W 85 G14
St. Austell *U.K.* 50°20N 4°47W 13 G3
St. Barbe *Canada* 51°12N 56°46W 73 B8
St-Barthélemy *W. Indies* 17°50N 62°50W 89 C7
St. Bees Hd. *U.K.* 54°31N 3°38W 12 C4
St. Bees I. *Australia* 20°56S 149°26E 62 b
St-Benoît *Réunion* 21°2S 55°43E 53 c
St. Bride's *Canada* 46°56N 54°10W 73 C9
St. Brides B. *U.K.* 51°49N 5°9W 13 F2
St-Brieuc *France* 48°30N 2°46W 20 B2
St. Catharines *Canada* 43°10N 79°15W 82 C5
St. Catherines I. *U.S.A.* 31°40N 81°10W 85 F14
St. Catherine's Pt. *U.K.* 50°34N 1°18W 13 G6
St-Chamond *France* 45°28N 4°31E 20 D6
St. Charles *Ill., U.S.A.* 41°54N 88°19W 80 E9
St. Charles *Md., U.S.A.* 38°36N 76°56W 81 F15
St. Charles *Mo., U.S.A.* 38°47N 90°29W 80 F8
St. Charles *Va., U.S.A.* 36°48N 83°4W 81 G12
St. Christopher-Nevis = St. Kitts & Nevis ■ *W. Indies* 17°20N 62°40W 89 C7
St. Clair *Mich., U.S.A.* 42°50N 82°30W 82 D2
St. Clair *Pa., U.S.A.* 40°43N 76°12W 83 F8
St. Clair → *U.S.A.* 42°38N 82°31W 82 D2
St. Clair, L. *N. Amer.* 42°27N 82°39W 82 D2
St. Clairsville *U.S.A.* 40°5N 80°54W 82 F4
St. Claude *Canada* 49°40N 98°20W 71 D9
St. Clears *U.K.* 51°49N 4°31W 13 F3
St-Clet *Canada* 45°21N 74°13W 83 A10
St. Cloud *Fla., U.S.A.* 28°15N 81°17W 85 G14
St. Cloud *Minn., U.S.A.* 45°34N 94°10W 80 C6
St. Cricq, C. *Australia* 25°17S 113°6E 61 E1
St. Croix *U.S. Virgin Is.* 17°45N 64°45W 89 C7
St. Croix → *U.S.A.* 44°45N 92°48W 80 C7
St. Croix Falls *U.S.A.* 45°24N 92°38W 80 C7
St. David's *Canada* 48°12N 58°52W 73 C8
St. David's *U.K.* 51°53N 5°16W 13 F2
St. David's Head *U.K.* 51°54N 5°19W 13 F2
St-Denis *France* 48°56N 2°20E 20 B5
St-Denis *Réunion* 20°52S 55°27E 53 c
St-Denis ✈ (RUN) *Réunion* 20°53S 55°32E 53 c
St-Dizier *France* 48°38N 4°56E 20 B6
St. Elias, Mt. *U.S.A.* 60°18N 140°56W 68 E3
St. Elias Mts. *N. Amer.* 60°33N 139°28W 70 A1
St-Étienne *France* 45°27N 4°22E 20 D6
St. Eugène *Canada* 45°30N 74°28W 83 A10
St. Eustatius *W. Indies* 17°20N 63°0W 89 C7
St-Félicien *Canada* 48°40N 72°25W 72 C5
St-Flour *France* 45°2N 3°6E 20 D5
St. Francis *U.S.A.* 39°47N 101°48W 80 F3
St. Francis → *U.S.A.* 34°38N 90°36W 85 D9
St. Francis, C. *S. Africa* 34°14S 24°49E 56 E3
St. Francisville *U.S.A.* 30°47N 91°23W 84 F9
St-François, L. *Canada* 45°10N 74°22W 83 A10
St-Gabriel *Canada* 46°17N 73°24W 72 C5
St. Gallen = Sankt Gallen *Switz.* 47°26N 9°22E 20 C8
St-Gaudens *France* 43°6N 0°44E 20 E4
St. George *Australia* 28°1S 148°30E 63 D4
St. George *N.B., Canada* 45°11N 66°50W 73 C6
St. George *Ont., Canada* 43°15N 80°15W 82 C4
St. George *S.C., U.S.A.* 33°11N 80°35W 85 E14
St. George *Utah, U.S.A.* 37°6N 113°35W 77 H7
St. George, C. *Canada* 48°30N 59°16W 73 C8
St. George, C. *U.S.A.* 29°40N 85°5W 85 G12
St. George I. *U.S.A.* 56°35N 169°35W 74 D6
St. George Ra. *Australia* 18°40S 125°0E 60 C4
St. George's *Canada* 48°26N 58°31W 73 C8
St-Georges *Canada* 46°8N 70°40W 73 C5
St. George's *Grenada* 12°5N 61°43W 89 D7
St. George's B. *Canada* 48°24N 58°53W 73 C8
St. Georges Basin *N.S.W., Australia* 35°7S 150°36E 63 F5
St. Georges Basin *W. Austral., Australia* 15°23S 125°2E 60 C4
St. George's Channel *Europe* 52°0N 6°0W 10 E6
St. Georges Hd. *Australia* 35°12S 150°42E 63 F5
St. Gotthard P. = San Gottardo, P. del *Switz.* 46°33N 8°33E 20 C8
St. Helena *Atl. Oc.* 15°58S 5°42W 48 H3
St. Helena *U.S.A.* 38°30N 122°28W 76 G2
St. Helena, Mt. *U.S.A.* 38°40N 122°36W 78 G4
St. Helena B. *S. Africa* 32°40S 18°10E 56 E2
St. Helens *Australia* 41°20S 148°15E 63 G4
St. Helens *U.K.* 53°27N 2°44W 12 D5
St. Helens *U.S.A.* 45°52N 122°48W 78 E4
St. Helens, Mt. *U.S.A.* 46°12N 122°12W 78 D4
St. Helier *U.K.* 49°10N 2°7W 13 H5
St-Hubert *Belgium* 50°2N 5°23E 15 D5
St-Hubert *Canada* 45°29N 73°25W 83 A11
St-Hyacinthe *Canada* 45°40N 72°58W 72 C5
St. Ignace *U.S.A.* 45°52N 84°44W 81 C11
St. Ignace I. *Canada* 48°45N 88°0W 72 C2
St. Ignatius *U.S.A.* 47°19N 114°6W 76 C6
St. Ives *Cambs., U.K.* 52°20N 0°4W 13 E7
St. Ives *Corn., U.K.* 50°12N 5°30W 13 G2
St. James *U.S.A.* 43°59N 94°38W 80 D6
St-Jean → *Canada* 50°17N 64°20W 73 B7
St-Jean, L. *Canada* 48°40N 72°0W 73 C5
St-Jean-Port-Joli *Canada* 47°15N 70°13W 73 C5
St-Jean-sur-Richelieu *Canada* 45°20N 73°20W 83 A11
St-Jérôme *Canada* 45°47N 74°0W 72 C5
St. John *Canada* 45°20N 66°8W 73 C6
St. John *U.S.A.* 38°0N 98°46W 80 F4
St. John → *N. Amer.* 45°12N 66°5W 81 C20
St. John, C. *Canada* 50°0N 55°32W 73 C8
St. John I. *U.S. Virgin Is.* 18°20N 64°42W 89 e
St. John's *Antigua & B.* 17°6N 61°51W 89 C7
St. John's *Canada* 47°35N 52°40W 73 C9
St. Johns *Ariz., U.S.A.* 34°30N 109°22W 77 J9
St. Johns *Mich., U.S.A.* 43°0N 84°33W 81 D11
St. Johns → *U.S.A.* 30°24N 81°24W 85 F14
St. John's Pt. *Ireland* 54°34N 8°27W 10 B3
St. Johnsbury *U.S.A.* 44°25N 72°1W 83 B12
St. Johnsville *U.S.A.* 43°0N 74°43W 83 C10
St. Joseph *Canada* 43°24N 81°42W 82 C3
St-Joseph *Martinique* 14°39N 61°4W 88 c
St-Joseph *Réunion* 21°22S 55°37E 53 c
St. Joseph *La., U.S.A.* 31°55N 91°14W 84 F9
St. Joseph *Mo., U.S.A.* 39°46N 94°50W 80 F6
St. Joseph → *U.S.A.* 42°7N 86°29W 80 D10
St. Joseph, I. *Canada* 46°12N 83°58W 72 C3
St. Joseph, L. *Canada* 51°10N 90°35W 72 B1
St. Kilda *U.K.* 57°49N 8°34W 14 C2
St. Kitts & Nevis ■ *W. Indies* 17°20N 62°40W 89 C7
St. Laurent *Canada* 50°25N 97°58W 71 C9
St. Lawrence *Australia* 22°16S 149°31E 62 C4
St. Lawrence *Canada* 46°54N 55°23W 73 C8
St. Lawrence → *Canada* 49°30N 66°0W 73 C6
St. Lawrence, Gulf of *Canada* 48°25N 62°0W 73 C7
St. Lawrence I. *U.S.A.* 63°30N 170°30W 74 C5
St. Lawrence Islands △ *Canada* 44°27N 75°52W 83 B9
St. Léonard *Canada* 47°12N 67°58W 73 C6
St-Leu *Réunion* 21°9S 55°18E 53 c
St. Lewis → *Canada* 52°26N 56°11W 73 B8
St-Lô *France* 49°7N 1°5W 20 B3
St-Louis *Guadeloupe* 15°56N 61°19W 88 b
St-Louis *Réunion* 21°16S 55°25E 53 c
St. Louis *Senegal* 16°8N 16°27W 50 E2
St. Louis *U.S.A.* 38°37N 90°11W 80 F8
St. Louis → *U.S.A.* 46°44N 92°9W 80 B7
St-Luc *Canada* 45°22N 73°18W 83 A11
St. Lucia ■ *W. Indies* 14°0N 60°57W 89 f
St. Lucia, L. *S. Africa* 28°5S 32°30E 57 D5
St. Lucia Channel *W. Indies* 14°15N 61°0W 89 D7
St. Maarten ☒ *W. Indies* 18°0N 63°5W 89 C7
St. Magnus B. *U.K.* 60°25N 1°35W 11 A7
St-Malo *France* 48°39N 2°1W 20 B2
St-Marc *Haiti* 19°10N 72°41W 89 C5
St. Maries *U.S.A.* 47°19N 116°35W 76 C5
St-Martin ☒ *W. Indies* 18°0N 63°0W 89 C7
St-Martin, C. *Martinique* 14°52N 61°14W 88 c
St. Martin, L. *Canada* 51°40N 98°30W 71 C9
St. Martins *Barbados* 13°5N 59°28W 89 g
St. Mary Pk. *Australia* 31°32S 138°34E 63 E2
St. Marys *Australia* 41°35S 148°11E 63 G4
St. Marys *Canada* 43°20N 81°10W 82 C3
St. Mary's *Corn., U.K.* 49°55N 6°18W 13 H1
St. Mary's *Orkney, U.K.* 58°54N 2°54W 11 C6
St. Marys *Ga., U.S.A.* 30°44N 81°33W 85 F14
St. Marys *Pa., U.S.A.* 41°26N 78°34W 82 E6
St. Mary's, C. *Canada* 46°50N 54°12W 73 C9
St. Mary's B. *Canada* 46°50N 53°50W 73 C9
St. Mary's Bay *Canada* 44°25N 66°10W 73 D6
St-Mathieu, Pte. *France* 48°20N 4°45W 20 B1
St. Matthew I. *U.S.A.* 60°24N 172°42W 74 C5
St-Maurice → *Canada* 46°21N 72°31W 72 C5
St. Mawes *U.K.* 50°10N 5°2W 13 G2
St-Nazaire *France* 47°17N 2°12W 20 C2
St. Neots *U.K.* 52°14N 0°15W 13 E7
St-Niklaas *Belgium* 51°10N 4°8E 15 C4
St-Omer *France* 50°45N 2°15E 20 A5
St-Pamphile *Canada* 46°58N 69°48W 73 C6
St-Pascal *Canada* 47°32N 69°48W 73 C6
St. Paul *Canada* 54°0N 111°17W 70 C6
St-Paul *Réunion* 20°59S 55°17E 53 c
St. Paul *Minn., U.S.A.* 44°56N 93°5W 80 C7
St. Paul *Nebr., U.S.A.* 41°13N 98°27W 80 E4
St-Paul → *Canada* 51°27N 57°42W 73 B8
St. Paul, Î. *Ind. Oc.* 38°55S 77°34E 3 F13
St. Paul I. *Canada* 47°12N 60°9W 73 C7
St. Paul I. *U.S.A.* 57°10N 170°15W 74 D5
St. Peter *U.S.A.* 44°20N 93°57W 80 C7
St. Peter Port *U.K.* 49°26N 2°33W 13 H5
St. Peters *N.S., Canada* 45°40N 60°53W 73 C7
St. Peters *P.E.I., Canada* 46°25N 62°35W 73 C7
St. Petersburg = Sankt-Peterburg *Russia* 59°55N 30°20E 9 G24
St. Petersburg *U.S.A.* 27°46N 82°40W 85 H13
St-Phillippe *Réunion* 21°21S 55°44E 53 c
St-Pie *Canada* 45°30N 72°54W 83 A12
St-Pierre *Martinique* 14°45N 61°10W 88 c
St-Pierre *Réunion* 21°19S 55°28E 53 c
St-Pierre *St-P. & M.* 46°46N 56°12W 73 C8
St-Pierre, L. *Canada* 46°12N 72°52W 72 C5
St-Pierre-et-Miquelon ☒ *N. Amer.* 46°55N 56°10W 73 C8
St-Quentin *Canada* 47°30N 67°23W 73 C6
St-Quentin *France* 49°50N 3°16E 20 B5
St. Regis *U.S.A.* 47°18N 115°6W 76 C6
St. Regis Falls *U.S.A.* 44°40N 74°32W 83 B10
St. Sebastien, Tanjon' i *Madag.* 12°26S 48°44E 57 A8
St-Siméon *Canada* 47°51N 69°54W 73 C6
St. Simons I. *U.S.A.* 31°12N 81°15W 85 F14
St. Simons Island *U.S.A.* 31°9N 81°22W 85 F14
St. Stephen *Canada* 45°16N 67°17W 73 C6
St. Thomas *Canada* 42°45N 81°10W 82 D3
St. Thomas I. *U.S. Virgin Is.* 18°20N 64°55W 89 e
St-Tite *Canada* 46°45N 72°34W 72 C5
St-Tropez *France* 43°17N 6°38E 20 E7
St. Troud = St. Truiden *Belgium* 50°48N 5°10E 15 D5
St. Truiden *Belgium* 50°48N 5°10E 15 D5
St. Vidgeon's ✪ *Australia* 14°47S 134°53E 62 A1
St. Vincent = São Vicente *C. Verde Is.* 17°0N 25°0W 50 b
St. Vincent, G. *Australia* 35°0S 138°0E 63 F2
St. Vincent & the Grenadines ■ *W. Indies* 13°0N 61°10W 89 D7
St. Vincent Passage *W. Indies* 13°30N 61°0W 89 D7
St-Vith *Belgium* 50°17N 6°9E 15 D6
St. Walburg *Canada* 53°39N 109°12W 71 C7
Ste-Agathe-des-Monts *Canada* 46°3N 74°17W 72 C5
Ste-Anne *Guadeloupe* 16°13N 61°24W 88 b
Ste-Anne *Seychelles* 4°36S 55°31E 53 b
Ste-Anne, L. *Canada* 50°0N 67°42W 73 B6
Ste-Anne-des-Monts *Canada* 49°8N 66°30W 73 C6
Ste. Genevieve *U.S.A.* 37°59N 90°2W 80 G8
Ste-Marguerite → *Canada* 50°9N 66°36W 73 B6
Ste-Marie *Canada* 46°26N 71°0W 73 C5
Ste-Marie *Martinique* 14°48N 61°1W 88 c
Ste-Marie *Réunion* 20°53S 55°33E 53 c
Ste-Marie, Ile = Nosy Boraha *Madag.* 16°50S 49°55E 57 B8
Ste-Rose *Guadeloupe* 16°20N 61°45W 88 b
Ste-Rose *Réunion* 21°8S 55°45E
Ste. Rose du Lac *Canada* 51°4N 99°30W
Saintes *France* 45°45N 0°37W
Saintes, Îs. des *Guadeloupe* 15°50N 61°35W
Saintfield *U.K.* 54°28N 5°49W
Saintonge *France* 45°40N 0°50W
Saipan *N. Marianas* 15°12N 145°45E
Sairang *India* 23°50N 92°45E 41
Sairecábur, Cerro *Bolivia* 22°43S 67°54W
Saitama *Japan* 35°54N 139°38E
Saitama □ *Japan* 36°25N 139°30E
Saiyid *Pakistan* 33°7N 73°2E
Sajama *Bolivia* 18°7S 69°0W
Sajószentpéter *Hungary* 48°12N 20°44E 17
Sajum *India* 33°20N 79°0E
Sak → *S. Africa* 30°52S 20°25E
Saka *Kenya* 0°9S 39°20E
Sakai *Japan* 34°34N 135°27E
Sakaide *Japan* 34°19N 133°50E
Sakaiminato *Japan* 35°38N 133°11E
Sakākah *Si. Arabia* 30°0N 40°8E
Sakakawea, L. *U.S.A.* 47°30N 101°25W
Sakami → *Canada* 53°40N 76°40W
Sakami, L. *Canada* 53°15N 77°0W
Sakania *Dem. Rep. of the Congo* 12°43S 28°30E
Sakaraha *Madag.* 22°55S 44°32E
Sakartvelo = Georgia ■ *Asia* 42°0N 43°0E
Sakarya *Turkey* 40°48N 30°25E
Sakashima-Guntō *Japan* 24°46N 124°0E
Sakata *Japan* 38°55N 139°50E
Sakchu *N. Korea* 40°23N 125°2E 35
Sakeny → *Madag.* 20°0S 45°25E
Sakha □ *Russia* 66°0N 130°0E 29
Sakhalin *Russia* 51°0N 143°0E 29
Sakhalinskiy Zaliv *Russia* 54°0N 141°0E 29
Šakiai *Lithuania* 54°59N 23°2E 9
Sakon Nakhon *Thailand* 17°10N 104°9E
Sakrand *Pakistan* 26°10N 68°15E
Sakri *India* 26°13N 86°5E 43
Sakrivier *S. Africa* 30°54S 20°28E
Sakti *India* 22°2N 82°58E 43
Sakuma *Japan* 35°3N 137°49E
Sakurai *Japan* 34°30N 135°51E
Sal *C. Verde Is.* 16°45N 22°55W
Sal Rei *C. Verde Is.* 16°11N 22°53W
Sala *Sweden* 59°58N 16°35E 9
Sala Consilina *Italy* 40°23N 15°36E
Sala-y-Gómez *Pac. Oc.* 26°28S 105°28W 65
Sala y Gómez Ridge *Pac. Oc.* 25°0S 98°0W 65
Salaberry-de-Valleyfield *Canada* 45°15N 74°8W 83
Salada, L. *Mexico* 32°20N 115°40W
Saladas *Argentina* 28°15S 58°40W
Saladillo *Argentina* 35°40S 59°55W
Salado → *B. Aires, Argentina* 35°44S 57°22W
Salado → *La Pampa, Argentina* 37°30S 67°0W
Salado → *Santa Fe, Argentina* 31°40S 60°41W
Salado → *Mexico* 26°52N 99°19W
Salaga *Ghana* 8°31N 0°31W
Sālah *Syria* 32°40N 36°45E
Şalāḩ ad Dīn □ *Iraq* 34°35N 43°35E
Salakos *Greece* 36°17N 27°57E
Salālah *Oman* 16°56N 53°59E
Salamanca *Chile* 31°46S 70°59W
Salamanca *Spain* 40°58N 5°39W
Salamanca *U.S.A.* 42°10N 78°43W
Salāmatābād *Iran* 35°39N 47°50E
Salamina *Greece* 37°56N 23°30E 23
Salamis *Cyprus* 35°11N 33°54E 25
Salar de Atacama *Chile* 23°30S 68°25W
Salar de Uyuni *Bolivia* 20°30S 67°45W
Salatiga *Indonesia* 7°19S 110°30E 37
Salavat *Russia* 53°21N 55°55E 18
Salaverry *Peru* 8°15S 79°0W
Salawati *Indonesia* 1°7S 130°52E
Salawin △ *Thailand* 18°18N 97°40E
Salaya *India* 22°19N 69°35E
Salayar *Indonesia* 6°7S 120°30E
S'Albufera *Spain* 39°47N 3°7E 24
Salcombe *U.K.* 50°14N 3°47W
Saldanha *S. Africa* 33°0S 17°58E
Saldanha B. *S. Africa* 33°6S 18°0E
Saldus *Latvia* 56°38N 22°30E 9
Sale *Australia* 38°6S 147°6E
Salé *Morocco* 34°3N 6°48W
Sale *U.K.* 53°26N 2°19W
Salekhard *Russia* 66°30N 66°35E
Salelologa *Samoa* 13°41S 172°11W
Salem *India* 11°40N 78°11E 40
Salem *Ill., U.S.A.* 38°38N 88°57W
Salem *Ind., U.S.A.* 38°36N 86°6W 80
Salem *Mass., U.S.A.* 42°31N 70°53W 83
Salem *Mo., U.S.A.* 37°39N 91°32W
Salem *N.H., U.S.A.* 42°45N 71°12W 83
Salem *N.J., U.S.A.* 39°34N 75°28W 81
Salem *N.Y., U.S.A.* 43°10N 73°20W 83
Salem *Ohio, U.S.A.* 40°54N 80°52W
Salem *Oreg., U.S.A.* 44°56N 123°2W
Salem *S. Dak., U.S.A.* 43°44N 97°23W
Salem *Va., U.S.A.* 37°18N 80°3W 81
Salerno *Italy* 40°41N 14°47E
Salford *U.K.* 53°30N 2°18W
Salgótarján *Hungary* 48°5N 19°47E 17
Salgueiro *Brazil* 8°4S 39°6W 93
Salibabu *Indonesia* 3°51N 126°40E
Salibea = Salybia *Trin. & Tob.* 10°43N 61°2W 93
Salida *U.S.A.* 38°32N 106°0W 75
Salihli *Turkey* 38°28N 28°8E 23
Salihorsk *Belarus* 52°51N 27°27E 17
Salima *Malawi* 13°47S 34°28E
Salina *Italy* 38°34N 14°50E

ina Kans., U.S.A. 38°50N 97°37W 80 F5
ina Utah, U.S.A. 38°58N 111°51W 76 G8
ina Cruz Mexico 16°10N 95°12W 87 D5
inas Brazil 16°10S 42°10W 93 G10
inas Ecuador 2°10S 80°58W 92 D2
inas U.S.A. 36°40N 121°39W 78 J5
inas → Guatemala 16°28N 90°31W 87 D6
inas → U.S.A. 36°45N 121°48W 78 J5
inas, B. de Nic. 11°4N 85°45W 88 D2
inas, Pampa de las Argentina 31°58S 66°42W 94 C2
inas Ambargasta Argentina 29°0S 65°0W 94 B3
inas de Hidalgo Mexico 22°38N 101°43W 86 C4
inas Grandes Argentina 30°0S 65°0W 94 C3
inas Pueblo Missions ⌂ U.S.A. 34°6N 106°4W 77 J10
inas Valley U.S.A. 36°15N 121°15W 78 J5
ine → Ark., U.S.A. 33°10N 92°8W 84 E8
ine → Kans., U.S.A. 38°52N 97°30W 80 F5
ines, C. de ses Spain 39°16N 3°4E 24 B10
inópolis Brazil 0°40S 47°20W 93 D9
isbury Australia 34°46S 138°38E 63 E2
isbury U.K. 51°4N 1°47W 13 F6
isbury Md., U.S.A. 38°22N 75°36W 81 F16
isbury N.C., U.S.A. 35°40N 80°29W 85 D14
isbury I. Canada 63°30N 77°0W 69 E16
isbury Plain U.K. 51°14N 1°55W 13 F6
khad Syria 32°29N 36°43E 46 C5
la Finland 66°50N 28°49E 8 C23
liq = Coral Harbour Canada 64°8N 83°10W 69 E15
lisaw U.S.A. 35°28N 94°47W 84 D7
luit Canada 62°14N 75°38W 69 E16
mās Iran 38°11N 44°47E 44 B5
mo Canada 49°10N 117°20W 70 D5
mon U.S.A. 45°11N 113°54W 76 D7
mon → Canada 54°3N 122°40W 70 C4
mon → U.S.A. 45°51N 116°47W 76 D5
mon Arm Canada 50°40N 119°15W 70 C5
mon Gums Australia 32°59S 121°38E 61 F3
mon Pt. Canada 43°52N 77°15W 82 C7
mon River Mts. U.S.A. 44°50N 115°30W 76 D6
mon River Res. U.S.A. 43°32N 75°55W 83 C9
o Finland 60°22N 23°10E 9 F20
ome U.S.A. 33°47N 113°37W 79 M13
on India 26°2N 81°27E 43 F9
on-de-Provence France 43°39N 5°6E 20 E6
onica = Thessaloniki Greece 40°38N 22°58E 23 D10
onta Romania 46°49N 21°42E 17 E11
pausselkä Finland 61°3N 26°15E 8 F22
sacate Argentina 31°20S 65°5W 94 C2
sk Russia 46°28N 41°30E 19 E7
so → Italy 37°6N 13°57E 22 F5
t → Canada 60°0N 112°25W 70 B6
t → U.S.A. 33°23N 112°19W 77 K7
t Fork Red → U.S.A. 34°27N 99°21W 84 D5
t L. Australia 30°6S 142°8E 63 E3
t Lake City U.S.A. 40°45N 111°53W 76 F8
t Range Pakistan 32°30N 72°25E 42 C5
ta Argentina 24°57S 65°25W 94 A2
ta □ Argentina 24°48S 65°30W 94 A2
tash U.K. 50°24N 4°14W 13 G3
tburn by the Sea U.K. 54°35N 0°58W 12 C7
tcoats U.K. 55°38N 4°47W 11 F4
tee Is. Ireland 52°7N 6°37W 10 D5
tfjellet Norway 66°40N 15°15E 8 C16
tfjorden Norway 67°15N 14°10E 8 C16
tillo Mexico 25°25N 101°0W 86 B4
to Argentina 34°20S 60°15W 94 C3
to Uruguay 31°27S 57°50W 94 C4
to → Italy 42°26N 12°25E 22 C5
to del Guaíra Paraguay 24°3S 54°17W 95 A5
ton City U.S.A. 33°18N 115°57W 79 M11
ton Sea U.S.A. 33°15N 115°45W 79 M11
tsburg U.S.A. 40°29N 79°27W 82 F5
uda → U.S.A. 34°1N 81°4W 85 D14
ûm Egypt 31°31N 25°7E 51 B11
ur India 18°27N 83°18E 41 K13
vador Brazil 13°0S 38°30W 93 F11
vador Canada 52°10N 109°32W 71 C7
vador, El ■ Cent. Amer. 13°50N 89°0W 88 D2
vador, L. U.S.A. 29°43N 90°15W 85 G9
ween → Burma 16°31N 97°37E 41 L20
yan Azerbaijan 39°33N 48°59E 45 B6
ybia Trin. & Tob. 10°43N 61°2W 93 K15
zach → Austria 48°12N 12°56E 16 D7
zburg Austria 47°48N 13°2E 16 E7
zgitter Germany 52°9N 10°19E 16 B6
zwedel Germany 52°52N 11°10E 16 B6
n India 26°50N 70°31E 42 F4
n Neua Laos 20°29N 104°5E 38 B5
n Ngao Thailand 17°18N 99°0E 38 D2
n Rayburn Res. U.S.A. 31°4N 94°5W 84 F7
n Son Vietnam 19°44N 105°54E 38 C5
na de Langreo = Langreo Spain 43°18N 5°40W 21 A3
nagaltay Russia 50°36N 95°3E 29 D10
nales Group Phil. 6°0N 122°0E 37 C6
nana India 30°10N 76°13E 42 D7
nana Cay Bahamas 23°3N 73°45W 89 B5
nanga Tanzania 8°20S 39°13E 55 D4
nangān □ Afghan. 36°15N 68°3E 40 B5
nangwa Dem. Rep. of the Congo 4°23S 24°10E 54 C1
nani Japan 42°7N 142°56E 30 C11
nar Phil. 12°0N 125°0E 37 B7
nara Russia 53°8N 50°6E 18 D9
naria = Shōmrōn West Bank 32°15N 35°13E 46 C4
naria △ Greece 35°17N 23°58E 25 D5
narinda Indonesia 0°30S 117°9E 36 E5
narkand = Samarqand Uzbekistan 39°40N 66°55E 28 F7
narqand Uzbekistan 39°40N 66°55E 28 F7
Sāmarrā' Iraq 34°12N 43°52E 44 C4
Samastipur India 25°50N 85°50E 43 G11
Samba Dem. Rep. of the Congo 4°38S 26°22E 54 C2
Samba India 32°32N 75°10E 43 C6
Sambalpur India 21°28N 84°4E 41 J14
Sambar, Tanjung Indonesia 2°59S 110°19E 36 E4
Sambas Indonesia 1°20N 109°20E 36 D3
Sambava Madag. 14°16S 50°10E 57 A9
Sambawizi Zimbabwe 18°24S 26°13E 55 F2
Sambhajinagar = Aurangabad India 19°50N 75°23E 40 K9
Sambhal India 28°35N 78°37E 43 E8
Sambhar India 26°52N 75°6E 42 F6
Sambhar L. India 26°55N 75°12E 42 F6
Sambiase Italy 38°58N 16°17E 22 E7
Sambir Ukraine 49°30N 23°10E 17 D12
Sambor Cambodia 12°46N 106°0E 38 F6
Sambor Prei Kuk Cambodia 12°46N 105°58E 38 F5
Samborombón, B. Argentina 36°5S 57°20W 94 D4
Samburu ⌂ Kenya 0°37N 37°31E 54 B4
Samcheok S. Korea 37°30N 129°10E 35 F15
Same Tanzania 4°2S 37°38E 54 C4
Samet, Ko Thailand 12°34N 101°27E 38 F3
Samfya Zambia 11°22S 29°31E 55 E2
Samnah Si. Arabia 25°10N 37°15E 44 E3
Samo Alto Chile 30°22S 71°0W 94 C1
Samoa ■ Pac. Oc. 14°0S 172°0W 59 b
Samoan Is. Pac. Oc. 14°0S 171°0W 59 b
Samokov Bulgaria 42°18N 23°35E 23 C10
Samos Greece 37°45N 26°50E 23 F12
Samothráki = Mathraki Greece 39°48N 19°31E 25 A3
Samothraki Greece 40°28N 25°28E 23 D11
Sampacho Argentina 33°20S 64°50W 94 C3
Sampalan Indonesia 8°41S 115°34E 37 K18
Sampang Indonesia 7°11S 113°13E 37 G15
Sampit Indonesia 2°34S 113°0E 36 E4
Sampit, Teluk Indonesia 3°5S 113°3E 36 E4
Samrong Cambodia 14°15N 103°30E 38 E4
Samsø Denmark 55°50N 10°35E 9 J14
Samsun Turkey 41°15N 36°22E 19 F6
Samui, Ko Thailand 9°30N 100°0E 39 b
Samusole Dem. Rep. of the Congo 10°2S 24°0E 55 E1
Samut Prakan Thailand 13°40N 100°36E 38 F3
Samut Sakhon Thailand 13°32N 100°17E 38 F3
Samut Songkhram Thailand 13°24N 100°1E 38 F3
Samwari Pakistan 28°30N 66°46E 42 E2
San Mali 13°15N 4°57W 50 F5
San → Cambodia 13°32N 105°57E 38 F5
San → Poland 50°45N 21°51E 17 C11
San Agustin, C. Phil. 6°20N 126°13E 37 C7
San Agustín de Valle Fértil Argentina 30°35S 67°30W 94 C2
San Ambrosio Pac. Oc. 26°28S 79°53W 90 F3
San Andreas U.S.A. 38°12N 120°41W 78 G6
San Andrés, I. de Caribbean 12°42N 81°46W 88 D3
San Andres Mts. U.S.A. 33°0N 106°30W 77 K10
San Andrés Tuxtla Mexico 18°27N 95°13W 87 D5
San Angelo U.S.A. 31°28N 100°26W 84 F4
San Anselmo U.S.A. 37°59N 122°34W 78 H4
San Antonio Belize 16°15N 89°2W 87 D7
San Antonio Chile 33°40S 71°40W 94 C1
San Antonio N. Mex., U.S.A. 33°55N 106°52W 77 K10
San Antonio Tex., U.S.A. 29°25N 98°29W 84 G5
San Antonio → U.S.A. 28°30N 96°54W 84 G6
San Antonio, C. Argentina 36°15S 56°40W 94 D4
San Antonio, C. de Cuba 21°50N 84°57W 88 B3
San Antonio, Mt. U.S.A. 34°17N 117°38W 79 L9
San Antonio de los Baños Cuba 22°54N 82°31W 88 B3
San Antonio de los Cobres Argentina 24°10S 66°17W 94 A2
San Antonio Oeste Argentina 40°40S 65°0W 96 E4
San Ardo U.S.A. 36°1N 120°54W 78 J6
San Augustín Canary Is. 27°47N 15°32W 24 G4
San Augustine U.S.A. 31°32N 94°7W 84 F7
San Bartolomé Canary Is. 28°59N 13°37W 24 F6
San Bartolomé de Tirajana Canary Is. 27°54N 15°34W 24 G4
San Benedetto del Tronto Italy 42°57N 13°53E 22 C5
San Benedicto, I. Mexico 19°18N 110°49W 86 D2
San Benito U.S.A. 26°8N 97°38W 84 H6
San Benito → U.S.A. 36°53N 121°34W 78 J5
San Benito Mt. U.S.A. 36°22N 120°37W 78 J6
San Bernardino U.S.A. 34°7N 117°19W 79 L9
San Bernardino Mts. U.S.A. 34°10N 116°45W 79 L10
San Bernardino Str. Phil. 12°30N 124°10E 37 B7
San Bernardo Chile 33°40S 70°50W 94 C1
San Bernardo, I. de Colombia 9°45N 75°50W 92 B3
San Blas Mexico 26°5N 108°46W 86 B3
San Blas, Arch. de Panama 9°50N 78°31W 88 E4
San Blas, C. U.S.A. 29°40N 85°21W 85 G12
San Borja Bolivia 14°50S 66°52W 92 F5
San Buenaventura = Ventura U.S.A. 34°17N 119°18W 79 L7
San Buenaventura Mexico 27°5N 101°32W 86 B4
San Carlos = Sant Carles Spain 39°3N 1°34E 24 B8
San Carlos Argentina 33°50S 69°0W 94 C2
San Carlos Chile 36°10S 72°0W 94 D1
San Carlos Baja Calif. S., Mexico 24°47N 112°7W 86 C2
San Carlos Coahuila, Mexico 29°1N 100°51W 86 B4
San Carlos Nic. 11°12N 84°50W 88 D3
San Carlos Phil. 10°29N 123°25E 37 B6
San Carlos Uruguay 34°46S 54°58W 95 C5
San Carlos U.S.A. 33°21N 110°27W 77 K8
San Carlos Venezuela 9°40N 68°36W 92 B5
San Carlos de Bariloche Argentina 41°10S 71°25W 96 E2
San Carlos de Bolívar Argentina 36°15S 61°6W 94 D3
San Carlos del Zulia Venezuela 9°1N 71°55W 92 B4
San Carlos L. U.S.A. 33°11N 110°32W 77 K8
San Clemente Chile 35°30S 71°29W 94 D1
San Clemente U.S.A. 33°26N 117°37W 79 M9
San Clemente I. U.S.A. 32°53N 118°29W 79 N8
San Cristóbal = Es Migjorn Gran Spain 39°57N 4°3E 24 B11
San Cristóbal Argentina 30°20S 61°10W 94 C3
San Cristóbal Dom. Rep. 18°25N 70°6W 89 C5
San Cristóbal Solomon Is. 10°30S 161°0E 58 C9
San Cristóbal Venezuela 7°46N 72°14W 92 B4
San Cristóbal de las Casas Mexico 16°45N 92°38W 87 D6
San Diego Calif., U.S.A. 32°42N 117°9W 79 N9
San Diego Tex., U.S.A. 27°46N 98°14W 84 H5
San Diego, C. Argentina 54°40S 65°10W 96 G3
San Diego de la Unión Mexico 21°28N 100°52W 86 C4
San Diego Int. ✕ (SAN) U.S.A. 32°44N 117°11W 79 N9
San Dimitri, Ras Malta 36°4N 14°11E 25 C1
San Dimitri Point = San Dimitri, Ras Malta 36°4N 14°11E 25 C1
San Estanislao Paraguay 24°39S 56°26W 94 A4
San Felipe Chile 32°43S 70°42W 94 C1
San Felipe Mexico 31°1N 114°52W 86 A2
San Felipe Venezuela 10°20N 68°44W 92 A5
San Felipe → U.S.A. 33°10N 115°49W 79 M11
San Félix Chile 28°56S 70°28W 94 B1
San Félix Pac. Oc. 26°23S 80°0W 90 F2
San Fernando = Sant Ferran Spain 38°42N 1°28E 24 C7
San Fernando Chile 34°30S 71°0W 94 C1
San Fernando Mexico 24°51N 98°10W 87 C5
San Fernando La Union, Phil. 16°40N 120°23E 37 A6
San Fernando Pampanga, Phil. 15°5N 120°37E 37 A6
San Fernando Spain 36°28N 6°17W 21 D2
San Fernando Trin. & Tob. 10°16N 61°28W 93 K15
San Fernando U.S.A. 34°17N 118°26W 79 L8
San Fernando de Apure Venezuela 7°54N 67°15W 92 B5
San Fernando de Atabapo Venezuela 4°3N 67°42W 92 C5
San Francisco Argentina 31°30S 62°5W 94 C3
San Francisco U.S.A. 37°46N 122°23W 78 H4
San Francisco → U.S.A. 32°59N 109°22W 77 K9
San Francisco, C. de Colombia 6°18N 77°29W 90 C3
San Francisco, Paso de S. Amer. 27°0S 68°0W 94 B2
San Francisco de Macorís Dom. Rep. 19°19N 70°15W 89 C5
San Francisco del Monte de Oro Argentina 32°36S 66°8W 94 C2
San Francisco del Oro Mexico 26°52N 105°51W 86 B3
San Francisco Int. ✕ (SFO) U.S.A. 37°37N 122°22W 78 H4
San Francisco Javier = Sant Francesc de Formentera Spain 38°42N 1°26E 24 C7
San Gabriel Chile 33°47S 70°15W 94 C1
San Gabriel Chilac Mexico 18°19N 97°21W 87 D5
San Gabriel Mts. U.S.A. 34°17N 117°38W 79 L9
San Gavino Monreale Italy 39°33N 8°47E 22 E3
San German Puerto Rico 18°4N 67°4W 89 d
San Gorgonio Mt. U.S.A. 34°6N 116°50W 79 L10
San Gottardo, P. del Switz. 46°33N 8°33E 20 C8
San Gregorio Uruguay 32°37S 55°40W 95 C4
San Gregorio U.S.A. 37°20N 122°23W 78 H4
San Guillermo △ Argentina 27°50S 69°45W 94 B2
San Ignacio Belize 17°10N 89°5W 87 D7
San Ignacio Bolivia 16°20S 60°55W 92 G6
San Ignacio Mexico 27°27N 112°51W 86 B2
San Ignacio Paraguay 26°52S 57°3W 94 B4
San Ignacio, L. Mexico 26°54N 113°13W 86 B2
San Ildefonso, C. Phil. 16°0N 122°1E 37 A6
San Isidro Argentina 34°29S 58°31W 94 C4
San Jacinto U.S.A. 33°47N 116°57W 79 M10
San Jaime = Sant Jaume Spain 39°54N 4°4E 24 B11
San Javier Misiones, Argentina 27°55S 55°5W 95 B4
San Javier Santa Fe, Argentina 30°40S 59°55W 94 C4
San Javier Bolivia 16°18S 62°30W 92 G6
San Javier Chile 35°40S 71°45W 94 D1
San Jerónimo Taviche Mexico 16°44N 96°35W 87 D5
San Joaquin U.S.A. 36°36N 120°11W 78 J6
San Joaquin → U.S.A. 38°4N 121°51W 78 G5
San Joaquin Valley U.S.A. 37°20N 121°0W 78 J6
San Jon U.S.A. 35°6N 103°20W 77 J12
San Jordi = Sant Jordi Spain 39°33N 2°46E 24 B9
San Jorge Argentina 31°54S 61°50W 94 C3
San Jorge, B. Mexico 31°20N 113°20W 86 A2
San Jorge, G. Argentina 46°0S 66°0W 96 F3
San José = San Josep Spain 38°55N 1°18E 24 C7
San José Argentina 32°12S 58°15W 94 C4
San José Costa Rica 9°55N 84°2W 88 E3
San José Guatemala 14°0N 90°50W 88 D1
San Jose Mind. Occ., Phil. 12°27N 121°4E 37 B6
San Jose Nueva Ecija, Phil. 15°45N 120°55E 37 A6
San Jose U.S.A. 37°20N 121°53W 78 H5
San Jose → U.S.A. 34°25N 106°45W 77 J10
San José, I. Mexico 25°0N 110°38W 86 C2
San Jose de Buenavista Phil. 10°45N 121°56E 37 B6
San José de Chiquitos Bolivia 17°53S 60°50W 92 G6
San José de Feliciano Argentina 30°26S 58°46W 94 C4
San José de Jáchal Argentina 30°15S 68°46W 94 C2
San José de Mayo Uruguay 34°27S 56°40W 94 C4
San José del Cabo Mexico 23°3N 109°41W 86 C3
San José del Guaviare Colombia 2°35N 72°38W 92 C4
San Josep Spain 38°55N 1°18E 24 C7
San Juan Argentina 31°30S 68°30W 94 C2
San Juan Puerto Rico 18°28N 66°7W 89 d
San Juan Trin. & Tob. 10°39N 61°29W 93 K15
San Juan □ Argentina 31°9S 69°0W 94 C2
San Juan → Argentina 32°20S 67°25W 94 C2
San Juan → Nic. 10°56N 83°42W 88 D3
San Juan → U.S.A. 37°16N 110°26W 77 H8
San Juan Bautista = Sant Joan de Labritja Spain 39°5N 1°31E 24 B8
San Juan Bautista Paraguay 26°37S 57°6W 94 B4
San Juan Bautista U.S.A. 36°51N 121°32W 78 J5
San Juan Capistrano U.S.A. 33°30N 117°40W 79 M9
San Juan Cr. → U.S.A. 35°40N 120°22W 78 J5
San Juan de Guadelupe Mexico 24°38N 102°44W 86 C4
San Juan de la Costa Mexico 24°20N 110°41W 86 C2
San Juan de los Lagos Mexico 21°15N 102°18W 86 C4
San Juan de los Morros Venezuela 9°55N 67°21W 92 B5
San Juan del Norte Nic. 10°58N 83°40W 88 D3
San Juan del Norte, B. de Nic. 11°0N 83°40W 88 D3
San Juan del Río Mexico 20°23N 100°0W 87 C5
San Juan del Sur Nic. 11°20N 85°51W 88 D2
San Juan I. U.S.A. 48°32N 123°5W 78 B3
San Juan Island ⌂ U.S.A. 48°35N 123°8W 78 B4
San Juan Mts. U.S.A. 37°30N 107°0W 77 H10
San Justo Argentina 30°47S 60°30W 94 C3
San Kamphaeng Thailand 18°45N 99°8E 38 C2
San Lázaro, C. Mexico 24°50N 112°18W 86 C2
San Leandro U.S.A. 37°42N 122°9W 78 H4
San Lorenzo = Sant Llorenç des Cardassar Spain 39°37N 3°17E 24 B10
San Lorenzo Argentina 32°45S 60°45W 94 C3
San Lorenzo Ecuador 1°15N 78°50W 92 C3
San Lorenzo Paraguay 25°20S 57°32W 94 B4
San Lorenzo → Mexico 24°15N 107°24W 86 C3
San Lorenzo, I. Mexico 28°38N 112°51W 86 B2
San Lorenzo, Mte. Argentina 47°40S 72°20W 96 F2
San Lucas = Cabo San Lucas Mexico 22°53N 109°54W 86 C3
San Lucas Bolivia 20°5S 65°7W 92 H5
San Lucas Mexico 27°10N 112°14W 86 B2
San Lucas U.S.A. 36°8N 121°1W 78 J5
San Lucas, C. Mexico 22°52N 109°53W 86 C3
San Luis Argentina 33°20S 66°20W 94 C2
San Luis Cuba 22°17N 83°46W 88 B3
San Luis Guatemala 16°14N 89°27W 88 C2
San Luis Ariz., U.S.A. 32°29N 114°47W 77 K6
San Luis Colo., U.S.A. 37°12N 105°25W 77 H11
San Luis □ Argentina 34°0S 66°0W 94 C2
San Luis, I. Mexico 29°58N 114°26W 86 B2
San Luis, Sierra de Argentina 32°30S 66°10W 94 C2
San Luis de la Paz Mexico 21°18N 100°31W 86 C4
San Luis Obispo U.S.A. 35°17N 120°40W 79 K6
San Luis Potosí Mexico 22°9N 100°59W 86 C4
San Luis Potosí □ Mexico 23°0N 101°0W 86 C4
San Luis Res. U.S.A. 37°4N 121°5W 78 H5
San Luis Rio Colorado Mexico 32°29N 114°58W 86 A2
San Manuel U.S.A. 32°36N 110°38W 77 K8
San Marcos Guatemala 14°59N 91°52W 88 D1
San Marcos Calif., U.S.A. 33°9N 117°10W 79 M9
San Marcos Tex., U.S.A. 29°53N 97°56W 84 G6
San Marcos, I. Mexico 27°13N 112°6W 86 B2
San Marino San Marino 43°55N 12°30E 16 G7
San Marino ■ Europe 43°56N 12°25E 22 C5
San Martín Antarctica 68°11S 67°0W 5 C17
San Martín Argentina 33°5S 68°28W 94 C2
San Martín → Bolivia 13°8S 63°43W 92 F6
San Martín, L. Argentina 48°50S 72°50W 96 F2
San Martín de los Andes Argentina 40°10S 71°20W 96 E2
San Mateo = Sant Mateu Spain 39°3N 1°23E 24 B7
San Mateo U.S.A. 37°34N 122°19W 78 H4
San Matías Bolivia 16°25S 58°20W 92 G7
San Matías, G. Argentina 41°30S 64°0W 96 E4
San Miguel = Sant Miquel Spain 39°3N 1°26E 24 B7
San Miguel El Salv. 13°30N 88°12W 88 D2
San Miguel Panama 8°27N 78°55W 88 E4
San Miguel U.S.A. 35°45N 120°42W 78 K6
San Miguel → Bolivia 13°52S 63°56W 92 F6
San Miguel de Tucumán Argentina 26°50S 65°20W 94 B2
San Miguel del Monte Argentina 35°23S 58°50W 94 D4
San Miguel I. U.S.A. 34°2N 120°23W 79 L6
San Nicolás Canary Is. 27°58N 15°47W 24 G4
San Nicolás de los Arroyos Argentina 33°25S 60°10W 94 C3
San Nicolas I. U.S.A. 33°15N 119°30W 79 M7
San Onofre U.S.A. 33°22N 117°34W 79 M9
San Pablo Bolivia 21°43S 66°38W 94 A2
San Pablo U.S.A. 37°58N 122°21W 78 H4
San Pedro B. Aires, Argentina 33°40S 59°40W 94 C4
San Pedro Misiones, Argentina 26°30S 54°10W 95 B5
San Pedro Chile 33°54S 71°28W 94 C1
San Pedro Ivory C. 4°50N 6°33W 50 H4
San Pedro Mexico 23°55N 110°17W 86 C2
San Pedro □ Paraguay 24°0S 57°0W 94 A4
San Pedro → Chihuahua, Mexico 28°20N 106°10W 86 B3
San Pedro → Nayarit, Mexico 21°45N 105°30W 86 C3
San Pedro → U.S.A. 32°59N 110°47W 77 K8
San Pedro, Pta. Chile 25°30S 70°38W 94 B1
San Pedro Channel U.S.A. 33°30N 118°25W 79 M8
San Pedro de Atacama Chile 22°55S 68°15W 94 A2
San Pedro de Jujuy Argentina 24°12S 64°55W 94 A3
San Pedro de las Colonias Mexico 25°45N 102°59W 86 B4
San Pedro de Macorís Dom. Rep. 18°30N 69°18W 89 C6
San Pedro del Norte Nic. 13°4N 84°33W 88 D3
San Pedro del Paraná Paraguay 26°43S 56°13W 94 B4
San Pedro Mártir, Sierra Mexico 31°0N 115°30W 86 A1
San Pedro Mártir △ Mexico 31°0N 115°21W 86 A1
San Pedro Ocampo = Melchor Ocampo Mexico 24°51N 101°39W 86 C4
San Pedro Pochutla Mexico 15°44N 96°28W 87 D5
San Pedro Sula Honduras 15°30N 88°0W 88 C2
San Pedro Tututepec Mexico 16°9N 97°38W 87 D5
San Pietro Italy 39°8N 8°17E 22 E3
San Rafael Argentina 34°40S 68°21W 94 C2
San Rafael Calif., U.S.A. 37°58N 122°32W 78 H4
San Rafael N. Mex., U.S.A. 35°7N 107°53W 77 J10
San Rafael Mt. U.S.A. 34°41N 119°52W 79 L7
San Rafael Mts. U.S.A. 34°40N 119°50W 79 L7
San Ramón de la Nueva Orán Argentina 23°10S 64°20W 94 A3
San Remo Italy 43°49N 7°46E 20 E7
San Roque Argentina 28°25S 58°45W 94 B4
San Roque Spain 36°17N 5°21W 21 D3
San Rosendo Chile 37°16S 72°43W 94 D1
San Saba U.S.A. 31°12N 98°43W 84 F5
San Salvador El Salv. 13°40N 89°10W 88 D2
San Salvador de Jujuy Argentina 24°10S 64°48W 94 A3
San Salvador I. Bahamas 24°0N 74°40W 89 B5
San Sebastián = Donostia-San Sebastián Spain 43°17N 1°58W 21 A5
San Sebastián Argentina 53°10S 68°30W 96 G3
San Sebastián Puerto Rico 18°20N 66°59W 89 d
San Sebastián de la Gomera Canary Is. 28°5N 17°7W 24 F2
San Serra = Son Serra Spain 39°43N 3°13E 24 B10
San Severo Italy 41°41N 15°23E 22 D6
San Simeon U.S.A. 35°39N 121°11W 78 K5
San Simon U.S.A. 32°16N 109°14W 77 K9
San Telmo = Sant Elm Spain 39°35N 2°21E 24 B9
San Telmo Mexico 30°58N 116°6W 86 A1
San Tiburcio Mexico 24°8N 101°32W 86 C4
San Valentin, Mte. Chile 46°30S 73°30W 96 F2
San Vicente de Cañete Peru 13°8S 76°30W 92 F3
San Vicente de la Barquera Spain 43°23N 4°29W 21 A3
San Vito Costa Rica 8°50N 82°58W 88 E3
Sana' Yemen 15°27N 44°12E 47 D3
Sana → Bos.-H. 45°3N 16°23E 16 F9
Sanae IV Antarctica 70°20S 9°0W 5 D2
Sanaga → Cameroon 3°35N 9°38E 52 D1
Sanak I. U.S.A. 54°25N 162°40W 74 E7
Sanaloa, Presa Mexico 24°50N 107°20W 86 C3
Sanana Indonesia 2°4S 125°58E 37 E7
Sanand India 22°59N 72°25E 42 H5
Sanandaj Iran 35°18N 47°1E 44 C5
Sanandita Bolivia 21°40S 63°45W 94 A3
Sanawad India 22°11N 76°5E 42 H7
Sancellas = Sencelles Spain 39°39N 2°54E 24 B9
Sanchahe China 44°50N 126°2E 35 B14
Sánchez Dom. Rep. 19°15N 69°36W 89 C6
Sanchor India 24°45N 71°55E 42 G4
Sancti Spíritus Cuba 21°52N 79°33W 88 B4
Sancy, Puy de France 45°32N 2°50E 20 D5
Sand = Polokwane → S. Africa 22°25S 30°5E 57 C5
Sand Hills U.S.A. 42°10N 101°30W 80 D3
Sand Lakes ⌂ Canada 57°51N 98°32W 71 B9
Sand Springs U.S.A. 36°9N 96°7W 84 C6
Sanda Japan 34°53N 135°14E 31 G7
Sandakan Malaysia 5°53N 118°4E 36 C5
Sandan = Sambor Cambodia 12°46N 106°0E 38 F6

Sandanski *Bulgaria* 41°35N 23°16E 23 D10
Sanday *U.K.* 59°16N 2°31W 11 B6
Sandefjord *Norway* 59°10N 10°15E 9 G14
Sanders *U.S.A.* 35°13N 109°20W 77 J9
Sanderson *U.S.A.* 30°9N 102°24W 84 F3
Sandersville *U.S.A.* 32°59N 82°48W 85 E13
Sandfire Roadhouse *Australia* 19°45S 121°15E 60 C3
Sandfly L. *Canada* 55°43N 106°6W 71 B7
Sandfontein *Namibia* 23°48S 19°1E 56 C2
Sandheads, The *India* 21°10N 88°20E 43 J13
Sandía *Peru* 14°10S 69°30W 92 F5
Sandila *India* 27°5N 80°31E 43 F9
Sandnes *Norway* 58°50N 5°45E 9 G11
Sandnessjøen *Norway* 66°2N 12°38E 8 C15
Sandoa *Dem. Rep. of the Congo* 9°41S 23°0E 52 F4
Sandomierz *Poland* 50°40N 21°43E 17 C11
Sandover → *Australia* 21°43S 136°32E 62 C2
Sandoway = Thandwe *Burma* 18°20N 94°30E 41 K19
Sandoy *Færoe Is.* 61°52N 6°46W 8 F9
Sandpoint *U.S.A.* 48°17N 116°33W 76 B5
Sandray *U.K.* 56°53N 7°31W 11 E1
Sandringham *U.K.* 52°51N 0°31E 12 E8
Sandstone *Australia* 27°59S 119°16E 61 E2
Sandusky *Mich., U.S.A.* 43°25N 82°50W 82 C2
Sandusky *Ohio, U.S.A.* 41°27N 82°42W 82 E2
Sandveld *Namibia* 21°25S 20°0E 56 C3
Sandviken *Sweden* 60°38N 16°46E 8 F17
Sandwich, C. *Australia* 18°14S 146°18E 62 B4
Sandwich B. *Canada* 53°40N 57°15W 73 B8
Sandwich B. *Namibia* 23°25S 14°20E 56 C1
Sandy *Oreg., U.S.A.* 45°24N 122°16W 78 E4
Sandy *Pa., U.S.A.* 41°6N 78°46W 82 E6
Sandy *Utah, U.S.A.* 40°32N 111°50W 76 F8
Sandy Bay *Canada* 55°31N 102°19W 71 B8
Sandy Bight *Australia* 33°50S 123°20E 61 F3
Sandy C. *Queens., Australia* 24°42S 153°15E 62 C5
Sandy C. *Tas., Australia* 41°25S 144°45E 63 G3
Sandy Cay *Bahamas* 23°13N 75°18W 89 B4
Sandy Cr. → *Australia* 31°54S 145°18E 63 B4
Sandy Cr. → *U.S.A.* 41°51N 109°47W 76 F9
Sandy Creek *U.S.A.* 43°38N 76°5W 83 C8
Sandy L. *Canada* 53°2N 93°0W 72 B1
Sandy Lake *Canada* 53°0N 93°15W 72 B1
Sandy Valley *U.S.A.* 35°49N 115°38W 79 K11
Sanford *Fla., U.S.A.* 28°48N 81°16W 85 G14
Sanford *Maine, U.S.A.* 43°27N 70°47W 83 C14
Sanford *N.C., U.S.A.* 35°29N 79°10W 85 D15
Sanford → *Australia* 27°22S 115°53E 61 E2
Sanford, Mt. *U.S.A.* 62°13N 144°8W 68 E3
Sang-i-Masha *Afghan.* 33°8N 67°27E 42 C2
Sanga *Mozam.* 12°22S 35°21E 55 E4
Sanga → *Congo* 1°5S 17°0E 52 E3
Sangamner *India* 19°37N 74°15E 40 K9
Sangān *Iran* 34°23N 60°15E 45 C9
Sangar *Afghan.* 32°56N 65°30E 42 C1
Sangar *Russia* 64°2N 127°31E 29 C13
Sangar Sarai *Afghan.* 34°27N 70°35E 42 B4
Sangarh → *Pakistan* 30°43N 70°44E 42 D4
Sangay *Ecuador* 2°0S 78°20W 92 D3
Sange *Dem. Rep. of the Congo* 6°58S 28°21E 54 D2
Sangeang *Indonesia* 8°12S 119°6E 37 F5
Sanger *U.S.A.* 36°42N 119°33W 78 J7
Sangerhausen *Germany* 51°28N 11°18E 16 C6
Sanggan He → *China* 38°12N 117°15E 34 E9
Sanggau *Indonesia* 0°5N 110°30E 36 D4
Sanghar *Pakistan* 26°2N 68°57E 42 F3
Sangihe, Kepulauan *Indonesia* 3°0N 125°30E 37 D7
Sangihe, Pulau *Indonesia* 3°35N 125°30E 37 D7
Sangiran △ *Indonesia* 7°27S 110°50E 37 G14
Sangju *S. Korea* 36°25N 128°10E 35 F15
Sangkapura *Indonesia* 5°52S 112°40E 36 F4
Sangkha *Thailand* 14°37N 103°52E 38 E4
Sangkhla Buri *Thailand* 14°57N 98°28E 38 E2
Sangkulirang *Indonesia* 0°59N 117°58E 36 D5
Sangla *Pakistan* 31°43N 73°23E 42 D5
Sangli *India* 16°55N 74°33E 40 L9
Sangmélima *Cameroon* 2°57N 12°1E 52 D2
Sango *Zimbabwe* 18°44S 31°44E 55 F3
Sangod *India* 24°55N 76°17E 42 G7
Sangre de Cristo Mts. *U.S.A.* 37°30N 105°20W 77 H11
Sangre Grande *Trin. & Tob.* 10°35N 61°8W 93 K15
Sangrur *India* 30°14N 75°50E 42 D6
Sangudo *Canada* 53°50N 114°54W 70 C6
Sangue → *Brazil* 11°1S 58°39W 92 F7
Sanibel *U.S.A.* 26°27N 82°1W 85 H13
Sanikluaq *Canada* 56°32N 79°14W 72 A4
Sanin-Kaigan △ *Japan* 35°39N 134°37E 31 G7
Sanirajak = Hall Beach *Canada* 68°46N 81°12W 69 D15
Sanjawi *Pakistan* 30°17N 68°21E 42 D3
Sanje *Uganda* 0°49S 31°30E 54 C3
Sanjo *Japan* 37°37N 138°57E 30 F9
Sankh → *India* 22°15N 84°48E 43 H11
Sankt Gallen *Switz.* 47°26N 9°22E 20 C8
Sankt Michel = Mikkeli *Finland* 61°43N 27°15E 8 F22
Sankt Moritz *Switz.* 46°30N 9°51E 20 C8
Sankt-Peterburg *Russia* 59°55N 30°20E 9 G24
Sankt Pölten *Austria* 48°12N 15°38E 16 D8
Sankuru → *Dem. Rep. of the Congo* 4°17S 20°25E 52 E4
Sanliurfa *Turkey* 37°12N 38°50E 44 B3
Sanlúcar de Barrameda *Spain* 36°46N 6°21W 21 D2
Sanmenxia *China* 34°47N 111°12E 34 G6
Sanming *China* 26°15N 117°40E 33 F12
Sannicandro Gargánico *Italy* 41°50N 15°34E 22 D6
Sânnicolau Mare *Romania* 46°5N 20°39E 17 E11
Sannieshof *S. Africa* 26°30S 25°47E 56 D4
Sannīn, J. *Lebanon* 33°57N 35°52E 46 B4
Sanniquellie *Liberia* 7°19N 8°38W 50 G4
Sano *Japan* 36°19N 139°35E 31 F9
Sanok *Poland* 49°35N 22°10E 17 D12
Sanquhar *U.K.* 55°22N 3°54W 11 F5
Sans Souci *Trin. & Tob.* 10°50N 61°0W 93 K16
Sant Antoni de Portmany *Spain* 38°59N 1°19E 24 C7
Sant Carles *Spain* 39°3N 1°34E 24 B8
Sant Elm *Spain* 39°35N 2°21E 24 B9
Sant Feliu de Guíxols *Spain* 41°45N 3°1E 21 B7
Sant Ferran *Spain* 38°42N 1°28E 24 C7
Sant Francesc de Formentera *Spain* 38°42N 1°26E 24 C7
Sant Jaume *Spain* 39°54N 4°4E 24 B11
Sant Joan *Spain* 39°36N 3°4E 24 B10
Sant Joan de Labritja *Spain* 39°5N 1°31E 24 B8
Sant Jordi *Ibiza, Spain* 38°53N 1°24E 24 C7
Sant Jordi *Mallorca, Spain* 39°33N 2°46E 24 B9
Sant Jordi, G. de *Spain* 40°53N 1°2E 21 B6
Sant Llorenç des Cardassar *Spain* 39°37N 3°17E 24 B10
Sant Mateu *Spain* 39°3N 1°23E 24 B7
Sant Miquel *Spain* 39°3N 1°26E 24 B7
Sant Salvador *Spain* 39°27N 3°11E 24 B10
Santa Agnés *Spain* 39°3N 1°21E 24 B7
Santa Ana *Bolivia* 13°50S 65°40W 92 F5
Santa Ana *El Salv.* 14°0N 89°31W 88 D2
Santa Ana *Mexico* 30°33N 111°7W 86 A2
Santa Ana *U.S.A.* 33°46N 117°52W 79 M9
Sant' Antíoco *Italy* 39°4N 8°27E 22 E3
Santa Bárbara *Chile* 37°40S 72°1W 94 D1
Santa Bárbara *Honduras* 14°53N 88°14W 88 D2
Santa Bárbara *Mexico* 26°48N 105°49W 86 B3
Santa Barbara *U.S.A.* 34°25N 119°42W 79 L7
Santa Barbara Channel *U.S.A.* 34°15N 120°0W 79 L7
Santa Barbara I. *U.S.A.* 33°29N 119°2W 79 M7
Santa Catalina, Gulf of *U.S.A.* 33°10N 117°50W 79 N9
Santa Catalina, I. *Mexico* 25°40N 110°47W 86 B2
Santa Catalina I. *U.S.A.* 33°23N 118°25W 79 M8
Santa Catarina □ *Brazil* 27°25S 48°30W 95 B6
Santa Catarina, I. de *Brazil* 27°30S 48°40W 95 B6
Santa Cecília *Brazil* 26°56S 50°18W 95 B5
Santa Clara *Cuba* 22°20N 80°0W 88 B4
Santa Clara *Calif., U.S.A.* 37°21N 121°57W 78 H5
Santa Clara *N.Y., U.S.A.* 44°38N 74°27W 83 B10
Santa Clara *Utah, U.S.A.* 37°8N 113°39W 77 H7
Santa Clara de Olimar *Uruguay* 32°50S 54°54W 95 C5
Santa Clara Valley *U.S.A.* 36°50N 121°30W 78 J5
Santa Clarita *U.S.A.* 34°24N 118°33W 79 L8
Santa Clotilde *Peru* 2°33S 73°45W 92 D4
Santa Coloma de Gramenet *Spain* 41°27N 2°13E 21 B7
Santa Cruz *Bolivia* 17°43S 63°10W 92 G6
Santa Cruz *Chile* 34°38S 71°27W 94 C1
Santa Cruz *Costa Rica* 10°15N 85°35W 88 D2
Santa Cruz *Madeira* 32°42N 16°46W 24 D3
Santa Cruz *Phil.* 14°20N 121°24E 37 B6
Santa Cruz *U.S.A.* 36°58N 122°1W 78 J4
Santa Cruz → *Argentina* 50°10S 68°20W 96 G3
Santa Cruz de la Palma *Canary Is.* 28°41N 17°46W 24 F2
Santa Cruz de la Palma ✕ (SPC) *Canary Is.* 28°40N 17°45W 24 F2
Santa Cruz de Tenerife *Canary Is.* 28°28N 16°15W 24 F3
Santa Cruz del Norte *Cuba* 23°9N 81°55W 88 B3
Santa Cruz del Sur *Cuba* 20°44N 78°0W 88 B4
Santa Cruz do Rio Pardo *Brazil* 22°54S 49°37W 95 A6
Santa Cruz do Sul *Brazil* 29°42S 52°25W 95 B5
Santa Cruz I. *U.S.A.* 34°1N 119°43W 79 M7
Santa Cruz Is. *Solomon Is.* 10°30S 166°0E 58 C9
Santa Cruz Mts. *Jamaica* 17°58N 77°43W 88 a
Santa Domingo, Cay *Bahamas* 21°25N 75°15W 88 B4
Santa Elena *Argentina* 30°58S 59°47W 94 C4
Santa Elena, C. *Costa Rica* 10°54N 85°56W 88 D2
Santa Eulària des Riu *Spain* 38°59N 1°32E 24 C8
Santa Fé *Argentina* 31°35S 60°41W 94 C3
Santa Fe *U.S.A.* 35°41N 105°57W 77 J11
Santa Fé □ *Argentina* 31°50S 60°55W 94 C3
Santa Fé do Sul *Brazil* 20°13S 50°56W 93 H8
Santa Filomena *Brazil* 9°6S 45°50W 93 E9
Santa Gertrudis *Spain* 39°0N 1°26E 24 C7
Santa Inês *Brazil* 13°17S 39°48W 93 F11
Santa Inés, I. *Chile* 54°0S 73°0W 96 G2
Santa Isabel *Argentina* 36°10S 66°54W 94 D2
Santa Isabel do Morro *Brazil* 11°34S 50°40W 93 F8
Santa Lucía *Corrientes, Argentina* 28°58S 59°5W 94 B4
Santa Lucía *San Juan, Argentina* 31°30S 68°30W 94 C2
Santa Lucía *Uruguay* 34°27S 56°24W 94 C4
Santa Lucia Range *U.S.A.* 36°0N 121°20W 78 K5
Santa Luzia *C. Verde Is.* 16°50N 24°35W 50 b
Santa Margalida *Spain* 39°42N 3°6E 24 B10
Santa Margarita *Argentina* 38°28S 61°35W 94 D3
Santa Margarita *U.S.A.* 35°23N 120°37W 78 K6
Santa Margarita → *U.S.A.* 33°13N 117°23W 79 M9
Santa Margarita, I. *Mexico* 24°27N 111°50W 86 C2
Santa María *Argentina* 26°40S 66°0W 94 B2
Santa Maria *Azores* 36°58N 25°6W 50 a
Santa Maria *Brazil* 29°40S 53°48W 95 B5
Santa Maria *C. Verde Is.* 16°31N 22°53W 50 b
Santa Maria *U.S.A.* 34°57N 120°26W 79 L6
Santa María → *Mexico* 31°0N 107°14W 86 A3
Santa María, B. de *Mexico* 25°4N 108°6W 86 B3
Santa Maria da Vitória *Brazil* 13°24S 44°12W 93 F10
Santa María del Camí *Spain* 39°38N 2°47E 24 B9
Santa Maria di Léuca, C. *Italy* 39°47N 18°22E 23 E8
Santa Marta *Colombia* 11°15N 74°13W 92 A4
Santa Marta, Sierra Nevada de *Colombia* 10°55N 73°50W 92 A4
Santa Marta Grande, C. *Brazil* 28°43S 48°50W 95 B6
Santa Maura = Lefkada *Greece* 38°40N 20°43E 23 E9
Santa Monica *U.S.A.* 34°1N 118°29W 79 M8
Santa Monica Mts. ⌂ *U.S.A.* 34°4N 118°44W 79 L8
Santa Paula *U.S.A.* 34°21N 119°4W 79 L7
Santa Ponça *Spain* 39°30N 2°28E 24 B9
Santa Rosa *La Pampa, Argentina* 36°40S 64°17W 94 D3
Santa Rosa *San Luis, Argentina* 32°21S 65°10W 94 C2
Santa Rosa *Brazil* 27°52S 54°29W 95 B5
Santa Rosa *Calif., U.S.A.* 38°26N 122°43W 78 G4
Santa Rosa *N. Mex., U.S.A.* 34°57N 104°41W 77 J11
Santa Rosa and San Jacinto Mts. ⌂ *U.S.A.* 33°28N 116°20W 79 M10
Santa Rosa de Copán *Honduras* 14°47N 88°46W 88 D2
Santa Rosa de Río Primero *Argentina* 31°8S 63°20W 94 C3
Santa Rosa del Sara *Bolivia* 17°7S 63°35W 92 G6
Santa Rosa I. *Calif., U.S.A.* 33°58N 120°6W 79 M6
Santa Rosa I. *Fla., U.S.A.* 30°20N 86°50W 85 F11
Santa Rosa Range *U.S.A.* 41°45N 117°40W 76 F5
Santa Rosalía *Mexico* 27°19N 112°17W 86 B2
Santa Sylvina *Argentina* 27°50S 61°10W 94 B3
Santa Tecla = Nueva San Salvador *El Salv.* 13°40N 89°18W 88 D2
Santa Teresa *Argentina* 33°25S 60°47W 94 C3
Santa Teresa *Australia* 24°8S 134°22E 62 C1
Santa Teresa *Mexico* 25°17N 97°51W 87 B5
Santa Teresa △ *Uruguay* 33°57S 53°31W 95 C5
Santa Teresita *Argentina* 36°32S 56°41W 94 D4
Santa Vitória do Palmar *Brazil* 33°32S 53°25W 95 C5
Santa Ynez → *U.S.A.* 34°41N 120°36W 79 L6
Santa Ynez Mts. *U.S.A.* 34°30N 120°0W 79 L6
Santa Ysabel *U.S.A.* 33°7N 116°40W 79 M10
Santai *China* 31°5N 104°58E 32 E9
Santana *Madeira* 32°48N 16°52W 24 D3
Santana, Coxilha de *Brazil* 30°50S 55°35W 95 C4
Santana do Livramento *Brazil* 30°55S 55°30W 95 C4
Santander *Spain* 43°27N 3°51W 21 A4
Santander Jiménez *Mexico* 24°13N 98°28W 87 C5
Santanilla, Is. *Honduras* 17°22N 83°57W 88 C3
Santanyí *Spain* 39°20N 3°5E 24 B10
Santaquin *U.S.A.* 39°59N 111°47W 76 G8
Santarém *Brazil* 2°25S 54°42W 93 D8
Santarém *Portugal* 39°12N 8°42W 21 C1
Santaren Channel *W. Indies* 24°0N 79°30W 88 B4
Santee *U.S.A.* 32°50N 116°58W 79 N10
Santee → *U.S.A.* 33°7N 79°17W 85 E15
Santiago = Río Grande de Santiago → *Mexico* 21°36N 105°26W 86 C3
Santiago = São Tiago *C. Verde Is.* 15°0N 23°40W 50 b
Santiago *Brazil* 29°11S 54°52W 95 B5
Santiago *Canary Is.* 28°2N 17°12W 24 F2
Santiago *Chile* 33°26S 70°40W 94 C1
Santiago *Panama* 8°0N 81°0W 88 E3
Santiago *Phil.* 16°41N 121°33E 37 A6
Santiago → *Peru* 4°27S 77°38W 92 D3
Santiago de Compostela *Spain* 42°52N 8°37W 21 A1
Santiago de Cuba *Cuba* 20°0N 75°49W 88 C4
Santiago de los Caballeros *Dom. Rep.* 19°30N 70°40W 89 C5
Santiago del Estero *Argentina* 27°50S 64°15W 94 B3
Santiago del Estero □ *Argentina* 27°40S 63°15W 94 B3
Santiago del Teide *Canary Is.* 28°17N 16°48W 24 F3
Santiago Ixcuintla *Mexico* 21°49N 105°13W 86 C3
Santiago Jamiltepec *Mexico* 16°17N 97°49W 87 D5
Santiago Papasquiaro *Mexico* 25°3N 105°25W 86 C3
Santiago Pinotepa Nacional *Mexico* 16°19N 98°1W 87 D5
Santiaguillo, L. de *Mexico* 24°48N 104°48W 86 C4
Santo Amaro *Brazil* 12°30S 38°43W 93 F11
Santo Anastácio *Brazil* 21°58S 51°39W 95 A5
Santo André *Brazil* 23°39S 46°29W 95 A6
Santo Ângelo *Brazil* 28°15S 54°15W 95 B5
Santo Antão *C. Verde Is.* 16°52N 25°10W 50 b
Santo Antônio do Içá *Brazil* 3°5S 67°57W 92 D5
Santo Antônio do Leverger *Brazil* 15°52S 56°5W 93 G7
Santo Domingo *Dom. Rep.* 18°30N 69°59W 89 C6
Santo Domingo *Baja Calif., Mexico* 30°43N 116°2W 86 A1
Santo Domingo *Baja Calif. S., Mexico* 25°29N 111°55W 86 B2
Santo Domingo *Nic.* 12°14N 84°59W 88 D3
Santo Domingo de los Colorados *Ecuador* 0°15S 79°9W 92 D3
Santo Domingo Pueblo *U.S.A.* 35°31N 106°22W 77 J10
Santo Tomás *Mexico* 31°33N 116°24W 86 A1
Santo Tomás *Peru* 14°26S 72°8W 92 F4
Santo Tomé *Argentina* 28°40S 56°5W 95 B4
Santo Tomé de Guayana = Ciudad Guayana *Venezuela* 8°0N 62°30W 92 B6
Santoña *Spain* 43°29N 3°27W 21 A4
Santorini *Greece* 36°23N 25°27E 23 F11
Santos *Brazil* 24°0S 46°20W 95 A6
Santos Dumont *Brazil* 22°55S 43°10W 95 A7
Santuario de Aves Laguna Colorada △ *Bolivia* 22°10S 67°45W 94 A2
Sanur *Indonesia* 8°41S 115°15E 37 K18
Sanwer *India* 22°59N 75°50E 42 H6
Sanxiang *China* 22°21N 113°25E 33 a
Sanya *China* 18°14N 109°29E 38 C7
Sanyuan *China* 34°35N 108°58E 34 G5
São Bernardo do Campo *Brazil* 23°45S 46°34W 95 A6
São Borja *Brazil* 28°39S 56°0W 95 B4
São Carlos *Brazil* 22°0S 47°50W 95 A6
São Cristóvão *Brazil* 11°1S 37°15W 93 F11
São Domingos *Brazil* 13°25S 46°19W 93 F9
São Filipe *C. Verde Is.* 15°2N 24°30W 50 b
São Francisco *Brazil* 16°0S 44°50W 93 G10
São Francisco → *Brazil* 10°30S 36°24W 93 F11
São Francisco do Sul *Brazil* 26°15S 48°36W 95 B6
São Gabriel *Brazil* 30°20S 54°20W 95 C5
São Gonçalo *Brazil* 22°48S 43°5W 95 A7
Sao Hill *Tanzania* 8°20S 35°12E 55 D4
São João da Boa Vista *Brazil* 22°0S 46°52W 95 A6
São João da Madeira *Portugal* 40°54N 8°30W 21 B1
São João del Rei *Brazil* 21°8S 44°15W 95 A7
São João do Araguaia *Brazil* 5°23S 48°46W 93 E9
São João do Piauí *Brazil* 8°21S 42°15W 93 E10
São Joaquim *Brazil* 28°18S 49°56W 95 B6
São Joaquim △ *Brazil* 28°12S 49°37W 95 B6
São Jorge *Azores* 38°38N 28°3W 50 a
São Jorge, Pta. de *Madeira* 32°50N 16°53W 24 D3
São José *Brazil* 27°38S 48°39W 95 B5
São José do Norte *Brazil* 32°1S 52°3W 95 C5
São José do Rio Preto *Brazil* 20°50S 49°20W 95 A6
São José dos Campos *Brazil* 23°7S 45°52W 95 A6
São Leopoldo *Brazil* 29°50S 51°10W 95 B5
São Lourenço *Brazil* 22°7S 45°3W 95 A6
São Lourenço → *Brazil* 17°53S 57°27W 93 G7
São Lourenço, Pta. de *Madeira* 32°44N 16°39W 24 D3
São Lourenço do Sul *Brazil* 31°22S 51°58W 95 C5
São Luís *Brazil* 2°39S 44°15W 93 D10
São Luís Gonzaga *Brazil* 28°25S 55°0W 95 B5
São Marcos → *Brazil* 18°15S 47°37W 93 G9
São Marcos, B. de *Brazil* 2°0S 44°0W 93 D10
São Mateus *Brazil* 18°44S 39°50W 93 G11
São Mateus do Sul *Brazil* 25°52S 50°23W 95 B5
São Miguel *Azores* 37°47N 25°30W 50 a
São Miguel do Oeste *Brazil* 26°45S 53°34W 95 B5
São Nicolau *C. Verde Is.* 16°20N 24°20W 50 b
São Paulo *Brazil* 23°32S 46°38W 95 A6
São Paulo □ *Brazil* 22°0S 49°0W 95 A6
São Paulo de Olivença *Brazil* 3°27S 68°48W 92 D5
São Roque *Madeira* 32°46N 16°48W 24 D3
São Roque, C. de *Brazil* 5°30S 35°16W 93 E11
São Sebastião, I. de *Brazil* 23°50S 45°18W 95 A6
São Sebastião do Paraíso *Brazil* 20°54S 46°59W 95 A6
São Tiago *C. Verde Is.* 15°0N 23°40W 50 b
São Tomé *São Tomé & Príncipe* 0°10N 6°39E 48 F4
São Tomé, C. de *Brazil* 22°0S 40°59W 95 A7
São Tomé & Príncipe ■ *Africa* 0°12N 6°39E 49 F4
São Vicente *Brazil* 23°57S 46°23W 95 A6
São Vicente *C. Verde Is.* 17°0N 25°0W 50 b
São Vicente *Madeira* 32°48N 17°3W 24 D2
São Vicente, C. de *Portugal* 37°0N 9°0W 21 D1
Saona, I. *Dom. Rep.* 18°10N 68°40W 89 C6
Saône → *France* 45°44N 4°50E 20 D6
Saonek *Indonesia* 0°22S 130°55E 37 E8
Sapam, Ao *Thailand* 8°0N 98°26E 39 a
Saparua *Indonesia* 3°33S 128°40E 37 E7
Sape *Indonesia* 8°34S 118°59E 37 F5
Sapele *Nigeria* 5°50N 5°40E 50 G7
Sapelo I. *U.S.A.* 31°25N 81°12W 85 F14
Sapi ⌂ *Zimbabwe* 15°48S 29°42E 55 F2
Saposoa *Peru* 6°55S 76°45W 92 E3
Sapphire *Australia* 23°28S 147°43E 62 C4
Sappho *U.S.A.* 48°4N 124°16W 78 B2
Sapporo *Japan* 43°0N 141°21E 30 C10
Sapudi *Indonesia* 7°6S 114°20E 37 G16
Sapulpa *U.S.A.* 35°59N 96°5W 84 D6
Saqqez *Iran* 36°15N 46°20E 44 B5
Sar Dasht *Āzarbāyjān-e Gharbī, Iran* 36°9N 45°28E 44 B5
Sar Dasht *Khuzestān, Iran* 32°32N 48°52E 45 C6
Sar-e Pol □ *Afghan.* 36°20N 65°50E 40 B4
Sar Gachīneh = Yāsūj *Iran* 30°31N 51°31E 45 D6
Sar Planina *Macedonia* 42°0N 21°0E 23 C9
Sara Buri = Saraburi *Thailand* 14°30N 100°55E 38
Sarāb *Iran* 37°55N 47°40E 44
Sarābādī *Iraq* 33°1N 44°48E 44
Saraburi *Thailand* 14°30N 100°55E 38
Saradiya *India* 21°34N 70°2E 42
Saragossa = Zaragoza *Spain* 41°39N 0°53W 21
Saraguro *Ecuador* 3°35S 79°16W 92
Sarahs *Turkmenistan* 36°32N 61°13E 45
Sarai Naurang *Pakistan* 32°50N 70°47E 42
Saraikela *India* 22°42N 85°56E 43 H
Sarajevo *Bos.-H.* 43°52N 18°26E 23
Sarakhs = Sarahs *Turkmenistan* 36°32N 61°13E 45
Saran, Gunung *Indonesia* 0°30S 111°25E 36
Saranac Lake *U.S.A.* 44°20N 74°10W 83 B
Saranac Lakes *U.S.A.* 44°20N 74°28W 83 B
Saranda *Tanzania* 5°45S 34°59E 54
Sarandí del Yí *Uruguay* 33°18S 55°38W 95
Sarandí Grande *Uruguay* 33°44S 56°20W 94
Sarangani B. *Phil.* 6°0N 125°13E 37
Sarangani Is. *Phil.* 5°25N 125°25E 37
Sarangarh *India* 21°30N 83°5E 41 J
Saransk *Russia* 54°10N 45°10E 18
Sarapul *Russia* 56°28N 53°48E 18
Sarasota *U.S.A.* 27°20N 82°32W 85 H
Saratoga *Calif., U.S.A.* 37°16N 122°2W 78
Saratoga *Wyo., U.S.A.* 41°27N 106°49W 76 F
Saratoga ⌂ *U.S.A.* 43°0N 73°38W 83 D
Saratoga L. *U.S.A.* 43°1N 73°44W 83 C
Saratoga Springs *U.S.A.* 43°5N 73°47W 83 C
Saratok *Malaysia* 1°55N 111°17E 36
Saratov *Russia* 51°30N 46°2E 19
Sarāvān *Iran* 27°25N 62°15E 45
Saravane *Laos* 15°43N 106°25E 38
Sarawak □ *Malaysia* 2°0N 113°0E 36
Saray *Turkey* 41°26N 27°55E 23 D
Saraykōy *Turkey* 37°55N 28°54E 23 F
Sarbāz *Iran* 26°38N 61°19E 45
Sarbīsheh *Iran* 32°30N 59°40E 45
Sarda → *India* 27°21N 81°23E 43
Sardarshahr *India* 28°30N 74°29E 42
Sardegna □ *Italy* 40°0N 9°0E 22
Sardhana *India* 29°9N 77°39E 42
Sardina, Pta. *Canary Is.* 28°9N 15°44W 24
Sardinia = Sardegna □ *Italy* 40°0N 9°0E 22
Sardis *Turkey* 38°28N 27°58E 23 E
Sārdūīyeh = Dar Mazār *Iran* 29°14N 57°20E 45
Saren *Indonesia* 8°26S 115°34E 37 J
S'Arenal *Spain* 39°30N 2°45E 24
Sarera, G. of *Indonesia* 2°0S 135°0E 58
Sargasso Sea *Atl. Oc.* 27°0N 72°0W 66 G
Sargodha *Pakistan* 32°10N 72°40E 42
Sarh *Chad* 9°5N 18°23E 51
Sārī *Iran* 36°30N 53°4E 45
Saria *India* 21°38N 83°22E 43 J
Sariab *Pakistan* 30°6N 66°59E 42
Sarıgöl *Turkey* 38°14N 28°41E 23 E
Sarikei *Malaysia* 2°8N 111°30E 36
Sarila *India* 25°46N 79°41E 43
Sarina *Australia* 21°22S 149°13E 62
Sarita *U.S.A.* 27°13N 97°47W 84
Sariwŏn *N. Korea* 38°31N 125°46E 35 E
Sarju → *India* 27°21N 81°23E 43
Sark *U.K.* 49°25N 2°22W 13
Sarkari Tala *India* 27°39N 70°52E 42
Şarköy *Turkey* 40°36N 27°6E 23 D
Sarlat-la-Canéda *France* 44°54N 1°13E 20
Sarmi *Indonesia* 1°49S 138°44E 37
Sarmiento *Argentina* 45°35S 69°5W 96
Särna *Sweden* 61°41N 13°8E 8 F
Sarnia *Canada* 42°58N 82°23W 82
Sarolangun *Indonesia* 2°19S 102°42E 36
Saronikos Kolpos *Greece* 37°45N 23°45E 23 F
Saros Körfezi *Turkey* 40°30N 26°15E 23 D
Sarpsborg *Norway* 59°16N 11°7E 9 G
Sarqan *Kazakhstan* 45°24N 79°55E 32
Sarre = Saar → *Europe* 49°41N 6°32E 15
Sarreguemines *France* 49°5N 7°4E 20
Sarthe → *France* 47°33N 0°31W 20
Saruna → *Pakistan* 26°31N 67°7E 42
Sarvar *India* 26°4N 75°0E 42
Sarvestān *Iran* 29°20N 53°10E 45
Sary-Tash *Kyrgyzstan* 39°44N 73°15E 28
Saryshaghan *Kazakhstan* 46°12N 73°38E 28
Sasan Gir *India* 21°10N 70°36E 42
Sasaram *India* 24°57N 84°5E 43 G
Sasebo *Japan* 33°10N 129°43E 31
Saser Kangri *India* 34°50N 77°50E 43
Saskatchewan □ *Canada* 54°40N 106°0W 71
Saskatchewan → *Canada* 53°37N 100°40W 71
Saskatoon *Canada* 52°10N 106°38W 71
Saskylakh *Russia* 71°55N 114°1E 29
Saslaya △ *Nic.* 13°45N 85°4W 88
Sasolburg *S. Africa* 26°46S 27°49E 57
Sasovo *Russia* 54°25N 41°55E 18
Sassandra *Ivory C.* 4°55N 6°8W 50
Sassandra → *Ivory C.* 4°58N 6°5W 50
Sássari *Italy* 40°43N 8°34E 22
Sassie *Australia* 10°2S 142°51E 6
Sassnitz *Germany* 54°29N 13°39E 16
Sassuolo *Italy* 44°33N 10°47E 22
Sastamala *Finland* 61°20N 22°54E 8
Sasyk, Ozero *Ukraine* 45°45N 29°40E 17
Sata-Misaki *Japan* 31°0N 130°40E 31
Satadougou *Mali* 12°25N 11°25W 50
Satakunda = Satakunta *Finland* 61°45N 23°0E 8
Satakunta *Finland* 61°45N 23°0E 8
Satara *India* 17°44N 73°58E 40
Satara *S. Africa* 24°29S 31°47E 57
Satbarwa *India* 23°55N 84°16E 43
Satevó *Mexico* 27°57N 106°7W 86

illa → *U.S.A.* 30°59N 81°29W **85** F14
ka *Russia* 55°3N 59°1E **18** C10
mala Hills *Andhra Pradesh, ndia* 19°45N 78°45E **40** K11
mala Hills *Maharashtra, ndia* 20°15N 74°40E **40** J9
na *India* 24°35N 80°50E **43** G9
oraljaújhely *Iungary* 48°25N 21°41E **17** D11
pura △ *India* 22°40N 78°15E **42** H8
pura Ra. *India* 21°25N 76°10E **40** J10
suna-Shotō *Japan* 30°0N 130°0E **31** K5
tahip *Thailand* 12°41N 100°54E **38** F3
u Mare *Romania* 47°46N 22°55E **17** E12
ui *Indonesia* 3°50S 115°27E **36** E5
un *Thailand* 6°43N 100°2E **39** J3
upa'itea *Samoa* 13°45S 172°18W **59** b
urnina → *Brazil* 12°15S 58°10W **92** F7
ıce *Argentina* 30°5S 58°46W **94** C4
ıceda *Mexico* 25°46N 101°19W **86** B4
ıcillo *Mexico* 28°1N 105°17W **86** B3
ıda *Norway* 59°40N 6°20E **9** G12
ıðarkrókur *Iceland* 65°45N 19°40W **8** D4
ıdi Arabia ■ *Asia* 26°0N 44°0E **44** B3
ıerland *Germany* 51°12N 7°59E **16** C4
ıgeen → *Canada* 44°30N 81°22W **82** B3
ıgerties *U.S.A.* 42°5N 73°57W **83** D11
ıgus *U.S.A.* 34°25N 118°32W **79** L8
ık Centre *U.S.A.* 45°44N 94°57W **80** C6
ık Rapids *U.S.A.* 45°35N 94°10W **80** C6
ılt Ste. Marie *Canada* 46°30N 84°20W **72** C3
ılt Ste. Marie *U.S.A.* 46°30N 84°21W **81** B11
ımlaki *Indonesia* 7°55S 131°20E **37** F8
ımur *France* 47°15N 0°5W **20** C3
ınders, C. *N.Z.* 45°53S 170°45E **59** F3
ınders I. *Antarctica* 57°48S 26°28W **5** B1
ınders Pt. *Australia* 27°52S 125°38E **61** E4
ırimo *Angola* 9°40S 20°12E **52** F4
ısalito *U.S.A.* 37°51N 122°29W **78** H4
á *Honduras* 15°32N 86°15W **88** C2
a → *Serbia* 44°50N 20°26E **23** B9
age *U.S.A.* 47°27N 104°21W **76** C11
age I. = Niue ⊠ *ac. Oc.* 19°2S 169°54W **65** J11
age River *Australia* 41°31S 145°14E **63** G4
ai'i *Samoa* 13°28S 172°24W **59** b
alou *Benin* 7°57N 1°58E **50** G6
ane *Mozam.* 19°37S 35°8E **55** F4
anna *U.S.A.* 42°5N 90°8W **80** D8
anna-la-Mar *Jamaica* 18°10N 78°10W **88** a
annah *Ga., U.S.A.* 32°5N 81°6W **85** E14
annah *Mo., U.S.A.* 39°56N 94°50W **80** F6
annah *Tenn., U.S.A.* 35°14N 88°15W **85** D10
annah → *U.S.A.* 32°2N 80°53W **85** E14
annakhet *Laos* 16°30N 104°49E **38** D5
ant L. *Canada* 50°16N 90°44W **72** B1
ant Lake *Canada* 50°14N 90°40W **72** B1
arkundla = Kundla *ıdia* 21°21N 71°25E **42** J4
e → *Mozam.* 21°16S 34°0E **57** C5
eh *Iran* 35°2N 50°20E **45** C6
elugu *Ghana* 9°38N 0°54W **50** G5
o *Finland* 62°45N 27°30E **8** E22
oie □ *France* 45°26N 6°25E **20** D7
olax = Savo *Finland* 62°45N 27°30E **8** E22
ona *Italy* 44°17N 8°30E **20** D8
ona *U.S.A.* 42°17N 77°13W **82** D7
onlinna *Finland* 61°52N 28°53E **8** F23
oy = Savoie □ *France* 45°26N 6°25E **20** D7
ur *Turkey* 37°34N 40°53E **44** B4
usavu *Fiji* 16°34S 179°15E **59** a
usavu B. *Fiji* 16°45S 179°15E **59** a
vahlunto *Indonesia* 0°40S 100°52E **36** E2
vai *Indonesia* 3°0S 129°5E **37** E7
vai Madhopur *India* 26°0N 76°25E **42** G7
valeke *Fiji* 17°59S 179°18E **59** a
vang Daen Din *hailand* 17°28N 103°28E **38** D4
vankhalok *Thailand* 17°19N 99°50E **38** D2
vara *Japan* 35°55N 140°30E **31** G10
vatch Range *U.S.A.* 39°0N 106°30W **76** G10
vel Mt. *U.K.* 54°50N 7°2W **10** B4
vi *Thailand* 10°14N 99°5E **39** G2
vmills *Zimbabwe* 19°30S 28°2E **55** F2
vtooth ⌂ *U.S.A.* 44°0N 114°50W **76** D6
vtooth Range *U.S.A.* 44°3N 114°58W **76** D6
vu *Indonesia* 10°35S 121°50E **37** F6
vu Sea *Indonesia* 9°30S 121°50E **37** F6
by → *Australia* 18°25S 140°53E **62** B3
mundham *U.K.* 52°13N 1°30E **13** E9
ony = Sachsen □ *ermany* 50°55N 13°10E **16** C7
ony, Lower = iedersachsen □ *ermany* 52°50N 9°0E **16** B5
ton *U.S.A.* 40°13N 78°15W **82** F6
abec *Canada* 48°35N 67°41W **73** C6
aboury *Laos* 19°15N 101°45E **38** C3
án *Peru* 11°8S 77°12W **92** F3
an, Vostochnyy *ussia* 54°0N 96°0E **29** D10
an, Zapadnyy *Russia* 52°30N 94°0E **32** A7
anogorsk *Russia* 53°5N 91°23E **29** D10
dā *Lebanon* 33°35N 35°25E **46** B4
handulaan = Öldziyt *Iongolia* 44°40N 109°1E **34** B5
ḥūt *Yemen* 15°12N 51°10E **47** D5
lac *Somali Rep.* 11°21N 43°30E **47** E3
nshand = Buyant-Uhaa *Iongolia* 44°55N 110°11E **34** B6
re *Okla., U.S.A.* 35°18N 99°38W **84** D5
re *Pa., U.S.A.* 41°59N 76°32W **83** E8
reville *U.S.A.* 40°28N 74°22W **83** F10
ula *Mexico* 19°52N 103°36W **86** D4
ward *Canada* 50°21N 125°55W **70** C3
anit *Albania* 40°30N 19°20E **23** D8
ava → *Czech Rep.* 49°53N 14°24E **16** D8
n *Pakistan* 35°35N 73°30E **43** B5
fell Pike *U.K.* 54°27N 3°14W **12** C4
lloway *U.K.* 60°9N 1°17W **11** A7

Scalpay *U.K.* 57°18N 6°0W **11** D3
Scandia *Canada* 50°20N 112°2W **70** C6
Scandicci *Italy* 43°45N 11°11E **22** C4
Scandinavia *Europe* 64°0N 12°0E **8** E15
Scapa Flow *U.K.* 58°53N 3°3W **11** C5
Scappoose *U.S.A.* 45°45N 122°53W **78** E4
Scarba *U.K.* 56°11N 5°43W **11** E3
Scarborough *Trin. & Tob.* 11°11N 60°42W **93** J16
Scarborough *U.K.* 54°17N 0°24W **12** C7
Scariff I. *Ireland* 51°44N 10°15W **10** E1
Scarp *U.K.* 58°1N 7°8W **11** C1
Scebeli = Shabeelle → *Somali Rep.* 2°0N 44°0E **47** G3
Schaffhausen *Switz.* 47°42N 8°39E **20** C8
Schagen *Neths.* 52°49N 4°48E **15** B4
Schaghticoke *U.S.A.* 42°54N 73°35W **83** D11
Schefferville *Canada* 54°48N 66°50W **73** B6
Schelde → *Belgium* 51°15N 4°16E **15** C4
Schell Creek Ra. *U.S.A.* 39°25N 114°40W **76** G6
Schellsburg *U.S.A.* 40°3N 78°39W **82** F6
Schenectady *U.S.A.* 42°49N 73°57W **83** D11
Schenevus *U.S.A.* 42°33N 74°50W **83** D10
Schiedam *Neths.* 51°55N 4°25E **15** C4
Schiermonnikoog *Neths.* 53°30N 6°15E **15** A6
Schio *Italy* 45°43N 11°21E **22** B4
Schiphol, Amsterdam ✈ **(AMS)** *Neths.* 52°18N 4°45E **15** B4
Schleswig *Germany* 54°31N 9°34E **16** A5
Schleswig-Holstein □ *Germany* 54°30N 9°30E **16** A5
Schœlcher *Martinique* 14°36N 61°7W **88** c
Schoharie *U.S.A.* 42°40N 74°19W **83** D10
Schoharie Cr. → *U.S.A.* 42°57N 74°18W **83** D10
Scholls *U.S.A.* 45°24N 122°56W **78** E4
Schouten I. *Australia* 42°20S 148°20E **63** G4
Schouten Is. = Supiori *Indonesia* 1°0S 136°0E **37** E9
Schouwen *Neths.* 51°43N 3°45E **15** C3
Schreiber *Canada* 48°45N 87°20W **72** C2
Schroffenstein *Namibia* 27°11S 18°42E **56** D2
Schroon Lake *U.S.A.* 43°50N 73°46W **83** C11
Schuler *Canada* 50°20N 110°6W **71** C6
Schumacher *Canada* 48°30N 81°16W **72** C3
Schurz *U.S.A.* 38°57N 118°49W **76** G4
Schuyler *U.S.A.* 41°27N 97°4W **80** E5
Schuylerville *U.S.A.* 43°6N 73°35W **83** C11
Schuylkill → *U.S.A.* 39°53N 75°12W **83** G9
Schuylkill Haven *U.S.A.* 40°37N 76°11W **83** F8
Schwäbische Alb *Germany* 48°20N 9°30E **16** D5
Schwaner, Pegunungan *Indonesia* 1°0S 112°30E **36** E4
Schwarzrand *Namibia* 25°37S 16°50E **56** D2
Schwarzwald *Germany* 48°30N 8°20E **16** D5
Schwatka Mts. *U.S.A.* 67°20N 156°30W **74** B8
Schwedt *Germany* 53°3N 14°16E **16** B8
Schweinfurt *Germany* 50°3N 10°14E **16** C6
Schweiz = Switzerland ■ *Europe* 46°30N 8°0E **20** C8
Schweizer-Reneke *S. Africa* 27°11S 25°18E **56** D4
Schwenningen = Villingen-Schwenningen *Germany* 48°3N 8°26E **16** D5
Schwerin *Germany* 53°36N 11°22E **16** B6
Schwyz *Switz.* 47°2N 8°39E **20** C8
Sciacca *Italy* 37°31N 13°3E **22** F5
Scilla *Italy* 38°15N 15°43E **22** E6
Scilly, Isles of *U.K.* 49°56N 6°22W **13** H1
Scioto → *U.S.A.* 38°44N 83°1W **81** F12
Scituate *U.S.A.* 42°12N 70°44W **83** D14
Scobey *U.S.A.* 48°47N 105°25W **76** B11
Scone *Australia* 32°5S 150°52E **63** E5
Scone *U.K.* 56°25N 3°24W **11** E5
Scoresbysund = Ittoqqortoormiit *Greenland* 70°20N 23°0W **4** B6
Scotia *Calif., U.S.A.* 40°29N 124°6W **76** F1
Scotia *N.Y., U.S.A.* 42°50N 73°58W **83** D11
Scotia Sea *Antarctica* 56°5S 56°0W **5** B18
Scotland *Canada* 43°1N 80°22W **82** C4
Scotland □ *U.K.* 57°0N 4°0W **11** E5
Scott *Antarctica* 77°50S 166°45E **5** D11
Scott, C. *Australia* 13°30S 129°49E **60** B4
Scott City *U.S.A.* 38°29N 100°54W **80** F3
Scott Glacier *Antarctica* 66°15S 100°5E **5** C8
Scott I. *Antarctica* 67°0S 179°0E **5** C11
Scott Is. *Canada* 50°48N 128°40W **70** C3
Scott L. *Canada* 59°55N 106°18W **71** B7
Scott Reef *Australia* 14°0S 121°50E **60** B3
Scottburgh *S. Africa* 30°15S 30°47E **57** E5
Scottdale *U.S.A.* 40°6N 79°35W **82** F5
Scottish Borders □ *U.K.* 55°35N 2°50W **11** F6
Scotts Valley *U.S.A.* 37°3N 122°1W **78** H4
Scottsbluff *U.S.A.* 41°52N 103°40W **80** E2
Scottsboro *U.S.A.* 34°40N 86°2W **85** D11
Scottsburg *U.S.A.* 38°41N 85°47W **81** F11
Scottsdale *Australia* 41°9S 147°31E **63** G4
Scottsdale *U.S.A.* 33°40N 111°53W **77** K8
Scottsville *Ky., U.S.A.* 36°45N 86°11W **80** G10
Scottsville *N.Y., U.S.A.* 43°2N 77°47W **82** C7
Scottville *U.S.A.* 43°58N 86°17W **80** D10
Scranton *U.S.A.* 41°25N 75°40W **83** E9
Scugog, L. *Canada* 44°10N 78°55W **82** B6
Scunthorpe *U.K.* 53°36N 0°39W **12** D7
Scutari = Shkodër *Albania* 42°4N 19°32E **23** C8
Sea Is. *U.S.A.* 31°30N 81°7W **85** F14
Seabrook, L. *Australia* 30°55S 119°40E **61** F2
Seaford *U.K.* 50°47N 0°7E **13** G8
Seaford *U.S.A.* 38°39N 75°37W **81** F16
Seaforth *Australia* 20°55S 148°57E **62** b
Seaforth *Canada* 43°35N 81°25W **82** C3
Seaforth, L. *U.K.* 57°52N 6°36W **11** D2
Seagraves *U.S.A.* 32°57N 102°34W **84** E3
Seaham *U.K.* 54°50N 1°20W **12** C6
Seal → *Canada* 59°4N 94°48W **71** B10
Seal L. *Canada* 54°20N 61°30W **73** B7
Sealy *U.S.A.* 29°47N 96°9W **84** G6
Searchlight *U.S.A.* 35°28N 114°55W **79** K12

Searcy *U.S.A.* 35°15N 91°44W **84** D9
Searles L. *U.S.A.* 35°44N 117°21W **79** K9
Seascale *U.K.* 54°24N 3°29W **12** C4
Seaside *Calif., U.S.A.* 36°37N 121°50W **78** J5
Seaside *Oreg., U.S.A.* 46°0N 123°56W **78** E3
Seaspray *Australia* 38°25S 147°15E **63** F4
Seattle *U.S.A.* 47°36N 122°19W **78** C4
Seattle-Tacoma Int. ✈ **(SEA)** *U.S.A.* 47°27N 122°18W **78** C4
Seba *Indonesia* 10°29S 121°50E **60** B3
Sebago L. *U.S.A.* 43°52N 70°34W **83** C14
Sebago Lake *U.S.A.* 43°51N 70°34W **83** C14
Sebastián Vizcaíno, B. *Mexico* 28°0N 114°30W **86** B2
Sebastopol = Sevastopol *Ukraine* 44°35N 33°30E **19** F5
Sebastopol *U.S.A.* 38°24N 122°49W **78** G4
Sebewaing *U.S.A.* 43°44N 83°27W **81** D12
Sebha = Sabhā *Libya* 27°9N 14°29E **51** C8
Şebinkarahisar *Turkey* 40°22N 38°28E **19** F6
Sebring *Fla., U.S.A.* 27°30N 81°27W **85** H14
Sebring *Ohio, U.S.A.* 40°55N 81°2W **82** F3
Sebringville *Canada* 43°24N 81°4W **82** C3
Sebta = Ceuta *N. Afr.* 35°52N 5°18W **21** E3
Sebuku *Indonesia* 3°30S 116°25E **36** E5
Sechelt *Canada* 49°25N 123°42W **70** D4
Sechura, Desierto de *Peru* 6°0S 80°30W **92** E2
Secretary I. *N.Z.* 45°15S 166°56E **59** F1
Secunderabad *India* 17°28N 78°30E **40** L11
Security *U.S.A.* 38°45N 104°45W **76** G11
Sedalia *U.S.A.* 38°42N 93°14W **80** F7
Sedan *France* 49°43N 4°57E **20** B6
Sedan *U.S.A.* 37°8N 96°11W **80** G5
Sedbergh *U.K.* 54°20N 2°31W **12** C5
Seddon *N.Z.* 41°40S 174°7E **59** D5
Seddonville *N.Z.* 41°33S 172°1E **59** D4
Sedé Boqér *Israel* 30°52N 34°47E **46** E3
Sedeh *Fārs, Iran* 30°45N 52°11E **45** D7
Sedeh *Khorāsān, Iran* 33°20N 59°14E **45** C8
Sederot *Israel* 31°32N 34°37E **46** D3
Sédhiou *Senegal* 12°44N 15°30W **50** F2
Sedley *Canada* 50°10N 104°0W **71** C8
Sedona *U.S.A.* 34°52N 111°46W **77** J8
Sedova, Pik *Russia* 73°29N 54°58E **28** B6
Sedro-Woolley *U.S.A.* 48°30N 122°14W **78** B4
Seeheim *Namibia* 26°50S 17°45E **56** D2
Seeis *Namibia* 22°29S 17°39E **56** C2
Seekoei → *S. Africa* 30°18S 25°1E **56** E4
Seeley's Bay *Canada* 44°29N 76°14W **83** B8
Seferihisar *Turkey* 38°10N 26°50E **23** E12
Sefophe *Botswana* 22°11S 27°58E **55** G2
Seg-ozero *Russia* 63°20N 33°46E **18** B5
Segamat *Malaysia* 2°30N 102°50E **39** L4
Segesta *Italy* 37°56N 12°50E **22** F5
Seget *Indonesia* 1°24S 130°58E **37** E8
Segezha *Russia* 63°44N 34°19E **18** B5
Ségou *Mali* 13°30N 6°16W **50** F4
Segovia = Coco → *Cent. Amer.* 15°0N 83°8W **88** D3
Segovia *Spain* 40°57N 4°10W **21** B3
Segre → *Spain* 41°40N 0°43E **21** B6
Seguam I. *U.S.A.* 52°19N 172°30W **74** E5
Séguéla *Ivory C.* 7°55N 6°40W **50** G4
Seguin *U.S.A.* 29°34N 97°58W **84** G6
Segundo → *Argentina* 30°53S 62°44W **94** C3
Segura → *Spain* 38°3N 0°44W **21** C5
Seh Konj, Kūh-e *Iran* 30°6N 57°30E **45** D8
Seh Qal'eh *Iran* 33°40N 58°24E **45** C8
Sehitwa *Botswana* 20°30S 22°30E **56** C3
Sehlabathebe △ *Lesotho* 29°53S 29°7E **57** D4
Sehore *India* 23°10N 77°5E **42** H7
Sehwan *Pakistan* 26°28N 67°53E **42** F2
Seikan Tunnel *Japan* 41°28N 140°10E **30** D10
Seil *U.K.* 56°18N 5°38W **11** E3
Seiland *Norway* 70°25N 23°15E **8** A20
Seiling *U.S.A.* 36°9N 98°56W **84** C5
Seinäjoki *Finland* 62°40N 22°51E **8** E20
Seine → *France* 49°26N 0°26E **20** B4
Seistan = Sīstān *Asia* 30°50N 61°0E **45** D9
Seistan, Daryācheh-ye = Sīstan, Daryācheh-ye *Iran* 31°0N 61°0E **45** D9
Sekayu *Indonesia* 2°51S 103°51E **36** E2
Seke *Tanzania* 3°20S 33°31E **54** C3
Sekenke *Tanzania* 4°18S 34°11E **54** C3
Sekondi-Takoradi *Ghana* 4°58N 1°45W **50** H5
Sekudai *Malaysia* 1°32N 103°39E **39** d
Sekuma *Botswana* 24°36S 23°50E **56** C3
Selah *U.S.A.* 46°39N 120°32W **76** C3
Selama *Malaysia* 5°12N 100°42E **39** K3
Selangor □ *Malaysia* 3°10N 101°30E **39** L3
Selaru *Indonesia* 8°9S 131°0E **37** F8
Selatan, Selat *Malaysia* 5°15N 100°20E **39** c
Selawik L. *U.S.A.* 66°30N 160°45W **74** B7
Selby *U.K.* 53°47N 1°5W **12** D6
Selby *U.S.A.* 45°31N 100°2W **80** C3
Selçuk *Turkey* 37°56N 27°22E **23** F12
Selden *U.S.A.* 39°33N 100°34W **80** F3
Sele → *Italy* 40°29N 14°56E **22** D6
Selebi-Phikwe *Botswana* 21°58S 27°48E **57** C4
Selemdzha → *Russia* 51°42N 128°53E **29** D13
Selenga = Selenge Mörön → *Asia* 52°16N 106°16E **32** A10
Selenge Mörön → *Asia* 52°16N 106°16E **32** A10
Seletan, Tanjung *Indonesia* 4°10S 114°40E **36** E4
Sélibabi *Mauritania* 15°10N 12°15W **50** E3
Seligman *U.S.A.* 35°20N 112°53W **77** J7
Selîma *Sudan* 21°22N 29°19E **51** D11
Selinda Spillway → *Botswana* 18°35S 23°10E **56** B3
Selinsgrove *U.S.A.* 40°48N 76°52W **82** F8
Selkirk *Man., Canada* 50°10N 96°55W **71** C9
Selkirk *Ont., Canada* 42°49N 79°56W **82** D5
Selkirk *U.K.* 55°33N 2°50W **11** F6
Selkirk I. = Horse I. *Canada* 53°20N 99°6W **71** C9
Selkirk Mts. *Canada* 51°15N 117°40W **68** G8
Sellafield *U.K.* 54°25N 3°29W **12** C4
Sellia *Greece* 35°12N 24°23E **25** D6

Sells *U.S.A.* 31°55N 111°53W **77** L8
Selma *Ala., U.S.A.* 32°25N 87°1W **85** E11
Selma *Calif., U.S.A.* 36°34N 119°37W **78** J7
Selma *N.C., U.S.A.* 35°32N 78°17W **85** D15
Selmer *U.S.A.* 35°10N 88°36W **85** D10
Selous ⌂ *Tanzania* 8°37S 37°42E **55** D4
Selowandoma Falls *Zimbabwe* 21°15S 31°50E **55** G3
Selpele *Indonesia* 0°1S 130°5E **37** E8
Selsey Bill *U.K.* 50°43N 0°47W **13** G7
Seltso *Russia* 53°22N 34°4E **18** D5
Selu *Indonesia* 7°32S 130°55E **37** F8
Selva *Argentina* 29°50S 62°0W **94** B3
Selva *Spain* 39°46N 2°54E **24** B9
Selva Lancandona = Montes Azules ⌂ *Mexico* 16°21N 91°3W **87** D6
Selvagens, Ilhas *Madeira* 30°5N 15°55W **50** B2
Selvas *Brazil* 6°30S 67°0W **92** E5
Selwyn L. *Canada* 60°0N 104°30W **71** B8
Selwyn Mts. *Canada* 63°0N 130°0W **68** E5
Selwyn Ra. *Australia* 21°10S 140°0E **62** C3
Seman → *Albania* 40°47N 19°30E **23** D8
Semarang *Indonesia* 7°0S 110°26E **37** G14
Semarapura = Klungkung *Indonesia* 8°32S 115°24E **37** K18
Sembabule *Uganda* 0°4S 31°25E **54** C3
Sembawang *Singapore* 1°27N 103°50E **39** d
Sembung *Indonesia* 8°28S 115°11E **37** J18
Semey *Kazakhstan* 50°30N 80°10E **32** A5
Seminoe Res. *U.S.A.* 42°9N 106°55W **76** E10
Seminole *Okla., U.S.A.* 35°14N 96°41W **84** D6
Seminole *Tex., U.S.A.* 32°43N 102°39W **84** E3
Seminole Draw → *U.S.A.* 32°27N 102°20W **84** E3
Semipalatinsk = Semey *Kazakhstan* 50°30N 80°10E **32** A5
Semirara Is. *Phil.* 12°0N 121°20E **37** B6
Semisopochnoi I. *U.S.A.* 51°55N 179°36E **74** E3
Semitau *Indonesia* 0°29N 111°57E **36** D4
Semliki △ *Uganda* 0°49N 30°4E **54** B3
Semmering P. *Austria* 47°41N 15°45E **16** E8
Semnān *Iran* 35°40N 53°23E **45** C7
Semnān □ *Iran* 36°0N 54°0E **45** C7
Semporna *Malaysia* 4°30N 118°33E **37** D5
Semuda *Indonesia* 2°51S 112°58E **36** E4
Sen → *Cambodia* 12°32N 104°28E **38** F5
Senā *Iran* 28°27N 51°36E **45** D6
Sena *Mozam.* 17°25S 35°0E **55** F4
Sena Madureira *Brazil* 9°5S 68°45W **92** E5
Senador Pompeu *Brazil* 5°40S 39°20W **93** E11
Senang, Pulau *Singapore* 1°11N 103°52E **39** d
Senanga *Zambia* 16°7S 23°16E **53** H4
Senatobia *U.S.A.* 34°37N 89°58W **85** D10
Sencelles *Spain* 39°39N 2°54E **24** B9
Sendai *Kagoshima, Japan* 31°50N 130°20E **31** J5
Sendai *Miyagi, Japan* 38°15N 140°53E **30** E10
Sendai-Wan *Japan* 38°15N 141°0E **30** E10
Sendhwa *India* 21°41N 75°6E **42** J6
Seneca *U.S.A.* 34°41N 82°57W **85** D13
Seneca Falls *U.S.A.* 42°55N 76°48W **83** D8
Seneca L. *U.S.A.* 42°40N 76°54W **82** D8
Senecaville L. *U.S.A.* 39°55N 81°25W **82** G3
Senegal ■ *W. Afr.* 14°30N 14°30W **50** F3
Sénégal → *W. Afr.* 15°48N 16°32W **50** E2
Senegambia *Africa* 12°45N 12°0W **48** E2
Senekal *S. Africa* 28°20S 27°36E **57** D4
Senga Hill *Zambia* 9°19S 31°11E **55** D3
Senge Khambab = Indus → *Pakistan* 24°20N 67°47E **42** G2
Sengerema *Tanzania* 2°42S 32°35E **54** C3
Senggigi *Indonesia* 8°29S 116°2E **37** J19
Sengua → *Zimbabwe* 17°7S 28°5E **55** F2
Senhor-do-Bonfim *Brazil* 10°30S 40°10W **93** F10
Senigállia *Italy* 43°43N 13°13E **22** C5
Senj *Croatia* 45°0N 14°58E **16** F8
Senja *Norway* 69°25N 17°30E **8** B17
Senkaku-Shotō *E. China Sea* 25°45N 123°30E **31** M1
Senlis *France* 49°13N 2°35E **20** B5
Senmonorom *Cambodia* 12°27N 107°12E **38** F6
Senneterre *Canada* 48°25N 77°15W **72** C4
Seno *Laos* 16°35N 104°50E **38** D5
Senqu = Orange → *S. Africa* 28°41S 16°28E **56** D2
Sens *France* 48°11N 3°15E **20** B5
Senta *Serbia* 45°55N 20°3E **23** B9
Sentery = Lubao *Dem. Rep. of the Congo* 5°17S 25°42E **54** D2
Sentinel *U.S.A.* 32°52N 113°13W **77** K7
Sentosa *Singapore* 1°16N 103°50E **39** d
Seo de Urgel = La Seu d'Urgell *Spain* 42°22N 1°23E **21** A6
Seogwipo *S. Korea* 33°13N 126°34E **35** H14
Seohara *India* 29°15N 78°33E **43** E8
Seonath → *India* 21°44N 82°28E **43** J10
Seondha *India* 26°9N 78°48E **43** F8
Seongnam *S. Korea* 37°26N 127°8E **35** F14
Seoni *India* 22°5N 79°30E **43** H8
Seoni Malwa *India* 22°27N 77°28E **42** H8
Seonsan *S. Korea* 36°14N 128°17E **35** F15
Seosan *S. Korea* 36°47N 126°27E **35** F14
Seoul *S. Korea* 37°31N 126°58E **35** F14
Sepīdān *Iran* 30°20N 52°5E **45** D7
Sepo-ri *N. Korea* 38°57N 127°25E **35** E14
Sepone *Laos* 16°45N 106°13E **38** D6
Sept-Îles *Canada* 50°13N 66°22W **73** B6
Sequim *U.S.A.* 48°5N 123°6W **78** B3
Sequoia △ *U.S.A.* 36°30N 118°30W **78** J8
Seraing *Belgium* 50°35N 5°32E **15** D5
Serakhis → *Cyprus* 35°13N 32°55E **25** D11
Seram *Indonesia* 3°10S 129°0E **37** E7
Seram Sea *Indonesia* 2°30S 128°30E **37** E7
Serampore = Shrirampur *India* 22°44N 88°21E **43** H13
Seranantsara *Madag.* 18°30S 49°5E **57** B8
Serang *Indonesia* 6°8S 106°10E **37** G12
Serangoon *Singapore* 1°23N 103°54E **39** d
Serasan *Indonesia* 2°29N 109°4E **36** D3

Serbia ■ *Europe* 43°20N 20°0E **23** B9
Serdar *Turkmenistan* 39°4N 56°23E **45** B8
Serdobsk *Russia* 52°28N 44°10E **18** D7
Seremban *Malaysia* 2°43N 101°53E **39** L3
Serengeti △ *Tanzania* 2°11S 35°0E **54** C3
Serengeti Plain *Tanzania* 2°40S 35°0E **54** C4
Serenje *Zambia* 13°14S 30°15E **55** E3
Sereth = Siret → *Romania* 45°24N 28°1E **17** F14
Sergeya Kirova, Ostrova *Russia* 77°30N 89°0E **29** B10
Sergino *Russia* 62°25N 65°12E **28** C7
Sergipe □ *Brazil* 10°30S 37°30W **93** F11
Sergiyev Posad *Russia* 56°20N 38°10E **18** C6
Serhetabat *Turkmenistan* 35°20N 62°18E **45** C9
Seria *Brunei* 4°37N 114°23E **36** D4
Serian *Malaysia* 1°10N 110°31E **36** D4
Sericho *Kenya* 1°5N 39°5E **54** B4
Serifos *Greece* 37°9N 24°30E **23** F11
Sérigny → *Canada* 56°47N 66°0W **73** A6
Seringapatam Reef *Australia* 13°38S 122°5E **60** B3
Seririt *Indonesia* 8°12S 114°56E **37** J17
Serkout *Algeria* 23°56N 6°47E **50** D7
Sermata *Indonesia* 8°15S 128°50E **37** F7
Sermersuaq *Greenland* 79°30N 62°0W **69** B19
Serov *Russia* 59°29N 60°35E **18** C11
Serowe *Botswana* 22°25S 26°43E **56** C4
Serpentine Lakes *Australia* 28°30S 129°10E **61** E4
Serpent's Mouth = Sierpe, Bocas de la *Venezuela* 10°0N 61°30W **93** L15
Serpukhov *Russia* 54°55N 37°28E **18** D6
Serra do Navio *Brazil* 0°59N 52°3W **93** C8
Serranía San Luís △ *Paraguay* 22°35S 57°22W **94** A4
Serranía San Rafael △ *Paraguay* 26°30S 56°0W **95** B4
Serres *Greece* 41°5N 23°31E **23** D10
Serrezuela *Argentina* 30°40S 65°20W **94** C2
Serrinha *Brazil* 11°39S 39°0W **93** F11
Sertanópolis *Brazil* 23°4S 51°2W **95** A5
Serua *Indonesia* 6°18S 130°1E **37** F8
Serui *Indonesia* 1°53S 136°10E **37** E9
Serule *Botswana* 21°57S 27°20E **56** C4
Ses Salines *Spain* 39°21N 3°3E **24** B10
Sese Is. *Uganda* 0°20S 32°20E **54** C3
Sesepe *Indonesia* 1°30S 127°59E **37** E7
Sesfontein *Namibia* 19°7S 13°39E **56** B1
Sesheke *Zambia* 17°29S 24°13E **56** B3
S'Espalmador *Spain* 38°47N 1°26E **24** C7
S'Espardell *Spain* 38°48N 1°29E **24** C7
S'Estanyol *Spain* 39°22N 2°54E **24** B9
Setana *Japan* 42°26N 139°51E **30** C9
Sète *France* 43°25N 3°42E **20** E5
Sete Lagoas *Brazil* 19°27S 44°16W **93** G10
Sétif *Algeria* 36°9N 5°26E **50** A7
Seto *Japan* 35°14N 137°6E **31** G8
Setonaikai *Japan* 34°20N 133°30E **31** G6
Setonaikai △ *Japan* 34°15N 133°15E **31** G6
Settat *Morocco* 33°0N 7°40W **50** B4
Setting L. *Canada* 55°0N 98°38W **71** C9
Settle *U.K.* 54°5N 2°16W **12** C5
Settlement, The *Br. Virgin Is.* 18°43N 64°22W **89** e
Settlers *S. Africa* 25°2S 28°30E **57** C4
Setúbal *Portugal* 38°30N 8°58W **21** C1
Setúbal, B. de *Portugal* 38°40N 8°56W **21** C1
Seul, Lac *Canada* 50°20N 92°30W **72** B1
Sevan, Ozero = Sevana Lich *Armenia* 40°30N 45°20E **44** A5
Sevana Lich *Armenia* 40°30N 45°20E **44** A5
Sevastopol *Ukraine* 44°35N 33°30E **19** F5
Seven Sisters *Canada* 54°56N 128°10W **70** C3
Severn → *Canada* 56°2N 87°36W **72** A2
Severn → *U.K.* 51°35N 2°40W **13** F5
Severnaya Zemlya *Russia* 79°0N 100°0E **29** B11
Severnyye Uvaly *Russia* 60°0N 50°0E **18** C8
Severo-Kurilsk *Russia* 50°40N 156°8E **29** D16
Severo-Yeniseyskiy *Russia* 60°22N 93°1E **29** C10
Severo-Zapadnyy □ *Russia* 65°0N 40°0E **28** C4
Severobaykalsk *Russia* 55°39N 109°19E **29** D11
Severodvinsk *Russia* 64°27N 39°58E **18** B6
Severomorsk *Russia* 69°5N 33°27E **8** B25
Severouralsk *Russia* 60°9N 59°57E **18** B10
Seversk *Russia* 56°36N 84°49E **28** D9
Sevier → *U.S.A.* 39°4N 113°6W **76** G7
Sevier Desert *U.S.A.* 39°40N 112°45W **76** G7
Sevier L. *U.S.A.* 38°54N 113°9W **76** G7
Sevilla *Spain* 37°23N 5°58W **21** D2
Seville = Sevilla *Spain* 37°23N 5°58W **21** D2
Sevlievo *Bulgaria* 43°2N 25°6E **23** C11
Sewani *India* 28°58N 75°39E **42** E6
Seward *Alaska, U.S.A.* 60°7N 149°27W **68** E2
Seward *Nebr., U.S.A.* 40°55N 97°6W **80** E5
Seward *Pa., U.S.A.* 40°25N 79°1W **82** F5
Seward Peninsula *U.S.A.* 65°30N 166°0W **74** B6
Sewell *Chile* 34°10S 70°23W **94** C1
Sewer *Indonesia* 5°53S 134°40E **37** F8
Sewickley *U.S.A.* 40°32N 80°12W **82** F4
Sexsmith *Canada* 55°21N 118°47W **70** B5
Seychelles ■ *Ind. Oc.* 5°0S 56°0E **53** b
Seyðisfjörður *Iceland* 65°16N 13°57W **8** D7
Seydişehir *Turkey* 37°25N 31°51E **19** G5
Seydvān *Iran* 38°34N 45°2E **44** B5
Seyhan → *Turkey* 36°43N 34°53E **44** B2
Seym → *Ukraine* 51°27N 32°34E **19** D5
Seymour *Australia* 37°2S 145°10E **63** F4
Seymour *S. Africa* 32°33S 26°46E **56** E4
Seymour *Conn., U.S.A.* 41°24N 73°4W **83** E11
Seymour *Ind., U.S.A.* 38°58N 85°53W **81** F11
Seymour *Tex., U.S.A.* 33°35N 99°16W **84** E5
Sfântu Gheorghe *Romania* 45°52N 25°48E **17** F13
Sfax *Tunisia* 34°49N 10°48E **51** B8
Sha Tau Kok *China* 22°33N 114°13E **33** a

Sha Tin *China* 22°23N 114°12E **33** a
Shaanxi □ *China* 35°0N 109°0E **34** G5
Shaba = Katanga □ *Dem. Rep. of the Congo* 8°0S 25°0E **54** D2
Shaba △ *Kenya* 0°38N 37°48E **54** B4
Shabeelle → *Somali Rep.* 2°0N 44°0E **47** G3
Shabogamo L. *Canada* 53°15N 66°30W **73** B6
Shabunda *Dem. Rep. of the Congo* 2°40S 27°16E **54** C2
Shache *China* 38°20N 77°10E **32** D4
Shackleton Fracture Zone *S. Ocean* 60°0S 60°0W **5** B18
Shackleton Ice Shelf *Antarctica* 66°0S 100°0E **5** C8
Shackleton Inlet *Antarctica* 83°0S 160°0E **5** E11
Shādegān *Iran* 30°40N 48°38E **45** D6
Shadi *India* 33°24N 77°14E **43** C7
Shadrinsk *Russia* 56°5N 63°32E **28** D7
Shadyside *U.S.A.* 39°58N 80°45W **82** G4
Shafter *U.S.A.* 35°30N 119°16W **79** K7
Shaftesbury *U.K.* 51°0N 2°11W **13** F5
Shaftsbury *U.S.A.* 43°0N 73°11W **83** D11
Shagram *Pakistan* 36°24N 72°20E **43** A5
Shah Alam *Malaysia* 3°5N 101°32E **39** L3
Shah Alizai *Pakistan* 29°25N 66°33E **42** E2
Shah Bunder *Pakistan* 24°13N 67°56E **42** G2
Shahabad *Punjab, India* 30°10N 76°55E **42** D7
Shahabad *Raj., India* 25°15N 77°11E **42** G7
Shahabad *Ut. P., India* 27°36N 79°56E **43** F8
Shahadpur *Pakistan* 25°55N 68°35E **42** G3
Shahbā' *Syria* 32°52N 36°38E **46** C5
Shahdād *Iran* 30°30N 57°40E **45** D8
Shahdād, Namakzār-e *Iran* 30°20N 58°20E **45** D8
Shahdadkot *Pakistan* 27°50N 67°55E **42** F2
Shahdol *India* 23°19N 81°26E **43** H9
Shahe *China* 37°0N 114°32E **34** F8
Shahganj *India* 26°3N 82°44E **43** F10
Shahgarh *India* 27°15N 69°50E **42** F3
Shahjahanpur *India* 27°54N 79°57E **43** F8
Shahpur = Salmās *Iran* 38°11N 44°47E **44** B5
Shahpur *India* 22°12N 77°58E **42** H7
Shahpur *Baluchistan, Pakistan* 28°46N 68°27E **42** E3
Shahpur *Punjab, Pakistan* 32°17N 72°26E **42** C5
Shahpur Chakar *Pakistan* 26°9N 68°39E **42** F3
Shahpura *Mad. P., India* 23°10N 80°45E **43** H9
Shahpura *Raj., India* 25°38N 74°56E **42** G6
Shahr-e Bābak *Iran* 30°7N 55°9E **45** D7
Shahr-e Kord *Iran* 32°15N 50°55E **45** C6
Shāhrakht *Iran* 33°38N 60°16E **45** C9
Shahrezā = Qomsheh *Iran* 32°0N 51°55E **45** D6
Shahrig *Pakistan* 30°15N 67°40E **42** D2
Shāhrud = Emāmrūd *Iran* 36°30N 55°0E **45** B7
Shahukou *China* 40°20N 112°18E **34** D7
Shaikhabad *Afghan.* 34°2N 68°45E **42** B3
Shajapur *India* 23°27N 76°21E **42** H7
Shajing *China* 22°44N 113°48E **33** a
Shakargarh *Pakistan* 32°17N 75°10E **42** C6
Shakawe *Botswana* 18°28S 21°49E **56** B3
Shaker Heights *U.S.A.* 41°28N 81°32W **82** E3
Shakhtersk *Russia* 49°10N 142°8E **29** E15
Shakhty *Russia* 47°40N 40°16E **19** E7
Shakhunya *Russia* 57°40N 46°46E **18** C8
Shaki *Nigeria* 8°41N 3°21E **50** G6
Shakotan-Hantō *Japan* 43°10N 140°30E **30** C10
Shaksam Valley *Asia* 36°0N 76°20E **43** A7
Shallow Lake *Canada* 44°36N 81°5W **82** B3
Shalqar *Kazakhstan* 47°48N 59°39E **28** E6
Shaluli Shan *China* 30°40N 99°55E **41** D21
Shām *Iran* 26°39N 57°21E **45** E8
Shām, Bādiyat ash *Asia* 32°0N 40°0E **44** C3
Shamâl Sînî □ *Egypt* 30°30N 33°30E **46** E2
Shamattawa *Canada* 55°51N 92°5W **72** A1
Shamattawa → *Canada* 55°1N 85°23W **72** A2
Shamīl *Iran* 27°30N 56°55E **45** E8
Shāmkūh *Iran* 35°47N 57°50E **45** C8
Shamli *India* 29°32N 77°18E **42** E7
Shammar, Jabal *Si. Arabia* 27°40N 41°0E **44** E4
Shamo = Gobi *Asia* 44°0N 110°0E **34** C6
Shamo, L. *Ethiopia* 5°45N 37°30E **47** F2
Shamokin *U.S.A.* 40°47N 76°34W **83** F8
Shamrock *Canada* 45°23N 76°50W **83** A8
Shamrock *U.S.A.* 35°13N 100°15W **84** D4
Shamva *Zimbabwe* 17°20S 31°32E **55** F3
Shan □ *Burma* 21°30N 98°30E **41** J21
Shan Xian *China* 34°50N 116°5E **34** G9
Shanchengzhen *China* 42°20N 125°20E **35** C13
Shāndak *Iran* 28°28N 60°27E **45** D9
Shandan *China* 38°45N 101°15E **32** D9
Shandon *U.S.A.* 35°39N 120°23W **78** K6
Shandong □ *China* 36°0N 118°0E **35** G10
Shandong Bandao *China* 37°0N 121°0E **35** F11
Shandur Pass *Pakistan* 36°4N 72°31E **43** A5
Shang Xian = Shangzhou *China* 33°50N 109°58E **34** H5
Shangalowe *Dem. Rep. of the Congo* 10°50S 26°30E **55** E2
Shangani *Zimbabwe* 19°41S 29°20E **57** B4
Shangani → *Zimbabwe* 18°41S 27°10E **55** F2
Shangbancheng *China* 40°50N 118°1E **35** D10
Shangdu *China* 41°30N 113°30E **34** D7
Shanghai *China* 31°15N 121°26E **33** E13
Shanghe *China* 37°20N 117°10E **35** F9
Shangnan *China* 33°32N 110°50E **34** H6
Shangqiu *China* 34°26N 115°36E **34** G8
Shangrao *China* 28°25N 117°59E **33** F12
Shangri-La = Zhongdian *China* 27°48N 99°42E **32** F8
Shangshui *China* 33°32N 114°35E **34** H8
Shangyi *China* 41°4N 113°57E **34** D7
Shangzhi *China* 45°22N 127°56E **35** B14
Shangzhou *China* 33°50N 109°58E **34** H5
Shanhetun *China* 44°33N 127°15E **35** B14
Shanklin *U.K.* 50°38N 1°11W **13** G6
Shannon *N.Z.* 40°33S 175°25E **59** D5
Shannon → *Ireland* 52°35N 9°30W **10** D2
Shannon ✈ (SNN) *Ireland* 52°42N 8°57W **10** D3
Shannon, Mouth of the *Ireland* 52°30N 9°55W **10** D2
Shannon △ *Australia* 34°35S 116°25E **61** F2
Shannonbridge *Ireland* 53°17N 8°3W **10** C3
Shansi = Shanxi □ *China* 37°0N 112°0E **34** F7
Shantar, Ostrov Bolshoy *Russia* 55°9N 137°40E **29** D14
Shantipur *India* 23°17N 88°25E **43** H13
Shantou *China* 23°18N 116°40E **33** G12
Shantung = Shandong □ *China* 36°0N 118°0E **35** G10
Shanxi □ *China* 37°0N 112°0E **34** F7
Shanyang *China* 33°31N 109°55E **34** H5
Shanyin *China* 39°25N 112°56E **34** E7
Shaoguan *China* 24°48N 113°35E **33** G11
Shaoyang *China* 27°14N 111°25E **33** F11
Shap *U.K.* 54°32N 2°40W **12** C5
Shapinsay *U.K.* 59°3N 2°51W **11** B6
Shaqra' *Si. Arabia* 25°15N 45°16E **44** E5
Shaqrā' *Yemen* 13°22N 45°44E **47** E4
Sharafkhāneh *Iran* 38°11N 45°29E **44** B5
Sharbot Lake *Canada* 44°46N 76°41W **83** B8
Shari *Japan* 43°55N 144°40E **30** C12
Sharjah = Ash Shāriqah *U.A.E.* 25°23N 55°26E **45** E7
Shark B. *Australia* 25°30S 113°32E **61** E1
Shark Bay △ *Australia* 25°30S 113°30E **61** E1
Sharm el Sheikh *Egypt* 27°53N 34°18E **51** C12
Sharon *Canada* 44°6N 79°26W **82** B5
Sharon *Mass., U.S.A.* 42°7N 71°11W **83** D13
Sharon *Pa., U.S.A.* 41°14N 80°31W **82** E4
Sharon Springs *Kans., U.S.A.* 38°54N 101°45W **80** F3
Sharon Springs *N.Y., U.S.A.* 42°48N 74°37W **83** D10
Sharp Pt. *Australia* 10°58S 142°43E **62** A3
Sharpe L. *Canada* 54°24N 93°40W **72** B1
Sharpsville *U.S.A.* 41°15N 80°29W **82** E4
Sharqi, Al Jabal ash *Lebanon* 33°40N 36°10E **46** B5
Sharya *Russia* 58°22N 45°20E **18** C8
Shashemene *Ethiopia* 7°13N 38°33E **47** F2
Shashi *Botswana* 21°15S 27°27E **57** C4
Shashi → *Africa* 21°14S 29°20E **55** G2
Shasta, Mt. *U.S.A.* 41°25N 122°12W **76** F2
Shasta L. *U.S.A.* 40°43N 122°25W **76** F2
Shatsky Rise *Pac. Oc.* 34°0N 157°0E **64** D7
Shatt al Arab *Asia* 29°57N 48°34E **45** D6
Shaunavon *Canada* 49°35N 108°25W **71** D7
Shaver L. *U.S.A.* 37°9N 119°18W **78** H7
Shaw → *Australia* 20°21S 119°17E **60** D2
Shaw I. *Australia* 20°30S 149°2E **62** b
Shawanaga *Canada* 45°31N 80°17W **82** A4
Shawangunk Mts. *U.S.A.* 41°35N 74°30W **83** E10
Shawano *U.S.A.* 44°47N 88°36W **80** C9
Shawinigan *Canada* 46°35N 72°50W **72** C5
Shawmari, J. ash *Jordan* 30°35N 36°35E **46** E5
Shawnee *U.S.A.* 35°20N 96°55W **84** D6
Shay Gap *Australia* 20°30S 120°10E **60** D3
Shaybārā *Si. Arabia* 25°26N 36°47E **44** E3
Shaykh, J. ash *Lebanon* 33°25N 35°50E **46** B4
Shaykh Miskīn *Syria* 32°49N 36°9E **46** C5
Shaykh Sa'd *Iraq* 32°34N 46°17E **44** C5
Shāzand *Iran* 33°56N 49°24E **45** C6
Shchūchīnsk *Kazakhstan* 52°56N 70°12E **28** D8
She Xian *China* 36°30N 113°40E **34** F7
Shebele = Shabeelle → *Somali Rep.* 2°0N 44°0E **47** G3
Sheboygan *U.S.A.* 43°46N 87°45W **80** D10
Shediac *Canada* 46°14N 64°32W **73** C7
Sheelin, L. *Ireland* 53°48N 7°20W **10** C4
Sheenjek → *U.S.A.* 66°45N 144°33W **74** B11
Sheep Haven *Ireland* 55°11N 7°52W **10** A4
Sheep Range *U.S.A.* 36°35N 115°15W **79** J11
Sheerness *U.K.* 51°26N 0°47E **13** F8
Sheet Harbour *Canada* 44°56N 62°31W **73** D7
Sheffield *U.K.* 53°23N 1°28W **12** D6
Sheffield *Ala., U.S.A.* 34°46N 87°41W **85** D11
Sheffield *Mass., U.S.A.* 42°5N 73°21W **83** D11
Sheffield *Pa., U.S.A.* 41°42N 79°3W **82** E5
Sheikhpura *India* 25°9N 85°53E **43** G11
Shekhupura *Pakistan* 31°42N 73°58E **42** D5
Shekou *China* 22°30N 113°55E **33** a
Shelburne *N.S., Canada* 43°47N 65°20W **73** D6
Shelburne *Ont., Canada* 44°4N 80°15W **82** B4
Shelburne *U.S.A.* 44°23N 73°14W **83** B11
Shelburne B. *Australia* 11°50S 142°50E **62** A3
Shelburne Falls *U.S.A.* 42°36N 72°45W **83** D12
Shelby *Mich., U.S.A.* 43°37N 86°22W **80** D10
Shelby *Miss., U.S.A.* 33°57N 90°46W **85** E9
Shelby *Mont., U.S.A.* 48°30N 111°51W **76** B8
Shelby *N.C., U.S.A.* 35°17N 81°32W **85** D14
Shelby *Ohio, U.S.A.* 40°53N 82°40W **82** F2
Shelbyville *Ill., U.S.A.* 39°24N 88°48W **80** F9
Shelbyville *Ind., U.S.A.* 39°31N 85°47W **81** F11
Shelbyville *Ky., U.S.A.* 38°13N 85°14W **81** F11
Shelbyville *Tenn., U.S.A.* 35°29N 86°28W **85** D11
Sheldon *U.S.A.* 43°11N 95°51W **80** D6
Sheldrake *Canada* 50°20N 64°51W **73** B7
Shelikhova, Zaliv *Russia* 59°30N 157°0E **29** D16
Shelikof Strait *U.S.A.* 57°30N 155°0W **74** D9
Shell Lakes *Australia* 29°20S 127°30E **61** E4
Shellbrook *Canada* 53°13N 106°24W **71** C7
Shellharbour *Australia* 34°31S 150°51E **63** E5
Shelter I. *U.S.A.* 41°4N 72°20W **83** E12
Shelton *Conn., U.S.A.* 41°19N 73°5W **83** E11
Shelton *Wash., U.S.A.* 47°13N 123°6W **78** C3
Shen Xian *China* 36°15N 115°40E **34** F8
Shenandoah *Iowa, U.S.A.* 40°46N 95°22W **80** E6
Shenandoah *Pa., U.S.A.* 40°49N 76°12W **83** F8
Shenandoah *Va., U.S.A.* 38°29N 78°37W **81** F14
Shenandoah → *U.S.A.* 39°19N 77°44W **81** F15
Shenandoah △ *U.S.A.* 38°35N 78°22W **81** F14
Shenchi *China* 39°8N 112°10E **34** E7
Shendam *Nigeria* 8°49N 9°30E **50** G7
Shendî *Sudan* 16°46N 33°22E **51** E12
Shengfang *China* 39°3N 116°42E **34** E9
Shenjingzi *China* 44°40N 124°30E **35** B13
Shenmu *China* 38°50N 110°29E **34** E6
Shenqiu *China* 33°25N 115°5E **34** H8
Shensi = Shaanxi □ *China* 35°0N 109°0E **34** G5
Shenyang *China* 41°48N 123°27E **35** D12
Shenzhen *China* 22°32N 114°5E **33** a
Shenzhen ✈ (SZX) *China* 22°41N 113°49E **33** a
Shenzhen Shuiku *China* 22°34N 114°8E **33** a
Shenzhen Wan *China* 22°27N 113°55E **33** a
Sheo *India* 26°11N 71°15E **42** F4
Sheopur Kalan *India* 25°40N 76°40E **42** G7
Shepetivka *Ukraine* 50°10N 27°10E **17** C14
Shepparton *Australia* 36°23S 145°26E **63** F4
Sheppey, I. of *U.K.* 51°25N 0°48E **13** F8
Shepton Mallet *U.K.* 51°11N 2°33W **13** F5
Sheqi *China* 33°12N 112°57E **34** H7
Sher Qila *Pakistan* 36°7N 74°2E **43** A6
Sherborne *U.K.* 50°57N 2°31W **13** G5
Sherbro I. *S. Leone* 7°30N 12°40W **50** G3
Sherbrooke *N.S., Canada* 45°8N 61°59W **73** C7
Sherbrooke *Qué., Canada* 45°28N 71°57W **83** A13
Sherburne *U.S.A.* 42°41N 75°30W **83** D9
Shergarh *India* 26°20N 72°18E **42** F5
Sherghati *India* 24°34N 84°47E **43** G11
Sheridan *Ark., U.S.A.* 34°19N 92°24W **84** D8
Sheridan *Wyo., U.S.A.* 44°48N 106°58W **76** D10
Sheringham *U.K.* 52°56N 1°13E **12** E9
Sherkin I. *Ireland* 51°28N 9°26W **10** E2
Sherkot *India* 29°22N 78°35E **43** E8
Sherlovaya Gora *Russia* 50°34N 116°15E **29** D12
Sherman *N.Y., U.S.A.* 42°9N 79°35W **82** D5
Sherman *Tex., U.S.A.* 33°38N 96°36W **84** E6
Sherpur *India* 25°34N 83°47E **43** G10
Sherridon *Canada* 55°8N 101°5W **71** B8
Sherwood Forest *U.K.* 53°6N 1°7W **12** D6
Sherwood Park *Canada* 53°31N 113°19W **70** C6
Sheslay → *Canada* 58°48N 132°5W **70** B2
Shethanei L. *Canada* 58°48N 97°50W **71** B9
Shetland □ *U.K.* 60°30N 1°30W **11** A7
Shetland Is. *U.K.* 60°30N 1°30W **11** A7
Shetrunji → *India* 21°19N 72°7E **42** J5
Sheung Shui *China* 22°31N 114°7E **33** a
Shey-Phoksundo △ *Nepal* 29°30N 82°45E **43** E10
Sheyang *China* 33°48N 120°29E **35** H11
Sheyenne → *U.S.A.* 47°2N 96°50W **80** B5
Shiashkotan, Ostrov *Russia* 48°49N 154°6E **29** E16
Shibām *Yemen* 15°59N 48°36E **47** D4
Shibata *Japan* 37°57N 139°20E **30** F9
Shibecha *Japan* 43°17N 144°36E **30** C12
Shibetsu *Japan* 43°30N 145°10E **30** C12
Shibetsu *Japan* 44°10N 142°23E **30** B11
Shibogama L. *Canada* 53°35N 88°15W **72** B2
Shibushi *Japan* 31°25N 131°8E **31** J5
Shickshinny *U.S.A.* 41°9N 76°9W **83** E8
Shickshock Mts. = Chic-Chocs, Mts. *Canada* 48°55N 66°0W **73** C6
Shidao *China* 36°50N 122°25E **35** F12
Shido *Japan* 34°19N 134°10E **31** G7
Shiel, L. *U.K.* 56°48N 5°34W **11** E3
Shield, C. *Australia* 13°20S 136°20E **62** A2
Shīeli *Kazakhstan* 44°20N 66°15E **28** E7
Shiga □ *Japan* 35°20N 136°0E **31** G8
Shiguaigou *China* 40°52N 110°15E **34** D6
Shihchiachuangi = Shijiazhuang *China* 38°2N 114°28E **34** E8
Shihezi *China* 44°15N 86°2E **32** C6
Shijiazhuang *China* 38°2N 114°28E **34** E8
Shikarpur *India* 28°17N 78°7E **42** E8
Shikarpur *Pakistan* 27°57N 68°39E **42** F3
Shikohabad *India* 27°6N 78°36E **43** F8
Shikoku □ *Japan* 33°30N 133°30E **31** H6
Shikoku-Sanchi *Japan* 33°30N 133°30E **31** H6
Shikotan, Ostrov *Asia* 43°47N 146°44E **29** E15
Shikotsu-Ko *Japan* 42°45N 141°25E **30** C10
Shikotsu-Tōya △ *Japan* 44°4N 145°8E **30** C10
Shiliguri *India* 26°45N 88°25E **41** F16
Shiliu = Changjiang *China* 19°20N 108°55E **38** C7
Shilka *Russia* 52°0N 115°55E **29** D12
Shilka → *Russia* 53°20N 121°26E **29** D13
Shillelagh *Ireland* 52°45N 6°32W **10** D5
Shillington *U.S.A.* 40°18N 75°58W **83** F9
Shillong *India* 25°35N 91°53E **41** G17
Shilo *West Bank* 32°4N 35°18E **46** C4
Shilou *China* 37°0N 110°48E **34** F6
Shimabara *Japan* 32°48N 130°20E **31** H5
Shimada *Japan* 34°49N 138°10E **31** G9
Shimane □ *Japan* 35°0N 132°30E **31** G6
Shimanovsk *Russia* 52°15N 127°30E **29** D13
Shimba Hills △ *Kenya* 4°14S 39°25E **54** C4
Shimbiris *Somali Rep.* 10°44N 47°14E **47** E4
Shimizu *Japan* 35°0N 138°30E **31** G9
Shimla *India* 31°2N 77°9E **42** D7
Shimodate *Japan* 36°20N 139°55E **31** F9
Shimoga *India* 13°57N 75°32E **40** N9
Shimokita-Hantō *Japan* 41°20N 141°0E **30** D10
Shimoni *Kenya* 4°38S 39°20E **54** C4
Shimonoseki *Japan* 33°58N 130°55E **31** H5
Shimpuru Rapids *Namibia* 17°45S 19°55E **56** B2
Shin, L. *U.K.* 58°5N 4°30W **11** C4
Shinano-Gawa → *Japan* 36°50N 138°30E **31** F9
Shināş *Oman* 24°46N 56°28E **45** E8
Shīndand *Afghan.* 33°12N 62°8E **40** C3
Shinglehouse *U.S.A.* 41°58N 78°12W **82** E6
Shingū *Japan* 33°40N 135°55E **31** H7
Shingwidzi *S. Africa* 23°5S 31°25E **57** C5
Shinjō *Japan* 38°46N 140°18E **30** E10
Shinkolobwe *Dem. Rep. of the Congo* 11°10S 26°40E **52** G5
Shinshār *Syria* 34°36N 36°43E **46** A5
Shinyanga *Tanzania* 3°45S 33°27E **54** C3
Shinyanga □ *Tanzania* 3°50S 34°0E **54** C3
Shio-no-Misaki *Japan* 33°25N 135°45E **31** H7
Shiogama *Japan* 38°19N 141°1E **30** E10
Shiojiri *Japan* 36°6N 137°58E **31** F8
Shipchenski Prokhod *Bulgaria* 42°45N 25°15E **23** C11
Shiping *China* 23°45N 102°23E **32** G9
Shippagan *Canada* 47°45N 64°45W **73** C7
Shippensburg *U.S.A.* 40°3N 77°31W **82** F7
Shippenville *U.S.A.* 41°15N 79°28W **82** E5
Shiprock *U.S.A.* 36°47N 108°41W **77** H9
Shiqma, N. → *Israel* 31°37N 34°30E **46** D3
Shiquan *China* 33°5N 108°15E **34** H5
Shiquan He = Indus → *Pakistan* 24°20N 67°47E **42** G2
Shīr Kūh *Iran* 31°39N 54°3E **45** D7
Shiragami-Misaki *Japan* 41°24N 140°12E **30** D10
Shirakawa *Fukushima, Japan* 37°7N 140°13E **31** F10
Shirakawa *Gifu, Japan* 36°17N 136°56E **31** F8
Shirane-San *Gumma, Japan* 36°48N 139°22E **31** F9
Shirane-San *Yamanashi, Japan* 35°42N 138°9E **31** G9
Shiraoi *Japan* 42°33N 141°21E **30** C10
Shīrāz *Iran* 29°42N 52°30E **45** D7
Shire → *Africa* 17°42S 35°19E **55** F4
Shiren *China* 41°57N 126°34E **35** D14
Shiretoko △ *Japan* 44°15N 145°15E **30** B12
Shiretoko-Misaki *Japan* 44°21N 145°20E **30** B12
Shirinab → *Pakistan* 30°15N 66°28E **42** D2
Shiriya-Zaki *Japan* 41°25N 141°30E **30** D10
Shiroishi *Japan* 38°0N 140°37E **30** F10
Shirshov Ridge *Pac. Oc.* 58°0N 170°0E **64** B8
Shīrvān *Iran* 37°30N 57°50E **45** B8
Shirwa, L. = Chilwa, L. *Malawi* 15°15S 35°40E **55** F4
Shishaldin Volcano *U.S.A.* 54°45N 163°58W **74** E7
Shivamogga = Shimoga *India* 13°57N 75°32E **40** N9
Shivpuri *India* 25°26N 77°42E **42** G7
Shixian *China* 43°5N 129°50E **35** C15
Shiyan *China* 22°42N 113°56E **33** a
Shiyan Shuiku *China* 22°42N 113°54E **33** a
Shizuishan *China* 39°15N 106°50E **34** E4
Shizuoka *Japan* 34°57N 138°24E **31** G9
Shizuoka □ *Japan* 35°15N 138°40E **31** G9
Shklov = Shklow *Belarus* 54°16N 30°15E **17** A16
Shklow *Belarus* 54°16N 30°15E **17** A16
Shkodër *Albania* 42°4N 19°32E **23** C8
Shkumbini → *Albania* 41°2N 19°31E **23** D8
Shmidta, Ostrov *Russia* 81°0N 91°0E **29** A10
Shō-Gawa → *Japan* 36°47N 137°4E **31** F8
Shoal L. *Canada* 49°33N 95°1W **71** D9
Shoal Lake *Canada* 50°30N 100°35W **71** C8
Shōdo-Shima *Japan* 34°30N 134°15E **31** G7
Sholapur = Solapur *India* 17°43N 75°56E **40** L9
Shōmrōn *West Bank* 32°15N 35°13E **46** C4
Shoreham *U.S.A.* 43°53N 73°18W **83** C11
Shoreham by Sea *U.K.* 50°50N 0°16W **13** G7
Shori → *Pakistan* 28°29N 69°44E **42** E3
Shorkot *Pakistan* 30°50N 72°0E **42** D4
Shorkot Road *Pakistan* 30°47N 72°15E **42** D5
Shoshone *Calif., U.S.A.* 35°58N 116°16W **79** K10
Shoshone *Idaho, U.S.A.* 42°56N 114°25W **76** E6
Shoshone L. *U.S.A.* 44°22N 110°43W **76** D8
Shoshone Mts. *U.S.A.* 39°20N 117°25W **76** G5
Shoshong *Botswana* 22°56S 26°31E **56** C4
Shoshoni *U.S.A.* 43°14N 108°7W **76** E9
Shouguang *China* 37°52N 118°45E **35** F10
Shouyang *China* 37°54N 113°8E **34** F7
Show Low *U.S.A.* 34°15N 110°2W **77** J8
Shqipëria = Albania ■ *Europe* 41°0N 20°0E **23** D9
Shreveport *U.S.A.* 32°31N 93°45W **84** E8
Shrewsbury *U.K.* 52°43N 2°45W **13** E5
Shri Mohangarh *India* 27°17N 71°18E **42** F4
Shrirampur *India* 22°44N 88°21E **43** H13
Shropshire □ *U.K.* 52°36N 2°45W **13** E5
Shū *Kazakhstan* 43°36N 73°42E **28** E8
Shuangcheng *China* 45°20N 126°15E **35** B14
Shuanggou *China* 34°2N 117°30E **35** G9
Shuangliao *China* 43°29N 123°30E **35** C12
Shuangshanzi *China* 40°20N 119°8E **35** D10
Shuangyang *China* 43°28N 125°40E **35** C13
Shuangyashan *China* 46°28N 131°5E **33** B15
Shuguri Falls *Tanzania* 8°33S 37°22E **55** D4
Shujalpur *India* 23°18N 76°46E **42** H7
Shukpa Kunzang *India* 34°22N 78°22E **43** B8
Shulan *China* 44°28N 127°0E **35** B14
Shule *China* 39°25N 76°3E **32** D4
Shule He → *China* 40°20N 92°50E **32** C7
Shumagin Is. *U.S.A.* 55°7N 160°30W **74** D7
Shumen *Bulgaria* 43°18N 26°55E **23** C12
Shumikha *Russia* 55°10N 63°15E **28** D7
Shungnak *U.S.A.* 66°52N 157°9W **74** B8
Shuo Xian = Shuozhou *China* 39°20N 112°33E **34** E7
Shuozhou *China* 39°20N 112°33E **34** E7
Shūr → *Fārs, Iran* 28°30N 55°0E **45** D7
Shūr → *Kermān, Iran* 30°52N 57°37E **45** D8
Shūr → *Yazd, Iran* 31°45N 55°15E **45** D7
Shūr Āb *Iran* 34°23N 51°11E **45** C6
Shūr Gaz *Iran* 29°10N 59°20E **45** D8
Shūrāb *Iran* 33°43N 56°29E **45** C8
Shūrjestān *Iran* 31°24N 52°25E **45** D7
Shurugwi *Zimbabwe* 19°40S 30°0E **55** F3
Shūsf *Iran* 31°50N 60°5E **45** D9
Shūshtar *Iran* 32°0N 48°50E **45** D6
Shuswap L. *Canada* 50°55N 119°3W **70** C5
Shute Harbour △ *Australia* 20°17S 148°47E **62** b
Shuyang *China* 34°10N 118°42E **35**
Shūzū *Iran* 29°52N 54°30E **45**
Shwebo *Burma* 22°30N 95°45E **41**
Shwegu *Burma* 24°15N 96°26E **41**
Shweli → *Burma* 23°45N 96°45E **41**
Shymkent *Kazakhstan* 42°18N 69°36E **28**
Shyok *India* 34°13N 78°12E **43**
Shyok → *Pakistan* 35°13N 75°53E **43**
Si Kiang = Xi Jiang → *China* 22°5N 113°20E **33**
Si Lanna △ *Thailand* 19°17N 99°12E **38**
Si Nakarin Res. *Thailand* 14°35N 99°0E **38**
Si-ngan = Xi'an *China* 34°15N 109°0E **34**
Si Phangnga *Thailand* 9°8N 98°29E **39**
Si Prachan *Thailand* 14°37N 100°9E **38**
Si Racha *Thailand* 13°10N 100°48E **38**
Si Sa Ket *Thailand* 15°8N 104°23E **38**
Siachen Glacier *Asia* 35°20N 77°30E **43**
Siahaf → *Pakistan* 29°3N 68°57E **42**
Siahan Range *Pakistan* 27°30N 64°40E **40**
Siak Sri Indrapura *Indonesia* 0°51N 102°0E **36**
Sialkot *Pakistan* 32°32N 74°30E **42**
Siam = Thailand ■ *Asia* 16°0N 102°0E **38**
Sian = Xi'an *China* 34°15N 109°0E **34**
Sian Ka'an △ *Mexico* 19°35N 87°40W **87**
Siantan *Indonesia* 3°10N 106°15E **36**
Siāreh *Iran* 28°5N 60°14E **45**
Siargao I. *Phil.* 9°52N 126°3E **37**
Siasi *Phil.* 5°34N 120°50E **37**
Siau *Indonesia* 2°50N 125°25E **37**
Šiauliai *Lithuania* 55°56N 23°15E **9**
Sibâi, Gebel el *Egypt* 25°45N 34°10E **44**
Sibang *Indonesia* 8°34S 115°13E **37**
Sibay *Russia* 52°42N 58°39E **18**
Sibayi, L. *S. Africa* 27°20S 32°45E **57**
Šibenik *Croatia* 43°48N 15°54E **22**
Siberia = Sibirskiy □ *Russia* 58°0N 90°0E **29**
Siberia *Russia* 60°0N 100°0E **4**
Siberut *Indonesia* 1°30S 99°0E **36**
Sibi *Pakistan* 29°30N 67°54E **42**
Sibiloi △ *Kenya* 4°0N 36°20E **54**
Sibirskiy □ *Russia* 58°0N 90°0E **29**
Sibirtsevo *Russia* 44°12N 132°26E **30**
Sibiti *Congo* 3°38S 13°19E **52**
Sibiu *Romania* 45°45N 24°9E **17**
Sibley *U.S.A.* 43°24N 95°45W **80**
Sibolga *Indonesia* 1°42N 98°45E **36**
Siborongborong *Indonesia* 2°13N 98°58E **39**
Sibsagar *India* 27°0N 94°36E **41**
Sibu *Malaysia* 2°18N 111°49E **36**
Sibuco *Phil.* 7°20N 122°10E **37**
Sibuguey B. *Phil.* 7°50N 122°45E **37**
Sibut *C.A.R.* 5°46N 19°10E **52**
Sibutu *Phil.* 4°45N 119°30E **37**
Sibutu Passage *E. Indies* 4°50N 120°0E **37**
Sibuyan I. *Phil.* 12°25N 122°40E **37**
Sibuyan Sea *Phil.* 12°30N 122°20E **37**
Sicamous *Canada* 50°49N 119°0W **70**
Siccus → *Australia* 31°55S 139°17E **63**
Sichon *Thailand* 9°0N 99°54E **39**
Sichuan □ *China* 30°30N 103°0E **32**
Sicilia *Italy* 37°30N 14°30E **22**
Sicily = Sicilia *Italy* 37°30N 14°30E **22**
Sicily, Str. of *Medit. S.* 37°35N 11°56E **22**
Sico → *Honduras* 15°58N 84°58W **88**
Sicuani *Peru* 14°21S 71°10W **92**
Sidari *Greece* 39°47N 19°41E **25**
Siddhapur *India* 23°56N 72°25E **42**
Siddipet *India* 18°5N 78°51E **40**
Sideros, Akra *Greece* 35°19N 26°19E **25**
Sidhauli *India* 27°17N 80°50E **43**
Sidhi *India* 24°25N 81°53E **43**
Sidi-bel-Abbès *Algeria* 35°13N 0°39W **50**
Sidi Ifni *Morocco* 29°29N 10°12W **50**
Sidikalang *Indonesia* 2°45N 98°19E **39**
Sidlaw Hills *U.K.* 56°32N 3°2W **11**
Sidley, Mt. *Antarctica* 77°2S 126°2W **5**
Sidmouth *U.K.* 50°40N 3°15W **13**
Sidmouth, C. *Australia* 13°25S 143°36E **62**
Sidney *Canada* 48°39N 123°24W **78**
Sidney *Mont., U.S.A.* 47°43N 104°9W **76**
Sidney *N.Y., U.S.A.* 42°19N 75°24W **83**
Sidney *Nebr., U.S.A.* 41°8N 102°59W **80**
Sidney *Ohio, U.S.A.* 40°17N 84°9W **81**
Sidney Lanier, L. *U.S.A.* 34°10N 84°4W **85**
Sidoarjo *Indonesia* 7°27S 112°43E **37**
Sidon = Saydā *Lebanon* 33°35N 35°25E **46**
Sidra, G. of = Surt, Khalīj *Libya* 31°40N 18°30E **51**
Siedlce *Poland* 52°10N 22°20E **17**
Sieg → *Germany* 50°46N 7°6E **16**
Siegen *Germany* 50°51N 8°0E **16**
Siem Pang *Cambodia* 14°7N 106°23E **38**
Siem Reap = Siemreab *Cambodia* 13°20N 103°52E **38**
Siemreab *Cambodia* 13°20N 103°52E **38**
Siena *Italy* 43°19N 11°21E **22**
Sieradz *Poland* 51°37N 18°41E **17**
Sierpe, Bocas de la *Venezuela* 10°0N 61°30W **93**
Sierra Blanca *U.S.A.* 31°11N 105°22W **84**
Sierra Blanca Peak *U.S.A.* 33°23N 105°49W **77**
Sierra City *U.S.A.* 39°34N 120°38W **78**
Sierra Colorada *Argentina* 40°35S 67°50W **96**
Sierra de Agalta △ *Honduras* 15°1N 85°48W **88**
Sierra de Bahoruco △ *Dom. Rep.* 18°10N 71°25W **89**
Sierra de La Culata △ *Venezuela* 8°45N 71°10W **89**
Sierra de Lancandón △ *Guatemala* 16°59N 90°23W **88**
Sierra de las Quijadas △ *Argentina* 32°29S 67°5W **94**
Sierra de San Luis △ *Venezuela* 11°20N 69°43W **89**

ra de San Pedro Mártir △
lexico 31°10N 115°30W 86 A1
ra Gorda Chile 22°50S 69°15W 94 A2
ra Leone ■ W. Afr. 9°0N 12°0W 50 G3
ra Madre Mexico 16°0N 93°0W 87 D6
ra Madre Occidental
lexico 27°0N 107°0W 86 B3
ra Madre Oriental
lexico 25°0N 100°0W 86 C5
ra Mojada Mexico 27°18N 103°41W 86 B4
ra Nevada Spain 37°3N 3°15W 21 D4
ra Nevada U.S.A. 39°0N 120°30W 78 H8
ra Nevada △
enezuela 8°35N 70°45W 89 E5
ra Nevada de Santa Marta △
olombia 10°56N 73°36W 89 D5
ra Vista U.S.A. 31°33N 110°18W 77 L8
raville U.S.A. 39°36N 120°22W 78 F6
os Greece 37°0N 24°45E 23 F11
n Canada 51°21N 100°8W 71 C8
n Pass Canada 57°52N 126°15W 70 B3
Algeria 35°32N 0°12W 50 A5
toka Fiji 18°8S 177°32E 59 a
etu-Marmației
omania 47°57N 23°52E 17 E12
ișoara Romania 46°12N 24°50E 17 E13
Indonesia 5°25N 96°0E 36 C1
ufjörður Iceland 66°12N 18°55W 8 C4
al de Botrang Belgium 50°29N 6°4E 15 D6
al Pk. U.S.A. 33°20N 114°2W 79 M12
y I. Antarctica 60°43S 45°36W 5 C18
ig Ecuador 3°0S 78°50W 92 D3
enza Spain 41°3N 2°40W 21 B4
iri Guinea 11°31N 9°10W 50 F4
lda Latvia 57°10N 24°55E 9 H21
ng China 33°27N 118°16E 35 H10
ra India 23°29N 80°6E 43 H9
ajoki → Finland 64°50N 24°43E 8 D21
njärvi Finland 63°4N 27°39E 8 E22
rira Ra. = Chizarira
mbabwe 17°36S 27°45E 55 F2
India 22°26N 69°47E 42 H3
nni Chief →
nada 57°47N 122°15W 70 B4
o Thailand 7°34N 99°21E 39 J2
r India 27°33N 75°10E 42 F6
sso Mali 11°18N 5°35W 50 F4
ston U.S.A. 36°53N 89°35W 80 G9
ote Alin, Khrebet
ussia 45°0N 136°0E 30 B8
ote Alin Ra. = Sikhote Alin,
hrebet Russia 45°0N 136°0E 30 B8
nos Greece 36°40N 25°8E 23 F11
im □ India 27°50N 88°30E 41 F16
→ Spain 42°27N 7°43W 21 A2
cayoapan Mexico 17°30N 98°9W 87 D5
wad India 21°54N 74°54E 42 J6
har India 24°49N 92°48E 41 G18
City U.S.A. 35°44N 79°28W 85 D15
ia = Śląsk Poland 51°0N 16°30E 16 C9
arhi Doti Nepal 29°15N 81°0E 43 E9
hat India 26°35N 93°0E 41 F18
ouette Seychelles 4°29S 55°12E 53 b
ke Turkey 36°22N 33°58E 44 B2
uri = Shiliguri India 26°45N 88°25E 41 F16
g Co China 31°50N 89°20E 32 E6
tra Bulgaria 44°6N 27°19E 23 B12
ri Turkey 41°4N 28°14E 23 D13
n Sweden 60°55N 14°45E 8 F16
eborg Denmark 56°10N 9°32E 9 H13
wood Australia 17°45S 146°2E 62 B4
jhuay, Cordillera
ile 19°46S 68°40W 92 G5
mäe Estonia 59°24N 27°45E 9 G22
th U.K. 54°52N 3°23W 12 C4
am Springs U.S.A. 36°11N 94°32W 84 C7
oi Turkey 37°15N 42°27E 44 B4
ee U.S.A. 30°21N 94°11W 84 F7
ė Lithuania 55°21N 21°33E 9 J19
an Turkey 38°7N 41°2E 44 B4
ani India 23°18N 78°25E 43 H8
r City U.S.A. 32°46N 108°17W 77 K9
r Cr. → U.S.A. 43°16N 119°13W 76 E4
r Creek U.S.A. 42°33N 79°10W 82 D5
r L. U.S.A. 38°39N 120°6W 78 G6
r Lake Calif.,
S.A. 35°21N 116°7W 79 K10
r Lake Oreg., U.S.A. 43°8N 121°3W 76 E3
rmine Mts. Ireland 52°47N 8°15W 10 D3
rton Colo., U.S.A. 37°49N 107°40W 77 H10
rton Tex., U.S.A. 34°28N 101°19W 84 D4
es → U.S.A. 43°34N 119°2W 76 E4
altala India 24°43N 86°33E 43 G12
anggang = Bandar Sri Aman
alaysia 1°15N 111°32E 36 D4
ao China 22°47N 101°5E 32 G9
ard, L. Canada 47°40N 78°40W 72 C4
av Turkey 39°4N 28°58E 23 E13
ba Tanzania 2°10S 37°36E 54 C4
birsk = Ulyanovsk
ussia 54°20N 48°25E 18 D8
bo Tanzania 4°51S 29°41E 54 C2
coe Canada 42°50N 80°23W 82 D4
coe, L. Canada 44°25N 79°20W 82 B5
dega India 22°37N 84°31E 43 H11
eria Romania 45°51N 23°1E 17 F12
eulue Indonesia 2°45N 95°45E 36 D1
feropol Ukraine 44°55N 34°3E 19 F5
Greece 36°35N 27°50E 23 F12
Valley U.S.A. 34°16N 118°47W 79 L8
kot Nepal 30°0N 81°50E 43 E9
a = Shimla India 31°2N 77°9E 42 D7
ipal △ India 21°45N 86°30E 43 J12
nie Canada 49°56N 108°6W 71 D7
nler U.S.A. 35°21N 119°59W 79 K7
älv = Simojoki →
nland 65°35N 25°1E 8 D21
ojoki → Finland 65°35N 25°1E 8 D21
ojovel Mexico 17°12N 92°38W 87 D6
onette → Canada 55°9N 118°15W 70 B5
Simonstown S. Africa 34°14S 18°26E 56 E2
Simpang Empat Malaysia 5°27N 100°29E 39 c
Simplonpass Switz. 46°15N 8°3E 20 C8
Simpson Desert Australia 25°0S 137°0E 62 D2
Simpson Desert △
Australia 24°59S 138°21E 62 C2
Simpson Pen. Canada 68°34N 88°45W 69 D14
Simrishamn Sweden 55°33N 14°22E 9 J16
Simsbury U.S.A. 41°53N 72°48W 83 E12
Simushir, Ostrov
Russia 46°50N 152°30E 29 E16
Sin Cowe I. S. China Sea 9°53N 114°19E 36 C4
Sina Dhago Somali Rep. 5°50N 47°0E 47 F4
Sinabang Indonesia 2°30N 96°24E 36 D1
Sinai = Es Sînâ' Egypt 29°0N 34°0E 46 F2
Sinai, Mt. = Mûsa, Gebel
Egypt 28°33N 33°59E 44 D2
Sinaloa □ Mexico 25°0N 107°30W 86 C3
Sinaloa de Leyva
Mexico 25°50N 108°14W 86 B3
Sinarades Greece 39°34N 19°51E 25 A3
Sincelejo Colombia 9°18N 75°24W 92 B3
Sinch'ang N. Korea 40°7N 128°28E 35 D15
Sinch'ŏn N. Korea 38°17N 125°21E 35 E13
Sinclair U.S.A. 41°47N 107°7W 76 F10
Sinclair Mills Canada 54°5N 121°40W 70 C4
Sinclair's B. U.K. 58°31N 3°5W 11 C5
Sinclairville U.S.A. 42°16N 79°16W 82 D5
Sincorá, Serra do Brazil 13°30S 41°0W 93 F10
Sind = Sindh □ Pakistan 26°0N 69°0E 42 G3
Sind → Jammu & Kashmir,
India 34°18N 74°45E 43 B6
Sind → Mad. P., India 26°26N 79°13E 43 F8
Sind Sagar Doab Pakistan 32°0N 71°30E 42 D4
Sindangan Phil. 8°10N 123°5E 37 C6
Sindangbarang Indonesia 7°27S 107°1E 37 G12
Sinde Zambia 17°28S 25°51E 55 F2
Sindh = Indus →
Pakistan 24°20N 67°47E 42 G2
Sindh □ Pakistan 26°0N 69°0E 42 G3
Sindri India 23°45N 86°42E 43 H12
Sines Portugal 37°56N 8°51W 21 D1
Sines, C. de Portugal 37°58N 8°53W 21 D1
Sineu Spain 39°38N 3°1E 24 B10
Sing Buri Thailand 14°53N 100°25E 38 E3
Singa Sudan 13°10N 33°57E 51 F12
Singalila △ India 27°10N 88°5E 43 F13
Singapore ■ Asia 1°17N 103°51E 39 d
Singapore, Straits of Asia 1°15N 104°0E 39 d
Singapore Changi ✈ (SIN)
Singapore 1°23N 103°59E 39 M4
Singaraja Indonesia 8°7S 115°6E 37 J18
Singatoka = Sigatoka Fiji 18°8S 177°32E 59 a
Singida Tanzania 4°49S 34°48E 54 C3
Singida □ Tanzania 6°0S 34°30E 54 D3
Singkaling Hkamti
Burma 26°0N 95°39E 41 G19
Singkang Indonesia 4°8S 120°1E 37 E6
Singkawang Indonesia 1°0N 108°57E 36 D3
Singkep Indonesia 0°30S 104°25E 36 E2
Singkil Indonesia 2°17N 97°49E 39 L1
Singkuang Indonesia 1°3N 98°55E 39 L2
Singleton Australia 32°33S 151°10E 63 E5
Singleton, Mt. N. Terr.,
Australia 22°0S 130°46E 60 D5
Singleton, Mt. W. Austral.,
Australia 29°27S 117°15E 61 E2
Singoli India 25°0N 75°22E 42 G6
Singora = Songkhla
Thailand 7°13N 100°37E 39 J3
Sinhŭng N. Korea 40°11N 127°34E 35 D14
Siniscóla Italy 40°34N 9°41E 22 D3
Sinjai Indonesia 5°7S 120°20E 37 F6
Sinjār Iraq 36°19N 41°52E 44 B4
Sinkat Sudan 18°55N 36°49E 51 E13
Sinkiang = Xinjiang Uygur
Zizhiqu □ China 42°0N 86°0E 32 C6
Sinmak N. Korea 38°25N 126°14E 35 E14
Sinmi-do N. Korea 39°33N 124°53E 35 E13
Sinnamary Fr. Guiana 5°25N 53°0W 93 B8
Sinni → Italy 40°8N 16°41E 22 D7
Sinop Turkey 42°1N 35°11E 19 F6
Sinor India 21°55N 73°20E 42 J5
Sinp'o N. Korea 40°0N 128°13E 35 E15
Sinsk Russia 61°8N 126°48E 29 C13
Sintang Indonesia 0°5N 111°35E 36 D4
Sinton U.S.A. 28°2N 97°31W 84 G6
Sintra Portugal 38°47N 9°25W 21 C1
Sinŭiju N. Korea 40°5N 124°24E 35 D13
Siocon Phil. 7°40N 122°10E 37 C6
Siófok Hungary 46°54N 18°3E 17 E10
Sion Switz. 46°14N 7°20E 20 C7
Sion Mills U.K. 54°48N 7°29W 10 B4
Sionainn = Shannon →
Ireland 52°35N 9°30W 10 D2
Sioux Center U.S.A. 43°5N 96°11W 80 D5
Sioux City U.S.A. 42°30N 96°24W 80 D5
Sioux Falls U.S.A. 43°33N 96°44W 80 D5
Sioux Lookout Canada 50°10N 91°50W 72 B1
Sioux Narrows Canada 49°25N 94°10W 71 D10
Sipadan Malaysia 4°6N 118°38E 37 D5
Siparia Trin. & Tob. 10°8N 61°31W 93 K15
Siping China 43°8N 124°21E 35 C13
Sipiwesk L. Canada 55°5N 97°35W 71 B9
Sipra → India 23°55N 75°28E 42 H6
Sipura Indonesia 2°18S 99°40E 36 E1
Siquia → Nic. 12°10N 84°20W 88 D3
Siquijor Phil. 9°12N 123°35E 37 C6
Siquirres Costa Rica 10°6N 83°30W 88 D3
Şīr Abu Nu'ayr U.A.E. 25°20N 54°20E 45 E7
Şīr Banī Yās U.A.E. 24°19N 52°37E 45 E7
Sir Edward Pellew Group
Australia 15°40S 137°10E 62 B2
Sir Graham Moore Is.
Australia 13°53S 126°34E 60 B4
Sir James MacBrien, Mt.
Canada 62°7N 127°41W 68 E6
Sira → Norway 58°23N 6°34E 9 G12
Siracusa Italy 37°4N 15°17E 22 F6
Sirajganj Bangla. 24°25N 89°47E 43 G13
Sirathu India 25°39N 81°19E 43 G9
Sīrdān Iran 36°39N 49°12E 45 B6
Siren U.S.A. 45°47N 92°24W 80 C7
Sirer Spain 38°56N 1°22E 24 C7
Siret → Romania 45°24N 28°1E 17 F14
Sirghāyā Syria 33°51N 36°8E 46 B5
Sirikit Res. Thailand 17°45N 100°34E 38 C3
Sirinat △ Thailand 8°6N 98°17E 39 a
Sīrjān Iran 29°30N 55°45E 45 D7
Sirmaur India 24°51N 81°23E 43 G9
Sirmilik △ Canada 72°50N 80°35W 69 C15
Sirohi India 24°52N 72°53E 42 G5
Sironj India 24°5N 77°39E 42 G7
Síros = Ermoupoli
Greece 37°28N 24°57E 23 F11
Sirr, Nafud as Si. Arabia 25°33N 44°35E 44 E5
Sirretta Pk. U.S.A. 35°56N 118°19W 79 K8
Sīrrī Iran 25°55N 54°32E 45 E7
Sirsa India 29°33N 75°4E 42 E6
Sirsa → India 26°51N 79°4E 43 F8
Sisak Croatia 45°30N 16°21E 16 F9
Sishen S. Africa 27°47S 22°59E 56 D3
Sishui China 35°42N 117°18E 35 G9
Sisipuk L. Canada 55°45N 101°50W 71 B8
Sisophon Cambodia 13°38N 102°59E 38 F4
Sisseton U.S.A. 45°40N 97°3W 80 C5
Sīstān Asia 30°50N 61°0E 45 D9
Sīstan, Daryācheh-ye Iran 31°0N 61°0E 45 D9
Sīstān va Balūchestān □
Iran 27°0N 62°0E 45 E9
Sisters U.S.A. 44°18N 121°33W 76 D3
Sisters, The Seychelles 4°16S 55°52E 53 b
Siswa Bazar India 27°9N 83°46E 43 F10
Sitamarhi India 26°37N 85°30E 43 F11
Sitampiky Madag. 16°41S 46°6E 57 B8
Sitapur India 27°38N 80°45E 43 F9
Siteki Swaziland 26°32S 31°58E 57 D5
Sitges Spain 41°17N 1°47E 21 B6
Sitia Greece 35°13N 26°6E 25 D8
Sitito-Ozima Ridge
Pac. Oc. 23°0N 143°0E 64 E6
Sitka U.S.A. 57°3N 135°20W 70 B1
Sittang Myit → Burma 17°20N 96°45E 41 L20
Sittard Neths. 51°0N 5°52E 15 C5
Sittingbourne U.K. 51°21N 0°45E 13 F8
Sittoung = Sittang Myit →
Burma 17°20N 96°45E 41 L20
Sittwe Burma 20°18N 92°45E 41 J18
Situbondo Indonesia 7°42S 113°56E 37 G16
Si'umu Samoa 14°1S 171°48W 59 b
Siuna Nic. 13°37N 84°45W 88 D3
Siuri India 23°50N 87°34E 43 H12
Sīvand Iran 30°5N 52°55E 45 D7
Sivas Turkey 39°43N 36°58E 44 B3
Siverek Turkey 37°50N 39°19E 44 B3
Sivomaskinskiy Russia 66°40N 62°35E 18 A11
Sivrihisar Turkey 39°30N 31°35E 19 G5
Sîwa Egypt 29°11N 25°31E 51 C11
Sîwa, El Wâhât es Egypt 29°10N 25°30E 48 D6
Siwa Oasis = Sîwa, El Wâhât es
Egypt 29°10N 25°30E 48 D6
Siwalik Range Nepal 28°0N 83°0E 43 F10
Siwan India 26°13N 84°21E 43 F11
Siwana India 25°38N 72°25E 42 G5
Six Cross Roads Barbados 13°7N 59°28W 89 g
Sixian China 33°30N 117°50E 35 H9
Sixmilebridge Ireland 52°44N 8°46W 10 D3
Sixth Cataract Sudan 16°20N 32°42E 51 E12
Siyah Kuh, Kavīr-e Iran 32°55N 53°55E 45 C7
Siyang China 33°42N 118°40E 35 H10
Siziwang Qi China 41°25N 111°40E 34 D6
Sjælland Denmark 55°30N 11°30E 9 J14
Sjumen = Shumen
Bulgaria 43°18N 26°55E 23 C12
Skadarsko Jezero Europe 42°10N 19°20E 23 C8
Skaftafell Iceland 64°1N 17°0W 8 D5
Skaftafell △ Iceland 64°9N 16°50W 8 D5
Skagafjörður Iceland 65°54N 19°35W 8 D4
Skagastølstindane
Norway 61°28N 7°52E 8 F12
Skagaströnd Iceland 65°50N 20°19W 8 D3
Skagen Denmark 57°43N 10°35E 9 H14
Skagerrak Denmark 57°30N 9°0E 9 H13
Skagit → U.S.A. 48°23N 122°22W 78 B4
Skagway U.S.A. 59°28N 135°19W 68 F4
Skala-Podilska Ukraine 48°50N 26°15E 17 D14
Skalat Ukraine 49°23N 25°55E 17 D13
Skåne Sweden 55°59N 13°30E 9 J15
Skaneateles U.S.A. 42°57N 76°26W 83 D8
Skaneateles L. U.S.A. 42°51N 76°22W 83 D8
Skardu Pakistan 35°20N 75°44E 43 B6
Skarżysko-Kamienna
Poland 51°7N 20°52E 17 C11
Skeena → Canada 54°9N 130°5W 70 C2
Skeena Mts. Canada 56°40N 128°30W 70 B3
Skegness U.K. 53°9N 0°20E 12 D8
Skeldon = Corriverton
Guyana 5°55N 57°20W 92 B7
Skeleton Coast Namibia 20°0S 13°0E 53 J2
Skeleton Coast ⌓ Namibia 20°0S 13°20E 56 C1
Skellefteå Sweden 64°45N 20°50E 8 D19
Skellefteälven →
Sweden 64°45N 21°10E 8 D19
Skelleftehamn Sweden 64°40N 21°9E 8 D19
Skellig Michael = Great Skellig
Ireland 51°47N 10°33W 10 E1
Skerries, The U.K. 53°25N 4°36W 12 D3
Ski Norway 59°43N 10°52E 9 G14
Skiathos Greece 39°12N 23°30E 23 E10
Skibbereen Ireland 51°33N 9°16W 10 E2
Skiddaw U.K. 54°39N 3°9W 12 C4
Skidegate Canada 53°15N 132°1W 70 C2
Skien Norway 59°12N 9°35E 9 G13
Skierniewice Poland 51°58N 20°10E 17 C11
Skihist, Mt. Canada 50°12N 121°54W 70 C4
Skikda Algeria 36°50N 6°58E 50 A7
Skilloura Cyprus 35°14N 33°10E 25 D12
Skipton U.K. 53°58N 2°3W 12 D5
Skirmish Pt. Australia 11°59S 134°17E 62 A1
Skiros Greece 38°55N 24°34E 23 E11
Skive Denmark 56°33N 9°2E 9 H13
Skjálfandafljót → Iceland 65°59N 17°25W 8 D5
Skjálfandi Iceland 66°5N 17°30W 8 C5
Skole Ukraine 49°3N 23°30E 17 D12
Skopelos Greece 39°9N 23°47E 23 E10
Skopi Greece 35°11N 26°2E 25 D8
Skopje Macedonia 42°1N 21°26E 23 C9
Skövde Sweden 58°24N 13°50E 9 G15
Skovorodino Russia 54°0N 124°0E 29 D13
Skowhegan U.S.A. 44°46N 69°43W 81 C19
Skull Ireland 51°32N 9°34W 10 E2
Skunk → U.S.A. 40°42N 91°7W 80 E8
Skuodas Lithuania 56°16N 21°33E 9 H19
Skvyra Ukraine 49°44N 29°40E 17 D15
Skye U.K. 57°15N 6°10W 11 D2
Skykomish U.S.A. 47°42N 121°22W 76 C3
Skyros = Skiros Greece 38°55N 24°34E 23 E11
Slade Point Australia 21°4S 149°13E 62 b
Slade Pt. Australia 21°5S 149°13E 62 b
Slættaratindur Færoe Is. 62°18N 7°1W 8 E9
Slagelse Denmark 55°23N 11°19E 9 J14
Slamet Indonesia 7°16S 109°8E 37 G13
Slane Ireland 53°42N 6°33W 10 C5
Slaney → Ireland 52°26N 6°33W 10 D5
Slangberge S. Africa 31°32S 20°48E 56 E3
Śląsk Poland 51°0N 16°30E 16 C9
Slate Is. Canada 48°40N 87°0W 72 C2
Slatina Romania 44°28N 24°22E 17 F13
Slatington U.S.A. 40°45N 75°37W 83 F9
Slaton U.S.A. 33°26N 101°39W 84 E4
Slave → Canada 61°18N 113°39W 70 A6
Slave Coast W. Afr. 6°0N 2°30E 50 G6
Slave Lake Canada 55°17N 114°43W 70 B6
Slave Pt. Canada 61°11N 115°56W 70 A5
Slavgorod Russia 53°1N 78°37E 28 D8
Slavonski Brod Croatia 45°11N 18°1E 23 B8
Slavuta Ukraine 50°15N 27°2E 17 C14
Slavyanka Russia 42°53N 131°21E 30 C5
Slavyansk = Slovyansk
Ukraine 48°55N 37°36E 19 E6
Slawharad Belarus 53°27N 31°0E 17 B16
Sleaford U.K. 53°0N 0°24W 12 D7
Sleaford B. Australia 34°55S 135°45E 63 E2
Sleat, Sd. of U.K. 57°5N 5°47W 11 D3
Sleeper Is. Canada 58°30N 81°0W 69 F15
Sleeping Bear Dunes ⌓
U.S.A. 44°50N 86°5W 80 C10
Sleepy Eye U.S.A. 44°18N 94°43W 80 C6
Slemon L. Canada 63°13N 116°4W 70 A5
Slide Mt. U.S.A. 42°0N 74°25W 83 E10
Slidell U.S.A. 30°17N 89°47W 85 F10
Sliema Malta 35°55N 14°30E 25 D2
Slieve Aughty Ireland 53°4N 8°30W 10 C3
Slieve Bloom Ireland 53°4N 7°40W 10 C4
Slieve Donard U.K. 54°11N 5°55W 10 B6
Slieve Gamph Ireland 54°6N 9°0W 10 B3
Slieve Gullion U.K. 54°7N 6°26W 10 B5
Slieve League Ireland 54°40N 8°42W 10 B3
Slieve Mish Ireland 52°12N 9°50W 10 D2
Slievenamon Ireland 52°25N 7°34W 10 D4
Sligeach = Sligo Ireland 54°16N 8°28W 10 B3
Sligo Ireland 54°16N 8°28W 10 B3
Sligo U.S.A. 41°6N 79°29W 82 E5
Sligo □ Ireland 54°8N 8°42W 10 B3
Sligo B. Ireland 54°18N 8°40W 10 B3
Slippery Rock U.S.A. 41°4N 80°3W 82 E4
Slite Sweden 57°42N 18°48E 9 H18
Sliven Bulgaria 42°42N 26°19E 23 C12
Sloan U.S.A. 35°57N 115°13W 79 K11
Sloansville U.S.A. 42°45N 74°22W 83 D10
Slobodskoy Russia 58°40N 50°6E 18 C9
Slobozia Romania 44°34N 27°23E 17 F14
Slocan Canada 49°48N 117°28W 70 D5
Slonim Belarus 53°4N 25°19E 17 B13
Slough U.K. 51°30N 0°36W 13 F7
Slough □ U.K. 51°30N 0°36W 13 F7
Sloughhouse U.S.A. 38°26N 121°12W 78 G5
Slovak Rep. ■ Europe 48°30N 20°0E 17 D10
Slovakia = Slovak Rep. ■
Europe 48°30N 20°0E 17 D10
Slovakian Ore Mts. = Slovenské
Rudohorie Slovak Rep. 48°45N 20°0E 17 D10
Slovenia ■ Europe 45°58N 14°30E 16 F8
Slovenija = Slovenia ■
Europe 45°58N 14°30E 16 F8
Slovenska = Slovak Rep. ■
Europe 48°30N 20°0E 17 D10
Slovenské Rudohorie
Slovak Rep. 48°45N 20°0E 17 D10
Slovyansk Ukraine 48°55N 37°36E 19 E6
Sluch → Ukraine 51°37N 26°38E 17 C14
Sluis Neths. 51°18N 3°23E 15 C3
Słupsk Poland 54°30N 17°3E 17 A9
Slurry S. Africa 25°49S 25°42E 56 D4
Slutsk Belarus 53°2N 27°31E 17 B14
Slyne Hd. Ireland 53°25N 10°10W 10 C1
Slyudyanka Russia 51°40N 103°40E 29 D11
Småland Sweden 57°15N 15°25E 9 H16
Smalltree L. Canada 61°0N 105°0W 71 A8
Smallwood Res. Canada 54°5N 64°30W 73 B7
Smarhon Belarus 54°20N 26°24E 17 A14
Smartt Syndicate Dam
S. Africa 30°45S 23°10E 56 E3
Smartville U.S.A. 39°13N 121°18W 78 F5
Smeaton Canada 53°30N 104°49W 71 C8
Smederevo Serbia 44°40N 20°57E 23 B9
Smerwick Harbour
Ireland 52°12N 10°23W 10 D1
Smethport U.S.A. 41°49N 78°27W 82 E6
Smidovich Russia 48°36N 133°49E 29 E14
Smith Canada 55°10N 114°0W 70 B6
Smith Center U.S.A. 39°47N 98°47W 80 F4
Smith I. Canada 60°45N 78°25W 69 E16
Smith Islands △ Australia 20°40S 149°9E 62 b
Smith River ⌓ U.S.A. 41°55N 124°0W 76 F2
Smithburne → Australia 17°3S 140°57E 62 B3
Smithers Canada 54°45N 127°10W 70 C3
Smithfield S. Africa 30°9S 26°30E 56 E4
Smithfield N.C., U.S.A. 35°31N 78°21W 85 D15
Smithfield Utah, U.S.A. 41°50N 111°50W 76 F8
Smiths Falls Canada 44°55N 76°0W 83 B9
Smithton Australia 40°53S 145°6E 63 G4
Smithville Canada 43°6N 79°33W 82 C5
Smithville U.S.A. 30°1N 97°10W 84 F6
Smoke Creek Desert
U.S.A. 40°30N 119°40W 76 F4
Smoky → Canada 56°10N 117°21W 70 B5
Smoky Bay Australia 32°22S 134°13E 63 E1
Smoky Hill → U.S.A. 39°4N 96°48W 80 F5
Smoky Hills U.S.A. 39°15N 99°30W 80 F4
Smoky Lake Canada 54°10N 112°30W 70 C6
Smøla Norway 63°23N 8°3E 8 E13
Smolensk Russia 54°45N 32°5E 18 D5
Smolikas, Oros Greece 40°9N 20°58E 23 D9
Smolyan Bulgaria 41°36N 24°38E 23 D11
Smooth Rock Falls
Canada 49°17N 81°37W 72 C3
Smoothstone L. Canada 54°40N 106°50W 71 C7
Smorgon = Smarhon
Belarus 54°20N 26°24E 17 A14
Smyrna = İzmir Turkey 38°25N 27°8E 23 E12
Smyrna U.S.A. 39°18N 75°36W 81 F16
Snæfell Iceland 64°48N 15°34W 8 D6
Snaefell I. of Man 54°16N 4°27W 12 C3
Snæfellsjökull Iceland 64°49N 23°46W 8 D2
Snake → U.S.A. 46°12N 119°2W 76 C4
Snake I. Australia 38°47S 146°33E 63 F4
Snake Range U.S.A. 39°0N 114°20W 76 G6
Snake River Plain
U.S.A. 42°50N 114°0W 76 E7
Snåsavatnet Norway 64°12N 12°0E 8 D15
Sneek Neths. 53°2N 5°40E 15 A5
Sneem Ireland 51°50N 9°54W 10 E2
Sneeuberge S. Africa 31°46S 24°20E 56 E3
Snelling U.S.A. 37°31N 120°26W 78 H6
Sněžka Europe 50°41N 15°50E 16 C8
Snizort, L. U.K. 57°33N 6°28W 11 D2
Snøhetta Norway 62°19N 9°16E 8 E13
Snohomish U.S.A. 47°55N 122°6W 78 C4
Snoqualmie Pass
U.S.A. 47°25N 121°25W 78 C5
Snoul Cambodia 12°4N 106°26E 39 F6
Snover U.S.A. 43°27N 82°58W 82 C2
Snow Hill U.S.A. 38°11N 75°24W 81 F16
Snow Lake Canada 54°52N 100°3W 71 C8
Snow Mt. Calif., U.S.A. 39°23N 122°45W 78 F4
Snow Mt. Maine, U.S.A. 45°18N 70°48W 83 A14
Snow Shoe U.S.A. 41°2N 77°57W 82 E7
Snowbird L. Canada 60°45N 103°0W 71 A8
Snowdon U.K. 53°4N 4°5W 12 D3
Snowdonia △ U.K. 53°7N 3°59W 12 D4
Snowdrift = Łutsel K'e
Canada 62°24N 110°44W 71 A6
Snowdrift → Canada 62°24N 110°44W 71 A6
Snowflake U.S.A. 34°30N 110°5W 77 J8
Snowshoe Pk. U.S.A. 48°13N 115°41W 76 B6
Snowtown Australia 33°46S 138°14E 63 E2
Snowville U.S.A. 41°58N 112°43W 76 F7
Snowy → Australia 37°46S 148°30E 63 F4
Snowy Mt. U.S.A. 43°42N 74°23W 83 C10
Snowy Mts. Australia 36°30S 148°20E 63 F4
Snug Corner Bahamas 22°33N 73°52W 89 B5
Snug Harbour Canada 45°22N 80°17W 82 A4
Snyatyn Ukraine 48°27N 25°38E 17 D13
Snyder Okla., U.S.A. 34°40N 98°57W 84 D5
Snyder Tex., U.S.A. 32°44N 100°55W 84 E4
Soahanina Madag. 18°42S 44°13E 57 B7
Soalala Madag. 16°6S 45°20E 57 B8
Soaloka Madag. 18°32S 45°15E 57 B8
Soamanonga Madag. 23°52S 44°47E 57 C7
Soan → Pakistan 33°1N 71°44E 42 C4
Soanierana-Ivongo
Madag. 16°55S 49°35E 57 B8
Soanindraniny Madag. 19°54S 47°14E 57 B8
Soavina Madag. 20°23S 46°56E 57 C8
Soavinandriana Madag. 19°9S 46°45E 57 B8
Sobat, Nahr → Sudan 9°22N 31°33E 51 G12
Sobhapur India 22°47N 78°17E 42 H8
Sobradinho, Represa de
Brazil 9°30S 42°0W 93 E10
Sobral Brazil 3°50S 40°20W 93 D10
Soc Giang Vietnam 22°54N 106°1E 38 A6
Soc Trang Vietnam 9°37N 105°50E 39 H5
Socastee U.S.A. 33°41N 78°59W 85 E15
Soch'e = Shache China 38°20N 77°10E 32 D4
Sochi Russia 43°35N 39°40E 19 F6
Société, Îs. de la
French Polynesia 17°0S 151°0W 65 J12
Society Is. = Société, Îs. de la
French Polynesia 17°0S 151°0W 65 J12
Socompa, Portezuelo de
Chile 24°27S 68°18W 94 A2
Socorro N. Mex., U.S.A. 34°4N 106°54W 77 J10
Socorro Tex., U.S.A. 31°39N 106°18W 84 F1
Socorro, I. Mexico 18°45N 110°58W 86 D2
Socotra Yemen 12°30N 54°0E 47 E5
Soda L. U.S.A. 35°10N 116°4W 77 J5
Soda Plains India 35°30N 79°0E 43 B8
Soda Springs U.S.A. 42°39N 111°36W 76 E8
Sodankylä Finland 67°29N 26°40E 8 C22
Soddy-Daisy U.S.A. 35°17N 85°10W 85 D12
Söderhamn Sweden 61°18N 17°10E 8 F17
Söderköping Sweden 58°31N 16°20E 9 G17
Södermanland Sweden 58°56N 16°55E 9 G17
Södertälje Sweden 59°12N 17°39E 9 G17
Sodiri Sudan 14°27N 29°0E 51 F11
Sodus U.S.A. 43°14N 77°4W 82 C7
Sodwana Bay △ S. Africa 27°35S 32°43E 57 D5
Soe Indonesia 9°52S 124°17E 60 A3
Soekmekaar = Morebeng
S. Africa 23°30S 29°55E 57 C4
Soest Neths. 52°9N 5°19E 15 B5
Sofala □ Mozam. 19°30S 34°30E 57 B5
Sofia = Sofiya Bulgaria 42°45N 23°20E 23 C10
Sofia → Madag. 15°27S 47°23E 57 B8
Sofiya Bulgaria 42°45N 23°20E 23 C10
Sōfu-Gan Japan 29°49N 140°21E 31 K10
Sogamoso Colombia 5°43N 72°56W 92 B4
Sogār Iran 25°53N 58°6E 45 E8
Sogndalsfjøra Norway 61°14N 7°5E 8 F12

Søgne *Norway* 58°5N 7°48E **9** G12
Sognefjorden *Norway* 61°10N 5°50E **8** F11
Soh *Iran* 33°26N 51°27E **45** C6
Sohâg *Egypt* 26°33N 31°43E **51** C12
Sohagpur *India* 22°42N 78°12E **42** H8
Soheuksando *S. Korea* 34°4N 125°7E **35** G13
Soignies *Belgium* 50°35N 4°5E **15** D4
Soissons *France* 49°25N 3°19E **20** B5
Sōja *Japan* 34°40N 133°45E **31** G6
Sojat *India* 25°55N 73°45E **42** G5
Sokal *Ukraine* 50°31N 24°15E **17** C13
Sokcho *S. Korea* 38°12N 128°36E **35** E15
Söke *Turkey* 37°48N 27°28E **23** F12
Sokelo
    Dem. Rep. of the Congo 9°55S 24°36E **55** D1
Sokhumi *Georgia* 43°0N 41°0E **19** F7
Soko Is. *China* 22°10N 113°54E **33** a
Sokodé *Togo* 9°0N 1°11E **50** G6
Sokol *Russia* 59°30N 40°5E **18** C7
Sokółka *Poland* 53°25N 23°30E **17** B12
Sokołów Podlaski
    Poland 52°25N 22°15E **17** B12
Sokoto *Nigeria* 13°2N 5°16E **50** F7
Sol Iletsk *Russia* 51°10N 55°0E **18** D10
Solai *Kenya* 0°2N 36°12E **54** B4
Solan *India* 30°55N 77°7E **42** D7
Solana Beach *U.S.A.* 32°58N 117°16W **79** N9
Solander I. *N.Z.* 46°34S 166°54E **59** G1
Solapur *India* 17°43N 75°56E **40** L9
Soldotna *U.S.A.* 60°29N 151°3W **68** C4
Soléa *Cyprus* 35°5N 33°4E **25** D12
Soledad *Colombia* 10°55N 74°46W **92** A4
Soledad *U.S.A.* 36°26N 121°20W **78** J5
Soledad *Venezuela* 8°10N 63°34W **92** B6
Solent, The *U.K.* 50°45N 1°25W **13** G6
Solfonntaggen *Norway* 60°2N 6°57E **9** F12
Solhan *Turkey* 38°57N 41°3E **44** B4
Soligalich *Russia* 59°5N 42°10E **18** C7
Soligorsk = Salihorsk
    Belarus 52°51N 27°27E **17** B14
Solihull *U.K.* 52°26N 1°47W **13** E6
Solikamsk *Russia* 59°38N 56°50E **18** C10
Solila *Madag.* 21°25S 46°37E **57** C8
Solimões = Amazonas →
    S. Amer. 0°5S 50°0W **93** D8
Solingen *Germany* 51°10N 7°5E **16** C4
Sollefteå *Sweden* 63°12N 17°20E **8** E17
Sóller *Spain* 39°46N 2°43E **24** B9
Solo → *Indonesia* 6°47S 112°22E **37** G15
Sologne *France* 47°40N 1°45E **20** C4
Solok *Indonesia* 0°45S 100°40E **36** E2
Sololá *Guatemala* 14°49N 91°10W **88** D1
Sololo *Kenya* 3°33N 38°39E **54** B4
Solomon, N. Fork →
    U.S.A. 39°29N 98°26W **80** F4
Solomon, S. Fork →
    U.S.A. 39°25N 99°12W **80** F4
Solomon Is. ■ *Pac. Oc.* 6°0S 155°0E **58** B8
Solomon Rise *Pac. Oc.* 1°0N 157°0E **64** G7
Solon *China* 46°32N 121°10E **33** B13
Solon Springs *U.S.A.* 46°22N 91°49W **80** B8
Solor *Indonesia* 8°27S 123°0E **37** F6
Solothurn *Switz.* 47°13N 7°32E **20** C7
Šolta *Croatia* 43°24N 16°15E **22** C7
Solţānābād *Khorāsān*,
    Iran 34°13N 59°58E **45** C8
Solţānābād *Khorāsān*, *Iran* 36°29N 58°5E **45** B8
Solunska Glava
    Macedonia 41°44N 21°31E **23** D9
Solvang *U.S.A.* 34°36N 120°8W **79** L6
Solvay *U.S.A.* 43°3N 76°13W **83** C8
Sölvesborg *Sweden* 56°5N 14°35E **9** H16
Solvychegodsk *Russia* 61°21N 46°56E **18** B8
Solway Firth *U.K.* 54°49N 3°35W **12** C4
Solwezi *Zambia* 12°11S 26°21E **55** E2
Sōma *Japan* 37°40N 140°50E **30** F10
Soma *Turkey* 39°10N 27°35E **23** E12
Somabhula *Zimbabwe* 19°42S 29°40E **57** B4
Somali Pen. *Africa* 7°0N 46°0E **48** F8
Somali Rep. ■ *Africa* 7°0N 47°0E **47** F4
Somalia = Somali Rep. ■
    Africa 7°0N 47°0E **47** F4
Somaliland □ *Somali Rep.* 9°0N 47°0E **47** F4
Sombor *Serbia* 45°46N 19°9E **23** B8
Sombra *Canada* 42°43N 82°29W **82** D2
Sombrerete *Mexico* 23°38N 103°39W **86** C4
Sombrero *Anguilla* 18°37N 63°30W **89** C7
Somdari *India* 25°47N 72°38E **42** G5
Somers *U.S.A.* 48°5N 114°13W **76** B6
Somerset *Ky., U.S.A.* 37°5N 84°36W **81** G11
Somerset *Mass., U.S.A.* 41°47N 71°8W **83** E13
Somerset *Pa., U.S.A.* 40°1N 79°5W **82** F5
Somerset □ *U.K.* 51°9N 3°0W **13** F5
Somerset East *S. Africa* 32°42S 25°35E **56** E4
Somerset I. *Canada* 73°30N 93°0W **68** C13
Somerset West *S. Africa* 34°8S 18°50E **56** E2
Somersworth *U.S.A.* 43°16N 70°52E **83** C14
Somerton *U.S.A.* 32°36N 114°43W **77** K6
Somerville *U.S.A.* 40°35N 74°38W **83** F10
Someş → *Romania* 47°49N 22°43E **17** D12
Somme → *France* 50°11N 1°38E **20** A4
Somnath *India* 20°53N 70°22E **42** J4
Somosierra, Puerto de
    Spain 41°4N 3°35W **21** B4
Somosomo *Fiji* 16°47S 179°58W **59** a
Somosomo Str. *Fiji* 16°0S 180°0E **59** a
Somoto *Nic.* 13°28N 86°37W **88** D2
Somport, Túnel de *Spain* 42°48N 0°31W **20** E3
Son → *India* 25°42N 84°52E **43** G11
Son Ha *Vietnam* 15°3N 108°34E **38** E7
Son La *Vietnam* 21°20N 103°50E **38** B4
Son Morrell *Spain* 39°44N 3°20E **24** B10
Son Rapinya *Spain* 39°35N 2°37E **24** B9
Son Sardina *Spain* 39°37N 2°45E **24** B9
Son Serra *Spain* 39°43N 3°13E **24** B10
Son Tay *Vietnam* 21°8N 105°30E **38** B5
Son Tinh *Vietnam* 15°10N 108°40E **38** E7
Soná *Panama* 8°0N 81°20W **88** E3
Sonamarg *India* 34°18N 75°21E **43** B6
Sonamukhi *India* 23°18N 87°27E **43** H12

Sonar → *India* 24°24N 79°56E **43** G8
Sŏnch'ŏn *N. Korea* 39°48N 124°55E **35** E13
Sondags → *S. Africa* 33°44S 25°51E **56** E4
Sønderborg *Denmark* 54°55N 9°49E **9** J13
Sóndrio *Italy* 46°10N 9°52E **20** C8
Sone *Mozam.* 17°23S 34°55E **55** F3
Sonepur *India* 20°55N 83°50E **41** J13
Song *Thailand* 18°28N 100°11E **38** C3
Song Cau *Vietnam* 13°27N 109°18E **38** F7
Song Ma *Vietnam* 20°56N 103°53E **38** B4
Song Xian *China* 34°12N 112°8E **34** G7
Songan *Indonesia* 8°13S 115°24E **37** J18
Songgang *China* 22°46N 113°50E **33** a
Songhua Hu *China* 43°35N 126°50E **35** C14
Songhua Jiang →
    China 47°45N 132°30E **33** B15
Songimvelo △ *S. Africa* 25°50S 31°2E **57** D5
Songjianghe *China* 42°10N 127°29E **35** C14
Sŏngjin = Kimch'aek
    N. Korea 40°40N 129°10E **35** D15
Songkhla *Thailand* 7°13N 100°37E **39** J3
Songnim *N. Korea* 38°45N 125°39E **35** E13
Songo *Mozam.* 15°34S 32°38E **55** F3
Songpan *China* 32°40N 103°30E **32** D5
Songwe
    Dem. Rep. of the Congo 3°20S 26°16E **54** C2
Songwe → *Africa* 9°44S 33°58E **55** D3
Sonhat *India* 23°29N 82°31E **43** H10
Sonid Youqi *China* 42°45N 112°48E **34** C7
Sonid Zuoqi *China* 43°50N 113°42E **34** C7
Sonipat *India* 29°0N 77°5E **42** E7
Sonkach *India* 22°59N 76°21E **42** H7
Sonmiani *Pakistan* 25°25N 66°40E **42** G2
Sonmiani B. *Pakistan* 25°15N 66°30E **42** G2
Sono → *Brazil* 9°58S 48°11W **93** E9
Sonoma *U.S.A.* 38°18N 122°28W **78** G4
Sonora *Calif., U.S.A.* 37°59N 120°23W **78** H6
Sonora *Tex., U.S.A.* 30°34N 100°39W **84** F4
Sonora □ *Mexico* 29°20N 110°40W **86** B2
Sonora → *Mexico* 29°5N 110°55W **86** B2
Sonoran Desert
    U.S.A. 33°40N 113°30W **79** L12
Sonoyta *Mexico* 31°51N 112°50W **86** A2
Sonsonate *El Salv.* 13°43N 89°44W **88** D2
Soochow = Suzhou
    China 31°19N 120°38E **33** E13
Sooke *Canada* 48°13N 123°43W **78** B3
Soomaaliya = Somali Rep. ■
    Africa 7°0N 47°0E **47** F4
Sop Hao *Laos* 20°33N 104°27E **38** B5
Sop Prap *Thailand* 17°53N 99°20E **38** D2
Sopi *Indonesia* 2°34N 128°28E **37** D7
Sopot *Poland* 54°27N 18°31E **17** A10
Sopron *Hungary* 47°45N 16°32E **17** E9
Sopur *India* 34°18N 74°27E **43** B6
Sør-Rondane *Antarctica* 72°0S 25°0E **5** D4
Sorah *Pakistan* 27°13N 68°56E **42** F3
Soraon *India* 25°37N 81°51E **43** G9
Sord = Swords *Ireland* 53°28N 6°13W **10** C5
Sorel-Tracy *Canada* 46°0N 73°10W **72** C5
Sórgono *Italy* 40°1N 9°6E **22** D3
Soria *Spain* 41°43N 2°32W **21** B4
Soriano *Uruguay* 33°24S 58°19W **94** C4
Sorkh, Kuh-e *Iran* 35°40N 58°30E **45** C8
Soroca *Moldova* 48°8N 28°12E **17** D15
Sorocaba *Brazil* 23°31S 47°27W **95** A6
Sorochinsk *Russia* 52°26N 53°10E **18** D9
Soroki = Soroca *Moldova* 48°8N 28°12E **17** D15
Soroni *Greece* 36°21N 28°1E **25** C10
Sorong *Indonesia* 0°55S 131°15E **37** E8
Soroti *Uganda* 1°43N 33°35E **54** B3
Sørøya *Norway* 70°40N 22°30E **8** A20
Sørøysundet *Norway* 70°25N 23°0E **8** A20
Sorrell *Australia* 42°47S 147°34E **63** G4
Sorrento *Italy* 40°37N 14°22E **22** D6
Sorsele *Sweden* 65°31N 17°30E **8** D17
Sorsogon *Phil.* 13°0N 124°0E **37** B6
Sortavala *Russia* 61°42N 30°41E **8** F24
Sortland *Norway* 68°42N 15°25E **8** B16
Soscumica, L. *Canada* 50°15N 77°27W **72** B4
Sosnogorsk *Russia* 63°37N 53°51E **18** B9
Sosnowiec *Poland* 50°20N 19°10E **17** C10
Sossus Vlei *Namibia* 24°40S 15°23E **56** C2
Sŏsura *N. Korea* 42°16N 130°36E **35** C16
Sot → *India* 27°27N 79°37E **43** F8
Sotavento *C. Verde Is.* 15°0N 25°0W **50** b
Sotik *Kenya* 0°41S 35°7E **54** C4
Sotkamo *Finland* 64°8N 28°23E **8** D23
Soto la Marina *Mexico* 23°46N 98°13W **87** C5
Soto la Marina →
    Mexico 23°45N 97°45W **87** C5
Sotuta *Mexico* 20°36N 89°1W **87** C7
Souanké *Congo* 2°10N 14°3E **52** D2
Souda *Greece* 35°29N 24°4E **25** D6
Soudas, Ormos *Greece* 35°25N 24°10E **25** D6
Souderton *U.S.A.* 40°19N 75°19W **83** F9
Soufrière *Guadeloupe* 16°5N 61°40W **88** b
Soufrière *St. Lucia* 13°51N 61°3W **89** f
Soufrière Bay *Dominica* 15°13N 61°22W **89** f
Soukhouma *Laos* 14°38N 105°48E **38** E5
Sŏul = Seoul *S. Korea* 37°31N 126°58E **35** F14
Sound, The *U.K.* 50°20N 4°10W **13** G3
Sources, Mt. aux *Lesotho* 28°45S 28°50E **57** D4
Soure *Brazil* 0°35S 48°30W **93** D9
Souris *Man., Canada* 49°40N 100°20W **71** D8
Souris *P.E.I., Canada* 46°21N 62°15W **73** C7
Souris → *N. Amer.* 49°40N 99°34W **80** A4
Sousa *Brazil* 6°45S 38°10W **93** E11
Sousse *Tunisia* 35°50N 10°38E **51** A8
Sout → *S. Africa* 31°35S 18°24E **56** E2
South Africa ■ *Africa* 32°0S 23°0E **56** E3
South America 10°0S 60°0W **90** E5
South Aulatsivik I.
    Canada 56°45N 61°30W **73** A7
South Australia □
    Australia 32°0S 139°0E **63** E2
South Australian Basin
    Ind. Oc. 38°0S 126°0E **64** L4
South Ayrshire □ *U.K.* 55°18N 4°41W **11** F4

South Baldy Pk. *U.S.A.* 34°6N 107°11W **77** J10
South Bass I. *U.S.A.* 41°39N 82°49W **82** E2
South Bend *Ind., U.S.A.* 41°41N 86°15W **80** E10
South Bend *Wash.,*
    U.S.A. 46°40N 123°48W **78** D3
South Boston *U.S.A.* 36°42N 78°54W **81** G14
South Branch *Canada* 47°55N 59°2W **73** C8
South Brook *Canada* 49°26N 56°5W **73** C8
South Bruny I. *Australia* 43°20S 147°15E **63** G4
South Carolina □ *U.S.A.* 34°0N 81°0W **85** E14
South Charleston
    U.S.A. 38°22N 81°44W **81** F13
South China Sea *Asia* 10°0N 113°0E **36** C4
South Cumberland Is. △
    Australia 20°42S 149°11E **62** b
South Dakota □ *U.S.A.* 44°15N 100°0W **80** C4
South Dayton *U.S.A.* 42°21N 79°3W **82** D5
South Deerfield *U.S.A.* 42°29N 72°37W **83** D12
South Downs *U.K.* 50°52N 0°25W **13** G7
South Downs △ *U.K.* 50°50N 0°50W **13** G7
South East C. *Australia* 43°40S 146°50E **63** G4
South East Is. *Australia* 34°17S 123°30E **61** F3
South Esk → *U.K.* 56°43N 2°31W **11** E6
South Fiji Basin *Pac. Oc.* 26°0S 175°0E **64** K9
South Foreland *U.K.* 51°8N 1°24E **13** F9
South Fork American →
    U.S.A. 38°57N 120°59W **78** G5
South Fork Feather →
    U.S.A. 39°17N 121°36W **78** F5
South Fork Grand →
    U.S.A. 45°43N 102°17W **80** C2
South Fork Milk →
    U.S.A. 48°4N 106°19W **76** B10
South Fork Republican →
    U.S.A. 40°3N 101°31W **80** E3
South Georgia *Antarctica* 54°30S 37°0W **96** G9
South Gloucestershire □
    U.K. 51°32N 2°28W **13** F5
South Hadley *U.S.A.* 42°16N 72°35W **83** D12
South Haven *U.S.A.* 42°24N 86°16W **80** D10
South Henik L. *Canada* 61°30N 97°20W **71** A9
South Horr *Kenya* 2°12N 36°56E **54** B4
South I. *Kenya* 2°35N 36°35E **54** B4
South I. *N.Z.* 44°0S 170°0E **59** F3
South Indian Lake
    Canada 56°47N 98°56W **71** B9
South Invercargill *N.Z.* 46°26S 168°23E **59** G2
South Island △ *Kenya* 2°39N 36°35E **54** B4
South Kitui △ *Kenya* 1°48S 38°46E **54** C4
South Knife → *Canada* 58°55N 94°37W **71** B10
South Koel → *India* 22°32N 85°14E **43** H11
South Korea ■ *Asia* 36°0N 128°0E **35** G15
South Lake Tahoe
    U.S.A. 38°57N 119°59W **78** G6
South Lanarkshire □
    U.K. 55°37N 3°53W **11** F5
South Loup → *U.S.A.* 41°4N 98°39W **80** E4
South Luangwa △ *Zambia* 13°0S 31°20E **55** E3
South Magnetic Pole
    Antarctica 64°8S 138°8E **5** C9
South Milwaukee
    U.S.A. 42°55N 87°52W **80** D10
South Molton *U.K.* 51°1N 3°51W **13** F4
South Moose L. *Canada* 53°49N 100°1W **71** C8
South Nahanni →
    Canada 61°3N 123°21W **70** A4
South Nation → *Canada* 45°34N 75°6W **83** A9
South Natuna Is. = Natuna
    Selatan, Kepulauan
    Indonesia 2°45N 109°0E **36** D3
South Negril Pt. *Jamaica* 18°16N 78°22W **88** a
South Orkney Is.
    Antarctica 63°0S 45°0W **5** C18
South Ossetia □ *Georgia* 42°21N 44°2E **19** F7
South Otselic *U.S.A.* 42°38N 75°46W **83** D9
South Pagai, I. = Pagai Selatan,
    Pulau *Indonesia* 3°0S 100°15E **36** E2
South Paris *U.S.A.* 44°14N 70°31W **83** B14
South Pittsburg *U.S.A.* 35°1N 85°42W **85** D12
South Platte → *U.S.A.* 41°7N 100°42W **80** E3
South Pole *Antarctica* 90°0S 0°0 **5** E
South Porcupine *Canada* 48°30N 81°12W **72** C3
South Portland *U.S.A.* 43°38N 70°15W **83** C14
South Pt. *Barbados* 13°2N 59°32W **89** g
South Pt. *U.S.A.* 42°52N 83°19W **82** D1
South River *Canada* 45°52N 79°23W **72** C4
South River *U.S.A.* 40°27N 74°23W **83** F10
South Ronaldsay *U.K.* 58°48N 2°58W **11** C6
South Rukuru → *Malawi* 10°44S 34°14E **55** E3
South Sandwich Is.
    Antarctica 57°0S 27°0W **5** B1
South Sandwich Trench
    Atl. Oc. 56°0S 24°0W **5** B1
South Saskatchewan →
    Canada 53°15N 105°5W **71** C7
South Seal → *Canada* 58°48N 98°8W **71** B9
South Shetland Is.
    Antarctica 62°0S 59°0W **5** C18
South Shields *U.K.* 55°0N 1°25W **12** C6
South Sioux City *U.S.A.* 42°28N 96°24W **80** D5
South Sister *U.S.A.* 44°4N 121°51W **76** D3
South Taranaki Bight
    N.Z. 39°40S 174°5E **59** C5
South Tasman Rise
    S. Ocean 48°0S 146°0E **5** A10
South Thompson →
    Canada 50°40N 120°20W **70** C4
South Twin I. *Canada* 53°7N 79°52W **72** B4
South Tyne → *U.K.* 54°59N 2°8W **12** C5
South Uist *U.K.* 57°20N 7°15W **11** D1
South Valley *U.S.A.* 35°1N 106°41W **77** J10
South Wellesley Is.
    Australia 16°58S 139°17E **62** B2
South West C. *Australia* 43°34S 146°3E **63** G4
South West C. *N.Z.* 47°17S 167°28E **59** G1
South Williamsport
    U.S.A. 41°13N 77°0W **82** E8
South Yorkshire □ *U.K.* 53°27N 1°36W **12** D6
Southampton *Canada* 44°30N 81°25W **82** B3
Southampton *U.K.* 50°54N 1°23W **13** G6

Southampton *U.S.A.* 40°53N 72°23W **83** F12
Southampton □ *U.K.* 50°54N 1°23W **13** G6
Southampton I. *Canada* 64°30N 84°0W **69** E15
Southaven *U.S.A.* 34°59N 90°0W **85** D9
Southbank *Canada* 54°2N 125°46W **70** C3
Southbridge *N.Z.* 43°48S 172°16E **59** E4
Southbridge *U.S.A.* 42°5N 72°2W **83** D12
Southend *Canada* 56°19N 103°22W **71** B8
Southend-on-Sea *U.K.* 51°32N 0°44E **13** F8
Southend-on-Sea □ *U.K.* 51°32N 0°44E **13** F8
Southern = Yuzhnyy □
    Russia 44°0N 40°0E **28** E5
Southern □ *Malawi* 15°0S 35°0E **55** F4
Southern □ *Zambia* 16°20S 26°20E **55** F2
Southern Alps *N.Z.* 43°41S 170°11E **59** E3
Southern Cross
    Australia 31°12S 119°15E **61** F2
Southern Indian L.
    Canada 57°10N 98°30W **71** B9
Southern Lau Group
    Fiji 18°40S 178°40W **59** a
Southern Ocean *Antarctica* 62°0S 60°0E **5** C6
Southern Pines *U.S.A.* 35°11N 79°24W **85** D15
Southern Uplands *U.K.* 55°28N 3°52W **11** F5
Southington *U.S.A.* 41°36N 72°53W **83** E12
Southland □ *N.Z.* 45°30S 168°0E **59** F1
Southold *U.S.A.* 41°4N 72°26W **83** E12
Southport *Australia* 27°58S 153°25E **63** D5
Southport *U.K.* 53°39N 3°0W **12** D4
Southport *U.S.A.* 42°3N 76°49W **82** D8
Southwest △ *Australia* 43°8S 146°5E **63** G4
Southwest Pacific Basin
    Pac. Oc. 40°0S 140°0W **5** A12
Southwold *U.K.* 52°20N 1°41E **13** E9
Southwood △ *Australia* 27°48S 150°8E **63** D5
Soutpansberg *S. Africa* 23°0S 29°30E **57** C4
Sovetsk *Kaliningrad, Russia* 55°6N 21°50E **9** J19
Sovetsk *Kirov, Russia* 57°38N 48°53E **18** C8
Sovetskaya Gavan
    Russia 48°58N 140°18E **29** E15
Soweto *S. Africa* 26°14S 27°52E **57** D4
Sowma'eh Sarā *Iran* 37°17N 49°18E **45** B6
Sōya-Kaikyō = La Perouse Str.
    Asia 45°40N 142°0E **30** B11
Sōya-Misaki *Japan* 45°30N 141°55E **30** B10
Soyo *Angola* 6°13S 12°20E **52** F2
Sozh → *Belarus* 51°57N 30°48E **17** B16
Spa *Belgium* 50°29N 5°53E **15** D5
Spain ■ *Europe* 39°0N 4°0W **21** B4
Spalding *Australia* 33°30S 138°37E **63** E2
Spalding *U.K.* 52°48N 0°9W **12** E7
Spanaway *U.S.A.* 47°6N 122°26W **78** C4
Spangler *U.S.A.* 40°39N 78°48W **82** F6
Spanish *Canada* 46°12N 82°20W **72** C3
Spanish Fork *U.S.A.* 40°7N 111°39W **76** F8
Spanish Town
    Br. Virgin Is. 17°43N 64°26W **89** e
Spanish Town *Jamaica* 18°0N 76°57W **88** a
Sparks *U.S.A.* 39°32N 119°45W **78** F7
Sparta = Sparti *Greece* 37°5N 22°25E **23** F10
Sparta *Mich., U.S.A.* 43°10N 85°42W **81** D11
Sparta *N.J., U.S.A.* 41°2N 74°38W **83** E10
Sparta *Wis., U.S.A.* 43°56N 90°49W **80** D8
Spartanburg *U.S.A.* 34°56N 81°57W **85** D14
Sparti *Greece* 37°5N 22°25E **23** F10
Spartivento, C. *Calabria,*
    Italy 37°55N 16°4E **22** F7
Spartivento, C. *Sard., Italy* 38°53N 8°50E **22** E3
Sparwood *Canada* 49°44N 114°53W **70** D6
Spassk Dalniy *Russia* 44°40N 132°48E **30** B6
Spatha, Akra *Greece* 35°42N 23°43E **25** D5
Spatsizi → *Canada* 57°42N 128°7W **70** B3
Spatsizi Plateau Wilderness △
    Canada 57°40N 128°0W **70** B3
Spean → *U.K.* 56°55N 4°59W **11** E4
Spearfish *U.S.A.* 44°30N 103°52W **80** C2
Spearman *U.S.A.* 36°12N 101°12W **84** C4
Speculator *U.S.A.* 43°30N 74°22W **83** C10
Speightstown *Barbados* 13°15N 59°39W **89** g
Speke Gulf *Tanzania* 2°20S 32°50E **54** C3
Spence Bay = Taloyoak
    Canada 69°32N 93°32W **68** D13
Spencer *Idaho, U.S.A.* 44°22N 112°11W **76** D7
Spencer *Iowa, U.S.A.* 43°9N 95°9W **80** D6
Spencer *N.Y., U.S.A.* 42°13N 76°30W **83** D8
Spencer *Nebr., U.S.A.* 42°53N 98°42W **80** D4
Spencer, C. *Australia* 35°20S 136°53E **63** F2
Spencer B. *Namibia* 25°30S 14°47E **56** D1
Spencer G. *Australia* 34°0S 137°20E **63** E2
Spencerville *Canada* 44°51N 75°33W **83** B9
Spences Bridge *Canada* 50°25N 121°20W **70** C4
Spennymoor *U.K.* 54°42N 1°36W **12** C6
Spenser Mts. *N.Z.* 42°15S 172°45E **59** E4
Sperrin Mts. *U.K.* 54°50N 7°0W **10** B5
Spey → *U.K.* 57°40N 3°6W **11** D5
Speyer *Germany* 49°29N 8°25E **16** D5
Spezand *Pakistan* 29°59N 67°0E **42** E2
Spiddle *Ireland* 53°15N 9°18W **10** C2
Spili *Greece* 35°13N 24°31E **25** D6
Spīn Būldak *Afghan.* 31°1N 66°25E **42** D2
Spinalonga *Greece* 35°18N 25°44E **25** D7
Spinifex △ *Australia* 28°30S 128°30E **61** E4
Spirit Lake *U.S.A.* 46°15N 122°9W **78** D4
Spirit River *Canada* 55°45N 118°50W **70** B5
Spiritwood *Canada* 53°24N 107°33W **71** C7
Spithead *U.K.* 50°45N 1°10W **13** G6
Spitzbergen = Svalbard
    Arctic 78°0N 17°0E **4** B8
Spjelkavik *Norway* 62°28N 6°22E **8** E12
Split *Croatia* 43°31N 16°26E **22** C7
Split L. *Canada* 56°8N 96°15W **71** B9
Split Lake *Canada* 56°8N 96°15W **71** B9
Spofford *U.S.A.* 29°10N 100°25W **84** G4
Spokane *U.S.A.* 47°40N 117°24W **76** C5
Spoleto *Italy* 42°44N 12°44E **22** C5
Spooner *U.S.A.* 45°50N 91°53W **80** C8
Sporyy Navolok, Mys
    Russia 75°50N 68°40E **28** B7
Sprague *U.S.A.* 47°18N 117°59W **76** C5
Spratly I. *S. China Sea* 8°38N 111°55E **36** C4

Spratly Is. *S. China Sea* 8°20N 112°0E **36**
Spray *U.S.A.* 44°50N 119°48W **76**
Spree → *Germany* 52°32N 13°12E **16**
Sprengisandur *Iceland* 64°52N 18°7W **8**
Spring City *U.S.A.* 40°11N 75°33W **8**
Spring Creek *U.S.A.* 40°44N 115°35W **7**
Spring Garden *U.S.A.* 39°52N 120°47W **7**
Spring Hall *Barbados* 13°18N 59°36W **8**
Spring Hill *U.S.A.* 28°27N 82°41W **8**
Spring Mts. *U.S.A.* 36°0N 115°45W **79**
Spring Valley *Calif.,*
    U.S.A. 32°44N 116°59W **79**
Spring Valley *Nev.,*
    U.S.A. 36°6N 115°14W **79**
Springbok *S. Africa* 29°42S 17°54E **5**
Springboro *U.S.A.* 41°48N 80°22W **8**
Springdale *Canada* 49°30N 56°6W **7**
Springdale *U.S.A.* 36°11N 94°8W **8**
Springer *U.S.A.* 36°22N 104°36W **77**
Springerville *U.S.A.* 34°8N 109°17W **7**
Springfield *Canada* 42°50N 80°56W **8**
Springfield *N.Z.* 43°19S 171°56E **5**
Springfield *Colo.,*
    U.S.A. 37°24N 102°37W **77**
Springfield *Ill., U.S.A.* 39°48N 89°39W **8**
Springfield *Mass., U.S.A.* 42°6N 72°35W **8**
Springfield *Mo., U.S.A.* 37°13N 93°17W **8**
Springfield *Ohio, U.S.A.* 39°55N 83°49W **81**
Springfield *Oreg., U.S.A.* 44°3N 123°1W **7**
Springfield *Tenn.,*
    U.S.A. 36°31N 86°53W **8**
Springfield *Vt., U.S.A.* 43°18N 72°29W **8**
Springfontein *S. Africa* 30°15S 25°40E **5**
Springhill *Canada* 45°40N 64°4W **7**
Springhill *U.S.A.* 33°0N 93°28W **8**
Springhouse *Canada* 51°56N 122°7W **7**
Springs *S. Africa* 26°13S 28°25E **5**
Springsure *Australia* 24°8S 148°6E **6**
Springvale *U.S.A.* 43°28N 70°48W **83**
Springville *Calif., U.S.A.* 36°8N 118°49W **7**
Springville *N.Y., U.S.A.* 42°31N 78°40W **8**
Springville *Utah, U.S.A.* 40°10N 111°37W **7**
Springwater *U.S.A.* 42°38N 77°35W **8**
Spruce-Creek *U.S.A.* 40°36N 78°9W **8**
Spruce Knob-Seneca Rocks △
    U.S.A. 38°50N 79°30W **81**
Spruce Mt. *U.S.A.* 44°12N 72°19W **83**
Sprucedale *Canada* 45°29N 79°28W **8**
Spur *U.S.A.* 33°28N 100°52W **8**
Spurn Hd. *U.K.* 53°35N 0°8E **1**
Spuzzum *Canada* 49°37N 121°23W **7**
Squam L. *U.S.A.* 43°45N 71°32W **83**
Squamish *Canada* 49°45N 123°10W **7**
Square Islands *Canada* 52°47N 55°47W **7**
Squires, Mt. *Australia* 26°14S 127°28E **6**
Srbija = Serbia ■ *Europe* 43°20N 20°0E **2**
Sre Ambel *Cambodia* 11°8N 103°46E **3**
Sre Khtum *Cambodia* 12°10N 106°52E **3**
Sre Umbell = Sre Ambel
    Cambodia 11°8N 103°46E **3**
Srebrenica *Bos.-H.* 44°6N 19°18E **2**
Sredinnyy Khrebet
    Russia 57°0N 160°0E **29**
Sredinnyy Ra. = Sredinnyy
    Khrebet *Russia* 57°0N 160°0E **29**
Srednekolymsk *Russia* 67°27N 153°40E **29**
Śrem *Poland* 52°6N 17°2E **1**
Sremska Mitrovica
    Serbia 44°59N 19°38E **2**
Srepok → *Cambodia* 13°33N 106°16E **3**
Sretensk *Russia* 52°10N 117°40E **29**
Sri Aman = Bandar Sri Aman
    Malaysia 1°15N 111°32E **3**
Sri Lanka ■ *Asia* 7°30N 80°50E **40**
Srikakulam *India* 18°14N 83°58E **41**
Srinagar *India* 34°5N 74°50E **4**
Srono *Indonesia* 8°24S 114°16E **37**
Ssese Is. = Sese Is. *Uganda* 0°20S 32°20E **5**
Staaten → *Australia* 16°24S 141°17E **6**
Staaten River △
    Australia 16°15S 142°40E **6**
Stade *Germany* 53°35N 9°29E **1**
Stadskanaal *Neths.* 53°4N 6°55E **1**
Staffa *U.K.* 56°27N 6°21W **1**
Stafford *U.K.* 52°49N 2°7W **1**
Stafford *U.S.A.* 37°58N 98°36W **8**
Stafford Springs *U.S.A.* 41°57N 72°18W **83**
Staffordshire □ *U.K.* 52°53N 2°10W **1**
Staines *U.K.* 51°26N 0°29W **1**
Stakhanov *Ukraine* 48°35N 38°40E **1**
Stalida *Greece* 35°17N 25°25E **2**
Stalowa Wola *Poland* 50°34N 22°3E **17**
Stalybridge *U.K.* 53°28N 2°3W **1**
Stamford *Australia* 21°15S 143°46E **6**
Stamford *U.K.* 52°39N 0°29W **1**
Stamford *Conn., U.S.A.* 41°3N 73°32W **83**
Stamford *N.Y., U.S.A.* 42°25N 74°38W **83**
Stamford *Tex., U.S.A.* 32°57N 99°48W **8**
Stampriet *Namibia* 24°20S 18°28E **5**
Stamps *U.S.A.* 33°22N 93°30W **8**
Standerton *S. Africa* 26°55S 29°7E **5**
Standish *U.S.A.* 43°59N 83°57W **8**
Stanford *S. Africa* 34°26S 19°29E **5**
Stanford *U.S.A.* 47°9N 110°13W **7**
Stanger *S. Africa* 29°27S 31°14E **5**
Stängselåsen = Salpausselkä
    Finland 61°3N 26°15E **8**
Stanislaus → *U.S.A.* 37°40N 121°14W **7**
Stanke Dimitrov
    Bulgaria 42°17N 23°9E **2**
Stanley *Australia* 40°46S 145°19E **6**
Stanley *China* 22°13N 114°12E **3**
Stanley *Falk. Is.* 51°40S 59°51W **9**
Stanley *U.K.* 54°53N 1°41W **1**
Stanley *Idaho, U.S.A.* 44°13N 114°56W **7**
Stanley *N. Dak., U.S.A.* 48°19N 102°23W **8**
Stanley *N.Y., U.S.A.* 42°48N 77°6W **8**
Stanley Mission
    Canada 55°25N 104°33W **7**
Stannards *U.S.A.* 42°5N 77°55W **8**

anovoy Khrebet
Russia 55°0N 130°0E 29 D14
anovoy Ra. = Stanovoy Khrebet
Russia 55°0N 130°0E 29 D14
ansmore Ra. Australia 21°23S 128°33E 60 D4
ansted, London ✈ (STN)
U.K. 51°54N 0°14E 13 F8
nthorpe Australia 28°36S 151°59E 63 D5
anton U.S.A. 32°8N 101°48W 84 E4
anwood U.S.A. 48°15N 122°23W 78 B4
aples Canada 42°10N 82°35W 82 D2
aples U.S.A. 46°21N 94°48W 80 B6
apleton U.S.A. 41°29N 100°31W 80 E3
ar City Canada 52°50N 104°20W 71 C8
ar Lake U.S.A. 44°10N 75°2W 83 B9
ara Planina Bulgaria 43°15N 23°0E 23 C10
ara Zagora Bulgaria 42°26N 25°39E 23 C11
arachowice Poland 51°3N 21°2E 17 C11
araya Russa Russia 57°58N 31°23E 18 C5
arbuck I. Kiribati 5°37S 155°55W 65 H12
arcke △ Australia 14°56S 145°2E 62 A4
argard Szczeciński
Poland 53°20N 15°0E 16 B8
aritsa Russia 56°33N 34°55E 18 C5
arke U.S.A. 29°57N 82°7W 85 G13
arkville U.S.A. 33°28N 88°49W 85 E10
arogard Gdański
Poland 53°59N 18°30E 17 B10
arokonstantinov =
Starokonstyantyniv
Ukraine 49°48N 27°10E 17 D14
arokonstyantyniv
Ukraine 49°48N 27°10E 17 D14
art Pt. U.K. 50°13N 3°39W 13 G4
aryy Chartoriysk
Ukraine 51°15N 25°54E 17 C13
aryy Oskol Russia 51°19N 37°55E 19 D6
ate College U.S.A. 40°48N 77°52W 82 F7
ateline U.S.A. 38°57N 119°56W 78 G7
aten, I. = Estados, I. de Los
Argentina 54°40S 64°30W 96 G4
aten I. U.S.A. 40°35N 74°9W 83 F10
atesboro U.S.A. 32°27N 81°47W 85 E14
atesville U.S.A. 35°47N 80°53W 85 D14
atia = St. Eustatius
W. Indies 17°20N 63°0W 89 C7
auffer U.S.A. 34°45N 119°3W 79 L7
aunton Ill., U.S.A. 39°1N 89°47W 80 F9
aunton Va., U.S.A. 38°9N 79°4W 81 F14
avanger Norway 58°57N 5°40E 9 G11
aveley N.Z. 43°40S 171°32E 59 E3
avelot Belgium 50°23N 5°55E 15 D5
avern Norway 59°0N 10°1E 9 G14
avoren Neths. 52°53N 5°22E 15 B5
avropol Russia 45°5N 42°0E 19 E7
avros Cyprus 35°1N 32°38E 25 D11
avros Greece 35°12N 24°45E 25 D6
avros, Akra Greece 35°26N 24°58E 25 D6
awell Australia 37°5S 142°47E 63 F3
awell → Australia 20°20S 142°55E 62 C3
ayner Canada 44°25N 80°5W 82 B4
ayton U.S.A. 44°48N 122°48W 76 D2
eamboat Springs
U.S.A. 40°29N 106°50W 76 F10
eele U.S.A. 46°51N 99°55W 80 B4
eelton U.S.A. 40°14N 76°50W 82 F8
een River Canada 59°40N 117°12W 70 B5
eens Mt. U.S.A. 42°35N 118°40W 76 E4
eenwijk Neths. 52°47N 6°7E 15 B6
eep Pt. Australia 26°8S 113°8E 61 E1
eep Rock Canada 51°30N 98°48W 71 C9
efanie L. = Chew Bahir
Ethiopia 4°40N 36°50E 47 G2
efansson Bay Antarctica 67°20S 59°8E 5 C5
efansson I. Canada 73°20N 105°45W 69 C10
eiermark □ Austria 47°26N 15°0E 16 E8
eilacoom U.S.A. 47°10N 122°36W 78 C4
eilrandberge Namibia 17°45S 13°20E 56 B1
einbach Canada 49°32N 96°40W 71 D9
einhausen Namibia 21°49S 18°20E 56 C2
einkjer Norway 64°1N 11°31E 8 D14
einkopf S. Africa 29°18S 17°43E 56 D2
ellarton Canada 45°32N 62°30W 73 C7
ellenbosch S. Africa 33°58S 18°50E 56 E2
endal Germany 52°36N 11°53E 16 B6
eornabhaigh = Stornoway
U.K. 58°13N 6°23W 11 C2
epanakert = Xankändi
Azerbaijan 39°52N 46°49E 44 B5
ephens Creek
Australia 31°50S 141°30E 63 E3
ephens I. Canada 54°10N 130°45W 70 C2
ephens L. Canada 56°32N 95°0W 71 B9
ephenville Canada 48°31N 58°35W 73 C8
ephenville U.S.A. 32°13N 98°12W 84 E5
eppe Asia 50°0N 50°0E 26 C7
erkstroom S. Africa 31°32S 26°32E 56 E4
erling Colo., U.S.A. 40°37N 103°13W 76 F12
erling Ill., U.S.A. 41°48N 89°42W 80 E9
erling Kans., U.S.A. 38°13N 98°12W 80 F4
erling City U.S.A. 31°51N 101°0W 84 F4
erling Heights U.S.A. 42°35N 83°2W 81 D12
erling Run U.S.A. 41°25N 78°12W 82 E6
erlitamak Russia 53°40N 56°0E 18 D10
ernes Greece 35°30N 24°9E 25 D6
ettin = Szczecin Poland 53°27N 14°27E 16 B8
ettiner Haff Germany 53°47N 14°15E 16 B8
ettler Canada 52°19N 112°40W 70 C6
eubenville U.S.A. 40°22N 80°37W 82 F4
evenage U.K. 51°55N 0°13W 13 F7
evens Point U.S.A. 44°31N 89°34W 80 C9
evenson U.S.A. 45°42N 121°53W 78 E5
evenson → Australia 27°6S 135°33E 63 D2
evenson L. Canada 53°55N 96°0W 71 C9
evensville U.S.A. 46°30N 114°5W 76 C6
ewart Canada 55°56N 129°57W 70 B3
ewart U.S.A. 39°5N 119°46W 78 F7
ewart → Canada 63°19N 139°26W 68 E4
ewart, C. Australia 11°57S 134°56E 62 A1
ewart, I. Chile 54°50S 71°15W 96 G2
Stewart I. N.Z. 46°58S 167°54E 59 G1
Stewarts Point U.S.A. 38°39N 123°24W 78 G3
Stewartville U.S.A. 43°51N 92°29W 80 D7
Stewiacke Canada 45°9N 63°22W 73 C7
Steynsburg S. Africa 31°15S 25°49E 56 E4
Steyr Austria 48°3N 14°25E 16 D8
Steytlerville S. Africa 33°17S 24°19E 56 E3
Stigler U.S.A. 35°15N 95°8W 84 D7
Stikine → Canada 56°40N 132°30W 70 B2
Stilfontein S. Africa 26°51S 26°50E 56 D4
Stillwater N.Z. 42°27S 171°20E 59 E3
Stillwater N.Y., U.S.A. 42°55N 73°41W 83 D11
Stillwater Okla., U.S.A. 36°7N 97°4W 84 C6
Stillwater Range U.S.A. 39°50N 118°5W 76 G4
Stillwater Res. U.S.A. 43°54N 75°3W 83 C9
Stilwell U.S.A. 35°49N 94°38W 84 D7
Stînga Nistrului □
Moldova 47°20N 29°15E 17 E15
Štip Macedonia 41°42N 22°10E 23 D10
Stirling Canada 44°18N 77°33W 82 B7
Stirling U.K. 56°8N 3°57W 11 E5
Stirling □ U.K. 56°12N 4°18W 11 E4
Stirling Ra. Australia 34°23S 118°0E 61 F2
Stirling Range △
Australia 34°26S 118°20E 61 F2
Stittsville Canada 45°15N 75°55W 83 A9
Stjernøya Norway 70°20N 22°40E 8 A20
Stjørdalshalsen Norway 63°29N 10°51E 8 E14
Stockerau Austria 48°24N 16°12E 16 D9
Stockholm Sweden 59°19N 18°4E 9 G18
Stockport U.K. 53°25N 2°9W 12 D5
Stocksbridge U.K. 53°29N 1°35W 12 D6
Stockton Calif., U.S.A. 37°58N 121°17W 78 H5
Stockton Kans., U.S.A. 39°26N 99°16W 80 F4
Stockton Mo., U.S.A. 37°42N 93°48W 80 G7
Stockton-on-Tees U.K. 54°35N 1°19W 12 C6
Stockton-on-Tees □
U.K. 54°35N 1°19W 12 C6
Stockton Plateau
U.S.A. 30°30N 102°30W 84 F3
Stoeng Treng Cambodia 13°31N 105°58E 38 F5
Stoer, Pt. of U.K. 58°16N 5°23W 11 C3
Stoke-on-Trent U.K. 53°1N 2°11W 12 D5
Stoke-on-Trent □ U.K. 53°1N 2°11W 12 D5
Stokes △ Australia 33°45S 121°11E 61 F3
Stokes Bay Canada 45°0N 81°28W 82 B3
Stokes Pt. Australia 40°10S 143°56E 63 G3
Stokes Ra. Australia 15°50S 130°50E 60 C5
Stokksnes Iceland 64°14N 14°58W 8 D6
Stokmarknes Norway 68°34N 14°54E 8 B16
Stolac Bos.-H. 43°5N 17°59E 23 C7
Stolbovoy, Ostrov
Russia 74°44N 135°14E 29 B14
Stolbtsy = Stowbtsy
Belarus 53°30N 26°43E 17 B14
Stolin Belarus 51°53N 26°50E 17 C14
Stomio Greece 35°21N 23°32E 25 D5
Stone U.K. 52°55N 2°9W 12 E5
Stoneboro U.S.A. 41°20N 80°7W 82 E4
Stonehaven U.K. 56°59N 2°12W 11 E6
Stonehenge Australia 24°22S 143°17E 62 C3
Stonehenge U.K. 51°9N 1°45W 13 F6
Stonewall Canada 50°10N 97°19W 71 C9
Stony I. U.S.A. 43°53N 76°19W 83 C8
Stony L. Man., Canada 58°51N 98°40W 71 B9
Stony L. Ont., Canada 44°30N 78°5W 82 B6
Stony Point U.S.A. 41°14N 73°59W 83 E11
Stony Pt. U.S.A. 43°50N 76°18W 83 C8
Stony Rapids Canada 59°16N 105°50W 71 B7
Stony Tunguska = Tunguska,
Podkamennaya →
Russia 61°50N 90°13E 29 C10
Stonyford U.S.A. 39°23N 122°33W 78 F4
Stora Lulevatten Sweden 67°10N 19°30E 8 C18
Storavan Sweden 65°45N 18°10E 8 D18
Stord Norway 59°52N 5°23E 9 G11
Store Bælt Denmark 55°20N 11°0E 9 J14
Storkerson Pen.
Canada 72°30N 106°30W 68 C10
Storm B. Australia 43°10S 147°30E 63 G4
Storm Lake U.S.A. 42°39N 95°13W 80 D6
Stormberge S. Africa 31°16S 26°17E 56 E4
Stormsrivier S. Africa 33°59S 23°52E 56 E3
Stornoway U.K. 58°13N 6°23W 11 C2
Storozhinets = Storozhynets
Ukraine 48°14N 25°45E 17 D13
Storozhynets Ukraine 48°14N 25°45E 17 D13
Storrs U.S.A. 41°49N 72°15W 83 E12
Storsjön Sweden 63°9N 14°30E 8 E16
Storuman Sweden 65°5N 17°10E 8 D17
Storuman, L. Sweden 65°13N 16°50E 8 D17
Stouffville Canada 43°58N 79°15W 82 C5
Stoughton Canada 49°40N 103°0W 71 D8
Stour → Dorset, U.K. 50°43N 1°47W 13 G6
Stour → Kent, U.K. 51°18N 1°22E 13 F9
Stour → Suffolk, U.K. 51°57N 1°4E 13 F9
Stourbridge U.K. 52°28N 2°8W 13 E5
Stout L. Canada 52°0N 94°40W 71 C10
Stove Pipe Wells Village
U.S.A. 36°35N 117°11W 79 J9
Stow U.S.A. 41°10N 81°27W 82 E3
Stowbtsy Belarus 53°30N 26°43E 17 B14
Stowmarket U.K. 52°12N 1°0E 13 E9
Strabane U.K. 54°50N 7°27W 10 B4
Strahan Australia 42°9S 145°20E 63 G4
Stralsund Germany 54°18N 13°4E 16 A7
Strand S. Africa 34°9S 18°48E 56 E2
Stranda Møre og Romsdal,
Norway 62°19N 6°58E 8 E12
Stranda Nord-Trøndelag,
Norway 63°33N 10°14E 8 E14
Strangford L. U.K. 54°30N 5°37W 10 B6
Stranorlar Ireland 54°48N 7°46W 10 B4
Stranraer U.K. 54°54N 5°1W 11 G3
Strasbourg Canada 51°4N 104°55W 71 C8
Strasbourg France 48°35N 7°42E 20 B7
Stratford Canada 43°23N 81°0W 82 C4
Stratford N.Z. 39°20S 174°19E 59 C5
Stratford Calif., U.S.A. 36°11N 119°49W 78 J7
Stratford Conn., U.S.A. 41°12N 73°8W 83 E11
Stratford Tex., U.S.A. 36°20N 102°4W 84 C3
Stratford-upon-Avon
U.K. 52°12N 1°42W 13 E6
Strath Spey U.K. 57°9N 3°49W 11 D5
Strathalbyn Australia 35°13S 138°53E 63 F2
Strathaven U.K. 55°40N 4°5W 11 F4
Strathcona △ Canada 49°38N 125°40W 70 D3
Strathmore Canada 51°5N 113°18W 70 C6
Strathmore U.K. 56°37N 3°7W 11 E5
Strathmore U.S.A. 36°9N 119°4W 78 J7
Strathnaver Canada 53°20N 122°33W 70 C4
Strathpeffer U.K. 57°35N 4°32W 11 D4
Strathroy Canada 42°58N 81°38W 82 D3
Strathy Pt. U.K. 58°36N 4°1W 11 C4
Strattanville U.S.A. 41°12N 79°19W 82 E5
Stratton U.S.A. 45°8N 70°26W 83 A14
Stratton Mt. U.S.A. 43°4N 72°55W 83 C12
Straubing Germany 48°52N 12°34E 16 D7
Straumnes Iceland 66°26N 23°8W 8 C2
Strawberry → U.S.A. 40°10N 110°24W 76 F8
Streaky B. Australia 32°48S 134°13E 63 E1
Streaky Bay Australia 32°51S 134°18E 63 E1
Streator U.S.A. 41°8N 88°50W 80 E9
Streetsboro U.S.A. 41°14N 81°21W 82 E3
Streetsville Canada 43°35N 79°42W 82 C5
Streng → Cambodia 13°12N 103°37E 38 F4
Streymoy Færoe Is. 62°8N 7°5W 8 E9
Strezhevoy Russia 60°42N 77°34E 28 C8
Strimonas → Greece 40°46N 23°51E 23 D10
Strokestown Ireland 53°46N 8°7W 10 C3
Stroma U.K. 58°41N 3°7W 11 C5
Strómboli Italy 38°47N 15°13E 22 E6
Stromeferry U.K. 57°21N 5°33W 11 D3
Stromness U.K. 58°58N 3°17W 11 C5
Stromsburg U.S.A. 41°7N 97°36W 80 E5
Strömstad Sweden 58°56N 11°10E 9 G14
Strömsund Sweden 63°51N 15°33E 8 E16
Strongsville U.S.A. 41°19N 81°50W 82 E3
Stronsay U.K. 59°7N 2°35W 11 B6
Stroud U.K. 51°45N 2°13W 13 F5
Stroud Road Australia 32°18S 151°57E 63 E5
Stroudsburg U.S.A. 40°59N 75°12W 83 F9
Stroumbi Cyprus 34°53N 32°29E 25 E11
Struer Denmark 56°30N 8°35E 9 H13
Strugi Krasnyye Russia 58°21N 29°1E 9 G23
Strumica Macedonia 41°28N 22°41E 23 D10
Struthers U.S.A. 41°4N 80°39W 82 E4
Stryker U.S.A. 48°41N 114°46W 76 B6
Stryy Ukraine 49°16N 23°48E 17 D12
Strzelecki Cr. →
Australia 29°37S 139°59E 63 D2
Strzelecki Desert
Australia 29°30S 140°0E 63 D3
Stuart Fla., U.S.A. 27°12N 80°15W 85 H14
Stuart Nebr., U.S.A. 42°36N 99°8W 80 D4
Stuart → Canada 54°0N 123°35W 70 C4
Stuart Bluff Ra.
Australia 22°50S 131°52E 60 D5
Stuart I. U.S.A. 63°35N 162°30W 74 C7
Stuart L. Canada 54°30N 124°30W 70 C4
Stuart Ra. Australia 29°10S 134°56E 63 D1
Stull L. Canada 54°24N 92°34W 72 B1
Stung Treng = Stoeng Treng
Cambodia 13°31N 105°58E 38 F5
Stupart → Canada 56°0N 93°25W 72 A1
Sturge I. Antarctica 67°26S 164°47E 5 C11
Sturgeon B. Canada 52°0N 97°50W 71 C9
Sturgeon Bay U.S.A. 44°50N 87°23W 80 C10
Sturgeon Falls Canada 46°25N 79°57W 72 C4
Sturgeon L. Alta., Canada 55°6N 117°32W 70 B5
Sturgeon L. Ont., Canada 50°0N 90°45W 72 C1
Sturgeon L. Ont., Canada 44°28N 78°43W 82 B6
Sturgis Canada 51°56N 102°36W 71 C8
Sturgis Mich., U.S.A. 41°48N 85°25W 81 E11
Sturgis S. Dak., U.S.A. 44°25N 103°31W 80 C2
Sturt △ Australia 27°17S 141°37E 63 D3
Sturt Cr. → Australia 19°8S 127°50E 60 C4
Sturt Stony Desert
Australia 28°30S 141°0E 63 D3
Stutterheim S. Africa 32°33S 27°28E 56 E4
Stuttgart Germany 48°48N 9°11E 16 D5
Stuttgart U.S.A. 34°30N 91°33W 84 D9
Stuyvesant U.S.A. 42°23N 73°45W 83 D11
Stykkishólmur Iceland 65°2N 22°40W 8 D2
Styria = Steiermark □
Austria 47°26N 15°0E 16 E8
Su Xian = Suzhou
China 33°41N 116°59E 34 H9
Suakin Sudan 19°8N 37°20E 51 E13
Suan N. Korea 38°42N 126°22E 35 E14
Suana Indonesia 8°44S 115°36E 37 K18
Suar India 29°2N 79°3E 43 E8
Şubā Si. Arabia 27°10N 35°40E 44 E2
Subang Indonesia 6°34S 107°45E 37 G12
Subansiri → India 26°48N 93°50E 41 F18
Subarnarekha → India 22°34N 87°24E 43 H12
Subayhah Si. Arabia 30°2N 38°50E 44 D3
Subi Indonesia 2°58N 108°50E 36 D3
Subotica Serbia 46°6N 19°39E 23 A8
Suceava Romania 47°38N 26°16E 17 E14
Suchan = Partizansk
Russia 43°8N 133°9E 29 E14
Suchil Mexico 23°39N 103°54W 86 C4
Suchou = Suzhou
China 31°19N 120°38E 33 E13
Süchow = Xuzhou
China 34°18N 117°10E 35 G9
Suck → Ireland 53°17N 8°3W 10 C3
Sucre Bolivia 19°0S 65°15W 92 G5
Sucuriú → Brazil 20°47S 51°38W 93 H8
Sud, Pte. du Canada 49°3N 62°14W 73 C7
Sud-Kivu □
Dem. Rep. of the Congo 3°0S 28°30E 54 C2
Sud-Ouest, Pte. Canada 49°23N 63°36W 73 C7
Sud Ouest, Pte. Mauritius 20°28S 57°18E 53 d
Sudan U.S.A. 34°4N 102°31W 84 D3
Sudan ■ Africa 15°0N 30°0E 51 E11
Sudbury Canada 46°30N 81°0W 72 C3
Sudbury U.K. 52°2N 0°45E 13 E8
Sûdd Sudan 8°20N 30°0E 51 G12
Sudeten Mts. = Sudety
Europe 50°20N 16°45E 17 C9
Sudety Europe 50°20N 16°45E 17 C9
Suðuroy Færoe Is. 61°32N 6°50W 8 F9
Sudi Tanzania 10°11S 39°57E 55 E4
Sudirman, Pegunungan
Indonesia 4°30S 137°0E 37 E9
Sudong, Pulau Singapore 1°13N 103°44E 39 d
Sueca Spain 39°12N 0°21W 21 C5
Suemez I. U.S.A. 55°15N 133°20W 70 B2
Suez = El Suweis Egypt 29°58N 32°31E 51 C12
Suez, G. of = Suweis, Khalîg el
Egypt 28°40N 33°0E 51 C12
Suez Canal = Suweis, Qanâ es
Egypt 31°0N 32°20E 51 B12
Suffield Canada 50°12N 111°10W 70 C6
Suffolk U.S.A. 36°44N 76°35W 81 G15
Suffolk □ U.K. 52°16N 1°0E 13 E9
Sugar Grove U.S.A. 41°59N 79°20W 82 E5
Sugarcreek U.S.A. 41°25N 79°52W 82 E5
Sugarive → India 26°16N 86°24E 43 F12
Sugluk = Salluit
Canada 62°14N 75°38W 69 E16
Şuḥār Oman 24°20N 56°40E 45 E8
Sühbaatar Mongolia 50°17N 106°10E 32 A10
Sühbaatar □ Mongolia 45°30N 114°0E 34 B8
Suhl Germany 50°36N 10°42E 16 C6
Sui Pakistan 28°37N 69°19E 42 E3
Sui Xian China 34°25N 115°2E 34 G8
Suide China 37°30N 110°12E 34 F6
Suifenhe China 44°25N 131°10E 35 B16
Suihua China 46°32N 126°55E 33 B14
Suining Jiangsu, China 33°56N 117°58E 35 H9
Suining Sichuan, China 30°26N 105°35E 32 E10
Suiping China 33°10N 113°59E 34 H7
Suir → Ireland 52°16N 7°9W 10 D4
Suisse = Switzerland ■
Europe 46°30N 8°0E 20 C8
Suisun City U.S.A. 38°15N 122°2W 78 G4
Suiyang China 44°30N 130°56E 35 B16
Suizhong China 40°21N 120°20E 35 D11
Sujangarh India 27°42N 74°31E 42 F6
Sujiatun China 41°39N 123°20E 35 D12
Sukabumi Indonesia 6°56S 106°50E 37 G12
Sukadana Indonesia 1°10S 110°0E 36 E4
Sukagawa Japan 37°17N 140°23E 31 F10
Sukamade Indonesia 8°30S 113°52E 37 H15
Sukaraja Indonesia 2°28S 110°25E 36 E4
Sukawati Indonesia 8°35S 115°17E 37 K18
Sukch'ŏn N. Korea 39°22N 125°35E 35 E13
Sukhona → Russia 61°15N 46°39E 18 C6
Sukhothai Thailand 17°1N 99°49E 38 D2
Sukhumi = Sokhumi
Georgia 43°0N 41°0E 19 F7
Sukkur Pakistan 27°42N 68°54E 42 F3
Sukkur Barrage Pakistan 27°40N 68°50E 42 F3
Sukri → India 25°4N 71°43E 42 G4
Sukumo Japan 32°56N 132°44E 31 H6
Sukunka → Canada 55°45N 121°15W 70 B4
Sula, Kepulauan Indonesia 1°45S 125°0E 37 E7
Sulaco → Honduras 15°2N 87°44W 88 C2
Sulaiman Range
Pakistan 30°30N 69°50E 42 D3
Sūlār Iran 31°53N 51°54E 45 D6
Sulawesi Indonesia 2°0S 120°0E 37 E6
Sulawesi Barat □ Indonesia 3°0S 119°0E 37 E5
Sulawesi Sea = Celebes Sea
Indonesia 3°0N 123°0E 37 D6
Sulawesi Selatan □
Indonesia 2°30S 120°0E 37 E6
Sulawesi Tengah □
Indonesia 1°30S 121°0E 37 E6
Sulawesi Tenggara □
Indonesia 3°50S 122°0E 37 E6
Sulawesi Utara □ Indonesia 1°0N 124°0E 37 D6
Sulima S. Leone 6°58N 11°32W 50 G3
Sulina Romania 45°10N 29°40E 17 F15
Sulitjelma Norway 67°9N 16°3E 8 C17
Sullana Peru 4°52S 80°39W 92 D2
Sullivan Ill., U.S.A. 39°36N 88°37W 80 F9
Sullivan Ind., U.S.A. 39°6N 87°24W 80 F10
Sullivan Mo., U.S.A. 38°13N 91°10W 80 F8
Sullivan Bay Canada 50°55N 126°50W 70 C3
Sullivan I. = Lanbi Kyun
Burma 10°50N 98°20E 39 G2
Sullom Voe U.K. 60°27N 1°20W 11 A7
Sulphur La., U.S.A. 30°14N 93°23W 84 F8
Sulphur Okla., U.S.A. 34°31N 96°58W 84 D6
Sulphur Pt. Canada 60°56N 114°48W 70 A6
Sulphur Springs U.S.A. 33°8N 95°36W 84 E7
Sultan Canada 47°36N 82°47W 72 C3
Sultan U.S.A. 47°52N 121°49W 78 C5
Sultan Hamud Kenya 2°1S 37°22E 54 C4
Sultanābād = Arāk Iran 34°0N 49°40E 45 C6
Sultanpur Mad. P., India 23°9N 77°56E 42 H8
Sultanpur Punjab, India 31°13N 75°11E 42 D6
Sultanpur Ut. P., India 26°18N 82°4E 43 F10
Sulu Arch. Phil. 6°0N 121°0E 37 C6
Sulu Sea E. Indies 8°0N 120°0E 37 C6
Sulūq Libya 31°44N 20°14E 51 B10
Sulzberger Ice Shelf
Antarctica 78°0S 150°0W 5 E13
Sumalata Indonesia 1°0N 122°31E 37 D6
Sumampa Argentina 29°25S 63°29W 94 B3
Sumatera Indonesia 0°40N 100°20E 36 D2
Sumatera Barat □
Indonesia 1°0S 101°0E 36 E2
Sumatera Selatan □
Indonesia 3°0S 104°0E 36 E2
Sumatera Utara □
Indonesia 2°30N 98°0E 36 D1
Sumatra = Sumatera
Indonesia 0°40N 100°20E 36 D2
Sumba Indonesia 9°45S 119°35E 37 F5
Sumba, Selat Indonesia 9°0S 118°40E 37 F5
Sumbawa Indonesia 8°26S 117°30E 36 F5
Sumbawa Besar
Indonesia 8°30S 117°26E 36 F5
Sumbawanga □ Tanzania 8°0S 31°30E 52 F6
Sumbe Angola 11°10S 13°48E 52 G2
Sumbu △ Zambia 8°43S 30°22E 55 D3
Sumburgh Hd. U.K. 59°52N 1°17W 11 B7
Sumdeo India 31°26N 78°44E 43 D8
Sumdo China 35°6N 78°41E 43 B8
Sumedang Indonesia 6°52S 107°55E 37 G12
Šumen = Shumen
Bulgaria 43°18N 26°55E 23 C12
Sumenep Indonesia 7°1S 113°52E 37 G15
Sumgait = Sumqayıt
Azerbaijan 40°34N 49°38E 19 F8
Summer L. U.S.A. 42°50N 120°45W 76 E3
Summerland Canada 49°32N 119°41W 70 D5
Summerside Canada 46°24N 63°47W 73 C7
Summersville U.S.A. 38°17N 80°51W 81 F13
Summerville Ga.,
U.S.A. 34°29N 85°21W 85 D12
Summerville S.C., U.S.A. 33°1N 80°11W 85 E14
Summit Lake Canada 54°20N 122°40W 70 C4
Summit Peak U.S.A. 37°21N 106°42W 77 H10
Sumner Iowa, U.S.A. 42°51N 92°6W 80 D7
Sumner Wash., U.S.A. 47°12N 122°14W 78 C4
Sumoto Japan 34°21N 134°54E 31 G7
Šumperk Czech Rep. 49°59N 16°59E 17 D9
Sumqayıt Azerbaijan 40°34N 49°38E 19 F8
Sumter U.S.A. 33°55N 80°21W 85 E14
Sumy Ukraine 50°57N 34°50E 19 D5
Sun City S. Africa 25°17S 27°3E 56 D4
Sun City Ariz., U.S.A. 33°35N 112°16W 77 K7
Sun City Calif., U.S.A. 33°42N 117°11W 79 M9
Sun City Center U.S.A. 27°43N 82°21W 85 H13
Sun Lakes U.S.A. 33°10N 111°52W 77 K8
Sun Valley U.S.A. 43°42N 114°21W 76 E6
Sunagawa Japan 43°29N 141°55E 30 C10
Sunan N. Korea 39°15N 125°40E 35 E13
Sunart, L. U.K. 56°42N 5°43W 11 E3
Sunburst U.S.A. 48°53N 111°55W 76 B8
Sunbury Australia 37°35S 144°44E 63 F3
Sunbury U.S.A. 40°52N 76°48W 83 F8
Sunchales Argentina 30°58S 61°35W 94 C3
Suncheon S. Korea 34°52N 127°31E 35 G14
Suncho Corral Argentina 27°55S 63°27W 94 B3
Sunch'on N. Korea 39°25N 125°56E 35 E13
Suncook U.S.A. 43°8N 71°27W 83 C13
Sunda, Selat Indonesia 6°20S 105°30E 36 F3
Sunda Is. Indonesia 5°0S 105°0E 64 H2
Sunda Str. = Sunda, Selat
Indonesia 6°20S 105°30E 36 F3
Sunda Trench = Java Trench
Ind. Oc. 9°0S 105°0E 36 F3
Sundance Canada 56°32N 94°4W 71 B10
Sundance U.S.A. 44°24N 104°23W 76 D11
Sundar Nagar India 31°32N 76°53E 42 D7
Sundarbans Asia 22°0N 89°0E 41 J16
Sundarbans △ India 22°0N 88°45E 43 J13
Sundargarh India 22°4N 84°5E 41 H14
Sundays = Sondags →
S. Africa 33°44S 25°51E 56 E4
Sunderland Canada 44°16N 79°4W 82 B5
Sunderland U.K. 54°55N 1°23W 12 C6
Sundown △ Australia 28°49S 151°38E 63 D5
Sundre Canada 51°49N 114°38W 70 C6
Sundsvall Sweden 62°23N 17°17E 8 E17
Sungai Acheh Malaysia 5°8N 100°30E 39 c
Sungai Kolok Thailand 6°2N 101°58E 39 J3
Sungai Lembing Malaysia 3°55N 103°3E 39 L4
Sungai Petani Malaysia 5°37N 100°30E 39 K3
Sungaigerong Indonesia 2°59S 104°52E 36 E2
Sungailiat Indonesia 1°51S 106°8E 36 E3
Sungaipenuh Indonesia 2°1S 101°20E 36 E2
Sungari = Songhua Jiang →
China 47°45N 132°30E 33 B15
Sunghua Chiang = Songhua
Jiang → China 47°45N 132°30E 33 B15
Sunland Park U.S.A. 31°50N 106°40W 77 L10
Sunndalsøra Norway 62°40N 8°33E 8 E13
Sunnyside U.S.A. 46°20N 120°0W 76 C4
Sunnyvale U.S.A. 37°23N 122°2W 78 H4
Sunrise Manor U.S.A. 36°12N 115°4W 79 J11
Suntar Russia 62°15N 117°30E 29 C12
Suomenselkä Finland 62°52N 24°0E 8 E21
Suomi = Finland ■ Europe 63°0N 27°0E 8 E22
Suomussalmi Finland 64°54N 29°10E 8 D23
Suoyarvi Russia 62°3N 32°20E 18 B5
Supai U.S.A. 36°15N 112°41W 77 H7
Supaul India 26°10N 86°40E 43 F12
Superior Ariz., U.S.A. 33°18N 111°6W 77 K8
Superior Mont., U.S.A. 47°12N 114°53W 76 C6
Superior Nebr., U.S.A. 40°1N 98°4W 80 E4
Superior Wis., U.S.A. 46°44N 92°6W 80 B7
Superior, L. N. Amer. 47°0N 87°0W 72 C2
Suphan Buri Thailand 14°14N 100°10E 38 E3
Suphan Dağı Turkey 38°54N 42°48E 44 B4
Supiori Indonesia 1°0S 136°0E 37 E9
Supung Shuiku China 40°35N 124°50E 35 D13
Sūq ash Shuyūkh Iraq 30°53N 46°28E 44 D5
Sūq Suwayq Si. Arabia 24°23N 38°27E 44 E3
Suqian China 33°54N 118°8E 35 H10
Suquṭra = Socotra Yemen 12°30N 54°0E 47 E5
Sūr Lebanon 33°19N 35°16E 46 B4
Şūr Oman 22°34N 59°32E 47 C6
Sur, Pt. U.S.A. 36°18N 121°54W 78 J5
Sura → Russia 56°6N 46°0E 18 C8
Surab Pakistan 28°25N 66°15E 42 E2
Surabaja = Surabaya
Indonesia 7°17S 112°45E 37 G15
Surabaya Indonesia 7°17S 112°45E 37 G15
Surakarta Indonesia 7°35S 110°48E 37 G14
Surat Australia 27°10S 149°6E 63 D4
Surat India 21°12N 72°55E 40 J8
Surat Thani Thailand 9°6N 99°20E 39 H2
Suratgarh India 29°18N 73°55E 42 E5
Şuraymilā Si. Arabia 25°7N 46°7E 44 E5
Surendranagar India 22°45N 71°40E 42 H4
Surf U.S.A. 34°41N 120°36W 79 L6
Surfers Paradise
Australia 28°0S 153°25E 63 D5
Surgut Russia 61°14N 73°20E 28 C8
Suriapet India 17°10N 79°40E 40 L11
Surigao Phil. 9°47N 125°29E 37 C7
Surin Thailand 14°50N 103°34E 38 E4

Surinam = Suriname ■ *S. Amer.* 4°0N 56°0W **93** C7
Suriname ■ *S. Amer.* 4°0N 56°0W **93** C7
Suriname → *Suriname* 5°50N 55°15W **93** B7
Sūrmaq *Iran* 31°3N 52°48E **45** D7
Surrey *Canada* 49°7N 122°45W **78** A4
Surrey □ *U.K.* 51°15N 0°31W **13** F7
Sursand *India* 26°39N 85°43E **43** F11
Sursar → *India* 26°14N 87°3E **43** F12
Surt *Libya* 31°11N 16°39E **51** B9
Surt, Khalīj *Libya* 31°40N 18°30E **51** B9
Surtanahu *Pakistan* 26°22N 70°0E **42** F4
Surtsey *Iceland* 63°20N 20°30W **8** E3
Suruga-Wan *Japan* 34°45N 138°30E **31** G9
Susaki *Japan* 33°22N 133°17E **31** H6
Sūsangerd *Iran* 31°35N 48°6E **45** D6
Susanville *U.S.A.* 40°25N 120°39W **76** F3
Susner *India* 23°57N 76°5E **42** H7
Susquehanna *U.S.A.* 41°57N 75°36W **83** E9
Susquehanna → *U.S.A.* 39°33N 76°5W **83** G8
Susques *Argentina* 23°35S 66°25W **94** A2
Sussex *Canada* 45°45N 65°37W **73** C6
Sussex *U.S.A.* 41°13N 74°37W **83** E10
Sussex, E. □ *U.K.* 51°0N 0°20E **13** G8
Sussex, W. □ *U.K.* 51°0N 0°30W **13** G7
Sustut → *Canada* 56°20N 127°30W **70** B3
Susuman *Russia* 62°47N 148°10E **29** C15
Susunu *Indonesia* 3°7S 133°39E **37** E8
Susurluk *Turkey* 39°54N 28°8E **23** E13
Sutay Uul *Asia* 46°35N 93°38E **32** B7
Sutherland *S. Africa* 32°24S 20°40E **56** E3
Sutherland *U.K.* 58°12N 4°50W **11** C4
Sutherland *U.S.A.* 41°10N 101°8W **80** E3
Sutherland Falls *N.Z.* 44°48S 167°46E **59** F1
Sutherlin *U.S.A.* 43°23N 123°19W **76** E2
Suthri *India* 23°3N 68°55E **42** H3
Sutlej → *Pakistan* 29°23N 71°3E **42** E4
Sutter *U.S.A.* 39°10N 121°45W **78** F5
Sutter Buttes *U.S.A.* 39°12N 121°49W **78** F5
Sutter Creek *U.S.A.* 38°24N 120°48W **78** G6
Sutton *Canada* 45°6N 72°37W **83** A12
Sutton *Nebr., U.S.A.* 40°36N 97°52W **80** E5
Sutton *W. Va., U.S.A.* 38°40N 80°43W **81** F13
Sutton → *Canada* 55°15N 83°45W **72** A3
Sutton Coldfield *U.K.* 52°35N 1°49W **13** E6
Sutton in Ashfield *U.K.* 53°8N 1°16W **12** D6
Sutton L. *Canada* 54°15N 84°42W **72** B3
Suttor → *Australia* 21°36S 147°2E **62** C4
Suttsu *Japan* 42°48N 140°14E **30** C10
Sutwik I. *U.S.A.* 56°34N 157°12W **74** D8
Suva *Fiji* 18°6S 178°30E **59** a
Suva Planina *Serbia* 43°10N 22°5E **23** C10
Suvorov Is. = Suwarrow Is. *Cook Is.* 13°15S 163°5W **65** J11
Suwałki *Poland* 54°8N 22°59E **17** A12
Suwannaphum *Thailand* 15°33N 103°47E **38** E4
Suwannee → *U.S.A.* 29°17N 83°10W **85** G13
Suwanose-Jima *Japan* 29°38N 129°43E **31** K4
Suwarrow Is. *Cook Is.* 13°15S 163°5W **65** J11
Suweis, Khalîg el *Egypt* 28°40N 33°0E **51** C12
Suweis, Qanâ es *Egypt* 31°0N 32°20E **51** B12
Suwon *S. Korea* 37°17N 127°1E **35** F14
Suzdal *Russia* 56°29N 40°26E **18** C7
Suzhou *Anhui, China* 33°41N 116°59E **34** H9
Suzhou *Jiangsu, China* 31°19N 120°38E **33** E13
Suzu *Japan* 37°25N 137°17E **31** F8
Suzu-Misaki *Japan* 37°31N 137°21E **31** F8
Suzuka *Japan* 34°55N 136°36E **31** G8
Svalbard *Arctic* 78°0N 17°0E **4** B8
Svalbard Radio = Longyearbyen *Svalbard* 78°13N 15°40E **4** B8
Svappavaara *Sweden* 67°40N 21°3E **8** C19
Svartisen *Norway* 66°40N 13°59E **8** C15
Svay Chek *Cambodia* 13°48N 102°58E **38** F4
Svay Rieng *Cambodia* 11°9N 105°45E **39** G5
Svealand *Sweden* 60°20N 15°0E **9** F16
Sveg *Sweden* 62°2N 14°21E **8** E16
Svendborg *Denmark* 55°4N 10°35E **9** J14
Sverdrup Chan. *Canada* 79°56N 96°25W **69** B12
Sverdrup Is. *Canada* 79°0N 97°0W **66** B10
Sverige = Sweden ■ *Europe* 57°0N 15°0E **9** H16
Svetlaya *Russia* 46°33N 138°18E **30** A9
Svetlogorsk = Svyetlahorsk *Belarus* 52°38N 29°46E **17** B15
Svir → *Russia* 60°30N 32°48E **18** B5
Svishtov *Bulgaria* 43°36N 25°23E **23** C11
Svislach *Belarus* 53°3N 24°2E **17** B13
Svizzera = Switzerland ■ *Europe* 46°30N 8°0E **20** C8
Svobodnyy *Russia* 51°20N 128°0E **29** D13
Svolvær *Norway* 68°15N 14°34E **8** B16
Svyetlahorsk *Belarus* 52°38N 29°46E **17** B15
Swabian Alps = Schwäbische Alb *Germany* 48°20N 9°30E **16** D5
Swaffham *U.K.* 52°39N 0°42E **13** E8
Swains I. *Amer. Samoa* 11°11S 171°4W **65** J11
Swainsboro *U.S.A.* 32°36N 82°20W **85** E13
Swakop → *Namibia* 22°38S 14°36E **56** C2
Swakopmund *Namibia* 22°37S 14°30E **56** C1
Swale → *U.K.* 54°5N 1°20W **12** C6
Swan → *Australia* 32°3S 115°45E **61** F2
Swan → *Canada* 52°30N 100°45W **71** C8
Swan Hill *Australia* 35°20S 143°33E **63** F3
Swan Hills *Canada* 54°43N 115°24W **70** C5
Swan Is. = Santanilla, Is. *Honduras* 17°22N 83°57W **88** C3
Swan L. *Man., Canada* 52°30N 100°40W **71** C8
Swan L. *Ont., Canada* 54°16N 91°11W **72** B1
Swan Ra. *U.S.A.* 48°0N 113°45W **76** C7
Swan River *Canada* 52°10N 101°16W **71** C8
Swanage *U.K.* 50°36N 1°58W **13** G6
Swansea *Australia* 42°8S 148°4E **63** G4
Swansea *Canada* 43°38N 79°28W **82** C5
Swansea *U.K.* 51°37N 3°57W **13** F4
Swansea □ *U.K.* 51°38N 4°3W **13** F3
Swartberge *S. Africa* 33°20S 22°0E **56** E3
Swartmodder *S. Africa* 28°1S 20°32E **56** D3
Swartnossob → *Namibia* 23°8S 18°42E **56** C2
Swartruggens *S. Africa* 25°39S 26°42E **56** D4
Swastika *Canada* 48°7N 80°6W **72** C3
Swat → *Pakistan* 34°40N 72°5E **43** B5
Swatow = Shantou *China* 23°18N 116°40E **33** G12
Swaziland ■ *Africa* 26°30S 31°30E **57** D5
Sweden ■ *Europe* 57°0N 15°0E **9** H16
Sweet Grass *U.S.A.* 48°59N 111°58W **76** B8
Sweet Home *U.S.A.* 44°24N 122°44W **76** D2
Sweetwater *Nev., U.S.A.* 38°27N 119°9W **78** G7
Sweetwater *Tenn., U.S.A.* 35°36N 84°28W **85** D12
Sweetwater *Tex., U.S.A.* 32°28N 100°25W **84** E4
Sweetwater → *U.S.A.* 42°31N 107°2W **76** E10
Swellendam *S. Africa* 34°1S 20°26E **56** E3
Świdnica *Poland* 50°50N 16°30E **17** C9
Świdnik *Poland* 51°13N 22°39E **17** C12
Świebodzin *Poland* 52°15N 15°31E **16** B8
Świecie *Poland* 53°25N 18°30E **17** B10
Swift Current *Canada* 50°20N 107°45W **71** C7
Swift Current → *Canada* 50°38N 107°44W **71** C7
Swilly, L. *Ireland* 55°12N 7°33W **10** A4
Swindon *U.K.* 51°34N 1°46W **13** F6
Swindon □ *U.K.* 51°34N 1°46W **13** F6
Swinemünde = Świnoujście *Poland* 53°54N 14°16E **16** B8
Swinford *Ireland* 53°57N 8°58W **10** C3
Świnoujście *Poland* 53°54N 14°16E **16** B8
Switzerland ■ *Europe* 46°30N 8°0E **20** C8
Swords *Ireland* 53°28N 6°13W **10** C5
Swoyerville *U.S.A.* 41°18N 75°53W **83** E9
Sydenham → *Canada* 42°33N 82°25W **82** D2
Sydney *Australia* 33°52S 151°12E **63** E5
Sydney *Canada* 46°7N 60°7W **73** C7
Sydney L. *Canada* 50°41N 94°25W **71** C10
Sydney Mines *Canada* 46°18N 60°15W **73** C7
Sydprøven = Alluitsup Paa *Greenland* 60°30N 45°35W **4** C5
Sydra, G. of = Surt, Khalīj *Libya* 31°40N 18°30E **51** B9
Sykesville *U.S.A.* 41°3N 78°50W **82** E6
Syktyvkar *Russia* 61°45N 50°40E **18** B9
Sylacauga *U.S.A.* 33°10N 86°15W **85** E11
Sylarna *Sweden* 63°2N 12°13E **8** E15
Sylhet *Bangla.* 24°54N 91°52E **41** G17
Sylhet □ *Bangla.* 24°50N 91°50E **41** G17
Sylt *Germany* 54°54N 8°22E **16** A5
Sylvan Beach *U.S.A.* 43°12N 75°44W **83** C9
Sylvan Lake *Canada* 52°20N 114°3W **70** C6
Sylvania *Ga., U.S.A.* 32°45N 81°38W **85** E14
Sylvania *Ohio, U.S.A.* 41°43N 83°42W **81** E12
Sylvester *U.S.A.* 31°32N 83°50W **85** F13
Sym *Russia* 60°20N 88°18E **28** C9
Synnot Ra. *Australia* 16°30S 125°20E **60** C4
Syowa *Antarctica* 68°50S 12°0E **5** C5
Syracuse *Kans., U.S.A.* 37°59N 101°45W **80** G3
Syracuse *N.Y., U.S.A.* 43°3N 76°9W **83** C8
Syracuse *Nebr., U.S.A.* 40°39N 96°11W **80** E5
Syrdarya → *Kazakhstan* 46°3N 61°0E **28** E7
Syria ■ *Asia* 35°0N 38°0E **44** C3
Syrian Desert = Shām, Bādiyat ash *Asia* 32°0N 40°0E **44** C3
Syros = Ermoupoli *Greece* 37°28N 24°57E **23** F11
Syzran *Russia* 53°12N 48°30E **18** D8
Szczecin *Poland* 53°27N 14°27E **16** B8
Szczecinek *Poland* 53°43N 16°41E **17** B9
Szczeciński, Zalew = Stettiner Haff *Germany* 53°47N 14°15E **16** B8
Szczytno *Poland* 53°33N 21°0E **17** B11
Szechwan = Sichuan □ *China* 30°30N 103°0E **32** E9
Szeged *Hungary* 46°16N 20°10E **17** E11
Székesfehérvár *Hungary* 47°15N 18°25E **17** E10
Szekszárd *Hungary* 46°22N 18°42E **17** E10
Szentes *Hungary* 46°39N 20°21E **17** E11
Szolnok *Hungary* 47°10N 20°15E **17** E11
Szombathely *Hungary* 47°14N 16°38E **17** E9

T

Ta Khli *Thailand* 15°15N 100°21E **38** E3
Ta Lai *Vietnam* 11°24N 107°23E **39** G6
Ta Phraya △ *Thailand* 14°11N 102°49E **38** E4
Tabacal *Argentina* 23°15S 64°15W **94** A3
Tabaco *Phil.* 13°22N 123°44E **37** B6
Ţābah *Si. Arabia* 26°55N 42°38E **44** E4
Tabanan *Indonesia* 8°32S 115°8E **37** K18
Ţabas *Khorāsān, Iran* 32°48N 60°12E **45** C9
Ţabas *Yazd, Iran* 33°35N 56°55E **45** C8
Tabasará, Serranía de *Panama* 8°35N 81°40W **88** E3
Tabasco □ *Mexico* 18°0N 92°40W **87** D6
Tabāsīn *Iran* 31°12N 57°54E **45** D8
Tabatinga, Serra da *Brazil* 10°30S 44°0W **93** F10
Taber *Canada* 49°47N 112°8W **70** D6
Tablas I. *Phil.* 12°25N 122°2E **37** B6
Table, Pte. de la *Réunion* 21°14S 55°48E **53** c
Table B. *Canada* 53°40N 56°25W **73** B8
Table B. *S. Africa* 33°35S 18°25E **56** E2
Table Mountain △ *S. Africa* 34°20S 18°28E **56** E2
Table Mt. *S. Africa* 33°58S 18°26E **56** E2
Table Rock L. *U.S.A.* 36°36N 93°19W **80** G7
Tabletop, Mt. *Australia* 23°24S 147°11E **62** C4
Tábor *Czech Rep.* 49°25N 14°39E **16** D8
Tabora *Tanzania* 5°2S 32°50E **54** D3
Tabora □ *Tanzania* 5°0S 33°0E **54** D3
Tabou *Ivory C.* 4°30N 7°20W **50** H4
Tabrīz *Iran* 38°7N 46°20E **44** B5
Tabuaeran *Kiribati* 3°51N 159°22W **65** G12
Tabūk *Si. Arabia* 28°23N 36°36E **44** D3
Tabūk □ *Si. Arabia* 27°40N 36°50E **44** E3
Tacámbaro de Codallos *Mexico* 19°14N 101°28W **86** D4
Tacheng *China* 46°40N 82°58E **32** B5
Tachilek *Burma* 20°26N 99°52E **38** B2
Tach'ing Shan = Daqing Shan *China* 40°40N 111°0E **34** D6
Tacloban *Phil.* 11°15N 124°58E **37** B6
Tacna *Peru* 18°0S 70°20W **92** G4
Tacoma *U.S.A.* 47°14N 122°26W **78** C4
Tacuarembó *Uruguay* 31°45S 56°0W **95** C4
Tademaït, Plateau du *Algeria* 28°30N 2°30E **50** C6
Tadjoura *Djibouti* 11°50N 42°55E **47** E3
Tadmor *N.Z.* 41°27S 172°45E **59** D4
Tadoule L. *Canada* 58°36N 98°20W **71** B9
Tadoussac *Canada* 48°11N 69°42W **73** C6
Tadzhikistan = Tajikistan ■ *Asia* 38°30N 70°0E **28** F8
Taegu = Daegu *S. Korea* 35°50N 128°37E **35** G15
Taegwan *N. Korea* 40°13N 125°12E **35** D13
Taejŏn = Daejeon *S. Korea* 36°20N 127°28E **35** F14
Taen, Ko *Thailand* 9°22N 99°57E **39** b
Tafalla *Spain* 42°30N 1°41W **21** A5
Tafelbaai = Table B. *S. Africa* 33°35S 18°25E **56** E2
Tafermaar *Indonesia* 6°47S 134°10E **37** F8
Tafí Viejo *Argentina* 26°43S 65°17W **94** B2
Tafīhān *Iran* 29°25N 52°39E **45** D7
Tafresh *Iran* 34°45N 49°57E **45** C6
Taft *Iran* 31°45N 54°14E **45** D7
Taft *Phil.* 11°57N 125°30E **37** B7
Taft *U.S.A.* 35°8N 119°28W **79** K7
Taftan *Pakistan* 29°0N 61°30E **40** E2
Taftān, Kūh-e *Iran* 28°40N 61°0E **45** D9
Taga *Samoa* 13°46S 172°28W **59** b
Taga Dzong *Bhutan* 27°5N 89°55E **41** F16
Taganrog *Russia* 47°12N 38°50E **19** E6
Tagbilaran *Phil.* 9°39N 123°51E **37** C6
Tagish *Canada* 60°19N 134°16W **70** A2
Tagish L. *Canada* 60°10N 134°20W **70** A2
Tagliamento → *Italy* 45°38N 13°6E **22** B5
Tagomago *Spain* 39°2N 1°39E **24** B8
Taguatinga *Brazil* 12°27S 46°22W **93** F10
Tagum *Phil.* 7°33N 125°53E **37** C7
Tagus = Tejo → *Europe* 38°40N 9°24W **21** C1
Tahakopa *N.Z.* 46°30S 169°23E **59** G2
Tahan, Gunung *Malaysia* 4°34N 102°17E **39** K4
Tahat *Algeria* 23°18N 5°33E **50** D7
Tāherī *Iran* 27°43N 52°20E **45** E7
Tahiti *French Polynesia* 17°37S 149°27W **59** d
Tahiti, I. *Tahiti* 17°37S 149°27W **59** d
Tahlequah *U.S.A.* 35°55N 94°58W **84** D7
Tahoe, L. *U.S.A.* 39°6N 120°2W **78** G6
Tahoe City *U.S.A.* 39°10N 120°9W **78** F6
Tahoka *U.S.A.* 33°10N 101°48W **84** E4
Taholah *U.S.A.* 47°21N 124°17W **78** C2
Tahoua *Niger* 14°57N 5°16E **50** F7
Tahrūd *Iran* 29°26N 57°49E **45** D8
Tahsis *Canada* 49°55N 126°40W **70** D3
Tahta *Egypt* 26°44N 31°32E **51** C12
Tahulandang *Indonesia* 2°27N 125°23E **37** D7
Tahuna *Indonesia* 3°38N 125°30E **37** D7
Tai Hu *China* 31°5N 120°10E **33** E13
Tai Mo Shan *China* 22°25N 114°7E **33** a
Tai O *China* 22°15N 113°52E **33** a
Tai Pang Wan *China* 22°33N 114°24E **33** a
Tai Po *China* 22°27N 114°10E **33** a
Tai Rom Yen △ *Thailand* 8°45N 99°30E **39** H2
Tai Shan *China* 36°25N 117°20E **35** F9
Tai Yue Shan = Lantau I. *China* 22°15N 113°56E **33** a
Tai'an *China* 36°12N 117°8E **35** F9
Taiarapu, Presqu'île de *Tahiti* 17°47S 149°14W **59** d
Taibai Shan *China* 33°57N 107°45E **34** H4
Taibei = T'aipei *Taiwan* 25°4N 121°29E **33** F13
Taibique *Canary Is.* 27°42N 17°58W **24** G2
Taibus Qi *China* 41°54N 115°22E **34** D8
T'aichung *Taiwan* 24°9N 120°37E **33** G13
Taieri → *N.Z.* 46°3S 170°12E **59** G3
Taigu *China* 37°28N 112°30E **34** F7
Taihang Shan *China* 36°0N 113°30E **34** G7
Taihape *N.Z.* 39°41S 175°48E **59** C5
Taihe *China* 33°20N 115°42E **34** H8
Taikang *China* 34°5N 114°50E **34** G8
Tailem Bend *Australia* 35°12S 139°29E **63** F2
Taimyr Peninsula = Taymyr, Poluostrov *Russia* 75°0N 100°0E **29** B11
Tain *U.K.* 57°49N 4°4W **11** D4
T'ainan *Taiwan* 23°0N 120°10E **33** G13
Taipa *China* 22°10N 113°35E **33** a
T'aipei *Taiwan* 25°4N 121°29E **33** F13
Taiping *Malaysia* 4°51N 100°44E **39** K3
Taipingchuan *China* 44°23N 123°11E **35** B12
Taipingzhen *China* 33°35N 111°42E **34** H6
Tairbeart = Tarbert *U.K.* 57°54N 6°49W **11** D2
Taita Hills *Kenya* 3°25S 38°15E **54** C4
Taitao, Pen. de *Chile* 46°30S 75°0W **96** F2
T'aitung *Taiwan* 22°43N 121°4E **33** G13
Taivalkoski *Finland* 65°33N 28°12E **8** D23
Taiwan ■ *Asia* 23°30N 121°0E **33** G13
Taiwan Strait *Asia* 24°40N 120°0E **33** G13
Taiyiba *Israel* 32°36N 35°27E **46** C4
Taiyuan *China* 37°52N 112°33E **34** F7
Taizhong = T'aichung *Taiwan* 24°9N 120°37E **33** G13
Taizhou *China* 32°28N 119°55E **33** E12
Ta'izz *Yemen* 13°35N 44°2E **47** E3
Tājābād *Iran* 30°2N 54°24E **45** D7
Tajikistan ■ *Asia* 38°30N 70°0E **28** F8
Tajima *Japan* 37°12N 139°46E **31** F9
Tajo = Tejo → *Europe* 38°40N 9°24W **21** C1
Tajrīsh *Iran* 35°48N 51°25E **45** C6
Tak *Thailand* 16°52N 99°8E **38** D2
Takāb *Iran* 36°24N 47°7E **44** B5
Takachiho *Japan* 32°42N 131°18E **31** H5
Takachu *Botswana* 22°37S 21°58E **56** C3
Takada *Japan* 37°7N 138°15E **31** F9
Takahagi *Japan* 36°43N 140°45E **31** F10
Takaka *N.Z.* 40°51S 172°50E **59** D4
Takamaka *Seychelles* 4°50S 55°30E **53** b
Takamatsu *Japan* 34°20N 134°5E **31** G7
Takaoka *Japan* 36°47N 137°0E **31** F8
Takapuna *N.Z.* 36°47S 174°47E **59** B5
Takasaki *Japan* 36°20N 139°0E **31** F9
Takatsuki *Japan* 34°51N 135°37E **31** G7
Takaungu *Kenya* 3°38S 39°52E **54** C4
Takayama *Japan* 36°18N 137°11E **31** F8
Take-Shima *Japan* 30°49N 130°26E **31** J5
Takefu *Japan* 35°50N 136°10E **31** G8
Takengon *Indonesia* 4°45N 96°50E **36** D1
Takeo *Cambodia* 10°59N 104°47E **39** G5
Takeo *Japan* 33°12N 130°1E **31** H5
Takeshima = Tokdo *Asia* 37°15N 131°52E **31** F5
Tākestān *Iran* 36°0N 49°40E **45** C6
Taketa *Japan* 32°58N 131°24E **31** H5
Takh *India* 33°6N 77°32E **43** C7
Takhmau *Cambodia* 11°29N 104°57E **39** G5
Takht-Sulaiman *Pakistan* 31°40N 69°58E **42** D3
Takikawa *Japan* 43°33N 141°54E **30** C10
Takla L. *Canada* 55°15N 125°45W **70** B3
Takla Landing *Canada* 55°30N 125°50W **70** B3
Takla Makan *China* 38°0N 83°0E **32** D5
Taklamakan Shamo = Takla Makan *China* 38°0N 83°0E **32** D5
Taksimo *Russia* 56°20N 114°52E **29** D12
Taku → *Canada* 58°30N 133°50W **70** B2
Takua Thung *Thailand* 8°24N 98°27E **39** a
Tal Halāl *Iran* 28°54N 55°1E **45** D7
Tala *Uruguay* 34°21S 55°46W **95** C4
Talagang *Pakistan* 32°55N 72°25E **42** C5
Talagante *Chile* 33°40S 70°50W **94** C1
Talamanca, Cordillera de *Cent. Amer.* 9°20N 83°20W **88** E3
Talampaya △ *Argentina* 29°43S 67°42W **94** B2
Talara *Peru* 4°38S 81°18W **92** D2
Talas *Kyrgyzstan* 42°30N 72°13E **28** E8
Talâta *Egypt* 30°36N 32°20E **46** E1
Talaud, Kepulauan *Indonesia* 4°30N 126°50E **37** D7
Talaud Is. = Talaud, Kepulauan *Indonesia* 4°30N 126°50E **37** D7
Talavera de la Reina *Spain* 39°55N 4°46W **21** C3
Talayan *Phil.* 6°52N 124°24E **37** C6
Talbandh *India* 22°3N 86°20E **43** H12
Talbot, C. *Australia* 13°48S 126°43E **60** B4
Talbragar → *Australia* 32°12S 148°37E **63** E4
Talca *Chile* 35°28S 71°40W **94** D1
Talcahuano *Chile* 36°40S 73°10W **94** D1
Talcher *India* 21°0N 85°18E **41** J14
Taldy Kurgan = Taldyqorghan *Kazakhstan* 45°10N 78°45E **32** B4
Taldyqorghan *Kazakhstan* 45°10N 78°45E **32** B4
Tālesh *Iran* 37°58N 48°58E **45** B6
Tālesh, Kūhhā-ye *Iran* 37°42N 48°55E **45** B6
Tali Post *Sudan* 5°55N 30°44E **51** G12
Taliabu *Indonesia* 1°50S 125°0E **37** E6
Talibon *Phil.* 10°9N 124°20E **37** B6
Talihina *U.S.A.* 34°45N 95°3W **84** D7
Taliwang *Indonesia* 8°50S 116°55E **36** F5
Tall 'Afar *Iraq* 36°22N 42°27E **44** B4
Tall Kalakh *Syria* 34°41N 36°15E **46** A5
Talladega *U.S.A.* 33°26N 86°6W **85** E11
Tallahassee *U.S.A.* 30°27N 84°17W **85** F12
Tallangatta *Australia* 36°15S 147°19E **63** F4
Tallering Pk. *Australia* 28°6S 115°37E **61** E2
Talli *Pakistan* 29°32N 68°8E **42** E3
Tallinn *Estonia* 59°22N 24°48E **9** G21
Tallmadge *U.S.A.* 41°6N 81°27W **82** E3
Tallulah *U.S.A.* 32°25N 91°11W **84** E9
Talnakh *Russia* 69°29N 88°22E **29** C9
Taloyoak *Canada* 69°32N 93°32W **68** D13
Talpa de Allende *Mexico* 20°23N 104°51W **86** C4
Talparo *Trin. & Tob.* 10°30N 61°17W **93** K15
Talsi *Latvia* 57°10N 22°30E **9** H20
Taltal *Chile* 25°23S 70°33W **94** B1
Taltson → *Canada* 61°24N 112°46W **70** A6
Talwood *Australia* 28°29S 149°29E **63** D4
Talyawalka Cr. → *Australia* 32°28S 142°22E **63** E3
Tam Dao △ *Vietnam* 21°45N 105°45E **38** B5
Tam Ky *Vietnam* 15°34N 108°29E **38** E7
Tam Quan *Vietnam* 14°35N 109°3E **38** E7
Tama *U.S.A.* 41°58N 92°35W **80** E7
Tama Abu, Banjaran *Malaysia* 3°50N 115°5E **36** D5
Tamale *Ghana* 9°22N 0°50W **50** G5
Taman Negara △ *Malaysia* 4°38N 102°26E **39** K4
Tamano *Japan* 34°29N 133°59E **31** G6
Tamanrasset *Algeria* 22°50N 5°30E **50** D7
Tamaqua *U.S.A.* 40°48N 75°58W **83** F9
Tamar → *U.K.* 50°27N 4°15W **13** G3
Tamarin *Mauritius* 20°19S 57°20E **53** d
Tamarinda *Spain* 39°55N 3°49E **24** B10
Tamashima *Japan* 34°32N 133°40E **31** G6
Tamatave = Toamasina *Madag.* 18°10S 49°25E **57** B8
Tamaulipas □ *Mexico* 24°0N 98°45W **87** C5
Tamaulipas, Sierra de *Mexico* 23°30N 98°20W **87** C5
Tamazula *Mexico* 24°57N 106°57W **86** C3
Tamazunchale *Mexico* 21°16N 98°47W **87** C5
Tambach *Kenya* 0°36N 35°31E **54** B4
Tambacounda *Senegal* 13°45N 13°40W **50** F3
Tambelan, Kepulauan *Indonesia* 1°0N 107°30E **36** D3
Tambellup *Australia* 34°4S 117°37E **61** F2
Tambo *Australia* 24°54S 146°14E **62** C4
Tambo de Mora *Peru* 13°30S 76°8W **92** F3
Tambohorano *Madag.* 17°30S 43°58E **57** B7
Tambora *Indonesia* 8°12S 118°5E **36** F5
Tamboura *C.A.R.* 5°10N 25°12E **54** A2
Tambov *Russia* 52°45N 41°28E **18** D7
Tambuku *Indonesia* 7°8S 113°40E **37** G15
Tâmega → *Portugal* 41°5N 8°21W **21**
Tamenglong *India* 25°0N 93°35E **41**
Tamiahua, L. de *Mexico* 21°35N 97°35W **87**
Tamil Nadu □ *India* 11°0N 77°0E **40**
Tamluk *India* 22°18N 87°58E **43**
Tammerfors = Tampere *Finland* 61°30N 23°50E **8**
Tampa *U.S.A.* 27°56N 82°27W **85**
Tampa, Tanjung *Indonesia* 8°55S 116°12E **37**
Tampa B. *U.S.A.* 27°50N 82°30W **85**
Tampere *Finland* 61°30N 23°50E **8**
Tampico *Mexico* 22°13N 97°51W **87**
Tampin *Malaysia* 2°28N 102°13E **39**
Tampoi *Malaysia* 1°30N 103°39E
Tamsagbulag *Mongolia* 47°14N 117°21E **33**
Tamu *Burma* 24°13N 94°12E **41**
Tamworth *Australia* 31°7S 150°58E **63**
Tamworth *Canada* 44°29N 77°0W **82**
Tamworth *U.K.* 52°39N 1°41W **13**
Tan An *Vietnam* 10°32N 106°25E **39**
Tan Chau *Vietnam* 10°48N 105°12E **39**
Tan Hiep *Vietnam* 10°27N 106°21E **39**
Tan-Tan *Morocco* 28°29N 11°1W **50**
Tan Yen *Vietnam* 22°4N 105°3E **38**
Tana → *Kenya* 2°32S 40°31E **54**
Tana → *Norway* 70°30N 28°14E **8**
Tana, L. *Ethiopia* 13°5N 37°30E **47**
Tana River Primate ⌓ *Kenya* 1°55S 40°7E **54**
Tanabe *Japan* 33°44N 135°22E **31**
Tanafjorden *Norway* 70°45N 28°25E **8**
Tanaga I. *U.S.A.* 51°48N 177°53W **74**
Tanah Merah *Malaysia* 5°48N 102°9E **39**
Tanahbala *Indonesia* 0°30S 98°30E **36**
Tanahgrogot *Indonesia* 1°55S 116°15E **36**
Tanahjampea *Indonesia* 7°10S 120°35E **37**
Tanahmasa *Indonesia* 0°12S 98°39E **36**
Tanahmerah *Indonesia* 6°5S 140°16E **37**
Tanakpur *India* 29°5N 80°7E **43**
Tanakura *Japan* 37°10N 140°20E **31**
Tanami Desert *Australia* 18°50S 132°0E **60**
Tanana *U.S.A.* 65°10N 152°4W **74**
Tanana → *U.S.A.* 65°10N 151°58W **68**
Tananarive = Antananarivo *Madag.* 18°55S 47°31E **57**
Tánaro → *Italy* 44°55N 8°40E **20**
Tancheng *China* 34°25N 118°20E **35**
Tanch'ŏn *N. Korea* 40°27N 128°54E **35**
Tanda *Ut. P., India* 26°33N 82°35E **43**
Tanda *Ut. P., India* 28°57N 78°56E **43**
Tandag *Phil.* 9°4N 126°9E **37**
Tandaia *Tanzania* 9°25S 34°15E **55**
Tandaué *Angola* 16°58S 18°5E **56**
Tandil *Argentina* 37°15S 59°6W **94**
Tandil, Sa. del *Argentina* 37°30S 59°0W **94**
Tandlianwala *Pakistan* 31°3N 73°9E **42**
Tando Adam *Pakistan* 25°45N 68°40E **42**
Tando Allahyar *Pakistan* 25°28N 68°43E **42**
Tando Bago *Pakistan* 24°47N 68°58E **42**
Tando Mohommed Khan *Pakistan* 25°8N 68°32E **42**
Tandou L. *Australia* 32°40S 142°5E **63**
Tandoureh △ *Iran* 37°50N 59°0E **45**
Tandragee *U.K.* 54°21N 6°24W **10**
Tane-ga-Shima *Japan* 30°30N 131°0E **31**
Taneatua *N.Z.* 38°4S 177°1E **59**
Tanen Tong Dan = Dawna Ra. *Burma* 16°30N 98°30E **38**
Tanezrouft *Algeria* 23°9N 0°11E **50**
Tang, Koh *Cambodia* 10°16N 103°7E **39**
Tang, Ra's-e *Iran* 25°21N 59°52E **45**
Tang Krasang *Cambodia* 12°34N 105°3E **38**
Tanga *Tanzania* 5°5S 39°2E **54**
Tanga □ *Tanzania* 5°20S 38°0E **54**
Tanganyika, L. *Africa* 6°40S 30°0E **54**
Tanger *Morocco* 35°50N 5°49W **50**
Tangerang *Indonesia* 6°11S 106°37E **37**
Tanggu *China* 39°2N 117°40E **35**
Tanggula Shan *China* 32°40N 92°10E **32**
Tanggula Shankou *China* 32°42N 92°27E **32**
Tanghe *China* 32°47N 112°50E **34**
Tanghla Range = Tanggula Shan *China* 32°40N 92°10E **32**
Tangier = Tanger *Morocco* 35°50N 5°49W **50**
Tangjia *China* 22°22N 113°35E
Tangjia Wan *China* 22°21N 113°36E
Tangorin *Australia* 21°47S 144°12E **62**
Tangorombohitr'i Makay *Madag.* 21°0S 45°15E **57**
Tangra Yumco *China* 31°0N 86°38E **32**
Tangshan *China* 39°38N 118°10E **35**
Tangtou *China* 35°28N 118°30E **35**
Tangyin *China* 35°54N 114°21E **34**
Tanimbar, Kepulauan *Indonesia* 7°30S 131°30E **37**
Tanimbar Is. = Tanimbar, Kepulauan *Indonesia* 7°30S 131°30E **37**
Taninthari = Tenasserim *Burma* 12°6N 99°3E **39**
Tanjay *Phil.* 9°30N 123°5E **37**
Tanjong Pelepas *Malaysia* 1°21N 103°33E
Tanjore = Thanjavur *India* 10°48N 79°12E **40**
Tanjung *Kalimantan Selatan, Indonesia* 2°10S 115°25E **36**
Tanjung *Nusa Tenggara Barat, Indonesia* 8°21S 116°9E **37**
Tanjung Malim *Malaysia* 3°42N 101°31E **39**
Tanjung Tokong *Malaysia* 5°28N 100°18E
Tanjungbalai *Indonesia* 2°55N 99°44E **36**
Tanjungbatu *Indonesia* 2°23N 118°3E **36**
Tanjungkarang Telukbetung = Bandar Lampung *Indonesia* 5°20S 105°10E **36**
Tanjungpandan *Indonesia* 2°43S 107°38E **36**

jungpinang Indonesia 1°5N 104°30E 36 D2
jungredeb Indonesia 2°9N 117°29E 36 D5
jungselor Indonesia 2°55N 117°25E 36 D5
k Pakistan 32°14N 70°25E 42 C4
khala India 21°58N 73°47E 42 J5
kwa-Karoo △
, Africa 32°14S 19°50E 56 E2
nersville U.S.A. 41°3N 75°18W 83 E9
nu Ola Asia 51°0N 94°0E 29 D10
num Sands
ustralia 23°57S 151°22E 62 C5
out Niger 14°50N 8°55E 50 F7
ta Egypt 30°45N 30°57E 51 B12
toyuca Mexico 21°21N 98°14W 87 C5
tung = Dandong
hina 40°10N 124°20E 35 D13
unda Australia 34°30S 139°0E 63 E2
zania ■ Africa 6°0S 34°0E 54 D3
zhou China 22°16N 113°28E 33 a
zilla → Canada 58°8N 130°43W 70 B2
, Ko Thailand 10°5N 99°52E 39 G2
'an = Taonan
hina 45°22N 122°40E 35 B12
'er He → China 45°45N 124°5E 35 B13
lanaro Madag. 25°2S 47°0E 57 D8
le China 38°48N 106°40E 34 E4
nan China 45°22N 122°40E 35 B12
s U.S.A. 36°24N 105°35W 77 H11
udenni Mali 22°40N 3°55W 50 D5
a Estonia 59°15N 25°50E 9 G21
a Shan = Daba Shan
hina 32°0N 109°0E 33 E10
achula Mexico 14°54N 92°17W 87 E6
ah Malaysia 4°12N 101°15E 39 K3
ajós → Brazil 2°24S 54°41W 93 D8
aktuan Indonesia 3°15N 97°10E 36 D1
anahoni → Suriname 4°20N 54°25W 93 C8
anui N.Z. 45°56S 169°18E 59 F2
auá → Brazil 5°40S 64°21W 92 E6
es Brazil 30°40S 51°23W 95 C5
eta Liberia 6°29N 8°52W 50 G4
han Hin Thailand 16°13N 100°26E 38 D3
irapecó, Serra
'enezuela 1°10N 65°0W 92 C6
o-Capara △ Venezuela 7°55N 71°15W 89 E5
ti → India 21°8N 72°41E 40 J8
uae-o-Uenuku N.Z. 42°0S 173°39E 59 E4
ul Group Phil. 5°35N 120°50E 37 C6
urucuará Brazil 0°24S 65°2W 92 D5
ţaq Iraq 35°53N 44°35E 44 C5
uara Brazil 29°36S 50°46W 95 B5
uari → Brazil 19°15S 57°17W 92 G7
a Australia 27°17S 150°31E 63 D5
a Canada 44°28N 81°9W 82 B3
a Russia 56°55N 74°24E 28 D8
a Zambia 16°58S 26°45E 55 F2
a → Montenegro 43°21N 18°51E 23 C8
ābulus Lebanon 34°31N 35°50E 46 A4
ābulus Libya 32°49N 13°7E 51 B8
ābulus N. Afr. 31°0N 13°0E 51 B8
adehi India 23°18N 79°21E 43 H8
ajalejo Canary Is. 28°12N 14°7W 24 F5
akan Indonesia 3°20N 117°35E 36 D5
akit, Mt. Kenya 2°2N 35°10E 54 B4
ama-Jima Japan 24°39N 124°42E 31 M2
an, Mys Russia 54°56N 19°59E 9 J18
anagar India 28°43N 74°50E 42 E6
anaki □ N.Z. 39°25S 174°30E 59 C5
anaki, Mt. N.Z. 39°17S 174°5E 59 C5
ancón Spain 40°1N 3°1W 21 B4
angire △ Tanzania 4°21S 36°7E 54 C4
ansay U.K. 57°54N 7°0W 11 D1
anto Italy 40°28N 17°14E 22 D7
anto, G. di Italy 40°8N 17°20E 22 D7
apacá Colombia 2°56S 69°46W 92 D5
apacá □ Chile 20°45S 69°30W 94 A2
apoto Peru 6°30S 76°20W 92 E3
arua Ra. N.Z. 40°45S 175°25E 59 D5
ashcha Ukraine 49°30N 30°31E 17 D16
auacá Brazil 8°6S 70°48W 92 E4
auacá → Brazil 6°42S 69°48W 92 E5
avao Tahiti 17°43S 149°18W 59 d
avao, Isthme de
ahiti 17°43S 149°19W 59 d
awa Kiribati 1°30N 173°0E 64 G9
awera N.Z. 39°2S 176°36E 59 C6
awera, L. N.Z. 38°13S 176°27E 59 C6
az Kazakhstan 42°54N 71°22E 32 C3
azona Spain 41°55N 1°43W 21 B5
bagatay, Khrebet
'azakhstan 48°0N 83°0E 32 B5
bat Ness U.K. 57°52N 3°47W 11 D5
bela Dam Pakistan 34°8N 72°52E 42 B5
bert Ireland 52°34N 9°22W 10 D2
bert Argyll & Bute,
.K. 55°52N 5°25W 11 F3
bert W. Isles, U.K. 57°54N 6°49W 11 D2
bes France 43°15N 0°3E 20 E4
boro U.S.A. 35°54N 77°32W 85 D16
coola Australia 30°44S 134°36E 63 E1
coon Australia 30°15S 146°43E 63 E4
ee Australia 31°50S 152°30E 63 E5
faya Morocco 27°55N 12°55W 50 C3
goviște Romania 44°55N 25°27E 17 F13
gu Jiu Romania 45°5N 23°19E 17 F12
gu Mureș Romania 46°31N 24°38E 17 E13
if U.A.E. 24°3N 53°46E 45 E7
ifa Spain 36°1N 5°36W 21 D3
ija Bolivia 21°30S 64°40W 94 A3
ija □ Bolivia 21°30S 63°30W 94 A3
iku → Indonesia 2°55S 138°26E 37 E9
im Basin = Tarim Pendi
'hina 40°0N 84°0E 32 C5
im He → China 39°30N 88°30E 32 D6
im Pendi China 40°0N 84°0E 32 C5
itatu → Indonesia 2°54S 138°27E 37 E9
ka → S. Africa 32°10S 26°0E 56 E4
ka La Bhutan 27°12N 89°44E 43 F13
kastad S. Africa 32°0S 26°16E 56 E4
khankut, Mys
'kraine 45°25N 32°30E 19 E5
Tarko Sale Russia 64°55N 77°50E 28 C8
Tarkwa Ghana 5°20N 2°0W 50 G5
Tarlac Phil. 15°29N 120°35E 37 A6
Tarma Peru 11°25S 75°45W 92 F3
Tarn → France 44°5N 1°6E 20 E4
Târnăveni Romania 46°19N 24°13E 17 E13
Tarnobrzeg Poland 50°35N 21°41E 17 C11
Tarnów Poland 50°3N 21°0E 17 C11
Tarnowskie Góry
Poland 50°27N 18°54E 17 C10
Ţārom Iran 28°11N 55°46E 45 D7
Taroom Australia 25°36S 149°48E 63 D4
Taroudannt Morocco 30°30N 8°52W 50 B4
Tarpon Springs U.S.A. 28°9N 82°45W 85 G13
Tarrafal C. Verde Is. 15°18N 23°39W 50 b
Tarragona Spain 41°5N 1°17E 21 B6
Tarraleah Australia 42°17S 146°26E 63 G4
Tarrasa = Terrassa Spain 41°34N 2°1E 21 B7
Tarrytown U.S.A. 41°4N 73°52W 83 E11
Tarshiha = Me'ona Israel 33°1N 35°15E 46 B4
Tarsus Turkey 36°58N 34°55E 44 B2
Tartagal Argentina 22°30S 63°50W 94 A3
Tartu Estonia 58°20N 26°44E 9 G22
Ţarţūs Syria 34°55N 35°55E 44 C2
Tarumizu Japan 31°29N 130°42E 31 J5
Tarutao = Ko Tarutao △
Thailand 6°31N 99°26E 39 J2
Tarutao, Ko Thailand 6°33N 99°40E 39 J2
Tarutung Indonesia 2°0N 98°54E 36 D1
Taseko → Canada 52°8N 123°45W 70 C4
Tash-Kömür Kyrgyzstan 41°40N 72°10E 28 E8
Tash-Kumyr = Tash-Kömür
Kyrgyzstan 41°40N 72°10E 28 E8
Tashauz = Dashoguz
Turkmenistan 41°49N 59°58E 28 E6
Tashi Chho Dzong = Thimphu
Bhutan 27°31N 89°45E 41 F16
Ţashk, Daryācheh-ye
Iran 29°45N 53°35E 45 D7
Tashkent = Toshkent
Uzbekistan 41°20N 69°10E 28 E7
Tashtagol Russia 52°47N 87°53E 28 D9
Tasiilaq Greenland 65°40N 37°20W 4 C6
Tasikmalaya Indonesia 7°18S 108°12E 37 G13
Tåsjön Sweden 64°15N 15°40E 8 D16
Taskan Russia 62°59N 150°20E 29 C16
Tasman B. N.Z. 40°59S 173°25E 59 D4
Tasman Basin Pac. Oc. 46°0S 158°0E 64 M7
Tasman Mts. N.Z. 41°3S 172°25E 59 D4
Tasman Pen. Australia 43°10S 148°0E 63 G4
Tasman Sea Pac. Oc. 36°0S 160°0E 58 E9
Tasmania □ Australia 42°0S 146°30E 63 G4
Tasmanian Wilderness World
Heritage Area △
Australia 43°0S 146°0E 63 G4
Tassili n'Ajjer Algeria 25°47N 8°1E 50 C7
Tassili-Oua-n-Ahaggar
Algeria 20°41N 5°30E 50 D7
Tat Ton △ Thailand 15°57N 102°2E 38 E4
Tata Morocco 29°46N 7°56W 50 C4
Tatabánya Hungary 47°32N 18°25E 17 E10
Tataouine Tunisia 32°57N 10°29E 51 B8
Tatar Republic = Tatarstan □
Russia 55°30N 51°30E 18 C9
Tatarbunary Ukraine 45°50N 29°39E 17 F15
Tatarsk Russia 55°14N 76°0E 28 D8
Tatarskiy Proliv Russia 50°0N 141°0E 29 E15
Tatarstan □ Russia 55°30N 51°30E 18 C9
Tatatua, Pte. Tahiti 17°44S 149°8W 59 d
Tateyama Japan 35°0N 139°50E 31 G9
Tathlina L. Canada 60°33N 117°39W 70 A5
Tathra Australia 36°44S 149°59E 63 F4
Tatinnai L. Canada 60°55N 97°40W 71 A9
Tatla Lake Canada 52°0N 124°20W 70 C4
Tatnam, C. Canada 57°16N 91°0W 71 B10
Tatra = Tatry Slovak Rep. 49°20N 20°0E 17 D11
Tatry Slovak Rep. 49°20N 20°0E 17 D11
Tatshenshini →
Canada 59°28N 137°45W 70 B1
Tatsuno Japan 34°52N 134°33E 31 G7
Tatta Pakistan 24°42N 67°55E 42 G2
Tatuí Brazil 23°25S 47°53W 95 A6
Tatum U.S.A. 33°16N 103°19W 77 K12
Tat'ung = Datong China 40°6N 113°18E 34 D7
Tatvan Turkey 38°31N 42°15E 44 B4
Ta'ū Amer. Samoa 14°15S 169°30W 59 b
Taubaté Brazil 23°0S 45°36W 95 A6
Tauern Austria 47°15N 12°40E 16 E7
Taumarunui N.Z. 38°53S 175°15E 59 C5
Taumaturgo Brazil 8°54S 72°51W 92 E4
Taung S. Africa 27°33S 24°47E 56 D3
Taungdwingyi Burma 20°1N 95°40E 41 J19
Taunggyi Burma 20°50N 97°0E 41 J20
Taungup Burma 18°51N 94°14E 41 K19
Taunsa Pakistan 30°42N 70°39E 42 D4
Taunsa Barrage Pakistan 30°42N 70°50E 42 D4
Taunton U.K. 51°1N 3°5W 13 F4
Taunton U.S.A. 41°54N 71°6W 83 E13
Taunus Germany 50°13N 8°34E 16 C5
Taupo N.Z. 38°41S 176°7E 59 C6
Taupo, L. N.Z. 38°46S 175°55E 59 C5
Tauragė Lithuania 55°14N 22°16E 9 J20
Tauranga N.Z. 37°42S 176°11E 59 B6
Tauranga Harb. N.Z. 37°30S 176°5E 59 B6
Taureau, Rés. Canada 46°46N 73°50W 72 C5
Taurianova Italy 38°21N 16°1E 22 E7
Taurus Mts. = Toros Dağları
Turkey 37°0N 32°30E 44 B2
Tautira Tahiti 17°44S 149°9W 59 d
Tauyskaya Guba
Russia 59°20N 150°20E 29 D16
Tavan Bogd Uul
Mongolia 49°10N 87°49E 32 B6
Tavastehus = Hämeenlinna
Finland 61°0N 24°28E 8 F21
Tavda Russia 58°7N 65°8E 28 D7
Tavda → Russia 57°47N 67°18E 28 D7
Taverner B. Canada 67°12N 72°25W 69 D17
Taveta Kenya 3°23S 37°37E 54 C4
Taveuni Fiji 16°51S 179°58W 59 a
Tavira Portugal 37°8N 7°40W 21 D2
Tavistock Canada 43°19N 80°50W 82 C4
Tavistock U.K. 50°33N 4°9W 13 G3
Tavoy Burma 14°2N 98°12E 38 E2
Tavoy Pt. Burma 13°32N 98°10E 38 F2
Tavrichanka Russia 43°18N 131°59E 30 C5
Tavua Fiji 17°37S 177°5E 59 a
Tavuki Fiji 19°7S 178°8E 59 a
Taw → U.K. 51°4N 4°4W 13 F3
Tawa → India 22°48N 77°48E 42 H8
Tawas City U.S.A. 44°16N 83°31W 81 C12
Tawau Malaysia 4°20N 117°55E 36 D5
Tawi-tawi I. Phil. 5°10N 120°15E 37 C6
Taxco Mexico 18°33N 99°36W 87 D5
Taxila Pakistan 33°42N 72°52E 42 C5
Taxkorgan Tajik Zizhixian
China 37°49N 75°14E 32 D4
Tay → U.K. 56°37N 3°38W 11 E5
Tay, Firth of U.K. 56°25N 3°8W 11 E5
Tay, L. Australia 32°55S 120°48E 61 F3
Tay, L. U.K. 56°32N 4°8W 11 E4
Tay ⌂ U.K. 56°43N 3°59W 11 E5
Tay Ninh Vietnam 11°20N 106°5E 39 G6
Tayabamba Peru 8°15S 77°16W 92 E3
Taygetos Oros Greece 37°0N 22°23E 23 F10
Taylakova Russia 59°13N 74°0E 28 D8
Taylor Canada 56°13N 120°40W 70 B4
Taylor Nebr., U.S.A. 41°46N 99°23W 80 E4
Taylor Pa., U.S.A. 41°23N 75°43W 83 E9
Taylor Tex., U.S.A. 30°34N 97°25W 84 F6
Taylor, Mt. U.S.A. 35°14N 107°37W 77 J10
Taylorville U.S.A. 39°33N 89°18W 80 F9
Taymā Si. Arabia 27°35N 38°45E 44 E3
Taymyr, Oz. Russia 74°20N 102°0E 29 B11
Taymyr, Poluostrov
Russia 75°0N 100°0E 29 B11
Taypaq Kazakhstan 49°0N 51°47E 19 E9
Tayport U.K. 56°27N 2°52W 11 E6
Tayrona △ Colombia 11°20N 74°2W 89 D5
Tayshet Russia 55°58N 98°1E 29 D10
Taytay Phil. 10°45N 119°30E 37 B5
Tāyyebād Iran 34°45N 60°45E 45 C9
Tāyyebād □ Iran 34°44N 60°44E 45 C9
Taz → Russia 67°32N 78°40E 28 C8
Taza Morocco 34°16N 4°6W 50 B5
Tāza Khurmātū Iraq 35°18N 44°20E 44 C5
Tazawa-Ko Japan 39°43N 140°40E 30 E10
Tazerbo Libya 25°45N 21°0E 51 C10
Tazin → Canada 59°48N 109°55W 71 B7
Tazin L. Canada 59°44N 108°42W 71 B7
Tazovskiy Russia 67°30N 78°44E 28 C8
Tbilisi Georgia 41°43N 44°50E 19 F7
Tchad = Chad ■ Africa 15°0N 17°15E 51 F8
Tchad, L. Chad 13°30N 14°30E 51 F8
Tch'eng-tou = Chengdu
China 30°38N 104°2E 32 E9
Tchentlo L. Canada 55°15N 125°0W 70 B4
Tchibanga Gabon 2°45S 11°0E 52 E2
Tch'ong-k'ing = Chongqing
China 29°35N 106°25E 32 F10
Tczew Poland 54°8N 18°50E 17 A10
Te Anau, L. N.Z. 45°15S 167°45E 59 F1
Te Aroha N.Z. 37°32S 175°44E 59 B5
Te Awamutu N.Z. 38°1S 175°20E 59 C5
Te Ika a Maui = North I.
N.Z. 38°0S 175°0E 59 C5
Te Kuiti N.Z. 38°20S 175°11E 59 C5
Te Puke N.Z. 37°46S 176°22E 59 B6
Te Waewae B. N.Z. 46°13S 167°33E 59 G1
Te Wai Pounamu = South I.
N.Z. 44°0S 170°0E 59 F3
Teague U.S.A. 31°38N 96°17W 84 F6
Teahupoo Tahiti 17°50S 149°16W 59 d
Teapa Mexico 17°33N 92°57W 87 D6
Tebakang Malaysia 1°6N 110°30E 36 D4
Teberau → Malaysia 1°30N 103°42E 39 d
Tébessa Algeria 35°22N 8°8E 50 A7
Tebicuary → Paraguay 26°36S 58°16W 94 B4
Tebingtinggi Riau,
Indonesia 1°0N 102°45E 36 E2
Tebingtinggi Sumatera Utara,
Indonesia 3°20N 99°9E 36 D1
Tecate Mexico 32°34N 116°38W 79 N10
Tecka Argentina 43°29S 70°48W 96 E2
Tecomán Mexico 18°55N 103°53W 86 D4
Tecopa U.S.A. 35°51N 116°13W 79 K10
Tecoripa Mexico 28°37N 109°57W 86 B3
Tecuala Mexico 22°23N 105°27W 86 C3
Tecuci Romania 45°51N 27°27E 17 F14
Tecumseh Canada 42°19N 82°54W 82 D2
Tecumseh Mich., U.S.A. 42°0N 83°57W 81 D12
Tecumseh Okla., U.S.A. 35°15N 96°56W 84 D6
Tedzhen = Tejen
Turkmenistan 37°23N 60°31E 28 F7
Tees → U.K. 54°37N 1°10W 12 C6
Tees B. U.K. 54°40N 1°9W 12 C6
Teeswater Canada 43°59N 81°17W 82 C3
Tefé Brazil 3°25S 64°50W 92 D6
Tegal Indonesia 6°52S 109°8E 37 G13
Tegallalang Indonesia 8°27S 115°17E 37 J18
Tegalsari Indonesia 8°25S 114°8E 37 J17
Tegid, L. = Bala, L. U.K. 52°53N 3°37W 12 E4
Tegucigalpa Honduras 14°5N 87°14W 88 D2
Tehachapi U.S.A. 35°8N 118°27W 79 K8
Tehachapi Mts. U.S.A. 35°0N 118°30W 79 L8
Teheran = Tehrān Iran 35°41N 51°25E 45 C6
Tehoru Indonesia 3°23S 129°30E 37 E7
Tehrān Iran 35°41N 51°25E 45 C6
Tehrān □ Iran 35°30N 51°30E 45 C6
Tehri India 30°23N 78°29E 43 D8
Tehuacán Mexico 18°27N 97°23W 87 D5
Tehuantepec Mexico 16°21N 95°13W 87 D5
Tehuantepec, G. de
Mexico 15°50N 95°12W 87 D5
Tehuantepec, Istmo de
Mexico 17°15N 94°30W 87 D6
Teide, Pico de Canary Is. 28°15N 16°38W 24 F3
Teifi → U.K. 52°5N 4°41W 13 E3
Teign → U.K. 50°32N 3°32W 13 G4
Teignmouth U.K. 50°33N 3°31W 13 G4
Tejakula Indonesia 8°8S 115°20E 37 J18
Tejam India 29°57N 80°11E 43 E9
Tejen Turkmenistan 37°23N 60°31E 28 F7
Tejen → Turkmenistan 37°24N 60°38E 45 B9
Tejo → Europe 38°40N 9°24W 21 C1
Tejon Pass U.S.A. 34°49N 118°53W 79 L8
Tekamah U.S.A. 41°47N 96°13W 80 E5
Tekapo, L. N.Z. 43°53S 170°33E 59 E3
Tekax Mexico 20°12N 89°17W 87 C7
Tekeli Kazakhstan 44°50N 79°0E 28 E8
Tekeze Dam Ethiopia 13°10N 38°45E 47 E2
Tekirdağ Turkey 40°58N 27°30E 23 D12
Tekkali India 18°37N 84°15E 41 K14
Tekoa U.S.A. 47°14N 117°4W 76 C5
Tekong Besar, Pulau
Singapore 1°25N 104°3E 39 d
Tel Aviv ✈ (TLV) Israel 32°5N 34°49E 46 C3
Tel Aviv-Yafo Israel 32°4N 34°48E 46 C3
Tel Lakhish Israel 31°34N 34°51E 46 D3
Tel Megiddo Israel 32°35N 35°11E 46 C4
Tela Honduras 15°40N 87°28W 88 C2
Telanaipura = Jambi
Indonesia 1°38S 103°30E 36 E2
Telavi Georgia 42°0N 45°30E 19 F8
Telde Canary Is. 27°59N 15°25W 24 G4
Telegraph Creek Canada 58°0N 131°10W 70 B2
Telekhany = Tsyelyakhany
Belarus 52°30N 25°46E 17 B13
Telemark Norway 59°15N 7°40E 9 G12
Telén Argentina 36°15S 65°31W 94 D2
Teleng Iran 25°47N 61°3E 45 E9
Teles Pires → Brazil 7°21S 58°3W 92 E7
Telescope Pk. U.S.A. 36°10N 117°5W 79 J9
Telford U.K. 52°40N 2°27W 13 E5
Telford and Wrekin □
U.K. 52°45N 2°27W 12 E5
Telkwa Canada 54°41N 127°5W 70 C3
Tell City U.S.A. 37°57N 86°46W 80 G10
Teller U.S.A. 65°16N 166°22W 74 B6
Tellicherry India 11°45N 75°30E 40 P9
Telluride U.S.A. 37°56N 107°49W 77 H10
Teloloapán Mexico 18°21N 99°51W 87 D5
Telpos Iz Russia 63°16N 59°13E 18 B10
Telsen Argentina 42°30S 66°50W 96 E3
Telšiai Lithuania 55°59N 22°14E 9 J20
Teluk Anson = Teluk Intan
Malaysia 4°3N 101°0E 39 K3
Teluk Bahang Malaysia 5°28N 100°13E 39 c
Teluk Betung = Bandar Lampung
Indonesia 5°20S 105°10E 36 F3
Teluk Intan Malaysia 4°3N 101°0E 39 K3
Teluk Kumbar Malaysia 5°18N 100°14E 39 c
Telukbutun Indonesia 4°13N 108°12E 36 D3
Telukdalam Indonesia 0°33N 97°50E 36 D1
Tema Ghana 5°41N 0°0 50 G5
Temagami, L. Canada 47°0N 80°10W 72 C3
Temax Mexico 21°9N 88°56W 87 C7
Temba S. Africa 25°20S 28°17E 57 D4
Tembagapura Indonesia 4°20S 137°0E 37 E9
Tembe
Dem. Rep. of the Congo 0°16S 28°14E 54 C2
Tembe ⌂ S. Africa 26°51S 32°24E 57 D5
Temblor Range U.S.A. 35°20N 119°50W 79 K7
Teme → U.K. 52°11N 2°13W 13 E5
Temecula U.S.A. 33°30N 117°9W 79 M9
Temengor, Tasik
Malaysia 5°24N 101°18E 39 K3
Temerloh Malaysia 3°27N 102°25E 39 L4
Teminabuan Indonesia 1°26S 132°1E 37 E8
Temir Kazakhstan 49°1N 57°14E 19 E10
Temirtau Kazakhstan 50°5N 72°56E 32 A3
Témiscamie → Canada 50°59N 73°5W 73 B5
Témiscaming Canada 46°44N 79°5W 72 C4
Témiscamingue, L.
Canada 47°10N 79°25W 72 C4
Temosachic Mexico 28°57N 107°51W 86 B3
Tempe U.S.A. 33°24N 111°54W 77 K8
Tempiute U.S.A. 37°39N 115°38W 78 H11
Temple U.S.A. 31°6N 97°21W 84 F6
Temple B. Australia 12°15S 143°3E 62 A3
Templemore Ireland 52°47N 7°51W 10 D4
Templeton U.S.A. 35°33N 120°42W 78 K6
Templeton → Australia 21°0S 138°40E 62 C2
Tempoal de Sánchez
Mexico 21°31N 98°23W 87 C5
Temuco Chile 38°45S 72°40W 96 D2
Temuka N.Z. 44°14S 171°17E 59 F3
Tenabo Mexico 20°3N 90°14W 87 C6
Tenaha U.S.A. 31°57N 94°15W 84 F7
Tenakee Springs
U.S.A. 57°47N 135°13W 70 B1
Tenali India 16°15N 80°35E 41 L12
Tenancingo de Degollado
Mexico 18°58N 99°36W 87 D5
Tenango del Valle
Mexico 19°7N 99°33W 87 D5
Tenaro, Akra Greece 36°22N 22°27E 23 F10
Tenasserim Burma 12°6N 99°3E 39 F2
Tenasserim □ Burma 14°0N 98°30E 38 F2
Tenasserim → Burma 12°24N 98°37E 38 F2
Tenby U.K. 51°40N 4°42W 13 F3
Tenda, Colle di France 44°7N 7°36E 20 D7
Tendaho Ethiopia 11°48N 40°54E 47 E3
Tendō Japan 38°20N 140°30E 30 E10
Tendukhera India 23°24N 79°33E 43 H8
Teneguía, Volcanes de
Canary Is. 28°28N 17°51W 24 F2
Ténéré, Erg du Niger 17°35N 10°55E 51 E8
Tenerife Canary Is. 28°15N 16°35W 24 F3
Tenerife, Pico Canary Is. 27°43N 18°1W 24 G1
Tenerife Norte ✈ (TFN)
Canary Is. 28°28N 16°17W 24 F3
Tenerife Sur ✈ (TFS)
Canary Is. 28°3N 16°33W 24 F3
Teng Xian China 35°5N 117°10E 35 G9
Tengah, Kepulauan
Indonesia 7°5S 118°15E 36 F5
Tengchong China 25°0N 98°28E 32 G8
Tengchowfu = Penglai
China 37°48N 120°42E 35 F11
Tenggarong Indonesia 0°24S 116°58E 36 E5
Tengger Shamo China 38°0N 104°10E 34 F3
Tenggol, Pulau Malaysia 4°48N 103°41E 39 K4
Tengiz Köli Kazakhstan 50°30N 69°0E 32 A2
Tengzhou China 35°4N 117°9E 35 G9
Teniente Enciso △
Paraguay 21°5S 61°8W 94 A3
Teniente Rodolfo Marsh
Antarctica 62°30S 58°0W 5 C18
Tenino U.S.A. 46°51N 122°51W 78 D4
Tenkasi India 8°55N 77°20E 40 Q10
Tenke Katanga,
Dem. Rep. of the Congo 11°22S 26°40E 55 E2
Tenke Katanga,
Dem. Rep. of the Congo 10°32S 26°7E 55 E2
Tennant Creek
Australia 19°30S 134°15E 62 B1
Tennessee □ U.S.A. 36°0N 86°30W 85 D11
Tennessee → U.S.A. 37°4N 88°34W 85 C10
Teno, Pta. de Canary Is. 28°21N 16°55W 24 F3
Tenojoki = Tana →
Norway 70°30N 28°14E 8 A23
Tenom Malaysia 5°4N 115°57E 36 C5
Tenosique Mexico 17°29N 91°26W 87 D6
Tenryū-Gawa → Japan 35°39N 137°48E 31 G8
Tenterden U.K. 51°4N 0°42E 13 F8
Tenterfield Australia 29°0S 152°0E 63 D5
Teófilo Otoni Brazil 17°50S 41°30W 93 G10
Tepa Indonesia 7°52S 129°31E 37 F7
Tepalcatepec →
Mexico 18°35N 101°59W 86 D4
Tepehuanes Mexico 25°21N 105°44W 86 B3
Tepetongo Mexico 22°28N 103°9W 86 C4
Tepic Mexico 21°30N 104°54W 86 C4
Teplice Czech Rep. 50°40N 13°48E 16 C7
Tepoca, C. Mexico 30°20N 112°25W 86 A2
Tequila Mexico 20°54N 103°47W 86 C4
Ter → Spain 42°2N 3°12E 21 A7
Ter Apel Neths. 52°53N 7°5E 15 B7
Téra Niger 14°0N 0°45E 50 F6
Teraina Kiribati 4°43N 160°25W 65 G12
Téramo Italy 42°39N 13°42E 22 C5
Terang Australia 38°15S 142°55E 63 F3
Terang, Teluk Indonesia 8°44S 116°0E 37 K19
Terceira Azores 38°43N 27°13W 50 a
Tercero → Argentina 32°58S 61°47W 94 C3
Terebovlya Ukraine 49°18N 25°44E 17 D13
Terek → Russia 44°0N 47°30E 19 F8
Terengganu □ Malaysia 4°55N 103°0E 39 K4
Terepaima △ Venezuela 9°58N 69°17W 89 E6
Teresina Brazil 5°9S 42°45W 93 E10
Terewah, L. Australia 29°52S 147°35E 63 D4
Teridgerie Cr. →
Australia 30°25S 148°50E 63 E4
Termas de Río Hondo
Argentina 27°29S 64°52W 94 B3
Termez = Termiz
Uzbekistan 37°15N 67°15E 28 F7
Términi Imerese Italy 37°59N 13°42E 22 F5
Términos, L. de Mexico 18°37N 91°33W 87 D6
Termiz Uzbekistan 37°15N 67°15E 28 F7
Térmoli Italy 42°0N 15°0E 22 C6
Ternate Indonesia 0°45N 127°25E 37 D7
Terneuzen Neths. 51°20N 3°50E 15 C3
Terney Russia 45°3N 136°37E 30 B8
Terni Italy 42°34N 12°38E 22 C5
Ternopil Ukraine 49°30N 25°40E 17 D13
Ternopil □ Ukraine 48°10N 25°10E 17 D13
Ternopol = Ternopil
Ukraine 49°30N 25°40E 17 D13
Terowie Australia 33°8S 138°55E 63 E2
Terra Bella U.S.A. 35°58N 119°3W 79 K7
Terra Nova △ Canada 48°33N 53°55W 73 C9
Terrace Canada 54°30N 128°35W 70 C3
Terrace Bay Canada 48°47N 87°5W 72 C2
Terracina Italy 41°17N 13°15E 22 D5
Terralba Italy 39°43N 8°39E 22 E3
Terrassa Spain 41°34N 2°1E 21 B7
Terre Haute U.S.A. 39°28N 87°25W 80 F10
Terrebonne B. U.S.A. 29°5N 90°35W 85 G9
Terrell U.S.A. 32°44N 96°17W 84 E6
Terrenceville Canada 47°40N 54°44W 73 C9
Terry U.S.A. 46°47N 105°19W 76 C11
Terryville U.S.A. 41°41N 73°3W 83 E11
Terschelling Neths. 53°25N 5°20E 15 A5
Teruel Spain 40°22N 1°8W 21 B5
Tervola Finland 66°6N 24°49E 8 C21
Teryaweynya L.
Australia 32°18S 143°22E 63 E3
Teshio Japan 44°53N 141°44E 30 B10
Teshio-Gawa → Japan 44°53N 141°45E 30 B10
Tesiyn Gol → Mongolia 50°40N 93°20E 32 A7
Teslin Canada 60°10N 132°43W 70 A2
Teslin → Canada 61°34N 134°35W 70 A2
Teslin L. Canada 60°15N 132°57W 70 A2
Tessalit Mali 20°12N 1°0E 50 D6
Tessaoua Niger 13°47N 7°56E 50 F7
Test → U.K. 50°56N 1°29W 13 G6
Testigos, Is. Los Venezuela 11°23N 63°7W 89 D7
Tetachuck L. Canada 53°18N 125°55W 70 C3
Tetas, Pta. Chile 23°31S 70°38W 94 A1
Tete Mozam. 16°13S 33°33E 55 F3
Tete □ Mozam. 15°15S 32°40E 55 F3
Teterev → Ukraine 51°1N 30°5E 17 C16
Teteven Bulgaria 42°58N 24°17E 23 C11
Tethul → Canada 60°35N 112°12W 70 A6
Tetiyev Ukraine 49°22N 29°38E 17 D15
Tetlit Zheh = Fort McPherson
Canada 67°30N 134°55W 68 D5
Teton → U.S.A. 47°56N 110°31W 76 C8
Tétouan Morocco 35°35N 5°21W 50 A4
Tetovo Macedonia 42°1N 20°59E 23 C9
Tetufera, Mt. Tahiti 17°40S 149°26W 59 d
Teuco → Argentina 25°35S 60°11W 94 B3
Teulon Canada 50°23N 97°16W 71 C9
Teun Indonesia 6°59S 129°8E 37 F7
Teuri-Tō Japan 44°26N 141°19E 30 B10
Teutoburger Wald
Germany 52°5N 8°22E 16 B5
Tevere → Italy 41°44N 12°14E 22 D5

Teverya Israel 32°47N 35°32E 46 C4
Teviot → U.K. 55°29N 2°38W 11 F6
Tewantin Australia 26°27S 153°3E 63 D5
Tewkesbury U.K. 51°59N 2°9W 13 F5
Texada I. Canada 49°40N 124°25W 70 D4
Texarkana Ark., U.S.A. 33°26N 94°2W 84 E7
Texarkana Tex., U.S.A. 33°26N 94°3W 84 E7
Texas Australia 28°49S 151°9E 63 D5
Texas □ U.S.A. 31°40N 98°30W 84 F5
Texas City U.S.A. 29°24N 94°54W 84 G7
Texel Neths. 53°5N 4°50E 15 A4
Texline U.S.A. 36°23N 103°2W 84 C3
Texoma, L. U.S.A. 33°50N 96°34W 84 E6
Teyateyaneng Lesotho 29°7S 27°34E 53 K5
Tezin Afghan. 34°24N 69°30E 42 B3
Teziutlán Mexico 19°49N 97°21W 87 D5
Tezpur India 26°40N 92°45E 41 F18
Tezzeron L. Canada 54°43N 124°30W 70 C4
Tha-anne → Canada 60°31N 94°37W 71 A10
Tha Deua Laos 17°57N 102°53E 38 D4
Tha Deua Laos 19°26N 101°50E 38 C3
Tha Li Thailand 17°37N 101°25E 38 D3
Tha Pla Thailand 17°48N 100°32E 38 D3
Tha Rua Thailand 14°34N 100°44E 38 E3
Tha Sala Thailand 8°40N 99°56E 39 H2
Tha Song Yang Thailand 17°34N 97°55E 38 D1
Thaba Putsoa Lesotho 29°45S 28°0E 57 D4
Thabana Ntlenyana Lesotho 29°30S 29°16E 57 D4
Thabazimbi S. Africa 24°40S 27°21E 57 C4
Thādiq Si. Arabia 25°18N 45°52E 44 E5
Thai Binh Vietnam 20°35N 106°1E 38 B6
Thai Muang Thailand 8°24N 98°16E 39 H2
Thai Nguyen Vietnam 21°35N 105°55E 38 B5
Thailand ■ Asia 16°0N 102°0E 38 E4
Thailand, G. of Asia 11°30N 101°0E 39 G3
Thakhek Laos 17°25N 104°45E 38 D5
Thal Pakistan 33°28N 70°33E 42 C4
Thal Desert Pakistan 31°10N 71°30E 42 D4
Thala La = Hkakabo Razi Burma 28°25N 97°23E 41 E20
Thalabarivat Cambodia 13°33N 105°57E 38 F5
Thalang = Amphoe Thalang Thailand 8°1N 98°20E 39 a
Thalang Thailand 8°2N 98°20E 39 H2
Thalassery = Tellicherry India 11°45N 75°30E 40 P9
Thale Sap Songkhla Thailand 7°12N 100°28E 39 J3
Thallon Australia 28°39S 148°49E 63 D4
Thalu, Ko Thailand 9°26N 99°54E 39 b
Tham Lot Thailand 19°34N 98°16E 38 C2
Thames N.Z. 37°7S 175°34E 59 B5
Thames → Canada 42°20N 82°25W 82 D2
Thames → U.K. 51°29N 0°34E 13 F8
Thames → U.S.A. 41°18N 72°5W 83 E12
Thames Estuary U.K. 51°29N 0°52E 13 F8
Thamesford Canada 43°4N 81°0W 82 C4
Thamesville Canada 42°33N 81°59W 82 D3
Than India 22°34N 71°11E 42 H4
Than Kyun Burma 9°49N 98°4E 39 H1
Than Uyen Vietnam 22°0N 103°54E 38 B4
Thana Gazi India 27°25N 76°19E 42 F7
Thandla India 23°0N 74°34E 42 H6
Thandwe Burma 18°20N 94°30E 41 K19
Thane India 19°12N 72°59E 40 K8
Thanesar India 30°1N 76°52E 42 D7
Thanet, I. of U.K. 51°21N 1°20E 13 F9
Thangool Australia 24°38S 150°42E 62 C5
Thanh Hoa Vietnam 19°48N 105°46E 38 C5
Thanh Hung Vietnam 9°55N 105°43E 39 H5
Thanh Pho Ho Chi Minh Vietnam 10°58N 106°40E 39 G6
Thanh Thuy Vietnam 22°55N 104°51E 38 A5
Thanjavur India 10°48N 79°12E 40 P11
Thanlwin = Salween → Burma 16°31N 97°37E 41 L20
Thano Bula Khan Pakistan 25°22N 67°50E 42 G2
Thanya Buri Thailand 14°1N 100°45E 38 E3
Thaolintoa L. Canada 61°30N 96°25W 71 A9
Thap Lan △ Thailand 14°15N 102°16E 38 E4
Thap Sakae Thailand 11°30N 99°37E 39 G2
Thap Than Thailand 15°27N 99°54E 38 E2
Thar Desert India 28°0N 72°0E 42 F5
Tharad India 24°30N 71°44E 42 G4
Thargomindah Australia 27°58S 143°46E 63 D3
Tharp Fracture Zone S. Ocean 54°0S 135°0W 5 B14
Tharrawaddy Burma 17°38N 95°48E 41 L19
Tharthār, Buḩayrat ath Iraq 34°0N 43°15E 44 C4
Tharthar, L. = Tharthār, Buḩayrat ath Iraq 34°0N 43°15E 44 C4
Tharthār, W. ath → Iraq 34°32N 43°4E 44 C4
Thasos Greece 40°40N 24°40E 23 D11
That Khe Vietnam 22°16N 106°28E 38 A6
Thatcher Ariz., U.S.A. 32°51N 109°46W 77 K9
Thatcher Colo., U.S.A. 37°33N 104°7W 77 H11
Thaton Burma 16°55N 97°22E 41 L20
Thaungdut Burma 24°30N 94°40E 41 G19
Thayawthadangy Kyun Burma 12°19N 98°0E 38 F1
Thayer U.S.A. 36°31N 91°33W 80 G8
Thayetmyo Burma 19°20N 95°10E 41 K19
Thazi Burma 21°0N 96°5E 41 J20
Thbeng Meanchey Cambodia 13°49N 104°58E 38 F5
The Broads U.K. 52°45N 1°30E 12 E9
The Everglades U.S.A. 25°50N 81°0W 85 J14
The Gulf = Persian Gulf Asia 27°0N 50°0E 45 E6
The Hague = 's-Gravenhage Neths. 52°7N 4°17E 15 B4
The Pas Canada 53°45N 101°15W 71 C8
The Wash U.K. 52°58N 0°20E 12 E8
The Weald U.K. 51°4N 0°20E 13 F8
Thebes = Thiva Greece 38°19N 23°19E 23 E10
Thebes Egypt 25°40N 32°35E 51 C12
Thedford U.S.A. 41°59N 100°35W 80 E3
Theebine Australia 25°57S 152°34E 63 D5
Thekulthili L. Canada 61°3N 110°0W 71 A7
Thelon → Canada 64°16N 96°4W 71 A8
Theodore Australia 24°55S 150°3E 62 C5
Theodore Canada 51°26N 102°55W 71 C8
Theodore U.S.A. 30°33N 88°10W 85 F10
Theodore Roosevelt △ U.S.A. 47°0N 103°25W 80 B2
Theodore Roosevelt L. U.S.A. 33°40N 111°10W 77 K8
Thepha Thailand 6°52N 100°58E 39 J3
Theresa U.S.A. 44°13N 75°48W 83 B9
Thermopolis U.S.A. 43°39N 108°13W 76 E9
Thermopylæ Greece 38°48N 22°35E 23 E10
Thessalon Canada 46°20N 83°30W 72 C3
Thessalonica, Gulf of = Thessaloniki Kolpos Greece 40°15N 22°45E 23 D10
Thessaloniki Greece 40°38N 22°58E 23 D10
Thessaloniki Kolpos Greece 40°15N 22°45E 23 D10
Thetford U.K. 52°25N 0°45E 13 E8
Thetford Mines Canada 46°8N 71°18W 73 C5
Theun → Laos 18°19N 104°0E 38 C5
Theunissen S. Africa 28°26S 26°43E 56 D4
Thevenard Australia 32°9S 133°38E 63 E1
Thibodaux U.S.A. 29°48N 90°49W 85 G9
Thicket Portage Canada 55°19N 97°42W 71 B9
Thief River Falls U.S.A. 48°7N 96°10W 80 A5
Thiel Mts. Antarctica 85°15S 91°0W 5 E16
Thiers France 45°52N 3°33E 20 D5
Thiès Senegal 14°50N 16°51W 50 F2
Thika Kenya 1°1S 37°5E 54 C4
Thimphu Bhutan 27°31N 89°45E 41 F16
Thina → S. Africa 31°18S 29°13E 57 E4
Þingvallavatn Iceland 64°11N 21°9W 8 D3
Thionville France 49°20N 6°10E 20 B7
Thira = Santorini Greece 36°23N 25°27E 23 F11
Third Cataract Sudan 19°42N 30°20E 51 E12
Thirsk U.K. 54°14N 1°19W 12 C6
Thiruvananthapuram = Trivandrum India 8°41N 77°0E 40 Q10
Thisted Denmark 56°58N 8°40E 9 H13
Thistle I. Australia 35°0S 136°8E 63 F2
Thithia = Cicia Fiji 17°45S 179°18W 59 a
Thiva Greece 38°19N 23°19E 23 E10
Þjórsá → Iceland 63°47N 20°48W 8 E3
Thlewiaza → Canada 60°29N 94°40W 71 A10
Thmar Puok Cambodia 13°57N 103°4E 38 F4
Tho Chu, Dao Vietnam 9°20N 103°28E 39 H4
Tho Vinh Vietnam 19°16N 105°42E 38 C5
Thoa → Canada 60°31N 109°47W 71 A7
Thoen Thailand 17°43N 99°12E 38 D2
Thoeng Thailand 19°41N 100°12E 38 C3
Thohoyandou S. Africa 22°58S 30°29E 53 J6
Thomas U.S.A. 35°45N 98°45W 84 D5
Thomas, L. Australia 26°4S 137°58E 63 D2
Thomas Hubbard, C. Canada 82°0N 94°25W 69 A13
Thomaston U.S.A. 32°53N 84°20W 85 E12
Thomasville Ala., U.S.A. 31°55N 87°44W 85 F11
Thomasville Ga., U.S.A. 30°50N 83°59W 85 F13
Thomasville N.C., U.S.A. 35°53N 80°5W 85 D14
Thompson Canada 55°45N 97°52W 71 B9
Thompson U.S.A. 41°52N 75°31W 83 E9
Thompson → Canada 50°15N 121°24W 70 C4
Thompson → U.S.A. 39°46N 93°37W 80 F7
Thompson Falls U.S.A. 47°36N 115°21W 76 C6
Thompson Pk. U.S.A. 41°0N 123°2W 76 F2
Thompson Springs U.S.A. 38°58N 109°43W 76 G9
Thompsontown U.S.A. 40°33N 77°14W 82 F7
Thomson U.S.A. 33°28N 82°30W 85 E13
Thomson → Australia 25°11S 142°53E 62 C3
Thon Buri Thailand 13°45N 100°29E 38 F3
Thong Yang Thailand 9°28N 99°56E 39 b
Thoothukudi = Tuticorin India 8°50N 78°12E 40 Q11
Þórisvatn Iceland 64°20N 18°55W 8 D4
Thornaby on Tees U.K. 54°33N 1°18W 12 C6
Thornbury Canada 44°34N 80°26W 82 B4
Thorne U.K. 53°37N 0°57W 12 D7
Thornton Canada 44°16N 79°43W 82 B5
Thornton U.S.A. 39°52N 104°58W 76 G11
Thorold Canada 43°7N 79°12W 82 C5
Þórshöfn Iceland 66°12N 15°20W 8 C6
Thouin, C. Australia 20°20S 118°10E 60 D2
Thousand Islands U.S.A. 44°20N 76°0W 83 B9
Thousand Oaks U.S.A. 34°10N 118°50W 79 L8
Thrace Turkey 41°0N 27°0E 23 D12
Three Forks U.S.A. 45°54N 111°33W 76 D8
Three Gorges Dam China 30°45N 111°15E 33 E11
Three Hills Canada 51°43N 113°15W 70 C6
Three Hummock I. Australia 40°25S 144°55E 63 G3
Three Pagodas Pass Asia 15°20N 98°30E 38 E2
Three Points, C. Ghana 4°42N 2°6W 50 H5
Three Rivers Calif., U.S.A. 36°26N 118°54W 78 J8
Three Rivers Tex., U.S.A. 28°28N 98°11W 84 G5
Three Springs Australia 29°32S 115°45E 61 E2
Thrissur = Trichur India 10°30N 76°18E 40 P10
Throssell, L. Australia 27°33S 124°10E 61 E3
Throssell Ra. Australia 22°3S 121°43E 60 D3
Thrushton △ Australia 27°47S 147°40E 63 D4
Thu Dau Mot Vietnam 10°58N 106°39E 39 G6
Thua → Kenya 1°31S 39°30E 54 C4
Thuan Hoa Vietnam 8°58N 105°30E 39 H5
Thubun Lakes Canada 61°30N 112°0W 71 A6
Thuin Belgium 50°20N 4°17E 15 D4
Thule = Qaanaaq Greenland 77°30N 69°10W 69 B18
Thule Air Base = Uummannaq Greenland 77°33N 68°52W 69 B18
Thun Switz. 46°45N 7°38E 20 C7
Thunder B. U.S.A. 45°0N 83°20W 82 B1
Thunder Bay Canada 48°20N 89°15W 72 C2
Thung Salaeng Luang △ Thailand 16°43N 100°50E 38 D3
Thung Song Thailand 8°10N 99°40E 39 H2
Thungyа △ Thailand 15°30N 98°40E 38 E2
Thunkar Bhutan 27°55N 91°0E 41 F17
Thuong Tra Vietnam 16°2N 107°42E 38 D6
Thüringer Wald Germany 50°35N 11°0E 16 C6
Thuringowa △ Australia 19°18S 146°43E 62 B4
Thurles Ireland 52°41N 7°49W 10 D4
Thurrock □ U.K. 51°31N 0°23E 13 F8
Thursday I. Australia 10°30S 142°3E 62 A3
Thurso Canada 45°36N 75°15W 72 C4
Thurso U.K. 58°36N 3°32W 11 C5
Thurso → U.K. 58°36N 3°32W 11 C5
Thurston I. Antarctica 72°0S 100°0W 5 D16
Thutade L. Canada 57°0N 126°55W 70 B3
Thyolo Malawi 16°7S 35°5E 55 F4
Ti-Tree Australia 22°5S 133°22E 62 C1
Tian Shan Asia 40°30N 76°0E 32 C4
Tianjin China 39°7N 117°12E 35 E9
Tianjin Shi □ China 39°8N 117°12E 35 E9
Tianjun China 38°0N 98°30E 32 D8
Tianmen China 30°39N 113°9E 33 E11
Tianshifu China 41°17N 124°22E 35 D13
Tianshui China 34°32N 105°40E 34 G3
Tianya China 18°18N 109°20E 38 C7
Tianyar Indonesia 8°12S 115°30E 37 J18
Tianzhen China 40°24N 114°5E 34 D8
Tianzhuangtai China 40°43N 122°5E 35 D12
Tiarei Tahiti 17°32S 149°20W 59 d
Tiaret Algeria 35°20N 1°21E 50 A6
Tibagi Brazil 24°30S 50°24W 95 A5
Tibagi → Brazil 22°47S 51°1W 95 A5
Tībastī, Sarīr Libya 22°50N 18°30E 51 D9
Tibati Cameroon 6°22N 12°30E 52 C2
Tiber = Tevere → Italy 41°44N 12°14E 22 D5
Tiber Res. U.S.A. 48°19N 111°6W 76 B8
Tiberias = Teverya Israel 32°47N 35°32E 46 C4
Tiberias, L. = Yam Kinneret Israel 32°45N 35°35E 46 C4
Tibesti Chad 21°0N 17°30E 51 D9
Tibet = Xizang Zizhiqu □ China 32°0N 88°0E 32 E6
Tibet, Plateau of Asia 32°0N 86°0E 26 E10
Tibnī Syria 35°36N 39°50E 44 C3
Tibooburra Australia 29°26S 142°1E 63 D3
Tiburón, I. Mexico 29°0N 112°25W 86 B2
Ticino → Italy 45°9N 9°14E 20 D8
Ticonderoga U.S.A. 43°51N 73°26W 83 C11
Ticul Mexico 20°24N 89°32W 87 C7
Tidaholm Sweden 58°12N 13°58E 9 G15
Tiddim Burma 23°28N 93°45E 41 H18
Tidioute U.S.A. 41°41N 79°24W 82 E5
Tidjikja Mauritania 18°29N 11°35W 50 E3
Tidore Indonesia 0°40N 127°25E 37 D7
Tiefa China 42°28N 123°26E 35 C12
Tiegang Shuiku China 22°37N 113°53E 33 a
Tiel Neths. 51°53N 5°26E 15 C5
Tieli China 46°57N 128°3E 33 B14
Tieling China 42°20N 123°55E 35 C12
Tielt Belgium 51°0N 3°20E 15 C3
Tien Shan = Tian Shan Asia 40°30N 76°0E 32 C4
Tien-tsin = Tianjin China 39°7N 117°12E 35 E9
Tien Yen Vietnam 21°20N 107°24E 38 B6
T'ienching = Tianjin China 39°7N 117°12E 35 E9
Tienen Belgium 50°48N 4°57E 15 D4
Tientsin = Tianjin China 39°7N 117°12E 35 E9
Tieri Australia 23°2S 148°21E 62 C4
Tierra Amarilla Chile 27°28S 70°18W 94 B1
Tierra Amarilla U.S.A. 36°42N 106°33W 77 H10
Tierra Colorada Mexico 17°11N 99°32W 87 D5
Tierra de Campos Spain 42°10N 4°50W 21 A3
Tierra del Fuego, I. Grande de Argentina 54°0S 69°0W 96 G3
Tiétar → Spain 39°50N 6°1W 21 C3
Tieté → Brazil 20°40S 51°35W 95 A5
Tiffin U.S.A. 41°7N 83°11W 81 E12
Tifton U.S.A. 31°27N 83°31W 85 F13
Tifu Indonesia 3°39S 126°24E 37 E7
Tighina Moldova 46°50N 29°30E 17 E15
Tigil Russia 57°49N 158°40E 29 D16
Tignish Canada 46°58N 64°2W 73 C7
Tigre → Peru 4°30S 74°10W 92 D4
Tigre → Venezuela 9°20N 62°30W 92 B6
Tigris = Dijlah, Nahr → Asia 31°0N 47°25E 44 D5
Tigyaing Burma 23°45N 96°10E 41 H20
Tīhāmah Asia 15°3N 41°55E 47 D3
Tijara India 27°56N 76°31E 42 F7
Tijuana Mexico 32°32N 117°1W 79 N9
Tikal Guatemala 17°13N 89°24W 88 C2
Tikal △ Guatemala 17°23N 89°34W 88 C2
Tikamgarh India 24°44N 78°50E 43 G8
Tikchik Lakes U.S.A. 60°0N 159°0W 74 D8
Tikhoretsk Russia 45°56N 40°5E 19 E7
Tikhvin Russia 59°35N 33°30E 18 C5
Tikirarjuaq = Whale Cove Canada 62°10N 92°34W 71 A10
Tikrīt Iraq 34°35N 43°37E 44 C4
Tiksi Russia 71°40N 128°45E 29 B13
Tilamuta Indonesia 0°32N 122°23E 37 D6
Tilburg Neths. 51°31N 5°6E 15 C5
Tilbury Canada 42°17N 82°23W 82 D2
Tilbury U.K. 51°27N 0°22E 13 F8
Tilcara Argentina 23°36S 65°23W 94 A2
Tilden U.S.A. 42°3N 97°50W 80 D5
Tilhar India 28°0N 79°45E 43 F8
Tilichiki Russia 60°27N 166°5E 29 C17
Tilissos Greece 35°20N 25°1E 25 D7
Till → U.K. 55°41N 2°13W 12 B5
Tillamook U.S.A. 45°27N 123°51W 76 D2
Tillsonburg Canada 42°53N 80°44W 82 D4
Tillyeria Cyprus 35°6N 32°40E 25 D11
Tilos Greece 36°27N 27°27E 23 F12
Tilpa Australia 30°57S 144°24E 63 E3
Tilt → U.K. 56°46N 3°51W 11 E5
Tilton U.S.A. 43°27N 71°36W 83 C13
Tiltonsville U.S.A. 40°10N 80°41W 82 F4
Timanfaya △ Canary Is. 29°0N 13°46W 24 F6
Timanskiy Kryazh Russia 65°58N 50°5E 18 A9
Timaru N.Z. 44°23S 171°14E 59 F3
Timau Kenya 0°4N 37°15E 54 B4
Timbaki Greece 35°4N 24°45E 25 D6
Timber Creek Australia 15°40S 130°29E 60 C5
Timber Lake U.S.A. 45°26N 101°5W 80 C3
Timber Mt. U.S.A. 37°6N 116°28W 78 H10
Timbué, Ponta Mozam. 18°47S 36°22E 55 F4
Timbuktu = Tombouctou Mali 16°50N 3°0W 50 E5
Timi Cyprus 34°44N 32°31E 25 E11
Timika Indonesia 4°47S 136°32E 37 E9
Timimoun Algeria 29°14N 0°16E 50 C6
Timiris, Râs Mauritania 19°21N 16°30W 50 E2
Timiskaming, L. = Témiscamingue, L. Canada 47°10N 79°25W 72 C4
Timişoara Romania 45°43N 21°15E 17 F11
Timmins Canada 48°28N 81°25W 72 C3
Timok → Serbia 44°10N 22°40E 23 B10
Timoleague Ireland 51°39N 8°46W 10 E3
Timor Asia 9°0S 125°0E 37 F7
Timor Leste = East Timor ■ Asia 8°50S 126°0E 37 F7
Timor Sea Ind. Oc. 12°0S 127°0E 60 B4
Tin Can Bay Australia 25°56S 153°0E 63 D5
Tin Mt. U.S.A. 36°50N 117°10W 78 J9
Tina = Thina → S. Africa 31°18S 29°13E 57 E4
Tîna, Khalîg el Egypt 31°10N 32°40E 46 D1
Tinaca Pt. Phil. 5°30N 125°25E 37 C7
Tinajo Canary Is. 29°4N 13°42W 24 E6
Tindal Australia 14°31S 132°22E 60 B5
Tindouf Algeria 27°42N 8°10W 50 C4
Tinfunque △ Paraguay 23°55S 60°17W 94 A3
Tinggi, Pulau Malaysia 2°18N 104°7E 39 L5
Tingo María Peru 9°10S 75°54W 92 E3
Tingrela Ivory C. 10°27N 6°25W 50 F4
Tingri China 28°34N 86°38E 32 F6
Tinh Bien Vietnam 10°36N 104°57E 39 G5
Tinian N. Marianas 15°0N 145°38E 64 F6
Tinnevelly = Tirunelveli India 8°45N 77°45E 40 Q10
Tinogasta Argentina 28°5S 67°32W 94 B2
Tinos Greece 37°33N 25°8E 23 F11
Tinpahar India 24°59N 87°44E 43 G12
Tintina Argentina 27°2S 62°45W 94 B3
Tintinara Australia 35°48S 140°2E 63 F3
Tiobraid Árran = Tipperary Ireland 52°28N 8°10W 10 D3
Tioga N. Dak., U.S.A. 48°24N 102°56W 80 A2
Tioga Pa., U.S.A. 41°55N 77°8W 82 E7
Tioman, Pulau Malaysia 2°50N 104°10E 39 L5
Tionesta U.S.A. 41°30N 79°28W 82 E5
Tipperary Ireland 52°28N 8°10W 10 D3
Tipperary □ Ireland 52°37N 7°55W 10 D4
Tipton Calif., U.S.A. 36°4N 119°19W 78 J7
Tipton Iowa, U.S.A. 41°46N 91°8W 80 E8
Tipton Mt. U.S.A. 35°32N 114°12W 79 K12
Tiptonville U.S.A. 36°23N 89°29W 85 C10
Tīrān Iran 32°45N 51°8E 45 C6
Tiranë Albania 41°18N 19°49E 23 D8
Tirari Desert Australia 28°22S 138°7E 63 D2
Tiraspol Moldova 46°55N 29°35E 17 E15
Tire Turkey 38°5N 27°45E 23 E12
Tirebolu Turkey 40°58N 38°45E 19 F6
Tiree U.K. 56°31N 6°55W 11 E2
Tiree, Passage of U.K. 56°30N 6°30W 11 E2
Tîrgovişte = Târgovişte Romania 44°55N 25°27E 17 F13
Tîrgu Jiu = Târgu Jiu Romania 45°5N 23°19E 17 F12
Tîrgu Mureş = Târgu Mureş Romania 46°31N 24°38E 17 E13
Tirich Mir Pakistan 36°15N 71°55E 40 A7
Tiritiri o te Moana = Southern Alps N.Z. 43°41S 170°11E 59 E3
Tirnavos Greece 39°45N 22°18E 23 E10
Tirodi India 21°40N 79°44E 40 J11
Tirol □ Austria 47°3N 10°43E 16 E6
Tirso → Italy 39°53N 8°32E 22 E3
Tirtagangga Indonesia 8°24S 115°35E 37 J18
Tiruchchirappalli India 10°45N 78°45E 40 P11
Tirunelveli India 8°45N 77°45E 40 Q10
Tirupati India 13°39N 79°25E 40 N11
Tiruppur India 11°5N 77°22E 40 P10
Tiruvannamalai India 12°15N 79°5E 40 N11
Tisa India 32°50N 76°9E 42 C7
Tisa → Serbia 45°15N 20°17E 23 B9
Tisdale Canada 52°50N 104°0W 71 C8
Tishomingo U.S.A. 34°14N 96°41W 84 D6
Tisza = Tisa → Serbia 45°15N 20°17E 23 B9
Tit-Ary Russia 71°55N 127°2E 29 B13
Tithwal Pakistan 34°21N 73°50E 43 B5
Titicaca, L. S. Amer. 15°30S 69°30W 92 G5
Titiwa Nigeria 12°14N 12°53E 51 F8
Titule Dem. Rep. of the Congo 3°15N 25°31E 54 B2
Titusville Fla., U.S.A. 28°37N 80°49W 85 G14
Titusville Pa., U.S.A. 41°38N 79°41W 82 E5
Tivaouane Senegal 14°56N 16°45W 50 F2
Tiverton Canada 44°16N 81°32W 82 B3
Tiverton U.K. 50°54N 3°29W 13 G4
Tivoli Italy 41°58N 12°45E 22 D5
Tiwi ✪ Australia 11°33S 130°42E 60 B5
Tizi-Ouzou Algeria 36°42N 4°3E 50 A6
Tizimín Mexico 21°9N 88°9W 87 C7
Tiznit Morocco 29°48N 9°45W 50 C4
Tjeggelvas Sweden 66°37N 17°45E 8 C17
Tjieggelvas = Tjeggelvas Sweden 66°37N 17°45E 8 C17
Tjirebon = Cirebon Indonesia 6°45S 108°32E 37 G13
Tjirrkarli ✪ Australia 26°0S 125°28E 61 E4
Tjluring Indonesia 8°25S 114°13E 37
Tjörn Sweden 58°0N 11°35E 9
Tjukayirla Roadhouse Australia 27°9S 124°34E 61
Tjurabalan ✪ Australia 19°50S 128°0E 6
Tlacotalpan Mexico 18°37N 95°40W 87
Tlahualilo de Zaragoza Mexico 26°7N 103°27W 8
Tlaquepaque Mexico 20°39N 103°19W 8
Tlaxcala Mexico 19°19N 98°14W 87
Tlaxcala □ Mexico 19°25N 98°10W 87
Tlaxiaco Mexico 17°25N 97°35W 87
Tlemcen Algeria 34°52N 1°21W 5
To Bong Vietnam 12°45N 109°16E 3
Toa Payoh Singapore 1°20N 103°51E
Toad → Canada 59°25N 124°57W 7
Toad River Canada 58°51N 125°14W 7
Toamasina Madag. 18°10S 49°25E 5
Toamasina □ Madag. 18°0S 49°0E 5
Toay Argentina 36°43S 64°38W 94
Toba Japan 34°30N 136°51E 31
Toba, Danau Indonesia 2°30N 98°30E 3
Toba Kakar Pakistan 31°30N 69°0E 42
Toba Tek Singh Pakistan 30°55N 72°25E 42
Tobago Trin. & Tob. 11°10N 60°30W 93
Tobelo Indonesia 1°45N 127°56E 37
Tobermory Canada 45°12N 81°40W 8
Tobermory U.K. 56°38N 6°5W 1
Tobi Palau 2°40N 131°10E 37
Tobi-Shima Japan 41°12N 139°34E 3
Tobin, L. Australia 21°45S 125°49E 6
Tobin L. Canada 53°35N 103°30W 7
Toboali Indonesia 3°0S 106°25E 3
Tobol → Russia 58°10N 68°12E 28
Tobol i Indonesia 0°38S 120°5E 3
Tobolsk Russia 58°15N 68°10E 28
Tobruk = Tubruq Libya 32°7N 23°55E 51
Tobyhanna U.S.A. 41°11N 75°25W 8
Tobyl = Tobol → Russia 58°10N 68°12E 28
Tocantinópolis Brazil 6°20S 47°25W 9
Tocantins □ Brazil 10°0S 48°0W 9
Tocantins → Brazil 1°45S 49°10W 9
Toccoa U.S.A. 34°35N 83°19W 85
Tochi → Pakistan 32°49N 70°41E 4
Tochigi Japan 36°25N 139°45E 3
Tochigi □ Japan 36°45N 139°45E 3
Toco Trin. & Tob. 10°49N 60°57W 93
Toconao Chile 23°11S 68°1W 9
Tocopilla Chile 22°5S 70°10W 9
Tocumwal Australia 35°51S 145°31E 6
Tocuyo → Venezuela 11°3N 68°23W 9
Todd → Australia 24°52S 135°48E 6
Todeli Indonesia 1°40S 124°29E 3
Todenyang Kenya 4°35N 35°56E 5
Todgarh India 25°42N 73°58E 4
Todos os Santos, B. de Brazil 12°48S 38°38W 93
Todos Santos Mexico 23°26N 110°13W 8
Toe Hd. U.K. 57°50N 7°8W 1
Tofield Canada 53°25N 112°40W 7
Tofino Canada 49°11N 125°55W 7
Tofua Tonga 19°45S 175°5W
Tōgane Japan 35°33N 140°22E 31
Togian, Kepulauan Indonesia 0°20S 121°50E 3
Togliatti Russia 53°32N 49°24E 1
Togo ■ W. Afr. 8°30N 1°35E 5
Togtoh China 40°15N 111°10E 3
Tohiea, Mt. Moorea 17°33S 149°49W
Tōhoku □ Japan 39°50N 141°45E 30
Töhöm Mongolia 44°27N 108°2E 3
Toinya Sudan 6°17N 29°46E 51
Toiyabe Range U.S.A. 39°30N 117°0W 7
Tojikiston = Tajikistan ■ Asia 38°30N 70°0E 2
Tojo Indonesia 1°20S 121°15E 3
Tōjō Japan 34°53N 133°16E 3
Tok U.S.A. 63°20N 142°59W 6
Tokachi-Dake Japan 43°17N 142°5E 30
Tokachi-Gawa → Japan 42°44N 143°42E 30
Tokala Indonesia 1°30S 121°40E 3
Tōkamachi Japan 37°8N 138°43E 3
Tokanui N.Z. 46°34S 168°56E 5
Tokara-Rettō Japan 29°37N 129°43E 3
Tokarahi N.Z. 44°56S 170°39E 5
Tokashiki-Shima Japan 26°11N 127°21E 3
Tokat Turkey 40°22N 36°35E 1
Tŏkch'ŏn N. Korea 39°45N 126°18E 35
Tokdo Asia 37°15N 131°52E 3
Tokeland U.S.A. 46°42N 123°59W 7
Tokelau Is. ⊠ Pac. Oc. 9°0S 171°45W 64
Tokmak Kyrgyzstan 42°49N 75°15E 2
Toko Ra. Australia 23°5S 138°20E 6
Tokoro-Gawa → Japan 44°7N 144°5E 30
Toku Tonga 18°10S 174°11W
Tokuno-Shima Japan 27°56N 128°55E 3
Tokushima Japan 34°4N 134°34E 3
Tokushima □ Japan 33°55N 134°0E 3
Tokuyama Japan 34°3N 131°50E 3
Tōkyō Japan 35°43N 139°45E 3
Tolaga Bay N.Z. 38°21S 178°20E 5
Tolbukhin = Dobrich Bulgaria 43°37N 27°49E 23
Toledo Brazil 24°44S 53°45W 9
Toledo Spain 39°50N 4°2W 2
Toledo Ohio, U.S.A. 41°39N 83°33W 81
Toledo Oreg., U.S.A. 44°37N 123°56W 7
Toledo Wash., U.S.A. 46°26N 122°51W 7
Toledo, Montes de Spain 39°33N 4°20W 2
Toledo Bend Res. U.S.A. 31°11N 93°34W 8
Tolga Australia 17°15S 145°29E 6
Toliara Madag. 23°21S 43°40E 5
Toliara □ Madag. 21°0S 45°0E 5
Tolima Colombia 4°40N 75°19W 9
Tolitoli Indonesia 1°5N 120°50E 37
Tollhouse U.S.A. 37°1N 119°24W 7
Tolmachevo Russia 58°56N 29°51E 9
Tolo, Teluk Indonesia 2°20S 122°10E 3
Tolo Harbour China 22°27N 114°12E 3

uca Mexico 19°17N 99°40W 87 D5
yatti = Togliatti
Russia 53°32N 49°24E 18 D8
n Burke S. Africa 23°55S 28°0E 57 C4
n Price Australia 22°40S 117°48E 60 D2
nah U.S.A. 43°59N 90°30W 80 D8
nahawk U.S.A. 45°28N 89°44W 80 C9
nakomai Japan 42°38N 141°36E 30 C10
nales U.S.A. 38°15N 122°53W 78 G4
nales B. U.S.A. 38°15N 123°58W 78 G3
nanivi Fiji 17°37S 178°1E 59 a
nar Portugal 39°36N 8°25W 21 C1
naszów Mazowiecki
Poland 51°30N 20°2E 17 C10
natlán Mexico 19°56N 105°15W 86 D3
nbador, Serra do Brazil 12°0S 58°0W 92 F7
nbigbee → U.S.A. 31°8N 87°57W 85 F11
nbouctou Mali 16°50N 3°0W 50 E5
nbstone U.S.A. 31°43N 110°4W 77 L8
nbua Angola 15°55S 11°55E 56 B1
né Chile 36°36S 72°57W 94 D1
nel Japan 38°40N 141°20E 30 E10
nelloso Spain 39°10N 3°2W 21 C4
nini Indonesia 0°30N 120°30E 37 D6
nini, Teluk Indonesia 0°10S 121°0E 37 E6
nintoul U.K. 57°15N 3°23W 11 D5
nkinson Ranges
Australia 26°11S 129°5E 61 E4
nmot Russia 59°4N 126°20E 29 D13
nnop Ta Suos
Cambodia 11°20N 104°15E 39 G5
no → Colombia 5°20N 67°48W 92 B5
ns Place U.S.A. 37°34N 118°41W 78 H8
ns River U.S.A. 39°58N 74°12W 83 G10
nsk Russia 56°30N 85°5E 28 D9
nalá Chiapas, Mexico 16°4N 93°45W 87 D6
nalá Jalisco, Mexico 20°37N 103°14W 86 C4
nantins Brazil 2°45S 67°45W 92 D5
nasket U.S.A. 48°42N 119°26W 76 B4
nawanda U.S.A. 43°1N 78°53W 82 D6
nb Iran 26°15N 55°15E 45 E7
nbridge U.K. 51°11N 0°17E 13 F8
ndano Indonesia 1°35N 124°54E 37 D6
ndoro Namibia 17°45S 18°50E 56 B2
ne → Australia 34°25S 116°25E 61 F2
ne-Gawa → Japan 35°44N 140°51E 31 F9
nekābon Iran 36°45N 51°12E 45 B6
ng Xian China 39°55N 116°35E 34 E9
ng-yeong S. Korea 34°50N 128°20E 35 G15
nga ■ Pac. Oc. 19°50S 174°30W 59 c
nga Trench Pac. Oc. 18°0S 173°0W 64 J10
ngaat S. Africa 29°33S 31°9E 57 D5
ngareva = Penrhyn
Cook Is. 9°0S 158°0W 65 H12
ngariro △ N.Z. 39°8S 175°33E 59 C5
ngatapu Tonga 21°10S 175°10W 59 c
ngatapu Group Tonga 21°0S 175°0W 59 c
ngchuan China 35°6N 109°3E 34 G5
ngeren Belgium 50°47N 5°28E 15 D5
nggu Jiao China 22°22N 113°37E 33 a
ngguan China 34°40N 110°25E 34 G6
nghua China 41°42N 125°58E 35 D13
ngjiang China 47°40N 132°27E 33 B15
ngjosŏn Man
N. Korea 39°30N 128°0E 35 E15
ngking, G. of = Tonkin, G. of
Asia 20°0N 108°0E 38 C7
ngliao China 43°38N 122°18E 35 C12
ngling China 30°55N 117°48E 33 E12
ngobory Madag. 23°32S 44°20E 57 C7
ngoy Chile 30°16S 71°31W 94 C1
ngren China 27°43N 109°11E 32 F10
ngres = Tongeren
Belgium 50°47N 5°28E 15 D5
ngsa Dzong Bhutan 27°31N 90°31E 41 F17
ngshi China 18°30N 109°20E 38 C7
ngue U.K. 58°29N 4°25W 11 C4
ngue → U.S.A. 46°25N 105°52W 76 C11
ngwei China 35°0N 105°5E 34 G3
ngxin China 36°59N 105°58E 34 F3
ngyang N. Korea 39°9N 126°53E 35 E14
ngyu China 44°45N 123°4E 35 B12
nj Sudan 7°20N 28°44E 51 G11
nk India 26°6N 75°54E 42 F6
nkawa U.S.A. 36°41N 97°18W 84 C6
nkin = Bac Phan
Vietnam 22°0N 105°0E 38 B5
nkin, G. of Asia 20°0N 108°0E 38 C7
nlé Sap Cambodia 13°0N 104°0E 38 F5
no Japan 39°19N 141°32E 30 E10
nopah U.S.A. 38°4N 117°14W 77 G5
nosí Panama 7°20N 80°20W 88 E3
ns → Haryana, India 30°30N 77°39E 42 D7
ns → Ut. P., India 26°1N 83°33E 43 F10
nsberg Norway 59°19N 10°25E 9 G14
nto △ U.S.A. 33°39N 111°7W 77 K8
numea Tonga 20°30S 174°30W 59 c
obanna Australia 18°42S 146°9E 62 B4
odyay Australia 31°34S 116°28E 61 F2
oele U.S.A. 40°32N 112°18W 76 F7
ompine Australia 27°15S 144°19E 63 D3
ora Australia 38°39S 146°23E 63 F4
ora-Khem Russia 52°28N 96°17E 29 D10
owoomba Australia 27°32S 151°56E 63 D5
p Springs Australia 16°37S 131°51E 60 C5
paz U.S.A. 38°41N 119°30W 78 G7
peka U.S.A. 39°3N 95°40W 80 F6
pley Canada 54°49N 126°18W 70 C3
pocalma, Pta. Chile 34°10S 72°2W 94 C1
pock U.S.A. 34°46N 114°29W 79 L12
pol'čany Slovak Rep. 48°35N 18°12E 17 D10
polobampo Mexico 25°36N 109°3W 86 B3
pozero, Ozero Russia 65°35N 32°0E 8 D25
ppenish U.S.A. 46°23N 120°19W 76 C3
raka Vestale Madag. 16°20S 43°58E 57 B7
rata Peru 17°23S 70°1W 92 G4
rbalı Turkey 38°10N 27°21E 23 E12
rbat-e Heydārīyeh
Iran 35°15N 59°12E 45 C8
rbat-e Jām Iran 35°16N 60°35E 45 C9
Torbay Canada 47°40N 52°42W 73 C9
Torbay □ U.K. 50°26N 3°31W 13 G4
Tordesillas Spain 41°30N 5°0W 21 B3
Torfaen □ U.K. 51°43N 3°3W 13 F4
Torgau Germany 51°34N 13°0E 16 C7
Torhout Belgium 51°5N 3°7E 15 C3
Tori-Shima Japan 30°29N 140°19E 31 J10
Torilla Pen. Australia 22°29S 150°5E 62 C5
Torino Italy 45°3N 7°40E 20 D7
Torit Sudan 4°27N 32°31E 51 H12
Torkamān Iran 37°35N 47°23E 44 B5
Tormes → Spain 41°18N 6°29W 21 B2
Tornado Mt. Canada 49°55N 114°40W 70 D6
Torneå = Tornio Finland 65°50N 24°12E 8 D21
Torneälven → Europe 65°50N 24°12E 8 D21
Torneträsk Sweden 68°24N 19°15E 8 B18
Torngat Mts. Canada 59°0N 63°40W 69 F19
Tornio Finland 65°50N 24°12E 8 D21
Tornionjoki = Torneälven →
Europe 65°50N 24°12E 8 D21
Tornquist Argentina 38°8S 62°15W 94 D3
Toro Spain 39°59N 4°8E 24 B11
Toro, Cerro del Chile 29°10S 69°50W 94 B2
Toro ⌂ Uganda 1°5N 30°22E 54 C3
Toro Pk. U.S.A. 33°34N 116°24W 79 M10
Toronto Canada 43°39N 79°20W 82 C5
Toronto U.S.A. 40°28N 80°36W 82 F4
Toronto Lester B. Pearson Int. ✈
(YYZ) Canada 43°46N 79°35W 82 C5
Toropets Russia 56°30N 31°40E 18 C5
Tororo Uganda 0°45N 34°12E 54 B3
Toros Dağları Turkey 37°0N 32°30E 44 B2
Torpa India 22°57N 85°6E 43 H11
Torquay U.K. 50°27N 3°32W 13 G4
Torrance U.S.A. 33°50N 118°20W 79 M8
Torre de Moncorvo
Portugal 41°12N 7°8W 21 B2
Torre del Greco Italy 40°47N 14°22E 22 D6
Torrejón de Ardoz Spain 40°27N 3°29W 21 B4
Torrelavega Spain 43°20N 4°5W 21 A3
Torremolinos Spain 36°38N 4°30W 21 D3
Torrens, L. Australia 31°0S 137°50E 63 E2
Torrens Cr. → Australia 22°23S 145°9E 62 C4
Torrens Creek Australia 20°48S 145°3E 62 C4
Torrent Spain 39°27N 0°28W 21 C5
Torreón Mexico 25°33N 103°26W 86 B4
Torres Brazil 29°21S 49°44W 95 B5
Torres Mexico 28°46N 110°47W 86 B2
Torres Strait Australia 9°50S 142°20E 62 a
Torres Vedras Portugal 39°5N 9°15W 21 C1
Torrevieja Spain 37°59N 0°42W 21 D5
Torrey U.S.A. 38°18N 111°25W 76 G8
Torridge → U.K. 51°0N 4°13W 13 G3
Torridon, L. U.K. 57°35N 5°50W 11 D3
Torrington Conn., U.S.A. 41°48N 73°7W 83 E11
Torrington Wyo.,
U.S.A. 42°4N 104°11W 76 E11
Tórshavn Færoe Is. 62°5N 6°56W 8 E9
Tortola Br. Virgin Is. 18°19N 64°45W 89 e
Tortosa Spain 40°49N 0°31E 21 B6
Tortosa, C. Spain 40°41N 0°52E 21 B6
Tortue, Î. de la Haiti 20°5N 72°57W 89 B5
Tortuguero △ Costa Rica 10°31N 83°29W 88 D3
Torūd Iran 35°25N 55°5E 45 C7
Toruń Poland 53°2N 18°39E 17 B10
Tory Hill Canada 44°58N 78°16W 82 B6
Tory I. Ireland 55°16N 8°14W 10 A3
Tosa Japan 33°24N 133°23E 31 H6
Tosa-Shimizu Japan 32°52N 132°58E 31 H6
Tosa-Wan Japan 33°15N 133°30E 31 H6
Toscana □ Italy 43°25N 11°0E 22 C4
Toshka Lakes Egypt 22°50N 31°0E 51 D12
Toshkent Uzbekistan 41°20N 69°10E 28 E7
Tostado Argentina 29°15S 61°50W 94 B3
Tostón, Pta. de Canary Is. 28°42N 14°2W 24 F5
Tosu Japan 33°22N 130°31E 31 H5
Toteng Botswana 20°22S 22°58E 56 C3
Totma Russia 60°0N 42°40E 18 C7
Totnes U.K. 50°26N 3°42W 13 G4
Totness Suriname 5°53N 56°19W 93 B7
Totonicapán Guatemala 14°58N 91°12W 88 D1
Totoya, I. Fiji 18°57S 179°50W 59 a
Totten Glacier Antarctica 66°45S 116°10E 5 C8
Tottenham Australia 32°14S 147°21E 63 E4
Tottenham Canada 44°1N 79°49W 82 B5
Tottori Japan 35°30N 134°15E 31 G7
Tottori □ Japan 35°30N 134°12E 31 G7
Toubkal, Djebel Morocco 31°0N 8°0W 50 B4
Tougan Burkina Faso 13°11N 2°58W 50 F5
Touggourt Algeria 33°6N 6°4E 50 B7
Toul France 48°40N 5°53E 20 B6
Toulon France 43°10N 5°55E 20 E6
Toulouse France 43°37N 1°27E 20 E4
Toummo Niger 22°45N 14°8E 51 D8
Toungoo Burma 19°0N 96°30E 41 K20
Touraine France 47°20N 0°30E 20 C4
Tourcoing France 50°42N 3°10E 20 A5
Touriñán, C. Spain 43°3N 9°18W 21 A1
Tournai Belgium 50°35N 3°25E 15 D3
Tournon-sur-Rhône
France 45°4N 4°50E 20 D6
Tours France 47°22N 0°40E 20 C4
Toussoro, Mt. C.A.R. 9°7N 23°14E 52 C4
Touws → S. Africa 33°45S 21°11E 56 E3
Touwsrivier S. Africa 33°20S 20°2E 56 E3
Towada Japan 40°37N 141°13E 30 D10
Towada-Hachimantai △
Japan 40°20N 140°55E 30 D10
Towada-Ko Japan 40°28N 140°55E 30 D10
Towanda U.S.A. 41°46N 76°27W 83 E8
Tower U.S.A. 47°48N 92°17W 80 B7
Towerhill Cr. →
Australia 22°28S 144°35E 62 C3
Towner U.S.A. 48°21N 100°25W 80 A3
Townsend U.S.A. 46°19N 111°31W 76 C8
Townshend I. Australia 22°10S 150°31E 62 C5
Townsville Australia 19°15S 146°45E 62 B4
Towot Sudan 6°12N 34°25E 54 A3
Towraghondī Afghan. 35°13N 62°16E 40 B3
Towson U.S.A. 39°24N 76°36W 81 F15
Towuti, Danau Indonesia 2°45S 121°32E 37 E6
Tōya-Ko Japan 42°35N 140°51E 30 C10
Toyama Japan 36°40N 137°15E 31 F8
Toyama □ Japan 36°45N 137°30E 31 F8
Toyama-Wan Japan 37°0N 137°30E 31 F8
Toyapakeh Indonesia 8°41S 115°29E 37 K18
Toyohashi Japan 34°45N 137°25E 31 G8
Toyokawa Japan 34°48N 137°27E 31 G8
Toyonaka Japan 34°46N 135°28E 31 G7
Toyooka Japan 35°35N 134°48E 31 G7
Toyota Japan 35°3N 137°7E 31 G8
Tozeur Tunisia 33°56N 8°8E 50 B7
Trá Li = Tralee Ireland 52°16N 9°42W 10 D2
Trá Mhór = Tramore
Ireland 52°10N 7°10W 10 D4
Tra On Vietnam 9°58N 105°55E 39 H5
Tra Vinh Vietnam 9°50N 106°15E 39 H6
Trabzon Turkey 41°0N 39°45E 19 F6
Tracadie Canada 47°30N 64°55W 73 C7
Tracy Canada 46°1N 73°9W 72 C5
Tracy Calif., U.S.A. 37°44N 121°26W 78 H5
Tracy Minn., U.S.A. 44°14N 95°37W 80 C6
Trafalgar, C. Spain 36°10N 6°2W 21 D2
Trail Canada 49°5N 117°40W 70 D5
Trainor L. Canada 60°24N 120°17W 70 A4
Trakai △ Lithuania 54°30N 25°10E 9 J21
Trákhonas Cyprus 35°12N 33°21E 25 D12
Tralee Ireland 52°16N 9°42W 10 D2
Tralee B. Ireland 52°17N 9°55W 10 D2
Tram Chim △ Vietnam 10°43N 105°31E 39 G5
Tramore Ireland 52°10N 7°10W 10 D4
Tramore B. Ireland 52°9N 7°10W 10 D4
Tramuntana, Serra de
Spain 39°48N 2°54E 24 B9
Tran Ninh, Cao Nguyen
Laos 19°30N 103°10E 38 C4
Tranås Sweden 58°3N 14°59E 9 G16
Trancas Argentina 26°11S 65°20W 94 B2
Trang Thailand 7°33N 99°38E 39 J2
Trangahy Madag. 19°7S 44°31E 57 B7
Trangan Indonesia 6°40S 134°20E 37 F8
Trangie Australia 32°4S 148°0E 63 E4
Trani Italy 41°17N 16°25E 22 D7
Tranoroa Madag. 24°42S 45°4E 57 C8
Tranqueras Uruguay 31°13S 55°45W 95 C4
Transantarctic Mts.
Antarctica 85°0S 170°0W 5 E12
Transilvania Romania 46°30N 24°0E 17 E12
Transilvanian Alps = Carpaţii
Meridionali Romania 45°30N 25°0E 17 F13
Transnistria = Stînga Nistrului □
Moldova 47°20N 29°15E 17 E15
Transylvania = Transilvania
Romania 46°30N 24°0E 17 E12
Trápani Italy 38°1N 12°29E 22 E5
Trapper Pk. U.S.A. 45°54N 114°18W 76 D6
Traralgon Australia 38°12S 146°34E 63 F4
Trasimeno, L. Italy 43°8N 12°6E 22 C5
Trat Thailand 12°14N 102°33E 39 F4
Tratani → Pakistan 29°19N 68°20E 42 E3
Traun Austria 48°14N 14°15E 16 D8
Travellers L. Australia 33°20S 142°0E 63 E3
Travemünde Germany 53°57N 10°52E 16 B6
Travers, Mt. N.Z. 42°1S 172°45E 59 E4
Traverse City U.S.A. 44°46N 85°38W 81 C11
Travis, L. U.S.A. 30°24N 97°55W 84 F6
Travnik Bos.-H. 44°17N 17°39E 23 B7
Trawbreaga B. Ireland 55°20N 7°25W 10 A4
Trébbia → Italy 45°4N 9°41E 20 D8
Třebíč Czech Rep. 49°14N 15°55E 16 D8
Trebinje Bos.-H. 42°44N 18°22E 23 C8
Trebonne Australia 18°37S 146°5E 62 B4
Tregaron U.K. 52°14N 3°56W 13 E4
Tregrosse Is. Australia 17°41S 150°43E 62 B5
Treherne Canada 49°38N 98°42W 71 D9
Treinta y Tres Uruguay 33°16S 54°17W 95 C5
Trelawney Zimbabwe 17°30S 30°30E 57 B5
Trelew Argentina 43°10S 65°20W 96 E3
Trelleborg Sweden 55°20N 13°10E 9 J15
Tremadog Bay U.K. 52°51N 4°18W 12 E3
Tremonton U.S.A. 41°43N 112°10W 76 F7
Tremp Spain 42°10N 0°52E 21 A6
Trenche → Canada 47°46N 72°53W 72 C5
Trenčín Slovak Rep. 48°52N 18°4E 17 D10
Trenggalek Indonesia 8°3S 111°43E 37 H14
Trenque Lauquen
Argentina 36°5S 62°45W 94 D3
Trent → Canada 44°6N 77°34W 82 B7
Trent → U.K. 53°41N 0°42W 12 D7
Trento Italy 46°4N 11°8E 22 A4
Trenton Canada 44°10N 77°34W 82 B7
Trenton Mo., U.S.A. 40°5N 93°37W 80 E7
Trenton N.J., U.S.A. 40°14N 74°46W 83 F10
Trenton Nebr., U.S.A. 40°11N 101°1W 80 E3
Trepassey Canada 46°43N 53°25W 73 C9
Tres Arroyos Argentina 38°26S 60°20W 94 D3
Três Corações Brazil 21°44S 45°15W 95 A6
Três Lagoas Brazil 20°50S 51°43W 93 H8
Tres Lomas Argentina 36°27S 62°51W 94 D3
Tres Montes, C. Chile 46°50S 75°30W 96 F1
Tres Pinos U.S.A. 36°48N 121°19W 78 J5
Três Pontas Brazil 21°23S 45°29W 95 A6
Tres Puentes Chile 27°50S 70°15W 94 B1
Tres Puntas, C. Argentina 47°0S 66°0W 96 F3
Três Rios Brazil 22°6S 43°15W 95 A7
Tres Valles Mexico 18°15N 96°8W 87 D5
Tresco U.K. 49°57N 6°20W 13 H1
Treviso Italy 45°40N 12°15E 22 B5
Triabunna Australia 42°30S 147°55E 63 G4
Trianda Greece 36°25N 28°10E 25 C10
Triangle Zimbabwe 21°2S 31°28E 57 C5
Tribal Areas □ Pakistan 33°0N 70°0E 42 C4
Tribulation, C. Australia 16°5S 145°29E 62 B4
Tribune U.S.A. 38°28N 101°45W 80 F3
Trichinopoly = Tiruchchirappalli
India 10°45N 78°45E 40 P11
Trichur India 10°30N 76°18E 40 P10
Trichy = Tiruchchirappalli
India 10°45N 78°45E 40 P11
Trida Australia 33°1S 145°1E 63 E4
Trier Germany 49°45N 6°38E 16 D4
Trieste Italy 45°40N 13°46E 22 B5
Triglav Slovenia 46°21N 13°50E 16 E7
Trikala Greece 39°34N 21°47E 23 E9
Trikomo Cyprus 35°17N 33°52E 25 D12
Trikora, Puncak
Indonesia 4°15S 138°45E 37 E9
Trim Ireland 53°33N 6°48W 10 C5
Trimmu Dam Pakistan 31°10N 72°8E 42 D5
Trincomalee Sri Lanka 8°38N 81°15E 40 Q12
Trindade Atl. Oc. 20°20S 29°50W 2 F8
Trindade Brazil 16°40S 49°30W 93 G9
Trinidad Bolivia 14°46S 64°50W 92 F6
Trinidad Cuba 21°48N 80°0W 88 B4
Trinidad Trin. & Tob. 10°30N 61°15W 93 K15
Trinidad Uruguay 33°30S 56°50W 94 C4
Trinidad U.S.A. 37°10N 104°31W 77 H11
Trinidad → Mexico 17°49N 95°9W 87 D5
Trinidad & Tobago ■
W. Indies 10°30N 61°20W 93 K15
Trinity Canada 48°59N 53°55W 73 C9
Trinity U.S.A. 30°57N 95°22W 84 F7
Trinity → Calif., U.S.A. 41°11N 123°42W 76 F2
Trinity → Tex., U.S.A. 29°45N 94°43W 84 G7
Trinity B. Canada 48°20N 53°10W 73 C9
Trinity Hills Trin. & Tob. 10°7N 61°7W 93 K15
Trinity Is. U.S.A. 56°33N 154°25W 74 D9
Trinity Range U.S.A. 40°10N 118°40W 76 F4
Trinkitat Sudan 18°45N 37°51E 51 E13
Trinway U.S.A. 40°9N 82°1W 82 F2
Triolet Mauritius 20°4S 57°32E 53 d
Tripoli = Tarābulus
Lebanon 34°31N 35°50E 46 A4
Tripoli = Tarābulus Libya 32°49N 13°7E 51 B8
Tripoli Greece 37°31N 22°25E 23 F10
Tripolitania = Tarābulus
N. Afr. 31°0N 13°0E 51 B8
Tripura □ India 24°0N 92°0E 41 H18
Tripylos Cyprus 34°59N 32°41E 25 E11
Tristan da Cunha Atl. Oc. 37°6S 12°20W 49 K2
Trisul India 30°19N 79°47E 43 D8
Trivandrum India 8°41N 77°0E 40 Q10
Trnava Slovak Rep. 48°23N 17°35E 17 D9
Trochu Canada 51°50N 113°13W 70 C6
Trodely I. Canada 52°15N 79°26W 72 B4
Troglav Croatia 43°56N 16°36E 22 C7
Troilus, L. Canada 50°50N 74°35W 72 B5
Trois-Pistoles Canada 48°5N 69°10W 73 C6
Trois-Rivières Canada 46°25N 72°34W 72 C5
Trois-Rivières Guadeloupe 15°57N 61°40W 88 b
Troitsk Russia 54°10N 61°35E 28 D7
Troitsko Pechorsk
Russia 62°40N 56°10E 18 B10
Trölladyngja Iceland 64°54N 17°16W 8 D5
Trollhättan Sweden 58°17N 12°20E 9 G15
Trollheimen Norway 62°46N 9°1E 8 E13
Trombetas → Brazil 1°55S 55°35W 93 D7
Tromsø Norway 69°40N 18°56E 8 B18
Tron Thailand 17°28N 100°7E 38 D3
Trona U.S.A. 35°46N 117°23W 79 K9
Tronador, Mte.
Argentina 41°10S 71°50W 96 E2
Trøndelag Norway 64°17N 11°50E 8 D14
Trondheim Norway 63°36N 10°25E 8 E14
Trondheimsfjorden
Norway 63°35N 10°30E 8 E14
Troodos Cyprus 34°55N 32°52E 25 E11
Troon U.K. 55°33N 4°39W 11 F4
Tropic U.S.A. 37°37N 112°5W 77 H7
Trostan U.K. 55°3N 6°10W 10 A5
Trou Gras Pt. St. Lucia 13°51N 60°53W 89 f
Trout → Canada 61°19N 119°51W 70 A5
Trout L. N.W.T., Canada 60°40N 121°14W 70 A4
Trout L. Ont., Canada 51°20N 93°15W 71 C10
Trout Lake Canada 56°30N 114°32W 70 B6
Trout Lake U.S.A. 46°0N 121°32W 78 E5
Trout River Canada 49°29N 58°8W 73 C8
Trout Run U.S.A. 41°23N 77°3W 82 E7
Trouville-sur-Mer France 49°21N 0°5E 20 B4
Trowbridge U.K. 51°18N 2°12W 13 F5
Troy Turkey 39°57N 26°12E 23 E12
Troy Ala., U.S.A. 31°48N 85°58W 85 F12
Troy Kans., U.S.A. 39°47N 95°5W 80 F6
Troy Mo., U.S.A. 38°59N 90°59W 80 F8
Troy Mont., U.S.A. 48°28N 115°53W 76 B6
Troy N.Y., U.S.A. 42°44N 73°41W 83 D11
Troy Ohio, U.S.A. 40°2N 84°12W 81 E11
Troy Pa., U.S.A. 41°47N 76°47W 83 E8
Troyes France 48°19N 4°3E 20 B6
Truchas Pk. U.S.A. 35°58N 105°39W 77 J11
Trucial States = United Arab
Emirates ■ Asia 23°50N 54°0E 45 F7
Truckee U.S.A. 39°20N 120°11W 78 F6
Trudovoye Russia 43°17N 132°5E 30 C6
Trujillo Honduras 16°0N 86°0W 88 C2
Trujillo Peru 8°6S 79°0W 92 E3
Trujillo Spain 39°28N 5°55W 21 C3
Trujillo U.S.A. 35°32N 104°42W 77 J11
Trujillo Venezuela 9°22N 70°38W 92 B4
Truk Micronesia 7°25N 151°46E 64 G7
Trumann U.S.A. 35°41N 90°31W 85 D9
Trumansburg U.S.A. 42°33N 76°40W 83 D8
Trumbull, Mt. U.S.A. 36°25N 113°19W 77 H7
Trundle Australia 32°53S 147°35E 63 E4
Trung Lon, Hon Vietnam 8°36N 106°8E 39 H6
Trung-Phan Vietnam 17°0N 109°0E 38 D6
Truro Canada 45°21N 63°14W 73 C7
Truro U.K. 50°16N 5°4W 13 G2
Truskavets Ukraine 49°17N 23°30E 17 D12
Trutch Canada 57°44N 122°57W 70 B4
Truth or Consequences
U.S.A. 33°8N 107°15W 77 K10
Trutnov Czech Rep. 50°37N 15°54E 16 C8
Tryonville U.S.A. 41°42N 79°48W 82 E5
Tsandi Namibia 17°42S 14°50E 56 B1
Tsaratanana Madag. 16°47S 47°39E 57 B8
Tsaratanana, Mt. de =
Maromokotro Madag. 14°0S 49°0E 57 A8
Tsaratanana ⌂ Madag. 13°57S 48°52E 57 A8
Tsau Botswana 20°8S 22°22E 56 C3
Tsavo Kenya 2°59S 38°28E 54 C4
Tsavo East △ Kenya 2°44S 38°47E 54 C4
Tsavo West △ Kenya 3°19S 37°57E 54 C4
Tsentralnyy □ Russia 52°0N 40°0E 28 D4
Tses Namibia 25°58S 18°8E 56 D2
Tsetserleg Mongolia 47°36N 101°32E 32 B9
Tseung Kwan China 22°20N 114°15E 33 a
Tsévié Togo 6°25N 1°20E 50 G6
Tshabong Botswana 26°2S 22°29E 56 D3
Tshane Botswana 24°5S 21°54E 56 C3
Tshela
Dem. Rep. of the Congo 4°57S 13°4E 52 E2
Tshesebe Botswana 21°51S 27°32E 57 C4
Tshibeke
Dem. Rep. of the Congo 2°40S 28°35E 54 C2
Tshibinda
Dem. Rep. of the Congo 2°23S 28°43E 54 C2
Tshikapa
Dem. Rep. of the Congo 6°28S 20°48E 52 F4
Tshilenge
Dem. Rep. of the Congo 6°17S 23°48E 54 D1
Tshinsenda
Dem. Rep. of the Congo 12°20S 28°0E 55 E2
Tshofa
Dem. Rep. of the Congo 5°13S 25°16E 54 D2
Tshwane = Pretoria
S. Africa 25°44S 28°12E 57 D4
Tshwane Botswana 22°24S 22°1E 56 C3
Tsigara Botswana 20°22S 25°54E 56 C4
Tsihombe Madag. 25°10S 45°41E 57 D8
Tsiigehtchic Canada 67°15N 134°0W 68 D5
Ts'il-os ⌂ Canada 51°9N 123°59W 70 C4
Tsimlyansk Res. = Tsimlyanskoye
Vdkhr. Russia 48°0N 43°0E 19 E7
Tsimlyanskoye Vdkhr.
Russia 48°0N 43°0E 19 E7
Tsinan = Jinan China 36°38N 117°1E 34 F9
Tsineng S. Africa 27°5S 23°5E 56 D3
Tsing Yi China 22°21N 114°6E 33 a
Tsinghai = Qinghai □
China 36°0N 98°0E 32 D8
Tsingtao = Qingdao
China 36°5N 120°20E 35 F11
Tsingy de Bemaraha ⌂
Madag. 18°35S 44°55E 57 B7
Tsingy de Namoroka ⌂
Madag. 16°29S 45°25E 57 B8
Tsinjoarivo Madag. 19°37S 47°40E 57 B8
Tsinjomitondraka Madag. 15°40S 47°8E 57 B8
Tsirigo = Kythira Greece 36°8N 23°0E 23 F10
Tsiroanomandidy Madag. 18°46S 46°2E 57 B8
Tsitondroina Madag. 21°19S 46°0E 57 C8
Tsitsikamma △ S. Africa 34°3S 23°40E 56 E3
Tsivory Madag. 24°4S 46°5E 57 C8
Tskhinvali Georgia 42°14N 44°1E 19 F7
Tsna → Russia 54°55N 41°58E 18 D7
Tso Moriri, L. India 32°50N 78°20E 43 C8
Tsobis Namibia 19°27S 17°30E 56 B2
Tsodilo Hill Botswana 18°49S 21°43E 56 B3
Tsogttsetsiy = Baruunsuu
Mongolia 43°43N 105°35E 34 C3
Tsolo S. Africa 31°18S 28°37E 57 E4
Tsomo S. Africa 32°0S 27°42E 57 E4
Tsu Japan 34°45N 136°25E 31 G8
Tsu L. Canada 60°40N 111°52W 70 A6
Tsuchiura Japan 36°5N 140°15E 31 F10
Tsuen Wan China 22°22N 114°6E 33 a
Tsugaru-Hantō Japan 41°0N 140°30E 30 D10
Tsugaru-Kaikyō Japan 41°35N 141°0E 30 D10
Tsumeb Namibia 19°9S 17°44E 56 B2
Tsumis Namibia 23°39S 17°29E 56 C2
Tsuruga Japan 35°45N 136°2E 31 G8
Tsurugi-San Japan 33°51N 134°6E 31 H7
Tsuruoka Japan 38°44N 139°50E 30 E9
Tsushima Gifu, Japan 35°10N 136°43E 31 G8
Tsushima Nagasaki,
Japan 34°20N 129°20E 31 G4
Tsuyama Japan 35°3N 134°0E 31 G7
Tsyelyakhany Belarus 52°30N 25°46E 17 B13
Tuam = Tuaim Ireland 53°31N 8°51W 10 C3
Tual Indonesia 5°38S 132°44E 37 F8
Tualatin U.S.A. 45°23N 122°45W 78 E4
Tuam Ireland 53°31N 8°51W 10 C3
Tuamotu, Îs.
French Polynesia 17°0S 144°0W 65 J13
Tuamotu Arch. = Tuamotu, Îs.
French Polynesia 17°0S 144°0W 65 J13
Tuamotu Ridge Pac. Oc. 20°0S 138°0W 65 K14
Tuan Giao Vietnam 21°35N 103°25E 38 B4
Tuao Phil. 17°45N 121°22E 37 A6
Tuapse Russia 44°5N 39°10E 19 F6
Tuas Singapore 1°19N 103°39E 39 d
Tuatapere N.Z. 46°8S 167°41E 59 G1
Tuba City U.S.A. 36°8N 111°14W 77 H8
Tuban Indonesia 6°54S 112°3E 37 G15
Tubani Botswana 24°46S 24°18E 56 C3
Tubarão Brazil 28°30S 49°0W 95 B6
Tūbās West Bank 32°20N 35°22E 46 C4
Tubas → Namibia 22°54S 14°35E 56 C2
Tübingen Germany 48°31N 9°4E 16 D5
Tubou Fiji 18°13S 178°48W 59 a
Tubruq Libya 32°7N 23°55E 51 B10
Tubuaï, Îs.
French Polynesia 25°0S 150°0W 65 K13
Tuc Trung Vietnam 11°1N 107°12E 39 G6
Tucacas Venezuela 10°48N 68°19W 92 A5
Tuchitua Canada 60°55N 129°13W 70 A3
Tuchodi → Canada 58°17N 123°42W 70 B4
Tuckanarra Australia 27°7S 118°5E 61 E2
Tucson U.S.A. 32°13N 110°58W 77 K8
Tucumán □ Argentina 26°48S 66°2W 94 B2
Tucumcari U.S.A. 35°10N 103°44W 77 J12
Tucupita Venezuela 9°2N 62°3W 92 B6
Tucuruí Brazil 3°42S 49°44W 93 D9
Tucuruí, Represa de
Brazil 4°0S 49°30W 93 D9
Tudela Spain 42°4N 1°39W 21 A5
Tudmur Syria 34°36N 38°15E 44 C3
Tudor, L. Canada 55°50N 65°25W 73 A6
Tuen Mun China 22°24N 113°59E 33 a

Tugela → S. Africa 29°14S 31°30E 57 D5
Tuguegarao Phil. 17°35N 121°42E 37 A6
Tugur Russia 53°44N 136°45E 29 D14
Tui Spain 42°3N 8°39W 21 A1
Tuineje Canary Is. 28°19N 14°3W 24 F5
Tukangbesi, Kepulauan
Indonesia 6°0S 124°0E 37 F6
Tukarak I. Canada 56°15N 78°45W 72 A4
Tukayyid Iraq 29°47N 45°36E 44 D5
Tuktoyaktuk Canada 69°27N 133°2W 68 D5
Tuktut Nogait △
Canada 69°15N 122°0W 68 D7
Tukums Latvia 56°58N 23°10E 9 H20
Tukuyu Tanzania 9°17S 33°35E 55 D3
Tula Hidalgo, Mexico 20°3N 99°21W 87 C5
Tula Tamaulipas, Mexico 23°0N 99°43W 87 C5
Tula Russia 54°13N 37°38E 18 D6
Tulach Mhór = Tullamore
Ireland 53°16N 7°31W 10 C4
Tulancingo Mexico 20°5N 98°22W 87 C5
Tulare U.S.A. 36°13N 119°21W 78 J7
Tulare Lake Bed U.S.A. 36°0N 119°48W 78 K7
Tularosa U.S.A. 33°5N 106°1W 77 K10
Tulbagh S. Africa 33°16S 19°6E 56 E2
Tulcán Ecuador 0°48N 77°43W 92 C3
Tulcea Romania 45°13N 28°46E 17 F15
Tulchyn Ukraine 48°41N 28°49E 17 D15
Tüleh Iran 34°35N 52°33E 45 C7
Tulemalu L. Canada 62°58N 99°25W 71 A9
Tuli Zimbabwe 21°58S 29°13E 55 G2
Tulia U.S.A. 34°32N 101°46W 84 D4
Tulita Canada 64°57N 125°30W 68 E6
Ţūlkarm West Bank 32°19N 35°2E 46 C4
Tulla Ireland 52°53N 8°46W 10 D3
Tullahoma U.S.A. 35°22N 86°13W 85 D11
Tullamore Australia 32°39S 147°36E 63 E4
Tullamore Ireland 53°16N 7°31W 10 C4
Tulle France 45°16N 1°46E 20 D4
Tullow Ireland 52°49N 6°45W 10 D5
Tully Australia 17°56S 145°55E 62 B4
Tully U.S.A. 42°48N 76°7W 83 D8
Tulsa U.S.A. 36°10N 95°55W 84 C7
Tulsequah Canada 58°39N 133°35W 70 B2
Tuluá Colombia 4°6N 76°11W 92 C3
Tulun Russia 54°32N 100°35E 29 D11
Tulungagung Indonesia 8°5S 111°54E 37 H14
Tuma → Nic. 13°6N 84°35W 88 D3
Tumacacori ⌓ U.S.A. 31°35N 111°6W 77 L8
Tumaco Colombia 1°50N 78°45W 92 C3
Tumakuru = Tumkur
India 13°18N 77°6E 40 N10
Tumatumari Guyana 5°20N 58°55W 92 B7
Tumba Sweden 59°12N 17°48E 9 G17
Tumba, L.
Dem. Rep. of the Congo 0°50S 18°0E 52 E3
Tumbarumba Australia 35°44S 148°0E 63 F4
Tumbaya Argentina 23°50S 65°26W 94 A2
Tumbes Peru 3°37S 80°27W 92 D2
Tumbler Ridge Canada 55°8N 121°0W 70 B4
Tumbwe
Dem. Rep. of the Congo 11°25S 27°15E 55 E2
Tumby Bay Australia 34°21S 136°8E 63 E2
Tumd Youqi China 40°30N 110°30E 34 D6
Tumen China 43°0N 129°50E 30 C4
Tumen Jiang → China 42°20N 130°35E 35 C16
Tumeremo Venezuela 7°18N 61°30W 92 B6
Tumkur India 13°18N 77°6E 40 N10
Tump Pakistan 26°7N 62°16E 40 F3
Tumpat Malaysia 6°11N 102°10E 39 J4
Tumu Ghana 10°56N 1°56W 50 F5
Tumucumaque, Serra
Brazil 2°0N 55°0W 93 C8
Tumut Australia 35°16S 148°13E 63 F4
Tumwater U.S.A. 47°1N 122°54W 78 C4
Tuna India 22°59N 70°5E 42 H4
Tunapuna Trin. & Tob. 10°38N 61°24W 93 K15
Tunas de Zaza Cuba 21°39N 79°34W 88 B4
Tunbridge Wells = Royal
Tunbridge Wells U.K. 51°7N 0°16E 13 F8
Tunchang China 19°21N 110°5E 38 C8
Tuncurry Australia 32°17S 152°29E 63 E5
Tundla India 27°12N 78°17E 42 F8
Tunduru Tanzania 11°8S 37°25E 55 E4
Tundzha → Bulgaria 41°40N 26°35E 23 C11
Tung Chung China 22°17N 113°57E 33 a
Tung Lung Chau China 22°15N 114°17E 33 a
Tungabhadra → India 15°57N 78°15E 40 M11
Tungla Nic. 13°24N 84°21W 88 D3
Tungsha Tao = Dongsha Dao
S. China Sea 20°45N 116°43E 33 G12
Tungsten Canada 61°57N 128°16W 70 A3
Tunguska, Nizhnyaya →
Russia 65°48N 88°4E 29 C9
Tunguska, Podkamennaya →
Russia 61°50N 90°13E 29 C10
Tunica U.S.A. 34°41N 90°23W 85 D9
Tunis Tunisia 36°50N 10°11E 51 A8
Tunisia ■ Africa 33°30N 9°10E 51 A7
Tunja Colombia 5°33N 73°25W 92 B4
Tunkhannock U.S.A. 41°32N 75°57W 83 E9
Tunliu China 36°13N 112°52E 34 F7
Tunnel Creek △
Australia 17°41S 125°18E 60 C4
Tunnsjøen Norway 64°45N 13°25E 8 D15
Tunungayualok I. Canada 56°0N 61°0W 73 A7
Tunuyán Argentina 33°35S 69°0W 94 C2
Tunuyán → Argentina 33°33S 67°30W 94 C2
Tuolumne U.S.A. 37°58N 120°15W 78 H6
Tuolumne → U.S.A. 37°36N 121°13W 78 H5
Tūp Āghāj Iran 36°3N 47°50E 44 B5
Tupã Brazil 21°57S 50°28W 95 A5
Tupelo U.S.A. 34°16N 88°43W 85 D10
Tupinambarana, I. Brazil 3°0S 58°0W 92 D7
Tupiza Bolivia 21°30S 65°40W 94 A2
Tupman U.S.A. 35°18N 119°21W 79 K7
Tupper Canada 55°32N 120°1W 70 B4
Tupper L. U.S.A. 44°10N 74°32W 83 B10
Tupper Lake U.S.A. 44°14N 74°28W 83 B10
Tupungato, Cerro
S. Amer. 33°15S 69°50W 94 C2
Tuquan China 45°18N 121°38E 35 B11
Túquerres Colombia 1°5N 77°37W 92 C3
Tura Russia 64°20N 100°17E 29 C11
Turabah Si. Arabia 28°20N 43°15E 44 D4
Tūrān Iran 35°39N 56°42E 45 C8
Turan Russia 51°55N 94°0E 29 D10
Ţurayf Si. Arabia 31°41N 38°39E 44 D3
Turda Romania 46°34N 23°47E 17 E12
Turek Poland 52°3N 18°30E 17 B10
Turén Venezuela 9°17N 69°6W 92 B5
Turfan = Turpan China 43°58N 89°10E 32 C6
Turfan Basin = Turpan Pendi
China 42°40N 89°25E 32 C6
Turfan Depression = Turpan
Pendi China 42°40N 89°25E 32 C6
Turgeon → Canada 50°0N 78°56W 72 C4
Tŭrgovishte Bulgaria 43°17N 26°38E 23 C12
Turgutlu Turkey 38°30N 27°43E 23 E12
Turgwe → Zimbabwe 21°31S 32°15E 57 C5
Turia → Spain 39°27N 0°19W 21 C5
Turiaçu Brazil 1°40S 45°19W 93 D9
Turiaçu → Brazil 1°36S 45°19W 93 D9
Turin = Torino Italy 45°3N 7°40E 20 D7
Turkana, L. Africa 3°30N 36°5E 54 B4
Turkestan = Türkistan
Kazakhstan 43°17N 68°16E 28 E7
Turkey ■ Eurasia 39°0N 36°0E 19 G6
Turkey Creek = Warmun
Australia 17°2S 128°12E 60 C4
Türkistan Kazakhstan 43°17N 68°16E 28 E7
Türkmenabat
Turkmenistan 39°6N 63°34E 45 B9
Türkmenbashi
Turkmenistan 40°5N 53°5E 45 A7
Turkmenistan ■ Asia 39°0N 59°0E 28 F6
Turks & Caicos Is. ⊠
W. Indies 21°20N 71°20W 89 B5
Turks Island Passage
W. Indies 21°30N 71°30W 89 B5
Turku Finland 60°30N 22°19E 9 F20
Turkwel → Kenya 3°6N 36°6E 54 B4
Turlock U.S.A. 37°30N 120°51W 78 H6
Turnagain Australia 9°34S 142°17E 62 a
Turnagain → Canada 59°12N 127°35W 70 B3
Turnagain, C. N.Z. 40°28S 176°38E 59 D6
Turneffe Is. Belize 17°20N 87°50W 87 D7
Turner U.S.A. 48°51N 108°24W 76 B9
Turner Pt. Australia 11°47S 133°32E 62 A1
Turner Valley Canada 50°40N 114°17W 70 C6
Turners Falls U.S.A. 42°36N 72°33W 83 D12
Turnhout Belgium 51°19N 4°57E 15 C4
Turnor L. Canada 56°35N 108°35W 71 B7
Tŭrnovo = Veliko Tŭrnovo
Bulgaria 43°5N 25°41E 23 C11
Turnu Măgurele
Romania 43°46N 24°56E 17 G13
Turnu Roşu, P.
Romania 45°33N 24°17E 17 F13
Turpan China 43°58N 89°10E 32 C6
Turpan Pendi China 42°40N 89°25E 32 C6
Turriff U.K. 57°32N 2°27W 11 D6
Tursāq Iraq 33°27N 45°47E 44 C5
Turtle Head I. Australia 10°56S 142°37E 62 A3
Turtle L. Canada 53°36N 108°38W 71 C7
Turtle Lake U.S.A. 47°31N 100°53W 80 B3
Turtleford Canada 53°23N 108°57W 71 C7
Turuépano △ Venezuela 10°34N 62°43W 89 D7
Turukhansk Russia 65°21N 88°5E 29 C9
Tuscaloosa U.S.A. 33°12N 87°34W 85 E11
Tuscany = Toscana □
Italy 43°25N 11°0E 22 C4
Tuscarawas → U.S.A. 40°24N 81°25W 82 F3
Tuscarora Mt. U.S.A. 40°55N 77°55W 82 F7
Tuscola Ill., U.S.A. 39°48N 88°17W 80 F9
Tuscola Tex., U.S.A. 32°12N 99°48W 84 E5
Tuscumbia U.S.A. 34°44N 87°42W 85 D11
Tuskegee U.S.A. 32°25N 85°42W 85 E12
Tuticorin India 8°50N 78°12E 40 Q11
Tutóia Brazil 2°45S 42°20W 93 D10
Tutong Brunei 4°47N 114°40E 36 D4
Tutrakan Bulgaria 44°2N 26°40E 23 B12
Tuttle Creek L. U.S.A. 39°15N 96°36W 80 F5
Tuttlingen Germany 47°58N 8°48E 16 E5
Tutuala E. Timor 8°25S 127°15E 37 F7
Tutuila Amer. Samoa 14°19S 170°50W 59 b
Tutume Botswana 20°30S 27°5E 53 J5
Tuul Gol → Mongolia 48°30N 104°25E 32 B9
Tuva □ Russia 51°30N 95°0E 29 D10
Tuvalu ■ Pac. Oc. 8°0S 178°0E 58 B10
Tuvuca Fiji 17°40S 178°48W 59 a
Tuxpan Mexico 20°57N 97°24W 87 C5
Tuxtla Gutiérrez Mexico 16°45N 93°7W 87 D6
Tuy = Tui Spain 42°3N 8°39W 21 A1
Tuy Duc Vietnam 12°15N 107°27E 39 F6
Tuy Phong Vietnam 11°14N 108°43E 39 G7
Tuya L. Canada 59°7N 130°35W 70 B2
Tuyen Hoa Vietnam 17°50N 106°10E 38 D6
Tuyen Quang Vietnam 21°50N 105°10E 38 B5
Tūysarkān Iran 34°33N 48°27E 45 C6
Tuz Gölü Turkey 38°42N 33°18E 19 G5
Ţūz Khurmātū Iraq 34°56N 44°38E 44 C5
Tuzigoot △ U.S.A. 34°46N 112°2W 77 J7
Tuzla Bos.-H. 44°34N 18°41E 23 B8
Tver Russia 56°55N 35°59E 18 C6
Twain U.S.A. 40°1N 121°3W 78 E5
Twain Harte U.S.A. 38°2N 120°14W 78 G6
Tweed Canada 44°29N 77°19W 82 B7
Tweed → U.K. 55°45N 2°0W 11 F6
Tweed Heads Australia 28°10S 153°31E 63 D5
Tweedsmuir ⌓ Canada 53°0N 126°20W 70 C3
Twentynine Palms
U.S.A. 34°8N 116°3W 79 L10
Twillingate Canada 49°42N 54°45W 73 C9
Twin Bridges U.S.A. 45°33N 112°20W 76 D7
Twin Falls Canada 53°30N 64°32W 73 B7
Twin Falls U.S.A. 42°34N 114°28W 76 E6
Twin Valley U.S.A. 47°16N 96°16W 80 B5
Twinsburg U.S.A. 41°19N 81°26W 82 E3
Twitchell Res. U.S.A. 34°59N 120°19W 79 L6
Two Harbors U.S.A. 47°2N 91°40W 80 B8
Two Hills Canada 53°43N 111°52W 70 C6
Two Rivers U.S.A. 44°9N 87°34W 80 C10
Two Rocks Australia 31°30S 115°35E 61 F2
Twofold B. Australia 37°8S 149°59E 63 F4
Tyachiv Ukraine 48°1N 23°35E 17 D12
Tychy Poland 50°9N 18°59E 17 C10
Tyler Minn., U.S.A. 44°17N 96°8W 80 C5
Tyler Tex., U.S.A. 32°21N 95°18W 84 E7
Tynda Russia 55°10N 124°43E 29 D13
Tyndall U.S.A. 43°0N 97°50W 80 D5
Tyne → U.K. 54°59N 1°32W 12 C6
Tyne & Wear □ U.K. 55°6N 1°17W 12 B6
Tynemouth U.K. 55°1N 1°26W 12 B6
Tyre = Sūr Lebanon 33°19N 35°16E 46 B4
Tyrifjorden Norway 60°2N 10°8E 9 F14
Tyrol = Tirol □ Austria 47°3N 10°43E 16 E6
Tyrone U.S.A. 40°40N 78°14W 82 F6
Tyrone □ U.K. 54°38N 7°11W 10 B4
Tyrrell → Australia 35°26S 142°51E 63 F3
Tyrrell, L. Australia 35°20S 142°50E 63 F3
Tyrrell L. Canada 63°7N 105°27W 71 A7
Tyrrhenian Sea Medit. S. 40°0N 12°30E 22 E5
Tysfjorden Norway 68°7N 16°25E 8 B17
Tyulgan Russia 52°22N 56°12E 18 D10
Tyumen Russia 57°11N 65°29E 28 D7
Tywi → U.K. 51°48N 4°21W 13 F3
Tywyn U.K. 52°35N 4°5W 13 E3
Tzaneen S. Africa 23°47S 30°9E 57 C5
Tzermiado Greece 35°12N 25°29E 25 D7
Tzukong = Zigong
China 29°15N 104°48E 32 F9

U

U.S.A. = United States of
America ■ N. Amer. 37°0N 96°0W 75 H20
U.S. Virgin Is. ⊠ W. Indies 18°20N 65°0W 89 e
Uanle Uen = Wanleweyne
Somali Rep. 2°37N 44°54E 47 G3
Uatumã → Brazil 2°26S 57°37W 92 D7
Uaupés Brazil 0°8S 67°5W 92 D5
Uaupés → Brazil 0°2N 67°16W 92 C5
Uaxactún Guatemala 17°25N 89°29W 88 C2
Ubá Brazil 21°8S 43°0W 95 A7
Ubaitaba Brazil 14°18S 39°20W 93 F11
Ubangi = Oubangi →
Dem. Rep. of the Congo 0°30S 17°50E 52 E3
Ubauro Pakistan 28°15N 69°45E 42 E3
Ubayyiḍ, W. al → Iraq 32°34N 43°48E 44 C4
Ube Japan 33°56N 131°15E 31 H5
Úbeda Spain 38°3N 3°23W 21 C4
Uberaba Brazil 19°50S 47°55W 93 G9
Uberlândia Brazil 19°0S 48°20W 93 G9
Ubin, Pulau Singapore 1°24N 103°57E 39 d
Ubly U.S.A. 43°42N 82°55W 82 C2
Ubolratna Res. Thailand 16°45N 102°30E 38 D4
Ubombo S. Africa 27°31S 32°4E 57 D5
Ubon Ratchathani
Thailand 15°15N 104°50E 38 E5
Ubondo
Dem. Rep. of the Congo 0°55S 25°42E 54 C2
Ubort → Belarus 52°6N 28°30E 17 B15
Ubud Indonesia 8°30S 115°16E 37 J18
Ubundu
Dem. Rep. of the Congo 0°22S 25°30E 54 C2
Ucayali → Peru 4°30S 73°30W 92 D4
Uchab Namibia 19°47S 17°42E 56 B2
Uchiura-Wan Japan 42°25N 140°40E 30 C10
Uchquduq Uzbekistan 41°50N 62°50E 28 E7
Uchur → Russia 58°48N 130°35E 29 D14
Ucluelet Canada 48°57N 125°32W 70 D3
Uda → Russia 54°42N 135°14E 29 D14
Udachnyy Russia 66°25N 112°24E 29 C12
Udagamandalam India 11°30N 76°44E 40 P10
Udainagar India 22°33N 76°13E 42 H7
Udaipur India 24°36N 73°44E 42 G5
Udaipur Garhi Nepal 27°0N 86°35E 43 F12
Udala India 21°35N 86°34E 43 J12
Uddevalla Sweden 58°21N 11°55E 9 G14
Uddjaure Sweden 65°56N 17°49E 8 D17
Uden Neths. 51°40N 5°37E 15 C5
Udgir India 18°25N 77°5E 40 K10
Udhampur India 33°0N 75°5E 43 C6
Údine Italy 46°3N 13°14E 22 A5
Udintsev Fracture Zone
S. Ocean 57°0S 145°0W 5 B13
Udmurtia □ Russia 57°30N 52°30E 18 C9
Udon Thani Thailand 17°29N 102°46E 38 D4
Udong Cambodia 11°48N 104°45E 39 G5
Udskaya Guba Russia 54°50N 135°45E 29 D14
Udu Pt. Fiji 16°9S 179°57W 59 a
Udupi India 13°25N 74°42E 40 N9
Udzungwa △ Tanzania 7°52S 36°35E 54 D4
Udzungwa Range
Tanzania 9°30S 35°10E 55 D4
Ueda Japan 36°24N 138°16E 31 F9
Uedineniya, Os. Russia 78°0N 85°0E 4 B12
Uele →
Dem. Rep. of the Congo 3°45N 24°45E 52 D4
Uelen Russia 66°10N 170°0W 29 C19
Uelzen Germany 52°57N 10°32E 16 B6
Ufa Russia 54°45N 55°55E 18 D10
Ufa → Russia 54°40N 56°0E 18 D10
Ugab → Namibia 20°55S 13°30E 56 C1
Ugalla → Tanzania 5°8S 30°42E 54 D3
Ugalla River ⌓ Tanzania 5°50S 31°54E 54 D3
Uganda ■ Africa 2°0N 32°0E 54 B3
Ugie S. Africa 31°10S 28°13E 57 E4
Uglegorsk Russia 49°5N 142°2E 29 E15
Ugljan Croatia 44°12N 15°10E 16 F8
Uglovoe Kopi Russia 64°44N 177°42E 29 C18
Uhlenhorst Namibia 23°45S 17°55E 56 C2
Uhrichsville U.S.A. 40°24N 81°21W 82 F3
Uibhist a Deas = South Uist
U.K. 57°20N 7°15W 11 D1
Uibhist a Tuath = North Uist
U.K. 57°40N 7°15W 11 D1
Uig U.K. 57°35N 6°21W 11 D2
Uíge Angola 7°30S 14°40E 52 F2
Uiha Tonga 19°54S 174°25W 59 c
Uijeongbu S. Korea 37°44N 127°2E 35 F14
Ŭiju N. Korea 40°15N 124°35E 35 D13
Uinta Mts. U.S.A. 40°45N 110°30W 76 F8
Uis Namibia 21°8S 14°49E 56 C1
Uiseong S. Korea 36°21N 128°45E 35 F15
Uitenhage S. Africa 33°40S 25°28E 56 E4
Uithuizen Neths. 53°24N 6°41E 15 A6
Ujh → India 32°10N 75°18E 42 C6
Ujhani India 28°0N 79°6E 43 F8
Uji-guntō Japan 31°15N 129°25E 31 J4
Ujjain India 23°9N 75°43E 42 H6
Ujung Kulon △
Indonesia 6°46S 105°19E 37 G11
Ujung Pandang = Makassar
Indonesia 5°10S 119°20E 37 F5
Uka Russia 57°50N 162°0E 29 D17
Ukara I. Tanzania 1°50S 33°0E 54 C3
Uke-Shima Japan 28°2N 129°14E 31 K4
Ukerewe I. Tanzania 2°0S 33°0E 54 C3
Ukhrul India 25°10N 94°25E 41 G19
Ukhta Russia 63°34N 53°41E 18 B9
Ukiah U.S.A. 39°9N 123°13W 78 F3
Ukmergė Lithuania 55°15N 24°45E 9 J21
Ukraine ■ Europe 49°0N 32°0E 19 E5
Ukwi Botswana 23°29S 20°30E 56 C3
Ulaan Nuur Mongolia 44°30N 103°40E 34 B2
Ulaan-Uul Mongolia 46°4N 100°49E 34 B6
Ulaanbaatar Mongolia 47°55N 106°53E 32 B10
Ulaangom Mongolia 50°5N 92°10E 32 A7
Ulaanjirem Mongolia 45°5N 105°30E 34 B3
Ulak I. U.S.A. 51°22N 178°57W 74 E4
Ulamba
Dem. Rep. of the Congo 9°3S 23°38E 55 D1
Ulan Bator = Ulaanbaatar
Mongolia 47°55N 106°53E 32 B10
Ulan Ude Russia 51°45N 107°40E 33 A10
Ulanhad = Chifeng
China 42°18N 118°58E 35 C10
Ulanhot = Horqin Youyi Qianqi
China 46°5N 122°3E 35 A12
Ulansuhai Nur China 40°53N 108°51E 34 D5
Ulaya Morogoro, Tanzania 7°3S 36°55E 54 D4
Ulaya Tabora, Tanzania 4°25S 33°30E 54 C3
Ulcinj Montenegro 41°58N 19°10E 23 D8
Ulco S. Africa 28°21S 24°15E 56 D3
Ule älv = Oulujoki →
Finland 65°1N 25°30E 8 D21
Ule träsk = Oulujärvi
Finland 64°25N 27°15E 8 D22
Uleåborg = Oulu Finland 65°1N 25°29E 8 D21
Ulefoss Norway 59°17N 9°16E 9 G13
Ulhasnagar India 19°15N 73°10E 40 K8
Uliastay Mongolia 47°56N 97°28E 32 B8
Uljin S. Korea 36°59N 129°24E 35 F15
Ulladulla Australia 35°21S 150°29E 63 F5
Ullapool U.K. 57°54N 5°9W 11 D3
Ulleungdo S. Korea 37°30N 130°30E 31 F5
Ullswater Canada 45°12N 79°29W 82 A5
Ullswater U.K. 54°34N 2°52W 12 C5
Ulm Germany 48°23N 9°58E 16 D5
Ulmarra Australia 29°37S 153°4E 63 D5
Ulonguè Mozam. 14°37S 34°19E 55 E3
Ulricehamn Sweden 57°46N 13°26E 9 H15
Ulsan S. Korea 35°20N 129°15E 35 G15
Ulsta U.K. 60°30N 1°9W 11 A7
Ulster □ U.K. 54°35N 6°30W 10 B5
Ulubat Gölü Turkey 40°9N 28°35E 23 D13
Uludağ Turkey 40°4N 29°13E 23 D13
Uluguru Mts. Tanzania 7°15S 37°40E 54 D4
Ulungur He → China 47°1N 87°24E 32 B6
Ulungur Hu China 47°20N 87°10E 32 B6
Uluru Australia 25°23S 131°5E 61 E5
Uluru-Kata Tjuta △
Australia 25°19S 131°1E 61 E5
Ulutau Kazakhstan 48°39N 67°1E 28 E7
Uluwatu Indonesia 8°50S 115°5E 37 K18
Ulva U.K. 56°29N 6°13W 11 E2
Ulverston U.K. 54°13N 3°5W 12 C4
Ulverstone Australia 41°11S 146°11E 63 G4
Ulya Russia 59°10N 142°0E 29 D15
Ulyanovsk Russia 54°20N 48°25E 18 D8
Ulyasutay = Uliastay
Mongolia 47°56N 97°28E 32 B8
Ulysses Kans., U.S.A. 37°35N 101°22W 80 G3
Ulysses Pa., U.S.A. 41°54N 77°46W 82 E7
Ulysses, Mt. Canada 57°20N 124°5W 70 B4
Umala Bolivia 17°25S 68°5W 92 G5
'Umān = Oman ■ Asia 23°0N 58°0E 47 C6
Uman Ukraine 48°40N 30°12E 17 D16
Umaria India 23°35N 80°50E 43 H9
Umarkot Pakistan 25°15N 69°40E 42 G3
Umarpada India 21°27N 73°30E 42 J5
Umatilla U.S.A. 45°55N 119°21W 76 D4
Umba Russia 66°42N 34°11E 18 A5
Umbagog L. U.S.A. 44°46N 71°3W 83 B13
Umbakumba Australia 13°47S 136°50E 62 A2
Umbrella Mts. N.Z. 45°35S 169°5E 59 F2
Umeå Sweden 63°45N 20°20E 8 E19
Umeälven → Sweden 63°45N 20°20E 8 E19
Umera Indonesia 0°12S 129°37E 37 E7
Umfuli → Zimbabwe 17°30S 29°23E 55 F2
Umfurudzi ⌓ Zimbabwe 17°6S 31°40E 55 F3
Umgusa Zimbabwe 19°29S 27°52E 55 F2
Umgwenya → Mozam. 25°14S 32°18E 57 D5
Umiujaq Canada 56°33N 76°33W 72 A4
Umkomaas S. Africa 30°13S 30°48E 57 E5
Umlazi S. Africa 29°59S 30°54E 53 L6
Umm ad Daraj, J. Jordan 32°18N 35°48E 46 C4
Umm al Qaywayn
U.A.E. 25°30N 55°35E 45 E7
Umm al Qittayn Jordan 32°18N 36°40E 46 C5
Umm Bāb Qatar 25°12N 50°48E 45 E6
Umm Durman = Omdurmân
Sudan 15°40N 32°28E 51 E12
Umm el Fahm Israel 32°31N 35°9E 46 C4
Umm Keddada Sudan 13°33N 26°35E 51 F11
Umm Lajj Si. Arabia 25°0N 37°23E 44 E3
Umm Qasr Iraq 30°1N 47°58E 44 D5
Umm Ruwaba Sudan 12°50N 31°20E 51 F12
Umnak I. U.S.A. 53°15N 168°20W 74
Umniati → Zimbabwe 16°49S 28°45E 55
Umpqua → U.S.A. 43°40N 124°12W 76
Umreth India 22°41N 73°4E 42
Umtata = Mthatha
S. Africa 31°36S 28°49E 57
Umuarama Brazil 23°45S 53°20W 95
Umvukwe Ra. Zimbabwe 16°45S 30°45E 55
Umzimvubu S. Africa 31°38S 29°33E 57
Umzingwane →
Zimbabwe 22°12S 29°56E 55
Umzinto = eMuziwezinto
S. Africa 30°15S 30°45E 57
Una India 20°46N 71°8E 42
Una → Bos.-H. 45°0N 16°20E 16
Unadilla U.S.A. 42°20N 75°19W 83
Unalakleet U.S.A. 63°52N 160°47W 74
Unalaska U.S.A. 53°53N 166°32W 74
Unalaska I. U.S.A. 53°35N 166°50W 74
'Unayzah Si. Arabia 26°6N 43°58E 44
'Unayzah, J. Asia 32°12N 39°18E 44
Uncía Bolivia 18°25S 66°40W 92
Uncompahgre Peak
U.S.A. 38°4N 107°28W 76
Uncompahgre Plateau
U.S.A. 38°20N 108°15W 76
Undara Volcanic △
Australia 18°14S 144°41E 62
Underbool Australia 35°10S 141°51E 63
Underwood Canada 44°18N 81°29W 82
Ungarie Australia 33°38S 146°56E 63
Ungarra Australia 34°12S 136°2E 63
Ungava, Pén. d' Canada 60°0N 74°0W 69
Ungava B. Canada 59°30N 67°30W 69
Ungeny = Ungheni
Moldova 47°11N 27°51E 17
Unggi N. Korea 42°16N 130°28E 35
Ungheni Moldova 47°11N 27°51E 17
Ungwana B. Kenya 2°45S 40°20E 54
União da Vitória Brazil 26°13S 51°5W 95
Unimak I. U.S.A. 54°45N 164°0W 74
Unimak Pass. U.S.A. 54°15N 164°30W 74
Union Miss., U.S.A. 32°34N 89°7W 85
Union Mo., U.S.A. 38°27N 91°0W 80
Union S.C., U.S.A. 34°43N 81°37W 85
Union City Calif., U.S.A. 37°36N 122°1W 78
Union City N.J., U.S.A. 40°45N 74°2W 83
Union City Pa., U.S.A. 41°54N 79°51W 82
Union City Tenn., U.S.A. 36°26N 89°3W 85
Union Dale U.S.A. 41°43N 75°29W 83
Union Gap U.S.A. 46°33N 120°28W 76
Union Springs Ala.,
U.S.A. 32°9N 85°43W 85
Union Springs N.Y.,
U.S.A. 42°50N 76°41W 83
Uniondale S. Africa 33°39S 23°7E 56
Uniontown U.S.A. 39°54N 79°44W 81
Unionville U.S.A. 40°29N 93°1W 80
United Arab Emirates ■
Asia 23°50N 54°0E 45
United Kingdom ■ Europe 53°0N 2°0W 14
United States of America ■
N. Amer. 37°0N 96°0W 75
Unity Canada 52°30N 109°5W 71
University Park
U.S.A. 32°17N 106°45W 77
University Place U.S.A. 47°14N 122°33W 78
Unjha India 23°46N 72°24E 42
Unnao India 26°35N 80°30E 43
Unsengedsi → Zimbabwe 15°43S 31°14E 55
Unst U.K. 60°44N 0°53W 11
Unuk → Canada 56°5N 131°3W 70
Unzen-Amakusa △
Japan 32°15N 130°10E 31
Uong Bi Vietnam 21°2N 106°47E 38
Uozu Japan 36°48N 137°24E 31
Upata Venezuela 8°1N 62°24W 92
Upemba, L.
Dem. Rep. of the Congo 8°30S 26°20E 55
Upemba △
Dem. Rep. of the Congo 9°0S 26°35E 55
Upernavik Greenland 72°49N 56°20W 4
Upington S. Africa 28°25S 21°15E 56
Upleta India 21°46N 70°16E 42
'Upolu Samoa 13°58S 172°0W 5
Upper Alkali L. U.S.A. 41°47N 120°8W 76
Upper Arrow L. Canada 50°30N 117°50W 70
Upper Daly ✪ Australia 14°26S 131°3E 60
Upper Darby U.S.A. 39°55N 75°16W 81
Upper Foster L. Canada 56°47N 105°20W 71
Upper Hutt N.Z. 41°8S 175°5E 59
Upper Klamath L.
U.S.A. 42°25N 121°55W 76
Upper Lake U.S.A. 39°10N 122°54W 78
Upper Liard Canada 60°3N 128°54W 70
Upper Manzanilla
Trin. & Tob. 10°31N 61°4W 93
Upper Missouri Breaks ⌓
U.S.A. 47°50N 109°55W 76
Upper Musquodoboit
Canada 45°10N 62°58W 73
Upper Red L. U.S.A. 48°8N 94°45W 80
Upper Sandusky
U.S.A. 40°50N 83°17W 81
Upper Volta = Burkina Faso ■
Africa 12°0N 1°0W 50
Uppland Sweden 59°59N 17°48E 9
Uppsala Sweden 59°53N 17°38E 9
Upshi India 33°48N 77°52E 43
Upton U.S.A. 44°6N 104°38W 76
Uqsuqtuuq = Gjoa Haven
Canada 68°38N 95°53W 68
Ur Iraq 30°55N 46°25E 44
Urad Qianqi China 40°40N 108°30E 34
Urahoro Japan 42°50N 143°40E 30
Urakawa Japan 42°9N 142°47E 30
Ural = Uralskiy □ Russia 64°0N 70°0E 28
Ural = Zhayyq →
Kazakhstan 47°0N 51°48E 19
Ural Australia 33°21S 146°12E 63

ral Mts. = Uralskie Gory Eurasia 60°0N 59°0E 18 C10
ralla Australia 30°37S 151°29E 63 E5
ralsk = Oral Kazakhstan 51°20N 51°20E 19 D9
ralskie Gory Eurasia 60°0N 59°0E 18 C10
ralskiy □ Russia 64°0N 70°0E 28 C7
rambo Tanzania 5°4S 32°0E 54 D3
randangi Australia 21°32S 138°14E 62 C2
ranium City Canada 59°34N 108°37W 71 B7
rapuntja = Utopia Australia 22°14S 134°33E 62 C1
raricoera → Brazil 3°2N 60°30W 92 C6
rasoe Japan 26°15N 127°43E 31 L3
rawa = Saitama Japan 35°54N 139°38E 31 G9
ray Russia 60°5N 65°15E 28 C7
ray'irah Si. Arabia 25°57N 48°53E 45 E6
rbana Ill., U.S.A. 40°7N 88°12W 80 E9
rbana Ohio, U.S.A. 40°7N 83°45W 81 E12
rbandale U.S.A. 41°38N 93°43W 80 E7
rbino Italy 43°43N 12°38E 22 C5
rbión, Picos de Spain 42°1N 2°52W 21 A4
rcos Peru 13°40S 71°38W 92 F4
rdinarrain Argentina 32°37S 58°52W 94 C4
re → U.K. 54°5N 1°20W 12 C6
rengoy Russia 65°58N 78°22E 28 C8
res Mexico 29°26N 110°24W 86 B2
rewera △ N.Z. 38°29S 177°7E 59 C6
rfa = Sanliurfa Turkey 37°12N 38°50E 44 B3
rganch Uzbekistan 41°40N 60°41E 28 E7
rgench = Urganch Uzbekistan 41°40N 60°41E 28 E7
rgüp Turkey 38°38N 34°56E 44 B2
ri India 34°8N 74°2E 43 B6
ribia Colombia 11°43N 72°16W 92 A4
riondo Bolivia 21°41S 64°41W 94 A3
rique Mexico 27°13N 107°55W 86 B3
rique → Mexico 26°29N 107°58W 86 B3
rk Neths. 52°39N 5°36E 15 B5
rla Turkey 38°20N 26°47E 23 E12
rmia = Orūmīyeh Iran 37°40N 45°0E 44 B5
roševac Kosovo 42°23N 21°10E 23 C9
royan, Montanas de Puerto Rico 18°12N 67°0W 89 d
rrampinyi Iltjiltarri ☼ Australia 24°20S 132°10E 60 D5
ruaçu Brazil 14°30S 49°10W 93 F9
ruapan Mexico 19°24N 102°3W 86 D4
rubamba → Peru 10°43S 73°48W 92 F4
rucará Brazil 2°32S 57°45W 92 D7
ruçuí Brazil 7°20S 44°28W 93 E10
ruguai → Brazil 26°0S 53°30W 95 B5
ruguaiana Brazil 29°50S 57°0W 94 B4
ruguay ■ S. Amer. 32°30S 56°30W 94 C4
ruguay → S. Amer. 34°12S 58°18W 94 C4
rumchi = Ürümqi China 43°45N 87°45E 32 C6
rümqi China 43°45N 87°45E 32 C6
runa ☼ Australia 24°3S 133°1E 60 D5
rup, Ostrov Russia 46°0N 151°0E 29 E16
rzhar Kazakhstan 47°5N 81°38E 28 E9
sa → Russia 66°16N 59°49E 18 A10
sak Turkey 38°43N 29°28E 19 G4
sakos Namibia 21°54S 15°31E 56 C2
sedom Germany 53°55N 14°2E 16 B8
seless Loop Australia 26°8S 113°23E 61 E1
shakova, Ostrov Russia 82°0N 80°0E 29 A8
shant = Ouessant, Î. d' France 48°28N 5°6W 20 B1
shashi Tanzania 1°59S 33°57E 54 C3
shayrah Si. Arabia 25°35N 45°47E 44 E5
shibuka Japan 32°11N 130°1E 31 H5
shtöbe Kazakhstan 45°16N 78°0E 28 E8
shuaia Argentina 54°50S 68°23W 96 G3
shumun Russia 52°47N 126°32E 29 D13
sinsk Russia 65°57N 57°27E 18 A10
k Canada 54°38N 128°26W 70 C3
k → U.K. 51°33N 2°58W 13 F5
ka India 27°12N 83°7E 43 F10
man Russia 52°5N 39°48E 18 D6
oke Tanzania 5°8S 32°24E 54 D3
solye-Sibirskoye Russia 52°48N 103°40E 29 D11
pallata, P. de Argentina 32°37S 69°22W 94 C2
suriysk Russia 43°48N 131°59E 30 C5
surka Russia 45°12N 133°31E 30 B6
t-Chaun Russia 68°47N 170°30E 29 C18
t-Ilimsk Russia 58°3N 102°39E 29 D11
t-Ishim Russia 57°45N 71°10E 28 D8
t-Kamchatsk Russia 56°10N 162°28E 29 D17
t-Kamenogorsk = Öskemen Kazakhstan 50°0N 82°36E 28 D9
t-Khayryuzovo Russia 57°15N 156°45E 29 D16
t-Kut Russia 56°50N 105°42E 29 D11
t-Kuyga Russia 70°1N 135°43E 29 B14
t-Maya Russia 60°30N 134°28E 29 C14
t-Mil Russia 59°40N 133°11E 29 D14
t-Nera Russia 64°35N 143°15E 29 C15
t-Nyukzha Russia 56°34N 121°37E 29 D13
t-Olenek Russia 73°0N 120°5E 29 B12
t-Omchug Russia 61°9N 149°38E 29 C15
t-Port Russia 69°40N 84°26E 28 C9
t-Tsilma Russia 65°28N 52°11E 18 A9
t-Urt = Ustyurt Plateau Asia 44°0N 55°0E 28 E6
t-Usa Russia 66°2N 56°57E 18 A10
t-Vorkuta Russia 67°24N 64°0E 18 A11
tí nad Labem Czech Rep. 50°41N 14°3E 16 C8
tica Italy 38°42N 13°11E 22 E5
tinov = Izhevsk Russia 56°51N 53°14E 18 C9
tyurt Plateau Asia 44°0N 55°0E 28 E6
uki Japan 33°8N 131°49E 31 H5
ulután El Salv. 13°25N 88°28W 88 D2
umacinta → Mexico 18°24N 92°38W 87 D6
umbura = Bujumbura Burundi 3°16S 29°18E 54 C2
ure Tanzania 4°40S 34°22E 54 C3
utuo → Mozam. 26°48S 32°7E 57 D5
Uta Indonesia 4°33S 136°0E 37 E9
Utah □ U.S.A. 39°20N 111°30W 76 G8
Utah L. U.S.A. 40°12N 111°48W 76 F8
Utara, Selat Malaysia 5°28N 100°20E 39 c
Utarni India 26°5N 71°58E 42 F4
Utatlan Guatemala 15°2N 91°11W 88 C1
Ute Creek → U.S.A. 35°21N 103°50W 77 J12
Utena Lithuania 55°27N 25°40E 9 J21
Utete Tanzania 8°0S 38°45E 54 D4
Uthai Thani Thailand 15°22N 100°3E 38 E3
Uthal Pakistan 25°44N 66°40E 42 G2
Utiariti Brazil 13°0S 58°10W 92 F7
Utica N.Y., U.S.A. 43°6N 75°14W 83 C9
Utica Ohio, U.S.A. 40°14N 82°27W 82 F2
Utikuma L. Canada 55°50N 115°30W 70 B5
Utila Honduras 16°6N 86°56W 88 C2
Utopia Australia 22°14S 134°33E 62 C1
Utraula India 27°19N 82°25E 43 F10
Utrecht Neths. 52°5N 5°8E 15 B5
Utrecht S. Africa 27°38S 30°20E 57 D5
Utrecht □ Neths. 52°6N 5°7E 15 B5
Utrera Spain 37°12N 5°48W 21 D3
Utsjoki → Finland 69°51N 26°59E 8 B22
Utsunomiya Japan 36°30N 139°50E 31 F9
Uttar Pradesh □ India 27°0N 80°0E 43 F9
Uttaradit Thailand 17°36N 100°5E 38 D3
Uttarakhand □ India 30°0N 79°30E 43 D8
Utterson Canada 45°12N 79°19W 82 A5
Uttoxeter U.K. 52°54N 1°52W 12 E6
Utuado Puerto Rico 18°16N 66°42W 89 d
Uummannaq Avannaarsua, Greenland 77°33N 68°52W 69 B18
Uummannaq Kitaa, Greenland 70°58N 52°17W 4 B5
Uummannarsuaq = Nunap Isua Greenland 59°48N 43°55W 66 D15
Uusikaarlepyy Finland 63°32N 22°31E 8 E20
Uusikaupunki Finland 60°47N 21°25E 8 F19
Uva Russia 56°59N 52°13E 18 C9
Uvalde U.S.A. 29°13N 99°47W 84 G5
Uvat Russia 59°5N 68°50E 28 D7
Uvinza Tanzania 5°5S 30°24E 54 D3
Uvira Dem. Rep. of the Congo 3°22S 29°3E 54 C2
Uvs Nuur Mongolia 50°20N 92°30E 32 A7
'Uwairidh, Ḩarrat al Si. Arabia 26°50N 38°0E 44 E3
Uwajima Japan 33°10N 132°35E 31 H6
Uwanda △ Tanzania 7°46S 32°0E 54 D3
Uweinat, Jebel Sudan 21°54N 24°58E 51 D10
Uxbridge Canada 44°6N 79°7W 82 B5
Uxin Qi China 38°50N 109°5E 34 E5
Uxmal Mexico 20°22N 89°46W 87 C7
Üydzin Mongolia 44°9N 107°0E 34 B4
Uyo Nigeria 5°1N 7°53E 50 G7
Uyûn Mûsa Egypt 29°53N 32°40E 46 F1
Uyuni Bolivia 20°28S 66°47W 92 H5
Uzbekistan ■ Asia 41°30N 65°0E 28 E7
Uzboy → Turkmenistan 39°30N 55°0E 45 B7
Uzen, Mal → Kazakhstan 49°4N 49°44E 19 E8
Uzerche France 45°25N 1°34E 20 D4
Uzh → Ukraine 51°15N 30°12E 17 C16
Uzhgorod = Uzhhorod Ukraine 48°36N 22°18E 17 D12
Uzhhorod Ukraine 48°36N 22°18E 17 D12
Užice Serbia 43°55N 19°50E 23 C8
Uzunköprü Turkey 41°16N 26°43E 23 D12

V

Vaal → S. Africa 29°4S 23°38E 56 D3
Vaal Dam S. Africa 27°0S 28°14E 57 D4
Vaalbos △ S. Africa 28°22S 24°20E 56 D3
Vaalwater S. Africa 24°15S 28°8E 57 C4
Vaasa Finland 63°6N 21°38E 8 E19
Vác Hungary 47°49N 19°10E 17 E10
Vacaria Brazil 28°31S 50°52W 95 B5
Vacata Fiji 17°15S 179°31W 59 a
Vacaville U.S.A. 38°21N 121°59W 78 G5
Vach = Vakh → Russia 60°45N 76°45E 28 C8
Vache, Î. à Haiti 18°2N 73°35W 89 C5
Vacoas Mauritius 20°18S 57°29E 53 d
Vadakara = Badagara India 11°35N 75°40E 40 P9
Vadnagar India 23°47N 72°40E 42 H5
Vadodara India 22°20N 73°10E 42 H5
Vadsø Norway 70°3N 29°50E 8 A23
Vaduz Liech. 47°8N 9°31E 20 C8
Værøy Norway 67°40N 12°40E 8 C15
Vágar Færoe Is. 62°5N 7°15W 8 E9
Vagia, Akra Greece 36°15N 28°11E 25 C10
Vågsfjorden Norway 68°50N 16°50E 8 B17
Váh → Slovak Rep. 47°43N 18°7E 17 D9
Vahsel B. Antarctica 75°0S 35°0W 5 D1
Váhtjer = Gällivare Sweden 67°9N 20°40E 8 C19
Vaï Greece 35°15N 26°18E 25 D8
Vaihiria, L. Tahiti 17°40S 149°25W 59 d
Vail U.S.A. 39°40N 106°20W 75 H18
Vairao Tahiti 17°47S 149°17W 59 d
Vaisali → India 26°28N 78°53E 43 F8
Vaitogi Amer. Samoa 14°24S 170°44W 59 b
Vakh → Russia 60°45N 76°45E 28 C8
Val-d'Or Canada 48°7N 77°47W 72 C4
Val Marie Canada 49°15N 107°45W 71 D7
Valaam Russia 61°22N 30°57E 8 F24
Valahia Romania 44°35N 25°0E 17 F13
Valandovo Macedonia 41°19N 22°34E 23 D10
Valatie U.S.A. 42°24N 73°40W 83 D11
Valcheta Argentina 40°40S 66°8W 96 E3
Valdai Hills = Valdayskaya Vozvyshennost Russia 57°0N 33°30E 18 C5
Valdayskaya Vozvyshennost Russia 57°0N 33°30E 18 C5
Valdepeñas Spain 38°43N 3°25W 21 C4
Valdés, Pen. Argentina 42°30S 63°45W 96 E4
Valdez U.S.A. 61°7N 146°16W 68 E2
Valdivia Chile 39°50S 73°14W 96 D2
Valdivia Abyssal Plain S. Ocean 62°30S 70°0E 5 C6
Valdosta U.S.A. 30°50N 83°17W 85 F13
Valdres Norway 61°5N 9°5E 8 F13
Vale U.S.A. 43°59N 117°15W 76 E5
Vale of Glamorgan □ U.K. 51°28N 3°25W 13 F4
Valemount Canada 52°50N 119°15W 70 C5
Valença Brazil 13°20S 39°5W 93 F11
Valença do Piauí Brazil 6°20S 41°45W 93 E10
Valence France 44°57N 4°54E 20 D6
Valencia Spain 39°27N 0°23W 21 C5
Valencia Trin. & Tob. 10°39N 61°11W 93 K15
Valencia U.S.A. 34°48N 106°43W 77 J10
Valencia Venezuela 10°11N 68°0W 92 A5
Valencia □ Spain 39°20N 0°40W 21 C5
Valencia, G. de Spain 39°30N 0°20E 21 C6
Valencia de Alcántara Spain 39°25N 7°14W 21 C2
Valencia I. Ireland 51°54N 10°22W 10 E1
Valenciennes France 50°20N 3°34E 20 A5
Valentim, Sa. do Brazil 6°0S 43°30W 93 E10
Valentin Russia 43°8N 134°17E 30 C7
Valentine Nebr., U.S.A. 42°52N 100°33W 80 D3
Valentine Tex., U.S.A. 30°35N 104°30W 84 F2
Valera Venezuela 9°19N 70°37W 92 B4
Valga Estonia 57°47N 26°2E 9 H22
Valier U.S.A. 48°18N 112°16W 76 B7
Valjevo Serbia 44°18N 19°53E 23 B8
Valka Latvia 57°46N 26°3E 9 H22
Valkeakoski Finland 61°16N 24°2E 8 F21
Valkenswaard Neths. 51°21N 5°29E 15 C5
Vall de Uxó = La Vall d'Uixó Spain 39°49N 0°15W 21 C5
Valladolid Mexico 20°41N 88°12W 87 C7
Valladolid Spain 41°38N 4°43W 21 B3
Valldemossa Spain 39°43N 2°37E 24 B9
Valle de la Pascua Venezuela 9°13N 66°0W 92 B5
Valle de las Palmas Mexico 32°20N 116°43W 79 N10
Valle de Santiago Mexico 20°23N 101°12W 86 C4
Valle de Zaragoza Mexico 27°25N 105°50W 86 B3
Valle Fértil, Sierra del Argentina 30°20S 68°0W 94 C2
Valle Gran Rey Canary Is. 28°5N 17°20W 24 F2
Valle Hermoso Mexico 25°35N 97°40W 87 B5
Valle Nacional Mexico 17°47N 96°18W 87 D5
Valledupar Colombia 10°29N 73°15W 92 A4
Vallehermoso Canary Is. 28°10N 17°15W 24 F2
Vallejo U.S.A. 38°7N 122°14W 78 G4
Vallenar Chile 28°30S 70°50W 94 B1
Valletta Malta 35°54N 14°31E 25 D2
Valley Center U.S.A. 33°13N 117°2W 79 M9
Valley City U.S.A. 46°55N 98°0W 80 B5
Valley Falls Oreg., U.S.A. 42°29N 120°17W 76 E3
Valley Falls R.I., U.S.A. 41°54N 71°24W 83 E13
Valley of Flowers △ India 30°50N 79°40E 43 D8
Valley Springs U.S.A. 38°12N 120°50W 78 G6
Valley View U.S.A. 40°39N 76°33W 83 F8
Valley Wells U.S.A. 35°27N 115°46W 79 K11
Valleyview Canada 55°5N 117°17W 70 B5
Vallimanca, Arroyo Argentina 35°40S 59°10W 94 D4
Valls Spain 41°18N 1°15E 21 B6
Valmiera Latvia 57°37N 25°29E 9 H21
Valognes France 49°30N 1°28W 20 B3
Valona = Vlorë Albania 40°32N 19°28E 23 D8
Valozhyn Belarus 54°3N 26°30E 17 A14
Valparaíso Chile 33°2S 71°40W 94 C1
Valparaíso Mexico 22°46N 103°34W 86 C4
Valparaiso U.S.A. 41°28N 87°4W 80 E10
Valparaíso □ Chile 33°2S 71°40W 94 C1
Vals → S. Africa 27°23S 26°30E 56 D4
Vals, Tanjung Indonesia 8°26S 137°25E 37 F9
Valsad India 20°40N 72°58E 40 J8
Valverde Canary Is. 27°48N 17°55W 24 G2
Valverde del Camino Spain 37°35N 6°47W 21 D2
Vamizi, I. Mozam. 11°2S 40°40E 55 E5
Vammala = Sastamala Finland 61°20N 22°54E 8 F20
Vamos Greece 35°24N 24°13E 25 D6
Van Turkey 38°30N 43°20E 44 B4
Van, L. = Van Gölü Turkey 38°30N 43°0E 44 B4
Van Alstyne U.S.A. 33°25N 96°35W 84 E6
Van Blommestein Meer Suriname 4°45N 55°5W 93 C7
Van Buren Canada 47°10N 67°55W 73 C6
Van Buren Ark., U.S.A. 35°26N 94°21W 84 D7
Van Buren Maine, U.S.A. 47°10N 67°58W 81 B20
Van Buren Mo., U.S.A. 37°0N 91°1W 80 G8
Van Canh Vietnam 13°37N 109°0E 38 F7
Van Diemen, C. N. Terr., Australia 11°9S 130°24E 60 B5
Van Diemen, C. Queens., Australia 16°30S 139°46E 62 B2
Van Diemen G. Australia 11°45S 132°0E 60 B5
Van Gölü Turkey 38°30N 43°0E 44 B4
Van Horn U.S.A. 31°3N 104°50W 84 F2
Van Ninh Vietnam 12°42N 109°14E 38 F7
Van Rees, Pegunungan Indonesia 2°35S 138°15E 37 E9
Van Wert U.S.A. 40°52N 84°35W 81 E11
Van Yen Vietnam 21°4N 104°42E 38 B5
Van Zylsrus S. Africa 26°52S 22°4E 56 D3
Vanadzor Armenia 40°48N 44°30E 19 F7
Vanavara Russia 60°22N 102°16E 29 C11
Vancouver Canada 49°15N 123°7W 78 A3
Vancouver U.S.A. 45°38N 122°40W 78 E4
Vancouver, C. Australia 35°2S 118°11E 61 G2
Vancouver I. Canada 49°50N 126°0W 78 B2
Vancouver Int. ✈ (YVR) Canada 49°10N 123°10W 78 A3
Vanda = Vantaa Finland 60°18N 24°56E 9 F21
Vandalia Ill., U.S.A. 38°58N 89°6W 80 F9
Vandalia Mo., U.S.A. 39°19N 91°29W 80 F8
Vandenberg Village U.S.A. 34°43N 120°28W 79 L6
Vandergrift U.S.A. 40°36N 79°34W 82 F5
Vanderhoof Canada 54°0N 124°0W 70 C4
Vanderkloof Dam S. Africa 30°4S 24°40E 56 E3
Vanderlin I. Australia 15°44S 137°2E 62 B2
Vänern Sweden 58°47N 13°30E 9 G15
Vänersborg Sweden 58°26N 12°19E 9 G15
Vang Vieng Laos 18°58N 102°32E 38 C4
Vanga Kenya 4°35S 39°12E 54 C4
Vangaindrano Madag. 23°21S 47°36E 57 C8
Vanguard Canada 49°55N 107°20W 71 D7
Vanino Russia 48°50N 140°5E 29 E15
Vankleek Hill Canada 45°32N 74°40W 83 A10
Vännäs Sweden 63°58N 19°48E 8 E18
Vannes France 47°40N 2°47W 20 C2
Vannøya Norway 70°6N 19°50E 8 A18
Vanrhynsdorp S. Africa 31°36S 18°44E 56 E2
Vansbro Sweden 60°32N 14°15E 9 F16
Vansittart B. Australia 14°3S 126°17E 60 B4
Vansittart I. Canada 65°50N 84°0W 69 D15
Vantaa Finland 60°18N 24°56E 9 F21
Vanua Balavu Fiji 17°12S 178°55W 59 a
Vanua Levu Fiji 16°33S 179°15E 59 a
Vanua Vatu Fiji 18°22S 179°15W 59 a
Vanuatu ■ Pac. Oc. 15°0S 168°0E 58 C9
Vanwyksvlei S. Africa 30°18S 21°49E 56 E3
Vapnyarka Ukraine 48°32N 28°45E 17 D15
Var → France 43°39N 7°12E 20 E7
Varāmīn Iran 35°20N 51°39E 45 C6
Varāmīn □ Iran 35°18N 51°35E 45 B6
Varanasi India 25°22N 83°0E 43 G10
Varangerfjorden Norway 70°3N 29°25E 8 A23
Varangerhalvøya Norway 70°25N 29°30E 8 A23
Varaždin Croatia 46°20N 16°20E 16 E9
Varberg Sweden 57°6N 12°20E 9 H15
Vardak □ Afghan. 34°0N 68°0E 40 B6
Vardar = Axios → Greece 40°57N 22°35E 23 D10
Varde Denmark 55°38N 8°29E 9 J13
Vardø Norway 70°23N 31°5E 8 A24
Varella, Mui = Nay, Mui Vietnam 12°54N 109°26E 38 F7
Varėna Lithuania 54°12N 24°30E 9 J21
Varese Italy 45°48N 8°50E 20 D8
Varginha Brazil 21°33S 45°25W 95 A6
Varillas Chile 24°0S 70°10W 94 A1
Varkaus Finland 62°19N 27°50E 8 E22
Varna Bulgaria 43°13N 27°56E 23 C12
Värnamo Sweden 57°10N 14°3E 9 H16
Várnjárga = Varangerhalvøya Norway 70°25N 29°30E 8 A23
Vars Canada 45°21N 75°21W 83 A9
Varysburg U.S.A. 42°46N 78°19W 82 D6
Varzaneh Iran 32°25N 52°40E 45 C7
Vasa = Vaasa Finland 63°6N 21°38E 8 E19
Vasa Barris → Brazil 11°10S 37°10W 93 F11
Vascongadas = País Vasco □ Spain 42°50N 2°45W 21 A4
Vasht = Khāsh Iran 28°15N 61°15E 45 D9
Vasilevichi Belarus 52°15N 29°50E 17 B15
Vasilkov = Vasylkiv Ukraine 50°7N 30°15E 17 C16
Vaslui Romania 46°38N 27°42E 17 E14
Vassar Canada 49°10N 95°55W 71 D9
Vassar U.S.A. 43°22N 83°35W 81 D12
Västerås Sweden 59°37N 16°38E 9 G17
Västerbotten Sweden 64°36N 20°4E 8 D19
Västerdalälven → Sweden 60°30N 15°7E 8 F16
Västervik Sweden 57°43N 16°33E 9 H17
Västmanland Sweden 59°45N 16°20E 9 G17
Vasto Italy 42°8N 14°40E 22 C6
Vasylkiv Ukraine 50°7N 30°15E 17 C16
Vat Phou Laos 14°51N 105°49E 38 E5
Vatersay U.K. 56°55N 7°32W 11 E1
Vati Greece 36°3N 27°53E 25 C9
Vatican City ■ Europe 41°54N 12°27E 22 D5
Vatili Cyprus 35°6N 33°40E 25 D12
Vatiu = Atiu Cook Is. 20°0S 158°10W 65 J12
Vatnajökull Iceland 64°30N 16°48W 8 D5
Vatolakkos Greece 35°27N 23°53E 25 D5
Vatoloha Madag. 17°52S 47°48E 57 B8
Vatomandry Madag. 19°20S 48°59E 57 B8
Vatra-Dornei Romania 47°22N 25°22E 17 E13
Vatrak → India 23°9N 73°2E 42 H5
Vättern Sweden 58°25N 14°30E 9 G16
Vatu Vara Fiji 17°26S 179°31W 59 a
Vatulele Fiji 18°33S 177°37E 59 a
Vaudreuil-Dorion Canada 45°23N 74°3W 83 A10
Vaughn Mont., U.S.A. 47°33N 111°33W 76 C8
Vaughn N. Mex., U.S.A. 34°36N 105°13W 77 J11
Vaujours L. Canada 55°27N 74°15W 72 A5
Vaupés = Uaupés → Brazil 0°2N 67°16W 92 C5
Vaupés □ Colombia 1°0N 71°0W 92 C4
Vauxhall Canada 50°5N 112°9W 70 C6
Vav India 24°22N 71°31E 42 G4
Vavatenina Madag. 17°28S 49°12E 57 B8
Vava'u Tonga 18°36S 174°0W 59 c
Vava'u Group Tonga 18°40S 174°0W 59 c
Vawkavysk Belarus 53°9N 24°30E 17 B13
Växjö Sweden 56°52N 14°50E 9 H16
Vaygach, Ostrov Russia 70°0N 60°0E 28 C7
Veaikevárri = Svappavaara Sweden 67°40N 21°3E 8 C19
Vechte → Neths. 52°34N 6°6E 15 B6
Vedea → Romania 43°42N 25°41E 17 G13
Vedia Argentina 34°30S 61°31W 94 C3
Veendam Neths. 53°5N 6°52E 15 A6
Veenendaal Neths. 52°2N 5°34E 15 B5
Vefsna → Norway 65°48N 13°10E 8 D15
Vega Norway 65°40N 11°55E 8 D14
Vega U.S.A. 35°15N 102°26W 84 D3
Vega Baja Puerto Rico 18°27N 66°23W 89 d
Vegreville Canada 53°30N 112°5W 70 C6
Vejer de la Frontera Spain 36°15N 5°59W 21 D3
Vejle Denmark 55°43N 9°30E 9 J13
Velas, C. Costa Rica 10°21N 85°52W 88 D2
Velasco, Sierra de Argentina 29°20S 67°10W 94 B2
Velddrif S. Africa 32°42S 18°11E 56 E2
Velebit Planina Croatia 44°50N 15°20E 16 F8
Veles Macedonia 41°46N 21°47E 23 D9
Vélez-Málaga Spain 36°48N 4°5W 21 D3
Vélez Rubio Spain 37°41N 2°5W 21 D4
Velhas → Brazil 17°13S 44°49W 93 G10
Velika Kapela Croatia 45°10N 15°5E 16 F8
Velikaya → Russia 57°48N 28°10E 18 C4
Velikaya Kema Russia 45°30N 137°12E 30 B8
Veliki Ustyug Russia 60°47N 46°20E 18 B8
Velikiye Luki Russia 56°25N 30°32E 18 C5
Veliko Tŭrnovo Bulgaria 43°5N 25°41E 23 C11
Velikonda Range India 14°45N 79°10E 40 M11
Velletri Italy 41°41N 12°47E 22 D5
Vellore India 12°57N 79°10E 40 N11
Velsk Russia 61°10N 42°5E 18 B7
Veluwezoom △ Neths. 52°5N 6°0E 15 B6
Velva U.S.A. 48°4N 100°56W 80 A3
Venado Mexico 22°56N 101°6W 86 C4
Venado Tuerto Argentina 33°50S 62°0W 94 C3
Vendée □ France 46°50N 1°35W 20 C3
Vendôme France 47°47N 1°3E 20 C4
Venézia Italy 45°27N 12°21E 22 B5
Venézia, G. di Italy 45°15N 13°0E 22 B5
Venezuela ■ S. Amer. 8°0N 66°0W 92 B5
Venezuela, G. de Venezuela 11°30N 71°0W 92 A4
Vengurla India 15°53N 73°45E 40 M8
Venice = Venézia Italy 45°27N 12°21E 22 B5
Venice U.S.A. 27°6N 82°27W 85 H13
Venkatapuram India 18°20N 80°30E 41 K12
Venlo Neths. 51°22N 6°11E 15 C6
Vennachar Junction Canada 45°5N 77°14W 82 A7
Vennesla Norway 58°15N 7°59E 9 G12
Venray Neths. 51°31N 6°0E 15 C6
Ventana, Punta de la Mexico 24°4N 109°48W 86 C3
Ventana, Sa. de la Argentina 38°0S 62°30W 94 D3
Ventersburg S. Africa 28°7S 27°9E 56 D4
Venterstad S. Africa 30°47S 25°48E 56 E4
Ventnor U.K. 50°36N 1°12W 13 G6
Ventotène Italy 40°47N 13°25E 22 D5
Ventoux, Mt. France 44°10N 5°17E 20 D6
Ventspils Latvia 57°25N 21°32E 9 H19
Ventuarí → Venezuela 3°58N 67°2W 92 C5
Ventucopa U.S.A. 34°50N 119°29W 79 L7
Ventura U.S.A. 34°17N 119°18W 79 L7
Vénus, Pte. Tahiti 17°29S 149°29W 59 d
Venus B. Australia 38°40S 145°42E 63 F4
Venustiano Carranza Mexico 30°25N 115°53W 86 A1
Vera Argentina 29°30S 60°20W 94 B3
Vera Spain 37°15N 1°51W 21 D5
Veracruz Mexico 19°11N 96°8W 87 D5
Veracruz □ Mexico 19°0N 96°15W 87 D5
Veraval India 20°53N 70°27E 42 J4
Verbánia Italy 45°56N 8°33E 20 D8
Vercelli Italy 45°19N 8°25E 20 D8
Verdalsøra Norway 63°48N 11°30E 8 E14
Verde → Argentina 41°56S 65°5W 96 E3
Verde → Goiás, Brazil 18°1S 50°14W 93 G8
Verde → Mato Grosso do Sul, Brazil 21°25S 52°20W 93 H8
Verde → Chihuahua, Mexico 26°29N 107°58W 86 B3
Verde → Oaxaca, Mexico 15°59N 97°50W 87 D5
Verde → Veracruz, Mexico 21°10N 102°50W 86 C4
Verde → Paraguay 23°9S 57°37W 94 A4
Verde → U.S.A. 33°33N 111°40W 74 J17
Verde, Cay Bahamas 23°0N 75°5W 88 B4
Verden Germany 52°55N 9°14E 16 B5
Verdi U.S.A. 39°31N 119°59W 78 F7
Verdun France 49°9N 5°24E 20 B6
Vereeniging S. Africa 26°38S 27°57E 57 D4
Verga, C. Guinea 10°30N 14°10W 50 F3
Vergara Uruguay 32°56S 53°57W 95 C5
Vergemont Cr. → Australia 24°16S 143°16E 62 C3
Vergennes U.S.A. 44°10N 73°15W 83 B11
Veria Greece 40°34N 22°12E 23 D10
Verín Spain 41°57N 7°27W 21 B2
Verkhnetulomskoye Vdkhr. Russia 68°36N 31°12E 8 B24
Verkhnevilyuysk Russia 63°27N 120°18E 29 C13
Verkhniy Baskunchak Russia 48°14N 46°44E 19 E8
Verkhoyansk Russia 67°35N 133°25E 29 C14
Verkhoyansk Ra. = Verkhoyanskiy Khrebet Russia 66°0N 129°0E 29 C13
Verkhoyanskiy Khrebet Russia 66°0N 129°0E 29 C13
Vermilion Canada 53°20N 110°50W 71 C6
Vermilion U.S.A. 41°25N 82°22W 82 E2
Vermilion → Canada 53°22N 110°51W 71 C6
Vermilion B. U.S.A. 29°42N 92°0W 84 G9
Vermilion Bay Canada 49°51N 93°34W 71 D10
Vermilion L. U.S.A. 47°53N 92°26W 80 B7
Vermillion U.S.A. 42°47N 96°56W 80 D5
Vermillion → Canada 47°38N 72°56W 72 C5
Vermont □ U.S.A. 44°0N 73°0W 83 C12
Vernadsky Antarctica 65°0S 64°0W 5 C17
Vernal U.S.A. 40°27N 109°32W 76 F9
Vernalis U.S.A. 37°36N 121°17W 78 H5
Verner Canada 46°25N 80°8W 72 C3
Verneukpan S. Africa 30°0S 21°0E 56 E3
Vernon Canada 50°20N 119°15W 70 C5
Vernon U.S.A. 34°9N 99°17W 84 D5
Vernonia U.S.A. 45°52N 123°11W 78 E3
Vero Beach U.S.A. 27°38N 80°24W 85 H14

Véroia = Veria Greece 40°34N 22°12E 23 D10
Verona Canada 44°29N 76°42W 83 B8
Verona Italy 45°27N 10°59E 22 B4
Versailles France 48°48N 2°7E 20 B5
Vert, C. Senegal 14°45N 17°30W 50 F2
Verulam S. Africa 29°38S 31°2E 57 D5
Verviers Belgium 50°37N 5°52E 15 D5
Veselovskoye Vdkhr. Russia 46°58N 41°25E 19 E7
Vesoul France 47°40N 6°11E 20 C7
Vesterålen Norway 68°45N 15°0E 8 B16
Vestfjorden Norway 67°55N 14°0E 8 C16
Vestmannaeyjar Iceland 63°27N 20°15W 8 E3
Vestspitsbergen Svalbard 78°40N 17°0E 4 B8
Vestvågøya Norway 68°18N 13°50E 8 B15
Vesuvio Italy 40°49N 14°26E 22 D6
Vesuvius, Mt. = Vesuvio Italy 40°49N 14°26E 22 D6
Veszprém Hungary 47°8N 17°57E 17 E9
Vetlanda Sweden 57°24N 15°3E 9 H16
Vetlugu → Russia 56°36N 46°4E 18 C8
Vettore, Mte. Italy 42°49N 13°16E 22 C5
Veurne Belgium 51°5N 2°40E 15 C2
Veys Iran 31°30N 49°0E 45 D6
Vezhen Bulgaria 42°50N 24°20E 23 C11
Vi Thanh Vietnam 9°42N 105°26E 39 H5
Viacha Bolivia 16°39S 68°18W 92 G5
Viamão Brazil 30°5S 51°0W 95 C5
Viana Brazil 3°13S 44°55W 93 D10
Viana do Alentejo Portugal 38°17N 7°59W 21 C2
Viana do Castelo Portugal 41°42N 8°50W 21 B1
Vianden Lux. 49°56N 6°12E 15 E6
Viangchan = Vientiane Laos 17°58N 102°36E 38 D4
Vianópolis Brazil 16°40S 48°35W 93 G9
Vianos Greece 35°2N 25°21E 25 D7
Viaréggio Italy 43°52N 10°14E 22 C4
Vibo Valéntia Italy 38°40N 16°6E 22 E7
Viborg Denmark 56°27N 9°23E 9 H13
Vic Spain 41°58N 2°19E 21 B7
Vicenza Italy 45°33N 11°33E 22 B4
Vich = Vic Spain 41°58N 2°19E 21 B7
Vichada → Colombia 4°55N 67°50W 92 C5
Vichy France 46°9N 3°26E 20 C5
Vicksburg Ariz., U.S.A. 33°45N 113°45W 79 M13
Vicksburg Miss., U.S.A. 32°21N 90°53W 85 E9
Victor India 21°0N 71°30E 42 J4
Victor Harbor Australia 35°30S 138°37E 63 F2
Victoria Argentina 32°40S 60°10W 94 C3
Victoria Canada 48°30N 123°25W 78 B3
Victoria Chile 38°13S 72°20W 96 D2
Victoria China 22°17N 114°9E 33 a
Victoria Malta 36°3N 14°14E 25 C1
Victoria Seychelles 4°38S 55°28E 53 b
Victoria Kans., U.S.A. 38°52N 99°9W 80 F4
Victoria Tex., U.S.A. 28°48N 97°0W 84 G6
Victoria □ Australia 37°0S 144°0E 63 F3
Victoria → Australia 15°10S 129°40E 60 C4
Victoria, Grand L. Canada 47°31N 77°30W 72 C4
Victoria, L. Africa 1°0S 33°0E 54 C3
Victoria, L. Australia 33°57S 141°15E 63 E3
Victoria, Mt. Burma 21°14N 93°55E 41 J18
Victoria Beach Canada 50°40N 96°35W 71 C9
Victoria de Durango = Durango Mexico 24°3N 104°39W 86 C4
Victoria de las Tunas = Las Tunas Cuba 20°58N 76°59W 88 B4
Victoria Falls Zimbabwe 17°58S 25°52E 55 F2
Victoria Harbour Canada 44°45N 79°45W 82 B5
Victoria I. Canada 71°0N 111°0W 68 C9
Victoria L. Canada 48°20N 57°27W 73 C8
Victoria Ld. Antarctica 75°0S 160°0E 5 D11
Victoria Nile → Uganda 2°14N 31°26E 54 B3
Victoria Pk. Belize 16°48N 88°37W 87 D7
Victoria River Australia 16°25S 131°0E 60 C5
Victoria Str. Canada 69°31N 100°30W 68 D11
Victoria West S. Africa 31°25S 23°4E 56 E3
Victoriaville Canada 46°4N 71°56W 73 C5
Victorica Argentina 36°20S 65°30W 94 D2
Victorville U.S.A. 34°32N 117°18W 79 L9
Vicuña Chile 30°0S 70°50W 94 C1
Vicuña Mackenna Argentina 33°53S 64°25W 94 C3
Vidal U.S.A. 34°7N 114°31W 79 L12
Vidalia U.S.A. 32°13N 82°25W 85 E13
Viddolatno = Vindelälven → Sweden 63°55N 19°50E 8 E18
Vidin Bulgaria 43°59N 22°50E 23 C10
Vidisha India 23°28N 77°53E 42 H7
Vidos Greece 39°38N 19°55E 25 A3
Vidzy Belarus 55°23N 26°37E 9 J22
Viedma Argentina 40°50S 63°0W 96 E4
Viedma, L. Argentina 49°30S 72°30W 96 F2
Vielsalm Belgium 50°17N 5°54E 15 D5
Vieng Pou Kha Laos 20°41N 101°4E 38 B3
Vienna = Wien Austria 48°12N 16°22E 16 D9
Vienna Canada 42°41N 80°48W 82 D4
Vienna Ill., U.S.A. 37°25N 88°54W 80 G9
Vienna Mo., U.S.A. 38°11N 91°57W 80 F8
Vienne France 45°31N 4°53E 20 D6
Vienne → France 47°13N 0°5E 20 C4
Vientiane Laos 17°58N 102°36E 38 D4
Vientos, Paso de los Caribbean 20°0N 74°0W 89 C5
Vieques Puerto Rico 18°8N 65°25W 89 d
Vierge Pt. St. Lucia 13°49N 60°53W 89 f
Vierzon France 47°13N 2°5E 20 C5
Viet Quang Vietnam 22°30N 104°48E 38 A5
Viet Tri Vietnam 21°18N 105°25E 38 B5
Vietnam ■ Asia 19°0N 106°0E 38 C6
Vieux Fort St. Lucia 13°46N 60°58W 89 f
Vigan Phil. 17°35N 120°28E 37 A6
Vigévano Italy 45°19N 8°51E 20 D8
Vigia Brazil 0°50S 48°5W 93 D9
Viglas, Akra Greece 35°54N 27°51E 25 D9
Vigo Spain 42°12N 8°41W 21 A1
Vihowa Pakistan 31°8N 70°30E 42 D4
Vihowa → Pakistan 31°8N 70°41E 42 D4
Vijayawada India 16°31N 80°39E 41 L12
Vijosë → Albania 40°37N 19°24E 23 D8
Vík Iceland 63°25N 19°1W 8 E4
Vikeke = Viqueque E. Timor 8°52S 126°23E 37 F7
Viking Canada 53°7N 111°50W 70 C6
Vikna Norway 64°55N 10°58E 8 D14
Vila da Maganja Mozam. 17°18S 37°30E 55 F4
Vila da Ribeira Brava C. Verde Is. 16°32N 24°25W 50 b
Vila do Bispo Portugal 37°5N 8°53W 21 D1
Vila Franca de Xira Portugal 38°57N 8°59W 21 C1
Vila Gamito Mozam. 14°12S 33°0E 55 E3
Vila Gomes da Costa Mozam. 24°20S 33°37E 57 C5
Vila Machado Mozam. 19°15S 34°14E 55 F3
Vila Mouzinho Mozam. 14°48S 34°25E 55 E3
Vila Nova de Gaia Portugal 41°8N 8°37W 21 B1
Vila Real Portugal 41°17N 7°48W 21 B2
Vila-Real Spain 39°55N 0°3W 21 C5
Vila Real de Santo António Portugal 37°10N 7°28W 21 D2
Vila Vasco da Gama Mozam. 14°54S 32°14E 55 E3
Vila Velha Brazil 20°20S 40°17W 95 A7
Vilagarcía de Arousa Spain 42°34N 8°46W 21 A1
Vilaine → France 47°30N 2°27W 20 C2
Vilanandro, Tanjona Madag. 16°11S 44°27E 57 B7
Vilanculos Mozam. 22°1S 35°17E 57 C6
Vilanova i la Geltrú Spain 41°13N 1°40E 21 B6
Vilcheka, Ostrov Russia 80°0N 58°31E 28 B6
Vilcheka, Zemlya Russia 80°30N 60°30E 4 A11
Vileyka Belarus 54°30N 26°53E 17 A14
Vilhelmina Sweden 64°35N 16°39E 8 D17
Vilhena Brazil 12°40S 60°5W 92 F6
Viliya = Neris → Lithuania 55°8N 24°16E 9 J21
Viljandi Estonia 58°28N 25°30E 9 G21
Vilkitskogo, Proliv Russia 78°0N 103°0E 29 B11
Vilkovo = Vylkove Ukraine 45°28N 29°32E 17 F15
Villa Abecia Bolivia 21°0S 68°18W 94 A2
Villa Ana Argentina 28°28S 59°40W 94 B4
Villa Angela Argentina 27°34S 60°45W 94 B3
Villa Bella Bolivia 10°25S 65°22W 92 F5
Villa Cañás Argentina 34°0S 61°35W 94 C3
Villa Constitución Argentina 33°15S 60°20W 94 C3
Villa de Arriaga Mexico 21°56N 101°20W 86 C4
Villa de María Argentina 29°55S 63°43W 94 B3
Villa de Méndez Mexico 25°7N 98°34W 87 B5
Villa Dolores Argentina 31°58S 65°15W 94 C2
Villa Frontera Mexico 26°56N 101°27W 86 B4
Villa Gesell Argentina 37°15S 56°55W 94 D4
Villa Guillermina Argentina 28°15S 59°29W 94 B4
Villa Hayes Paraguay 25°5S 57°20W 94 B4
Villa Hidalgo Mexico 24°15N 99°26W 87 C5
Villa Iris Argentina 38°12S 63°12W 94 D3
Villa María Argentina 32°20S 63°10W 94 C3
Villa Mazán Argentina 28°40S 66°30W 94 B2
Villa Montes Bolivia 21°10S 63°30W 94 A3
Villa Ocampo Argentina 28°30S 59°20W 94 B4
Villa Ocampo Mexico 26°27N 105°31W 86 B3
Villa Ojo de Agua Argentina 29°30S 63°44W 94 B3
Villa San Martín Argentina 28°15S 64°9W 94 B3
Villa Unión Mexico 23°12N 106°14W 86 C3
Villacarlos Spain 39°53N 4°17E 24 B11
Villacarrillo Spain 38°7N 3°3W 21 C4
Villach Austria 46°37N 13°51E 16 E7
Villafranca de Bonany Spain 39°34N 3°5E 24 B10
Villagrán Mexico 24°29N 99°29W 87 C5
Villaguay Argentina 32°0S 59°0W 94 C4
Villahermosa Mexico 17°59N 92°55W 87 D6
Villajoyosa Spain 38°30N 0°12W 21 C5
Villalba Spain 43°26N 7°40W 21 A2
Villanueva U.S.A. 35°16N 105°22W 77 J11
Villanueva de la Serena Spain 38°59N 5°50W 21 C3
Villanueva y Geltrú = Vilanova i la Geltrú Spain 41°13N 1°40E 21 B6
Villarreal = Vila-Real Spain 39°55N 0°3W 21 C5
Villarrica Chile 39°15S 72°15W 96 D2
Villarrica Paraguay 25°40S 56°30W 94 B4
Villarrobledo Spain 39°18N 2°36W 21 C4
Villavicencio Argentina 32°28S 69°0W 94 C2
Villavicencio Colombia 4°9N 73°37W 92 C4
Villaviciosa Spain 43°32N 5°27W 21 A3
Villazón Bolivia 22°0S 65°35W 94 A2
Ville-Marie Canada 47°20N 79°30W 72 C4
Ville Platte U.S.A. 30°41N 92°17W 84 F8
Villena Spain 38°39N 0°52W 21 C5
Villeneuve-d'Ascq France 50°38N 3°9E 20 A5
Villeneuve-sur-Lot France 44°24N 0°42E 20 D4
Villiers S. Africa 27°2S 28°36E 57 D4
Villingen-Schwenningen Germany 48°3N 8°26E 16 D5
Villmanstrand = Lappeenranta Finland 61°3N 28°12E 8 F23
Vilna Canada 54°7N 111°55W 70 C6
Vilnius Lithuania 54°38N 25°19E 9 J21
Vilvoorde Belgium 50°56N 4°26E 15 D4
Vilyuchinsk Russia 52°55N 158°24E 29 D16
Vilyuy → Russia 64°24N 126°26E 29 C13
Vilyuysk Russia 63°40N 121°35E 29 C13
Viña del Mar Chile 33°0S 71°30W 94 C1
Vinarós Spain 40°30N 0°27E 21 B6
Vincennes U.S.A. 38°41N 87°32W 80 F10
Vincennes Bay S. Ocean 66°0S 109°0E 5 C8
Vincent U.S.A. 34°33N 118°11W 79 L8
Vinchina Argentina 28°45S 68°15W 94 B2
Vindelälven → Sweden 63°55N 19°50E 8 E18
Vindeln Sweden 64°12N 19°43E 8 D18
Vindhya Ra. India 22°50N 77°0E 42 H7
Vineland U.S.A. 39°29N 75°2W 81 F16
Vinh Vietnam 18°45N 105°38E 38 C5
Vinh Long Vietnam 10°16N 105°57E 39 G5
Vinh Moc Vietnam 21°18N 105°1E 38 B5
Vinh Yen Vietnam 21°21N 105°35E 38 B5
Vinita U.S.A. 36°39N 95°9W 84 C7
Vinkovci Croatia 45°19N 18°48E 23 B8
Vinnitsa = Vinnytsya Ukraine 49°15N 28°30E 17 D15
Vinnytsya Ukraine 49°15N 28°30E 17 D15
Vinnytsya □ Ukraine 49°20N 28°15E 17 D15
Vinson Massif Antarctica 78°35S 85°25W 5 D16
Vinton Calif., U.S.A. 39°48N 120°10W 78 F6
Vinton Iowa, U.S.A. 42°10N 92°1W 80 D7
Vinton La., U.S.A. 30°11N 93°35W 84 F8
Violet Valley ✪ Australia 17°10S 127°55E 60 C4
Viqueque E. Timor 8°52S 126°23E 37 F7
Virac Phil. 13°30N 124°20E 37 B6
Virachey Cambodia 13°59N 106°49E 38 F6
Virachey △ Cambodia 14°14N 106°55E 38 E6
Virago Sd. Canada 54°0N 132°30W 70 C2
Viramgam India 23°5N 72°0E 42 H5
Viranşehir Turkey 37°13N 39°45E 44 B3
Virawah Pakistan 24°31N 70°46E 42 G4
Virden Canada 49°50N 100°56W 71 D8
Vire France 48°50N 0°53W 20 B3
Vírgenes, C. Argentina 52°19S 68°21W 96 G3
Virgin → U.S.A. 36°28N 114°21W 77 H6
Virgin Gorda Br. Virgin Is. 18°30N 64°26W 89 e
Virgin Is. (British) ☑ W. Indies 18°30N 64°30W 89 e
Virgin Is. (U.S.) ☑ W. Indies 18°20N 65°0W 89 e
Virgin Islands △ U.S. Virgin Is. 18°21N 64°43W 89 C7
Virginia S. Africa 28°8S 26°55E 56 D4
Virginia U.S.A. 47°31N 92°32W 80 B7
Virginia □ U.S.A. 37°30N 78°45W 81 G14
Virginia Beach U.S.A. 36°49N 76°9W 81 F16
Virginia City Mont., U.S.A. 45°18N 111°56W 76 D8
Virginia City Nev., U.S.A. 39°19N 119°39W 78 F7
Virginia Falls Canada 61°38N 125°42W 70 A3
Virginiatown Canada 48°9N 79°36W 72 C4
Viroqua U.S.A. 43°34N 90°53W 80 D8
Virovitica Croatia 45°51N 17°21E 22 B7
Virpur India 21°51N 70°42E 42 J4
Virton Belgium 49°35N 5°32E 15 E5
Virudunagar India 9°30N 77°58E 40 Q10
Virunga △ Dem. Rep. of the Congo 0°5N 29°38E 54 B2
Vis Croatia 43°4N 16°10E 22 C7
Visalia U.S.A. 36°20N 119°18W 78 J7
Visayan Sea Phil. 11°30N 123°30E 37 B6
Visby Sweden 57°37N 18°18E 9 H18
Viscount Melville Sd. Canada 74°10N 108°0W 69 C10
Visé Belgium 50°44N 5°41E 15 D5
Vise, Ostrov Russia 79°33N 76°50E 28 B8
Višegrad Bos.-H. 43°47N 19°17E 23 C8
Viseu Brazil 1°10S 46°5W 93 D9
Viseu Portugal 40°40N 7°55W 21 B2
Vishakhapatnam India 17°45N 83°20E 41 L13
Visnagar India 23°45N 72°32E 42 H5
Viso, Mte. Italy 44°38N 7°5E 20 D7
Visokoi I. Antarctica 56°43S 27°15W 5 B1
Vista U.S.A. 33°12N 117°14W 79 M9
Vistula = Wisła → Poland 54°22N 18°55E 17 A10
Vitebsk = Vitsyebsk Belarus 55°10N 30°15E 18 C5
Viterbo Italy 42°25N 12°6E 22 C5
Viti Levu Fiji 17°30S 177°30E 59 a
Vitigudino Spain 41°1N 6°26W 21 B2
Vitim Russia 59°28N 112°35E 29 D12
Vitim → Russia 59°26N 112°34E 29 D12
Vitória Brazil 20°20S 40°22W 93 H10
Vitória da Conquista Brazil 14°51S 40°51W 93 F10
Vitória de Santo Antão Brazil 8°10S 35°20W 93 E11
Vitoria-Gasteiz Spain 42°50N 2°41W 21 A4
Vitsyebsk Belarus 55°10N 30°15E 18 C5
Vittória Italy 36°57N 14°32E 22 F6
Vittório Véneto Italy 45°59N 12°18E 22 B5
Viveiro Spain 43°39N 7°38W 21 A2
Vivian U.S.A. 32°53N 93°59W 84 E8
Viwa Fiji 17°10S 177°58E 59 a
Vizcaíno, Desierto de Mexico 27°30N 113°45W 86 B2
Vizcaíno, Sierra Mexico 27°30N 114°0W 86 B2
Vize Turkey 41°34N 27°45E 23 D12
Vizianagaram India 18°6N 83°30E 41 K13
Vlaardingen Neths. 51°55N 4°21E 15 C4
Vladikavkaz Russia 43°0N 44°35E 19 F7
Vladimir Russia 56°15N 40°30E 18 C7
Vladimir Volynskiy = Volodymyr-Volynskyy Ukraine 50°50N 24°18E 17 C13
Vladivostok Russia 43°10N 131°53E 30 C5
Vlieland Neths. 53°16N 4°55E 15 A4
Vlissingen Neths. 51°26N 3°34E 15 C3
Vlorë Albania 40°32N 19°28E 23 D8
Vltava → Czech Rep. 50°21N 14°30E 16 D8
Vo Dat Vietnam 11°9N 107°31E 39 G6
Voe U.K. 60°21N 1°16W 11 A7
Vogelkop = Doberai, Jazirah Indonesia 1°25S 133°0E 37 E8
Vogelsberg Germany 50°31N 9°12E 16 C5
Voghera Italy 44°59N 9°1E 20 D8
Vohibinany Madag. 18°49S 49°4E 57 B8
Vohilava Madag. 21°4S 48°0E 57 C8
Vohimarina = Iharana Madag. 13°25S 50°0E 57 A9
Vohimena, Tanjon' i Madag. 25°36S 45°8E 57 D8
Vohipeno Madag. 22°22S 47°51E 57 C8
Voi Kenya 3°25S 38°32E 54 C4
Voiron France 45°22N 5°35E 20 D6
Voisey B. Canada 56°15N 61°50W 73 A7
Vojmsjön Sweden 65°0N 16°24E 8 D17
Vojvodina □ Serbia 45°20N 20°0E 23 B9
Volborg U.S.A. 45°51N 105°41W 76 D11
Volcán de Colima △ Mexico 19°30N 103°40W 86 D4
Volcano Is. = Kazan-Rettō Pac. Oc. 25°0N 141°0E 64 E6
Volcans △ Rwanda 1°30S 29°26E 54 C2
Volda Norway 62°9N 6°5E 8 E12
Volga = Privolzhskiy □ Russia 56°0N 50°0E 28 D6
Volga → Russia 46°0N 48°30E 19 E8
Volga Hts. = Privolzhskaya Vozvyshennost Russia 51°0N 46°0E 19 D8
Volgodonsk Russia 47°33N 42°5E 19 E7
Volgograd Russia 48°40N 44°25E 19 E7
Volgogradskoye Vdkhr. Russia 50°0N 45°20E 19 D8
Volkhov → Russia 60°8N 32°20E 18 B5
Volkovysk = Vawkavysk Belarus 53°9N 24°30E 17 B13
Volksrust S. Africa 27°24S 29°53E 57 D4
Volochanka Russia 71°0N 94°28E 29 B10
Volodymyr-Volynskyy Ukraine 50°50N 24°18E 17 C13
Vologda Russia 59°10N 39°45E 18 C6
Volos Greece 39°24N 22°59E 23 E10
Volosovo Russia 59°27N 29°32E 9 G23
Volovets Ukraine 48°43N 23°11E 17 D12
Volozhin = Valozhyn Belarus 54°3N 26°30E 17 A14
Volsk Russia 52°5N 47°22E 18 D8
Volta → Ghana 5°46N 0°41E 48 F4
Volta, L. Ghana 7°30N 0°0 50 G6
Volta Redonda Brazil 22°31S 44°5W 95 A7
Voltaire, C. Australia 14°16S 125°35E 60 B4
Volterra Italy 43°24N 10°51E 22 C4
Volturno → Italy 41°1N 13°55E 22 D5
Volyn □ Ukraine 51°15N 24°30E 17 C13
Volzhskiy Russia 48°56N 44°46E 19 E7
Vomo Fiji 17°30S 177°15E 59 a
Vondrozo Madag. 22°49S 47°20E 57 C8
Vopnafjörður Iceland 65°45N 14°50W 8 D6
Vorkuta Russia 67°48N 64°20E 18 A11
Vormsi Estonia 59°1N 23°13E 9 G20
Voronezh Russia 51°40N 39°10E 19 D6
Võrts järv Estonia 58°16N 26°3E 9 G22
Võru Estonia 57°48N 26°54E 9 H22
Vosges France 48°20N 7°10E 20 B7
Voss Norway 60°38N 6°26E 8 F12
Vostok Antarctica 78°30S 106°50E 5 D8
Vostok I. Kiribati 10°5S 152°23W 65 J12
Votkinsk Russia 57°0N 53°55E 18 C9
Votkinskoye Vdkhr. Russia 57°22N 55°12E 18 C10
Votsuri-Shima Japan 25°45N 123°29E 31 M1
Vouga → Portugal 40°41N 8°40W 21 B1
Vouxa, Akra Greece 35°37N 23°32E 25 D5
Voyageurs △ U.S.A. 48°32N 93°0W 80 A7
Voynitsa Russia 65°10N 30°20E 8 D24
Voyvozh Russia 62°56N 54°56E 28 C6
Vozhe, Ozero Russia 60°45N 39°0E 18 B6
Voznesensk Ukraine 47°35N 31°21E 19 E5
Voznesenye Russia 61°0N 35°28E 18 B6
Vrangel Russia 42°43N 133°5E 30 C6
Vrangelya, Ostrov Russia 71°0N 180°0E 29 B18
Vranje Serbia 42°34N 21°54E 23 C9
Vratsa Bulgaria 43°15N 23°30E 23 C10
Vrbas → Bos.-H. 45°8N 17°29E 22 B7
Vrede S. Africa 27°24S 29°6E 57 D4
Vredefort S. Africa 27°0S 27°22E 56 D4
Vredenburg S. Africa 32°56S 18°0E 56 E2
Vredendal S. Africa 31°41S 18°35E 56 E2
Vrindavan India 27°37N 77°40E 42 F7
Vrisses Greece 35°23N 24°13E 25 D6
Vršac Serbia 45°8N 21°20E 23 B9
Vryburg S. Africa 26°55S 24°45E 56 D3
Vryheid S. Africa 27°45S 30°47E 57 D5
Vu Liet Vietnam 18°43N 105°23E 38 C5
Vukovar Croatia 45°21N 18°59E 23 B8
Vuktyl Russia 63°52N 57°20E 28 C6
Vulcan Canada 50°25N 113°15W 70 C6
Vulcan Romania 45°23N 23°17E 17 F12
Vulcaneşti Moldova 45°41N 28°18E 17 F15
Vulcano Italy 38°24N 14°58E 22 E6
Vulkaneshty = Vulcaneşti Moldova 45°41N 28°18E 17 F15
Vunduzi → Mozam. 18°56S 34°1E 55 F3
Vung Tau Vietnam 10°21N 107°4E 39 G6
Vunidawa Fiji 17°50S 178°21E 59 a
Vunisea Fiji 19°3S 178°10E 59 a
Vwaza △ Malawi 10°58S 33°25E 55 E3
Vyartsilya Russia 62°8N 30°45E 8 E24
Vyatka = Kirov Russia 58°35N 49°40E 18 C8
Vyatka → Russia 55°37N 51°28E 18 C9
Vyatskiye Polyany Russia 56°14N 51°5E 18 C9
Vyazemskiy Russia 47°32N 134°45E 29 E14
Vyazma Russia 55°10N 34°15E 18 C5
Vyborg Russia 60°43N 28°47E 8 F23
Vychegda → Russia 61°18N 46°36E 18 B8
Vychodné Beskydy Europe 49°20N 22°0E 17 D11
Vyg-ozero Russia 63°47N 34°29E 18 B5
Vylkove Ukraine 45°28N 29°32E 17 F15
Vynohradiv Ukraine 48°9N 23°2E 17 D12
Vyrnwy, L. U.K. 52°48N 3°31W 12 E4
Vyshniy Volochek Russia 57°30N 34°30E 18 C5
Vyshzha = imeni 26 Bakinskikh Komissarov Turkmenistan 39°22N 54°10E 45 B7
Vyškov Czech Rep. 49°17N 17°0E 17 D9
Vytegra Russia 61°0N 36°27E 18 B6

W

W.A.C. Bennett Dam Canada 56°2N 122°6W 70 B4
Wa Ghana 10°7N 2°25W 50 F5
Waal → Neths. 51°37N 5°0E 15 C5
Waalwijk Neths. 51°42N 5°4E 15 C5
Waanyi-Garawa ✪ Australia 18°2S 137°33E 62 B2
Wabakimi △ Canada 50°43N 89°29W 72 B
Wabana Canada 47°40N 53°0W 73 C
Wabasca → Canada 58°22N 115°20W 70 B
Wabasca-Desmarais Canada 55°57N 113°56W 70 B
Wabash U.S.A. 40°48N 85°49W 81 E1
Wabash → U.S.A. 37°48N 88°2W 80 G
Wabigoon L. Canada 49°44N 92°44W 71 D1
Wabowden Canada 54°55N 98°38W 71 C
Wabuk Pt. Canada 55°20N 85°5W 72 A
Wabush Canada 52°55N 66°52W 73 B
Waco U.S.A. 31°33N 97°9W 84 F
Waconichi, L. Canada 50°8N 74°0W 72 B
Wad Hamid Sudan 16°30N 32°45E 51 E1
Wad Medanî Sudan 14°28N 33°30E 51 F1
Wad Thana Pakistan 27°22N 66°23E 42 F
Wadai Africa 12°0N 19°0E 48 E
Wadayama Japan 35°19N 134°52E 31 G
Waddeneilanden Neths. 53°25N 5°10E 15 A
Waddenzee Neths. 53°6N 5°10E 15 A
Waddington U.S.A. 44°52N 75°12W 83 B
Waddington, Mt. Canada 51°23N 125°15W 70 C
Waddy Pt. Australia 24°58S 153°21E 63 C
Wadebridge U.K. 50°31N 4°51W 13 G
Wadena Canada 51°57N 103°47W 71 C
Wadena U.S.A. 46°26N 95°8W 80 B
Wadeye Australia 14°28S 129°52E 60 B
Wadhams Canada 51°30N 127°30W 70 C
Wadhwan India 22°42N 71°41E 42 H
Wādī as Sīr Jordan 31°56N 35°49E 46 D
Wadi Halfa Sudan 21°53N 31°19E 51 D1
Wadi Rum △ Jordan 29°30N 35°20E 46 F
Wadsworth Nev., U.S.A. 39°38N 119°17W 76 G
Wadsworth Ohio, U.S.A. 41°2N 81°44W 82 E
Waegwan S. Korea 35°59N 128°23E 35 G1
Wafangdian China 39°38N 121°58E 35 E1
Wafrah Kuwait 28°33N 47°56E 44 D
Wagait ✪ Australia 13°1S 130°5E 60 B
Wageningen Neths. 51°58N 5°40E 15 C
Wager B. Canada 65°26N 88°40W 69 D1
Wagga Wagga Australia 35°7S 147°24E 63 F
Waghete Indonesia 4°10S 135°50E 37 E
Wagin Australia 33°17S 117°25E 61 F
Wagner U.S.A. 43°5N 98°18W 80 D
Wagon Mound U.S.A. 36°1N 104°42W 77 H1
Wagoner U.S.A. 35°58N 95°22W 84 D
Wah Pakistan 33°45N 72°40E 42 C
Wahai Indonesia 2°48S 129°35E 37 E
Wâḥid Egypt 30°48N 32°21E 46 E
Wahnai Afghan. 32°40N 65°50E 42 C
Wahoo U.S.A. 41°13N 96°37W 80 E
Wahpeton U.S.A. 46°16N 96°36W 80 E
Waialua U.S.A. 21°34N 158°8W 75
Waiau → N.Z. 42°47S 173°22E 59
Waibeem Indonesia 0°30S 132°59E 37 E
Waigeo Indonesia 0°20S 130°40E 37 E
Waihi N.Z. 37°23S 175°52E 59 E
Waihou → N.Z. 37°15S 175°40E 59 E
Waika Dem. Rep. of the Congo 2°22S 25°42E 54 C
Waikabubak Indonesia 9°45S 119°25E 37
Waikaremoana, L. N.Z. 38°49S 177°9E 59
Waikari N.Z. 42°58S 172°41E 59
Waikato → N.Z. 37°23S 174°43E 59
Waikelo Indonesia 9°24S 119°19E 60
Waikerie Australia 34°9S 140°0E 63
Waikokopu N.Z. 39°3S 177°52E 59
Waikouaiti N.Z. 45°36S 170°41E 59
Waimakariri → N.Z. 43°24S 172°42E 59
Waimate N.Z. 44°45S 171°3E 59
Wainganga → India 18°50N 79°55E 40 K
Waingapu Indonesia 9°35S 120°11E 37
Waini → Guyana 8°20N 59°50W 92
Wainwright Canada 52°50N 110°50W 71
Wainwright U.S.A. 70°38N 160°2W 74
Waiouru N.Z. 39°28S 175°41E 59
Waipara N.Z. 43°3S 172°46E 59
Waipawa N.Z. 39°56S 176°38E 59
Waipiro N.Z. 38°2S 178°22E 59
Waipoua Forest N.Z. 35°39S 173°33E 59
Waipu N.Z. 35°59S 174°29E 59
Waipukurau N.Z. 40°1S 176°33E 59
Wairakei N.Z. 38°37S 176°6E 59
Wairarapa, L. N.Z. 41°14S 175°15E 59
Wairoa N.Z. 39°3S 177°25E 59
Waitaki → N.Z. 44°56S 171°7E 59
Waitangi N.Z. 35°16S 174°5E 59
Waitara N.Z. 38°59S 174°15E 59
Waitomo Caves N.Z. 38°16S 175°7E 59
Waitsburg U.S.A. 46°16N 118°9W 76
Waiuku N.Z. 37°15S 174°45E 59
Wajima Japan 37°30N 137°0E 31
Wajir Kenya 1°42N 40°5E 54
Wakasa Japan 35°20N 134°24E 31
Wakasa-Wan Japan 35°40N 135°30E 31
Wakatipu, L. N.Z. 45°5S 168°33E 59
Wakaw Canada 52°39N 105°44W 71
Wakaya Fiji 17°37S 179°0E 5
Wakayama Japan 34°15N 135°15E 31
Wakayama □ Japan 33°50N 135°30E 31
Wake Forest U.S.A. 35°59N 78°30W 85 D
Wake I. Pac. Oc. 19°18N 166°36E 64
WaKeeney U.S.A. 39°1N 99°53W 80
Wakefield Jamaica 18°26N 77°42W 8
Wakefield N.Z. 41°24S 173°5E 59
Wakefield U.K. 53°41N 1°29W 12
Wakefield Mass., U.S.A. 42°30N 71°5W 83 D
Wakefield Mich., U.S.A. 46°29N 89°56W 80

kkanai Japan 45°28N 141°35E 30 B10
kkerstroom S. Africa 27°24S 30°10E 57 D5
kool Australia 35°28S 144°23E 63 F3
kool → Australia 35°55S 143°33E 63 F3
kre Indonesia 0°19S 131°5E 37 E8
kuach, L. Canada 55°34N 67°32W 73 A6
agunya ✪ Australia 23°10S 120°50E 60 D3
amba Zambia 13°30S 28°42E 55 E2
brzych Poland 50°45N 16°18E 16 C9
bury Hill U.K. 51°21N 1°28W 13 F6
cha Australia 30°55S 151°31E 63 E5
cheren Neths. 51°30N 3°35E 15 C3
cott U.S.A. 41°46N 106°51W 76 F10
cz Poland 53°17N 16°27E 16 B9
dburg Ra. Australia 24°40S 117°35E 61 D2
den Colo., U.S.A. 40°44N 106°17W 76 F10
den N.Y., U.S.A. 41°34N 74°11W 83 E10
dport U.S.A. 44°26N 124°4W 76 D1
dron U.S.A. 34°54N 94°5W 84 D7
ebing Australia 30°41S 116°13E 61 F2
es □ U.K. 52°19N 4°43W 13 E3
es I. Canada 68°1N 86°40W 69 D14
gett Australia 30°0S 148°5E 63 E4
green Coast
ntarctica 75°15S 105°0W 5 D15
lker U.S.A. 47°6N 94°35W 80 B6
lker, L. Canada 50°20N 67°11W 73 B6
lker L. Canada 54°42N 95°57W 71 C9
lker L. U.S.A. 38°42N 118°43W 76 G4
lkerston Australia 21°11S 149°8E 62 b
lkerton Canada 44°10N 81°10W 82 B3
ll U.S.A. 44°0N 102°8W 80 C2
lla Walla U.S.A. 46°4N 118°20W 76 C4
llace Idaho, U.S.A. 47°28N 115°56W 76 C6
llace N.C., U.S.A. 34°44N 77°59W 85 D16
llaceburg Canada 42°34N 82°23W 82 D2
llachia = Valahia
omania 44°35N 25°0E 17 F13
llal Australia 26°32S 146°7E 63 D4
llam Cr. →
ustralia 28°40S 147°20E 63 D4
llambin, L. Australia 30°57S 117°35E 61 F2
llangarra Australia 28°56S 151°58E 63 D5
llaroo Australia 33°56S 137°39E 63 E2
llenpaupack, L.
.S.A. 41°25N 75°15W 83 E9
llingford Conn.,
.S.A. 41°27N 72°50W 83 E12
llingford Vt., U.S.A. 43°28N 72°58W 83 C12
llis & Futuna, Is. ☑
ac. Oc. 13°18S 176°10W 58 C11
llowa U.S.A. 45°34N 117°32W 76 D5
llowa Mts. U.S.A. 45°20N 117°30W 76 D5
lls U.K. 60°14N 1°33W 11 A7
lls of Jerusalem △
ustralia 41°56S 146°15E 63 G4
llula U.S.A. 46°5N 118°54W 76 C4
llumbilla Australia 26°33S 149°9E 63 D4
lmsley L. Canada 63°25N 108°36W 71 A7
lney, I. of U.K. 54°6N 3°15W 12 C4
lnut Canyon △
.S.A. 35°15N 111°20W 77 J8
lnut Creek U.S.A. 37°54N 122°4W 78 H4
lnut Ridge U.S.A. 36°4N 90°57W 85 C9
lpole Australia 34°58S 116°44E 61 F2
lpole U.S.A. 42°9N 71°15W 83 D13
lpole-Nornalup △
ustralia 35°0S 116°45E 61 G2
lsall U.K. 52°35N 1°58W 13 E6
lsenburg U.S.A. 37°38N 104°47W 77 H11
lsh U.S.A. 37°23N 102°17W 77 H12
lsh → Australia 16°31S 143°42E 62 B3
lsingham Canada 42°40N 80°31W 82 D4
lter F. George Res.
.S.A. 31°38N 85°4W 85 F12
lterboro U.S.A. 32°55N 80°40W 85 E14
lters U.S.A. 34°22N 98°19W 84 D5
ltham U.S.A. 42°23N 71°14W 83 D13
ltman U.S.A. 43°4N 107°12W 76 E10
lton U.S.A. 42°10N 75°8W 83 D9
lton-on-the-Naze U.K. 51°51N 1°17E 13 F9
lvis Bay Namibia 23°0S 14°28E 56 C1
lvisbaai = Walvis Bay
amibia 23°0S 14°28E 56 C1
mba
em. Rep. of the Congo 2°10N 27°57E 54 B2
mba Kenya 0°58N 37°19E 54 B4
mbardi ✪ Australia 16°2S 130°55E 60 C5
mego U.S.A. 39°12N 96°18W 80 F5
mena Indonesia 4°4S 138°57E 37 E9
mpana Karlantijpa ✪
ustralia 17°45S 132°10E 60 C5
mpaya ✪ Australia 17°20S 135°0E 62 B2
mpsville U.S.A. 43°4N 75°42W 83 C9
msutter U.S.A. 41°40N 107°58W 76 F10
mulan Indonesia 3°27S 126°7E 37 E7
n Xian China 38°47N 115°7E 34 E8
na Pakistan 32°20N 69°32E 42 C3
naaring Australia 29°38S 144°9E 63 D3
naka N.Z. 44°42S 169°9E 59 F2
naka, L. N.Z. 44°33S 169°7E 59 F2
napitei L. Canada 46°45N 80°40W 72 C3
ndel Sea = McKinley Sea
rctic 82°0N 0°0 4 A7
ndhari Pakistan 27°42N 66°48E 42 F2
ndoan Australia 26°5S 149°55E 63 D4
nfu China 40°8N 122°38E 35 D12
ng → Thailand 17°8N 99°2E 38 D2
ng Noi Thailand 14°13N 100°44E 38 E3
ng Saphung
hailand 17°18N 101°46E 38 D3
ng Thong Thailand 16°50N 100°26E 38 D3
nga
em. Rep. of the Congo 2°58N 29°12E 54 B2
ngal Indonesia 6°8S 134°9E 37 F8
nganella Australia 35°6S 144°49E 63 F3
nganui N.Z. 39°56S 175°3E 59 C5
ngaratta Australia 36°21S 146°19E 63 F4
ngary Australia 34°35S 135°29E 63 E2
ngcun China 36°41N 117°41E 35 F9
Wangdu China 38°40N 115°7E 34 E8
Wangerooge Germany 53°47N 7°54E 16 B4
Wangiwangi Indonesia 5°22S 123°37E 37 F6
Wangqing China 43°12N 129°42E 35 C15
Wanimiyn ✪ Australia 15°55S 130°49E 60 C5
Wanjina Wunggurr Wilinggin ✪
Australia 16°0S 127°0E 60 C4
Wankaner India 22°35N 71°0E 42 H4
Wanless Canada 54°11N 101°21W 71 C8
Wanleweyne Somali Rep. 2°37N 44°54E 47 G3
Wanneroo Australia 31°42S 115°46E 61 F2
Wanning China 18°48N 110°22E 38 C8
Wannoo Billabong Roadhouse
Australia 27°25S 115°49E 61 E2
Wanon Niwat Thailand 17°38N 103°46E 38 D4
Wanqinsha China 22°43N 113°33E 33 a
Wanquan China 40°50N 114°40E 34 D8
Wanrong China 35°25N 110°50E 34 G6
Wanshan Qundao China 21°57N 113°45E 33 a
Wantage U.K. 51°35N 1°25W 13 F6
Wanxian China 30°42N 108°20E 32 E10
Wanzai China 22°12N 113°31E 33 a
Wapakoneta U.S.A. 40°34N 84°12W 81 E11
Wapato U.S.A. 46°27N 120°25W 76 C3
Wapawekka L. Canada 54°55N 104°40W 71 C8
Wapikopa L. Canada 52°56N 87°53W 72 B2
Wapiti → Canada 55°5N 118°18W 70 B5
Wappingers Falls
U.S.A. 41°36N 73°55W 83 E11
Wapsipinicon → U.S.A. 41°44N 90°19W 80 E8
Wapusk △ Canada 57°46N 93°22W 71 B10
Warakurna Australia 24°59S 128°17E 61 D4
Warangal India 17°58N 79°35E 40 L11
Waraseoni India 21°45N 80°2E 43 J9
Waratah Australia 41°30S 145°30E 63 G4
Waratah B. Australia 38°54S 146°5E 63 F4
Warburton Vic.,
Australia 37°47S 145°42E 63 F4
Warburton W. Austral.,
Australia 26°8S 126°35E 61 E4
Warburton ✪ Australia 26°7S 126°34E 61 E4
Warburton → Australia 28°4S 137°28E 63 D2
Warburton Groove
Australia 28°18S 137°8E 63 A2
Warburton Ra.
Australia 25°55S 126°28E 61 E4
Ward N.Z. 41°49S 174°11E 59 D5
Ward → Australia 26°28S 146°6E 63 D4
Ward Mt. U.S.A. 37°12N 118°54W 78 H8
Warden S. Africa 27°50S 29°0E 57 D4
Wardha India 20°45N 78°39E 40 J11
Wardha → India 19°57N 79°11E 40 K11
Wardsville Canada 42°39N 81°45W 82 D3
Ware U.K. 51°49N 0°0 13 F8
Ware U.S.A. 42°16N 72°14W 83 D12
Waregem Belgium 50°53N 3°27E 15 D3
Wareham U.S.A. 41°46N 70°43W 83 E14
Waremme Belgium 50°43N 5°15E 15 D5
Warialda Australia 29°29S 150°33E 63 D5
Wariap Indonesia 1°30S 134°5E 37 E8
Warin Chamrap
Thailand 15°12N 104°53E 38 E5
Warkopi Indonesia 1°12S 134°9E 37 E8
Warm Springs U.S.A. 38°10N 116°20W 77 G5
Warman Canada 52°19N 106°30W 71 C7
Warmbad = Bela Bela
S. Africa 24°51S 28°19E 57 C4
Warmbad Namibia 28°25S 18°42E 56 D2
Warminster U.K. 51°12N 2°10W 13 F5
Warminster U.S.A. 40°12N 75°6W 83 F9
Warmun Australia 17°2S 128°12E 60 C4
Warner Mts. U.S.A. 41°40N 120°15W 76 F3
Warner Robins U.S.A. 32°37N 83°36W 85 E13
Waroona Australia 32°50S 115°58E 61 F2
Warrabri ✪ Australia 21°1S 134°19E 62 C1
Warracknabeal Australia 36°9S 142°26E 63 F3
Warragul Australia 38°10S 145°58E 63 F4
Warrego → Australia 30°24S 145°21E 63 E4
Warrego Ra. Australia 24°58S 146°0E 62 C4
Warren Australia 31°42S 147°51E 63 E4
Warren Ark., U.S.A. 33°37N 92°4W 84 E8
Warren Mich., U.S.A. 42°28N 83°1W 81 D12
Warren Minn., U.S.A. 48°12N 96°46W 80 A5
Warren Ohio, U.S.A. 41°14N 80°49W 82 E4
Warren Pa., U.S.A. 41°51N 79°9W 82 E5
Warrenpoint U.K. 54°6N 6°15W 10 B5
Warrensburg Mo.,
U.S.A. 38°46N 93°44W 80 F7
Warrensburg N.Y.,
U.S.A. 43°29N 73°46W 83 C11
Warrenton S. Africa 28°9S 24°47E 56 D3
Warrenton U.S.A. 46°10N 123°56W 78 D3
Warri Nigeria 5°30N 5°41E 50 G7
Warrington U.K. 53°24N 2°35W 12 D5
Warrington U.S.A. 30°23N 87°17W 85 F11
Warrington □ U.K. 53°24N 2°35W 12 D5
Warrnambool Australia 38°25S 142°30E 63 F3
Warroad U.S.A. 48°54N 95°19W 80 A6
Warruwi Australia 11°36S 133°20E 62 A1
Warsa Indonesia 0°47S 135°55E 37 E9
Warsak Dam Pakistan 34°11N 71°19E 42 B4
Warsaw = Warszawa
Poland 52°14N 21°0E 17 B11
Warsaw Ind., U.S.A. 41°14N 85°51W 81 E11
Warsaw N.Y., U.S.A. 42°45N 78°8W 82 D6
Warsaw Ohio, U.S.A. 40°20N 82°0W 82 F3
Warszawa Poland 52°14N 21°0E 17 B11
Warta → Poland 52°35N 14°39E 16 B8
Warthe = Warta →
Poland 52°35N 14°39E 16 B8
Waru Indonesia 3°30S 130°36E 37 E8
Warumungu ✪
Australia 19°15S 134°44E 62 B1
Warwick Australia 28°10S 152°1E 63 D5
Warwick U.K. 52°18N 1°35W 13 E6
Warwick N.Y., U.S.A. 41°16N 74°22W 83 E10
Warwick R.I., U.S.A. 41°42N 71°28W 83 E13
Warwickshire □ U.K. 52°14N 1°38W 13 E6
Wasaga Beach Canada 44°31N 80°1W 82 B4
Wasagaming Canada 50°39N 99°58W 71 C9
Wasatch Ra. U.S.A. 40°0N 111°30W 76 F8
Wasbank S. Africa 28°15S 30°9E 57 D5
Wasco Calif., U.S.A. 35°36N 119°20W 79 K7
Wasco Oreg., U.S.A. 45°36N 120°42W 76 D3
Waseca U.S.A. 44°5N 93°30W 80 C7
Wasekamio L. Canada 56°45N 108°45W 71 B7
Wash, The U.K. 52°58N 0°20E 12 E8
Washago Canada 44°45N 79°20W 82 B5
Washburn N. Dak.,
U.S.A. 47°17N 101°2W 80 B3
Washburn Wis., U.S.A. 46°40N 90°54W 80 B8
Washim India 20°3N 77°0E 40 J10
Washington U.K. 54°55N 1°30W 12 C6
Washington D.C., U.S.A. 38°53N 77°2W 81 F15
Washington Ga.,
U.S.A. 33°44N 82°44W 85 E13
Washington Ind.,
U.S.A. 38°40N 87°10W 80 F10
Washington Iowa,
U.S.A. 41°18N 91°42W 80 E8
Washington Mo., U.S.A. 38°33N 91°1W 80 F8
Washington N.C.,
U.S.A. 35°33N 77°3W 85 D16
Washington N.J.,
U.S.A. 40°46N 74°59W 83 F10
Washington Pa., U.S.A. 40°10N 80°15W 82 F4
Washington Utah,
U.S.A. 37°8N 113°31W 77 H7
Washington □ U.S.A. 47°30N 120°30W 76 C3
Washington, Mt.
U.S.A. 44°16N 71°18W 83 B13
Washington Court House
U.S.A. 39°32N 83°26W 81 F12
Washington I. U.S.A. 45°23N 86°54W 80 C10
Washougal U.S.A. 45°35N 122°21W 78 E4
Washpool △ Australia 29°22S 152°20E 63 D5
Wasian Indonesia 1°47S 133°19E 37 E8
Wasilla U.S.A. 61°35N 149°26W 68 E2
Wasini Marine ⌂ Kenya 4°33S 39°14E 54 C4
Wasior Indonesia 2°43S 134°30E 37 E8
Wāsiţ □ Iraq 32°50N 45°50E 44 C5
Waskaganish Canada 51°30N 78°40W 72 B4
Waskaiowaka L.
Canada 56°33N 96°23W 71 B9
Waskesiu Lake Canada 53°55N 106°5W 71 C7
Wasserkuppe Germany 50°29N 9°55E 16 C5
Waswanipi Canada 49°40N 76°29W 72 C4
Waswanipi, L. Canada 49°35N 76°40W 72 C4
Watampone Indonesia 4°29S 120°25E 37 E6
Watamu Kenya 3°23S 40°0E 54 C5
Watarrka △ Australia 24°20S 131°30E 60 D5
Water Park Pt.
Australia 22°56S 150°47E 62 C5
Water Valley U.S.A. 34°10N 89°38W 85 D10
Waterberg Plateau ⌂
Namibia 20°25S 17°18E 56 C2
Waterberge S. Africa 24°10S 28°0E 57 C4
Waterbury Conn., U.S.A. 41°33N 73°3W 83 E11
Waterbury Vt., U.S.A. 44°20N 72°46W 83 B12
Waterbury L. Canada 58°10N 104°22W 71 B8
Waterdown Canada 43°20N 79°53W 82 C5
Waterford Canada 42°56N 80°17W 82 D4
Waterford Ireland 52°15N 7°8W 10 D4
Waterford Calif., U.S.A. 37°38N 120°46W 78 H6
Waterford Pa., U.S.A. 41°57N 79°59W 82 E5
Waterford □ Ireland 52°10N 7°40W 10 D4
Waterford Harbour
Ireland 52°8N 6°58W 10 D5
Waterhen L. Canada 52°10N 99°40W 71 C9
Waterloo Belgium 50°43N 4°25E 15 D4
Waterloo Ont., Canada 43°30N 80°32W 82 C4
Waterloo Qué., Canada 45°22N 72°32W 83 A12
Waterloo Ill., U.S.A. 38°20N 90°9W 80 F8
Waterloo Iowa, U.S.A. 42°30N 92°21W 80 D7
Waterloo N.Y., U.S.A. 42°54N 76°52W 82 D8
Watersmeet U.S.A. 46°16N 89°11W 80 B9
Waterton Lakes △
Canada 48°45N 115°0W 70 D6
Watertown Conn.,
U.S.A. 41°36N 73°7W 83 E11
Watertown N.Y., U.S.A. 43°59N 75°55W 83 C9
Watertown S. Dak.,
U.S.A. 44°54N 97°7W 80 C5
Watertown Wis., U.S.A. 43°12N 88°43W 80 D9
Waterval-Boven S. Africa 25°40S 30°18E 57 D5
Waterville Canada 45°16N 71°54W 83 A13
Waterville Maine,
U.S.A. 44°33N 69°38W 81 C19
Waterville N.Y., U.S.A. 42°56N 75°23W 83 D9
Waterville Pa., U.S.A. 41°19N 77°21W 82 E7
Waterville Wash., U.S.A. 47°39N 120°4W 76 C3
Watervliet U.S.A. 42°44N 73°42W 83 D11
Wates Indonesia 7°51S 110°10E 37 G14
Watford Canada 42°57N 81°53W 82 D3
Watford U.K. 51°40N 0°24W 13 F7
Watford City U.S.A. 47°48N 103°17W 80 B2
Wathaman → Canada 57°16N 102°59W 71 B8
Wathaman L. Canada 56°58N 103°44W 71 B8
Watheroo Australia 30°15S 116°5E 61 F2
Watheroo △ Australia 30°19S 115°48E 61 F2
Wating China 35°40N 106°38E 34 G4
Watkins Glen U.S.A. 42°23N 76°52W 82 D8
Watling I. = San Salvador I.
Bahamas 24°0N 74°40W 89 B5
Watonga U.S.A. 35°51N 98°25W 84 D5
Watrous Canada 51°40N 105°25W 71 C7
Watrous U.S.A. 35°48N 104°59W 77 J11
Watsa
Dem. Rep. of the Congo 3°4N 29°30E 54 B2
Watseka U.S.A. 40°47N 87°44W 80 E10
Watson Canada 52°10N 104°30W 71 C8
Watson Lake Canada 60°6N 128°49W 70 A3
Watsontown U.S.A. 41°5N 76°52W 82 E8
Watsonville U.S.A. 36°55N 121°45W 78 J5
Wattiwarriganna Cr. →
Australia 28°57S 136°10E 63 D2
Watuata = Batuata
Indonesia 6°12S 122°42E 37 F6
Watubela, Kepulauan
Indonesia 4°28S 131°35E 37 E8
Watubela Is. = Watubela,
Kepulauan Indonesia 4°28S 131°35E 37 E8
Wau = Wâw Sudan 7°45N 28°1E 51 G11
Waubamik Canada 45°27N 80°1W 82 A4
Waubay U.S.A. 45°20N 97°18W 80 C5
Wauchope N.S.W.,
Australia 31°28S 152°45E 63 E5
Wauchope N. Terr.,
Australia 20°36S 134°15E 62 C1
Wauchula U.S.A. 27°33N 81°49W 85 H14
Waukarlycarly, L.
Australia 21°18S 121°56E 60 D3
Waukegan U.S.A. 42°22N 87°50W 80 D10
Waukesha U.S.A. 43°1N 88°14W 80 D9
Waukon U.S.A. 43°16N 91°29W 80 D8
Waupaca U.S.A. 44°21N 89°5W 80 C9
Waupun U.S.A. 43°38N 88°44W 80 D9
Waurika U.S.A. 34°10N 98°0W 84 D6
Wausau U.S.A. 44°58N 89°38W 80 C9
Wautoma U.S.A. 44°4N 89°18W 80 C9
Wauwatosa U.S.A. 43°2N 88°0W 80 D9
Wave Hill = Kalkarindji
Australia 17°30S 130°47E 60 C5
Wave Rock ⌂ Australia 32°26S 118°53E 61 F2
Waveney → U.K. 52°35N 1°39E 13 E9
Waverley N.Z. 39°46S 174°37E 59 C5
Waverly Iowa, U.S.A. 42°44N 92°29W 80 D7
Waverly N.Y., U.S.A. 42°1N 76°32W 83 E8
Wavre Belgium 50°43N 4°38E 15 D4
Wâw Sudan 7°45N 28°1E 51 G11
Wāw al Kabīr Libya 25°20N 16°43E 51 C9
Wawa Canada 47°59N 84°47W 72 C3
Wawanesa Canada 49°36N 99°40W 71 D9
Wawona U.S.A. 37°32N 119°39W 78 H7
Waxahachie U.S.A. 32°24N 96°51W 84 E6
Way, L. Australia 26°45S 120°16E 61 E3
Waya Fiji 17°19S 177°10E 59 a
Waycross U.S.A. 31°13N 82°21W 85 F13
Wayland U.S.A. 42°34N 77°35W 82 D7
Wayne Nebr., U.S.A. 42°14N 97°1W 80 D5
Wayne W. Va., U.S.A. 38°13N 82°27W 81 F12
Waynesboro Ga., U.S.A. 33°6N 82°1W 85 E13
Waynesboro Miss.,
U.S.A. 31°40N 88°39W 85 F10
Waynesboro Pa.,
U.S.A. 39°45N 77°35W 81 F15
Waynesboro Va., U.S.A. 38°4N 78°53W 81 F14
Waynesburg U.S.A. 39°54N 80°11W 81 F13
Waynesville U.S.A. 35°28N 82°58W 85 D13
Waynoka U.S.A. 36°35N 98°53W 84 C5
Wazirabad Pakistan 32°30N 74°8E 42 C6
Waziristan Pakistan 33°0N 70°0E 42 C4
We Indonesia 5°51N 95°18E 36 C1
Weald, The U.K. 51°4N 0°20E 13 F8
Wear → U.K. 54°55N 1°23W 12 C6
Weatherford Okla.,
U.S.A. 35°32N 98°43W 84 D5
Weatherford Tex., U.S.A. 32°46N 97°48W 84 E6
Weaverville U.S.A. 40°44N 122°56W 76 F2
Webb City U.S.A. 37°9N 94°28W 80 G6
Webequie Canada 52°59N 87°21W 72 B2
Webster Mass., U.S.A. 42°3N 71°53W 83 D13
Webster N.Y., U.S.A. 43°13N 77°26W 82 C7
Webster S. Dak., U.S.A. 45°20N 97°31W 80 C5
Webster City U.S.A. 42°28N 93°49W 80 D7
Webster Springs U.S.A. 38°29N 80°25W 81 F13
Webuye Kenya 0°37N 34°46E 54 B3
Weda Indonesia 0°21N 127°50E 37 D7
Weda, Teluk Indonesia 0°20N 128°0E 37 D7
Weddell Abyssal Plain
S. Ocean 65°0S 20°0W 5 C2
Weddell I. Falk. Is. 51°50S 61°0W 96 G4
Weddell Sea Antarctica 72°30S 40°0W 5 D1
Wedderburn Australia 36°26S 143°33E 63 F3
Wedgeport Canada 43°44N 65°59W 73 D6
Wedza Zimbabwe 18°40S 31°33E 55 F3
Wee Waa Australia 30°11S 149°26E 63 E4
Weed U.S.A. 41°25N 122°23W 76 F2
Weed Heights U.S.A. 38°59N 119°13W 78 G7
Weedsport U.S.A. 43°2N 76°33W 83 C8
Weedville U.S.A. 41°17N 78°30W 82 E6
Weenen S. Africa 28°48S 30°7E 57 D5
Weert Neths. 51°15N 5°43E 15 C5
Wei He → Hebei, China 36°10N 115°45E 34 F8
Wei He → Shaanxi,
China 34°38N 110°15E 34 G6
Weichang China 41°58N 117°49E 35 D9
Weichuan China 34°20N 113°59E 34 G7
Weiden Germany 49°41N 12°10E 16 D7
Weifang China 36°44N 119°7E 35 F10
Weihai China 37°30N 122°6E 35 F12
Weihui China 35°25N 114°3E 34 G8
Weimar Germany 50°58N 11°19E 16 C6
Weinan China 34°31N 109°29E 34 G5
Weipa Australia 12°40S 141°50E 62 A3
Weir → Australia 28°20S 149°50E 63 D4
Weir → Canada 56°54N 93°21W 71 B10
Weir River Canada 56°49N 94°6W 71 B10
Weirton U.S.A. 40°24N 80°35W 82 F4
Weiser U.S.A. 44°15N 116°58W 76 D5
Weishan China 34°47N 117°5E 35 G9
Weishan Hu China 34°35N 117°14E 35 G9
Weiyuan China 35°7N 104°10E 34 G3
Wejherowo Poland 54°35N 18°12E 17 A10
Wekusko L. Canada 54°40N 99°50W 71 C9
Welch U.S.A. 37°26N 81°35W 81 G13
Welford △ Australia 25°5S 143°16E 62 D3
Welkom S. Africa 28°0S 26°46E 56 D4
Welland Canada 43°0N 79°15W 82 D5
Welland → U.K. 52°51N 0°5W 13 E7
Wellesley Is. Australia 16°42S 139°30E 62 B2
Wellesley Islands ✪
Australia 16°32S 139°23E 62 B2
Wellingborough U.K. 52°19N 0°41W 13 E7
Wellington Australia 32°35S 148°59E 63 E4
Wellington Canada 43°57N 77°20W 82 C7
Wellington N.Z. 41°19S 174°46E 59 D5
Wellington S. Africa 33°38S 19°1E 56 E2
Wellington Somst., U.K. 50°58N 3°13W 13 G4
Wellington Telford & Wrekin,
U.K. 52°42N 2°30W 13 E5
Wellington Colo.,
U.S.A. 40°42N 105°0W 76 F11
Wellington Kans., U.S.A. 37°16N 97°24W 80 G5
Wellington Nev., U.S.A. 38°45N 119°23W 78 G7
Wellington Ohio, U.S.A. 41°10N 82°13W 82 E2
Wellington Tex., U.S.A. 34°51N 100°13W 84 D4
Wellington, I. Chile 49°30S 75°0W 96 F2
Wellington, L. Australia 38°6S 147°20E 63 F4
Wellington Chan. Canada 75°0N 93°0W 69 C13
Wells U.K. 51°13N 2°39W 13 F5
Wells Maine, U.S.A. 43°20N 70°35W 83 C14
Wells N.Y., U.S.A. 43°24N 74°17W 83 C10
Wells Nev., U.S.A. 41°7N 114°58W 76 F6
Wells, L. Australia 26°44S 123°15E 61 E3
Wells, Mt. Australia 17°25S 127°8E 60 C4
Wells Gray ⌂ Canada 52°30N 120°15W 70 C4
Wells-next-the-Sea U.K. 52°57N 0°51E 12 E8
Wells River U.S.A. 44°9N 72°4W 83 B12
Wellsboro U.S.A. 41°45N 77°18W 82 E7
Wellsburg U.S.A. 40°16N 80°37W 82 F4
Wellsville N.Y., U.S.A. 42°7N 77°57W 82 D7
Wellsville Ohio, U.S.A. 40°36N 80°39W 82 F4
Wellsville Utah, U.S.A. 41°38N 111°56W 76 F8
Wellton U.S.A. 32°40N 114°8W 77 K6
Wels Austria 48°9N 14°1E 16 D8
Welshpool U.K. 52°39N 3°8W 13 E4
Welwyn Garden City
U.K. 51°48N 0°12W 13 F7
Wem U.K. 52°52N 2°44W 12 E5
Wembere → Tanzania 4°10S 34°15E 54 C3
Wemindji Canada 53°0N 78°49W 72 B4
Wemyss Canada 44°52N 76°23W 83 B8
Wen Xian Gansu, China 32°43N 104°36E 34 H3
Wen Xian Henan, China 34°55N 113°5E 34 G7
Wenatchee U.S.A. 47°25N 120°19W 76 C3
Wenchang China 19°38N 110°42E 38 C8
Wenchi Ghana 7°46N 2°8W 50 G5
Wenchow = Wenzhou
China 28°0N 120°38E 33 F13
Wenden U.S.A. 33°49N 113°33W 79 M13
Wendeng China 37°15N 122°5E 35 F12
Wendesi Indonesia 2°30S 134°17E 37 E8
Wendover U.S.A. 40°44N 114°2W 76 F6
Wenlock → Australia 12°2S 141°55E 62 A3
Wenshan China 23°20N 104°18E 32 G9
Wenshang China 35°45N 116°30E 34 G9
Wenshui China 37°26N 112°1E 34 F7
Wensleydale U.K. 54°17N 2°0W 12 C6
Wensu China 41°15N 80°10E 32 C5
Wensum → U.K. 52°40N 1°15E 12 E8
Wentworth Australia 34°2S 141°54E 63 E3
Wentzel L. Canada 59°2N 114°28W 70 B6
Wenut Indonesia 3°11S 133°19E 37 E8
Wenxi China 35°20N 111°10E 34 G6
Wenzhou China 28°0N 120°38E 33 F13
Weott U.S.A. 40°20N 123°55W 76 F2
Wepener S. Africa 29°42S 27°3E 56 D4
Werda Botswana 25°24S 23°15E 56 D3
Weri Indonesia 3°10S 132°38E 37 E8
Werra → Germany 51°24N 9°39E 16 C5
Werrimull Australia 34°25S 141°38E 63 E3
Werris Creek Australia 31°18S 150°38E 63 E5
Weser → Germany 53°36N 8°28E 16 B5
Wesiri Indonesia 7°30S 126°30E 37 F7
Weslaco U.S.A. 26°10N 97°58W 84 H6
Weslemkoon L. Canada 45°2N 77°25W 82 A7
Wesleyville Canada 49°9N 53°33W 73 C9
Wesleyville U.S.A. 42°9N 80°1W 82 D4
Wessel, C. Australia 10°59S 136°46E 62 A2
Wessel Is. Australia 11°10S 136°45E 62 A2
Wessington Springs
U.S.A. 44°5N 98°34W 80 C4
West U.S.A. 31°48N 97°6W 84 F6
West → U.S.A. 42°52N 72°33W 83 D12
West Allis U.S.A. 43°1N 88°0W 80 D9
West Antarctica
Antarctica 80°0S 90°0W 5 D15
West Baines →
Australia 15°38S 129°59E 60 C4
West Bank ■ Asia 32°6N 35°13E 46 C4
West Bend U.S.A. 43°25N 88°11W 80 D9
West Bengal □ India 23°0N 88°0E 43 H13
West Berkshire □ U.K. 51°25N 1°17W 13 F6
West Beskids = Západné Beskydy
Europe 49°30N 19°0E 17 D10
West Branch U.S.A. 44°17N 84°14W 81 C11
West Branch Susquehanna →
U.S.A. 40°53N 76°48W 83 F8
West Bromwich U.K. 52°32N 1°59W 13 E6
West Burra U.K. 60°5N 1°21W 11 A7
West Canada Cr. →
U.S.A. 43°1N 74°58W 83 C10
West Caroline Basin
Pac. Oc. 4°0N 138°0E 64 G5
West Chazy U.S.A. 44°49N 73°28W 83 B11
West Chester U.S.A. 39°58N 75°36W 83 G9
West Coast ⌂ Namibia 21°53S 14°14E 56 C1
West Coast △ S. Africa 33°13S 18°0E 56 E2
West Columbia U.S.A. 29°9N 95°39W 84 G7
West Covina U.S.A. 34°4N 117°54W 79 L9
West Des Moines U.S.A. 41°35N 93°43W 80 E7
West Dunbartonshire □
U.K. 55°59N 4°30W 11 F4
West End Bahamas 26°41N 78°58W 88 A4
West Falkland Falk. Is. 51°40S 60°0W 96 G4
West Fargo U.S.A. 46°52N 96°54W 80 B5
West Fiji Basin Pac. Oc. 17°0S 173°0E 64 J9
West Fjord = Vestfjorden
Norway 67°55N 14°0E 8 C16
West Fork Trinity →
U.S.A. 32°48N 96°54W 84 E6
West Frankfort U.S.A. 37°54N 88°55W 80 G9
West Grand L. U.S.A. 45°14N 67°51W 81 C20
West Hartford U.S.A. 41°45N 72°44W 83 E12
West Haven U.S.A. 41°17N 72°57W 83 E12
West Hazleton U.S.A. 40°58N 76°0W 83 F9
West Helena U.S.A. 34°33N 90°38W 85 D9
West Hurley U.S.A. 41°59N 74°7W 83 E10
West Ice Shelf Antarctica 67°0S 85°0E 5 C7
West Indies Cent. Amer. 15°0N 65°0W 89 D7

West Jordan U.S.A. 40°36N 111°56W 76 F8
West Lamma Channel China 22°14N 114°4E 33 a
West Linn U.S.A. 45°21N 122°36W 78 E4
West Lorne Canada 42°36N 81°36W 82 D3
West Lothian □ U.K. 55°54N 3°36W 11 F5
West Lunga → Zambia 13°6S 24°39E 55 E1
West MacDonnell △ Australia 23°38S 132°59E 60 D5
West Mariana Basin Pac. Oc. 15°0N 137°0E 64 F5
West Memphis U.S.A. 35°8N 90°10W 85 D9
West Midlands □ U.K. 52°26N 2°0W 13 E6
West Mifflin U.S.A. 40°21N 79°52W 82 F5
West Milford U.S.A. 41°8N 74°22W 83 E10
West Milton U.S.A. 41°1N 76°50W 82 E8
West Monroe U.S.A. 32°31N 92°9W 84 E8
West Newton U.S.A. 40°14N 79°46W 82 F5
West Nicholson Zimbabwe 21°2S 29°20E 55 G2
West Odessa U.S.A. 31°50N 102°30W 84 F3
West Palm Beach U.S.A. 26°43N 80°3W 85 H14
West Plains U.S.A. 36°44N 91°51W 80 G8
West Point Miss., U.S.A. 33°36N 88°39W 85 E10
West Point N.Y., U.S.A. 41°24N 73°58W 83 E11
West Point Nebr., U.S.A. 41°51N 96°43W 80 E5
West Point Va., U.S.A. 37°32N 76°48W 81 G15
West Point L. U.S.A. 33°8N 85°0W 85 E12
West Pt. = Ouest, Pte. de l' Canada 49°52N 64°40W 73 C7
West Pt. Australia 35°1S 135°56E 63 F2
West Road → Canada 53°18N 122°53W 70 C4
West Rutland U.S.A. 43°36N 73°3W 83 C11
West Schelde = Westerschelde → Neths. 51°25N 3°25E 15 C3
West Seneca U.S.A. 42°51N 78°48W 82 D6
West Siberian Plain Russia 62°0N 75°0E 26 B9
West Sussex □ U.K. 50°55N 0°30W 13 G7
West-Terschelling Neths. 53°22N 5°13E 15 A5
West Valley City U.S.A. 40°42N 111°58W 76 F8
West Virginia □ U.S.A. 38°45N 80°30W 81 F13
West-Vlaanderen □ Belgium 51°0N 3°0E 15 D2
West Walker → U.S.A. 38°54N 119°9W 78 G7
West Wyalong Australia 33°56S 147°10E 63 E4
West Yellowstone U.S.A. 44°40N 111°6W 76 D8
West Yorkshire □ U.K. 53°45N 1°40W 12 D6
Westall, Pt. Australia 32°55S 134°4E 63 E1
Westbrook U.S.A. 43°41N 70°22W 81 D18
Westbury Australia 41°30S 146°51E 63 G4
Westby U.S.A. 48°52N 104°3W 76 B11
Westend U.S.A. 35°42N 117°24W 79 K9
Westerland Germany 54°54N 8°17E 16 A5
Westerly U.S.A. 41°22N 71°50W 83 E13
Western □ Kenya 0°30N 34°30E 54 B3
Western □ Zambia 15°0S 24°4E 55 F1
Western Australia □ Australia 25°0S 118°0E 61 E2
Western Cape □ S. Africa 34°0S 20°0E 56 E3
Western Dvina = Daugava → Latvia 57°4N 24°3E 9 H21
Western Ghats India 14°0N 75°0E 40 N9
Western Isles □ U.K. 57°30N 7°10W 11 D1
Western Sahara ■ Africa 25°0N 13°0W 50 D3
Western Samoa = Samoa ■ Pac. Oc. 14°0S 172°0W 59 b
Western Sierra Madre = Madre Occidental, Sierra Mexico 27°0N 107°0W 86 B3
Westernport U.S.A. 39°29N 79°3W 81 F14
Westerschelde → Neths. 51°25N 3°25E 15 C3
Westerwald Germany 50°38N 7°56E 16 C4
Westfield Mass., U.S.A. 42°7N 72°45W 83 D12
Westfield N.Y., U.S.A. 42°20N 79°35W 82 D5
Westfield Pa., U.S.A. 41°55N 77°32W 82 E7
Westhill U.K. 57°9N 2°19W 11 D6
Westhope U.S.A. 48°55N 101°1W 80 A3
Westland △ N.Z. 43°16S 170°16E 59 E2
Westland Bight N.Z. 42°55S 170°5E 59 E3
Westlock Canada 54°9N 113°55W 70 C6
Westmar Australia 27°55S 149°44E 63 D4
Westmeath □ Ireland 53°33N 7°34W 10 C4
Westminster Calif., U.S.A. 33°45N 118°0W 79 M8
Westminster Colo., U.S.A. 39°50N 105°2W 76 G11
Westminster Md., U.S.A. 39°34N 76°59W 81 F15
Westmont U.S.A. 40°19N 78°58W 82 F6
Westmoreland Barbados 13°13N 59°37W 89 g
Westmorland U.S.A. 33°2N 115°37W 79 M11
Weston Oreg., U.S.A. 45°49N 118°26W 76 D4
Weston W. Va., U.S.A. 39°2N 80°28W 81 F13
Weston I. Canada 52°33N 79°36W 72 B4
Weston-super-Mare U.K. 51°21N 2°58W 13 F5
Westover U.S.A. 40°45N 78°40W 82 F6
Westport Canada 44°40N 76°25W 83 B8
Westport Ireland 53°48N 9°31W 10 C2
Westport N.Z. 41°46S 171°37E 59 D3
Westport N.Y., U.S.A. 44°11N 73°26W 83 B11
Westport Oreg., U.S.A. 46°8N 123°23W 78 D3
Westport Wash., U.S.A. 46°53N 124°6W 78 D2
Westray Canada 53°36N 101°24W 71 C8
Westray U.K. 59°18N 3°0W 11 B5
Westree Canada 47°26N 81°34W 72 C3
Westville U.S.A. 39°8N 120°42W 78 F6
Westwood U.S.A. 40°18N 121°0W 76 F3
Wetar Indonesia 7°48S 126°30E 37 F7
Wetaskiwin Canada 52°55N 113°24W 70 C6
Wete Tanzania 5°4S 39°43E 52 F7
Wetherby U.K. 53°56N 1°23W 12 D6
Wethersfield U.S.A. 41°42N 72°40W 83 E12
Wetteren Belgium 51°0N 3°53E 15 D3
Wetzlar Germany 50°32N 8°31E 16 C5
Wewoka U.S.A. 35°9N 96°30W 84 D6
Wexford Ireland 52°20N 6°28W 10 D5
Wexford □ Ireland 52°20N 6°25W 10 D5
Wexford Harbour Ireland 52°20N 6°25W 10 D5
Weyburn Canada 49°40N 103°50W 71 D8
Weymouth Canada 44°30N 66°1W 73 D6
Weymouth U.K. 50°37N 2°28W 13 G5
Weymouth U.S.A. 42°13N 70°58W 83 D14
Weymouth, C. Australia 12°37S 143°27E 62 A3
Wha Ti Canada 63°8N 117°16W 68 E8
Whakaari N.Z. 37°30S 177°13E 59 B6
Whakatane N.Z. 37°57S 177°1E 59 B6
Whale = Baleine → Canada 58°15N 67°40W 73 A6
Whale B. Burma 11°37N 98°38E 39 G2
Whale Cove Canada 62°10N 92°34W 71 A10
Whales, B. of Antarctica 78°0S 160°0W 5 D12
Whalsay U.K. 60°22N 0°59W 11 A8
Whangamata N.Z. 37°12S 175°53E 59 B5
Whangamomona N.Z. 39°8S 174°44E 59 C5
Whanganui △ N.Z. 39°17S 174°53E 59 C5
Whangarei N.Z. 35°43S 174°21E 59 A5
Whangarei Harb. N.Z. 35°45S 174°28E 59 A5
Wharekauri = Chatham Is. Pac. Oc. 44°0S 176°40W 64 M10
Wharfe → U.K. 53°51N 1°9W 12 D6
Wharfedale U.K. 54°6N 2°1W 12 C5
Wharton N.J., U.S.A. 40°54N 74°35W 83 F10
Wharton Pa., U.S.A. 41°31N 78°1W 82 E6
Wharton Tex., U.S.A. 29°19N 96°6W 84 G6
Wharton Basin Ind. Oc. 22°0S 92°0E 64 K1
Wheatland Calif., U.S.A. 39°1N 121°25W 78 F5
Wheatland Wyo., U.S.A. 42°3N 104°58W 76 E11
Wheatley Canada 42°6N 82°27W 82 D2
Wheaton Md., U.S.A. 39°3N 77°3W 81 F15
Wheaton Minn., U.S.A. 45°48N 96°30W 80 C5
Wheelbarrow Pk. U.S.A. 37°26N 116°5W 78 H10
Wheeler Oreg., U.S.A. 45°41N 123°53W 76 D2
Wheeler Tex., U.S.A. 35°27N 100°16W 84 D4
Wheeler → Canada 57°2N 67°13W 73 A6
Wheeler L. U.S.A. 34°48N 87°23W 85 D11
Wheeler Pk. N. Mex., U.S.A. 36°34N 105°25W 77 H11
Wheeler Pk. Nev., U.S.A. 38°57N 114°15W 76 G6
Wheeler Ridge U.S.A. 35°0N 118°57W 79 L8
Wheelersburg U.S.A. 38°44N 82°51W 81 F12
Wheeling U.S.A. 40°4N 80°43W 82 F4
Whernside U.K. 54°14N 2°24W 12 C5
Whiddy I. Ireland 51°41N 9°31W 10 E2
Whim Creek Australia 20°50S 117°49E 60 D2
Whiskey Jack L. Canada 58°23N 101°55W 71 B8
Whiskeytown-Shasta-Trinity ⌂ U.S.A. 40°45N 122°15W 76 F2
Whistleduck Cr. → Australia 20°15S 135°18E 62 C2
Whistler Canada 50°7N 122°58W 70 C4
Whitby Canada 43°52N 78°56W 82 C6
Whitby U.K. 54°29N 0°37W 12 C7
White → Ark., U.S.A. 33°57N 91°5W 84 E9
White → Ind., U.S.A. 38°25N 87°45W 80 F10
White → S. Dak., U.S.A. 43°42N 99°27W 80 D4
White → Tex., U.S.A. 33°14N 100°56W 84 E4
White → Utah, U.S.A. 40°4N 109°41W 76 F9
White → Vt., U.S.A. 43°37N 72°20W 83 C12
White → Wash., U.S.A. 47°12N 122°15W 78 C4
White, L. Australia 21°9S 128°56E 60 D4
White B. Canada 50°0N 56°35W 73 C8
White Bird U.S.A. 45°46N 116°18W 76 D5
White Butte U.S.A. 46°23N 103°18W 80 B2
White City U.S.A. 42°26N 122°51W 76 E2
White Cliffs Australia 30°50S 143°10E 63 E3
White Hall U.S.A. 39°26N 90°24W 80 F8
White Haven U.S.A. 41°4N 75°47W 83 E9
White Horse, Vale of U.K. 51°37N 1°30W 13 F6
White I. = Whakaari N.Z. 37°30S 177°13E 59 B6
White L. Canada 45°18N 76°31W 83 A8
White L. U.S.A. 29°44N 92°30W 84 G8
White Lake Canada 45°21N 76°29W 83 A8
White Mountain Peak U.S.A. 37°38N 118°15W 77 H4
White Mts. Calif., U.S.A. 37°30N 118°15W 78 H8
White Mts. N.H., U.S.A. 44°15N 71°15W 83 B13
White Mts. △ Australia 20°43S 145°12E 62 C4
White Nile = Nîl el Abyad → Sudan 15°38N 32°31E 51 E12
White Otter L. Canada 49°5N 91°55W 72 C1
White Pass U.S.A. 46°38N 121°24W 78 D5
White Plains U.S.A. 41°2N 73°46W 83 E11
White River Canada 48°35N 85°20W 72 C2
White River U.S.A. 43°34N 100°45W 80 D3
White Rock Canada 49°2N 122°48W 78 A4
White Rock U.S.A. 35°50N 106°12W 77 J10
White Russia = Belarus ■ Europe 53°30N 27°0E 17 B14
White Sands ⌂ U.S.A. 32°46N 106°20W 77 K10
White Sea = Beloye More Russia 66°30N 38°0E 18 A6
White Sulphur Springs Mont., U.S.A. 46°33N 110°54W 76 C8
White Sulphur Springs W. Va., U.S.A. 37°48N 80°18W 81 G13
White Swan U.S.A. 46°23N 120°44W 78 D6
Whitecliffs N.Z. 43°26S 171°55E 59 E3
Whitecourt Canada 54°10N 115°45W 70 C5
Whiteface Mt. U.S.A. 44°22N 73°54W 83 B11
Whitefield U.S.A. 44°23N 71°37W 83 B13
Whitefish U.S.A. 48°25N 114°20W 76 B6
Whitefish B. U.S.A. 46°40N 84°55W 72 C3
Whitefish L. Canada 62°41N 106°48W 71 A7
Whitefish Pt. U.S.A. 46°45N 84°59W 81 B11
Whitegull, L. = Goélands, L. aux Canada 55°27N 64°17W 73 A7
Whitehall Mich., U.S.A. 43°24N 86°21W 80 D10
Whitehall Mont., U.S.A. 45°52N 112°6W 76 D7
Whitehall N.Y., U.S.A. 43°33N 73°24W 83 C11
Whitehall Wis., U.S.A. 44°22N 91°19W 80 C8
Whitehaven U.K. 54°33N 3°35W 12 C4
Whitehorse Canada 60°43N 135°3W 70 A1
Whitemark Australia 40°7S 148°3E 63 G4
Whiteriver U.S.A. 33°50N 109°58W 77 K9
Whitesand → Canada 60°9N 115°45W 70 A5
Whitesands S. Africa 34°23S 20°50E 56 E3
Whitesboro N.Y., U.S.A. 43°7N 75°18W 83 C9
Whitesboro Tex., U.S.A. 33°39N 96°54W 84 E6
Whiteshell ⌂ Canada 50°0N 95°40W 71 D9
Whiteville U.S.A. 34°20N 78°42W 85 D15
Whitewater U.S.A. 42°50N 88°44W 80 D9
Whitewater Baldy U.S.A. 33°20N 108°39W 77 K9
Whitewater L. Canada 50°50N 89°10W 72 B2
Whitewood Australia 21°28S 143°30E 62 C3
Whitewood Canada 50°20N 102°20W 71 C8
Whithorn U.K. 54°44N 4°26W 11 G4
Whitianga N.Z. 36°47S 175°41E 59 B5
Whitman U.S.A. 42°5N 70°56W 83 D14
Whitney Canada 45°31N 78°14W 82 A6
Whitney, Mt. U.S.A. 36°35N 118°18W 78 J8
Whitney Point U.S.A. 42°20N 75°58W 83 D9
Whitstable U.K. 51°21N 1°3E 13 F9
Whitsunday I. Australia 20°15S 149°4E 62 b
Whitsunday Islands △ Australia 20°15S 149°0E 62 b
Whittier Alaska, U.S.A. 60°47N 148°41W 74 C10
Whittier Calif., U.S.A. 33°58N 118°2W 79 M8
Whittlesea Australia 37°27S 145°9E 63 F4
Wholdaia L. Canada 60°43N 104°20W 71 A8
Whyalla Australia 33°2S 137°30E 63 E2
Wiang Kosai △ Thailand 17°54N 99°29E 38 D2
Wiang Sa Thailand 18°34N 100°45E 38 C3
Wiarton Canada 44°40N 81°10W 82 B3
Wiay U.K. 57°24N 7°13W 11 D1
Wibaux U.S.A. 46°59N 104°11W 76 C11
Wichian Buri Thailand 15°39N 101°7E 38 E3
Wichita U.S.A. 37°42N 97°20W 80 G5
Wichita Falls U.S.A. 33°54N 98°30W 84 E5
Wick U.K. 58°26N 3°5W 11 C5
Wickenburg U.S.A. 33°58N 112°44W 77 K7
Wickepin Australia 32°50S 117°30E 61 F2
Wickham Australia 20°42S 117°11E 60 D2
Wickham, C. Australia 39°35S 143°57E 63 F3
Wickliffe U.S.A. 41°36N 81°28W 82 E3
Wicklow Ireland 52°59N 6°3W 10 D5
Wicklow □ Ireland 52°57N 6°25W 10 D5
Wicklow Hd. Ireland 52°58N 6°0W 10 D6
Wicklow Mts. Ireland 52°58N 6°26W 10 C5
Wicklow Mts. △ Ireland 53°6N 6°21W 10 C5
Widgeegoara Cr. → Australia 28°51S 146°34E 63 D4
Widgiemooltha Australia 31°30S 121°34E 61 F3
Widnes U.K. 53°23N 2°45W 12 D5
Wieluń Poland 51°15N 18°34E 17 C10
Wien Austria 48°12N 16°22E 16 D9
Wiener Neustadt Austria 47°49N 16°16E 16 E9
Wiesbaden Germany 50°4N 8°14E 16 C5
Wigan U.K. 53°33N 2°38W 12 D5
Wiggins Colo., U.S.A. 40°14N 104°4W 76 F11
Wiggins Miss., U.S.A. 30°51N 89°8W 85 F10
Wight, I. of U.K. 50°41N 1°17W 13 G6
Wigston U.K. 52°35N 1°6W 13 E6
Wigton U.K. 54°50N 3°10W 12 C4
Wigtown U.K. 54°53N 4°27W 11 G4
Wigtown B. U.K. 54°46N 4°15W 11 G4
Wik and Wikway People ❂ Australia 13°55S 142°28E 62 A3
Wilber U.S.A. 40°29N 96°58W 80 E5
Wilberforce Canada 45°2N 78°13W 82 A6
Wilberforce, C. Australia 11°54S 136°35E 62 A2
Wilburton U.S.A. 34°55N 95°19W 84 D7
Wilcannia Australia 31°30S 143°26E 63 E3
Wilcox U.S.A. 41°35N 78°41W 82 E6
Wildrose U.S.A. 36°14N 117°11W 79 J9
Wildspitze Austria 46°53N 10°53E 16 E6
Wilge → S. Africa 27°3S 28°20E 57 D4
Wilhelm II Coast Antarctica 68°0S 90°0E 5 C7
Wilhelmshaven Germany 53°31N 8°7E 16 B5
Wilhelmstal Namibia 21°58S 16°21E 56 C2
Wilis Indonesia 7°51S 111°48E 37 G15
Wilkes-Barre U.S.A. 41°15N 75°53W 83 E9
Wilkes Land Antarctica 69°0S 120°0E 5 D9
Wilkie Canada 52°27N 108°42W 71 C7
Wilkinsburg U.S.A. 40°26N 79°52W 82 F5
Wilkinson Lakes Australia 29°40S 132°39E 61 E5
Willandra Creek → Australia 33°22S 145°52E 63 E4
Willapa B. U.S.A. 46°40N 124°0W 76 C1
Willapa Hills U.S.A. 46°35N 123°25W 78 D3
Willard U.S.A. 41°3N 82°44W 82 E2
Willare Bridge Roadhouse Australia 17°43S 123°38E 60 C3
Willcox U.S.A. 32°15N 109°50W 77 K9
Willemstad Curaçao 12°5N 68°55W 89 D6
William → Canada 59°8N 109°19W 71 B7
William 'Bill' Dannelly Res. U.S.A. 32°6N 87°24W 85 E11
William Creek Australia 28°58S 136°22E 63 D2
Williams Australia 33°2S 116°52E 61 F2
Williams Ariz., U.S.A. 35°15N 112°11W 77 J7
Williams Calif., U.S.A. 39°9N 122°9W 78 F4
Williams Harbour Canada 52°33N 55°47W 73 B8
Williams Lake Canada 52°10N 122°10W 70 C4
Williamsburg Ky., U.S.A. 36°44N 84°10W 81 G11
Williamsburg Pa., U.S.A. 40°28N 78°12W 82 F6
Williamsburg Va., U.S.A. 37°16N 76°43W 81 G15
Williamson N.Y., U.S.A. 43°14N 77°11W 82 C7
Williamson W. Va., U.S.A. 37°41N 82°17W 81 G12
Williamsport U.S.A. 41°15N 77°1W 82 E7
Williamston U.S.A. 35°51N 77°4W 85 D16
Williamstown Australia 37°51S 144°52E 63 F3
Williamstown Ky., U.S.A. 38°38N 84°34W 81 F11
Williamstown Mass., U.S.A. 42°43N 73°12W 83 D11
Williamstown N.Y., U.S.A. 43°26N 75°53W 83 C9
Willimantic U.S.A. 41°43N 72°13W 83 E12
Willingboro U.S.A. 40°3N 74°54W 81 E16
Williston S. Africa 31°20S 20°53E 56 E3
Williston Fla., U.S.A. 29°23N 82°27W 85 G13
Williston N. Dak., U.S.A. 48°9N 103°37W 80 A2
Williston L. Canada 56°0N 124°0W 70 B4
Willits U.S.A. 39°25N 123°21W 76 G2
Willmar U.S.A. 45°7N 95°3W 80 C6
Willmore Wilderness ⌂ Canada 53°45N 119°30W 70 C5
Willoughby U.S.A. 41°39N 81°24W 82 E3
Willow Bunch Canada 49°20N 105°35W 71 D7
Willow L. Canada 62°10N 119°8W 70 A5
Willow Wall, The China 42°10N 122°0E 35 C12
Willowick U.S.A. 41°38N 81°28W 82 E3
Willowlake → Canada 62°42N 123°8W 70 A4
Willowmore S. Africa 33°15S 23°30E 56 E3
Willows U.S.A. 39°31N 122°12W 78 F4
Willowvale = Gatyana S. Africa 32°16S 28°31E 57 E4
Wills, L. Australia 21°25S 128°51E 60 D4
Wills Cr. → Australia 22°43S 140°2E 62 C3
Willsboro U.S.A. 44°21N 73°24W 83 B11
Willunga Australia 35°15S 138°30E 63 F2
Wilmette U.S.A. 42°4N 87°42W 80 D10
Wilmington Australia 32°39S 138°7E 63 E2
Wilmington Del., U.S.A. 39°45N 75°33W 81 F16
Wilmington N.C., U.S.A. 34°14N 77°55W 85 D16
Wilmington Ohio, U.S.A. 39°27N 83°50W 81 F12
Wilmington Vt., U.S.A. 42°52N 72°52W 83 D12
Wilmslow U.K. 53°19N 2°13W 12 D5
Wilpena Cr. → Australia 31°25S 139°29E 63 E2
Wilsall U.S.A. 45°59N 110°38W 76 D8
Wilson N.C., U.S.A. 35°44N 77°55W 85 D16
Wilson N.Y., U.S.A. 43°19N 78°50W 82 C6
Wilson Pa., U.S.A. 40°41N 75°15W 83 F9
Wilson → Australia 16°48S 128°16E 60 C4
Wilson Bluff Australia 31°41S 129°0E 61 F4
Wilson Inlet Australia 35°0S 117°22E 61 G2
Wilsons Promontory Australia 38°59S 146°23E 63 F4
Wilton U.S.A. 47°10N 100°47W 80 B3
Wilton → Australia 14°45S 134°33E 62 A1
Wiltshire □ U.K. 51°18N 1°53W 13 F6
Wiltz Lux. 49°57N 5°55E 15 E5
Wiluna Australia 26°36S 120°14E 61 E3
Wimborne Minster U.K. 50°48N 1°59W 13 G6
Wimmera → Australia 36°8S 141°56E 63 F3
Winam G. Kenya 0°20S 34°15E 54 C3
Winburg S. Africa 28°30S 27°2E 56 D4
Winchendon U.S.A. 42°41N 72°3W 83 D12
Winchester U.K. 51°4N 1°18W 13 F6
Winchester Conn., U.S.A. 41°53N 73°9W 83 E11
Winchester Idaho, U.S.A. 46°14N 116°38W 76 C5
Winchester Ind., U.S.A. 40°10N 84°59W 81 E11
Winchester Ky., U.S.A. 37°59N 84°11W 81 G11
Winchester N.H., U.S.A. 42°46N 72°23W 83 D12
Winchester Nev., U.S.A. 36°7N 115°7W 79 J11
Winchester Tenn., U.S.A. 35°11N 86°7W 85 D11
Winchester Va., U.S.A. 39°11N 78°10W 81 F14
Wind → U.S.A. 43°12N 108°12W 76 E9
Wind Cave △ U.S.A. 43°32N 103°17W 80 D2
Wind River Range U.S.A. 43°0N 109°30W 76 E9
Windamere, L. Australia 32°48S 149°51E 63 B4
Windau = Ventspils Latvia 57°25N 21°32E 9 H19
Windber U.S.A. 40°14N 78°50W 82 F6
Winder U.S.A. 34°0N 83°45W 85 D13
Windermere U.K. 54°23N 2°55W 12 C5
Windhoek Namibia 22°35S 17°4E 56 C2
Windidda ❂ Australia 26°23S 122°12E 61 E3
Windjana Gorge △ Australia 17°51S 125°0E 60 C3
Windom U.S.A. 43°52N 95°7W 80 D6
Windorah Australia 25°24S 142°36E 62 D3
Window Rock U.S.A. 35°41N 109°3W 77 J9
Windrush → U.K. 51°43N 1°24W 13 F6
Windsor Australia 33°37S 150°50E 63 E5
Windsor N.S., Canada 44°59N 64°5W 73 D7
Windsor Ont., Canada 42°18N 83°0W 82 D2
Windsor U.K. 51°29N 0°36W 13 F7
Windsor Calif., U.S.A. 38°33N 122°49W 76 G2
Windsor Colo., U.S.A. 40°29N 104°54W 76 F11
Windsor Conn., U.S.A. 41°50N 72°39W 83 E12
Windsor Mo., U.S.A. 38°32N 93°31W 80 F7
Windsor N.Y., U.S.A. 42°5N 75°37W 83 D9
Windsor Vt., U.S.A. 43°29N 72°24W 83 C12
Windsor & Maidenhead □ U.K. 51°29N 0°40W 13 F7
Windsorton S. Africa 28°16S 24°44E 56 D3
Windward Is. W. Indies 13°0N 61°0W 89 D7
Windward Passage = Vientos, Paso de los Caribbean 20°0N 74°0W 89 C5
Winefred L. Canada 55°30N 110°30W 71 B6
Winfield U.S.A. 37°15N 96°59W 80 G5
Wingate Mts. Australia 14°25S 130°40E 60 B5
Wingellina = Irrunytju Australia 26°3S 128°56E 61 E4
Wingham Australia 31°48S 152°22E 63 E5
Wingham Canada 43°55N 81°20W 82 C3
Winisk → Canada 55°17N 85°5W 72 A2
Winisk L. Canada 52°55N 87°22W 72 B2
Wink U.S.A. 31°45N 103°9W 84 F3
Winkler Canada 49°10N 97°56W 71 D9
Winlock U.S.A. 46°30N 122°56W 78 D4
Winneba Ghana 5°25N 0°36W 50 G5
Winnebago, L. U.S.A. 44°0N 88°26W 80 D9
Winnecke Cr. → Australia 18°35S 131°34E 60
Winnemucca U.S.A. 40°58N 117°44W 7
Winnemucca L. U.S.A. 40°7N 119°21W 7
Winner U.S.A. 43°22N 99°52W 80
Winnett U.S.A. 47°0N 108°21W 76
Winnfield U.S.A. 31°56N 92°38W 84
Winnibigoshish, L. U.S.A. 47°27N 94°13W 80
Winnipeg Canada 49°54N 97°9W 71
Winnipeg → Canada 50°38N 96°19W 71
Winnipeg, L. Canada 52°0N 97°0W 71
Winnipeg Beach Canada 50°30N 96°58W 71
Winnipegosis Canada 51°39N 99°55W 71
Winnipegosis L. Canada 52°30N 100°0W 71
Winnipesaukee, L. U.S.A. 43°38N 71°21W 83
Winnisquam L. U.S.A. 43°33N 71°31W 83
Winnsboro La., U.S.A. 32°10N 91°43W 84
Winnsboro S.C., U.S.A. 34°23N 81°5W 85
Winnsboro Tex., U.S.A. 32°58N 95°17W 84
Winokapau, L. Canada 53°15N 62°50W 73
Winona Minn., U.S.A. 44°3N 91°39W 80
Winona Miss., U.S.A. 33°29N 89°44W 85
Winooski U.S.A. 44°29N 73°11W 83
Winooski → U.S.A. 44°32N 73°17W 83
Winschoten Neths. 53°9N 7°3E 15
Winsford U.K. 53°12N 2°31W 12
Winslow = Bainbridge Island U.S.A. 47°38N 122°32W 78
Winslow U.S.A. 35°2N 110°42W 77
Winsted U.S.A. 41°55N 73°4W 83
Winston-Salem U.S.A. 36°6N 80°15W 85
Winter Garden U.S.A. 28°34N 81°35W 85
Winter Haven U.S.A. 28°1N 81°44W 85
Winter Park U.S.A. 28°36N 81°20W 85
Winterhaven U.S.A. 32°44N 114°38W 79
Winters U.S.A. 38°32N 121°58W 78
Winterset U.S.A. 41°20N 94°1W 80
Wintersville U.S.A. 40°23N 80°42W 82
Winterswijk Neths. 51°58N 6°43E 15
Winterthur Switz. 47°30N 8°44E 20
Winthrop U.S.A. 48°28N 120°10W 76
Winton Australia 22°24S 143°3E 62
Winton N.Z. 46°8S 168°20E 59
Wirlyajarrayi ❂ Australia 21°45S 132°35E 60
Wirrulla Australia 32°24S 134°31E 63
Wisbech U.K. 52°41N 0°9E 13
Wisconsin □ U.S.A. 44°45N 89°30W 80
Wisconsin → U.S.A. 43°0N 91°15W 80
Wisconsin Rapids U.S.A. 44°23N 89°49W 80
Wisdom U.S.A. 45°37N 113°27W 76
Wiseman U.S.A. 67°25N 150°6W 74
Wishaw U.K. 55°46N 3°54W 11
Wishek U.S.A. 46°16N 99°33W 80
Wisła → Poland 54°22N 18°55E 17
Wismar Germany 53°54N 11°29E 16
Wisner U.S.A. 41°59N 96°55W 80
Witbank = eMalahleni S. Africa 25°51S 29°14E 57
Witdraai S. Africa 26°58S 20°48E 56
Witham U.K. 51°48N 0°40E 13
Witham → U.K. 52°59N 0°2W 12
Withernsea U.K. 53°44N 0°1E 12
Witjira △ Australia 26°22S 135°37E 63
Witless Bay Canada 47°17N 52°50W 73
Witney U.K. 51°48N 1°28W 13
Witnossob → Namibia 23°55S 18°45E 56
Witriver S. Africa 25°20S 31°0E 57
Wittenberge Germany 53°0N 11°45E 16
Wittenoom Australia 22°15S 118°20E 60
Witu Kenya 2°23S 40°26E 54
Witvlei Namibia 22°23S 18°32E 56
Wiwon N. Korea 40°54N 126°3E 35
Wkra → Poland 52°27N 20°44E 17
Wlingi Indonesia 8°5S 112°25E 37
Włocławek Poland 52°40N 19°3E 17
Włodawa Poland 51°33N 23°31E 17
Woburn U.S.A. 42°29N 71°9W 83
Wodian China 32°50N 112°35E 34
Wodonga Australia 36°5S 146°50E 63
Wokam Indonesia 5°45S 134°28E 37
Woking U.K. 51°19N 0°34W 13
Wokingham U.K. 51°24N 0°49W 13
Wokingham □ U.K. 51°25N 0°51W 13
Wolf → Canada 60°17N 132°33W 70
Wolf Creek U.S.A. 47°0N 112°4W 76
Wolf L. Canada 60°24N 131°40W 70
Wolf Point U.S.A. 48°5N 105°39W 76
Wolfe Creek Crater △ Australia 19°10S 127°47E 60
Wolfe I. Canada 44°7N 76°20W 83
Wolfeboro U.S.A. 43°35N 71°13W 83
Wolfsberg Austria 46°50N 14°52E 16
Wolfsburg Germany 52°25N 10°48E 16
Wolin Poland 53°50N 14°37E 16
Wollaston, Is. Chile 55°40S 67°30W 96
Wollaston L. Canada 58°7N 103°10W 71
Wollaston Lake Canada 58°3N 103°33W 71
Wollaston Pen. Canada 69°30N 115°0W 68
Wollongong Australia 34°25S 150°54E 63
Wolmaransstad S. Africa 27°12S 25°59E 56
Wolseley S. Africa 33°26S 19°7E 56
Wolsey U.S.A. 44°25N 98°28W 80
Wolstenholme, C. Canada 62°35N 77°30W 66
Wolvega Neths. 52°52N 6°0E 15
Wolverhampton U.K. 52°35N 2°7W 13
Wondai Australia 26°20S 151°49E 63
Wongalarroo L. Australia 31°32S 144°0E 63
Wongan Hills Australia 30°51S 116°37E 61
Wonju S. Korea 37°22N 127°58E 35
Wonosari Indonesia 7°58S 110°36E 37
Wonosobo Indonesia 7°22S 109°54E 37
Wonowon Canada 56°44N 121°48W 70
Wŏnsan N. Korea 39°11N 127°27E 35
Wonthaggi Australia 38°37S 145°37E 63
Wood Buffalo △ Canada 59°0N 113°41W 70
Wood Is. Australia 16°24S 123°19E 60

od L. *Canada* 55°17N 103°17W **71** B8
odah, I. *Australia* 13°27S 136°10E **62** A2
odbourne *U.S.A.* 41°46N 74°36W **83** E10
odbridge *Canada* 43°47N 79°36W **82** C5
odbridge *U.K.* 52°6N 1°20E **13** E9
odburn *Australia* 29°6S 153°23E **63** A5
odburn *U.S.A.* 45°9N 122°51W **76** D2
odenbong *Australia* 28°24S 152°39E **63** D5
odend *Australia* 37°20S 144°33E **63** F3
odford *Australia* 26°58S 152°47E **63** D5
odfords *U.S.A.* 38°47N 119°50W **78** G7
odlake *U.S.A.* 36°25N 119°6W **78** J7
odland *Calif., U.S.A.* 38°41N 121°46W **78** G5
odland *Maine, U.S.A.* 45°9N 67°25W **81** C20
odland *Pa., U.S.A.* 41°0N 78°21W **82** F6
odland *Wash., U.S.A.* 45°54N 122°45W **78** E4
odland Caribou △ *Canada* 51°0N 94°45W **71** C10
odlands *Singapore* 1°26N 103°46E **39** d
odlands, The *U.S.A.* 30°9N 95°29W **84** F7
odonga *Australia* 36°10S 146°50E **63** F4
odridge *Canada* 49°20N 96°9W **71** D9
odroffe, Mt. *Australia* 26°20S 131°45E **61** E5
ods, L. *Australia* 17°50S 133°30E **62** B1
ods, L. of the *Amer.* 49°15N 94°45W **71** D10
ods Bay *Canada* 45°8N 79°59W **82** A5
odstock *Australia* 19°35S 146°50E **62** B4
odstock *N.B., Canada* 46°11N 67°37W **73** C6
odstock *Ont., Canada* 43°10N 80°45W **82** C4
odstock *U.K.* 51°51N 1°20W **13** F6
odstock *Ill., U.S.A.* 42°19N 88°27W **80** D9
odstock *N.Y., U.S.A.* 42°2N 74°7W **83** D10
odstock *Vt., U.S.A.* 43°37N 72°31W **83** C12
odsville *U.S.A.* 44°9N 72°2W **83** B13
odview *Canada* 44°35N 78°8W **82** B6
odville *N.Z.* 40°20S 175°53E **59** D5
odville *Miss., U.S.A.* 31°6N 91°18W **84** F9
odville *Tex., U.S.A.* 30°47N 94°25W **84** F7
odward *U.S.A.* 36°26N 99°24W **84** C5
dy *U.S.A.* 35°42N 118°50W **79** K8
dy → *Canada* 52°31N 100°51W **71** C8
lacombe *U.K.* 51°10N 4°13W **13** F3
lamai, C. *Australia* 38°30S 145°23E **63** F4
ler *U.K.* 55°33N 2°1W **12** B5
lgoolga *Australia* 30°6S 153°11E **63** E5
mera *Australia* 31°5S 136°50E **63** E2
nsocket *R.I., U.S.A.* 42°0N 71°31W **83** E13
nsocket *S. Dak., U.S.A.* 44°3N 98°17W **80** C4
ramel → *Australia* 25°47S 114°10E **61** E1
ramel Roadhouse *Australia* 25°45S 114°17E **61** E1
roonooran △ *Australia* 16°25S 146°1E **62** B4
ster *U.S.A.* 40°48N 81°56W **82** F3
cester *S. Africa* 33°39S 19°27E **56** E2
cester *U.K.* 52°11N 2°12W **13** E5
cester *Mass., U.S.A.* 42°16N 71°48W **83** D13
cester *N.Y., U.S.A.* 42°36N 74°45W **83** D10
cestershire □ *U.K.* 52°13N 2°10W **13** E5
kington *U.K.* 54°39N 3°33W **12** C4
ksop *U.K.* 53°18N 1°7W **12** D6
kum *Neths.* 52°59N 5°26E **15** B5
land *U.S.A.* 44°1N 107°57W **76** D10
ms *Germany* 49°37N 8°21E **16** D5
sley *Canada* 56°31N 119°8W **70** B5
tham *U.S.A.* 31°47N 96°28W **84** F6
thing *Barbados* 13°5N 59°35W **89** g
thing *U.K.* 50°49N 0°21W **13** G7
thington *Minn., U.S.A.* 43°37N 95°36W **80** D6
thington *Pa., U.S.A.* 40°50N 79°38W **82** F5
i *Indonesia* 0°15S 128°0E **37** E7
e *Kenya* 1°47S 37°38E **54** C4
alum ✪ *Australia* 16°30S 123°45E **60** C3
obaluk ✪ *Australia* 37°0S 142°0E **63** C3
-han = Wuhan *China* 30°31N 114°18E **33** E11
si = Wuxi *China* 31°33N 120°18E **33** E13
voni *Indonesia* 4°5S 123°5E **37** E6
ngel I. = Vrangelya, Ostrov *Russia* 71°0N 180°0E **29** B18
ngell *U.S.A.* 56°28N 132°23W **70** B2
ngell Mts. *U.S.A.* 61°30N 142°0W **68** E3
th, C. *U.K.* 58°38N 5°1W **11** C3
U.S.A. 40°5N 102°13W **76** F12
kin, The *U.K.* 52°41N 2°32W **13** E5
ns *U.S.A.* 33°12N 82°23W **85** E13
xham *U.K.* 53°3N 3°0W **12** D4
xham □ *U.K.* 53°1N 2°58W **12** D5
ght *Fla., U.S.A.* 30°27N 86°38W **85** F11
ght *Wyo., U.S.A.* 43°45N 105°28W **76** E11
ght Pt. *Canada* 43°48N 81°44W **82** C3
ghtson, Mt. *U.S.A.* 31°42N 110°51W **77** L8
ghtwood *U.S.A.* 34°22N 117°38W **79** L9
gley *Canada* 63°16N 123°37W **68** E7
claw *Poland* 51°5N 17°5E **17** C9
śnia *Poland* 52°21N 17°36E **17** B9
iang → *China* 29°40N 107°20E **32** F10
Kau Tang *China* 22°30N 114°14E **33** a
an *China* 36°40N 114°15E **34** F8
alawun ✪ *Australia* 15°28S 133°1E **60** C5
in *Australia* 30°6S 116°37E **61** F2
u *China* 37°28N 110°42E **34** F6
hang *China* 44°55N 127°5E **35** B14
heng *China* 37°12N 116°20E **34** F9
huan *China* 41°5N 111°28E **34** D6
a *China* 39°29N 106°42E **34** E4
i *China* 37°40N 117°35E **35** F9
ing He → *China* 37°2N 110°23E **34** F6
inna *Australia* 33°0S 135°22E **63** E2
u *China* 33°22N 104°54E **34** H3
uishan *China* 22°25N 113°25E **33** a
ullar = Beswick *Australia* 14°34S 132°53E **60** B5
ai *China* 39°39N 106°48E **34** E4
an *China* 30°31N 114°18E **33** E11
Wuhe *China* 33°10N 117°50E **35** H9
Wuhsi = Wuxi *China* 31°33N 120°18E **33** E13
Wuhu *China* 31°22N 118°21E **33** E12
Wukari *Nigeria* 7°51N 9°42E **50** G7
Wulajie *China* 44°6N 126°33E **35** B14
Wulanbulang *China* 41°5N 110°55E **34** D6
Wular L. *India* 34°20N 74°30E **43** B6
Wulian *China* 35°40N 119°12E **35** G10
Wuliaru *Indonesia* 7°27S 131°0E **37** F8
Wulumuchi = Ürümqi *China* 43°45N 87°45E **32** C6
Wundanyi *Kenya* 3°24S 38°22E **54** C4
Wunnummin L. *Canada* 52°55N 89°10W **72** B2
Wuntho *Burma* 23°55N 95°45E **41** H19
Wupatki ⌂ *U.S.A.* 35°35N 111°20W **77** J8
Wuppertal *Germany* 51°16N 7°12E **16** C4
Wuppertal *S. Africa* 32°13S 19°12E **56** E2
Wuqing *China* 39°23N 117°4E **35** E9
Wurralibi ✪ *Australia* 15°43S 137°1E **62** B2
Wurtsboro *U.S.A.* 41°35N 74°29W **83** E10
Würzburg *Germany* 49°46N 9°55E **16** D5
Wushan *China* 34°43N 104°53E **34** G3
Wushi *China* 41°9N 79°13E **32** C4
Wutai *China* 38°40N 113°12E **34** E7
Wutai Shan *China* 39°3N 113°32E **34** E7
Wuting = Huimin *China* 37°27N 117°28E **35** F9
Wutonghaolai *China* 42°50N 120°5E **35** C11
Wutongqiao *China* 29°22N 103°50E **32** F9
Wuwei *China* 37°57N 102°34E **32** D9
Wuxi *China* 31°33N 120°18E **33** E13
Wuxiang *China* 36°49N 112°50E **34** F7
Wuyang *China* 33°25N 113°35E **34** H7
Wuyi *China* 37°46N 115°56E **34** F8
Wuyi Shan *China* 27°0N 117°0E **33** F12
Wuyuan *China* 41°2N 108°20E **34** D5
Wuzhai *China* 38°54N 111°48E **34** E6
Wuzhi Shan *China* 18°45N 109°45E **38** C7
Wuzhong *China* 38°2N 106°12E **34** E4
Wuzhou *China* 23°30N 111°18E **33** G11
Wuzishan = Tongshi *China* 18°30N 109°20E **38** C7
Wyaaba Cr. → *Australia* 16°27S 141°35E **62** B3
Wyalkatchem *Australia* 31°8S 117°22E **61** F2
Wyalusing *U.S.A.* 41°40N 76°16W **83** E8
Wyandotte *U.S.A.* 42°12N 83°9W **81** D12
Wyandra *Australia* 27°12S 145°56E **63** D4
Wyangala, L. *Australia* 33°54S 149°0E **63** E4
Wyara, L. *Australia* 28°42S 144°14E **63** D3
Wycheproof *Australia* 36°5S 143°17E **63** F3
Wycliffe Well *Australia* 20°48S 134°14E **62** C1
Wye → *U.K.* 51°38N 2°40W **13** F5
Wyemandoo *Australia* 28°28S 118°29E **61** E2
Wymondham *U.K.* 52°35N 1°7E **13** E9
Wymore *U.S.A.* 40°7N 96°40W **80** E5
Wyndham *Australia* 15°33S 128°3E **60** C4
Wyndham *N.Z.* 46°20S 168°51E **59** G2
Wynne *U.S.A.* 35°14N 90°47W **85** D9
Wynyard *Australia* 41°5S 145°44E **63** G4
Wynyard *Canada* 51°45N 104°10W **71** C8
Wyola L. *Australia* 29°8S 130°17E **61** E5
Wyoming *Canada* 42°57N 82°7W **82** D2
Wyoming □ *U.S.A.* 43°0N 107°30W **76** E10
Wyomissing *U.S.A.* 40°20N 75°59W **83** F9
Wyong *Australia* 33°14S 151°24E **63** E5
Wytheville *U.S.A.* 36°57N 81°5W **81** G13

X

Xaafuun *Somali Rep.* 10°25N 51°16E **47** E5
Xaafuun, Ras *Somali Rep.* 10°27N 51°24E **47** E5
Xaçmaz *Azerbaijan* 41°31N 48°42E **19** F8
Xai-Xai *Mozam.* 25°6S 33°31E **57** D5
Xaignabouri = Sayaboury *Laos* 19°15N 101°45E **38** C3
Xainza *China* 30°58N 88°35E **32** E6
Xalapa *Mexico* 19°32N 96°55W **87** D5
Xangongo *Angola* 16°45S 15°5E **56** B2
Xankändi *Azerbaijan* 39°52N 46°49E **44** B5
Xanthi *Greece* 41°10N 24°58E **23** D11
Xanxerê *Brazil* 26°53S 52°23W **95** B5
Xapuri *Brazil* 10°35S 68°35W **92** F5
Xar Moron He → *China* 43°25N 120°35E **35** C11
Xátiva *Spain* 38°59N 0°32W **21** C5
Xau, L. *Botswana* 21°15S 24°44E **56** C3
Xavantina *Brazil* 21°15S 52°48W **95** A5
Xebert *China* 44°2N 122°0E **35** B12
Xenia *U.S.A.* 39°41N 83°56W **81** F12
Xeropotamos → *Cyprus* 34°42N 32°33E **25** E11
Xhora *S. Africa* 31°55S 28°38E **57** E4
Xhumo *Botswana* 21°7S 24°35E **56** C3
Xi Jiang → *China* 22°5N 113°20E **33** G11
Xi Ujimqin Qi *China* 44°32N 117°40E **35** B9
Xi Xian *China* 36°41N 110°58E **34** F6
Xia Xian *China* 35°8N 111°12E **34** G6
Xiachengzi *China* 44°40N 130°18E **35** B16
Xiajin *China* 36°56N 116°0E **34** F9
Xiamen *China* 24°25N 118°4E **33** G12
Xi'an *China* 34°15N 109°0E **34** G5
Xian Xian *China* 38°12N 116°6E **34** E9
Xiang Jiang → *China* 28°55N 112°50E **33** F11
Xiang Khouang *Laos* 19°17N 103°25E **38** C4
Xiangcheng *China* 33°50N 113°27E **34** H7
Xiangfan *China* 32°2N 112°8E **33** E11
Xianggang = Hong Kong □ *China* 22°11N 114°14E **33** a
Xianghuang Qi *China* 42°2N 113°50E **34** C7
Xiangning *China* 35°58N 110°50E **34** G6
Xiangquan *China* 36°30N 113°1E **34** F7
Xiangquan He = Sutlej → *Pakistan* 29°23N 71°3E **42** E4
Xiangshui *China* 34°12N 119°33E **35** G10
Xiangtan *China* 27°51N 112°54E **33** F11
Xiantao *China* 30°25N 113°25E **33** E11
Xianyang *China* 34°20N 108°40E **34** G5
Xiao Hinggan Ling *China* 49°0N 127°0E **33** B14
Xiao Xian *China* 34°15N 116°55E **34** G9
Xiaoshan *China* 30°12N 120°18E **33** E13
Xiaowutai Shan *China* 39°51N 114°59E **34** E8
Xiaoyi *China* 37°8N 111°48E **34** F6
Xiashan = Zhanjiang *China* 21°15N 110°20E **33** G11
Xiawa *China* 42°35N 120°38E **35** C11
Xiayi *China* 34°15N 116°10E **34** G9
Xichang *China* 27°51N 102°19E **32** F9
Xichuan *China* 33°0N 111°30E **34** H6
Xifei He → *China* 32°45N 116°40E **34** H9
Xifeng *Gansu, China* 35°40N 107°40E **34** G4
Xifeng *Liaoning, China* 42°42N 124°45E **35** C13
Xifengzhen = Xifeng *China* 35°40N 107°40E **34** G4
Xigazê *China* 29°5N 88°45E **32** F6
Xihe *China* 34°2N 105°20E **34** G3
Xihua *China* 33°45N 114°30E **34** H8
Xili Shuiku *China* 22°36N 113°57E **33** a
Xiliao He → *China* 43°32N 123°35E **35** C12
Xilinhot *China* 43°52N 116°2E **34** C9
Ximana *Mozam.* 19°24S 33°58E **55** F3
Ximiao *China* 40°59N 100°12E **32** C9
Xin Xian = Xinzhou *China* 38°22N 112°46E **34** E7
Xinavane *Mozam.* 25°2S 32°47E **57** D5
Xinbin *China* 41°40N 125°2E **35** D13
Xing Xian *China* 38°27N 111°7E **34** E6
Xingcheng *China* 40°40N 120°45E **35** D11
Xinghe *China* 40°55N 113°55E **34** D7
Xinghua *China* 32°58N 119°48E **35** H10
Xinglong *China* 40°25N 117°30E **35** D9
Xingping *China* 34°20N 108°28E **34** G5
Xingtai *China* 37°3N 114°32E **34** F8
Xingu → *Brazil* 1°30S 51°53W **93** D8
Xingxingxia *China* 41°47N 95°0E **32** C8
Xingyang *China* 34°45N 112°52E **34** G7
Xingyi *China* 25°3N 104°59E **32** F9
Xinhe *China* 37°30N 115°15E **34** F8
Xining *China* 36°34N 101°40E **32** D9
Xinjiang *China* 35°34N 111°11E **34** G6
Xinjiang Uygur Zizhiqu □ *China* 42°0N 86°0E **32** C6
Xinjin = Pulandian *China* 39°25N 121°58E **35** E11
Xinkai He → *China* 43°32N 123°35E **35** C12
Xinken *China* 22°39N 113°36E **33** a
Xinle *China* 38°25N 114°40E **34** E8
Xinlitun *China* 42°0N 122°8E **35** D12
Xinmin *China* 41°59N 122°50E **35** D12
Xintai *China* 35°55N 117°45E **35** G9
Xinwan *China* 22°47N 113°40E **33** a
Xinxiang *China* 35°18N 113°50E **34** G7
Xinyang *China* 32°6N 114°3E **33** E11
Xinyi *China* 34°23N 118°21E **35** G10
Xinzhan *China* 43°50N 127°18E **35** C14
Xinzheng *China* 34°20N 113°45E **34** G7
Xinzhou *Hainan, China* 19°43N 109°17E **38** C7
Xinzhou *Shanxi, China* 38°22N 112°46E **34** E7
Xiongyuecheng *China* 40°12N 122°5E **35** D12
Xiping *Henan, China* 33°22N 114°5E **34** H8
Xiping *Henan, China* 33°25N 111°8E **34** H6
Xique-Xique *Brazil* 10°50S 42°40W **93** F10
Xisha Qundao = Paracel Is. *S. China Sea* 15°50N 112°0E **36** A4
Xiuyan *China* 40°18N 123°11E **35** D12
Xixabangma Feng *China* 28°20N 85°40E **43** E11
Xixia *China* 33°25N 111°29E **34** H6
Xixiang *Guangdong, China* 22°34N 113°52E **33** a
Xixiang *Shaanxi, China* 33°0N 107°44E **34** H4
Xixón = Gijón *Spain* 43°32N 5°42W **21** A3
Xiyang *China* 37°38N 113°38E **34** F7
Xizang Zizhiqu □ *China* 32°0N 88°0E **32** E6
Xlendi *Malta* 36°1N 14°12E **25** C1
Xochob *Mexico* 19°21N 89°48W **87** D7
Xuan Loc *Vietnam* 10°56N 107°14E **39** G6
Xuanhua *China* 40°40N 115°2E **34** D8
Xuanwei *China* 26°15N 103°59E **32** F9
Xuchang *China* 34°2N 113°48E **34** G7
Xun Xian *China* 35°42N 114°33E **34** G8
Xunyang *China* 32°48N 109°22E **34** H5
Xunyi *China* 35°8N 108°20E **34** G5
Xúquer → *Spain* 39°5N 0°10W **21** C5
Xushui *China* 39°2N 115°40E **34** E8
Xuwen *China* 20°20N 110°10E **33** G11
Xuyi *China* 32°55N 118°32E **35** H10
Xuzhou *China* 34°18N 117°10E **35** G9
Xylophagou *Cyprus* 34°54N 33°51E **25** E12

Y

Ya Xian = Sanya *China* 18°14N 109°29E **38** C7
Yaamba *Australia* 23°8S 150°22E **62** C5
Ya'an *China* 29°58N 103°5E **32** F9
Yaapeet *Australia* 35°45S 142°3E **63** F3
Yablonovyy Khrebet *Russia* 53°0N 114°0E **29** D12
Yablonovyy Ra. = Yablonovyy Khrebet *Russia* 53°0N 114°0E **29** D12
Yabrai Shan *China* 39°40N 103°0E **34** E2
Yabrūd *Syria* 33°58N 36°39E **46** B5
Yabucoa *Puerto Rico* 18°3N 65°53W **89** d
Yacambú △ *Venezuela* 9°42N 69°27W **89** E6
Yacheng *China* 18°22N 109°6E **38** C7
Yacuiba *Bolivia* 22°0S 63°43W **94** A3
Yacuma → *Bolivia* 13°38S 65°23W **92** F5
Yadgir *India* 16°45N 77°5E **40** L10
Yadkin → *U.S.A.* 35°23N 80°4W **85** D14
Yadua *Fiji* 16°49S 178°18E **59** a
Yaeyama-Rettō *Japan* 24°30N 123°40E **31** M1
Yagasa Cluster *Fiji* 18°57S 178°28W **59** a
Yagodnoye *Russia* 62°33N 149°40E **29** C15
Yahila *Dem. Rep. of the Congo* 0°13N 24°28E **54** B1
Yahk *Canada* 49°6N 116°10W **70** D5
Yahuma *Dem. Rep. of the Congo* 1°0N 23°10E **52** D4
Yaita *Japan* 36°48N 139°56E **31** F9
Yaiza *Canary Is.* 28°57N 13°46W **24** F6
Yakeshi *China* 49°17N 120°44E **33** B13
Yakima *U.S.A.* 46°36N 120°31W **76** C3
Yakima → *U.S.A.* 46°15N 119°14W **76** C4
Yakishiri-Jima *Japan* 44°26N 141°25E **30** B10
Yakobi I. *U.S.A.* 58°0N 136°30W **70** B1
Yakovlevka *Russia* 44°26N 133°28E **30** B6
Yaku-Shima *Japan* 30°20N 130°30E **31** J5
Yakumo *Japan* 42°15N 140°16E **30** C10
Yakutat *U.S.A.* 59°33N 139°44W **68** F4
Yakutat B. *U.S.A.* 59°45N 140°45W **74** D11
Yakutia = Sakha □ *Russia* 66°0N 130°0E **29** C14
Yakutsk *Russia* 62°5N 129°50E **29** C13
Yala *Thailand* 6°33N 101°18E **39** J3
Yalata *Australia* 31°59S 132°26E **61** F5
Yalata ✪ *Australia* 31°35S 132°7E **61** F5
Yalboroo *Australia* 20°50S 148°40E **62** b
Yale *U.S.A.* 43°8N 82°48W **82** C2
Yalgorup △ *Australia* 32°39S 115°38E **61** F2
Yalinga *C.A.R.* 6°33N 23°10E **52** C4
Yalkabul, Pta. *Mexico* 21°32N 88°37W **87** C7
Yalleroi *Australia* 24°3S 145°42E **62** C4
Yalobusha → *U.S.A.* 33°33N 90°10W **85** E9
Yalong *China* 18°12N 109°42E **38** C7
Yalong Jiang → *China* 26°40N 101°55E **32** F9
Yalova *Turkey* 40°41N 29°15E **23** D13
Yalpirakinu ✪ *Australia* 22°24S 132°15E **60** D5
Yalta *Ukraine* 44°30N 34°10E **19** F5
Yalu Jiang → *China* 39°55N 124°19E **35** E13
Yam *Australia* 9°54S 142°46E **62** a
Yam Ha Melah = Dead Sea *Asia* 31°30N 35°30E **46** D4
Yam Kinneret *Israel* 32°45N 35°35E **46** C4
Yamada *Japan* 33°33N 130°49E **31** H5
Yamagata *Japan* 38°15N 140°15E **30** E10
Yamagata □ *Japan* 38°30N 140°0E **30** E10
Yamaguchi *Japan* 34°10N 131°32E **31** G5
Yamaguchi □ *Japan* 34°20N 131°40E **31** G5
Yamal, Poluostrov *Russia* 71°0N 70°0E **28** B8
Yamal Pen. = Yamal, Poluostrov *Russia* 71°0N 70°0E **28** B8
Yamanashi □ *Japan* 35°40N 138°40E **31** G9
Yamantau, Gora *Russia* 54°15N 58°6E **18** D10
Yamato Ridge *Sea of Japan* 39°20N 135°0E **30** E7
Yamba *Australia* 29°26S 153°23E **63** D5
Yambarran Ra. *Australia* 15°10S 130°25E **60** C5
Yâmbiô *Sudan* 4°35N 28°16E **51** H11
Yambol *Bulgaria* 42°30N 26°30E **23** C12
Yamburg *Russia* 68°21N 77°8E **28** C8
Yamdena *Indonesia* 7°45S 131°20E **37** F8
Yame *Japan* 33°13N 130°35E **31** H5
Yamethin *Burma* 20°29N 96°18E **41** J20
Yamma Yamma, L. *Australia* 26°16S 141°20E **63** D3
Yamoussoukro *Ivory C.* 6°49N 5°17W **50** G4
Yampa → *U.S.A.* 40°32N 108°59W **76** F9
Yampi Sd. *Australia* 16°8S 123°38E **60** C3
Yampil *Moldova* 48°15N 28°15E **17** D15
Yampol = Yampil *Moldova* 48°15N 28°15E **17** D15
Yamuna → *India* 25°30N 81°53E **43** G9
Yamunanagar *India* 30°7N 77°17E **42** D7
Yamzho Yumco *China* 28°48N 90°35E **32** F7
Yana → *Russia* 71°30N 136°0E **29** B14
Yanagawa *Japan* 33°10N 130°24E **31** H5
Yanai *Japan* 33°58N 132°7E **31** H6
Yan'an *China* 36°35N 109°26E **34** F5
Yanaul *Russia* 56°25N 55°0E **18** C10
Yanbu 'al Baḩr *Si. Arabia* 24°0N 38°5E **44** F3
Yanchang *China* 36°43N 110°1E **34** F6
Yancheng *Henan, China* 33°35N 114°0E **34** H8
Yancheng *Jiangsu, China* 33°23N 120°8E **35** H11
Yanchep *Australia* 31°33S 115°37E **61** F2
Yanchi *China* 37°48N 107°20E **34** F4
Yanchuan *China* 36°51N 110°10E **34** F6
Yanco Cr. → *Australia* 35°14S 145°35E **63** F4
Yandeyarra ✪ *Australia* 21°17S 118°24E **60** D2
Yandicoogina *Australia* 22°49S 119°12E **60** D2
Yandoon *Burma* 17°0N 95°40E **41** L19
Yang Xian *China* 33°15N 107°30E **34** H4
Yang-yang *S. Korea* 38°4N 128°38E **35** E15
Yangambi *Dem. Rep. of the Congo* 0°47N 24°24E **54** B1
Yangcheng *China* 35°28N 112°22E **34** G7
Yangch'ü = Taiyuan *China* 37°52N 112°33E **34** F7
Yanggao *China* 40°21N 113°55E **34** D7
Yanggu *China* 36°8N 115°43E **34** F8
Yangjiang *China* 21°50N 111°59E **33** G11
Yangliuqing *China* 39°2N 117°5E **35** E9
Yangon = Rangoon *Burma* 16°45N 96°20E **41** L20
Yangpingguan *China* 32°58N 106°5E **34** H4
Yangquan *China* 37°58N 113°31E **34** F7
Yangtse = Chang Jiang → *China* 31°48N 121°10E **33** E13
Yangtze Kiang = Chang Jiang → *China* 31°48N 121°10E **33** E13
Yangyuan *China* 40°1N 114°10E **34** D8
Yangzhou *China* 32°21N 119°26E **33** E12
Yanji *China* 42°59N 129°30E **35** C15
Yankton *U.S.A.* 42°53N 97°23W **80** D5
Yankunytjatjara-Antakirinja ✪ *Australia* 27°20S 134°30E **63** A1
Yanonge *Dem. Rep. of the Congo* 0°35N 24°38E **54** B1
Yanqi *China* 42°5N 86°35E **32** C6
Yanqing *China* 40°30N 115°58E **34** D8
Yanshan *China* 38°4N 117°22E **35** E9
Yanshou *China* 45°28N 128°22E **35** B15
Yantabulla *Australia* 29°21S 145°0E **63** D4
Yantai *China* 37°34N 121°22E **35** F11
Yantian *China* 22°35N 114°16E **33** a
Yantongshan *China* 43°17N 126°0E **35** C14
Yanuca *Fiji* 18°24S 178°0E **59** a
Yanzhou *China* 35°35N 116°49E **34** G9
Yao Noi, Ko *Thailand* 8°7N 98°37E **39** a
Yao Xian *China* 34°55N 108°59E **34** G5
Yao Yai, Ko *Thailand* 8°0N 98°35E **39** a
Yaoundé *Cameroon* 3°50N 11°35E **52** D2
Yaowan *China* 34°15N 118°3E **35** G10
Yap *Pac. Oc.* 9°30N 138°10E **64** G5
Yapen *Indonesia* 1°50S 136°0E **37** E9
Yapen, Selat *Indonesia* 1°20S 136°10E **37** E9
Yapero *Indonesia* 4°59S 137°11E **37** E9
Yappar → *Australia* 18°22S 141°16E **62** B3
Yapuparra ✪ *Australia* 27°15S 126°20E **61** E4
Yaqaga *Fiji* 16°35S 178°36E **59** a
Yaqui → *Mexico* 27°37N 110°39W **86** B2
Yar-Sale *Russia* 66°50N 70°50E **28** C8
Yaraka *Australia* 24°53S 144°3E **62** C3
Yaransk *Russia* 57°22N 47°49E **18** C8
Yare → *U.K.* 52°35N 1°38E **13** E9
Yaremcha *Ukraine* 48°27N 24°33E **17** D13
Yarensk *Russia* 62°11N 49°15E **18** B8
Yarí → *Colombia* 0°20S 72°20W **92** D4
Yarkand = Shache *China* 38°20N 77°10E **32** D4
Yarkant He → *China* 40°26N 80°59E **32** C5
Yarker *Canada* 44°23N 76°46W **83** B8
Yarkhun → *Pakistan* 36°17N 72°30E **43** A5
Yarlung Ziangbo Jiang = Brahmaputra → *Asia* 23°40N 90°35E **43** H13
Yarmouth *Canada* 43°50N 66°7W **73** D6
Yarmūk → *Syria* 32°42N 35°40E **46** C4
Yaroslavl *Russia* 57°35N 39°55E **18** C6
Yarqa, W. → *Egypt* 30°0N 33°49E **46** F2
Yarra Yarra Lakes *Australia* 29°40S 115°45E **61** E2
Yarram *Australia* 38°29S 146°39E **63** F4
Yarraman *Australia* 26°50S 152°0E **63** D5
Yarras *Australia* 31°25S 152°20E **63** E5
Yarrie *Australia* 20°40S 120°12E **60** D3
Yartsevo *Russia* 60°20N 90°0E **29** C10
Yarumal *Colombia* 6°58N 75°24W **92** B3
Yasawa *Fiji* 16°47S 177°31E **59** a
Yasawa Group *Fiji* 17°0S 177°23E **59** a
Yaselda *Belarus* 52°7N 26°28E **17** B14
Yasin *Pakistan* 36°24N 73°23E **43** A5
Yasinski, L. *Canada* 53°16N 77°35W **72** B4
Yasinya *Ukraine* 48°16N 24°21E **17** D13
Yasothon *Thailand* 15°50N 104°10E **38** E5
Yass *Australia* 34°49S 148°54E **63** E4
Yāsūj *Iran* 30°31N 51°31E **45** D6
Yatağan *Turkey* 37°20N 28°10E **23** F13
Yates Center *U.S.A.* 37°53N 95°44W **80** G6
Yathkyed L. *Canada* 62°40N 98°0W **71** A9
Yatsushiro *Japan* 32°30N 130°40E **31** H5
Yatta Plateau *Kenya* 2°0S 38°0E **54** C4
Yauco *Puerto Rico* 18°2N 66°51W **89** d
Yavarí → *Peru* 4°21S 70°2W **92** D4
Yávaros *Mexico* 26°42N 109°31W **86** B3
Yavatmal *India* 20°20N 78°15E **40** J11
Yavne *Israel* 31°52N 34°45E **46** D3
Yavoriv *Ukraine* 49°55N 23°20E **17** D12
Yavorov = Yavoriv *Ukraine* 49°55N 23°20E **17** D12
Yawatahama *Japan* 33°27N 132°24E **31** H6
Yawri B. *S. Leone* 8°22N 13°0W **50** G3
Yaxian = Sanya *China* 18°14N 109°29E **38** C7
Yazd *Iran* 31°55N 54°27E **45** D7
Yazd □ *Iran* 32°0N 55°0E **45** D7
Yazd-e Khvāst *Iran* 31°31N 52°7E **45** D7
Yazman *Pakistan* 29°8N 71°45E **42** E4
Yazoo → *U.S.A.* 32°22N 90°54W **85** E9
Yazoo City *U.S.A.* 32°51N 90°25W **85** E9
Ybycuí △ *Paraguay* 26°5S 56°46W **94** B4
Ybytyruzú △ *Paraguay* 25°51S 56°11W **95** B4
Ye *Burma* 15°15N 97°15E **38** E1
Ye Xian *China* 33°35N 113°25E **34** H7
Yebyu *Burma* 14°15N 98°13E **38** E2
Yecheng *China* 37°54N 77°26E **32** D4
Yecheon *S. Korea* 36°39N 128°27E **35** F15
Yecla *Spain* 38°35N 1°5W **21** C5
Yécora *Mexico* 28°20N 108°58W **86** B3
Yedintsy = Edineț *Moldova* 48°9N 27°18E **17** D14
Yegros *Paraguay* 26°20S 56°25W **94** B4
Yehbuah *Indonesia* 8°23S 114°45E **37** J17
Yehuda, Midbar *Israel* 31°35N 35°15E **46** D4
Yei *Sudan* 4°9N 30°40E **51** H12
Yekaterinburg *Russia* 56°50N 60°30E **28** D7
Yelarbon *Australia* 28°33S 150°38E **63** D5
Yelets *Russia* 52°40N 38°30E **18** D6
Yelin = Lingshui *China* 18°27N 110°0E **38** C8
Yelizavetgrad = Kirovohrad *Ukraine* 48°35N 32°20E **19** E5
Yelizovo *Russia* 53°11N 158°23E **29** D16
Yell *U.K.* 60°35N 1°5W **11** A7
Yell Sd. *U.K.* 60°33N 1°15W **11** A7
Yellow = Huang He → *China* 37°55N 118°50E **35** F10
Yellow Sea *China* 35°0N 123°0E **35** G12
Yellowhead Pass *Canada* 52°53N 118°25W **70** C5
Yellowknife *Canada* 62°27N 114°29W **70** A6
Yellowknife → *Canada* 62°31N 114°19W **70** A6
Yellowstone → *U.S.A.* 47°59N 103°59W **76** C12
Yellowstone △ *U.S.A.* 44°40N 110°30W **76** D8
Yellowstone L. *U.S.A.* 44°27N 110°22W **76** D8
Yelsk *Belarus* 51°50N 29°10E **17** C15
Yemen ■ *Asia* 15°0N 44°0E **47** E3
Yen Bai *Vietnam* 21°42N 104°52E **38** B5
Yenangyaung *Burma* 20°30N 95°0E **41** J19
Yenbo = Yanbu 'al Baḩr *Si. Arabia* 24°0N 38°5E **44** F3
Yenda *Australia* 34°13S 146°14E **63** E4
Yeni Erenköy = Yialousa *Cyprus* 35°32N 34°10E **25** D13
Yenice *Turkey* 39°55N 27°17E **23** E12
Yenisey → *Russia* 71°50N 82°40E **28** B9
Yeniseysk *Russia* 58°27N 92°13E **29** D10
Yeniseyskiy Zaliv *Russia* 72°20N 81°0E **28** B9
Yenyuka *Russia* 57°57N 121°15E **29** D13
Yeo → *U.K.* 51°2N 2°49W **13** G5
Yeo, L. *Australia* 28°0S 124°30E **61** E3
Yeo I. *Canada* 45°24N 81°48W **82** A3
Yeola *India* 20°2N 74°30E **40** J9

Yeong-wol S. Korea 37°11N 128°28E 35 F15
Yeongcheon S. Korea 35°58N 128°56E 35 G15
Yeongdeok S. Korea 36°24N 129°22E 35 F15
Yeongdong S. Korea 36°10N 127°46E 35 F14
Yeongju S. Korea 36°50N 128°40E 35 F15
Yeosu S. Korea 34°47N 127°45E 35 G14
Yeotmal = Yavatmal India 20°20N 78°15E 40 J11
Yeovil U.K. 50°57N 2°38W 13 G5
Yeppoon Australia 23°5S 150°47E 62 C5
Yerbent Turkmenistan 39°30N 58°50E 28 F6
Yerbogachen Russia 61°16N 108°0E 29 C11
Yerevan Armenia 40°10N 44°31E 44 A5
Yerington U.S.A. 38°59N 119°10W 76 G4
Yermo U.S.A. 34°54N 116°50W 79 L10
Yerólakkos Cyprus 35°11N 33°15E 25 D12
Yeropol Russia 65°15N 168°40E 29 C17
Yeroskipos Cyprus 34°46N 32°28E 25 E11
Yershov Russia 51°23N 48°27E 19 D8
Yerushalayim = Jerusalem Israel/West Bank 31°47N 35°10E 46 D4
Yes Tor U.K. 50°41N 4°0W 13 G4
Yesan S. Korea 36°41N 126°51E 35 F14
Yeso U.S.A. 34°26N 104°37W 77 J11
Yessey Russia 68°29N 102°10E 29 C11
Yetman Australia 28°56S 150°48E 63 D5
Yeu, Î. d' France 46°42N 2°20W 20 C2
Yevpatoriya Ukraine 45°15N 33°20E 19 E5
Yeysk Russia 46°40N 38°12E 19 E6
Yezd = Yazd Iran 31°55N 54°27E 45 D7
Ygatimi Paraguay 24°5S 55°40W 95 A4
Yhati Paraguay 25°45S 56°35W 94 B4
Yhú Paraguay 25°0S 56°0W 95 B4
Yí → Uruguay 33°7S 57°8W 94 C4
Yi 'Allaq, G. Egypt 30°21N 33°31E 46 E2
Yi He → China 34°10N 118°8E 35 G10
Yi Xian Hebei, China 39°20N 115°30E 34 E8
Yi Xian Liaoning, China 41°30N 121°22E 35 D11
Yialiás → Cyprus 35°9N 33°44E 25 D12
Yialousa Cyprus 35°32N 34°10E 25 D13
Yibin China 28°45N 104°32E 32 F9
Yichang China 30°40N 111°20E 33 E11
Yicheng China 35°42N 111°40E 34 G6
Yichuan China 36°2N 110°10E 34 F6
Yichun China 47°44N 128°52E 33 B14
Yidu China 36°43N 118°28E 35 F10
Yijun China 35°28N 109°8E 34 G5
Yıldız Dağları Turkey 41°48N 27°36E 23 D12
Yilehuli Shan China 51°20N 124°20E 33 A13
Yima China 34°44N 111°53E 34 G6
Yimianpo China 45°7N 128°2E 35 B15
Yin Xu China 36°8N 114°10E 34 F8
Yi'nan China 35°31N 118°24E 35 G10
Yinchuan China 38°30N 106°15E 34 E4
Yindarlgooda, L. Australia 30°40S 121°52E 61 F3
Yindjibarndi ⚙ Australia 22°0S 118°35E 60 D2
Ying He → China 32°30N 116°30E 34 H9
Ying Xian China 39°32N 113°10E 34 E7
Yingkou China 40°37N 122°18E 35 D12
Yingpanshui China 37°26N 104°18E 34 F3
Yingualyalya ⚙ Australia 18°49S 129°12E 60 C4
Yining China 43°58N 81°10E 32 C5
Yiningarra ⚙ Australia 20°53S 129°27E 60 D4
Yinmabin Burma 22°10N 94°55E 41 H19
Yirga Alem Ethiopia 6°48N 38°22E 47 F2
Yirrkala Australia 12°14S 136°56E 62 A2
Yishan China 24°28N 108°38E 32 G10
Yishui China 35°47N 118°30E 35 G10
Yishun Singapore 1°26N 103°51E 39 d
Yitong China 43°13N 125°20E 35 C13
Yixing China 31°21N 119°48E 33 E12
Yiyang Henan, China 34°27N 112°10E 34 G7
Yiyang Hunan, China 28°35N 112°18E 33 F11
Yli-Kitka Finland 66°8N 28°30E 8 C23
Ylitornio Finland 66°19N 23°39E 8 C20
Ylivieska Finland 64°4N 24°28E 8 D21
Yoakum U.S.A. 29°17N 97°9W 84 G6
Yog Pt. Phil. 14°6N 124°12E 37 B6
Yogyakarta Indonesia 7°49S 110°22E 37 G14
Yogyakarta □ Indonesia 7°48S 110°22E 37 G14
Yoho △ Canada 51°25N 116°30W 70 C5
Yojoa, L. de Honduras 14°53N 88°0W 88 D2
Yok Don △ Vietnam 12°50N 107°40E 38 F6
Yokadouma Cameroon 3°26N 14°55E 52 D2
Yokkaichi Japan 34°55N 136°38E 31 G8
Yoko Cameroon 5°32N 12°20E 52 C2
Yokohama Japan 35°27N 139°28E 31 G9
Yokosuka Japan 35°20N 139°40E 31 G9
Yokote Japan 39°20N 140°30E 30 E10
Yola Nigeria 9°10N 12°29E 51 G8
Yolaina, Cordillera de Nic. 11°30N 84°0W 88 D3
Yŏlöten Turkmenistan 37°18N 62°21E 45 B9
Yom → Thailand 15°35N 100°1E 38 E3
Yonago Japan 35°25N 133°19E 31 G6
Yonaguni-Jima Japan 24°27N 123°0E 31 M1
Yŏnan N. Korea 37°55N 126°11E 35 F14
Yonezawa Japan 37°57N 140°4E 30 F10
Yong-in S. Korea 37°14N 127°12E 35 F14
Yong Peng Malaysia 2°0N 103°3E 39 M4
Yong Sata Thailand 7°8N 99°41E 39 J2
Yongamp'o N. Korea 39°56N 124°23E 35 E13
Yong'an China 25°59N 117°25E 33 F12
Yongchang China 38°17N 102°7E 32 D9
Yongcheng China 33°55N 116°20E 34 H9
Yongdeng China 36°38N 103°25E 34 F2
Yonghe China 36°46N 110°38E 34 F6
Yŏnghŭng N. Korea 39°31N 127°18E 35 E14
Yongji Jilin, China 43°33N 126°13E 35 C14
Yongji Shanxi, China 34°52N 110°28E 34 G6
Yongnian China 36°47N 114°29E 34 F8
Yongning China 38°15N 106°14E 34 E4
Yongzhou China 26°17N 111°37E 33 F11
Yonibana S. Leone 8°30N 12°19W 50 G3
Yonkers U.S.A. 40°56N 73°52W 83 F11
Yonne → France 48°23N 2°58E 20 B5
York Australia 31°52S 116°47E 61 F2
York U.K. 53°58N 1°6W 12 D6
York Ala., U.S.A. 32°29N 88°18W 85 E10
York Nebr., U.S.A. 40°52N 97°36W 80 E5
York Pa., U.S.A. 39°58N 76°44W 81 F15
York, C. Australia 10°42S 142°31E 62 A3
York, City of □ U.K. 53°58N 1°6W 12 D6
York, Kap Greenland 75°55N 66°25W 69 B18
York, Vale of U.K. 54°15N 1°25W 12 C6
York Sd. Australia 15°0S 125°5E 60 C4
Yorke Pen. Australia 34°50S 137°40E 63 E2
Yorkshire Dales △ U.K. 54°12N 2°10W 12 C5
Yorkshire Wolds U.K. 54°8N 0°31W 12 C7
Yorkton Canada 51°11N 102°28W 71 C8
Yorkville U.S.A. 38°52N 123°13W 78 G3
Yoro Honduras 15°9N 87°7W 88 C2
Yoron-Jima Japan 27°2N 128°26E 31 L4
Yos Sudarso, Pulau = Dolak, Pulau Indonesia 8°0S 138°30E 37 F9
Yosemite △ U.S.A. 37°45N 119°40W 78 H7
Yosemite Village U.S.A. 37°45N 119°35W 78 H7
Yoshino-Kumano △ Japan 34°12N 135°55E 31 H8
Yoshkar Ola Russia 56°38N 47°55E 18 C8
Yotvata Israel 29°55N 35°2E 46 F4
Youbou Canada 48°53N 124°13W 78 B2
Youghal Ireland 51°56N 7°52W 10 E4
Youghal B. Ireland 51°55N 7°49W 10 E4
Young Australia 34°19S 148°18E 63 E4
Young Canada 51°47N 105°45W 71 C7
Young Uruguay 32°44S 57°36W 94 C4
Young I. Antarctica 66°25S 162°24E 5 C11
Younghusband, L. Australia 30°50S 136°5E 63 E2
Younghusband Pen. Australia 36°0S 139°25E 63 F2
Youngstown Canada 51°35N 111°10W 71 C6
Youngstown N.Y., U.S.A. 43°15N 79°3W 82 C5
Youngstown Ohio, U.S.A. 41°6N 80°39W 82 E4
Youngsville U.S.A. 41°51N 79°19W 82 E5
Youngwood U.S.A. 40°14N 79°34W 82 F5
Youyu China 40°10N 112°20E 34 D7
Yozgat Turkey 39°51N 34°47E 19 G5
Ypacaraí △ Paraguay 25°18S 57°19W 94 B4
Ypané → Paraguay 23°29S 57°19W 94 A4
Ypres = Ieper Belgium 50°51N 2°53E 15 D2
Yreka U.S.A. 41°44N 122°38W 76 F2
Ystad Sweden 55°26N 13°50E 9 J15
Ysyk-Köl = Balykchy Kyrgyzstan 42°26N 76°12E 32 C4
Ysyk-Köl Kyrgyzstan 42°25N 77°15E 28 E8
Ythan → U.K. 57°19N 1°59W 11 D7
Ytyk-Kyuyel Russia 62°30N 133°45E 29 C14
Yu Jiang → China 23°22N 110°3E 33 G11
Yu Shan Taiwan 23°25N 120°52E 33 G13
Yu Xian = Yuzhou China 34°10N 113°28E 34 G7
Yu Xian Hebei, China 39°50N 114°35E 34 E8
Yu Xian Shanxi, China 38°5N 113°20E 34 E7
Yuan Jiang → China 28°55N 111°50E 33 F11
Yuanping China 38°42N 112°46E 34 E7
Yuanqu China 35°18N 111°40E 34 G6
Yuanyang China 35°3N 113°58E 34 G7
Yuba → U.S.A. 39°8N 121°36W 78 F5
Yuba City U.S.A. 39°8N 121°37W 78 F5
Yūbari Japan 43°4N 141°59E 30 C10
Yūbetsu Japan 44°13N 143°50E 30 B11
Yucatán □ Mexico 20°50N 89°0W 87 C7
Yucatán, Canal de Caribbean 22°0N 86°30W 88 B2
Yucatán, Península de Mexico 19°30N 89°0W 66 H11
Yucatan Basin Cent. Amer. 19°0N 86°0W 87 D7
Yucatan Channel = Yucatán, Canal de Caribbean 22°0N 86°30W 88 B2
Yucca U.S.A. 34°52N 114°9W 79 L12
Yucca Valley U.S.A. 34°8N 116°27W 79 L10
Yucheng China 36°55N 116°32E 34 F9
Yuci = Jinzhong China 37°42N 112°46E 34 F7
Yuen Long China 22°26N 114°2E 33 a
Yuendumu Australia 22°16S 131°49E 60 D5
Yuendumu ⚙ Australia 22°21S 131°40E 60 D5
Yueyang China 29°21N 113°5E 33 F11
Yugorenok Russia 59°47N 137°40E 29 D14
Yukon U.S.A. 35°31N 97°45W 84 D6
Yukon → U.S.A. 62°32N 163°54W 74 C7
Yukon Flats U.S.A. 66°40N 145°45W 74 B10
Yukon Territory □ Canada 63°0N 135°0W 68 E5
Yukta Russia 63°26N 105°42E 29 C11
Yukuhashi Japan 33°44N 130°59E 31 H5
Yulara Australia 25°10S 130°55E 61 E5
Yule → Australia 20°41S 118°17E 60 D2
Yuleba Australia 26°37S 149°24E 63 D4
Yulin Hainan, China 18°10N 109°31E 38 C7
Yulin Shaanxi, China 38°20N 109°30E 34 E5
Yuma Ariz., U.S.A. 32°43N 114°37W 79 N12
Yuma Colo., U.S.A. 40°8N 102°43W 76 F12
Yuma, B. de Dom. Rep. 18°20N 68°35W 89 C6
Yumbe Uganda 3°28N 31°15E 54 B3
Yumbi Dem. Rep. of the Congo 1°12S 26°15E 54 C2
Yumen China 39°50N 97°30E 32 D8
Yun Xian China 32°50N 110°46E 34 H6
Yuna Australia 28°20S 115°0E 61 E2
Yuncheng Henan, China 35°36N 115°57E 34 G8
Yuncheng Shanxi, China 35°2N 111°0E 34 G6
Yungang Shiku China 40°5N 113°0E 34 D7
Yungas Bolivia 17°0S 66°0W 92 G5
Yungay Chile 37°10S 72°5W 94 D1
Yunkanjini ⚙ Australia 22°33S 131°6E 60 D5
Yunling Shan China 28°30N 98°50E 32 F8
Yunnan □ China 25°0N 102°0E 32 G9
Yunta Australia 32°34S 139°36E 63 E2
Yunxi China 33°0N 110°22E 34 H6
Yunyang China 33°26N 112°42E 34 H7
Yupanqui Basin Pac. Oc. 19°0S 101°0W 65 J17
Yuraygir △ Australia 29°45S 153°15E 63 D5
Yurga Russia 55°42N 84°51E 28 D9
Yurimaguas Peru 5°55S 76°7W 92 E3
Yurubí △ Venezuela 10°26N 68°42W 89 D6
Yuscarán Honduras 13°58N 86°45W 88 D2
Yushe China 37°4N 112°58E 34 F7
Yushu Jilin, China 44°43N 126°38E 35 B14
Yushu Qinghai, China 33°5N 96°55E 32 E8
Yutai China 35°0N 116°45E 34 G9
Yutian Hebei, China 39°53N 117°45E 35 E9
Yutian Sinkiang-Uigur, China 36°52N 81°42E 32 D5
Yuxarı Qarabağ = Nagorno-Karabakh □ Azerbaijan 39°55N 46°45E 44 B5
Yuxi China 24°30N 102°35E 32 G9
Yuzawa Japan 39°10N 140°30E 30 E10
Yuzhno-Kurilsk Russia 44°1N 145°51E 29 E15
Yuzhno-Sakhalinsk Russia 46°58N 142°45E 33 B17
Yuzhnyy □ Russia 44°0N 40°0E 28 E5
Yuzhou China 34°10N 113°28E 34 G7
Yvetot France 49°37N 0°44E 20 B4

Z

Zaanstad Neths. 52°27N 4°50E 15 B4
Zāb al Kabīr → Iraq 36°1N 43°24E 44 B4
Zāb aş Şaghīr → Iraq 35°17N 43°29E 44 C4
Zabaykalsk Russia 49°40N 117°25E 29 E12
Zabol Iran 31°0N 61°32E 45 D9
Zābol □ Afghan. 32°0N 67°0E 40 D5
Zābolī Iran 27°10N 61°35E 45 E9
Zabrze Poland 50°18N 18°50E 17 C10
Zacapa Guatemala 14°59N 89°31W 88 D2
Zacapu Mexico 19°50N 101°43W 86 D4
Zacatecas Mexico 22°47N 102°35W 86 C4
Zacatecas □ Mexico 23°0N 103°0W 86 C4
Zacatecoluca El Salv. 13°29N 88°51W 88 D2
Zachary U.S.A. 30°39N 91°9W 84 F9
Zacoalco de Torres Mexico 20°14N 103°35W 86 C4
Zacualtipán Mexico 20°39N 98°36W 87 C5
Zadar Croatia 44°8N 15°14E 16 F8
Zadetkyi Kyun Burma 10°0N 98°25E 39 G2
Zafarqand Iran 33°11N 52°29E 45 C7
Zafra Spain 38°26N 6°30W 21 C2
Żagań Poland 51°39N 15°22E 16 C8
Zagaoua Chad 15°30N 22°24E 51 E10
Zagazig Egypt 30°40N 31°30E 51 B12
Zāgheh Iran 33°30N 48°42E 45 C6
Zagreb Croatia 45°50N 15°58E 16 F9
Zāgros, Kūhhā-ye Iran 33°45N 48°5E 45 C6
Zagros Mts. = Zāgros, Kūhhā-ye Iran 33°45N 48°5E 45 C6
Zahamena ⌂ Madag. 17°37S 48°49E 57 B8
Zāhedān Fārs, Iran 28°46N 53°52E 45 D7
Zāhedān Sīstān va Balūchestān, Iran 29°30N 60°50E 45 D9
Zahlah Lebanon 33°52N 35°50E 46 B4
Zaïre = Congo → Africa 6°4S 12°24E 52 F2
Zaječar Serbia 43°53N 22°18E 23 C10
Zaka Zimbabwe 20°20S 31°29E 57 C5
Zakamensk Russia 50°23N 103°17E 29 D11
Zakhodnaya Dzvina = Daugava → Latvia 57°4N 24°3E 9 H21
Zākhū Iraq 37°10N 42°50E 44 B4
Zakinthos = Zakynthos Greece 37°47N 20°54E 23 F9
Zakopane Poland 49°18N 19°57E 17 D10
Zakros Greece 35°6N 26°10E 25 D8
Zakynthos Greece 37°47N 20°54E 23 F9
Zalaegerszeg Hungary 46°53N 16°47E 17 E9
Zalantun China 48°0N 122°43E 33 B13
Zalari Russia 53°33N 102°30E 29 D11
Zalău Romania 47°12N 23°3E 17 E12
Zaleshchiki = Zalishchyky Ukraine 48°45N 25°45E 17 D13
Zalew Wiślany Poland 54°20N 19°50E 17 A10
Zalingei Sudan 12°51N 23°29E 51 F10
Zalishchyky Ukraine 48°45N 25°45E 17 D13
Zama L. Canada 58°45N 119°5W 70 B5
Zambeke Dem. Rep. of the Congo 2°8N 25°17E 54 B2
Zambeze → Africa 18°35S 36°20E 55 F4
Zambezi = Zambeze → Africa 18°35S 36°20E 55 F4
Zambezi Zambia 13°30S 23°15E 53 G4
Zambezi △ Zimbabwe 17°54S 25°41E 55 F2
Zambezia □ Mozam. 16°15S 37°30E 55 F4
Zambia ■ Africa 15°0S 28°0E 55 F2
Zamboanga Phil. 6°59N 122°3E 37 C6
Zamora Mexico 19°59N 102°16W 86 D4
Zamora Spain 41°30N 5°45W 21 B3
Zamość Poland 50°43N 23°15E 17 C12
Zanda China 31°32N 79°50E 32 E4
Zandvoort Neths. 52°22N 4°32E 15 B4
Zanesville U.S.A. 39°56N 82°1W 82 G2
Zangābād Iran 38°26N 46°44E 44 B5
Zangue → Mozam. 17°50S 35°21E 55 F4
Zanjān Iran 36°40N 48°35E 45 B6
Zanjān □ Iran 37°20N 49°30E 45 B6
Zanjān → Iran 37°8N 47°47E 45 B6
Zante = Zakynthos Greece 37°47N 20°54E 23 F9
Zanthus Australia 31°2S 123°34E 61 F3
Zanzibar Tanzania 6°12S 39°12E 54 D4
Zaouiet El-Kala = Bordj Omar Driss Algeria 28°10N 6°40E 50 C7
Zaouiet Reggâne Algeria 26°32N 0°3E 50 C6
Zaoyang China 32°10N 112°45E 33 E11
Zaozhuang China 34°50N 117°35E 35 G9
Zap Suyu = Zāb al Kabīr → Iraq 36°1N 43°24E 44 B4
Zapadnaya Dvina = Daugava → Latvia 57°4N 24°3E 9 H21
Západné Beskydy Europe 49°30N 19°0E 17 D10
Zapala Argentina 39°0S 70°5W 96 D2
Zapaleri, Cerro Bolivia 22°49S 67°11W 94 A2
Zapata U.S.A. 26°55N 99°16W 84 H5
Zapolyarnyy Russia 69°26N 30°51E 8 B24
Zapopán Mexico 20°43N 103°24W 86 C4
Zaporizhzhya Ukraine 47°50N 35°10E 19 E6
Zaporozhye = Zaporizhzhya Ukraine 47°50N 35°10E 19 E6
Zara Turkey 39°58N 37°43E 44 B3
Zaragoza Coahuila, Mexico 28°29N 100°55W 86 B4
Zaragoza Nuevo León, Mexico 23°58N 99°46W 87 C5
Zaragoza Spain 41°39N 0°53W 21 B5
Zarand Kermān, Iran 30°46N 56°34E 45 D8
Zarand Markazī, Iran 35°18N 50°25E 45 C6
Zaranj Afghan. 30°55N 61°55E 40 D2
Zarasai Lithuania 55°40N 26°20E 9 J22
Zárate Argentina 34°7S 59°0W 94 C4
Zard, Kūh-e Iran 32°22N 50°4E 45 C6
Zāreh Iran 35°7N 49°9E 45 C6
Zaria Nigeria 11°0N 7°40E 50 F7
Zarneh Iran 33°55N 46°10E 44 C5
Zaros Greece 35°8N 24°54E 25 D6
Zarqā', Nahr az → Jordan 32°10N 35°37E 46 C4
Zarrīn Iran 32°46N 54°37E 45 C7
Zaruma Ecuador 3°40S 79°38W 92 D3
Żary Poland 51°37N 15°10E 16 C8
Zarzis Tunisia 33°31N 11°2E 51 B8
Zaskar → India 34°13N 77°20E 43 B7
Zaskar Mts. India 33°15N 77°30E 43 C7
Zastron S. Africa 30°18S 27°7E 56 E4
Zavāreh Iran 33°29N 52°28E 45 C7
Zave Zimbabwe 17°6S 30°1E 57 B5
Zavitinsk Russia 50°10N 129°20E 29 D13
Zavodovski I. Antarctica 56°0S 27°45W 5 B1
Zawiercie Poland 50°30N 19°24E 17 C10
Zāwiyat al Bayḑā = Al Bayḑā Libya 32°50N 21°44E 51 B10
Zāyā Iraq 33°33N 44°13E 44 C5
Zāyandeh → Iran 32°35N 52°0E 45 C7
Zaysan Kazakhstan 47°28N 84°52E 32 B5
Zaysan Köli Kazakhstan 48°0N 83°0E 28 E9
Zayū China 28°48N 97°27E 41 E20
Zazafotsy Madag. 21°11S 46°21E 57 C8
Zbarazh Ukraine 49°43N 25°44E 17 D13
Zdolbuniv Ukraine 50°30N 26°15E 17 C14
Zduńska Wola Poland 51°37N 18°59E 17 C10
Zeballos Canada 49°59N 126°50W 70 D3
Zebediela S. Africa 24°20S 29°17E 57 C4
Zeebrugge Belgium 51°19N 3°12E 15 C3
Zeehan Australia 41°52S 145°25E 63 G4
Zeeland □ Neths. 51°30N 3°50E 15 C3
Zeerust S. Africa 25°31S 26°4E 56 D4
Ẕefat Israel 32°58N 35°29E 46 C4
Zehak Iran 30°53N 61°42E 45 D9
Zeil, Mt. Australia 23°30S 132°23E 60 D5
Zeila = Saylac Somali Rep. 11°21N 43°30E 47 E3
Zeist Neths. 52°5N 5°15E 15 B5
Zeitz Germany 51°2N 12°7E 16 C7
Zelenogorsk Russia 60°12N 29°43E 8 F23
Zelenograd Russia 56°1N 37°12E 18 C6
Zelienople U.S.A. 40°48N 80°8W 82 F4
Zémio C.A.R. 5°2N 25°5E 54 A2
Zempoala Mexico 19°27N 96°23W 87 D5
Zemun Serbia 44°51N 20°25E 23 B9
Zenica Bos.-H. 44°10N 17°57E 23 B7
Žepče Bos.-H. 44°28N 18°2E 23 B8
Zevenaar Neths. 51°56N 6°5E 15 C6
Zeya Russia 53°48N 127°14E 29 D13
Zeya → Russia 51°42N 128°53E 29 D13
Zêzere → Portugal 39°28N 8°20W 21 C1
Zghartā Lebanon 34°21N 35°53E 46 A4
Zgorzelec Poland 51°10N 15°0E 16 C8
Zhabinka Belarus 52°13N 24°2E 17 B13
Zhambyl = Taraz Kazakhstan 42°54N 71°22E 32 C3
Zhangaözen Kazakhstan 43°18N 52°48E 19 F9
Zhangaqazaly Kazakhstan 45°48N 62°6E 28 E7
Zhangbei China 41°10N 114°45E 34 D8
Zhangguangcai Ling China 45°0N 129°0E 35 B15
Zhangjiabian China 22°33N 113°28E 33 a
Zhangjiakou China 40°48N 114°55E 34 D8
Zhangye China 38°50N 100°23E 32 D9
Zhangzhou China 24°30N 117°35E 33 G12
Zhanhua China 37°40N 118°8E 35 F10
Zhanjiang China 21°15N 110°20E 33 G11
Zhannetty, Ostrov Russia 76°43N 158°0E 29 B16
Zhanyu China 44°30N 122°30E 35 B12
Zhao Xian China 37°43N 114°45E 34 F8
Zhaocheng China 36°22N 111°38E 34 F6
Zhaoqing China 23°0N 112°20E 33 G11
Zhaotong China 27°20N 103°44E 32 F9
Zhaoyang Hu China 35°4N 116°47E 34 G9
Zhaoyuan Heilongjiang, China 45°27N 125°0E 35 B13
Zhaoyuan Shandong, China 37°20N 120°23E 35 F11
Zhari Namco China 31°6N 85°36E 32 E6
Zhashkiv Ukraine 49°15N 30°5E 17 D16
Zhashui China 33°40N 109°8E 34 H5
Zhayylma Kazakhstan 51°37N 61°33E 28 D7
Zhayyq → Kazakhstan 47°0N 51°48E 19 E9
Zhdanov = Mariupol Ukraine 47°5N 37°31E 19 E6
Zhecheng China 34°7N 115°20E 34 G8
Zhejiang □ China 29°0N 120°0E 33 F13
Zheleznodorozhnyy Russia 62°35N 50°55E 18 B9
Zheleznogorsk-Ilimskiy Russia 56°34N 104°8E 29 D11
Zhen'an China 33°27N 109°9E 34 H5
Zhengding China 38°8N 114°32E 34 E8
Zhengzhou China 34°45N 113°34E 34 G7
Zhenlai China 45°50N 123°5E 35 B12
Zhenping China 33°10N 112°16E 34 H7
Zhenyuan China 35°35N 107°30E 34 G4
Zhetiqara Kazakhstan 52°11N 61°12E 28 D7
Zhezqazghan Kazakhstan 47°44N 67°40E 28 E7
Zhidan China 36°48N 108°48E 34 F5
Zhigansk Russia 66°48N 123°27E 29 C13
Zhilinda Russia 70°0N 114°20E 29 C12
Zhitomir = Zhytomyr Ukraine 50°20N 28°40E 17 C15
Zhlobin Belarus 52°55N 30°0E 17 B16
Zhmerynka Ukraine 49°2N 28°2E 17
Zhob Pakistan 31°20N 69°31E
Zhob → Pakistan 32°4N 69°50E
Zhodzina Belarus 54°5N 28°17E 17
Zhokhova, Ostrov Russia 76°4N 152°40E 29
Zhongba China 29°39N 84°10E
Zhongdian China 27°48N 99°42E
Zhongning China 37°29N 105°40E
Zhongshan Antarctica 69°0S 39°50E
Zhongshan China 22°26N 113°20E
Zhongshankong China 22°35N 113°29E
Zhongtiao Shan China 35°0N 111°10E
Zhongwei China 37°30N 105°12E
Zhongyang China 37°20N 111°11E
Zhosaly Kazakhstan 45°29N 64°4E
Zhoucun China 36°47N 117°48E
Zhoukou China 33°38N 114°38E
Zhoukoudian China 39°41N 115°56E
Zhouzhi China 34°10N 108°12E
Zhovkva Ukraine 50°4N 23°58E 17
Zhuanghe China 39°40N 123°0E 35
Zhucheng China 36°0N 119°27E 35
Zhugqu China 33°40N 104°30E
Zhuhai China 22°17N 113°34E
Zhujiang Kou China 22°20N 113°45E
Zhumadian China 32°59N 114°2E 33
Zhuo Xian = Zhuozhou China 39°28N 115°58E
Zhuolu China 40°20N 115°12E
Zhuozhou China 39°28N 115°58E
Zhuozi China 41°0N 112°25E
Zhuzhou China 27°49N 113°12E 33
Zhytomyr Ukraine 50°20N 28°40E 17
Zhytomyr □ Ukraine 50°50N 28°10E 17
Ziarat Pakistan 30°25N 67°49E
Zibo China 36°47N 118°3E 35
Zichang China 37°18N 109°40E
Zidi = Wandhari Pakistan 27°42N 66°48E
Zielona Góra Poland 51°57N 15°31E
Zierikzee Neths. 51°40N 3°55E
Zigong China 29°15N 104°48E
Ziguéy Chad 14°43N 15°50E
Ziguinchor Senegal 12°35N 16°20W
Zihuatanejo Mexico 17°39N 101°33W
Žilina Slovak Rep. 49°12N 18°42E 17
Zillah Libya 28°30N 17°33E
Zima Russia 54°0N 102°5E 29
Zimapán Mexico 20°45N 99°21W
Zimba Zambia 17°20S 26°11E
Zimbabwe Zimbabwe 20°16S 30°54E
Zimbabwe ■ Africa 19°0S 30°0E
Zimnicea Romania 43°40N 25°22E 17
Zinave △ Mozam. 21°35S 33°40E
Zinder Niger 13°48N 9°0E
Zinga Tanzania 9°16S 38°49E
Zion △ U.S.A. 37°15N 113°5W
Ziros Greece 35°5N 26°8E
Zirreh, Gowd-e Afghan. 29°45N 62°0E
Zitácuaro Mexico 19°24N 100°22W
Zitundo Mozam. 26°48S 32°47E
Ziwa Maghariba = Kagera □ Tanzania 2°0S 31°30E
Ziway, L. Ethiopia 8°0N 38°50E
Ziyang China 32°32N 108°31E
Zlatograd Bulgaria 41°22N 25°7E 23
Zlatoust Russia 55°10N 59°40E 18
Zlín Czech Rep. 49°14N 17°40E
Zmeinogorsk Kazakhstan 51°10N 82°13E
Znojmo Czech Rep. 48°50N 16°2E
Zobeyrī Iran 34°10N 46°40E
Zobia Dem. Rep. of the Congo 3°0N 25°59E
Zoetermeer Neths. 52°3N 4°30E
Zohreh → Iran 30°16N 51°15E
Zolochiv Ukraine 49°45N 24°51E 17
Zomba Malawi 15°22S 35°19E
Zongo Dem. Rep. of the Congo 4°20N 18°35E
Zonguldak Turkey 41°28N 31°50E
Zonqor Pt. Malta 35°52N 14°34E
Zorritos Peru 3°43S 80°40W
Zou Xiang China 35°30N 116°58E
Zouar Chad 20°30N 16°32E
Zouérate = Zouîrât Mauritania 22°44N 12°21W
Zouîrât Mauritania 22°44N 12°21W
Zoutkamp Neths. 53°20N 6°18E
Zrenjanin Serbia 45°22N 20°23E
Zufār Oman 17°40N 54°0E
Zug Switz. 47°10N 8°31E
Zugspitze Germany 47°25N 10°59E
Zuid-Holland □ Neths. 52°0N 4°35E
Zuidbeveland Neths. 51°30N 3°50E
Zuidhorn Neths. 53°15N 6°23E
Zula Eritrea 15°17N 39°40E
Zumbo Mozam. 15°35S 30°26E
Zumpango Mexico 19°48N 99°6W
Zunhua China 40°18N 117°58E
Zuni Pueblo U.S.A. 35°4N 108°51W
Zunyi China 27°42N 106°53E 32
Zuoquan China 37°1N 113°27E
Zurbātīyah Iraq 33°9N 46°3E
Zürich Switz. 47°22N 8°32E
Zutphen Neths. 52°9N 6°12E
Zuurberg △ S. Africa 33°12S 25°32E
Zuwārah Libya 32°58N 12°1E
Zūzan Iran 34°22N 59°53E
Zvishavane Zimbabwe 20°17S 30°2E
Zvolen Slovak Rep. 48°33N 19°10E 17
Zwettl Austria 48°35N 15°9E
Zwickau Germany 50°44N 12°30E
Zwolle Neths. 52°31N 6°6E
Zwolle U.S.A. 31°38N 93°39W
Żyrardów Poland 52°3N 20°28E 17
Zyryan Kazakhstan 49°43N 84°20E
Zyryanka Russia 65°45N 150°51E 29
Zyryanovsk = Zyryan Kazakhstan 49°43N 84°20E
Żywiec Poland 49°42N 19°10E 17